PERIODIC TABLE OF THE ELEMENTS

Legend:
- Metals
- Nonmetals
- Metalloids

IA (1)	IIA (2)	IIIB (3)	IVB (4)	VB (5)	VIB (6)	VIIB (7)	(8)	VIIIB (9)	(10)	IB (11)	IIB (12)	IIIA (13)	IVA (14)	VA (15)	VIA (16)	VIIA (17)	VIIIA (18)
1 **H** 1.0079																1 **H** 1.0079	2 **He** 4.0026
3 **Li** 6.941	4 **Be** 9.0122											5 **B** 10.811	6 **C** 12.011	7 **N** 14.0067	8 **O** 15.9994	9 **F** 18.9984	10 **Ne** 20.1797
11 **Na** 22.9898	12 **Mg** 24.3050											13 **Al** 26.9815	14 **Si** 28.0855	15 **P** 30.9738	16 **S** 32.066	17 **Cl** 35.4527	18 **Ar** 39.948
19 **K** 39.0983	20 **Ca** 40.078	21 **Sc** 44.9559	22 **Ti** 47.88	23 **V** 50.9415	24 **Cr** 51.9961	25 **Mn** 54.9380	26 **Fe** 55.847	27 **Co** 58.9332	28 **Ni** 58.69	29 **Cu** 63.546	30 **Zn** 65.39	31 **Ga** 69.723	32 **Ge** 72.61	33 **As** 74.9216	34 **Se** 78.96	35 **Br** 79.904	36 **Kr** 83.80
37 **Rb** 85.4678	38 **Sr** 87.62	39 **Y** 88.9059	40 **Zr** 91.224	41 **Nb** 92.9064	42 **Mo** 95.94	43 **Tc** (98)	44 **Ru** 101.07	45 **Rh** 102.9055	46 **Pd** 106.42	47 **Ag** 107.8682	48 **Cd** 112.411	49 **In** 114.82	50 **Sn** 118.710	51 **Sb** 121.75	52 **Te** 127.60	53 **I** 126.9045	54 **Xe** 131.29
55 **Cs** 132.9054	56 **Ba** 137.327	57 **La** 138.9055 *	72 **Hf** 178.49	73 **Ta** 180.9479	74 **W** 183.85	75 **Re** 186.207	76 **Os** 190.2	77 **Ir** 192.22	78 **Pt** 195.08	79 **Au** 196.9665	80 **Hg** 200.59	81 **Tl** 204.3833	82 **Pb** 207.2	83 **Bi** 208.9804	84 **Po** (209)	85 **At** (210)	86 **Rn** (222)
87 **Fr** (223)	88 **Ra** (226)	89 **Ac** (227) **	104 **Rf** (261)	105 **Db** (262)	106 **Sg** (266)	107 **Bh** (264)	108 **Hs** (277)	109 **Mt** (268)	110 **Uun** (281)	111 **Uuu** (272)	112 **Uub** (285)		114 **Uuq** (289)				

*Lanthanide Series

58 **Ce** 140.115	59 **Pr** 140.9...	60 **Nd**	61 **Pm** (145)	62 **Sm** 150.36	63 **Eu** 151.965	64 **Gd** 157.25	65 **Tb** 158.9253	66 **Dy** 162.50	67 **Ho** 164.9303	68 **Er** 167.26	69 **Tm** 168.9342	70 **Yb** 173.04	71 **Lu** 174.967

**Actinide Series

90 **Th** 232.0381	91 **Pa** 231.0359	**U** 238.0289	93 **Np** (237)	94 **Pu** (244)	95 **Am** (243)	96 **Cm** (247)	97 **Bk** (247)	98 **Cf** (251)	99 **Es** (252)	100 **Fm** (257)	101 **Md** (258)	102 **No** (259)	103 **Lr** (260)

Note: Atomic masses are IUPAC values (up to four decimal places). More accurate values for some elements are given in the International Table of Atomic Weights at the back of the book.

We dedicate this book to the memory of
our longtime editor, publisher, mentor, and friend,
John Vondeling (1933–2001)

CONTENTS OVERVIEW

CONTENTS

CHAPTER 13: Liquids and Solids   477

THE LIQUID STATE   486

THE SOLID STATE   497

CHAPTER 14: Solutions   533

THE DISSOLUTION PROCESS   535

COLLIGATIVE PROPERTIES OF SOLUTIONS   548

COLLOIDS   567

CHAPTER 15: Chemical Thermodynamics   *582*

HEAT CHANGES AND THERMOCHEMISTRY   584

SPONTANEITY OF PHYSICAL AND CHEMICAL CHANGES   612

CHAPTER 16: Chemical Kinetics   *638*

FACTORS THAT AFFECT REACTION RATES   646

ENRICHMENT
Calculus Derivation of Integrated Rate Equations...662

ENRICHMENT
Using Integrated Rate Equations to Determine Reaction Order...664

CHEMISTRY IN USE
Ozone...686

CHAPTER 17: Chemical Equilibrium   *699*

CHAPTER 27: Organic Chemistry I: Formulas, Names, and Properties *1033*

CHAPTER 28: Organic Chemistry II: Shapes, Selected Reactions, and Biopolymers *1100*

The active metal potassium, K, reacts violently with water:

$$2K(s) + 2H_2O(\ell) \longrightarrow$$
$$H_2(g) + 2KOH(aq)$$

Heat released by the reaction causes unreacted potassium to give off a characteristic purple glow and ignites the hydrogen gas that is formed.

*G*eneral Chemistry *and* General Chemistry *with* A Qualitative Analysis Supplement, *seventh edition, are intended for use in the introductory chemistry course taken by students of chemistry, biology, geology, physics, engineering, and related subjects. Although some background in high school science is helpful, no specific knowledge of topics in chemistry is presupposed. These books are self-contained presentations of the fundamentals of chemistry. Their aim is to convey to students the dynamic and changing aspects of chemistry in the modern world.*

This text provides students with an understanding of fundamental concepts of chemistry; their ability to solve problems is based on this understanding. Our goal in this revision is to provide students with the best possible tool for learning chemistry by incorporating and amplifying features that enhance their understanding of concepts and guide them through the more challenging aspects of learning chemistry.

CHANGES TO THE SEVENTH EDITION

In revising this edition of *General Chemistry*, we have stayed true to the original vision and ideas of the book but have added some new and unique features and made some other modifications based on reviewer feedback.

First and foremost in our list of changes is the addition of our new co-author George Stanley, Cyril & Tutta Vetter Alumni Professor at Louisiana State University. His extensive research in inorganic chemistry and his knowledge of the General Chemistry curriculum have greatly enhanced this revision. George has won numerous awards and accolades, both nationally and locally, including the NSF *Special Creativity Award* in 1994, the LSU *University Excellence in Teaching Award* in 1995, the LSU College of Basic Sciences *Center for Excellence in Science Teaching* each year since 1997, and the Baton Rouge-ACS *Charles E. Coates Award* in 1999. George is thus a superb addition to a team of authors highly recognized for their teaching. Ken Whitten, Professor Emeritus at the University of Georgia, received many teaching awards, including the *Northeast Georgia ACS Section Outstanding Undergraduate Teaching Award* and the *General Sandy Beaver Teaching Award.* Teaching awards to Larry Peck, Professor and Director of First Year Chemistry Programs at Texas A&M University, include that University's *Association of Former Students Distinguished Achievement Award in Chemistry Teaching* in 2002 and the *Catalyst Award* (a national award for excellence in Chemistry Teaching) presented by the Chemical Manufacturers Association in 2000. Ray Davis is University Distinguished Teaching Professor at the University of Texas at Austin; his teaching awards include the *Minnie Stevens Piper Professorship* in 1992, the *Jean Holloway Award for Excellence in Teaching* in 1996, and (five times) the *Outstanding Teacher Award* given by campus freshman honor societies. He was an inaugural member of the University's Academy of Distinguished Teachers in 1995.

We have gone over the entire text in the book and edited the narrative for even better clarity.

Inspired by George Stanley, a new margin note outlines *common mistakes* students taking General Chemistry tend to make and how to correct them. These notes, associated with both the narrative and the examples, gently remind students of possible misconceptions about a topic or procedure and emphasize points often overlooked by students. They are identified by the icon in the margin.

SCi
LINKS.

The art program has been revised to include more molecular art. Charge distributions within many molecules are illustrated with colorful Electrostatic Charge Potential (ECP) figures, to help students visualize the effects of these charges on molecular properties and intermolecular interactions.

Margin references to additional reading and study on the World Wide Web via SciLinks have been included. Associated with the National Science Teachers Association (NSTA), SciLinks is an enormous database of science-related websites. Each site in the database has been found and checked by chemical educators and screened by the NSTA to be relevant and appropriate. The sites are intended to help students see molecular properties in chemistry they otherwise wouldn't be able to visualize. We have hand picked over 75 keywords corresponding to several hundred SciLinks sites on which students can find more information about a given topic.

New general and conceptual questions have been written for the end-of-chapter exercise sets.

Beyond the Textbook is a new subset of media-related exercises in this edition. Housed on the book's companion website at **http://www.brookscole.com/chemistry/whitten**, the *Beyond the Textbook* feature contains additional General Chemistry Interactive CD-ROM, Version 3.0 exercises and also directs students to the SciLinks database, other relevant sites, or library resources for information to use in solving these problems.

We also have made some changes to the narrative and have rearranged sections within some chapters. We have modified and reorganized some parts of Chapter 4, "Some Types of Chemical Reactions," so that the discussion on naming inorganic compounds precedes the sections on classifying chemical reactions. A new section in Chapter 5, The Structure of Atoms, covers paramagnetism and diamagnetism, material that was in an Enrichment section in the previous edition. A new section in Chapter 7, Chemical Bonding, includes information on formal charges.

PROVEN FEATURES

We *also* have continued to employ and amplify features that were well received in earlier editions of the text:

- Margin callouts direct students to specific topic screens on the **General Chemistry Interactive CD-ROM, Version 3.0,** by William J. Vining, University of Massachusetts, Amherst; John C. Kotz, SUNY—Oneonta; and Patrick Harman.

- **Conceptual problems** in the end-of-chapter problem sets emphasize conceptual understanding rather than computation.

- **Macro–micro and molecular art** has been utilized wherever appropriate to improve the students' ability to visualize molecular-level aspects of chemical properties and concepts.

- The book's **companion website** at **http://www.brookscole.com/chemistry/whitten** provides a wide variety of material to enhance student learning, including many aspects specific to this text. Pointers to this and other websites are provided in this text and in appropriate ancillaries.

- A chapter **outline** and a list of **objectives** are provided at the beginning of each chapter. These allow students to preview the chapter prior to reading it and help them to know the expectations of the chapter.

- **Margin notes** are used to point out historical facts, provide additional information, further emphasize some important points, relate information to ideas developed earlier, and note the relevance of discussions.

- **Key terms** are boldfaced in the text and are defined at the end of each chapter, immediately reinforcing terminology and concepts.

- **Figures** have been redrawn as necessary to improve appearance and clarity, and new **photographs** have been added to illustrate important points and provide visual interest.

- **Problem-solving tips** are found in almost every chapter. These highlighted helpful hints guide students through more complex subject areas. These tips are based on the authors' experiences and sensitivity to difficulties that the students face and work in tandem with the new *common mistakes* feature.

- **"Chemistry in Use" boxes,** a successful feature from previous editions, has been retained, and several interesting new topics have been introduced. "Chemistry in Use" boxes are identified by the special icon shown in the margin.

- **Enrichment** sections provide more insight into selected topics for better-prepared students, but they can be easily omitted without any loss of continuity.

- **Titles appear on each Example** so that students can see more clearly what concept or skill the Example is explaining. This is also useful for review purposes before exams. A note at the end of most Examples, "You should now work Exercise X," encourages students to practice the appropriate end-of-chapter Exercise and more closely ties illustrative Examples to related Exercises, thereby reinforcing concepts. Each Example also contains a Plan that explains the logic used to solve the problem.

- The end-of-chapter **Exercises** have been carefully examined and revised: More than one third of the problems are new or modified. All Exercises have been carefully reviewed for accuracy. The *Building Your Knowledge* category of end-of-chapter Exercises asks students to apply knowledge they learned in previous chapters to the current chapter. These questions help students retain previously learned information and show them that chemistry is an integrated science.

- A **Glossary** is included in the index so that students can look up a term at the back of the book as well as in the Key Terms at the end of the chapter.

We have also continued to use many ideas and teaching philosophies developed over the seven editions of this text:

We have kept in mind that chemistry is an experimental science and have emphasized the important role of theory in science. We have presented many of the classic experiments followed by interpretations and explanations of these milestones in the development of scientific thought.

We have defined each new term as accurately as possible and illustrated its meaning as early as was practical. We begin each chapter at a very fundamental level and then progress through carefully graded steps to a reasonable level of sophistication. Numerous illustrative Examples are provided throughout the text and keyed to end-of-chapter Exercises. The first Examples in each section are quite simple; the last is considerably more complex. The unit-factor method has been emphasized where appropriate.

We believe that the central concepts of chemical change are best understood in the sequence of chemical thermodynamics (*Is the forward or the reverse reaction favored?*), followed by chemical kinetics (*How fast does the reaction go?*), and then by chemical equilibrium (*How far does the reaction go?*). Our presentation in Chapters 15 through 17 reflects this belief.

We have used color extensively to make it easier to read the text and comprehend its organization. A detailed description of our pedagogical use of color starts on page xxxi in the "To the Student" section. Pedagogical use of color makes the text clearer, more accurate, and easier to understand.

We have used a blend of SI and more traditional metric units, because many students plan careers in areas in which SI units are not yet widely used. The health-care fields, the biological sciences, textiles, and agriculture are typical examples. We have used the joule rather than the calorie in nearly all energy calculations. We have emphasized the use of natural logarithms in mathematical relationships and problems, except where common practice retains the use of base-10 logarithms, such as in pH and related calculations and in the Nernst equation.

ORGANIZATION

There are 28 chapters in *General Chemistry*, and eight additional chapters in *A Qualitative Analysis Supplement.*

We present stoichiometry (Chapters 2 and 3) before atomic structure and bonding (Chapters 5–9) to establish a sound foundation for a laboratory program as early as possible. These chapters are virtually self-contained to provide flexibility to those who wish to cover structure and bonding before stoichiometry.

Because much of chemistry involves chemical reactions, we have introduced chemical reactions in a simplified, systematic way early in the text (Chapter 4). A logical, orderly introduction to formula unit, total ionic, and net ionic equations is included so that this information can be used throughout the remainder of the text. Solubility guidelines are clarified in this chapter so that students can use them in writing chemical equations in their laboratory work. Finally, naming inorganic compounds gives students early exposure to systematic nomenclature.

Many students have difficulty systematizing and using information, so we have done our utmost to assist them. At many points throughout the text we summarize the results of recent discussions or illustrative examples in tabular form to help students see the "big picture." The basic ideas of chemical periodicity are introduced early (Chapters 4 and 6) and are used throughout the text. The simplified classification of acids and bases introduced in Chapter 4 is expanded in Chapter 10, Acids, Bases, and Salts, after the appropriate background on structure and bonding. References are made to the classification of acids and bases and to the solubility guidelines throughout the text to emphasize the importance of systematizing and using previously covered information. Chapter 11 covers solution stoichiometry for both acid–base and redox reactions, emphasizing the mole method.

After our excursion through Gases and the Kinetic–Molecular Theory (Chapter 12), Liquids and Solids (Chapter 13), and Solutions (Chapter 14), students have the appropriate background for a wide variety of laboratory experiments.

Comprehensive chapters are presented on Chemical Thermodynamics (Chapter 15) and Chemical Kinetics (Chapter 16). The distinction between the roles of standard and non-standard Gibbs free energy change in predicting reaction spontaneity is clearly discussed. Chapter 16, Chemical Kinetics, provides early and consistent emphasis on the experimental basis of kinetics.

These chapters provide the necessary background for a strong introduction to Chemical Equilibrium in Chapter 17. This is followed by three chapters on Equilibria in Aqueous Solutions. A chapter on Electrochemistry (Chapter 21) completes the common core of the text except for Nuclear Chemistry (Chapter 26), which is self-contained and may be studied at any point in the course.

A group of basically descriptive chapters follow. Chapter 22, Metals I: Metallurgy, and Chapter 23, Metals II: Properties and Reactions, give broad coverage to the chemistry of metals. Chapter 24 covers Some Nonmetals and Metalloids, and Chapter 25 is a sound introduction to Coordination Compounds. Throughout these chapters, we have been careful to include appropriate applications of the principles that have been developed in the first part of the text to explain descriptive chemistry.

Organic chemistry is discussed in Chapters 27 and 28. Chapter 27 (Organic Chemistry I: Formulas, Names, and Properties) presents the classes of organic compounds, their structures and nomenclature (with major emphasis on the principal functional groups), and some fundamental classes of organic reactions. Chapter 28 (Organic Chemistry II: Shapes, Selected Reactions, and Biopolymers) presents isomerism and geometries of organic molecules, selected specific types, and an introduction to biopolymers.

Eight additional chapters are included in *A Qualitative Analysis Supplement*. In Chapter 29, important properties of the metals of the cation groups are tabulated, their properties are discussed, the sources of the elements are listed, their metallurgies are described, and a few uses of each metal are given.

Chapter 30 is a detailed introduction to the laboratory procedures used in semimicro qualitative analysis.

Chapters 31–35 cover the analysis of the groups of cations. (Cations that create serious disposal problems are no longer used in the qualitative analysis chapters. Mercury, silver, lead, and most chromium have been removed.) Each chapter includes a discussion of the important oxidation states of the metals, an introduction to the analytical procedures, and comprehensive discussions of the chemistry of each cation group. Detailed laboratory instructions, set off in color, follow. Students are alerted to pitfalls in advance, and alternate confirmatory tests and "clean-up" procedures are described for troublesome cations. A set of Exercises accompanies each chapter.

In Chapter 31, the traditional Group I has been replaced by the traditional Group IIA (minus lead). Traditional Group IIB (minus mercury) constitutes the first part of Chapter 32; then Groups I and II (traditional IIA + IIB minus lead and mercury) make up the last part of Chapter 32. Chapter 33 includes all of the usual Group III elements. Chapter 34 covers Group IV and Chapter 35 discusses Group V.

Chapter 36 contains a discussion of some of the more sophisticated ionic equilibria of qualitative analysis. The material is presented in a single chapter for the convenience of the instructor.

A FLEXIBLE PRESENTATION

We have exerted great effort to make the presentation as flexible as possible to give instructors the freedom to choose the order in which they teach topics. Some examples follow:

1. As in previous editions, we have clearly delineated the parts of Chapter 15, Chemical Thermodynamics, that can easily be moved forward by instructors who wish to cover thermochemistry (Sections 15-1 through 15-8) after stoichiometry (Chapters 2 and 3).

2. Chapter 4, Some Types of Chemical Reactions, is based on the periodic table and introduces chemical reactions just after stoichiometry. Reactions are classified as (a) oxidation–reduction reactions, (b) combination reactions, (c) displacement reactions, and (d) metathesis reactions (two types).

3. Some instructors prefer to discuss gases (Chapter 12) after stoichiometry (Chapters 2 and 3). Chapter 12 can be moved to that position.

4. Chapter 5 (The Structure of Atoms), Chapter 6 (Chemical Periodicity), and Chapter 7 (Chemical Bonding) provide comprehensive coverage of these key topics.

5. As in earlier editions, Molecular Structure and Covalent Bonding Theories (Chapter 8) includes parallel comprehensive VSEPR and VB descriptions of simple molecules. This approach has been widely accepted. However, some instructors prefer separate presentations of these theories of covalent bonding. The chapter has been carefully organized into numbered subdivisions to accommodate these professors; detailed suggestions are included at the beginning of the chapter.

6. Chapter 9 (Molecular Orbitals in Chemical Bonding) is a "stand-alone chapter" that may be omitted or moved with no loss in continuity.

7. Chapter 10 (Reactions in Aqueous Solutions I: Acids, Bases, and Salts) and Chapter 11 (Reactions in Aqueous Solution II: Calculations) include comprehensive discussions of acid–base and redox reactions in aqueous solutions, and solution stoichiometry calculations for acid–base and redox reactions.

The following ancillaries are available to qualified adopters. Please consult your local Brooks/Cole • Thomson sales representative for details.

INSTRUCTOR ANCILLARIES

Instructor's Solutions Manual by Vickie Williamson, Texas A&M University. Contains answers and solutions to all odd-numbered end-of-chapter exercises.

Test Bank by Nancy Faulk, Blinn College—Bryan Campus. Contains approximately 100 questions in every chapter, including five conceptual types of questions and multiple-choice questions.

Computerized Test Bank ExamView Computerized Testing . . . including online testing! Create, deliver, and customize tests and study guides (both print and online) in minutes with this assessment and tutorial system. ExamView offers both a Quick Test Wizard and an Online Test Wizard that guide you step by step through the process of creating tests.

General Chemistry Interactive CD-ROM, Version 3.0, by William J. Vining, University of Massachusetts, Amherst; John C. Kotz, SUNY— Oneonta; and Patrick Harman. Automatically included FREE with every new copy of this book, the General Chemistry Interactive CD-ROM, Version 3.0, works hand in hand and chapter by chapter with this book—giving students many opportunities to reinforce visually and interactively their understanding of general chemistry's important concepts and principles. The CD-ROM includes numerous guided simulations, tutorials with specific and targeted feedback, and media-based exercises; more than 100 video segments; hundreds of interactive models; a plotting tool, and molecular mass and molarity calculators; and an extensive database of compounds with their thermodynamic properties. Throughout the book, notes and icons appear in the margins and direct students to videos, animations, and other references that will help them master concepts.

OWL: Online Web-based Learning System Developed over the past 15 years at the University of Massachusetts and class-tested by thousands of students, OWL is a fully customizable and flexible teaching and learning system. With both numerical and chemical parameterization and useful, specific feedback built right in, OWL produces countless general chemistry questions correlated to Brooks/Cole general chemistry textbooks. OWL is the only system specifically designed to support mastery learning, where students work as long as they need to master each chemical concept and skill. OWL provides the framework for a Web-based virtual-learning environment that can improve your students' performance, automatically assess student homework, reduce your faculty workload, and save you time as you manage your general chemistry course.

Multimedia Manager for General Chemistry, Seventh Edition This one-stop digital library and presentation tool—a cross-platform CD-ROM—includes text, art, and tables in a variety of electronic formats that are easily exported into other software packages. This enhanced CD-ROM also contains simulations, molecular models, and QuickTime® movies to supplement your lectures. You also can customize your presentation by importing your personal lecture slides or other material you choose.

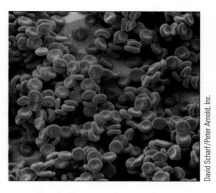

Red blood cells (1200x). The red blood cells that transport O_2 throughout our bodies contain hemoglobin, a coordination compound.

David Scharf/Peter Arnold, Inc.

The Brooks/Cole Chemistry Resource Center at http://www.brookscole.com/chemistry When you purchase this text, you have access to a rich array of learning resources that you won't find anywhere else. This outstanding site features everything from online quizzing to tutorials and more. The text-specific site—resident at the Resource Center—includes online quizzing, laboratory experiments, links to other useful sites, and other resources.

MyCourse 2.0 Whether you want only the easy-to-use tools to build it or the content to furnish it, MyCourse 2.0 offers you the simple solution for a custom course website that allows you to assign, track, and report on student progress, load your syllabus, and more. For a demo, visit http://mycourse.thomsonlearning.com.

Overhead Transparencies A set of 125 full-color overhead transparency acetates enhances the lecture presentation.

InfoTrac® College Edition Adopters and their students automatically receive a four-month subscription to InfoTrac College Edition with every new copy of this book! Newly improved, this extensive online library opens the door to the full text (not just abstracts) of countless articles from thousands of publications, going back over 20 years. An icon at the end of each chapter indicates exercises that require students to access and use InfoTrac College Edition.

STUDENT ANCILLARIES

Student Solutions Manual by Wendy Keeney-Kennicutt, Texas A&M University, and Yi-Noo Tang, Texas A&M University. This manual includes worked-out solutions to all of the even-numbered problems in the text. The solutions are worked in a manner consistent with the problem-solving approach of the book. However, when appropriate, alternate methods of solving the same problem are included, and problems are tied back to specific numbered examples in the text.

Student Study Guide by Raymond E. Davis, University of Texas at Austin, and James Petrich, San Antonio College. This study guide includes chapter summaries that highlight the main themes, study goals with section references, innovative tools for mastering important terms and concepts, a preliminary test for each chapter that provides an average of 80 drill and concept questions, and answers to the preliminary tests. The Study Guide has a new section called "Quotefalls" (word puzzles) in addition to the sections already listed. This section will come just before the preliminary test in each chapter.

Problem Solving in General Chemistry by Leslie N. Kinsland, Cornell University. This valuable study tool sharpens students' problem-solving skills. It contains a brief discussion of topics, a variety of examples and exercises, and answers to all the questions.

PowerPoint™ Slides with Lecture Outline by Charles Atwood. This outline helps students organize the material, prepare for class, and reduce the burden of note-taking in class. It provides great flexibility for the professor and makes more time available for other activities. It also comes with a special **PowerPoint™ CD-ROM** developed by Charles Atwood and Joel Caughran, both of the University of Georgia.

ACKNOWLEDGMENTS

With a mixture of sadness at his passing in January 2001 and fondness for his memory, we acknowledge above all the many roles of John Vondeling in the development and continued success of this book. John was our longtime editor, publisher, and friend, and his contributions to science publishing in general and this book in particular are beyond measure. In John's many years at Saunders College Publishing, we, like his many other authors, enjoyed his unwavering enthusiasm for publishing, his limitless knowledge of all aspects of the business, and his insights into the myriad details of authorship and production that make a book successful. One of John's final contributions to this book was his insightful recommendation of George G. Stanley as a new member of this author team; the other authors remain in his debt for that suggestion. John was our good friend, and we miss him.

Courtesy of Dr. Andre Geim, Manchester University

A live frog suspended in an extremely strong magnetic field (16 Tesla) demonstrating the diamagnetic repulsion effect.

The list of other individuals who contributed to the evolution of this book is long indeed. First, we would like to express our appreciation to the professors who contributed so much to our scientific education: Professors Arnold Gilbert, M. L. Bryant, the late W. N. Pirkle and Alta Sproull, C. N. Jones, S. F. Clark, R. S. Drago (KWW); the late Dorothy Vaughn, the late David Harker, and Calvin Vanderwerf, Professors Ralph N. Adams, F. S. Rowland, A. Tulinsky, and William von E. Doering (RED); Professors R. O'Connor, G. L. Baker, W. B. Cook, G. J. Hunt, A. E. Martell, and the late M. Passer (MLP); and Professors Richard Eisenberg, F. Albert Cotton, the late John A. Osborn, and Dr. Jerry Unruh (GGS).

As preparations for this edition were beginning, the ownership of Saunders College Publishing was transferred to the Brooks/Cole group of Thomson Learning. We are grateful to Angus McDonald for his guidance through this transition process, and for getting us off to a good start in this new publishing environment. The staff at Brooks/Cole Publishers has contributed immeasurably to the evolution of this book. As Chemistry Acquisitions Editor, John Holdcroft provided the authors a firm guiding hand and unstinting support throughout an often hectic development and production schedule. Jay Campbell, our Developmental Editor, coordinated innumerable details of reviewer comments, scheduling, and manuscript preparation and submission; we are grateful to Jay for his many unseen contributions and his expert guidance through many aspects of the process. As with prior editions (though with a far more constrained schedule!) Dena Digilio-Betz, our photo researcher, has gathered many excellent photographs with ingenuity, persistence, and patience. Lisa Weber's work as Project Manager for Editorial Production and her experienced hand in weaving together the many threads of the production process have been nothing short of amazing; we are grateful for her many contributions to the appearance, consistency, and quality of the book. As Project Managers, Doris Bruey and Anne Gibby of Sparkpoint Communications handled countless details of the production process thoroughly and efficiently, and helped us not only to keep to the schedule but to improve the consistency of the presentation. Stephen Rapley, Creative Director, oversaw development and execution of high-quality design and artwork that enhance both the appearance and the substance of the book. Greg Gambino's expert artwork is a major contribution to this edition. We also thank Karoliina Tuovinen, Assistant Editor, for coordinating the preparation of the print ancillaries.

Jim Morgenthaler (Athens), Charles Steele (Austin), and Charles D. Winters (Oneonta) did the original photography for this book.

Marcia Gillette (University of Indiana—Kokomo) and Alvin Holder (Virginia Polytechnical Institute and State University) wrote new conceptual exercises for the end-of-chapter Exercise sets.

Our thanks go to Marcy Whitney (University of Alabama) and Gary Riley (St. Louis College of Pharmacy) for checking the accuracy of the text and the end-of-chapter Exercises.

The end-of-chapter Exercises were considerably improved by the careful checking and numerous suggestions of Wendy Keeney-Kennicutt and Vickie Williamson.

Finally, we are deeply indebted to our families, Betty, Andy, and Kathryn Whitten; Sharon and Brian Davis; Angela Loera, and Laura Kane; and Sandy Peck, Molly Levine, and Marci Culp. They have supported us during the many years we have worked on this project. Their understanding, encouragement, and moral support have "kept us going."

REVIEWERS OF THE SEVENTH EDITION

The following individuals performed a pre-revision review of the seventh edition, and their valuable comments helped guide the development of this edition.

Ann Cartwright, *San Jacinto College Central*
Mark Draganjac, *Arkansas State University*
Dr. Lucio Gelmini, *Grant MacEwan College*
Jack D. Hefley, *Blinn College—Bryan Campus*
Wendy Keeney-Kennicutt, *Texas A&M University*
Larry Krannich, *University of Alabama, Birmingham*
Peter Krieger, *Palm Beach Community College*
Stephanie Morris, *Pellissippi State Technical Community College*
James M. Schlegal, *Rutgers University—Newark*
Cheryl Snyder, *Schoolcraft College*
John Thompson, *Lane Community College*
Mona Wahby, *Macomb Community College*

REVIEWERS OF THE FIRST SIX EDITIONS OF GENERAL CHEMISTRY

Edwin Abbott, Montana State University; Ed Acheson, Millikin University; David R. Adams, North Shore Community College; Carolyn Albrecht; Steven Albrecht, Ball State University; Dolores Aquino, San Jacinto College Central; Ale Arrington, South Dakota School of Mines; George Atkinson, Syracuse University; Charles Atwood, University of Georgia; Jerry Atwood, University of Alabama; William G. Bailey, Broward Community College; Major Charles Bass, United States Military Academy; J. M. Bellama, University of Maryland; Carl B. Bishop, Clemson University; Muriel B. Bishop, Clemson University; James R. Blanton, The Citadel; George Bodner, Purdue University; Joseph Branch, Central Alabama Community College; Greg Brewer, The Citadel; Clark Bricker, University of Kansas; Robert Broman, University of Missouri; William Brown, Beloit College; Robert F. Bryan, University of Virginia; Barbara Burke, California State Polytechnic, Pomona; L. A. Burns, St. Clair County Community College; James Carr, University of Nebraska, Lincoln; Elaine Carter, Los Angeles City College; Ann Cartwright, San Jacinto College Central; Thomas Cassen, University of North Carolina; Martin Chin, San Jose State University; Evelyn A. Clarke, Community College of Philadelphia; Kent Clinger, David Lipscomb University; Lawrence Conroy, University of Minnesota; Mark Cracolice, University of Montana; Julian Davies, University of Toledo; John DeKorte, Glendale Community

College (Arizona); George Eastland, Jr., Saginaw Valley State University; Harry Eick, Michigan State University; Mohammed El-Mikki, University of Toledo; Dale Ensor, Tennessee Technological University; Lawrence Epstein, University of Pittsburgh; Sandra Etheridge, Gulf Coast Community College; Darrell Eyman, University of Iowa; Nancy Faulk, Blinn College; Wade A. Freeman, University of Illinois, Chicago Circle; Mark Freilich, Memphis State University; Richard Gaver, San Jose State University; Gary Gray, University of Alabama, Birmingham; Robert Hanrahan, University of Florida; Alton Hassell, Baylor University; Henry Heikkinen, University of Maryland; Forrest C. Hentz, North Carolina State University; R. K. Hill, University of Georgia; Bruce Hoffman, Lewis and Clark College; Larry Houck, Memphis State University; Arthur Hufnagel, Erie Community College, North Campus; Wilbert Hutton, Iowa State University; Albert Jache, Marquette University; William Jensen, South Dakota State University; M. D. Joeston, Vanderbilt University; Stephen W. John, Lane Community College; Andrew Jorgensen, University of Toledo; Margaret Kastner, Bucknell University; Philip Kinsey, University of Evansville; Leslie N. Kinsland, University of Southwestern Louisiana; Donald Kleinfelter, University of Memphis; Marlene Kolz; Bob Kowerski, College of San Mateo; Larry Krannich, University of Alabama at Birmingham; Peter Krieger, Palm Beach Community College; Charles Kriley, Grove City College; Charles Kriley, Purdue University, Calumet; James Krueger, Oregon State University; Norman Kulevsky, University of North Dakota; Robert Lamb, Ohio Northern University; Alfred Lee, City College of San Francisco; Patricia Lee, Bakersfield College; William Litchman, University of New Mexico; Ramon Lopez de la Vega, Florida International University; Joyce Maddox, Tennessee State University; Gilbert J. Mains, Oklahoma State University; Ronald Marks, Indiana University of Pennsylvania; William Masterton, University of Connecticut; William E. McMullen, Texas A&M University; Clinton Medbery, The Citadel; Joyce Miller, San Jacinto College; Richard Mitchell, Arkansas State University; Kathleen Murphy, Daemen College; Joyce Neiburger, Purdue University; Deborah Nycz, Broward Community College; Barbara O'Brien, Texas A&M University; Christopher Ott, Assumption College; James L. Pauley, Pittsburgh State University; John Phillips, Purdue University, Calumet; Richard A. Pierce, Jefferson College; William Pietro, University of Wisconsin, Madison; Ronald O. Ragsdale, University of Utah; Susan Raynor, Rutgers University; Randal Remmel; Gary F. Riley, St. Loui College of Pharmacy; Don Roach, Miami Dade Community College; Eugene Rochow, Harvard University; Roland R. Roskos, University of Wisconsin, La Crosse; John Ruff, University of Georgia; George Schenk, Wayne State University; Mary Jane Schultz, Tufts University; William Scroggins, El Camino College; Curtis Sears, Georgia State University; Diane Sedney, George Washington University; Mahesh Sharma, Columbus College; C. H. Stammer, University of Georgia; Yi- Noo Tang, Texas A&M University; Margaret Tierney, Prince George's Community College; Henry Tracy, University of Southern Maine; Janice Turner, Augusta College; James Valentini, University of California, Irvine; Douglas Vaughan; Victor Viola, Indiana University; W. H. Waggoner, University of Georgia; Susan Weiner, West Valley College; Donald Williams, Hope College; Vickie Williamson, Texas A&M University; David Winters, Tidewater Community College; Wendy S. Wolbach, Illinois Wesleyan University; Kevin L. Wolf, Texas A&M University; Marie Wolff, Joliet Jr. College; James Wood, Palm Beach Community College; Robert Zellmer, Ohio State University; and Steve Zumdahl, University of Illinois.

Kenneth W. Whitten
Raymond E. Davis
M. Larry Peck
George G. Stanley

TO THE STUDENT

The earth is a huge chemical system. Many different chemicals can be found in the earth's atmosphere. It is our responsibility to keep harmful chemicals from polluting that system. We depend upon it for the oxygen we breathe, the carbon dioxide and nitrogen plants consume, and the water clouds transport. Maintaining life on the planet requires understanding and intelligent use of these resources. Scientists can provide important information about the processes, but each of us must share in the responsibility for our environment.

We have written this text to assist you as you study chemistry. Chemistry is a fundamental science—some call it the central science. As you and your classmates pursue diverse career goals, you will find that the vocabulary and ideas presented in this text will be useful in more places and in more ways than you may imagine now.

We begin with the most basic vocabulary and ideas. We then carefully evolve increasingly sophisticated ideas that are necessary and useful in all the other physical sciences, the biological sciences, and the applied sciences such as medicine, dentistry, engineering, agriculture, and home economics.

We have made the early chapters as nearly self-contained as possible. The material can be presented in the order considered most appropriate by your professor. Some professors will cover chapters in a different sequence or will omit some chapters completely—the text was designed to accommodate this.

Early in each section we have attempted to provide the experimental basis for the ideas we evolve. By experimental basis we mean the observations and experiments on the phenomena that have been most important in developing concepts. We then present an explanation of the experimental observations.

Chemistry is an experimental science. We know what we know because we (literally thousands of scientists) have observed it to be true. Theories have been evolved to explain experimental observations (facts). Successful theories explain observations fully and accurately. More importantly, they enable us to predict the results of experiments that have not yet been performed. Thus, we should always keep in mind the fact that experiment and theory go hand in hand. They are intimately related parts of our attempt to understand and explain natural phenomena.

"What is the best way to study chemistry?" is a question we are asked often by our students. While there is no single answer to this question, the following suggestions may be helpful. Your professor may provide additional suggestions. A number of supplementary materials accompany this text. All are designed to assist you as you study chemistry. Your professor may suggest that you use some of them.

Students often underestimate the importance of the act of *writing* as a tool for learning. Whenever you read, do not just highlight passages in the text, but also *take notes*. Whenever you work problems or answer questions *write yourself explanations* of why each step was done or how you reasoned out the answer. Keep a special section of your notebook for working out problems or answering questions. The very act of writing forces you to concentrate more on what you are doing, and you learn more. This is true even if you never go back to review what you wrote earlier. Of course, these notes will also help you to review for an examination.

You should always read over the assigned material before it is covered in class. This helps you to recognize the ideas as your professor discusses them. Take careful class notes. *At the first opportunity*, and certainly the same day, you should recopy your class notes. As you do this, fill in more detail where you can. Try to work the illustrative examples that your professor solved in class, without looking at the solution in your notes. If you must look at the solution, look at only one line (step), and then try to figure out the next step. Read the assigned material again and take notes, integrating these with your class notes. Reading should be much more informative the second time.

Review the "key terms" at the end of the chapter to be sure that you know the exact meaning of each term. Work the illustrative examples in the text while covering the solutions with a sheet of paper. If you find it necessary to look at the solutions, look at only one line at a time and try to figure out the next step. Answers to illustrative examples are displayed on blue backgrounds. At the end of most examples, we suggest related questions from the end-of-chapter exercises. You should work these suggested exercises as you come to them. Make sure that you read the Problem-Solving Tips; these will help you avoid common mistakes and understand more complex ideas.

This is a good time to work through the appropriate chapter in the STUDY GUIDE TO GENERAL CHEMISTRY. This will help you to see an overview of the chapter, to set specific study goals, and then to check and improve your grasp of basic vocabulary, concepts, and skills. Next, work the assigned exercises at the end of the chapter.

The appendices contain much useful information. You should become familiar with them and their contents so that you may use them whenever necessary. Answers to selected even-numbered numerical exercises are given at the end of the text so that you may check your work.

Ice is slightly less dense than liquid water, so ice floats in water.

The World Wide Web (Internet) is an increasingly important source of many kinds of information. The extensive website at **http://www.brookscole.com/chemistry/whitten** provides a wide variety of material to enhance your learning, including many features specific to this text. One new feature that you will find at that website is a new type of problems called "Beyond the Textbook." These problems direct you to the CD-ROM, to other sites on the Web, or to library resources for additional information to solve the problems.

Pointers to this and other websites are provided in this text and in appropriate ancillaries. We have incorporated references to SciLinks, a huge database of websites made available by the National Science Teachers Association (NSTA), into the margin notes in the text. These sites have plenty of useful supplementary information on many of the topics we cover in the book. We would appreciate hearing about other chemistry-related websites that you have found interesting or useful. You will also find many references throughout this text to the **General Chemistry Interactive CD-ROM, Version 3.0,** by William J. Vining, University of Massachusetts, Amherst; John C. Kotz, SUNY—Oneonta; and Patrick Harman. This multimedia presentation contains more than 600 screens including full-motion videos showing chemical reactions in progress, animations and other presentations of key concepts and experiments, an interactive periodic table of the elements, and narrated problem-solving hints and suggestions.

We heartily recommend the **Study Guide to General Chemistry** by Raymond E. Davis and James Petrich, the **Solutions Manual** by Wendy Keeney-Kennicutt and Yi-Noo Tang, **Problem Solving for General Chemistry** by Leslie Kinsland, and **The Student Lecture Outline** by Charles Atwood, Kenneth W. Whitten, and Richard Hedges, all of which were written to accompany this text. The STUDY GUIDE provides an overview of each chapter and emphasizes the threads of continuity that run through chemistry. It lists study goals, tells you which ideas are most important and why they are important, and provides many forward and backward references. Additionally, the STUDY GUIDE contains many easy to moderately difficult questions that enable you to gauge your progress. These short questions provide excellent practice in preparing for examinations. Answers are provided for all questions, and many have explanations or references to appropriate sections in the text.

The SOLUTIONS MANUAL contains detailed solutions and answers to all even-numbered end-of-chapter exercises. It also has many helpful references to appropriate sections and illustrative examples in the text.

PROBLEM SOLVING FOR GENERAL CHEMISTRY provides a valuable study tool if you want more practice with problem solving. It contains a brief discussion of appropriate topics, a variety of examples to illustrate the topics, a set of exercises coded by topic, miscellaneous exercises, and answers to all exercises.

The STUDENT LECTURE OUTLINE helps you organize material in the text and serves as a helpful classroom note-taking supplement so that you can pay more attention to the lecture. It is packaged with a PowerPoint CD-ROM.

If you have suggestions for improving this text, please write to us and tell us about them. You can contact us by email via **http://www.brookscole.com/chemistry** (the Brooks/ Cole College website).

MOLECULAR ART

This edition contains extensive new molecular art, most of it computer generated. Some examples of the ways in which molecular art is used in this edition are

1. **Structures or reactions.** Molecular art is used to give a molecular-level view of a concept being discussed, as in the following interpretation of a balanced chemical equation.

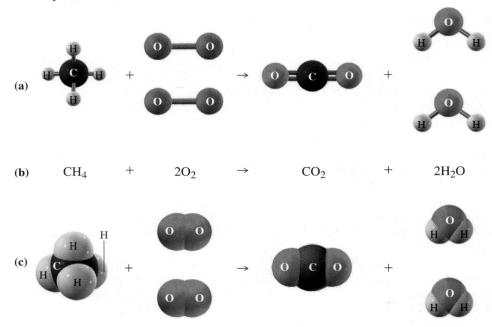

Figure 2-6 on page 53 uses this art to illustrate two common representations of molecules—ball-and-stick and space-filling models.

2. **Macro–micro art.** Molecular art presented together with a photograph of a sample or an experiment clarifies the molecular behavior.

3. **Electrostatic charge potential (ECP) figures.** Charge distributions within molecules are illustrated with colorful ECP figures. These will help you visualize their effects on molecular properties and intermolecular interactions.

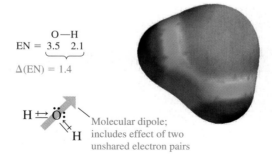

KEYS FOR COLOR CODES

In addition to full-color photography and art, we have used color to help you identify and organize important ideas, techniques, and concepts as you study this book.

1. Important ideas, mathematical relationships, and summaries are displayed on blue screens, the width of the text.

> There is no observable change in the quantity of matter during a chemical reaction or during a physical change.

2. Answers to examples are shown on blue screens. Intermediate steps (logic, guidance, and so on) are shown on gold screens.

EXAMPLE 1-3   *Unit Factors*

Express 1.47 mi in inches.

Plan

First we write down the units of what we wish to know, preceded by a question mark. Then we set it equal to whatever we are given:

$$? \text{ inches} = 1.47 \text{ miles}$$

Then we choose unit factors to convert the given units (miles) to the desired units (inches):

$$\text{miles} \longrightarrow \text{feet} \longrightarrow \text{inches}$$

Solution

$$? \text{ in.} = 1.47 \text{ mi} \times \frac{5280 \text{ ft}}{1 \text{ mi}} \times \frac{12 \text{ in.}}{1 \text{ ft}} = 9.31 \times 10^4 \text{ in.} \quad \text{(calculator gives 93139.2)}$$

Note that both miles and feet cancel, leaving only inches, the desired unit. Thus, there is no ambiguity as to how the unit factors should be written. The answer contains three significant figures because there are three significant figures in 1.47 miles.

---

3. Acidic and basic properties are contrasted by using pink and blue, respectively. Neutral solutions are indicated in pale purple.

TABLE 19-5	*Titration Data for 100.0 mL of 0.100 M HCl versus NaOH*		
mL of 0.100 M NaOH Added	**mmol NaOH Added**	**mmol Excess Acid or Base**	**pH**
0.0	0.00	10.0 H_3O^+	1.00
20.0	2.00	8.0	1.18
50.0	5.00	5.0	1.48
90.0	9.00	1.0	2.28
99.0	9.90	0.10	3.30
99.5	9.95	0.05	3.60
100.0	10.00	0.00 (eq. pt.)	7.00
100.5	10.05	0.05 OH^-	10.40
110.0	11.00	1.00	11.68
120.0	12.00	2.00	11.96

4. Red and blue are used in oxidation–reduction reactions and electrochemistry.
   (a) Oxidation numbers are shown in red circles to avoid confusion with ionic charges. Oxidation is indicated by blue and reduction is indicated by red.

$$2[\text{Ag}^+(aq) + \text{NO}_3^-(aq)] + \text{Cu}(s) \longrightarrow [\text{Cu}^{2+}(aq) + 2\text{NO}_3^-(aq)] + 2\text{Ag}(s)$$

(+1)   (+5)(−2)   (0)   (+2)   (+5)(−2)   (0)

+2

−1

The nitrate ions, NO_3^-, are spectator ions. Canceling them from both sides gives the net ionic equation:

$$\overset{+1}{2Ag^+(aq)} + \overset{0}{Cu(s)} \longrightarrow \overset{+2}{Cu^{2+}(aq)} + \overset{0}{2Ag(s)}$$

This is a redox equation. The oxidation number of silver decreases from +1 to zero; silver ion is reduced and is the oxidizing agent. The oxidation number of copper increases from zero to +2; copper is oxidized and is the reducing agent.

(b) In electrochemistry (Chapter 21), we learn that oxidation occurs at the *anode;* we use blue to indicate the anode and its half-reaction. Similarly, reduction occurs at the *cathode;* we use red to indicate the cathode and its half-reaction.

$$
\begin{array}{ll}
2Cl^- \longrightarrow Cl_2(g) + 2e^- & \text{(oxidation, anode half-reaction)} \\
2[Na^+ + e^- \longrightarrow Na(\ell)] & \text{(reduction, cathode half-reaction)} \\
\hline
\underbrace{2Na^+ + 2Cl^-}_{2NaCl(\ell)} \longrightarrow 2Na(\ell) + Cl_2(g) & \text{(overall cell reaction)}
\end{array}
$$

5. Atomic orbitals are shown in yellow. Bonding and antibonding molecular orbitals are shown in blue and red, respectively.

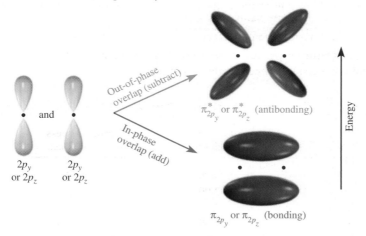

6. Hybridization schemes and hybrid orbitals are emphasized in green.

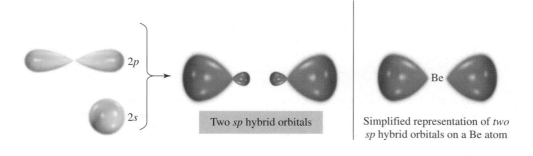

7. Electrostatic Charge Potential (ECP) representations emphasize the distribution of charge in a molecule. In these drawings, the charge is shown on a color scale ranging from red (most negative) through green (neutral) to blue (most positive).

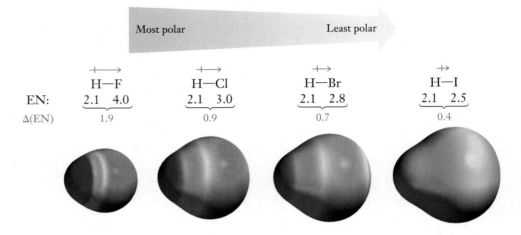

8. Color-coded periodic tables emphasize the classification of the elements as metals (blue), nonmetals (yellow), and metalloids (green). Please study the periodic table inside the front cover carefully so that you recognize this color scheme.

The Foundations of Chemistry

1

Charles D. Winters

The earth is a huge chemical system. Many different chemicals can be found in the earth's atmosphere. It is our responsibility to keep harmful chemicals from polluting that system. We depend upon it for the oxygen we breathe, the carbon dioxide and nitrogen plants consume, and the water clouds transport. Maintaining life on the planet requires understanding and intelligent use of these resources. Scientists can provide important information about the processes, but each of us must share in the responsibility for our environment.

OUTLINE

OBJECTIVES

After you have studied this chapter, you should be able to

- Use the basic vocabulary of matter and energy
- Distinguish between chemical and physical properties and between chemical and physical changes
- Recognize various forms of matter: homogeneous and heterogeneous mixtures, substances, compounds, and elements
- Apply the concept of significant figures
- Apply appropriate units to describe the results of measurement
- Use the unit factor method to carry out conversions among units
- Describe temperature measurements on various common scales, and convert between these scales
- Carry out calculations relating temperature change to heat absorbed or liberated

Thousands of practical questions are studied by chemists. A few of them are

How can we modify a useful drug so as to improve its effectiveness while minimizing any harmful or unpleasant side effects?

How can we develop better materials to be used as synthetic organs for replacement surgery?

Which substances could help to avoid rejection of foreign tissue in organ transplants?

What improvements in fertilizers or pesticides can increase agricultural yields? How can this be done with minimal environmental danger?

How can we get the maximum work from a fuel while producing the least harmful emissions possible?

Which really poses the greater environmental threat—the burning of fossil fuels and the resulting contribution to the greenhouse effect and climatic change, or the use of nuclear power and the related radiation and disposal problems?

How can we develop suitable materials for the semiconductor and microelectronics industry? Can we develop a battery that is cheaper, lighter, and more powerful?

What changes in structural materials could help to make aircraft lighter and more economical, yet at the same time stronger and safer?

What relationship is there between the substances we eat, drink, or breathe and the possibility of developing cancer? How can we develop substances that are effective in killing cancer cells preferentially over normal cells?

Can we economically produce fresh water from sea water for irrigation or consumption?

How can we slow down unfavorable reactions, such as the corrosion of metals, while speeding up favorable ones, such as the growth of foodstuffs?

Chemistry touches almost every aspect of our lives, our culture, and our environment. Its scope encompasses the air we breathe, the food we eat, the fluids we drink, the clothing we wear, the dwellings we live in, and the transportation and fuel supplies we use, as well as our fellow creatures.

> Chemistry is the science that describes matter—its properties, the changes it undergoes, and the energy changes that accompany those processes.

Enormous numbers of chemical reactions are necessary to produce a human being.

Matter includes everything that is tangible, from our bodies and the stuff of our everyday lives to the grandest objects in the universe. Some call chemistry the central science. It rests on the foundation of mathematics and physics and in turn underlies the life sciences—biology and medicine. To understand living systems fully, we must first understand the chemical reactions and chemical influences that operate within them. The chemicals of our bodies profoundly affect even the personal world of our thoughts and emotions.

No one can be expert in all aspects of such a broad science as chemistry. Sometimes we arbitrarily divide the study of chemistry into various branches. Carbon is very versatile in its bonding and behavior and is a key element in many substances that are essential to life. All living matter contains compounds with carbon combined with hydrogen and sometimes with a few other elements such as oxygen, nitrogen, and sulfur. **Organic chemistry** is the study of all such compounds. **Inorganic chemistry** is the study of all other compounds, but also includes some of the simpler carbon-containing compounds such as carbon monoxide, carbon dioxide, carbonates, and bicarbonates. (In the early days of chemistry, living matter and inanimate matter were believed to be entirely different. We now know that many of the compounds found in living matter can be made from nonliving, or "inorganic," sources. Thus, the terms "organic" and "inorganic" have different meanings than they did originally.) The branch of chemistry that is concerned with the detection or identification of substances present in a sample (*qualitative analysis*) or with the amount of each that is present (*quantitative analysis*) is called **analytical chemistry. Physical chemistry** applies the mathematical theories and methods of physics to the properties of matter and to the study of chemical processes and the accompanying energy changes. As its name suggests, **biochemistry** is the study of the chemistry of processes in living organisms. Such divisions are arbitrary, and most chemical studies involve more than one of these traditional areas of chemistry. The principles you will learn in a general chemistry course are the foundation of all branches of chemistry.

We understand simple chemical systems well; they lie near chemistry's fuzzy boundary with physics. They can often be described exactly by mathematical equations. We fare less well with more complicated systems. Even where our understanding is fairly thorough, we must make approximations, and often our knowledge is far from complete. Each year researchers provide new insights into the nature of matter and its interactions. As chemists find answers to old questions, they learn to ask new ones. Our scientific knowledge has been described as an expanding sphere that, as it grows, encounters an ever-enlarging frontier.

In our search for understanding, we eventually must ask fundamental questions, such as the following:

How do substances combine to form other substances? How much energy is involved in changes that we observe?

How is matter constructed in its intimate detail? How are atoms and the ways that they combine related to the properties of the matter that we can measure, such as color, hardness, chemical reactivity, and electrical conductivity?

What fundamental factors influence the stability of a substance? How can we force a desired (but energetically unfavorable) change to take place? What factors control the rate at which a chemical change takes place?

In your study of chemistry, you will learn about these and many other basic ideas that chemists have developed to help them describe and understand the behavior of matter. Along the way, we hope that you come to appreciate the development of this science, one of the grandest intellectual achievements of human endeavor. You will also learn how to apply these fundamental principles to solve real problems. One of your major goals in the study of chemistry should be to develop your ability to think critically and to solve problems (not just do numerical calculations!). In other words, you need to learn to manipulate not only numbers, but also quantitative ideas, words, and concepts.

In the first chapter, our main goals are (1) to begin to get an idea of what chemistry is about and the ways in which chemists view and describe the material world and (2) to acquire some skills that are useful and necessary in the understanding of chemistry, its contribution to science and engineering, and its role in our daily lives.

1-1 MATTER AND ENERGY

Matter is anything that has mass and occupies space. **Mass** is a measure of the quantity of matter in a sample of any material. The more massive an object is, the more force is required to put it in motion. All bodies consist of matter. Our senses of sight and touch usually tell us that an object occupies space. In the case of colorless, odorless, tasteless gases (such as air), our senses may fail us.

Energy is defined as the capacity to do work or to transfer heat. We are familiar with many forms of energy, including mechanical energy, light energy, electrical energy, and heat energy. Light energy from the sun is used by plants as they grow, electrical energy allows us to light a room by flicking a switch, and heat energy cooks our food and warms our homes. Energy can be classified into two principal types: kinetic energy and potential energy.

A body in motion, such as a rolling boulder, possesses energy because of its motion. Such energy is called **kinetic energy.** Kinetic energy represents the capacity for doing work directly. It is easily transferred between objects. **Potential energy** is the energy an object possesses because of its position, condition, or composition. Coal, for example, possesses chemical energy, a form of potential energy, because of its composition. Many electrical generating plants burn coal, producing heat, which is converted to electrical energy. A

We might say that we can "touch" air when it blows in our faces, but we depend on other evidence to show that a still body of air fits our definition of matter.

The term "kinetic" comes from the Greek word *kinein,* meaning "to move." The word "cinema" is derived from the same Greek word.

Nuclear energy is an important kind of potential energy.

boulder located atop a mountain possesses potential energy because of its height. It can roll down the mountainside and convert its potential energy into kinetic energy. We discuss energy because all chemical processes are accompanied by energy changes. As some processes occur, energy is released to the surroundings, usually as heat energy. We call such processes **exothermic.** Any combustion (burning) reaction is exothermic. Some chemical reactions and physical changes, however, are **endothermic;** that is, they absorb energy from their surroundings. An example of a physical change that is endothermic is the melting of ice.

The Law of Conservation of Matter

When we burn a sample of metallic magnesium in oxygen, the magnesium combines with the oxygen (Figure 1-1) to form magnesium oxide, a white powder. This chemical reaction is accompanied by the release of large amounts of heat energy and light energy. When we weigh the product of the reaction, magnesium oxide, we find that it is heavier than the original piece of magnesium. The increase in the mass of a solid is due to the combination of oxygen with magnesium to form magnesium oxide. Many experiments have shown that the mass of the magnesium oxide is exactly the sum of the masses of magnesium and oxygen that combined to form it. Similar statements can be made for all chemical reactions. These observations are summarized in the **Law of Conservation of Matter:**

> There is no observable change in the quantity of matter during a chemical reaction or during a physical change.

This statement is an example of a **scientific (natural) law,** a general statement based on the observed behavior of matter to which no exceptions are known. A nuclear reaction is *not* a chemical reaction.

The Law of Conservation of Energy

In exothermic chemical reactions, *chemical energy* is usually converted into *heat energy.* Some exothermic processes involve other kinds of energy changes. For example, some liberate light energy without heat, and others produce electrical energy without heat or light. In *endothermic* reactions, heat energy, light energy, or electrical energy is converted into chemical energy. Although chemical changes always involve energy changes, some energy transformations do not involve chemical changes at all. For example, heat energy may be converted into electrical energy or into mechanical energy without any simultaneous chemical changes. Many experiments have demonstrated that all the energy involved in any chemical or physical change appears in some form after the change. These observations are summarized in the **Law of Conservation of Energy:**

> Energy cannot be created or destroyed in a chemical reaction or in a physical change. It can only be converted from one form to another.

The Law of Conservation of Matter and Energy

With the dawn of the nuclear age in the 1940s, scientists, and then the world, became aware that matter can be converted into energy. In nuclear reactions (Chapter 26), matter

Figure 1-1 Magnesium burns in oxygen to form magnesium oxide, a white solid. The mass of magnesium oxide formed is equal to the sum of the masses of oxygen and magnesium that formed it.

Charles D. Winters

Electricity is produced in hydroelectric plants by the conversion of mechanical energy (from flowing water) into electrical energy.

is transformed into energy. The relationship between matter and energy is given by Albert Einstein's now famous equation

$$E = mc^2$$

This equation tells us that the amount of energy released when matter is transformed into energy is the product of the mass of matter transformed and the speed of light squared. At the present time, we have not (knowingly) observed the transformation of energy into matter on a large scale. It does, however, happen on an extremely small scale in "atom smashers," or particle accelerators, used to induce nuclear reactions. Now that the equivalence of matter and energy is recognized, the **Law of Conservation of Matter and Energy** can be stated in a single sentence:

The combined amount of matter and energy in the universe is fixed.

1-2 STATES OF MATTER

Matter can be classified into three states (Figure 1-2), although most of us can think of examples that do not fit neatly into any of the three categories. In the **solid** state, substances are rigid and have definite shapes. Volumes of solids do not vary much with changes in temperature and pressure. In many solids, called crystalline solids, the individual particles that make up the solid occupy definite positions in the crystal structure. The strengths of interaction between the individual particles determine how hard and how strong the crystals are. In the **liquid** state, the individual particles are confined to a given volume. A liquid flows and assumes the shape of its container up to the volume of the liquid. Liquids are very hard to compress. **Gases** are much less dense than liquids and solids. They occupy all parts of any vessel in which they are confined. Gases are capable of infinite expansion and are compressed easily. We conclude that they consist primarily of empty space, meaning that the individual particles are quite far apart.

1-3 CHEMICAL AND PHYSICAL PROPERTIES

To distinguish among samples of different kinds of matter, we determine and compare their **properties.** We recognize different kinds of matter by their properties, which are broadly classified into chemical properties and physical properties.

Chemical properties are exhibited by matter as it undergoes changes in composition. These properties of substances are related to the kinds of chemical changes that the substances undergo. For instance, we have already described the combination of metallic magnesium with gaseous oxygen to form magnesium oxide, a white powder. A chemical property of magnesium is that it can combine with oxygen, releasing energy in the process. A chemical property of oxygen is that it can combine with magnesium.

All substances also exhibit **physical properties** that can be observed in the *absence of any change in composition.* Color, density, hardness, melting point, boiling point, and electrical and thermal conductivities are physical properties. Some physical properties of a substance depend on the conditions, such as temperature and pressure, under which they are measured. For instance, water is a solid (ice) at low temperatures but is a liquid at higher

Einstein formulated this equation in 1905 as a part of his theory of relativity. Its validity was demonstrated in 1939 with the first controlled nuclear reaction.

sciLINKS. **TOPIC:** Matter and Energy
GO TO: www.scilinks.org
sciLINKS **CODE:** WCH0110

See the General Chemistry Interactive CD-ROM, Version 3.0, *Screen 1.3, States of Matter.*

The properties of a person include height, weight, sex, skin and hair color, and the many subtle features that constitute that person's general appearance.

See the General Chemistry Interactive CD-ROM, Version 3.0, *Screen 1.2, Physical Properties of Matter.*

Charles Steele

Property	Solid	Liquid	Gas
Rigidity	Rigid	Flows and assumes shape of container	Fills any container completely
Expansion on heating	Slight	Slight	Expands infinitely
Compressibility	Slight	Slight	Easily compressed

Figure 1-2 A comparison of some physical properties of the three states of matter. (*left*) Iodine, a solid element. (*center*) Bromine, a liquid element. (*right*) Chlorine, a gaseous element.

temperatures. At still higher temperatures, it is a gas (steam). As water is converted from one state to another, its composition is constant. Its chemical properties change very little. On the other hand, the physical properties of ice, liquid water, and steam are different (Figure 1-3).

Properties of matter can be further classified according to whether or not they depend on the *amount* of substance present. The volume and the mass of a sample depend on, and are directly proportional to, the amount of matter in that sample. Such properties, which depend on the amount of material examined, are called **extensive properties.** By contrast, the color and the melting point of a substance are the same for a small sample and for a large one. Properties such as these, which are independent of the amount of material examined, are called **intensive properties.** All chemical properties are intensive properties.

Because no two different substances have identical sets of chemical and physical properties under the same conditions, we are able to identify and distinguish among different substances. For instance, water is the only clear, colorless liquid that freezes at 0°C, boils at 100°C at one atmosphere of pressure, dissolves a wide variety of substances (e.g., copper(II)

www Many compilations of chemical and physical properties of matter can be found on the World Wide Web. One site is maintained by the U.S. National Institute of Standards and Technology (NIST) at **http://webbook.nist.gov** Perhaps you can find other sites.

SCiLINKS.
TOPIC: Phys/Chem Properties
GO TO: www.scilinks.org
*sci*LINKS CODE: WCH0120

Figure 1-3 Physical changes that occur among the three states of matter. *Sublimation* is the conversion of a solid directly to a gas without passing through the liquid state; the reverse of that process is called *deposition*. The changes shown in blue are endothermic (absorb heat); those shown in red are exothermic (release heat). Water is a substance that is familiar to us in all three physical states. The molecules are close together in the solid and the liquid but far apart in the gas. The molecules in the solid are relatively fixed in position, but those in the liquid and gas can flow around each other.

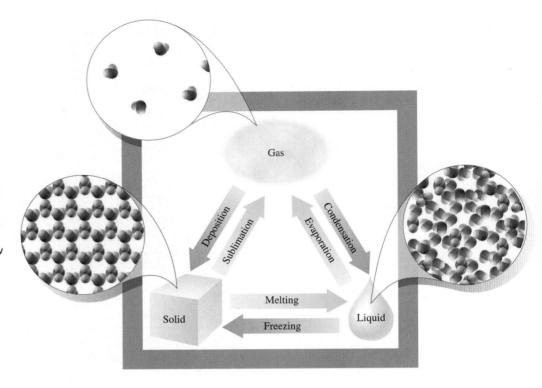

(a)   **(b)**   **(c)**   **(d)**

Figure 1-4 Some physical and chemical properties of water. *Physical:* (a) It melts at 0°C; (b) it boils at 100°C (at normal atmospheric pressure); (c) it dissolves a wide range of substances, including copper(II) sulfate, a blue solid. *Chemical:* (d) It reacts violently with sodium to form hydrogen gas and sodium hydroxide.

TABLE 1-1	Physical Properties of a Few Common Substances (at 1 atm pressure)				
			Solubility at 25°C (g/100 g)		
Substance	**Melting Point (°C)**	**Boiling Point (°C)**	**In water**	**In ethyl alcohol**	**Density (g/cm³)**
acetic acid	16.6	118.1	infinite	infinite	1.05
benzene	5.5	80.1	0.07	infinite	0.879
bromine	−7.1	58.8	3.51	infinite	3.12
iron	1530	3000	insoluble	insoluble	7.86
methane	−182.5	−161.5	0.0022	0.033	6.67×10^{-4}
oxygen	−218.8	−183.0	0.0040	0.037	1.33×10^{-3}
sodium chloride	801	1473	36.5	0.065	2.16
water	0	100	—	infinite	1.00

sulfate), and reacts violently with sodium (Figure 1-4). Table 1-1 compares several physical properties of a few substances. A sample of any of these substances can be distinguished from the others by observing their properties.

One atmosphere of pressure is the average atmospheric pressure at sea level.

1-4 CHEMICAL AND PHYSICAL CHANGES

We described the reaction of magnesium as it burns in the oxygen of the air (see Figure 1-1). This reaction is a *chemical change*. In any **chemical change,** (1) one or more substances are used up (at least partially), (2) one or more new substances are formed, and (3) energy is absorbed or released. As substances undergo chemical changes, they demonstrate their chemical properties. A **physical change,** on the other hand, occurs with *no change in chemical composition*. Physical properties are usually altered significantly as matter undergoes physical changes (Figure 1-3). In addition, a physical change *may* suggest that a chemical change has also taken place. For instance, a color change, a warming, or the formation of a solid when two solutions are mixed could indicate a chemical change.

Energy is always released or absorbed when chemical or physical changes occur. Energy is required to melt ice, and energy is required to boil water. Conversely, the condensation of steam to form liquid water always liberates energy, as does the freezing of liquid water to form ice. The changes in energy that accompany these physical changes for water are shown in Figure 1-5. At a pressure of one atmosphere, ice always melts at the same temperature (0°C), and pure water always boils at the same temperature (100°C).

CD-ROM Screens 1.11, Chemical Change, and 1.12, Chemical Change on the Molecular Scale.

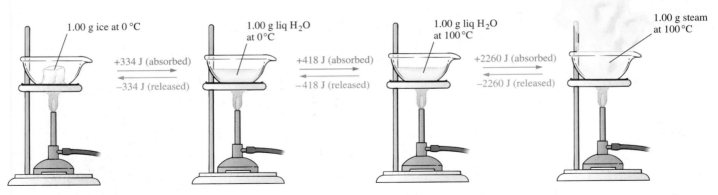

Figure 1-5   Changes in energy that accompany some physical changes for water. The energy unit joules (J) is defined in Section 1-13.

1-5   MIXTURES, SUBSTANCES, COMPOUNDS, AND ELEMENTS

By "composition of a mixture," we mean both the identities of the substances present and their relative amounts in the mixture.

The blue copper(II) sulfate solution in Figure 1-4c is a homogeneous mixture.

A heterogeneous mixture of two minerals: galena (black) and quartz (white).

Mixtures are combinations of two or more pure substances in which each substance retains its own composition and properties. Almost every sample of matter that we ordinarily encounter is a mixture. The most easily recognized type of mixture is one in which different portions of the sample have recognizably different properties. Such a mixture, which is not uniform throughout, is called **heterogeneous.** Examples include mixtures of salt and charcoal (in which two components with different colors can be distinguished readily from each other by sight), foggy air (which includes a suspended mist of water droplets), and vegetable soup. Another kind of mixture has uniform properties throughout; such a mixture is described as a **homogeneous mixture** and is also called a **solution.** Examples include salt water; some *alloys*, which are homogeneous mixtures of metals in the solid state; and air (free of particulate matter or mists). Air is a mixture of gases. It is mainly nitrogen, oxygen, argon, carbon dioxide, and water vapor. There are only trace amounts of other substances in the atmosphere.

An important characteristic of all mixtures is that they can have variable composition. (For instance, we can make an infinite number of different mixtures of salt and sugar by varying the relative amounts of the two components used.) Consequently, repeating the same experiment on mixtures from different sources may give different results, whereas the same treatment of a pure sample will always give the same results. When the distinction between homogeneous mixtures and pure substances was realized and methods were developed (in the late 1700s) for separating mixtures and studying pure substances, consistent results could be obtained. This resulted in reproducible chemical properties, which formed the basis of real progress in the development of chemical theory.

Mixtures can be separated by physical means because each component retains its properties (Figures 1-6 and 1-7). For example, a mixture of salt and water can be separated by evaporating the water and leaving the solid salt behind. To separate a mixture of sand and salt, we could treat it with water to dissolve the salt, collect the sand by filtration, and then evaporate the water to reclaim the solid salt. Very fine iron powder can be mixed with powdered sulfur to give what appears to the naked eye to be a homogeneous mixture of

CHEMISTRY IN USE

The Development of Science

The Resources of the Ocean

As is apparent to anyone who has swum in the ocean, sea water is not pure water but contains a large amount of dissolved solids. In fact, each cubic kilometer of sea water contains about 3.6×10^{10} kilograms of dissolved solids. Nearly 71% of the earth's surface is covered with water. The oceans cover an area of 361 million square kilometers at an average depth of 3729 meters and hold approximately 1.35 billion cubic kilometers of water. This means that the oceans contain a total of more than 4.8×10^{21} kilograms of dissolved material (or more than 100,000,000,000,000,000,000 pounds). Rivers flowing into the oceans and submarine volcanoes constantly add to this storehouse of minerals. The formation of sediment and the biological demands of organisms constantly remove a similar amount.

Sea water is a very complicated solution of many substances. The main dissolved component of sea water is sodium chloride, common salt. Besides sodium and chlorine, the main elements in sea water are magnesium, sulfur, calcium, potassium, bromine, carbon, nitrogen, and strontium. Together these 10 elements make up more than 99% of the dissolved materials in the oceans. In addition to sodium chloride, they combine to form such compounds as magnesium chloride, potassium sulfate, and calcium carbonate (lime). Animals absorb the latter from the sea and build it into bones and shells.

Many other substances exist in smaller amounts in sea water. In fact, most of the 92 naturally occurring elements have been measured or detected in sea water, and the remainder will probably be found as more sensitive analytical techniques become available. There are staggering amounts of valuable metals in sea water, including approximately 1.3×10^{11} kilograms of copper, 4.2×10^{12} kilograms of uranium, 5.3×10^{9} kilograms of gold, 2.6×10^{9} kilograms of silver, and 6.6×10^{8} kilograms of lead. Other elements of economic importance include 2.6×10^{12} kilograms of aluminum, 1.3×10^{10} kilograms of tin, 26×10^{11} kilograms of manganese, and 4.0×10^{10} kilograms of mercury.

One would think that with such a large reservoir of dissolved solids, considerable "chemical mining" of the ocean would occur. At present, only four elements are commercially extracted in large quantities. They are sodium and chlorine, which are produced from the sea by solar evaporation; magnesium; and bromine. In fact, most of the U.S. pro-

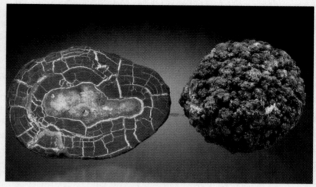

© Charles D. Winters/Photo Researchers, Inc.

duction of magnesium is derived from sea water, and the ocean is one of the principal sources of bromine. Most of the other elements are so thinly scattered through the ocean that the cost of their recovery would be much higher than their economic value. However, it is probable that as resources become more and more depleted from the continents, and as recovery techniques become more efficient, mining of sea water will become a much more desirable and feasible prospect.

One promising method of extracting elements from sea water uses marine organisms. Many marine animals concentrate certain elements in their bodies at levels many times higher than the levels in sea water. Vanadium, for example, is taken up by the mucus of certain tunicates and can be concentrated in these animals to more than 280,000 times its concentration in sea water. Other marine organisms can concentrate copper and zinc by a factor of about 1 million. If these animals could be cultivated in large quantities without endangering the ocean ecosystem, they could become a valuable source of trace metals.

In addition to dissolved materials, sea water holds a great store of suspended particulate matter that floats through the water. Some 15% of the manganese in sea water is present in particulate form, as are appreciable amounts of lead and iron. Similarly, most of the gold in sea water is thought to adhere to the surfaces of clay minerals in suspension. As in the case of dissolved solids, the economics of filtering these very fine particles from sea water are not favorable at present. However, because many of the particles suspended in sea water carry an electric charge, ion exchange techniques and modifications of electrostatic processes may someday provide important methods for the recovery of trace metals.

(a)   **(b)**

Figure 1-6   (a) A mixture of iron and sulfur is a *heterogeneous* mixture. (b) Like any mixture, it can be separated by physical means, such as removing the iron with a magnet.

the two. Separation of the components of this mixture is easy, however. The iron may be removed by a magnet, or the sulfur may be dissolved in carbon disulfide, which does not dissolve iron (Figure 1-6).

In *any* mixture, (1) the composition can be varied and (2) each component of the mixture retains its own properties.

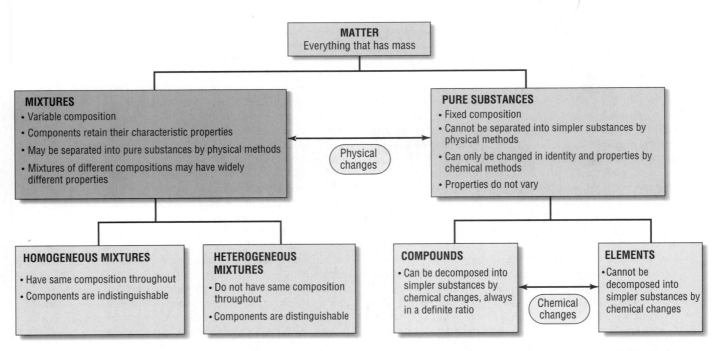

Figure 1-7   One scheme for classification of matter. Arrows indicate the general means by which matter can be separated.

Imagine that we have a sample of muddy river water (a heterogeneous mixture). We might first separate the suspended dirt from the liquid by filtration. Then we could remove dissolved air by warming the water. Dissolved solids might be removed by cooling the sample until some of it freezes, pouring off the liquid, and then melting the ice. Other dissolved components might be separated by distillation or other methods. Eventually we would obtain a sample of pure water that could not be further separated by any physical separation methods. No matter what the original source of the impure water—the ocean, the Mississippi River, a can of tomato juice—water samples obtained by purification all have identical composition, and, under identical conditions, they all have identical properties. Any such sample is called a substance, or sometimes a pure substance.

> A **substance** cannot be further broken down or purified by physical means. A substance is matter of a particular kind. Each substance has its own characteristic properties that are different from the set of properties of any other substance.

Now suppose we decompose some water by passing electricity through it (Figure 1-8). (An *electrolysis* process is a chemical reaction.) We find that the water is converted into two simpler substances, hydrogen and oxygen; more significantly, hydrogen and oxygen are

The first ice that forms is quite pure. The dissolved solids tend to stay behind in the remaining liquid.

If we use the definition given here of a *substance,* the phrase *pure substance* may appear to be redundant.

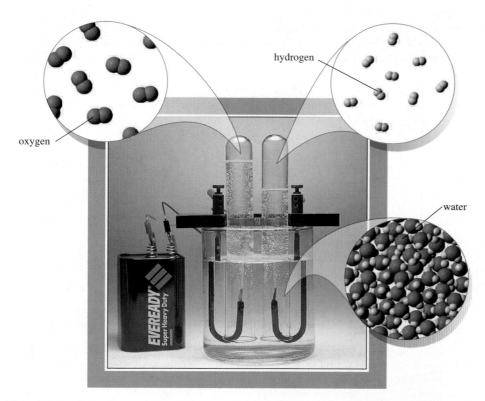

Figure 1-8 Electrolysis apparatus for small-scale chemical decomposition of water by electrical energy. The volume of hydrogen produced (*right*) is twice that of oxygen (*left*). Some dilute sulfuric acid is added to increase the conductivity.

always present in the same ratio by mass, 11.1% to 88.9%. These observations allow us to identify water as a compound.

> A **compound** is a substance that can be decomposed by chemical means into simpler substances, always in the same ratio by mass.

As we continue this process, starting with any substance, we eventually reach a stage at which the new substances formed cannot be further broken down by chemical means. The substances at the end of this chain are called elements.

> An **element** is a substance that cannot be decomposed into simpler substances by chemical changes.

SC*LINKS*.   TOPIC: Mixtures
              GO TO: www.scilinks.org
              *sci*LINKS CODE: WCH0130

For instance, neither of the two gases obtained by the electrolysis of water—hydrogen and oxygen—can be further decomposed, so we know that they are elements.

As another illustration (Figure 1-9), pure calcium carbonate (a white solid present in limestone and seashells) can be broken down by heating to give another white solid (call it A) and a gas (call it B) in the mass ratio 56.0:44.0. This observation tells us that calcium carbonate is a compound. The white solid A obtained from calcium carbonate can be further broken down into a solid and a gas in a definite ratio by mass, 71.5:28.5. But neither of these can be further decomposed, so they must be elements. The gas is identical to the oxygen obtained from the electrolysis of water; the solid is a metallic element called calcium. Similarly, the gas B, originally obtained from calcium carbonate, can be decomposed into two elements, carbon and oxygen, in a fixed mass ratio, 27.3:72.7. This sequence illustrates that a compound can be broken apart into simpler substances at a fixed mass ratio; those simpler substances may be either elements or simpler compounds.

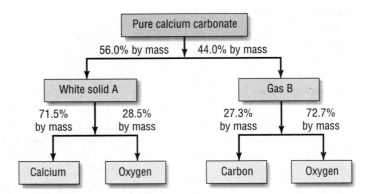

Figure 1-9 Diagram of the decomposition of calcium carbonate to give a white solid A (56.0% by mass) and a gas B (44.0% by mass). This decomposition into simpler substances at a fixed ratio proves that calcium carbonate is a compound. The white solid A further decomposes to give the elements calcium (71.5% by mass) and oxygen (28.5% by mass). This proves that the white solid A is a compound; it is known as calcium oxide. The gas B also can be broken down to give the elements carbon (27.3% by mass) and oxygen (72.7% by mass). This establishes that gas B is a compound; it is known as carbon dioxide.

Furthermore, we may say that *a compound is a pure substance consisting of two or more different elements in a fixed ratio.* Water is 11.1% hydrogen and 88.9% oxygen by mass. Similarly, carbon dioxide is 27.3% carbon and 72.7% oxygen by mass, and calcium oxide (the white solid A in the previous discussion) is 71.5% calcium and 28.5% oxygen by mass. We could also combine the numbers in the previous paragraph to show that calcium carbonate is 40.1% calcium, 12.0% carbon, and 47.9% oxygen by mass. Observations such as these on innumerable pure compounds led to the statement of the **Law of Definite Proportions** (also known as the **Law of Constant Composition**):

> Different samples of any pure compound contain the same elements in the same proportions by mass.

The physical and chemical properties of a compound are different from the properties of its constituent elements. Sodium chloride is a white solid that we ordinarily use as table salt (Figure 1-10). This compound is formed by the combination of the element sodium (a soft, silvery white metal that reacts violently with water; see Figure 1-4d) and the element chlorine (a pale green, corrosive, poisonous gas; see Figure 1-2c).

Recall that elements are substances that cannot be decomposed into simpler substances by chemical changes. Nitrogen, silver, aluminum, copper, gold, and sulfur are other examples of elements.

We use a set of **symbols** to represent the elements. These symbols can be written more quickly than names, and they occupy less space. The symbols for the first 109 elements consist of either a capital letter *or* a capital letter and a lowercase letter, such as C (carbon) or Ca (calcium). A list of the known elements and their symbols is given inside the front cover.

In the past, the discoverers of elements claimed the right to name them although the question of who had actually discovered the elements first was sometimes disputed. In modern times, new elements are given temporary names and three-letter symbols based on a numerical system. These designations are used until the question of the right to name the newly discovered elements is resolved. Decisions resolving the names of elements 104 through 109 were announced in 1997 by the International Union of Pure and Applied Chemistry (IUPAC), an international organization that represents chemical societies from 40 countries. IUPAC makes recommendations regarding many matters of convention and terminology in chemistry. These recommendations carry no legal force, but they are normally viewed as authoritative throughout the world.

A short list of symbols of common elements is given in Table 1-2. Many symbols consist of the first one or two letters of the element's English name. Some are derived from the element's Latin name (indicated in parentheses in Table 1-2) and one, W for tungsten, is from the German *Wolfram.* You should learn the list in Table 1-2. Names and symbols for additional elements should be learned as they are encountered.

Most of the earth's crust is made up of a relatively small number of elements. Only 10 of the 88 naturally occurring elements make up more than 99% by mass of the earth's crust, oceans, and atmosphere (Table 1-3). Oxygen accounts for roughly half. Relatively few elements, approximately one fourth of the naturally occurring ones, occur in nature as free elements. The rest are always found chemically combined with other elements.

A very small amount of the matter in the earth's crust, oceans, and atmosphere is involved in living matter. The main element in living matter is carbon, but only a tiny fraction of the carbon in the environment occurs in living organisms. More than one quarter

Charles D. Winters

Figure 1-10 The reaction of sodium, a solid element, and chlorine, a gaseous element, to produce sodium chloride (table salt). This reaction gives off considerable energy in the form of heat and light.

See the essay "The Names of Elements" on page 68.

The other known elements have been made artificially in laboratories, as described in Chapter 26.

Mercury is the only metal that is a liquid at room temperature.

The stable form of sulfur at room temperature is a solid.

TABLE 1-2	Some Common Elements and Their Symbols				
Symbol	**Element**	**Symbol**	**Element**	**Symbol**	**Element**
Ag	silver (*argentum*)	F	fluorine	Ni	nickel
Al	aluminum	Fe	iron (*ferrum*)	O	oxygen
Au	gold (*aurum*)	H	hydrogen	P	phosphorus
B	boron	He	helium	Pb	lead (*plumbum*)
Ba	barium	Hg	mercury (*hydrargyrum*)	Pt	platinum
Bi	bismuth	I	iodine	S	sulfur
Br	bromine	K	potassium (*kalium*)	Sb	antimony (*stibium*)
C	carbon	Kr	krypton	Si	silicon
Ca	calcium	Li	lithium	Sn	tin (*stannum*)
Cd	cadmium	Mg	magnesium	Sr	strontium
Cl	chlorine	Mn	manganese	Ti	titanium
Co	cobalt	N	nitrogen	U	uranium
Cr	chromium	Na	sodium (*natrium*)	W	tungsten (*Wolfram*)
Cu	copper (*cuprum*)	Ne	neon	Zn	zinc

TABLE 1-3	Abundance of Elements in the Earth's Crust, Oceans, and Atmosphere						
Element	**Symbol**	**% by Mass**		**Element**	**Symbol**	**% by Mass**	
oxygen	O	49.5%		chlorine	Cl	0.19%	
silicon	Si	25.7		phosphorus	P	0.12	
aluminum	Al	7.5		manganese	Mn	0.09	
iron	Fe	4.7		carbon	C	0.08	
calcium	Ca	3.4	99.2%	sulfur	S	0.06	0.7%
sodium	Na	2.6		barium	Ba	0.04	
potassium	K	2.4		chromium	Cr	0.033	
magnesium	Mg	1.9		nitrogen	N	0.030	
hydrogen	H	0.87		fluorine	F	0.027	
titanium	Ti	0.58		zirconium	Zr	0.023	
				All others combined		≈0.1%	

of the total mass of the earth's crust, oceans, and atmosphere is made up of silicon, yet it has almost no biological role.

1-6 MEASUREMENTS IN CHEMISTRY

In the next section, we introduce the standards for basic units of measurement. These standards were selected because they are reproducible and unchanging and because they allow us to make precise measurements. The values of fundamental units are arbitrary.[1] In

[1]*Prior to the establishment of the National Bureau of Standards in 1901, at least 50 different distances had been used as "1 foot" in measuring land within New York City. Thus the size of a 100-ft by 200-ft lot in New York City depended on the generosity of the seller and did not necessarily represent the expected dimensions.*

TABLE 1-4	The Seven Fundamental Units of Measurement (SI)	
Physical Property	**Name of Unit**	**Symbol**
length	meter	m
mass	kilogram	kg
time	second	s
electric current	ampere	A
temperature	kelvin	K
luminous intensity	candela	cd
amount of substance	mole	mol

SC/LINKS.

TOPIC: SI Units
GO TO: www.scilinks.org
sciLINKS CODE: WCH0140

the United States, all units of measure are set by the National Institute of Standards and Technology, NIST (formerly the National Bureau of Standards, NBS). Measurements in the scientific world are usually expressed in the units of the metric system or its modernized successor, the International System of Units (SI). The SI, adopted by the National Bureau of Standards in 1964, is based on the seven fundamental units listed in Table 1-4. All other units of measurement are derived from them.

In this text we shall use both metric units and SI units. Conversions between non-SI and SI units are usually straightforward. Appendix C lists some important units of measurement and their relationships to one another. Appendix D lists several useful physical constants. The most frequently used of these appear on the inside back cover.

The metric and SI systems are *decimal systems, in which prefixes are used to indicate fractions and multiples of ten.* The same prefixes are used with all units of measurement. The distances and masses in Table 1-5 illustrate the use of some common prefixes and the relationships among them.

The abbreviation SI comes from the French *le Système International.*

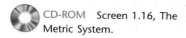 CD-ROM Screen 1.16, The Metric System.

TABLE 1-5	Common Prefixes Used in the SI and Metric Systems		
Prefix	**Abbreviation**	**Meaning**	**Example**
mega-	M	10^6	1 megameter (Mm) = 1×10^6 m
kilo-*	k	10^3	1 kilometer (km) = 1×10^3 m
deci-	d	10^{-1}	1 decimeter (dm) = 1×10^{-1} m
centi-*	c	10^{-2}	1 centimeter (cm) = 1×10^{-2} m
milli-*	m	10^{-3}	1 milligram (mg) = 1×10^{-3} g
micro-*	μ[†]	10^{-6}	1 microgram (μg) = 1×10^{-6} g
nano-*	n	10^{-9}	1 nanogram (ng) = 1×10^{-9} g
pico-	p	10^{-12}	1 picogram (pg) = 1×10^{-12} g

*These prefixes are commonly used in chemistry.
[†]This is the Greek letter μ (pronounced "mew").

The prefixes used in the SI and metric systems may be thought of as *multipliers.* For example, the prefix *kilo-* indicates multiplication by 1000 or 10^3, and *milli-* indicates multiplication by 0.001 or 10^{-3}.

1-7 UNITS OF MEASUREMENT

Mass and Weight

We distinguish between mass and weight. **Mass** is the measure of the quantity of matter a body contains (see Section 1-1). The mass of a body does not vary as its position changes. On the other hand, the **weight** of a body is a measure of the gravitational attraction of the earth for the body, and this varies with distance from the center of the earth. An object weighs very slightly less high up on a mountain than at the bottom of a deep valley. Because the mass of a body does not vary with its position, the mass of a body is a more fundamental property than its weight. We have become accustomed, however, to using the term "weight" when we mean mass, because weighing is one way of measuring mass (Figure 1-11). Because we usually discuss chemical reactions at constant gravity, weight relationships are just as valid as mass relationships. Nevertheless, we should keep in mind that the two are not identical.

The basic unit of mass in the SI system is the **kilogram** (Table 1-6). The kilogram is defined as the mass of a platinum–iridium cylinder stored in a vault in Sèvres, near Paris, France. A 1-lb object has a mass of 0.4536 kg. The basic mass unit in the earlier *metric system* was the gram. A U.S. five-cent coin (a "nickel") has a mass of about 5 g.

TABLE 1-6	Some SI Units of Mass
*kilo*gram, kg	base unit
gram, g	1,000 g = 1 kg
*milli*gram, mg	1,000 mg = 1 g
*micro*gram, μg	1,000,000 μg = 1 g

Length

The **meter** is the standard unit of length (distance) in both SI and metric systems. The meter is defined as the distance light travels in a vacuum in 1/299,792,468 second. It is approximately 39.37 inches. In situations in which the English system would use inches, the metric centimeter (1/100 meter) is convenient. The relationship between inches and centimeters is shown in Figure 1-12.

The meter was originally defined (1791) as one ten-millionth of the distance between the North Pole and the equator.

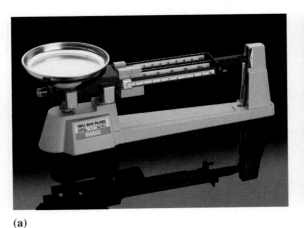

(a)

(b)

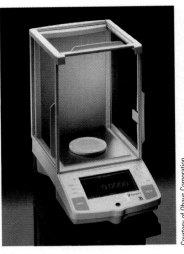

Courtesy of Ohaus Corporation

(c)

Figure 1-11 Three types of laboratory balances. (a) A triple-beam balance used for determining mass to about ±0.01 g. (b) A modern electronic top-loading balance that gives a direct readout of mass to ±0.001 g. (c) A modern analytical balance that can be used to determine mass to ±0.0001 g. Analytical balances are used when masses must be determined as precisely as possible.

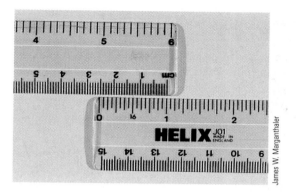

Figure 1-12 The relationship between inches and centimeters: 1 in. = 2.54 cm (exactly).

Volume

Volumes are often measured in liters or milliliters in the metric system. One liter (1 L) is one cubic decimeter (1 dm^3), or 1000 cubic centimeters (1000 cm^3). One milliliter (1 mL) is 1 cm^3. In medical laboratories, the cubic centimeter (cm^3) is often abbreviated cc. In the SI, the cubic meter is the basic volume unit and the cubic decimeter replaces the metric unit, liter. Different kinds of glassware are used to measure the volume of liquids. The one we choose depends on the accuracy we desire. For example, the volume of a liquid dispensed can be measured more accurately with a buret than with a small graduated cylinder (Figure 1-13). Equivalences between common English units and metric units are summarized in Table 1-7.

Sometimes we must combine two or more units to describe a quantity. For instance, we might express the speed of a car as 60 mi/h (also mph). Recall that the algebraic notation x^{-1} means $1/x$; applying this notation to units, we see that h^{-1} means 1/h, or "per hour." So the unit of speed could also be expressed as mi·h^{-1}.

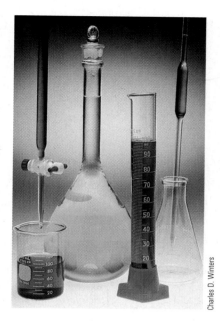

Figure 1-13 Some laboratory apparatus used to measure volumes of liquids: 150-mL beaker (*bottom left,* green liquid); 25-mL buret (*top left,* red); 1000-mL volumetric flask (*center,* yellow); 100-mL graduated cylinder (*right front,* blue); and 10-mL volumetric pipet (*right rear,* green).

TABLE 1-7	Conversion Factors Relating Length, Volume, and Mass (weight) Units					
	Metric		**English**		**Metric–English Equivalents**	
Length	1 km	$= 10^3$ m	1 ft	= 12 in.	2.54 cm	= 1 in.
	1 cm	$= 10^{-2}$ m	1 yd	= 3 ft	39.37 in.*	= 1 m
	1 mm	$= 10^{-3}$ m	1 mile	= 5280 ft	1.609 km*	= 1 mile
	1 nm	$= 10^{-9}$ m				
	1 Å	$= 10^{-10}$ m				
Volume	1 mL	$= 1$ cm$^3 = 10^{-3}$ L	1 gal	= 4 qt = 8 pt	1 L	= 1.057 qt*
	1 m^3	$= 10^6$ cm$^3 = 10^3$ L	1 qt	= 57.75 in.^{3}*	28.32 L	= 1 ft^3
Mass	1 kg	$= 10^3$ g	1 lb	= 16 oz	453.6 g*	= 1 lb
	1 mg	$= 10^{-3}$ g			1 g	= 0.03527 oz*
	1 metric tonne	$= 10^3$ kg	1 short ton	= 2000 lb	1 metric tonne	= 1.102 short ton*

These conversion factors, unlike the others listed, are inexact. They are quoted to four significant figures, which is ordinarily more than sufficient.

1-8 USE OF NUMBERS

CD-ROM   Screen 1.17, Using Numerical Information.

In chemistry, we measure and calculate many things, so we must be sure we understand how to use numbers. In this section we discuss two aspects of the use of numbers: (1) the notation of very large and very small numbers and (2) an indication of how well we actually know the numbers we are using. You will carry out many calculations with calculators. Please refer to Appendix A for some instructions about the use of electronic calculators.

Scientific Notation

We use **scientific notation** when we deal with very large and very small numbers. For example, 197 grams of gold contains approximately

$$602,000,000,000,000,000,000,000 \text{ gold atoms}$$

The mass of one gold atom is approximately

$$0.000\ 000\ 000\ 000\ 000\ 000\ 000\ 327 \text{ gram}$$

In scientific notation, these numbers are

6.02×10^{23} gold atoms

3.27×10^{-22} gram

In using such large and small numbers, it is inconvenient to write down all the zeroes. In scientific (exponential) notation, we place one nonzero digit to the left of the decimal.

$$4,300,000. = 4.3 \times 10^6$$

6 places to the left, $\therefore$ exponent of 10 is 6

$$0.000348 = 3.48 \times 10^{-4}$$

4 places to the right, $\therefore$ exponent of 10 is -4

The reverse process converts numbers from exponential to decimal form. See Appendix A for more detail, if necessary.

✓ Problem-Solving Tip: *Know How to Use Your Calculator*

Students sometimes make mistakes when they try to enter numbers into their calculators in scientific notation. Suppose you want to enter the number 4.36×10^{-2}. On most calculators, you would

1. Press 4.36
2. Press EE or EXP, which stands for "times ten to the"
3. Press 2 (the magnitude of the exponent) and then ± or CHS (to change its sign)

The calculator display might show the value as $\boxed{4.36 \qquad -02}$ or as $\boxed{0.0436}$. Different calculators show different numbers of digits, which can sometimes be adjusted.
   If you wished to enter -4.36×10^2, you would

1. Press 4.36, then press ± or CHS to change its sign,
2. Press EE or EXP, and then press 2

The calculator would then show $\boxed{-4.36 \qquad 02}$ or $\boxed{-436.0}$.

Caution: Be sure you remember that the EE or EXP button *includes* the "times 10" operation. An error that beginners often make is to enter " × 10" explicitly when trying to enter a number in scientific notation. Suppose you mistakenly enter 3.7×10^2 as follows:

1. Enter 3.7
2. Press × and then enter 10
3. Press EXP or EE and then enter 2

The calculator then shows the result as 3.7×10^3 or 3700—why? This sequence is processed by the calculator as follows: Step 1 enters the number 3.7; step 2 multiplies by 10, to give 37; step 3 multiplies this by 10^2, to give 37×10^2 or 3.7×10^3.

Other common errors include changing the sign of the exponent when the intent was to change the sign of the entire number (e.g., -3.48×10^4 entered as 3.48×10^{-4}).

When in doubt, carry out a trial calculation for which you already know the answer. For instance, multiply 300 by 2 by entering the first value as 3.00×10^2 and then multiplying by 2; you know the answer should be 600, and if you get any other answer, you know you have done something wrong. If you cannot find (or understand) the printed instructions for *your calculator*, your instructor or a classmate might be able to help.

Significant Figures

There are two kinds of numbers. *Numbers obtained by counting or from definitions are* **exact numbers.** They are known to be absolutely accurate. For example, the exact number of people in a closed room can be counted, and there is no doubt about the number of people. A dozen eggs is defined as exactly 12 eggs, no more, no fewer (Figure 1-14).

An *exact* number may be thought of as containing an *infinite* number of significant figures.

(a)

(b)

Figure 1-14   (a) A dozen eggs is exactly 12 eggs. (b) A specific swarm of honeybees contains an *exact* number of live bees (but it would be difficult to count them, and any two *swarms* would be unlikely to contain the same exact number of bees).

There is some uncertainty in all measurements.

Significant figures indicate the *uncertainty* in measurements.

SC*L*INKS.

TOPIC: Significant
Figures
GO TO: www.scilinks.org
sciLINKS CODE: WCH0150

Numbers obtained from measurements are not exact. Every measurement involves an estimate. For example, suppose you are asked to measure the length of this page to the nearest 0.1 mm. How do you do it? The smallest divisions (calibration lines) on a meter stick are 1 mm apart (see Figure 1-12). An attempt to measure to 0.1 mm requires estimation. If three different people measure the length of the page to 0.1 mm, will they get the same answer? Probably not. We deal with this problem by using significant figures.

Significant figures are digits believed to be correct by the person who makes a measurement. We assume that the person is competent to use the measuring device. Suppose one measures a distance with a meter stick and reports the distance as 343.5 mm. What does this number mean? In this person's judgment, the distance is greater than 343.4 mm but less than 343.6 mm, and the best estimate is 343.5 mm. The number 343.5 mm contains four significant figures. The last digit, 5, is a *best estimate* and is therefore doubtful, but it is considered to be a significant figure. In reporting numbers obtained from measurements, *we report one estimated digit, and no more.* Because the person making the measurement is not certain that the 5 is correct, it would be meaningless to report the distance as 343.53 mm.

To see more clearly the part significant figures play in reporting the results of measurements, consider Figure 1-15a. Graduated cylinders are used to measure volumes of liquids when a high degree of accuracy is not necessary. The calibration lines on a 50-mL graduated cylinder represent 1-mL increments. Estimation of the volume of liquid in a 50-mL cylinder to within 0.2 mL ($\frac{1}{5}$ of one calibration increment) with reasonable certainty is possible. We might measure a volume of liquid in such a cylinder and report the volume as 38.6 mL, that is, to three significant figures.

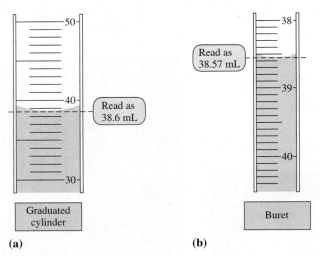

Figure 1-15 Measurement of the volume of water using two types of volumetric glassware. For consistency, we always read the bottom of the meniscus (the curved surface of the water). (a) A graduated cylinder is used to measure the amount of liquid *contained* in the glassware, so the scale increases from bottom to top. The level in a 50-mL graduated cylinder can usually be estimated to within 0.2 mL. The level here is 38.6 mL (three significant figures). (b) We use a buret to measure the amount of liquid *delivered* from the glassware, by taking the difference between an initial and a final volume reading. The level in a 50-mL buret can be read to within 0.02 mL. The level here is 38.57 mL (four significant figures).

Burets are used to measure volumes of liquids when higher accuracy is required. The calibration lines on a 50-mL buret represent 0.1-mL increments, allowing us to make estimates to within 0.02 mL ($\frac{1}{5}$ of one calibration increment) with reasonable certainty (Figure 1-15b). Experienced individuals estimate volumes in 50-mL burets to 0.01 mL with considerable reproducibility. For example, using a 50-mL buret, we can measure out 38.57 mL (four significant figures) of liquid with reasonable accuracy.

Accuracy refers to how closely a measured value agrees with the correct value. **Precision** refers to how closely individual measurements agree with one another. Ideally, all measurements should be both accurate and precise. Measurements may be quite precise yet quite inaccurate because of some *systematic error*, which is an error repeated in each measurement. (A faulty balance, for example, might produce a systematic error.) Very accurate measurements are seldom imprecise.

Measurements are frequently repeated to improve accuracy and precision. Average values obtained from several measurements are usually more reliable than individual measurements. Significant figures indicate how precisely measurements have been made (assuming the person who made the measurements was competent).

Some simple rules govern the use of significant figures.

1. Nonzero digits are always significant.

For example, 38.57 mL has four significant figures; 288 g has three significant figures.

2. Zeroes are sometimes significant, and sometimes they are not.
   a. Zeroes at the *beginning* of a number (used just to position the decimal point) are never significant.

For example, 0.052 g has two significant figures; 0.00364 m has three significant figures. These could also be reported in scientific notation (Appendix A) as 5.2×10^{-2} g and 3.64×10^{-3} m, respectively.

   b. Zeroes *between* nonzero digits are always significant.

For example, 2007 g has four significant figures; 6.08 km has three significant figures.

   c. Zeroes at the *end* of a number that contains a decimal point are always significant.

For example, 38.0 cm has three significant figures; 440.0 m has four significant figures. These could also be reported as 3.80×10^1 cm and 4.400×10^2 m, respectively.

   d. Zeroes at the *end* of a number that does not contain a decimal point may or may not be significant.

When we wish to specify that all of the zeroes in such a number *are* significant, we may indicate this by placing a decimal point after the number. For instance, 130. grams can represent a mass known to *three* significant figures, that is, 130 ± 1 gram.

For example, the quantity 24,300 km could represent three, four, or five significant figures. We are given insufficient information to answer the question. If both of the zeroes are used just to place the decimal point, the number should appear as 2.43×10^4 km (three significant figures). If only one of the zeroes is used to place the decimal point (i.e., the number was measured ±10), the number is 2.430×10^4 km (four significant figures). If the number is actually known to be 24,300 ± 1, it should be written as 2.4300×10^4 km (five significant figures).

> **3.** Exact numbers can be considered as having an unlimited number of significant figures. This applies to defined quantities.

For example, in the equivalence 1 yard = 3 feet, the numbers 1 and 3 are exact, and we do not apply the rules of significant figures to them. The equivalence 1 inch = 2.54 centimeters is an exact one.

A calculated number can never be more precise than the numbers used to calculate it. The following rules show how to get the number of significant figures in a calculated number.

> **4.** In addition and subtraction, the last digit retained in the sum or difference is determined by the position of the first doubtful digit.

EXAMPLE 1-1 *Significant Figures (Addition and Subtraction)*

(a) Add 37.24 mL and 10.3 mL. (b) Subtract 21.2342 g from 27.87 g.

Plan

We first check to see that the quantities to be added or subtracted are expressed in the same units. We carry out the addition or subtraction. Then we follow Rule 4 for significant figures to express the answer to the correct number of significant figures.

Solution

Doubtful digits are underlined in this example.

(a)
$$
\begin{array}{r}
37.2\underline{4} \text{ mL} \\
+10.\underline{3} \text{ mL} \\
\hline
47.\underline{54} \text{ mL}
\end{array}
$$
is reported as 47.5 mL (calculator gives 47.54)

(b)
$$
\begin{array}{r}
27.8\underline{7} \quad \text{g} \\
-21.234\underline{2} \text{ g} \\
\hline
6.6\underline{358} \text{ g}
\end{array}
$$
is reported as 6.64 g (calculator gives 6.6358)

> **5.** In multiplication and division, an answer contains no more significant figures than the least number of significant figures used in the operation.

EXAMPLE 1-2 *Significant Figures (Multiplication)*

What is the area of a rectangle 1.23 cm wide and 12.34 cm long?

Plan

The area of a rectangle is its length times its width. We must first check to see that the width and length are expressed in the same units. (They are, but if they were not, we must first convert one to the units of the other.) Then we multiply the width by the length. We then follow Rule 5 for significant figures to find the correct number of significant figures. The units for the result are equal to the product of the units for the individual terms in the multiplication.

Solution

$$A = \ell \times w = (12.34 \text{ cm})(1.23 \text{ cm}) = \boxed{15.2 \text{ cm}^2}$$

$$(\text{calculator result} = 15.1782)$$

Because three is the smallest number of significant figures used, the answer should contain only three significant figures. The number generated by an electronic calculator (15.1782) implies more accuracy than is justified; the result cannot be more accurate than the information that led to it. Calculators have no judgment, so you must exercise yours.

You should now work Exercise 32.

With many examples we suggest selected exercises from the end of the chapter. These exercises use the skills or concepts from that example. Now you should work Exercise 32 from the end of this chapter.

The step-by-step calculation in the margin demonstrates why the area is reported as 15.2 cm² rather than 15.1782 cm². The length, 12.34 cm, contains four significant figures, whereas the width, 1.23 cm, contains only three. If we underline each uncertain figure, as well as each figure obtained from an uncertain figure, the step-by-step multiplication gives the result reported in Example 1-2. We see that there are only two certain figures (15) in the result. We report the first doubtful figure (.2), but no more. Division is just the reverse of multiplication, and the same rules apply.

In the three simple arithmetic operations we have performed, the number combination generated by an electronic calculator is not the "answer" in a single case! The correct result of each calculation, however, can be obtained by "rounding." The rules of significant figures tell us where to round.

In rounding off, certain conventions have been adopted. When the number to be dropped is less than 5, the preceding number is left unchanged (e.g., 7.34 rounds to 7.3). When it is more than 5, the preceding number is increased by 1 (e.g., 7.37 rounds to 7.4). When the number to be dropped is 5, the preceding number is set to the nearest *even* number (e.g., 7.45 rounds to 7.4, and 7.35 rounds to 7.4).

```
   12.34 cm
×   1.23 cm
   3702
  2468
 1234
15.1782 cm² = 15.2 cm²
```

Rounding to an even number is intended to reduce the accumulation of errors in chains of calculations.

✓ **Problem-Solving Tip:** *When Do We Round?*

When a calculation involves several steps, we often show the answer to each step to the correct number of significant figures. *But we carry all digits in the calculator to the end of the calculation; then we round the final answer to the appropriate number of significant figures.* When carrying out such a calculation, it is safest to carry extra figures through all steps and then to round the final answer appropriately.

1-9 THE UNIT FACTOR METHOD (DIMENSIONAL ANALYSIS)

Many chemical and physical processes can be described by numerical relationships. In fact, many of the most useful ideas in science must be treated mathematically. In this section, we review some problem-solving skills.

It would be nonsense to say that the length of a piece of cloth is 4.7. We must specify units with the number—4.7 inches, 4.7 feet, or 4.7 meters, for instance.

> The units must *always* accompany the numeric value of a measurement, whether we are writing about the quantity, talking about it, or using it in calculations.

Multiplication by unity (by one) does not change the value of an expression. If we represent "one" in a useful way, we can do many conversions by just "multiplying by one." This method of performing calculations is known as **dimensional analysis,** the **factor-label method,** or the **unit factor method.** Regardless of the name chosen, it is a powerful mathematical tool that is almost foolproof.

Unit factors may be constructed from any two terms that describe the same or equivalent "amounts" of whatever we may consider. For example, 1 foot is equal to exactly 12 inches, by definition. We may write an equation to describe this equality:

$$1 \text{ ft} = 12 \text{ in.}$$

SC*LINKS.*

TOPIC: Measurements
GO TO: www.scilinks.org
*sci*LINKS CODE: WCH0160

Dividing both sides of the equation by 1 ft gives

$$\frac{1 \text{ ft}}{1 \text{ ft}} = \frac{12 \text{ in.}}{1 \text{ ft}} \qquad \text{or} \qquad 1 = \frac{12 \text{ in.}}{1 \text{ ft}}$$

The factor (fraction) 12 in./1 ft is a unit factor because the numerator and denominator describe the same distance. Dividing both sides of the original equation by 12 in. gives 1 = 1 ft/12 in., a second unit factor that is the reciprocal of the first. *The reciprocal of any unit factor is also a unit factor.* Stated differently, division of an amount by the same amount always yields one!

In the English system, we can write many unit factors, such as

Unless otherwise indicated, a "ton" refers to a "short ton," 2000 lb. There are also the "long ton," which is 2240 lb, and the metric tonne, which is 1000 kg.

$$\frac{1 \text{ yd}}{3 \text{ ft}}, \quad \frac{1 \text{ yd}}{36 \text{ in.}}, \quad \frac{1 \text{ mi}}{5280 \text{ ft}}, \quad \frac{4 \text{ qt}}{1 \text{ gal}}, \quad \frac{2000 \text{ lb}}{1 \text{ ton}}$$

The reciprocal of each of these is also a unit factor. Items in retail stores are frequently priced with unit factors, such as 39¢/lb and $3.98/gal. When all the quantities in a unit factor come from definitions, the unit is known to an unlimited (infinite) number of significant figures. For instance, if you bought eight 1-gallon jugs of something priced at $3.98/gal, the total cost would be 8 × $3.98, or $31.84; the merchant would not round this to $31.80, let alone to $30.

In science, nearly all numbers have units. What does 12 mean? Usually we must supply appropriate units, such as 12 eggs or 12 people. In the unit factor method, the units guide us through calculations in a step-by-step process, because all units except those in the desired result cancel.

EXAMPLE 1-3   Unit Factors

Express 1.47 mi in inches.

Plan

First we write down the units of what we wish to know, preceded by a question mark. Then we set it equal to whatever we are given:

$$\underline{?} \text{ inches} = 1.47 \text{ miles}$$

Then we choose unit factors to convert the given units (miles) to the desired units (inches):

$$\text{miles} \longrightarrow \text{feet} \longrightarrow \boxed{\text{inches}}$$

We relate (a) miles to feet and then (b) feet to inches.

Solution

$$\underline{?} \text{ in.} = 1.47 \text{ mi} \times \frac{5280 \text{ ft}}{1 \text{ mi}} \times \frac{12 \text{ in.}}{1 \text{ ft}} = \boxed{9.31 \times 10^4 \text{ in.}} \quad \text{(calculator gives 93139.2)}$$

Note that both miles and feet cancel, leaving only inches, the desired unit. Thus, there is no ambiguity as to how the unit factors should be written. The answer contains three significant figures because there are three significant figures in 1.47 miles.

In the interest of clarity, cancellation of units will be omitted in the remainder of this book. You may find it useful to continue the cancellation of units.

✓ **Problem-Solving Tip:** *Significant Figures*

"How do defined quantities affect significant figures?" Any quantity that comes from a *definition* is exact, that is, it is known to an unlimited number of significant figures. In Example 1-3, the quantities 5280 ft, 1 mile, 12 in., and 1 ft all come from definitions, so they do not limit the significant figures in the answer.

✓ **Problem-Solving Tip:** *Think About Your Answer!*

It is often helpful to ask yourself, "Does the answer make sense?" In Example 1-3, the distance involved is more than a mile. We expect this distance to be many inches, so a large answer is not surprising. Suppose we had mistakenly multiplied by the unit factor $\frac{1 \text{ mile}}{5280 \text{ feet}}$ (and not noticed that the units did not cancel properly); we would have gotten the answer 3.34×10^{-3} in. (0.00334 in.), which we should have immediately recognized as nonsense!

Within the SI and metric systems, many measurements are related to one another by powers of ten.

EXAMPLE 1-4   Unit Conversions

The Ångstrom (Å) is a unit of length, 1×10^{-10} m, that provides a convenient scale on which to express the radii of atoms. Radii of atoms are often expressed in nanometers. The radius of a phosphorus atom is 1.10 Å. What is the distance expressed in centimeters and nanometers?

$\text{Å} \rightarrow \text{m} \rightarrow \boxed{\text{cm}}$

$\text{Å} \rightarrow \text{m} \rightarrow \boxed{\text{nm}}$

Plan

We use the equalities $1\ \text{Å} = 1 \times 10^{-10}\ \text{m}$, $1\ \text{cm} = 1 \times 10^{-2}\ \text{m}$, and $1\ \text{nm} = 1 \times 10^{-9}\ \text{m}$ to construct the unit factors that convert $1.10\ \text{Å}$ to the desired units.

Solution

$$\underline{?}\ \text{cm} = 1.10\ \text{Å} \times \frac{1 \times 10^{-10}\ \text{m}}{1\ \text{Å}} \times \frac{1\ \text{cm}}{1 \times 10^{-2}\ \text{m}} = \boxed{1.10 \times 10^{-8}\ \text{cm}}$$

$$\underline{?}\ \text{nm} = 1.10\ \text{Å} \times \frac{1.0 \times 10^{-10}\ \text{m}}{1\ \text{Å}} \times \frac{1\ \text{nm}}{1 \times 10^{-9}\ \text{m}} = \boxed{1.10 \times 10^{-1}\ \text{nm}}$$

All the unit factors used in this example contain only exact numbers.

You should now work Exercise 34.

EXAMPLE 1-5 *Volume Calculation*

Assuming a phosphorus atom is spherical, calculate its volume in Å^3, cm^3, and nm^3. The formula for the volume of a sphere is $V = (\frac{4}{3})\pi r^3$. Refer to Example 1-4.

Plan

We use the results of Example 1-4 to calculate the volume in each of the desired units.

Solution

$$\underline{?}\ \text{Å}^3 = (\tfrac{4}{3})\pi(1.10\ \text{Å})^3 = \boxed{5.58\ \text{Å}^3}$$

$1\ \text{Å} = 10^{-10}\ \text{m} = 10^{-8}\ \text{cm}$

$$\underline{?}\ \text{cm}^3 = (\tfrac{4}{3})\pi(1.10 \times 10^{-8}\ \text{cm})^3 = \boxed{5.58 \times 10^{-24}\ \text{cm}^3}$$

$$\underline{?}\ \text{nm}^3 = (\tfrac{4}{3})\pi(1.10 \times 10^{-1}\ \text{nm})^3 = \boxed{5.58 \times 10^{-3}\ \text{nm}^3}$$

You should now work Exercise 38.

EXAMPLE 1-6 *Mass Conversion*

A sample of gold has a mass of 0.234 mg. What is its mass in g? in kg?

Plan

We use the relationships $1\ \text{g} = 1000\ \text{mg}$ and $1\ \text{kg} = 1000\ \text{g}$ to write the required unit factors.

Solution

$$\underline{?}\ \text{g} = 0.234\ \text{mg} \times \frac{1\ \text{g}}{1000\ \text{mg}} = \boxed{2.34 \times 10^{-4}\ \text{g}}$$

$$\underline{?}\ \text{kg} = 2.34 \times 10^{-4}\ \text{g} \times \frac{1\ \text{kg}}{1000\ \text{g}} = \boxed{2.34 \times 10^{-7}\ \text{kg}}$$

Again, this example includes unit factors that contain only exact numbers.

✓ **Problem-Solving Tip:** *Conversions Within the Metric or SI System*

The SI and metric systems of units are based on powers of ten. This means that many unit conversions *within* these systems can be carried out just by shifting the decimal point. For instance, the conversion from milligrams to grams in Example 1-6 just involves shifting the decimal point to the *left* by three places. How do we know to move it to the left? We know that the gram is a larger unit of mass than the milligram, so the number of grams in a given mass must be a *smaller* number than the number of milligrams. After you carry out many such conversions using unit factors, you will probably begin to take such shortcuts. Always think about the answer, to see whether it should be larger or smaller than the quantity was before conversion.

Unity raised to *any* power is 1. *Any* unit factor raised to a power is still a unit factor, as the next example shows.

EXAMPLE 1-7   *Volume Conversion*

One liter is exactly 1000 cm³. How many cubic inches are there in 1000 cm³?

Plan

We would multiply by the unit factor $\dfrac{1 \text{ in.}}{2.54 \text{ cm}}$ to convert cm to in. Here we require the *cube* of this unit factor.

Solution

$$\underline{?} \text{ in.}^3 = 1000 \text{ cm}^3 \times \left(\frac{1 \text{ in.}}{2.54 \text{ cm}}\right)^3 = 1000 \text{ cm}^3 \times \frac{1 \text{ in.}}{16.4 \text{ cm}^3} = \boxed{61.0 \text{ in.}^3}$$

Suppose we start with the equality

$$1 \text{ in.} = 2.54 \text{ cm}$$

We can perform the same operation on both sides of the equation. Let's cube both sides:

$$(1 \text{ in.})^3 = (2.54 \text{ cm})^3 = 16.4 \text{ cm}^3$$

so the quantity

$$\frac{(1 \text{ in.})^3}{(2.54 \text{ cm})^3} = 1$$

is a unit factor.

Example 1-7 shows that a unit factor *cubed* is still a unit factor.

EXAMPLE 1-8   *Energy Conversion*

A common unit of energy is the erg. Convert 3.74×10^{-2} erg to the SI units of energy, joules, and kilojoules. One erg is exactly 1×10^{-7} joule (J).

Plan

The definition that relates ergs and joules is used to generate the needed unit factor. The second conversion uses a unit factor that is based on the definition of the prefix *kilo-*.

Solution

$$\underline{?} \text{ J} = 3.74 \times 10^{-2} \text{ erg} \times \frac{1 \times 10^{-7} \text{ J}}{1 \text{ erg}} = \boxed{3.74 \times 10^{-9} \text{ J}}$$

$$\underline{?} \text{ kJ} = 3.74 \times 10^{-9} \text{ J} \times \frac{1 \text{ kJ}}{1000 \text{ J}} = \boxed{3.74 \times 10^{-12} \text{ kJ}}$$

Conversions between the English and SI (metric) systems are conveniently made by the unit factor method. Several conversion factors are listed in Table 1-7. It may be helpful to remember one each for

length	1 in. = 2.54 cm (exact)
mass and weight	1 lb = 454 g (near sea level)
volume	1 qt = 0.946 L or 1 L = 1.06 qt

We relate
(a) gallons to quarts, then
(b) quarts to liters, and then
(c) liters to milliliters.

EXAMPLE 1-9   *English–Metric Conversion*

Express 1.0 gallon in milliliters.

Plan

We ask $\underline{?}$ mL = 1.0 gal and multiply by the appropriate factors.

$$\text{gallons} \longrightarrow \text{quarts} \longrightarrow \text{liters} \longrightarrow \text{milliliters}$$

Solution

$$\underline{?} \text{ mL} = 1.0 \text{ gal} \times \frac{4 \text{ qt}}{1 \text{ gal}} \times \frac{1 \text{ L}}{1.06 \text{ qt}} \times \frac{1000 \text{ mL}}{1 \text{ L}} = 3.8 \times 10^3 \text{ mL}$$

You should now work Exercise 36.

The fact that all other units cancel to give the desired unit, milliliters, shows that we used the correct unit factors. The factors 4 qt/gal and 1000 mL/L contain only exact numbers. The factor 1 L/1.06 qt contains three significant figures. Because 1.0 gal contains only two, the answer contains only two significant figures.

Examples 1-1 through 1-9 show that multiplication by one or more unit factors changes the units and the number of units, but not the amount of whatever we are calculating.

1-10   PERCENTAGE

We often use percentages to describe quantitatively how a total is made up of its parts. In Table 1-3, we described the amounts of elements present in terms of the percentage of each element.

Percentages can be treated as unit factors. For any mixture containing substance A,

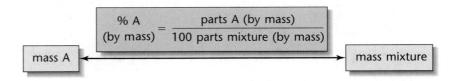

If we say that a sample is 24.4% carbon by mass, we mean that out of every 100 parts (exactly) by mass of sample, 24.4 parts by mass are carbon. This relationship can be represented by whichever of the two unit factors we find useful:

$$\frac{24.4 \text{ parts carbon}}{100 \text{ parts sample}} \quad \text{or} \quad \frac{100 \text{ parts sample}}{24.4 \text{ parts carbon}}$$

This ratio can be expressed in terms of grams of carbon for every 100 grams of sample, pounds of carbon for every 100 pounds of sample, or any other mass or weight unit. The next example illustrates the use of dimensional analysis involving percentage.

EXAMPLE 1-10 *Percentage*

U.S. pennies made since 1982 consist of 97.6% zinc and 2.4% copper. The mass of a particular penny is measured to be 1.494 grams. How many grams of zinc does this penny contain?

Plan

From the percentage information given, we may write the required unit factor

$$\frac{97.6 \text{ g zinc}}{100 \text{ g sample}}$$

Solution

$$\underline{?} \text{ g zinc} = 1.494 \text{ g sample} \times \frac{97.6 \text{ g zinc}}{100 \text{ g sample}} = \boxed{1.46 \text{ g zinc}}$$

The number of significant figures in the result is limited by the three significant figures in 97.6%. Because the definition of percentage involves *exactly* 100 parts, the number 100 is known to an infinite number of significant figures.

You should now work Exercises 63 and 64.

Six materials with different densities. The liquid layers are gasoline (*top*), water (*middle*), and mercury (*bottom*). A cork floats on gasoline. A piece of oak wood sinks in gasoline but floats on water. Brass sinks in water but floats on mercury.

1-11 DENSITY AND SPECIFIC GRAVITY

In science, we use many terms that involve combinations of different units. Such quantities may be thought of as unit factors that can be used to convert among these units. The **density** of a sample of matter is defined as the mass per unit volume:

$$\text{density} = \frac{\text{mass}}{\text{volume}} \quad \text{or} \quad D = \frac{m}{V}$$

Densities may be used to distinguish between two substances or to assist in identifying a particular substance. They are usually expressed as g/cm^3 or g/mL for liquids and solids and as g/L for gases. These units can also be expressed as g·cm^{-3}, g·mL^{-1}, and g·L^{-1}, respectively. Densities of several substances are listed in Table 1-8.

CD-ROM Screen 1.8, Density.

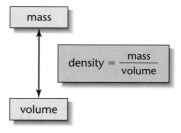

The intensive property *density* relates the two extensive properties: *mass* and *volume*.

EXAMPLE 1-11 *Density, Mass, Volume*

A 47.3-mL sample of ethyl alcohol (ethanol) has a mass of 37.32 g. What is its density?

Plan

We use the definition of density.

Solution

$$D = \frac{m}{V} = \frac{37.32 \text{ g}}{47.3 \text{ mL}} = \boxed{0.789 \text{ g/mL}}$$

You should now work Exercise 40.

TABLE 1-8	*Densities of Common Substances**		
Substance	**Density (g/cm³)**	**Substance**	**Density (g/cm³)**
hydrogen (gas)	0.000089	sand*	2.32
carbon dioxide (gas)	0.0019	aluminum	2.70
cork*	0.21	iron	7.86
oak wood*	0.71	copper	8.92
ethyl alcohol	0.789	silver	10.50
water	1.00	lead	11.34
magnesium	1.74	mercury	13.59
table salt	2.16	gold	19.30

Cork, oak wood, and sand are common materials that have been included to provide familiar reference points. They are not pure elements or compounds as are the other substances listed.

These densities are given at room temperature and *one atmosphere* pressure, the average atmospheric pressure at sea level. Densities of solids and liquids change only slightly, but densities of gases change greatly, with changes in temperature and pressure.

EXAMPLE 1-12   *Density, Mass, Volume*

If 116 g of ethanol is needed for a chemical reaction, what volume of liquid would you use?

Plan

We determined the density of ethanol in Example 1-11. Here we are given the mass, m, of a sample of ethanol. So we know values for D and m in the relationship

$$D = \frac{m}{V}$$

We rearrange this relationship to solve for V, put in the known values, and carry out the calculation. Alternatively, we can use the unit factor method to solve the problem.

Solution

The density of ethanol is 0.789 g/mL (Table 1-8).

$$D = \frac{m}{V}, \quad \text{so} \quad V = \frac{m}{D} = \frac{116 \text{ g}}{0.789 \text{ g/mL}} = \boxed{147 \text{ mL}}$$

Observe that density gives two unit factors. In this case, they are $\frac{0.789\,\text{g}}{1\ \text{mL}}$ and $\frac{1\ \text{mL}}{0.789\,\text{g}}$

Alternatively,

$$\underline{?}\text{ mL} = 116 \text{ g} \times \frac{1 \text{ mL}}{0.789 \text{ g}} = \boxed{147 \text{ mL}}$$

You should now work Exercise 42.

EXAMPLE 1-13   *Unit Conversion*

Express the density of mercury in lb/ft³.

Plan

The density of mercury is 13.59 g/cm³ (see Table 1-8). To convert this value to the desired units, we can use unit factors constructed from the conversion factors in Table 1-7.

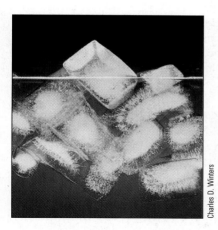

Ice is slightly less dense than liquid water, so ice floats in water.

Solid ethyl alcohol is more dense than liquid ethyl alcohol. This is true of nearly every known substance.

Solution

$$\underline{?}\ \frac{lb}{ft^3} = 13.59\ \frac{g}{cm^3} \times \frac{1\ lb}{453.6\ g} \times \left(\frac{2.54\ cm}{1\ in.}\right)^3 \times \left(\frac{12\ in.}{1\ ft}\right)^3 = \boxed{848.4\ lb/ft^3}$$

It would take a very strong person to lift a cubic foot of mercury!

The **specific gravity** (Sp. Gr.) of a substance is the ratio of its density to the density of water, both at the same temperature.

$$Sp.\ Gr. = \frac{D_{substance}}{D_{water}}$$

Density and specific gravity are both intensive properties; that is, they do not depend on the size of the sample. Specific gravities are dimensionless numbers.

The density of water is 1.000 g/mL at 3.98°C, the temperature at which the density of water is greatest. Variations in the density of water with changes in temperature, however, are small enough that we can use 1.00 g/mL up to 25°C without introducing significant errors into our calculations.

EXAMPLE 1-14   *Density, Specific Gravity*

The density of table salt is 2.16 g/mL at 20°C. What is its specific gravity?

Plan

We use the preceding definition of specific gravity. The numerator and denominator have the same units, so the result is dimensionless.

Solution

$$Sp.\ Gr. = \frac{D_{salt}}{D_{water}} = \frac{2.16\ g/mL}{1.00\ g/mL} = \boxed{2.16}$$

This example also demonstrates that the density and specific gravity of a substance are numerically equal near room temperature if density is expressed in g/mL (g/cm³).

Labels on commercial solutions of acids give specific gravities and the percentage by mass of the acid present in the solution. From this information, the amount of acid present in a given volume of solution can be calculated.

EXAMPLE 1-15   *Specific Gravity, Volume, Percentage by Mass*

Battery acid is 40.0% sulfuric acid, H_2SO_4, and 60.0% water by mass. Its specific gravity is 1.31. Calculate the mass of pure H_2SO_4 in 100.0 mL of battery acid.

Plan

The percentages are given on a mass basis, so we must first convert the 100.0 mL of acid solution (soln) to mass. To do this, we need a value for the density. We have demonstrated that density and specific gravity are numerically equal at 20°C because the density of water is 1.00 g/mL. We can use the density as a unit factor to convert the given volume of the solution to mass of the solution. Then we use the percentage by mass to convert the mass of the solution to the mass of the acid.

Solution

From the given value for specific gravity, we may write

$$\text{Density} = 1.31 \text{ g/mL}$$

The solution is 40.0% H_2SO_4 and 60.0% H_2O by mass. From this information we may construct the desired unit factor:

$$\frac{40.0 \text{ g } H_2SO_4}{100 \text{ g soln}} \longrightarrow \quad \text{because 100 g of solution contains 40.0 g of } H_2SO_4$$

We can now solve the problem:

$$\underline{?} \ H_2SO_4 = 100.0 \text{ mL soln} \times \frac{1.31 \text{ g soln}}{1 \text{ mL soln}} \times \frac{40.0 \text{ g } H_2SO_4}{100 \text{ g soln}} = \boxed{52.4 \text{ g } H_2SO_4}$$

You should now work Exercise 48.

1-12 HEAT AND TEMPERATURE

CD-ROM   Screen 1.10, Temperature.

In Section 1-1 you learned that heat is one form of energy. You also learned that the many forms of energy can be interconverted and that in chemical processes, chemical energy is converted to heat energy or vice versa. The amount of heat a process uses (*endothermic*) or gives off (*exothermic*) can tell us a great deal about that process. For this reason, it is important for us to be able to measure the intensity of heat.

Temperature measures the intensity of heat, the "hotness" or "coldness" of a body. A piece of metal at 100°C feels hot to the touch, whereas an ice cube at 0°C feels cold. Why? Because the temperature of the metal is higher, and that of the ice cube lower, than body temperature. **Heat** is a form of energy that *always flows spontaneously from a hotter body to a colder body*—it never flows in the reverse direction.

Temperatures can be measured with mercury-in-glass thermometers. A mercury thermometer consists of a reservoir of mercury at the base of a glass tube, open to a very thin (capillary) column extending upward. Mercury expands more than most other liquids as its temperature rises. As it expands, its movement up into the evacuated column can be seen.

Anders Celsius (1701–1744), a Swedish astronomer, developed the Celsius temperature scale, formerly called the centigrade temperature scale. When we place a Celsius thermometer in a beaker of crushed ice and water, the mercury level stands at exactly 0°C, the lower reference point. In a beaker of water boiling at one atmosphere pressure, the mercury level stands at exactly 100°C, the higher reference point. There are 100 equal steps between these two mercury levels. They correspond to an interval of 100 degrees between the melting point of ice and the boiling point of water at one atmosphere. Figure 1-16 shows how temperature marks between the reference points are established.

In the United States, temperatures are frequently measured on the temperature scale devised by Gabriel Fahrenheit (1686–1736), a German instrument maker. On this scale, the freezing and boiling points of water are now defined as 32°F and 212°F, respectively. In scientific work, temperatures are often expressed on the **Kelvin** (absolute) temperature scale. As we shall see in Section 12-5, the zero point of the Kelvin temperature scale is *derived* from the observed behavior of all matter.

Relationships among the three temperature scales are illustrated in Figure 1-17. Between the freezing point of water and the boiling point of water, there are 100 steps (°C or kelvins, respectively) on the Celsius and Kelvin scales. Thus the "degree" is the same size on the Celsius and Kelvin scales. But every Kelvin temperature is 273.15 units above the corresponding Celsius temperature. The relationship between these two scales is as follows:

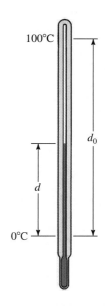

Figure 1-16 At 45°C, as read on a mercury-in-glass thermometer, d equals $0.45d_0$ where d_0 is the distance from the mercury level at 0°C to the level at 100°C.

$$? \text{ K} = °\text{C} + 273.15° \qquad \text{or} \qquad ? °\text{C} = \text{K} - 273.15°$$

We shall usually round 273.15 to 273.

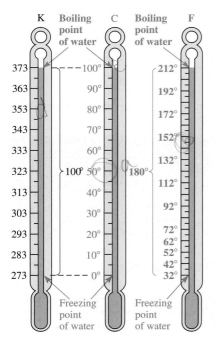

Figure 1-17 The relationships among the Kelvin, Celsius (centigrade), and Fahrenheit temperature scales.

In the SI system, "degrees Kelvin" are abbreviated simply as K rather than °K and are called **kelvins.**

Any temperature *change* has the same numerical value whether expressed on the Celsius scale or on the Kelvin scale. For example, a change from 25°C to 59°C represents a *change* of 34 Celsius degrees. Converting these to the Kelvin scale, the same change is expressed as (273 + 25) = 298 K to (59 + 273) = 332 K, or a *change* of 34 kelvins.

Comparing the Fahrenheit and Celsius scales, we find that the intervals between the same reference points are 180 Fahrenheit degrees and 100 Celsius degrees, respectively. Thus a Fahrenheit degree must be smaller than a Celsius degree. It takes 180 Fahrenheit degrees to cover the same temperature *interval* as 100 Celsius degrees. From this information, we can construct the unit factors for temperature *changes:*

The numbers in these ratios are *exact* numbers, so they do not affect the number of significant figures in the calculated result.

$$\frac{180°F}{100°C} \quad \text{or} \quad \frac{1.8°F}{1.0°C} \quad \text{and} \quad \frac{100°C}{180°F} \quad \text{or} \quad \frac{1.0°C}{1.8°F}$$

But the starting points of the two scales are different, so we *cannot convert* a temperature on one scale to a temperature on the other just by multiplying by the unit factor. In converting from °F to °C, we must subtract 32 Fahrenheit degrees to reach the zero point on the Celsius scale (Figure 1-17).

These are often remembered in abbreviated form:

°F = 1.8°C + 32°

$$°C = \frac{(°F - 32°)}{1.8}$$

Either of these equations can be rearranged to obtain the other one, so you need to learn only one of them.

$$\underline{?}\,°F = \left(x°C \times \frac{1.8°F}{1.0°C}\right) + 32°F \quad \text{and} \quad \underline{?}\,°C = \frac{1.0°C}{1.8°F}(x°F - 32°F)$$

EXAMPLE 1-16   *Temperature Conversion*

When the temperature reaches "100.°F in the shade," it's hot. What is this temperature on the Celsius scale?

Plan

We use the relationship $\underline{?}\,°C = \frac{1.0°C}{1.8°F}(x°F - 32°F)$ to carry out the desired conversion.

Solution

A temperature of 100.°F is 38°C.

$$\underline{?}\,°C = \frac{1.0°C}{1.8°F}(100.°F - 32°F) = \frac{1.0°C}{1.8°F}(68°F) = \boxed{38°C}$$

EXAMPLE 1-17   *Temperature Conversion*

When the absolute temperature is 400 K, what is the Fahrenheit temperature?

Plan

We first use the relationship $\underline{?}\,°C = K - 273°$ to convert from kelvins to degrees Celsius; then we carry out the further conversion from degrees Celsius to degrees Fahrenheit.

Solution

$$\underline{?}\,°C = (400 \text{ K} - 273 \text{ K})\frac{1.0°C}{1.0 \text{ K}} = 127°C$$

$$\underline{?}\,°F = \left(127°C \times \frac{1.8°F}{1.0°C}\right) + 32°F = \boxed{261°F}$$

You should now work Exercise 50.

1-13 HEAT TRANSFER AND THE MEASUREMENT OF HEAT

Chemical reactions and physical changes occur with either the simultaneous evolution of heat **(exothermic processes)** or the absorption of heat **(endothermic processes)**. The amount of heat transferred in a process is usually expressed in joules or in calories.

The SI unit of energy and work is the **joule (J)**, which is defined as $1 \text{ kg·m}^2/\text{s}^2$. The kinetic energy (KE) of a body of mass m moving at speed v is given by $\frac{1}{2}mv^2$. A 2-kg object moving at one meter per second has KE $= \frac{1}{2}(2 \text{ kg})(1 \text{ m/s})^2 = 1 \text{ kg·m}^2/\text{s}^2 = 1 \text{ J}$. You may find it more convenient to think in terms of the amount of heat required to raise the temperature of one gram of water from 14.5°C to 15.5°C, which is 4.184 J.

One **calorie** is defined as exactly 4.184 J. The so-called "large calorie," used to indicate the energy content of foods, is really one kilocalorie, that is, 1000 calories. We shall do most calculations in joules.

The **specific heat** of a substance is the amount of heat required to raise the temperature of one gram of the substance one degree Celsius (also one kelvin) with no change in phase. Changes in phase (physical state) absorb or liberate relatively large amounts of energy (see Figure 1-5). The specific heat of each substance, a physical property, is different for the solid, liquid, and gaseous phases of the substance. For example, the specific heat of ice is 2.09 J/g·°C near 0°C; for liquid water it is 4.18 J/g·°C; and for steam it is 2.03 J/g·°C near 100°C. The specific heat for water is quite high. A table of specific heats is provided in Appendix E.

$$\text{specific heat} = \frac{(\text{amount of heat in J})}{(\text{mass of substance in g})(\text{temperature change in °C})}$$

The units of specific heat are $\dfrac{\text{J}}{\text{g·°C}}$ or $\text{J·g}^{-1}\text{·°C}^{-1}$.

The **heat capacity** of a body is the amount of heat required to raise its temperature 1°C. The heat capacity of a body is its mass in grams times its specific heat. The heat capacity refers to the mass of that particular body, so its units do not include mass. The units are J/°C or J·°C^{-1}.

> In English units, this corresponds to a 4.4-pound object moving at 197 feet per minute, or 2.2 miles per hour. In terms of electrical energy, one joule is equal to one watt · second. Thus, one joule is enough energy to operate a 10-watt light bulb for $\frac{1}{10}$ second.

> The calorie was originally defined as the amount of heat necessary to raise the temperature of one gram of water at one atmosphere from 14.5°C to 15.5°C.

> The specific heat of a substance varies *slightly* with temperature and pressure. These variations can be ignored for calculations in this text.

EXAMPLE 1-18 Specific Heat

How much heat, in joules, is required to raise the temperature of 205 g of water from 21.2°C to 91.4°C?

Plan

The specific heat of a substance is the amount of heat required to raise the temperature of 1 g of substance 1°C:

$$\text{specific heat} = \frac{(\text{amount of heat in J})}{(\text{mass of substance in g})(\text{temperature change in °C})}$$

We can rearrange the equation so that

(amount of heat) = (mass of substance) (specific heat) (temperature change)

Alternatively, we can use the unit factor approach.

Solution

$$\text{amount of heat} = (205 \text{ g}) (4.18 \text{ J/g·°C}) (70.2°C) = \boxed{6.02 \times 10^4 \text{ J}}$$

> In this example, we calculate the amount of heat needed to prepare a cup of hot tea.

By the unit factor approach,

$$\text{amount of heat} = (205 \text{ g})\left(\frac{4.18 \text{ J}}{1 \text{ g·°C}}\right)(70.2°C) = \boxed{6.02 \times 10^4 \text{ J}} \quad \text{or} \quad \boxed{60.2 \text{ kJ}}$$

All units except joules cancel. To cool 205 g of water from 91.4°C to 21.2°C, it would be necessary to remove exactly the same amount of heat, 60.2 kJ.

You should now work Exercises 58 and 59.

When two objects at different temperatures are brought into contact, heat flows from the hotter to the colder body (Figure 1-18); this continues until the two are at the same temperature. We say that the two objects are then in *thermal equilibrium*. The temperature change that occurs for each object depends on the initial temperatures and the relative masses and specific heats of the two materials.

EXAMPLE 1-19 *Specific Heat*

A 588-gram chunk of iron is heated to 97.5°C. Then it is immersed in 247 grams of water originally at 20.7°C. When thermal equilibrium has been reached, the water and iron are both at 36.2°C. Calculate the specific heat of iron.

Plan

The amount of heat gained by the water as it is warmed from 20.7°C to 36.2°C is the same as the amount of heat lost by the iron as it cools from 97.5°C to 36.2°C. We can equate these two amounts of heat and solve for the unknown specific heat.

> In specific heat calculations, we use the *magnitude* of the temperature change (i.e., a positive number), so we subtract the lower temperature from the higher one in both cases.

Solution

$$\text{temperature change of water} = 36.2°C - 20.7°C = 15.5°C$$
$$\text{temperature change of iron} = 97.5°C - 36.2°C = 61.3°C$$

(a)

(b)

Figure 1-18 A hot object, such as a heated piece of metal (a), is placed into cooler water. Heat is transferred from the hotter metal bar to the cooler water until the two reach the same temperature (b). We say that they are then at *thermal equilibrium*.

$$\text{number of joules gained by water} = (247\text{ g})\left(4.18\ \frac{\text{J}}{\text{g}\cdot°\text{C}}\right)(15.5°\text{C})$$

Let x = specific heat of iron

$$\text{number of joules lost by iron} = (588\text{ g})\left(x\ \frac{\text{J}}{\text{g}\cdot°\text{C}}\right)(61.3°\text{C})$$

We set these two quantities equal to one another and solve for x.

$$(247\text{g})\left(4.18\ \frac{\text{J}}{\text{g}\cdot°\text{C}}\right)(15.5°\text{C}) = (588\text{ g})\left(x\ \frac{\text{J}}{\text{g}\cdot°\text{C}}\right)(61.3°\text{C})$$

$$x = \frac{(247\text{ g})\left(4.18\ \dfrac{\text{J}}{\text{g}\cdot°\text{C}}\right)(15.5°\text{C})}{(588\text{ g})(61.3°\text{C})} = 0.444\ \frac{\text{J}}{\text{g}\cdot°\text{C}}$$

You should now work Exercise 62.

The specific heat of iron is much smaller than the specific heat of water.

$$\frac{\text{specific heat of iron}}{\text{specific heat of water}} = \frac{0.444\text{ J/g}\cdot°\text{C}}{4.18\text{ J/g}\cdot°\text{C}} = 0.106$$

The amount of heat required to raise the temperature of 205 g of iron by 70.2°C (as we calculated for water in Example 1-18) is

$$\text{amount of heat} = (205\text{ g})\left(\frac{0.444\text{ J}}{\text{g}\cdot°\text{C}}\right)(70.2°\text{C}) = 6.39 \times 10^3\text{ J, or 6.39 kJ}$$

We see that the amount of heat required to accomplish a given change in temperature for a given quantity of iron is less than that for the same quantity of water, by the same ratio.

$$\frac{\text{number of joules required to warm 205 g of iron by 70.2°C}}{\text{number of joules required to warm 205 g of water by 70.2°C}} = \frac{6.39\text{ kJ}}{60.2\text{ kJ}} = 0.106$$

It might not be necessary to carry out explicit calculations when we are looking only for qualitative comparisons.

EXAMPLE 1-20 *Comparing Specific Heats*

We add the same amount of heat to 10.0 grams of each of the following substances starting at 20.0°C: liquid water, $H_2O(\ell)$; liquid mercury; $Hg(\ell)$, liquid benzene, $C_6H_6(\ell)$; and solid aluminum, $Al(s)$. Rank the samples from lowest to highest final temperature. Refer to Appendix E for required data.

Plan

We can obtain the values of specific heats (Sp. Ht.) for these substances from Appendix E. The higher the specific heat for a substance, the more heat is required to raise a given mass of sample by a given temperature change, so the less its temperature changes by a given amount of heat. The substance with the lowest specific heat undergoes the largest temperature change, and the one with the highest specific heat undergoes the smallest temperature change. *It is not necessary to calculate the amount of heat required to answer this question.*

Solution

The specific heats obtained from Appendix E are as follows:

Substance	Sp. Ht. $\left(\dfrac{J}{g \cdot °C}\right)$
$H_2O(\ell)$	4.18
$Hg(\ell)$	0.138
$C_6H_6(\ell)$	1.74
$Al(s)$	0.900

Ranked from highest to lowest specific heats: $H_2O(\ell) > C_6H_6(\ell) > Al(s) > Hg(\ell)$. Adding the same amount of heat to the same size sample of these substances changes the temperature of $H_2O(\ell)$ the least and that of $Hg(\ell)$ the most. The ranking from lowest to highest final temperature is

$$H_2O(\ell) < C_6H_6(\ell) < Al(s) < Hg(\ell)$$

You should now work Exercise 71.

Key Terms

Accuracy   How closely a measured value agrees with the correct value.

Calorie   Defined as exactly 4.184 joules. Originally defined as the amount of heat required to raise the temperature of one gram of water from 14.5°C to 15.5°C.

Chemical change   A change in which one or more new substances are formed.

Chemical property   See *Properties*.

Compound   A substance composed of two or more elements in fixed proportions. Compounds can be decomposed into their constituent elements.

Density   Mass per unit volume, $D = m/V$.

Element   A substance that cannot be decomposed into simpler substances by chemical means.

Endothermic   Describes processes that absorb heat energy.

Energy   The capacity to do work or transfer heat.

Exothermic   Describes processes that release heat energy.

Extensive property   A property that depends on the amount of material in a sample.

Heat   A form of energy that flows between two samples of matter because of their difference in temperature.

Heat capacity   The amount of heat required to raise the temperature of a body (of whatever mass) one degree Celsius.

Heterogeneous mixture   A mixture that does not have uniform composition and properties throughout.

Homogeneous mixture   A mixture that has uniform composition and properties throughout.

Intensive property   A property that is independent of the amount of material in a sample.

Joule   A unit of energy in the SI system. One joule is $1 \text{ kg} \cdot \text{m}^2/\text{s}^2$, which is also 0.2390 cal.

Kinetic energy   Energy that matter possesses by virtue of its motion.

Law of Conservation of Energy   Energy cannot be created or destroyed in a chemical reaction or in a physical change; it may be changed from one form to another.

Law of Conservation of Matter   No detectable change occurs in the total quantity of matter during a chemical reaction or during a physical change.

Law of Conservation of Matter and Energy   The combined amount of matter and energy available in the universe is fixed.

Law of Constant Composition   See *Law of Definite Proportions*.

Law of Definite Proportions   Different samples of any pure compound contain the same elements in the same proportions by mass; also known as the *Law of Constant Composition*.

Mass   A measure of the amount of matter in an object. Mass is usually measured in grams or kilograms.

Matter   Anything that has mass and occupies space.

Mixture   A sample of matter composed of variable amounts of two or more substances, each of which retains its identity and properties.

Physical change   A change in which a substance changes from one physical state to another, but no substances with different compositions are formed.

Physical property   See *Properties*.

Potential energy   Energy that matter possesses by virtue of its position, condition, or composition.

Precision How closely repeated measurements of the same quantity agree with one another.

Properties Characteristics that describe samples of matter. Chemical properties are exhibited as matter undergoes chemical changes. Physical properties are exhibited by matter with no changes in chemical composition.

Scientific (natural) law A general statement based on the observed behavior of matter, to which no exceptions are known.

Significant figures Digits that indicate the precision of measurements—digits of a measured number that have uncertainty only in the last digit.

Specific gravity The ratio of the density of a substance to the density of water at the same temperature.

Specific heat The amount of heat required to raise the temperature of one gram of a substance one degree Celsius.

Substance Any kind of matter all specimens of which have the same chemical composition and physical properties.

Symbol (of an element) A letter or group of letters that represents (identifies) an element.

Temperature A measure of the intensity of heat, that is, the hotness or coldness of a sample or object.

Unit factor A factor in which the numerator and denominator are expressed in different units but represent the same or equivalent amounts. Multiplying by a unit factor is the same as multiplying by one.

Weight A measure of the gravitational attraction of the earth for a body.

Exercises

Asterisks are used to denote some of the more challenging exercises.

Matter and Energy

1. Define the following subdivisions of chemistry: (a) biochemistry; (b) analytical chemistry; (c) geochemistry; (d) nuclear chemistry. (Hint: You may need to consult a dictionary for answers to this question.)
2. Define the following subdivisions of chemistry: (a) organic chemistry; (b) forensic chemistry; (c) physical chemistry; (d) medicinal chemistry. (Hint: You may need to consult a dictionary for answers to this question.)
3. Define the following terms, and illustrate each with a specific example: (a) matter; (b) kinetic energy; (c) mass; (d) exothermic process; (e) intensive property.
4. Define the following terms, and illustrate each with a specific example: (a) weight; (b) potential energy; (c) temperature; (d) endothermic process; (e) extensive property.
5. State the Law of Conservation of Matter and Energy and explain how it differs from the Law of Conservation of Matter and the Law of Conservation of Energy.
6. Write Einstein's equation and describe how it can be used to relate the mass change in a nuclear reaction to energy.
7. State the following laws, and illustrate each: (a) the Law of Conservation of Matter; (b) the Law of Conservation of Energy; (c) the Law of Conservation of Matter and Energy.
8. All electrical motors are less than 100% efficient in converting electrical energy into useable work. How can the efficiency of electrical motors be less than 100% and the Law of Conservation of Energy still be valid?
9. An incandescent light bulb functions because of the flow of electric current. Does the incandescent light bulb convert all of the electrical energy to light? Observe a functioning incandescent light bulb, and explain what occurs with reference to the Law of Conservation of Energy.

States of Matter

10. List the three states of matter and some characteristics of each. How are they alike? How are they different?
11. What is a homogeneous mixture? Which of the following are pure substances? Which of the following are homogeneous mixtures? Explain your answers. (a) Sugar dissolved in water; (b) tea and ice; (c) french onion soup; (d) mud; (e) gasoline; (f) carbon dioxide; (g) melted chocolate-chip mint ice cream.

12. Define the following terms clearly and concisely. Give two illustrations of each: (a) substance; (b) mixture; (c) element; (d) compound.
13. Classify each of the following as an element, a compound, or a mixture. Justify your classification: (a) a soft drink;

(b) seawater; (c) air; (d) chicken noodle soup; (e) table salt; (f) copper wire; (g) popcorn; (h) ice cream.

14. Classify each of the following as an element, a compound, or a mixture. Justify your classification: (a) gasoline; (b) tap water; (c) calcium carbonate; (d) ink from a ballpoint pen; (e) toothpaste; (f) aluminum foil.

15. Sand, candle wax, and table sugar are placed in a beaker and stirred. (a) Is the resulting combination a mixture? If so, what kind of mixture? (b) Design an experiment in which the sand, candle wax, and table sugar can be separated.

16. A $10 gold piece minted in the early part of the 1900s appeared to have a dirty area. The dirty appearance could not be removed by careful cleaning. Close examination of the coin revealed that the "dirty" area was really pure copper. Is the mixture of gold and copper in this coin a heterogeneous or homogeneous mixture?

Chemical and Physical Properties

17. Distinguish between the following pairs of terms and give two specific examples of each: (a) chemical properties and physical properties; (b) intensive properties and extensive properties; (c) chemical changes and physical changes; (d) mass and weight.

18. Which of the following are chemical properties, and which are physical properties? (a) Striking a match causes it to burst into flames. (b) A particular type of steel is very hard and consists of 95% iron, 4% carbon, and 1% miscellaneous other elements. (c) The density of gold is 19.3 g/mL. (d) Iron dissolves in hydrochloric acid with the evolution of hydrogen gas. (e) Fine steel wool burns in air. (f) Refrigeration slows the rate at which fruit ripens.

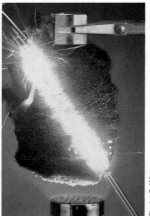

Charles D. Winters

19. Label each of the following as either a physical process or a chemical process: (a) rusting of an iron bridge; (b) melting of ice; (c) burning of a wooden stick; (d) digestion of a baked potato.

20. Describe each of the following as a chemical change, a physical change, or both. (a) A wet towel dries in the sun. (b) Lemon juice is added to tea, causing its color to change. (c) Hot air rises over a radiator. (d) Coffee is brewed by passing hot water through ground coffee. (e) An antacid tablet decreases the acidity of an over acid stomach.

21. Which of the following illustrate the concept of potential energy and which illustrate kinetic energy? (a) water stored in a water tower; (b) a rubber band stretched around a newspaper; (c) a frozen pint of ice cream; (d) a comet moving through space; (e) a basketball dropping through a net; (f) the roof of a house.

22. Which of the following illustrate the concept of potential energy and which illustrate kinetic energy? (a) a moving car; (b) an inflated balloon; (c) a pitched baseball just as it is being released by the pitcher; (d) a flashlight battery; (e) a frozen lake; (f) a car as it moves along the highway.

23. Which of the following processes are exothermic? endothermic? How can you tell? (a) burning gasoline; (b) freezing ice cream; (c) melting chocolate; (d) cooling hot water; (e) condensing water vapor; (f) burning a match.

24. Which of the following processes are exothermic? endothermic? How can you tell? (a) combustion; (b) freezing water; (c) melting ice; (d) boiling water; (e) condensing steam; (f) burning paper.

25. A sample of yellow sulfur powder is placed in a sealed flask with the air removed and replaced with an inert gas. Heat is applied by means of a flame from a Bunsen burner until the sulfur melts and begins to boil. After cooling, the material in the flask is reddish and has the consistency of used chewing gum. Careful chemical analysis tells us that the substance is pure sulfur. Is this a chemical or physical change? Propose an explanation for the change.

26. A weighed sample of yellow sulfur is placed in a flask. The flask is gently heated using a Bunsen burner. Observation indicates that nothing appears to happen to the sulfur during the heating, but the mass of sulfur is less than before the heating, and there is a sharp odor that was not present before the heating. Propose an explanation of what caused the change in the mass of the sulfur. Is your hypothesis of the mass change a chemical or physical change?

Charles Steele

sulfur

Measurements and Calculations

27. Express the following numbers in scientific notation: (a) 650.; (b) 0.0630; (c) 8600 (assume that this number is measured to 10); (d) 8600 (assume that this number is measured to 1); (e) 16,000; (f) 0.100010.

28. For each of the following quantities underline the zeros that are significant figures, determine the number of significant figures in the quantity, and rewrite the quantity using scientific notation. (a) 423.06 mL; (b) 0.0001073040 g; (c) 1,081.02 pounds.

29. Which of the following are likely to be exact numbers? Why? (a) 128 students; (b) 7 railroad cars; (c) $20,355.47; (d) 25 lb of sugar; (e) 12.5 gal of diesel fuel; (f) 5446 ants.

30. Express the following exponentials as ordinary numbers: (a) 5.06×10^3; (b) 4.0010×10^{-3}; (c) 16.10×10^3; (d) 0.206×10^{-4}; (e) 9.000×10^3; (f) 9.000×10^{-3}.

31. The circumference of a circle is given by πd, where d is the diameter of the circle. Calculate the circumference of a circle with a diameter of 7.41 cm. Use the value of 3.141593 for π. (Show your answer with the correct number of significant figures.)

32. A box is 252.56 cm wide, 19.23 cm deep and 6.5 cm tall. Calculate the volume of the box. (Show your answer with the correct number of significant figures.)

33. Indicate the multiple or fraction of 10 by which a quantity is multiplied when it is preceded by each of the following prefixes. (a) M; (b) m; (c) c; (d) d; (e) k; (f) n.

34. Carry out each of the following conversions. (a) 28.5 m to km; (b) 36.3 km to m; (c) 447 kg to g; (d) 1.32 L to mL; (e) 55.9 dL to L; (f) 6251 L to cm^3.

35. Express 5.31 centimeters in meters, millimeters, kilometers, and micrometers.

36. Express (a) 1.00 cubic foot in units of liters; (b) 1.00 liter in units of pints; (c) miles per gallon in kilometers per liter.

37. The screen of a laptop computer measures 8.25 in. wide and 6.25 in. tall. If this computer were being sold in Europe, what would be the metric size, in cm, of the screen used in the specifications for the computer?

38. If the price of gasoline is $1.389/gal, what is its price in cents per liter?

39. Suppose your automobile gas tank holds 14 gal and the price of gasoline is $0.325/L. How much would it cost to fill your gas tank?

40. What is the density of silicon, if 50.6 g occupies 21.72 mL?

41. What is the mass of a rectangular piece of copper 24.4 cm $\times$ 11.4 cm $\times$ 7.9 cm? The density of copper is 8.92 g/cm^3.

42. A small crystal of sucrose (table sugar) had a mass of 5.536 mg. The dimensions of the box-like crystal were 2.20 mm $\times$ 1.36 mm $\times$ 1.12 mm. What is the density of sucrose expressed in g/cm^3?

43. Vinegar has a density of 1.0056 g/cm^3. What is the mass of 3 L of vinegar?

44. The density of silver is 10.5 g/cm^3. (a) What is the volume, in cm^3, of an ingot of silver with mass 0.615 kg? (b) If this sample of silver is a cube, how long is each edge in cm? (c) How long is the edge of this cube in inches?

*45. A container has a mass of 78.91 g when empty and 92.44 g when filled with water. The density of water is 1.0000 g/cm^3. (a) Calculate the volume of the container. (b) When filled with an unknown liquid, the container had a mass of 88.42 g. Calculate the density of the unknown liquid.

*46. The mass of an empty container is 77.664 g. The mass of the container filled with water is 99.646 g. (a) Calculate the volume of the container, using a density of 1.0000 g/cm^3 for water. (b) A piece of metal was added to the empty container, and the combined mass was 85.308 g. Calculate the mass of the metal. (c) The container with the metal was filled with water, and the mass of the entire system was 106.442 g. What mass of water was added? (d) What volume of water was added? (e) What is the volume of the piece of metal? (f) Calculate the density of the metal.

47. A solution is 40.0% acetic acid (the characteristic component in vinegar) by mass. The density of this solution is 1.049 g/mL at 20°C. Calculate the mass of pure acetic acid in 250.0 mL of this solution at 20°C.

48. A solution that is 40.0% iron(III) chloride by mass has a density of 1.149 g/mL. What mass of iron(III) chloride, in g, is present in 3.50 L of this solution?

Heat Transfer and Temperature Measurement

49. Express (a) 245°C in K; (b) 25.25 K in °C; (c) −42.0°C in °F; (d) 110.0°F in K.

50. Express (a) 10°F in °C; (b) 34.6°F in K; (c) 348 K in °F; (d) 10.3°C in °F.

51. Make each of the following temperature conversions: (a) 27°C to °F; (b) −27°C to °F; (c) 100°F to °C.

*52. On the Réamur scale, which is no longer used, water freezes at 0°R and boils at 80°R. (a) Derive an equation that relates this to the Celsius scale. (b) Derive an equation that relates this to the Fahrenheit scale. (c) Mercury is a liquid

metal at room temperature. It boils at 356.6°C (673.9°F). What is the boiling point of mercury on the Réamur scale?

53. Liquefied gases have boiling points well below room temperature. On the Kelvin scale the boiling points of the following gases are: He, 4.2 K; N_2, 77.4 K. Convert these temperatures to the Celsius and the Fahrenheit scales.

54. Convert the temperatures at which the following metals melt to the Celsius and Fahrenheit scales: Al, 933.6 K; Ag, 1235.1 K.

55. What is the melting point of lead in °F (mp 327.5°C)?

56. The average temperature of a healthy German shepherd is 101.5°F. Express this temperature in degrees Celsius. Express this temperature in kelvins.

Taxi/GETTY

57. Calculate the amount of heat required to raise the temperature of 78.2 g of water from 10.0°C to 32.0°C. The specific heat of water is 4.184 J/g·°C.

58. The specific heat of aluminum is 0.895 J/g·°C. Calculate the amount of heat required to raise the temperature of 25.1 g of aluminum from 27.0°C to 65.5°C.

59. How much heat must be removed from 15.5 g of water at 90.0°C to cool it to 38.2°C?

*** 60.** In some solar-heated homes, heat from the sun is stored in rocks during the day and then released during the cooler night. (a) Calculate the amount of heat required to raise the temperature of 78.7 kg of rocks from 25.0°C to 43.0°C. Assume that the rocks are limestone, which is essentially pure calcium carbonate. The specific heat of calcium carbonate is 0.818 J/g·°C. (b) Suppose that when the rocks in part (a) cool to 30.0°C, all the heat released goes to warm the 10,000 ft^3 (2.83×10^5 L) of air in the house, originally at 10.0°C. To what final temperature would the air be heated? The specific heat of air is 1.004 J/g·°C, and its density is 1.20×10^{-3} g/mL.

*** 61.** A small immersion heater is used to heat water for a cup of coffee. We wish to use it to heat 245 mL of water (about a teacupful) from 25°C to 85°C in 2.00 min. What must be the heat rating of the heater, in kJ/min, to accomplish

this? Ignore the heat that goes to heat the cup itself. The density of water is 0.997 g/mL.

62. When 75.0 grams of metal at 75.0°C is added to 150 grams of water at 15.0°C, the temperature of the water rises to 18.3°C. Assume that no heat is lost to the surroundings. What is the specific heat of the metal?

Mixed Exercises

63. A sample is marked as containing 25.8% calcium carbonate by mass. (a) How many grams of calcium carbonate are contained in 75.45 g of the sample? (b) How many grams of the sample would contain 18.8 g of calcium carbonate?

64. An iron ore is found to contain 9.24% hematite (a compound that contains iron). (a) How many tons of this ore would contain 8.40 tons of hematite? (b) How many kilograms of this ore would contain 9.40 kg of hematite?

*** 65.** The radius of a hydrogen atom is about 0.37 Å, and the average radius of the earth's orbit around the sun is about 1.5×10^8 km. Find the ratio of the average radius of the earth's orbit to the radius of the hydrogen atom.

66. A notice on a bridge informs drivers that the height of the bridge is 25.5 ft. How tall in meters is an 18-wheel tractor-trailer combination if it just touches the bridge?

67. Some American car manufacturers install speedometers that indicate speed in the English system and in the metric system (mi/h and km/h). What is the metric speed if the car is traveling at 70 mi/h?

*** 68.** The lethal dose of potassium cyanide (KCN) taken orally is 1.6 mg/kg of body weight. Calculate the lethal dose of potassium cyanide taken orally by a 165-lb person.

69. Suppose you ran a mile in 4.50 min. (a) What would be your average speed in km/h? (b) What would be your average speed in cm/s? (c) What would be your time (in minutes: seconds) for 1500 m?

CONCEPTUAL EXERCISES

70. Suggest a means of using physical properties or physical changes to separate and recover the components of the following mixtures: (a) oil and vinegar; (b) salt and pepper.

71. If you were given the job of choosing the materials from which pots and pans were to be made, what kinds of materials would you choose on the basis of specific heat? Why?

72. When she discovered that a piece of shiny gray zinc was too large to fit through the opening of an Erlenmeyer flask, a student cut the zinc into smaller pieces so they would fit. She then poured enough blue copper chloride solution into the flask to cover the zinc pieces. After 20 min the solution became colorless, the bottom of the flask was slightly warm to the touch, the zinc pieces were visibly reduced in size, and brown granular material appeared in the mixture. List the physical properties, physical changes, and chemical changes the student should have noted and recorded in her laboratory notebook.

73. Which is more dense at 0°C, ice or water? How do you know?

74. Which has the higher temperature, a sample of water at 65°C or a sample of iron at 65°F?

75. The drawing in the circle (below) is a greatly expanded representation of the molecules in the liquid of the thermometer on the left. The thermometer registers 20°C. Which of the figures (a–d) is the best representation of the liquid in this same thermometer at 10°C? (Assume that the same volume of liquid is shown in each expanded representation.)

76. Household ammonia is 5% ammonia by mass and has a density of 1.006 g/mL. What volume of this solution must a person purchase to obtain 17.4 g of ammonia?

77. Although newly minted pennies look as though they are composed of copper, they actually contain only 2.7% copper. The remainder of the metal is zinc. If the densities of copper and zinc are 8.72 g/cm³ and 7.14 g/cm³, respectively, what is the density of a newly minted penny?

78. During the past several years, you have gained chemical vocabulary and understanding from a variety of academic and entertainment venues. List three events that occurred early in the development of your current chemical knowledge.

*79. At what temperature will a Fahrenheit thermometer give: (a) the same reading as a Celsius thermometer? (b) a reading that is twice that on the Celsius thermometer? (c) a

reading that is numerically the same but opposite in sign from that on the Celsius thermometer?

*80. Cesium atoms are the largest naturally occurring atoms. The radius of a cesium atom is 2.62 Å. How many cesium atoms would have to be laid side by side to give a row of cesium atoms 1.00 in. long? Assume that the atoms are spherical.

81. Based on what you have learned during the study of this chapter, write a question that requires you to use chemical information but no mathematical calculations.

82. As you write out the answer to an end-of-chapter exercise, what chemical changes occur? Did your answer involve knowledge not covered in Chapter 1?

83. *Combustion* is discussed later in this textbook; however, you probably already know what the term means. (Look it up to be sure.) List two other chemical terms that were in your vocabulary before you read Chapter 1.

BEYOND THE TEXTBOOK

Go to the textbook website at

http://www.brookscole.com/chemistry/whitten

for additional activities and exercises based on the General Chemistry Interactive CD-ROM, the World Wide Web, and library resources.

InfoTrac College Edition

For additional readings, go to InfoTrac College Edition, your online research library at:

http://infotrac.thomsonlearning.com

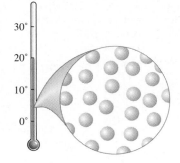

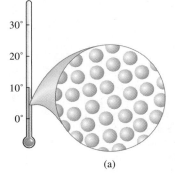 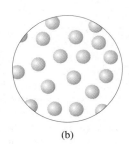 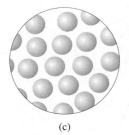 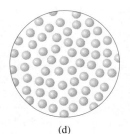

(a)　　　　(b)　　　　(c)　　　　(d)

Exercise 75

2 Chemical Formulas and Composition Stoichiometry

Charles D. Winters

Citrine, a form of quartz (SiO_2) in which traces of iron oxide are present.

OBJECTIVES

After you have studied this chapter, you should be able to

- Understand some early concepts of atoms
- Use chemical formulas to solve various kinds of chemical problems
- Relate names to formulas and charges of simple ions
- Combine simple ions to write formulas and names of some ionic compounds
- Recognize and use formula weights and mole relationships
- Interconvert masses, moles, and formulas
- Determine percent compositions in compounds
- Determine formulas from composition
- Perform calculations of purity of substances

The language we use to describe the forms of matter and the changes in its composition is not limited to use in chemistry courses; it appears throughout the scientific world. Chemical symbols, formulas, and equations are used in such diverse areas as agriculture, home economics, engineering, geology, physics, biology, medicine, and dentistry. In this chapter we describe the simplest atomic theory. We shall use it as we represent the chemical formulas of elements and compounds. Later, after additional facts have been introduced, this theory will be expanded.

The word "stoichiometry" is derived from the Greek *stoicheion*, which means "first principle or element," and *metron*, which means "measure." **Stoichiometry** describes the quantitative relationships among elements in compounds (composition stoichiometry) and among substances as they undergo chemical changes (reaction stoichiometry). In this chapter we are concerned with chemical formulas and **composition stoichiometry**. In Chapter 3 we shall discuss chemical equations and **reaction stoichiometry**.

It is important to learn this fundamental material well as it is the basis of all chemistry.

2-1 ATOMS AND MOLECULES

The term "atom" comes from the Greek language and means "not divided" or "indivisible."

The Greek philosopher Democritus (470–400 BC) suggested that all matter is composed of tiny, discrete, indivisible particles that he called atoms. His ideas, based entirely on philosophical speculation rather than experimental evidence, were rejected for 2000 years. By the late 1700s, scientists began to realize that the concept of atoms provided an explanation for many experimental observations about the nature of matter.

By the early 1800s, the Law of Conservation of Matter (Section 1-1) and the Law of Definite Proportions (Section 1-5) were both accepted as general descriptions of how matter behaves. John Dalton (1766–1844), an English schoolteacher, tried to explain why matter behaves in such systematic ways as those expressed here. In 1808, he published the first "modern" ideas about the existence and nature of atoms. Dalton's explanation summarized and expanded the nebulous concepts of early philosophers and scientists; more importantly, his ideas were based on *reproducible experimental results* of measurements by many scientists. These ideas form the core of **Dalton's Atomic Theory,** one of the highlights in the history of scientific thought. In condensed form, Dalton's ideas may be stated as follows:

The radius of a calcium atom is only 1.97×10^{-8} cm, and its mass is 6.66×10^{-23} g.

Statement 3 is true for *chemical* reactions. It is not true, however, for *nuclear* reactions (Chapter 26).

1. An element is composed of extremely small, indivisible particles called atoms.
2. All atoms of a given element have identical properties that differ from those of other elements.
3. Atoms cannot be created, destroyed, or transformed into atoms of another element.
4. Compounds are formed when atoms of different elements combine with one another in small whole-number ratios.
5. The relative numbers and kinds of atoms are constant in a given compound.

Dalton believed that atoms were solid, indivisible spheres, an idea we now reject. But he showed remarkable insight into the nature of matter and its interactions. Some of his ideas could not be verified (or refuted) experimentally at the time. They were based on the limited experimental observations of his day. Even with their shortcomings, Dalton's ideas provided a framework that could be modified and expanded by later scientists. Thus John Dalton is often considered to be the father of modern atomic theory.

The smallest particle of an element that maintains its chemical identity through all chemical and physical changes is called an **atom** (Figure 2-1). In Chapter 5, we shall study the structure of the atom in detail; let us simply summarize here the main features of atomic composition. Atoms, and therefore *all* matter, consist principally of three **fundamental**

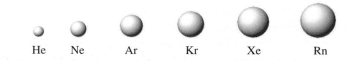

He   Ne   Ar   Kr   Xe   Rn

Figure 2-1 Relative sizes for atoms of the noble gases.

TABLE 2-1	*Fundamental Particles of Matter*	
Particle (symbol)	Approximate Mass (amu)*	Charge (relative scale)
electron (e^-)	0.0	$1-$
proton (p or p^+)	1.0	$1+$
neutron (n or n^0)	1.0	none

$1\ amu = 1.6605 \times 10^{-24}$ g

particles: *electrons, protons,* and *neutrons.* These are the basic building blocks of atoms. The masses and charges of the three fundamental particles are shown in Table 2-1. The masses of protons and neutrons are nearly equal, but the mass of an electron is much smaller. Neutrons carry no charge. The charge on a proton is equal in magnitude, but opposite in sign, to the charge on an electron. Because any atom is electrically neutral it contains an equal number of electrons and protons.

The **atomic number** (symbol is **Z**) of an element is defined as the number of protons in the nucleus. In the periodic table, elements are arranged in order of increasing atomic numbers. These are the red numbers above the symbols for the elements in the periodic table on the inside front cover. For example, the atomic number of silver is 47.

A **molecule** is the smallest particle of an element or compound that can have a stable independent existence. In nearly all molecules, two or more atoms are bonded together in very small, discrete units (particles) that are electrically neutral.

Individual oxygen atoms are not stable at room temperature and atmospheric pressure. Single atoms of oxygen mixed under these conditions quickly combine to form pairs. The oxygen with which we are all familiar is made up of two atoms of oxygen; it is a *diatomic* molecule, O_2. Hydrogen, nitrogen, fluorine, chlorine, bromine, and iodine are other examples of diatomic molecules (Figure 2-2).

Some other elements exist as more complex molecules. One form of phosphorus molecules consists of four atoms, and sulfur exists as eight-atom ring-shaped molecules at ordinary temperatures and pressures. Molecules that contain two or more atoms are called *polyatomic* molecules (Figure 2-3).

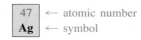

47 ← atomic number
Ag ← symbol

For Group VIIIA elements, the noble gases, a molecule contains only one atom, and so an atom and a molecule are the same (see Figure 2-1).

You should remember the common elements that occur as diatomic molecules: H_2, N_2, O_2, F_2, Cl_2, Br_2, I_2.

Some common prefixes:
  di = two
  tri = three
  tetra = four
  penta = five
  hexa = six
  poly = more than one

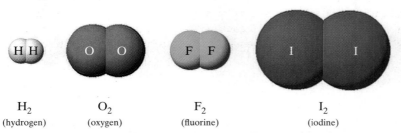

H_2	O_2	F_2	I_2
(hydrogen)	(oxygen)	(fluorine)	(iodine)

Figure 2-2 Models of diatomic molecules of some elements, approximately to scale. These are called space-filling models because they show the atoms with their approximate relative sizes.

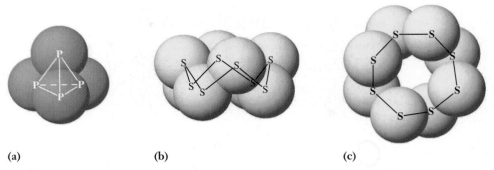

(a)   (b)   (c)

Figure 2-3 (a) A model of the P_4 molecule of white phosphorus. (b) A model of the S_8 ring found in rhombic sulfur. (c) Top view of the S_8 ring in rhombic sulfur.

In modern terminology, O_2 is named dioxygen, H_2 is dihydrogen, P_4 is tetraphosphorus, and so on. Even though such terminology is officially preferred, it has not yet gained wide acceptance. Most chemists still refer to O_2 as oxygen, H_2 as hydrogen, P_4 as phosphorus, and so on.

Molecules of compounds are composed of more than one kind of atom. A water molecule consists of two atoms of hydrogen and one atom of oxygen. A molecule of methane consists of one carbon atom and four hydrogen atoms. The shapes of a few molecules are shown in Figure 2-4 as ball-and-stick models.

Atoms are the building blocks of molecules, and molecules are the stable forms of many elements and compounds. We are able to study samples of compounds and elements that consist of large numbers of atoms and molecules. With the scanning tunneling microscope it is now possible to "see" atoms (Figure 2-5). It would take millions of atoms to make a row as long as the diameter of the period at the end of this sentence.

Methane is the principal component of natural gas.

CD-ROM Screens 1.5, Electrons and Atoms; 1.6, Compounds and Molecules; 3.2, Elements that Exist as Molecules; and 3.3, Molecular Compounds.

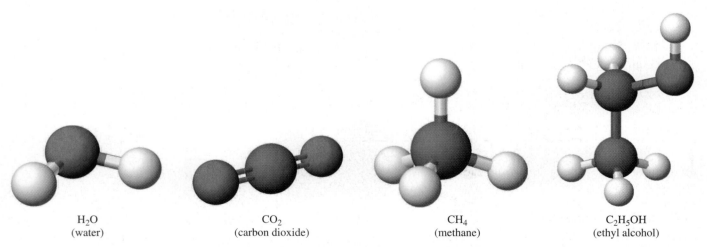

| H_2O | CO_2 | CH_4 | C_2H_5OH |
| (water) | (carbon dioxide) | (methane) | (ethyl alcohol) |

Figure 2-4 Formulas and ball-and-stick models for molecules of some compounds. Ball-and-stick models represent the atoms as smaller spheres than in space-filling models, in order to show the chemical bonds between the atoms as "sticks."

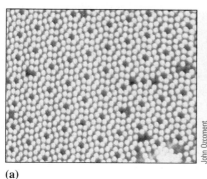
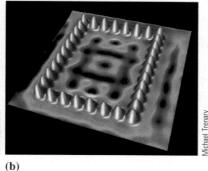

(a)   (b)

John Ozcoment

Michael Trenary

Figure 2-5   (a) A computer reconstruction of the surface of a sample of silicon, as observed with a scanning tunneling electron microscope (STM), reveals the regular pattern of individual silicon atoms. The larger darker areas represent missing surface silicon atoms. The distance between adjacent "touching" Si atoms is about 1.7 Å, or 1.7 × 10^{-10} m. Many important reactions occur on the surfaces of solids. Observations of the atomic arrangements on surfaces help chemists understand such reactions. (b) 34 iron atoms (blue cones) arranged on a copper surface.

2-2 CHEMICAL FORMULAS

The **chemical formula** for a substance shows its chemical composition. This represents the elements present as well as the ratio in which the atoms of the elements occur. The formula for a single atom is the same as the symbol for the element. Thus, Na can represent a single sodium atom. It is unusual to find such isolated atoms in nature, with the exception of the noble gases (He, Ne, Ar, Kr, Xe, and Rn). A subscript following the symbol of an element indicates the number of atoms in a molecule. For instance, F_2 indicates a molecule containing two fluorine atoms, and P_4 a molecule containing four phosphorus atoms.

Some elements exist in more than one form. Familiar examples include (1) oxygen, found as O_2 molecules, and ozone, found as O_3 molecules, and (2) two crystalline forms of carbon—diamond and graphite (Figure 13-32). Different forms of the same element in the same physical state are called **allotropic modifications,** or **allotropes.**

Compounds contain two or more elements in chemical combination in fixed proportions. Many compounds exist as molecules (Table 2-2). Hence, each molecule of hydrogen

An O_2 molecule.

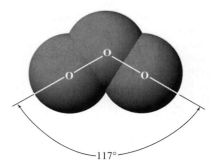

117°

An O_3 molecule.

TABLE 2-2	*Names and Formulas of Some Common Molecular Compounds*				
Name	**Formula**	**Name**	**Formula**	**Name**	**Formula**
water	H_2O	sulfur dioxide	SO_2	butane	C_4H_{10}
hydrogen peroxide	H_2O_2	sulfur trioxide	SO_3	pentane	C_5H_{12}
hydrogen chloride*	HCl	carbon monoxide	CO	benzene	C_6H_6
sulfuric acid	H_2SO_4	carbon dioxide	CO_2	methanol (methyl alcohol)	CH_3OH
nitric acid	HNO_3	methane	CH_4	ethanol (ethyl alcohol)	CH_3CH_2OH
acetic acid	CH_3COOH	ethane	C_2H_6	acetone	CH_3COCH_3
ammonia	NH_3	propane	C_3H_8	diethyl ether (ether)	$CH_3CH_2OCH_2CH_3$

Called hydrochloric acid if dissolved in water.

chloride, HCl, contains one atom of hydrogen and one atom of chlorine; each molecule of acetic acid, CH_3COOH, contains two carbon atoms, four hydrogen atoms, and two oxygen atoms. An aspirin molecule, $C_9H_8O_4$, contains nine carbon atoms, eight hydrogen atoms, and four oxygen atoms.

Many of the molecules found in nature are organic compounds. **Organic compounds** contain C—C or C—H bonds or both, often in combination with nitrogen, oxygen, sulfur, and other elements. Eleven of the compounds listed in Table 2-2 are organic compounds (acetic acid and the last ten entries). All of the other compounds in the table are **inorganic compounds** (compounds that do not contain C—H bonds).

Some groups of atoms behave chemically as single entities. For instance, an oxygen atom that is bonded to a hydrogen atom and also to a carbon atom that is bonded to three other atoms forms the reactive combination of atoms known as the alcohol group or molecule. In formulas of compounds containing two or more of the same group, the group formula is enclosed in parentheses. Thus, ethylene glycol contains two *alcohol groups* and its formula is $C_2H_4(OH)_2$ (see structure in the margin). When you count the number of atoms in this molecule from its formula, you must multiply the numbers of hydrogen and oxygen atoms in the OH group by 2. There are *two* carbon atoms, *six* hydrogen atoms and *two* oxygen atoms in a molecule of ethylene glycol (i.e., $C_2H_6O_2$).

Compounds were first recognized as distinct substances because of their different physical properties and because they could be separated from one another by physical methods. Once the concept of atoms and molecules was established, the reason for these differences in properties could be understood: Two compounds differ from each other because their molecules are different. Conversely, if two molecules contain the same number of the same kinds of atoms, arranged the same way, then both are molecules of the same compound. Thus, the atomic theory explains the **Law of Definite Proportions** (see Section 1-5).

This law, also known as the **Law of Constant Composition,** can now be extended to include its interpretation in terms of atoms. It is so important for performing the calculations in this chapter that we restate it here:

> Different pure samples of a compound always contain the same elements in the same proportion by mass; this corresponds to atoms of these elements combined in fixed numerical ratios.

So we see that for a substance composed of molecules, the **chemical formula** gives the number of atoms of each type in the molecule. But this formula does not express the order in which the atoms in the molecules are bonded together. The **structural formula** shows the order in which atoms are connected. The lines connecting atomic symbols represent chemical bonds between atoms. The bonds are actually forces that tend to hold atoms at certain distances and angles from one another. For instance, the structural formula of propane shows that the three C atoms are linked in a chain, with three H atoms bonded to each of the end C atoms and two H atoms bonded to the center C. **Ball-and-stick** molecular models and **space-filling** molecular models help us to see the shapes and relative sizes of molecules. These four representations are shown in Figure 2-6. The ball-and-stick and space-filling models show (1) the *bonding sequence*, that is the order in which the atoms are connected to each other, and (2) the *geometrical arrangements* of the atoms in the molecule. As we shall see later, both are extremely important because they determine the properties of compounds.

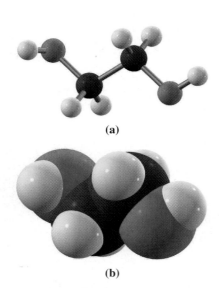

(a)

(b)

(a) Ball-and-stick and (b) space-filling models of ethylene glycol.

Chemical Formula	Structural Formula	Ball-and-Stick Model	Space-Filling Model
H_2O, water	H—O—H		
H_2O_2, hydrogen peroxide	H—O—O—H		
CCl_4, carbon tetrachloride	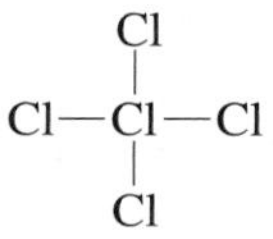 Cl—Cl—Cl with Cl above and Cl below	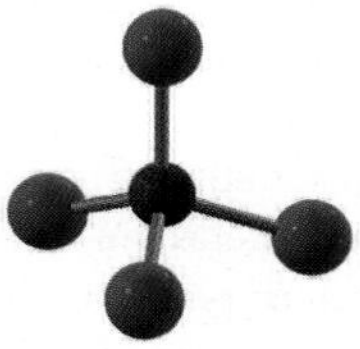	
C_2H_5OH, ethanol	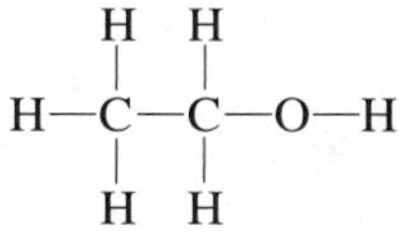 H—C—C—O—H with H atoms above and below the carbons	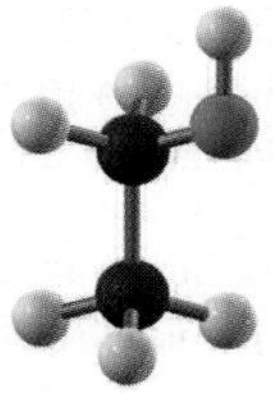	

Figure 2-6 Formulas and models for some molecules. Structural formulas show the order in which atoms are connected but do not represent true molecular shapes. Ball-and-stick models use balls of different colors to represent atoms and sticks to represent bonds; they show the three-dimensional shapes of molecules. Space-filling models show the (approximate) relative sizes of atoms and the shapes of molecules, but the bonds between the atoms are hidden by the overlapping spheres that represent the atoms.

2-3 IONS AND IONIC COMPOUNDS

So far we have discussed only compounds that exist as discrete molecules. Some compounds, such as sodium chloride, NaCl, consist of collections of large numbers of ions. An **ion** is an atom or group of atoms that carries an electric charge. Ions that possess a *positive* charge, such as the sodium ion, Na^+, are called **cations.** Those carrying a *negative* charge, such as the chloride ion, Cl^-, are called **anions.** The charge on an ion *must* be included as

The words "cation" (kat′-i-on) and "anion" (an′-i-on) and their relationship to cathode and anode will be described in Chapter 21.

Figure 2-7 The arrangement of ions in NaCl. (a) A crystal of sodium chloride consists of an extended array that contains equal numbers of sodium ions (*small spheres*) and chloride ions (*large spheres*). Within the crystal, (b) each chloride ion is surrounded by six sodium ions, and (c) each sodium ion is surrounded by six chloride ions.

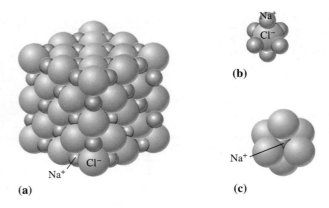

(a)   (b)   (c)

CD-ROM   Screens 3.5, Ions: Cations and Anions and 3.6, Polyatomic Ions.

The general term "formula unit" applies to molecular or ionic compounds, whereas the more specific term "molecule" applies only to elements and compounds that exist as discrete molecules.

In this text, we use the standard convention of representing multiple charges with the number before the sign, e.g., Ca^{2+}, *not* Ca^{+2} and SO_4^{2-}, *not* SO_4^{-2}.

a superscript on the right side of the chemical symbol(s) when we write the formula for the individual ion.

As discussed in detail in Chapter 5, an atom consists of a very small, very dense, positively charged *nucleus* surrounded by a diffuse distribution of negatively charged particles called *electrons*. The number of positive charges in the nucleus defines the identity of the element to which the atom corresponds. Electrically neutral atoms contain the same number of electrons outside the nucleus as positive charges (protons) within the nucleus. Ions are formed when neutral atoms lose or gain electrons. An Na^+ ion is formed when a sodium atom loses one electron, and a Cl^- ion is formed when a chlorine atom gains one electron.

The compound NaCl consists of an extended array of Na^+ and Cl^- ions (Figure 2-7). Within the crystal (though not on the surface) each Na^+ ion is surrounded at equal distances by six Cl^- ions, and each Cl^- ion is similarly surrounded by six Na^+ ions. *Any* compound, whether ionic or molecular, is electrically neutral; that is, it has no net charge. In NaCl this means that the Na^+ and Cl^- ions are present in a 1:1 ratio, and this is indicated by the formula NaCl.

Because there are no "molecules" of ionic substances, we should not refer to "a molecule of sodium chloride, NaCl," for example. Instead, we refer to a **formula unit** of NaCl, which consists of one Na^+ ion and one Cl^- ion. Likewise, one formula unit of $CaCl_2$ consists of one Ca^{2+} ion and two Cl^- ions. As you will see in the next section, we speak of the formula unit of all ionic compounds as the smallest, whole-number ratios of ions that yield neutral representations. It is also acceptable to refer to a formula unit of a molecular compound. One formula unit of propane, C_3H_8, is the same as one molecule of C_3H_8; it contains three C atoms and eight H atoms bonded together into a group. For the present, we shall tell you which substances are ionic and which are molecular when it is important to know. Later you will learn to make the distinction yourself.

Polyatomic ions are groups of atoms that bear an electric charge. The first atom in the formula is usually the central atom to which the other atoms are bonded to make a stable combination. Examples include the ammonium ion, NH_4^+; the sulfate ion, SO_4^{2-}; and the nitrate ion, NO_3^-. Table 2-3 shows the formulas, ionic charges, and names of some common ions. When writing the formula of a polyatomic compound, we show groups in parentheses when they appear more than once. For example, $(NH_4)_2SO_4$ represents a compound that has two NH_4^+ ions for each SO_4^{2-} ion.

TABLE 2-3	*Formulas, Ionic Charges, and Names of Some Common Ions*					
Common Cations (positive ions)				**Common Anions (negative ions)**		
Formula	*Charge*	*Name*		*Formula*	*Charge*	*Name*
Na^+	1+	sodium		F^-	1−	fluoride
K^+	1+	potassium		Cl^-	1−	chloride
NH_4^+	1+	ammonium		Br^-	1−	bromide
Ag^+	1+	silver		OH^-	1−	hydroxide
				CH_3COO^-	1−	acetate
Mg^{2+}	2+	magnesium		NO_3^-	1−	nitrate
Ca^{2+}	2+	calcium				
Zn^{2+}	2+	zinc		O^{2-}	2−	oxide
Cu^+	1+	copper(I)		S^{2-}	2−	sulfide
Cu^{2+}	2+	copper(II)		SO_4^{2-}	2−	sulfate
Fe^{2+}	2+	iron(II)		SO_3^{2-}	2−	sulfite
				CO_3^{2-}	2−	carbonate
Fe^{3+}	3+	iron(III)				
Al^{3+}	3+	aluminum		PO_4^{3-}	3−	phosphate

As we shall see, some metals can form more than one kind of ion with a positive charge. For such metals, we specify which ion we mean with a Roman numeral, for example, iron(II) or iron(III). Because zinc forms no stable ions other than Zn^{2+}, we do not need to use Roman numerals in its name.

2-4 NAMES AND FORMULAS OF SOME IONIC COMPOUNDS

During your study of chemistry you will have many occasions to refer to compounds by name. In this section, we shall see how a few compounds should be named. More comprehensive rules for naming compounds are presented at the appropriate places later in the text.

Table 2-2 includes examples of names for a few common molecular compounds. You should learn that short list before proceeding much farther in this textbook. We shall name many more molecular compounds as we encounter them in later chapters.

Ionic compounds (*clockwise, from top*): salt (sodium chloride, NaCl), calcite (calcium carbonate, $CaCO_3$), cobalt(II) chloride hexahydrate, ($CoCl_2 \cdot 6H_2O$), fluorite (calcium fluoride, CaF_2).

Charles D. Winters

CD-ROM   Screen 3.11, Naming Ionic Compounds.

The names of some common ions appear in Table 2-3. You will need to know the names and formulas of these frequently encountered ions. They can be used to write the formulas and names of many ionic compounds. We write the formula of an ionic compound by adjusting the relative numbers of positive and negative ions so their total charges cancel (i.e., add to zero). The name of an ionic compound is formed by giving the names of the ions, with the positive ion named first.

✓ **Problem-Solving Tip:** *Where to Start in Learning to Name Compounds*

You may not be sure of the best point to start learning the naming of compounds. It has been found that before rules for naming can make much sense or before we can expand our knowledge to more complex compounds, we need to know the names and formulas in Tables 2-2 and 2-3. If you are unsure of your ability to recall a name or a formula in Tables 2-2 and 2-3 when given the other, prepare flash cards, lists, and so on that you can use to learn these tables.

EXAMPLE 2-1   *Formulas for Ionic Compounds*

Write the formulas for the following ionic compounds: (a) sodium fluoride, (b) calcium fluoride, (c) iron(II) sulfate, (d) zinc phosphate.

Plan

In each case, we identify the chemical formulas of the ions from Table 2-3. These ions must be present in the simplest whole-number ratio that gives the compound *no net charge*. Recall that the formulas and names of ionic compounds are written by giving the positively charged ion first.

Solution

(a) The formula for the sodium ion is Na^+, and the formula for the fluoride ion is F^- (Table 2-3). Because the charges on these two ions are equal in magnitude, the ions must be present in equal numbers, or in a 1:1 ratio. Thus, the formula for sodium fluoride is NaF.

(b) The formula for the calcium ion is Ca^{2+} and the formula for the fluoride ion is F^-. Now each positive ion (Ca^{2+}) provides twice as much charge as each negative ion (F^-). So there must be twice as many F^- ions as Ca^{2+} ions to equalize the charge. This means that the ratio of calcium to fluoride ions is 1:2. So the formula for calcium fluoride is CaF_2.

(c) The iron(II) ion is Fe^{2+}, and the sulfate ion is SO_4^{2-}. As in part (a), the equal magnitudes of positive and negative charges tell us that the ions must be present in equal numbers, or in a 1:1 ratio. The formula for iron(II) sulfate is $FeSO_4$.

(d) The zinc ion is Zn^{2+}, and the phosphate ion is PO_4^{3-}. Now it will take *three* Zn^{2+} ions to account for as much charge (6+ total) as would be present in *two* PO_4^{3-} ions (6− total). So the formula for zinc phosphate is $Zn_3(PO_4)_2$.

You should now work Exercises 16 and 23.

EXAMPLE 2-2   *Names for Ionic Compounds*

Name the following ionic compounds: (a) $(NH_4)_2S$, (b) $Cu(NO_3)_2$, (c) $ZnCl_2$, (d) $Fe_2(CO_3)_3$.

Plan

In naming ionic compounds, it is helpful to inspect the formula for atoms or groups of atoms that we recognize as representing familiar ions.

Solution

(a) The presence of the polyatomic grouping NH_4 in the formula suggests to us the presence of the ammonium ion, NH_4^+. There are two of these, each accounting for 1+ in charge. To balance this, the single S must account for 2− in charge, or S^{2-}, which we recognize as the sulfide ion. Thus, the name of the compound is ammonium sulfide.

(b) The NO_3 grouping in the formula tells us that the nitrate ion, NO_3^-, is present. Two of these nitrate ions account for $2 \times 1- = 2-$ in negative charge. To balance this, copper must account for 2+ charge and be the copper(II) ion. The name of the compound is copper(II) nitrate.

(c) The positive ion present is zinc ion, Zn^{2+}, and the negative ion is chloride, Cl^-. The name of the compound is zinc chloride.

(d) Each CO_3 grouping in the formula must represent the carbonate ion, CO_3^{2-}. The presence of *three* such ions accounts for a total of 6− in negative charge, so there must be a total of 6+ present in positive charge to balance this. It takes *two* iron ions to provide this 6+, so each ion must have a charge of 3+ and be Fe^{3+}, the iron(III) ion, or ferric ion. The name of the compound is iron(III) carbonate.

You should now work Exercises 15 and 22.

> We use the information that the carbonate ion has a 2− charge to find the charge on the iron ions. The total charges must add up to zero.

A more extensive discussion on naming compounds appears in Chapter 4.

2-5 ATOMIC WEIGHTS

As the chemists of the eighteenth and nineteenth centuries painstakingly sought information about the compositions of compounds and tried to systematize their knowledge, it became apparent that each element has a characteristic mass relative to every other element. Although these early scientists did not have the experimental means to measure the mass of each kind of atom, they succeeded in defining a *relative* scale of atomic masses.

An early observation was that carbon and hydrogen have relative atomic masses, also traditionally called **atomic weights (AW),** of approximately 12 and 1, respectively. Thousands of experiments on the compositions of compounds have resulted in the establishment of a scale of relative atomic weights based on the **atomic mass unit (amu),** which is defined as *exactly $\frac{1}{12}$ of the mass of an atom of a particular kind of carbon atom, called carbon-12.*

On this scale, the atomic weight of hydrogen (H) is 1.00794 amu, that of sodium (Na) is 22.989768 amu, and that of magnesium (Mg) is 24.3050 amu. This tells us that Na atoms have nearly 23 times the mass of H atoms, and Mg atoms are about 24 times heavier than H atoms.

When you need values of atomic weights, consult the periodic table or the alphabetical listing of elements, both found on facing pages inside the front cover.

> The term "atomic weight" is widely accepted because of its traditional use, although it is properly a mass rather than a weight. "Atomic mass" is often used and is technically more accurate.

2-6 THE MOLE

Even the smallest bit of matter that can be handled reliably contains an enormous number of atoms. So we must deal with large numbers of atoms in any real situation, and some unit for conveniently describing a large number of atoms is desirable. The idea of using a unit to describe a particular number (amount) of objects has been around for a long time. You are already familiar with the dozen (12 items) and the gross (144 items).

SCI*LINKS.*

TOPIC: The Mole
GO TO: www.scilinks.org
*sci***LINKS CODE:** WCH0210

"Mole" is derived from the Latin word *moles,* which means "a mass." "Molecule" is the diminutive form of this word and means "a small mass."

The SI unit for amount is the **mole,** abbreviated mol. It is *defined* as the amount of substance that contains as many entities (atoms, molecules, or other particles) as there are atoms in exactly 0.012 kg of pure carbon-12 atoms. Many experiments have refined the number, and the currently accepted value is

$$1 \text{ mole} = 6.0221367 \times 10^{23} \text{ particles}$$

This number, often rounded to 6.022×10^{23}, is called **Avogadro's number** in honor of Amedeo Avogadro (1776–1856), whose contributions to chemistry are discussed in Section 12-8.

According to its definition, the mole unit refers to a fixed number of items, the identities of which must be specified. Just as we speak of a dozen eggs or a pair of aces, we refer to a mole of atoms or a mole of molecules (or a mole of ions, electrons, or other particles). We could even think about a mole of eggs, although the size of the required carton staggers the imagination! Helium exists as discrete He atoms, so one mole of helium consists of 6.022×10^{23} He *atoms.* Hydrogen commonly exists as diatomic (two-atom) molecules, so one mole of hydrogen is 6.022×10^{23} H_2 *molecules* and $2(6.022 \times 10^{23})$ H atoms.

Every kind of atom, molecule, or ion has a definite characteristic mass. It follows that one mole of a given pure substance also has a definite mass, regardless of the source of the sample. This idea is of central importance in many calculations throughout the study of chemistry and the related sciences.

Because the mole is defined as the number of atoms in 0.012 kg (or 12 g) of carbon-12, and the atomic mass unit is defined as $\frac{1}{12}$ of the mass of a carbon-12 atom, the following convenient relationship is true:

The mass of one mole of atoms of a pure element in grams is numerically equal to the atomic weight of that element in atomic mass units. This is also called the **molar mass** of the element; its units are grams/mole, also written as g/mol or g·mol^{-1}.

CD-ROM   Screen 3.14, Compounds, Molecules, and the Mole: Molar Mass.

For instance, if you obtain a pure sample of the metallic element titanium (Ti), whose atomic weight is 47.88 amu, and measure out 47.88 g of it, you will have one mole, or 6.022×10^{23} titanium atoms.

The symbol for an element can be used to (1) identify the element, (2) represent one atom of the element, or (3) represent one mole of atoms of the element. The last interpretation will be extremely useful in calculations in the next chapter.

The atomic weight of iron (Fe) is 55.847 amu. Assume that one dozen large eggs weighs 24 oz.

A quantity of a substance may be expressed in a variety of ways. For example, consider a dozen eggs and 55.847 grams (or one mole) of iron (Figure 2-8). We can express the amount of eggs or iron present in any of several units. We can then construct unit factors to relate an amount of the substance expressed in one kind of unit to the same amount expressed in another unit.

Unit Factors for Eggs	Unit Factors for Iron
$\dfrac{12 \text{ eggs}}{1 \text{ doz eggs}}$	$\dfrac{6.022 \times 10^{23} \text{ Fe atoms}}{1 \text{ mol Fe atoms}}$
$\dfrac{12 \text{ eggs}}{24 \text{ oz eggs}}$	$\dfrac{6.022 \times 10^{23} \text{ Fe atoms}}{55.847 \text{ g Fe}}$
and so on	and so on

12 large eggs
or
1 dozen eggs
or
24 ounces of eggs

6.022×10^{23} Fe atoms
or
1 mole of Fe atoms
or
55.847 grams of iron

Charles Steele

Figure 2-8 Three ways of representing amounts.

As Table 2-4 suggests, the concept of a mole as applied to atoms is especially useful. It provides a convenient basis for comparing the masses of equal numbers of atoms of different elements.

Figure 2-9 shows what one mole of atoms of each of some common elements looks like. Each of the examples in Figure 2-9 represents 6.022×10^{23} *atoms* of the element.

The relationship between the mass of a sample of an element and the number of moles of atoms in the sample is illustrated in Example 2-3.

In this textbook we usually work problems involving atomic weights (masses) or formula weights (masses) rounded to only one decimal place. We round the answer further if initial data do not support the number of significant figures obtained using the rounded atomic weights. Similarly, if the initial data indicate that more significant figures are justified, we will rework such problems using atomic weights and formula weights containing values beyond the tenths place.

TABLE 2-4	*Mass of One Mole of Atoms of Some Common Elements*	

Element	A Sample with a Mass of	Contains
carbon	12.0 g C	6.02×10^{23} C atoms or 1 mol of C atoms
titanium	47.9 g Ti	6.02×10^{23} Ti atoms or 1 mol of Ti atoms
gold	197.0 g Au	6.02×10^{23} Au atoms or 1 mol of Au atoms
hydrogen	1.0 g H_2	6.02×10^{23} H atoms or 1 mol of H atoms (3.01×10^{23} H_2 molecules or $\frac{1}{2}$ mol of H_2 molecules)
sulfur	32.1 g S_8	6.02×10^{23} S atoms or 1 mol of S atoms (0.753×10^{23} S_8 molecules or $\frac{1}{8}$ mol of S_8 molecules)

Figure 2-9 One mole of atoms of some common elements. Back row (*left to right*): bromine (79.9 g), aluminum (27.0 g), mercury (200.6 g), copper (63.5 g). Front row (*left to right*): sulfur (32.1 g), zinc (65.4 g), iron (55.8 g).

To the required four significant figures, 1 mol Fe atoms = 55.85 g Fe.

EXAMPLE 2-3　*Moles of Atoms*

How many moles of atoms does 136.9 g of iron metal contain?

Plan

The atomic weight of iron is 55.85 amu. This tells us that the molar mass of iron is 55.85 g/mol, or that one mole of iron atoms is 55.85 g of iron. We can express this as either of two unit factors:

$$\frac{1 \text{ mol Fe atoms}}{55.85 \text{ g Fe}} \quad \text{or} \quad \frac{55.85 \text{ g Fe}}{1 \text{ mol Fe atoms}}$$

Because one mole of iron has a mass of 55.85 g, we expect that 136.9 g will be a fairly small number of moles (greater than 1, but less than 10).

Solution

$$\underline{\ ?\ } \text{ mol Fe atoms} = 136.9 \text{ g Fe} \times \frac{1 \text{ mol Fe atoms}}{55.85 \text{ g Fe}} = \boxed{2.451 \text{ mol Fe atoms}}$$

You should now work Exercises 32 and 40.

Once the number of moles of atoms of an element is known, the number of atoms in the sample can be calculated, as Example 2-4 illustrates.

EXAMPLE 2-4　*Numbers of Atoms*

How many atoms are contained in 2.451 mol of iron?

Plan

One mole of atoms of an element contains Avogadro's number of atoms, or 6.022×10^{23} atoms. This lets us generate the two unit factors

$$\frac{6.022 \times 10^{23} \text{ atoms}}{1 \text{ mol atoms}} \quad \text{and} \quad \frac{1 \text{ mol atoms}}{6.022 \times 10^{23} \text{ atoms}}$$

Solution

$$\underline{?} \text{ Fe atoms} = 2.451 \text{ mol Fe atoms} \times \frac{6.022 \times 10^{23} \text{ Fe atoms}}{1 \text{ mol Fe atoms}} = 1.476 \times 10^{24} \text{ Fe atoms}$$

We expected the number of atoms in more than two moles of atoms to be a very large number. Written in nonscientific notation, the answer to this example is: 1,476,000,000,000,000,000,000,000.

You should now work Exercise 42.

Try to name this number with its many zeroes.

If we know the atomic weight of an element on the carbon-12 scale, we can use the mole concept and Avogadro's number to calculate the *average* mass of one atom of that element in grams (or any other mass unit we choose).

CHEMISTRY IN USE

The Development of Science

Avogadro's Number

If you think that the value of Avogadro's number, 6×10^{23}, is too large to be useful to anyone but chemists, look up into the sky on a clear night. You may be able to see about 3000 stars with the naked eye, but the total number of stars swirling around you in the known universe is approximately equal to Avogadro's number. Just think, the known universe contains approximately one mole of stars! You don't have to leave earth to encounter such large numbers. The water in the Pacific Ocean has a volume of about 6×10^{23} mL and a mass of about 6×10^{23} g.

Avogadro's number is almost incomprehensibly large. For example, if one mole of dollars given away at the rate of a million per second beginning when the earth first formed some 4.5 billion years ago, would any remain today? Surprisingly, about three fourths of the original mole of dollars would be left today; it would take about 14,500,000,000 more years to give away the remaining money at $1 million per second.

Computers can be used to provide another illustration of the magnitude of Avogadro's number. If a computer can count up to one billion in one second, it would take that computer about 20 million years to count up to 6×10^{23}. In contrast, recorded human history goes back only a few thousand years.

The impressively large size of Avogadro's number can give us very important insights into the very small sizes of individual molecules. Suppose one drop of water evaporates in one hour. There are about 20 drops in one milliliter of water, which weighs one gram. So one drop of water is about 0.05 g of water. How many H_2O molecules evaporate per second?

$$\frac{\underline{?} \ H_2O \text{ molecules}}{1 \text{ s}} = \frac{0.05 \text{ g } H_2O}{1 \text{ h}} \times \frac{1 \text{ mol } H_2O}{18 \text{ g } H_2O} \times$$
$$\frac{6 \times 10^{23} \ H_2O \text{ molecules}}{1 \text{ mol } H_2O} \times \frac{1 \text{ h}}{60 \text{ min}} \times \frac{1 \text{ min}}{60 \text{ s}}$$
$$= 5 \times 10^{17} \ H_2O \text{ molecules/s}$$

5×10^{17} H_2O molecules evaporating per second is five hundred million billion H_2O molecules evaporating per second —a number that is beyond our comprehension! This calculation helps us to recognize that water molecules are incredibly small. There are approximately 1.7×10^{21} water molecules in a single drop of water.

By gaining some appreciation of the vastness of Avogadro's number, we gain a greater appreciation of the extremely tiny volumes occupied by individual atoms, molecules, and ions.

Ronald DeLorenzo
Middle Georgia College
Original concept by Larry Nordell

EXAMPLE 2-5   *Masses of Atoms*

Calculate the average mass of one iron atom in grams.

Plan

We expect that the mass of a single atom in grams would be a *very* small number. We know that one mole of Fe atoms has a mass of 55.85 g and contains 6.022×10^{23} Fe atoms. We use this information to generate unit factors to carry out the desired conversion.

Solution

$$\frac{?\ \text{g Fe}}{\text{Fe atom}} = \frac{55.85\ \text{g Fe}}{1\ \text{mol Fe atoms}} \times \frac{1\ \text{mol Fe atoms}}{6.022 \times 10^{23}\ \text{Fe atoms}} = 9.274 \times 10^{-23}\ \text{g Fe/Fe atom}$$

Thus, we see that the average mass of one Fe atom is only 9.274×10^{-23} g, that is, 0.00000000000000000000009274 g.

Example 2-5 demonstrates how small atoms are and why it is necessary to use large numbers of atoms in practical work. The next example will help you to realize how large Avogadro's number is.

EXAMPLE 2-6   *Avogadro's Number*

A stack of 500 sheets of typing paper is 1.9 inches thick. Calculate the thickness, in inches and in miles, of a stack of typing paper that contains one mole (Avogadro's number) of sheets.

Plan

We construct unit factors from the data given, from conversion factors in Table 1-7, and from Avogadro's number.

Solution

$$?\ \text{in.} = 1\ \text{mol sheets} \times \frac{6.022 \times 10^{23}\ \text{sheets}}{1\ \text{mol sheets}} \times \frac{1.9\ \text{in.}}{500\ \text{sheets}} = 2.3 \times 10^{21}\ \text{in.}$$

$$?\ \text{mi} = 2.3 \times 10^{21}\ \text{in.} \times \frac{1\ \text{ft}}{12\ \text{in.}} \times \frac{1\ \text{mi}}{5280\ \text{ft}} = 3.6 \times 10^{16}\ \text{mi}$$

By comparison, the sun is about 93 million miles from the earth. This stack of paper would make 390 million stacks that reach from the earth to the sun.

Imagine the number of trees required to make this much paper! This would weigh about 5×10^{20} kg, which far exceeds the mass of all the trees on earth. The mass of the earth is about 6×10^{24} kg.

✓ **Problem-Solving Tip:** *When Do We Round?*

Even though the number 1.9 has two significant figures, we carry the other numbers in Example 2-6 to more significant figures. Then we round at the end to the appropriate number of significant figures. The numbers in the distance conversions are exact numbers.

2-7 FORMULA WEIGHTS, MOLECULAR WEIGHTS, AND MOLES

The **formula weight (FW)** of a substance is the sum of the atomic weights (AW) of the elements in the formula, each taken the number of times the element occurs. Hence a formula weight gives the mass of one formula unit in atomic mass units.

Formula weights, like the atomic weights on which they are based, are relative masses. The formula weight for sodium hydroxide, NaOH (rounded off to the nearest 0.1 amu) is found as follows.

Number of Atoms of Stated Kind		× Mass of One Atom	= Mass Due to Element
$1 \times Na =$	1	× 23.0 amu	= 23.0 amu of Na
$1 \times H =$	1	× 1.0 amu	= 1.0 amu of H
$1 \times O =$	1	× 16.0 amu	= 16.0 amu of O
		Formula weight of NaOH	= 40.0 amu

The term "formula weight" is correctly used for either ionic or molecular substances. When we refer specifically to molecular (non-ionic) substances, that is, substances that exist as discrete molecules, we often substitute the term **molecular weight (MW)**.

EXAMPLE 2-7 *Formula Weights*

Calculate the formula weight (molecular weight) of acetic acid (vinegar), CH_3COOH, using rounded values for atomic weights given in the International Table of Atomic Weights inside the front cover of the text.

Plan

We add the atomic weights of the elements in the formula, each multiplied by the number of times the element occurs.

Solution

Number of Atoms of Stated Kind		× Mass of One Atom	= Mass Due to Element
$2 \times C =$	2	× 12.0 amu	= 24.0 amu of C
$4 \times H =$	4	× 1.0 amu	= 4.0 amu of H
$2 \times O =$	2	× 16.0 amu	= 32.0 amu of O
		Formula weight (molecular weight) of acetic acid (vinegar) =	60.0 amu

You should now work Exercise 28.

CD-ROM Screen 3.15, Using Molar Mass

SC*i*LINKS. **TOPIC:** Molar Masses
GO TO: www.scilinks.org
*sci*LINKS CODE: WCH0220

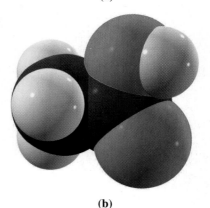

(a)

(b)

(a) Ball-and-stick and (b) space-filling model of an acetic acid (vinegar) molecule, CH_3COOH.

TABLE 2-5	*One Mole of Some Common Molecular Substances*		

Substance	Molecular Weight	A Sample with a Mass of	Contains
oxygen	32.0	32.0 g O_2	1 mol of O_2 molecules 6.02×10^{23} molecules O_2 ($2 \times 6.02 \times 10^{23}$ atoms of O)
water	18.0	18.0 g H_2O	1 mol of H_2O molecules 6.02×10^{23} molecules of H_2O ($2 \times 6.02 \times 10^{23}$ atoms of H and 6.02×10^{23} atoms of O)
methane	16.0	16.0 g CH_4	1 mol of CH_4 molecules 6.02×10^{23} molecules of CH_4 ($4 \times 6.02 \times 10^{23}$ atoms of H and 6.02×10^{23} atoms of C)
sucrose (sugar)	342.3	342.3 g $C_{12}H_{22}O_{11}$	1 mol of $C_{12}H_{22}O_{11}$ molecules 6.02×10^{23} molecules of sucrose ($12 \times 6.02 \times 10^{23}$ atoms of C, $22 \times 6.02 \times 10^{23}$ atoms of H, and $11 \times 6.02 \times 10^{23}$ atoms of O)

The amount of substance that contains the mass in grams numerically equal to its formula weight in amu contains 6.022×10^{23} formula units, or *one mole* of the substance. This is sometimes called the **molar mass** of the substance. Molar mass is *numerically equal* to the formula weight of the substance (the atomic weight for atoms of elements) and has the units grams/mole (g/mol).

One mole of sodium hydroxide is 40.0 g of NaOH, and one mole of acetic acid is 60.0 g of CH_3COOH. One mole of any molecular substance contains 6.02×10^{23} molecules of that substance, as Table 2-5 illustrates.

Because no simple NaCl molecules exist at ordinary temperatures, it is inappropriate to refer to the "molecular weight" of NaCl or any ionic compound. One mole of an ionic compound contains 6.02×10^{23} *formula units* of the substance. Recall that one formula unit of sodium chloride consists of one sodium ion, Na^+, and one chloride ion, Cl^-. One mole, or 58.4 g, of NaCl contains 6.02×10^{23} Na^+ ions and 6.02×10^{23} Cl^- ions (Table 2-6).

The mole concept, together with Avogadro's number, provides important connections among the extensive properties mass of substance, number of moles of substance, and number of molecules or ions. These are summarized as follows.

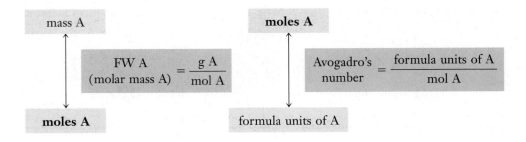

TABLE 2-6	One Mole of Some Ionic Compounds		
Compound	Formula Weight	A Sample with a Mass of 1 Mol	Contains
sodium chloride	58.4	58.4 g NaCl	6.02×10^{23} Na$^+$ ions or 1 mol of Na$^+$ ions 6.02×10^{23} Cl$^-$ ions or 1 mol of Cl$^-$ ions
calcium chloride	111.0	111.0 g CaCl$_2$	6.02×10^{23} Ca^{2+} ions or 1 mol of Ca^{2+} ions $2(6.02 \times 10^{23})$ Cl$^-$ ions or 2 mol of Cl$^-$ ions
aluminum sulfate	342.1	342.1 g Al$_2$(SO$_4$)$_3$	$2(6.02 \times 10^{23})$ Al^{3+} ions or 2 mol of Al^{3+} ions $3(6.02 \times 10^{23})$ SO$_4{}^{2-}$ ions or 3 mol of SO$_4{}^{2-}$ ions

The following examples show the relations between numbers of molecules, atoms, or formula units and their masses.

EXAMPLE 2-8 Masses of Molecules

What is the mass in grams of 10.0 million SO$_2$ molecules?

Plan

One mole of SO$_2$ contains 6.02×10^{23} SO$_2$ molecules and has a mass of 64.1 grams.

Solution

$$\underline{?}\ \text{g SO}_2 = 10.0 \times 10^6\ \text{SO}_2\ \text{molecules} \times \frac{64.1\ \text{g SO}_2}{6.02 \times 10^{23}\ \text{SO}_2\ \text{molecules}}$$

$$= 1.06 \times 10^{-15}\ \text{g SO}_2$$

When fewer than four significant figures are needed in calculations, Avogadro's number may be rounded to 6.02×10^{23}.

Ten million SO$_2$ molecules have a mass of only 0.00000000000000106 g. Commonly used analytical balances are capable of weighing to ±0.0001 g.

You should now work Exercise 44.

EXAMPLE 2-9 Moles

How many (a) moles of O$_2$, (b) O$_2$ molecules, and (c) O atoms are contained in 40.0 g of oxygen gas (dioxygen) at 25°C?

Plan

We construct the needed unit factors from the following equalities: (a) the mass of one mole of O$_2$ is 32.0 g (molar mass O$_2$ = 32.0 g/mol); (b) one mole of O$_2$ contains 6.02×10^{23} O$_2$ molecules; (c) one O$_2$ molecule contains two O atoms.

Solution

One mole of O_2 contains 6.02×10^{23} O_2 molecules, and its mass is 32.0 g.

(a)
$$\underline{?}\text{ mol } O_2 = 40.0 \text{ g } O_2 \times \frac{1 \text{ mol } O_2}{32.0 \text{ g } O_2} = \boxed{1.25 \text{ mol } O_2}$$

(b)
$$\underline{?}\text{ } O_2 \text{ molecules} = 40.0 \text{ g } O_2 \times \frac{6.02 \times 10^{23} \text{ } O_2 \text{ molecules}}{32.0 \text{ g } O_2}$$
$$= \boxed{7.52 \times 10^{23} \text{ molecules}}$$

Or, we can use the number of moles of O_2 calculated in part (a) to find the number of O_2 molecules.

$$\underline{?}\text{ } O_2 \text{ molecules} = 1.25 \text{ mol } O_2 \times \frac{6.02 \times 10^{23} \text{ } O_2 \text{ molecules}}{1 \text{ mol } O_2} = \boxed{7.52 \times 10^{23} \text{ } O_2 \text{ molecules}}$$

(c)
$$\underline{?}\text{ O atoms} = 40.0 \text{ g } O_2 \times \frac{6.02 \times 10^{23} \text{ } O_2 \text{ molecules}}{32.0 \text{ g } O_2} \times \frac{2 \text{ O atoms}}{1 \text{ } O_2 \text{ molecule}}$$
$$= \boxed{1.50 \times 10^{24} \text{ O atoms}}$$

You should now work Exercise 36.

EXAMPLE 2-10   *Numbers of Atoms*

Calculate the number of hydrogen atoms in 39.6 g of ammonium sulfate, $(NH_4)_2SO_4$.

Plan

One mole of $(NH_4)_2SO_4$ is 6.02×10^{23} formula units and has a mass of 132.1 g.

| g of $(NH_4)_2SO_4$ | $\longrightarrow$ | mol of $(NH_4)_2SO_4$ | $\longrightarrow$ | formula units of $(NH_4)_2SO_4$ | $\longrightarrow$ | H atoms |

In Example 2-10, we relate (a) grams to moles, (b) moles to formula units, and (c) formula units to H atoms.

Solution

$$\underline{?}\text{ H atoms} = 39.6 \text{ g } (NH_4)_2SO_4 \times \frac{1 \text{ mol } (NH_4)_2SO_4}{132.1 \text{ g } (NH_4)_2SO_4} \times$$

$$\frac{6.02 \times 10^{23} \text{ formula units } (NH_4)_2SO_4}{1 \text{ mol } (NH_4)_2SO_4} \times \frac{8 \text{ H atoms}}{1 \text{ formula units } (NH_4)_2SO_4}$$

$$= \boxed{1.44 \times 10^{24} \text{ H atoms}}$$

You should now work Exercise 34.

2-8 PERCENT COMPOSITION AND FORMULAS OF COMPOUNDS

If the formula of a compound is known, its chemical composition can be expressed as the mass percent of each element in the compound (percent composition). For example, one carbon dioxide molecule, CO_2, contains one C atom and two O atoms. Percentage is the part divided by the whole times 100 percent (or simply parts per 100), so we can represent the percent composition of carbon dioxide as follows:

AW = atomic weight (mass)
MW = molecular weight (mass)

$$\% \ C = \frac{\text{mass of C}}{\text{mass of CO}_2} \times 100\% = \frac{\text{AW of C}}{\text{MW of CO}_2} \times 100\% = \frac{12.0 \text{ amu}}{44.0 \text{ amu}} \times 100\% = \boxed{27.3\%}$$

$$\% \ O = \frac{\text{mass of O}}{\text{mass of CO}_2} \times 100\% = \frac{2 \times \text{AW of O}}{\text{MW of CO}_2} \times 100\% = \frac{2(16.0 \text{ amu})}{44.0 \text{ amu}} \times 100\% = \boxed{72.7\% \ O}$$

One *mole* of CO_2 (44.0 g) contains one *mole* of C atoms (12.0 g) and two *moles* of O atoms (32.0 g). We could therefore have used these masses in the preceding calculation. These numbers are the same as the ones used—only the units are different. In Example 2-11 we shall base our calculation on one *mole* rather than one *molecule*.

As a check, we see that the percentages add to 100%.

EXAMPLE 2-11 *Percent Composition*

Calculate the percent composition by mass of HNO_3.

CD-ROM Screen 3.16, Percent Composition.

Plan

We first calculate the mass of one mole as in Example 2-7. Then we express the mass of each element as a percent of the total.

Solution

The molar mass of HNO_3 is calculated first.

Number of Mol of Atoms		× Mass of One Mol of Atoms	= Mass Due to Element
$1 \times H =$	1	× 1.0 g	= 1.0 g of H
$1 \times N =$	1	× 14.0 g	= 14.0 g of N
$3 \times O =$	3	× 16.0 g	= 48.0 g of O

$$\text{Mass of 1 mol of HNO}_3 = 63.0 \text{ g}$$

Now, its percent composition is

$$\% \ H = \frac{\text{mass of H}}{\text{mass of HNO}_3} \times 100\% = \frac{1.0 \text{ g}}{63.0 \text{ g}} \times 100\% = \boxed{1.6\% \ H}$$

$$\% \ N = \frac{\text{mass of N}}{\text{mass of HNO}_3} \times 100\% = \frac{14.0 \text{ g}}{63.0 \text{ g}} \times 100\% = \boxed{22.2\% \ N}$$

$$\% \ O = \frac{\text{mass of O}}{\text{mass of HNO}_3} \times 100\% = \frac{48.0 \text{ g}}{63.0 \text{ g}} \times 100\% = \boxed{76.2\% \ O}$$

$$\text{Total} = 100.0\%$$

When chemists use the % notation, they mean percent by mass unless they specify otherwise.

You should now work Exercise 62.

Nitric acid is 1.6% H, 22.2% N, and 76.2% O by mass. All samples of pure HNO_3 have this composition, according to the Law of Definite Proportions.

CHEMISTRY IN USE

The Development of Science

Names of the Elements

If you were to discover a new element, how would you name it? Throughout history, scientists have answered this question in different ways. Most have chosen to honor a person or place or to describe the new substance.

Until the Middle Ages only nine elements were known: gold, silver, tin, mercury, copper, lead, iron, sulfur, and carbon. The metals' chemical symbols are taken from descriptive Latin names: *aurum* ("yellow"), *argentum* ("shining"), *stannum* ("dripping" or "easily melted"), *hydrargyrum* ("silvery water"), *cuprum* ("Cyprus," where many copper mines were located), *plumbum* (exact meaning unknown—possibly "heavy"), and *ferrum* (also unknown). Mercury is named after the planet, one reminder that the ancients associated metals with gods and celestial bodies. In turn, both the planet, which moves rapidly across the sky, and the element, which is the only metal that is liquid at room temperature and thus flows rapidly, are named for the fleet god of messengers in Roman mythology. In English, mercury is nicknamed "quicksilver."

Prior to the reforms of Antoine Lavoisier (1743–1794), chemistry was a largely nonquantitative, unsystematic science in which experimenters had little contact with each other. In 1787, Lavoisier published his *Methode de Nomenclature Chimique*, which proposed, among other changes, that all new elements be named descriptively. For the next 125 years, most elements were given names that corresponded to their properties. Greek roots were one popular source, as evidenced by hydrogen (*hydros-gen*, "water-producing"), oxygen (*oksys-gen*, "acid-producing"), nitrogen (*nitron-gen*, "soda-producing"), bromine (*bromos*, "stink"), and argon (*a-er-gon*, "no reaction"). The discoverers of argon, Sir William Ramsay (1852–1916) and Baron Rayleigh (1842–1919), originally proposed the name *aeron* (from *aer* or "air"), but critics thought it was too close to the biblical name Aaron! Latin roots, such as *radius* ("ray"), were also used (radium and radon are both naturally radioactive elements that emit "rays"). Color was often the determining property, especially after the invention of the spectroscope in 1859, because different elements (or the light that they emit) have prominent characteristic colors. Cesium, indium, iodine, rubidium, and thallium were all named in this manner. Their respective Greek and Latin roots denote blue-gray, indigo, violet, red, and green (*thallus* means "tree sprout"). Because of the great variety of colors of its compounds, iridium takes its name from the Latin *iris*, meaning "rainbow." Alternatively, an element name might suggest a mineral or the ore that contained it. One example is Wolfram or tungsten (W), which was isolated from wolframite. Two other "inconsistent" elemental symbols, K and Na, arose from occurrence as well. *Kalium* was first obtained from the saltwort plant, *Salsola kali*, and *natrium* from niter. Their English names, potassium and sodium, are derived from the ores potash and soda.

Other elements, contrary to Lavoisier's suggestion, were named after planets, mythological figures, places, or superstitions. "Celestial elements" include helium ("sun"), tellurium ("earth"), selenium ("moon"—the element was discovered in close proximity to tellurium), cerium (the asteroid Ceres, which was discovered only two years before the element), and uranium (the planet Uranus, discovered a few years earlier). The first two transuranium elements (those *beyond* uranium) to be produced were named neptunium and plutonium for the next two planets, Neptune and Pluto. The names promethium (Prometheus, who stole fire from heaven), vanadium (Scandinavian goddess Vanadis), titanium (Titans, the first sons of the earth), tantalum (Tantalos, father of the Greek goddess Niobe), and thorium (Thor, Scandinavian god of war) all arise from Greek or Norse mythology.

"Geographical elements," shown on the map, sometimes honored the discoverer's native country or workplace. The Latin names for Russia (*ruthenium*), France (*gallium*), Paris (*lutetium*), and Germany (*germanium*) were among those used. Marie Sklodowska Curie named one of the elements that she discovered polonium, after her native Poland. Often the locale of discovery lends its name to the element; the record holder is certainly the Swedish village Ytterby, the site of ores from which the four elements terbium, erbium, ytterbium, and yttrium were isolated. Elements honoring important scientists include curium, einsteinium, nobelium, fermium, and lawrencium.

Most of the elements now known were given titles peacefully, but a few were not. Niobium, isolated in 1803 by Ekeberg from an ore that also contained tantalum, and named after Niobe (daughter of Tantalos), was later found to be identical to an 1802 discovery of C. Hatchett, columbium. (Interestingly, Hatchett first found the element in an ore sample that had been sent to England more than a century

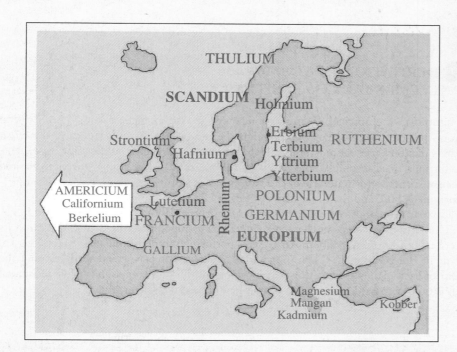

THULIUM

SCANDIUM Holmium

Strontium
Erbium
Terbium RUTHENIUM
Hafnium
Yttrium
Ytterbium

AMERICIUM
Californium POLONIUM
Berkelium Lutetium
GERMANIUM
FRANCIUM Rhenium
EUROPIUM

GALLIUM

Magnesium
Mangan
Kadmium Kobber

Many chemical elements were named after places. From Vivi Ringnes, "Origin and Names of Chemical Elements," *Journal of Chemical Education*, 66, 1989, 731–737. Reprinted by permission

earlier by John Winthrop, the first governor of Connecticut.) Although "niobium" became the accepted designation in Europe, the Americans, not surprisingly, chose "columbium." It was not until 1949—when the International Union of Pure and Applied Chemistry (IUPAC) ended more than a century of controversy by ruling in favor of mythology—that element 41 received a unique name.

In 1978, the IUPAC recommended that elements beyond 103 be known temporarily by systematic names based on numerical roots; element 104 is unnilquadium (*un* for 1, *nil* for 0, *quad* for 4, plus the *-ium* ending), followed by unnilpentium, unnilhexium, and so on. Arguments over the names of elements 104 and 105 prompted the IUPAC to begin hearing claims of priority to numbers 104 to 109. The IUPAC's final recommendations for these element names were announced in 1997. The names and symbols recommended by that report are: element 104, rutherfordium, Rf; element 105, dubnium, Db; element 106, seaborgium, Sg; element 107, bohrium, Bh; element 108, hassium, Hs; and element 109, meitnerium, Mt. Some of these (Rf and Bh) are derived from the names of scientists prominent in the development of atomic theory; others (Sg, Hs, and Mt) are named for scientists who were involved in the discovery of heavy elements. Dubnium is named in honor of the Dubna laboratory in the former Soviet Union, where important contributions to the creation of heavy elements have originated.

Lisa Saunders Boffa
Senior Chemist
Exxon Corporation

Problem-Solving Tip: *The Whole Is Equal to the Sum of Its Parts*

Percentages must add to 100%. Roundoff errors may not cancel, however, and totals such as 99.9% or 100.1% may be obtained in calculations. As an alternative method of calculation, if we know all of the percentages except one, we can subtract their sum from 100% to obtain the remaining value.

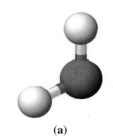

(a)

(b)

A ball-and-stick model of (a) a molecule of water, H_2O, and (b) a molecule of hydrogen peroxide, H_2O_2.

2-9 DERIVATION OF FORMULAS FROM ELEMENTAL COMPOSITION

The **simplest,** or **empirical, formula** for a compound is the smallest whole-number ratio of atoms present. For molecular compounds the **molecular formula** indicates the *actual* numbers of atoms present in a molecule of the compound. It may be the same as the simplest formula or else some whole-number multiple of it. For example, the simplest and molecular formulas for water are both H_2O; however, for hydrogen peroxide, the empirical formula is HO, and the molecular formula is H_2O_2.

Each year thousands of new compounds are made in laboratories or discovered in nature. One of the first steps in characterizing a new compound is the determination of its percent composition. A *qualitative* analysis is performed to determine *which* elements are present in the compound. Then a *quantitative* analysis is performed to determine the *amount* of each element. Once the percent composition of a compound (or its elemental composition by mass) is known, the simplest formula can be determined.

EXAMPLE 2-12   *Simplest Formulas*

Compounds containing sulfur and oxygen are serious air pollutants; they represent the major cause of acid rain. Analysis of a sample of a pure compound reveals that it contains 50.1% sulfur and 49.9% oxygen by mass. What is the simplest formula of the compound?

Plan

One mole of atoms of any element is 6.02×10^{23} atoms, so the ratio of moles of atoms in any sample of a compound is the same as the ratio of atoms in that compound. This calculation is carried out in two steps.

Step 1: Let's consider 100.0 g of compound, which contains 50.1 g of S and 49.9 g of O. We calculate the number of moles of atoms of each.

Step 2: We then obtain a whole-number ratio between these numbers that gives the ratio of atoms in the sample and hence in the simplest formula for the compound.

Remember that *percent* means "parts per hundred."

Solution

Step 1: $\underline{?}$ mol S atoms $= 50.1 \text{ g S} \times \dfrac{1 \text{ mol S atoms}}{32.1 \text{ g S}} = 1.56$ mol S atoms

$\underline{?}$ mol O atoms $= 49.9 \text{ g O} \times \dfrac{1 \text{ mol O atoms}}{16.0 \text{ g O}} = 3.12$ mol O atoms

CD-ROM   Screen 3.18, Determining Empirical Formulas

Step 2: Now we know that 100.0 g of the compound contains 1.56 mol of S atoms and 3.12 mol of O atoms. We obtain a whole-number ratio between these numbers that gives the ratio of atoms in the simplest formula.

$$\frac{1.56}{1.56} = 1.00 \ S$$

$$\frac{3.12}{1.56} = 2.00 \ O$$

SO_2

You should now work Exercise 54.

> A simple and useful way to obtain whole-number ratios among several numbers follows: (a) Divide each number by the smallest number, and then, (b) if necessary, multiply all the resulting numbers by the smallest whole number that will eliminate fractions.

The solution for Example 2-12 can be set up in tabular form.

Element	Relative Mass of Element	Relative Number of Atoms (divide mass by AW)	Divide by Smaller Number	Smallest Whole-Number Ratio of Atoms
S	50.1	$\dfrac{50.1}{32.1} = 1.56$	$\dfrac{1.56}{1.56} = 1.00 \ S$	
O	49.9	$\dfrac{49.9}{16.0} = 3.12$	$\dfrac{3.12}{1.56} = 2.00 \ O$	SO_2

> The "Relative Mass of Element" column is proportional to the mass of each element in grams. With this interpretation, the next column could be headed "Relative Number of *Moles* of Atoms." Then the last column would represent the smallest whole-number ratios of *moles* of atoms. But because a mole is always the same number of items (atoms), that ratio is the same as the smallest whole-number ratio of atoms.

This tabular format provides a convenient way to solve simplest-formula problems, as the next example illustrates.

EXAMPLE 2-13 *Simplest Formula*

A 20.882-g sample of an ionic compound is found to contain 6.072 g of Na, 8.474 g of S, and 6.336 g of O. What is its simplest formula?

Plan

We reason as in Example 2-12, calculating the number of moles of each element and the ratio among them. Here we use the tabular format that was introduced earlier.

Solution

Element	Relative Mass of Element	Relative Number of Atoms (divide mass by AW)	Divide by Smallest Number	Convert Fractions to Whole Numbers (multiply by integer)	Smallest Whole-Number Ratio of Atoms
Na	6.072	$\dfrac{6.072}{23.0} = 0.264$	$\dfrac{0.264}{0.264} = 1.00$	$1.00 \times 2 = 2 \ Na$	
S	8.474	$\dfrac{8.474}{32.1} = 0.264$	$\dfrac{0.264}{0.264} = 1.00$	$1.00 \times 2 = 2 \ S$	$Na_2S_2O_3$
O	6.336	$\dfrac{6.336}{16.0} = 0.396$	$\dfrac{0.396}{0.264} = 1.50$	$1.50 \times 2 = 3 \ O$	

The ratio of atoms in the simplest formula *must be a whole-number ratio* (by definition). To convert the ratio $1:1:1.5$ to a whole-number ratio, each number in the ratio was multiplied by 2, which gave the simplest formula $Na_2S_2O_3$.

You should now work Exercise 56.

 Don't forget to multiply all calculated coefficients by this common factor to obtain the final integer coefficients!

✔ **Problem-Solving Tip:** *Know Common Fractions in Decimal Form*

As Example 2-13 illustrates, sometimes we must convert a fraction to a whole number by multiplying the fraction by the correct integer. But we must first recognize which fraction is represented by a nonzero part of a number. The decimal equivalents of the following fractions may be useful.

Decimal Equivalent (to two places)	Fraction	To Convert to Integer, Multiply by
0.50	$\frac{1}{2}$	2
0.33	$\frac{1}{3}$	3
0.67	$\frac{2}{3}$	3
0.25	$\frac{1}{4}$	4
0.75	$\frac{3}{4}$	4
0.20	$\frac{1}{5}$	5

The fractions $\frac{2}{5}$, $\frac{3}{5}$, and $\frac{4}{5}$ are equal to 0.40, 0.60, and 0.80, respectively; these should be multiplied by 5.

When we use the procedure given in this section, we often obtain numbers such as 0.99 and 1.52. Because the results obtained by analysis of samples usually contain some error (as well as roundoff errors), we would interpret 0.99 as 1.0 and 1.52 as 1.5.

Millions of compounds are composed of carbon, hydrogen, and oxygen. Analyses for C and H can be performed in a C-H combustion system (Figure 2-10). An accurately known

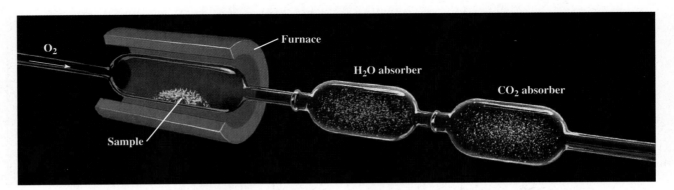

Figure 2-10 A combustion train used for carbon–hydrogen analysis. The absorbent for water is magnesium perchlorate, $Mg(ClO_4)_2$. Carbon dioxide is absorbed by finely divided sodium hydroxide supported on glass wool. Only a few milligrams of sample is needed for analysis. The increase in weight of the H_2O absorber can be converted into the amount of hydrogen present, while the increase in the CO_2 absorber can be used to calculate the amount of carbon present.

mass of a compound is burned in a furnace in a stream of oxygen. The carbon and hydrogen in the sample are converted to carbon dioxide and water vapor, respectively. The resulting increases in masses of the CO_2 and H_2O absorbers can then be related to the masses and percentages of carbon and hydrogen in the original sample.

EXAMPLE 2-14 *Percent Composition*

Hydrocarbons are organic compounds composed entirely of hydrogen and carbon. A 0.1647-g sample of a pure hydrocarbon was burned in a C-H combustion train to produce 0.4931 g of CO_2 and 0.2691 g of H_2O. Determine the masses of C and H in the sample and the percentages of these elements in this hydrocarbon.

Plan

Step 1: We use the observed mass of CO_2, 0.4931 g, to determine the mass of carbon in the original sample. There is one mole of carbon atoms, 12.01 g, in each mole of CO_2, 44.01 g; we use this information to construct the unit factor

$$\frac{12.01 \text{ g C}}{44.01 \text{ g } CO_2}$$

Step 2: Likewise, we can use the observed mass of H_2O, 0.2691 g, to calculate the amount of hydrogen in the original sample. We use the fact that there are two moles of hydrogen atoms, 2.016 g, in each mole of H_2O, 18.02 g, to construct the unit factor

$$\frac{2.016 \text{ g H}}{18.02 \text{ g } H_2O}$$

Step 3: Then we calculate the percentages by mass of each element in turn, using the relationship

$$\% \text{ element} = \frac{\text{g element}}{\text{g sample}} \times 100\%$$

Solution

Step 1: $\underline{?}$ g C = 0.4931 g $CO_2 \times \dfrac{12.01 \text{ g C}}{44.01 \text{ g } CO_2}$ = $\boxed{0.1346 \text{ g C}}$

Step 2: $\underline{?}$ g H = 0.2691 g $H_2O \times \dfrac{2.016 \text{ g H}}{18.02 \text{ g } H_2O}$ = $\boxed{0.03010 \text{ g H}}$

Step 3:   % C = $\dfrac{0.1346 \text{ g C}}{0.1647 \text{ g sample}} \times 100\%$ = $\boxed{81.72\% \text{ C}}$

              %H = $\dfrac{0.03010 \text{ g H}}{0.1647 \text{ g sample}} \times 100\%$ = $\boxed{18.28\% \text{ H}}$

Total = 100.00%

You should now work Exercise 66.

Hydrocarbons are obtained from coal and coal tar and from oil and gas wells. The main use of hydrocarbons is as fuels. The simplest hydrocarbons are

methane	CH_4
ethane	C_2H_6
propane	C_3H_8
butane	C_4H_{10}

We could calculate the mass of H by subtracting the mass of C from the mass of the sample. It is good experimental practice, however, when possible, to base both on experimental measurements, as we have done here. This would help to check for errors in the analysis or calculation.

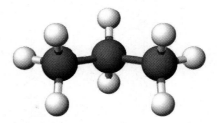

Can you show that the hydrocarbon in Example 2-14 is propane, C_3H_8?

When the compound to be analyzed contains oxygen, the calculation of the amount or percentage of oxygen in the sample is somewhat different. Part of the oxygen that goes to form CO_2 and H_2O comes from the sample, and part comes from the oxygen stream supplied. For that reason, we cannot directly determine the amount of oxygen already in the

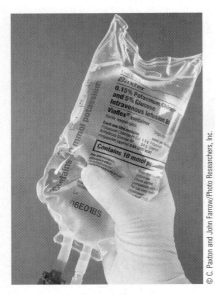

Glucose, a simple sugar, is the main component of intravenous feeding liquids. Its common name is dextrose. It is also one of the products of carbohydrate metabolism.

sample. The approach is to analyze as we did in Example 2-14 for all elements *except* oxygen. Then we subtract the sum of their masses from the mass of the original sample to obtain the mass of oxygen. The next example illustrates such a calculation.

EXAMPLE 2-15   *Percent Composition*

A 0.1014-g sample of purified glucose was burned in a C-H combustion train to produce 0.1486 g of CO_2 and 0.0609 g of H_2O. An elemental analysis showed that glucose contains only carbon, hydrogen, and oxygen. Determine the masses of C, H, and O in the sample and the percentages of these elements in glucose.

Plan

Steps 1 and 2: We first calculate the masses of carbon and hydrogen as we did in Example 2-14.

Step 3: The rest of the sample must be oxygen because glucose has been shown to contain only C, H, and O. So we subtract the masses of C and H from the total mass of sample.

Step 4: Then we calculate the percentage by mass for each element.

Solution

We say that the mass of O in the sample is calculated by *difference*.

Step 1: $\text{? g C} = 0.1486 \text{ g } CO_2 \times \dfrac{12.01 \text{ g C}}{44.01 \text{ g } CO_2} = \boxed{0.04055 \text{ g C}}$

Step 2: $\text{? g H} = 0.0609 \text{ g } H_2O \times \dfrac{2.016 \text{ g H}}{18.02 \text{ g } H_2O} = \boxed{0.00681 \text{ g H}}$

Step 3: $\text{? g O} = 0.1014 \text{ g sample} - [0.04055 \text{ g C} + 0.00681 \text{ g H}] = \boxed{0.0540 \text{ g O}}$

Step 4: Now we can calculate the percentages by mass for each element:

$$\% \text{ C} = \dfrac{0.04055 \text{ g C}}{0.1014 \text{ g}} \times 100\% = \boxed{39.99\% \text{ C}}$$

$$\% \text{ H} = \dfrac{0.00681 \text{ g H}}{0.1014 \text{ g}} \times 100\% = \boxed{6.72\% \text{ H}}$$

$$\% \text{ O} = \dfrac{0.0540 \text{ g O}}{0.1014 \text{ g}} \times 100\% = \boxed{53.2\% \text{ O}}$$

Total $= \ 99.9\%$

You should now work Exercise 68.

2-10   DETERMINATION OF MOLECULAR FORMULAS

Percent composition data yield only simplest formulas. To determine the molecular formula for a molecular compound, *both* its *empirical* formula and its molecular weight must be known. Some methods for experimental determination of molecular weights are introduced in Chapters 12 and 14.

© C. Paxton and John Farrow/Photo Researchers, Inc.

For many compounds the molecular formula is a multiple of the simplest formula. Consider butane, C_4H_{10}. The *empirical* formula for butane is C_2H_5, but the molecular formula contains twice as many atoms; that is, $2 \times (C_2H_5) = C_4H_{10}$. Benzene, C_6H_6, is another example. The simplest formula for benzene is CH, but the molecular formula contains six times as many atoms; that is, $6 \times (CH) = C_6H_6$.

The molecular formula for a compound is either the same as, or an *integer* multiple of, the (*empirical*) formula.

$$\text{molecular formula} = n \times \text{simplest formula}$$

So we can write

$$\text{molecular weight} = n \times \text{simplest formula weight}$$

$$n = \frac{\text{molecular weight}}{\text{simplest formula weight}}$$

The molecular formula is then obtained by multiplying the *empirical* formula by the integer, n.

CD-ROM   Screen 3.18,
Determining Molecular Formulas

EXAMPLE 2-16   *Molecular Formula*

In Example 2-15, we found the elemental composition of glucose. Other experiments show that its molecular weight is approximately 180 amu. Determine the simplest formula and the molecular formula of glucose.

Plan

Step 1: We first use the masses of C, H, and O found in Example 2-15 to determine the simplest formula.

Step 2: We can use the simplest formula to calculate the simplest formula weight. Because the molecular weight of glucose is known (approximately 180 amu), we can determine the molecular formula by dividing the molecular weight by the simplest formula weight.

$$n = \frac{\text{molecular weight}}{\text{simplest formula weight}}$$

The molecular weight is n times the *empirical* formula weight, so the molecular *formula* of glucose is n times the *empirical formula*.

As an alternative, we could have used the percentages by mass from Example 2-15. Using the earliest available numbers helps to minimize the effects of rounding errors.

Solution

Step 1:

Element	Relative Mass of Element	Relative Number of Atoms (divide mass by AW)	Divide by Smallest	Smallest Whole-Number Ratio of Atoms
C	0.04055	$\dfrac{0.04055}{12.01} = 0.003376$	$\dfrac{0.003376}{0.003376} = 1.00$ C	
H	0.00681	$\dfrac{0.00681}{1.008} = 0.00676$	$\dfrac{0.00676}{0.003376} = 2.00$ H	 CH_2O
O	0.0540	$\dfrac{0.0540}{16.00} = 0.00338$	$\dfrac{0.00338}{0.003376} = 1.00$ O	

Many sugars are rich sources in our diet. The most familiar is ordinary table sugar, which is sucrose, $C_{12}H_{22}O_{11}$. An enzyme in our saliva readily splits sucrose into two simple sugars, glucose and fructose. The simplest formula for both glucose and fructose is $C_6H_{12}O_6$. They have different structures and different properties, however, so they are different compounds.

Step 2: The simplest formula is CH_2O, which has a formula weight of 30.03 amu. Because the molecular weight of glucose is approximately 180 amu, we can determine the molecular formula by dividing the molecular weight by the *empirical* formula weight.

$$n = \frac{180 \text{ amu}}{30.03 \text{ amu}} = 6.00$$

The molecular weight is six times the simplest formula weight, $6 \times (CH_2O) = C_6H_{12}O_6$, so the molecular formula of glucose is $C_6H_{12}O_6$.

You should now work Exercises 49 and 50.

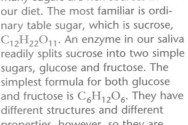

As we shall see when we discuss the composition of compounds in some detail, two (and sometimes more) elements may form more than one compound. The **Law of Multiple Proportions** summarizes many experiments on such compounds. It is usually stated: When two elements, A and B, form more than one compound, the ratio of the masses of element B that combine with a given mass of element A in each of the compounds can be expressed by small whole numbers. Water, H_2O, and hydrogen peroxide, H_2O_2, provide an example. The ratio of masses of oxygen that combine with a given mass of hydrogen in H_2O and H_2O_2 is 1:2. In H_2O, one mole of oxygen combines with two moles of hydrogen atoms, while in hydrogen peroxide, H_2O_2, two moles of oxygen combine with two moles of hydrogen atoms. Thus the ratio of oxygen atoms in the two compounds compared to a given number of hydrogen atoms is 1:2. Many similar examples, such as CO and CO_2 (1:2 oxygen ratio) and SO_2 and SO_3 (2:3 oxygen ratio), are known. The Law of Multiple Proportions had been recognized from studies of elemental composition before the time of Dalton. It provided additional support for his atomic theory.

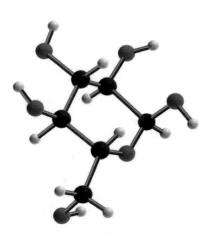

EXAMPLE 2-17 ·*Law of Multiple Proportions*

What is the ratio of the numbers of oxygen atoms that are combined with a given number of nitrogen atoms in the compounds N_2O_3 and NO?

Plan

To compare the number of oxygen atoms, we need to have *equal numbers of nitrogen atoms.*

Solution

Because NO has half as many nitrogen atoms in its formula relative to N_2O_3, we must multiply it by a factor of 2 to compare the two elements on the basis of an equal number of nitrogen atoms. We could, of course, also divide N_2O_3 by a factor of two, but that would produce fractional subscripts that most chemists like to avoid. (This approach would, however, give us the same answer.) Note that we show the number of each element present, cancel out the equal amounts of nitrogen atoms, leaving the ratio of oxygen atoms.

$$\text{oxygen ratio} = \frac{N_2O_3}{2(NO)} = \frac{3O/2\cancel{N}}{2O/2\cancel{N}} = \frac{3O}{2O} = \frac{3}{2}$$

You should now work Exercises 71 and 72.

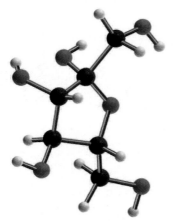

Ball and stick models of glucose (*above*) and fructose (*below*). Gray atoms = carbon, white atoms = hydrogen, and red atoms = oxygen.

2-11 SOME OTHER INTERPRETATIONS OF CHEMICAL FORMULAS

Once we master the mole concept and the meaning of chemical formulas, we can use them in many other ways. The examples in this section illustrate a few additional kinds of information we can get from a chemical formula and the mole concept.

EXAMPLE 2-18 *Composition of Compounds*

What mass of chromium is contained in 35.8 g of $(NH_4)_2Cr_2O_7$?

Plan

Let us first solve the problem in several steps.

Step 1: The formula tells us that each mole of $(NH_4)_2Cr_2O_7$ contains two moles of Cr atoms, so we first find the number of moles of $(NH_4)_2Cr_2O_7$, using the unit factor

$$\frac{1 \text{ mol } (NH_4)_2Cr_2O_7}{252.0 \text{ g } (NH_4)_2Cr_2O_7}$$

Step 2: Then we convert the number of moles of $(NH_4)_2Cr_2O_7$ into the number of moles of Cr atoms it contains, using the unit factor

$$\frac{2 \text{ mol Cr atoms}}{1 \text{ mol } (NH_4)_2Cr_2O_7}$$

Step 3: We then use the atomic weight of Cr to convert the number of moles of chromium atoms to the mass of chromium.

$$\text{mass } (NH_4)_2Cr_2O_7 \longrightarrow \text{mol } (NH_4)_2Cr_2O_7 \longrightarrow \text{mol Cr} \longrightarrow \boxed{\text{mass Cr}}$$

Solution

Step 1: $\underline{?}$ mol $(NH_4)_2Cr_2O_7 = 35.8 \text{ g } (NH_4)_2Cr_2O_7 \times \dfrac{1 \text{ mol } (NH_4)_2Cr_2O_7}{252.0 \text{ g } (NH_4)_2Cr_2O_7}$

$= \boxed{0.142 \text{ mol } (NH_4)_2Cr_2O_7}$

Step 2: $\underline{?}$ mol Cr atoms $= 0.142 \text{ mol } (NH_4)_2Cr_2O_7 \times \dfrac{2 \text{ mol Cr atoms}}{1 \text{ mol } (NH_4)_2Cr_2O_7}$

$= \boxed{0.284 \text{ mol Cr atoms}}$

Step 3: $\underline{?}$ g Cr $= 0.284 \text{ mol Cr atoms} \times \dfrac{52.0 \text{ g Cr}}{1 \text{ mol Cr atoms}} = \boxed{14.8 \text{ g Cr}}$

If you understand the reasoning in these conversions, you should be able to solve this problem in a single setup:

$\underline{?}$ g Cr $= 35.8 \text{ g } (NH_4)_2Cr_2O_7 \times \dfrac{1 \text{ mol } (NH_4)_2Cr_2O_7}{252.0 \text{ g } (NH_4)_2Cr_2O_7} \times \dfrac{2 \text{ mol Cr atoms}}{1 \text{ mol } (NH_4)_2Cr_2O_7} \times \dfrac{52.0 \text{ g Cr}}{1 \text{ mol Cr}} = \boxed{14.8 \text{ g Cr}}$

You should now work Exercise 76.

EXAMPLE 2-19   Composition of Compounds

What mass of potassium chlorate, $KClO_3$, would contain 40.0 g of oxygen?

Plan

The formula $KClO_3$ tells us that each mole of $KClO_3$ contains three moles of oxygen atoms. Each mole of oxygen atoms weighs 16.0 g. So we can set up the solution to convert:

$$\text{mass O} \longrightarrow \text{mol O} \longrightarrow \text{mol } KClO_3 \longrightarrow \text{mass } KClO_3$$

Solution

$$\underline{?}\ g\ KClO_3 = 40.0\ g\ O \times \frac{1\ \text{mol O atoms}}{16.0\ g\ O\ \text{atoms}} \times \frac{1\ \text{mol } KClO_3}{3\ \text{mol O atoms}} \times \frac{122.6\ g\ KClO_3}{1\ \text{mol } KClO_3}$$

$$= \boxed{102\ g\ KClO_3}$$

You should now work Exercise 78.

✓ Problem-Solving Tip: *How Do We Know When . . . ?*

How do we know when to represent oxygen as O and when as O_2? A *compound* that contains oxygen *does not* contain O_2 molecules. So we solve problems such as Example 2-19 in terms of moles of O atoms. Thus, we must use the formula weight for O, which is 16.0 g O atoms/1 mol O atoms. Similar reasoning applies to compounds containing other elements that are polyatomic molecules in *pure elemental form*, such as H_2, Cl_2, or P_4.

EXAMPLE 2-20   Composition of Compounds

(a) What mass of sulfur dioxide, SO_2, would contain the same mass of oxygen as is contained in 33.7 g of arsenic pentoxide, As_2O_5?
(b) What mass of calcium chloride, $CaCl_2$, would contain the same number of chloride ions as are contained in 48.6 g of sodium chloride, NaCl?

Plan

(a) We could find explicitly the number of grams of O in 33.7 g of As_2O_5, and then find the mass of SO_2 that contains that same number of grams of O. But this method includes some unnecessary calculation. We need only convert to *moles* of O (because this is the same amount of O regardless of its environment) and then to moles of SO_2 to obtain mass of SO_2.

$$\text{mass } As_2O_5 \longrightarrow \text{mol } As_2O_5 \longrightarrow \text{mol O atoms} \longrightarrow \text{mol } SO_2 \longrightarrow \text{mass } SO_2$$

(b) Because one mole always consists of the same number (Avogadro's number) of items, we can reason in terms of *moles* of Cl^- ions and solve as in part (a).

$$\text{mass NaCl} \longrightarrow \text{mol NaCl} \longrightarrow \text{mol } Cl^- \text{ ions} \longrightarrow \text{mol } CaCl_2 \longrightarrow \text{mass } CaCl_2$$

Solution

(a)

$$\underline{?}\ g\ SO_2 = 33.7\ g\ As_2O_5 \times \frac{1\ \text{mol } As_2O_5}{229.8\ g\ As_2O_5} \times \frac{5\ \text{mol O atoms}}{1\ \text{mol } As_2O_5}$$

$$\times \frac{1\ \text{mol } SO_2}{2\ \text{mol O atoms}} \times \frac{64.1\ g\ SO_2}{1\ \text{mol } SO_2} = \boxed{23.5\ SO_2}$$

(b) $\underline{?} \text{ g CaCl}_2 = 48.6 \text{ NaCl} \times \dfrac{1 \text{ mol NaCl}}{58.4 \text{ g NaCl}} \times \dfrac{1 \text{ mol Cl}^-}{1 \text{ mol NaCl}}$

$\times \dfrac{1 \text{ mol CaCl}_2}{2 \text{ mol Cl}^-} \times \dfrac{111.0 \text{ g CaCl}_2}{1 \text{ mol CaCl}_2} = \boxed{46.2 \text{ g CaCl}_2}$

You should now work Exercise 80.

Charles Steele

Figure 2-11 One mole of some compounds. The colorless liquid is water, H_2O (1 mol = 18.0 g = 18.0 mL). The white solid (*front left*) is *anhydrous* oxalic acid, $(COOH)_2$ (1 mol = 90.0 g). The second white solid (*front right*) is *hydrated* oxalic acid, $(COOH)_2 \cdot 2H_2O$ (1 mol = 126.0 g). The blue solid is hydrated copper(II) sulfate, $CuSO_4 \cdot 5H_2O$ (1 mol = 249.7 g). The red solid is mercury(II) oxide (1 mol = 216.6 g).

The physical appearance of one mole of each of some compounds is illustrated in Figure 2-11. Two different forms of oxalic acid are shown. The formula unit (molecule) of oxalic acid is $(COOH)_2$ (FW = 90.0 amu; molar mass = 90.0 g/mol). When oxalic acid is obtained by crystallization from a water solution, however, two molecules of water are present for each molecule of oxalic acid, even though it appears dry. The formula of this **hydrate** is $(COOH)_2 \cdot 2H_2O$ (FW = 126.1 amu; molar mass = 126.1 g/mol). The dot shows that the crystals contain two H_2O molecules per $(COOH)_2$ molecule. The water can be driven out of the crystals by heating to leave **anhydrous** oxalic acid, $(COOH)_2$. Anhydrous means "without water." Copper(II) sulfate, an *ionic* compound, shows similar behavior. Anhydrous copper(II) sulfate ($CuSO_4$; FW = 159.6 amu; molar mass = 159.6 g/mol) is almost white. Hydrated copper(II) sulfate ($CuSO_4 \cdot 5H_2O$; FW = 249.7 amu; molar mass = 249.7 g/mol) is deep blue. The following example illustrates how we might find and use the formula of a hydrate.

Charles D. Winters

Heating blue $CuSO_4 \cdot 5H_2O$ forms anhydrous $CuSO_4$, which is gray. Some blue $CuSO_4 \cdot 5H_2O$ is visible in the cooler center portion of the crucible.

EXAMPLE 2-21   *Composition of Compounds*

A reaction requires pure anhydrous calcium sulfate, $CaSO_4$. Only an unidentified hydrate of calcium sulfate, $CaSO_4 \cdot xH_2O$, is available.

(a) We heat 67.5 g of unknown hydrate until all the water has been driven off. The resulting mass of pure $CaSO_4$ is 53.4 g. What is the formula of the hydrate, and what is its formula weight?

(b) Suppose we wish to obtain enough of this hydrate to supply 95.5 g of $CaSO_4$ after heating. How many grams should we weigh out?

Plan

(a) To determine the formula of the hydrate, we must find the value of x in the formula $CaSO_4 \cdot xH_2O$. The mass of water removed from the sample is equal to the difference in the two masses given. The value of x is the number of moles of H_2O per mole of $CaSO_4$ in the hydrate.

(b) The formula weights of $CaSO_4$, 136.2 g/mol, and of $CaSO_4 \cdot xH_2O$, $(136.2 + x18.0)$ g/mol, allow us to write the conversion factor required for the calculation.

Solution

(a) ? g water driven off = 67.5 g $CaSO_4 \cdot xH_2O$ − 53.4 g $CaSO_4$ = 14.1 g H_2O

$$x = \frac{?\ \text{mol } H_2O}{\text{mol } CaSO_4} = \frac{14.1\ \text{g } H_2O}{53.4\ \text{g } CaSO_4} \times \frac{1\ \text{mol } H_2O}{18.0\ \text{g } H_2O} \times \frac{136.2\ \text{g } CaSO_4}{1\ \text{mol } CaSO_4} = \frac{2.00\ \text{mol } H_2O}{\text{mol } CaSO_4}$$

Thus, the formula of the hydrate is $CaSO_4 \cdot 2H_2O$. Its formula weight is

$$FW = 1 \times (\text{formula weight } CaSO_4) + 2 \times (\text{formula weight } H_2O)$$

$$= 136.2\ \text{g/mol} + 2(18.0\ \text{g/mol}) = 172.2\ \text{g/mol}$$

(b) The formula weights of $CaSO_4$ (136.2 g/mol) and of $CaSO_4 \cdot 2H_2O$ (172.2 g/mol) allow us to write the unit factor

$$\frac{172.2\ \text{g } CaSO_4 \cdot 2H_2O}{136.2\ \text{g } CaSO_4}$$

We use this factor to perform the required conversion:

$$?\ \text{g } CaSO_4 \cdot 2H_2O = 95.5\ \text{g } CaSO_4 \text{ desired} \times \frac{172.2\ \text{g } CaSO_4 \cdot 2H_2O}{136.2\ \text{g } CaSO_4}$$

$$= 121\ \text{g } CaSO_4 \cdot 2H_2O$$

You should now work Exercise 82.

CD-ROM   Screen 3.19, Hydrated Compounds

Impurities are not necessarily bad. For example, inclusion of 0.02% KI, potassium iodide, in ordinary table salt has nearly eliminated goiter in the United States. Goiter is a disorder of the thyroid gland caused by a deficiency of iodine. Mineral water tastes better than purer, distilled water.

2-12 PURITY OF SAMPLES

Most substances obtained from laboratory reagent shelves are not 100% pure. The **percent purity** is the mass percentage of a specified substance in an impure sample. When impure samples are used for precise work, account must be taken of impurities. The photo in the margin shows the label from reagent-grade sodium hydroxide, NaOH, which is 98.2% pure by mass. From this information we know that total impurities represent 1.8% of the mass of this material. We can write several unit factors:

$$\frac{98.2\ \text{g NaOH}}{100\ \text{g sample}}, \quad \frac{1.8\ \text{g impurities}}{100\ \text{g sample}}, \quad \text{and} \quad \frac{1.8\ \text{g impurities}}{98.2\ \text{g NaOH}}$$

The inverse of each of these gives us a total of six unit factors.

EXAMPLE 2-22   *Percent Purity*

Calculate the masses of NaOH and impurities in 45.2 g of 98.2% pure NaOH.

Plan

The percentage of NaOH in the sample gives the unit factor $\dfrac{98.2\ \text{g NaOH}}{100\ \text{g sample}}$. The remainder of the sample is 100% − 98.2% = 1.8% impurities; this gives the unit factor $\dfrac{1.8 \text{ g impurities}}{100 \text{ g sample}}$.

Solution

$$\underline{?} \text{ g NaOH} = 45.2 \text{ g sample} \times \frac{98.2 \text{ g NaOH}}{100 \text{ g sample}} = \boxed{44.4 \text{ g NaOH}}$$

$$\underline{?} \text{ g impurities} = 45.2 \text{ g sample} \times \frac{1.8 \text{ g impurities}}{100 \text{ g sample}} = \boxed{0.81 \text{ g impurities}}$$

You should now work Exercises 86 and 87.

ACTUAL ANALYSIS, LOT G22931

Meets A.C.S. Specifications

Assay (NaOH) (by acidimetry)	98.2	%
Sodium Carbonate (Na₂CO₃)	0.2	%
Chloride (Cl)	< 0.0005	%
Ammonium Hydroxide Precipitate	< 0.01	%
Heavy Metals (as Ag)	< 0.0005	%
Copper (Cu)	0.0003	%
Potassium (K) (by FES)	0.002	%
Trace Impurities (in ppm):		
Nitrogen Compounds (as N)	< 2	
Phosphate (PO₄)	< 1	
Sulfate (SO₄)	< 5	
Iron (Fe)	< 2	
Mercury (Hg) (by AAS)	< 0.003	
Nickel (Ni)	< 2	

A label from a bottle of sodium hydroxide, NaOH.

✓ Problem-Solving Tip: *Utility of the Unit Factor Method*

Observe the beauty of the unit factor approach to problem solving! Such questions as "do we multiply by 0.982 or divide by 0.982?" never arise. The units always point toward the correct answer because we use unit factors constructed so that units *always* cancel out so that we arrive at the desired unit.

Many important relationships have been introduced in this chapter. Some of the most important transformations you have seen in Chapters 1 and 2 are summarized in Figure 2-12.

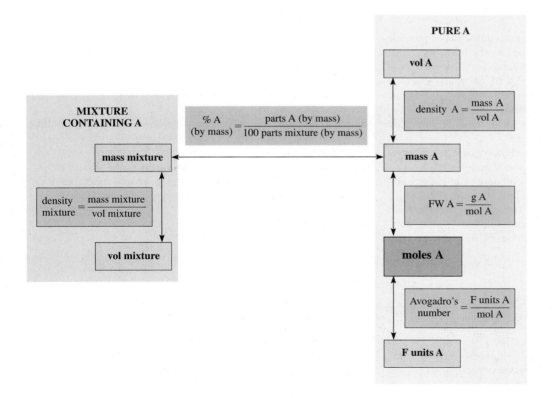

Figure 2-12 Some important relationships from Chapters 1 and 2. The relationships that provide unit factors are enclosed in green boxes. F units = formula units (molecules for molecular systems).

Key Terms

Allotropic modifications (allotropes)   Different forms of the same element in the same physical state.

Anhydrous   Without water.

Anion   An ion with a negative charge.

Atom   The smallest particle of an element that maintains its chemical identity through all chemical and physical changes.

Atomic mass unit (amu)   One twelfth of the mass of an atom of the carbon-12 isotope; a unit used for stating atomic and formula weights.

Atomic number   The number of protons in the nucleus of an atom.

Atomic weight   Weighted average of the masses of the constituent isotopes of an element; the relative mass of atoms of different elements.

Avogadro's number 6.022×10^{23} units of a specified item. See *Mole*.

Cation   An ion with a positive charge.

Chemical formula   Combination of element symbols that indicates the chemical composition of a substance.

Composition stoichiometry   Describes the quantitative (mass) relationships among elements in compounds.

Empirical formula   See *Simplest formula*.

Formula   Combination of element symbols that indicates the chemical composition of a substance.

Formula unit   The smallest repeating unit of a substance—for non-ionic substances, the molecule.

Formula weight   The mass, in atomic mass units, of one formula unit of a substance. Numerically equal to the mass, in grams, of one mole of the substance (see *Molar mass*). This number is obtained by adding the atomic weights of the atoms specified in the formula.

Hydrate   A crystalline sample that contains water, H_2O, and another compound in a fixed mole ratio. Examples include $CuSO_4 \cdot 5H_2O$ and $(COOH)_2 \cdot 2H_2O$.

Ion   An atom or group of atoms that carries an electric charge. A positive ion is a *cation*; a negative ion is an *anion*.

Ionic compound   A compound that is composed of cations and anions. An example is sodium chloride, NaCl.

Law of Constant Composition   See *Law of Definite Proportions*.

Law of Definite Proportions   Different samples of a pure compound always contain the same elements in the same proportions by mass; this corresponds to atoms of these elements in fixed numerical ratios. Also known as the Law of Constant Composition.

Law of Multiple Proportions   When two elements, A and B, form more than one compound, the ratio of the masses of element B that combine with a given mass of element A in each of the compounds can be expressed by small whole numbers.

Molar mass   The mass, in grams, of one mole of a substance; numerically equal to the formula weight of the substance. See *Formula weight*; see *Molecular weight*.

Mole 6.022×10^{23} (Avogadro's number of) formula units (or molecules, for a molecular substance) of a substance. The mass, in grams, of one mole is numerically equal to the formula (molecular) weight of the substance.

Molecular formula   A formula that indicates the actual number of atoms present in a molecule of a molecular substance. Compare with *Simplest (empirical) formula*.

Molecular weight   The mass, in atomic mass units, of one molecule of a non-ionic (molecular) substance. Numerically equal to the mass, in grams, of one mole of such a substance. This number is obtained by adding the atomic weights of the atoms specified in the formula.

Molecule   The smallest particle of an element or compound that can have a stable independent existence.

Percent composition   The mass percentage of each element in a compound.

Percent purity   The mass percentage of a specified compound or element in an impure sample.

Polyatomic   Consisting of more than one atom. Elements such as Cl_2, P_4, and S_8 exist as polyatomic molecules. Examples of polyatomic ions are the ammonium ion, NH_4^+, and the sulfate ion, SO_4^{2-}.

Simplest formula   The smallest whole-number ratio of atoms present in a compound; also called *empirical formula*. Compare with *Molecular formula*.

Stoichiometry   Description of the quantitative relationships among elements in compounds (composition stoichiometry) and among substances as they undergo chemical changes (reaction stoichiometry).

Structural formula   A representation that shows how atoms are connected in a compound.

Exercises

Basic Ideas

1. (a) Define "stoichiometry". (b) Distinguish between composition stoichiometry and reaction stoichiometry.
2. Define "atomic number". Distinguish between atom and molecule.
3. For each of the following elements, give the number of protons in the nucleus, the charge of the nucleus, and the number of electrons surrounding the nucleus: (a) lithium; (b) chlorine; (c) radon; (d) iron; (e) carbon.
4. Give two examples of molecules that represent allotropes. Which of the compounds you selected are diatomic?

5. Write formulas for: (a) hydrogen peroxide; (b) sulfur trioxide; (c) butane; (d) acetic acid.

6. What structural feature distinguishes organic compounds from inorganic compounds?

7. Define the following terms: (a) formula weight; (b) molecular weight; (c) structural formula; (d) ion.

8. Give examples of molecules that contain (a) three atoms of oxygen; (b) two atoms of hydrogen; (c) four atoms total; (d) eight atoms total.

9. Which of the compounds in Table 2-2 are organic compounds?

10. Which of the compounds in Table 2-2 are inorganic and have "acid" in their name?

Names and Formulas

11. Write formulas for the following compounds: (a) pentane; (b) ethyl alcohol; (c) carbon dioxide; (d) acetone; (e) carbon tetrachloride.

12. Name the following: (a) HNO_3; (b) C_4H_{10}; (c) NH_4^+; (d) CH_3OH.

13. Name each of the following ions. Classify each as a monatomic or polyatomic ion. Classify each as a cation or an anion. (a) Na^+; (b) OH^-; (c) SO_4^{2-}; (d) S^{2-}; (e) Zn^{2+}; (f) Fe^{2+}.

14. Write the chemical symbol for each of the following ions. Classify each as a monatomic or polyatomic ion. Classify each as a cation or an anion. (a) magnesium ion; (b) sulfite ion; (c) copper(I) ion; (d) ammonium ion; (e) bromide ion.

15. Name each of the following compounds: (a) $CaCl_2$; (b) $Cu(NO_3)_2$; (c) Na_2SO_4; (d) $Ca(OH)_2$; (e) $FeSO_4$.

16. Write the chemical formula and name for each of the ionic compounds that could be formed by the combination of two of the following ions: Ca^{2+}; Na^+; Cl^-; and SO_4^{2-}.

17. Write the chemical formula for the ionic compound formed between each of the following pairs of ions. Name each compound. (a) Na^+ and S^{2-}, (b) Al^{3+} and SO_4^{2-}; (c) Na^+ and PO_4^{3-}; (d) Mg^{2+} and NO_2^-; (e) Fe^{2+} and CO_3^{2-}.

18. Write the chemical formula for the ionic compound formed between each of the following pairs of ions. Name each compound. (a) Cu^{2+} and CO_3^{2-}; (b) Sr^{2+} and Cl^-; (c) NH_4^+ and CO_3^{2-}; (d) Zn^{2+} and O^{2-}; (e) Fe^{3+} and CH_3COO^-.

19. A student was asked to write the chemical formulas for the following compounds. If the formulas are correct, say so. Explain why any that are incorrect are in error, and correct them. (a) potassium iodide: PI; (b) copper(I) nitrate; $CuNO_3$; (c) silver sulfite: $AgSO_4$.

20. Convert each of the following into a correct formula represented with correct notation. (a) $AlCO_3$; (b) Mg_2Cl; (c) $Zn(OH)_3$; (d) $(NH_4)_3S$; (e) $Na_2(I)_2$.

21. Convert each of the following into a correct formula represented with correct notation. (a) $AlOH_3$; (b) Mg_3CO_3; (c) $Zn(CO_3)_2$; (d) $(NH_4)_3SO_4$; (e) $Mg_2(SO_4)_2$.

22. Write the formula of the compound produced by the combination of each of the following pairs of elements. Name each compound. (a) potassium and chlorine; (b) magnesium and chlorine; (c) sulfur and oxygen; (d) calcium and oxygen; (e) sodium and sulfur; (f) aluminum and sulfur.

23. Write the chemical formula of each of the following: (a) calcium carbonate—major component of coral, seashells, and limestone—found in antacid preparations; (b) magnesium sulfate—found in Epsom salts; (c) acetic acid—the acid in vinegar; (d) sodium hydroxide—common name is lye; (e) zinc oxide—used to protect from sunlight's UV rays when blended in an ointment.

Zinc oxide used as a sunscreen.

Atomic and Formula Weights

24. What is the mass ratio (four significant figures) of one atom of Rb to one atom of Cl?

25. An atom of an element has a mass ever so slightly greater than twice the mass of a Ni atom. Identify the element.

26. (a) What is the atomic weight of an element? (b) Why can atomic weights be referred to as relative numbers?

27. (a) What is the atomic mass unit (amu)? (b) The atomic weight of vanadium is 50.942 amu, and the atomic weight of ruthenium is 101.07 amu. What can we say about the relative masses of V and Ru atoms?

28. Determine the formula weight of each of the following substances: (a) bromine, Br_2; (b) hydrogen peroxide, H_2O_2; (c) saccharin, $C_7H_5NSO_3$; (d) potassium chromate, K_2CrO_4.

29. Determine the formula weight of each of the following substances: (a) calcium sulfate, $CaSO_4$; (b) propane, C_3H_8; (c) the sulfa drug sulfanilamide, $C_6H_4SO_2(NH_2)_2$; (d) uranyl phosphate, $(UO_2)_3(PO_4)_2$.

30. Determine the formula weight of each of the following common acids: (a) hydrochloric acid, HCl; (b) acetic acid, CH_2COOH; (c) phosphoric acid, H_3PO_4; (d) sulfuric acid, H_2SO_4.

31. A sample of 6.68 g of calcium combines exactly with 6.33 g of fluorine, forming calcium fluoride, CaF_2. Find the relative masses of the atoms of calcium and fluorine. Check your answer using a table of atomic weights. If the formula were not known, could you still do this calculation?

The Mole Concept

32. Calculate the mass in grams and kilograms of 1.788 moles of gold.

33. What mass, in grams, should be weighed for an experiment that requires 1.26 mol of $(NH_4)_2HPO_4$?

34. How many hydrogen atoms are contained in 175 grams of propane, C_3H_8?

35. (a) How many formula units are contained in 154.3 g of K_2CrO_4? (b) How many potassium ions? (c) How many CrO_4^{2-} ions? (d) How many atoms of all kinds?

36. How many moles of substance are contained in each of the following samples? (a) 12.3 g of NH_3; (b) 3.32 g of ammonium bromide; (c) 5.6 g of PCl_5; (d) 115 g of Sn.

37. A large neon sign is to be filled with a mixture of gases, including 6.348 g neon. What number of moles is this?

38. How many molecules are in 12.5 g of each of the following substances? (a) CO_2; (b) N_2; (c) P_4; (d) P_2. (e) Do parts (c) and (d) contain the same number of atoms of phosphorus?

39. Sulfur molecules exist under various conditions as S_8, S_6, S_4, S_2, and S. (a) Is the mass of one mole of each of these molecules the same? (b) Is the number of molecules in one mole of each of these molecules the same? (c) Is the mass of sulfur in one mole of each of these molecules the same? (d) Is the number of atoms of sulfur in one mole of each of these molecules the same?

40. Complete the following table. Refer to a table of atomic weights.

Element	Atomic Weight	Mass of One Mole of Atoms
(a) Sn		
(b)	79.904 g	
(c) Ca		
(d)		51.9961 g

41. Complete the following table. Refer to a table of atomic weights.

Element	Formula	Mass of One Mole of Molecules
(a) Br	Br_2	
(b)	H_2	
(c)	P_4	
(d)		20.1797 g
(e) S		256.528 g
(f) O		

42. Complete the following table.

Moles of Compound	Moles of Cations	Moles of Anions
1 mol $KClO_4$		
2 mol Na_2SO_4		
0.2 mol calcium sulfate		
	0.50 mol NH_4^+	0.25 mol SO_4^{2-}

43. Calculate the number of Ni atoms in 1.0 trillionth of a gram of nickel.

44. What is the mass of 10.0 million methane, CH_4, molecules?

45. A sample of propane, C_3H_8, has the same mass as 10.0 million molecules of methane, CH_4. How many C_3H_8 molecules does the sample contain?

Composition Stoichiometry

46. Draw structural, ball-and-stick and space-filling representations of the following chemical formulas: (a) H_2O_2, hydrogen peroxide (b) Cl_2, chlorine (c) CO_2, carbon dioxide.

47. What percent by mass of iron(II) phosphate is iron?

48. Calculate the percent by mass of silver found in a particular mineral that is determined to be silver carbonate.

49. An alcohol is 64.81% C, 13.60% H, and 21.59% O by mass. Another experiment shows that its molecular weight is approximately 74 amu. What is the molecular formula of the alcohol?

50. Skatole is found in coal tar and in human feces. It contains three elements: C, H, and N. It is 82.40% C and 6.92% H by mass. Its simplest formula is its molecular formula. What are (a) the formula and (b) the molecular weight of skatole?

51. Testosterone, the male sex hormone, contains only C, H, and O. It is 79.12% C and 9.79% H by mass. Each molecule contains two O atoms. What are (a) the molecular weight and (b) the molecular formula for testosterone?

*52. The beta-blocker drug, timolol, is expected to reduce the need for heart bypass surgery. Its composition by mass is 49.4% C, 7.64% H, 17.7% N, 15.2% O, and 10.1% S. The mass of 0.0100 mol of timolol is 3.16 g. (a) What is the simplest formula of timolol? (b) What is the molecular formula of timolol?

53. Determine the simplest formula for each of the following compounds: (a) copper(II) tartrate: 30.03% Cu, 22.70% C, 1.91% H, 45.37% O. (b) nitrosyl fluoroborate: 11.99% N, 13.70% O, 9.25% B, 65.06% F.

54. The hormone norepinephrine is released in the human body during stress and increases the body's metabolic rate. Like many biochemical compounds, norepinephrine is composed of carbon, hydrogen, oxygen, and nitrogen. The percent composition of this hormone is 56.8% C, 6.56% H, 28.4% O, and 8.28% N. What is the simplest formula of norepinephrine?

55. (a) A sample of a compound is found to contain 5.60 g N, 14.2 g Cl, and 0.800 g H. What is the simplest formula of this compound? (b) A sample of another compound containing the same elements is found to be 26.2% N, 66.4% Cl, and 7.5% H. What is the simplest formula of this compound?

56. A common product found in nearly every kitchen contains 27.37% sodium, 1.20% hydrogen, 14.30% carbon, and 57.14% oxygen. The simplest formula is the same as the formula of the compound. Find the formula of this compound.

57. Bupropion is present in a medication that is an antidepressant and is also used to aid in quitting smoking. The composition of bupropion is 65.13% carbon, 7.57% hydrogen, 14.79% chlorine, 5.84% nitrogen, and 6.67% oxygen. The simplest formula is the same as the molecular formula of this compound. Determine the formula.

58. Lysine is an essential amino acid. One experiment showed that each molecule of lysine contains two nitrogen atoms. Another experiment showed that lysine contains 19.2% N, 9.64% H, 49.3% C, and 21.9% O by mass. What is the molecular formula for lysine?

59. Cocaine has the following percent composition by mass: 67.30% C, 6.930% H, 21.15% O, and 4.62% N. What is the simplest formula of cocaine?

60. A compound with the molecular weight of 56.0 g was found as a component of photochemical smog. The compound is composed of carbon and oxygen, 42.9% and 57.1%, respectively. What is the formula of this compound?

61. Calculate the percent composition of each of the following compounds: (a) aspartame, $C_{14}H_{18}N_2O_5$; (b) carborundum, SiC; (c) aspirin, $C_9H_8O_4$.

62. Calculate the percent composition of each of the following compounds: (a) dopa, $C_9H_{11}NO_4$; (b) vitamin E, $C_{29}H_{50}O_2$; (c) vanillin, $C_8H_8O_3$.

Determination of Simplest and Molecular Formulas

63. A green compound forms when chromium metal is burned. The percent composition of the compound is: 68.42% Cr and 31.58% O. Is the compound CrO_2, CrO, Cr_2O_3, CrO_3 or some other combination of chromium and oxygen?

*64. Copper is obtained from ores containing the following minerals: azurite, $Cu_3(CO_3)_2(OH)_2$; chalcocite, Cu_2S; chalcopyrite, $CuFeS_2$; covelite, CuS; cuprite, Cu_2O; and malachite, $Cu_2CO_3(OH)_2$. Which mineral has the highest copper content as a percent by mass?

A sample of malachite

65. A 1.20-g sample of a compound gave 2.92 g of CO_2 and 1.22 g of H_2O on combustion in oxygen. The compound is known to contain only C, H, and O. What is its simplest formula?

66. A 0.1153-gram sample of a pure hydrocarbon was burned in a C-H combustion train to produce 0.3986 gram of CO_2 and 0.0578 gram of H_2O. Determine the masses of C and H in the sample and the percentages of these elements in this hydrocarbon.

67. Naphthalene is a hydrocarbon that is used for mothballs. A 0.3204-gram sample of naphthalene was burned in a C-H combustion train to produce 1.100 grams of carbon dioxide and 0.1802 grams of water. What masses and percentages of C and H are present in naphthalene?

68. Combustion of 0.3710 mg of a hydrocarbon produces 1.164 mg of CO_2. What is the simplest formula of the hydrocarbon?

*69. Complicated chemical reactions occur at hot springs on the ocean floor. One compound obtained from such a hot spring consists of Mg, Si, H, and O. From a 0.301-g sample, the Mg is recovered as 0.104 g of MgO; H is recovered as 23.1 mg of H_2O; and Si is recovered as 0.155 g of SiO_2. What is the simplest formula of this compound?

70. A 1.000-gram sample of an alcohol was burned in oxygen to produce 1.913 g of CO_2 and 1.174 g of H_2O. The alcohol contained only C, H, and O. What is the simplest formula of the alcohol?

The Law of Multiple Proportions

71. Show that the compounds water, H_2O, and hydrogen peroxide, H_2O_2, obey the Law of Multiple Proportions.

72. Nitric oxide, NO, is produced in internal combustion engines. When NO comes in contact with air, it is quickly converted into nitrogen dioxide, NO_2, a very poisonous, corrosive gas. What mass of O is combined with 3.00 g of N in (a) NO and (b) NO_2? Show that NO and NO_2 obey the Law of Multiple Proportions.

73. Sulfur forms two chlorides. A 45.00-gram sample of one chloride decomposes to give 8.30 g of S and 36.71 g of Cl. A 45.00-gram sample of the other chloride decomposes to give 5.90 g of S and 39.11 g of Cl. Show that these compounds obey the Law of Multiple Proportions.

74. What mass of oxygen is combined with 5.54 g of sulfur in (a) sulfur dioxide, SO_2, and in (b) sulfur trioxide, SO_3?

Interpretation of Chemical Formulas

75. One prominent ore of copper contains chalcopyrite, $CuFeS_2$. How many pounds of copper are contained in 4.63 pounds of pure $CuFeS_2$?

76. Mercury occurs as a sulfide ore called *cinnabar*, HgS. How

many grams of mercury are contained in 925 g of pure HgS?

A sample of cinnabar

77. (a) How many grams of copper are contained in 523 g of $CuSO_4$? (b) How many grams of copper are contained in 523 g of $CuSO_4 \cdot 5H_2O$?

78. What mass of $KMnO_4$ would contain 22.0 g of manganese?

79. What mass of azurite, $Cu_3(CO_3)_2(OH)_2$, would contain 815 g of copper?

A sample of azurite

80. Two minerals that contain copper are chalcopyrite, $CuFeS_2$, and chalcocite, Cu_2S. What mass of chalcocite would contain the same mass of copper as is contained in 250 pounds of chalcopyrite?

81. Tungsten is a very dense metal (19.3 g/cm^3) with extremely high melting and boiling points (3370°C and 5900°C). When a small amount of it is included in steel, the resulting alloy is far harder and stronger than ordinary steel. Two important ores of tungsten are FeWO4 and $CaWO_4$. How many grams of $CaWO_4$ would contain the same mass of tungsten that is present in 835 g of $FeWO_4$?

*82. When a mole of $CuSO_4 \cdot 5H_2O$ is heated to 110°C, it loses four moles of H_2O to form $CuSO_4 \cdot H_2O$. When it is heated to temperatures above 150°C, the other mole of H_2O is lost. (a) How many grams of $CuSO_4 \cdot H_2O$ could be obtained by heating 782 g of $CuSO_4 \cdot 5H_2O$ to 110°C? (b) How many grams of anhydrous $CuSO_4$ could be obtained by heating 782 g of $CuSO_4 \cdot 5H_2O$ to 180°C?

Percent Purity

83. A particular ore of lead, galena, is 10.0% lead sulfide, PbS, and 90.0% impurities by weight. What mass of lead is contained in 50.0 grams of this ore?

84. What mass of chromium is present in 150. grams of an ore of chromium that is 55.0% iron(II) dichromate, $FeCr_2O_7$, and 45.0% impurities by mass? If 90.0% of the chromium can be recovered from 100.0 grams of the ore, what mass of pure chromium is obtained?

85. What masses of (a) Sr and (b) N are contained in 276.7 g of 88.2% pure $Sr(NO_3)_2$? Assume that the impurities do not contain the elements mentioned.

86. (a) What weight of magnesium carbonate is contained in 375 pounds of an ore that is 26.7% magnesium carbonate by weight? (b) What weight of impurities is contained in the sample? (c) What weight of magnesium is contained in the sample? (Assume that no magnesium is present in the impurities.)

87. Vinegar is 5.0% acetic acid, $C_2H_4O_2$, by mass. (a) How many grams of acetic acid are contained in 84.0 g of vinegar? (b) How many pounds of acetic acid are contained in 84.0 pounds of vinegar? (c) How many grams of sodium chloride, NaCl, are contained in 34.0 g of saline solution that is 5.0% NaCl by mass?

*88. What is the percent by mass of copper sulfate, $CuSO_4$, in a sample of copper sulfate pentahydrate, $CuSO_4 \cdot 5H_2O$? (b) What is the percent by mass of $CuSO_4$ in a sample that is 74.4% $CuSO_4 \cdot 5H_2O$ by mass?

Mixed Examples

89. Ammonium nitrate, NH_4NO_3, and urea, CH_4N_2O, are both commonly used as sources of nitrogen in commercial fertilizers. If ammonium nitrate sells for $2.95/lb and urea for $3.65, which has the more nitrogen for the dollar?

90. (a) How many moles of ozone molecules are contained in 96.0 g of ozone, O_3? (b) How many moles of oxygen atoms are contained in 96.0 g of ozone? (c) What mass of O_2 would contain the same number of oxygen atoms as 96.0 g of ozone? (d) What mass of oxygen gas, O_2, would contain the same number of molecules as 96.0 g of ozone?

91. The recommended daily allowance of calcium is 1200 mg. Calcium carbonate is an inexpensive source of calcium and useful as a dietary supplement as long as it is taken along with Vitamin D which is essential to calcium absorption. How many grams of calcium carbonate must an individual take per day to provide for his/her recommended daily allowance of calcium?

92. Vitamin E is an antioxidant that plays an especially important role protecting cellular structures in the lungs. Combustion of a 0.497 g sample of Vitamin E produced 1.47 g of carbon dioxide and 0.518 g of water. Determine the empirical formula of Vitamin E.

93. A metal, M, forms an oxide having the simplest formula M_2O_3. This oxide contains 52.9% of the metal by mass. (a) Calculate the atomic weight of the metal. (b) Identify the metal.

94. Three samples of magnesium oxide were analyzed to determine the mass ratios O/Mg, giving the following results:

$$\frac{1.60 \text{ g O}}{2.43 \text{ g Mg}}, \quad \frac{0.658 \text{ g O}}{1.00 \text{ g Mg}}, \quad \frac{2.29 \text{ g O}}{3.48 \text{ g Mg}}$$

Which law of chemical combination is illustrated by these data?

*95. The molecular weight of hemoglobin is about 65,000 g/mol. Hemoglobin contains 0.35% Fe by mass. How many iron atoms are in a hemoglobin molecule?

*96. More than 1 billion pounds of adipic acid (MW 146.1 g/mol) is manufactured in the United States each year. Most of it is used to make synthetic fabrics. Adipic acid contains only C, H, and O. Combustion of a 1.6380-g sample of adipic acid gives 2.960 g of CO_2 and 1.010 g of H_2O. (a) What is the simplest formula for adipic acid? (b) What is its molecular formula?

97. Crystals of hydroxyapatite, $Ca_{10}(PO_4)_6(OH)_2$, provide the hardness associated with bones. In hydroxyapatite crystals, what is (a) the percent calcium and (b) the percent phosphorus?

CONCEPTUAL EXERCISES

98. If a 2.0-mole sample of each of the following compounds is completely burned, which one would produce the most moles of water? Which would produce the fewest? (a) CH_3CH_2OH; (b) CH_3OH; (c) CH_3OCH_3.

99. When a sample is burned in a combustion train, the percent oxygen in the sample cannot be determined directly from the mass of water and carbon dioxide formed. Why?

100. What mass of NaCl would contain the same *total* number of ions as 245 g of $MgCl_2$?

101. Two deposits of minerals containing silver are found. One of the deposits contains silver oxide, and the other contains silver sulfide. The deposits can be mined at the same price per ton of the original silver-containing compound, but only one deposit can be mined by your company. Which of the deposits would you recommend and why?

102. A decision is to be made as to the least expensive source of zinc. One source of zinc is zinc sulfate, $ZnSO_4$, and another is zinc acetate dihydrate, $Zn(CH_3COO)_2 \cdot 2H_2O$. These two sources of zinc can be purchased at the same price per kilogram of compound. Which is the most economical source of zinc and by how much?

103. Assume that a penny is $\frac{1}{16}$ in. thick and that the moon is 222,000 mi at its closest approach to the earth (perigee). Show by calculation whether or not a picomole of pen-

nies stacked on their faces would reach from the earth to the moon.

104. Find the number of moles of Ag needed to form each of the following: (a) 0.652 mol Ag_2S; (b) 0.652 mol Ag_2O; (c) 0.652 g Ag_2S; (d) 6.52×10^{22} formula units of Ag_2S.

BUILDING YOUR KNOWLEDGE

NOTE *Beginning with this chapter, exercises under the "Building Your Knowledge" heading will often require that you use skills, concepts, or information that you should have mastered in earlier chapters. This provides you an excellent opportunity to "tie things together" as you study.*

105. Vegetarians sometimes suffer from the lack of vitamin B_{12}. Each molecule of vitamin B_{12} contains a single atom of cobalt and is 4.35% cobalt by mass. What is the molecular weight of vitamin B_{12}?

106. A student wants to determine the empirical and molecular formulas of a compound containing only carbon, hydrogen, and oxygen. To do so, he combusted a 0.625 g sample of the compound and formed 1.114 g of CO_2 and 0.455 g of water. An independent analysis indicated that the molar mass of the compound is 74.1 g/mol. What are the empirical and molecular formulas of this compound?

107. Elemental lead is needed in the construction of storage batteries. The lead is obtained by first roasting galena (PbS) in limited air to produce sulfur dioxide and lead oxide. The lead oxide formed is 92.83% lead. The lead is then obtained from the lead oxide in a process that is nearly 100% efficient. What is the simplest formula of the lead oxide formed?

108. Near room temperature, the density of water is 1.00 g/mL, and the density of ethanol (grain alcohol) is 0.789 g/mL. What volume of ethanol contains the same number of molecules as are present in 175 mL of H_2O?

BEYOND THE TEXTBOOK

Go to the textbook website at

http://www.brookscole.com/chemistry/whitten

for additional activities and exercises based on the General Chemistry Interactive CD-ROM, the World Wide Web, and library resources.

InfoTrac College Edition

For additional readings, go to InfoTrac College Edition, your online research library at:

http://infotrac.thomsonlearning.com

3

Chemical Equations and Reaction Stoichiometry

OUTLINE

Charles D. Winters

Steel wool burns in pure oxygen gas.

$$4Fe + 3O_2 \longrightarrow 2Fe_2O_3$$

OBJECTIVES

After you have studied this chapter, you should be able to

- Write balanced chemical equations to describe chemical reactions
- Interpret balanced chemical equations to calculate the moles of reactants and products involved in each of the reactions
- Interpret balanced chemical equations to calculate the masses of reactants and products involved in each of the reactions
- Determine which reactant is the limiting reactant in reactions
- Use the limiting reactant concept in calculations recording chemical equations
- Compare the amount of substance actually formed in a reaction (actual yield) with the predicted amount (theoretical yield), and determine the percent yield
- Work with sequential reactions
- Use the terminology of solutions—solute, solvent, concentration
- Calculate concentrations of solutions when they are diluted
- Carry out calculations related to the use of solutions in chemical reactions

I n Chapter 2 we studied composition stoichiometry, the quantitative relationships among elements in compounds. In this chapter we study **reaction stoichiometry**—the quantitative relationships among substances as they participate in chemical reactions. In this study we will ask several important questions. *How* can we describe the reaction of one substance with another? *How much* of one substance reacts with a given amount of another substance? *Which reactant* determines the amounts of products formed in a chemical reaction? *How* can we describe reactions in aqueous solutions?

Whether we are concerned with describing a reaction used in a chemical analysis, one used industrially in the production of some useful material, or one that occurs during metabolism in the body, we must describe it accurately. Chemical equations represent a very precise, yet a very versatile, language that describes chemical changes. We begin our study by examining chemical equations.

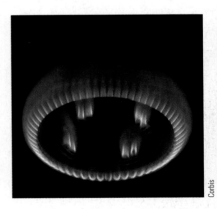

Corbis

Methane, CH_4, is the main component of natural gas.

3-1 CHEMICAL EQUATIONS

Chemical reactions always involve changing one or more substances into one or more different substances. In other words, chemical reactions rearrange atoms or ions to form other substances.

CD-ROM Screen 4.2, Chemical Equations.

Chemical equations are used to describe chemical reactions, and they show (1) *the substances that react*, called **reactants**; (2) *the substances formed*, called **products**; and (3) *the relative amounts of the substances involved*. We write the reactants to the *left* of an arrow and the products to the *right* of the arrow. As a typical example, let's consider the combustion (burning) of natural gas, a reaction used to heat buildings and cook food. Natural gas is a mixture of several substances, but the principal component is methane, CH_4. The equation that describes the reaction of methane with excess oxygen is

$$\underbrace{CH_4 + 2O_2}_{\text{reactants}} \longrightarrow \underbrace{CO_2 + 2H_2O}_{\text{products}}$$

What does this equation tell us? In the simplest terms, it tells us that methane reacts with oxygen to produce carbon dioxide, CO_2, and water. More specifically, it says that for every CH_4 molecule that reacts, two molecules of O_2 also react, and that one CO_2 molecule and two H_2O molecules are formed. That is,

$$\underset{\text{1 molecule}}{CH_4} + \underset{\text{2 molecules}}{2O_2} \xrightarrow{\text{heat}} \underset{\text{1 molecule}}{CO_2} + \underset{\text{2 molecules}}{2H_2O}$$

This description of the reaction of CH_4 with O_2 is based on *experimental observations*. By this we mean experiments have shown that when one CH_4 molecule reacts with two O_2 molecules, one CO_2 molecule and two H_2O molecules are formed. *Chemical equations are based on experimental observations*. Special conditions required for some reactions are indicated by notation over the arrow. Figure 3-1 is a pictorial representation of the rearrangement of atoms described by this equation.

As we pointed out in Section 1-1, *there is no detectable change in the quantity of matter during an ordinary chemical reaction*. This guiding principle, the **Law of Conservation of Matter,** provides the basis for "balancing" chemical equations and for performing calculations based on those equations. Because matter is neither created nor destroyed during a chemical reaction,

Sometimes it is not possible to represent a chemical change with a single chemical equation. For example, when too little O_2 is present, both CO_2 and CO are found as products, and a second chemical equation must be used to describe the process. In the present case (excess oxygen), only one equation is required.

The arrow may be read "yields." The capital Greek letter delta (Δ) is sometimes used in place of the word "heat."

CD-ROM   Screen 4.3, Law of Conservation of Matter.

SCiLINKS.

TOPIC: Chemical Equations
GO TO: www.scilinks.org
*sci*LINKS CODE: WCH0310

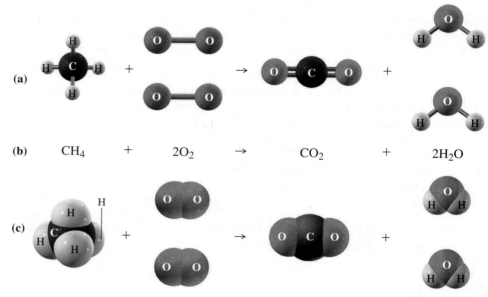

Figure 3-1 Three representations of the reaction of methane with oxygen to form carbon dioxide and water. Chemical bonds are broken and new ones are formed in each representation. Part (a) illustrates the reaction using ball-and-stick models, (b) uses chemical formulas, and (c) uses space-filling models.

a balanced chemical equation must always include the same number of each kind of atom on both sides of the equation.

Chemists usually write equations with the smallest possible whole-number coefficients.

Before we attempt to balance an equation, all substances must be represented by formulas that describe them *as they exist*. For instance, we must write H_2 to represent diatomic hydrogen molecules—not H, which represents hydrogen atoms. Once the correct formulas are written, the subscripts in the formulas may not be changed. Different subscripts in formulas specify different compounds, so changing the formulas would mean that the equation would no longer describe the same reaction.

Dimethyl ether, C_2H_6O, burns in an excess of oxygen to give carbon dioxide and water. Let's balance the equation for this reaction. In unbalanced form, the equation is

$$C_2H_6O + O_2 \longrightarrow CO_2 + H_2O$$

Carbon appears in only one compound on each side, and the same is true for hydrogen. We begin by balancing these elements:

$$C_2H_6O + O_2 \longrightarrow 2CO_2 + 3H_2O$$

Now we have an odd number of atoms of O on each side. The single O in C_2H_6O balances one of the atoms of O on the right. We balance the other six by placing a coefficient of 3 before O_2 on the left.

$$C_2H_6O + 3O_2 \longrightarrow 2CO_2 + 3H_2O$$

When we think we have finished the balancing, we should *always* do a complete check for each element, as shown in red in the margin.

Let's generate the balanced equation for the reaction of aluminum metal with hydrochloric acid to produce aluminum chloride and hydrogen. The unbalanced "equation" is

$$Al + HCl \longrightarrow AlCl_3 + H_2$$

As it now stands, the "equation" does not satisfy the Law of Conservation of Matter because there are two H atoms in the H_2 molecule and three Cl atoms in one formula unit of $AlCl_3$ (right side), but only one H atom and one Cl atom in the HCl molecule (left side).

Let us first balance chlorine by putting a coefficient of 3 in front of HCl.

$$Al + 3HCl \longrightarrow AlCl_3 + H_2$$

Now there are 3H on the left and 2H on the right. The least common multiple of 3 and 2 is 6; to balance H, we multiply the 3HCl by 2 and the H_2 by 3.

$$Al + 6HCl \longrightarrow AlCl_3 + 3H_2$$

Now Cl is again unbalanced (6Cl on the left, 3 on the right), but we can fix this by putting a coefficient of 2 in front of $AlCl_3$ on the right.

$$Al + 6HCl \longrightarrow 2AlCl_3 + 3H_2$$

Now all elements except Al are balanced (1 on the left, 2 on the right); we complete the balancing by putting a coefficient of 2 in front of Al on the left.

$$\underset{\text{aluminum}}{2Al} + \underset{\text{hydrochloric acid}}{6HCl} \longrightarrow \underset{\text{aluminum chloride}}{2AlCl_3} + \underset{\text{hydrogen}}{3H_2}$$

Balancing chemical equations "by inspection" is a *trial-and-error* approach. It requires a great deal of practice, but it is *very important!* Remember that we use the smallest whole-number coefficients. Some chemical equations are difficult to balance by inspection or "trial and error." In Chapter 11 we will learn methods for balancing complex equations.

In Reactants   **In Products**
2C, 6H, 3O   1C, 2H, 3O
C, H are not balanced

In Reactants   **In Products**
2C, 6H, 3O   2C, 6H, 7O
Now O is not balanced

In Reactants   **In Products**
2C, 6H, 7O   2C, 6H, 7O
Now the equation is balanced

In Reactants   **In Products**
1Al, 1H, 1Cl   1Al, 2H, 3Cl
H, Cl are not balanced

In Reactants   **In Products**
1Al, 3H, 3Cl   1Al, 2H, 3Cl
Now H is not balanced

In Reactants   **In Products**
1Al, 6H, 6Cl   1Al, 6H, 3Cl
Now Cl is not balanced

In Reactants   **In Products**
1Al, 6H, 6Cl   2Al, 6H, 6Cl
Now Al is not balanced

In Reactants   **In Products**
2Al, 6H, 6Cl   2Al, 6H, 6Cl
Now the equation is balanced

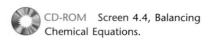
CD-ROM Screen 4.4, Balancing Chemical Equations.

✔ **Problem-Solving Tip:** *Balancing Chemical Equations*

There is no one best place to start when balancing a chemical equation, but the following suggestions might be helpful:

1. Look for elements that appear in only one place on each side of the equation (in only one reactant and in only one product), and balance those elements first.

2. If free, uncombined elements appear on either side, balance them last.

Notice how these suggestions worked in the procedures illustrated in this section. Above all, remember that we should *never* change subscripts in formulas, because doing so would describe different substances. We only adjust the coefficients to balance the equation.

EXAMPLE 3-1   *Balancing Chemical Equations*

Balance the following chemical equations:
(a) $P_4 + Cl_2 \longrightarrow PCl_5$
(b) $RbOH + SO_2 \longrightarrow Rb_2SO_3 + H_2O$
(c) $P_4O_{10} + Ca(OH)_2 \longrightarrow Ca_3(PO_4)_2 + H_2O$

Plan

For each equation, we balance one element at a time by inspection, making sure that there are the same numbers of each type of element on the reactant and product sides of the equations.

Solution

(a) Let's first balance P.

$$P_4 + Cl_2 \longrightarrow 4PCl_5$$

Now we balance Cl:

$$P_4 + 10Cl_2 \longrightarrow 4PCl_5$$

As a check, we see that there are 4P and 20Cl on each side, so the equation is balanced.

(b) We can first balance Rb.

$$2RbOH + SO_2 \longrightarrow Rb_2SO_3 + H_2O$$

We see that each side now has 2Rb, 1S, 2H, and 4O, so the equations is balanced.

(c) Let's balance P first. There are two P on the right and 4P on the left, so we multiply $Ca_3(PO_4)_2$ by 2.

$$P_4O_{10} + Ca(OH)_2 \longrightarrow 2Ca_3(PO_4)_2 + H_2O$$

Next we can balance Ca by multiplying $Ca(OH)_2$ by 6.

$$P_4O_{10} + 6Ca(OH)_2 \longrightarrow 2Ca_3(PO_4)_2 + H_2O$$

Now we see that there are 12H on the left and only 2H on the right, so we multiply H_2O by 6.

$$P_4O_{10} + 6Ca(OH)_2 \longrightarrow 2Ca_3(PO_4)_2 + 6H_2O$$

Checking the O balance we see that there are $(10 + (2 \times 6)) = 22O$ on the left and $((2 \times 4 \times 2) + (6 \times 1)) = 22O$ on the right. All other elements are also balanced so the equation is balanced.

You should now work Exercises 8 and 10.

3-2 CALCULATIONS BASED ON CHEMICAL EQUATIONS

We are now ready to use chemical equations to calculate the relative *amounts* of substances involved in chemical reactions. Let us again consider the combustion of methane in excess oxygen. The balanced chemical equation for that reaction is

$$CH_4 + 2O_2 \longrightarrow CO_2 + 2H_2O$$

On a quantitative basis, at the molecular level, the equation says

$$CH_4 \quad + \quad 2O_2 \quad \longrightarrow \quad CO_2 \quad + \quad 2H_2O$$

| 1 molecule of methane | 2 molecules of oxygen | 1 molecule of carbon dioxide | 2 molecules of water |

CD-ROM Screens 4.5, Weight Relations, and 4.6, Calculations in Stoichiometry.

EXAMPLE 3-2 Number of Molecules

How many O_2 molecules react with 47 CH_4 molecules according to the preceding equation?

Plan

The *balanced* equation tells us that *one* CH_4 molecule reacts with *two* O_2 molecules. We can construct two unit factors from this fact:

$$\frac{1\ CH_4\ \text{molecule}}{2\ O_2\ \text{molecules}} \quad \text{and} \quad \frac{2\ O_2\ \text{molecules}}{1\ CH_4\ \text{molecule}}$$

These expressions are unit factors for *this* reaction because the numerator and denominator are *chemically equivalent*. In other words, the numerator and the denominator represent the same amount of reaction. To convert CH_4 molecules to O_2 molecules, we multiply by the second of the two factors.

Solution

$$\underline{?}\ O_2\ \text{molecules} = 47\ CH_4\ \text{molecules} \times \frac{2\ O_2\ \text{molecules}}{1\ CH_4\ \text{molecule}} = \boxed{94\ O_2\ \text{molecules}}$$

You should now work Exercise 12.

A balanced chemical equation may be interpreted on a *molecular* basis.

A chemical equation also indicates the relative amounts of each reactant and product in a given chemical reaction. We showed earlier that formulas can represent moles of substances. Suppose Avogadro's number of CH_4 molecules, rather than just one CH_4 molecule, undergo this reaction. Then the equation can be written

$$CH_4 \quad + \quad 2O_2 \quad \longrightarrow \quad CO_2 \quad + \quad 2H_2O$$

| 6.02×10^{23} molecules = 1 mol | $2(6.02 \times 10^{23}$ molecules) = 2 mol | 6.02×10^{23} molecules = 1 mol | $2(6.02 \times 10^{23}$ molecules) = 2 mol |

This interpretation tells us that *one* mole of methane reacts with *two* moles of oxygen to produce *one* mole of carbon dioxide and *two* moles of water.

We usually cannot work with individual molecules; a mole of a substance is an amount we might use in a laboratory experiment.

A balanced chemical equation may be interpreted in terms of *moles* of reactants and products.

EXAMPLE 3-3 Number of Moles Formed

How many moles of water could be produced by the reaction of 3.5 mol of methane with excess oxygen (i.e., more than a sufficient amount of oxygen is present)?

Plan

The equation for the combustion of methane

$$CH_4 + 2O_2 \longrightarrow CO_2 + 2H_2O$$

| 1 mol | 2 mol | 1 mol | 2 mol |

Please don't try to memorize unit factors for chemical reactions; instead, learn the general *method* for constructing them from balanced chemical equations.

shows that 1 mol of methane reacts with 2 mol of oxygen to produce 2 mol of water. From this information we construct two *unit factors*:

$$\frac{1 \text{ mol CH}_4}{2 \text{ mol H}_2\text{O}} \quad \text{and} \quad \frac{2 \text{ mol H}_2\text{O}}{1 \text{ mol CH}_4}$$

We use the second factor in this calculation.

Solution

$$\underline{?} \text{ mol H}_2\text{O} = 3.5 \text{ mol CH}_4 \times \frac{2 \text{ mol H}_2\text{O}}{1 \text{ mol CH}_4} = \boxed{7.0 \text{ mol H}_2\text{O}}$$

You should now work Exercises 14 and 18.

We know the mass of 1 mol of each of these substances, so we can also write

$$\text{CH}_4 \; + \; 2\text{O}_2 \; \longrightarrow \; \text{CO}_2 \; + \; 2\text{H}_2\text{O}$$

1 mol	2 mol	1 mol	2 mol
16.0 g	2(32.0 g)	44.0 g	2(18.0 g)
16.0 g	64.0 g	44.0 g	36.0 g

80.0 g reactants   80.0 g products

A balanced equation may be interpreted on a *mass* basis.

The equation now tells us that 16.0 g of CH_4 reacts with 64.0 g of O_2 to form 44.0 g of CO_2 and 36.0 g of H_2O. The Law of Conservation of Matter is satisfied. Chemical equations describe **reaction ratios,** that is, the *mole ratios* of reactants and products as well as the *relative masses* of reactants and products.

> ✓ **Problem-Solving Tip:** *Use the Reaction Ratio in Calculations with Balanced Chemical Equations*
>
> The most important way of interpreting the balanced chemical equation is in terms of moles. We use the coefficients to get the reaction ratio (in moles) of any two substances we want to relate. Then we apply it as
>
> $$\begin{pmatrix} \text{moles of} \\ \text{desired} \\ \text{substance} \end{pmatrix} = \begin{pmatrix} \text{moles of} \\ \text{substance} \\ \text{given} \end{pmatrix} \times \begin{pmatrix} \text{reaction ratio} \\ \text{from balanced} \\ \text{chemical equation} \end{pmatrix}$$
>
> It is important to include the substance formulas as part of the units; this can help us decide how to set up the unit factors. Notice that in Example 3-3 we want to cancel the term mol CH_4, so we know that mol CH_4 must be in the denominator of the mole ratio by which we multiply; we want mol O_2 as the answer, so mol O_2 must appear in the numerator of the mole ratio. In other words, do not just write $\frac{\text{mol}}{\text{mol}}$; write $\frac{\text{mol of something}}{\text{mol of something else}}$, giving the formulas of the two substances involved.

SCLINKS.

TOPIC: The Mole
GO TO: www.scilinks.org
*sci*LINKS CODE: WCH0210

EXAMPLE 3-4   *Mass of a Reactant Required*

What mass of oxygen is required to react completely with 1.20 mol of CH_4?

Plan

The balanced equation

$$\text{CH}_4 \; + \; 2\text{O}_2 \; \longrightarrow \; \text{CO}_2 \; + \; 2\text{H}_2\text{O}$$

1 mol	2 mol	1 mol	2 mol
16.0 g	2(32.0 g)	44.0 g	2(18.0 g)

gives the relationships among moles and grams of reactants and products.

$$\text{mol CH}_4 \; \longrightarrow \; \boxed{\text{mol O}_2} \; \longrightarrow \; \boxed{\text{g O}_2}$$

Solution

$$\underline{?}\text{ g O}_2 = 1.20\text{ mol CH}_4 \times \frac{2\text{ mol O}_2}{1\text{ mol CH}_4} \times \frac{32.0\text{ g O}_2}{1\text{ mol O}_2} = \boxed{76.8\text{ g O}_2}$$

EXAMPLE 3-5 *Mass of a Reactant Required*

What mass of oxygen is required to react completely with 24.0 g of CH_4?

Plan

Recall the balanced equation in Example 3-4.

$$CH_4 + 2O_2 \longrightarrow CO_2 + 2H_2O$$

| 1 mol | 2 mol | 1 mol | 2 mol |
| 16.0 g | 2(32.0) g | 44.0 g | 2(18.0) g |

This shows that 1 mol of CH_4 reacts with 2 mol of O_2. These two quantities are chemically equivalent, so we can construct *unit factors*.

Solution

$$CH_4 + 2O_2 \longrightarrow CO_2 + 2H_2O$$

| 1 mol | 2 mol | 1 mol | 2 mol |

$$\underline{?}\text{ mol CH}_4 = 24.0\text{ g CH}_4 \times \frac{1\text{ mol CH}_4}{16.0\text{ g CH}_4} = 1.50\text{ mol CH}_4$$

$$\underline{?}\text{ mol O}_2 = 1.50\text{ mol CH}_4 \times \frac{2\text{ mol O}_2}{1\text{ mol CH}_4} = 3.00\text{ mol O}_2$$

$$\underline{?}\text{ g O}_2 = 3.00\text{ mol O}_2 \times \frac{32.0\text{ g O}_2}{1\text{ mol O}_2} = \boxed{96.0\text{ g O}_2}$$

Here we solve the problem in three steps; we convert

1. $g\,CH_4 \rightarrow mol\,CH_4$

2. $mol\,CH_4 \rightarrow mol\,O_2$

3. $mol\,O_2 \rightarrow g\,O_2$

All these steps could be combined into one setup in which we convert

$$g\text{ of CH}_4 \longrightarrow \text{mol of CH}_4 \longrightarrow \text{mol of O}_2 \longrightarrow g\text{ of O}_2$$

$$\underline{?}\text{ g O}_2 = 24.0\text{ g CH}_4 \times \frac{1\text{ mol CH}_4}{16.0\text{ g CH}_4} \times \frac{2\text{ mol O}_2}{1\text{ mol CH}_4} \times \frac{32.0\text{ g O}_2}{1\text{ mol O}_2} = \boxed{96.0\text{ g O}_2}$$

The same answer, 96.0 g of O_2, is obtained by both methods.

You should now work Exercise 22.

The question posed in Example 3-5 may be reversed, as in Example 3-6.

EXAMPLE 3-6 *Mass of a Reactant Required*

What mass of CH_4, in grams, is required to react with 96.0 g of O_2?

Plan

We recall that 1 mole of CH_4 reacts with 2 moles of O_2.

Solution

$$\underline{?}\text{ g CH}_4 = 96.0\text{ g O}_2 \times \frac{1\text{ mol O}_2}{32.0\text{ g O}_2} \times \frac{1\text{ mol CH}_4}{2\text{ mol O}_2} \times \frac{16.0\text{ g CH}_4}{1\text{ mol CH}_4} = \boxed{24.0\text{ g CH}_4}$$

These unit factors are the reciprocals of those used in Example 3-5.

You should now work Exercise 24.

This is the amount of CH_4 in Example 3-5 that reacted with 96.0 g of O_2.

EXAMPLE 3-7   *Mass of a Product Formed*

It is important to recognize that the reaction must stop when the 6.00 mol of CH_4 has been used up. Some O_2 will remain unreacted.

Most combustion reactions occur in excess O_2, that is, more than enough O_2 to burn the substance completely. Calculate the mass of CO_2, in grams, that can be produced by burning 6.00 mol of CH_4 in excess O_2.

Plan

The balanced equation tells us that 1 mol CH_4 produces 1 mol CO_2.

$$CH_4 \; + \quad 2O_2 \quad \longrightarrow \quad CO_2 \; + \quad 2H_2O$$
$$\text{1 mol} \qquad \text{2 mol} \qquad \qquad \text{1 mol} \qquad \text{2 mol}$$
$$\text{16.0 g} \qquad \text{2(32.0 g)} \qquad \text{44.0 g} \qquad \text{2(18.0 g)}$$

Solution

$$\underline{?} \text{ g } CO_2 = 6.00 \text{ mol } CH_4 \times \frac{1 \text{ mol } CO_2}{1 \text{ mol } CH_4} \times \frac{44.0 \text{ g } CO_2}{1 \text{ mol } CO_2} = \boxed{2.64 \times 10^2 \text{ g } CO_2}$$

You should now work Exercise 26.

Reaction stoichiometry usually involves interpreting a balanced chemical equation to relate a *given* bit of information to the *desired* bit of information.

EXAMPLE 3-8   *Mass of a Reactant Required*

Never start a calculation involving a chemical reaction without first checking that the equation is balanced.

Phosphorus, P_4, burns with excess oxygen to form tetraphosphorus decoxide, P_4O_{10}. In this reaction, what mass of P_4 reacts with 1.50 moles of O_2?

Plan

The balanced equation tells us that 1 mole of P_4 reacts with 5 moles of O_2.

$$P_4 \; + \; 5O_2 \; \longrightarrow \; P_4O_{10}$$
$$\text{1 mol} \quad \text{5 mol} \qquad \text{1 mol}$$

$$\text{mol } O_2 \; \longrightarrow \; \text{mol } P_4 \; \longrightarrow \; \boxed{\text{mass } P_4}$$

Solution

$$\underline{?} \text{ g } P_4 = 1.50 \text{ mol } O_2 \times \frac{1 \text{ mol } P_4}{5 \text{ mol } O_2} \times \frac{124.0 \text{ g } P_4}{\text{mol } P_4} = \boxed{37.2 \text{ g } P_4}$$

You should now work Exercise 28.

The possibilities for this kind of problem solving are limitless. Before you continue, you should work Exercises 12–29 at the end of the chapter.

3-3   THE LIMITING REACTANT CONCEPT

 CD-ROM   Screens 4.7, Reactions Controlled by the Supply of One Reactant, and 4.8, Limiting Reactants: The Details.

In the problems we have worked thus far, the presence of an excess of one reactant was stated or implied. The calculations were based on the substance that was used up first, called the **limiting reactant.** Before we study the concept of the limiting reactant in stoichiometry, let's develop the basic idea by considering a simple but analogous nonchemical example.

Suppose you have four slices of ham and six slices of bread and you wish to make as many ham sandwiches as possible using only one slice of ham and two slices of bread per sandwich. Obviously, you can make only three sandwiches, at which point you run out of bread. (In a chemical reaction this would correspond to one of the reactants being used up—so the reaction would stop.) The bread is therefore the *limiting reactant*, and the extra slice of ham is the *excess reactant*. The amount of product, ham sandwiches, is determined by the amount of the limiting reactant, bread in this case. The limiting reactant is not necessarily the reactant present in the smallest amount. We have four slices of ham, the smallest amount, and six slices of bread, but the *reaction ratio* is two slices of bread to one piece of ham, and so bread is the limiting reactant.

EXAMPLE 3-9   *Limiting Reactant*

What mass of CO_2 could be formed by the reaction of 16.0 g of CH_4 with 48.0 g of O_2?

Plan

The balanced equation tells us that 1 mol CH_4 reacts with 2 mol O_2.

$$CH_4 + 2O_2 \longrightarrow CO_2 + 2H_2O$$

| 1 mol | 2 mol | 1 mol | 2 mol |
| 16.0 g | 2(32.0 g) | 44.0 g | 2(18.0 g) |

We are given masses of both CH_4 and O_2, so we calculate the number of moles of each reactant, and then determine the number of moles of each reactant required to react with the other. From these calculations we identify the limiting reactant. Then we base the calculation on it.

Solution

$$\underline{?}\ mol\ CH_4 = 16.0\ g\ CH_4 \times \frac{1\ mol\ CH_4}{16.0\ g\ CH_4} = 1.00\ mol\ CH_4$$

$$\underline{?}\ mol\ O_2 = 48.0\ g\ O_2 \times \frac{1\ mol\ O_2}{32.0\ g\ O_2} = 1.50\ mol\ O_2$$

Now we return to the balanced equation. First we calculate the number of moles of O_2 that would be required to react with 1.00 mole of CH_4.

$$\underline{?}\ mol\ O_2 = 1.00\ mol\ CH_4 \times \frac{2\ mol\ O_2}{1\ mol\ CH_4} = 2.00\ mol\ O_2$$

We see that 2.00 moles of O_2 are required, but we have only 1.50 moles of O_2, so O_2 is the limiting reactant. Alternatively, we can calculate the number of moles of CH_4 that would react with 1.50 moles of O_2.

$$\underline{?}\ mol\ CH_4 = 1.50\ mol\ O_2 \times \frac{1\ mol\ CH_4}{2\ mol\ O_2} = 0.750\ mol\ CH_4$$

This tells us that only 0.750 mole of CH_4 would be required to react with 1.50 moles of O_2. But we have 1.00 mole of CH_4, so we see again that O_2 is the limiting reactant. The reaction must stop when the limiting reactant, O_2, is used up, so we base the calculation on O_2.

$$g\ of\ O_2 \longrightarrow mol\ of\ O_2 \longrightarrow mol\ of\ CO_2 \longrightarrow g\ of\ CO_2$$

$$\underline{?}\ g\ CO_2 = 48.0\ g\ O_2 \times \frac{1\ mol\ O_2}{32.0\ O_2} \times \frac{1\ mol\ CO_2}{2\ mol\ O_2} \times \frac{44.0\ g\ CO_2}{1\ mol\ CO_2} = 33.0\ g\ CO_2$$

You should now work Exercise 30.

Thus, 33.0 g of CO_2 is the most CO_2 that can be produced from 16.0 g of CH_4 and 48.0 g of O_2. If we had based our calculation on CH_4 rather than O_2, our answer would be too big (44.0 g) and *wrong* because more O_2 than we have would be required.

Another approach to problems like Example 3-9 is to calculate the number of moles of each reactant:

$$\underline{?}\ \text{mol}\ CH_4 = 16.0\ \text{g}\ CH_4 \times \frac{1\ \text{mol}\ CH_4}{16.0\ \text{g}\ CH_4} = \underline{1.00\ \text{mol}\ CH_4}$$

$$\underline{?}\ \text{mol}\ O_2 = 48.0\ \text{g}\ O_2 \times \frac{1\ \text{mol}\ O_2}{32.0\ \text{g}\ O_2} = \underline{1.50\ \text{mol}\ O_2}$$

Then we return to the balanced equation. We first calculate the *required ratio* of reactants as indicated by the balanced chemical equation. We then calculate the *available ratio* of reactants and compare the two:

Required Ratio	Available Ratio
$\dfrac{1\ \text{mol}\ CH_4}{2\ \text{mol}\ O_2} = \dfrac{0.500\ \text{mol}\ CH_4}{1.00\ \text{mol}\ O_2}$	$\dfrac{1.00\ \text{mol}\ CH_4}{1.50\ \text{mol}\ O_2} = \dfrac{0.667\ \text{mol}\ CH_4}{1.00\ \text{mol}\ O_2}$

We see that each mole of O_2 requires exactly 0.500 mol of CH_4 to be completely used up. We have 0.667 mol of CH_4 for each mole of O_2, so there is more than enough CH_4 to react with the O_2 present. That means that there is *insufficient* O_2 to react with all the available CH_4. The reaction must stop when the O_2 is gone; O_2 is the limiting reactant, and we must base the calculation on it. (Suppose the available ratio of CH_4 to O_2 had been *smaller than* the required ratio. Then we would have concluded that there is not enough CH_4 to react with all the O_2, so CH_4 would have been the limiting reactant.)

✓ **Problem-Solving Tip:** *Choosing the Limiting Reactant*

Students often wonder how to know which ratio to calculate to help find the limiting reactant.

1. The ratio must involve the two reactants with amounts given in the problem.
2. It doesn't matter which way you calculate the ratio, as long as you calculate both required ratio and available ratio in the same order. For example, we could calculate the required and available ratio of $\dfrac{\text{mol}\ O_2}{\text{mol}\ CH_4}$ in the approach we just illustrated.

If you can't decide how to solve a limiting reactant problem, as a last resort, do the entire calculation twice—once based on each reactant amount given. The *smaller* answer is the correct one.

EXAMPLE 3-10 *Limiting Reactant*

What is the maximum mass of $Ni(OH)_2$ that could be prepared by mixing two solutions that contain 25.9 g of $NiCl_2$ and 10.0 g of NaOH, respectively?

$$NiCl_2 + 2NaOH \longrightarrow Ni(OH)_2 + 2NaCl$$

Plan

Interpreting the balanced equation as usual, we have

$$NiCl_2 + 2NaOH \longrightarrow Ni(OH)_2 + 2NaCl$$

1 mol	2 mol	1 mol	2 mol
129.6 g	2(40.0 g)	92.7 g	2(58.4 g)

We determine the number of moles of $NiCl_2$ and NaOH present. Then we find the number of moles of each reactant required to react with the other reactant. These calculations identify the limiting reactant. We base the calculation on it.

Solution

$$\underline{?}\ mol\ NiCl_2 = 25.9\ g\ NiCl_2 \times \frac{1\ mol\ NiCl_2}{129.6\ g\ NiCl_2} = 0.200\ mol\ NiCl_2$$

$$\underline{?}\ mol\ NaOH = 10.0\ g\ NaOH \times \frac{1\ mol\ NaOH}{40.0\ g\ NaOH} = 0.250\ mol\ NaOH$$

We return to the balanced equation and calculate the number of moles of NaOH required to react with 0.200 mol of $NiCl_2$.

$$\underline{?}\ mol\ NaOH = 0.200\ mol\ NiCl_2 \times \frac{2\ mol\ NaOH}{1\ mol\ NiCl_2} = 0.400\ mol\ NaOH$$

But we have only 0.250 moles of NaOH, so NaOH is the limiting reactant.

$$g\ of\ NaOH \longrightarrow mol\ of\ NaOH \longrightarrow mol\ Ni(OH)_2 \longrightarrow g\ of\ Ni(OH)_2$$

$$\underline{?}\ g\ Ni(OH)_2 = 10.0\ g\ NaOH \times \frac{1\ mol\ NaOH}{40.0\ g\ NaOH} \times \frac{1\ mol\ Ni(OH)_2}{2\ mol\ NaOH} \times \frac{92.7\ g\ Ni(OH)_2}{1\ mol\ Ni(OH)_2}$$

$$= \boxed{11.6\ g\ Ni(OH)_2}$$

You should now work Exercises 34 and 36.

> When we are given the amounts of *two (or more)* reactants, we should suspect that we are dealing with a limiting reagent problem. It is very unlikely that *exactly* the stoichiometric amounts of both reactants are present in a reaction mixture.

Even though the reaction occurs in aqueous solution, this calculation is similar to earlier examples because we are given the amounts of pure reactants.

A precipitate of solid $Ni(OH)_2$ forms when colorless NaOH solution is added to green $NiCl_2$ solution. (Example 3–10).

3-4 PERCENT YIELDS FROM CHEMICAL REACTIONS

The **theoretical yield** from a chemical reaction is the yield calculated by assuming that the reaction goes to completion. In practice we often do not obtain as much product from a reaction mixture as is theoretically possible. This is true for several reasons. (1) Many reactions do not go to completion; that is, the reactants are not completely converted to products. (2) In some cases, a particular set of reactants undergoes two or more reactions simultaneously, forming undesired products as well as desired products. Reactions other than the desired one are called *side reactions*. (3) In some cases, separation of the desired product from the reaction mixture is so difficult that not all of the product formed is successfully isolated. The **actual yield** is the amount of a specified pure product actually obtained from a given reaction.

In the examples we have worked to this point, the amounts of products that we calculated were theoretical yields.

CD-ROM   Screen 4.9,
Percent Yield.

The term **percent yield** is used to indicate how much of a desired product is obtained from a reaction.

$$\text{percent yield} = \frac{\text{actual yield of product}}{\text{theoretical yield of product}} \times 100\%$$

Consider the preparation of nitrobenzene, $C_6H_5NO_2$, by the reaction of a limited amount of benzene, C_6H_6, with excess nitric acid, HNO_3. The balanced equation for the reaction may be written as

$$C_6H_6 + HNO_3 \longrightarrow C_6H_5NO_2 + H_2O$$

| 1 mol | 1 mol | 1 mol | 1 mol |
| 78.1 g | 63.0 g | 123.1 g | 18.0 g |

EXAMPLE 3-11   *Percent Yield*

A 15.6-g sample of C_6H_6 is mixed with excess HNO_3. We isolate 18.0 g of $C_6H_5NO_2$. What is the percent yield of $C_6H_5NO_2$ in this reaction?

Plan

First we interpret the balanced chemical equation to calculate the theoretical yield of $C_6H_5NO_2$. Then we use the actual (isolated) yield and the previous definition to calculate the percent yield.

Solution

We calculate the theoretical yield of $C_6H_5NO_2$.

It is not necessary to know the mass of one mole of HNO_3 to solve this problem.

$$\underline{?}\text{ g } C_6H_5NO_2 = 15.6\text{ g } C_6H_6 \times \frac{1\text{ mol } C_6H_6}{78.1\text{ g } C_6H_6} \times \frac{1\text{ mol } C_6H_5NO_2}{1\text{ mol } C_6H_6} \times \frac{123.1\text{ g } C_6H_5NO_2}{1\text{ mol } C_6H_5NO_2}$$

$$= 24.6\text{ g } C_6H_5NO_2 \leftarrow \text{theoretical yield}$$

This tells us that if *all* the C_6H_6 were converted to $C_6H_5NO_2$ and isolated, we should obtain 24.6 g of $C_6H_5NO_2$ (100% yield). We isolate only 18.0 g of $C_6H_5NO_2$, however.

$$\text{percent yield} = \frac{\text{actual yield of product}}{\text{theoretical yield of product}} \times 100\% = \frac{18.0\text{ g}}{24.6\text{ g}} \times 100\%$$

$$= 73.2\%$$

You should now work Exercise 44.

The amount of nitrobenzene obtained *in this experiment* is 73.2% of the amount that would be expected *if* the reaction had gone to completion, *if* there were no side reactions, and *if* we could have recovered all the product as pure substance.

3-5 SEQUENTIAL REACTIONS

Often more than one reaction is required to change starting materials into the desired product. This is true for many reactions that we carry out in the laboratory and for many industrial processes. These are called **sequential reactions.** The amount of desired product from each reaction is taken as the starting material for the next reaction.

EXAMPLE 3-12 *Sequential Reactions*

At high temperatures, carbon reacts with water to produce a mixture of carbon monoxide, CO, and hydrogen, H_2.

$$C + H_2O \xrightarrow{\text{heat}} CO + H_2$$

Carbon monoxide is separated from H_2 and then used to separate nickel from cobalt by forming a gaseous compound, nickel tetracarbonyl, $Ni(CO)_4$.

$$Ni + 4CO \longrightarrow Ni(CO)_4$$

What mass of $Ni(CO)_4$ could be obtained from the CO produced by the reaction of 75.0 g of carbon? Assume 100% yield.

Plan

We interpret both chemical equations in the usual way, and solve the problem in two steps. They tell us that one mole of C produces one mole of CO and that four moles of CO is required to produce one mole of $Ni(CO)_4$.

1. We determine the number of moles of CO formed in the first reaction.
2. From the number of moles of CO produced in the first reaction, we calculate the number of grams of $Ni(CO)_4$ formed in the second reaction.

Solution

1.
$$C + H_2O \longrightarrow CO + H_2$$
$$\text{1 mol} \quad \text{1 mol} \quad \text{1 mol} \quad \text{1 mol}$$
$$\text{12.0 g}$$

$$\underline{?} \text{ mol CO} = 75.0 \text{ g C} \times \frac{1 \text{ mol C}}{12.0 \text{ g C}} \times \frac{1 \text{ mol CO}}{1 \text{ mol C}} = 6.25 \text{ mol CO}$$

2.
$$Ni + 4CO \longrightarrow Ni(CO)_4$$
$$\text{1 mol} \quad \text{4 mol} \quad \text{1 mol}$$
$$\text{171 g}$$

$$\underline{?} \text{ g Ni(CO)}_4 = 6.25 \text{ mol CO} \times \frac{1 \text{ mol Ni(CO)}_4}{4 \text{ mol CO}} \times \frac{171 \text{ g Ni(CO)}_4}{1 \text{ mol Ni(CO)}_4} = \boxed{267 \text{ g Ni(CO)}_4}$$

Alternatively, we can set up a series of unit factors based on the conversions in the reaction sequence and solve the problem in one setup.

$$\text{g C} \longrightarrow \text{mol C} \longrightarrow \text{mol CO} \longrightarrow \text{mol Ni(CO)}_4 \longrightarrow \text{g Ni(CO)}_4$$

$$\underline{?} \text{ g Ni(CO)}_4 = 75.0 \text{ g C} \times \frac{1 \text{ mol C}}{12.0 \text{ g C}} \times \frac{1 \text{ mol CO}}{1 \text{ mol C}} \times \frac{1 \text{ mol Ni(CO)}_4}{4 \text{ mol CO}} \times \frac{171 \text{ g Ni(CO)}_4}{1 \text{ mol Ni(CO)}_4} = \boxed{267 \text{ g Ni(CO)}_4}$$

You should now work Exercise 50.

Two large uses for H_3PO_4 are in fertilizers and cola drinks. Approximately 100 lb of H_3PO_4– based fertilizer are used per year per person in America. The unit factors that account for less than 100% reaction and less than 100% recovery are included in parentheses.

EXAMPLE 3-13   *Sequential Reactions*

Phosphoric acid, H_3PO_4, is a very important compound used to make fertilizers. It is also present in cola drinks.

H_3PO_4 can be prepared in a two-step process.

$$\text{Reaction 1:} \quad P_4 + 5O_2 \longrightarrow P_4O_{10}$$

$$\text{Reaction 2:} \quad P_4O_{10} + 6H_2O \longrightarrow 4H_3PO_4$$

We allow 272 g of phosphorus to react with excess oxygen, which forms tetraphosphorus decoxide, P_4O_{10}, in 89.5% yield. In the second step reaction, a 96.8% yield of H_3PO_4 is obtained. What mass of H_3PO_4 is obtained?

Plan

1. We interpret the first equation as usual and calculate the amount of P_4O_{10} *obtained.*

$$
\begin{array}{ccc}
P_4 & + \quad 5O_2 & \longrightarrow \quad P_4O_{10} \\
1\ \text{mol} & 5\ \text{mol} & 1\ \text{mol} \\
124\ \text{g} & 5(32.0\ \text{g}) & 284\ \text{g}
\end{array}
$$

$$\text{g } P_4 \longrightarrow \text{mol } P_4 \longrightarrow \text{mol } P_4O_{10} \longrightarrow \boxed{\text{g } P_4O_{10}}$$

2. Then we interpret the second equation and calculate the amount of H_3PO_4 *obtained* from the P_4O_{10} from the first step.

$$
\begin{array}{ccc}
P_4O_{10} & + \quad 6H_2O & \longrightarrow \quad 4H_3PO_4 \\
1\ \text{mol} & 6\ \text{mol} & 4\ \text{mol} \\
284\ \text{g} & 6(18.0\ \text{g}) & 4(98.0\ \text{g})
\end{array}
$$

$$\text{g } P_4O_{10} \longrightarrow \text{mol } P_4O_{10} \longrightarrow \text{mol } H_3PO_4 \longrightarrow \boxed{\text{g } H_3PO_4}$$

Solution

1. $\underline{?}\ \text{g } P_4O_{10} = 272\ \text{g } P_4 \times \dfrac{1\ \text{mol } P_4}{124\ \text{g } P_4} \times \dfrac{1\ \text{mol } P_4O_{10}}{1\ \text{mol } P_4} \times \dfrac{284\ \text{g } P_4O_{10}\ \text{theoretical}}{1\ \text{mol } P_4O_{10}\ \text{theoretical}}$

$$\times \left(\dfrac{89.5\ \text{g } P_4O_{10}\ \text{actual}}{100\ \text{g } P_4O_{10}\ \text{theoretical}} \right) = \boxed{558\ \text{g } P_4O_{10}}$$

The unit factors that account for less than 100% reaction and less than 100% recovery are included in parentheses.

2. $\underline{?}\ \text{g } H_3PO_4 = 558\ \text{g } P_4O_{10} \times \dfrac{1\ \text{mol } P_4O_{10}}{284\ \text{g } P_4O_{10}} \times \dfrac{4\ \text{mol } H_3PO_4}{1\ \text{mol } P_4O_{10}}$

$$\times \dfrac{98.0\ \text{g } H_3PO_4\ \text{theoretical}}{1\ \text{mol } H_3PO_4\ \text{theoretical}} \times \left(\dfrac{96.8\ \text{g } H_3PO_4\ \text{actual}}{100\ \text{g } H_3PO_4\ \text{theoretical}} \right) = \boxed{746\ \text{g } H_3PO_4}$$

You should now work Exercises 52 and 54.

3-6   CONCENTRATIONS OF SOLUTIONS

Many chemical reactions are more conveniently carried out with the reactants mixed in solution rather than as pure substances. A **solution** is a homogeneous mixture, at the molecular level, of two or more substances. Simple solutions usually consist of one substance, the **solute,** dissolved in another substance, the **solvent.** The solutions used in the laboratory are usually liquids, and the solvent is often water. These are called **aqueous solutions.** For example, solutions of hydrochloric acid can be prepared by dissolving hydrogen chloride (HCl, a gas at room temperature and atmospheric pressure) in water. Solutions of sodium hydroxide are prepared by dissolving solid NaOH in water.

We often use solutions to supply the reactants for chemical reactions. Solutions allow the most intimate mixing of the reacting substances at the molecular level, much more than would be possible in solid form. (A practical example is drain cleaner, shown in the photo.) We sometimes adjust the concentrations of solutions to speed up or slow down the rate of a reaction. In this section we study methods for expressing the quantities of the various components present in a given amount of solution.

Concentrations of solutions are expressed in terms of *either* the amount of solute present in a given mass or volume of *solution*, or the amount of solute dissolved in a given mass or volume of *solvent*.

In some solutions, such as a nearly equal mixture of ethyl alcohol and water, the distinction between solute and solvent is arbitrary.

Percent by Mass

Concentrations of solutions may be expressed in terms of **percent by mass** of solute, which gives the mass of solute per 100 mass units of solution. The gram is the usual mass unit.

$$\text{percent solute} = \frac{\text{mass of solute}}{\text{mass of solution}} \times 100\%$$

$$\text{percent} = \frac{\text{mass of solute}}{\text{mass of solute} + \text{mass of solvent}} \times 100\%$$

Thus, a solution that is 10.0% calcium gluconate, $Ca(C_6H_{11}O_7)_2$, by mass contains 10.0 grams of calcium gluconate in 100.0 grams of *solution*. This could be described as 10.0 grams of calcium gluconate in 90.0 grams of water. The density of a 10.0% solution of calcium gluconate is 1.07 g/mL, so 100 mL of a 10.0% solution of calcium gluconate has a mass of 107 grams. Observe that 100 grams of a solution usually does *not* occupy 100 mL. Unless otherwise specified, percent means percent *by mass*, and water is the solvent.

The sodium hydroxide and aluminum in some drain cleaners do not react while they are stored in solid form. When water is added, the NaOH dissolves and begins to act on trapped grease. At the same time, NaOH and Al react to produce H_2 gas; the resulting turbulence helps to dislodge the blockage. Do you see why the container should be kept tightly closed?

EXAMPLE 3-14   *Percent of Solute*

Calculate the mass of nickel(II) sulfate, $NiSO_4$, contained in 200. g of a 6.00% solution of $NiSO_4$.

Plan

The percentage information tells us that the solution contains 6.00 g of $NiSO_4$ per 100. g of solution. The desired information is the mass of $NiSO_4$ in 200. g of solution. A unit factor is constructed by placing 6.00 g $NiSO_4$ over 100. g of solution. Multiplication of the mass of the solution, 200. g, by this unit factor gives the mass of $NiSO_4$ in the solution.

Solution

$$\underline{?}\text{ g NiSO}_4 = 200.\text{ g soln} \times \frac{6.00\text{ g NiSO}_4}{100.\text{ g soln}} = \boxed{12.0\text{ g NiSO}_4}$$

A 10.0% solution of $Ca(C_6H_{11}O_7)_2$ is sometimes administered intravenously in emergency treatment for black widow spider bites.

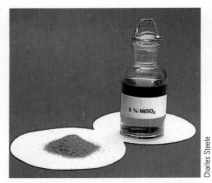

Solid $NiSO_4$ and a 6% solution of $NiSO_4$ in water (Examples 3-14, 3-15, and 3-16).

EXAMPLE 3-15   Mass of Solution

A 6.00% $NiSO_4$ solution contains 40.0 g of $NiSO_4$. Calculate the mass of the solution.

Plan

Placing 100. g of solution over 6.00 g of $NiSO_4$ gives the desired unit factor.

Solution

$$\text{? g soln} = 40.0 \text{ g NiSO}_4 \times \frac{100. \text{ g soln}}{6.00 \text{ g NiSO}_4} = \boxed{667 \text{ g soln}}$$

EXAMPLE 3-16   Mass of Solute

Calculate the mass of $NiSO_4$ present in 200. mL of a 6.00% solution of $NiSO_4$. The density of the solution is 1.06 g/mL at 25°C.

Plan

The volume of a solution multiplied by its density gives the mass of the solution (see Section 1-11). The mass of the solution is then multiplied by the mass fraction due to $NiSO_4$ (6.00 g $NiSO_4$/100. g soln) to give the mass of $NiSO_4$ in 200. mL of solution.

Solution

$$\text{? g NiSO}_4 = \underbrace{200. \text{ mL soln} \times \frac{1.06 \text{ g soln}}{1.00 \text{ mL soln}}}_{212 \text{ g soln}} \times \frac{6.00 \text{ g NiSO}_4}{100. \text{ g soln}} = \boxed{12.7 \text{ g NiSO}_4}$$

You should now work Exercise 58.

EXAMPLE 3-17   Percent Solute and Density

What volume of a solution that is 15.0% iron(III) nitrate contains 30.0 g of $Fe(NO_3)_3$? The density of the solution is 1.16 g/mL at 25°C.

Plan

Two unit factors relate mass of $Fe(NO_3)_3$ and mass of solution, 15.0 g $Fe(NO_3)_3$/100 g and 100 g/15.0 g $Fe(NO_3)_3$. The second factor converts grams of $Fe(NO_3)_3$ to grams of solution.

Solution

$$\text{? mL soln} = \underbrace{30.0 \text{ g Fe(NO}_3)_3 \times \frac{100. \text{ g soln}}{15.0 \text{ g Fe(NO}_3)_3}}_{200 \text{ g soln}} \times \frac{1.00 \text{ mL soln}}{1.16 \text{ g soln}} = \boxed{172 \text{ mL}}$$

Note that the answer is not 200. mL but considerably less because 1.00 mL of solution has a mass of 1.16 g; however, 172 mL of the solution has a mass of 200 g.

You should now work Exercise 60.

Molarity

Molarity (*M*), or molar concentration, is a common unit for expressing the concentrations of solutions. **Molarity** is defined as the number of moles of solute per liter of solution:

$$\text{molarity} = \frac{\text{number of moles of solute}}{\text{number of liters of solution}}$$

To prepare one liter of a one molar solution, one mole of solute is placed in a one liter volumetric flask, enough solvent is added to dissolve the solute, and solvent is then added until the volume of the solution is exactly one liter. A 0.100 *M* solution contains 0.100 mol of solute per liter of solution, and a 0.0100 *M* solution contains 0.0100 mol of solute per liter of solution (Figure 3-2).

Water is the solvent in *most* of the solutions that we encounter. Unless otherwise indicated, we assume that water is the solvent. When the solvent is other than water, we state this explicitly.

 CD-ROM   Screen 5.15, Preparing Solutions, Direct Addition.

 Students sometimes make the mistake of assuming that one molar solution contains one mole of solute in a liter of solvent. This is not the case; one liter of solvent plus one mole of solute usually has a total volume of more than one liter.

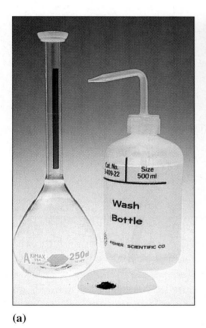

(a)

(b)

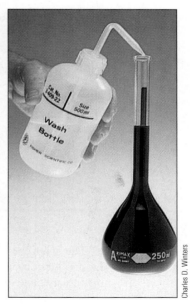

(c)

Charles D. Winters

Figure 3-2   Preparation of 0.0100 *M* solution of KMnO$_4$, potassium permanganate. A 250.-mL sample of 0.0100 *M* KMnO$_4$ solution contains 0.395 g of KMnO$_4$ (1 mol = 158 g). (a) 0.395 g of KMnO$_4$ (0.00250 mol) is weighed out carefully and transferred into a 250.-mL volumetric flask. (b) The KMnO$_4$ is dissolved in water. (c) Distilled H$_2$O is added to the volumetric flask until the volume of solution is 250. mL. The flask is then stoppered, and its contents are mixed thoroughly to give a homogeneous solution.

EXAMPLE 3-18   *Molarity*

Calculate the molarity (M) of a solution that contains 3.65 g of HCl in 2.00 L of solution.

Plan

We are given the number of grams of HCl in 2.00 L of solution. We apply the definition of molarity, remembering to convert grams of HCl to moles of HCl.

Solution

We place 3.65 g HCl over 2.00 L of solution, and then convert g HCl to mol HCl.

$$\frac{?\ \text{mol HCl}}{\text{L soln}} = \frac{3.65\ \text{g HCl}}{2.00\ \text{L soln}} \times \frac{1\ \text{mol HCl}}{36.5\ \text{g HCl}} = \boxed{0.0500\ \text{mol HCl/L soln}}$$

The concentration of the HCl solution is 0.0500 molar, and the solution is called 0.0500 M hydrochloric acid. One liter of the solution contains 0.0500 mol of HCl.

You should now work Exercise 62.

EXAMPLE 3-19   *Mass of Solute*

Calculate the mass of $Ba(OH)_2$ required to prepare 2.50 L of a 0.0600 M solution of barium hydroxide.

Plan

The volume of the solution, 2.50 L, is multiplied by the concentration, 0.0600 mol $Ba(OH)_2$/L, to give the number of moles of $Ba(OH)_2$. The number of moles of $Ba(OH)_2$ is then multiplied by the mass of $Ba(OH)_2$ in one mole, 171.3 g $Ba(OH)_2$/mol $Ba(OH)_2$, to give the mass of $Ba(OH)_2$ in the solution.

Solution

$$?\ \text{g Ba(OH)}_2 = 2.50\ \text{L soln} \times \frac{0.0600\ \text{mol Ba(OH)}_2}{1\ \text{L soln}} \times \frac{171.3\ \text{g Ba(OH)}_2}{1\ \text{mol Ba(OH)}_2}$$

$$= \boxed{25.7\ \text{g Ba(OH)}_2}$$

You should now work Exercise 64.

The solutions of acids and bases that are sold commercially are too concentrated for most laboratory uses. We often dilute these solutions before we use them. We must know the molar concentration of a stock solution before it is diluted. This can be calculated from the specific gravity and the percentage data given on the label of the bottle.

EXAMPLE 3-20   *Molarity*

A sample of commercial sulfuric acid is 96.4% H_2SO_4 by mass, and its specific gravity is 1.84. Calculate the molarity of this sulfuric acid solution.

Plan

The density of a solution, grams per milliliter, is numerically equal to its specific gravity, so the density of the solution is 1.84 g/mL. The solution is 96.4% H_2SO_4 by mass; therefore, 100. g of solution contains 96.4 g of *pure* H_2SO_4. From this information, we can find the molarity of the solution. First, we calculate the mass of 1 L of solution.

Solution

$$\frac{? \text{ g soln}}{\text{L soln}} = \frac{1.84 \text{ g soln}}{\text{mL soln}} \times \frac{1000 \text{ mL soln}}{\text{L soln}} = 1.84 \times 10^3 \text{ g soln/L soln}$$

The solution is 96.4% H_2SO_4 by mass, so the mass of H_2SO_4 in 1 L is

$$\frac{? \text{ g } H_2SO_4}{\text{L soln}} = \frac{1.84 \times 10^3 \text{ g soln}}{\text{L soln}} \times \frac{96.4 \text{ g } H_2SO_4}{100. \text{ g soln}} = 1.77 \times 10^3 \text{ g } H_2SO_4/\text{L soln}$$

The molarity is the number of moles of H_2SO_4 per liter of solution.

$$\frac{? \text{ mol } H_2SO_4}{\text{L soln}} = \frac{1.77 \times 10^3 \text{ g } H_2SO_4}{\text{L soln}} \times \frac{1 \text{ mol } H_2SO_4}{98.1 \text{ g } H_2SO_4} = \boxed{18.0 \text{ mol } H_2SO_4/\text{L soln}}$$

Thus, the solution is an 18.0 M H_2SO_4 solution. This problem can also be solved by using a series of three unit factors.

$$\frac{? \text{ mol } H_2SO_4}{\text{L soln}} = \frac{1.84 \text{ g soln}}{\text{mL soln}} \times \frac{1000 \text{ mL soln}}{\text{L soln}} \times \frac{96.4 \text{ g } H_2SO_4}{100. \text{ g soln}} \times \frac{1 \text{ mol } H_2SO_4}{98.1 \text{ g } H_2SO_4}$$

$$= 18.1 \text{ mol } H_2SO_4/\text{L soln} = \boxed{18.1 \, M \, H_2SO_4}$$

You should now work Exercise 70.

✔ **Problem-Solving Tip: *Write Complete Units***

A common pitfall is to write units that are not complete enough to be helpful. For instance writing the density in Example 3-20 as just $\frac{1.84 \text{ g}}{\text{mL}}$ doesn't help us figure out the required conversions. It is much safer to write $\frac{1.84 \text{ g soln}}{\text{mL soln}}$, $\frac{1000 \text{ mL soln}}{\text{L soln}}$, $\frac{96.4 \text{ g } H_2SO_4}{100 \text{ g soln}}$, and so on. In Example 3-20, we have written complete units to help guide us through the problem.

3-7 DILUTION OF SOLUTIONS

Recall that the definition of molarity is the number of moles of solute divided by the volume of the solution in liters:

$$\text{molarity} = \frac{\text{number of moles of solute}}{\text{number of liters of solution}}$$

Multiplying both sides of the equation by the volume, we obtain

$$\text{volume (in L)} \times \text{molarity} = \text{number of moles of solute}$$

ANALYSIS

Assay (H_2SO_4) W/W...Min. 95.0%--Max. 98.0%

MAXIMUM LIMITS OF IMPURITIES

Appearance........Passes A.C.S. Test
Color (APHA).....................10 Max.
Residue after Ignition............4 ppm
Chloride (Cl)....................0.2 ppm
Nitrate (NO_3)...................0.5 ppm
Ammonium (NH_4).................1 ppm
Substances Reducing $KMnO_4$ (limit about 2ppm as SO_2).........Passes A.C.S. Test
Arsenic (As)..................0.004 ppm
Heavy Metals (as Pb)..........0.8 ppm
Iron (Fe)........................0.2 ppm
Mercury (Hg).......................5 ppb
Specific Gravity...................~1.84
Normality.............................~36

Suitable for Mercury Determinations

A label that shows the analysis of sulfuric acid.

The small difference between these two answers is due to rounding.

CD-ROM Screen 5.16, Preparing Solutions, Dilution.

Multiplication of the volume of a solution, in liters, by its molar concentration gives the amount of solute in the solution.

A can of frozen orange juice contains a certain mass (or moles) of vitamin C. After the frozen contents of the can are diluted by addition of water, the amount of vitamin C in the resulting total amount of solution will be unchanged. The concentration, or amount per a selected volume, will be less in the final solution, however.

When we dilute a solution by mixing it with more solvent, the amount of solute present does not change. But the volume and the concentration of the solution *do* change. Because the same number of moles of solute is divided by a larger number of liters of solution, the molarity decreases. Using a subscript 1 to represent the original concentrated solution and a subscript 2 to represent the dilute solution, we obtain

$$\text{volume}_1 \times \text{molarity}_1 = \text{number of moles of solute} = \text{volume}_2 \times \text{molarity}_2$$

or

⚠ Never use this equation to relate two different substances in a chemical reaction. It only applies to dilution calculations

$$V_1 M_1 = V_2 M_2 \qquad \text{(for dilution only)}$$

We could use any volume unit as long as we use the same unit on both sides of the equation. This relationship also applies when the concentration is changed by evaporating some solvent.

This expression can be used to calculate any one of four quantities when the other three are known (Figure 3-3). We frequently need a certain volume of dilute solution of a given molarity for use in the laboratory, and we know the concentration of the initial solution available. Then we can calculate the amount of initial solution that must be used to make the dilute solution.

CAUTION!   Dilution of a concentrated solution, especially of a strong acid or base, frequently liberates a great deal of heat. This can vaporize drops of water as they hit the concentrated solution and can cause dangerous spattering. As a safety precaution, *concentrated solutions of acids or bases are always poured slowly into water*, allowing the heat to be absorbed by the larger quantity of water. Calculations are usually simpler to visualize, however, by assuming that the water is added to the concentrated solution.

EXAMPLE 3-21   *Dilution*

How many milliliters of 18.0 M H_2SO_4 are required to prepare 1.00 L of a 0.900 M solution of H_2SO_4?

Plan

The volume (1.00 L) and molarity (0.900 M) of the final solution and the molarity (18.0 M) of the original solution are given. We can use the relationship $V_1 M_1 = V_2 M_2$ with subscript 1 for the initial acid solution and subscript 2 for the dilute solution. We solve

$$V_1 M_1 = V_2 M_2 \qquad \text{for } V_1$$

Solution

$$V_1 = \frac{V_2 M_2}{M_1} = \frac{1.00 \text{ L} \times 0.900 \ M}{18.0 \ M} = 0.0500 \text{ L} = \boxed{50.0 \text{ mL}}$$

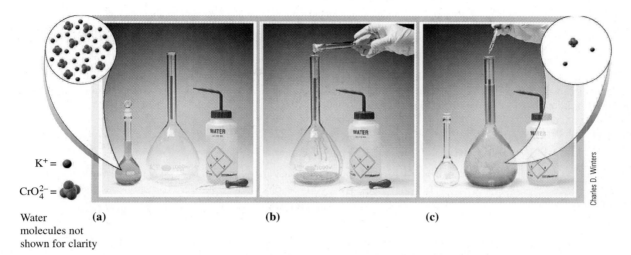

$K^+ = $
$CrO_4^{2-} = $
Water **(a)** **(b)** **(c)**
molecules not
shown for clarity

Figure 3-3 Dilution of a solution. (a) A 100.-mL volumetric flask is filled to the calibration line with a 0.100 M potassium chromate, K_2CrO_4 solution. (b) The 0.100 M K_2CrO_4 solution is transferred into a 1.00-L volumetric flask. The smaller flask is rinsed with a small amount of distilled H_2O. The rinse solution is added to the solution in the larger flask. To make sure that all the original K_2CrO_4 solution is transferred to the larger flask, the smaller flask is rinsed twice more and each rinse is added to the solution in the larger flask. (c) Distilled water is added to the 1.00-L flask until the liquid level coincides with its calibration line. The flask is stoppered and its contents are mixed thoroughly. The new solution is 0.0100 M K_2CrO_4. (100. mL of 0.100 M K_2CrO_4 solution has been diluted to 1000. mL.) The 100. mL of original solution and the 1000. mL of final solution both contain the amount of K_2CrO_4 dissolved in the original 100. mL of 0.100 M K_2CrO_4.

The dilute solution contains 1.00 L $\times$ 0.900 M = 0.900 mol of H_2SO_4, so 0.900 mol of H_2SO_4 must also be present in the original concentrated solution. Indeed, 0.0500 L $\times$ 18.0 M = 0.900 mol of H_2SO_4.

You should now work Exercises 72 and 74.

3-8 USING SOLUTIONS IN CHEMICAL REACTIONS

If we plan to carry out a reaction in a solution, we must calculate the amounts of solutions that we need. If we know the molarity of a solution, we can calculate the amount of solute contained in a specified volume of that solution. This procedure is illustrated in Example 3-22.

CD-ROM   Screen 5.18,
Stoichiometry of Reactions
in Solution.

EXAMPLE 3-22 *Amount of Solute*

Calculate (a) the number of moles of H_2SO_4 and (b) the number of grams of H_2SO_4 in 500. mL of 0.324 M H_2SO_4 solution.

Plan

Because we have two parallel calculations in this example, we state the plan for each step just before the calculation is done.

Solution

500. mL is more conveniently expressed as 0.500 L in this problem. By now, you should be able to convert mL to L (and the reverse) without writing out the conversion.

(a) The volume of a solution in liters multiplied by its molarity gives the number of moles of solute, H_2SO_4 in this case.

$$\underline{?}\text{ mol }H_2SO_4 = 0.500\text{ L soln} \times \frac{0.324\text{ mol }H_2SO_4}{\text{L soln}} = \boxed{0.162\text{ mol }H_2SO_4}$$

(b) We may use the results of part (a) to calculate the mass of H_2SO_4 in the solution.

A mole of H_2SO_4 is 98.1 g.

$$\underline{?}\text{ g }H_2SO_4 = 0.162\text{ mol }H_2SO_4 \times \frac{98.1\text{ g }H_2SO_4}{1\text{ mol }H_2SO_4} = \boxed{15.9\text{ g }H_2SO_4}$$

The mass of H_2SO_4 in the solution can be calculated without solving explicitly for the number of moles of H_2SO_4.

$$\underline{?}\text{ g }H_2SO_4 = 0.500\text{ L soln} \times \frac{0.324\text{ mol }H_2SO_4}{\text{L soln}} \times \frac{98.1\text{ g }H_2SO_4}{1\text{ mol }H_2SO_4} = \boxed{15.9\text{ g }H_2SO_4}$$

Molarity can be used as a unit factor, that is, $\dfrac{\text{mol solute}}{\text{L soln}}$.

One of the most important uses of molarity relates the volume of a solution of known concentration of one reactant to the mass of the other reactant.

EXAMPLE 3-23   *Solution Stoichiometry*

Calculate the volume in liters and in milliliters of a 0.324 *M* solution of sulfuric acid required to react completely with 2.792 g of Na_2CO_3 according to the equation

$$H_2SO_4 + Na_2CO_3 \longrightarrow Na_2SO_4 + CO_2 + H_2O$$

Plan

The balanced equation tells us that 1 mol H_2SO_4 reacts with 1 mol Na_2CO_3, and we can write

$$H_2SO_4 + Na_2CO_3 \longrightarrow Na_2SO_4 + CO_2 + H_2O$$
$$\text{1 mol} \qquad \text{1 mol} \qquad\quad \text{1 mol} \quad \text{1 mol} \quad \text{1 mol}$$
$$\qquad\qquad \text{106.0 g}$$

We convert (1) grams of Na_2CO_3 to moles of Na_2CO_3, (2) moles of Na_2CO_3 to moles of H_2SO_4, and (3) moles of H_2SO_4 to liters of H_2SO_4 solution.

$$\text{g }Na_2CO_3 \longrightarrow \text{mol }Na_2CO_3 \longrightarrow \text{mol }H_2SO_4 \longrightarrow \boxed{\text{L }H_2SO_4\text{ soln}}$$

Solution

$$\underline{?}\text{ L }H_2SO_4 = 2.792\text{ g }Na_2CO_3 \times \frac{1\text{ mol }Na_2CO_3}{106.0\text{ g }Na_2CO_3} \times \frac{1\text{ mol }H_2SO_4}{1\text{ mol }Na_2CO_3} \times \frac{1\text{ L }H_2SO_4\text{ soln}}{0.324\text{ mol }H_2SO_4}$$

$$= \boxed{0.0813\text{ L }H_2SO_4\text{ soln}} \quad \text{or} \quad \boxed{81.3\text{ mL }H_2SO_4\text{ soln}}$$

You should now work Exercise 78.

Charles Steele

The indicator methyl orange changes from yellow, its color in basic solutions, to orange, its color in acidic solutions, when the reaction in Example 3-23 reaches completion.

Often we must calculate the volume of solution of known molarity that is required to react with a specified volume of another solution. We always examine the balanced chemical equation for the reaction to determine the *reaction ratio*, that is, the relative numbers of moles of reactants.

EXAMPLE 3-24   *Volume of Solution Required*

Find the volume in liters and in milliliters of a 0.505 M NaOH solution required to react with 40.0 mL of 0.505 M H_2SO_4 solution according to the reaction

$$H_2SO_4 + 2NaOH \longrightarrow Na_2SO_4 + 2H_2O$$

Plan

We shall work this example in several steps, stating the "plan," or reasoning, just before each step in the calculation. Then we shall use a single setup to solve the problem.

Solution

The balanced equation tells us that the reaction ratio is 1 mol H_2SO_4 to 2 mol NaOH.

$$H_2SO_4 + 2NaOH \longrightarrow Na_2SO_4 + 2H_2O$$
$$\text{1 mol} \qquad \text{2 mol} \qquad \quad \text{1 mol} \qquad \text{2 mol}$$

From the volume and the molarity of the H_2SO_4 solution, we can calculate the number of moles of H_2SO_4.

$$\underline{?}\ \text{mol}\ H_2SO_4 = 0.0400\ \text{L}\ H_2SO_4\ \text{soln} \times \frac{0.505\ \text{mol}\ H_2SO_4}{\text{L soln}} = 0.0202\ \text{mol}\ H_2SO_4$$

The number of moles of H_2SO_4 is related to the number of moles of NaOH by the reaction ratio, 1 mol H_2SO_4/2 mol NaOH:

$$\underline{?}\ \text{mol NaOH} = 0.0202\ \text{mol}\ H_2SO_4 \times \frac{2\ \text{mol NaOH}}{1\ \text{mol}\ H_2SO_4} = 0.0404\ \text{mol NaOH}$$

⚠️ A common mistake is to use the dilution equation $V_1M_1 = V_2M_2$ to solve problems involving reactions. Though it can sometimes give the correct answer, this is coincidental; for Example 3-24 it would give the wrong answer.

The volume of H_2SO_4 solution is expressed as 0.0400 L rather than 40.0 mL.

Now we can calculate the volume of 0.505 M NaOH solution that contains 0.0404 mol of NaOH:

Again we see that molarity is a unit factor. In this case,

$$\frac{1.00 \text{ L NaOH soln}}{0.505 \text{ mol NaOH}}$$

$$\underline{?} \text{ L NaOH soln} = 0.0404 \text{ mol NaOH} \times \frac{1.00 \text{ L NaOH soln}}{0.505 \text{ mol NaOH}} = \boxed{0.0800 \text{ L NaOH soln}}$$

which we usually call $\boxed{80.0 \text{ mL of NaOH solution.}}$

We have worked through the problem stepwise; let us solve it in a single setup.

$$\boxed{\begin{array}{c} \text{L H}_2\text{SO}_4 \text{ soln} \\ \text{available} \end{array}} \longrightarrow \boxed{\begin{array}{c} \text{mol H}_2\text{SO}_4 \\ \text{available} \end{array}} \longrightarrow \boxed{\begin{array}{c} \text{mol NaOH} \\ \text{soln needed} \end{array}} \longrightarrow \boxed{\begin{array}{c} \text{L NaOH} \\ \text{soln needed} \end{array}}$$

$$\underline{?} \text{ L NaOH soln} = 0.0400 \text{ L H}_2\text{SO}_4 \text{ soln} \times \frac{0.505 \text{ mol H}_2\text{SO}_4}{\text{L H}_2\text{SO}_4 \text{ soln}} \times \frac{2 \text{ mol NaOH}}{1 \text{ mol H}_2\text{SO}_4}$$

$$\times \frac{1.00 \text{ L NaOH soln}}{0.505 \text{ mol NaOH}}$$

$$= \boxed{0.0800 \text{ L NaOH soln or } 80.0 \text{ mL NaOH soln}}$$

You should now work Exercise 80.

Key Terms

Actual yield The amount of a specified pure product actually obtained from a given reaction. Compare with *Theoretical yield*.

Aqueous solution A solution in which the solvent is water.

Chemical equation Description of a chemical reaction by placing the formulas of reactants on the left and the formulas of products on the right of an arrow. A chemical equation must be balanced; that is, it must have the same number of each kind of atom on both sides.

Concentration The amount of solute per unit volume or mass of solvent or of solution.

Dilution The process of reducing the concentration of a solute in solution, usually simply by adding more solvent.

Limiting reactant A substance that stoichiometrically limits the amount of product(s) that can be formed.

Molarity (*M*) The number of moles of solute per liter of solution.

Percent by mass 100% multiplied by the mass of a solute divided by the mass of the solution in which it is contained.

Percent yield 100% times actual yield divided by theoretical yield.

Products Substances produced in a chemical reaction.

Reactants Substances consumed in a chemical reaction.

Reaction ratio The relative amounts of reactants and products involved in a reaction; may be the ratio of moles, or masses.

Reaction stoichiometry Description of the quantitative relationships among substances as they participate in chemical reactions.

Sequential reaction A chemical process in which several reaction steps are required to convert starting materials into products.

Solute The dispersed (dissolved) phase of a solution.

Solution A homogeneous mixture of two or more substances.

Solvent The dispersing medium of a solution.

Stoichiometry Description of the quantitative relationships among elements and compounds as they undergo chemical changes.

Theoretical yield The maximum amount of a specified product that could be obtained from specified amounts of reactants, assuming complete consumption of the limiting reactant according to only one reaction and complete recovery of the product. Compare with *Actual yield*.

Exercises

Chemical Equations

1. In a chemical equation, what are the reactants? What information must an unbalanced chemical equation contain?

2. In a balanced chemical equation there must be the same number of atoms of each element on both sides of the equation. What scientific (natural) law requires that there be equal numbers of atoms of each element in both the products and the reactants?

3. Write a chemical equation for the following reaction: In sunlight hydrogen gas reacts explosively with chlorine to produce hydrogen chloride.

4. Write a chemical equation for the following reaction: In the presence of a spark, hydrogen gas reacts explosively with oxygen to produce gaseous water.

5. Use words to explicitly state the relationships among numbers of molecules of reactants and products in the equation for the combustion of butane, C_4H_{10}.

$$2C_4H_{10} + 13O_2 \longrightarrow 8CO_2 + 10H_2O$$

A butane-fueled lighter

6. Use words to explicitly state the information given in the following equation.

$$S(s) + O_2(g) \longrightarrow SO_2(g)$$

7. Write the chemical equation for the production of magnesium oxide by the reaction of magnesium and oxygen.

Balance each "equation" in Exercises 8-11 by inspection.

8. (a) $Li + O_2 \longrightarrow Li_2O$
   (b) $Mg_3N_2 + H_2O \longrightarrow NH_3 + Mg(OH)_2$
   (c) $NaCl + Pb(NO_3)_2 \longrightarrow PbCl_2 + NaNO_3$
   (d) $H_2O + KO_2 \longrightarrow KOH + O_2$
   (e) $H_2SO_4 + NH_3 \longrightarrow (NH_4)_2SO_4$

9. (a) $Na + O_2 \longrightarrow Na_2O_2$
   (b) $P_4 + O_2 \longrightarrow P_4O_{10}$
   (c) $Ca(HCO_3)_2 + Na_2CO_3 \longrightarrow CaCO_3 + NaHCO_3$
   (d) $NH_3 + O_2 \longrightarrow NO + H_2O$
   (e) $Rb + H_2O \longrightarrow RbOH + H_2$

10. (a) $Fe_2O_3 + CO \longrightarrow Fe + CO_2$
    (b) $K + H_2O \longrightarrow KOH + H_2$
    (c) $K + KNO_3 \longrightarrow K_2O + N_2$
    (d) $(NH_4)_2Cr_2O_7 \longrightarrow N_2 + H_2O + Cr_2O_3$
    (e) $Al + Cr_2O_3 \longrightarrow Al_2O_3 + Cr$

11. (a) $PbO + PbS \longrightarrow Pb + SO_2$
    (b) $NaCl + H_2O + SiO_2 \longrightarrow HCl + Na_2SiO_3$
    (c) $K_2CO_3 + Al_2Cl_6 \longrightarrow Al_2(CO_3)_3 + KCl$
    (d) $KClO_3 + C_{12}H_{22}O_{11} \longrightarrow KCl + CO_2 + H_2O$
    (e) $PCl_3 + O_2 \longrightarrow POCl_3$

Calculations Based on Chemical Equations

In Exercises 12-15, (a) write the balanced chemical equation that represents the reaction described by words, and then perform calculations to answer parts (b) and (c).

12. (a) Nitrogen, N_2, combines with hydrogen, H_2, to form ammonia, NH_3.
    (b) How many hydrogen molecules are required to react with 400 nitrogen molecules?
    (c) How many ammonia molecules are formed in part (b)?

13. (a) Sulfur, S_8, combines with oxygen at elevated temperatures to form sulfur dioxide.
    (b) If 400 oxygen molecules are used up in this reaction, how many sulfur molecules react?
    (c) How many sulfur dioxide molecules are formed in part (b)?

14. (a) Limestone, $CaCO_3$, dissolves in muriatic acid, HCl, to form calcium chloride, $CaCl_2$, carbon dioxide, and water.
    (b) How many moles of HCl are required to dissolve 5.7 mol of $CaCO_3$?
    (c) How many moles of water are formed in part (b)?

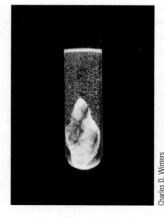

Limestone dissolving in HCl

15. (a) Aluminum building materials have a hard, transparent, protective coating of aluminum oxide, Al_2O_3, formed by reaction with oxygen in the air. The sulfuric acid,

H_2SO_4, in acid rain dissolves this protective coating and forms aluminum sulfate, $Al_2(SO_4)_3$, and water.

(b) How many moles of H_2SO_4 are required to react with 3.8 mol of Al_2O_3?

(c) How many moles of $Al_2(SO_4)_3$ are formed in part (b)?

16. Calculate the number of grams of baking soda, $NaHCO_3$, that contain 17.5 moles of carbon.

Baking soda

17. Limestone, coral, and seashells are composed primarily of calcium carbonate. The test for the identification of a carbonate is to use a few drops of hydrochloric acid. The unbalanced equation is

$$CaCO_3 + HCl \longrightarrow CaCl_2 + CO_2 + H_2O.$$

(a) Balance the equation.

(b) How many atoms are in 0.150 moles of calcium carbonate?

(c) What number of carbon dioxide molecules is released on the reaction of 0.150 moles of calcium carbonate?

18. How many moles of oxygen can be obtained by the decomposition of 6.5 mol of reactant in each of the following reactions?

(a) $2KClO_3 \longrightarrow 2KCl + 3O_2$

(b) $2H_2O_2 \longrightarrow 2H_2O + O_2$

(c) $2HgO \longrightarrow 2Hg + O_2$

(d) $2NaNO_3 \longrightarrow 2NaNO_2 + O_2$

(e) $KClO_4 \longrightarrow KCl + 2O_2$

19. For the formation of 6.5 mol of water, which reaction uses the most nitric acid?

(a) $3Cu + 8HNO_3 \longrightarrow 3Cu(NO_3)_2 + 2NO + 4H_2O$

(b) $Al_2O_3 + 6HNO_3 \longrightarrow 2Al(NO_3)_3 + 3H_2O$

(c) $4Zn + 10HNO_3 \longrightarrow$
    $\qquad 4Zn(NO_3)_2 + NH_4NO_3 + 3H_2O$

20. Consider the reaction

$$NH_3 + O_2 \xrightarrow{\text{not balanced}} NO + H_2O$$

For every 12.00 mol of NH_3, (a) how many moles of O_2 are required, (b) how many moles of NO are produced, and (c) how many moles of H_2O are produced?

21. Consider the reaction

$$2NO + Br_2 \longrightarrow 2NOBr$$

For every 6.25 mol of bromine that reacts, how many moles of (a) NO react and (b) NOBr are produced?

22. We allow 32.0 g of methane, CH_4, to react as completely as possible with excess oxygen, O_2, to form CO_2 and water. Write the balanced equation for this reaction. What mass of oxygen reacts?

23. Iron(III) oxide, Fe_2O_3, is a result of the reaction of iron with the oxygen in air.

(a) What is the balanced equation for this reaction?

(b) What number of moles of iron react with 15.25 mol of oxygen from the air?

(c) What mass of iron is required to react with 15.25 mol of oxygen?

24. A sample of magnetic iron oxide, Fe_3O_4, reacts completely with hydrogen at red heat. The water vapor formed by the reaction

$$Fe_3O_4 + 4H_2 \xrightarrow{\text{heat}} 3Fe + 4H_2O$$

is condensed and found to weigh 28.15 g. Calculate the mass of Fe_3O_4 that reacted.

25. What masses of cobalt(II) chloride and of hydrogen fluoride are needed to prepare 5.25 mol of cobalt(II) fluoride by the following reaction?

$$CoCl_2 + 2HF \longrightarrow CoF_2 + 2HCl$$

*26. Sodium iodide, NaI, is a source of iodine used to produce iodized salt. (a) Write the balanced chemical equation for the reaction of sodium and iodine. (b) How many grams of sodium iodide are produced by the reaction of 46.63 grams of iodine?

Iodized salt

*27. Calculate the mass of calcium required to react with 1.885 g of carbon during the production of calcium carbide, CaC_2.

28. Calculate the mass of propane, C_3H_8, that will produce 2.50 moles of water when burned in excess oxygen.

$$C_3H_8 + O_2 \xrightarrow{\text{not balanced}} CO_2 + H_2O$$

29. What mass of pentane, C_5H_{12}, produces 9.033×10^{22} CO_2 molecules when burned in excess oxygen?

Limiting Reactant

30. How many grams of NH_3 can be prepared from 89.78 g of N_2 and 18.17 g of H_2?

$$N_2 + 3H_2 \longrightarrow 2NH_3$$

*31. Silver nitrate solution reacts with calcium chloride solution according to the equation

$$2AgNO_3 + CaCl_2 \longrightarrow Ca(NO_3)_2 + 2AgCl$$

All of the substances involved in this reaction are soluble in water except silver chloride, AgCl, which forms a solid (precipitate) at the bottom of the flask. Suppose we mix together a solution containing 6.30 g of $AgNO_3$ and one containing 4.20 g of $CaCl_2$. What mass of AgCl is formed?

*32. "Superphosphate," a water-soluble fertilizer, is sometimes marketed as "triple phosphate." It is a mixture of $Ca(H_2PO_4)_2$ and $CaSO_4$ on a 1:2 *mole* basis. It is formed by the reaction

$$Ca_3(PO_4)_2 + 2H_2SO_4 \longrightarrow Ca(H_2PO_4)_2 + 2CaSO_2$$

We treat 400. g of $Ca_3(PO_4)_2$ with 267 g of H_2SO_4. How many grams of superphosphate could be formed?

Superphosphate fertilizer

33. Gasoline is produced from crude oil, a nonrenewable resource. Ethanol is mixed with gasoline to produce a fuel called gasohol. Calculate the mass of water produced when 100.0 g of ethanol, C_2H_5OH, is burned in 82.82 g of oxygen.

34. What mass of potassium can be produced by the reaction of 175.0 g of Na with 175.0 g of KCl?

$$Na + KCl \xrightarrow{\text{heat}} NaCl + K$$

35. Silicon carbide, an abrasive, is made by the reaction of silicon dioxide with graphite.

$$SiO_2 + C \xrightarrow{\text{heat}} SiC + CO \quad \text{(balanced?)}$$

We mix 300. g of SiO_2 and 203 g of C. If the reaction proceeds as far as possible, which reactant is left over? How much of this reactant remains?

36. What is the maximum amount of $Ca_3(PO_4)_2$ that can be prepared from 9.25 g of $Ca(OH)_2$ and 12.25 g of H_3PO_4?

$$3Ca(OH)_2 + 2H_3PO_4 \longrightarrow Ca_3(PO_4)_2 + 6H_2O$$

37. A reaction mixture contains 14.3 g of PCl_3 and 9.10 g of PbF_2. What mass of $PbCl_2$ can be obtained from the following reaction?

$$3PbF_2 + 2PCl_3 \longrightarrow 2PF_3 + 3PbCl_2$$

How much of which reactant is left unchanged?

38. The following equation represents a reaction between aqueous solutions of silver nitrate and barium chloride.

$$2AgNO_3(aq) + BaCl_2(aq) \longrightarrow$$
$$2AgCl(s) + Ba(NO_3)_2(aq)$$

According to this equation, if a solution that contains 41.6 g $AgNO_3$ is mixed with a solution that contains 35.4 g $BaCl_2$, (a) which is the limiting reactant? (b) How many grams of which reactant will be left over? (c) How many grams of AgCl will be formed?

39. The following reaction takes place at high temperatures.

$$Cr_2O_3(s) + 2Al(\ell) \longrightarrow 2Cr(\ell) + Al_2O_3(\ell)$$

If 42.7 g Cr_2O_3 and 9.8 g Al are mixed and reacted until one of the reactants is used up, (a) which reactant will be left over? (b) How much will be left? (c) How many grams of chromium will be formed?

Percent Yield from Chemical Reactions

40. The percent yield for the reaction

$$PCl_3 + Cl_2 \longrightarrow PCl_5$$

is 77.5%. What mass of PCl_5 is expected from the reaction of 96.7 g of PCl_3 with excess chlorine?

41. The percent yield for the following reaction carried out in carbon tetrachloride solution is 66.0%.

$$Br_2 + Cl_2 \longrightarrow 2BrCl$$

(a) What amount of BrCl is formed from the reaction of 0.0350 mol Br_2 with 0.0350 mol Cl_2?
(b) What amount of Br_2 is left unchanged?

42. When heated, potassium chlorate, $KClO_3$, melts and decomposes to potassium chloride and diatomic oxygen.
(a) What is the theoretical yield of O_2 from 3.75 g $KClO_3$?
(b) If 1.10 g of O_2 is obtained, what is the percent yield?

43. The percent yield of the following reaction is consistently 92%.

$$CH_4(g) + 4S(g) \longrightarrow CS_2(g) + 2H_2S(g)$$

How many grams of sulfur would be needed to obtain 80.0 g of CS_2?

44. Solid silver nitrate undergoes thermal decomposition to form silver metal, nitrogen dioxide, and oxygen. Write the chemical equation for this reaction. A 0.665-g sample of silver metal is obtained from the decomposition of a 1.076-g sample of $AgNO_3$. What is the percent yield of the reaction?

45. Tin(IV) chloride is produced in 81.1% yield by the reaction of tin with chlorine. How much tin is required to produce a kilogram of tin(IV) chloride?

46. Give the percent yield if 108 mg SO_2 is obtained from the combustion of 78.1 mg of carbon disulfide according to the reaction:

$$CS_2 + 3O_2 \longrightarrow CO_2 + 2SO_2$$

47. From a 45.0-g sample of an iron ore containing Fe_3O_4, 1.56 g of Fe is obtained by the reaction:

$$Fe_3O_4 + 2C \longrightarrow 3Fe + 2CO_2$$

What is the percent of Fe_3O_4 in the ore?

48. The reaction of finely divided aluminum and iron(III) oxide, Fe_2O_3, is called the thermite reaction. It produces a tremendous amount of heat, making the welding of railroad track possible. The reaction of 500. grams of aluminum and 500. grams of iron(III) oxide produces 166.5 grams of iron. (a) Calculate the mass of iron that should be released by this reaction. (b) What is the percent yield of iron?

$$Fe_2O_3 + 2Al \longrightarrow 2Fe + Al_2O_3 + heat$$

Charles D. Winters

Thermite reaction

49. Lime, $Ca(OH)_2$, can be used to neutralize an acid spill. A 5.57-g sample of $Ca(OH)_2$ reacts with an excess of hy-drochloric acid; 7.41 g of calcium chloride is collected. What is the percent yield of this experiment?

$$Ca(OH)_2 + 2HCl \longrightarrow CaCl_2 + 2H_2O$$

Sequential Reactions

50. Consider the two-step process for the formation of tellurous acid described by the following equations:

$$TeO_2 + 2OH^- \longrightarrow TeO_3^{2-} + H_2O$$
$$TeO_3^{2-} + 2H^+ \longrightarrow H_2TeO_3$$

What mass of H_2TeO_3 is formed from 64.2 g of TeO_2, assuming 100% yield?

51. Consider the formation of cyanogen, C_2N_2, and its subsequent decomposition in water given by the equations

$$2Cu^{2+} + 6CN^- \longrightarrow 2[Cu(CN)_2]^- + C_2N_2$$
$$C_2N_2 + H_2O \longrightarrow HCN + HOCN$$

How much hydrocyanic acid, HCN, can be produced from 85.77 g of KCN, assuming 100% yield?

52. What mass of potassium chlorate is required to supply the proper amount of oxygen needed to burn 66.3 g of methane, CH_4?

$$2KClO_3 \longrightarrow 2KCl + 3O_2$$
$$CH_4 + 2O_2 \longrightarrow CO_2 + 2H_2O$$

53. Hydrogen, obtained by the electrical decomposition of water, is combined with chlorine to produce 444.2 g of hydrogen chloride. Calculate the mass of water decomposed.

$$2H_2O \longrightarrow 2H_2 + O_2$$
$$H_2 + Cl_2 \longrightarrow 2HCl$$

54. Ammonium nitrate, known for its use in agriculture, can be produced from ammonia by the following sequence of reactions:

$$NH_3(g) + O_2(g) \longrightarrow NO(g) + H_2O(g)$$
$$NO(g) + O_2(g) \longrightarrow NO_2(g)$$
$$NO_2(g) + H_2O(\ell) \longrightarrow HNO_3(aq) + NO(g)$$
$$HNO_3(aq) + NH_3(g) \longrightarrow NH_4NO_3(aq)$$

(a) Balance each equation.
(b) How many moles of nitrogen atoms are required for every mole of ammonium nitrate (NH_4NO_3)?
(c) How much ammonia is needed to prepare 500.0 grams of ammonium nitrate (NH_4NO_3)?

55. Calcium sulfate is the essential component of plaster and sheet rock. Waste calcium sulfate can be converted into quicklime, CaO, by reaction with carbon at high temperatures. The following two reactions represent a sequence of reactions that might take place:

$$CaSO_3(s) + 4C(s) \longrightarrow CaS(\ell) + 4CO(g)$$
$$CaS(\ell) + 3CaSO_4(s) \longrightarrow 4CaO(s) + 4SO_2(g)$$

What weight of sulfur dioxide (in grams) could be obtained from 1.500 kg of calcium sulfate?

*56. The chief ore of zinc is the sulfide, ZnS. The ore is concentrated by flotation and then heated in air, which converts the ZnS to ZnO.

$$2ZnS + 3O_2 \longrightarrow 2ZnO + 2SO_2$$

The ZnO is then treated with dilute H_2SO_4

$$ZnO + H_2SO_4 \longrightarrow ZnSO_4 + H_2O$$

to produce an aqueous solution containing the zinc as $ZnSO_4$. An electric current is passed through the solution to produce the metal (electrolysis).

$$2ZnSO_4 + 2H_2O \longrightarrow 2Zn + 2H_2SO_4 + O_2$$

What mass of Zn is obtained from an ore containing 345 kg of ZnS? Assume the flotation process to be 89.6% efficient, the electrolysis step to be 92.2% efficient, and the other steps to be 100% efficient.

Concentrations of Solutions—Percent by Mass

57. (a) How many moles of solute are contained in 500. g of a 15.00% aqueous solution of $K_2Cr_2O_7$?
(b) How many grams of solute are contained in the solution of part (a)?
(c) How many grams of water (the solvent) are contained in the solution of part (a)?

58. The density of an 18.0% solution of ammonium sulfate, $(NH_4)_2SO_4$, is 1.10 g/mL. What mass of $(NH_4)_2SO_4$ is required to prepare 450. mL of this solution?

59. The density of an 18.0% solution of ammonium chloride, NH_4Cl, solution is 1.05 g/mL. What mass of NH_4Cl does 450 mL of this solution contain?

60. What volume of the solution of $(NH_4)_2SO_4$ described in Exercise 58 contains 125.0 g of $(NH_4)_2SO_4$?

*61. A reaction requires 65.6 g of NH_4Cl. What volume of the solution described in Exercise 59 do you need if you want to use a 25.0% excess of NH_4Cl?

Concentrations of Solutions—Molarity

62. What is the molarity of a solution prepared by dissolving 355 g of sodium phosphate, Na_3PO_4, in water and diluting to 3.50 L of solution?

63. What is the molarity of a solution prepared by dissolving 4.49 g of sodium chloride in water and diluting to 30.0 mL of solution?

64. How many grams of Na_2HPO_4 (a) are needed to prepare 250 mL of 0.50 M solution, and (b) are in 250 mL of 0.50 M solution?

65. How many kilograms of ethylene glycol, $C_2H_6O_2$, are needed to prepare a 9.00 M solution to protect a 12.0-L

car radiator against freezing? What is the mass of $C_2H_6O_2$ in 12.0 L of 9.00 M solution?

Antifreeze being poured into water

66. A solution made by dissolving 16.0 g of $CaCl_2$ in 64.0 g of water has a density of 1.180 g/mL at 20°C. (a) What is the percent by mass of $CaCl_2$ in the solution? (b) What is the molarity of $CaCl_2$ in the solution?

67. A solution contains 0.100 mol/L of each of the following acids: HCl, H_2SO_4, H_3PO_4. (a) Is the molarity the same for each acid? (b) Is the number of molecules per liter the same for each acid? (c) Is the mass per liter the same for each acid?

68. What is the molarity of a barium chloride solution prepared by dissolving 1.50 g of $BaCl_2 \cdot 2H_2O$ in enough water to make 550. mL of solution?

69. How many grams of potassium benzoate trihydrate, $KC_7H_5O_2 \cdot 3H_2O$, are needed to prepare 1 L of a 0.150 M solution of potassium benzoate?

70. Stock hydrofluoric acid solution is 49.0% HF and has a specific gravity of 1.17. What is the molarity of the solution?

71. Stock phosphoric acid solution is 85.0% H_3PO_4 and has a specific gravity of 1.70. What is the molarity of the solution?

Dilution of Solutions

72. Commercial concentrated hydrochloric acid is 12.0 M HCl. What volume of concentrated hydrochloric acid is required to prepare 2.50 L of 2.25 M HCl solution?

73. Commercially available concentrated sulfuric acid is 18.0 M H_2SO_4. Calculate the volume of concentrated sulfuric acid required to prepare 2.50 L of 2.25 M H_2SO_4 solution.

74. Calculate the volume of 0.0550 M $Ba(OH)_2$ solution that contains the same number of moles of $Ba(OH)_2$ as 135 mL of 0.0900 M $Ba(OH)_2$ solution.

75. Calculate the volume of 4.00 M NaOH solution required to prepare 200. mL of a 0.600 M solution of NaOH.

76. Calculate the final volume of solution obtained if 100. mL of 12.0 M NaOH is diluted to 4.20 M.

77. In a laboratory preparation room one may find a reagent bottle containing 5.0 L of 12 M NaOH. Write a set of instructions for the production of 250. mL of 3.50 M NaOH from such a solution.

Using Solutions in Chemical Reactions

78. Calculate the volume of a 0.225 M solution of potassium hydroxide, KOH, required to react with 0.385 g of acetic acid, CH_3COOH, according to the following reaction.

$$KOH + CH_3COOH \longrightarrow KCH_3COO + H_2O$$

79. Calculate the number of grams of carbon dioxide, CO_2 that can react with 43.44 mL of a 0.957 M solution of potassium hydroxide, KOH, according to the following reaction.

$$2KOH + CO_2 \longrightarrow K_2CO_3 + H_2O$$

80. What volume of 0.446 M HNO_3 solution is required to react completely with 35.62 mL of 0.0515 M $Ba(OH)_2$?

$$Ba(OH)_2 + 2HNO_3 \longrightarrow Ba(NO_3)_2 + 2H_2O$$

81. What volume of 0.750 M HBr is required to react completely with 0.800 mol of $Ca(OH)_2$?

$$2HBr + Ca(OH)_2 \longrightarrow CaBr_2 + 2H_2O$$

82. An excess of $AgNO_3$ reacts with 115.5 mL of an $AlCl_3$ solution to give 0.215 g of AgCl. What is the concentration, in moles per liter, of the $AlCl_3$ solution?

$$AlCl_3 + 3AgNO_3 \longrightarrow 3AgCl + Al(NO_3)_3$$

83. An impure sample of solid Na_2CO_3 is allowed to react with 0.1755 M HCl.

$$Na_2CO_3 + 2HCl \longrightarrow 2NaCl + CO_2 + H_2O$$

A 0.2337-g sample of sodium carbonate requires 15.55 mL of HCl solution. What is the purity of the sodium carbonate?

Mixed Exercises

*84. An iron ore that contains Fe_3O_4 reacts according to the reaction

$$Fe_3O_4 + 2C \longrightarrow 3Fe + 2CO_2$$

We obtain 3.49 g of Fe from the reaction of 75.0 g of the ore. What is the percent Fe_3O_4 in the ore?

*85. If 86.3% of the iron can be recovered from an ore that is 43.2% magnetic iron oxide, Fe_3O_4, what mass of iron could be recovered from 2.50 kg of this ore? The reduction of magnetic iron oxide is a complex process that can be represented in simplified form as:

$$Fe_3O_4 + 4CO \longrightarrow 3Fe + 4CO_2$$

86. Gaseous chlorine will displace bromide ion from an aqueous solution of potassium bromide to form aqueous potassium chloride and aqueous bromine. Write the chemical equation for this reaction. What mass of bromine is produced if 0.631 g of chlorine undergoes reaction?

87. Calculate the volume of 2.25 M phosphoric acid solution necessary to react with 45.0 mL of 0.150 M $Mg(OH)_2$.

$$2H_3PO_4 + 3Mg(OH)_2 \longrightarrow Mg_3(PO_4)_2 + 6H_2O$$

CONCEPTUAL EXERCISES

88. Using your own words, give a definition of a chemical reaction.

89. Write directions for the preparation of 1.000 L of 0.0500 M NaCl solution from a 0.350 M solution of NaCl.

90. Without making any calculations, arrange the following three solutions in order of increasing molarity. (a) 1.0% (by mass) NaCl; (b) 1.0% (by mass) $SnCl_2$; (c) 1.0% (by mass) $AlCl_3$. (Assume that these solutions have nearly equal densities.)

91. To what volume must a student dilute 50 mL of a solution containing 25 mg of $AlCl_3$/mL so that the Al^{3+} concentration in the new solution is 0.022 M?

92. Bismuth dissolves in nitric acid according to the reaction below. What volume of 30.0% HNO_3 by mass (density = 1.182 g/mL) would be required to dissolve 20.0 g of Bi?

$$Bi + 4HNO_3 + 3H_2O \longrightarrow Bi(NO_3)_3 \cdot 5H_2O + NO$$

93. How would you prepare 1 L of 1.25×10^{-6} M NaCl (molecular weight 58.44 g/mol) solution by using a balance that can measure mass only to 0.01 g?

94. The drawings shown below represent beakers of aqueous solutions. Each sphere represents a dissolved solute particle.
   (a) Which solution is most concentrated?
   (b) Which solution is least concentrated?
   (c) Which two solutions have the same concentration?
   (d) When solutions E and F are combined, the resulting solution has the same concentration as solution ——.

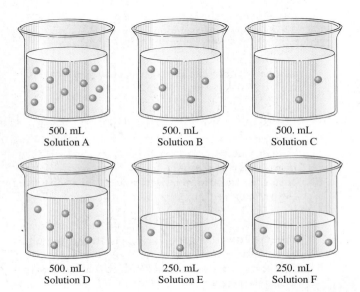

500. mL Solution A	500. mL Solution B	500. mL Solution C
500. mL Solution D	250. mL Solution E	250. mL Solution F

95. You prepared a NaCl solution by adding 58.44 g of NaCl to a 1-L volumetric flask and then adding water to dissolve it. When finished, the final volume in your flask looked like the illustration.

The solution you prepared is
(a) greater than 1 M because you added more solvent than necessary.
(b) less than 1 M because you added less solvent than necessary.
(c) greater than 1 M because you added less solvent than necessary.
(d) less than 1 M because you added more solvent than necessary.
(e) is 1 M because the amount of solute, not solvent, determines the concentration.

96. Zinc is more active chemically than is silver; it can be used to remove ionic silver from solution.

$$Zn(s) + 2AgNO_3(aq) \longrightarrow Zn(NO_3)_2(aq) + 2Ag(s)$$

The concentration of a silver nitrate solution is determined to be 1.330 mol/L. Pieces of zinc totaling 100.0 g are added to 1.000 L of the solution; 90.0 g of silver is collected.
(a) Calculate the percent yield of silver.
(b) Suggest a reason why the yield is less than 100.0%.

97. Ammonia is formed in a direct reaction of nitrogen and hydrogen.

$$N_2(g) + 3H_2(g) \longrightarrow 2NH_3(g)$$

A tiny portion of the starting mixture is represented by the top diagram, where the blue spheres represent N and the white spheres represent H. Which of the others represents the product mixture?

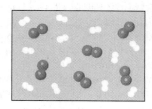

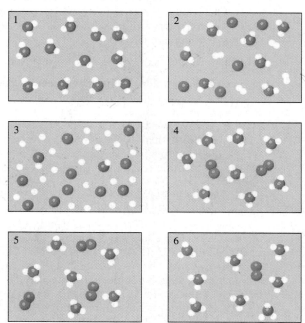

For the reaction of the given sample, which of the following is true?
(a) N_2 is the limiting reactant.
(b) H_2 is the limiting reactant.
(c) NH_3 is the limiting reactant.
(d) No reactant is limiting; they are present in the correct stoichiometric ratio.

BUILDING YOUR KNOWLEDGE

98. Acetic acid, CH_3COOH, reacts with ethanol, CH_3CH_2OH, to form ethyl acetate, $CH_3COOCH_2CH_3$, (density 0.902 g/mL) by the following reaction:

$$CH_3COOH + CH_3CH_2OH \longrightarrow$$
$$CH_3COOCH_2CH_3 + H_2O$$

We combine 20.2 mL of acetic acid with 20.1 mL of ethanol.
(a) Which compound is the limiting reactant?
(b) If 27.5 mL of pure ethyl acetate is produced, what is the percent yield? [*Hint:* See Tables 1-1 and 1-8.]

99. The label on a bottle of concentrated ammonia says that the contents are 28.0% (by mass) NH_3 and have a density of 0.898 g/mL. What is the molarity of the concentrated ammonia?

100. What is the molarity of a solution prepared by mixing 25.0 mL of 0.375 M NaCl solution with 42.0 mL of a 0.632 M NaCl solution?

101. Concentrated hydrochloric acid solution is 37.0% HCl and has a density of 1.19 g/mL. A dilute solution of HCl is prepared by diluting 4.50 mL of this concentrated HCl solution to 100.00 mL with water. Then 10.0 mL of this dilute HCl solution reacts with an $AgNO_3$ solution according to the following reaction:

$$HCl(aq) + AgNO_3(aq) \longrightarrow HNO_3(aq) + AgCl(s)$$

How many milliliters of 0.108 M $AgNO_3$ solution is required to precipitate all of the chloride as $AgCl(s)$?

102. In a particular experiment, 225 g of phosphorus, P_4, reacted with excess oxygen to form tetraphosphorus decoxide, P_4O_{10}, in 89.5% yield. In the second step reaction, a 97.8% yield of H_3PO_4 was obtained.
  (a) Write the balanced equations for these two reaction steps.
  (b) What mass of H_3PO_4 was obtained?

103. A student measured 50.00 mL of a NaCl solution labeled "5.00% NaCl (Sp. Gr. = 1.036)" into a 250-mL volumetric flask and diluted the solution to volume. The student labeled the flask "0.273 M NaCl". Does the label on the flask reflect the true molarity of the solution? If not, what is the molarity of the solution the student prepared?

104. What mass of AgCl could be formed by mixing 10.0 mL of a 1.20% NaCl by mass solution (d = 1.02 g/mL) with 50.0 mL of 1.21×10^{-2} M $AgNO_3$? The chemical equation for the reaction of these two solutions is:

$$NaCl(aq) + AgNO_3(aq) \longrightarrow AgCl(s) + NaNO_3(aq)$$

105. Magnesium displaces copper from a dilute solution of copper(II) sulfate; the pure copper will settle out of the solution.

$$Mg(s) + CuSO_4(aq) \longrightarrow MgSO_4(aq) + Cu(s)$$

A copper(II) sulfate solution is mixed by dissolving 25.000 g of copper(II) sulfate, and then it is treated with an excess of magnesium metal. The mass of copper collected is 8.786 g after drying. Calculate the percent yield of copper.

106. Suppose you are designing an experiment for the preparation of hydrogen. For the production of equal amounts of hydrogen, which metal, Zn or Al, is less expensive if Zn costs about half as much as Al on a mass basis?

$$Zn + 2HCl \longrightarrow ZnCl_2 + H_2$$

$$2Al + 6HCl \longrightarrow 2AlCl_3 + 3H_2$$

107. Gaseous chlorine and gaseous fluorine undergo a combination reaction to form the interhalogen compound ClF.
  (a) Write the chemical equation for this reaction.
  (b) Calculate the mass of fluorine needed to react with 3.47 g of Cl_2.
  (c) How many grams of ClF are formed?

BEYOND THE TEXTBOOK

Go to the textbook website at

http://www.brookscole.com/chemistry/whitten

for additional activities and exercises based on the General Chemistry Interactive CD-ROM, the World Wide Web, and library resources.

InfoTrac College Edition

For additional readings, go to InfoTrac College Edition, your online research library at:

http://infotrac.thomsonlearning.com

Some Types of Chemical Reactions

4

Charles D. Winters

The active metal potassium, K, reacts violently with water:

$$2K(s) + 2H_2O(\ell) \longrightarrow H_2(g) + 2KOH(aq)$$

Heat released by the reaction causes unreacted potassium to give off a characteristic purple glow and ignites the hydrogen gas that is formed.

OBJECTIVES

After you have studied this chapter, you should be able to

- Describe the periodic table and some of the relationships that it summarizes
- Recognize and describe nonelectrolytes, strong electrolytes, and weak electrolytes
- Recognize and classify acids (strong, weak), bases (strong, weak, insoluble), and salts (soluble, insoluble); use the solubility guidelines
- Describe reactions in aqueous solutions by writing formula unit equations, total ionic equations, and net ionic equations
- Assign oxidation numbers to elements when they are free, in compounds, or in ions
- Name and write formulas for common binary and ternary inorganic compounds
- Recognize oxidation–reduction reactions and identify which species are oxidized, reduced, oxidizing agents, and reducing agents
- Recognize and describe classes of reactions: decomposition reactions, displacement reactions, and various types of metathesis reactions

In this chapter we examine some types of chemical reactions. Millions of reactions are known, so it is useful to group them into classes, or types, so that we can deal systematically with these massive amounts of information. We describe how some compounds behave in aqueous solution, including how well their solutions conduct electricity and whether or not the compounds dissolve in water. We introduce several ways to represent chemical reactions in aqueous solution—formula unit equations, total ionic equations, and net ionic equations—and the advantages and disadvantages of these methods. We also introduce systematic ways to describe chemical reactions, as well as their reactants and their products.

Let's first take a brief look at the periodic table, which helps us to organize many properties of the elements, including their chemical reactions.

4-1 THE PERIODIC TABLE: METALS, NONMETALS, AND METALLOIDS

In 1869, the Russian chemist Dmitri Mendeleev (1834–1907) and the German chemist Lothar Meyer (1830–1895) independently published arrangements of known elements that are much like the periodic table in use today. Mendeleev's classification was based primarily on chemical properties of the elements, whereas Meyer's classification was based largely on physical properties. The tabulations were surprisingly similar. Both emphasized the *periodicity*, or regular periodic repetition, of properties with increasing atomic weight.

Mendeleev arranged the known elements in order of increasing atomic weight in successive sequences so that elements with similar chemical properties fell into the same column. He noted that both physical and chemical properties of the elements vary in a periodic fashion with atomic weight. His periodic table of 1872 contained the 62 known elements (Figure 4-1). Mendeleev placed H, Li, Na, and K in his table as "Gruppe I." These were

Pronounced "men-del-*lay*-ev."

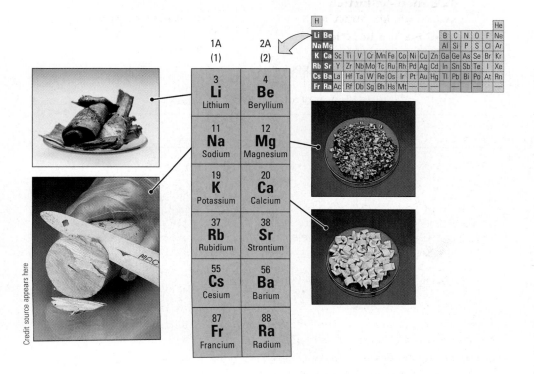

Credit source appears here

Lithium (*top left*) and sodium (*bottom left*) from Group IA. Magnesium (*top right*) and calcium (*bottom right*) from Group IIA.

known to combine with F, Cl, Br, and I of "Gruppe VII" to produce compounds that have similar formulas such as HF, LiCl, NaCl, and KI. All these compounds dissolve in water to produce solutions that conduct electricity. The "Gruppe II" elements were known to form compounds such as $BeCl_2$, $MgBr_2$, and $CaCl_2$, as well as compounds with O and S from

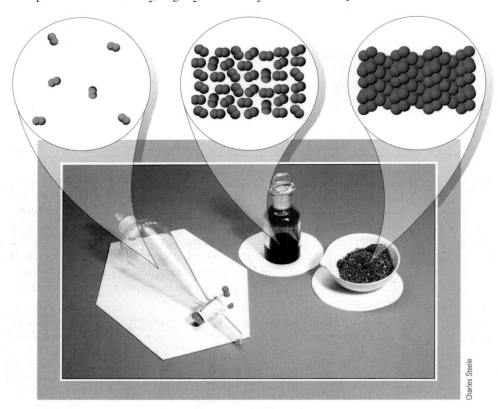

Three of the halogens, elements from Group VIIA (*left to right*): chlorine, bromine, iodine.

Charles Steele

REIHEN	GRUPPE I – R^2O	GRUPPE II – RO	GRUPPE III – R^2O^3	GRUPPE IV RH^4 RO^2	GRUPPE V RH^3 R^2O^5	GRUPPE VI RH^2 RO^3	GRUPPE VII RH R^2O^7	GRUPPE VIII – RO^4
1	H = 1							
2	Li = 7	Be = 9,4	B = 11	C = 12	N = 14	O = 16	F = 19	
3	Na = 23	Mg = 24	Al = 27,3	Si = 28	P = 31	S = 32	Cl = 35,5	
4	K = 39	Ca = 40	– = 44	Ti = 48	V = 51	Cr = 52	Mn = 55	Fe = 56, Co = 59, Ni = 59, Cu = 63.
5	(Cu = 63)	Zn = 65	– = 68	– = 72	As = 75	Se = 78	Br = 80	
6	Rb = 85	Sr = 87	?Yt = 88	Zr = 90	Nb = 94	Mo = 96	– = 100	Ru = 104, Rh = 104, Pd = 106, Ag = 108.
7	(Ag = 108)	Cd = 112	In = 113	Sn = 118	Sb = 122	Te = 125	J = 127	
8	Cs = 133	Ba = 137	?Di = 138	?Ce = 140	–	–	–	– – – –
9	(–)	–	–	–	–	–	–	
10	–	–	?Er = 178	?La = 180	Ta = 182	W = 184	–	Os = 195, Ir = 197, Pt = 198, Au = 199.
11	(Au = 199)	Hg = 200	Tl = 204	Pb = 207	Bi = 208	–	–	
12	–	–	–	Th = 231	–	U = 240	–	– – – –

Figure 4-1   Mendeleev's early periodic table (1872). "J" is the German symbol for iodine.

"Gruppe VI" such as MgO, CaO, MgS, and CaS. These and other chemical properties led him to devise a table in which the elements were arranged by increasing atomic weights and grouped into vertical families.

In most areas of human endeavor progress is slow and faltering. Occasionally, however, an individual develops concepts and techniques that clarify confused situations. Mendeleev was such an individual. One of the brilliant successes of his periodic table was that it provided for elements that were unknown at the time. When he encountered "missing" elements, Mendeleev left blank spaces. Some appreciation of his genius in constructing the table as he did can be gained by comparing the predicted (1871) and observed properties of germanium, which was not discovered until 1886. Mendeleev called the undiscovered element eka-silicon because it fell below silicon in his table. He was familiar with the properties of germanium's neighboring elements. They served as the basis for his predictions of properties of germanium (Table 4-1). Some modern values for properties of germanium differ significantly from those reported in 1886. But many of the values on which Mendeleev based his predictions were also inaccurate.

Because Mendeleev's arrangement of the elements was based on increasing *atomic weights*, several elements would have been out of place in his table. Mendeleev put the controversial elements (Te and I, Co and Ni) in locations consistent with their properties, however. He thought the apparent reversal of atomic weights was due to inaccurate values for those weights. Careful redetermination showed that the values were correct. Explanation of the locations of these "out-of-place" elements had to await the development of the concept of *atomic number*, approximately 50 years after Mendeleev's work. The **atomic number** (Section 5-5) of an element is the number of protons in the nucleus of its atoms. (It is also the

Charles D. Winters

Not all properties of the elements can be predicted from the trends in the periodic table. For example, gallium's melting point being low enough to melt from the heat of the hand is inconsistent with the properties of the elements above and below it.

TABLE 4-1	Predicted and Observed Properties of Germanium		
Property	**Eka-Silicon Predicted, 1871**	**Germanium Reported, 1886**	**Modern Values**
Atomic weight	72	72.32	72.61
Atomic volume	13 cm³	13.22 cm³	13.5 cm³
Specific gravity	5.5	5.47	5.35
Specific heat	0.073 cal/g°C	0.076 cal/g°C	0.074 cal/g°C
Maximum valence*	4	4	4
Color	Dark gray	Grayish white	Grayish white
Reaction with water	Will decompose steam with difficulty	Does not decompose water	Does not decompose water
Reactions with acids and alkalis	Slight with acids; more pronounced with alkalis	Not attacked by HCl or dilute aqueous NaOH; reacts vigorously with molten NaOH	Not dissolved by HCl or H_2SO_4 or dilute NaOH; dissolved by concentrated NaOH
Formula of oxide	EsO_2	GeO_2	GeO_2
Specific gravity of oxide	4.7	4.703	4.228
Specific gravity of tetrachloride	1.9 at 0°C	1.887 at 18°C	1.8443 at 30°C
Boiling point of tetrachloride	100°C	86°C	84°C
Boiling point of tetraethyl derivative	160°C	160°C	186°C

*"Valence" refers to the combining power of a specific element.

CD-ROM Screens 2.13, The Periodic Table, and 2.14, Chemical Periodicity

number of electrons in a neutral atom of an element.) This quantity is fundamental to the identity of each element because it is related to the electrical makeup of atoms. Elements are arranged in the periodic table in order of increasing atomic number. With the development of this concept, the **periodic law** attained essentially its present form:

> The properties of the elements are periodic functions of their atomic numbers.

The periodic law tells us that if we arrange the elements in order of increasing atomic number, we periodically encounter elements that have similar chemical and physical properties. The presently used "long form" of the periodic table (Table 4-2 and inside the front cover) is such an arrangement. The vertical columns are referred to as **groups** or **families,** and the horizontal rows are called **periods.** Elements in a *group* have similar chemical and physical properties, and those within a *period* have properties that change progressively across the table. Several groups of elements have common names that are used so frequently they should be learned. The Group IA elements, except H, are referred to as **alkali metals,** and the Group IIA elements are called the **alkaline earth metals.** The Group VIIA

Alkaline means basic. The character of basic compounds is described in Section 10-4.

About 80% of the elements are metals.

TABLE 4-2   *The Periodic Table*

There are other systems for numbering the groups in the periodic table. We number the groups by the standard American system of A and B groups. An alternative system in which the groups are numbered 1 through 18 is shown in parentheses.

TABLE 4-3	*Some Physical Properties of Metals and Nonmetals*

Metals	Nonmetals
1. High electrical conductivity that decreases with increasing temperature	1. Poor electrical conductivity (except carbon in the form of graphite)
2. High thermal conductivity	2. Good heat insulators (except carbon in the form of diamond)
3. Metallic gray or silver luster*	3. No metallic luster
4. Almost all are solids†	4. Solids, liquids, or gases
5. Malleable (can be hammered into sheets)	5. Brittle in solid state
6. Ductile (can be drawn into wires)	6. Nonductile

*Except copper and gold.
†Except mercury; cesium and gallium melt in a protected hand.

TABLE 4-4	*Some Chemical Properties of Metals and Nonmetals*

Metals	Nonmetals
1. Outer shells contain few electrons—usually three or fewer	1. Outer shells contain four or more electrons*
2. Form cations (positive ions) by losing electrons	2. Form anions (negative ions) by gaining electrons†
3. Form ionic compounds with nonmetals	3. Form ionic compounds with metals† and molecular (covalent) other compounds with nonmetals
4. Solid state characterized by metallic bonding	4. Covalently bonded molecules; noble gases are monatomic

*Except hydrogen and helium.
†Except the noble gases.

elements are called **halogens,** which means "salt formers," and the Group VIIIA elements are called **noble (or rare) gases.**

The general properties of metals and nonmetals are distinct. Physical and chemical properties that distinguish metals from nonmetals are summarized in Tables 4-3 and 4-4. Not all metals and nonmetals possess all these properties, but they share most of them to varying degrees. The physical properties of metals can be explained on the basis of metallic bonding in solids (Section 13-17).

Table 4-2, The Periodic Table, shows how we classify the known elements as *metals* (shown in blue), *nonmetals* (tan), and *metalloids* (green). The elements to the left of those touching the heavy stairstep line are metals (except hydrogen), and those to the right are nonmetals. Such a classification is somewhat arbitrary, and several elements do not fit neatly into either class. Most elements adjacent to the heavy line are often called **metalloids** (or semimetals), because they are metallic (or nonmetallic) only to a limited degree.

Metallic character increases from top to bottom and decreases from left to right with respect to position in the periodic table.

Cesium, atomic number 55, is the most active naturally occurring metal. Francium and radium are radioactive and do not occur in nature in appreciable amounts. Noble gases seldom bond with other elements. They are unreactive, monatomic gases. The most active nonmetal is fluorine, atomic number 9.

Nonmetallic character decreases from top to bottom and increases from left to right in the periodic table.

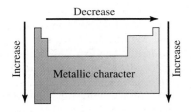

General trends in metallic character of A group elements with position in the periodic table.

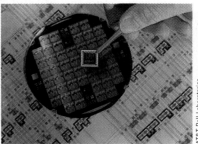

Silicon, a metalloid, is widely used in the manufacture of electronic chips.

CHEMISTRY IN USE

The Development of Science

The Discovery of Phosphorus

Technology and its impact on society have always been intriguing subjects for artists. This was particularly true during the Industrial Revolution, when chemistry was on the verge of transforming itself from alchemical "magic" into a scientific discipline. It is easy to see how the scientist, toiling away in a laboratory full of strange equipment and trying to make sense of the natural world, held a certain heroic appeal to artists.

One of the more romantic accounts of chemical activity during that period is *The Alchymist in Search of the Philosophers' Stone Discovers Phosphorus* (1771) by the English painter Joseph Wright of Derby (1734–1797). In Wright's depiction, a stately, bearded alchemist has just isolated a new element by distillation. As the substance collects in the flask it begins to glow in the dark, illuminating the laboratory with an eerie white light and bringing the imaginary scientist to his knees in wonder. The element phosphorus was in fact named for this property—*phosphorescence*—with both words deriving from the Greek *phosphoros*, or "giving light."

The actual discovery of elemental phosphorus was probably not quite as dramatic as Joseph Wright envisioned. It was first isolated from urine by the German chemist Henning Brand in 1669, by a much more laborious process than the one represented by the tidy distillation apparatus in Wright's painting. The first step of the preparation, as described in a 1726 treatise entitled "Phosphoros Elementalis," in fact involved steeping 50 or 60 pails of urine in tubs for two weeks ". . . till it putrify and breed Worms"—hardly a fitting subject for eighteenth century artwork!

The glowing material was of such novelty that two of Brand's scientific contemporaries offered to find a royal buyer for his process. Expecting a bigger reward at a later date, Brand gave the two the recipe for phosphorus in exchange for some small gifts. However, one man instead claimed the discovery for himself after repeating Brand's work in his own laboratory. Through the other, Brand did receive a contract with the Duke of Hanover for the preparation of phosphorus; however, he was dissatisfied with his pay, and it was only after writing a number of complaint letters (and enlisting his wife to do the same) that he finally received what he felt was fair compensation for his discovery.

A number of other eighteenth-century scientific tableaux were immortalized by Wright. He was particularly fascinated

Derby Museums and Art Gallery

The Alchymist in Search of the Philosophers' Stone Discovers Phosphorus, *by Joseph Wright (1771).*

by light and shadow effects. This, combined with his interest in technological subjects (the town of Derby played an important part in the beginnings of the Industrial Revolution), led him to use other unusual objects, such as glowing iron ingots (*Iron Forge*, 1772) and laboratory candles (*Experiment on a Bird in an Air Pump*, 1786), as focal points in paintings of industrial or scientific scenes.

Lisa S. Boffa
Senior Chemist
Exxon Corporation

Metalloids show some properties that are characteristic of both metals and nonmetals. Many of the metalloids, such as silicon, germanium, and antimony, act as semiconductors, which are important in solid-state electronic circuits. **Semiconductors** are insulators at lower temperatures but become conductors at higher temperatures (Section 13-17). The conductivities of metals, by contrast, decrease with increasing temperature.

Aluminum is the least metallic of the metals and is sometimes classified as a metalloid. It is metallic in appearance and an excellent conductor of electricity.

> Aluminum is the most abundant metal in the earth's crust (7.5% by mass).

In this and later chapters we will study some chemical reactions of elements and their compounds and relate the reactions to the locations of the elements in the periodic table. First, we will describe some important properties of solutions and what they tell us about the nature and behavior of the dissolved substances, the solutes.

4-2 AQUEOUS SOLUTIONS: AN INTRODUCTION

Approximately three fourths of the earth's surface is covered with water. The body fluids of all plants and animals are mainly water. Thus we can see that many important chemical reactions occur in aqueous (water) solutions, or in contact with water. In Chapter 3, we introduced solutions and methods of expressing concentrations of solutions. It is useful to know the kinds of substances that are soluble in water, and the forms in which they exist, before we begin our systematic study of chemical reactions.

> CD-ROM Screen 5.3, Compounds in Aqueous Solution

1 Electrolytes and Extent of Ionization

Solutes that are water-soluble can be classified as either electrolytes or nonelectrolytes. **Electrolytes** are substances whose aqueous solutions conduct electric current. **Strong electrolytes** are substances that conduct electricity well in dilute aqueous solution. **Weak electrolytes** conduct electricity poorly in dilute aqueous solution. Aqueous solutions of **nonelectrolytes** do not conduct electricity. Electric current is carried through aqueous solution by the movement of ions. The strength of an electrolyte depends on the number of ions in solution and also on the charges on these ions (Figure 4-2).

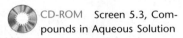

> TOPIC: Electrolytes/ Nonelectrolytes
> GO TO: www.scilinks.org
> *sci*LINKS CODE: WCH0420

Dissociation refers to the process in which a solid *ionic compound*, such as NaCl, separates into its ions in solution:

$$NaCl(s) \xrightarrow{H_2O} Na^+(aq) + Cl^-(aq)$$

> Recall that *ions* are charged particles. The movement of charged particles conducts electricity.

Molecular compounds, for example *pure* HCl, exist as discrete molecules and do not contain ions; however, many such compounds form ions in solution. **Ionization** refers to the process in which a *molecular compound* separates or reacts with water to form ions in solution:

$$HCl(g) \xrightarrow{H_2O} H^+(aq) + Cl^-(aq)$$

> In Chapter 10, we will see that it is appropriate to represent $H^+(aq)$ as H_3O^+ to emphasize its interaction with water.

> Three major classes of solutes are strong electrolytes: (1) strong acids, (2) strong bases, and (3) most soluble salts. *These compounds are completely or nearly completely ionized (or dissociated) in dilute aqueous solutions*, so they are strong electrolytes.

> Acids and bases are further identified in Subsections 2, 3, and 4. They are discussed in more detail in Chapter 10.

An **acid** can be defined as a substance that produces hydrogen ions, H^+, in aqueous solutions. We usually write the formulas of inorganic acids with hydrogen written first. Organic acids can often be recognized by the presence of the COOH group in the formula. A **base** is a substance that produces hydroxide ions, OH^-, in aqueous solutions. A **salt** is a compound that contains a cation other than H^+ and an anion other than hydroxide ion,

> Positively charged ions are called *cations*, and negatively charged ions are called *anions* (Section 2-3). The formula for a salt may include H or OH, but it *must* contain another cation *and* another anion. For example, $NaHSO_4$ and $Al(OH)_2Cl$ are salts.

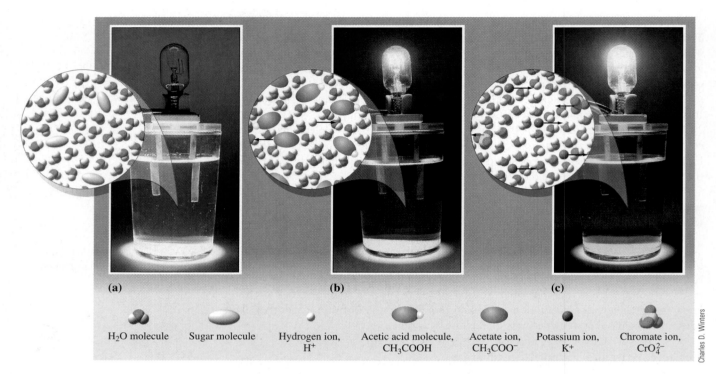

(a)   **(b)**   **(c)**

H₂O molecule   Sugar molecule   Hydrogen ion,   Acetic acid molecule,   Acetate ion,   Potassium ion,   Chromate ion,
                                    H⁺   CH₃COOH   CH₃COO⁻   K⁺   CrO₄²⁻

Figure 4-2 An experiment to demonstrate the presence of ions in solution. Two copper electrodes are dipped into a liquid in a beaker. When the liquid contains significant concentrations of ions, the ions move between the electrodes to complete the circuit (which includes a light bulb). (a) Pure water and sugar are nonelectrolytes. (b) A solution of a weak electrolyte, acetic acid (CH₃COOH); it contains low concentrations of ions, and so the bulb glows dimly. (c) A solution of a strong electrolyte, potassium chromate (K₂CrO₄); it contains a high concentration of ions, and so the bulb glows brightly.

OH^-, or oxide ion, O^{2-} (see Table 2-3). As we will see later in this chapter, salts are formed when acids react with bases.

2 Strong and Weak Acids

Many properties of aqueous solutions of acids are due to H⁺(aq) ions. These are described in Section 10-4.

To give a more complete description of reactions, we indicate the physical states of reactants and products: (g) for gases, (ℓ) for liquids, and (s) for solids. The notation (aq) following ions indicates that they are hydrated in aqueous solution; that is, they interact with water molecules in solution. The complete ionization of a strong electrolyte is indicated by a single arrow (→).

As a matter of convenience we place acids into two classes: strong acids and weak acids. **Strong acids** ionize (separate into hydrogen ions and stable anions) completely, or very nearly completely, in dilute aqueous solution. Seven strong acids and their anions are listed in Table 4-5. Please learn this short list; you can then assume that other acids you encounter are weak.

Because strong acids ionize completely or very nearly completely in dilute solutions, their solutions contain predominantly ions rather than acid molecules. Consider the ionization of hydrochloric acid. Pure hydrogen chloride, HCl, is a molecular compound that is a gas at room temperature and atmospheric pressure. When it dissolves in water, it reacts nearly 100% to produce a solution that contains hydrogen ions and chloride ions:

$$HCl(g) \xrightarrow{H_2O} H^+(aq) + Cl^-(aq) \qquad \text{(to completion)}$$

Similar equations can be written for all strong acids.

TABLE 4-5   *Some Strong Acids and Their Anions*

Common Strong Acids		Anions of These Strong Acids	
Formula	*Name*	*Formula*	*Name*
HCl	hydrochloric acid	Cl^-	chloride ion
HBr	hydrobromic acid	Br^-	bromide ion
HI	hydroiodic acid	$I-$	iodide ion
HNO_3	nitric acid	NO_3^-	nitrate ion
$HClO_4$	perchloric acid	ClO_4^-	perchlorate ion
$(HClO_3$	chloric acid$)$	ClO_3^-	chlorate ion
H_2SO_4	sulfuric acid	$\begin{cases} HSO_4^- \\ SO_4^{2-} \end{cases}$	hydrogen sulfate ion sulfate ion

Chloric acid is sometimes not listed with the common strong acids since it is not commonly encountered. However, its anion is much more common. So we have included both the acid and its anion in this table.

Weak acids ionize only slightly (usually less than 5%) in dilute aqueous solution. Some common weak acids are listed in Appendix F. Several of them and their anions are given in Table 4-6.

The equation for the ionization of acetic acid, CH_3COOH, in water is typical of weak acids:

$$CH_3COOH(aq) \rightleftharpoons H^+(aq) + CH_3COO^-(aq) \quad \text{(reversible)}$$

The double arrow $\rightleftharpoons$ generally signifies that the reaction occurs in *both* directions and that the forward reaction does not go to completion. All of us are familiar with solutions of acetic acid. Vinegar is 5% acetic acid by mass. Our use of oil and vinegar as a salad dressing suggests that acetic acid is a weak acid; we could not safely drink a 5% solution of any strong acid. To be specific, acetic acid is 0.5% ionized (and 99.5% nonionized) in 5% solution.

 CD-ROM   Screen 5.8, Acids

Acetic acid is the most familiar organic acid.

Our stomachs have linings that are much more resistant to attack by acids than are our other tissues.

TABLE 4-6   *Some Common Weak Acids and Their Anions*

Common Weak Acids		Anions of These Weak Acids	
Formula	*Name*	*Formula*	*Name*
HF^*	hydrofluoric acid	F^-	fluoride ion
CH_3COOH	acetic acid	CH_3COO^-	acetate ion
HCN	hydrocyanic acid	CN^-	cyanide ion
$HNO_2{}^\dagger$	nitrous acid	NO_2^-	nitrite ion
$H_2CO_3{}^\dagger$	carbonic acid	$\begin{cases} HCO_3^- \\ CO_3^{2-} \end{cases}$	hydrogen carbonate ion carbonate ion
$H_2SO_3{}^\dagger$	sulfurous acid	$\begin{cases} HSO_3^- \\ SO_3^{2-} \end{cases}$	hydrogen sulfite ion sulfite ion
H_3PO_4	phosphoric acid	$\begin{cases} H_2PO_4^- \\ HPO_4^{2-} \\ PO_4^{3-} \end{cases}$	dihydrogen phosphate ion hydrogen phosphate ion phosphate ion
$(COOH)_2$	oxalic acid	$\begin{cases} H(COO)_2^- \\ (COO)_2^{2-} \end{cases}$	hydrogen oxalate ion oxalate ion

The names given here correspond to the aqueous solutions.

*HF is a weak acid, whereas HCl, HBr, and HI are strong acids.

†Free acid molecules exist only in dilute aqueous solution or not at all. Many salts of these acids are common, stable compounds, however.

Charles Steele

Citrus fruits contain citric acid, so their juices are acidic. This is shown here by the red colors on the indicator paper. Acids taste sour.

The carboxylate group —COOH is

$$-\overset{\displaystyle O}{\underset{\displaystyle O-H}{C}}$$

Other organic acids have other groups in the position of the H_3C— group in acetic acid.

Inorganic acids may be strong or weak.

A multitude of organic acids occur in living systems. Organic acids contain the carboxylate grouping of atoms, —COOH. Most common organic acids are weak. They can ionize slightly by breaking the O—H bond, as shown on the following page for acetic acid:

$$H_3C-\overset{\displaystyle O}{\underset{\displaystyle O-H}{C}}\ (aq) \rightleftharpoons H_3C-\overset{\displaystyle O}{\underset{\displaystyle O^-}{C}}\ (aq) + H^+(aq)$$

Organic acids are discussed in Chapter 27. Some naturally occurring organic weak acids are tartaric acid (grapes), lactic acid (sour milk), and formic acid (ants). Carbonic acid, H_2CO_3, and hydrocyanic acid, HCN(aq), are two common acids that contain carbon but that are considered to be *inorganic* acids. Inorganic acids are often called **mineral acids** because they are obtained primarily from nonliving sources.

Many common food and household products are acidic (orange juice, vinegar, soft drink, citrus fruits) or basic (cleaning preparations, baking soda).

EXAMPLE 4-1   *Strong and Weak Acids*

In the following lists of common acids, which are strong and which are weak? (a) H_3PO_4, HCl, H_2CO_3, HNO_3; (b) $HClO_4$, H_2SO_4, HClO, HF.

Plan

We recall that Table 4-5 lists some common strong acids. Other *common* acids are assumed to be weak.

Solution

(a) HCl and HNO_3 are strong acids; H_3PO_4 and H_2CO_3 are weak acids.
(b) $HClO_4$ and H_2SO_4 are strong acids; HClO and HF are weak acids.

You should now work Exercises 19 and 21.

3 Reversible Reactions

Reactions that can occur in both directions are **reversible reactions.** We use a double arrow ($\rightleftharpoons$) to indicate that a reaction is *reversible*. What is the fundamental difference between reactions that go to completion and those that are reversible? We have seen that the ionization of HCl in water is nearly complete. Suppose we dissolve some table salt, NaCl, in water and then add some dilute nitric acid to it. The resulting solution contains Na^+ and Cl^- ions (from the dissociation of NaCl) as well as H^+ and NO_3^- (from the ionization of HNO_3). The H^+ and Cl^- ions do *not* react significantly to form nonionized HCl molecules; this would be the reverse of the ionization of HCl.

$$H^+(aq) + Cl^-(aq) \longrightarrow \text{no reaction}$$

Na^+ and NO_3^- ions do not combine because $NaNO_3$ is a soluble ionic compound.

In contrast, when a sample of sodium acetate, $NaCH_3COO$, is dissolved in H_2O and mixed with nitric acid, the resulting solution initially contains Na^+, CH_3COO^-, H^+, and NO_3^- ions. But most of the H^+ and CH_3COO^- ions combine to produce nonionized molecules of acetic acid, the reverse of the ionization of the acid. Thus, the ionization of acetic acid, like that of any other weak electrolyte, is reversible.

$$H^+(aq) + CH_3COO^-(aq) \rightleftharpoons CH_3COOH(aq) \qquad \text{(reversible)}$$

4 Strong Bases, Insoluble Bases, and Weak Bases

Solutions of bases have a set of common properties due to the OH^- ion. These are described in Section 10-4.

Most common bases are *ionic* metal hydroxides. **Strong bases** are soluble in water and are dissociated completely in dilute aqueous solution. The common strong bases are listed in

TABLE 4-7	*Common Strong Bases*	
Group IA		**Group IIA**
LiOH	lithium hydroxide	
NaOH	sodium hydroxide	
KOH	potassium hydroxide	$Ca(OH)_2$ calcium hydroxide
RbOH	rubidium hydroxide	$Sr(OH)_2$ strontium hydroxide
CsOH	cesium hydroxide	$Ba(OH)_2$ barium hydroxide

CD-ROM Screen 5.9, Bases

Table 4-7. They are the hydroxides of the Group IA metals and the heavier members of Group IIA. The equation for the dissociation of sodium hydroxide in water is typical. Similar equations can be written for other strong bases.

$$NaOH(s) \xrightarrow{H_2O} Na^+(aq) + OH^-(aq) \quad \text{(to completion)}$$

Strong bases are ionic compounds in the solid state.

Other metals form ionic hydroxides, but these are so sparingly soluble in water that they cannot produce strongly basic solutions. They are called **insoluble bases** or sometimes sparingly soluble bases. Typical examples include $Cu(OH)_2$, $Zn(OH)_2$, $Fe(OH)_2$, and $Fe(OH)_3$.

Common **weak bases** are molecular substances that are soluble in water but form only low concentrations of ions in solution. The most common weak base is ammonia, NH_3.

$$NH_3(aq) + H_2O(\ell) \rightleftharpoons NH_4^+(aq) + OH^-(aq) \quad \text{(reversible)}$$

The weak bases are *molecular* substances that dissolve in water to give slightly basic solutions; they are sometimes called molecular bases.

Closely related N-containing compounds, the *amines*, such as methylamine, CH_3NH_2, and aniline, $C_6H_5NH_2$, are also weak bases. Nicotine (found in tobacco) and caffeine (found in coffee, tea, and cola drinks) are naturally occurring amines.

EXAMPLE 4-2 *Classifying Bases*

From the following lists, choose (i) the strong bases, (ii) the insoluble bases, and (iii) the weak bases. (a) NaOH, $Cu(OH)_2$, $Pb(OH)_2$, $Ba(OH)_2$; (b) $Fe(OH)_3$, KOH, $Mg(OH)_2$, $Sr(OH)_2$, NH_3.

Plan

(i) We recall that Table 4-7 lists the *common strong bases*. (ii) Other common metal hydroxides are assumed to be *insoluble bases*. (iii) Ammonia and closely related nitrogen-containing compounds, the amines, are the common *weak bases*.

Solution

(a) (i) The strong bases are NaOH and $Ba(OH)_2$, so
    (ii) the insoluble bases are $Cu(OH)_2$ and $Pb(OH)_2$.
(b) (i) The strong bases are KOH and $Sr(OH)_2$, so
    (ii) the insoluble bases are $Fe(OH)_3$ and $Mg(OH)_2$, and
    (iii) the weak base is NH_3.

You should now work Exercises 22 and 24.

SCi
LINKS.
TOPIC: Acids/Bases
GO TO: www.scilinks.org
*sci*LINKS **CODE:** WCH0430

5 Solubility Guidelines for Compounds in Aqueous Solution

Solubility is a complex phenomenon, and it is not possible to give a complete summary of all our observations. The following brief summary for solutes in aqueous solutions will be very useful. These generalizations are often called the *solubility guidelines*. Compounds whose solubility in water is less than about 0.02 mol/L are usually classified as insoluble

There is no sharp dividing line between "soluble" and "insoluble"
compounds. Compounds whose
solubilities fall near the arbitrary
division are called "moderately soluble" compounds.

CD-ROM   Screens 3.10,
and 5.4, Solubility of Ionic
Compounds

*sci*LINKS.
TOPIC: Solubility
GO TO: www.scilinks.org
*sci*LINKS CODE: WCH0440

compounds, whereas those that are more soluble are classified as soluble compounds.
No gaseous or solid substances are infinitely soluble in water. You may wish to review
Tables 2-3, 4-5, and 4-6. They list some common ions. Table 4-11 contains a more compre-
hensive list.

1. The common inorganic acids are soluble in water. Low-molecular-weight organic
   acids are also soluble.
2. All common compounds of the Group IA metal ions (Li^+, Na^+, K^+, Rb^+, Cs^+) and
   the ammonium ion, NH_4^+, are soluble in water.
3. The common nitrates, NO_3^-; acetates, CH_3COO^-; chlorates, ClO_3^-; and perchlo-
   rates, ClO_4^-, are soluble in water.
4. (a) The common chlorides, Cl^-, are soluble in water except $AgCl$, Hg_2Cl_2, and
       $PbCl_2$.
   (b) The common bromides, Br^-, and iodides, I^-, show approximately the same sol-
       ubility behavior as chlorides, but there are some exceptions. As these halide ions
       (Cl^-, Br^-, I^-) increase in size, the solubilities of their slightly soluble com-
       pounds decrease.
   (c) The common fluorides, F^-, are soluble in water except MgF_2, CaF_2, SrF_2, BaF_2,
       and PbF_2.
5. The common sulfates, SO_4^{2-}, are soluble in water except $PbSO_4$, $BaSO_4$, and $HgSO_4$;
   $CaSO_4$, $SrSO_4$, and Ag_2SO_4 are moderately soluble.
6. The common metal hydroxides, OH^-, are *insoluble* in water except those of the
   Group IA metals and the heavier members of the Group IIA metals, beginning with
   $Ca(OH)_2$.
7. The common carbonates, CO_3^{2-}, phosphates, PO_4^{3-}, and arsenates, AsO_4^{3-}, are *insoluble* in water except those of the Group IA metals and NH_4^+. $MgCO_3$ is moderately soluble.
8. The common sulfides, S^{2-}, are *insoluble* in water except those of the Group IA and
   Group IIA metals and the ammonium ion.

Table 4-8 summarizes much of the information about the solubility guidelines.
   We have distinguished between strong and weak electrolytes and between soluble and
insoluble compounds. Let us now see how we can describe chemical reactions in aqueous
solutions.

EXAMPLE 4-3   *Solubility of Some Common Ionic Salts*

From the following compounds, choose (a) those that are likely to be soluble in water and
(b) those that are likely to be insoluble: $NaBr$, $Cu(OH)_2$, $PbCl_2$, AgI, Fe_2O_3, $Mg(NO_3)_2$,
$(NH_4)_2SO_4$.

Plan

We recall from the guidelines and Table 4-8 that all Na^+, K^+, and NH_4^+ salts are soluble.
Therefore, we predict that $NaBr$ and $(NH_4)_2SO_4$ will be soluble. Similarly, NO_3^- salts are
soluble, so $Mg(NO_3)_2$ should also be soluble. All other compounds listed in this example
should be insoluble.

Solution

(a) The soluble compounds are $NaBr$, $(NH_4)_2SO_4$, and $Mg(NO_3)_2$.
(b) The insoluble compounds are $Cu(OH)_2$, $PbCl_2$, AgI, and Fe_2O_3.

You should now work Exercises 27 and 30.

TABLE 4-8	*Solubility Guidelines for Common Ionic Compounds in Water*

Generally Soluble	**Exceptions**
Na^+, K^+, NH_4^+ compounds	No common exceptions
fluorides (F^-)	Insoluble: MgF_2, CaF_2, SrF_2, BaF_2, PbF_2
chlorides (Cl^-)	Insoluble: $AgCl$, Hg_2Cl_2
	Soluble in hot water: $PbCl_2$
bromides (Br^-)	Insoluble: $AgBr$, Hg_2Br_2, $PbBr_2$
	Moderately soluble: $HgBr_2$
iodides (I^-)	Insoluble: many heavy-metal iodides
sulfates (SO_4^{2-})	Insoluble: $BaSO_4$, $PbSO_4$, $HgSO_4$
	Moderately soluble: $CaSO_4$, $SrSO_4$, Ag_2SO_4
nitrates (NO_3^-), nitrites (NO_2^-)	Moderately soluble: $AgNO_2$
chlorates (ClO_3^-), perchlorates (ClO_4^-)	No common exceptions
acetates (CH_3COO^-)	Moderately soluble: $AgCH_3COO$

Generally Insoluble	**Exceptions**
sulfides (S^{2-})†	Soluble: those of NH_4^+, Na^+, K^+, Mg^{2+}, Ca^{2+}
oxides (O^{2-}), hydroxides (OH^-)	Soluble: Li_2O^*, $LiOH$, Na_2O^*, $NaOH$, K_2O^*, KOH, BaO^*, $Ba(OH)_2$
	Moderately soluble: CaO^*, $Ca(OH)_2$, SrO^*, $Sr(OH)_2$
carbonates (CO_3^{2-}), phosphates (PO_4^{3-}), arsenates (AsO_4^{3-})	Soluble: those of NH_4^+, Na^+, K^+

*Dissolves with evolution of heat and formation of hydroxides.
†Dissolves with formation of HS^- and H_2S.

4-3 REACTIONS IN AQUEOUS SOLUTIONS

Many important chemical reactions occur in aqueous solutions. In this chapter you should learn to describe such aqueous reactions and to predict the products of many reactions.

First, let's look at how we write chemical equations that describe reactions in aqueous solutions. We use three kinds of chemical equations. Table 4-9 shows the kinds of information about each substance that we use in writing equations for reactions in aqueous solutions. Some typical examples are included. Refer to Table 4-9 often as you study the following sections.

1. In **formula unit equations,** we show complete formulas for all compounds. When metallic copper is added to a solution of (colorless) silver nitrate, the more active metal —copper—displaces silver ions from the solution. The resulting solution contains blue copper(II) nitrate, and metallic silver forms as a finely divided solid (Figure 4-3):

$$2AgNO_3(aq) + Cu(s) \longrightarrow 2Ag(s) + Cu(NO_3)_2(aq)$$

Both silver nitrate and copper(II) nitrate are soluble ionic compounds (for solubility guidelines see Table 4-8 and Section 4-2 Part 5).

2. In **total ionic equations,** formulas are written to show the (predominant) form in which each substance exists when it is in contact with aqueous solution. We often use brackets in total ionic equations to show ions that have a common source or that remain in solution after the reaction is complete. The total ionic equation for this reaction is

$$2[Ag^+(aq) + NO_3^-(aq)] + Cu(s) \longrightarrow 2Ag(s) + [Cu^{2+}(aq) + 2NO_3^-(aq)]$$

The White Cliffs of Dover, England, are composed mainly of calcium carbonate ($CaCO_3$).

Because we have not studied periodic trends in properties of transition metals, it would be difficult for you to predict that·Cu is more active than Ag. The fact that this reaction occurs (see Figure 4-3) shows that it is.

TABLE 4-9 *Bonding, Solubility, Electrolyte Characteristics, and Predominant Forms of Solutes in Contact with Water*

	Acids		Bases			Salts	
	Strong acids	**Weak acids**	**Strong bases**	**Insoluble bases**	**Weak bases**	**Soluble salts**	**Insoluble salts**
Examples	HCl HNO$_3$	CH$_3$COOH HF	NaOH Ca(OH)$_2$	Mg(OH)$_2$ Al(OH)$_3$	NH$_3$ CH$_3$NH$_2$	KCl, NaNO$_3$, NH$_4$Br	BaSO$_4$, AgCl, Ca$_3$(PO$_4$)$_2$
Pure compound ionic or molecular?	Molecular	Molecular	Ionic	Ionic	Molecular	Ionic	Ionic
Water-soluble or insoluble?	Soluble*	Soluble*	Soluble	Insoluble	Soluble†	Soluble	Insoluble
≈ 100% ionized or dissociated in dilute aqueous solution?	Yes	No	Yes	(footnote‡)	No	Yes§	(footnote‡)
Written in ionic equations as	Separate ions	Molecules	Separate ions	Complete formulas	Molecules	Separate ions	Complete formulas

*Most common inorganic acids and the low-molecular-weight organic acids (—COOH) are water-soluble.

†The low-molecular-weight amines are water-soluble.

‡The very small concentrations of "insoluble" metal hydroxides and insoluble salts in saturated aqueous solutions are nearly completely dissociated.

§There are a few exceptions. A few soluble salts are molecular (and not ionic) compounds.

Ions that appear in solution on both sides of the total ionic equation are called **spectator ions;** they undergo no change in the chemical reaction.

Brackets are not used in net ionic equations.

This is why it is important to know how and when to construct net ionic equations from formula unit equations.

Examination of the total ionic equation shows that NO_3^- ions do not participate in the reaction. Because they do not change, they are often called "spectator" ions.

3. In **net ionic equations,** we show only the species that react. The net ionic equation is obtained by eliminating the spectator ions and the brackets from the total ionic equation.

$$2Ag^+(aq) + Cu(s) \longrightarrow 2Ag(s) + Cu^{2+}(aq)$$

Net ionic equations allow us to focus on the *essence* of a chemical reaction in aqueous solutions. On the other hand, if we are dealing with stoichiometric calculations, we frequently must deal with formula weights and therefore with the *complete* formulas of all species. In

Figure 4-3 (a) Copper wire and a silver nitrate solution. (b) The copper wire has been placed in the solution and some finely divided silver has deposited on the wire. The solution is blue because it contains copper(II) nitrate.

(a)   (b)

Charles Steele

✓ Problem-Solving Tip: *Writing Ionic Equations*

The following chart will help in deciding which formula units are to be written as separate ions in the total ionic equation and which ones are to be written as unchanged formula units. You must answer two questions about a substance to determine whether it should be written in ionic form or as a formula unit in the total and net ionic equations.

1. Does it dissolve in water? If not, write the full formula.
2. (a) If it dissolves, does it ionize (a strong acid)?
   (b) If it dissolves, does it dissociate (a strong base or a soluble salt)?

If the answer to *either part* of the second question is yes, the substance is a soluble strong electrolyte, and its formula is written in ionic form.

 The only common substances that should be written as ions in ionic equations are (1) strong acids, (2) strong bases, and (3) soluble ionic salts.

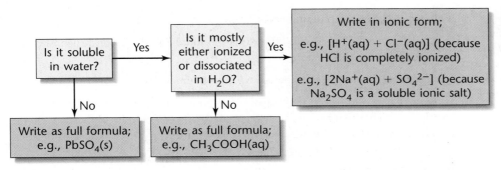

CD-ROM Screen 5.5, Equations of Reaction in Aqueous Solution: Net Ionic Equations

such cases, formula unit equations are more useful. Total ionic equations provide the bridge between the two.

Recall the lists of strong acids (Table 4-5) and strong bases (Table 4-7). These acids and bases are completely or almost completely ionized or dissociated in dilute aqueous solutions. Other common acids and bases are either insoluble or only slightly ionized or dissociated. In addition, the solubility guidelines (Table 4-8 and Section 4-2 Part 5) allow you to determine which salts are soluble in water. Most salts that are soluble in water are also strong electrolytes. Exceptions such as lead acetate, $Pb(CH_3COO)_2$, which is soluble but does not ionize appreciably, will be noted as they are encountered.

4-4 OXIDATION NUMBERS

The many reactions that involve the transfer of electrons from one species to another are called **oxidation–reduction reactions,** or simply, **redox reactions.** We use oxidation numbers to keep track of electron transfers. The systematic naming of compounds (Sections 4-5 and 4-6) also makes use of oxidation numbers.

The **oxidation number,** or **oxidation state,** of an element in a simple *binary* ionic compound is the number of electrons gained or lost by an atom of that element when it forms the compound. In the case of a single-atom ion, it corresponds to the actual charge on the ion. In molecular compounds, oxidation numbers do not have the same significance they

Oxidation–reduction and displacement reactions are discussed in Sections 4-7 and 4-10.

Binary means two. Binary compounds contain two elements.

have in binary ionic compounds. Oxidation numbers, however, are very useful aids in writing formulas and in balancing equations. In molecular species, the oxidation numbers are assigned according to an arbitrary set of rules. The element farther to the right and higher up in the periodic table is assigned a negative oxidation number, and the element farther to the left and lower down in the periodic table is assigned a positive oxidation number.

Some rules for assigning oxidation numbers follow. These rules are not comprehensive, but they cover most cases. In applying these rules, keep in mind two important points. First, oxidation numbers are always assigned on a per atom basis; second, treat the rules in order of *decreasing* importance—the first rule that applies takes precedence over any subsequent rules that seem to apply.

Polyatomic elements have two or more atoms per molecule.

CD-ROM   Screen 5.13, Oxidation Numbers

1. The oxidation number of the atoms in any free, uncombined element is zero. This includes polyatomic elements such as H_2, O_2, O_3, P_4, and S_8.

2. The oxidation number of an element in a simple (monatomic) ion is equal to the charge on the ion.

3. The *sum* of the oxidation numbers of all atoms in a compound is zero.

4. In a polyatomic ion, the *sum* of the oxidation numbers of the constituent atoms is equal to the charge on the ion.

5. Fluorine has an oxidation number of -1 in its compounds.

6. Hydrogen has an oxidation number of $+1$ in compounds unless it is combined with metals, in which case it has an oxidation number of -1. Examples of these exceptions are NaH and CaH_2.

7. Oxygen usually has an oxidation number of -2 in its compounds. There are some exceptions:

   a. Oxygen has an oxidation number of -1 in hydrogen peroxide, H_2O_2, and in peroxides, which contain the O_2^{2-} ion; examples are CaO_2 and Na_2O_2.

   b. Oxygen has an oxidation number of $-\frac{1}{2}$ in superoxides, which contain the O_2^- ion; examples are KO_2 and RbO_2.

   c. When combined with fluorine in OF_2, oxygen has an oxidation number of $+2$.

8. The position of the element in the periodic table helps to assign its oxidation number:

   a. Group IA elements have oxidation numbers of $+1$ in all of their compounds.

   b. Group IIA elements have oxidation numbers of $+2$ in all of their compounds.

   c. Group IIIA elements have oxidation numbers of $+3$ in all of their compounds, with a few rare exceptions.

   d. Group VA elements have oxidation numbers of -3 in *binary* compounds with metals, with H, or with NH_4^+. Exceptions are compounds with a Group VA element combined with an element to its right in the periodic table; in this case, their oxidation numbers can be found by using rules 3 and 4.

   e. Group VIA elements below oxygen have oxidation numbers of -2 in *binary* compounds with metals, with H, or with NH_4^+. When these elements are combined with oxygen or with a lighter halogen, their oxidation numbers can be found by using rules 3 and 4.

TABLE 4-10 *Common Oxidation Numbers (States) for Group A Elements in Compounds and Ions*

Element(s)	Common Ox. Nos.	Examples	Other Ox. Nos.
H	+1	H_2O, CH_4, NH_4Cl	−1 in metal hydrides, e.g., NaH, CaH_2
Group IA	+1	KCl, NaH, $RbNO_3$, K_2SO_4	None
Group IIA	+2	$CaCl_2$, MgH_2, $Ba(NO_3)_2$, $SrSO_4$	None
Group IIIA	+3	$AlCl_3$, BF_3, $Al(NO_3)_3$, GaI_3	None in common compounds
Group IVA	+2 +4	CO, PbO, $SnCl_2$, $Pb(NO_3)_2$ CCl_4, SiO_2, SiO_3^{2-}, $SnCl_4$	Many others are also seen for C and Si
Group VA	−3 in binary compounds with metals −3 in NH_4^+, binary compounds with H	Mg_3N_2, Na_3P, Cs_3As NH_3, PH_3, AsH_3, NH_4^+	+3, e.g., NO_2^-, PCl_3 +5, e.g., NO_3^-, PO_4^{3-}, AsF_5, P_4O_{10}
O	−2	H_2O, P_4O_{10}, Fe_2O_3, CaO, ClO_3^-	+2 in OF_2 −1 in peroxides, e.g., H_2O_2, Na_2O_2 $-\frac{1}{2}$ in superoxides, e.g., KO_2, RbO_2
Group VIA (other than O)	−2 in binary compounds with metals and H −2 in binary compounds with NH_4^+	H_2S, CaS, Fe_2S_3, Na_2Se $(NH_4)_2S$, $(NH_4)_2Se$	+4 with O and the lighter halogens, e.g., SO_2, SeO_2, Na_2SO_3, SO_3^{2-}, SF_4 +6 with O and the lighter halogens, e.g., SO_3, TeO_3, H_2SO_4, SO_4^{2-}, SF_6
Group VIIA	−1 in binary compounds with metals and H −1 in binary compounds with NH_4^+	MgF_2, KI, $ZnCl_2$, $FeBr_3$ NH_4Cl, NH_4Br	Cl, Br, or I with O or with a lighter halogen +1, e.g., BrF, ClO^-, BrO^- +3, e.g., ICl_3, ClO_2^-, BrO_2^- +5, e.g., BrF_5, ClO_3^-, BrO_3^- +7, e.g., IF_7, ClO_4^-, BrO_4^-

f. Group VIIA elements have oxidation numbers of −1 in *binary* compounds with metals, with H, with NH_4^+, or with a heavier halogen. When these elements except fluorine (i.e., Cl, Br, I) are combined with oxygen or with a lighter halogen, their oxidation numbers can be found by using rules 3 and 4.

Table 4-10 summarizes rules 5 through 8, with many examples.

EXAMPLE 4-4 *Oxidation Numbers*

Determine the oxidation numbers of nitrogen in the following species: (a) N_2O_4, (b) NH_3, (c) HNO_3, (d) NO_3^-, (e) N_2.

Plan

We first assign oxidation numbers to elements that exhibit a single common oxidation number (see Table 4-10). We recall that oxidation numbers are represented *per atom*, that the sum of the oxidation numbers in a compound is zero, and that the sum of the oxidation numbers in an ion equals the charge on the ion.

Aqueous solutions of some compounds that contain chromium. *Left to right:* chromium(II) chloride, $CrCl_2$, is blue; chromium(III) chloride, $CrCl_3$, is green; potassium chromate, K_2CrO_4, is yellowish; potassium dichromate, $K_2Cr_2O_7$, is orangish.

By convention, *oxidation numbers* are represented as $+n$ and $-n$, but ionic charges are represented as $n+$ and $n-$. We shall circle oxidation numbers associated with formulas and show them in red. Both oxidation numbers and ionic charges can be combined algebraically.

Usually the element with the positive oxidation number is written first. For historic reasons, however, in compounds containing nitrogen and hydrogen, such as NH_3, and many compounds containing carbon and hydrogen, such as CH_4, hydrogen is written last, although it has a positive oxidation number.

Solution

(a) The oxidation number of O is -2. The sum of the oxidation numbers for all atoms in a compound must be zero:

ox. no./atom: $\overset{x}{N_2}\overset{-2}{O_4}$

total ox. no.: $2x + 4(-2) = 0$ or $x = \boxed{+4}$

(b) The oxidation number of H is $+1$:

ox. no./atom: $\overset{x}{N}\overset{+1}{H_3}$

total ox. no.: $x + 3(1) = 0$ or $x = \boxed{-3}$

(c) The oxidation number of H is $+1$ and the oxidation number of O is -2.

ox. no./atom: $\overset{+1}{H}\overset{x}{N}\overset{-2}{O_3}$

total ox. no.: $1 + x + 3(-2) = 0$ or $x = \boxed{+5}$

(d) The sum of the oxidation numbers for all atoms in an ion equals the charge on the ion:

ox. no./atom: $\overset{x}{N}\overset{-2}{O_3}{}^-$

total ox. no.: $x + 3(-2) = -1$ or $x = \boxed{+5}$

(e) The oxidation number of any free element is $\boxed{\text{zero.}}$

You should now work Exercise 42.

NAMING SOME INORGANIC COMPOUNDS

CD-ROM Screen 3.11, Naming Ionic Compounds

The rules for naming inorganic compounds are set down by the Committee on Inorganic Nomenclature of the International Union of Pure and Applied Chemistry (IUPAC). The names and formulas of a few organic compounds were given in Table 2-2, and more systematic rules for naming them will appear in Chapter 27.

4-5 NAMING BINARY COMPOUNDS

Millions of compounds are known, so it is important to be able to associate names and formulas in a systematic way.

Binary compounds consist of two elements; they may be either ionic or molecular. The rule is to name the more metallic element first and the less metallic element second. The less metallic element is named by adding an "-ide" suffix to the element's *unambiguous* stem. Stems for the nonmetals follow.

The stem for each element is derived from the name of the element.

IIIA		IVA		VA		VIA		VIIA	
B	bor	C	carb	N	nitr	O	ox	H	hydr
		Si	silic	P	phosph	S	sulf	F	fluor
				As	arsen	Se	selen	Cl	chlor
				Sb	antimon	Te	tellur	Br	brom
								I	iod

Binary ionic compounds contain metal cations and nonmetal anions. The cation is named first and the anion second.

Formula	Name	Formula	Name
KBr	potassium bromide	Rb_2S	rubidium sulfide
$CaCl_2$	calcium chloride	Ba_3N_2	barium nitride
NaH	sodium hydride	SrO	strontium oxide

Notice that there is a space between the name of the cation and the name of the anion.

The preceding method is sufficient for naming binary ionic compounds containing metals that exhibit *only one oxidation number* other than zero (Section 4-4). Most transition metals and the metals of Groups IIIA (except Al), IVA, and VA exhibit more than one oxidation number. These metals may form two or more binary compounds with the same nonmetal. To distinguish among all the possibilities, the oxidation number of the metal is indicated by a Roman numeral in parentheses following its name. This method can be applied to any binary compound of a metal and a nonmetal.

Roman numerals are *not* necessary for metals that commonly exhibit only one oxidation number in their compounds.

Formula	Ox. No. of Metal	Name	Formula	Ox. No. of Metal	Name
Cu_2O	+1	copper(I) oxide	$SnCl_2$	+2	tin(II) chloride
CuF_2	+2	copper(II) fluoride	$SnCl_4$	+4	tin(IV) chloride
FeS	+2	iron(II) sulfide	PbO	+2	lead(II) oxide
Fe_2O_3	+3	iron(III) oxide	PbO_2	+4	lead(IV) oxide

Notice that prefixes are not used to indicate the number of ions in the formula. For example, the correct name is tin(IV) chloride not tetratin monochloride.

The advantage of the IUPAC system is that if you know the formula, you can write the exact and unambiguous name; if you are given the name, you can write the formula at once. An older method, still in use but not recommended by the IUPAC, uses "-ous" and "-ic" suffixes to indicate lower and higher oxidation numbers, respectively. This system can distinguish between only two different oxidation numbers for a metal. It is therefore not as useful as the Roman numeral system.

Familiarity with the older system is still necessary. It is still widely used in many scientific, engineering, and medical fields.

Formula	Ox. No. of Metal	Name	Formula	Ox. No. of Metal	Name
CuCl	+1	cuprous chloride	SnF_2	+2	stannous fluoride
$CuCl_2$	+2	cupric chloride	SnF_4	+4	stannic fluoride
FeO	+2	ferrous oxide	Hg_2Cl_2	+1	mercurous chloride
$FeBr_3$	+3	ferric bromide	$HgCl_2$	+2	mercuric chloride

Some compounds contain polyatomic ions that behave much like monatomic anions. Compounds that contain these ions are called **pseudobinary ionic compounds.** The prefix "pseudo-" means "false"; these compounds are named as though they were binary compounds. The common examples of such polyatomic anions are the hydroxide ion, OH^-, and the cyanide ion, CN^-. The ammonium ion, NH_4^+, is the common cation that behaves like a simple metal cation.

Formula	Name	Formula	Name
NH_4I	ammonium iodide	NH_4CN	ammonium cyanide
$Ca(CN)_2$	calcium cyanide	$Cu(OH)_2$	copper(II) hydroxide or cupric hydroxide
NaOH	sodium hydroxide	$Fe(OH)_3$	iron(III) hydroxide or ferric hydroxide

A list of common cations and anions appears in Table 4-11. It will enable you to name many of the ionic compounds you encounter.

Nearly all **binary molecular compounds** involve two *nonmetals* bonded together.

TABLE 4-11 *Formulas, Ionic Charges, and Names for Some Common Ions*

Common Cations			Common Anions		
Formula	Charge	Name	Formula	Charge	Name
Li^+	1+	lithium ion	F^-	1−	fluoride ion
Na^+	1+	sodium ion	Cl^-	1−	chloride ion
K^+	1+	potassium ion	Br^-	1−	bromide ion
NH_4^+	1+	ammonium ion	I^-	1−	iodide ion
Ag^+	1+	silver ion	OH^-	1−	hydroxide ion
			CN^-	1−	cyanide ion
Mg^{2+}	2+	magnesium ion	ClO^-	1−	hypochlorite ion
Ca^{2+}	2+	calcium ion	ClO_2^-	1−	chlorite ion
Ba^{2+}	2+	barium ion	ClO_3^-	1−	chlorate ion
Cd^{2+}	2+	cadmium ion	ClO_4^-	1−	perchlorate ion
Zn^{2+}	2+	zinc ion	CH_3COO^-	1−	acetate ion
Cu^{2+}	2+	copper(II) ion or cupric ion	MnO_4^-	1−	permanganate ion
Hg_2^{2+}	2+	mercury(I) ion or mercurous ion	NO_2^-	1−	nitrite ion
Hg^{2+}	2+	mercury(II) ion or mercuric ion	NO_3^-	1−	nitrate ion
Mn^{2+}	2+	manganese(II) ion or manganous ion	SCN^-	1−	thiocyanate ion
Co^{2+}	2+	cobalt(II) ion or cobaltous ion			
Ni^{2+}	2+	nickel(II) ion or nickelous ion	O^{2-}	2−	oxide ion
Pb^{2+}	2+	lead(II) ion or plumbous ion	S^{2-}	2−	sulfide ion
Sn^{2+}	2+	tin(II) ion or stannous ion	HSO_3^-	1−	hydrogen sulfite ion or bisulfite ion
Fe^{2+}	2+	iron(II) ion or ferrous ion	SO_3^{2-}	2−	sulfite ion
			HSO_4^-	1−	hydrogen sulfate ion or bisulfate ion
Fe^{3+}	3+	iron(III) ion or ferric ion	SO_4^{2-}	2−	sulfate ion
Al^{3+}	3+	aluminum ion	HCO_3^-	1−	hydrogen carbonate ion or bicarbonate ion
Cr^{3+}	3+	chromium(III) ion or chromic ion	CO_3^{2-}	2−	carbonate ion
			CrO_4^{2-}	2−	chromate ion
			$Cr_2O_7^{2-}$	2−	dichromate ion
			PO_4^{3-}	3−	phosphate ion
			AsO_4^{3-}	3−	arsenate ion

If you don't already know them, you should learn these common prefixes.

Number	Prefix
2	di
3	tri
4	tetra
5	penta
6	hexa
7	hepta
8	octa
9	nona
10	deca

Although many nonmetals can exhibit different oxidation numbers, their oxidation numbers are *not* properly indicated by Roman numerals or suffixes. Instead, elemental proportions in binary covalent compounds are indicated by using a *prefix* system for both elements. The Greek and Latin prefixes for one through ten are *mono-, di-, tri-, tetra-, penta-, hexa-, hepta, octa-, nona-,* and *deca-*. The prefix "mono-" is omitted for both elements except in the common name for CO, carbon monoxide. We use the minimum number of prefixes needed to name a compound unambiguously. The final "a" in a prefix is omitted when the nonmetal stem begins with the letter "o"; we write "heptoxide," not "heptaoxide."

Formula	Name	Formula	Name
SO_2	sulfur dioxide	Cl_2O_7	dichlorine heptoxide
SO_3	sulfur trioxide	CS_2	carbon disulfide
N_2O_4	dinitrogen tetroxide	SF_4	sulfur tetrafluoride
As_4O_6	tetraarsenic hexoxide	SF_6	sulfur hexafluoride

Binary acids are compounds in which H is bonded to a Group VIA element other than O or to a Group VIIA element; they act as acids when dissolved in water. *The pure compounds are named as typical binary compounds.* Their aqueous solutions are named by modifying the characteristic stem of the nonmetal with the prefix "hydro-" and the suffix "-ic" followed by the word "acid." The stem for sulfur in this instance is "sulfur" rather than "sulf."

Formula	Name of Compound	Name of Aqueous Solution
HCl	hydrogen chloride	hydrochloric acid, HCl(aq)
HF	hydrogen fluoride	hydrofluoric acid, HF(aq)
H_2S	hydrogen sulfide	hydrosulfuric acid, H_2S(aq)
HCN	hydrogen cyanide	hydrocyanic acid, HCN(aq)

In later chapters we will learn additional systematic rules for naming more complex compounds.

SC/LINKS. **TOPIC:** Naming Compounds **GO TO:** www.scilinks.org *sci*LINKS **CODE:** WCH0450

4-6 NAMING TERNARY ACIDS AND THEIR SALTS

A ternary compound consists of three elements. **Ternary acids (oxoacids)** are compounds of hydrogen, oxygen, and (usually) a nonmetal. Nonmetals that exhibit more than one oxidation state form more than one ternary acid. These ternary acids differ in the number of oxygen atoms they contain. The suffixes "-ous" and "-ic" following the stem name of the central element indicate lower and higher oxidation states, respectively. One common ternary acid of each nonmetal is (somewhat arbitrarily) designated as the "-ic" acid. That is, it is named by putting the element stem before the "-ic" suffix. The common ternary "-ic acids" are shown in Table 4-12. It is important to learn the names and formulas of these acids because the names of all other ternary acids and salts are derived from them. There are no common "-ic" ternary acids for the omitted nonmetals.

The oxoacid with the central element in the highest oxidation state usually contains more O atoms. Oxoacids with their central elements in lower oxidation states usually have fewer O atoms.

TABLE 4-12	*Formulas of Some "-ic" Acids*

Periodic Group of Central Elements				
IIA	*IVA*	*VA*	*VIA*	*VIIA*
(+3) H_3BO_3 boric acid	(+4) H_2CO_3 carbonic acid	(+5) HNO_3 nitric acid		
	(+4) H_4SiO_4 silicic acid	(+5) H_3PO_4 phosphoric acid	(+6) H_2SO_4 sulfuric acid	(+5) $HClO_3$ chloric acid
		(+5) H_3AsO_4 arsenic acid	(+6) H_2SeO_4 selenic acid	(+5) $HBrO_3$ bromic acid
			(+6) H_6TeO_6 telluric acid	(+5) HIO_3 iodic acid

Note that the oxidation state of the central atom is equal to its periodic group number, except for the halogens.

Acids containing *one fewer oxygen atom* per central atom are named in the same way except that the "-ic" suffix is changed to "-ous." The oxidation number of the central element is *lower by 2* in the "-ous" acid than in the "-ic" acid.

Formula	Ox. No.	Name	Formula	Ox. No.	Name
H_2SO_3	+4	sulfur*ous* acid	H_2SO_4	+6	sulfur*ic* acid
HNO_2	+3	nitr*ous* acid	HNO_3	+5	nitr*ic* acid
H_2SeO_3	+4	selen*ous* acid	H_2SeO_4	+6	selen*ic* acid
$HBrO_2$	+3	brom*ous* acid	$HBrO_3$	+5	brom*ic* acid

Ternary acids that have one fewer O atom than the "-ous" acids (two fewer O atoms than the "-ic" acids) are named using the prefix "hypo-" and the suffix "-ous." These are acids in which the oxidation state of the central nonmetal is *lower by 2* than that of the central nonmetal in the "-ous acids."

Formula	Ox. No.	Name
$HClO$	+1	*hypo*chlor*ous* acid
H_3PO_2	+1	*hypo*phosphor*ous* acid
HIO	+1	*hypo*iod*ous* acid
$H_2N_2O_2$	+1	*hypo*nitr*ous* acid

Notice that $H_2N_2O_2$ has a 1:1 ratio of nitrogen to oxygen, as would the hypothetical HNO.

Acids containing *one more oxygen atom* per central nonmetal atom than the normal "-ic acid" are named "*per*stem*ic*" acids.

Formula	Ox. No.	Name
$HClO_4$	+7	*per*chlor*ic* acid
$HBrO_4$	+7	*per*brom*ic* acid
HIO_4	+7	*per*iod*ic* acid

The oxoacids of chlorine can be summarized as follows:

Formula	Ox. No.	Name
$HClO$	+1	*hypo*chlor*ous* acid
$HClO_2$	+3	chlor*ous* acid
$HClO_3$	+5	chlor*ic* acid
$HClO_4$	+7	*per*chlor*ic* acid

Ternary salts are compounds that result from replacing the hydrogen in a ternary acid with another ion. They usually contain metal cations or the ammonium ion. As with binary compounds, the cation is named first. The name of the anion is based on the name of the ternary acid from which it is derived.

An anion derived from a ternary acid with an "-ic" ending is named by dropping the "-ic acid" and replacing it with "-ate." An anion derived from an "-ous acid" is named by replacing the suffix "-ous acid" with "-ite." The "per-" and "hypo-" prefixes are retained.

Formula	Name
$(NH_4)_2SO_4$	ammonium sulfate (SO_4^{2-}, from sulfuric acid, H_2SO_4)
KNO_3	potassium nitrate (NO_3^-, from nitric acid, HNO_3)
$Ca(NO_2)_2$	calcium nitrite (NO_2^-, from nitrous acid, HNO_2)
$LiClO_4$	lithium perchlorate (ClO_4^-, from perchloric acid, $HClO_4$)
$FePO_4$	iron(III) phosphate (PO_4^{3-}, from phosphoric acid, H_3PO_4)
$NaClO$	sodium hypochlorite (ClO^-, from hypochlorous acid, $HClO$)

Acidic salts contain anions derived from ternary polyprotic acids in which one or more acidic hydrogen atoms remain. These salts are named as if they were the usual type of ternary salt, with the word "hydrogen" or "dihydrogen" inserted after the name of the cation to show the number of acidic hydrogen atoms.

Formula	Name	Formula	Name
$NaHSO_4$	sodium hydrogen sulfate	KH_2PO_4	potassium dihydrogen phosphate
$NaHSO_3$	sodium hydrogen sulfite	K_2HPO_4	potassium hydrogen phosphate
		$NaHCO_3$	sodium hydrogen carbonate

An older, commonly used method (which is not recommended by the IUPAC, but which is widely used in commerce) involves the use of the prefix "bi-" attached to the name of the anion to indicate the presence of an acidic hydrogen. According to this system, $NaHSO_4$ is called sodium bisulfate and $NaHCO_3$ is called sodium bicarbonate.

✓ **Problem-Solving Tip:** *Naming Ternary Acids and Their Anions*

The following table might help you to remember the names of the ternary acids and their ions. First learn the formulas of the acids mentioned earlier that end with "-ic acid." Then relate possible other acids to the following table. The stem (**XXX**) represents the stem of the name, for example, "nitr," "sulfur," or "chlor."

Decreasing oxidation number of central atom	Ternary Acid	Anion	Decreasing number of oxygen atoms on central atom
	*per***XXX***ic* acid	*per***XXX***ate*	
	XXX*ic* acid	**XXX***ate*	
	XXX*ous* acid	**XXX***ite*	
	*hypo***XXX***ous* acid	*hypo***XXX***ite*	

CLASSIFYING CHEMICAL REACTIONS

We now discuss chemical reactions in further detail. We classify them as oxidation–reduction reactions, combination reactions, decompositions reactions, displacement reactions, and metathesis reactions. The last type can be further described as precipitation reactions and acid–base (neutralization) reactions. We will see that many reactions, especially oxidation–reduction reactions, fit into more than one category, and that some reactions do not fit neatly into any of them. As we study different kinds of chemical reactions, we will learn to predict the products of other similar reactions. In Chapter 6 we will describe typical

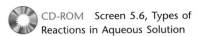

CD-ROM Screen 5.6, Types of Reactions in Aqueous Solution

reactions of hydrogen, oxygen, and their compounds. These reactions will illustrate periodic relationships with respect to chemical properties. It should be emphasized that our system is not an attempt to transform nature so that it fits into small categories but rather an effort to give some order to our many observations of nature.

4-7   OXIDATION–REDUCTION REACTIONS: INTRODUCTION

The term "oxidation" originally referred to the combination of a substance with oxygen. This results in an increase in the oxidation number of an element in that substance. According to the original definition, the following reactions involve oxidation of the substance shown on the far left of each equation. Oxidation numbers are shown for *one* atom of the indicated kind.

1. The formation of rust, Fe_2O_3, iron(III) oxide: oxidation state of Fe

$$4Fe(s) + 3O_2(g) \longrightarrow 2Fe_2O_3(s) \qquad\qquad 0 \longrightarrow +3$$

2. Combustion reactions: oxidation state of C

$$C(s) + O_2(g) \longrightarrow CO_2(g) \qquad\qquad 0 \longrightarrow +4$$
$$2CO(g) + O_2(g) \longrightarrow 2CO_2(g) \qquad\qquad +2 \longrightarrow +4$$
$$C_3H_8(g) + 5O_2(g) \longrightarrow 3CO_2(g) + 4H_2O(g) \qquad -8/3 \longrightarrow +4$$

Originally, *reduction* described the removal of oxygen from a compound. Oxide ores are reduced to metals (a very real reduction in mass). For example, tungsten for use in light bulb filaments can be prepared by reduction of tungsten(VI) oxide with hydrogen at 1200°C:

oxidation number of W

$$WO_3(s) + 3H_2(g) \longrightarrow W(s) + 3H_2O(g) \qquad +6 \longrightarrow 0$$

Tungsten is reduced, and its oxidation state decreases from +6 to zero. Hydrogen is oxidized from zero to the +1 oxidation state. The terms "oxidation" and "reduction" are now applied much more broadly.

> **Oxidation** is an increase in oxidation number and corresponds to the loss, or apparent loss, of electrons. **Reduction** is a decrease in oxidation number and corresponds to a gain, or apparent gain, of electrons.

Electrons are neither created nor destroyed in chemical reactions. So oxidation and reduction always occur simultaneously, and to the same extent, in ordinary chemical reactions. In the four equations cited previously as *examples of oxidation*, the oxidation numbers of iron and carbon atoms increase as they are oxidized. In each case oxygen is reduced as its oxidation number decreases from zero to −2.

Because oxidation and reduction occur simultaneously in all of these reactions, they are referred to as oxidation–reduction reactions. For brevity, we usually call them **redox** reactions. Redox reactions occur in nearly every area of chemistry and biochemistry. We need to be able to identify oxidizing agents and reducing agents and to balance oxidation–reduction equations. These skills are necessary for the study of electrochemistry in Chapter 21. Electrochemistry involves electron transfer between physically separated oxidizing and

---

Oxidation number is a formal concept adopted for our convenience. The numbers are determined by relying on rules. These rules can result in a fractional oxidation number, as shown here. This does not mean that electronic charges are split.

The terms "oxidation number" and "oxidation state" are used interchangeably.

In biological systems, *reduction* often corresponds to the addition of hydrogen to molecules or polyatomic ions and *oxidation* often corresponds to the removal of hydrogen.

CD-ROM   Screens 5.13, Redox Reactions and Electron Transfer, and 5.14, Recognizing Oxidation–Reduction Reactions

reducing agents and interconversions between chemical energy and electrical energy. These skills are also fundamental to the study of biology, biochemistry, environmental science, and materials science.

> **Oxidizing agents** are species that (1) oxidize other substances, (2) contain atoms that are reduced, and (3) gain (or appear to gain) electrons. **Reducing agents** are species that (1) reduce other substances, (2) contain atoms that are oxidized, and (3) lose (or appear to lose) electrons.

The following abbreviations are widely used:

ox. no. = oxidation number
ox. agt. = oxidizing agent
red. agt. = reducing agent

The following equations represent examples of redox reactions. Oxidation numbers are shown above the formulas, and oxidizing and reducing agents are indicated:

$$\overset{0}{2Fe(s)} + \overset{0}{3Cl_2(g)} \longrightarrow \overset{+3\;-1}{2FeCl_3(s)}$$
$$\text{red. agt.} \qquad \text{ox. agt.}$$

$$\overset{+3\;-1}{2FeBr_3(aq)} + \overset{0}{3Cl_2(g)} \longrightarrow \overset{+3\;-1}{2FeCl_3(aq)} + \overset{0}{3Br_2(\ell)}$$
$$\text{red. agt.} \qquad \text{ox. agt.}$$

Equations for redox reactions can also be written as total ionic and net ionic equations. For example, the second equation may also be written as:

$$2[Fe^{3+}(aq) + 3Br^-(aq)] + 3Cl_2(g) \longrightarrow 2[Fe^{3+}(aq) + 3Cl^-(aq)] + 3Br_2(\ell)$$

We distinguish between oxidation numbers and actual charges on ions by denoting oxidation numbers as $+n$ or $-n$ in red circles *just above the symbols of the elements*, and actual charges as $n+$ or $n-$ above and to the right of formulas of ions. The spectator ions, Fe^{3+}, do not participate in electron transfer. Their cancellation allows us to focus on the oxidizing agent, $Cl_2(g)$, and the reducing agent, $Br^-(aq)$.

$$2Br^-(aq) + Cl_2(g) \longrightarrow 2Cl^-(aq) + Br_2(\ell)$$

A **disproportionation reaction** is a redox reaction in which the same element is oxidized and reduced. An example is:

$$\overset{0}{Cl_2} + H_2O \longrightarrow \overset{-1}{HCl} + \overset{+1}{HClO}$$

Iron reacting with chlorine to form iron(III) chloride.

EXAMPLE 4-5 *Redox Reactions*

Write each of the following formula unit equations as a net ionic equation if the two differ. Which ones are redox reactions? For the redox reactions, identify the oxidizing agent, the reducing agent, the species oxidized, and the species reduced.

(a) $2AgNO_3(aq) + Cu(s) \longrightarrow Cu(NO_3)_2(aq) + 2Ag(s)$

(b) $4KClO_3(s) \xrightarrow{\text{heat}} KCl(s) + 3KClO_4(s)$

(c) $3AgNO_3(aq) + K_3PO_4(aq) \longrightarrow Ag_3PO_4(s) + 3KNO_3(aq)$

Plan

To write ionic equations, we must recognize compounds that are (1) soluble in water and (2) ionized or dissociated in aqueous solutions. To determine which are oxidation–reduction reactions, we should assign an oxidation number to each element.

Metallic silver formed by immersing a spiral of copper wire in a silver nitrate solution (see Example 4-5a).

Solution

(a) According to the solubility guidelines (Section 4-2 Part 5), both silver nitrate, $AgNO_3$, and copper(II) nitrate, $Cu(NO_3)_2$, are water-soluble ionic compounds. The total ionic equation and oxidation numbers are

$$2[\overset{+1}{Ag^+}(aq) + \overset{+5}{N}\overset{-2}{O_3}^-(aq)] + \overset{0}{Cu}(s) \longrightarrow [\overset{+2}{Cu^{2+}}(aq) + 2\overset{+5}{N}\overset{-2}{O_3}^-(aq)] + 2\overset{0}{Ag}(s)$$

$$\underline{+2} \uparrow$$
$$\underline{-1}$$

The nitrate ions, NO_3^-, are spectator ions. Canceling them from both sides gives the net ionic equation:

$$2\overset{+1}{Ag^+}(aq) + \overset{0}{Cu}(s) \longrightarrow \overset{+2}{Cu^{2+}}(aq) + 2\overset{0}{Ag}(s)$$

This is a redox equation. The oxidation number of silver decreases from +1 to zero; silver ion is reduced and is the oxidizing agent. The oxidation number of copper increases from zero to +2; copper is oxidized and is the reducing agent.

(b) This reaction involves only solids, so there are no ions in solution, and the formula unit and net ionic equations are identical. It is a redox reaction:

$$4\overset{+1}{K}\overset{+5}{Cl}\overset{-2}{O_3}(s) \longrightarrow \overset{+1}{K}\overset{-1}{Cl}(s) + 3\overset{+1}{K}\overset{+7}{Cl}\overset{-2}{O_4}(s)$$

$$\underline{-6} \uparrow$$
$$\underline{+2}$$

Chlorine is reduced from +5 in $KClO_3$ to the −1 oxidation state in KCl; the oxidizing agent is $KClO_3$. Chlorine is oxidized from +5 in $KClO_3$ to the +7 oxidation state in $KClO_4$. $KClO_3$ is also the reducing agent. This is a disproportionation reaction. We see that $KClO_3$ is both the oxidizing agent and the reducing agent.

(c) The solubility guidelines indicate that all these salts are soluble except for silver phosphate, Ag_3PO_4. The total ionic equation is

$$3[Ag^+(aq) + NO_3^-(aq)] + [3K^+(aq) + PO_4^{3-}(aq)] \longrightarrow Ag_3PO_4(s) + 3[K^+(aq) + NO_3^-(aq)]$$

Eliminating the spectator ions gives the net ionic equation:

$$3\overset{+1}{Ag^+}(aq) + \overset{+5}{P}\overset{-2}{O_4}^{3-}(aq) \longrightarrow \overset{+1}{Ag_3}\overset{+5}{P}\overset{-2}{O_4}(s)$$

There are no changes in oxidation numbers; this is not a redox reaction.

You should now work Exercises 66 and 69.

The reaction of $AgNO_3$(aq) and K_3PO_4(aq) is a precipitation reaction (see Example 4-5c).

James W. Morganthaler

✓ Problem-Solving Tip: *A Foolproof Way to Recognize a Redox Reaction*

You can always recognize a redox reaction by analyzing oxidation numbers. First determine the oxidation number of each element wherever it appears in the reaction. If no elements change in oxidation numbers, the reaction is not an oxidation–reduction reaction. If changes do occur, the reaction is an oxidation–reduction reaction. Remember that oxidation and reduction must always occur together; if some atoms increase in oxidation numbers, then others must decrease.

In Chapter 11 we will learn to balance redox equations and to carry out stoichiometric calculations using the balanced equations.

4-8 COMBINATION REACTIONS

Reactions in which two or more substances combine to form a compound are called **combination reactions.**

They may involve (1) the combination of two elements to form a compound, (2) the combination of an element and a compound to form a single new compound, or (3) the combination of two compounds to form a single new compound. Let's examine some of these reactions.

1 Element + Element → Compound

For this type of combination reaction, each element goes from an uncombined state, where its oxidation state is zero, to a combined state in a compound, where its oxidation state is not zero. Thus reactions of this type are oxidation–reduction reactions (see Section 4-7).

Metal + Nonmetal → Binary Ionic Compound

Most metals react with most nonmetals to form binary ionic compounds. The Group IA metals combine with the Group VIIA nonmetals to form binary *ionic* compounds with the general formula MX (Section 7-2):

$$2M(s) + X_2 \longrightarrow 2(M^+X^-)(s) \qquad M = Li, Na, K, Rb, Cs$$
$$X = F, Cl, Br, I$$

This general equation thus represents the 20 combination reactions that form the ionic compounds listed in Table 4-13. Sodium, a silvery-white metal, combines with chlorine, a pale green gas, to form sodium chloride, or ordinary table salt. All members of both families undergo similar reactions.

$$2Na(s) + Cl_2(g) \longrightarrow 2NaCl(s) \qquad \text{sodium chloride (mp } 801°C)$$

As we might expect, the Group IIA metals also combine with the Group VIIA nonmetals to form binary compounds. Except for $BeCl_2$, $BeBr_2$, and BeI_2, these are ionic compounds. In general terms these combination reactions may be represented as:

$$M(s) + X_2 \longrightarrow MX_2(s) \qquad M = Be, Mg, Ca, Sr, Ba$$
$$X = F, Cl, Br, I$$

Consider the reaction of magnesium with fluorine to form magnesium fluoride:

$$Mg(s) + F_2(g) \longrightarrow MgF_2(s) \qquad \text{magnesium fluoride (mp } 1266°C)$$

Because all the IIA and VIIA elements undergo similar reactions, the general equation, written above, represents 20 reactions. We omit radium and astatine, the rare and highly radioactive members of the families.

Nonmetal + Nonmetal → Binary Covalent Compound

When two nonmetals combine with each other, they form binary *covalent* compounds. In such reactions, the oxidation number of the element with the more positive oxidation

Potassium, a metal, reacts with chlorine, a nonmetal, to form potassium chloride, KCl. The reaction releases energy in the form of heat and light.

Another important reaction of this kind is the formation of metal oxides (Section 6-8).

TABLE 4-13	Alkali Metal Halides: Compounds Formed by Group IA and VIIA Elements		
LiF	LiCl	LiBr	LiI
NaF	NaCl	NaBr	NaI
KF	KCl	KBr	KI
RbF	RbCl	RbBr	RbI
CsF	CsCl	CsBr	CsI

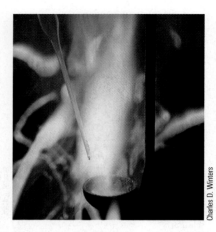

Phosphorus and chlorine, two non-metals, react to form phosphorus pentachloride, PCl_5.

Nonmetals in odd-numbered periodic groups favor odd oxidation numbers, whereas those in even-numbered groups favor even oxidation numbers in their compounds. The *maximum* oxidation number for a representative element is equal to its periodic group number. For example, sulfur (Group VIA) can form both SF_4 and SF_6.

number is often variable, depending on reaction conditions. For example, phosphorus (Group VA) combines with a *limited amount* of chlorine to form phosphorus trichloride, in which phosphorus exhibits the +3 oxidation state.

$$P_4(s) + 6Cl_2(g) \longrightarrow 4\overset{+3}{P}Cl_3(\ell) \quad \text{(with limited } Cl_2) \quad \text{(mp } -112°C)$$

With an excess of chlorine, the product is phosphorus pentachloride, which contains phosphorus in the +5 oxidation state:

$$P_4(s) + 10Cl_2(g) \longrightarrow 4\overset{+5}{P}Cl_5(s) \quad \text{(with excess } Cl_2) \quad \text{(decomposes at } 167°C)$$

In general, *a higher oxidation state of a nonmetal is formed when it reacts with an excess of another nonmetal.* There are many more reactions in which two elements combine to form a compound (see Sections 6-7 and 6-8).

2 Compound + Element → Compound

Phosphorus in the +3 oxidation state in PCl_3 molecules can be converted to the +5 state in PCl_5 by combination with chlorine:

$$\overset{+3}{P}Cl_3(\ell) + Cl_2(g) \longrightarrow \overset{+5}{P}Cl_5(s)$$

Likewise, sulfur in the +4 state is converted to the +6 state when SF_4 reacts with fluorine to form SF_6:

$$\overset{+4}{S}F_4(g) + F_2(g) \longrightarrow \overset{+6}{S}F_6(g) \quad \text{sulfur hexafluoride (mp } -50.5°C)$$

Combination reactions of this type are also oxidation–reduction reactions.

3 Compound + Compound → Compound

An example of reactions in this category is the combination of calcium oxide with carbon dioxide to produce calcium carbonate:

$$CaO(s) + CO_2(g) \longrightarrow CaCO_3(s)$$

Pyrosulfuric acid is produced by dissolving sulfur trioxide in concentrated sulfuric acid:

$$SO_3(g) + H_2SO_4(\ell) \longrightarrow H_2S_2O_7(\ell)$$

Pyrosulfuric acid, $H_2S_2O_7$, is then diluted with water to make H_2SO_4:

$$H_2S_2O_7(\ell) + H_2O(\ell) \longrightarrow 2H_2SO_4(\ell)$$

Oxides of the Group IA and IIA metals react with water to form metal hydroxides, e.g.:

$$CaO(s) + H_2O(\ell) \longrightarrow Ca(OH)_2(aq)$$

4-9 DECOMPOSITION REACTIONS

Decomposition reactions can be considered as the opposite of combination reactions.

Decomposition reactions are those in which a compound decomposes to produce (1) two elements, (2) one or more elements *and* one or more compounds, or (3) two or more compounds.

Examples of each type follow.

1 Compound → Element + Element

The electrolysis of water produces two elements by the decomposition of a compound. A compound that ionizes, such as H_2SO_4, is added to increase the conductivity of water and the rate of the reaction (Figure 1-8), but it does not participate in the reaction:

$$2H_2O(\ell) \xrightarrow{\text{electrolysis}} 2H_2(g) + O_2(g)$$

Small amounts of oxygen can be prepared by the thermal decomposition of certain oxygen-containing compounds. Some metal oxides, such as mercury(II) oxide, HgO, decompose on heating to produce oxygen:

$$\underset{\text{mercury(II) oxide}}{2HgO(s)} \xrightarrow{\text{heat}} 2Hg(\ell) + O_2(g)$$

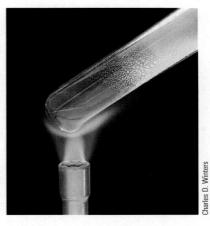

Mercury(II) oxide, a red compound, decomposes when heated into the two elements: mercury (a metal) and oxygen (a nonmetal). Mercury vapor condenses on the cooler upper portion of the test tube.

2 Compound → Compound + Element

The alkali metal chlorates, such as $KClO_3$, decompose when heated to produce the corresponding chlorides and liberate oxygen. Potassium chlorate is a common laboratory source of small amounts of oxygen:

$$\underset{\text{potassium chlorate}}{2KClO_3(s)} \xrightarrow[\text{MnO}_2]{\text{heat}} \underset{\text{potassium chloride}}{2KCl(s)} + 3O_2(g)$$

Nitrate salts of alkali metals or alkaline earth metals decompose to form metal nitrites and oxygen gas.

$$2NaNO_3(s) \longrightarrow 2NaNO_2(s) + O_2(g)$$

Hydrogen peroxide decomposes to form water and oxygen.

$$H_2O_2(\ell) \longrightarrow 2H_2O(\ell) + O_2(g)$$

Manganese dioxide, MnO_2, is used as a catalyst, a substance that speeds up a chemical reaction but is not consumed. Here it allows the decomposition to occur at a lower temperature.

3 Compound → Compound + Compound

The thermal decomposition of calcium carbonate (limestone) and other carbonates produces two compounds, a metal oxide and carbon dioxide:

$$CaCO_3(s) \xrightarrow{\text{heat}} CaO(s) + CO_2(g)$$

This is an important reaction in the production of cement. Calcium oxide is also used as a base in industrial processes.

When some solid hydroxides are heated, they decompose to form a metal oxide and water vapor.

$$Mg(OH)_2(s) \xrightarrow{\text{heat}} MgO(s) + H_2O(g)$$

Magnesium oxide, MgO, is pressed into sheets for use as a thermal insulating material in oven walls.

Ammonium salts lose ammonia.

$$(NH_4)_2SO_4(s) \xrightarrow{\text{heat}} 2NH_3(g) + H_2SO_4(\ell)$$

Alkali metal carbonates do not decompose when heated.

A decomposition reaction may or may not also be an oxidation–reduction reaction. You can always identify a redox reaction by determining the oxidation state of each element in each occurrence in the reaction (see the Problem-Solving Tip in Section 4-7).

Charles D. Winters

Solid ammonium dichromate, [(NH$_4$)$_2$Cr$_2$O$_7$, *orange*] decomposes when heated into chromium(II) oxide, (Cr$_2$O$_3$, *green*), nitrogen, and steam (water vapor). This reaction is sometimes demonstrated as the "classroom volcano," but it must be done with extreme caution due to the carcinogenic (cancer-causing) nature of (NH$_4$)$_2$Cr$_2$O$_7$.

If the ammonium salt contains an anion that is a strong oxidizing agent (e.g., nitrate, nitrite, or dichromate), its decomposition reaction produces an oxide, water (as vapor at high temperatures), and nitrogen gas. Such a reaction is a redox reaction.

$$(NH_4)_2Cr_2O_7(s) \xrightarrow{\text{heat}} Cr_2O_3(s) + 4H_2O(g) + N_2(g)$$

4-10 DISPLACEMENT REACTIONS

Reactions in which one element displaces another from a compound are called **displacement reactions.**

These reactions are always redox reactions. The more readily a metal undergoes oxidation, the more active we say it is.

Active metals displace less active metals or hydrogen from their compounds in aqueous solution to form the oxidized form of the more active metal and the reduced (free metal) form of the other metal or hydrogen.

In Table 4-14, the most active metals are listed at the top of the first column. These metals tend to react to form their oxidized forms (cations). Elements at the bottom of the activity series (the first column of Table 4-14) tend to remain in their reduced form. They are easily converted from their oxidized forms to their reduced forms.

$$1\begin{bmatrix}\textbf{More Active Metal +}\\ \textbf{Salt of Less Active Metal}\end{bmatrix} \longrightarrow \begin{bmatrix}\textbf{Less Active Metal +}\\ \textbf{Salt of More Active Metal}\end{bmatrix}$$

The reaction of copper with silver nitrate that was described in detail in Section 4-3 is typical. Please refer to it.

EXAMPLE 4-6 *Displacement Reaction*

A large piece of zinc metal is placed in a copper(II) sulfate, CuSO$_4$, solution. The blue solution becomes colorless as copper metal falls to the bottom of the container. The resulting solution contains zinc sulfate, ZnSO$_4$. Write balanced formula unit, total ionic, and net ionic equations for the reaction.

Plan

The metals zinc and copper are *not* ionized or dissociated in contact with H$_2$O. Both CuSO$_4$ and ZnSO$_4$ are soluble salts (solubility guideline 5), so they are written in ionic form.

Solution

$$Zn(s) + CuSO_4(aq) \longrightarrow Cu(s) + ZnSO_4(aq)$$
$$Zn(s) + [Cu^{2+}(aq) + SO_4{}^{2-}(aq)] \longrightarrow Cu(s) + [Zn^{2+}(aq) + SO_4{}^{2-}(aq)]$$
$$Zn(s) + Cu^{2+}(aq) \longrightarrow Cu(s) + Zn^{2+}(aq)$$

A strip of zinc metal was placed in a blue solution of copper(II) sulfate, CuSO$_4$. The copper has been displaced from solution and has fallen to the bottom of the beaker. The resulting zinc sulfate, ZnSO$_4$, solution is colorless.

In this *displacement reaction*, the more active metal, zinc, displaces the ions of the less active metal, copper, from aqueous solution.

You should now work Exercise 78.

TABLE 4-14	*Activity Series of Some Elements*		
Element		**Common Reduced Form**	**Common Oxidized Forms**
Li		Li	Li$^+$
K		K	K$^+$
Ca		Ca	Ca^{2+}
Na		Na	Na$^+$
Mg		Mg	Mg^{2+}
Al		Al	Al^{3+}
Mn		Mn	Mn^{2+}
Zn		Zn	Zn^{2+}
Cr		Cr	Cr^{3+}, Cr^{6+}
Fe		Fe	Fe^{2+}, Fe^{3+}
Cd		Cd	Cd^{2+}
Co		Co	Co^{2+}
Ni		Ni	Ni^{2+}
Sn		Sn	Sn^{2+}, Sn^{4+}
Pb		Pb	Pb^{2+}, Pb^{4+}
H (a nonmetal)		H$_2$	H$^+$
Sb (a metalloid)		Sb	Sb^{3+}
Cu		Cu	Cu$^+$, Cu^{2+}
Hg		Hg	Hg$_2$$^{2+}$, Hg^{2+}
Ag		Ag	Ag$^+$
Pt		Pt	Pt^{2+}, Pt^{4+}
Au		Au	Au$^+$, Au^{3+}

Displace hydrogen from nonoxidizing acids
Displace hydrogen from steam
Displace hydrogen from cold water

SC*LINKS*.
TOPIC: Activity Series
GO TO: www.scilinks.org
*sci*LINKS CODE: WCH0470

H$_2$SO$_4$ can function as an oxidizing agent with other substances, but it is not an oxidizing agent in its reaction with active metals.

2 [Active Metal + Nonoxidizing Acid] ⟶ [Hydrogen + Salt of Acid]

A common method for preparing small amounts of hydrogen involves the reaction of active metals with nonoxidizing acids, such as HCl and H$_2$SO$_4$. For example, when zinc is dissolved in H$_2$SO$_4$, the reaction produces zinc sulfate; hydrogen is displaced from the acid, and it bubbles off as gaseous H$_2$. The formula unit equation for this reaction is

$$Zn(s) + H_2SO_4(aq) \longrightarrow ZnSO_4(aq) + H_2(g)$$

strong acid   soluble salt

Both sulfuric acid (in very dilute solution) and zinc sulfate exist primarily as ions, so the total ionic equation is

$$Zn(s) + [2H^+(aq) + SO_4{}^{2-}(aq)] \longrightarrow [Zn^{2+}(aq) + SO_4{}^{2-}(aq)] + H_2(g)$$

Elimination of spectator ions from the total ionic equation gives the net ionic equation:

$$Zn(s) + 2H^+(aq) \longrightarrow Zn^{2+}(aq) + H_2(g)$$

Table 4-14 lists the **activity series.** When any metal listed above hydrogen in this series is added to a solution of a *nonoxidizing* acid such as hydrochloric acid, HCl, and sulfuric acid, H$_2$SO$_4$, the metal dissolves to produce hydrogen, and a salt is formed. HNO$_3$ is the common *oxidizing acid.* It reacts with active metals to produce oxides of nitrogen, but *not* hydrogen, H$_2$.

Charles Steele

Zinc reacts with dilute H$_2$SO$_4$ to produce H$_2$ and a solution that contains ZnSO$_4$. This is a displacement reaction.

Troublesome Displacement Reactions

The deterioration of the Statue of Liberty and the damage done at the Three Mile Island and Chernobyl nuclear facilities are just a few of the major problems that have resulted from ignorance about chemical reactivity.

When originally constructed over one hundred years ago the Statue of Liberty had a 200,000-pound outer copper skin supported by a framework of 2000 iron bars. First, oxygen in the air oxidized the copper skin to form copper oxide. In a series of reactions, iron (the more active metal) then reduced the Cu^{2+} ions in copper oxide.

$$2Fe + 3Cu^{2+} \longrightarrow 2Fe^{3+} + 3Cu$$

Over the years, the supporting iron frame was reduced to less than half its original thickness; this made necessary the repairs done to the statue before the celebration of its 100th birthday on July 4, 1986.

Two major nuclear power plant accidents, one at Three Mile Island near Harrisburg, Pennsylvania, in 1979 and the other at Chernobyl in Ukraine in 1986, were also unexpected consequences of chemical reactivity. In each case, cooling pump failures sent temperatures soaring above 340°C. Like aluminum, zirconium (used in building the reactors) forms an oxide coating that shields it from further reactions. However, that protective coating breaks down at high temperatures. Without its protective coating, zirconium reacts with steam.

$$Zr(s) + 2H_2O(g) \longrightarrow ZrO_2(s) + 2H_2(g)$$

At Three Mile Island, this displacement reaction produced a 1000-cubic foot bubble of hydrogen gas. Because hydrogen is easily ignited by a spark, the nuclear power plant was in real danger of a complete meltdown until the hydrogen could be removed.

During the Middle Ages (~AD 400–1400), another displacement reaction completely misled alchemists into foolishly pursuing a philosophers' stone that was believed to have the power to turn base metals such as iron and lead into more precious metals such as silver and gold. The alchemists' ignorance of relative activities of metals led them to believe that they had turned iron into a more precious metal when they inserted an iron rod into a blue copper(II) sulfate solution. In fact, the following displacement reaction had occurred, plating shiny copper metal onto the iron rod:

$$Fe(s) + 3Cu^{2+}(aq) \longrightarrow 2Fe^{3+}(aq) + Cu(s)$$

In the 1960s and 1970s, some automobile manufacturers showed their ignorance of chemical reactivity by building cars with aluminum water pumps and aluminum engine heads

Patricia Caufield/Photo Researchers, Inc.

attached to cast-iron engine blocks. These water pumps often leaked and the engine heads quickly deteriorated. These problems occurred as the more active aluminum reacted with iron(II) oxide (formed when the iron engine reacted with atmospheric oxygen).

$$Al + Fe^{3+} \longrightarrow Al^{3+} + Fe$$

Some dentists have made similar mistakes by placing gold caps over teeth that are adjacent to existing fillings. The slightly oxidized gold can react with a dental amalgam filling (an alloy of silver, tin, copper, and mercury). As the dental amalgam is oxidized, it dissolves in saliva to produce a persistent metallic taste in the patient's mouth.

When plumbers connect galvanized pipes (iron pipes coated with zinc) to copper pipes, copper ions oxidize the zinc coating and expose the underlying iron, allowing it to rust. The displacement reaction that occurs is

$$Zn + Cu^{2+} \longrightarrow Zn^{2+} + Cu$$

Once the zinc coating has been punctured on an iron pipe, oxidation of the iron pipes occurs rapidly because iron is a more active metal than copper.

It is important to keep in mind that a variety of reactions other than the displacement reactions discussed here probably take place. For example, less active metals (such as copper) can conduct electrons from the metals being oxidized to oxidizing agents (such as oxygen or the oxide of nitrogen and sulfur) that are present in the atmosphere. Oxygen plays an important role in all these displacement examples.

Ronald DeLorenzo
Middle Georgia College

EXAMPLE 4-7   *Displacement Reaction*

Which of the following metals can displace hydrogen from hydrochloric acid solution? Write balanced formula unit, total ionic, and net ionic equations for reactions that can occur.

$$Al, \quad Cu, \quad Ag$$

Plan

The activity series of the metals, Table 4-14, tells us that copper and silver *do not* displace hydrogen from solutions of nonoxidizing acids. Aluminum is an active metal that can displace H_2 from HCl and form aluminum chloride.

Solution

$$2Al(s) + 6HCl(aq) \longrightarrow 3H_2(g) + 2AlCl_3(aq)$$

$$2Al(s) + 6[H^+(aq) + Cl^-(aq)] \longrightarrow 3H_2(g) + 2[Al^{3+}(aq) + 3Cl^-(aq)]$$

$$2Al(s) + 6H^+(aq) \longrightarrow 3H_2(g) + 2Al^{3+}(aq)$$

You should now work Exercises 77 and 79.

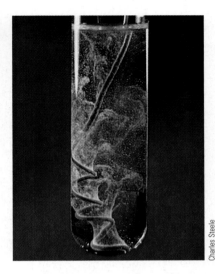

Aluminum displaces H_2 from a hydrochloric acid solution.

Very active metals can even displace hydrogen from water. The reaction of potassium, or another metal of Group IA, with water is also a *displacement reaction*:

$$2K(s) + 2H_2O(\ell) \longrightarrow 2[K^+(aq) + OH^-(aq)] + H_2(g)$$

Such reactions of very active metals of Group IA are dangerous, however, because they generate enough heat to cause explosive ignition of the hydrogen (Figure 4-4).

EXAMPLE 4-8   *Displacement Reaction*

Which of the following metals can displace hydrogen from water at room temperature? Write balanced formula unit, total ionic, and net ionic equations for reactions that can occur.

$$Sn, \quad Ca, \quad Hg$$

Plan

The activity series, Table 4-14, tells us that tin and mercury *cannot* displace hydrogen from water. Calcium is a very active metal (see Table 4-14) that displaces hydrogen from cold water and forms calcium hydroxide, a strong base.

Solution

$$Ca(s) + 2H_2O(\ell) \longrightarrow H_2(g) + Ca(OH)_2(aq)$$

$$Ca(s) + 2H_2O(\ell) \longrightarrow H_2(g) + [Ca^{2+}(aq) + 2OH^-(aq)]$$

$$Ca(s) + 2H_2O(\ell) \longrightarrow H_2(g) + Ca^{2+}(aq) + 2OH^-(aq)$$

You should now work Exercise 85.

Figure 4-4   Potassium, like other Group IA metals, reacts vigorously with water. For this photograph, the room was completely dark, and all the light you see here was produced by dropping a small piece of potassium into a beaker of water.

3 $\begin{bmatrix} \text{Active Nonmetal +} \\ \text{Salt of Less Active Nonmetal} \end{bmatrix} \longrightarrow \begin{bmatrix} \text{Less Active Nonmetal +} \\ \text{Salt of More Active Nonmetal} \end{bmatrix}$

Many *nonmetals* displace less active nonmetals from combination with a metal or other cation. For example, when chlorine is bubbled through a solution containing bromide ions

The displacement reaction of calcium with water at room temperature produces bubbles of hydrogen.

Activity of the halogens decreases going down the group in the periodic table.

(derived from a soluble ionic salt such as sodium bromide, NaBr), chlorine displaces bromide ions to form elemental bromine and chloride ions (as aqueous sodium chloride):

$$Cl_2(g) + 2[Na^+(aq) + Br^-(aq)] \longrightarrow 2[Na^+(aq) + Cl^-(aq)] + Br_2(\ell)$$

chlorine   sodium bromide   sodium chloride   bromine

Similarly, when bromine is added to a solution containing iodide ions, the iodide ions are displaced by bromine to form iodine and bromide ions:

$$Br_2(\ell) + 2[Na^+(aq) + I^-(aq)] \longrightarrow 2[Na^+(aq) + Br^-(aq)] + I_2(s)$$

bromine   sodium iodide   sodium bromide   iodine

Each halogen will displace less active (heavier) halogens from their binary salts; that is, the order of decreasing activities is

$$F_2 > Cl_2 > Br_2 > I_2$$

Conversely, a halogen will *not* displace more active (lighter) members from their salts:

$$I_2(s) + 2F^- \longrightarrow \text{no reaction}$$

EXAMPLE 4-9 *Displacement Reactions*

Which of the following combinations would result in a displacement reaction? Write balanced formula unit, total ionic, and net ionic equations for reactions that occur.

(a) $I_2(s)$ $+ NaBr(aq) \longrightarrow$

(b) $Cl_2(g) + NaI(aq) \longrightarrow$

(c) $Br_2(\ell) + NaCl(aq) \longrightarrow$

(a) Bromine, Br$_2$, in water (*pale orange*) is poured into an aqueous solution of NaI, the top layer in the cylinder. (b) Br$_2$ displaces I$^-$ from solution and forms solid iodine, I$_2$. The I$_2$ dissolves in water to give a brown solution but is more soluble in many organic liquids (*purple bottom layer*).

(a)

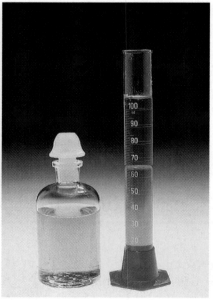

(b)

Plan

The activity of the halogens decreases from top to bottom in the periodic table. We see (a) that Br is above I and (c) that Cl is above Br in the periodic table; therefore, neither combination (a) nor combination (c) could result in reaction. Cl is above I in the periodic table, and so combination (b) results in a displacement reaction.

Solution

The more active halogen, Cl_2, displaces the less active halogen, I_2, from its compounds.

$$Cl_2(g) + 2NaI(aq) \longrightarrow I_2(s) + 2NaCl(aq)$$

$$Cl_2(g) + 2[Na^+(aq) + I^-(aq)] \longrightarrow I_2(s) + 2[Na^+(aq) + Cl^-(aq)]$$

$$Cl_2(g) + 2I^-(aq) \longrightarrow I_2(s) + 2Cl^-(aq)$$

You should now work Exercise 86.

4-11 METATHESIS REACTIONS

In many reactions between two compounds in aqueous solution, the positive and negative ions appear to "change partners" to form two new compounds, with no change in oxidation numbers. Such reactions are called **metathesis reactions.**

Pronounced "meh-*tath*-uh-sis." Metathesis reactions are also sometimes referred to as **double displacement reactions.**

We can represent such reactions by the following general equation, where A and B represent positive ions (cations) and X and Y represent negative ions (anions):

$$AX + BY \longrightarrow AY + BX$$

For example, when we mix silver nitrate and sodium chloride solutions, solid silver chloride is formed and sodium nitrate remains dissolved in water:

$$AgNO_3(aq) + NaCl(aq) \longrightarrow AgCl(s) + NaNO_3(aq)$$

Metathesis reactions result in the removal of ions from solution; this removal of ions can be thought of as the *driving force* for the reaction—the reason it occurs. The removal of ions can occur in three ways, which can be used to classify two types of metathesis reactions:

1. Formation of predominantly nonionized molecules (weak or nonelectrolytes) in solution; the most common such nonelectrolyte product is water

2. Formation of an insoluble solid, called a *precipitate* (which separates from the solution)

1 Acid–Base (Neutralization) Reactions: Formation of a Nonelectrolyte

Acid–base reactions are among the most important kinds of chemical reactions. Many acid–base reactions occur in nature in both plants and animals. Many acids and bases are essential compounds in an industrialized society (see Table 4-15). For example, approximately 300 pounds of sulfuric acid, H_2SO_4, and approximately 100 pounds of ammonia, NH_3, is required to support the lifestyle of an average American for one year.

The reaction of an acid with a metal hydroxide base produces a salt and water. Such reactions are called **neutralization reactions** because the typical properties of acids and bases are neutralized.

The manufacture of fertilizers consumes more H_2SO_4 *and* more NH_3 than any other single use.

TABLE 4-15	2001 Production of Inorganic Acids, Bases, and Salts in the United States		
Formula	**Name**	**Billions of Pounds**	**Major Uses**
H_2SO_4	sulfuric acid	80.11	Manufacture of fertilizers and other chemicals
CaO, $Ca(OH)_2$	lime (calcium oxide and calcium hydroxide)	41.23	Manufacture of other chemicals, steelmaking, water treatment
NH_3	ammonia	26.09	Fertilizer; manufacture of fertilizers and other chemicals
H_3PO_4	phosphoric acid	23.21	Manufacture of fertilizers
Na_2CO_3	sodium carbonate (soda ash)	22.71	Manufacture of glass, other chemicals, detergents, pulp, and paper
NaOH	sodium hydroxide	22.37	Manufacture of other chemicals, pulp and paper, soap and detergents, aluminum, textiles
HNO_3	nitric acid	15.65	Manufacture of fertilizers, explosives, plastics, and lacquers
NH_4NO_3	ammonium nitrate	14.20	Fertilizer and explosives
HCl	hydrochloric acid	8.84	Manufacture of other chemicals and rubber; metal cleaning
$(NH_4)_2SO_4$	ammonium sulfate	5.12	Fertilizer
TiO_2	titanium dioxide	2.93	Paints and coatings
KOH, K_2CO_3	potash	2.65	Manufacture of fertilizers
$Al_2(SO_4)_2$	aluminum sulfate	2.33	Water treatment, dyeing textiles
$NaClO_3$	sodium chlorate	1.81	Manufacture of other chemicals, explosives, plastics
Na_2SO_4	sodium sulfate	1.13	Manufacture of paper, glass, and detergents

When a base such as ammonia or an amine reacts with an acid, a salt, but no water, is formed. This is still called an acid–base, or neutralization, reaction.

In nearly all neutralization reactions, the driving force is the combination of $H^+(aq)$ from an acid and $OH^-(aq)$ from a base (or a base plus water) to form water molecules.

 CD-ROM Screen 5.10, Acid–Base Reactions

Consider the reaction of hydrochloric acid, HCl(aq), with aqueous sodium hydroxide, NaOH. Table 4-5 tells us that HCl is a strong acid, and Table 4-7 tells us that NaOH is a strong base. The salt sodium chloride, NaCl, is formed in this reaction. It contains the cation of its parent base, Na^+, and the anion of its parent acid, Cl^-. Solubility guidelines 2 and 4 tell us that NaCl is a soluble salt.

$$HCl(aq) + NaOH(aq) \longrightarrow H_2O(\ell) + NaCl(aq)$$
$$[H^+(aq) + Cl^-(aq)] + [Na^+(aq) + OH^-(aq)] \longrightarrow H_2O(\ell) + [Na^+(aq) + Cl^-(aq)]$$
$$H^+(aq) + OH^-(aq) \longrightarrow H_2O(\ell)$$

The net ionic equation for *all* reactions of strong acids with strong bases that form soluble salts and water is

$$H^+(aq) + OH^-(aq) \longrightarrow H_2O(\ell)$$

✓ **Problem-Solving Tip:** *Salt Formation*

The salt that is formed in a neutralization reaction is composed of the cation of the base and the anion of the acid. The salt may be soluble or insoluble. If our goal were to obtain the salt from the reaction of aqueous HCl with aqueous NaOH, we could evaporate the water and obtain solid NaCl.

EXAMPLE 4-10 *Neutralization Reactions*

Predict the products of the reaction between $HI(aq)$ and $Ca(OH)_2(aq)$. Write balanced formula unit, total ionic, and net ionic equations.

Plan

This is an acid–base neutralization reaction; the products are H_2O and the salt that contains the cation of the base, Ca^{2+}, and the anion of the acid, I^-; CaI_2 is a soluble salt (solubility guideline 4). HI is a strong acid (see Table 4-5), $Ca(OH)_2$ is a strong base (see Table 4-7), and CaI_2 is a soluble ionic salt, so all are written in ionic form.

Solution

$$2HI(aq) + Ca(OH)_2(aq) \longrightarrow CaI_2(aq) + 2H_2O(\ell)$$
$$2[H^+(aq) + I^-(aq)] + [Ca^{2+}(aq) + 2OH^-(aq)] \longrightarrow [Ca^{2+}(aq) + 2I^-(aq)] + 2H_2O(\ell)$$

We cancel the spectator ions.

$$2H^+(aq) + 2OH^-(aq) \longrightarrow 2H_2O(\ell)$$

Dividing by 2 gives the net ionic equation:

$$H^+(aq) + OH^-(aq) \longrightarrow H_2O(\ell)$$

You should now work Exercise 91.

Recall that in balanced equations we show the smallest whole-number coefficients possible.

Reactions of *weak* acids with strong bases also produce salts and water, but there is a significant difference in the balanced ionic equations because weak acids are only *slightly* ionized.

EXAMPLE 4-11 *Neutralization Reactions*

Write balanced formula unit, total ionic, and net ionic equations for the reaction of acetic acid with potassium hydroxide.

Plan

Neutralization reactions involving metal hydroxide bases produce a salt and water. CH_3COOH is a weak acid (see Table 4-6), so it is written as formula units. KOH is a strong base (see Table 4–7) and KCH_3COO is a soluble salt (solubility guidelines 2 and 3), so both are written in ionic form.

Solution

$$CH_3COOH(aq) + KOH(aq) \longrightarrow KCH_3COO(aq) + H_2O(\ell)$$
$$CH_3COOH(aq) + [K^+(aq) + OH^-(aq)] \longrightarrow [K^+(aq) + CH_3COO^-(aq)] + H_2O(\ell)$$

The spectator ion is K^+, the cation of the strong base, KOH.

$$CH_3COOH(aq) + OH^-(aq) \longrightarrow CH_3COO^-(aq) + H_2O(\ell)$$

Thus, we see that *this* net ionic equation includes *molecules* of the weak acid and *anions* of the weak acid.

You should now work Exercise 92.

A monoprotic acid contains one acidic H per formula unit.

> The reactions of *weak monoprotic acids* with *strong bases* that form *soluble salts* can be represented in general terms as
>
> $$HA(aq) + OH^-(aq) \longrightarrow A^-(aq) + H_2O(\ell)$$
>
> where HA represents the weak acid and A^- represents its anion.

EXAMPLE 4-12 *Salt Formation*

Write balanced formula unit, total ionic, and net ionic equations for an acid–base reaction that will produce the salt, barium chloride.

Plan

Neutralization reactions produce a salt. The salt contains the cation from the base and the anion from the acid. The base must therefore contain Ba^{2+}, that is, $Ba(OH)_2$, and the acid must contain Cl^-, that is, HCl. We write equations that represent the reaction between the strong base, $Ba(OH)_2$, and the strong acid, HCl.

Solution

$$2HCl(aq) + Ba(OH)_2(aq) \longrightarrow BaCl_2(aq) + 2H_2O(\ell)$$

$$2[H^+(aq) + Cl^-(aq)] + [Ba^{2+}(aq) + 2OH^-(aq)] \longrightarrow [Ba^{2+}(aq) + 2Cl^-(aq)] + 2H_2O(\ell)$$

We cancel the spectator ions.

$$2H^+(aq) + 2OH^-(aq) \longrightarrow 2H_2O(\ell)$$

Dividing by 2 gives the net ionic equation:

$$H^+(aq) + OH^-(aq) \longrightarrow H_2O(\ell)$$

The net ionic equation shows the driving force for this reaction. The formula unit equation shows the salt formed or that could be isolated if the water were evaporated.

You should now work Exercise 100.

---

2 Precipitation Reactions

To understand the discussion of precipitation reactions, you must know the solubility guidelines, Table 4-8.

In **precipitation reactions** an insoluble solid, a **precipitate,** forms and then settles out of solution. The driving force for these reactions is the strong attraction between cations and anions. This results in the removal of ions from solution by the formation of a precipitate. Our teeth and bones were formed by very slow precipitation reactions in which mostly calcium phosphate $Ca_3(PO_4)_2$ was deposited in the correct geometric arrangements.

An example of a precipitation reaction is the formation of bright yellow insoluble lead(II) chromate when we mix solutions of the soluble ionic compounds lead(II) nitrate and potassium chromate (Figure 4-5). The other product of the reaction is KNO_3, a soluble ionic salt.

 CD-ROM   Screen 5.7, Precipitation Reactions

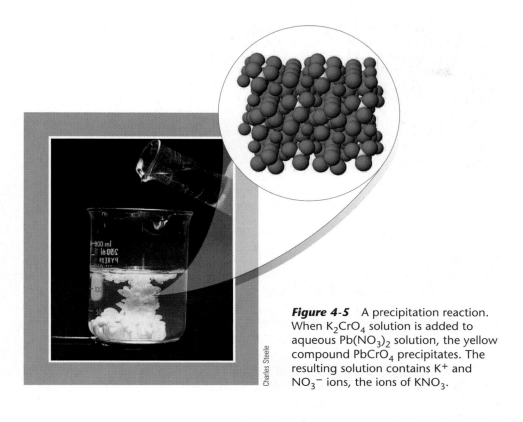

Figure 4-5 A precipitation reaction. When K_2CrO_4 solution is added to aqueous $Pb(NO_3)_2$ solution, the yellow compound $PbCrO_4$ precipitates. The resulting solution contains K^+ and NO_3^- ions, the ions of KNO_3.

The balanced formula unit, total ionic, and net ionic equations for this reaction follow:

$$Pb(NO_3)_2(aq) + K_2CrO_4(aq) \longrightarrow PbCrO_4(s) + 2KNO_3(aq)$$

$$[Pb^{2+}(aq) + 2\,NO_3^-(aq)] + [2K^+(aq) + CrO_4^{2-}(aq)] \longrightarrow$$
$$PbCrO_4(s) + 2[K^+(aq) + NO_3^-(aq)]$$

$$Pb^{2+}(aq) + CrO_4^{2-}(aq) \longrightarrow PbCrO_4(s)$$

Another important precipitation reaction involves the formation of insoluble carbonates (solubility guideline 7). Limestone deposits are mostly calcium carbonate, $CaCO_3$, although many also contain significant amounts of magnesium carbonate, $MgCO_3$.

Suppose we mix together aqueous solutions of sodium carbonate, Na_2CO_3, and calcium chloride, $CaCl_2$. We recognize that *both* Na_2CO_3 and $CaCl_2$ (solubility guidelines 2, 4a, and 7) are soluble ionic compounds. At the instant of mixing, the resulting solution contains four ions:

$$Na^+(aq), \quad CO_3^{2-}(aq), \quad Ca^{2+}(aq), \quad Cl^-(aq)$$

One pair of ions, Na^+ and Cl^-, *cannot* form an insoluble compound (solubility guidelines 2 and 4). We look for a pair of ions that could form an insoluble compound. Ca^{2+} ions and CO_3^{2-} ions are such a combination; they form insoluble $CaCO_3$ (solubility guideline 7). The equations for the reaction follow:

$$CaCl_2(aq) + Na_2CO_3(aq) \longrightarrow CaCO_3(s) + 2\,NaCl(aq)$$

$$[Ca^{2+}(aq) + 2\,Cl^-(aq)] + [2Na^+(aq) + CO_3^{2-}(aq)] \longrightarrow$$
$$CaCO_3(s) + 2[Na^+(aq) + Cl^-(aq)]$$

$$Ca^{2+}(aq) + CO_3^{2-}(aq) \longrightarrow CaCO_3(s)$$

Seashells, which are formed in very slow precipitation reactions, are mostly calcium carbonate ($CaCO_3$), a white compound. Traces of transition metal ions give them color.

SC*LINKS*.

TOPIC: Precipitation
Reactions
GO TO: www.scilinks.org
*sci*LINKS **CODE:** WCH0480

EXAMPLE 4-13   *Solubility Guidelines and Precipitation Reactions*

Will a precipitate form when aqueous solutions of $Ca(NO_3)_2$ and NaCl are mixed in reasonable concentrations? Write balanced formula unit, total ionic, and net ionic equations for any reaction.

Plan

We recognize that both $Ca(NO_3)_2$ (solubility guideline 3) and NaCl (solubility guidelines 2 and 4) are soluble compounds. We use the solubility guidelines to determine whether any of the possible products are insoluble.

Solution

At the instant of mixing, the resulting solution contains four ions:

$$Ca^{2+}(aq), \qquad NO_3^-(aq), \qquad Na^+(aq), \qquad Cl^-(aq)$$

New combinations of ions *could* be $CaCl_2$ and $NaNO_3$. But solubility guideline 4 tells us that $CaCl_2$ is a soluble compound, and solubility guidelines 2 and 3 tell us that $NaNO_3$ is a soluble compound. Therefore, no precipitate forms in this solution.

You should now work Exercise 104.

EXAMPLE 4-14   *Solubility Guidelines and Precipitation Reactions*

Will a precipitate form when aqueous solutions of $CaCl_2$ and K_3PO_4 are mixed in reasonable concentrations? Write balanced formula unit, total ionic, and net ionic equations for any reaction.

Plan

We recognize that both $CaCl_2$ (solubility guideline 4) and K_3PO_4 (solubility guideline 2) are soluble compounds. We use the solubility guidelines to determine whether any of the possible products are insoluble.

Solution

At the instant of mixing, the resulting solution contains four ions:

$$Ca^{2+}(aq), \qquad Cl^-(aq), \qquad K^+(aq), \qquad PO_4^{3-}(aq)$$

New combinations of ions *could* be KCl and $Ca_3(PO_4)_2$. Solubility guidelines 2 and 4 tell us that KCl is a soluble compound; solubility guideline 7 tells us that $Ca_3(PO_4)_2$ is an insoluble compound, so a precipitate of $Ca_3(PO_4)_2$ forms in this solution.

The equations for the formation of calcium phosphate follow:

$$3CaCl_2(aq) + 2K_3PO_4(aq) \longrightarrow Ca_3(PO_4)_2(s) + 6KCl(aq)$$

$$3[Ca^{2+}(aq) + 2Cl^-(aq)] + 2[3K^+(aq) + PO_4^{3-}(aq)] \longrightarrow$$
$$Ca_3(PO_4)_2(s) + 6[K^+(aq) + Cl^-(aq)]$$

$$3Ca^{2+}(aq) + 2PO_4^{3-}(aq) \longrightarrow Ca_3(PO_4)_2(s)$$

You should now work Exercise 110.

4-12 SUMMARY OF REACTION TYPES

Table 4-16 summarizes the reaction types we have presented. Remember that a reaction might be classified in more than one category.

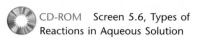

CD-ROM   Screen 5.6, Types of Reactions in Aqueous Solution

EXAMPLE 4-15   *Classifying Reactions*

Classify each of the following reactions.

(a) $Zn(s) + AgNO_3(aq) \longrightarrow Zn(NO_3)_2(aq) + 2Ag(s)$

(b) $Ca(OH)_2(s) \xrightarrow{\text{heat}} CaO(s) + H_2O(g)$

(c) $2HI(g) \xrightarrow{\text{heat}} H_2(g) + I_2(g)$

(d) $Cu(NO_3)_2(aq) + Na_2S(aq) \longrightarrow CuS(s) + 2NaNO_3(aq)$

(e) $SO_2(g) + H_2O(\ell) \longrightarrow H_2SO_3(aq)$

(f) $H_2SO_3(aq) + 2KOH(aq) \longrightarrow K_2SO_3(aq) + 2H_2O(\ell)$

Plan

We identify each reaction type by its characteristics, using Table 4-16 and the appropriate sections as a guide.

Solution

(a) One element, Zn, displaces another, Ag, from a compound; this is a displacement reaction.

(b) A single compound breaks apart into two compounds; this is a decomposition reaction.

(c) A single compound breaks apart into two elements; this is another decomposition reaction. However, now there are changes in oxidation numbers; H changes from +1 in HI to 0 in H_2, and I changes from −1 in HI to 0 in I_2. So this is also an oxidation–reduction (redox) reaction.

(d) The positive and negative ions in the two reactant compounds change partners; this is a metathesis reaction. An insoluble product, CuS(s), is formed, so the reaction is a precipitation reaction.

(e) Two compounds combine to form a single product; this is a combination reaction.

(f) The positive and negative ions change partners; this is a metathesis reaction. An acid and a base react to form a salt and water; this is an acid–base (neutralization) reaction.

You should now work Exercises 113 through 121.

TABLE 4-16	*Summary and Examples of Reaction Types*

Section	Reaction Type, Examples	Characteristics
4-7	Oxidation–Reduction (redox)	Oxidation numbers (Section 4-4) of some elements change; at least one element must increase and at least one must decrease in oxidation number
4-8	Combination 1. element + element → compound $2Al(s) + 3Cl_2(g) → 2AlCl_3(s)^*$ $P_4(s) + 10Cl_2(g) → 4PCl_5(s)^*$ 2. compound + element → compound $SF_4(g) + F_2(g) → SF_6(g)^*$ $2SO_2(g) + O_2(g) → 2SO_3(\ell)^*$ 3. compound + compound → compound $CaO(s) + CO_2(g) → CaCO_3(s)$ $Na_2O(s) + H_2O(\ell) → 2NaOH(aq)$	More than one reactant, single product
4-9	Decomposition 1. compound → element + element $2HgO(s) → 2Hg(g) + O_2(g)^*$ $2H_2O(\ell) → 2H_2(g) + O_2(g)^*$ 2. compound → compound + element $2NaNO_3(s) → 2NaNO_2(s) + O_2(g)^*$ $2H_2O_2(\ell) → 2H_2O(\ell) + O_2(g)^*$ 3. compound → compound + compound $CaCO_3(s) → CaO(s) + CO_2(g)$ $Mg(OH)_2(s) → MgO(s) + H_2O(\ell)$	Single reactant, more than one product
4-10	Displacement $Zn(s) + CuSO_4(aq) → Cu(s) + ZnSO_4(aq)^*$ $Zn(s) + H_2SO_4(aq) → H_2(g) + ZnSO_4(aq)^*$ $Cl_2(g) + 2NaI(aq) → I_2(s) + 2NaCl(aq)^*$	One element displaces another from a compound: Element + compound → element + compound Activity series (Table 4-14) summarizes metals and hydrogen; halogen activities (Group VIIA) decrease going down the group
4-11	Metathesis	Positive and negative ions in two compounds appear to "change partners" to form two new compounds; no change in oxidation numbers
	1. acid–base neutralization $HCl(aq) + NaOH(aq) → NaCl(aq) + H_2O(\ell)$ $CH_3COOH(aq) + KOH(aq) → KCH_3COO(aq) + H_2O(\ell)$ $HCl(aq) + NH_3(aq) → NH_4Cl(aq)$ $2H_3PO_4(aq) + 3Ca(OH)_2(aq) → Ca_3(PO_4)_2(s) + 6H_2O(\ell)^†$	Product is a salt; water is often formed
	2. precipitation $CaCl_2(aq) + Na_2CO_3(aq) → CaCO_3(s) + 2NaCl(aq)$ $Pb(NO_3)_2(aq) + K_2CrO_4(aq) → PbCrO_4(s) + 2KNO_3(aq)$ $2H_3PO_4(aq) + 3Ca(OH)_2(aq) → Ca_3(PO_4)_2(s) + 6H_2O(\ell)^†$	Products include an insoluble substance, which precipitates from solution as a solid; solubility guidelines assist in predicting, recognizing

*These examples are also oxidation–reduction (redox) reactions.

†This reaction is both an acid–base neutralization reaction and a precipitation reaction.

Key Terms

Acid A substance that produces $H^+(aq)$ ions in aqueous solution. Strong acids ionize completely or almost completely in dilute aqueous solution. Weak acids ionize only slightly.

Acid–base reaction See *Neutralization reaction.*

Active metal A metal that readily loses electrons to form cations.

Activity series A listing of metals (and hydrogen) in order of decreasing activity.

Alkali metals Elements of Group IA in the periodic table, except hydrogen.

Alkaline earth metals Group IIA elements in the periodic table.

Atomic number The number of protons in the nucleus of an atom of an element.

Base A substance that produces $OH^-(aq)$ ions in aqueous solution. Strong bases are soluble in water and are completely *dissociated*. Weak bases ionize only slightly.

Binary acid A binary compound in which H is bonded to a nonmetal in Group VIIA or a nonmetal other than oxygen in Group VIA.

Binary compound A compound consisting of two elements; may be ionic or molecular.

Chemical periodicity The variation in properties of elements with their positions in the periodic table.

Combination reaction Reaction in which two substances (elements or compounds) combine to form one compound.

Decomposition reaction Reaction in which a compound decomposes to form two or more products (elements, compounds, or some combination of these).

Displacement reaction A reaction in which one element displaces another from a compound.

Disproportionation reaction A redox reaction in which the oxidizing agent and the reducing agent are the same element.

Dissociation In aqueous solution, the process in which a solid *ionic compound* separates into its ions.

Electrolyte A substance whose aqueous solutions conduct electricity.

Formula unit equation An equation for a chemical reaction in which all formulas are written as complete formulas.

Group (family) The elements in a vertical column of the periodic table.

Halogens Group VIIA elements in the periodic table.

Ionization In aqueous solution, the process in which a *molecular compound* separates to form ions.

Metal An element below and to the left of the stepwise division (metalloids) of the periodic table; about 80% of the known elements are metals.

Metalloids Elements with properties intermediate between metals and nonmetals: B, Si, Ge, As, Sb, Te, Po, and At.

Metathesis reaction A reaction in which the positive and negative ions in two compounds "change partners," with no change in oxidation numbers, to form two new compounds.

Net ionic equation An equation that results from canceling spectator ions from a total ionic equation.

Neutralization reaction The reaction of an acid with a base to form a salt. Often, the reaction of hydrogen ions with hydroxide ions to form water molecules.

Noble (rare) gases Elements of Group VIIIA in the periodic table.

Nonelectrolyte A substance whose aqueous solutions do not conduct electricity.

Nonmetals Elements above and to the right of the metalloids in the periodic table.

Oxidation An increase in oxidation number; corresponds to a loss of electrons.

Oxidation numbers Arbitrary numbers that can be used as mechanical aids in writing formulas and balancing equations; for single-atom ions they correspond to the charge on the ion; less metallic atoms are assigned negative oxidation numbers in compounds and polyatomic ions.

Oxidation–reduction reaction A reaction in which oxidation and reduction occur; also called a redox reaction.

Oxidation states See *Oxidation numbers.*

Oxidizing agent The substance that oxidizes another substance and is reduced.

Oxoacid See *Ternary acid.*

Period The elements in a horizontal row of the periodic table.

Periodic law The properties of the elements are periodic functions of their atomic numbers.

Periodic table An arrangement of elements in order of increasing atomic number that also emphasizes periodicity.

Periodicity Regular periodic variations of properties of elements with atomic number (and position in the periodic table).

Precipitate An insoluble solid that forms and separates from a solution.

Precipitation reaction A reaction in which a solid (precipitate) forms.

Pseudobinary ionic compound A compound that contains more than two elements but is named like a binary compound.

Redox reaction See *Oxidation–reduction reaction.*

Reducing agent The substance that reduces another substance and is oxidized.

Reduction A decrease in oxidation number; corresponds to a gain of electrons.

Reversible reaction A reaction that occurs in both directions; described with double arrows ($\rightleftharpoons$).

Salt A compound that contains a cation other than H^+ and an anion other than OH^- or O^{2-}.

Semiconductor A substance that does not conduct electricity at low temperatures but does so at higher temperatures.

Spectator ions Ions that appear in solution on both sides of the total ionic equation; they undergo no change in the chemical reaction.

Strong acid An acid that ionizes (separates into ions) completely, or very nearly completely, in dilute aqueous solution.

Strong electrolyte A substance that conducts electricity well in dilute aqueous solution.

Strong base Metal hydroxide that is soluble in water and dissociates completely in dilute aqueous solution.

Ternary acid A ternary compound containing H, O, and another element, usually a nonmetal.

Ternary compound A compound consisting of three elements; may be ionic or molecular.

Ternary salt A salt resulting from replacing the hydrogen in a ternary acid with another ion.

Total ionic equation An equation for a chemical reaction written to show the predominant form of all species in aqueous solution or in contact with water.

Weak acid An acid that ionizes only slightly in dilute aqueous solution.

Weak base A molecular substance that ionizes only slightly in water to produce an alkaline (base) solution.

Weak electrolyte A substance that conducts electricity poorly in dilute aqueous solution.

Exercises

The Periodic Table

1. State the periodic law. What does it mean?

2. What was Mendeleev's contribution to the construction of the modern periodic table?

3. Consult a handbook of chemistry, and look up melting points of the elements of Periods 2 and 3. Show that melting point is a property that varies periodically for these elements.

*4. Mendeleev's periodic table was based on increasing atomic weight. Argon has a higher atomic weight than potassium, yet in the modern table argon appears before potassium. Explain how this can be.

5. Estimate the density of antimony from the following densities (g/cm^3): As, 5.72; Bi, 9.8; Sn, 7.30; Te, 6.24. Show how you arrived at your answer. Using a reference other than your textbook, look up the density of antimony. How does your predicted value compare with the reported value?

Product made from antimony

6. Estimate the density of selenium from the following densities (g/cm^3): S, 2.07; Te, 6.24; As, 5.72; Br, 3.12. Show how you arrived at your answer. Using a reference other than your textbook, look up the density of selenium. How does your predicted value compare with the reported value?

7. Estimate the specific heat of antimony from the following specific heats (J/g°C): As, 0.34; Bi, 0.14; Sn, 0.23; Te, 0.20. Show how you arrived at your answer.

8. Given the following melting points in °C, estimate the value for CBr$_4$: CF$_4$, −184; CCl$_4$, −23; CI$_4$, 171 (decomposes). Using a reference other than your textbook, look up the melting point of CBr$_4$. How does your predicted value compare with the reported value?

9. Calcium and magnesium form the following compounds: CaCl$_2$, MgCl$_2$, CaO, MgO, Ca$_3$N$_2$, and Mg$_3$N$_2$. Predict the formula for a compound of (a) magnesium and sulfur, (b) barium and iodine.

10. The formulas of some hydrides of second-period representative elements are as follows: BeH$_2$, BH$_3$, CH$_4$, NH$_3$, H$_2$O, HF. A famous test in criminology laboratories for the presence of arsenic (As) involves the formation of arsine, the hydride of arsenic. Predict the formula of arsine.

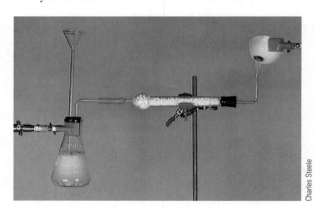

Arsine burns to form a dark spot.

11. Clearly distinguish between the following terms and provide specific examples of each: groups (families) of elements, and periods of elements.

12. Write names and symbols for (a) the alkaline earth metals, (b) the Group IVA elements, (c) the Group VIB elements.

13. Write names and symbols for (a) the alkali metals, (b) the noble gases, (c) the Group IIIA elements.

14. Clearly and concisely define the following terms and

provide examples of each: (a) metals, (b) nonmetals, (c) halogens.

Aqueous Solutions

15. Define and distinguish among (a) strong electrolytes, (b) weak electrolytes, and (c) nonelectrolytes.
16. Three common classes of compounds are electrolytes. Name them and give an example of each.
17. Define (a) acids, (b) bases, (c) salts, and (d) molecular compounds.
18. How can a salt be related to a particular acid and a particular base?
19. List the names and formulas of seven strong acids.
20. Write equations for the ionization of the following acids: (a) hydrochloric acid, (b) nitric acid, (c) perchloric acid.
21. List the names and formulas of five weak acids.
22. List the names and formulas of the common soluble strong bases.
23. Write equations for the ionization of the following acids. Which ones ionize only slightly? (a) HF, (b) H_2SO_4, (c) CH_3COOH, (d) HNO_2.
24. The most common weak base is present in a common household chemical. Write the equation for the ionization of this weak base.
25. Summarize the electrical properties of strong electrolytes, weak electrolytes, and nonelectrolytes.
26. What is the difference between ionization and dissociation in aqueous solution?
27. List the names and formulas of five insoluble bases.
28. Which of the following are strong electrolytes? Weak electrolytes? Nonelectrolytes? (a) Na_2S, (b) $Ba(OH)_2$, (c) CH_3OH, (d) HCN, (e) $Al(NO_3)_3$.
29. Classify the following as strong electrolytes, weak electrolytes, or nonelectrolytes: (a) $NaClO_4$, (b) $HClO_2$, (c) CH_3CH_2OH, (d) CH_3COOH, (e) HNO_3.
30. Write the formulas of two soluble and two insoluble chlorides, sulfates, and hydroxides.
31. Describe an experiment for classifying each of these compounds as a strong electrolyte, a weak electrolyte, or a nonelectrolyte: Na_2CO_3, HCN, CH_3COOH, H_2S, H_2SO_4, NH_3. Predict and explain the expected results.
32. (a) Which of these are acids? HI, NH_3, H_2SeO_4, BF_3, $Fe(OH)_3$, H_2S, C_6H_6, CsOH, H_3PO_3, HCN. (b) Which of these are bases? NaOH, H_2Se, BCl_3, NH_3.
*33. Classify each substance as either an electrolyte or a nonelectrolyte: NH_4Cl, HI, C_6H_6, $Zn(CH_3COO)_2$, $Cu(NO_3)_2$, CH_3CH_2OH, $C_{12}H_{22}O_{11}$ (sugar), LiOH, $KHCO_3$, CCl4, $La_2(SO_4)_3$, I_2.
*34. Classify each substance as either a strong or weak electrolyte, and then list (a) the strong acids, (b) the strong bases, (c) the weak acids, and (d) the weak bases. NaBr, $MgSO_4$, HCl, $H_2C_2O_4$, $Ba(NO_3)_2$, H_3PO_4, CsOH, HNO_3, HI, $Ba(OH)_2$, LiOH, H_2O, NH_3, KOH, $Mg(CH_3COO)_2$, HCN, $HClO_4$.

35. Classify each substance as soluble, moderately soluble, or insoluble. Ag_2SO_4, $(NH_4)_2CO_3$, AgCl, $HgBr_2$
36. Classify each substance as soluble, moderately soluble, or insoluble. $Ca(CH_3COO)_2$, NH_4Cl, $AgNO_3$, $PbCl_2$
37. What are reversible reactions? Give some examples.
38. Many household "chemicals" are acidic or basic. List a few of each kind.
39. Some chemical reactions reach an equilibrium rather than going to completion. What is "equal" in such an equilibrium?
40. Vinegar is 5% acetic acid, an organic acid, by mass. Many organic acids occur in living systems. What conclusion can be drawn from this information as to the strengths of organic acids?

Oxidation Numbers

41. Assign oxidation numbers to the element specified in each group of compounds. (a) N in NO, N_2O_3, N_2O_4, NH_4Cl, N_2H_4, NH_2OH, HNO_2, HNO_3 (b) C in CO, CO_2, CH_2O, CH_4O, C_2H_6O, $(COOH)_2$, Na_2CO_3, C_6H_6 (c) S in S^{2-}, SO_3^{2-}, SO_4^{2-}, $S_2O_3^{2-}$, $S_4O_6^{2-}$, HS
42. Assign oxidation numbers to the element specified in each group of compounds. (a) P in PCl_5, P_4O_6, P_4O_{10}, HPO_3, H_3PO_3, $POCl_3$, $H_4P_2O_7$, $Mg_3(PO_4)_2$ (b) Br in Br^-, BrO^-, BrO_2^-, BrO_3^-, BrO_4^- (c) Mn in MnO, MnO_2, $Mn(OH)_2$, K_2MnO_4, $KMnO_4$, Mn_2O_7 (d) O in OF_2, Na_2O, Na_2O_2, KO_2.
43. Assign oxidation numbers to the element specified in each group of ions. (a) S in S_8, H_2S, SO_2, SO_3, Na_2SO_3, H_2SO_4, K_2SO_4 (b) Cr in CrO_2^-, $Cr(OH)_4^-$, CrO_4^-, $Cr_2O_7^{2-}$ (c) B in BO_2^-, BO_3^{3-}, $B_4O_7^{2-}$.
44. Assign oxidation numbers to the element specified in each group of ions. (a) N in N^{3-}, NO_2^-, NO_3^-, N_3^-, NH_4^- (b) Cl in Cl_2, HCl, HClO, $HClO_2$, $KClO_3$, Cl_2O_7, $Ca(ClO_4)_2$, PCl_5^-.

Naming Compounds

45. Name the following common anions using the IUPAC system of nomenclature: (a) NO_3^-; (b) SO_4^{2-}; (c) ClO_3^-; (d) CH_3COO^-; (e) PO_4^{2-}.
46. Name the following monatomic cations using the IUPAC system of nomenclature: (a) Na^+, (b) Au^{3+}, (c) Ca^{2+}, (d) Zn^{2+}, (e) Ag^+.
47. Write the chemical symbol for each of the following: (a) sodium ion, (b) iron(II) ion, (c) silver ion, (d) mercury(II) ion, (e) bismuth(III) ion.
48. Write the chemical formula for each of the following: (a) chloride ion, (b) hydrogen sulfide ion, (c) telluride ion, (d) oxide ion, (e) nitrite ion.
49. Name the following ionic compounds: (a) Li_2S, (b) SnO_2, (c) RbBr, (d) K_2O, (e) Ba_3N_2.
50. Name the following ionic compounds: (a) CuI_2, (b) Hg_2Cl_2, (c) Li_3N, (d) $MnCl_2$, (e) $CuCO_3$, (f) FeO.

51. Write the chemical formula for each of the following compounds: (a) lithium fluoride, (b) zinc oxide, (c) barium oxide, (d) magnesium bromide, (e) hydrogen cyanide, (f) copper(I) chloride.

52. Write the chemical formula for each of the following compounds: (a) copper(II) chlorite, (b) potassium nitrate, (c) barium phosphate, (d) copper(I) sulfate, (e) sodium sulfite.

53. What is the name of the acid with the formula H_2CO_3? Write the formulas of the two anions derived from it and name these ions.

54. What is the name of the acid with the formula H_3PO_3? What is the name of the HPO_3^{2-} ion?

55. Name the following binary molecular compounds: (a) NO_2, (b) CO_2, (c) SF_6, (d) $SiCl_4$, (e) IF.

56. Name the following binary molecular compounds: (a) AsF_3, (b) Br_2O, (c) BrF_5, (d) CSe_2, (e) Cl_2O_7.

57. Write the chemical formula for each of the following compounds: (a) iodine bromide, (b) silicon dioxide, (c) phosphorus trichloride, (d) tetrasulfur dinitride, (e) bromine trifluoride, (f) hydrogen telluride, (g) xenon tetrafluoride.

58. Write the chemical formula for each of the following compounds: (a) diboron trioxide, (b) dinitrogen pentasulfide, (c) phosphorus triiodide, (d) sulfur tetrachloride, (e) silicon sulfide, (f) hydrogen sulfide, (g) tetraphosphorus hexoxide.

59. Write formulas for the compounds that are expected to be formed by the following pairs of ions:

	A. Cl^-	B. OH^-	C. SO_4^{2-}	D. PO_4^{3-}	E. NO_3^-
1. NH_4^+		Omit – see note			
2. Na^+					
3. Mg^{2+}					
4. Ni^{2+}					
5. Fe^{3+}					
6. Ag^+					

NOTE: The compound NH_4OH does not exist. The solution commonly labeled "NH_4OH" is aqueous ammonia, $NH_3(aq)$.

60. Write the names for t he compounds of Exercise 59.

61. Write balanced chemical equations for each of the following processes: (a) Calcium phosphate reacts with sulfuric acid to produce calcium sulfate and phosphoric acid. (b) Calcium phosphate reacts with water containing dissolved carbon dioxide to produce calcium hydrogen carbonate and calcium hydrogen phosphate.

62. Write balanced chemical equations for each of the following processes: (a) When heated, nitrogen and oxygen combine to form nitrogen oxide. (b) Heating a mixture of lead(II) sulfide and lead(II) sulfate produces metallic lead and sulfur dioxide.

Oxidation–Reduction Reactions

63. Define and provide examples of the following terms: (a) oxidation, (b) reduction, (c) oxidizing agent, (d) reducing agent.

64. Why must oxidation and reduction always occur simultaneously in chemical reactions?

65. Determine which of the following are oxidation–reduction reactions. For those that are, identify the oxidizing and reducing agents.
    (a) $3Zn(s) + 2CoCl_3(aq) \rightarrow 3ZnCl_2(aq) + 2Co(s)$
    (b) $ICl(s) + H_2O(\ell) \rightarrow HCl(aq) + HIO(aq)$
    (c) $3HCl(aq) + HNO_3(aq) \rightarrow Cl_2(g) + NOCl(g) + 2H_2O(\ell)$
    (d) $Fe_2O_3(s) + 3CO(g) \xrightarrow{heat} 2Fe(s) + 3CO_2(g)$

66. Determine which of the following are oxidation–reduction reactions. For those that are, identify the oxidizing and reducing agents.
    (a) $HgCl_2(aq) + 2KI(aq) \rightarrow HgI_2(s) + 2KCl(aq)$
    (b) $4NH_3(g) + 3O_2(g) \rightarrow 2N_2(g) + 6H_2O(g)$
    (c) $CaCO_3(s) + 2HNO_3(aq) \rightarrow$
    $ Ca(NO_3)_2(aq) + CO_2(g) + H_2O(\ell)$
    (d) $PCl_3(\ell) + 3H_2O(\ell) \rightarrow 3HCl(aq) + H_3PO_3(aq)$

67. Write balanced formula unit equations for the following redox reactions:
    (a) aluminum reacts with sulfuric acid, H_2SO_4, to produce aluminum sulfate, $Al_2(SO_4)_3$, and hydrogen
    (b) nitrogen, N_2, reacts with hydrogen, H_2, to form ammonia, NH_3
    (c) zinc sulfide, ZnS, reacts with oxygen, O_2, to form zinc oxide, ZnO, and sulfur dioxide, SO_2
    (d) carbon reacts with nitric acid, HNO_3, to produce nitrogen dioxide, NO_2, carbon dioxide, CO_2, and water
    (e) sulfuric acid reacts with hydrogen iodide, HI, to produce sulfur dioxide, SO_2, iodine, I_2, and water.

68. Identify the oxidizing agents and reducing agents in the oxidation-reduction reactions given in Exercise 67.

69. Write total ionic and net ionic equations for the following redox reactions occurring in aqueous solution or in contact with water:
    (a) $Fe + 2HCl \rightarrow FeCl_2 + H_2$
    (b) $2KMnO_4 + 16HCl \rightarrow$
    $ 2MnCl_2 + 2KCl + 5Cl_2 + 8H_2O$
    (*Note*: $MnCl_2$ is water-soluble.)
    (c) $4Zn + 10HNO_3 \rightarrow 4Zn(NO_3)_2 + NH_4NO_2 + 3H_2O$.

70. Write total ionic and net ionic equations for the following unbalanced redox reactions occurring in aqueous solution or in contact with water:
    (a) $H_2(g) + Al_2O_3(s) \rightarrow Al(s) + H_2O(\ell)$
    (b) $Zn(s) + CoCl_3(aq) \rightarrow ZnCl_2(aq) + Co(s)$
    (c) $Fe_2O_3(s) + CO(g) \rightarrow Fe(s) + CO_2(g)$.

Combination Reactions

71. Write balanced equations that show the combination reactions of the following Group IA metals combining with the Group VIIA nonmetals. (a) Li and Cl_2; (b) K and F_2; (c) Na and I_2

72. Write balanced equations that show the combination reactions of the following Group IIA metals and Group VIIA nonmetals. (a) Be and F_2, (b) Ca and Br_2, (c) Ba and Cl_2

In Exercises 73 and 74, some combination reactions are described by words. Write the balanced chemical equation for each, and assign oxidation numbers to elements other than H and O.

73. (a) Antimony reacts with a limited amount of chlorine to form antimony(III) chloride. (b) Antimony(III) chloride reacts with excess chlorine to form antimony(V) chloride. (c) Carbon burns in a limited amount of oxygen to form carbon monoxide.

74. (a) Sulfur trioxide reacts with aluminum oxide to form aluminum sulfate. (b) Dichlorine heptoxide reacts with water to form perchloric acid. (c) When cement "sets," the main reaction is the combination of calcium oxide with silicon dioxide to form calcium silicate, $CaSiO_3$.

Decomposition Reactions

In Exercises 75 and 76, write balanced formula unit equations for the reactions described by words. Assign oxidation numbers to all elements.

75. (a) Hydrogen peroxide, H_2O_2, is used as an antiseptic. Blood causes it to decompose into water and oxygen; (b) When heated, ammonium nitrate can decompose explosively to form nitrogen oxide and steam.

A model of hydrogen peroxide

76. (a) A "classroom volcano" is made by heating solid ammonium dichromate, $(NH_4)_2Cr_2O_7$, which decomposes into nitrogen, chromium(III) oxide, and steam. (b) At high temperatures, sodium nitrate (a fertilizer) forms sodium nitrite and oxygen.

Ammonium Dichromate Volcano

Displacement Reactions

77. Which of the following would displace hydrogen when a piece of the metal is dropped into dilute H_2SO_4 solution? Write balanced total ionic and net ionic equations for the reactions: Zn, Cu, Sn, Al.

78. Which of the following metals would displace copper from an aqueous solution of copper(II) sulfate? Write balanced total ionic and net ionic equations for the reactions: Hg, Zn, Fe, Ni.

79. Arrange the metals listed in Exercise 77 in order of increasing activity.

80. Arrange the metals listed in Exercise 78 in order of increasing activity.

81. Which of the following metals would displace hydrogen from cold water? Write balanced net ionic equations for the reactions: Ag, Na, Ca, Cr.

82. Arrange the metals listed in Exercise 81 in order of increasing activity.

83. What is the order of decreasing activity of the halogens?

84. Name five elements that will react with steam but not with cold water.

85. (a) Name two common metals: one that *does not* displace hydrogen from water, and one that *does not* displace hydrogen from water or acid solutions. (b) Name two common metals: one that *does* displace hydrogen from water, and one that displaces hydrogen from acid solutions but not from water. Write net ionic equations for the reactions that occur.

86. Of the possible displacement reactions shown, which one(s) could occur?
(a) $2Cl^-(aq) + Br_2(\ell) \rightarrow 2Br^-(aq) + Cl_2(g)$
(b) $2Br_2(aq) + F_2(g) \rightarrow 2F^-(aq) + Br_2(\ell)$
(c) $2I^-(aq) + Cl_2(g) \rightarrow 2Cl^-(aq) + I_2(s)$
(d) $2Br^-(aq) + I_2(s) \rightarrow 2I^-(aq) + Br_2(\ell)$

87. Use the activity series to predict whether or not the following reactions will occur:
(a) $Cu(s) + Mg^{2+} \rightarrow Mg(s) + Cu^{2+}$

(b) $Ni(s) + Cu^{2+} \rightarrow Ni^{2+} + Cu(s)$
(c) $Cu(s) + 2H^+ \rightarrow Cu^{2+} + H_2(g)$
(d) $Mg(s) + H_2O(g) \rightarrow MgO(s) + H_2(g)$

88. Repeat Exercise 87 for
(a) $Sn(s) + Ca^{2+} \rightarrow Sn^{2+} + Ca(s)$
(b) $Al_2O_3(s) + 3H_2(g) \xrightarrow{\text{heat}} 2Al(s) + 3H_2O(g)$
(c) $Cu(s) + 2H^+ \rightarrow Cu^{2+} + H_2(g)$
(d) $Cu(s) + Pb^{2+} \rightarrow Cu^{2+} + Pb(s)$

Metathesis Reactions

Exercises 89 and 90 describe precipitation reactions *in aqueous solutions*. For each, write balanced (i) formula unit, (ii) total ionic, and (iii) net ionic equations. Refer to the solubility guidelines as necessary.

89. (a) Black-and-white photographic film contains some silver bromide, which can be formed by the reaction of sodium bromide with silver nitrate. (b) Barium sulfate is used when x-rays of the gastrointestinal tract are made. Barium sulfate can be prepared by reacting barium chloride with dilute sulfuric acid. (c) In water purification systems small solid particles are often "trapped" as aluminum hydroxide precipitates and fall to the bottom of the sedimentation pool. Aluminum sulfate reacts with calcium hydroxide (from lime) to form aluminum hydroxide and calcium sulfate.

90. (a) Our bones are mostly calcium phosphate. Calcium chloride reacts with potassium phosphate to form calcium phosphate and potassium chloride. (b) Mercury compounds are very poisonous. Mercury(II) nitrate reacts with sodium sulfide to form mercury(II) sulfide, which is very insoluble, and sodium nitrate. (c) Chromium(III) ions are very poisonous. They can be removed from solution by precipitating very insoluble chromium(III) hydroxide. Chromium(III) chloride reacts with calcium hydroxide to form chromium-(III) hydroxide and calcium chloride.

In Exercises 91 through 94, write balanced (i) formula unit, (ii) total ionic, and (iii) net ionic equations for the reactions that occur between the acid and the base. Assume that all reactions occur in water or in contact with water.

91. (a) hydrochloric acid + calcium hydroxide
(b) dilute sulfuric acid + strontium hydroxide
(c) perchloric acid + aqueous ammonia

92. (a) acetic acid + potassium hydroxide
(b) sulfurous acid + sodium hydroxide
(c) hydrofluoric acid + sodium hydroxide

*93. (a) potassium hydroxide + hydrosulfuric acid
(b) potassium hydroxide + hydrochloric acid
(c) lead(II) hydroxide + hydrosulfuric acid

94. (a) lithium hydroxide + sulfuric acid
(b) calcium hydroxide + phosphoric acid
(c) copper(II) hydroxide + nitric acid

In Exercises 95 through 98, write balanced (i) formula unit, (ii) total ionic, and (iii) net ionic equations for the reaction of an acid and a base that will produce the indicated salts.

95. (a) sodium chloride, (b) sodium phosphate, (c) barium nitrate

96. (a) calcium perchlorate, (b) ammonium sulfate, (c) copper(II) acetate

*97. (a) sodium carbonate, (b) barium carbonate, (c) nickel(II) acetate

*98. (a) sodium sulfide, (b) aluminum phosphate, (c) lead(II) acetate

99. (a) Propose a definition for salts, as a class of compounds, on the basis of how they are formed. (b) Provide an example, in the form of a chemical reaction, to illustrate your definition of salts.

100. We can tell from the formula of a salt how it can be produced. Write a balanced chemical equation for the production of each of the following salts: (a) magnesium nitrate, (b) aluminum sulfite, (c) potassium carbonate, (d) zinc chlorate, (e) lithium acetate.

101. Magnesium hydroxide is a gelatinous material that forms during the water purification process in some water treatment plants because of magnesium ions in the water. (a) Write the chemical equation for the reaction of hydrochloric acid with magnesium hydroxide. (b) Explain what drives this reaction to completion.

Precipitation Reactions

102. A common test for the presence of chloride ions is the formation of a heavy, white precipitate when a solution of silver nitrate is added. (a) Write the balanced chemical equation for the production of silver chloride from silver nitrate solution and calcium chloride solution. (b) Explain why this reaction goes to completion.

103. Based on the solubility guidelines given in Table 4-8, how would you write the formulas for the following substances in a total ionic equation? (a) $PbSO_4$, (b) $Na(CH_3COO)$, (c) Na_2CO_3, (d) MnS, (e) $BaCl_2$.

104. Repeat Exercise 103 for the following: (a) $(NH_4)_2SO_4$, (b) $NaBr$, (c) $SrCl_2$, (d) MgF_2, (e) Na_2CO_3.

Refer to the solubility guidelines given in Table 4-8. Classify the compounds in Exercises 105 through 108 as soluble, moderately soluble, or insoluble in water.

105. (a) $NaClO_4$, (b) $AgCl$, (c) $Pb(NO_3)_2$, (d) KOH, (e) $CaSO_4$

106. (a) $BaSO_4$, (b) $Al(NO_3)_3$, (c) CuS, (d) Na_3AsO_4, (e) $Ca(CH_3COO)_2$

107. (a) $Fe(NO_3)_3$, (b) $Hg(CH_3COO)_2$, (c) $BeCl_2$, (d) $CuSO_4$, (e) $CaCO_3$

108. (a) $KClO_3$, (b) NH_4Cl, (c) NH_3, (d) HNO_2, (e) PbS

In Exercises 109 and 110, write balanced (i) formula unit, (ii) total ionic, and (iii) net ionic equations for the reactions that occur when *aqueous solutions* of the compounds are mixed.

109. (a) $Ba(NO_3)_2 + K_2CO_3 \rightarrow$
     (b) $NaOH + NiCl_2 \rightarrow$
     (c) $Al_2(SO_4)_3 + NaOH \rightarrow$
110. (a) $Cu(NO_3)_2 + Na_2S \rightarrow$
     (b) $CdSO_4 + H_2S \rightarrow$
     (c) $Bi_2(SO_4)_3 + (NH_4)_2S \rightarrow$
111. In each of the following, both compounds are water-soluble. Predict whether a precipitate will form when solutions of the two are mixed, and, if so, identify the compound that precipitates. (a) $Pb(NO_3)_2$, NaI; (b) $Ba(NO_3)_2$, KCl; (c) $(NH_4)_2S$, $AgNO_3$
112. In each of the following, both compounds are water-soluble. Predict whether a precipitate will form when solutions of the two are mixed, and, if so, identify the compound that precipitates. (a) NH_4Br, $Hg_2(NO_3)_2$; (b) KOH, Na_2S; (c) Cs_2SO_4, $MgCl_2$

Identifying Reaction Types

The following reactions apply to Exercises 113 through 121.

(a) $H_2SO_4(aq) + 2KOH(aq) \rightarrow K_2SO_4(aq) + 2H_2O(\ell)$

(b) $2Rb(s) + Br_2(\ell) \xrightarrow{heat} 2RbBr(s)$

(c) $2KI(aq) + F_2(g) \rightarrow 2KF(aq) + I_2(s)$

(d) $CaO(s) + SiO_2(s) \xrightarrow{heat} CaSiO_3(s)$

(e) $S(s) + O_2(g) \xrightarrow{heat} SO_2(g)$

(f) $BaCO_3(s) \xrightarrow{heat} BaO(s) + CO_2(g)$

(g) $HgS(s) + O_2(g) \xrightarrow{heat} Hg(\ell) + SO_2(g)$

(h) $AgNO_3(aq) + HCl(aq) \rightarrow AgCl(s) + HNO_3(aq)$

(i) $Pb(s) + 2HBr(aq) \rightarrow PbBr_2(s) + H_2(g)$

(j) $2HI(aq) + H_2O_2(aq) \rightarrow I_2(s) + 2H_2O(\ell)$

(k) $RbOH(aq) + HNO_3(aq) \rightarrow RbNO_3(aq) + H_2O(\ell)$

(l) $N_2O_5(s) + H_2O(\ell) \rightarrow 2HNO_3(aq)$

(m) $H_2O(g) + CO(g) \xrightarrow{heat} H_2(g) + CO_2(g)$

(n) $MgO(s) + H_2O(\ell) \rightarrow Mg(OH)_2(s)$

(o) $PbSO_4(s) + PbS(s) \xrightarrow{heat} 2Pb(s) + 2SO_2(g)$

113. Identify the precipitation reactions.
114. Identify the acid–base reactions.
115. Identify the oxidation–reduction reactions.
116. Identify the oxidizing agent and reducing agent for each oxidation–reduction reaction.
117. Identify the displacement reactions.
118. Identify the metathesis reactions.
119. Identify the combination reactions.
120. Identify the decomposition reactions.
121. (a) Do any of these reactions fit into more than one class? Which ones? (b) Do any of these reactions not fit into any of our classes of reactions? Which ones?
122. Predict whether or not a solid is formed when we mix the following; identify any solid product by name and identify the reaction type: (a) copper(II) nitrate solution and magnesium metal, (b) barium nitrate and sodium phosphate solutions, (c) calcium acetate solution with aluminum metal, (d) silver nitrate and sodium iodide solutions.
123. Predict whether or not a solid is formed when we mix the following; identify any solid product by formula and by name: (a) potassium permanganate and sodium phosphate solutions, (b) lithium carbonate and cadmium nitrate, (c) stannous fluoride and bismuth chloride, (d) strontium sulfate with barium chloride solutions.

CONCEPTUAL EXERCISES

124. The following statements are a collection of incorrect statements. Explain why each is wrong. (a) A 1.2 M CH_3COOH solution is a stronger acid than a 0.12 M CH_3COOH solution. (b) The salt produced during the neutralization of nitric acid with potassium hydroxide is KNO_4. (c) Nickel will react with a HCl solution but not with steam to produce hydrogen and nickel ion solution. Magnesium will react with steam to produce hydrogen and magnesium ions. Therefore, nickel is more reactive than magnesium.

As we have seen, two substances may react to form different products when they are mixed in different proportions under different conditions. In Exercises 125 and 126, write balanced equations for the reactions described by words. Assign oxidation numbers.

125. (a) Ethane burns in excess air to form carbon dioxide and water. (b) Ethane burns in a limited amount of air to form carbon monoxide and water. (c) Ethane burns (poorly) in a very limited amount of air to form elemental carbon and water.
126. (a) Butane (C_4H_{10}) burns in excess air to form carbon dioxide and water. (b) Butane burns in a limited amount of air to form carbon monoxide and water. (c) When heated in the presence of *very little* air, butane "cracks" to form acetylene, C_2H_2, carbon monoxide, and hydrogen.
127. Calcium phosphate is the component of human bone that provides rigidity. Fallout from a nuclear bomb can contain radioactive strontium-90. These two facts are closely tied together when one considers human health. Explain.
128. Limestone consists mainly of the mineral calcite, which is calcium carbonate. A very similar deposit called dolostone is composed primarily of the mineral dolomite, an ionic substance that contains carbonate ions and a mixture of magnesium and calcium ions. (a) Is this a surprising mixture of ions? Explain, based on the periodic table. (b) A test for limestone is to apply cold dilute hydrochloric acid, which causes the rapid formation of bubbles. What causes these bubbles?

The Dolomite Alps of Italy.

129. Chemical equations can be interpreted on either a particulate level (atoms, molecules, ions) or a mole level (moles of reactants and products). Write word statements to describe the combustion of propane on a particulate level and a mole level.

$$2C_3H_8(g) + 10O_2(g) \rightarrow 6CO_2(g) + 8H_2O(\ell)$$

130. Write word statements to describe the following reaction on a particulate level and a mole level.

$$P_4(s) + 6Cl_2(g) \rightarrow 4PCl_3(\ell)$$

131. When each of the following pairs of reactants are combined in a beaker: (a) describe in words what the contents of the beaker would look like before and after any reaction that might occur, (b) use different circles for atoms, molecules, and ions to draw a nanoscale (particulate-level) diagram of what the contents would look like, and (c) write a chemical equation for any reactions that might occur.

$$LiCl(aq) \text{ and } AgNO_3(aq)$$
$$NaOH(aq) \text{ and } HCl(aq)$$
$$CaCO_3(s) \text{ and } HCl(aq)$$
$$Na_2CO_3(aq) \text{ and } Ca(OH)_2(aq)$$

132. Explain how you could prepare barium sulfate by (a) an acid-base reaction, (b) a precipitation reaction, and (c) a gas-forming reaction. The materials you have to start with are $BaCO_3$, $Ba(OH)_2$, Na_2SO_4, and H_2SO_4.

BUILDING YOUR KNOWLEDGE

133. Magnesium oxide, marketed as tablets or as an aqueous slurry called "milk of magnesia," is a common commercial antacid. What volume, in milliliters, of fresh gastric juice,

Milk of Magnesia

corresponding in acidity to 0.17 M HCl, could be neutralized by 104 mg of magnesium oxide?

$$MgO(s) + 2HCl(aq) \rightarrow MgCl_2(aq) + H_2O(\ell)$$

134. How many moles of oxygen can be obtained by the decomposition of 15.0 grams of reactant in each of the following reactions?

(a) $2KClO_3(s) \xrightarrow{\text{heat}} 2KCl(s) + 3O_2(g)$

(b) $2H_2O_2(aq) \rightarrow 2H_2O(\ell) + O_2(g)$

(c) $2HgO(s) \xrightarrow{\text{heat}} 2Hg(\ell) + O2(g)$

BEYOND THE TEXTBOOK

Go to the textbook website at

http://www.brookscole.com/chemistry/whitten

for additional activities and exercises based on the General Chemistry Interactive CD-ROM, the World Wide Web, and library resources.

InfoTrac College Edition

For additional readings, go to InfoTrac College Edition, your online research library at:

http://infotrac.thomsonlearning.com

Metal salts often produce distinctive flame colors when burned (here with methanol): barium (green), strontium (red), and calcium (orange).

Charles D. Winters

OBJECTIVES

After you have studied this chapter, you should be able to

- Describe the evidence for the existence and properties of electrons, protons, and neutrons
- Predict the arrangements of the particles in atoms
- Describe isotopes and their composition
- Calculate atomic weights from isotopic abundance
- Describe the wave properties of light and how wavelength, frequency, and speed are related
- Use the particle description of light, and explain how it is related to the wave description
- Relate atomic emission and absorption spectra to important advances in atomic theory
- Describe the main features of the quantum mechanical picture of the atom
- Describe the four quantum numbers, and give possible combinations of their values for specific atomic orbitals
- Describe the shapes of orbitals and recall the usual order of their relative energies
- Write the electron configurations of atoms
- Relate the electron configuration of an atom to its position in the periodic table

The Dalton theory of the atom and related ideas were the basis for our study of *composition stoichiometry* (Chapter 2) and *reaction stoichiometry* (Chapter 3), but that level of atomic theory leaves many questions unanswered. *Why* do atoms combine to form compounds? *Why* do they combine only in simple numerical ratios? *Why* are particular numerical ratios of atoms observed in compounds? *Why* do different elements have different properties? *Why* are they gases, liquids, solids, metals, nonmetals, and so on? *Why* do some groups of elements have similar properties and form compounds with similar formulas? The answers to these and many other fascinating questions in chemistry are supplied by our modern understanding of the nature of atoms. But how can we study something as small as an atom?

Much of the development of modern atomic theory was based on two broad types of research carried out by dozens of scientists just before and after 1900. The first type dealt with the electrical nature of matter. These studies led scientists to recognize that atoms are composed of more fundamental particles and helped them to describe the approximate arrangements of these particles in atoms. The second broad area of research dealt with the interaction of matter with energy in the form of light. Such research included studies of the colors of the light that substances give off or absorb. These studies led to a much more detailed understanding of the arrangements of particles in atoms. It became clear that the arrangement of the particles determines the chemical and physical properties of each element. As we learn more about the structures of atoms, we are able to organize chemical facts in ways that help us to understand the behavior of matter.

We will first study the particles that make up atoms and the basic structure of atoms. Then we will take a look at the quantum mechanical theory of atoms and see how this theory describes the arrangement of the electrons in atoms. Current atomic theory is considerably less than complete. Even so, it is a powerful tool that helps us describe the forces holding atoms in chemical combination with one another.

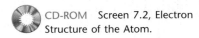 CD-ROM   Screen 7.2, Electron Structure of the Atom.

www Accounts of some important developments in atomic theory appear on the World Wide Web; for example, The Science Museum has an interesting on-line history of atomic structure discoveries at **http://www.sciencemuseum .org.uk/on-line/electron/**

SUBATOMIC PARTICLES

5-1 FUNDAMENTAL PARTICLES

In our study of atomic structure, we look first at the **fundamental particles.** These are the basic building blocks of all atoms. Atoms, and hence *all* matter, consist principally of three fundamental particles: *electrons, protons,* and *neutrons.* Knowledge of the nature and functions of these particles is essential to understanding chemical interactions. The relative masses and charges of the three fundamental particles are shown in Table 5-1. The

Many other subatomic particles, such as quarks, positrons, neutrinos, pions, and muons, have also been discovered. It is not necessary to study their characteristics to learn the fundamentals of atomic structure that are important in chemical reactions.

TABLE 5-1	*Fundamental Particles of Matter*	
Particle	Mass	Charge (relative scale)
electron (e^-)	0.00054858 amu	1−
proton (p or p^+)	1.0073 amu	1+
neutron (n or n^0)	1.0087 amu	none

mass of an electron is very small compared with the mass of either a proton or a neutron. The charge on a proton is equal in magnitude, but opposite in sign, to the charge on an electron. Let's examine these particles in more detail.

5-2 THE DISCOVERY OF ELECTRONS

The process is called chemical electrolysis. *Lysis* means "splitting apart."

Some of the earliest evidence about atomic structure was supplied in the early 1800s by the English chemist Humphry Davy (1778–1829). He found that when he passed electric current through some substances, the substances decomposed. He therefore suggested that the elements of a chemical compound are held together by electrical forces. In 1832–1833, Michael Faraday (1791–1867), Davy's student, determined the quantitative relationship between the amount of electricity used in electrolysis and the amount of chemical reaction that occurs. Studies of Faraday's work by George Stoney (1826–1911) led him to suggest in 1874 that units of electric charge are associated with atoms. In 1891, he suggested that they be named *electrons*.

Study Figures 5-1 and 5-2 carefully as you read this section.

The most convincing evidence for the existence of electrons came from experiments using **cathode-ray tubes** (Figure 5-1). Two electrodes are sealed in a glass tube containing

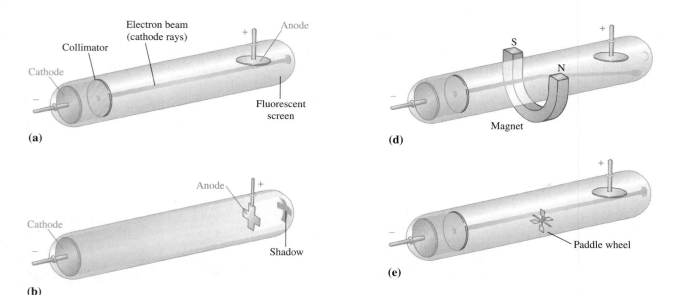

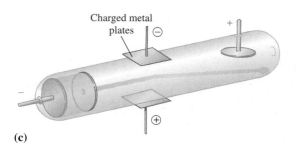

Figure 5-1 Some experiments with cathode-ray tubes that show the nature of cathode rays. (a) A cathode-ray (discharge) tube, showing the production of a beam of electrons (cathode rays). The beam is detected by observing the glow of a fluorescent screen. (b) A small object placed in a beam of cathode rays casts a shadow. This shows that cathode rays travel in straight lines. (c) Cathode rays have negative electric charge, as demonstrated by their deflection in an electric field. (The electrically charged plates produce an electric field.) (d) Interaction of cathode rays with a magnetic field is also consistent with negative charge. The magnetic field goes from one pole to the other. (e) Cathode rays have mass, as shown by their ability to turn a small paddle wheel in their path.

gas at a very low pressure. When a high voltage is applied, current flows and rays are given off by the cathode (negative electrode). These rays travel in straight lines toward the anode (positive electrode) and cause the walls opposite the cathode to glow. An object placed in the path of the cathode rays casts a shadow on a zinc sulfide fluorescent screen placed near the anode. The shadow shows that the rays travel from the negatively charged cathode toward the positively charged anode. The rays must therefore be negatively charged. Furthermore, they are deflected by electric and magnetic fields in the directions expected for negatively charged particles.

In 1897 J. J. Thomson (1856–1940) studied these negatively charged particles more carefully. He called them **electrons,** the name Stoney had suggested in 1891. By studying the degree of deflections of cathode rays in different strengths of electric and magnetic fields, Thomson determined the ratio of the charge (e) of the electron to its mass (m). The modern value for this ratio is

$$e/m = 1.75882 \times 10^8 \text{ coulomb (C)/gram}$$

This ratio is the same regardless of the type of gas in the tube, the composition of the electrodes, or the nature of the electric power source. The clear implication of Thomson's work was that electrons are fundamental particles present in all atoms. We now know that this is true and that all atoms contain integral (whole) numbers of electrons.

After the charge-to-mass ratio for the electron had been determined, additional experiments were necessary to determine the value of either its mass or its charge so that the other could be calculated. In 1909, Robert Millikan (1868–1953) solved this dilemma with the famous "oil-drop experiment," in which he determined the charge of the electron. This experiment is described in Figure 5-2. All of the charges measured by Millikan turned out

CD-ROM   Screens 2.6, Electrons, and 2.7, Mass of the Electrons

The coulomb (C) is the standard unit of *quantity* of electric charge. It is defined as the quantity of electricity transported in one second by a current of one ampere. It corresponds to the amount of electricity that will deposit 0.00111798 g of silver in an apparatus set up for plating silver.

X-rays are a form of radiation of much shorter wavelength than visible light (see Section 5-10). They are sufficiently energetic to knock electrons out of the atoms in the air. In Millikan's experiment these free electrons became attached to some of the oil droplets.

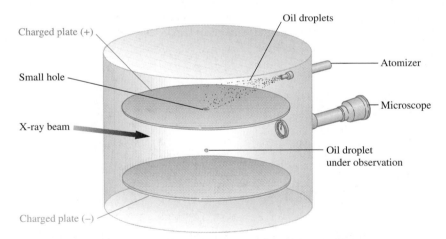

Figure 5-2   The Millikan oil-drop experiment. Tiny spherical oil droplets are produced by an atomizer. The mass of the spherical drop can be calculated from its volume (obtained from a measurement of the radius of the drop with a microscope) and the known density of the oil. A few droplets fall through the hole in the upper plate. Irradiation with X-rays gives some of these oil droplets a negative charge. When the voltage between the plates is increased, a negatively charged drop falls more slowly because it is attracted by the positively charged upper plate and repelled by the negatively charged lower plate. At one particular voltage, the electrical force (up) and the gravitational force (down) on the drop are exactly balanced, and the drop remains stationary. Knowing this voltage and the mass of the drop, we can calculate the charge on the drop.

Robert A. Millikan (*left*) was an American physicist who was a professor at the University of Chicago and later director of the physics laboratory at the California Institute of Technology. He won the 1923 Nobel Prize in physics.

to be integral multiples of the same number. He assumed that this smallest charge was the charge on one electron. This value is 1.60218×10^{-19} coulomb (modern value).

The charge-to-mass ratio, $e/m = 1.75882 \times 10^8$ C/g, can be used in inverse form to calculate the mass of the electron:

$$m = \frac{1 \text{ g}}{1.75882 \times 10^8 \text{ C}} \times 1.60218 \times 10^{-19} \text{ C}$$

$$= 9.10940 \times 10^{-28} \text{ g}$$

This is only about 1/1836 the mass of a hydrogen atom, the lightest of all atoms. Millikan's simple oil-drop experiment stands as one of the cleverest, yet most fundamental, of all classic scientific experiments. It was the first experiment to suggest that atoms contain integral numbers of electrons; we now know this to be true.

5-3 CANAL RAYS AND PROTONS

CD-ROM   Screen 2.8, Protons.

In 1886 Eugen Goldstein (1850–1930) first observed that a cathode-ray tube also generates a stream of positively charged particles that moves toward the cathode. These were called **canal rays** because they were observed occasionally to pass through a channel, or "canal," drilled in the negative electrode (Figure 5-3). These *positive rays*, or *positive ions*, are created when the gaseous atoms in the tube lose electrons. Positive ions are formed by the process

$$\text{Atom} \longrightarrow \text{cation}^+ + e^- \quad \text{or} \quad X \longrightarrow X^+ + e^- \quad \text{(energy absorbed)}$$

Different elements give positive ions with different e/m ratios. The regularity of the e/m values for different ions led to the idea that there is a unit of positive charge and that it resides in the **proton**. The proton is a fundamental particle with a charge equal in magnitude but opposite in sign to the charge on the electron. Its mass is almost 1836 times that of the electron.

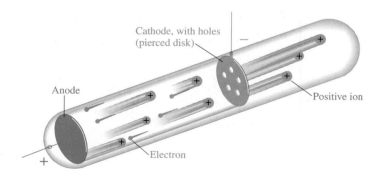

Figure 5-3 A cathode-ray tube with a different design and with a perforated cathode. Such a tube was used to produce canal rays and to demonstrate that they travel toward the cathode. Like cathode rays, these *positive* rays are deflected by electric or magnetic fields, but in the opposite direction from cathode rays (electron beam). Canal ray particles have e/m ratios many times smaller than those of electrons because of their much greater masses. When different elements are in the tube, positive ions with different e/m ratios are observed.

5-4 RUTHERFORD AND THE NUCLEAR ATOM

By the early 1900s, it was clear that each atom contains regions of both positive and negative charge. The question was, how are these charges distributed? The dominant view of that time was summarized in J. J. Thomson's model of the atom; the positive charge was assumed to be distributed evenly throughout the atom. The negative charges were pictured as being imbedded in the atom like plums in a pudding (the "plum pudding model").

Soon after Thomson developed his model, tremendous insight into atomic structure was provided by one of Thomson's former students, Ernest Rutherford (1871–1937), who was the outstanding experimental physicist of his time.

By 1909, Ernest Rutherford had established that alpha (α) particles are positively charged particles. They are emitted at high kinetic energies by some radioactive atoms, that is, atoms that disintegrate spontaneously. In 1910 Rutherford's research group carried out a series of experiments that had an enormous impact on the scientific world. They bombarded a very thin piece of gold foil with α-particles from a radioactive source. A fluorescent zinc sulfide screen was placed behind the foil to indicate the scattering of the α-particles by the gold foil (Figure 5-4). Scintillations (flashes) on the screen, caused by the individual α-particles, were counted to determine the relative numbers of α-particles deflected at various angles. At the time, alpha particles were believed to be extremely dense, much denser than the gold atom.

If the Thomson model of the atom were correct, any α-particles passing through the foil would have been deflected by very small angles. Quite unexpectedly, nearly all of the α-particles passed through the foil with little or no deflection. A few, however, were deflected

CD-ROM Screens 2.5, Evidence of Subatomic Particles: Radioactivity; and 2.9, The Nucleus of the Atom.

α-Particles are now known to be He^{2+} ions, that is, helium atoms without their two electrons. (See Chapter 26.)

Radioactivity is contrary to the Daltonian idea of the indivisibility of atoms.

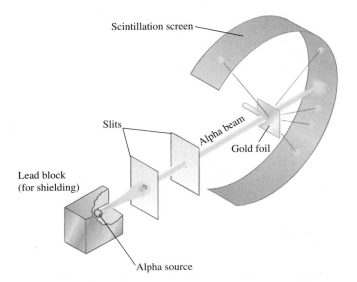

SC/*LINKS.*

TOPIC: Subatomic Particles
GO TO: www.scilinks.org
*sci*LINKS **CODE:** WCH0510

Figure 5-4 The Rutherford scattering experiment. A narrow beam of α-particles from a radioactive source was directed at a very thin sheet of gold foil. Most of the particles passed right through the gold foil (*brown*). Many were deflected through moderate angles (*shown in red*). These deflections were surprising, but the 0.001% of the total that were reflected at acute angles (*shown in blue*) were totally unexpected. Similar results were observed using foils of other metals.

Ernest Rutherford was one of the giants in the development of our understanding of atomic structure. While working with J. J. Thomson at Cambridge University, he discovered α and β radiation. He spent the years 1899–1907 at McGill University in Canada where he proved the nature of these two radiations, for which he received the Nobel Prize in chemistry in 1908. He returned to England in 1908, and it was there, at Manchester University, that he and his coworkers Hans Geiger and Ernst Marsden performed the famous gold foil experiments that led to the nuclear model of the atom. Not only did he perform much important research in physics and chemistry, but he also guided the work of ten future recipients of the Nobel Prize.

This representation is *not* to scale. If nuclei were as large as the black dots that represent them, each white region, which represents the size of an atom, would have a diameter of more than 30 feet!

through large angles, and a very few α-particles even returned from the gold foil in the direction from which they had come! Rutherford was astounded. In his own words,

> *It was quite the most incredible event that has ever happened to me in my life. It was almost as if you fired a 15-inch shell into a piece of tissue paper and it came back and hit you.*

Rutherford's mathematical analysis of his results showed that the scattering of positively charged α-particles was caused by repulsion from very dense regions of positive charge in the gold foil. He concluded that the mass of one of these regions is nearly equal to that of a gold atom, but that the diameter is no more than 1/10,000 that of an atom. Many experiments with foils of different metals yielded similar results. Realizing that these observations were inconsistent with previous theories about atomic structure, Rutherford discarded the old theory and proposed a better one. He suggested that each atom contains a *tiny, positively charged, massive center* that he called an atomic **nucleus.** Most α-particles pass through metal foils undeflected because atoms are *primarily* empty space populated only by the very light electrons. The few particles that are deflected are the ones that come close to the heavy, highly charged metal nuclei (Figure 5-5).

Rutherford was able to determine the magnitudes of the positive charges on the atomic nuclei. The picture of atomic structure that he developed is called the Rutherford model of the atom.

Atoms consist of very small, very dense positively charged nuclei surrounded by clouds of electrons at relatively large distances from the nuclei.

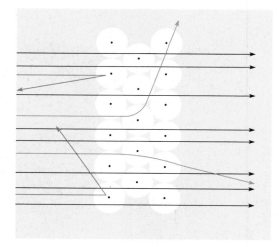

Figure 5-5 An interpretation of the Rutherford scattering experiment. The atom is pictured as consisting mostly of "open" space. At the center is a tiny and extremely dense nucleus that contains all of the atom's positive charge and nearly all of its mass. The electrons are thinly distributed throughout the "open" space. Most of the positively charged α-particles (*black arrows*) pass through the open space undeflected, not coming near any gold nuclei. The few that pass fairly close to a nucleus (*red arrows*) are repelled by electrostatic forces and thereby deflected. The very few particles that are on a path to collide with gold nuclei are repelled backward at acute angles (*blue arrows*). Calculations based on the results of the experiment indicated that the diameter of the open-space portion of the atom is from 10,000 to 100,000 times greater than the diameter of the nucleus.

5-5 ATOMIC NUMBER

Only a few years after Rutherford's scattering experiments, H. G. J. Moseley (1887–1915) studied X-rays given off by various elements. Max von Laue (1879–1960) had shown that X-rays could be diffracted by crystals into a spectrum in much the same way that visible light can be separated into its component colors. Moseley generated X-rays by aiming a beam of high-energy electrons at a solid target made of a single pure element (Figure 5-6a).

The spectra of X-rays produced by targets of different elements were recorded photographically. Each photograph consisted of a series of lines representing X-rays at various wavelengths; each element produced its own distinct set of wavelengths. Comparison of results from different elements revealed that corresponding lines were displaced toward shorter wavelengths as atomic weights of the target materials increased, with a few exceptions. But in 1913 Moseley showed that the X-ray wavelengths could be better correlated with the atomic number. A plot illustrating this interpretation of Moseley's data appears in Figure 5-6b. On the basis of his mathematical analysis of these X-ray data, he concluded that

> each element differs from the preceding element by having one more positive charge in its nucleus.

For the first time it was possible to arrange all known elements in order of increasing nuclear charge. This discovery led to the realization that atomic number, related to the electrical

In a modern technique known as "X-ray fluorescence spectroscopy," the wavelengths of X-rays given off by a sample target indicate which elements are present in the sample.

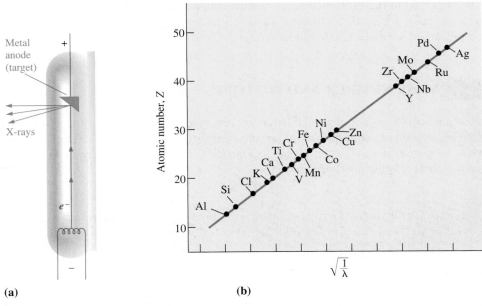

(a)   **(b)**

H. G. J. Moseley was one of the many remarkable scientists who worked with Ernest Rutherford. Moseley's scientific career was very short. He enlisted in the British army during World War I and died in battle in the Gallipoli campaign in 1915.

Figure 5-6 (a) A simplified representation of the production of X-rays by bombardment of a solid target with a high-energy beam of electrons. (b) A plot of some of Moseley's X-ray data. The atomic number of an element is found to be directly proportional to the square root of the reciprocal of the wavelength of a particular X-ray spectral line. Wavelength (Section 5-10) is represented by λ.

properties of the atom, was more fundamental to determining the properties of the elements than atomic weight was. This discovery put the ideas of the periodic table on a more fundamental footing.

We now know that every nucleus contains an integral number of protons exactly equal to the number of electrons in a neutral atom of the element. Every hydrogen atom contains one proton, every helium atom contains two protons, and every lithium atom contains three protons. The number of protons in the nucleus of an atom determines its identity; this number is known as the **atomic number** of that element.

5-6 NEUTRONS

The third fundamental particle, the neutron, eluded discovery until 1932. James Chadwick (1891–1974) correctly interpreted experiments on the bombardment of beryllium with high-energy α-particles. Later experiments showed that nearly all elements up to potassium, element 19, produce neutrons when they are bombarded with high-energy α-particles. The **neutron** is an uncharged particle with a mass slightly greater than that of the proton.

This does not mean that elements above number 19 do not have neutrons, only that neutrons are not generally knocked out of atoms of higher atomic number by α-particle bombardment.

> Atoms consist of very small, very dense nuclei surrounded by clouds of electrons at relatively great distances from the nuclei. All nuclei contain protons; nuclei of all atoms except the common form of hydrogen also contain neutrons.

Nuclear diameters are about 10^{-5} nanometers (nm); atomic diameters are about 10^{-1} nm. To put this difference in perspective, suppose that you wish to build a model of an atom using a basketball (diameter about 9.5 inches) as the nucleus; on this scale, the atomic model would be nearly 6 miles across!

5-7 MASS NUMBER AND ISOTOPES

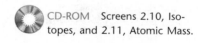 CD-ROM   Screens 2.10, Isotopes, and 2.11, Atomic Mass.

Most elements consist of atoms of different masses, called **isotopes.** The isotopes of a given element contain the same number of protons (and also the same number of electrons) because they are atoms of the same element. They differ in mass because they contain different numbers of neutrons in their nuclei.

> Isotopes are atoms of the same element with different masses; they are atoms containing the same number of protons but different numbers of neutrons.

For example, there are three distinct kinds of hydrogen atoms, commonly called hydrogen, deuterium, and tritium. (This is the only element for which we give each isotope a different name.) Each contains one proton in the atomic nucleus. The predominant form of hydrogen contains no neutrons, but each deuterium atom contains one neutron and each tritium atom contains two neutrons in its nucleus (Table 5-2). All three forms of hydrogen display very similar chemical properties.

| | | | | Atomic | | | No. of Electrons |
Name	Symbol	Nuclide Symbol	Mass (amu)	Abundance in Nature	No. of Protons	No. of Neutrons	(in neutral atoms)
hydrogen	H	^1_1H	1.007825	99.985%	1	0	1
deuterium	D	^2_1H	2.01400	0.015%	1	1	1
tritium*	T	^3_1H	3.01605	0.000%	1	2	1

TABLE 5-2 *The Three Isotopes of Hydrogen*

No known natural sources; produced by decomposition of artificial isotopes.

The **mass number** of an atom is the sum of the number of protons and the number of neutrons in its nucleus; that is

$$\text{Mass number} = \text{number of protons} + \text{number of neutrons}$$
$$= \text{atomic number} + \text{neutron number}$$

The mass number for normal hydrogen atoms is 1; for deuterium, 2; and for tritium, 3. The composition of a nucleus is indicated by its **nuclide symbol.** This consists of the symbol for the element (E), with the atomic number (Z) written as a subscript at the lower left and the mass number (A) as a superscript at the upper left, $^A_Z E$. By this system, the three isotopes of hydrogen are designated as ^1_1H, ^2_1H, and ^3_1H.

A mass number is a count of the *number* of protons plus neutrons present, so it must be a whole number. Because the masses of the proton and the neutron are both about 1 amu, the mass number is *approximately* equal to the actual mass of the isotope (which is not a whole number).

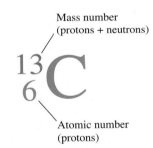

Figure 5-7 Interpretation of a nuclide symbol. Chemists often omit the subscripted atomic number because the element symbol implies the atomic number.

EXAMPLE 5-1 *Determination of Atomic Makeup*

Determine the number of protons, neutrons, and electrons in each of the following species. Are the members within each pair isotopes?

(a) $^{35}_{17}Cl$ and $^{37}_{17}Cl$ (b) $^{63}_{29}Cu$ and $^{65}_{29}Cu$

Plan

Knowing that the number at the bottom left of the nuclide symbol is the atomic number or number of protons, we can verify the identity of the element in addition to knowing the number or protons per nuclide. From the mass number at the top left, we know the number of protons plus neutrons. The number of protons (atomic number) minus the number of electrons must equal the charge, if any, shown at the top right. From these data one can determine if two nuclides have the same number of protons and are therefore the same element. If they are the same element, they will be isotopes only if their mass numbers differ.

Solution

(a) For $^{35}_{17}Cl$:   Atomic number = 17. There are therefore 17 protons per nucleus.
Mass number = 35. There are therefore 35 protons plus neutrons or, because we know that there are 17 protons, there are 18 neutrons.
Because no charge is indicated, there must be equal numbers of protons and electrons, or 17 electrons.

For $^{37}_{17}Cl$:   There are 17 protons, 20 neutrons, and 17 electrons per atom.

These are isotopes of the same element. Both have 17 protons, but they differ in their numbers of neutrons: one has 18 neutrons and the other has 20.

(b) For $^{63}_{29}Cu$:   Atomic number = 29. There are 29 protons per nucleus.
Mass number = 63. There are 29 protons plus 34 neutrons.
Because no charge is indicated, there must be equal numbers of protons and electrons, or 29 electrons.

For $^{65}_{29}Cu$:   There are 29 protons, 36 neutrons, and 29 electrons per atom.

These are isotopes. Both have 29 protons, but they differ in their numbers of neutrons: one isotope has 34 neutrons and the other has 36.

You should now work Exercises 18 and 20.

5-8 MASS SPECTROMETRY AND ISOTOPIC ABUNDANCE

Mass spectrometers are instruments that measure the charge-to-mass ratio of charged particles (Figures 5-8). A gas sample at very low pressure is bombarded with high-energy electrons. This causes electrons to be ejected from some of the gas molecules, creating positive ions. The positive ions are then focused into a very narrow beam and accelerated by an electric field toward a magnetic field. The magnetic field deflects the ions from their straight-line path. The extent to which the beam of ions is deflected depends on four factors:

1. *Magnitude of the accelerating voltage (electric field strength).* Higher voltages result in beams of more rapidly moving particles that are deflected less than the beams of the more slowly moving particles produced by lower voltages.

2. *Magnetic field strength.* Stronger fields deflect a given beam more than weaker fields.

3. *Masses of the particles.* Because of their inertia, heavier particles are deflected less than lighter particles that carry the same charge.

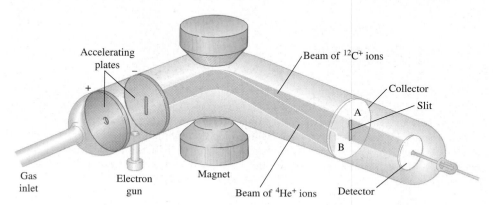

Figure 5-8   The mass spectrometer. In the mass spectrometer, gas molecules at low pressure are ionized and accelerated by an electric field. The ion beam is then passed through a magnetic field. In that field the beam is resolved into components, each containing particles of equal charge-to-mass ratio. Lighter particles are deflected more strongly than heavy ones with the same charge. In a beam containing $^{12}C^+$ and $^{4}He^+$ ions, the lighter $^{4}He^+$ ions would be deflected more than the heavier $^{12}C^+$ ions. The spectrometer shown is adjusted to detect the $^{12}C^+$ ions. By changing the magnitude of the magnetic or electric field, we can move the beam of $^{4}He^+$ ions striking the collector from B to A, where it would be detected. The relative masses of the ions are calculated from the changes required to refocus the beam.

4. *Charges on the particles.* Particles with higher charges interact more strongly with magnetic fields and are thus deflected more than particles of equal mass with smaller charges.

The mass spectrometer is used to measure masses of isotopes as well as isotopic abundances, that is, the relative amounts of the isotopes. Helium occurs in nature almost exclusively as ^{4}He. Its atomic mass can be determined in an experiment such as that illustrated in Figure 5-8.

A beam of Ne$^+$ ions in the mass spectrometer is split into three segments. The mass spectrum of these ions (a graph of the relative numbers of ions of each mass) is shown in Figure 5-9. This indicates that neon occurs in nature as three isotopes: ^{20}Ne, ^{21}Ne, and ^{22}Ne. In Figure 5-9 we see that the isotope ^{20}Ne, mass 19.99244 amu, is the most abundant isotope (has the tallest peak). It accounts for 90.48% of the atoms. ^{22}Ne, mass 21.99138, accounts for 9.25%, and ^{21}Ne, mass 20.99384, for only 0.27% of the atoms.

Figure 5-10 shows a modern mass spectrometer. In nature, some elements, such as fluorine and phosphorus, exist in only one form, but most elements occur as isotopic mixtures.

We can omit the atomic number subscript 2 for He, because all helium atoms have atomic number 2.

Isotopes are two or more forms of atoms of the same element with different masses; the atoms contain the same number of protons but different numbers of neutrons.

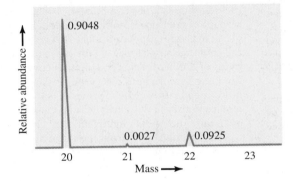

Figure 5-9 Mass spectrum of neon (1+ ions only). Neon consists of three isotopes, of which neon-20 is by far the most abundant (90.48%). The mass of that isotope, to five decimal places, is 19.99244 amu on the carbon-12 (^{12}C) scale. The number by each peak corresponds to the fraction of all Ne$^+$ ions represented by the isotope with that mass.

(a)

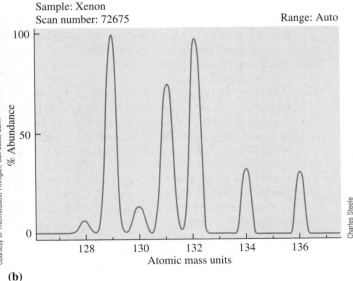

(b)

Figure 5-10 (a) A modern mass spectrometer. (b) The mass spectrum of Xe$^+$ ions. The isotope ^{126}Xe is at too low an abundance (0.090%) to appear in this experiment.

CHEMISTRY IN USE

The Development of Science

Stable Isotope Ratio Analysis

Many elements exist as two or more stable isotopes, although one isotope is usually present in far greater abundance. For example, there are two stable isotopes of carbon, ^{13}C and ^{12}C, of which ^{12}C is the more abundant, constituting 98.89% of all carbon. Similarly, there are two stable isotopes of nitrogen, ^{14}N and ^{15}N, of which ^{14}N makes up 99.63% of all nitrogen.

Differences in chemical and physical properties that arise from differences in atomic mass of an element are known as isotope effects. We know that the extranuclear structure of an element (the number of electrons and their arrangement) essentially determines its chemical behavior, whereas the nucleus has more influence on many of the physical properties of the element. Because all isotopes of a given element contain the same number and arrangement of electrons, it was assumed for a long time that isotopes would behave identically in chemical reactions. In reality, although isotopes behave very similarly in chemical reactions, the correspondence is not perfect. The mass differences between different isotopes of the same element cause them to have slightly different physical and chemical properties. For example, the presence of only one additional neutron in the nucleus of the heavier isotope can cause it to react a little more slowly than its lighter counterpart. Such an effect often results in a ratio of heavy isotope to light isotope in the product of a reaction that is different from the ratio found in the reactant.

Stable isotope ratio analysis (SIRA) is an analytical technique that takes advantage of the chemical and physical properties of isotopes. In SIRA the isotopic composition of a sample is measured using a mass spectrometer. This composition is then expressed as the relative ratios of two or more of the stable isotopes of a specific element. For instance, the ratio of ^{13}C to ^{12}C in a sample can be determined. This ratio is then compared with the isotope ratio of a defined standard. Because mass differences are most pronounced among the lightest elements, those elements experience the greatest isotope effects. Thus, the isotopes of the elements H, C, N, O, and S are used most frequently for SIRA. These elements have further significance because they are among the most abundant elements in biological systems.

The isotopic composition of a sample is usually expressed as a "del" value (∂), defined as

$$\partial X_{\text{sample}} (^0/_{00}) = \frac{(R_{\text{sample}} - R_{\text{standard}})}{R_{\text{standard}}} \times 1000$$

where $\partial X_{\text{sample}}$ is the isotope ratio relative to a standard, and R_{sample} and R_{standard} are the absolute isotope ratios of the sample and standard, respectively. Multiplying by 1000 allows the values to be expressed in parts per thousand ($^0/_{00}$). If the del value is a positive number, the sample has a greater amount of the heavier isotope than does the standard.

An interesting application of SIRA is the determination of the adulteration of food. As already mentioned, the isotope ratios of different plants and animals have been determined. For instance, corn has a $\partial^{13}C$ value of about $-12^0/_{00}$ and most flowering plants have $\partial^{13}C$ values of about $-26^0/_{00}$. The difference in these $\partial^{13}C$ values arises because these plants carry out photosynthesis by slightly different chemical reactions. In the first reaction of photosynthesis, corn produces a molecule that contains four carbons, whereas flowering plants produce a molecule that has only three carbons. High-fructose corn syrup (HFCS) is thus derived from a "C_4" plant, whereas the nectar that bees gather comes from "C_3" plants. The slight differences in the photosynthetic pathways of C_3 and C_4 plants create the major differences in their $\partial^{13}C$ values. Brokers who buy and sell huge quantities of "sweet" products are able to monitor HFCS adulteration of honey, maple syrup, apple juice, and so on by taking advantage of the SIRA technique. If the $\partial^{13}C$ value of one of these products is not appropriate, then the product obviously has had other substances added to it; that is, it, has been adulterated. The U.S. Department of Agriculture conducts routine isotope analyses to ensure the purity of those products submitted for subsidy programs. Similarly, the honey industry monitors itself with the SIRA technique.

Stable isotope ratio analysis is a powerful tool; many of its potential uses are only slowly being recognized by researchers. In the meantime, the use of stable isotope methods in research is becoming increasingly common, and through these methods scientists are attaining new levels of understanding of chemical, biological, and geological processes.

Beth A. Trust

TABLE 5-3 *Some Naturally Occurring Isotopic Abundances*

Element	Atomic Weight (amu)	Isotope	% Natural Abundance	Mass (amu)
boron	10.811	^{10}B	19.91	10.01294
		^{11}B	80.09	11.00931
oxygen	15.9994	^{16}O	99.762	15.99492
		^{17}O	0.038	16.99913
		^{18}O	0.200	17.99916
chlorine	35.4527	^{35}Cl	75.770	34.96885
		^{37}Cl	24.230	36.96590
uranium	238.0289	^{234}U	0.0055	234.0409
		^{235}U	0.720	235.0439
		^{238}U	99.2745	238.0508

The 20 elements that have only one naturally occurring isotope are ^9Be, ^{19}F, ^{23}Na, ^{27}Al, ^{31}P, ^{45}Sc, ^{55}Mn, ^{59}Co, ^{75}As, ^{89}Y, ^{93}Nb, ^{103}Rh, ^{127}I, ^{133}Cs, ^{141}Pr, ^{159}Tb, ^{165}Ho, ^{169}Tm, ^{197}Au, and ^{209}Bi. There are, however, other artificially produced isotopes of these elements.

Some examples of natural isotopic abundances are given in Table 5-3. The percentages are based on the numbers of naturally occurring atoms of each isotope, *not* on their masses.

The distribution of isotopic masses, although nearly constant, does vary somewhat depending on the source of the element. For example, the abundance of ^{13}C in atmospheric CO_2 is slightly different from that in seashells. The chemical history of a compound can be inferred from small differences in isotope ratios.

5-9 THE ATOMIC WEIGHT SCALE AND ATOMIC WEIGHTS

We said in Section 2-5 that the **atomic weight scale** is based on the mass of the carbon-12 (^{12}C) isotope. As a result of action taken by the International Union of Pure and Applied Chemistry in 1962,

> one **amu** is exactly 1/12 of the mass of a carbon-12 atom.

Described another way, the mass of one atom of ^{12}C is *exactly* 12 amu.

This is approximately the mass of one atom of ^1H, the lightest isotope of the element with lowest mass.

In Section 2-6 we said that one mole of atoms contains 6.022×10^{23} atoms. The mass of one mole of atoms of any element, in grams, is numerically equal to the atomic weight of the element. Because the mass of one ^{12}C atom is exactly 12 amu, the mass of one mole of carbon-12 atoms is exactly 12 grams.

To show the relationship between atomic mass units and grams, let's calculate the mass, in amu, of 1.000 gram of ^{12}C atoms.

$$\text{? amu} = 1.000 \text{ g } ^{12}\text{C atoms} \times \frac{1 \text{ mol } ^{12}\text{C}}{12 \text{ g } ^{12}\text{C atoms}} \times \frac{6.022 \times 10^{23} \text{ } ^{12}\text{C atoms}}{1 \text{ mol } ^{12}\text{C atoms}} \times \frac{12 \text{ amu}}{^{12}\text{C atom}}$$

$$= 6.022 \times 10^{23} \text{ amu (in 1 g)}$$

Thus,

We saw in Chapter 2 that Avogadro's number is the number of particles of a substance in one mole of that substance. We now see that Avogadro's number also represents the number of amu in one gram. You may wish to verify that the same result is obtained regardless of the element or isotope chosen.

$$1 \text{ g} = 6.022 \times 10^{23} \text{ amu} \quad \textit{or} \quad 1 \text{ amu} = 1.660 \times 10^{-24} \text{ g}$$

At this point, we emphasize the following:

1. The *atomic number, Z,* is an integer equal to the number of protons in the nucleus of an atom of the element. It is also equal to the number of electrons in a neutral atom. It is the same for all atoms of an element.

2. The *mass number, A,* is an integer equal to the *sum* of the number of protons and the number of neutrons in the nucleus of an atom of a *particular isotope* of an element. It is different for different isotopes of the same element.

3. Many elements occur in nature as mixtures of isotopes. The *atomic weight* of such an element is the weighted average of the masses of its isotopes. Atomic weights are fractional numbers, not integers.

The atomic weight that we determine experimentally (for an element that consists of more than one isotope) is such a weighted average. The following example shows how an atomic weight can be calculated from measured isotopic abundances.

EXAMPLE 5-2   *Calculation of Atomic Weight*

Three isotopes of magnesium occur in nature. Their abundances and masses, determined by mass spectrometry, are listed in the following table. Use this information to calculate the atomic weight of magnesium.

Isotope	% Abundance	Mass (amu)
^{24}Mg	78.99	23.98504
^{25}Mg	10.00	24.98584
^{26}Mg	11.01	25.98259

Plan

We multiply the fraction (percent divided by 100) of each isotope by its mass and add these numbers to obtain the atomic weight of magnesium.

Solution

$$\text{Atomic weight} = 0.7899(23.98504 \text{ amu}) + 0.1000(24.98584 \text{ amu}) + 0.1101(25.98259 \text{ amu})$$

$$= 18.946 \text{ amu} \qquad + 2.4986 \text{ amu} \qquad + 2.8607 \text{ amu}$$

$$= \boxed{24.30 \text{ amu}} \quad \text{(to four significant figures)}$$

The two heavier isotopes make small contributions to the atomic weight of magnesium because most magnesium atoms are the lightest isotope.

You should now work Exercises 28 and 30.

✓ Problem-Solving Tip: *"Weighted" Averages*

Consider the following analogy to the calculation of atomic weights. Suppose you want to calculate the average weight of your classmates. Imagine that one half of them weigh 100 pounds each, and the other half weigh 200 pounds each. The average weight would be

$$\text{Average weight} = \frac{1}{2}(100 \text{ lb}) + \frac{1}{2}(200 \text{ lb}) = 150 \text{ lb}$$

Imagine, however, that three quarters of the class members weigh 100 pounds each, and the other quarter weigh 200 pounds each. Now, the average weight would be

$$\text{Average weight} = \frac{3}{4}(100 \text{ lb}) + \frac{1}{4}(200 \text{ lb}) = 125 \text{ lb}$$

We can express the fractions in this calculation in decimal form:

$$\text{Average weight} = 0.750(100 \text{ lb}) + 0.250(200 \text{ lb}) = 125 \text{ lb}$$

In such a calculation, the value (in this case, the weight) of each thing (people, atoms) is multiplied by the fraction of things that have that value. In Example 5-2 we expressed each percentage as a decimal fraction, such as

$$78.99\% = \frac{78.99 \text{ parts}}{100 \text{ parts total}} = 0.7899$$

Example 5-3 shows how the process can be reversed. Isotopic abundances can be calculated from isotopic masses and from the atomic weight of an element that occurs in nature as a mixture of only two isotopes.

EXAMPLE 5-3   *Calculation of Isotopic Abundance*

The atomic weight of gallium is 69.72 amu. The masses of the naturally occurring isotopes are 68.9257 amu for ^{69}Ga and 70.9249 amu for ^{71}Ga. Calculate the percent abundance of each isotope.

Plan

We represent the fraction of each isotope algebraically. Atomic weight is the weighted average of the masses of the constituent isotopes. So the fraction of each isotope is multiplied by its mass, and the sum of the results is equal to the atomic weight.

Solution

Let x = fraction of ^{69}Ga. Then $(1 - x)$ = fraction of ^{71}Ga.

$$x(68.9257 \text{ amu}) + (1 - x)(70.9249 \text{ amu}) = 69.72 \text{ amu}$$
$$68.9257x + 70.9249 - 70.9249x = 69.72$$
$$-1.9992x = -1.20$$
$$x = 0.600$$

When a quantity is represented by fractions, the sum of the fractions must always be unity. In this case, $x + (1 - x) = 1$.

$$x = 0.600 = \text{fraction of } {}^{69}\text{Ga} \quad \therefore \quad \boxed{60.0\% \ {}^{69}\text{Ga}}$$

$$(1 - x) = 0.400 = \text{fraction of } {}^{71}\text{Ga} \quad \therefore \quad \boxed{40.0\% \ {}^{71}\text{Ga}}$$

You should now work Exercise 32.

THE ELECTRONIC STRUCTURES OF ATOMS

The Rutherford model of the atom is consistent with the evidence presented so far, but it has some serious limitations. It does not answer such important questions as: *Why* do different elements have such different chemical and physical properties? *Why* does chemical bonding occur at all? *Why* does each element form compounds with characteristic formulas? *How* can atoms of different elements give off or absorb light only of characteristic colors (as was known long before 1900)?

To improve our understanding, we must first learn more about the arrangements of electrons in atoms. The theory of these arrangements is based largely on the study of the light given off and absorbed by atoms. Then we will develop a detailed picture of the *electron configurations* of different elements. A knowledge of these arrangements will help us to understand the periodic table and chemical bonding.

Alfred Pasieka/Peter Arnold, Inc.

White light is dispersed by a prism into a *continuous* spectrum.

CD-ROM Screens 7.3, Electromagnetic Radiation, and 7.4, The Electromagnetic Spectrum.

SCi LINKS. TOPIC: Wave Properties
GO TO: www.scilinks.org
*sci*LINKS CODE: WCH0520

One cycle per second is also called one *hertz* (Hz), after Heinrich Hertz (1857–1894). In 1887, Hertz discovered electromagnetic radiation outside the visible range and measured its speed and wavelengths.

5-10 ELECTROMAGNETIC RADIATION

Our ideas about the arrangements of electrons in atoms have evolved slowly. Much of the information has been derived from atomic **emission spectra.** These are the bright lines, or bands, produced on photographic film by radiation that has passed through a refracting glass prism after being emitted from electrically or thermally excited atoms. To help us understand the nature of atomic spectra, we first describe electromagnetic radiation.

All types of electromagnetic radiation, or radiant energy, can be described in the terminology of waves. To help characterize any wave, we specify its *wavelength* (or its *frequency*). Let's use a familiar kind of wave, that on the surface of water (Figure 5-11), to illustrate these terms. The significant feature of wave motion is its repetitive nature. The **wavelength,** λ (Greek letter "lambda"), is the distance between any two adjacent identical points of the wave, for instance, two adjacent crests. The **frequency** is the number of wave crests passing a given point per unit time; it is represented by the symbol ν (Greek letter "nu") and is usually expressed in cycles per second or, more commonly, simply as 1/s or s^{-1} with "cycles" understood. For a wave that is "traveling" at some speed, the wavelength and the frequency are related to each other by

$$\lambda\nu = \text{speed of propagation of the wave} \quad \text{or} \quad \lambda\nu = c$$

Thus, wavelength and frequency are inversely proportional to each other; for the same wave speed, shorter wavelengths correspond to higher frequencies.

For water waves, it is the surface of the water that changes repetitively; for a vibrating violin string, it is the displacement of any point on the string. Electromagnetic radiation is

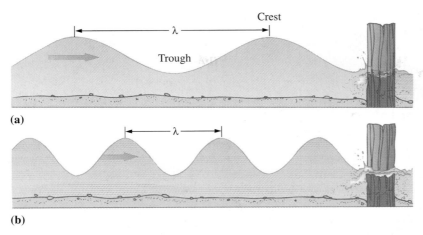

(a)

(b)

Figure 5-11 Illustrations of the wavelength and frequency of water waves. The distance between any two identical points, such as crests, is the wavelength, λ. We could measure the frequency, ν, of the wave by observing how often the level rises and falls at a fixed point in its path—for instance, at the post—or how often crests hit the post. (a) and (b) represent two waves that are traveling at the same speed. In (a) the wave has long wavelength and low frequency; in (b) the wave has shorter wavelength and higher frequency.

The diffraction of white light by the closely spaced grooves of a compact disk spreads the light into its component colors. Diffraction is described as the constructive and destructive interference of light waves.

a form of energy that consists of electric and magnetic fields that vary repetitively. The electromagnetic radiation most obvious to us is visible light. It has wavelengths ranging from about 4.0×10^{-7} m (violet) to about 7.5×10^{-7} m (red). Expressed in frequencies, this range is about 7.5×10^{14} Hz (violet) to about 4.0×10^{14} Hz (red).

Isaac Newton (1642–1727) first recorded the separation of sunlight into its component colors by allowing it to pass through a glass prism. Because sunlight (white light) contains all wavelengths of visible light, it gives the *continuous spectrum* observed in a rainbow (Figure 5-12a). Visible light represents only a tiny segment of the electromagnetic radiation spectrum (Figure 5-12b). In addition to all wavelengths of visible light, sunlight also contains shorter wavelength (ultraviolet) radiation as well as longer wavelength (infrared) radiation. Neither of these can be detected by the human eye. Both may be detected and recorded photographically or by detectors designed for that purpose. Many other familiar kinds of radiation are simply electromagnetic radiation of longer or shorter wavelengths.

In a vacuum, the speed of electromagnetic radiation, c, is the same for all wavelengths, 2.99792458×10^8 m/s. The relationship between the wavelength and frequency of electromagnetic radiation, with c rounded to three significant figures, is

$$\lambda \nu = c = 3.00 \times 10^8 \text{ m/s (or 186,000 miles/s)}$$

Sir Isaac Newton, one of the giants of science. You probably know of him from his theory of gravitation. In addition, he made enormous contributions to the understanding of many other aspects of physics, including the nature and behavior of light, optics, and the laws of motion. He is credited with the discoveries of differential calculus and of expansions into infinite series.

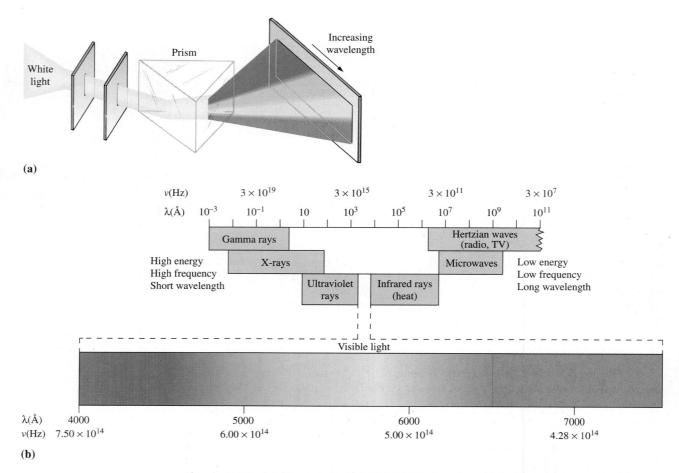

Figure 5-12 (a) Dispersion of visible light by a prism. Light from a source of white light is passed through a slit and then through a prism. It is spread into a continuous spectrum of all wavelengths of visible light. (b) Visible light is only a very small portion of the electromagnetic spectrum. Some radiant energy has longer or shorter wavelengths than our eyes can detect. The upper part shows the approximate ranges of the electromagnetic spectrum on a logarithmic scale. The lower part shows the visible region on an expanded scale. Note that wavelength increases as frequency decreases.

EXAMPLE 5-4 *Wavelength of Light*

Light near the middle of the ultraviolet region of the electromagnetic radiation spectrum has a frequency of 2.73×10^{16} s^{-1}. Yellow light near the middle of the visible region of the spectrum has a frequency of 5.26×10^{14} s^{-1}. Calculate the wavelength that corresponds to each of these two frequencies of light.

Plan

Wavelength and frequency are inversely proportional to each other, $\lambda \nu = c$. We solve this relationship for λ and calculate the wavelengths.

Solution

(ultraviolet light) $\lambda = \dfrac{c}{\nu} = \dfrac{3.00 \times 10^8 \text{ m}\cdot\text{s}^{-1}}{2.73 \times 10^{16} \text{ s}^{-1}} = \boxed{1.10 \times 10^{-8} \text{ m } (1.10 \times 10^2 \text{ Å})}$

(yellow light) $\lambda = \dfrac{c}{\nu} = \dfrac{3.00 \times 10^8 \text{ m}\cdot\text{s}^{-1}}{5.26 \times 10^{14} \text{ s}^{-1}} = \boxed{5.70 \times 10^{-7} \text{ m } (5.70 \times 10^3 \text{ Å})}$

You should now work Exercise 40.

We have described light in terms of wave behavior. Under certain conditions, it is also possible to describe light as composed of *particles*, or **photons**. According to the ideas presented by Max Planck (1858–1947) in 1900, each photon of light has a particular amount (a **quantum**) of energy. The amount of energy possessed by a photon depends on the frequency of the light. The energy of a photon of light is given by Planck's equation

$$E = h\nu \quad \text{or} \quad E = \frac{hc}{\lambda}$$

where h is Planck's constant, $6.6260755 \times 10^{-34}$ J·s, and ν is the frequency of the light. Thus, energy is directly proportional to frequency. Planck's equation is used in Example 5-5 to show that a photon of ultraviolet light has more energy than a photon of yellow light.

 CD-ROM   Screen 7.5, Planck's Equation

 SCLINKS **TOPIC:** Electromagnetic Spectrum
GO TO: www.scilinks.org
*sci*LINKS CODE: WCH0530

EXAMPLE 5-5   *Energy of Light*

In Example 5-4 we calculated the wavelengths of ultraviolet light of frequency 2.73×10^{16} s^{-1} and of yellow light of frequency 5.26×10^{14} s^{-1}. Calculate the energy, in joules, of an individual photon of each. Compare these photons by calculating the ratio of their energies.

Plan

We use each frequency to calculate the photon energy from the relationship $E = h\nu$. Then we calculate the required ratio.

Solution

(ultraviolet light) $E = h\nu = (6.626 \times 10^{-34} \text{ J}\cdot\text{s})(2.73 \times 10^{16} \text{ s}^{-1}) = \boxed{1.81 \times 10^{-17} \text{ J}}$

(yellow light) $E = h\nu = (6.626 \times 10^{-34} \text{ J}\cdot\text{s})(5.26 \times 10^{14} \text{ s}^{-1}) = \boxed{3.49 \times 10^{-19} \text{ J}}$

(You can check these answers by calculating the energies directly from the wavelengths, using the equation $E = hc/\lambda$.)

Now, we compare the energies of these two photons.

$$\frac{E_{\text{uv}}}{E_{\text{yellow}}} = \frac{1.81 \times 10^{-17} \text{ J}}{3.49 \times 10^{-19} \text{ J}} = \boxed{51.9}$$

A photon of light near the middle of the ultraviolet region is more than 51 times more energetic than a photon of light near the middle of the visible region.

You should now work Exercise 41.

This is one reason why ultraviolet (UV) light damages your skin much more rapidly than visible light. Another reason is that many of the organic compounds in the skin absorb UV light more readily than visible light. For many biologically important molecules, a single photon of ultraviolet light is enough energy to break a chemical bond.

5-11 THE PHOTOELECTRIC EFFECT

One experiment that had not been satisfactorily explained with the wave theory of light was the **photoelectric effect.** The apparatus for the photoelectric effect is shown in Figure 5-13. The negative electrode in the evacuated tube is made of a pure metal such as cesium. When light of a sufficiently high energy strikes the metal, electrons are knocked off its surface. They then travel to the positive electrode and form a current flowing through the circuit. The important observations follow.

1. Electrons are ejected only if the light is of sufficiently short wavelength (has sufficiently high energy), no matter how long or how brightly the light shines. This wavelength limit is different for different metals.

2. The number of electrons emitted per second (the current) increases as the brightness (intensity) of the light increases, once the photon energy of the light is high enough to start the photoelectric effect (point 1). The amount of current does not depend on the wavelength (color) of the light used after the minimum photon energy needed to initiate the photoelectric effect is reached.

Classical theory said that even "low-" energy light should cause current to flow if the metal is irradiated long enough. Electrons should accumulate energy and be released when they have enough energy to escape from the metal atoms. According to the old theory, if the light is made more energetic, then the current should increase even though the light intensity remains the same. Such is *not* the case.

The answer to the puzzle was provided by Albert Einstein (1879–1955). In 1905 he extended Planck's idea that light behaves as though it were composed of *photons*, each with a particular amount (a quantum) of energy. According to Einstein, each photon can transfer its energy to a single electron during a collision. When we say that the intensity of light is increased, we mean that the number of photons striking a given area per second is increased. The picture is now one of a particle of light striking an electron near the surface of the metal and giving up its energy to the electron. If that energy is equal to or greater than the amount needed to liberate the electron, it can escape to join the photoelectric current. For this explanation, Einstein received the 1921 Nobel Prize in physics.

The intensity of light is the brightness of the light. In wave terms, it is related to the amplitude of the light waves. In photon terms, it is the number of photons hitting the target.

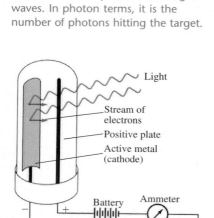

Figure 5-13 The photoelectric effect. When electromagnetic radiation of sufficient minimum energy strikes the surface of a metal (negative electrode) inside an evacuated tube, electrons are stripped off the metal to create an electric current. The current increases with increasing radiation intensity.

5-12 ATOMIC SPECTRA AND THE BOHR ATOM

Incandescent ("red hot" or "white hot") solids, liquids, and high-pressure gases give continuous spectra. When an electric current is passed through a gas in a vacuum tube at very low pressures, however, the light that the gas emits can be dispersed by a prism into distinct lines (Figure 5-14a). Such an **emission spectrum** is described as a *bright line spectrum.* The lines can be recorded photographically, and the wavelength of light that produced each line can be calculated from the position of that line on the photograph.

Similarly, we can shine a beam of white light (containing a continuous distribution of wavelengths) through a gas and analyze the beam that emerges. We find that only certain wavelengths have been absorbed (Figure 5-14b). The wavelengths that are absorbed in this **absorption spectrum** are also given off in the emission experiment. Each spectral line corresponds to a specific wavelength of light and thus to a specific amount of energy that is

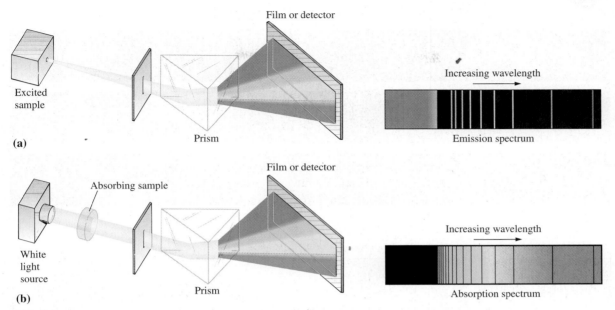

Figure 5-14 (a) *Atomic emission.* The light emitted by a sample of excited hydrogen atoms (or any other element) can be passed through a prism and separated into certain discrete wavelengths. Thus, an emission spectrum, which is a photographic recording of the separated wavelengths, is called a line spectrum. Any sample of reasonable size contains an enormous number of atoms. Although a single atom can be in only one excited state at a time, the collection of atoms contains all possible excited states. The light emitted as these atoms fall to lower energy states is responsible for the spectrum. (b) *Atomic absorption.* When white light is passed through unexcited hydrogen and then through a slit and a prism, the transmitted light is lacking in intensity at the same wavelengths as are emitted in part (a). The recorded absorption spectrum is also a line spectrum and the photographic negative of the emission spectrum.

either absorbed or emitted. An atom of each element displays its own characteristic set of lines in its emission or absorption spectrum (Figure 5-15). These spectra can serve as "fingerprints" that allow us to identify different elements present in a sample, even in trace amounts.

The sensor for each "pixel" in a digital camera is a tiny photoelectric device. The photoelectric sensors that open some supermarket and elevator doors also utilize this effect.

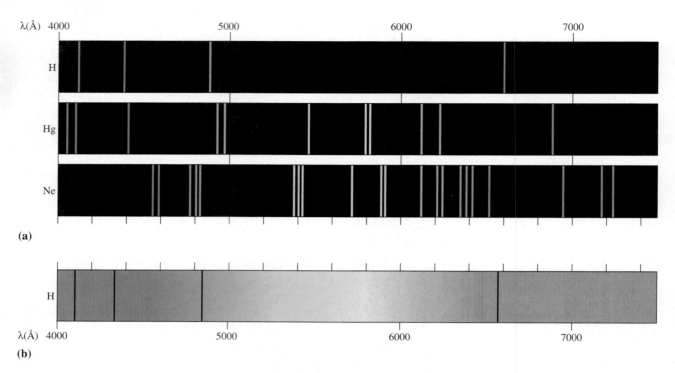

Figure 5-15 Atomic spectra in the visible region for some elements. Figure 5-14a shows how such spectra are produced. (a) Emission spectra for some elements. (b) Absorption spectrum for hydrogen. Compare the positions of these lines with those in the emission spectrum for H in (a).

EXAMPLE 5-6 *Energy of Light*

A green line of wavelength 4.86×10^{-7} m is observed in the emission spectrum of hydrogen. Calculate the energy of one photon of this green light.

Plan

We know the wavelength of the light, and we calculate its frequency so that we can then calculate the energy of each photon.

Solution

$$E = \frac{hc}{\lambda} = \frac{(6.626 \times 10^{-34}\,\text{J}\cdot\text{s})(3.00 \times 10^{8}\,\text{m/s})}{(4.86 \times 10^{-7}\,\text{m})} = \boxed{4.09 \times 10^{-19}\,\text{J/photon}}$$

To gain a better appreciation of the amount of energy involved, let's calculate the total energy, in kilojoules, emitted by one mole of atoms. (Each atom emits one photon.)

$$\frac{?\ \text{kJ}}{\text{mol}} = 4.09 \times 10^{-19}\ \frac{\text{J}}{\text{atom}} \times \frac{1\ \text{kJ}}{1 \times 10^{3}\ \text{J}} \times \frac{6.02 \times 10^{23}\ \text{atoms}}{\text{mol}} = \boxed{2.46 \times 10^{2}\ \text{kJ/mol}}$$

This calculation shows that when each atom in one mole of hydrogen atoms emits light of wavelength 4.86×10^{-7} m, the mole of atoms loses 246 kJ of energy as green light. (This would be enough energy to operate a 100-watt light bulb for more than 40 minutes.)

You should now work Exercises 42 and 44.

When an electric current is passed through hydrogen gas at very low pressures, several series of lines in the spectrum of hydrogen are produced. These lines were studied intensely by many scientists. In the late 19th century, Johann Balmer (1825–1898) and Johannes Rydberg (1854–1919) showed that the wavelengths of the various lines in the hydrogen spectrum can be related by a mathematical equation:

$$\frac{1}{\lambda} = R\left(\frac{1}{n_1^2} - \frac{1}{n_2^2}\right)$$

Here R is $1.097 \times 10^7 \text{ m}^{-1}$ and is known as the Rydberg constant. The n's are positive integers, and n_1 is smaller than n_2. The Balmer–Rydberg equation was derived from numerous observations, not theory. It is thus an empirical equation.

In 1913 Niels Bohr (1885–1962), a Danish physicist, provided an explanation for Balmer and Rydberg's observations. He wrote equations that described the electron of a hydrogen atom as revolving around its nucleus in circular orbits. He included the assumption that the electronic energy is *quantized*; that is, only certain values of electronic energy are possible. This led him to suggest that electrons can only be in certain discrete orbits, and that they absorb or emit energy in discrete amounts as they move from one orbit to

The lightning flashes produced in electrical storms and the light produced by neon gas in neon signs are two familiar examples of visible light produced by electronic transitions.

CD-ROM   Screens 7.6, Atomic Line Spectra, and 7.7, Bohr's Model of the Hydrogen Atom.

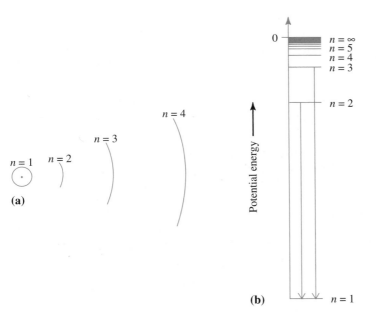

Figure 5-16 (a) The radii of the first four Bohr orbits for a hydrogen atom. The dot at the center represents the nuclear position. The radius of each orbit is proportional to n^2, so the orbits are more widely spaced as the n value increases. These four radii are in the ratio 1:4:9:16. (b) Relative values for the energies associated with some Bohr energy levels in a hydrogen atom. By convention, the potential energy of the electron is defined as zero when it is at an infinite distance from the nucleus. Any more stable arrangement would have a lower potential energy (more stable). The energy spacing between orbits gets smaller as the n value increases. For very large values of n, the energy levels are so close together that they form a continuum. Some possible electronic transitions corresponding to lines in the hydrogen emission spectrum are indicated by arrows. Transitions in the opposite direction account for lines in the absorption spectrum. The biggest energy *change* occurs when an electron jumps between $n = 1$ and $n = 2$; a considerably smaller energy change occurs when the electron jumps between $n = 3$ and $n = 4$.

AIP Emilio Segrè Visual Archives

The Danish physicist Niels Bohr was one of the most influential scientists of the 20th century. Like many other now-famous physicists of his time, he worked for a time in England with J. J. Thomson and later with Ernest Rutherford. During this period, he began to develop the ideas that led to the publication of his explanation of atomic spectra and his theory of atomic structure, for which he received the Nobel Prize in 1922. After escaping from German-occupied Denmark to Sweden in 1943, he helped to arrange the escape of hundreds of Danish Jews from the Hitler regime. He later went to the United States, where, until 1945, he worked with other scientists at Los Alamos, New Mexico, on the development of the atomic bomb. From then until his death in 1962, he worked for the development and use of atomic energy for peaceful purposes.

another. Each orbit thus corresponds to a definite *energy level* for the electron. When an electron is promoted from a lower energy level to a higher one, it absorbs a definite (or quantized) amount of energy. When the electron falls back to the original energy level, it emits exactly the same amount of energy it absorbed in moving from the lower to the higher energy level. Figure 5-16 illustrates these transactions schematically. The values of n_1 and n_2 in the Balmer–Rydberg equation identify the lower and higher levels, respectively, of these electronic transitions.

Enrichment   The Bohr Theory and the Balmer–Rydberg Equation

From mathematical equations describing the orbits for the hydrogen atom, together with the assumption of quantization of energy, Bohr was able to determine two significant aspects of each allowed orbit:

1. *Where* the electron can be with respect to the nucleus—that is, the radius, r, of the circular orbit. This is given by

$$r = n^2 a_0$$

Note: r is proportional to n^2.

where n is a positive integer $(1, 2, 3, \ldots)$ that tells which orbit is being described and a_0 is the *Bohr radius*. Bohr was able to calculate the value of a_0 from a combination of Planck's constant, the charge of the electron, and the mass of the electron as

$$a_0 = 5.292 \times 10^{-11} \text{ m} \stackrel{m}{=} 0.5292 \text{ Å}$$

2. *How stable* the electron would be in that orbit—that is, its potential energy, E. This is given by

$$E = -\frac{1}{n^2}\left(\frac{h^2}{8\pi^2 m a_0^2}\right) = -\frac{2.180 \times 10^{-18} \text{ J}}{n^2}$$

Note: E is proportional to $-\dfrac{1}{n^2}$.

where h = Planck's constant, m = the mass of the electron, and the other symbols have the same meaning as before. E is always negative when the electron is in the atom; $E = 0$ when the electron is completely removed from the atom (n = infinity).

We define the potential energy of a set of charged particles to be zero when the particles are infinitely far apart.

Results of evaluating these equations for some of the possible values of n $(1, 2, 3, \ldots)$ are shown in Figure 5-17. The larger the value of n, the farther from the nucleus is the orbit being described, and the radius of this orbit increases as the *square of n* increases. As n increases, n^2 increases, $1/n^2$ decreases, and thus the electronic energy increases (becomes less negative and smaller in magnitude). For orbits farther from the nucleus, the

(Continued on page 200)

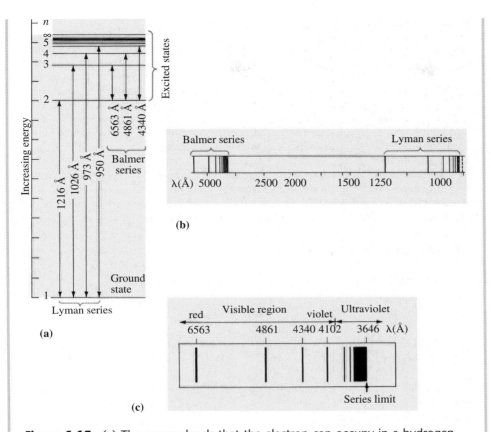

Figure 5-17 (a) The energy levels that the electron can occupy in a hydrogen atom and a few of the transitions that cause the emission spectrum of hydrogen. The numbers on the vertical lines show the wavelengths of light emitted when the electron falls to a lower energy level. (Light of the same wavelength is absorbed when the electron is promoted to the higher energy level.) The difference in energy between two given levels is exactly the same for all hydrogen atoms, so it corresponds to a specific wavelength and to a specific line in the emission spectrum of hydrogen. In a given sample, some hydrogen atoms could have their electrons excited to the $n = 2$ level. Some of these electrons could then fall to the $n = 1$ energy level, giving off the *difference* in energy in the form of light (the 1216-Å transition). Other hydrogen atoms might have their electrons excited to the $n = 3$ level; subsequently some could fall to the $n = 1$ level (the 1026-Å transition). Because higher energy levels become closer and closer in energy, *differences* in energy between successive transitions become smaller and smaller. The corresponding lines in the emission spectrum become closer together and eventually result in a continuum, a series of lines so close together that they are indistinguishable. (b) The emission spectrum of hydrogen. The series of lines produced by the electron falling to the $n = 1$ level is known as the *Lyman series*; it is in the ultraviolet region. A transition in which the electron falls to the $n = 2$ level gives rise to a similar set of lines in the visible region of the spectrum, known as the *Balmer series*. Not shown are series involving transitions to energy levels with higher values of n. (c) The Balmer series shown on an expanded scale. The line at 6563 Å (the $n = 3 \rightarrow n = 2$ transition) is much more intense than the line at 4861 Å (the $n = 4 \rightarrow n = 2$ transition) because the first transition occurs much more frequently than the second. Successive lines in the spectrum become less intense as the series limit is approached because the transitions that correspond to these lines are less probable.

(Enrichment, continued)

electronic potential energy is higher (less negative—the electron is in a *higher* energy level or in a less stable state). Going away from the nucleus, the allowable orbits are farther apart in distance, but closer together in energy. Consider the two possible limits of these equations. One limit is when $n = 1$; this describes the electron at the smallest possible distance from the nucleus and at its lowest (most negative) energy, that is, at its most stable location. The other limit is for very large values of n, that is, as n approaches infinity. As this limit is approached, the electron is very far from the nucleus, or effectively removed from the atom; the potential energy is as high as possible, approaching zero, that is, at its least stable location.

Each line in the emission spectrum represents the *difference in energies* between two allowed energy levels for the electron. When the electron goes from energy level n_2 to energy level n_1, the difference in energy is given off as a single photon. The energy of this photon can be calculated from Bohr's equation for the energy, as follows.

$$E \text{ of photon} = E_2 - E_1 = \left(-\frac{2.180 \times 10^{-18}\,\text{J}}{n_2^2}\right) - \left(-\frac{2.180 \times 10^{-18}\,\text{J}}{n_1^2}\right)$$

Factoring out the constant 2.180×10^{-18} J and rearranging, we get

$$E \text{ of photon} = 2.180 \times 10^{-18}\,\text{J}\left(\frac{1}{n_1^2} - \frac{1}{n_2^2}\right)$$

The Planck equation, $E = hc/\lambda$, relates the energy of the photon to the wavelength of the light, so

$$\frac{hc}{\lambda} = 2.180 \times 10^{-18}\,\text{J}\left(\frac{1}{n_1^2} - \frac{1}{n_2^2}\right)$$

Rearranging for $1/\lambda$, we obtain

$$\frac{1}{\lambda} = \frac{2.180 \times 10^{-18}\,\text{J}}{hc}\left(\frac{1}{n_1^2} - \frac{1}{n_2^2}\right)$$

Comparing this to the Balmer–Rydberg equation, Bohr showed that the Rydberg constant is equivalent to 2.180×10^{-18} J/hc. We can use the values for h and c to obtain the same value, 1.097×10^7 m^{-1}, that was obtained by Rydberg on a solely empirical basis. Furthermore, Bohr showed the physical meaning of the two whole numbers n_1 and n_2; they represent the two energy states between which the transition takes place. Using this approach, Bohr was able to use fundamental constants to calculate the wavelengths of the observed lines in the hydrogen emission spectrum. Thus, Bohr's application of the idea of quantization of energy to the electron in an atom provided the answer to a half-century-old puzzle concerning the discrete colors given off in the spectrum.

We now accept the fact that electrons occupy only certain energy levels in atoms. In most atoms, some of the energy differences between levels correspond to the energy of visible light. Thus, colors associated with electronic transitions in such elements can be observed by the human eye.

Although the Bohr theory satisfactorily explained the spectra of hydrogen and of other species containing one electron (He$^+$, Li^{2+}, etc.) the wavelengths in the observed spectra of more complex species could not be calculated. Bohr's assumption of circular orbits was modified in 1916 by Arnold Sommerfeld (1868–1951), who assumed elliptical orbits. Even so, the Bohr approach was doomed to failure, because it modified classical mechanics to solve a problem that could not be solved by classical mechanics. It was a contrived solution.

Various metals emit distinctive colors of visible light when heated to a high enough temperature (flame test). This is the basis for all fireworks, which use the salts of different metals like strontium (red), barium (green), and copper (blue) to produce the beautiful colors.

This failure of classical mechanics set the stage for the development of a new physics, quantum mechanics, to deal with small particles. The Bohr theory, however, did introduce the ideas that only certain energy levels are possible, that these energy levels are described by quantum numbers that can have only certain allowed values, and that the quantum numbers indicate something about where and how stable the electrons are in these energy levels. The ideas of modern atomic theory have replaced Bohr's original theory. But his achievement in showing a link between electronic arrangements and Balmer and Rydberg's empirical description of light absorption, and in establishing the quantization of electronic energy, was a very important step toward an understanding of atomic structure.

Two big questions remained about electrons in atoms: (1) How are electrons arranged in atoms? (2) How do these electrons behave? We now have the background to consider how modern atomic theory answers these questions.

Materials scientists study electron diffraction patterns to learn about the surfaces of solids.

5-13 THE WAVE NATURE OF THE ELECTRON

Einstein's idea that light can exhibit both wave properties and particle properties suggested to Louis de Broglie (1892–1987) that very small particles, such as electrons, might also display wave properties under the proper circumstances. In his doctoral thesis in 1925, de Broglie predicted that a particle with a mass m and velocity v should have the wavelength associated with it. The numerical value of this de Broglie wavelength is given by

$$\lambda = h/mv \quad \text{(where } h = \text{Planck's constant)}$$

Two years after de Broglie's prediction, C. Davisson (1882–1958) and L. H. Germer (1896–1971) at the Bell Telephone Laboratories demonstrated diffraction of electrons by a crystal of nickel. This behavior is an important characteristic of waves. It shows conclusively that electrons do have wave properties. Davisson and Germer found that the wavelength associated with electrons of known energy is exactly that predicted by de Broglie. Similar diffraction experiments have been successfully performed with other particles, such as neutrons.

Be careful to distinguish between the letter v, which represents velocity, and the Greek letter nu, ν, which represents frequency. (See Section 5-10.)

CD-ROM   Screen 7.8, Wave Properties of the Electron.

EXAMPLE 5-7 de Broglie Equation

(a) Calculate the wavelength in meters of an electron traveling at 1.24×10^7 m/s. The mass of an electron is 9.11×10^{-28} g. (b) Calculate the wavelength of a baseball of mass 5.25 oz traveling at 92.5 mph. Recall that $1 \text{ J} = 1 \text{ kg} \cdot \text{m}^2/\text{s}^2$.

Plan

For each calculation, we use the de Broglie equation

$$\lambda = \frac{h}{mv}$$

where

$$h \text{ (Planck's constant)} = 6.626 \times 10^{-34} \text{ J} \cdot \text{s} \times \frac{1\dfrac{\text{kg} \cdot \text{m}^2}{\text{s}^2}}{1 \text{ J}}$$

$$= 6.626 \times 10^{-34} \frac{\text{kg} \cdot \text{m}^2}{\text{s}}$$

For consistency of units, mass must be expressed in kilograms. In part (b), we must also convert the speed to meters per second.

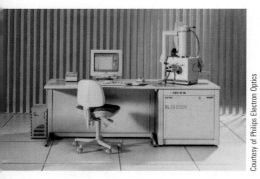

A modern electron microscope.

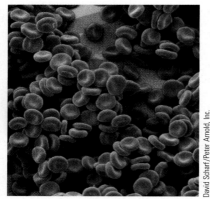

A color-enhanced scanning electron micrograph of human red blood cells, magnified 1200×.

Solution

(a)
$$m = 9.11 \times 10^{-28} \text{ g} \times \frac{1 \text{ kg}}{1000 \text{ g}} = 9.11 \times 10^{-31} \text{ kg}$$

Substituting into the de Broglie equation,

$$\lambda = \frac{h}{mv} = \frac{6.626 \times 10^{-34} \frac{\text{kg} \cdot \text{m}^2}{\text{s}}}{(9.11 \times 10^{-31} \text{ kg})\left(1.24 \times 10^7 \frac{\text{m}}{\text{s}}\right)} = 5.87 \times 10^{-11} \text{ m}$$

Though this seems like a very short wavelength, it is similar to the spacing between atoms in many crystals. A stream of such electrons hitting a crystal gives measurable diffraction patterns.

(b)
$$m = 5.25 \text{ oz} \times \frac{1 \text{ lb}}{16 \text{ oz}} \times \frac{1 \text{ kg}}{2.205 \text{ lb}} = 0.149 \text{ kg}$$

$$v = \frac{92.5 \text{ miles}}{\text{h}} \times \frac{1 \text{ h}}{3600 \text{ s}} \times \frac{1.609 \text{ km}}{1 \text{ mile}} \times \frac{1000 \text{ m}}{1 \text{ km}} = 41.3 \frac{\text{m}}{\text{s}}$$

Now, we substitute into the de Broglie equation.

$$\lambda = \frac{h}{mv} = \frac{6.626 \times 10^{-34} \frac{\text{kg} \cdot \text{m}^2}{\text{s}}}{(0.149 \text{ kg})\left(41.3 \frac{\text{m}}{\text{s}}\right)} = 1.08 \times 10^{-34} \text{ m}$$

This wavelength is far too short to give any measurable effects. Recall that atomic diameters are in the order of 10^{-10} m, which is 24 powers of 10 greater than the baseball "wavelength."

You should now work Exercise 62.

As you can see from the results of Example 5-7, the particles of the subatomic world behave very differently from the macroscopic objects with which we are familiar. To talk about the behavior of atoms and their particles, we must give up many of our long-held views about the behavior of matter. We must be willing to visualize a world of new and unfamiliar properties, such as the ability to act in some ways like a particle and in other ways like a wave.

The wave behavior of electrons is exploited in the electron microscope. This instrument allows magnification of objects far too small to be seen with an ordinary light microscope.

5-14 THE QUANTUM MECHANICAL PICTURE OF THE ATOM

Through the work of de Broglie, Davisson and Germer, and others, we now know that electrons in atoms can be treated as waves more effectively than as small compact particles traveling in circular or elliptical orbits. Large objects such as golf balls and moving automobiles obey the laws of classical mechanics (Isaac Newton's laws), but very small particles such as electrons, atoms, and molecules do not. A different kind of mechanics, called

quantum mechanics, which is based on the *wave* properties of matter, describes the behavior of very small particles much better. Quantization of energy is a consequence of these properties.

One of the underlying principles of quantum mechanics is that we cannot determine precisely the paths that electrons follow as they move about atomic nuclei. The **Heisenberg Uncertainty Principle,** stated in 1927 by Werner Heisenberg (1901–1976), is a theoretical assertion that is consistent with all experimental observations.

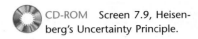 CD-ROM   Screen 7.9, Heisenberg's Uncertainty Principle.

> It is impossible to determine accurately both the momentum and the position of an electron (or any other very small particle) simultaneously.

Momentum is mass times velocity, *mv.* Because electrons are so small and move so rapidly, their motion is usually detected by electromagnetic radiation. Photons that interact with electrons have about the same energies as the electrons. Consequently, the interaction of a photon with an electron severely disturbs the motion of the electron. It is not possible to determine simultaneously both the position and the velocity of an electron, so we resort to a statistical approach and speak of the probability of finding an electron within specified regions in space.

This is like trying to locate the position of a moving automobile by driving another automobile into it.

With these ideas in mind, we list some basic ideas of quantum mechanics.

1. Atoms and molecules can exist only in certain energy states. In each energy state, the atom or molecule has a definite energy. When an atom or molecule changes its energy state, it must emit or absorb just enough energy to bring it to the new energy state (the quantum condition).

Atoms and molecules possess various forms of energy. Let's focus our attention on their *electronic energies.*

2. When atoms or molecules emit or absorb radiation (light), they change their energies. The energy change in the atom or molecule is related to the frequency or wavelength of the light emitted or absorbed by the equations:

$$\Delta E = h\nu \qquad \text{or} \qquad \Delta E = hc/\lambda$$

Recall that $\lambda\nu = c$, so $\nu = c/\lambda$.

This gives a relationship between the energy change, ΔE, and the wavelength, λ, of the radiation emitted or absorbed. *The energy lost (or gained) by an atom as it goes from higher to lower (or lower to higher) energy states is equal to the energy of the photon emitted (or absorbed) during the transition.*

3. The allowed energy states of atoms and molecules can be described by sets of numbers called *quantum numbers.*

The mathematical approach of quantum mechanics involves treating the electron in an atom as a *standing wave.* A standing wave is a wave that does not travel and therefore has at least one point at which it has zero amplitude, called a node. As an example, consider the various ways that a guitar string can vibrate when it is plucked (Figure 5-18). Because both ends are fixed (nodes), the string can vibrate only in ways in which there is a whole number of *half-wavelengths* in the length of the string (Figure 5-18a). Any possible motion of the string can be described as some combination of these allowed vibrations. In a similar way, we can imagine that the electron in the hydrogen atom behaves as a wave (recall the

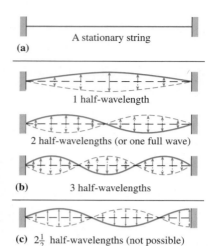

(a) A stationary string

1 half-wavelength

2 half-wavelengths (or one full wave)

(b) 3 half-wavelengths

(c) $2\frac{1}{2}$ half-wavelengths (not possible)

Figure 5-18 When a string that is fixed at both ends—such as (a) a guitar string—is plucked, it has a number of natural patterns of vibration, called normal modes. Because the string is fixed at both ends, the ends must be stationary. Each different possible vibration is a standing wave and can be described by a wave function. The only waves that are possible are those in which a whole number of half-wavelengths fits into the string length. These allowed waves constitute a harmonic series. Any total motion of the string is some combination of these allowed harmonics. (b) Some of the ways in which a plucked guitar string can vibrate. The position of the string at one extreme of each vibration is shown as a solid line, and at the other extreme as a dashed line. (c) An example of vibration that is *not* possible for a plucked string. In such a vibration, an end of the string would move; this is not possible because the ends are fixed.

de Broglie relationship in the last section). The electron can be described by the same kind of standing-wave mathematics that is applied to the vibrating guitar string. In this approach, the electron is characterized by a three-dimensional wave function, ψ. In a given space around the nucleus, only certain "waves" can exist. Each "allowed wave" corresponds to a stable energy state for the electron and is described by a particular set of quantum numbers.

The quantum mechanical treatment of atoms and molecules is highly mathematical. The important point is that each solution of the Schrödinger wave equation (see the following Enrichment section) describes a possible energy state for the electrons in the atom. Each solution is described by a set of **quantum numbers.** These numbers are in accord with those deduced from experiment and from empirical equations such as the Balmer–Rydberg equation. Solutions of the Schrödinger equation also tell us about the shapes and orientations of the probability distributions of the electrons. (The Heisenberg Principle implies that this is how we must describe the positions of the electrons.) These *atomic orbitals* (which are described in Section 5-16) are deduced from the solutions of the Schrödinger equation. The orbitals are defined by the quantum numbers.

Enrichment   The Schrödinger Equation

In 1926 Erwin Schrödinger (1887–1961) modified an existing equation that described a three-dimensional standing wave by imposing wavelength restrictions suggested by de Broglie's ideas. The modified equation allowed him to calculate the energy levels in the hydrogen atom. It is a differential equation that need not be memorized or even understood to read this book. A knowledge of differential calculus would be necessary to solve it.

$$-\frac{h^2}{8\pi^2 m}\left(\frac{\partial^2 \psi}{\partial x^2} + \frac{\partial^2 \psi}{\partial y^2} + \frac{\partial^2 \psi}{\partial z^2}\right) + V\psi = E\psi$$

This equation has been solved exactly only for one-electron species such as the hydrogen atom and the ions He^+ and Li^{2+}. Simplifying assumptions are necessary to solve the equation for more complex atoms and molecules. Chemists and physicists have used their intuition and ingenuity (and modern computers), however, to apply this equation to more complex systems.

In 1928 Paul A. M. Dirac (1902–1984) reformulated electron quantum mechanics to take into account the effects of relativity. This gave rise to a fourth quantum number.

5-15 QUANTUM NUMBERS

The solutions of the Schrödinger and Dirac equations for hydrogen atoms give wave functions, ψ, that describe the various states available to hydrogen's single electron. Each of these possible states is described by four quantum numbers. We can use these quantum numbers to designate the electronic arrangements in all atoms, their so-called **electron configurations.** These quantum numbers play important roles in describing the energy levels of electrons and the shapes of the orbitals that describe distributions of electrons in space. The interpretation will become clearer when we discuss atomic orbitals in the following section. For now, let's say that

> an **atomic orbital** is a region of space in which the probability of finding an electron is high.

We define each quantum number and describe the range of values it may take.

1. The **principal quantum number,** n, describes the *main energy level,* or shell, that an electron occupies. It may be any positive integer:

$$n = 1, 2, 3, 4, \ldots$$

2. The **angular momentum quantum number,** ℓ, designates the *shape of the region* in space that an electron occupies. Within a shell (defined by the value of n, the principal quantum number), different sublevels or subshells are possible, each with a characteristic shape. The angular momentum quantum number designates a *sublevel,* or specific *shape* of atomic orbital that an electron may occupy. This number, ℓ, may take integral values from 0 up to and including $(n - 1)$:

$$\ell = 0, 1, 2, \ldots, (n - 1)$$

Thus, the maximum value of ℓ is $(n - 1)$. We give a letter notation to each value of ℓ. Each letter corresponds to a different sublevel (subshell) and a differently shaped orbital:

$$\ell = 0, 1, 2, 3, \ldots, (n - 1)$$
$$s \quad p \quad d \quad f$$

The *s, p, d, f* designations arise from the characteristics of spectral emission lines produced by electrons occupying the orbitals: *s* (sharp), *p* (principal), *d* (diffuse), and *f* (fundamental).

In the first shell, the maximum value of ℓ is zero, which tells us that there is only an s subshell and no p subshell. In the second shell, the permissible values of ℓ are 0 and 1, which tells us that there are only s and p subshells.

3. The **magnetic quantum number,** m_ℓ, designates the specific orbital within a subshell. Orbitals within a given subshell differ in their orientations in space, but not in their energies. Within each subshell, m_ℓ may take any integral values from $-\ell$ through zero up to and including $+\ell$:

$$m_\ell = (-\ell), \ldots, 0, \ldots, (+\ell)$$

The maximum value of m_ℓ depends on the value of ℓ. For example, when $\ell = 1$, which designates the p subshell, there are three permissible values of m_ℓ: -1, 0, and $+1$. Thus, three distinct regions of space, called atomic orbitals, are associated with a p subshell. We refer to these orbitals as the p_x, p_y, and p_z orbitals (see Section 5-16).

TABLE 5-4	*Permissible Values of the Quantum Numbers Through* $n = 4$				
n	ℓ	m_ℓ	m_s	Electron Capacity of Subshell = $4\ell + 2$	Electron Capacity of Shell = $2n^2$
1	0 (1*s*)	0	$+\frac{1}{2}, -\frac{1}{2}$	2	2
2	0 (2*s*)	0	$+\frac{1}{2}, -\frac{1}{2}$	2	8
	1 (2*p*)	$-1, 0, +1$	$\pm\frac{1}{2}$ for each value of m_ℓ	6	
3	0 (3*s*)	0	$+\frac{1}{2}, -\frac{1}{2}$	2	18
	1 (3*p*)	$-1, 0, +1$	$\pm\frac{1}{2}$ for each value of m_ℓ	6	
	2 (3*d*)	$-2, -1, 0, +1, +2$	$\pm\frac{1}{2}$ for each value of m_ℓ	10	
4	0 (4*s*)	0	$+\frac{1}{2}, -\frac{1}{2}$	2	32
	1 (4*p*)	$-1, 0, +1$	$\pm\frac{1}{2}$ for each value of m_ℓ	6	
	2 (4*d*)	$-2, -1, 0, +1, +2$	$\pm\frac{1}{2}$ for each value of m_ℓ	10	
	3 (4*f*)	$-3, -2, -1, 0, +1, +2, +3$	$\pm\frac{1}{2}$ for each value of m_ℓ	14	

CD-ROM Screens 7.10, Schrödinger's Equation and Wave Functions; 7.11, Shells, Subshells and Orbitals; and 7.12, Quantum Numbers and Orbitals.

4. The **spin quantum number,** m_s, refers to the spin of an electron and the orientation of the magnetic field produced by this spin. For every set of n, ℓ, and m_ℓ values, m_s can take the value $+\frac{1}{2}$ or $-\frac{1}{2}$:

$$m_s = \pm\frac{1}{2}$$

The values of n, ℓ, and m_ℓ describe a particular atomic orbital. Each atomic orbital can accommodate no more than two electrons, one with $m_s = +\frac{1}{2}$ and another with $m_s = -\frac{1}{2}$.

Table 5-4 summarizes some permissible values for the four quantum numbers. Spectroscopic evidence confirms the quantum mechanical predictions about the number of atomic orbitals in each shell.

5-16 ATOMIC ORBITALS

As you study the next two sections, keep in mind that the wave function, ψ, for an orbital characterizes two features of an electron in that orbital: (1) *where* (the region in space) the probability of finding the electron is high and (2) *how stable* that electron is (its energy).

Let's now describe the distributions of electrons in atoms. For each neutral atom, we must account for a number of electrons equal to the number of protons in the nucleus, that is, the atomic number of the atom. Each electron is said to occupy an atomic orbital defined by a set of quantum numbers n, ℓ, and m_ℓ. In any atom, each orbital can hold a maximum of two electrons. Within each atom, these atomic orbitals, taken together, can be represented as a diffuse cloud of electrons (Figure 5-19).

The main shell of each atomic orbital in an atom is indicated by the principal quantum number n (from the Schrödinger equation). As we have seen, the principal quantum number takes integral values: $n = 1, 2, 3, 4, \ldots$. The value $n = 1$ describes the first, or innermost, shell. These shells have been referred to as electron energy levels. Successive shells are at increasingly greater distances from the nucleus. For example, the $n = 2$ shell is farther from the nucleus than the $n = 1$ shell. The electron capacity of each shell is indicated in the right-hand column of Table 5-4. For a given n, the capacity is $2n^2$.

By the rules of Section 5-15, each shell has an s subshell (defined by $\ell = 0$) consisting of one s atomic orbital (defined by $m_\ell = 0$). We distinguish among orbitals in different principal shells (main energy levels) by using the principal quantum number as a coefficient; 1*s*

SC*L*INKS. **TOPIC:** Atomic Orbitals
GO TO: www.scilinks.org
*sci*LINKS **CODE:** WCH0550

Figure 5-19 An electron cloud surrounding an atomic nucleus. The electron density drops off rapidly but smoothly as distance from the nucleus increases.

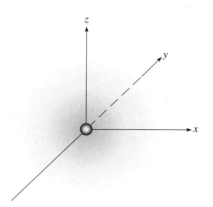

indicates the *s* orbital in the first shell, *2s* is the *s* orbital in the second shell, *2p* is a *p* orbital in the second shell, and so on (Table 5-4).

For each solution to the quantum mechanical equation, we can calculate the electron probability density (sometimes just called the electron density) at each point in the atom. This is the probability of finding an electron at that point. It can be shown that this electron density is proportional to $r^2\psi^2$, where r is the distance from the nucleus.

In the graphs in Figure 5-20, the electron probability density at a given distance from the nucleus is plotted against distance from the nucleus, for *s* orbitals. It is found that the

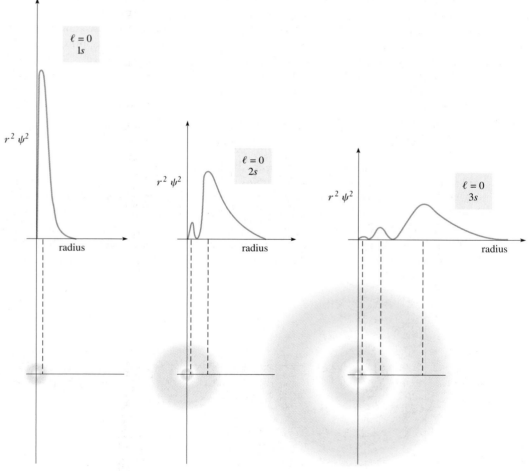

Figure 5-20 Plots of the electron density distributions associated with *s* orbitals. For any *s* orbital, this plot is the same in any direction (orbital is spherically symmetrical). The sketch below each plot shows a cross section, in the plane of the atomic nucleus, of the electron cloud associated with that orbital. Electron density is proportional to $r^2\psi^2$. *s* orbitals with $n > 1$ have $n - 1$ regions where the electron density drops to zero. These are indicated on the plots by the electron density dropping to zero near the nucleus and on the cross sections by the white region.

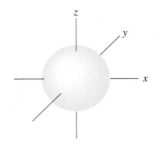

Figure 5-21   The shape of an s orbital.

CD-ROM   Screens 7.13, Shapes of Atomic Orbitals, and 7.14, Orbital Shapes and Chemistry.

electron probability density curve is the same regardless of the direction in the atom. We describe an *s* orbital as *spherically symmetrical*; that is, it is round like a basketball (Figure 5-21). The electron clouds (electron densities) associated with the 1*s*, 2*s*, and 3*s* atomic orbitals are shown just below the plots in Figure 5-20. The electron clouds are three-dimensional, and only cross sections are shown here. The regions shown in some figures (Figures 5-21 through 5-25) appear to have surfaces or skins only because they are arbitrarily "cut off" so that there is a 90% probability of finding an electron occupying the orbital somewhere within the volume defined by the surface.

Beginning with the second shell, each shell also contains a *p* subshell, defined by $\ell = 1$. Each of these subshells consists of a set of *three p* atomic orbitals, corresponding to the three allowed values of m_ℓ (-1, 0, and $+1$) when $\ell = 1$. The sets are referred to as 2*p*, 3*p*, 4*p*, 5*p*, . . . orbitals to indicate the main shells in which they are found. Each set of atomic *p* orbitals resembles three mutually perpendicular equal-arm dumbbells (see Figure 5-22). The nucleus defines the origin of a set of Cartesian coordinates with the usual *x*, *y*, and *z* axes (see Figure 5-23a). The subscript *x*, *y*, or *z* indicates the axis along which each of the orbitals is directed. A set of three *p* atomic orbitals may be represented as in Figure 5-23b.

Beginning at the third shell, each shell also contains a third subshell ($\ell = 2$) composed of a set of *five d* atomic orbitals ($m_\ell = -2, -1, 0, +1, +2$). They are designated 3*d*, 4*d*, 5*d*, . . . to indicate the shell in which they are found. The shapes of the members of a set are indicated in Figure 5-24.

In each of the fourth and larger shells, there is also a fourth subshell, containing a set of *seven f* atomic orbitals ($\ell = 3$, $m_\ell = -3, -2, -1, 0, +1, +2, +3$). These are shown in Figure 5-25.

Thus, we see the first shell contains only the 1*s* orbital; the second shell contains the 2*s* and three 2*p* orbitals; the third shell contains the 3*s*, three 3*p*, and five 3*d* orbitals; and the fourth shell consists of a 4*s*, three 4*p*, five 4*d*, and seven 4*f* orbitals. All subsequent shells contain *s*, *p*, *d*, and *f* subshells as well as others that are not occupied in any presently known elements in their lowest energy states.

The sizes of orbitals increase with increasing *n*, and the true shapes of *p* orbitals are "diffuse," as shown in Figure 5-26. The directions of *p*, *d*, and *f* orbitals, however, are easier to

Figure 5-22   Three representations of the shape of a *p* orbital. The top view is a probability diagram, while the middle view shows how most chemists draw a *p* orbital. The plot at the bottom is along the axis of maximum electron density for this orbital. A plot along any other direction would be different, because a *p* orbital is *not* spherically symmetrical like an *s* orbital.

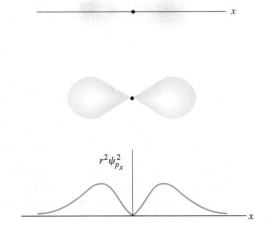

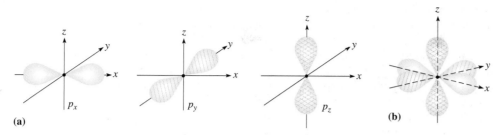

(a)

(b)

Figure 5-23 (a) The relative directional character of a set of *p* orbitals. (b) A model of three *p* orbitals (p_x, p_y, and p_z) of a single set of orbitals. The nucleus is at the center. The lobes are actually more diffuse ("fatter") than depicted. See Figure 5-26.

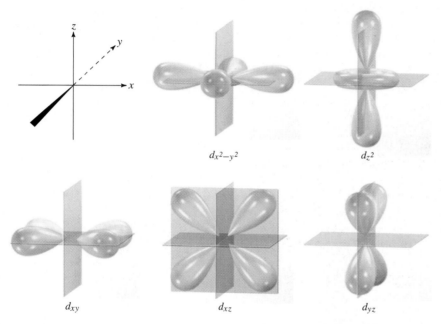

$d_{x^2-y^2}$

d_{z^2}

d_{xy}

d_{xz}

d_{yz}

Figure 5-24 Spatial orientation of *d* orbitals. Note that the lobes of the $d_{x^2-y^2}$ and d_{z^2} orbitals lie along the axes, whereas the lobes of the others lie along diagonals between the axes.

www Some excellent visual representations of orbital shapes can be seen at **http://www.shef.ac.uk/chemistry/orbitron/index.html**

visualize in drawings such as those in Figures 5-23, 5-24, and 5-25; therefore, these "slender" representations are usually used.

In this section, we haven't yet discussed the fourth quantum number, the spin quantum number, m_s. Because m_s has two possible values, $+\frac{1}{2}$ and $-\frac{1}{2}$, each atomic orbital, defined by the values of n, ℓ, and m_ℓ, has a capacity of two electrons. Electrons are negatively charged, and they behave as though they were spinning about axes through their centers, so they act like tiny magnets. The motions of electrons produce magnetic fields, and these can interact with one another. Two electrons in the same orbital having opposite m_s values are said to be **spin-paired,** or simply **paired** (Figure 5-27).

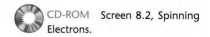 CD-ROM Screen 8.2, Spinning Electrons.

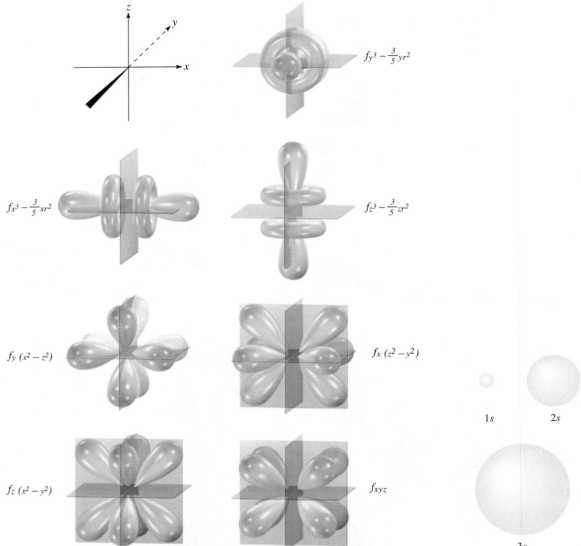

$$f_{y^3 - \frac{3}{5}yr^2}$$

$$f_{x^3 - \frac{3}{5}xr^2}$$

$$f_{z^3 - \frac{3}{5}zr^2}$$

$$f_{y\,(x^2 - z^2)}$$

$$f_{x\,(z^2 - y^2)}$$

$$f_{z\,(x^2 - y^2)}$$

$$f_{xyz}$$

Figure 5-25 Relative directional character of *f* orbitals. The seven orbitals are shown within cubes as an aid to visualization.

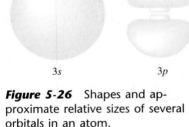

1s   2s   2p

3s   3p

Figure 5-26 Shapes and approximate relative sizes of several orbitals in an atom.

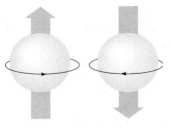

One electron has $m_s = +\frac{1}{2}$;

the other has $m_s = -\frac{1}{2}$.

Figure 5-27 Electron spin. Electrons act as though they spin about an axis through their centers. Because electrons may spin in two directions, the spin quantum number has two possible values, $+\frac{1}{2}$ and $-\frac{1}{2}$, sometimes referred to as "spin up" or "spin down." Each electron spin produces a magnetic field. When two electrons have opposite spins, the attraction due to their opposite magnetic fields (*gray arrows*) helps to overcome the repulsion of their like charges. This permits two electrons to occupy the same region (orbital).

Let's summarize, in tabular form, some of the information we have developed to this point. The principal quantum number n indicates the main shell. The number of subshells per shell is equal to n, the number of atomic orbitals per shell is n^2, and the maximum number of electrons per shell is $2n^2$, because each atomic orbital can hold two electrons.

Shell n	Number of Subshells per Shell n	Number of Atomic Orbitals n^2	Maximum Number of Electrons $2n^2$
1	1	1 ($1s$)	2
2	2	4 ($2s$, $2p_x$, $2p_y$, $2p_z$)	8
3	3	9 ($3s$, three $3p$'s, five $3d$'s)	18
4	4	16	32
5	5	25	50

5-17 ELECTRON CONFIGURATIONS

The wave function for an atom simultaneously depends on (describes) all the electrons in the atom. The Schrödinger equation is much more complicated for atoms with more than one electron than for a one-electron species such as a hydrogen atom, and an explicit solution to this equation is not possible even for helium, let alone for more complicated atoms. We must therefore rely on approximations to solutions of the many-electron Schrödinger equation. We shall use one of the most common and useful, called the *orbital approximation.* In this approximation, the electron cloud of an atom is assumed to be the superposition of charge clouds, or orbitals, arising from the individual electrons; these orbitals resemble the atomic orbitals of hydrogen (for which exact solutions are known), which we described in some detail in the previous section. Each electron is described by the same allowed combinations of quantum numbers (n, ℓ, m_ℓ, and m_s) that we used for the hydrogen atom; however, the order of energies of the orbitals is often different from that in hydrogen.

The great power of modern computers has allowed scientists to make numerical approximations to this solution to very high accuracy for simple atoms such as helium. As the number of electrons increases, however, even such numerical approaches become quite difficult to apply and interpret. For multielectron atoms, more quantitative approximations are used.

Let us now examine the electronic structures of atoms of different elements. The electronic arrangement that we will describe for each atom is called the **ground state electron configuration.** This corresponds to the isolated atom in its lowest energy, or unexcited, state. Electron configurations for the elements, as determined by experiment, are given in Appendix B. We will consider the elements in order of increasing atomic number, using as our guide the periodic table on the inside front cover of this text.

In describing ground state electron configuration, the guiding idea is that the *total energy* of the atom is as low as possible. To determine these configurations, we use the **Aufbau Principle** as a guide:

The German verb *aufbauen* means "to build up."

> Each atom is "built up" by (1) adding the appropriate numbers of protons and neutrons in the nucleus as specified by the atomic number and the mass number, and (2) adding the necessary number of electrons into orbitals in the way that gives the lowest *total* energy for the atom.

As we apply this principle, we will focus on the difference in electronic arrangement between a given element and the element with an atomic number that is one lower. In doing this, we emphasize the particular electron that distinguishes each element from the previous one; however, we should remember that this distinction is artificial because electrons are not really distinguishable. Though we do not always point it out, we *must* keep in mind that the atomic number (the charge on the nucleus) also differs from one element to the next.

The orbitals increase in energy with increasing value of the quantum number n. For a given value of n, energy increases with increasing value of ℓ. In other words, within a particular main shell, the s subshell is lowest in energy, the p subshell is the next lowest, then the d, then the f, and so on. As a result of changes in the nuclear charge and interactions among the electrons in the atom, the order of energies of the orbitals can vary somewhat from atom to atom.

Two general rules help us to predict electron configurations.

1. Electrons are assigned to orbitals in order of increasing value of $(n + \ell)$.
2. For subshells with the same value of $(n + \ell)$, electrons are assigned first to the subshell with lower n.

For example, the $2s$ subshell has ($n + \ell = 2 + 0 = 2$), and the $2p$ subshell has ($n + \ell = 2 + 1 = 3$), so we would expect to fill the $2s$ subshell before the $2p$ subshell (Rule 1). This rule also predicts that the $4s$ subshell ($n + \ell = 4 + 0 = 4$) will fill before the $3d$ subshell ($n + \ell = 3 + 2 = 5$). Rule 2 reminds us to fill $2p$ ($n + \ell = 2 + 1 = 3$) before $3s$ ($n + \ell = 3 + 0 = 3$) because $2p$ has a lower value of n. The *usual* order of energies of orbitals of an atom and a helpful device for remembering this order are shown in Figures 5-28 and 5-29.

But we should consider these only as a *guide* to predicting electron arrangements. The observed electron configurations of lowest total energy do not always match those predicted by the Aufbau guide, and we will see a number of exceptions, especially for elements in the B groups (transition metals) of the periodic table.

The electronic structures of atoms are governed by the **Pauli Exclusion Principle:**

No two electrons in an atom may have identical sets of four quantum numbers.

An orbital is described by a particular allowed set of values for n, ℓ, and m_ℓ. Two electrons can occupy the same orbital only if they have opposite spins, m_s. Two such electrons in the same orbital are said to be *spin-paired*, or simply *paired*. A single electron that occupies an orbital by itself is said to be *unpaired*. For simplicity, we shall indicate atomic orbitals as ___ and show an unpaired electron as ↑ and spin-paired electrons as ↑↓.

Row 1. The first shell consists of only one atomic orbital, $1s$. This can hold a maximum of two electrons. Hydrogen, as we have already noted, contains just one electron. Helium, a noble gas, has a filled first main shell (two electrons) and is so stable that no chemical reactions of helium are known.

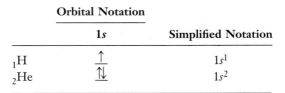

	Orbital Notation	Simplified Notation
	$1s$	
$_1$H	↑	$1s^1$
$_2$He	↑↓	$1s^2$

Row 2. Elements of atomic numbers 3 through 10 occupy the second period, or horizontal row, in the periodic table. In neon atoms the second main shell is filled completely. Neon, a noble gas, is extremely stable. No reactions of it are known.

CD-ROM Screens 8.4, The Pauli Exclusion Principle, and 8.5, Atomic Subshell Energies.

Helium's electrons can be displaced only by very strong forces, as in excitation by high-voltage discharges.

In the simplified notation, we indicate with superscripts the number of electrons in each subshell.

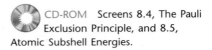

principal quantum number n

$1s^2$

number of electrons in orbital or set of equivalent orbitals

angular momentum quantum number ℓ

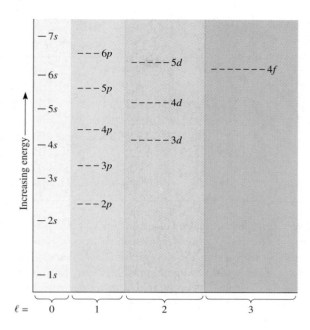

Figure 5-28 The usual order of filling (Aufbau order) of the orbitals of an atom. The relative energies are different for different elements, but the following main features should be noted: (1) The largest energy gap is between the 1s and 2s orbitals. (2) The energies of orbitals are generally closer together at higher energies. (3) The gap between np and $(n + 1)s$ (e.g., between 2p and 3s or between 3p and 4s) is fairly large. (4) The gap between $(n - 1)d$ and ns (e.g., between 3d and 4s) is quite small. (5) The gap between $(n - 2)f$ and ns (e.g., between 4f and 6s) is even smaller.

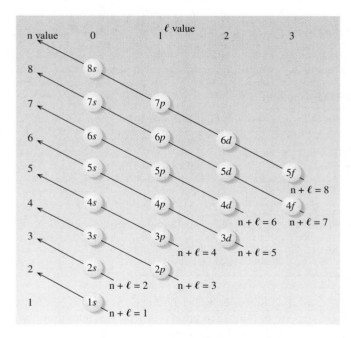

Figure 5-29 An aid to remembering the usual order of filling of atomic orbitals. Write each shell (value of n) on one horizontal line, starting with $n = 1$ at the bottom. Write all like subshells (same ℓ values) in the same vertical column. Subshells are filled in order of increasing $(n + \ell)$. When subshells have the same $(n + \ell)$, the subshell with the lower n fills first. To use the diagram, we follow the diagonal arrows in order, reading bottom to top.

In writing electron configurations of atoms, we frequently simplify the notations. The abbreviation [He] indicates that the $1s$ orbital is completely filled, $1s^2$, as in helium.

As with helium, neon's electrons can be displaced by high-voltage electric discharge, as is observed in neon signs.

	Orbital Notation			Simplified Notation		
	$1s$	$2s$	$2p$			
$_3$Li	⇅	↑		$1s^2 2s^1$	or	[He] $2s^1$
$_4$Be	⇅	⇅		$1s^2 2s^2$		[He] $2s^2$
$_5$B	⇅	⇅	↑ __ __	$1s^2 2s^2 2p^1$		[He] $2s^2 2p^1$
$_6$C	⇅	⇅	↑ ↑ __	$1s^2 2s^2 2p^2$		[He] $2s^2 2p^2$
$_7$N	⇅	⇅	↑ ↑ ↑	$1s^2 2s^2 2p^3$		[He] $2s^2 2p^3$
$_8$O	⇅	⇅	⇅ ↑ ↑	$1s^2 2s^2 2p^4$		[He] $2s^2 2p^4$
$_9$F	⇅	⇅	⇅ ⇅ ↑	$1s^2 2s^2 2p^5$		[He] $2s^2 2p^5$
$_{10}$Ne	⇅	⇅	⇅ ⇅ ⇅	$1s^2 2s^2 2p^6$		[He] $2s^2 2p^6$

We see that some atoms have unpaired electrons in the same set of energetically equivalent, or **degenerate,** orbitals. We have already seen that two electrons can occupy a given atomic orbital (with the same values of n, ℓ, and m_ℓ) *only* if their spins are paired (have opposite values of m_s). Even with pairing of spins, however, two electrons that are in the same orbital repel each other more strongly than do two electrons in different (but equal-energy) orbitals. Thus, both theory and experimental observations (see Section 5-18) lead to **Hund's Rule:**

> Electrons occupy all the orbitals of a given subshell singly before pairing begins. These unpaired electrons have parallel spins.

Thus, carbon has two unpaired electrons in its $2p$ orbitals, and nitrogen has three.

Row 3. The next element beyond neon is sodium. Here we begin to add electrons to the third shell. Elements 11 through 18 occupy the third period in the periodic table.

	Orbital Notation		Simplified Notation
	$3s$	$3p$	
$_{11}$Na	[Ne] ↑		[Ne] $3s^1$
$_{12}$Mg	[Ne] ⇅		[Ne] $3s^2$
$_{13}$Al	[Ne] ⇅	↑ __ __	[Ne] $3s^2 3p^1$
$_{14}$Si	[Ne] ⇅	↑ ↑ __	[Ne] $3s^2 3p^2$
$_{15}$P	[Ne] ⇅	↑ ↑ ↑	[Ne] $3s^2 3p^3$
$_{16}$S	[Ne] ⇅	⇅ ↑ ↑	[Ne] $3s^2 3p^4$
$_{17}$Cl	[Ne] ⇅	⇅ ⇅ ↑	[Ne] $3s^2 3p^5$
$_{18}$Ar	[Ne] ⇅	⇅ ⇅ ⇅	[Ne] $3s^2 3p^6$

Although the third shell is not yet filled (the d orbitals are still empty), argon is a noble gas. All noble gases except helium have $ns^2 np^6$ electron configurations (where n indicates the largest occupied shell). The noble gases are quite unreactive.

Rows 4 and 5. It is an experimentally observed fact that *an electron occupies the available orbital that gives the atom the lowest total energy.* It is observed that filling the $4s$ orbitals before electrons enter the $3d$ orbitals *usually* leads to a lower total energy for the atom than some other arrangements. We therefore fill the orbitals in this order (see Figure 5-28). According to the Aufbau order (recall Figures 5-28 and 5-29), $4s$ fills before $3d$. In general, *the $(n + 1)s$ orbital fills before the nd orbital.* This is sometimes referred to as the $(n + 1)$ *rule.*

After the $3d$ sublevel is filled to its capacity of 10 electrons, the $4p$ orbitals fill next, taking us to the noble gas krypton. Then the $5s$ orbital, the five $4d$ orbitals, and the three $5p$ orbitals fill to take us to xenon, a noble gas.

Let's now examine the electronic structure of the 18 elements in the fourth period in some detail. Some of these have electrons in d orbitals.

Orbital Notation

		3d	4s	4p	Simplified Notation
$_{19}$K	[Ar]		↑		[Ar] $4s^1$
$_{20}$Ca	[Ar]		↑↓		[Ar] $4s^2$
$_{21}$Sc	[Ar]	↑ _ _ _ _	↑↓		[Ar] $3d^14s^2$
$_{22}$Ti	[Ar]	↑ ↑ _ _ _	↑↓		[Ar] $3d^24s^2$
$_{23}$V	[Ar]	↑ ↑ ↑ _ _	↑↓		[Ar] $3d^34s^2$
$_{24}$Cr	[Ar]	↑ ↑ ↑ ↑ ↑	↑		[Ar] $3d^54s^1$
$_{25}$Mn	[Ar]	↑ ↑ ↑ ↑ ↑	↑↓		[Ar] $3d^54s^2$
$_{26}$Fe	[Ar]	↑↓ ↑ ↑ ↑ ↑	↑↓		[Ar] $3d^64s^2$
$_{27}$Co	[Ar]	↑↓ ↑↓ ↑ ↑ ↑	↑↓		[Ar] $3d^74s^2$
$_{28}$Ni	[Ar]	↑↓ ↑↓ ↑↓ ↑ ↑	↑↓		[Ar] $3d^84s^2$
$_{29}$Cu	[Ar]	↑↓ ↑↓ ↑↓ ↑↓ ↑↓	↑		[Ar] $3d^{10}4s^1$
$_{30}$Zn	[Ar]	↑↓ ↑↓ ↑↓ ↑↓ ↑↓	↑↓		[Ar] $3d^{10}4s^2$
$_{31}$Ga	[Ar]	↑↓ ↑↓ ↑↓ ↑↓ ↑↓	↑↓	↑ _ _	[Ar] $3d^{10}4s^24p^1$
$_{32}$Ge	[Ar]	↑↓ ↑↓ ↑↓ ↑↓ ↑↓	↑↓	↑ ↑ _	[Ar] $3d^{10}4s^24p^2$
$_{33}$As	[Ar]	↑↓ ↑↓ ↑↓ ↑↓ ↑↓	↑↓	↑ ↑ ↑	[Ar] $3d^{10}4s^24p^3$
$_{34}$Se	[Ar]	↑↓ ↑↓ ↑↓ ↑↓ ↑↓	↑↓	↑↓ ↑ ↑	[Ar] $3d^{10}4s^24p^4$
$_{35}$Br	[Ar]	↑↓ ↑↓ ↑↓ ↑↓ ↑↓	↑↓	↑↓ ↑↓ ↑	[Ar] $3d^{10}4s^24p^5$
$_{36}$Kr	[Ar]	↑↓ ↑↓ ↑↓ ↑↓ ↑↓	↑↓	↑↓ ↑↓ ↑↓	[Ar] $3d^{10}4s^24p^6$

As you study these electron configurations, you should be able to see how most of them are predicted from the Aufbau order. However, as we fill the $3d$ set of orbitals, from Sc to Zn, we see that these orbitals are not filled quite regularly. As the $3d$ orbitals are filled, their energies get closer to that of the $4s$ orbital and eventually become lower. If the order of filling of electrons on chromium gave the expected configuration, it would be: [Ar] $4s^23d^4$. Chemical and spectroscopic evidence indicates, however, that the configuration of Cr has only one electron in the $4s$ orbital, [Ar] $4s^13d^5$. For this element, the $4s$ and $3d$ orbitals are nearly equal in energy. Six electrons in these six orbitals of nearly the same energy are more stable with the electrons all unpaired, [Ar] $3d$ ↑ ↑ ↑ ↑ ↑ $4s$ ↑ rather than the predicted order [Ar] $3d$ ↑ ↑ ↑ ↑ _ $4s$ ↑↓.

The next elements, Mn to Ni, have configurations as predicted by the Aufbau order, presumably because forming a pair of electrons in the larger $4s$ orbital is easier than in a smaller, less diffuse $3d$ orbital. By the time Cu is reached, the energy of $3d$ is sufficiently lower than that of $4s$ so that the total energy of the configuration [Ar] $4s^13d^{10}$ is lower than that of [Ar] $4s^23d^9$.

We notice that the exceptions for Cr and Cu give half-filled or filled sets of equivalent orbitals (d^5 and d^{10}, respectively), and this is also true for several other exceptions to the Aufbau order. You may wonder why such an exception does not occur in, for example, Si

End-of-chapter Exercises 79–109 provide much valuable practice in writing electron configurations.

or Ge, where we could have an s^1p^3 configuration that would have half-filled sets of s and p orbitals. It does not occur because of the very large energy gap between ns and np orbitals. There is some evidence that does, however, suggest an enhanced stability of half-filled sets of p orbitals.

    The electron configurations for the transition metals discussed here and in Appendix B are for individual metal atoms in the gas phase. Most chemists work with the transition metals either in the metallic state or as coordination compounds (see Chapter 25). A solid transition metal has a band structure of overlapping d and s orbital levels (Section 13-7). When transition metal atoms have other types of atoms or molecules bonded to them, however, the electronic configuration becomes simpler in that the d orbitals fill first, followed by the next higher s orbital. This is illustrated by Cr, which has a $4s^13d^5$ electronic configuration as a free atom in the gas phase. But in the compound $Cr(CO)_6$, chromium hexacarbonyl, which contains a central Cr atom surrounded by six neutral carbon monoxide (or carbonyl) groups, the chromium atom has a $3d^6$ electronic configuration.

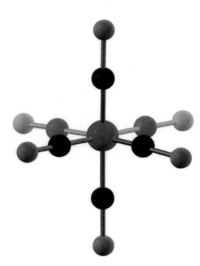

Ball-and-stick model of $Cr(CO)_6$ (Cr = purple, C = black/gray, O = red)

The electron configurations of elements 1 through 109 are given in Appendix B.

✓ **Problem-Solving Tip:** *Exceptions to the Aufbau Order*

In Appendix B, you will find a number of exceptions to the electron configurations predicted from the Aufbau Principle. You should realize that statements such as the Aufbau Principle and the $(n + 1)$ rule merely represent general guidelines and should not be viewed as hard-and-fast rules; exceptions occur to make the *total energy* of the atom as low as possible. Some of the reasons for exceptions are

1. The Aufbau order of orbital energies is based on calculations for the hydrogen atom, which contains only one electron. The orbital energies also depend on additional factors such as the nuclear charge and interactions of electrons in different occupied orbitals.

2. The energy scale varies with the atomic number.

3. Some orbitals are very close together, so their order can change, depending on the occupancies of other orbitals.

Some types of exceptions to the Aufbau order are general enough to remember easily, for example, those based on the special stability of filled or half-filled sets of orbitals. Other exceptions are quite unpredictable. Your instructor may expect you to remember some of the exceptions.

    Let us now write the quantum numbers to describe each electron in an atom of nitrogen. Keep in mind the fact that Hund's Rule must be obeyed. Thus, there is only one (unpaired) electron in each $2p$ orbital in a nitrogen atom.

EXAMPLE 5-8 *Electron Configurations and Quantum Numbers*

Write an acceptable set of four quantum numbers for each electron in a nitrogen atom.

Plan

Nitrogen has seven electrons, which occupy the lowest energy orbitals available. Two electrons can occupy the first shell, $n = 1$, in which there is only one s orbital; when $n = 1$, then ℓ must be zero, and therefore $m_\ell = 0$. The two electrons differ only in spin quantum number, m_s. The next five electrons can all fit into the second shell, for which $n = 2$ and ℓ may be either 0 or 1. The $\ell = 0$ (s) subshell fills first, and the $\ell = 1$ (p) subshell is occupied next.

Solution

Electron	n	ℓ	m_ℓ	m_s	e^- Configuration
1, 2	$\begin{cases}1\\1\end{cases}$	0 0	0 0	$+\frac{1}{2}$ $-\frac{1}{2}$	$1s^2$
3, 4	$\begin{cases}2\\2\end{cases}$	0 0	0 0	$+\frac{1}{2}$ $-\frac{1}{2}$	$2s^2$
5, 6, 7	$\begin{cases}2\\2\\2\end{cases}$	1 1 1	-1 0 $+1$	$+\frac{1}{2}$ or $-\frac{1}{2}$ $+\frac{1}{2}$ or $-\frac{1}{2}$ $+\frac{1}{2}$ or $-\frac{1}{2}$	$2p_x^{\ 1}$ $2p_y^{\ 1}$ or $2p^3$ $2p_z^{\ 1}$

Electrons are indistinguishable. We have numbered them 1, 2, 3, and so on as an aid to counting them.

In the lowest energy configurations, the three $2p$ electrons either have $m_s = +\frac{1}{2}$ or all have $m_s = -\frac{1}{2}$.

EXAMPLE 5-9 *Electron Configurations and Quantum Numbers*

Write an acceptable set of four quantum numbers for each electron in a chlorine atom.

Plan

Chlorine is element number 17. Its first seven electrons have the same quantum numbers as those of nitrogen in Example 5-8. Electrons 8, 9, and 10 complete the filling of the $2p$ subshell ($n = 2$, $\ell = 1$) and therefore also the second energy level. Electrons 11 through 17 fill the $3s$ subshell ($n = 3$, $\ell = 0$) and partially fill the $3p$ subshell ($n = 3$, $\ell = 1$).

Solution

Electron	n	ℓ	m_ℓ	m_s	e^- Configuration
1, 2	1	0	0	$\pm\frac{1}{2}$	$1s^2$
3, 4	2	0	0	$\pm\frac{1}{2}$	$2s^2$
5–10	$\begin{cases}2\\2\\2\end{cases}$	1 1 1	-1 0 $+1$	$\pm\frac{1}{2}$ $\pm\frac{1}{2}$ $\pm\frac{1}{2}$	$2p^6$
11, 12	3	0	0	$\pm\frac{1}{2}$	$3s^2$
13–17	$\begin{cases}3\\3\\3\end{cases}$	1 1 1	-1 0 $+1$	$\pm\frac{1}{2}$ $\pm\frac{1}{2}$ $+\frac{1}{2}$ or $-\frac{1}{2}*$	$3p^5$

The 3p orbital with only a single electron can be any one of the set, not necessarily the one with $m_\ell = +1$.

You should now work Exercises 98 and 102.

5-18 PARAMAGNETISM AND DIAMAGNETISM

Substances that contain unpaired electrons are weakly *attracted* into magnetic fields and are said to be **paramagnetic.** By contrast, those in which all electrons are paired are very weakly repelled by magnetic fields and are called **diamagnetic.** The magnetic effect can be measured by hanging a test tube full of a substance on a balance by a long thread and suspending it above the gap of an electromagnet (Figure 5-30). When the current is switched on, a paramagnetic substance such as copper(II) sulfate is pulled into the strong field. The

Both paramagnetism and diamagnetism are hundreds to thousands of times weaker than *ferromagnetism,* the effect seen in iron bar magnets.

Figure 5-30 Diagram of an apparatus for measuring the paramagnetism of a substance. The tube contains a measured amount of the substance, often in solution. (a) Before the magnetic field is turned on, the position and mass of the sample are determined. (b) When the field is on, a paramagnetic substance is attracted *into* the field. (c) A diamagnetic substance would be repelled *far more weakly* by the field.

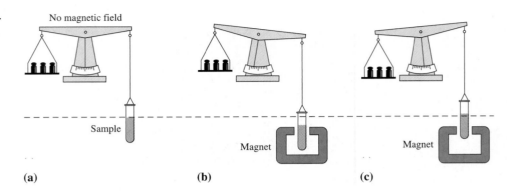

No magnetic field

Sample

Magnet   Magnet

(a)   **(b)**   **(c)**

CD-ROM   Screen 8.3, Spinning Electrons and Magnetism.

A live frog suspended in an extremely strong magnetic field (16 Tesla) demonstrating the diamagnetic repulsion effect.

Courtesy of Dr. Andre Geim, Manchester University

A newer IUPAC system numbers the columns in the periodic table from 1 to 18 for the *s*, *d*, and *p*- block elements. This is shown in most of the periodic tables in this book in parentheses under the older Roman numeral plus A or B naming system.

SCiLINKS.
TOPIC: Periodic Table
GO TO: www.scilinks.org
*sci***LINKS CODE:** WCH0410

paramagnetic attraction per mole of substance can be measured by weighing the sample before and after energizing the magnet. The paramagnetism per mole increases with increasing number of unpaired electrons per formula unit. Many transition metals and ions have one or more unpaired electrons and are paramagnetic.

Iron, cobalt, and nickel are the only *free* elements that exhibit **ferromagnetism.** This property is much stronger than paramagnetism; it allows a substance to become permanently magnetized when placed in a magnetic field. This happens as randomly oriented electron spins align themselves with an applied field. To exhibit ferromagnetism, the atoms must be within the proper range of sizes so that unpaired electrons on adjacent atoms can interact cooperatively with one another, but not to the extent that they pair. Experimental evidence suggests that in ferromagnets, atoms cluster together into *domains* that contain large numbers of atoms in fairly small volumes. The atoms within each domain interact cooperatively with one another to generate ferromagnetism.

5-19 THE PERIODIC TABLE AND ELECTRON CONFIGURATIONS

In this section, we view the *periodic table* (see inside front cover and Section 4-1) from a modern, much more useful perspective—as a systematic representation of the electron configurations of the elements. In the periodic table, elements are arranged in blocks based on the kinds of atomic orbitals that are being filled (Figure 5-31). The periodic tables in this text are divided into "A" and "B" groups. The A groups contain elements in which *s* and *p* orbitals are being filled. Elements within any particular A group have similar electron configurations and chemical properties, as we shall see in the next chapter. The B groups include the transition metals in which there are one or two electrons in the *s* orbital of the outermost occupied shell, and the *d* orbitals, one shell smaller, are being filled.

Lithium, sodium, and potassium, elements of the leftmost column of the periodic table (Group IA), have a single electron in their outermost *s* orbital (ns^1). Beryllium and magnesium, of Group IIA, have two electrons in their outermost shell, ns^2, while boron and aluminum (Group IIIA) have three electrons in their outermost shell, ns^2np^1. Similar observations can be made for other A group elements.

The electron configurations of the A group elements and the noble gases can be predicted reliably from Figures 5-28 and 5-29. However, there are some more pronounced irregularities in the B groups below the fourth period. In the heavier B group elements, the higher energy subshells in different principal shells have energies that are very nearly equal

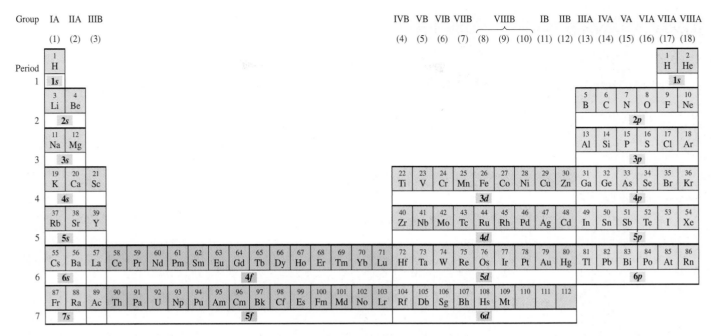

Figure 5-31 A periodic table colored to show the kinds of atomic orbitals (subshells) being filled and the symbols of blocks of elements. The electronic structures of the A group elements are quite regular and can be predicted from their positions in the periodic table, but many exceptions occur in the d and f blocks. The colors in this figure are the same as those in Figure 5-28.

Hydrogen and helium are shown here in their usual positions in the periodic table. These may seem somewhat unusual based just on their electron configurations. We should remember, however, that the first shell ($n = 1$) can hold a maximum of only two electrons. This shell is entirely filled in helium, so He behaves as a noble gas, and we put it in the column with the other noble gases (Group VIIIA). Hydrogen has one electron that is easily lost, like the metals in Group IA, so we put it in Group IA even though it is not a metal. Furthermore, hydrogen is one electron short of a noble gas configuration (He), so we could also place it with the other such elements in Group VIIA.

(Figure 5-29). It is easy for an electron to jump from one orbital to another of nearly the same energy, even in a different set. This is because the orbital energies are *perturbed* (change slightly) as the nuclear charge changes, and an extra electron is added in going from one element to the next. This phenomenon gives rise to other irregularities that are analogous to those seen for Cr and Cu, described earlier.

We can extend the information in Figure 5-31 to indicate the electron configurations that are represented by each *group* (column) of the periodic table. Table 5-5 shows this interpretation of the periodic table, along with the most important exceptions. We can use this interpretation of the periodic table to write, quickly and reliably, the electron configurations for elements.

A more complete listing of electron configurations is given in Appendix B.

TABLE 5-5 *The s, p, d, and f Blocks of the Periodic Table**

s orbital block

GROUPS

p orbital block

d orbital block

f orbital block

	IA	IIA	IIIB			IVB	VB	VIB	VIIB		VIIIB		IB	IIB	IIIA	IVA	VA	VIA	VIIA	VIIIA
	(1)	(2)	(3)			(4)	(5)	(6)	(7)	(8)	(9)	(10)	(11)	(12)	(13)	(14)	(15)	(16)	(17)	(18)
	s^1	s^2																		s^2
$n=1$	1 H																			2 He
			d^1s^2			d^2s^2	d^3s^2	d^5s^1	d^5s^2	d^6s^2	d^7s^2	d^8s^2	$d^{10}s^1$	$d^{10}s^2$	s^2p^1	s^2p^2	s^2p^3	s^2p^4	s^2p^5	s^2p^6
$n=2$	3 Li	4 Be													5 B	6 C	7 N	8 O	9 F	10 Ne
$n=3$	11 Na	12 Mg													13 Al	14 Si	15 P	16 S	17 Cl	18 Ar
$n=4$	19 K	20 Ca	21 Sc			22 Ti	23 V	24 Cr	25 Mn	26 Fe	27 Co	28 Ni	29 Cu	30 Zn	31 Ga	32 Ge	33 As	34 Se	35 Br	36 Kr
$n=5$	37 Rb	38 Sr	39 Y			40 Zr	41 Nb d^4s^1	42 Mo	43 Tc	44 Ru d^7s^1	45 Rh d^8s^1	46 Pd $d^{10}s^0$	47 Ag	48 Cd	49 In	50 Sn	51 Sb	52 Te	53 I	54 Xe
$n=6$	55 Cs	56 Ba	57 La	58 Ce→71 Lu		72 Hf	73 Ta	74 W d^4s^2	75 Re	76 Os	77 Ir	78 Pt d^9s^1	79 Au	80 Hg	81 Tl	82 Pb	83 Bi	84 Po	85 At	86 Rn
$n=7$	87 Fr	88 Ra	89 Ac	90 Th→103 Lr		104 Rf	105 Db	106 Sg	107 Bh	108 Hs	109 Mt	110	111	112						

*n is the principal quantum number. The d^1s^2, d^2s^2, . . . designations represent known configurations. They refer to $(n-1)d$ and ns orbitals. Several exceptions to the configurations indicated above each group are shown in gray.

EXAMPLE 5-10 *Electron Configurations*

Use Table 5-5 to determine the electron configurations of (a) magnesium, Mg; (b) germanium, Ge; and (c) molybdenum, Mo.

Plan

We will use the electron configurations indicated in Table 5-5 for each group. Each *period* (row) begins filling a new shell (new value of n). Elements to the right of the d orbital block have the d orbitals in the $(n-1)$ shell already filled. We often find it convenient to collect all sets of orbitals with the same value of n together, to emphasize the number of electrons in the *outermost* shell, that is, the shell with the highest value of n.

Solution

Although the 4s orbital usually fills before the 3d, most chemists will understand and accept either of the answers shown for the electron configurations of Ge and Mo. The same holds for the configurations of other elements.

(a) Magnesium, Mg, is in Group IIA, which has the general configuration s^2; it is in Period 3 (third row). The last filled noble gas configuration is that of neon, or [Ne]. The electron configuration of Mg is [Ne] $3s^2$.

(b) Germanium, Ge, is in Group IVA, for which Table 5-5 shows the general configuration s^2p^2. It is in Period 4 (the fourth row), so we interpret this as $4s^24p^2$. The last filled noble gas configuration is that of argon, Ar, accounting for 18 electrons. In addition, Ge lies beyond the d orbital block, so we know that the 3d orbitals are completely filled. The electron configuration of Ge is [Ar] $4s^23d^{10}4p^2$ or [Ar] $3d^{10}4s^24p^2$.

(c) Molybdenum, Mo, is in Group VIB, with the general configuration d^5s^1; it is in Period 5, which begins with $5s$ and is beyond the noble gas krypton. The electron configuration of Mo is [Kr] $5s^1 4d^5$ or [Kr] $4d^5 5s^1$. The electron configuration of molybdenum is analogous to that of chromium, Cr, the element just above it. The configuration of Cr was discussed in Section 5-17 as one of the exceptions to the Aufbau order of filling.

You should now work Exercise 100.

EXAMPLE 5-11   *Unpaired Electrons*

Determine the number of unpaired electrons in an atom of tellurium, Te.

Plan

Te is in Group VIA in the periodic table, which tells us that its configuration is s^2p^4. All other shells are completely filled, so they contain only paired electrons. We need only to find out how many unpaired electrons are represented by s^2p^4.

Solution

The notation s^2p^4 is a short representation for s ⇅ p ⇅ ↑ ↑. This shows that an atom of Te contains two unpaired electrons.

You should now work Exercises 106 and 108.

The periodic table has been described as "the chemist's best friend." Chemical reactions involve loss, gain, or sharing of electrons. In this chapter, we have seen that the fundamental basis of the periodic table is that it reflects similarities and trends in electron configurations. It is easy to use the periodic table to determine many important aspects of electron configurations of atoms. Practice until you can use the periodic table with confidence to answer many questions about electron configurations. As we continue our study, we will learn many other useful ways to interpret the periodic table. We should always keep in mind that the many trends in chemical and physical properties that we correlate with the periodic table are ultimately based on the trends in electron configurations.

Key Terms

Absorption spectrum The spectrum associated with absorption of electromagnetic radiation by atoms (or other species) resulting from transitions from lower to higher electronic energy states.

Alpha (α) particle A helium ion with a 2+ charge; an assembly of two protons and two neutrons.

amu See *Atomic mass unit.*

Angular momentum quantum number (ℓ) The quantum mechanical solution to a wave equation that designates the subshell, or set of orbitals (s, p, d, f), within a given main shell in which an electron resides.

Anode In a cathode-ray tube, the positive electrode.

Atomic mass unit An arbitrary mass unit defined to be exactly one twelfth the mass of the carbon-12 isotope.

Atomic number The integral number of protons in the nucleus; defines the identity of an element.

Atomic orbital The region or volume in space in which the probability of finding electrons is highest.

Aufbau ("building up") Principle A guide for predicting the order in which electrons fill subshells and shells in atoms.

Balmer–Rydberg equation An empirical equation that relates wavelengths in the hydrogen emission spectrum to simple integers.

Canal ray A stream of positively charged particles (cations) that moves toward the negative electrode in a cathode-ray tube; observed to pass through canals (holes) in the negative electrode.

Cathode In a cathode-ray tube, the negative electrode.

Cathode ray The beam of electrons going from the negative electrode toward the positive electrode in a cathode-ray tube.

Cathode-ray tube A closed glass tube containing a gas under low pressure, with electrodes near the ends and a luminescent screen at the end near the positive electrode; produces cathode rays when high voltage is applied.

Continuous spectrum A spectrum that contains all wavelengths in a specified region of the electromagnetic spectrum.

d orbitals Beginning in the third shell, a set of five degenerate orbitals per shell, higher in energy than s and p orbitals in the same shell.

Degenerate orbitals   Two or more orbitals that have the same energy.

Diamagnetism   *Weak* repulsion by a magnetic field; associated with all electrons in an atom, molecule, or substance being paired.

Electromagnetic radiation   Energy that is propagated by means of electric and magnetic fields that oscillate in directions perpendicular to the direction of travel of the energy.

Electron   A subatomic particle having a mass of 0.00054858 amu and a charge of 1−.

Electron configuration   The specific distribution of electrons in the atomic orbitals of atoms and ions.

Electron transition   The transfer of an electron from one energy level to another.

Emission spectrum   The spectrum associated with emission of electromagnetic radiation by atoms (or other species) resulting from electron transitions from higher to lower energy states.

Excited state   Any energy state other than the ground state of an atom, ion, or molecule.

f **orbitals**   Beginning in the fourth shell, a set of seven degenerate orbitals per shell, higher in energy than s, p, and d orbitals in the same shell.

Ferromagnetism   The property that allows a substance to become permanently magnetized when placed in a magnetic field; exhibited by iron, cobalt, and nickel and some of their alloys.

Frequency (ν)   The number of crests of a wave that pass a given point per unit time.

Fundamental particles   Subatomic particles of which all matter is composed; protons, electrons, and neutrons are fundamental particles.

Ground state   The lowest energy state or most stable state of an atom, molecule, or ion.

Group   A vertical column in the periodic table; also called a family.

Heisenberg Uncertainty Principle   It is impossible to determine accurately both the momentum and position of an electron simultaneously.

Hund's Rule   Each orbital of a given subshell is occupied by a single electron before pairing begins. See *Aufbau Principle*.

Isotopes   Two or more forms of atoms of the same element with different masses; that is, atoms containing the same number of protons but different numbers of neutrons.

Line spectrum   An atomic emission or absorption spectrum.

Magnetic quantum number (m_ℓ)   Quantum mechanical solution to a wave equation that designates the particular orbital within a given subshell (s, p, d, f) in which an electron resides. The p_x, p_y, and p_z orbitals have different magnetic quantum numbers.

Mass number   The integral sum of the numbers of protons and neutrons in an atom.

Mass spectrometer   An instrument that measures the charge-to-mass ratios of charged particles.

Natural radioactivity   Spontaneous decomposition of an atom.

Neutron   A subatomic nuclear particle having a mass of 1.0087 amu and no charge.

Nucleus   The very small, very dense, positively charged center of an atom containing protons and neutrons, except for $_1^1H$.

Nuclide symbol   The symbol for an atom, $_Z^A E$, in which E is the symbol for an element, Z is its atomic number, and A is its mass number.

p **orbitals**   Beginning with the second shell, a set of three degenerate mutually perpendicular, equal-arm, dumbbell-shaped atomic orbitals per shell.

Pairing of electrons   Interaction of two electrons with opposite m_s values in the same orbital ($\uparrow\downarrow$).

Paramagnetism   Attraction toward a magnetic field, stronger than diamagnetism, but still very weak compared with ferromagnetism; due to presence of unpaired electrons.

Pauli Exclusion Principle   No two electrons in the same atom may have identical sets of four quantum numbers.

Period   A horizontal row in the periodic table.

Photoelectric effect   Emission of an electron from the surface of a metal, caused by impinging electromagnetic radiation of certain minimum energy; the resulting current increases with increasing intensity of radiation.

Photon   A "packet" of light or electromagnetic radiation; also called a quantum of light.

Principal quantum number (n)   The quantum mechanical solution to a wave equation that designates the main shell, or energy level, in which an electron resides.

Proton   A subatomic particle having a mass of 1.0073 amu and a charge of 1+, found in the nuclei of atoms.

Quantum   A "packet" of energy. See *Photon*.

Quantum mechanics   A mathematical method of treating particles on the basis of quantum theory, which assumes that energy (of small particles) is not infinitely divisible.

Quantum numbers   Numbers that describe the energies of electrons in atoms; they are derived from quantum mechanical treatment.

Radiant energy   See *Electromagnetic radiation*.

s **orbital**   A spherically symmetrical atomic orbital; one per shell.

Spectral line   Any of a number of lines corresponding to definite wavelengths in an atomic emission or absorption spectrum; these lines represent the energy difference between two energy levels.

Spectrum   Display of component wavelengths of electromagnetic radiation.

Spin quantum number (m_s)   The quantum mechanical solution to a wave equation that indicates the relative spins of electrons ("spin up" and "spin down").

Wavelength (λ)   The distance between two identical points of a wave.

Exercises

Particles and the Nuclear Atom

1. List the three fundamental particles of matter, and indicate the mass and charge associated with each.
2. In the oil-drop experiment, how did Millikan know that none of the oil droplets he observed were ones that had a deficiency of electrons rather than excess?
3. How many electrons carry a total charge of 1.00 coulomb?
4. (a) How do we know that canal rays have charges opposite in sign to cathode rays? What are canal rays? (b) Why are cathode rays from all samples of gases identical, whereas canal rays are not?
*5. The following data are measurements of the charges on oil droplets using an apparatus similar to that used by Millikan:

13.458×10^{-19} C	17.308×10^{-19} C
15.373×10^{-19} C	28.844×10^{-19} C
17.303×10^{-19} C	11.545×10^{-19} C
15.378×10^{-19} C	19.214×10^{-19} C

Each should be a whole-number ratio of some fundamental charge. Using these data, determine the value of the fundamental charge.
*6. Suppose we discover a new positively charged particle, which we call the "whizatron." We want to determine its charge. (a) What modifications would we have to make to the Millikan oil-drop apparatus to carry out the corresponding experiment on whizatrons? (b) In such an experiment, we observe the following charges on five different droplets:

6.52×10^{-19} C	1.14×10^{-18} C
8.16×10^{-19} C	9.78×10^{-19} C
3.26×10^{-19} C	

What is the charge on the whizatron?
7. Outline Rutherford's contribution to understanding the nature of atoms.
8. Why was Rutherford so surprised that some of the α-particles were scattered backward in the gold foil experiment?
9. Summarize Moseley's contribution to our knowledge of the structure of atoms.
10. The approximate radius of a hydrogen atom is 0.0529 nm, and that of a proton is 1.5×10^{-15} m. Assuming both the hydrogen atom and the proton to be spherical, calculate the fraction of the space in an atom of hydrogen that is occupied by the nucleus. $V = (\frac{4}{3})\pi r^3$ for a sphere.
11. The approximate radius of a neutron is 1.5×10^{-15} m, and the mass is 1.675×10^{-27} kg. Calculate the density of a neutron. $V = (\frac{4}{3})\pi r^3$ for a sphere.

Atom Composition, Isotopes, and Atomic Weights

12. Arrange the following in order of increasing ratio of charge to mass: $^{12}C^+$, $^{12}C^{2+}$, $^{13}C^+$, $^{13}C^{2+}$.
13. Refer to Exercise 12. Suppose all of these high-energy ions are present in a mass spectrometer. For which one will its path be changed (a) the most and (b) the least by increasing the external magnetic field?
14. Estimate the percentage of the total mass of a ^{197}Au atom that is due to (a) electrons, (b) protons, and (c) neutrons by *assuming* that the mass of the atom is simply the sum of the masses of the appropriate numbers of subatomic particles.

Charles D. Winters

Native gold

15. (a) How are isotopic abundances determined experimentally? (b) How do the isotopes of a given element differ?
16. Clearly define and provide examples that illustrate the meaning of each. (a) atomic number, (b) isotope, (c) mass number, (d) nuclear charge.
17. Write the composition of one atom of each of the three isotopes of neon: ^{20}Ne, ^{21}Ne, ^{22}Ne.
18. Write the composition of one atom of each of the four isotopes of strontium: ^{84}Sr, ^{86}Sr, ^{87}Sr, ^{88}Sr.
19. Complete Chart A for neutral atoms.

Chart A

Kind of Atom	Atomic Number	Mass Number	Isotope	Number of Protons	Number of Electrons	Number of Neutrons
			^{40}Ca			
potassium		39				
	14	28				
		202		80		

Chart B

Kind of Atom	Atomic Number	Mass Number	Isotope	Number of Protons	Number of Electrons	Number of Neutrons
cobalt	_____	_____		_____	_____	32
_____	_____	_____	^{193}Ir	_____	_____	_____
_____	_____	_____		_____	25	30
_____	_____	182	_____	_____	78	_____

20. Complete Chart B for neutral atoms.

*21. Prior to 1962, the atomic weight scale was based on the assignment of an atomic weight of exactly 16 amu to the *naturally occurring* mixture of oxygen. The atomic weight of cobalt is 58.9332 amu on the carbon-12 scale. What was it on the older scale?

22. Determine the number of protons, neutrons, and electrons in each of the following species: (a) ^{24}Mg; (b) ^{45}Sc; (c) ^{91}Zr; (d) ^{27}Al; (e) ^{65}Zn^{2+}; (f) ^{108}Ag^{+}.

23. Determine the number of protons, neutrons, and electrons in each of the following species: (a) ^{52}Cr; (b) ^{93}Nb; (c) ^{137}Ba^{2+}; (d) ^{63}Cu^{+}; (e) ^{56}Fe^{2+}; (f) ^{56}Fe^{3+}.

24. What is the symbol of the species composed of each of the following sets of subatomic particles? (a) $25p$, $30n$, $25e$; (b) $20p$, $20n$, $18e$; (c) $33p$, $42n$, $33e$; (d) $53p$, $74n$, $54e$.

25. What is the symbol of the species composed of each of the following sets of subatomic particles? (a) $94p$, $150n$, $94e$; (b) $79p$, $118n$, $76e$; (c) $34p$, $45n$, $36e$; (d) $54p$, $77n$, $54e$

26. The atomic weight of lithium is 6.941 amu. The two naturally occurring isotopes of lithium have the following masses: ^{6}Li, 6.01512 amu; ^{7}Li, 7.01600 amu. Calculate the percent of ^{6}Li in naturally occurring lithium.

27. The atomic weight of rubidium is 85.4678 amu. The two naturally occurring isotopes of rubidium have the following masses: ^{85}Rb, 84.9118 amu; ^{87}Rb, 86.9092 amu. Calculate the percent of ^{85}Rb in naturally occurring rubidium.

28. Bromine is composed of ^{79}Br, 78.9183 amu, and ^{81}Br, 80.9163 amu. The percent composition of a sample is 50.69% Br-79 and 49.31% Br-81. Based on this sample, calculate the atomic weight of bromine.

29. What is the atomic weight of a hypothetical element that consists of the following isotopes in the indicated relative abundances?

Isotope	Isotopic Mass (amu)	% Natural Abundance
1	94.9	12.4
2	95.9	73.6
3	97.9	14.0

30. Naturally occurring iron consists of four isotopes with the abundances indicated. From the following masses and relative abundances of these isotopes, calculate the atomic weight of naturally occurring iron.

Isotope	Isotopic Mass (amu)	% Natural Abundance
^{54}Fe	53.9396	5.82
^{56}Fe	55.9349	91.66
^{57}Fe	56.9354	2.19
^{58}Fe	57.9333	0.33

31. Calculate the atomic weight of nickel from the following information.

Isotope	Isotopic Mass (amu)	% Natural Abundance
^{58}Ni	57.9353	67.88
^{60}Ni	59.9332	26.23
^{61}Ni	60.9310	1.19
^{62}Ni	61.9283	3.66
^{64}Ni	63.9280	1.08

32. The atomic weight of copper is 63.546 amu. The two naturally occurring isotopes of copper have the following masses: ^{63}Cu, 62.9298 amu; ^{65}Cu, 64.9278 amu. Calculate the percent of ^{63}Cu in naturally occurring copper.

Charles D. Winters

Native copper

33. Silver consists of two naturally occurring isotopes: ^{107}Ag, which has a mass of 106.90509 amu, and ^{109}Ag, which has a mass of 108.9047 amu. The atomic weight of silver is

107.8682 amu. Determine the percent abundance of each isotope in naturally occurring silver.

34. Refer to Table 5-3 *only* and calculate the atomic weights of oxygen and chlorine. Do your answers agree with the atomic weights given in that table?

35. The following is a mass spectrum of the 1+ charged ions of an element. Calculate the atomic weight of the element. What is the element?

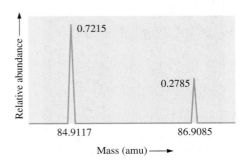

36. Suppose you measure the mass spectrum of the 1+ charged ions of germanium, atomic weight 72.61 amu. Unfortunately, the recorder on the mass spectrometer jams at the beginning and again at the end of your experiment. You obtain only the following partial spectrum, which *may or may not be complete*. From the information given here, can you tell whether one of the germanium isotopes is missing? If one is missing, at which end of the plot should it appear?

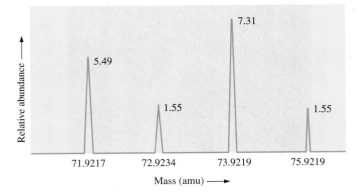

37. Calculate the atomic weight of silicon using the following data for the percent natural abundance and mass of each isotope: (a) 92.23% ^{28}Si (27.9769 amu); (b) 4.67% ^{29}Si (28.9765 amu); (c) 3.10% ^{30}Si (29.9738 amu).

38. Calculate the atomic weight of chromium using the following data for the percent natural abundance and mass of each isotope: (a) 4.35% ^{50}Cr (49.9461 amu); (b) 83.79% ^{52}Cr (51.9405 amu); (c) 9.50% ^{53}Cr (52.9406 amu); (d) 2.36% ^{54}Cr (53.9389 amu).

Electromagnetic Radiation

39. Calculate the wavelengths, in meters, of radiation of the following frequencies: (a) 4.80×10^{15} s^{-1}; (b) 1.18×10^{14} s^{-1}; (c) 5.44×10^{12} s^{-1}.

40. Calculate the frequency of radiation of each of the following wavelengths: (a) 9774 Å; (b) 442 nm; (c) 4.92 cm; (d) 4.92×10^{-9} cm.

41. What is the energy of a photon of each of the radiations in Exercise 39? Express your answer in joules per photon. In which regions of the electromagnetic spectrum do these radiations fall?

42. Excited lithium ions emit radiation at a wavelength of 670.8 nm in the visible range of the spectrum. (This characteristic color is often used as a qualitative analysis test for the presence of Li$^+$.) Calculate (a) the frequency and (b) the energy of a photon of this radiation. (c) What color is this light?

43. Calculate the energy, in joules per photon, of the red line, 6573 Å, in the discharge spectrum of atomic calcium.

44. Ozone in the upper atmosphere absorbs ultraviolet radiation, which induces the following chemical reaction:

$$O_3(g) \longrightarrow O_2(g) + O(g)$$

What is the energy of a 3400-Å photon that is absorbed? What is the energy of a mole of these photons?

*45. During photosynthesis, chlorophyll-α absorbs light of wavelength 440 nm and emits light of wavelength 670 nm. What is the energy available for photosynthesis from the absorption–emission of a mole of photons?

Photosynthesis

46. Alpha Centauri is the star closest to our solar system. It is 4.3 light years away. How many miles is this? A light year is the distance that light travels (in a vacuum) in one year. Assume that space is essentially a vacuum.

The Photoelectric Effect

47. What evidence supports the idea that electromagnetic radiation is (a) wave-like; (b) particle-like?

48. Describe the influence of frequency and intensity of electromagnetic radiation on the current in the photoelectric effect.

*49. Cesium is often used in "electric eyes" for self-opening doors in an application of the photoelectric effect. The amount of energy required to ionize (remove an electron from) a cesium atom is 3.89 electron volts (1 eV = 1.60×10^{-19} J). Show by calculation whether a beam of yellow light with wavelength 5830 Å would ionize a cesium atom.

*50. Refer to Exercise 49. What would be the wavelength, in nanometers, of light with just sufficient energy to ionize a cesium atom? What color would this light be?

Atomic Spectra and the Bohr Theory

51. (a) Distinguish between an atomic emission spectrum and an atomic absorption spectrum. (b) Distinguish between a continuous spectrum and a line spectrum.

52. Prepare a sketch similar to Figure 5-16b that shows a ground energy state and three excited energy states. Using vertical arrows, indicate the transitions that would correspond to the absorption spectrum for this system.

53. Why is the Bohr model of the hydrogen atom referred to as the solar system model?

*54. If each atom in one mole of atoms emits a photon of wavelength 5.55×10^3 Å, how much energy is lost? Express the answer in kJ/mol. As a reference point, burning one mole (16 g) of CH_4 produces 819 kJ of heat.

55. What is the Balmer–Rydberg equation? Why is it called an empirical equation?

56. Hydrogen atoms absorb energy so that the electrons are excited to the energy level $n = 7$. Electrons then undergo these transitions: (1) $n = 7 \rightarrow n = 1$; (2) $n = 7 \rightarrow n = 2$; (3) $n = 2 \rightarrow n = 1$. Which of these transitions will produce the photon with (a) the smallest energy; (b) the highest frequency; (c) the shortest wavelength? (d) What is the frequency of a photon resulting from the transition $n = 6 \rightarrow n = 1$?

*57. Five energy levels of the He atom are given in joules per atom above an *arbitrary* reference energy: (1) 6.000×10^{-19}; (2) 8.812×10^{-19}; (3) 9.381×10^{-19}; (4) 10.443×10^{-19}; (5) 10.934×10^{-19}. Construct an energy level diagram for He and find the energy of the photon (a) absorbed for the electron transition from level 1 to level 5 and (b) emitted for the electron transition from level 4 to level 1.

58. The following are prominent lines in the visible region of the emission spectra of the elements listed. The lines can be used to identify the elements. What color is the light responsible for each line? (a) lithium, 4603 Å; (b) neon, 540.0 nm; (c) calcium, 6573 Å; (d) potassium, $\nu = 3.90 \times 10^{14}$ Hz.

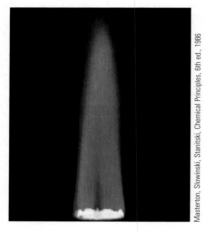

Masterton, Slowinski, Stanitski, Chemical Principles, 6th ed., 1986

59. Hydrogen atoms have an absorption line at 1026 Å. What is the frequency of the photons absorbed, and what is the energy difference, in joules, between the ground state and this excited state of the atom?

60. An argon laser emits blue light with a wavelength of 488.0 nm. How many photons are emitted by this laser in 2.00 seconds, operating at a power of 515 milliwatts? One watt (a unit of power) is equal to 1 joule/second.

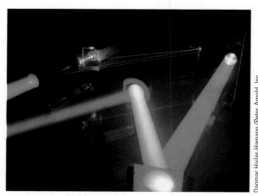

Dagmar Hailer-Hamann/Peter Arnold, Inc.

Lasers

The Wave-Particle View of Matter

61. (a) What evidence supports the idea that electrons are particle-like? (b) What evidence supports the idea that electrons are wave-like?

62. (a) What is the de Broglie wavelength of a proton moving at a speed of 2.50×10^7 m/s? The proton mass is 1.67×10^{-24} g. (b) What is the de Broglie wavelength of

a stone with a mass of 30.0 g moving at 2.00×10^3 m/h ($\approx$100 mph)? (c) How do the wavelengths in parts (a) and (b) compare with the typical radii of atoms? (See the atomic radii in Figure 6-1).

63. What is the wavelength corresponding to a neutron of mass 1.67×10^{-27} kg moving at 2360 m/s?

64. What is the velocity of an α-particle (a helium nucleus) that has a de Broglie wavelength of 0.529 Å?

Quantum Numbers and Atomic Orbitals

65. (a) What is a quantum number? What is an atomic orbital? (b) How many quantum numbers are required to specify a single atomic orbital? What are they?

66. How are the possible values for the angular momentum quantum number for a given electron restricted by the value of n?

67. Without giving the ranges of possible values of the four quantum numbers, n, ℓ, m_ℓ, and m_s, describe briefly what information each one gives.

68. An electron is in one of the $3p$ orbitals. What are the possible values of the quantum numbers n, ℓ, m_ℓ, and m_s for the electron?

69. What is the maximum number of electrons in an atom that can have the following quantum numbers? (a) $n = 2$; (b) $n = 3$ and $\ell = 1$; (c) $n = 3$, $\ell = 0$, and $m_\ell = 0$; (d) $n = 3$, $\ell = 1$, $m_\ell = -1$, and $m_s = -\frac{1}{2}$.

70. What is the maximum number of electrons in an atom that can have the following quantum numbers? (a) $n = 2$ and $\ell = 1$; (b) $n = 3$ and $\ell = 1$; (c) $n = 2$, $\ell = 1$ and $m_\ell = -1$; (d) $n = 3$, $\ell = 1$, and $m_\ell = -1$; (e) $n = 3$, $\ell = 1$, $m_\ell = 0$, and $m_s = -\frac{1}{2}$.

71. What are the values of n and ℓ for the following subshells? (a) $1s$; (b) $3s$; (c) $5p$; (d) $3d$; (e) $4f$.

72. How many individual orbitals are there in the third shell? Write out n, ℓ, and m_ℓ quantum numbers for each one, and label each set by the s, p, d, f designations.

73. (a) Write the possible values of ℓ when $n = 4$. (b) Write the allowed number of orbitals (1) with the quantum numbers $n = 3$, $\ell = 1$; (2) with the quantum number $n = 2$, $\ell = 1$; (3) with the quantum numbers $n = 3$, $\ell = 1$, $m_\ell = -1$; (4) with the quantum numbers $n = 1$.

74. Write the subshell notations that correspond to (a) $n = 3$, $\ell = 0$; (b) $n = 3$, $\ell = 1$; (c) $n = 6$, $\ell = 1$; (d) $n = 3$, $\ell = 2$.

75. What values can m_ℓ take for (a) a $4d$ orbital, (b) a $3s$ orbital, and (c) a $3p$ orbital?

76. How many orbitals in any atom can have the given quantum number or designation? (a) $4p$; (b) $3s$; (c) $3p_x$; (d) $n = 4$; (e) $6d$; (f) $5d$; (g) $5f$; (h) $6s$.

77. The following incorrect sets of quantum numbers in the order n, ℓ, m_ℓ, m_s are written for paired electrons or for one electron in an orbital. Correct them, assuming n values are correct. (a) 1, 0, 0, $+\frac{1}{2}$, $+\frac{1}{2}$; (b) 2, 0, 1, $\pm\frac{1}{2}$; (c) 3, 2, 3, $\pm\frac{1}{2}$; (d) 3, 1, 2, $+\frac{1}{2}$; (e) 2, 1, -1, 0; (f) 3, -1, -1, $-\frac{1}{2}$.

78. (a) How are a $2s$ orbital and a $3s$ orbital in an atom similar? How do they differ? (b) How are a $3p_x$ orbital and a $3d_{yz}$ orbital in an atom similar? How do they differ?

Electron Configurations and the Periodic Table

You should be able to use the positions of elements in the periodic table to answer the exercises in this section.

79. Draw representations of ground state electron configurations using the orbital notation ($\uparrow\downarrow$) for the following elements. (a) F; (b) V; (c) Br; (d) Rh.

80. Draw representations of ground state electron configurations using the orbital notation ($\uparrow\downarrow$) for the following elements. (a) K; (b) Sc; (c) Ga; (d) Cd.

81. Determine the number of electrons in the outer occupied shell of each of the following elements, and indicate the principal quantum number of that shell. (a) Na; (b) S; (c) Si; (d) Sr; (e) Ba; (f) Br.

82. With the help of Appendix B, list the symbols for the first five elements, by atomic number, that have an unpaired electron in an s orbital. Identify the group in which most of these are found in the periodic table.

83. List the elements having an atomic number of 18 or less that have two unpaired p orbital electrons. Indicate the group to which each of these elements belongs in the periodic table.

84. Identify the element, or elements possible, given only the number of electrons in the outermost shell and the principal quantum number of that shell. (a) 2 electrons, first shell; (b) 2 electrons, second shell; (c) 3 electrons, third shell; (d) 2 electrons, seventh shell; (e) 4 electrons, third shell; (f) 8 electrons, fifth shell.

85. Give the ground state electron configurations for the elements of Exercise 79 using shorthand notation—that is, $1s^2 2s^2 2p^6$, and so on.

86. Give the ground state electron configurations for the elements of Exercise 80 using shorthand notation—that is, $1s^2 2s^2 2p^6$, and so on.

87. State the Pauli Exclusion Principle. Would any of the following electron configurations violate this rule: (a) $1s^3$, (b) $1s^2 2s^2 2p_x^3$; (c) $1s^2 2s^2 2p_x^2$? Explain.

88. State Hund's Rule. Would any of the following electron configurations violate this rule: (a) $1s^2$; (b) $1s^2 2s^2 2p_x^2$; (c) $1s^2 2s^2 2p_x^1 2p_y^1$; (d) $1s^2 2s^2 2p_x^2 2p_z^1$; (e) $1s^2 2s^1 2p_x^2 2p_y^1 2p_z^1$? Explain.

*89. Classify each of the following atomic electron configurations as (i) a ground state, (ii) an excited state, or (iii) a forbidden state: (a) $1s^2 2s^0 2p^3$; (b) [Kr] $4d^{10} 5s^3$; (c) $1s^2 2s^2 2p^6 3s^2 3p^6 3d^{12} 4s^2$; (d) $1s^2 2s^2 2p^6 3s^2 3p^6 2d^1$; (e) $1s^2 2s^2 2p^8 3s^2 3p^5$.

90. Which of the elements with atomic numbers of 10 or less are paramagnetic when in the atomic state?

91. Semiconductor industries depend on such elements as Si, Ga, As, Ge, B, Cd, and S. Write the predicted electron configuration of each element.

92. The manufacture of high-temperature ceramic supercon-
ductors depends on such elements as Cu, O, La, Y, Ba, Tl,
and Bi. Write the predicted electron configuration of
each element. (Consult Appendix B if necessary.)

93. In nature, potassium and sodium are often found together. (a) Write the electron configurations for potassium and for sodium. (b) How are they similar? (c) How
do they differ?

Metallic potassium

Metallic sodium

94. Which elements are represented by the following electron configurations?
(a) $1s^2 2s^2 2p^6 3s^2 3p^6 3d^{10} 4s^2 4p^3$
(b) [Kr] $4d^{10} 4f^{14} 5s^2 5p^6 5d^{10} 6s^2 6p^3$
(c) [Kr] $4d^{10} 4f^{14} 5s^2 5p^6 5d^{10} 6s^2 6p^6 7s^2$
(d) [Kr] $4d^5 5s^2$
(e) $1s^2 2s^2 2p^6 3s^2 3p^6 3d^3 4s^1$

95. Repeat Exercise 94 for
(a) $1s^2 2s^2 2p^6 3s^2 3p^6 3d^{10} 4s^1$
(b) [Kr] $4d^{10} 4f^{14} 5s^2 5p^6 5d^{10} 6s^2 6p^4$
(c) $1s^2 2s^2 2p^6 3s^2 3p^5$
(d) [Kr] $4d^{10} 4f^{14} 5s^2 5p^6 5d^{10} 6s^2 6p^6 7s^2$

96. Find the total number of s, p, and d electrons in each of
the following: (a) P; (b) Kr; (c) Ni; (d) Zn; (e) Ti.

97. Write the electron configurations of the Group IIA elements Be, Mg, and Ca (see inside front cover). What similarity do you observe?

98. Construct a table in which you list a possible set of values for the four quantum numbers for each electron in
the following atoms in their ground states. (a) Na; (b) O;
(c) Ca.

99. Construct a table in which you list a possible set of values for the four quantum numbers for each electron in
the following atoms in their ground states. (a) Mg; (b) S;
(c) Sc.

100. Draw general electron structures for the A group ele-

ments using the ⇅ notation, where n is the principal quantum number for the highest occupied energy level.

	ns	np
IA	—	— — —
IIA	—	— — —

and so on

101. Repeat Exercise 100 using $ns^x np^y$ notation.

102. List n, ℓ, and m_ℓ quantum numbers for the highest energy
electron (or one of the highest energy electrons if there
are more than one) in the following atoms in their ground
states. (a) S; (b) Ac; (c) Cl; (d) Pr.

103. List n, ℓ, and m_ℓ quantum numbers for the highest energy
electron (or one of the highest energy electrons if there
are more than one) in the following atoms in their ground
states. (a) Se; (b) Zn; (c) Ca; (d) U.

104. Write the ground state electron configurations for elements A–E.

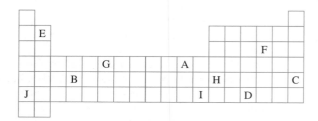

105. Repeat Exercise 104 for elements F–J.

106. How many unpaired electrons are in atoms of Na, Ne, Al,
Be, Br, and Ti?

107. (a) Distinguish between the terms "diamagnetic" and
"paramagnetic," and provide an example that illustrates
the meaning of each. (b) How is paramagnetism measured
experimentally?

108. Which of the following ions or atoms possess paramagnetic properties? (a) Br; (b) Kr; (c) Ne$^+$; (d) Fe; (e) Br$^-$.

109. Which of the following ions or atoms possess paramagnetic properties? (a) Cl$^-$; (b) Ca^{2+}; (c) Ca; (d) Ar$^-$; (e) Si.

CONCEPTUAL EXERCISES

110. The atomic mass of chlorine is reported to be 35.5, yet no
atom of chlorine has the mass of 35.5 amu. Explain.

111. Chemists often use the terms "atomic weight" and "atomic
mass" interchangeably. Explain why it would be more accurate if, in place of either of these terms, we used the
phrase "average atomic mass".

112. Draw a three-dimensional representation of each of the
following orbitals. (a) $3p_z$; (b) $2s$; (c) $3d_{xy}$; (d) $3d_{z^2}$.

113. We often show the shapes of orbitals as drawings. What
are some of the limitations of these drawings?

114. An atom in its ground state contains 18 electrons. How

many of these electrons are in orbitals with $\ell = 0$ values?

115. Suppose that scientists were to discover a new element, one that has the chemical properties of the noble gases, and it was positioned directly below radon on the periodic table. Assuming that the g orbitals of the elements preceding it in the period had not yet begun to fill, what would be the atomic number and ground state electron configuration of this new element?

116. For a lithium atom, give: (a) its ground state electron configuration; (b) the electron configuration for one of its lowest energy excited states; and (c) an electron configuration for a forbidden or impossible state.

117. Suppose we could excite all of the electrons in a sample of hydrogen atoms to the $n = 6$ level. They would then emit light as they relaxed to lower energy states. Some atoms might undergo the transition $n = 6$ to $n = 1$, and others might go from $n = 6$ to $n = 5$, then from $n = 5$ to $n = 4$, and so on. How many lines would we expect to observe in the resulting emission spectrum?

BUILDING YOUR KNOWLEDGE

118. Two isotopes of hydrogen occur naturally (^{1}H, >99%, and ^{2}H, <1%) and two of chlorine occur naturally (^{35}Cl, 76%, and ^{37}Cl, 24%). (a) How many different masses of HCl molecules can be formed from these isotopes? (b) What is the approximate mass of each of the molecules, expressed in atomic mass units? (Use atomic weights rounded to the nearest whole number.) (c) List these HCl molecules in order of decreasing relative abundance.

119. CH_4 is methane. If ^{1}H, ^{2}H, ^{12}C, and ^{13}C were the only isotopes present in a given sample of methane, show the different formulas and formula weights that might exist in that sample. (Use atomic weights rounded to the nearest whole number.)

120. Sodium is easily identified in a solution by its strong emission at $\lambda = 589$ nm. According to Einstein's equation, $E = mc^2$ (where m is mass) this amount of energy can be converted into mass. What is the mass equivalent of one photon emitted by an excited sodium atom? ($1\ \text{J} = 1\ \text{kg} \cdot \text{m}^2/\text{s}^2$)

121. A student was asked to calculate the wavelength and frequency of light emitted for an electron making the following transitions: (a) n = 6 → n = 2, and (b) n = 6 → n = 3. She was asked to determine whether she would be able to visually detect either of these electron transitions. Are her responses below correct? If not, make the necessary corrections.
    (a) $1/\lambda = (1.097 \times 10^7/\text{m})(1/2^2 - 1/6^2) = 2.44 \times 10^6/\text{m}$; $\lambda = 244$ nm
    (b) $1/\lambda = (1.097 \times 10^7/\text{m})(1/3^2 - 1/6^2) = 9.14 \times 10^5/\text{m}$; $\lambda = 1090$ nm
    She concluded that she couldn't see either of the transitions because neither is in the visible region of the spectrum.

122. When compounds of barium are heated in a flame, green light of wavelength 554 nm is emitted. How much energy is lost when one mole of barium atoms each emit one photon of this wavelength?

123. A 60-watt light bulb consumes energy at the rate of $60\ \text{J} \cdot \text{s}^{-1}$. Much of the light is emitted in the infrared region, and less than 5% of the energy appears as visible light. Calculate the number of visible photons emitted per second. Make the simplifying assumptions that 5.0% of the light is visible and that all visible light has a wavelength of 550 nm (yellow-green).

Light Bulb

124. Classical music radio station KMFA in Austin broadcasts at a frequency of 89.5 MHz. What is the wavelength of its signal in meters?

BEYOND THE TEXTBOOK

Go to the textbook website at

http://www.brookscole.com/chemistry/whitten

for additional activities and exercises based on the General Chemistry Interactive CD-ROM, the World Wide Web, and library resources.

InfoTrac College Edition

For additional readings, go to InfoTrac College Edition, your online research library at:

http://infotrac.thomsonlearning.com

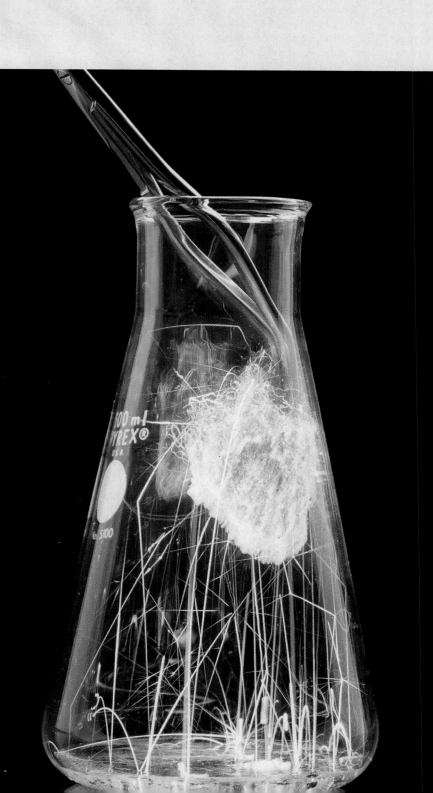

OUTLINE

Steel wool (iron) burns brightly in pure oxygen.

OBJECTIVES

After you have studied this chapter, you should be able to

- More effectively use the periodic table
- Discuss chemical periodicity of the following physical properties:
    Atomic radii
    Ionization energy
    Electron affinity
    Ionic radii
    Electronegativity
- Describe chemical periodicity in the reactions of
    Hydrogen
    Oxygen
- Describe chemical periodicity in the compounds of
    Hydrogen
    Oxygen

6-1 MORE ABOUT THE PERIODIC TABLE

In Chapter 4 we described the development of the periodic table, some terminology for it, and its guiding principle, the *periodic law*.

The properties of the elements are periodic functions of their atomic numbers.

In Chapter 5 we described electron configurations of the elements. In the long form of the periodic table, elements are arranged in blocks based on the kinds of atomic orbitals being filled. (Please review Table 5-5 and Figure 5-31 carefully.) We saw that electron

AT&T Bell Laboratories

The properties of elements are correlated with their positions in the periodic table. Chemists use the periodic table as an invaluable guide in their search for new, useful materials. A barium sodium niobate crystal can convert infrared laser light into visible green light. This harmonic generation or "frequency doubling" is very important in chemical research using lasers and in the telecommunications industry.

Some transition metals (*left to right*): Ti, V, Cr, Mn, Fe, Co, Ni, Cu.

Charles D. Winters

$[He] = 1s^2$

CD-ROM   Screen 8.15, Chemical Reactions and Periodic Properties.

SCiLINKS.   **TOPIC:** Periodic Table
GO TO: www.scilinks.org
*sci*LINKS **CODE:** WCH0410

configurations of elements in the A groups are entirely predictable from their positions in the periodic table. We also noted, however, that some irregularities occur within the B groups.

Now we classify the elements according to their electron configurations, which is a very useful system.

Noble Gases. For many years the Group VIIIA elements—the noble gases—were called inert gases because no chemical reactions were known for them. We now know that the heavier members do form compounds, mostly with fluorine and oxygen. Except for helium, each of these elements has eight electrons in its outermost occupied shell. Their outer shell may be represented as having the electron configuration . . . ns^2np^6.

Representative Elements. The A group elements in the periodic table are called representative elements. Their "last" electron is assigned to an outer shell s or p orbital. These elements show distinct and fairly regular variations in their properties with changes in atomic number.

d-**Transition Elements.** Elements in the B groups in the periodic table are known as the *d*-transition elements or, more simply, as transition elements or transition metals. The elements of the four transition series are all metals and are characterized by electrons being assigned to *d* orbitals. Stated differently, the *d*-transition elements contain electrons in both the *ns* and $(n - 1)d$ orbitals, but not in the *np* orbitals. The first transition series, Sc through Zn, has electrons in the 4*s* and 3*d* orbitals, but not in the 4*p* orbitals. They are referred to as

First transition series:	$_{21}$Sc through $_{30}$Zn
Second transition series:	$_{39}$Y through $_{48}$Cd
Third transition series:	$_{57}$La and $_{72}$Hf through $_{80}$Hg
Fourth transition series:	$_{89}$Ac and $_{104}$Rf through element 112

 HEMISTRY IN USE   *The Development of Science*

The Periodic Table

The periodic table is one of the first things a student of chemistry encounters. It appears invariably in textbooks, in lecture halls, and in laboratories. Scientists consider it an indispensable reference. And yet, less than 150 years ago, the idea of arranging the elements by atomic weight or number was considered absurd. At an 1866 meeting of the Chemical Society at Burlington House, England, J. A. R. Newlands (1837–1898) presented a theory he called the law of octaves. It stated that when the known elements were listed by increasing atomic weights, those that were eight places apart would be similar, much like notes on a piano keyboard. His colleagues' reactions are probably summed up best by the remark of a Professor Foster: "Have you thought of arranging the elements according to their initial letters? Maybe some better connections would come to light that way."

It is not surprising that poor Newlands was not taken seriously. In the 1860s, little information was available to illustrate relationships among the elements. Only 62 of them had been distinguished from more complex substances when Mendeleev first announced his discovery of the periodic law in 1869. As advances in atomic theory were made, however, and as new experiments contributed to the understanding of chemical behavior, some scientists had begun to see similarities and patterns among the elements. In 1869, Lothar Meyer and Dmitri Mendeleev independently published similar versions of the now-famous periodic table.

Mendeleev's discovery was the result of many years of hard work. He gathered information on the elements from all corners of the earth—by corresponding with colleagues, studying books and papers, and redoing experiments to confirm data. He put the statistics of each element on a small card and pinned the cards to his laboratory wall, where he arranged and rearranged them many times until he was sure that they were in the right order. One especially farsighted feature of Mendeleev's accomplishment was his realization that some elements were missing from the table. He predicted the properties of these substances (gallium, scandium, and germanium). (It is important to remember that Mendeleev's periodic table organization was devised more than 50 years before the discovery and characterization of subatomic particles.)

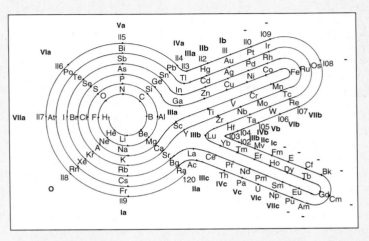

An alternative representation of the periodic table, as proposed by Charles Janet, 1928.

Since its birth in 1869, the periodic table has been discussed and revised many times. Spectroscopic and other discoveries have filled in the blanks left by Mendeleev and added a new column consisting of the noble gases. As scientists learned more about atomic structure, the basis for ordering was changed from atomic weight to atomic number. The perplexing *f*-transition elements were sorted out and given a special place, along with many of the radioactive elements created by atomic bombardment. Even the form of the table has been experimented with, resulting in everything from spiral and circular tables to exotic shapes such as the one suggested by Charles Janet (shown here). A three-dimensional periodic table that takes into account valence-shell energies has been proposed by Professor Leland C. Allen of Princeton University.

During the past century, chemistry has become a fast-moving science in which methods and instruments are often outdated within a few years. But it is doubtful that our old friend, the periodic table, will ever become obsolete. It may be modified, but it will always stand as a statement of basic relationships in chemistry and as a monument to the wisdom and insight of its creator, Dmitri Mendeleev.

Lisa Saunders Boffa
Senior Chemist
Exxon Corporation

CHEMISTRY IN USE

The Development of Science

Glenn Seaborg: A Human Side to the Modern Periodic Table

If it were possible to associate a human face with the modern periodic table, that face would most likely belong to Glenn Seaborg (1912–1999), the codiscoverer of ten transuranium elements and the name behind Element 106. Seaborg's contributions to heavy-element chemistry began in 1940, when he and co-workers at the University of California at Berkeley produced the first sample of plutonium by bombarding uranium with deuterons (^{2}H nuclei) in a particle accelerator. They found that the isotope plutonium-239 undergoes nuclear fission (see Chapter 26), making it a potential energy source for nuclear power or nuclear weapons.

As American involvement in World War II grew, President Franklin Roosevelt called Seaborg and other eminent scientists to the Wartime Metallurgical Laboratory at the University of Chicago, where they figured out how to prepare and purify plutonium-239 in useful quantities for the Manhattan Project, the making of the atom bomb. In 1945, Seaborg was one of the signers of the Franck Report, a document recommending that a safe demonstration test of the atomic bomb might persuade Japan to surrender without the bomb actually being used. Professor Seaborg served as a scientific advisor for nine other presidents following Roosevelt and was chairman of the U.S. Atomic Energy Commission under Kennedy, Johnson, and Nixon.

Seaborg's contributions illustrate how certain areas of science can be highly influenced by a particular institution or even a national tradition over time. His discovery of plutonium at the University of California at Berkeley followed the 1940 synthesis of neptunium at the same site by Edwin McMillan, who shared the 1951 Nobel Prize in chemistry

with Seaborg for these accomplishments. Since that time, Seaborg and other teams involving Berkeley researchers at the University's Lawrence Berkeley Laboratory have prepared nine more heavy elements. He and co-workers hold the world's only patents on chemical elements, for americium and curium. The original location of the first transuranium laboratory on the Berkeley campus (a few yards from the later site of Professor Seaborg's reserved "Nobel Laureate" parking space) is now a national historic landmark.

Laboratories in the United States, Russia, and Germany have been the most active in the synthesis of new elements. In 1994, nationalistic feelings invaded what should have been impartial decisions by the International Union of Pure and Applied Chemistry (IUPAC) regarding the official names for elements 101 and 109. In some cases, researchers from different countries had proposed different names for these elements based on where credit for their discovery was felt to be deserved. For example, the name "hahnium" was proposed for element 105 by American researchers, while the Russians preferred the more snappy "nielsbohrium." The American Chemical Society proposed to name element 106 seaborgium (Sg), but the IUPAC's nomenclature committee rejected the choice, objecting to the fact that Seaborg was still alive ("and they can prove it," he quipped). Outrage at this and some of the other naming decisions prompted many scientists to ignore the IUPAC's recommended name for 106, rutherfordium, and to continue to use seaborgium. In 1997, the IUPAC reversed its decision and endorsed Sg, saving the chemical literature from future confusion caused by different naming practices in the scientific journals and conferences of different countries.

If Professor Seaborg had been nominated for a different honor—appearance on a U.S. postage stamp—the story would have had an unhappier ending. Although surely not as rare a commodity as the names of new chemical elements, United States stamps are not permitted to honor living individuals. Seaborg would have been the only person in the world who could have received mail addressed entirely in elements: Seaborgium, Lawrencium (for the Lawrence Berkeley Laboratory), Berkelium, Californium, Americium—and don't forget the ZIP code, 94720.

Lisa Saunders Boffa
Senior Chemist
Exxon Corporation

The elements of Period 3. Properties progress (*left to right*) from solids (Na, Mg, Al, Si, P, S) to gases (Cl, Ar) and from the most metallic (Na) to the most nonmetallic (Ar).

***f*-Transition Elements.** Sometimes known as *inner transition elements*, these are elements in which electrons are being added to *f* orbitals. In these elements, the second from the outermost occupied shell is building from 18 to 32 electrons. All are metals. The *f*-transition elements are located between Groups IIIB and IVB in the periodic table. They are

First *f*-transition series (lanthanides): $_{58}Ce$ through $_{71}Lu$

Second *f*-transition series (actinides): $_{90}Th$ through $_{103}Lr$

The A and B designations for groups of elements in the periodic table are somewhat arbitrary, and they are reversed in some periodic tables. In another standard designation, the groups are numbered 1 through 18. The system used in this text is the one commonly used in the United States. Elements with the same group numbers, but with different letters, have relatively few similar properties. The origin of the A and B designations is the fact that some compounds of elements with the same group numbers have similar formulas but quite different properties, for example, NaCl (IA) and AgCl (IB), $MgCl_2$ (IIA) and $ZnCl_2$ (IIB). As we shall see, variations in the properties of the B groups across a row are not nearly as regular and dramatic as the variations observed across a row of A group elements.

The *outermost* electrons have the greatest influence on the properties of elements. Adding an electron to an *inner d* orbital results in less striking changes in properties than adding an electron to an *outer s* or *p* orbital.

In any atom the *outermost* electrons are those that have the highest value of the principal quantum number, *n*.

PERIODIC PROPERTIES OF THE ELEMENTS

Now we investigate the nature of periodicity. Knowledge of periodicity is valuable in understanding bonding in simple compounds. Many physical properties, such as melting points, boiling points, and atomic volumes, show periodic variations. For now, we describe the variations that are most useful in predicting chemical behavior. The variations in these properties depend on electron configurations, especially the configurations in the outermost occupied shell, and on how far away that shell is from the nucleus.

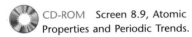 http://webbook.nist.gov

CD-ROM   Screen 8.9, Atomic Properties and Periodic Trends.

6-2 ATOMIC RADII

CD-ROM   Screen 8.10, Atomic Size.

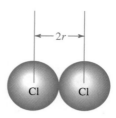

The radius of an atom, *r*, is taken as half of the distance between nuclei in *homonuclear* molecules such as Cl_2.

In Section 5-16 we described individual atomic orbitals in terms of probabilities of distributions of electrons over certain regions in space. Similarly, we can visualize the total electron cloud that surrounds an atomic nucleus as somewhat indefinite. We cannot isolate a single atom and measure its diameter the way we can measure the diameter of a golf ball. For all practical purposes, the size of an individual atom cannot be uniquely defined. An indirect approach is required. The size of an atom is determined by its immediate environment, especially its interaction with surrounding atoms. By analogy, suppose we arrange some golf balls in an orderly array in a box. If we know how the balls are positioned, the number of balls, and the dimensions of the box, we can calculate the diameter of an individual ball. Application of this reasoning to solids and their densities leads us to values for the atomic sizes of many elements. In other cases, we derive atomic radii from the observed distances between atoms that are combined with one another. For example, the distance between atomic centers (nuclei) in the Cl_2 molecule is measured to be 2.00 Å. This suggests that the radius of *each* Cl atom is half the interatomic distance, or 1.00 Å. We collect the data obtained from many such measurements to indicate the *relative* sizes of individual atoms.

The top of Figure 6-1 displays the relative sizes of atoms of the representative elements and the noble gases. It shows the periodicity in atomic radii. (The ionic radii at the bottom of Figure 6-1 are discussed in Section 6-5.)

The **effective nuclear charge,** Z_{eff}, experienced by an electron in an outer shell is less than the actual nuclear charge, Z. This is because the *attraction* of outer-shell electrons by the nucleus is partly counterbalanced by the repulsion of these outer-shell electrons by electrons in inner shells. We say that the electrons in inner shells *screen*, or *shield*, electrons in outer shells from the full effect of the nuclear charge. This concept of a **screening,** or **shielding, effect** helps us understand many periodic trends in atomic properties.

Consider an atom of lithium; it has two electrons in a filled shell, $1s^2$, and one electron in the $2s$ orbital, $2s^1$. The electron in the $2s$ orbital is fairly effectively screened from the nucleus by the two electrons in the filled $1s$ orbital, so the $2s$ electron does not "feel" the full 3+ charge of the nucleus. The effective nuclear charge, Z_{eff}, experienced by the electron in the $2s$ orbital, however, is not 1 (3 minus 2) either. The electron in the outer shell of lithium has some probability of being found close to the nucleus (see Figure 5-20). We say that, to some extent, it *penetrates* the region of the $1s$ electrons; that is, the $1s$ electrons do not completely shield the outer-shell electrons from the nucleus. The electron in the $2s$ shell "feels" an effective nuclear charge a little larger than 1+. Sodium, element number 11, has ten electrons in inner shells, $1s^22s^22p^6$, and one electron in an outer shell, $3s^1$. The ten inner-shell electrons of the sodium atom screen (shield) the outer-shell electron somewhat from the nucleus, counteracting some of the 11+ nuclear charge. But the $3s$ electron of sodium penetrates the inner shells to a significant extent, so the effective nuclear charge felt by the outermost ($3s$) electron is actually greater than it is for lithium ($2s$). The somewhat increased attraction for the outermost electron in sodium is outweighed, however, by the fact that the "outer" electron in a sodium atom is in the third shell, whereas in lithium it is in the second shell. Recall from Chapter 5 that the third shell ($n = 3$) is farther from the nucleus than the second shell ($n = 2$). Thus, we see why sodium atoms are larger than lithium atoms. Similar reasoning explains why potassium atoms are larger than sodium atoms and why the sizes of the elements in each column of the periodic table are related in a similar way.

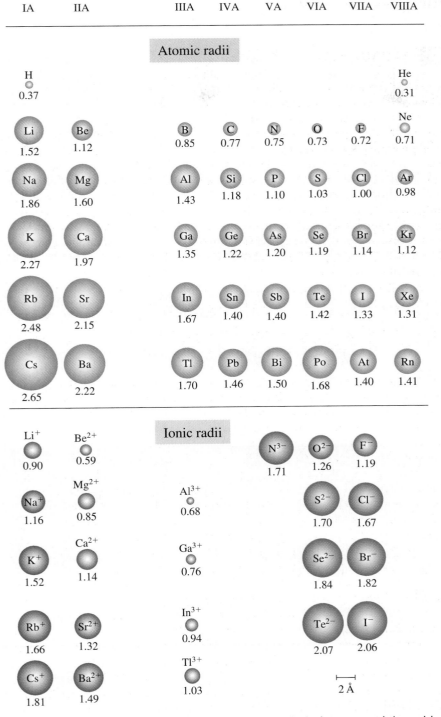

Figure 6-1 (*Top*) Atomic radii of the A group (representative) elements and the noble gases, in angstroms, Å (Section 6-2). Atomic radii *increase going down a group* because electrons are being added to shells farther from the nucleus. Atomic radii *decrease from left to right within a given period* owing to increasing effective nuclear charge. Hydrogen atoms are the smallest and cesium atoms are the largest naturally occurring atoms.

(*Bottom*) Sizes of ions of the A group elements, in angstroms (Section 6-5). Positive ions (cations, blue) are always *smaller* than the neutral atoms from which they are formed. Negative ions (anions, green) are always *larger* than the neutral atoms from which they are formed.

237

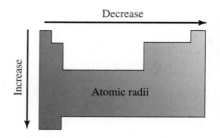

General trends in atomic radii of A group elements with position in the periodic table.

Within a family (vertical group on the periodic table) of representative elements, atomic radii increase from top to bottom as electrons are added to shells farther from the nucleus.

As we move *across* the periodic table, atoms become smaller due to increasing effective nuclear charges. Consider the elements B ($Z = 5$, $1s^2 2s^2 2p^1$) to F ($Z = 9$, $1s^2 2s^2 2p^5$). In B there are two electrons in a noble gas configuration, $1s^2$, and three electrons in the second shell, $2s^2 2p^1$. The two electrons in the noble gas configuration fairly effectively screen out the effect of two protons in the nucleus. So the electrons in the second shell of B "feel" a greater effective nuclear charge than do those of Be. By similar arguments, we see that in carbon ($Z = 6$, $1s^2 2s^2 2p^2$) the electrons in the second shell "feel" an effective nuclear charge greater than those of B. So we expect C atoms to be smaller than B atoms, and they are. In nitrogen ($Z = 7$, $1s^2 2s^2 2p^3$), the electrons in the second shell "feel" an even greater effective nuclear charge, and so N atoms are smaller than C atoms.

As we move from left to right *across a period* in the periodic table, atomic radii of representative elements *decrease* as a proton is added to the nucleus and an electron is added to a particular shell.

CD-ROM   Screen 8.14, Properties of Transition Metals.

For the transition elements, the variations are not so regular because electrons are being added to an inner shell. All transition elements have smaller radii than the preceding Group IA and IIA elements in the same period.

EXAMPLE 6-1   *Trends in Atomic Radii*

Arrange the following elements in order of increasing atomic radii. Justify your order.

$$Cs, \quad F, \quad K, \quad Cl$$

Plan

Both K and Cs are Group IA metals, whereas F and Cl are halogens (VIIA nonmetals). Figure 6-1 shows that atomic radii increase as we descend a group, so K < Cs and F < Cl. Atomic radii decrease from left to right.

Solution

The order of increasing atomic radii is F < Cl < K < Cs.

You should now work Exercise 16.

6-3   IONIZATION ENERGY

CD-ROM   Screen 8.12, Ionization Energy.

The first **ionization energy (IE$_1$)**, also called *first ionization potential*, is

the minimum amount of energy required to remove the most loosely bound electron from an isolated gaseous atom to form an ion with a 1+ charge.

TABLE 6-1 *First Ionization Energies (kJ/mol of atoms) of Some Elements*

H																	He
1312																	2372
Li	Be											B	C	N	O	F	Ne
520	899											801	1086	1402	1314	1681	2081
Na	Mg											Al	Si	P	S	Cl	Ar
496	738											578	786	1012	1000	1251	1521
K	Ca	Sc	Ti	V	Cr	Mn	Fe	Co	Ni	Cu	Zn	Ga	Ge	As	Se	Br	Kr
419	599	631	658	650	652	717	759	758	757	745	906	579	762	947	941	1140	1351
Rb	Sr	Y	Zr	Nb	Mo	Tc	Ru	Rh	Pd	Ag	Cd	In	Sn	Sb	Te	I	Xe
403	550	617	661	664	685	702	711	720	804	731	868	558	709	834	869	1008	1170
Cs	Ba	La	Hf	Ta	W	Re	Os	Ir	Pt	Au	Hg	Tl	Pb	Bi	Po	At	Rn
377	503	538	681	761	770	760	840	880	870	890	1007	589	715	703	812	890	1037

For calcium, for example, the first ionization energy, IE_1, is 590 kJ/mol:

$$Ca(g) + 590 \text{ kJ} \longrightarrow Ca^+(g) + e^-$$

The second **ionization energy (IE_2)** is the amount of energy required to remove the second electron. For calcium, it may be represented as

$$Ca^+(g) + 1145 \text{ kJ} \longrightarrow Ca^{2+}(g) + e^-$$

For a given element, IE_2 *is always greater than* IE_1 because it is always more difficult to remove a negatively charged electron from a positively charged ion than from the corresponding neutral atom. Table 6-1 gives first ionization energies.

Ionization energies measure how tightly electrons are bound to atoms. Ionization always requires energy to remove an electron from the attractive force of the nucleus. Low ionization energies indicate ease of removal of electrons, and hence ease of positive ion (cation) formation. Figure 6-2 shows a plot of first ionization energy versus atomic number for several elements.

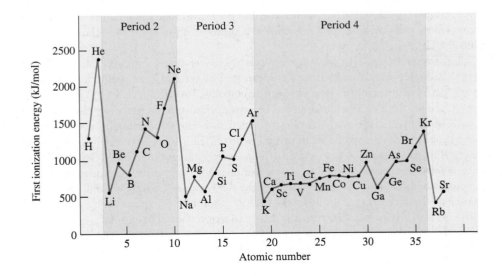

Figure 6-2 A plot of first ionization energies for the first 38 elements versus atomic number. The noble gases have very high first ionization energies, and the IA metals have low first ionization energies. Note the similarities in the variations for the Period 2 elements, 3 through 10, to those for the Period 3 elements, 11 through 18, as well as for the later A group elements. Variations for B group elements are not nearly so pronounced as those for A group elements.

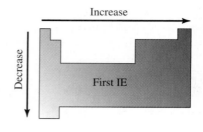

General trends in first ionization energies of A group elements with position in the periodic table. Exceptions occur at Groups IIIA and VIA.

By Coulomb's Law, $F \propto \dfrac{(q^+)(q^-)}{d^2}$, the attraction for the outer shell electrons is directly proportional to the *effective* charges and inversely proportional to the *square* of the distance between the charges. Even though the effective nuclear charge increases going down a group, the greatly increased size results in a weaker net attraction for the outer electrons and thus in a lower first ionization energy.

Elements with low ionization energies (IE) lose electrons easily to form cations.

We see that in each period of Figure 6-2, the noble gases have the highest first ionization energies. This should not be surprising because the noble gases are known to be very unreactive elements. It requires more energy to remove an electron from a helium atom (slightly less than 4.0×10^{-18} J/atom, or 2372 kJ/mol) than to remove one from a neutral atom of any other element.

$$He(g) + 2372 \text{ kJ} \longrightarrow He^+(g) + e^-$$

The Group IA metals (Li, Na, K, Rb, Cs) have very low first ionization energies. Each of these elements has only one electron in its outermost shell ($\ldots ns^1$), and they are the largest atoms in their periods. The first electron added to a shell is easily removed to form a noble gas configuration. As we move down the group, the first ionization energies become smaller. The force of attraction of the positively charged nucleus for electrons decreases as the square of the distance between them increases. So as atomic radii increase in a given group, first ionization energies decrease because the outermost electrons are farther from the nucleus.

Effective nuclear charge, Z_{eff}, increases going from left to right across a period. The increase in effective nuclear charge causes the outermost electrons to be held more tightly, making them harder to remove. The first ionization energies therefore generally *increase* from left to right across the periodic table. The reason for the trend in first ionization energies is the same as that used in Section 6-2 to explain trends in atomic radii. The first ionization energies of the Group IIA elements (Be, Mg, Ca, Sr, Ba) are significantly higher than those of the Group IA elements in the same periods. This is because the Group IIA elements have higher Z_{eff} values and smaller atomic radii. Thus, their outermost electrons are held more tightly than those of the neighboring IA metals. It is harder to remove an electron from a pair in the filled outermost s orbitals of the Group IIA elements than to remove the single electron from the half-filled outermost s orbitals of the Group IA elements.

The first ionization energies for the Group IIIA elements (B, Al, Ga, In, Tl) are exceptions to the general horizontal trends. They are *lower* than those of the IIA elements in the same periods because the IIIA elements have only a single electron in their outermost p orbitals. Less energy is required to remove the first p electron than the second s electron from the outermost shell because the p orbital is at a higher energy (less stable) than an s orbital within the same shell (n value).

Going from Groups IIIA to VA, electrons are going singly into separate np orbitals, where they do not shield one another significantly. The general left-to-right increase in IE_1 for each period is interrupted by a dip between Groups VA (N, P, As, Sb, Bi) and VIA elements (O, S, Se, Te, Po). Presumably, this behavior is because the fourth np electron in the Group VIA elements is paired with another electron in the same orbital, so it experiences greater repulsion than it would in an orbital by itself. This increased repulsion apparently outweighs the increase in Z_{eff}, so the fourth np electron in an outer shell (Group VIA elements) is somewhat easier to remove (lower ionization energy) than is the third np electron in an outer shell (Group VA elements). After the dip between Groups VA and VIA, the importance of the increasing Z_{eff} outweighs the repulsion of electrons needing to be paired, and the general left-to-right increases in first ionization energies resume.

Knowledge of the relative values of ionization energies assists us in predicting whether an element is likely to form ionic or molecular (covalent) compounds. Elements with low ionization energies form ionic compounds by losing electrons to form *cations* (positively

As one goes across a period on the periodic table, the slight breaks in the increasing ionization energies occur between Groups IIA and IIIA (electrons first enter the np subshell) and again between Groups VA and VIA (electrons are first paired in the np subshell).

charged ions). Elements with intermediate ionization energies generally form molecular compounds by sharing electrons with other elements. Elements with very high ionization energies, such as Groups VIA and VIIA, often gain electrons to form *anions* (negatively charged ions) with closed shell (noble gas) electron configurations.

One factor that favors an atom of a *representative* element forming a monatomic ion in a compound is the formation of a stable noble gas electron configuration. Energy considerations are consistent with this observation. For example, as 1 mol Li from Group IA forms 1 mol Li^+ ions, it absorbs 520 kJ/mol of Li atoms. The IE_2 value is 14 times greater, 7298 kJ/mol, and is prohibitively large for the formation of Li^{2+} ions under ordinary conditions. For Li^{2+} ions to form, an electron would have to be removed from the filled first shell, which is very unlikely. The other alkali metals behave in the same way, for similar reasons.

The first two ionization energies of Be (Group IIA) are 899 and 1757 kJ/mol, but IE_3 is more than eight times larger, 14,849 kJ/mol. So Be forms Be^{2+} ions, but not Be^{3+} ions. The other alkaline earth metals—Mg, Ca, Sr, Ba, and Ra—behave in a similar way. Only the lower members of Group IIIA, beginning with Al, form 3+ ions. Bi and some *d*- and *f*-transition metals do so, too. We see that the magnitudes of successive ionization energies support the ideas of electron configurations discussed in Chapter 5.

> Due to the high energy required, *simple monatomic cations with charges greater than 3+ do not form under ordinary circumstances.*

EXAMPLE 6-2   *Trends in First IEs*

Arrange the following elements in order of increasing first ionization energy. Justify your order.

Na,   Mg,   Al,   Si

Plan

The first ionization energies generally increase from left to right in the same period, but there are exceptions at Groups IIIA and VIA. Al is a IIIA element with only one electron in its outer *p* orbitals, $1s^2 2s^2 2p^6 3s^2 3p^1$, which makes the first ionization energy lower than one might normally expect.

Solution

There is a slight dip at Group IIIA in the plot of first IE versus atomic number (see Figure 6-2). The order of increasing first ionization energy is Na < Al < Mg < Si.

You should now work Exercise 24.

6-4   ELECTRON AFFINITY

The **electron affinity (EA)** of an element may be defined as

> the amount of energy *absorbed* when an electron is added to an isolated gaseous atom to form an ion with a 1− charge.

The convention is to assign a positive value when energy is absorbed and a negative value when energy is released. Most elements have no affinity for an additional electron and thus

Here is one reason why trends in ionization energies are important.

Noble gas electron configurations are stable only for ions in compounds. In fact, $Li^+(g)$ is less stable than $Li(g)$ by 520 kJ/mol.

 CD-ROM   Screen 8.13, Electron Affinity.

This is consistent with thermodynamic convention.

The value of EA for Cl can also be represented as -5.79×10^{-19} J/atom or -3.61 eV/atom. The electron volt (eV) is a unit of energy (1 eV $= 1.6022 \times 10^{-19}$ J).

have an electron affinity equal to zero. We can represent the electron affinities of helium and chlorine as

$$\text{He}(g) + e^- \xrightarrow{\quad\times\quad} \text{He}^-(g) \qquad \text{EA} = \quad 0 \text{ kJ/mol}$$

$$\text{Cl}(g) + e^- \longrightarrow \text{Cl}^-(g) + 349 \text{ kJ} \qquad \text{EA} = -349 \text{ kJ/mol}$$

The first equation tells us that helium will not add an electron. The second equation tells us that when one mole of gaseous chlorine atoms gain one electron each to form gaseous chloride ions, 349 kJ of energy is *released (exothermic)*. Figure 6-3 shows a plot of electron affinity versus atomic number for several elements.

Electron affinity involves the *addition* of an electron to a neutral gaseous atom. The process by which a neutral atom X gains an electron

$$\text{X}(g) + e^- \longrightarrow \text{X}^-(g) \qquad \text{(EA)}$$

is *not* the reverse of the ionization process

$$\text{X}^+(g) + e^- \longrightarrow \text{X}(g) \qquad \text{(reverse of IE}_1)$$

The first process begins with a neutral atom, whereas the second begins with a positive ion. Thus, IE_1 and EA are *not* simply equal in value with the signs reversed. We see from Figure 6-3 that electron affinities generally become more negative from left to right across a row in the periodic table (excluding the noble gases). This means that most representative elements in Groups IA to VIIA show a greater attraction for an extra electron from left to right. Halogen atoms, which have the outer electron configuration ns^2np^5, have the most negative electron affinities. They form stable anions with noble gas configurations, $\ldots ns^2np^6$, by gaining one electron.

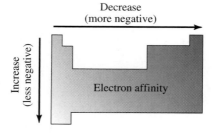

General trends in electron affinities of A group elements with position in the periodic table. There are many exceptions.

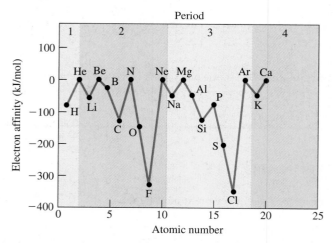

Figure 6-3 A plot of electron affinity versus atomic number for the first 20 elements. The *general* horizontal trend is that electron affinities become more negative (more energy is released as an extra electron is added) from Group IA through Group VIIA for a given period. Exceptions occur at the IIA and VA elements.

TABLE 6-2 *Electron Affinity Values (kJ/mol) of Some Elements**

Period	IA	IIA		IIIA	IVA	VA	VIA	VIIA	VIIIA
1	H −73								He 0
2	Li −60	Be (~0)		B −29	C −122	N 0	O −141	F −328	Ne 0
3	Na −53	Mg (~0)		Al −43	Si −134	P −72	S −200	Cl −349	Ar 0
4	K −48	Ca (~0)	Cu −118	Ga −29	Ge −119	As −78	Se −195	Br −324	Kr 0
5	Rb −47	Sr (~0)	Ag −125	In −29	Sn −107	Sb −101	Te −190	I −295	Xe 0
6	Cs −45	Ba (~0)	Au −282	Tl −19	Pb −35	Bi −91			

*Estimated values are in parentheses.

Elements with very negative electron affinities gain electrons easily to form negative ions (anions).

"Electron affinity" is a precise and quantitative term, like "ionization energy," but it is difficult to measure. Table 6-2 shows electron affinities for several elements.

For many reasons, the variations in electron affinities are not regular across a period. The general trend is: the electron affinities of the elements become more negative from left to right in each period. Noteworthy exceptions are the elements of Groups IIA and VA, which have less negative values than the trends suggest (see Figure 6-3). It is very difficult to add an electron to a IIA metal atom because its outer s subshell is filled. The values for the VA elements are slightly less negative than expected because they apply to the addition of an electron to a half-filled set of np orbitals ($ns^2np^3 \rightarrow ns^2np^4$), which requires pairing. The resulting repulsion overcomes the increased attractive force of the nucleus.

Energy is always required to bring a negative charge (electron) closer to another negative charge (anion). So the addition of a second electron to a 1− anion to form an ion with a 2− charge is always endothermic. Thus, electron affinities of *anions* are always positive.

This reasoning is similar to that used to explain the low IE_1 values for Group VIA elements.

EXAMPLE 6-3 *Trends in EAs*

Arrange the following elements from most negative to least negative electron affinity.

$$K, \quad Br, \quad Cs, \quad Cl$$

Plan

Electron affinity values generally become more negative from left to right across a period, with major exceptions at Groups IIA (Be) and VA (N). They generally become more negative from bottom to top.

In fact, K and Cs readily form 1+ cations.

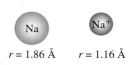

r = 1.52 Å   r = 0.90 Å

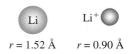

r = 1.86 Å   r = 1.16 Å

The nuclear charge remains constant when the ion is formed.

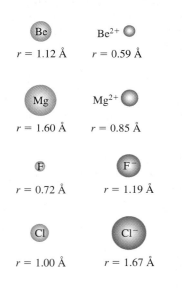

Solution

The order of electron affinities from most negative to least negative is

(most negative EA) $Cl < Br < K < Cs$ (least negative EA)

This means that Cl most readily forms a $1-$ ion, while Cs is least able to do so.

You should now work Exercises 30 and 31.

6-5 IONIC RADII

Many elements on the left side of the periodic table react with other elements by *losing* electrons to form positively charged ions. Each of the Group IA elements (Li, Na, K, Rb, Cs) has only one electron in its outermost shell (electron configuration . . . ns^1). These elements react with other elements by losing one electron to form the ions Li^+, Na^+, K^+, Rb^+, and Cs^+, thus attaining noble gas configurations. A neutral lithium atom, Li, contains three protons in its nucleus and three electrons, with its outermost electron in the $2s$ orbital. A lithium ion, Li^+, however, contains three protons in its nucleus but only two electrons, both in the $1s$ orbital. So a Li^+ ion is much smaller than a neutral Li atom (see figure in the margin). Likewise, a sodium ion, Na^+, is considerably smaller than a sodium atom, Na. The relative sizes of atoms and common ions of some representative elements are shown in Figure 6-1.

Isoelectronic species have the same number of electrons. We see that the ions formed by the Group IIA elements (Be^{2+}, Mg^{2+}, Ca^{2+}, Sr^{2+}, Ba^{2+}) are significantly smaller than the *isoelectronic* ions formed by the Group IA elements in the same period. The radius of the Li^+ ion is 0.90 Å, whereas the radius of the Be^{2+} ion is only 0.59 Å. This is what we should expect. A beryllium ion, Be^{2+}, is formed when a beryllium atom, Be, loses both of its $2s$ electrons, while the 4+ nuclear charge remains constant. We expect the 4+ nuclear charge in Be^{2+} to attract the remaining two electrons quite strongly. Comparison of the ionic radii of the IIA elements with their atomic radii indicates the validity of our reasoning. Similar reasoning indicates that the ions of the Group IIIA metals (Al^{3+}, Ga^{3+}, In^{3+}, Tl^{3+}) should be even smaller than the ions of Group IA and Group IIA elements in the same periods.

Now consider the Group VIIA elements (F, Cl, Br, I). These have the outermost electron configuration . . . ns^2np^5. These elements can completely fill their outermost p orbitals by *gaining* one electron to attain noble gas configurations. Thus, when a fluorine atom (with seven electrons in its outer shell) gains one electron, it becomes a fluoride ion, F^-, with eight electrons in its outer shell. These eight electrons repel one another more strongly than the original seven, so the electron cloud expands. The F^- ion is, therefore, much larger than the neutral F atom (see figure in the margin). Similar reasoning indicates that a chloride ion, Cl^-, should be larger than a neutral chlorine atom, Cl. Observed ionic radii (see Figure 6-1) verify these predictions.

Comparing the sizes of an oxygen atom (Group VIA) and an oxide ion, O^{2-}, again we find that the negatively charged ion is larger than the neutral atom. The oxide ion is also larger than the isoelectronic fluoride ion because the oxide ion contains ten electrons held by a nuclear charge of only 8+, whereas the fluoride ion has ten electrons held by a nuclear charge of 9+. Comparison of radii is not a simple issue when we try to compare atoms, positive and negative ions, and ions with varying charge. Sometimes we compare atoms

r = 1.12 Å   r = 0.59 Å

r = 1.60 Å   r = 0.85 Å

r = 0.72 Å   r = 1.19 Å

r = 1.00 Å   r = 1.67 Å

Decrease →

Increase ↓

Ionic radii

General trends in ionic radii of A group elements with position in the periodic table.

to their ions, atoms or ions that are vertically or horizontally positioned on the periodic table, or isoelectronic species. The following guidelines are often considered in the order given.

1. Simple positively charged ions (cations) are always smaller than the neutral atoms from which they are formed.

2. Simple negatively charged ions (anions) are always larger than the neutral atoms from which they are formed.

3. The sizes of cations decrease from left to right across a period.

4. The sizes of anions decrease from left to right across a period.

5. Within an isoelectronic series, radii decrease with increasing atomic number because of increasing nuclear charge.

6. Both cation and anion sizes increase going down a group.

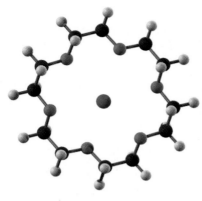

Large ring compounds can selectively trap ions based on the ability of various ions to fit within a cavity in the large compound. This ball-and-stick model is an example of this with a Na cation in the middle (magenta atom). This results in selectivity such as occurs in transport across membranes in biological systems.

An isoelectronic series of ions

	N^{3-}	O^{2-}	F^-	Na^+	Mg^{2+}	Al^{3+}
Ionic radius (Å)	1.71	1.26	1.19	1.16	0.85	0.68
No. of electrons	10	10	10	10	10	10
Nuclear charge	+7	+8	+9	+11	+12	+13

 CD-ROM   Screen 8.11, Ion Sizes.

EXAMPLE 6-4   *Trends in Ionic Radii*

Arrange the following ions in order of increasing ionic radii: (a) Ca^{2+}, K^+, Al^{3+}; (b) Se^{2-}, Br^-, Te^{2-}.

Plan

Some of the pairs of ions are isoelectronic, so we can compare their sizes on the basis of nuclear charges. Other comparisons can be made based on the outermost occupied shell (highest value of n).

Solution

(a) Ca^{2+} and K^+ are isoelectronic (18 electrons each) with an outer-shell electron configuration of $3s^2 3p^6$. Because Ca^{2+} has a higher nuclear charge (20+) than K^+ (19+), Ca^{2+} holds its 18 electrons more tightly, and Ca^{2+} is smaller than K^+. Al^{3+} has electrons only in the second main shell (outer-shell electron configuration of $2s^2 2p^6$), so it is smaller than either of the other two ions.

$$Al^{3+} < Ca^{2+} < K^+$$

(b) Br^- and Se^{2-} are isoelectronic (36 electrons each) with an outer-shell electron configuration of $4s^2 4p^6$. Because Br^- has a higher nuclear charge (35+) than Se^{2-} (34+), Br^- holds its

36 electrons more tightly, and Br^- is smaller than Se^{2-}. Te^{2-} has electrons in the fifth main shell (outer configuration of $5s^25p^6$), so it is larger than either of the other two ions.

$$Br^- < Se^{2-} < Te^{2-}$$

You should now work Exercises 36 and 38.

6-6 ELECTRONEGATIVITY

Because the noble gases form few compounds, they are not included in this discussion.

The **electronegativity (EN)** of an element is a measure of the relative tendency of an atom to attract electrons to itself *when it is chemically combined with another atom.*

Elements with high electronegativities (nonmetals) often gain electrons to form anions. Elements with low electronegativities (metals) often lose electrons to form cations.

Electronegativities of the elements are expressed on a somewhat arbitrary scale, called the Pauling scale (Table 6-3). The electronegativity of fluorine (4.0) is higher than that of any other element. This tells us that when fluorine is chemically bonded to other elements, it

TABLE 6-3 *Electronegativity Values of the Elements* [a]

IA																		VIIIA
1	1 H 2.1	IIA			Metals								IIIA	IVA	VA	VIA	VIIA	2 He
2	3 Li 1.0	4 Be 1.5			Nonmetals								5 B 2.0	6 C 2.5	7 N 3.0	8 O 3.5	9 F 4.0	10 Ne
3	11 Na 1.0	12 Mg 1.2	IIIB		Metalloids				VIIIB			IB	IIB	13 Al 1.5	14 Si 1.8	15 P 2.1	16 S 2.5	17 Cl 3.0	18 Ar
4	19 K 0.9	20 Ca 1.0	21 Sc 1.3	22 Ti 1.4	23 V 1.5	24 Cr 1.6	25 Mn 1.6	26 Fe 1.7	27 Co 1.7	28 Ni 1.8	29 Cu 1.8	30 Zn 1.6	31 Ga 1.7	32 Ge 1.9	33 As 2.1	34 Se 2.4	35 Br 2.8	36 Kr
5	37 Rb 0.9	38 Sr 1.0	39 Y 1.2	40 Zr 1.3	41 Nb 1.5	42 Mo 1.6	43 Tc 1.7	44 Ru 1.8	45 Rh 1.8	46 Pd 1.8	47 Ag 1.6	48 Cd 1.6	49 In 1.6	50 Sn 1.8	51 Sb 1.9	52 Te 2.1	53 I 2.5	54 Xe
6	55 Cs 0.8	56 Ba 1.0	57 * La 1.1	72 Hf 1.3	73 Ta 1.4	74 W 1.5	75 Re 1.7	76 Os 1.9	77 Ir 1.9	78 Pt 1.8	79 Au 1.9	80 Hg 1.7	81 Tl 1.6	82 Pb 1.7	83 Bi 1.8	84 Po 1.9	85 At 2.1	86 Rn
7	87 Fr 0.8	88 Ra 1.0	89 † Ac 1.1															

*	58 Ce 1.1	59 Pr 1.1	60 Nd 1.1	61 Pm 1.1	62 Sm 1.1	63 Eu 1.1	64 Gd 1.1	65 Tb 1.1	66 Dy 1.1	67 Ho 1.1	68 Er 1.1	69 Tm 1.1	70 Yb 1.0	71 Lu 1.2
†	90 Th 1.2	91 Pa 1.3	92 U 1.5	93 Np 1.3	94 Pu 1.3	95 Am 1.3	96 Cm 1.3	97 Bk 1.3	98 Cf 1.3	99 Es 1.3	100 Fm 1.3	101 Md 1.3	102 No 1.3	103 Lr 1.5

[a] Electronegativity values are given at the bottoms of the boxes.

has a greater tendency to attract electron density to itself than does any other element. Oxygen is the second most electronegative element.

> For the representative elements, electronegativities usually increase from left to right across periods and decrease from top to bottom within groups.

Variations among the transition metals are not as regular. In general, both ionization energies and electronegativities are low for elements at the lower left of the periodic table and high for those at the upper right.

General trends in electronegativities of A group elements with position in the periodic table.

EXAMPLE 6-5 *Trends in ENs*

Arrange the following elements in order of increasing electronegativity.

B,   Na,   F,   O

Plan

Electronegativities increase from left to right across a period and decrease from top to bottom within a group.

Solution

The order of increasing electronegativity is Na < B < O < F.

You should now work Exercise 42.

Although the electronegativity scale is somewhat arbitrary, we can use it with reasonable confidence to make predictions about bonding. Two elements with quite different electronegativities (a metal and a nonmetal) tend to react with each other to form ionic compounds. The less electronegative element gives up its electron(s) to the more electronegative element. Two nonmetals with similar electronegativities tend to form covalent bonds with each other. That is, they share their electrons. In this sharing, the more electronegative element attains a greater share. This is discussed in detail in Chapters 7 and 8.

Ionization energy (Section 6-3) and electron affinity (Section 6-4) are precise quantitative concepts. We find, however, that the more qualitative concept of *electronegativity* is more useful in describing chemical bonding.

CHEMICAL REACTIONS AND PERIODICITY

Now we will illustrate the periodicity of chemical properties by considering some reactions of hydrogen, oxygen, and their compounds. We choose to discuss hydrogen and oxygen because, of all the elements, they form the most kinds of compounds with other elements. Additionally, compounds of hydrogen and oxygen are very important in such diverse phenomena as all life processes and most corrosion processes.

6-7 HYDROGEN AND THE HYDRIDES

Hydrogen

Elemental hydrogen is a colorless, odorless, tasteless diatomic gas with the lowest molecular weight and density of any known substance. Discovery of the element is attributed to the Englishman Henry Cavendish (1731–1810), who prepared it in 1766 by passing steam through a red-hot gun barrel (mostly iron) and by the reaction of acids with active metals.

The name "hydrogen" means "water former."

Can you write the net ionic equation for the reaction of Zn with HCl(aq)?

The latter is still the method commonly used for the preparation of small amounts of H_2 in the laboratory. In each case, H_2 is liberated by a displacement (and redox) reaction, of the kind described in Section 4-8. (See the activity series, Table 4-14.)

$$3Fe(s) + 4H_2O(g) \xrightarrow{\text{heat}} Fe_3O_4(s) + 4H_2(g)$$

$$Zn(s) + 2HCl(aq) \longrightarrow ZnCl_2(aq) + H_2(g)$$

Hydrogen also can be prepared by electrolysis of water.

$$2H_2O(\ell) \xrightarrow{\text{electricity}} 2H_2(g) + O_2(g)$$

This is the reverse of the decomposition of H_2O.

In the future, if it becomes economical to convert solar energy into electrical energy that can be used to electrolyze water, H_2 could become an important fuel (although the dangers of storage and transportation would have to be overcome). The *combustion* of H_2 liberates a great deal of heat. **Combustion** is the highly exothermic combination of a substance with oxygen, usually with a flame. (See Section 6-8, Combustion Reactions.)

$$2H_2(g) + O_2(g) \xrightarrow[\text{or heat}]{\text{spark}} 2H_2O(\ell) + \text{energy}$$

Hydrogen is no longer used in blimps and dirigibles. It has been replaced by helium, which is slightly denser, non-flammable, and much safer.

Hydrogen is very flammable; it was responsible for the *Hindenburg* airship disaster in 1937. A spark is all it takes to initiate the **combustion reaction,** which is exothermic enough to provide the heat necessary to sustain the reaction.

Hydrogen is prepared by the "water gas reaction," which results from the passage of steam over white-hot coke (impure carbon, a nonmetal) at 1500°C. The mixture of products commonly called "water gas" is used industrially as a fuel. Both components, CO and H_2, undergo combustion.

$$\underset{\text{in coke}}{C(s)} + \underset{\text{steam}}{H_2O(g)} \longrightarrow \underset{\text{"water gas"}}{\underbrace{CO(g) + H_2(g)}}$$

NECAR 5, Courtesy of Mercedes-Benz, USA

Vast quantities of hydrogen are produced commercially each year by a process called *steam cracking*. Methane reacts with steam at 830°C in the presence of a nickel catalyst.

$$CH_4(g) + H_2O(g) \xrightarrow[\text{Ni}]{\text{heat}} CO(g) + 3H_2(g)$$

In 2002 a prototype car from DaimlerChrysler that uses fuel-cell technology completed a 16 day cross-country trip. A small chemical reactor first converts methanol (CH_3OH), H_2O, and O_2 into H_2 and CO_2. The H_2 then reacts further with O_2 in a fuel cell to produce electricity that powers the car. The advantage of using methanol is that it is easier and safer to store than H_2 gas. The fuel cell has considerably higher efficiency compared to direct combustion of methanol in an internal combustion engine.

The mixture of H_2 and CO gases is also referred to as "synthesis gas." It can be used to produce a wide variety of organic chemicals like methanol (CH_3OH) and hydrocarbon mixtures for gasoline, kerosene, and related fuels.

Reactions of Hydrogen and Hydrides

Atomic hydrogen has the $1s^1$ electron configuration. It reacts with metals and with other nonmetals to form binary compounds called **hydrides.** These can be (a) **ionic hydrides,** which contain hydride ions, H^-, formed when hydrogen gains one electron per atom from an active metal; or (2) **molecular hydrides,** in which hydrogen shares electrons with an atom of another nonmetal.

The ionic or molecular character of the binary compounds of hydrogen depends on the position of the other element in the periodic table (Figure 6-4). The reactions of H_2 with the *alkali* (IA) and the heavier (more active) *alkaline earth* (IIA) *metals* result in solid *ionic hydrides.* The reaction with the molten (liquid) IA metals may be represented in general terms as

The use of the term "hydride" does not necessarily imply the presence of the hydride ion, H^-.

$$2M(\ell) + H_2(g) \xrightarrow[\text{high pressures}]{\text{high temperatures}} 2(M^+, H^-)(s) \qquad M = Li, Na, K, Rb, Cs$$

IA	IIA	IIIA	IVA	VA	VIA	VIIA
LiH	BeH_2	B_2H_6	CH_4	NH_3	H_2O	HF
NaH	MgH_2	$(AlH_3)_x$	SiH_4	PH_3	H_2S	HCl
KH	CaH_2	Ga_2H_6	GeH_4	AsH_3	H_2Se	HBr
RbH	SrH_2	InH_3	SnH_4	SbH_3	H_2Te	HI
CsH	BaH_2	TlH	PbH_4	BiH_3	H_2Po	HAt

Figure 6-4 Common hydrides of the representative elements. The ionic hydrides are shaded blue, molecular hydrides are shaded red, and those of intermediate character are shaded purple. The Group VIIA hydrogen halides are acids and do not include "hydride" in their names.

When the other element has similar or higher electronegativity than hydrogen, the hydride has more molecular character. When the other element is considerably less electronegative than hydrogen, ionic hydrides containing H^- are formed (CsH).

Thus, hydrogen combines with lithium to form lithium hydride and with sodium to form sodium hydride.

$$2Li(\ell) + H_2(g) \longrightarrow 2LiH(s) \qquad \text{lithium hydride (mp 680°C)}$$

$$2Na(\ell) + H_2(g) \longrightarrow 2NaH(s) \qquad \text{sodium hydride (mp 800°C)}$$

The ionic hydrides are named by naming the metal first, followed by "hydride."

In general terms, the reactions of the heavier (more active) IIA metals may be represented as

$$M(\ell) + H_2(g) \longrightarrow (M^{2+}, 2H^-)(s) \qquad M = Ca, Sr, Ba$$

Thus, calcium combines with hydrogen to form calcium hydride:

$$Ca(\ell) + H_2(g) \longrightarrow CaH_2(s) \qquad \text{calcium hydride (mp 816°C)}$$

These *ionic hydrides are all basic* because they react with water to form hydroxide ions. When water is added by drops to lithium hydride, for example, lithium hydroxide and hydrogen are produced. The reaction of calcium hydride is similar.

$$LiH(s) + H_2O(\ell) \longrightarrow LiOH(s) + H_2(g)$$

$$CaH_2(s) + 2H_2O(\ell) \longrightarrow Ca(OH)_2(s) + 2H_2(g)$$

Ionic hydrides can serve as sources of hydrogen. They must be stored in environments free of moisture and O_2.

We show LiOH and $Ca(OH)_2$ as solids here because not enough water is available to act as a solvent.

Hydrogen reacts with *nonmetals* to form binary *molecular hydrides*. For example, H_2 combines with the halogens to form colorless, gaseous hydrogen halides (Figure 6-5):

$$H_2(g) + X_2 \longrightarrow 2HX(g) \qquad X = F, Cl, Br, I$$
$$\text{hydrogen halides}$$

Specifically, hydrogen reacts with fluorine to form hydrogen fluoride and with chlorine to form hydrogen chloride:

$$H_2(g) + F_2(g) \longrightarrow 2HF(g) \qquad \text{hydrogen fluoride}$$

$$H_2(g) + Cl_2(g) \longrightarrow 2HCl(g) \qquad \text{hydrogen chloride}$$

The hydrogen halides are named by the word "hydrogen" followed by the stem for the halogen with an "-ide" ending.

Hydrogen combines with Group VIA elements to form molecular compounds:

$$2H_2(g) + O_2(g) \xrightarrow{\text{heat}} 2H_2O(g)$$

The heavier members of this family also combine with hydrogen to form binary compounds that are gases at room temperature. Their formulas resemble that of water.

These compounds are named:
  H_2O, hydrogen oxide (water)
  H_2S, hydrogen sulfide
  H_2Se, hydrogen selenide
  H_2Te, hydrogen telluride
All except H_2O are *very* toxic.

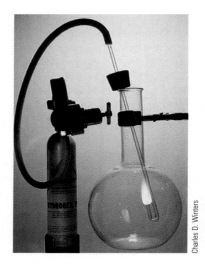

Figure 6-5 Hydrogen, H_2, burns in an atmosphere of pure chlorine, Cl_2, to produce hydrogen chloride.

$$H_2 + Cl_2 \longrightarrow 2HCl$$

Figure 6-6 Ammonia may be applied directly to the soil as a fertilizer.

The primary industrial use of H_2 is in the synthesis of ammonia, a molecular hydride, by the Haber process (Section 17-7). Most of the NH_3 is used in liquid form as a fertilizer (Figure 6-6) or to make other fertilizers, such as ammonium nitrate, NH_4NO_3, and ammonium sulfate, $(NH_4)_2SO_4$:

$$N_2(g) + 3H_2(g) \xrightarrow[\text{heat, high pressure}]{\text{catalysts}} 2NH_3(g)$$

Many of the molecular (nonmetal) hydrides are acidic; their aqueous solutions contain hydrogen cations (H^+). These include HF, HCl, HBr, HI, H_2S, H_2Se, and H_2Te.

EXAMPLE 6-6   *Predicting Products of Reactions*

Predict the products of the reactions involving the reactants shown. Write a balanced formula unit equation for each.

(a) $H_2(g) + I_2(g) \xrightarrow{\text{heat}}$

(b) $K(\ell) + H_2(g) \xrightarrow{\text{heat}}$

(c) $NaH(s) + H_2O(\ell)$ (excess) $\longrightarrow$

Plan

(a) Hydrogen reacts with the halogens (Group VIIA) to form hydrogen halides—in this example, HI.

(b) Hydrogen reacts with active metals to produce hydrides—in this case, KH.

(c) Active metal hydrides react with water to produce a metal hydroxide and H_2.

Solution

(a) $H_2(g) + I_2(g) \xrightarrow{\text{heat}} 2HI(g)$

(b) $2K(\ell) + H_2(g) \xrightarrow{\text{heat}} 2KH(s)$

(c) $NaH(s) + H_2O(\ell) \longrightarrow NaOH(aq) + H_2(g)$

Remember that hydride ions, H^-, react with (reduce) water to produce OH^- ions and $H_2(g)$.

EXAMPLE 6-7 *Ionic and Molecular Properties*

Predict the ionic or molecular character of the products in Example 6-6.

Plan

We refer to Figure 6-4, which displays the nature of hydrides.

Solution

Reaction (a) is a reaction between hydrogen and another nonmetal. The product, HI, must be molecular. Reaction (b) is the reaction of hydrogen with an active Group IA metal. Thus, KH must be ionic. The products of reaction (c) are molecular $H_2(g)$ and the strong base, NaOH, which is ionic.

You should now work Exercises 54 and 55.

6-8 OXYGEN AND THE OXIDES

Oxygen and Ozone

Oxygen was discovered in 1774 by an English minister and scientist, Joseph Priestley (1733–1804). He observed the thermal decomposition of mercury(II) oxide, a red powder, to form liquid Hg and a colorless gas:

$$2HgO(s) \xrightarrow{heat} 2Hg(\ell) + O_2(g)$$

That part of the earth we see—land, water, and air—is approximately 50% oxygen by mass. About two thirds of the mass of the human body is due to oxygen in H_2O. Elemental oxygen, O_2, is an odorless and colorless gas that makes up about 21% by volume of dry air. In the liquid (bp = $-183°C$) and solid states (mp = $-218°C$), it is pale blue. Oxygen is only slightly soluble in water; only about 0.04 g dissolves in 1 L water at 25°C (0.001 M solutions). This is sufficient to sustain fish and other marine organisms. Oxygen is obtained commercially by the fractional distillation of liquid air. The greatest single industrial use of O_2 is for oxygen-enrichment in blast furnaces for the conversion of pig iron (reduced, high-carbon iron) to steel.

Oxygen also exists in a second allotropic form, ozone, O_3. Ozone is an unstable, pale blue gas at room temperature. It is formed by passing an electrical discharge through gaseous oxygen. Its unique, pungent odor is often noticed during electrical storms and in the vicinity of electrical equipment. Not surprisingly, its density is about $1\frac{1}{2}$ times that of O_2. At $-112°C$, it condenses to a deep blue liquid. It is a very strong oxidizing agent. As a concentrated gas or a liquid, ozone can easily decompose explosively:

$$2O_3(g) \longrightarrow 3O_2(g)$$

Oxygen atoms, or **radicals,** are intermediates in this exothermic decomposition of O_3 to O_2. They act as strong oxidizing agents in such applications as destroying bacteria in water purification.

The ozone molecule is angular (page 51). The two oxygen–oxygen bond lengths (1.28 Å) are identical and are intermediate between typical single and double bond lengths.

Reactions of Oxygen and the Oxides

Oxygen forms oxides by direct combination with all other elements except the noble gases and noble (unreactive) metals (Au, Pd, Pt). **Oxides** are binary compounds that contain oxygen. Although such reactions are generally very exothermic, many proceed quite slowly

The name "oxygen" means "acid former."

Liquid O_2 is used as an oxidizer for rocket fuels. O_2 also is used in the health fields for oxygen-enriched air.

Allotropes are different forms of the same element in the same physical state (Section 2-2).

A *radical* is a species containing one or more unpaired electrons; many radicals are very reactive.

TOPIC: Oxygen
GO TO: www.scilinks.org
*sci*LINKS **CODE:** WCH0630

and require heating to supply the energy necessary to break the strong bonds in O_2 molecules. After these reactions are initiated, most release more than enough energy to be self-sustaining, and some become "red hot."

Reactions of O_2 with Metals

In general, metallic oxides (including peroxides and superoxides) are ionic solids. The Group IA metals combine with oxygen to form three kinds of solid ionic products called oxides, peroxides, and superoxides. Lithium combines with oxygen to form lithium oxide.

$$4Li(s) + O_2(g) \longrightarrow 2Li_2O(s) \qquad \text{lithium oxide (mp > 1700°C)}$$

By contrast, sodium reacts with an excess of oxygen to form sodium peroxide, Na_2O_2, as the major product, rather than sodium oxide, Na_2O.

$$2Na(s) + O_2(g) \longrightarrow Na_2O_2(g) \qquad \text{sodium peroxide (decomposes at 460°C)}$$

Peroxides contain the $O-O^{2-}$ ion (O_2^{2-}), in which the oxidation number of each oxygen is -1, whereas **normal oxides** such as lithium oxide, Li_2O, contain oxide ions, O^{2-}. The heavier members of the family (K, Rb, Cs) react with excess oxygen to form **superoxides.** These contain the superoxide ion, O_2^-, in which the oxidation number of each oxygen is $-\frac{1}{2}$. The reaction with K is

$$K(s) + O_2(g) \longrightarrow KO_2(s) \qquad \text{potassium superoxide (mp 430°C)}$$

The larger K^+, Rb^+, and Cs^+ cations prefer the larger O^{2-} and O_2^{2-} anions owing to better solid-state packing, which results from better size matching of the ions.

The tendency of the Group IA metals to form oxygen-rich compounds increases going down the group. This is because cation radii increase going down the group. You can recognize these classes of compounds as

Class	Contains Ions	Oxidation No. of Oxygen
normal oxides	O^{2-}	-2
peroxides	O_2^{2-}	-1
superoxides	O_2^-	$-\frac{1}{2}$

Beryllium reacts with oxygen only at elevated temperatures and forms only the normal oxide, BeO. The other Group IIA metals form normal oxides at moderate temperatures.

The Group IIA metals react with oxygen to form normal ionic oxides, MO, but at high pressures of oxygen the heavier ones form ionic peroxides, MO_2 (Table 6-4).

$$2M(s) + O_2(g) \longrightarrow 2(M^{2+}, O^{2-})(s) \qquad \text{M = Be, Mg, Ca, Sr, Ba}$$
$$M(s) + O_2(g) \longrightarrow (M^{2+}, O_2^{2-})(s) \qquad \text{M = Ca, Sr, Ba}$$

TABLE 6-4 | *Oxygen Compounds of the IA and IIA Metals**

	IA					IIA				
	Li	Na	K	Rb	Cs	Be	Mg	Ca	Sr	Ba
normal oxides	Li_2O	Na_2O	K_2O	Rb_2O	Cs_2O	BeO	MgO	CaO	SrO	BaO
peroxides	Li_2O_2	Na_2O_2	K_2O_2	Rb_2O_2	Cs_2O_2			CaO_2	SrO_2	BaO_2
superoxides		NaO_2	KO_2	RbO_2	CsO_2					

The shaded compounds represent the principal products of the direct reaction of the metal with oxygen.

For example, the equations for the reactions of calcium and oxygen are

$$2Ca(s) + O_2(g) \longrightarrow 2CaO(s) \qquad \text{calcium oxide (mp 2580°C)}$$

$$Ca(s) + O_2(g) \longrightarrow CaO_2(s) \qquad \text{calcium peroxide (decomposes at 275°C)}$$

The other metals, with the exceptions noted previously (Au, Pd, and Pt), react with oxygen to form solid metal oxides. Many metals to the right of Group IIA show variable oxidation states, so they may form several oxides. For example, iron combines with oxygen in the following series of reactions to form three different oxides (Figure 6-7).

$$2Fe(s) + O_2(g) \xrightarrow{\text{heat}} 2FeO(s) \qquad \text{iron(II) oxide}$$

$$6FeO(s) + O_2(g) \xrightarrow{\text{heat}} 2Fe_3O_4(s) \qquad \text{magnetic iron oxide (a mixed oxide)}$$

$$4Fe_3O_4(s) + O_2(g) \xrightarrow{\text{heat}} 6Fe_2O_3(s) \qquad \text{iron(III) oxide}$$

Copper reacts with a limited amount of oxygen to form red Cu_2O, whereas with excess oxygen it forms black CuO.

$$4Cu(s) + O_2(g) \xrightarrow{\text{heat}} 2Cu_2O(s) \qquad \text{copper(I) oxide}$$

$$2Cu(s) + O_2(g) \xrightarrow{\text{heat}} 2CuO(s) \qquad \text{copper(II) oxide}$$

Figure 6-7 Steel wool burns brilliantly to form iron(III) oxide, Fe_2O_3.

Metals that exhibit variable oxidation states react with a limited amount of oxygen to give oxides with lower oxidation states (such as FeO and Cu_2O). They react with an excess of oxygen to give oxides with higher oxidation states (such as Fe_2O_3 and CuO).

Reactions of Metal Oxides with Water

Oxides of metals are called **basic anhydrides** (or **basic oxides**) because many of them combine with water to form bases with no change in oxidation state of the metal (Figure 6-8). "Anhydride" means "without water," and in a sense, the metal oxide is a hydroxide base with the water "removed." Metal oxides that are soluble in water react to produce the corresponding hydroxides.

Increasing acidic character ⟶

	IA	IIA	IIIA	IVA	VA	VIA	VIIA
	Li_2O	BeO	B_2O_3	CO_2	N_2O_5		OF_2
	Na_2O	MgO	Al_2O_3	SiO_2	P_4O_{10}	SO_3	Cl_2O_7
	K_2O	CaO	Ga_2O_3	GeO_2	As_2O_5	SeO_3	Br_2O_7
	Rb_2O	SrO	In_2O_3	SnO_2	Sb_2O_5	TeO_3	I_2O_7
	Cs_2O	BaO	Tl_2O_3	PbO_2	Bi_2O_5	PoO_3	At_2O_7

Increasing base character (vertical, downward)

Figure 6-8 The normal oxides of the representative elements in their maximum oxidation states. Acidic oxides (acid anhydrides) are shaded red, amphoteric oxides are shaded purple, and basic oxides (basic anhydrides) are shaded blue. An **amphoteric oxide** is one that shows some acidic and some basic properties.

Metal Oxide + Water $\longrightarrow$ Metal Hydroxide (base)				
sodium oxide	$Na_2O(s)$	$+ H_2O(\ell) \longrightarrow$	$2\ NaOH(aq)$	sodium hydroxide
calcium oxide	$CaO(s)$	$+ H_2O(\ell) \longrightarrow$	$Ca(OH)_2(aq)$	calcium hydroxide
barium oxide	$BaO(s)$	$+ H_2O(\ell) \longrightarrow$	$Ba(OH)_2(aq)$	barium hydroxide

The oxides of the Group IA metals and the heavier Group IIA metals dissolve in water to give solutions of strong bases. Most other metal oxides are relatively insoluble in water.

Reactions of O_2 with Nonmetals

Oxygen combines with many nonmetals to form molecular oxides. For example, carbon burns in oxygen to form carbon monoxide or carbon dioxide, depending on the relative amounts of carbon and oxygen.

Recall that oxidation states are indicated by red numbers in red circles.

$$2C(s) + O_2(g) \longrightarrow 2\overset{(+2)}{C}O(s) \qquad \text{(excess C and limited } O_2\text{)}$$

$$C(s) + O_2(g) \longrightarrow \overset{(+4)}{C}O_2(g) \qquad \text{(limited C and excess } O_2\text{)}$$

Carbon monoxide is a very poisonous gas because it forms a stronger bond with the iron atom in hemoglobin than does an oxygen molecule. Attachment of the CO molecule to the iron atom destroys the ability of hemoglobin to pick up oxygen in the lungs and carry it to the brain and muscle tissue. Carbon monoxide poisoning is particularly insidious because the gas has no odor and because the victim first becomes drowsy.

Unlike carbon monoxide, carbon dioxide is not toxic. It is one of the products of the respiratory process. It is used to make carbonated beverages, which are mostly saturated solutions of carbon dioxide in water; a small amount of the carbon dioxide combines with the water to form carbonic acid (H_2CO_3), a very weak acid.

Phosphorus reacts with a limited amount of oxygen to form tetraphosphorus hexoxide, P_4O_6:

$$P_4(s) + 3O_2(g) \longrightarrow \overset{(+3)}{P_4}O_6(s) \qquad \text{tetraphosphorus hexoxide}$$

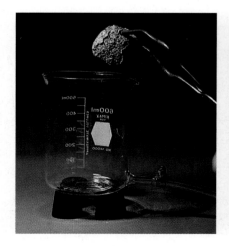 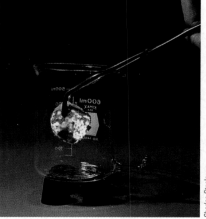

Carbon burns brilliantly in pure O_2 to form CO_2.

Charles Steele

whereas reaction with an excess of oxygen gives tetraphosphorus decoxide, P_4O_{10}:

$$P_4(s) + 5O_2(g) \longrightarrow \overset{(+5)}{P_4}O_{10}(s) \qquad \text{tetraphosphorus decoxide}$$

Sulfur burns in oxygen to form primarily sulfur dioxide (Figure 6-9) and only very small amounts of sulfur trioxide.

$$S_8(s) + 8O_2(g) \longrightarrow 8\overset{(+4)}{S}O_2(g) \qquad \text{sulfur dioxide (bp } -10.0°\text{C)}$$

$$S_8(s) + 12O_2(g) \longrightarrow 8\overset{(+6)}{S}O_3(\ell) \qquad \text{sulfur trioxide (bp } 43.4°\text{C)}$$

The production of SO_3 at a reasonable rate requires the presence of a catalyst.

OXIDATION STATES OF NONMETALS   Nearly all nonmetals exhibit more than one oxidation state in their compounds. In general, the *most common* oxidation states of a nonmetal are (1) its periodic group number, (2) its periodic group number minus two, and (3) its periodic group number minus eight. The reactions of nonmetals with a limited amount of oxygen usually give products that contain the nonmetals (other than oxygen) in lower oxidation states, usually case (2). Reactions with excess oxygen give products in which the nonmetals exhibit higher oxidation states, case (1). The examples we have cited are CO and CO_2, P_4O_6 and P_4O_{10}, and SO_2 and SO_3. The molecular formulas of the oxides are sometimes not easily predictable, but the *simplest* formulas are. For example, the two most common oxidation states of phosphorus in molecular compounds are +3 and +5. The simplest formulas for the corresponding phosphorus oxides therefore are P_2O_3 and P_2O_5, respectively. The molecular (true) formulas are twice these, P_4O_6 and P_4O_{10}.

Reactions of Nonmetal Oxides with Water

Nonmetal oxides are called **acid anhydrides** (or **acidic oxides**) because many of them dissolve in water to form acids *with no change in oxidation state of the nonmetal* (see Figure 6-8). Several **ternary acids** can be prepared by reaction of the appropriate nonmetal oxides with water. Ternary acids contain three elements, usually H, O, and another nonmetal.

	Nonmetal Oxide	+ Water	⟶	Ternary Acid	
carbon dioxide	$\overset{(+4)}{C}O_2(g)$	$+ H_2O(\ell)$	⟶	$\overset{(+4)}{H_2C}O_3(aq)$	carbonic acid
sulfur dioxide	$\overset{(+4)}{S}O_2(g)$	$+ H_2O(\ell)$	⟶	$\overset{(+4)}{H_2S}O_3(aq)$	sulfurous acid
sulfur trioxide	$\overset{(+6)}{S}O_3(\ell)$	$+ H_2O(\ell)$	⟶	$\overset{(+6)}{H_2S}O_4(aq)$	sulfuric acid
dinitrogen pentoxide	$\overset{(+5)}{N_2}O_5(s)$	$+ H_2O(\ell)$	⟶	$2\overset{(+5)}{H}NO_3(aq)$	nitric acid
tetraphosphorus decoxide	$\overset{(+5)}{P_4}O_{10}(s)$	$+ 6H_2O(\ell)$	⟶	$4\overset{(+5)}{H_3P}O_4(aq)$	phosphoric acid

Nearly all oxides of nonmetals react with water to give solutions of ternary acids. The oxides of boron and silicon, which are insoluble, are two exceptions.

Figure 6-9 Sulfur burns in oxygen to form sulfur dioxide.

Reactions of Metal Oxides with Nonmetal Oxides

Another common kind of reaction of oxides is the *combination of metal oxides (basic anhydrides) with nonmetal oxides (acid anhydrides), with no change in oxidation states, to form salts.*

	Metal Oxide	+	Nonmetal Oxide	$\longrightarrow$	Salt	
calcium oxide + sulfur trioxide	$\overset{+2}{\text{CaO}}(s)$	+	$\overset{+6}{\text{SO}_3}(\ell)$	$\longrightarrow$	$\overset{+2}{\text{Ca}}\overset{+6}{\text{SO}_4}(s)$	calcium sulfate
magnesium oxide + carbon dioxide	$\overset{+2}{\text{MgO}}(s)$	+	$\overset{+4}{\text{CO}_2}(g)$	$\longrightarrow$	$\overset{+2}{\text{Mg}}\overset{+4}{\text{CO}_3}(s)$	magnesium carbonate
sodium oxide + tetraphosphorus decoxide	$6\overset{+1}{\text{Na}_2\text{O}}(s)$	+	$\overset{+5}{\text{P}_4\text{O}_{10}}(s)$	$\longrightarrow$	$4\overset{+1}{\text{Na}_3}\overset{+5}{\text{PO}_4}(s)$	sodium phosphate

EXAMPLE 6-8   Acidic Character of Oxides

Arrange the following oxides in order of increasing molecular (acidic) character: SO_3, Cl_2O_7, CaO, and PbO_2.

Plan

Molecular (acidic) character of oxides increases as nonmetallic character of the element that is combined with oxygen increases (see Figure 6-8).

increasing nonmetallic character $\longrightarrow$

Ca < Pb < S < Cl

Periodic group: IIA   IVA   VIA   VIIA

Solution

increasing molecular character $\longrightarrow$

Thus, the order is $CaO < PbO_2 < SO_3 < Cl_2O_7$

EXAMPLE 6-9   Basic Character of Oxides

Arrange the oxides in Example 6-8 in order of increasing basicity.

Plan

The greater the molecular character of an oxide, the more acidic it is. Thus, the most basic oxides have the least molecular (most ionic) character (see Figure 6-8).

Solution

increasing basic character $\longrightarrow$

molecular $Cl_2O_7 < SO_3 < PbO_2 < CaO$ ionic

EXAMPLE 6-10   Predicting Reaction Products

Predict the products of the following pairs of reactants. Write a balanced equation for each reaction.

(a) $Cl_2O_7(\ell) + H_2O(\ell) \longrightarrow$

(b) $As_4(s) + O_2(g)$ (excess) $\xrightarrow{\text{heat}}$

(c) $Mg(s) + O_2(g) \xrightarrow{\text{heat}}$

Plan

(a) The reaction of a nonmetal oxide (acid anhydride) with water forms a ternary acid in which the nonmetal (Cl) has the same oxidation state ($+7$) as in the oxide. Thus, the acid is perchloric acid, $HClO_4$.

(b) Arsenic, a Group VA nonmetal, exhibits common oxidation states of $+5$ and $+5 - 2 = +3$. Reaction of arsenic with *excess* oxygen produces the higher-oxidation-state oxide, As_2O_5. By analogy with the oxide of phosphorus in the $+5$ oxidation state, P_4O_{10}, we might write the formula as As_4O_{10}, but this oxide exists as As_2O_5.

(c) The reaction of a Group IIA metal with oxygen produces the normal metal oxide—MgO in this case.

Solution

(a) $Cl_2O_7(\ell) + H_2O(\ell) \longrightarrow 2HClO_4(aq)$

(b) $As_4(s) + 5O_2(g) \xrightarrow{\text{heat}} 2As_2O_5(s)$

(c) $2Mg(s) + O_2(g) \xrightarrow{\text{heat}} 2MgO(s)$

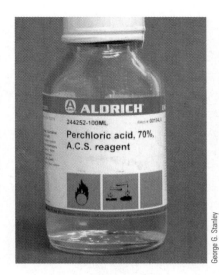

George G. Stanley

EXAMPLE 6-11 *Predicting Reaction Products*

Predict the products of the following pairs of reactants. Write a balanced equation for each reaction.

(a) $CaO(s) + H_2O(\ell) \longrightarrow$

(b) $Li_2O(s) + SO_3(\ell) \longrightarrow$

Plan

(a) The reaction of a metal oxide with water produces the metal hydroxide.

(b) The reaction of a metal oxide with a nonmetal oxide produces a salt containing the cation of the metal oxide and the anion of the acid for which the nonmetal oxide is the anhydride. SO_3 is the acid anhydride of sulfuric acid, H_2SO_4.

Solution

(a) Calcium oxide reacts with water to form calcium hydroxide.

$$CaO(s) + H_2O(\ell) \longrightarrow Ca(OH)_2(aq)$$

(b) Lithium oxide reacts with sulfur trioxide to form lithium sulfate.

$$Li_2O(s) + SO_3(\ell) \longrightarrow Li_2SO_4(s)$$

You should now work Exercises 66–69.

CaO is called quicklime. $Ca(OH)_2$ is called slaked lime.

Combustion Reactions

Combustion, or burning, is an oxidation–reduction reaction in which oxygen combines rapidly with oxidizable materials in highly exothermic reactions, usually with a visible flame. The complete combustion of *hydrocarbons*, in fossil fuels for example, produces carbon dioxide and water (steam) as the major products:

Hydrocarbons are compounds that contain only hydrogen and carbon.

$$\overset{\text{(-4)(+1)}}{CH_4(g)} + \overset{\text{(0)}}{2O_2(g)} \xrightarrow{\text{heat}} \overset{\text{(+4)(-2)}}{CO_2(g)} + \overset{\text{(+1)(-2)}}{2H_2O(g)} + \text{heat}$$
$$\text{excess}$$

$$\overset{\text{(-2)(+1)}}{C_6H_{12}(g)} + \overset{\text{(0)}}{9O_2(g)} \xrightarrow{\text{heat}} \overset{\text{(+4)(-2)}}{6CO_2(g)} + \overset{\text{(+1)(-2)}}{6H_2O(g)} + \text{heat}$$
$$\text{cyclohexane} \qquad \text{excess}$$

As we have seen, the origin of the term "oxidation" lies in just such reactions, in which oxygen "oxidizes" another species.

Combustion of Fossil Fuels and Air Pollution

Fossil fuels are mixtures of variable composition that consist primarily of hydrocarbons. We burn them to use the energy that is released rather than to obtain chemical products (Figure 6-10). For example, burning octane, C_8H_{18}, in an excess of oxygen (plenty of air) produces carbon dioxide and water. There are many similar compounds in gasoline and diesel fuels.

$$2C_8H_{18}(\ell) + 25O_2(g) \longrightarrow 16CO_2(g) + 18H_2O(\ell)$$
$$\text{excess}$$

Carbon monoxide is produced by the incomplete burning of carbon-containing compounds in a limited amount of oxygen.

$$2C_8H_{18}(\ell) + 17O_2(g) \longrightarrow 16CO(g) + 18H_2O(\ell)$$
$$\text{limited amount}$$

In very limited oxygen, carbon (soot) is produced by partially burning hydrocarbons. For octane, the reaction is

$$2C_8H_{18}(\ell) + 9O_2(g) \longrightarrow 16C(s) + 18H_2O(\ell)$$
$$\text{very limited amount}$$

When you see blue or black smoke (carbon) coming from an internal combustion engine (or smell unburned fuel in the air), you may be quite sure that lots of toxic carbon monoxide is also being produced and released into the air.

We see that the incomplete combustion of hydrocarbons yields undesirable products—carbon monoxide and elemental carbon (soot), which pollute the air. Unfortunately, all fossil fuels—natural gas, coal, gasoline, kerosene, oil, and so on—also have undesirable nonhydrocarbon impurities that burn to produce oxides that act as additional air pollutants. At this time it is not economically feasible to remove all of these impurities from the fuels before burning them.

Fossil fuels result from the decay of animal and vegetable matter (Figure 6-11). All living matter contains some sulfur and nitrogen, so fossil fuels also contain sulfur and nitrogen impurities to varying degrees. Table 6-5 gives composition data for some common kinds of coal.

Combustion of sulfur produces sulfur dioxide, SO_2, probably the single most harmful pollutant.

$$\overset{\text{(0)}}{S_8(s)} + 8O_2(g) \xrightarrow{\text{heat}} \overset{\text{(+4)}}{8SO_2(g)}$$

Large amounts of SO_2 are produced by the burning of sulfur-containing coal.

Figure 6-10 Georgia Power Company's Plant Bowen at Taylorsville, Georgia. Plants such as this one burn more than 10^7 tons of coal and produce over 2×10^8 megawatt-hours of electricity each year.

"White smoke" from auto exhaust systems is tiny droplets of water condensing in the cooler air.

Carbon in the form of soot is one of the many kinds of *particulate matter* in polluted air.

TABLE 6-5	*Some Typical Coal Compositions in Percent (dry, ash-free)*				
	C	**H**	**O**	**N**	**S**
lignite	70.59	4.47	23.13	1.04	0.74
subbituminous	77.2	5.01	15.92	1.30	0.51
bituminous	80.2	5.80	7.53	1.39	5.11
anthracite	92.7	2.80	2.70	1.00	0.90

Figure 6-11 The luxuriant growth of vegetation that occurred during the carboniferous age is the source of our coal deposits.

Many metals occur in nature as sulfides. The process of extracting the free (elemental) metals involves **roasting**—heating an ore in the presence of air. For many metal sulfides this produces a metal oxide and SO_2. The metal oxides are then reduced to the free metals. Consider lead sulfide, PbS, as an example:

$$2PbS(s) + 3O_2(g) \longrightarrow 2PbO(s) + 2SO_2(g)$$

Sulfur dioxide is corrosive; it damages plants, structural materials, and humans. It is a nasal, throat, and lung irritant. Sulfur dioxide is slowly oxidized to sulfur trioxide, SO_3, by oxygen in air:

$$2SO_2(g) + O_2(g) \longrightarrow 2SO_3(\ell)$$

Sulfur trioxide combines with moisture in the air to form the strong, corrosive acid, sulfuric acid:

$$SO_3(\ell) + H_2O(\ell) \longrightarrow H_2SO_4(\ell)$$

Oxides of sulfur are the main cause of acid rain.

Compounds of nitrogen are also impurities in fossil fuels; they burn to form nitric oxide, NO. Most of the nitrogen in the NO in exhaust gases from furnaces, automobiles, airplanes, and so on, however, comes from the air that is mixed with the fuel.

$$\overset{0}{N_2}(g) + O_2(g) \longrightarrow 2\overset{+2}{N}O(g)$$

Remember that "clean air" is *about* 80% N_2 and 20% O_2 by mass. This reaction does *not* occur at room temperature but does occur at the high temperatures of furnaces, internal combustion engines, and jet engines.

The Blue Ridge Mountain next to Palmerton, Pennsylvania, was decimated by SO_2 pollution from the zinc refinery located there. The roasting of the ZnS ore to produce ZnO (used to produce Zn metal) also made SO_2, most of which was captured to generate sulfuric acid. But cumulative SO_2 releases over many years killed most of the vegetation on the mountain next to the refinery. The refinery is now shut down, and the Environmental Protection Agency (EPA) is doing extensive remediation of the mountain with acid- and zinc-resistant plants.

NO can be further oxidized by oxygen to nitrogen dioxide, NO_2; this reaction is enhanced in the presence of ultraviolet light from the sun.

$$2\overset{+2}{N}O(g) + O_2(g) \xrightarrow[\text{light}]{\text{UV}} 2\overset{+4}{N}O_2(g) \qquad \text{(a reddish-brown gas)}$$

NO_2 is responsible for the reddish-brown haze that hangs over many cities on sunny afternoons (Figure 6-12) and probably for most of the respiratory problems associated with this kind of air pollution. It can react to produce other oxides of nitrogen and other secondary pollutants.

In addition to being a pollutant itself, nitrogen dioxide reacts with water in the air to form nitric acid, another major contributor to acid rain:

$$3NO_2(g) + H_2O(\ell) \longrightarrow 2HNO_3(aq) + NO(g)$$

Figure 6-12 Photochemical pollution (a brown haze) enveloping a city.

Key Terms

Acid anhydride A nonmetal oxide that reacts with water to form an acid.

Acidic oxide See *Acid anhydride*.

Actinides Elements 90 through 103 (after *actinium*).

Amphoteric oxide An oxide that shows some acidic and some basic properties.

Amphoterism The ability of a substance to react with both acids and bases.

Angstrom (Å) 10^{-10} meter, 10^{-1} nm, or 100 pm.

Atomic radius The radius of an atom.

Basic anhydride A metal oxide that reacts with water to form a base.

Basic oxide See *Basic anhydride*.

Catalyst A substance that speeds up a chemical reaction without itself being consumed in the reaction.

Combustion reaction The reaction of a substance with oxygen in a highly exothermic reaction, usually with a visible flame.

***d*-Transition elements (metals)** The B group elements in the periodic table; sometimes called simply transition elements.

Effective nuclear charge (Z_{eff}) The nuclear charge experienced by the outermost electrons of an atom; the actual nuclear charge minus the effects of shielding due to inner shell electrons.

Electron affinity The amount of energy absorbed in the process in which an electron is added to a neutral isolated gaseous atom to form a gaseous ion with a $1-$ charge; has a negative value if energy is released.

Electronegativity A measure of the relative tendency of an atom to attract electrons to itself when chemically combined with another atom.

***f*-Transition elements (metals)** Elements 58 through 71 and 90 through 103; also called inner transition elements (metals).

Hydride A binary compound of hydrogen.

Inner transition elements See *f-Transition elements*.

Ionic hydride An ionic compound that contains the hydride ion, H^-.

Ionic radius The radius of an ion.

Ionization energy The amount of energy required to remove the most loosely held electron of an isolated gaseous atom or ion.

Isoelectronic Having the same number of electrons.

Lanthanides Elements 58 through 71 (after *lanthanum*).

Molecular hydride A compound in which hydrogen shares electrons with an atom of another nonmetal.

Noble gas configuration The stable electron configuration of a noble gas.

Noble gases Elements of periodic Group VIIIA; also called rare gases; formerly called inert gases.

Normal oxide A metal oxide containing the oxide ion, O^{2-} (oxygen in the -2 oxidation state).

Oxide A binary compound of oxygen.

Periodicity Regular periodic variations of properties of elements with atomic number and position in the periodic table.

Periodic law The properties of the elements are periodic functions of their atomic numbers.

Peroxide A compound containing oxygen in the -1 oxidation state. Metal peroxides contain the peroxide ion, O_2^{2-}.

Radical A species containing one or more unpaired electrons; many radicals are very reactive.

Representative elements   The A group elements in the periodic table.

Roasting   Heating an ore of an element in the presence of air.

Shielding effect   Electrons in filled sets of s and p orbitals between the nucleus and outer shell electrons shield the outer shell electrons somewhat from the effect of protons in the nucleus; also called screening effect.

Superoxide   A compound containing the superoxide ion, O_2^- (oxygen in the $-\frac{1}{2}$ oxidation state).

Ternary acid   An acid containing three elements: H, O, and (usually) another nonmetal.

Exercises

Classification of the Elements

1. Define and illustrate the following terms clearly and concisely: (a) representative elements; (b) d-transition elements; (c) inner transition elements.

2. The third shell ($n = 3$) has s, p, and d subshells. Why does Period 3 contain only eight elements?

3. Account for the number of elements in Period 6.

*4. What would be the atomic number of the as-yet undiscovered alkali metal of Period 8?

5. Identify the group, family, or other periodic table location of each element with the outer electron configuration (a) ns^2np^3; (b) ns^1; (c) $ns^2(n-1)d^{0-2}(n-2)f^{1-14}$

6. Repeat Exercise 5 for (a) ns^2np^5; (b) ns^2; (c) $ns^2(n-1)d^{1-10}$; (d) ns^2np^1

7. Write the outer electron configurations for the (a) alkaline earth metals; (b) tenth column of d-transition metals; and (c) halogens.

8. Which of the elements in the following periodic table is (are) (a) alkali metals; (b) an element with the outer configuration of d^7s^2; (c) lanthanides; (d) p-block representative elements; (e) elements with partially filled f-subshells; (f) halogens; (g) s-block representative elements; (h) actinides; (i) d-transition elements; (j) noble gases; (k) alkaline earth elements?

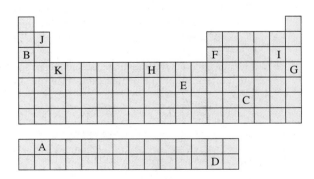

9. Identify the elements and the part of the periodic table in which the elements with the following configurations are found. (a) $1s^2 2s^2 2p^6 3s^2 3p^6 4s^1$; (b) [Kr] $4d^8 5s^2$; (c) [Xe] $4f^{14} 5d^5 6s^1$; (d) [Xe] $4f^{12} 6s^2$; (e) [Kr] $4d^{10} 5s^2 5p^4$; (f) [Kr] $4d^{10} 4f^{14} 5s^2 5p^6 5d^{10} 6s^2 6p^3$

10. Which of the following species are isoelectronic? O^{2-}, F^-, Ne, Na, Ca^{2+}, Al^{3+}

11. Which of the following species are isoelectronic? P^{3-}, S^{2-}, Cl^-, Ar, K, Ca^{2+}

Atomic Radii

12. What is meant by nuclear shielding? What effect does it have on trends in atomic radii?

13. Why do atomic radii decrease from left to right within a period in the periodic table?

14. Consider the elements in Group VIA. Even though it has not been characterized, what can be predicted about the ionic radius of element number 116?

15. Variations in the atomic radii of the transition elements are not so pronounced as those of the representative elements. Why?

16. Arrange each of the following sets of atoms in order of increasing atomic radii: (a) the alkaline earth elements; (b) the noble gases; (c) the representative elements in the third period; (d) C, Si, Sn, and Pb.

Charles D. Winters

Carbon (right), silicon (bottom), tin (left) and lead (top)

17. Arrange the following species in order of decreasing radius: Br, I^-, Se, Li^+. Explain your answer.

Ionization Energy

18. Define (a) first ionization energy and (b) second ionization energy.
19. Why is the second ionization energy for a given element always greater than the first ionization energy?
20. What is the usual relationship between atomic radius and first ionization energy, other factors being equal?
21. What is the usual relationship between nuclear charge and first ionization energy, other factors being equal?
22. Going across a period on the periodic table, what is the relationship between shielding and first ionization energy?
23. Within a group on the periodic table, what is the relationship between shielding and first ionization energy?
24. Arrange the members of each of the following sets of elements in order of increasing first ionization energies: (a) the alkali metals; (b) the halogens; (c) the elements in the second period; (d) Br, Cl, B, Ga, Cs, and H.
25. The following series of five ionization energies pertains to an element in the second period: $IE_1 = 1.33 \times 10^{-21}$ kJ/atom; $IE_2 = 4.03 \times 10^{-21}$ kJ/atom; $IE_3 = 6.08 \times 10^{-21}$ kJ/atom; $IE_4 = 4.16 \times 10^{-20}$ kJ/atom; $IE_5 = 5.45 \times 10^{-20}$ kJ/atom. Identify the element and explain why you selected that element.
26. What is the general relationship between the sizes of the atoms of Period 2 and their first ionization energies? Rationalize the relationship.
27. In a plot of first ionization energy versus atomic number for Periods 2 and 3, "dips" occur at the IIIA and VIA elements. Account for these dips.
28. On the basis of electron configurations, would you expect a Na^{2+} ion to exist in compounds? Why or why not? How about Mg^{2+}?
29. How much energy, in kilojoules, must be absorbed by 1.25 mol of gaseous lithium atoms to convert all of them to gaseous Li^+ ions?

Electron Affinity

30. Arrange the following elements in order of increasing negative values of electron affinity: P, S, Cl, and I.
31. Arrange the members of each of the following sets of elements in order of increasingly negative electron affinities: (a) the Group IA metals; (b) the Group IVA elements; (c) the elements in the second period; (d) Li, K, C, F, and Cl.
32. The electron affinities of the halogens are much more negative than those of the Group VIA elements. Why is this so?
33. The addition of a second electron to form an ion with a 2− charge is always endothermic. Why is this so?
34. Write the equation for the change described by each of the following, and write the electron configuration for each atom or ion shown: (a) the electron affinity of oxygen; (b) the electron affinity of chlorine; (c) the electron affinity of magnesium.

Ionic Radii

35. Compare the sizes of cations and the neutral atoms from which they are formed by citing three specific examples.
36. Arrange the members of each of the following sets of cations in order of increasing ionic radii: (a) K^+, Ca^{2+}, Ga^{3+}; (b) Ca^{2+}, Be^{2+}, Ba^{2+}, Mg^{2+}; (c) Al^{3+}, Sr^{2+}, Rb^+, K^+; (d) K^+, Ca^{2+}, Rb^+.
37. Compare the sizes of anions and the neutral atoms from which they are formed by citing three specific examples.
38. Arrange the following sets of anions in order of increasing ionic radii: (a) Cl^-, S^{2-}, P^{3-}; (b) O^{2-}, S^{2-}, Te^{2-}; (c) N^{3-}, S^{2-}, Br^-, P^{3-}; (d) Cl^-, Br^-, I^-.
39. Explain the trend in size of either the atom or ion as one moves down a group.
40. Many transition metals can form more than one simple positive ion. For example, iron forms both Fe^{2+} and Fe^{3+} ions, and tin forms both Sn^{2+} and Sn^{4+} ions. Which is the smaller ion of each pair, and why?

Electronegativity

41. What is electronegativity?
42. Arrange the members of each of the following sets of elements in order of increasing electronegativities: (a) Pb, C, Sn, Ge; (b) S, Na, Mg, Cl; (c) P, N, Sb, Bi; (d) Se, Ba, F, Si, Sc.
43. Which of the following statements is better? Why? (a) Magnesium has a weak attraction for electrons in a chemical bond because it has a low electronegativity. (b) The electronegativity of magnesium is low because magnesium has a weak attraction for electrons in a chemical bond.
44. Compare the electronegativity values of the metalloids.
45. One element takes on only a negative oxidation number when combined with other elements. From the table of electronegativity values, determine which element this is.

Additional Exercises on the Periodic Table

46. The bond lengths in F_2 and Cl_2 molecules are 1.42 Å and 1.98 Å, respectively. Calculate the atomic radii for these elements. Predict the Cl—F bond length. (The actual Cl—F bond length is 1.64 Å.)
*47. The atoms in crystalline nickel are arranged so that they touch one another in a plane as shown in the sketch:

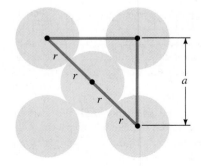

From plane geometry, we can see that $4r = a\sqrt{2}$. Calculate the radius of a nickel atom given that $a = 3.5238$ Å.

48. Compare the respective values of the first ionization energy (see Table 6-1) and electron affinity (see Table 6-2) for several elements. Which energy is greater? Why?

49. Compare the respective values of the first ionization energy (see Table 6-1) and electron affinity (see Table 6-2) for nitrogen to those for carbon and oxygen. Explain why the nitrogen values are considerably different.

*50. Based on general trends, the electron affinity of fluorine would be expected to be greater than that of chlorine; however, the value is less and is similar to the value for bromine. Explain.

51. Atomic number 91 is the element protactinium, an extremely rare element. There is very little known about protactinium; even the boiling point is unknown. Thorium, atomic number 90, has the boiling point of 4788°C and uranium's boiling point is 4131°C. Predict the boiling point of protactinium, and discuss how close the prediction is likely to be.

Hydrogen and the Hydrides

52. Summarize the physical properties of hydrogen.

53. Write balanced formula unit equations for (a) the reaction of iron with steam, (b) the reaction of calcium with hydrochloric acid, (c) the electrolysis of water, and (d) the "water gas" reaction.

54. Write a balanced formula unit equation for the preparation of (a) an ionic hydride and (b) a molecular hydride.

55. Classify the following hydrides as molecular or ionic: (a) NaH, (b) H_2S, (c) AlH_3, (d) KH, (e) NH_3.

56. Explain why NaH and H_2S are different kinds of hydrides.

57. Write formula unit equations for the reactions of (a) CaH_2 and (b) AlH_3 with water.

Charles D. Winters

Reaction of CaH_2 and H_2O

58. Name the following (pure) compounds: (a) H_2S, (b) HCl, (c) NaH, (d) NH_3, (e) H_2Se, (f) MgH_2, (g) AlH_3.

59. Write a formula unit equation for the reaction of the an ionic hydride with water. What products are always formed during such a reaction?

Oxygen and the Oxides

60. Briefly compare and contrast the properties of oxygen with those of hydrogen.

61. Write molecular equations to show how oxygen can be prepared from (a) mercury(II) oxide, HgO, (b) hydrogen peroxide, H_2O_2, and (c) potassium chlorate, $KClO_3$

*62. Which of the following elements form normal oxides as the *major* products of reactions with oxygen? (a) Li, (b) Ba, (c) Rb, (d) Mg, (e) Zn (exhibits only one common oxidation state), (f) Al.

63. Oxygen has a positive oxidation number when combined with which element? Compare the electronegativity values of oxygen and that element.

64. Write formula unit equations for the reactions of the following elements with a *limited* amount of oxygen: (a) C, (b) As_4, (c) Ge.

65. Write formula unit equations for the reactions of the following elements with an *excess* of oxygen: (a) C, (b) As_4, (c) Ge.

66. Distinguish among normal oxides, peroxides, and superoxides. What is the oxidation state of oxygen in each case?

*67. Which of the following can be classified as basic anhydrides? (a) CO_2, (b) Li_2O, (c) SeO_3, (d) CaO, (e) N_2O_5.

68. Write balanced formula unit equations for the following reactions and name the products: (a) carbon dioxide, CO_2, with water (b) sulfur trioxide, SO_3, with water (c) selenium trioxide, SeO_3, with water (d) dinitrogen pentoxide, N_2O_5, with water (e) dichlorine heptoxide, Cl_2O_7, with water.

69. Write balanced formula unit equations for the following reactions and name the products: (a) sodium oxide, Na_2O, with water (b) calcium oxide, CaO, with water (c) lithium oxide, Li_2O, with water (d) magnesium oxide, MgO, with sulfur dioxide, SO_2 (e) calcium oxide, CaO, with carbon dioxide, CO_2.

*70. Identify the acid anhydrides of the following ternary acids: (a) H_2SO_4, (b) H_2CO_3, (c) H_2SO_3, (d) H_3PO_4, (e) HNO_2.

71. Identify the basic anhydrides of the following metal hydroxides: (a) $NaOH$, (b) $Ca(OH)_2$, (c) $Fe(OH)_2$, (d) $Al(OH)_3$.

Combustion Reactions

72. Define combustion. Why are all combustion reactions also redox reactions?

73. Write equations for the complete combustion of the following compounds: (a) methane, $CH_4(g)$; (b) propane, $C_3H_8(g)$; (c) methanol, $CH_3OH(\ell)$.

74. Write equations for the *incomplete* combustion of the following compounds to produce carbon monoxide: (a) methane, $CH_4(g)$; (b) propane, $C_3H_8(g)$.

As we have seen, two substances may react to form different products when they are mixed in different proportions under different conditions. In Exercises 75 and 76, write balanced equations for the reactions described. Assign oxidation numbers.

75. (a) Ethane burns in excess air to form carbon dioxide and water. (b) Ethane burns in a limited amount of air to form

carbon monoxide and water. (c) Ethane burns (poorly) in a very limited amount of air to form elemental carbon and water.

76. (a) Butane (C_4H_{10}) burns in excess air to form carbon dioxide and water. (b) Butane burns in a limited amount of air to form carbon monoxide and water. (c) When heated in the presence of *very little* air, butane "cracks" to form acetylene, C_2H_2; carbon monoxide; and hydrogen.

*77. (a) How much SO_2 would be formed by burning 1.00 kg of bituminous coal that is 5.15% sulfur by mass? Assume that all of the sulfur is converted to SO_2. (b) If 19.0% of the SO_2 escaped into the atmosphere and 75.0% of the escaped SO_2 were converted to H_2SO_4, how many grams of H_2SO_4 would be produced in the atmosphere?

*78. Write equations for the complete combustion of the following compounds. Assume that sulfur is converted to SO_2 and nitrogen is converted to NO. (a) $C_6H_5NH_2(\ell)$, (b) $C_3H_7SH(\ell)$, (c) $C_7H_{10}NO_2S(\ell)$.

CONCEPTUAL EXERCISES

79. Write the electron configuration for the product of the second ionization of the third largest alkaline earth metal.

80. You are given the atomic radii of 110 pm, 118 pm, 120 pm, 122 pm, and 135 pm, but do not know to which element (As, Ga, Ge, P, and Si) these values correspond. Which must be the value of Ge?

81. What compound will most likely form between potassium and element X, if element X has the electronic configuration $1s^22s^22p^63s^23p^4$?

82. Write the electron configurations of beryllium and magnesium. What similarities in their chemical properties can you predict on the basis of their electron configurations?

83. Dolostone is often more porous than limestone. One explanation of the origin of dolostone is that it results from partial replacement of calcium by magnesium in an original limestone sediment. Is this explanation reasonable, given what you know of the ionic radii of magnesium and calcium ions?

BUILDING YOUR KNOWLEDGE

84. Hydrogen can be obtained from water by electrolysis. Hydrogen may someday be an important replacement for current fuels. Describe some of the problems that

you would predict if hydrogen were used in today's motor vehicles.

85. The only chemically stable ion of rubidium is Rb⁺. The most stable monatomic ion of bromine is Br⁻. Krypton (Kr) is among the least reactive of all elements. Compare the electron configurations of Rb⁺, Br⁻, and Kr. Then predict the most stable monatomic ions of strontium (Sr) and selenium (Se).

86. The first ionization energy of potassium, K, is 419 kJ/mol. What is the minimum frequency of light required to ionize gaseous potassium atoms?

87. Potassium and argon would be anomalies in a periodic table in which elements are arranged in order of increasing atomic weights. Identify two other elements among the transition elements whose positions in the periodic table would have been reversed in a "weight-sequence" arrangement. Which pair of elements would most obviously be out of place on the basis of their chemical behavior? Explain your answer in terms of the current atomic model, showing electron configurations for these elements.

88. The second ionization energy for magnesium is 1451 kJ/mol. How much energy, in kilojoules, must be absorbed by 1.50 g of gaseous magnesium atoms to convert them to gaseous Mg^{2+} ions?

89. The chemical reactivities of carbon and lead are similar, but there are also major differences. Using their electron configurations, explain why these similarities and differences may exist.

BEYOND THE TEXTBOOK

Go to the textbook website at

http://www.brookscole.com/chemistry/whitten

for additional activities and exercises based on the General Chemistry Interactive CD-ROM, the World Wide Web, and library resources.

InfoTrac College Edition

For additional readings, go to InfoTrac College Edition, your online research library at:

http://infotrac.thomsonlearning.com

Chemical Bonding

Charles D. Winters

Carbon atoms are covalently bonded together in a three-dimensional array to make diamond, the hardest substance known.

OBJECTIVES

After you have studied this chapter, you should be able to

- Write Lewis dot representations of atoms
- Predict whether bonding between specified elements will be primarily ionic, covalent, or polar covalent
- Compare and contrast characteristics of ionic and covalent compounds
- Describe how the properties of compounds depend on their bonding
- Describe how the elements bond by electron transfer (ionic bonding)
- Describe energy relationships in ionic compounds
- Predict the formulas of ionic compounds
- Describe how elements bond by sharing electrons (covalent bonding)
- Write Lewis dot and dash formulas for molecules and polyatomic ions
- Recognize exceptions to the octet rule
- Write formal charges for atoms in covalent structures
- Describe resonance, and know when to write resonance structures and how to do so
- Relate the nature of bonding to electronegativity differences

Chemical bonding refers to the attractive forces that hold atoms together in compounds. There are two major classes of bonding. (1) **Ionic bonding** results from electrostatic interactions among ions, which often results from the net *transfer* of one or more electrons from one atom or group of atoms to another. (2) **Covalent bonding** results from *sharing* one or more electron pairs between two atoms. These two classes represent two extremes; all bonds between atoms of different elements have at least some degree of both ionic and covalent character. Compounds containing predominantly ionic bonding are called **ionic compounds.** Those that are held together mainly by covalent bonds are called **covalent compounds.** Some nonmetallic elements, such as H_2, Cl_2, N_2, and P_4, also involve covalent bonding. Some properties usually associated with many simple ionic and covalent compounds are summarized in the following list. The differences in these properties can be accounted for by the differences in bonding between the atoms or ions.

As you study Chapters 7 and 8, keep in mind the periodic similarities that you learned in Chapters 4 and 6. What you learn about the bonding of an element usually applies to the other elements in the same column of the periodic table, with minor variations.

Ionic Compounds	Covalent Compounds
1. They are solids with high melting points (typically >400°C).	1. They are gases, liquids, or solids with low melting points (typically <300°C).
2. Many are soluble in polar solvents such as water.	2. Many are insoluble in polar solvents.
3. Most are insoluble in nonpolar solvents, such as hexane, C_6H_{14}, and carbon tetrachloride, CCl_4.	3. Most are soluble in nonpolar solvents, such as hexane, C_6H_{14}, and carbon tetrachloride, CCl_4.
4. Molten compounds conduct electricity well because they contain mobile charged particles (ions).	4. Liquid and molten compounds do not conduct electricity.
5. Aqueous solutions conduct electricity well because they contain mobile charged particles (ions).	5. Aqueous solutions are *usually* poor conductors of electricity because most do not contain charged particles.
6. They are often formed between two elements with quite different electronegativities, usually a metal and a nonmetal.	6. They are often formed between two elements with similar electronegativities, usually nonmetals.

The distinction between polar and nonpolar molecules is made in Section 7-9.

As we saw in Section 4-2, aqueous solutions of some covalent compounds do conduct electricity, because they react with water to some extent to form ions.

7-1 LEWIS DOT FORMULAS OF ATOMS

The number and arrangements of electrons in the outermost shells of atoms determine the chemical and physical properties of the elements as well as the kinds of chemical bonds they form. We write **Lewis formulas** (sometimes called **Lewis dot formulas** or **Lewis dot representations**) as a convenient bookkeeping method for keeping track of these "chemically important electrons." We now introduce this method for atoms of elements; in our discussion of chemical bonding in subsequent sections, we will frequently use such formulas for atoms, molecules, and ions.

Chemical bonding involves only the **valence electrons,** which are usually the electrons in the outermost occupied shells. In Lewis dot representations, only the electrons in the outermost occupied *s* and *p* orbitals are shown as dots. Table 7-1 shows Lewis dot formulas for atoms of the representative elements. All elements in a given group have the same outer-shell electron configuration. It is somewhat arbitrary on which side of the atom symbol we write the electron dots. We do, however, represent an electron pair as a pair of dots and an unpaired electron as a single dot. Because of the large numbers of dots, such formulas are not as useful for the transition and inner transition elements.

 CD-ROM   Screen 9.4, Valence Electrons

For example, Al has the electron configuration [Ar] $3s^2 3p^1$. The three dots in the Lewis dot formula for Al represent the two *s* electrons (the pair of dots) and the *p* electron (the single dot) beyond the noble gas configuration.

TABLE 7-1	*Lewis Dot Formulas for Representative Elements*							
Group	IA	IIA	IIIA	IVA	VA	VIA	VIIA	VIIIA
Number of electrons in valence shell	1	2	3	4	5	6	7	8 (except He)
Period 1	H ·							He :
Period 2	Li ·	Be :	B·	C·	·N·	·O:	·F:	: Ne :
Period 3	Na ·	Mg :	Al·	Si·	·P·	·S:	·Cl:	: Ar :
Period 4	K ·	Ca :	Ga·	Ge·	·As·	·Se:	·Br:	: Kr :
Period 5	Rb ·	Sr :	In·	Sn·	·Sb·	·Te:	·I:	: Xe :
Period 6	Cs ·	Ba :	Tl·	Pb·	·Bi·	·Po:	·At:	: Rn :
Period 7	Fr ·	Ra :						

For the Group A elements, the number of valence electrons (dots in the Lewis formula) for the *neutral* atom is equal to the group number. Exceptions: H (one valence electron) and He (two valence electrons).

IONIC BONDING

7-2 FORMATION OF IONIC COMPOUNDS

The first kind of chemical bonding we shall describe is **ionic bonding.** We recall (Section 2-3) that an **ion** is an atom or a group of atoms that carries an electrical charge. An ion in which the atom or group of atoms has fewer electrons than protons is positively charged and is called a **cation;** one that has more electrons than protons is negatively charged and is called an **anion.** An ion that consists of only one atom is described as a **monatomic ion.** Examples include the chloride ion, Cl^-, and the magnesium ion, Mg^{2+}. An ion that contains more than one atom is called a **polyatomic ion.** Examples include the ammonium ion, NH_4^+; the hydroxide ion, OH^-; and the sulfate ion, SO_4^{2-}. The atoms of a polyatomic ion are held together by covalent bonds. In this section we shall discuss how ions can be formed from individual atoms; polyatomic ions will be discussed along with other covalently bonded species.

The chemical and physical properties of an ion are quite different from those of the atom from which the ion is derived. For example, an atom of Na and an Na^+ ion are quite different.

SCiLINKS.

TOPIC: Ionic Bonding
GO TO: www.scilinks.org
*sci*LINKS CODE: WCH0710

Ionic bonding is the attraction of oppositely charged ions (cations and anions) in large numbers to form a solid. Such a solid compound is called an *ionic solid.*

As our previous discussions of ionization energy, electronegativity, and electron affinity would suggest, ionic bonding can occur easily when elements that have low electronegativities and low ionization energies (metals) react with elements that have high electronegativities and very negative electron affinities (nonmetals). Many metals are easily *oxidized*—that is, they lose electrons to form cations; and many nonmetals are readily *reduced*—that is, they gain electrons to form anions.

> When the electronegativity difference, $\Delta(EN)$, between two elements is large, as between a metal and a nonmetal, the elements are likely to form a compound by ionic bonding (transfer of electrons).

Let's describe some combinations of metals with nonmetals to form ionic compounds.

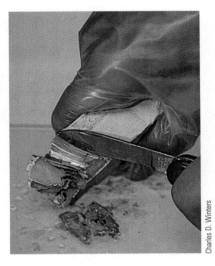

Freshly cut sodium has a metallic luster. A little while after being cut, the sodium metal surface turns white as it reacts with the air.

Group IA Metals and Group VIIA Nonmetals

Consider the reaction of sodium (a Group IA metal) with chlorine (a Group VIIA nonmetal). Sodium is a soft silvery metal (mp 98°C), and chlorine is a yellowish green corrosive gas at room temperature. Both sodium and chlorine react with water, sodium vigorously. By contrast, sodium chloride is a white solid (mp 801°C) that dissolves in water with no reaction and with the absorption of just a little heat. We can represent the reaction for its formation as

$$2Na(s) + Cl_2(g) \longrightarrow 2NaCl(s)$$
$$\text{sodium} \qquad \text{chlorine} \qquad \text{sodium chloride}$$

We can understand this reaction better by showing electron configurations for all species. We represent chlorine as individual atoms rather than molecules, for simplicity.

Na [Ne] $\underset{3s}{\uparrow}$

Cl [Ne] $\underset{3s}{\uparrow\downarrow}$ $\underset{3p}{\uparrow\downarrow \; \uparrow\downarrow \; \uparrow}$

$\longrightarrow$

Na$^+$ [Ne] $1e^-$ lost

Cl$^-$ [Ne] $\underset{3s}{\uparrow\downarrow}$ $\underset{3p}{\uparrow\downarrow \; \uparrow\downarrow \; \uparrow\downarrow}$ $1e^-$ gained

In this reaction, Na atoms lose one electron each to form Na$^+$ ions, which contain only ten electrons, the same number as the *preceding* noble gas, neon. We say that sodium ions have the noble gas electronic structure of neon: Na$^+$ is *isoelectronic* with Ne (Section 6-5). In contrast, Cl atoms gain one electron each to form Cl$^-$ ions, which contain 18 electrons.

Isoelectronic species have the same number of electrons (see Section 6-5). Some isoelectronic species:

O^{2-}	8 protons	10 electrons
F$^-$	9 protons	10 electrons
Ne	10 protons	10 electrons
Na$^+$	11 protons	10 electrons
Mg^{2+}	12 protons	10 electrons

Some ionic compounds. Clockwise from front right: sodium chloride (NaCl, *white*); copper(II) sulfate pentahydrate (CuSO$_4$·5H$_2$O, *blue*); nickel(II) chloride hexahydrate (NiCl$_2$·6H$_2$O, *green*); potassium dichromate (K$_2$Cr$_2$O$_7$, *orange*); and cobalt(II) chloride hexahydrate (CoCl$_2$·6H$_2$O, *red*). One mole of each substance is shown.

The loss of electrons is *oxidation* (Section 4-7). Na atoms are *oxidized* to form Na$^+$ ions.

This is the same number as the *following* noble gas, argon; Cl$^-$ is *isoelectronic* with Ar. These processes can be represented compactly as

$$Na \longrightarrow Na^+ + e^- \quad \text{and} \quad Cl + e^- \longrightarrow Cl^-$$

The gain of electrons is *reduction* (Section 4-7). Cl atoms are *reduced* to form Cl$^-$ ions.

Similar observations apply to most ionic compounds formed by reactions between *representative metals and representative nonmetals*.

We can use Lewis dot formulas (Section 7-1) to represent the reaction.

$$Na \cdot + \; : \overset{\cdot\cdot}{Cl} \cdot \longrightarrow Na^+ \Big[: \overset{\cdot\cdot}{\underset{\cdot\cdot}{Cl}} : \Big]^-$$

The Na$^+$ and Cl$^-$ ions can be present only in a 1:1 ratio in the compound sodium chloride, so the formula must be NaCl. We predict the same formula based on the fact that each Na atom can lose only one electron from its outermost occupied shell and each Cl atom needs only one electron to fill completely its outermost *p* orbitals.

The chemical formula NaCl does not explicitly indicate the ionic nature of the compound, only the ratio of ions. Furthermore, values of electronegativities are not always available. So we must learn to recognize, from positions of elements in the periodic table and known trends in electronegativity, when the difference in electronegativity is large enough to favor ionic bonding.

The noble gases are excluded from this generalization.

> The farther apart across the periodic table two Group A elements are, the more ionic their bonding will be.

The greatest difference in electronegativity occurs from the lower left to upper right, so CsF ($\Delta[\text{EN}] = 3.2$) is more ionic than LiI ($\Delta[\text{EN}] = 1.5$).

All the Group IA metals (Li, Na, K, Rb, Cs) will react with the Group VIIA elements (F, Cl, Br, I) to form ionic compounds of the same general formula, MX. All the resulting ions, M$^+$ and X$^-$, have noble gas configurations. Once we understand the bonding of one member of a group (column) in the periodic table, we know a great deal about the others in the family. Combining each of the five common alkali metals with each of the four common halogens gives $5 \times 4 = 20$ possible compounds. The discussion of NaCl presented here applies also to the other 19 such compounds.

Formulas for some of these are

LiF	NaF	KF
LiCl	NaCl	KCl
LiBr	NaBr	KBr
LiI	NaI	KI

The collection of isolated positive and negative ions occurs at a higher energy than the elements from which they are formed. The ion formation alone is not sufficient to account for the formation of ionic compounds. Some other favorable factor must account for the observed stability of these compounds. Because of the opposite charges on Na$^+$ and Cl$^-$, an attractive force is developed. According to Coulomb's Law, the force of attraction, F, between two oppositely charged particles of charge magnitudes q^+ and q^- is directly proportional to the product of the charges and inversely proportional to the square of the distance separating their centers, d. Thus, the greater the charges on the ions and the smaller the ions are, the stronger the resulting ionic bonding. Of course, like-charged ions repel each other, so the distances separating the ions in ionic solids are those at which the attractions exceed the repulsions by the greatest amount.

Coulomb's Law is $F \propto \dfrac{q^+ q^-}{d^2}$. The symbol $\propto$ means "is proportional to."

The energy associated with the attraction of separated gaseous positive and negative ions to form an ionic solid is the *crystal lattice energy* of the solid. For NaCl, this energy is -789 kJ/mol; that is, one mole of NaCl solid is 789 kJ lower in energy (*more stable*) than one mole of isolated Na$^+$ ions and one mole of isolated Cl$^-$ ions. We could also say that it would require 789 kJ of energy to separate one mole of NaCl solid into isolated gaseous ions. The stability of ionic compounds is thus due to the interplay of the energy cost of ion formation and the energy repaid by the crystal lattice energy. The best trade-off usually comes when the monatomic ions of representative elements have noble gas configurations. For more on these ideas, see the Enrichment feature "Introduction to Energy Relationships in Ionic Bonding."

The structure of common table salt, sodium chloride (NaCl), is shown in Figure 7-1. Like other simple ionic compounds, NaCl(s) exists in a regular, extended array of positive and negative ions, Na^+ and Cl^-. Distinct molecules of solid ionic substances do not exist, so we must refer to *formula units* (Section 2-3) instead of molecules. The forces that hold all the particles together in an ionic solid are quite strong. This explains why such substances have quite high melting and boiling points (a topic that we will discuss more fully in Chapter 13). When an ionic compound is melted or dissolved in water, its charged particles are free to move in an electric field, so such a liquid shows high electrical conductivity (Section 4-2, Part 1).

We can represent the general reaction of the IA metals with the VIIA elements as follows:

$$2M(s) + X_2 \longrightarrow 2MX(s) \qquad M = Li,\ Na,\ K,\ Rb,\ Cs;\ X = F,\ Cl,\ Br,\ I$$

The Lewis dot representation for the generalized reaction is

$$2M\cdot\ +\ :\overset{\cdot\cdot}{\underset{\cdot\cdot}{X}}:\overset{\cdot\cdot}{\underset{\cdot\cdot}{X}}:\ \longrightarrow\ 2\,(M^+\Big[:\overset{\cdot\cdot}{\underset{\cdot\cdot}{X}}:\Big]^-)$$

Group IA Metals and Group VIA Nonmetals

Next, consider the reaction of lithium (Group IA) with oxygen (Group VIA) to form lithium oxide, a solid ionic compound (mp > 1700°C) (see next page). We may represent the reaction as

$$4Li(s) + O_2(g) \longrightarrow 2Li_2O(s)$$
lithium   oxygen   lithium oxide

The formula for lithium oxide, Li_2O, indicates that two atoms of lithium combine with one atom of oxygen. If we examine the structures of the atoms before reaction, we can see the reason for this ratio.

In a compact representation,

$$2[Li \longrightarrow Li^+ + e^-] \qquad \text{and} \qquad O + 2e^- \longrightarrow O^{2-}$$

The Lewis dot formulas for the atoms and ions are

$$2Li\cdot\ +\ :\overset{\cdot\cdot}{O}\cdot\ \longrightarrow\ 2Li^+\Big[:\overset{\cdot\cdot}{\underset{\cdot\cdot}{O}}:\Big]^{2-}$$

Lithium ions, Li^+, are isoelectronic with helium atoms ($2\ e^-$). Oxide ions, O^{2-}, are isoelectronic with neon atoms ($10\ e^-$).

The very small size of the Li^+ ion gives it a much higher *charge density* (ratio of charge to size) than that of the larger Na^+ ion (Figure 6-1). Similarly, the O^{2-} ion is smaller than the Cl^- ion, so its smaller size and double negative charge give it a much higher charge density. These more concentrated charges and smaller sizes bring the Li^+ and O^{2-} ions closer together in Li_2O than the Na^+ and Cl^- ions are in NaCl. Consequently, the q^+q^- product in the numerator of Coulomb's Law is greater in Li_2O, and the d^2 term in the denominator is smaller. The net result is that the ionic bonding is much stronger (the lattice energy

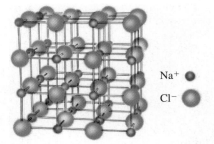

Figure 7-1 A representation of the crystal structure of NaCl. Each Cl^- ion (*green*) is surrounded by six sodium ions, and each Na^+ ion (*gray*) is surrounded by six chloride ions. Any NaCl crystal includes billions of ions in the pattern shown. Adjacent ions actually are in contact with one another; in this drawing, the structure has been expanded to show the spatial arrangement of ions. The lines *do not* represent covalent bonds. Compare with Figure 2-7, a space-filling drawing of the NaCl structure.

Although the oxides of the other Group IA metals are prepared by different methods, similar descriptions apply to compounds between the Group IA metals (Li, Na, K, Rb, Cs) and the Group VIA nonmetals (O, S, Se, Te, Po).

Each Li atom has 1 e^- in its valence shell, one more e^- than a noble gas configuration, [He]. Each O atom has 6 e^- in its valence shell and needs 2 e^- more to attain a noble gas configuration [Ne]. The Li^+ ions are formed by oxidation of Li atoms, and the O^{2-} ions are formed by reduction of O atoms.

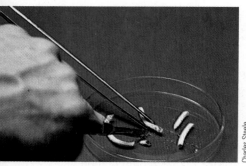

Lithium is a metal, as the shiny surface of freshly cut Li shows. Where it has been exposed to air, the surface is covered with lithium oxide.

This discussion also applies to other ionic compounds between any Group IIA metal (Be, Mg, Ca, Sr, Ba) and any Group VIA nonmetal (O, S, Se, Te).

is much more negative) in Li_2O than in NaCl. This is consistent with the higher melting temperature of Li_2O (>1700°C) compared to NaCl (801°C).

Group IIA Metals and Group VIA Nonmetals

As our final illustration of ionic bonding, consider the reaction of calcium (Group IIA) with oxygen (Group VIA). This reaction forms calcium oxide, a white solid ionic compound with a very high melting point, 2580°C.

$$2Ca(s) + O_2(g) \longrightarrow 2CaO(s)$$

calcium   oxygen   calcium oxide

Again, we show the electronic structure of the atoms and ions, representing the inner electrons by the symbol of the preceding noble gas.

$$\left. \begin{array}{l} \text{Ca} \quad [\text{Ar}] \; \underset{4s}{\uparrow\downarrow} \\[2em] \text{O} \quad [\text{He}] \; \underset{2s}{\uparrow\downarrow} \; \underset{2p}{\uparrow\downarrow \; \uparrow \; \uparrow} \end{array} \right\} \longrightarrow \left\{ \begin{array}{ll} \text{Ca}^{2+} \; [\text{Ar}] \; \underset{4s}{\underline{}} & 2\,e^- \text{ lost} \\[2em] \text{O}^{2-} \; [\text{He}] \; \underset{2s}{\uparrow\downarrow} \; \underset{2p}{\uparrow\downarrow \; \uparrow\downarrow \; \uparrow\downarrow} & 2\,e^- \text{ gained} \end{array} \right.$$

The Lewis dot notation for the reacting atoms and the resulting ions is

$$\text{Ca} : + \; : \overset{..}{\text{O}} \cdot \; \longrightarrow \; \text{Ca}^{2+} \left[: \overset{..}{\underset{..}{\text{O}}} : \right]^{2-}$$

Calcium ions, Ca^{2+}, are isoelectronic with argon (18 e^-), the preceding noble gas. Oxide ions, O^{2-}, are isoelectronic with neon (10 e^-), the following noble gas.

Ca^{2+} is about the same size as Na^+ (see Figure 6-1) but carries twice the charge, so its charge density is higher. Because the attraction between the two small, highly charged ions Ca^{2+} and O^{2-} is quite high, the ionic bonding is very strong, accounting for the very high melting point of CaO, 2580°C.

d-Transition Metal Ions

Chemists do not ordinarily represent transition metals or their ions as dot formulas.

Electron configurations of the *d*-transition metal atoms include the *s* electrons in the outermost occupied shell and the *d* electrons one energy level lower (e.g., $3d4s$ for the first transition series in Period 4). The outer *s* electrons lie outside the *d* electrons and are *always* the first ones lost when transition metals form simple ions. In the first transition series, scandium and zinc each form a single positive ion. Scandium loses its two $4s$ electrons and its only $3d$ electron to form Sc^{3+}. Zinc loses its two $4s$ electrons to form Zn^{2+}.

$$\left. \begin{array}{l} \quad\quad\quad \overset{\textstyle 3d}{\rule{2.5cm}{0.4pt}} \quad \overset{\textstyle 4s}{\rule{0.6cm}{0.4pt}} \\ \text{Sc} \; [\text{Ar}] \; \uparrow \, \_\_\_\_\_ \; \uparrow\downarrow \\[1em] \text{Zn} \; [\text{Ar}] \; \uparrow\downarrow \; \uparrow\downarrow \; \uparrow\downarrow \; \uparrow\downarrow \; \uparrow\downarrow \; \uparrow\downarrow \end{array} \right\} \longrightarrow \left\{ \begin{array}{ll} \quad\quad \overset{\textstyle 3d}{\rule{2.5cm}{0.4pt}} \quad \overset{\textstyle 4s}{\rule{0.6cm}{0.4pt}} \\ \text{Sc}^{3+} \; [\text{Ar}] \; \_\_\_\_\_ \; \_ & 3\,e^- \text{ lost} \\[1em] \text{Zn}^{2+} \; [\text{Ar}] \; \uparrow\downarrow \; \uparrow\downarrow \; \uparrow\downarrow \; \uparrow\downarrow \; \uparrow\downarrow \; \_ & 2\,e^- \text{ lost} \end{array} \right.$$

Most other $3d$-transition metals can form at least two cations in their compounds. For example, cobalt can form Co^{2+} and Co^{3+} ions.

Many transition metal ions are highly colored. These flasks contain (*left to right*), aqueous solutions of $Fe(NO_3)_3$, $Co(NO_3)_2$, $Ni(NO_3)_2$, $Cu(NO_3)_2$, and $Zn(NO_3)_2$. Colorless Zn^{2+} ions differ from the others by having completely filled $3d$ orbitals.

$$\left. \begin{array}{l} \quad\quad\quad \overset{\textstyle 3d}{\rule{2.5cm}{0.4pt}} \quad \overset{\textstyle 4s}{\rule{0.6cm}{0.4pt}} \\ \text{Co} \; [\text{Ar}] \; \uparrow\downarrow \; \uparrow\downarrow \; \uparrow \; \uparrow \; \uparrow \; \uparrow\downarrow \\[1em] \text{Co} \; [\text{Ar}] \; \uparrow\downarrow \; \uparrow\downarrow \; \uparrow \; \uparrow \; \uparrow \; \uparrow\downarrow \end{array} \right\} \longrightarrow \left\{ \begin{array}{ll} \quad\quad \overset{\textstyle 3d}{\rule{2.5cm}{0.4pt}} \quad \overset{\textstyle 4s}{\rule{0.6cm}{0.4pt}} \\ \text{Co}^{2+} \; [\text{Ar}] \; \uparrow\downarrow \; \uparrow\downarrow \; \uparrow \; \uparrow \; \uparrow \; \_ & 2\,e^- \text{ lost} \\[1em] \text{Co}^{3+} \; [\text{Ar}] \; \uparrow\downarrow \; \uparrow \; \uparrow \; \uparrow \; \uparrow \; \_ & 3\,e^- \text{ lost} \end{array} \right.$$

TABLE 7-2	*Simple Binary Ionic Compounds*					
Metal		**Nonmetal**		**General Formula**	**Ions Present**	**Example**	**mp (°C)**

Metal		Nonmetal		General Formula	Ions Present	Example	mp (°C)
IA*	+	VIIA	$\longrightarrow$	MX	(M^+, X^-)	LiBr	547
IIA	+	VIIA	$\longrightarrow$	MX_2	$(M^{2+}, 2X^-)$	$MgCl_2$	708
IIIA	+	VIIA	$\longrightarrow$	MX_3	$(M^{3+}, 3X^-)$	GaF_3	800 (subl)
IA*†	+	VIA	$\longrightarrow$	M_2X	$(2M^+, X^{2-})$	Li_2O	>1700
IIA	+	VIA	$\longrightarrow$	MX	(M^{2+}, X^{2-})	CaO	2580
IIA	+	VIA	$\longrightarrow$	M_2X_3	$(2M^{3+}, 3X^{2-})$	Al_2O_3	2045
IA*	+	VA	$\longrightarrow$	M_3X	$(3M^+, X^{3-})$	Li_3N	840
IIA	+	VA	$\longrightarrow$	M_3X_2	$(3M^{2+}, 2X^{3-})$	Ca_3P_2	≈1600
IIIA	+	VA	$\longrightarrow$	MX	(M^{3+}, X^{3-})	AlP	

*Hydrogen is a nonmetal. All binary compounds of hydrogen are covalent, except for certain metal hydrides such as NaH and CaH_2, which contain hydride, H^-, ions.

†As we saw in Section 6-8, the metals in Groups IA and IIA also commonly form peroxides (containing the O_2^{2-} ion) or superoxides (containing the O_2^- ion). See Table 6-4. The peroxide and superoxide ions contain atoms that are covalently bonded to one another.

Binary Ionic Compounds: A Summary

Table 7-2 summarizes the general formulas of binary ionic compounds formed by the representative elements. "M" represents metals, and "X" represents nonmetals from the indicated groups. In these examples of ionic bonding, each of the metal atoms has lost one, two, or three electrons, and each of the nonmetal atoms has gained one, two, or three electrons. *Simple (monatomic) ions rarely have charges greater than 3+ or 3−.* Ions with greater charges would interact so strongly with the electron clouds of other ions that the electron clouds would be very distorted, and considerable covalent character in the bonds would result. Distinct molecules of solid ionic substances do not exist. The sum of the attractive forces of all the interactions in an ionic solid is substantial. Therefore, binary ionic compounds have high melting and boiling points.

The *d*- and *f*-transition elements form many compounds that are essentially ionic in character. Many simple ions of transition metals do not have noble gas configurations. All common monatomic anions have noble gas configurations. Most monatomic cations of the representative elements (A groups) have noble gas configurations.

Binary compounds contain *two* elements.

 CD-ROM   Screen 8.8, Electron Configurations in Ions.

The distortion of the electron cloud of an anion by a small, highly charged cation is called *polarization*.

Enrichment   Introduction to Energy Relationships in Ionic Bonding

The following discussion may help you to understand why ionic bonding occurs between elements with low ionization energies and those with high electronegativities. There is a general tendency in nature to achieve stability. One way to do this is by lowering potential energy; *lower* energies generally represent *more stable* arrangements.

Let's use energy relationships to describe why the ionic solid NaCl is more stable than a mixture of individual Na and Cl atoms. Consider a gaseous mixture of one mole of sodium atoms and one mole of chlorine atoms, $Na(g) + Cl(g)$. The energy change associated with the loss of one mole of electrons by one mole of Na atoms to form one mole of Na^+ ions (Step 1 in Figure 7-2) is given by the *first ionization energy* of Na (see Section 6-3).

$$Na(g) \longrightarrow Na^+(g) + e^-   \text{first ionization energy} = 496 \text{ kJ/mol}$$

(Enrichment, continued)

This is a positive value, so the mixture $Na^+(g) + e^- + Cl(g)$ is 496 kJ/mol higher in energy than the original mixture of atoms (the mixture $Na^+ + e^- + Cl$ is *less stable* than the mixture of atoms).

The energy change for the gain of one mole of electrons by one mole of Cl atoms to form one mole of Cl^- ions (Step 2) is given by the *electron affinity* of Cl (see Section 6-4).

$$Cl(g) + e^- \longrightarrow Cl^-(g) \qquad \text{electron affinity} = -349 \text{ kJ/mol}$$

This negative value, -349 kJ/mol, lowers the energy of the mixture, but the mixture of separated ions, $Na^+ + Cl^-$, is still *higher* in energy (*less stable*) by $(496 - 349)$ kJ/mol = 147 kJ/mol than the original mixture of atoms (the red arrow in Figure 7-2). Thus, just the formation of ions does not explain why the process occurs. The strong attractive force between ions of opposite charge draws the ions together into the regular array shown in Figure 7-1. The energy associated with this attraction (Step 3) is the *crystal lattice energy* of NaCl, -789 kJ/mol.

$$Na^+(g) + Cl^-(g) \longrightarrow NaCl(s) \qquad \text{crystal lattice energy} = -789 \text{ kJ/mol}$$

The crystal (solid) formation thus further *lowers* the energy to $(147 - 789)$ kJ/mol = -642 kJ/mol. The overall result is that one mole of NaCl(s) is 642 kJ/mol lower in energy (more stable) than the original mixture of atoms (the blue arrow in Figure 7-2). Thus, we see that a major driving force for the formation of ionic compounds is the large electrostatic stabilization due to the attraction of the ionic charges (Step 3).

CD-ROM   Screen 9.3, Lattice Energy.

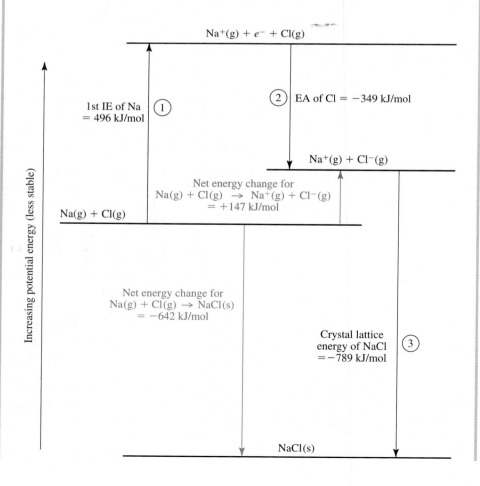

Figure 7-2 A schematic representation of the energy changes that accompany the process

Na(g) + Cl(g) → NaCl(s).

The red arrow represents the *positive* energy change (unfavorable) for the process of ion formation,

Na(g) + Cl(g) → Na⁺(g) + Cl⁻(g).

The blue arrow represents the *negative* energy change (favorable) for the overall process, including the formation of the ionic solid.

In this discussion we have not taken into account the fact that sodium is a solid metal or that chlorine actually exists as diatomic molecules. The additional energy changes involved when these are changed to gaseous Na and Cl atoms, respectively, are small enough that the overall energy change starting from $Na(s)$ and $Cl_2(g)$ is still negative.

COVALENT BONDING

Ionic bonding cannot result from a reaction between two nonmetals, because their electronegativity difference is not great enough for electron transfer to take place. Instead, reactions between two nonmetals result in *covalent bonding*.

A **covalent bond** is formed when two atoms share one or more pairs of electrons. Covalent bonding occurs when the electronegativity difference, $\Delta(EN)$, between elements (atoms) is zero or relatively small.

In predominantly covalent compounds the bonds between atoms *within* a molecule (*intra*molecular bonds) are relatively strong, but the forces of attraction *between* molecules (*inter*molecular forces) are relatively weak. As a result, covalent compounds have lower melting and boiling points than ionic compounds. The relation of bonding types to physical properties of liquids and solids will be developed more fully in Chapter 13.

7-3 FORMATION OF COVALENT BONDS

Let's look at the simplest case of covalent bonding, the reaction of two hydrogen atoms to form the diatomic molecule H_2. As you recall, an isolated hydrogen atom has the ground state electron configuration $1s^1$, with the probability density for this one electron spherically distributed about the hydrogen nucleus (Figure 7-3a). As two hydrogen atoms approach each other from large distances, the electron of each hydrogen atom is attracted by the nucleus of the *other* hydrogen atom as well as by its own nucleus (Figure 7-3b). If these two electrons have opposite spins so that they can occupy the same region (orbital), both electrons can now preferentially occupy the region *between* the two nuclei (Figure 7-3c), because they are attracted by both nuclei. The electrons are *shared* between the two hydrogen atoms, and a single covalent bond is formed. We say that the $1s$ orbitals *overlap* so that both electrons are now in the orbitals of both hydrogen atoms. The closer together the atoms come, the more nearly this is true. In that sense, each hydrogen atom then has the helium configuration $1s^2$.

The bonded atoms are at lower energy (more stable) than the separated atoms. This is shown in the plot of energy versus distance in Figure 7-4. As the two atoms get closer together, however, the two nuclei, being positively charged, exert an increasing repulsion on each other. At some distance, a minimum energy, -435 kJ/mol, is reached; it corresponds to the most stable arrangement and occurs at 0.74 Å, the actual distance between two hydrogen nuclei in an H_2 molecule. At distances greater than this, the attractive forces exceed the repulsive ones, so the atoms are drawn closer together. At smaller separations, repulsive forces increase more rapidly than attractive ones, so repulsive forces dominate and the arrangement is less stable. The magnitude of this stabilization is called the H—H *bond energy*. Bond energies for covalent bonds will be discussed more extensively in Section 15-9.

Other pairs of nonmetal atoms share electron pairs to form covalent bonds. The result of this sharing is that each atom attains a more stable electron configuration—frequently

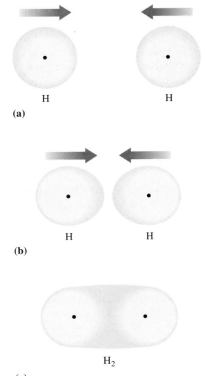

(a)

(b)

(c)

Figure 7-3 A representation of the formation of a covalent bond between two hydrogen atoms. The position of each positively charged nucleus is represented by a black dot. Electron density is indicated by the depth of shading. (a) Two hydrogen atoms separated by a large distance. (b) As the atoms approach each other, the electron of each atom is attracted by the positively charged nucleus of the other atom, so the electron density begins to shift. (c) The two electrons can both be in the region where the two $1s$ orbitals overlap; the electron density is highest in the region between the nuclei of the two atoms.

Figure 7-4 The potential energy of the H_2 molecule as a function of the distance between the two nuclei. The lowest point in the curve, -435 kJ/mol, corresponds to the internuclear distance actually observed in the H_2 molecule, 0.74 Å. At distances longer than this, attractive forces dominate; at distances shorter than this, repulsive forces dominate. (The minimum potential energy, -435 kJ/mol, corresponds to the value of -7.23×10^{-19} J per H_2 molecule.) Energy is compared with that of two separated hydrogen atoms.

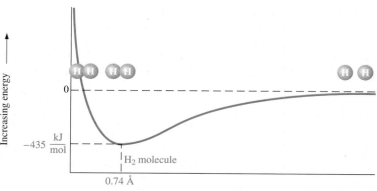

SciLINKS

TOPIC: Covalent Bonding
GO TO: www.scilinks.org
*sci*LINKS CODE: WCH0720

CD-ROM   Screens 9.5, Chemical Bond Formation: Covalent Bonding; 9.11, Bond Properties; 10.2, Models of Chemical Bonding; and 10.3, Valence Bond Theory: Bond Formation.

The bonding in the other halogens, Cl_2, Br_2, and I_2, is analogous to that in F_2.

The bonding in gaseous HCl, HBr, and HI is analogous to that in HF.

the same as that of the nearest noble gas. This results in a more stable arrangement for the bonded atoms. (This is discussed in Section 7-5.) Most covalent bonds involve sharing of two, four, or six electrons—that is, one, two, or three *pairs* of electrons. Two atoms form a **single covalent bond** when they share one pair of electrons, a **double covalent bond** when they share two electron pairs, and a **triple covalent bond** when they share three electron pairs. These are usually called simply *single*, *double*, and *triple* bonds. Covalent bonds that involve sharing of one and three electrons are known but are relatively rare.

In a Lewis formula, we represent a covalent bond by writing *each shared electron pair* either as a pair of two dots between the two atom symbols or as a dash connecting them. Thus, the formation of H_2 from H atoms could be represented as

$$H\cdot \; + \; \cdot H \longrightarrow H:H \qquad or \qquad H{-}H$$

where the dash represents a single bond. Similarly, the combination of two fluorine atoms to form a molecule of fluorine, F_2, can be shown as

$$:\!\ddot{F}\cdot \; + \; \cdot\ddot{F}\!: \longrightarrow :\!\ddot{F}\!:\!\ddot{F}\!: \qquad or \qquad :\!\ddot{F}\!{-}\!\ddot{F}\!:$$

The formation of a hydrogen fluoride, HF, molecule from a hydrogen atom and a fluorine atom can be shown as

$$H\cdot \; + \; \cdot\ddot{F}\!: \longrightarrow H\!:\!\ddot{F}\!: \qquad or \qquad H{-}\ddot{F}\!:$$

We will see many more examples of this representation.

In our discussion, we have postulated that bonds form by the **overlap** of two atomic orbitals. This is the essence of the **valence bond theory**, which we will describe in more detail in the next chapter. Another theory, **molecular orbital theory**, is discussed in Chapter 9. For now, let us concentrate on the *number* of electron pairs shared and defer the discussion of *which* orbitals are involved in the sharing until the next chapter.

CD-ROM   Screen 9.3, Lewis Electron Dot Structures.

7-4 LEWIS FORMULAS FOR MOLECULES AND POLYATOMIC IONS

In Sections 7-1 and 7-2 we drew *Lewis formulas* for atoms and monatomic ions. In Section 7-3, we used Lewis formulas to show the *valence electrons* in three simple molecules. A water molecule can be represented by either of the following diagrams.

In H_2O, two of the six valence electrons of an oxygen atom are used in covalent bonding; the one valence electron of each hydrogen atom is used in covalent bonding.

An H_2O molecule has two shared electron pairs, that is, two single covalent bonds. The O atom also has two unshared pairs.

dot formula   dash formula

In *dash formulas*, a shared pair of electrons is indicated by a dash. The following diagrams show the two *double* bonds in carbon dioxide and its Lewis formula.

$$:O::C::O:$$ or $$:O=C=O:$$

dot formula   dash formula

A CO_2 molecule has four shared electron pairs in two double bonds. The central atom (C) has no unshared pairs.

In CO_2, the four valence electrons of a carbon atom are used in covalent bonding; two of the six valence electrons of each oxygen atom are used in covalent bonding.

The covalent bonds in a polyatomic ion can be represented in the same way. The Lewis formula for the ammonium ion, NH_4^+, shows only eight electrons, even though the N atom has five electrons in its valence shell and each H atom has one, for a total of $5 + 4(1) = 9$ electrons. The NH_4^+ ion, with a charge of $1+$, has one less electron than the original atoms.

A polyatomic ion is an ion that contains more than one atom.

$$\left[\begin{array}{c} H \\ H:N:H \\ H \end{array} \right]^+ \quad \left[\begin{array}{c} H \\ | \\ H-N-H \\ | \\ H \end{array} \right]^+$$

dot formula   dash formula

The NH_3 molecule, like the NH_4^+ ion, has eight valence electrons about the N atom.

$$H:N:H \quad \text{or} \quad H-N-H$$
$$\ \ \ H \qquad\qquad\qquad H$$

The writing of Lewis formulas is an electron bookkeeping method that is useful as a first approximation to suggest bonding schemes. It is important to remember that Lewis dot formulas only show the number of valence electrons, the number and kinds of bonds, and the order in which the atoms are connected. *They are not intended to show the three-dimensional shapes of molecules and polyatomic ions.* We will see in Chapter 8, however, that the three-dimensional geometry of a molecule can be predicted from its Lewis formula.

7-5 WRITING LEWIS FORMULAS: THE OCTET RULE

Representative elements usually attain stable noble gas electron configurations when they share electrons. In the water molecule eight electrons are in the outer shell of the O atom, and it has the neon electron configuration; two electrons are in the valence shell of each H atom, and each has the helium electron configuration. Likewise, the C and O of CO_2 and the N of NH_3 and the NH_4^+ ion each have a share in eight electrons in their outer shells. The H atoms in NH_3 and NH_4^+ each share two electrons. Many Lewis formulas are based on the idea that

> in *most* of their compounds, the representative elements achieve noble gas configurations.

In some compounds, the central atom does not achieve a noble gas configuration. Such exceptions to the octet rule are discussed in Section 7-8.

This statement is usually called the **octet rule** because the noble gas configurations have $8\ e^-$ in their outermost shells (except for He, which has $2\ e^-$).

For now, we restrict our discussion to compounds of the *representative elements*. The octet rule alone does not let us write Lewis formulas. We still must decide how to place the electrons around the bonded atoms—that is, how many of the available valence electrons are **bonding electrons** (shared) and how many are **unshared electrons** (associated with only one atom). A pair of unshared electrons in the same orbital is called a **lone pair.** A simple mathematical relationship is helpful here:

The representative elements are those in the A groups of the periodic table.

$$S = N - A$$

S is the total number of electrons *shared* in the molecule or polyatomic ion.

N is the total number of valence shell electrons *needed* by all the atoms in the molecule or ion to achieve noble gas configurations ($N = 8 \times$ number of atoms that are not H, plus $2 \times$ number of H atoms).

> A is the number of electrons *available* in the valence shells of all of the (representative) atoms. This is equal to the sum of their periodic group numbers. We must adjust A, if necessary, for ionic charges. We add electrons to account for negative charges and subtract electrons to account for positive charges.

Let's see how this relationship applies to some species whose Lewis formulas we have already shown.

For F_2,

$$N = 2 \times 8 \text{ (for two F atoms)} = 16 \ e^- \text{ needed}$$
$$A = 2 \times 7 \text{ (for two F atoms)} = 14 \ e^- \text{ available}$$
$$S = N - A = 16 - 14 = 2 \ e^- \text{ shared}$$

$:\!\ddot{F}\!:\!\ddot{F}\!:$

The Lewis formula for F_2 shows 14 valence electrons total, with 2 e^- shared in a single bond.

For HF,

$$N = 1 \times 2 \text{ (for one H atom)} + 1 \times 8 \text{ (for one F atom)} = 10 \ e^- \text{ needed}$$
$$A = 1 \times 1 \text{ (for one H atom)} + 1 \times 7 \text{ (for one F atom)} = 8 \ e^- \text{ available}$$
$$S = N - A = 10 - 8 = 2 \ e^- \text{ shared}$$

$H\!:\!\ddot{F}\!:$

The Lewis formula for HF shows 8 valence electrons total, with 2 e^- shared in a single bond.

For H_2O,

$$N = 2 \times 2 \text{ (for two H atoms)} + 1 \times 8 \text{ (for one O atom)} = 12 \ e^- \text{ needed}$$
$$A = 2 \times 1 \text{ (for two H atoms)} + 1 \times 6 \text{ (for one O atom)} = 8 \ e^- \text{ available}$$
$$S = N - A = 12 - 8 = 4 \ e^- \text{ shared}$$

$H\!:\!\ddot{O}\!:$
H

The Lewis formula for H_2O shows 8 valence electrons total, with a total of 4 e^- shared, 2 e^- in each single bond.

For CO_2,

$$N = 1 \times 8 \text{ (for one C atom)} + 2 \times 8 \text{ (for two O atoms)} = 24 \ e^- \text{ needed}$$
$$A = 1 \times 4 \text{ (for one C atom)} + 2 \times 6 \text{ (for two O atoms)} = 16 \ e^- \text{ available}$$
$$S = N - A = 24 - 16 = 8 \ e^- \text{ shared}$$

$:\!\ddot{O}\!:\!:\!C\!:\!:\!\ddot{O}\!:$

The Lewis formula for CO_2 shows 16 valence electrons total, with a total of 8 e^- shared, 4 e^- in each double bond.

For NH_4^+,

$$N = 1 \times 8 \text{ (for one N atom)} + 4 \times 2 \text{ (for four H atoms)} = 16 \ e^- \text{ needed}$$
$$A = 1 \times 5 \text{ (for one N atom)} + 4 \times 1 \text{ (for four H atoms)} - 1 \text{ (for 1+ charge)} = 8 \ e^- \text{ available}$$
$$S = N - A = 16 - 8 = 8 \ e^- \text{ shared}$$

The 1+ ionic charge is due to a *deficiency* of one e^- relative to the neutral atoms.

$$\left[\begin{array}{c} H \\ H:N:H \\ H \end{array} \right]^+$$

The Lewis formula for NH_4^+ shows 8 valence electrons total, with all 8 e^- shared, 2 e^- in each single bond.

The following general steps describe the use of the $S = N - A$ relationship in constructing dot formulas for molecules and polyatomic ions.

A Guide to Writing Lewis Formulas

1. Select a reasonable (symmetrical) "skeleton" for the molecule or polyatomic ion.

   a. The *least electronegative element* is usually the central element, except that H is never the central element, because it forms only one bond. The least electronegative element is usually the one that needs the most electrons to fill its octet. Example: CS_2 has the skeleton S   C   S.

   b. Carbon bonds to two, three, or four atoms, but *never* more than four. Nitrogen bonds to one (rarely), two, three (most commonly), or four atoms. Oxygen bonds to one, two (most commonly), or three atoms.

   c. Oxygen atoms do not bond to each other except in (1) O_2 and O_3 molecules; (2) hydrogen peroxide, H_2O_2, and its derivatives, the peroxides, which contain the O_2^{2-} group; and (3) the rare superoxides, which contain the O_2^- group. Example: The nitrate ion, NO_3^-, has the skeleton

$$\begin{bmatrix} & O & \\ & N & \\ O & & O \end{bmatrix}^-$$

   d. In *ternary oxoacids*, hydrogen usually bonds to an O atom, *not* to the central atom. Example: nitrous acid, HNO_2, has the skeleton H   O   N   O. There are a few exceptions to this guideline, such as H_3PO_3 and H_3PO_2.

   > A ternary oxoacid contains *three* elements—H, O, and another element, often a nonmetal.

   e. For ions or molecules that have more than one central atom, the most symmetrical skeletons possible are used. Examples: C_2H_4 and $P_2O_7^{4-}$ have the following skeletons:

$$\begin{matrix} H & H \\ C & C \\ H & H \end{matrix} \quad \text{and} \quad \begin{bmatrix} & O & & O & \\ O & P & O & P & O \\ & O & & O & \end{bmatrix}^{4-}$$

2. Calculate N, *the number of valence (outer) shell electrons* needed by all atoms in the molecule or ion to achieve noble gas configurations. Examples:

   > For compounds containing only representative elements, N is equal to 8 × number of atoms that are *not* H, plus 2 × number of H atoms.

   For PF_3,

$$N = 1 \times 8 \text{ (P atom)} + 3 \times 8 \text{ (F atoms)} = 32 \ e^- \text{ needed}$$

   For CH_3OH,

$$N = 1 \times 8 \text{ (C atom)} + 4 \times 2 \text{ (H atoms)} + 1 \times 8 \text{ (O atom)} = 24 \ e^- \text{ needed}$$

   For NO_3^-,

$$N = 1 \times 8 \text{ (N atom)} + 3 \times 8 \text{ (O atoms)} = 32 \ e^- \text{ needed}$$

   Calculate A, *the number of electrons available* in the valence (outer) shells of all the atoms. For negatively charged ions, add to this total the number of electrons equal to the charge on the anion; for positively charged ions, subtract the number of electrons equal to the charge on the cation. Examples:

   > For the representative elements, the number of valence shell electrons in an atom is equal to its periodic group number. Exceptions: 1 for an H atom and 2 for He.

   For PF_3,

$$A = 1 \times 5 \text{ (P atom)} + 3 \times 7 \text{ (F atoms)} = 26 \ e^- \text{ available}$$

   For CH_3OH,

$$A = 1 \times 4 \text{ (C atom)} + 4 \times 1 \text{ (H atoms)} + 1 \times 6 \text{ (O atom)} = 14 \ e^- \text{ available}$$

For NO_3^-,

$$A = 1 \times 5 \text{ (N atom)} + 3 \times 6 \text{ (O atoms)} + 1 \text{ (for 1- charge)} = 5 + 18 + 1 = 24 \ e^- \text{ available}$$

Calculate S, *total number of electrons shared* in the molecule or ion, using the relationship $S = N - A$. Examples:

For PF_3,

$$S = N - A = 32 - 26$$
$$= 6 \text{ electrons shared (3 pairs of } e^- \text{ shared)}$$

For CH_3OH,

$$S = N - A = 24 - 14$$
$$= 10 \text{ electrons shared (5 pairs of } e^- \text{ shared)}$$

For NO_3^-,

$$S = N - A = 32 - 24 = 8 \ e^- \text{ shared (4 pairs of } e^- \text{ shared)}$$

C, N, and O often form double and triple bonds. S can form double bonds with C, N, and O.

3. Place the S electrons into the skeleton as *shared pairs*. Use double and triple bonds only when necessary. Lewis formulas may be shown as either dot formulas or dash formulas.

Formula	Skeleton	Dot Formula ("bonds" in place, but incomplete)	Dash Formula ("bonds" in place, but incomplete)
PF_3	F P F F	F : P : F F	F —P— F | F
CH_3OH	H H C O H H	H H : C : O : H H	H | H— C — O —H | H
NO_3^-	$\begin{bmatrix} & O & \\ & N & \\ O & & O \end{bmatrix}^-$	$\begin{bmatrix} & O & \\ & N & \\ O & & O \end{bmatrix}^-$	$\begin{bmatrix} & O & \\ & N & \\ O & & O \end{bmatrix}^-$

Formulas are sometimes written to give clues about which atoms are bonded together. In this format, the H atoms are bonded to the atom preceding them; in CH_3OH the first three H atoms are bonded to C, and the last H atom is bonded to O.

4. Place the additional electrons into the skeleton as *unshared (lone) pairs* to fill the octet of every A group element (except H, which can share only 2 e^-). Check that the total number of electrons is equal to A, from Step 2. Examples:

For PF_3,

 A very common error in writing Lewis formulas is showing the wrong number of electrons. Always make a final check to be sure that the Lewis formula you write shows the same number of electrons you calculated as A.

$$: \overset{..}{\underset{..}{F}} : \overset{..}{\underset{..}{P}} : \overset{..}{\underset{..}{F}} : \quad \text{or} \quad : \overset{..}{\underset{..}{F}} — \overset{}{\underset{}{P}} — \overset{..}{\underset{..}{F}} :$$
$$: \overset{}{\underset{..}{F}} : \qquad\qquad\qquad\qquad : \overset{}{\underset{..}{F}} :$$

Check: 13 pairs of e^- have been used and all octets are satisfied. $2 \times 13 = 26 \ e^-$ available.

For CH$_3$OH,

$$
\begin{array}{ccc}
\text{H} & & \text{H} \\
\text{H}:\overset{\cdot\cdot}{\underset{\text{H}}{\text{C}}}:\overset{\cdot\cdot}{\underset{\cdot\cdot}{\text{O}}}:\text{H} & \quad\text{or}\quad & \text{H}-\overset{\text{H}}{\underset{\text{H}}{\text{C}}}-\overset{\cdot\cdot}{\underset{\cdot\cdot}{\text{O}}}-\text{H}
\end{array}
$$

CD-ROM For a slightly different approach than the one used in this text, see the CD-ROM, Screen 9.7, Drawing Lewis Structures.

Check: 7 pairs of e^- have been used and all octets are satisfied. $2 \times 7 = 14\ e^-$ available.

For NO$_3^-$,

$$
\left[\ \overset{\overset{\displaystyle \cdot\cdot}{\text{O}}}{\underset{\overset{\displaystyle \cdot\cdot}{\text{O}}\quad\overset{\displaystyle \cdot\cdot}{\text{O}}}{:::\ \ \text{N}\ \ }}\ \right]^- \quad\text{or}\quad \left[\ \overset{\overset{\displaystyle \cdot\cdot}{\text{O}}}{\underset{\overset{\displaystyle \cdot\cdot}{\text{O}}\quad\overset{\displaystyle \cdot\cdot}{\text{O}}}{\ \ \text{N}\ \ }}\ \right]^-
$$

Check: 12 pairs of e^- have been used and all octets are satisfied. $2 \times 12 = 24\ e^-$ available.

EXAMPLE 7-1 *Writing Lewis Formulas*

Write the Lewis formula for the nitrogen molecule, N$_2$.

Plan

We follow the stepwise procedure that was just presented for writing Lewis formulas.

Solution

Step 1:   The skeleton is N   N.

Step 2: $N = 2 \times 8 = 16\ e^-$ needed (total) by both atoms

   $A = 2 \times 5 = 10\ e^-$ available (total) for both atoms

   $S = N - A = 16\ e^- - 10\ e^- = 6\ e^-$ shared

Step 3:   N$:::$N $6\ e^-$ (3 pairs) are shared; a *triple* bond.

Step 4:   The additional $4\ e^-$ are accounted for by a lone pair on each N. The complete Lewis formula is

$$
:\text{N}:::\text{N}: \quad\text{or}\quad :\text{N}\equiv\text{N}:
$$

Check: $10\ e^-$ (5 pairs) have been used.

EXAMPLE 7-2 *Writing Lewis Formulas*

Write the Lewis formula for carbon disulfide, CS$_2$, a foul-smelling liquid.

Plan

Again, we follow the stepwise procedure to apply the relationship $S = N - A$.

Solution

Step 1:   The skeleton is S   C   S.

Step 2: $N = 1 \times 8$ (for C) $+ 2 \times 8$ (for S) $= 24\ e^-$ needed by all atoms

   $A = 1 \times 4$ (for C) $+ 2 \times 6$ (for S) $= 16\ e^-$ available

   $S = N - A = 24\ e^- - 16\ e^- = 8\ e^-$ shared

Step 3:   S$::$C$::$S $8\ e^-$ (4 pairs) are shared; two *double* bonds.

C is the central atom, or the element in the middle of the molecule. It needs four more electrons to acquire an octet, and each S atom needs only two more electrons.

Calcium carbonate, CaCO$_3$, occurs in several forms in nature. The mineral calcite forms very large clear crystals (*right*); in microcrystalline form it is the main constituent of limestone (*top left*). Seashells (*bottom*) are largely calcium carbonate.

Step 4: C already has an octet, so the remaining 8 e^- are distributed as lone pairs on the S atoms to give each S an octet. The complete Lewis formula is

$$\ddot{S}::C::\ddot{S} \quad \text{or} \quad \ddot{S}=C=\ddot{S}$$

Check: 16 e^- (8 pairs) have been used. The bonding picture is similar to that of CO$_2$; this is not surprising, because S is below O in Group VIA.

You should now work Exercise 27.

EXAMPLE 7-3 *Writing Lewis Formulas*

Write the Lewis formula for the carbonate ion, CO$_3^{2-}$.

Plan

The same stepwise procedure can be applied to ions. We must remember to adjust A, the total number of electrons, to account for the charge shown on the ion.

Solution

Step 1: The skeleton is O   C   O$\qquad$O^{2-}

Step 2: $N = 1 \times 8$ (for C) $+ 3 \times 8$ (for O) $= 8 + 24 = 32\ e^-$ needed by all atoms

$A = 1 \times 4$ (for C) $+ 3 \times 6$ (for O) $+ 2$ (for the 2$-$ charge)

$\qquad = 4 + 18 + 2 = 24\ e^-$ available

$S = N - A = 32\ e^- - 24\ e^- = 8\ e^-$ (4 pairs) shared

Step 3: $O:\ddot{C}::O$ (Four pairs are shared. At this point it doesn't matter which O is doubly bonded.)

Step 4: The Lewis formula is

   or

Check: 24 e^- (12 pairs) have been used.

You should now work Exercises 28 and 36.

Acetylene burns in oxygen with a flame so hot that it is used to weld metals.

For practice, apply the methods of this section to write these Lewis formulas.

You should practice writing many Lewis formulas. A few common types of organic compounds and their Lewis formulas are shown here. Each follows the octet rule. Methane, CH$_4$, is the simplest of a huge family of organic compounds called *hydrocarbons* (composed solely of hydrogen and carbon). Ethane, C$_2$H$_6$, is another hydrocarbon that contains only single bonds. Ethylene, C$_2$H$_4$, has a carbon–carbon double bond, and acetylene, C$_2$H$_2$, contains a carbon–carbon triple bond.

methane, CH$_4$$\qquad$ethane, C$_2H_6$$\qquad$ethylene, C$_2H_4$$\qquad$acetylene, C$_2H_2$

Halogen atoms can appear in place of hydrogen atoms in many organic compounds because hydrogen and halogen atoms each need one more electron to attain noble gas configurations. An example is chloroform, $CHCl_3$. Alcohols contain the group C—O—H; the most common alcohol is ethanol, CH_3CH_2OH. An organic compound that contains a carbon–oxygen double bond is formaldehyde, H_2CO.

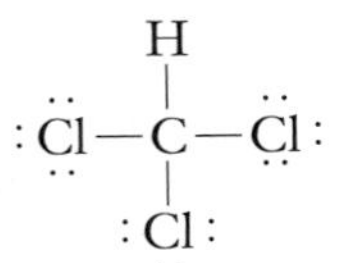

chloroform, $CHCl_3$ 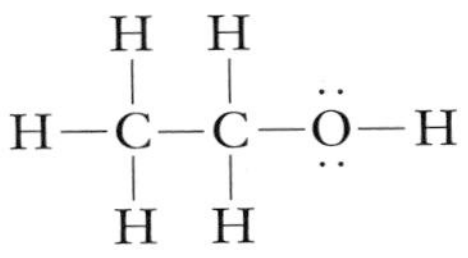   ethanol, CH_3CH_2OH 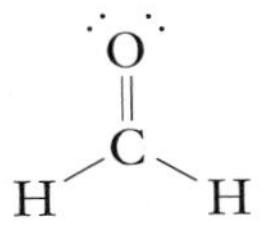   formaldehyde, H_2CO

✔ **Problem-Solving Tip:** *Drawing Lewis Formulas*

The following guidelines might help you draw Lewis formulas.

1. In most of their covalent compounds, the representative elements follow the octet rule, except that hydrogen always shares only two electrons.

2. Hydrogen forms only one bond to another element; thus hydrogen can never be a central atom.

3. Carbon always forms four bonds. This can be accomplished as:
   a. four single bonds
   b. two double bonds
   c. two single bonds and one double bond or
   d. one single bond and one triple bond

4. In neutral (uncharged) species, nitrogen forms three bonds, and oxygen forms two bonds.

5. Nonmetals can form single, double, or triple bonds, but never quadruple bonds.

6. Carbon forms double or triple bonds to C, N, O, or S atoms; oxygen can form double bonds with many other elements.

C, N, and O often form double and triple bonds. S can form double bonds with C, N, and O.

7-6 RESONANCE

In addition to the Lewis formula shown in Example 7-3, two other Lewis formulas with the same skeleton for the CO_3^{2-} ion are equally acceptable. In these formulas, the double bond could be between the carbon atom and either of the other two oxygen atoms.

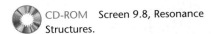 CD-ROM Screen 9.8, Resonance Structures.

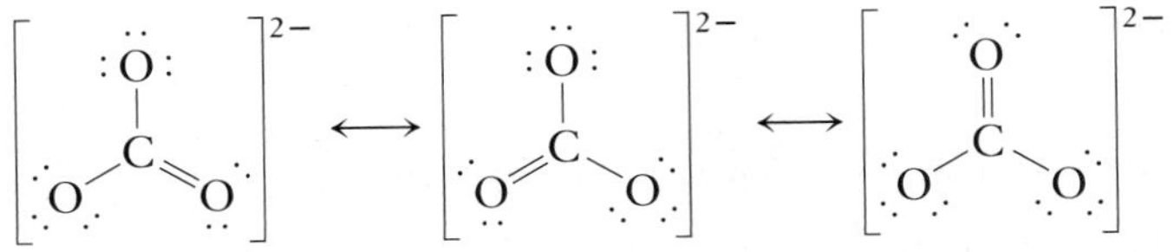

A molecule or polyatomic ion for which two or more Lewis formulas with the same arrangements of atoms can be drawn to describe the bonding is said to exhibit **resonance.** The three structures shown here are **resonance structures** of the carbonate ion. The relationship among them is indicated by the double-headed arrows, ↔. This symbol *does not mean* that the ion flips back and forth among these three structures. The true structure can be described as an average, or hybrid, of the three.

When electrons are shared among more than two atoms, the electrons are said to be *delocalized*. The concept of delocalization is important in molecular orbital theory (Chapter 9).

Experiments show that the C—O bonds in CO_3^{2-} are *neither* double nor single bonds but are intermediate in bond length and strength. Based on measurements in many compounds, the typical C—O single bond length is 1.43 Å, and the typical C=O double bond length is 1.22 Å. The C—O bond length for each bond in the CO_3^{2-} ion is intermediate at 1.29 Å. Another way to represent this situation is by **delocalization** of bonding electrons:

$$\begin{bmatrix} & O & \\ & \| & \\ O \!\!\! - \!\!\! & C & \!\!\! - \!\!\! O \end{bmatrix}^{2-} \qquad \text{(lone pairs on O atoms not shown)}$$

The dashed lines indicate that some of the electrons shared between C and O atoms are *delocalized* among all four atoms; that is, the four pairs of shared electrons are equally distributed among three C—O bonds.

Charles D. Winters

The rose on the left is in an atmosphere of sulfur dioxide, SO_2. Gaseous SO_2 and its aqueous solutions are used as bleaching agents. A similar process is used to bleach wood pulp before it is converted to paper.

EXAMPLE 7-4   *Lewis Formulas, Resonance*

Draw two resonance structures for the sulfur dioxide molecule, SO_2.

Plan

The stepwise procedure (Section 7-5) can be used to write each resonance structure.

Solution

$$\begin{array}{cc} S & O \\ \searrow & \searrow \end{array}$$
$$N = 1(8) + 2(8) = 24 \ e^-$$
$$\underline{A = 1(6) + 2(6) = 18 \ e^-}$$
$$S = \quad N - A \quad = 6 \ e^- \text{ shared}$$

The resonance structures are

$$:\!\ddot{O}:\;:\!\ddot{S}:\;\ddot{O}\!: \quad\longleftrightarrow\quad :\!\ddot{O}:\;\ddot{S}:\;:\!\ddot{O}\!: \qquad \text{or} \qquad :\!\ddot{O}\!=\!\ddot{S}\!-\!\ddot{O}\!: \quad\longleftrightarrow\quad :\!\ddot{O}\!-\!\ddot{S}\!=\!\ddot{O}\!:$$

We could show delocalization of electrons as follows:

$$O \!=\!\!=\! \ddot{S} \!=\!\!=\! O \qquad \text{(lone pairs on O not shown)}$$

Remember that Lewis formulas *do not necessarily show shapes.* SO_2 molecules are angular, not linear.

You should now work Exercises 48 and 50.

✓ Problem-Solving Tip: *Some Guidelines for Resonance Structures*

Resonance involves several different acceptable Lewis formulas with the same arrangement of atoms. They differ only in the arrangements of electron pairs, never in atom positions. The actual structure of such a molecule or ion is the average, or composite, of its resonance structures, but this does not mean that the electrons are moving from one place to another. This average structure is more stable than any of the individual resonance structures.

7-7 FORMAL CHARGES

An experimental determination of the structure of a molecule or polyatomic ion is necessary to establish unequivocally its correct structure. We often do not have these results available, however. **Formal charge** is the hypothetical charge on an atom *in a molecule or polyatomic ion*; to find the formal charge, we count bonding electrons as though they were equally shared between the two bonded atoms. The concept of formal charges helps us to write correct Lewis formulas in most cases. The most energetically favorable formula for a molecule is usually one in which the formal charge on each atom is zero or as near zero as possible.

Consider the reaction of NH_3 with hydrogen ion, H^+, to form the ammonium ion, NH_4^+.

$$\text{H}-\overset{..}{\underset{\underset{\text{H}}{|}}{\text{N}}}-\text{H} + \text{H}^+ \longrightarrow \left[\text{H}-\overset{\overset{\text{H}}{|}}{\underset{\underset{\text{H}}{|}}{\text{N}}}-\text{H}\right]^+$$

The previously unshared pair of electrons on the N atom in the NH_3 molecule is shared with the H^+ ion to form the NH_4^+ ion, in which the N atom has four covalent bonds. Because N is a Group VA element, we expect it to form three covalent bonds to complete its octet. How can we describe the fact that N has four covalent bonds in species like NH_4^+? The answer is obtained by calculating the *formal charge* on each atom in NH_4^+ by the following rules:

The H^+ ion has a vacant orbital, which accepts a share in the lone pair on nitrogen. The formation of a covalent bond by the sharing of an electron pair that is provided by one atom is called *coordinate covalent bond formation*. This type of bond formation is discussed again in Chapters 10 and 25.

Rules for Assigning Formal Charges to Atoms of Group A Elements

1. The formal charge, abbreviated FC, on an atom in a Lewis formula is given by the relationship

   FC = (group number) − [(number of bonds) + (number of unshared e^-)]

   Formal charges are represented by $\oplus$ and $\ominus$ to distinguish them from real charges on ions.

2. In a Lewis formula, an atom that has the same number of bonds as its periodic group number has a formal charge of zero.

3. **a.** In a molecule, the sum of the formal charges is zero.
   b. In a polyatomic ion, the sum of the formal charges is equal to the charge.

Let us apply these rules to the ammonia molecule, NH_3, and to the ammonium ion, NH_4^+. Because N is a Group VA element, its group number is 5.

$$\text{H}-\overset{..}{\underset{\underset{\text{H}}{|}}{\text{N}}}-\text{H} \qquad \left[\text{H}-\overset{\overset{\text{H}}{|}}{\underset{\underset{\text{H}}{|}}{\text{N}}}-\text{H}\right]^+$$

In NH_3 the N atom has 3 bonds and 2 unshared e^-, and so for N,

FC = (group number) − [(number of bonds) + (number of unshared e^-)]

   = 5 − (3 + 2) = 0 (for N)

For H,

$$FC = \text{(group number)} - [\text{(number of bonds)} + \text{(number of unshared } e^-)]$$
$$= 1 - (1 + 0) = 0 \text{ (for H)}$$

The formal charges of N and H are both zero in NH_3, so the sum of the formal charges is $0 + 3(0) = 0$, consistent with Rule 3a.

In NH_4^+ the atom has four bonds and no unshared e^-, and so for N,

$$FC = \text{(group number)} - [\text{(number of bonds)} + \text{(number of unshared } e^-)]$$
$$= 5 - (4 + 0) = +1 \text{ (for N)}$$

Calculation of the FC for H atoms gives zero, as shown previously. The sum of the formal charges in NH_4^+ is $(+1) + 4(0) = +1$. This is consistent with Rule 3b.

$$\left[\begin{array}{c} \text{H} \\ | \\ \text{H}-\overset{\oplus}{\text{N}}-\text{H} \\ | \\ \text{H} \end{array} \right]^{+}$$

FCs are indicated by $\oplus$ and $\ominus$. The sum of the formal charges in a polyatomic ion is equal to the charge on the ion: +1 in NH_4^+.

Thus, we see the octet rule is obeyed in both NH_3 and NH_4^+. The sum of the formal charges in each case is that predicted by Rule 3, even though nitrogen has four covalent bonds in the NH_4^+ ion.

This bookkeeping system helps us to choose among various Lewis formulas for a molecule or ion, according to the following guidelines.

CD-ROM   Screen 9.15, Formal Charge.

> **a.** The most likely formula for a molecule or ion is usually one in which the formal charge on each atom is zero or as near zero as possible.
>
> **b.** Negative formal charges are more likely to occur on the more electronegative elements.
>
> **c.** Lewis formulas in which adjacent atoms have formal charges of the same sign are usually *not* accurate representations (the *adjacent charge rule*).

Now let's write some Lewis formulas for, and assign formal charges to, the atoms in nitrosyl chloride, NOCl, a compound often used in organic synthesis. The Cl atom and the O atom are both bonded to the N atom. Two Lewis formulas that satisfy the octet rule are

(i) $:\!\overset{\oplus}{\text{Cl}}\!=\!\text{N}\!-\!\overset{..}{\underset{..}{\text{O}}}\!:^{\ominus}$ (ii) $:\!\overset{..}{\underset{..}{\text{Cl}}}\!-\!\text{N}\!=\!\overset{..}{\underset{..}{\text{O}}}\!:$

For Cl, $FC = 7 - (2 + 4) = +1$	For Cl, $FC = 7 - (1 + 6) = 0$
For N, $FC = 5 - (3 + 2) = 0$	For N, $FC = 5 - (3 + 2) = 0$
For O, $FC = 6 - (1 + 6) = -1$	For O, $FC = 6 - (2 + 4) = 0$

We believe that (ii) is a preferable Lewis formula because it has smaller formal charges than (i). We see that a double-bonded terminal Cl atom would have its electrons arranged as $:\!\overset{..}{\text{Cl}}\!=\!\text{X}$, with the formal charge of Cl equal to $7 - (2 + 4) = +1$. A positive formal charge on such an electronegative element is quite unlikely, and double bonding to chlorine does not occur. The same reasoning would apply to the other halogens.

7-8 WRITING LEWIS FORMULAS: LIMITATIONS OF THE OCTET RULE

Recall that representative elements achieve noble gas electron configurations in *most of* their compounds. But when the octet rule is not applicable, the relationship $S = N - A$ is not valid without modification. The following are general cases for which the procedure in Section 7-5 *must be modified*—that is, there are four types of limitations of the octet rule.

A. Most covalent compounds of beryllium, Be. Because Be contains only two valence shell electrons, it usually forms only two covalent bonds when it bonds to two other atoms. We therefore use *four electrons* as the number *needed* by Be in Step 2. In Steps 3 and 4 we use only two pairs of electrons for Be.

B. Most covalent compounds of the Group IIIA elements, especially boron, B. The IIIA elements contain only three valence shell electrons, so they often form three covalent bonds when they bond to three other atoms. We therefore use *six electrons* as the number *needed* by the IIIA elements in Step 2; and in Steps 3 and 4 we use only three pairs of electrons for the IIIA elements.

C. Compounds or ions containing an odd number of electrons. Examples are NO, with 11 valence shell electrons, and NO_2, with 17 valence shell electrons.

D. Compounds or ions in which the central element needs a share in more than eight valence shell electrons to hold all the available electrons, A. We say that the central atom in such species has an **expanded valence shell**. Extra rules are added to Steps 2 and 4 when this is encountered.

Step 2a: If S, the number of electrons shared, is less than the number needed to bond all atoms to the central atom, then S is increased to the number of electrons needed.

Step 4a: If S must be increased in Step 2a, then the octets of all the atoms might be satisfied before all of the electrons (A) have been added. Place the extra electrons on the central element.

Many species that violate the octet rule are quite reactive. For instance, compounds containing atoms with only four valence shell electrons (limitation type A) or six valence shell electrons (limitation type B) frequently react with other species that supply electron pairs. A compound that accepts a share in a pair of electrons are called is called a **Lewis acid;** a **Lewis base** is a species that makes available a share in a pair of electrons. (This kind of behavior will be discussed in detail in Section 10-10.) Molecules with an odd number of electrons often *dimerize* (combine in pairs) to give products that do satisfy the octet rule. Examples are the dimerization of NO to form N_2O_2 (Section 24-15) and NO_2 to form N_2O_4 (Section 24-15). Examples 7-5 through 7-9 illustrate some limitations and show how such Lewis formulas are constructed.

CD-ROM Screen 9.10, Free Radicals—Exceptions to the Octet Rule.

Lewis formulas are not normally written for compounds containing *d*- and *f*-transition metals. The *d*- and *f*-transition metals utilize *d* or *f* orbitals (or both) in bonding as well as *s* and *p* orbitals. Thus, they can accommodate more than eight valence electrons.

EXAMPLE 7-5   *Limitation A of the Octet Rule*

Write the Lewis formula for gaseous beryllium chloride, $BeCl_2$, a covalent compound.

Plan

This is an example of limitation type A. So, as we follow the steps in writing the Lewis formula, we must remember to use *four electrons* as the number *needed* by Be in Step 2. Steps 3 and 4 should show only two pairs of electrons for Be.

Solution

Step 1:   The skeleton is Cl   Be   Cl

see limitation type A
↓

Step 2: $N = 2 \times 8$ (for Cl) $+ 1 \times 4$ (for Be) $= 20\ e^-$ needed

$A = 2 \times 7$ (for Cl) $+ 1 \times 2$ (for Be) $= 16\ e^-$ available

$S = N - A = 20\ e^- - 16\ e^- = 4\ e^-$ shared

Step 3:   Cl : Be : Cl

Step 4: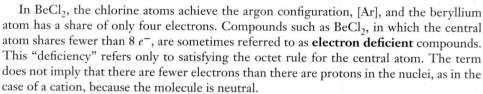

Calculation of formal charges shows that

for Be, FC $= 2 - (2 + 0) = 0$ and   for Cl, FC $= 7 - (1 + 6) = 0$

CD-ROM   Screen 9.9, Electron-Deficient Compounds.

In $BeCl_2$, the chlorine atoms achieve the argon configuration, [Ar], and the beryllium atom has a share of only four electrons. Compounds such as $BeCl_2$, in which the central atom shares fewer than 8 e^-, are sometimes referred to as **electron deficient** compounds. This "deficiency" refers only to satisfying the octet rule for the central atom. The term does not imply that there are fewer electrons than there are protons in the nuclei, as in the case of a cation, because the molecule is neutral.

A Lewis formula can be written for $BeCl_2$ that *does* satisfy the octet rule (see structure shown in the margin). Let us evaluate the formal charges for that formula:

$\overset{..}{\underset{..}{Cl}}{=}Be{=}\overset{..}{\underset{..}{Cl}}:$

for Be, FC $= 2 - (4 + 0) = -2$ and   for Cl, FC $= 7 - (2 + 4) = +1$

We have said that the most favorable structure for a molecule is one in which the formal charge on each atom is zero, if possible. In case some atoms did have nonzero formal charges, we would expect that the more electronegative atoms (Cl) would be the ones with lowest formal charge. Thus, we prefer the Lewis structure shown in Example 7-5 over the one in the margin.

One might expect a similar situation for compounds of the other IIA metals, Mg, Ca, Sr, Ba, and Ra. These elements, however, have *lower ionization energies* and *larger radii* than Be, so they usually form ions by losing two electrons.

Charles D. Winters

BF_3 and BCl_3 are gases at room temperature. Liquid BBr_3 and solid BI_3 are shown here.

EXAMPLE 7-6   *Limitation B of the Octet Rule*

Write the Lewis formula for boron trichloride, BCl_3, a covalent compound.

Plan

This covalent compound of boron is an example of limitation type B. As we follow the steps in writing the Lewis formula, we use *six electrons* as the number *needed* by boron in Step 2. Steps 3 and 4 should show only three pairs of electrons for boron.

Solution

Cl

Step 1:   The skeleton is Cl   B   Cl

see limitation type B
↓

Step 2: $N = 3 \times 8$ (for Cl) $+ 1 \times 6$ (for B) $= 30\ e^-$ needed

$A = 3 \times 7$ (for Cl) $+ 1 \times 3$ (for B) $= 24\ e^-$ available

$S = N - A = 30\ e^- - 24\ e^- = 6\ e^-$ shared

Step 3:
$$
\begin{array}{c}
\text{Cl} \\
\text{Cl} : \text{B} : \text{Cl}
\end{array}
$$

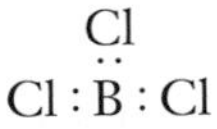

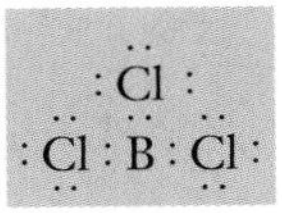

Step 4: $:\ddot{\text{Cl}}:\ddot{\text{B}}:\ddot{\text{Cl}}:$ or $:\ddot{\text{Cl}}\!-\!\text{B}\!-\!\ddot{\text{Cl}}:$ with $:\ddot{\text{Cl}}:$ above B

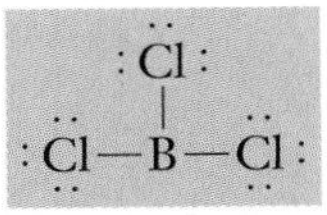

Each chlorine atom achieves the Ne configuration. The boron (central) atom acquires a share of only six valence shell electrons. Calculation of formal charges shows that

for B, FC $= 3 - (3 + 0) = 0$ and   for Cl, FC $= 7 - (1 + 6) = 0$

You should now work Exercise 52.

EXAMPLE 7-7   *Limitation D of the Octet Rule*

Write the Lewis formula for phosphorus pentafluoride, PF_5, a covalent compound.

Plan

We apply the usual stepwise procedure to write the Lewis formula. In PF_5, all five F atoms are bonded to P. This requires the sharing of a minimum of 10 e^-, so this is an example of limitation type D. We therefore add Step 2a, and increase S from the calculated value of 8 e^- to 10 e^-.

Solution

Step 1:   The skeleton is
$$
\begin{array}{ccc}
 & \text{F} & \\
\text{F} & \text{P} & \text{F} \\
\text{F} & & \text{F}
\end{array}
$$

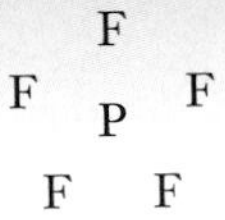

Step 2: $N = 5 \times 8$ (for F) $+ 1 \times 8$ (for P) $= 48\ e^-$ needed

$A = 5 \times 7$ (for F) $+ 1 \times 5$ (for P) $= 40\ e^-$ available

$S = N - A = 8\ e^-$ shared

Five F atoms are bonded to P. This requires the sharing of a minimum of 10 e^-. But only 8 e^- have been calculated. This is therefore an example of limitation type D.

Step 2a: Increase S from 8 e^- to 10 e^-, the number required to bond five F atoms to one P atom. The number of electrons available, 40, does not change.

Step 3:
$$
\begin{array}{ccc}
 & \text{F} & \\
\text{F} : & \text{P} & : \text{F} \\
\text{F} & & \text{F}
\end{array}
$$

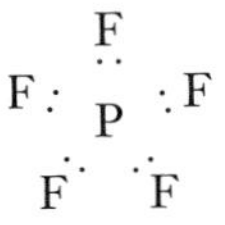

Step 4:   (two Lewis structures shown)   or   (structure with bonds)

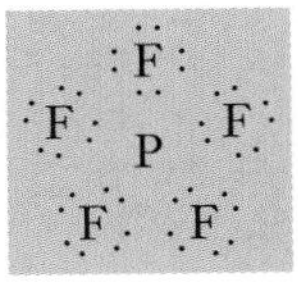

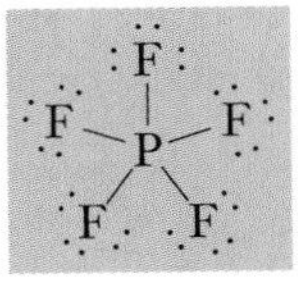

When the octets of the five F atoms have been satisfied, all 40 of the available electrons have been added. The phosphorus (central) atom has a share of 10 electrons.

Calculation of formal charges shows that

$$\text{for P, FC} = 5 - (5 + 0) = 0 \qquad \text{and} \qquad \text{for F, FC} = 7 - (1 + 6) = 0$$

An atom that exhibits an expanded valence shell is said to be hypervalent.

We see that P in PF_5 exhibits an **expanded valence shell.** The electronic basis of the octet rule is that one *s* and three *p* orbitals in the valence shell of an atom can accommodate a maximum of eight electrons. The valence shell of phosphorus has $n = 3$, so it also has *3d* orbitals available that can be involved in bonding. It is for this reason that phosphorus (and many other representative elements of Period 3 and beyond) can exhibit an expanded valence shell. By contrast, elements in the *second row* of the periodic table can *never* exceed eight electrons in their valence shells because each atom has only one *s* and three *p* orbitals in that shell. Thus, we understand why PF_5 can exist but NF_5 cannot.

EXAMPLE 7-8 *Limitation D of the Octet Rule*

Write the Lewis formula for sulfur tetrafluoride, SF_4.

Plan

We apply the usual stepwise procedure. The calculation of $S = N - A$ in Step 2 shows only 6 e^- shared, but a minimum of 8 e^- are required to bond four F atoms to the central S atom. Limitation type D applies, and we proceed accordingly.

Solution

Step 1:   The skeleton is
$$\begin{array}{ccc} F & & F \\ & S & \\ F & & F \end{array}$$

Step 2: $N = 1 \times 8$ (S atom) $+ 4 \times 8$ (F atoms) $= 40\ e^-$ needed

$A = 1 \times 6$ (S atom) $+ 4 \times 7$ (F atoms) $= 34\ e^-$ available

$S = N - A = 40 - 34 = 6\ e^-$ shared. Four F atoms are bonded to the central S. This requires a minimum of 8 e^-, but only 6 e^- have been calculated in Step 2. This is therefore an example of limitation type D.

Step 2a: We increase *S* from 6 e^- to 8 e^-.

Step 3:   F : S : F
          F·   ·F

Step 4:   :F: S :F: We first complete octets on the atoms around the central atom.
          :F: :F:

Step 4a: Now we have satisfied the octet rule, but we have used only 32 of the 34 e^- available. We place the other two electrons on the central S atom.

In SF_4, sulfur has an expanded valence shell.

Calculation of the formal charge shows that

$$\text{for S, FC} = 6 - (4 + 2) = 0$$
$$\text{for F, FC} = 7 - (1 + 6) = 0$$

EXAMPLE 7-9 *Limitation D of the Octet Rule*

Write the Lewis formula for the triiodide ion, I_3^-.

Plan

We apply the usual stepwise procedure. The calculation of $S = N - A$ in Step 2 shows only $2\ e^-$ shared, but a minimum of $4\ e^-$ are required to bond two I atoms to the central I. Limitation type D applies, and we proceed accordingly.

Solution

Step 1: The skeleton is $[\text{I}\quad\text{I}\quad\text{I}]^-$

Step 2: $N = 3 \times 8\ \text{(for I)} = 24\ e^-$ needed

$A = 3 \times 7\ \text{(for I)} + 1\ \text{(for the 1− charge)} = 22\ e^-$ available

$S = N - A = 2\ e^-$ shared. Two I atoms are bonded to the central I. This requires a minimum of $4\ e^-$, but only $2\ e^-$ have been calculated in Step 4. This is therefore an example of limitation type D.

Step 2a: Increase S from $2\ e^-$ to $4\ e^-$.

Step 3: $[\text{I}:\text{I}:\text{I}]^-$

Step 4: $\left[:\overset{..}{\underset{..}{\text{I}}}:\overset{..}{\underset{..}{\text{I}}}:\overset{..}{\underset{..}{\text{I}}}:\right]^-$

Step 4a: Now we have satisfied the octets of all atoms using only 20 of the $22\ e^-$ available. We place the other two electrons on the central I atom.

$$\left[:\overset{..}{\underset{..}{\text{I}}}:\overset{..}{\underset{..}{\text{I}}}:\overset{..}{\underset{..}{\text{I}}}:\right]^- \quad\text{or}\quad \left[:\overset{..}{\underset{..}{\text{I}}}\!-\!\overset{..}{\underset{..}{\text{I}}}\!-\!\overset{..}{\underset{..}{\text{I}}}:\right]^-$$

The central iodine atom in I_3^- has an expanded valence shell.

Calculation of the formal charge shows that

$$\text{for I on ends, FC} = 7 - (1 + 6) = 0$$
$$\text{for I in middle, FC} = 7 - (2 + 6) = -1$$

You should now work Exercise 56.

A Lewis formula for sulfuric acid, H_2SO_4, can be written that follows the octet rule. This Lewis formula and the formal charges are

$$\begin{array}{c}:\overset{..}{\text{O}}:\\ |\\ \text{H}-\overset{..}{\underset{..}{\text{O}}}-\text{S}-\overset{..}{\underset{..}{\text{O}}}-\text{H}\\ |\\ :\underset{..}{\text{O}}:\end{array}$$

for S, FC $= 6 - (4 + 0) = +2$

for H, FC $= 1 - (1 + 0) = 0$

for O (in $-\text{OH}$), FC $= 6 - (2 + 4) = 0$

for other O, FC $= 6 - (1 + 6) = -1$

As we saw in Example 7-8, the availability of *d* orbitals in its valence shell permits sulfur to have an expanded valence shell. A different Lewis formula that does *not* follow the octet rule is

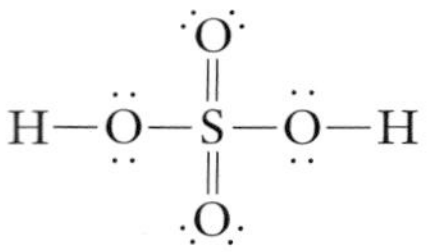

$$
\begin{array}{l}
\ddot{\text{O}} \\
\parallel \\
\ddot{\text{H}}-\ddot{\text{O}}-\text{S}-\ddot{\text{O}}-\text{H} \\
\parallel \\
\ddot{\text{O}}
\end{array}
$$

for S, FC $= 6 - (6 + 0) = 0$
for H, FC $= 1 - (1 + 0) = 0$
for O (in $-$OH), FC $= 6 - (2 + 4) = 0$
for other O, FC $= 6 - (2 + 4) = 0$

Many chemists prefer this formula because it has more favorable formal charges.

We have seen that *atoms attached to the central atom nearly always attain noble gas configurations,* even when the central atom does not.

7-9 POLAR AND NONPOLAR COVALENT BONDS

CD-ROM   Screen 9.13, Bond Polarity and Electronegativity.

Covalent bonds may be either *polar* or *nonpolar.* In a **nonpolar bond** such as that in the hydrogen molecule, H_2 (H : H or H—H), the electron pair is *shared equally* between the two hydrogen nuclei. We defined electronegativity as the tendency of an atom to attract electrons to itself in a chemical bond (see Section 6-6). Both H atoms have the same electronegativity. This means that the shared electrons are equally attracted to both hydrogen nuclei and therefore spend equal amounts of time near each nucleus. In this nonpolar covalent bond, the **electron density** is symmetrical about a plane that is perpendicular to a line between the two nuclei. This is true for all homonuclear *diatomic molecules,* such as H_2, O_2, N_2, F_2, and Cl_2, because the two identical atoms have identical electronegativities. We can generalize:

A **homonuclear** molecule contains only one kind of atom. A molecule that contains two or more kinds of atoms is described as **heteronuclear.**

Remember that ionic compounds are solids at room temperature.

The covalent bonds in all homonuclear diatomic molecules must be nonpolar.

Let us now consider *heteronuclear diatomic molecules.* Start with the fact that hydrogen fluoride, HF, is a gas at room temperature. This tells us that it is a covalent compound. We also know that the H—F bond has some degree of polarity because H and F are not identical atoms and therefore do not attract the electrons equally. But how polar will this bond be?

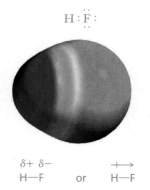

H : F̈ :

$\delta+$ $\delta-$

H—F   or $\xrightarrow{\quad\quad}$ H—F

The electronegativity of hydrogen is 2.1, and that of fluorine is 4.0 (see Table 6-3). Clearly, the F atom, with its higher electronegativity, attracts the shared electron pair much more strongly than does the H atom. We can represent the structure of HF as shown in the margin. The electron density is distorted in the direction of the more electronegative F atom. This small shift of electron density leaves H somewhat positive.

Covalent bonds, such as the one in HF, in which the *electron pairs are shared unequally* are called **polar covalent bonds.** Two kinds of notation used to indicate polar bonds are shown in the margin.

The word "dipole" means "two poles." Here it refers to the positive and negative poles that result from the separation of charge within a molecule.

The $\delta-$ over the F atom indicates a "partial negative charge." This means that the F end of the molecule is somewhat more negative than the H end. The $\delta+$ over the H atom indicates a "partial positive charge," or that the H end of the molecule is positive *with respect to* the F end. We are *not* saying that H has a charge of $+1$ or that F has a charge of -1! A second way to indicate the polarity is to draw an arrow so that the head points toward the negative end (F) of the bond and the crossed tail indicates the positive end (H).

The separation of charge in a polar covalent bond creates an electric **dipole.** We expect the dipoles in the covalent molecules HF, HCl, HBr, and HI to be different because F, Cl, Br, and I have different electronegativities. This tells us that atoms of these elements have different tendencies to attract an electron pair that they share with hydrogen. We indicate this difference as shown here, where Δ(EN) is the difference in electronegativity between two atoms that are bonded together.

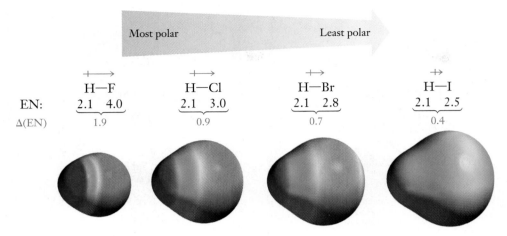

The values of electronegativity are obtained from Table 6-3.

The longest arrow indicates the largest dipole, or greatest separation of electron density in the molecule (see Table 7-3). For comparison, the Δ(EN) values for some typical 1:1 ionic compounds are RbCl, 2.1; NaF, 3.0; and KCl, 2.1

TABLE 7-3 Δ(EN) Values and Dipole Moments for Some Pure (Gaseous) Substances

Substance	Δ(EN)	Dipole Moment (μ)*
HF	1.9	1.91 D
HCl	0.9	1.03 D
HBr	0.7	0.79 D
HI	0.4	0.38 D
H—H	0	0 D

*The magnitude of a dipole moment is given by the product of charge × distance of separation. Molecular dipole moments are usually expressed in debyes (D).

EXAMPLE 7-10   Polar Bonds

Each halogen can form single covalent bonds with other halogens, to form compounds called *interhalogens;* some examples are ClF and BrF. Use the electronegativity values in Table 6-3 to rank the following single bonds from most polar to least polar: F—F, F—Cl, F—Br, Cl—Br, Cl—I, and Cl—Cl.

Plan

The bond polarity decreases as the electronegativity difference between the two atoms decreases. We can calculate Δ(EN) for each bond, and then arrange them according to decreasing Δ(EN) value.

Solution

We know that two F atoms have the same electronegativity, so Δ(EN) for F—F must be zero, and the F—F bond is nonpolar; the same reasoning applies to Cl—Cl. We use the values from Table 6-3 to calculate Δ(EN) for each of the other pairs, always subtracting the smaller from the larger value:

Elements	Δ(EN)
F, Cl	$4.0 - 3.0 = 1.0$
F, Br	$4.0 - 2.8 = 1.2$
Cl, Br	$3.0 - 2.8 = 0.2$
Cl, I	$3.0 - 2.5 = 0.5$

The bonds, arranged from most polar to least polar, are

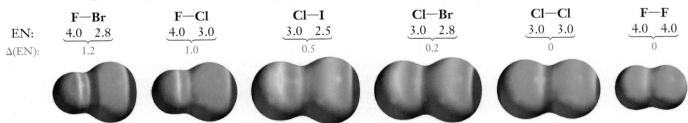

You should now work Exercise 72.

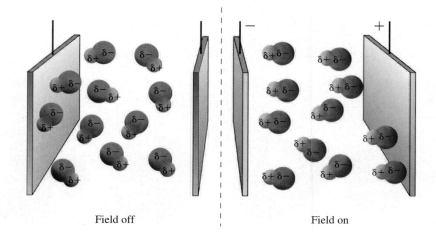

Figure 7-5 If polar molecules, such as HF, are subjected to an electric field, they tend to line up very slightly in a direction opposite to that of the field. This minimizes the electrostatic energy of the molecules. Nonpolar molecules are not oriented by an electric field. The effect is greatly exaggerated in this drawing.

Field off   Field on

$$\mu = d \times q$$

7-10 DIPOLE MOMENTS

It is convenient to express bond polarities on a numerical scale. We indicate the polarity of a molecule by its dipole moment, which measures the separation of charge within the molecule. The **dipole moment**, μ, is defined as the product of the distance, d, separating charges of equal magnitude and opposite sign, and the magnitude of the charge, q. A dipole moment is measured by placing a sample of the substance between two plates and applying a voltage. This causes a small shift in electron density of any molecule, so the applied voltage is diminished very slightly. Diatomic molecules that contain polar bonds, however, such as HF, HCl, and CO, tend to orient themselves in the electric field (Figure 7-5). This causes the measured voltage between the plates to decrease more markedly for these substances, and we say that these molecules are *polar*. Molecules such as F_2 or N_2 do not reorient, so the change in voltage between the plates remains slight; we say that these molecules are *nonpolar*.

Generally, as electronegativity differences increase in diatomic molecules, the measured dipole moments increase. This can be seen clearly from the data for the hydrogen halides (see Table 7-3).

Unfortunately, the dipole moments associated with *individual bonds* can be measured only in simple diatomic molecules. *Entire molecules* rather than selected pairs of atoms must be subjected to measurement. Measured values of dipole moments reflect the *overall* polarities of molecules. For polyatomic molecules they are the result of all the bond dipoles in the molecules. In Chapter 8, we will see that structural features, such as molecular geometry and the presence of lone (unshared) pairs of electrons, also affect the polarity of a molecule.

7-11 THE CONTINUOUS RANGE OF BONDING TYPES

Now let's now clarify our classification of bonding types. The degree of electron sharing or transfer depends on the electronegativity difference between the bonding atoms. Nonpolar covalent bonding (involving *equal sharing* of electron pairs) is one extreme, occurring when the atoms are identical (Δ(EN) is zero). Ionic bonding (involving *complete transfer* of electrons) represents the other extreme, and occurs when two elements with very different electronegativities interact (Δ(EN) is large).

Polar covalent bonds may be thought of as intermediate between pure (nonpolar) covalent bonds and pure ionic bonds. In fact, bond polarity is sometimes described in terms of *partial ionic character*. This usually increases with increasing difference in electronegativity between bonded atoms. Calculations based on the measured dipole moment of gaseous HCl indicate about 17% "ionic character."

When cations and anions interact strongly, some amount of electron sharing takes place; in such cases we can consider the ionic compound as having some *partial covalent character*. For instance, the high charge density of the very small Li^+ ion causes it to distort large anions that it approaches. The distortion attracts electron density from the anion to the region between it and the Li^+ ion, giving lithium compounds a higher degree of covalent character than in other alkali metal compounds.

Almost all bonds have both ionic and covalent character. By experimental means, a given type of bond can usually be identified as being "closer" to one or the other extreme type. We find it useful and convenient to use the labels for the major classes of bonds to describe simple substances, keeping in mind that they represent ranges of behavior.

Above all, we must recognize that any classification of a compound that we might suggest based on electronic properties *must* be consistent with the physical properties of ionic and covalent substances described at the beginning of the chapter. For instance, HCl has a rather large electronegativity difference (0.9), and its aqueous solutions conduct electricity. But we know that we cannot view it as an ionic compound because it is a gas, and not a solid, at room temperature. Liquid HCl is a nonconductor.

Let us point out another aspect of the classification of compounds as ionic or covalent. Not all ions consist of single charged atoms. Many are small groups of atoms that are covalently bonded together, yet they still have excess positive or negative charge. Examples of such *polyatomic ions* are ammonium ion, NH_4^+, sulfate ion, SO_4^{2-}, and nitrate ion, NO_3^-. A compound such as potassium sulfate, K_2SO_4, contains potassium ions, K^+, and sulfate ions, SO_4^{2-}, in a 2:1 ratio. We should recognize that this compound contains both covalent bonding (electron sharing *within* each sulfate ion) and ionic bonding (electrostatic attractions *between* potassium and sulfate ions). We classify this compound as *ionic*, however, because it is a high-melting solid (mp 1069°C), it conducts electricity both in molten form and in aqueous solution, and it displays other properties that we generally associate with ionic compounds. Put another way, covalent bonding holds each sulfate ion together, but the forces that hold the *entire* substance together are ionic.

In summary, we can describe chemical bonding as a continuum that may be represented as

HCl *ionizes* in aqueous solution. We will study more about this behavior in Chapter 10.

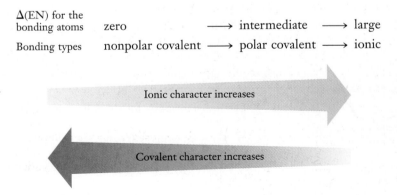

| Δ(EN) for the bonding atoms | zero | $\longrightarrow$ intermediate $\longrightarrow$ large |
| Bonding types | nonpolar covalent $\longrightarrow$ polar covalent $\longrightarrow$ ionic |

Ionic character increases

Covalent character increases

Key Terms

Anion A negatively charged ion; that is, an ion in which the atom or group of atoms has more electrons than protons.

Binary compound A compound consisting of two elements; may be ionic or covalent.

Bonding pair   A pair of electrons involved in a covalent bond. Also called shared pair.

Cation   A positively charged ion; that is, an ion in which the atom or group of atoms has fewer electrons than protons.

Chemical bonds   Attractive forces that hold atoms together in elements and compounds.

Covalent bond   A chemical bond formed by the sharing of one or more electron pairs between two atoms.

Covalent compound   A compound containing predominantly covalent bonds.

Debye   The unit used to express dipole moments.

Delocalization of electrons   Refers to bonding electrons distributed among more than two atoms that are bonded together; occurs in species that exhibit resonance.

Dipole   Refers to the separation of charge between two covalently bonded atoms.

Dipole moment (μ)   The product of the distance separating opposite charges of equal magnitude and the magnitude of the charge; a measure of the polarity of a bond or molecule. A measured dipole moment refers to the dipole moment of an entire molecule.

Double bond   A covalent bond resulting from the sharing of four electrons (two pairs) between two atoms.

Electron-deficient compound   A compound containing at least one atom (other than H) that has fewer than eight valence shell electrons.

Expanded valence shell   Describes an atom that contains more than eight valence shell electrons.

Formal charge   The hypothetical charge on an atom in a covalently bonded molecule or ion; bonding electrons are counted as though they were shared equally between the two bonded atoms.

Heteronuclear   Consisting of different elements.

Homonuclear   Consisting of only one element.

Hypervalent   See *Expanded valence shell.*

Ion   An atom or a group of atoms that carries an electrical charge.

Ionic bonding   The attraction of oppositely charged ions (cations and anions) in large numbers to form a solid. Ions result from the transfer of one or more electrons from one atom or group of atoms to another.

Ionic compound   A compound containing predominantly ionic bonding.

Isoelectronic   Having the same number of electrons.

Lewis acid   A substance that accepts a share in a pair of electrons from another species.

Lewis base   A substance that makes available a share in an electron pair.

Lewis formula   The representation of a molecule, ion, or formula unit by showing atomic symbols and only outer-shell electrons; does not represent the shape of the molecule or ion. Each bonding electron pair can be represented by a pair of dots (the Lewis dot formula) or by a dash (the Lewis dash formula).

Lone pair   A pair of electrons residing on one atom and not shared by other atoms; unshared pair.

Monatomic ion   An ion that consists of only one atom.

Nonpolar bond   A covalent bond between two atoms with the same electronegativity, so that the electron density is symmetrically distributed.

Octet rule   Many representative elements attain at least a share of eight electrons in their valence shells when they form molecular or ionic compounds; there are some limitations.

Polar bond   A covalent bond between two atoms with different electronegativities, so that the electron density is unsymmetrically distributed.

Polyatomic ion   An ion that consists of more than one atom.

Resonance   A concept in which two or more Lewis formulas for the same arrangement of atoms (resonance structures) are used to describe the bonding in a molecule or ion.

Single bond   A covalent bond resulting from the sharing of two electrons (one pair) between two atoms.

Triple bond   A covalent bond resulting from the sharing of six electrons (three pairs) between two atoms.

Unshared pair   See *Lone pair.*

Valence electrons   The s and p electrons in the outermost shell of an atom.

Exercises

Chemical Bonding: Basic Ideas

1. What type of force is responsible for chemical bonding? Explain the differences between ionic bonding and covalent bonding.

2. What kind of bonding (ionic or covalent) would you predict for the products resulting from the following combination of elements? (a) Na + Br_2; (b) C + S; (c) N_2 + O_2; (d) S + O_2.

3. Why are covalent bonds called directional bonds, whereas ionic bonding is termed non-directional?

4. (a) What do Lewis dot formulas for atoms show? (b) Write Lewis dot formulas for the following atoms: He; S; N; Ne; Mg; Br.

5. Write Lewis dot formulas for the following atoms: Li; B; As; K; Ne; Al.

6. Describe the types of bonding in sodium chlorate, $NaClO_3$.

7. Describe the types of bonding in ammonium chloride, NH_4Cl.

$$\left[\begin{array}{c} H \\ | \\ H-N-H \\ | \\ H \end{array} \right]^+ \qquad :\ddot{Cl}:^-$$

8. Based on the positions in the periodic table of the following pairs of elements, predict whether bonding between the two would be primarily ionic or covalent. Justify your answers. (a) Ca and Cl; (b) P and O; (c) Cl and I; (d) Na and I; (e) Si and Br; (f) Ba and F.

9. Predict whether the bonding between the following pairs of elements would be primarily ionic or covalent. Justify your answers. (a) Rb and Cl; (b) P and O; (c) Ca and F; (d) N and S; (e) C and F; (f) K and O.

10. Classify the following compounds as ionic or covalent: (a) $Ca(ClO_3)_2$; (b) H_2Se; (c) KNO_3; (d) $CaCl_2$; (e) H_2CO_3; (f) NCl_3; (g) Li_2O; (h) N_2H_4; (i) $SOCl_2$.

Ionic Bonding

11. Describe what happens to the valence electron(s) as a metal atom and a nonmetal atom combine to form an ionic compound.

*12. Describe an ionic crystal. What factors might determine the geometrical arrangement of the ions?

13. Why are solid ionic compounds rather poor conductors of electricity? Why does conductivity increase when an ionic compound is melted or dissolved in water?

14. Write the formula for the ionic compound that forms between each of the following pairs of elements: (a) Mg and Cl_2; (b) Ba and Cl_2; (c) Na and Cl_2.

15. Write the formula for the ionic compound that forms between each of the following pairs of elements: (a) Cs and Cl_2; (b) Sr and S; (c) Li and Se.

16. When a d-transition metal ionizes, it loses its outer s electrons before it loses any d electrons. Using [noble gas] $(n-1)d^x$ representations, write the outer-electron configurations for the following ions: (a) Cr^{3+}; (b) Mn^{2+}; (c) Ag^+; (d) Fe^{2+}; (e) Cu^{2+}; (f) Sc^{2+}; (g) Fe^{3+}.

17. Which of the following do not accurately represent stable binary ionic compounds? Why? $BaCl_2$; NaS; AlF_4; SrS_2; Ca_2O_3; $NaBr_2$; NCl_4; $LiSe_2$.

18. Which of the following do not accurately represent stable binary ionic compounds? Why? MgI; $Al(OH)_2$; InF_2; CO_3; $RbCl_2$; CsS; Be_3O.

19. (a) Write Lewis formulas for the positive and negative ions in these compounds: $CaBr_2$; Li_2O; Ca_3P_2; $PbCl_2$; Bi_2O_3. (b) Which ions do not have a noble gas configuration?

20. Write formulas for two cations that have each of the following electron configurations in their *highest* occupied energy level: (a) $3s^2 3p^6$; (b) $5s^2 5p^6$.

21. Write formulas for two anions that have each of the electron configurations listed in Exercise 20.

Covalent Bonding: General Concepts

22. What does Figure 7-4 tell us about the attractive and repulsive forces in a hydrogen molecule?

23. Distinguish between heteronuclear and homonuclear diatomic molecules.

24. How many electrons are shared between two atoms in (a) a single covalent bond, (b) a double covalent bond, and (c) a triple covalent bond?

25. What is the maximum number of covalent bonds that a second-period element could form? How can the representative elements beyond the second period form more than this number of covalent bonds?

Lewis Formulas for Molecules and Polyatomic Ions

26. What information about chemical bonding can a Lewis formula give for a compound or ion? What information about bonding is not directly represented by a Lewis formula?

27. Write Lewis formulas for the following: H_2; N_2; Br_2; HCl; HBr.

28. Write Lewis formulas for the following: H_2S; NH_3; OH^-; F^-.

29. Use Lewis formulas to represent the covalent molecules formed by these pairs of elements. Write only structures that satisfy the octet rule. (a) P and H; (b) Se and Br; (c) P and Cl; (d) Si and Cl.

30. Use Lewis formulas to represent the covalent molecules formed by these pairs of elements. Write only structures that satisfy the octet rule. (a) S and Cl; (b) As and F; (c) I and Cl; (d) N and Cl.

31. Find the total number of valence electrons in each of the following molecules or ions. (a) NH_2^-; (b) ClO_4^-; (c) HCN; (d) $SnCl_4$.

32. How many valence electrons does each of these molecules or ions have? (a) H_2Se; (b) PCl_3; (c) NOCl; (d) OH^-

33. Write Lewis structures for the molecules or ions in Exercise 31.

34. Write Lewis structures for the molecules or ions in Exercise 32.

35. Write Lewis structures for the following molecules or ions: (a) ClO_3^-; (b) C_2H_6O (two possibilities); (c) HOCl; (d) SO_3^{2-}.

36. Write Lewis structures for the following molecules or ions: (a) H_2CO; (b) ClF; (c) BF_4^-; (d) PO_4^{3-}; (e) ClO_2^-.

*37. (a) Write the Lewis formula for $AlCl_3$, a molecular compound. Note that in $AlCl_3$, the aluminum atom is an exception to the octet rule. (b) In the gaseous phase, two molecules of $AlCl_3$ join together (dimerize) to form Al_2Cl_6. (The two molecules are joined by two "bridging" Al—Cl—Al bonds.) Write the Lewis formula for this molecule.

*38. Write the Lewis formula for molecular ClO_2. There is a single unshared electron on the chlorine atom in this molecule.

39. Write Lewis formulas for CH_4 and SiH_4; explain the similarity.

40. Write Lewis formulas for CCl_4, SiF_4, and SnI_4; explain the similarity.

41. Write Lewis formulas for the fluorine molecule and for sodium fluoride. Describe the nature of the chemical bonding involved in each substance.

42. Write Lewis formulas for butane, $CH_3CH_2CH_2CH_3$, and propane, $CH_3CH_2CH_3$. Describe the nature of the bond indicated: CH_3CH_2—CH_2CH_3 and CH_3—CH_2CH_3.

43. Write Lewis structures for the following molecules: (a) Formic acid, HCOOH, in which the atomic arrangement is

O

H C O H

(b) Acetonitrile, CH_3CN; (c) Vinyl chloride, CH_2CHCl, the molecule from which PVC plastics are made.

44. Write Lewis structures for the following molecules: (a) Tetrafluoroethylene, C_2F_4, the molecule from which Teflon is made; (b) Acrylonitrile, CH_2CHCN, the molecule from which Orlon is made.

45. What do we mean by the term "resonance"? Do the resonance structures that we draw actually represent the bonding in the substance? Explain your answer.

46. Careful examination of the ozone molecule indicates that the two outer oxygens are the same distance from the central oxygen. Write Lewis formulas or resonance structures that are consistent with this finding.

47. Draw resonance structures for the nitric acid molecule.

*48. We can write two resonance structures for toluene, $C_6H_5CH_3$:

How would you expect the carbon-carbon bond lengths in the six-membered ring to compare with the carbon-carbon bond length between the CH_3 group and the carbon atom on the ring?

49. Write resonance structures for the formate ion, $HCOO^-$.

50. Write resonance structures for each of the following ions. (a) NO_2^-; (b) BrO_3^-; (c) PO_2^{3-}. (*Hint*: Consider both the Br and the P to share more than an octet of electrons in order for them to also form at least one double bond.)

51. Write Lewis formulas for (a) H_2NOH (i.e., one H bonded to O); (b) S_8 (a ring of eight atoms); (c) PCl_3; (d) F_2O_2 (O atoms in center, F atoms on outside); (e) CO; (f) $SeCl_6$.

52. Write the Lewis formula for each of the following covalent compounds. Which ones contain at least one atom with a share in less than an octet of valence electrons? (a) $BeBr_2$; (b) BBr_3; (c) PCl_3; (d) $AlCl_3$.

53. Which of the following species contain at least one atom that violates the octet rule?

54. Write the Lewis formula for each of the following molecules or ions. Which ones contain at least one atom with a share in less than an octet of valence electrons? (a) CBr_2Cl_2; (b) BF_3; (c) BCl_4^-; (d) AlF_4^-.

*55. None of the following is known to exist. What is wrong with each one?

56. Write the Lewis formula for each of the following molecules or ions. Which ones contain at least one atom with a share in more than an octet of valence electrons? (a) AsF_6^-; (b) $AsCl_5$; (c) ICl_3; (d) C_2H_6.

*57. Suppose that "El" is the general symbol for a representative element. In each case, in which periodic group is El located? Justify your answers and cite a specific example for each one.

*58. Suppose that "El" is the general symbol for a representative element. In each case, in which periodic group is El located? Justify your answers and cite a specific example for each one.

(a) $:\!\overset{..}{\underset{..}{Br}}\!-\!\overset{..}{\underset{\underset{\displaystyle :\overset{..}{\underset{..}{Br}}:}{|}}{El}}\!-\!\overset{..}{\underset{..}{Br}}\!:$

(b) $:\!\overset{..}{\underset{..}{O}}\!::\!El\!::\!\overset{..}{\underset{..}{O}}\!:$

(c) $\left[\,\overset{\displaystyle \overset{\textstyle H}{|}}{H\!-\!\underset{\underset{\textstyle H}{|}}{El}\!-\!H}\,\right]^{+}$

(d) $\left[\,H:\!\overset{..}{\underset{\underset{\textstyle H}{|}}{El}}\!:H\,\right]^{+}$

59. Many common stains, such as those of chocolate and other fatty foods, can be removed by dry-cleaning solvents such as tetrachloroethylene, C_2Cl_4. Is C_2Cl_4 ionic or covalent? Write its Lewis formula.

*60. Write acceptable Lewis formulas for the following common air pollutants: (a) SO_2; (b) NO_2; (c) CO; (d) O_3 (ozone); (e) SO_3; (f) $(NH_4)_2SO_4$. Which one is a solid? Which ones exhibit resonance? Which ones violate the octet rule?

Formal Charges

61. Assign a formal charge to each atom in the following:

(a) $:\!\overset{..}{\underset{..}{Cl}}\!-\!\overset{..}{\underset{\underset{\displaystyle :\overset{..}{\underset{..}{Cl}}:}{|}}{O}}\!:$

(b) $:\!\overset{..}{\underset{..}{O}}\!-\!\overset{..}{\underset{..}{S}}\!=\!\overset{..}{\underset{..}{O}}\!:$

(c) $:\!\overset{..}{\underset{..}{O}}\!-\!\overset{\overset{\textstyle :\overset{..}{O}:}{|}}{\underset{\underset{\textstyle :\overset{..}{\underset{..}{O}}:}{|}}{Cl}}\!-\!\overset{..}{\underset{..}{O}}\!-\!\overset{\overset{\textstyle :\overset{..}{O}:}{|}}{\underset{\underset{\textstyle :\overset{..}{\underset{..}{O}}:}{|}}{Cl}}\!-\!\overset{..}{\underset{..}{O}}\!:$

(d) $\left[\,\overset{..}{\underset{..}{O}}\!=\!C\!-\!\overset{..}{\underset{\underset{\textstyle :\overset{..}{\underset{..}{O}}:}{|}}{O}}\!:\,\right]^{2-}$

(e) $\left[\,:\!\overset{..}{\underset{..}{O}}\!-\!\overset{\overset{\textstyle :\overset{..}{O}:}{|}}{\underset{\underset{\textstyle :\overset{..}{\underset{..}{O}}:}{|}}{Cl}}\!-\!\overset{..}{\underset{..}{O}}\!:\,\right]^{-}$

62. Assign a formal charge to each atom in the following:

(a) $:\!\overset{..}{\underset{..}{F}}\!-\!\overset{..}{\underset{\underset{\displaystyle :\overset{..}{\underset{..}{F}}:}{|}}{As}}\!-\!\overset{..}{\underset{..}{F}}\!:$

(b) $\overset{..}{\underset{..}{F}}\!\diagdown\!\overset{\overset{\textstyle \overset{..}{F}\ \ \overset{..}{F}\!\cdot}{}}{\underset{\underset{\textstyle :\overset{..}{\underset{..}{F}}:}{|}}{P}}\!\diagup\!\overset{..}{\underset{..}{F}}$

(c) $\overset{..}{\underset{..}{O}}\!=\!C\!=\!\overset{..}{\underset{..}{O}}\!:$

(d) $\left[\,\overset{..}{\underset{..}{O}}\!=\!N\!=\!\overset{..}{\underset{..}{O}}\!:\,\right]^{+}$

(e) $\left[\,:\!\overset{..}{\underset{..}{Cl}}\!-\!\overset{\overset{\textstyle :\overset{..}{Cl}:}{|}}{\underset{\underset{\textstyle :\overset{..}{\underset{..}{Cl}}:}{|}}{Al}}\!-\!\overset{..}{\underset{..}{Cl}}\!:\,\right]^{-}$

*63. With the aid of formal charges, explain which Lewis formula is more likely to be correct for each given molecule.

(a) For Cl_2O, $\quad :\!\overset{..}{\underset{..}{Cl}}\!-\!\overset{..}{\underset{..}{O}}\!-\!\overset{..}{\underset{..}{Cl}}\!:\quad$ or $\quad :\!\overset{..}{\underset{..}{Cl}}\!-\!\overset{..}{\underset{..}{Cl}}\!-\!\overset{..}{\underset{..}{O}}\!:$

(b) For HN_3,

$H\!-\!\overset{..}{\underset{..}{N}}\!=\!N\!=\!\overset{..}{\underset{..}{N}}\!:\quad$ or $\quad H\!-\!\overset{..}{\underset{..}{N}}\!\equiv\!N\!-\!\overset{..}{\underset{..}{N}}\!:$

(c) For N_2O, $\quad :\!\overset{..}{\underset{..}{N}}\!=\!O\!=\!\overset{..}{\underset{..}{N}}\!:\quad$ or $\quad :\!N\!\equiv\!N\!-\!\overset{..}{\underset{..}{O}}\!:$

64. Write Lewis formulas for six different resonance forms of the sulfate ion, SO_4^{2-}. Indicate all formal charges. Predict which arrangement is likely to be the least stable and justify your selection. (*Hint:* The answer must include resonance structures that have more than an octet of electrons about the sulfur atom.)

Ionic Versus Covalent Character and Bond Polarities

65. Distinguish between polar and nonpolar covalent bonds.

66. Why is an HCl molecule polar but a Cl_2 molecule is nonpolar?

67. How does one predict that the chemical bonding between two elements is likely to be ionic?

68. How does one predict that a covalent bond between two atoms is polar?

69. Ionic compounds generally have a higher melting point than covalent compounds. What is the major difference in the structures of ionic and covalent compounds that explains the difference in melting points?

70. Explain why the electrons in the carbon-chlorine covalent bond tend to move more toward the halogen atom than do the electrons in the carbon-bromine covalent bond.

71. Why do we show only partial charges, and not full charges, on the atoms of a polar molecule?

72. In each pair of bonds, indicate the more polar bond, and use $\delta+$ and $\delta-$ to show the direction of polarity in each bond. (a) C—O and C—N; (b) B—O and P—S; (c) P—H; and P—N; (d) B—H and B—I

73. The molecule below is urea, a compound used in plastics and fertilizers.

$$\overset{\displaystyle \quad\quad :\overset{..}{\underset{..}{O}}:}{\underset{\displaystyle H\quad\quad\quad\quad\quad H}{H\diagdown_{\textstyle N}\diagup^{\textstyle \|}_{\textstyle C}\diagdown_{\textstyle N}\diagup H}}$$

(a) Which bonds in this molecule are polar and which are nonpolar? (b) Which is the most polar bond in the molecule? Which atom is the partial negative end of this bond?

74. (a) Which two of the following pairs of elements are most likely to form ionic bonds? Te and H; C and F; Ba and F; N and F; K and O. (b) Of the remaining three pairs, which one forms the least polar, and which the most polar, covalent bond?

75. (a) List three reasonably nonpolar covalent bonds between dissimilar atoms. (b) List three pairs of elements whose compounds should exhibit extreme ionic character.

76. Classify the bonding between the following pairs of atoms as ionic, polar covalent, or nonpolar covalent. (a) Li and O; (b) Br and I; (c) Na and H; (d) O and O; (e) H and O.

77. For each of the following, tell whether the bonding is primarily ionic or covalent. (a) potassium and iodine in potassium iodide; (b) beryllium and the nitrate ion in beryllium nitrate; (c) the carbon-carbon bond in CH_3CH_3;

(d) carbon and oxygen in carbon monoxide; (e) phosphorus and oxygen in the phosphate ion.

78. Identify the bond in each of the following bonded pairs that is likely to have the greater proportion of "ionic character." (a) Na—Cl or Mg—Cl; (b) Ca—Cl or Fe—Cl; (c) Al—Br or O—Br; (d) Ra—H or C—H.

79. Identify the bond in each of the following bonded pairs that is likely to have the greater proportion of "covalent character." (a) K—Br or Rb—Br; (b) Li—O or H—O; (c) Al—Cl or C—Cl; (d) As—S or O—S.

CONCEPTUAL EXERCISES

80. When asked to give an example of resonance structures, a student drew the following. Why is this example incorrect?

$$\underset{\underset{H}{|}}{\overset{\overset{O}{\|}}{H-C-C-H}} \longleftrightarrow H-\overset{OH}{\underset{}{C}}=C\overset{H}{\underset{H}{<}}$$

81. Why is the following not an example of resonance structures?

$$:\!\ddot{S}\!-\!C\!\equiv\!N\!: \longleftrightarrow :\!\ddot{S}\!-\!N\!\equiv\!C\!:$$

82. Describe the circumstances under which one would expect a bond to exhibit 100% "covalent character" and 0% "ionic character." Give an example of two bonded atoms that would be predicted to exhibit 0% "ionic character."

83. The following properties can be found in a handbook of chemistry:

camphor, $C_{10}H_{16}O$—colorless crystals; specific gravity, 0.990 at 25°C; sublimes, 204°C; insoluble in water; very soluble in alcohol and ether.

praseodymium chloride, $PrCl_3$—blue-green needle crystals; specific gravity, 4.02; melting point, 786°C; boiling point, 1700°C; solubility in cold water, 103.9 g/100 mL H_2O; very soluble in hot water. Would you describe each of these as ionic or covalent? Why?

BUILDING YOUR KNOWLEDGE

84. Look up the properties of NaCl and PCl_3 in a handbook of chemistry. Why do we describe NaCl as an ionic compound and PCl_3 as a covalent compound?

85. (a) How many moles of electrons are transferred when 10.0 g of magnesium react as completely as possible with 10.0 g of fluorine to form MgF_2? (b) How many electrons is this? (c) Look up the charge on the electron in coulombs. What is the total charge, in coulombs, that is transferred?

86. Write the formula for the compound that forms between (a) calcium and nitrogen, (b) aluminum and oxygen, (c) potassium and selenium, and (d) strontium and chlorine. Classify each compound as covalent or ionic.

*87. Write the Lewis formulas for the nitric acid molecule (HNO_3) that are consistent with the following bond length data: 1.405 Å for the bond between the nitrogen atom and the oxygen atom that is attached to the hydrogen atom; 1.206 Å for the bonds between the nitrogen atom and each of the other oxygen atoms.

88. Write the total ionic and net ionic equations for the reaction between each of the following pairs of compounds in aqueous solution. Then give the Lewis formula for each species in these equations. (a) HCN and NaOH; (b) HCl and NaOH; (c) $CaCl_2$ and Na_2CO_3.

89. Sketch a portion of an aqueous solution of NaCl. Show the Lewis formulas of the solute and solvent species. Suggest the relative location of each species with respect to the others.

90. Sketch a portion of an aqueous solution of CH_3COOH. Show the Lewis formulas of the solute and solvent species. Suggest the relative location of each species with respect to the others.

BEYOND THE TEXTBOOK

Go to the textbook website at

http://www.brookscole.com/chemistry/whitten

for additional activities and exercises based on the General Chemistry Interactive CD-ROM, the World Wide Web, and library resources.

InfoTrac College Edition

For additional readings, go to InfoTrac College Edition, your online research library at:

http://infotrac.thomsonlearning.com

Charles Steele

Camphor sublimes readily.

Molecular Structure and Covalent Bonding Theories

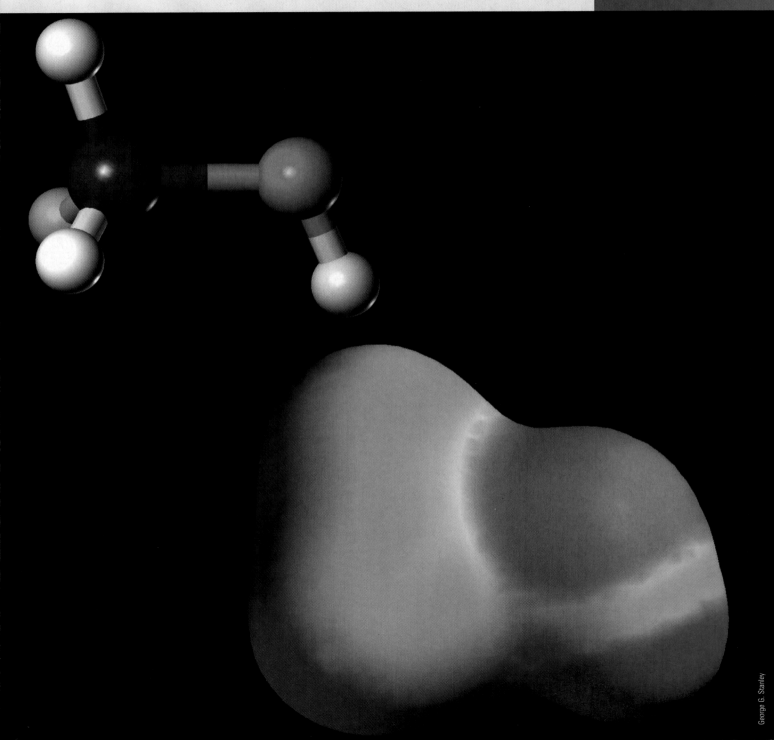

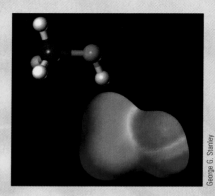

A ball-and-stick model of methanol, CH₃OH (also known as wood alcohol), is shown in the upper corner, while an electrostatic charge potential plot (surface) around the methanol molecule is shown to the lower right. This surface is color-coded showing the relative charges on the atoms that range from red (most negative) through green (neutral) to dark blue (most positive).

George G. Stanley

OUTLINE

OBJECTIVES

After you have studied this chapter, you should be able to

- Describe the basic ideas of the valence shell electron pair repulsion (VSEPR) theory

- Use the VSEPR theory to predict the electronic geometry and the molecular geometry of polyatomic molecules and ions

- Describe the relationships between molecular shapes and polarities

- Predict whether a molecule is polar or nonpolar

- Describe the basic ideas of the valence bond (VB) theory

- Analyze the hybrid orbitals used in bonding in polyatomic molecules and ions

- Use the theory of hybrid orbitals to describe the bonding in double and triple bonds

We know a great deal about the molecular structures of millions of compounds, all based on reliable experiments. In our discussion of theories of covalent bonding, we must keep in mind that the theories represent *an attempt to explain and organize* what we know. For bonding theories to be valid, they must be consistent with the large body of experimental observations about molecular structure. In this chapter we will study two theories of covalent bonding. These theories allow us to predict correct structures and properties. Like any simplified theories, they are not entirely satisfactory in describing *every* known structure; however, their successful application to the vast majority of structures studied justifies their continued use.

8-1 A PREVIEW OF THE CHAPTER

The electrons in the outer shell, or **valence shell,** of an atom are the electrons involved in bonding. In most of our discussion of covalent bonding, we will focus attention on these electrons. Valence shell electrons are those that were not present in the *preceding* noble gas orbitals. Lewis formulas show the number of valence shell electrons in a polyatomic molecule or ion (Sections 7-4 through 7-8). We will write Lewis formulas for each molecule or polyatomic ion we discuss. The theories introduced in this chapter apply equally well to polyatomic molecules and to ions.

Two theories go hand in hand in a discussion of covalent bonding. The *valence shell electron pair repulsion (VSEPR) theory* helps us to predict the spatial arrangement of atoms in a polyatomic molecule or ion. It does not, however, explain *how* bonding occurs, just *where* it occurs and where unshared pairs of valence shell electrons are directed. The *valence bond (VB) theory* describes *how* the bonding takes place, in terms of *overlapping atomic orbitals.* In this theory, the atomic orbitals discussed in Chapter 5 are often "mixed," or *hybridized,* to form new orbitals with different spatial orientations. Used together, these two simple ideas enable us to understand the bonding, molecular shapes, and properties of a wide variety of polyatomic molecules and ions.

We will first discuss the basic ideas and application of these two theories. Then we will learn how an important molecular property, *polarity,* depends on molecular shape. Most of this chapter will then be devoted to studying how these ideas are applied to various types of polyatomic molecules and ions.

In Chapter 7 we used valence bond terminology to discuss the bonding in H_2, although we did not name the theory there.

IMPORTANT NOTE   Different instructors prefer to cover these two theories in different ways. Your instructor will tell you the order in which you should study the material in this chapter. Regardless of how you study this chapter, Tables 8-1, 8-2, 8-3, and 8-4 are important summaries, and you should refer to them often.

1. One approach is to discuss both the VSEPR theory and the VB theory together, emphasizing how they complement each other. If your instructor prefers this parallel approach, you should study the chapter in the order in which it is presented.

2. An alternative approach is to first master the VSEPR theory and the related topic of molecular polarity for different structures, and then learn how the VB theory describes the overlap of bonding orbitals in these structures. If your instructor takes this approach, you should study this chapter in the following order:
   a. Read the summary material under the main heading "Molecular Shapes and Bonding" preceding Section 8-5.
   b. *VSEPR theory, molecular polarity.* Study Sections 8-2 and 8-3; then in Sections 8-5 through 8-12, study only the subsections marked A and B.
   c. *VB theory.* Study Section 8-4; then in Sections 8-5 through 8-12, study the Valence Bond Theory subsections, marked C; then study Sections 8-13 and 8-14.

CD-ROM   You can see and rotate many molecular models on the CD-ROM. Go to any chapter and then select "Molecular Models" from the Tools menu.

No matter which order your instructor prefers, the following procedure will help you analyze the structure and bonding in any compound.

Never skip to Step 4 until you have done Step 3. The electronic geometry and the molecular geometry may or may not be the same; knowing the electronic geometry first will enable you to find the correct molecular geometry.

Many chemists use the terms *lone pair* and *unshared pair* interchangeably, as we will do throughout this discussion.

You should pay attention to these procedures as you study this chapter.

1. Write the Lewis formula for the molecule or polyatomic ion, and identify a *central atom*—an atom that is bonded to more than one other atom (Section 8-2).

2. Count the *number of regions of high electron density* on the central atom (Section 8-2).

3. Apply the VSEPR theory to determine the arrangement of the *regions of high electron density* (the *electronic geometry*) about the central atom (Section 8-2; Tables 8-1 and 8-4).

4. Using the Lewis formula as a guide, determine the arrangement of the *bonded atoms* (the *molecular geometry*) about the central atom, as well as the location of the unshared valence electron pairs on that atom (parts B of Sections 8-5 through 8-12; Tables 8-3 and 8-4). This description includes ideal bond angles.

5. If there are lone (unshared) pairs of valence shell electrons on the central atom, consider how their presence might modify somewhat the *ideal* molecular geometry and bond angles deduced in Step 4 (Section 8-2; of Sections 8-8 through 8-12).

6. Use the VB theory to determine the *hybrid orbitals* utilized by the central atom; describe the overlap of these orbitals to form bonds; describe the orbitals that contain unshared pairs of valence shell electrons on the central atom (Parts C of Sections 8-5 through 8-12; Sections 8-13 and 8-14; Tables 8-2 and 8-4).

7. If more than one atom can be identified as a central atom (as in many organic molecules), repeat Steps 2 through 6 for each central atom, to build up a picture of the geometry and bonding in the entire molecule or ion.

8. When all central atoms in the molecule or ion have been accounted for, use the entire molecular geometry, electronegativity differences, and the presence of lone pairs of valence shell electrons on the central atom to predict *molecular polarity* (Section 8-3; Parts B of Sections 8-5 through 8-12).

The following diagram summarizes this procedure.

Learn this procedure, and use it as a mental "checklist." Trying to do this reasoning in a different order often leads to confusion or wrong answers.

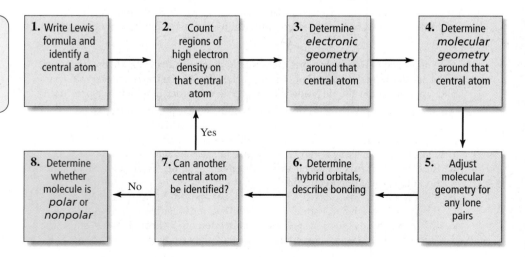

In Section 7-4 we showed that Lewis formulas of polyatomic ions can be constructed in the same way as those of neutral molecules. Once the Lewis formula of an ion is known, we use the VSEPR and VB theories to deduce its electronic geometry, shape, and hybridization, just as for neutral molecules.

Recall that we must take into account the "extra" electrons on anions and the "missing" electrons of cations.

8-2 VALENCE SHELL ELECTRON PAIR REPULSION THEORY

The basic ideas of the **valence shell electron pair repulsion (VSEPR) theory** are:

> Each set of valence shell electrons on a central atom is significant. The sets of valence shell electrons on the *central atom* repel one another. They are arranged about the *central atom* so that repulsions among them are as small as possible.

SC/INKS.

TOPIC: VSEPR Theory
GO TO: www.scilinks.org
*sci*LINKS CODE: WCH0810

This results in maximum separation of the regions of high electron density about the central atom.

A **central atom** is any atom that is bonded to more than one other atom. In some molecules, more than one central atom may be present. In such cases, we determine the arrangement around each in turn, to build up a picture of the overall shape of the entire molecule or ion. We first count the number of **regions of high electron density** around the *central atom*, as follows:

> 1. Each bonded atom is counted as *one* region of high electron density, *whether the bonding is single, double, or triple.*
> 2. Each unshared pair of valence electrons on the central atom is counted as *one* region of high electron density.

The VSEPR theory assumes that regions of high electron density (electron pairs) on the central atom will be as far from one another as possible.

Consider the following molecules and polyatomic ions as examples.

Formula:	CO_2	NH_3	CH_4	NO_3^-
Lewis dot formula:	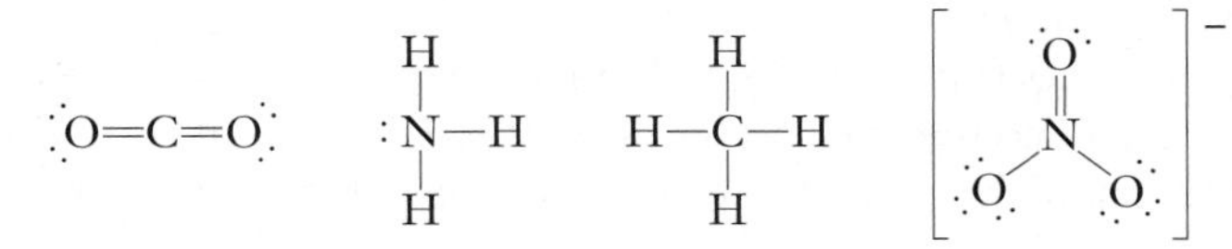 $\ddot{O}=C=\ddot{O}$	H $:N-H$ H	H $H-C-H$ H	$\left[\begin{array}{c} \ddot{O} \\ N \\ \ddot{O} \quad \ddot{O} \end{array}\right]^-$
Central atom:	C	N	C	N
Number of atoms bonded to *central atom:*	2	3	4	3
Number of unshared pairs on *central atom:*	0	1	0	0
Total number of regions of high electron density on *central atom:*	2	4	4	3

Although the terminology is not as precise as we might wish, we use "molecular geometry" to describe the arrangement of atoms in *polyatomic ions* as well as in molecules.

According to VSEPR theory, the molecule or ion is most stable when the regions of high electron density on the central atom are as far apart as possible. The arrangement of these *regions of high electron density* around the central atom is referred to as the **electronic geometry** of the central atom.

For instance, two regions of high electron density are most stable on opposite sides of the central atom (the linear arrangement). Three regions are most stable when they are arranged at the corners of an equilateral triangle. Each different number of regions of high electron density corresponds to a most stable arrangement of those regions. Table 8-1 shows the relationship between the common numbers or regions of high electron density and their corresponding electronic geometries. After we know the electronic geometry (and *only then*), we consider how many of these regions of high electron density connect (bond) the central atom to other atoms. This lets us deduce the arrangement of *atoms* around the central atom, called the **molecular geometry.** If necessary, we repeat this procedure for each central atom in the molecule or ion. These procedures are illustrated in Parts B of Sections 8-5 through 8-12.

8-3 POLAR MOLECULES: THE INFLUENCE OF MOLECULAR GEOMETRY

In Chapter 7 we saw that the unequal sharing of electrons between two atoms with different electronegativities, $\Delta(EN) > 0$, results in a *polar bond*. For heteronuclear diatomic molecules such as HF, this bond polarity results in a *polar molecule*. Then the entire molecule acts as a dipole, and we would find that the molecule has a measurable *dipole moment*, that is, it is greater than zero.

When a molecule consists of more than two atoms joined by polar bonds, we must also take into account the *arrangement* of the resulting bond dipoles in deciding whether or not a molecule is polar. For such a case, we first use the VSEPR theory to deduce the molecular geometry (arrangement of atoms), as described in the preceding section and exemplified in Parts A and B of Sections 8-5 through 8-12. Then we determine whether the bond dipoles are arranged in such a way that they cancel (so that the resulting molecule is *nonpolar*) or do not cancel (so that the resulting molecule is *polar*).

In this section we will discuss the ideas of cancellation of dipoles in general terms, using general atomic symbols A and B. Then we will apply these ideas to specific molecular geometries and molecular polarities in Parts B of Sections 8-5 through 8-12.

Let us consider a heteronuclear triatomic molecule with the formula AB_2 (A is the central atom). Such a molecule must have one of the following two molecular geometries:

The angular form could have different angles, but either the molecule is linear or it is not. The angular arrangement is sometimes called *V-shaped* or *bent*.

$$\underset{\text{linear}}{B—A—B} \qquad \text{or} \qquad \underset{\text{angular}}{B—A\diagdown_B}$$

Suppose that atom B has a higher electronegativity than atom A. Then each A—B bond is polar, with the negative end of the bond dipole pointing toward B. We can view each bond dipole as an *electronic vector*, with a *magnitude* and a *direction*. In the linear AB_2 arrangement, the two bond dipoles are *equal* in magnitude and *opposite* in direction. They therefore cancel to give a nonpolar molecule (dipole moment equal to zero).

TABLE 8-1	Number of Regions of High Electron Density About a Central Atom		
No. Regions of High Electron Density	Electronic Geometry*		
	Description; Angles†	Line Drawing‡	Ball-and-Stick Model
2	linear; 180°	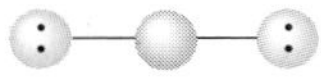	
3	trigonal planar; 120°	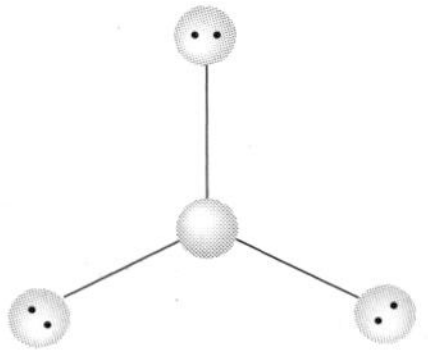	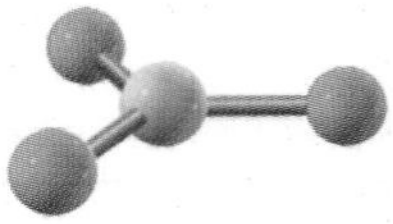
4	tetrahedral; 109.5°	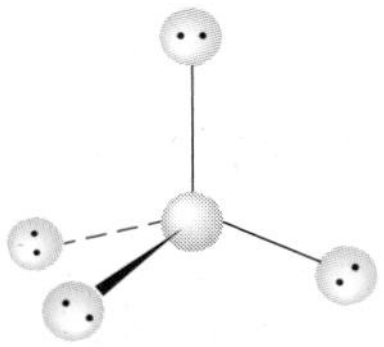	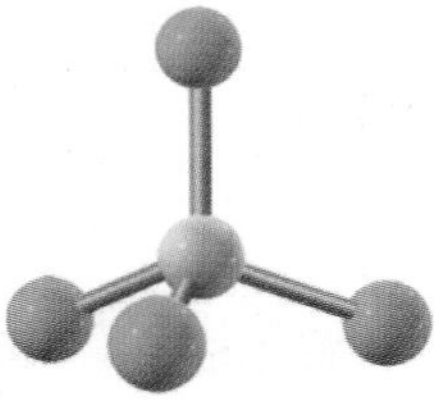
5	trigonal bipyramidal; 90°, 120°, 180°	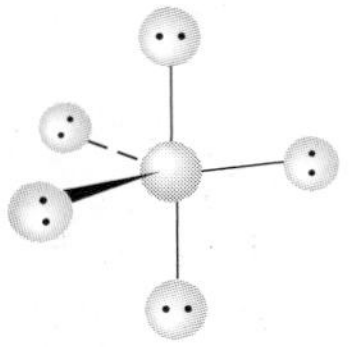	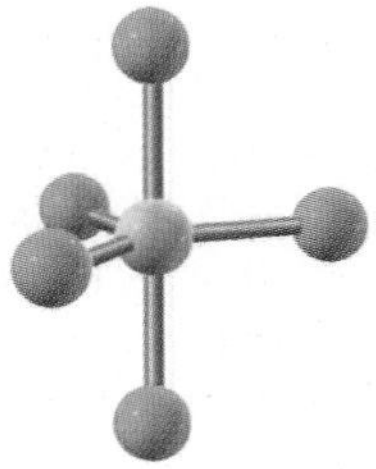
6	octahedral; 90°, 180°	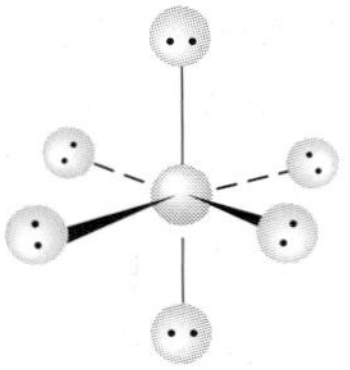	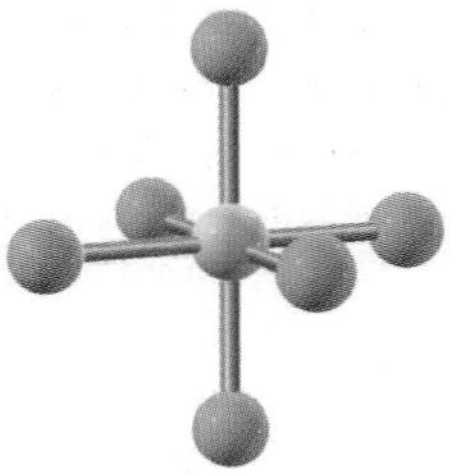

*Electronic geometries are illustrated here using only single pairs of electrons as regions of high electron density. Each orange sphere represents a region of high electron density about the central atom (gray sphere). Each may represent either an atom bonded to the central atom or a lone pair on the central atom.

†Angles made by imaginary lines through the nucleus and the centers of regions of high electron density.

‡By convention, a line in the plane of the drawing is represented by a solid line ——, a line behind this plane is shown as a dashed line -----, and a line in front of this plane is shown as a wedge ➤ with the fat end of the wedge nearest the viewer.

$$\overset{\longleftarrow\;+\;+\longrightarrow}{\text{B—A—B}}$$

net dipole = 0
(nonpolar molecule)

In the case of the angular arrangement, the two equal dipoles *do not cancel*, but add to give a dipole moment greater than zero. The angular molecular arrangement represents a polar molecule.

In this chapter the direction of the net dipole is indicated by a shaded arrow:

net dipole > 0
(polar molecule)

If the electronegativity differences were reversed in this B—A—B molecule—that is, if A were more electronegative than B—the directions of all bond polarities would be reversed. But the bond polarities would still cancel in the linear arrangement to give a nonpolar molecule. In the angular arrangement, bond polarities would still add to give a polar molecule, but with the net dipole pointing in the opposite direction from that described earlier.

We can make similar arguments based on addition of bond dipoles for other arrangements. As we will see in Section 8-8, lone pairs on the central atom can also affect the direction and the magnitude of the net molecular dipole, so the presence of lone pairs on the central atom must always be taken into account.

> For a molecule to be polar, *two* conditions must *both* be met:
>
> 1. There must be at least one polar bond or one lone (unshared) pair on the central atom.
>   *and*
> 2. **a.** The polar bonds, if there are more than one, must not be arranged so that their polarities (bond dipoles) cancel.
>   *or*
>   **b.** If there are two or more lone (unshared) pairs on the central atom, they must not be arranged so that their polarities cancel.

CD-ROM   Screen 9.19, Molecular Polarity.

$$\overset{\longleftarrow\;+\;+\longrightarrow}{\overset{..}{O}=C=\overset{..}{O}}$$

linear molecule;
bond dipoles cancel;
molecule is nonpolar

Put another way, if there are no polar bonds or unshared pairs of electrons on the central atom, the molecule *cannot* be polar. Even if polar bonds or unshared pairs are present, they may be arranged so that their polarities cancel one another, resulting in a nonpolar molecule.

Carbon dioxide, CO_2, is a three-atom molecule in which each carbon–oxygen bond is *polar* because of the electronegativity difference between C and O. But the molecule *as a whole* is shown by experiment (dipole moment measurement) to be nonpolar. This tells us that the polar bonds are arranged in such a way that the bond polarities cancel. Water, H_2O, on the other hand, is a very polar molecule; this tells us that the H—O bond polarities do not cancel one another. Molecular shapes clearly play a crucial role in determining molecular dipole moments. We will develop a better understanding of molecular shapes in order to understand molecular polarities.

The logic used in deducing whether a molecule is polar or nonpolar is outlined in Figure 8-1. The approach described in this section will be applied to various electronic and molecular geometries in Parts B of Sections 8-5 through 8-12.

H—Ö—H

angular molecule;
bond dipoles do not cancel;
molecule is polar

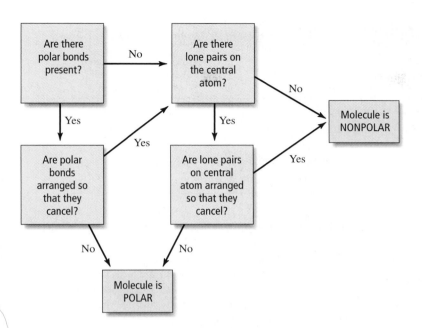

Figure 8-1 A guide to determining whether a polyatomic molecule is polar or nonpolar. Study the more detailed presentation in the text.

8-4 VALENCE BOND THEORY

In Chapter 7 we described covalent bonding as electron pair sharing that results from the overlap of orbitals from two atoms. This is the basic idea of the **valence bond (VB) theory**—it describes *how* bonding occurs. In many examples throughout this chapter, we first use the VSEPR theory to describe the *orientations* of the regions of high electron density. Then we use the VB theory to describe the atomic orbitals that overlap to produce the bonding with that geometry. We also assume that each lone pair occupies a separate orbital. Thus, the two theories work together to give a fuller description of the bonding.

We learned in Chapter 5 that an isolated atom has its electrons arranged in orbitals in the way that leads to the lowest total energy for the atom. Usually, however, these "pure atomic" orbitals do not have the correct energies or orientations to describe where the electrons are when an atom is bonded to other atoms. When other atoms are nearby as in a molecule or ion, an atom can combine its valence shell orbitals to form a new set of orbitals that is at a lower total energy in the presence of the other atoms than the pure atomic orbitals would be. This process is called **hybridization,** and the new orbitals that are formed are called **hybrid orbitals.** These hybrid orbitals can overlap with orbitals on other atoms to share electrons and form bonds. Such hybrid orbitals *usually* give an improved description of the experimentally observed geometry of the molecule or ion.

The designation (label) given to a set of hybridized orbitals reflects the *number and kind* of atomic orbitals that hybridize to produce the new set (Table 8-2). Further details about hybridization and hybrid orbitals appear in the following sections. Throughout the text, hybrid orbitals are shaded in green.

We can describe hybridization as the mathematical combination of the waves that represent the orbitals of the atom. This is analogous to the formation of new waves on the surface of water when different waves interact.

CD-ROM Screens 10.2, Models of Chemical Bonding, and 10.3, Valence Bond Theory.

MOLECULAR SHAPES AND BONDING

We are now ready to study the structures of some simple molecules. We often refer to generalized chemical formulas in which "A" represents the central atom and "B" represents an atom bonded to A. We follow the eight steps of analysis outlined in Section 8-1. We first

TABLE 8-2	*Relation Between Electronic Geometries and Hybridization*		
Regions of High Electron Density	Electronic Geometry	Atomic Orbitals Mixed from Valence Shell of Central Atom	Hybridization
2	linear	one *s*, one *p*	*sp*
3	trigonal planar	one *s*, two *p*'s	sp^2
4	tetrahedral	one *s*, three *p*'s	sp^3
5	trigonal bipyramidal	one *s*, three *p*'s, one *d*	sp^3d
6	octahedral	one *s*, three *p*'s, two *d*'s	sp^3d^2

CD-ROM Screens 10.4, Hybrid Orbitals, and 10.6, Determining Hybrid Orbitals.

See the "Important Note" in Section 8-1; consult your instructor for guidance on the order in which you should study Sections 8-5 through 8-12.

give the known (experimentally determined) facts about the polarity and the shape of the molecule. We then write the Lewis formula (Part A of each section). Then we explain these facts in terms of the VSEPR and VB theories. The simpler VSEPR theory will be used to explain (or predict) first the *electronic geometry* and then the *molecular geometry* in the molecule (Part B). We then show how the molecular polarity of a molecule is a result of bond polarities, lone pairs, and molecular geometry. Finally, we use the VB theory to describe the bonding in molecules in more detail, usually using hybrid orbitals (Part C). As you study each section, refer frequently to the summaries that appear in Table 8-4.

8-5 LINEAR ELECTRONIC GEOMETRY: AB₂ SPECIES (NO LONE PAIRS OF ELECTRONS ON A)

A. Experimental Facts and Lewis Formulas

Several linear molecules consist of a central atom plus two atoms of another element, abbreviated as AB_2. These compounds include $BeCl_2$, $BeBr_2$, and BeI_2 (in their gaseous state), as well as CdX_2 and HgX_2, where X = Cl, Br, or I. All of these are known to be linear (bond angle = 180°), nonpolar, covalent compounds.

Let's focus on *gaseous* $BeCl_2$ molecules (mp 405°C). We wrote the Lewis formula for $BeCl_2$ in Example 7-5. It shows two single covalent bonds, with Be and Cl each contributing one electron to each bond.

In many of its compounds, Be does not satisfy the octet rule (see Section 7-8).

$$\ddot{\text{C}}\text{l} : \text{Be} : \ddot{\text{C}}\text{l} \quad \text{or} \quad : \ddot{\text{C}}\text{l} - \text{Be} - \ddot{\text{C}}\text{l} :$$

The high melting point of $BeCl_2$ is due to its polymeric nature in the solid state.

B. VSEPR Theory

Valence shell electron pair repulsion theory places the two electron pairs on Be 180° apart, that is, with **linear** *electronic geometry*. Both electron pairs are bonding pairs, so VSEPR also predicts a linear atomic arrangement, or *linear molecular geometry*, for $BeCl_2$.

The VSEPR theory describes the locations of bonded atoms around the central atom, as well as where its lone pairs of valence shell electrons are directed.

$$: \ddot{\text{C}}\text{l} \overset{180°}{\underset{\frown}{-\text{Be}-}} \ddot{\text{C}}\text{l} :$$

If we examine the bond dipoles, we see that the electronegativity difference (see Table 6-3) is large (1.5 units) and each bond is quite polar:

A model of $BeCl_2$, a linear AB_2 molecule.

$$\begin{array}{ccc} & \text{Cl} - \text{Be} - \text{Cl} & \longleftarrow + \longrightarrow \\ \text{EN} = & 3.0 \quad 1.5 \quad 3.0 & : \ddot{\text{C}}\text{l} - \text{Be} - \ddot{\text{C}}\text{l} : \\ \Delta(\text{EN}) = & \underbrace{}_{1.5} \underbrace{}_{1.5} & \text{net dipole} = 0 \end{array}$$

The two bond dipoles are *equal* in magnitude and *opposite* in direction. They therefore cancel to give a nonpolar molecule.

The difference in electronegativity between Be and Cl is so large that we might expect ionic bonding. The radius of Be^{2+} is so small (0.59 Å) and its **charge density** (ratio of charge to size) is so high, however, that most simple beryllium compounds are covalent rather than ionic. The high charge density of Be^{2+} causes it to attract and distort the electron cloud of monatomic anions of all but the most electronegative elements. As a result, the bonds in $BeCl_2$ are polar covalent rather than ionic. Two exceptions are BeF_2 and BeO. They are ionic compounds because they contain the two most electronegative elements bonded to Be.

> It is important to distinguish between *nonpolar bonds* and *nonpolar molecules*.

> We say that the Be^{2+} ion *polarizes* the anions, Cl^-.

C. Valence Bond Theory

Consider the ground state electron configuration of Be. There are two electrons in the $1s$ orbital, but these nonvalence (inner) electrons are *not* involved in bonding. Two more electrons are *paired* in the $2s$ orbital. How, then, will a Be atom bond to two Cl atoms? The Be atom must somehow make available one orbital for each bonding Cl electron (the unpaired p electrons). The following *ground state* electron configuration for Be is the configuration for an isolated Be atom. Another configuration may be more stable when the Be atom is covalently bonded. Suppose that the Be atom "promoted" one of the paired $2s$ electrons to one of the $2p$ orbitals, the next higher energy orbitals.

> Cl ground state configuration:
>
> $$[He] \quad \underset{3s}{\uparrow\downarrow} \quad \underset{3p}{\uparrow\downarrow \;\; \uparrow\downarrow \;\; \uparrow}$$

$$\text{Be [He]} \quad \underset{2s}{\uparrow\downarrow} \quad \underset{2p}{\underline{\;\;}\;\underline{\;\;}\;\underline{\;\;}} \quad \xrightarrow{\text{promote}} \quad \text{Be [He]} \quad \underset{2s}{\uparrow} \quad \underset{2p}{\underline{\uparrow}\;\underline{\;\;}\;\underline{\;\;}}$$

Then there would be two Be orbitals available for bonding. This description, however, is still not fully consistent with experimental fact. The Be $2s$ and $2p$ orbitals could not overlap a Cl $3p$ orbital with equal effectiveness; that is, this "promoted pure atomic" arrangement would predict two *nonequivalent* Be—Cl bonds. Yet we observe experimentally that the Be—Cl bonds are *identical* in bond length and strength.

For these two orbitals on Be to become equivalent, they must *hybridize* to give two orbitals that reflect the equal contribution of the s and p orbitals. These are called **sp hybrid orbitals.** Consistent with Hund's Rule, each of these equivalent hybrid orbitals on Be would contain one electron.

> Hund's Rule is discussed in Section 5-17.

$$\text{Be [He]} \quad \underset{2s}{\uparrow\downarrow} \quad \underset{2p}{\underline{\;\;}\;\underline{\;\;}\;\underline{\;\;}} \quad \xrightarrow{\text{hybridize}} \quad \text{Be [He]} \quad \underset{sp}{\underline{\uparrow}\;\underline{\uparrow}} \quad \underset{2p}{\underline{\;\;}\;\underline{\;\;}}$$

The sp hybrid orbitals are described as *linear orbitals*, and we say that Be has *linear electronic geometry*.

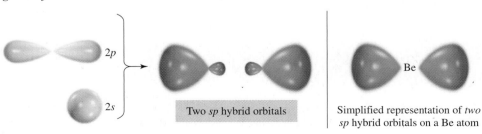

Two *sp* hybrid orbitals   Simplified representation of *two* sp hybrid orbitals on a Be atom

> As we did for pure atomic orbitals, we often draw hybrid orbitals more slender than they actually are. Such drawings are intended to remind us of the orientations and general shapes of orbitals.

We can imagine that there is one electron in each of these hybrid orbitals on the Be atom. Recall that each Cl atom has a half-filled $3p$ orbital that can overlap with a half-filled sp hybrid of Be. We picture the bonding in $BeCl_2$ in the following diagram, in which only the bonding electrons are represented.

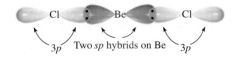

Unshared pairs of electrons on Cl atoms are not shown. The hybrid orbitals on the central atom are shown in green in this and subsequent drawings.

Thus, the Be and two Cl nuclei would lie on a straight line. *This is consistent with the experimental observation that the molecule is linear.*

The structures of beryllium bromide, $BeBr_2$, and beryllium iodide, BeI_2, are similar to that of $BeCl_2$. The chlorides, bromides, and iodides of cadmium, CdX_2, and mercury, HgX_2, are also linear, covalent molecules (where X = Cl, Br, or I).

The two X's within one structure are identical.

> sp Hybridization occurs at the central atom whenever there are two regions of high electron density around the central atom. AB_2 molecules and ions with no lone pairs on the central atom have linear electronic geometry, linear molecular geometry, and sp hybridization on the central atom.

> ✔ **Problem-Solving Tip:** *Number and Kind of Hybrid Orbitals*
>
> One additional idea about hybridization is worth special emphasis:
>
> **The number of hybrid orbitals is always equal to the number of atomic orbitals that hybridize.**
>
> Hybrid orbitals are named by indicating the *number and kind* of atomic orbitals hybridized. Hybridization of *one s* orbital and *one p* orbital gives *two sp hybrid orbitals*. We shall see presently that hybridization of *one s* and *two p* orbitals gives *three sp²* hybrid orbitals; hybridization of *one s* orbital and *three p* orbitals gives *four sp³* hybrids, and so on (see Table 8-2).

Hybridization usually involves orbitals from the same main shell (same n).

8-6 TRIGONAL PLANAR ELECTRONIC GEOMETRY: AB₃ SPECIES (NO LONE PAIRS OF ELECTRONS ON A)

A. Experimental Facts and Lewis Formulas

Boron is a Group IIIA element that forms many covalent compounds by bonding to three other atoms. Typical examples include boron trifluoride, BF_3 (mp $-127°C$); boron trichloride, BCl_3 (mp $-107°C$); boron tribromide, BBr_3 (mp $-46°C$); and boron triiodide, BI_3 (mp $50°C$). All are trigonal planar nonpolar molecules.

A trigonal planar molecule is a flat molecule in which all three bond angles are 120°.

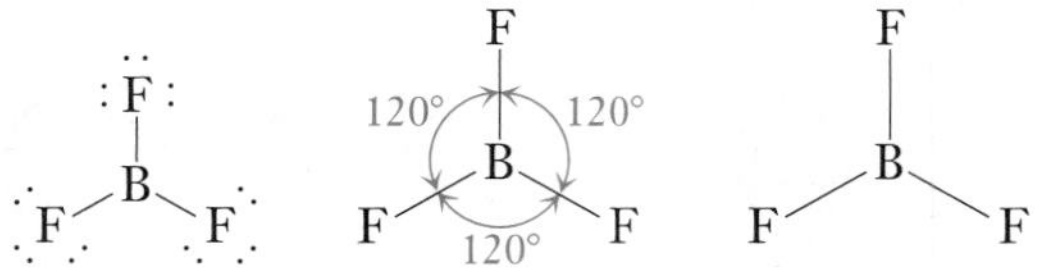

The solid lines represent bonds between B and F atoms.

The Lewis formula for BF_3 is derived from the following: (a) each B atom has three electrons in its valence shell and (b) each B atom is bonded to three F (or Cl, Br, I) atoms. In

Example 7-6 we wrote the Lewis formula for BCl$_3$. Both F and Cl are members of Group VIIA, and so the Lewis formulas for BF$_3$ and BCl$_3$ should be similar.

We see that BF$_3$ and other similar molecules have a central element that does *not* satisfy the octet rule. Boron shares only six electrons.

A model of BF$_3$, a trigonal planar AB$_3$ molecule.

B. VSEPR Theory

Boron, the central atom, has three regions of high electron density (three bonded atoms, no lone pairs on B). The VSEPR theory predicts **trigonal planar** *electronic geometry* for molecules such as BF$_3$ because this structure gives maximum separation among the three regions of high electron density. There are no lone pairs of electrons associated with the boron atom, so a fluorine atom is at each corner of the equilateral triangle, and the *molecular geometry* is also trigonal planar. The maximum separation of any three items (electron pairs) around a fourth item (B atom) is at 120° angles in a single plane. All four atoms are in the same plane. The three F atoms are at the corners of an equilateral triangle, with the B atom in the center. The structures of BCl$_3$, BBr$_3$, and BI$_3$ are similar.

Examination of the bond dipoles of BF$_3$ shows that the electronegativity difference (see Table 6-3) is very large (2.0 units) and that the bonds are very polar.

Trigonal planar geometry is sometimes called *plane triangular* or, simply, *triangular*.

The B^{3+} ion is so small (radius = 0.20 Å) that boron does not form simple ionic compounds.

$$\begin{array}{c} \text{B—F} \\ \text{EN} = \underbrace{2.0 \quad 4.0} \\ \Delta(\text{EN}) = \quad 2.0 \end{array}$$

$$\begin{array}{c} \text{F} \\ \updownarrow \\ \text{F} \overset{\nwarrow}{\underset{\nearrow}{\text{B}}} \text{F} \end{array}$$

net molecular dipole = 0

However, the three bond dipoles are symmetrical, so they cancel to give a nonpolar molecule.

C. Valence Bond Theory

To be consistent with experimental findings and the predictions of the VSEPR theory, the VB theory must explain three *equivalent* B—F bonds. Again we use the idea of hybridization. Now the 2s orbital and two of the 2p orbitals of B hybridize to form a set of three equivalent *sp^2* **hybrid orbitals.**

> Students often forget that the notation for hybridized orbitals indicates both the number of orbitals used in the hybridization and the number of hybrid orbitals obtained. For example, sp^3 indicates four orbitals.

Three *sp^2* hybrid orbitals point toward the corners of an equilateral triangle:

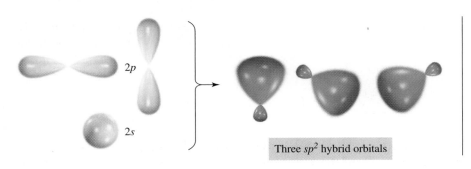

Three *sp^2* hybrid orbitals

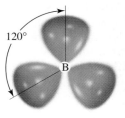

Simplified representation of *three sp^2* hybrid orbitals on a B atom

We can imagine that there is one electron in each of these hybrid orbitals. Each of the three F atoms has a $2p$ orbital with one unpaired electron. The $2p$ orbitals can overlap the three sp^2 hybrid orbitals on B. Three electron pairs are shared among one B and three F atoms:

Lone pairs of electrons are not shown for the F atoms.

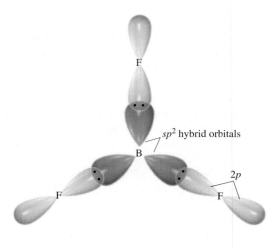

sp^2 Hybridization occurs at the central atom whenever there are three regions of high electron density around the central atom. AB_3 molecules and ions with no unshared pairs on the central atom have trigonal planar electronic geometry, trigonal planar molecular geometry, and sp^2 hybridization on the central atom.

8-7   TETRAHEDRAL ELECTRONIC GEOMETRY: AB_4 SPECIES (NO LONE PAIRS OF ELECTRONS ON A)

A. Experimental Facts and Lewis Formulas

Each Group IVA element has four electrons in its highest occupied energy level. The Group IVA elements form many covalent compounds by sharing those four electrons with four other atoms. Typical examples include CH_4 (mp $-182°C$), CF_4 (mp $-184°C$), CCl_4 (mp $-23°C$), SiH_4 (mp $-185°C$), and SiF_4 (mp $-90°C$). All are tetrahedral, nonpolar molecules (bond angles = 109.5°). In each, the Group IVA atom is located in the center of a regular tetrahedron. The other four atoms are located at the four corners of the tetrahedron.

The Group IVA atom contributes four electrons to the bonding in a tetrahedral AB_4 molecule, and the other four atoms contribute one electron each. The Lewis formulas for methane, CH_4, and carbon tetrafluoride, CF_4, are typical.

The names of many solid figures are based on the numbers of plane faces they have. A *regular* tetrahedron is a three-dimensional figure with four equal-sized equilateral triangular faces (the prefix *tetra*- means "four").

$$\begin{array}{cc} \mathrm{H} & \ddot{\mathrm{F}}\!: \\ | & | \\ \mathrm{H}\!-\!\overset{\displaystyle |}{\underset{\displaystyle |}{\mathrm{C}}}\!-\!\mathrm{H} & :\!\ddot{\mathrm{F}}\!-\!\overset{\displaystyle |}{\underset{\displaystyle |}{\mathrm{C}}}\!-\!\ddot{\mathrm{F}}\!: \\ \mathrm{H} & :\!\ddot{\mathrm{F}}\!: \end{array}$$

CH₄, methane   CF₄, carbon tetrafluoride

The ammonium ion, NH_4^+, is a familiar example of polyatomic ion of this type. In an ammonium ion, the central atom is located at the center of a regular tetrahedron with the other atoms at the corners (H—N—H bond angles = 109.5°).

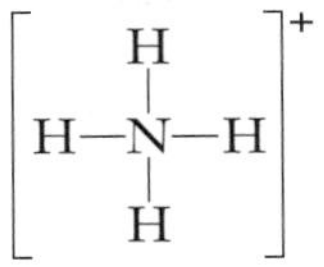

$$
\left[\begin{array}{c} H \\ | \\ H-N-H \\ | \\ H \end{array} \right]^+
$$

NH_4^+, ammonium ion

B. VSEPR Theory

VSEPR theory predicts that four valence shell electron pairs are directed toward the corners of a regular tetrahedron. That shape gives the maximum separation for four electron pairs around one atom. Thus, VSEPR theory predicts **tetrahedral** *electronic geometry* for an AB$_4$ molecule that has no unshared electrons on A. There are no lone pairs of electrons on the central atom, so another atom is at each corner of the tetrahedron. VSEPR theory predicts a *tetrahedral molecular geometry* for each of these molecules.

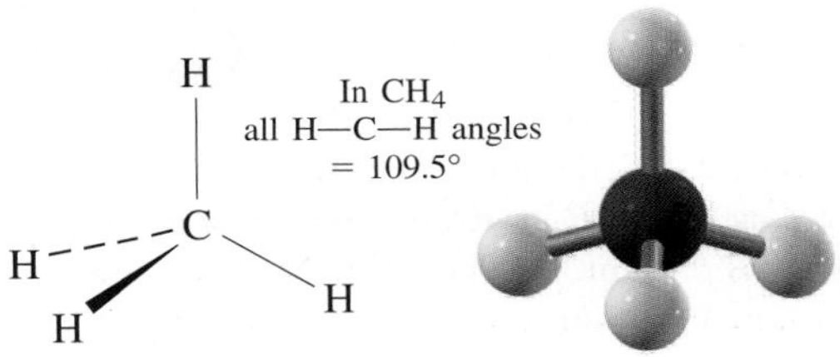

In CH$_4$ all H—C—H angles = 109.5°

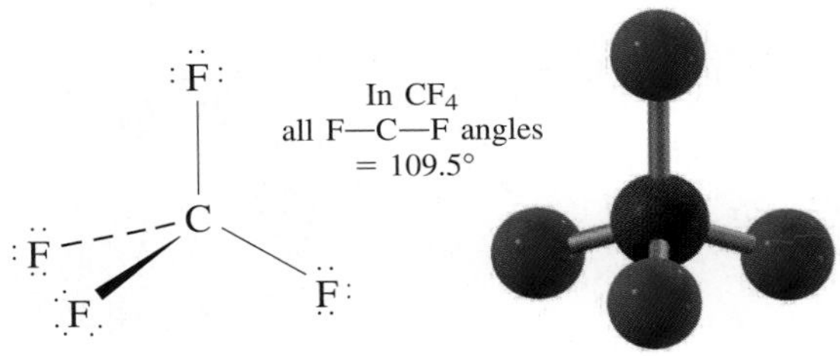

In CF$_4$ all F—C—F angles = 109.5°

The results that we discussed in Sections 8-5 (BeX$_2$, CdX$_2$, and HgX$_2$, where X = Cl, Br, or I), and 8-6 (BX$_3$, where X = F, Cl, Br, or I), and in this section (CH$_4$, CF$_4$, CCl$_4$, SiH$_4$, SiF$_4$, and NH$_4^+$) illustrate an important generalization:

When a molecule or polyatomic ion has no unshared pairs of valence electrons on the central atom, the *electronic geometry* and the *molecular geometry* are the same.

 This is a good generalization to remember.

Examination of bond dipoles shows that in CH$_4$ the individual bonds are only slightly polar, whereas in CF$_4$ the bonds are quite polar. In CH$_4$ the bond dipoles are directed toward carbon, but in CF_4 they are directed away from carbon. Both molecules are very symmetrical, so the bond dipoles cancel, and both molecules are nonpolar. This is true for all AB_4 molecules in which there are *no unshared electron pairs on the central element* and all four B atoms are identical.

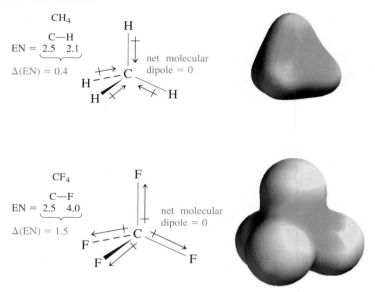

In some tetrahedral molecules, the atoms bonded to the central atom are not all the same. Such molecules are usually polar, with the degree of polarity depending on the relative sizes of the bond dipoles present. In CH_3F or CH_2F_2, for example, the addition of unequal dipoles makes the molecule polar.

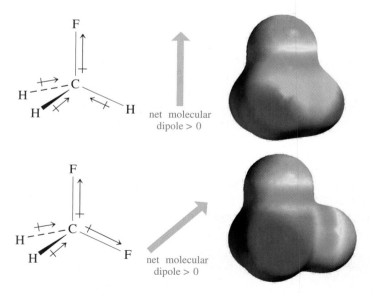

The VSEPR theory also predicts that the NH_4^+ ion has tetrahedral electronic geometry. Each region of high electron density bonds the central atom to another atom (H in NH_4^+) at the corner of the tetrahedral arrangement. We describe the molecular geometry of this ion as tetrahedral.

You may wonder whether square planar AB_4 molecules exist. We will discuss some examples of square planar AB_4 species in Section 8-12. The bond angles in square planar molecules are only 90°. Most AB_4 molecules are tetrahedral, however, with larger bond angles (109.5°) and greater separation of valence electron pairs around A.

C. Valence Bond Theory

According to VB theory, each Group IVA atom (C in our example) must make four equivalent orbitals available for bonding. To do this, C forms four *sp³* **hybrid orbitals** by mixing the *s* and all three *p* orbitals in its outer ($n = 2$) shell. This results in four unpaired electrons.

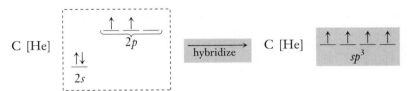

CD-ROM   Screen 10.4, Hybrid Orbitals. Be sure to see the animation of *sp³* orbital formation on that screen.

These *sp³* hybrid orbitals are directed toward the corners of a regular tetrahedron, which has a 109.5° angle from any corner to the center to any other corner.

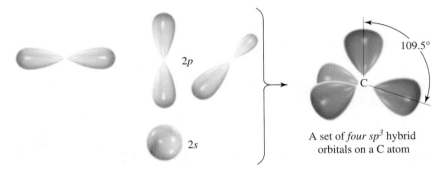

A set of *four sp³* hybrid orbitals on a C atom

Each of the four atoms that bond to C has a half-filled atomic orbital; these can overlap the half-filled *sp³* hybrid orbitals, as is illustrated for CH_4 and CF_4.

CD-ROM   Screen 10.5, Sigma Bonding.

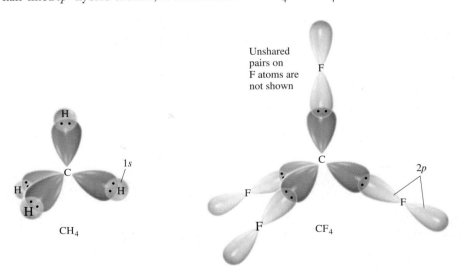

Unshared pairs on F atoms are not shown

CH_4

CF_4

✓ **Problem-Solving Tip:** *Use Periodic Relationships*

Because F, Cl, Br, and I are all in the same group of the periodic table, we know that their compounds will be similar. We expect that the detailed descriptions we have seen for CF_4 will also apply to CCl_4, CBr_4, and CI_4, and we do not need to go through the entire reasoning for each one. Thus, we can say that each of the CX_4 molecules (X = F, Cl, Br, or I) also has tetrahedral electronic geometry, tetrahedral molecular geometry, sp^3 hybridization on the carbon atom, zero dipole moment, and so on.

We can give the same VB description for the hybridization of the central atoms in polyatomic ions. In NH_4^+ the N atoms forms four sp^3 hybrid orbitals directed toward the corners of a regular tetrahedron. Each of these sp^3 hybrid orbitals overlaps with an orbital on a neighboring atom (H in NH_4^+) to form a bond.

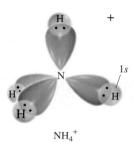

NH_4^+

We see that sp^3 hybridization occurs at the central atom whenever there are four regions of high electron density around the central atom. AB_4 molecules and ions with no unshared pairs on the central atom have tetrahedral electronic geometry, tetrahedral molecular geometry, and sp^3 hybridization on the central atom.

✓ **Problem-Solving Tip:** *Often There Is More Than One Central Atom*

Many molecules contain more than one central atom, that is, there is more than one atom that is bonded to several other atoms. We can analyze such molecules one central atom at a time, to build up a picture of the three-dimensional aspects of the molecule. An example is ethane, C_2H_6; its Lewis formula is

$$
\begin{array}{c}
\text{H}\text{H} \\
| | \\
\text{H}-\text{C}-\text{C}-\text{H} \\
| | \\
\text{H}\text{H}
\end{array}
$$

Let us first consider the left-hand carbon atom. The arrangements of its regions of high electron density allows us to locate the other C and three H atoms with respect to that C atom (the atoms outlined in red). Then we carry out a similar analysis for the right-hand C atom to deduce the arrangements of *its* neighbors (outlined in blue).

Each C atom in C_2H_6 has four regions of high electron density. The VSEPR theory tells us that each C atom has tetrahedral electronic geometry; the resulting atomic arrangement around each C atom has one C and three H atoms at the corners of this tetrahedral arrangement. The VB interpretation is that each C atom is sp^3 hybridized. The C—C bond is formed by overlap of a half-filled sp^3 hybrid orbital of one C atom with a half-filled sp^3 hybrid orbital of the other C atom. Each C—H bond is formed by the overlap of a half-filled sp^3 hybrid orbital on C with the half-filled $1s$ orbital of an H atom.

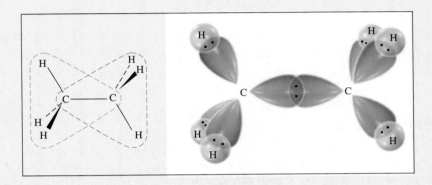

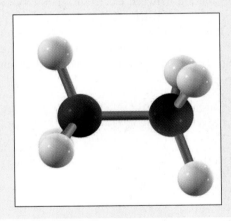

8-8 TETRAHEDRAL ELECTRONIC GEOMETRY: AB₃U SPECIES (ONE LONE PAIR OF ELECTRONS ON A)

We are now ready to study the structures of some simple molecules with unshared valence electron pairs (lone pairs) on the central atom. In this and subsequent sections, we use generalized chemical formulas in which "A" represents the central atom, "B" represents an atom bonded to A, and "U" represents an unshared valence shell electron pair (lone pair) on the central atom A. For instance, AB_3U would represent any molecule with three B atoms bonded to a central atom A, with one unshared valence pair on A.

A. Experimental Facts and Lewis Formulas

Some Group VA elements also form covalent compounds by sharing all five valence electrons (Sections 7-8 and 8-11).

Each Group VA element has five electrons in its valence shell. The Group VA elements form some covalent compounds by sharing three of those electrons with three other atoms. Let us describe two examples: ammonia, NH_3, and nitrogen trifluoride, NF_3. Each is a trigonal pyramidal, polar molecule with an unshared pair on the nitrogen atom. Each has a nitrogen atom at the apex and the other three atoms at the corners of the triangular base of the pyramid.

The Lewis formulas for NH_3 and NF_3 are

$$H-\overset{\cdot\cdot}{\underset{H}{N}}-H \qquad :\overset{\cdot\cdot}{\underset{\cdot\cdot}{F}}-\overset{\cdot\cdot}{\underset{\cdot\cdot}{\underset{:F:}{N}}}-\overset{\cdot\cdot}{F}:$$

$$NH_3 \qquad\qquad NF_3$$

B. VSEPR Theory

As in Section 8-7, VSEPR theory predicts that the *four* regions of high electron density around a central atom will be directed toward the corners of a tetrahedron because this gives maximum separation. So N has tetrahedral electronic geometry in NH_3 and NF_3. We must reemphasize the distinction between electronic geometry and molecular geometry. *Electronic geometry* refers to the geometric arrangement of the *regions of electron density* around the central atom. But the *molecular geometry* excludes the unshared pairs on the central atom, and describes only the arrangement of *atoms* (i.e., nuclei) around the central atom. We can represent the tetrahedral electronic geometry around N in NH_3 or NF_3 as follows.

We then use the Lewis formula as a guide to put the bonded atoms and the lone pairs in these tetrahedral sites around the nitrogen atom.

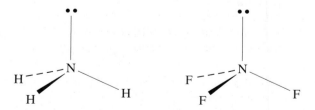

Then we describe the *molecular geometry* as the arrangement of the *atoms*. In each of these molecules, the N atom is at the apex of a (shallow) trigonal pyramidal arrangement and the other three atoms are at the corners of the triangular base of the pyramid. Thus, the molecular geometry of each molecule is described as *trigonal pyramidal.*

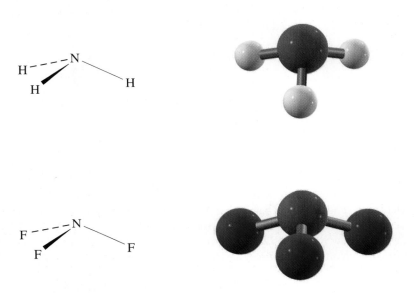

We have seen that CH_4, CF_4, NH_3, and NF_3 all have tetrahedral electronic geometry. But CH_4 and CF_4 (AB_4) have tetrahedral molecular geometry, whereas NH_3 and NF_3 (AB_3U) have trigonal pyramidal molecular geometry.

> In molecules or polyatomic ions that contain lone (unshared) pairs of valence electrons on the central atom, the *electronic geometry* and the *molecular geometry* cannot be the same.

Because this trigonal pyramidal molecular geometry is a fragment of tetrahedral electronic geometry, we expect that the H—N—H angle would be close to the tetrahedral value, 109.5°. In CH_4 (a tetrahedral AB_4 molecule), all H—C—H bond angles are observed to be this ideal value, 109.5°. In NH_3, however, the H—N—H bond angles are observed to be less than this, 107.3°. How can we explain this deviation?

A lone pair is a pair of valence electrons that is associated with only one nucleus in contrast to a bonded pair, which is associated with two nuclei. The known geometries of many molecules and polyatomic ions, based on measurements of bond angles, show that *lone pairs of electrons occupy more space than bonding pairs.* A lone pair has only one atom exerting strong attractive forces on it, so it resides closer to the nucleus than do bonding electrons. The relative magnitudes of the repulsive forces between pairs of electrons on an atom are

$$lp/lp \gg lp/bp > bp/bp$$

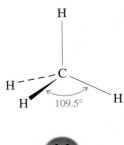

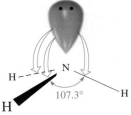

where *lp* refers to lone (unshared) pairs and *bp* refers to bonding pairs of valence shell electrons. We are most concerned with the repulsions among the electrons in the valence shell of the *central atom* of a molecule or polyatomic ion. The angles at which repulsive forces among valence shell electron pairs are minimized are the angles at which the bonding pairs and unshared pairs (and therefore nuclei) are found in covalently bonded molecules and polyatomic ions. Due to *lp/bp* repulsions in NH_3 and NF_3, their bond angles are *less* than the angles of 109.5° we observed in CH_4 and CF_4 molecules.

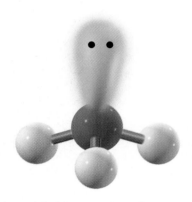

A model of NH_3, a trigonal pyramidal AB_3U molecule, showing the lone pair.

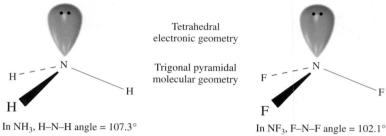

Tetrahedral electronic geometry

Trigonal pyramidal molecular geometry

In NH_3, H–N–H angle = 107.3°   In NF_3, F–N–F angle = 102.1°

The formulas are frequently written as $:NH_3$ and $:NF_3$ to emphasize the unshared pairs of electrons. The unshared pairs must be considered as the polarities of these molecules are examined; they are extremely important in chemical reactions. This is why NH_3 is a base, as we saw in Section 4-2.4 and as we shall discuss more fully in Chapter 10. The contribution of each unshared pair to polarity can be depicted as shown below and discussed in the margin.

The electronegativity differences in NH_3 and NF_3 are nearly equal, *but* the resulting nearly equal bond polarities are in opposite directions.

$$\begin{array}{cc} N-H & N-F \\ EN = 3.0 \quad 2.1 \qquad \overset{\longleftarrow +}{N-H} & EN = 3.0 \quad 4.0 \qquad \overset{+\longrightarrow}{N-F} \\ \Delta(EN) = \quad 0.9 & \Delta(EN) = \quad 1.0 \end{array}$$

Thus, we have

In NH_3 the bond dipoles *reinforce* the effect of the unshared pair, so NH_3 is very polar (μ = 1.47 D). In NF_3 the bond dipoles *oppose* the effect of the unshared pair, so NF_3 is only slightly polar (μ = 0.23 D).

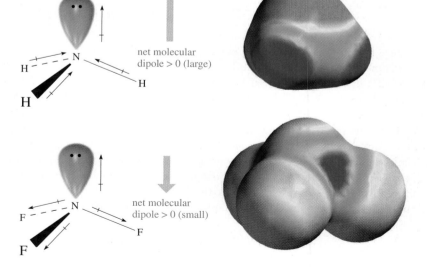

net molecular dipole > 0 (large)

net molecular dipole > 0 (small)

We can now use this information to explain the bond angles observed in NF_3 and NH_3. Because of the direction of the bond dipoles in NH_3, the electron-rich end of each N—H bond is at the central atom, N. On the other hand, the fluorine end of each bond in NF_3 is the electron-rich end. As a result, the lone pair can more closely approach the N in NF_3 than in NH_3. In NF_3 the lone pair therefore exerts greater repulsion toward the bonded pairs than in NH_3. In addition, the longer N—F bond length makes the *bp–bp* distance greater in NF_3 than in NH_3, so that the *bp/bp* repulsion in NF_3 is less than that in NH_3. The net effect is that the bond angles are reduced more in NF_3. We can represent this situation as:

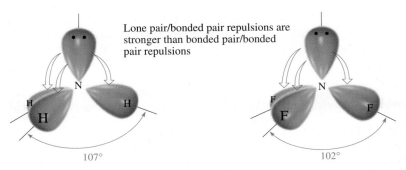

Lone pair/bonded pair repulsions are stronger than bonded pair/bonded pair repulsions

107°   102°

Bonded pair/bonded pair repulsions are weaker in NF_3 than in NH_3 due to the longer N—F bond

We might expect the larger F atoms ($r = 0.72$ Å) to repel each other more strongly than the H atoms ($r = 0.37$ Å), leading to larger bond angles in NF_3 than in NH_3. This is not the case, however, because the N—F bond is longer than the N—H bond. The N—F bond density is farther from the N than the N—H bond density.

C. Valence Bond Theory

Experimental results suggest four nearly equivalent orbitals (three involved in bonding, a fourth to accommodate the lone pair), so we again need four sp^3 hybrid orbitals.

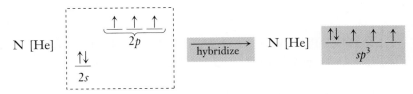

In both NH_3 and NF_3 the lone pair of electrons occupies one of the sp^3 hybrid orbitals. Each of the other three sp^3 orbitals participates in bonding by sharing electrons with another atom. They overlap with half-filled H 1s orbitals and F 2p orbitals in NH_3 and NF_3, respectively.

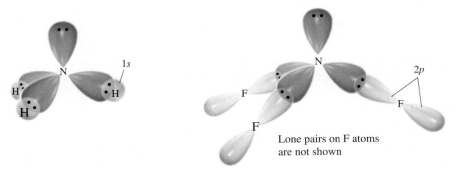

Lone pairs on F atoms are not shown

> AB$_3$U molecules and ions, each having four regions of high electron density around the central atom, *usually* have tetrahedral electronic geometry, trigonal pyramidal molecular geometry, and *sp^3* hybridization on the central atom.

We must remember that *theory* (and its application) depends on fact, not the other way around. Sometimes the experimental facts are not consistent with the existence of hybrid orbitals. In PH$_3$ and AsH$_3$, each H—P—H bond angle is 93.7°, and each H—As—H bond angle is 91.8°. These angles very nearly correspond to three *p* orbitals at 90° to each other. Thus, there appears to be no need to use the VSEPR theory or hybridization to describe the bonding in these molecules. In such cases, we just use the "pure" atomic orbitals rather than hybrid orbitals to describe the bonding.

✓ **Problem-Solving Tip:** *When Do We Not Need To Describe Hybrid Orbitals?*

Students often wonder how to recognize when hybridization does not apply, as in the PH$_3$ and AsH$_3$ cases just described and in H$_2$S (see Section 8-9). Remember that models such as hybridization are our attempts to explain observations such as bond angles. At the level of our studies, we will be given information about the observed molecular geometry, such as measured bond angles. If these are near the angles of pure (unhybridized) orbitals, then hybridization is not needed; if they are near the predicted angles for hybridized orbitals, then hybridization should be used in our explanation. If no information about observed molecular geometry (molecular shape, bond angles, etc.) is supplied, you should assume that the VSEPR and hybridization approaches presented in this chapter should be used.

8-9   TETRAHEDRAL ELECTRONIC GEOMETRY: AB$_2$U$_2$ SPECIES (TWO LONE PAIRS OF ELECTRONS ON A)

A. Experimental Facts and Lewis Formulas

Each Group VIA element has six electrons in its valence shell. The Group VIA elements form many covalent compounds by sharing a pair of electrons with each of two other atoms. Typical examples are H$_2$O, H$_2$S, and Cl$_2$O. The Lewis formulas for these molecules are

$$\text{H}_2\text{O} \qquad \text{H}\!-\!\ddot{\text{O}}\!: \qquad \text{H}_2\text{S} \qquad \text{H}\!-\!\ddot{\text{S}}\!: \qquad \text{Cl}_2\text{O} \qquad :\!\ddot{\text{Cl}}\!-\!\ddot{\text{O}}\!:$$
$$\qquad\quad\; \overset{|}{\text{H}} \qquad\qquad\qquad\quad \overset{|}{\text{H}} \qquad\qquad\qquad\quad\;\; \overset{|}{\underset{\cdot\cdot}{:\text{Cl}}}\!:$$

All are angular, polar molecules. The bond angle in water, for example, is 104.5°, and the molecule is very polar with a dipole moment of 1.85 D.

B. VSEPR Theory

The VSEPR theory predicts that the four electron pairs around the oxygen atom in H$_2$O should be 109.5° apart in a tetrahedral arrangement. The observed H—O—H bond angle is 104.5°. The two lone (unshared) pairs strongly repel each other and the bonding pairs of electrons. These repulsions force the bonding pairs closer together and result in the decreased bond angle. The decrease in the H—O—H bond angle (from 109.5° to 104.5°) is greater than the corresponding decrease in the H—N—H bond angles in ammonia (from 109.5° to 107.3°) because of the *lp/lp* repulsion in H_2O.

The electronegativity difference is large (1.4 units), and so the bonds are quite polar. Additionally, the bond dipoles *reinforce* the effect of the two unshared pairs, so the H_2O molecule is very polar. Its dipole moment is 1.8 D. Water has unusual properties, which can be explained in large part by its high polarity.

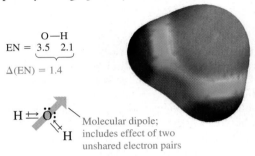

$$EN = \underbrace{\begin{matrix} O—H \\ 3.5 \quad 2.1 \end{matrix}}$$

$$\Delta(EN) = 1.4$$

H⇌Ö:
H

Molecular dipole; includes effect of two unshared electron pairs

C. Valence Bond Theory

The bond angle in H_2O (104.5°) is closer to the tetrahedral value (109.5°) than to the 90° angle that would result from bonding by pure $2p$ atomic orbitals on O. Valence bond theory therefore postulates four sp^3 hybrid orbitals centered on the O atom: two to participate in bonding and two to hold the two unshared pairs.

AB_2U_2 molecules and ions, each having four regions of high electron density around the central atom, *usually* have tetrahedral electronic geometry, angular molecular geometry, and sp^3 hybridization on the central atom.

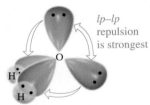

lp–lp repulsion is strongest

There are now *two* lone pairs that repel the bonded pairs

Hydrogen sulfide, H_2S, is also an angular molecule, but the H—S—H bond angle is 92.2°. This is very close to the 90° angles between two unhybridized $3p$ orbitals of S. We therefore *do not* propose hybrid orbitals to describe the bonding in H_2S. The two H atoms are able to exist at approximately right angles to each other when they are bonded to the larger S atom. The bond angles in H_2Se and H_2Te are 91° and 89.5°, respectively.

Sulfur is located directly below oxygen in Group VIA.

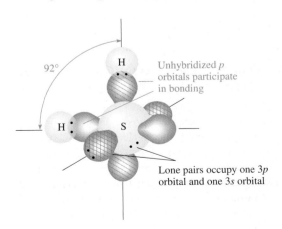

92°

H

Unhybridized *p* orbitals participate in bonding

H

S

Lone pairs occupy one $3p$ orbital and one $3s$ orbital

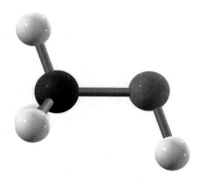

> ✓ **Problem-Solving Tip:** *Some Molecules Have Two Central Atoms and Lone Pairs*
>
> Methanol, CH_3OH, is a simple molecule that has more than one central atom *and* two lone pairs. It is the simplest member of a family of organic molecules called *alcohols*; all alcohols contain the atom grouping C—O—H. The Lewis formula for methanol is
>
> $$\begin{array}{c} H \\ | \\ H-C-\ddot{O}-H \\ | \\ H \end{array}$$
>
> Again, we consider the arrangements around two central atoms in sequence. The carbon atom (outlined here in red) has four regions of high electron density. VSEPR theory tells us that this atom has tetrahedral *electronic* geometry; the resulting atomic arrangement around the C atom has four atoms (O and three H) at the corners of this tetrahedral arrangement, so the molecular geometry about the C atom is tetrahedral. The oxygen atom (outlined in blue) has four regions of high electron density, so it, too, has tetrahedral *electronic* geometry. Thus, the C—O—H arrangement is angular, and there are two unshared pairs on O. The VB interpretation is that both atoms are sp^3 hybridized. The C—O bond is formed by overlap of a half-filled sp^3 hybrid orbital on the C atom with a half-filled sp^3 hybrid orbital on the O atom. Each covalent bond to an H is formed by the overlap of a half-filled sp^3 hybrid orbital with the half-filled $1s$ orbital on an H atom. Each unshared pair of electrons on O is in an sp^3 hybrid orbital.

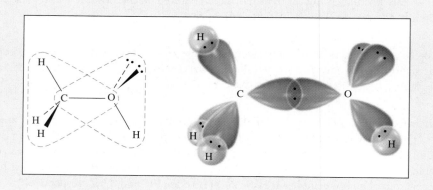

8-10 TETRAHEDRAL ELECTRONIC GEOMETRY: ABU_3 SPECIES (THREE LONE PAIRS OF ELECTRONS ON A)

We represent the halogen as X.

$$HX, \; H:\ddot{\underset{\cdot\cdot}{X}}: \qquad X_2, \; :\ddot{\underset{\cdot\cdot}{X}}:\ddot{\underset{\cdot\cdot}{X}}:$$

In latter case, either halogen may be considered the A atom of AB.

Each Group VIIA element has seven electrons in its highest occupied energy level. The Group VIIA elements form molecules such as H—F, H—Cl, Cl—Cl, and I—I by sharing one of their electrons with another atom. The other atom contributes one electron to the bonding. Lewis dot formulas for these molecules are shown in the margin. Any diatomic molecule must be linear. Neither VSEPR theory nor VB theory adds anything to what we already know about the molecular geometry of such molecules.

8-11 TRIGONAL BIPYRAMIDAL ELECTRONIC GEOMETRY: AB_5, AB_4U, AB_3U_2, AND AB_2U_3

A. Experimental Facts and Lewis Formulas

In Section 8-8 we saw that the Group VA elements have five electrons in their outermost occupied shells and form some molecules by sharing only three of these electrons with other atoms (e.g., NH_3, NF_3, and PCl_3). Group VA elements (P, As, and Sb) beyond the second period also form some covalent compounds by sharing all five of their valence electrons with five other atoms (see Section 7-8). Phosphorus pentafluoride, PF_5 (mp $-83°C$), is such a compound. Each P atom has five valence electrons to share with five F atoms. The Lewis formula for PF_5 (see Example 7-7) is shown in the margin. PF_5 molecules are *trigonal bipyramidal* nonpolar molecules. A **trigonal bipyramid** is a six-sided polyhedron consisting of two pyramids joined at a common triangular (trigonal) base.

B. VSEPR Theory

The VSEPR theory predicts that the five regions of high electron density around the phosphorus atom in PF_5 should be as far apart as possible. Maximum separation of five items around a sixth item is achieved when the five items (bonding pairs) are placed at the corners and the sixth item (P atom) is placed in the center of a trigonal (or triangular) bipyramid. This is in agreement with experimental observation.

A trigonal bipyramid.

The three F atoms marked *e* are at the corners of the common base, in the same plane as the P atom. These are called *equatorial* F atoms (*e*). The other two F atoms, one above and one below the plane, are called *axial* F atoms (*a*). The F—P—F bond angles are 90° (axial to equatorial), 120° (equatorial to equatorial), and 180° (axial to axial).

The large electronegativity difference between P and F (1.9) suggests very polar bonds. Let's consider the bond dipoles in two groups, because there are two different kinds of P—F bonds in PF_5 molecules, axial and equatorial.

$$
\begin{array}{c}
\text{P—F} \\
EN = \underbrace{2.1 \quad 4.0} \\
\Delta(EN) = \quad 1.9
\end{array}
$$

axial bonds   equatorial bonds

As an exercise in geometry, in how many different ways can five fluorine atoms be arranged *symmetrically* around a phosphorus atom? Compare the hypothetical bond angles in such arrangements with those in a trigonal bipyramidal arrangement.

The two axial bonds are in a linear arrangement, like the two bonds in $BeCl_2$ (see Section 8-5); the three equatorial bonds are in a trigonal planar arrangement, like the three bonds in BF_3 (see Section 8-6).

The two axial bond dipoles cancel each other, and the three equatorial bond dipoles cancel, so PF_5 molecules are nonpolar.

C. Valence Bond Theory

Because phosphorus is the central atom in a PF_5 molecule, it must have available five half-filled orbitals to form bonds with five F atoms. Hybridization involves one d orbital from the vacant set of $3d$ orbitals along with the $3s$ and $3p$ orbitals of the P atom.

The five sp^3d **hybrid orbitals** on P point toward the corners of a trigonal bipyramid. Each is overlapped by a singly occupied $2p$ orbital of an F atom. The resulting pairing of P and F electrons forms five covalent bonds.

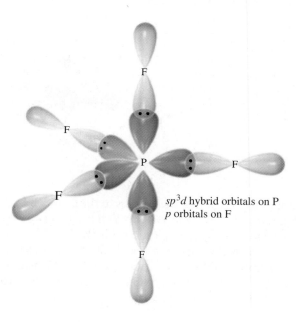

sp^3d hybrid orbitals on P
p orbitals on F

sp^3d Hybridization occurs at the central atom whenever there are five regions of high electron density around the central atom. AB_5 molecules and ions with no unshared pairs on the central atom have trigonal bipyramidal electronic geometry, trigonal bipyramidal molecular geometry, and sp^3d hybridization on the central atom.

We see that sp^3d hybridization uses an available d orbital in the outermost occupied shell of the central atom. The heavier Group VA elements—P, As, and Sb—can form five covalent bonds using this hybridization. But nitrogen, also in Group VA, cannot form five covalent bonds because the valence shell of N has only one s and three p orbitals (and no d orbitals). The set of s and p orbitals in a given energy level (and therefore any set of hybrids composed only of s and p orbitals) can accommodate a *maximum* of eight electrons and participate in a *maximum* of four covalent bonds. The same is true of all elements of the second period because they have only s and p orbitals in their valence shells. No atoms in the first and second periods can exhibit expanded valence.

> The P atom is said to have an *expanded valence shell* (see Section 7-8).

D. Unshared Valence Electron Pairs in Trigonal Bipyramidal Electronic Geometry

As we saw in Sections 8-8 and 8-9, lone pairs of electrons occupy more space than bonding pairs, resulting in increased repulsions from lone pairs. What happens when one or more of the five regions of high electron density on the central atom are lone pairs? Let us first consider a molecule such as SF_4, for which the Lewis formula is

> Reminder: The relative magnitudes of repulsive forces are:
>
> $$lp/lp \gg lp/bp > bp/bp$$

$$\ddot{F} \diagdown \quad \ddot{F}$$
$$S$$
$$\ddot{F} \quad \diagup \ddot{F}$$

The central atom, S, is bonded to four atoms and has one lone pair. This is an example of the general formula AB_4U. Sulfur has five regions of high electron density, so we know that the electronic geometry is trigonal bipyramidal and that the bonding orbitals are sp^3d hybrids. But now a new question arises: Is the lone (unshared) pair more stable in an axial (a) or in an equatorial (e) position? If it were in an axial position, it would be 90° from

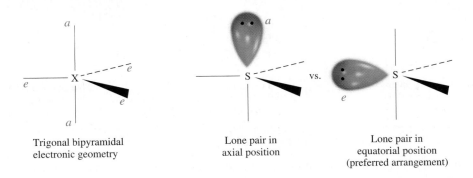

Trigonal bipyramidal electronic geometry

Lone pair in axial position

Lone pair in equatorial position (preferred arrangement)

the *three* closest other pairs (the pairs bonding three F atoms in equatorial positions) and 180° from the other axial pair. If it were in an equatorial position, only the *two* axial pairs would be at 90°, and the other two equatorial pairs would be less crowded at 120° apart. The lone pair would be less crowded in an *equatorial* position. The four F atoms then occupy the remaining four positions. We describe the resulting arrangement of *atoms* as a **seesaw arrangement.**

Imagine rotating the arrangement so that the line joining the two axial positions is the board on which the two seesaw riders sit, and the two bonded equatorial positions are the pivot of the seesaw.

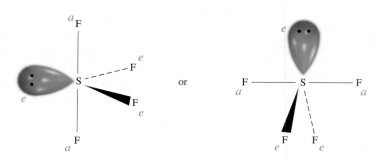

As we saw in Sections 8-8 and 8-9, the differing magnitudes of repulsions involving lone pairs and bonding pairs often result in observed bond angles that are slightly different from idealized values. For instance, *lp/bp* repulsion in the seesaw molecule SF_4 causes distortion of the axial S—F bonds away from the lone pair, to an angle of 177°; the two equatorial S—F bonds, ideally at 120°, move much closer together to an angle of 101.6°.

By the same reasoning, we understand why additional lone pairs also take equatorial positions (AB_3U_2 with both lone pairs equatorial or AB_2U_3 with all three lone pairs equatorial). These arrangements are summarized in Figure 8-2.

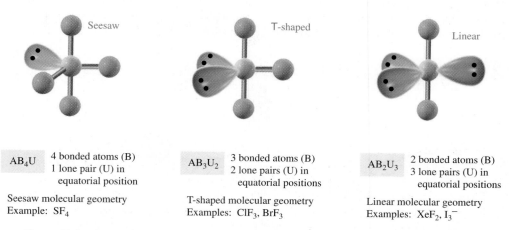

AB_4U	4 bonded atoms (B) 1 lone pair (U) in equatorial position

Seesaw molecular geometry
Example: SF_4

AB_3U_2	3 bonded atoms (B) 2 lone pairs (U) in equatorial positions

T-shaped molecular geometry
Examples: ClF_3, BrF_3

AB_2U_3	2 bonded atoms (B) 3 lone pairs (U) in equatorial positions

Linear molecular geometry
Examples: XeF_2, I_3^-

Figure 8-2 Arrangements of bonded atoms and lone pairs (five regions of high electron density—trigonal pyramidal electronic geometry).

8-12 OCTAHEDRAL ELECTRONIC GEOMETRY: AB$_6$, AB$_5$U, AND AB$_4$U$_2$

A. Experimental Facts and Lewis Formulas

The Group VIA elements below oxygen form some covalent compounds of the AB$_6$ type by sharing their six valence electrons with six other atoms. Sulfur hexafluoride, SF$_6$ (mp −51°C), an unreactive gas, is an example. Sulfur hexafluoride molecules are nonpolar octahedral molecules. The hexafluorophosphate ion, PF$_6^-$, is an example of a polyatomic ion of the type AB$_6$.

B. VSEPR Theory

In an SF$_6$ molecule we have six valence shell electron pairs and six F atoms surrounding one S atom. Because the valence shell of sulfur contains no lone pairs, the electronic and molecular geometries in SF$_6$ are identical. The maximum separation possible for six electron pairs around one S atom is achieved when the electron pairs are at the corners and the S atom is at the center of a regular octahedron. Thus, VSEPR theory is consistent with the observation that SF$_6$ molecules are **octahedral.**

In a regular octahedron, each of the eight faces is an equilateral triangle.

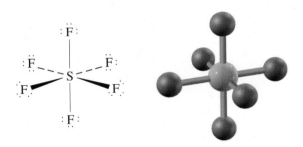

In this octahedral molecule the F—S—F bond angles are 90° and 180°. Each S—F bond is quite polar, but each bond dipole is canceled by an equal dipole at 180° from it. So the large bond dipoles cancel and the SF$_6$ molecule is nonpolar.

By similar reasoning, VSEPR theory predicts octahedral electronic geometry and octahedral molecular geometry for the PF$_6^-$ ion, which has six valence shell electron pairs and six F atoms surrounding one P atom.

C. Valence Bond Theory

Sulfur atoms can use one 3s, three 3p, and two 3d orbitals to form six hybrid orbitals that accommodate six electron pairs:

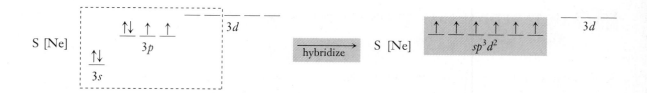

The six sp^3d^2 **hybrid orbitals** are directed toward the corners of a regular octahedron. Each sp^3d^2 hybrid orbital is overlapped by a half-filled $2p$ orbital from fluorine to form a total of six covalent bonds.

Se and Te, in the same group, form analogous compounds. O cannot do so, for the same reasons as discussed earlier for N (see Section 8-11).

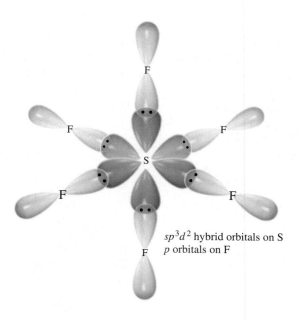

sp^3d^2 hybrid orbitals on S
p orbitals on F

An analogous picture could be drawn for the PF_6^- ion.

> sp^3d^2 Hybridization occurs at the central atom whenever there are six regions of high electron density around the central atom. AB_6 molecules and ions with no lone pairs on the central atom have octahedral electronic geometry, octahedral molecular geometry, and sp^3d^2 hybridization on the central atom.

D. Unshared Valence Electron Pairs in Octahedral Electronic Geometry

We can reason along the lines used in Part D of Section 8-11 to predict the preferred locations of lone pairs on the central atom in octahedral electronic geometry. Because of the high symmetry of the octahedral arrangement, all six positions are equivalent, so it does not matter in which position in the drawing we put the first lone pair. AB_5U molecules and ions are described as having **square pyramidal** molecular geometry. When a second lone pair is present, the most stable arrangement has the two lone pairs in two octahedral positions at 180° angles from each other. This leads to a **square planar** molecular geometry for AB_4U_2 species. These arrangements are shown in Figure 8-3. Table 8-3 summarizes a great deal of information—study this table carefully.

Two lone pairs at 90° from each other would be much more crowded.

✔ **Problem-Solving Tip:** *Placing Lone Pairs on the Central Atom*

Remember that lone pairs occupy more space than bonded pairs, so the lone pairs are always put in positions where they will be least crowded.

If the Lewis formula for a molecule or ion shows only one lone pair: In linear, trigonal planar, tetrahedral, or octahedral electronic geometry, all positions are equivalent, so it doesn't matter where we place the lone pair. In trigonal bipyramidal electronic geometry, place the lone pair in the equatorial position where it is least crowded, and put the bonded atoms in the other positions.

If the Lewis formula shows two lone pairs: In trigonal planar or tetrahedral electronic geometry, we can place the lone pairs in any two positions and the bonded atoms in the other position(s). In trigonal bipyramidal electronic geometry, place the two lone pairs in two equatorial positions (120° apart) where they are least crowded, and put the bonded atoms in the other positions. In octahedral electronic geometry, place the two lone pairs in two positions *across* (180°) from each other, and put the bonded atoms in the other positions.

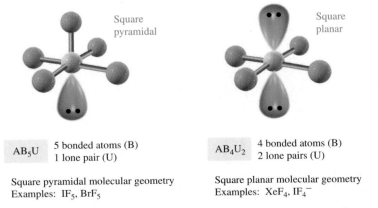

Square pyramidal

| AB_5U | 5 bonded atoms (B) |
| | 1 lone pair (U) |

Square pyramidal molecular geometry
Examples: IF_5, BrF_5

Square planar

| AB_4U_2 | 4 bonded atoms (B) |
| | 2 lone pairs (U) |

Square planar molecular geometry
Examples: XeF_4, IF_4^-

Figure 8-3 Arrangements of bonded atoms and lone pairs (six regions of high electron density—octahedral electronic geometry).

TABLE 8-3   *Molecular Geometry of Species with Lone Pairs (U) on the Central Atom*

General Formula	Regions of High Electron Density	Electronic Geometry	Hybridization at Central Atom	Lone Pairs	Molecular Geometry	Examples
AB_2U	3	trigonal planar	sp^2	1	Angular	O_3, NO_2^-, SO_2
AB_3U	4	tetrahedral	sp^3	1	Trigonal pyramidal	NH_3, SO_3^{2-}
AB_2U_2	4	tetrahedral	sp^3	2	Angular	H_2O, NH_2^-
AB_4U	5	trigonal bipyramidal	sp^3d	1	Seesaw	SF_4

TABLE 8-3	*(continued)*					
General Formula	Regions of High Electron Density	Electronic Geometry	Hybridization at Central Atom	Lone Pairs	Molecular Geometry	Examples
AB_3U_2	5	trigonal bipyramidal	sp^3d	2	T-shaped	ICl_3, ClF_3
AB_2U_3	5	trigonal bipyramidal	sp^3d	3	Linear	XeF_2, I_3^-
AB_5U	6	octahedral	sp^3d^2	1	Square pyramidal	IF_5, BrF_5
AB_4U_2	6	octahedral	sp^3d^2	2	Square planar	XeF_4, IF_4^-

8-13 COMPOUNDS CONTAINING DOUBLE BONDS

In Chapter 7 we constructed Lewis formulas for some molecules and polyatomic ions that contain double and triple bonds. We have not yet considered bonding and shapes for such species. Let us consider ethylene (ethene), C_2H_4, as a specific example. Its dot formula is

$$S = N - A$$
$$= 24 - 12 = \underline{12e^- \text{ shared}}$$

$$\overset{H}{\underset{H}{\diagup}} C = C \overset{H}{\underset{H}{\diagdown}}$$

Here each C atom is considered a central atom. Remember that each bonded atom counts as *one* region of high electron density.

Each atom has three regions of high electron density. The VSEPR theory tells us that each C atom is at the center of a trigonal plane.

Valence bond theory pictures each doubly bonded carbon atom as sp^2 hybridized, with one electron in each sp^2 hybrid orbital and one electron in the unhybridized $2p$ orbital. This $2p$ orbital is perpendicular to the plane of the three sp^2 hybrid orbitals:

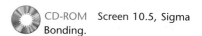

Recall that sp^2 hybrid orbitals are directed toward the corners of an equilateral triangle. Figure 8-4 shows top and side views of these hybrid orbitals.

The two C atoms interact by head-on (end-to-end) overlap of sp^2 hybrids pointing toward each other to form a *sigma (σ) bond* and by side-on overlap of the unhybridized $2p$ orbitals to form a *pi (π) bond*.

CD-ROM   Screen 10.5, Sigma Bonding.

> A **sigma bond** is a bond resulting from head-on overlap of atomic orbitals. *The region of electron sharing is along and cylindrically around an imaginary line connecting the bonded atoms.*

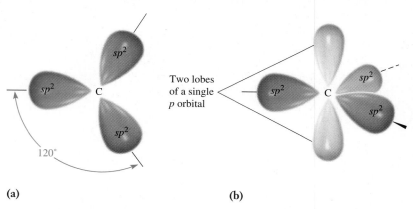

(a)   (b)

Figure 8-4 (a) A top view of three sp^2 hybrid orbitals (*green*). The remaining unhybridized *p* orbital (not shown in this view) is perpendicular to the plane of the drawing. (b) A side view of a carbon atom in a trigonal planar (sp^2-hybridized) environment, showing the remaining *p* orbital (*yellow*). This *p* orbital is perpendicular to the plane of the three sp^2 hybrid orbitals.

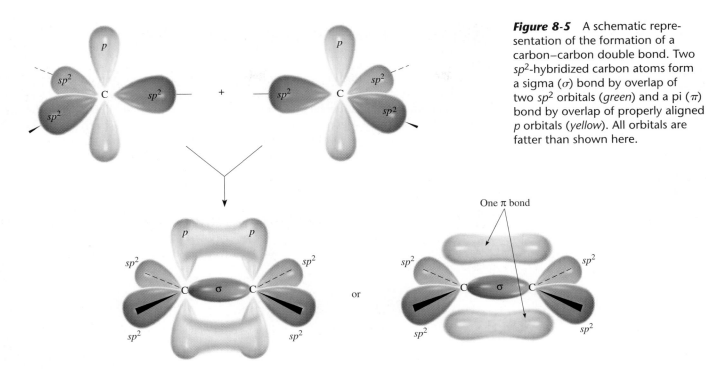

Figure 8-5 A schematic representation of the formation of a carbon–carbon double bond. Two sp^2-hybridized carbon atoms form a sigma (σ) bond by overlap of two sp^2 orbitals (*green*) and a pi (π) bond by overlap of properly aligned p orbitals (*yellow*). All orbitals are fatter than shown here.

All single bonds are sigma bonds. Many kinds of pure atomic orbitals and hybridized orbitals can be involved in sigma bond formation.

> **A pi bond** is a bond resulting from side-on overlap of atomic orbitals. *The regions of electron sharing are on opposite sides of an imaginary line connecting the bonded atoms and parallel to this line.*

CD-ROM Screen 10.7, Multiple Bonding.

A pi bond can form *only* if there is *also* a sigma bond between the same two atoms. The sigma and pi bonds together make a double bond (Figure 8-5). The 1s orbitals (with one e^- each) of four hydrogen atoms overlap the remaining four sp^2 orbitals (with one e^- each) on the carbon atoms to form four C—H sigma bonds (Figure 8-6).

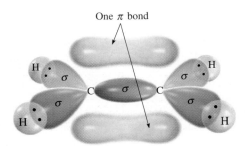

Figure 8-6 Four C—H σ bonds, one C—C σ bond (*green*), and one C—C π bond (*yellow*) in the planar C_2H_4 molecule.

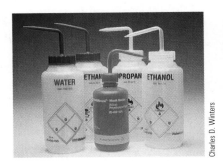

Charles D. Winters

An important use of ethylene, C_2H_4, is in the manufacture of polyethylene, a nonbreakable, nonreactive plastic.

CD-ROM   Screen 10.8, Molecular Fluxionality: Bond Rotations.

As a consequence of the sp^2 hybridization of C atoms in carbon–carbon double bonds, each carbon atom is at the center of a trigonal plane. The p orbitals that overlap to form the π bond must be parallel to each other for effective overlap to occur. This adds the further restriction that these trigonal planes (sharing a common corner) must also be *coplanar*. Thus, all four atoms attached to the doubly bonded C atoms lie in the same plane (see Figure 8-6). Many other important organic compounds contain carbon–carbon double bonds. Several are described in Chapter 27.

8-14   COMPOUNDS CONTAINING TRIPLE BONDS

One compound that contains a triple bond is ethyne (acetylene), C_2H_2. Its Lewis formula is

$$S = N - A$$
$$= 20 - 10 = \underline{10e^- \text{ shared}}$$

$$H : C :: C : H \qquad H—C≡C—H$$

The VSEPR theory predicts that the two regions of high electron density around each carbon atom are 180° apart.

Each triple-bonded carbon atom has two regions of high electron density, so valence bond theory postulates that each is sp hybridized (see Section 8-5). Let us designate the p_x orbitals as the ones involved in hybridization. Carbon has one electron in each sp hybrid orbital and one electron in each of the $2p_y$ and $2p_z$ orbitals (before bonding is considered). See Figure 8-7.

The three p orbitals in a set are indistinguishable. We can label the one involved in hybridization as "p_x" to help us visualize the orientations of the two unhybridized p orbitals on carbon.

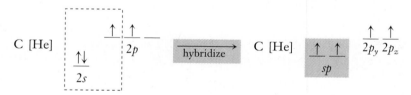

The two carbon atoms form one sigma bond by head-on overlap of the sp hybrid orbitals; each C atom also forms a sigma bond with one H atom. The sp hybrids on each atom are 180° apart. Thus, the entire molecule must be linear.

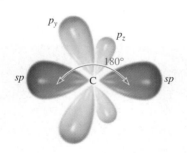

Figure 8-7   Diagram of the two linear hybridized sp orbitals (*green*) of an atom. These lie in a straight line, and the two unhybridized p orbitals p_y and p_z (*yellow*) lie in the perpendicular plane and are perpendicular to each other.

The unhybridized atomic $2p_y$ and $2p_z$ orbitals are perpendicular to each other and to the line through the centers of the two sp hybrid orbitals (Figure 8-8). The side-on overlap of the $2p_y$ orbitals on the two C atoms forms one pi bond; the side-on overlap of the $2p_z$ orbitals forms another pi bond.

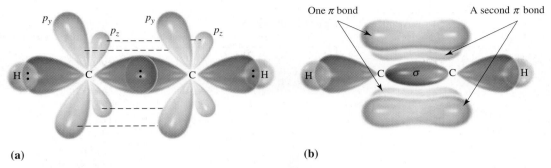

(a)

(b)

Figure 8-8 The acetylene molecule, C_2H_2. (a) The overlap diagram of two *sp*-hybridized carbon atoms and two *s* orbitals from two hydrogen atoms. The hybridized *sp* orbitals on each C are shown in green, and the unhybridized *p* orbitals are shown in tan. The dashed lines, each connecting two lobes, indicate the side-by-side overlap of the four unhybridized *p* orbitals to form two π bonds. There are two C—H σ bonds, one C—C σ bond (*green*), and two C—C π bonds. This makes the net carbon–carbon bond a triple bond. (b) The π-bonding orbitals (*yellow*) are positioned with one above and below the line of the σ bonds (*green*) and the other behind and in front of the line of the σ bonds.

Some other molecules containing triply bonded atoms are nitrogen, :N≡N:, hydrogen cyanide, H—C≡N:, and propyne, CH_3—C≡C—H. In each case, both atoms involved in the triple bonds are *sp* hybridized. In the triple bond, each atom participates in one sigma and two pi bonds. The C atom in carbon dioxide, :O=C=O:, must participate in two pi bonds (to two different O atoms) and also participates in two sigma bonds, so it is also *sp* hybridized, and the molecule is linear.

In propyne, the C atom in the CH_3 group is *sp^3* hybridized and at the center of a tetrahedral arrangement.

8-15 A SUMMARY OF ELECTRONIC AND MOLECULAR GEOMETRIES

We have discussed several common types of polyatomic molecules and ions, and provided a reasonable explanation for the observed structures and polarities of these species. Table 8-4 provides a summation.

Our discussion of covalent bonding illustrates two important points:

1. Molecules and polyatomic ions have definite shapes.
2. The properties of molecules and polyatomic ions are determined to a great extent by their shapes. Incompletely filled electron shells and unshared pairs of electrons on the central element are very important.

Our ideas about chemical bonding have developed over many years. As experimental techniques for determining the *structures* of molecules have improved, our understanding of chemical bonding has improved also. Experimental observations on molecular geometry support our ideas about chemical bonding. The ultimate test for any theory is this: Can it

continued on page 341

TABLE 8-4 *A Summary of Electronic and Molecular Geometries of Polyatomic Molecules and Ions*

Regions of High Electron Density[a]	Electronic Geometry	Hybridization at Central Atom (Angles)	Hybridized Orbital Orientation	Examples	Molecular Geometry
2	linear	sp (180°)		$BeCl_2$ $HgBr_2$ CdI_2 CO_2[b] C_2H_2[c]	linear linear linear linear linear
3	trigonal planar	sp^2 (120°)		BF_3 BCl_3 NO_3^-[e] SO_2[d,e] NO_2^-[d,e] C_2H_4[f]	trigonal planar trigonal planar trigonal planar angular (AB_2U) angular (AB_2U) planar (trig. planar at each C)
4	tetrahedral	sp^3 (109.5°)		CH_4 CCl_4 NH_4^+ SO_4^{2-} $CHCl_3$ NH_3[d] SO_3^{2-}[d] H_3O^+[d] H_2O[d]	tetrahedral tetrahedral tetrahedral tetrahedral distorted tet. pyramidal (AB_3U) pyramidal (AB_3U) pyramidal (AB_3U) angular (AB_2U_2)
5	trigonal bipyramidal	sp^3d (90°, 120°, 180°)		PF_5 $SbCl_5$ SF_4[d] ClF_3[d] XeF_2[d] I_3^-[d]	trigonal bipyramidal trigonal bipyramidal seesaw (AB_4U) T-shaped (AB_3U_2) linear (AB_2U_3) linear (AB_2U_3)
6	octahedral	sp^3d^2 (90°, 180°)		SF_6 SeF_6 PF_6^- BrF_5[d] XeF_4[d]	octahedral octahedral octahedral square pyramidal (AB_5U) square planar (AB_4U_2)

[a] The number of locations of high electron density around the central atom. A region of high electron density may be a single bond, a double bond, a triple bond, or an unshared pair. These determine the electronic geometry, and thus the hybridization of the central atom.

[b] Contains two double bonds.

[c] Contains a triple bond.

[d] Central atom in molecule or ion has unshared pair(s) of electrons.

[e] Bonding involves resonance.

[f] Contains one double bond.

correctly predict the results of experiments before they are performed? When the answer is *yes*, we have confidence in the theory. When the answer is *no*, the theory must be modified. Current theories of chemical bonding enable us to make predictions that are usually accurate.

Key Terms

Angular A term used to describe the molecular geometry of a molecule that has two atoms bonded to a central atom and one or more unshared pairs on the central atom (AB_2U or AB_2U_2). Also called *V-shaped* or *bent*.

Central atom An atom in a molecule or polyatomic ion that is bonded to more than one other atom.

Electronic geometry The geometric arrangement of orbitals containing the shared and unshared electron pairs surrounding the central atom of a molecule or polyatomic ion.

Hybridization The mixing of a set of atomic orbitals on an atom to form a new set of hybrid orbitals with the same total electron capacity and with properties and energies intermediate between those of the original unhybridized orbitals.

Hybrid orbitals Orbitals formed on an atom by the process of hybridization.

Lewis formula A method of representing a molecule or formula unit by showing atoms and only outer-shell electrons; does not show shape.

Linear A term used to describe the electronic geometry around a central atom that has two regions of high electron density. Also used to describe the molecular geometry of a molecule or polyatomic ion that has one atom in the center bonded to two atoms on opposite sides ($180°$) of the central atom (AB_2 or AB_2U_3).

Molecular geometry The arrangement of atoms (*not* unshared pairs of electrons) around a central atom of a molecule or polyatomic ion.

Octahedral A term used to describe the electronic geometry around a central atom that has six regions of high electron density. Also used to describe the molecular geometry of a molecule or polyatomic ion that has one atom in the center bonded to six atoms at the corners of an octahedron (AB_6).

Octahedron A polyhedron with eight equal-sized, equilateral triangular faces and six apices (corners).

Overlap of orbitals The interaction of orbitals on different atoms in the same region of space.

Pi (π) bond A bond resulting from the side-on overlap of atomic orbitals, in which the regions of electron sharing are on opposite sides of and parallel to an imaginary line connecting the bonded atoms.

Seesaw A term used to describe the molecular geometry of a molecule or polyatomic ion that has four atoms bonded to a central atom and one unshared pair on the central atom (AB_4U).

Sigma (σ) bond A bond resulting from the head-on overlap of atomic orbitals, in which the region of electron sharing is along and (cylindrically) symmetrical to an imaginary line connecting the bonded atoms.

Square planar A term used to describe molecules and polyatomic ions that have one atom in the center and four atoms at the corners of a square.

Square pyramidal A term used to describe the molecular geometry of a molecule or polyatomic ion that has five atoms bonded to a central atom and one unshared pair on the central atom (AB_5U).

Tetrahedral A term used to describe the electronic geometry around a central atom that has four regions of high electron density. Also used to describe the molecular geometry of a molecule or polyatomic ion that has one atom in the center bonded to four atoms at the corners of a tetrahedron (AB_4).

Tetrahedron A polyhedron with four equal-sized, equilateral triangular faces and four apices (corners).

Trigonal bipyramid A six-sided polyhedron with five apices (corners), consisting of two pyramids sharing a common triangular base.

Trigonal bipyramidal A term used to describe the electronic geometry around a central atom that has five regions of high electron density. Also used to describe the molecular geometry of a molecule or polyatomic ion that has one atom in the center bonded to five atoms at the corners of a trigonal bipyramid (AB_5).

Trigonal planar (also plane triangular) A term used to describe the electronic geometry around a central atom that has three regions of high electron density. Also used to describe the molecular geometry of a molecule or polyatomic ion that has one atom in the center bonded to three atoms at the corners of an equilateral triangle (AB_3).

Trigonal pyramidal A term used to describe the molecular geometry of a molecule or polyatomic ion that has three atoms bonded to a central atom and one unshared pair on the central atom (AB_3U).

T-shaped A term used to describe the molecular geometry of a molecule or polyatomic ion that has three atoms bonded to a central atom and two unshared pairs on the central atom (AB_3U_2).

Valence bond (VB) theory Assumes that covalent bonds are formed when atomic orbitals on different atoms overlap and electrons are shared.

Valence shell The outermost occupied electron shell of an atom.

Valence shell electron pair repulsion (VSEPR) theory Assumes that valence electron pairs are arranged around the central element of a molecule or polyatomic ion so that there is maximum separation (and minimum repulsion) among regions of high electron density.

Exercises

VSEPR Theory: General Concepts

1. State in your own words the basic idea of the VSEPR theory.
2. (a) Distinguish between "lone pairs" and "bonding pairs" of electrons. (b) Which has the greater spatial requirement? How do we know this? (c) Indicate the order of increasing repulsions among lone pairs and bonding pairs of electrons.
3. What two shapes could a triatomic species have? How would the electronic geometries for the two shapes differ?
4. How are double and triple bonds treated when the VSEPR theory is used to predict molecular geometry? How is a single unshared electron treated?
5. Sketch the three different possible arrangements of the three B atoms around the central atom A for the molecule AB_3U_2. Which of these structures correctly describes the molecular geometry? Why? What are the predicted ideal bond angles? How would observed bond angles deviate from these values?
6. Sketch the three different possible arrangements of the two B atoms around the central atom A for the molecule AB_2U_3. Which of these structures correctly describes the molecular geometry? Why?

Valence Bond Theory: General Concepts

7. What are hybridized atomic orbitals? How is the theory of hybridized orbitals useful?
8. (a) What is the relationship between the number of regions of high electron density on an atom and the number of its pure atomic orbitals that hybridize? (b) What is the relationship between the number of atomic orbitals that hybridize and the number of hybrid orbitals formed?
9. What hybridization is associated with these electronic geometries: trigonal planar; linear; tetrahedral; octahedral; trigonal bipyramidal?
10. Prepare sketches of the orbitals around atoms that are (a) sp, (b) sp^2, (c) sp^3, (d) sp^3d, and (e) sp^3d^2 hybridized. Show in the sketches any unhybridized p orbitals that might participate in multiple bonding.
11. What types of hybridization would you predict for molecules having the following general formulas? (a) AB_4; (b) AB_2U_3; (c) AB_3U; (d) ABU_4; (e) ABU_3.
12. Repeat Exercise 11 for (a) ABU_5; (b) AB_2U_4; (c) AB_3; (d) AB_3U_2; (e) AB_5.
13. (a) What is the maximum number of bonds that an atom can form without expanding its valence shell? (b) What must be true of the electron configuration of an element for it to be able to expand its valence shell? (c) Tell which of the following elements can expand its valence shell: N, O, F, P, S, Cl, Xe.
14. What angles are associated with orbitals in the following sets of hybrid orbitals? (a) sp; (b) sp^2; (c) sp^3; (d) sp^3d; (e) sp^3d^2. Sketch each.
15. What are the primary factors on which we base a decision on whether the bonding in a molecule is better described in terms of simple orbital overlap or overlap involving hybridized atomic orbitals?
16. The elements in Group IIA form compounds, such as Cl—Be—Cl, that are linear and, therefore, nonpolar. What is the hybridization at the central atoms?
17. The elements in Group IIIA form compounds, such as $AlCl_3$, that are planar and, therefore, nonpolar. What is the hybridization at the central atoms?

Electronic and Molecular Geometry

18. Under what conditions is molecular (or ionic) geometry identical to electronic geometry about a central atom?
19. Distinguish between electronic geometry and molecular geometry.
20. Identify the central atom (or atoms) in each of the following compounds or ions: (a) HCO_3^-; (b) SiO_2; (c) SO_2; (d) $Al(OH)_4^-$; (e) $BeBr_2$; (f) $(CH_3)_4Pb$.
21. Identify the central atom (or atoms) in each of the following compounds or ions: (a) H_2SO_4; (b) NH_3; (c) NH_4^+; (d) BCl_3; (e) CH_3NH_2; (f) $CdCl_2$.
22. Write a Lewis formula for each of the following species. Indicate the number of regions of high electron density and the electronic and molecular or ionic geometries. (a) $BeCl_2$; (b) $SnCl_4$; (c) BrF_3; (d) SbF_6^-.
23. Write a Lewis formula for each of the following species. Indicate the number of regions of high electron density and the electronic and molecular or ionic geometries. (a) BF_3; (b) SO_2; (c) IO_3^-; (d) $SiCl_4$; (e) SeF_6.
24. (a) What would be the ideal bond angles in each molecule or ion in Exercise 22, ignoring lone pair effects? (b) How do these differ, if at all, from the actual values? Why?
25. (a) What would be the ideal bond angles in each molecule or ion in Exercise 23, ignoring lone pair effects? (b) Are these values greater than, less than, or equal to the actual values? Why?
26. Write the Lewis formula and identify the electronic geometry and molecular geometry for each polyatomic species in the following equations:
   (a) $H^+ + H_2O \rightarrow H_3O^+$
   (b) $NH_3 + H^+ \rightarrow NH_4^+$.
27. Pick the member of each pair that you would expect to have the smaller bond angles, if different, and explain why. (a) SF_2 and SO_2; (b) BF_3 and BCl_3; (c) SiF_4 and SF_4; (d) NF_3 and OF_2.
28. Draw a Lewis formula, sketch the three-dimensional shape, and name the electronic and ionic geometries for the following polyatomic ions. (a) H_3O^+; (b) PCl_6^-; (c) PCl_4^-; (d) $SbCl_4^+$.

29. As the name implies, the interhalogens are compounds that contain two different halogens. Write Lewis formulas and three dimensional structures for the following. Name the electronic and molecular geometries of each. (a) BrF_3; (b) BrF; (c) BrF_5.

*30. (a) Write a Lewis formula for each of the following molecules: SiF_4; SF_4; XeF_4. (b) Contrast the molecular geometries of these three molecules. Account for differences in terms of the VSEPR theory.

31. Write the Lewis formulas and predict the shapes of these very reactive carbon-containing species: H_3C^+(a carbocation); $H_3C:^-$ (a carbanion); and $:CH_2$ (a carbene whose unshared electrons are paired).

32. Write the Lewis formulas and predict the shapes of (a) I_3^-; (b) $TeCl_4$; (c) XeO_3; (d) NOBr (N is the central atom); (e) NO_2Cl (N is the central atom); (f) $SOCl_2$ (S is the central atom).

33. Describe the shapes of these polyatomic ions: (a) AlO_3^{3-}; (b) AsO_4^{3-}; (c) SO_3^{2-}; (d) NO_3^-.

34. Would you predict a nitrogen–oxygen bond to have the same magnitude of bond polarity as a hydrogen–oxygen bond? Explain your answer.

35. Which of the following molecules are polar? Why? (a) CH_4; (b) CH_3Br; (c) CH_2Br_2; (d) CHI_3; (e) CI_4.

36. Which of the following molecules are polar? Why? (a) CdI_2; (b) BCl_3; (c) $AsCl_3$; (d) H_2S; (e) SF_6.

37. Which of the following molecules are nonpolar? Justify your answer. (a) SO_3; (b) IF; (c) Cl_2O; (d) NF_3; (e) $CHCl_3$.

38. The PF_2Cl_3 molecule is nonpolar. Use this information to sketch its three-dimensional shape. Justify your choice.

*39. In what two major ways does the presence of unshared pairs of valence electrons affect the polarity of a molecule? Describe two molecules for which the presence of unshared pairs on the central atom helps to make the molecules polar. Can you think of a bonding arrangement that has unshared pairs of valence electrons on the central atom but that is nonpolar?

40. Is the phosphorus–chlorine bond in phosphorus trichloride a polar bond? Is phosphorus trichloride a polar molecule? Explain.

41. Is the phosphorus–chlorine bond in phosphorus pentachloride a polar bond? Is phosphorus pentachloride a polar molecule? Explain.

42. Write the Lewis formula for each of the following. Indicate which bonds are polar. (See Table 6-3.) Indicate which molecules are polar. (a) CS_2; (b) AlF_3; (c) H_2S; (d) SnF_2.

43. Write the Lewis formula for each of the following. Indicate which bonds are polar. (See Table 6-3.) Indicate which molecules are polar. (a) OF_2; (b) CH_4; (c) H_2SO_4; (d) SnF_4.

Valence Bond Theory

44. Describe the orbital overlap model of covalent bonding.
45. Briefly summarize the reasoning by which we might have

predicted that the formula of the simplest stable hydrocarbon would be CH_2, if we did not consider hybridization. Would this species satisfy the octet rule?

46. What is the hybridization of the central atom in each of the following? (a) PCl_5; (b) molecular $AlCl_3$; (c) CF_4; (d) SF_6; (e) ClO_4^-.

47. What is the hybridization of the central atom in each of the following? (a) IF_4^-; (b) SiO_4^{4-}; (c) AlH_4^-; (d) PH_4^+; (e) PBr_3; (f) ClO_3^-.

48. (a) Describe the hybridization of the central atom in each of these covalent species. (1) $CHCl_3$; (2) CH_2Br_2; (3) NF_3; (4) PO_4^{3-}; (5) IF_6^+; (6) SiF_6^{2-}. (b) Give the shape of each species.

49. Describe the hybridization of the underlined atoms in $\underline{C}_2F_6$, $\underline{C}_2F_4$, $\underline{N}_2F_4$, and $(H_2N)_2\underline{C}O$.

50. Prepare a sketch of the molecule $CH_3CCl{=}CH_2$ showing orbital overlaps. Identify the type of hybridization of atomic orbitals on each carbon atom.

*51. After comparing experimental and calculated dipole moments, Charles A. Coulson suggested that the Cl atom in HCl is sp hybridized. (a) Give the orbital electronic structure for an sp hybridized Cl atom. (b) Which HCl molecule would have a larger dipole moment—one in which the chlorine uses pure p orbitals for bonding with the H atom or one in which sp hybrid orbitals are used?

*52. Predict the hybridization at each carbon atom in each of the following molecules.

(a) acetone (a common solvent)

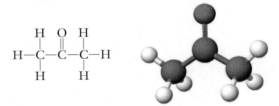

(b) glycine (an amino acid)

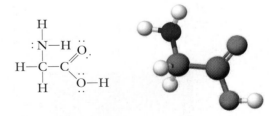

(c) nitrobenzene

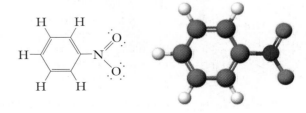

(d) chloroprene (used to make neoprene, a synthetic rubber)

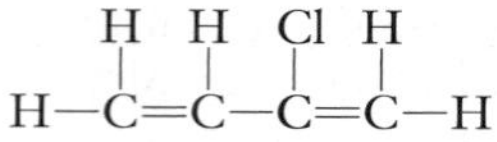

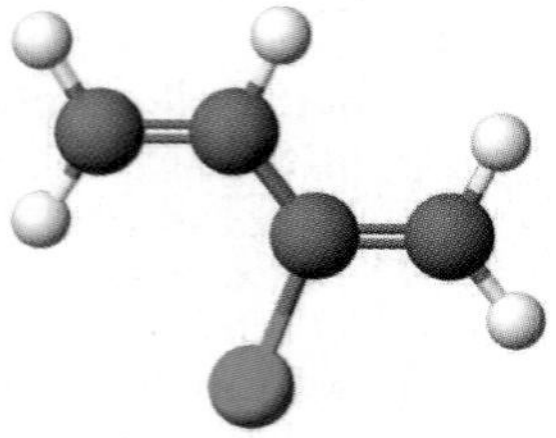

$$H-\overset{H}{\underset{}{C}}=\overset{H}{\underset{}{C}}-\overset{Cl}{\underset{}{C}}=\overset{H}{\underset{}{C}}-H$$

(e) 4-penten-1-yne

$$H-\overset{H}{\underset{}{C}}=\overset{H}{\underset{}{C}}-\overset{H}{\underset{H}{C}}-C\equiv C-H$$

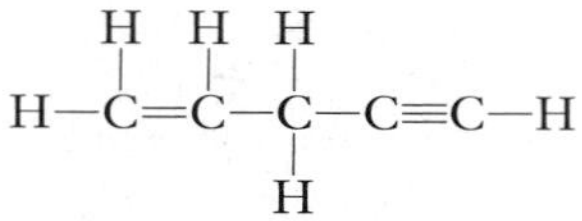

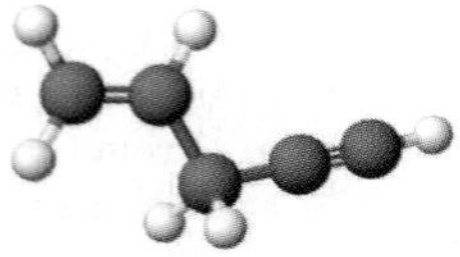

*53. Predict the hybridization at the numbered atoms (①, ②, and so on) in the following molecules and predict the approximate bond angles at those atoms.

(a) diethyl ether, an anesthetic

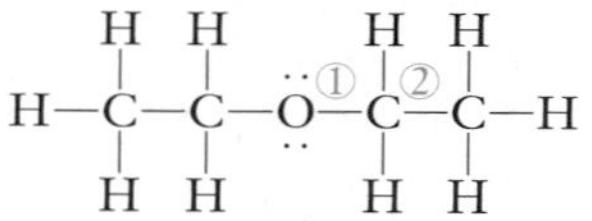

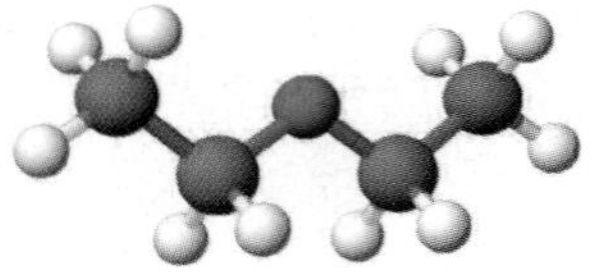

$$H-\overset{H}{\underset{H}{C}}-\overset{H}{\underset{H}{C}}-\overset{..}{\underset{..}{O}}①-\overset{H}{\underset{H}{C}}②-\overset{H}{\underset{H}{C}}-H$$

(b) caffeine, a stimulant in coffee and in many over-the-counter medicinals [1]

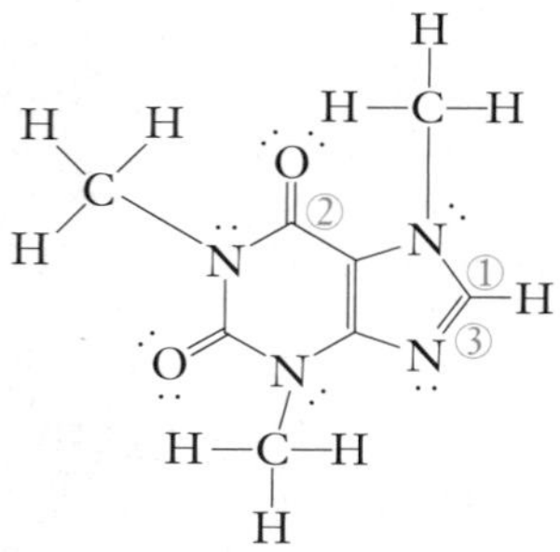

Coffee

(c) acetylsalicylic acid (aspirin) [1]

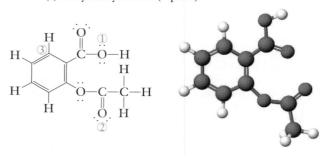

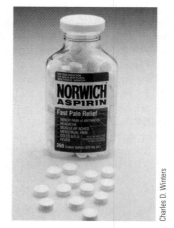

Aspirin

(d) nicotine [1]

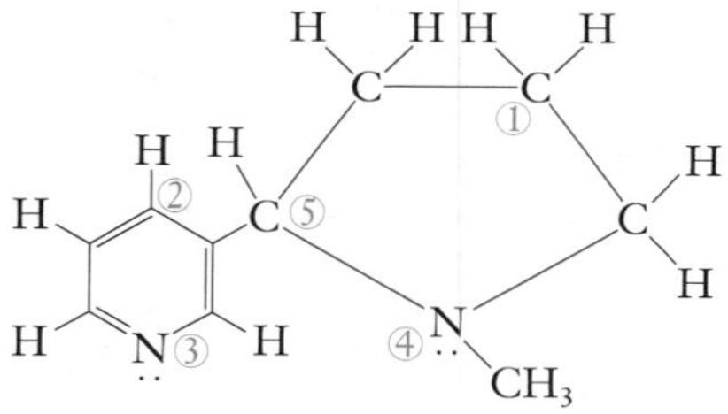

(e) ephedrine, a nasal decongestant [1]

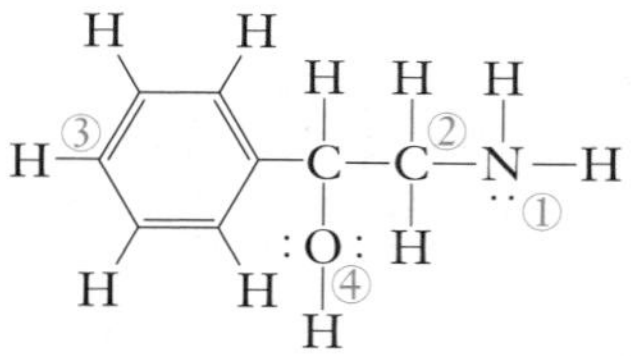

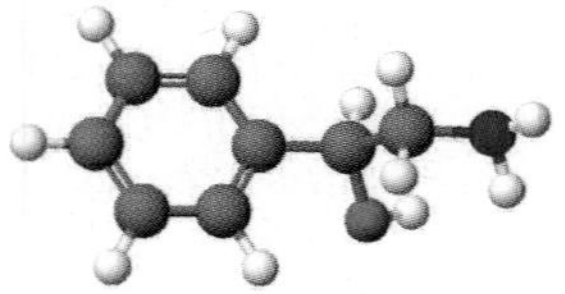

54. Prepare sketches of the overlaps of the following atomic orbitals: (a) *s* with *s*; (b) *s* with *p* along the bond axis; (c) *p* with *p* along the bond axis (head-on overlap); (d) *p* with *p* perpendicular to the bond axis (side-on overlap).

55. Prepare a sketch of the cross-section (through the atomic

[1] *In these kinds of structural drawings, each intersection of lines represents a C atom.*

centers) taken between two atoms that have formed (a) a single σ bond, (b) a double bond consisting of a σ bond and a π bond, and (c) a triple bond consisting of a σ bond and two π bonds.

56. How many sigma and how many pi bonds are there in each of the following molecules?

(a) H—C—C=C—C—H   (c) H—C—C

(b) C=C=C

*57. How many sigma and how many pi bonds are there in each of the following molecules?

(a) H—C—C—C=C   (c) C=C

(b) H—C—C—C   (d) $CH_2CHCH_2OCH_2CH_3$

58. Describe the bonding in the carbide ion, C_2^{2-}, with a three-dimensional VB structure. Show the orbital overlap, and label the orbitals.

59. Write Lewis formulas for molecular oxygen and ozone. Assuming that all of the valence electrons in the oxygen atoms are in hybrid orbitals, what would be the hybridization of the oxygen atoms in each substance? Prepare sketches of the molecules.

*60. Draw a Lewis formula and a three-dimensional structure for each of the following polycentered molecules. Indicate hybridizations and bond angles at each carbon atom. (a) butane, C_4H_{10}; (b) propene, $H_2C{=}CHCH_3$; (c) 1-butyne, $HC{\equiv}CCH_2CH_3$; (d) acetaldehyde, CH_3CHO.

61. How many σ bonds and how many π bonds are there in each of the molecules of Exercise 60?

62. (a) Describe the hybridization of N in each of these species.
(i) $:NH_3$; (ii) NH_4^+ (iii) $HN{=}NH$; (iv) $HC{\equiv}N:$;
(v) $H_2N{-}NH_2$.
(b) Give an orbital description for each species, specifying the location of any unshared pairs and the orbitals used for the multiple bonds.

63. Write the Lewis formulas and predict the hybrid orbitals

and the shapes of these polyatomic ions and covalent molecules: (a) $HgCl_2$; (b) BF_3; (c) BF_4^-; (d) $SbCl_5$; (e) SbF_6^-.

64. (a) What is the hybridization of each C in these molecules? (i) $H_2C{=}O$; (ii) $HC{\equiv}N$; (iii) $CH_3CH_2CH_3$; (iv) ketene, $H_2C{=}C{=}O$. (b) Describe the shape of each molecule.

*65. The following fluorides of xenon have been well characterized: XeF_2, XeF_4, and XeF_6. (a) Write Lewis formulas for these substances and decide what type of hybridization of the Xe atomic orbitals has taken place. (b) Draw all of the possible atomic arrangements of XeF_2, and discuss your choice of molecular geometry. (c) What shape do you predict for XeF_4?

*66. Iodine and fluorine form a series of interhalogen molecules and ions. Among these are IF (minute quantities observed spectroscopically), IF_3, IF_4^-, and IF_5. (a) Write Lewis formulas for each of these species. (b) Identify the type of hybridization that the orbitals of the iodine atom have undergone in each substance. (c) Identify the shape of the molecule or ion.

Mixed Exercises

67. In the pyrophosphate ion, $P_2O_7^{4-}$, one oxygen atom is bonded to both phosphorus atoms.

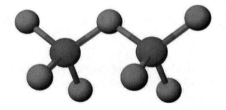

Write a Lewis formula, and sketch the three-dimensional shape of the ion. Describe the electronic and molecular geometries with respect to the central O atom and with respect to each P atom.

68. Briefly discuss the bond angles in the hydroxylamine molecule in terms of the ideal geometry and the small changes caused by electron-pair repulsions.

H—N—O—H

69. Repeat Exercise 68 for the nitrite ion.

*70. The methyl free radical $\cdot CH_3$ has bond angles of about 120°, whereas the methyl carbanion $:CH_3^-$ has bond angles of about 109°. What can you infer from these facts about the repulsive force exerted by an unpaired, unshared electron as compared with that exerted by an unshared pair of electrons?

*71. Two Lewis structures can be written for the square planar molecule $PtCl_2Br_2$:

Show how a difference in dipole moments can distinguish between these two possible structures.

72. (a) Describe the hybridization of N in NO_2^+ and NO_2^-.
(b) Predict the bond angle in each case.

*73. The skeleton and a ball-and-stick model for the nitrous acid molecule, HNO_2, are shown here. Draw the Lewis formula. What are the hybridizations at the middle O and N atoms?

74. Describe the change in hybridization that occurs at the central atom of the reactant at the left in each of the following reactions.
(a) $PF_5 + F^- \rightarrow PF_6^-$
(b) $2CO + O_2 \rightarrow 2CO_2$
(c) $AlI_3 + I^- \rightarrow AlI_4^-$
(d) What change in hybridization occurs in the following reaction? $:NH_3 + BF_3 \rightarrow H_3N:BF_3$

*75. Consider the following proposed Lewis formulas for ozone (O_3):

(i) $:\overset{..}{O}-\overset{..}{O}=\overset{..}{O}: \longleftrightarrow \overset{..}{O}=\overset{..}{O}-\overset{..}{O}:$ (ii) $:\overset{..}{O}-\overset{..}{O}:$

(iii) $:\overset{..}{O}-\overset{..}{O}-\overset{..}{O}:$

(a) Which of these correspond to a polar molecule? (b) Which of these predict covalent bonds of equal lengths and strengths? (c) Which of these predict a diamagnetic molecule? (d) The properties listed in parts (a), (b), and (c) are those observed for ozone. Which structure correctly predicts all three? (e) Which of these contain a considerable amount of "strain"? Explain.

76. What hybridizations are predicted for the central atoms in molecules having the formulas AB_2U_2 and AB_3U? What

are the predicted bond angles for these molecules? The observed bond angles for representative substances are

H_2O	104.5°	NH_3	106.7°
H_2S	92.2°	PH_3	93.7°
H_2Se	91.0°	AsH_3	91.8°
H_2Te	89.5°	SbH_3	91.3°

What would be the predicted bond angle if no hybridization occurred? What conclusion can you draw concerning the importance of hybridization for molecules of compounds involving elements with higher atomic numbers?

CONCEPTUAL EXERCISES

77. Draw and explain the difference in the three-dimensional shapes of $CH_3CH_2CH_3$ and CH_3COCH_3.

78. Complete the following table.

Molecule or Ion	Electronic Geometry	Molecular Geometry	Hybridization of the Sulfur Atom
SO_2			
SCl_2			
SO_3			
SO_3^{2-}			
SF_4			
SO_4^{2-}			
SF_5^+			
SF_6			

79. Compare the shapes of the following pairs of molecules: (a) H_2CO and CH_4; (b) $PbCl_4$ and PbO_2; (c) CH_4 and $PbCl_4$.

80. (a) Write two Lewis formulas for CO_3^{2-}. Do the two formulas represent resonance forms? (b) Do the same for HSO_4^-.

81. What advantages does the VSEPR model of chemical bonding have compared with Lewis formulas?

82. What evidence could you present to show that two carbon atoms joined by a single sigma bond are able to rotate about an axis that coincides with the bond, but two carbon atoms bonded by a double bond cannot rotate about an axis along the double bond?

BUILDING YOUR KNOWLEDGE

83. Sketch three-dimensional representations of the following molecules and indicate the direction of any net dipole for each molecule: (a) CH_4; (b) CH_3Cl; (c) CH_2Cl_2; (d) $CHCl_3$; (e) CCl_4.

84. Carbon forms two common oxides, CO and CO_2. It also forms a third (very uncommon) oxide, carbon suboxide, C_3O_2, which is linear. The structure has terminal oxygen atoms on both ends. Write the Lewis formula for C_3O_2. How many regions of high electron density are there about each of the three carbon atoms?

85. Draw the Lewis formula of an ammonium ion. Describe the formation of the ammonium ion from ammonia plus H^+. Does the hybridization of orbitals on nitrogen change during the formation of the ammonium ion? Do the bond angles change?

86. The following is an incomplete Lewis formula for a molecule. This formula has all the atoms at the correct places, but it is missing several valence electrons. Complete this Lewis formula, including lone pairs.

$$\begin{array}{ccc} Cl & H & Br \\ | & | & | \\ H-C & C-C & O \end{array}$$

87. The following is an incomplete Lewis formula for a molecule. This formula has all the atoms at the correct places, but it is missing several valence electrons. Complete this Lewis formula, including lone pairs.

$$\begin{array}{cccccc} H-O & H \\ | \\ O & C & C & C & C & N-H \\ | & | \\ H & Br & H & H \end{array}$$

BEYOND THE TEXTBOOK

Go to the textbook website at

http://www.brookscole.com/chemistry/whitten

for additional activities and exercises based on the General Chemistry Interactive CD-ROM, the World Wide Web, and library resources.

InfoTrac College Edition

For additional readings, go to InfoTrac College Edition, your online research library at:

http://infotrac.thomsonlearning.com

Molecular Orbitals in Chemical Bonding

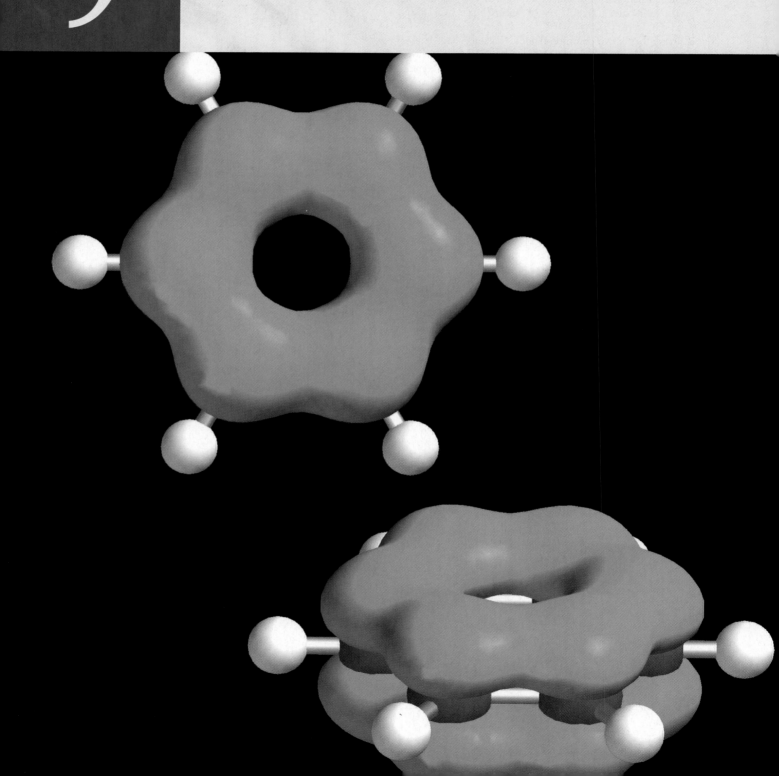

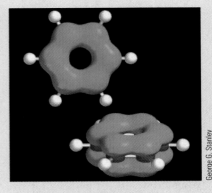

A computer representation of one of the π molecular orbitals of benzene.

OBJECTIVES

After you have studied this chapter, you should be able to

- Describe the basic concepts of molecular orbital theory
- Relate the shapes and overlap of atomic orbitals to the shapes and energies of the resulting molecular orbitals
- Distinguish among bonding, antibonding, and nonbonding orbitals
- Apply the Aufbau Principle to find molecular orbital descriptions for homonuclear diatomic molecules and ions
- Apply the Aufbau Principle to find molecular orbital descriptions for heteronuclear diatomic molecules and ions with small $\Delta(EN)$ values
- Find the bond order in diatomic molecules and ions
- Relate bond order to bond stability
- Use the MO concept of delocalization for molecules in which valence bond theory would postulate resonance

We have described bonding and molecular geometry in terms of valence bond theory. In valence bond theory, we postulate that bonds result from the sharing of electrons in overlapping orbitals of different atoms. We describe electrons in overlapping orbitals of different atoms as being localized in the bonds between the two atoms involved. We use hybridization to help account for the geometry of a molecule. In valence bond theory, however, we view each orbital as belonging to an individual atom.

In **molecular orbital theory,** we postulate that

the combination of atomic orbitals on different atoms forms **molecular orbitals** (MOs), so that electrons in them belong to the molecule as a whole.

In some polyatomic molecules, a molecular orbital may extend over only a fraction of the molecule.

Polyatomic ions such as CO_3^{2-}, SO_4^{2-}, and NH_4^+ can be described by the molecular orbital approach.

Valence bond and molecular orbital theories are alternative descriptions of chemical bonding. They have strengths and weaknesses, so they are complementary. Valence bond theory is descriptively attractive, and it lends itself well to visualization. Molecular orbital theory gives better descriptions of electron cloud distributions, bond energies, and magnetic properties, but its results are not as easy to visualize.

The valence bond picture of bonding in the O_2 molecule involves a double bond.

$$\ddot{O} :: \ddot{O}$$

This shows no unpaired electrons, so it predicts that O_2 is diamagnetic. Experiments show, however, that O_2 is paramagnetic; therefore, it has unpaired electrons. Thus, the valence bond structure is inconsistent with experiment and cannot be accepted as a description of the bonding. Molecular orbital theory accounts for the fact that O_2 has two unpaired electrons. This ability of MO theory to explain the paramagnetism of O_2 gave it credibility as a major theory of bonding. We shall develop some of the ideas of MO theory and apply them to some molecules and polyatomic ions.

An early triumph of molecular orbital theory was its ability to account for the observed paramagnetism of oxygen, O_2. According to earlier theories, O_2 was expected to be diamagnetic, that is, to have only paired electrons.

9-1 MOLECULAR ORBITALS

We learned in Chapter 5 that each solution to the Schrödinger equation, called a wave function, represents an atomic orbital. The mathematical pictures of hybrid orbitals in valence bond theory can be generated by combining the wave functions that describe two or more atomic orbitals on a *single* atom. Similarly, combining wave functions that describe atomic orbitals on *separate* atoms generates mathematical descriptions of molecular orbitals.

An orbital has physical meaning only when we square its wave function to describe the electron density. Thus, the overall sign on the wave function that describes an atomic orbital is not important, but when we *combine* two orbitals, the signs of the wave functions become very significant. When waves are combined, they may interact either constructively or destructively (Figure 9-1). Likewise, when two atomic orbitals overlap, they can be in phase (added) or out of phase (subtracted). When they overlap in phase, constructive interaction occurs in the region between the nuclei, and a **bonding orbital** is produced. The energy of the bonding orbital is always lower (more stable) than the energies of the combining orbitals. When the orbitals overlap out of phase, destructive interaction reduces the probability of finding electrons in the region between the nuclei, and an **antibonding orbital** is produced. This is higher in energy (less stable) than the original atomic orbitals, leading to a repulsion between the two atoms. The overlap of two atomic orbitals always produces two MOs: one bonding and one antibonding.

Another way of viewing the relative stabilities of these orbitals follows. In a bonding molecular orbital, the electron density is high *between* the two atoms, where it stabilizes the arrangement by attracting both nuclei. By contrast, an antibonding orbital has a node (a region of zero electron density) between the nuclei; this allows the nuclei to repel one another more strongly, which makes the arrangement less stable. Electrons are *more* stable (have lower energy) in bonding molecular orbitals than in the individual atoms. Placing electrons in antibonding orbitals, on the other hand, requires an increase in their energy, which makes them *less* stable than in the individual atoms.

To summarize, electrons in *bonding* orbitals tend to *stabilize* the molecule or ion; electrons in *antibonding* orbitals tend to *destabilize* the molecule or ion. The relative number of electrons in bonding versus antibonding orbitals determines the *overall stability* of the molecule or ion.

We can illustrate this basic principle by considering the combination of the $1s$ atomic orbitals *on two different atoms* (Figure 9-2). When these orbitals are occupied by electrons,

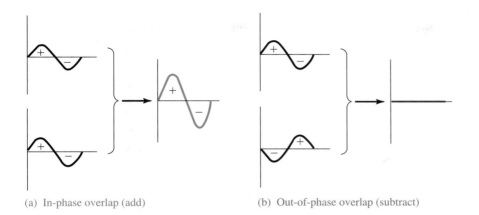

(a) In-phase overlap (add)   (b) Out-of-phase overlap (subtract)

Figure 9-1 An illustration of constructive and destructive interference of waves. (a) If the two identical waves shown at the left are added, they interfere constructively to produce the more intense wave at the right. (b) Conversely, if they are subtracted, it is as if the phases (signs) of one wave were reversed and added to the first wave. This causes destructive interference, resulting in the wave at the right with zero amplitude; that is, a straight line.

the shapes of the orbitals are plots of electron density. These plots show the regions in molecules where the probabilities of finding electrons are the greatest.

In the bonding orbital, the two $1s$ orbitals have reinforced each other in the region between the two nuclei by in-phase overlap, or addition of their electron waves. In the antibonding orbital, they have canceled each other in this region by out-of-phase overlap, or subtraction of their electron waves. We designate both molecular orbitals as **sigma (σ) molecular orbitals** (which indicates that they are cylindrically symmetrical about the internuclear axis). We indicate with subscripts the atomic orbitals that have been combined. The asterisk (*) denotes an antibonding orbital. Thus, two $1s$ orbitals produce a σ_{1s} (read "sigma-1s") bonding orbital and a σ_{1s}^* (read "sigma-1s-star") antibonding orbital. The right-hand side of Figure 9-2 shows the relative energy levels of these orbitals. All sigma antibonding orbitals have nodal planes bisecting the internuclear axis. A **node,** or **nodal plane,** is a region in which the probability of finding electrons is zero.

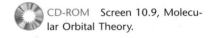 CD-ROM Screen 10.9, Molecular Orbital Theory.

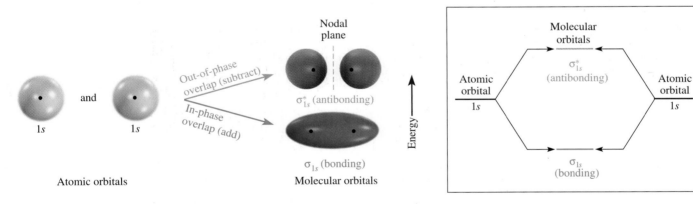

Figure 9-2 Molecular orbital diagram for the combination of the $1s$ atomic orbitals on two identical atoms (*at the left*) to form two MOs. One is a *bonding* orbital, σ_{1s} (*blue*), resulting from addition of the wave functions of the $1s$ orbitals. The other is an *antibonding* orbital, σ_{1s}^* (*red*), at higher energy resulting from subtraction of the waves that describe the combining $1s$ orbitals. In all σ-type MOs, the electron density is symmetrical about an imaginary line connecting the two nuclei. The terms "subtraction of waves," "out of phase," and "destructive interference in the region between the nuclei" all refer to the formation of an antibonding MO. Nuclei are represented by dots.

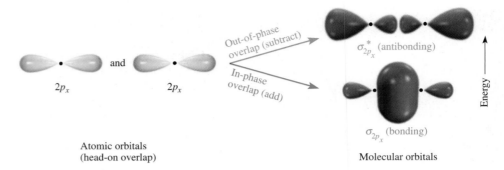

Figure 9-3 Production of σ_{2p_x} and $\sigma_{2p_x}^*$ molecular orbitals by overlap of $2p_x$ orbitals on two atoms.

How we name the axes is arbitrary. We designate the internuclear axis as the x direction.

This would involve rotating Figures 9-2, 9-3, and 9-4 by 90° so that the internuclear axes are perpendicular to the plane of the pages.

For any two sets of p orbitals on two different atoms, corresponding orbitals such as p_x orbitals can overlap *head-on*. This gives σ_p and σ_p^* orbitals, as shown in Figure 9-3 for the head-on overlap of $2p_x$ orbitals on the two atoms. If the remaining p orbitals overlap (p_y with p_y and p_z with p_z), they must do so sideways, or *side-on*, forming **pi (π) molecular orbitals.** Depending on whether all p orbitals overlap, there can be as many as two π_p and two π_p^* orbitals. Figure 9-4 illustrates the overlap of two corresponding $2p$ orbitals on two atoms to form π_{2p} and π_{2p}^* molecular orbitals. There is a nodal plane along the internuclear axis for all pi molecular orbitals. If one views a sigma molecular orbital along the internuclear axis, it appears to be symmetrical around the axis like a pure s atomic orbital. A similar cross-sectional view of a pi molecular orbital looks like a pure p atomic orbital, with a node along the internuclear axis.

> The number of molecular orbitals formed is equal to the number of atomic orbitals that are combined. When two atomic orbitals are combined, one of the resulting MOs is at a *lower* energy than the original atomic orbitals; this is a *bonding* orbital. The other MO is at a *higher* energy than the original atomic orbitals; this is an *antibonding* orbital.

If we had chosen the z axis as the axis of head-on overlap of the 2p orbitals in Figure 9-3, side-on overlap of the $2p_x–2p_x$ and $2p_y–2p_y$ orbitals would form the π-type molecular orbitals.

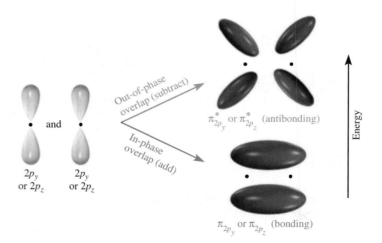

Figure 9-4 The π_{2p} and π_{2p}^* molecular orbitals from overlap of one pair of $2p$ atomic orbitals (for instance, $2p_y$ orbitals). There can be an identical pair of molecular orbitals at right angles to these, formed by another pair of p orbitals on the same two atoms (in this case, $2p_z$ orbitals).

9-2 MOLECULAR ORBITAL ENERGY LEVEL DIAGRAMS

Figure 9-5 shows molecular orbital energy level diagrams for homonuclear diatomic molecules of elements in the first and second periods. Each diagram is an extension of the right-hand diagram in Figure 9-2, to which we have added the molecular orbitals formed from $2s$ and $2p$ atomic orbitals.

For the diatomic species shown in Figure 9-5a, the two π_{2p} orbitals are lower in energy than the σ_{2p} orbital. Molecular orbital calculations indicate, however, that for O_2, F_2, and hypothetical Ne_2 molecules, the σ_{2p} orbital is lower in energy than the π_{2p} orbitals (see Figure 9-5b).

"Homonuclear" means consisting only of atoms of the same element. "Diatomic" means consisting of two atoms.

Spectroscopic data support these orders.

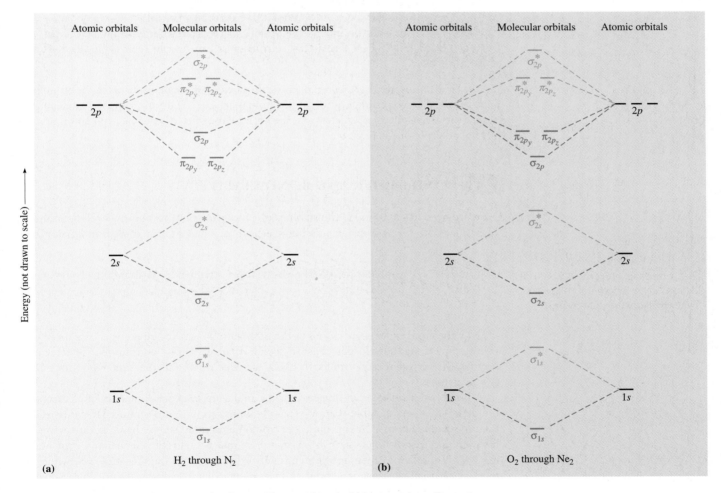

Figure 9-5 Energy level diagrams for first- and second-period homonuclear diatomic molecules and ions (not drawn to scale). The solid lines represent the relative energies of the indicated atomic and molecular orbitals. (a) The diagram for H_2, He_2, Li_2, Be_2, B_2, C_2, and N_2 molecules and their ions. (b) The diagram for O_2, F_2 and Ne_2 molecules and their ions.

Diagrams such as these are used to describe the bonding in a molecule in MO terms. Electrons occupy MOs according to the same rules developed for atomic orbitals; they follow the Aufbau Principle, the Pauli Exclusion Principle, and Hund's Rule. (See Section 5-17.) To obtain the molecular orbital description of the bonding in a molecule or ion, follow these steps:

CD-ROM   Screen 10.10, Molecular Electron Configurations.

1. Draw (or select) the appropriate molecular orbital energy level diagram.
2. Determine the *total* number of electrons in the molecule. Note that in applying MO theory, we will account for *all* electrons. This includes both the inner-shell electrons and the valence electrons.
3. Add these electrons to the energy level diagram, putting each electron into the lowest energy level available.
   a. A maximum of *two* electrons can occupy any given molecular orbital, and then only if they have opposite spin (Pauli Exclusion Principle).
   b. Electrons must occupy all the orbitals of the same energy singly before pairing begins. These unpaired electrons must have parallel spins (Hund's Rule).

For isolated atoms, we represented the resulting arrangement of electrons in the atomic orbitals as the electron configuration of the atom (e.g., $1s^2 2s^2$). In the same way, we can represent the arrangement of electrons in the molecular orbitals as the electron configuration of the molecule or ion (e.g., $\sigma_{1s}^2 \sigma_{1s}^{*2} \sigma_{2s}^2 \sigma_{2s}^{*2}$).

9-3 BOND ORDER AND BOND STABILITY

Now we need a way to judge the stability of a molecule once its energy level diagram has been filled with the appropriate number of electrons. This criterion is the **bond order** (bo):

Electrons in bonding orbitals are often called **bonding electrons,** and electrons in antibonding orbitals are called **antibonding electrons.**

$$\text{bond order} = \frac{(\text{number of bonding electrons}) - (\text{number of antibonding electrons})}{2}$$

Usually the bond order corresponds to the number of bonds described by the valence bond theory. Fractional bond orders exist in species that contain an odd number of electrons, such as the nitrogen oxide molecule, NO, (15 electrons) and the superoxide ion, O_2^-, (17 electrons).

A bond order *equal to zero* means that the molecule has equal numbers of electrons in bonding MOs (more stable than in separate atoms) and in antibonding MOs (less stable than in separate atoms). Such a molecule would be no more stable than separate atoms, so it would not exist. A bond order *greater than zero* means that more electrons occupy bonding MOs (stabilizing) than antibonding MOs (destabilizing). Such a molecule would be more stable than the separate atoms, and we predict that its existence is possible. But such a molecule could be quite reactive.

The greater the bond order of a diatomic molecule or ion, the more stable we predict it to be. Likewise, for a bond between two given atoms, the greater the bond order, the shorter is the bond length and the greater is the bond energy.

The **bond energy** is the amount of energy necessary to break a mole of bonds (Section 15-9); therefore, bond energy is a measure of bond strength.

> ### ✓ Problem-Solving Tip: *Working with MO Theory*
>
> MO theory is often the best model to predict the bond order, bond stability, or magnetic properties of a molecule or ion. The procedure is as follows:
>
> 1. Draw (or select) the appropriate MO energy level diagram.
> 2. Count the total number of electrons in the molecule or ion.
> 3. Follow the Pauli Exclusion Principle and Hund's Rule to add the electrons to the MO diagram.
> 4. Calculate the bond order: $\text{bond order} = \left(\dfrac{\text{bonding } e\text{'s} - \text{antibonding } e\text{'s}}{2} \right)$.
> 5. Use the bond order to evaluate stability.
> 6. Look for the presence of unpaired electrons to determine if a species is paramagnetic.

 CD-ROM   Screen 10.11, Homonuclear Diatomic Molecules.

9-4 HOMONUCLEAR DIATOMIC MOLECULES

The electron configurations for the homonuclear diatomic molecules of the first and second periods are shown in Table 9-1 together with their bond orders, bond lengths, and bond energies. You can derive these electron configurations by putting the required numbers of electrons into the appropriate diagram of Figure 9-5. This procedure is illustrated by the following cases.

This is just like the Aufbau process we used in Chapter 5 to develop the electron configurations of isolated atoms. There we put electrons into the atomic orbitals of isolated atoms; here we put them into the molecular orbitals of a collection of atoms.

The Hydrogen Molecule, H_2

The overlap of the $1s$ orbitals of two hydrogen atoms produces σ_{1s} and σ_{1s}^* molecular orbitals. The two electrons of the molecule occupy the lower energy σ_{1s} orbital (Figure 9-6a).

Because the two electrons in an H_2 molecule are in a bonding orbital, the bond order is one. We conclude that the H_2 molecule would be stable, and we know it is. The energy associated with two electrons in the H_2 molecule is lower than that associated with the same two electrons in separate $1s$ atomic orbitals. The lower the energy of a system, the more stable it is. As the energy of a system decreases, its stability increases.

H_2 bond order $= \dfrac{2 - 0}{2} = 1$

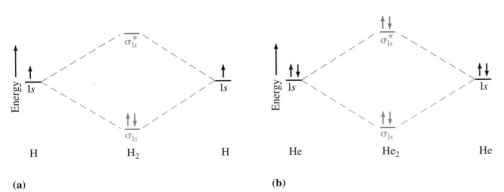

Figure 9-6 Molecular orbital diagrams for (a) H_2 and (b) He_2.

The Helium Molecule (Hypothetical), He_2

$$He_2 \text{ bond order} = \frac{2-2}{2} = 0$$

The energy level diagram for He_2 is similar to that for H_2 except that it has two more electrons. These occupy the antibonding σ_{1s}^* orbital (see Figures 9-5a and 9-6b and Table 9-1), giving He_2 a bond order of zero. That is, the two electrons in the bonding orbital of He_2 would make the molecule *more stable* than the separate atoms. But the two electrons in the antibonding orbital would make the molecule *less stable* than the separate atoms. These effects cancel, so the molecule would be no more stable than the separate atoms. The bond order is zero, and the molecule would not exist. In fact, He_2 is unknown.

The Boron Molecule, B_2

$$B_2 \text{ bond order} = \frac{6-4}{2} = 1$$

The boron atom has the configuration $1s^2 2s^2 2p^1$. Here p electrons participate in the bonding. Figure 9-5a and Table 9-1 show that the π_{p_y} and π_{p_z} molecular orbitals are lower in energy than the σ_{2p} for B_2. Thus, the electron configuration is

$$\sigma_{1s}^2 \quad \sigma_{1s}^{*2} \quad \sigma_{2s}^2 \quad \sigma_{2s}^{*2} \quad \pi_{2p_y}^1 \quad \pi_{2p_z}^1$$

Orbitals of equal energy are called *degenerate* orbitals. Hund's Rule for filling degenerate orbitals was discussed in Section 5-17.

The unpaired electrons are consistent with the observed paramagnetism of B_2. Here we illustrate Hund's Rule in molecular orbital theory. The π_{2p_y} and π_{2p_z} orbitals are equal in energy and contain a total of two electrons. Accordingly, one electron occupies each orbital. The bond order is one. Experiments verify that B_2 molecules exist in the vapor state.

The Nitrogen Molecule, N_2

$$N_2 \text{ bond order} = \frac{10-4}{2} = 3$$

In the valence bond representation, N_2 is shown as $:N \equiv N:$, with a triple bond.

Experimental thermodynamic data show that the N_2 molecule is stable, is diamagnetic, and has a very high bond energy, 946 kJ/mol. This is consistent with molecular orbital theory. Each nitrogen atom has 7 electrons, so the diamagnetic N_2 molecule has 14 electrons.

$$\sigma_{1s}^2 \quad \sigma_{1s}^{*2} \quad \sigma_{2s}^2 \quad \sigma_{2s}^{*2} \quad \pi_{2p_y}^2 \quad \pi_{2p_z}^2 \quad \sigma_{2p}^2$$

Six more electrons occur in bonding orbitals than in antibonding orbitals, so the bond order is three. We see (Table 9-1) that N_2 has a very short bond length, only 1.09 Å, the shortest of any diatomic species except H_2.

The Oxygen Molecule, O_2

Among the homonuclear diatomic molecules, only N_2 and the very small H_2 have shorter bond lengths than O_2, 1.21 Å. Recall that VB theory predicts that O_2 is diamagnetic. Experiments show, however, that it is paramagnetic, with two unpaired electrons. MO theory predicts a structure consistent with this observation. For O_2, the σ_{2p} orbital is lower in energy than the π_{2p_y} and π_{2p_z} orbitals. Each oxygen atom has 8 electrons, so the O_2 molecule has 16 electrons.

$$\sigma_{1s}^2 \quad \sigma_{1s}^{*2} \quad \sigma_{2s}^2 \quad \sigma_{2s}^{*2} \quad \sigma_{2p}^2 \quad \pi_{2p_y}^2 \quad \pi_{2p_z}^2 \quad \pi_{2p_y}^{*1} \quad \pi_{2p_z}^{*1}$$

$$O_2 \text{ bond order} = \frac{10-6}{2} = 2$$

The two unpaired electrons reside in the *degenerate* antibonding orbitals, $\pi_{2p_y}^*$ and $\pi_{2p_z}^*$. Because there are four more electrons in bonding orbitals than in antibonding orbitals, the bond order is two (see Figure 9-5b and Table 9-1). We see why the molecule is much more stable than two free O atoms.

TABLE 9-1 *Molecular Orbitals for First- and Second-Period (Row) Diatomic Molecules*[a]

	H_2	He_2[c]	Li_2[b]	Be_2[c]	B_2[b]	C_2[b]	N_2		O_2	F_2	Ne_2[c]
σ_{2p}^*	—	—	—	—	—	—	—		—	—	⇅
$\pi_{2p_y}^*,\ \pi_{2p_z}^*$	— —	— —	— —	— —	— —	— —	— —		↑ ↑	⇅ ⇅	⇅ ⇅
σ_{2p}	—	—	—	—	—	—	⇅	$\pi_{2p_y},\ \pi_{2p_z}$	⇅ ⇅	⇅ ⇅	⇅ ⇅
$\pi_{2p_y},\ \pi_{2p_z}$	— —	— —	— —	— —	↑ ↑	⇅ ⇅	⇅ ⇅	σ_{2p}	⇅	⇅	⇅
σ_{2s}^*	—	—	—	⇅	⇅	⇅	⇅		⇅	⇅	⇅
σ_{2s}	—	—	⇅	⇅	⇅	⇅	⇅		⇅	⇅	⇅
σ_{1s}^*	—	⇅	⇅	⇅	⇅	⇅	⇅		⇅	⇅	⇅
σ_{1s}	⇅	⇅	⇅	⇅	⇅	⇅	⇅		⇅	⇅	⇅
Paramagnetic?	no	no	no	no	yes	no	no		yes	no	no
Bond order	1	0	1	0	1	2	3		2	1	0
Observed bond length (Å)	0.74	—	2.67	—	1.59	1.31	1.09		1.21	1.43	—
Observed bond energy (kJ/mol)	436	—	110	9	≈270	602	945		498	155	—

Increasing energy (not to scale)

[a] *Electron distribution in molecular orbitals, bond order, bond length, and bond energy of homonuclear diatomic molecules of the first- and second-period elements. Note that nitrogen molecules, N_2, have the highest bond energies listed; they have a bond order of three. The species C_2 and O_2, with a bond order of two, have the next highest bond energies.*

[b] *Exists only in the vapor state at elevated temperatures.*

[c] *Unknown species.*

Similarly, MO theory can be used to predict the structures and stabilities of ions, as Example 9-1 shows.

EXAMPLE 9-1 *Predicting Stabilities and Bond Orders*

Predict the stabilities and bond orders of the ions (a) O_2^+ and (b) O_2^-.

Plan

(a) The O_2^+ ion is formed by removing one electron from the O_2 molecule. The electrons that are withdrawn most easily are those in the highest energy orbitals. (b) The superoxide ion, O_2^-, results from adding an electron to the O_2 molecule.

Solution

(a) We remove one of the π_{2p}^* electrons of O_2 to find the configuration of O_2^+:

$$\sigma_{1s}^2\ \ \sigma_{1s}^{*2}\ \ \sigma_{2s}^2\ \ \sigma_{2s}^{*2}\ \ \sigma_{2p}^2\ \ \pi_{2p_y}^2\ \ \pi_{2p_z}^2\ \ \pi_{2p_y}^{*1}$$

There are five more electrons in bonding orbitals than in antibonding orbitals, so the bond order is 2.5. We conclude that the ion would be reasonably stable relative to other diatomic ions, and it does exist.

In fact, the unusual ionic compound $[O_2^+][PtF_6^-]$ played an important role in the discovery of the first noble gas compound, $XePtF_6$ (Section 24-2).

(b) We add one electron to the appropriate orbital of O_2 to find the configuration of O_2^-. Following Hund's Rule, we add this electron into the $\pi_{2p_y}^*$ orbital to form a pair:

$$\sigma_{1s}^2\ \ \sigma_{1s}^{*2}\ \ \sigma_{2s}^2\ \ \sigma_{2s}^{*2}\ \ \sigma_{2p}^2\ \ \pi_{2p_y}^2\ \ \pi_{2p_z}^2\ \ \pi_{2p_y}^{*2}\ \ \pi_{2p_y}^{*1}$$

> There are three more bonding electrons than antibonding electrons, so the bond order is 1.5. We conclude that the ion should exist but be less stable than O_2.

The known superoxides of the heavier Group IA elements—KO_2, RbO_2, and CsO_2—contain the superoxide ion, O_2^-. These compounds are formed by combination of the free metals with oxygen (Section 6-8, second subsection).

You should now work Exercises 21 and 22.

The Fluorine Molecule, F_2

Each fluorine atom has 9 electrons, so there are 18 electrons in F_2.

$$\sigma_{1s}^2 \quad \sigma_{1s}^{*2} \quad \sigma_{2s}^2 \quad \sigma_{2s}^{*2} \quad \sigma_{2p}^2 \quad \pi_{2p_y}^2 \quad \pi_{2p_z}^2 \quad \pi_{2p_y}^{*2} \quad \pi_{2p_z}^{*2}$$

$$F_2 \text{ bond order} = \frac{10 - 8}{2} = 1$$

The bond order is one. As you know, F_2 exists. The F—F bond distance is longer (1.43 Å) than the bond distances in O_2 (1.21 Å) or N_2 (1.09 Å) molecules. The bond order in F_2 (one) is less than that in O_2 (two) or N_2 (three). The bond energy of the F_2 molecules is lower than that of either O_2 or N_2 (see Table 9-1). As a result, F_2 molecules are the most reactive of the three.

Heavier Homonuclear Diatomic Molecules

It might appear reasonable to use the same types of molecular orbital diagrams to predict the stability or existence of homonuclear diatomic molecules of the third and subsequent periods. However, the heavier halogens, Cl_2, Br_2, and I_2, which contain only sigma (single) bonds, are the only well-characterized examples at room temperature. We would predict from both molecular orbital theory and valence bond theory that the other (nonhalogen) homonuclear diatomic molecules from below the second period would exhibit pi bonding and therefore multiple bonding.

Some heavier elements, such as S_2, exist as diatomic species in the vapor phase at elevated temperatures. These species are neither common nor very stable. The instability is related to the inability of atoms of the heavier elements to form strong pi bonds *with each other*. For larger atoms, the sigma bond length is too great to allow the atomic *p* orbitals on different atoms to overlap very effectively. The strength of pi bonding therefore decreases rapidly with increasing atomic size. For example, N_2 is *much* more stable than P_2. This is because the 3*p* orbitals on one P atom do not overlap side by side in a pi-bonding manner with corresponding 3*p* orbitals on another P atom nearly as effectively as do the corresponding 2*p* orbitals on the smaller N atoms. MO theory does not predict multiple bonding for Cl_2, Br_2, or I_2, each of which has a bond order of one.

9-5 HETERONUCLEAR DIATOMIC MOLECULES

Heteronuclear Diatomic Molecules of Second-Period Elements

Corresponding atomic orbitals of two different elements, such as the 2*s* orbitals of nitrogen and oxygen atoms, have different energies because their nuclei have different charges and therefore different attractions for electrons. Atomic orbitals of the *more electronegative element* are *lower* in energy than the corresponding orbitals of the less electronegative element. Accordingly, a molecular orbital diagram such as Figure 9-5 is inappropriate for *heteronuclear* diatomic molecules. If the two elements are similar (as in NO or CN molecules, for example), we can modify the diagram of Figure 9-5 by skewing it slightly. Figure 9-7 shows the energy level diagram and electron configuration for nitrogen oxide, NO, also known as nitric oxide.

The closer the energy of a molecular orbital is to the energy of one of the atomic orbitals from which it is formed, the more of the character of that atomic orbital it shows. Thus, as we see in Figure 9-7, the bonding MOs in the NO molecule have more oxygen-like atomic orbital character, and the antibonding orbitals have more nitrogen-like atomic orbital character.

In general the energy differences ΔE_1, ΔE_2, and ΔE_3 (*green backgrounds* in Figure 9-7) depend on the difference in electronegativities between the two atoms. The greater these energy differences, the more polar is the bond joining the atoms and the greater is its ionic character. On the other hand, the energy differences reflect the degree of overlap between atomic orbitals; the smaller these differences are, the more the orbitals can overlap, and the greater the covalent character of the bond is.

We see that NO has a total of 15 electrons, making it isoelectronic with the N_2^- ion. The distribution of electrons is therefore the same in NO as in N_2^-, although we expect the energy levels of the MOs to be different. In accord with our predictions, nitrogen oxide is a stable molecule. It has a bond order of 2.5, a short nitrogen–oxygen bond length of 1.15 Å, a low dipole moment of 0.15 D, and a high bond energy of 891 kJ/mol.

Note: CN is a reactive molecule, not the stable cyanide ion, CN^-.

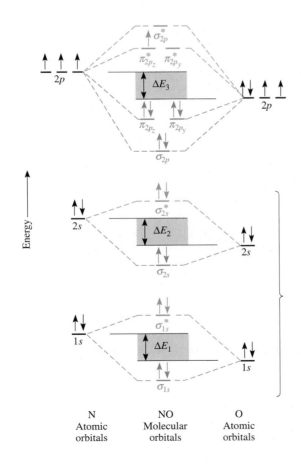

Figure 9-7 MO energy level diagram for nitrogen oxide, NO, a slightly polar heteronuclear diatomic molecule ($\mu = 0.15$ D). The atomic orbitals of oxygen, the more electronegative element, are a little lower in energy than the corresponding atomic orbitals of nitrogen, the less electronegative element. For this molecule, the energy differences ΔE_1, ΔE_2, and ΔE_3 are not very large; the molecule is not very polar.

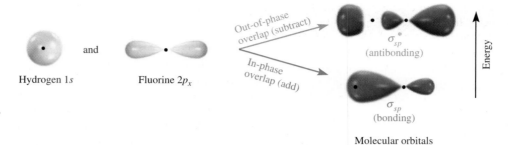

Figure 9-8 Formation of σ_{sp} and σ_{sp}^* molecular orbitals in HF by overlap of the 1s orbital of H with a 2p orbital of F.

The Hydrogen Fluoride Molecule, HF

The electronegativity difference between hydrogen (EN = 2.1) and fluorine (EN = 4.0) is very large (Δ(EN) = 1.9). The hydrogen fluoride molecule contains a very polar bond (μ = 1.91 D). The bond in HF involves the 1s electron of H and an unpaired electron from a 2p orbital of F. Figure 9-8 shows the overlap of the 1s orbital of H with a 2p orbital of F to form σ_{sp} and σ_{sp}^* molecular orbitals. The remaining two 2p orbitals of F have no net overlap with H orbitals. They are called **nonbonding orbitals.** The same is true for the F 2s and 1s orbitals. These nonbonding orbitals retain the characteristics of the F atomic orbitals from which they are formed. The MO diagram of HF is shown in Figure 9-9.

Other Diatomic Species with Large Δ(EN) Values

If the energies of the atomic orbitals of the two atoms of a diatomic molecule or ion are quite different, the MO diagram may be unlike that known for any homonuclear species. Its unique MO diagram is constructed by combining the Schrödinger equations for the two atoms. Construction of the MO diagram for CO is a complex case, beyond the coverage in this textbook.

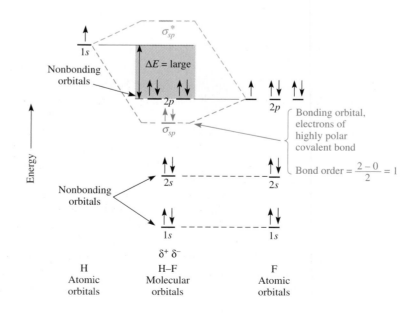

Figure 9-9 MO energy level diagram for hydrogen fluoride, HF, a very polar molecule (μ = 1.91 D). ΔE is large because the electronegativity difference is large.

9-6 DELOCALIZATION AND THE SHAPES OF MOLECULAR ORBITALS

In Section 7-6 we described resonance formulas for molecules and polyatomic ions. Resonance is said to exist when two or more equivalent Lewis formulas can be written for the same species and a single such formula does not account for the properties of a substance. In molecular orbital terminology, a more appropriate description involves **delocalization** of electrons. The shapes of molecular orbitals for species in which electron delocalization occurs can be predicted by combining all the contributing atomic orbitals.

This section includes pictorial representations of molecular orbitals. The details of the corresponding molecular orbital diagrams are too complicated to present here.

The Carbonate Ion, CO_3^{2-}

Consider the trigonal planar carbonate ion, CO_3^{2-}, as an example. All the carbon–oxygen bonds in the ion have the same bond length and the same energy, intermediate between those of typical C—O and C=O bonds. Valence bond theory describes the ion in terms of three contributing resonance structures (Figure 9-10a). No one of the three resonance forms adequately describes the bonding.

According to valence bond theory, the C atom is described as sp^2 hybridized, and it forms one sigma bond with each of the three O atoms. This leaves one unhybridized $2p$ atomic orbital on the C atom, say the $2p_z$ orbital. This orbital is capable of overlapping and mixing with the $2p_z$ orbital of any of the three O atoms. The sharing of two electrons in the resulting localized pi orbital would form a pi bond. Thus, three equivalent resonance structures can be drawn in valence bond terms (Figure 9-10b). We emphasize that there is *no evidence* for the existence of these separate resonance structures.

The average carbon–oxygen bond order in the CO_3^{2-} ion is $1\frac{1}{3}$.

The MO description of the pi bonding involves the simultaneous overlap and mixing of the carbon $2p_z$ orbital with the $2p_z$ orbitals of all three oxygen atoms. This forms a delocalized bonding pi molecular orbital system extending above and below the plane of the sigma system, as well as an antibonding pi orbital system. Electrons are said to occupy the entire set of bonding pi MOs, as depicted in Figure 9-10c. The shape is obtained by averaging the contributing valence bond resonance structures. The bonding in such species as nitrate ion, NO_3^-, and ozone, O_3, can be described similarly.

The Benzene Molecule, C_6H_6

Now let us consider the benzene molecule, C_6H_6, whose two valence bond resonance forms are shown in Figure 9-11a. The valence bond description involves sp^2 hybridization at each C atom. Each C atom is at the center of a trigonal plane, and the entire molecule is known to be planar. There are sigma bonds from each C atom to the two adjacent C atoms and to one H atom. This leaves one unhybridized $2p_z$ orbital on each C atom and one remaining valence electron for each. According to valence bond theory, adjacent pairs of $2p_z$ orbitals and the six remaining electrons occupy the regions of overlap to form a total of three pi bonds in either of the two ways shown in Figure 9-11b.

There is no evidence for the existence of either of these forms of benzene. The MO description of benzene is far better than the valence bond description.

Experimental studies of the C_6H_6 structure prove that it does *not* contain alternating single and double carbon–carbon bonds. The usual C—C single bond length is 1.54 Å, and the usual C=C double bond length is 1.34 Å. All six of the carbon–carbon bonds in benzene are the same length, 1.39 Å, intermediate between those of single and double bonds.

This is well explained by the MO theory, which predicts that the six $2p_z$ orbitals of the C atoms overlap and mix to form three pi-bonding and three pi-antibonding molecular

continued on page 364

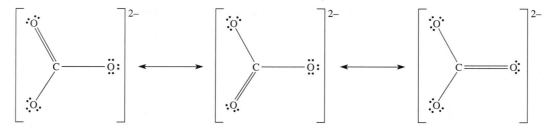

(a) Lewis formulas for valence bond resonance structures

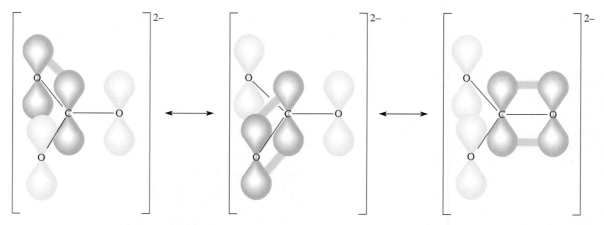

(b) *p*-Orbital overlap in valence bond resonance

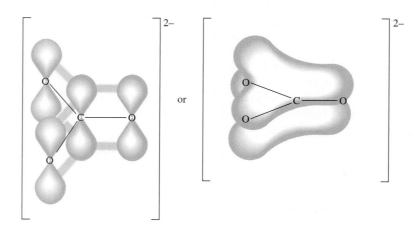

or

(c) Delocalized MO representation

Figure 9-10   Alternative representations of the bonding in the carbonate ion, CO_3^{2-}.
(a) Lewis formulas of the three valence bond resonance structures. (b) Representation of
the *p* orbital overlap in the valence bond resonance structures. In each resonance form,
the *p* orbitals on two atoms would overlap to form the π components of the hypothetical
double bonds. Each O atom has two additional sp^2 orbitals (not shown) in the plane of
the nuclei. Each of these additional sp^2 orbitals contains an oxygen unshared pair. (c) In
the MO description, the electrons in the π-bonded region are spread out, or *delocalized,*
over all four atoms of the CO_3^{2-} ion. This MO description is more consistent with the experimental observation of equal bond lengths and energies than are the valence bond
pictures in parts (a) and (b).

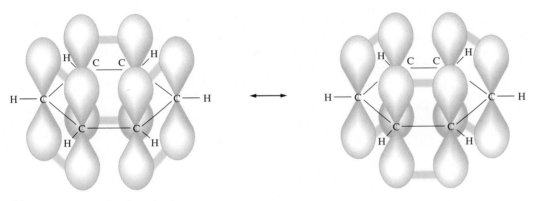

(a) Lewis formulas for valence bond resonance structures

(b) *p*-Orbital overlap in valence bond resonance structures

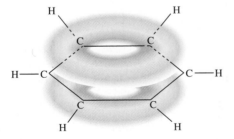

(c) Delocalized MO representation

Figure 9-11   Representations of the bonding in the benzene molecule, C_6H_6. (a) Lewis formulas of the two valence bond resonance structures. (b) The six *p* orbitals of the benzene ring, shown overlapping to form the (hypothetical) double bonds of the two resonance forms of valence bond theory. (c) In the MO description the six electrons in the pi-bonded region are *delocalized*, meaning they occupy an extended pi-bonding region above and below the plane of the six C atoms.

CD-ROM   Screen 10.13,
Molecular Orbitals and Vision.

orbitals. For instance, the most strongly bonding pi molecular orbital in the benzene pi–MO system is that in Figure 9-11c. The six pi electrons occupy three bonding MOs of this extended (delocalized) system. Thus, they are distributed throughout the molecule as a whole, above and below the plane of the sigma-bonded framework. This results in identical character for all carbon–carbon bonds in benzene. Each carbon–carbon bond has a bond order of 1.5. The MO representation of the extended pi system is the same as that obtained by averaging the two contributing valence bond resonance structures.

Key Terms

Antibonding electrons   Electrons in antibonding orbitals.

Antibonding orbital   A molecular orbital higher in energy than any of the atomic orbitals from which it is derived; when populated with electrons, lends instability to a molecule or ion. Denoted with an asterisk (*) superscript on its symbol.

Bond energy   The amount of energy necessary to break one mole of bonds of a given kind (in the gas phase).

Bond order   Half the number of electrons in bonding orbitals minus half the number of electrons in antibonding orbitals.

Bonding electrons   Electrons in bonding orbitals

Bonding molecular orbital   A molecular orbital lower in energy than any of the atomic orbitals from which it is derived; when populated with electrons, lends stability to a molecule or ion.

Degenerate orbitals   Orbitals of the same energy.

Delocalization   The formation of a set of molecular orbitals that extend over more than two atoms; important in species that valence bond theory describes in terms of resonance (Section 7-6).

Heteronuclear   Consisting of different elements.

Homonuclear   Consisting of only one element.

Molecular orbital (MO)   An orbital resulting from overlap and mixing of atomic orbitals on different atoms. An MO belongs to the molecule as a whole.

Molecular orbital theory   A theory of chemical bonding based on the postulated existence of molecular orbitals.

Nodal plane (node)   A region in which the probability of finding an electron is zero.

Nonbonding molecular orbital   A molecular orbital derived only from an atomic orbital of one atom; lends neither stability nor instability to a molecule or ion when populated with electrons.

Pi (π) bond   A bond resulting from electron occupation of a pi molecular orbital.

Pi (π) molecular orbital   A molecular orbital resulting from side-on overlap of atomic orbitals.

Sigma (σ) bond   A bond resulting from electron occupation of a sigma molecular orbital.

Sigma (σ) molecular orbital   A molecular orbital resulting from head-on overlap of two atomic orbitals.

Exercises

MO Theory: General Concepts

1. Describe the main differences between the valence bond theory and the molecular orbital theory.

2. In molecular orbital theory, what is a molecular orbital? What two types of information can be obtained from molecular orbital calculations? How do we use such information to describe the bonding within a molecule?

3. What is the relationship between the maximum number of electrons that can be accommodated by a set of molecular orbitals and the maximum number that can be accommodated by the atomic orbitals from which the MOs are formed? What is the maximum number of electrons that one MO can hold?

4. Answer Exercise 3 after replacing "molecular orbitals" with "hybridized atomic orbitals."

5. What differences and similarities exist among (a) atomic

orbitals, (b) localized hybridized atomic orbitals according to valence bond theory, and (c) molecular orbitals?

6. Describe the shapes, including the locations of the nuclei, of σ and σ^* orbitals.

7. Describe the shapes, including the locations of the nuclei, of π and π^* orbitals.

8. State the three rules for placing electrons in molecular orbitals.

9. What is meant by the term "bond order"? How is the value of the bond order calculated?

10. Compare and illustrate the differences between (a) atomic orbitals and molecular orbitals, (b) bonding and antibonding molecular orbitals, (c) σ bonds and π bonds, and (d) localized and delocalized molecular orbitals.

11. Is it possible for a molecule or polyatomic ion in its ground state to have a negative bond order? Why?

12. What is the relationship between the energy of a bonding

molecular orbital and the energies of the original atomic orbitals? What is the relationship between the energy of an antibonding molecular orbital and the energies of the original atomic orbitals?

13. Compare and contrast the following three concepts: (a) bonding orbitals; (b) antibonding orbitals; (c) nonbonding orbitals.

Homonuclear Diatomic Species

14. What do we mean when we say that a molecule or ion is (a) homonuclear, (b) heteronuclear, or (c) diatomic?
15. Use the appropriate molecular orbital energy diagram to write the electron configurations of the following molecules and ions: (a) Be_2, Be_2^+, Be_2^-; (b) B_2, B_2^+, B_2^-.
16. What is the bond order of each of the species in Exercise 15?
17. Which of the species in Exercise 15 are diamagnetic and which are paramagnetic?
18. Use MO theory to predict relative stabilities of the species in Exercise 15. Comment on the validity of these predictions. What else *must* be considered in addition to electron occupancy of MOs?
19. Use the appropriate molecular orbital energy diagram to write the electron configuration for each of the following; calculate the bond order of each, and predict which would exist. (a) H_2^+; (b) H_2; (c) H_2^-; (d) H_2^{2-}.
20. Repeat Exercise 19 for (a) He_2^+; (b) He_2; (c) He_2^{2+}.
21. Repeat Exercise 19 for (a) N_2; (b) Ne_2; (c) C_2^{2-}.
22. Repeat Exercise 19 for (a) Li_2; (b) Li_2^+; (c) O_2^{2-}.
*23. Which homonuclear diatomic molecules or ions of the second period have the following electron distributions in MOs? In other words, identify X in each.
   (a) X_2 $\sigma_{1s}^2 \sigma_{1s}^{*2} \sigma_{2s}^2 \sigma_{2s}^{*2} \pi_{2p_y}^2 \pi_{2p_z}^2 \sigma_{2p}^2$
   (b) X_2 $\sigma_{1s}^2 \sigma_{1s}^{*2} \sigma_{2s}^2 \sigma_{2s}^{*2} \sigma_{2p}^2 \pi_{2p_y}^2 \pi_{2p_z}^2 \pi_{2p_y}^{*1} \pi_{2p_z}^{*1}$
   (c) X_2^- $\sigma_{1s}^2 \sigma_{1s}^{*2} \sigma_{2s}^2 \sigma_{2s}^{*2} \pi_{2p_y}^2 \pi_{2p_z}^2 \sigma_{2p}^2 \pi_{2p_y}^{*1}$
24. What is the bond order of each of the species in Exercise 23?
25. Assuming that the σ_{2p} MO is lower in energy than the π_{2p_y} and π_{2p_z} MOs for the following species, write out electron configurations for (a) F_2, F_2^-, F_2^+; (b) C_2, C_2^+, C_2^-.
26. (a) What is the bond order of each species in Exercise 25? (b) Are they diamagnetic or paramagnetic? (c) What would MO theory predict about the stabilities of these species?
27. (a) Give the MO designations for O_2, O_2^-, O_2^{2-}, O_2^+, and O_2^{2+}. (b) Give the bond order in each case. (c) Match these species with the following observed bond lengths: 1.04 Å; 1.12 Å; 1.21 Å; 1.33 Å; and 1.49 Å.
28. (a) Give the MO designations for N_2, N_2^-, and N_2^+. (b) Give the bond order in each case. (c) Rank these three species by increasing predicted bond length.

Heteronuclear Diatomic Species

The following is a molecular orbital energy level diagram for a heteronuclear diatomic molecule, XY, in which both X and Y are from Period 2 and Y is slightly more electronegative. This diagram may be used in answering questions in this section.

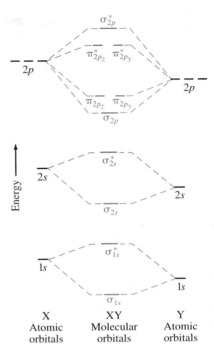

	X	XY	Y
	Atomic orbitals	Molecular orbitals	Atomic orbitals

29. Use the preceding diagram to fill in an MO diagram for NO^-. What is the bond order of NO^-? Is it paramagnetic? How would you assess its stability?
30. Repeat Exercise 29 for NO^+.
31. Repeat Exercise 29 for CN^+. Refer to the preceding diagram but assume that the π_{2p_y} and π_{2p_z} MOs are lower in energy than the σ_{2p} MO.
32. Compare the MO descriptions for CN, CN^-, and CN^{2-}. Refer to the preceding diagram but assume that the π_{2p_y} and π_{2p_z} MOs are lower in energy than the σ_{2p} MO. Which would be most stable? Why?
33. For each of the two species OF and OF^-: (a) Draw MO energy level diagrams. (b) Write out electron configurations. (c) Determine bond orders and predict relative stabilities. (d) Predict diamagnetism or paramagnetism.

O_2 is paramagnetic

34. For each of the two species NF and NF^+: (a) Draw MO energy level diagrams. (b) Write out electron configurations. (c) Determine bond orders and predict relative stabilities. (d) Predict diamagnetism or paramagnetism.

35. Considering the shapes of MO energy level diagrams for nonpolar covalent and polar covalent molecules, what would you predict about MO diagrams, and therefore about overlap of atomic orbitals, for ionic compounds?

36. To increase the strength of the bonding in the hypothetical compound BC, would you add or subtract an electron? Explain your answer with the aid of an MO electron structure.

Delocalization

37. Use Lewis formulas to depict the resonance structures of the following species from the valence bond point of view, and then sketch MOs for the delocalized π systems. (a) NO_3^-, nitrate ion; (b) HCO_3^-, hydrogen carbonate ion (H is bonded to O); (c) NO_2^-, nitrite ion.

38. Use Lewis formulas to depict the resonance structures of the following species from the valence bond point of view, and then sketch MOs for the delocalized π systems: (a) SO_2, sulfur dioxide; (b) O_3, ozone; (c) HCO_2^-, formate ion (H is bonded to C).

Mixed Exercises

39. Draw and label the complete MO energy level diagrams for the following species. For each, determine the bond order, predict the stability of the species, and predict whether the species will be paramagnetic or diamagnetic. (a) He_2^+; (b) CN; (c) HeH^+.

40. Draw and label the complete MO energy level diagrams for the following species. For each, determine the bond order, predict the stability of the species, and predict whether the species will be paramagnetic or diamagnetic. (a) O_2^{2+}; (b) HO^-; (c) HF.

41. Which of these species would you expect to be paramagnetic or diamagnetic? (a) He_2^-; (b) N_2; (c) NO^+; (d) N_2^{2+}; (e) F_2^+.

CONCEPTUAL EXERCISES

42. Refer to the diagrams in Figure 9-5 as needed. Can the bond order of a diatomic species having 20 or fewer electrons be greater than three? Can the bond order be a value that is not divisible by 0.5? Why?

43. As NO ionizes to form NO^+, does the nitrogen–oxygen bond become stronger or weaker?

$$NO \longrightarrow NO^+ + e^-$$

44. Which of the homonuclear diatomic molecules of the second row of the periodic table (Li_2 to Ne_2) are predicted by MO theory to be paramagnetic? Which ones are predicted to have a bond order of one? Which ones are predicted to have a bond order of two? Which one is predicted to have the highest bond order?

*45. Use valence bond and molecular orbital theory to describe the change in Cl—Cl bond length when Cl_2 loses an electron to form Cl_2^+. Would the cation be diamagnetic or paramagnetic? Explain.

*46. Briefly explain the finding that the N—O bond is longer in NO^+ than in NO. Should NO^- exist, and if so, how would its bond length compare with those in NO^+ and NO?

BUILDING YOUR KNOWLEDGE

47. When carbon vaporizes at extremely high temperatures, among the species present in the vapor is the diatomic molecule C_2. Write a Lewis formula for C_2. Does your Lewis formula of C_2 obey the octet rule? (C_2 does not contain a quadruple bond.) Does C_2 contain a single, a double, or a triple bond? Is it paramagnetic or diamagnetic? Show how molecular orbital theory can be used to predict the answers to questions left unanswered by valence bond theory.

48. Rationalize the following observations in terms of the stabilities of σ and π bonds: (a) The most common form of nitrogen is N_2, whereas the most common form of phosphorus is P_4 (see the structure in Figure 2-3); (b) The most common forms of oxygen are O_2 and (less common) O_3, whereas the most common form of sulfur is S_8.

BEYOND THE TEXTBOOK

Go to the textbook website at

http://www.brookscole.com/chemistry/whitten

for additional activities and exercises based on the General Chemistry Interactive CD-ROM, the World Wide Web, and library resources.

InfoTrac College Edition

For additional readings, go to InfoTrac College Edition, your online research library at:

http://infotrac.thomsonlearning.com

Reactions in Aqueous Solutions I: Acids, Bases, and Salts

10

D. Winters

Charles D. Winters

Ammonium chloride (NH₄Cl), is the white salt formed from the reaction of the base ammonia (NH₃) and hydrochloric acid (HCl). Heating NH₄Cl in a spoon produces some gaseous HCl and NH₃ that react again above the spoon to form white NH₄Cl "smoke."

OUTLINE

OBJECTIVES

After you have studied this chapter, you should be able to

- Describe the Arrhenius theory of acids and bases
- Describe hydrated hydrogen ions
- Describe the Brønsted–Lowry theory of acids and bases
- List properties of aqueous solutions of acids
- List properties of aqueous solutions of bases
- Arrange binary acids in order of increasing strength
- Arrange ternary acids in order of increasing strength
- Describe the Lewis theory of acids and bases
- Complete and balance equations for acid–base reactions
- Define acidic and basic salts
- Explain amphoterism
- Describe methods for preparing acids

You will encounter many of these in your laboratory work.

In technological societies, acids, bases, and salts are indispensable compounds. Table 4-15 lists the 15 such compounds that were included in the top 50 chemicals produced in the United States in 2001. The production of H_2SO_4 (number 1) was almost twice as great as the production of lime (number 2). Sixty-five percent of the H_2SO_4 is used in the production of fertilizers.

Many acids, bases, and salts occur in nature and serve a wide variety of purposes. For instance, your stomach "digestive juice" contains approximately 0.10 mole of hydrochloric acid per liter. Human blood and the aqueous components of most cells are slightly basic. The liquid in your car battery is approximately 40% H_2SO_4 by mass. Baking soda is a salt of carbonic acid. Sodium hydroxide, a base, is used in the manufacture of soaps, paper, and many other chemicals. "Drāno" is solid NaOH that contains some aluminum chips. Sodium chloride is used to season food and as a food preservative. Calcium chloride is used to melt ice on highways and in the emergency treatment of cardiac arrest. Several ammonium salts are used as fertilizers. Many organic acids (carboxylic acids) and their derivatives occur

(a)

(b)

Mama G. Clarke

Left, Many common household liquids are acidic, including soft drinks, vinegar, and fruit juices. *Right,* Most cleaning materials are basic.

in nature. Acetic acid is present in vinegar; the sting of an ant bite is due to formic acid. Amino acids are the building blocks of proteins, which are important materials in the bodies of animals, including humans. Amino acids are carboxylic acids that also contain basic groups derived from ammonia. The pleasant odors and flavors of ripe fruit are due in large part to the presence of esters (Chapter 27), which are formed from the acids in unripe fruit.

SciLINKS.

TOPIC: Acids/Bases
GO TO: www.scilinks.org
*sci*LINKS CODE: WCH0430

10-1 PROPERTIES OF AQUEOUS SOLUTIONS OF ACIDS AND BASES

Aqueous solutions of most **protic acids** (those containing acidic hydrogen atoms) exhibit certain properties, which are properties of hydrated hydrogen ions in aqueous solution.

Chemists also use "protonic" to describe these acids.

1. They have a sour taste. Pickles are usually preserved in vinegar, a 5% solution of acetic acid. Many pickled condiments contain large amounts of sugar so that the taste of acetic acid is partially masked by the sweet taste of sugar. Lemons contain citric acid, which is responsible for their characteristic sour taste.

2. They change the colors of many indicators (highly colored dyes whose colors depend on the acidic or basic character of the solution). Acids turn blue litmus red, and cause bromthymol blue to change from blue to yellow.

3. Nonoxidizing acids react with metals above hydrogen in the activity series (Section 4-10, Part 2) to liberate hydrogen gas, H_2. (HNO_3, a common oxidizing acid, reacts with metals to produce primarily nitrogen oxides.)

4. They react with (neutralize) metal oxides and metal hydroxides to form salts and water (Section 4-11, Part 1).

5. They react with salts of weaker acids to form the weaker acid and the salt of the stronger acid; for example,

$$3HCl(aq) + Na_3PO_4(aq) \longrightarrow H_3PO_4(aq) + 3NaCl(aq).$$

6. Aqueous solutions of acids conduct an electric current because they are totally or partly ionized.

Caution: We should *never* try to identify a substance in the laboratory by taste. You have, however, probably experienced the sour taste of acetic acid in vinegar or citric acid in foods that contain citrus fruits.

The indicator bromthymol blue is yellow in acidic solution and blue in basic solution.

Aqueous solutions of most bases also exhibit certain properties, which are due to the hydrated hydroxide ions present in aqueous solutions of bases.

1. They have a bitter taste.
2. They have a slippery feeling. Soaps are common examples that are mildly basic. A solution of household bleach feels very slippery because it is quite basic.
3. They change the colors of many indicators: bases turn litmus from red to blue, and bromthymol blue changes from yellow to blue.
4. They react with (neutralize) acids to form salts and, in most cases, water.
5. Their aqueous solutions conduct an electric current because they are dissociated or ionized.

10-2 THE ARRHENIUS THEORY

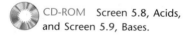

This is an extremely important concept.

CD-ROM   Screen 5.8, Acids, and Screen 5.9, Bases.

In 1680, Robert Boyle noted that acids (1) dissolve many substances, (2) change the colors of some natural dyes (indicators), and (3) lose their characteristic properties when mixed with alkalis (bases). By 1814, J. Gay-Lussac concluded that acids *neutralize* bases and that the two classes of substances should be defined in terms of their reactions with each other.

In 1884, Svante Arrhenius (1859–1927) presented his theory of electrolytic dissociation, which resulted in the Arrhenius theory of acid–base reactions. In his view,

> an **acid** is a substance that contains hydrogen and produces H^+ in aqueous solution. A **base** is a substance that contains the OH (hydroxyl) group and produces hydroxide ions, OH^-, in aqueous solution.

Neutralization is defined as the combination of H^+ ions with OH^- ions to form H_2O molecules.

We now know that all ions are hydrated in aqueous solution; they interact weakly with water molecules.

$$H^+(aq) + OH^-(aq) \longrightarrow H_2O(\ell) \qquad \text{(neutralization)}$$

Review Sections 4-2, 4-11, part 1, 6-7, and 6-8.

The Arrhenius theory of acid–base behavior satisfactorily explained reactions of *protic acids* with metal hydroxides (hydroxy bases). It was a significant contribution to chemical thought and theory in the latter part of the nineteenth century. The Arrhenius model of acids and bases, although limited in scope, led to the development of more general theories of acid–base behavior. They will be considered in later sections.

10-3 THE HYDRONIUM ION (HYDRATED HYDROGEN ION)

The most common isotope of hydrogen, ^1H, has no neutrons. Thus, $^1H^+$ is a bare proton. In discussions of acids and bases, we use the terms "hydrogen ion," "proton," and "H^+" interchangeably.

Although Arrhenius described H^+ ions in water as bare ions (protons), we now know that they are hydrated in aqueous solution and exist as $H^+(H_2O)_n$, in which n is some small integer. This is due to the attraction of the H^+ ions, or protons, for the oxygen end ($\delta-$) of water molecules. Although we do not know the extent of hydration of H^+ in most solutions, we usually represent the hydrated hydrogen ion as the **hydronium ion**, H_3O^+, or $H^+(H_2O)_n$, in which $n = 1$.

> The hydrated hydrogen ion is the species that gives aqueous solutions of acids their characteristic acidic properties.

Whether we use the designation $H^+(aq)$ or H_3O^+, we always mean the hydrated hydrogen ion.

$$H^+ + :\overset{..}{\underset{|}{O}}{-}H \longrightarrow \left[H{-}\overset{..}{\underset{|}{O}}{-}H\right]^+$$

SCiLINKS.

TOPIC: Hydronium Ion
GO TO: www.scilinks.org
sciLINKS CODE: WCH1020

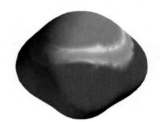

Hydronium ion, H_3O^+.

10-4 THE BRØNSTED–LOWRY THEORY

In 1923, J. N. Brønsted (1879–1947) and T. M. Lowry (1874–1936) independently presented logical extensions of the Arrhenius theory. Brønsted's contribution was more thorough than Lowry's, and the result is known as the **Brønsted theory** or the **Brønsted–Lowry theory.**

> An **acid** is defined as a *proton donor* (H^+), and a **base** is defined as a *proton acceptor.*

These definitions are sufficiently broad that any hydrogen-containing molecule or ion capable of releasing a proton, H^+, is an acid, whereas any molecule or ion that can accept a proton is a base. (In the Arrhenius theory of acids and bases, only substances that contain the OH^- group would be called bases.)

> An acid–base reaction is the transfer of a proton from an acid to a base.

Thus, the complete ionization of hydrogen chloride, HCl, a *strong* acid, in water is an acid–base reaction in which water acts as a base or proton acceptor.

Step 1: $HCl(aq) \longrightarrow H^+(aq) + Cl^-(aq)$ (Arrhenius description)
Step 2: $H_2O(\ell) + H^+(aq) \longrightarrow H_3O^+$
Overall: $H_2O(\ell) + HCl(aq) \longrightarrow H_3O^+ + Cl^-(aq)$ (Brønsted–Lowry description)

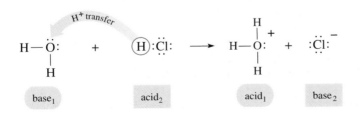

The Brønsted–Lowry theory is especially useful for reactions in aqueous solutions. It is widely used in medicine and in the biological sciences.

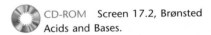

CD-ROM Screen 17.2, Brønsted Acids and Bases.

Remember that in this text we use red to indicate acids and blue to indicate bases.

The ionization of hydrogen fluoride, a *weak* acid, is similar, but it occurs to only a slight extent, so we use a double arrow to indicate that it is reversible.

Various measurements (electrical conductivity, freezing point depression, etc.) indicate that HF is only *slightly* ionized in water.

The double arrow is used to indicate that the reaction occurs in both the forward and the reverse directions. It does not mean that there are equal amounts of reactants and products present.

We use rectangles to indicate one conjugate acid–base pair and ovals to indicate the other pair.

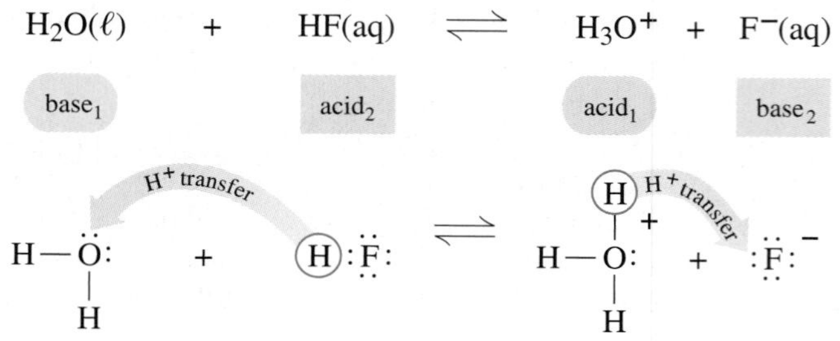

It makes no difference which conjugate acid–base pair, HF and F^- or H_3O^+ and H_2O, is assigned the subscripts 1 and 2.

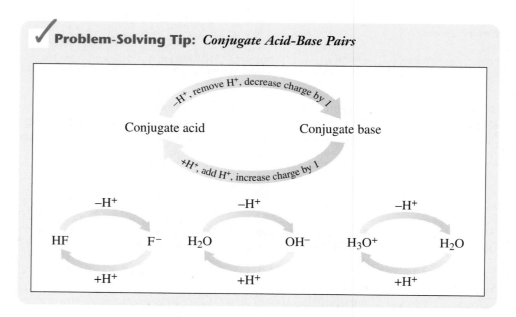

We can describe Brønsted–Lowry acid–base reactions in terms of **conjugate acid–base pairs.** These are two species that differ by a proton. In the preceding equation, HF ($acid_2$) and F^- ($base_2$) are one conjugate acid–base pair, and H_2O ($base_1$) and H_3O^+ ($acid_1$) are the other pair. The members of each conjugate pair are designated by the same numerical subscript. In the forward reaction, HF and H_2O act as acid and base, respectively. In the reverse reaction, H_3O^+ acts as the acid, or proton donor, and F^- acts as the base, or proton acceptor.

✔ **Problem-Solving Tip:** *Conjugate Acid-Base Pairs*

When the *weak* acid, HF, dissolves in water, the HF molecules give up some H^+ ions that can be accepted by either of two bases, F^- or H_2O. The fact that HF is only slightly ionized tells us that F^- is a stronger base than H_2O. When the *strong* acid, HCl, dissolves in water, the HCl molecules give up H^+ ions that can be accepted by either of two bases, Cl^- or H_2O. The fact that HCl is completely ionized in dilute aqueous solution tells us that Cl^- is a weaker base than H_2O. Thus, the weaker acid, HF, has the stronger conjugate base, F^-, while the stronger acid, HCl, has the weaker conjugate base, Cl^-. We can generalize:

F^- is a stronger base than H_2O. H_2O is a stronger base than Cl^-. F^- is therefore a stronger base than Cl^-.

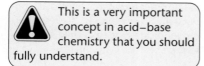 This is a very important concept in acid–base chemistry that you should fully understand.

> The stronger the acid, the weaker is its conjugate base; the weaker the acid, the stronger is its conjugate base.

"Strong" and "weak," like many adjectives, are used here in a relative sense. When we say that the fluoride anion, F^-, is a stronger base than Cl^- ion we do *not* necessarily mean that it is a strong base in an absolute sense. For example, hydroxide ion, OH^-, is also a stronger base than Cl^- ion. That does not mean that hydroxide and fluoride ions have similar base strengths. In fact, the aqueous hydroxide ion is a far stronger base than the fluoride ion. "Stronger" and "weaker," therefore, are simply comparative terms that we use to compare the basicity (or acidity) of two (or more) ions or molecules. We often compare base strengths either to the hydroxide ion, which is a very strong base, or to the *anions of the strong acids*, which are very weak bases (e.g., Cl^-, Br^-, I^-, NO_3^-).

Ammonia acts as a weak Brønsted–Lowry base, and water acts as an acid in the ionization of aqueous ammonia.

$$NH_3(aq) \;+\; H_2O(\ell) \;\rightleftharpoons\; NH_4^+(aq) \;+\; OH^-(aq)$$

base$_1$ acid$_2$ acid$_1$ base$_2$

Be careful to avoid confusing *solubility in water* and *extent of ionization*. They are not necessarily related. Ammonia is very *soluble* in water (≈ 15 mol/L at 25°C). In a 0.10 M solution, NH_3 is only 1.3% ionized and 98.7% nonionized.

ammonia   water   ammonium ion   hydroxide ion

As we see in the reverse reaction, ammonium ion, NH_4^+, is the conjugate acid of NH_3. The hydroxide ion, OH^-, is the conjugate base of water. In three dimensions, the molecular structures are

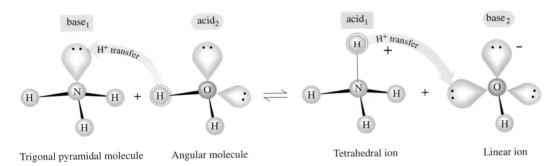

base$_1$ acid$_2$ acid$_1$ base$_2$

Trigonal pyramidal molecule   Angular molecule   Tetrahedral ion   Linear ion

Water acts as an acid (H^+ donor) in its reaction with NH_3, whereas it acts as a base (H^+ acceptor) in its reactions with HCl and with HF.

Whether water acts as an acid or as a base depends on the other species present.

EXAMPLE 10-1   *Identifying Conjugate Acid–Base Pairs*

(a) Write the equation for the ionization of HNO_3 in water. (b) Identify the conjugate acid–base pairs in the equation for part (a). (c) What is the conjugate base of H_3PO_4? (d) What is the conjugate acid of SO_4^{2-}?

Plan

When an acid ionizes (loses a H^+) the deprotonated product (often anionic) is the conjugate base, while a base that reacts with H^+ forms a protonated product that is the conjugate acid.

(a) HNO_3 is a strong acid and ionizes in water to produce $NO_3^-(aq)$ and H_3O^+ (H^+).

(b) HNO_3 is an acid, and its conjugate base is NO_3^-; H_2O is a base, and H_3O^+ is the conjugate acid.

(c) H_3PO_4 is an acid that reacts with water via the following equation:

$$H_3PO_4(aq) + H_2O(\ell) \rightleftharpoons H_2PO_4^-(aq) + H_3O^+(aq)$$

In this reaction the $H_2PO_4^-(aq)$ is the conjugate base to the $H_3PO_4(aq)$ acid.

(d) $SO_4^{2-}(aq)$ can react as a base with H^+ to form the conjugate acid $HSO_4^-(aq)$.

Solution

(a) $\overset{\text{acid}_1}{HNO_3(aq)} + \overset{\text{base}_2}{H_2O(\ell)} \longrightarrow \overset{\text{base}_1}{NO_3^-(aq)} + \overset{\text{acid}_2}{H_3O^+(aq)}$

(b) Conjugate acid/base pairs for the preceding reaction:

HNO_3 (acid) and NO_3^- (conjugate base)

H_2O (base) and H_3O^+ (conjugate acid)

(c) Conjugate base of H_3PO_4: $H_2PO_4^-$

(d) Conjugate acid of SO_4^{2-}: HSO_4^-

B—H, C—H, and N—H bonds in neutral molecules are usually too strong to ionize to produce H^+. Most acids have H bonded to an electronegative atom such as O (especially when the O atom is bonded to an electronegative element in a high oxidation state) or to a halogen (F, Cl, Br, I). N—H bonds to cationic nitrogen atoms are also weakly acidic (for example, NH_4^+).

Bases must have at least one lone pair on an atom to be able to form a new bond to H^+. Lone pairs on nitrogen and oxygen are the most common basic sites in molecules (e.g., NH_3 and H_2O). Anions, whether monatomic or polyatomic, typically have lone pairs that are basic; examples are F^-, OH^-, and PO_4^{3-}.

10-5 THE AUTOIONIZATION OF WATER

Careful measurements show that pure water ionizes ever so slightly to produce equal numbers of hydrated hydrogen ions and hydroxide ions.

Because an equal number of H^+ and OH^- ions are produced, the solution is neither acidic nor basic. This is referred to in acid–base terms as a *neutral* solution.

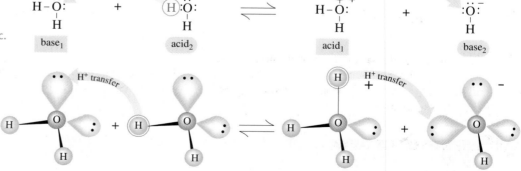

In simplified notation, we represent this reaction as

$$H_2O(\ell) \rightleftharpoons H^+(aq) + OH^-(aq)$$

CD-ROM   Screen 17.3, The Acid–Base Properties of Water: Autoionization

This **autoionization** (self-ionization) of water is an acid–base reaction according to the Brønsted–Lowry theory. One H_2O molecule (the acid) donates a proton to another H_2O molecule (the base). The H_2O molecule that donates a proton becomes an OH^- ion, the conjugate base of water. The H_2O molecule that accepts a proton becomes an H_3O^+ ion. Examination of the reverse reaction (right to left) shows that H_3O^+ (an acid) donates a proton to OH^- (a base) to form two H_2O molecules. One H_2O molecule behaves as an acid and the other acts as a base in the autoionization of water. Water is said to be **amphiprotic**; that is, H_2O molecules can both donate and accept protons.

As we saw in Section 4-11, Part 1, H_3O^+ and OH^- ions combine to form nonionized water molecules when strong acids and strong soluble bases react to form soluble salts and water. The reverse reaction, the autoionization of water, occurs only slightly, as expected.

The prefix "amphi-" means "of both kinds." "Amphiprotism" refers to amphoterism by accepting and donating a proton in different reactions (see Section 10-6).

10-6 AMPHOTERISM

As we have seen, whether a particular substance behaves as an acid or as a base depends on its environment. Earlier we described the amphiprotic nature of water. **Amphoterism** is a more general term that describes the ability of a substance to react either as an acid or as a base. *Amphiprotic behavior* describes the cases in which substances exhibit amphoterism by accepting and by donating a proton, H^+. Several *insoluble* metal hydroxides are amphoteric; that is, they react with acids to form salts and water, but they also dissolve in and react with excess strong bases.

Aluminum hydroxide is a typical amphoteric metal hydroxide. Its behavior as a *base* is illustrated by its reaction with nitric acid to form a *normal salt*. The balanced formula unit, total ionic, and net ionic equations for this reaction are, respectively:

A normal salt (see Section 10-9) contains no ionizable H atoms or OH groups.

$$Al(OH)_3(s) + 3HNO_3(aq) \longrightarrow Al(NO_3)_3(aq) + 3H_2O(\ell)$$

formula unit equation

$$Al(OH)_3(s) + 3[H^+(aq) + NO_3^-(aq)] \longrightarrow [Al^{3+}(aq) + 3NO_3^-(aq)] + 3H_2O(\ell)$$

total ionic equation

$$Al(OH)_3(s) + 3H^+(aq) \longrightarrow Al^{3+}(aq) + 3H_2O(\ell)$$

net ionic equation

When an excess of a solution of any strong base, such as NaOH, is added to solid aluminum hydroxide, the $Al(OH)_3$ acts like an acid and dissolves. The equation for the reaction is usually written

All hydroxides containing small, highly charged metal ions are insoluble in water.

$$Al(OH)_3(s) + NaOH(aq) \longrightarrow NaAl(OH)_4(aq)$$
$$\quad \text{an acid} \qquad \text{a base} \qquad \text{sodium aluminate,} \atop \text{a soluble compound}$$

The total ionic and net ionic equations are

$$Al(OH)_3(s) + [Na^+(aq) + OH^-(aq)] \longrightarrow [Na^+(aq) + Al(OH)_4^-(aq)]$$

$$Al(OH)_3(s) + OH^-(aq) \longrightarrow Al(OH)_4^-(aq)$$

Other amphoteric metal hydroxides undergo similar reactions.

Table 10-1 contains lists of the common amphoteric hydroxides. Three are hydroxides of metalloids, As, Sb, and Si, which are located along the line that divides metals and nonmetals in the periodic table.

Generally, elements of intermediate electronegativity form amphoteric hydroxides. Those of high and low electronegativity form acidic and basic "hydroxides," respectively.

10-7 STRENGTHS OF ACIDS

Binary Acids

The ease of ionization of binary protic acids depends on both (1) the ease of breaking H—X bonds and (2) the stability of the resulting ions in solution. Let us consider the relative

TABLE 10-1	Some Amphoteric Hydroxides	
Metal or Metalloid ions	Insoluble Amphoteric Hydroxide	Complex Ion Formed in an Excess of a Strong Base
Be^{2+}	$Be(OH)_2$	$[Be(OH)_4]^{2-}$
Al^{3+}	$Al(OH)_3$	$[Al(OH)_4]^-$
Cr^{3+}	$Cr(OH)_3$	$[Cr(OH)_4]^-$
Zn^{2+}	$Zn(OH)_2$	$[Zn(OH)_4]^{2-}$
Sn^{2+}	$Sn(OH)_2$	$[Sn(OH)_3]^-$
Sn^{4+}	$Sn(OH)_4$	$[Sn(OH)_6]^{2-}$
Pb^{2+}	$Pb(OH)_2$	$[Pb(OH)_4]^{2-}$
As^{3+}	$As(OH)_3$	$[As(OH)_4]^-$
Sb^{3+}	$Sb(OH)_3$	$[Sb(OH)_4]^-$
Si^{4+}	$Si(OH)_4$	SiO_4^{4-} and SiO_3^{2-}
Co^{2+}	$Co(OH)_2$	$[Co(OH)_4]^{2-}$
Cu^{2+}	$Cu(OH)_2$	$[Cu(OH)_4]^{2-}$

 A *weak* acid may be very reactive. For example, HF dissolves sand and glass. The equation for its reaction with sand is

$$SiO_2(s) + 4HF(g) \longrightarrow SiF_4(g) + 2H_2O(\ell)$$

The reaction with glass and other silicates is similar. These reactions are *not* related to acid strength; none of the three *strong* hydrohalic acids—HCl, HBr, or HI—undergoes such a reaction.

Bond strength is shown by the bond energies introduced in Chapter 7 and tabulated in Section 15-9. The strength of the H—F bond is due largely to the very small size of the F atom and the strong ionic attraction between the F^- and the H^+ ions.

strengths of the Group VIIA hydrohalic acids. Hydrogen fluoride ionizes only slightly in dilute aqueous solutions.

$$HF(aq) + H_2O(\ell) \rightleftharpoons H_3O^+(aq) + F^-(aq)$$

HCl, HBr, and HI, however, ionize completely or nearly completely in dilute aqueous solutions because the H—X bonds are much weaker.

$$HX(aq) + H_2O(\ell) \longrightarrow H_3O^+(aq) + X^-(aq) \qquad X = Cl, Br, I$$

The order of *bond strengths* for the hydrogen halides is

(strongest bonds) $HF \gg HCl > HBr > HI$ (weakest bonds)

To understand why HF is so much weaker an acid than the other hydrogen halides, let us consider the following factors.

1. In HF the electronegativity difference is 1.9, compared with 0.9 in HCl, 0.7 in HBr, and 0.4 in HI (Section 7-9). We might expect the very polar H—F bond in HF to ionize easily. The fact that HF is the *weakest* of these acids suggests that this effect must be of minor importance.

2. The bond strength is considerably greater in HF than in the other three molecules. This tells us that the H—F bond is harder to break than the H—Cl, H—Br, and H—I bonds.

3. The small, highly charged F^- ion, formed when HF ionizes, causes increased ordering of the water molecules. This increase is unfavorable to the process of ionization.

The net result of all factors is that HF is a much weaker acid than the other hydrohalic acids: HCl, HBr, and HI.

In dilute aqueous solutions, hydrochloric, hydrobromic, and hydroiodic acids are completely ionized, and all show the same apparent acid strength. Water is sufficiently basic that it does not distinguish among the acid strengths of HCl, HBr, and HI, and therefore it is referred to as a **leveling solvent** for these acids. It is not possible to determine the order of the strengths of these three acids in water because they are so nearly completely ionized.

When these compounds dissolve in anhydrous acetic acid or other solvents less basic than water, however, they exhibit significant differences in their acid strengths. The observed order of acid strengths is

$$HCl < HBr < HI$$

We observe that

> the hydronium ion is the strongest acid that can exist in aqueous solution. All acids stronger than $H_3O^+(aq)$ react completely with water to produce $H_3O^+(aq)$ and their conjugate bases.

CD-ROM Screen 17.13, The Acid–Base Properties of Water: Autoionization.

This is called the **leveling effect** of water. For example $HClO_4$ (Table 10-2) reacts completely with H_2O to form $H^+(aq)$ and $ClO_4^-(aq)$.

$$HClO_4(aq) + H_2O(\ell) \longrightarrow H_3O^+(aq) + ClO_4^-(aq)$$

Similar observations have been made for aqueous solutions of strong bases such as NaOH and KOH. Both are completely dissociated in dilute aqueous solutions.

$$NaOH(s) \xrightarrow{H_2O} Na^+(aq) + OH^-(aq)$$

> The hydroxide ion is the strongest base that can exist in aqueous solution. Bases stronger than OH^- react completely with H_2O to produce OH^- and their conjugate acids.

When metal amides such as sodium amide, $NaNH_2$, are placed in H_2O, the amide ion, NH_2^-, reacts with H_2O completely.

$$NH_2^-(aq) + H_2O(\ell) \longrightarrow NH_3(aq) + OH^-(aq)$$

Thus, we see that H_2O is a leveling solvent for all bases stronger than OH^-.

The amide ion, NH_2^-, is a stronger base than OH^-.

TABLE 10-2 *Relative Strengths of Conjugate Acid–Base Pairs*

Acid		Base
$HClO_4$ \\ HI \\ HBr \\ HCl \\ HNO_3	100% ionized in dilute aq. soln. No molecules of nonionized acid.	Negligible base strength in water. $\quad$ ClO_4^- \\ I^- \\ Br^- \\ Cl^- \\ NO_3^-

acid loses H^+ ⇌ base gains H^+

| H_3O^+ \\ HF \\ CH_3COOH \\ HCN \\ NH_4^+ \\ H_2O \\ NH_3 | Equilibrium mixture of nonionized molecules of acid, conjugate base, and $H^+(aq)$. Reacts completely with H_2O to form OH^-; cannot exist in aqueous solution. | H_2O \\ F^- \\ CH_3COO^- \\ CN^- \\ NH_3 \\ OH^- \\ NH_2^- |

Acid strength increases

Base strength increases

Acid strengths for binary acids of elements in other groups of the periodic table vary in the same way as those of the VIIA elements. The order of bond strengths for the VIA hydrides is

$$\text{(strongest bonds)} \quad H_2O \gg H_2S > H_2Se > H_2Te \quad \text{(weakest bonds)}$$

H—O bonds are much stronger than the bonds in the other Group VI hydrides. As we might expect, the order of acid strengths for these hydrides is just the reverse of the order of bond strengths.

$$\text{(weakest acid)} \quad H_2O \ll H_2S < H_2Se < H_2Te \quad \text{(strongest acid)}$$

Table 10-2 displays relative acid and base strengths of a number of conjugate acid–base pairs.

> The trends in binary acid strengths *across* a period (e.g., $CH_4 < NH_3 < H_2O < HF$) are *not* those predicted from trends in bond energies and electronegativity differences. The correlations used for *vertical* trends cannot be used for *horizontal* trends. This is because a "horizontal" series of compounds has different stoichiometries and different numbers of unshared pairs of electrons on the central atoms.

Ternary Acids

Most ternary acids (containing three different elements) are *hydroxyl compounds of nonmetals* (oxoacids) that ionize to produce $H^+(aq)$. The formula for nitric acid is commonly written HNO_3 to emphasize the presence of an acidic hydrogen atom. The hydrogen atom is bonded to one of the oxygen atoms so it could also be written as $HONO_2$ (see margin).

In most ternary acids the hydroxyl oxygen is bonded to a fairly electronegative nonmetal. In nitric acid the nitrogen draws the electrons of the N—O (hydroxyl) bond closer to itself than would a less electronegative element such as sodium. The oxygen pulls the electrons of the O—H bond close enough so that the hydrogen atom ionizes as H^+, leaving NO_3^-. The regions of positive charge in HNO_3 are shown in blue in the electrostatic charge potential plot (margin).

$$HNO_3(aq) \longrightarrow H^+(aq) + NO_3^-(aq)$$

Let us consider the hydroxyl compounds of metals. We call these compounds *hydroxides* because they can produce hydroxide ions in water to give basic solutions. Oxygen is much more electronegative than most metals, such as sodium. It draws the electrons of the sodium–oxygen bond in NaOH (a strong base) so close to itself that the bonding is ionic. NaOH therefore exists as Na^+ and OH^- ions, even in the solid state, and dissociates into Na^+ and OH^- ions when it dissolves in H_2O.

$$NaOH(s) \xrightarrow{\text{H}_2\text{O}} Na^+(aq) + OH^-(aq)$$

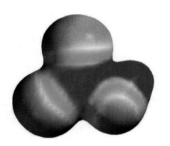

Bond that breaks to form H^+ and NO_3^-

Hydroxyl group

Returning to our consideration of ternary acids, we usually write the formula for sulfuric acid as H_2SO_4 to emphasize that it is a polyprotic acid. The formula can also be written as $(HO)_2SO_2$, however, because the structure of sulfuric acid (see margin) shows clearly that H_2SO_4 contains two O—H groups bound to a sulfur atom. Because the O—H bonds are easier to break than the S—O bonds, sulfuric acid ionizes as an acid.

$$\text{Step 1:} \quad H_2SO_4(aq) \longrightarrow H^+(aq) + HSO_4^-(aq)$$
$$\text{Step 2:} \quad HSO_4^-(aq) \rightleftharpoons H^+(aq) + SO_4^{2-}(aq)$$

The first step in the ionization of H_2SO_4 is complete in dilute aqueous solution. The second step is nearly complete in very dilute aqueous solutions. The first step in the ionization of a polyprotic acid always occurs to a greater extent than the second step because it is easier to remove a proton from a neutral acid molecule than from a negatively charged anion.

Sulfurous acid, H_2SO_3, is a polyprotic acid that contains the same elements as H_2SO_4. H_2SO_3 is a weak acid, however, which tells us that the H—O bonds in H_2SO_3 are stronger than those in H_2SO_4.

> Sulfuric acid is called a **polyprotic acid** because it has more than one ionizable hydrogen atom per molecule. It is the only *common* polyprotic acid that is also a strong acid.

Comparison of the acid strengths of nitric acid, HNO_3, and nitrous acid, HNO_2, shows that HNO_3 is a much stronger acid than HNO_2.

> Acid strengths of most ternary acids containing the same central element increase with increasing oxidation state of the central element and with increasing numbers of oxygen atoms.

The following orders of increasing acid strength are typical.

$$H_2SO_3 < H_2SO_4$$
$$HNO_2 < HNO_3$$
$$HClO < HClO_2 < HClO_3 < HClO_4$$

(strongest acids are on the right side)

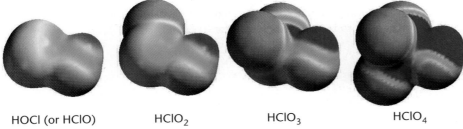

HOCl (or HClO) HClO₂ HClO₃ HClO₄

Electrostatic charge potential plots of H_2SO_4 (top) and HSO_4^- (bottom) showing the decrease in positive charge (blue) on the hydrogen atom in the anionic HSO_4^-. This directly relates to the reduced acidity of HSO_4^- compared to H_2SO_4.

> For most ternary acids containing different elements in the same oxidation state from the same group in the periodic table, acid strengths increase with increasing electronegativity of the central element.

$$H_2SeO_4 < H_2SO_4 \qquad H_2SeO_3 < H_2SO_3$$
$$H_3PO_4 < HNO_3$$
$$HBrO_4 < HClO_4 \qquad HBrO_3 < HClO_3$$

Contrary to what we might expect, H_3PO_3 is a stronger acid than HNO_2. Care must be exercised to compare acids that have *similar structures*. For example, H_3PO_2, which has two H atoms bonded to the P atom, is a stronger acid than H_3PO_3, which has one H atom bonded to the P atom. H_3PO_3 is a stronger acid than H_3PO_4, which has no H atoms bonded to the P atom.

> ⚠ In *most* ternary inorganic acids, all H atoms are bonded to O. But it is very important to be sure of their structures before comparing them.

10-8 ACID–BASE REACTIONS IN AQUEOUS SOLUTIONS

In Section 4-11 we introduced classical acid–base reactions. We defined neutralization as the reaction of an acid with a base to form a salt and (in most cases) water. Most *salts* are ionic compounds that contain a cation other than H^+ and an anion other than OH^- or O^{2-}. The common *strong acids* and common *strong bases* are listed in the margin. All other common acids may be assumed to be weak. The other common metal hydroxides (bases) are insoluble in water.

Arrhenius and Brønsted–Lowry acid–base neutralization reactions all have one thing in common.

Strong Acids	
Binary	**Ternary**
HCl	$HClO_4$
HBr	$HClO_3$
HI	HNO_3
	H_2SO_4

Strong Bases	
LiOH	
NaOH	
KOH	$Ca(OH)_2$
RbOH	$Sr(OH)_2$
CsOH	$Ba(OH)_2$

Neutralization reactions involve the reaction of an acid with a base to form a salt that contains the cation characteristic of the base and the anion characteristic of the acid. Water is also usually formed.

This is indicated in the formula unit equation. The general form of the net ionic equation, however, is different for different acid–base reactions. The net ionic equations depend on the solubility and extent of ionization or dissociation of each reactant and product.

In writing ionic equations, we always write the formulas of the predominant forms of the compounds in, or in contact with, aqueous solution. Writing ionic equations from formula unit equations requires a knowledge of the lists of strong acids and strong bases, as well as of the generalizations on solubilities of inorganic compounds. Please review carefully all of Sections 4-2 and 4-3. Study Tables 4-8 and 4-9 carefully because they summarize much information that you are about to use again.

In Section 4-2 we examined some reactions of strong acids with strong bases to form soluble salts. Let us illustrate one additional example. Perchloric acid, $HClO_4$, reacts with sodium hydroxide to produce sodium perchlorate, $NaClO_4$, a soluble ionic salt.

$$HClO_4(aq) + NaOH(aq) \longrightarrow NaClO_4(aq) + H_2O(\ell)$$

The total ionic equation for this reaction is

$$[H^+(aq) + ClO_4^-(aq)] + [Na^+(aq) + OH^-(aq)] \longrightarrow [Na^+(aq) + ClO_4^-(aq)] + H_2O(\ell)$$

Eliminating the spectator ions, Na^+ and ClO_4^-, gives the net ionic equation

$$H^+(aq) + OH^-(aq) \longrightarrow H_2O(\ell)$$

This is the net ionic equation for the reaction of all strong acids with strong bases to form soluble salts and water.

Many weak acids react with strong bases to form soluble salts and water. For example, acetic acid, CH_3COOH, reacts with sodium hydroxide, NaOH, to produce sodium acetate, $NaCH_3COO$.

$$CH_3COOH(aq) + NaOH(aq) \longrightarrow NaCH_3COO(aq) + H_2O(\ell)$$

The total ionic equation for this reaction is

$$CH_3COOH(aq) + [Na^+(aq) + OH^-(aq)] \longrightarrow [Na^+(aq) + CH_3COO^-(aq)] + H_2O(\ell)$$

Elimination of Na^+ from both sides gives the net ionic equation

$$CH_3COOH(aq) + OH^-(aq) \longrightarrow CH_3COO^-(aq) + H_2O(\ell)$$

In general terms, the reaction of a *weak monoprotic acid* with a *strong base* to form a *soluble salt* may be represented as

$$HA(aq) + OH^-(aq) \longrightarrow A^-(aq) + H_2O(\ell) \quad \text{(net ionic equation)}$$

CD-ROM   Screen 5.10, Acid–Base Reactions.

This is considered to be the same as

$$H_3O^+ + OH^- \rightarrow 2H_2O$$

Monoprotic acids contain one, diprotic acids contain two, and triprotic acids contain three acidic (ionizable) hydrogen atoms per formula unit. Polyprotic acids (those that contain more than one ionizable hydrogen atom) are discussed in detail in Chapter 18.

When all the H^+ from the acid and all the OH^- of the base have reacted, we say that there has been complete neutralization.

EXAMPLE 10-2   *Equations for Acid–Base Reactions*

Write (a) formula unit, (b) total ionic, and (c) net ionic equations for the complete neutralization of phosphoric acid, H_3PO_4, with potassium hydroxide, KOH.

CHEMISTRY IN USE

Our Daily Lives

Everyday Salts of Ternary Acids

You may have encountered some salts of ternary acids without even being aware of them. For example, the iron in many of your breakfast cereals and breads may have been added in the form of iron(II) sulfate, $FeSO_4$, or iron(II) phosphate, $Fe_3(PO_4)_2$; the calcium in these foods often comes from the addition of calcium carbonate, $CaCO_3$. Fruits and vegetables keep fresh longer after an application of sodium sulfite, Na_2SO_3, and sodium hydrogen sulfite, $NaHSO_3$. Restaurants also use these two sulfites to keep their salad bars more appetizing. The red color of fresh meat is maintained for much longer by the additives sodium nitrate, $NaNO_3$, and sodium nitrite, $NaNO_2$. Sodium phosphate, Na_3PO_4, is used to prevent metal ion flavors and to control acidity in some canned goods.

Many salts of ternary acids are used in medicine. Lithium carbonate, Li_2CO_3, has been used successfully to combat severe jet lag. Lithium carbonate is also useful in the treatment of mania, depression, alcoholism, and schizophrenia. Magnesium sulfate, $MgSO_4$, sometimes helps to prevent convulsions during pregnancy and to reduce the solubility of toxic barium sulfate in internally administered preparations consumed before gastrointestinal X-ray films are taken.

Other salts of ternary acids that you may find in your home include potassium chlorate, $KClO_3$, in matches as an oxidizing agent and oxygen source; sodium hypochlorite, $NaClO$, in bleaches and mildew removers; and ammonium carbonate, $(NH_4)_2CO_3$, which is the primary ingredient in smelling salts. Limestone and marble are calcium carbonate; gypsum and plaster of Paris are primarily calcium sulfate, $CaSO_4$. Fire-

Charles Steele

The tips of "strike anywhere" matches contain tetraphosphorus trisulfide, red phosphorus, and potassium chlorate. Friction converts kinetic energy into heat, which initiates a spontaneous reaction.

$$P_4S_3(s) + 8O_2 \longrightarrow P_4O_{10}(s) + 3SO_2(g)$$

The thermal decomposition of $KClO_3$ provides additional oxygen for this reaction.

works get their brilliant colors from salts such as barium nitrate, $Ba(NO_3)_2$, which imparts a green color; strontium carbonate, $SrCO_3$, which gives a red color; and copper(II) sulfate, $CuSO_4$, which produces a blue color. Should your fireworks get out of hand and accidentally start a fire, the ammonium phosphate, $(NH_4)_3PO_4$, sodium hydrogen carbonate, $NaHCO_3$, and potassium hydrogen carbonate, $KHCO_3$, in your ABC dry fire extinguisher will come in handy.

An unexpected place to find ternary acid salts is in your long-distance phone bills; nitrates, $NO_3{}^-$, are cheaper than day rates.

Ronald DeLorenzo
Middle Georgia College

Plan

(a) The salt produced in the reaction contains the cation of the base, K^+, and the $PO_4{}^{3-}$ anion of the acid. The salt is K_3PO_4.

(b) H_3PO_4 is a weak acid—it is not written in ionic form. KOH is a strong base, and so it is written in ionic form. K_3PO_4 is a *soluble salt*, and so it is written in ionic form.

(c) The spectator ions are canceled to give the net ionic equation.

Solution

(a) $H_3PO_4(aq) + 3KOH(aq) \longrightarrow K_3PO_4(aq) + 3H_2O(\ell)$

(b) $H_3PO_4(aq) + 3[K^+(aq) + OH^-(aq)] \longrightarrow [3K^+(aq) + PO_4{}^{3-}(aq)] + 3H_2O(\ell)$

(c) $H_3PO_4(aq) + 3OH^-(aq) \longrightarrow PO_4{}^{3-}(aq) + 3H_2O(\ell)$

You should now work Exercise 56.

Neutralization reactions of ammonia, NH_3, form salts but do not form water.

EXAMPLE 10-3   *Equations for Acid–Base Reactions*

Write (a) formula unit, (b) total ionic, and (c) net ionic equations for the neutralization of aqueous ammonia with nitric acid.

Plan

(a) The salt produced in the reaction contains the cation of the base, NH_4^+, and the anion of the acid, NO_3^-. The salt is NH_4NO_3.

(b) HNO_3 is a strong acid—we write it in ionic form. Ammonia is a weak base. NH_4NO_3 is a soluble salt that is completely dissociated—we write it in ionic form.

(c) We cancel the spectator ions, NO_3^-, and obtain the net ionic equation.

Solution

(a) $HNO_3(aq) + NH_3(aq) \longrightarrow NH_4NO_3(aq)$

(b) $[H^+(aq) + NO_3^-(aq)] + NH_3(aq) \longrightarrow [NH_4^+(aq) + NO_3^-(aq)]$

(c) $H^+(aq) + NH_3(aq) \longrightarrow NH_4^+(aq)$

You should now work Exercise 58.

EXAMPLE 10-4   *Preparation of Salts*

Write the formula unit equation for the reaction of an acid and a base that will produce each of the following salts: (a) Na_3PO_4, (b) $Ca(ClO_3)_2$, (c) $MgSO_4$.

Plan

(a) The salt contains the ions, Na^+ and PO_4^{3-}. Na^+ is the cation in the strong base, $NaOH$. PO_4^{3-} is the anion in the weak acid, H_3PO_4. The reaction of $NaOH$ with H_3PO_4 should therefore produce the desired salt plus water.

(b) The cation, Ca^{2+}, is from the strong base, $Ca(OH)_2$. The anion, ClO_3^-, is from the strong acid, $HClO_3$. The reaction of $Ca(OH)_2$ with $HClO_3$ will produce water and the desired salt.

(c) $MgSO_4$ is the salt produced in the reaction of $Mg(OH)_2(s)$ and $H_2SO_4(aq)$.

Solution

(a) $H_3PO_4(aq) + 3NaOH(aq) \longrightarrow Na_3PO_4(aq) + 3H_2O(\ell)$

(b) $2HClO_3(aq) + Ca(OH)_2(aq) \longrightarrow Ca(ClO_3)_2(aq) + 2H_2O(\ell)$

(c) $H_2SO_4(aq) + Mg(OH)_2(s) \longrightarrow MgSO_4(aq) + 2H_2O(\ell)$

You should now work Exercise 60.

10-9 ACIDIC SALTS AND BASIC SALTS

To this point we have examined acid–base reactions in which stoichiometric amounts of Arrhenius acids and bases were mixed. Those reactions form *normal salts*. As the name implies, **normal salts** contain no ionizable H atoms or OH groups. The *complete* neutralization of phosphoric acid, H_3PO_4, with sodium hydroxide, $NaOH$, produces the normal salt, Na_3PO_4. The equation for this complete neutralization is

$$H_3PO_4(aq) + 3NaOH(aq) \longrightarrow Na_3PO_4(aq) + 3H_2O(\ell)$$

$\qquad$ 1 mole $\qquad\qquad$ 3 moles $\qquad\qquad$ sodium phosphate,
$\qquad\qquad\qquad\qquad\qquad\qquad\qquad\qquad$ a normal salt

Charles D. Winters

Sodium hydrogen carbonate, baking soda, is the most familiar example of an acidic salt. It can neutralize strong bases, but its aqueous solutions are slightly basic, as the blue color of the indicator bromthymol blue shows.

$NaHCO_3$ (baking soda) is very handy for neutralizing chemical spills of acids or bases due to its ability to react with and neutralize both strong acids and strong bases. Containers of $NaHCO_3$ are kept in many chemistry labs for this purpose.

If less than stoichiometric amounts of bases react with *polyprotic* acids, the resulting salts are known as **acidic salts** because they can neutralize additional base.

$$H_3PO_4(aq) + NaOH(aq) \longrightarrow NaH_2PO_4(aq) + H_2O(\ell)$$
1 mole   1 mole   sodium dihydrogen phosphate, an acidic salt

$$H_3PO_4(aq) + 2NaOH(aq) \longrightarrow Na_2HPO_4(aq) + 2H_2O(\ell)$$
1 mole   2 moles   sodium hydrogen phosphate, an acidic salt

The reaction of phosphoric acid, H_3PO_4, a weak acid, with strong bases can produce the three salts shown in the three preceding equations, depending on the relative amounts of acid and base used. The acidic salts, NaH_2PO_4 and Na_2HPO_4, can react further with bases such as NaOH.

$$NaH_2PO_4(aq) + 2NaOH(aq) \longrightarrow Na_3PO_4(aq) + 2H_2O(\ell)$$
$$Na_2HPO_4(aq) + NaOH(aq) \longrightarrow Na_3PO_4(aq) + H_2O(\ell)$$

There are many additional examples of acidic salts. Sodium hydrogen carbonate, $NaHCO_3$, commonly called sodium bicarbonate, is classified as an acidic salt. It is, however, the acidic salt of an extremely weak acid—carbonic acid, H_2CO_3—and solutions of sodium bicarbonate are slightly basic, as are solutions of salts of other extremely weak acids.

Polyhydroxy bases (bases that contain more than one OH per formula unit) react with stoichiometric amounts of acids to form normal salts.

> Remember that acidic and basic salts do not necessarily form solutions that are acidic or basic, respectively. The acidic and basic names refer to their ability to react with strong acids (basic salts) and bases (acidic salts). They can also often react with both acids and bases (e.g., $NaHCO_3$), but are usually named for their primary acid or base reaction chemistry.

$$Al(OH)_3(s) + 3HCl(aq) \longrightarrow AlCl_3(aq) + 3H_2O(\ell)$$
1 mole   3 moles   aluminum chloride, a normal salt

The reaction of polyhydroxy bases with less than stoichiometric amounts of acids forms **basic salts,** that is, salts that contain unreacted OH groups. For example, the reaction of aluminum hydroxide with hydrochloric acid can produce two different basic salts:

These basic aluminum salts are called "aluminum chlorohydrate." They are components of some deodorants.

$$Al(OH)_3(s) + HCl(aq) \longrightarrow Al(OH)_2Cl(s) + H_2O(\ell)$$
1 mole   1 mole   aluminum dihydroxide chloride, a basic salt

$$Al(OH)_3(s) + 2HCl(aq) \longrightarrow Al(OH)Cl_2(s) + 2H_2O(\ell)$$
1 mole   2 moles   aluminum hydroxide dichloride, a basic salt

Aqueous solutions of basic salts are not necessarily basic, but they can neutralize acids, such as

$$Al(OH)_2Cl + 2HCl \longrightarrow AlCl_3 + 2H_2O$$

Most basic salts are rather insoluble in water.

10-10 THE LEWIS THEORY

This is the same Lewis who made many contributions to our understanding of chemical bonding.

CD-ROM   Screen 17.12, Lewis Acids and Bases, Screen 17.13, Cationic Lewis Acids, and Screen 17.14, Neutral Lewis Acids.

In 1923, Professor G. N. Lewis (1875–1946) presented the most comprehensive of the classic acid–base theories. The Lewis definitions follow.

> An **acid** is any species that can accept a share in an electron pair. A **base** is any species that can make available, or "donate," a share in an electron pair.

These definitions do *not* specify that an electron pair must be transferred from one atom to another—only that an electron pair, residing originally on one atom, must be shared between two atoms. *Neutralization* is defined as **coordinate covalent bond formation.** This results in a covalent bond in which both electrons were furnished by one atom or ion.

The reaction of boron trichloride with ammonia is a typical Lewis acid–base reaction.

In BCl_3 boron has an empty orbital and only 6 valence electrons.

The Lewis theory is sufficiently general that it covers *all* acid–base reactions that the other theories include, plus many additional reactions such as complex formation (Chapter 25).

The autoionization of water (Section 10-5) was described in terms of Brønsted–Lowry theory. In Lewis theory terminology, this is also an acid–base reaction. The acceptance of a proton, H^+, by a base involves the formation of a coordinate covalent bond.

base   acid

Theoretically, any species that contains an unshared electron pair could act as a base. In fact, most ions and molecules that contain unshared electron pairs undergo some reactions by sharing their electron pairs with atoms or molecules that have low energy vacant orbitals. Conversely, many Lewis acids contain only six electrons in the highest occupied energy level of the central element. They react by accepting a share in an additional pair of electrons. Many compounds of the Group IIIA elements are Lewis acids, as illustrated by the reaction of boron trichloride with ammonia, presented earlier.

Anhydrous aluminum chloride, $AlCl_3$, is a common Lewis acid that is used to catalyze many organic reactions. $AlCl_3$ acts as a Lewis acid when it dissolves in hydrochloric acid to give a solution that contains $AlCl_4^-$ ions.

Almost all transition metal complexes (see Chapter 25) form bonds by the donation of available electron pairs from surrounding atoms or molecules to vacant orbitals on the metal atom.

$$AlCl_3(s) + Cl^-(aq) \longrightarrow AlCl_4^-(aq)$$

acid   base   product

Other ions and molecules behave as Lewis acids by expansion of the valence shell of the central element. Anhydrous tin(IV) chloride is a colorless liquid that also is frequently used as a Lewis acid catalyst. The tin atom (Group IVA) can expand its valence shell by utilizing vacant d orbitals. It can accept shares in two additional electron pairs, as its reaction with hydrochloric acid illustrates.

$$\underset{\text{acid}}{SnCl_4(\ell)} + \underset{\text{base}}{2Cl^-(aq)} \longrightarrow SnCl_6^{2-}(aq)$$

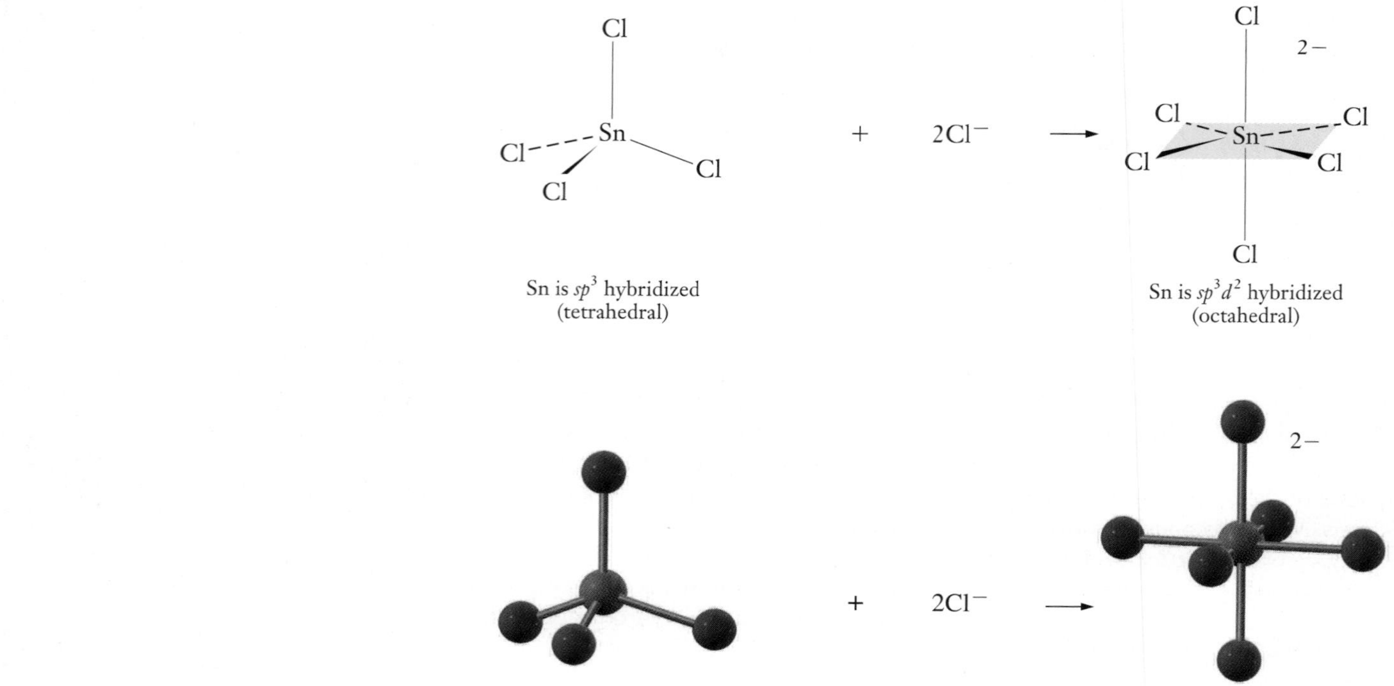
Cl
Sn
Cl
Cl
Cl
+
2Cl−
Cl
2−
Cl
Sn
Cl
Cl
Cl
Cl
Sn is sp3 hybridized (tetrahedral)
Sn is sp3d2 hybridized (octahedral)
+
2Cl−
2−

Many organic and biological reactions are acid–base reactions that do not fit within the Arrhenius or Brønsted–Lowry theories. Experienced chemists find the Lewis theory to be very useful because so many other chemical reactions are covered by it. The less experienced sometimes find the theory less useful, but as their knowledge expands so does its utility.

✓ Problem-Solving Tip: *Which Acid–Base Theory Should You Use?*

Remember the following:

1. Arrhenius acids and bases are also Brønsted–Lowry acids and bases; the reverse is not true.
2. Brønsted–Lowry acids and bases are also Lewis acids and bases; the reverse is not true.
3. We usually prefer the Arrhenius or the Brønsted–Lowry theory when water or another protic solvent is present.
4. Although the Lewis theory can be used to explain the acidic or basic property of some species in protic solvents, the most important use of the Lewis theory is for acid-base reactions in many nonaqueous solvents and with transition metal complexes.

10-11 THE PREPARATION OF ACIDS

Binary acids may be prepared by combination of appropriate elements with hydrogen (Section 6-7, Part 2).

Small quantities of the hydrogen halides (their solutions are called hydrohalic acids) and other *volatile acids* are usually prepared by adding concentrated nonvolatile acids to the appropriate salts. (Sulfuric and phosphoric acids are classified as *nonvolatile acids* because they have much higher boiling points than other common acids.) The reactions of concentrated sulfuric acid with solid sodium fluoride and sodium chloride produce gaseous hydrogen fluoride and hydrogen chloride, respectively.

$$H_2SO_4(\ell) + NaF(s) \longrightarrow NaHSO_4(s) + HF(g)$$

sulfuric acid — sodium fluoride — sodium hydrogen sulfate — hydrogen fluoride
bp = 336°C — — — bp = 19.6°C

$$H_2SO_4(\ell) + NaCl(s) \longrightarrow NaHSO_4(s) + HCl(g)$$

— sodium chloride — — hydrogen chloride
— — — bp = −84.9°C

The volatile acid HCl can be made by dropping concentrated H_2SO_4 onto solid NaCl. Gaseous HCl is liberated. HCl(g) dissolves in the water on a piece of filter paper. The indicator methyl red on the paper turns red, its color in acidic solution.

Because concentrated sulfuric acid is a fairly strong oxidizing agent, it cannot be used to prepare hydrogen bromide or hydrogen iodide; instead, the free halogens are produced. Phosphoric acid, a nonoxidizing acid, is dropped onto solid sodium bromide or sodium iodide to produce hydrogen bromide or hydrogen iodide, as the following equations show:

$$H_3PO_4(\ell) + NaBr(s) \xrightarrow{heat} NaH_2PO_4(s) + HBr(g)$$

phosphoric acid — sodium bromide — sodium dihydrogen phosphate — hydrogen bromide
bp = 213°C — — — bp = −67.0°C

$$H_3PO_4(\ell) + NaI(s) \xrightarrow{heat} NaH_2PO_4(s) + HI(g)$$

— sodium iodide — — hydrogen iodide
— — — bp = −35°C

This kind of reaction may be generalized as

$$\text{nonvolatile acid} + \text{salt of volatile acid} \longrightarrow \text{salt of nonvolatile acid} + \text{volatile acid}$$

Dissolving each of the gaseous hydrogen halides in water gives the corresponding hydrohalic acid.

In Section 6-8, Part 2 we saw that many nonmetal oxides, called acid anhydrides, react with water to form *ternary acids* with no changes in oxidation numbers. For example, dichlorine heptoxide, Cl_2O_7, forms perchloric acid when it dissolves in water.

$$\overset{+7}{Cl_2}O_7(\ell) + H_2O(\ell) \longrightarrow 2[H^+(aq) + \overset{+7}{Cl}O_4^-(aq)]$$

Some *high oxidation state transition metal oxides* are acidic oxides; that is, they dissolve in water to give solutions of ternary acids. Manganese(VII) oxide, Mn_2O_7, and chromium(VI) oxide, CrO_3, are the most common examples.

A solution of dichromic acid, $H_2Cr_2O_7$, is deep red.

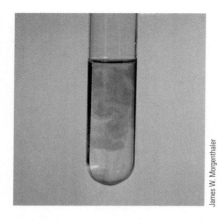

A drop of PCl_3 is added to water that contains the indicator methyl orange. As PCl_3 reacts with water to form HCl and H_3PO_3, the indicator turns red, its color in acidic solution.

$$\underset{\text{manganese(VII) oxide}}{\overset{+7}{Mn_2O_7}(\ell)} + H_2O(\ell) \longrightarrow 2[H^+(aq) + \underset{\text{permanganic acid}}{\overset{+7}{MnO_4^-}}(aq)]$$

$$\underset{\text{chromium(VI) oxide}}{2\overset{+6}{CrO_3}(s)} + H_2O(\ell) \longrightarrow [2H^+(aq) + \underset{\text{dichromic acid}}{\overset{+6}{Cr_2O_7^{2-}}}(aq)]$$

Neither permanganic acid nor dichromic acid has been isolated in pure form. Many stable salts of both are well known.

The halides and oxyhalides of some nonmetals hydrolyze (react with water) to produce two acids: a (binary) hydrohalic acid and a (ternary) oxyacid of the nonmetal. Phosphorus trihalides react with water to produce the corresponding hydrohalic acids and phosphorous acid, a weak diprotic acid, whereas phosphorus pentahalides give phosphoric acid and the corresponding hydrohalic acid.

$$\overset{+3}{PX_3} + 3H_2O(\ell) \longrightarrow \overset{+3}{H_3PO_3}(aq) + 3HX(aq)$$

$$\overset{+5}{PX_5} + 4H_2O(\ell) \longrightarrow \overset{+5}{H_3PO_4}(aq) + 5HX(aq)$$

There are no changes in oxidation numbers in these reactions. Examples include the reactions of PCl_3 and PCl_5 with H_2O.

$$\underset{\substack{\text{phosphorus}\\\text{trichloride}}}{\overset{+3}{PCl_3}(\ell)} + 3H_2O(\ell) \longrightarrow \underset{\text{phosphorous acid}}{\overset{+3}{H_3PO_3}}(aq) + 3[H^+(aq) + Cl^-(aq)]$$

$$\underset{\substack{\text{phosphorus}\\\text{pentachloride}}}{\overset{+5}{PCl_5}(s)} + 4H_2O(\ell) \longrightarrow \underset{\text{phosphoric acid}}{\overset{+5}{H_3PO_4}}(aq) + 5[H^+(aq) + Cl^-(aq)]$$

Key Terms

Acid (Arrhenius or Brønsted–Lowry)   A substance that produces $H^+(aq)$ ions in aqueous solution. Strong acids ionize completely or almost completely in dilute aqueous solution; weak acids ionize only slightly.

Acid anhydride   The oxide of a nonmetal that reacts with water to form an acid.

Acidic salt   A salt that contains an ionizable hydrogen atom; does not necessarily produce acidic solutions.

Amphiprotism   The ability of a substance to exhibit amphoterism by accepting or donating protons.

Amphoterism   Ability of a substance to act as either an acid or a base.

Anhydrous   Without water.

Autoionization   An ionization reaction between identical molecules.

Base (Arrhenius)   A substance that produces $OH^-(aq)$ ions in aqueous solution. Strong bases are soluble in water and are completely *dissociated*. Weak bases ionize only slightly.

Basic anhydride   The oxide of a metal that reacts with water to form a base.

Basic salt   A salt containing a basic OH group.

Brønsted–Lowry acid   A proton donor.

Brønsted–Lowry base   A proton acceptor.

Conjugate acid–base pair   In Brønsted–Lowry terminology, a reactant and product that differ by a proton, H^+.

Coordinate covalent bond   A covalent bond in which both shared electrons are furnished by the same species; the bond between a Lewis acid and a Lewis base.

Dissociation   In aqueous solution, the process in which a *solid ionic compound* separates into its ions.

Electrolyte A substance whose aqueous solutions conduct electricity.

Formula unit equation A chemical equation in which all compounds are represented by complete formulas.

Hydride A binary compound of hydrogen.

Hydronium ion H_3O^+, the usual representation of the hydrated hydrogen ion.

Ionization In aqueous solution, the process in which a *molecular* compound reacts with water to form ions.

Leveling effect The effect by which all acids stronger than the acid that is characteristic of the solvent react with the solvent to produce that acid; a similar statement applies to bases. The strongest acid (base) that can exist in a given solvent is the acid (base) characteristic of that solvent.

Lewis acid Any species that can accept a share in an electron pair to form a coordinate covalent bond.

Lewis base Any species that can make available a share in an electron pair to form a coordinate covalent bond.

Net ionic equation The equation that results from canceling spectator ions and eliminating brackets from a total ionic equation.

Neutralization The reaction of an acid with a base to form a salt and (usually) water; usually, the reaction of hydrogen ions with hydroxide ions to form water molecules.

Nonelectrolyte A substance whose aqueous solutions do not conduct electricity.

Normal salt A salt containing no ionizable H atoms or OH groups.

Polyprotic acid An acid that contains more than one ionizable hydrogen atom per formula unit.

Protic acid An Arrhenius acid, or a Brønsted–Lowry acid.

Salt A compound that contains a cation other than H^+ and an anion other than OH^- or O^{2-}; also called protonic acid.

Spectator ions Ions in solution that do not participate in a chemical reaction.

Strong electrolyte A substance that conducts electricity well in dilute aqueous solution.

Ternary acid An acid that contains three elements—usually H, O, and another nonmetal.

Ternary compound A compound that contains three different elements.

Total ionic equation The equation for a chemical reaction written to show the predominant form of all species in aqueous solution or in contact with water.

Weak electrolyte A substance that conducts electricity poorly in dilute aqueous solution.

Exercises

Basic Ideas

1. Which properties of acids did Robert Boyle observe?
2. Gay-Lussac reached an important conclusion about acids and bases. What was it?
3. Define the following terms. You may wish to refer to Chapter 4 to check the definitions. (a) acid; (b) neutralization; (c) ionization; (d) dissociation; (e) salt.

The Arrhenius Theory

4. Outline Arrhenius' ideas about acids and bases. (a) How did he define the following terms: acid, base, neutralization? (b) Give an example that illustrates each term.
5. Define and illustrate the following terms clearly and concisely. Give an example of each. (a) strong electrolyte; (b) weak electrolyte; (c) nonelectrolyte; (d) strong acid; (e) strong base; (f) weak acid; (g) weak base; (h) insoluble base.
6. Describe an experiment for classifying compounds as strong electrolytes, weak electrolytes, or nonelectrolytes. Tell what would be observed for each of the following compounds and classify each. Na_2SO_4; HCN; CH_3COOH; CH_3OH; HF; $HClO_4$; HCOOH; NH_3.
7. Summarize the electrical properties of strong electrolytes, weak electrolytes, and nonelectrolytes.

8. Distinguish between the following pairs of terms, and provide a specific example of each. (a) strong acid and weak acid; (b) strong base and weak base; (c) strong base and insoluble base.
9. Write formulas and names for (a) the common strong acids; (b) three weak acids; (c) the common strong bases; (d) the most common weak base; (e) four soluble ionic salts; (f) four insoluble salts.

The Hydrated Hydrogen Ion

10. Write the formula of a hydrated hydrogen ion that contains only one water of hydration. Give another name for the hydrated hydrogen ion.
11. Why is the hydrated hydrogen ion important?
12. Criticize the following statement: "The hydrated hydrogen ion should always be represented as H_3O^+."

H_3O^+

Brønsted–Lowry Theory

13. State the basic ideas of the Brønsted–Lowry theory.

14. Use Brønsted–Lowry terminology to define the following terms. Illustrate each with a specific example. (a) acid; (b) conjugate base; (c) base; (d) conjugate acid; (e) conjugate acid–base pair.

15. Write balanced equations that describe the ionization of the following acids in dilute aqueous solution. Use a single arrow ($\rightarrow$) to represent complete, or nearly complete, ionization and a double arrow ($\rightleftharpoons$) to represent a small extent of ionization. (a) HCl; (b) C_2H_5COOH; (c) H_2S; (d) HCN; (e) HF; (f) $HClO_4$.

16. Use words and equations to describe how ammonia can act as a base in (a) aqueous solution and (b) the pure state, that is, as gaseous ammonia molecules when it reacts with gaseous hydrogen chloride or a similar anhydrous acid.

17. What does autoionization mean? How can the autoionization of water be described as an acid–base reaction? What structural features must a compound have to be able to undergo autoionization?

18. Illustrate, with appropriate equations, the fact that these species are bases in water: NH_3; HS^-; CH_3COO^-; O^{2-}.

19. In terms of Brønsted–Lowry theory, state the differences between (a) a strong and a weak base and (b) a strong and a weak acid.

20. Give the products in the following acid–base reactions. Identify the conjugate acid–base pairs. (a) $NH_4^+ + CN^-$; (b) $HS^- + H_2SO_4$; (c) $HClO_4 + [H_2NNH_3]^+$; (d) $NH_2^- + H_2O$

21. Give the conjugate acids of H_2O, OH^-, I^-, AsO_4^{3-}, NH_2^-, HPO_4^{2-}, and NO_2^-.

22. Give the conjugate bases of H_2O, HS^-, HCl, PH_4^+, and $HOCH_3$.

23. Identify the Brønsted–Lowry acids and bases in these reactions and group them into conjugate acid–base pairs.
(a) $NH_3 + HBr \rightleftharpoons NH_4^+ + Br^-$
(b) $NH_4^+ + HS^- \rightleftharpoons NH_3 + H_2S$
(c) $H_3O^+ + PO_4^{3-} \rightleftharpoons HPO_4^{2-} + H_2O$
(d) $HSO_3^- + CN^- \rightleftharpoons HCN + SO_3^{2-}$
(e) $O^{2-} + H_2O \rightleftharpoons OH^- + OH^-$

24. Identify each species in the following reactions as either an acid or a base, in the Brønsted–Lowry sense.
(a) $CN^- + H_2O \rightleftharpoons HCN + OH^-$
(b) $HCO_3^- + H_2SO_4 \rightleftharpoons HSO_4^- + H_2CO_3$
(c) $CH_3COOH + NO_2^- \rightleftharpoons HNO_2 + CH_3COO^-$
(d) $NH_2^- + H_2O \rightleftharpoons NH_3 + OH^-$

25. Identify each reactant and product in the following chemical reactions as a Brønsted–Lowry acid, a Brønsted–Lowry base, or neither. Arrange the species in each reaction as conjugate acid–base pairs.
(a) $H_2SeO_4 + H_2O \rightleftharpoons H_3O^+ + HSeO_4^-$
(b) $HPO_4^{2-} + H_2O \rightleftharpoons H_3O^+ + PO_4^{3-}$
(c) $NH_3 + H^- \rightleftharpoons H_2 + NH_2^-$
(d) $HCl + NH_3 \rightleftharpoons NH_4Cl$
(e) $NH_4^+ + HSO_3^- \rightleftharpoons NH_3 + H_2SO_3$

26. Identify each reactant and product in the following chemical reactions as a Brønsted–Lowry acid, a Brønsted–Lowry base, or neither. Arrange the species in each reaction as conjugate acid–base pairs.
(a) $H_2CO_3 + H_2O \rightleftharpoons H_3O^+ + HCO_3^-$
(b) $HSO_4^- + H_2O \rightleftharpoons H_3O^+ + SO_4^{2-}$
(c) $H_3PO_4 + CN^- \rightleftharpoons HCN + H_2PO_4^-$
(d) $HS^- + OH^- \rightleftharpoons H_2O + S^{2-}$

27. Arrange the species in the reactions of Exercise 24 as Brønsted–Lowry conjugate pairs.

Properties of Aqueous Solutions of Acids and Bases

28. Write equations and designate conjugate pairs for the stepwise reactions in water of (a) H_2SO_4 and (b) H_2SO_3.

29. We say that strong acids, weak acids, and weak bases *ionize* in water, but strong bases *dissociate* in water. What is the difference between ionization and dissociation?

30. Distinguish between solubility in water and extent of ionization in water. Provide specific examples that illustrate the meanings of both terms.

31. Write three general statements that describe the extents to which acids, bases, and salts are ionized in dilute aqueous solutions.

32. List five properties of bases in aqueous solution. Does aqueous ammonia exhibit these properties? Why?

33. List six properties of aqueous solutions of protic acids.

Amphoterism

34. Use chemical equations to illustrate the hydroxides of beryllium, zinc, arsenic, and antimony reacting (a) as acids; (b) as bases.

35. Draw the Lewis formula of aluminum hydroxide, and explain the features that enable it to possess amphoteric properties.

Strengths of Acids

36. Briefly explain how the following reaction illustrates the leveling effect of water.

$$K_2O(s) + H_2O(\ell) \rightleftharpoons 2K^+(aq) + 2OH^-(aq)$$

37. What property is characteristic of all strong acids and strong bases but not weak acids and weak bases?

38. What does "base strength" mean? What does "acid strength" mean?

39. Classify each of the following substances as (a) a strong base, (b) an insoluble base, (c) a strong acid, or (d) a weak acid: LiOH; HCl; $Ba(OH)_2$; $Cu(OH)_2$; H_2S; H_2CO_3; H_2SO_4; $Zn(OH)_2$.

40. (a) What are binary protic acids? (b) Write names and formulas for four binary protic acids.

41. (a) How can the order of increasing acid strength in a series of similar binary protic acids be explained? (b) Illustrate your answer for the series HF, HCl, HBr, and HI. (c) What is the order of increasing base strength of the conjugate bases of the acids in (b)? Why? (d) Is your explanation applicable to the series H_2O, H_2S, H_2Se, and H_2Te? Why?

42. Arrange the members of each group in order of decreasing acidity: (a) H_2O, H_2Se, H_2S; (b) HI, HCl, HF, HBr; (c) H_2S, S_2^-, HS^-.

43. (a) Which is the stronger acid of each pair? (1) NH_4^+, NH_3; (2) H_2O, H_3O^+; (3) HS^-, H_2S; (4) HSO_3^-, H_2SO_3. (b) How are acidity and charge related?

44. Classify each of the hydrides NaH, BeH_2, BH_3, CH_4, NH_3, H_2O, and HF as a Brønsted–Lowry base, a Brønsted–Lowry acid, or neither.

45. Illustrate the leveling effect of water by writing equations for the reactions of HCl and HNO_3 with water.

Ternary Acids

46. In what sense can we describe nitric and sulfuric acids as hydroxyl compounds of nonmetals?

47. What are ternary acids? Write names and formulas for four of them.

48. Among the compounds formed by phosphorous are phosphoric acid (H_3PO_4) and phosphorous acid (H_3PO_3). H_3PO_4 is a triprotic acid, while H_3PO_3 is only diprotic. Draw Lewis formulas of these two compounds that would account for the triprotic nature of H_3PO_4 and the diprotic nature of H_3PO_3, and explain how your structures show this behavior. Suggest an alternative means of writing the formula of phosphorous acid that would better represent its acid–base behavior.

49. Explain the order of increasing acid strength for the following groups of acids and the order of increasing base strength for their conjugate bases. (a) H_2SO_3, H_2SO_4; (b) HNO_2, HNO_3; (c) H_3PO_3, H_3PO_4; (d) HClO, $HClO_2$, $HClO_3$, $HClO_4$.

50. (a) Write a generalization that describes the order of acid strengths for a series of ternary acids that contain different elements in the same oxidation state from the same group in the periodic table. (b) Indicate the order of acid strengths for the following: (1) HNO_3, H_3PO_4; (2) H_3PO_4, H_3AsO_4; (3) $HClO_3$, H_2SeO_4; (4) $HClO_3$, $HBrO_3$, HIO_3.

*51. List the following acids in order of increasing strength: (a) sulfuric, phosphoric, and perchloric; (b) HIO_3, HIO_2, HIO, and HIO_4; (c) selenous, sulfurous, and tellurous acids; (d) hydrosulfuric, hydroselenic, and hydrotelluric acids; (e) H_2CrO_4, H_2CrO_2, $HCrO_3$, and H_3CrO_3.

Reactions of Acids and Bases

52. Why are acid–base reactions described as neutralization reactions?

53. Distinguish among (a) formula unit equations, (b) total ionic equations, and (c) net ionic equations. What are the advantages and limitations of each?

54. Classify each substance as either an electrolyte or a nonelectrolyte: NH_4Cl; HI; C_6H_6; RaF_2; $Zn(CH_3COO)_2$; $Cu(NO_3)_2$; CH_3COOH; $C_{12}H_{22}O_{11}$ (table sugar); LiOH; $KHCO_3$; $NaClO_4$; $La_2(SO_4)_3$; I_2.

55. Classify each substance as either a strong or a weak electrolyte, and then list (a) the strong acids, (b) the strong bases, (c) the weak acids, and (d) the weak bases. NaCl; $MgSO_4$; HCl; CH_3COOH; $Ba(NO_3)_2$; H_3PO_4; $Sr(OH)_2$; HNO_3; HI; $Ba(OH)_2$; LiOH; C_3H_5COOH; NH_3; CH_3NH_2; KOH; HCN; $HClO_4$.

For Exercises 56–58, write balanced (1) formula unit, (2) total ionic, and (3) net ionic equations for reactions between the acid–base pairs. Name all compounds except water. Assume complete neutralization.

56. (a) HNO_2 + NaOH →
    (b) H_2SO_4 + KOH →
    (c) HCl + NH_3 →
    (d) CH_3COOH + KOH →
    (e) HI + NaOH →

57. (a) H_2CO_3 + $Ba(OH)_2$ →
    (b) H_2SO_4 + $Ca(OH)_2$ →
    (c) H_3PO_4 + $Ba(OH)_2$ →
    (d) HBr + NaOH →
    (e) H_3AsO_4 + KOH →

58. (a) $HClO_4$ + $Ba(OH)_2$ →
    (b) HI + $Ca(OH)_2$ →
    (c) H_2SO_4 + NH_3 →
    (d) H_2SO_4 + $Fe(OH)_3$ →
    (e) H_2SO_4 + $Ba(OH)_2$ →

59. Complete these equations by writing the formulas of the omitted compounds.
    (a) $Ba(OH)_2$ + ? → $BaSO_4$(s) + $2H_2O$
    (b) FeO(s) + ? → $FeCl_2$(aq) + H_2O
    (c) HCl(aq) + ? → AgCl(s) + ?
    (d) Na_2O + ? → 2NaOH(aq)
    (e) NaOH + ? → Na_2HPO_4(aq) + ?
       (two possible answers)

60. Although many salts may be formed by a variety of reactions, salts are usually thought of as being derived from the reaction of an acid with a base. For each of the salts listed here, choose the acid and base that would react with each other to form the salt. Write the (i) formula unit, (ii) total ionic, and (iii) net ionic equations for the formation of each salt. (a) $Pb(NO_3)_2$; (b) $AlCl_3$; (c) $(NH_4)_2SO_4$; (d) $Ca(ClO_4)_2$; (e) $Al_2(SO_4)_3$.

61. (a) Which of the following compounds are salts? $CaCO_3$; Li_2O; $U(NO_3)_5$; $AgNO_3$; $Sr(CH_3COO)_2$. (b) Write an acid–base equation that accounts for the formation of those identified as being salts.

62. Repeat Exercise 61 for: $KMnO_4$; $CaSO_3$; P_4O_{10}; SnF_2; K_3PO_4.

Acidic and Basic Salts

63. What are polyprotic acids? Write names and formulas for five polyprotic acids.

64. What are acidic salts? Write balanced equations to show how the following acidic salts can be prepared from the appropriate acid and base: $NaHSO_3$; $KHCO_3$; NaH_2PO_4; Na_2HPO_4; NaHS.

65. Indicate the mole ratio of acid and base required in each case in Exercise 64.

66. The following salts are components of fertilizers. They are made by reacting gaseous NH_3 with concentrated solutions of acids. The heat produced by the reactions evaporates most of the water. Write balanced formula unit equations that show the formation of each. (a) NH_4NO_3; (b) $NH_4H_2PO_4$; (c) $(NH_4)_2HPO_4$; (d) $(NH_4)_3PO_4$; (e) $(NH_4)_2SO_4$.

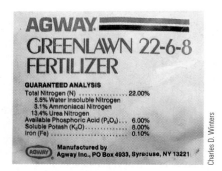

Common lawn fertilizer.

67. What are polyhydroxy bases? Write names and formulas for five polyhydroxy bases.

68. What are basic salts? (a) Write balanced equations to show how each of the following basic salts can be prepared from the appropriate acid and base: $Ca(OH)Cl$; $Al(OH)_2Cl$; $Al(OH)Cl_2$. (b) Indicate the mole ratio of acid and base required in each case.

69. What are amphoteric metal hydroxides? (a) Are they bases? (b) Write the names and formulas for four amphoteric metal hydroxides.

70. Chromium(III) hydroxide and lead(II) hydroxide are typical amphoteric hydroxides. (a) Write the formula unit, total ionic, and net ionic equations for the complete reaction of each hydroxide with nitric acid. (b) Write the same kinds of equations for the reaction of each hydroxide with an excess of potassium hydroxide solution. Reference to Table 10-1 may be helpful.

71. Write the chemical equations for the stepwise ionization of oxalic acid, $(COOH)_2$, a diprotic acid.

72. Write the chemical equations for the stepwise ionization of citric acid, $C_3H_5O(COOH)_3$, a triprotic acid.

The Lewis Theory

73. Define and illustrate the following terms clearly and concisely. Write an equation to illustrate the meaning of each term. (a) Lewis acid; (b) Lewis base; (c) neutralization according to Lewis theory.

74. Write a Lewis formula for each species in the following equations. Label the acids and bases using Lewis theory terminology.

(a) $H_2O + H_2O \rightleftharpoons H_3O^+ + OH^-$
(b) $HCl(g) + H_2O \rightarrow H_3O^+ + Cl^-$
(c) $NH_3(g) + H_2O \rightleftharpoons NH_4^+ + OH^-$
(d) $NH_3(g) + HBr(g) \rightarrow NH_4Br(s)$

75. Explain the differences between the Brønsted–Lowry and the Lewis acid–base theories, using the formation of the ammonium ion from ammonia and water to illustrate your points.

76. What are the advantages and limitations of the Brønsted–Lowry theory?

77. Iodine, I_2, is much more soluble in a water solution of potassium iodide, KI, than it is in H_2O. The anion found in the solution is I_3^-. Write an equation for the reaction that forms I_3^-, indicating the Lewis acid and the Lewis base.

78. Identify the Lewis acid and base and the donor and acceptor atoms in each of the following reactions.

(a) $H{-}\overset{\cdot\cdot}{\underset{}{O}}{:} + H^+ \rightarrow \left[H{-}\overset{H}{\underset{}{O}}{-}H \right]^+$

(b) $6\left[:\overset{\cdot\cdot}{\underset{\cdot\cdot}{Cl}}: \right]^- + Pt^{4+} \rightarrow \left[\begin{matrix} & Cl & Cl & \\ :Cl{-}Pt{-}Cl: \\ & Cl & Cl & \end{matrix} \right]^{2-}$

79. What is the term for a single covalent bond in which both electrons in the shared pair come from the same atom? Identify the Lewis acid and base and the donor and acceptor atoms in the following reaction.

$$H{-}\overset{H}{\underset{H}{N}}{:} + \overset{:F:}{\underset{:F:}{B}}{-}F: \rightarrow H{-}\overset{H}{\underset{H}{N}}{-}\overset{:F:}{\underset{:F:}{B}}{-}F:$$

80. A group of very strong acids are the fluoroacids, H_mXF_n. Two such acids are formed by Lewis acid–base reactions. (a) Identify each Lewis acid and Lewis base.

$HF + SbF_5 \longrightarrow H(SbF_6)$ (called a "super" acid, hexafluoroantimonic acid)

$HF + BF_3 \longrightarrow H(BF_4)$ (tetrafluoroboric acid)

(b) To which atom is the H of the product bonded? How is the H bonded?

Preparation of Acids

81. A volatile acid such as nitric acid, HNO_3, can be prepared by adding concentrated H_2SO_4 to a salt of the acid. (a) Write the chemical equation for the reaction of H_2SO_4 with sodium nitrate (called Chile saltpeter). (b) A dilute aqueous solution of H_2SO_4 cannot be used. Why?

82. Outline a method of preparing each of the following acids

and write appropriate balanced equations for each preparation: (a) H_2S; (b) HCl; (c) CH_3COOH.

83. Repeat Exercise 82 for (a) carbonic acid, (b) chloric acid, (c) permanganic acid, and (d) phosphoric acid (two methods).

Mixed Exercises

84. Give the formula for an example chosen from the representative elements for (a) an acidic oxide, (b) an amphoteric oxide, and (c) a basic oxide.

85. Identify each of the following as (i) acidic, (ii) basic, or (iii) amphoteric. Assume all oxides are dissolved in or are in contact with water. Do not be intimidated by the way in which the formula of the compound is written. (a) Rb_2O; (b) Cl_2O_5; (c) HCl; (d) $SO_2(OH)_2$; (e) HNO_2, (f) Al_2O_3; (g) CaO; (h) H_2O; (i) CO_2; (j) SO_2.

86. Indicate which of the following substances—(a) H_2S; (b) $PO(OH)_3$; (c) H_2CaO_2; (d) $ClO_3(OH)$; (e) $Sb(OH)_3$—can act as (i) an acid, (ii) a base, or (iii) both according to the Arrhenius (classical) theory or the Brønsted–Lowry theory. Do not be confused by the unusual way in which the formulas are written.

87. (a) Write equations for the reactions of HCO_3^- with H_3O^+ and HCO_3^- with OH^-, and indicate the conjugate acid–base pairs in each case. (b) A substance such as HCO_3^- that reacts with both H_3O^+ and OH^- is said to be _____. (Fill in the missing word.)

88. (a) List the conjugate bases of H_3PO_4, NH_4^+, and OH^- and the conjugate acids of HSO_4^-, PH_3, and PO_4^{3-}. (b) Given that NO_2^- is a stronger base than NO_3^-, which is the stronger acid—nitric acid, HNO_3, or nitrous acid, HNO_2?

*89. A 0.1 M solution of copper(II) chloride, $CuCl_2$, causes the light bulb in Figure 4-2 to glow brightly. When hydrogen sulfide, H_2S, a very weak acid, is added to the solution, a black precipitate of copper(II) sulfide, CuS, forms, and the bulb still glows brightly. The experiment is repeated with a 0.1 M solution of copper(II) acetate, $Cu(CH_3COO)_2$, which also causes the bulb to glow brightly. Again, CuS forms, but this time the bulb glows dimly. With the aid of ionic equations, explain the difference in behavior between the $CuCl_2$ and $Cu(CH_3COO)_2$ solutions.

90. Referring again to Figure 4-2, explain the following results of a conductivity experiment (use ionic equations). (a) Individual solutions of NaOH and HCl cause the bulb to glow brightly. When the solutions are mixed, the bulb still glows brightly but not as brightly as before. (b) Individual solutions of NH_3 and CH_3COOH cause the bulb to glow dimly. When the solutions are mixed, the bulb glows brightly.

91. When a 0.1 M aqueous ammonia solution is tested with a conductivity apparatus (Figure 4.2), the bulb glows dimly. When a 0.1 M hydrochloric acid solution is tested, the bulb glows brightly. Would you expect the bulb to glow more brightly, stop glowing, or stay the same as water is added to each of the solutions? Explain your reasoning.

CONCEPTUAL EXERCISES

92. The following diagrams are nanoscale representations of different acids in aqueous solution; the water molecules are not shown. The small, dark circles are hydrogen atoms

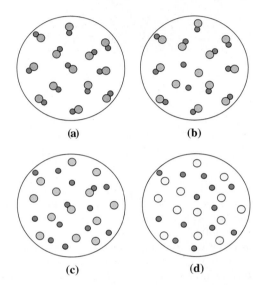

(a)   (b)

(c)   (d)

or ions. The larger, lighter circles represent the anions. (a) Which diagram best represents hydrochloric acid? (b) Which diagram best represents acetic acid?

93. One of the chemical products of muscle contraction is lactic acid ($CH_3CH(OH)COOH$), a monoprotic acid whose structure is

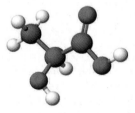

Prolonged exercise can temporarily overload the body's capacity for elimination of this substance, and the resulting increase in lactic acid concentration in the muscles causes pain and stiffness. (a) Lactic acid has six H atoms, yet it acts as a monoprotic acid in an aqueous environment. Which of the H atoms is ionizable? (b) Draw the structural formula of the conjugate base. (c) Write a net ionic equation that illustrates the ionization of lactic acid in water. (d) Describe the geometry around each of the carbon atoms in lactic acid.

Runners risk suffering from lactic acid buildup.

Cave formation

94. On the planet Baseacidopolous, the major solvent is liquid ammonia, not water. Ammonia autoionizes much like water $(2NH_3 \rightleftharpoons NH_4^+ + NH_2^-)$. If instead of water, ammonia is used as a solvent: (a) What is the formula of the cation that would indicate that a compound is an acid? (b) What is the formula of the anion produced if a compound is a base? (c) Look at the way that NaCl is formed from an acid–base reaction on earth and determine if NaCl can be a salt on Baseacidopolous.

95. Some of the acid formed in tissues is excreted through the kidneys. One of the bases removing the acid is HPO_4^{2-}. Write the equation for the reaction. Could Cl^- serve this function?

BUILDING YOUR KNOWLEDGE

96. Autoionization can occur when an ion other than an H^+ is transferred, as exemplified by the transfer of a Cl^- ion from one PCl_5 molecule to another. Write the equation for this reaction. What are the shapes of the two ions that are formed?

97. Limestone, $CaCO_3$, is a water-insoluble material, whereas $Ca(HCO_3)_2$ is soluble. Caves are formed when rainwater containing dissolved CO_2 passes over limestone for long periods of time. Write a chemical equation for the acid–base reaction.

98. Acids react with metal carbonates and hydrogen carbonates to form carbon dioxide and water. (a) Write the balanced equation for the reaction that occurs when baking soda, $NaHCO_3$, and vinegar, 5% acetic acid, are mixed. What causes the "fizz"? (b) Lactic acid, $CH_3CH(OH)COOH$, is found in sour milk and in buttermilk. Many of its reactions are very similar to those of acetic acid. Write the balanced equation for the reaction of baking soda, $NaHCO_3$, with lactic acid. Explain why bread "rises" during the baking process.

BEYOND THE TEXTBOOK

Go to the textbook website at

http://www.brookscole.com/chemistry/whitten

for additional activities and exercises based on the General Chemistry Interactive CD-ROM, the World Wide Web, and library resources.

InfoTrac College Edition

For additional readings, go to InfoTrac College Edition, your online research library at:

http://infotrac.thomsonlearning.com

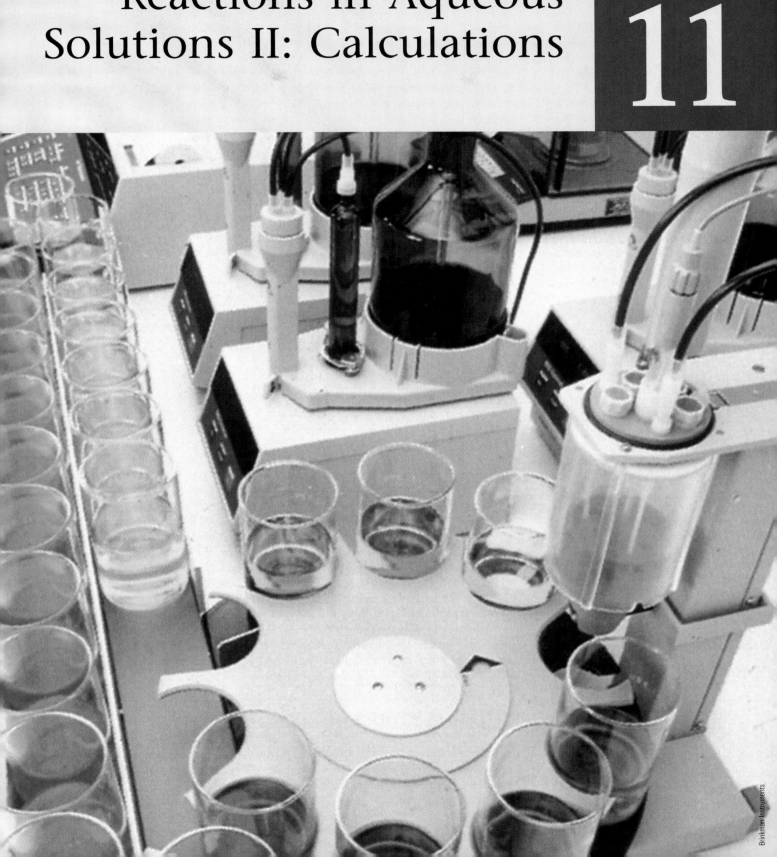

Automatic titrators are used in modern analytical laboratories. Such titrators rely on electrical properties of the solutions. Methyl red indicator changes from yellow to red at the end point of the titration.

OBJECTIVES

After you have studied this chapter, you should be able to

- Perform molarity calculations
- Solve acid–base stoichiometry calculations
- Describe titration and standardization
- Use the mole method and molarity in acid–base titration reactions
- Perform calculations involving equivalent weights and normality of acid and base solutions
- Balance oxidation–reduction equations
- Perform calculations associated with redox reactions

AQUEOUS ACID–BASE REACTIONS

Digestive juice is the acidic fluid secreted by glands in the lining of the stomach.

Hydrochloric acid, HCl, is called "stomach acid" because it is the main acid ($\approx 0.10\ M$) in our digestive juices. When the concentration of HCl is too high in humans, problems result. These problems may range from "heartburn" to ulcers that can eat through the lining of the stomach wall. Snakes have very high concentrations of HCl in their digestive juices so that they can digest whole small animals and birds.

Automobile batteries contain 40% H_2SO_4 by mass. When the battery has "run down," the concentration of H_2SO_4 is significantly lower than 40%. A technician checks an automobile battery by drawing some battery acid into a hydrometer, which indicates the density of the solution. This density is related to the concentration of H_2SO_4.

There are many practical applications of acid–base chemistry in which we must know the concentration of a solution of an acid or a base.

11-1 CALCULATIONS INVOLVING MOLARITY

In Sections 3-6 through 3-8 we introduced methods for expressing concentrations of solutions and discussed some related calculations. Review of those sections will be helpful as we learn more about acid–base reactions in solutions.

In *some cases*, one mole of an acid reacts with one mole of a base to yield neutralization.

$$HCl + NaOH \longrightarrow NaCl + H_2O$$
$$HNO_3 + KOH \longrightarrow KNO_3 + H_2O$$

Because one mole of each acid reacts with one mole of each base in these cases, *one liter of a one-molar solution of either of these acids* reacts with *one liter of a one-molar solution of either of these bases*. These acids have only one acidic hydrogen per formula unit, and these bases have one hydroxide ion per formula unit, so one formula unit of base reacts with one formula unit of acid.

The *reaction ratio* is the relative numbers of moles of reactants and products shown in the balanced equation.

EXAMPLE 11-1 *Acid–Base Reactions*

If 100. mL of 0.100 *M* HCl solution and 100. mL of 0.100 *M* NaOH are mixed, what is the molarity of the salt in the resulting solution? Assume that the volumes are additive.

Plan

We first write the balanced equation for the acid–base reaction and then construct the reaction summary that shows the amounts (moles) of HCl and NaOH. We determine the amount of salt formed from the reaction summary. The final (total) volume is the sum of the volumes mixed. Then we calculate the molarity of the salt.

Experiments have shown that volumes of dilute aqueous solutions are very nearly additive. No significant error is introduced by making this assumption. 0.100 L of dilute NaOH solution mixed with 0.100 L of dilute HCl solution gives 0.200 L of solution.

Solution

The following tabulation shows that equal numbers of moles of HCl and NaOH are mixed and, therefore, all of the HCl and NaOH react. The resulting solution contains only NaCl, the salt formed by the reaction, and water.

	HCl	+	NaOH	$\longrightarrow$	NaCl	+ H$_2$O
Rxn ratio:	1 mol		1 mol		1 mol	1 mol
Start:	$\left[0.100\ L\left(\dfrac{0.100\ \text{mol}}{L}\right)\right]$		$\left[0.100\ L\left(\dfrac{0.100\ \text{mol}}{L}\right)\right]$		0 mol	
	0.0100 mol HCl		0.0100 mol NaOH			
Change:	−0.0100 mol		−0.0100 mol		+0.0100 mol	
After rxn:	0 mol		0 mol		0.0100 mol	

The HCl and NaOH neutralize each other exactly, and the resulting solution contains 0.0100 mol of NaCl in 0.200 L of solution. Its molarity is

$$\underset{?}{\text{—}}\frac{\text{mol NaCl}}{L} = \frac{0.0100\ \text{mol NaCl}}{0.200\ L} = \boxed{0.0500\ M\ \text{NaCl}}$$

The amount of water produced by the reaction is negligible.

We often express the volume of a solution in milliliters rather than in liters. Likewise, we may express the amount of solute in millimoles (mmol) rather than in moles. Because one milliliter is 1/1000 of a liter and one **millimole** is 1/1000 of a mole, molarity also may be expressed as the number of millimoles of solute per milliliter of solution:

1 mol = 1000 mmol
1 L = 1000 mL

$$\text{molarity} = \frac{\text{no. mol}}{L} = \frac{\text{no. mmol}}{mL}$$

$$\text{molarity} = \frac{\text{number of millimoles of solute}}{\text{number of milliliters of solution}}$$

For volumes and concentrations that are commonly used in laboratory experiments, solving problems in terms of millimoles and milliliters often involves more convenient numbers than using moles and liters. We should note also that the reaction ratio that we obtain from any balanced chemical equation is exactly the same whether we express all quantities in moles or in millimoles. We will work many problems in this chapter using millimoles and milliliters. Let us see how we might solve Example 11-1 in these terms.

As in Example 11-1 we first write the balanced equation for the acid–base reaction, and then construct the reaction summary that shows the amounts (millimoles) of NaOH and HCl. We determine the amount of salt formed from the reaction summary. The final (total) volume is the sum of the volumes mixed. Then we can calculate the molarity of the salt.

The tabulation for the solution is

	HCl	+	NaOH	$\longrightarrow$	NaCl	+	H_2O
Rxn ratio:	1 mmol		1 mmol		1 mmol		1 mmol
Start:	$\left[100.\ mL\left(\dfrac{0.100\ mmol}{mL}\right)\right]$		$\left[100.\ mL\left(\dfrac{0.100\ mmol}{mL}\right)\right]$		0 mmol		
	= 10.0 mmol HCl		= 10.0 mmol NaOH				
Change:	−10.0 mmol		−10.0 mmol		+10.0 mmol		
After rxn:	0 mmol		0 mmol		10.0 mmol		

$$\underline{\quad?\quad}\ \frac{\text{mmol NaCl}}{mL} = \frac{10.0\ \text{mmol NaCl}}{200.\ mL} = \boxed{0.0500\ M\ \text{NaCl}}$$

EXAMPLE 11-2   Acid–Base Reactions

If 100. mL of 1.00 M HCl and 100. mL of 0.80 M NaOH solutions are mixed, what are the molarities of the solutes in the resulting solution?

Plan

We proceed as we did in Example 11-1. This reaction summary shows that NaOH is the limiting reactant and that we have excess HCl.

Solution

CD-ROM   Screen 5.15, Solution
Concentration—Molarity.

	HCl	+	NaOH	$\longrightarrow$	NaCl	+	H_2O
Rxn ratio:	1 mmol		1 mmol		1 mmol		1 mmol
Start:	100. mmol		80. mmol		0 mmol		
Change:	−80. mmol		−80. mmol		+80. mmol		
After rxn:	20. mmol		0 mmol		80. mmol		

Because two solutes are present in the solution after reaction, we must calculate the concentrations of both.

$$\underset{?}{\frac{\text{mmol HCl}}{\text{mL}}} = \frac{20.\ \text{mmol HCl}}{200.\ \text{mL}} = \boxed{0.10\ M\ \text{HCl}}$$

$$\underset{?}{\frac{\text{mmol NaCl}}{\text{mL}}} = \frac{80.\ \text{mmol NaCl}}{200.\ \text{mL}} = \boxed{0.40\ M\ \text{NaCl}}$$

Both HCl and NaCl are strong electrolytes, so the solution is 0.10 M in $H^+(aq)$, (0.10 + 0.40) M = 0.50 M in Cl^-, and 0.40 M in Na^+ ions.

You should now work Exercise 14.

 You should master these initial examples before attempting the following more complex ones.

✓ Problem-Solving Tip: *Review Limiting Reactant Calculations*

To solve many of the problems in this chapter, you will need to apply the limiting reactant concept (Section 3-3). In Example 11-1, we confirm that the two reactants are initially present in the mole ratio required by the balanced chemical equation; they both react completely, so there is no excess of either one. In Example 11-2, we need to determine which reactant limits the reaction. Before you proceed, be sure you understand how the ideas of Section 3-3 are used in these examples.

In many cases more than one mole of a base will be required to neutralize completely one mole of an acid, or more than one mole of an acid will be required to neutralize completely one mole of a base.

$$\underset{\substack{\text{1 mol}}}{H_2SO_4} + \underset{\substack{\text{2 mol}}}{2NaOH} \longrightarrow \underset{\substack{\text{1 mol}}}{Na_2SO_4} + 2H_2O$$

$$\underset{\substack{\text{2 mol}}}{2HCl} + \underset{\substack{\text{1 mol}}}{Ca(OH)_2} \longrightarrow \underset{\substack{\text{1 mol}}}{CaCl_2} + 2H_2O$$

The first equation shows that one mole of H_2SO_4 reacts with two moles of NaOH. Thus, *two* liters of 1 M NaOH solution are required to neutralize one liter of 1 M H_2SO_4 solution. The second equation shows that two moles of HCl react with one mole of $Ca(OH)_2$. Thus, *two* liters of HCl solution are required to neutralize one liter of $Ca(OH)_2$ solution of equal molarity.

 Students often mistakenly try to use the relation

$$V_1M_1 = V_2M_2$$

to solve problems about solution stoichiometry. This equation is intended *only* for dilution problems, and not for problems involving reactions.

EXAMPLE 11-3 *Volume of Acid to Neutralize Base*

What volume of 0.00300 M HCl solution would just neutralize 30.0 mL of 0.00100 M $Ca(OH)_2$ solution?

Plan

We write the balanced equation for the reaction to determine the reaction ratio. Then we (1) convert milliliters of $Ca(OH)_2$ solution to millimoles of $Ca(OH)_2$ using molarity in the reaction ratio, 0.00100 mmol $Ca_2(OH)/1.00$ mL $Ca(OH)_2$ solution; (2) convert millimoles of $Ca(OH)_2$ to millimoles of HCl using the reaction ratio, 2 mmol HCl/1 mmol Ca(OH) (from the balanced equation); and (3) convert millimoles of HCl to milliliters of HCl solution using the reaction ratio, 1.00 mL HCl/0.00300 mmol HCl, that is, molarity inverted.

CD-ROM Screen 5.18, Stoichiometry of Reactions in Solution.

$$\boxed{\begin{array}{c}\text{mL Ca(OH)}_2\\ \text{soln}\end{array}} \longrightarrow \boxed{\begin{array}{c}\text{mmol Ca(OH)}_2\\ \text{present}\end{array}} \longrightarrow \boxed{\begin{array}{c}\text{mmol HCl}\\ \text{needed}\end{array}} \longrightarrow \boxed{\begin{array}{c}\text{mL HCl(aq)}\\ \text{needed}\end{array}}$$

Solution

The balanced equation for the reaction is

$$2HCl + Ca(OH)_2 \longrightarrow CaCl_2 + 2H_2O$$
$$\text{2 mmol} \qquad \text{1 mmol} \qquad \text{1 mmol} \quad \text{2 mmol}$$

The balanced chemical equation allows us to construct the reaction ratio as either a mole ratio or a milli-mole ratio.

$$\frac{2 \text{ mol HCl}}{1 \text{ mol Ca(OH)}_2} \text{ or } \frac{2 \text{ mmol HCl}}{1 \text{ mmol Ca(OH)}_2}$$

$$\underline{?} \text{ mL HCl} =$$

$$30.0 \text{ mL Ca(OH)}_2 \times \frac{0.00100 \text{ mmol Ca(OH)}_2}{1.00 \text{ mL Ca(OH)}_2} \times \frac{2 \text{ mmol HCl}}{1 \text{ mmol Ca(OH)}_2} \times \frac{1.00 \text{ mL HCl}}{0.00300 \text{ mmol HCl}}$$

$$= \boxed{20.0 \text{ mL HCl}}$$

You should now work Exercise 16.

In the preceding example we used the unit factor, 2 mol HCl/1 mol Ca(OH)$_2$, to convert moles of Ca(OH)$_2$ to moles of HCl because the balanced equation shows that two moles of HCl are required to neutralize one mole of Ca(OH)$_2$. We must always write the balanced equation and determine the *reaction ratio*.

> ✓ **Problem-Solving Tip:** *There Is More Than One Way to Solve Some Problems*
>
> In many problems more than one "plan" can be followed. In Example 11-3 a particular plan was used successfully. Many students can more easily visualize the solution by following a plan like that in Examples 11-1 and 11-4. We suggest that you use the plan that you find most understandable.

EXAMPLE 11-4   *Acid–Base Reactions*

If 100. mL of 1.00 *M* H$_2$SO$_4$ solution is mixed with 200. mL of 1.00 *M* KOH, what salt is produced, and what is its molarity?

Plan

We proceed as we did in Example 11-2. We note that the reaction ratio is 1 mmol of H$_2$SO$_4$ to 2 mmol of KOH to 1 mmol of K$_2$SO$_4$.

Solution

	H$_2$SO$_4$	+	2KOH	$\longrightarrow$	K$_2$SO$_4$	+ H$_2$O
Rxn ratio:	1 mmol		2 mmol		1 mmol	
Start:	$\left[100.\text{ mL}\left(\dfrac{1.00 \text{ mmol}}{\text{mL}}\right)\right]$		$\left[200.\text{ mL}\left(\dfrac{1.00 \text{ mmol}}{\text{mL}}\right)\right]$			
	= 100. mmol		= 200. mmol		0 mmol	
Change:	−100. mmol		−200. mmol		+100. mmol	
After rxn:	0 mmol		0 mmol		100. mmol	

 Don't forget to add the solution volumes to find the volume of the final solution.

The reaction produces 100. mmol of potassium sulfate. This is contained in 300. mL of solution, and so the concentration is

$$\frac{?\ mmol\ K_2SO_4}{mL} = \frac{100.\ mmol\ K_2SO_4}{300.\ mL} = 0.333\ M\ K_2SO_4$$

You should now work Exercise 12.

Because K_2SO_4 is a soluble salt, this corresponds to 0.666 M K^+ and 0.333 M SO_4^{2-}.

11-2 TITRATIONS

In Examples 3-24 and 11-3, we calculated the volume of one solution that is required to react with a given volume of another solution, with the concentrations of *both* solutions given. In the laboratory we often measure the volume of one solution that is required to react with a given volume of another solution of known concentration. Then we calculate the concentration of the first solution. The process is called **titration** (Figure 11-1).

SC*LINKS*. TOPIC: Titration
GO TO: www.scilinks.org
*sci*LINKS CODE: WCH1110

(a)

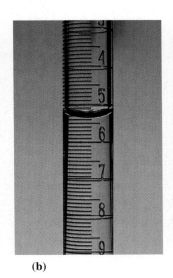
(b)

(c)

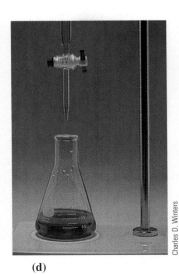
(d)

Charles D. Winters

Figure 11-1 The titration process. (a) A typical setup for titration in a teaching laboratory. The solution to be titrated is placed in an Erlenmeyer flask, and a few drops of indicator are added. The buret is filled with a standard solution (or the solution to be standardized). The volume of solution in the buret is read carefully. (b) The meniscus describes the surface of the liquid in the buret. Aqueous solutions wet glass, so the meniscus of an aqueous solution is always concave. The position of the *bottom* of the meniscus is read and recorded. (c) The solution in the buret is added (dropwise near the end point), with stirring, to the Erlenmeyer flask until the end point is reached. (d) The end point is signaled by the appearance (or change) of color *throughout* the solution being titrated. (A very large excess of indicator was used to make this photograph.) The volume of the liquid is read again—the difference between the final and initial buret readings is the volume of the solution used.

Titration is the process in which a solution of one reactant, the titrant, is carefully added to a solution of another reactant, and the volume of titrant required for complete reaction is measured.

How does one know when to stop a titration—that is, when is the chemical reaction just complete? In one method, a few drops of an indicator solution are added to the solution to be titrated. An acid–base **indicator** is a substance that can exist in different forms, with different colors that depend on the concentration of H^+ in the solution. At least one of these forms must be very intensely colored so that even very small amounts of it can be seen.

We can titrate an acid solution of unknown concentration by adding a standardized solution of sodium hydroxide dropwise from a **buret** (see Figure 11-1). A common buret is graduated in large intervals of 1 mL and in smaller intervals of 0.1 mL so that it is possible to estimate the volume of a solution dispensed to within at least ±0.02 mL. (Experienced individuals can often read a buret to ±0.01 mL.) The analyst tries to choose an indicator that changes color clearly at the point at which stoichiometrically equivalent amounts of acid and base have reacted, the **equivalence point.** The point at which the indicator changes color and the titration is stopped is called the **end point.** Ideally, the end point should coincide with the equivalence point. Phenolphthalein is colorless in acidic solution and reddish violet in basic solution. In a titration in which a base is added to an acid, phenolphthalein is often used as an indicator. The end point is signaled by the first appearance of a faint pink coloration that persists for at least 15 seconds as the solution is swirled.

The choice of indicators will be discussed in Section 19.4.

CD-ROM   Screen 5.15, Titrations; view the simulation in that screen.

EXAMPLE 11-5   *Titration*

What is the molarity of a hydrochloric acid solution if 36.7 mL of the HCl solution is required to react with 43.2 mL of 0.236 M sodium hydroxide solution?

$$HCl + NaOH \longrightarrow NaCl + H_2O$$

Plan

The balanced equation tells us that the reaction ratio is one millimole of HCl to one millimole of NaOH, which gives the conversion factor, 1 mmol HCl/1 mmol NaOH.

$$HCl \ + NaOH \longrightarrow NaCl \ + \ H_2O$$
$$1 \text{ mmol} \quad 1 \text{ mmol} \qquad 1 \text{ mmol} \quad 1 \text{ mmol}$$

First we find the number of millimoles of NaOH. The reaction ratio is one millimole of HCl to one millimole of NaOH, so the HCl solution must contain the same number of millimoles of HCl. Then we can calculate the molarity of the HCl solution because we know its volume.

Solution

The volume of a solution (in milliliters) multiplied by its molarity gives the number of millimoles of solute.

$$\underline{?} \text{ mmol NaOH} = 43.2 \text{ mL NaOH soln} \times \frac{0.236 \text{ mmol NaOH}}{1 \text{ mL NaOH soln}} = 10.2 \text{ mmol NaOH}$$

Because the reaction ratio is one millimole of NaOH to one millimole of HCl, the HCl solution must contain 10.2 millimole HCl.

$$\underline{?} \text{ mol HCl} = 10.2 \text{ mmol NaOH} \times \frac{1 \text{ mmol HCl}}{1 \text{ mmol NaOH}} = 10.2 \text{ mmol HCl}$$

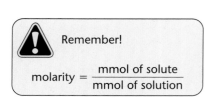

Remember!

$$\text{molarity} = \frac{\text{mmol of solute}}{\text{mmol of solution}}$$

The indicator phenolphthalein changes from colorless, its color in acidic solutions, to pink, its color in basic solutions, when the reaction in Example 11-5 reaches completion. Note the first appearance of a faint pink coloration in the middle beaker; this signals that the end point is near.

We know the volume of the HCl solution, so we can calculate its molarity.

$$\frac{?\ \text{mmol HCl}}{\text{mL HCl soln}} = \frac{10.2\ \text{mmol HCl}}{36.7\ \text{mL HCl soln}} = 0.278\ M\ \text{HCl}$$

EXAMPLE 11-6   Titration

A 43.2-mL sample of 0.236 M sodium hydroxide solution reacts completely with 36.7 mL of a sulfuric acid solution. What is the molarity of the H_2SO_4 solution?

$$H_2SO_4 + 2NaOH \longrightarrow Na_2SO_4 + 2H_2O$$

Plan

The balanced equation tells us that the reaction ratio is one millimole of H_2SO_4 to two millimoles of NaOH, which gives the conversion factor, 1 mmol H_2SO_4/2 mmol NaOH.

$$H_2SO_4 + 2NaOH \longrightarrow Na_2SO_4 + 2H_2O$$
$$\text{1 mmol} \quad \text{2 mmol} \qquad\qquad \text{1 mmol} \quad \text{2 mmol}$$

First we find the number of millimoles of NaOH. The reaction ratio is one millimole of H_2SO_4 to two millimoles of NaOH, so the number of millimoles of H_2SO_4 must be one half of the number of millimoles of NaOH. Then we can calculate the molarity of the H_2SO_4 solution because we know its volume.

Solution

The volume of a solution (in milliliters) multiplied by its molarity gives the number of millimoles of solute.

$$?\ \text{mmol NaOH} = 43.2\ \text{mL NaOH soln} \times \frac{0.236\ \text{mmol NaOH}}{1\ \text{mL NaOH soln}} = 10.2\ \text{mmol NaOH}$$

Because the reaction ratio is two millimoles of NaOH to one millimole of H_2SO_4, the H_2SO_4 solution must contain 5.10 millimoles of H_2SO_4.

$$\underline{?}\ \text{mmol}\ H_2SO_4 = 10.2\ \text{mmol NaOH} \times \frac{1\ \text{mmol}\ H_2SO_4}{2\ \text{mmol NaOH}} = 5.10\ \text{mmol}\ H_2SO_4$$

We know the volume of the H_2SO_4 solution, so we can calculate its molarity.

$$\frac{\underline{?}\ \text{mmol}\ H_2SO_4}{\text{mL}\ H_2SO_4\ \text{soln}} = \frac{5.10\ \text{mmol}\ H_2SO_4}{36.7\ \text{mL}\ H_2SO_4\ \text{soln}} = \boxed{0.139\ M\ H_2SO_4}$$

You should now work Exercise 38.

Solutions with accurately known concentrations are called **standard solutions.** Often we prepare a solution of a substance and then determine its concentration by titration with a standard solution.

Standardization is the process by which one determines the concentration of a solution by measuring accurately the volume of the solution required to react with an accurately known amount of a **primary standard.** The standardized solution is then known as a **secondary standard** and is used in the analysis of unknowns.

The properties of an ideal *primary standard* include the following:

1. It must not react with or absorb the components of the atmosphere, such as water vapor, oxygen, and carbon dioxide.
2. It must react according to one invariable reaction.
3. It must have a high percentage purity.
4. It should have a high formula weight to minimize the effect of error in weighing.
5. It must be soluble in the solvent of interest.
6. It should be nontoxic.
7. It should be readily available (inexpensive).
8. It should be environmentally friendly.

The first five of these characteristics are essential to minimize the errors involved in analytical methods. The last three characteristics are just as important as the first five in most analytical laboratories. Because primary standards are often costly and difficult to prepare, secondary standards are often used in day-to-day work.

11-3 THE MOLE METHOD AND MOLARITY

Let us now describe the use of a few primary standards for acids and bases. One primary standard for solutions of acids is sodium carbonate, Na_2CO_3, a solid compound.

$$H_2SO_4 + Na_2CO_3 \longrightarrow Na_2SO_4 + CO_2 + H_2O$$
$$\text{1 mol} \qquad \text{1 mol} \qquad \text{1 mol} \quad \text{1 mol} \quad \text{1 mol}$$

$$\text{1 mol}\ Na_2CO_3 = 106.0\ \text{g} \qquad \text{and} \qquad \text{1 mmol}\ Na_2CO_3 = 0.1060\ \text{g}$$

Sodium carbonate is a salt. Because a base can be broadly defined as a substance that reacts with hydrogen ions, in *this* reaction Na_2CO_3 can be thought of as a base.

Notice the similarity between Examples 11-5 and 11-6 in which 43.2 mL of 0.236 *M* NaOH solution is used. In Example 11-5 the reaction ratio is 1 mmol acid/1 mmol base, whereas in Example 11-6 the reaction ratio is 1 mmol acid/2 mmol base, and so the molarity of the HCl solution (0.278 *M*) is twice the molarity of the H_2SO_4 solution (0.139 *M*).

CO_2, H_2O, and O_2 are present in the atmosphere. They react with many substances.

CD-ROM   Screen 5.19, Titrations.

Refer to the Brønsted–Lowry theory. (Section 10-4).

EXAMPLE 11-7 *Standardization of an Acid Solution*

Calculate the molarity of a solution of H_2SO_4 if 40.0 mL of the solution neutralizes 0.364 g of Na_2CO_3.

Plan

We know from the balanced equation that 1 mol of H_2SO_4 reacts with 1 mol of Na_2CO_3, 106.0 g. This provides the reaction ratio that converts 0.364 g of Na_2CO_3 to the corresponding number of moles of H_2SO_4, from which we can calculate molarity.

$$\boxed{\begin{array}{c} g\ Na_2CO_3 \\ available \end{array}} \longrightarrow \boxed{\begin{array}{c} mol\ Na_2CO_3 \\ present \end{array}} \longrightarrow \boxed{\begin{array}{c} mol\ H_2SO_4 \\ used \end{array}} \longrightarrow \boxed{\begin{array}{c} molarity \\ of\ H_2SO_4 \end{array}}$$

Solution

$$\underline{?}\ mol\ H_2SO_4 = 0.364\ g\ Na_2CO_3 \times \frac{1\ mol\ Na_2CO_3}{106.0\ g\ Na_2CO_3} \times \frac{1\ mol\ H_2SO_4}{1\ mol\ Na_2CO_3}$$

$$= 0.00343\ mol\ H_2SO_4 \quad \text{(present in 40.0 mL of solution)}$$

Now we calculate the molarity of the H_2SO_4 solution

$$\frac{\underline{?}\ mol\ H_2SO_4}{L} = \frac{0.00343\ mol\ H_2SO_4}{0.0400\ L} = \boxed{0.0858\ M\ H_2SO_4}$$

You should now work Exercise 26.

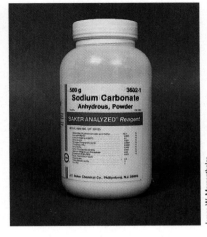

Sodium carbonate is often used as a primary standard for acids.

Most inorganic bases are metal hydroxides, all of which are solids. Even in the solid state, however, most inorganic bases react rapidly with CO_2 (an acid anhydride) from the atmosphere. Most metal hydroxides also absorb H_2O from the air. These properties make it *very* difficult to weigh out samples of pure metal hydroxides accurately. Chemists obtain solutions of bases of accurately known concentration by standardizing the solutions against an acidic salt, potassium hydrogen phthalate, $KC_6H_4(COO)(COOH)$. This is produced by neutralization of one of the two ionizable hydrogens of an organic acid, phthalic acid.

The "ph" in *phthalate* is silent. Phthalate is pronounced "thalate."

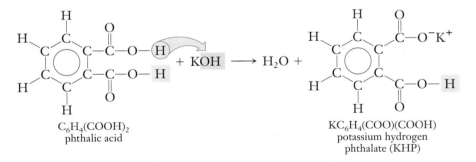

$C_6H_4(COOH)_2$
phthalic acid

$KC_6H_4(COO)(COOH)$
potassium hydrogen
phthalate (KHP)

This acidic salt, known simply as KHP, has one acidic hydrogen (highlighted) that reacts with bases. KHP is easily obtained in a high state of purity and is soluble in water. It is used as a primary standard for bases.

The P in KHP stands for the phthalate ion, $C_6H_4(COO)_2^{2-}$, *not* phosphorus.

EXAMPLE 11-8 *Standardization of Base Solution*

A 20.00-mL sample of a solution of NaOH reacts with 0.3641 g of KHP. Calculate the molarity of the NaOH solution.

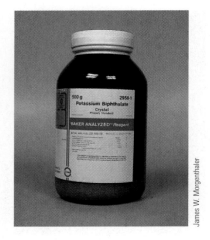

Very pure KHP is readily available.

Plan

We first write the balanced equation for the reaction between NaOH and KHP. We then calculate the number of moles of NaOH in 20.00 mL of solution from the amount of KHP that reacts with it. Then we can calculate the molarity of the NaOH solution.

$$\boxed{\begin{array}{c} \text{g KHP} \\ \text{available} \end{array}} \longrightarrow \boxed{\begin{array}{c} \text{mol KHP} \\ \text{available} \end{array}} \longrightarrow \boxed{\begin{array}{c} \text{mol NaOH} \\ \text{required} \end{array}} \longrightarrow \boxed{\begin{array}{c} \text{molarity} \\ \text{of NaOH} \end{array}}$$

Solution

$$\text{NaOH} + \text{KHP} \longrightarrow \text{NaKP} + \text{H}_2\text{O}$$
$$\text{1 mol} \quad \text{1 mol} \quad\quad \text{1 mol} \quad \text{1 mol}$$

We see that NaOH and KHP react in a 1:1 reaction ratio. One mole of KHP is 204.2 g.

$$\underline{?}\ \text{mol NaOH} = 0.3641\ \text{g KHP} \times \frac{1\ \text{mol KHP}}{204.2\ \text{g KHP}} \times \frac{1\ \text{mol NaOH}}{1\ \text{mol KHP}} = 0.001783\ \text{mol NaOH}$$

Then we calculate the molarity of the NaOH solution.

$$\frac{\underline{?}\ \text{mol NaOH}}{\text{L}} = \frac{0.001783\ \text{mol NaOH}}{0.02000\ \text{L}} = \boxed{0.08915\ M\ \text{NaOH}}$$

You should now work Exercise 28.

Impure samples of acids can be titrated with standard solutions of bases. The results can be used to determine percentage purity of the samples.

EXAMPLE 11-9   *Determination of Percent Acid*

Oxalic acid, $(\text{COOH})_2$, is used to remove rust stains and some ink stains from fabrics. A 0.1743-g sample of *impure* oxalic acid required 39.82 mL of 0.08915 M NaOH solution for complete neutralization. No acidic impurities were present. Calculate the percentage purity of the $(\text{COOH})_2$.

Plan

We write the balanced equation for the reaction and calculate the number of moles of NaOH in the standard solution. Then we calculate the mass of $(\text{COOH})_2$ in the sample, which gives us the information we need to calculate percentage purity.

Solution

The equation for the complete neutralization of $(\text{COOH})_2$ with NaOH is

$$2\text{NaOH} + (\text{COOH})_2 \longrightarrow \text{Na}_2(\text{COO})_2 + 2\text{H}_2\text{O}$$
$$\text{2 mol} \quad\quad \text{1 mol} \quad\quad\quad \text{1 mol} \quad\quad \text{2 mol}$$

Each molecule of $(\text{COOH})_2$ contains two acidic H's.

$$\begin{array}{cc} \text{O} & \text{O} \\ \parallel & \parallel \\ \text{H}-\text{O}-\text{C}-\text{C}-\text{O}-\text{H} \end{array}$$
$$\text{1 mol} = 90.04\ \text{g}$$

Two moles of NaOH neutralizes completely one mole of $(\text{COOH})_2$. The number of moles of NaOH that react is the volume times the molarity of the solution.

$$\underline{?}\ \text{mol NaOH} = 0.03982\ \text{L} \times \frac{0.08915\ \text{mol NaOH}}{\text{L}} = 0.003550\ \text{mol NaOH}$$

Now we calculate the mass of $(\text{COOH})_2$ that reacts with 0.003550 mol NaOH.

$$\underline{?}\ g\ (COOH)_2 = 0.003550\ mol\ NaOH \times \frac{1\ mol\ (COOH)_2}{2\ mol\ NaOH} \times \frac{90.04\ g\ (COOH)_2}{1\ mol\ (COOH)_2}$$

$$= 0.1598\ g\ (COOH)_2$$

The 0.1743-g sample contained 0.1598 g of $(COOH)_2$, so its percentage purity was

$$\%\ purity = \frac{0.1598\ g\ (COOH)_2}{0.1743\ g\ sample} \times 100\% = \boxed{91.68\%\ pure\ (COOH)_2}$$

You should now work Exercise 34.

11-4 EQUIVALENT WEIGHTS AND NORMALITY

In acid–base reactions, the mass of the acid (expressed in grams) that could furnish 6.022×10^{23} hydrogen ions (1 mol) or that could react with 6.022×10^{23} hydroxide ions (1 mol) is defined as one **equivalent weight,** or **equivalent (eq), of acid**. One mole of an acid contains 6.022×10^{23} formula units of the acid. Consider hydrochloric acid as a typical monoprotic acid.

$$HCl \xrightarrow{H_2O} H^+(aq) + Cl^-(aq)$$

1 mol	1 mol	1 mol
36.46 g	1.008 g	35.45 g
6.022×10^{23} formula units	6.022×10^{23} formula units	6.022×10^{23} formula units

We see that one mole of HCl can produce 6.022×10^{23} H$^+$ ions, and so *one mole of HCl is one equivalent.* The same is true for all monoprotic acids.

Sulfuric acid is a diprotic acid. One molecule of H_2SO_4 can furnish 2H$^+$ ions.

$$H_2SO_4 \xrightarrow{H_2O} 2H^+(aq) + SO_4{}^{2-}(aq)$$

1 mol	2 mol	1 mol
98.08 g	2(1.008 g)	96.06 g
6.022×10^{23} formula units	$2(6.022 \times 10^{23})$ formula units	6.022×10^{23} formula units

This equation shows that one mole of H_2SO_4 can produce $2(6.022 \times 10^{23})$ H$^+$; therefore, one mole of H_2SO_4 is *two* equivalent weights in all reactions in which *both* acidic hydrogen atoms react.

One **equivalent weight of a base** is defined as the mass of the base (expressed in grams) that will furnish 6.022×10^{23} hydroxide ions or the mass of the base that will react with 6.022×10^{23} hydrogen ions.

The equivalent weight of an *acid* is obtained by dividing its formula weight in grams either by the number of acidic hydrogens that could be furnished by one formula unit of the acid *or* by the number of hydroxide ions with which one formula unit of the acid reacts. The equivalent weight of a *base* is obtained by dividing its formula weight in grams either by the number of hydroxide ions furnished by one formula unit *or* by the number of hydrogen ions with which one formula unit of the base reacts. Equivalent weights of some common acids and bases are given in Table 11-1.

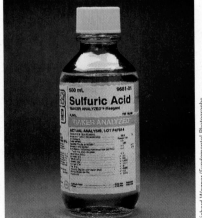

One mole of H_2SO_4 is two equivalent weights of H_2SO_4.

Richard Wagner/Fundamental Photographs

TABLE 11-1	Equivalent Weights* of Some Acids and Bases				
Acids			**Bases**		
Symbolic representation	One equivalent		Symbolic representation	One equivalent	
$\dfrac{HNO_3}{1}$	$= \dfrac{63.02 \text{ g}}{1}$	$= 63.02 \text{ g } HNO_3$	$\dfrac{NaOH}{1}$	$= \dfrac{40.00 \text{ g}}{1}$	$= 40.00 \text{ g NaOH}$
$\dfrac{CH_3COOH}{1}$	$= \dfrac{60.03 \text{ g}}{1}$	$= 60.03 \text{ g } CH_3COOH$	$\dfrac{NH_3}{1}$	$= \dfrac{17.04 \text{ g}}{1}$	$= 17.04 \text{ g } NH_3$
$\dfrac{KHP}{1}$	$= \dfrac{204.2 \text{ g}}{1}$	$= 204.2 \text{ g KHP}$	$\dfrac{Ca(OH)_2}{2}$	$= \dfrac{74.10 \text{ g}}{2}$	$= 37.05 \text{ g } Ca(OH)_2$
$\dfrac{H_2SO_4}{2}$	$= \dfrac{98.08 \text{ g}}{2}$	$= 49.04 \text{ g } H_2SO_4$	$\dfrac{Ba(OH)_2}{2}$	$= \dfrac{171.36 \text{ g}}{2}$	$= 85.68 \text{ g } Ba(OH)_2$

Complete neutralization is assumed.

Any calculation that can be carried out with equivalent weights and normality can also be done by the mole method using molarity. The methods of this section are widely used, however, in health-related fields and in many industrial laboratories.

An **equivalent weight** is often referred to simply as an **equivalent** (eq).

A **milliequivalent weight** is often referred to simply as a **milliequivalent** (meq).

Because one mole of an acid does not necessarily neutralize one mole of a base, some chemists prefer a method of expressing concentration other than molarity to retain a one-to-one relationship. Concentrations of solutions of acids and bases are frequently expressed as *normality* (N). The **normality** of a solution is defined as the number of equivalent weights, or simply equivalents (eq), of solute per liter of solution. Normality may be represented symbolically as

$$\text{normality} = \frac{\text{number of equivalent weights of solute}}{\text{liter of solution}} = \frac{\text{no. eq}}{L}$$

By definition there are 1000 milliequivalent weights (meq) in one equivalent weight of an acid or base. Normality may also be represented as

$$\text{normality} = \frac{\text{number of milliequivalent weights of solute}}{\text{milliliter of solution}} = \frac{\text{no. meq}}{mL}$$

EXAMPLE 11-10   Concentration of a Solution

Calculate the normality of a solution of 4.202 grams of HNO_3 in 600. mL of solution.

Plan

We convert grams of HNO_3 to moles of HNO_3 and then to equivalents of HNO_3, which lets us calculate the normality.

$$\frac{g \; HNO_3}{L} \longrightarrow \frac{mol \; HNO_3}{L} \longrightarrow \frac{eq \; HNO_3}{L} = N \; HNO_3$$

Solution

$$N = \frac{\text{no. eq } HNO_3}{L}$$

$$\underset{M_{HNO_3}}{\underbrace{\frac{? \text{ eq } HNO_3}{L} = \frac{4.202 \text{ g } HNO_3}{0.600 \text{ L}} \times \frac{1 \text{ mol } HNO_3}{63.02 \text{ g } HNO_3}}} \times \frac{1 \text{ eq } HNO_3}{\text{mol } HNO_3} = \boxed{0.111 \text{ } N \text{ } HNO_3}$$

Because normality is equal to molarity times the number of equivalents per mole of solute, a solution's normality is always equal to or greater than its molarity.

$$\text{normality} = \text{molarity} \times \frac{\text{no. eq}}{\text{mol}} \quad \text{or} \quad N = M \times \frac{\text{no. eq}}{\text{mol}}$$

EXAMPLE 11-11 *Concentration of a Solution*

Calculate (a) the molarity and (b) the normality of a solution that contains 9.50 grams of barium hydroxide in 2000. mL of solution.

Plan

(a) We use the same kind of logic we used in Example 11-10.

(b) Because each mole of $Ba(OH)_2$ produces 2 moles of OH^- ions, 1 mole of $Ba(OH)_2$ is 2 equivalents. Thus,

$$N = M \times \frac{2 \text{ eq}}{\text{mol}} \quad \text{or} \quad M = \frac{N}{2 \text{ eq/mol}}$$

Solution

(a) $\dfrac{? \text{ mol } Ba(OH)_2}{L} = \dfrac{9.50 \text{ g } Ba(OH)_2}{2.00 \text{ L}} \times \dfrac{1 \text{ mol } Ba (OH)_2}{171.36 \text{ g } Ba(OH)_2} = \boxed{0.0277 \text{ } M \text{ } Ba(OH)_2}$

(b) $\dfrac{? \text{ eq } Ba(OH)_2}{L} = \dfrac{0.0277 \text{ mol } Ba(OH)_2}{L} \times \dfrac{2 \text{ eq } Ba(OH)_2}{1 \text{ mol } Ba(OH)_2} = \boxed{0.0554 \text{ } N \text{ } Ba(OH)_2}$

Because each formula unit of $Ba(OH)_2$ contains two OH^- ions,

1 mol $Ba(OH)_2$ = 2 eq $Ba(OH)_2$

Thus, molarity is one half of normality for $Ba(OH)_2$ solutions.

You should now work Exercises 40 through 42.

From the definitions of one equivalent of an acid and of a base, we see that *one equivalent of an acid reacts with one equivalent of any base.* It is *not* true that one mole of any acid reacts with one mole of any base in any specific chemical reaction that goes to completion. As a consequence of the definition of equivalents, 1 eq acid ≅ 1 eq base. We may write the following for *all* acid–base reactions that go to completion.

The notation ≅ is read "is equivalent to."

> number of equivalents of acid = number of equivalents of base

The product of the volume of a solution, in liters, and its normality is equal to the number of equivalents of solute contained in the solution. For a solution of an acid,

$$L_{acid} \times N_{acid} = L_{acid} \times \frac{\text{eq acid}}{L_{acid}} = \text{eq acid}$$

Remember that the product of volume and concentration equals the amount of solute.

Similar relationships can be written for a solution of a base. Because 1 eq of acid *always* reacts with 1 eq of base, we may write

$$\text{number of equivalents of acid} = \text{number of equivalents of base}$$

so

The conversion factors needed to convert liters to milliliters on each side of this equation will cancel.

$$L_{acid} \times N_{acid} = L_{base} \times N_{base} \quad \text{or} \quad mL_{acid} \times N_{acid} = mL_{base} \times N_{base}$$

EXAMPLE 11-12   *Volume Required for Neutralization*

What volume of $0.100\ N$ HNO_3 solution is required to neutralize completely 50.0 mL of a 0.150 N solution of $Ba(OH)_2$?

Plan

We know three of the four variables in the relationship

$$mL_{acid} \times N_{acid} = mL_{base} \times N_{base}$$

and so we solve for mL acid.

Solution

$$\underline{?\ mL_{acid}} = \frac{mL_{base} \times N_{base}}{N_{acid}} = \frac{50.0\ mL \times 0.150\ N}{0.100\ N} = \boxed{75.0\ mL\ of\ HNO_3\ solution}$$

You should now work Exercise 49.

In Example 11-13 let's again solve Example 11-7, this time using normality rather than molarity. The balanced equation for the reaction of H_2SO_4 with Na_2CO_3, interpreted in terms of equivalent weights, is

By definition, there must be equal numbers of equivalents of reactants in a balanced chemical equation.

$$\begin{array}{ccccccc} H_2SO_4 & + & Na_2CO_3 & \longrightarrow & Na_2SO_4 & + & CO_2 & + & H_2O \\ 1\ mol & & 1\ mol & & 1\ mol & & 1\ mol & & 1\ mol \\ 2\ eq & & 2\ eq & & & & & & \\ 98.08\ g & & 106.0\ g & & & & & & \end{array}$$

So, 1 eq $Na_2CO_3 = 53.0$ g

EXAMPLE 11-13   *Standardization of Acid Solution*

Calculate the normality of a solution of H_2SO_4 if 40.0 mL of the solution reacts completely with 0.364 gram of Na_2CO_3.

Plan

We refer to the balanced equation. We are given the mass of Na_2CO_3, so we convert grams of Na_2CO_3 to equivalents of Na_2CO_3, then to equivalents of H_2SO_4, which lets us calculate the normality of the H_2SO_4 solution.

$$\boxed{\begin{array}{c} g\ Na_2CO_3 \\ present \end{array}} \longrightarrow \boxed{\begin{array}{c} eq\ Na_2CO_3 \\ present \end{array}} \longrightarrow \boxed{\begin{array}{c} eq\ H_2SO_4 \\ needed \end{array}} \longrightarrow \boxed{\dfrac{eq\ H_2SO_4}{L}}$$

Solution

First we calculate the number of equivalents of Na_2CO_3 in the sample.

$$\text{no. eq } Na_2CO_3 = 0.364 \text{ g } Na_2CO_3 \times \frac{1 \text{ eq } Na_2CO_3}{53.0 \text{ g } Na_2CO_3} = 6.87 \times 10^{-3} \text{ eq } Na_2CO_3$$

Because no. eq $H_2SO_4 = $ no. eq Na_2CO_3, we can write

$$L_{H_2SO_4} \times N_{H_2SO_4} = 6.87 \times 10^{-3} \text{ eq } H_2SO_4$$

$$N_{H_2SO_4} = \frac{6.87 \times 10^{-3} \text{ eq } H_2SO_4}{L_{H_2SO_4}} = \frac{6.87 \times 10^{-3} \text{ eq } H_2SO_4}{0.040 \text{ L}} = \boxed{0.172 \ N \ H_2SO_4}$$

You should now work Exercise 46.

The starting values in this example are the same as those in Example 11-7. The normality of this H_2SO_4 solution is twice the molarity obtained in Example 11-7 because 1 mol of H_2SO_4 is 2 eq.

OXIDATION–REDUCTION REACTIONS

Our rules for assigning oxidation numbers are constructed so that in all redox reactions

the total increase in oxidation numbers must equal the total decrease in oxidation numbers.

This equivalence provides the basis for balancing redox equations. Although there is no single "best method" for balancing all redox equations, the half-reaction method is the most widely used. Many redox equations can be balanced by simple inspection, but you should master the half-reaction method so it can be used to balance difficult equations.

This method is used extensively in electrochemistry (Chapter 21).

SC*i*LINKS.

TOPIC: Redox Reactions
GO TO: www.scilinks.org
*sci***LINKS CODE:** WCH0460

All balanced equations must satisfy two criteria.

1. There must be mass balance. That is, the same number of atoms of each kind must appear in reactants and products.
2. There must be charge balance. The sums of actual charges on the left and right sides of the equation must equal each other. In a balanced *formula unit equation*, the total charge on each side will be equal to zero. In a balanced *net ionic equation*, the total charge on each side might not be zero, but it still must be equal on the two sides of the equation.

11-5 THE HALF-REACTION METHOD

In the half-reaction method we separate and completely balance equations describing oxidation and reduction **half-reactions.** Then we equalize the numbers of electrons gained and lost in each. Finally, we add the resulting half-reactions to give the overall balanced equation. The general procedure follows.

1. Write as much of the overall unbalanced equation as possible, omitting spectator ions.

2. Construct unbalanced oxidation and reduction half-reactions (these are usually incomplete as well as unbalanced). Show complete formulas for polyatomic ions and molecules.

3. Balance by inspection all elements in each half-reaction, except H and O. Then use the chart in Section 11-6 to balance H and O in each half-reaction.

4. Balance the charge in each half-reaction by adding electrons as "products" or "reactants."

5. Balance the electron transfer by multiplying the balanced half-reactions by appropriate integers.

6. Add the resulting half-reactions and eliminate any common terms.

CD-ROM   Screen 20.3, Balancing Equations for Redox Reactions.

EXAMPLE 11-14   *Balancing Redox Equations*

A useful analytical procedure involves the oxidation of iodide ions to free iodine. The free iodine is then titrated with a standard solution of sodium thiosulfate, $Na_2S_2O_3$. Iodine oxidizes $S_2O_3^{2-}$ ions to tetrathionate ions, $S_4O_6^{2-}$, and is reduced to I^- ions. Write the balanced net ionic equation for this reaction.

Plan

We are given the formulas for two reactants and two products. We use these to write as much of the equations as possible. We construct and balance the appropriate half-reactions using the rules just described. Then we add the half-reactions and eliminate common terms.

Solution

$$I_2 + S_2O_3^{2-} \longrightarrow I^- + S_4O_6^{2-}$$

$$I_2 \longrightarrow I^- \qquad \text{(red. half-reaction)}$$

$$I_2 \longrightarrow 2I^-$$

Each I_2 gains $2e^-$. I_2 is reduced; it is the oxidizing agent.

$$I_2 + 2e^- \longrightarrow 2I^- \qquad \text{(balanced red. half-reaction)}$$

$$S_2O_3^{2-} \longrightarrow S_4O_6^{2-} \qquad \text{(ox. half-reaction)}$$

$$2S_2O_3^{2-} \longrightarrow S_4O_6^{2-}$$

Each $S_2O_3^{2-}$ ion loses an e^-. $S_2O_3^{2-}$ is oxidized; it is the reducing agent.

$$2S_2O_3^{2-} \longrightarrow S_4O_6^{2-} + 2e^- \qquad \text{(balanced ox. half-reaction)}$$

Each balanced half-reaction involves a transfer of two electrons. We add these half-reactions and cancel the electrons.

Oxidizing and reducing agents were defined in Section 4-7.

$$I_2 + 2e^- \longrightarrow 2I^-$$

$$2S_2O_3^{2-} \longrightarrow S_4O_6^{2-} + 2e^-$$

$$\overline{I_2(s) + 2S_2O_3^{2-}(aq) \longrightarrow 2I^-(aq) + S_4O_6^{2-}(aq)}$$

11-6 ADDING H$^+$, OH$^-$, OR H$_2$O TO BALANCE OXYGEN OR HYDROGEN

Frequently we need more oxygen or hydrogen to complete the mass balance for a reaction or half-reaction in aqueous solution. We must be careful, however, not to introduce other changes in oxidation number or to use species that could not actually be present in the solution. We cannot add H$_2$ or O$_2$ to equations because these species are not present in aqueous solutions. Acidic solutions do not contain significant concentrations of OH$^-$ ions. Basic solutions do not contain significant concentrations of H$^+$ ions.

In acidic solution:	We add only H$^+$ or H$_2$O (*not* OH$^-$ in acidic solution).
In basic solution:	We add only OH$^-$ or H$_2$O (*not* H$^+$ in basic solution).

The following chart shows how to balance hydrogen and oxygen.

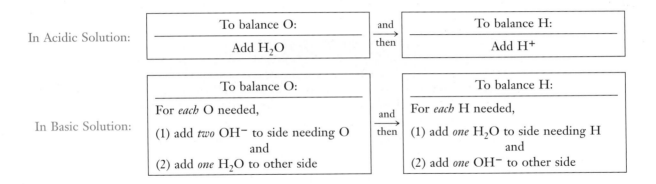

When balancing redox equations, we often find it convenient to omit the spectator ions (Section 4-3) so that we can focus on the oxidation and reduction processes. We use the methods presented in this chapter to balance the net ionic equation. If necessary we add the spectator ions and combine species to write the balanced formula unit equation. Examples 11-15 and 11-16 illustrate this approach.

In Section 4-3, we first wrote the *formula unit equation*. We separated any ionized or dissociated species into ions to obtain the *total ionic equation*. Then we eliminated the spectator ions to obtain the *net ionic equation*. In Examples 11-15 and 11-16, we reverse the procedure.

EXAMPLE 11-15   *Net Ionic Equations*

Permanganate ions oxidize iron(II) to iron(III) in sulfuric acid solution. Permanganate ions are reduced to manganese(II) ions. Write the balanced net ionic equation for this reaction.

Plan

We use the given information to write as much of the equation as possible. Then we follow Steps 2 through 6 in Section 11-5. The reaction occurs in H$_2$SO$_4$ solution; we can add H$^+$ and H$_2$O as needed to balance H and O in the half-reactions (Step 3).

Potassium permanganate, $KMnO_4$, is a commonly used oxidizing agent in the laboratory

Solution

$$Fe^{2+} + MnO_4^- \longrightarrow Fe^{3+} + Mn^{2+}$$

$$Fe^{2+} \longrightarrow Fe^{3+} \qquad \text{(ox. half-reaction)}$$

$$Fe^{2+} \longrightarrow Fe^{3+} + 1e^- \qquad \text{(balanced ox. half-reaction)}$$

- -

$$MnO_4^- \longrightarrow Mn^{2+} \qquad \text{(red. half-reaction)}$$

$$MnO_4^- + 8H^+ \longrightarrow Mn^{2+} + 4H_2O$$

$$MnO_4^- + 8H^+ + 5e^- \longrightarrow Mn^{2+} + 4H_2O \qquad \text{(balanced red. half-reaction)}$$

The oxidation half-reaction involves one electron, and the reduction half-reaction involves five electrons. Now we balance the electron transfer and then add the two equations term by term. This gives the balanced net ionic equation.

$$5(Fe^{2+} \longrightarrow Fe^{3+} + 1e^-)$$

$$1(MnO_4^- + 8H^+ + 5e^- \longrightarrow Mn^{2+} + 4H_2O)$$

$$5Fe^{2+}(aq) + MnO_4^-(aq) + 8H^+(aq) \longrightarrow 5Fe^{3+}(aq) + Mn^{2+}(aq) + 4H_2O(\ell)$$

EXAMPLE 11-16 *Total Ionic and Formula Unit Equations*

Write the balanced total ionic and the formula unit equations for the reaction in Example 11-15, given that the reactants were $KMnO_4$, $FeSO_4$, and H_2SO_4.

Plan

The K^+ is the cationic spectator ion, and the anionic spectator ion is SO_4^{2-}. The Fe^{3+} ion will need to occur twice in the product $Fe_2(SO_4)_3$, so there must be an even number of Fe atoms. So the net ionic equation is multiplied by two. It now becomes

$$10Fe^{2+}(aq) + 2MnO_4^-(aq) + 16H^+(aq) \longrightarrow 10Fe^{3+}(aq) + 2Mn^{2+}(aq) + 8H_2O(\ell)$$

Based on the $10Fe^{2+}$ and the $16H^+$, we add $18SO_4^{2-}$ to the reactant side of the equation; we must also add them to the product side to keep the equation balanced. Based on the $2MnO_4^-$, we add $2K^+$ to each side of the equation.

Solution

Total ionic equation

$$10[Fe^{2+}(aq) + SO_4^{2-}(aq)] + 2[K^+(aq) + MnO_4^-(aq)] + 8[2H^+(aq) + SO_4^{2-}(aq)] \longrightarrow$$
$$5[2Fe^{3+}(aq) + 3SO_4^{2-}(aq)] + 2[Mn^{2+}(aq) + SO_4^{2-}(aq)] + 8H_2O(\ell)$$
$$+ [2K^+(aq) + SO_4^{2-}(aq)]$$

Balanced formula unit equation

$$10FeSO_4(aq) + 2KMnO_4(aq) + 8H_2SO_4(aq) \longrightarrow$$
$$5Fe_2(SO_4)_3(aq) + 2MnSO_4(aq) + K_2SO_4(aq) + 8H_2O(\ell)$$

You should now work Exercise 60.

Caution: These common household chemicals, ammonia and bleach, should never be mixed because they react to form chloramine (NH_2Cl), a very poisonous volatile compound.

$$NH_3(aq) + ClO^-(aq) \longrightarrow NH_2Cl(aq) + OH^-(aq)$$

Bleaches sold under trade names such as Clorox and Purex are 5% solutions of sodium hypochlorite. The hypochlorite ion is a very strong oxidizing agent in basic solution. It oxidizes many stains to colorless substances.

EXAMPLE 11-17   *Balancing Redox Equations*

In basic solution, hypochlorite ions, ClO^-, oxidize chromite ions, CrO_2^-, to chromate ions, CrO_4^{2-}, and are reduced to chloride ions. Write the balanced net ionic equation for this reaction.

Plan

We are given the formulas for two reactants and two products; we write as much of the equations as possible. The reaction occurs in basic solution; we can add OH^- and H_2O as needed. We construct and balance the appropriate half-reactions, equalize the electron transfer, add the half-reactions, and eliminate common terms.

Solution

$$CrO_2^- + ClO^- \longrightarrow CrO_4^{2-} + Cl^-$$

$$CrO_2^- \longrightarrow CrO_4^{2-} \qquad\qquad \text{(ox. half-rxn)}$$

$$CrO_2^- + 4OH^- \longrightarrow CrO_4^{2-} + 2H_2O$$

$$CrO_2^- + 4OH^- \longrightarrow CrO_4^{2-} + 2H_2O + 3e^- \qquad \text{(balanced ox. half-rxn)}$$

- -

$$ClO^- \longrightarrow Cl^- \qquad\qquad \text{(red. half-rxn)}$$

$$ClO^- + H_2O \longrightarrow Cl^- + 2OH^-$$

$$ClO^- + H_2O + 2e^- \longrightarrow Cl^- + 2OH^- \qquad \text{(balanced red. half-rxn)}$$

The oxidation half-reaction involves three electrons, and the reduction half-reaction involves two electrons. We balance the electron transfer and add the half-reactions term by term.

$$2(CrO_2^- + 4OH^- \longrightarrow CrO_4^{2-} + 2H_2O + 3e^-)$$

$$3(ClO^- + H_2O + 2e^- \longrightarrow Cl^- + 2OH^-)$$

$$\overline{2CrO_2^- + 8OH^- + 3ClO^- + 3H_2O \longrightarrow 2CrO_4^{2-} + 4H_2O + 3Cl^- + 6OH^-}$$

We see that 6 OH^- and 3 H_2O can be eliminated from both sides to give the balanced net ionic equation.

$$2CrO_2^-(aq) + 2OH^-(aq) + 3ClO^-(aq) \longrightarrow 2CrO_4^{2-}(aq) + H_2O(\ell) + 3Cl^-(aq)$$

You should now work Exercise 56.

Caution: These common household chemicals, vinegar and bleach, should never be mixed because they react to form chlorine, a very poisonous gas.

$$2H^+(aq) + ClO^-(aq) + Cl^-(aq) \longrightarrow Cl_2(g) + H_2O(\ell)$$

James W. Morgenthaler

> ✓ **Problem-Solving Tip:** *Converting Ionic to Formula Unit Equations*
>
> We learned in Section 4-3 how to convert the formula unit equation to the net ionic equation. To do this, we convert the formulas for all *strong electrolytes* into their ions, and then cancel *spectator ions* from both sides of the equation. In Example 11-16 we reverse this procedure. To balance this excess charge, we must add negatively charged spectator ions to combine with the positively charged reactants, and we add positively charged spectator ions to combine with the negatively charged reactants. Any spectator ions added to the reactant side of the equation must also be added to the product side. Then we combine species to give complete formula units. Now we can write total ionic and formula unit equations for Exercise 11-17 if we know what spectator ions are present. Just for practice, consider the spectator ions to be Na^+.

> ⚠ In a balanced ionic equation, the total charge on the left side of the equation must equal the total charge on the right side.

Every balanced equation must have both mass balance and charge balance. Once the redox part of an equation has been balanced, we must next count *either* atoms or charges. Suppose we had balanced a redox equation in ionic form to give

$$H^+ + 3C_2H_5OH + Cr_2O_7{}^{2-} \longrightarrow 2Cr^{3+} + 3C_2H_4O + H_2O$$

The net charge on the left side is $(1 + 2-) = 1-$. On the right, it is $2(3+) = 6+$. Because H^+ is the *only charged species whose coefficient isn't known*, we add 7 *more* H^+ to give a net charge of $6+$ on both sides.

$$8H^+ + 3C_2H_5OH + Cr_2O_7{}^{2-} \longrightarrow 2Cr^{3+} + 3C_2H_4O + H_2O$$

Now we have 10 O on the left and only 4 O on the right. We add six *more* H_2O molecules to give the balanced net ionic equation.

$$8H^+(aq) + 3C_2H_5OH(\ell) + Cr_2O_7{}^{2-}(aq) \longrightarrow 2Cr^{3+}(aq) + 3C_2H_4O(\ell) + 7H_2O(\ell)$$

How can you tell whether to balance atoms or charges first? Look at the equation *after you have balanced the redox part*. Decide which is simpler, and do that. In the preceding equation, it is easier to balance charges than to balance atoms.

11-7 STOICHIOMETRY OF REDOX REACTIONS

One method of analyzing samples quantitatively for the presence of *oxidizable* or *reducible* substances is by **redox titration.** In such analyses, the concentration of a solution is determined by allowing it to react with a carefully measured amount of a *standard* solution of an oxidizing or reducing agent.

As in other kinds of chemical reactions, we must pay particular attention to the mole ratio in which oxidizing agents and reducing agents react.

Potassium permanganate, $KMnO_4$, is a strong oxidizing agent. Through the years it has been the "workhorse" of redox titrations. For example, in acidic solution, $KMnO_4$ reacts with iron(II) sulfate, $FeSO_4$, according to the balanced equation in the following example. A strong acid, such as H_2SO_4, is used in such titrations (Example 11-15).

A word about terminology. The reaction involves $MnO_4{}^-$ ions and Fe^{2+} ions in acidic solution. The source of $MnO_4{}^-$ ions usually is the soluble ionic compound $KMnO_4$. We often refer to "permanganate solutions." Such solutions also contain cations—in this case, K^+. Likewise, we often refer to "iron(II) solutions" without specifying what the anion is.

Because it has an intense purple color, $KMnO_4$ acts as its own indicator. One drop of 0.020 M $KMnO_4$ solution imparts a pink color to a liter of pure water. When $KMnO_4$ solution is added to a solution of a reducing agent, the end point in the titration is taken as the point at which a pale pink color appears in the solution being titrated and persists for at least 30 seconds.

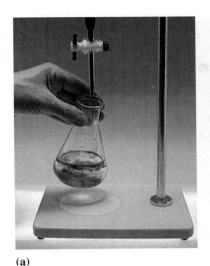

(a)   (b)

Charles D. Winters

Figure 11-2 (a) Nearly colorless FeSO₄ solution is titrated with deep-purple KMnO₄. (b) The end point is the point at which the solution becomes pink, owing to a *very small* excess of KMnO₄. Here a considerable excess of KMnO₄ was added so that the pink color could be reproduced photographically.

EXAMPLE 11-18 *Redox Titration*

What volume of 0.0200 M KMnO₄ solution is required to oxidize 40.0 mL of 0.100 M FeSO₄ in sulfuric acid solution (Figure 11-2)?

Plan

The balanced equation in Example 11-15 gives the reaction ratio, 1 mol MnO₄⁻/5 mol Fe²⁺. Then we calculate the number of moles of Fe²⁺ to be titrated, which lets us find the number of moles of MnO₄⁻ required *and* the volume in which this number of moles of KMnO₄ is contained.

One mole of KMnO₄ contains one mole of MnO₄⁻ ions. The number of moles of KMnO₄ is therefore *always* equal to the number of moles of MnO₄⁻ ions required in a reaction. Similarly, one mole of FeSO₄ contains 1 mole of Fe²⁺ ions.

Solution

The reaction ratio is

$$MnO_4^-(aq) + 8H^+(aq) + 5Fe^{2+}(aq) \longrightarrow 5Fe^{3+}(aq) + Mn^{2+}(aq) + 4H_2O(\ell)$$

rxn ratio: 1 mol   5 mol

The number of moles of Fe²⁺ to be titrated is

$$\underline{?}\ mol\ Fe^{2+} = 40.0\ mL \times \frac{0.100\ mol\ Fe^{2+}}{1000\ mL} = 4.00 \times 10^{-3}\ mol\ Fe^{2+}$$

We use the balanced equation to find the number of moles of MnO₄⁻ required.

$$\underline{?}\ mol\ MnO_4^- = 4.00 \times 10^{-3}\ mol\ Fe^{2+} \times \frac{1\ mol\ MnO_4^-}{5\ mol\ Fe^{2+}} = 8.00 \times 10^{-4}\ mol\ MnO_4^-$$

Each formula unit of KMnO₄ contains one MnO₄⁻ ion, and so

$$1\ mol\ KMnO_4 \cong 1\ mol\ MnO_4^-$$

The volume of 0.0200 M KMnO₄ solution that contains 8.00×10^{-4} mol of KMnO₄ is

$$\underline{?}\ mL\ KMnO_4\ soln = 8.00 \times 10^{-4}\ mol\ KMnO_4 \times \frac{1000\ mL\ KMnO_4\ soln}{0.0200\ mol\ KMnO_4}$$

$$= 40.0\ mL\ KMnO_4\ soln$$

You should now work Exercises 64 and 66.

Potassium dichromate, $K_2Cr_2O_7$, is another frequently used oxidizing agent. However, an indicator must be used when reducing agents are titrated with dichromate solutions. $K_2Cr_2O_7$ is orange, and its reduction product, Cr^{3+}, is green.

Consider the oxidation of sulfite ions, SO_3^{2-}, to sulfate ions, SO_4^{2-}, by $Cr_2O_7^{2-}$ ions in the presence of a strong acid such as sulfuric acid. We shall balance the equation by the half-reaction method.

$$Cr_2O_7^{2-} \longrightarrow Cr^{3+} \qquad \text{(red. half-rxn)}$$

$$Cr_2O_7^{2-} \longrightarrow 2Cr^{3+}$$

$$14H^+ + Cr_2O_7^{2-} \longrightarrow 2Cr^{3+} + 7H_2O$$

$$6e^- + 14H^+ + Cr_2O_7^{2-} \longrightarrow 2Cr^{3+} + 7H_2O \qquad \text{(balanced red. half-rxn)}$$

- -

$$SO_3^{2-} \longrightarrow SO_4^{2-} \qquad \text{(ox. half-rxn)}$$

$$SO_3^{2-} + H_2O \longrightarrow SO_4^{2-} + 2H^+$$

$$SO_3^{2-} + H_2O \longrightarrow SO_4^{2-} + 2H^+ + 2e^- \qquad \text{(balanced ox. half-rxn)}$$

We now equalize the electron transfer, add the balanced half-reactions, and eliminate common terms.

$$(6e^- + 14H^+ + Cr_2O_7^{2-} \longrightarrow 2Cr^{3+} + 7H_2O) \qquad \text{(reduction)}$$

$$\underline{3(SO_3^{2-} + H_2O \longrightarrow SO_4^{2-} + 2H^+ + 2e^-)} \qquad \text{(oxidation)}$$

$$8H^+(aq) + Cr_2O_7^{2-}(aq) + 3SO_3^{2-}(aq) \longrightarrow 2Cr^{3+}(aq) + 3SO_4^{2-}(aq) + 4H_2O(\ell)$$

The balanced equation tells us that the reaction ratio is 3 mol SO_3^{2-}/mol $Cr_2O_7^{2-}$ or 1 mol $Cr_2O_7^{2-}$/3 mol SO_3^{2-}. Potassium dichromate is the usual source of $Cr_2O_7^{2-}$ ions, and Na_2SO_3 is the usual source of SO_3^{2-} ions. Thus, the preceding reaction ratio could also be expressed as 1 mol $K_2Cr_2O_7$/3 mol Na_2SO_3.

EXAMPLE 11-19   *Redox Titration*

A 20.00-mL sample of Na_2SO_3 was titrated with 36.30 mL of 0.05130 M $K_2Cr_2O_7$ solution in the presence of H_2SO_4. Calculate the molarity of the Na_2SO_3 solution.

Plan

We can calculate the number of millimoles of $Cr_2O_7^{2-}$ in the standard solution. Then we refer to the balanced equation in the preceding discussion, which gives us the reaction ratio, 3 mmol SO_3^{2-}/1 mmol $Cr_2O_7^{2-}$. The reaction ratio lets us calculate the number of millimoles of SO_3^{2-} (Na_2SO_3) that reacted and the molarity of the solution.

$$\text{mL } Cr_2O_7^{2-} \text{ soln} \longrightarrow \text{mmol } Cr_2O_7^{2-} \longrightarrow \text{mmol } SO_3^{2-} \longrightarrow M \text{ } SO_3^{2-} \text{ soln}$$

Solution

From the preceding discussion we know the balanced equation and the reaction ratio.

$$3SO_3^{2-} + Cr_2O_7^{2-} + 8H^+ \longrightarrow 3SO_4^{2-} + 2Cr^{3+} + 4H_2O$$
$$\quad\; \text{3 mmol} \quad\;\; \text{1 mmol}$$

The number of millimoles of $Cr_2O_7^{2-}$ used is

$$\underline{?} \text{ mmol } Cr_2O_7^{2-} = 36.30 \text{ mL} \times \frac{0.05130 \text{ mmol } Cr_2O_7^{2-}}{\text{mL}} = 1.862 \text{ mmol } Cr_2O_7^{2-}$$

$Cr_2(SO_4)_3$ is green in acidic solution. $K_2Cr_2O_7$ is orange in acidic solution.

Charles Steele

The number of millimoles of SO_3^{2-} that reacted with 1.862 mmol of $Cr_2O_7^{2-}$ is

$$\underline{?}\ mmol\ SO_3^{2-} = 1.862\ mmol\ Cr_2O_7^{2-} \times \frac{3\ mmol\ SO_3^{2-}}{1\ mmol\ Cr_2O_7^{2-}} = 5.586\ mmol\ SO_3^{2-}$$

The Na_2SO_3 solution contained 5.586 mmol of SO_3^{2-} (or 5.586 mmol of Na_2SO_3). Its molarity is

$$\underline{?}\ \frac{mmol\ Na_2SO_3}{mL} = \frac{5.586\ mmol\ Na_2SO_3}{20.00\ mL} = 0.2793\ M\ Na_2SO_3$$

You should now work Exercise 68.

Key Terms

Buret A piece of volumetric glassware, usually graduated in 0.1-mL intervals, that is used in titrations to deliver solutions in a quantitative (dropwise) manner.

End point The point at which an indicator changes color and a titration is stopped.

Equivalence point The point at which chemically equivalent amounts of reactants have reacted.

Equivalent weight in acid–base reactions The mass of an acid or base that furnishes or reacts with 6.022×10^{23} H_3O^+ or OH^- ions.

Half-reaction Either the oxidation part or the reduction part of a redox reaction.

Indicator For acid–base titrations, an organic compound that exhibits its different colors in solutions of different acidities; used to determine the point at which the reaction between two solutes is complete.

Milliequivalent 1/1000 equivalent.

Millimole 1/1000 mole.

Molarity (M) The number of moles of solute per liter of solution or the number of millimoles of solute per milliliter of solution.

Normality (N) The number of equivalent weights (equivalents) of solute per liter of solution.

Oxidation An algebraic increase in oxidation number; may correspond to a loss of electrons.

Oxidation–reduction reaction A reaction in which oxidation and reduction occur; also called redox reaction.

Oxidizing agent The substance that oxidizes another substance and is reduced.

Primary standard A substance of a known high degree of purity that undergoes one invariable reaction with the other reactant of interest.

Redox reaction An oxidation-reduction reaction.

Redox titration The quantitative analysis of the amount or concentration of an oxidizing or reducing agent in a sample by observing its reaction with a known amount or concentration of a reducing or oxidizing agent.

Reducing agent The substance that reduces another substance and is oxidized.

Reduction An algebraic decrease in oxidation number; may correspond to a gain of electrons.

Secondary standard A solution that has been titrated against a primary standard. A standard solution in a secondary standard.

Standard solution A solution of accurately known concentration.

Standardization The process by which the concentration of a solution is accurately determined by titrating it against an accurately known amount of a primary standard.

Titration The process by which the volume of a standard solution required to react with a specific amount of a substance is determined.

Exercises

You may assume that all species shown in chemical equations are *in aqueous solutions* unless otherwise indicated.

Molarity

1. Why can we describe molarity as a "method of convenience" for expressing concentrations of solutions?
2. Why is the molarity of a solution the same number whether we describe it in mol/L or in mmol/mL?
3. Calculate the molarities of solutions that contain the following masses of solute in the indicated volumes: (a) 35.5 g of H_3AsO_4 in 500. mL of solution; (b) 8.33 g of $(COOH)_2$ in 600. mL of solution; (c) 8.25 g of $(COOH)_2 \cdot 2H_2O$ in 750. mL of solution.
4. What is the molarity of a solution made by dissolving 66.3 g of magnesium sulfate in sufficient water to produce a total of 3.50 L?

5. There are 85.0 g of iron(II) nitrate present in 850. mL of a solution. Calculate the molarity of that solution.

6. Calculate the molarity of a solution that is 39.77% H_2SO_4 by mass. The specific gravity of the solution is 1.305.

7. Calculate the molarity of a solution that is 19.0% HNO_3 by mass. The specific gravity of the solution is 1.11.

8. If 225. mL of 4.32 M HCl solution is added to 450. mL of 2.16 M NaOH solution, the resulting solution will be _____ molar in NaCl.

9. What is the molarity of the salt solution produced when 750. mL of 3.00 M HCl and 750. mL of 3.00 M LiOH are mixed? (Assume that the volumes are additive.) Give the name and formula of the salt formed.

10. Potassium iodide is sometimes used as a sodium chloride replacement for those people who cannot tolerate table salt. Calculate the molarity of potassium iodide solution produced when 35.5 mL of 9.00 M HI and 35.5 mL of 9.00 M KOH are mixed.

11. What is the salt concentration produced if we mix 8.00 mL of 4.50 M HCl with 4.50 mL of 4.00 M $Ba(OH)_2$? Give the name and formula of the salt formed.

12. What is the concentration of barium iodide produced by mixing 7.50 mL of 0.125 M $Ba(OH)_2$ with 18.0 mL of 0.0650 M HI?

13. What is the concentration of the ammonium chloride produced when 22.0 mL of 12.0 M HCl and 18.5 mL of 8.00 M NH_3 are mixed?

14. If 225 mL of 5.52 M H_3PO_4 solution is added to 775 mL of 5.52 M NaOH solution, the resulting solution will be _____ molar in Na_3PO_4 and _____ molar in _____.

15. If 100. mL of 0.200 M HCl solution is added to 200. mL of 0.0400 M $Ba(OH)_2$ solution, the resulting solution will be _____ molar in $BaCl_2$ and _____ molar in _____.

16. What volume of 0.0125 M acetic acid solution would completely neutralize 15.58 mL of 0.0105 M $Ba(OH)_2$ solution?

17. What volume of 0.150 M potassium hydroxide solution would completely neutralize 17.5 mL of 0.100 M H_2SO_4 solution?

18. A vinegar solution is 5.11% acetic acid. Its density is 1.007 g/mL. What is its molarity?

19. A household ammonia solution is 5.03% ammonia. Its density is 0.979 g/mL. What is its molarity?

20. (a) What volumes of 2.25 M NaOH and 4.50 M H_3PO_4 solutions would be required to form 1.00 mol of Na_3PO_4? (b) What volumes of the solutions would be required to form 1.00 mol of Na_2HPO_4?

Standardization and Acid–Base Titrations: Mole Method

21. Define and illustrate the following terms clearly and concisely: (a) standard solution; (b) titration; (c) primary standard; (d) secondary standard.

22. Describe the preparation of a standard solution of NaOH, a compound that absorbs both CO_2 and H_2O from the air.

23. Distinguish between the *net ionic equation* and the *formula unit equation*.

24. (a) What is potassium hydrogen phthalate, KHP? (b) For what is it used?

25. Why can sodium carbonate be used as a primary standard for solutions of acids?

26. Calculate the molarity of a solution of HNO_3 if 35.72 mL of the solution neutralizes 0.4040 g of Na_2CO_3.

27. If 35.38 mL of a sulfuric acid solution reacts completely with 0.3545 g of Na_2CO_3, what is the molarity of the sulfuric acid solution?

28. A solution of sodium hydroxide is standardized against potassium hydrogen phthalate. From the following data, calculate the molarity of the NaOH solution.

mass of KHP used	0.4536 g
buret reading before titration	0.23 mL
buret reading after titration	31.26 mL

29. Calculate the molarity of a KOH solution if 30.68 mL of the KOH solution reacted with 0.4084 g of potassium hydrogen phthalate, KHP.

30. Calcium carbonate tablets can be used as an antacid and a source of dietary calcium. A bottle of generic antacid tablets states that each tablet contains 600. mg calcium carbonate. What volume of 6.0 M HNO_3 could be neutralized by the calcium carbonate in one tablet?

31. What volume of 18.0 M H_2SO_4 is required to react with 65.5 mL of 6.00 M NaOH to produce a Na_2SO_4 solution? What volume of water must be added to the resulting solution to obtain a 1.25 M Na_2SO_4 solution?

32. (a) What are the properties of an ideal primary standard? (b) What is the importance of each property?

33. The secondary standard solution of NaOH of Exercise 28 was used to titrate a solution of unknown concentration of HCl. A 30.00-mL sample of the HCl solution required 34.21 mL of the NaOH solution for complete neutralization. What is the molarity of the HCl solution?

*34. An impure sample of $(COOH)_2 \cdot 2H_2O$ that had a mass of 1.00 g was dissolved in water and titrated with standard NaOH solution. The titration required 19.16 mL of 0.198 M NaOH solution. Calculate the percent $(COOH)_2 \cdot 2H_2O$ in the sample. Assume that the sample contains no acidic impurities.

*35. A 25.0-mL sample of 0.0500 M $Ca(OH)_2$ is added to 10.0 mL of 0.100 M HNO_3. (a) Is the resulting solution acidic or basic? (b) How many moles of excess acid or base are present? (c) How many additional mL of 0.0500 M $Ca(OH)_2$ or 0.100 M HNO_3 would be required to completely neutralize the solution?

*36. An antacid tablet containing calcium carbonate as an active ingredient required 22.3 mL of 0.0932 M HCl for complete neutralization. What mass of $CaCO_3$ did the tablet contain?

*37. Butyric acid, whose empirical formula is C_2H_4O, is the acid responsible for the odor of rancid butter. The acid has one ionizable hydrogen per molecule. A 1.000-g sample of butyric acid is neutralized by 36.28 mL of 0.3132 M NaOH solution. What are (a) the molecular weight and (b) the molecular formula of butyric acid?

38. What is the molarity of a solution of sodium hydroxide, NaOH, if 36.2 mL of this solution is required to react with 37.5 mL of 0.0342 M nitric acid solution according to the following reaction?

$$HNO_3 + NaOH \longrightarrow NaNO_3 + H_2O(\ell)$$

39. What is the molarity of a solution of sodium hydroxide, NaOH, if 18.45 mL of this solution is required to react with 17.60 mL of 0.101 M hydrochloric acid solution according to the following reaction?

$$HCl + NaOH \longrightarrow NaCl + H_2O(\ell)$$

Standardization and Acid–Base Titrations: Equivalent Weight Method

In answering Exercises 40–49, assume that the acids and bases will be completely neutralized.

40. What is the normality of each of the following acid or base solutions? (a) 0.55 M HCl; (b) 0.55 M H_2SO_4; (c) 0.55 M H_3PO_4; (d) 0.55 M NaOH.

41. What is the normality of each of the following acid or base solutions? (a) 0.215 M $Ca(OH)_2$; (b) 0.215 M $Al(OH)_3$; (c) 0.215 M HNO_3; (d) 0.105 M H_2Se.

42. What is the normality of a solution that contains 9.78 g of H_3PO_4 in 185 mL of solution?

43. What are the molarity and normality of a sulfuric acid solution that is 19.6% H_2SO_4 by mass? The density of the solution is 1.14 g/mL.

44. Calculate the molarity and the normality of a solution that contains 16.6 g of arsenic acid, H_3AsO_4, in enough water to make 470. mL of solution.

45. Calculate the normality and molarity of an H_2SO_4 solution if 44.3 mL of the solution reacts with 0.484 g of Na_2CO_3.

$$H_2SO_4 + Na_2CO_3 \longrightarrow Na_2SO_4 + CO_2(g) + H_2O(\ell)$$

46. Calculate the normality and molarity of an HCl solution if 38.1 mL of the solution reacts with 0.438 g of Na_2CO_3.

$$2HCl + Na_2CO_3 \longrightarrow 2NaCl + CO_2(g) + H_2O(\ell)$$

47. To minimize the effect of buret reading errors, titrations performed using a 50-mL buret are most accurate when titrant volumes are in the range of 35–45 mL. Suggest a range of sample weights that would yield a 35–45 mL titration range for the standardization of solutions of the following approximate concentrations. (a) 0.0325 M NaOH using potassium hydrogen phthalate ($C_8H_5KO_4$), a monoprotic acid. (b) 0.060 M KOH using primary standard benzoic acid (C_6H_5COOH) a monoprotic acid.

48. Magnesium hydroxide, $Mg(OH)_2$, is commonly used as the active ingredient in antacid tablets. A student analyzed an antacid tablet for mass percent $Mg(OH)_2$ by dissolving a tablet weighing 1.462 g in 25.00 mL of 0.953 M HCl, and neutralizing the unreacted HCl. That neutral-

ization required 12.29 mL of 0.602 M NaOH. Calculate the mass percent of $Mg(OH)_2$ in the antacid tablet.

49. Vinegar is an aqueous solution of acetic acid, CH_3COOH. Suppose you titrate a 25.00-mL sample of vinegar with 17.62 mL of a standardized 0.1045 N solution of NaOH. (a) What is the normality of acetic acid in this vinegar? (b) What is the mass of acetic acid contained in 1.000 L of vinegar?

Balancing Redox Equations

In Exercises 50 and 51, write balanced formula unit equations for the reactions described by words.

50. (a) Iron reacts with hydrochloric acid to form aqueous iron(II) chloride and gaseous hydrogen. (b) Chromium reacts with sulfuric acid to form aqueous chromium(III) sulfate and gaseous hydrogen. (c) Tin reacts with concentrated nitric acid to form tin(IV) oxide, nitrogen dioxide, and water.

51. (a) Carbon reacts with hot concentrated nitric acid to form carbon dioxide, nitrogen dioxide, and water. (b) Sodium reacts with water to form aqueous sodium hydroxide and gaseous hydrogen. (c) Zinc reacts with sodium hydroxide solution to form aqueous sodium tetrahydroxozincate and gaseous hydrogen. (The tetrahydroxozincate ion is $[Zn(OH)_4]^{2-}$.)

52. Copper is a widely used metal. Before it is welded (brazed), copper is cleaned by dipping it into nitric acid. HNO_3 oxidizes Cu to Cu^{2+} ions and is reduced to NO. The other product is H_2O. Write the balanced net ionic and formula unit equations for the reaction. Excess HNO_3 is present.

Copper is cleaned by dipping it into nitric acid.

53. Balance the following equations. For each equation tell what is oxidized, what is reduced, what is the oxidizing agent, and what is the reducing agent.
    (a) $Cu(NO_3)_2(s) \xrightarrow{\text{heat}} CuO(s) + NO_2(g) + O_2(g)$
    (b) $Hg_2Cl_2(s) + NH_3(aq) \longrightarrow$
        $Hg(\ell) + HgNH_2Cl(s) + NH_4^+(aq) + Cl^-(aq)$
    (c) $Ba(s) + H_2O(\ell) \longrightarrow Ba(OH)_2(aq) + H_2(g)$

54. Balance the following equations. For each equation tell what is oxidized, what is reduced, what is the oxidizing agent, and what is the reducing agent.
    (a) $MnO_4^-(aq) + H^+(aq) + Br^-(aq) \longrightarrow$
    $$Mn^{2+}(aq) + Br_2(\ell) + H_2O(\ell)$$
    (b) $Cr_2O_7^{2-}(aq) + H^+(aq) + I^-(aq) \longrightarrow$
    $$Cr^{3+}(aq) + I_2(s) + H_2O(\ell)$$
    (c) $MnO_4^-(aq) + SO_3^{2-}(aq) + H^+(aq) \longrightarrow$
    $$Mn^{2+}(aq) + SO_4^{2-}(aq) + H_2O(\ell)$$
    (d) $Cr_2O_7^{2-}(aq) + Fe^{2+}(aq) + H^+(aq) \longrightarrow$
    $$Cr^{3+}(aq) + Fe^{3+}(aq) + H_2O(\ell)$$

55. Balance the following ionic equations. For each equation tell what is oxidized, what is reduced, what is the oxidizing agent, and what is the reducing agent.
    (a) $C_2H_4(g) + MnO_4^-(aq) + H^+(aq) \longrightarrow$
    $$CO_2(g) + Mn^{2+}(aq) + H_2O(\ell)$$
    (b) $H_2S(aq) + H^+(aq) + Cr_2O_7^{2-}(aq) \longrightarrow$
    $$Cr^{3+}(aq) + S(s) + H_2O(\ell)$$
    (c) $ClO_3^-(aq) + H_2O(\ell) + I_2(s) \longrightarrow$
    $$IO_3^-(aq) + Cl^-(aq) + H^+(aq)$$
    (d) $Cu(s) + H^+(aq) + SO_4^{2-}(aq) \longrightarrow$
    $$Cu^{2+}(aq) + H_2O(\ell) + SO_2(g)$$

56. Drāno drain cleaner is solid sodium hydroxide that contains some aluminum chips. When Drāno is added to water, the NaOH dissolves rapidly with the evolution of a lot of heat. The Al reduces H_2O in the basic solution to produce $[Al(OH)_4]^-$ ions and H_2 gas, which gives the bubbling action. Write the balanced net ionic and formula unit equations for this reaction.

The Drāno reaction.

57. Balance the following ionic equations. For each equation tell what is oxidized, what is reduced, what is the oxidizing agent, and what is the reducing agent.
    (a) $Cr(OH)_4^-(aq) + OH^-(aq) + H_2O_2(aq) \longrightarrow$
    $$CrO_4^{2-}(aq) + H_2O(\ell)$$
    (b) $MnO_2(s) + H^+(aq) + NO_2^-(aq) \longrightarrow$
    $$NO_3^-(aq) + Mn^{2+}(aq) + H_2O(\ell)$$
    (c) $Sn(OH)_3^-(aq) + Bi(OH)_3(s) + OH^-(aq) \longrightarrow$
    $$Sn(OH)_6^{2-}(aq) + Bi(s)$$

(d) $CrO_4^{2-}(aq) + H_2O(\ell) + HSnO_2^-(aq) \longrightarrow$
$$CrO_2^-(aq) + OH^-(aq) + HSnO_3^-(aq)$$

58. Balance the following ionic equations for reactions in acidic solution. H^+ or H_2O (but not OH^-) may be added as necessary.
    (a) $Fe^{2+}(aq) + MnO_4^-(aq) \longrightarrow Fe^{3+}(aq) + Mn^{2+}(aq)$
    (b) $Br_2(\ell) + SO_2(g) \longrightarrow Br^-(aq) + SO_4^{2-}(aq)$
    (c) $Cu(s) + NO_3^-(aq) \longrightarrow Cu^{2+}(aq) + NO_2(g)$
    (d) $PbO_2(s) + Cl^-(aq) \longrightarrow PbCl_2(s) + Cl_2(g)$
    (e) $Zn(s) + NO_3^-(aq) \longrightarrow Zn^{2+}(aq) + N_2(g)$

59. Balance the following ionic equations for reactions in acidic solution. H^+ or H_2O (but not OH^-) may be added as necessary.
    (a) $P_4(s) + NO_3^-(aq) \longrightarrow H_3PO_4(aq) + NO(g)$
    (b) $H_2O_2(aq) + MnO_4^-(aq) \longrightarrow Mn^{2+}(aq) + O_2(g)$
    (c) $HgS(s) + Cl^-(aq) + NO_3^-(aq) \longrightarrow$
    $$HgCl_4^{2-}(aq) + NO_2(g) + S(s)$$
    (d) $HBrO(aq) \longrightarrow Br^-(aq) + O_2(g)$

60. Write the balanced net ionic equations for the reactions given. Then, using the reactants shown in parentheses convert each balanced net ionic equation to a balanced formula unit equation.
    (a) $MnO_4^- + C_2O_4^{2-} + H^+ \longrightarrow$
    $$Mn^{2+} + CO_2(g) + H_2O(\ell)$$
    $$(KMnO_4, HCl, \text{ and } K_2C_2O_4)$$
    (b) $Zn + NO_3^- + H^+ \longrightarrow Zn^{2+} + NH_4^+ + H_2O(\ell)$
    $$(Zn(s) \text{ and } HNO_3)$$

61. Write the balanced net ionic equations for the reactions given. Then, using the reactants shown in parentheses convert each balanced net ionic equation to a balanced formula unit equation.
    (a) $I_2 + S_2O_3^{2-} \longrightarrow I^- + S_4O_6^{2-}$ (I_2 and $Na_2S_2O_3$)
    (b) $IO_3^- + N_2H_4 + Cl^- + H^+ \longrightarrow$
    $$N_2(g) + ICl_2^- + H_2O(\ell)$$
    $$(NaIO_3 + N_2H_4, \text{ and } HCl)$$

62. Write the balanced net ionic equations for the reactions given. Then, using the reactants shown in parentheses convert each balanced net ionic equation to a balanced formula unit equation.
    (a) $Zn(s) + Cu^{2+} \longrightarrow Cu(s) + Zn^{2+}$ (Zn and $CuSO_4$)
    (b) $Cr(s) + H^+ \longrightarrow Cr^{3+} + H_2(g)$ (Cr and H_2SO_4)

63. Write the balanced net ionic equations for the reactions given. Then, using the reactants shown in parentheses convert each balanced net ionic equation to a balanced formula unit equation.
    (a) $Cl_2 + OH^- \longrightarrow ClO_3^- + Cl^- + H_2O(\ell)$
    $$(Cl_2 \text{ and hot NaOH})$$
    (b) $Pb(s) + H^+ + Br^- \longrightarrow PbBr_2(s) + H_2(g)$
    $$(Pb(s) \text{ and HBr})$$

Redox Titrations: Mole Method and Molarity

64. What volume of $0.142\ M$ $KMnO_4$ would be required to oxidize 25.0 mL of $0.100\ M$ $FeSO_4$ in acidic solution? Refer to Example 11-18.

65. What volume of $0.142\ M$ $K_2Cr_2O_7$ would be required

to oxidize 70.0 mL of 0.100 M Na_2SO_3 in acidic solution? The products include Cr^{3+} and SO_4^{2-} ions. Refer to Example 11-19.

66. What volume of 0.190 M $KMnO_4$ would be required to oxidize 40.0 mL of 0.100 M KI in acidic solution? Products include Mn^{2+} and I_2.

67. What volume of 0.190 M $K_2Cr_2O_7$ would be required to oxidize 50.0 mL of 0.150 M KI in acidic solution? Products include Cr^{3+} and I_2.

68. (a) A solution of sodium thiosulfate, $Na_2S_2O_3$, is 0.1442 M. 44.00 mL of this solution reacts with 26.85 mL of I_2 solution. Calculate the molarity of the I_2 solution.

$$2Na_2S_2O_3 + I_2 \longrightarrow Na_2S_4O_6 + 2NaI$$

(b) 25.32 mL of the I_2 solution is required to titrate a sample containing As_2O_3. Calculate the mass of As_2O_3 (197.8 g/mol) in the sample.

$$As_2O_3 + 5H_2O(\ell) + 2I_2 \longrightarrow 2H_3AsO_4 + 4HI$$

69. Copper(II) ions, Cu^{2+}, can be determined by the net reaction

$$2Cu^{2+} + 2I^- + 2S_2O_3^{2-} \longrightarrow 2CuI(s) + S_4O_6^{2-}$$

A 2.115-g sample containing $CuSO_4$ and excess KI is titrated with 32.55 mL of 0.1214 M solution of $Na_2S_2O_3$. What is the percent $CuSO_4$ (159.6 g/mol) in the sample?

70. What volume of 3.0 M nitrate ion solution would be required to react with 25. mL of 0.75 M sulfide ion solution? (*Hint:* The equation is not balanced.)

$$NO_3^- + S^{2-} \longrightarrow NO + S(s) \quad \text{(acidic solution)}$$

*71. The iron in a 5.675-g sample containing some Fe_2O_3 is reduced to Fe^{2+}. The Fe^{2+} is titrated with 12.42 mL of 0.1467 M $K_2Cr_2O_7$ in an acid solution.

$$6Fe^{2+} + Cr_2O_7^{2-} + 14H^+ \longrightarrow$$
$$6Fe^{3+} + 2Cr^{3+} + 7H_2O(\ell)$$

Find (a) the mass of Fe and (b) the percentage of Fe in the sample.

72. Calculate the molarity of a solution that contains 12.6 g of $KMnO_4$ in 500. mL of solution to be used in the reaction that produces MnO_4^{2-} ions as the reduction product.

*73. A 0.823-g sample of an ore of iron is dissolved in acid and converted to Fe(II). The sample is oxidized by 38.50 mL of 0.161 M ceric sulfate, $Ce(SO_4)_2$, solution; the cerium(IV) ion, Ce^{4+}, is reduced to Ce^{3+} ion. (a) Write a balanced equation for the reaction. (b) What is the percent iron in the ore?

Mixed Exercises

74. Calculate the molarity of a hydrochloric acid solution if 32.75 mL of it reacts with 0.3811 g of sodium carbonate.

75. Calculate the molarity and the normality of a sulfuric acid solution if 32.75 mL of it reacts with 0.3811 g of sodium carbonate.

76. Find the number of mmol of HCl that reacts with 25.5 mL of 0.220 M NaOH. What volume of 0.606 M HCl is needed to furnish this amount of HCl?

77. What is the composition of the final solution when 25.5 mL of 0.220 M NaOH and 25.5 mL of 0.410 M HCl solutions are mixed?

78. What volume of 0.1123 M HCl is needed to completely neutralize 1.79 g of $Ca(OH)_2$?

79. What mass of NaOH is needed to neutralize 33.50 mL of 0.1036 M HCl? If the NaOH is available as a 0.1533 M aqueous solution, what volume will be required?

80. What volume of 0.246 M H_2SO_4 solution would be required to completely neutralize 34.4 mL of 0.302 M KOH solution?

81. What volume of 0.388 N H_2SO_4 solution would be required to completely neutralize 34.4 mL of 0.302 N KOH solution?

82. What volume of 0.1945 normal sodium hydroxide would be required to neutralize completely 34.38 mL of 0.1023 normal H_2SO_4 solution?

83. Benzoic acid, C_6H_5COOH, is sometimes used as a primary standard for the standardization of solutions of bases. A 1.862-g sample of this acid is neutralized by 29.00 mL of NaOH solution. What is the molarity of the base solution?

$$C_6H_5COOH(s) + NaOH(aq) \longrightarrow$$
$$C_6H_5COONa(aq) + H_2O(\ell)$$

84. Find the volume of 0.225 M HI solution required to titrate
(a) 25.0 mL of 0.100 M NaOH
(b) 5.03 g of $AgNO_3$ ($Ag^+ + I^- \longrightarrow AgI(s)$)
(c) 0.621 g $CuSO_4$ ($2Cu^{2+} + 4I^- \longrightarrow 2CuI(s) + I_2(s)$)

CONCEPTUAL EXERCISES

85. Describe how you could prepare 1.00 L of 1.00×10^{-6} M NaCl solution by using a balance that can measure masses to only 0.01 g.

86. Ascorbic acid (vitamin C), along with many other reputed properties, acts as an antioxidant. The following equation illustrates its antioxidant properties.

$$H_2C_6H_6O_6 \longrightarrow C_6H_6O_6 + H_2(g)$$

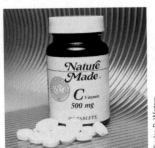

What is an antioxidant? Assign oxidation numbers. Is vitamin C oxidized or reduced in this reaction?

BUILDING YOUR KNOWLEDGE

87. For the formation of 1.00 mol of water, which reaction uses the most nitric acid?

    (a) $3Cu(s) + 8HNO_3(aq) \longrightarrow$
$$3Cu(NO_3)_2(aq) + 2NO(g) + 4H_2O(\ell)$$

    (b) $Al_2O_3(s) + 6HNO_3(aq) \longrightarrow$
$$2Al(NO_3)_3(aq) + 3H_2O(\ell)$$

    (c) $4Zn(s) + 10HNO_3(aq) \longrightarrow$
$$4Zn(NO_3)_2(aq) + NH_4NO_3(aq) + 3H_2O(\ell)$$

88. Limonite is an ore of iron that contains $2Fe_2O_3 \cdot 3H_2O$. A 0.5166-g sample of limonite is dissolved in acid and treated so that all the iron is converted to ferrous ions, Fe^{2+}. This sample requires 42.96 mL of 0.02130 M sodium dichromate solution, $Na_2Cr_2O_7$, for titration. Fe^{2+} is oxidized to Fe^{3+}, and $Cr_2O_7{}^{2-}$ is reduced to Cr^{3+}. What is the percent iron in the limonite? If your answer had been over 100% limonite, what conclusion could you make, presuming that the analytical data are correct?

89. One of the troublesome products of a water treatment plant in some areas of the country is $Mg(OH)_2$, a gelatinous precipitate formed during water softening. A suggestion was made that instead of shoveling the precipitate out of the pool during cleaning, the $Mg(OH)_2$ could be neutralized with hydrochloric acid to produce a soluble compound, $MgCl_2$. Then the pool could be flushed out with fresh water. Calculate the volume of 12.0 M HCl necessary to neutralize 4750 L of solution containing 1.50 g of $Mg(OH)_2$ per liter.

90. Silver nitrate and calcium chloride solutions produce a heavy, white precipitate when mixed. Chemical analysis indicates that the precipitate is silver chloride. What mass of silver chloride would be produced if 45 mL of 6.0 M silver nitrate is mixed with 45 mL of 6.0 M calcium chloride?

91. A 0.500-g sample of a crystalline monoprotic acid was dissolved in sufficient water to produce 100. mL of solution. Neutralization of the resulting solution required 75.0 mL of 0.150 M NaOH. How many moles of the acid were present in the initial acid solution?

92. The typical concentration of HCl in stomach acid (digestive juice) is a concentration of about 8.0×10^{-2} M. One experiences "acid stomach" when the stomach contents reach about 1.0×10^{-1} M HCl. One antacid tablet contains 334 mg of active ingredient, $NaAl(OH)_2CO_3$. Assume that you have acid stomach and that your stomach contains 800. mL of 1.0×10^{-1} M HCl. Calculate the number of mmol of HCl in the stomach and the number of mmol of HCl that the tablet *can* neutralize. Which is

greater? (The neutralization reaction produces NaCl, $AlCl_3$, CO_2, and H_2O.)

93. Refer to Exercises 18 and 19. Notice that the percent by mass of solute is nearly the same for both solutions. How many moles of solute are present per liter of each solution? Are the moles of solute per liter also nearly equal? Why or why not?

94. The etching of glass by hydrofluoric acid may be represented by the simplified reaction of silica with HF.

$$SiO_2(s) + HF(aq) \longrightarrow H_2SiF_6(aq) + H_2O(\ell)$$

This is an acid–base reaction in which a weak acid is used to produce an even weaker acid. Is it also an oxidation–reduction reaction? Balance the equation.

95. Write a Lewis formula for the anion $SiF_6{}^{2-}$ that would be produced from the weak acid H_2SiF_6. Use the VSEPR theory to predict the shape of $SiF_6{}^{2-}$.

96. Baking soda, $NaHCO_3$, used to be a common remedy for "acid stomach." What weight of baking soda would be required to neutralize 85 mL of digestive juice, corresponding in acidity to 0.17 M HCl?

97. Oxalic acid, a poisonous compound, is found in certain vegetables such as spinach and rhubarb, but in concentrations well below toxic limits. The manufacturers of a spinach juice concentrate routinely test their product using an oxalic acid analysis to avoid any problems from an unexpectedly high concentration of this chemical. A titration with potassium permanganate is used for the oxalic acid assay, according to the following net equation.

$$5H_2C_2O_4 + 2MnO_4{}^- + 6H^+ \longrightarrow$$
$$10CO_2 + 2Mn^{2+} + 8H_2O(\ell)$$

Calculate the molarity of an oxalic acid solution requiring 23.2 mL of 0.127 M permanganate for a 25.0 mL portion of the solution.

BEYOND THE TEXTBOOK

Go to the textbook website at

http://www.brookscole.com/chemistry/whitten

for additional activities and exercises based on the General Chemistry Interactive CD-ROM, the World Wide Web, and library resources.

InfoTrac College Edition

For additional readings, go to InfoTrac College Edition, your online research library at:

http://infotrac.thomsonlearning.com

Gases and the Kinetic–Molecular Theory

12

The lower density of the hot air trapped in these balloons causes them to rise into the more dense atmosphere.

Digital Vision /Getty

OUTLINE

OBJECTIVES

After you have studied this chapter, you should be able to

- List the properties of gases and compare gases, liquids, and solids

- Describe how pressure is measured

- Use and understand the absolute (Kelvin) temperature scale

- Describe the relationships among pressure, volume, temperature, and amount of gas (Boyle's Law, Charles's Law, Avogadro's Law, and the Combined Gas Law), and the limitations of each

- Use Boyle's Law, Charles's Law, Avogadro's Law, and the Combined Gas Law, as appropriate, to calculate changes in pressure, volume, temperature, and amount of gas

- Calculate gas densities and the standard molar volume

- Use the ideal gas equation to do pressure, volume, temperature, and mole calculations as related to gas samples

- Determine molecular weights and formulas of gaseous substances from measured properties of gases

- Describe how mixtures of gases behave and predict their properties (Dalton's Law of Partial Pressures)

- Carry out calculations about the gases involved in chemical reactions

- Apply the kinetic–molecular theory of gases and describe how this theory is consistent with the observed gas laws

- Describe molecular motion, diffusion, and effusion of gases

- Describe the molecular features that are responsible for nonideal behavior of real gases and explain when this nonideal behavior is important

TABLE 12-1	*Densities and Molar Volumes of Three Substances at Atmospheric Pressure**					
Substance	Solid		Liquid (20°C)		Gas (100°C)	
	Density (g/mL)	Molar volume (mL/mol)	Density (g/mL)	Molar volume (mL/mol)	Density (g/mL)	Molar volume (mL/mol)
water (H_2O)	0.917 (0°C)	19.6	0.998	18.0	0.000588	30,600
benzene (C_6H_6)	0.899 (0°C)	86.9	0.876	89.2	0.00255	30,600
carbon tetrachloride (CCl_4)	1.70 (−25°C)	90.5	1.59	96.8	0.00503	30,600

*The molar volume of a substance is the volume occupied by one mole of that substance.

12-1 COMPARISON OF SOLIDS, LIQUIDS, AND GASES

Matter exists on earth in three physical states: solids, liquids, and gases. In the solid state H_2O is known as ice, in the liquid state it is called water, and in the gaseous state it is known as steam or water vapor. Most, but not all, substances can exist in all three states. Most solids change to liquids and most liquids turn into gases as they are heated. Liquids and gases are known as **fluids** because they flow freely. Solids and liquids are referred to as **condensed states** because they have much higher densities than gases. Table 12-1 displays the densities of a few common substances in different physical states.

As the data in Table 12-1 indicate, solids and liquids are many times denser than gases. The molecules must be very far apart in gases and much closer together in liquids and solids. For example, the volume of one mole of liquid water is about 18 milliliters, whereas one mole of steam occupies about 30,600 milliliters at 100°C and atmospheric pressure. Gases are easily compressed, and they completely fill any container in which they are present. This tells us that the molecules in a gas are far apart relative to their sizes and that interactions among them are weak. The possibilities for interaction among gaseous molecules would be minimal (because they are so far apart) were it not for their rapid motion.

All substances that are gases at room temperature may be liquefied by cooling and compressing them. Volatile liquids are easily converted to gases at room temperature or slightly above. The term **vapor** refers to a gas that is formed by evaporation of a liquid or sublimation of a solid. We often use this term when some of the liquid or solid remains in contact with the gas.

12-2 COMPOSITION OF THE ATMOSPHERE AND SOME COMMON PROPERTIES OF GASES

Many important chemical substances are gases at ambient conditions. The earth's atmosphere is a mixture of gases and particles of liquids and solids (Table 12-2). The major gaseous components are N_2 (bp −195.79°C) and O_2 (bp −182.98°C), with smaller concentrations of other gases. All gases are *miscible*; that is, they mix completely *unless* they react with one another.

Several scientists, notably Torricelli (1643), Boyle (1660), Charles (1787), and Graham (1831), laid an experimental foundation on which our present understanding of gases is based. For example, their investigations showed that

Some compounds decompose before melting or boiling.

Ice is less dense than liquid water. This behavior is quite unusual; most substances are denser in the solid state than in the liquid state.

 CD-ROM Screen 1.3, States of Matter.

Volatile liquids evaporate readily. They have low boiling points.

TABLE 12-2	*Composition of Dry Air*
Gas	% by Volume
N_2	78.09
O_2	20.94
Ar	0.93
CO_2	0.03*
He, Ne, Kr, Xe	0.002
CH_4	0.00015*
H_2	0.00005
All others combined[†]	< 0.00004

*Variable.

[†]Atmospheric moisture varies considerably.

Br₂ (gas)

Br₂ (liquid)

Diffusion of bromine vapor in air. Some liquid bromine (*dark reddish brown*) was placed in the small inner bottle. As the liquid evaporated, the resulting reddish brown gas diffused.

1. Gases can be compressed into smaller volumes; that is, their densities can be increased by applying increased pressure.

2. Gases exert pressure on their surroundings; in turn, pressure must be exerted to confine gases.

3. Gases expand without limits, and so gas samples completely and uniformly occupy the volume of any container.

4. Gases diffuse into one another, and so samples of gas placed in the same container mix completely. Conversely, different gases in a mixture do not separate on standing.

5. The amounts and properties of gases are described in terms of temperature, pressure, the volume occupied, and the number of molecules present. For example, a sample of gas occupies a greater volume when hot than it does when cold at the same pressure, but the number of molecules does not change.

Investigating four variables at once is difficult. In Sections 12-4 through 12-8 we shall see how to study these variables two at a time. Section 12-9 will consolidate these descriptions into a single relationship, the ideal gas equation.

12-3 PRESSURE

Pressure is defined as force per unit area—for example, pounds per square inch (lb/in.²), commonly known as *psi*. Pressure may be expressed in many different units, as we shall see. The mercury **barometer** is a simple device for measuring atmospheric pressures. Figure 12-1a illustrates the "heart" of the mercury barometer. A glass tube (about 800 mm long) is sealed at one end, filled with mercury, and then carefully inverted into a dish of mercury without air being allowed to enter. The mercury in the tube falls to the level at which the pressure of the air on the surface of the mercury in the dish equals the gravitational pull downward on the mercury in the tube. The air pressure is measured in terms of the height

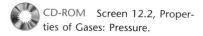

CD-ROM Screen 12.2, Properties of Gases: Pressure.

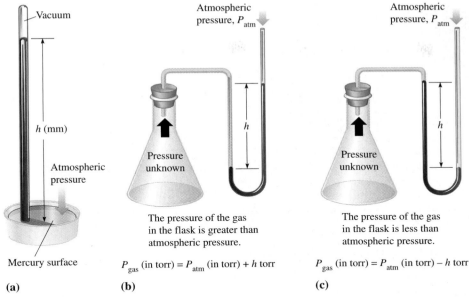

Vacuum

h (mm)

Atmospheric pressure

Mercury surface

(a)

Atmospheric pressure, P_{atm}

h

Pressure unknown

The pressure of the gas in the flask is greater than atmospheric pressure.

P_{gas} (in torr) = P_{atm} (in torr) + h torr

(b)

Atmospheric pressure, P_{atm}

h

Pressure unknown

The pressure of the gas in the flask is less than atmospheric pressure.

P_{gas} (in torr) = P_{atm} (in torr) − h torr

(c)

Figure 12-1 Some laboratory devices for measuring pressure. (a) Schematic diagram of a closed-end barometer. At the level of the lower mercury surface, the pressure both inside and outside the tube must be equal to that of the atmosphere. There is no air inside the tube, so the pressure is exerted only by the mercury column h mm high. Hence, the atmospheric pressure must equal the pressure exerted by h mm Hg, or h torr. (b) The two-arm mercury barometer is called a manometer. In this sample, the pressure of the gas inside the flask is *greater than* the external atmospheric pressure. At the level of the lower mercury surface, the total pressure on the mercury in the left arm must equal the total pressure on the mercury in the right arm. The pressure exerted by the gas is equal to the external pressure *plus* the pressure exerted by the mercury column of height h mm, or P_{gas} (in torr) = P_{atm} (in torr) + h torr. (c) When the gas pressure measured by the manometer is *less than* the external atmospheric pressure, the pressure exerted by the atmosphere is equal to the gas pressure *plus* the pressure exerted by the mercury column, or P_{atm} = P_{gas} + h. We can rearrange this to write P_{gas} (in torr) = P_{atm} (in torr) − h torr.

of the mercury column, that is, the vertical distance between the surface of the mercury in the open dish and that inside the closed tube. The pressure exerted by the atmosphere is equal to the pressure exerted by the column of mercury.

Mercury barometers are simple and well known, so gas pressures are frequently expressed in terms of millimeters of mercury (mm Hg, or just mm). In recent years the unit **torr** has been used to indicate pressure; it is defined as 1 torr = 1 mm Hg.

A mercury **manometer** consists of a glass U-tube partially filled with mercury. One arm is open to the atmosphere, and the other is connected to a container of gas (see Figure 12-1b,c).

Atmospheric pressure varies with atmospheric conditions and distance above sea level. The atmospheric pressure decreases with increasing elevation because there is a decreasing mass of air above it. Approximately one half of the matter in the atmosphere is less than 20,000 feet above sea level. Thus, atmospheric pressure is only about one half as great at

The unit *torr* was named for Evangelista Torricelli (1608–1647), who invented the mercury barometer.

(text continues on page 432)

CHEMISTRY IN USE

The Environment

The Greenhouse Effect

During the 20th century, the great increase in our use of fossil fuels caused a significant rise in the concentration of carbon dioxide, CO_2, in the atmosphere. Scientists believe that the concentration of atmospheric CO_2 could double by early in the 21st century, compared with its level just before the Industrial Revolution. During the last 100 to 200 years, the CO_2 concentration has increased by 25%. The curve in Figure (a) shows the recent steady rise in atmospheric CO_2 concentration.

Energy from the sun reaches the earth in the form of light. Neither CO_2 nor H_2O vapor absorbs the visible light in sunlight, so they do not prevent it from reaching the surface of the earth. The energy given off by the earth in the form of lower-energy infrared (heat) radiation, however, is readily absorbed by both CO_2 and H_2O (as it is by the glass or plastic of greenhouses). Thus, some of the heat the earth must lose to stay in thermal equilibrium can become trapped in the atmosphere, causing the temperature to rise (Figure b). This phenomenon, called the **greenhouse effect,** has been the subject of much discussion among scientists and the topic of many articles in the popular press. The anticipated rise in average global temperature by the year 2050 due to increased CO_2 concentration is predicted to be 2 to 5°C.

An increase of 2 to 5°C may not seem like much. However, this is thought to be enough to cause a dramatic change in climate, transforming now productive land into desert and altering the habitats of many animals and plants beyond their ability to adapt. Another drastic consequence of even this small temperature rise would be the partial melting of the polar ice caps. The resulting rise in sea level, though only a few feet, would mean that water would inundate coastal cities such as Los Angeles, New York, and Houston, and low-lying coastal areas such as southern Florida and Louisiana. On a global scale, the effects would be devastating.

The earth's forests and jungles play a crucial role in maintaining the balance of gases in the atmosphere, removing CO_2 and supplying O_2. The massive destruction, for economic reasons, of heavily forested areas such as the Amazon rain forest in South America is cited as another long-term contributor to global environmental problems. Worldwide, more than 3 million square miles of once-forested land is now barren for some reason; at least 60% of this land is now unused. Envi-

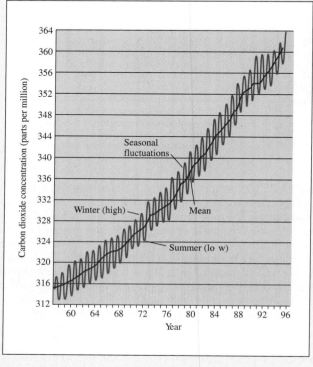

(a) A plot of the monthly average CO_2 concentration in parts per million (ppm), measured at Mauna Loa Observatory, Hawaii, far from significant sources of CO_2 from human activities. Annual fluctuations occur because plants in the Northern Hemisphere absorb CO_2 in the spring and release it as they decay in the fall.

ronmental scientists estimate that if even one quarter of this land could be reforested, the vegetation would absorb 1.1 billion tons of CO_2 annually.

Some scientists are more skeptical than others about the role of human-produced CO_2 in climate changes and, indeed,

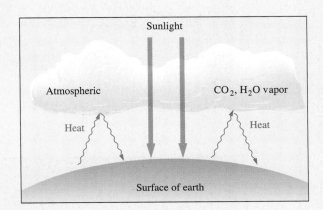

(b) The greenhouse effect. Visible light passes through atmospheric H_2O and CO_2, but heat radiated from the surface of the earth is absorbed by these gases.

about whether global warming is a significant phenomenon or simply another of the recognized warm–cold cycles that have occurred throughout the earth's history. Such skeptics point out an unexplained increase in atmospheric CO_2 during

an extended period in the 17th century, and an even higher and more prolonged peak about 130,000 years ago. Even the most skeptical observers, however, seem to agree that responsible stewardship of the planet requires that we do something in a reasoned fashion to reduce production of greenhouse gases, primarily CO_2, and that this will involve decreasing our dependence on energy from fossil fuels. Despite the technical and political problems of waste disposal, an essentially all-electric economy based on nuclear power may someday be a usable solution.

Much CO_2 is eventually absorbed by the vast amount of water in the oceans, where the carbonate–bicarbonate buffer system almost entirely counteracts any adverse effects of ocean water acidity. Ironically, there is also evidence to suggest that other types of air pollution in the form of particulate matter may partially counteract the greenhouse effect. The particles reflect visible (sun) radiation rather than absorbing it, blocking some light from entering the atmosphere. It seems foolish, however, to depend on one form of pollution to help rescue us from the effects of another! Real solutions to current environmental problems such as the greenhouse effect are not subject to quick fixes, but depend on long-term cooperative international efforts that are based on the firm knowledge resulting from scientific research.

Jacques Jangoux/Peter Arnold, Inc.

Tropical rain forests are important in maintaining the balance of CO_2 and O_2 in the earth's atmosphere. In recent years a portion of the South American forests (by far the world's largest) larger than France has been destroyed, either by flooding caused by hydroelectric dams or by clearing of forest land for agricultural or ranching use. Such destruction continues at a rate of more than 20,000 square kilometers per year. If current trends continue, many of the world's rain forests will be severely reduced or even obliterated in the next few years. The fundamental question— "What are the long-term consequences of the destruction of tropical rain forests?"—remains unanswered.

Figure 12-2 Some commercial pressure-measuring devices. (a) A commercial mercury barometer. (b) Portable barometers. This type is called an *aneroid* ("not wet") barometer. Some of the air has been removed from the airtight box, which is made of thin, flexible metal. When the pressure of the atmosphere changes, the remaining air in the box expands or contracts (Boyle's Law), moving the flexible box surface and an attached pointer along a scale. (c) A tire gauge. This kind of gauge registers "relative" pressure, that is, the *difference* between internal pressure and the external atmospheric pressure. For instance, when the gauge reads 30 psi (pounds per square inch), the total gas pressure in the tire is 30 psi + 1 atm, or about 45 psi. In engineering terminology, this is termed "psig" (g = gauge).

(a) **(b)** **(c)**

A common pressure unit used in the rest of the world is the bar, which is nearly equal to an atmosphere of pressure:
1.00 bar = 100. kPa
1.00 atm = 1.01 bar

Acceleration is the change in velocity (m/s) per unit time (s), m/s².

20,000 feet as it is at sea level. Mountain climbers and pilots use portable barometers to determine their altitudes (Figure 12-2). At sea level, at a latitude of 45°, the average atmospheric pressure supports a column of mercury 760 mm high in a simple mercury barometer when the mercury and air are at 0°C. This average sea-level pressure of 760 mm Hg is called one **atmosphere** of pressure.

> one atmosphere (atm) = 760 mm Hg at 0°C = 760 torr

The SI unit of pressure is the **pascal** (Pa), defined as the pressure exerted by a force of one newton acting on an area of one square meter. By definition, one *newton* (N) is the force required to give a mass of one kilogram an acceleration of one meter per second per second. Symbolically we represent one newton as

$$1 \text{ N} = \frac{1 \text{ kg} \cdot \text{m}}{\text{s}^2} \qquad \text{so} \qquad 1 \text{ Pa} = \frac{1 \text{ N}}{\text{m}^2} = \frac{1 \text{ kg}}{\text{m} \cdot \text{s}^2}$$

One atmosphere of pressure = 1.01325×10^5 Pa, or 101.325 kPa.

12-4 BOYLE'S LAW: THE VOLUME–PRESSURE RELATIONSHIP

Early experiments on the behavior of gases were carried out by Robert Boyle (1627–1691) in the 17th century. In a typical experiment (Figure 12-3), a sample of a gas was trapped in a U-tube and allowed to come to constant temperature. Then its volume and the difference in the heights of the two mercury columns were recorded. This difference in height plus the pressure of the atmosphere represents the pressure on the gas. Addition of more mercury to the tube increases the pressure by changing the height of the mercury column. As a result, the gas volume decreases. The results of several such experiments are tabulated in Figure 12-4a.

Boyle showed that for a given sample of gas at constant temperature, the product of pressure and volume, $P \times V$, was always the same number.

> At a given temperature, the product of pressure and volume of a definite mass of gas is constant.
>
> $$PV = k \qquad \text{(constant } n, T\text{)}$$

This relationship is **Boyle's Law.** The value of k depends on the amount (number of moles, n) of gas present and on the temperature, T. Units for k are determined by the units used to express the volume (V) and pressure (P).

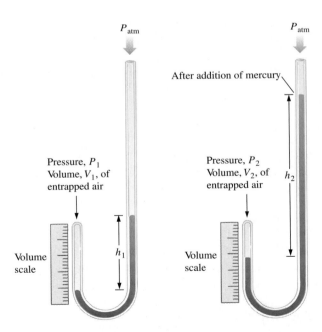

Figure 12-3 A representation of Boyle's experiment. A sample of air is trapped in a tube in such a way that the pressure on the air can be changed and the volume of the air measured. P_{atm} is the atmospheric pressure, measured with a barometer. $P_1 = h_1 + P_{atm}$, $P_2 = h_2 + P_{atm}$.

P	V	P × V	1/P
5.0	40.0	200	0.20
10.0	20.0	200	0.10
15.0	13.3	200	0.0667
17.0	11.8	201	0.0588
20.0	10.0	200	0.0500
22.0	9.10	200	0.0455
30.0	6.70	201	0.0333
40.0	5.00	200	0.0250

(a)

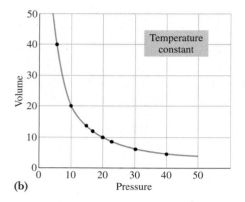
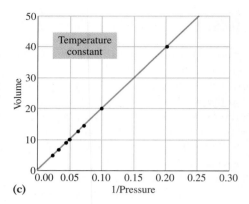

(b)   (c)

Figure 12-4 (a) Some typical data from an experiment such as that shown in Figure 12-3. Measured values of *P* and *V* are presented in the first two columns, on an arbitrary scale. (b, c) Graphical representations of Boyle's Law, using the data of part (a). (b) *V* versus *P*. (c) *V* versus 1/*P*.

When the volume of a gas is plotted against its pressure at constant temperature, the resulting curve is one branch of a hyperbola. Figure 12-4b is a graphic illustration of this inverse relationship. When volume is plotted versus the reciprocal of the pressure, 1/*P*, a straight line results (Figure 12-4c). In 1662, Boyle summarized the results of his experiments on various samples of gases in an alternative statement of Boyle's Law:

> At constant temperature the volume, *V*, occupied by a definite mass of a gas is inversely proportional to the applied pressure, *P*.
>
> $$V \propto \frac{1}{P} \quad \text{or} \quad V = k\left(\frac{1}{P}\right) \quad \text{(constant } n, T\text{)}$$

The symbol ∝ reads "is proportional to." A proportionality is converted into an equality by introducing a proportionality constant, *k*.

At normal temperatures and pressure, most gases obey Boyle's Law rather well. We call this *ideal behavior*. Deviations from ideality are discussed in Section 12-15.

Let us think about a fixed mass of gas at constant temperature, but at two different conditions of pressure and volume (see Figure 12-3). For the first condition we can write

$$P_1V_1 = k \quad \text{(constant } n, T\text{)}$$

and for the second condition we can write

$$P_2V_2 = k \quad \text{(constant } n, T\text{)}$$

Because the right-hand sides of these two equations are the same, the left-hand sides must be equal, or

$$P_1V_1 = P_2V_2 \quad \text{(for a given amount of a gas at constant temperature)}$$

This form of Boyle's Law is useful for calculations involving pressure and volume changes, as the following examples demonstrate.

EXAMPLE 12-1 *Boyle's Law Calculation*

A sample of gas occupies 12 L under a pressure of 1.2 atm. What would its volume be if the pressure were increased to 2.4 atm?

Plan

We know the volume at one pressure and want to find the volume at another pressure (constant temperature). This suggests that we use Boyle's Law. We tabulate what is known and what is asked for, and then solve the Boyle's Law equation for the unknown quantity, V_2.

Pressure and volume are inversely proportional. Doubling the pressure halves the volume of a sample of gas at constant temperature.

Solution

We have

$$V_1 = 12 \text{ L} \qquad P_1 = 1.2 \text{ atm}$$
$$V_2 = \underline{?} \qquad P_2 = 2.4 \text{ atm}$$

It is often helpful to tabulate what is given and what is asked for in a problem.

Solving Boyle's Law, $P_1V_1 = P_2V_2$, for V_2 and substituting gives

$$V_2 = \frac{P_1V_1}{P_2} = \frac{(1.2 \text{ atm})(12 \text{ L})}{2.4 \text{ atm}} = \boxed{6.0 \text{ L}}$$

✓ Problem-Solving Tip: *Units in Boyle's Law Calculations*

Which units for volume and pressure are appropriate for Boyle's Law calculations? Boyle's Law in the form $P_1V_1 = P_2V_2$ can also be rearranged and written as $V_1/V_2 = P_2/P_1$. This involves a ratio of volumes, so they can be expressed in any volume units—liters, milliliters, cubic feet—as long as the *same* units are used for both volumes. Likewise, because Boyle's Law involves a ratio of pressures, you can use any units for pressures—atmospheres, torr, pascals—as long as the *same* units are used for both pressures.

EXAMPLE 12-2 *Boyle's Law Calculation*

A sample of oxygen occupies 10.0 L under a pressure of 790. torr (105 kPa). At what pressure would it occupy 13.4 L if the temperature did not change?

Plan

We know the pressure at one volume and wish to find the pressure at another volume (at constant temperature). We can solve Boyle's Law for the second pressure and substitute.

Solution

We have $P_1 = 790.$ torr; $V_1 = 10.0$ L; $P_2 = \underline{?}$; $V_2 = 13.4$ L. Solving Boyle's Law, $P_1V_1 = P_2V_2$, for P_2 and substituting yields

$$P_2 = \frac{P_1V_1}{V_2} = \frac{(790. \text{ torr})(10.0 \text{ L})}{13.4 \text{ L}} = \boxed{590. \text{ torr}} \qquad \left(\times \frac{101.3 \text{ kPa}}{760. \text{ torr}} = \boxed{78.6 \text{ kPa}} \right)$$

You should now work Exercises 16 and 17.

 It is a very good idea to think qualitatively about what the answer should be before doing the numerical calculation. For example, since the volume is *increasing*, we should reason qualitatively from Boyle's Law that the pressure must be *decreasing*. This provides a good check on the final numerical answer you calculate.

> ✔ **Problem-Solving Tip:** *Use What You Can Predict About the Answer*
>
> In Example 12-1 the calculated volume decrease is consistent with the increase in pressure. We can use this reasoning in another method for solving that problem, that is, by setting up a "Boyle's Law factor" to change the volume in the direction required by the pressure change. We reason that the pressure increases from 1.2 atm to 2.4 atm, so the volume *decreases* by the factor (1.2 atm/2.4 atm). The solution then becomes
>
> $$\underline{?}\ L = 12\ L \times (\text{Boyle's Law factor that would decrease the volume})$$
>
> $$= 12\ L \times \left(\frac{1.2\ \text{atm}}{2.4\ \text{atm}}\right) = 6.0\ L$$
>
> Now solve Example 12-2 using a Boyle's Law factor.

An artist's representation of Jacques Charles's first ascent in a hydrogen balloon at the Tuileries, Paris, December 1, 1783.

Lord Kelvin (1824–1907) was born William Thompson. At the age of ten he was admitted to Glasgow University. Because its new appliance was based on Kelvin's theories, a refrigerator company named its product the Kelvinator.

Recall that temperatures on the Kelvin scale are expressed in kelvins (not degrees Kelvin) and represented by K, not °K.

Absolute zero may be thought of as the limit of thermal contraction for an ideal gas.

12-5 CHARLES'S LAW: THE VOLUME–TEMPERATURE RELATIONSHIP; THE ABSOLUTE TEMPERATURE SCALE

In his pressure–volume studies on gases, Robert Boyle noticed that heating a sample of gas caused some volume change, but he did not follow up on this observation. About 1800, two French scientists—Jacques Charles (1746–1823) and Joseph Gay-Lussac (1778–1850), pioneer balloonists at the time—began studying the expansion of gases with increasing temperature. Their studies showed that the rate of expansion with increased temperature was constant and was the same for all the gases they studied as long as the pressure remained constant. The implications of their discovery were not fully recognized until nearly a century later. Then scientists used this behavior of gases as the basis of a new temperature scale, the absolute temperature scale.

The change of volume with temperature, at constant pressure, is illustrated in Figure 12-5. From the table of typical data in Figure 12-5b, we see that volume (V, mL) increases as temperature (t,°C) increases, but the quantitative relationship is not yet obvious. These data are plotted in Figure 12-5c (line A), together with similar data for the same gas sample at different pressures (lines B and C).

Lord Kelvin, a British physicist, noticed that an extension of the different temperature–volume lines back to zero volume (dashed line) yields a common intercept at −273.15°C on the temperature axis. Kelvin named this temperature **absolute zero.** The degrees are the same size over the entire scale, so 0°C becomes 273.15 degrees above absolute zero. In honor of Lord Kelvin's work, this scale is called the Kelvin temperature scale. As pointed out in Section 1-12, the relationship between the Celsius and Kelvin temperature scales is K = °C + 273.15°.

If we convert temperatures (°C) to absolute temperatures (K), the green scale in Figure 12-5c, the volume–temperature relationship becomes obvious. This relationship is known as **Charles's Law.**

> At constant pressure, the volume occupied by a definite mass of a gas is directly proportional to its absolute temperature.

We can express Charles's Law in mathematical terms as

$$V \propto T \qquad \text{or} \qquad V = kT \qquad (\text{constant } n, P)$$

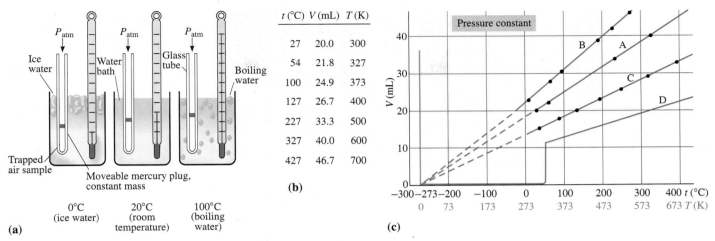

Figure 12-5 An experiment showing that the volume of an ideal gas increases as the temperature is increased at constant pressure. (a) A mercury plug of constant weight, plus atmospheric pressure, maintains a constant pressure on the trapped air. (b) Some representative volume–temperature data at constant pressure. The relationship becomes clear when t (°C) is converted to T (K) by adding 273°C. (c) A graph in which volume is plotted versus temperature on two different scales. Lines A, B, and C represent the same mass of the same ideal gas at different pressures. Line A represents the data tabulated in part (b). Graph D shows the behavior of a gas that condenses to form a liquid (in this case, at 50°C) as it is cooled. The volume does not really drop to zero when the gas forms a liquid, but it does become much smaller than the gaseous volume.

Rearranging the expression gives $V/T = k$, a concise statement of Charles's Law. As the temperature increases, the volume must increase proportionally. If we let subscripts 1 and 2 represent values for the same sample of gas at two different temperatures, we obtain

$$\frac{V_1}{T_1} = \frac{V_2}{T_2} \quad \text{(for a definite mass of gas at constant pressure)}$$

which is the more useful form of Charles's Law. This relationship is valid *only* when temperature, T, is expressed on an absolute (usually the Kelvin) scale.

EXAMPLE 12-3 *Charles's Law Calculation*

A sample of nitrogen occupies 117 mL at 100.°C. At what temperature in °C would it occupy 234 mL if the pressure did not change?

Plan

We know the volume of the sample at one temperature and wish to know its temperature corresponding to a second volume (constant pressure). We can solve Charles's Law for the second temperature. We must remember to carry out calculations with all temperatures expressed on the Kelvin scale, converting to or from Celsius as necessary.

 It is very important to remember to convert the temperatures into kelvins for all gas law calculations. Don't forget to convert the temperature back to the proper units at the end of the problem!

Solution

$$V_1 = 117 \text{ mL} \qquad\qquad V_2 = 234 \text{ mL}$$
$$T_1 = 100.°\text{C} + 273° = 373 \text{ K} \qquad T_2 = \underline{?}$$

Charles D. Winters

When balloons filled with air are cooled in liquid nitrogen (bp −196°C), each shrinks to a small fraction of its original volume. Because the boiling points of the other components of air, except He and Ne, are higher than −196°C, they condense to form liquids. When the balloons are removed from the liquid nitrogen, the liquids vaporize to form gases again. As the air warms to room temperature, the balloons expand to their original volume (Charles's Law).

$$\frac{V_1}{T_1} = \frac{V_2}{T_2} \quad \text{and} \quad T_2 = \frac{V_2 T_1}{V_1} = \frac{(234 \text{ mL})(373 \text{ K})}{(117 \text{ mL})} = \boxed{746 \text{ K}}$$

$$°C = 746 \text{ K} - 273° = \boxed{473°C}$$

The temperature doubles on the Kelvin scale, from 373 K to 746 K, so the volume doubles.

You should now work Exercise 24.

✓ **Problem-Solving Tip:** *Be Careful of Units in Charles's Law Calculations*

Which units for volume and temperature are appropriate for Charles's Law calculations? The equation $\frac{V_1}{T_1} = \frac{V_2}{T_2}$ can be written as $\frac{V_1}{V_2} = \frac{T_1}{T_2}$. This involves a ratio of volumes, so they can be expressed in any volume units—liters, milliliters, cubic feet—as long as the *same* units are used for both volumes. But the relationship *does not apply at all* unless the temperatures are both expressed on an absolute scale. Remember to express all temperatures in kelvins for Charles's Law calculations.

12-6 STANDARD TEMPERATURE AND PRESSURE

We have seen that both temperature and pressure affect the volumes (and therefore the densities) of gases. It is often convenient to choose some "standard" temperature and pressure as a reference point for discussing gases. **Standard temperature and pressure (STP)** are, by international agreement, exactly 0°C (273.15 K) and one atmosphere of pressure (760. torr).

12-7 THE COMBINED GAS LAW EQUATION

Boyle's Law relates the pressures and volumes of a sample of gas at constant temperature, $P_1 V_1 = P_2 V_2$. Charles's Law relates the volumes and temperatures at constant pressure,

$V_1/T_1 = V_2/T_2$. Combination of Boyle's Law and Charles's Law into a single expression gives the **Combined Gas Law equation.**

$$\frac{P_1V_1}{T_1} = \frac{P_2V_2}{T_2} \qquad \text{(constant amount of gas)}$$

When any five of the variables in the equation are known, the sixth variable can be calculated.

Notice that the Combined Gas Law equation becomes

1. $P_1V_1 = P_2V_2$ (Boyle's Law) when T is constant;

2. $\dfrac{V_1}{T_1} = \dfrac{V_2}{T_2}$ (Charles's Law) when P is constant; and

3. $\dfrac{P_1}{T_1} = \dfrac{P_2}{T_2}$ when V is constant.

EXAMPLE 12-4 Combined Gas Law Calculation

A sample of neon occupies 105 liters at 27°C under a pressure of 985 torr. What volume would it occupy at standard temperature and pressure (STP)?

Plan

A sample of gas is changing in all three quantities P, V, and T. This suggests that we use the Combined Gas Law equation. We tabulate what is known and what is asked for, solve the Combined Gas Law equation for the unknown quantity, V_2, and substitute known values.

Solution

$$V_1 = 105 \text{ L} \qquad P_1 = 985 \text{ torr} \qquad T_1 = 27°C + 273° = 300.\text{ K}$$
$$V_2 = \underline{\ ?\ } \qquad P_2 = 760.\text{ torr} \qquad T_2 = 273 \text{ K}$$

Solving for V_2,

$$\frac{P_1V_1}{T_1} = \frac{P_2V_2}{T_2} \qquad \text{so} \qquad V_2 = \frac{P_1V_1T_2}{P_2T_1} = \frac{(985 \text{ torr})(105 \text{ L})(273 \text{ K})}{(760.\text{ torr})(300.\text{ K})} = \boxed{124 \text{ L}}$$

Alternatively, we can multiply the original volume by a Boyle's Law factor and a Charles's Law factor. As the pressure decreases from 985 torr to 760. torr, the volume increases, so the Boyle's Law factor is 985 torr/760. torr. As the temperature decreases from 300. K to 273 K, the volume decreases, so the Charles's Law factor is 273 K/300. K. Multiplication of the original volume by these factors gives the same result.

$$\underline{\ ?\ } \text{ L} = 105 \text{ L} \times \frac{985 \text{ torr}}{760.\text{ torr}} \times \frac{273 \text{ K}}{300.\text{ K}} = \boxed{124 \text{ L}}$$

The temperature decrease (from 300. K to 273 K) alone would give only a small *decrease* in the volume of neon, by a factor of 273 K/300. K, or 0.910. The pressure decrease (from 985 torr to 760. torr) alone would result in a greater *increase* in the volume, by a factor of 985 torr/760. torr, or 1.30. The result of the two changes is that the volume increases from 105 liters to 124 liters.

EXAMPLE 12-5 Combined Gas Law Calculation

A sample of gas occupies 12.0 liters at 240.°C under a pressure of 80.0 kPa. At what temperature would the gas occupy 15.0 liters if the pressure were increased to 107 kPa?

Plan

The approach is the same as for Example 12-4 except that the unknown quantity is now the temperature, T_2.

Solution

$$V_1 = 12.0 \text{ L} \qquad P_1 = 80.0 \text{ kPa} \qquad T_1 = 240.°\text{C} + 273° = 513 \text{ K}$$
$$V_2 = 15.0 \text{ L} \qquad P_2 = 107 \text{ kPa} \qquad T_2 = \underline{?}$$

We solve the Combined Gas Law equation for T_2.

$$\frac{P_1 V_1}{T_1} = \frac{P_2 V_2}{T_2} \quad \text{so} \quad T_2 = \frac{P_2 V_2 T_1}{P_1 V_1} = \frac{(107 \text{ kPa})(15.0 \text{ L})(513 \text{ K})}{(80.0 \text{ kPa})(12.0 \text{ L})} = \boxed{858 \text{ K}}$$

$$\text{K} = °\text{C} + 273° \quad \text{so} \quad °\text{C} = 858 \text{ K} - 273° = \boxed{585°\text{C}}$$

You should now work Exercises 32 and 33.

✓ **Problem-Solving Tip:** *Units in Combined Gas Law Calculations*

The Combined Gas Law equation is derived by combining Boyle's and Charles's Laws, so the comments in earlier Problem-Solving Tips also apply to this equation. Remember to express all temperatures in kelvins. Volumes can be expressed in any units as long as both are in the same units. Similarly, any pressure units can be used, so long as both are in the same units. Example 12-4 uses torr for both pressures; Example 12-5 uses kPa for both pressures.

12-8 AVOGADRO'S LAW AND THE STANDARD MOLAR VOLUME

In 1811, Amedeo Avogadro postulated that

CD-ROM   Screen 12.3, Gas Laws.

at the same temperature and pressure, equal volumes of all gases contain the same number of molecules.

Many experiments have demonstrated that Avogadro's hypothesis is accurate to within about ±2%, and the statement is now known as **Avogadro's Law.**

Avogadro's Law can also be stated as follows.

At constant temperature and pressure, the volume, V, occupied by a gas sample is directly proportional to the number of moles, n, of gas.

$$V \propto n \quad \text{or} \quad V = kn \quad \text{or} \quad \frac{V}{n} = k \quad (\text{constant } P, T)$$

For two samples of gas at the same temperature and pressure, the relation between volumes and numbers of moles can be represented as

$$\frac{V_1}{n_1} = \frac{V_2}{n_2} \quad (\text{constant } T, P)$$

TABLE 12-3	Standard Molar Volumes and Densities of Some Gases (0°C)			
Gas	Formula	(g/mol)	Standard Molar Volume (L/mol)	Density at STP (g/L)
hydrogen	H_2	2.02	22.428	0.090
helium	He	4.003	22.426	0.178
neon	Ne	20.18	22.425	0.900
nitrogen	N_2	28.01	22.404	1.250
oxygen	O_2	32.00	22.394	1.429
argon	Ar	39.95	22.393	1.784
carbon dioxide	CO_2	44.01	22.256	1.977
ammonia	NH_3	17.03	22.094	0.771
chlorine	Cl_2	70.91	22.063	3.214

Deviations in standard molar volume indicate that gases do not behave ideally.

The volume occupied by a mole of gas at *standard temperature and pressure*, STP, is referred to as the standard molar volume. It is nearly constant for all gases (Table 12-3).

The volume percentages given in Table 12-2 are also equal to mole percentages.

The **standard molar volume** of an ideal gas is taken to be 22.414 liters per mole at STP.

Gas densities depend on pressure and temperature; however, the number of moles of gas in a given sample does not change with temperature or pressure. Pressure changes affect volumes of gases according to Boyle's Law, and temperature changes affect volumes of gases according to Charles's Law. We can use these laws to convert gas densities at various temperatures and pressures to *standard temperature and pressure*. Table 12-3 gives the experimentally determined densities of several gases at standard temperature and pressure.

Density is defined as mass per unit volume. For solids and liquids this is usually expressed in g/mL, but for gases g/L is more convenient.

EXAMPLE 12-6 *Molecular Weight, Density*

One (1.00) mole of a gas occupies 27.0 liters, and its density is 1.41 g/L at a particular temperature and pressure. What is its molecular weight? What is the density of the gas at STP?

Plan

We can use dimensional analysis to convert the density, 1.41 g/L, to molecular weight, g/mol. To calculate the density at STP, we recall that the volume occupied by one mole would be 22.4 L.

Solution

We multiply the density under the original conditions by the unit factor 27.0 L/1.00 mol to generate the appropriate units, g/mol.

$$\frac{?\ g}{mol} = \frac{1.41\ g}{L} \times \frac{27.0\ L}{mol} = \boxed{38.1\ g/mol}$$

At STP, 1.00 mol of the gas, 38.1 g, would occupy 22.4 L, and its density would be

$$Density = \frac{38.1\ g}{1\ mol} \times \frac{1\ mol}{22.4\ L} = \boxed{1.70\ g/L}\ at\ STP$$

You should now work Exercises 38 and 40.

12-9   SUMMARY OF GAS LAWS: THE IDEAL GAS EQUATION

CD-ROM   Screen 12.4, The Ideal Gas Law.

Let us summarize what we have learned about gases. Any sample of gas can be described in terms of its pressure, temperature (in kelvins), volume, and the number of moles, n, present. Any three of these variables determine the fourth. The gas laws we have studied give several relationships among these variables. An **ideal gas** is one that exactly obeys these gas laws. Many real gases show slight deviations from ideality, but at normal temperatures and pressures the deviations are usually small enough to be ignored. We will do so for the present and discuss deviations from ideal behavior later.

We can summarize the behavior of ideal gases as follows.

TOPIC: Gas Laws
GO TO: www.scilinks.org
sciLINKS CODE: WCH1210

Boyle's Law	$V \propto \dfrac{1}{P}$	(at constant T and n)
Charles's Law	$V \propto T$	(at constant P and n)
Avogadro's Law	$V \propto n$	(at constant T and P)
Summarizing	$V \propto \dfrac{nT}{P}$	(no restrictions)

As before, a proportionality can be written as an equality by introducing a proportionality constant, for which we'll use the symbol R. This gives

$$V = R\left(\frac{nT}{P}\right) \qquad \text{or, rearranging,} \qquad \boxed{PV = nRT}$$

This equation takes into account the values of n, T, P, and V. Restrictions that apply to the individual gas laws are therefore not needed for the ideal gas equation.

This relationship is called the **ideal gas equation** or the *Ideal Gas Law*. The numerical value of R, the **universal gas constant,** depends on the choices of the units for P, V, and T. One mole of an ideal gas occupies 22.414 liters at 1.0000 atmosphere and 273.15 K (STP). Solving the ideal gas equation for R gives

$$R = \frac{PV}{nT} = \frac{(1.0000 \text{ atm})(22.414 \text{ L})}{(1.0000 \text{ mol})(273.15 \text{ K})} = 0.082057 \; \frac{\text{L} \cdot \text{atm}}{\text{mol} \cdot \text{K}}$$

In working problems, we often round R to 0.0821 L·atm/mol·K. We can express R in other units, as shown inside the back cover of this text.

> It is important to use the R value whose units match the pressure and volume units in the problem. We will most commonly use atm and liters as units, but remember to pay attention to what is being used in the problem at hand. As mentioned before, one should always use kelvins for the temperature.

EXAMPLE 12-7   *Units of R*

R can have any *energy* units per mole per kelvin. Calculate R in terms of joules per mole per kelvin and in SI units of kPa·dm³/mol·K.

Plan

We apply dimensional analysis to convert to the required units.

Solution

Appendix C shows that 1 L · atm = 101.325 joules.

$$R = \frac{0.082057 \text{ L·atm}}{\text{mol} \cdot \text{K}} \times \frac{101.325 \text{ J}}{1 \text{ L·atm}} = \boxed{8.3144 \text{ J/mol·K}}$$

Now evaluate R in SI units. One atmosphere pressure is 101.325 kilopascals, and the molar volume at STP is 22.414 dm³.

Recall that 1 dm³ = 1 L.

$$R = \frac{PV}{nT} = \frac{101.325 \text{ kPa} \times 22.414 \text{ dm}^3}{1 \text{ mol} \times 273.15 \text{ K}} = 8.3145 \frac{\text{kPa} \cdot \text{dm}^3}{\text{mol} \cdot \text{K}}$$

You should now work Exercise 41.

We can now express R, the universal gas constant, to four digits in three different sets of units.

$$R = 0.08206 \frac{\text{L} \cdot \text{atm}}{\text{mol} \cdot \text{K}} = 8.314 \frac{\text{J}}{\text{mol} \cdot \text{K}} = 8.314 \frac{\text{kPa} \cdot \text{dm}^3}{\text{mol} \cdot \text{K}}$$

The usefulness of the ideal gas equation is that it relates the four variables, P, V, n, and T, that describe a sample of gas at *one set of conditions*. If any three of these variables are known, the fourth can be calculated.

EXAMPLE 12-8 *Ideal Gas Equation*

What pressure, in atm, is exerted by 54.0 grams of Xe in a 1.00-liter flask at 20.°C?

Plan

We list the variables with the proper units. Then we solve the ideal gas equation for P and substitute values.

Solution

$$V = 1.00 \text{ L} \qquad n = 54.0 \text{ g Xe} \times \frac{1 \text{ mol}}{131.3 \text{ g Xe}} = 0.411 \text{ mol}$$

$$T = 20.°\text{C} + 273° = 293 \text{ K} \qquad P = \underline{?}$$

Solving $PV = nRT$ for P and substituting gives

$$P = \frac{nRT}{V} = \frac{(0.411 \text{ mol})\left(\dfrac{0.0821 \text{ L} \cdot \text{atm}}{\text{mol} \cdot \text{K}}\right)(293 \text{ K})}{1.00 \text{ L}} = 9.89 \text{ atm}$$

We are rounding the value of R to three significant figures.

EXAMPLE 12-9 *Ideal Gas Equation*

What is the volume of a gas balloon filled with 4.00 moles of He when the atmospheric pressure is 748 torr and the temperature is 30.°C?

Plan

We first list the variables with the proper units. Then we solve the ideal gas equation for V and substitute known values.

Solution

$$P = 748 \text{ torr} \times \frac{1 \text{ atm}}{760 \text{ torr}} = 0.984 \text{ atm} \qquad n = 4.00 \text{ mol}$$

$$T = 30.°\text{C} + 273° = 303 \text{ K} \qquad V = \underline{?}$$

Solving $PV = nRT$ for V and substituting gives

$$V = \frac{nRT}{P} = \frac{(4.00 \text{ mol})\left(0.0821 \dfrac{\text{L·atm}}{\text{mol·K}}\right)(303 \text{ K})}{0.984 \text{ atm}} = \boxed{101 \text{ L}}$$

You should now work Exercise 44.

You may wonder why pressures are given in torr or mm Hg and temperatures in °C. This is because pressures are often measured with mercury barometers, and temperatures are measured with Celsius thermometers.

A helium-filled weather balloon.

National Center for Atmospheric Research/National Science Foundation

EXAMPLE 12-10 *Ideal Gas Equation*

A helium-filled weather balloon has a volume of 7240 cubic feet. How many grams of helium would be required to inflate this balloon to a pressure of 745 torr at 21°C? (1 ft³ = 28.3 L)

Plan

We use the ideal gas equation to find n, the number of moles required, and then convert to grams. We must convert each quantity to one of the units stated for R. ($R = 0.0821$ L·atm/mol·K)

Solution

$$P = 745 \text{ torr} \times \frac{1 \text{ atm}}{760 \text{ torr}} = 0.980 \text{ atm} \qquad T = 21°C + 273° = 294 \text{ K}$$

$$V = 7240 \text{ ft}^3 \times \frac{28.3 \text{ L}}{1 \text{ ft}^3} = 2.05 \times 10^5 \text{ L} \qquad n = \underline{\ ?\ }$$

Solving $PV = nRT$ for n and substituting gives

$$n = \frac{PV}{RT} = \frac{(0.980 \text{ atm})(2.05 \times 10^5 \text{ L})}{\left(0.0821 \dfrac{\text{L·atm}}{\text{mol·K}}\right)(294 \text{ K})} = 8.32 \times 10^3 \text{ mol He}$$

$$\underline{\ ?\ } \text{ g He} = (8.32 \times 10^3 \text{ mol He})\left(4.00 \frac{\text{g}}{\text{mol}}\right) = \boxed{3.33 \times 10^4 \text{ g He}}$$

You should now work Exercise 45.

✔ Problem-Solving Tip: *Watch Out for Units in Ideal Gas Law Calculations*

The units of R that are appropriate for Ideal Gas Law calculations are those that involve units of volume, pressure, moles, and temperature. When you use the value $R = 0.0821$ L·atm/mol·K, remember to express all quantities in a calculation in these units. Pressures should be expressed in atmospheres, volumes in liters, temperature in kelvins, and amount of gas in moles. In Examples 12-9 and 12-10 we converted pressures from torr to atm. In Example 12-10 the volume was converted from ft³ to L.

Each of the individual gas laws can be derived from the ideal gas equation.

Summary of the Ideal Gas Laws

1. The individual gas laws are usually used to calculate the *changes* in conditions for a sample of gas (subscripts can be thought of as "before" and "after").

    Boyle's Law $\qquad P_1V_1 = P_2V_2$ $\qquad$ (for a given amount of a gas at constant temperature)

    Charles's Law $\qquad \dfrac{V_1}{T_1} = \dfrac{V_2}{T_2}$ $\qquad$ (for a given amount of a gas at constant pressure)

    Combined Gas Law $\qquad \dfrac{P_1V_1}{T_1} = \dfrac{P_2V_2}{T_2}$ $\qquad$ (for a given amount of a gas)

    Avogadro's Law $\qquad \dfrac{V_1}{n_1} = \dfrac{V_2}{n_2}$ $\qquad$ (for gas samples at the same temperature and pressure)

2. The ideal gas equation is used to calculate one of the four variables P, V, n, and T, which describe a sample of gas at *any single set of conditions*.

    $$PV = nRT$$

 When volume, temperature, or pressure change, it is more convenient to use one of the individual gas laws. When the calculation involves the amount of gas (g or moles), the ideal gas equation is usually more useful.

The ideal gas equation can also be used to calculate the densities of gases.

EXAMPLE 12-11 *Ideal Gas Equation*

Nitric acid, a very important industrial chemical, is made by dissolving the gas nitrogen dioxide, NO_2, in water. Calculate the density of NO_2 gas, in g/L, at 1.24 atm and 50.°C.

CD-ROM Screen 12.5, Gas Density.

Plan

We use the ideal gas equation to find the number of moles, n, in any volume, V, at the specified pressure and temperature. Then we convert moles to grams. Because we want to express density in g/L, we choose a volume of one liter.

Solution

$$V = 1.00 \text{ L} \qquad\qquad n = \underline{\ ?\ }$$
$$T = 50.°\text{C} + 273° = 323 \text{ K} \qquad P = 1.24 \text{ atm}$$

Solving $PV = nRT$ for n and substituting gives

$$n = \frac{PV}{RT} = \frac{(1.24 \text{ atm})(1.00 \text{ L})}{\left(0.0821 \dfrac{\text{L·atm}}{\text{mol·K}}\right)(323 \text{ K})} = 0.0468 \text{ mol}$$

So there is 0.0468 mol NO_2/L at the specified P and T. Converting this to grams of NO_2 per liter, we obtain

$$\text{Density} = \frac{\underline{\ ?\ } \text{ g}}{\text{L}} = \frac{0.0468 \text{ mol } NO_2}{\text{L}} \times \frac{46.0 \text{ g } NO_2}{\text{mol } NO_2} = \boxed{2.15 \text{ g/L}}$$

You should now work Exercise 39.

12-10 DETERMINATION OF MOLECULAR WEIGHTS AND MOLECULAR FORMULAS OF GASEOUS SUBSTANCES

CD-ROM   Screen 12.6, Using Gas Laws: Determining Molar Mass.

In Section 2-10 we distinguished between simplest and molecular formulas of compounds. We showed how simplest formulas can be calculated from percent compositions of compounds. The molecular weight must be known to determine the molecular formula of a compound. For compounds that are gases at convenient temperatures and pressures, the ideal gas law provides a basis for determining molecular weights.

> ⚠ Remember to convert °C into K and torr into atm. One must then rearrange the ideal gas equation to solve for *n*, the number of moles.

EXAMPLE 12-12   Molecular Weight

A 0.109-gram sample of a pure gaseous compound occupies 112 mL at 100.°C and 750. torr. What is the molecular weight of the compound?

Plan

We first use the ideal gas equation, $PV = nRT$, to find the number of moles of gas. Then, knowing the mass of that number of moles of gas, we calculate the mass of one mole, the molecular weight.

Solution

$$V = 0.112 \text{ L} \qquad T = 100.°C + 273° = 373 \text{ K} \qquad P = 750. \text{ torr} \times \frac{1 \text{ atm}}{760. \text{ torr}} = 0.987 \text{ atm}$$

$$n = \frac{PV}{RT} = \frac{(0.987 \text{ atm})(0.112 \text{ L})}{\left(0.0821 \dfrac{\text{L·atm}}{\text{mol·K}}\right)(373 \text{ K})} = 0.00361 \text{ mol}$$

The mass of 0.00361 mole of this gas is 0.109 g, so the mass of one mole is

$$\frac{? \text{ g}}{\text{mol}} = \frac{0.109 \text{ g}}{0.00361 \text{ mol}} = \boxed{30.2 \text{ g/mol}}$$

The molecular weight of the gas is 30.2 amu. The gas could be ethane, C_2H_6, MW = 30.1 amu. Can you think of other possibilities? Could the gas have been NO, CH_3OH, O_2, or CH_3NH_2?

EXAMPLE 12-13   Molecular Weight

A 120.-mL flask contained 0.345 gram of a gaseous compound at 100.°C and 1.00 atm pressure. What is the molecular weight of the compound?

Plan

We use the ideal gas equation, $PV = nRT$, to determine the number of moles of gas that filled the flask. Then, knowing the mass of this number of moles, we can calculate the mass of one mole.

Solution

$$V = 0.120 \text{ L} \qquad P = 1.00 \text{ atm} \qquad T = 100.°C + 273° = 373 \text{ K}$$

$$n = \frac{PV}{RT} = \frac{(1.00 \text{ atm})(0.120 \text{ L})}{\left(0.0821 \dfrac{\text{L·atm}}{\text{mol·K}}\right)(373 \text{ K})} = 0.00392 \text{ mol}$$

The mass of 0.00392 mol of gas is 0.345 g, so the mass of one mole is

$$\frac{? \text{ g}}{\text{mol}} = \frac{0.345 \text{ g}}{0.00392 \text{ mol}} = \boxed{88.0 \text{ g/mol}}$$

You should now work Exercises 49 and 52.

Let's carry the calculation one step further in the next example.

EXAMPLE 12-14   *Molecular Formula*

Additional analysis of the gaseous compound in Example 12-13 showed that it contained 54.5% carbon, 9.10% hydrogen, and 36.4% oxygen by mass. What is its molecular formula?

Plan

We first find the simplest formula for the compound as we did in Section 2-9 (Examples 2-12 and 2-13). Then we use the molecular weight that we determined in Example 12-13 to find the molecular formula. To find the molecular formula, we reason as in Example 2-16. We use the experimentally known molecular weight to find the ratio

$$n = \frac{\text{molecular weight}}{\text{simplest-formula weight}}$$

The molecular weight is n times the simplest-formula weight, so the molecular formula is n times the simplest formula.

Solution

Element	Relative Mass of Element	Relative Number of Atoms (divide mass by AW)	Divide by Smallest Number	Smallest Whole-Number Ratio of Atoms
C	54.5	$\frac{54.5}{12.0} = 4.54$	$\frac{4.54}{2.28} = 1.99$	2
H	9.10	$\frac{9.10}{1.01} = 9.01$	$\frac{9.01}{2.28} = 3.95$	4 C_2H_4O
O	36.4	$\frac{36.4}{16.0} = 2.28$	$\frac{2.28}{2.28} = 1.00$	1

The simplest formula is C_2H_4O and the simplest-formula weight is 44.0 amu.
    Division of the molecular weight by the simplest-formula weight gives

$$\frac{\text{molecular weight}}{\text{simplest-formula weight}} = \frac{88.0 \text{ amu}}{44.0 \text{ amu}} = 2$$

The molecular formula is therefore $2 \times (C_2H_4O) = \boxed{C_4H_8O_2}$.
    The gas could be either ethyl acetate or butyric acid. Both have the formula $C_4H_8O_2$. They have very different odors, however. Ethyl acetate has the odor of nail polish remover. Butyric acid has the foul odor of rancid butter.

You should now work Exercise 50.

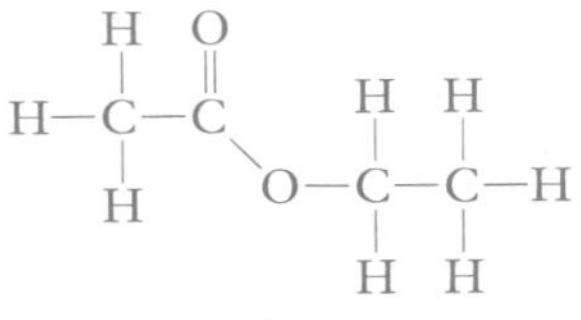

ethyl acetate

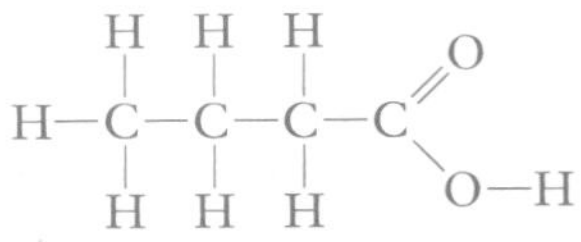

butyric acid

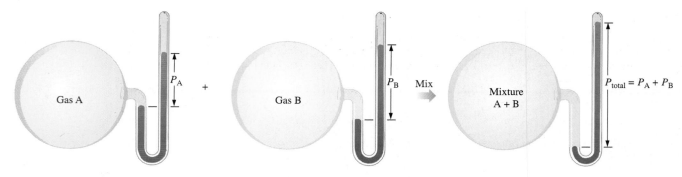

Figure 12-6 An illustration of Dalton's Law. When the two gases A and B are mixed in the same container at the same temperature, they exert a total pressure equal to the sum of their partial pressures.

12-11 DALTON'S LAW OF PARTIAL PRESSURES

CD-ROM   Screen 12.8, Gas Mixtures and Partial Pressures.

Many gas samples, including our atmosphere, are mixtures that consist of different kinds of gases. The total number of moles in a mixture of gases is

$$n_{total} = n_A + n_B + n_C + \cdots$$

where n_A, n_B, and so on represent the number of moles of each kind of gas present. Rearranging the ideal gas equation, $P_{total}V = n_{total}RT$, for the total pressure, P_{total}, and then substituting for n_{total} gives

$$P_{total} = \frac{n_{total}RT}{V} = \frac{(n_A + n_B + n_C + \cdots)\,RT}{V}$$

Multiplying out the right-hand side gives

$$P_{total} = \frac{n_ART}{V} + \frac{n_BRT}{V} + \frac{n_CRT}{V} + \cdots$$

John Dalton was the first to notice this effect. He did so in 1807 while studying the compositions of moist and dry air. The pressure that each gas exerts in a mixture is called its **partial pressure.** No way has been devised to measure the pressure of an individual gas in a mixture; it must be calculated from other quantities.

Now n_ART/V is the *partial pressure* P_A that the n_A moles of gas A alone would exert in the container at temperature T; similarly, $n_BRT/V = P_B$, and so on. Substituting these into the equation for P_{total}, we obtain **Dalton's Law of Partial Pressures** (Figure 12-6).

$$P_{total} = P_A + P_B + P_C + \cdots \qquad \text{(constant } V, T\text{)}$$

The total pressure exerted by a mixture of ideal gases is the sum of the partial pressures of those gases.

Dalton's Law is useful in describing real gaseous mixtures at moderate pressures because it allows us to relate total measured pressures to the composition of mixtures.

EXAMPLE 12-15   *Mixture of Gases*

A 10.0-liter flask contains 0.200 mole of methane, 0.300 mole of hydrogen, and 0.400 mole of nitrogen at 25°C. (a) What is the pressure, in atmospheres, inside the flask? (b) What is the partial pressure of each component of the mixture of gases?

Plan

(a) We are given the number of moles of each component. The ideal gas law is then used to calculate the total pressure from the total number of moles. (b) The partial pressure of each gas in the mixture can be calculated by substituting the number of moles of each gas individually into $PV = nRT$.

Solution

(a) $n = 0.200$ mol $CH_4 + 0.300$ mol $H_2 + 0.400$ mol $N_2 = 0.900$ mol of gas

$V = 10.0$ L $T = 25°C + 273° = 298$ K

Solving $PV = nRT$ for P gives $P = nRT/V$. Substitution gives

$$P = \frac{(0.900 \text{ mol})\left(0.0821 \dfrac{\text{L·atm}}{\text{mol·K}}\right)(298 \text{ K})}{10.0 \text{ L}} = 2.20 \text{ atm}$$

(b) Now we find the partial pressures. For CH_4, $n = 0.200$ mol, and the values for V and T are the same as in Part (a).

$$P_{CH_4} = \frac{(n_{CH_4})RT}{V} = \frac{(0.200 \text{ mol})\left(0.0821 \dfrac{\text{L·atm}}{\text{mol·K}}\right)(298 \text{ K})}{10.0 \text{ L}} = 0.489 \text{ atm}$$

Similar calculations for the partial pressures of hydrogen and nitrogen give

$$P_{H_2} = 0.734 \text{ atm} \quad \text{and} \quad P_{N_2} = 0.979 \text{ atm}$$

As a check, we use Dalton's Law: $P_{total} = P_A + P_B + P_C + \cdots$. Addition of the partial pressures in this mixture gives the total pressure we calculated in Part (a).

$$P_{total} = P_{CH_4} + P_{H_2} + P_{N_2} = (0.489 + 0.734 + 0.979) \text{ atm} = 2.20 \text{ atm}$$

You should now work Exercises 56 and 57.

✓ **Problem-Solving Tip:** *Amounts of Gases in Mixtures Can Be Expressed in Various Units*

In Example 12-15 we were given the number of moles of each gas. Sometimes the amount of a gas is expressed in other units that can be converted to number of moles. For instance, if we know the formula weight (or the formula), we can convert a given mass of gas to number of moles.

We can describe the composition of any mixture in terms of the mole fraction of each component. The **mole fraction,** X_A, of component A in a mixture is defined as

$$X_A = \frac{\text{no. mol A}}{\text{total no. mol of all components}}$$

Like any other fraction, mole fraction is a dimensionless quantity. For each component in a mixture, the mole fraction is

The sum of all mole fractions in a mixture is equal to 1.

$X_A + X_B + \cdots = 1$ for any mixture

We can use this relationship to check mole fraction calculations or to find a remaining mole fraction if we know all the others.

$$X_A = \frac{\text{no. mol A}}{\text{no. mol A} + \text{no. mol B} + \cdots},$$

$$X_B = \frac{\text{no. mol B}}{\text{no. mol A} + \text{no. mol B} + \cdots}, \quad \text{and so on}$$

For a gaseous mixture, we can relate the mole fraction of each component to its partial pressure as follows. From the ideal gas equation, the number of moles of each component can be written as

$$n_A = P_A V/RT, \qquad n_B = P_B V/RT, \qquad \text{and so on}$$

and the total number of moles is

$$n_{total} = P_{total} V/RT$$

Substituting into the definition of X_A,

$$X_A = \frac{n_A}{n_A + n_B + \cdots} = \frac{P_A V/RT}{P_{total} V/RT}$$

The quantities V, R, and T cancel to give

$$X_A = \frac{P_A}{P_{total}}; \qquad \text{similarly, } X_B = \frac{P_B}{P_{total}}; \qquad \text{and so on}$$

We can rearrange these equations to give another statement of Dalton's Law of Partial Pressures.

$$P_A = X_A \times P_{total}; \qquad P_B = X_B \times P_{total}; \qquad \text{and so on}$$

The partial pressure of each gas is equal to its mole fraction in the gaseous mixture times the total pressure of the mixture.

EXAMPLE 12-16   *Mole Fraction, Partial Pressure*

Calculate the mole fractions of the three gases in Example 12-15.

Plan

One way to solve this problem is to use the numbers of moles given in the problem. Alternatively we could use the partial pressures and the total pressure from Example 12-15.

Solution

Using the moles given in Example 12-15,

Example 12-16 shows that for a gas mixture the relative numbers of moles of components are the same as relative pressures of the components. This is true for all ideal gas mixtures.

$$X_{CH_4} = \frac{n_{CH_4}}{n_{total}} = \frac{0.200 \text{ mol}}{0.900 \text{ mol}} = \boxed{0.222}$$

$$X_{H_2} = \frac{n_{H_2}}{n_{total}} = \frac{0.300 \text{ mol}}{0.900 \text{ mol}} = \boxed{0.333}$$

$$X_{N_2} = \frac{n_{N_2}}{n_{total}} = \frac{0.400 \text{ mol}}{0.900 \text{ mol}} = \boxed{0.444}$$

Using the partial and total pressures calculated in Example 12-15,

$$X_{CH_4} = \frac{P_{CH_4}}{P_{total}} = \frac{0.489}{2.20 \text{ atm}} = 0.222$$

$$X_{H_2} = \frac{P_{H_2}}{P_{total}} = \frac{0.734 \text{ atm}}{2.20 \text{ atm}} = 0.334$$

The difference between the two calculated results is due to rounding.

$$X_{N_2} = \frac{P_{N_2}}{P_{total}} = \frac{0.979 \text{ atm}}{2.20 \text{ atm}} = 0.445$$

You should now work Exercise 58.

EXAMPLE 12-17   *Partial Pressure, Mole Fraction*

The mole fraction of oxygen in the atmosphere is 0.2094. Calculate the partial pressure of O_2 in air when the atmospheric pressure is 760. torr.

Plan

The partial pressure of each gas in a mixture is equal to its mole fraction in the mixture times the total pressure of the mixture.

Solution

$$P_{O_2} = X_{O_2} \times P_{total}$$
$$= 0.2094 \times 760. \text{ torr} = 159 \text{ torr}$$

Dalton's Law can be used in combination with other gas laws, as the following example shows.

EXAMPLE 12-18   *Mixture of Gases*

Two tanks are connected by a closed valve. Each tank is filled with gas as shown, and both tanks are held at the same temperature. We open the valve and allow the gases to mix.

(a) After the gases mix, what is the partial pressure of each gas, and what is the total pressure?

(b) What is the mole fraction of each gas in the mixture?

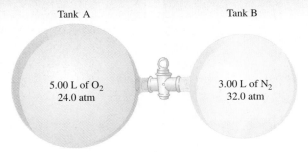

Tank A

Tank B

5.00 L of O_2
24.0 atm

3.00 L of N_2
32.0 atm

Plan

(a) Each gas expands to fill the available volume, 5.00 liters plus 3.00 liters or a total volume of 8.00 liters. We can use Boyle's Law to calculate the partial pressure that each gas would exert after it expands to fill 8.00 L. The total pressure is equal to the sum of the partial pressures of the two gases. (b) The mole fractions can be calculated from the ratio of the partial pressure of each gas to the total pressure.

Solution

(a) For O_2,

$$P_1V_1 = P_2V_2 \quad \text{or} \quad P_{2,O_2} = \frac{P_1V_1}{V_2} = \frac{24.0 \text{ atm} \times 5.00 \text{ L}}{8.00 \text{ L}} = \boxed{15.0 \text{ atm}}$$

For N_2,

$$P_1V_1 = P_2V_2 \quad \text{or} \quad P_{2,N_2} = \frac{P_1V_1}{V_2} = \frac{32.0 \text{ atm} \times 3.00 \text{ L}}{8.00 \text{ L}} = \boxed{12.0 \text{ atm}}$$

The total pressure is the sum of the partial pressures.

$$P_{\text{total}} = P_{2,O_2} + P_{2,N_2} = 15.0 \text{ atm} + 12.0 \text{ atm} = \boxed{27.0 \text{ atm}}$$

(b)

$$X_{O_2} = \frac{P_{2,O_2}}{P_{\text{total}}} = \frac{15.0 \text{ atm}}{27.0 \text{ atm}} = \boxed{0.556}$$

$$X_{N_2} = \frac{P_{2,N_2}}{P_{\text{total}}} = \frac{12.0 \text{ atm}}{27.0 \text{ atm}} = \boxed{0.444}$$

As a check, the sum of the mole fractions is 1.

You should now work Exercise 60.

Some gases can be collected over water. Figure 12-7 illustrates the collection of a sample of hydrogen by displacement of water. A gas produced in a reaction displaces the denser water from the inverted water-filled container. The pressure on the gas inside the collection container could be made equal to atmospheric pressure by collecting gas until the water level inside is the same as that outside.

One complication arises, however. A gas in contact with water soon becomes saturated with water vapor. The pressure inside the container is the sum of the partial pressure of the gas itself *plus* the partial pressure exerted by the water vapor in the gas mixture (the **vapor pressure** of water). Every liquid shows a characteristic vapor pressure that varies only with

Notice that this problem has been solved without calculating the number of moles of either gas.

Gases that are very soluble in water or that react with water cannot be collected by this method. Other liquids can be used.

The partial pressure exerted by the vapor above a liquid is called the vapor pressure of that liquid. A more extensive table of the vapor pressure of water appears in Appendix E.

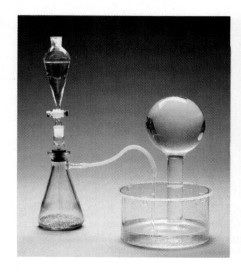

Figure 12-7 Apparatus for preparing hydrogen from zinc and sulfuric acid.

$$Zn(s) + 2H^+(aq) \longrightarrow Zn^{2+}(aq) + H_2(g)$$

The hydrogen is collected by displacement of water.

Marna G. Clarke

temperature, and *not* with the volume of vapor present, so long as both liquid and vapor are present. Table 12-4 displays the vapor pressure of water near room temperature.

The relevant point here is that a gas collected over water is "moist"; that is, it is saturated with water vapor. Measuring the atmospheric pressure at which the gas is collected, we can write

$$P_{atm} = P_{gas} + P_{H_2O} \quad \text{or} \quad P_{gas} = P_{atm} - P_{H_2O}$$

Example 12-19 provides a detailed illustration.

TABLE 12-4	Vapor Pressure of Water Near Room Temperature
Temperature (°C)	Vapor Pressure of Water (torr)
19	16.48
20	17.54
21	18.65
22	19.83
23	21.07
24	22.38
25	23.76
26	25.21
27	26.74
28	28.35

EXAMPLE 12-19 Gas Collected Over Water

Hydrogen was collected over water (Figure 12-7) at 21°C on a day when the atmospheric pressure was 748 torr. The volume of the gas sample collected was 300. mL. (a) How many moles of H_2 were present? (b) How many moles of water vapor were present in the moist gas mixture? (c) What is the mole fraction of hydrogen in the moist gas mixture? (d) What would be the mass of the gas sample if it were dry?

Plan

(a) The vapor pressure of H_2O, $P_{H_2O} = 19$ torr at 21°C, is obtained from Table 12-4. Applying Dalton's Law, $P_{H_2} = P_{atm} - P_{H_2O}$. We then use the partial pressure of H_2 in the ideal gas equation to find the number of moles of H_2 present. (b) The partial pressure of water vapor (the vapor pressure of water at the stated temperature) is used in the ideal gas equation to find the number of moles of water vapor present. (c) The mole fraction of H_2 is the ratio of its partial pressure to the total pressure. (d) The number of moles found in Part (a) can be converted to mass of H_2.

Solution

(a) $P_{H_2} = P_{atm} - P_{H_2O} = (748 - 19) \text{ torr} = 729 \text{ torr} \times \dfrac{1 \text{ atm}}{760 \text{ torr}} = 0.959 \text{ atm}$

We also need to convert the volume from mL to L and the temperature to K

$$V = 300. \text{ mL} = 0.300 \text{ L} \quad \text{and} \quad T = 21°C + 273° = 294 \text{ K}$$

Solving the ideal gas equation for n_{H_2} gives

$$n_{H_2} = \frac{P_{H_2}V}{RT} = \frac{(0.959 \text{ atm})(0.300 \text{ L})}{\left(0.0821 \dfrac{\text{L·atm}}{\text{mol·K}}\right)(294 \text{ K})} = 1.19 \times 10^{-2} \text{ mol } H_2$$

(b) $P_{H_2O} = 19 \text{ torr} \times \dfrac{1 \text{ atm}}{760 \text{ torr}} = 0.025 \text{ atm}$

V and T have the same values as in Part (a).

$$n_{H_2O} = \frac{P_{H_2O}V}{RT} = \frac{(0.025 \text{ atm})(0.300 \text{ L})}{\left(0.0821 \dfrac{\text{L·atm}}{\text{mol·K}}\right)(294 \text{ K})} = 3.1 \times 10^{-4} \text{ mol } H_2O \text{ vapor}$$

(c) $X_{H_2} = \dfrac{P_{H_2}}{P_{total}} = \dfrac{729 \text{ torr}}{748 \text{ torr}} = 0.974$

(d) $? \text{ g } H_2 = 1.19 \times 10^{-2} \text{ mol} \times \dfrac{2.02 \text{ g}}{1 \text{ mol}} = 2.40 \times 10^{-2} \text{ g } H_2$

You should now work Exercise 62.

Remember that *each* gas occupies the *total* volume of the container.

At STP, this dry hydrogen would occupy 267 mL. Can you calculate this?

The nitrogen gas formed in the rapid reaction

$$2NaN_3(s) \longrightarrow 2Na(s) + 3N_2(g)$$

fills an automobile air bag during a collision. The air bag fills within 1/20th of a second after a front collision.

12-12 MASS–VOLUME RELATIONSHIPS IN REACTIONS INVOLVING GASES

Many chemical reactions produce gases. For instance, the combustion of hydrocarbon in excess oxygen at high temperatures produces both carbon dioxide and water as gases, as illustrated for octane.

$$2C_8H_{18}(g) + 25O_2(g) \longrightarrow 16CO_2(g) + 18H_2O(g)$$

The N_2 gas produced by the very rapid decomposition of sodium azide, $NaN_3(s)$, inflates air bags used as safety devices in automobiles.

We know that one mole of gas, measured at STP, occupies 22.4 liters; we can use the ideal gas equation to find the volume of a mole of gas at any other conditions. This information can be utilized in stoichiometry calculations (Section 3-2).

Small amounts of oxygen can be produced in the laboratory by heating solid potassium chlorate, $KClO_3$, in the presence of a catalyst, manganese(IV) oxide, MnO_2. Solid potassium chloride, KCl, is also produced. (CAUTION: Heating $KClO_3$ can be dangerous.)

$$2KClO_3(s) \xrightarrow[\text{heat}]{MnO_2} 2KCl(s) + 3O_2(g)$$

| 2 mol | 2 mol | 3 mol |
| 2(122.6 g) | 2(74.6 g) | 3(22.4 L_{STP}) |

Unit factors can be constructed using any two of these quantities.

Production of a gas by a reaction.

$$2NaOH(aq) + 2Al(s) + 6H_2O(\ell) \longrightarrow$$
$$2Na[Al(OH)_4](aq) + 3H_2(g)$$

This reaction is used in some solid drain cleaners.

EXAMPLE 12-20 *Gas Volume in a Chemical Reaction*

What volume of O_2 (STP) can be produced by heating 112 grams of $KClO_3$?

Plan

The preceding equation shows that two moles of $KClO_3$ produce three moles of O_2. We construct appropriate unit factors from the balanced equation and the standard molar volume of oxygen to solve the problem.

Solution

$$\underline{?}\,L_{STP}\,O_2 = 112\text{ g }KClO_3 \times \frac{1\text{ mol }KClO_3}{122.6\text{ g }KClO_3} \times \frac{3\text{ mol }O_2}{2\text{ mol }KClO_3} \times \frac{22.4\text{ }L_{STP}\text{ }O_2}{1\text{ mol }O_2}$$

$$= \boxed{30.7\text{ }L_{STP}\text{ }O_2}$$

This calculation shows that the thermal decomposition of 112 grams of $KClO_3$ produces 30.7 liters of oxygen measured at standard conditions.

You should now work Exercise 74.

EXAMPLE 12-21 *Gas Volume in a Chemical Reaction*

A 1.80-gram mixture of potassium chlorate, $KClO_3$, and potassium chloride, KCl, was heated until all of the $KClO_3$ had decomposed. After being dried, the liberated oxygen occupied 405 mL at 25°C when the barometric pressure was 745 torr. (a) How many moles of O_2 were produced? (b) What percentage of the mixture was $KClO_3$?

Plan

(a) The number of moles of O_2 produced can be calculated from the ideal gas equation. (b) Then we use the balanced chemical equation to relate the known number of moles of O_2 formed and the mass of $KClO_3$ that decomposed to produce it.

Solution

(a) $V = 405$ mL $= 0.405$ L; $P = 745$ torr $\times \dfrac{1 \text{ atm}}{760 \text{ torr}} = 0.980$ atm

$T = 25°C + 273° = 298$ K

Solving the ideal gas equation for n and evaluating gives

$$n = \frac{PV}{RT} = \frac{(0.980 \text{ atm})(0.405 \text{ L})}{\left(0.0821 \dfrac{\text{L·atm}}{\text{mol·K}} (298 \text{ K})\right)} = 0.0162 \text{ mol } O_2$$

(b) $\underline{?}$ g $KClO_3 = 0.0162$ mol $O_2 \times \dfrac{2 \text{ mol } KClO_3}{3 \text{ mol } O_2} \times \dfrac{122.6 \text{ g } KClO_3}{1 \text{ mol } KClO_3} = 1.32$ g $KClO_3$

The sample contained 1.32 grams of $KClO_3$. The percent of $KClO_3$ in the sample is

$$\% \ KClO_3 = \frac{\text{g } KClO_3}{\text{g sample}} \times 100\% = \frac{1.32 \text{ g}}{1.80 \text{ g}} \times 100\% = 73.3\% \ KClO_3$$

You should now work Exercise 78.

Our study of stochiometry has shown that substances react in definite mole and mass proportions. Using previously discussed gas laws, we can show that gases also react in simple, definite proportions by volume. For example, *one* volume of hydrogen always combines (reacts) with *one* volume of chlorine to form *two* volumes of hydrogen chloride, if all volumes are measured at the same temperature and pressure

$$H_2(g) \ + \ Cl_2(g) \ \longrightarrow \ 2HCl(g)$$
$$1 \text{ volume} + 1 \text{ volume} \longrightarrow 2 \text{ volumes}$$

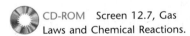 CD-ROM   Screen 12.7, Gas Laws and Chemical Reactions.

Volumes may be expressed in any units as long as the same unit is used for all. Gay-Lussac summarized several experimental observations on combining volumes of gases. The summary is known as **Gay-Lussac's Law of Combining Volumes:**

> At constant temperature and pressure, the volumes of reacting gases can be expressed as a ratio of simple whole numbers.

The ratio is obtained from the coefficients in the balanced equation for the reaction. Clearly, the law applies only to *gaseous* substances at the same temperature and pressure. No generalizations can be made about the volumes of solids and liquids as they undergo chemical reactions. Consider the following examples, based on experimental observations at constant temperature and pressure. Hundreds more could be cited.

1. One volume of nitrogen can react with three volumes of hydrogen to form two volumes of ammonia

$$N_2(g) \ + \ 3H_2(g) \ \longrightarrow \ 2NH_3(g)$$
1 volume + 3 volumes ⟶ 2 volumes

2. One volume of methane reacts with (burns in) two volumes of oxygen to give one volume of carbon dioxide and two volumes of steam

$$CH_4(g) \ + \ 2O_2(g) \ \longrightarrow \ CO_2(g) + 2H_2O(g)$$
1 volume + 2 volumes ⟶ 1 volume + 2 volumes

3. Sulfur (a solid) reacts with one volume of oxygen to form one volume of sulfur dioxide

$$S(s) + \ O_2(g) \ \longrightarrow \ SO_2(g)$$
1 volume ⟶ 1 volume

4. Four volumes of ammonia burn in five volumes of oxygen to produce four volumes of nitric oxide and six volumes of steam

$$4NH_3(g) \ + \ 5O_2(g) \ \longrightarrow \ 4\ NO(g) + 6H_2O(g)$$
4 volumes   5 volumes ⟶ 4 volumes   6 volumes

12-13 THE KINETIC–MOLECULAR THEORY

As early as 1738, Daniel Bernoulli (1700–1782) envisioned gaseous molecules in ceaseless motion striking the walls of their container and thereby exerting pressure. In 1857, Rudolf Clausius (1822–1888) published a theory that attempted to explain various experimental observations that had been summarized by Boyle's, Dalton's, Charles's, and Avogadro's laws. The basic assumptions of the **kinetic–molecular theory** for an ideal gas follow.

1. Gases consist of discrete molecules. The individual molecules are very small and are very far apart relative to their own sizes.

2. The gas molecules are in continuous, random, straight-line motion with varying velocities (see Figure 12-8).

3. The collisions between gas molecules and with the walls of the container are elastic; the total energy is conserved during a collision; that is, there is no net energy gain or loss.

4. Between collisions, the molecules exert no attractive or repulsive forces on one another; instead, each molecule travels in a straight line with a constant velocity.

Kinetic energy is the energy a body possesses by virtue of its motion. It is $\frac{1}{2}mu^2$, where m, the body's mass, can be expressed in grams and u, its velocity, can be expressed in meters per second (m/s). The assumptions of the kinetic–molecular theory can be used to relate temperature and molecular kinetic energy (see the Enrichment section, pages 459–461).

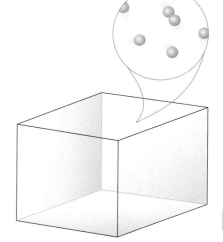

Figure 12-8 A representation of molecular motion. Gaseous molecules, in constant motion, undergo collisions with one another and with the walls of the container.

1. The observation that gases can be easily compressed indicates that the molecules are far apart. At ordinary temperatures and pressures, the gas molecules themselves occupy an insignificant fraction of the total volume of the container.

2. Near temperatures and pressures at which a gas liquefies, the gas does not behave ideally (Section 12-15) and attractions or repulsions among gas molecules *are* significant.

3. At any given instant, only a small fraction of the molecules are involved in collisions.

The average kinetic energy of gaseous molecules is directly proportional to the absolute temperature of the sample. The average kinetic energies of molecules of different gases are equal at a given temperature.

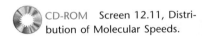

TOPIC: Kinetic Theory
GO TO: www.scilinks.org
*sci***LINKS CODE:** WCH1220

For instance, in samples of H_2, He, CO_2, and SO_2 at the same temperature, all the molecules have the same average kinetic energies. But the lighter molecules, H_2 and He, have much higher average velocities than do the heavier molecules, CO_2 and SO_2, at the same temperature.

We can summarize this very important result from the kinetic–molecular theory.

$$\text{average molecular } KE = \overline{KE} \propto T$$

or

$$\text{average molecular speed} = \bar{u} \propto \sqrt{\frac{T}{\text{molecular weight}}}$$

A bar over a quantity denotes an *average* of that quantity.

Molecular kinetic energies of gases increase with increasing temperature and decrease with decreasing temperature. We have referred only to the *average* kinetic energy; in a given sample, some molecules may be moving quite rapidly while others are moving more slowly. Figure 12-9 shows the distribution of speeds of gaseous molecules at two temperatures.

The kinetic–molecular theory satisfactorily explains most of the observed behavior of gases in terms of molecular behavior. Let's look at the gas laws in terms of the kinetic–molecular theory.

CD-ROM Screen 12.11, Distribution of Molecular Speeds.

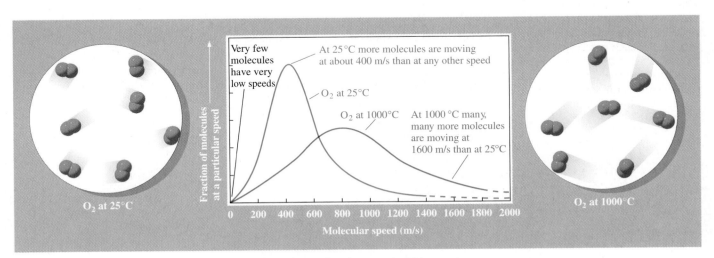

Figure 12-9 The Maxwellian distribution function for molecular speeds. This graph shows the relative numbers of O_2 molecules having a given speed at 25°C and at 1000°C. At 25°C, most O_2 molecules have speeds between 200 and 600 m/s (450–1350 miles per hour). Some of the molecules have very high speeds, so the distribution curve never reaches the horizontal axis. The average molecular speed is higher at 1000°C than at 25°C.

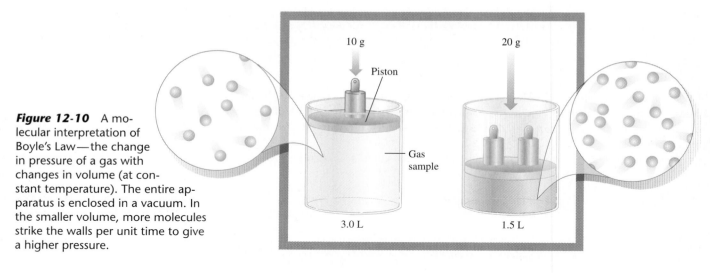

Figure 12-10 A molecular interpretation of Boyle's Law—the change in pressure of a gas with changes in volume (at constant temperature). The entire apparatus is enclosed in a vacuum. In the smaller volume, more molecules strike the walls per unit time to give a higher pressure.

Boyle's Law

The pressure exerted by a gas on the walls of its container is caused by gas molecules striking the walls. Clearly, pressure depends on two factors: (1) the number of molecules striking the walls per unit time and (2) how vigorously the molecules strike the walls. If the temperature is held constant, the average speed and the force of the collisions remain the same. But halving the volume of a sample of gas doubles the pressure because twice as many molecules strike a given area on the walls per unit time. Likewise, doubling the volume of a sample of gas halves the pressure because only half as many gas molecules strike a given area on the walls per unit time (Figure 12-10).

CD-ROM Screen 12.10, Gas Laws and the Kinetic–Molecular Theory.

Dalton's Law

Figure 12-11 A molecular interpretation of Dalton's Law. The molecules act independently in the mixture, so each gas exerts its own partial pressure due to its molecular collisions with the walls. The total gas pressure is the sum of the partial pressures of the component gases.

In a gas sample the molecules are very far apart and do not attract one another significantly. Each kind of gas molecule acts independently of the presence of the other kind. The molecules of each gas thus collide with the walls with a frequency and vigor that do not change even if other molecules are present (Figure 12-11). As a result, each gas exerts a partial

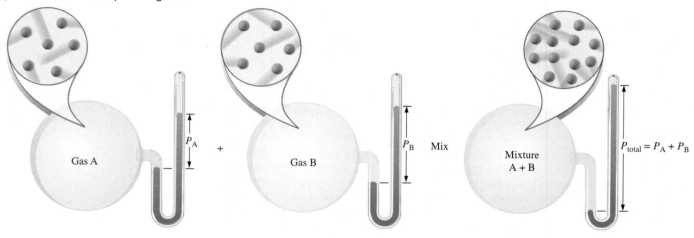

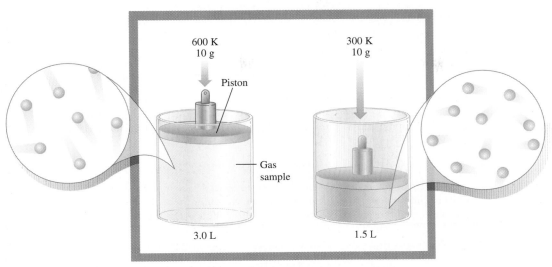

Figure 12-12 A molecular interpretation of Charles's Law—the change in volume of a gas with changes in temperature (at constant pressure). At the lower temperature, molecules strike the walls less often and less vigorously. Thus, the volume must be less to maintain the same pressure.

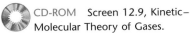

CD-ROM Screen 12.9, Kinetic–Molecular Theory of Gases.

pressure that is independent of the presence of the other gas, and the total pressure is due to the sum of all the molecule–wall collisions.

Charles's Law

Recall that average kinetic energy is directly proportional to the absolute temperature. Doubling the *absolute* temperature of a sample of gas doubles the average kinetic energy of the gaseous molecules, and the increased force of the collisions of molecules with the walls doubles the volume at constant pressure. Similarly, halving the absolute temperature decreases kinetic energy to half its original value; at constant pressure, the volume decreases by half because of the reduced vigor of the collision of gaseous molecules with the container walls (Figure 12-12).

E n r i c h m e n t   **Kinetic–Molecular Theory, the Ideal Gas Equation, and Molecular Speeds**

In 1738, Daniel Bernoulli derived Boyle's Law from Newton's laws of motion applied to gas molecules. This derivation was the basis for an extensive mathematical development of the kinetic–molecular theory more than a century later by Clausius, Maxwell, Boltzmann, and others. Although we do not need to study the detailed mathematical presentation of this theory, we can gain some insight into its concepts from the reasoning behind Bernoulli's derivation. Here we present that reasoning based on proportionality arguments.

In the kinetic–molecular theory pressure is viewed as the result of collisions of gas molecules with the walls of the container. As each molecule strikes a wall, it exerts a small impulse. The pressure is the total force thus exerted on the walls divided by the area of the walls. The total force on the walls (and thus the pressure) is proportional to two factors:

(Enrichment, continued)

(1) the impulse exerted by each collision and (2) the rate of collisions (number of collisions in a given time interval).

$$P \propto (\text{impulse per collision}) \times (\text{rate of collisions})$$

Recall that momentum is mass × speed.

Let us represent the mass of an individual molecule by m and its speed by u. The heavier the molecule is (greater m) and the faster it is moving (greater u), the harder it pushes on the wall when it collides. The impulse due to each molecule is proportional to its *momentum, mu*.

$$\text{impulse per collision} \propto mu$$

The rate of collisions, in turn, is proportional to two factors. First, the rate of collision must be proportional to the molecular speed; the faster the molecules move, the more often they reach the wall to collide. Second, this collision rate must be proportional to the number of molecules per unit volume, N/V. The greater the number of molecules, N, in a given volume, the more molecules collide in a given time interval.

$$\text{rate of collisions} \propto (\text{molecular speed}) \times (\text{molecules per unit volume})$$

or

$$\text{rate of collisions} \propto (u) \times \left(\frac{N}{V}\right)$$

We can introduce these proportionalities into the one describing pressure, to conclude that

$$P \propto (mu) \times u \times \frac{N}{V} \quad \text{or} \quad P \propto \frac{Nmu^2}{V} \quad \text{or} \quad PV \propto Nmu^2$$

$\overline{u^2}$ is the average of the squares of the molecular speeds. It is proportional to the square of the average speed, but the two quantities are not equal.

At any instant not all molecules are moving at the same speed, u. We should reason in terms of the *average* behavior of the molecules, and express the quantity u^2 in average terms as $\overline{u^2}$, the **mean-square speed**.

$$PV \propto Nm\overline{u^2}$$

Not all molecules collide with the walls at right angles, so we must average (using calculus) over all the trajectories. This gives a proportionality constant of $\frac{1}{3}$, and

$$PV = \tfrac{1}{3}Nm\overline{u^2}$$

This describes the quantity PV (pressure × volume) in terms of *molecular quantities*—number of molecules, molecular masses, and molecular speeds. The number of molecules, N, is given by the number of moles, n, times Avogadro's number, N_{Av}, or $N = nN_{Av}$. Making this substitution, we obtain

$$PV = \tfrac{1}{3}nN_{Av}m\overline{u^2}$$

The ideal gas equation describes (pressure × volume) in terms of *measurable quantities*—number of moles and absolute temperature.

$$PV = nRT$$

So we see that the ideas of the kinetic–molecular theory lead to an equation of the same form as the macroscopic ideal gas equation. Thus, the molecular picture of the theory is consistent with the ideal gas equation and gives support to the theory. Equating the right-hand sides of these last two equations and canceling n gives

$$\tfrac{1}{3}N_{Av}m\overline{u^2} = RT$$

This equation can also be written as

$$\tfrac{1}{3}N_{Av} \times (2 \times \tfrac{1}{2}m\overline{u^2}) = RT$$

From physics we know that the *kinetic energy* of a particle of mass m moving at speed u is $\tfrac{1}{2}mu^2$. So we can write

$$\tfrac{2}{3}N_{Av} \times (\text{avg } KE \text{ per molecule}) = RT$$

or

$$N_{Av} \times (\text{avg } KE \text{ per molecule}) = \tfrac{3}{2}RT$$

This equation shows that the absolute temperature is directly proportional to the average molecular kinetic energy, as postulated by the kinetic–molecular theory. Because there are N_{Av} molecules in a mole, the left-hand side of this equation is equal to the total kinetic energy of a mole of molecules.

$$\text{total kinetic energy per mole of gas} = \tfrac{3}{2}RT$$

With this interpretation, the total molecular–kinetic energy of a mole of gas depends *only* on the temperature, and not on the mass of the molecules or the gas density.

We can also obtain some useful equations for molecular speeds from the previous reasoning. Solving the equation

$$\tfrac{1}{3}N_{Av}m\overline{u^2} = RT$$

for **root-mean-square speed, $u_{rms} = \sqrt{\overline{u^2}}$,** we obtain

$$u_{rms} = \sqrt{\dfrac{3RT}{N_{Av}m}}$$

We recall that m is the mass of a single molecule. So $N_{Av}m$ is the mass of Avogadro's number of molecules, or one mole of substance; this is equal to the *molecular weight, M,* of the gas.

$$u_{rms} = \sqrt{\dfrac{3RT}{M}}$$

EXAMPLE 12-22   *Molecular Speed*

Calculate the root-mean-square speed of H_2 molecules in meters per second at 20°C. Recall that

$$1 \text{ J} = 1 \, \dfrac{\text{kg} \cdot \text{m}^2}{\text{s}^2}$$

Plan

We substitute the appropriate values into the equation relating u_{rms} to temperature and molecular weight. Remember that R must be expressed in the appropriate units.

$$R = 8.314 \, \dfrac{\text{J}}{\text{mol} \cdot \text{K}} = 8.314 \, \dfrac{\text{kg} \cdot \text{m}^2}{\text{mol} \cdot \text{K} \cdot \text{s}^2}$$

Solution

$$u_{rms} = \sqrt{\frac{3RT}{M}} = \sqrt{\frac{3 \times 8.314 \dfrac{kg \cdot m^2}{mol \cdot K \cdot s^2} \times 293\ K}{2.016 \dfrac{g}{mol} \times \dfrac{1\ kg}{1000\ g}}}$$

$$u_{rms} = \sqrt{3.62 \times 10^6\ m^2/s^2} = \boxed{1.90 \times 10^3\ m/s} \qquad \text{(about 4250 mph)}$$

You should now work Exercise 86.

12-14   DIFFUSION AND EFFUSION OF GASES

Effusion is the escape of a gas through a tiny hole. **Diffusion** is the movement of a substance (for example, a gas) into a space or the mixing of one substance (for example, a gas) with another.

Because gas molecules are in constant, rapid, random motion, they diffuse quickly throughout any container (Figure 12-13). For example, if hydrogen sulfide (the smell of rotten eggs) is released in a large room, the odor can eventually be detected throughout the room. If a mixture of gases is placed in a container with thin porous walls, the molecules effuse through the walls. Because they move faster, lighter gas molecules effuse through the tiny openings of porous materials faster than heavier molecules (Figure 12-14).

Although they are the most abundant elements in the universe, hydrogen and helium occur as gases only in trace amounts in our atmosphere. This is due to the high average molecular speeds resulting from their low molecular weights. At temperatures in our atmosphere, these molecules reach speeds exceeding the escape velocity required for them to break out of the earth's gravitational pull and diffuse into interplanetary space. Thus, most of the gaseous hydrogen and helium that were probably present in large concentrations in the earth's early atmosphere have long since diffused away. The same is true for the abundance of these gases on other small planets in our solar system, especially those with higher average temperatures than ours (Mercury and Venus). The Mariner 10 spacecraft in 1974 revealed measurable amounts of He in the atmosphere of Mercury; the source of this helium is unknown. Massive bodies such as stars (including our own sun) are mainly composed of H and He.

CD-ROM   Screen 12.12, Applications of the Kinetic–Molecular Theory.

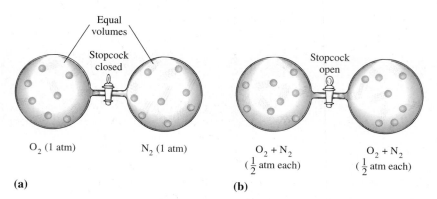

(a)   **(b)**

Figure 12-13 A representation of diffusion of gases. The space between the molecules allows for ease of mixing one gas with another. Collisions of molecules with the walls of the container are responsible for the pressure of the gas.

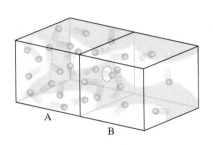

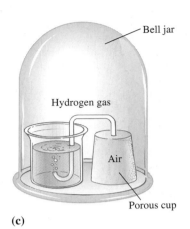

(a)

(b)

(c)

Figure 12-14 Effusion of gases. (a) A molecular interpretation of effusion. Molecules are in constant motion; occasionally they strike the opening and escape. (b) Latex balloons were filled with the same volume of He (*yellow*), N_2 (*blue*), and O_2 (*red*). Lighter molecules, such as He, effuse through the tiny pores of the latex balloons more rapidly than does N_2 or O_2. The silver party balloon is made of a metal-coated polymer with pores that are too small to allow rapid He effusion. (c) If a bell jar full of hydrogen is brought down over a porous cup full of air, rapidly moving hydrogen effuses into the cup faster than the oxygen and nitrogen in the air can effuse out of the cup. This causes an increase in pressure in the cup sufficient to produce bubbles in the water in the beaker.

12-15 REAL GASES: DEVIATIONS FROM IDEALITY

Until now our discussions have dealt with *ideal* behavior of gases. By this we mean that the identity of a gas does not affect how it behaves, and the same equations should work equally well for all gases. Under ordinary conditions most *real* gases do behave ideally; their P and V are predicted by the ideal gas laws, so they do obey the postulates of the kinetic–molecular theory. According to the kinetic–molecular model, (1) all but a negligible volume of a gas sample is empty space, and (2) the molecules of *ideal* gases do not attract one another because they are so far apart relative to their own sizes.

Under some conditions, however, most gases can have pressures and/or volumes that are *not* accurately predicted by the ideal gas laws. Figure 12-15 illustrates the nonideal behavior of several gases. The ratio of PV/nRT versus pressure should equal 1.0 for an ideal gas. Most gases, however, show marked deviations from ideal behavior at high pressures. This tells us that they are not behaving entirely as postulated by the kinetic–molecular theory.

Nonideal gas behavior (deviation from the predictions of the ideal gas laws) is most significant at *high pressures* and/or *low temperatures*, that is, near the conditions under which the gas liquefies.

Johannes van der Waals (1837–1923) studied deviations of real gases from ideal behavior. In 1867, he empirically adjusted the ideal gas equation

$$P_{ideal}V_{ideal} = nRT$$

to take into account two complicating factors.

NH$_3$ gas (*left*) and HCl gas (*right*) escape from concentrated aqueous solutions. The white smoke (solid NH$_4$Cl) shows where the gases mix and react.

$$NH_3(g) + HCl(g) \longrightarrow NH_4Cl(s)$$

Figure 12-15 The nonideal behavior of real gases compared with ideal behavior. For a gas that behaves ideally, $PV = nRT$ at all pressures, so $PV/nRT = 1$ at all pressures (horizontal line). We can test a gas for ideal behavior by measuring P, V, n, and T for a sample of the gas at various pressures and then calculating PV/nRT. This plot shows that different gases deviate differently from ideal behavior, and that the deviations from ideality become more pronounced at higher pressures.

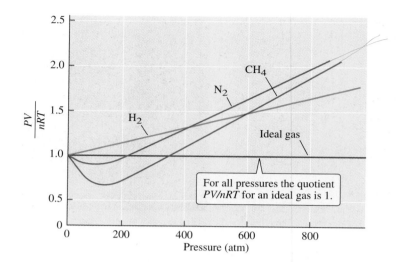

1. According to the kinetic–molecular theory, the molecules are so small, relative to the total volume of the gas, that each molecule can move through virtually the entire *measured volume* of the container, $V_{measured}$ (Figure 12-16a). But under high pressures, a gas is compressed so that the volume of the molecules themselves becomes a significant fraction of the total volume occupied by the gas. As a result, the *available volume*, $V_{available}$, for any molecule to move in is less than the *measured volume* by an amount that depends on the volume excluded by the presence of the other molecules (Figure 12-16b). To account for this, we subtract a correction factor, nb.

$$V_{ideally\ available} = V_{measured} - nb$$

The factor nb corrects for the volume occupied by the molecules themselves. Larger molecules have greater values of b, and the greater the number of molecules in a sample (higher n), the larger is the volume correction. The correction term becomes negligibly small, however, when the volume is large (or the pressure is low).

2. The kinetic–molecular theory describes pressure as resulting from molecular collisions with the walls of the container; this theory assumes that attractive forces between molecules are insignificant. For any real gas, the molecules can attract one

Figure 12-16 A molecular interpretation of deviations from ideal behavior. (a) A sample of gas at a low temperature. Each sphere represents a molecule. Because of their low kinetic energies, attractive forces between molecules can now cause a few molecules to "stick together." (b) A sample of gas under high pressure. The molecules are quite close together. The free volume is now a much smaller fraction of the total volume.

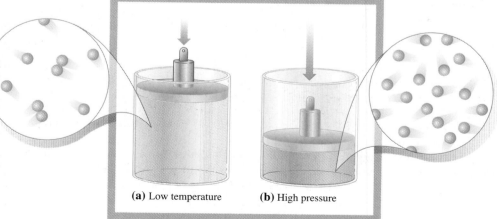

(a) Low temperature   (b) High pressure

another. But at higher temperatures, the potential energy due to intermolecular attractions is negligibly small compared with the high kinetic energy due to the rapid motion of the molecules and to the great distances between them. When the temperature is quite low (low kinetic energy), the molecules move so slowly that the potential energy due to even small attractive forces *does* become important. This perturbation becomes even more important when the molecules are very close together (at high pressure). As a result, the molecules deviate from their straight-line paths and take longer to reach the walls, so fewer collisions take place in a given time interval. Furthermore for a molecule about to collide with the wall, the attraction by its neighbors causes the collision to be less energetic than it would otherwise be (Figure 12-17). As a consequence, the pressure that the gas exerts, P_{measured}, is less than the pressure it would exert if attractions were truly negligible, $P_{\text{ideally exerted}}$. To correct for this, we subtract a correction factor, n^2a/V^2, from the ideal pressure.

$$P_{\text{measured}} = P_{\text{ideally exerted}} - \frac{n^2a}{V^2_{\text{measured}}}$$

or

$$P_{\text{ideally exerted}} = P_{\text{measured}} + \frac{n^2a}{V^2_{\text{measured}}}$$

In this correction term, large values of a indicate strong attractive forces. When more molecules are present (greater n) and when the molecules are close together (smaller V^2 in the denominator), the correction term becomes larger. The correction term becomes negligibly small, however, when the volume is large.

When we substitute these two expressions for corrections into the ideal gas equation, we obtain the equation

$$\left(P_{\text{measured}} + \frac{n^2a}{V^2_{\text{measured}}}\right)(V_{\text{measured}} - nb) = nRT$$

or

$$\left(P + \frac{n^2a}{V^2}\right)(V - nb) = nRT$$

This is the **van der Waals equation.** In this equation, P, V, T, and n represent the *measured* values of pressure, volume, temperature (expressed on the absolute scale), and number of moles, respectively, just as in the ideal gas equation. The quantities a and b are experimentally derived constants that differ for different gases (Table 12-5). When a and b are both zero, the van der Waals equation reduces to the ideal gas equation.

We can understand the relative values of a and b in Table 12-5 in terms of molecular properties. Note that a for helium is very small. This is the case for all noble gases and many other nonpolar molecules, because only very weak attractive forces, called dispersion forces, exist between them. **Dispersion forces** result from short-lived electrical dipoles produced by the attraction of one atom's nucleus for an adjacent atom's electrons. These forces exist for all molecules but are especially important for nonpolar molecules, which would never liquefy if dispersion forces did not exist. Polar molecules such as ammonia, NH_3, have permanent charge separations (dipoles), so they exhibit greater forces of attraction for one another. This explains the high value of a for ammonia. Dispersion forces and permanent dipole forces of attraction are discussed in more detail in Chapter 13.

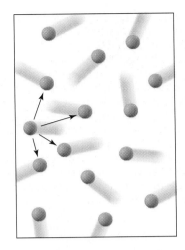

Figure 12-17 A gas molecule strikes the walls of a container with diminished force. The attractive forces between a molecule and its neighbors are significant.

The van der Waals equation, like the ideal gas equation, is known as an *equation of state,* that is, an equation that describes a state of matter.

TABLE 12-5	van der Waals Constants	
Gas	a (L²·atm/mol²)	b (L/mol)
H_2	0.244	0.0266
He	0.034	0.0237
N_2	1.39	0.0391
NH_3	4.17	0.0371
CO_2	3.59	0.0427
CH_4	2.25	0.0428

Larger molecules have greater values of b. For instance, H_2, a first-row diatomic molecule, has a greater b value than the first-row monatomic He. The b value for CO_2, which contains three second-row atoms, is greater than that for N_2, which contains only two second-row atoms.

The following example illustrates the deviation of methane, CH_4, from ideal gas behavior under high pressure.

EXAMPLE 12-23   van der Waals Equation

Calculate the pressure exerted by 1.00 mole of methane, CH_4, in a 500.-mL vessel at 25.0°C assuming (a) ideal behavior and (b) nonideal behavior.

Plan

(a) Ideal gases obey the ideal gas equation. We can solve this equation for P.

(b) To describe methane as a nonideal gas, we use the van der Waals equation and solve for P.

Solution

(a) Using the ideal gas equation to describe ideal gas behavior,

$$PV = nRT$$

$$P = \frac{nRT}{V} = \frac{(1.00 \text{ mol})\left(\dfrac{0.0821 \text{ L} \cdot \text{atm}}{\text{mol} \cdot \text{K}}\right)(298 \text{ K})}{0.500 \text{ L}} = \boxed{48.9 \text{ atm}}$$

(b) Using the van der Waals equation to describe nonideal gas behavior,

$$\left(P + \frac{n^2 a}{V^2}\right)(V - nb) = nRT$$

For CH_4, $a = 2.25 \text{ L}^2 \cdot \text{atm/mol}^2$ and $b = 0.0428 \text{ L/mol}$ (see Table 12-5).

$$\left[P + \frac{(1.00 \text{ mol})^2(2.25 \text{ L}^2 \cdot \text{atm/mol}^2)}{(0.500 \text{ L})^2}\right]\left[0.500 \text{ L} - (1.00 \text{ mol})\left(0.0428 \frac{\text{L}}{\text{mol}}\right)\right]$$

$$= (1.00 \text{ mol})\left(\frac{0.0821 \text{ L} \cdot \text{atm}}{\text{mol} \cdot \text{K}}\right)(298 \text{ K})$$

Combining terms and canceling units, we get

$$P + 9.00 \text{ atm} = \frac{24.5 \text{ L} \cdot \text{atm}}{0.457 \text{ L}} = 53.6 \text{ atm}$$

$$P = \boxed{44.6 \text{ atm}}$$

You should now work Exercises 94 and 95.

The pressure is 4.3 atm (8.8%) less than that calculated from the ideal gas law. A significant error would be introduced by assuming ideal behavior at this high pressure.

Repeating the calculations of Example 12-23 with the volume twenty times higher ($V = 10.0$ L) gives ideal and nonideal pressures, respectively, of 2.45 and 2.44 atm, a difference of only 0.4%.

Many other equations have been developed to describe the behavior of real gases. Each of these contains quantities that must be empirically derived for each gas.

Key Terms

Absolute zero   The zero point on the absolute temperature scale; −273.15°C or 0 K; theoretically, the temperature at which molecular motion is a minimum.

Atmosphere (atm)   A unit of pressure; the pressure that will support a column of mercury 760 mm high at 0°C; 760 torr.

Avogadro's Law   At the same temperature and pressure, equal volumes of all gases contain the same number of molecules.

Bar   A unit of pressure; 1.00 bar is equal to 100. kPa (or 0.987 atm).

Barometer   A device for measuring atmospheric pressure. See Figures 12-1 and 12-2. The liquid is usually mercury.

Boyle's Law   At constant temperature, the volume occupied by a given mass of a gas is inversely proportional to the applied pressure.

Charles's Law   At constant pressure, the volume occupied by a definite mass of a gas is directly proportional to its absolute temperature.

Condensed states   The solid and liquid states.

Dalton's Law of Partial Pressures   The total pressure exerted by a mixture of gases is the sum of the partial pressures of the individual gases.

Diffusion   The movement of a substance (e.g., a gas) into a space or the mixing of one substance (e.g., a gas) with another.

Dispersion forces   Weak, short-range attractive forces between short-lived temporary dipoles.

Effusion   The escape of a gas through a tiny hole or a thin porous wall.

Fluids   Substances that flow freely; gases and liquids.

Gay-Lussac's Law of Combining Volumes   At constant temperature and pressure, the volumes of reacting gases (and any gaseous products) can be expressed as ratios of small whole numbers.

Ideal gas   A hypothetical gas that obeys exactly all postulates of the kinetic–molecular theory.

Ideal Gas Equation   The product of the pressure and volume of an ideal gas is directly proportional to the number of moles of the gas and the absolute temperature.

Kinetic–molecular theory   A theory that attempts to explain macroscopic observations on gases in microscopic or molecular terms.

Manometer   A two-armed barometer. See Figure 12-1.

Mole fraction   The number of moles of a component of a mixture divided by the total number of moles in the mixture.

Partial pressure   The pressure exerted by one gas in a mixture of gases.

Pascal (Pa)   The SI unit of pressure; it is defined as the pressure exerted by a force of one newton acting on an area of one square meter.

Pressure   Force per unit area.

Real gases   Gases that deviate from ideal gas behavior.

Root-mean-square speed, u_{rms}   The square root of the mean-square speed, $\sqrt{\overline{u^2}}$. This is equal to $\sqrt{\dfrac{3RT}{M}}$ for an ideal gas. The root-mean-square speed is slightly different from the average speed, but the two quantities are proportional.

Standard molar volume   The volume occupied by one mole of an ideal gas under standard conditions; 22.414 liters.

Standard temperature and pressure (STP)   Standard temperature 0°C (273.15 K), and standard pressure, one atmosphere, are standard conditions for gases.

Torr   A unit of pressure; the pressure that will support a column of mercury 1 mm high at 0°C.

Universal gas constant R, the proportionality constant in the ideal gas equation, $PV = nRT$.

van der Waals equation   An equation of state that extends the ideal gas law to real gases by inclusion of two empirically determined parameters, which are different for different gases.

Vapor   A gas formed by boiling or evaporation of a liquid or sublimation of a solid; a term commonly used when some of the liquid or solid remains in contact with the gas.

Vapor pressure   The pressure exerted by a vapor in equilibrium with its liquid or solid.

Exercises

You may assume *ideal gas behavior* unless otherwise indicated.

Basic Ideas

1. Define pressure. Give a precise scientific definition—one that can be understood by someone without any scientific training.

2. State whether each property is characteristic of all gases, some gases, or no gas: (a) transparent to light; (b) colorless; (c) unable to pass through filter paper; (d) more difficult to compress than liquid water; (e) odorless; (f) settles on standing.

3. Describe the mercury barometer. How does it work?

Charles D. Winters

4. What is a manometer? How does it work?
5. Express a pressure of 675 torr in the following units: (a) mm Hg; (b) atm; (c) Pa; (d) kPa.
6. A typical laboratory atmospheric pressure reading is 752 torr. Convert this value to (a) psi, (b) cm Hg, (c) inches Hg, (d) kPa, (e) atm, and (f) ft H_2O.
7. Complete the following table.

	atm	torr	Pa	kPa
Standard atmosphere	1			
Partial pressure of nitrogen in the atmosphere		593		
A tank of compressed hydrogen			1.61×10^5	
Atmospheric pressure at the summit of Mt. Everest				33.7

8. State whether each of the following samples of matter is a gas. If the information is insufficient for you to decide, write "insufficient information." (a) A material is in a steel tank at 100. atm pressure. When the tank is opened to the atmosphere, the material immediately expands, increasing its volume many-fold. (b) A material, on being emitted from an industrial smokestack, rises about 10 m into the air. Viewed against a clear sky, it has a white appearance. (c) 1.0 mL of material weighs 8.2 g. (d) When a material is released from a point 30 ft below the level of a lake at sea level (equivalent in water pressure to about 76 cm of mercury), it rises rapidly to the surface, at the same time doubling its volume. (e) A material is transparent and pale green in color. (f) One cubic meter of a material contains

as many molecules as 1 m^3 of air at the same temperature and pressure.
*9. The densities of mercury and corn oil are 13.5 g/mL and 0.92 g/mL, respectively. If corn oil were used in a barometer, what would be the height of the column, in meters, at standard atmospheric pressure? (The vapor pressure of the oil is negligible.)
10. Steel tanks for storage of gases are capable of withstanding pressures greater than 125 atm. Express this pressure in psi.
11. Automobile tires are normally inflated to a pressure of 33 psi as measured by a tire gauge. (a) Express this pressure in atmospheres. (b) Assuming standard atmospheric pressure, calculate the internal pressure of the tire.

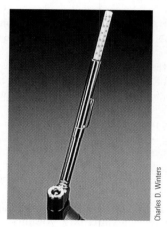

Charles D. Winters

Boyle's Law: The Pressure–Volume Relationship

12. (a) On what kinds of observations (measurements) is Boyle's Law based? State the law. (b) Use the statement of Boyle's Law to derive a simple mathematical expression for Boyle's Law.
13. Could the words "a fixed number of moles" be substituted for "a definite mass" in the statement of Boyle's Law? Explain.
14. A gas sample contained in a cylinder equipped with a moveable piston occupied 300. mL at a pressure of 2.00 atm. What would be the final pressure if the volume were increased to 550. mL at constant temperature?
15. A balloon that contains 1.50 liters of air at 1.00 atm is taken under water to a depth at which the pressure is 2.50 atm. Calculate the new volume of the balloon. Assume that the temperature remains constant.
16. A 50.-L sample of gas collected in the upper atmosphere at a pressure of 58.3 torr is compressed into a 150.-mL container at the same temperature. (a) What is the new pressure, in atmospheres? (b) To what volume would the original sample have had to be compressed to exert a pressure of 10.0 atm?

17. A sample of krypton gas occupies 75.0 mL at 0.400 atm. If the temperature remained constant, what volume would the krypton occupy at (a) 4.00 atm, (b) 0.0400 atm, (c) 765 torr, (d) 4.00 torr, and (e) 3.5×10^{-2} torr?

*18. A cylinder containing 15 L of helium gas at a pressure of 165 atm is to be used to fill toy balloons to a pressure of 1.1 atm. Each inflated balloon has a volume of 2.5 L. What is the maximum number of balloons that can be inflated? (Remember that 15 L of helium at 1.1 atm will remain in the "exhausted" cylinder.)

19. (a) Can an absolute temperature scale based on Fahrenheit rather than Celsius degrees be developed? Why? (b) Can an absolute temperature scale that is based on a "degree" twice as large as a Celsius degree be developed? Why?

20. (a) What does "absolute temperature scale" mean? (b) Describe the experiments that led to the evolution of the absolute temperature scale. What is the relationship between the Celsius and Kelvin temperature scales? (c) What does "absolute zero" mean?

21. Complete the table by making the required temperature conversions. Pay attention to significant figures.

	Temperature	
	K	**°C**
Normal boiling point of water		100
Reference for thermodynamic data	298.15	
Dry ice becomes a gas at atmospheric pressure		−78.5
The center of the sun (estimated)	1.53×10^7	

Charles's Law: The Volume–Temperature Relationship

22. (a) Why is a plot of volume versus temperature at constant pressure a straight line (see Figure 12-5)? (b) On what kind of observations (measurements) is Charles's Law based? State the law.

23. A gas occupies a volume of 31.0 L at 19.0°C. If the gas temperature rises to 38.0°C at constant pressure, (a) would you expect the volume to double to 62.0 L? Explain. Calculate the new volume (b) at 38.0°C, (c) at 400. K, and (d) at 0.00°C.

24. Several balloons are inflated with helium to a volume of 0.85 L at 27°C. One of the balloons was found several hours later; the temperature had dropped to 22°C. What would be the volume of the balloon when found, if no helium has escaped?

Charles D. Winters

25. Which of the following statements are true? Which are false? Why is each true or false? *Assume constant pressure* in each case. (a) If a sample of gas is heated from 100°C to 200°C, the volume will double. (b) If a sample of gas is heated from 0.°C to 273°C, the volume will double. (c) If a sample of gas is cooled from 1273°C to 500.°C, the volume will decrease by a factor of 2. (d) If a sample of gas is cooled from 1000.°C to 200.°C, the volume will decrease by a factor of 5. (e) If a sample of gas is heated from 473°C to 1219°C, the volume will increase by a factor of 2.

*26. The device shown here is a gas thermometer. (a) At the ice point, the gas volume is 1.400 L. What would be the new volume if the gas temperature were raised from the ice point to 8.0°C? (b) Assume the cross-sectional area of the graduated arm is 1.0 cm². What would be the difference in height if the gas temperature changed from 0°C to 8.0°C? (c) What modifications could be made to increase the sensitivity of the thermometer?

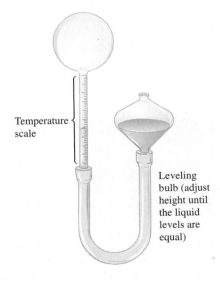

Temperature scale

Leveling bulb (adjust height until the liquid levels are equal)

27. A weather balloon is filled to the volume of 150. L on a day when the temperature is 19°C. If no gases escaped, what would be the volume of the weather balloon after it rises to an altitude where the temperature is −8°C?

28. Calculate the volume of an ideal gas at the temperatures of dry ice (−78.5°C), liquid N_2 (−195.8°C), and liquid He (−268.9°C) if it occupies 5.00 L at 25.0°C. Assume constant pressure. Plot your results, and extrapolate to zero volume. At what temperature would zero volume be theoretically reached?

The Combined Gas Law

29. Classify the relationship between the variables (a) P and V, (b) V and T, and (c) P and T as either (i) directly proportional or (ii) inversely proportional.

30. Prepare sketches of plots of (a) P vs. V, (b) P vs. $1/V$, (c) V vs. T, and (d) P vs. T for an ideal gas.

31. A sample of gas occupies 300. mL at STP. Under what pressure would this sample occupy 200. mL if the temperature were increased to 819°C?

32. A 380.-mL sample of neon exerts a pressure of 670. torr at 26°C. At what temperature in °C would it exert a pressure of 940. torr in a volume of 440. mL?

33. A 247-mL sample of a gas exerts a pressure of 2.75 atm at 16.0°C. What volume would it occupy at 100.°C and 1.00 atm?

34. Show how Boyle's and Charles's gas laws can be obtained from the combined gas law equation.

STP, Standard Molar Volume, and Gas Densities

35. How many molecules of an ideal gas are contained in a 2.50-L flask at STP?

36. (a) What is Avogadro's Law? What does it mean? (b) What does "standard molar volume" mean? (c) Are there conditions other than STP at which 1 mole of an ideal gas would occupy 22.4 L? Explain.

37. Using a ground-based telescope and a spectrometer, sodium vapor has been detected as a major component of the thin atmosphere of Mercury. Its concentration is

NASA/JPL

The surface of the planet Mercury.

estimated to be about 1.0×10^5 atoms per cm³. (a) Express this in moles per liter. (b) The maximum temperature of the atmosphere was measured by Mariner 10 to be about 970.°C. What is the approximate partial pressure of sodium vapor at that temperature?

38. Ethylene dibromide (EDB) was formerly used as a fumigant for fruits and grains, but now it is banned because it is a potential health hazard. EDB is a liquid that boils at 109°C. Its molecular weight is 188 g/mol. Calculate the density of its vapor at 165°C and 1.00 atm.

39. Nitrogen is slightly less dense than is a sample of air at the same temperature and pressure. Calculate the density of N_2, in g/L, at 1.25 atm and 35°C. If the average molecular weight of the air is 29.2, what is the density of air at the same conditions?

40. A laboratory technician forgot what the color coding on some commercial cylinders of gas meant, but remembered that each of two specific tanks contained one of the following gases: He, Ne, Ar, or Kr. Measurements at STP made on samples of the gases from the two cylinders showed the gas densities to be 3.74 g/L and 0.178 g/L. (a) Determine by calculation which of these gases was present in each tank. (b) Could this determination be made if the densities had been at a temperature and pressure different from STP?

The Ideal Gas Equation

41. Calculate R in L·atm/mol·K, in kPa·dm³/mol·K, in J/mol·K, and in kJ/mol·K.

42. (a) What is an ideal gas? (b) What is the ideal gas equation? (c) Outline the logic used to obtain the ideal gas equation. (d) What is R? How is it obtained?

43. (a) A chemist is preparing to carry out a reaction at high pressure that requires 36.0 mol of hydrogen gas. The chemist pumps the hydrogen into a 12.3-L rigid steel container at 25°C. To what pressure (in atmospheres) must the hydrogen be compressed? (b) What would be the density of the high-pressure hydrogen?

44. Calculate the pressure needed to contain 2.54 mol of an ideal gas at 45°C in a volume of 3.75 L.

45. (a) How many molecules are in a 500.-mL container of gaseous oxygen if the pressure is 2.50×10^{-9} torr and the temperature is 1225 K? (b) How many grams of oxygen are in the container?

*46. A barge containing 580. tons of liquid chlorine was involved in an accident. (a) What volume would this amount of chlorine occupy if it were all converted to a gas at 750. torr and 18°C? (b) Assume that the chlorine is confined to a width of 0.500 mile and an average depth of 60. ft. What would be the length, in feet, of this chlorine "cloud"?

Molecular Weights and Formulas for Gaseous Compounds

47. Analysis of a volatile liquid shows that it contains 37.23% carbon, 7.81% hydrogen, and 54.96% chlorine by mass. At 150.°C and 1.00 atm, 500. mL of the vapor has a mass of 0.922 g. (a) What is the molecular weight of the compound? (b) What is its molecular formula?

*48. A student was given a container of ethane, C_2H_6, that had been closed at STP. By making appropriate measurements, the student found that the mass of the sample of ethane was 0.244 g and the volume of the container was 185 mL. Use the student's data to calculate the molecular weight of ethane. What percent error is obtained? Suggest some possible sources of the error.

49. Calculate the molecular weight of a gaseous element if 0.480 g of the gas occupies 367 mL at 365 torr and 45°C. Suggest the identity of the element.

50. A cylinder was found in a storeroom of a manufacturing plant. The label on the cylinder was gone and no one remembered what the cylinder held. A 0.00500-gram sample was found to occupy 4.13 mL at 23°C and 745 torr. The sample was also found to be composed of only carbon and hydrogen. Identify the gas.

51. A sample of porous rock was brought back from the planet Farout on the other side of the galaxy. Trapped in the rock was a carbon-oxygen gas. The unknown gas was extracted and evaluated. A volume of 3.70 mL of the gas was collected under the conditions of STP. The mass of the gas sample was determined to be 0.00726 grams. Additional analysis proved that there was only one compound present in the gas sample. What was the most probable identity of the gas based on these analyses?

*52. A highly volatile liquid was allowed to vaporize completely into a 250.-mL flask immersed in boiling water. From the following data, calculate the molecular weight (in amu per molecule) of the liquid. Mass of empty flask = 65.347 g; mass of flask filled with water at room temperature = 327.4 g; mass of flask and condensed liquid = 65.739 g; atmospheric pressure = 743.3 torr; temperature of boiling water = 99.8°C; density of water at room temperature = 0.997 g/mL.

53. A pure gas contains 85.63% carbon and 14.37% hydrogen by mass. Its density is 2.50 g/L at STP. What is its molecular formula?

Gas Mixtures and Dalton's Law

54. (a) What are partial pressures of gases? (b) State Dalton's Law. Express it symbolically.

55. A sample of oxygen of mass 30.0 g is confined in a vessel at 0°C and 1000. torr. Then 8.00 g of hydrogen is pumped into the vessel at constant temperature. What will be the final pressure in the vessel (assuming only mixing with no reaction)?

56. A gaseous mixture contains 5.23 g of chloroform, $CHCl_3$, and 1.66 g of methane, CH_4. What pressure is exerted by the mixture inside a 50.0-mL metal container at 275°C? What pressure is contributed by the $CHCl_3$?

57. A cyclopropane-oxygen mixture can be used as an anesthetic. If the partial pressures of cyclopropane and oxygen are 140. torr and 560. torr, respectively, what is the ratio of the number of moles of cyclopropane to the number of moles of oxygen in this mixture? What is the corresponding ratio of molecules?

58. What is the mole fraction of each gas in a mixture having the partial pressures of 0.467 atm of He, 0.317 atm of Ar, and 0.277 atm of Xe?

*59. Assume that unpolluted air has the composition shown in Table 12-2. (a) Calculate the number of molecules of N_2, of O_2, and of Ar in 1.00 L of air at 21°C and 1.00 atm. (b) Calculate the mole fractions of N_2, O_2, and Ar in the air.

60. Individual samples of O_2, N_2, and He are present in three 2.25-L vessels. Each exerts a pressure of 1.50 atm. (a) If all three gases are forced into the same 1.00-L container with no change in temperature, what will be the resulting pressure? (b) What is the partial pressure of O_2 in the mixture? (c) What are the partial pressures of N_2 and He?

61. Hydrogen was collected over water at 20°C and 757 torr. The volume of this gas sample was 35.3 mL. What volume would the dry hydrogen occupy at STP?

Charles D. Winters

62. A nitrogen sample occupies 417 mL at STP. If the same sample was collected over water at 25°C and 750. torr, what would be the volume of the gas sample?

*63. A study of climbers who reached the summit of Mt. Everest without supplemental oxygen revealed that the partial pressures of O_2 and CO_2 in their lungs were 35 torr and 7.5 torr, respectively. The barometric pressure at the summit was 253 torr. Assume that the lung gases are saturated with moisture at a body temperature of 37°C. Calculate the partial pressure of inert gas (mostly nitrogen) in the climbers' lungs.

64. A 4.00-L flask containing He at 6.00 atm is connected to a 3.00-L flask containing N_2 at 3.00 atm and the gases are allowed to mix. (a) Find the partial pressures of each gas after they are allowed to mix. (b) Find the total pressure of the mixture. (c) What is the mole fraction of helium?

65. A 3.46-liter sample of a gas was collected over water on a day when the temperature was 21°C and the barometric pressure was 718 torr. The dry sample of gas had a mass of 4.20 g. What is the molecular weight of the gas? At 21°C the vapor pressure of water is 18.65 torr.

Stoichiometry in Reactions Involving Gases

66. During a collision, automobile air bags are inflated by the N_2 gas formed by the explosive decomposition of sodium azide, NaN_3.

$$2NaN_3 \longrightarrow 2Na + 3N_2$$

What mass of sodium azide would be needed to inflate a 30.0-L bag to a pressure of 1.40 atm at 25°C?

67. Assuming the volumes of all gases in the reaction are measured at the same temperature and pressure, calculate the volume of water vapor obtainable by the explosive reaction of a mixture of 625 mL of hydrogen gas and 325 mL of oxygen gas.

*68. One liter of sulfur vapor, $S_8(g)$, at 600°C and 1.00 atm is burned in excess pure oxygen to give sulfur dioxide gas, SO_2, measured at the same temperature and pressure. What mass of SO_2 gas is obtained?

69. Calculate the volume of methane, CH_4, measured at 300. K and 825 torr, that can be produced by the bacterial breakdown of 1.00 kg of a simple sugar.

$$C_6H_{12}O_6 \longrightarrow 3CH_4 + 3CO_2$$

*70. A common laboratory preparation of oxygen is

$$2KClO_3(s) \xrightarrow{\text{heat/MnO}_2} 2KCl(s) + 3O_2(g)$$

If you were designing an experiment to generate four bottles (each containing 250. mL) of O_2 at 25°C and 762 torr and allowing for 25% waste, what mass of potassium chlorate would be required?

71. Many campers use small propane stoves to cook meals. What volume of air (see Table 12-2) will be required to burn 10.5 L of propane, C_3H_8? Assume all gas volumes are measured at the same temperature and pressure.

$$C_3H_8(g) + 5O_2(g) \longrightarrow 3CO_2(g) + 4H_2O(g)$$

Charles D. Winters

*72. If 2.00 L of nitrogen and 5.00 L of hydrogen were allowed to react, how many liters of $NH_3(g)$ could form? Assume all gases are at the same temperature and pressure, and that the limiting reactant is used up.

$$N_2(g) + 3H_2(g) \longrightarrow 2NH_3(g)$$

*73. We burn 15.00 L of ammonia in 20.00 L of oxygen at 500.°C. What volume of nitric oxide, NO, gas can form? What volume of steam, $H_2O(g)$, is formed? Assume all gases are at the same temperature and pressure, and that the limiting reactant is used up.

$$4NH_3(g) + 5O_2(g) \longrightarrow 4NO(g) + 6H_2O(g)$$

74. What mass of KNO_3 would have to be decomposed to produce 12.1 L of oxygen measured at STP?

$$2KNO_3(s) \xrightarrow{\text{heat}} 2KNO_2(s) + O_2(g)$$

75. Refer to Exercise 74. An impure sample of KNO_3 that had a mass of 55.8 g was heated until all of the KNO_3 had decomposed. The liberated oxygen occupied 4.22 L at STP. What percentage of the sample was KNO_3? Assume that no impurities decompose to produce oxygen.

*76. Heating a 5.913-g sample of an ore containing a metal sulfide, in the presence of excess oxygen, produces 1.177 L of dry SO_2, measured at 35.0°C and 755 torr. Calculate the percentage by mass of sulfur in the ore.

*77. The following reactions occur in a gas mask (a self contained breathing apparatus) sometimes used by underground miners. The H_2O and CO_2 come from exhaled air, and O_2 is inhaled as it is produced. KO_2 is potassium superoxide. The CO_2 is converted to the solid salt $KHCO_3$, potassium hydrogen carbonate, so that CO_2 is not inhaled in significant amounts.

$$4KO_2(s) + 2H_2O(\ell) \longrightarrow 4KOH(s) + 3O_2(g)$$
$$CO_2(g) + KOH(s) \longrightarrow KHCO_3(s)$$

(a) What volume of O_2, measured at STP, is produced by the complete reaction of 1.25 g of KO_2? (b) What is this volume at body temperature, 37°C, and 1.00 atm? (c) What mass of KOH is produced in part (a)? (d) What volume of CO_2, measured at STP, will react with the mass of KOH of part (c)? (e) What is the volume of CO_2 in part (d) measured at 37°C and 1.00 atm?

*78. Let us represent gasoline as octane, C_8H_{18}. When hydrocarbon fuels burn in the presence of sufficient oxygen, CO_2 is formed.

Reaction A: $2C_8H_{18} + 25O_2 \longrightarrow 16CO_2 + 18H_2O$

But when the supply of oxygen is limited, the poisonous gas carbon monoxide, CO, is formed.

Reaction B: $2C_8H_{18} + 17O_2 \longrightarrow 16CO + 18H_2O$

Any automobile engine, no matter how well tuned, burns its fuel by some combination of these two reactions.

Suppose an automobile engine is running at idle speed in a closed garage with air volume 97.5 m³. This engine burns 95.0% of its fuel by reaction A, and the remainder by reaction B. (a) How many liters of octane, density 0.702 g/mL, must be burned for the CO to reach a concentration of 2.00 g/m³? (b) If the engine running at idle speed burns fuel at the rate of 1.00 gal/h (0.0631 L/min), how long does it take to reach the CO concentration in (a)?

The Kinetic-Molecular Theory and Molecular Speeds

79. Outline the kinetic-molecular theory.
80. The radius of a typical molecule of a gas is 2.00 Å. (a) Find the volume of a molecule assuming it to be spherical. For a sphere, $V = 4/3 \, \pi r^3$. (b) Calculate the volume actually occupied by 1.00 mol of these molecules. (c) If 1.0 mol of this gas occupies 22.4 L, find the fraction of the volume actually occupied by the molecules. (d) Comment on your answer to part (c) in view of the first statement summarizing the kinetic-molecular theory of an ideal gas.
81. How does the kinetic-molecular theory explain (a) Boyle's Law? (b) Dalton's Law? (c) Charles's Law?
82. SiH_4 molecules are heavier than CH_4 molecules; yet, according to kinetic-molecular theory, the average kinetic energies of the two gases at the same temperature are equal. How can this be?
*83. At 22°C, Cl_2 molecules have some rms speed (which we need not calculate). At what temperature would the rms speed of F_2 molecules be the same?
*84. (a) How do average speeds of gaseous molecules vary with temperature? (b) Calculate the ratio of the rms speed of N_2 molecules at 100.°C to the rms speed of the same molecules at 0.0°C.
85. How do the average kinetic energies and average speeds of each gas in a mixture compare?
86. (a) If you heat a gaseous sample in a fixed volume container, the pressure increases. Use the kinetic-molecular theory to explain the increased pressure. (b) If the volume of a gaseous sample is reduced at constant temperature, the pressure increases. Use the kinetic-molecular theory to explain the increase in pressure.

Real Gases and Deviations from Ideality

87. What is the van der Waals equation? How does it differ from the ideal gas equation?
88. Which of the following gases would be expected to behave most nearly ideally under the same conditions? H_2, F_2, HF. Which one would be expected to deviate from ideal behavior the most? Explain both answers.
89. Does the effect of intermolecular attraction on the properties of a gas become more significant or less significant if (a) the gas is compressed to a smaller volume at constant temperature? (b) more gas is forced into the same volume at the same temperature? (c) the temperature of the gas is raised at constant pressure?

90. Does the effect of molecular volume on the properties of a gas become more significant or less significant if (a) the gas is compressed to a smaller volume at constant temperature? (b) more gas is forced into the same volume at the same temperature? (c) the temperature of the gas is raised at constant pressure?
91. A sample of gas has a molar volume of 10.1 L at a pressure of 745 torr and a temperature of −138°C. Is the gas behaving ideally?
92. Calculate the compressibility factor, $(P_{real})(V_{real})/RT$, for a 1.00-mol sample of NH_3 under the following conditions: in a 500.-mL vessel at −10.0°C it exerts a pressure of 30.0 atm. What would be the *ideal* pressure for 1.00 mol of NH_3 at −10.0°C in a 500.-mL vessel? Compare this with the real pressure and account for the difference.
93. (a) How do "real" and "ideal" gases differ? (b) Under what kinds of conditions are deviations from ideality most important? Why?
94. Find the pressure of a sample of carbon tetrachloride, CCl_4, if 1.00 mol occupies 35.0 L at 77.0°C (slightly above its normal boiling point). Assume that CCl_4 obeys (a) the ideal gas law; (b) the van der Waals equation. The van der Waals constants for CCl_4 are $a = 20.39 \; L^2 \cdot atm/mol^2$ and $b = 0.1383 \; L/mol$.
95. Repeat the calculations of Exercise 94 using a 3.10-mol gas sample confined to 5.75 L at 135°C.

Mixed Exercises

96. A student is to perform a laboratory experiment that requires the evolution and collection of 75 mL of dry oxygen gas at one atmosphere and 25°C. What is the minimum mass of water required to generate the oxygen by electrolysis of water?
97. A tilting McLeod gauge is used to measure very low pressures of gases in glass vacuum lines in the laboratory. It operates by compressing a large volume of gas at low pressure to a much smaller volume so that the pressure is more easily measured. What is the pressure of a gas in a vacuum line if a 53.3-mL volume of the gas, when compressed to 0.235 mL, supports a 16.9-mm column of mercury?

A McLeod gauge.

98. Imagine that you live in a cabin with an interior volume of 175 m³. On a cold morning your indoor air temperature is 10.°C, but by the afternoon the sun has warmed the cabin air to 18°C. The cabin is not sealed; therefore, the pressure inside is the same as it is outdoors. Assume that the pressure remains constant during the day. How many cubic meters of air would have been forced out of the cabin by the sun's warming? How many liters?

99. A particular tank can safely hold gas up to a pressure of 44.3 atm. When the tank contains 38.1 g of N_2 at 25°C, the gas exerts a pressure of 10.1 atm. What is the highest temperature to which the gas sample can be heated safely?

100. Find the molecular weight of Freon-12 (a chlorofluorocarbon) if 8.29 L of vapor at 200.°C and 790. torr has a mass of 26.8 g.

101. A flask of unknown volume was filled with air to a pressure of 3.25 atm. This flask was then attached to an evacuated flask with a known volume of 5.00 L, and the air was allowed to expand into the flask. The final pressure of the air (in both flasks) was 2.40 atm. Calculate the volume of the first flask.

*102. Relative humidity is the ratio of the pressure of water vapor in the air to the pressure of water vapor in air that is saturated with water vapor at the same temperature.

Relative humidity

$$= \frac{\text{actual partial pressure of } H_2O}{\text{partial pressure of } H_2O \text{ vapor if sat'd}}$$

Often this quantity is multiplied by 100 to give the percent relative humidity. Suppose the percent relative humidity is 80.0% at 91.4°F (33.0°C) in a house with volume 245 m³. Then an air conditioner is turned on. Due to the condensation of water vapor on the cold coils of the air conditioner, water vapor is also removed from the air as it cools. After the air temperature has reached 77.0°F (25.0°C), the percent relative humidity is measured to be 15.0%. (a) What mass of water has been removed from the air in the house? (*Reminder:* Take into account the difference in saturated water vapor pressure at the two temperatures.) (b) What volume would this liquid water occupy at 25°C? (Density of liquid water at 25.0°C = 0.997 g/cm³.)

103. A 450.-mL flask contains 0.500 g of nitrogen gas at a pressure of 744 torr. Are these data sufficient to allow you to calculate the temperature of the gas? If not, what is missing? If so, what is the temperature in °C?

104. Use both the ideal gas law and the van der Waals equation to calculate the pressure exerted by a 10.0-mol sample of ammonia in a 60.0-L container at 100.°C. By what percentage do the two results differ?

105. What volume of hydrogen fluoride at 743 torr and 24°C will be released by the reaction of 38.3 g of xenon difluoride with a stoichiometric amount of water? The *unbalanced* equation is

$$XeF_2(s) + H_2O(\ell) \longrightarrow Xe(g) + O_2(g) + HF(g)$$

What volumes of oxygen and xenon will be released under these conditions?

106. Cyanogen is 46.2% carbon and 53.8% nitrogen by mass. At a temperature of 25°C and a pressure of 750. torr, 1.00 g of cyanogen gas occupies 0.476 L. Determine the empirical formula and the molecular formula of cyanogen.

107. Incandescent light bulbs contain noble gases, such as argon, so that the filament will last longer. The approximate volume of a 100.-watt bulb is 130. cm³, and the bulb contains 0.125 g of argon. How many grams of argon would be contained in a 150.-watt bulb under the same pressure and temperature conditions if the volume of the larger wattage bulb is 180. cm³?

CONCEPTUAL EXERCISES

108. An attempt was made to collect carbon dioxide, isolated from the decomposition of a carbonate-containing mineral, by first bubbling the gas through pure liquid acetic acid. The experiment yielded 500. mL of a gaseous mixture of acetic acid and carbon dioxide at 1.00 atm and 16.0°C. The vapor pressure of pure acetic acid at 16.0°C is 400. torr. What should be the total mass of the collected sample?

109. Acetylene (C_2H_2), the gas used in welders' torches, is produced by the reaction of calcium carbide (CaC_2) with water. The other reaction product is calcium hydroxide. (a) Write the chemical equation for the production of C_2H_2. (b) What volume of C_2H_2, measured at 22°C and 965 mm Hg, would be produced from the complete reaction of 10.2 g of CaC_2 with 5.23 g of H_2O?

Charles D. Winters

110. Redraw Figure 12-12 so that it more accurately depicts the decrease in space between molecules as the system is cooled from 600. K to 300. K but also emphasizes the change in kinetic energy.

111. Suppose you were asked to supply a particular mass of a specified gas in a container of fixed volume at a specified pressure and temperature. Is it likely that you could fulfill the request? Explain.

112. The gas molecules in the box undergo a reaction at constant temperature and pressure.

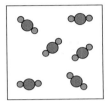

If the initial volume is 2.5 L and the final volume is 1.25 L, which of the following boxes could be the products of the reaction? Explain your reasoning.

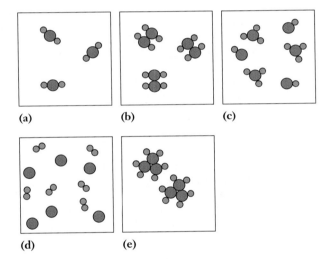

(a)   (b)   (c)

(d)   (e)

113. A 503-mL flask contains 0.0243 mol of an ideal gas at a given temperature and pressure. Another flask contains 0.0388 mol of the gas at the same temperature and pressure. What is the volume of the second flask?

114. Mole fraction is one of the few quantities that is expressed as a fraction. Frequently, percent is used instead. Cite an example earlier in this textbook where percent was used instead of a fraction. Write an equation that relates mole fraction and mole percent.

115. Explain why gases at high pressure or low temperature do not act as ideal gases.

BUILDING YOUR KNOWLEDGE

116. A 5.00-L reaction vessel contains hydrogen at a partial pressure of 0.588 atm and oxygen gas at a partial pressure of 0.302 atm. Which element is the limiting reactant in the following reaction? $2H_2(g) + O_2(g) \longrightarrow 2H_2O(g)$

117. Suppose the gas mixture in Exercise 116 is ignited and the reaction produces the theoretical yield of the product. What would be the partial pressure of each substance present in the final mixture?

118. A sample of carbon was burned in pure oxygen, forming carbon dioxide. The CO_2 was then bubbled into 3.50 L of 0.437 M NaOH, forming sodium carbonate, as shown by the reaction below.

$$CO_2 + 2\,NaOH \longrightarrow Na_2CO_3 + H_2O$$

Following the reaction, the excess NaOH was exactly neutralized with 1.71 L of 0.350 M HCl. What volume of O_2, measured at 8.6 atm and 20°C was consumed in the process?

119. When magnesium carbonate, $MgCO_3$, is heated to a high temperature, it decomposes.

$$MgCO_3(s) \xrightarrow{heat} MgO(s) + CO_2(g)$$

A 20.29-gram sample of impure magnesium carbonate is completely decomposed at 1000.°C in a previously evacuated 2.00-L reaction vessel. After the reaction was complete, the solid residue (consisting only of MgO and the original impurities) had a mass of 15.90 grams. Assume that no other constituent of the sample produced a gas and that the volume of any solid was negligible compared with the gas volume. (a) How many grams of CO_2 were produced? (b) What was the pressure of the CO_2 produced? (c) What percent of the original sample was magnesium carbonate?

120. One natural source of atmospheric carbon dioxide is precipitation reactions such as the precipitation of silicates in the oceans.

$$Mg^{2+}(aq) + SiO_2(dispersed) + 2HCO_3^{-}(aq) \longrightarrow$$
$$MgSiO_3(s) + 2CO_2(g) + H_2O(\ell)$$

How many grams of magnesium silicate would be precipitated during the formation of 100. L of carbon dioxide at 30.°C and 775 torr?

121. Table 12-2 states that dry air is 20.94% (by volume) oxygen. What is the partial pressure of oxygen under the conditions of STP? *Hint:* For gaseous samples the mole ratios are equal to the volume ratios.

BEYOND THE TEXTBOOK

Go to the textbook website at

http://www.brookscole.com/chemistry/whitten

for additional activities and exercises based on the General Chemistry Interactive CD-ROM, the World Wide Web, and library resources.

InfoTrac College Edition

For additional readings, go to InfoTrac College Edition, your online research library at:

http://infotrac.thomsonlearning.com

Liquids and Solids

13

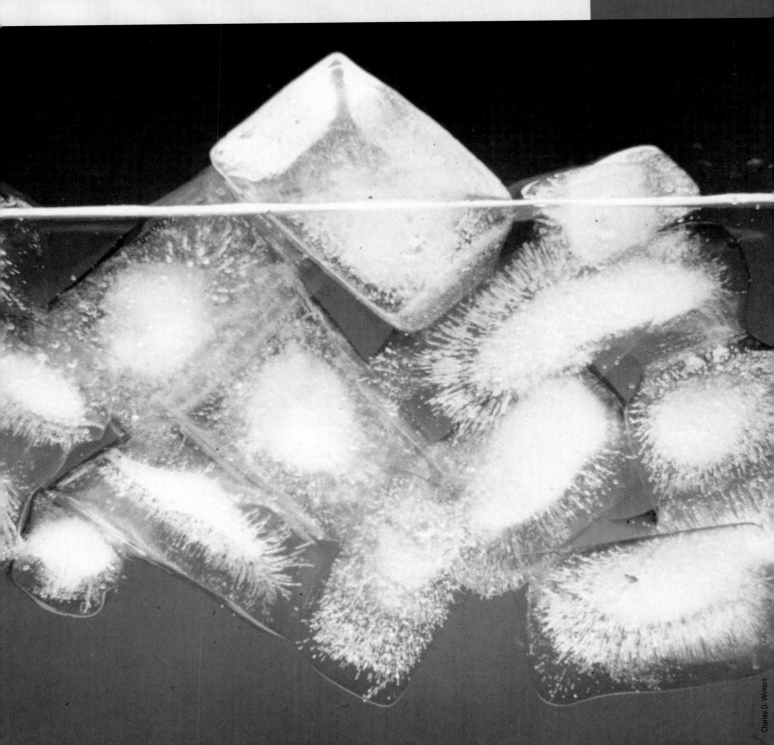

At the melting point of ice the solid and liquid phases of water are in equilibrium.

Charles D. Winters

OUTLINE

OBJECTIVES

After you have studied this chapter, you should be able to

- Describe the properties of liquids and solids and how they differ from gases
- Understand the kinetic–molecular description of liquids and solids, and show how this description differs from that for gases
- Use the terminology of phase changes
- Understand various kinds of intermolecular attractions and how they are related to physical properties such as vapor pressure, viscosity, melting point, and boiling point
- Describe evaporation, condensation, and boiling in molecular terms
- Calculate the heat transfer involved in warming or cooling without change of phase
- Calculate the heat transfer involved in phase changes
- Describe melting, solidification, sublimation, and deposition in molecular terms
- Interpret P versus T phase diagrams
- Describe the regular structure of crystalline solids
- Describe various types of solids
- Relate the properties of different types of solids to the bonding or interactions among particles in these solids
- Visualize some common simple arrangements of atoms in solids
- Carry out calculations relating atomic arrangement, density, unit cell size, and ionic or atomic radii in some simple crystalline arrangements
- Describe the bonding in metals
- Explain why some substances are conductors, some are insulators, and others are semiconductors

he molecules of most gases are so widely separated at ordinary temperatures and pressures that they do not interact with one another significantly. The physical properties of gases are reasonably well described by the simple relationships in Chapter 12. In liquids and solids, the so-called **condensed phases,** the particles are close together so they interact much more strongly. Although the properties of liquids and solids can be described, they cannot be adequately explained by simple mathematical relationships. Table 13-1 and Figure 13-1 summarize some of the characteristics of gases, liquids, and solids.

CD-ROM   Screen 13.2, Phases of Matter; the Kinetic–Molecular Theory. This screen contains an animated version of Figure 13-1.

TABLE 13-1	*Some Characteristics of Solids, Liquids, and Gases*	
Solids	**Liquids**	**Gases**
1. Have definite shape (resist deformation)	1. Have no definite shape (assume shapes of containers)	1. Have no definite shape (fill containers completely)
2. Are nearly incompressible	2. Have definite volume (are only very slightly compressible)	2. Are compressible
3. Usually have higher density than liquids	3. Have high density	3. Have low density
4. Are not fluid	4. Are fluid	4. Are fluid
5. Diffuse only very slowly through solids	5. Diffuse through other liquids	5. Diffuse rapidly
6. Have an ordered arrangement of particles that are very close together; particles usually have only vibrational motion	6. Consist of disordered clusters of particles that are quite close together; particles have random motion in three dimensions	6. Consist of extremely disordered particles with much empty space between them; particles have rapid, random motion in three dimensions

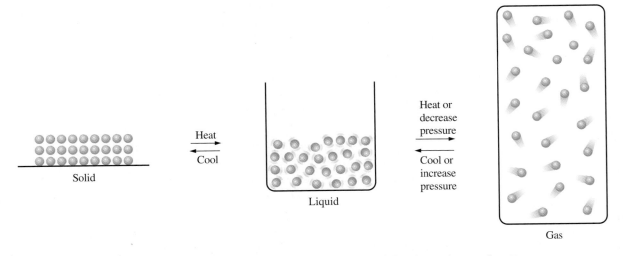

Figure 13-1   Representations of the kinetic–molecular description of the three phases of matter.

13-1 KINETIC–MOLECULAR DESCRIPTION OF LIQUIDS AND SOLIDS

The properties listed in Table 13-1 can be qualitatively explained in terms of the kinetic–molecular theory of Chapter 12. We saw in Section 12-13 that the average kinetic energy of a collection of gas molecules decreases as the temperature is lowered. As a sample of gas is cooled and compressed, the rapid, random motion of gaseous molecules decreases. The molecules approach one another, and the intermolecular attractions increase. Eventually, these increasing intermolecular attractions overcome the reduced kinetic energies. At this point condensation (liquefaction) occurs. The temperatures and pressures required for condensation vary from gas to gas, because different kinds of molecules have different attractive forces.

In the liquid state the forces of attraction among particles are great enough that disordered clustering occurs. The particles are so close together that very little of the volume occupied by a liquid is empty space. As a result, it is very hard to compress a liquid. Particles in liquids have sufficient energy of motion to overcome partially the attractive forces among them. They are able to slide past one another so that liquids assume the shapes of their containers up to the volume of the liquid.

Liquids diffuse into other liquids with which they are *miscible*. For example, when a drop of red food coloring is added to a glass of water, the water becomes red throughout after diffusion is complete. The natural diffusion rate is slow at normal temperatures. Because the average separations among particles in liquids are far less than those in gases, the densities of liquids are much higher than the densities of gases (Table 12-1).

Cooling a liquid lowers its molecular kinetic energy and causes its molecules to slow down even more. If the temperature is lowered sufficiently, at ordinary pressures, stronger but shorter-range attractive interactions overcome the reduced kinetic energies of the molecules to cause *solidification*. The temperature required for *crystallization* at a given pressure depends on the nature of short-range interactions among the particles and is characteristic of each substance.

Most solids have ordered arrangements of particles with a very restricted range of motion. Particles in the solid state cannot move freely past one another so they only vibrate about fixed positions. Consequently, solids have definite shapes and volumes. Because the particles are so close together, solids are nearly incompressible and are very dense relative to gases. Solid particles do not diffuse readily into other solids. However, analysis of two

Intermolecular attractions are those between different molecules or ions. Intramolecular attractions are those between atoms within a single molecule or ion.

The miscibility of two liquids refers to their ability to mix and produce a homogeneous solution.

Solidification and crystallization refer to the process in which a liquid changes to a solid. Crystallization is a more specific term, referring to the formation of a very ordered solid material.

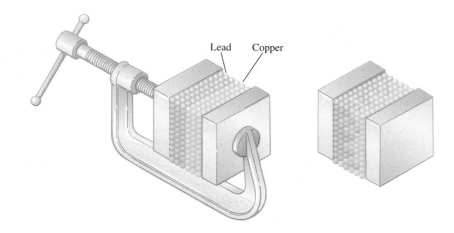

Lead   Copper

Figure 13-2   A representation of diffusion in solids. When blocks of two different metals are clamped together for a long time, a few atoms of each metal diffuse into the other metal.

blocks of different solids, such as copper and lead, that have been pressed together for a period of years shows that each block contains some atoms of the other element. This demonstrates that solids do diffuse, but very slowly (Figure 13-2).

13-2 INTERMOLECULAR ATTRACTIONS AND PHASE CHANGES

We have seen (Section 12-15) how the presence of strong attractive forces between gas molecules can cause gas behavior to become nonideal when the molecules get close together. In liquids and solids the molecules are much closer together than in gases. As a result, properties of liquids, such as boiling point, vapor pressure, viscosity, and heat of vaporization, depend markedly on the strengths of the intermolecular attractive forces. These forces are also directly related to the properties of solids, such as melting point and heat of fusion. Let us preface our study of these condensed phases with a discussion of the types of attractive forces that can exist between molecules and ions.

*Inter*molecular forces refer to the forces *between* individual particles (atoms, molecules, ions) of a substance. These forces are quite weak relative to *intra*molecular forces, that is, covalent and ionic bonds *within* compounds. For example, 927 kJ of energy is required to decompose one mole of water vapor into H and O atoms. This reflects the strength of intramolecular forces (covalent bonds).

$$H-\overset{..}{\underset{..}{O}}-H(g) \longrightarrow 2H\cdot(g) + \cdot\overset{..}{\underset{..}{O}}\cdot(g) \qquad \text{(absorbs 927 kJ/mol)}$$

Only 40.7 kJ is required to convert one mole of liquid water into steam at 100°C.

$$H_2O(\ell) \longrightarrow H_2O(g) \qquad \text{(absorbs 40.7 kJ/mol)}$$

This reflects the lower strength of the intermolecular forces of attraction between the water molecules, compared to the covalent bonds within the water molecules. The attractive forces between water molecules are mainly due to *hydrogen bonding*.

If it were not for the existence of intermolecular attractions, condensed phases (liquids and solids) could not exist. These are the forces that hold the particles close to one another in liquids and solids. As we shall see, the effects of these attractions on melting points of solids parallel those on boiling points of liquids. High boiling points are associated with compounds that have strong intermolecular attractions. Let us consider the effects of the general types of forces that exist among ionic, covalent, and monatomic species.

Ion–Ion Interactions

According to Coulomb's Law, the *force of attraction* between two oppositely charged ions is directly proportional to the charges on the ions, q^+ and q^-, and inversely proportional to the square of the distance between them, d.

$$F \propto \frac{q^+ q^-}{d^2}$$

Energy has the units of force × distance, $F \times d$, so the *energy of attraction* between two oppositely charged ions is directly proportional to the charges on the ions and inversely proportional to the distance of separation.

$$E \propto \frac{q^+ q^-}{d}$$

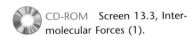 CD-ROM Screen 13.3, Intermolecular Forces (1).

It is important to be able to tell whether a substance is ionic, nonpolar covalent, or polar covalent. You should review the discussion of bonding in Chapters 7 and 8.

When oppositely charged ions are close together, d (the denominator) is small, so F, the attractive force between them, is large.

The covalent bonding *within* a polyatomic ion such as NH_4^+ or SO_4^{2-} is very strong, but the forces that hold the *entire substance* together are ionic. Thus a compound that contains a polyatomic ion is an ionic compound (see Section 7-11).

Ionic compounds such as NaCl, $CaBr_2$, and K_2SO_4 exist as extended arrays of discrete ions in the solid state. As we shall see in Section 13-16, the oppositely charged ions in these arrays are quite close together. As a result of these small distances, d, the energies of attraction in these solids are substantial. Most ionic bonding is strong, and as a result most ionic compounds have relatively high melting points (Table 13-2). At high enough temperatures, ionic solids melt as the added heat energy overcomes the potential energy associated with the attraction of oppositely charged ions. The ions in the resulting liquid samples are free to move about, which accounts for the excellent electrical conductivity of molten ionic compounds.

For most substances, the liquid is less dense than the solid. Melting a solid nearly always produces greater average separations among the particles. This means that the forces (and energies) of attractions among the ions in an ionic liquid are less than in the solid state because the average d is greater in the melt. However, these energies of attraction are still much greater in magnitude than the energies of attraction among neutral species (molecules or atoms).

⚠ The simple concepts of ion size and magnitudes of the ionic charges combined with Coulomb's Law can give you very good guidance on understanding a number of solid-state properties ranging from melting point trends to solubility.

The product q^+q^- increases as the charges on ions increase. Ionic substances containing multiply charged ions, such as Al^{3+}, Mg^{2+}, O^{2-}, and S^{2-} ions, *usually* have higher melting and boiling points than ionic compounds containing only singly charged ions, such as Na^+, K^+, F^-, and Cl^-. For a series of ions of similar charges, the closer approach of smaller ions results in stronger interionic attractive forces and higher melting points (compare NaF, NaCl, and NaBr in Table 13-2).

Dipole–Dipole Interactions

CD-ROM   Screen 13.4, Intermolecular Forces (2).

> Permanent dipole–dipole interactions occur between polar covalent molecules because of the attraction of the $\delta+$ atoms of one molecule to the $\delta-$ atoms of another molecule (Section 7-10).

Electrostatic forces between two ions decrease by the factor $1/d^2$ as their separation, d, increases. But dipole–dipole forces vary as $1/d^4$. Because of the higher power of d in the denominator, $1/d^4$ diminishes with increasing d much more rapidly than does $1/d^2$. As a result, dipole forces are effective only over very short distances. Furthermore, for dipole–dipole forces, $q+$ and q^- represent only "partial charges," so these forces are weaker than ion–ion forces. Average dipole–dipole interaction energies are approximately 4 kJ per mole of bonds. They are much weaker than ionic and covalent bonds, which have typical energies of about 400 kJ per mole of bonds. Substances in which permanent dipole–dipole interactions affect physical properties include bromine fluoride, BrF, and sulfur dioxide, SO_2.

TABLE 13-2	*Melting Points of Some Ionic Compounds*				
Compound	**mp (°C)**	**Compound**	**mp (°C)**	**Compound**	**mp (°C)**
NaF	993	CaF_2	1423	MgO	2800
NaCl	801	Na_2S	1180	CaO	2580
NaBr	747	K_2S	840	BaO	1923
KCl	770				

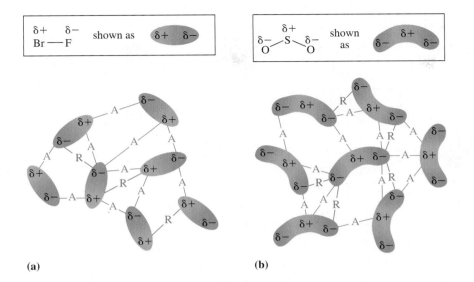

(a)   (b)

Figure 13-3 Dipole–dipole interactions among polar molecules. (a) Bromine fluoride, BrF. (b) Sulfur dioxide, SO_2. Each polar molecule is shaded with regions of highest negative charge ($\delta-$) in red and regions of highest positive charge ($\delta+$) colored blue. Attractive forces are shown as —A—, and repulsive forces are shown as —R—. Molecules tend to arrange themselves to maximize attractions by bringing regions of opposite charge together while minimizing repulsions by separating regions of like charge.

Dipole–dipole interactions are illustrated in Figure 13-3. All dipole–dipole interactions, including hydrogen bonding (discussed in the following section), are somewhat directional. An increase in temperature causes an increase in translational, rotational, and vibrational motion of molecules. This produces more random orientations of molecules relative to one another. Consequently, dipole–dipole interactions become less important as temperature increases. All these factors make compounds having only dipole–dipole interactions more volatile than ionic compounds.

Hydrogen Bonding

Hydrogen bonds are a special case of very strong dipole–dipole interaction. They are not really chemical bonds in the formal sense.

> Strong hydrogen bonding occurs among polar covalent molecules containing H and one of the three small, highly electronegative elements—F, O, or N.

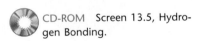 CD-ROM Screen 13.5, Hydrogen Bonding.

Like ordinary dipole–dipole interactions, hydrogen bonds result from the attractions between $\delta+$ atoms of one molecule, in this case H atoms, and the $\delta-$ atoms of another molecule. The small sizes of the F, O, and N atoms, combined with their high electronegativities, concentrate the electrons of these molecules around these $\delta-$ atoms. This causes an H atom bonded to one of these highly electronegative elements to become quite positive. The $\delta+$ H atom is attracted to a lone pair of electrons on an F, O, or N atom other than the atom to which it is covalently bonded (Figure 13-4). The molecule that contains the hydrogen-bonding $\delta+$ H atom is often referred to as the *hydrogen-bond donor*; the $\delta-$ atom to which it is attracted is called the *hydrogen-bond acceptor*.

Recently, careful studies of light absorption and magnetic properties in solution and of the arrangements of molecules in solids have led to the conclusion that the same kind of attraction occurs (although more weakly) when H is bonded to carbon. In some cases, very weak C—H---O "hydrogen bonds" exist. Similar observations suggest the existence of weak hydrogen bonds to chlorine atoms, such as O—H---Cl. However, most chemists usually

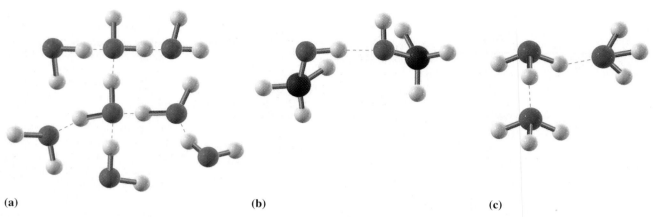

(a)   **(b)**   **(c)**

Figure 13-4 Hydrogen bonding (indicated by dashed lines) in (a) water, H_2O; (b) methyl alcohol, CH_3OH; and (c) ammonia, NH_3. Hydrogen bonding is a special case of very strong dipole interaction.

restrict usage of the term "hydrogen bonding" to compounds in which H is covalently bonded to F, O, or N, and we will do likewise throughout this text.

Typical hydrogen-bond energies are in the range 15 to 20 kJ/mol, which is four to five times greater than the energies of other dipole–dipole interactions. As a result, hydrogen bonds exert a considerable influence on the properties of substances. Hydrogen bonding is responsible for the unusually high melting and boiling points of compounds such as water, methyl alcohol, and ammonia compared with other compounds of similar molecular weight and molecular geometry (Figure 13-5). Hydrogen bonding between amino acid subunits is very important in establishing the three-dimensional structures of proteins.

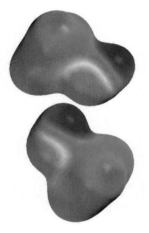

Electrostatic charge potential plots of two hydrogen bonded methyl alcohol molecules. The polar O-H bonds are colored red (oxygen atom, partial negative charge) and darker blue (hydrogen atom, partial positive charge). The hydrogen bonding is due to the electrostatic attraction between the $\delta+$ charged hydrogen of one methanol to the $\delta-$ charged oxygen atom of another.

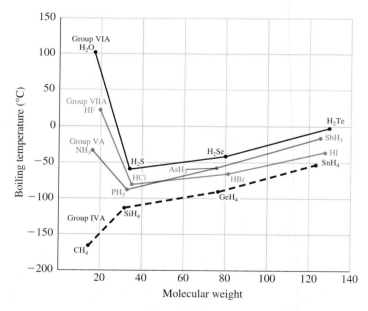

Figure 13-5 Boiling points of some hydrides as a function of molecular weight. The unusually high boiling points of NH_3, H_2O, and HF compared with those of other hydrides of the same groups are due to hydrogen bonding. The electronegativity difference between H and C is small, and there are no unshared pairs on C; thus, CH_4 is not hydrogen bonded. Increasing molecular weight corresponds to increasing number of electrons; this makes the electron clouds easier to deform and causes increased dispersion forces, accounting for the increase in boiling points for the nonhydrogen-bonded members of each series.

Dispersion Forces

Dispersion forces are weak attractive forces that are important only over *extremely* short distances because they vary as $1/d^7$. They are present between all types of molecules in condensed phases but are weak for small molecules. Dispersion forces are the only kind of intermolecular forces present among symmetrical nonpolar substances such as SO_3, CO_2, O_2, N_2, Br_2, H_2, and monatomic species such as the noble gases. Without dispersion forces, such substances could not condense to form liquids or solidify to form solids. Condensation of some substances occurs only at very low temperatures and/or high pressures.

Dispersion forces result from the attraction of the positively charged nucleus of one atom for the electron cloud of an atom in nearby molecules. This induces *temporary* dipoles in neighboring atoms or molecules. As electron clouds become larger and more diffuse, they are attracted less strongly by their own (positively charged) nuclei. Thus, they are more easily distorted, or *polarized*, by adjacent nuclei.

Polarizability increases with increasing numbers of electrons and therefore with increasing sizes of molecules. Therefore, dispersion forces are generally stronger for molecules that are larger or that have more electrons. For molecules that are large or quite polarizable the total effect of dispersion forces can be even higher than dipole–dipole interactions or hydrogen bonding.

Dispersion forces are depicted in Figure 13-6. They exist in all substances.

Figure 13-5 shows that polar covalent compounds with hydrogen bonding (H_2O, HF, NH_3) boil at higher temperatures than analogous polar compounds without hydrogen

Dispersion forces are often called London forces, after the German-born physicist Fritz London (1900–1954). He initially postulated their existence in 1930, on the basis of quantum theory.

Although the term "van der Waals forces" usually refers to all intermolecular attractions, it is also often used interchangeably with "dispersion forces," as are the terms "London forces" and "dipole-induced dipole forces."

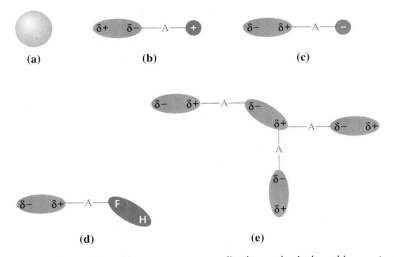

(a)   (b)   (c)

(d)   (e)

Figure 13-6 An illustration of how a temporary dipole can be induced in an atom. (a) An isolated argon atom, with spherical charge distribution (no dipole). (b) When a cation approaches the argon atom, the outer portion of the electron cloud is weakly attracted by the ion's positive charge. This induces a weak *temporary* dipole in the argon atom. (c) A temporary dipole can also be induced if the argon atom is approached by an anion. (d) The approach of a molecule with a permanent dipole (for instance, HF) could also temporarily polarize the argon atom. (e) Even in pure argon, the close approach of one argon atom to another results in temporary dipole formation in both atoms as each atom's electron cloud is attracted by the nucleus of the other atom or is repelled by the other atom's electron cloud. The resulting temporary dipoles cause weak attractions among the argon atoms. Molecules are even more easily polarized than isolated atoms.

 CD-ROM   Screen 13.7, Intermolecular Forces (3).

For a physical analogy to dispersion forces think about Velcro. Each little hook and loop in Velcro represents a very weak interaction. But adding them all together can make the total attraction very strong.

George Semple

TABLE 13-3 *Approximate Contributions to the Total Energy of Interaction Between Molecules, in kJ/mol*

Molecule	Permanent Dipole Moment (D)	Permanent Dipole–Dipole Energy	Dispersion Energy	Total Energy	Molar Heat of Vaporization (kJ/mol)
Ar	0	0	8.5	8.5	6.7
CO	0.1	≈0	8.7	8.7	8.0
HCl	1.03	3.3	17.8	21	16.2
NH_3	1.47	13*	16.3	29	27.4
H_2O	1.85	36*	10.9	47	40.7

*Hydrogen-bonded.

The ability of the Tokay gecko to climb on walls and ceilings is due to the dispersion forces between the tiny hairs on the gecko's foot and the surface.

Minden Pictures

bonding (H_2S, HCl, PH_3). Symmetrical, nonpolar compounds (CH_4, SiH_4) of comparable molecular weight boil at lower temperatures. In the absence of hydrogen bonding, boiling points of analogous substances (CH_4, SiH_4, GeH_4, SnH_4) increase fairly regularly with increasing number of electrons and molecular size (molecular weight). This is due to increasing effectiveness of dispersion forces of attraction in the larger molecules and occurs even in the case of some polar covalent molecules. The increasing effectiveness of dispersion forces, for example, accounts for the increase in boiling points in the sequences HCl < HBr < HI and H_2S < H_2Se < H_2Te, which involve nonhydrogen-bonded polar covalent molecules. The differences in electronegativities between hydrogen and other nonmetals *decrease* in these sequences, and the increasing dispersion forces override the decreasing permanent dipole–dipole forces. The *permanent* dipole–dipole interactions therefore have very little effect on the boiling point trend of these compounds.

Let's compare the magnitudes of the various contributions to the total energy of interactions in some simple molecules. Table 13-3 shows the permanent dipole moments and the energy contributions for five simple molecules. The contribution from dispersion forces is substantial in all cases. The permanent dipole–dipole energy is greatest for substances in which hydrogen bonding occurs. The variations of these total energies of interaction are closely related to molar heats of vaporization. As we shall see in Section 13-9, the heat of vaporization measures the amount of energy required to overcome the attractive forces that hold the molecules together in a liquid.

The properties of a liquid or a solid are often the result of many forces. The properties of an ionic compound are determined mainly by the very strong ion–ion interactions, even though other forces may also be present. In a polar covalent compound that contains N—H, O—H, or F—H bonds, strong hydrogen bonding is often the strongest force present. If hydrogen bonding is absent in a polar covalent compound, dispersion forces are likely to be the most important forces. In a slightly polar or nonpolar covalent compound or a monatomic nonmetal, the dispersion forces, though weak, are still the strongest ones present, so they determine the forces. For large molecules, even the very weak dispersion forces can total up to a considerable interactive force.

THE LIQUID STATE

Honey is a very viscous liquid.

Charles D. Winters

We shall briefly describe several properties of the liquid state. These properties vary markedly among various liquids, depending on the nature and strength of the attractive forces among the particles (atoms, molecules, ions) making up the liquid.

13-3 VISCOSITY

Viscosity is the resistance to flow of a liquid. Honey has a high viscosity at room temperature, and freely flowing gasoline has a low viscosity. The viscosity of a liquid can be measured with a viscometer such as the one in Figure 13-7.

For a liquid to flow, the molecules must be able to slide past one another. In general, the stronger the intermolecular forces of attraction, the more viscous the liquid is. Substances that have a great ability to form hydrogen bonds, especially involving several hydrogen-bonding sites per molecule, such as glycerine (margin), usually have high viscosities. Increasing the size and surface area of molecules generally results in increased viscosity, due to the increased dispersion forces. For example, the shorter-chain hydrocarbon pentane (a free-flowing liquid at room temperature) is less viscous than dodecane (an oily liquid at room temperature). The longer the molecules are, the more they attract one another, and the harder it is for them to flow.

The *poise* is the unit used to express viscosity. The viscosity of water at 25°C is 0.89 centipoise.

🔵 CD-ROM   Screen 13.11, Properties of Liquids (4): Surface Tension/Capillary Action/Viscosity.

pentane, C_5H_{12}
viscosity = 0.215
   centipoise at 25°C

$$H-\underset{\underset{H}{|}}{\overset{\overset{H}{|}}{C}}-\underset{\underset{H}{|}}{\overset{\overset{H}{|}}{C}}-\underset{\underset{H}{|}}{\overset{\overset{H}{|}}{C}}-\underset{\underset{H}{|}}{\overset{\overset{H}{|}}{C}}-\underset{\underset{H}{|}}{\overset{\overset{H}{|}}{C}}-H$$

dodecane, $C_{12}H_{26}$
viscosity = 1.38
   centipoise at 25°C

$$H-\overset{H}{\underset{H}{C}}-\overset{H}{\underset{H}{C}}-\overset{H}{\underset{H}{C}}-\overset{H}{\underset{H}{C}}-\overset{H}{\underset{H}{C}}-\overset{H}{\underset{H}{C}}-\overset{H}{\underset{H}{C}}-\overset{H}{\underset{H}{C}}-\overset{H}{\underset{H}{C}}-\overset{H}{\underset{H}{C}}-\overset{H}{\underset{H}{C}}-\overset{H}{\underset{H}{C}}-H$$

As temperature increases and the molecules move more rapidly, their kinetic energies are better able to overcome intermolecular attractions. Thus, viscosity decreases with increasing temperature, as long as no changes in composition occur.

$$H-\overset{OH}{\underset{H}{C}}-\overset{OH}{\underset{H}{C}}-\overset{OH}{\underset{H}{C}}-H$$
glycerine

Viscosity = 945 centipoise at 25°C

$\left(\begin{array}{c}\text{stronger}\\\text{attractive}\\\text{forces}\end{array}\right) \longleftrightarrow \left(\begin{array}{c}\text{higher}\\\text{viscosity}\end{array}\right)$

$\left(\begin{array}{c}\text{increasing}\\\text{temperature}\end{array}\right) \longleftrightarrow \left(\begin{array}{c}\text{lower}\\\text{viscosity}\end{array}\right)$

13-4 SURFACE TENSION

Molecules below the surface of a liquid are influenced by intermolecular attractions from all directions. Those on the surface are attracted only toward the interior (Figure 13-8); these attractions pull the surface layer toward the center. The most stable situation is one

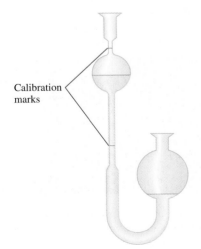

Calibration marks

Figure 13-7 The Ostwald viscometer, a device used to measure viscosity of liquids. The time it takes for a known volume of a liquid to flow through a small neck of known size is measured. Liquids with low viscosities flow rapidly.

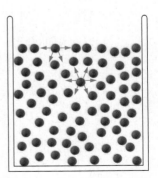

Figure 13-8 A molecular-level view of the attractive forces experienced by molecules at and below the surface of a liquid.

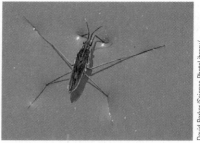

The surface tension of water supports this water strider. The non-polar surfaces of its feet also help to repel the water.

Droplets of mercury lying on a glass surface. The small droplets are almost spherical, whereas the larger droplets are flattened due to the effects of gravity. This shows that surface tension has more influence on the shape of the small (lighter) droplets.

(*Right*) The shape of a soap bubble is due to the inward force (surface tension) that acts to minimize the surface area.

in which the surface area is minimal. For a given volume, a sphere has the least possible surface area, so drops of liquid tend to assume spherical shapes. **Surface tension** is a measure of the inward forces that must be overcome to expand the surface area of a liquid.

13-5 CAPILLARY ACTION

All forces holding a liquid together are called **cohesive forces.** The forces of attraction between a liquid and another surface are **adhesive forces.** The partial positive charges on the H atoms of water hydrogen bond strongly to the partial negative charges on the oxygen atoms at the surface of the glass. As a result, water *adheres* to glass, or is said to *wet* glass. As water creeps up the side of the glass tube, its favorable area of contact with the glass increases. The surface of the water, its **meniscus,** has a concave shape (Figure 13-9). On the other hand, mercury does not wet glass because its cohesive forces (due to dispersion forces) are much stronger than its attraction to glass. Thus, its meniscus is convex. **Capillary action** occurs when one end of a capillary tube, a glass tube with a small bore (inside diameter), is immersed in a liquid. If adhesive forces exceed cohesive forces, the liquid creeps up the sides of the tube until a balance is reached between adhesive forces and the weight of liquid. The smaller the bore, the higher the liquid climbs. Capillary action helps plant roots take up water and dissolved nutrients from the soil and transmit them up the

Figure 13-9   The meniscus, as observed in glass tubes with water and with mercury.

Coating glass with a silicone polymer greatly reduces the adhesion of water to the glass. The left side of each glass has been treated with Rain-X, which contains a silicone polymer. Water on the treated side forms droplets that are easily swept away.

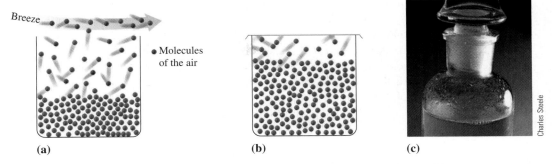

(a)   **(b)**   **(c)**

Figure 13-10 (a) Liquid continuously evaporates from an open vessel. (b) Equilibrium between liquid and vapor is established in a closed container in which molecules return to the liquid at the same rate as they leave it. (c) A bottle in which liquid–vapor equilibrium has been established. Note that droplets have condensed.

stems. The roots, like glass, exhibit strong adhesive forces for water. Osmotic pressure (Section 14-15) also plays a major role in this process.

13-6   EVAPORATION

Evaporation, or **vaporization,** is the process by which molecules on the surface of a liquid break away and go into the gas phase (Figure 13-10). Kinetic energies of molecules in liquids depend on temperature in the same way as they do in gases. The distribution of kinetic energies among liquid molecules at two different temperatures is shown in Figure 13-11. To break away, the molecules must possess at least some minimum kinetic energy. Figure 13-11 shows that at a higher temperature, a greater fraction of molecules possess at least that minimum energy. The rate of evaporation increases as temperature increases.

Only the higher-energy molecules can escape from the liquid phase. The average molecular kinetic energy of the molecules remaining in the liquid state is thereby lowered, resulting in a lower temperature in the liquid. The liquid would then be cooler than its surroundings, so it absorbs heat from its surroundings. The cooling of your body by evaporation of perspiration is a familiar example of the cooling of the surroundings by evaporation of a liquid. This is called "cooling by evaporation."

The dew on this spider web was formed by condensation of water vapor from the air.

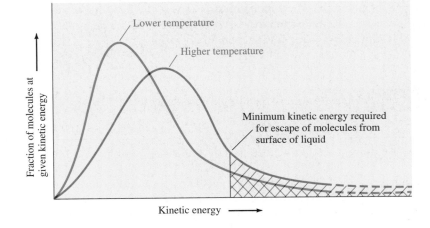

Figure 13-11 Distribution of kinetic energies of molecules in a liquid at different temperatures. At the lower temperature, a smaller fraction of the molecules have the energy required to escape from the liquid, so evaporation is slower and the equilibrium vapor pressure (Section 13-7) is lower.

A molecule in the vapor may strike the liquid surface and be captured there. This process, the reverse of evaporation, is called **condensation.** As evaporation occurs in a closed container, the volume of liquid decreases and the number of gas molecules above the surface increases. Because more gas phase molecules can collide with the surface, the rate of condensation increases. The system composed of the liquid and gas molecules of the same substance eventually achieves a **dynamic equilibrium** in which the rate of evaporation equals the rate of condensation in the closed container.

As an analogy, suppose that 30 students per minute leave a classroom, moving into the closed hallway outside, and 30 students per minute enter it. The total number of students in the room would remain constant, as would the total number of students outside the room.

$$\text{liquid} \underset{\text{condensation}}{\overset{\text{evaporation}}{\rightleftharpoons}} \text{vapor}$$

The two opposing rates are not zero, but are equal to each other—hence we call this "dynamic," rather than "static," equilibrium. Even though evaporation and condensation are both continuously occurring, *no net change occurs* because the rates are equal.

However, if the vessel were left open to the air, this equilibrium could not be reached. Molecules would diffuse away, and slight air currents would also sweep some gas molecules away from the liquid surface. This would allow more evaporation to occur to replace the lost vapor molecules. Consequently, a liquid can eventually evaporate entirely if it is left uncovered. This situation illustrates **LeChatelier's Principle:**

This is one of the guiding principles that allows us to understand chemical equilibrium. It is discussed further in Chapter 17.

> A system at equilibrium, or changing toward equilibrium, responds in the way that tends to relieve or "undo" any stress placed on it.

In this example the stress is the removal of molecules in the vapor phase. The response is the continued evaporation of the liquid.

13-7 VAPOR PRESSURE

CD-ROM   Screen 13.9, Properties of Liquids (2): Vapor Pressure; this screen contains an animation of the vaporization process.

Vapor molecules cannot escape when vaporization of a liquid occurs in a closed container. As more molecules leave the liquid, more gaseous molecules collide with the walls of the container, with one another, and with the liquid surface, so more condensation occurs. This is responsible for the formation of liquid droplets that adhere to the sides of the vessel above a liquid surface and for the eventual establishment of equilibrium between liquid and vapor (see Figure 13-10b and c).

As long as some liquid remains in contact with the vapor, the pressure does not depend on the volume or surface area of the liquid.

> The partial pressure of vapor molecules above the surface of a liquid at equilibrium at a given temperature is the **vapor pressure (vp)** of the liquid at that temperature. Because the rate of evaporation increases with increasing temperature, vapor pressures of liquids *always* increase as temperature increases.

TABLE 13-4	Vapor Pressures (in torr) of Some Liquids					
	0°C	25°C	50°C	75°C	100°C	125°C
water	4.6	23.8	92.5	300	760	1741
benzene	27.1	94.4	271	644	1360	
methyl alcohol	29.7	122	404	1126		
diethyl ether	185	470	1325	2680	4859	

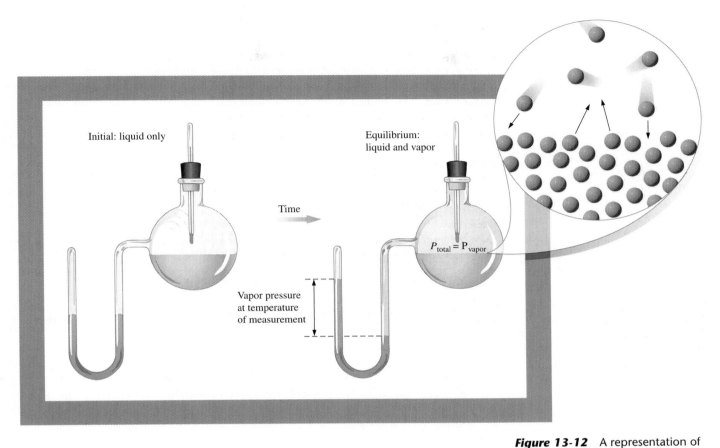

Initial: liquid only

Time

Equilibrium:
liquid and vapor

$P_{total} = P_{vapor}$

Vapor pressure
at temperature
of measurement

Easily vaporized liquids are called **volatile** liquids, and they have relatively high vapor pressures. The most volatile liquid in Table 13-4 is diethyl ether, whereas water is the least volatile. Vapor pressures can be measured with manometers (Figure 13-12).

Stronger cohesive forces tend to hold molecules in the liquid state. Methyl alcohol molecules are strongly linked by hydrogen bonding, whereas diethyl ether molecules are not, so methyl alcohol has a lower vapor pressure than diethyl ether. The very strong hydrogen bonding in water accounts for its unusually low vapor pressure (see Table 13-4) and high

Figure 13-12 A representation of the measurement of vapor pressure of a liquid at a given temperature. The container is evacuated before the liquid is added. At the instant the liquid is added to the container, there are no molecules in the gas phase so the pressure is zero. Some of the liquid then vaporizes until equilibrium is established. The difference in heights of the mercury column is a measure of the vapor pressure of the liquid at that temperature.

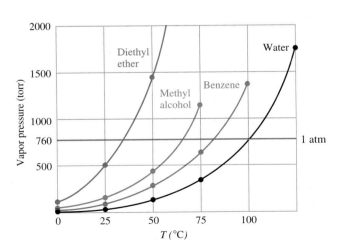

Figure 13-13 Plots of the vapor pressures of the liquids in Table 13-4. The *normal* boiling point of a liquid is the temperature at which its vapor pressure is equal to one atmosphere. Normal boiling points are: water, 100°C; benzene; 80.1°C; methyl alcohol, 65.0°C; and diethyl ether, 34.6°C. Notice that the increase in vapor pressure is *not* linear with temperature.

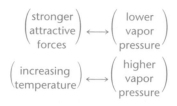

CD-ROM   Screen 13.10, Properties of Liquids (3): Boiling Point.

As water is being heated, but before it boils, small bubbles may appear in the container. This is not boiling, but rather the formation of bubbles of dissolved gases such as CO_2 and O_2 whose solubilities in water decrease with increasing temperature.

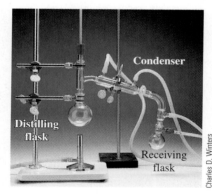

Figure 13-14 A laboratory setup for distillation. During distillation of an impure liquid, nonvolatile substances remain in the distilling flask. The liquid is vaporized and condensed before being collected in the receiving flask. If any of the substances are flammable, using an open flame would be dangerous; in such a case, another source of heat, such as an electric heater, should be used.

boiling point. Dispersion forces generally increase with increasing molecular size, so substances composed of larger molecules have lower vapor pressures.

We can understand the order of vapor pressures of the four liquids cited in Table 13-4 and Figure 13-13 by considering the strengths of their intermolecular attractions. Water has the lowest vapor pressure (strongest cohesive forces) because each molecule has two hydrogen atoms to act as hydrogen-bond donors and each molecule can accept hydrogen bonds from two other molecules. Methyl alcohol has only one potential hydrogen-bond donor, so its average cohesive forces are weaker than those in water and its vapor pressure is higher. In benzene and diethyl ether, the hydrogen atoms are all bonded to carbon, so strong hydrogen bonds are not possible. Electrons can move easily throughout the delocalized π-bonding orbitals of benzene, however, so benzene is quite polarizable and exhibits significant dispersion forces. In addition, the hydrogen atoms of benzene are more positive than most hydrogens that are bonded to carbon. The H atoms of benzene are attracted to the electron-rich π-bonding regions of nearby molecules. The accumulation of these forces gives benzene rather strong cohesive forces, resulting in a lower vapor pressure than we might expect for a hydrocarbon. The diethyl ether molecule is only slightly polar, resulting in weak dipole–dipole forces and a high vapor pressure.

13-8   BOILING POINTS AND DISTILLATION

When heat energy is added to a liquid, it increases the kinetic energy of the molecules, and the temperature of the liquid increases. Heating a liquid always increases its vapor pressure. When a liquid is heated to a sufficiently high temperature under a given applied (usually atmospheric) pressure, bubbles of vapor begin to form below the surface. If the vapor pressure inside the bubbles is less than the applied pressure on the surface of the liquid, the bubbles collapse as soon as they form. If the temperature is raised sufficiently, the vapor pressure is high enough that the bubbles can persist, rise to the surface, and burst, releasing the vapor into the air. This process is called *boiling* and is different from evaporation. The **boiling point** of a liquid is the temperature at which its vapor pressure equals the external pressure. The **normal boiling point** is the temperature at which the vapor pressure of a liquid is equal to exactly one atmosphere (760 torr). The vapor pressure of water is 760 torr at 100°C, its normal boiling point. As heat energy is added to a pure liquid *at its boiling point*, the temperature remains constant, because the energy is used to overcome the cohesive forces in the liquid to form vapor.

If the applied pressure is lower than 760 torr, say on the top of a mountain, water boils below 100°C. The chemical reactions involved in cooking food occur more slowly at the lower temperature, so it takes longer to cook food in boiling water at high altitudes than at sea level. A pressure cooker cooks food rapidly because water boils at higher temperatures under increased pressures. The higher temperature of the boiling water increases the rate of cooking.

Different liquids have different cohesive forces, so they have different vapor pressures and boil at different temperatures. A mixture of liquids with sufficiently different boiling points can often be separated into its components by **distillation**. In this process the mixture is heated slowly until the temperature reaches the point at which the most volatile liquid boils off. If this component is a liquid under ordinary conditions, it is subsequently recondensed in a water-cooled condensing column (Figure 13-14) and collected as a distillate. After enough heat has been added to vaporize all of the most volatile liquid, the temperature again rises slowly until the boiling point of the next substance is reached, and the process continues. Any nonvolatile substances dissolved in the liquid do not boil, but remain in the distilling flask. Impure water can be purified and separated from its dissolved salts by distillation. Compounds with similar boiling points, especially those that interact very strongly with one another, are not effectively separated by simple distillation but require a modification called fractional distillation (Section 14-10).

13-9 HEAT TRANSFER INVOLVING LIQUIDS

Heat must be added to a liquid to raise its temperature (Section 1-13). The **specific heat** ($J/g \cdot °C$) or **molar heat capacity** ($J/mol \cdot °C$) of a liquid is the amount of heat that must be added to the stated mass of liquid to raise its temperature by one degree Celsius. If heat is added to a liquid under constant pressure, the temperature rises until its boiling point is reached. Then the temperature remains constant until enough heat has been added to boil away all the liquid. The **molar heat** (or **enthalpy**) **of vaporization** (ΔH_{vap}) of a liquid is the amount of heat that must be added to one mole of the liquid at its boiling point to convert it to vapor with no change in temperature. Heats of vaporization can also be expressed as energy per gram. For example, the heat of vaporization for water at its boiling point is 40.7 kJ/mol, or 2.26×10^3 J/g.

$$\frac{? \text{ J}}{g} = \frac{40.7 \text{ kJ}}{\text{mol}} \times \frac{1000 \text{ J}}{\text{kJ}} \times \frac{1 \text{ mol}}{18.0 \text{ g}} = 2.26 \times 10^3 \text{ J/g}$$

Like many other properties of liquids, heats of vaporization reflect the strengths of intermolecular forces. Heats of vaporization generally increase as boiling points and intermolecular forces increase and as vapor pressures decrease. Table 13-5 illustrates this. The high heats of vaporization of water, ethylene glycol, and ethyl alcohol are due mainly to the strong hydrogen-bonding interactions in these liquids (see Section 13-2). The very high value for water makes it very effective as a coolant and, in the form of steam, as a source of heat.

Liquids can evaporate even below their boiling points. The water in perspiration is an effective coolant for our bodies. Each gram of water that evaporates absorbs 2.41 kJ of heat from the body. We feel even cooler in a breeze because perspiration evaporates faster, so heat is removed more rapidly.

The specific heat and heat capacity of a substance change somewhat with its temperature. For most substances, this variation is small enough to ignore.

Molar heats of vaporization (also called molar *enthalpies* of vaporization) are often expressed in kilojoules rather than joules. The units of heat of vaporization do *not* include temperature. This is because boiling occurs with *no change in temperature.*

CD-ROM Screen 13.8, Properties of Liquids (1): Enthalpy of Vaporization.

The heat of vaporization of water is higher at 37°C (normal body temperature) than at 100°C (2.41 kJ/g compared to 2.26 kJ/g).

TABLE 13-5 *Heats of Vaporization, Boiling Points, and Vapor Pressures of Some Common Liquids*

Liquid	Vapor Pressure (torr at 20°C)	Boiling Point at 1 atm (°C)	Heat of Vaporization at Boiling Point	
			J/g	*kJ/mol*
water, H_2O	17.5	100.	2260	40.7
ethyl alcohol, CH_3CH_2OH	43.9	78.3	855	39.3
benzene, C_6H_6	74.6	80.1	395	30.8
diethyl ether, $CH_3CH_2OCH_2CH_3$	442.	34.6	351	26.0
carbon tetrachloride, CCl_4	85.6	76.8	213	32.8
ethylene glycol, CH_2OHCH_2OH	0.1	197.3	984	58.9

Condensation is the reverse of evaporation. The amount of heat that must be removed from a vapor to condense it (without change in temperature) is called the **heat of condensation.**

$$\text{liquid} + \text{heat} \xrightleftharpoons[\text{condensation}]{\text{evaporation}} \text{vapor}$$

The heat of condensation of a liquid is equal in magnitude to the heat of vaporization. It is released by the vapor during condensation.

Because of the large amount of heat released by steam as it condenses, burns caused by steam at 100°C are much more severe than burns caused by liquid water at 100°C.

Because 2.26 kJ must be absorbed to vaporize one gram of water at 100°C, that same amount of heat must be released to the environment when one gram of steam at 100°C condenses to form liquid water at 100°C. In steam-heated radiators, steam condenses and releases 2.26 kJ of heat per gram as its molecules collide with the cooler radiator walls and condense there. The metallic walls conduct heat well. They transfer the heat to the air in contact with the outside walls of the radiator. The heats of condensation and vaporization of non–hydrogen-bonded liquids, such as benzene, have smaller magnitudes than those of hydrogen-bonded liquids (see Table 13-5). They are, therefore, much less effective as heating and cooling agents.

EXAMPLE 13-1   *Heat of Vaporization*

Calculate the amount of heat, in joules, required to convert 180. grams of water at 10.0°C to steam at 105.0°C.

Plan

The total amount of heat absorbed is the sum of the amounts required to (1) warm the liquid water from 10.0°C to 100.0°C, (2) convert the liquid water to steam at 100.0°C, and (3) warm the steam from 100.0°C to 105.0°C.

180. g H$_2$O(ℓ) at 10.0°C	Step 1: warm the liquid (temp. change)	180. g H$_2$O(ℓ) at 100.0°C	Step 2: boil the liquid (phase change)	180. g H$_2$O(g) at 100.0°C	Step 3: warm the steam (temp. change)	180. g H$_2$O(g) at 105.0°C

Steps 1 and 3 of this example involve warming with *no* phase change. Such calculations were introduced in Section 1-13.

Steps 1 and 3 involve the specific heats of water and steam, 4.18 J/g·°C and 2.03 J/g·°C, respectively (Appendix E), whereas step 2 involves the heat of vaporization of water (2.26×10^3 J/g).

Solution

Step 1: Temperature change only

1. $? \text{ J} = 180. \text{ g} \times \dfrac{4.18 \text{ J}}{\text{g} \cdot \text{°C}} \times (100.0\text{°C} - 10.0\text{°C}) = 6.77 \times 10^4 \text{ J} = 0.677 \times 10^5 \text{ J}$

Step 2: Phase change only

2. $? \text{ J} = 180. \text{ g} \times \dfrac{2.26 \times 10^3 \text{ J}}{\text{g}} \qquad\qquad = 4.07 \times 10^5 \text{ J}$

Step 3: Temperature change only

3. $? \text{ J} = 180. \text{ g} \times \dfrac{2.03 \text{ J}}{\text{g} \cdot \text{°C}} \times (105.0\text{°C} - 100.0\text{°C}) = 1.8 \times 10^3 \text{ J} = 0.018 \times 10^5 \text{ J}$

Total amount of heat absorbed = 4.76×10^5 J

You should now work Exercises 50 and 51.

Distillation is not an economical way to purify large quantities of water for public water supplies. The high heat of vaporization of water makes it too expensive to vaporize large volumes of water.

✔ **Problem-Solving Tip:** *Temperature Change or Phase Change?*

A problem such as Example 13-1 can be broken down into steps so that each involves *either* a temperature change *or* a phase change, but not both. A temperature change calculation uses the specific heat of the substance (steps 1 and 3 of Example 13-1); remember that each different phase has its own specific heat. A phase change always takes place with *no change* in temperature, so that calculation does not involve temperature (step 2 of Example 13-1).

EXAMPLE 13-2   *Heat of Vaporization*

Compare the amount of "cooling" experienced by an individual who drinks 400. mL of ice water (0.0°C) with the amount of "cooling" experienced by an individual who "sweats out" 400. mL of water. Assume that the sweat is essentially pure water and that all of it evaporates. The density of water is very nearly 1.00 g/mL at both 0.0°C and 37.0°C, average body temperature. The heat of vaporization of water is 2.41 kJ/g at 37.0°C.

Plan

In the case of drinking ice water, the body is cooled by the amount of heat required to raise the temperature of 400. mL (400. g) of water from 0.0°C to 37.0°C. The amount of heat lost by perspiration is equal to the amount of heat required to vaporize 400. g of water at 37.0°C.

Solution

Raising the temperature of 400. g of water from 0.0°C to 37.0°C requires

$$\underline{?}\ J = (400.\ g)(4.18\ J/g\cdot°C)(37.0°C) = 6.19 \times 10^4\ J,\ \text{or}\ \boxed{61.9\ kJ}$$

Evaporating (i.e., "sweating out") 400. mL of water at 37°C requires

$$\underline{?}\ J = (400.\ g)(2.41 \times 10^3\ J/g) = 9.64 \times 10^5\ J,\ \text{or}\ \boxed{964\ kJ}$$

Thus, we see that "sweating out" 400. mL of water removes 964 kJ of heat from one's body, whereas drinking 400. mL of ice water cools it by only 61.9 kJ. Stated differently, sweating removes (964/61.9) = 15.6 times more heat than drinking ice water!

You should now work Exercise 55.

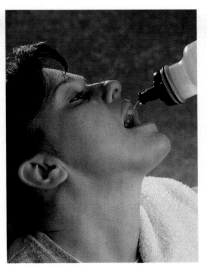

For health reasons, it is important to replace the water lost by perspiration.

Enrichment   The Clausius–Clapeyron Equation

We have seen (Figure 13-13) that vapor pressure increases with increasing temperature. Let us now discuss the quantitative expression of this relationship.

When the temperature of a liquid is changed from T_1 to T_2, the vapor pressure of the liquid changes from P_1 to P_2. These changes are related to the molar heat of vaporization, ΔH_{vap}, for the liquid by the **Clausius–Clapeyron equation.**

$$\ln\left(\frac{P_2}{P_1}\right) = \frac{\Delta H_{vap}}{R}\left(\frac{1}{T_1} - \frac{1}{T_2}\right)$$

Although ΔH_{vap} changes somewhat with temperature, it is usually adequate to use the value tabulated at the normal boiling point of the liquid (Appendix E) unless more precise values are available. The units of R must be consistent with those of ΔH_{vap}.

The Clausius–Clapeyron equation is used for three types of calculations: (1) to predict the vapor pressure of a liquid at a specified temperature, as in Example 13-3; (2) to determine the temperature at which a liquid has a specified vapor pressure; and (3) to calculate ΔH_{vap} from measurement of vapor pressures at different temperatures.

(Enrichment, continued)

EXAMPLE 13-3   *Vapor Pressure Versus Temperature*

The normal boiling point of ethanol, C_2H_5OH, is 78.3°C, and its molar heat of vaporization is 39.3 kJ/mol (Appendix E). What would be the vapor pressure, in torr, of ethanol at 50.0°C?

Plan

The normal boiling point of a liquid is the temperature at which its vapor pressure is 760 torr, so we designate this as one of the conditions (subscript 1). We wish to find the vapor pressure at another temperature (subscript 2), and we know the molar heat of vaporization. We use the Clausius–Clapeyron equation to solve for P_2.

Solution

$$P_1 = 760 \text{ torr} \quad \text{at} \quad T_1 = 78.3°C + 273.2 = 351.5 \text{ K}$$
$$P_2 = \underline{?} \quad \text{at} \quad T_2 = 50.0°C + 273.2 = 323.2 \text{ K}$$
$$\Delta H_{vap} = 39.3 \text{ kJ/mol} \quad \text{or} \quad 3.93 \times 10^4 \text{ J/mol}$$

We solve for P_2.

$$\ln\left(\frac{P_2}{760 \text{ torr}}\right) = \frac{3.93 \times 10^4 \text{ J/mol}}{\left(8.314 \dfrac{\text{J}}{\text{mol·K}}\right)}\left(\frac{1}{351.5 \text{ K}} - \frac{1}{323.2 \text{ K}}\right)$$

$$\ln\left(\frac{P_2}{760 \text{ torr}}\right) = -1.18 \quad \text{so} \quad \left(\frac{P_2}{760 \text{ torr}}\right) = e^{-1.18} = 0.307$$

$$P_2 = 0.307(760 \text{ torr}) = \boxed{233 \text{ torr}} \quad \text{(lower vapor pressure at lower temperature)}$$

You should now work Exercise 58.

We have described many properties of liquids and discussed how they depend on intermolecular forces of attraction. The general effects of these attractions on the physical properties of liquids are summarized in Table 13-6. "High" and "low" are relative terms. Table 13-6 is intended to show only very general trends. Example 13-4 illustrates the use of intermolecular attractions to predict boiling points.

TABLE 13-6	*General Effects of Intermolecular Attractions on Physical Properties of Liquids*	
Property	Volatile Liquids (weak intermolecular attractions)	Nonvolatile Liquids (strong intermolecular attractions)
cohesive forces	low	high
viscosity	low	high
surface tension	low	high
specific heat	low	high
vapor pressure	high	low
rate of evaporation	high	low
boiling point	low	high
heat of vaporization	low	high

> **EXAMPLE 13-4** *Boiling Points Versus Intermolecular Forces*
>
> Predict the order of increasing boiling points for the following: H_2S; H_2O; CH_4; H_2; KBr.
>
> **Plan**
>
> We analyze the polarity and size of each substance to determine the kinds of intermolecular forces that are present. In general, the stronger the intermolecular forces, the higher is the boiling point of the substance.
>
> **Solution**
>
> KBr is ionic, so it boils at the highest temperature. Water exhibits strong hydrogen bonding and boils at the next highest temperature. Hydrogen sulfide is the only other polar covalent substance in the list, so it boils below H_2O but above the other two substances. Both CH_4 and H_2 are nonpolar. The larger CH_4 molecule is more easily polarized than the very small H_2 molecule, so dispersion forces are stronger in CH_4. Thus, CH_4 boils at a higher temperature than H_2.
>
> $$H_2 < CH_4 < H_2S < H_2O < KBr$$
> $$\xrightarrow{\hspace{4cm}}$$
> increasing boiling points

You should now work Exercise 24.

THE SOLID STATE

13-10 MELTING POINT

The **melting point (freezing point)** of a substance is the temperature at which its solid and liquid phases coexist in equilibrium.

$$\text{solid} \underset{\text{freezing}}{\overset{\text{melting}}{\rightleftharpoons}} \text{liquid}$$

The *melting point* of a solid is the same as the *freezing point* of its liquid. It is the temperature at which the rate of melting of a solid is the same as the rate of freezing of its liquid under a given applied pressure.

The **normal melting point** of a substance is its melting point at one atmosphere pressure. Changes in pressure have very small effects on melting points, but they have large effects on boiling points.

13-11 HEAT TRANSFER INVOLVING SOLIDS

When heat is added to a solid below its melting point, its temperature rises. After enough heat has been added to bring the solid to its melting point, additional heat is required to convert the solid to liquid. During this melting process, the temperature remains constant at the melting point until all of the substance has melted. After melting is complete, the continued addition of heat results in an increase in the temperature of the liquid, until the boiling point is reached. This is illustrated graphically in the first three segments of the heating curve in Figure 13-15.

The **molar heat** (or **enthalpy**) **of fusion** (ΔH_{fus}; kJ/mol) is the amount of heat required to melt one mole of a solid at its melting point. Heats of fusion can also be expressed on a

During any phase change,
liquid ⇌ gas
and
solid ⇌ liquid
the temperature remains constant.

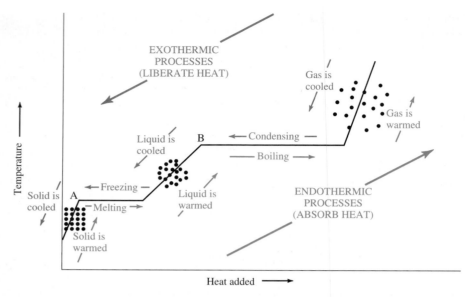

Charles D. Winters

Figure 13-15 A typical heating curve at constant pressure. When heat energy is added to a solid below its melting point, the temperature of the solid rises until its melting point is reached (point A). In this region of the plot, the slope is rather steep because of the low specific heats of solids [e.g., 2.09 J/g·°C for $H_2O(s)$]. If the solid is heated at its melting point (A), its temperature remains constant until the solid has melted, because the melting process requires energy. The length of this horizontal line is proportional to the heat of fusion of the substance—the higher the heat of fusion, the longer the line. When all of the solid has melted, heating the liquid raises its temperature until its boiling point is reached (point B). The slope of this line is less steep than that for warming the solid, because the specific heat of the liquid phase [e.g., 4.18 J/g·°C for $H_2O(\ell)$] is usually greater than that of the corresponding solid. If heat is added to the liquid at its boiling point (B), the added heat energy is absorbed as the liquid boils. This horizontal line is longer than the previous one, because the heat of vaporization of a substance is always higher than its heat of fusion. When all of the liquid has been converted to a gas (vapor), the addition of more heat raises the temperature of the gas. This segment of the plot has a steep slope because of the relatively low specific heat of the gas phase [e.g., 2.03 J/g·°C for $H_2O(g)$]. Each step in the process can be reversed by removing the same amount of heat.

Melting is always endothermic. The term "fusion" means "melting."

per gram basis. The heat of fusion depends on the *inter*molecular forces of attraction in the solid state. These forces "hold the molecules together" as a solid. Heats of fusion are *usually* higher for substances with higher melting points. Values for some common compounds are shown in Table 13-7. Appendix E has more values.

The **heat** (or **enthalpy**) **of solidification** of a liquid is equal in magnitude to the heat of fusion. It represents removal of a sufficient amount of heat from a given amount (1 mol or 1 g) of liquid to solidify the liquid at its freezing point. For water,

$$\text{ice} \underset{\substack{\text{6.02 kJ/mol} \\ \text{or 334 J/g released}}}{\overset{\substack{\text{6.02 kJ/mol} \\ \text{or 334 J/g absorbed}}}{\rightleftharpoons}} \text{water} \qquad \text{(at 0°C)}$$

TABLE 13-7 *Some Melting Points and Heats of Fusion*

Substance	Melting Point (°C)	Heat of Fusion	
		J/g	kJ/mol
methane, CH_4	−182	58.6	0.92
ethyl alcohol, CH_3CH_2OH	−117	109	5.02
water, H_2O	0	334	6.02
naphthalene, $C_{10}H_8$	80.2	147	18.8
silver nitrate, $AgNO_3$	209	67.8	11.5
aluminum, Al	658	395	10.6
sodium chloride, NaCl	801	519	30.3

EXAMPLE 13-5 *Heat of Fusion*

The molar heat of fusion, ΔH_{fus}, of Na is 2.6 kJ/mol at its melting point, 97.5°C. How much heat must be absorbed by 5.0 g of solid Na at 97.5°C to melt it?

Plan

Melting takes place at a constant temperature. The molar heat of fusion tells us that every mole of Na, 23 grams, absorbs 2.6 kJ of heat at 97.5°C during the melting process. We want to know the amount of heat that 5.0 grams would absorb. We use the appropriate unit factors, constructed from the atomic weight and ΔH_{fus}, to find the amount of heat absorbed.

Solution

$$\underline{?}\ kJ = 5.0\ g\ Na \times \frac{1\ mol\ Na}{23\ g\ Na} \times \frac{2.6\ kJ}{1\ mol\ Na} = \boxed{0.57\ kJ}$$

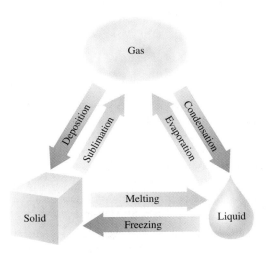

Transitions among the three states of matter. The transitions shown in blue are endothermic (absorb heat); those shown in red are exothermic (release heat).

Ice is very efficient for cooling because considerable heat is required to melt a given mass of it. However, ΔH_{vap} is generally much greater than ΔH_{fusion}, so evaporative cooling is preferable when possible.

 It is important to know the melting and boiling point temperatures for the material in question. We must always treat the phase change as a separate part of the energy calculation.

Figure 13-16 Sublimation can be used to purify volatile solids. The high vapor pressure of the solid substance causes it to sublime when heated. Crystals of purified substance are formed when the vapor is deposited to form solid on the cooler (upper) portion of the apparatus. Iodine, I_2, sublimes readily. I_2 vapor is purple.

Charles D. Winters

EXAMPLE 13-6   *Heat of Fusion*

Calculate the amount of heat that must be absorbed by 50.0 grams of ice at $-12.0°C$ to convert it to water at $20.0°C$. Refer to Appendix E.

Plan

We must determine the amount of heat absorbed during three steps: (1) warming 50.0 g of ice from $-12.0°C$ to its melting point, $0.0°C$ (we use the specific heat of ice, 2.09 J/g·°C); (2) melting the ice with no change in temperature (we use the heat of fusion of ice at $0.0°C$, 334 J/g; and (3) warming the resulting liquid from $0.0°C$ to $20.0°C$ (we use the specific heat of water, 4.18 J/g·°C).

$$50.0 \text{ g } H_2O(s) \text{ at } -12.0°C \xrightarrow[\text{(temp. change)}]{\text{Step 1: warm the ice}} 50.0 \text{ g } H_2O(s) \text{ at } 0.0°C \xrightarrow[\text{(phase change)}]{\text{Step 2: melt the ice}}$$

$$50.0 \text{ g } H_2O(\ell) \text{ at } 0.0°C \xrightarrow[\text{(temp. change)}]{\text{Step 3: warm the liquid}} 50.0 \text{ g } H_2O(\ell) \text{ at } 20°C$$

Solution

1. $50.0 \text{ g} \times \dfrac{2.09 \text{ J}}{\text{g·°C}} \times [0.0 - (-12.0)]°C = 1.25 \times 10^3 \text{ J} = 0.125 \times 10^4 \text{ J}$

2. $50.0 \text{ g} \times \dfrac{334 \text{ J}}{\text{g}} = 1.67 \times 10^4 \text{ J}$

3. $50.0 \text{ g} \times \dfrac{4.18 \text{ J}}{\text{g·°C}} \times (20.0 - 0.0)°C = 4.18 \times 10^3 \text{ J} = 0.418 \times 10^4 \text{ J}$

Total amount of heat absorbed $= 2.21 \times 10^4 \text{ J} = 22.1 \text{ kJ}$

Note that most of the heat was absorbed in step 2, melting the ice.

You should now work Exercise 56.

13-12   SUBLIMATION AND THE VAPOR PRESSURE OF SOLIDS

Some solids, such as iodine and carbon dioxide, vaporize at atmospheric pressure without passing through the liquid state. This process is known as **sublimation.** Solids exhibit vapor pressures just as liquids do, but they generally have much lower vapor pressures. Solids with high vapor pressures sublime easily. The characteristic odor of a common household solid, *para*-dichlorobenzene (moth repellent), is due to sublimation. The reverse process, by which a vapor solidifies without passing through the liquid phase, is called **deposition.**

$$\text{solid} \underset{\text{deposition}}{\overset{\text{sublimation}}{\rightleftharpoons}} \text{gas}$$

Some impure solids can be purified by sublimation and subsequent deposition of the vapor (as a solid) onto a cooler surface. Purification of iodine by sublimation is illustrated in Figure 13-16.

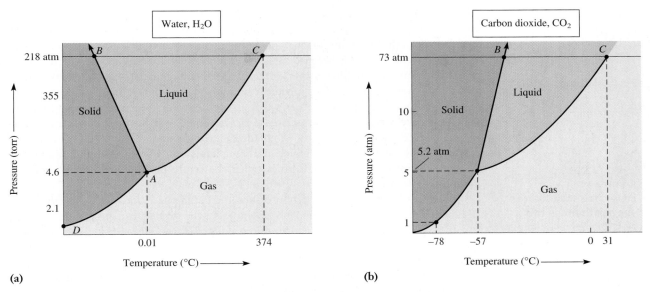

Figure 13-17 Phase diagrams (not to scale). (a) Diagram for water. For water and a few other substances for which the solid is less dense than the liquid, the solid–liquid equilibrium line (*AB*) has negative slope, that is, up and to the left. (b) Diagram for carbon dioxide, a substance for which the solid is denser than the liquid. Note that the solid–liquid equilibrium line has positive slope, that is, up and to the right. This is true for most substances.

13-13 PHASE DIAGRAMS (*P* VERSUS *T*)

We have discussed the general properties of the three phases of matter. Now we can describe **phase diagrams** that show the equilibrium pressure–temperature relationships among the different phases of a given pure substance in a closed system. Our discussion of phase diagrams applies only to *closed systems* (e.g., a sample in a sealed container), in which matter does not escape into the surroundings. This limitation is especially important when the vapor phase is involved. Figure 13-17 shows a portion of the phase diagrams for water and carbon dioxide. The curves are not drawn to scale. The distortion allows us to describe the changes of state over wide ranges of pressure or temperature using one diagram.

The curved line from *A* to *C* in Figure 13-17a is a vapor pressure curve obtained experimentally by measuring the vapor pressures of water at various temperatures (Table 13-8). Points along this curve represent the temperature–pressure combinations for which liquid and gas (vapor) coexist in equilibrium. At points above *AC*, the stable form of water is liquid; below the curve, it is vapor.

CD-ROM   Screen 13.17, Phase Changes: Phase Diagrams.

TABLE 13-8	Points on the Vapor Pressure Curve for Water								
temperature (°C)	−10	0	20	30	50	70	90	95	100	101
vapor pressure (torr)	2.1	4.6	17.5	31.8	92.5	234	526	634	760	788

 CD-ROM   Screen 13.6, The
Weird Properties of Water.

Benzene is *denser* as a solid than as
a liquid, so the solid sinks in the liquid (*left*). This is the behavior shown
by nearly all known substances except water (*right*).

The CO_2 in common fire extinguishers is liquid. As you can see from Figure 13-17b, the liquid must be at
some pressure greater than 10 atm
for temperatures above 0°C. It is ordinarily at about 65 atm (more than
900 lb/in.²), so these cylinders must
be *handled with care*.

Phase diagrams are obtained by combining the results of heating curves
measured experimentally at different
pressures.

Water is one of the few compounds
that expands when it freezes. This
expansion on freezing can lead to
the breaking of sealed containers
that are too full of water.

Line *AB* represents the liquid–solid equilibrium conditions. We see that it has a negative slope. Water is one of the very few substances for which this is the case. The negative
slope (up and to the left) indicates that increasing the pressure sufficiently on the surface of
ice causes it to melt. This is because ice is *less dense* than liquid water in the vicinity of the
liquid–solid equilibrium. The network of hydrogen bonding in ice is more extensive than
that in liquid water and requires a greater separation of H_2O molecules. This causes ice to
float in liquid water. Almost all other solids are denser than their corresponding liquids;
they would have positive slopes associated with line *AB*. The stable form of water at points
to the left of *AB* is solid (ice). Thus *AB* is called a *melting curve*.

There is only one point, *A*, at which all three phases of a substance—solid, liquid, and
gas—can coexist at equilibrium. This is called the **triple point.** For water it occurs at
4.6 torr and 0.01°C.

At pressures below the triple-point pressure, the liquid phase does not exist; rather, the
substance goes directly from solid to gas (sublimes) or the reverse happens (crystals deposit
from the gas). At pressures and temperatures along *AD*, the *sublimation curve*, solid and
vapor are in equilibrium.

Consider CO_2 (Figure 13-17b). The triple point is at 5.2 atmospheres and −57°C. This
pressure is *above* normal atmospheric pressure, so liquid CO_2 cannot exist at atmospheric
pressure. Dry ice (solid CO_2) sublimes and does not melt at atmospheric pressure.

The **critical temperature** is the temperature above which a gas cannot be liquefied, that
is, the temperature above which the liquid and gas do not exist as distinct phases. A substance at a temperature above its critical temperature is called a *supercritical fluid*. The **critical pressure** is the pressure required to liquefy a gas (vapor) *at* its critical temperature. The
combination of critical temperature and critical pressure is called the **critical point** (*C* in
Figure 13-17). For H_2O, the critical point is at 374°C and 218 atmospheres; for CO_2, it is
at 31°C and 73 atmospheres. There is no such upper limit to the solid–liquid line, however,
as emphasized by the arrowhead at the top of that line.

To illustrate the use of a phase diagram in determining the physical state or states of
a system under different sets of pressures and temperatures, let's consider a sample of water
at point *E* in Figure 13-18a (355 torr and −10°C). At this point all the water is in the
form of ice, $H_2O(s)$. Suppose that we hold the pressure constant and gradually increase the
temperature—in other words, trace a path from left to right along *EG*. At the temperature at which *EG* intersects *AB*, the melting curve, some of the ice melts. If we stopped
here, equilibrium between solid and liquid water would eventually be established, and both
phases would be present. If we added more heat, all the solid would melt with no temperature change. Remember that all phase changes of pure substances occur at constant
temperature.

Once the solid is completely melted, additional heat causes the temperature to rise.
Eventually, at point *F* (355 torr and 80°C), some of the liquid begins to boil; liquid, $H_2O(\ell)$,
and vapor, $H_2O(g)$, are in equilibrium. Adding more heat at constant pressure vaporizes the
rest of the water with no temperature change. Adding still more heat warms the vapor (gas)
from *F* to *G*. Complete vaporization would also occur if, at point *F* and before all the liquid had vaporized, the temperature were held constant and the pressure were decreased to,
say, 234 torr at point *H*. If we wished to hold the pressure at 234 torr and condense some
of the vapor, it would be necessary to cool the vapor to 70°C, point *I*, which lies on the vapor pressure curve, *AC*. To state this in another way, the vapor pressure of water at 70°C is
234 torr.

Suppose we move back to solid at point *E* (355 torr and −10°C). If we now hold the
temperature at −10°C and reduce the pressure, we move vertically down along *EJ*. At a
pressure of 2.1 torr we reach the sublimation curve, at which point the solid passes directly

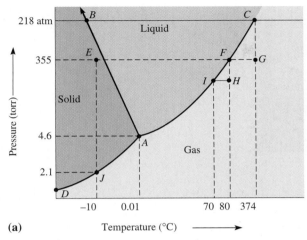

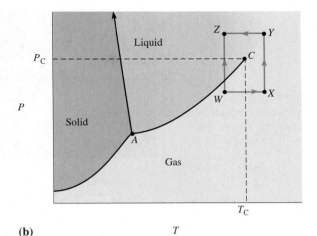

(a)   Temperature (°C) ⟶   **(b)**   T

Figure 13-18 Some interpretations of phase diagrams. (a) The phase diagram of water. Phase relationships at various points in this diagram are described in the text. (b) Two paths by which a gas can be liquefied. (1) Below the critical temperature. Compressing the sample at *constant* temperature is represented by the vertical line *WZ*. Where this line crosses the vapor pressure curve *AC*, the gas liquefies; at that set of conditions, *two distinct phases,* gas and liquid, are present in equilibrium with each other. These two phases have different properties, for example, different densities. Raising the pressure further results in a completely liquid sample at point *Z*. (2) Above the critical temperature. Suppose that we instead first warm the gas at constant pressure from *W* to *X,* a temperature above its critical temperature. Then, holding the temperature constant, we increase the pressure to point *Y*. Along this path, the sample increases *smoothly* in density, with no sharp transition between phases. From *Y,* we then decrease the temperature to reach final point *Z,* where the sample is clearly a liquid.

into the gas phase (sublimes) until all the ice has sublimed. An important application of this phenomenon is in the freeze-drying of foods. In this process a water-containing food is cooled below the freezing point of water to form ice, which is then removed as a vapor by decreasing the pressure.

Let's clarify the nature of the fluid phases (liquid and gas) and of the critical point by describing two different ways that a gas can be liquefied. A sample at point *W* in the phase diagram of Figure 13-18b is in the vapor (gas) phase, below its critical temperature. Suppose we compress the sample at constant *T* from point *W* to point *Z*. We can identify a definite pressure (the intersection of line *WZ* with the vapor pressure curve *AC*) where the transition from gas to liquid takes place. If we go *around* the critical point by the path *WXYZ,* however, no such clear-cut transition takes place. By this second path, the density and other properties of the sample vary in a continuous manner; there is no definite point at which we can say that the sample changes from gas to liquid.

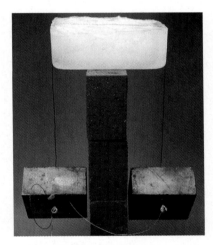

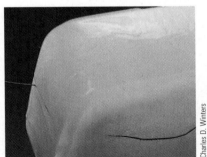

A weighted wire cuts through a block of ice. The ice melts under the high pressure of the wire, and then refreezes behind the wire.

A fluid *below* its critical temperature may properly be identified as a liquid or as a gas. *Above* the critical temperature, we should use the term "fluid."

13-14 AMORPHOUS SOLIDS AND CRYSTALLINE SOLIDS

We have already seen that solids have definite shapes and volumes, are not very compressible, are dense, and diffuse only very slowly into other solids. They are generally characterized by compact, ordered arrangements of particles that vibrate about fixed positions in their structures.

Charles D. Winters

The regular external shape of a crystal is the result of regular internal arrangements of atoms, molecules, or ions. The crystal shown is quartz.

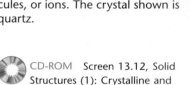 CD-ROM   Screen 13.12, Solid Structures (1): Crystalline and Amorphous Solids.

One test for the purity of a crystalline solid is the sharpness of its melting point. Impurities disrupt the intermolecular forces and cause melting to occur over a considerable temperature range.

The lattice planes are planes within the crystal containing ordered arrangements of particles.

Some noncrystalline solids, called **amorphous solids,** have no well-defined, ordered structure. Examples include rubber, some kinds of plastics, and amorphous sulfur.

Some amorphous solids are called glasses because, like liquids, they flow, although *very* slowly. The irregular structures of glasses are intermediate between those of freely flowing liquids and those of crystalline solids; there is only short-range order. Crystalline solids such as ice and sodium chloride have well-defined, sharp melting temperatures. Particles in amorphous solids are irregularly arranged, so intermolecular forces among their particles vary in strength within a sample. Melting occurs at different temperatures for various portions of the same sample as the intermolecular forces are overcome. Unlike crystalline solids, glasses and other amorphous solids do not exhibit sharp melting points, but soften over a temperature range.

The shattering of a crystalline solid produces fragments having the same (or related) interfacial angles and structural characteristics as the original sample. The shattering of a cube of rock salt produces several smaller cubes of rock salt. This cleaving occurs preferentially along crystal lattice planes between which the interionic or intermolecular forces of attraction are weakest. Amorphous solids with irregular structures, such as glasses, shatter irregularly to yield pieces with curved edges and irregular angles.

William and Lawrence Bragg are the only father and son to receive the Nobel Prize, which they shared in physics in 1915.

Enrichment   X-Ray Diffraction

Atoms, molecules, and ions are much too small to be seen with the eye. The arrangements of particles in crystalline solids are determined indirectly by X-ray diffraction (scattering). In 1912, the German physicist Max von Laue (1879–1960) showed that any crystal could serve as a three-dimensional diffraction grating for incident electromagnetic radiation with wavelengths approximating the internuclear separations of atoms in the crystal. Such radiation is in the X-ray region of the electromagnetic spectrum. Using an apparatus such as that shown in Figure 13-19, a monochromatic (single-wavelength) X-ray beam is defined by a system of slits and directed onto a crystal. The crystal is rotated to vary the angle of incidence θ. At various angles, strong beams of deflected X-rays hit a photographic plate. Upon development, the plate shows a set of symmetrically arranged spots due to deflected X-rays. Different crystals produce different arrangements of spots.

In 1913, the English scientists William (1862–1942) and Lawrence (1890–1971) Bragg found that diffraction photographs are more easily interpreted by considering the crystal as a reflection grating rather than a diffraction grating. The analysis of the spots is somewhat complicated, but an experienced crystallographer can determine the separations between atoms within identical layers and the distances between layers of atoms. The more electrons an atom has, the more strongly it scatters X-rays, so it is also possible to determine the identities of individual atoms using this technique.

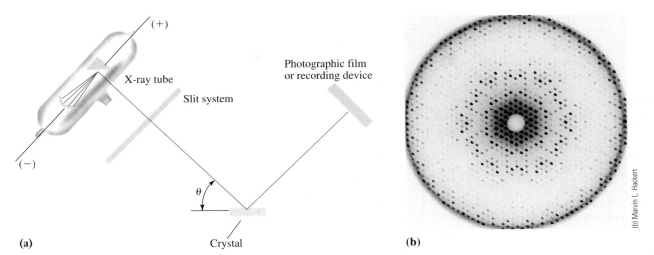

(a)

X-ray tube

Slit system

Photographic film or recording device

θ

Crystal

(b)

(b) Marvin L. Hackert

Figure 13-19 (a) X-ray diffraction by crystals (schematic). (b) A photograph of the X-ray diffraction pattern from a crystal of the enzyme histidine decarboxylase (MW $\approx$ 37,000 amu). The crystal was rotated so that many different lattice planes with different spacings were moved in succession into diffracting position (see Figure 13-20).

Figure 13-20 illustrates the determination of spacings between layers of atoms. The X-ray beam strikes parallel layers of atoms in the crystal at an angle θ. Those rays colliding with atoms in the first layer are reflected at the same angle θ. Those passing through the first layer may be reflected from the second layer, third layer, and so forth. A reflected beam results only if all rays are in phase.

For the waves to be in phase (interact constructively), the difference in path length must be equal to the wavelength, λ, times an integer, n. This leads to the condition known as the **Bragg equation.**

$$n\lambda = 2d \sin \theta \qquad \text{or} \qquad \sin \theta = \frac{n\lambda}{2d}$$

It tells us that for X-rays of a given wavelength λ, atoms in planes separated by distances d give rise to reflections at angles of incidence θ. The reflection angle increases with increasing order, $n = 1, 2, 3, \ldots$.

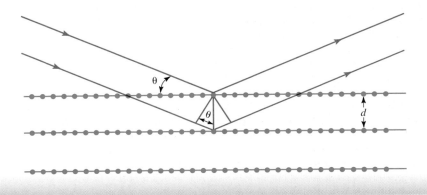

Figure 13-20 Reflection of a monochromatic beam of X-rays by two lattice planes (layers of atoms) of a crystal.

Figure 13-21 Patterns that repeat in two dimensions. Such patterns might be used to make wallpaper. We must imagine that the pattern extends indefinitely (to the end of the wall). In each pattern two of the many possible choices of unit cells are outlined. Once we identify a unit cell and its contents, repetition by translating this unit generates the entire pattern. In (a) the unit cell contains only one cat. In (b) each cell contains two cats related to one another by a 180° rotation. Any crystal is an analogous pattern in which the contents of the three-dimensional unit cell consist of atoms, molecules, or ions. The pattern extends in *three* dimensions to the boundaries of the crystal, usually including many thousands of unit cells.

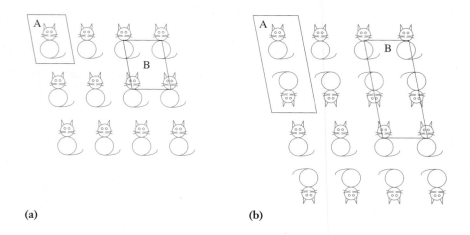

(a)   (b)

13-15 STRUCTURES OF CRYSTALS

All crystals contain regularly repeating arrangements of atoms, molecules, or ions. They are analogous (but in three dimensions) to a wallpaper pattern (Figure 13-21). Once we discover the pattern of a wallpaper, we can repeat it in two dimensions to cover a wall. To describe such a repeating pattern we must specify two things: (1) the size and shape of the repeating unit and (2) the contents of this unit. In the wallpaper pattern of Figure 13-21a, two different choices of the repeating unit are outlined. Repeating unit A contains one complete cat; unit B, with the same area, contains parts of four different cats, but these still add up to one complete cat. From whichever unit we choose, we can obtain the entire pattern by repeatedly translating the contents of that unit in two dimensions.

In a crystal the repeating unit is three-dimensional; its contents consist of atoms, molecules, or ions. The smallest unit of volume of a crystal that shows all the characteristics of the crystal's pattern is a **unit cell.** We note that the unit cell is just the fundamental *box* that describes the arrangement. The unit cell is described by the lengths of its edges—*a, b, c* (which are related to the spacings between layers, *d*)—and the angles between the edges—α, β, γ (Figure 13-22). Unit cells are stacked in three dimensions to build a lattice, the three-dimensional *arrangement* corresponding to the crystal. It can be proven that unit cells must fit into one of the seven crystal systems (Table 13-9). Each crystal system is

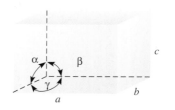

Figure 13-22 A representation of a unit cell.

TABLE 13-9	*The Unit Cell Relationships for the Seven Crystal Systems**		
	Unit Cell		
System	*Lengths*	*Angles*	**Example (common name)**
cubic	$a = b = c$	$\alpha = \beta = \gamma = 90°$	NaCl (rock salt)
tetragonal	$a = b \neq c$	$\alpha = \beta = \gamma = 90°$	TiO_2 (rutile)
orthorhombic	$a \neq b \neq c$	$\alpha = \beta = \gamma = 90°$	$MgSO_4 \cdot 7H_2O$ (epsomite)
monoclinic	$a \neq b \neq c$	$\alpha = \gamma = 90°; \beta \neq 90°$	$CaSO_4 \cdot 2H_2O$ (gypsum)
triclinic	$a \neq b \neq c$	$\alpha \neq \beta \neq \gamma \neq 90°$	$K_2Cr_2O_7$ (potassium dichromate)
hexagonal	$a = b \neq c$	$\alpha = \beta = 90°; \gamma = 120°$	SiO_2 (silica)
rhombohedral	$a = b = c$	$\alpha = \beta = \gamma \neq 90°$	$CaCO_3$ (calcite)

In these definitions, the sign ≠ means "is not necessarily equal to."

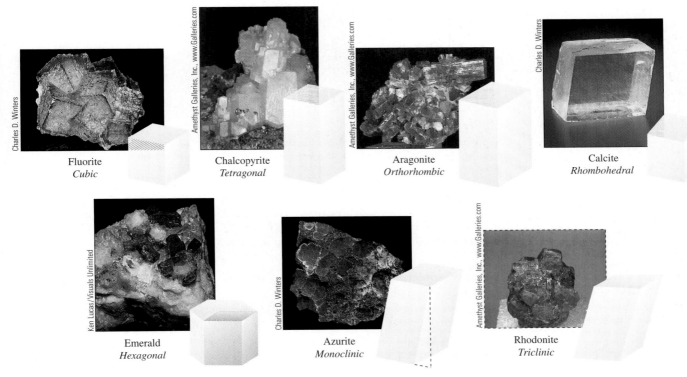

Fluorite
Cubic

Chalcopyrite
Tetragonal

Aragonite
Orthorhombic

Calcite
Rhombohedral

Emerald
Hexagonal

Azurite
Monoclinic

Rhodonite
Triclinic

Figure 13-23 Shapes of unit cells for the seven crystal systems and a representative mineral of each system.

distinguished by the relations between the unit cell lengths and angles *and* by the symmetry of the resulting three-dimensional patterns. Crystals have the same symmetry as their constituent unit cells because all crystals are repetitive multiples of such cells.

Let's replace each repeat unit in the crystal by a point (called a *lattice point*) placed at the same place in the unit. All such points have the same environment and are indistinguishable from one another. The resulting three-dimensional array of points is called a **lattice**. It is a simple but complete description of the way in which a crystal structure is built up.

The unit cells shown in Figure 13-23a are the simple, or primitive, unit cells corresponding to the seven crystal systems listed in Table 13-9. Each of these unit cells corresponds to *one* lattice point. As a two-dimensional representation of the reasoning behind this statement, look at the unit cell marked "B" in Figure 13-21a. Each corner of the unit cell is a lattice point, and can be imagined to represent one cat. The cat at each corner is shared among four unit cells (remember—we are working in two dimensions here). The unit cell has four corners, and in the corners of the unit cell are enough pieces to make one complete cat. Thus, unit cell B contains one cat, the same as the alternative unit cell choice marked "A." Now imagine that each lattice point in a three-dimensional crystal represents an object (a molecule, an atom, and so on). Such an object at a corner (Figure 13-24a) is shared by the eight unit cells that meet at that corner. Each unit cell has eight corners, so it contains eight "pieces" of the object, so it contains $8 \times \frac{1}{8} = 1$ object. Similarly, an object on an edge, but not at a corner, is shared by four unit cells (Figure 13-24b), and an object on a face is shared by two unit cells (Figure 13-24c).

Each unit cell contains atoms, molecules, or ions in a definite arrangement. Often the unit cell contents are related by some additional symmetry. (For instance, the unit cell in Figure 13-21b contains *two* cats, related to one another by a rotation of 180°.) Different substances that crystallize in the same type of lattice with the same atomic arrangement are

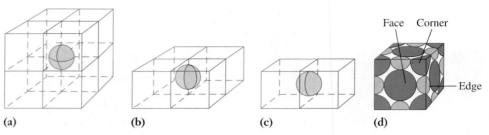

Figure 13-24 Representation of the sharing of an object (an atom, ion, or molecule) among unit cells. The fraction of each sphere that "belongs" to a single unit cell is shown in red. (a) The sharing of an object at a corner by eight unit cells. (b) The sharing of an object on an edge by four unit cells. (c) The sharing of an object in a face by two unit cells. (d) A representation of a unit cell that illustrates the portions of atoms presented in more detail in Figure 13-28. The green ion at each corner is shared by eight unit cells, as in part (a). The gray ion at each edge is shared by four unit cells, as in Part (b). The blue ion in each face is shared by two unit cells, as in Part (c).

said to be **isomorphous.** A single substance that can crystallize in more than one arrangement is said to be **polymorphous.**

In a *simple*, or *primitive*, lattice, only the eight corners of the unit cell are equivalent. In other types of crystals, objects equivalent to those forming the outline of the unit cell may occupy extra positions within the unit cell. (In this context, "equivalent" means that the same atoms, molecules, or ions appear in *identical environments and orientations* at the eight corners of the cell and, when applicable, at other locations in the unit cell.) This results in additional lattices besides the simple ones in Figure 13-23. Two of these are shown in Figure 13-25b, c. A *body-centered* lattice has equivalent points at the eight unit cell corners *and* at the center of the unit cell (see Figure 13-25). Iron, chromium, and many other metals crystallize in a body-centered cubic (bcc) arrangement. The unit cell of such a metal contains $8 \times \frac{1}{8} = 1$ atom at the corners of the cell *plus* one atom at the center of the cell (and

A crystal of one form of polonium metal has Po atoms at the corners of a simple cubic unit cell that is 3.35 Å on edge (Example 13-7).

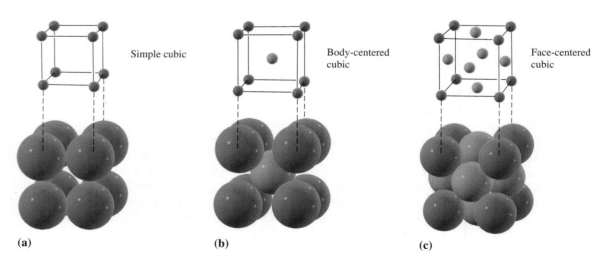

Figure 13-25 Unit cells for (a) simple cubic, (b) body-centered cubic, and (c) face-centered cubic. The spheres in each figure represent *identical* atoms or ions; different colors are shown *only* to help you visualize the spheres in the center of the cube in body-centered cubic (b) and in face-centered cubic (c) forms.

therefore entirely in this cell); this makes a total of *two* atoms per unit cell. A *face-centered* structure involves the eight points at the corners and six more equivalent points, one in the middle of each of the six square faces of the cell. A metal (calcium and silver are cubic examples) that crystallizes in this arrangement has $8 \times \frac{1}{8} = 1$ atom at the corners *plus* $6 \times \frac{1}{2} = 3$ more in the faces, for a total of *four* atoms per unit cell. In more complicated crystals, each lattice site may represent several atoms or an entire molecule.

We have discussed some simple structures that are easy to visualize. More complex compounds crystallize in structures with unit cells that can be more difficult to describe. Experimental determination of the crystal structures of such solids is correspondingly more complex. Modern computer-controlled instrumentation can collect and analyze the large amounts of X-ray diffraction data used in such studies. This now allows analysis of structures ranging from simple metals to complex biological molecules such as proteins and nucleic acids. Most of our knowledge about the three-dimensional arrangements of atoms depends on crystal structure studies.

Each object in a face is shared between two unit cells, so it is counted $\frac{1}{2}$ in each; there are six faces in each unit cell.

13-16 BONDING IN SOLIDS

We classify crystalline solids into categories according to the types of particles in the crystal and the bonding or interactions among them. The four categories are (1) metallic solids, (2) ionic solids, (3) molecular solids, and (4) covalent solids. Table 13-10 summarizes these categories of solids and their typical properties.

Metallic Solids

Metals crystallize as solids in which metal ions may be thought to occupy the lattice sites and are embedded in a cloud of delocalized valence electrons. Nearly all metals crystallize in one of three types of lattices: (1) body-centered cubic (bcc), (2) face-centered cubic (fcc; also called cubic close-packed), and (3) hexagonal close-packed. The latter two types are called close-packed structures because the particles (in this case metal atoms) are packed together as closely as possible. The differences between the two close-packed structures are

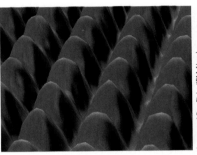

Scanning tunneling microscope image of nickel atoms on the surface of nickel metal.

Courtesy of Don Eigler/IBM Almaden Research Center

TABLE 13-10	*Characteristics of Types of Solids*			
	Metallic	**Ionic**	**Molecular**	**Covalent**
Particles of unit cell	Metal ions in "electron cloud"	Anions, cations	Molecules (or atoms)	Atoms
Strongest interparticle forces	Metallic bonds attraction between cations and e^-'s)	Electrostatic	Dispersion, dipole–dipole, and/or hydrogen bonds	Covalent bonds
Properties	Soft to very hard; good thermal and electrical conductors; wide range of melting points (-39 to $3400°C$)	Hard; brittle; poor thermal and electrical conductors; high melting points (400 to $3000°C$)	Soft; poor thermal and electrical conductors; low melting points (-272 to $400°C$)	Very hard; poor thermal and electrical conductors;* high melting points (1200 to $4000°C$)
Examples	Li, K, Ca, Cu, Cr, Ni (metals)	$NaCl$, $CaBr_2$, K_2SO_4 (typical salts)	CH_4 (methane), P_4, O_2, Ar, CO_2, H_2O, S_8	C (diamond), SiO_2 (quartz)

Exceptions: Diamond is a good conductor of heat; graphite is soft and conducts electricity well.

Figure 13-26 (a) Spheres in the same plane, packed as closely as possible. Each sphere touches six others. (b) Spheres in two planes, packed as closely as possible. All spheres represent *identical* atoms or ions; different colors are shown *only* to help you visualize the layers. Real crystals have many more than two planes. Each sphere touches six others in its own layer, three in the layer below it, and three in the layer above it; that is, it contacts a total of 12 other spheres (has a coordination number of 12).

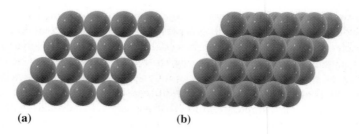

(a)   **(b)**

illustrated in Figures 13-26 and 13-27. Let spheres of equal size represent identical metal atoms, or any other particles, that form close-packed structures. Consider a layer of spheres packed in a plane, *A*, as closely as possible (Figure 13-27a). An identical plane of spheres, *B*, is placed in the depressions of plane *A*. If the third plane is placed with its spheres directly above those in plane *A*, the *ABA* arrangement results. This is the hexagonal close-packed

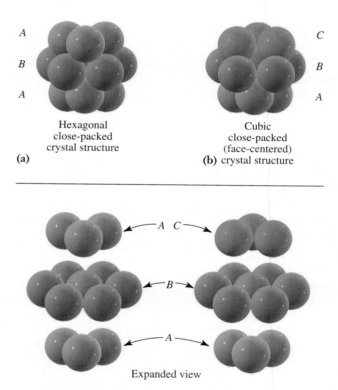

The most efficient solid-state packing is usually preferred in order to minimize the amount of "empty" space in a structure. The exact crystal structure type, however, can vary depending on the shapes and sizes of the ions, atoms, or molecules making up the solid.

Hexagonal close-packed crystal structure
(a)

Cubic close-packed (face-centered) **(b)** crystal structure

Expanded view

Figure 13-27 There are two crystal structures in which atoms are packed together as compactly as possible. The diagrams show the structures expanded to clarify the difference between them. (a) In the hexagonal close-packed structure, the first and third layers are oriented in the same direction, so that each atom in the third layer (*A*) lies directly above an atom in the first layer (*A*). (b) In the cubic close-packed structure, the first and third layers are oriented in opposite directions, so that no atom in the third layer (*C*) is directly above an atom in either of the first two layers (*A* and *B*). In both cases, every atom is surrounded by 12 other atoms if the structure is extended indefinitely, so each atom has a coordination number of 12. Although it is not obvious from this figure, the cubic close-packed structure is face-centered cubic. To see this, we would have to include additional atoms and tilt the resulting cluster of atoms. All spheres represent identical atoms or ions; different colors are shown only to help you visualize the layers.

structure (Figure 13-27a). The extended pattern of arrangement of planes is *ABABAB* .... If the third layer is placed in the alternate set of depressions in the second layer so that spheres in the first and third layers are not directly above and below each other, the cubic close-packed structure, *ABCABCABC* ... , results (Figure 13-27b). In close-packed structures each sphere has a *coordination number* of 12, that is, 12 nearest neighbors. In ideal close-packed structures 74% of a given volume is due to spheres and 26% is empty space. The body-centered cubic structure is less efficient in packing; each sphere has only eight nearest neighbors, and there is more empty space.

The term "coordination number" is used in crystallography in a somewhat different sense from that in coordination chemistry (Section 25-3). Here it refers to the number of nearest neighbors.

✓ **Problem-Solving Tip:** *The Locations of the Nearest Neighbors in Cubic Crystals*

The distance from an atom to one of its nearest neighbors in any crystal structure depends on the arrangement of atoms and on the size of the unit cell. For structures (such as metals) that contain only one kind of atom, the nearest neighbors of each atom can be visualized as follows. (Recall that for a cubic structure, the unit cell edge is *a*). In a simple cubic structure, the nearest neighbors are along the cell edge (i). In a face-centered cubic structure, the nearest neighbors are along the face diagonal (ii). In a body-centered cubic structure, they are along the body diagonal (iii).

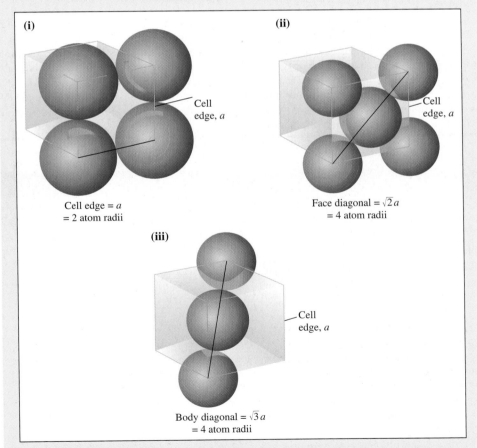

(i)

Cell edge, *a*

Cell edge = *a*
= 2 atom radii

(ii)

Cell edge, *a*

Face diagonal = $\sqrt{2}\,a$
= 4 atom radii

(iii)

Cell edge, *a*

Body diagonal = $\sqrt{3}\,a$
= 4 atom radii

The relationships just described hold only for structures composed of a single kind of atom. For other structures, the relationships are more complex.

EXAMPLE 13-7   Nearest Neighbors

In the simple cubic form of polonium there are Po atoms at the corners of a simple cubic unit cell that is 3.35 Å on edge. (a) What is the shortest distance between centers of neighboring Po atoms? (b) How many nearest neighbors does each atom have?

Plan

We visualize the simple cubic cell.

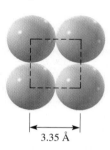

3.35 Å

One face of the simple
cubic unit cell

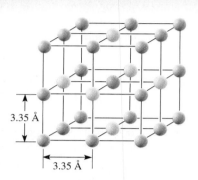

3.35 Å

Several unit cells
(atoms shown smaller for clarity)

Solution

(a) One face of the cubic unit cell is shown in the left-hand drawing, with the atoms touching. The centers of the nearest neighbor atoms are separated by one unit cell edge, at the distance 3.35 Å. (b) A three-dimensional representation of eight unit cells is also shown. In that drawing the atoms are shown smaller for clarity. Some atoms are represented with different colors to aid in visualizing the arrangement, but *all atoms are identical.* Consider the atom shown in red at the center (at the intersection of the eight unit cells). Its nearest neighbors in all of the unit cell directions are shown as light red atoms. As we can see, there are six nearest neighbors. The same would be true of any atom in the structure.

EXAMPLE 13-8   Nearest Neighbors

Silver crystals are face-centered cubic, with a cell edge of 4.086 Å. (a) What is the distance between centers of the two closest Ag atoms? (b) What is the atomic radius of silver in this crystal? (c) How many nearest neighbors does each atom have?

Plan

We reason as in Example 13-7, except that now the two atoms closest to each other are those along the face diagonal.

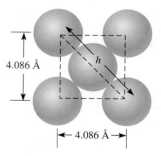

4.086 Å

h

4.086 Å

One face of the face-centered
cubic unit cell

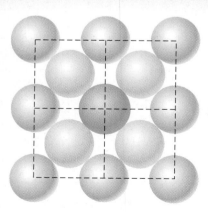

One face of four adjacent unit cells
(*x–y* plane)

Solution

(a) One face of the face-centered cubic unit cell is shown in the left-hand drawing, with the atoms touching. The nearest neighbor atoms are the ones along the diagonal of the face of the cube. We may visualize the face as consisting of two right isosceles triangles sharing a common hypotenuse, h, and having sides of length $a = 4.086$ Å. The hypotenuse is equal to *twice* the center-to-center distance. The hypotenuse can be calculated from the Pythagorean theorem, $h^2 = a^2 + a^2$. The length of the hypotenuse equals the square root of the sum of the squares of the sides.

$$h = \sqrt{a^2 + a^2} = \sqrt{2a^2} = \sqrt{2(4.086 \text{ Å})^2} = 5.778 \text{ Å}$$

The distance between centers of adjacent silver atoms is one half of h, so

$$\text{Distance} = \frac{5.778 \text{ Å}}{2} = \boxed{2.889 \text{ Å}}$$

(b) The hypotenuse of the unit cell face is four times the radius of the silver atom.

$$\text{Atom radius} = \frac{5.778 \text{ Å}}{4} = \boxed{1.444 \text{ Å}}$$

(c) To see the number of nearest neighbors, we expand the left-hand drawing to include several unit cells, as shown in the right-hand drawing. Suppose that this is the x–y plane. The atom shown in red has four nearest neighbors in this plane. There are four more such neighbors in the x–z plane (perpendicular to the x–y plane), and four additional neighbors in the y–z plane (also perpendicular to the x–y plane). This gives a total of $\boxed{12 \text{ nearest neighbors.}}$

EXAMPLE 13-9 *Density and Cell Volume*

From data in Example 13-8, calculate the density of metallic silver.

Plan

We first determine the mass of a unit cell, that is, the mass of four atoms of silver. The density of the unit cell, and therefore of silver, is its mass divided by its volume.

Solution

$$\underline{?} \text{ g Ag per unit cell} = \frac{4 \text{ Ag atoms}}{\text{unit cell}} \times \frac{1 \text{ mol Ag}}{6.022 \times 10^{23} \text{ Ag atoms}} \times \frac{107.87 \text{ g Ag}}{1 \text{ mol Ag}}$$

$$= 7.165 \times 10^{-22} \text{ g Ag/unit cell}$$

$$V_{\text{unit cell}} = (4.086 \text{ Å})^3 = 68.22 \text{ Å}^3 \times \left(\frac{10^{-8} \text{ cm}}{\text{Å}}\right)^3 = 6.822 \times 10^{-23} \text{ cm}^3/\text{unit cell}$$

$$\text{Density} = \frac{7.165 \times 10^{-22} \text{ g Ag/unit cell}}{6.822 \times 10^{-23} \text{ cm}^3/\text{unit cell}} = \boxed{10.50 \text{ g/cm}^3}$$

A handbook gives the density of silver as 10.5 g/cm^3 at 20°C.

You should now work Exercises 90 and 92.

Data obtained from crystal structures and observed densities give us information from which we can calculate the value of Avogadro's number. The next example illustrates these calculations.

EXAMPLE 13-10 Density, Cell Volume, and Avogadro's Number

Titanium crystallizes in a body-centered cubic unit cell with an edge length of 3.306 Å. The density of titanium is 4.401 g/cm³. Use these data to calculate Avogadro's number.

Plan

We relate the density and the volume of the unit cell to find the total mass contained in one unit cell. Knowing the number of atoms per unit cell, we can then find the mass of one atom. Comparing this to the known atomic weight, which is the mass of one mole (Avogadro's number) of atoms, we can evaluate Avogadro's number.

Solution

We first determine the volume of the unit cell.

$$V_{cell} = (3.306 \text{ Å})^3 = 36.13 \text{ Å}^3$$

We now convert Å³ to cm³.

$$\underline{?} \text{ cm}^3 = 36.13 \text{ Å}^3 \times \left(\frac{10^{-8} \text{ cm}}{\text{Å}}\right)^3 = 3.613 \times 10^{-23} \text{ cm}^3$$

The mass of the unit cell is its volume times the observed density.

$$\text{Mass of unit cell} = 3.613 \times 10^{-23} \text{ cm}^3 \times \frac{4.401 \text{ g}}{\text{cm}^3} = 1.590 \times 10^{-22} \text{ g}$$

The bcc unit cell contains $8 \times \frac{1}{8} + 1 = 2$ Ti atoms, so this represents the mass of two Ti atoms. The mass of a single Ti atom is

$$\text{Mass of atom} = \frac{1.590 \times 10^{-22} \text{ g}}{2 \text{ atoms}} = 7.950 \times 10^{-23} \text{ g/atom}$$

From the known atomic weight of Ti (47.88), we know that the mass of one mole of Ti is 47.88 g/mol. Avogadro's number represents the number of atoms per mole, and can be calculated as

$$N_{Av} = \frac{47.88 \text{ g}}{\text{mol}} \times \frac{1 \text{ atom}}{7.950 \times 10^{-23} \text{ g}} = 6.023 \times 10^{23} \text{ atoms/mol}$$

You should now work Exercise 96.

Ionic Solids

Most salts crystallize as ionic solids with ions occupying the unit cell. Sodium chloride (Figure 13-28) is an example. Many other salts crystallize in the sodium chloride (face-centered cubic) arrangement. Examples are the halides of Li^+, K^+, and Rb^+, and $M^{2+}X^{2-}$ oxides and sulfides such as MgO, CaO, CaS, and MnO. Two other common ionic structures are those of cesium chloride, CsCl (simple cubic lattice), and zincblende, ZnS (face-centered cubic lattice), shown in Figure 13-29. Salts that are isomorphous with the CsCl structure include CsBr, CsI, NH_4Cl, TlCl, TlBr, and TlI. The sulfides of Be^{2+}, Cd^{2+}, and Hg^{2+}, together with CuBr, CuI, AgI, and ZnO, are isomorphous with the zincblende structure (Figure 13-29c).

Ionic solids are usually poor electrical and thermal conductors. Liquid (molten) ionic compounds are excellent electrical conductors, however, because their ions are freely mobile.

In certain types of solids, including ionic crystals, particles *different* from those at the corners of the unit cell occupy extra positions within the unit cell. For example, the face-centered cubic unit cell of sodium chloride can be visualized as having chloride ions at the corners and middles of the faces; sodium ions are on the edges between the chloride ions and in the center (see Figures 13-28 and 13-29b). Thus, a unit cell of NaCl contains the following.

CD-ROM Screen 13.13, Solid Structures (2): Ionic Solids.

Cl⁻: $\dfrac{\text{eight at}}{\text{corners}}$ + $\dfrac{\text{six in middles}}{\text{of faces}}$ = 1 + 3 = 4 Cl⁻ ions/unit cell

    $8 \times \frac{1}{8}$ + $6 \times \frac{1}{2}$

Na⁺: $\dfrac{\text{twelve on}}{\text{edges}}$ + one in center = 3 + 1 = 4 Na⁺ ions/unit cell

    $12 \times \frac{1}{4}$ +   1

The unit cell contains equal numbers of Na⁺ and Cl⁻ ions, as required by its chemical formula. Alternatively, we could translate the unit cell by half its length in any axial direction within the lattice and visualize the unit cell in which sodium and chloride ions have exchanged positions. Such an exchange is not always possible. You should confirm that this alternative description also gives four chloride ions and four sodium ions per unit cell.

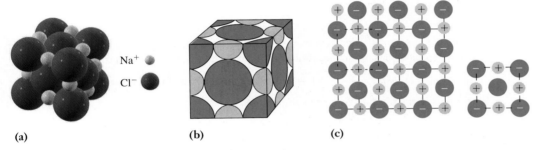

(a)   (b)   (c)

Na⁺ •
Cl⁻ ●

Figure 13-28 Some representations of the crystal structure of sodium chloride, NaCl. Sodium ions are shown in gray and chloride ions are shown in green. (a) One unit cell of the crystal structure of sodium chloride. (b) A representation of the unit cell of sodium chloride that indicates the relative sizes of the Na⁺ and Cl⁻ ions as well as how ions are shared between unit cells. Particles at the corners, edges, and faces of unit cells are shared by other unit cells. Remember that there is an additional Na⁺ ion at the center of the cube. (c) A cross-section of the structure of NaCl, showing the repeating pattern of its unit cell at the right.

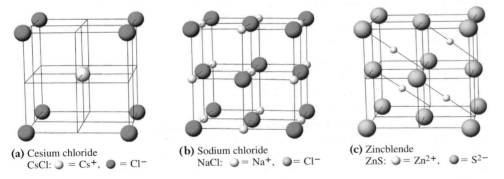

(a) Cesium chloride
    CsCl: ○ = Cs⁺, ● = Cl⁻

(b) Sodium chloride
    NaCl: ○ = Na⁺, ● = Cl⁻

(c) Zincblende
    ZnS: ○ = Zn²⁺, ● = S²⁻

Figure 13-29 Crystal structures of some ionic compounds of the MX type. The gray circles represent cations. One unit cell of each structure is shown. (a) The structure of cesium chloride, CsCl, is simple cubic. It is *not* body-centered, because the point at the center of the cell (Cs⁺, gray) is not the same as the point at a corner of the cell (Cl⁻, green). (b) Sodium chloride, NaCl, is face-centered cubic. (c) Zincblende, ZnS, is face-centered cubic, with four Zn²⁺ (gray) and four S²⁻ (yellow) ions per unit cell. The Zn²⁺ ions are related by the same translations as the S²⁻ ions.

Ionic radii such as those in Figure 6-1 and Table 14-1 are obtained from X-ray crystallographic determinations of unit cell dimensions, assuming that adjacent ions are in contact with each other.

EXAMPLE 13-11   *Ionic Radii from Crystal Data*

Lithium bromide, LiBr, crystallizes in the NaCl face-centered cubic structure with a unit cell edge length of $a = b = c = 5.501$ Å. Assume that the Br^- ions at the corners of the unit cell are in contact with those at the centers of the faces. Determine the ionic radius of the Br^- ion. One face of the unit cell is depicted in Figure 13-30.

Plan

We may visualize the face as consisting of two right isosceles triangles sharing a common hypotenuse, h, and having sides of length $a = 5.501$ Å. The hypotenuse is equal to four times the radius of the bromide ion, $h = 4r_{Br^-}$.

Solution

The hypotenuse can be calculated from the Pythagorean theorem, $h^2 = a^2 + a^2$. The length of the hypotenuse equals the square root of the sum of the squares of the sides.

$$h = \sqrt{a^2 + a^2} = \sqrt{2a^2} = \sqrt{2(5.501\ \text{Å})^2} = 7.780\ \text{Å}$$

The radius of the bromide ion is one fourth of h, so

$$r_{Br^-} = \frac{7.780\ \text{Å}}{4} = \boxed{1.945\ \text{Å}}$$

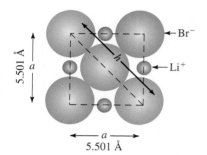

Figure 13-30   One face of the face-centered cubic unit cell of lithium bromide (Example 13-11).

EXAMPLE 13-12   *Ionic Radii from Crystal Data*

Refer to Example 13-11. Calculate the ionic radius of Li^+ in LiBr, assuming anion–cation contact along an edge of the unit cell.

Plan

The edge length, $a = 5.501$ Å, is twice the radius of the Br^- ion plus twice the radius of the Li^+ ion. We know from Example 13-11 that the radius for the Br^- ion is 1.945 Å.

Solution

$$5.501\ \text{Å} = 2\ r_{Br^-} + 2\ r_{Li^+}$$

$$2\ r_{Li^+} = 5.501\ \text{Å} - 2(1.945\ \text{Å}) = 1.611\ \text{Å}$$

$$r_{Li^+} = \boxed{0.806\ \text{Å}}$$

You should now work Exercise 88.

The value of 1.945 Å for the Br^- radius calculated in Example 13-11 is a little different from the value of 1.82 Å given in Figure 6-1; the Li^+ value of 0.806 Å from Example 13-12 also differs somewhat from the value of 0.90 Å given in Figure 6-1. We should remember that the tabulated value in Figure 6-1 is the *average* value obtained from a number of crystal structures of compounds containing the specified ion. Calculations of ionic radii usually assume that anion–anion contact exists, but this assumption is not always true. Calculated radii therefore vary from structure to structure, and we should not place too much emphasis on a value of an ionic radius obtained from any *single* structure determination. We now see that there is some difficulty in determining precise values of ionic radii. Similar difficulties can arise in the determination of atomic radii from molecular and covalent solids or of metallic radii from solid metals.

Molecular Solids

The lattice positions that describe unit cells of molecular solids represent molecules or monatomic elements (sometimes referred to as monatomic molecules). Figure 13-31 shows the unit cells of two simple molecular crystals. Although the bonds *within* molecules are covalent and strong, the forces of attraction *between* molecules are much weaker. They range from hydrogen bonds and weaker dipole–dipole interactions in polar molecules such as H_2O and SO_2 to very weak dispersion forces in symmetrical, nonpolar molecules such as CH_4, CO_2, and O_2 and monatomic elements, such as the noble gases. Because of the relatively weak intermolecular forces of attraction, molecules can be easily displaced. Thus, molecular solids are usually soft substances with low melting points. Because electrons do not move from one molecule to another under ordinary conditions, molecular solids are poor electrical conductors and good insulators.

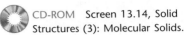

CD-ROM   Screen 13.14, Solid Structures (3): Molecular Solids.

Dispersion forces are also present among polar molecules.

Covalent Solids

Covalent solids (or "network solids") can be considered giant molecules that consist of covalently bonded atoms in an extended, rigid crystalline network. Diamond (one crystalline form of carbon) and quartz are examples of covalent solids (Figure 13-32). Because of their rigid, strongly bonded structures, *most* covalent solids are very hard and melt at high temperatures. Because electrons are localized in covalent bonds, they are not freely mobile. As a result, covalent solids are *usually* poor thermal and electrical conductors at ordinary temperatures. (Diamond, however, is a good conductor of heat; jewelers use this property to distinguish diamonds from imitations.)

An important exception to these generalizations about properties is *graphite*, an allotropic form of carbon. It has the layer structure shown in Figure 13-32b. The overlap of an extended π-electron network in each plane makes graphite an excellent conductor. The very weak attraction between layers allows these layers to slide over one another easily. Graphite is used as a lubricant, as an additive for motor oil, and in pencil "lead" (combined with clay and other fillers to control hardness).

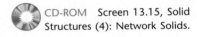

CD-ROM   Screen 13.15, Solid Structures (4): Network Solids.

It is interesting to note that these two allotropes of carbon include one very hard substance and one very soft substance. They differ only in the arrangement and bonding of the C atoms.

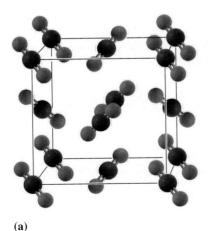
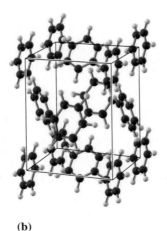

(a)   (b)

Figure 13-31 The packing arrangement in a molecular crystal depends on the shape of the molecule as well as on the electrostatic interactions of any regions of excess positive and negative charge in the molecules. The arrangements in some molecular crystals are shown here: (a) carbon dioxide, CO_2; (b) benzene, C_6H_6.

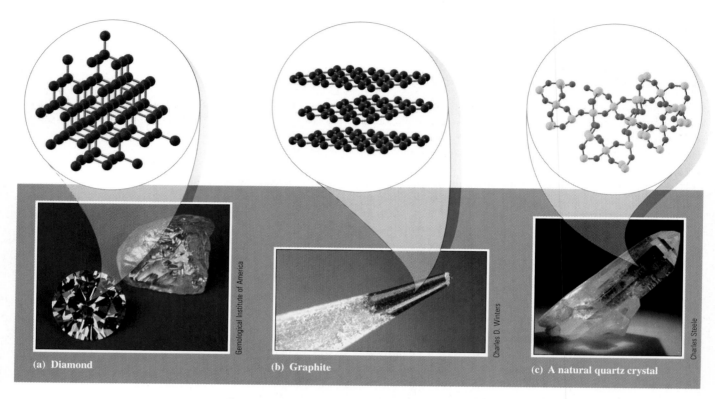

(a) Diamond

(b) Graphite

(c) A natural quartz crystal

Figure 13-32 Portions of the atomic arrangements in three covalent solids. (a) Diamond. Each C is bonded tetrahedrally to four others through sp^3-sp^3 σ-bonds. (1.54 Å). (b) Graphite. C atoms are linked in planes by sp^2-sp^2 σ- and π-bonds (1.42 Å). The crystal is soft, owing to the weakness of the attractions between planes (3.40 Å). Electrons move freely through the delocalized π-bonding network in these planes, but they do not jump between planes easily. (c) Quartz (SiO_2). Each Si atom (*gray*) is bonded tetrahedrally to four O atoms (*red*).

13-17 BAND THEORY OF METALS

As described in the previous section, most metals crystallize in close-packed structures. The ability of metals to conduct electricity and heat must result from strong electronic interactions of an atom with its 8 to 12 nearest neighbors. This might be surprising at first if we recall that each Group IA and Group IIA metal atom has only one or two valence electrons available for bonding. This is too few to participate in bonds localized between it and each of its nearest neighbors.

Bonding in metals is called **metallic bonding.** It results from the electrical attractions among positively charged metal ions and mobile, delocalized electrons belonging to the crystal as a whole. The properties associated with metals—metallic luster, high thermal and electrical conductivity, and so on—can be explained by the **band theory** of metals, which we now describe.

The overlap interaction of two atomic orbitals, say the 3s orbitals of two sodium atoms, produces two molecular orbitals, one bonding orbital and one antibonding orbital (Chapter 9). If N atomic orbitals interact, N molecular orbitals are formed. In a single metallic crystal containing one mole of sodium atoms, for example, the interaction (overlap) of

Metals can be formed into many shapes because of their malleability and ductility.

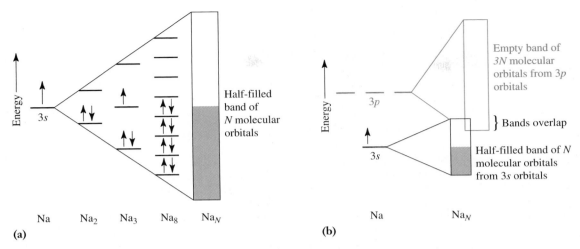

Figure 13-33 (a) The band of orbitals resulting from interaction of the 3*s* orbitals in a crystal of sodium. (b) Overlapping of a half-filled "3*s*" band (*black*) with an empty "3*p*" band (*red*) of Na$_N$ crystal.

6.022×10^{23} 3*s* atomic orbitals produces 6.022×10^{23} molecular orbitals. Atoms interact more strongly with nearby atoms than with those farther away. The energy that separates bonding and antibonding molecular orbitals resulting from two given atomic orbitals decreases as the overlap between the atomic orbitals decreases (Chapter 9). The interactions among the mole of Na atoms result in a series of very closely spaced molecular orbitals (formally σ_{3s} and σ_{3s}^*). These constitute a nearly continuous **band** of orbitals that belongs to the crystal as a whole. One mole of Na atoms contributes 6.022×10^{23} valence electrons (Figure 13-33a), so the 6.022×10^{23} orbitals in the band are half-filled.

The ability of metallic Na to conduct electricity is due to the ability of any of the highest energy electrons in the "3*s*" band to jump to a slightly higher-energy vacant orbital in the same band when an electric field is applied. The resulting net flow of electrons through the crystal is in the direction of the applied field.

The empty 3*p* atomic orbitals of the Na atoms also interact to form a wide band of 3 × 6.022×10^{23} orbitals. The 3*s* and 3*p* atomic orbitals are quite close in energy, so the fanned-out bands of molecular orbitals overlap, as shown in Figure 13-33b. The two overlapping bands contain $4 \times 6.022 \times 10^{23}$ orbitals and only 6.022×10^{23} electrons. Because each orbital can hold two electrons, the resulting combination of 3*s* and 3*p* bands is only one-eighth full.

The alkali metals are those of Group IA; the alkaline earth metals are those of Group IIA.

Overlap of "3*s*" and "3*p*" bands is not necessary to explain the ability of Na or of any other Group IA metal to conduct electricity, as the half-filled "3*s*" band is sufficient for this. In the Group IIA metals, however, such overlap is important. Consider a crystal of magnesium as an example. The 3*s* atomic orbital of an isolated Mg atom is filled with two electrons. Thus, without this overlap, the "3*s*" band in a crystal of Mg is also filled. Mg is a good conductor at room temperature because the highest energy electrons are able to move readily into vacant orbitals in the "3*p*" band (Figure 13-34).

Figure 13-34 Overlapping of a filled "3*s*" band (*blue*) with an empty "3*p*" band of Mg$_N$ crystal. The higher-energy electrons are able to move into the "3*p*" band (*red*) as a result of this overlap. There are now empty orbitals immediately above the filled orbitals, leading to conductivity.

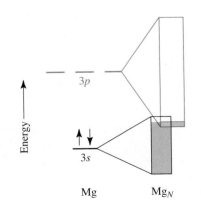

CHEMISTRY IN USE

Research & Technology

Semiconductors

A **semiconductor** is an element or a compound with filled bands that are only slightly below, but do not overlap with, empty bands. The difference between an insulator and a semiconductor is only the size of the energy gap, and there is no sharp distinction between them. An **intrinsic** semiconductor (i.e., a semiconductor in its pure form) is a much poorer conductor of electricity than a metal because, for conduction to occur in a semiconductor, electrons must be excited from bonding orbitals in the filled *valence band* into the empty *conduction band*. Figure (a) shows how this happens. An electron that is given an excitation energy greater than or equal to the **band gap (E_g)** enters the conduction band and leaves behind a positively charged **hole** (h^+, the absence of a bonding electron) in the valence band. Both the electron and the hole reside in *delocalized* orbitals, and both can move in an electric field, much as electrons move in a metal. (Holes migrate when an electron in a nearby orbital moves to fill in the hole, thereby creating a new hole in the nearby orbital.) Electrons and holes move in opposite directions in an electric field.

Silicon, a semiconductor of great importance in electronics, has a band gap of 1.94×10^{-22} kJ, or 1.21 *electron volts* (eV). This is the energy needed to create one electron and one hole or, put another way, the energy needed to break one Si—Si bond. This energy can be supplied either thermally or by using light with a photon energy greater than the band gap. To excite one *mole* of electrons from the valence band to the conduction band, an energy of

$$\frac{6.022 \times 10^{23} \text{ electrons}}{\text{mol}} \times \frac{1.94 \times 10^{-22} \text{ kJ}}{\text{electron}} = 117 \text{ kJ/mol}$$

is required. For silicon, a large amount of energy is required, so there are very few mobile electrons and holes (about one electron in a trillion—i.e., 1 in 10^{12}—is excited thermally at room temperature); the conductivity of pure silicon is therefore about 10^{11} times lower than that of highly conductive metals such as silver. The number of electrons excited thermally is proportional to $e^{-E_g/2RT}$. Increasing the temperature or decreasing the band gap energy leads to higher conductivity for an intrinsic semiconductor. Insulators such as diamond and silicon dioxide (quartz), which have very large values of E_g, have conductivities 10^{15} to 10^{20} times lower than most metals.

The electrical conductivity of a semiconductor can be greatly increased by **doping** with impurities. For example, silicon, a Group IVA element, can be doped by adding small amounts of a Group VA element, such as phosphorus, or a Group IIIA element, such as boron. Figure (b) shows the effect of substituting phosphorus for silicon in the crystal structure (silicon has the same structure as diamond, Figure 13-31a). There are exactly enough valence band orbitals to accommodate four of the valence electrons from the phosphorus atom. However, a phosphorus atom has one more electron (and one more proton in its nucleus) than does silicon. The fifth electron enters a higher energy orbital that is localized in the lattice near the phosphorus atom; the energy of this orbital, called a **donor level**, is just below the conduction band, within the energy gap. An electron in this orbital can easily become *delocalized* when a small amount of thermal energy promotes it into the conduction band. Because the phosphorus-doped silicon contains mobile, *negatively* charged carriers (electrons), it is said to be doped **n-type**. Doping the silicon crystal with boron produces a related, but opposite, effect. Each boron atom contributes only three valence electrons to bonding orbitals in the valence band, and therefore a *hole* is localized near each boron atom. Thermal energy is enough to separate the negatively charged boron atom from the hole, de-

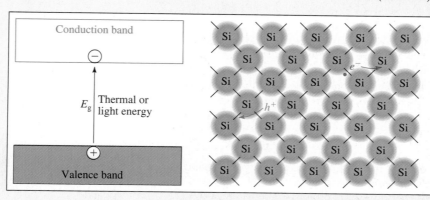

(a) Generation of an electron–hole pair in silicon, an intrinsic semiconductor. The electron (e^-) and hole (h^+) have opposite charges, and so move in opposite directions in an electric field.

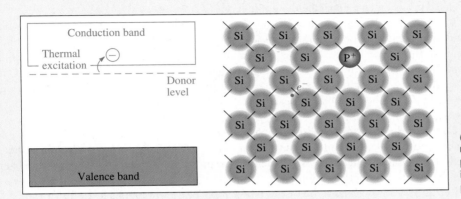

(b) n-type doping of silicon by phosphorus. The extra valence electron from a phosphorus atom is thermally excited into the conduction band, leaving a fixed positive charge on the phosphorus atom.

localizing the latter. In this case the charge carriers are the holes, which are *positive*, and the crystal is doped **p-type.** In both p- and n-type doping, an extremely small concentration of dopants (as little as one part per billion) is enough to cause a significant increase in conductivity. For this reason, great

The colors of semiconductors are determined by the band gap energy E_g. Only photons with energy greater than E_g can be absorbed. From the Planck radiation formula ($E = h\nu$) and $\lambda\nu = c$, we calculate that the wavelength, λ, of an absorbed photon must be less than hc/E_g. Gallium arsenide (GaAs; $E_g = 1.4$ eV) absorbs photons of wavelengths shorter than 890 nm, which is in the near infrared region. Because it absorbs all wavelengths of visible light, gallium arsenide appears black to the eye. Iron oxide (Fe_2O_3; $E_g = 2.2$ eV) absorbs light of wavelengths shorter than 570 nm; it absorbs both yellow and blue light, and therefore appears red. Cadmium sulfide (CdS; $E_g = 2.6$ eV), which absorbs blue light ($\lambda \leq 470$ nm), appears yellow. Strontium titanate ($SrTiO_3$; $E_g = 3.2$ eV) absorbs only in the ultraviolet ($\lambda \leq 390$ nm). It appears white to the eye because visible light of all colors is reflected by the fine particles.

pains are taken to purify the semiconductors used in electronic devices.

Even in a doped semiconductor, mobile electrons and holes are both present, although one carrier type is predominant. For example, in a sample of silicon doped with arsenic (n-type doping), the concentrations of mobile electrons are slightly less than the concentration of arsenic atoms (usually expressed in terms of atoms/cm³), and the concentrations of mobile holes are extremely low. Interestingly, the concentrations of electrons and holes always follow an equilibrium expression that is entirely analogous to that for the autodissociation of water into H^+ and OH^- ions (Chapter 18); that is,

$$[e^-][h^+] = K_{eq}$$

where the equilibrium constant K_{eq} depends only on the identity of the semiconductor and the absolute temperature. For silicon at room temperature, $K_{eq} = 4.9 \times 10^{19}$ carriers²/cm⁶.

Doped semiconductors are extremely important in electronic applications. A **p–n junction** is formed by joining p- and n-type semiconductors. At the junction, free electrons and holes combine, annihilating each other and leaving positively and negatively charged dopant atoms on opposite sides. The unequal charge distribution on the two sides of the junction causes an electric field to develop and gives rise to current rectification (electrons can flow, with a small applied voltage, only from the n side to the p side of the junction; holes flow only in the reverse direction). Devices such as **diodes** and **transistors,** which form the bases of most analog and digital electronic circuits, are composed of p–n junctions.

Professor Thomas A. Mallouk
Pennsylvania State University

According to band theory, the highest energy electrons of metallic crystals occupy either a partially filled band or a filled band that overlaps an empty band. A band within which (or into which) electrons move to allow electrical conduction is called a **conduction band**. The electrical conductivity of a metal decreases as temperature increases. The increase in temperature causes thermal agitation of the metal ions. This impedes the flow of electrons when an electric field is applied.

Crystalline nonmetals, such as diamond and phosphorus, are **insulators**—they do not conduct electricity. The reason for this is that their highest energy electrons occupy filled bands of molecular orbitals that are separated from the lowest empty band (conduction band) by an energy difference called the **band gap**. In an insulator, this band gap is an energy difference that is too large for electrons to jump to get to the conduction band (Figure 13-35). Many ionic solids are also insulators, but are good conductors in their molten (liquid) state.

Elements that are **semiconductors** have filled bands that are only slightly below, but do not overlap with, empty bands. They do not conduct electricity at low temperatures, but a small increase in temperature is sufficient to excite some of the highest-energy electrons into the empty conduction band.

Let us now summarize some of the physical properties of metals in terms of the band theory of metallic bonding.

1. We have just accounted for the *ability of metals to conduct electricity*.

2. Metals are also *conductors of heat*. They can absorb heat as electrons become thermally excited to low-lying vacant orbitals in a conduction band. The reverse process accompanies the release of heat.

3. Metals have a *lustrous appearance* because the mobile electrons can absorb a wide range of wavelengths of radiant energy as they jump to higher energy levels. Then they emit photons of visible light and fall back to lower levels within the conduction band.

4. Metals are *malleable or ductile* (or both). A crystal of a metal is easily deformed when a mechanical stress is applied to it. All of the metal ions are identical, and they are imbedded in a "sea of electrons." As bonds are broken, new ones are readily formed

Various samples of elemental silicon. The circle at the lower right is a disk of ultrapure silicon on which many electronic circuits have been etched.

Charles D. Winters

A **malleable** substance can be rolled or pounded into sheets. A **ductile** substance can be drawn into wires. Gold is the most malleable metal known.

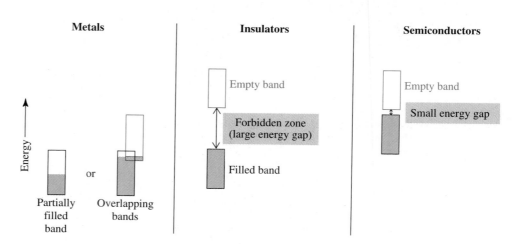

Figure 13-35 Distinction among metals, insulators, and semiconductors. In each case an unshaded area represents a conduction band.

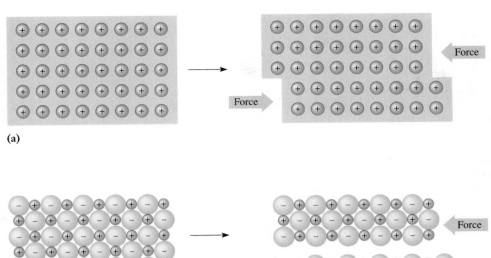

(a)

(b)

A crystal can be cleaved into smaller crystals that have the same appearance as the larger crystal.

Figure 13-36 (a) In a metal, the positively charged metal ions are immersed in a delocalized "cloud of electrons." When the metal is distorted (e.g., rolled into sheets or drawn into wires), the environment around the metal atoms is essentially unchanged, and no new repulsive forces occur. This explains why metal sheets and wires remain intact. (b) By contrast, when an ionic crystal is subjected to a force that causes it to slip along a plane, the increased repulsive forces between like-charged ions cause the crystal to break or cleave along a crystal plane.

with adjacent metal ions. The features of the arrangement remain unchanged, and the environment of each metal ion is the same as before the deformation occurred (Figure 13-36). The breakage of bonds involves the promotion of electrons to higher-energy levels. The formation of bonds is accompanied by the return of the electrons to the original energy levels.

Key Terms

Adhesive force   Force of attraction between a liquid and another surface.

Allotropes   Different forms of the same element in the same physical state.

Amorphous solid   A noncrystalline solid with no well-defined, ordered structure.

Band   A series of very closely spaced, nearly continuous molecular orbitals that belong to the material as a whole.

Band gap   An energy separation between an insulator's highest filled electron energy band and the next higher-energy vacant band.

Band theory of metals   A theory that accounts for the bonding and properties of metallic solids.

Boiling point   The temperature at which the vapor pressure of a liquid is equal to the external pressure; also the condensation point.

Capillary action   The drawing of a liquid up the inside of a small-bore tube when adhesive forces exceed cohesive forces, or the depression of the surface of the liquid when cohesive forces exceed adhesive forces.

Cohesive forces   All the forces of attraction among particles of a liquid.

Condensation   Liquefaction of vapor.

Condensed phases   The liquid and solid phases; phases in which particles interact strongly.

Conduction band   A partially filled band or a band of vacant energy levels just higher in energy than a filled band; a band within which, or into which, electrons must be promoted to allow electrical conduction to occur in a solid.

Coordination number   In describing crystals, the number of nearest neighbors of an atom or ion.

Critical point   The combination of critical temperature and critical pressure of a substance.

Critical pressure   The pressure required to liquefy a gas (vapor) at its critical temperature.

Critical temperature   The temperature above which a gas cannot be liquefied; the temperature above which a substance cannot exhibit distinct gas and liquid phases.

Crystal lattice   The pattern of arrangement of particles in a crystal.

Crystalline solid   A solid characterized by a regular, ordered arrangement of particles.

Deposition   The direct solidification of a vapor by cooling; the reverse of sublimation.

Dipole–dipole interactions   Interactions between polar molecules, that is, between molecules with permanent dipoles.

Dipole-induced dipole interaction   See *Dispersion forces*.

Dispersion forces   Very weak and very short-range attractive forces between short-lived temporary (induced) dipoles; also called London forces.

Distillation   The separation of a liquid mixture into its components on the basis of differences in boiling points.

Dynamic equilibrium   A situation in which two (or more) processes occur at the same rate so that no net change occurs.

Enthalpy of fusion   See *Heat of fusion*.

Enthalpy of solidification   See *Heat of solidification*.

Evaporation   Vaporization of a liquid below its boiling point.

Freezing point   See *Melting point*.

Heat of condensation   The amount of heat that must be removed from a specific amount of a vapor at its condensation point to condense the vapor with no change in temperature; usually expressed in J/g or kJ/mol; in the latter case it is called the *molar heat of condensation*.

Heat of fusion   The amount of heat required to melt a specific amount of a solid at its melting point with no change in temperature; usually expressed in J/g or kJ/mol; in the latter case it is called the *molar heat of fusion*.

Heat of solidification   The amount of heat that must be removed from a specific amount of a liquid at its freezing point to freeze it with no change in temperature; usually expressed in J/g or kJ/mol; in the latter case it is called the *molar heat of solidification*.

Heat of vaporization   The amount of heat required to vaporize a specific amount of a liquid at its boiling point with no change in temperature; usually expressed in J/g or kJ/mol; in the latter case it is called the *molar heat of vaporization*.

Hydrogen bond   A fairly strong dipole–dipole interaction (but still considerably weaker than covalent or ionic bonds) between molecules containing hydrogen directly bonded to a small, highly electronegative atom, such as N, O, or F.

Insulator   A poor conductor of electricity and heat.

Intermolecular forces   Forces *between* individual particles (atoms, molecules, ions) of a substance.

Intramolecular forces   Forces between atoms (or ions) *within* molecules (or formula units).

Isomorphous   Refers to crystals having the same atomic arrangement.

LeChatelier's Principle   A system at equilibrium, or striving to attain equilibrium, responds in such a way as to counteract any stress placed upon it.

London forces   See *Dispersion forces*.

Melting point   The temperature at which liquid and solid coexist in equilibrium; also the freezing point.

Meniscus   The upper surface of a liquid in a cylindrical container.

Metallic bonding   Bonding within metals due to the electrical attraction of positively charged metal ions for mobile electrons that belong to the crystal as a whole.

Molar enthalpy of vaporization   See *Molar heat of vaporization*.

Molar heat capacity   The amount of heat necessary to raise the temperature of one mole of a substance one degree Celsius with no change in state; usually expressed in kJ/mol·°C. See *Specific heat*.

Molar heat of condensation   The amount of heat that must be removed from one mole of a vapor at its condensation point to condense the vapor with no change in temperature; usually expressed in kJ/mol. See *Heat of condensation*.

Molar heat of fusion   The amount of heat required to melt one mole of a solid at its melting point with no change in temperature; usually expressed in kJ/mol. See *Heat of fusion*.

Molar heat of vaporization   The amount of heat required to vaporize one mole of a liquid at its boiling point with no change in temperature; usually expressed in kJ/mol. See *Heat of vaporization*.

Normal boiling point   The temperature at which the vapor pressure of a liquid is equal to one atmosphere pressure.

Normal melting point   The melting (freezing) point at one atmosphere pressure.

Phase diagram   A diagram that shows equilibrium temperature–pressure relationships for different phases of a substance.

Polymorphous   Refers to substances that crystallize in more than one crystalline arrangement.

Semiconductor   A substance that does not conduct electricity well at low temperatures but that does at higher temperatures.

Specific heat   The amount of heat necessary to raise the temperature of a specific amount of a substance one degree Celsius with no change in state; usually expressed in J/g·°C. See *Molar heat capacity*.

Sublimation   The direct vaporization of a solid by heating without passing through the liquid state.

Supercritical fluid   A substance at a temperature above its critical temperature. A supercritical fluid cannot be described as either a liquid or gas, but has the properties of both.

Surface tension   The result of inward intermolecular forces of attraction among liquid particles that must be overcome to expand the surface area.

Triple point The point on a phase diagram that corresponds to the only pressure and temperature at which three phases (usually solid, liquid, and gas) of a substance can coexist at equilibrium.

Unit cell The smallest repeating unit showing all the structural characteristics of a crystal.

Vapor pressure The partial pressure of a vapor in equilibrium with its parent liquid or solid.

Viscosity The tendency of a liquid to resist flow; the inverse of its fluidity.

Volatility The ease with which a liquid vaporizes.

Exercises

General Concepts

1. What causes dispersion forces? What factors determine the strengths of dispersion forces between molecules?

2. What is hydrogen bonding? Under what conditions can strong hydrogen bonds be formed?

3. Which of the following substances have permanent dipole-dipole forces? (a) GeH_4; (b) molecular $MgCl_2$; (c) NBr_3; (d) F_2O.

4. Which of the following substances have permanent dipole-dipole forces? (a) molecular $AlCl_3$; (b) PCl_5; (c) CO; (d) SeF_4.

5. For which of the substances in Exercise 3 are dispersion forces the only important forces in determining boiling points?

6. For which of the substances in Exercise 4 are dispersion forces the only important forces in determining boiling points?

7. For each of the following pairs of compounds, predict which compound would exhibit strong hydrogen bonding. Justify your prediction. It may help to write a Lewis formula for each. (a) water, H_2O, or hydrogen sulfide, H_2S; (b) dichloromethane, CH_2Cl_2 or fluoroamine, NH_2F; (c) acetone, C_3H_6O (contains a C=O double bond) or ethyl alcohol, C_2H_6O (contains one C—O single bond).

8. Hydrogen bonding is a very strong dipole-dipole interaction. Why is hydrogen bonding so strong in comparison with other dipole-dipole interactions?

9. Give an example of each of the six types of phase changes.

10. Which of the following substances exhibits strong hydrogen bonding in the liquid and solid states? (a) CH_3OH (methyl alcohol); (b) PH_3; (c) CH_4; (d) $(CH_3)_2NH$; (e) CH_3NH_2.

11. Which of the following substances exhibits strong hydrogen bonding in the liquid and solid states? (a) H_2S; (b) HBr; (c) SiH_4; (d) HF; (e) HCl.

12. The molecular weights of CH_4 and NH_3 are nearly the same. Account for the fact that the melting and boiling points of NH_3 ($-77.7°C$ and $-33.3°C$) are higher than those of CH_4 ($-184°C$ and $-161.5°C$).

*13. Imagine replacing one H atom of a methane molecule, CH_4, with another atom or group of atoms. Account for the order in the normal boiling points of the resulting compounds: CH_4($-161°C$); CH_3Cl ($-24.2°C$); CH_3F ($-78°C$); CH_3OH ($65°C$).

14. For each of the following pairs of compounds, predict which would exhibit strong hydrogen bonding. Justify your prediction. It may help to write a Lewis formula for each. (a) ammonia, NH_3, or phosphine, PH_3; (b) ethylene, C_2H_4, or hydrazine, N_2H_4; (c) hydrogen fluoride, HF, or hydrogen chloride, HCl.

15. Describe the intermolecular forces that are present in each of the following compounds. Which kind of force would have the greatest influence on the properties of each compound? (a) bromine pentafluoride, BrF_5; (b) acetone, C_3H_6O (contains a central C=O double bond); (c) formaldehyde, H_2CO

16. Describe the intermolecular forces that are present in each of the following compounds. Which kind of force would have the greatest influence on the properties of each compound? (a) ethyl alcohol, C_2H_6O (contains one C—O single bond); (b) phosphine, PH_3; (c) sulfur hexafluoride, SF_6.

17. Account for the fact that ethylene glycol ($HOCH_2CH_2OH$) is less viscous than glycerine ($HOCH_2CHOHCH_2OH$), but more viscous than ethyl alcohol (CH_3CH_2OH).

18. Water forms a concave meniscus in a 10-mL glass graduated cylinder but the meniscus it forms in a 10-mL plastic graduated cylinder is flat. Why is the appearance of the water surface in these two measuring devices different?

19. Why does HF have a lower boiling point and lower heat of vaporization than H_2O, even though their molecular weights are nearly the same and the hydrogen bonds between molecules of HF are stronger?

*20. Many carboxylic acids form dimers in which two molecules "stick together." These dimers result from the formation of *two* hydrogen bonds between the two molecules. Use acetic acid to draw a likely structure for this kind of hydrogen-bonded dimer.

$$CH_3-\overset{\overset{\displaystyle \cdot\cdot O \cdot\cdot}{\|}}{C}-\overset{\cdot\cdot}{\underset{\cdot\cdot}{O}}-H$$

The Liquid State

21. Use the kinetic-molecular theory to describe the behavior of liquids with changing temperature. Why are liquids more dense than gases?

22. Distinguish between evaporation and boiling. Use the kinetic-molecular theory to explain the dependence of rate of evaporation on temperature.

23. Support or criticize the statement that liquids with high normal boiling points have low vapor pressures. Give examples of three common liquids that have relatively high vapor pressures at 25°C and three that have low vapor pressures at 25°C.

24. Within each group, assign each of the boiling points to the appropriate substance on the basis of intermolecular forces. (a) Ne, Ar, Kr: −246°C, −186°C, −152°C; (b) NH_3, H_2O, HF: −33°C, 20°C, 100°C.

25. Within each group, assign each of the boiling points to the respective substances on the basis of intermolecular forces. (a) N_2, HCN, C_2H_6: −196°C, −89°C, 26°C; (b) H_2, HCl, Cl_2: −35°C, −259°C, −85°C.

26. (a) What is the definition of the normal boiling point? (b) Why is it necessary to specify the atmospheric pressure over a liquid when measuring a boiling point?

27. What factors determine how viscous a liquid is? How does viscosity change with increasing temperature?

28. What is the surface tension of a liquid? What causes this property? How does surface tension change with increasing temperature?

29. What happens inside a capillary tube when a liquid "wets" the tube? What happens when a liquid does not "wet" the tube?

30. What are some of the similarities of the molecular-level descriptions of the viscosity, surface tension, vapor pressure and the rate of evaporation of a liquid?

31. Dispersion forces are extremely weak in comparison to the other intermolecular attractions. Explain why this is so.

32. Choose from each pair the substance that, in the liquid state, would have the greater vapor pressure at a given temperature. Base your choice on predicted strengths of intermolecular forces. (a) $BiBr_3$ or $BiCl_3$; (b) CO or CO_2; (c) N_2 or NO; (d) CH_3COOH or $HCOOCH_3$.

33. Repeat Exercise 32 for (a) C_6H_6 or C_6Cl_6; (b) $F_2C=O$ or CH_3OH; (c) He or H_2.

34. The temperatures at which the vapor pressures of the following liquids are all 100 torr are given. Predict the order of increasing boiling points of the liquids. Butane, C_4H_{10}, −44.2°C; 1-butanol, $C_4H_{10}O$, 70.1°C; diethyl ether, $C_4H_{10}O$, −11.5°C.

35. Plot a vapor pressure curve for $GaCl_3$ from the following vapor pressures. Estimate the boiling point of $GaCl_3$ under a pressure of 250 torr from the plot:

t (°C)	91	108	118	132	153	176	200
vp (torr)	20	40	60	100	200	400	760

36. Plot a vapor pressure curve for Cl_2O_7 from the following vapor pressures. Estimate the boiling point of Cl_2O_7 under a pressure of 125 torr from the plot:

t (°C)	−24	−13	−2	10	29	45	62	79
vp (torr)	5	10	20	40	100	200	400	760

37. The vapor pressure of liquid bromine at room temperature is 168 torr. Suppose that bromine is introduced drop by drop into a closed system containing air at 775 torr and room temperature. (The volume of liquid bromine is negligible compared to the gas volume.) If the bromine is added until no more vaporizes and a few drops of liquid are present in the flask, what would be the total pressure? What would be the total pressure if the volume of this closed system were decreased to one half its original value at the same temperature?

38. A closed flask contains water at 75.0°C. The total pressure of the air-and-water-vapor mixture is 633.5 torr. The vapor pressure of water at this temperature is given in Appendix E as 289.1 torr. What is the partial pressure of the air in the flask?

*39. ΔH_{vap} is usually greater than ΔH_{fus} for a substance, yet the *nature* of interactions that must be overcome in the vaporization and fusion processes are similar. Why is ΔH_{vap} greater?

*40. The heat of vaporization of water at 100°C is 2.26 kJ/g; at 37°C (body temperature) it is 2.41 kJ/g. (a) Convert the latter value to standard molar heat of vaporization, $\Delta H°_{vap}$, at 37°C. (b) Why is the heat of vaporization greater at 37°C than at 100°C?

41. Plot a vapor pressure curve for $C_2Cl_2F_4$ from the following vapor pressures. Estimate the boiling point of $C_2Cl_2F_4$ under a pressure of 300 torr from the plot:

t (°C)	−95.4	−72.3	−53.7	−39.1	−12.0	3.5
vp (torr)	1	10	40	100	400	760

42. Plot a vapor pressure curve for $C_2H_4F_2$ from the following vapor pressures. From the plot, estimate the boiling point of $C_2H_4F_2$ under a pressure of 200 torr.

t (°C)	−77.2	−51.2	−31.1	−15.0	14.8	31.7
vp (torr)	1	10	40	100	400	760

Phase Changes and Associated Heat Transfer

The following values will be useful in some exercises in this section:

Specific heat of ice	2.09 J/g·°C
Heat of fusion of ice at 0°C	334 J/g
Specific heat of liquid H_2O	4.184 J/g·°C
Heat of vaporization of liquid H_2O at 100°C	2.26×10^3 J/g
Specific heat of steam	2.03 J/g·°C

43. What amount of heat energy, in joules, must be removed to condense 25.4 g of water vapor at 122.5°C to liquid at 23.1°C?

44. Is the equilibrium that is established between two physical states of matter an example of static or dynamic equilibrium? How could one demonstrate the type of equilibrium established? Explain your answer.

45. Which of the following changes of state are exothermic? (a) fusion; (b) liquefaction; (c) sublimation; (d) deposition. Explain.

46. Suppose that heat was added to a 21.8-g sample of solid zinc at the rate of 9.84 J/s. After the temperature reached the normal melting point of zinc, 420.°C, it remained constant for 3.60 minutes. Calculate ΔH_{fus} at 420.°C, in J/mol, for zinc.

47. The specific heat of silver is 0.237 J/g·°C. Its melting point is 961°C. Its heat of fusion is 11 J/g. How much heat is needed to change 10.72 g of silver from solid at 23°C to liquid at 961°C?

48. The heat of fusion of thallium is 21 J/g, and its heat of vaporization is 795 J/g. The melting and boiling points are 304°C and 1457°C. The specific heat of liquid thallium is 0.13 J/g·°C. How much heat is needed to change 236 g of solid thallium at 304°C to vapor at 1457°C and 1 atm?

49. Calculate the amount of heat required to convert 65.0 g of ice at 0°C to liquid water at 100.°C.

50. Calculate the amount of heat required to convert 65.0 g of ice at −15.0°C to steam at 125.0°C.

51. Use data in Appendix E to calculate the amount of heat required to warm 165 g of mercury from 25°C to its boiling point and then to vaporize it.

52. If 275 g of liquid water at 100.°C and 525 g of water at 30.0°C are mixed in an insulated container, what is the final temperature?

53. If 10.0 g of ice at −10.0°C and 20.0 g of liquid water at 100.°C are mixed in an insulated container, what will the final temperature be?

54. If 185 g of liquid water at 0°C and 18.5 g of steam at 110.°C are mixed in an insulated container, what will the final temperature be?

55. Water can be cooled in hot climates by the evaporation of water from the surfaces of canvas bags. What mass of water can be cooled from 35.0°C to 25.0°C by the evaporation of one gram of water? Assume that ΔH_{vap} does not change with temperature.

56. (a) How much heat must be removed to prepare 14.0 g of ice at 0.0°C from 14.0 g of water at 25.0°C? (b) Calculate the mass of water at 100.0°C that could be cooled to 25.5°C by the same amount of heat as that calculated in part (a).

Clausius-Clapeyron Equation

57. Toluene, $C_6H_5CH_3$, is a liquid used in the manufacture of TNT. Its normal boiling point is 111.0°C, and its molar heat of vaporization is 35.9 kJ/mol. What would be the vapor pressure (torr) of toluene at 85.00°C?

58. At their normal boiling points, the heat of vaporization of water (100°C) is 40,656 J/mol and that of heavy water (101.41°C) is 41,606 J/mol. Use these data to calculate the vapor pressure of each liquid at 85.00°C.

59. (a) Use the Clausius-Clapeyron equation to calculate the temperature (°C) at which pure water would boil at a pressure of 400.0 torr. (b) Compare this result with the temperature read from Figure 13-13. (c) Compare the results of (a) and (b) with a value obtained from Appendix E.

*60. Show that the Clausius-Clapeyron equation can be written at a single temperature as

$$\ln P = \frac{-\Delta H_{vap}}{RT} + B$$

where B is a constant that has different values for different substances. This is an equation for a straight line. (a) What is the expression for the slope of this line? (b) Using the following vapor pressure data, plot $\ln P$ vs. $1/T$ for ethyl acetate, $CH_3COOC_2H_5$, a common organic solvent used in nail polish removers.

t (°C)	−43.4	−23.5	−13.5	−3.0	+9.1
vp (torr)	1	5	10.	20.	40.

t (°C)	16.6	27.0	42.0	59.3	
vp (torr)	60.	100.	200.	400.	

(c) From the plot, estimate ΔH_{vap} for ethyl acetate. (d) From the plot, estimate the normal boiling point of ethyl acetate.

Products containing ethyl acetate

*61. Repeat Exercise 60(b) and 60(c) for mercury, using the following data for liquid mercury. Then compare this value with the one in Appendix E.

t (°C)	126.2	184.0	228.8	261.7	323.0
vp (torr)	1	10.	40.	100.	400.

62. Isopropyl alcohol, C_3H_8O, is marketed as "rubbing alcohol." Its vapor pressure is 100. torr at 39.5°C and 400. torr at 67.8°C. Estimate the molar heat of vaporization of isopropyl alcohol.

63. Using data from Exercise 62, predict the normal boiling point of isopropyl alcohol.

64. Boiling mercury is often used in diffusion pumps to attain a very high vacuum; pressures down to 10^{-10} atm can be readily attained with such a system. Mercury vapor is very toxic to inhale, however. The normal boiling point of liquid mercury is 357°C. What would be the vapor pressure of mercury at 25°C?

Phase Diagrams

65. How many phases exist at a triple point? Describe what would happen if a small amount of heat were added under constant-volume conditions to a sample of water at the triple point. Assume a negligible volume change during fusion.

66. What is the critical point? Will a substance always be a liquid below the critical temperature? Why or why not?

Refer to the phase diagram of CO_2 in Figure 13-17b to answer Exercises 67–70.

67. What phase of CO_2 exists at 1.5 atm pressure and a temperature of: (a) −90°C; (b) −60°C; (c) 0°C?

68. What phases of CO_2 are present (a) at a temperature of −78°C and a pressure of 1 atm? (b) at −57°C and a pressure of 5.2 atm?

69. List the phases that would be observed if a sample of CO_2 at 8 atm pressure were heated from −80°C to 40°C.

70. How does the melting point of CO_2 change with pressure? What does this indicate about the relative density of solid CO_2 versus liquid CO_2?

*71. You are given the following data for ethanol, C_2H_5OH.

Normal melting point	−117°C
Normal boiling point	78.0°C
Critical temperature	243°C
Critical pressure	63.0 atm

Assume that the triple point is slightly lower in temperature than the melting point and that the vapor pressure at the triple point is about 10^{-5} torr. (a) Sketch a phase diagram for ethanol. (b) Ethanol at 1 atm and 140.°C is compressed to 70. atm. Are two phases present at any time during this process? (c) Ethanol at 1 atm and 270.°C is compressed to 70. atm. Are two phases present at any time during this process?

*72. You are given the following data for butane, C_4H_{10}.

Normal melting point	−138°C
Normal boiling point	0°C
Critical temperature	152°C
Critical pressure	38 atm

Assume that the triple point is slightly lower in temperature than the melting point and that the vapor pressure at the triple point is 3×10^{-5} torr. (a) Sketch a phase diagram for butane. (b) Butane at 1 atm and 140°C is compressed to 40 atm. Are two phases present at any time during this process? (c) Butane at 1 atm and 200°C is compressed to 40 atm. Are two phases present at any time during this process?

Exercises 73 and 74 refer to the following phase diagram for sulfur. (The vertical axis is on a logarithmic scale.) Sulfur has two *solid* forms, monoclinic and rhombic.

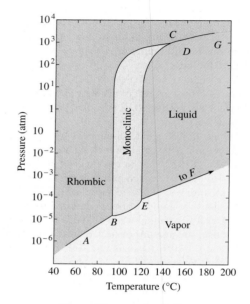

Phase Diagram for Sulfur

*73. (a) How many triple points are there for sulfur? (b) Indicate the approximate pressure and temperature at each triple point. (c) Which phases are in equilibrium at each triple point?

74. Which physical states should be present at equilibrium under the following conditions? (a) 10^{-1} atm and 110°C; (b) 10^{-5} atm and 80°C; (c) 5×10^3 atm and 160°C; (d) 10^{-1} atm and 80°C; (e) 10^{-5} atm and 140°C; (f) 1 atm and 140°C.

The Solid State

75. Comment on the following statement: "The only perfectly ordered state of matter is the crystalline state."

76. Ice floats in water. Why? Would you expect solid mercury to float in liquid mercury at its freezing point? Explain.

77. Distinguish among and compare the characteristics of molecular, covalent, ionic, and metallic solids. Give two examples of each kind of solid.

78. Classify each of the following substances, in the solid state, as molecular, ionic, covalent (network), or metallic solids:

	Melting Point (°C)	Boiling Point (°C)	Electrical Conductor	
			Solid	Liquid
MoF_6	17.5 (at 406 torr)	35	no	no
BN	3000 (sublimes)	—	no	no
Se_8	217	684	poor	poor
Pt	1769	3827	yes	yes
RbI	642	1300	no	yes

79. Classify each of the following substances, in the solid state, as molecular, ionic, covalent (network), or metallic solids:

	Melting Point (°C)	Boiling Point (°C)	Electrical Conductor	
			Solid	Liquid
$CeCl_3$	848	1727	no	yes
Ti	1675	3260	yes	yes
$TiCl_4$	−25	136	no	no
NO_3F	−175	−45.9	no	no
B	2300	2550	no	no

80. Based only on their formulas, classify each of the following in the solid state as a molecular, ionic, covalent (network), or metallic solid: (a) SO_2F; (b) MgF_2; (c) W; (d) Pb; (e) PF_5.

81. Based only on their formulas, classify each of the following in the solid state as a molecular, ionic, covalent (network), or metallic solid: (a) Au; (b) NO_2; (c) CaF_2; (d) SF_4; (e) $C_{diamond}$.

82. Arrange the following solids in order of increasing melting points and account for the order: NaF, MgF_2, AlF_3.

83. Arrange the following solids in order of increasing melting points and account for the order: MgO, CaO, SrO, BaO.

Unit Cell Data: Atomic and Ionic Sizes

84. Distinguish among and sketch simple cubic, body-centered cubic (bcc), and face-centered cubic (fcc) lattices. Use CsCl, sodium, and nickel as examples of solids existing in simple cubic, bcc, and fcc lattices, respectively.

85. Describe a unit cell as precisely as you can.

86. Determine the number of ions of each type present in each unit cell shown in Figure 13-29.

87. Refer to Figure 13-29a. (a) If the unit cell edge is represented as a, what is the distance (center to center) from Cs^+ to its nearest neighbor? (b) How many equidistant nearest neighbors does each Cs^+ ion have? What are the identities of these nearest neighbors? (c) What is the distance (center to center), in terms of a, from a Cs^+ ion to the nearest Cs^+ ion? (d) How many equidistant nearest neighbors does each Cl^- ion have? What are their identities?

88. Refer to Figure 13-28. (a) If the unit cell edge is represented as a, what is the distance (center to center) from Na^+ to its nearest neighbor? (b) How many equidistant nearest neighbors does each Na^+ ion have? What are the identities of these nearest neighbors? (c) What is the distance (center to center), in terms of a, from a Na^+ ion to the nearest Na^+ ion? (d) How many equidistant nearest neighbors does each Cl^- ion have? What are their identities?

89. Polonium crystallizes in a simple cubic unit cell with an edge length of 3.36 Å. (a) What is the mass of the unit cell? (b) What is the volume of the unit cell? (c) What is the theoretical density of Po?

90. Calculate the density of Na metal. The length of the body-centered cubic unit cell is 4.24 Å.

91. Tungsten has a density of 19.3 g/cm^3 and crystallizes in a cubic lattice whose unit cell edge length is 3.16 Å. Which type of cubic unit cell is it?

92. The atomic radius of iridium is 1.36 Å. The unit cell of iridium is a face-centered cube. Calculate the density of iridium.

93. A certain metal has a specific gravity of 10.200 at 25°C. It crystallizes in a body-centered cubic arrangement with a unit cell edge length of 3.147 Å. Determine the atomic weight and identify the metal.

94. The structure of diamond is shown below, with each sphere representing a carbon atom. (a) How many carbon atoms are there per unit cell in the diamond structure? (b) Verify, by extending the drawing if necessary, that each carbon atom has four nearest neighbors. What is the arrangement of these nearest neighbors? (c) What is the distance (center to center) from any carbon atom to its nearest neighbor, expressed in terms of a, the unit cell edge? (d) The observed unit cell edge length in diamond is 3.567 Å. What is the C—C single bond length in diamond? (e) Calculate the density of diamond.

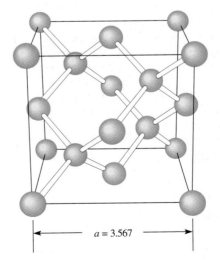

$\longleftarrow a = 3.567 \longrightarrow$

95. The crystal structure of CO_2 is cubic, with a cell edge length of 5.540 Å. A diagram of the cell is shown in Figure 13-31a. (a) What is the number of molecules of CO_2 per unit cell? (b) Is this structure face-centered cubic? How can you tell? (c) What is the density of solid CO_2 at this temperature?

96. A Group IVA element with a density of 11.35 g/cm³ crystallizes in a face-centered cubic lattice whose unit cell edge length is 4.95 Å. Calculate its atomic weight. What is the element?

97. Crystalline silicon has the same structure as diamond, with a unit cell edge length of 5.430 Å. (a) What is the Si–Si distance in this crystal? (b) Calculate the density of crystalline silicon.

*98. (a) What types of electromagnetic radiation are suitable for diffraction studies of crystals? (b) Describe the X-ray diffraction experiment. (c) What must be the relationship between the wavelength of incident radiation and the spacing of the particles in a crystal for diffraction to occur?

*99. (a) Write the Bragg equation. Identify each symbol. (b) X-rays from a palladium source ($\lambda = 0.576$ Å) were reflected by a sample of copper at an angle of 9.40°. This reflection corresponds to the unit cell length ($d = a$) with $n = 2$ in the Bragg equation. Calculate the length of the copper unit cell.

100. The spacing between successive planes of platinum atoms parallel to the cubic unit cell face is 2.256 Å. When X-radiation emitted by copper strikes a crystal of platinum metal, the minimum diffraction angle of X-rays is 19.98°. What is the wavelength of the Cu radiation?

101. Gold crystallizes in a fcc structure. When X-radiation of 0.70926 Å wavelength from molybdenum is used to determine the structure of metallic gold, the minimum diffraction angle of X-rays by the gold is 8.683°. Calculate the spacing between parallel layers of gold atoms.

Metallic Bonding and Semiconductors

102. In general, metallic solids are ductile and malleable, whereas ionic salts are brittle and shatter readily (although they are hard). Explain this observation.

103. What single factor accounts for the ability of metals to conduct both heat and electricity in the solid state? Why are ionic solids poor conductors of heat and electricity although they are composed of charged particles?

104. Compare the temperature dependence of electrical conductivity of a metal with that of a typical metalloid. Explain the difference.

Mixed Exercises

105. Benzene, C_6H_6, boils at 80.1°C. How much energy, in joules, would be required to change 450.0 g of liquid benzene at 21.5°C to a vapor at its boiling point? (The specific heat of liquid benzene is 1.74 J/g·°C and its heat of vaporization is 395 J/g.)

106. The three major components of air are N_2 (bp −196°C), O_2 (bp −183°C), and Ar (bp −186°C). Suppose we have a sample of liquid air at −200°C. In what order will these gases evaporate as the temperature is raised?

*107. A 20.0-g sample of liquid ethanol, C_2H_5OH, absorbs 6.84×10^3 J of heat at its normal boiling point, 78.0°C. The molar enthalpy of vaporization of ethanol, ΔH_{vap}, is 39.3 kJ/mol. (a) What volume of C_2H_5OH vapor is produced? The volume is measured at 78.0°C and 1.00 atm pressure. (b) What mass of C_2H_5OH remains in the liquid state?

*108. What is the pressure predicted by the ideal gas law for one mole of steam in 31.0 L at 100°C? What is the pressure predicted by the van der Waals equation (Section 12-15) given that $a = 5.464$ L²·atm/mol² and $b = 0.03049$ L/mol? What is the percent difference between these values? Does steam deviate from ideality significantly at 100°C? Why?

*109. The boiling points of HCl, HBr, and HI increase with increasing molecular weight. Yet the melting and boiling points of the sodium halides, NaCl, NaBr, and NaI, decrease with increasing formula weight. Explain why the trends are opposite. Describe the intermolecular forces present in each of these compounds and predict which has the lowest boiling point.

110. The structures for three molecules having the formula $C_2H_2Cl_2$ are

Describe the intermolecular forces present in each of these compounds and predict which has the lowest boiling point.

111. Are the following statements true or false? Indicate why if a statement is false. (a) The vapor pressure of a liquid will decrease if the volume of liquid decreases. (b) The normal boiling point of a liquid is the temperature at which the external pressure equals the vapor pressure of the liquid. (c) The vapor pressures of liquids in a similar series tend to increase with increasing molecular weight.

112. Are the following statements true or false? Indicate why if a statement is false. (a) The equilibrium vapor pressure of a liquid is independent of the volume occupied by the vapor above the liquid. (b) The normal boiling point of a liquid changes with changing atmospheric pressure. (c) The vapor pressure of a liquid will increase if the mass of liquid is increased.

113. The following are vapor pressures at 20°C. Predict the order of increasing normal boiling points of the liquids, acetone, 185 torr; ethanol, 44 torr; carbon disulfide, CS_2, 309 torr.

114. Refer to Exercise 113. What is the expected order of increasing molar heats of vaporization, ΔH_{vap}, of these liquids at their boiling points? Account for the order.

115. Refer to the sulfur phase diagram on page 528. (a) Can rhombic sulfur be sublimed? If so, under what conditions? (b) Can monoclinic sulfur be sublimed? If so, under what conditions? (c) Describe what happens if rhombic sulfur is slowly heated from 80°C to 140°C at constant 1-atm pressure. (d) What happens if rhombic sulfur is heated from 80°C to 140°C under constant pressure of 5×10^{-6} atm?

116. The normal boiling point of ammonia, NH_3, is $-33°C$, and its freezing point is $-78°C$. fill in the blanks. (a) At STP (0°C, 1 atm pressure), NH_3 is a _____. (b) If the temperature drops to $-40°C$, the ammonia will _____ and become a _____. (c) If the temperature drops further to $-80°C$ and the molecules arrange themselves in an orderly pattern, the ammonia will _____ and become a _____. (d) If crystals of ammonia are left on the planet Mars at a temperature of $-100°C$, they will gradually disappear by the process of _____ and form a _____.

117. Give the correct names for these changes in state: (a) Crystals of *para*-dichlorobenzene, used as a moth repellent, gradually become vapor without passing through the liquid phase. (b) As you enter a warm room from the outdoors on a cold winter day, your eyeglasses become fogged with a film of moisture. (c) On the same (windy) winter day, a pan of water is left outdoors. Some of it turns to vapor, the rest to ice.

118. The normal boiling point of trichlorofluoromethane, CCl_3F, is 24°C, and its freezing point is $-111°C$. Complete these sentences by supplying the proper terms that describe a state of matter or a change in state. (a) At standard temperature and pressure, CCl_3F is a _____. (b) In an arctic winter at $-40°C$ and 1-atm pressure, CCl_3F is a _____. If it is cooled to $-120°C$, the molecules arrange themselves in an orderly lattice, the CCl_3F _____ and becomes a _____. (c) If crystalline CCl_3F is held at a temperature of $-120°C$ while a stream of helium gas is blown over it, the crystals will gradually disappear by the process of _____. If liquid CCl_3F is boiled at atmospheric pressure, it is converted to a _____ at a temperature of _____.

119. The van der Waals constants (Section 12-15) are $a = 19.01 \text{ L}^2 \cdot \text{atm/mol}^2$, $b = 0.1460$ L/mol for pentane, and $a = 18.05 \text{ L}^2 \cdot \text{atm/mol}^2$, $b = 0.1417$ L/mol for isopentane.

$$CH_3-CH_2-CH_2-CH_2-CH_3 \qquad CH_3-\underset{\underset{H}{|}}{\overset{\overset{CH_3}{|}}{C}}-CH_2-CH_3$$

pentane   isopentane

(a) Basing your reasoning on intermolecular forces, why would you expect a for pentane to be greater? (b) Basing your reasoning on molecular size, why would you expect b for pentane to be greater?

CONCEPTUAL EXERCISES

*120. Iodine sublimes at room temperature and pressure; water does not. Explain the differences you would expect to observe at room temperature if 10.0 grams of iodine crystals were sealed in a 10.-mL container and 5.0 mL of water were sealed in a similar container.

Charles D. Winters

121. Referring to the phase diagram for carbon dioxide shown in figure 13-17b for approximate values, draw a heating curve similar to that in Figure 13-15 for carbon dioxide at 1 atmosphere pressure. Draw a second heating curve for carbon dioxide at 5 atm pressure. Estimate the transition temperatures.

122. A popular misconception is that "hot water freezes more quickly than cold water." In an experiment two 100.-mL samples of water, in identical containers, were placed far apart in a freezer at $-25°C$. One sample had an initial temperature of 78°C, while the other was at 24°C. The second sample took 151 minutes to freeze, and the warmer sample took 166 minutes. The warmer sample took more time but not much. How can you explain their taking about the same length of time to freeze?

123. Consider the portions of the heating curves shown here. Which compound has the highest specific heat capacity?

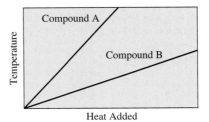

BUILDING YOUR KNOWLEDGE

124. More than 150 years ago Pierre Dulong and A. T. Petit discovered a *rule of thumb* that the heat capacity of one mole of a pure solid element is about 6.0 calories per °C (i.e., about 25 J/°C). A 100.2-g sample of an unknown metal at 99.9°C is placed in 50.6 g of water at 24.8°C. The temperature is 36.6°C when the system comes to equilibrium. Assume all heat lost by the metal is absorbed by the water. What is the likely identity of this metal?

125. In a flask containing dry air and some *liquid* silicon tetrachloride, $SiCl_4$, the total pressure was 988 torr at 225°C. Halving the volume of the flask increased the pressure to 1742 torr at constant temperature. What is the vapor pressure of $SiCl_4$ in this flask at 225°C?

126. A friend comes to you with this problem: "I looked up the vapor pressure of water in a table; it is 26.7 torr at 300 K and 92,826 torr at 600 K. That means that the vapor pressure increases by a factor of 3477 when the absolute temperature doubles over this temperature range. But I thought the pressure was proportional to the absolute temperature, $P = nRT/V$. The pressure doesn't just double. Why?" How would you help the friend?

127. Using as many as six drawings (frames) for each, depict the changes that occur at the molecular level during each of the following physical changes: (a) melting an ice cube; (b) sublimation of an ice cube below 4.6°C and 4.6 torr, and (c) evaporation of a droplet of water at room temperature and pressure.

128. Write the Lewis formula of each member of each of the following pairs. Then use VSEPR theory to predict the geometry about each central atom, and describe any features that lead you to decide which member of each pair would have the lower boiling point. (a) CH_3COOH and $HCOCH_3$; (b) NHF_2 and BH_2Cl; (c) CH_3CH_2OH and CH_3OCH_3.

129. At its normal melting point of 271.3°C, solid bismuth has a density of 9.73 g/cm³ and liquid bismuth has a density of 10.05 g/cm³. A mixture of liquid and solid bismuth is in equilibrium at 271.3°C. If the pressure were increased from 1 atm to 10 atm, would more solid bismuth melt or would more liquid bismuth freeze? What unusual property does bismuth share with water?

130. The doping of silicon with boron to an atomic concentration of 0.0010% boron atoms vs. silicon atoms increases its conductivity by a factor of 10^3 at room temperature. How many atoms of boron would be needed to dope 15.0 g of silicon? What mass of boron is this?

BEYOND THE TEXTBOOK

Go to the textbook website at

http://www.brookscole.com/chemistry/whitten

for additional activities and exercises based on the General Chemistry Interactive CD-ROM, the World Wide Web, and library resources.

InfoTrac College Edition

For additional readings, go to InfoTrac College Edition, your online research library at:

http://infotrac.thomsonlearning.com

Solutions

The ocean is a solution of many different dissolved compounds. In this chapter we look at solution formation and solution properties.

OUTLINE

OBJECTIVES

After you have studied this chapter, you should be able to

- Describe the factors that favor the dissolution process
- Describe the dissolution of solids in liquids, liquids in liquids, and gases in liquids
- Describe how temperature and pressure affect solubility
- Express concentrations of solutions in terms of molality and mole fractions
- Describe the four colligative properties of solutions and some of their applications
- Carry out calculations involving the four colligative properties of solutions: lowering of vapor pressure (Raoult's Law), boiling point elevation, freezing point depression, and osmotic pressure
- Use colligative properties to determine molecular weights of compounds
- Describe dissociation and ionization of compounds, and the associated effects on colligative properties
- Recognize and describe colloids: the Tyndall effect, the adsorption phenomenon, hydrophilic and hydrophobic colloids

A solution is defined as a *homogeneous mixture*, at the molecular level, of two or more substances in which settling does not occur. A solution consists of a solvent and one or more solutes, whose proportions can vary from one solution to another. By contrast, a pure substance has fixed composition. The *solvent* is the medium in which the *solutes* are dissolved. The fundamental units of solutes are usually ions or molecules.

Solutions include different combinations in which a solid, liquid, or gas acts as either solvent or solute. Usually the solvent is a liquid. For instance, sea water is an aqueous solution of many salts and some gases such as carbon dioxide and oxygen. Carbonated water is a saturated solution of carbon dioxide in water. Solutions are common in nature and are extremely important in all life processes, in all scientific areas, and in many industrial processes. The body fluids of all forms of life are solutions. Variations in concentrations of our bodily fluids, especially those of blood and urine, give physicians valuable clues about a person's health. Solutions in which the solvent is not a liquid are also common. Air is a solution of gases with variable composition. Dental fillings are solid amalgams, or solutions of liquid mercury dissolved in solid metals. Alloys are solid solutions of solids dissolved in a metal.

It is usually obvious which of the components of a solution is the solvent and which is (are) the solute(s): The solvent is usually the most abundant species present. In a cup of instant coffee, the coffee and any added sugar are considered solutes, and the hot water is the solvent. If we mix 10 grams of alcohol with 90 grams of water, alcohol is the solute. If we mix 10 grams of water with 90 grams of alcohol, water is the solute. But which is the solute and which is the solvent in a solution of 50 grams of water and 50 grams of alcohol? In such cases, the terminology is arbitrary and, in fact, unimportant.

Solutions were defined in Section 3-6.

Many naturally occurring fluids contain particulate matter suspended in a solution. For example, blood contains a solution (plasma) with suspended blood cells. Sea water contains dissolved substances as well as suspended solids.

THE DISSOLUTION PROCESS

14-1 SPONTANEITY OF THE DISSOLUTION PROCESS

In Section 4-2, part 5, we listed the solubility guidelines for aqueous solutions. Now we investigate the major factors that influence solubility.

A substance may dissolve with or without reaction with the solvent. For example, when metallic sodium reacts with water, there is the evolution of bubbles of hydrogen and a great deal of heat. A chemical change occurs in which H_2 and soluble ionic sodium hydroxide, NaOH, are produced.

$$2Na(s) + 2H_2O(\ell) \longrightarrow 2Na^+(aq) + 2OH^-(aq) + H_2(g)$$

If the resulting solution is evaporated to dryness, solid sodium hydroxide, NaOH, is obtained rather than metallic sodium. This, along with the production of bubbles of hydrogen, is evidence of a reaction with the solvent. Reactions that involve oxidation state changes are usually considered as chemical reactions and not as dissolution.

Solid sodium chloride, NaCl, on the other hand, dissolves in water with no evidence of chemical reaction.

$$NaCl(s) \xrightarrow{\;H_2O\;} Na^+(aq) + Cl^-(aq)$$

Evaporation of the water from the sodium chloride solution yields the original NaCl. In this chapter we focus on dissolution processes of this type, in which no irreversible reaction occurs between components.

The ease of dissolution of a solute depends on two factors: (1) the change in energy and (2) the change in disorder (called entropy change) that accompanies the process. In the next

Sodium reacts with water to form hydrogen gas and a solution of sodium hydroxide. A small amount of phenolphthalein indicator added to this solution turns pink in the presence of sodium hydroxide.

Ionic solutes that do not react with the solvent undergo solvation. This is a kind of reaction in which molecules of solvent are attached in oriented clusters to the solute particles.

chapter we will study both of these factors in detail for many kinds of physical and chemical changes. For now, we point out that a process is *favored* by (1) a *decrease in the energy* of the system, which corresponds to an *exothermic process*, and (2) an *increase in the disorder*, or randomness, of the system.

Let us look at the first of these factors. If a solution gets hotter as a substance dissolves, energy is being released in the form of heat. The energy change that accompanies a dissolution process is called the **heat of solution, $\Delta H_{solution}$.** It depends mainly on how strongly solute and solvent particles interact. A negative value of $\Delta H_{solution}$ designates the release of heat. More negative (less positive) values of $\Delta H_{solution}$ favor the dissolution process.

In a pure liquid all the intermolecular forces are between like molecules; when the liquid and a solute are mixed, each molecule then interacts with molecules (or ions) unlike it as well as with like molecules. The relative strengths of these interactions help to determine the extent of solubility of a solute in a solvent. The main interactions that affect the dissolution of a solute in a solvent follow.

CD-ROM   Screen 14.3, The Solution Process.

---

   a. Weak solute–solute attractions favor solubility.

   b. Weak solvent–solvent attractions favor solubility.

   c. Strong solvent–solute attractions favor solubility.

Figure 14-1 illustrates the interplay of these factors. The intermolecular or interionic attractions among solute particles in the pure solute must be overcome (Step a) to dissolve the solute. This part of the process requires an *input* of energy (endothermic). Separating the solvent molecules from one another (Step b) to "make room" for the solute particles also requires the *input* of energy (endothermic). Energy is *released*, however, as the solute particles and solvent molecules interact in the solution (Step c, exothermic). The overall dissolution process is exothermic (and favored) if the amount of heat absorbed in hypothetical Steps a and b is less than the amount of energy released in Step c. The process is endothermic (and disfavored) if the amount of energy absorbed in Steps a and b is greater than the amount of heat released in Step c.

We can consider the energy changes separately, even though the actual process cannot be carried out in these separate steps.

Many solids do dissolve in liquids by *endothermic* processes, however. The reason such processes can occur is that the endothermicity can be outweighed by a large increase in disorder of the solute during the dissolution process. The solute particles are highly ordered in a solid crystal, but are free to move about randomly in liquid solutions. Likewise, the

Figure 14-1 A diagram representing the changes in energy content associated with the hypothetical three-step sequence in a dissolution process—in this case, for a solid solute dissolving in a liquid solvent. (Similar considerations would apply to other combinations.) The process depicted here is exothermic. The amount of heat absorbed in Steps a and b is less than the amount released in Step c. In an *endothermic* process (not shown), the heat content of the solution would be *higher* than that of the original solvent plus solute. Thus, the amount of energy absorbed in Steps a and b would be *greater* than the amount of heat released in Step c.

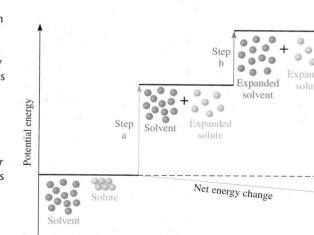

degree of disorder in the solvent increases as the solution is formed, because solvent molecules are then in a more random environment. They are surrounded by a mixture of solvent and solute particles.

Most dissolving processes are accompanied by an overall increase in disorder. Thus, the disorder factor is usually *favorable* to solubility. The determining factor, then, is whether the heat of solution (energy) also favors dissolution or, if it does not, whether it is small enough to be outweighed by the favorable effects of the increasing disorder. In gases, for instance, the molecules are so far apart that intermolecular forces are quite weak. Thus, when gases are mixed, changes in the intermolecular forces are very slight. So the very favorable increase in disorder that accompanies mixing is always more important than possible changes in intermolecular attractions (energy). Hence, gases that do not react with one another can always be mixed in any proportion.

The most common types of solutions are those in which the solvent is a liquid. In the next several sections we consider liquid solutions in more detail.

The dissolution of NaF in water is one of the few cases in which the dissolving process is not accompanied by an overall increase in disorder. The water molecules become more ordered around the small F^- ions. This is due to the strong hydrogen bonding between H_2O molecules and F^- ions.

$$O-H \cdots F^- \cdots H-O$$
$$| \qquad\qquad\qquad |$$
$$H \qquad\qquad\qquad H$$

The amount of heat released on mixing, however, outweighs the disadvantage of this ordering.

14-2 DISSOLUTION OF SOLIDS IN LIQUIDS

The ability of a solid to go into solution depends most strongly on its crystal lattice energy, or the strength of attractions among the particles making up the solid. The **crystal lattice energy** is defined as the energy change accompanying the formation of one mole of formula units in the crystalline state from constituent particles in the gaseous state. This process is always exothermic; that is, crystal lattice energies are always *negative*. For an ionic solid, the process is written as

A very negative crystal lattice energy indicates very strong attractions within the solid.

$$M^+(g) + X^-(g) \longrightarrow MX(s) + energy$$

The amount of energy involved in this process depends on the attraction between ions in the solid. When these attractions are strong, a large amount of energy is released as the solid forms, and so the solid is very stable.

The reverse of the crystal formation reaction is the separation of the crystal into ions.

$$MX(s) + energy \longrightarrow M^+(g) + X^-(g)$$

This process can be considered the hypothetical first step (Step a in Figure 14-1) in forming a solution of a solid in a liquid. It is always endothermic. The smaller the magnitude of the crystal lattice energy (a measure of the solute–solute interactions), the more readily dissolution occurs. Less energy must be supplied to start the dissolution process.

If the solvent is water, the energy that must be supplied to expand the solvent (Step b in Figure 14-1) includes that required to break up some of the hydrogen bonding between water molecules.

The third major factor contributing to the heat of solution is the extent to which solvent molecules interact with particles of the solid. The process in which solvent molecules surround and interact with solute ions or molecules is called **solvation.** When the solvent is water, the more specific term is **hydration. Hydration energy** (equal to the sum of Steps b and c in Figure 14-1) is defined as the energy change involved in the (exothermic) hydration of one mole of gaseous ions.

Hydration energy is also referred to as the **heat of hydration.**

$$M^{n+}(g) + xH_2O(\ell) \longrightarrow M(OH_2)_x^{n+} + energy \qquad \text{(for cation)}$$
$$X^{y-}(g) + rH_2O(\ell) \longrightarrow X(H_2O)_r^{y-} + energy \qquad \text{(for anion)}$$

Hydration is usually highly exothermic for ionic or polar covalent compounds, because the polar water molecules interact very strongly with ions and polar molecules. In fact, the only

CD-ROM   Screen 14.4, Energetics of Solution Formation.

solutes that are appreciably soluble in water either undergo dissociation or ionization or are able to form hydrogen bonds with water.

The overall heat of solution for a solid dissolving in a liquid is equal to the heat of solvation minus the crystal lattic energy.

$$\Delta H_{solution} = (\text{heat of solvation}) - (\text{crystal lattice energy})$$

Remember that both terms on the right are always negative.

Nonpolar solids such as naphthalene, $C_{10}H_8$, do not dissolve appreciably in polar solvents such as water because the two substances do not attract each other significantly. This is true despite the fact that crystal lattice energies of solids consisting of nonpolar molecules are much less negative (smaller in magnitude) than those of ionic solids. Naphthalene dissolves readily in nonpolar solvents such as benzene because there are no strong attractive forces between solute molecules or between solvent molecules. In such cases, the increase in disorder controls the process. These facts help explain the observation that "*like dissolves like.*"

> The statement "like dissolves like" means that polar solvents dissolve ionic and polar molecular solutes, and nonpolar solvents dissolve nonpolar molecular solutes.

Consider what happens when a piece of sodium chloride, a typical ionic solid, is placed in water. The $\delta+$ parts of water molecules (H atoms) attract the negative chloride ions on the surface of the solid NaCl, as shown in Figure 14-2. Likewise, the $\delta-$ parts of H_2O molecules (O atoms) orient themselves toward the Na^+ ions and solvate them. These at-

A cube of sugar is slowly lifted through a solution.

Glucose, like all sugars, is quite polar and dissolves in water, which is also very polar. Naphthalene is nonpolar and insoluble in water.

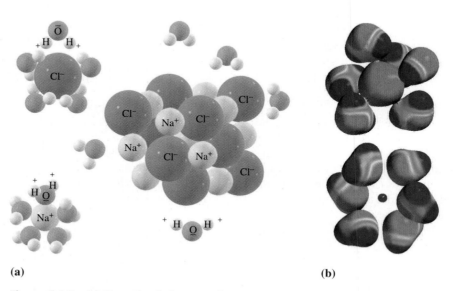

(a)   **(b)**

Figure 14-2   (a) The role of electrostatic attractions in the dissolution of NaCl in water. The $\delta+$ H of the polar H_2O molecule helps to attract Cl^- away from the crystal. Likewise, Na^+ is attracted by the $\delta-$ O. Once they are separated from the crystal, both kinds of ions are surrounded by water molecules, to complete the hydration process. (b) Electrostatic plots illustrating the attraction of the positive portions (blue) of the water molecules to the negative ions, and the attraction of the negative portions (red) of the water molecules to the positive ions.

tractions help to overcome the forces holding the ions in the crystal, and NaCl dissolves in the H_2O.

$$NaCl(s) \xrightarrow{H_2O} Na^+(aq) + Cl^-(aq)$$

When we write $Na^+(aq)$ and $Cl^-(aq)$, we refer to hydrated ions. The number of H_2O molecules attached to an ion differs with different ions. Sodium ions are thought to be hexahydrated; that is, $Na^+(aq)$ probably represents $[Na(OH_2)_6]^+$. Most cations in aqueous solution are surrounded by four to nine H_2O molecules, with six being the most common. Generally, larger cations can accommodate more H_2O molecules than smaller cations.

Many solids that are appreciably soluble in water are ionic compounds. Magnitudes of crystal lattice energies generally increase with increasing charge and decreasing size of ions. That is, the size of the lattice energy increases as the ionic charge densities increase and, therefore, as the strength of electrostatic attractions within the crystal increases. Hydration energies vary in the same order (Table 14-1). As we indicated earlier, crystal lattice energies and hydration energies are generally much smaller in magnitude for molecular solids than for ionic solids.

Hydration and the effects of attractions in a crystal oppose each other in the dissolution process. Hydration energies and lattice energies are usually of about the same magnitude for low-charge species, so they often nearly cancel each other. As a result, the dissolution process is slightly endothermic for many ionic substances. Ammonium nitrate, NH_4NO_3, is an example of a salt that dissolves endothermically. This property is used in the "instant cold packs" used to treat sprains and other minor injuries. Ammonium nitrate and water are packaged in a plastic bag in which they are kept separate by a partition that is easily broken when squeezed. As the NH_4NO_3 dissolves in the H_2O, the mixture absorbs heat from its surroundings and the bag becomes cold to the touch.

Ionic solids, if they are quite soluble, usually dissolve with the release of heat. Examples are anhydrous sodium sulfate, Na_2SO_4; calcium acetate, $Ca(CH_3COO)_2$; calcium chloride, $CaCl_2$; and lithium sulfate hydrate, $Li_2SO_4 \cdot H_2O$.

As the charge-to-size ratio (charge density) increases for ions in ionic solids, the magnitude of the crystal lattice energy usually increases more than the hydration energy. This

For simplicity, we often omit the (aq) designations from dissolved ions. Remember that all ions are hydrated in aqueous solution, whether this is indicated or not.

Review the sizes of ions in Figure 6-1 carefully.

The charge/radius ratio (charge density) is the ionic charge divided by the ionic radius in angstroms. This is a measure of the *charge density* around the ion. A negative value for heat of hydration indicates that heat is *released* during hydration.

Charles Steele

Solid ammonium nitrate, NH_4NO_3, dissolves in water in a very endothermic process, absorbing heat from its surroundings. It is used in instant cold packs for early treatment of injuries, such as sprains and bruises, to minimize swelling.

TABLE 14-1	Ionic Radii, Charge/Radius Ratios, and Hydration Energies for Some Cations		
Ion	Ionic Radius (Å)	Charge/Radius Ratio	Hydration Energy (kJ/mol)
K^+	1.52	0.66	−351
Na^+	1.16	0.86	−435
Li^+	0.90	1.11	−544
Ca^{2+}	1.14	1.75	−1650
Fe^{2+}	0.76	2.63	−1980
Zn^{2+}	0.74	2.70	−2100
Cu^{2+}	0.72	2.78	−2160
Fe^{3+}	0.64	4.69	−4340
Cr^{3+}	0.62	4.84	−4370
Al^{3+}	0.68	4.41	−4750

makes dissolution of solids that contain highly charged ions—such as aluminum fluoride, AlF_3; magnesium oxide, MgO; and chromium (III) oxide, Cr_2O_3—too endothermic to be very soluble in water.

14-3 DISSOLUTION OF LIQUIDS IN LIQUIDS (MISCIBILITY)

In science, **miscibility** is used to describe the ability of one liquid to dissolve in another. The three kinds of attractive interactions (solute–solute, solvent–solvent, and solvent–solute) must be considered for liquid–liquid solutions just as they were for solid–liquid solutions. Because solute–solute attractions are usually much weaker for liquid solutes than for solids, this factor is less important and so the mixing process is often exothermic for miscible liquids. Polar liquids tend to interact strongly with and dissolve readily in other polar liquids. Methanol, CH_3OH; ethanol, CH_3CH_2OH; acetonitrile, CH_3CN; and sulfuric acid, H_2SO_4, are all polar liquids that are soluble in most polar solvents (such as water). The hydrogen bonding between methanol and water molecules and the dipolar interaction between acetonitrile and water molecules are depicted in Figure 14-3.

Because hydrogen bonding is so strong between sulfuric acid, H_2SO_4, and water, a large amount of heat is released when concentrated H_2SO_4 is diluted with water (Figure 14-4). This can cause the solution to boil and spatter. If the major component of the mixture is water, this heat can be absorbed with less increase in temperature because of the unusually high specific heat of H_2O. For this reason, *sulfuric acid (as well as other mineral acids) is always diluted by adding the acid slowly and carefully to water. Water should never be added to the acid.* If spattering does occur when the acid is added to water, it is mainly water that spatters, not the corrosive concentrated acid.

Nonpolar liquids that do not react with the solvent generally are not very soluble in polar liquids because of the mismatch of forces of interaction. They are said to be *immiscible.* Nonpolar liquids are, however, usually quite soluble (*miscible*) in other nonpolar liquids. Between nonpolar molecules (whether alike or different) there are only dispersion forces,

Hydrogen bonding and dipolar interactions were discussed in Section 13-2.

When water is added to concentrated acid, the danger is due more to the spattering of the acid itself than to the steam from boiling water.

The nonpolar molecules in oil do not attract polar water molecules, so oil and water are immiscible. The polar water molecules attract one another strongly—they "squeeze out" the nonpolar molecules in the oil. Oil is less dense than water, so it floats on water.

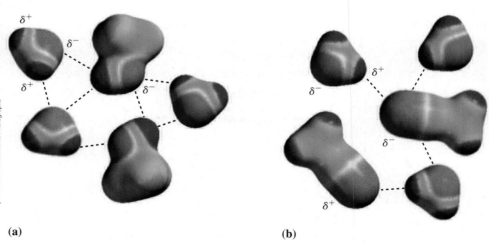

Kip Peticolas Fundamenta Photographs

(a)   (b)

Figure 14-3 (a) Hydrogen bonding in a solution of methanol (CH_3OH) and water. (b) Dipolar interactions in a solution of acetonitrile (CH_3CN) and water. In this representation, colors for each molecule range from blue (most positive regions) to red (most negative regions). Molecules tend to arrange themselves to maximize attractions by bringing regions of opposite charge together while minimizing repulsions by separating regions of like charge.

Charles Steele

Figure 14-4 The heat released by pouring 50 mL of sulfuric acid, H_2SO_4, into 50 mL of water increases the temperature by 100°C (from 21°C to 121°C)!

which are weak and easily overcome. As a result, when two nonpolar liquids are mixed, their molecules just "slide between" one another.

14-4 DISSOLUTION OF GASES IN LIQUIDS

Based on Section 13-2 and the foregoing discussion, we expect that polar gases are most soluble in polar solvents and nonpolar gases are most soluble in nonpolar liquids. Although carbon dioxide and oxygen are nonpolar gases, they do dissolve slightly in water. CO_2 is somewhat more soluble because it reacts with water to some extent to form carbonic acid, H_2CO_3. This in turn ionizes slightly in two steps to give hydrogen ions, bicarbonate ions, and carbonate ions.

$$CO_2(g) + H_2O(\ell) \rightleftharpoons H_2CO_3(aq) \qquad \text{carbonic acid (exists only in solution)}$$

$$H_2CO_3(aq) \rightleftharpoons H^+(aq) + HCO_3^-(aq)$$

$$HCO_3^-(aq) \rightleftharpoons H^+(aq) + CO_3^{2-}(aq)$$

Approximately 1.45 grams of CO_2 (0.0329 mole) dissolves in a liter of water at 25°C and one atmosphere pressure.

Oxygen, O_2, is less soluble in water than CO_2, but it does dissolve to a noticeable extent due to dispersion forces (induced dipoles, Section 13-2). Only about 0.041 gram of O_2 (1.3×10^{-3} mole) dissolves in a liter of water at 25°C and 1 atm pressure. This is sufficient to support aquatic life.

The hydrogen halides, HF, HCl, HBr, and HI, are all polar covalent gases. In the gas phase the interactions among the widely separated molecules are not very strong, so

Carbon dioxide is called an acid anhydride, that is, an "acid without water." As noted in Section 6-8, many other oxides of nonmetals, such as N_2O_5, SO_3, and P_4O_{10}, are also acid anhydrides.

Aqueous HCl, HBr, and HI are strong acids (Section 4-2 , Part 2, and Section 10-7). Aqueous HF is a weak acid.

Immersed in water, this green plant oxidizes the water to form gaseous oxygen. A small amount of the oxygen dissolves in water, and the excess oxygen forms bubbles of gas. The gaseous oxygen in the bubbles and the dissolved oxygen are in equilibrium with one another.

Concentrated hydrochloric acid, HCl(aq) is an approximately 40% solution of HCl gas in water.

solute–solute attractions are minimal, and the dissolution processes in water are exothermic. The resulting solutions, called hydrohalic acids, contain predominantly ionized HX (X = Cl, Br, I). The ionization involves *protonation* of a water molecule by HX to form a hydrated hydrogen ion and halide ion X^- (also hydrated). HCl is used as an example.

$$\overset{H^+ \text{ transfer}}{\textcircled{H} : \overset{..}{\underset{..}{Cl}} : \; + \; H\!-\!\overset{..}{\underset{|}{O}} : \longrightarrow \; : \overset{..}{\underset{..}{Cl}} : ^- \; + \; \left[H\!-\!\overset{\overset{\displaystyle H}{|}}{\underset{|}{\overset{..}{O}}} : \right]^+ }$$

HF is only slightly ionized in aqueous solution because of the strong covalent bond. In addition, the polar bond between H and the small F atoms in HF causes very strong hydrogen bonding between H_2O and the largely intact HF molecules.

$$\overset{\delta+ \quad \delta- \qquad \delta+ \quad \delta-}{H\!-\!\overset{..}{\underset{\underset{\displaystyle H}{|}}{O}} : \text{---} H\!-\!\overset{..}{\underset{..}{F}} :} \quad \text{as well as} \quad \overset{\delta+ \quad \delta- \qquad \delta+ \quad \delta-}{H\!-\!\overset{..}{\underset{..}{F}} : \text{---} H\!-\!\overset{..}{\underset{\underset{\displaystyle H}{|}}{O}} :}$$

$$\qquad\qquad \delta+ \qquad\qquad\qquad\qquad\qquad\qquad\qquad \delta+$$

> The only gases that dissolve appreciably in water are those that are capable of hydrogen bonding (such as HF), those that ionize extensively (such as HCl, HBr, and HI), and those that react with water (such as CO_2).

14-5 RATES OF DISSOLUTION AND SATURATION

At a given temperature, the rate of dissolution of a solid increases if large crystals are ground to a powder. Grinding increases the surface area, which in turn increases the number of solute ions or molecules in contact with the solvent. When a solid is placed in water, some of its particles solvate and dissolve. The rate of this process slows as time passes because the surface area of the crystals gets smaller and smaller. At the same time, the number of solute

A saturated solution of copper(II) sulfate, $CuSO_4$, in water. As H_2O evaporates, blue $CuSO_4\cdot5H_2O$ crystals form. They are in dynamic equilibrium with the saturated solution.

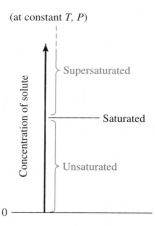

(at constant *T, P*)

Concentration of solute

Supersaturated

Saturated

Unsaturated

0

A solution that contains more than the amount of solute normally contained at saturation is said to be supersaturated. A solution that contains less than the amount of solute necessary for saturation is said to be unsaturated.

particles in solution increases, so they collide with the solid more frequently. Some of these collisions result in recrystallization. The rates of the two opposing processes become equal after some time. The solid and dissolved ions are then in equilibrium with each other.

$$\text{solid} \underset{\text{crystallization}}{\overset{\text{dissolution}}{\rightleftharpoons}} \text{dissolved particles}$$

Such a solution is said to be **saturated.** Saturation occurs at very low concentrations of dissolved species for slightly soluble substances and at high concentrations for very soluble substances. When imperfect crystals are placed in saturated solutions of their ions, surface defects on the crystals are slowly "patched" with no net increase in mass of the solid. Often, after some time has passed, we see fewer but larger crystals. These observations provide evidence of the dynamic nature of the solubility equilibrium. After equilibrium is established, no more solid dissolves without the simultaneous crystallization of an equal mass of dissolved ions.

The solubilities of many solids increase at higher temperatures. **Supersaturated solutions** contain higher-than-saturated concentrations of solute. They can sometimes be prepared by saturating a solution at a high temperature. The saturated solution is cooled slowly, without agitation, to a temperature at which the solute is less soluble. At this point, the resulting supersaturated solution is *metastable* (temporarily stable). This may be thought of as a state of pseudoequilibrium in which the system is at a higher energy than in its most stable state. In such a case, the solute has not yet become sufficiently organized for crystallization to begin. A supersaturated solution produces crystals rapidly if it is slightly disturbed or if it is "seeded" with a dust particle or a tiny crystal. Under such conditions enough solid crystallizes to leave a saturated solution (Figure 14-5).

 CD-ROM   Screen 14.2, Solubility.

Dynamic equilibria occur in all saturated solutions; for instance, there is a continuous exchange of oxygen molecules across the surface of water in an open container. This is fortunate for fish, which "breathe" dissolved oxygen.

Charles Steele

Figure 14-5 Another method of seeding a supersaturated solution is by pouring it very slowly onto a seed crystal. A supersaturated sodium acetate solution was used in these photographs.

Robert Metz

A tiny crystal of sodium acetate, NaCH₃COO, was added to a clear, colorless, supersaturated solution of NaCH₃COO. This photo shows solid NaCH₃COO just beginning to crystallize in a very rapid process.

14-6 EFFECT OF TEMPERATURE ON SOLUBILITY

In Section 13-6 we introduced LeChatelier's Principle, which states that *when a stress is applied to a system at equilibrium, the system responds in a way that best relieves the stress.* Recall that exothermic processes release heat and endothermic processes absorb heat.

$$\text{Exothermic:} \qquad \text{reactants} \longrightarrow \text{products} + \text{heat}$$
$$\text{Endothermic:} \qquad \text{reactants} + \text{heat} \longrightarrow \text{products}$$

Many ionic solids dissolve by endothermic processes. Their solubilities in water usually *increase* as heat is added and the temperature increases. For example, KCl dissolves endothermically.

$$KCl(s) + 17.2 \text{ kJ} \xrightarrow{\;H_2O\;} K^+(aq) + Cl^-(aq)$$

Figure 14-6 shows that the solubility of KCl increases as the temperature increases because more heat is available to increase the dissolving process. Raising the temperature (adding *heat*) causes a stress on the solubility equilibrium. This stress favors the process that *consumes* heat. In this case, more KCl dissolves.

Solid iodine, I_2, dissolves to a limited extent in water to give an orange solution. This aqueous solution does not mix with nonpolar carbon tetrachloride, CCl_4 (*left*). After the funnel is shaken and the liquids are allowed to separate (*right*), the upper aqueous phase is lighter orange and the lower CCl_4 layer is much more highly colored. This is because iodine is much more soluble in the nonpolar carbon tetrachloride than in water; much of the iodine dissolves preferentially in the lower (CCl_4) phase. The design of the separatory funnel allows the lower (denser) layer to be drained off. Fresh CCl_4 could be added and the process repeated. This method of separation is called *extraction.* It takes advantage of the different solubilities of a solute in two immiscible liquids.

Leon Lewandowski

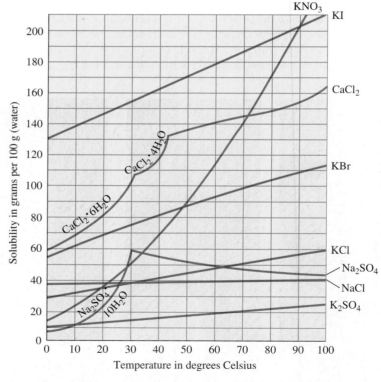

Figure 14-6 A graph that illustrates the effect of temperature on the solubilities of some salts. Some compounds exist either as nonhydrated crystalline substances or as hydrated crystals. Hydrated and nonhydrated crystal forms of the same compounds often have different solubilities because of the different total forces of attraction in the solids. The discontinuities in the solubility curves for $CaCl_2$ and Na_2SO_4 are due to transitions between hydrated and nonhydrated crystal forms.

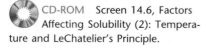

CD-ROM Screen 14.6, Factors Affecting Solubility (2): Temperature and LeChatelier's Principle.

Calcium acetate, $Ca(CH_3COO)_2$, is more soluble in cold water than in hot water. When a cold, concentrated solution of calcium acetate is heated, solid calcium acetate precipitates.

Some solids, such as anhydrous Na_2SO_4, and many liquids and gases dissolve by exothermic processes. Their solubilities usually decrease as temperature increases. The solubility of O_2 in water decreases (by 34%) from 0.041 gram per liter of water at 25°C to 0.027 gram per liter at 50°C. Raising the temperature of rivers and lakes by dumping heated waste water from industrial plants and nuclear power plants is called **thermal pollution.** A slight increase in the temperature of the water causes a small but significant decrease in the concentration of dissolved oxygen. As a result, the water can no longer support the marine life it ordinarily could.

CD-ROM   Screen 14.5, Factors Affecting Solubility (1): Henry's Law and Gas Pressure.

The dissolution of anhydrous calcium chloride, $CaCl_2$, in water is quite exothermic. Here the temperature increases from 21°C to 88°C. This dissolution process is utilized in commercial instant hot packs for quick treatment of injuries requiring heat.

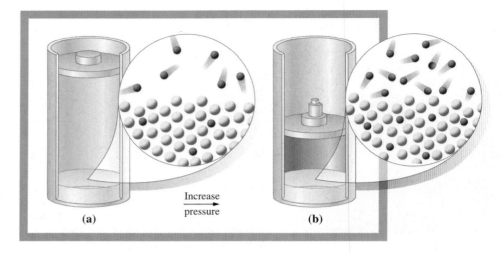

Figure 14-7 An illustration of Henry's Law. The solubility of a gas (red) that does not react completely with the solvent (yellow) increases with increasing pressure of the gas above the solution.

(a)   Increase pressure →   (b)

Carbonated beverages can be used to illustrate Henry's Law. When the bottle is opened, the equilibrium is disturbed and bubbles of CO_2 form within the liquid and rise to the surface. After some time, an equilibrium between dissolved CO_2 and atmospheric CO_2 is reached.

14-7 EFFECT OF PRESSURE ON SOLUBILITY

Changing the pressure has no appreciable effect on the solubilities of either solids or liquids in liquids. The solubilities of gases in all solvents increase, however, as the partial pressures of the gases increase (Figure 14-7). Carbonated water is a saturated solution of carbon dioxide in water under pressure. When a can or bottle of a carbonated beverage is opened, the pressure on the surface of the beverage is reduced to atmospheric pressure, and much of the CO_2 bubbles out of solution. If the container is left open, the beverage becomes "flat" because the released CO_2 escapes.

Henry's Law applies to gases that do not react with the solvent in which they dissolve (or, in some cases, gases that react incompletely). It is usually stated as follows.

> The pressure of a gas above the surface of a solution is proportional to the concentration of the gas in the solution. Henry's Law can be represented symbolically as
>
> $$P_{gas} = kC_{gas}$$

P_{gas} is the pressure of the gas above the solution, and k is a constant for a particular gas and solvent at a particular temperature. C_{gas} represents the concentration of dissolved gas; it is usually expressed either as molarity (Section 3-6) or as mole fraction (Section 14-8). The relationship is valid at low concentrations and low pressures.

14-8 MOLALITY AND MOLE FRACTION

We saw in Section 3-6 that concentrations of solutions are often expressed as percent by mass of solute or as molarity. Discussion of many physical properties of solutions is often made easier by expressing concentrations either in molality units or as mole fractions (Sections 14-9 to 14-14).

Molality

The **molality**, m, of a solute in solution is the number of moles of solute *per kilogram of solvent.*

$$\text{molality} = \frac{\text{number of moles solute}}{\text{number of kilograms solvent}}$$

Molality is based on the amount of *solvent not solution.*

EXAMPLE 14-1 *Molality*

What is the molality of a solution that contains 128 g of CH_3OH in 108 g of water?

Plan

We convert the amount of solute (CH_3OH) to moles, express the amount of solvent (water) in kilograms, and apply the definition of molality.

$$g\ CH_3OH \longrightarrow mol\ CH_3OH$$
$$\searrow$$
$$molality$$
$$g\ H_2O \longrightarrow kg\ H_2O \nearrow$$

Solution

$$\frac{?\ mol\ CH_3OH}{kg\ H_2O} = \frac{128\ g\ CH_3OH}{0.108\ kg\ H_2O} \times \frac{1\ mol\ CH_3OH}{32.0\ g\ CH_3OH} = \frac{37.0\ mol\ CH_3OH}{kg\ H_2O}$$

$$= 37.0\ m\ CH_3OH$$

You should now work Exercise 28.

Each beaker holds the amount of a crystalline ionic compound that will dissolve in 100. grams of water at 100°C. The compounds are (*top row, left to right*) 39 grams of sodium chloride (NaCl, *white*), 102 grams of potassium dichromate ($K_2Cr_2O_7$, *red-orange*), 341 grams of nickel sulfate hexahydrate ($NiSO_4 \cdot 6H_2O$, *green*); (*bottom row, left to right*) 79 grams of potassium chromate (K_2CrO_4, *yellow*), 191 grams of cobalt(II) chloride hexahydrate ($CoCl_2 \cdot 6H_2O$, *dark red*), and 203 grams of copper sulfate pentahydrate ($CuSO_4 \cdot 5H_2O$, *blue*).

EXAMPLE 14-2 *Molality*

How many grams of H_2O must be used to dissolve 50.0 grams of sucrose to prepare a 1.25 *m* solution of sucrose, $C_{12}H_{22}O_{11}$?

Plan

We convert the amount of solute ($C_{12}H_{22}O_{11}$) to moles, solve the molality expression for kilograms of solvent (water), and then express the result in grams.

Solution

$$?\ mol\ C_{12}H_{22}O_{11} = 50.0\ g\ C_{12}H_{22}O_{11} \times \frac{1\ mol\ C_{12}H_{22}O_{11}}{342\ g\ C_{12}H_{22}O_{11}} = 0.146\ mol\ C_{12}H_{22}O_{11}$$

$$\text{molality of solution} = \frac{mol\ C_{12}H_{22}O_{11}}{kg\ H_2O}$$

Rearranging gives

$$kg\ H_2O = \frac{mol\ C_{12}H_{22}O_{11}}{\text{molality of solution}} = \frac{0.146\ mol\ C_{12}H_{22}O_{11}}{1.25\ mol\ C_{12}H_{22}O_{11}/kg\ H_2O}$$

$$= 0.117\ kg\ H_2O = \boxed{117\ g\ H_2O}$$

 It is very easy to confuse molality (*m*) with molarity (*M*). Pay close attention to what is being asked.

In other examples later in this chapter we will calculate several properties of this solution.

Mole Fraction

Recall that in Chapter 12 the **mole fractions,** X_A and X_B, of each component in a mixture containing components A and B were defined as

$$X_A = \frac{\text{no. mol A}}{\text{no. mol A + no. mol. B}} \quad \text{and} \quad X_B = \frac{\text{no. mol B}}{\text{no. mol A + no. mol B}}$$

Mole fraction is a dimensionless quantity, that is, it has no units.

EXAMPLE 14-3   *Mole Fraction*

What are the mole fractions of CH_3OH and H_2O in the solution described in Example 14-1? It contains 128 grams of CH_3OH and 108 grams of H_2O.

Plan

We express the amount of both components in moles, and then apply the definition of mole fraction.

Solution

$$\underline{?}\ \text{mol } CH_3OH = 128\ \text{g } CH_3OH \times \frac{1\ \text{mol } CH_3OH}{32.0\ \text{g } CH_3OH} = 4.00\ \text{mol } CH_3OH$$

$$\underline{?}\ \text{mol } H_2O = 108\ \text{g } H_2O \times \frac{1\ \text{mol } H_2O}{18.0\ \text{g } H_2O} = 6.00\ \text{mol } H_2O$$

Now we calculate the mole fraction of each component.

$$X_{CH_3OH} = \frac{\text{no. mol } CH_3OH}{\text{no. mol } CH_3OH + \text{no. mol } H_2O} = \frac{4.00\ \text{mol}}{(4.00 + 6.00)\ \text{mol}} = \boxed{0.400}$$

$$X_{H_2O} = \frac{\text{no. mol } H_2O}{\text{no. mol } CH_3OH + \text{no. mol } H_2O} = \frac{6.00\ \text{mol}}{(4.00 + 6.00)\ \text{mol}} = \boxed{0.600}$$

You should now work Exercise 30.

In any mixture the sum of the mole fractions must be 1:

$$0.400 + 0.600 = 1$$

COLLIGATIVE PROPERTIES OF SOLUTIONS

Colligative means "tied together."

Physical properties of solutions that depend on the *number*, but not the *kind*, of solute particles in a given amount of solvent are called **colligative properties.** There are four important colligative properties of a solution that are directly proportional to the number of solute particles present. They are (1) vapor pressure lowering, (2) boiling point elevation, (3) freezing point depression, and (4) osmotic pressure. These properties of a solution depend on the *total concentration of all solute particles*, regardless of their ionic or molecular nature, charge, or size. For most of this chapter, we will consider *nonelectrolyte* solutes (Section 4-2, Part 1); these substances dissolve to give one mole of dissolved particles for each mole of solute. In Section 14-14 we will learn to modify our predictions of colligative properties to account for ion formation in electrolyte solutions.

14-9 LOWERING OF VAPOR PRESSURE AND RAOULT'S LAW

Many experiments have shown that a solution containing a *nonvolatile* liquid or a solid as a solute always has a lower vapor pressure than the pure solvent (Figure 14-8). The vapor pressure of a liquid depends on the ease with which the molecules are able to escape from the surface of the liquid. When a solute is dissolved in a liquid, some of the total volume of the solution is occupied by solute molecules, and so there are fewer solvent molecules *per unit area* at the surface. As a result, solvent molecules vaporize at a slower rate than if no solute were present. The increase in disorder that accompanies evaporation is also a significant factor. Because a solution is already more disordered ("mixed up") than a pure solvent, the evaporation of the pure solvent involves a larger increase in disorder, and is thus more favorable. Hence, the pure solvent exhibits a higher vapor pressure than does the solution. The lowering of the vapor pressure of the solution is a colligative property. It is a function of the number, and not the kind, of solute particles in solution. We emphasize that solutions of gases or low-boiling (volatile) liquid solutes can have *higher* total vapor pressures than the pure solvents, so this discussion does not apply to them.

The lowering of the vapor pressure of a solvent due to the presence of *nonvolatile, nonionizing* solutes is summarized by **Raoult's Law.**

A *vapor* is a gas formed by the boiling or evaporation of a liquid or sublimation of a solid. The *vapor pressure* of a liquid is the pressure (partial pressure) exerted by a vapor in equilibrium with its liquid (see Section 13-7).

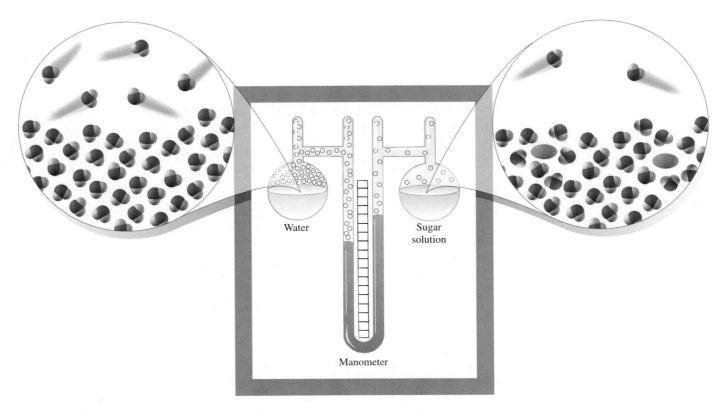

Water   Sugar solution   Manometer

Figure 14-8 Lowering of vapor pressure. If no air is present in the apparatus, the pressure above each liquid is due to water vapor. This pressure is less over the solution of sugar and water.

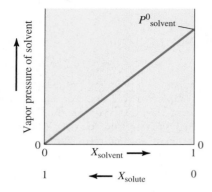

Figure 14-9 Raoult's Law for an ideal solution of a solute in a volatile liquid. The vapor pressure exerted by the liquid is proportional to its mole fraction in the solution.

CD-ROM   Screen 14.7, Colligative Properties (1): Vapor Pressure and Raoult's Law.

The vapor pressure of water in the solution is $(23.8 - 0.524)$ torr $= 23.3$ torr. We could calculate this vapor pressure directly from the mole fraction of the solvent (water) in the solution, using the relationship

$$P_{solvent} = X_{solvent}P^0_{solvent}.$$

> The vapor pressure of a solvent in an ideal solution is directly proportional to the mole fraction of the solvent in the solution.

The relationship can be expressed mathematically as

$$P_{solvent} = X_{solvent}P^0_{solvent}$$

where $X_{solvent}$ represents the mole fraction of the solvent in a solution, $P^0_{solvent}$ is the vapor pressure of the *pure* solvent, and $P_{solvent}$ is the vapor pressure of the solvent *in the solution* (see Figure 14-9). If the solute is nonvolatile, the vapor pressure of the solution is entirely due to the vapor pressure of the solvent, $P_{solution} = P_{solvent}$.

The *lowering* of the vapor pressure, $\Delta P_{solvent}$, is defined as

$$\Delta P_{solvent} = P^0_{solvent} - P_{solvent}$$

Thus,

$$\Delta P_{solvent} = P^0_{solvent} - (X_{solvent}P^0_{solvent}) = (1 - X_{solvent})P^0_{solvent}$$

Now $X_{solvent} + X_{solute} = 1$, so $1 - X_{solvent} = X_{solute}$. We can express the *lowering* of the vapor pressure in terms of the mole fraction of solute.

$$\Delta P_{solvent} = X_{solute}P^0_{solvent}$$

Solutions that obey this relationship exactly are called **ideal solutions.** The vapor pressures of many solutions, however, do not behave ideally.

EXAMPLE 14-4   *Vapor Pressure of a Solution of Nonvolatile Solute*

Sucrose is a nonvolatile, nonionizing solute in water. Determine the vapor pressure lowering, at 25°C, of the 1.25 *m* sucrose solution in Example 14-2. Assume that the solution behaves ideally. The vapor pressure of pure water at 25°C is 23.8 torr (Appendix E).

Plan

The solution in Example 14-2 was made by dissolving 50.0 grams of sucrose (0.146 mol) in 117 grams of water (6.50 mol). We calculate the mole fraction of solute in the solution. Then we apply Raoult's Law to find the vapor pressure lowering, $\Delta P_{solvent}$.

Solution

$$X_{sucrose} = \frac{0.146 \text{ mol}}{0.146 \text{ mol} + 6.50 \text{ mol}} = 0.0220$$

Applying Raoult's Law in terms of the vapor pressure lowering,

$$\Delta P_{solvent} = (X_{solute})(P^0_{solvent}) = (0.0220)(23.8 \text{ torr}) = \boxed{0.524 \text{ torr}}$$

You should now work Exercise 38.

When a solution consists of two components that are very similar, each component behaves essentially as it would if it were pure. For example, the two liquids heptane, C_7H_{16}, and octane, C_8H_{18}, are so similar that each heptane molecule experiences nearly the same intermolecular forces whether it is near another heptane molecule or near an octane molecule, and similarly for each octane molecule. The properties of such a solution can be predicted from a knowledge of its composition and the properties of each component. Such a solution is very nearly ideal.

Consider an ideal solution of two volatile components, A and B. The vapor pressure of each component above the solution is proportional to its mole fraction in the solution.

$$P_A = X_A P^0{}_A \quad \text{and} \quad P_B = X_B P^0{}_B$$

The total vapor pressure of the solution is, by Dalton's Law of Partial Pressures (Section 12-11), equal to the sum of the vapor pressures of the two components.

$$P_{total} = P_A + P_B \quad \text{or} \quad P_{total} = X_A P^0{}_A + X_B P^0{}_B$$

This is shown graphically in Figure 14-10. We can use these relationships to predict the vapor pressures of an ideal solution, as Example 14-5 illustrates.

> If component B were nonvolatile, then $P^0{}_B$ would be zero, and this description would be the same as that given earlier for a solution of a nonvolatile, nonionizing solute in a volatile solvent, $P_{total} = P_{solvent}$.

EXAMPLE 14-5 Vapor Pressure of a Solution of Volatile Components

At 40°C, the vapor pressure of pure heptane is 92.0 torr and the vapor pressure of pure octane is 31.0 torr. Consider a solution that contains 1.00 mole of heptane and 4.00 moles of octane. Calculate the vapor pressure of each component and the total vapor pressure above the solution.

Plan

We first calculate the mole fraction of each component in the liquid solution. Then we apply Raoult's Law to each of the two volatile components. The total vapor pressure is the sum of the vapor pressures of the components.

Solution

We first calculate the mole fraction of each component in the liquid solution.

$$X_{heptane} = \frac{1.00 \text{ mol heptane}}{(1.00 \text{ mol heptane}) + (4.00 \text{ mol octane})} = 0.200$$

$$X_{octane} = 1 - X_{heptane} = 0.800$$

Then, applying Raoult's Law for volatile components,

$$P_{heptane} = X_{heptane} P^0{}_{heptane} = (0.200)(92.0 \text{ torr}) = \boxed{18.4 \text{ torr}}$$

$$P_{octane} = X_{octane} P^0{}_{octane} = (0.800)(31.0 \text{ torr}) = \boxed{24.8 \text{ torr}}$$

$$P_{total} = P_{heptane} + P_{octane} = 18.4 \text{ torr} + 24.8 \text{ torr} = \boxed{43.2 \text{ torr}}$$

You should now work Exercise 40.

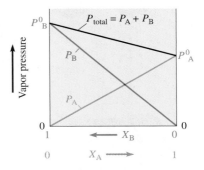

Figure 14-10 Raoult's Law for an ideal solution of two volatile components. The left-hand side of the plot corresponds to pure B ($X_A = 0$, $X_B = 1$), and the right-hand side corresponds to pure A ($X_A = 1$, $X_B = 0$). Of these hypothetical liquids, B is more volatile than A ($P^0{}_B > P^0{}_A$).

The vapor in equilibrium with a liquid solution of two or more volatile components has a higher mole fraction of the more volatile component than does the liquid solution.

EXAMPLE 14-6 Composition of Vapor

Calculate the mole fractions of heptane and octane in the vapor that is in equilibrium with the solution in Example 14-5.

Plan

We learned in Section 12-11 that the mole fraction of a component in a gaseous mixture equals the ratio of its partial pressure to the total pressure. In Example 14-5 we calculated the partial pressure of each component in the vapor and the total vapor pressure.

Heptane (pure vapor pressure = 92.0 torr at 40°C) is a more volatile liquid than octane (pure vapor pressure = 31.0 torr at 40°C). Its mole fraction in the vapor, 0.426, is higher than its mole fraction in the liquid, 0.200.

$$CH_3-\overset{\overset{\displaystyle \cdot\cdot O \cdot\cdot}{\|}}{C}-CH_3$$

acetone

$$\cdot\cdot S = C = S \cdot\cdot$$

carbon disulfide

Solution

In the *vapor*

$$X_{\text{heptane}} = \frac{P_{\text{heptane}}}{P_{\text{total}}} = \frac{18.4 \text{ torr}}{43.2 \text{ torr}} = \boxed{0.426}$$

$$X_{\text{octane}} = \frac{P_{\text{octane}}}{P_{\text{total}}} = \frac{24.8 \text{ torr}}{43.2 \text{ torr}} = \boxed{0.574}$$

or

$$X_{\text{octane}} = 1.000 - 0.426 = \boxed{0.574}$$

You should now work Exercise 42.

Many dilute solutions behave ideally. Some solutions do not behave ideally over the entire concentration range. For some solutions, the observed vapor pressure is greater than that predicted by Raoult's Law (Figure 14-11a). This kind of deviation, known as a *positive deviation*, is due to differences in polarity of the two components. On the molecular level, the two substances do not mix entirely randomly, so there is self-association of each component with local regions enriched in one type of molecule or the other. In a region enriched in A molecules, substance A acts as though its mole fraction were greater than it is in the solution as a whole, and the vapor pressure due to A is greater than if the solution were ideal. A similar description applies to component B. The total vapor pressure is then greater than it would be if the solution were behaving ideally. A solution of acetone (polar) and carbon disulfide (nonpolar) is an example of a solution that shows a positive deviation from Raoult's Law.

Another, more common type of deviation occurs when the total vapor pressure is less than that predicted (Figure 14-11b). This is called a *negative deviation*. Such an effect is due to unusually strong attractions (such as hydrogen bonding) between *different polar*

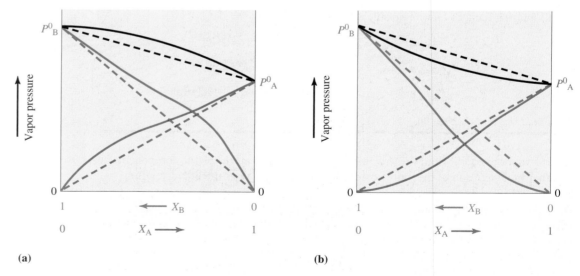

(a)   (b)

Figure 14-11 Deviations (*solid lines*) from Raoult's Law for two volatile components. Ideal behavior (Raoult's Law) is shown by the dashed lines. (a) Positive deviation. (b) Negative deviation.

molecules. As a result, different polar molecules are strongly attracted to one another, so fewer molecules escape to the vapor phase. The observed vapor pressure of each component is thus less than ideally predicted. An acetone–chloroform solution and an ethanol–water solution are two polar combinations that show negative deviations from Raoult's Law.

14-10 FRACTIONAL DISTILLATION

In Section 13-8 we described *simple* distillation as a process in which a liquid solution can be separated into volatile and nonvolatile components. But separation of volatile components is not very efficient by this method. Consider the simple distillation of a liquid solution consisting of two volatile components. If the temperature is slowly raised, the solution begins to boil when the sum of the vapor pressures of the components reaches the applied pressure on the surface of the solution. Both components exert vapor pressures, so both are carried away as a vapor. The resulting distillate is richer than the original liquid in the more volatile component (Example 14-6).

As a mixture of volatile liquids is distilled, the compositions of both the liquid and the vapor, as well as the boiling point of the solution, change continuously. *At constant pressure*, we can represent these quantities in a **boiling point diagram**, Figure 14-12. In such a diagram the lower curve represents the boiling point of a liquid mixture with the indicated composition. The upper curve represents the composition of the *vapor* in equilibrium with

$CHCl_3$
chloroform

CH_3CH_2OH
ethanol

The applied pressure is usually atmospheric pressure.

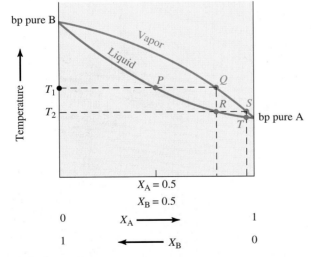

Figure 14-12 A boiling point diagram for a solution of two volatile liquids, A and B. The lower curve represents the boiling point of a liquid mixture with the indicated composition. The upper curve represents the composition of the *vapor* in equilibrium with the boiling liquid mixture at the indicated temperature. Pure liquid A boils at a lower temperature than pure liquid B; hence, A is the more volatile liquid in this illustration. Suppose we begin with an ideal equimolar mixture ($X_A = X_B = 0.5$) of liquids A and B. The point *P* represents the temperature at which this solution boils, T_1. The vapor that is present at this equilibrium is indicated by point *Q* ($X_A \approx 0.8$). Condensation of that vapor at temperature T_2 gives a liquid of the same composition (point *R*). At this point we have described one step of simple distillation. The boiling liquid at point *R* is in equilibrium with the vapor of composition indicated by point *S* ($X_A > 0.95$), and so on.

Distillation under vacuum lowers the applied pressure. This allows boiling at lower temperatures than under atmospheric pressure. This technique allows distillation of some substances that would decompose at higher temperatures.

the boiling liquid mixture at the indicated temperature. The intercepts at the two vertical axes show the boiling points of the two pure liquids. The distillation of the two liquids is described in the legend to Figure 14-12.

From the boiling point diagram in Figure 14-12, we see that two or more volatile liquids cannot be completely separated from each other by a single distillation step. The vapor collected at any boiling temperature is always enriched in the more volatile component (A); however, at any temperature the vapor still contains both components. A *series* of simple distillations would provide distillates increasingly richer in the more volatile component, but the repeated distillations would be very tedious.

Repeated distillations may be avoided by using **fractional distillation.** A *fractionating column* is inserted above the solution and attached to the condenser, as shown in Figure 14-13. The column is constructed so that it has a large surface area or is packed with many small glass beads or another material with a large surface area. These provide surfaces on which condensation can occur. Contact between the vapor and the packing favors condensation of the less volatile component. The column is cooler at the top than at the bottom. By the time the vapor reaches the top of the column, practically all of the less volatile component has condensed and fallen back down the column. The more volatile component goes into the condenser, where it is liquefied and delivered as a highly enriched distillate into the collection flask. The longer the column or the greater the packing, the more efficient is the separation.

Many fractions, such as gasoline, kerosene, fuel oil, paraffin, and asphalt, are separated from crude oil by fractional distillation.

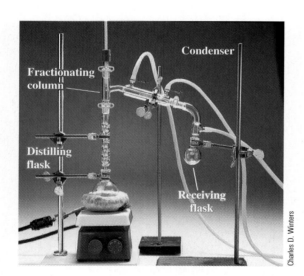

Charles D. Winters

Figure 14-13 A fractional distillation apparatus. The vapor phase rising in the column is in equilibrium with the liquid phase that has condensed out and is flowing slowly back down the column.

Union carbide industrial Gases, Linde Division

Fractional distillation is used for separations in many industrial processes. In this Pennsylvania plant, atmospheric air is liquefied by cooling and compression and then is separated by distillation in towers. This plant produces daily more than 1000 tons of gases from air (nitrogen, oxygen, and argon).

14-11 BOILING POINT ELEVATION

Recall that the boiling point of a liquid is the temperature at which its vapor pressure equals the applied pressure on its surface (see Section 13-8). For liquids in open containers, this is atmospheric pressure. We have seen that the vapor pressure of a solvent at a given temperature is lowered by the presence in it of a *nonvolatile* solute. Such a solution must be heated to a higher temperature than the pure solvent to cause the vapor pressure of the solvent to equal atmospheric pressure (Figure 14-14). In accord with Raoult's Law, the elevation of the boiling point of a solvent caused by the presence of a nonvolatile, nonionized solute is proportional to the number of moles of solute dissolved in a given mass of solvent. Mathematically, this is expressed as

> When the solute is nonvolatile, only the *solvent* distills from the solution.

$$\Delta T_b = K_b m$$

The term ΔT_b represents the elevation of the boiling point of the solvent, that is, the boiling point of the solution minus the boiling point of the pure solvent. The m is the molality of the solute, and K_b is a proportionality constant called the molal **boiling point elevation constant.** This constant is different for different solvents and does not depend on the solute (Table 14-2).

> ⚠ $\Delta T_b = T_{b(soln)} - T_{b(solvent)}$. The boiling points of solutions that contain nonvolatile solutes are always higher than the boiling points of the pure solvents. So ΔT_b is always positive.

K_b corresponds to the change in boiling point produced by a one-molal *ideal* solution of a nonvolatile nonelectrolyte. The units of K_b are °C/m.

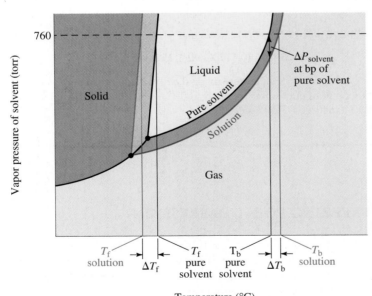

Figure 14-14 Because a *nonvolatile* solute lowers the vapor pressure of a solvent, the boiling point of a solution is higher and the freezing point lower than the corresponding points for the pure solvent. The magnitude of the boiling point elevation, ΔT_b, is less than the magnitude of the freezing point depression, ΔT_f.

TABLE 14-2	*Some Properties of Common Solvents*				
Solvent	bp (pure)	K_b (°C/m)		fp (pure)	K_f (°C/m)
water	100*	0.512		0*	1.86
benzene	80.1	2.53		5.48	5.12
acetic acid	118.1	3.07		16.6	3.90
nitrobenzene	210.88	5.24		5.7	7.00
phenol	182	3.56		43	7.40
camphor	207.42	5.61		178.40	40.0

Exact values.

Elevations of boiling points and depressions of freezing points, which will be discussed later, are usually quite small for solutions of typical concentrations. They can be measured, however, with specially constructed differential thermometers that measure small temperature changes accurately to the nearest 0.001°C.

EXAMPLE 14-7 *Boiling Point Elevation*

Predict the boiling point of the 1.25 m sucrose solution in Example 14-2.

Plan

We first find the *increase* in boiling point from the relationship $\Delta T_b = K_b m$. The boiling point is *higher* by this amount than the normal boiling point of pure water.

Solution

From Table 14-2, K_b for H_2O = 0.512°C/m, so

$$\Delta T_b = (0.512°C/m)(1.25\ m) = 0.640°C$$

The solution would boil at a temperature that is 0.640°C higher than pure water would boil. The normal boiling point of pure water is exactly 100°C, so at 1.00 atm this solution is predicted to boil at 100°C + 0.640°C = 100.640°C.

You should now work Exercise 46.

14-12 FREEZING POINT DEPRESSION

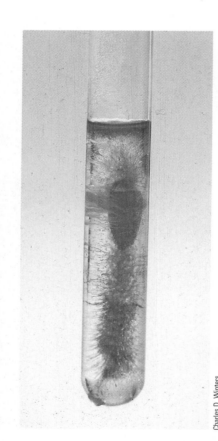

When a solution freezes, the solvent solidifies as the pure substance. For this photo a dye has been added. As the solute freezes along the wall of the test tube, the dye concentration increases near the center.

Charles D. Winters

Molecules of liquids move more slowly and approach one another more closely as the temperature is lowered. The freezing point of a liquid is the temperature at which the forces of attraction among molecules are just great enough to overcome their kinetic energies and thus cause a phase change from the liquid to the solid state. Strictly speaking, the freezing (melting) point of a substance is the temperature at which the liquid and solid phases are in equilibrium. When a dilute solution freezes, it is the *solvent* that begins to solidify first, leaving the solute in a more concentrated solution. Solvent molecules in a solution are somewhat more separated from one another (because of solute particles) than they are in the pure solvent. Consequently, the temperature of a solution must be lowered below the freezing point of the pure solvent to freeze it.

The freezing point depressions of solutions of nonelectrolytes have been found to be equal to the molality of the solute times a proportionality constant called the molal **freezing point depression constant, K_f.**

$$\Delta T_f = K_f m$$

ΔT_f is the *depression* of freezing point. It is defined as

$\Delta T_f = T_{f(solvent)} - T_{f(soln)}$

so it is always *positive*.

The values of K_f for a few solvents are given in Table 14-2. Each is numerically equal to the freezing point depression of a one-molal *ideal* solution of a nonelectrolyte in that solvent.

EXAMPLE 14-8 *Freezing Point Depression*

When 15.0 grams of ethyl alcohol, C_2H_5OH, is dissolved in 750. grams of formic acid, the freezing point of the solution is 7.20°C. The freezing point of pure formic acid is 8.40°C. Evaluate K_f for formic acid.

CD-ROM Screen 14.8, Colligative Properties (2): Boiling Point and Freezing Point.

Plan

The molality and the depression of the freezing point are calculated first. Then we solve the equation $\Delta T_f = K_f m$ for K_f and substitute values for m and ΔT_f.

$$K_f = \frac{\Delta T_f}{m}$$

Remember to use molality (m) and *not* molarity (M) when calculating boiling point elevation or freezing point depression.

Solution

$$\frac{?\ \text{mol } C_2H_5OH}{\text{kg formic acid}} = \frac{15.0\ \text{g } C_2H_5OH}{0.750\ \text{kg formic acid}} \times \frac{1\ \text{mol } C_2H_5OH}{46.0\ \text{g } C_2H_5OH} = 0.435\ m$$

$$\Delta T_f = (T_{f[\text{formic acid}]}) - (T_{f[\text{solution}]}) = 8.40°C - 7.20°C = 1.20°C \quad (\text{depression})$$

Then $K_f = \dfrac{\Delta T_f}{m} = \dfrac{1.20°C}{0.435\ m} = \boxed{2.76°C/m}$ for formic acid.

You should now work Exercise 48.

EXAMPLE 14-9 *Freezing Point Depression*

Calculate the freezing point of the 1.25 m sucrose solution in Example 14-2.

Plan

We first find the *decrease* in freezing point from the relationship $\Delta T_f = K_f m$. The temperature at which the solution freezes is *lower* than the freezing point of pure water by this amount.

Solution

From Table 14-2, K_f for H_2O = 1.86°C/m, so

$$\Delta T_f = (1.86°C/m)(1.25\ m) = 2.32°C$$

The solution freezes at a temperature 2.32°C *below* the freezing point of pure water, or

$$T_{f(\text{solution})} = 0.00°C - 2.32°C = \boxed{-2.32°C}$$

You should now work Exercise 56.

Lime, CaO, is added to molten iron ore during the manufacture of pig iron. It lowers the melting point of the mixture. The metallurgy of iron is discussed in more detail in Chapter 22.

Courtesy of bethlehem Steel

The total concentration of all dissolved solute species determines the colligative properties. As we will emphasize in Section 14-14, we must take into account the extent of ion formation in solutions of ionic solutes.

You may be familiar with several examples of the effects we have studied. Sea water does not freeze on some days when fresh water does, because sea water contains higher concentrations of solutes, mostly ionic solutes. Spreading soluble salts such as sodium chloride, NaCl, or calcium chloride, $CaCl_2$, on an icy road lowers the freezing point of the ice, causing the ice to melt.

A familiar application is the addition of "permanent" antifreeze, mostly ethylene glycol, $HOCH_2CH_2OH$, to the water in an automobile radiator. Because the boiling point of the solution is elevated, addition of a solute as a winter antifreeze also helps protect against loss of the coolant by summer "boil-over." The amounts by which the freezing and boiling points change depend on the concentration of the ethylene glycol solution. The addition of too much ethylene glycol is counterproductive, however. The freezing point of pure ethylene glycol is about $-12°C$. A solution that is mostly ethylene glycol would have a somewhat lower freezing point due to the presence of water as a solute. Suppose you graph the freezing point depression of water below $0°C$ as ethylene glycol is added, and also graph the freezing point depression of ethylene glycol below $-12°C$ as water is added. These two curves would intersect at some temperature, indicating the limit of lowering that can occur. (At these high concentrations, the solutions do not behave ideally, so the temperatures could not be accurately predicted by the equations we have introduced in this chapter, but the main ideas still apply.) Most antifreeze labels recommend a 50:50 mixture by volume (fp $-34°F$, bp $265°F$ with a 15-psi pressure cap on the radiator), and cite the limit of possible protection with a 70:30 mixture by volume of antifreeze:water (fp $-84°F$, bp $276°F$ with a 15-psi pressure cap).

14-13 DETERMINATION OF MOLECULAR WEIGHT BY FREEZING POINT DEPRESSION OR BOILING POINT ELEVATION

The colligative properties of freezing point depression and, to a lesser extent, boiling point elevation are useful in the determination of molecular weights of solutes. The solutes *must* be nonvolatile in the temperature range of the investigation if boiling point elevations are to be determined. We will restrict our discussion of determination of molecular weight to nonelectrolytes.

EXAMPLE 14-10   *Molecular Weight from a Colligative Property*

A 1.20-gram sample of an unknown covalent compound is dissolved in 50.0 grams of benzene. The solution freezes at 4.92°C. Calculate the molecular weight of the compound.

Plan

To calculate the molecular weight of the unknown compound, we find the number of moles that is represented by the 1.20 grams of unknown compound. We first use the freezing point data to find the molality of the solution. The molality relates the number of moles of solute and the mass of solvent (known), so this allows us to calculate the number of moles of unknown.

Ethylene glycol, $HOCH_2CH_2OH$, is the major component of "permanent" antifreeze. It depresses the freezing point of water in an automobile radiator and also raises its boiling point. The solution remains in the liquid phase over a wider temperature range than does pure water. This protects against both freezing and boil-over.

Charles D. Winters

Solution

From Table 14-2, the freezing point of pure benzene is 5.48°C and K_f is 5.12°C/m.

$$\Delta T_f = 5.48°C - 4.92°C = 0.56°C$$

$$m = \frac{\Delta T_f}{K_f} = \frac{0.56°C}{5.12°C/m} = 0.11\ m$$

The molality is the number of moles of solute per kilogram of benzene, so the number of moles of solute in 50.0 g (0.0500 kg) of benzene can be calculated.

$$0.11\ m = \frac{?\ \text{mol solute}}{0.0500\ \text{kg benzene}}$$

$$?\ \text{mol solute} = (0.11\ m)(0.0500\ \text{kg}) = 0.0055\ \text{mol solute}$$

$$\text{mass of 1.0 mol} = \frac{\text{no. of g solute}}{\text{no. of mol solute}} = \frac{1.20\ \text{g solute}}{0.0055\ \text{mol solute}} = 2.2 \times 10^2\ \text{g/mol}$$

$$\text{molecular weight} = \boxed{2.2 \times 10^2\ \text{amu}}$$

You should now work Exercise 61.

EXAMPLE 14-11 *Molecular Weight from a Colligative Property*

Either camphor ($C_{10}H_{16}O$, molecular weight = 152 g/mol) or naphthalene ($C_{10}H_8$, molecular weight = 128 g/mol) can be used to make mothballs. A 5.2-gram sample of mothballs was dissolved in 100. grams of ethyl alcohol, and the resulting solution had a boiling point of 78.90°C. Were the mothballs made of camphor or naphthalene? Pure ethyl alcohol has a boiling point of 78.41°C; its K_b = 1.22°C/m.

Plan

We can distinguish between the two possibilities by determining the molecular weight of the unknown solute. We do this by the method shown in Example 14-10, except that now we use the observed boiling point data.

Solution

The observed boiling point elevation is

$$\Delta T_b = T_{b(\text{solution})} - T_{b(\text{solvent})} = (78.90 - 78.41)°C = 0.49°C$$

Using $\Delta T_b = 0.49°C$ and $K_b = 1.22°C/m$, we can find the molality of the solution.

$$\text{molality} = \frac{\Delta T_b}{K_b} = \frac{0.49°C}{1.22°C/m} = 0.40\ m$$

The number of moles of solute in the 100. g (0.1000 kg) of solvent used is

$$\left(0.40\ \frac{\text{mol solute}}{\text{kg solvent}}\right)(0.100\ \text{kg solvent}) = 0.040\ \text{mol solute}$$

The molecular weight of the solute is its mass divided by the number of moles.

$$\frac{?\ \text{g}}{\text{mol}} = \frac{5.2\ \text{g}}{0.040\ \text{mol}} = 130\ \text{g/mol}$$

The value 130 g/mol for the molecular weight indicates that naphthalene was used to make these mothballs.

You should now work Exercise 59.

14-14   COLLIGATIVE PROPERTIES AND DISSOCIATION OF ELECTROLYTES

As we have emphasized, colligative properties depend on the *number* of solute particles in a given mass of solvent. A 0.100 molal *aqueous* solution of a covalent compound that does not ionize gives a freezing point depression of 0.186°C. If dissociation were complete, 0.100 *m* KBr would have an *effective* molality of 0.200 *m* (i.e., 0.100 *m* K$^+$ + 0.100 *m* Br$^-$). So we might predict that a 0.100 molal solution of this 1 : 1 strong electrolyte would have a freezing point depression of 2 × 0.186°C, or 0.372°C. In fact, the *observed* depression is only 0.349°C. This value for ΔT_f is about 6% less than we would expect for an effective molarity of 0.200 *m*.

In an ionic solution the solute particles are not randomly distributed. Rather, each positive ion has more negative than positive ions near it. The resulting electrical interactions cause the solution to behave nonideally. Some of the ions undergo **association** in solution (Figure 14-15). At any given instant, some K$^+$ and Br$^-$ ions collide and "stick together." During the brief time that they are in contact, they behave as a single particle. This tends to reduce the effective molality. The freezing point depression (ΔT_f) is therefore reduced (as well as the boiling point elevation (ΔT_b) and the lowering of vapor pressure).

A (more concentrated) 1.00 *m* solution of KBr might be expected to have a freezing point depression of 2 × 1.86°C = 3.72°C, but the observed depression is only 3.29°C. This value for ΔT_f is about 11% less than we would expect. We see a greater deviation from the depression predicted (ignoring ionic association) in the more concentrated solution. This is because the ions are closer together and collide more often in the more concentrated solution. Consequently, the ionic association is greater.

One measure of the extent of dissociation (or ionization) of an electrolyte in water is the **van't Hoff factor, *i*,** for the solution. This is the ratio of the *actual* colligative property to the value that *would* be observed *if no dissociation occurred.*

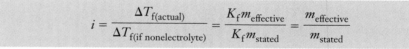

$$i = \frac{\Delta T_{f(actual)}}{\Delta T_{f(if\ nonelectrolyte)}} = \frac{K_f m_{effective}}{K_f m_{stated}} = \frac{m_{effective}}{m_{stated}}$$

The ideal, or limiting, value of *i* for a solution of KBr would be 2, and the value for a 2 : 1 electrolyte such as Na$_2$SO$_4$ would be 3; these values would apply to infinitely dilute solutions in which no appreciable ion association occurs. For 0.10 *m* and 1.0 *m* solutions of KBr, *i* is *less than* 2.

For 0.10 *m*: $i = \dfrac{0.349°C}{0.186°C} = 1.88$ For 1.0 *m*: $i = \dfrac{3.29°C}{1.86°C} = 1.77$

Table 14-3 lists actual and ideal values of *i* for solutions of some strong electrolytes, based on measurements of freezing point depressions.

Many weak electrolytes are quite soluble in water, but they ionize only slightly. The percent ionization and *i* value for a weak electrolyte in solution can also be determined from freezing point depression data (Example 14-12).

$\Delta T_f = K_f m = (1.86°C/m)(0.100\ m)$
$\quad = 0.186°C$

Ionic solutions are elegantly described by the Debye–Hückel theory, which is beyond the scope of this text.

Figure 14-15 Diagrammatic representation of the various species thought to be present in a solution of KBr in water. This would explain unexpected values for its colligative properties, such as freezing point depression.

Weak acids and weak bases (Section 4-2) are weak electrolytes.

TABLE 14-3	*Actual and Ideal van't Hoff Factors, i, for Aqueous Solutions of Nonelectrolytes and Strong Electrolytes*	
Compound	*i* for 1.00 *m* Solution	*i* for 0.100 *m* Solution
nonelectrolytes	1.00 (ideal)	1.00 (ideal)
sucrose, $C_{12}H_{22}O_{11}$	1.00	1.00
If 2 ions in solution/formula unit	2.00 (ideal)	2.00 (ideal)
KBr	1.77	1.88
NaCl	1.83	1.87
If 3 ions in solution/formula unit	3.00 (ideal)	3.00 (ideal)
K_2CO_3	2.39	2.45
K_2CrO_4	1.95	2.39
If 4 ions in solution/formula unit	4.00 (ideal)	4.00 (ideal)
$K_3[Fe(CN)_6]$	—	2.85

EXAMPLE 14-12   *Colligative Property and Weak Electrolytes*

Lactic acid, $C_2H_4(OH)COOH$, is found in sour milk. It is also formed in muscles during intense physical activity and is responsible for the pain felt during strenuous exercise. It is a weak monoprotic acid and therefore a weak electrolyte. The freezing point of a 0.0100 *m* aqueous solution of lactic acid is $-0.0206°C$. Calculate (a) the *i* value and (b) the percent ionization in the solution.

Plan for (a)

To evaluate the van't Hoff factor, *i*, we first calculate $m_{effective}$ from the observed freezing point depression and K_f for water; we then compare $m_{effective}$ and m_{stated} to find *i*.

Solution for (a)

$$m_{effective} = \frac{\Delta T_f}{K_f} = \frac{0.0206°C}{1.86°C/m} = 0.0111 \ m$$

$$i = \frac{m_{effective}}{m_{stated}} = \frac{0.0111 \ m}{0.0100 \ m} = \boxed{1.11}$$

Plan for (b)

The percent ionization is given by

$$\% \ \text{ionization} = \frac{m_{ionized}}{m_{original}} \times 100\% \quad (\text{where } m_{original} = m_{stated} = 0.0100 \ m)$$

The freezing point depression is caused by the $m_{effective}$, the total *concentration of all dissolved species*—in this case, the sum of the concentrations of HA, H^+, and A^-. We know the value of $m_{effective}$ from part (a). Thus, we need to construct an expression for the effective molality in terms of the amount of lactic acid that ionizes. We represent the molality of lactic acid that ionizes as an unknown, *x*, and write the concentrations of all species in terms of this unknown.

Solution for (b)

In many calculations, it is helpful to write down (1) the values, or symbols for the values, of initial concentrations; (2) changes in concentrations due to reaction; and (3) final concentrations, as shown here. The coefficients of the equation are all ones, so the reaction ratio must be $1:1:1$.

To simplify the notation, we denote the weak acid as HA and its anion as A^-. The reaction summary used here to analyze the extent of reaction was introduced in Chapter 11.

Let x = molality of lactic acid that ionizes; then

     x = molality of H^+ and lactate ions that have been formed

	HA	$\longrightarrow$	H^+	+	A^-
Start	0.0100 m		0		0
Change	$-x\ m$		$+x\ m$		$+x\ m$
Final	$(0.0100 - x)\ m$		$x\ m$		$x\ m$

The $m_{effective}$ is equal to the sum of the molalities of all the solute particles.

$$m_{effective} = m_{HA} + m_{H^+} + m_{A^-}$$
$$= (0.0100 - x)\ m + x\ m + x\ m = (0.0100 + x)\ m$$

This must equal the value for $m_{effective}$ calculated earlier, 0.0111 m.

$$0.0111\ m = (0.0100 + x)\ m$$

$$x = 0.0011\ m = \text{molality of the acid that ionizes}$$

We can now calculate the percent ionization.

$$\% \text{ ionization} = \frac{m_{ionized}}{m_{original}} \times 100\% = \frac{0.0011\ m}{0.0100\ m} \times 100\% = \boxed{11\%}$$

This experiment shows that in 0.0100 m solutions, only 11% of the lactic acid has been converted into H^+ and $C_2H_4(OH)COO^-$ ions. The remainder, 89%, exists as nonionized molecules.

You should now work Exercises 78 and 80.

✓ **Problem-Solving Tip:** *Selection of a van't Hoff Factor*

Use an ideal value for the van't Hoff factor unless the question clearly indicates to do otherwise, as in the previous example and in some of the end-of-chapter exercises. For a strong electrolyte dissolved in water, the ideal value for its van't Hoff factor is listed in Table 14-3. For nonelectrolytes dissolved in water or any solute dissolved in common nonaqueous solvents, the van't Hoff factor is considered to be 1. For weak electrolytes dissolved in water, the van't Hoff factor is a little greater than 1.

Osmosis is one of the main ways in which water molecules move into and out of living cells. The membranes and cell walls in living organisms allow solvent to pass through. Some of these also selectively permit passage of ions and other small solute particles.

14-15   OSMOTIC PRESSURE

Osmosis is the spontaneous process by which the solvent molecules pass through a semipermeable membrane from a solution of lower concentration of solute into a solution of higher concentration of solute. A **semipermeable membrane** (e.g., cellophane) separates two solutions. Solvent molecules may pass through the membrane in either direction, but the rate at which they pass into the more concentrated solution is found to be greater than the rate in the opposite direction. The initial difference between the two rates is directly proportional to the difference in concentration between the two solutions. Solvent particles continue to pass through the membrane (Figure 14-16a). The column of liquid continues to rise until the hydrostatic pressure due to the weight of the solution in the column is sufficient to force solvent molecules back through the membrane at the same rate at which they enter from the dilute side. The pressure exerted under this condition is called the **osmotic pressure** of the solution.

CD-ROM   Screen 14.9, Colligative Properties (3): Osmosis.

SC*L*INKS.   **TOPIC:** Osmosis
GO TO: www.scilinks.org
*sci***LINKS CODE:** WCH1410

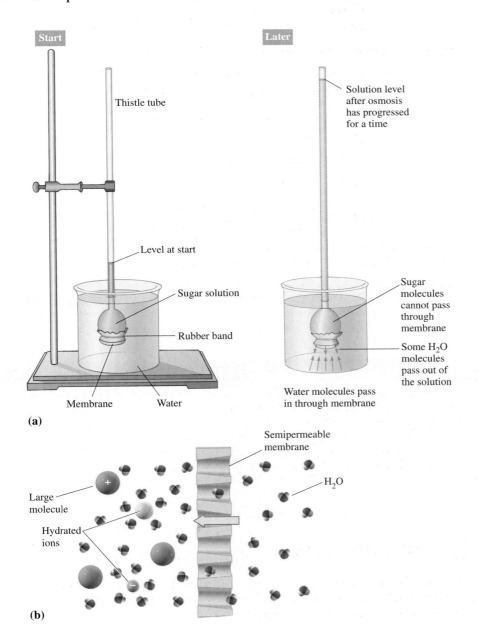

(a)

(b)

Figure 14-16 (a) Laboratory apparatus for demonstrating osmosis. The picture at the right gives some details of the process. (b) A simplified representation of the function of a semipermeable membrane.

> Osmotic pressure depends on the number, and not the kind, of solute particles in solution; it is therefore a colligative property.

The osmotic pressure of a given aqueous solution can be measured with an apparatus such as that depicted in Figure 14-16a. The solution of interest is placed inside an inverted glass (thistle) tube that has a membrane firmly fastened across the bottom. This part of the thistle tube and its membrane are then immersed in a container of pure water. As time passes, the height of the solution in the neck rises until the pressure it exerts just counterbalances the osmotic pressure.

The greater the number of solute particles, the greater the height to which the column rises, and the greater the osmotic pressure.

Alternatively, we can view osmotic pressure as the external pressure exactly sufficient to prevent osmosis. The pressure required (Figure 14-17) is equal to the osmotic pressure of the solution.

Like molecules of an ideal gas, solute particles are widely separated in very dilute solutions and do not interact significantly with one another. For very dilute solutions, osmotic pressure, π, is found to follow the equation

$$\pi = \frac{nRT}{V}$$

In this equation n is the number of moles of solute in volume, V, (in liters) of the solution. The other quantities have the same meaning as in the ideal gas law. The term n/V is a concentration term. In terms of molarity, M,

$$\pi = MRT$$

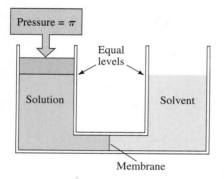

Figure 14-17 The pressure that is just sufficient to prevent solvent flow from the pure solvent side through the semipermeable membrane to the solution side is a measure of the osmotic pressure of the solution.

For a solution of an electrolyte, $\pi = M_{effective}RT$.

For *dilute aqueous solutions,* the molarity is approximately equal to the molality (because the density of the solution is nearly 1 g/mL, kg/L), so

$$\pi = mRT$$

(dilute aqueous solutions)

Osmotic pressure increases with increasing temperature because T affects the number of solvent–membrane collisions per unit time. It also increases with increasing molarity because M affects the difference in the numbers of solvent molecules hitting the membrane from the two sides, and because a higher M leads to a stronger drive to equalize the concentration difference by dilution and to increase disorder in the solution.

EXAMPLE 14-13   *Osmotic Pressure Calculation*

What osmotic pressure would the 1.25 m sucrose solution in Example 14-2 exhibit at 25°C? The density of this solution is 1.34 g/mL.

Plan

We note that the approximation $M \approx m$ is not very good for this solution, because the density of this solution is quite different from 1 g/mL or kg/L. Thus, we must first find the molarity of sucrose, and then use the relationship $\pi = MRT$.

Solution

Recall from Example 14-2 that there is 50.0 g of sucrose (0.146 mol) in 117 g of H_2O which gives 167 g of solution. The volume of this solution is

$$\text{vol solution} = 167 \text{ g} \times \frac{1 \text{ mL}}{1.34 \text{ g}} = 125 \text{ mL, or } 0.125 \text{ L}$$

Thus, the molarity of sucrose in the solution is

$$M_{sucrose} = \frac{0.146 \text{ mol}}{0.125 \text{ L}} = 1.17 \text{ mol/L}$$

Now we can calculate the osmotic pressure.

$$\pi = MRT = (1.17 \text{ mol/L})\left(0.0821 \ \frac{\text{L·atm}}{\text{mol·K}}\right)(298 \text{ K}) = \boxed{28.6 \text{ atm}}$$

You should now work Exercise 84.

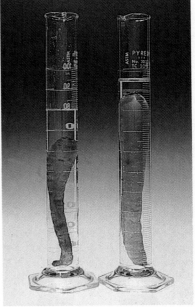

Let's compare the calculated values of the four colligative properties for the 1.25 *m* sucrose solution.

vapor pressure lowering	= 0.524 torr	(Example 14-4)
boiling point elevation	= 0.640°C	(Example 14-7)
freezing point depression	= 2.32°C	(Example 14-9)
osmotic pressure	= 28.6 atm	(Example 14-13)

The first of these is so small that it would be hard to measure precisely. Even this small lowering of the vapor pressure is sufficient to raise the boiling point by an amount that could be measured, although with difficulty. The freezing point depression is greater, but still could not be measured very precisely without a special apparatus. The osmotic pressure, on the other hand, is so large that it could be measured much more precisely. Thus, osmotic pressure is often the most easily measured of the four colligative properties, especially when very dilute solutions are used.

Osmotic pressures represent very significant forces. For example, a 1.00 molar solution of a nonelectrolyte in water at 0°C produces an equilibrium osmotic pressure of approximately 22.4 atmospheres (≈330 psi).

The use of measurements of osmotic pressure for the determination of molecular weights has several advantages. Even very dilute solutions give easily measurable osmotic pressures. This method therefore is useful in determination of the molecular weights of (1) very expensive substances, (2) substances that can be prepared only in very small amounts, and (3) substances of very high molecular weight that are not very soluble. Because high-molecular-weight materials are often difficult, and in some cases impossible, to obtain in a high state of purity, determinations of their molecular weights are not as accurate as we might like. Nonetheless, osmotic pressures provide a very useful method of estimating molecular weights.

An illustration of osmosis. When a carrot is soaked in a concentrated salt solution, water flows out of the plant cells by osmosis. A carrot soaked overnight in salt solution (*left*) has lost much water and become limp. A carrot soaked overnight in pure water (*right*) is little affected.

EXAMPLE 14-14 *Molecular Weight from Osmotic Pressure*

Pepsin is an enzyme present in the human digestive tract. A solution of a 0.500-gram sample of purified pepsin in 30.0 mL of aqueous solution exhibits an osmotic pressure of 8.92 torr at 27.0°C. Estimate the molecular weight of pepsin.

An enzyme is a protein that acts as a biological catalyst. Pepsin catalyzes the metabolic cleavage of amino acid chains (called peptide chains) in other proteins.

Plan

As we did in earlier molecular weight determinations (Section 14-13), we must first find *n*, the number of moles that 0.500 grams of pepsin represents. We start with the equation $\pi = MRT$. The molarity of pepsin is equal to the number of moles of pepsin per liter of solution, n/V. We substitute this for *M* and solve for *n*.

Solution

$$\pi = MRT = \left(\frac{n}{V}\right)RT \qquad \text{or} \qquad n = \frac{\pi V}{RT}$$

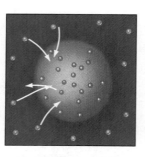
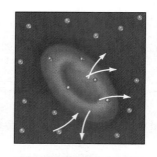
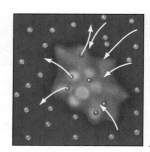

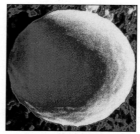
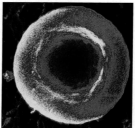
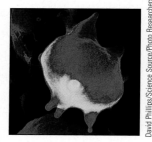

Cells expand in solution of
lower solute concentration
(a hypotonic solution)

Normal cells in isotonic solution

Cells shrink in solution of
greater solute concentration
(a hypertonic solution)

David Phillips/Science Source/Photo Researchers, Inc.

Living cells contain solutions. When living cells are put in contact with solutions having different total solute concentrations, the resulting osmotic pressure can cause solvent to flow into (*left*) or out of (*right*) the cells.

The freezing point depression of this very dilute solution would be only about 0.0009°C, which would be difficult to measure accurately. The osmotic pressure of 8.92 torr, on the other hand, is easily measured.

We convert 8.92 torr to atmospheres to be consistent with the units of R.

$$n = \frac{\pi V}{RT} = \frac{\left(8.92 \text{ torr} \times \dfrac{1 \text{ atm}}{760 \text{ torr}}\right)(0.0300 \text{ L})}{\left(0.0821 \dfrac{\text{L} \cdot \text{atm}}{\text{mol} \cdot \text{K}}\right)(300 \text{ K})} = 1.43 \times 10^{-5} \text{ mol pepsin}$$

Thus, 0.500 g of pepsin is 1.43×10^{-5} mol. We now estimate its molecular weight.

$$\underline{?} \text{ g/mol} = \frac{0.500}{1.43 \times 10^{-5} \text{ mol}} = \boxed{3.50 \times 10^4 \text{ g/mol}}$$

The molecular weight of pepsin is approximately 35,000 amu. This is typical for medium-sized proteins.

You should now work Exercise 90.

George Semple

Sports drinks were developed to counteract dehydration while maintaining electrolyte balance. Such solutions also help to prevent cell damage due to the osmotic pressure that would be caused by drinking large amounts of water.

✓ **Problem-Solving Tip:** *Units in Osmotic Pressure Calculations*

Strictly speaking, the equation for osmotic pressure is presented in terms of molarity, $\pi = MRT$. Osmotic pressure, π, has the units of pressure (atmospheres); M (mol/L); and T (kelvins). Therefore, the appropriate value of R is $0.0821 \dfrac{\text{L} \cdot \text{atm}}{\text{mol} \cdot \text{K}}$. We can balance the units in this equation as follows.

$$\text{atm} = \left(\frac{\text{mol}}{\text{L}}\right)\left(\frac{\text{L} \cdot \text{atm}}{\text{mol} \cdot \text{K}}\right)(\text{K})$$

When we use the approximation that $M \approx m$, we might think that the units do not balance. For very dilute aqueous solutions, we can think of M as being *numerically* equal to m, but having the units mol/L, as shown above.

TABLE 14-4		Types of Colloids		
Dispersed (solute-like) Phase		**Dispersing (solvent-like) Medium**	**Common Name**	**Examples**
solid	in	solid	solid sol	Many alloys (e.g., steel and duralumin), some colored gems, reinforced rubber, porcelain, pigmented plastics
liquid	in	solid	solid emulsion	Cheese, butter, jellies
gas	in	solid	solid foam	Sponge, rubber, pumice, Styrofoam
solid	in	liquid	sols and gels	Milk of magnesia, paints, mud, puddings
liquid	in	liquid	emulsion	Milk, face cream, salad dressings, mayonnaise
gas	in	liquid	foam	Shaving cream, whipped cream, foam on beer
solid	in	gas	solid aerosol	Smoke, airborne viruses and particulate matter, auto exhaust
liquid	in	gas	liquid aerosol	Fog, mist, aerosol spray, clouds

Freshly made wines are often cloudy because of colloidal particles. Removing these colloidal particles clarifies the wine.

COLLOIDS

A solution is a homogeneous mixture in which no settling occurs and in which solute particles are at the molecular or ionic state of subdivision. This represents one extreme of mixtures. The other extreme is a suspension, a clearly heterogeneous mixture in which solute-like particles settle out after mixing with a solvent-like phase. Such a situation results when a handful of sand is stirred into water. **Colloids (colloidal dispersions)** represent an intermediate kind of mixture in which the solute-like particles, or **dispersed phase,** are suspended in the solvent-like phase, or **dispersing medium.** The particles of the dispersed phase are so small that settling is negligible. They are large enough, however, to make the mixture appear cloudy or even opaque, because light is scattered as it passes through the colloid.

Table 14-4 indicates that all combinations of solids, liquids, and gases can form colloids except mixtures of nonreacting gases (all of which are homogeneous and, therefore, true solutions). Whether a given mixture forms a solution, a colloidal dispersion, or a suspension depends on the size of the solute-like particles (Table 14-5), as well as solubility and miscibility.

CD-ROM   Screen 14.10, Colloids.

SCiLINKS.

TOPIC: Colloids
GO TO: www.scilinks.org
*sci*LINKS CODE: WCH1420

TABLE 14-5	Approximate Sizes of Dispersed Particles	
Mixture	**Example**	**Approximate Particle Size**
suspension	sand in water	larger than 10,000 Å
colloidal dispersion	starch in water	10–10,000 Å
solution	sugar in water	1–10 Å

CHEMISTRY IN USE

Our Daily Lives

Water Purification and Hemodialysis

Semipermeable membranes play important roles in the normal functioning of many living systems. In addition, they are used in a wide variety of industrial and medical applications. Membranes with different permeability characteristics have been developed for many different purposes. One of these is the purification of water by reverse osmosis.

Suppose we place a semipermeable membrane between a saline (salt) solution and pure water. If the saline solution is pressurized under a greater pressure than its osmotic pressure, the direction of flow can be reversed. That is, the net flow of water molecules will be from the saline solution through the membrane into the pure water. This process is called **reverse osmosis.** The membrane usually consists of cellulose acetate or hollow fibers of a material structurally similar to nylon. This method has been used for the purification of brackish (mildly saline) water. It has the economic advantages of low cost, ease of apparatus construction, and simplicity of operation. Because this method of water purification requires no heat, it has a great advantage over distillation.

The city of Sarasota, Florida, has built a large reverse osmosis plant to purify drinking water. It processes more than 4 million gallons of water per day from local wells. Total dissolved solids are reduced in concentration from 1744 parts per million (ppm) (0.1744% by mass) to 90 ppm. This water is mixed with additional well water purified by an ion exchange system. The final product is more than 10 million gal-

A reverse osmosis unit used to provide all the fresh water (82,500 gallons per day) for the steamship *Norway.*

lons of water per day containing less than 500 ppm of total dissolved solids, the standard for drinking water set by the World Health Organization. The Kuwaiti and Saudi water purification plants that were of strategic concern in the Persian Gulf war use reverse osmosis in one of their primary stages.

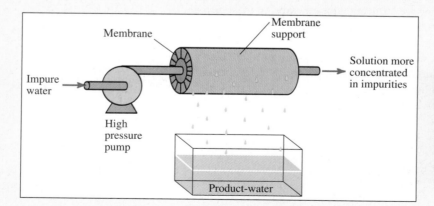

The reverse osmosis method of water purification.

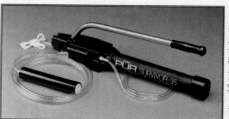

A portable reverse osmosis unit designed for individual use.

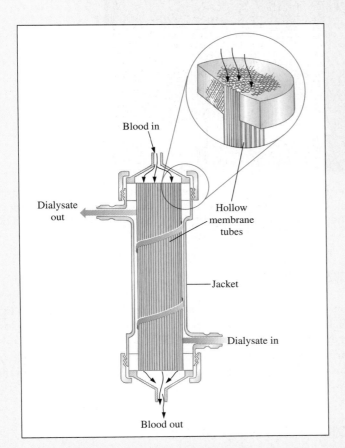

A schematic diagram of the hollow fiber (or capillary) dialyzer, the most commonly used artificial kidney. The blood flows through many small tubes constructed of semipermeable membrane; these tubes are bathed in the dialyzing solution.

Human kidneys carry out many important functions. One of the most crucial is the removal of metabolic waste products (e.g., creatinine, urea, and uric acid) from the blood without removal of substances needed by the body (e.g., glucose, electrolytes, and amino acids). The process by which this is accomplished in the kidney involves *dialysis*, a phenomenon in which the membrane allows transfer of both solvent molecules *and* certain solute molecules and ions, usually small ones. Many patients whose kidneys have failed can have this dialysis performed by an artificial kidney machine. In this mechanical procedure, called *hemodialysis*, the blood is withdrawn

from the body and passed in contact with a semipermeable membrane.

The membrane separates the blood from a dialyzing solution, or *dialysate*, that is similar to blood plasma in its concentration of needed substances (e.g., electrolytes and amino acids) but contains none of the waste products. Because the concentrations of undesirable substances are thus higher in the blood than in the dialysate, they flow preferentially out of the blood and are washed away. The concentrations of *needed* substances are the same on both sides of the membrane, so these substances are maintained at the proper concentrations in the blood. The small pore size of the membrane prevents passage of blood cells. However, Na^+ and Cl^- ions and some small molecules do pass through the membrane. A patient with total kidney failure may require up to four hemodialysis sessions per week, at 3 to 4 hours per session. To help hold down the cost of such treatment, the dialysate solution is later purified by a combination of filtration, distillation, and reverse osmosis and is then reused.

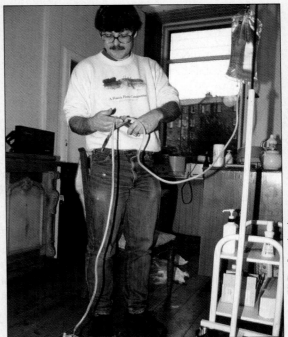

A portable dialysis unit.

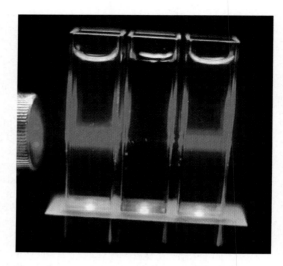

Figure 14-18   The dispersion of a beam of light by colloidal particles (*left and right tubes*) is called the Tyndall effect. The liquid in the middle tube is a true solution, so it does not scatter the light beam. The presence of colloidal particles is easily detected with the aid of a light beam.

14-16   THE TYNDALL EFFECT

The scattering of light by colloidal particles is called the **Tyndall effect** (Figure 14-18). Particles cannot scatter light if they are too small. Solute particles in solutions are below this limit. The maximum dimension of colloidal particles is about 10,000 Å.

The scattering of light from automobile headlights by fogs and mists is an example of the Tyndall effect, as is the scattering of a light beam in a laser show by dust particles in the air in a darkened room.

14-17   THE ADSORPTION PHENOMENON

Much of the chemistry of everyday life is the chemistry of colloids, as one can tell from the examples in Table 14-4. Because colloidal particles are so finely divided, they have tremendously high total surface area in relation to their volume. It is not surprising, therefore, that an understanding of colloidal behavior requires an understanding of surface phenomena.

Atoms on the surface of a colloidal particle are bonded only to other atoms of the particle on and below the surface. These atoms interact with whatever comes in contact with the surface. Colloidal particles often adsorb ions or other charged particles, as well as gases and liquids. The process of **adsorption** involves adhesion of any such species onto the surfaces of particles. For example, a bright red **sol** (solid dispersed in liquid) is formed by mixing hot water with a concentrated aqueous solution of iron(III) chloride (Figure 14-19).

$$2x[Fe^{3+}(aq) + 3Cl^-(aq)] + x(3 + y)H_2O \longrightarrow [Fe_2O_3 \cdot yH_2O]_x(s) + 6x[H^+ + Cl^-]$$
$$\text{yellow solution} \hspace{5cm} \text{bright red sol}$$

Each colloidal particle of this sol is a cluster of many formula units of hydrated Fe_2O_3. Each attracts positively charged Fe^{3+} ions to its surface. Because each particle is then surrounded by a shell of positively charged ions, the particles repel one another and cannot combine to the extent necessary to cause precipitation.

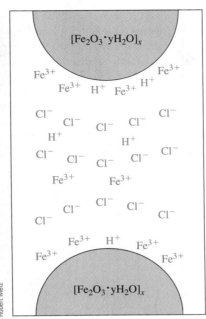

Figure 14-19 Stabilization of a colloid (Fe$_2$O$_3$ sol) by electrostatic forces. Each colloidal particle of this red sol is a cluster of many formula units of hydrated Fe$_2$O$_3$. Each attracts positively charged Fe^{3+} ions to its surface. (Fe^{3+} ions fit readily into the crystal structure, so they are preferentially adsorbed rather than the Cl$^-$ ions.) Each particle is then surrounded by a shell of positively charged ions, so the particles repel one another and cannot combine to the extent necessary to cause actual precipitation. The suspended particles scatter light, making the path of the light beam through the suspension visible.

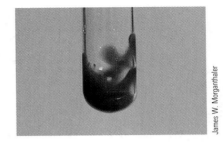

Fe(OH)$_3$ is a gelatinous precipitate (a gel).

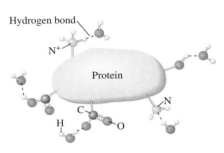

Figure 14-20 Examples of hydrophilic groups at the surface of a giant molecule (macromolecule) that help keep the macromolecule suspended in water.

14-18 HYDROPHILIC AND HYDROPHOBIC COLLOIDS

Colloids are classified as **hydrophilic** ("water loving") or **hydrophobic** ("water hating") based on the surface characteristics of the dispersed particles.

Hydrophilic Colloids

Proteins such as the oxygen-carrier hemoglobin form hydrophilic sols when they are suspended in saline aqueous body fluids such as blood plasma. Such proteins are macromolecules (giant molecules) that fold and twist in an aqueous environment so that polar groups are exposed to the fluid, whereas nonpolar groups are encased (Figure 14-20). Protoplasm and human cells are examples of **gels,** which are special types of sols in which the solid particles (in this case mainly proteins and carbohydrates) join together in a semirigid network structure that encloses the dispersing medium. Other examples of gels are gelatin, jellies, and gelatinous precipitates such as Al(OH)$_3$ and Fe(OH)$_3$.

Hydrophobic Colloids

Hydrophobic colloids cannot exist in polar solvents without the presence of **emulsifying agents,** or **emulsifiers.** These agents coat the particles of the dispersed phase to prevent their coagulation into a separate phase. Milk and mayonnaise are examples of hydrophobic

Some edible colloids.

TOPIC: Emulsions
GO TO: www.scilinks.org
*sci*LINKS **CODE:** WCH1430

"Dry" cleaning does not involve water. The solvents that are used in dry cleaning dissolve grease to form true solutions.

colloids (milk fat in milk, vegetable oil in mayonnaise) that stay suspended with the aid of emulsifying agents (casein in milk and egg yolk in mayonnaise).

Consider the mixture resulting from vigorous shaking of salad oil (nonpolar) and vinegar (polar). Droplets of hydrophobic oil are temporarily suspended in the water. In a short time, however, the very polar water molecules, which attract one another strongly, squeeze out the nonpolar oil molecules. The oil then coalesces and floats to the top. If we add an emulsifying agent, such as egg yolk, and shake or beat the mixture, a stable emulsion (mayonnaise) results.

Oil and grease are mostly long-chain hydrocarbons that are very nearly nonpolar. Our most common solvent is water, a polar substance that does not dissolve nonpolar substances. To use water to wash soiled fabrics, greasy dishes, or our bodies, we must enable the water to suspend and remove nonpolar substances. Soaps and detergents are emulsifying agents that accomplish this. Their function is controlled by the intermolecular interactions that result from their structures.

Solid soaps are usually sodium salts of long-chain organic acids called fatty acids. They have a polar "head" and a nonpolar "hydrocarbon tail." Sodium stearate, a typical soap, is shown here.

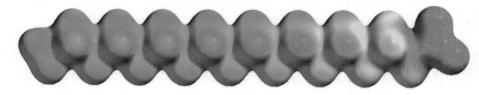

hydrocarbon tail (soluble in oil)   polar head
(soluble in H_2O)

sodium stearate (a soap)

(*Left*) Powdered sulfur floats on pure water because of the high surface tension of water. (*Right*) When a drop of detergent solution is added to the water, its surface tension is lowered and sulfur sinks. This lowering of surface tension enhances the cleaning action of detergent solutions.

Charles Steele

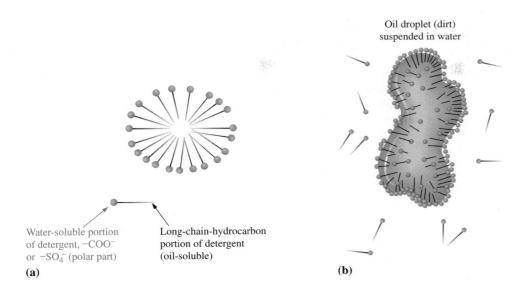

Oil droplet (dirt)
suspended in water

Water-soluble portion
of detergent, $-COO^-$
or $-SO_4^-$ (polar part)

Long-chain-hydrocarbon
portion of detergent
(oil-soluble)

(a)

(b)

Figure 14-21 (a) A representation of a micelle. The nonpolar tails "dissolve" in one another in the center of the cluster and the polar heads on the outside interact favorably with the polar water molecules. (b) Attachment of soap or detergent molecules to a droplet of oily dirt to suspend it in water.

The stearate ion is typical of the anions in soaps. It has a polar carboxylate head,

$$-\overset{\overset{\displaystyle O}{\|}}{C}-O^-,$$ and a long nonpolar tail, $CH_3(CH_2)_{16}-$. The head of the stearate ion is compatible with ("soluble in") water, whereas the hydrocarbon tail is compatible with ("soluble in") oil and grease. Groups of such ions can be dispersed in water because they form **micelles** (Figure 14-21a). Their "water-insoluble" tails are in the interior of a micelle and their polar heads on the outside where they can interact with the polar water molecules. When sodium stearate is stirred into water, the result is not a true solution. Instead it contains negatively charged micelles of stearate ions, surrounded by the positively charged Na^+ ions. The result is a suspension of micelles in water. These micelles are large enough to scatter light, so a soap–water mixture appears cloudy. Oil and grease "dissolve" in soapy water because the nonpolar oil and grease are taken into the nonpolar interior of micelles (Figure 14-21b). Micelles form a true emulsion in water, so the oil and grease can be washed away. Sodium stearate is called a **surfactant** (meaning "surface-active agent") or wetting agent because it has the ability to suspend and wash away oil and grease. Other soaps and detergents behave similarly.

"**Hard**" **water** contains Fe^{3+}, Ca^{2+}, and/or Mg^{2+} ions, all of which displace Na^+ from soaps to form precipitates. This removes the soap from the water and puts an undesirable coating on the bathtub or on the fabric being laundered. Synthetic **detergents** are soaplike emulsifiers that contain sulfonate, $-SO_3^-$, or sulfate, $-OSO_3^-$, instead of carboxylate groups, $-COO^-$. They do not precipitate the ions of hard water, so they can be used in hard water as soap substitutes without forming undesirable scum.

Phosphates were added to commercial detergents for various purposes. They complexed the metal ions that contribute to water hardness and kept them dissolved, controlled acid-

Sodium stearate is also a major component of some stick deodorants.

 CD-ROM Screen 14.11, Surfactants.

Phosphates and nonbiodegradable detergents are responsible for the devastation of plant life in and along this river.

Chinch Gryniewicz/Ecoscene/CORBIS

ity, and influenced micelle formation. The use of detergents containing phosphates is now discouraged because they cause **eutrophication** in rivers and streams that receive sewage. This is a condition (not related to colloids) in which an overgrowth of vegetation is caused by the high concentration of phosphorus, a plant nutrient. This overgrowth and the subsequent decay of the dead plants lead to decreased dissolved O_2 in the water, which causes the gradual elimination of marine life. There is also a foaming problem associated with branched alkylbenzenesulfonate (ABS) detergents in streams and in pipes, tanks, and pumps of sewage treatment plants. Such detergents are not **biodegradable;** that is, they cannot be broken down by bacteria.

a sodium branched alkylbenzenesulfonate (ABS)—a nonbiodegradable detergent

The two detergents shown here each have a $C_{12}H_{25}$ tail but the branched one is not biodegradable.

Currently used linear-chain alkylbenzenesulfonate (LAS) detergents are biodegradable and do not cause such foaming.

$$\left[CH_3CH_2CH_2CH_2CH_2CH_2CH_2CH_2CH_2CH_2CH_2CH_2 \text{—} \bigcirc \text{—} SO_3 \right]^{-} Na^+$$

sodium lauryl benzenesulfonate
a linear alkylbenzenesulfonate (LAS)—a biodegradable detergent

---

CHEMISTRY IN USE   *Our Daily Lives*

Why Does Red Wine Go with Red Meat?

Choosing the appropriate wine to go with dinner is a problem for some diners. Experts, however, have offered a simple rule for generations, "serve red wine with red meat and white wine with fish and chicken." Are these cuisine choices just traditions, or are there fundamental reasons for them?

Red wine is usually served with red meat because of a desirable matching of the chemicals found in each. The most influential ingredient in red meat is fat; it gives red meats their desirable flavor. As you chew a piece of red meat, the fat from the meat coats your tongue and palate, which desensitizes your taste buds. As a result, your second bite of red meat is less tasty than the first. Your steak would taste better if you washed your mouth between mouthfuls; there is an easy way to wash away the fat deposits.

Red wine contains a surfactant that cleanses your mouth, removing fat deposits, re-exposing your taste buds, and allowing you to savor the next bite of red meat almost as well as the first bite. The tannic acid (also called tannin) in red wine provides a soap-like action. Like soap, tannic acid consists of both

a nonpolar complex hydrocarbon part as well as a polar one. The polar part of tannic acid dissolves in polar saliva, while the nonpolar part dissolves in the fat film that coats your palate. When you sip red wine, a suspension of micelles forms in the saliva. This micelle emulsion has the fat molecules in its interior; the fat is washed away by swallowing the red wine.

White wines go poorly with red meats because they lack the tannic acid needed to cleanse the palate. In fact, the presence or absence of tannic acid distinguishes red wines from white wines. Grapes fermented with their skins produce red wines; grapes fermented without their skins produce white wines.

Because fish and chicken have less fat than red meats, they can be enjoyed without surfactants to cleanse the palate. Also, tannic acid has a strong flavor that can overpower the delicate flavor of many fish. The absence of tannic acid in white wines gives them a lighter flavor than red wines, and many people prefer this lighter flavor with their fish or chicken dinners.

Ronald DeLorenzo
Middle Georgia College

Key Terms

The following terms were defined at the end of Chapter 3: **concentration, dilution, molarity, percent by mass, solute, solution,** and **solvent**. The following terms were defined at the end of Chapter 13: **condensation, condensed phases, evaporation, phase diagram,** and **vapor pressure**.

Adsorption   Adhesion of species onto surfaces of particles.

Associated ions   Short-lived species formed by the collision of dissolved ions of opposite charge.

Biodegradability   The ability of a substance to be broken down into simpler substances by bacteria.

Boiling point elevation   The increase in the boiling point of a solvent caused by dissolution of a nonvolatile solute.

Boiling point elevation constant, K_b   A constant that corresponds to the change (increase) in boiling point produced by a one-molal *ideal* solution of a nonvolatile nonelectrolyte.

Colligative properties   Physical properties of solutions that depend on the number but not the kind of solute particles present.

Colloid   A heterogeneous mixture in which solute-like particles do not settle out; also called *colloidal dispersion*.

Crystal lattice energy   The energy change when one mole of formula units of a crystalline solid is formed from its ions, atoms, or molecules in the gas phase; always negative.

Detergent   A soap-like emulsifier that contains a sulfonate, $-SO_3^-$, or sulfate, $-OSO_3^-$, group instead of a carboxylate, $-COO^-$, group.

Dispersed phase   The solute-like species in a colloid.

Dispersing medium   The solvent-like phase in a colloid.

Dispersion   See *Colloid*.

Distillation   The process in which components of a mixture are separated by boiling away the more volatile liquid.

Effective molality   The sum of the molalities of all solute particles in solution.

Emulsifier   See *Emulsifying agent*.

Emulsifying agent   A substance that coats the particles of a dispersed phase and prevents coagulation of colloidal particles; an emulsifier.

Emulsion   A colloidal dispersion of a liquid in a liquid.

Eutrophication   The undesirable overgrowth of vegetation caused by high concentrations of plant nutrients in bodies of water.

Foam   A colloidal dispersion of a gas in a liquid.

Fractional distillation   The process in which a fractionating column is used in a distillation apparatus to separate components of a liquid mixture that have different boiling points.

Freezing point depression   The decrease in the freezing point of a solvent caused by the presence of a solute.

Freezing point depression constant, K_f   A constant that corresponds to the change in freezing point produced by a one-molal *ideal* solution of a nonvolatile nonelectrolyte.

Gel   A colloidal dispersion of a solid in a liquid; a semirigid sol.

Hard water   Water containing Fe^{3+}, Ca^{2+}, or Mg^{2+} ions, which form precipitates with soaps.

Heat of solution (molar)   The amount of heat absorbed in the formation of a solution that contains one mole of solute; the value is positive if heat is absorbed (endothermic) and negative if heat is released (exothermic).

Henry's Law   The pressure of the gas above a solution is proportional to the concentration of the gas in the solution.

Hydration   The interaction (surrounding) of solute particles with water molecules.

Hydration energy (molar) of an ion   The energy change accompanying the hydration of a mole of gaseous ions.

Hydrophilic colloids   Colloidal particles that attract water molecules.

Hydrophobic colloids   Colloidal particles that repel water molecules.

Ideal solution   A solution that obeys Raoult's Law exactly.

Liquid aerosol   A colloidal dispersion of a liquid in a gas.

Micelle   A cluster of a large number of soap or detergent molecules or ions, assembled with their hydrophobic tails directed toward the center and their hydrophilic heads directed outward.

Miscibility   The ability of one liquid to mix with (dissolve in) another liquid.

Molality (m)   Concentration expressed as number of moles of solute per kilogram of solvent.

Mole fraction of a component in solution   The number of moles of the component divided by the total number of moles of all components.

Osmosis   The process by which solvent molecules pass through a semipermeable membrane from a dilute solution into a more concentrated solution.

Osmotic pressure   The hydrostatic pressure produced on the surface of a semipermeable membrane by osmosis.

Percent ionization of weak electrolytes   The percent of the weak electrolyte that ionizes in a solution of a given concentration.

Raoult's Law   The vapor pressure of a solvent in an ideal solution is directly proportional to the mole fraction of the solvent in the solution.

Reverse osmosis   The forced flow of solvent molecules through a semipermeable membrane from a concentrated solution into a dilute solution. This is accomplished by application of hydrostatic pressure on the concentrated side greater than the osmotic pressure that is opposing it.

Saturated solution   A solution in which no more solute will dissolve at a given temperature.

Semipermeable membrane   A thin partition between two solutions through which certain molecules can pass but others cannot.

Soap   An emulsifier that can disperse nonpolar substances in water; the sodium salt of a long-chain organic acid; consists of a long hydrocarbon chain attached to a carboxylate group, $-CO_2^-Na^+$.

Sol   A colloidal dispersion of a solid in a liquid.

Solid aerosol   A colloidal dispersion of a solid in a gas.

Solid emulsion   A colloidal dispersion of a liquid in a solid.

Solid foam   A colloidal dispersion of a gas in a solid.

Solid sol   A colloidal dispersion of a solid in a solid.

Solvation   The process by which solvent molecules surround and interact with solute ions or molecules.

Supersaturated solution   A (metastable) solution that contains a higher-than-saturation concentration of solute; slight disturbance or seeding causes crystallization of excess solute.

Surfactant   A "surface-active agent"; a substance that has the abil-ity to emulsify and wash away oil and grease in an aqueous suspension.

Thermal pollution   Introduction of heated waste water into nat-ural waters.

Tyndall effect   The scattering of light by colloidal particles.

van't Hoff factor, *i*   A number that indicates the extent of disso-ciation or ionization of a solute; equal to the actual colligative property divided by the colligative property calculated assuming no ionization or dissociation.

Exercises

General Concepts: The Dissolving Process

1. Support or criticize the statement "Solutions and mix-tures are the same thing."

2. Give an example of a solution that contains each of the following: (a) a solid dissolved in a liquid; (b) a gas dis-solved in a gas; (c) a gas dissolved in a liquid; (d) a liquid dissolved in a liquid; (e) a solid dissolved in a solid. Iden-tify the solvent and the solute in each case.

3. There are no *true* solutions in which the solvent is gaseous and the solute is either liquid or solid. Why?

4. Explain why (a) solute–solute, (b) solvent–solvent, and (c) solute–solvent interactions are important in determin-ing the extent to which a solute dissolves in a solvent.

5. Define and distinguish between dissolution, solvation, and hydration.

*6. The amount of heat released or absorbed in the dissolu-tion process is important in determining whether the dis-solution process is spontaneous, meaning, whether it can occur. What is the other important factor? How does it influence solubility?

7. An old saying is that "oil and water don't mix." Explain the molecular basis for this saying.

*8. Two liquids, A and B, do not react chemically and are completely miscible. What would be observed as one is poured into the other? What would be observed in the case of two completely immiscible liquids and in the case of two partially miscible liquids?

9. Consider the following solutions. In each case, predict whether the solubility of the solute should be high or low. Justify your answers. (a) LiCl in hexane, C_6H_{14}; (b) $BaCl_2$ in H_2O; (c) C_6H_{14} in H_2O; (d) $CHCl_3$ in C_6H_{14}; (e) C_6H_{14} in CCl_4.

10. Consider the following solutions. In each case predict whether the solubility of the solute should be high or low. Justify your answers. (a) HCl in H_2O; (b) HF in H_2O; (c) Al_2O_3 in H_2O; (d) S_8 in H_2O; (e) Na_2SO_4 in hexane, C_6H_{14}.

11. For those solutions in Exercise 9 that can be prepared in "reasonable" concentrations, classify the solutes as non-electrolytes, weak electrolytes, or strong electrolytes.

12. For those solutions in Exercise 10 that can be prepared in "reasonable" concentrations, classify the solutes as non-electrolytes, weak electrolytes, or strong electrolytes.

13. Both methanol, CH_3OH, and ethanol, CH_3CH_2OH, are completely miscible with water at room temperature be-cause of strong solvent-solute intermolecular attractions. Predict the trend in solubility in water for 1-propanol, $CH_3CH_2CH_2OH$; 1-butanol, $CH_3CH_2CH_2CH_2OH$; and 1-pentanol, $CH_3CH_2CH_2CH_2CH_2OH$.

14. (a) Does the solubility of a solid in a liquid exhibit an ap-preciable dependence on pressure? (b) Is the same true for the solubility of a liquid in a liquid? Why?

15. Describe a technique for determining whether or not a solution contains an electrolyte.

16. A reagent bottle in the storeroom is labeled as containing a saturated sodium chloride solution. How can one deter-mine whether or not the solution is saturated?

*17. A handbook lists the value of the Henry's Law constant as 3.02×10^4 atm for ethane, C_2H_6, dissolved in water at 25°C. The absence of concentration units on k means that the constant is meant to be used with concentration ex-pressed as a mole fraction. Calculate the mole fraction of ethane in water at an ethane pressure of 0.15 atm.

*18. The mole fraction of methane, CH_4, dissolved in wa-ter can be calculated from the Henry's Law constants of 4.13×10^4 atm at 25°C and 5.77×10^4 atm at 50.°C. Cal-culate the solubility of methane at these temperatures for a methane pressure of 10. atm above the solution. Does the solubility increase or decrease with increasing temper-ature? (See Exercise 17 for interpretation of units.)

19. Choose the ionic compound from each pair for which the crystal lattice energy should be the most negative. Justify your choice. (a) LiF or LiBr; (b) KF or CaF_2; (c) FeF_2 or FeF_3; (d) NaF or KF.

20. Choose the ion from each pair that should be more strongly hydrated in aqueous solution. Justify your choice. (a) Na^+ or Rb^+; (b) Cl^- or Br^-; (c) Fe^{3+} or Fe^{2+}; (d) Na^+ or Mg^{2+}.

*21. The crystal lattice energy for LiBr(s) is -818.6 kJ/mol at 25°C. The hydration energy of the ions of LiBr is -867.4 kJ/mol at 25°C (for infinite dilution). (a) What is the heat of solution of LiBr(s) at 25°C (for infinite dilution)? (b) The hydration energy of $Li^+(g)$ is -544 kJ/mol at 25°C. What is the hydration energy for $Br^-(g)$ at 25°C?

22. Why is the dissolving of many ionic solids in water an endothermic process, whereas the mixing of most miscible liquids is an exothermic process?

Concentrations of Solutions

23. Under what conditions are the molarity and molality of a solution nearly the same? Which concentration unit is more useful when measuring volume with burets, pipets, and volumetric flasks in the laboratory? Why?

24. Many handbooks list solubilities in units of (g solute/ 100. g H_2O). How would you convert from this unit to mass percent?

25. The density of a 15.00% by mass aqueous solution of acetic acid, CH_3COOH, is 1.0187g/mL. What is (a) the molarity? (b) the molality? (c) the mole fraction of each component?

*26. Describe how to prepare 1.000 L of 0.245 m NaCl. The density of this solution is 1.01 g/mL.

*27. Urea, $(NH_2)_2CO$, is a product of metabolism of proteins. An aqueous solution is 32.0% urea by mass and has a density of 1.087 g/mL. Calculate the molality of urea in the solution.

28. Calculate the molality of a solution that contains 56.5 g of benzoic acid, C_6H_5COOH, in 325 mL of ethanol, C_2H_5OH. The density of ethanol is 0.789 g/mL.

29. Sodium fluoride has a solubility of 4.22 g in 100.0 g of water at 18°C. Express this solute concentration in terms of (a) mass percent, (b) mole fraction, and (c) molality.

30. What are the mole fractions of ethanol, C_2H_5OH, and water in a solution prepared by mixing 75.0 g of ethanol with 25.0 g of water?

31. What are the mole fractions of ethanol, C_2H_5OH, and water in a solution prepared by mixing 75.0 mL of ethanol with 25.0 mL of water at 25°C? The density of ethanol is 0.789 g/mL, and that of water is 1.00 g/mL.

32. The density of an aqueous solution containing 12.50 g K_2SO_4 in 100.00 g solution is 1.083 g/mL. Calculate the concentration of this solution in terms of molarity, molality, percent of K_2SO_4, and mole fraction of solvent.

*33. A piece of jewelry is marked "14 carat gold," meaning that on a mass basis the jewelry is 14/24 pure gold. What is the molality of this alloy—considering the other metal as the solvent?

*34. A solution that is 21.0% fructose, $C_6H_{12}O_6$, in water has a density of 1.10 g/mL at 20°C. (a) What is the molality of fructose in this solution? (b) At a higher temperature, the density is lower. Would the molality be less than, greater than, or the same as the molality at 20°C? Explain.

35. The density of a sulfuric acid solution taken from a car battery is 1.225 g/cm³. This corresponds to a 3.75 M solution. Express the concentration of this solution in terms of molality, mole fraction of H_2SO_4, and percentage of water by mass.

Car battery.

Raoult's Law and Vapor Pressure

36. In your own words, explain briefly *why* the vapor pressure of a solvent is lowered by dissolving a nonvolatile solute in it.

37. (a) Calculate the vapor pressure lowering associated with dissolving 25.5 g of table sugar, $C_{12}H_{22}O_{11}$, in 400. g of water at 25.0°C. (b) What is the vapor pressure of the solution? Assume that the solution is ideal. The vapor pressure of pure water at 25°C is 23.76 torr. (c) What is the vapor pressure of the solution at 100.°C?

38. Calculate (a) the lowering of the vapor pressure and (b) the vapor pressure of a solution prepared by dissolving 25.5 g of naphthalene, $C_{10}H_8$ (a nonvolatile nonelectrolyte), in 150.0 g of benzene, C_6H_6, at 20.°C. Assume that the solution is ideal. The vapor pressure of pure benzene is 74.6 torr at 20.°C.

39. At $-100.$°C ethane, CH_3CH_3, and propane, $CH_3CH_2CH_3$, are liquids. At that temperature, the vapor pressure of pure ethane is 394 torr and that of pure propane is 22 torr. What is the vapor pressure at $-100.$°C over a solution containing equal molar amounts of these substances?

40. Using Raoult's Law, predict the partial pressures in the vapor above a solution containing 0.350 mol acetone $(P^0 = 345$ torr) and 0.350 mol chloroform $(P^0 = 295$ torr).

41. What is the composition of the vapor above the solution described in Exercise 39?

42. What is the composition of the vapor above the solution described in Exercise 40?

43. Use the following vapor pressure diagram to estimate (a) the partial pressure of chloroform, (b) the partial pressure of acetone, and (c) the total vapor pressure of a solution in which the mole fraction of $CHCl_3$ is 0.30, assuming *ideal* behavior.

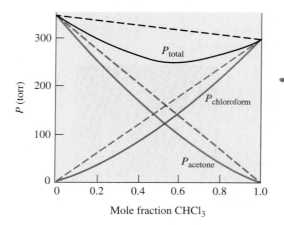

Mole fraction CHCl$_3$

44. Answer Exercise 43 for the *real* solution of acetone and chloroform.

45. A solution is prepared by mixing 50.0 g of dichloromethane, CH$_2$Cl$_2$, and 30.0 g of dibromomethane, CH$_2$Br$_2$, at 0°C. The vapor pressure at 0°C of pure CH$_2$Cl$_2$ is 0.175 atm, and that of CH$_2$Br$_2$ is 0.0150 atm. (a) Assuming ideal behavior, calculate the total vapor pressure of the solution. (b) Calculate the mole fractions of CH$_2$Cl$_2$ and CH$_2$Br$_2$ in the *vapor* above the liquid. Assume that both the vapor and the liquid behave ideally.

Boiling Point Elevation and Freezing Point Depression: Solutions of Nonelectrolytes

46. Calculate the boiling point of a 2.25 *m* aqueous solution of ethylene glycol, a nonvolatile nonelectrolyte.

47. A solution is prepared by dissolving 6.85 g of ordinary sugar (sucrose, C$_{12}$H$_{22}$O$_{11}$, 342 g/mol) in 34.0 g of water. Calculate the boiling point of the solution. Sucrose is a nonvolatile nonelectrolyte.

48. What is the freezing point of the solution described in Exercise 46?

49. What is the freezing point of the solution described in Exercise 47?

50. Refer to Table 14-2. Suppose you had a 0.175 *m* solution of a nonvolatile nonelectrolyte in each of the solvents listed there. Which one should have (a) the greatest freezing point depression, (b) the lowest freezing point, (c) the greatest boiling point elevation, and (d) the highest boiling point?

51. A 5.00% solution of dextrose, C$_6$H$_{12}$O$_6$, in water is referred to as DW$_5$. This solution can be used as a source of nourishment when introduced by intravenous injection. Calculate the freezing point and the boiling point of a DW$_5$ solution.

52. Lemon juice is a complex solution that does not freeze in a home freezer at temperatures as low as −11°C. At what temperature will the lemon juice freeze if its effective molality is the equivalent of a 10.0 *m* glucose solution?

*53. The "proof" of an alcoholic beverage is twice the volume percent of ethyl alcohol, C$_2$H$_5$OH, in water. The density of ethyl alcohol is 0.789 g/mL and that of water is 1.00 g/mL. A bottle of 100-proof rum is left outside on a cold winter day. (a) Will the rum freeze if the temperature drops to −18°C? (b) Rum is used in cooking and baking. At what temperature does 100-proof rum boil?

54. A 3.0-g sample of a nonelectrolyte was isolated from beef fat. The molecular weight of the compound was determined to be 137 amu. The sample was dissolved in 250.0 mL of ethyl alcohol. At what temperature should the solution boil? For ethyl alcohol: boiling point −78.41°C; K_b = 1.22°C/*m*, density = 0.789 g/mL.

55. The normal boiling point of benzene is 80.1°C. A 0.85-gram sample of a nonvolatile compound with the molar mass of 185 g/mol is dissolved in 2.75 g of benzene. What is the expected boiling point of this solution?

56. Pure copper melts at 1083°C. Its molal freezing point depression constant is 23°C/*m*. What will be the melting point of a brass made of 12% Zn and 88% Cu by mass?

57. How many grams of the nonelectrolyte sucrose, C$_{12}$H$_{22}$O$_{11}$, should be dissolved in 575 g of water to produce a solution that freezes at −1.85°C?

58. What mass of naphthalene, C$_{10}$H$_8$, should be dissolved in 375 g of nitrobenzene, C$_6$H$_5$NO$_2$, to produce a solution that boils at 214.20°C? See Table 14-2.

59. A solution was made by dissolving 3.75 g of a nonvolatile solute in 108.7 g of acetone. The solution boiled at 56.58°C. The boiling point of pure acetone is 55.95°C, and K_b = 1.71°C/*m*. Calculate the molecular weight of the solute.

60. The molecular weight of an organic compound was determined by measuring the freezing point depression of a benzene solution. A 0.500-g sample was dissolved in 50.0 g of benzene, and the resulting depression was 0.42°C. What is the approximate molecular weight? The compound gave the following elemental analysis: 40.0% C, 6.67% H, 53.3% O by mass. Determine the formula and exact molecular weight of the substance.

61. When 0.154 g of sulfur is finely ground and melted with 4.38 g of camphor, the freezing point of the camphor is lowered by 5.47°C. What is the molecular weight of sulfur? What is its molecular formula?

*62. (a) Suppose we dissolve a 6.00-g sample of a mixture of naphthalene, C$_{10}$H$_8$, and anthracene, C$_{14}$H$_{10}$, in 360. g of benzene. The solution is observed to freeze at 4.85°C. Find the percent composition (by mass) of the sample. (b) At what temperature should the solution boil? Assume that naphthalene and anthracene are nonvolatile nonelectrolytes.

Boiling Point Elevation and Freezing Point Depression: Solutions of Electrolytes

63. What is ion association in solution? Can you suggest why the term "ion pairing" is sometimes used to describe this phenomenon?

64. You have separate 0.10 M aqueous solutions of the following salts: $LiNO_3$, $Ca(NO_3)_2$, and $Al(NO_3)_3$. In which one would you expect to find the highest particle concentration? Which solution would you expect to conduct electricity most strongly? Explain your reasoning.

65. What is the significance of the van't Hoff factor, i?

66. What is the value of the van't Hoff factor, i, for the following strong electrolytes at infinite dilution? (a) Na_2SO_4; (b) KOH; (c) $Al_2(SO_4)_3$; (d) $SrSO_4$.

67. Compare the number of solute particles that are present in equal volumes of solutions of equal concentrations of strong electrolytes, weak electrolytes, and non-electrolytes.

68. Four beakers contain 0.010 m aqueous solutions of CH_3OH, $KClO_3$, $CaCl_2$, and CH_3COOH, respectively. Without calculating the actual freezing points of each of these solutions, arrange them from lowest to highest freezing point.

*69. A 0.050 m aqueous solution of $K_3[Fe(CN)_6]$ has a freezing point of $-0.2800°C$. Calculate the total concentration of solute particles in this solution and interpret your results.

*70. A solution is made by dissolving 1.50 gram each of $NaCl$, $NaBr$, and NaI in 150. g of water. What is the vapor pressure above this solution at 100°C? Assume complete dissociation of the three salts.

71. The ice fish lives under the polar ice cap where the water temperature is $-4°C$. This fish does not freeze at that temperature due to the solutes in its blood. The solute concentration of the fish's blood can be related to a sodium chloride solution that would have the same freezing point. What is the minimum concentration of a sodium chloride solution that would not freeze at $-4°C$?

72. Which solution would freeze at a lower temperature, 0.150 m sodium sulfate or 0.150 m calcium sulfate? Explain.

73. Solutions are produced by dissolving 20.0 grams of sodium nitrate in 0.500 kg of water and by dissolving 20.0 grams of calcium nitrate in 0.500 kg of water. Which solution would freeze at a lower temperature? Explain.

74. A series of 1.1 m aqueous solutions is produced. Predict which solution of each pair would boil at the higher temperature. (a) $NaCl$ or $LiCl$; (b) $LiCl$ or Li_2SO_4; (c) Na_2SO_4 or HCl; (d) HCl or $C_6H_{12}O_6$; (e) $C_6H_{12}O_6$ or CH_3OH; (f) CH_3OH or CH_3COOH; (g) CH_3COOH or KCl.

75. Identify which of the following solutions, each prepared by dissolving 65.0 g of the solute in 150. mL of water, would display the highest boiling point and explain your choice. (a) $NaCl$ or $LiCl$; (b) $LiCl$ or Li_2SO_4; (c) Na_2SO_4 or HCl; (d) HCl or $C_6H_{12}O_6$; (e) $C_6H_{12}O_6$ or CH_3OH; (f) CH_3OH or CH_3COOH; (g) CH_3COOH or KCl.

76. Synthetic ocean water can be produced by dissolving 36.0 g of table salt in 1000. mL of water. What is the boiling point of this solution? (Assume that the density of the water is 0.998 g/mL and that table salt is pure sodium chloride.)

77. Formic acid, $HCOOH$, ionizes slightly in water.

$$HCOOH(aq) \rightleftharpoons H^+(aq) + HCOO^-(aq)$$

A 0.0100 m formic acid solution freezes at $-0.0209°C$. Calculate the percent ionization of $HCOOH$ in this solution.

78. A 0.100 m acetic acid solution in water freezes at $-0.188°C$. Calculate the percent ionization of CH_3COOH in this solution.

*79. In a home ice cream freezer, we lower the freezing point of the water bath surrounding the ice cream can by dissolving $NaCl$ in water to make a brine solution. A 15.0% brine solution is observed to freeze at $-10.89°C$. What is the van't Hoff factor, i, for this solution?

80. $CsCl$ dissolves in water according to

$$CsCl(s) \xrightarrow{H_2O} Cs^+(aq) + Cl^-(aq)$$

A 0.121 m solution of $CsCl$ freezes at $-0.403°C$. Calculate i and the apparent percent dissociation of $CsCl$ in this solution.

Osmotic Pressure

81. What are osmosis and osmotic pressure?

*82. Show numerically that the molality and molarity of $1.00 \times 10^{-4} M$ aqueous sodium chloride are nearly equal. Why is this true? Would this be true if another solvent, say acetonitrile, CH_3CN, replaced water? Why or why not? The density of CH_3CN is 0.786 g/mL at 20.0°C.

83. Show how the expression $\pi = MRT$, where π is osmotic pressure, is similar to the ideal gas law. Rationalize qualitatively why this should be so.

84. What is the osmotic pressure associated with a 0.0151 M aqueous solution of a nonvolatile nonelectrolyte solute at 75°C?

85. The osmotic pressure of an aqueous solution of a nonvolatile nonelectrolyte solute is 1.27 atm at 0.0°C. What is the molarity of the solution?

86. Calculate the freezing point depression and boiling point elevation associated with the solution in Exercise 85.

87. Estimate the osmotic pressure associated with 32.5 g of an enzyme of molecular weight 4.21×10^6 dissolved in 1740. mL of ethyl acetate solution at 38.0°C.

88. Calculate the osmotic pressure at 25°C of 1.00 m K_2CrO_4 in water, taking ion association into account. Refer to Table 14-3. The density of a 1.00 m K_2CrO_4 solution is 1.14 g/mL.

89. Estimate the osmotic pressure at 25°C of 1.00 m K_2CrO_4 in water, assuming no ion association. The density of a 1.00 m K_2CrO_4 solution is 1.14 g/mL.

*90. Many biological compounds are isolated and purified in very small amounts. We dissolve 11.0 mg of a biological macromolecule with molecular weight of 2.00×10^4 in 10.0 g of water. (a) Calculate the freezing point of the solution. (b) Calculate the osmotic pressure of the solution at 25°C. (c) Suppose we are trying to use freezing point measurements to *determine* the molecular weight of this substance and that we make an error of only 0.001°C in the temperature measurement. What percent error would this cause in the calculated molecular weight? (d) Suppose we could measure the osmotic pressure with an error of only 0.1 torr (not a very difficult experiment). What percent error would this cause in the calculated molecular weight?

Colloids

91. How does a colloidal dispersion differ from a true solution?

92. Distinguish among (a) sol, (b) gel, (c) emulsion, (d) foam, (e) solid sol, (f) solid emulsion, (g) solid foam, (h) solid aerosol, and (i) liquid aerosol. Try to give an example of each that is not listed in Table 14-4.

93. What is the Tyndall effect, and how is it caused?

94. Distinguish between hydrophilic and hydrophobic colloids.

95. What is an emulsifier?

96. Distinguish between soaps and detergents. How do they interact with hard water? Write an equation to show the interaction between a soap and hard water that contains Ca^{2+} ions.

97. What is the disadvantage of branched alkylbenzenesulfonate (ABS) detergents compared with linear alkylbenzenesulfonate (LAS) detergents?

Mixed Exercises

*98. An aqueous solution prepared by dissolving 1.56 g of anhydrous $AlCl_3$ in 50.0 g of water has a freezing point of −1.61°C. Calculate the boiling point and osmotic pressure at 25°C of this solution. The density of the solution at 25°C is 1.002 g/mL. K_f and K_b for water are 1.86°C/m and 0.512°C/m, respectively.

99. Dry air contains 20.94% O_2 by volume. The solubility of O_2 in water at 25°C is 0.041 gram O_2 per liter of water. How many liters of water would dissolve the O_2 in one liter of dry air at 25°C and 1.00 atm?

100. (a) The freezing point of a 1.00% aqueous solution of acetic acid, CH_3COOH, is −0.310°C. What is the approximate formula weight of acetic acid in water? (b) A 1.00% solution of acetic acid in benzene has a freezing point depression of 0.441°C. What is the formula weight of acetic acid in this solvent? Explain the difference.

101. An aqueous ammonium chloride solution contains 5.75 mass % NH_4Cl. The density of the solution is 1.0195 g/mL. Express the concentration of this solution in molarity, molality, and mole fraction of solute.

102. Starch contains C—C, C—H, C—O, and O—H bonds. Hydrocarbons contain only C—C and C—H bonds. Both starch and hydrocarbon oils can form colloidal dispersions in water. (a) Which dispersion is classified as

hydrophobic? (b) Which is hydrophilic? (c) Which dispersion would be easier to make and maintain?

*103. Suppose we put some one-celled microorganisms in various aqueous NaCl solutions. We observe that the cells remain unperturbed in 0.7% NaCl, whereas they shrink in more concentrated solutions and expand in more dilute solutions. Assume that 0.7% NaCl behaves as an *ideal* 1:1 electrolyte. Calculate the osmotic pressure of the aqueous fluid within the cells at 25°C.

*104. A sample of a drug ($C_{21}H_{23}O_5N$, molecular weight = 369 g/mol) mixed with lactose (a sugar, $C_{12}H_{22}O_{11}$, molecular weight = 342 g/mol) was analyzed by osmotic pressure to determine the amount of sugar present. If 100. mL of solution containing 1.00 g of the drug–sugar mixture has an osmotic pressure of 519 torr at 25°C, what is the percent sugar present?

105. A solution containing 4.52 g of a nonelectrolyte polymer per liter of benzene solution has an osmotic pressure of 0.786 torr at 20.0°C. (a) Calculate the molecular weight of the polymer. (b) Assume that the density of the dilute solution is the same as that of benzene, 0.879 g/mL. What would be the freezing point depression for this solution? (c) Why are boiling point elevations and freezing point depressions difficult to use to measure molecular weights of polymers?

106. On what basis would you choose the components to prepare an ideal solution of a molecular solute? Which of the following combinations would you expect to act most nearly ideally? (a) $CH_4(\ell)$ and $CH_3OH(\ell)$; (b) $CH_3OH(\ell)$ and NaCl(s); (c) $CH_4(\ell)$ and $CH_3CH_3(\ell)$.

107. Physiological saline (normal saline) is a 0.90% NaCl solution. This solution is isotonic with human blood. Calculate the freezing point and boiling point of physiological saline.

CONCEPTUAL EXERCISES

108. In the first five sections of this chapter the term "dissolution" was used to describe the process by which a solute is dispersed by a solvent to form a solution. A popular dictionary defines dissolution as "decomposition into fragments or parts." Using either of the definitions, compare the following two uses of the term. "It was ruled that there must be dissolution of the estate." "The ease of dissolution of a solute depends on two factors. . . ."

109. Consider two nonelectrolytes A and B; A has a higher molecular weight than B, and both are soluble in solvent C. A solution is made by dissolving x grams of A in 100 grams of C; another solution is made by dissolving the same number of grams, x, of B in 100 grams of C. Assume that the two solutions have the same density. (a) Which solution has the higher molality? (b) Which solution has the higher mole fraction? (c) Which solution has the higher percent by mass? (d) Which solution has the higher molarity?

110. The actual value for the van't Hoff factor, i, is often observed to be less than the ideal value. Less commonly, it is greater than the ideal value. Give molecular explanations for the cases where i is observed to be greater than the ideal value.

*111. When the van't Hoff factor, i, is included, the boiling point elevation equation becomes $\Delta T_b = i K_b m$. The van't Hoff factor can be inserted in a similar fashion in the freezing point depression equation and the osmotic pressure equation. The van't Hoff factor, however, cannot be inserted in the same way into the vapor pressure-lowering equation. Show algebraically that $\Delta P_{solvent} = iX_{solute}P^0_{solvent}$ is not equivalent to $\Delta P_{solvent} = \{(i \times moles_{solute})/[(i \times moles_{solute}) + (moles_{solvent})]\}P^0_{solvent}$. Which equation for $\Delta P_{solvent}$ is correct?

112. The two solutions shown were both prepared by dissolving 194 g of K_2CrO_4 (1.00 mol) in a 1.00-L volumetric flask. The solution on the right was diluted to the 1-L mark on the neck of the flask, and the solution on the left was diluted by adding 1.00 L of water. Which solution, the one on the left or the one on the right, is: (a) more concentrated; (b) a 1.00 m solution; (c) a 1.00 M solution?

Charles D. Winters

113. Would you expect lowering the freezing point or elevating the boiling point to be the better method to obtain the approximate molecular weight of an unknown? Explain your choice.

114. Rock candy consists of crystals of sugar on a string or stick. Propose a method of making rock candy, and explain each step.

115. Concentrations expressed in units of parts per million and parts per billion often have no meaning for people until they relate these small and large numbers to their own experiences. How many seconds are equal to 1 ppm of a year?

BUILDING YOUR KNOWLEDGE

*116. DDT is a toxin still found in the fatty tissues of some animals. DDT was transported into our lakes and streams as runoff from agricultural operations where it was originally used several years ago as an insecticide. In the lakes and streams it did not dissolve to any great extent; it collected in the lake and stream bottoms. It entered the bodies of animals via fatty tissues in their diet; microorganisms collected the DDT, the fish ate the microorganisms, and so on. Fortunately, much of the once large quantities of DDT in lakes and streams has biodegraded. Based on this information, what can you conclude regarding the intermolecular forces present in DDT?

117. Draw Figure 14-1, but instead of using colored circles to represent the solvent and solute molecules, use Lewis formulas to represent water as the solvent and acetone, CH_3COCH_3, as the solute. Use dashed lines to show hydrogen bonds. Twelve water molecules and two acetone molecules should be sufficient to illustrate the interaction between these two kinds of molecules.

118. A sugar maple tree grows to a height of about 45 feet, and its roots are in contact with water in the soil. What must be the concentration of the sugar in its sap if osmotic pressure is solely responsible for forcing the sap to the top of the tree at 15°C? The density of mercury is 13.6 g/cm^3, and the density of the sap can be considered to be 1.10 g/cm^3.

119. Many metal ions become hydrated in solution by forming coordinate covalent bonds with the unshared pair of electrons from the water molecules to form "AB$_6$" ions. Because of their sizes, these hydrated ions are unable to pass through the semipermeable membrane described in Section 14-15, whereas water as a trimer, $(H_2O)_3$, or a tetramer, $(H_2O)_4$, can pass through. Anions tend to be less hydrated. Using the VSEPR theory, prepare three-dimensional drawings of $Cu(H_2O)_6^{2+}$ and a possible $(H_2O)_3$ that show their relative shapes and sizes.

BEYOND THE TEXTBOOK

Go to the textbook website at

http://www.brookscole.com/chemistry/whitten

for additional activities and exercises based on the General Chemistry Interactive CD-ROM, the World Wide Web, and library resources.

InfoTrac College Edition

For additional readings, go to InfoTrac College Edition, your online research library at:

http://infotrac.thomsonlearning.com

15 Chemical Thermodynamics

OUTLINE

The thermite reaction is a highly exothermic reaction of aluminum powder and iron oxide to produce molten steel (iron) and aluminum oxide. It is being used here to weld steel rails together for a new light rail line in Salt Lake City.

© Steve Griffin/The Salt Lake Tribune

OBJECTIVES

After you have studied this chapter, you should be able to

- Understand the terminology of thermodynamics, and the meaning of the signs of changes
- Use the concept of state functions
- Carry out calculations of calorimetry to determine changes in energy and enthalpy
- Use Hess's Law to find the enthalpy change, ΔH, for a reaction by combining thermochemical equations with known ΔH values
- Use Hess's Law to find the enthalpy change, ΔH, for a reaction by using tabulated values of standard molar enthalpies of formation
- Use Hess's Law to find the enthalpy of formation given ΔH for a reaction and the known enthalpies of formation of the other substances in the reaction
- Use the First Law of Thermodynamics to relate heat, work, and energy changes
- Relate the work done on or by a system to changes in its volume
- Use bond energies to estimate heats of reaction for gas phase reactions; use ΔH values for gas phase reactions to find bond energies
- Understand what is meant by a product-favored process; by a reactant-favored process
- Understand the relationship of entropy to the order or disorder of a system
- Understand how the spontaneity of a process is related to entropy changes— the Second Law of Thermodynamics
- Use tabulated values of absolute entropies to calculate the entropy change, ΔS
- Calculate changes in Gibbs free energy, ΔG, by two methods: (a) from values of ΔH and ΔS and (b) from tabulated values of standard molar free energies of formation; know when to use each type of calculation
- Use ΔG to predict whether a process is product-favored at constant T and P
- Understand how changes in temperature can affect the spontaneity of a process
- Predict the temperature range of spontaneity of a chemical or physical process

E nergy is very important in every aspect of our daily lives. The food we eat supplies the energy to sustain life with all of its activities and concerns. The availability of relatively inexpensive energy is an important factor in our technological society. This is seen in the costs of fuel, heating and cooling our homes and workplaces, and the electricity to power our lights, appliances, and computers. It is also seen in the costs of the goods and services we purchase, because a substantial part of the cost of production is for energy in one form or another. We must understand the storage and use of energy on a scientific basis to learn how to decrease our dependence on consumable oil and natural gas as our main energy sources. Such understanding has profound ramifications, ranging from our daily lifestyles to international relations.

The concept of energy is at the very heart of science. All physical and chemical processes are accompanied by the transfer of energy. Because energy cannot be created or destroyed, we must understand how to do the "accounting" of energy transfers from one body or one substance to another or from one form of energy to another.

In **thermodynamics** we study the energy changes that accompany physical and chemical processes. Usually these energy changes involve *heat*—hence the "thermo-" part of the term. In this chapter we study the two main aspects of thermodynamics. The first is **thermochemistry.** This practical subject is concerned with how we *observe, measure,* and *predict* energy changes for both physical changes and chemical reactions. The second part of the chapter addresses a more fundamental aspect of thermodynamics. There we will learn to use energy changes to tell whether or not a given process can occur under specified conditions to give predominantly products (or reactants) and how to make a process more (or less) favorable.

Some forms of energy are potential, kinetic, electrical, nuclear, heat, and light.

 CD-ROM   Screen 6.4, Energy, and Screen 6.5, Forms of Energy.

HEAT CHANGES AND THERMOCHEMISTRY

15-1 THE FIRST LAW OF THERMODYNAMICS

We can define energy as follows.

> Energy is the capacity to do work or to transfer heat.

We classify energy into two general types: kinetic and potential. **Kinetic energy** is the energy of motion. The kinetic energy of an object is equal to one half its mass, *m*, times the square of its velocity, *v*.

$$E_{\text{kinetic}} = \tfrac{1}{2}mv^2$$

The heavier a hammer is and the more rapidly it moves, the greater its kinetic energy and the more work it can accomplish.

Potential energy is the energy that a system possesses by virtue of its position or composition. The work that we do to lift an object is stored in the object as energy; we describe this as potential energy. If we drop a hammer, its potential energy is converted into kinetic energy as it falls, and it could do work on something it hits—for example, drive a nail or break a piece of glass. Similarly, an electron in an atom has potential energy because of the electrostatic force on it that is due to the positively charged nucleus and the other electrons in that atom and surrounding atoms. Energy can take many other forms: electrical energy,

© Royalty-free /CORBIS

As matter falls from a higher to a lower level, its gravitational potential energy is converted into kinetic energy. A hydroelectric power plant converts the kinetic energy of falling water into electrical (potential) energy.

radiant energy (light), nuclear energy, and chemical energy. At the atomic or molecular level, we can think of each of these as either kinetic or potential energy.

The chemical energy in a fuel or food comes from potential energy stored in atoms due to their arrangements in the molecules. This stored chemical energy can be released when compounds undergo chemical changes, such as those that occur in combustion and metabolism.

Reactions that release energy in the form of heat are called **exothermic** reactions. Combustion reactions of fossil fuels are familiar examples of exothermic reactions. Hydrocarbons—including methane, the main component of natural gas, and octane, a minor component of gasoline—undergo combustion with an excess of O_2 to yield CO_2 and H_2O. These reactions release heat energy. The amounts of heat energy released at constant pressure are shown for the reactions of one mole of methane and of two moles of octane.

$$CH_4(g) + 2O_2(g) \longrightarrow CO_2(g) + 2H_2O(\ell) + 890 \text{ kJ}$$

$$2C_8H_{18}(\ell) + 25O_2(g) \longrightarrow 16CO_2(g) + 18H_2O(\ell) + 1.090 \times 10^4 \text{ kJ}$$

In such reactions, the total energy of the products is lower than that of the reactants by the amount of energy released, most of which is heat. Some initial activation (e.g., by heat) is needed to get these reactions started. This is shown for CH_4 in Figure 15-1. This activation energy *plus* 890 kJ is released as one mole of $CO_2(g)$ and two moles of $H_2O(\ell)$ are formed.

A process that absorbs energy from its surroundings is called **endothermic.** One such process is shown in Figure 15-2.

Energy changes accompany physical changes, too (Chapter 13). For example, the melting of one mole of ice at 0°C at constant pressure must be accompanied by the absorption of 6.02 kJ of energy.

$$H_2O(s) + 6.02 \text{ kJ} \longrightarrow H_2O(\ell)$$

This tells us that the total energy of the water is raised by 6.02 kJ in the form of heat during the phase change even though the temperature remains constant.

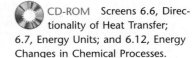 CD-ROM   Screens 6.6, Directionality of Heat Transfer; 6.7, Energy Units; and 6.12, Energy Changes in Chemical Processes.

A hydrocarbon is a binary compound of only hydrogen and carbon. Hydrocarbons may be gaseous, liquid, or solid. All burn.

The amount of heat shown in such an equation always refers to the reaction for the number of moles of reactants and products specified by the coefficients. We call this *one mole of reaction.*

> ⚠ It is important to specify the physical states of all substances, because different physical states have different energy contents.

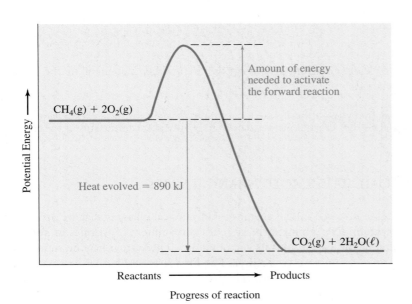

Amount of energy needed to activate the forward reaction

$CH_4(g) + 2O_2(g)$

Potential Energy

Heat evolved = 890 kJ

$CO_2(g) + 2H_2O(\ell)$

Reactants ⟶ Products

Progress of reaction

Figure 15-1 The difference between the potential energy of the reactants—one mole of $CH_4(g)$ and two moles of $O_2(g)$—and that of the products—one mole of $CO_2(g)$ and two moles of $H_2O(\ell)$—is the amount of heat evolved in this *exothermic* reaction at constant pressure. For this reaction, it is 890 kJ/mol of reaction. In this chapter, we see how to measure the heat absorbed or released and how to calculate it from other known heat changes. Some initial activation, for example by heat, is needed to get the reaction started (see Section 16-6). In the absence of such activation energy, a mixture of CH_4 and O_2 can be kept at room temperature for a long time without reacting. For an *endothermic* reaction, the final level is higher than the initial level.

(a)   (b)

Figure 15-2 An endothermic process. (a) When solid hydrated barium hydroxide, $Ba(OH)_2 \cdot 8H_2O$, and *excess* solid ammonium nitrate, NH_4NO_3, are mixed, an endothermic reaction occurs.

$$Ba(OH)_2 \cdot 8H_2O(s) + 2NH_4NO_3(s) \longrightarrow Ba(NO_3)_2(s) + 2NH_3(g) + 10H_2O(\ell)$$

The excess ammonium nitrate dissolves in the water produced in the reaction. (b) The dissolution process is also very endothermic. If the flask is placed on a wet wooden block, the water freezes and attaches the block to the flask.

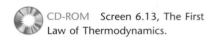 CD-ROM   Screen 6.13, The First Law of Thermodynamics.

Some important ideas about energy are summarized in the **First Law of Thermodynamics.**

> The combined amount of matter and energy in the universe is constant.

The **Law of Conservation of Energy** is just another statement of the First Law of Thermodynamics.

> Energy is neither created nor destroyed in chemical reactions and physical changes.

15-2 SOME THERMODYNAMIC TERMS

The substances involved in the chemical and physical changes that we are studying are called the **system.** Everything in the system's environment constitutes its **surroundings.** The **universe** is the system plus its surroundings. The system may be thought of as the part of the universe under investigation. The First Law of Thermodynamics tells us that energy is neither created nor destroyed; it is only transferred between the system and its surroundings.

The **thermodynamic state of a system** is defined by a set of conditions that completely specifies all the properties of the system. This set commonly includes the temperature, pressure, composition (identity and number of moles of each component), and physical state (gas, liquid, or solid) of each part of the system. Once the state has been specified, all other properties—both physical and chemical—are fixed.

The properties of a system—such as P, V, T—are called **state functions.** The *value* of a state function depends *only* on the state of the system and not on the way in which the system came to be in that state. A *change* in a state function describes a *difference* between the two states. It is independent of the process or pathway by which the change occurs.

For instance, consider a sample of one mole of pure liquid water at 30°C and 1 atm pressure. If at some later time the temperature of the sample is 22°C at the same pressure, then it is in a different thermodynamic state. We can tell that the *net* temperature change is −8°C. It does not matter whether (1) the cooling took place directly (either slowly or rapidly) from 30°C to 22°C, or (2) the sample was first heated to 36°C, then cooled to 10°C, and finally warmed to 22°C, or (3) any other conceivable path was followed from the initial state to the final state. The change in other properties (e.g., the pressure) of the sample is likewise independent of path.

The most important use of state functions in thermodynamics is to describe *changes*. We describe the difference in any quantity, X, as

$$\Delta X = X_{\text{final}} - X_{\text{initial}}$$

When X increases, the final value is greater than the initial value, so ΔX is *positive;* a decrease in X makes ΔX a *negative* value.

You can consider a state function as analogous to a bank account. With a bank account, at any time you can measure the amount of money in your account (your balance) in convenient terms—dollars and cents. Changes in this balance can occur for several reasons, such as deposit of your paycheck, writing of checks, or service charges assessed by the bank. In our analogy these transactions are *not* state functions, but they do cause *changes in* the state function (the balance in the account). You can think of the bank balance on a vertical scale; a deposit of $150 changes the balance by +$150, no matter what it was at the start, just as a withdrawal of $150 would change the balance by −$150. Similarly, we shall see that the energy of a system is a state function that can be changed—for instance, by an energy "deposit" of heat absorbed or work done on the system, or by an energy "withdrawal" of heat given off or work done by the system.

We can describe *differences* between levels of a state function, regardless of where the zero level is located. In the case of a bank balance, the "natural" zero level is obviously the point at which we open the account, before any deposits or withdrawals. In contrast, the zero levels on most temperature scales are set arbitrarily. When we say that the temperature of an ice–water mixture is "zero degrees Celsius," we are not saying that the mixture contains no temperature! We have simply chosen to describe this point on the temperature scale by the number *zero;* conditions of higher temperature are described by positive temperature values, and those of lower temperature have negative values, "below zero." The phrase "15 degrees cooler" has the same meaning anywhere on the scale. Many of the scales that we use in thermodynamics are arbitrarily defined in this way. Arbitrary scales are useful when we are interested only in *changes* in the quantity being described.

Any property of a system that depends only on the values of its state functions is also a state function. For instance, the volume of a given sample of water depends only on temperature, pressure, and physical state; volume is a state function. We shall encounter other thermodynamic state functions.

State functions are represented by capital letters. Here P refers to pressure, V to volume, and T to absolute temperature.

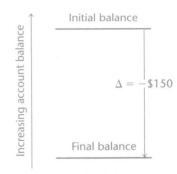

Here is a graphical representation of a $150 decrease in your bank balance. We express the change in your bank balance as $\Delta\$ = \$_{\text{final}} - \$_{\text{initial}}$. Your final balance is *less* than your initial balance, so the result is *negative,* indicating a *decrease.* There are many ways to get this same net change—one large withdrawal or some combination of deposits, withdrawals, interest earned, and service charges. All of the Δ values we will see in this chapter can be thought of in this way.

15-3 ENTHALPY CHANGES

Most chemical reactions and physical changes occur at constant (usually atmospheric) pressure.

We use q to represent the amount of heat absorbed by the system. The subscript p indicates a constant-pressure process.

> The quantity of heat transferred into or out of a system as it undergoes a chemical or physical change at constant pressure, q_p, is defined as the **enthalpy change, ΔH,** of the process.

An enthalpy change is sometimes loosely referred to as a *heat change* or a *heat of reaction*. The enthalpy change is equal to the enthalpy or "heat content," H, of the substances produced minus the enthalpy of the substances consumed.

CD-ROM   Screen 6.14, Enthalpy Change and ΔH.

$$\Delta H = H_{final} - H_{initial} \qquad \text{or} \qquad \Delta H = H_{substances\ produced} - H_{substances\ consumed}$$

It is impossible to know the absolute enthalpy (heat content) of a system. *Enthalpy is a state function*, however, and it is the *change in enthalpy* in which we are interested; this can be measured for many processes. In the next several sections we focus on chemical reactions and the enthalpy changes that occur in these processes. We first discuss the experimental determination of enthalpy changes.

15-4 CALORIMETRY

CD-ROM   Screen 6.11, Heat Transfer Between Substances, and Screen 6.16, Measuring Heats of Reaction: Calorimetry.

We can determine the energy change associated with a chemical or physical process by using an experimental technique called **calorimetry.** This technique is based on observing the temperature change when a system absorbs or releases energy in the form of heat. The experiment is carried out in a device called a **calorimeter,** in which the temperature change of a known amount of substance (often water) of known specific heat is measured. The temperature change is caused by the absorption or release of heat by the chemical or physical process under study. A review of calculations involved with heat transfer (Sections 1-13, 13-9, and 13-11) may be helpful for understanding this section.

A "coffee-cup" calorimeter (Figure 15-3) is often used in laboratory classes to measure "heats of reaction" at constant pressure, q_p, in aqueous solutions. Reactions are chosen so that there are no gaseous reactants or products. Thus, all reactants and products remain in the vessel throughout the experiment. Such a calorimeter could be used to measure the amount of heat absorbed or released when a reaction takes place in aqueous solution. We can consider the reactants and products as the system and the calorimeter plus the solution (mostly water) as the surroundings. For an exothermic reaction, the amount of heat evolved by the reaction can be calculated from the amount by which it causes the temperature of the calorimeter and the solution to rise. The heat can be visualized as divided into two parts.

The polystyrene insulation of the simple coffee-cup calorimeter ensures that little heat escapes from or enters the container.

$$\begin{pmatrix} \text{amount of heat} \\ \text{released by reaction} \end{pmatrix} = \begin{pmatrix} \text{amount of heat absorbed} \\ \text{by calorimeter} \end{pmatrix} + \begin{pmatrix} \text{amount of heat} \\ \text{absorbed by solution} \end{pmatrix}$$

The amount of heat absorbed by a calorimeter is sometimes expressed as the *heat capacity* of the calorimeter, in joules per degree.

The heat capacity of a calorimeter is determined by adding a known amount of heat and measuring the rise in temperature of the calorimeter and of the solution it contains. This heat capacity of a calorimeter is sometimes called its *calorimeter constant.*

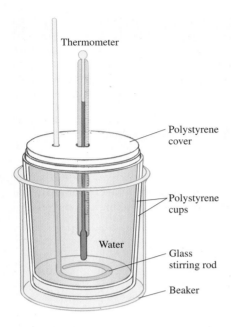

Thermometer

Polystyrene cover

Polystyrene cups

Water

Glass stirring rod

Beaker

Figure 15-3 A coffee-cup calorimeter. The stirring rod is moved up and down to ensure thorough mixing and uniform heating of the solution during reaction. The polystyrene walls and top provide insulation so that very little heat escapes. This kind of calorimeter measures q_p, the heat transfer due to a reaction occurring at constant *pressure*.

EXAMPLE 15-1 *Heat Capacity of a Calorimeter*

We add 3.358 kJ of heat to a calorimeter that contains 50.00 g of water. The temperature of the water and the calorimeter, originally at 22.34°C, increases to 36.74°C. Calculate the heat capacity of the calorimeter in J/°C. The specific heat of water is 4.184 J/g·°C.

One way to add heat is to use an electric heater.

Plan

We first calculate the amount of heat gained by the water in the calorimeter. The rest of the heat must have been gained by the calorimeter, so we can determine the heat capacity of the calorimeter.

Solution

$$50.00 \text{ g } H_2O(\ell) \text{ at } 22.34°C \longrightarrow 50.00 \text{ g } H_2O(\ell) \text{ at } 36.74°C$$

The temperature change is $(36.74 - 22.34)°C = 14.40°C$.

$$\underline{?} \text{ J} = 50.00 \text{ g} \times \frac{4.184 \text{ J}}{\text{g·°C}} \times 14.40°C = 3.012 \times 10^3 \text{ J}$$

The total amount of heat added was 3.358 kJ or 3.358×10^3 J. The difference between these heat values is the amount of heat absorbed by the calorimeter.

$$\underline{?} \text{ J} = 3.358 \times 10^3 \text{ J} - 3.012 \times 10^3 \text{ J} = 0.346 \times 10^3 \text{ J, or } 346 \text{ J absorbed by calorimeter}$$

To obtain the heat capacity of the calorimeter, we divide the amount of heat absorbed by the calorimeter, 346 J, by its temperature change.

$$\underline{?} \frac{\text{J}}{°C} = \frac{346 \text{ J}}{14.40°C} = \boxed{24.0 \text{ J/°C}}$$

The calorimeter absorbs 24.0 J of heat for each degree Celsius increase in its temperature.

You should now work Exercise 58.

 Note that, because we are using the *change* in temperature (ΔT) in this example, it is fine to use temperatures as °C or K. That is because the magnitude of a change in 1°C is equal to 1 K. In this problem $\Delta T = 14.40°C = 14.40$ K. But, when working with mathematical equations that use absolute temperatures T (not ΔT!), it is important to express temperatures in kelvins. We will see mathematical formulas in this chapter that use T and ΔT, so it is important to keep track of which is being used.

EXAMPLE 15-2 *Heat Measurements Using a Calorimeter*

A 50.0-mL sample of 0.400 M copper(II) sulfate solution at 23.35°C is mixed with 50.0 mL of 0.600 M sodium hydroxide solution, also at 23.35°C, in the coffee-cup calorimeter of Example 15-1. After the reaction occurs, the temperature of the resulting mixture is measured to be 25.23°C. The density of the final solution is 1.02 g/mL. Calculate the amount of heat evolved. Assume that the specific heat of the solution is the same as that of pure water, 4.184 J/g·°C.

$$CuSO_4(aq) + 2NaOH(aq) \longrightarrow Cu(OH)_2(s) + Na_2SO_4(aq)$$

Plan

When *dilute aqueous solutions* are mixed, their volumes are very nearly additive.

The amount of heat released by the reaction is absorbed by the calorimeter *and* by the solution. To find the amount of heat absorbed by the solution, we must know the mass of solution; to find that, we assume that the volume of the reaction mixture is the sum of volumes of the original solutions.

Solution

The mass of solution is

$$\underline{?}\text{ g soln} = (50.0 + 50.0)\text{ mL} \times \frac{1.02\text{ g soln}}{\text{mL}} = 102\text{ g soln}$$

The amount of heat absorbed by the calorimeter *plus* the amount absorbed by the solution is

$$\underline{?}\text{ J} = \overbrace{\frac{24.0\text{ J}}{°\text{C}} \times (25.23 - 23.35)°\text{C}}^{\substack{\text{amount of heat} \\ \text{absorbed by calorimeter}}} + \overbrace{102\text{ g} \times \frac{4.18\text{ J}}{\text{g}·°\text{C}} \times (25.23 - 23.35)°\text{C}}^{\substack{\text{amount of heat} \\ \text{absorbed by solution}}}$$

$$= 45\text{ J} + 801\text{ J} = 846\text{ J absorbed by solution plus calorimeter}$$

Thus, the reaction must have liberated 846 J, or 0.846 kJ, of heat.

You should now work Exercise 62(a).

The heat released by the reaction of HCl(aq) with NaOH(aq) causes the temperature of the solution to rise.

15-5 THERMOCHEMICAL EQUATIONS

A balanced chemical equation, together with its value of ΔH, is called a **thermochemical equation.** For example,

$$\underset{\text{1 mol}}{C_2H_5OH(\ell)} + \underset{\text{3 mol}}{3O_2(g)} \longrightarrow \underset{\text{2 mol}}{2CO_2(g)} + \underset{\text{3 mol}}{3H_2O(\ell)} + 1367\text{ kJ}$$

is a thermochemical equation that describes the combustion (burning) of one mole of liquid ethanol at a particular temperature and pressure. The coefficients in such a description *must* be interpreted as *numbers of moles.* Thus, 1367 kJ of heat is released when *one* mole of $C_2H_5OH(\ell)$ reacts with *three* moles of $O_2(g)$ to give *two* moles of $CO_2(g)$ and *three* moles of $H_2O(\ell)$. We can refer to this amount of reaction as one **mole of reaction,** which we abbreviate "mol rxn." This interpretation allows us to write various unit factors as desired.

$$\frac{1\text{ mol }C_2H_5OH(\ell)}{1\text{ mol rxn}}, \quad \frac{2\text{ mol }CO_2(g)}{1\text{ mol rxn}}, \quad \text{and so on}$$

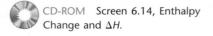

CD-ROM Screen 6.14, Enthalpy Change and ΔH.

We can also write the thermochemical equation as

$$C_2H_5OH(\ell) + 3O_2(g) \longrightarrow 2CO_2(g) + 3H_2O(\ell) \qquad \Delta H = -1367 \text{ kJ/mol rxn}$$

The negative sign indicates that this is an *exothermic* reaction (i.e., it gives *off* heat).

> We always interpret ΔH as the enthalpy change for the reaction as written; that is, as (enthalpy change)/(mole of reaction), where the denominator means "for the number of moles of each substance shown in the balanced equation."

We can then use several unit factors to interpret this thermochemical equation.

$$\frac{1367 \text{ kJ given off}}{\text{mol of reaction}} = \frac{1367 \text{ kJ given off}}{\text{mol } C_2H_5OH(\ell) \text{ consumed}} = \frac{1367 \text{ kJ given off}}{3 \text{ mol } O_2(g) \text{ consumed}}$$

$$= \frac{1367 \text{ kJ given off}}{2 \text{ mol } CO_2(g) \text{ formed}} = \frac{1367 \text{ kJ given off}}{3 \text{ mol } H_2O(\ell) \text{ formed}}$$

The reverse reaction would require the absorption of 1367 kJ under the same conditions;

$$1367 \text{ kJ} + 2CO_2(g) + 3H_2O(\ell) \longrightarrow C_2H_5OH(\ell) + 3O_2(g)$$

That is, it is *endothermic*, with $\Delta H = +1367$ kJ.

$$2CO_2(g) + 3H_2O(\ell) \longrightarrow C_2H_5OH(\ell) + 3O_2(g) \qquad \Delta H = +1367 \text{ kJ/mol rxn}$$

It is important to remember the following conventions regarding thermochemical equations:

1. The coefficients in a balanced thermochemical equation refer to the numbers of *moles* of reactants and products involved. In the thermodynamic interpretation of equations we *never* interpret the coefficients as *numbers of molecules*. Thus, it is acceptable to write coefficients as fractions rather than as integers, when necessary.

2. The numerical value of ΔH (or any other thermodynamic change) refers to the *number of moles* of substances specified by the equation. This amount of change of substances is called *one mole of reaction*, so we can express ΔH in units of energy/mol rxn. For brevity, the units of ΔH are sometimes written kJ/mol or even just kJ. No matter what units are used, be sure that you interpret the thermodynamic change *per mole of reaction for the balanced chemical equation to which it refers*. If a different amount of material is involved in the reaction, then the ΔH (or other change) must be scaled accordingly.

3. The physical states of all species are important and must be specified. Heat is given off or absorbed when phase changes occur, so different amounts of heat could be involved in a reaction depending on the phases of reactants and products.

4. The value of ΔH usually does not change significantly with moderate changes in temperature.

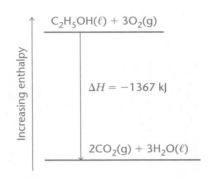

Before we know what a value of ΔH means, we must know the balanced chemical equation to which it refers.

CD-ROM Screen 6.15, Enthalpy Changes for Chemical Reactions.

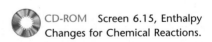

TOPIC: Thermochemistry
GO TO: www.scilinks.org
*sci*LINKS CODE: WCH1510

The launch of the space shuttle requires about 2×10^{10} kilojoules of energy. About one sixth of this comes from the reaction of hydrogen, H_2, and oxygen, O_2. The rest comes from the explosive decomposition of ammonium perchlorate, NH_4ClO_4, in solid-fuel rockets.

EXAMPLE 15-3   *Thermochemical Equations*

When 2.61 grams of dimethyl ether, CH_3OCH_3, is burned at constant pressure, 82.5 kJ of heat is given off. Find ΔH for the reaction

$$CH_3OCH_3(\ell) + 3O_2(g) \longrightarrow 2CO_2(g) + 3H_2O(\ell)$$

Plan

We scale the amount of heat given off in the experiment to correspond to the amount of CH_3OCH_3 shown in the balanced equation.

Solution

$$\frac{? \text{ kJ given off}}{\text{mol rxn}} = \frac{82.5 \text{ kJ given off}}{2.61 \text{ g } CH_3OCH_3} \times \frac{46.0 \text{ g } CH_3OCH_3}{\text{mol } CH_3OCH_3} \times \frac{1 \text{ mol } CH_3OCH_3}{\text{mol rxn}}$$

$$= 1450 \text{ kJ/mol rxn}$$

Because heat is given off, we know that the reaction is exothermic and the value of ΔH is negative, so

$$\Delta H = -1450 \text{ kJ/mol rxn}$$

You should now work Exercise 19.

EXAMPLE 15-4   *Thermochemical Equations*

Write the thermochemical equation for the reaction in Example 15-2.

Plan

We must determine *how much* reaction occurred—that is, how many moles of reactants were consumed. We first multiply the volume, in liters, of each solution by its concentration in mol/L (molarity) to determine the number of moles of each reactant mixed. Then we identify the limiting reactant. We scale the amount of heat released in the experiment to correspond to the number of moles of that reactant shown in the balanced equation.

Solution

Using the data from Example 15-2,

$$\underline{?} \text{ mol } CuSO_4 = 0.0500 \text{ L} \times \frac{0.400 \text{ mol } CuSO_4}{1.00 \text{ L}} = 0.0200 \text{ mol } CuSO_4$$

$$\underline{?} \text{ mol } NaOH = 0.0500 \text{ L} \times \frac{0.600 \text{ mol } NaOH}{1.00 \text{ L}} = 0.0300 \text{ mol } NaOH$$

We determine which is the limiting reactant (review Section 3-3).

Required Ratio	**Available Ratio**
$\dfrac{1 \text{ mol } CuSO_4}{2 \text{ mol } NaOH} = \dfrac{0.50 \text{ mol } CuSO_4}{1.00 \text{ mol } NaOH}$	$\dfrac{0.0200 \text{ mol } CuSO_4}{0.0300 \text{ mol } NaOH} = \dfrac{0.667 \text{ mol } CuSO_4}{1.00 \text{ mol } NaOH}$

NaOH is the limiting reactant.

More $CuSO_4$ is available than is required to react with the NaOH. Thus, 0.846 kJ of heat was given off during the consumption of 0.0300 mol of NaOH. The amount of heat given off per "mole of reaction" is

$$\frac{? \text{ kJ released}}{\text{mol rxn}} = \frac{0.846 \text{ kJ given off}}{0.0300 \text{ mol } NaOH} \times \frac{2 \text{ mol } NaOH}{\text{mol rxn}} = \frac{56.4 \text{ kJ given off}}{\text{mol rxn}}$$

Thus, when the reaction occurs *to the extent indicated by the balanced chemical equation*, 56.4 kJ is released. Remembering that exothermic reactions have negative values of ΔH_{rxn}, we write

$$CuSO_4(aq) + 2NaOH(aq) \longrightarrow Cu(OH)_2(s) + Na_2SO_4(aq) \qquad \Delta H_{rxn} = -56.4 \text{ kJ/mol rxn}$$

Heat is released, so this is an exothermic reaction.

You should now work Exercise 62(b).

EXAMPLE 15-5 *Amount of Heat Produced*

When aluminum metal is exposed to atmospheric oxygen (as in aluminum doors and windows), it is oxidized to form aluminum oxide. How much heat is released by the complete oxidation of 24.2 grams of aluminum at 25°C and 1 atm? The thermochemical equation is

$$4Al(s) + 3O_2(g) \longrightarrow 2Al_2O_3(s) \qquad \Delta H = -3352 \text{ kJ/mol rxn}$$

Plan

The thermochemical equation tells us that 3352 kJ of heat is released for every mole of reaction, that is, for every 4 moles of Al that reacts. We convert 24.2 g of Al to moles, and then calculate the number of kilojoules corresponding to that number of moles of Al, using the unit factors

$$\frac{-3352 \text{ kJ}}{\text{mol rxn}} \quad \text{and} \quad \frac{1 \text{ mol rxn}}{4 \text{ mol Al}}$$

Solution

For 24.2 g Al,

$$\underline{?} \text{ kJ} = 24.2 \text{ g Al} \times \frac{1 \text{ mol Al}}{27.0 \text{ g Al}} \times \frac{1 \text{ mol rxn}}{4 \text{ mol Al}} \times \frac{-3352 \text{ kJ}}{\text{mol rxn}} = -751 \text{ kJ}$$

This tells us that 751 kJ of heat is released to the surroundings during the oxidation of 24.2 grams of aluminum.

You should now work Exercises 14 and 15.

 The *sign* tells us that heat was released, but it would be grammatical nonsense to say in words that "−751 kJ of heat was released." As an analogy, suppose you give your friend $5. Your $\Delta$$ is −$5, but in describing the transaction you would not say "I gave her minus five dollars," but rather "I gave her five dollars."

✓ **Problem-Solving Tip:** *Mole of Reaction*

Remember that a thermochemical equation can imply *different* numbers of moles of *different* reactants or products. In Example 15-5 one mole of reaction also corresponds to 3 moles of $O_2(g)$ and to 2 moles of $Al_2O_3(s)$.

15-6 STANDARD STATES AND STANDARD ENTHALPY CHANGES

The **thermodynamic standard state** of a substance is its most stable pure form under standard pressure (one atmosphere)* and at some specific temperature (25°C or 298 K unless otherwise specified). Examples of elements in their standard states at 25°C are hydrogen, gaseous diatomic molecules, $H_2(g)$; mercury, a silver-colored liquid metal, $Hg(\ell)$;

A temperature of 25°C is 77°F. This is slightly above typical room temperature. Notice that these thermodynamic "standard conditions" are not the same as the "standard temperature and pressure (STP)" that we used in gas calculations involving standard molar volume (Chapter 12).

*IUPAC has changed the standard pressure from 1 atm to 1 bar. Because 1 bar is equal to 0.987 atm, the differences in thermodynamic calculations are negligible except in work of very high precision. Many tables of thermodynamic data are still based on a standard pressure of 1 atm, so we will use it in this book.

sodium, a silvery white solid metal, Na(s); and carbon, a grayish black solid called graphite, C(graphite). We use C(graphite) instead of C(s) to distinguish it from other solid forms of carbon, such as C(diamond). The reaction C(diamond) → C(graphite) would be *exothermic* by 1.897 kJ/mol rxn; C(graphite) is thus more stable than C(diamond). Examples of standard states of compounds include ethanol (ethyl alcohol or grain alcohol), a liquid, $C_2H_5OH(\ell)$; water, a liquid, $H_2O(\ell)$; calcium carbonate, a solid, $CaCO_3(s)$; and carbon dioxide, a gas, $CO_2(g)$. Keep in mind the following conventions for thermochemical standard states.

1. For a *pure* substance in the liquid or solid phase, the standard state is the pure liquid or solid.
2. For a gas, the standard state is the gas at a pressure of *one atmosphere;* in a mixture of gases, its partial pressure must be one atmosphere.
3. For a substance in solution, the standard state refers to *one-molar* concentration.

For ease of comparison and tabulation, we often refer to thermochemical or thermodynamic changes "at standard states" or, more simply, to a *standard change*. To indicate a change at standard pressure, we add a superscript zero. If some temperature other than standard temperature of 25°C (298 K) is specified, we indicate it with a subscript; if no subscript appears, a temperature of 25°C (298 K) is implied.

The **standard enthalpy change, ΔH^0_{rxn},** for reaction

$$\text{reactants} \longrightarrow \text{products}$$

refers to the ΔH when the specified number of moles of reactants, all at standard states, are converted *completely* to the specified number of moles of products, all at standard states.

We allow a reaction to take place, with changes in temperature or pressure if necessary; when the reaction is complete, we return the products to the same conditions of temperature and pressure that we started with, *keeping track of energy or enthalpy changes* as we do so. When we describe a process as taking place "at constant T and P," we mean that the initial and final conditions are the same. Because we are dealing with changes in state functions, the net change is the same as the change we would have obtained hypothetically with T and P actually held constant.

15-7 STANDARD MOLAR ENTHALPIES OF FORMATION, ΔH^0_f

It is not possible to determine the total enthalpy content of a substance on an absolute scale. We need to describe only *changes* in this state function, however, so we can define an *arbitrary scale* as follows.

The **standard molar enthalpy of formation, ΔH^0_f,** of a substance is the enthalpy change for the reaction in which *one mole* of the substance in a specified state is formed from its elements in their standard states. By convention, the ΔH^0_f value for any *element in its standard state* is zero.

If the substance exists in several different forms, the form that is most stable at 25°C and 1 atm is the standard state.

For gas laws (Chapter 12) standard temperature is taken as 0°C. For thermodynamics it is taken as 25°C.

This is sometimes referred to as the *standard heat of reaction.*

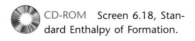 CD-ROM Screen 6.18, Standard Enthalpy of Formation.

We can think of ΔH^0_f as the enthalpy content of each substance, in its standard state, relative to the enthalpy content of the elements, in their standard states. This is why ΔH^0_f for an element in its standard state is zero.

TABLE 15-1 *Selected Standard Molar Enthalpies of Formation at 298 K*

Substance	ΔH_f^0 (kJ/mol)	Substance	ΔH_f^0 (kJ/mol)
$Br_2(\ell)$	0	HgS(s) red	−58.2
$Br_2(g)$	30.91	$H_2(g)$	0
C(diamond)	1.897	HBr(g)	−36.4
C(graphite)	0	$H_2O(\ell)$	−285.8
$CH_4(g)$	−74.81	$H_2O(g)$	−241.8
$C_2H_4(g)$	52.26	NO(g)	90.25
$C_6H_6(\ell)$	49.03	Na(s)	0
$C_2H_5OH(\ell)$	−277.7	NaCl(s)	−411.0
CO(g)	−110.5	$O_2(g)$	0
$CO_2(g)$	−393.5	$SO_2(g)$	−296.8
CaO(s)	−635.5	$SiH_4(g)$	34.0
$CaCO_3(s)$	−1207.0	$SiCl_4(g)$	−657.0
$Cl_2(g)$	0	$SiO_2(s)$	−910.9

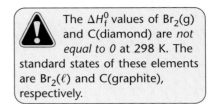

The ΔH_f^0 values of $Br_2(g)$ and C(diamond) are *not equal to 0* at 298 K. The standard states of these elements are $Br_2(\ell)$ and C(graphite), respectively.

Standard molar enthalpy of formation is often called **standard molar heat of formation** or, more simply, **heat of formation.** The superscript zero in ΔH_f^0 signifies standard pressure, 1 atmosphere. Negative values for ΔH_f^0 describe exothermic formation reactions, whereas positive values for ΔH_f^0 describe endothermic formation reactions.

The enthalpy change for a balanced equation that gives a compound from its elements does not necessarily give a molar enthalpy of formation for the compound. Consider the following exothermic reaction at standard conditions.

$$H_2(g) + Br_2(\ell) \longrightarrow 2HBr(g) \qquad \Delta H_{rxn}^0 = -72.8 \text{ kJ/mol rxn}$$

We see that *two* moles of HBr(g) are formed in the reaction as written. Half as much energy, 36.4 kJ, is liberated when *one mole* of HBr(g) is produced from its constituent elements in their standard states. For HBr(g), $\Delta H_f^0 = -36.4$ kJ/mol. This can be shown by dividing all coefficients in the balanced equation by 2.

$$\tfrac{1}{2}H_2(g) + \tfrac{1}{2}Br_2(\ell) \longrightarrow HBr(g) \qquad \Delta H_{rxn}^0 = -36.4 \text{ kJ/mol rxn}$$

$$\Delta H_{f\ HBr(g)}^0 = -36.4 \text{ kJ/mol HBr(g)}$$

Standard heats of formation of some common substances are tabulated in Table 15-1. Appendix K contains a larger listing.

When referring to a thermodynamic quantity for a *substance,* we often omit the description of the substance from the units. Units for tabulated ΔH_f^0 values are given as "kJ/mol"; we must interpret this as "per mole of the substance in the specified state." For instance, for HBr(g) the tabulated ΔH_f^0 value of −36.4 kJ/mol should be interpreted as $\dfrac{-36.4 \text{ kJ}}{\text{mol HBr(g)}}$.

The coefficients $\tfrac{1}{2}$ preceding $H_2(g)$ and $Br_2(\ell)$ do *not* imply half a molecule of each. In thermochemical equations, the coefficients always refer to the number of *moles* under consideration.

EXAMPLE 15-6 *Interpretation of ΔH_f^0*

The standard molar enthalpy of formation of ethanol, $C_2H_5OH(\ell)$, is −277.7 kJ/mol. Write the thermochemical equation for the reaction for which $\Delta H_{rxn}^0 = -277.7$ kJ/mol rxn.

Plan

The definition of ΔH_f^0 of a substance refers to a reaction in which *one mole* of the substance is formed. We put one mole of $C_2H_5OH(\ell)$ on the right side of the chemical equation and put the appropriate elements in their standard states on the left. We balance the equation *without changing the coefficient of the product*, even if we must use fractional coefficients on the left.

Solution

$$2C(\text{graphite}) + 3H_2(g) + \tfrac{1}{2}O_2(g) \longrightarrow C_2H_5OH(\ell) \qquad \Delta H = -277.7 \text{ kJ/mol rxn}$$

You should now work Exercise 26.

✓ **Problem-Solving Tip:** *How Do We Interpret Fractional Coefficients?*

Remember that we *always* interpret the coefficients in thermochemical equations as numbers of *moles* of reactants or products. The $\tfrac{1}{2}O_2(g)$ in the answer to Example 15-6 refers to $\tfrac{1}{2}$ *mole* of O_2 molecules, or

$$\tfrac{1}{2} \text{ mol } O_2 \times \frac{32.0 \text{ g } O_2}{\text{mol } O_2} = 16.0 \text{ g } O_2$$

It is important to realize that this is *not* the same as one mole of O atoms (though that would also weigh 16.0 g).

Similarly, the fractional coefficients in

$$\tfrac{1}{2}H_2(g) + \tfrac{1}{2}Br_2(\ell) \longrightarrow HBr(g)$$

refer to

$$\tfrac{1}{2} \text{ mol } H_2 \times \frac{2.0 \text{ g } H_2}{\text{mol } H_2} = 1.0 \text{ g } H_2$$

and

$$\tfrac{1}{2} \text{ mol } Br_2 \times \frac{159.8 \text{ g } Br_2}{\text{mol } Br_2} = 79.9 \text{ g } Br_2$$

respectively.

CD-ROM　Screen 6.17, Hess's Law.

15-8 HESS'S LAW

In 1840, G. H. Hess (1802–1850) published his **law of heat summation,** which he derived on the basis of numerous thermochemical observations.

> The enthalpy change for a reaction is the same whether it occurs by one step or by any series of steps.

As an analogy, consider traveling from Kansas City (elevation 884 ft above sea level) to Denver (elevation 5280 ft). The change in elevation is $(5280 - 884)$ ft = 4396 ft, regardless of the route taken.

Enthalpy is a state function. Its *change* is therefore independent of the pathway by which a reaction occurs. We do not need to know whether the reaction *does*, or even *can*, occur by the series of steps used in the calculation. The steps must (if only "on paper") result in the overall reaction. Hess's Law lets us calculate enthalpy changes for reactions for which the changes could be measured only with difficulty, if at all. In general terms, Hess's Law of heat summation may be represented as

$$\Delta H^0_{rxn} = \Delta H^0_a + \Delta H^0_b + \Delta H^0_c + \cdots$$

Here a, b, c, . . . refer to balanced thermochemical equations that can be summed to give the equation for the desired reaction.

Consider the following reaction.

$$C(graphite) + \tfrac{1}{2}O_2(g) \longrightarrow CO(g) \qquad \Delta H^0_{rxn} = \underline{?}$$

The enthalpy change for this reaction cannot be measured directly. Even though $CO(g)$ is the predominant product of the reaction of graphite with a *limited* amount of $O_2(g)$, some $CO_2(g)$ is always produced as well. The following reactions do go to completion with excess $O_2(g)$; therefore, ΔH^0 values have been measured experimentally for them. [Pure $CO(g)$ is readily available.]

$$C(graphite) + O_2(g) \longrightarrow CO_2(g) \qquad \Delta H^0_{rxn} = -393.5 \text{ kJ/mol rxn} \qquad (1)$$
$$CO(g) + \tfrac{1}{2}O_2(g) \longrightarrow CO_2(g) \qquad \Delta H^0_{rxn} = -283.0 \text{ kJ/mol rxn} \qquad (2)$$

We can "work backward" to find out how to combine these two known equations to obtain the desired equation. We want one mole of CO on the right, so we reverse equation (2) [designated below as (−2)]; heat is then absorbed instead of released, so we must change the sign of its ΔH^0 value. Then we add it to equation (1), canceling equal numbers of moles of the same species on each side. This gives the equation for the reaction we want. Adding the corresponding enthalpy changes gives the enthalpy change we seek.

You are familiar with the addition and subtraction of algebraic equations. This method of combining thermochemical equations is analogous.

$$\Delta H^0$$

$$\begin{array}{lll}
C(graphite) + O_2(g) \longrightarrow \cancel{CO_2(g)} & -393.5 \text{ kJ/mol rxn} & (1) \\
\cancel{CO_2(g)} \longrightarrow CO(g) + \tfrac{1}{2}O_2(g) & -(-283.0 \text{ kJ/mol rxn}) & (-2) \\
\hline
C(graphite) + \tfrac{1}{2}O_2(g) \longrightarrow CO(g) & \Delta H^0_{rxn} = -110.5 \text{ kJ/mol rxn} &
\end{array}$$

This equation shows the formation of one mole of $CO(g)$ in its standard state from the elements in their standard states. In this way, we determine that ΔH^0_f for $CO(g)$ is −110.5 kJ/mol.

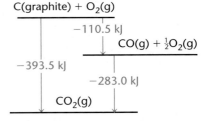

Above is a schematic representation of the enthalpy changes for the reaction $C(graphite) + \tfrac{1}{2}O_2(g) \to CO(g)$. The ΔH value for each step is based on the number of moles of each substance indicated.

EXAMPLE 15-7 *Combining Thermochemical Equations: Hess's Law*

Use the thermochemical equations shown here to determine ΔH^0_{rxn} at 25°C for the following reaction.

$$C(graphite) + 2H_2(g) \longrightarrow CH_4(g)$$

$$\Delta H^0$$

$$\begin{array}{lll}
C(graphite) + O_2(g) \longrightarrow CO_2(g) & -393.5 \text{ kJ/mol rxn} & (1) \\
H_2(g) + \tfrac{1}{2}O_2(g) \longrightarrow H_2O(\ell) & -285.8 \text{ kJ/mol rxn} & (2) \\
CH_4(g) + 2O_2(g) \longrightarrow CO_2(g) + 2H_2O(\ell) & -890.3 \text{ kJ/mol rxn} & (3)
\end{array}$$

These are combustion reactions, for which ΔH^0_{rxn} values can be readily determined from calorimetry experiments.

Plan

(i) We want one mole of C(graphite) as reactant, so we write down equation (1).

(ii) We want two moles of $H_2(g)$ as reactants, so we multiply equation (2) by 2 [designated below as $2 \times (2)$].

(iii) We want one mole of $CH_4(g)$ as product, so we reverse equation (3) to give (-3).

(iv) We do the same operations on each ΔH^0 value.

(v) Then we add these equations term by term. The result is the desired thermochemical equation, with all unwanted substances canceling. The sum of the ΔH^0 values is the ΔH^0 for the desired reaction.

Solution

We have used a series of reactions for which ΔH^0 values can be easily measured to calculate ΔH^0 for a reaction that cannot be carried out.

$$\Delta H^0$$

$$C(graphite) + O_2(g) \longrightarrow CO_2(g) \qquad -393.5 \text{ kJ/mol rxn} \qquad (1)$$
$$2H_2(g) + O_2(g) \longrightarrow 2H_2O(\ell) \qquad 2(-285.8 \text{ kJ/mol rxn}) \qquad 2 \times (2)$$
$$CO_2(g) + 2H_2O(\ell) \longrightarrow CH_4(g) + 2O_2(g) \qquad +890.3 \text{ kJ/mol rxn} \qquad (-3)$$

$$C(graphite) + 2H_2(g) \longrightarrow CH_4(g) \qquad \Delta H^0_{rxn} = -74.8 \text{ kJ/mol rxn}$$

$CH_4(g)$ cannot be formed directly from C(graphite) and $H_2(g)$, so its ΔH^0_f value cannot be measured directly. The result of this example tells us that this value is -74.8 kJ/mol.

EXAMPLE 15-8   Combining Thermochemical Equations: Hess's Law

Given the following thermochemical equations, calculate the heat of reaction at 298 K for the reaction of ethylene with water to form ethanol.

$$C_2H_4(g) + H_2O(\ell) \longrightarrow C_2H_5OH(\ell)$$

$$\Delta H^0$$

$$C_2H_5OH(\ell) + 3O_2(g) \longrightarrow 2CO_2(g) + 3H_2O(\ell) \qquad -1367 \text{ kJ/mol rxn} \qquad (1)$$
$$C_2H_4(g) + 3O_2(g) \longrightarrow 2CO_2(g) + 2H_2O(\ell) \qquad -1411 \text{ kJ/mol rxn} \qquad (2)$$

Plan

We reverse equation (1) to give (-1); when the equation is reversed, the sign of ΔH^0 is changed because the reverse of an exothermic reaction is endothermic. Then we add it to equation (2).

Solution

If you reverse the chemical equation, don't forget to switch the sign of ΔH^0_{rxn}.

$$\Delta H^0$$

$$2CO_2(g) + 3H_2O(\ell) \longrightarrow C_2H_5OH(\ell) + 3O_2(g) \qquad +1367 \text{ kJ/mol rxn} \qquad (-1)$$
$$C_2H_4(g) + 3O_2(g) \longrightarrow 2CO_2(g) + 2H_2O(\ell) \qquad -1411 \text{ kJ/mol rxn} \qquad (2)$$

$$C_2H_4(g) + H_2O(\ell) \longrightarrow C_2H_5OH(\ell) \qquad \Delta H^0_{rxn} = -44 \text{ kJ/mol rxn}$$

You should now work Exercises 30 and 32.

✔ Problem-Solving Tip: ΔH_f^0 *Refers to a Specific Reaction*

The ΔH^0 for the reaction in Example 15-8 is -44 kJ for each mole of $C_2H_5OH(\ell)$ formed. This reaction, however, does not involve formation of $C_2H_5OH(\ell)$ from its constituent elements; therefore, ΔH_{rxn}^0 is *not* ΔH_f^0 for $C_2H_5OH(\ell)$. We have seen the reaction for ΔH_f^0 of $C_2H_5OH(\ell)$ in Example 15-6.

Similarly, the ΔH_{rxn}^0 for

$$CO(g) + \tfrac{1}{2}O_2(g) \longrightarrow CO_2(g)$$

is *not* ΔH_f^0 for $CO_2(g)$.

Another interpretation of Hess's Law lets us use tables of ΔH_f^0 values to calculate the enthalpy change for a reaction. Let us consider again the reaction of Example 15-8.

$$C_2H_4(g) + H_2O(\ell) \longrightarrow C_2H_5OH(\ell)$$

A table of ΔH_f^0 values (Appendix K) gives $\Delta H_{f\ C_2H_5OH(\ell)}^0 = -277.7$ kJ/mol, $\Delta H_{f\ C_2H_4(g)}^0 = 52.3$ kJ/mol, and $\Delta H_{f\ H_2O(\ell)}^0 = -285.8$ kJ/mol. We may express this information in the form of the following thermochemical equations.

$$\Delta H^0$$

$2C(\text{graphite}) + 3H_2(g) + \tfrac{1}{2}O_2(g) \longrightarrow C_2H_5OH(\ell)$	-277.7 kJ/mol rxn	(1)
$2C(\text{graphite}) + 2H_2(g) \longrightarrow C_2H_4(g)$	52.3 kJ/mol rxn	(2)
$H_2(g) + \tfrac{1}{2}O_2(g) \longrightarrow H_2O(\ell)$	-285.8 kJ/mol rxn	(3)

We may generate the equation for the desired net reaction by adding equation (1) to the reverse of equations (2) and (3). The value of ΔH^0 for the desired reaction is then the sum of the corresponding ΔH^0 values.

$$\Delta H^0$$

$2C(\text{graphite}) + 3H_2(g) + \tfrac{1}{2}O_2(g) \longrightarrow C_2H_5OH(\ell)$	-277.7 kJ/mol rxn	(1)
$C_2H_4(g) \longrightarrow 2C(\text{graphite}) + 2H_2(g)$	-52.3 kJ/mol rxn	(-2)
$H_2O(\ell) \longrightarrow H_2(g) + \tfrac{1}{2}O_2(g)$	$+285.8$ kJ/mol rxn	(-3)

net rxn: $C_2H_4(g) + H_2O(\ell) \longrightarrow C_2H_5OH(\ell)$ $\Delta H_{rxn}^0 = -44.2$ kJ/mol rxn

We see that ΔH^0 for this reaction is given by

$$\Delta H_{rxn}^0 = \Delta H_{(1)}^0 + \Delta H_{(-2)}^0 + \Delta H_{(-3)}^0$$

or by

$$\Delta H_{rxn}^0 = \underset{\underset{\text{product}}{\uparrow}}{\Delta H_{f\ C_2H_5OH(\ell)}^0} - [\underset{\underset{\text{reactants}}{\uparrow}}{\Delta H_{f\ C_2H_4(g)}^0} + \Delta H_{f\ H_2O(\ell)}^0]$$

In general terms this is a very useful form of Hess's Law.

$$\Delta H_{rxn}^0 = \Sigma\, n\, \Delta H_{f\ \text{products}}^0 - \Sigma\, n\, \Delta H_{f\ \text{reactants}}^0$$

The standard enthalpy change of a reaction is equal to the sum of the standard molar enthalpies of formation of the products, each multiplied by its coefficient, n, in the *balanced equation*, minus the corresponding sum of the standard molar enthalpies of formation of the reactants.

The capital Greek letter sigma (Σ) is read "the sum of." The $\Sigma\, n$ means that the ΔH_f^0 value of each product and reactant must be multiplied by its coefficient, n, in the balanced equation. The resulting values are then added.

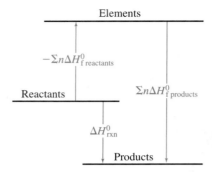

Figure 15-4   A schematic representation of Hess's Law. The red arrow represents the *direct* path from reactants to products. The series of blue arrows is a path (hypothetical) in which reactants are converted to elements, and they in turn are converted to products—all in their standard states.

$O_2(g)$ is an element in its standard state, so its ΔH_f^0 is zero.

In effect this form of Hess's Law supposes that the reaction occurs by converting reactants to the elements in their standard states, then converting these to products (Figure 15-4). Few, if any, reactions actually occur by such a pathway. Nevertheless, the ΔH^0 for this *hypothetical* pathway for *reactants → products* would be the same as that for any other pathway—including the one by which the reaction actually occurs.

EXAMPLE 15-9   *Using ΔH_f^0 Values: Hess's Law*

Calculate ΔH_{rxn}^0 for the following reaction at 298 K.

$$SiH_4(g) + 2O_2(g) \longrightarrow SiO_2(s) + 2H_2O(\ell)$$

Plan

We apply Hess's Law in the form $\Delta H_{rxn}^0 = \Sigma\, n\, \Delta H_{f\,products}^0 - \Sigma\, n\, \Delta H_{f\,reactants}^0$, so we use the ΔH_f^0 values tabulated in Appendix K.

Solution

We can first list the ΔH_f^0 values we obtain from Appendix K:

	$SiH_4(g)$	$O_2(g)$	$SiO_2(s)$	$H_2O(\ell)$
ΔH_f^0, kJ/mol:	34.3	0	−910.9	−285.8

$$\Delta H_{rxn}^0 = \Sigma\, n\, \Delta H_{f\,products}^0 - \Sigma\, n\, \Delta H_{f\,reactants}^0$$

$$\Delta H_{rxn}^0 = [\Delta H_{f\,SiO_2(s)}^0 + 2\,\Delta H_{f\,H_2O(\ell)}^0] - [\Delta H_{f\,SiH_4(g)}^0 + 2\,\Delta H_{f\,O_2(g)}^0]$$

$$\Delta H_{rxn}^0 = \left[\frac{1 \text{ mol } SiO_2(s)}{\text{mol rxn}} \times \frac{-910.9 \text{ kJ}}{\text{mol } SiO_2(s)} + \frac{2 \text{ mol } H_2O(\ell)}{\text{mol rxn}} \times \frac{-285.8 \text{ kJ}}{\text{mol } H_2O(\ell)}\right]$$

$$- \left[\frac{1 \text{ mol } SiH_4(g)}{\text{mol rxn}} \times \frac{+34.3 \text{ kJ}}{\text{mol } SiH_4(g)} + \frac{2 \text{ mol } O_2(g)}{\text{mol rxn}} \times \frac{0 \text{ kJ}}{\text{mol } O_2(g)}\right]$$

$$\Delta H_{rxn}^0 = \boxed{-1515.7 \text{ kJ/mol rxn}}$$

You should now work Exercise 36.

Each term in the sums on the right-hand side of the solution in Example 15-9 has the units

$$\frac{\text{mol substance}}{\text{mol rxn}} \times \frac{\text{kJ}}{\text{mol substance}} \quad \text{or} \quad \frac{\text{kJ}}{\text{mol rxn}}$$

For brevity, we shall omit units in the intermediate steps of calculations of this type, and just assign the proper units to the answer. Be sure that you understand how these units arise.

Suppose we measure ΔH_{rxn}^0 at 298 K and know all but one of the ΔH_f^0 values for reactants and products. We can then calculate the unknown ΔH_f^0 value.

EXAMPLE 15-10   Using ΔH_f^0 Values: Hess's Law

Use the following information to determine ΔH_f^0 for PbO(s, yellow).

$$PbO(s, yellow) + CO(g) \longrightarrow Pb(s) + CO_2(g) \qquad \Delta H_{rxn}^0 = -65.69 \text{ kJ}$$

$$\Delta H_f^0 \text{ for } CO_2(g) = -393.5 \text{ kJ/mol} \quad \text{and} \quad \Delta H_f^0 \text{ for } CO(g) = -110.5 \text{ kJ/mol}$$

We will consult Appendix K, only after working the problem, to check the answer.

Plan

We again use Hess's Law in the form $\Delta H_{rxn}^0 = \Sigma n \Delta H_{f \text{ products}}^0 - \Sigma n \Delta H_{f \text{ reactants}}^0$. The standard state of lead is Pb(s), so $\Delta H_{f \text{ Pb(s)}}^0 = 0$ kJ/mol. Now we are given ΔH_{rxn}^0 and the ΔH_f^0 values for all substances *except* PbO(s, yellow). We can solve for this unknown.

Solution

We list the known ΔH_f^0 values:

	PbO(s, yellow)	CO(g)	Pb(s)	$CO_2(g)$,
ΔH_f^0, kJ/mol:	$\Delta H_{f \text{ PbO}_2(s, \text{ yellow})}^0$	-110.5	0	-393.5

$$\Delta H_{rxn}^0 = \Sigma n \Delta H_{f \text{ products}}^0 \qquad - \Sigma n \Delta H_{f \text{ reactants}}^0$$

$$\Delta H_{rxn}^0 = \Delta H_{f \text{ Pb(s)}}^0 + \Delta H_{f \text{ CO}_2(g)}^0 - [\Delta H_{f \text{ PbO(s, yellow)}}^0 + \Delta H_{f \text{ CO(g)}}^0]$$

$$-65.69 = 0 \qquad + (-393.5) - [\Delta H_{f \text{ PbO(s, yellow)}}^0 + (-110.5)]$$

Rearranging to solve for $\Delta H_{f \text{ PbO(s, yellow)}}^0$, we have

$$\Delta H_{f \text{ PbO(s, yellow)}}^0 = 65.69 - 393.5 + 110.5 = \boxed{-217.3 \text{ kJ/mol of PbO}}$$

You should now work Exercise 42.

✓ Problem-Solving Tip: *Remember the Values of ΔH_f^0 for Elements*

In Example 15-10, we were not given the value of ΔH_f^0 for Pb(s). We should know without reference to tables that ΔH_f^0 for an *element* in its most stable form is exactly 0 kJ/mol. But the element *must* be in its most stable form. Thus, ΔH_f^0 for $O_2(g)$ is zero, because ordinary oxygen is gaseous and diatomic. We would *not* assume that ΔH_f^0 would be zero for oxygen atoms, O(g), or for ozone, $O_3(g)$. Similarly, ΔH_f^0 is zero for $Cl_2(g)$ and for $Br_2(\ell)$, but not for $Br_2(g)$. Recall that bromine is one of the few elements that is liquid at room temperature and 1 atm pressure.

15-9   BOND ENERGIES

Chemical reactions involve the breaking and making of chemical bonds. Energy is always required to break a chemical bond. Often this energy is supplied in the form of heat.

The **bond energy (B.E.)** is the amount of energy necessary to break *one mole* of bonds in a gaseous covalent substance to form products in the gaseous state at constant temperature and pressure.

For all practical purposes, the bond energy is the same as bond enthalpy. Tabulated values of average bond energies are actually average bond enthalpies. We use the term "bond energy" rather than "bond enthalpy" because it is common practice to do so.

The greater the bond energy, the more stable (stronger) the bond is, and the harder it is to break. Thus bond energy is a measure of bond strengths.

TABLE 15-2			*Some Average Single Bond Energies (kJ/mol of bonds)*								
H	**C**	**N**	**O**	**F**	**Si**	**P**	**S**	**Cl**	**Br**	**I**	
436	413	391	463	565	318	322	347	432	366	299	H
	346	305	358	485			272	339	285	213	C
		163	201	283				192			N
			146		452	335		218	201	201	O
				155	565	490	284	253	249	278	F
					222		293	381	310	234	Si
						201		326		184	P
							226	255			S
								242	216	208	Cl
									193	175	Br
										151	I

We have discussed these changes in terms of absorption or release of heat. Another way of breaking bonds is by absorption of light energy (Chapter 5). Bond energies can be determined from the energies of the photons that cause bond dissociation.

Consider the following reaction.

$$H_2(g) \longrightarrow 2H(g) \qquad \Delta H^0_{rxn} = \Delta H_{H-H} = +436 \text{ kJ/mol H—H bonds}$$

The bond energy of the hydrogen–hydrogen bond is 436 kJ/mol of bonds. In other words, 436 kJ of energy must be absorbed for every mole of H—H bonds that are broken. This endothermic reaction (ΔH^0_{rxn} is positive) can be written

$$H_2(g) + 436 \text{ kJ} \longrightarrow 2H(g)$$

Some average bond energies are listed in Tables 15-2 and 15-3. We see from Table 15-3 that for any combination of elements, a triple bond is stronger than a double bond, which in turn is stronger than a single bond. Bond energies for double and triple bonds are *not* simply two or three times those for the corresponding single bonds. A single bond is a σ bond, whereas double and triple bonds involve a combination of σ and π bonding. The bond energy measures the difficulty of overcoming the orbital overlap, and we should not expect the strength of a π bond to be the same as that of a σ bond between the same two atoms.

We should keep in mind that each of the values listed is the average bond energy from a variety of compounds. The *average C—H bond energy* is 413 kJ/mol of bonds. Average

TABLE 15-3	*Comparison of Some Average Single and Multiple Bond Energies (kJ/mol of bonds)*				
Single Bonds		**Double Bonds**		**Triple Bonds**	
C—C	346	C=C	602	C≡C	835
N—N	163	N=N	418	N≡N	945
O—O	146	O=O	498		
C—N	305	C=N	615	C≡N	887
C—O	358	C=O	732*	C≡O	1072

Except in CO_2, where it is 799 kJ/mol.

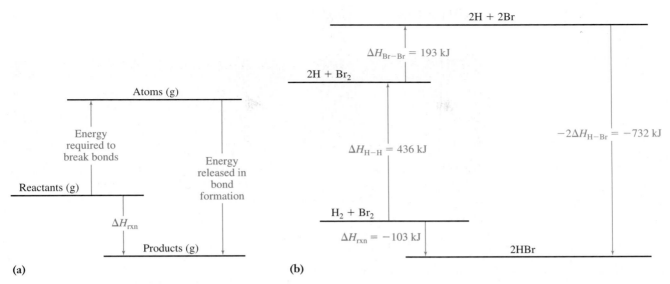

Figure 15-5 A schematic representation of the relationship between bond energies and ΔH_{rxn} for gas phase reactions. (a) For a general reaction (exothermic). (b) For the gas phase reaction

$$H_2(g) + Br_2(g) \longrightarrow 2HBr(g)$$

As usual for such diagrams, the value shown for each change refers to the number of moles of substances or bonds indicated in the diagram.

C—H bond energies differ slightly from compound to compound, as in CH_4, CH_3Cl, CH_3NO_2, and so on. Nevertheless, they are sufficiently constant to be useful in estimating thermodynamic data that are not readily available by another approach. Values of ΔH_{rxn}^0 estimated in this way are not as reliable as those obtained from ΔH_f^0 values for the substances involved in the reaction.

A special case of Hess's Law involves the use of bond energies to *estimate* heats of reaction. Consider the enthalpy diagrams in Figure 15-5. In general terms, ΔH_{rxn}^0 is related to the bond energies of the reactants and products in *gas phase reactions* by the following version of Hess's Law.

$$\Delta H_{rxn}^0 = \Sigma \text{ B.E.}_{\text{reactants}} - \Sigma \text{ B.E.}_{\text{products}} \qquad \text{in gas phase reactions only}$$

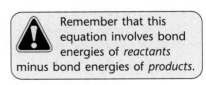

Remember that this equation involves bond energies of *reactants* minus bond energies of *products*.

The net enthalpy change of a reaction is the amount of energy required to break all the bonds in reactant molecules *minus* the amount of energy required to break all the bonds in product molecules. Stated in another way, the amount of energy released when a bond is formed is equal to the amount absorbed when the same bond is broken. The heat of reaction for a gas phase reaction can be described as the amount of energy released in forming all the bonds in the products minus the amount of energy released in forming all the bonds in the reactants (see Figure 15-5). This heat of reaction can be estimated using the average bond energies in Tables 15-2 and 15-3.

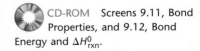

CD-ROM Screens 9.11, Bond Properties, and 9.12, Bond Energy and ΔH_{rxn}^0.

The definition of bond energies is limited to the bond-breaking process *only*, and does not include any provision for changes of state. Thus, it is valid only for substances in the gaseous state. The calculations of this section therefore apply *only* when all substances in the reaction are gases. If liquids or solids were involved, then additional information such as heats of vaporization and fusion would be needed to account for phase changes.

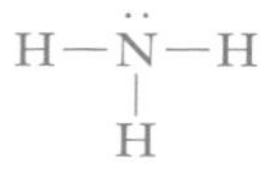

H—N̈—H
|
H

For each term in the sum, the units are

$$\frac{\text{mol bonds}}{\text{mol rxn}} \times \frac{\text{kJ}}{\text{mol bonds}}$$

EXAMPLE 15-11   Bond Energies

Use the bond energies listed in Table 15-2 to estimate the heat of reaction at 298 K for the following reaction.

$$N_2(g) + 3H_2(g) \longrightarrow 2NH_3(g)$$

Plan

Each NH_3 molecule contains three N—H bonds, so two moles of NH_3 contain six moles of N—H bonds. Three moles of H_2 contain a total of three moles of H—H bonds, and one mole of N_2 contains one mole of N≡N bonds. From this we can estimate the heat of reaction.

Solution

Using the bond energy form of Hess's Law,

$$\Delta H^0_{rxn} = [\Delta H_{N\equiv N} + 3\Delta H_{H-H}] - [6\Delta H_{N-H}]$$

$$= 945 + 3(436) - 6(391) = \boxed{-93 \text{ kJ/mol rxn}}$$

You should now work Exercise 46.

EXAMPLE 15-12   Bond Energies

Use the bond energies listed in Table 15-2 to estimate the heat of reaction at 298 K for the following reaction.

$$C_3H_8(g) \qquad + Cl_2(g) \longrightarrow \qquad C_3H_7Cl(g) \qquad + HCl(g)$$

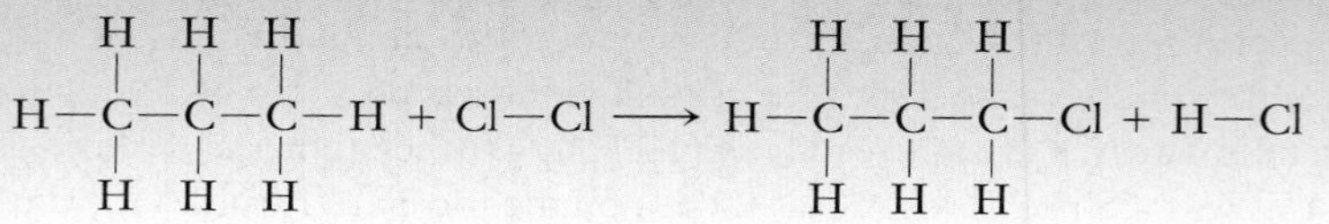

Plan

Two moles of C—C bonds and seven moles of C—H bonds are the same before and after reaction, so we do not need to include them in the bond energy calculation. The only reactant bonds that are broken are one mole of C—H bonds and one mole of Cl—Cl bonds. On the product side, the only new bonds formed are one mole of C—Cl bonds and one mole of H—Cl bonds. We need to take into account only the bonds that are different on the two sides of the equation. As before, we add and subtract the appropriate bond energies, using values from Table 15-2.

We would get the same value for ΔH^0_{rxn} if we used the full bond energy form of Hess's Law and assumed that *all* bonds in reactants were broken and then *all* bonds in products were formed. In such a calculation the bond energies for the unchanged bonds would cancel. Why? Try it!

Solution

$$\Delta H^0_{rxn} = [\Delta H_{C-H} + \Delta H_{Cl-Cl}] - [\Delta H_{C-Cl} + \Delta H_{H-Cl}]$$

$$= [413 + 242] - [339 + 432] = \boxed{-116 \text{ kJ/mol rxn}}$$

You should now work Exercises 48 and 50.

15-10 CHANGES IN INTERNAL ENERGY, ΔE

Internal energy is a state function, so it is represented by a capital letter.

The **internal energy, E,** of a specific amount of a substance represents all the energy contained within the substance. It includes such forms as kinetic energies of the molecules; energies of attraction and repulsion among subatomic particles, atoms, ions, or molecules; and other forms of energy. The internal energy of a collection of molecules is a state function. The difference between the internal energy of the products and the internal energy of the reactants of a chemical reaction or physical change, ΔE, is given by the equation

$$\Delta E = E_{final} - E_{initial} = E_{products} - E_{reactants} = q + w$$

The terms q and w represent heat and work, respectively. These are two ways in which energy can flow into or out of a system. **Work** involves a change of energy in which a body is moved through a distance, d, against some force, f; that is, $w = fd$.

ΔE = (amount of heat absorbed by system) + (amount of work done on system)

The following conventions apply to the signs of q and w.

q is positive:	Heat is *absorbed* by the system from the surroundings.
q is negative:	Heat is *released* by the system to the surroundings.
w is positive:	Work is done *on* the system by the surroundings.
w is negative:	Work is done *by* the system on the surroundings.

Whenever a given amount of energy is added to or removed from a system, either as heat or as work, the energy of the system changes by that same amount. Thus the equation $\Delta E = q + w$ is another way of expressing the First Law of Thermodynamics (see Section 15-1).

The only type of work involved in most chemical and physical changes is pressure–volume work. From dimensional analysis we can see that the product of pressure and volume is work. Pressure is the force exerted per unit area, where area is distance squared, d^2; volume is distance cubed, d^3. Thus, the product of pressure and volume is force times distance, which is work. An example of a physical change (a phase change) in which the system expands and thus does work as it absorbs heat is shown in Figure 15-6. Even if the weight of the book had not been present, the expanding system pushing against the atmosphere would have done work for the expansion.

When energy is released by a reacting system, ΔE is negative; energy can be written as a product in the equation for the reaction. When the system absorbs energy from the surroundings, ΔE is positive; energy can be written as a reactant in the equation.

For example, the complete combustion of CH_4 at constant volume at 25°C *releases* energy.

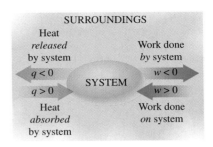

Sign conventions for q and w.

CD-ROM   Screen 6.13, The First Law of Thermodynamics.

$$\frac{F}{d^2} \times d^3 = Fd = w$$

Figure 15-6 A system that absorbs heat and does work. (a) Some powdered dry ice (solid CO_2) is placed into a flexible bag, which is then sealed. (b) As the dry ice absorbs heat from the surroundings, some solid CO_2 sublimes to form gaseous CO_2. The larger volume of the gas causes the bag to expand. The expanding gas does the work of raising a book that has been placed on the bag. Work would be done by the expansion, even if the book were not present, as the bag pushes against the surrounding atmosphere. The heat absorbed by such a process at constant pressure, q_p, is equal to ΔH for the process.

Charles D. Winters

(a)

Charles D. Winters

(b)

At 25°C the change in internal energy for the combustion of methane is -887 kJ/mol CH_4. The change in heat content is -890 kJ/mol CH_4 (see Section 15-1). The small difference is due to work done on the system as it is compressed by the atmosphere.

$$CH_4(g) + 2O_2(g) \longrightarrow CO_2(g) + 2H_2O(\ell) + 887 \text{ kJ}$$

indicates release of energy

We can write the *change in energy* that accompanies this reaction as

$$CH_4(g) + 2O_2(g) \longrightarrow CO_2(g) + 2H_2O(\ell) \qquad \Delta E = -887 \text{ kJ/mol rxn}$$

As discussed in Section 15-2, the negative sign indicates a *decrease* in energy of the system, or a *release* of energy by the system.

The reverse of this reaction *absorbs* energy. It can be written as

$$CO_2(g) + 2H_2O(\ell) + 887 \text{ kJ} \longrightarrow CH_4(g) + 2O_2(g)$$

indicates absorption of energy

or

$$CO_2(g) + 2H_2O(\ell) \longrightarrow CH_4(g) + 2O_2(g) \qquad \Delta E = +887 \text{ kJ/mol rxn}$$

If the latter reaction could be forced to occur, the system would have to absorb 887 kJ of energy per mole of reaction from its surroundings.

When a gas is produced against constant external pressure, such as in an open vessel at atmospheric pressure, the gas does work as it expands against the pressure of the atmosphere. If no heat is absorbed during the expansion, the result is a decrease in the internal energy of the system. On the other hand, when a gas is consumed in a process, the atmosphere does work on the reacting system.

Let us illustrate the latter case. Consider the complete reaction of a 2:1 mole ratio of H_2 and O_2 to produce steam at some constant temperature above 100°C and at one atmosphere pressure (Figure 15-7).

$$2H_2(g) + O_2(g) \longrightarrow 2H_2O(g) + \text{heat}$$

Assume that the constant-temperature bath surrounding the reaction vessel completely absorbs all the evolved heat so that the temperature of the gases does not change. The volume of the system decreases by one third (3 mol gaseous reactants → 2 mol gaseous prod-

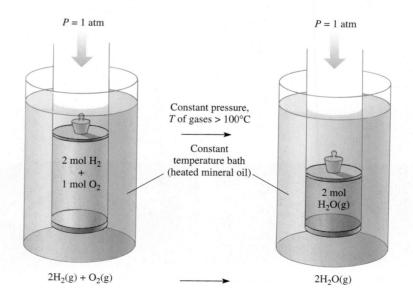

Figure 15-7 An illustration of the one-third decrease in volume that accompanies the reaction of H_2 with O_2 at constant temperature. The temperature is above 100°C.

ucts). The surroundings exert a constant pressure of one atmosphere and do work on the system by compressing it. The internal energy of the system increases by an amount equal to the amount of work done on it.

The work done on or by a system depends on the *external* pressure and the volume. When the external pressure is constant during a change, the amount of work done is equal to this pressure times the change in volume. The work done *on* a system equals $-P\,\Delta V$ or $-P(V_2 - V_1)$.

V_2 is the final volume, and V_1 is the initial volume.

Compression (volume decreases)	Expansion (volume increases)
Work is done *by* the surroundings *on* the system, so the sign of w is positive	Work is done *by* the system *on* the surroundings, so the sign of w is negative
V_2 is less than V_1, so $\Delta V = (V_2 - V_1)$ is negative	V_2 is greater than V_1, so $\Delta V = (V_2 - V_1)$ is positive
$w = -P\,\Delta V$ is positive $$(-) \times (+) \times (-) = +$$	$w = -P\,\Delta V$ is negative $$(-) \times (+) \times (+) = -$$
Can be due to a *decrease* in number of moles of gas (Δn negative)	Can be due to an *increase* in number of moles of gas (Δn positive)

We substitute $-P\,\Delta V$ for w in the equation $\Delta E = q + w$ to obtain

$$\Delta E = q - P\,\Delta V$$

In constant-volume reactions, no $P\,\Delta V$ work is done. Volume does not change, so nothing "moves through a distance," and $d = 0$ and $fd = 0$. The change in internal energy of the system is just the amount of heat absorbed or released at constant volume, q_v.

$$\Delta E = q_v$$

Do not make the error of setting work equal to $V\,\Delta P$.

A subscript v indicates a constant-volume process; a subscript p indicates a constant-pressure process.

Figure 15-8 shows the same phase change process as in Figure 15-6, but at constant volume condition, so no work is done.

Solids and liquids do not expand or contract significantly when the pressure changes ($\Delta V \approx 0$). In reactions in which equal numbers of moles of gases are produced and consumed at constant temperature and pressure, essentially no work is done. By the ideal gas equation, $P\,\Delta V = (\Delta n)RT$ and $\Delta n = 0$, where Δn equals the number of moles of gaseous products minus the number of moles of gaseous reactants. Thus, the work term w has a significant value at constant pressure only when there are different numbers of moles of gaseous products and reactants so that the volume of the system changes.

Charles D. Winters

Figure 15-8 A system that absorbs heat at constant volume. Some dry ice [$CO_2(s)$] is placed into a rigid flask, which is then sealed. As the dry ice absorbs heat from the surroundings, some $CO_2(s)$ sublimes to form $CO_2(g)$. In contrast to the case in Figure 15-6, this system cannot expand ($\Delta V = 0$), so no work is done, and the pressure in the flask increases. Thus, the heat absorbed at constant volume, q_v, is equal to ΔE for the process.

EXAMPLE 15-13 *Predicting the Sign of Work*

For each of the following chemical reactions carried out at constant temperature and constant pressure, predict the sign of w and tell whether work is done *on* or *by* the system. Consider the reaction mixture to be the system.

(a) Ammonium nitrate, commonly used as a fertilizer, decomposes explosively.

$$2NH_4NO_3(s) \longrightarrow 2N_2(g) + 4H_2O(g) + O_2(g)$$

This reaction was responsible for an explosion in 1947 that destroyed nearly the entire port of Texas City, Texas, and killed 576 people.

(b) Hydrogen and chlorine combine to form hydrogen chloride gas.

$$H_2(g) + Cl_2(g) \longrightarrow 2HCl(g)$$

The decomposition of NH_4NO_3 produces large amounts of gas, which expands rapidly as the very fast reaction occurs. This explosive reaction was the main cause of the destruction of the Federal Building in Oklahoma City in 1995.

CORBIS-Bettmann

(c) Sulfur dioxide is oxidized to sulfur trioxide, one step in the production of sulfuric acid.

$$2SO_2(g) + O_2(g) \longrightarrow 2SO_3(g)$$

Plan

Δn refers to the balanced equation.

For a process at constant pressure, $w = -P\,\Delta V = -(\Delta n)RT$. For each reaction, we evaluate Δn, the change in the number of moles of *gaseous* substances in the reaction.

$$\Delta n = \text{(no. of moles of gaseous products)} - \text{(no. of moles of gaseous reactants)}$$

Because both R and T (on the Kelvin scale) are positive quantities, the sign of w is opposite from that of Δn; it tells us whether the work is done *on* ($w = +$) or *by* ($w = -$) the system.

Solution

Here there are no gaseous reactants.

(a) $\Delta n = [2 \text{ mol } N_2(g) + 4 \text{ mol } H_2O(g) + 1 \text{ mol } O_2(g)] - 0 \text{ mol}$
    $= 7 \text{ mol} - 0 \text{ mol} = +7 \text{ mol}$

Δn is positive, so w is negative. This tells us that work is done *by* the system. The large amount of gas formed by the reaction pushes against the surroundings (as happened with devastating effect in the Texas City disaster).

(b) $\Delta n = [2 \text{ mol } HCl(g)] - [1 \text{ mol } H_2(g) + 1 \text{ mol } Cl_2(g)]$
    $= 2 \text{ mol} - 2 \text{ mol} = 0 \text{ mol}$

Thus, $w = 0$, and no work is done as the reaction proceeds. We can see from the balanced equation that for every two moles (total) of gas that react, two moles of gas are formed, so the volume neither expands nor contracts as the reaction occurs.

(c) $\Delta n = [2 \text{ mol } SO_3(g)] - [2 \text{ mol } SO_2(g) + 1 \text{ mol } O_2(g)]$
    $= 2 \text{ mol} - 3 \text{ mol} = -1 \text{ mol}$

Δn is negative, so w is positive. This tells us that work is done *on the system* as the reaction proceeds. The surroundings push against the diminishing volume of gas.

You should now work Exercises 75 and 76.

The "calorie content" of a food can be determined by burning it in excess oxygen inside a bomb calorimeter and determining the heat released. 1 "nutritional Calorie" = 1 kcal = 4.184 kJ.

A **bomb calorimeter** is a device that measures the amount of heat evolved or absorbed by a reaction occurring at constant volume (Figure 15-9). A strong steel vessel (the bomb) is immersed in a large volume of water. As heat is produced or absorbed by a reaction inside the steel vessel, the heat is transferred to or from the large volume of water. Thus, only rather small temperature changes occur. For all practical purposes, the energy changes associated with the reactions are measured at constant volume and constant temperature. No work is done when a reaction is carried out in a bomb calorimeter, even if gases are involved, because $\Delta V = 0$. Therefore,

$$\Delta E = q_v \qquad \text{(constant volume)}$$

CD-ROM   You should look again at Screen 6.16, Measuring Heats of Reaction: Calorimetry.

EXAMPLE 15-14   *Bomb Calorimeter*

A 1.000-gram sample of ethanol, C_2H_5OH, was burned in a bomb calorimeter whose heat capacity had been determined to be 2.71 kJ/°C. The temperature of 3000 grams of water rose from 24.284°C to 26.225°C. Determine ΔE for the reaction in joules per gram of ethanol, and then in kilojoules per mole of ethanol. The specific heat of water is 4.184 J/g·°C. The combustion reaction is

$$C_2H_5OH(\ell) + 3O_2(g) \longrightarrow 2CO_2(g) + 3H_2O(\ell)$$

Plan

The amount of heat given off by the system (in the sealed compartment) raises the temperature of the calorimeter and its water. The amount of heat absorbed by the water can be calculated using the specific heat of water; similarly, we use the heat capacity of the calorimeter to find the amount of heat absorbed by the calorimeter. The sum of these two amounts of heat is the total amount of heat released by the combustion of 1.000 gram of ethanol. We must then scale that result to correspond to one mole of ethanol.

Solution

The increase in temperature is

$$\underline{?}\ °C = 26.225°C - 24.284°C = 1.941°C \text{ rise}$$

The amount of heat responsible for this increase in temperature of 3000 grams of water is

$$\text{heat to warm water} = 1.941°C \times \frac{4.184\ J}{g \cdot °C} \times 3000\ g = 2.436 \times 10^4\ J = 24.36\ kJ$$

The amount of heat responsible for the warming of the calorimeter is

$$\text{heat to warm calorimeter} = 1.941°C \times \frac{2.71\ kJ}{°C} = 5.26\ kJ$$

The total amount of heat absorbed by the calorimeter *and* by the water is

$$\text{total amount of heat} = 24.36\ kJ + 5.26\ kJ = 29.62\ kJ$$

Combustion of one gram of C_2H_5OH liberates 29.62 kJ of energy in the form of heat, that is

$$\Delta E = q_v = \boxed{-29.62\ kJ/g \text{ ethanol}}$$

The negative sign indicates that energy is released by the system to the surroundings. Now we may evaluate ΔE in kJ/mol of ethanol by converting grams of C_2H_5OH to moles.

$$\frac{\underline{?}\ kJ}{\text{mol ethanol}} = \frac{-29.62\ kJ}{g} \times \frac{46.07\ g\ C_2H_5OH}{1\ mol\ C_2H_5OH} = -1365\ kJ/mol \text{ ethanol}$$

$$\Delta E = \boxed{-1365\ kJ/mol \text{ ethanol}}$$

This calculation shows that for the combustion of ethanol at constant temperature and constant volume, the change in internal energy is -1365 kJ/mol ethanol.

You should now work Exercises 64 and 65.

Benzoic acid, C_6H_5COOH, is often used to determine the heat capacity of a calorimeter. It is a solid that can be compressed into pellets. Its heat of combustion is accurately known: 3227 kJ/mol benzoic acid, or 26.46 kJ/g benzoic acid. Another way to measure the heat capacity of a calorimeter is to add a known amount of heat electrically.

Charles D. Winters

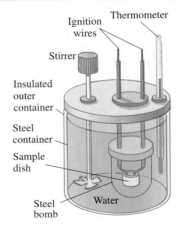

Figure 15-9   A bomb calorimeter measures q_v, the amount of heat given off or absorbed by a reaction occurring at constant *volume*. The amount of energy introduced via the ignition wires is measured and taken into account.

(a)

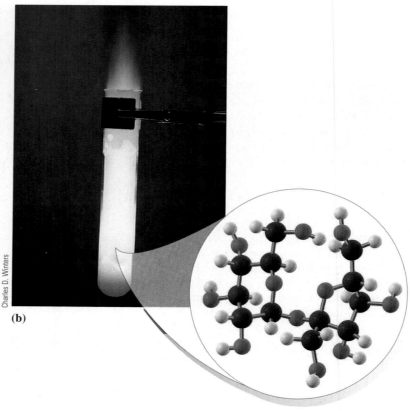

(b)

(a) This small piece of candy is mostly sucrose, $C_{12}H_{22}O_{11}$, a sugar. (b) When the piece of candy is heated together with potassium chlorate, $KClO_3$ (a good oxidizing agent), a highly product-favored reaction occurs.

$$C_{12}H_{22}O_{11}(s) + 12O_2(g) \longrightarrow$$
$$12CO_2(g) + 11H_2O(g)$$

If that amount of sucrose is completely metabolized to carbon dioxide and water vapor in your body, the same amount of energy is released, though more slowly.

The balanced chemical equation involves one mole of ethanol, so we can write the unit factor $\dfrac{1 \text{ mol ethanol}}{1 \text{ mol rxn}}$. Then we express the result of Example 15-14 as

$$\Delta E = \frac{-1365 \text{ kJ}}{\text{mol ethanol}} \times \frac{1 \text{ mol ethanol}}{1 \text{ mol rxn}} = -1365 \text{ kJ/mol rxn}$$

15-11 RELATIONSHIP BETWEEN ΔH AND ΔE

The fundamental definition of enthalpy, H, is

$$H = E + PV$$

For a process at constant temperature and pressure,

$$\Delta H = \Delta E + P\,\Delta V \qquad \text{(constant } T \text{ and } P\text{)}$$

From Section 15-10, we know that $\Delta E = q + w$, so

$$\Delta H = q + w + P\,\Delta V \qquad \text{(constant } T \text{ and } P\text{)}$$

At constant pressure, $w = -P\,\Delta V$, so

$$\Delta H = q + (-P\,\Delta V) + P\,\Delta V$$

$$\Delta H = q_p \qquad \text{(constant } T \text{ and } P\text{)}$$

The difference between ΔE and ΔH is the amount of expansion work ($P\,\Delta V$ work) that the system can do. Unless there is a change in the number of moles of gas present, this difference is extremely small and can usually be neglected. For an ideal gas, $PV = nRT$. At constant temperature and constant pressure, $P\,\Delta V = (\Delta n)RT$, a work term. Substituting gives

$$\Delta H = \Delta E + (\Delta n)RT \qquad \text{or} \qquad \Delta E = \Delta H - (\Delta n)RT \qquad \text{(constant } T \text{ and } P\text{)}$$

> ⚠ As usual, Δn refers to the number of moles of *gaseous products* minus the number of moles of *gaseous reactants* in the *balanced chemical equation*.

> ✓ **Problem-Solving Tip:** *Two Equations Relate ΔH and ΔE—Which One Should Be Used?*
>
> The relationship $\Delta H = \Delta E + P\,\Delta V$ is valid for *any* process that takes place at constant temperature and pressure. It is very useful for physical changes that involve volume changes, such as expansion or compression of a gas. When a chemical reaction occurs and causes a change in the number of moles of gas, it is more convenient to use the relationship in the form $\Delta H = \Delta E + (\Delta n)RT$. You should always remember that Δn refers to the change in number of moles of *gas* in the balanced chemical equation.

In Example 15-14 we found that the change in internal energy, ΔE, for the combustion of ethanol is -1365 kJ/mol ethanol at 298 K. Combustion of one mole of ethanol at 298 K and constant pressure releases 1367 kJ of heat. Therefore (see Section 15-5)

$$\Delta H = -1367 \; \frac{\text{kJ}}{\text{mol ethanol}}$$

The difference between ΔH and ΔE is due to the work term, $-P\,\Delta V$ or $-(\Delta n)RT$. In this balanced equation there are fewer moles of gaseous products than of gaseous reactants: $\Delta n = 2 - 3 = -1$.

$$C_2H_5OH(\ell) + 3O_2(g) \longrightarrow 2CO_2(g) + 3H_2O(\ell)$$

Thus, the atmosphere does work on the system (compresses it). Let us find the work done on the system per mole of reaction.

$$w = -P\,\Delta V = -(\Delta n)RT$$

$$= -(-1 \text{ mol})\left(\frac{8.314 \text{ J}}{\text{mol}\cdot\text{K}}\right)(298 \text{ K}) = +2.48 \times 10^3 \text{ J}$$

$$w = +2.48 \text{ kJ} \qquad \text{or} \qquad (\Delta n)RT = -2.48 \text{ kJ}$$

We can now calculate ΔE for the reaction from ΔH and $(\Delta n)RT$ values.

$$\Delta E = \Delta H - (\Delta n)RT = [-1367 - (-2.48)] = -1365 \text{ kJ/mol rxn}$$

This value agrees with the result that we obtained in Example 15-14. The size of the work term ($+2.48$ kJ) is very small compared with ΔH (-1367 kJ/mol rxn). This is true for many

The positive sign is consistent with the fact that work is done on the system. The balanced equation involves one mole of ethanol, so this is the amount of work done when one mole of ethanol undergoes combustion.

reactions. Of course, if $\Delta n = 0$, then $\Delta H = \Delta E$, and the same amount of heat would be absorbed or given off by the reaction whether it is carried out at constant pressure or at constant volume.

SPONTANEITY OF PHYSICAL AND CHEMICAL CHANGES

CD-ROM   Screens 6.2,
Product-Favored Systems;
6.3, Control of Chemical Reactions; and
19.3, Directionality of Reactions: Matter
and Energy Dispersal.

Another major concern of thermodynamics is predicting *whether* a particular process can occur under specified conditions to give predominantly products. We may summarize this concern in the question "Which would be more stable at the given conditions—the reactants or the products?" A change for which the collection of products is thermodynamically *more stable* than the collection of reactants under the given conditions is said to be **product-favored,** or **spontaneous,** under those conditions. A change for which the products are thermodynamically *less stable* than the reactants under the given conditions is described as **reactant-favored,** or **nonspontaneous,** under those conditions. Some changes are spontaneous under all conditions; others are nonspontaneous under all conditions. The great majority of changes, however, are spontaneous under some conditions but not under others. We use thermodynamics to predict conditions for which the latter type of reactions can occur to give predominantly products.

The concept of spontaneity has a very specific interpretation in thermodynamics. A spontaneous chemical reaction or physical change is one that can happen without any continuing outside influence. Any spontaneous change has a natural direction, like the rusting of a piece of iron, the burning of a piece of paper, or the melting of ice at room temperature. We can think of a spontaneous process as one for which products are favored over reactants at the specified conditions. Although a spontaneous reaction *might* occur rapidly, thermodynamic spontaneity is not related to speed. The fact that a process is spontaneous does not mean that it will occur at an observable rate. It may occur rapidly, at a moderate rate, or very slowly. The rate at which a spontaneous reaction occurs is addressed by kinetics (Chapter 16). We now study the factors that influence spontaneity of a physical or chemical change.

15-12   THE TWO ASPECTS OF SPONTANEITY

Many product-favored reactions are exothermic. For instance, the combustion (burning) reactions of hydrocarbons such as methane and octane are all exothermic and highly product-favored (spontaneous). The enthalpy contents of the products are lower than those of the reactants. Not all exothermic changes are spontaneous, however, nor are all spontaneous changes exothermic. As an example, consider the freezing of water, which is an exothermic process (heat is released). This process is spontaneous at temperatures below 0°C, but it certainly is not spontaneous at temperatures above 0°C. Likewise, we can find conditions at which the melting of ice, an endothermic process, is spontaneous. Spontaneity is *favored* but not required when heat is released during a chemical reaction or a physical change.

Another factor, related to the disorder of reactants and products, also plays a role in determining spontaneity. The dissolution of ammonium nitrate, NH_4NO_3, in water is spontaneous. Yet a beaker in which this process occurs becomes colder (see Figure 15-2). The system (consisting of the water, the solid NH_4NO_3, and the resulting hydrated NH_4^+ and NO_3^- ions) absorbs heat from the surroundings as the endothermic process occurs. Nevertheless, the process is spontaneous because the system becomes more disordered as the regularly arranged ions of crystalline ammonium nitrate become more randomly distributed hydrated ions in solution (Figure 15-10). An increase in disorder in the system favors the spontaneity of a reaction. In this particular case, the increase in disorder overrides the effect of endothermicity.

> Two factors affect the spontaneity of any physical or chemical change:
>
> 1. Spontaneity is *favored* when *heat is released* during the change (exothermic).
> 2. Spontaneity is *favored* when the change causes an *increase in disorder*.

The balance of these two effects is considered in Section 15-15.

Figure 15-10 As particles leave a crystal to go into solution, they become more disordered. This increase in disorder favors the dissolution of the crystal.

15-13 THE SECOND LAW OF THERMODYNAMICS

We now know that two factors determine whether a reaction is spontaneous under a given set of conditions. The effect of one factor, the enthalpy change, is that spontaneity is favored (but not required) by exothermicity, and nonspontaneity is favored (but not required) by endothermicity. The effect of the other factor is summarized by the **Second Law of Thermodynamics.**

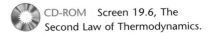

CD-ROM   Screen 19.6, The Second Law of Thermodynamics.

> In spontaneous changes, the universe tends toward a state of greater disorder.

The Second Law of Thermodynamics is based on our experiences. Some examples illustrate this law in the macroscopic world. When a mirror is dropped, it can shatter. When a drop of food coloring is added to a glass of water, it diffuses until a homogeneously colored solution results. When a truck is driven down the street, it consumes fuel and oxygen, producing carbon dioxide, water vapor, and other emitted substances.

The reverse of any spontaneous change is nonspontaneous, because if it did occur, the universe would tend toward a state of greater order. This is contrary to our experience. We would be very surprised if we dropped some pieces of silvered glass on the floor and a mirror spontaneously assembled. A truck cannot be driven along the street, even in reverse gear, so that it sucks up CO_2, water vapor, and other substances and produces fuel and oxygen.

15-14 ENTROPY, *S*

The thermodynamic state function **entropy, *S*,** is a measure of the disorder of the system. The greater the disorder of a system, the higher is its entropy. For any substance, the particles are more highly ordered in the solid state than in the liquid state. These, in turn, are

CD-ROM   Screen 19.4, Entropy: Matter Dispersal or Disorder.

Figure 15-11   As a sample changes from solid to liquid to gas, its particles become increasingly less ordered (more disordered), so its entropy increases.

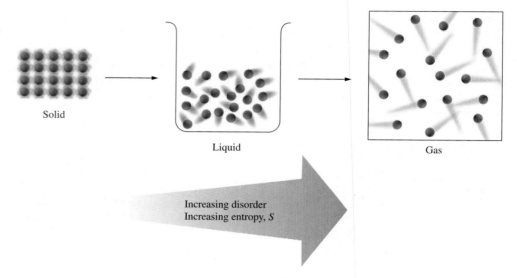

Solid

Liquid

Gas

Increasing disorder
Increasing entropy, S

(a) Stopcock closed

(b) Stopcock open

Figure 15-12   (a) A sample of gas in which all molecules of one gas are in one bulb and all molecules of the other gas are in the other bulb. (b) A sample of gas that contains the same number of each kind of molecule as in (a), but with the two kinds randomly mixed in the two bulbs. Sample (b) has greater disorder (higher entropy), and is thus more probable.

more highly ordered than in the gaseous state. Thus, the entropy of any substance increases as the substance goes from solid to liquid to gas (Figure 15-11).

If the entropy of a system increases during a process, the spontaneity of the process is favored but not required. The Second Law of Thermodynamics says that the entropy of the *universe* (not the system) increases during a spontaneous process, that is,

$$\Delta S_{universe} = \Delta S_{system} + \Delta S_{surroundings} > 0 \qquad \text{(spontaneous process)}$$

Of the two ideal gas samples in Figure 15-12, the more ordered arrangement (Figure 15-12a) has lower entropy than the randomly mixed arrangement with the same volume (Figure 15-12b). Because these ideal gas samples mix without absorbing or releasing heat and without a change in total volume, they do not interact with the surroundings, so the entropy of the surroundings does not change. In this case

$$\Delta S_{universe} = \Delta S_{system}$$

If we open the stopcock between the two bulbs in Figure 15-12a, we expect the gases to mix spontaneously, with an increase in the disorder of the system, that is, ΔS_{system} is positive.

$$\text{unmixed gases} \longrightarrow \text{mixed gases} \qquad \Delta S_{universe} = \Delta S_{system} > 0$$

We do not expect the more homogeneous sample in Figure 15-12b to spontaneously "un-mix" to give the arrangement in Figure 15-12a (which would correspond to a decrease in ΔS_{system}).

$$\text{mixed gases} \longrightarrow \text{unmixed gases} \qquad \Delta S_{universe} = \Delta S_{system} < 0$$

The ideas of entropy, order, and disorder are related to probability. The more ways an event can happen, the more probable that event is. In Figure 15-12b any individual red molecule is equally likely to be in either container, as is any individual blue molecule. As a result, there are many ways in which the mixed arrangement of Figure 15-12b can occur, so the probability of its occurrence is high, and so its entropy is high. In contrast, there is only one way the unmixed arrangement in Figure 15-12a can occur. The resulting probability is extremely low, and the entropy of this arrangement is low.

The entropy of a system can decrease during a spontaneous process or increase during a nonspontaneous process, depending on the accompanying ΔS_{surr}. If ΔS_{sys} is negative (decrease in disorder), then ΔS_{univ} may still be positive (overall increase in disorder) *if* ΔS_{surr} is more positive than ΔS_{sys} is negative. A refrigerator provides an illustration. It removes heat from inside the box (the system) and ejects that heat, *plus* the heat generated by the compressor, into the room (the surroundings). The entropy of the system decreases because the air molecules inside the box move more slowly. The increase in the entropy of the surroundings more than makes up for that, however, so the entropy of the universe (refrigerator + room) increases.

Similarly, if ΔS_{sys} is positive but ΔS_{surr} is even more negative, then ΔS_{univ} is still negative. Such a process will be nonspontaneous.

Let's consider the entropy changes that occur when a liquid solidifies at a temperature *below* its freezing (melting) point (Figure 15-13a). ΔS_{sys} is negative because a solid forms from its liquid, yet we know that this is a spontaneous process. A liquid releases heat to its surroundings (atmosphere) as it crystallizes. The released heat increases the motion (disorder) of the molecules of the surroundings, so ΔS_{surr} is positive. As the temperature decreases, the ΔS_{surr} contribution becomes more important. When the temperature is low enough (below the freezing point), the positive ΔS_{surr} outweighs the negative ΔS_{sys}. Then ΔS_{univ} becomes positive, and the freezing process becomes spontaneous.

The situation is reversed when a liquid is boiled or a solid is melted (Figure 15-13b). For example, at temperatures above its melting point, a solid spontaneously melts, and ΔS_{sys} is positive. The heat absorbed when the solid (system) melts comes from its surroundings. This decreases the motion of the molecules of the surroundings. Thus, ΔS_{surr} is negative (the surroundings become less disordered). The positive ΔS_{sys} is greater in magnitude than the negative ΔS_{surr}, however, so ΔS_{univ} is positive and the process is spontaneous.

Above the melting point, ΔS_{univ} is positive for melting. Below the melting point, ΔS_{univ} is positive for freezing. At the melting point, ΔS_{surr} is equal in magnitude and opposite in sign to ΔS_{sys}. Then ΔS_{univ} is zero for both melting and freezing; the system is at *equilibrium*. Table 15-4 lists the entropy effects for these changes of physical state.

We have said that ΔS_{univ} is positive for all spontaneous (product-favored) processes. Unfortunately, it is not possible to make direct measurements of ΔS_{univ}. Consequently, entropy changes accompanying physical and chemical changes are reported in terms of ΔS_{sys}. The subscript "sys" for system is usually omitted. The symbol ΔS refers to the change in entropy of the reacting system, just as ΔH refers to the change in enthalpy of the reacting system.

We abbreviate these subscripts as follows: system = sys, surroundings = surr, and universe = univ.

Can you develop a comparable table for boiling (liquid → gas) and condensation (gas → liquid)? (Study Table 15-4 carefully.)

(a) Freezing below mp

(b) Melting above mp

Figure 15-13 A schematic representation of heat flow and entropy changes for (a) freezing and (b) melting of a pure substance.

| | TABLE 15-4 | *Entropy Effects Associated with Melting and Freezing* | | | | | |

Change	Temperature	Sign of		(Magnitude of ΔS_{sys}) Compared with (Magnitude of ΔS_{surr})	$\Delta S_{univ} =$ $\Delta S_{sys} + \Delta S_{surr}$	Spontaneity
		ΔS_{sys}	ΔS_{surr}			
1. Melting	(a) > mp	+	−	>	> 0	Spontaneous
(solid → liquid)	(b) = mp	+	−	=	= 0	Equilibrium
	(c) < mp	+	−	<	< 0	Nonspontaneous
2. Freezing	(a) > mp	−	+	>	< 0	Nonspontaneous
(liquid → solid)	(b) = mp	−	+	=	= 0	Equilibrium
	(c) < mp	−	+	<	> 0	Spontaneous

The **Third Law of Thermodynamics** establishes the zero of the entropy scale.

> The entropy of a pure, perfect crystalline substance (perfectly ordered) is zero at absolute zero (0 K).

Enthalpies are measured only as *differences* with respect to an arbitrary standard state. Entropies, in contrast, are defined relative to an absolute zero level. In either case, the *per mole* designation means *per mole of substance in the specified state.*

As the temperature of a substance increases, the particles vibrate more vigorously, so the entropy increases (Figure 15-14). Further heat input causes either increased temperature (still higher entropy) or phase transitions (melting, sublimation, or boiling) that also result in higher entropy. The entropy of a substance at any condition is its **absolute entropy,** also called **standard molar entropy.** Consider the absolute entropies at 298 K listed in Table 15-5. At 298 K, *any* substance is more disordered than if it were in a perfect crystalline state at absolute zero, so tabulated S^0_{298} values for compounds and elements are *always positive.* Notice especially that S^0_{298} of an element, unlike its ΔH^0_f, is *not* equal to zero.

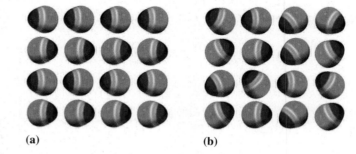

| | TABLE 15-5 | *Absolute Entropies at 298 K for a Few Common Substances* |

Substance	S^0 (J/mol · K)
C(diamond)	2.38
C(g)	158.0
$H_2O(\ell)$	69.91
$H_2O(g)$	188.7
$I_2(s)$	116.1
$I_2(g)$	260.6

Figure 15-14 (a) A simplified representation of a side view of a "perfect" crystal of a polar substance of 0 K. Note the perfect alignment of the dipoles in all molecules in a perfect crystal. This causes its entropy to be zero at 0 K. There are no perfect crystals, however, because even the purest substances that scientists have prepared are contaminated by traces of impurities that occupy a few of the positions in the crystal structure. Additionally, there are some vacancies in the crystal structures of even very highly purified substances such as those used in semiconductors (see Section 13-17). (b) A simplified representation of the same "perfect" crystal at a temperature above 0 K. Vibrations of the individual molecules within the crystal cause some dipoles to be oriented in directions other than those in a perfect arrangement. The entropy of such a crystalline solid is greater than zero, because there is disorder in the crystal.

The reference state for absolute entropy is specified by the Third Law of Thermodynamics. It is different from the reference state for ΔH_f^0 (see Section 15-7). The absolute entropies, S_{298}^0, of various substances under standard conditions are tabulated in Appendix K.

The **standard entropy change, ΔS^0**, of a reaction can be determined from the absolute entropies of reactants and products. The relationship is analogous to Hess's Law.

$$\Delta S_{rxn}^0 = \Sigma\, n\, S_{products}^0 - \Sigma\, n\, S_{reactants}^0$$

The $\Sigma\, n$ means that each S^0 value must be multiplied by the appropriate coefficient, n, from the balanced equation. These values are then added.

S^0 values are tabulated in units of $J/mol \cdot K$ rather than the larger units involving kilojoules that are used for enthalpy changes. The "mol" term in the units for a *substance* refers to a mole of the substance, whereas for a *reaction* it refers to a mole of reaction. Each term in the sums on the right-hand side of the equation has the units

$$\frac{mol\ substance}{mol\ rxn} \times \frac{J}{(mol\ substance) \cdot K} = \frac{J}{(mol\ rxn) \cdot K}$$

The result is usually abbreviated as $J/mol \cdot K$, or sometimes even as J/K. As before, we will usually omit units in intermediate steps and then apply appropriate units to the result.

CD-ROM   Screen 20.5, Calculating ΔS for a Chemical Reaction.

EXAMPLE 15-15   *Calculation of ΔS_{rxn}^0*

Use the values of standard molar entropies in Appendix K to calculate the entropy change at 25°C and one atmosphere pressure for the reaction of hydrazine with hydrogen peroxide. This explosive reaction has been used for rocket propulsion. Do you think the reaction is spontaneous? The balanced equation for the reaction is

$$N_2H_4(\ell) + 2H_2O_2(\ell) \longrightarrow N_2(g) + 4H_2O(g) \qquad \Delta H_{rxn}^0 = -642.2 \text{ kJ/mol reaction}$$

Plan

We use the equation for standard entropy change to calculate ΔS_{rxn}^0 from the tabulated values of standard molar entropies, S_{298}^0, for the substances in the reaction.

Solution

We can list the S_{298}^0 values that we obtain from Appendix K for each substance:

	$N_2H_4(\ell)$	$H_2O_2(\ell)$	$N_2(g)$	$H_2O(g)$
S^0, J/mol · K:	121.2	109.6	191.5	188.7

$$\Delta S_{rxn}^0 = \Sigma\, n\, S_{products}^0 - \Sigma\, n\, S_{reactants}^0$$
$$= [S_{N_2(g)}^0 + 4S_{H_2O(g)}^0] - [S_{N_2H_4(\ell)}^0 + 2S_{H_2O_2(\ell)}^0]$$
$$= [1\,(191.5) + 4\,(188.7)] - [1\,(121.2) + 2\,(109.6)]$$

$$\Delta S_{rxn}^0 = +605.9 \text{ J/mol} \cdot \text{K}$$

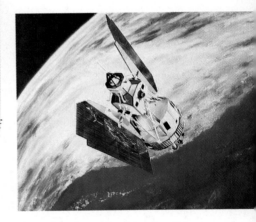

Small booster rockets adjust the course of a satellite in orbit. Some of these small rockets are powered by the N_2H_4–H_2O_2 reaction.

The "mol" designation for ΔS_{rxn}^0 refers to a mole of reaction, that is, one mole of $N_2H_4(\ell)$, two moles of $H_2O_2(\ell)$, and so on. Although it may not appear to be, $+605.9$ J/mol · K is a relatively large value of ΔS_{sys}^0. The positive entropy change favors spontaneity. This reaction is also exothermic (ΔH^0 is negative). As we shall see, this reaction *must* be spontaneous, because both factors are favorable: the reaction is exothermic (ΔH_{rxn}^0 is negative) and the disorder of the system increases (ΔS_{rxn}^0 is positive).

You should now work Exercise 92.

Because changes in the thermodynamic quantity *entropy* may be understood in terms of changes in *molecular disorder*, we can often predict the sign of ΔS_{sys}. The following illustrations emphasize several common types of processes that result in predictable entropy changes for the system.

1. *Phase changes.* When melting occurs, the molecules or ions are taken from their quite ordered crystalline arrangement to a more disordered one in which they are able to move past one another in the liquid. Thus, a melting process is always accompanied by an entropy increase ($\Delta S_{sys} > 0$). Likewise, vaporization and sublimation both take place with large increases in disorder, and hence with increases in entropy. For the reverse processes of freezing, condensation, and deposition, entropy decreases because order increases.

2. *Temperature changes*—for example, warming a gas from 25°C to 50°C. As any sample is warmed, the molecules undergo more (random) motion; hence entropy increases ($\Delta S_{sys} > 0$) as temperature increases. Likewise, as we raise the temperature of a solid, the particles vibrate more vigorously about their positions in the crystal, so that at any instant there is a larger average displacement from their mean positions; this results in an increase in entropy.

3. *Volume changes.* When the volume of a sample of gas increases, the molecules can occupy more positions, and hence are more randomly arranged than when they are closer together in a smaller volume. Hence, an expansion is accompanied by an increase in entropy ($\Delta S_{sys} > 0$). Conversely, as a sample is compressed, the molecules are more restricted in their locations, and a situation of greater order (lower entropy) results.

4. *Mixing of substances,* even without chemical reaction. Situations in which the molecules are more "mixed up" are more disordered, and hence are at higher entropy. We pointed out that the mixed gases of Figure 15-12b were more disordered than the separated gases of Figure 15-12a, and that the former was a situation of higher entropy. We see that mixing of gases by diffusion is a process for which $\Delta S_{sys} > 0$; we know from experience that it is always spontaneous. We have already pointed

The vaporization of bromine, $Br_2(\ell) \rightarrow Br_2(g)$ (*left*) and the sublimation of iodine, $I_2(s) \rightarrow I_2(g)$ (*right*) both lead to an increase in disorder, so $\Delta S_{sys} > 0$ for each process. Which do you think results in the more positive ΔS? Carry out the calculation using values from Appendix K to check whether your prediction was correct.

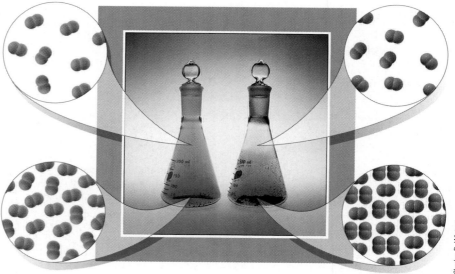

Charles D. Winters

When water, H_2O, and propyl alcohol, $CH_3CH_2CH_2OH$ (*left*) are mixed to form a solution (*right*), disorder increases. $\Delta S > 0$ for the mixing of any two molecular substances.

out (Section 14-2) that the increase in disorder (entropy increase) that accompanies mixing often provides the driving force for solubility of one substance in another. For example, when one mole of solid NaCl dissolves in water, NaCl(s) → NaCl(aq), the entropy (Appendix K) increases from 72.4 J/mol·K to 115.5 J/mol·K, or $\Delta S^0 = +43.1$ J/mol·K. The term "mixing" can be interpreted rather liberally. For example, the reaction $H_2(g) + Cl_2(g) \rightarrow 2HCl(g)$ has $\Delta S^0 > 0$; in the reactants, each atom is bonded to an identical atom, a less "mixed-up" situation than in the products, where unlike atoms are bonded together.

5. *Increase in the number of particles*, as in the dissociation of a diatomic gas such as $F_2(g) \rightarrow 2F(g)$. Any process in which the number of particles increases results in an increase in entropy, $\Delta S_{sys} > 0$. Values of ΔS^0 calculated for several reactions of this type are given in Table 15-6. As you can see, the ΔS^0 values for the dissociation process $X_2 \rightarrow 2X$ are all similar for X = H, F, Cl, and N. Why is the value given in Table 15-6 so much larger for X = Br? This process starts with *liquid* Br_2. The total process $Br_2(\ell) \rightarrow 2Br(g)$, for which $\Delta S^0 = 197.5$ J/mol·K, can be treated as the result of *two* processes. The first of these is *vaporization*, $Br_2(\ell) \rightarrow Br_2(g)$, for which $\Delta S^0 = 93.1$ J/mol·K. The second step is the dissociation of gaseous bromine, $Br_2(g) \rightarrow 2Br(g)$, for which $\Delta S^0 = 104.4$ J/mol·K; this entropy increase is about the same as for the other processes that involve *only* dissociation of a gaseous diatomic species. Can you rationalize the even higher value given in the table for the process $I_2(s) \rightarrow 2I(g)$?

TABLE 15-6	Entropy Changes for Some Processes $X_2 \rightarrow 2X$

Reaction	ΔS^0 (J/mol·K)
$H_2(g) \longrightarrow 2H(g)$	98.0
$N_2(g) \longrightarrow 2N(g)$	114.9
$O_2(g) \longrightarrow 2O(g)$	117.0
$F_2(g) \longrightarrow 2F(g)$	114.5
$Cl_2(g) \longrightarrow 2Cl(g)$	107.2
$Br_2(\ell) \longrightarrow 2Br(g)$	197.5
$I_2(s) \longrightarrow 2I(g)$	245.3

Charles D. Winters

Do you think that the reaction

$$2H_2(g) + O_2(g) \longrightarrow 2H_2O(\ell)$$

would have a higher or lower value of ΔS^0 than when the water is in the gas phase? Confirm by calculation.

6. *Changes in the number of moles of gaseous substances.* Processes that result in an increase in the number of moles of gaseous substances have $\Delta S_{sys} > 0$. Example 15-15 illustrates this. There are no gaseous reactants, but the products include five moles of gas. Conversely, we would predict that the process $2H_2(g) + O_2(g) \rightarrow 2H_2O(g)$ has a negative ΔS^0 value; here, three moles of gas is consumed while only two moles is produced, for a net decrease in the number of moles in the gas phase. You should be able to calculate the value of ΔS^0 for this reaction from the values in Appendix K.

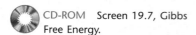

CD-ROM   Screen 19.7, Gibbs Free Energy.

15-15   FREE ENERGY CHANGE, ΔG, AND SPONTANEITY

Energy is the capacity to do work. If heat is released in a chemical reaction (ΔH is negative), *some* of the heat may be converted into useful work. Some of it may be expended to increase the order of the system (if ΔS is negative). If a system becomes more disordered ($\Delta S > 0$), however, more useful energy becomes available than indicated by ΔH alone. J. Willard Gibbs (1839–1903), a prominent 19th-century American professor of mathematics and physics, formulated the relationship between enthalpy and entropy in terms of another state function that we now call the **Gibbs free energy, G.** It is defined as

$$G = H - TS$$

The **Gibbs free energy change, ΔG,** at constant temperature and pressure, is

This is often called simply the *Gibbs energy change* or the *free energy change.*

$$\Delta G = \Delta H - T\,\Delta S \qquad \text{(constant } T \text{ and } P)$$

The amount by which the Gibbs free energy decreases is the *maximum useful energy* obtainable in the form of work from a given process at constant temperature and pressure. It is also the *indicator of spontaneity of a reaction or physical change* at constant T and P. If there is a net decrease of useful energy, ΔG is negative and the process is spontaneous (product-favored). We see from the equation that ΔG becomes more negative as (1) ΔH becomes

(a) The entropy of an organism decreases (unfavorable) when new cells are formed. The energy to sustain animal life is provided by the metabolism of food. This energy is released when the chemical bonds in the food are broken. Exhalation of gases and excretion of waste materials increase the entropy of the surroundings enough so that the entropy of the universe increases and the overall process can occur. (b) Stored chemical energy can later be transformed by the organism to the mechanical energy for muscle contraction, to the electrical energy for brain function, or to another needed form.

(a)

(b)

more negative (the process gives off more heat) and (2) ΔS becomes more positive (the process results in greater disorder). If there is a net increase in free energy of the system during a process, ΔG is positive and the process is nonspontaneous (reactant-favored). This means that the reverse process is spontaneous under the given conditions. When $\Delta G = 0$, there is no net transfer of free energy; both the forward and reverse processes are equally favorable. Thus, $\Delta G = 0$ describes a system at *equilibrium*.

The relationship between ΔG and spontaneity may be summarized as follows.

ΔG	Spontaneity of Reaction (constant T and P)
ΔG is positive	Reaction is nonspontaneous (reactant-favored)
ΔG is zero	System is at equilibrium
ΔG is negative	Reaction is spontaneous (product-favored)

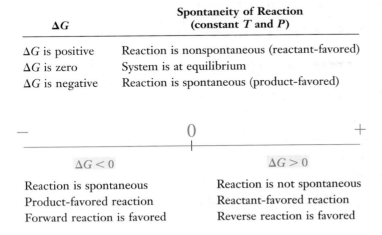

The free energy content of a system depends on temperature and pressure (and, for mixtures, on concentrations). The value of ΔG for a process depends on the states and the concentrations of the various substances involved. It also depends strongly on temperature, because the equation $\Delta G = \Delta H - T\,\Delta S$ includes temperature. Just as for other thermodynamic variables, we choose some set of conditions as a standard state reference. The standard state for ΔG^0 is the same as for ΔH^0—1 atm and the specified temperature, usually 25°C (298 K). Values of standard molar free energy of formation, ΔG_f^0, for many substances are tabulated in Appendix K. For *elements* in their standard states, $\Delta G_f^0 = 0$. The values of ΔG_f^0 may be used to calculate the standard free energy change of a reaction *at 298 K* by using the following relationship.

$$\Delta G_{rxn}^0 = \Sigma\, n\, \Delta G_{f\ products}^0 - \Sigma\, n\, \Delta G_{f\ reactants}^0 \qquad \text{(1 atm and 298 K \textit{only})}$$

The value of ΔG_{rxn}^0 allows us to predict the spontaneity of a very special hypothetical reaction that we call the *standard reaction*.

In the **standard reaction,** the numbers of moles of reactants shown in the balanced equation, all at standard conditions, are *completely* converted to the numbers of moles of products shown in the balanced equation, all at standard conditions.

We want to know which are more stable *at standard conditions*—the *reactants* or the *products.*

We must remember that it is ΔG, and not ΔG^0, that is the general criterion for spontaneity. ΔG depends on concentrations of reactants and products in the mixture. For most reactions, there is an *equilibrium mixture* of reactants and products that is more stable than either all reactants or all products. In Chapter 17 we will study the concept of equilibrium and see how to find ΔG for mixtures.

EXAMPLE 15-16　*Spontaneity of Standard Reaction*

Diatomic nitrogen and oxygen molecules make up about 99% of all the molecules in reasonably "unpolluted" dry air. Evaluate ΔG^0 for the following reaction at 298 K, using ΔG_f^0 values from Appendix K. Is the standard reaction spontaneous?

$$N_2(g) + O_2(g) \longrightarrow 2NO(g) \qquad \text{(nitrogen oxide)}$$

Plan

The reaction conditions are 1 atm and 298 K, so we can use the tabulated values of ΔG_f^0 for each substance in Appendix K to evaluate ΔG_{rxn}^0 in the preceding equation. The treatment of units for calculation of ΔG^0 is the same as that for ΔH^0 in Example 15-9.

Solution

We obtain the following values of ΔG_f^0 from Appendix K:

	$N_2(g)$	$O_2(g)$	$NO(g)$
ΔG_f^0, kJ/mol:	0	0	86.57

$$
\begin{aligned}
\Delta G_{rxn}^0 &= \Sigma \, n \, \Delta G_{f\ products}^0 - \Sigma \, n \, \Delta G_{f\ reactants}^0 \\
&= 2 \, \Delta G_{f\ NO(g)}^0 \quad - [\Delta G_{f\ N_2(g)}^0 + \Delta G_{f\ O_2(g)}^0] \\
&= 2(86.57) \qquad - [0 + 0]
\end{aligned}
$$

$$\boxed{\Delta G_{rxn}^0 = +173.1 \text{ kJ/mol rxn}} \qquad \text{for the reaction as written}$$

Because ΔG^0 is positive, the reaction is nonspontaneous at 298 K under standard state conditions.

You should now work Exercise 99.

For the reverse reaction at 298 K, $\Delta G_{rxn}^0 = -173.1$ kJ/mol. It is product-favored but very slow at room temperature. The NO formed in automobile engines is oxidized to even more harmful NO_2 much more rapidly than it decomposes to N_2 and O_2. Thermodynamic spontaneity does not guarantee that a process occurs at an observable rate. The oxides of nitrogen in the atmosphere represent a major environmental problem.

The value of ΔG^0 can also be calculated by the equation

$$\Delta G^0 = \Delta H^0 - T \Delta S^0 \qquad \text{(constant } T \text{ and } P\text{)}$$

Strictly, this last equation applies to standard conditions; however, ΔH^0 and ΔS^0 often do not vary much with temperature, so the equation can often be used to *estimate* free energy changes at other temperatures.

✓ Problem-Solving Tip: *Some Common Pitfalls in Calculating* ΔG_{rxn}^0

Be careful of these points when you carry out calculations that involve ΔG^0:

1. The calculation of ΔG_{rxn}^0 from tabulated values of ΔG_f^0 is valid *only* if the reaction is at 25°C (298 K) and one atmosphere.

2. Calculations with the equation $\Delta G^0 = \Delta H^0 - T \Delta S^0$ must be carried out with the temperature in kelvins.

3. The energy term in ΔS^0 is usually in joules, whereas that in ΔH^0 is usually in kilojoules; remember to convert one of these so that units are consistent before you combine them.

EXAMPLE 15-17 *Spontaneity of Standard Reaction*

Make the same determination as in Example 15-16, using heats of formation and absolute entropies rather than free energies of formation.

Plan

First we calculate ΔH^0_{rxn} and ΔS^0_{rxn}. We use the relationship $\Delta G^0 = \Delta H^0 - T\Delta S^0$ to evaluate the free energy change under standard state conditions at 298 K.

Solution

The values we obtain from Appendix K are

	$N_2(g)$	$O_2(g)$	$NO(g)$
ΔH^0_f, kJ/mol:	0	0	90.25
S^0, J/mol·K:	191.5	205.0	210.7

$$\Delta H^0_{rxn} = \Sigma\, n\, \Delta H^0_{f\,products} \quad - \Sigma\, n\, \Delta H^0_{f\,reactants}$$
$$= 2\, \Delta H^0_{f\,NO(g)} \quad - [\Delta H^0_{f\,N_2(g)} + \Delta H^0_{f\,O_2(g)}]$$
$$= [2(90.25)] \quad - (0 + 0) = 180.5 \text{ kJ/mol}$$

$$\Delta S^0_{rxn} = \Sigma\, nS^0_{products} \quad - \Sigma\, nS^0_{reactants}$$
$$= 2S^0_{NO(g)} \quad - [S^0_{N_2(g)} + S^0_{O_2(g)}]$$
$$= [2\,(210.7)] \quad - (191.5 + 205.0)] = 24.9 \text{ J/mol·K} = 0.0249 \text{ kJ/mol·K}$$

Now we use the relationship $\Delta G^0 = \Delta H^0 - T\Delta S^0$, with $T = 298$ K, to evaluate the free energy change under standard state conditions at 298 K.

$$\Delta G^0_{rxn} = \Delta H^0_{rxn} \quad - T\Delta S^0_{rxn}$$
$$= 180.5 \text{ kJ/mol} - (298 \text{ K})(0.0249 \text{ kJ/mol·K})$$
$$= 180.5 \text{ kJ/mol} - 7.42 \text{ kJ/mol}$$

$$\boxed{\Delta G^0_{rxn} = +173.1 \text{ kJ/mol rxn,}} \quad \text{the same value obtained in Example 15-16.}$$

You should now work Exercise 100.

 Remember to use the same energy units in ΔS^0 and ΔH^0!

15-16 THE TEMPERATURE DEPENDENCE OF SPONTANEITY

The methods developed in Section 15-15 can also be used to estimate the temperature at which a process is in equilibrium. When a system is at equilibrium, $\Delta G = 0$. Thus,

$$\Delta G_{rxn} = \Delta H_{rxn} - T\Delta S_{rxn} \quad \text{or} \quad 0 = \Delta H_{rxn} - T\Delta S_{rxn}$$

so

$$\Delta H_{rxn} = T\Delta S_{rxn} \quad \text{or} \quad T = \frac{\Delta H_{rxn}}{\Delta S_{rxn}} \quad \text{(at equilibrium)}$$

 CD-ROM Screen 19.8, Free Energy and Temperature.

EXAMPLE 15-18 *Estimation of Boiling Point*

Use the thermodynamic data in Appendix K to estimate the normal boiling point of bromine, Br_2. Assume that ΔH and ΔS do not change with temperature.

Actually, both ΔH^0_{rxn} and ΔS^0_{rxn} vary with temperature, but usually not enough to introduce significant errors for modest temperature changes. The value of ΔG^0_{rxn}, on the other hand, is strongly dependent on the temperature.

Plan

The process we must consider is

$$Br_2(\ell) \longrightarrow Br_2(g)$$

By definition, the normal boiling point of a liquid is the temperature at which pure liquid and pure gas coexist in equilibrium at 1 atm. Therefore, $\Delta G = 0$. We assume that $\Delta H_{rxn} = \Delta H^0_{rxn}$ and $\Delta S_{rxn} = \Delta S^0_{rxn}$. We can evaluate these two quantities, substitute them in the relationship $\Delta G = \Delta H - T\,\Delta S$, and then solve for T.

Solution

The required values (Appendix K) are as follows:

	$Br_2(\ell)$	$Br_2(g)$
ΔH^0_f, kJ/mol:	0	30.91
S^0, J/mol·K:	152.2	245.4

$$\Delta H_{rxn} = \Delta H^0_{f\,Br_2(g)} - \Delta H^0_{f\,Br_2(\ell)}$$
$$= 30.91 \quad - 0 = 30.91 \text{ kJ/mol}$$
$$\Delta S_{rxn} = S^0_{Br2(g)} \quad - S^0_{Br2(\ell)}$$
$$= (245.4 \quad - 152.2) = 93.2 \text{ J/mol·K} = 0.0932 \text{ kJ/mol·K}$$

We can now solve for the temperature at which the system is in equilibrium, that is, the boiling point of Br_2.

$$\Delta G_{rxn} = \Delta H_{rxn} - T\,\Delta S_{rxn} = 0 \quad \text{so} \quad \Delta H_{rxn} = T\,\Delta S_{rxn}$$

$$T = \frac{\Delta H_{rxn}}{\Delta S_{rxn}} = \frac{30.91 \text{ kJ/mol}}{0.0932 \text{ kJ/mol·K}} = \boxed{332 \text{ K } (59°C)}$$

This is the temperature at which the system is in equilibrium, that is, the boiling point of Br_2. The value listed in a handbook of chemistry and physics is 58.78°C.

You should now work Exercise 110.

The free energy change and spontaneity of a reaction depend on both enthalpy and entropy changes. Both ΔH and ΔS may be either positive or negative, so we can group reactions in four classes with respect to spontaneity (Figure 15-15).

$\Delta G = \Delta H - T\,\Delta S$		(constant temperature and pressure)
1. $\Delta H = -$ (favorable)	$\Delta S = +$ (favorable)	Reactions are product-favored at all temperatures
2. $\Delta H = -$ (favorable)	$\Delta S = -$ (unfavorable)	Reactions become product-favored below a definite temperature
3. $\Delta H = +$ (unfavorable)	$\Delta S = +$ (favorable)	Reactions become product-favored above a definite temperature
4. $\Delta H = +$ (unfavorable)	$\Delta S = -$ (unfavorable)	Reactions are reactant-favored at all temperatures

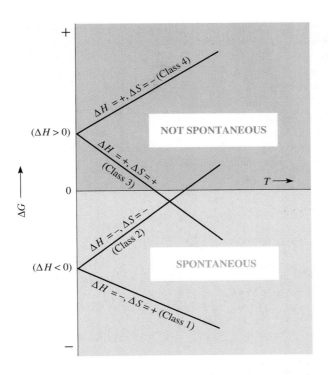

Figure 15-15 A graphical representation of the dependence of ΔG and spontaneity on temperature for each of the four classes of reactions listed in the text and in Table 15-7.

When ΔH and ΔS have opposite signs (classes 1 and 4), they act in the same direction, so the direction of spontaneous change does not depend on temperature. When ΔH and ΔS have the same signs (classes 2 and 3), their effects oppose one another, so changes in temperature can cause one factor or the other to dominate, and spontaneity depends on temperature. For class 2, decreasing the temperature decreases the importance of the *unfavorable* $T\,\Delta S$ term, so the reaction becomes product-favored at lower temperatures. For class 3, increasing the temperature increases the importance of the *favorable* $T\,\Delta S$ term, so the reaction becomes product-favored at higher temperatures.

We can estimate the temperature range over which a chemical reaction in class 2 or 3 is spontaneous by evaluating ΔH^0_{rxn} and ΔS^0_{rxn} from tabulated data. The temperature at which $\Delta G^0_{rxn} = 0$ is the temperature limit of spontaneity. The sign of ΔS^0_{rxn} tells us whether the reaction is spontaneous *below* or *above* this limit (Table 15-7).

TABLE 15-7 *Thermodynamic Classes of Reactions*

Class	Examples	ΔH (kJ/mol)	ΔS (J/mol·K)	Temperature Range of Spontaneity
1	$2H_2O_2(\ell) \longrightarrow 2H_2O(\ell) + O_2(g)$	-196	$+126$	All temperatures
	$H_2(g) + Br_2(\ell) \longrightarrow 2HBr(g)$	-72.8	$+114$	All temperatures
2	$NH_3(g) + HCl(g) \longrightarrow NH_4Cl(s)$	-176	-285	Lower temperatures (<619 K)
	$2H_2S(g) + SO_2(g) \longrightarrow 3S(s) + 2H_2O(\ell)$	-233	-424	Lower temperatures (<550 K)
3	$NH_4Cl(s) \longrightarrow NH_3(g) + HCl(g)$	$+176$	$+285$	Higher temperatures (>619 K)
	$CCl_4(\ell) \longrightarrow C(graphite) + 2Cl_2(g)$	$+135$	$+235$	Higher temperatures (>517 K)
4	$2H_2O(\ell) + O_2(g) \longrightarrow 2H_2O_2(\ell)$	$+196$	-126	Nonspontaneous, all temperatures
	$3O_2(g) \longrightarrow 2O_3(g)$	$+285$	-137	Nonspontaneous, all temperatures

Heating red HgS in air produces liquid Hg. Gaseous SO_2 escapes. Cinnabar, an important ore of mercury, contains HgS.

EXAMPLE 15-19 *Temperature Range of Spontaneity*

Mercury(II) sulfide is a dark red mineral called cinnabar. Metallic mercury is obtained by roasting the sulfide in a limited amount of air. Estimate the temperature range in which the *standard* reaction is product-favored.

$$HgS(s) + O_2(g) \longrightarrow Hg(\ell) + SO_2(g)$$

Plan

We evaluate ΔH_{rxn}^0 and ΔS_{rxn}^0 and assume that their values are independent of temperature. We find that both factors are favorable to spontaneity.

Solution

From Appendix K:

	HgS(s)	O_2	Hg(ℓ)	SO_2(g)
ΔH_f^0, kJ/mol:	−58.2	0	0	−296.8
S^0, J/mol·K:	82.4	205.0	76.0	248.1

$$\Delta H_{rxn}^0 = \Delta H_{f\,Hg(\ell)}^0 + \Delta H_{f\,SO_2(g)}^0 - [\Delta H_{f\,HgS(s)}^0 + \Delta H_{f\,O_2(g)}^0]$$
$$= 0 - 296.8 \qquad\qquad - [-58.2 + 0] = -238.6 \text{ kJ/mol}$$

$$\Delta S_{rxn}^0 = S_{Hg(\ell)}^0 + S_{SO_2(g)}^0 - [S_{HgS(s)}^0 + S_{O_2(g)}^0]$$
$$= 76.02 + 248.1 \quad - [82.4 + 205.0] = +36.7 \text{ J/mol·K}$$

ΔH_{rxn}^0 is negative and ΔS_{rxn}^0 is positive, so the reaction is product-favored at all temperatures. The reverse reaction is, therefore, nonspontaneous at all temperatures.

The fact that a reaction is product-favored at all temperatures does not mean that the reaction occurs rapidly enough to be useful at all temperatures. As a matter of fact, Hg(ℓ) can be obtained from HgS(s) by this reaction at a reasonable rate only at high temperatures.

EXAMPLE 15-20 *Temperature Range of Spontaneity*

Estimate the temperature range for which the following standard reaction is product-favored.

$$SiO_2(s) + 2C(\text{graphite}) + 2Cl_2(g) \longrightarrow SiCl_4(g) + 2CO(g)$$

Plan

When we proceed as in Example 15-19, we find that ΔS_{rxn}^0 is favorable to spontaneity, whereas ΔH_{rxn}^0 is unfavorable. Thus, we know that the reaction becomes product-favored *above* some temperature. We can set ΔG^0 equal to zero in the equation $\Delta G^0 = \Delta H^0 - T\Delta S^0$ and solve for the temperature at which the system is *at equilibrium*. This will represent the temperature above which the reaction would be product-favored.

Solution

From Appendix K:

	SiO_2(s)	C(graphite)	Cl_2(g)	$SiCl_4$(g)	CO(g)
ΔH_f^0, kJ/mol:	−910.9	0	0	−657.0	−110.5
S^0, J/mol·K:	41.84	5.740	223.0	330.6	197.6

$$\Delta H^0_{rxn} = [\Delta H^0_{f\ SiCl_4(g)} + 2\ \Delta H^0_{f\ CO(g)}] - [\Delta H^0_{f\ SiO_2(s)} + 2\ \Delta H^0_{f\ C(graphite)} + 2\ \Delta H^0_{f\ Cl_2(g)}]$$

$$= [(-657.0) + 2(-110.5)] - [(-910.9) + 2(0) + 2(0)]$$

$$= +32.9\ kJ/mol$$

$$\Delta S^0_{rxn} = S^0_{SiCl_4(g)} + 2S^0_{CO(g)} - [S^0_{SiO_2(s)} + 2S^0_{C(graphite)} + 2S^0_{Cl_2(g)}]$$

$$= [330.6 + 2(197.6)] - [41.84 + 2(5.740) + 2(223.0)]$$

$$= 226.5\ J/mol \cdot K = 0.2265\ kJ/mol \cdot K$$

When $\Delta G^0 = 0$, neither the forward nor the reverse reaction is favored. Let's find the temperature at which $\Delta G^0 = 0$ and the system is at equilibrium.

$$0 = \Delta G^0 = \Delta H^0 - T\ \Delta S^0$$

$$\Delta H^0 = T\ \Delta S^0$$

$$T = \frac{\Delta H^0}{\Delta S^0} = \frac{+32.9\ kJ/mol}{+0.2265\ kJ/mol \cdot K} = 145\ K$$

At temperatures above 145 K, the $T\ \Delta S^0$ term would be greater ($-T\ \Delta S^0$ would be more negative) than the ΔH^0 term, which would make ΔG^0 negative; so the reaction would be product-favored above 145 K. At temperatures below 145 K, the $T\ \Delta S^0$ term would be smaller than the ΔH^0 term, which would make ΔG^0 positive; so the reaction would be reactant-favored below 145 K.

However, 145 K ($-128°C$) is a very low temperature. For all practical purposes, the reaction is product-favored at all but very low temperatures. In practice, it is carried out at 800°C to 1000°C because of the greater reaction rate at these higher temperatures. This gives a useful and economical rate of production of $SiCl_4$, an important industrial chemical.

You should now work Exercises 104 and 108.

Key Terms

Absolute entropy (of a substance)   The entropy of a substance relative to its entropy in a perfectly ordered crystalline form at 0 K (where its entropy is zero).

Bomb calorimeter   A device used to measure the heat transfer between system and surroundings at constant volume.

Bond energy   The amount of energy necessary to break one mole of bonds in a gaseous substance, to form gaseous products at the same temperature and pressure.

Calorimeter   A device used to measure the heat transfer that accompanies a physical or chemical change.

Endothermic process   A process that absorbs heat.

Enthalpy change, ΔH   The quantity of heat transferred into or out of a system as it undergoes a chemical or physical change at constant temperature and pressure.

Entropy, S   A thermodynamic state property that measures the degree of disorder or randomness of a system.

Equilibrium   A state of dynamic balance in which the rates of forward and reverse processes (reactions) are equal; the state of a system when neither the forward nor the reverse process is thermodynamically favored.

Exothermic process   A process that gives off (releases) heat.

First Law of Thermodynamics   The total amount of energy in the universe is constant (also known as the Law of Conservation of Energy); energy is neither created nor destroyed in ordinary chemical reactions and physical changes.

Gibbs free energy, G   The thermodynamic state function of a system that indicates the amount of energy available for the system to do useful work at constant T and P. It is defined as $G = H - TS$.

Gibbs free energy change, ΔG   The indicator of spontaneity of a process at constant T and P. $\Delta G = \Delta H - T\ \Delta S$. If ΔG is negative, the process is product-favored (spontaneous).

Hess's Law of Heat Summation   The enthalpy change for a reaction is the same whether it occurs in one step or a series of steps.

Internal energy, E   All forms of energy associated with a specific amount of a substance.

Law of Conservation of Energy   Energy cannot be created or destroyed in a chemical reaction or in a physical change; it may be changed from one form to another; see *First Law of Thermodynamics*.

Mole of reaction (mol rxn)   The amount of reaction that corresponds to the number of moles of each substance shown in the balanced equation.

Nonspontaneous change   See *Reactant-favored change.*

Pressure–volume work   Work done by a gas when it expands against an external pressure or work done on a system as gases are compressed or consumed in the presence of an external pressure.

Product-favored change   A change for which the collection of products is more stable than the collection of reactants under the given conditions; also called spontaneous change.

Reactant-favored change   A change for which the collection of reactants is more stable than the collection of products under the given conditions; also called nonspontaneous change.

Second Law of Thermodynamics   The universe tends toward a state of greater disorder in spontaneous processes.

Spontaneous change   See *Product-favored change.*

Standard enthalpy change, ΔH^0   The enthalpy change in which the number of moles of reactants specified in the balanced chemical equation, all at standard states, is converted completely to the specified number of moles of products, all at standard states.

Standard entropy change, ΔS^0   The entropy change in which the number of moles of reactants specified in the balanced chemical equation, all at standard states, is converted completely to the specified number of moles of products, all at standard states.

Standard molar enthalpy of formation, ΔH_f^0 (of a substance)   The enthalpy change for the formation of one mole of a substance in a specified state from its elements in their standard states; also called *standard molar heat of formation* or just *heat of formation.*

Standard molar entropy, S^0 (of a substance)   The absolute entropy of a substance in its standard state at 298 K.

Standard reaction   A reaction in which the numbers of moles of reactants shown in the balanced equation, all in their standard states, are *completely* converted to the numbers of moles of products shown in the balanced equation, also all at their standard states.

Standard state (of a substance)   See *Thermodynamic standard state of a substance.*

State function   A variable that defines the state of a system; a function that is independent of the pathway by which a process occurs.

Surroundings   Everything in the environment of the system.

System   The substances of interest in a process; the part of the universe under investigation.

Thermochemical equation   A balanced chemical equation together with a designation of the corresponding value of ΔH_{rxn}. Sometimes used with changes in other thermodynamic quantities.

Thermochemistry   The observation, measurement, and prediction of energy changes for both physical changes and chemical reactions.

Thermodynamics   The study of the energy transfers accompanying physical and chemical processes.

Thermodynamic state of a system   A set of conditions that completely specifies all of the properties of the system.

Thermodynamic standard state of a substance   The most stable state of the substance at one atmosphere pressure and at some specific temperature (25°C unless otherwise specified).

Third Law of Thermodynamics   The entropy of a hypothetical pure, perfect, crystalline substance at absolute zero temperature is zero.

Universe   The system plus the surroundings.

Work   The application of a force through a distance; for physical changes or chemical reactions at constant external pressure, the work done on the system is $-P\,\Delta V$; for chemical reactions that involve gases, the work done on the system can be expressed as $-(\Delta n)RT$.

Exercises

General Concepts

1. State precisely the meaning of each of the following terms. You may need to review Chapter 1 to refresh your memory concerning terms introduced there. (a) energy; (b) kinetic energy; (c) potential energy; (d) joule.

2. State precisely the meaning of each of the following terms. You may need to review Chapter 1 to refresh your memory about terms introduced there. (a) heat; (b) temperature; (c) system; (d) surroundings; (e) thermodynamic state of system; (f) work.

3. (a) Give an example of the conversion of heat into work. (b) Give an example of the conversion of work into heat.

4. Distinguish between endothermic and exothermic processes. If we know that a reaction is endothermic in one direction, what can be said about the reaction in the reverse direction?

5. According to the First Law of Thermodynamics, the total amount of energy in the universe is constant. Why, then, do we say that we are experiencing a declining supply of energy?

6. Use the First Law of Thermodynamics to describe what occurs when an incandescent light is turned on.

7. Define enthalpy and give an example of a reaction that has a negative enthalpy change.

8. Which of the following are examples of state functions? (a) your bank balance; (b) your mass; (c) your weight; (d) the heat lost by perspiration during a climb up a mountain along a fixed path.

9. What is a state function? Would Hess's Law be a law if enthalpy were not a state function?

Enthalpy and Changes in Enthalpy

10. (a) Distinguish between ΔH and ΔH^0 for a reaction. (b) Distinguish between ΔH^0_{rxn} and ΔH^0_f.

11. A reaction is characterized by $\Delta H_{rxn} = -500$ kJ/mol. Does the reaction mixture absorb heat from the surroundings or release heat to them?

12. For each of the following reactions, (a) does the enthalpy increase or decrease; (b) is $H_{reactant} > H_{product}$ or is $H_{product} > H_{reactant}$; (c) is ΔH positive or negative?
    (i) $Al_2O_3(s) \longrightarrow 2Al(s) + \frac{3}{2}O_2(g)$ (endothermic)
    (ii) $Sn(s) + Cl_2(g) \longrightarrow SnCl_2(s)$ (exothermic)

13. (a) The combustion of 0.0222 g of isooctane vapor, $C_8H_{18}(g)$, at constant pressure raises the temperature of a calorimeter 0.400°C. The heat capacity of the calorimeter and water combined is 2.48 kJ/°C. Find the molar heat of combustion of gaseous isooctane.

    $$C_8H_{18}(g) + 12\frac{1}{2}O_2(g) \longrightarrow 8CO_2(g) + 9H_2O(\ell)$$

    (b) How many grams of $C_8H_{18}(g)$ must be burned to obtain 295 kJ of heat energy?

14. Methanol, CH_3OH, is an efficient fuel with a high octane rating that can be produced from coal and hydrogen.

    $$CH_3OH(g) + \frac{3}{2}O_2(g) \longrightarrow CO_2(g) + 2H_2O(\ell)$$
    $$\Delta H = -764 \text{ kJ/mol rxn}$$

    (a) Find the heat evolved when 110.0 g $CH_3OH(g)$ burns in excess oxygen. (b) What mass of O_2 is consumed when 975 kJ of heat is given out?

15. How much heat is liberated when 0.143 mole of sodium reacts with excess water according to the following equation?

    $$2Na(s) + 2H_2O(\ell) \longrightarrow H_2(g) + 2NaOH(aq)$$
    $$\Delta H = -368 \text{ kJ/mol rxn}$$

16. What is ΔH for the reaction

    $$PbO(s) + C(s) \longrightarrow Pb(s) + CO(g)$$

    if 5.95 kJ must be supplied to convert 13.43 g lead(II) oxide to lead?

17. The standard molar enthalpy of formation, ΔH^0_f, listed in Appendix K is zero for almost all elements. A few entries are not zero; find two examples and tell why they are not zeros.

18. Why is the standard molar enthalpy of formation, ΔH^0_f, for liquid water different than is ΔH^0_f for water vapor, both at 25°C. Which formation reaction is more exothermic? Does your answer indicate that $H_2O(\ell)$ is at a higher or lower enthalpy than $H_2O(g)$?

19. Methylhydrazine is burned with dinitrogen tetroxide in the attitude-control engines of the space shuttles.

    $$CH_6N_2(\ell) + \tfrac{5}{4}N_2O_4(\ell) \longrightarrow$$
    $$CO_2(g) + 3H_2O(\ell) + \tfrac{9}{4}N_2(g)$$

    The two substances ignite instantly on contact, producing a flame temperature of 3000 K. The energy liberated per 0.100 g of CH_6N_2 at constant atmospheric pressure after the products are cooled back to 25°C is 750 J. (a) Find ΔH for the reaction as written. (b) How many kilojoules are liberated when 62.5 g of N_2 is produced?

A space shuttle.

20. Which is more exothermic, the combustion of one mole of methane to form $CO_2(g)$ and liquid water or the combustion of one mole of methane to form $CO_2(g)$ and steam? Why? (No calculations are necessary.)

21. Which is more exothermic, the combustion of one mole of gaseous benzene, C_6H_6, or the combustion of one mole of liquid benzene? Why? (No calculations are necessary.)

Thermochemical Equations, ΔH^0_f, and Hess's Law

22. Explain the meaning of each word in the term "thermodynamic standard state of a substance."

23. Explain the meaning of each word in the term "standard molar enthalpy of formation." Give an example.

24. From the data in Appendix K, determine the form that represents the standard state for each of the following elements: (a) chlorine; (b) chromium; (c) bromine; (d) iodine; (e) sulfur.

25. From the data in Appendix K, determine the form that represents the standard state for each of the following elements: (a) oxygen; (b) carbon; (c) phosphorus; (d) rubidium; (e) mercury.

26. Write the balanced chemical equation whose ΔH^0_{rxn} value is equal to ΔH^0_f for each of the following substances: (a) calcium hydroxide, $Ca(OH)_2(s)$; (b) benzene, $C_6H_6(\ell)$; (c) sodium bicarbonate, $NaHCO_3(s)$; (d) calcium fluoride, $CaF_2(s)$; (e) phosphine, $PH_3(g)$; (f) propane, $C_3H_8(g)$; (g) atomic sulfur, $S(g)$.

27. Write the balanced chemical equation whose ΔH^0_{rxn} value is equal to ΔH^0_f for each of the following substances: (a) hydrogen sulfide, $H_2S(g)$; (b) lead(II) chloride, $PbCl_2(s)$; (c) atomic oxygen, $O(g)$; (d) benzoic acid, $C_6H_5COOH(s)$; (e) hydrogen peroxide, $H_2O_2(\ell)$; (f) dinitrogen pentoxide, $N_2O_5(g)$.

*28. We burn 2.88 g of lithium in excess oxygen at constant atmospheric pressure to form Li_2O. Then we bring the reaction mixture back to 25°C. In this process 121 kJ of heat is given off. What is the standard molar enthalpy of formation of Li_2O?

*29. We burn 7.20 g of magnesium in excess nitrogen at constant atmospheric pressure to form Mg_3N_2. Then we bring the reaction mixture back to 25°C. In this process 68.35 kJ of heat is given off. What is the standard molar enthalpy of formation of Mg_3N_2?

30. From the following enthalpies of reaction,

$$4HCl(g) + O_2(g) \longrightarrow 2H_2O(\ell) + 2Cl_2(g)$$
$$\Delta H = -202.4 \text{ kJ/mol rxn}$$

$$\tfrac{1}{2}H_2(g) + \tfrac{1}{2}F_2(g) \longrightarrow HF(\ell)$$
$$\Delta H = -600.0 \text{ kJ/mol rxn}$$

$$H_2(g) + \tfrac{1}{2}O_2(g) \longrightarrow H_2O(\ell)$$
$$\Delta H = -285.8 \text{ kJ/mol rxn}$$

find ΔH_{rxn} for $2HCl(g) + F_2(g) \longrightarrow 2HF(\ell) + Cl_2(g)$.

31. From the following enthalpies of reaction,

$$CaCO_3(s) \longrightarrow CaO(s) + CO_2(g)$$
$$\Delta H = -178.1 \text{ kJ/mol rxn}$$

$$CaO(s) + H_2O(\ell) \longrightarrow Ca(OH)_2(s)$$
$$\Delta H = -65.3 \text{ kJ/mol rxn}$$

$$Ca(OH)_2(s) \longrightarrow Ca^{2+}(aq) + 2OH^-(aq)$$
$$\Delta H = -16.2 \text{ kJ/mol rxn}$$

calculate ΔH_{rxn} for

$$Ca^{2+}(aq) + 2OH^-(aq) + CO_2(g) \longrightarrow$$
$$CaCO_3(s) + H_2O(\ell)$$

32. Given that

$$S(s) + O_2(g) \longrightarrow SO_2(g) \qquad \Delta H = -296.8 \text{ kJ/mol}$$
$$S(s) + \tfrac{3}{2}O_2(g) \longrightarrow SO_3(g) \qquad \Delta H = -395.6 \text{ kJ/mol}$$

determine the enthalpy change for the decomposition reaction

$$2SO_3(g) \longrightarrow 2SO_2(g) + O_2(g).$$

33. Aluminum reacts vigorously with many oxidizing agents. For example,

$$4Al(s) + 3O_2(g) \longrightarrow 2Al_2O_3(s)$$
$$\Delta H = -3352 \text{ kJ/mol}$$

$$4Al(s) + 3MnO_2(s) \longrightarrow 3Mn(s) + 2Al_2O_3(s)$$
$$\Delta H = -1792 \text{ kJ/mol}$$

Use this information to determine the molar enthalpy of formation of $MnO_2(s)$.

34. Given that

$$2H_2(g) + O_2(g) \longrightarrow 2H_2O(\ell)$$
$$\Delta H = -571.6 \text{ kJ/mol}$$

$$C_3H_4(g) + 4O_2(g) \longrightarrow 3CO_2(g) + 2H_2O(\ell)$$
$$\Delta H = -1937 \text{ kJ/mol}$$

$$C_3H_8(g) + 5O_2(g) \longrightarrow 3CO_2(g) + 4H_2O(\ell)$$
$$\Delta H = -2220 \text{ kJ/mol}$$

determine the heat of the hydrogenation reaction

$$C_3H_4(g) + 2H_2(g) \longrightarrow C_3H_8(g)$$

35. Determine the molar heat of formation of liquid hydrogen peroxide at 25°C from the following thermochemical equations.

$$H_2(g) + \tfrac{1}{2}O_2(g) \longrightarrow H_2O(g)$$
$$\Delta H^0 = -241.82 \text{ kJ/mol}$$

$$2H(g) + O(g) \longrightarrow H_2O(g)$$
$$\Delta H^0 = -926.92 \text{ kJ/mol}$$

$$2H(g) + 2O(g) \longrightarrow H_2O_2(g)$$
$$\Delta H^0 = -1070.60 \text{ kJ/mol}$$

$$2O(g) \longrightarrow O_2(g)$$
$$\Delta H^0 = -498.34 \text{ kJ/mol}$$

$$H_2O_2(\ell) \longrightarrow H_2O_2(g)$$
$$\Delta H^0 = 51.46 \text{ kJ/mol}$$

36. Use data in Appendix K to find the enthalpy of reaction for
   (a) $NH_4NO_3(s) \longrightarrow N_2O(g) + 2H_2O(\ell)$
   (b) $2FeS_2(s) + \tfrac{11}{2}O_2(g) \longrightarrow Fe_2O_3(s) + 4SO_2(g)$
   (c) $SiO_2(s) + 3C(s) \longrightarrow SiC(s) + 2CO(g)$

37. Repeat Exercise 36 for
   (a) $CaCO_3(s) \longrightarrow CaO(s) + CO_2(g)$
   (b) $2HI(g) + F_2(g) \longrightarrow 2HF(g) + I_2(s)$
   (c) $SF_6(g) + 3H_2O(\ell) \longrightarrow 6HF(g) + SO_3(g)$

38. The internal combustion engine uses heat produced during the burning of a fuel. Propane, $C_3H_8(g)$, is sometimes used as the fuel. Gasoline is the most commonly used fuel. Assume that the gasoline is pure octane, $C_8H_{18}(\ell)$ and the fuel and oxygen are completely converted into $CO_2(g)$ and $H_2O(g)$. For each of these fuels, determine the heat released per gram of fuel burned.

39. Propane, $C_3H_8(g)$, is used as the fuel for some modern internal combustion engines. Methane, $CH_4(g)$, has been proposed by the movie industry as the post-Apocalypse

fuel when gasoline and propane are supposedly no longer available. Assume that the fuel and oxygen are completely converted into $CO_2(g)$ and $H_2O(g)$. For each of these fuels determine the heat released per gram of fuel burned. Compare your answers to the answers for Exercise 38.

40. The thermite reaction, used for welding iron, is the reaction of Fe_3O_4 with Al.

$$8Al(s) + 3Fe_3O_4(s) \longrightarrow 4Al_2O_3(s) + 9Fe(s)$$
$$\Delta H^0 = -3350 \text{ kJ/mol rxn}$$

Because this large amount of heat cannot be rapidly dissipated to the surroundings, the reacting mass may reach temperatures near 3000°C. How much heat is released by the reaction of 27.6 g of Al with 69.12 g of Fe_3O_4?

The thermite reaction.

41. When a welder uses an acetylene torch, the combustion of acetylene liberates the intense heat needed for welding metals together. The equation for this combustion reaction is

$$2C_2H_2(g) + 5O_2(g) \longrightarrow 4CO_2(g) + 2H_2O(g)$$

The heat of combustion of acetylene is -1255.5 kJ/mol of C_2H_2. How much heat is liberated when 1.731 kg of C_2H_2 is burned?

42. Silicon carbide, or carborundum, SiC, is one of the hardest substances known and is used as an abrasive. It has the structure of diamond with half of the carbons replaced by silicon. It is prepared industrially by reduction of sand (SiO_2) with carbon in an electric furnace.

$$SiO_2(s) + 3C(graphite) \longrightarrow SiC(s) + 2CO(g)$$

ΔH^0 for this reaction is 624.6 kJ, and the ΔH_f^0 for $SiO_2(s)$ and $CO(g)$ are -910.9 kJ/mol and -110.5 kJ/mol, respectively. Calculate ΔH_f^0 for silicon carbide.

43. Natural gas is mainly methane, $CH_4(g)$. Assume that gasoline is octane, $C_8H_{18}(\ell)$, and that kerosene is $C_{10}H_{22}(\ell)$.

(a) Write the balanced equations for the combustion of each of these three hydrocarbons in excess O_2. The products are $CO_2(g)$ and $H_2O(\ell)$. (b) Calculate ΔH_{rxn}^0 at 25°C for each combustion reaction. ΔH_f^0 for $C_{10}H_{22}$ is -300.9 kJ/mol. (c) When burned at standard conditions, which of these three fuels would produce the most heat per mole? (d) When burned at standard conditions, which of the three would produce the most heat per gram?

Bond Energies

44. (a) How is the heat released or absorbed in a *gas phase reaction* related to bond energies of products and reactants? (b) Hess's Law states that

$$\Delta H_{rxn}^0 = \Sigma\, n\, \Delta H_{f\,products}^0 - \Sigma\, n\, \Delta H_{f\,reactants}^0$$

The relationship between ΔH_{rxn}^0 and bond energies for a *gas phase reaction* is

$$\Delta H_{rxn}^0 =$$
$$\Sigma \text{ bond energies}_{reactants} - \Sigma \text{ bond energies}_{products}$$

It is *not* true, in general, that ΔH_f^0 for a substance is equal to the negative of the sum of the bond energies of the substance. Why?

45. (a) Suggest a reason for the fact that different amounts of energy are required for the successive removal of the three hydrogen atoms of an ammonia molecule, even though all N—H bonds in ammonia are equivalent. (b) Suggest why the N—H bonds in different compounds such as ammonia, NH_3; methylamine, CH_3NH_2; and ethylamine, $C_2H_5NH_2$, have slightly different bond energies.

46. Use tabulated bond energies to estimate the enthalpy of reaction for each of the following gas phase reactions.
(a) $H_2C=CH_2 + Br_2 \longrightarrow BrH_2C-CH_2Br$
(b) $H_2O_2 \longrightarrow H_2O + \frac{1}{2}O_2$

47. Use tabulated bond energies to estimate the enthalpy of reaction for each of the following gas phase reactions.
(a) $N_2 + 3H_2 \longrightarrow 2NH_3$
(b) $CH_4 + Cl_2 \longrightarrow CH_3Cl + HCl$
(c) $CO + H_2O \longrightarrow CO_2 + H_2$

48. Use the bond energies listed in Table 15-2 to estimate the heat of reaction for

$$\underset{\overset{|}{F}}{\overset{\overset{Cl}{|}}{Cl-C-F}}(g) + F-F(g) \longrightarrow \underset{\overset{|}{F}}{\overset{\overset{F}{|}}{F-C-F}}(g) + Cl-Cl(g)$$

49. Estimate ΔH for the burning of one mole of butane, using the bond energies listed in Tables 15-2 and 15-3.

$$\underset{\overset{|}{H}\ \overset{|}{H}\ \overset{|}{H}\ \overset{|}{H}}{\overset{\overset{H}{|}\ \overset{H}{|}\ \overset{H}{|}\ \overset{H}{|}}{H-C-C-C-C-H}}(g) + \tfrac{13}{2}O=O(g) \longrightarrow$$
$$4O=C=O(g) + 5H-O-H(g)$$

50. (a) Use the bond energies listed in Table 15-2 to estimate the heats of formation of HCl(g) and HF(g). (b) Compare your answers to the standard heats of formation in Appendix K.

51. (a) Use the bond energies listed in Table 15-2 to estimate the heats of formation of $H_2O(g)$ and $O_3(g)$. (b) Compare your answers to the standard heats of formation in Appendix K.

52. Using data in Appendix K, calculate the average P—Cl bond energy in $PCl_3(g)$.

53. Using data in Appendix K, calculate the average P—H bond energy in $PH_3(g)$.

54. Using data in Appendix K, calculate the average P—Cl bond energy in $PCl_5(g)$. Compare your answer with the value calculated in Exercise 52.

*55. Methane undergoes several different exothermic reactions with gaseous chlorine. One of these forms chloroform, $CHCl_3(g)$.

$$CH_4(g) + 3Cl_2(g) \longrightarrow CHCl_3(g) + 3HCl(g)$$
$$\Delta H^0_{rxn} = -305.2 \text{ kJ/mol rxn}$$

Average bond energies per mole of bonds are: C—H = 413 kJ; Cl—Cl = 242 kJ; H—Cl = 432 kJ. Use these to calculate the average C—Cl bond energy in chloroform. Compare this with the value in Table 15-2.

*56. Ethylamine undergoes an endothermic gas phase dissociation to produce ethylene (or ethene) and ammonia.

$$\Delta H^0_{rxn} = +53.6 \text{ kJ/mol rxn}$$

The following average bond energies per mole of bonds are given: C—H = 413 kJ; C—C = 346 kJ; C=C = 602 kJ; N—H = 391 kJ. Calculate the C—N bond energy in ethylamine. Compare this with the value in Table 15-2.

Calorimetry

57. What is a coffee-cup calorimeter? How do coffee-cup calorimeters give us useful information?

58. A calorimeter contained 75.0 g of water at 16.95°C. A 93.3-g sample of iron at 65.58°C was placed in it, giving a final temperature of 19.68°C for the system. Calculate the heat capacity of the calorimeter. Specific heats are 4.184 J/g · °C for H_2O and 0.444 J/g · °C for Fe.

59. A student wishes to determine the heat capacity of a coffee-cup calorimeter. After she mixes 100.0 g of water at 58.5°C with 100.0 g of water, already in the calorimeter, at 22.8°C, the final temperature of the water is 39.7°C. (a) Calculate the heat capacity of the calorimeter in J/°C. Use 4.184 J/g · °C as the specific heat of water. (b) Why is it more useful to express the value in J/°C rather than units of J/(g calorimeter · °C)?

60. A coffee-cup calorimeter is used to determine the specific heat of a metallic sample. The calorimeter is filled with 50.0 mL of water at 25.0°C (density = 0.997 g/mL). A 36.5-gram sample of the metallic material is taken from water boiling at 100.0°C and placed in the calorimeter. The equilibrium temperature of the water and sample is 32.5°C. The calorimeter constant is known to be 1.87 J/°C. Calculate the specific heat of the metallic material.

61. A 5.1-gram piece of gold jewelry is removed from water at 100.0°C and placed in a coffee-cup calorimeter containing 16.9 g of water at 22.5°C. The equilibrium temperature of the water and jewelry is 23.2°C. The calorimeter constant is known from calibration experiments to be 1.54 J/°C. What is the specific heat of this piece of jewelry? The specific heat of pure gold is 0.129 J/g · °C. Is the jewelry pure gold?

62. A coffee-cup calorimeter having a heat capacity of 472 J/°C is used to measure the heat evolved when the following aqueous solutions, both initially at 22.6°C, are mixed: 100. g of solution containing 6.62 g of lead(II) nitrate, $Pb(NO_3)_2$, and 100. g of solution containing 6.00 g of sodium iodide, NaI. The final temperature is 24.2°C. Assume that the specific heat of the mixture is the same as that for water, 4.184 J/g · °C. The reaction is

$$Pb(NO_3)_2(aq) + 2NaI(aq) \longrightarrow PbI_2(s) + 2NaNO_3(aq)$$

(a) Calculate the heat evolved in the reaction. (b) Calculate the ΔH for the reaction under the conditions of the experiment.

63. A coffee-cup calorimeter is used to determine the heat of reaction for the acid-base neutralization

$$CH_3COOH(aq) + NaOH(aq) \longrightarrow$$
$$NaCH_3COO(aq) + H_2O(\ell)$$

When we add 20.00 mL of 0.625 M NaOH at 21.400°C to 30.00 mL of 0.500 M CH_3COOH already in the calorimeter at the same temperature, the resulting temperature is observed to be 24.347°C. The heat capacity of the calorimeter has previously been determined to be 27.8 J/°C. Assume that the specific heat of the mixture is the same as that of water, 4.184 J/g · °C, and that the density of the mixture is 1.02 g/mL. (a) Calculate the amount of heat given off in the reaction. (b) Determine ΔH for the reaction under the conditions of the experiment.

64. In a bomb calorimeter compartment surrounded by 945 g of water, the combustion of 1.048 g of benzene, $C_6H_6(\ell)$, raised the temperature of the water from 23.640°C to 32.692°C. The heat capacity of the calorimeter is 891 J/°C. (a) Write the balanced equation for the combustion reaction, assuming that $CO_2(g)$ and $H_2O(\ell)$ are the only products. (b) Use the calorimetric data to calculate ΔE for the combustion of benzene in kJ/g and in kJ/mol.

65. A 2.00-g sample of hydrazine, N_2H_4, is burned in a bomb

calorimeter that contains 6.40×10^3 g of H_2O, and the temperature increases from 25.00°C to 26.17°C. The heat capacity of the calorimeter is 3.76 kJ/°C. Calculate ΔE for the combustion of N_2H_4 in kJ/g and in kJ/mol.

*66. A strip of magnesium metal having a mass of 1.22 g dissolves in 100. mL of 6.02 M HCl, which has a specific gravity of 1.10. The hydrochloric acid is initially at 23.0°C, and the resulting solution reaches a final temperature of 45.5°C. The heat capacity of the calorimeter in which the reaction occurs is 562 J/°C. Calculate ΔH for the reaction under the conditions of the experiment, assuming the specific heat of the final solution is the same as that for water, 4.184 J/g·°C.

$$Mg(s) + 2HCl(aq) \longrightarrow MgCl_2(aq) + H_2(g)$$

67. When 3.16 g of salicylic acid, $C_7H_6O_3$, is burned in a bomb calorimeter containing 5.00 kg of water originally at 23.00°C, 69.3 kJ of heat is evolved. The calorimeter constant is 3255 J/°C. Calculate the final temperature.

68. A 6.620-gram sample of decane, $C_{10}H_{22}(\ell)$, was burned in a bomb calorimeter whose heat capacity had been determined to be 2.45 kJ/°C. The temperature of 1250.0 grams of water rose from 24.6°C to 26.4°C. Calculate ΔE for the reaction in joules per gram of decane and in kilojoules per mole of decane. The specific heat of water is 4.184 J/g·°C.

69. A nutritionist determines the caloric value of a 10.00-gram sample of beef fat by burning it in a bomb calorimeter. The calorimeter held 2.500 kg of water, the heat capacity of the bomb is 1.360 kJ/°C, and the temperature of the calorimeter increased from 25.0°C to 56.9°C. (a) Calculate the number of joules released per gram of beef fat. (b) One nutritional Calorie is 1 kcal or 4184 joules. What is the dietary, caloric value of beef fat, in nutritional Calories per gram?

Internal Energy and Changes in Internal Energy

70. (a) What are the sign conventions for q, the amount of heat added to or removed from a system? (b) What are the sign conventions for w, the amount of work done on or by a system?

71. What happens to ΔE for a system during a process in which (a) $q < 0$ and $w < 0$, (b) $q = 0$ and $w > 0$, and (c) $q > 0$ and $w < 0$?

72. What happens to ΔE for a system during a process in which (a) $q > 0$ and $w > 0$, (b) $q = w = 0$, and (c) $q < 0$ and $w > 0$?

73. A system performs 720 L·atm of pressure-volume work (1 L·atm = 101.325 J) on its surroundings and absorbs 5500. J of heat from its surroundings. What is the change in internal energy of the system?

74. A system receives 93 J of electrical work, performs 227 J of pressure-volume work, and releases 155 J of heat. What is the change in internal energy of the system?

75. For each of the following chemical and physical changes carried out at constant pressure, state whether work is done by the system on the surroundings or by the surroundings on the system, or whether the amount of work is negligible.
(a) $C_6H_6(\ell) \longrightarrow C_6H_6(g)$
(b) $\frac{1}{2}N_2(g) + \frac{3}{2}H_2(g) \longrightarrow NH_3(g)$
(c) $SiO_2(s) + 3C(s) \longrightarrow SiC(s) + 2CO(g)$

76. Repeat Exercise 75 for
(a) $2SO_2(g) + O_2(g) \longrightarrow 2SO_3(g)$
(b) $CaCO_3(s) \longrightarrow CaO(s) + CO_2(g)$
(c) $CO_2(g) + H_2O(\ell) + CaCO_3(s) \longrightarrow$
$\qquad\qquad\qquad\qquad Ca^{2+}(aq) + 2HCO_3^-(aq)$

77. Assuming that the gases are ideal, calculate the amount of work done (in joules) in each of the following reactions. In each case, is the work done *on* or *by* the system? (a) A reaction in the Mond process for purifying nickel that involves formation of the gas nickel(0) tetracarbonyl at 50–100°C. Assume one mole of nickel is used and a constant temperature of 75°C is maintained.

$$Ni(s) + 4CO(g) \longrightarrow Ni(CO)_4(g)$$

(b) The conversion of one mole of brown nitrogen dioxide into colorless dinitrogen tetroxide at 10.0°C.

$$2NO_2(g) \longrightarrow N_2O_4(g)$$

78. Assuming that the gases are ideal, calculate the amount of work done (in joules) in each of the following reactions. In each case, is the work done *on* or *by* the system? (a) The oxidation of one mole of HCl(g) at 200°C.

$$4HCl(g) + O_2(g) \longrightarrow 2Cl_2(g) + 2H_2O(g)$$

(b) The decomposition of one mole of nitric oxide (an air pollutant) at 300°C.

$$2NO(g) \longrightarrow N_2(g) + O_2(g)$$

*79. When an ideal gas expands at *constant temperature*, there is no change in molecular kinetic energy (kinetic energy is proportional to temperature), and there is no change in potential energy due to intermolecular attractions (these are zero for an ideal gas). Thus, for the isothermal (constant temperature) expansion of an ideal gas, $\Delta E = 0$. Suppose we allow an ideal gas to expand isothermally from 2.00 L to 5.00 L in two steps: (a) against a constant external pressure of 3.00 atm until equilibrium is reached, then (b) against a constant external pressure of 2.00 atm until equilibrium is reached. Calculate q and w for this two-step expansion.

Entropy and Entropy Changes

80. A car uses gasoline as a fuel. Describe the burning of the fuel in terms of chemical and physical changes. Relate your answer to the Second Law of Thermodynamics.

81. State the Second Law of Thermodynamics. We cannot

use ΔS_{univ} directly as a measure of the spontaneity of a reaction. Why?

82. State the Third Law of Thermodynamics. What does it mean?

83. Explain why ΔS may be referred to as a contributor to spontaneity.

84. For each of the following processes, tell whether the entropy of the system increases, decreases, or remains constant:
    (a) Melting one mole of ice to water at 0°C
    (b) Freezing one mole of water to ice at 0°C
    (c) Freezing one mole of water to ice at -10°C
    (d) Freezing one mole of water to ice at 0°C and then cooling it to -10°C

85. When solid sodium chloride is cooled from 25°C to 0°C, the entropy change is -4.4 J/mol · K. Is this an increase or decrease in randomness? Explain this entropy change in terms of what happens in the solid at the molecular level.

86. When a one-mole sample of argon gas at 0°C is compressed to one half its original volume, the entropy change is -5.76 J/mol · K. Is this an increase or a decrease in randomness? Explain this entropy change in terms of what happens in the gas at the molecular level.

87. Which of the following processes are accompanied by an increase in entropy of the system? (No calculation is necessary.) (a) A building is constructed from bricks, mortar, lumber, and nails. (b) A building collapses into bricks, mortar, lumber, and nails. (c) Iodine sublimes, $I_2(s) \longrightarrow I_2(g)$. (d) White silver sulfate, Ag_2SO_4, precipitates from a solution containing silver ions and sulfate ions. (e) A marching band is gathered into formation. (f) A partition is removed to allow two gases to mix.

88. Which of the following processes are accompanied by an increase in entropy of the system? (No calculation is necessary.) (a) Thirty-five pennies are removed from a bag and placed heads-up on a table. (b) The pennies of Part (a) are swept off the table and back into the bag. (c) Water freezes. (d) Carbon tetrachloride, CCl_4, evaporates. (e) The reaction $PCl_5(g) \longrightarrow PCl_3(g) + Cl_2(g)$ occurs. (f) The reaction $PCl_3(g) + Cl_2(g) \longrightarrow PCl_5(g)$ occurs.

89. For each of the following processes, tell whether the entropy of the *universe* increases, decreases, or remains constant. (a) melting one mole of ice to water at 0°C; (b) freezing one mole of water to ice at 0°C; (c) freezing one mole of water to ice at -15°C; (d) freezing one mole of water to ice at 0°C and then cooling it to -15°C.

*90. Consider the boiling of a pure liquid at constant pressure. Is each of the following greater than, less than, or equal to zero? (a) ΔS_{sys}; (b) ΔH_{sys}; (c) ΔT_{sys}.

91. Use S^0 data from Appendix K to calculate the value of ΔS^0_{298} for each of the following reactions. Compare the signs and magnitudes for these ΔS^0_{298} values and explain your observations.
    (a) $2NO(g) + H_2(g) \longrightarrow N_2O(g) + H_2O(g)$

    (b) $2N_2O_5(g) \longrightarrow 4NO_2(g) + O_2(g)$
    (c) $NH_4NO_3(s) \longrightarrow N_2O(g) + 2H_2O(g)$

92. Use S^0 data from Appendix K to calculate the value of ΔS^0_{298} for each of the following reactions. Compare the signs and magnitudes for these ΔS^0_{298} values and explain your observations.
    (a) $4HCl(g) + O_2(g) \longrightarrow 2Cl_2(g) + 2H_2O(g)$
    (b) $PCl_3(g) + Cl_2(g) \longrightarrow PCl_5(g)$
    (c) $2N_2O(g) \longrightarrow 2N_2(g) + O_2(g)$

Gibbs Free Energy Changes and Spontaneity

93. (a) What are the two factors that favor spontaneity of a process? (b) What is Gibbs free energy? What is change in Gibbs free energy? (c) Most spontaneous reactions are exothermic, but some are not. Explain. (d) Explain how the signs and magnitudes of ΔH and ΔS are related to the spontaneity of a process and how they affect it.

94. Which of the following conditions would predict a process that is (a) always spontaneous, (b) always nonspontaneous, or (c) spontaneous or nonspontaneous depending on the temperature and magnitudes of ΔH and ΔS? (i) $\Delta H > 0$, $\Delta S > 0$; (ii) $\Delta H > 0$, $\Delta S < 0$; (iii) $\Delta H < 0$, $\Delta S > 0$; (iv) $\Delta H < 0$, $\Delta S < 0$.

95. For the decomposition of $O_3(g)$ to $O_2(g)$

$$2O_3(g) \longrightarrow 3O_2(g)$$

$\Delta H^0 = -285.4$ kJ/mol and $\Delta S^0 = 137.55$ J/mol · K at 25°C. Calculate ΔG^0 for the reaction. Is the reaction spontaneous? Is either or both of the driving forces (ΔH^0 and ΔS^0) for the reaction favorable?

96. Calculate ΔG^0 at 25°C for the reaction

$$2NO_2(g) \longrightarrow N_2O_4(g)$$

given $\Delta H^0 = -57.20$ kJ/mol and $\Delta S^0 = -175.83$ J/mol·K. Is this reaction spontaneous? What is the driving force for spontaneity?

97. The standard Gibbs free energy of formation is -286.06 kJ/mol for $NaI(s)$, -261.90 kJ/mol for $Na^+(aq)$, and -51.57 kJ/mol for $I^-(aq)$ at 25°C. Calculate ΔG^0 for the reaction

$$NaI(s) \xrightarrow{H_2O} Na^+(aq) + I^-(aq)$$

*98. Use the following equations to find ΔG^0_f for HBr(g) at 25°C.

$Br_2(\ell) \longrightarrow Br_2(g)$	$\Delta G^0 = \quad 3.14$ kJ/mol
$HBr(g) \longrightarrow H(g) + Br(g)$	$\Delta G^0 = 339.09$ kJ/mol
$Br_2(g) \longrightarrow 2Br(g)$	$\Delta G^0 = 161.7$ kJ/mol
$H_2(g) \longrightarrow 2H(g)$	$\Delta G^0 = 406.494$ kJ/mol

99. Use values of standard free energy of formation, ΔG^0_f, from Appendix K, to calculate the standard free energy

change for each of the following reactions at 25°C and 1 atm.

(a) $3NO_2(g) + H_2O(\ell) \longrightarrow 2HNO_3(\ell) + NO(g)$
(b) $SnO_2(s) + 2CO(g) \longrightarrow 2CO_2(g) + Sn(s)$
(c) $2Na(s) + 2H_2O(\ell) \longrightarrow 2NaOH(aq) + H_2(g)$

100. Make the same calculations as in Exercise 99, using values of standard enthalpy of formation and absolute entropy instead of values of ΔG_f^0.

101. Calculate ΔG^0 for the reduction of the oxides of iron and copper by carbon at 650 K represented by the equations

(a) $2Fe_2O_3(s) + 3C(\text{graphite}) \longrightarrow 4Fe(s) + 3CO_2(g)$

(b) $2CuO(s) + C(\text{graphite}) \longrightarrow 2Cu(s) + CO_2(g)$

Values of ΔG^0_f at 700K are -92 kJ/mol for CuO(s), -632 kJ/mol for $Fe_2O_3(s)$, and -395 kJ/mol for $CO_2(g)$. (c) Which oxide can be reduced using carbon in a wood fire (which has a temperature of about 700 K), assuming standard state conditions?

Temperature Range of Spontaneity

102. Are the following statements true or false? Justify your answers. (a) An exothermic reaction is spontaneous. (b) If ΔH and ΔS are both positive, then ΔG will decrease when the temperature increases. (c) A reaction for which ΔS_{sys} is positive is spontaneous.

103. For the reaction

$$C(s) + O_2(g) \longrightarrow CO_2(g)$$

$\Delta H^0 = -393.51$ kJ/mol and $\Delta S^0 = 2.86$ J/mol·K at 25°C. (a) Does this reaction become more or less favorable as the temperature increases? (b) For the reaction

$$C(s) + \tfrac{1}{2}O_2(g) \longrightarrow CO(g)$$

$\Delta H^0 = -110.52$ kJ/mol and $\Delta S^0 = 89.36$ J/mol·K at 25°C. Does this reaction become more or less favorable as the temperature increases? (c) Compare the temperature dependencies of these reactions.

104. (a) Calculate ΔH^0, ΔG^0, and ΔS^0 for the reaction

$$2H_2O_2(\ell) \longrightarrow 2H_2O(\ell) + O_2(g) \text{ at } 25°C.$$

(b) Is there any temperature at which $H_2O_2(\ell)$ is stable at 1 atm?

105. When is it true that $\Delta S = \dfrac{\Delta H}{T}$?

106. Dissociation reactions are those in which molecules break apart. Why do high temperatures favor the spontaneity of most dissociation reactions?

107. Estimate the temperature range over which each of the following standard reactions is spontaneous.

(a) $2Al(s) + 3Cl_2(g) \longrightarrow 2AlCl_3(s)$
(b) $2NOCl(g) \longrightarrow 2NO(g) + Cl_2(g)$
(c) $4NO(g) + 6H_2O(g) \longrightarrow 4NH_3(g) + 5O_2(g)$
(d) $2PH_3(g) \longrightarrow 3H_2(g) + 2P(g)$

108. Estimate the temperature range over which each of the following standard reactions is spontaneous. (a) The reac-

tion by which sulfuric acid droplets from polluted air convert water-insoluble limestone or marble (calcium carbonate) to slightly soluble calcium sulfate, which is slowly washed away by rain:

$$CaCO_3(s) + H_2SO_4(\ell) \longrightarrow$$
$$CaSO_4(s) + H_2O(\ell) + CO_2(g)$$

(b) The reaction by which Antoine Lavoisier achieved the first laboratory preparation of oxygen in the late eighteenth century: the thermal decomposition of the red-orange powder, mercury(II) oxide, to oxygen and the silvery liquid metal, mercury:

$$2HgO(s) \longrightarrow 2Hg(\ell) + O_2(g)$$

(c) The reaction of coke (carbon) with carbon dioxide to form the reducing agent, carbon monoxide, which is used to reduce some metal ores to metals:

$$CO_2(g) + C(s) \longrightarrow 2CO(g)$$

(d) The reverse of the reaction by which iron rusts:

$$2Fe_2O_3(s) \longrightarrow 4Fe(s) + 3O_2(g)$$

109. Estimate the normal boiling point of pentacarbonyl-iron(0), $Fe(CO)_5$, at 1 atm pressure, using Appendix K.

110. (a) Estimate the normal boiling point of water, at 1 atm pressure, using Appendix K. (b) Compare the temperature obtained with the known boiling point of water. Can you explain the discrepancy?

111. Sublimation and subsequent deposition onto a cold surface are a common method of purification of I_2 and other solids that sublime readily. Estimate the sublimation temperature (solid to vapor) of the dark violet solid iodine, I_2, at 1 atm pressure, using the data of Appendix K.

Sublimation and deposition of I_2.

*112. (a) Is the reaction C(diamond) $\longrightarrow$ C(graphite) spontaneous at 25°C and 1 atm? (b) Now are you worried about your diamonds turning to graphite? Why or why not? (c) Is there a temperature at which diamond and graphite are in equilibrium? If so, what is this temperature? (d) How do you account for the formation of diamonds in nature? (*Hint:* Diamond has a higher density than graphite.)

Mixed Exercises

*113. An ice calorimeter, shown here, can be used to measure the amount of heat released or absorbed by a reaction that is carried out at a constant temperature of 0°C. If heat is transferred from the system to the bath, some of the ice melts. A given mass of liquid water has a smaller volume than the same mass of ice, so the total volume of the ice and water mixture decreases. Measuring the volume decrease using the scale at the left indicates the amount of heat released by the reacting system. As long as some ice remains in the bath, the temperature remains at 0°C. In Example 15-2 we saw that the reaction

$$CuSO_4(aq) + 2NaOH(aq) \longrightarrow$$
$$Cu(OH)_2(s) + Na_2SO_4(aq)$$

releases 846 J of heat at constant temperature and pressure when 50.0 mL of 0.400 M CuSO4 solution and 50.0 mL of 0.600 M NaOH solution are allowed to react. (Because no gases are involved in the reaction, the volume change of the reaction mixture is negligible.) Calculate the change in volume of the ice and water mixture that would be observed if we carried out the same experiment in an ice calorimeter. The density of $H_2O(\ell)$ at 0°C is 0.99987 g/mL and that of ice is 0.917 g/mL. The heat of fusion of ice at 0°C is 334 J/g.

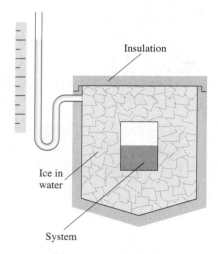

Insulation

Ice in water

System

114. It is difficult to prepare many compounds directly from their elements, so ΔH_f^0 values for these compounds cannot be measured directly. For many organic compounds, it is easier to measure the standard enthalpy of combustion by reaction of the compound with excess $O_2(g)$ to form $CO_2(g)$ and $H_2O(\ell)$. From the following standard enthalpies of combustion at 25°C, determine ΔH_f^0 for the compound. (a) cyclohexane, $C_6H_{12}(\ell)$, a useful organic solvent: $\Delta H_{combustion}^0 = -3920$ kJ/mol; (b) phenol, $C_6H_5OH(s)$, used as a disinfectant and in the production of thermo-setting plastics: $\Delta H_{combustion}^0 = -3053$ kJ/mol.

*115. Standard entropy changes cannot be measured directly in the laboratory. They are calculated from experimentally obtained values of ΔG^0 and ΔH^0. From the data given here, calculate ΔS^0 at 298 K for each of the following reactions.

(a) $OF_2(g) + H_2O(g) \longrightarrow O_2(g) + 2HF(g)$

$\Delta H^0 = -323.2$ kJ/mol $\Delta G^0 = -358.4$ kJ/mol

(b) $CaC_2(s) + 2H_2O(\ell) \longrightarrow Ca(OH)_2(s) + C_2H_2(g)$

$\Delta H^0 = -125.4$ kJ/mol $\Delta G^0 = -145.4$ kJ/mol

(c) $CaO(s) + H_2O(\ell) \longrightarrow Ca(OH)_2(aq)$

$\Delta H^0 = 81.5$ kJ/mol $\Delta G^0 = -26.20$ kJ/mol

*116. Calculate q, w, and ΔE for the vaporization of 12.5 g of liquid ethanol (C_2H_5OH) at 1.00 atm at 78.0°C, to form gaseous ethanol at 1.00 atm at 78.0°C. Make the following simplifying assumptions: (a) the density of liquid ethanol at 78.0°C is 0.789 g/mL, and (b) gaseous ethanol is adequately described by the ideal gas equation. The heat of vaporization of ethanol is 855 J/g.

117. We add 0.100 g of CaO(s) to 125 g H_2O at 23.6°C in a coffee-cup calorimeter. The following reaction occurs. What will be the final temperature of the solution?

$$CaO(s) + H_2O(\ell) \longrightarrow Ca(OH)_2(aq)$$
$$\Delta H^0 = 81.5 \text{ kJ/mol rxn}$$

118. (a) The accurately known molar heat of combustion of naphthalene, $C_{10}H_8(s)$, $\Delta H = -5156.8$ kJ/mol $C_{10}H_8$, is used to calibrate calorimeters. The complete combustion of 0.01520 g of $C_{10}H_8$ at constant pressure raises the temperature of a calorimeter by 0.212°C. Find the heat capacity of the calorimeter. (b) The initial temperature of the calorimeter (Part a) is 22.102°C; 0.1040 g of $C_8H_{18}(\ell)$, octane (molar heat of combustion $\Delta H = -5451.4$ kJ/mol C_8H_{18}), is completely burned in the calorimeter. Find the final temperature of the calorimeter.

CONCEPTUAL EXERCISES

119. When a gas expands suddenly, it may not have time to absorb a significant amount of heat: $q = 0$. Assume that 1.00 mol N_2 expands suddenly, doing 3000 J of work. (a) What is ΔE for the process? (b) The heat capacity of N_2 is 20.9 J/mol · °C. How much does its temperature fall during this expansion? (This is the principle of most snow-making machines, which use compressed air mixed with water vapor.)

120. As a rubber band is stretched, it gets warmer; when released, it gets cooler. To obtain the more nearly linear arrangement of the rubber band's polymeric material from the more random relaxed rubber band requires that there be rotation about carbon-carbon single bonds. Based on these data, give the sign of ΔG, ΔH, and ΔS for the stretching of a rubber band and for the relaxing of a stretched rubber band. What drives the spontaneous process?

121. (a) The decomposition of mercury(II) oxide has been used as a method for producing oxygen, but this is not a recommended method. Why not? (b) Write the balanced equation for the decomposition of mercury(II) oxide. (c) Calculate the ΔH^0, ΔS^0, and ΔG^0 for the reaction. (d) Is the reaction spontaneous at room temperature?

*122. (a) A student heated a sample of a metal weighing 32.6 g to 99.83°C and put it into 100.0 g of water at 23.62°C in a calorimeter. The final temperature was 24.41°C. The student calculated the specific heat of the metal, but neglected to use the heat capacity of the calorimeter. The specific heat of water is 4.184 J/g · °C. What was his answer? The metal was known to be chromium, molybdenum, or tungsten. By comparing the value of the specific heat to those of the metals (Cr, 0.460; Mo, 0.250; W, 0.135 J/g · °C), the student identified the metal. What was the metal? (b) A student at the next laboratory bench did the same experiment, obtained the same data, and used the heat capacity of the calorimeter in his calculations. The heat capacity of the calorimeter was 410 J/°C. Was his identification of the metal different?

*123. A sugar cube dissolves in a cup of coffee in an endothermic process. (a) Is the entropy change of the system (sugar plus coffee) greater than, less than, or equal to zero? (b) Is the entropy change of the universe greater than, less than, or equal to zero? (c) Is the entropy change of the surroundings greater than, less than, or equal to zero?

BUILDING YOUR KNOWLEDGE

*124. Energy to power muscular work is produced from stored carbohydrate (glycogen) or fat (triglycerides). Metabolic consumption and production of energy are described with the nutritional "Calorie, " which is equal to 1 kilocalorie. Average energy output per minute for various activities follows: sitting, 1.7 kcal; walking, level, 3.5 mph, 5.5 kcal; cycling, level, 13 mph, 10 kcal; swimming, 8.4 kcal; running, 10 mph, 19 kcal. Approximate energy values of some common foods are also given: large apple, 100 kcal; 8-oz cola drink, 105 kcal; malted milkshake, 8 oz milk, 500 kcal; $\frac{3}{4}$ cup pasta with tomato sauce and cheese, 195 kcal; hamburger on bun with sauce, 350 kcal; 10-oz sirloin steak, including fat, 1000 kcal. To maintain body weight, fuel intake should balance energy output. Prepare a table showing (a) each given food, (b) its fuel value, and (c) the minutes of each activity that would balance the kcal of each food.

125. From its heat of fusion, calculate the entropy change associated with the melting of one mole of ice at its melting point. From its heat of vaporization, calculate the entropy change associated with the boiling of one mole of water at its boiling point. Are your calculated values consistent with the simple model that we use to describe order in solids, liquids, and gases?

126. The energy content of dietary fat is 39 kJ/g, and for pro-

tein and carbohydrate it is 17 and 16 kJ/g, respectively. A 70.0-kg (155-lb) person utilizes 335 kJ/hr while resting and 1250 kJ/h while walking 6 km/h. How many hours would the person need to walk per day instead of resting if he or she consumed 100 g (about $\frac{1}{4}$ lb) of fat instead of 100 g of protein?

127. The enthalpy change for melting one mole of water at 273 K is $\Delta H_{273}^0 = 6010$ J/mol, whereas that for vaporizing a mole of water at 373 K is $\Delta H_{273}^0 = 40{,}660$ J/mol. Why is the second value so much larger?

128. A 436-g chunk of lead was removed from a beaker of boiling water, quickly dried, and dropped into a Styrofoam cup containing 50.0 g of water at 25.0°C. As the system reached equilibrium, the water temperature rose to 40.8°C. Calculate the heat capacity and the specific heat of the lead.

129. Methane, $CH_4(g)$, is the main constituent of natural gas. In excess oxygen, methane burns to $CO_2(g)$ and $H_2O(\ell)$, whereas in limited oxygen, the products are $CO(g)$ and $H_2O(\ell)$. Which would result in a higher temperature: a gas–air flame or a gas–oxygen flame? How can you tell?

Atlanta Gas & Light Company

A methane flame.

130. A 0.483-g sample of butter was burned in a bomb calorimeter whose heat capacity was 4572 J/°C, and the temperature was observed to rise from 24.76 to 27.93°C. Calculate the fuel value of butter in (a) kJ/g; (b) nutritional Calories/g (one nutritional Calorie is equal to one kilocalorie); (c) nutritional Calories/5-gram pat.

BEYOND THE TEXTBOOK

Go to the textbook website at

http://www.brookscole.com/chemistry/whitten

for additional activities and exercises based on the General Chemistry Interactive CD-ROM, the World Wide Web, and library resources.

InfoTrac College Edition

For additional readings, go to InfoTrac College Edition, your online research library at:

http://infotrac.thomsonlearning.com

16 Chemical Kinetics

OUTLINE

Lincoln Potter/Tony Stone Images

A burning building is an example of a rapid, highly exothermic reaction. Firefighters use basic principles of chemical kinetics to battle the fire. When water is sprayed onto a fire, its evaporation absorbs a large amount of energy; this lowers the temperature and slows the reaction. Other common methods for extinguishing fires include covering them with CO_2 (as with most household extinguishers), which decreases the supply of oxygen, and backburning (for grass and forest fires), which removes combustible material. In both cases, the removal of a reactant slows (or stops) the reaction.

OBJECTIVES

After you have studied this chapter, you should be able to

- Express the rate of a chemical reaction in terms of changes in concentrations of reactants and products with time
- Describe the experimental factors that affect the rates of chemical reactions
- Use the rate-law expression for a reaction—the relationship between concentration and rate
- Use the concept of order of a reaction
- Apply the method of initial rates to find the rate-law expression for a reaction
- Use the integrated rate-law expression for a reaction—the relationship between concentration and time
- Analyze concentration-versus-time data to determine the order of a reaction
- Describe the collision theory of reaction rates
- Describe the main aspects of transition state theory and the role of activation energy in determining the rate of a reaction
- Explain how the mechanism of a reaction is related to its rate-law expression
- Predict the rate-law expression that would result from a proposed reaction mechanism
- Identify reactants, products, intermediates, and catalysts in a multistep reaction mechanism
- Explain how temperature affects rates of reactions
- Use the Arrhenius equation to relate the activation energy for a reaction to changes in its rate constant with changing temperature
- Explain how a catalyst changes the rate of a reaction
- Describe homogeneous catalysis and heterogeneous catalysis

CD-ROM Screens 6.3, Control of Chemical Reactions: Thermodynamics and Kinetics; and 19.2, Reaction Favorability: Thermodynamics and Kinetics.

⚠ Remember that *spontaneous* is a thermodynamic term for a reaction that releases free energy, which could potentially be used to do work. It does not mean that a reaction is *instantaneous*. Only by studying the kinetics of a spontaneous reaction can we determine how rapidly it will proceed.

This is one of the reactions that occurs in a human digestive system when an antacid containing relatively insoluble magnesium hydroxide neutralizes excess stomach acid.

Recall that the units kJ/mol refer to the numbers of moles of reactants and products in the balanced equation.

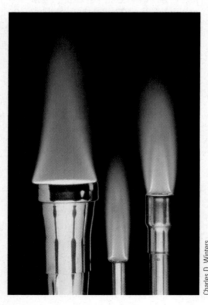

Propane burning is a very rapid reaction.

We are all familiar with processes in which some quantity changes with time—a car travels at 40 miles/hour, a faucet delivers water at 3 gallons/minute, or a factory produces 32,000 tires/day. Each of these ratios is called a rate. The **rate of a reaction** describes how fast reactants are used up and products are formed. **Chemical kinetics** is the study of rates of chemical reactions, the factors that affect reaction rates, and the mechanisms (the series of steps) by which reactions occur.

Our experience tells us that different chemical reactions occur at very different rates. For instance, combustion reactions—such as the burning of methane, CH_4, in natural gas and the combustion of isooctane, C_8H_{18}, in gasoline—proceed very rapidly, sometimes even explosively.

$$CH_4(g) + 2O_2(g) \longrightarrow CO_2(g) + 2H_2O(g)$$
$$2C_8H_{18}(g) + 25O_2(g) \longrightarrow 16CO_2(g) + 18H_2O(g)$$

On the other hand, the rusting of iron occurs only very slowly.

In our study of thermodynamics, we learned to assess whether a particular reaction was spontaneous or not and how much energy was released or absorbed. The question of how rapidly a reaction proceeds is addressed by kinetics. Even though a reaction is thermodynamically spontaneous, it might not occur at a measurable rate.

The reactions of strong acids with strong bases are thermodynamically favored and occur at very rapid rates. Consider, for example, the reaction of hydrochloric acid solution with solid magnesium hydroxide. It is thermodynamically spontaneous at standard state conditions, as indicated by the negative ΔG_{rxn}^0 value. It also occurs rapidly.

$$2HCl(aq) + Mg(OH)_2(s) \longrightarrow MgCl_2(aq) + 2H_2O(\ell) \qquad \Delta G_{rxn}^0 = -97 \text{ kJ/mol}$$

The reaction of diamond with oxygen is also thermodynamically spontaneous.

$$C(diamond) + O_2(g) \longrightarrow CO_2(g) \qquad \Delta G_{rxn}^0 = -397 \text{ kJ/mol}$$

However, we know from experience that diamonds exposed to air, even over long periods, do not react to form carbon dioxide. The reaction does not occur at an observable rate near room temperature.

The reaction of graphite with oxygen is also spontaneous, with a similar value of ΔG_{rxn}^0, -394 kJ/mol. Once it is started, this reaction occurs rapidly. These observations of reaction speeds are explained by kinetics, not thermodynamics.

16-1 THE RATE OF A REACTION

Rates of reactions are usually expressed in units of moles per liter per unit time. If we know the chemical equation for a reaction, its rate can be determined by following the change in concentration of any product or reactant that can be detected quantitatively.

To describe the rate of a reaction, we must determine the concentration of a reactant or product at various times as the reaction proceeds. Devising effective methods for this is a continuing challenge for chemists who study chemical kinetics. If a reaction is slow enough, we can take samples from the reaction mixture after successive time intervals and then analyze them. For instance, if one reaction product is an acid, its concentration can be determined by titration (Section 11-2) after each time interval. The reaction of ethyl acetate with water in the presence of a small amount of strong acid produces acetic acid. The extent of the reaction at any time can be determined by titration of the acetic acid.

Charles D. Winters

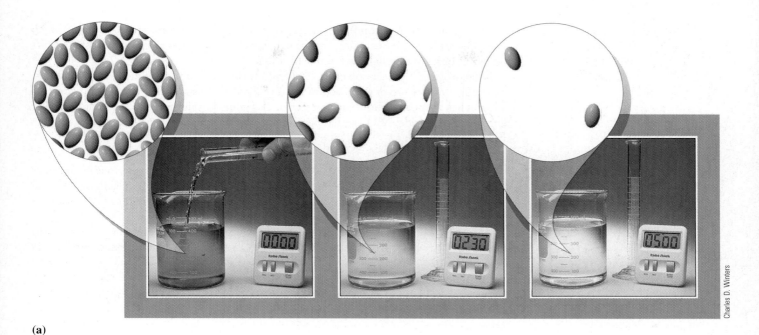

(a)

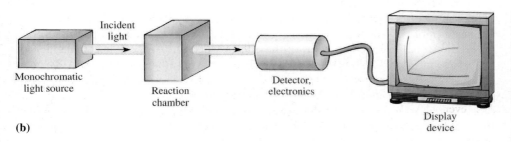

(b)

$$\underset{\text{ethyl acetate}}{CH_3\overset{\displaystyle O}{\overset{\|}{C}}\!-\!OCH_2CH_3(aq)} + H_2O(\ell) \xrightarrow{H^+} \underset{\text{acetic acid}}{CH_3\overset{\displaystyle O}{\overset{\|}{C}}\!-\!OH(aq)} + \underset{\text{ethanol}}{CH_3CH_2OH(aq)}$$

Figure 16-1 (a) Blue dye is reacting with bleach, which converts it into a colorless product. The color decreases and eventually disappears. The rate of the reaction could be determined by repeatedly measuring both the color intensity and the elapsed time. The concentration of dye could be calculated from the intensity of the blue color. (Only the dye molecules are shown for clarity.) (b) A spectroscopic method for determining reaction rates. Light of a wavelength that is absorbed by some substance whose concentration is changing is passed through a reaction chamber. Recording the change in light intensity gives a measure of the changing concentration of a reactant or product as the reaction progresses.

This approach is suitable only if the reaction is sufficiently slow that the time elapsed during withdrawal and analysis of the sample is negligible. Sometimes the sample is withdrawn and then quickly cooled ("quenched"). This slows the reaction (Section 16-8) so much that the desired concentration does not change significantly while the analysis is performed.

It is more convenient, especially when a reaction is rapid, to use a technique that continually monitors the change in some physical property of the system. If one of the reactants or products is colored, the increase (or decrease) in intensity of its color might be used to measure a decrease or increase in its concentration. Such an experiment is a special case of *spectroscopic* methods. These methods involve passing light (visible, infrared, or ultraviolet) through the sample. The light should have a wavelength that is absorbed by some substance whose concentration is changing (Figure 16-1). An appropriate light-sensing apparatus provides a signal that depends on the concentration of the absorbing substance. Modern techniques that use computer-controlled pulsing and sensing of lasers have enabled scientists to sample concentrations at very frequent intervals on the order of picoseconds

 CD-ROM Screen 15.2, Rates of Chemical Reactions.

(1 picosecond $= 10^{-12}$ second) or even femtoseconds (1 femtosecond $= 10^{-15}$ second). Such studies have yielded information about very fast reactions, such as energy transfer resulting from absorption of light in photosynthesis.

If the progress of a reaction causes a change in the total number of moles of gas present, the change in pressure of the reaction mixture (held at constant temperature and constant volume) lets us measure how far the reaction has gone. For instance, the decomposition of dinitrogen pentoxide, $N_2O_5(g)$, has been studied by this method.

$$2N_2O_5(g) \longrightarrow 4NO_2(g) + O_2(g)$$

For every two moles of N_2O_5 gas that react, a total of five moles of gas is formed (four moles of NO_2 and one mole of O_2). The resulting increase in pressure can be related by the ideal gas equation to the total number of moles of gas present. This indicates the extent to which the reaction has occurred.

Once we have measured the changes in concentrations of reactants or products with time, how do we describe the rate of a reaction? Consider a hypothetical reaction.

$$aA + bB \longrightarrow cC + dD$$

In this generalized representation, a represents the coefficient of substance A in the balanced chemical equation, b is the coefficient of substance B, and so on. For example, in an earlier equation given for the decomposition of N_2O_5, $a = 2$, A represents N_2O_5, $c = 4$, C represents NO_2, and so on.

The amount of each substance present can be given by its concentration, usually expressed as molarity (mol/L) and designated by brackets. The rate at which the reaction proceeds can be described in terms of the rate at which one of the reactants disappears, $-\Delta[A]/\Delta t$ or $-\Delta[B]/\Delta t$, or the rate at which one of the products appears, $\Delta[C]/\Delta t$ or $\Delta[D]/\Delta t$. The reaction rate must be positive because it describes the forward (left-to-right) reaction, which consumes A and B. The concentrations of reactants A and B decrease in the time interval Δt. Thus, $\Delta[A]/\Delta t$ and $\Delta[B]/\Delta t$ would be *negative* quantities. The purpose of the negative sign in the definition when using a reactant is to make the rate a positive quantity.

If no other reaction takes place, the changes in concentration are related to one another. For every a mol/L that [A] decreases, [B] must decrease by b mol/L, [C] must increase by c mol/L, and so on. We wish to describe the rate of reaction on a basis that is the same regardless of which reactant or product we choose to measure. To do this, we can describe the number of *moles of reaction* that occur per liter in a given time. For instance, this is accomplished for reactant A as

$$\left(\frac{1 \text{ mol rxn}}{a \text{ mol A}}\right)\left(\begin{array}{c}\text{rate of decrease}\\\text{in [A]}\end{array}\right) = -\left(\frac{1 \text{ mol rxn}}{a \text{ mol A}}\right)\left(\frac{\Delta[A]}{\Delta t}\right)$$

The units for rate of reaction are $\dfrac{\text{mol rxn}}{\text{L·time}}$, which we usually shorten to $\dfrac{\text{mol}}{\text{L·time}}$ or $\text{mol·L}^{-1}\text{·time}^{-1}$. The units $\dfrac{\text{mol}}{\text{L}}$ represent molarity, M, so the units for rate of reaction can also be written as $\dfrac{M}{\text{time}}$ or $M\text{·time}^{-1}$. Similarly, we can divide each concentration change by its coefficient in the balanced equation. Bringing signs to the beginning of each term, we write the rate of reaction based on the rate of change of concentration of each species.

In reactions involving gases, rates of reactions may be related to rates of change of partial pressures. Pressures of gases and concentrations of gases are directly proportional.

$$PV = nRT \quad \text{or} \quad P = \frac{n}{V}RT = MRT$$

where M is molarity.

The Greek "delta," Δ, stands for "change in," just as it did in Chapter 15.

$$\text{rate of reaction} = \frac{1}{a}\left(\begin{array}{c}\text{rate of} \\ \text{decrease} \\ \text{in [A]}\end{array}\right) = \frac{1}{b}\left(\begin{array}{c}\text{rate of} \\ \text{decrease} \\ \text{in [B]}\end{array}\right) = \frac{1}{c}\left(\begin{array}{c}\text{rate of} \\ \text{increase} \\ \text{in [C]}\end{array}\right) = \frac{1}{d}\left(\begin{array}{c}\text{rate of} \\ \text{increase} \\ \text{in [D]}\end{array}\right)$$

$$\text{rate of reaction} = -\frac{1}{a}\left(\frac{\Delta[A]}{\Delta t}\right) = -\frac{1}{b}\left(\frac{\Delta[B]}{\Delta t}\right) = \frac{1}{c}\left(\frac{\Delta[C]}{\Delta t}\right) = \frac{1}{d}\left(\frac{\Delta[D]}{\Delta t}\right)$$

This representation gives several equalities, any one of which can be used to relate changes in observed concentrations to the rate of reaction.

The expressions just given describe the *average* rate over a time period Δt. The rigorous expressions for the rate at any instant involve the derivatives of concentrations with respect to time.

$$-\frac{1}{a}\left(\frac{d[A]}{dt}\right), \frac{1}{c}\left(\frac{d[C]}{dt}\right), \text{ and so on.}$$

The shorter the time period, the closer $\dfrac{\Delta(\text{concentration})}{\Delta t}$ is to the corresponding derivative.

✓ **Problem-Solving Tip:** *Signs and Divisors in Expressions for Rate*

As an analogy to these chemical reaction rate expressions, suppose we make sardine sandwiches by the following procedure:

$$2 \text{ bread slices} + 3 \text{ sardines} + 1 \text{ pickle} \longrightarrow 1 \text{ sandwich}$$

As time goes by, the number of sandwiches increases, so $\Delta(\text{sandwiches})$ is positive; the rate of the process is given by $\Delta(\text{sandwiches})/\Delta(\text{time})$. Alternatively, we would count the decreasing number of pickles at various times. Because $\Delta(\text{pickles})$ is negative, we must multiply by (-1) to make the rate positive; rate $= -\Delta(\text{pickles})/\Delta(\text{time})$. If we measure the rate by counting slices of bread, we must also take into account that bread slices are consumed *twice as fast* as sandwiches are produced, so rate $= -\frac{1}{2}(\Delta(\text{bread})/\Delta(\text{time}))$. Four different ways of describing the rate all have the same numerical value.

$$\text{rate} = \left(\frac{\Delta(\text{sandwiches})}{\Delta t}\right) = -\frac{1}{2}\left(\frac{\Delta(\text{bread})}{\Delta t}\right) = -\frac{1}{3}\left(\frac{\Delta(\text{sardines})}{\Delta t}\right) = -\left(\frac{\Delta(\text{pickles})}{\Delta t}\right)$$

Consider as a specific chemical example the gas-phase reaction that occurs when we mix 1.000 mole of hydrogen and 2.000 moles of iodine chloride at 230°C in a closed 1.000-liter container.

$$H_2(g) + 2ICl(g) \longrightarrow I_2(g) + 2HCl(g)$$

The coefficients tell us that one mole of H_2 disappears for every two moles of ICl that disappear and for every one mole of I_2 and two moles of HCl that are formed. In other terms, the rate of disappearance of moles of H_2 is one-half the rate of disappearance of moles of ICl, and so on. So we write the rate of reaction as

$$\text{rate of reaction} = \begin{pmatrix} \text{rate of} \\ \text{decrease} \\ \text{in } [H_2] \end{pmatrix} = \frac{1}{2}\begin{pmatrix} \text{rate of} \\ \text{decrease} \\ \text{in } [ICl] \end{pmatrix} = \begin{pmatrix} \text{rate of} \\ \text{increase} \\ \text{in } [I_2] \end{pmatrix} = \frac{1}{2}\begin{pmatrix} \text{rate of} \\ \text{increase} \\ \text{in } [HCl] \end{pmatrix}$$

$$\text{rate of reaction} = -\left(\frac{\Delta[H_2]}{\Delta t}\right) = -\frac{1}{2}\left(\frac{\Delta[ICl]}{\Delta t}\right) = \left(\frac{\Delta[I_2]}{\Delta t}\right) = \frac{1}{2}\left(\frac{\Delta[HCl]}{\Delta t}\right)$$

Table 16-1 lists the concentrations of reactants remaining at 1-second intervals, beginning with the time of mixing ($t = 0$ seconds). The *average* rate of reaction over each 1-second interval is indicated in terms of the rate of decrease in concentration of hydrogen. Verify for yourself that the rate of loss of ICl is twice that of H_2. Therefore, the rate of reaction could also be expressed as rate $= -\frac{1}{2}(\Delta[ICl]/\Delta t)$. Increases in concentration of products could be used instead. Figure 16-2 shows graphically the rates of change of concentrations of all reactants and products.

TABLE 16-1	*Concentration and Rate Data for Reaction of 2.000 M ICl and 1.000 M H_2 at 230°C*		
		Average Rate During	
		One Time Interval $= -\dfrac{\Delta[H_2]}{\Delta t}$	
[ICl] (mol/L)	**[H$_2$]** (mol/L)	**($M \cdot s^{-1}$)**	**Time (t)** (seconds)
2.000	1.000		0
1.348	0.674	0.326	1
1.052	0.526	0.148	2
0.872	0.436	0.090	3
0.748	0.374	0.062	4
0.656	0.328	0.046	5
0.586	0.293	0.035	6
0.530	0.265	0.028	7
0.484	0.242	0.023	8

For example, the *average* rate over the interval from 1 to 2 seconds can be calculated as

$$-\frac{\Delta[H_2]}{\Delta t} =$$

$$-\frac{(0.526 - 0.674)\ \text{mol} \cdot L^{-1}}{(2 - 1)\ s}$$

$$= 0.148\ \text{mol} \cdot L^{-1}s^{-1}$$

$$= 0.148\ M \cdot s^{-1}$$

This does *not* mean that the reaction proceeds at this rate during the entire interval.

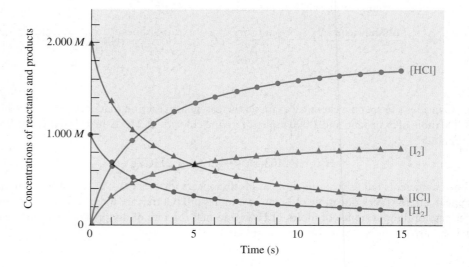

Figure 16-2 Plot of concentrations of all reactants and products versus time in the reaction of 1.000 *M* H_2 with 2.000 *M* ICl at 230°C, from data in Table 16-1 (and a few more points).

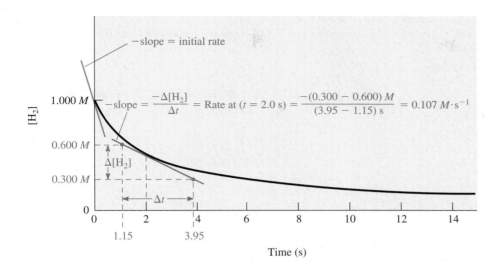

Figure 16-3 Plot of H_2 concentration versus time for the reaction of 1.000 M H_2 with 2.000 M ICl. The instantaneous rate of reaction at any time, t, equals the negative of the slope of the tangent to this curve at time t. The initial rate of the reaction is equal to the negative of the initial slope ($t = 0$). We take the negative of the slope in order to get a positive rate. The determination of the instantaneous rate at $t = 2$ seconds is illustrated. (If you do not recall how to find the slope of a straight line, refer to Figure 16-5.)

Figure 16-3 is a plot of the hydrogen concentration versus time, using data of Table 16-1. The initial rate, or the rate at the instant of mixing the reactants, is the negative of the slope at $t = 0$. The *instantaneous* rate of reaction at time t (2.0 seconds, for example) is the negative of the slope of the tangent to the curve at time t. We see that the rate decreases with time; lower concentrations of H_2 and ICl result in slower reaction. Had we plotted concentration of a product versus time, the rate would have been related to the *positive* slope of the tangent at time t.

Suppose a driver travels 40 miles in an hour; we describe his average speed (rate) as 40 mi/h. This does not necessarily mean that he drove at a steady speed. He might have stopped at a few traffic signals, made a fuel stop, driven sometimes faster, sometimes slower—his *instantaneous rate* (the rate at which he was traveling at any instant) was quite changeable.

EXAMPLE 16-1 *Rate of Reaction*

At some time, we observe that the reaction $2N_2O_5(g) \rightarrow 4NO_2(g) + O_2(g)$ is forming NO_2 at the rate of $0.0072 \dfrac{mol}{L \cdot s}$.

(a) What is the rate of change of $[O_2]$, $\dfrac{\Delta[O_2]}{\Delta t}$, in $\dfrac{mol}{L \cdot s}$, at this time?

(b) What is the rate of change of $[N_2O_5]$, $\dfrac{\Delta[N_2O_5]}{\Delta t}$, in $\dfrac{mol}{L \cdot s}$, at this time?

(c) What is the rate of reaction at this time?

Plan

We can use the mole ratios from the balanced equation to determine the rates of change of other products and reactants. The rate of reaction can then be derived from any one of these individual rates.

Solution

(a) The balanced equation gives the reaction ratio $\dfrac{1 \text{ mol } O_2}{4 \text{ mol } NO_2}$.

$$\text{rate of change of } [O_2] = \frac{\Delta[O_2]}{\Delta t} = \frac{0.0072 \text{ mol } NO_2}{L \cdot s} \times \frac{1 \text{ mol } O_2}{4 \text{ mol } NO_2} = 0.0018 \frac{\text{mol } O_2}{L \cdot s}$$

(b) The balanced equation shows that 2 mol N_2O_5 is *consumed* for every 4 mol NO_2 that is *formed*. Because $[N_2O_5]$ is decreasing as $[NO_2]$ increases, we should write the reaction ratio as $\dfrac{-2 \text{ mol } N_2O_5}{4 \text{ mol } NO_2}$.

$$\text{rate of change of } [N_2O_5] = \frac{\Delta[N_2O_5]}{\Delta t} = \frac{0.0072 \text{ mol } NO_2}{L \cdot s} \times \frac{-2 \text{ mol } N_2O_5}{4 \text{ mol } NO_2}$$

$$= \boxed{-0.0036 \; \frac{\text{mol } N_2O_5}{L \cdot s}}$$

The rate of *change* of $[N_2O_5]$ with time, $\dfrac{\Delta[N_2O_5]}{\Delta t}$, is $-0.0036 \; \dfrac{\text{mol } N_2O_5}{L \cdot s}$, a *negative* number, because N_2O_5, a reactant, is being used up.

(c) The rate of reaction can be calculated from the rate of decrease of any reactant concentration or the rate of increase of any product concentration.

$$\text{rate of reaction} = -\frac{1}{2}\left(\frac{\Delta[N_2O_5]}{\Delta t}\right) = -\frac{1}{2}\left(-0.0036 \; \frac{\text{mol}}{L \cdot s}\right) = \boxed{0.0018 \; \frac{\text{mol}}{L \cdot s}}$$

$$\text{rate of reaction} = \frac{1}{4}\left(\frac{\Delta[NO_2]}{\Delta t}\right) = \frac{1}{4}\left(0.0072 \; \frac{\text{mol}}{L \cdot s}\right) = \boxed{0.0018 \; \frac{\text{mol}}{L \cdot s}}$$

$$\text{rate of reaction} = \frac{1}{1}\left(\frac{\Delta[O_2]}{\Delta t}\right) = \boxed{0.0018 \; \frac{\text{mol}}{L \cdot s}}$$

We see that the rate of reaction is the same, no matter which reactant or product we use to determine it. Remember that the mol in these units is interpreted as "moles of reaction."

You should now work Exercise 10.

> **⚠ Always remember to divide by the balanced reactant or product coefficients in order to convert the individual reactant and product rates into an overall reaction rate. The overall reaction rate is always positive and expressed on a per mole basis.**

FACTORS THAT AFFECT REACTION RATES

SC*i*LINKS. TOPIC: Reaction Rate
GO TO: www.scilinks.org
*sci*LINKS CODE: WCH1610

Often we want a reaction to take place rapidly enough to be practical but not so rapidly as to be dangerous. The controlled burning of fuel in an internal combustion engine is an example of such a process. On the other hand, we want some undesirable reactions, such as the spoiling of food, to take place more slowly.

Four factors have marked effects on the rates of chemical reactions. They are (1) nature of the reactants, (2) concentrations of the reactants, (3) temperature, and (4) the presence of a catalyst. Understanding their effects can help us control the rates of reactions in desirable ways. The study of these factors gives important insight into the details of the processes by which a reaction occurs. This kind of study is the basis for developing theories of chemical kinetics. Now we study these factors and the related theories—collision theory and transition state theory.

16-2 NATURE OF THE REACTANTS

The physical states of reacting substances are important in determining their reactivities. A puddle of liquid gasoline can burn smoothly, but gasoline vapors can burn explosively. Two immiscible liquids may react slowly at their interface, but if they are intimately mixed to provide better contact, the reaction speeds up. White phosphorus and red phosphorus are different solid forms (allotropes) of elemental phosphorus. White phosphorus ignites when exposed to oxygen in the air. By contrast, red phosphorus can be kept in open containers for long periods of time without noticeable reaction.

Samples of dry solid potassium sulfate, K_2SO_4, and dry solid barium nitrate, $Ba(NO_3)_2$, can be mixed with no appreciable reaction occurring for several years. But if aqueous solutions of the two are mixed, a reaction occurs rapidly, forming a white precipitate of barium sulfate.

$$Ba^{2+}(aq) + SO_4^{2-}(aq) \longrightarrow BaSO_4(s) \qquad \text{(net ionic equation)}$$

Chemical identities of elements and compounds affect reaction rates. Metallic sodium, with its low ionization energy, reacts rapidly with water at room temperature; metallic calcium has a higher ionization energy and reacts only slowly with water at room temperature. Solutions of a strong acid and a strong base react rapidly when they are mixed because the interactions involve mainly electrostatic attractions between ions in solution. Reactions that involve the breaking of covalent bonds are usually slower.

The extent of subdivision of solids or liquids can be crucial in determining reaction rates. Large chunks of most metals do not burn. But many powdered metals, with larger surface areas and hence more atoms exposed to the oxygen of the air, burn easily. One pound of fine iron wire rusts much more rapidly than a solid one-pound chunk of iron. Violent explosions sometimes occur in grain elevators, coal mines, and chemical plants in which large amounts of powdered substances are produced. These explosions are examples of the effect of large surface areas on rates of reaction. The rate of reaction depends on the surface area or degree of subdivision. The ultimate degree of subdivision would make all reactant molecules (or ions or atoms) accessible to react at any given time. This situation can be achieved when the reactants are in the gaseous state or in solution.

Two allotropes of phosphorus. White phosphorus (*above*) ignites and burns rapidly when exposed to oxygen in the air, so it is stored under water. Red phosphorus (*below*) reacts with air much more slowly, and can be stored in contact with air.

CD-ROM   Screen 15.3, Control of Reaction Rates (1): Surface Area.

Powdered chalk (mostly calcium carbonate, $CaCO_3$) reacts rapidly with dilute hydrochloric acid because it has a large total surface area. A stick of chalk has a much smaller surface area, so it reacts much more slowly.

Powdered iron burns very rapidly when heated in a flame. Iron oxide is formed.

16-3 CONCENTRATIONS OF REACTANTS: THE RATE-LAW EXPRESSION

CD-ROM   Screen 15.4, Control of Reaction Rates (2): Concentration Dependence.

As the concentrations of reactants change at constant temperature, the rate of reaction changes. We write the **rate-law expression** (often called simply the **rate law**) for a reaction to describe how its rate depends on concentrations; this rate law is experimentally deduced for each reaction from a study of how its rate varies with concentration.

> The rate-law expression for a reaction in which A, B, . . . are reactants has the general form
>
> $$\text{rate} = k[A]^x[B]^y \dots$$
>
> The constant k is called the **specific rate constant** (or just the **rate constant**) for the reaction at a particular temperature. The values of the exponents, x and y, and of the rate constant, k bear no necessary relationship to the coefficients in the *balanced chemical equation* for the overall reaction and must be determined *experimentally*.

The order of the reaction with respect to a reactant is usually called simply the order of that reactant. The word *order* is used in kinetics in its mathematical meaning. This use is unrelated to the order–disorder discussion of entropy (Chapter 15).

The powers to which the concentrations are raised, x and y, are usually small integers or zero but are occasionally fractional or even negative. A power of *one* means that the rate is directly proportional to the concentration of that reactant. A power of *two* means that the rate is directly proportional to the *square* of that concentration. A power of *zero* means that the rate does not depend on the concentration of that reactant, *so long as some of the reactant is present*. The value of x is said to be the order of the reaction with respect to A, and y is the order of the reaction with respect to B. The overall <u>order of the reaction</u> is $x + y$. Examples of observed rate laws for some reactions follow.

1. $3NO(g) \longrightarrow N_2O(g) + NO_2(g)$

$\text{rate} = k[NO]^2$
second order in NO; second order overall

2. $2NO_2(g) + F_2(g) \longrightarrow 2NO_2F(g)$

$\text{rate} = k[NO_2][F_2]$
first order in NO_2 and first order in F_2;
second order overall

3. $2NO_2(g) \longrightarrow 2NO(g) + O_2(g)$

$\text{rate} = k[NO_2]^2$
second order in NO_2;
second order overall

Any number raised to the zero power is one. Here $[H^+]^0 = 1$.

4. $H_2O_2(aq) + 3I^-(aq) + 2H^+(aq) \longrightarrow 2H_2O(\ell) + I_3^-(aq)$

$\text{rate} = k[H_2O_2][I^-]$
first order in H_2O_2 and first order in I^-;
zero order in H^+; second order overall

> We see that the orders (exponents) in the rate law expression *may* or *may not* match the coefficients in the balanced equation. There is *no* way to predict reaction orders from the balanced overall chemical equation. The orders must be determined experimentally.

More details about values and units of k will be discussed in later sections.

It is important to remember the following points about this specific rate constant, k.

1. Once the reaction orders are known, experimental data must be used to determine the value of k for the reaction at appropriate conditions.

2. The value we determine is for a *specific reaction*, represented by a balanced equation.

3. The units of k depend on the *overall order* of the reaction.

4. The value we determine does not change with concentrations of either reactants or products.

5. The value we determine does not change with time (Section 16-4).

6. The value we determine refers to the reaction *at a particular temperature* and changes if we change the temperature (Section 16-8).

7. The value we determine depends on whether a *catalyst* is present (Section 16-9).

EXAMPLE 16-2 *Interpretation of the Rate Law*

For a hypothetical reaction

$$A + B + C \longrightarrow products$$

the rate law is determined to be

$$rate = k[A][B]^2$$

What happens to the reaction rate when we make each of the following concentration changes? (a) We double the concentration of A without changing the concentration of B or C. (b) We double the concentration of B without changing the concentration of A or C. (c) We double the concentration of C without changing the concentration of A or B. (d) We double all three concentrations simultaneously.

Plan

We interpret the rate law to predict the changes in reaction rate. We remember that changing concentrations does not change the value of k.

Solution

(a) We see that rate is directly proportional to the *first power* of [A]. We do not change [B] or [C]. Doubling [A] (i.e., increasing [A] by a factor of 2) causes the reaction rate to increase by a factor of $2^1 = 2$ so the reaction rate doubles.

(b) We see that rate is directly proportional to the *second power* of [B], that is $[B]^2$. We do not change [A] or [C]. Doubling [B] (i.e., increasing [B] by a factor of 2) causes the reaction rate to increase by a factor of $2^2 = 4$.

(c) The reaction rate is independent of [C], so changing [C] causes no change in reaction rate.

(d) Doubling all concentrations would cause the changes described in (a), (b), and (c) simultaneously. The rate would increase by a factor of 2 due to the change in [A], by a factor of 4 due to the change in [B], and be unaffected by the change in [C]. The result is that the reaction rate increases by a factor of $2^1 \times 2^2 = 8$.

You should now work Exercises 14 and 15.

CD-ROM   Screen 15.5, Determination of Rate Equations (1): Method of Initial Rates.

We can use the **method of initial rates** to deduce the rate law from experimentally measured rate data. Usually we know the concentrations of all reactants at the start of the reaction. We can then measure the initial rate of the reaction, corresponding to these initial concentrations. The following tabulated data refer to the hypothetical reaction

$$A + 2B \longrightarrow C$$

at a specific temperature. The brackets indicate the concentrations of the reacting species *at the beginning* of each experimental run listed in the first column—that is, the initial concentrations for each experiment.

In such an experiment we often keep some initial concentrations the same and vary others by simple factors, such as 2 or 3. This makes it easier to access the effect of each change on the rate.

Experiment	Initial [A]	Initial [B]	Initial Rate of Formation of C
1	$1.0 \times 10^{-2}\ M$	$1.0 \times 10^{-2}\ M$	$1.5 \times 10^{-6}\ M \cdot s^{-1}$
2	$1.0 \times 10^{-2}\ M$	$2.0 \times 10^{-2}\ M$	$3.0 \times 10^{-6}\ M \cdot s^{-1}$
3	$2.0 \times 10^{-2}\ M$	$1.0 \times 10^{-2}\ M$	$6.0 \times 10^{-6}\ M \cdot s^{-1}$

Because we are describing the same reaction in each experiment, each is governed by the same rate-law expression. This expression has the form

$$\text{rate} = k[A]^x[B]^y$$

Let's compare the initial rates of formation of product (reaction rates) for different experimental runs to see how changes in concentrations of reactants affect the rate of reaction. This lets us calculate the values of *x* and *y*, and then *k*.

We see that the initial concentration of A is the same in experiments 1 and 2; for these trials, any change in reaction rate would be due to different initial concentrations of B. Comparing these two experiments, we see that [B] has been changed by a factor of

$$\frac{2.0 \times 10^{-2}}{1.0 \times 10^{-2}} = 2.0 = [B]\ \text{ratio}$$

Whenever possible pick two experiments where only one reactant has different initial concentrations. We can calculate the order for this reactant from the effect of the concentration change on the initial rate of reaction.

The rate changes by a factor of

$$\frac{3.0 \times 10^{-6}}{1.5 \times 10^{-6}} = 2.0 = \text{rate ratio}$$

The exponent *y* can be deduced from

$$\text{rate ratio} = ([B]\ \text{ratio})^y$$

$$2.0 = (2.0)^y \quad \text{so} \quad y = 1$$

The reaction is first order in [B]. Thus far we know that the rate expression is

$$\text{rate} = k[A]^x[B]^1$$

To evaluate *x*, we observe that the concentrations of [A] are different in experiments 1 and 3. For these two trials, the initial concentration of B is the same, so any change in reaction rate could only be due to different initial concentrations of A. Comparing these two experiments, we see that [A] has been increased by a factor of

$$\frac{2.0 \times 10^{-2}}{1.0 \times 10^{-2}} = 2.0 = [A]\ \text{ratio}$$

The rate increases by a factor of

$$\frac{6.0 \times 10^{-6}}{1.5 \times 10^{-6}} = 4.0 = \text{rate ratio}$$

The exponent x can be deduced from

$$\text{rate ratio} = (\text{[A] ratio})^x$$

$$4.0 = (2.0)^x \qquad \text{so} \qquad x = 2$$

The reaction is second order in [A]. We can now write its rate-law expression as

$$\text{rate} = k[A]^2[B]$$

Now that we know the orders, the specific rate constant, k, can be evaluated by substituting any of the three sets of data into the rate-law expression. Using the data from experiment 1 gives

$$\text{rate}_1 = k[A]_1{}^2[B]_1 \qquad \text{or} \qquad k = \frac{\text{rate}_1}{[A]_1{}^2[B]_1}$$

$$k = \frac{1.5 \times 10^{-6}\ M \cdot s^{-1}}{(1.0 \times 10^{-2}\ M)^2(1.0 \times 10^{-2}\ M)} = 1.5\ M^{-2} \cdot s^{-1}$$

At the temperature at which the measurements were made, the rate-law expression for this reaction is

$$\text{rate} = k[A]^2[B] \qquad \text{or} \qquad \text{rate} = 1.5\ M^{-2} \cdot s^{-1}\ [A]^2[B]$$

We can check our result by evaluating k from one of the other sets of data.

 Remember that the specific rate constant k does *not* change with concentration. Only a temperature change or the introduction of a catalyst can change the value of k.

The units of k depend on the overall order of the reaction, consistent with converting the product of concentrations on the right to concentration/time on the left. For any reaction that is third order overall, the units of k are $M^{-2} \cdot \text{time}^{-1}$.

✓ Problem-Solving Tip: *Be Sure to Use the Rate of Reaction*

The rate-law expression should always give the dependence of the *rate of reaction* on concentrations. The data for the preceding calculation describe the rate of formation of the product C; the coefficient of C in the balanced equation is one, so the rate of reaction is equal to the rate of formation of C. If the coefficient of the measured substance had been 2, then before we began the analysis we should have divided each value of the "initial rate of formation" by 2 to obtain the initial rate of reaction. For instance, suppose we measure the rate of formation of AB in the reaction

$$A_2 + B_2 \longrightarrow 2AB$$

Then

$$\text{rate of increase} = \frac{1}{2}\left(\frac{\Delta[AB]}{\Delta t}\right) = \frac{1}{2}\ (\text{rate of formation of AB})$$

Then we would analyze how this reaction rate changes as we change the concentrations of reactants.

An Alternative Method

We can also use a simple algebraic approach to find the exponents in a rate-law expression. Consider the set of rate data given earlier for the hypothetical reaction

$$A + 2B \longrightarrow C$$

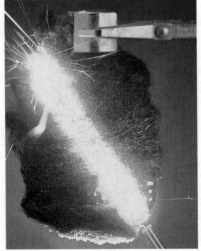

Charles D. Winters

When heated in air, steel wool glows but does not burn rapidly, due to the low O_2 concentration in air (about 21%). When pure oxygen is passed through the center of the steel wool via a porous tube, the steel wool burns vigorously because of the much greater accessibility of O_2 reactant molecules.

Experiment	Initial [A] (M)	Initial [B] (M)	Initial Rate of Formation of C (M·s^{-1})
1	1.0×10^{-2}	1.0×10^{-2}	1.5×10^{-6}
2	1.0×10^{-2}	2.0×10^{-2}	3.0×10^{-6}
3	2.0×10^{-2}	1.0×10^{-2}	6.0×10^{-6}

Because we are describing the same reaction in each experiment, all the experiments are governed by the same rate-law expression,

$$\text{rate} = k[A]^x[B]^y$$

The initial concentration of A is the same in experiments 1 and 2, so any change in the initial rates for these experiments would be due to different initial concentrations of B. To evaluate y, we solve the ratio of the rate-law expressions of these two experiments for y. We can divide the first rate-law expression by the corresponding terms in the second rate-law expression.

$$\frac{\text{rate}_1}{\text{rate}_2} = \frac{k[A]_1{}^x[B]_1{}^y}{k[A]_2{}^x[B]_2{}^y}$$

The value of k always cancels from such a ratio because it is constant at a particular temperature. The initial concentrations of A are equal, so they too cancel. Thus, the expression simplifies to

$$\frac{\text{rate}_1}{\text{rate}_2} = \left(\frac{[B]_1}{[B]_2}\right)^y$$

The only unknown in this equation is y. We substitute data from experiments 1 and 2 into the equation, which gives us

$$\frac{1.5 \times 10^{-6}\,M \cdot s^{-1}}{3.0 \times 10^{-6}\,M \cdot s^{-1}} = \left(\frac{1.0 \times 10^{-2}\,M}{2.0 \times 10^{-2}\,M}\right)^y$$

$$0.5 = (0.5)^y \qquad \text{so} \qquad y = 1$$

Thus far, we know that the rate-law expression is

$$\text{rate} = k[A]^x[B]^1$$

Next we evaluate x. In experiments 1 and 3, the initial concentration of B is the same, so any change in the initial rates for these experiments would be due to the different initial concentrations of A. We solve the ratio of the rate-law expressions of these two experiments for x. We divide the third rate-law expression by the corresponding terms in the first rate-law expression.

$$\frac{\text{rate}_3}{\text{rate}_1} = \frac{k[A]_3{}^x[B]_3{}^1}{k[A]_1{}^x[B]_1{}^1}$$

The value k cancels, and so do the concentrations of B because they are equal. Thus, the expression simplifies to

$$\frac{\text{rate}_3}{\text{rate}_1} = \frac{[A]_3{}^x}{[A]_1{}^x} = \left(\frac{[A]_3}{[A]_1}\right)^x$$

It does not matter which way we take the ratio. We would get the same value for y if we divided the second rate-law expression by the first— try it!

$$\frac{6.0 \times 10^{-6} \, M \cdot s^{-1}}{1.5 \times 10^{-6} \, M \cdot s^{-1}} = \left(\frac{2.0 \times 10^{-2} \, M}{1.0 \times 10^{-2} \, M} \right)^x$$

$$4.0 = (2.0)^x \quad \text{so} \quad x = 2$$

The power to which [A] is raised in the rate-law expression is 2, so the rate-law expression for this reaction is the same as that obtained earlier.

$$\text{rate} = k[A]^2[B]^1 \quad \text{or} \quad \text{rate} = k[A]^2[B]$$

EXAMPLE 16-3 *Method of Initial Rates*

Given the following data, determine the rate-law expression and the value of the rate constant for the reaction

$$2A + B + C \longrightarrow D + E$$

Experiment	Initial [A]	Initial [B]	Initial [C]	Initial Rate of Formation of E
1	0.20 M	0.20 M	0.20 M	$2.4 \times 10^{-6} \, M \cdot \text{min}^{-1}$
2	0.40 M	0.30 M	0.20 M	$9.6 \times 10^{-6} \, M \cdot \text{min}^{-1}$
3	0.20 M	0.30 M	0.20 M	$2.4 \times 10^{-6} \, M \cdot \text{min}^{-1}$
4	0.20 M	0.40 M	0.60 M	$7.2 \times 10^{-6} \, M \cdot \text{min}^{-1}$

The coefficient of E in the balanced equation is 1, so the rate of reaction is equal to the rate of formation of E.

Plan

The rate law is of the form rate = $k[A]^x[B]^y[C]^z$. We must evaluate x, y, z, and k. We use the reasoning outlined earlier; in this presentation the first method is used.

The alternative algebraic method outlined previously can also be used.

Solution

Dependence on [B]: In experiments 1 and 3, the initial concentrations of A and C are unchanged. Thus, any change in the rate would be due to the change in concentration of B. But we see that the rate is the same in experiments 1 and 3, even though the concentration of B is different. Thus, the reaction rate is independent of [B], so $y = 0$. We can neglect changes in [B] in the subsequent reasoning. The rate law must be

$[B]^0 = 1$

$$\text{rate} = k[A]^x[C]^z$$

Dependence on [C]: Experiments 1 and 4 involve the same initial concentration of A; thus the observed change in rate must be due entirely to the changed [C]. So we compare experiments 1 and 4 to find z.

$$[C] \text{ has increased by a factor of } \frac{0.60}{0.20} = 3.0 = [C] \text{ ratio}$$

For convenience we usually set up the ratios with the larger concentration on top to give a ratio > 1.

The rate changes by a factor of

$$\frac{7.2 \times 10^{-6}}{2.4 \times 10^{-6}} = 3.0 = \text{rate ratio}$$

The exponent z can be deduced from

$$\text{rate ratio} = ([C] \text{ ratio})^z$$

$$3.0 = (3.0)^z \quad \text{so} \quad z = 1 \qquad \text{The reaction is first order in [C].}$$

Now we know that the rate law is of the form

$$\text{rate} = k[A]^x[C]$$

Dependence on [A]: We use experiments 1 and 2 to evaluate x, because [A] is changed, [B] does not matter, and [C] is unaltered. The observed rate change is due *only* to the changed [A].

Note that the coefficient of 2 for reactant A cancels out in the ratio, so it does not play a role here.

$$[A] \text{ has increased by a factor of } \frac{0.40}{0.20} = 2.0 = [A] \text{ ratio}$$

The rate increases by a factor of

$$\frac{9.6 \times 10^{-6}}{2.4 \times 10^{-6}} = 4.0 = \text{rate ratio}$$

The exponent x can be deduced from

$$\text{rate ratio} = ([A] \text{ ratio})^x$$

$$4.0 = (2.0)^x \quad \text{so} \quad \boxed{x = 2} \quad \text{The reaction is second order in [A].}$$

From these results we can write the complete rate-law expression.

$$\text{rate} = k[A]^2[B]^0[C]^1 \quad \text{or} \quad \boxed{\text{rate} = k[A]^2[C]}$$

We can evaluate the specific rate constant, k, by substituting any of the four sets of data into the rate-law expression we have just derived. Data from experiment 2 give

$$\text{rate}_2 = k[A]_2{}^2[C]_2$$

$$k = \frac{\text{rate}_2}{[A]_2{}^2[C]_2} = \frac{9.6 \times 10^{-6} \, M \cdot \text{min}^{-1}}{(0.40 \, M)^2(0.20 \, M)} = \boxed{3.0 \times 10^{-4} \, M^{-2} \cdot \text{min}^{-1}}$$

The rate-law expression can also be written with the value of k incorporated.

$$\boxed{\text{rate} = 3.0 \times 10^{-4} \, M^{-2} \cdot \text{min}^{-1} \, [A]^2[C]}$$

This expression allows us to calculate the rate at which this reaction occurs with any known concentrations of A and C (provided some B is present). As we shall see presently, changes in temperature change reaction rates. This value of k is valid *only* at the temperature at which the data were collected.

You should now work Exercises 17 and 18.

✓ **Problem-Solving Tip:** *Check the Rate Law You Have Derived*

If the rate law that you deduce from initial rate data is correct, it will not matter which set of data you use to calculate k. As a check, you can calculate k several times, once from each set of experimental concentration and rate data. If the reaction orders in your derived rate law are correct, then all sets of experimental data will give the same value of k (within rounding error); but if the orders are wrong, then the k values will vary considerably.

EXAMPLE 16-4 *Method of Initial Rates*

Use the following initial rate data to determine the form of the rate-law expression for the reaction

$$3A + 2B \longrightarrow 2C + D$$

Experiment	Initial [A]	Initial [B]	Initial Rate of Formation of D
1	$1.00 \times 10^{-2} M$	$1.00 \times 10^{-2} M$	$6.00 \times 10^{-3} M\cdot min^{-1}$
2	$2.00 \times 10^{-2} M$	$3.00 \times 10^{-2} M$	$1.44 \times 10^{-1} M\cdot min^{-1}$
3	$1.00 \times 10^{-2} M$	$2.00 \times 10^{-2} M$	$1.20 \times 10^{-2} M\cdot min^{-1}$

Plan

The rate law is of the form rate = $k[A]^x[B]^y$. No two experiments have the same initial [B], so let's use the alternative method presented earlier to evaluate x and y.

Solution

The initial concentration of A is the same in experiments 1 and 3. We divide the third rate-law expression by the corresponding terms in the first one

$$\frac{rate_3}{rate_1} = \frac{k[A]_3{}^x[B]_3{}^y}{k[A]_1{}^x[B]_1{}^y}$$

The initial concentrations of A are equal, so they cancel, as does k. Simplifying and then substituting known values of rates and [B],

$$\frac{rate_3}{rate_1} = \frac{[B]_3{}^y}{[B]_1{}^y} \quad \text{or} \quad \frac{1.20 \times 10^{-2}\, M\cdot min}{6.00 \times 10^{-3}\, M\cdot min} = \left(\frac{2.00 \times 10^{-2}\, M}{1.00 \times 10^{-2}\, M}\right)^y$$

$$2.0 = (2.0)^y \quad \text{or} \quad \boxed{y = 1} \quad \text{The reaction is first order in [B].}$$

No two of the experimental runs have the same concentrations of B, so we must proceed somewhat differently. Let's compare experiments 1 and 2. The observed change in rate must be due to the *combination* of the changes in [A] and [B]. We can divide the second rate-law expression by the corresponding terms in the first one, cancel the equal k values, and collect terms.

This is solvable because we have already determined the order of B.

$$\frac{rate_2}{rate_1} = \frac{k[A]_2{}^x[B]_2{}^y}{k[A]_1{}^x[B]_1{}^y} = \left(\frac{[A]_2}{[A]_1}\right)^x\left(\frac{[B]_2}{[B]_1}\right)^y$$

Now let's insert the known values for rates and concentrations and the known [B] exponent of 1.

$$\frac{1.44 \times 10^{-1}\, M\cdot min^{-1}}{6.00 \times 10^{-3}\, M\cdot min^{-1}} = \left(\frac{2.00 \times 10^{-2}\, M}{1.00 \times 10^{-2}\, M}\right)^x\left(\frac{3.00 \times 10^{-2}\, M}{1.00 \times 10^{-2}\, M}\right)^1$$

$$24.0 = (2.00)^x(3.00)$$

$$8.00 = (2.00)^x \quad \text{or} \quad \boxed{x = 3} \quad \text{The reaction is third order in [A].}$$

The rate-law expression has the form $\boxed{\text{rate} = k[A]^3[B].}$

You should now work Exercise 22.

16-4 CONCENTRATION VERSUS TIME: THE INTEGRATED RATE EQUATION

Often we want to know the concentration of a reactant that would remain after some specified time, or how long it would take for some amount of the reactants to be used up.

> The equation that relates *concentration* and *time* is the **integrated rate equation.** We can also use it to calculate the **half-life,** $t_{1/2}$, of a reactant—the time it takes for half of that reactant to be converted into product. The integrated rate equation and the half-life are different for reactions of different order.

CD-ROM Screen 15.6, Concentration–Time Relationships.

We will look at relationships for some simple cases. If you know calculus, you may be interested in the derivation of the integrated rate equations. This development is presented in the Enrichments at the end of this section.

First-Order Reactions

For reactions involving $aA \rightarrow$ products that are *first order in* A and *first order overall*, the integrated rate equation is

a represents the coefficient of reactant A in the balanced overall equation.

$$\ln\left(\frac{[A]_0}{[A]}\right) = akt \qquad \text{(first order)}$$

$[A]_0$ is the initial concentration of reactant A, and $[A]$ is its concentration at some time, t, after the reaction begins. Solving this relationship for t gives

$$t = \frac{1}{ak} \ln\left(\frac{[A]_0}{[A]}\right)$$

When time $t_{1/2}$ has elapsed, half of the original $[A]_0$ has reacted, so half of it remains.

By definition, $[A] = \frac{1}{2}[A]_0$ at $t = t_{1/2}$. Thus

$$t_{1/2} = \frac{1}{ak} \ln\frac{[A]_0}{\frac{1}{2}[A]_0} = \frac{1}{ak} \ln 2$$

$$t_{1/2} = \frac{\ln 2}{ak} = \frac{0.693}{ak} \qquad \text{(first order)}$$

Nuclear decay (Chapter 26) is a very important first-order process. Exercises at the end of that chapter involve calculations of nuclear decay rates.

This relates the half-life of a reactant in a *first-order reaction* and its rate constant, k. In such reactions, the half-life *does not depend* on the initial concentration of A. This is not true for reactions having overall orders other than first order.

EXAMPLE 16-5 Half-Life: First-Order Reaction

Compound A decomposes to form B and C in a reaction that is first order with respect to A and first order overall. At 25°C, the specific rate constant for the reaction is 0.0450 s⁻¹. What is the half-life of A at 25°C?

$$A \longrightarrow B + C$$

Plan

We use the equation given earlier for $t_{1/2}$ for a first-order reaction. The value of k is given in the problem; the coefficient of reactant A is $a = 1$.

CD-ROM   Screen 15.8, Half-Life: First-Order Reactions.

Solution

$$t_{1/2} = \frac{\ln 2}{ak} = \frac{0.693}{1(0.0450 \text{ s}^{-1})} = 15.4 \text{ s}$$

After 15.4 seconds of reaction, half of the original reactant remains, so that $[A] = \frac{1}{2}[A]_0$.

You should now work Exercise 39.

EXAMPLE 16-6   *Concentration Versus Time: First-Order Reaction*

The reaction $2N_2O_5(g) \rightarrow 2N_2O_4(g) + O_2(g)$ obeys the rate law: rate $= k[N_2O_5]$, in which the specific rate constant is 0.00840 s^{-1} at a certain temperature. (a) If 2.50 moles of N_2O_5 were placed in a 5.00-liter container at that temperature, how many moles of N_2O_5 would remain after 1.00 minute? (b) How long would it take for 90% of the original N_2O_5 to react?

If the problem you are solving asks for or gives *time*, use the integrated rate equation. If it asks for or gives *rate*, use the rate-law expression.

Plan

We apply the integrated first-order rate equation.

$$\ln \left(\frac{[N_2O_5]_0}{[N_2O_5]} \right) = akt$$

(a) First, we must determine $[N_2O_5]_0$, the original molar concentration of N_2O_5. Then we solve for $[N_2O_5]$, the molar concentration after 1.00 minute. We must remember to express k and t using the same time units. Finally, we convert molar concentration of N_2O_5 to moles remaining. (b) We solve the integrated first-order equation for the required time.

Solution

The original concentration of N_2O_5 is

$$[N_2O_5]_0 = \frac{2.50 \text{ mol}}{5.00 \text{ L}} = 0.500 \ M$$

The other quantities are

$$a = 2 \quad k = 0.00840 \text{ s}^{-1} \quad t = 1.00 \text{ min} = 60.0 \text{ s} \quad [N_2O_5] = \underline{?}$$

The only unknown in the integrated rate equation is $[N_2O_5]$ after 1.00 minute. Let us solve for the unknown. Because $\ln x/y = \ln x - \ln y$,

Make sure the units of time are the same— chemists often use seconds (s) in specific rate constants.

$$\ln \frac{[N_2O_5]_0}{[N_2O_5]} = \ln [N_2O_5]_0 - \ln [N_2O_5] = akt$$

$$\ln [N_2O_5] = \ln [N_2O_5]_0 - akt$$

$$= \ln (0.500) - (2)(0.00840 \text{ s}^{-1})(60.0 \text{ s}) = -0.693 - 1.008$$

$$\ln [N_2O_5] = -1.701$$

Taking the inverse natural logarithm of both sides gives

$$[N_2O_5] = 1.82 \times 10^{-1} \ M$$

inv $\ln x = e^{\ln x}$

See Appendix A for more detail about the relation between e and $\ln$.

Thus, after 1.00 minute of reaction, the concentration of N_2O_5 is 0.182 M. The number of moles of N_2O_5 left in the 5.00-L container is

$$\underline{?}\text{ mol } N_2O_5 = 5.00 \text{ L} \times \frac{0.182 \text{ mol}}{\text{L}} = \boxed{0.910 \text{ mol } N_2O_5}$$

(b) Because the integrated *first-order* rate equation involves a *ratio* of concentrations, we do not need to obtain the numerical value of the required concentration. When 90.0% of the original N_2O_5 has reacted, 10.0% remains, or

$$[N_2O_5] = (0.100)[N_2O_5]_0$$

We make this substitution into the integrated rate equation and solve for the elapsed time, t.

$$\ln \frac{[N_2O_5]_0}{[N_2O_5]} = akt$$

$$\ln \frac{[N_2O_5]_0}{(0.100)[N_2O_5]_0} = (2)(0.00840 \text{ s}^{-1})t$$

$$\ln (10.0) = (0.0168 \text{ s}^{-1})t$$

$$2.302 = (0.0168 \text{ s}^{-1})t \quad \text{or} \quad t = \frac{2.302}{0.0168 \text{ s}^{-1}} = \boxed{137 \text{ seconds}}$$

You should now work Exercises 36 and 40.

✓ Problem-Solving Tip: *Does Your Answer Make Sense?*

We know that the amount of N_2O_5 in Example 16-6 must be decreasing. The calculated result, 0.910 mol N_2O_5 after 1.00 minute, is less than the initial amount, 2.50 mol N_2O_5, which is a reasonable result. If our solution had given a result that was *larger* than the original, we should recognize that we must have made some error. For example, if we had *incorrectly* written the equation as

$$\ln\frac{[N_2O_5]}{[N_2O_5]_0} = akt$$

we would have obtained $[N_2O_5] = 1.37$ M, corresponding to 6.85 mol N_2O_5. This would be more N_2O_5 than was originally present, which we should immediately recognize as an impossible answer.

Second-Order Reactions

For reactions involving $aA \rightarrow$ products that are *second order with respect to A and second order overall*, the integrated rate equation is

$$\frac{1}{[A]} - \frac{1}{[A]_0} = akt \qquad \left(\begin{array}{l}\text{second order in A,}\\\text{second order overall}\end{array}\right)$$

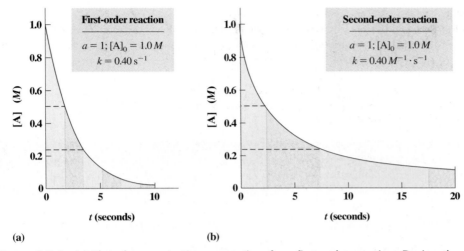

(a)   (b)

Figure 16-4 (a) Plot of concentration versus time for a first-order reaction. During the first half-life, 1.73 seconds, the concentration of A falls from 1.00 M to 0.50 M. An additional 1.73 seconds is required for the concentration to fall by half again, from 0.50 M to 0.25 M, and so on. For a first-order reaction, $t_{1/2} = \dfrac{\ln 2}{ak} = \dfrac{0.693}{ak}$; $t_{1/2}$ does not depend on the concentration at the beginning of that time period. (b) Plot of concentration versus time for a second-order reaction. The same values are used for a, $[A]_0$, and k as in Part (a). During the first half-life, 2.50 seconds, the concentration of A falls from 1.00 M to 0.50 M. The concentration falls by half again from 2.50 to 7.50 seconds, so the second half-life is 5.00 seconds. The half-life beginning at 0.25 M is 10.00 seconds. For a second-order reaction, $t_{1/2} = \dfrac{1}{ak[A]_0}$; $t_{1/2}$ is inversely proportional to the concentration at the beginning of that time period.

For $t = t_{1/2}$, we have $[A] = \frac{1}{2}[A]_0$, so

$$\frac{1}{\frac{1}{2}[A]_0} - \frac{1}{[A]_0} = akt_{1/2}$$

Simplifying and solving for $t_{1/2}$, we obtain the relationship between the rate constant and $t_{1/2}$.

$$t_{1/2} = \frac{1}{ak[A]_0} \qquad \left(\begin{array}{l} \text{second order in A,} \\ \text{second order overall} \end{array} \right)$$

In this case $t_{1/2}$ *depends on the initial concentration of* A. Figure 16-4 illustrates the different behavior of half-life for first- and second-order reactions.

You should carry out the algebraic steps to solve for $t_{1/2}$.

EXAMPLE 16-7 *Half-Life: Second-Order Reaction*

Compounds A and B react to form C and D in a reaction that was found to be second order in A and second order overall. The rate constant at 30°C is 0.622 liter per mole per minute. What is the half-life of A when 4.10×10^{-2} M A is mixed with excess B?

$$A + B \longrightarrow C + D \qquad \text{rate} = k[A]^2$$

Plan

As long as some B is present, only the concentration of A affects the rate. The reaction is second order in [A] and second order overall, so we use the appropriate equation for the half-life.

Solution

$$t_{1/2} = \frac{1}{ak[A]_0} = \frac{1}{(1)(0.622 \; M^{-1} \cdot min^{-1})(4.10 \times 10^{-2} \; M)} = \boxed{39.2 \; min}$$

EXAMPLE 16-8   Concentration Versus Time: Second-Order Reaction

The gas-phase decomposition of NOBr is second order in [NOBr], with $k = 0.810 \; M^{-1} \cdot s^{-1}$ at 10°C. We start with $4.00 \times 10^{-3} \; M$ NOBr in a flask at 10°C. How many seconds does it take to use up $1.50 \times 10^{-3} \; M$ of this NOBr?

$$2NOBr(g) \longrightarrow 2NO(g) + Br_2(g) \qquad rate = k[NOBr]^2$$

Plan

We first determine the concentration of NOBr that remains after $1.50 \times 10^{-3} \; M$ is used up. Then we use the second-order integrated rate equation to determine the time required to reach that concentration.

Solution

$$\underline{?} \; M \; NOBr \; remaining = (0.00400 - 0.00150) \; M = 0.00250 \; M = [NOBr]$$

We solve the integrated rate equation $\dfrac{1}{[NOBr]} - \dfrac{1}{[NOBr]_0} = akt$ for t.

The coefficient of NOBr is $a = 2$.

$$t = \frac{1}{ak}\left(\frac{1}{[NOBr]} - \frac{1}{[NOBr]_0}\right) = \frac{1}{(2)(0.810 \; M^{-1} \cdot s^{-1})}\left(\frac{1}{0.00250 \; M} - \frac{1}{0.00400 \; M}\right)$$

$$= \frac{1}{1.62 \; M^{-1} \cdot s^{-1}}(400 \; M^{-1} - 250 \; M^{-1})$$

$$= \boxed{92.6 \; s}$$

You should now work Exercise 34.

EXAMPLE 16-9   Concentration Versus Time: Second-Order Reaction

Consider the reaction of Example 16-8 at 10°C. If we start with $2.40 \times 10^{-3} \; M$ NOBr, what concentration of NOBr will remain after 5.00 minutes of reaction?

Plan

We use the integrated second-order rate equation to solve for the concentration of NOBr remaining at $t = 5.00$ minutes.

Solution

Again, we start with the expression $\dfrac{1}{[NOBr]} - \dfrac{1}{[NOBr]_0} = akt$. Then we put in the known values and solve for [NOBr].

$$\frac{1}{[NOBr]} - \frac{1}{2.40 \times 10^{-3} \, M} = (2)(0.810 \, M^{-1} \cdot s^{-1})(5.00 \text{ min})\left(\frac{60 \text{ s}}{1 \text{ min}}\right)$$

$$\frac{1}{[NOBr]} - 4.17 \times 10^2 \, M^{-1} = 486 \, M^{-1}$$

$$\frac{1}{[NOBr]} = 486 \, M^{-1} + 417 \, M^{-1} = 903 \, M^{-1}$$

$$[NOBr] = \frac{1}{903 \, M^{-1}} = 1.11 \times 10^{-3} \, M \qquad \text{(46.2\% remains unreacted)}$$

Thus, 53.8% of the original concentration of NOBr reacts within the first 5 minutes. This is reasonable because, as you can easily verify, the reaction has an initial half-life of 257 seconds, or 4.29 minutes.

You should now work Exercises 35 and 38.

Zero-Order Reaction

For a reaction $a\text{A} \rightarrow$ products that is zero order, the reaction rate is independent of concentrations. We can write the rate-law expression as

$$\text{rate} = -\frac{1}{a}\left(\frac{\Delta[A]}{\Delta t}\right) = k$$

The corresponding integrated rate equation is

$$[A] = [A]_0 - akt \qquad \text{(zero order)}$$

and the half-life is

$$t_{1/2} = \frac{[A]_0}{2ak} \qquad \text{(zero order)}$$

Table 16-2 summarizes the relationships that we have presented in Sections 16-3 and 16-4.

TABLE 16-2 *Summary of Relationships for Various Orders of the Reaction $a\text{A} \rightarrow$ Products*

	Order		
	Zero	**First**	**Second**
Rate-law expression	rate $= k$	rate $= k[A]$	rate $= k[A]^2$
Integrated rate equation	$[A] = [A]_0 - akt$	$\ln\dfrac{[A]_0}{[A]} = akt$ or $\log\dfrac{[A]_0}{[A]} = \dfrac{akt}{2.303}$	$\dfrac{1}{[A]} - \dfrac{1}{[A]_0} = akt$
Half-life, $t_{1/2}$	$\dfrac{[A]_0}{2ak}$	$\dfrac{\ln 2}{ak} = \dfrac{0.693}{ak}$	$\dfrac{1}{ak[A]_0}$

✓ **Problem-Solving Tip:** *Which Equation Should Be Used?*

How can you tell which equation to use to solve a particular problem?

1. You must decide whether to use the rate-law expression or the integrated rate equation. Remember that

   the *rate-law expression* relates *rate and concentration*

   whereas

   the *integrated rate equation* relates *time and concentration*.

   When you need to find the *rate* that corresponds to particular concentrations, or the concentrations needed to give a desired rate, you should use the rate-law expression. When *time* is involved in the problem, you should use the integrated rate equation.

2. You must choose the form of the rate-law expression or the integrated rate equation— zero, first, or second order—that is appropriate to the order of the reaction. These are summarized in Table 16-2. One of the following usually helps you decide.
   a. The statement of the problem may state explicitly what the order of the reaction is.
   b. The rate-law expression may be given, so that you can tell the order of the reaction from the exponents in that expression.
   c. The units of the specific rate constant, k, may be given; you can interpret these stated units to tell you the order of the reaction.

Order	Units of k
0	$M \cdot \text{time}^{-1}$
1	time^{-1}
2	$M^{-1} \cdot \text{time}^{-1}$

You should test this method using the concentration-versus-time data of Example 16-10, plotted in Figure 16-8b.

One method of assessing reaction order is based on comparing successive half-lives. As we have seen, $t_{1/2}$ for a first-order reaction does not depend on initial concentration. We can measure the time required for different concentrations of a reactant to fall to half of their original values. If this time remains constant, it is an indication that the reaction is first order for that reactant and first order overall (see Figure 16-4a). By contrast, for other orders of reaction, $t_{1/2}$ would change depending on initial concentration. For a second-order reaction, successively measured $t_{1/2}$ values would increase by a factor of 2 as $[A]_0$ decreases by a factor of 2 (see Figure 16-4b). $[A]_0$ is measured at the *beginning of each particular measurement period*.

Enrichment   **Calculus Derivation of Integrated Rate Equations**

The derivation of the integrated rate equation is an example of the use of calculus in chemistry. The following derivation is for a reaction that is assumed to be first order in a reactant A and first order overall. If you do not know calculus, you can still use the results of this derivation, as we have already shown in this section. For the reaction

$$aA \longrightarrow products$$

the rate is expressed as

$$rate = -\frac{1}{a}\left(\frac{\Delta[A]}{\Delta t}\right)$$

For a first-order reaction, the rate is proportional to the first power of [A].

$$-\frac{1}{a}\left(\frac{\Delta[A]}{\Delta t}\right) = k[A]$$

In calculus terms, we express the change during an infinitesimally short time dt as the derivative of [A] with respect to time.

$$-\frac{1}{a}\frac{d[A]}{dt} = k[A]$$

Separating variables, we obtain

$$-\frac{d[A]}{[A]} = (ak)dt$$

We integrate this equation with limits: As the reaction progresses from time = 0 (the start of the reaction) to time = t elapsed, the concentration of A goes from $[A]_0$, its starting value, to [A], the concentration remaining after time t:

$$-\int_{[A]_0}^{[A]} \frac{d[A]}{[A]} = ak \int_0^t dt$$

The result of the integration is

$$-(\ln [A] - \ln [A]_0) = ak(t - 0) \qquad \text{or} \qquad \ln [A]_0 - \ln [A] = akt$$

Remembering that $\ln(x) - \ln(y) = \ln(x/y)$, we obtain

$$\ln\frac{[A]_0}{[A]} = akt \qquad \text{(first order)}$$

This is the integrated rate equation for a reaction that is first order in reactant A and first order overall.

Integrated rate equations can be derived similarly from other simple rate laws. For a reaction $aA \rightarrow$ products that is second order in reactant A and second order overall, we can write the rate equation as

$$-\frac{d[A]}{adt} = k[A]^2$$

Again, using the methods of calculus, we can separate variables, integrate, and rearrange to obtain the corresponding integrated second-order rate equation.

$$\frac{1}{[A]} - \frac{1}{[A]_0} = akt \qquad \text{(second order)}$$

$-\frac{1}{a}\left(\frac{\Delta[A]}{\Delta t}\right)$ represents the average rate over a finite time interval Δt.

$-\frac{1}{a}\left(\frac{d[A]}{dt}\right)$ involves a change over an infinitesimally short time interval dt, so it represents the *instantaneous* rate.

(Enrichment, continued)

For a reaction $a\text{A} \rightarrow$ products that is zero order overall, we can write the rate equation as

$$-\frac{d[\text{A}]}{a\,dt} = k$$

In this case, the calculus derivation already described leads to the integrated zero-order rate equation

$$[\text{A}] = [\text{A}]_0 - akt \qquad \text{(zero order)}$$

Enrichment   Using Integrated Rate Equations to Determine Reaction Order

The integrated rate equation can help us to analyze concentration-versus-time data to determine reaction order. A graphical approach is often used. We can rearrange the integrated first-order rate equation

$$\ln \frac{[\text{A}]_0}{[\text{A}]} = akt$$

as follows. The logarithm of a quotient, $\ln (x/y)$, is equal to the difference of the logarithms, $\ln x - \ln y$, so we can write

$$\ln [\text{A}]_0 - \ln [\text{A}] = akt \qquad \text{or} \qquad \ln [\text{A}] = -akt + \ln [\text{A}]_0$$

Recall that the equation for a straight line may be written as

$$y = mx + b$$

where y is the variable plotted along the ordinate (vertical axis), x is the variable plotted along the abscissa (horizontal axis), m is the slope of the line, and b is the intercept of the line with the y axis (Figure 16-5). If we compare the last two equations, we find that $\ln [\text{A}]$ can be interpreted as y, and t as x.

$$\underbrace{\ln [\text{A}]}_{\downarrow \atop y} = \underbrace{-akt}_{\downarrow\downarrow \atop m\,x\,+} + \underbrace{\ln [\text{A}]_0}_{\downarrow \atop b}$$

CD-ROM   Screen 15.7, Determination of Rate Equations (2): Graphical Methods.

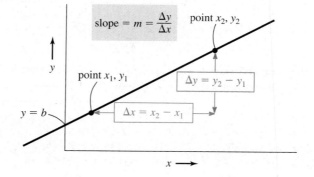

Figure 16-5 Plot of the equation $y = mx + b$, where m and b are constant. The slope of the line (positive in this case) is equal to m; the intercept on the y axis is equal to b.

slope $= m = \dfrac{\Delta y}{\Delta x}$ point x_2, y_2

point x_1, y_1

$\Delta y = y_2 - y_1$

$y = b$

$\Delta x = x_2 - x_1$

Figure 16-6 Plot of ln [A] versus time for a reaction $aA \longrightarrow$ products that follows first-order kinetics. The observation that such a plot gives a straight line would confirm that the reaction is first order in [A] and first order overall, that is, rate = k[A]. The slope is equal to $-ak$. Because a and k are positive numbers, the slope of the line is always negative. Logarithms are dimensionless, so the slope has the units $(\text{time})^{-1}$. The logarithm of a quantity less than 1 is negative, so data points for concentrations less than 1 molar would have negative values and appear below the time axis.

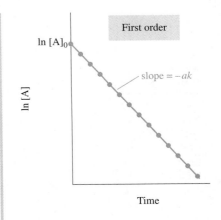

The quantity $-ak$ is a constant as the reaction proceeds, so it can be interpreted as m. The initial concentration of A is fixed, so ln $[A]_0$ is a constant for each experiment, and ln $[A]_0$ can be interpreted as b. Thus, a plot of ln [A] versus time for a first-order reaction would be expected to give a straight line (Figure 16-6) with the slope of the line equal to $-ak$ and the intercept equal to ln $[A]_0$.

We can proceed in a similar fashion with the integrated rate equation for a reaction that is second order in A and second order overall. We rearrange

$$\frac{1}{[A]} - \frac{1}{[A]_0} = akt \qquad \text{to read} \qquad \frac{1}{[A]} = akt + \frac{1}{[A]_0}$$

Again comparing this with the equation for a straight line, we see that a plot of 1/[A] versus time would be expected to give a straight line (Figure 16-7). The line would have a slope equal to ak and an intercept equal to $1/[A]_0$.

For a zero-order reaction, we can rearrange the integrated rate equation

$$[A]_0 - [A] = akt \qquad \text{to} \qquad [A] = -akt + [A]_0$$

Comparing this with the equation for a straight line, we see that a straight-line plot would be obtained by plotting concentration versus time, [A] versus t. The slope of this line is $-ak$, and the intercept is $[A]_0$.

This discussion suggests another way to deduce an unknown rate-law expression from experimental concentration data. The following approach is particularly useful for any decomposition reaction, one that involves only one reactant.

$$aA \longrightarrow \text{products}$$

We plot the data in various ways as suggested above. *If* the reaction followed zero-order kinetics, *then* a plot of [A] versus t would give a straight line. But *if* the reaction followed first-order kinetics, *then* a plot of ln [A] versus t would give a straight line whose slope could be interpreted to derive a value of k. *If* the reaction were second order in A and second order overall, *then* neither of these plots would give a straight line, but a plot of 1/[A] versus t would. If none of these plots gave a straight line (within expected scatter due to experimental error), we would know that none of these is the correct order (rate law) for the reaction. Plots to test for other orders can be devised, as can graphical tests for rate-law expressions involving more than one reactant, but those are subjects for more advanced texts. The graphical approach that we have described is illustrated in the following example.

It is not possible for *all* of the plots suggested here to yield straight lines for a given reaction. The nonlinearity of the plots may not become obvious, however, if the reaction times used are too short. In practice, all three lines might seem to be straight; we should then suspect that we need to observe the reaction for a longer time.

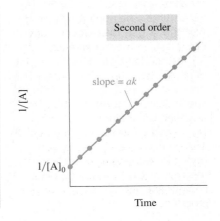

Figure 16-7 Plot of 1/[A] versus time for a reaction $aA \rightarrow$ products that follows a second-order kinetics. The observation that such a plot gives a straight line would confirm that the reaction is second order in [A] and second order overall, that is, rate = $k[A]^2$. The slope is equal to ak. Because a and k are positive numbers, the slope of the line is always positive. Because concentrations cannot be negative, 1/[A] is always positive, and the line is always above the time axis.

Time (min)	[A] (mol/L)
0.00	2.000
2.00	1.107
4.00	0.612
6.00	0.338
8.00	0.187
10.00	0.103

(Enrichment, continued)

EXAMPLE 16-10　*Graphical Determination of Reaction Order*

We carry out the reaction $A \rightarrow B + C$ at a particular temperature. As the reaction proceeds, we measure the molarity of the reactant, [A], at various times. The observed data are tabulated in the margin. (a) Plot [A] versus time. (b) Plot ln [A] versus time. (c) Plot 1/[A] versus time. (d) What is the order of the reaction? (e) Write the rate-law expression for the reaction. (f) What is the value of k at this temperature?

Plan

For Parts (a)–(c), we use the observed data to make the required plots, calculating related values as necessary. (d) We can determine the order of the reaction by observing which of these plots gives a straight line. (e) Knowing the order of the reaction, we can write the rate-law expression. (f) The value of k can be determined from the slope of the straight-line plot.

Solution

(a) The plot of [A] versus time is given in Figure 16-8b.

(b) We first use the given data to calculate the ln [A] column in Figure 16-8a. These data are then used to plot ln [A] versus time, as shown in Figure 16-8c.

(c) The given data are used to calculate the 1/[A] column in Figure 16-8a. Then we plot 1/[A] versus time, as shown in Figure 16-8d.

(d) It is clear from the answer to Part (b) that the plot of ln [A] versus time gives a straight line. This tells us that the reaction is first order in [A].

Figure 16-8 Data conversion and plots for Example 16-10. (a) The data are used to calculate the two columns ln [A] and 1/[A]. (b) Test for zero-order kinetics: a plot of [A] versus time. The nonlinearity of this plot shows that the reaction does not follow zero-order kinetics. (c) Test for first-order kinetics: a plot of ln [A] versus time. The observation that this plot gives a straight line indicates that the reaction follows first-order kinetics. (d) Test for second-order kinetics: a plot of 1/[A] versus time. If the reaction had followed second-order kinetics, this plot would have resulted in a straight line and the plot in Part (c) would not.

Time (min)	[A]	ln [A]	1/[A]
0.00	2.000	0.693	0.5000
2.00	1.107	0.102	0.9033
4.00	0.612	−0.491	1.63
6.00	0.338	−1.085	2.95
8.00	0.187	−1.677	5.35
10.00	0.103	−2.273	9.71

(a) Data for Example 16-10.

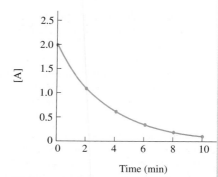

(b) Example 16-10(a).

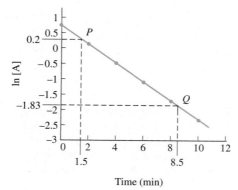

(c) Example 16-10(b).

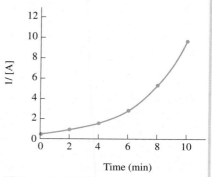

(d) Example 16-10(c).

(e) In the form of a rate-law expression, the answer to Part (d) gives rate $= k[A]$.

(f) We use the straight-line plot in Figure 16-8c to find the value of the rate constant for this first-order reaction from the relationship

$$\text{slope} = -ak \qquad \text{or} \qquad k = -\frac{\text{slope}}{a}$$

To determine the slope of the straight line, we pick any two points, such as P and Q, on the line. From their coordinates, we calculate

$$\text{slope} = \frac{\text{change in ordinate}}{\text{change in abscissa}} = \frac{(-1.83) - (0.27)}{(8.50 - 1.50)\ \text{min}} = -0.300\ \text{min}^{-1}$$

$$k = -\frac{\text{slope}}{a} = -\frac{-0.300\ \text{min}^{-1}}{1} = 0.300\ \text{min}^{-1}$$

You should now work Exercises 44 and 45.

The graphical interpretations of concentration-versus-time data for some common reaction orders are summarized in Table 16-3.

✓ **Problem-Solving Tip:** *Some Warnings About the Graphical Method for Determining Reaction Order*

1. If we were dealing with real experimental data, there would be some error in each of the data points on the plot. For this reason, we should *not* use experimental data points to determine the slope. (Random experimental errors of only 10% can introduce errors of more than 100% in slopes based on only two points.) Rather we should draw the best straight line and then use points on that line to find its slope. Errors are further minimized by choosing points that are widely separated.

2. Remember that the ordinate is the vertical axis and the abscissa is the horizontal one. If you are not careful to keep the points in the same order in the numerator and denominator, you will get the wrong sign for the slope.

TABLE 16-3 *Graphical Interpretations for Various Orders of the Reaction* $aA \rightarrow Products$

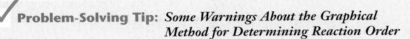

	Order		
	Zero	**First**	**Second**
Plot that gives straight line	$[A]$ vs. t	$\ln [A]$ vs. t	$\dfrac{1}{[A]}$ vs. t
Direction of straight-line slope	down with time	down with time	up with time
Interpretation of slope	$-ak$	$-ak$	ak
Interpretation of intercept	$[A]_0$	$\ln [A]_0$	$\dfrac{1}{[A]_0}$

16-5 COLLISION THEORY OF REACTION RATES

The fundamental notion of the **collision theory of reaction rates** is that for a reaction to occur, molecules, atoms, or ions must first collide. Increased concentrations of reacting species result in greater numbers of collisions per unit time. However, not all collisions result in reaction; that is, not all collisions are **effective collisions.** For a collision to be effective, the reacting species must (1) possess at least a certain minimum energy necessary to rearrange outer electrons in breaking bonds and forming new ones and (2) have the proper orientations toward one another at the time of collision.

CD-ROM   Screen 15.9, Microscopic View of Reactions (1): Collision Theory.

> Collisions must occur in order for a chemical reaction to proceed, but they do not guarantee that a reaction will occur.

A collision between atoms, molecules, or ions is not like one between two hard billiard balls. Whether or not chemical species "collide" depends on the distance at which they can interact with one another. For instance, the gas-phase ion–molecule reaction $CH_4^+ + CH_4 \rightarrow CH_5^+ + CH_3$ can occur with a fairly long-range contact. This is because the interactions between ions and induced dipoles are effective over a relatively long distance. By contrast, the reacting species in the gas reaction $CH_3 + CH_3 \rightarrow C_2H_6$ are both neutral. They interact appreciably only through very short-range forces between induced dipoles, so they must approach one another very closely before we could say that they "collide."

Recall (Chapter 12) that the average kinetic energy of a collection of molecules is proportional to the absolute temperature. At higher temperatures, more of the molecules possess sufficient energy to react (Section 16-8).

The colliding molecules must have the proper orientation relative to one another *and* have sufficient energy to react. If colliding molecules have improper orientations, they do not react even though they may possess sufficient energy. Figure 16-9 depicts some possible collisions between molecules of NO and N_2O, which can react to form NO_2 and N_2.

$$NO + N_2O \longrightarrow NO_2 + N_2$$

$Zn(s) + 2H^+(aq) \longrightarrow$
$\qquad Zn^{2+}(aq) + H_2(g)$

Dilute sulfuric acid reacts slowly with zinc metal (*left*), whereas more concentrated acid reacts rapidly (*right*). The $H^+(aq)$ concentration is higher in the more concentrated acid, and so more $H^+(aq)$ ions collide with Zn per unit time.

Charles Steele

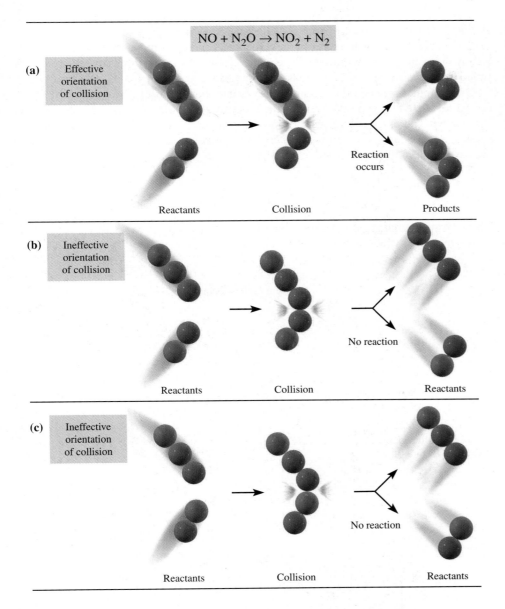

$$NO + N_2O \rightarrow NO_2 + N_2$$

(a) Effective orientation of collision

Reactants   Collision   Reaction occurs   Products

(b) Ineffective orientation of collision

Reactants   Collision   No reaction   Reactants

(c) Ineffective orientation of collision

Reactants   Collision   No reaction   Reactants

Figure 16-9 Some possible collisions between N_2O and NO molecules in the gas phase. (a) A collision that could be effective in producing the reaction. (b, c) Collisions that would be ineffective. The molecules must have the proper orientations relative to one another *and* have sufficient energy to react.

Only the collision in Figure 16-9a is in the correct orientation to transfer an oxygen atom from the linear N_2O molecule to form the angular NO_2 molecule. For some reactions, the presence of a heterogeneous catalyst (Section 16-9) can increase the fraction of colliding molecules that have the proper orientations.

16-6 TRANSITION STATE THEORY

Chemical reactions involve the making and breaking of chemical bonds. The energy associated with a chemical bond is a form of potential energy. Reactions are accompanied by changes in potential energy. Consider the following hypothetical, one-step reaction at a certain temperature.

$$A + B_2 \longrightarrow AB + B$$

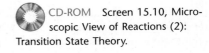

CD-ROM   Screen 15.10, Microscopic View of Reactions (2): Transition State Theory.

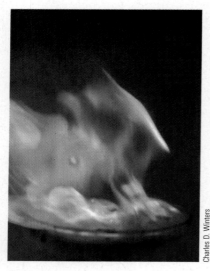

The reaction of potassium metal with water is spontaneous and has a very low activation energy (E_a). The small E_a means that the reaction will be very fast.

The **progress of reaction** represents how far the reaction has proceeded *along the pathway* leading from reactants to products. This coordinate is sometimes labeled the **reaction coordinate**.

Figure 16-10 shows plots of potential energy versus the progress of the reaction. In Figure 16-10a the ground state energy of the reactants, A and B_2, is higher than the ground state energy of the products, AB and B. The energy released in the reaction is the difference between these two energies, ΔE. It is related to the change in enthalpy, ΔH^0_{rxn} (Section 15-11).

Quite often, for reaction to occur, some covalent bonds must be broken so that others can be formed. This can occur only if the molecules collide *with enough kinetic energy* to overcome the potential energy stabilization of the bonds. According to the **transition state theory**, the reactants pass through a short-lived, high-energy intermediate state, called a **transition state**, before the products are formed.

$$A + B\!-\!B \longrightarrow A\text{-}\text{-}\text{-}B\text{-}\text{-}\text{-}B \longrightarrow A\!-\!B + B$$

reactants	transition state	products
A + B_2	AB_2	AB + B

The **activation energy** (or **activation barrier**), E_a, is the kinetic energy that reactant molecules must have to allow them to reach the transition state. If A and B_2 molecules do not

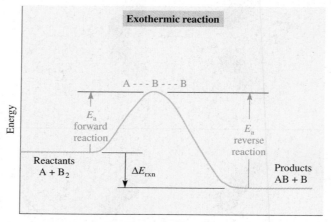

(a)

Figure 16-10 A potential energy diagram. (a) A reaction that releases energy (exothermic). An example of an exothermic gas-phase reaction is

$$H + I_2 \longrightarrow HI + I$$

(b) A reaction that absorbs energy (endothermic). An example of an endothermic gas-phase reaction is

$$I + H_2 \longrightarrow HI + H$$

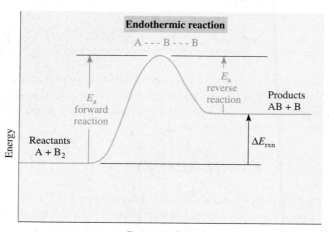

(b)

possess the necessary amount of energy, E_a, when they collide, reaction cannot occur. If they do possess sufficient energy to "climb the energy barrier" to reach the transition state, the reaction can proceed. When the atoms go from the transition state arrangement to the product molecules, energy is *released*. If the reaction results in a net *release* of energy (see Figure 16-10a), *more* energy than the activation energy is returned to the surroundings and the reaction is exothermic. If the reaction results in a *net absorption* of energy (see Figure 16-10b), an amount less than E_a is given off when the transition state is converted to products and the reaction is endothermic. The *net* release of energy is ΔE_{rxn}.

For the reverse reaction to occur, some molecules on the right (AB) must have kinetic energy equal to the reverse activation energy, $E_{a\ reverse}$, to allow them to reach the transition state. As you can see from the potential energy diagrams in Figure 16-10,

$$E_{a\ forward} - E_{a\ reverse} = \Delta E_{reaction}$$

As we shall see, increasing the temperature changes the rate by altering the fraction of molecules that can get over a given energy barrier (Section 16-8). Introducing a catalyst increases the rate by providing a different pathway that has a lower activation energy (Section 16-9).

As a specific example that illustrates the ideas of collision theory and transition state theory, consider the reaction of iodide ions with methyl chloride.

$$I^- + CH_3Cl \longrightarrow CH_3I + Cl^-$$

Many studies have established that this reaction proceeds as shown in Figure 16-11a. The I^- ion must approach the CH_3Cl from the "back side" of the C—Cl bond, through the middle of the three hydrogen atoms. A collision of an I^- ion with a CH_3Cl molecule from any other direction would not lead to reaction. But a sufficiently energetic collision with

Remember that ΔE_{rxn} relates product energy to reactant energy, regardless of the pathway. ΔE_{rxn} is negative when energy is given off; ΔE_{rxn} is positive when energy is absorbed from the surroundings.

The CH_3Cl and CH_3I molecules each have tetrahedral molecular geometry.

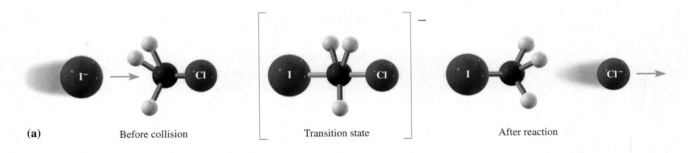

(a)　　Before collision　　　　Transition state　　　　After reaction

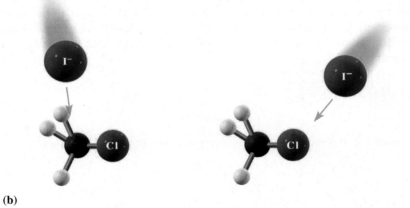

(b)

Figure 16-11 (a) A collision that could lead to reaction of $I^- + CH_3Cl$ to give $CH_3I + Cl^-$. The I^- must approach along the "back side" of the C—Cl bond. (b) Two collisions that are not in the "correct" orientation to cause a reaction.

the appropriate orientation could allow the new I—C bond to form at the same time that the C—Cl bond is breaking. This collection of atoms, which we represent as

$$
\begin{array}{ccc}
 & H \quad H \\
 & \backslash \;\; \nearrow \\
I \cdots & C & \cdots Cl \\
 & | \\
 & H
\end{array}
$$

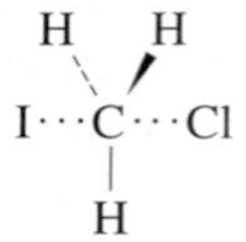

is what we call the transition state of this reaction (Figure 16-11a). From this state, either of two things could happen: (1) the I—C bond could finish forming and the C—Cl bond could finish breaking with Cl⁻ leaving, leading to products, or (2) the I—C bond could fall apart with I⁻ leaving and the C—Cl bond re-forming, leading back to reactants.

16-7 REACTION MECHANISMS AND THE RATE-LAW EXPRESSION

Some reactions take place in a single step, but most reactions occur in a series of **elementary** or **fundamental steps.** The step-by-step pathway by which a reaction occurs is called the **reaction mechanism.**

> The reaction orders *for any single elementary step* are equal to the coefficients for that step.

In many mechanisms, however, one step is much slower than the others.

> A overall reaction can never occur faster than its slowest elementary reaction step.

This slowest step is called the **rate-determining step.** The speed at which the slowest step occurs limits the rate at which the overall reaction occurs.

As an analogy, suppose you often drive a distance of 120 miles at the speed limit of 60 mi/h, requiring 2 hours. But one day there is an accident along the route, causing a slow-down for several hours. After passing the accident scene, you resume the posted speed of 60 mi/h. If the total time for this trip was 4 hours, then the *average* speed would be only 120 miles/4 hours, or 30 mi/h. Even though you drove for many miles at the same high speed, 60 mi/h, the overall rate was limited by the slow step, passing the accident scene.

The balanced equation for the overall reaction is equal to the sum of *all* the individual fundamental steps, including any that might follow the rate-determining step. We emphasize again that the rate-law exponents *do not necessarily match* the coefficients of the *overall* balanced equation.

> For the general overall reaction
>
> $$aA + bB \longrightarrow cC + dD$$
>
> the experimentally determined rate-law expression has the form
>
> $$\text{rate} = k[A]^x[B]^y$$
>
> The values of x and y are related to the coefficients of the reactants in the slowest (rate-determining) step, influenced in some cases by earlier steps.

We can view this transition state as though carbon is only partially bonded to I and only partially bonded to Cl.

 CD-ROM Screen 15.12, Reaction Mechanisms.

⚠ A fundamental reaction step can often look similar to an overall balanced reaction equation, so they are not easy to distinguish. In general, you will be told if a given reaction is fundamental or elementary in nature. If you are not told that a reaction is a fundamental step, you need to be given additional experimental information in order to determine the kinetic orders.

Using a combination of experimental data and chemical intuition, we can *postulate* a mechanism by which a reaction could occur. We can never prove that a proposed mechanism is correct. All we can do is postulate a mechanism that is *consistent* with experimental data. We might later detect reaction-intermediate species that are not explained by the proposed mechanism. We must then modify the mechanism or discard it and propose a new one.

As an example, the reaction of nitrogen dioxide and carbon monoxide has been found to be second order with respect to NO_2 and zero order with respect to CO below 225°C.

$$NO_2(g) + CO(g) \longrightarrow NO(g) + CO_2(g) \qquad rate = k[NO_2]^2$$

The balanced equation for the overall reaction shows the stoichiometry but *does not necessarily mean* that the reaction simply occurs by one molecule of NO_2 colliding with one molecule of CO. If the reaction really took place in *that* one step, then the rate would be first order in NO_2 and first order in CO, or rate = $k[NO_2][CO]$. The fact that the experimentally determined orders do not match the coefficients in the overall balanced equation tells us that *the reaction does not take place in one step.*

The following proposed two-step mechanism is consistent with the observed rate-law expression.

(1) $NO_2 + NO_2 \longrightarrow N_2O_4$ (slow)

(2) $N_2O_4 + CO \longrightarrow NO + CO_2 + NO_2$ (fast)

$NO_2 + CO \longrightarrow NO + CO_2$ (overall)

The rate-determining step of this mechanism involves a *bimolecular* collision between two NO_2 molecules. This is consistent with the rate expression involving $[NO_2]^2$. Because the CO is involved only after the slow step has occurred, the reaction rate would not depend on [CO] (i.e., the reaction would be zero order in CO) if this were the actual mechanism. In this proposed mechanism, N_2O_4 is formed in one step and is completely consumed in a later step. Such a species is called a **reaction intermediate.**

In other studies of this reaction, however, nitrogen trioxide, NO_3, has been detected as a transient (short-lived) intermediate. The mechanism now thought to be correct is

(1) $NO_2 + NO_2 \longrightarrow NO_3 + NO$ (slow)

(2) $NO_3 + CO \longrightarrow NO_2 + CO_2$ (fast)

$NO_2 + CO \longrightarrow NO + CO_2$ (overall)

In this proposed mechanism two molecules of NO_2 collide to produce one molecule each of NO_3 and NO. The reaction intermediate NO_3 then collides with one molecule of CO and reacts very rapidly to produce one molecule each of NO_2 and CO_2. Even though two NO_2 molecules are consumed in the first step, one is produced in the second step. The net result is that only one NO_2 molecule is consumed in the overall reaction.

Each of these proposed mechanisms meets both criteria for a plausible mechanism: (1) The steps add to give the equation for the overall reaction, and (2) the mechanism is consistent with the experimentally determined rate-law expression (in that two NO_2 molecules and no CO molecules are reactants in the slowest reaction step). The NO_3 that has been detected is evidence in favor of the second mechanism, but this does not unequivocally *prove* that mechanism; it may be possible to think of other mechanisms that would involve NO_3 as an intermediate and would also be consistent with the observed rate law.

You should be able to distinguish among various species that can appear in a reaction mechanism. So far, we have seen three such species:

CD-ROM   Screen 15.13, Reaction Mechanisms and Rate Equations.

The *rate* of a reaction involves only the steps up to and including the rate-determining step. The *overall stoichiometry* includes all steps in a reaction.

Some reaction intermediates are so unstable that it is very difficult to prove experimentally that they exist.

Nonproductive side reactions can produce observable intermediates (such as NO_3) that may not be part of the overall reaction to make products. This is one of the factors that can make kinetics very complicated.

1. *Reactant:* More is consumed than is formed.

2. *Product:* More is formed than is consumed.

3. *Reaction intermediate:* Formed in earlier steps, then consumed in an equal amount in later steps.

The gas-phase reaction of NO and Br_2 is known to be second order in NO and first order in Br_2.

$$2NO(g) + Br_2(g) \longrightarrow 2NOBr(g) \qquad rate = k[NO]^2[Br_2]$$

A one-step collision involving two NO molecules and one Br_2 molecule would be consistent with the experimentally determined rate-law expression. However, the likelihood of all *three* molecules colliding simultaneously is far less than the likelihood of two colliding. *Routes involving only bimolecular collisions or unimolecular decompositions are thought to be far more favorable in reaction mechanisms.* The mechanism is believed to be

> Think how unlikely it is for three moving billiard balls to collide *simultaneously*.

> Any fast step that precedes a slow step reaches equilibrium.

$$
\begin{array}{lll}
(1) & NO + Br_2 \rightleftharpoons NOBr_2 & \text{(fast, equilibrium)} \\
(2) & NOBr_2 + NO \longrightarrow 2NOBr & \text{(slow)} \\
\hline
& 2NO + Br_2 \longrightarrow 2NOBr & \text{(overall)}
\end{array}
$$

The first step involves the collision of one NO molecule (reactant) and one Br_2 molecule (reactant) to produce the intermediate species $NOBr_2$. The $NOBr_2$ can react rapidly, however, to re-form NO and Br_2. We say that this is an *equilibrium step*. Eventually another NO molecule (reactant) can collide with a short-lived $NOBr_2$ molecule and react to produce two NOBr molecules (product).

To analyze the rate law that would be consistent with this proposed mechanism, we again start with the slow (rate-determining) Step 2. Denoting the rate constant for this step as k_2, we could express the rate of this step as

> The rate-law expression of Step 2 (the rate-determining step) determines the rate law for the overall reaction. The overall rate law must not include the concentrations of any intermediate species formed in elementary reaction steps.

$$rate = k_2[NOBr_2][NO]$$

However, $NOBr_2$ is a reaction intermediate, so its concentration at the beginning of the second step may not be easy to measure directly. Because $NOBr_2$ is formed in a fast equilibrium step, we can relate its concentration to the concentrations of the original reactants. When a reaction or reaction step is at *equilibrium*, its forward (f) and reverse (r) rates are equal.

$$rate_{1f} = rate_{1r}$$

Because this is an elementary step, we can write the rate expression for both directions from the equation for the elementary step

$$k_{1f}[NO][Br_2] = k_{1r}[NOBr_2]$$

and then rearrange for $[NOBr_2]$.

$$[NOBr_2] = \frac{k_{1f}}{k_{1r}}[NO][Br_2]$$

When we substitute the right side of this equation for $[NOBr_2]$ in the rate expression for the rate-determining step, $rate = k_2[NOBr_2][NO]$, we arrive at the experimentally determined rate-law expression.

> The product and quotient of constants k_2, k_{1f}, and k_{1r} is another constant, k.

$$rate = k_2\left(\frac{k_{1f}}{k_{1r}}[NO][Br_2]\right)[NO] \qquad or \qquad rate = k[NO]^2[Br_2]$$

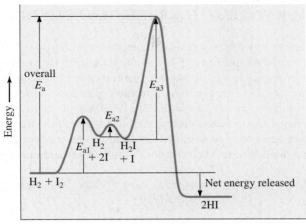

Figure 16-12 A graphical representation of the relative energies of activation for a postulated mechanism for the gas-phase reaction

$$H_2 + I_2 \longrightarrow 2HI$$

Similar interpretations apply to most other overall third- or higher-order reactions, as well as many lower-order reactions. When several steps are about equally slow, however, the analysis of experimental data is more complex. Fractional or negative reaction orders can result from complex multistep mechanisms.

One of the earliest kinetic studies involved the gas-phase reaction of hydrogen and iodine to form hydrogen iodide. The reaction was found to be first order in both hydrogen and iodine.

$$H_2(g) + I_2(g) \longrightarrow 2HI(g) \qquad rate = k[H_2][I_2]$$

The mechanism that was accepted for many years involved collision of single molecules of H_2 and I_2 in a simple one-step reaction. Current evidence indicates a more complex process, however. Most kineticists now accept the following mechanism.

(1) I_2		$\rightleftharpoons$	$2I$	(fast, equilibrium)
(2) I	$+ H_2$	$\rightleftharpoons$	H_2I	(fast, equilibrium)
(3) $H_2I + I$		$\longrightarrow$	$2HI$	(slow)
$H_2 + I_2$		$\longrightarrow$	$2HI$	(overall)

Apply the algebraic approach described earlier to show that this mechanism is consistent with the observed rate-law expression.

In this case neither original reactant appears in the rate-determining step, but both appear in the rate-law expression. Each step is a fundamental reaction in itself. Transition state theory tells us that each step has its own activation energy. Because Step 3 is the slowest, we know that its activation energy is the highest, as shown in Figure 16-12.

In summary

> the experimentally determined reaction orders of reactants indicate the number of molecules of those reactants involved in (1) the slow step only, if it occurs first, or (2) the slow step *and* any fast equilibrium steps preceding the slow step.

16-8 TEMPERATURE: THE ARRHENIUS EQUATION

The average kinetic energy of a collection of molecules is proportional to the absolute temperature. At a particular temperature, T_1, a definite fraction of the reactant molecules have sufficient kinetic energy, KE $> E_a$, to react to form product molecules on collision. At a higher temperature, T_2, a greater fraction of the molecules possess the necessary activation energy, and the reaction proceeds at a faster rate. This is depicted in Figure 16-13a.

From experimental observations, Svante Arrhenius developed the mathematical relationship among activation energy, absolute temperature, and the specific rate constant of a reaction, k, at that temperature. The relationship, called the **Arrhenius equation,** is

$e = 2.718$ is the base of *natural* logarithms (ln).

$$k = Ae^{-E_a/RT}$$

or, in logarithmic form,

$$\ln k = \ln A - \frac{E_a}{RT}$$

In this expression, A is a constant having the same units as the rate constant. It is equal to the fraction of collisions with the proper orientations when all reactant concentrations are one molar. R is the universal gas constant, expressed with the same energy units in its numerator as are used for E_a. For instance, when E_a is known in J/mol, the value $R = 8.314 \, \text{J/mol} \cdot \text{K}$ is appropriate. Here the unit "mol" is interpreted as "mole of reaction," as described in Chapter 15. One important point is the following: The greater the value of E_a,

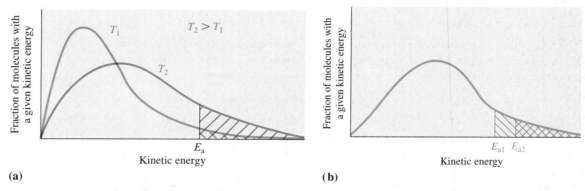

(a)   **(b)**

Figure 16-13   (a) The effect of temperature on the number of molecules that have kinetic energies greater than E_a. At T_2, a higher fraction of molecules possess at least E_a, the activation energy. The area between the distribution curve and the horizontal axis is proportional to the total number of molecules present. The total area is the same at T_1 and T_2. The shaded areas represent the number of particles that exceed the energy of activation, E_a. (b) Consider two hypothetical reactions 1 and 2, where the activation energy of reaction 1 is less than that of reaction 2—that is, $E_{a1} < E_{a2}$. At any given temperature, a larger fraction of the molecules have energies that exceed E_{a1} than that exceed E_{a2}, so reaction 1 would have a higher specific rate constant, k, than reaction 2 at the same reactant concentrations.

Antimony powder reacts with bromine more rapidly at 75°C (*left*) than at 25°C (*right*).

the smaller the value of k and the slower the reaction rate (other factors being equal). This is because fewer collisions take place with sufficient energy to get over a high-energy barrier (see Figure 16-13b).

The Arrhenius equation predicts that increasing T results in a faster reaction for the same E_a and concentrations.

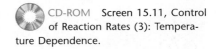

CD-ROM   Screen 15.11, Control of Reaction Rates (3): Temperature Dependence.

| **If T increases** | $\Rightarrow$ | E_a/RT decreases | $\Rightarrow$ | $-E_a/RT$ increases | $\Rightarrow$ | $e^{-E_a/RT}$ increases | $\Rightarrow$ | k increases | $\Rightarrow$ | **Reaction speeds up** |

Let's look at how the rate constant varies with temperature for a given single reaction. Assume that the activation energy and the factor A do not vary with temperature. We can write the Arrhenius equation for two different temperatures. Then we subtract one equation from the other and rearrange the result to obtain

$$\ln \frac{k_2}{k_1} = \frac{E_a}{R}\left(\frac{1}{T_1} - \frac{1}{T_2}\right)$$

Let's substitute some typical values into this equation. The activation energy for many reactions that occur near room temperature is about 50 kJ/mol. For such a reaction, a temperature increase from 300 K to 310 K would result in

$$\ln \frac{k_2}{k_1} = \frac{50,000 \text{ J/mol}}{(8.314 \text{ J/mol·K})}\left(\frac{1}{300 \text{ K}} - \frac{1}{310 \text{ K}}\right) = 0.647$$

$$\frac{k_2}{k_1} = 1.91 \approx 2$$

Chemists sometimes use the rule of thumb that near room temperature the rate of a reaction approximately doubles with a 10°C rise in temperature. Such a "rule" must be used with care, however, because it obviously depends on the activation energy.

A fever is your body's natural way of speeding up the immune system (and the rest of your metabolism) to more effectively fight an infection. Part of the old saying "feed a fever" makes a lot of sense from a kinetics viewpoint. Why—in terms of what you have learned in this chapter?

The reaction rate of chemilumines-
cent light sticks is strongly affected
by temperature. In the top photo
the light sticks immersed in cold
water (0.2°C) are barely glowing
relative to those sitting outside the
beaker at room temperature (22°C).
In the bottom photo the light sticks
in the beaker are at 59°C and are
glowing very brightly compared to
the room temperature sticks. The
light sticks at 59°C will "burn out",
that is, consume all the reactants
that produce the light much sooner
due to the faster reaction at the
higher temperature.

EXAMPLE 16-11 *Arrhenius Equation*

The specific rate constant, k, for the following first-order reaction is 9.16×10^{-3} s^{-1} at 0.0°C.
The activation energy of this reaction is 88.0 kJ/mol. Determine the value of k at 2.0°C.

$$N_2O_5 \longrightarrow NO_2 + NO_3$$

Plan

First we tabulate the values, remembering to convert temperature to the Kelvin scale.

$$E_a = 88,000 \text{ J/mol} \qquad\qquad R = 8.314 \text{ J/mol·K}$$
$$k_1 = 9.16 \times 10^{-3} \text{ s}^{-1} \quad \text{at} \quad T_1 = 0.0°C + 273 = 273 \text{ K}$$
$$k_2 = \underline{?} \quad \text{at} \quad T_2 = 2.0°C + 273 = 275 \text{ K}$$

We use these values in the "two-temperature" form of the Arrhenius equation.

Solution

$$\ln \frac{k_2}{k_1} = \frac{E_a}{R}\left(\frac{1}{T_1} - \frac{1}{T_2}\right)$$

$$\ln\left(\frac{k_2}{9.16 \times 10^{-3} \text{ s}^{-1}}\right) = \frac{88,000 \text{ J/mol}}{8.314 \dfrac{\text{J}}{\text{mol·K}}}\left(\frac{1}{273 \text{ K}} - \frac{1}{275 \text{ K}}\right) = 0.282$$

Taking inverse (natural) logarithms of both sides,

$$\frac{k_2}{9.16 \times 10^{-3} \text{ s}^{-1}} = 1.32$$

$$k_2 = 1.32(9.16 \times 10^{-3} \text{ s}^{-1}) = \boxed{1.21 \times 10^{-2} \text{ s}^{-1}}$$

We see that a very small temperature difference, only 2°C, causes an increase in the rate constant (and hence in the reaction rate for the same concentrations) of about 32%. Such sensitivity of rate to temperature change makes the control and measurement of temperature extremely
important in chemical reactions.

You should now work Exercise 53.

EXAMPLE 16-12 *Activation Energy*

The gas-phase decomposition of ethyl iodide to give ethylene and hydrogen iodide is a first-
order reaction.

$$C_2H_5I \longrightarrow C_2H_4 + HI$$

At 600. K, the value of k is 1.60×10^{-5} s^{-1}. When the temperature is raised to 700. K, the value
of k increases to 6.36×10^{-3} s^{-1}. What is the activation energy for this reaction?

Plan

We know k at two different temperatures. We solve the two-temperature forms of the Arrhenius
equation for E_a and evaluate.

Solution

$$k_1 = 1.60 \times 10^{-5} \text{ s}^{-1} \text{ at } T_1 = 600. \text{ K} \qquad k_2 = 6.36 \times 10^{-3} \text{ s}^{-1} \text{ at } T_2 = 700. \text{ K}$$
$$R = 8.314 \text{ J/mol·K} \qquad\qquad E_a = \underline{?}$$

We arrange the Arrhenius equation for E_a.

$$\ln \frac{k_2}{k_1} = \frac{E_a}{R}\left(\frac{1}{T_1} - \frac{1}{T_2}\right) \quad \text{so} \quad E_a = \frac{R \ln \frac{k_2}{k_1}}{\left(\frac{1}{T_1} - \frac{1}{T_2}\right)}$$

Substituting,

$$E_a = \frac{\left(8.314 \frac{J}{mol \cdot K}\right)\ln\left(\frac{6.36 \times 10^{-3}\,s^{-1}}{1.60 \times 10^{-5}\,s^{-1}}\right)}{\left(\frac{1}{600.K} - \frac{1}{700.K}\right)} = \frac{\left(8.314 \frac{J}{mol \cdot K}\right)(5.98)}{2.38 \times 10^{-4}\,K^{-1}} = \boxed{\begin{array}{l} 2.09 \times 10^5\,J/mol \\ or \quad 209\,kJ/mol \end{array}}$$

You should now work Exercise 54.

The determination of E_a in the manner illustrated in Example 16-12 may be subject to considerable error because it depends on the measurement of k at only two temperatures. Any error in either of these k values would greatly affect the resulting value of E_a. A more reliable method that uses many measured values for the same reaction is based on a graphical approach. Let us rearrange the single-temperature logarithmic form of the Arrhenius equation and compare it with the equation for a straight line.

Compare this approach to that described in the earlier Enrichment section for determining k.

$$\underbrace{\ln k}_{y} = \underbrace{-\left(\frac{E_a}{R}\right)}_{m}\underbrace{\left(\frac{1}{T}\right)}_{x} + \underbrace{\ln A}_{b}$$

The value of the collision frequency factor, A, is very nearly constant over moderate temperature changes. Thus, $\ln A$ can be interpreted as the constant term in the equation (the intercept). The slope of the straight line obtained by plotting $\ln k$ versus $1/T$ equals $-E_a/R$. This allows us to determine the value of the activation energy from the slope (Figure 16-14). Exercises 57 and 58 use this method.

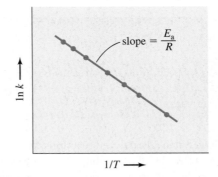

Figure 16-14 A graphical method for determining activation energy, E_a. At each of several different temperatures, the rate constant, k, is determined by methods such as those in Sections 16-3 and 16-4. A plot of $\ln k$ versus $1/T$ gives a straight line with negative slope. The slope of this straight line is $-E_a/R$. Use of this graphical method is often desirable, because it partially compensates for experimental errors in individual k and T values.

Figure 16-15 Potential energy diagrams showing the effect of a catalyst. The catalyst provides a different mechanism, corresponding to a lower-energy pathway, for the formation of the products. A catalyzed reaction typically occurs in several steps, each with its own barrier, but the overall energy barrier for the net reaction, E_a', is lower than that for the uncatalyzed reaction, E_a. The value of ΔE_{rxn} depends only on the states of the reactants and products, so it is the same for either path.

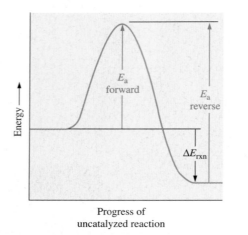

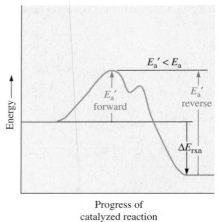

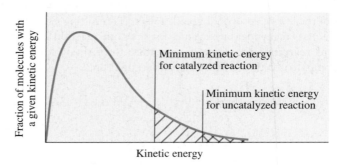

Figure 16-16 When a catalyst is present, the energy barrier is lowered. Thus, more molecules possess the minimum kinetic energy necessary for reaction. This is analogous to allowing more students to pass a course by lowering the requirements.

16-9 CATALYSTS

Catalysts are substances that can be added to reacting systems to increase the rate of reaction. They allow reactions to occur via alternative pathways that increase reaction rates by lowering activation energies.

The activation energy is lowered in all catalyzed reactions, as depicted in Figures 16-15 and 16-16. A catalyst does take part in the reaction, but all of it is re-formed in later steps. Thus, a catalyst does not appear in the balanced equation for the reaction.

For constant T and the same concentrations,

CD-ROM Screen 15.14, Catalysis and Reaction Rate.

| If E_a decreases | $\Rightarrow$ | E_a/RT decreases | $\Rightarrow$ | $-E_a/RT$ increases | $\Rightarrow$ | $e^{-E_a/RT}$ increases | $\Rightarrow$ | k increases | $\Rightarrow$ | **Reaction speeds up** |

We can describe two categories of catalysts: (1) homogeneous catalysts and (2) heterogeneous catalysts.

Homogeneous Catalysis

A **homogeneous catalyst** exists in the same phase as the reactants. Ceric ion, Ce^{4+}, at one time was an important laboratory oxidizing agent that was used in many redox titrations (Section 11.7). For example, Ce^{4+} oxidizes thallium(I) ions in solution; this reaction is catalyzed by the addition of a very small amount of a soluble salt containing manganese(II) ions, Mn^{2+}. The Mn^{2+} acts as a homogeneous catalyst.

$$2Ce^{4+} + Tl^+ \xrightarrow{Mn^{2+}} 2Ce^{3+} + Tl^{3+}$$

This reaction is thought to proceed by the following sequence of elementary steps.

$$
\begin{array}{ll}
Ce^{4+} + Mn^{2+} \longrightarrow Ce^{3+} + Mn^{3+} & \text{step 1} \\
Ce^{4+} + Mn^{3+} \longrightarrow Ce^{3+} + Mn^{4+} & \text{step 2} \\
\underline{Mn^{4+} + Tl^+ \longrightarrow Mn^{2+} + Tl^{3+}} & \text{step 3} \\
2Ce^{4+} + Tl^+ \longrightarrow 2Ce^{3+} + Tl^{3+} & \text{overall}
\end{array}
$$

Some of the Mn^{2+} catalyst reacts in Step 1, but an equal amount is regenerated in Step 3 and is thus available to react again. The two ions shown in blue, Mn^{3+} and Mn^{4+}, are *reaction intermediates*. Mn^{3+} ions are formed in Step 1 and consumed in an equal amount in Step 2; similarly, Mn^{4+} ions are formed in Step 2 and consumed in an equal amount in Step 3.

We can now summarize the species that can appear in a reaction mechanism.

1. *Reactant:* More is consumed than is formed.
2. *Product:* More is formed than is consumed.
3. *Reaction intermediate:* Formed in earlier steps, then consumed in an equal amount in later steps.
4. *Catalyst:* Consumed in earlier steps, then regenerated in an equal amount in later steps.

Strong acids function as homogeneous catalysts in the acid-catalyzed hydrolysis of esters (a class of organic compounds—Section 27-14). Using ethyl acetate (a component of nail polish removers) as an example of an ester, we can write the overall reaction as follows.

"Hydrolysis" means reaction with water.

$$\underset{\text{ethyl acetate}}{CH_3\overset{\displaystyle O}{\overset{\|}{C}}{-}OCH_2CH_3(aq)} + H_2O \xrightarrow{H^+} \underset{\text{acetic acid}}{CH_3\overset{\displaystyle O}{\overset{\|}{C}}{-}OH(aq)} + \underset{\text{ethanol}}{CH_3CH_2OH(aq)}$$

This is a thermodynamically favored reaction, but because of its high energy of activation, it occurs only very, very slowly when no catalyst is present. In the presence of strong acids, however, the reaction occurs more rapidly. In this acid-catalyzed hydrolysis, different intermediates with lower activation energies are formed. The sequence of steps in the *proposed* mechanism follows.

Groups of atoms that are involved in the change in each step are shown in blue. The catalyst, H$^+$, is shown in red.

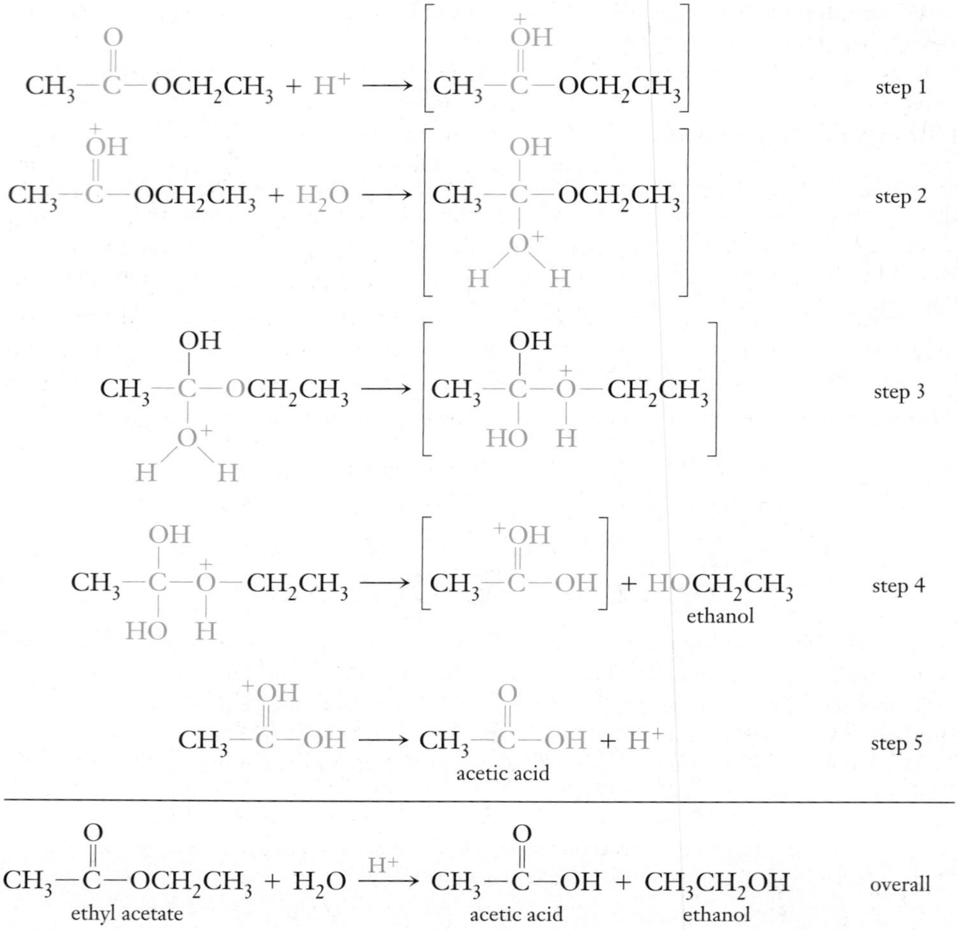

$$CH_3-\overset{O}{\overset{\|}{C}}-OCH_2CH_3 + H^+ \longrightarrow \left[CH_3-\overset{\overset{+}{O}H}{\overset{\|}{C}}-OCH_2CH_3\right] \qquad \text{step 1}$$

$$CH_3-\overset{\overset{+}{O}H}{\overset{\|}{C}}-OCH_2CH_3 + H_2O \longrightarrow \left[CH_3-\overset{OH}{\underset{\underset{H}{O^+}\underset{}{}}{\overset{|}{C}}}-OCH_2CH_3\right] \qquad \text{step 2}$$

$$CH_3-\overset{OH}{\underset{\underset{H}{\overset{|}{O^+}}\ \ H}{\overset{|}{C}}}-OCH_2CH_3 \longrightarrow \left[CH_3-\overset{OH}{\underset{HO\ \ H}{\overset{|}{C}}}-\overset{+}{O}-CH_2CH_3\right] \qquad \text{step 3}$$

$$CH_3-\overset{OH}{\underset{HO\ \ H}{\overset{|}{C}}}-\overset{+}{O}-CH_2CH_3 \longrightarrow \left[CH_3-\overset{+OH}{\overset{\|}{C}}-OH\right] + HOCH_2CH_3 \qquad \text{step 4}$$
$$\text{ethanol}$$

$$CH_3-\overset{+OH}{\overset{\|}{C}}-OH \longrightarrow CH_3-\overset{O}{\overset{\|}{C}}-OH + H^+ \qquad \text{step 5}$$
$$\text{acetic acid}$$

$$CH_3-\overset{O}{\overset{\|}{C}}-OCH_2CH_3 + H_2O \xrightarrow{H^+} CH_3-\overset{O}{\overset{\|}{C}}-OH + CH_3CH_2OH \qquad \text{overall}$$
$$\text{ethyl acetate} \qquad\qquad\quad \text{acetic acid} \qquad \text{ethanol}$$

We see that H$^+$ is a reactant in Step 1, but it is completely regenerated in Step 5. H$^+$ is therefore a catalyst. The species shown in brackets in Steps 1 through 4 are *reaction intermediates*. Ethyl acetate and water are the reactants, and acetic acid and ethanol are the products of the overall catalyzed reaction.

All intermediates in this sequence of elementary steps are charged species, but this is not always the case.

Heterogeneous Catalysis

A **heterogeneous catalyst** (also known as a **contact catalyst**) is present in a different phase from the reactants. Such catalysts are usually solids, and they lower activation energies by providing surfaces on which reactions can occur. The first step in the catalytic process is usually *adsorption*, in which one or more of the reactants become attached to the solid surface. Some reactant molecules may be held in particular orientations, or some bonds may be weakened; in other molecules, some bonds may be broken to form atoms or smaller

The petroleum industry uses numerous heterogeneous catalysts. Many of them contain highly colored compounds of transition metal ions. Several are shown here.

molecular fragments. This causes *activation* of the reactants. As a result, *reaction* occurs more readily than would otherwise be possible. In a final step, *desorption*, the product molecules leave the surface, freeing reaction sites to be used again. Most contact catalysts are more effective as small particles, because they have relatively large surface areas.

Transition metals and their compounds function as effective catalysts in many homogeneous and heterogeneous reactions. Vacant *d* orbitals in many transition metal ions can accept electrons from reactants to form intermediates. These subsequently decompose to form products. Three transition metals, Pt, Pd, and Ni, are often used as finely divided solids to provide surfaces on which heterogeneous reactions can occur.

The catalytic converters (Figure 16-17) built into automobile exhaust systems contain two types of heterogeneous catalysts, powdered noble metals and powdered transition metal oxides. They catalyze the oxidation of unburned hydrocarbon fuel (reaction 1) and of partial combustion products such as carbon monoxide (reaction 2, shown in Figure 16-18).

> Approximately 98% of all catalysts used in industry are heterogeneous. Therefore, when industrial chemists talk about a "catalyst" they usually mean a heterogeneous catalyst.

1.
$$2C_8H_{18}(g) + 25O_2(g) \xrightarrow[\text{NiO}]{\text{Pt}} 16CO_2(g) + 18H_2O(g)$$
$\uparrow$
isooctane (a component of gasoline)

2.
$$2CO(g) + O_2(g) \xrightarrow[\text{NiO}]{\text{Pt}} 2CO_2(g)$$

It is desirable to carry out these reactions in automobile exhaust systems. Carbon monoxide is very poisonous. The latter reaction is so slow that a mixture of CO and O_2 gas at the exhaust temperature would remain unreacted for thousands of years in the absence of a catalyst! Yet the addition of only a small amount of a solid, finely divided transition metal catalyst promotes the production of up to a mole of CO_2 per minute. Because this reaction is a very simple but important one, it has been studied extensively by surface chemists. It is one of the best understood heterogeneously catalyzed reactions. The major features of the catalytic process are shown in Figure 16-18.

The same catalysts also catalyze another reaction, the decomposition of nitrogen oxide, NO, into harmless N_2 and O_2.

$$2NO(g) \xrightarrow[\text{NiO}]{\text{Pt}} N_2(g) + O_2(g)$$

> At the high temperatures of the combustion of any fuel in air, nitrogen and oxygen combine to form nitrogen oxide.

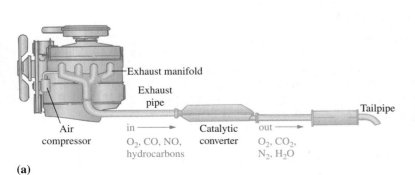

Exhaust manifold
Exhaust pipe
Tailpipe
Air compressor
in ⟶
O_2, CO, NO, hydrocarbons
Catalytic converter
out ⟶
O_2, CO_2, N_2, H_2O

(a)

(b)

Courtesy of General Motors

Figure 16-17 (a) The arrangement of a catalytic converter in an automobile. (b) A cutaway view of a catalytic converter, showing the pellets of catalyst.

(a) *Adsorption*: CO and O_2 reactant molecules become bound to the surface:

$$CO(g) \longrightarrow CO(surface) \quad \text{and} \quad O_2(g) \longrightarrow O_2(surface)$$

The CO molecules are linked through their C atoms to one or more metal atoms on the surface. The O_2 molecules are more weakly bound.

(b) *Activation*: The O_2 molecules dissociate into O atoms, which are held in place more tightly:

$$O_2(surface) \longrightarrow 2O(surface)$$

The CO molecules stick to the surface, but they migrate easily across the surface.

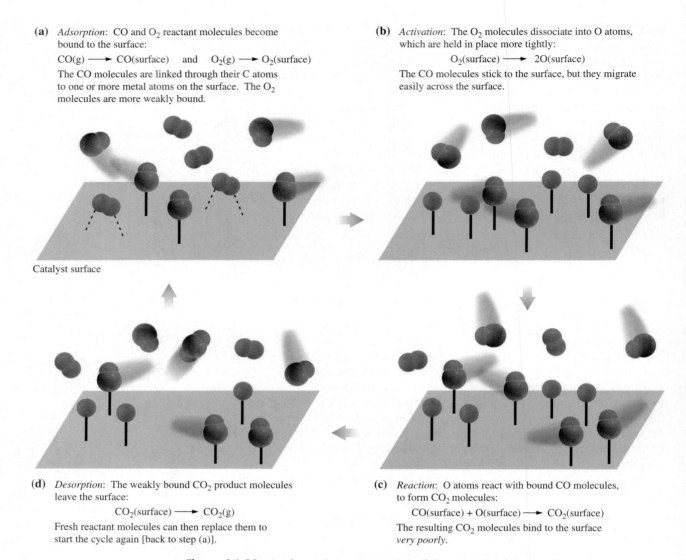

Catalyst surface

(d) *Desorption*: The weakly bound CO_2 product molecules leave the surface:

$$CO_2(surface) \longrightarrow CO_2(g)$$

Fresh reactant molecules can then replace them to start the cycle again [back to step (a)].

(c) *Reaction*: O atoms react with bound CO molecules, to form CO_2 molecules:

$$CO(surface) + O(surface) \longrightarrow CO_2(surface)$$

The resulting CO_2 molecules bind to the surface *very poorly*.

Figure 16-18 A schematic representation of the catalysis of the reaction

$$2CO(g) + O_2(g) \longrightarrow 2CO_2(g)$$

on a metallic surface.

See the Chemistry in Use essay "Nitrogen Oxides and Photochemical Smog" in Chapter 24.

Nitrogen oxide is a serious air pollutant because it is oxidized to nitrogen dioxide, NO_2, which reacts with water to form nitric acid and with other products of the incomplete combustion of hydrocarbons to form nitrates. The latter are eye irritants in photochemical smog.

These three reactions, catalyzed in catalytic converters, are all exothermic and thermodynamically favored. Unfortunately, other energetically favored reactions are also accelerated by the mixed catalysts. All fossil fuels contain sulfur compounds, which are oxidized to sulfur dioxide during combustion. Sulfur dioxide, itself an air pollutant, undergoes further oxidation to form sulfur trioxide as it passes through the catalytic bed.

(a)   (b)   (c)

A 30% hydrogen peroxide, H_2O_2, solution at room temperature decomposes very slowly to form O_2 and H_2O. (a) This reaction takes place more rapidly if the solution is heated. (b) A very small amount of a transition metal oxide is added. (c) This oxide catalyzes the decomposition reaction. The catalyzed reaction is rapid, so the exothermic reaction quickly heats the solution to the boiling point of water, forming steam. The temperature increase further accelerates the decomposition. Never use a syringe with a metal tip to withdraw a sample from a 30% hydrogen peroxide solution, because the metal can dangerously catalyze the rapid decomposition. (d) The bombardier beetle uses a catalyzed decomposition of hydrogen peroxide as a means of defense. An enzyme produced by the beetle catalyzes the rapid exothermic reaction. The resulting steam, along with other irritating chemicals, is ejected.

(d)

$$2SO_2(g) + O_2(g) \xrightarrow[\text{NiO}]{\text{Pt}} 2SO_3(g)$$

Sulfur trioxide is probably a worse pollutant than sulfur dioxide, because SO_3 is the acid anhydride of strong, corrosive sulfuric acid. Sulfur trioxide reacts with water vapor in the air, as well as in auto exhausts, to form sulfuric acid droplets. This problem must be overcome if the current type of catalytic converter is to see continued use. These same catalysts also suffer from the problem of being "poisoned"—that is, made inactive—by lead. Leaded fuels contain tetraethyl lead, $Pb(C_2H_5)_4$, and tetramethyl lead, $Pb(CH_3)_4$. Such fuels are not suitable for automobiles equipped with catalytic converters and are excluded by U. S. law from use in such cars.

Reactions that occur in the presence of a solid catalyst, as on a metal surface (heterogeneous catalysis) often follow zero-order kinetics. For instance, the rate of decomposition of $NO_2(g)$ at high pressures on a platinum metal surface does not change if we add more NO_2. This is because only the NO_2 molecules on the surface can react. If the metal surface is completely covered with NO_2 molecules, no additional molecules can be *adsorbed* until the ones already there have reacted and the products have *desorbed*. Thus, the rate of the reaction is controlled only by the availability of reaction sites on the Pt surface, and not by the total number of NO_2 molecules available.

Maintaining the continued efficiency of all three reactions in a "three-way" catalytic converter is a delicate matter. It requires control of such factors as the O_2 supply pressure and the order in which the reactants reach the catalyst. Modern automobile engines use microcomputer chips, based on an O_2 sensor in the exhaust stream, to control air valves.

(text continues on page 688)

Ozone

Ozone, O_3, is such a powerful oxidizing agent that in significant concentrations it degrades many plastics, metals, and rubber, as well as both plant and animal tissues. We therefore try to minimize exposure to ozone in our immediate environment. In the upper atmosphere, however, ozone plays a very important role in the absorption of harmful radiation from the sun. Maintaining appropriate concentrations of ozone—minimizing its production where ozone is harmful and preventing its destruction where ozone is helpful—is an important challenge in environmental chemistry.

Ozone is formed in the upper atmosphere as some O_2 molecules absorb high-energy UV radiation from the sun and dissociate into oxygen atoms; these then combine with other O_2 molecules to form ozone.

$$O_2(g) + UV \text{ radiation} \longrightarrow 2O(g) \quad \text{(step 1—occurs once)}$$
$$O_2(g) + O(g) \longrightarrow O_3(g) \quad \text{(step 2—occurs twice)}$$

$$3O_2(g) + UV \text{ radiation} \longrightarrow 2O_3(g) \quad \text{(net reaction)}$$

Although it also decomposes in the upper atmosphere, the ozone supply is continuously replenished by this process. Its concentration in the stratosphere ($\approx 7–31$ miles above the earth's surface) is about 10 ppm (parts per million), whereas it is only about 0.04 ppm near the earth's surface.

The high-altitude ozone layer is responsible for absorbing much of the dangerous ultraviolet light from the sun in the 20–30 Å wavelength range.

$$O_3(g) + UV \text{ radiation} \longrightarrow O_2(g) + O(g)$$
$$\text{(step 1—occurs once)}$$
$$O_2(g) + O(g) \longrightarrow O_3(g) \quad \text{(step 2—occurs once)}$$

$$\text{No net reaction}$$

We see that each time this sequence takes place, it absorbs one photon of ultraviolet light; however, the process regenerates as much ozone as it uses up. Each stratospheric ozone molecule can thus absorb a significant amount of ultraviolet light. If this high-energy radiation reached the surface of the earth in higher intensity, it would be very harmful to plants and animals (including humans). It has been estimated that the incidence of skin cancer would increase by 2% for every 1% decrease in the concentration of ozone in the stratosphere.

Chlorofluorocarbons (CFCs) are chemically inert, nonflammable, nontoxic compounds that are superb solvents and have been used in many industrial processes; they are excellent coolants for air conditioners and refrigerators. Two CFCs that have been widely used are Freon-11 and Freon-12 (Freon is a DuPont trade name).

$$\begin{array}{cc} & \text{Cl} \\ & | \\ \text{Cl}-\text{C}-\text{F} \\ & | \\ & \text{Cl} \end{array} \qquad \begin{array}{cc} & \text{Cl} \\ & | \\ \text{Cl}-\text{C}-\text{F} \\ & | \\ & \text{F} \end{array}$$

Freon-11 is CCl_3F Freon-12 is CCl_2F_2

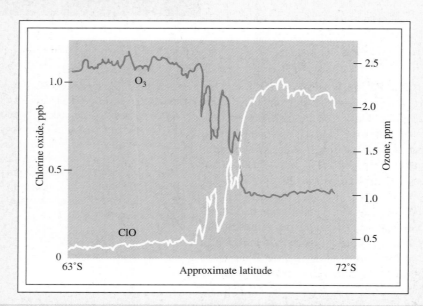

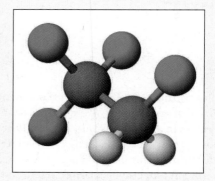

The compound known as HCFC-134, a fluorocarbon currently used in home and automobile air conditioners.

A plot that shows the decrease in [O_3] as [ClO] increases over Antarctica.

The CFCs are so unreactive that they do not readily decompose, that is, break down into simpler compounds, when they are released into the atmosphere. Over time the CFCs are carried into the stratosphere by air currents, where they are exposed to large amounts of ultraviolet radiation.

In 1974, Mario Molina and Sherwood Rowland of the University of California–Irvine demonstrated in their laboratory that when CFCs are exposed to high energy ultraviolet radiation they break down to form chlorine *radicals* ($:\overset{..}{\underset{..}{Cl}}\cdot$).

$$F-\overset{\displaystyle F}{\underset{\displaystyle Cl}{\overset{|}{\underset{|}{C}}}}-Cl \xrightarrow[\text{radiation}]{\text{UV}} F-\overset{\displaystyle F}{\underset{\displaystyle Cl}{\overset{|}{\underset{|}{C}}}}\cdot \ + \ :\overset{..}{\underset{..}{Cl}}\cdot \quad \text{(a chlorine radical)}$$

Molina and Rowland predicted that these very reactive radicals could cause problems by catalyzing the destruction of ozone in the stratosphere.

Each spring since 1979, researchers have observed a thinning of the ozone layer over Antarctica. Each spring (autumn in the Northern Hemisphere) beginning in 1983, satellite images have shown a "hole" in the ozone layer over the South Pole. During August and September 1987, a NASA research team flew a plane equipped with sophisticated analytical instruments into the ozone hole 25 times. Their measurements demonstrated that as the concentration of the chlorine oxide radicals, Cl—O increased, the concentration of ozone decreased.

By September 1992, this **ozone hole** was nearly three times the area of the United States. In December 1994, three years of data from NASA's Upper Atmosphere Research Satellite (UARS) provided conclusive evidence that CFCs are primarily responsible for this destruction of the ozone layer. Considerable thinning of the ozone layer in the Northern Hemisphere has also been observed.

The following is a simplified representation of the **chain reaction** that is now believed to account for most of the ozone destruction in the stratosphere.

$$:\overset{..}{\underset{..}{Cl}}\cdot \ + \ O_3 \longrightarrow Cl-\overset{..}{\underset{..}{O}}\cdot \ + \ O_2 \quad \text{(step 1)}$$
$$Cl-\overset{..}{\underset{..}{O}}\cdot \ + \ O \longrightarrow :\overset{..}{\underset{..}{Cl}}\cdot \ + \ O_2 \quad \text{(step 2)}$$
$$\overline{O_3(g) \ + \ O \longrightarrow 2O_2(g) \qquad \text{(net reaction)}}$$

A sufficient supply of oxygen atoms, O, is available in the upper atmosphere for the second step to occur. The net reaction results in the destruction of a molecule of ozone. The chlorine radical that initiates the first step of this reaction sequence is regenerated in the second step, however, and so a single chlorine radical can act as a catalyst to destroy many

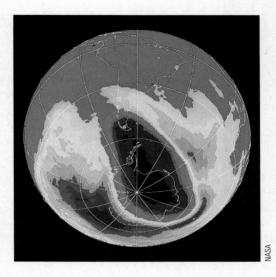

A computer-generated image of part of the Southern Hemisphere on October 17, 1994, reveals the ozone "hole" (black and purple areas) over Antarctica and the tip of South America. Relatively low ozone levels (blue and green areas) extend into much of South America as well as Central America. Normal ozone levels are shown in yellow, orange, and red. The ozone hole is not stationary but moves about as a result of air currents.

thousands of O_3 molecules. Other well-known reactions also destroy ozone in the stratosphere, but the evidence shows conclusively that the CFCs are the principal culprits.

Since January 1978, the use of CFCs in aerosol cans in the United States has been banned; increasingly strict laws prohibit the release into the atmosphere of CFCs from sources such as automobile air conditioners and discarded refrigerators. The Montreal Protocol, signed by 24 countries in 1989, called for reductions in production and use of many CFCs. International agreements have since called for a complete ban on CFC production. Efforts to develop suitable replacement substances and controls for existing CFCs continue. The good news is that scientists expect the ozone hole to decrease and possibly disappear during the 21st century *if* current international treaties remain in effect and *if* they are implemented throughout the world. These are two very large *if*s.

Additional information on stratospheric ozone can be found at the EPA Web site, http://www.epa.gov/docs/ozone. The Web site for this textbook will direct you to additional information about ozone.

When heated, a sugar cube (sucrose, melting point 185°C) melts but does not burn. A sugar cube rubbed in cigarette ash burns before it melts. The cigarette ash contains trace amounts of metal compounds that catalyze the combustion of sugar.

Some other important reactions that are catalyzed by transition metals and their oxides follow.

1. The Haber process for the production of ammonia (Section 17-7).

$$N_2 + 3H_2 \xrightarrow[\text{high T, P}]{\text{Fe, Fe oxides}} 2NH_3$$

2. The contact process for the production of sulfur trioxide in the manufacture of sulfuric acid (Section 24-11).

$$2SO_2 + O_2 \xrightarrow[\text{400°C}]{V_2O_5} 2SO_3$$

3. The chlorination of benzene (Section 27-16).

$$\underset{\text{benzene}}{C_6H_6} + Cl_2 \xrightarrow{FeBr_3} \underset{\text{chlorobenzene}}{C_6H_5Cl} + HCl$$

4. The hydrogenation of unsaturated hydrocarbons (Section 27-17).

$$RCH{=}CH_2 + H_2 \xrightarrow{Pt} RCH_2CH_3 \qquad R = \text{organic groups}$$

Enzymes as Biological Catalysts

Enzymes are proteins that act as catalysts for specific biochemical reactions in living systems. The reactants in enzyme-catalyzed reactions are called **substrates.** Thousands of vital processes in our bodies are catalyzed by many distinct enzymes. For instance, the enzyme carbonic anhydrase catalyzes the combination of CO_2 and water (the substrates), facilitating most of the transport of carbon dioxide in the blood. This combination reaction, ordinarily quite slow, proceeds rapidly in the presence of carbonic anhydrase; a single molecule of this enzyme can promote the conversion of more than 1 million molecules of carbon dioxide each second. Each enzyme is extremely specific, catalyzing only a few closely related reactions—or, in many cases, only one particular reaction—for only certain substrates. Modern theories of enzyme action attribute this to the requirement of very specific matching of shapes (molecular geometries) for a particular substrate to bind to a particular enzyme (Figure 16-19).

Enzyme-catalyzed reactions are important examples of zero-order reactions; that is, the rate of such a reaction is independent of the concentration of the substrate (provided *some* substrate is present).

$$\text{rate} = k$$

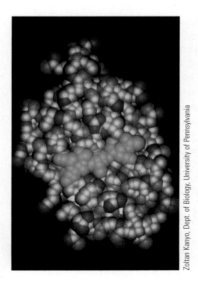

A space-filling model of the enzyme lysozyme. This enzyme catalyzes the hydrolysis of polysaccharides (complex carbohydrates) found in bacterial cell walls, that is, it breaks the link between two adjacent sugar units in the polysaccharide. The intact polysaccharide, shown here in green, fits into the active site, which is a cleft in the surface of the lysozyme molecule. The arrangement of hydrogen-bonding groups in the active site of the enzyme surface matches that on the polysaccharide, but other molecules with different hydrogen-bonding groups do not fit as well, so the enzyme exhibits very specific bonding for polysaccharides. After the hydrolysis reaction has taken place, the product molecules do not fit the active site as well, so they are released, making way for another polysaccharide to be bound to the enzyme active site.

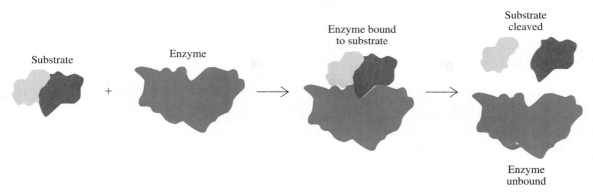

Figure 16-19 A schematic representation of a simplified mechanism (lock-and-key) for enzyme reaction. The substrates (reactants) fit the active sites of the enzyme molecule much as keys fit locks. When the reaction is complete, the products do not fit the active sites as well as the reactants did. They separate from the enzyme, leaving it free to catalyze the reaction of additional reactant molecules. The enzyme is not permanently changed by the process. The illustration here is for a process in which a complex reactant molecule is split to form two simpler product molecules. The formation of simple sugars from complex carbohydrates is a similar reaction. Some enzymes catalyze the combination of simple molecules to form more complex ones.

The active site on an enzyme can bind to only one substrate molecule at a time (or one pair, if the reaction links two reactant molecules), no matter how many other substrate molecules are available in the vicinity.

Ammonia is a very important industrial chemical that is used as a fertilizer and in the manufacture of many other chemicals. The reaction of nitrogen with hydrogen is a thermodynamically spontaneous reaction (product-favored), but without a catalyst it is very slow, even at high temperatures. The Haber process for its preparation involves the use of iron as a catalyst at 450°C to 500°C and high pressures.

$$N_2(g) + 3H_2(g) \xrightarrow{\text{Fe}} 2NH_3(g) \qquad \Delta G^0 = -194.7 \text{ kJ/mol (at 500°C)}$$

Even so, iron is not a very effective catalyst.

In contrast, the reaction between N_2 and H_2 to form NH_3 is catalyzed at room temperature and atmospheric pressure by a class of enzymes, called nitrogenases, that are present in some bacteria. Legumes are plants that support these bacteria; they are able to obtain nitrogen as N_2 from the atmosphere and convert it to ammonia.

In comparison with manufactured catalysts, most enzymes are tremendously efficient under very mild conditions. If chemists and biochemists could develop catalysts with a small fraction of the efficiency of enzymes, such catalysts could be a great boon to the world's health and economy. One of the most active areas of current chemical research involves attempts to discover or synthesize catalysts that can mimic the efficiency of naturally occurring enzymes such as nitrogenases. Such a development would be important in industry. It would eliminate the costs of the high temperature and high pressure that are necessary in the Haber process. This could decrease the cost of food grown with the aid of ammonia-based fertilizers. Ultimately this would help greatly to feed the world's growing population.

Transition metal ions are present in the active sites of some enzymes.

The process is called nitrogen fixation. The ammonia can be used in the synthesis of many nitrogen-containing biological compounds such as proteins and nucleic acids.

Key Terms

Activation energy (or **activation barrier**)   The kinetic energy that reactant molecules must have to allow them to reach the transition state so that a reaction can occur.

Arrhenius equation   An equation that relates the specific rate constant to activation energy and temperature.

Catalyst   A substance that increases the rate at which a reaction occurs. It remains unchanged when the reaction is complete.

Chemical kinetics   The study of rates and mechanisms of chemical reactions and of the factors on which they depend.

Collision theory   A theory of reaction rates that states that effective collisions between reactant molecules must take place for reaction to occur.

Contact catalyst   See *Heterogeneous catalyst.*

Effective collision   A collision between molecules that results in reaction; one in which molecules collide with proper orientations and with sufficient energy to react.

Elementary step   An individual step in the mechanism by which a reaction occurs. For each elementary step, the reaction orders *do* match the reactant coefficients in that step.

Enzyme   A protein that acts as a catalyst in a biological system.

Fundamental step   See *Elementary step.*

Half-life of a reactant   The time required for half of that reactant to be converted into product(s).

Heterogeneous catalyst   A catalyst that exists in a different phase (solid, liquid, or gas) from the reactants; a contact catalyst.

Homogeneous catalyst   A catalyst that exists in the same phase (liquid or gas) as the reactants.

Integrated rate equation   An equation that relates the concentration of a reactant remaining to the time elapsed; has different mathematical forms for different orders of reaction.

Method of initial rates   A method of determining the rate-law expression by carrying out a reaction with different initial concentrations and analyzing the resulting changes in initial rates.

Order of a reactant   The power to which the reactant's concentration is raised in the rate-law expression.

Order of a reaction   The sum of the powers to which all concentrations are raised in the rate-law expression; also called overall order of a reaction.

Progress of reaction (or **reaction coordinate**)   The progress along the pathway from reactants to products; sometimes called "progress of reaction".

Rate constant (also called **specific rate constant**)   An experimentally determined proportionality constant that is different for different reactions and that, for a given reaction, changes only with temperature or the presence of a catalyst; k in the rate-law expression, rate = $k[A]^x[B]^y$.

Rate-determining step   The slowest elementary step in a reaction mechanism; the step that limits the overall rate of reaction.

Rate-law expression (also called **rate law**)   An equation that relates the rate of a reaction to the concentrations of the reactants and the specific rate constant; rate = $k[A]^x[B]^y$. The exponents of reactant concentrations *do not necessarily* match the coefficients in the overall balanced chemical equation. The rate-law expression must be determined from experimental data.

Rate of reaction   The change in concentration of a reactant or product per unit time.

Reaction intermediate   A species that is produced and then entirely consumed during a multistep reaction; usually short-lived.

Reaction mechanism   The sequence of fundamental steps by which reactants are converted into products.

Substrate   A reactant in an enzyme-catalyzed reaction.

Thermodynamically favorable (spontaneous) reaction   A reaction that occurs with a net release of free energy, G; a reaction for which ΔG is negative (see Section 15-15).

Transition state   A relatively high-energy state in which bonds in reactant molecules are partially broken and new ones are partially formed.

Transition state theory   A theory of reaction rates that states that reactants pass through high-energy transition states before forming products.

Exercises

General Concepts

1. Briefly summarize the effects of each of the four factors that affect rates of reactions.

2. Describe the basic features of collision theory and transition state theory.

3. What is a rate-law expression? Describe how it is determined for a particular reaction.

4. Distinguish between reactions that are thermodynamically favorable and reactions that are kinetically favorable. What can be said about relationships between the two?

5. What is meant by the mechanism of a reaction? How does the mechanism relate to the order of the reaction?

6. What, if anything, can be said about the relationship between the coefficients of the balanced *overall* equation for a reaction and the powers to which concentrations are raised in the rate-law expression? To what are these powers related?

7. Express the rate of reaction in terms of the rate of change of each reactant and product in the following reactions.

   (a) $H_2O_2(aq) + 2H^+(aq) + 2I^-(aq) \longrightarrow$
   $$I_2(aq) + 2H_2O(\ell)$$

   (b) $2NO(g) + Br_2(g) \longrightarrow 2NOBr(g)$

   (c) $CH_3COOH(aq) + OH^-(aq) \longrightarrow$
   $$CH_3COO^-(aq) + H_2O(\ell)$$

   (d) $N_2O_5(g) \longrightarrow NO_2(g) + NO_3(g)$

8. Express the rate of reaction in terms of the rate of change of each reactant and each product in the following.
   (a) $2N_2O_5(g) \longrightarrow 4NO_2(g) + O_2(g)$
   (b) $2SO_2(g) + O_2(g) \longrightarrow 2SO_3(g)$
   (c) $C_2H_4(g) + Br_2(g) \longrightarrow C_2H_4Br_2(g)$
   (d) $(C_2H_5)_2(NH)_2 + I_2 \longrightarrow (C_2H_5)_2N_2 + 2HI$

9. At a given time, N_2 is reacting with H_2 at a rate of 0.25 M/min to produce NH_3. At that same time, what is the rate at which the other reactant is changing and the rate at which the product is changing?

$$N_2 + 3H_2 \longrightarrow 2NH_3$$

10. The following equation shows the production of NO and H_2O by oxidation of ammonia. At a given time, NH_3 is reacting at a rate of 1.10 M/min. At that same time, what is the rate at which the other reactant is changing and the rate at which each product is changing?

$$4NH_3 + 5O_2 \longrightarrow 4NO + 6H_2O$$

11. Why do large crystals of sugar burn more slowly than finely ground sugar?

12. Some fireworks are bright because of the burning of magnesium. Speculate on how fireworks might be constructed using magnesium. How might the sizes of the pieces of magnesium be important? What would you expect to occur if pieces that were too large were used? Too small?

Charles Steele

Rate-Law Expression

13. If doubling the initial concentration of a reactant doubles the initial rate of reaction, what is the order of the reaction with respect to the reactant? If this concentration change causes the rate to increase by a factor of 8, what is the order? If the concentration changes and the rate remains the same, what is the order?

14. The rate expression for the following reaction at a certain temperature is rate = $k[NO]^2[O_2]$. Two experiments involving this reaction are carried out at the same tempera-

ture. In the second experiment the initial concentration of NO is halved while the initial concentration of O_2 is doubled. The initial rate in the second experiment will be _____ times that of the first.

$$2NO + O_2 \longrightarrow 2NO_2$$

15. The rate-law expression for the following reaction is found to be rate = k $[N_2O_5]$. What is the overall reaction order?

$$2N_2O_5(g) \longrightarrow 4NO_2(g) + O_2(g)$$

16. Use times expressed in seconds to give the units of the rate constant for reactions that are overall (a) first order; (b) second order; (c) third order; (d) of order $1\frac{1}{2}$.

17. Rate data were obtained at 25°C for the following reaction. What is the rate-law expression for this reaction?

$$A + 2B \longrightarrow C + 2D$$

Expt.	Initial [A] (mol/L)	Initial [B] (mol/L)	Initial Rate of Formation of C ($M \cdot min^{-1}$)
1	0.10	0.10	3.0×10^{-4}
2	0.30	0.30	9.0×10^{-4}
3	0.30	0.10	3.0×10^{-4}
4	0.40	0.20	6.0×10^{-4}

18. Rate data were obtained for the following reaction at 25°C. What is the rate-law expression for the reaction?

$$2A + B + 2C \longrightarrow D + 2E$$

Expt.	Initial [A] (M)	Initial [B] (M)	Initial [C] (M)	Initial Rate of Formation of D ($M \cdot min^{-1}$)
1	0.10	0.20	0.10	5.0×10^{-4}
2	0.20	0.20	0.30	1.5×10^{-3}
3	0.30	0.20	0.10	5.0×10^{-4}
4	0.40	0.60	0.30	4.5×10^{-3}

19. (a) A certain reaction is zero order in reactant A and second order in reactant B. If the concentrations of both reactants are doubled, what happens to the reaction rate? (b) What would happen to the reaction rate if the reaction in part (a) were first order in A and first order in B?

20. The rate expression for the following reaction is shown to be rate = $k[A]^2[B_2]$. If, during a reaction, the concentrations of both A and B_2 were suddenly halved, the rate of the reaction would _____ by a factor of _____.

$$A + B_2 \longrightarrow products$$

21. Rate data were collected for the following reaction at a particular temperature.

$$A + B \longrightarrow products$$

Expt.	Initial [A] (mol/L)	Initial [B] (mol/L)	Initial Rate of Reaction $(M \cdot s^{-1})$
1	0.10	0.10	0.0090
2	0.20	0.10	0.036
3	0.10	0.20	0.018
4	0.10	0.30	0.027

(a) What is the rate-law expression for this reaction?
(b) Describe the order of the reaction with respect to each reactant and to the overall order.

22. Rate data were collected for the following reaction at a particular temperature.

$$2ClO_2(aq) + 2OH^-(aq) \longrightarrow$$
$$ClO_3^-(aq) + ClO_2^-(aq) + H_2O(\ell)$$

Expt.	Initial $[ClO_2]$ (mol/L)	Initial $[OH^-]$ (mol/L)	Initial Rate of Reaction $(M \cdot s^{-1})$
1	0.012	0.012	2.07×10^{-4}
2	0.024	0.012	8.28×10^{-4}
3	0.012	0.024	4.14×10^{-4}
4	0.024	0.024	1.66×10^{-3}

(a) What is the rate-law expression for this reaction?
(b) Describe the order of the reaction with respect to each reactant and to the overall order.

23. The reaction

$$(C_2H_5)_2(NH)_2 + I_2 \longrightarrow (C_2H_5)_2N_2 + 2HI$$

gives the following initial rates.

Expt.	$[(C_2H_5)_2(NH)_2]_0$ (mol/L)	$[I_2]_0$ (mol/L)	Initial Rate of Formation of $(C_2H_5)_2N_2$
1	0.015	0.015	$3.15\ M \cdot s^{-1}$
2	0.015	0.045	$9.45\ M \cdot s^{-1}$
3	0.030	0.045	$18.9\ M \cdot s^{-1}$

Write the rate-law expression.

*24. Given these data for the reaction $A + B \longrightarrow C$, write the rate-law expression.

Expt.	Initial [A] (M)	Initial [B] (M)	Initial Rate of Formation of C $(M \cdot s^{-1})$
1	0.20	0.10	5.0×10^{-6}
2	0.30	0.10	7.5×10^{-6}
3	0.40	0.20	4.0×10^{-5}

*25. Given these data for the reaction $A + B \longrightarrow C$, write the rate-law expression.

Expt.	Initial [A] (M)	Initial [B] (M)	Initial Rate of Formation of C $(M \cdot s^{-1})$
1	0.25	0.15	8.0×10^{-5}
2	0.25	0.30	3.2×10^{-4}
3	0.50	0.60	5.12×10^{-3}

*26. Given these data for the reaction $A + B \longrightarrow C$, write the rate-law expression.

Expt.	Initial [A] (M)	Initial [B] (M)	Initial Rate of Formation of C $(M \cdot s^{-1})$
1	0.10	0.10	2.0×10^{-4}
2	0.20	0.10	8.0×10^{-4}
3	0.40	0.20	2.56×10^{-2}

*27. Consider a chemical reaction between compounds A and B that is first order in A and first order in B. From the information shown here, fill in the blanks.

Expt.	Rate $(M \cdot s^{-1})$	[A]	[B]
1	0.24	0.20 M	0.050 M
2	0.20	___ M	0.030 M
3	0.80	0.40 M	___ M

*28. Consider a chemical reaction of compounds A and B that was found to be first order in A and second order in B. From the following information, fill in the blanks.

Expt.	Rate $(M \cdot s^{-1})$	[A]	[B]
1	0.150	1.00 M	0.200 M
2	___	2.00 M	0.200 M
3	___	2.00 M	0.400 M

29. The following rearrangement reaction is first order:

$$CH_3NC \longrightarrow CH_3CN$$

In a table of kinetics data we find the following values listed for this reaction: $A = 3.98 \times 10^{13}$ s^{-1}, $E_a = 160$ kJ/mol. (a) Calculate the value of the specific rate constant at room temperature, 25°C. (b) Calculate the value of the specific rate constant at 115°C.

30. The following gas-phase decomposition reaction is first order:

$$C_2H_5Cl \longrightarrow C_2H_4 + HCl$$

In a table of kinetics data we find the following values listed for this reaction: $A = 1.58 \times 10^{13}$ s^{-1}, $E_a = 237$ kJ/mol. (a) Calculate the value of the specific rate constant at room temperature, 25°C. (b) Calculate the value of the specific rate constant at 275°C.

*31. The rate of decomposition of NO_2 by the following reaction at some temperature is 5.4×10^{-5} mol NO_2/L·s when $[NO_2] = 0.0110$ mol/L.

$$2NO_2(g) \longrightarrow 2NO(g) + O_2(g)$$

(a) Assume that the rate law is rate $= k[NO_2]$. What rate of disappearance of NO_2 would be predicted when $[NO_2] = 0.00550$ mol/L? (b) Now assume that the rate law is rate $= k[NO_2]^2$. What rate of disappearance of NO_2 would be predicted when $[NO_2] = 0.00550$ mol/L? (c) The rate when $[NO_2] = 0.00550$ mol/L is observed to be 1.4×10^{-5} mol NO_2/L·s. Which rate law is correct? (d) Calculate the rate constant. (*Reminder*: Express the rate of reaction in terms of rate of disappearance of NO_2.)

Integrated Rate Equations and Half-Life

32. What is meant by the half-life of a reactant?
33. The rate law for the reaction of sucrose in water,

$$C_{12}H_{22}O_{11} + H_2O \longrightarrow 2C_6H_{12}O_6$$

is rate $= k[C_{12}H_{22}O_{11}]$. After 2.57 hours at 25°C, 6.00 g/L of $C_{12}H_{22}O_{11}$ has decreased to 5.40 g/L. Evaluate k for this reaction at 25°C.

34. The rate constant for the decomposition of nitrogen dioxide

$$2NO_2 \longrightarrow 2NO + O_2$$

with a laser beam is 1.70 M^{-1}·min^{-1}. Find the time, in seconds, needed to decrease 2.00 mol/L of NO_2 to 1.25 mol/L.

35. The second-order rate constant for the following gas-phase reaction is 0.0442 M^{-1}·s^{-1}. We start with 0.135 mol C_2F_4 in a 2.00-liter container, with no C_4F_8 initially present.

$$2C_2F_4 \longrightarrow C_4F_8$$

(a) What will be the concentration of C_2F_4 after 1.00 hour? (b) What will be the concentration of C_4F_8 after 1.00 hour? (c) What is the half-life of the reaction for the initial C_2F_4 concentration given in part (a)?

(d) How long will it take for half of the C_2F_4 that remains after 1.00 hour to disappear?

36. The decomposition reaction of carbon disulfide, CS_2, to carbon monosulfide, CS, and sulfur is first order with $k = 2.8 \times 10^{-7}$ s^{-1} at 1000°C.

$$CS_2 \longrightarrow CS + S$$

(a) What is the half-life of this reaction at 1000°C? (b) How many days would pass before a 2.00-gram sample of CS_2 had decomposed to the extent that 0.75 gram of CS_2 remained? (c) Refer to part (b). How many grams of CS would be present after this length of time? (d) How much of a 2.00-gram sample of CS_2 would remain after 45.0 days?

*37. The first-order rate constant for the conversion of cyclobutane to ethylene at 1000°C is 87 s^{-1}. (a) What is the half-life of this reaction at 1000°C? (b) If you started with 4.00 g of cyclobutane, how long would it take to consume 2.50 g of it? (*Hint:* Write the ratio of concentrations, $[A]_0/[A]$, in terms of mass, molecular weight, and volume.) (c) How much of an initial 1.00-g sample of cyclobutane would remain after 1.00 s?

cyclobutane   ethylene

*38. For the reaction

$$2NO_2 \longrightarrow 2NO + O_2$$

the rate equation is

$$\text{rate} = 1.4 \times 10^{-10} \ M^{-1}\cdot s^{-1} \ [NO_2]^2 \ \text{at } 25°C$$

(a) If 3.00 mol of NO_2 is initially present in a sealed 2.00-L vessel at 25°C, what is the half-life of the reaction? (b) Refer to part (a). What concentration and how many grams of NO_2 remain after 115 years? (c) Refer to part (b). What concentration of NO would have been produced during the same period of time?

39. The first-order rate constant for the radioactive decay of radium-223 is 0.0606 day^{-1}. What is the half-life of radium-223?

40. Cyclopropane rearranges to form propene

cyclopropane   propene

in a reaction that follows first-order kinetics. At 800 K, the specific rate constant for this reaction is $2.74 \times 10^{-3} \text{ s}^{-1}$. Suppose we start with a cyclopropane concentration of 0.375 M. How long will it take for 99.0% of the cyclopropane to disappear according to this reaction?

41. The rate constant for the first-order reaction

$$N_2O_5 \longrightarrow 2NO_2 + \tfrac{1}{2}O_2$$

is $1.20 + 10^{-2} \text{ s}^{-1}$ at 45°C, and the initial concentration of N_2O_5 is 0.01500 M. (a) How long will it take for the concentration to decrease to 0.00110 M? (b) How much longer will it take for a further decrease to 0.000900 M?

42. It is found that 54.0 minutes is required for the concentration of substance A to decrease from 0.75 M to 0.20 M. What is the rate constant for this first-order decomposition?

$$A \longrightarrow B + C$$

43. The thermal decomposition of ammonia at high temperatures was studied in the presence of inert gases. Data at 2000 K are given for a single experiment.

$$NH_3 \longrightarrow NH_2 + H$$

t (hours)	$[NH_3]$ (mol/L)
0	8.000×10^{-7}
25	6.75×10^{-7}
50	5.84×10^{-7}
75	5.15×10^{-7}

Plot the appropriate concentration expressions against time to find the order of the reaction. Find the rate constant of the reaction from the slope of the line. Use the given data and the appropriate integrated rate equation to check your answer.

44. The following data were obtained from a study of the decomposition of a sample of HI on the surface of a gold wire. (a) Plot the data to find the order of the reaction, the rate constant, and the rate equation. (b) Calculate the HI concentration in mmol/L at 600. s.

t (seconds)	$[HI]$ (mmol/L)
0.	5.46
250.	4.10
500.	2.73
750.	1.37

45. The decomposition of SO_2Cl_2 in the gas phase,

$$SO_2Cl_2 \longrightarrow SO_2 + Cl_2$$

can be studied by measuring the concentration of Cl_2 as the reaction proceeds. We begin with $[SO_2Cl_2]_0 = 0.250$ M.

Holding the temperature constant at 320°C, we monitor the Cl_2 concentration, with the following results.

t (hours)	$[Cl_2]$ (mol/L)
0.00	0.000
2.00	0.037
4.00	0.068
6.00	0.095
8.00	0.117
10.00	0.137
12.00	0.153
14.00	0.168
16.00	0.180
18.00	0.190
20.00	0.199

(a) Plot $[Cl_2]$ versus t. (b) Plot $[SO_2Cl_2]$ versus t. (c) Determine the rate law for this reaction. (d) What is the value, with units, for the specific rate constant at 320°C? (e) How long would it take for 95% of the original SO_2Cl_2 to react?

*46. At some temperature, the rate constant for the decomposition of HI on a gold surface is 0.080 $M \cdot \text{s}^{-1}$.

$$2HI(g) \longrightarrow H_2(g) + I_2(g)$$

(a) What is the order of the reaction? (b) How long will it take for the concentration of HI to drop from 1.00 M to 0.30 M?

Activation Energy, Temperature, and Catalysts

47. Draw typical reaction energy diagrams for one-step reactions that release energy and that absorb energy. Distinguish between the net energy change, ΔE, for each kind of reaction and the activation energy. Indicate potential energies of products and reactants for both kinds of reactions.

48. Use graphs to illustrate how the presence of a catalyst can affect the rate of a reaction.

49. How do homogeneous catalysts and heterogeneous catalysts differ?

50. (a) Why should one expect an increase in temperature to increase the initial rate of reaction? (b) Why should one expect a reaction in the gaseous state to be faster than the same reaction in the solid state?

51. Assume that the activation energy for a certain reaction is 173 kJ/mol and the reactions are started with equal initial concentrations of reactants. How many times faster will the reaction occur at 40°C than at 10°C?

52. What is the activation energy for a reaction if its rate constant is found to triple when the temperature is raised from 600 K to 610 K?

53. For a gas-phase reaction, $E_a = 103$ kJ/mol, and the rate constant is 0.0850 min^{-1} at 273 K. Find the rate constant at 323 K.

54. The rate constant of a reaction is tripled when the temperature is increased from 298 K to 308 K. Find E_a.

55. The rate constant for the decomposition of N_2O

$$2N_2O(g) \longrightarrow 2N_2(g) + O_2(g)$$

is 2.6×10^{-11} s^{-1} at 300°C and 2.1×10^{-10} s^{-1} at 330°C. Calculate the activation energy for this reaction. Prepare a reaction coordinate diagram like Figure 16-10 using -164.1 kJ/mol as the ΔE_{rxn}.

56. For a particular reaction, $\Delta E^0 = 51.51$ kJ/mol, $k = 8.0 \times 10^{-7}$ s^{-1} at 0.0°C, and $k = 8.9 \times 10^{-4}$ s^{-1} at 50.0°C. Prepare a reaction coordinate diagram like Figure 16-10 for this reaction.

*57. You are given the rate constant as a function of temperature for the exchange reaction

$$Mn(CO)_5(CH_3CN)^+ + NC_5H_5 \longrightarrow$$
$$Mn(CO)_5(NC_5H_5)^+ + CH_3CN$$

T (K)	k (min^{-1})
298	0.0409
308	0.0818
318	0.157

(a) Calculate E_a from a plot of log k versus $1/T$. (b) Use the graph to predict the value of k at 311 K. (c) What is the numerical value of the collision frequency factor, A, in the Arrhenius equation?

*58. The rearrangement of cyclopropane to propene described in Exercise 40 has been studied at various temperatures. The following values for the specific rate constant have been determined experimentally.

T (K)	k (s^{-1})
600.	3.30×10^{-9}
650.	2.19×10^{-7}
700.	7.96×10^{-6}
750.	1.80×10^{-4}
800.	2.74×10^{-3}
850.	3.04×10^{-2}
900.	2.58×10^{-1}

(a) From the appropriate plot of these data, determine the value of the activation energy for this reaction. (b) Use the graph to estimate the value of k at 500 K. (c) Use the graph to estimate the temperature at which the value of k would be equal to 5.00×10^{-5} s^{-1}.

59. Biological reactions nearly always occur in the presence of enzymes as catalysts. The enzyme catalase, which acts on peroxides, reduces the E_a for the reaction from 72 kJ/mol (uncatalyzed) to 28 kJ/mol (catalyzed). By what factor does the reaction rate increase at normal body temperature, 37.0°C, for the same reactant (peroxide) concentration? Assume that the collision factor, A, remains constant.

*60. The enzyme carbonic anhydrase catalyzes the hydration of carbon dioxide.

$$CO_2 + H_2O \longrightarrow H_2CO_3$$

This reaction is involved in the transfer of CO_2 from tissues to the lungs via the bloodstream. One enzyme molecule hydrates 10^6 molecules of CO_2 per second. How many grams of CO_2 are hydrated in one minute in 1 L by 1.0×10^{-6} M enzyme?

61. The following gas-phase reaction follows first-order kinetics.

$$ClO_2F \longrightarrow ClOF + O$$

The activation energy of this reaction is 186 kJ/mol. The value of k at 322°C is 6.76×10^{-4} s^{-1}. (a) What would be the value of k for this reaction at 25°C? (b) At what temperature would this reaction have a k value of 3.00×10^{-2} s^{-1}?

62. The following gas-phase reaction is first order.

$$N_2O_5 \longrightarrow NO_2 + NO_3$$

The activation energy of this reaction is 88 kJ/mol. The value of k at 0°C is 9.16×10^{-3} s^{-1}. (a) What would be the value of k for this reaction at room temperature, 25°C? (b) At what temperature would this reaction have a k value of 3.00×10^{-2} s^{-1}?

Reaction Mechanisms

63. Define *reaction mechanism*. Why do we believe that only bimolecular collisions and unimolecular decompositions are important in most reaction mechanisms?

64. The rate equation for the reaction

$$Cl_2(aq) + H_2S(aq) \longrightarrow S(s) + 2HCl(aq)$$

is found to be rate = $k[Cl_2][H_2S]$. Which of the following mechanisms are consistent with the rate law?

(a)

$Cl_2 \longrightarrow Cl^+ + Cl^-$	(slow)
$Cl^- + H_2S \longrightarrow HCl + HS^-$	(fast)
$Cl^+ + HS^- \longrightarrow HCl + S$	(fast)
$Cl_2 + H_2S \longrightarrow S + 2HCl$	(overall)

(b)

$Cl_2 + H_2S \longrightarrow HCl + Cl^+ + HS^-$	(slow)
$Cl^+ + HS^- \longrightarrow HCl + S$	(fast)
$Cl_2 + H_2S \longrightarrow S + 2HCl$	(overall)

(c) $Cl_2 \rightleftharpoons Cl + Cl$ (fast, equilibrium)
       $Cl + H_2S \rightleftharpoons HCl + HS$ (fast, equilibrium)
       $HS + Cl \longrightarrow HCl + S$ (slow)

 ———————————————————————

       $Cl_2 + H_2S \longrightarrow S + 2HCl$ (overall)

65. Write the overall reaction and the rate expressions that correspond to the following reaction mechanisms. Be sure to eliminate intermediates from the answers:

(a) $A + B \rightleftharpoons C + D$ (fast, equilibrium)
     $C + E \longrightarrow F$ (slow)

(b) $A \rightleftharpoons B + C$ (fast, equilibrium)
     $C + D \rightleftharpoons E$ (fast, equilibrium)
       $E \longrightarrow F$ (slow)

66. Write the overall reaction and the rate expressions that correspond to the following mechanisms. Be sure to eliminate intermediates from the answers:

(a) $2A + B \rightleftharpoons D$ (fast, equilibrium)
     $D + B \longrightarrow E + F$ (slow)
       $F \longrightarrow G$ (fast)

(b) $A + B \rightleftharpoons C$ (fast, equilibrium)
     $C + D \rightleftharpoons F$ (fast, equilibrium)
       $F \longrightarrow G$ (slow)

67. The ozone, O_3, of the stratosphere can be decomposed by reaction with nitrogen oxide (commonly called nitric oxide), NO, from high-flying jet aircraft.

$$O_3(g) + NO(g) \longrightarrow NO_2(g) + O_2(g)$$

The rate expression is rate = $k[O_3][NO]$. Which of the following mechanisms are consistent with the observed rate expression?

(a) $NO + O_3 \longrightarrow NO_3 + O$ (slow)
     $NO_3 + O \longrightarrow NO_2 + O_2$ (fast)

 ———————————————————————

     $O_3 + NO \longrightarrow NO_2 + O_2$ (overall)

(b) $NO + O_3 \longrightarrow NO_2 + O_2$ (slow)
                                (one step)

(c) $O_3 \longrightarrow O_2 + O$ (slow)
       $O + NO \longrightarrow NO_2$ (fast)

 ———————————————————————

     $O_3 + NO \longrightarrow NO_2 + O_2$ (overall)

(d) $NO \longrightarrow N + O$ (slow)
       $O + O_3 \longrightarrow 2O_2$ (fast)
       $O_2 + N \longrightarrow NO_2$ (fast)

 ———————————————————————

     $O_3 + NO \longrightarrow NO_2 + O_2$ (overall)

(e) $NO \rightleftharpoons N + O$ (fast, equilibrium)
       $O + O_3 \longrightarrow 2O_2$ (slow)
       $O_2 + N \longrightarrow NO_2$ (fast)

 ———————————————————————

     $O_3 + NO \longrightarrow NO_2 + O_2$ (overall)

68. A proposed mechanism for the decomposition of ozone, $2O_3 \longrightarrow 3O_2$, is

$$O_3 \rightleftharpoons O_2 + O \qquad \text{(fast, equilibrium)}$$
$$O + O_3 \longrightarrow 2O_2 \qquad \text{(slow)}$$

Derive the rate equation for the net reaction.

69. A mechanism for the gas-phase reaction

$$H_2 + I_2 \longrightarrow 2HI$$

was discussed in the chapter. (a) Show that this mechanism predicts the correct rate law, rate = $k[H_2][I_2]$.

$$I_2 \rightleftharpoons 2I \qquad \text{(fast, equilibrium)}$$
$$I + H_2 \rightleftharpoons H_2I \qquad \text{(fast, equilibrium)}$$
$$H_2I + I \longrightarrow 2HI \qquad \text{(slow)}$$

(b) Identify any reaction intermediates in this proposed mechanism.

70. The combination of Cl atoms is catalyzed by $N_2(g)$. The following mechanism is suggested.

$$N_2 + Cl \rightleftharpoons N_2Cl \qquad \text{(fast, equilibrium)}$$
$$N_2Cl + Cl \longrightarrow Cl_2 + N_2 \qquad \text{(slow)}$$

(a) Identify any reaction intermediates in this proposed mechanism. (b) Is this mechanism consistent with the experimental rate law, rate = $k[N_2][Cl]_2$?

71. The reaction between NO and Br_2 was discussed in Section 16-7. The following mechanism has also been proposed.

$$2NO \rightleftharpoons N_2O_2 \qquad \text{(fast, equilibrium)}$$
$$N_2O_2 + Br_2 \longrightarrow 2NOBr \qquad \text{(slow)}$$

Is this mechanism consistent with the observation that the reaction is second order in NO and first order in Br_2?

*72. The following mechanism for the reaction between H_2 and CO to form formaldehyde, H_2CO, has been proposed.

$$H_2 \rightleftharpoons 2H \qquad \text{(fast, equilibrium)}$$
$$H + CO \longrightarrow HCO \qquad \text{(slow)}$$
$$H + HCO \longrightarrow H_2CO \qquad \text{(fast)}$$

(a) Write the balanced equation for the overall reaction. (b) The observed rate dependence is found to be one-half order in H_2 and first order in CO. Is this proposed reaction mechanism consistent with the observed rate dependence?

73. The reaction between nitrogen dioxide and ozone,

$$2NO_2 + O_3 \longrightarrow N_2O_5 + O_2$$

has been studied at 231 K. The experimental rate equation is rate = $k[NO_2][O_3]$. (a) What is the order of the reaction? (b) Is either of the following proposed mechanisms consistent with the given kinetic data? Show how you arrived at your answer.

(a) $NO_2 + NO_2 \rightleftharpoons N_2O_4$ (fast, equilibrium)
$N_2O_4 + O_3 \longrightarrow N_2O_5 + O_2$ (slow)

(b) $NO_2 + O_3 \longrightarrow NO_3 + O_2$ (slow)
$NO_3 + NO_2 \longrightarrow N_2O_5$ (fast)

Mixed Exercises

74. (a) What is the transition state in a reaction mechanism? (b) Are the energy of activation and the transition state related concepts? Explain. (c) How does the activation energy affect the rate of reaction?

75. Refer to the reaction and data in Exercise 61. We begin with 2.80 mol of ClO_2F in a 3.00-L container. (a) How many moles of ClO_2F would remain after 2.00 min at 25°C? (b) How much time would be required for 99.0% of the ClO_2F to decompose at 25°C?

76. Refer to the reaction and data in Exercise 62. We begin with 2.80 mol of N_2O_5 in a 3.00-L container. (a) How many moles of N_2O_5 would remain after 2.00 min at 25°C? (b) How much time would be required for 99.0% of the N_2O_5 to decompose at 25°C?

77. The decomposition of gaseous dimethyl ether

$$CH_3OCH_3 \longrightarrow CH_4 + CO + H_2$$

follows first-order kinetics. Its half-life is 25.0 min at 500°C. (a) Starting with 12.00 g of dimethyl ether at 500°C, how many grams would remain after 150 min? (b) In part (a), how many grams would remain after 180 min? (c) In part (b), what fraction remains and what fraction reacts? (d) Calculate the time, in minutes, required to decrease 24.0 mg of dimethyl ether to 2.40 mg.

78. The rate of the hemoglobin (Hb)–carbon monoxide reaction,

$$4Hb + 3CO \longrightarrow Hb_4(CO)_3$$

has been studied at 20°C. Concentrations are expressed in micromoles per liter (μmol/L).

Concentration (μmol/L)		Rate of Disappearance of Hb (μmol · L^{-1} · s^{-1})
[Hb]	[CO]	
3.36	1.00	0.941
6.72	1.00	1.88
6.72	3.00	5.64

(a) Write the rate equation for the reaction. (b) Calculate the rate constant for the reaction. (c) Calculate the rate, at the instant when [Hb] = 1.50 and [CO] = 0.600 μmol/L.

79. How does an enzyme change the speed with which a reaction reaches equilibrium? Can an enzyme change the final equilibrium concentrations? Explain.

CONCEPTUAL EXERCISES

80. Some reactions occur faster than others due to differences in the shapes of the reactants. Use the collision theory to explain these observations.

81. How it is possible for two reactant molecules to collide with the correct orientation and still not react?

82. Write the net ionic equation for the following reaction. Construct a potential energy diagram, like Figure 16-10, for this reaction.

$$HCl(aq) + NaOH(aq) \longrightarrow NaCl(aq) + H_2O(\ell)$$

83. Starting with only two molecules of each reactant in a reaction that is first order in each reactant, show how the collision theory predicts that the rate of reaction will double if the amount of either reactant is doubled.

84. A sentence in an introductory chemistry textbook reads, "Dioxygen reacts with itself to form trioxygen, ozone, according to the following equation, $3O_2 \longrightarrow 2O_3$." As a student of chemistry, what would you write to criticize this sentence?

85. A stream of gaseous H_2 is directed onto finely divided platinum powder in the open air. The metal immediately glows white-hot and continues to do so as long as the stream continues. Explain.

86. Is the activation energy of a reaction expected to be higher or lower when the same reactants are in the gaseous state rather than the liquid or solid state? Explain.

BUILDING YOUR KNOWLEDGE

87. The following explanation of the operation of a pressure cooker appears in a cookbook: "Boiling water in the presence of air can never produce a temperature higher than 212°F, no matter how high the heat source. But in a pressure cooker, the air is withdrawn first, so the boiling water can be maintained at higher temperatures." Support or criticize this explanation.

*88. A cookbook gives the following general guideline for use of a pressure cooker. "For steaming vegetables, cooking time at a gauge pressure of 15 pounds per square inch (psi) is $\frac{1}{3}$ that at atmospheric pressure." Remember that gauge pressure is measured relative to the external atmospheric pressure, which is 15 psi at sea level. From this information, estimate the activation energy for the process of steaming vegetables. (*Hint:* Clausius and Clapeyron may be able to help you.)

89. For most reactions that involve an enzyme, the rate of product formation versus reactant concentration increases as reactant concentration increases until a maximum value is obtained, after which further increases do not yield increased rates. Using a description like that in Figure 16-19, describe how the reaction may be first order with respect to substrate but the amount of enzyme can also be a determining factor.

90. Using the mechanism and relative energy values shown in Figure 16-12, prepare Lewis formulas that illustrate the species that are likely to be present at each of the peaks and troughs in the graphical representation given in Figure 16-12. (*Hint:* You may need to label some bonds as being weaker, stretched, in the process of being formed, and so on.)

91. The activation energy for the reaction

$$2HI(g) \longrightarrow H_2(g) + I_2(g)$$

is 179 kJ/mol. Construct a diagram similar to Figure 16-10 for this reaction. (*Hint:* Calculate ΔH^0 from values in Appendix K. How does ΔH^0 compare with ΔE^0 for this reaction?)

*92. The activation energy for the reaction between O_3 and NO is 9.6 kJ/mol.

$$O_3(g) + NO(g) \longrightarrow NO_2(g) + O_2(g)$$

(a) Use the thermodynamic quantities in Appendix K to calculate ΔH^0 for this reaction. (b) Prepare an activation energy plot similar to Figure 16-10 for this reaction. (*Hint:* How does ΔH^0 compare with ΔE^0 for this reaction?)

BEYOND THE TEXTBOOK

Go to the textbook website at

http://www.brookscole.com/chemistry/whitten

for additional activities and exercises based on the General Chemistry Interactive CD-ROM, the World Wide Web, and library resources.

InfoTrac College Edition

For additional readings, go to InfoTrac College Edition, your online research library at:

http://infotrac.thomsonlearning.com

Chemical Equilibrium

Digital Vision/Getty

Chemical equilibrium plays an important role in the operation of chemical plants where manufacturers want to optimize the amount of product. This can be done by adjusting reaction conditions such as temperature, pressure, and concentrations of reactants used, all key factors that affect chemical equilibrium.

OUTLINE

OBJECTIVES

After you have studied this chapter, you should be able to

- Explain the basic ideas of chemical equilibrium
- Explain what an equilibrium constant is and what it tells us
- Explain what a reaction quotient is and what it tells us
- Use equilibrium constants to describe systems at equilibrium
- Recognize the factors that affect equilibria and predict the resulting effects
- Use the equilibrium constant expressed in terms of partial pressures (K_p) and relate K_p to K_c
- Describe heterogeneous equilibria and write their equilibrium constants
- Use the relationships between thermodynamics and equilibrium
- Estimate equilibrium constants at different temperatures

17-1 BASIC CONCEPTS

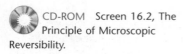

CD-ROM Screen 16.2, The Principle of Microscopic Reversibility.

Chemical reactions that can occur in either direction are called **reversible reactions.** Most reversible reactions do not go to completion. That is, even when reactants are mixed in stoichiometric quantities, they are not completely converted to products.

Reversible reactions can be represented in general terms as follows, where the capital letters represent formulas and the lowercase letters represent the stoichiometric coefficients in the balanced equation.

$$a\text{A} + b\text{B} \rightleftharpoons c\text{C} + d\text{D}$$

The double arrow ($\rightleftharpoons$) indicates that the reaction is reversible—that is, both the forward and reverse reactions occur simultaneously. In discussions of chemical equilibrium, the substances that appear on the left side of the balanced chemical equation are called the "reactants," and those on the right side are called the "products." In fact, the reaction can proceed in *either direction*. When A and B react to form C and D at the same rate at which C and D react to form A and B, the system is at *equilibrium*.

> **Chemical equilibrium** exists when two opposing reactions occur simultaneously at the same rate.

Chemical equilibria are **dynamic equilibria;** that is, individual molecules are continually reacting, even though the overall composition of the reaction mixture does not change. In a system at equilibrium, the equilibrium is said to lie toward the right if more C and D are present than A and B, and to lie toward the left if more A and B are present.

Consider a case in which the coefficients in the equation for a reaction are all 1. When substances A and B react, the rate of the forward reaction decreases as time passes because the concentration of A and B decrease.

$$A + B \longrightarrow C + D \tag{1}$$

As the concentrations of C and D build up, they start to re-form A and B.

$$C + D \longrightarrow A + B \tag{2}$$

As more C and D molecules are formed, more can react, and so the rate of reaction between C and D increases with time. Eventually, the two reactions occur at the same rate, and the system is at equilibrium (Figure 17-1).

$$A + B \rightleftharpoons C + D$$

If a reaction begins with only C and D present, the rate of reaction (2) decreases with time, and the rate of reaction (1) increases with time until the two rates are equal.

The dynamic nature of chemical equilibrium can be proved experimentally by inserting radioactive atoms into a small percentage of reactant or product molecules. Even when the initial mixture is at equilibrium, radioactive atoms eventually appear in both reactant and product molecules.

CD-ROM   Screen 16.3, The Equilibrium State.

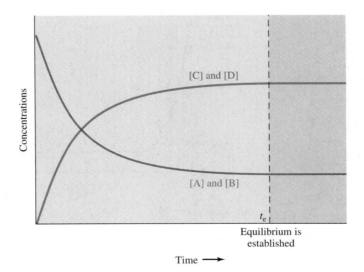

Figure 17-1 Variation in the concentrations of species present in the A + B $\rightleftharpoons$ C + D system as equilibrium is approached, beginning with equal concentrations of A and B only. For this reaction, production of products is favored. As we will see, this corresponds to a value of the equilibrium constant greater than 1. Brackets, [ ], represent the concentration, in moles per liter, of the species enclosed within them. The time axis may be in any convenient units—seconds, minutes, hours, and so on.

The SO_2–O_2–SO_3 System

Consider the reversible reaction of sulfur dioxide with oxygen to form sulfur trioxide at 1500 K.

$$2SO_2(g) + O_2(g) \rightleftharpoons 2SO_3(g)$$

The numbers in this discussion were determined experimentally.

Suppose 0.400 mole of SO_2 and 0.200 mole of O_2 are injected into a closed 1.00-liter container. When equilibrium is established (at time t_e, Figure 17-2a), we find that 0.056 mole of SO_3 has formed and that 0.344 mole of SO_2 and 0.172 mole of O_2 remain unreacted. The amount of product at equilibrium does not increase further if the reaction is given additional time. These changes are summarized in the following reaction summary, using molarity units rather than moles. (They are numerically identical here because the volume of the reaction vessel is 1.00 liter.) The changes due to reaction are represented by the *changes* in concentrations.

A setup such as this is called a "reaction summary." The ratio in the "change due to rxn" line is determined by the coefficients in the balanced equation.

	$2SO_2(g)$	+	$O_2(g)$	$\rightleftharpoons$	$2SO_3(g)$
initial conc'n	0.400 *M*		0.200 *M*		0
change due to rxn	−0.056 *M*		−0.028 *M*		+0.056 *M*
equilibrium conc'n	0.344 *M*		0.172 *M*		0.056 *M*

In a similar experiment, 0.500 mole of SO_3 is introduced alone into a closed 1.00-liter container. When equilibrium is established (at time t_e, Figure 17-2b), 0.076 mole of SO_3, 0.212 mole of O_2, and 0.424 mole of SO_2 are present. These equilibrium amounts differ from those in the previous case, but they are related in an important way, as we will see in the next section. This time the reaction proceeds from *right to left* as the equation is written. The *changes in concentration* are in the same 2:1:2 ratio as in the previous case, as required by the coefficients of the balanced equation. The time required to reach equilibrium may be longer or shorter.

	$2SO_2(g)$	+	$O_2(g)$	$\rightleftharpoons$	$2SO_3(g)$
initial conc'n	0		0		0.500 *M*
change due to rxn	+0.424 *M*		+0.212 *M*		−0.424 *M*
equilibrium conc'n	0.424 *M*		0.212 *M*		0.076 *M*

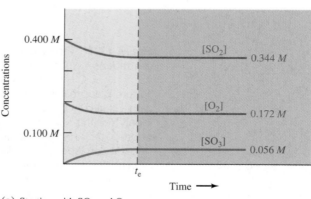

(a) Starting with SO_2 and O_2.

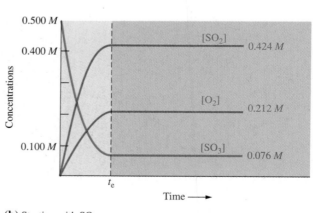

(b) Starting with SO_3.

Figure 17-2 Establishment of equilibrium in the $2SO_2 + O_2 \rightleftharpoons 2SO_3$ system. (a) Beginning with stoichiometric amounts of SO_2 and O_2 and no SO_3. (b) Beginning with only SO_3 and no SO_2 or O_2. Greater changes in concentrations occur to establish equilibrium when starting with SO_3 than when starting with SO_2 and O_2. The equilibrium favors SO_2 and O_2.

The results of these experiments are summarized in the following table and in Figure 17-2.

	Initial Concentrations			Equilibrium Concentrations		
	$[SO_2]$	$[O_2]$	$[SO_3]$	$[SO_2]$	$[O_2]$	$[SO_3]$
Experiment 1	0.400 M	0.200 M	0 M	0.344 M	0.172 M	0.056 M
Experiment 2	0 M	0 M	0.500 M	0.424 M	0.212 M	0.076 M

17-2 THE EQUILIBRIUM CONSTANT

Suppose a reversible reaction occurs by a single elementary reaction step:

$$2A + B \rightleftharpoons A_2B$$

The rate of the forward reaction is $rate_f = k_f[A]^2[B]$; the rate of the reverse reaction is $rate_r = k_r[A_2B]$. In these expressions, k_f and k_r are the *specific rate constants* of the forward and reverse reactions, respectively. By definition, the two rates are equal *at equilibrium* ($rate_f = rate_r$). So we write

$$k_f[A]^2[B] = k_r[A_2B] \qquad \text{(at equilibrium)}$$

Dividing both sides of this equation by k_r and by $[A]^2[B]$ gives

$$\frac{k_f}{k_r} = \frac{[A_2B]}{[A]^2[B]}$$

At any specific temperature, both k_f and k_r are constants, so k_f/k_r is also a constant. This ratio is given a special name and symbol, the *equilibrium constant*, K_c or simply K.

$$K_c = \frac{[A_2B]}{[A]^2[B]} \qquad \text{(at equilibrium)}$$

Even if the overall reaction occurs by a multistep mechanism, we can show that the equilibrium constant is the product and ratio of the rate constants for each step of the mechanism. Regardless of the mechanism by which this reaction occurs, the concentrations of reaction intermediates cancel out and the equilibrium constant expression has the same form. For a reaction in general terms, the equilibrium constant can always be written as follows:

For $\underbrace{aA + bB}_{\text{reactants}} \rightleftharpoons \underbrace{cC + dD}_{\text{products}}$, $K_c = \dfrac{\overbrace{[C]_{eq}{}^c[D]_{eq}{}^d}^{\substack{\text{product} \\ \text{concentrations}}}}{\underbrace{[A]_{eq}{}^a[B]_{eq}{}^b}_{\substack{\text{reactant} \\ \text{concentrations}}}}$

The **equilibrium constant K_c**, is defined as the product of the *equilibrium concentrations* (in moles per liter) of the products, each raised to the power that corresponds to its coefficient in the balanced equation, divided by the product of the *equilibrium concentrations* of reactants, each raised to the power that corresponds to its coefficient in the balanced equation.

Remember that for elementary (or fundamental) reactions the coefficients do correspond to the orders of the reactant and product concentrations.

The subscript c refers to concentrations. The brackets, [ ], in this expression indicate *equilibrium* concentrations in moles per liter.

CD-ROM   Screen 16.4, The Equilibrium Constant.

Moles per liter will also be referred to as molarity (*M*) or molar concentration.

CD-ROM   Screen 16.6, Writing Equilibrium Expressions.

We emphasize that the concentrations in the equilibrium constant expression are those at equilibrium.

Numerical values for K_c can come only from experiments. Some equilibrium constant expressions and their numerical values at 25°C are

$$N_2(g) + O_2(g) \rightleftharpoons 2NO(g) \qquad K_c = \frac{[NO]^2}{[N_2][O_2]} = 4.5 \times 10^{-31}$$

$$CH_4(g) + Cl_2(g) \rightleftharpoons CH_3Cl(g) + HCl(g) \qquad K_c = \frac{[CH_3Cl][HCl]}{[CH_4][Cl_2]} = 1.2 \times 10^{18}$$

$$N_2(g) + 3H_2(g) \rightleftharpoons 2NH_3(g) \qquad K_c = \frac{[NH_3]^2}{[N_2][H_2]^3} = 3.6 \times 10^8$$

Remember that calculations with K_c values always involve *equilibrium values* of concentrations.

The thermodynamic definition of the equilibrium constant involves activities rather than concentrations. The **activity** of a component of an ideal mixture is the ratio of its concentration or partial pressure to a standard concentration (1 M) or pressure (1 atm). For now, we can consider the activity of each species to be a dimensionless quantity whose numerical value can be determined as follows.

1. For any pure liquid or pure solid, the activity is taken as 1.

2. For components of ideal solutions, the activity of each component is taken to be the ratio of its molar concentration to a standard concentration of 1 M, so the units cancel.

3. For gases in an ideal mixture, the activity of each component is taken to be the ratio of its partial pressure to a standard pressure of 1 atm, so again the units cancel.

Because of the use of activities, *the equilibrium constant has no units; the values we put into K_c are numerically equal to molar concentrations, but are dimensionless,* that is, they have no units. In this text, calculations have usually included units along with numbers. Calculations involving equilibrium are frequently carried out without units; we will follow that practice in this text.

These are three very important ideas.

The magnitude of K_c is a measure of the extent to which reaction occurs. For any balanced chemical equation, the value of K_c

1. is constant at a given temperature;

2. changes if the temperature changes;

3. does not depend on the initial concentrations.

CD-ROM   Screen 16.5, The Meaning of the Equilibrium Constant.

A value of K_c *much* greater than 1 indicates that the "numerator concentrations" (products) would be much greater than the "denominator concentrations" (reactants); this means that at equilibrium most of the reactants would be converted into products. For example, the reaction $CH_4(g) + Cl_2(g) \rightleftharpoons CH_3Cl(g) + HCl(g)$ shown earlier goes nearly to completion; in Chapter 15, we called such a reaction "product-favored." On the other hand, if K_c is quite small, equilibrium is established when most of the reactants remain unreacted and only small amounts of products are formed. The reaction $N_2(g) + O_2(g) \rightleftharpoons 2NO(g)$ shown earlier reaches equilibrium with only a tiny amount of NO present; in Chapter 15, we called such a reaction "reactant-favored."

For a given chemical reaction at a specific temperature, the product of the concentrations of the products formed by the reaction, each raised to the appropriate power, divided by the product of the concentrations of the reactants, each raised to the appropriate power, always has the same value, that is, K_c. This does *not* mean that the individual equilibrium concentrations for a given reaction are always the same, but it does mean that this particular numerical combination of their values (K_c) is constant.

Consider again the SO_2–O_2–SO_3 equilibrium described in Section 17-1. We can use the *equilibrium* concentrations from either experiment to calculate the value of the equilibrium constant for this reaction at 1500 K.

From experiment 1: $2SO_2(g) + O_2(g) \rightleftharpoons 2SO_3(g)$
equilibrium conc'n   0.344 *M*   0.172 *M*   0.056 *M*

Substituting the numerical values (without units) into the equilibrium expression gives the value of the equilibrium constant.

Alternatively, $$K_c = \frac{[SO_3]^2}{[SO_2]^2[O_2]} = \frac{(0.056)^2}{(0.344)^2(0.172)} = 0.15$$

From experiment 2: $2SO_2(g) + O_2(g) \rightleftharpoons 2SO_3(g)$
equilibrium conc'n   0.424 *M*   0.212 *M*   0.076 *M*

$$K_c = \frac{(0.076)^2}{(0.424)^2(0.212)} = 0.15$$

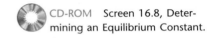
CD-ROM   Screen 16.8, Determining an Equilibrium Constant.

No matter what combinations of reactant and product concentrations we start with, the resulting *equilibrium* concentrations at 1500 K for this reversible reaction would always give the same value of K_c, 0.15.

For the reversible reaction written *as it is* with SO_2 and O_2 as reactants and SO_3 as the product, K_c is 0.15 at 1500 K.

EXAMPLE 17-1   *Calculation of K_c*

Some nitrogen and hydrogen are placed in an empty 5.00-liter container at 500°C. When equilibrium is established, 3.01 mol of N_2, 2.10 mol of H_2, and 0.565 mol of NH_3 are present. Evaluate K_c for the following reaction at 500°C.

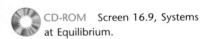
CD-ROM   Screen 16.9, Systems at Equilibrium.

$$N_2(g) + 3H_2(g) \rightleftharpoons 2NH_3(g)$$

Plan

The *equilibrium concentrations* are obtained by dividing the number of moles of each reactant and product by the volume, 5.00 liters. Then we substitute these equilibrium concentrations into the equilibrium constant expression.

Solution

The equilibrium concentrations are

$$[N_2] = 3.01 \text{ mol}/5.00 \text{ L} = 0.602 \text{ } M$$
$$[H_2] = 2.10 \text{ mol}/5.00 \text{ L} = 0.420 \text{ } M$$
$$[NH_3] = 0.565 \text{ mol}/5.00 \text{ L} = 0.113 \text{ } M$$

We substitute these numerical values into the expression for K_c.

$$K_c = \frac{[NH_3]^2}{[N_2][H_2]^3} = \frac{(0.113)^2}{(0.602)(0.420)^3} = 0.286$$

Remember that the concentrations in K_c calculations are *equilibrium* values of *molar* concentration.

Thus, for the reaction of N_2 and H_2 to form NH_3 at 500°C, we can write

$$K_c = \frac{[NH_3]^2}{[N_2][H_2]^3} = 0.286$$

The small value of K_c indicates that the equilibrium lies to the left, that is, the reaction is reactant-favored.

The value of K_c from Example 17-1 is much different from the value given earlier for the same reaction at 25°C. For this reaction, products are favored at the lower temperature ($K_c = 3.6 \times 10^8$ at 25°C), whereas reactants are favored at the higher temperature ($K_c = 0.286$ at 500°C). The dependence of K_c on temperature will be discussed later in this chapter; for now, we note that it can depend strongly on temperature.

EXAMPLE 17-2   *Calculation of K_c*

We put 10.0 moles of N_2O into a 2.00-L container at some temperature, where it decomposes according to

$$2N_2O(g) \rightleftharpoons 2N_2(g) + O_2(g)$$

At equilibrium, 2.20 moles of N_2O remain. Calculate the value of K_c for the reaction.

Plan

We express all concentrations in moles per liter. The mole ratio from the balanced chemical equation allows us to find the changes in concentrations of the other substances in the reaction. We use the reaction summary to find the equilibrium concentrations to use in the K_c expression.

Solution

At equilibrium, 2.20 mol N_2O remain, so

$$\underline{?}\ mol\ N_2O\ reacting = 10.00\ mol\ N_2O\ initial - 2.20\ mol\ N_2O\ remaining$$

$$= 7.80\ mol\ N_2O\ reacting$$

The initial $[N_2O] = (10.00\ mol)/(2.00\ L) = 5.00\ M$; the concentration of N_2O that reacts is $(7.80\ mol)/(2.00\ L) = 3.90\ M$. From the balanced chemical equation, each 2 mol N_2O that react produces 2 mol N_2 and 1 mol O_2, or a reaction ratio of

$$1\ mol\ N_2O\ reacting : 1\ mol\ N_2\ formed : \tfrac{1}{2}\ mol\ O_2\ formed$$

We can now write the reaction summary.

	$2N_2O(g) \rightleftharpoons$	$2N_2(g)$	$+$	$O_2(g)$
initial	5.00 M	0		0
change due to rxn	−3.90 M	+3.90 M		$+\tfrac{1}{2}(3.90) = 1.95\ M$
equilibrium	1.10 M	3.90 M		1.95 M

We put these equilibrium concentrations into the equilibrium constant expression and evaluate K_c.

$$K_c = \frac{[N_2]^2[O_2]}{[N_2O]^2} = \frac{(3.90)^2(1.95)}{(1.10)^2} = 24.5$$

You should now work Exercises 24 and 28.

17-3 VARIATION OF K_c WITH THE FORM OF THE BALANCED EQUATION

The value of K_c depends on how we write the balanced equation for the reaction. We wrote the equation for the reaction of SO_2 and O_2 to produce SO_3, and its equilibrium constant expression as

$$2SO_2(g) + O_2(g) \rightleftharpoons 2SO_3(g) \qquad \text{and} \qquad K_c = \frac{[SO_3]^2}{[SO_2]^2[O_2]} = 0.15$$

Suppose we write the equation for the same reaction in reverse. The equation and its equilibrium constant, written this way, are

$$2SO_3(g) \rightleftharpoons 2SO_2(g) + O_2(g) \qquad \text{and} \qquad K_c' = \frac{[SO_2]^2[O_2]}{[SO_3]^2} = \frac{1}{K_c} = \frac{1}{0.15} = 6.7$$

We see that K_c', the equilibrium constant for the equation written in reverse, is the *reciprocal* of K_c, the equilibrium constant for the original equation.

If the equation for the reaction were written as

$$SO_2(g) + \tfrac{1}{2}O_2(g) \rightleftharpoons SO_3(g) \qquad K_c'' = \frac{[SO_3]}{[SO_2][O_2]^{1/2}} = K_c^{1/2} = 0.39$$

If an equation for a reaction is multiplied by any factor, n, then the original value of K_c is raised to the nth power. Thus, we must always write the balanced chemical equation to accompany the value of K_c for a chemical reaction.

 Reversing an equation is the same as multiplying all coefficients by -1. This reverses the roles of "reactants" and "products."

We see that K_c'' is the square root of K_c. $K_c^{1/2}$ means the square root of K_c.

 CD-ROM   Screen 16.7, Manipulating Equilibrium Expressions.

EXAMPLE 17-3 *Variation of the Form of K_c*

You are given the following chemical equation and its equilibrium constant at a given temperature.

$$2HBr(g) + Cl_2(g) \rightleftharpoons 2HCl(g) + Br_2(g) \qquad K_c = 4.0 \times 10^4$$

Write the expression for, and calculate the numerical value of, the equilibrium constant for each of the following at the same temperature.

(a) $\qquad\qquad 4HBr(g) + 2Cl_2(g) \rightleftharpoons 4HCl(g) + 2Br_2(g)$

(b) $\qquad\qquad HBr(g) + \tfrac{1}{2}Cl_2(g) \rightleftharpoons HCl(g) + \tfrac{1}{2}Br_2(g)$

A coefficient of $\tfrac{1}{2}$ refers to $\tfrac{1}{2}$ of a mole, *not* $\tfrac{1}{2}$ of a molecule.

Plan

We recall the definition of the equilibrium constant. For the original equation,

$$K_c = \frac{[HCl]^2[Br_2]}{[HBr]^2[Cl_2]} = 4.0 \times 10^4$$

Solution

(a) The original equation has been multiplied by 2, so K_c must be squared.

$$K_c' = \frac{[HCl]^4[Br_2]^2}{[HBr]^4[Cl_2]^2} \qquad K_c' = (K_c)^2 = (4.0 \times 10^4)^2 = 1.6 \times 10^9$$

(b) The original equation has been multiplied by $\tfrac{1}{2}$ (divided by 2), so K_c must be raised to the $\tfrac{1}{2}$ power. The value of K_c'' is the square root of the original K_c value.

$$K_c'' = \frac{[HCl][Br_2]^{1/2}}{[HBr][Cl_2]^{1/2}} = \sqrt{K_c} = \sqrt{4.0 \times 10^4} = 2.0 \times 10^2$$

You should now work Exercise 30.

17-4 THE REACTION QUOTIENT

The reaction quotient is sometimes called the **mass action expression**.

The **reaction quotient, Q,** for the general reaction is given as follows.

$$\text{For } aA + bB \rightleftharpoons cC + dD, \qquad Q = \frac{[C]^c[D]^d}{[A]^a[B]^b} \quad \longleftarrow \begin{array}{l} \text{not necessarily} \\ \text{equilibrium} \\ \text{concentrations} \end{array}$$

The reaction quotient has the same *form* as the equilibrium constant, but it involves specific values that are not *necessarily* equilibrium concentrations. If they *are* equilibrium concentrations, then $Q = K_c$. The concept of the reaction quotient is very useful. We can compare the magnitude of Q with that of K_c for a reaction under given conditions to decide whether the forward or the reverse reaction must occur to a greater extent to establish equilibrium.

When the forward reaction occurs to a greater extent than the reverse reaction, we say that a *net* forward reaction has occurred.

We can think of the reaction quotient, Q, as a measure of the progress of the reaction. When the mixture contains *only* reactants, the concentrations in the numerator are zero, so $Q = 0$. As the reaction proceeds to the right, the product concentrations (numerator) increase and the reactant concentrations (denominator) decrease, so Q increases to an infinitely large value when all reactants have been consumed and only products remain. The value of K_c is a particular value of Q that represents equilibrium mixtures for the reaction.

If at any time $Q < K_c$, the forward reaction must occur to a greater extent than the reverse reaction for equilibrium to be established. This is because when $Q < K_c$, the numerator of Q is too small and the denominator is too large. To increase the numerator and to reduce the denominator, A and B must react to produce C and D. Conversely, if $Q > K_c$, the reverse reaction must occur to a greater extent than the forward reaction for equilibrium to be reached. When the value of Q reaches the value of K_c, the system is at equilibrium, so no further *net* reaction occurs.

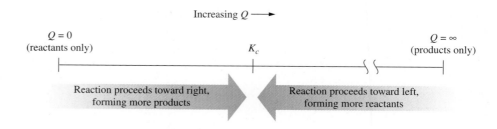

$Q < K_c$ Forward reaction predominates until equilibrium is established.

$Q = K_c$ System is at equilibrium.

$Q > K_c$ Reverse reaction predominates until equilibrium is established.

In Example 17-4 we calculate the value for Q and compare it with the *known* value of K_c to predict the direction of the reaction that leads to equilibrium.

EXAMPLE 17-4 *The Reaction Quotient*

At a very high temperature, $K_c = 65.0$ for the following reaction.

$$2HI(g) \rightleftharpoons H_2(g) + I_2(g)$$

The following concentrations were detected in a mixture. Is the system at equilibrium? If not, in which direction must the reaction proceed for equilibrium to be established?

$$[HI] = 0.500\ M, \qquad [H_2] = 2.80\ M, \qquad \text{and} \qquad [I_2] = 3.40\ M$$

These concentrations could be present if we started with a mixture of HI, H_2, and I_2.

Plan

We substitute these concentrations into the expression for the reaction quotient to calculate Q. Then we compare Q with K_c to see whether the system is at equilibrium.

Solution

$$Q = \frac{[H_2][I_2]}{[HI]^2} = \frac{(2.80)(3.40)}{(0.500)^2} = 38.1$$

But $K_c = 65.0$, so $Q < K_c$. The system is *not* at equilibrium. For equilibrium to be established, the value of Q must *increase* until it equals K_c. This can occur only if the numerator *increases* and the denominator *decreases*. Thus, the forward (left-to-right) reaction must occur to a greater extent than the reverse reaction; that is, some HI must react to form more H_2 and I_2 to reach equilibrium.

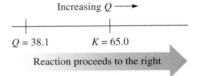

Increasing $Q \longrightarrow$

$Q = 38.1 \qquad K = 65.0$

Reaction proceeds to the right

You should now work Exercises 38, 40, and 42.

17-5 USES OF THE EQUILIBRIUM CONSTANT, K_c

We have seen (Section 17-2) how to calculate the value of K_c from one set of equilibrium concentrations. Once that value has been obtained, the process can be turned around to calculate equilibrium *concentrations* from the equilibrium *constant*.

The equilibrium constant is a "constant" only if the temperature does not change.

EXAMPLE 17-5 *Finding Equilibrium Concentrations*

The equation for the following reaction and the value of K_c at a given temperature are given. An equilibrium mixture in a 1.00-liter container contains 0.25 mol of PCl_5 and 0.16 mol of PCl_3. What equilibrium concentration of Cl_2 must be present?

$$PCl_3(g) + Cl_2(g) \rightleftharpoons PCl_5(g) \qquad K_c = 1.9$$

CD-ROM   Screen 16.10, Estimating Equilibrium Concentrations.

Plan

We write the equilibrium constant expression and its value. Only one term, $[Cl_2]$, is unknown. We solve for it.

Solution

Because the volume of the container is 1.00 liter, the molar concentration (mol/L) of each substance is numerically equal to the number of moles. The equilibrium constant expression and its numeric value are

$$K_c = \frac{[PCl_5]}{[PCl_3][Cl_2]} = 1.9$$

$$[Cl_2] = \frac{[PCl_5]}{K_c[PCl_3]} = \frac{(0.25)}{(1.9)(0.16)} = \boxed{0.82 \ M}$$

Often we know the starting concentrations and want to know how much of each reactant and each product would be present at equilibrium. The next two examples illustrate this important kind of calculation.

EXAMPLE 17-6   *Finding Equilibrium Concentrations*

For the following reaction, the equilibrium constant is 49.0 at a certain temperature. If 0.400 mol each of A and B are placed in a 2.00-liter container at that temperature, what concentrations of all species are present at equilibrium?

$$A + B \rightleftharpoons C + D$$

We have represented chemical formulas by single letters to simplify the notation in these calculations.

Plan

First we find the initial concentrations. Then we write the reaction summary and represent the equilibrium concentrations algebraically. Finally we substitute the algebraic representations of equilibrium concentrations into the K_c expression and find the equilibrium concentrations.

Solution

The initial concentrations are

$$[A] = \frac{0.400 \ mol}{2.00 \ L} = 0.200 \ M \qquad [C] = 0 \ M$$

$$[B] = \frac{0.400 \ mol}{2.00 \ L} = 0.200 \ M \qquad [D] = 0 \ M$$

We know that the reaction can only proceed to the right because only "reactants" are present. The reaction summary includes the values, or symbols for the values, of (1) initial concentrations, (2) changes in concentrations, and (3) concentrations at equilibrium.

Let x = moles per liter of A that react; then x = moles per liter of B that react and x = moles per liter of C and D that are formed.

The coefficients in the equation are all 1's, so the reaction ratio must be 1:1:1:1.

	A	+	B	$\rightleftharpoons$	C	+	D
initial	0.200 *M*		0.200 *M*		0 *M*		0 *M*
change due to rxn	$-x$ *M*		$-x$ *M*		$+x$ *M*		$+x$ *M*
at equilibrium	$(0.200 - x)M$		$(0.200 - x)M$		x *M*		x *M*

Now K_c is known, but concentrations are not. But the equilibrium concentrations have all been expressed in terms of the single variable x. We substitute the equilibrium concentrations (*not* the initial ones) into the K_c expression and solve for x.

$$K_c = \frac{[C][D]}{[A][B]} = 49.0$$

$$\frac{(x)(x)}{(0.200 - x)(0.200 - x)} = \frac{x^2}{(0.200 - x)^2} = 49.0$$

This quadratic equation has a perfect square on both sides. We solve it by taking the square roots of both sides of the equation and then rearranging for x.

$$\frac{x}{0.200 - x} = 7.00$$

$$x = 1.40 - 7.00x \qquad 8.00x = 1.40 \qquad x = \frac{1.40}{8.00} = 0.175$$

Now we know the value of x, so the equilibrium concentrations are

$$[A] = (0.200 - x)\, M = \boxed{0.025\, M}\,; \qquad [C] = x\, M = \boxed{0.175\, M}$$

$$[B] = (0.200 - x)\, M = \boxed{0.025\, M}\,; \qquad [D] = x\, M = \boxed{0.175\, M}$$

We see that the equilibrium concentrations of products are much greater than those of reactants because K_c is much greater than 1.

To check our answers we use the equilibrium concentrations to calculate Q and verify that its value is equal to K_c.

$$Q = \frac{[C][D]}{[A][B]} = \frac{(0.175)(0.175)}{(0.025)(0.025)} = 49 \qquad \text{Recall that } K_c = 49.0$$

The ideas developed in Example 17-6 may be applied to cases in which the reactants are mixed in nonstoichiometric amounts. This is shown in Example 17-7.

EXAMPLE 17-7 Finding Equilibrium Concentrations

Consider the same system as in Example 17-6 at the same temperature. If 0.600 mol of A and 0.200 mol of B are mixed in a 2.00-liter container and allowed to reach equilibrium, what are the equilibrium concentrations of all species?

Plan

We proceed as we did in Example 17-6. The only difference is that now we have *nonstoichiometric* starting amounts of reactants.

Solution

As in Example 17-6, we let $x = $ mol/L of A that react; then $x = $ mol/L of B that react, and $x = $ mol/L of C and D formed.

	A	+	B	$\rightleftharpoons$	C	+	D
initial	0.300 M		0.100 M		0 M		0 M
change due to rxn	$-x\, M$		$-x\, M$		$+x\, M$		$+x\, M$
equilibrium	$(0.300 - x)\, M$		$(0.100 - x)\, M$		$x\, M$		$x\, M$

The initial concentrations are governed by the amounts of reactants mixed together. But *changes in concentrations* due to reaction must still occur in the 1:1:1:1 ratio required by the coefficients in the balanced equation.

$$K_c = \frac{[C][D]}{[A][B]} = 49.0 \qquad \text{so} \qquad \frac{(x)(x)}{(0.300 - x)(0.100 - x)} = 49.0$$

The left side of this equation is *not a perfect square.*

We can arrange this quadratic equation into the standard form.

$$\frac{x^2}{0.0300 - 0.400x + x^2} = 49.0$$

$$x^2 = 1.47 - 19.6x + 49.0x^2$$

$$48.0x^2 - 19.6x + 1.47 = 0$$

Quadratic equations can be solved by use of the quadratic formula.

$$x = \frac{-b \pm \sqrt{b^2 - 4ac}}{2a}$$

In this case $a = 48.0$, $b = -19.6$, and $c = 1.47$. Substituting these values gives

$$x = \frac{-(-19.6) \pm \sqrt{(-19.6)^2 - 4(48.0)(1.47)}}{2(48.0)}$$

$$= \frac{19.6 \pm 10.1}{96.0} = 0.309 \quad \text{or} \quad 0.099$$

Evaluating the quadratic formula always yields two roots. One root (the answer) has physical meaning. The other root, while mathematically correct, is extraneous; that is, it has no physical meaning. The value of x is defined as the number of moles of A per liter that react and the number of moles of B per liter that react. No more B can be consumed than was initially present (0.100 M), so $x = 0.309$ is the extraneous root. Thus, $x = 0.099$ is the root that has physical meaning, and the extraneous root is 0.309. The equilibrium concentrations are

$$[A] = (0.300 - x)\, M = \boxed{0.201\ M}\ ; \qquad [B] = (0.100 - x)\, M = \boxed{0.001\ M}\ ;$$

$$[C] = [D] = x\, M \quad = \boxed{0.099\ M}$$

You should now work Exercises 34, 44, and 46.

✓ **Problem-Solving Tip:** *Solving Quadratic Equations— Which Root Shall We Use?*

Quadratic equations can be rearranged into standard form.

$$ax^2 + bx + c = 0$$

All can be solved by the quadratic formula, which is

$$x = \frac{-b \pm \sqrt{b^2 - 4ac}}{2a} \quad \text{(Appendix A)}$$

This formula gives *two* roots, both of which are *mathematically* correct. A foolproof way to determine which root of the equation has physical meaning is to substitute the value of the variable into the expressions for the equilibrium concentrations. For the extraneous root, one or more of these substitutions will lead to a negative concentration, which is physically impossible (there cannot be *less than none* of a substance present!). The correct root will give all positive concentrations. In Example 17-7, substitution of the extraneous root $x = 0.309$ would give $[A] = (0.300 - 0.309)\ M = -0.009\ M$ and $[B] = (0.100 - 0.309)\ M = -0.209\ M$. Either of these concentration values is impossible, so we would know that 0.309 is an extraneous root. You should apply this check to subsequent calculations that involve solving a quadratic equation.

Check Example 17-6:

$$Q = \frac{[C][D]}{[A][B]} = \frac{(0.175)(0.175)}{(0.025)(0.025)}$$

$$Q = 49 = K_c$$

Check Example 17-7:

$$Q = \frac{(0.099)(0.099)}{(0.201)(0.001)}$$

$$Q = 49 = K_c$$

The following table summarizes Examples 17-6 and 17-7.

	Initial Concentrations (M)				Equilibrium Concentrations (M)			
	[A]	[B]	[C]	[D]	[A]	[B]	[C]	[D]
Example 17-6	0.200	0.200	0	0	0.025	0.025	0.175	0.175
Example 17-7	0.300	0.100	0	0	0.201	0.001	0.099	0.099

The data from the table can be substituted into the reaction quotient expression, Q, as a check. Even though the reaction is initiated by different relative amounts of reactants in the two cases, the ratios of equilibrium concentrations of products to reactants (each raised to the first power) agree within roundoff error.

17-6 DISTURBING A SYSTEM AT EQUILIBRIUM: PREDICTIONS

Once a reacting system has reached equilibrium, it remains at equilibrium until it is disturbed by some change of conditions. The guiding principle is known as **LeChatelier's Principle** (Section 13-6).

LeChatelier is pronounced "le-SHOT-lee-ay."

> If a system at equilibrium is disturbed by changing its conditions (applying a stress), the system shifts in the direction that reduces the stress. If given sufficient time, a new state of equilibrium is established.

Remember that the *value* of an equilibrium constant changes only with temperature.

The reaction quotient, Q, helps us predict the direction of this shift. Three types of changes can disturb the equilibrium of a reaction.

1. Changes in concentration
2. Changes in pressure or volume (for reactions that involve gases)
3. Changes in temperature

We now study the effects of these types of stresses from a qualitative, or descriptive, point of view. In Section 17-8 we expand our discussion with quantitative examples.

For reactions involving gases at constant temperature, changes in volume cause changes in pressure, and vice versa.

Changes in Concentration

Consider the following system *starting at equilibrium*.

$$A + B \rightleftharpoons C + D \qquad K_c = \frac{[C][D]}{[A][B]}$$

When more of any reactant or product is *added* to the system, the value of Q changes, so it no longer matches K_c, and the reaction is no longer at equilibrium. The stress due to the added substance is relieved by shifting the equilibrium in the direction that consumes some of the added substance, moving the value of Q back toward K_c. Let us compare the mass action expressions for Q and K_c. If more A or B is added, then $Q < K_c$, and the forward reaction occurs more rapidly and to a greater extent than the reverse reaction until equilibrium is reestablished. If more C or D is added, $Q > K_c$, and the reverse reaction occurs more rapidly and to a greater extent until equilibrium is reestablished.

CD-ROM Screen 16.11, Disturbing a Chemical Equilibrium (1): LeChatelier's Principle.

> Adding or removing reactants or products changes the value of Q. It *does not change* the value of K_c.

CD-ROM Screen 16.13, Disturbing a Chemical Equilibrium (3): Addition or Removal of a Reagent.

We view LeChatelier's Principle in the kinetic terms we used to introduce equilibrium. The rate of the forward reaction is proportional to the reactant concentrations raised to some powers,

$$\text{rate}_f = k_f[A]^x[B]^y$$

Effects of changes in concentration on the equilibrium

$$[Co(OH_2)_6]^{2+} + 4Cl^- \rightleftharpoons [CoCl_4]^{2-} + 6H_2O$$

A solution of $CoCl_2 \cdot 6H_2O$ in isopropyl alcohol and water is purple (not shown) due to the mixture of $[Co(OH_2)_6]^{2+}$ (*pink*) and $[CoCl_4]^{2-}$ (*blue*). When we add concentrated HCl, the excess Cl^- shifts the reaction to the right (*blue, right*). Adding $AgNO_3(aq)$ removes some Cl^- by precipitation of AgCl(s) and favors the reaction to the left (produces more $[Co(OH_2)_6]^{2+}$; the resulting solution is pink (*left*). Each insert shows the structure of the cobalt complex species present in higher concentration; other ions and solvent molecules are not shown.

The terminology used here is not as precise as we might like, but it is widely used. When we say that the equilibrium is "shifted to the left," we mean that the reaction to the left occurs to a greater extent than the reaction to the right.

When we add more A to an equilibrium mixture, this rate increases so that it no longer matches the rate of the reverse reaction. As the reaction proceeds to the right, consuming some A and B and forming more C and D, the forward rate diminishes and the reverse rate increases until they are again equal. At that point, a new equilibrium condition has been reached, with more C and D than were present in the *original* equilibrium mixture. Not all of the added A has been consumed when the new equilibrium is reached, however.

If a reactant or product is *removed* from a system at equilibrium, the reaction that produces *that* substance occurs more rapidly and to a greater extent than its reverse. If some C or D is removed, then $Q < K_c$, and the forward reaction is favored until equilibrium is reestablished. If some A or B is removed, the reverse reaction is favored.

 This tabulation summarizes a lot of useful information. Study it carefully.

Stress	Q	Direction of Shift of $A + B \rightleftharpoons C + D$
Increase concentration of A or B	$Q < K$	$\longrightarrow$ right
Increase concentration of C or D	$Q > K$	left $\longleftarrow$
Decrease concentration of A or B	$Q > K$	left $\longleftarrow$
Decrease concentration of C or D	$Q < K$	$\longrightarrow$ right

When a new equilibrium condition is established, (1) the rates of the forward and reverse reactions are again equal; and (2) K_c is again satisfied by the new concentrations of reactants and products.

Practical applications of changes of this type are of great economic importance. Removing a product of a reversible reaction forces the reaction to produce more product than could be obtained if the reaction were simply allowed to reach equilibrium.

EXAMPLE 17-8 Disturbing a System at Equilibrium: Predictions

Given the following reaction at equilibrium in a closed container at 500°C, predict the effect of each of the following changes on the amount of NH_3 present at equilibrium: (a) forcing more H_2 into the system; (b) removing some NH_3 from the system.

$$N_2(g) + 3H_2(g) \rightleftharpoons 2NH_3(g)$$

Plan
We apply LeChatelier's Principle to each part of the question individually.

Solution
(a) Adding a substance favors the reaction that uses up that substance (forward in this case).

More NH_3 is formed.

(b) Removing a substance favors the reaction that produces that substance (forward in this case).

More NH_3 is formed.

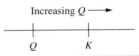

(a) Reaction proceeds to the right

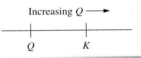

(b) Reaction proceeds to the right

At the new equilibrium, the concentration of any reactant or product that was added will still be greater than it was in the original equilibrium mixture. The concentration of any reactant or product that was removed will be less in the new equilibrium even though the reaction will have replaced part of the removed material.

Changes in Volume and Pressure

Changes in pressure have little effect on the concentrations of solids or liquids because they are only slightly compressible. Changes in pressure do cause significant changes in concentrations of gases, however. Such changes therefore affect the value of Q for reactions in which the number of moles of gaseous reactants differs from the number of moles of gaseous products. For an ideal gas,

$$PV = nRT \qquad \text{or} \qquad P = \left(\frac{n}{V}\right)(RT)$$

The term (n/V) represents concentration, that is, mol/L. At constant temperature, n, R, and T are constants. Thus, if the volume occupied by a gas decreases, its partial pressure increases and its concentration (n/V) increases. If the volume of a gas increases, both its partial pressure and its concentration decrease.
    Consider the following gaseous system at equilibrium.

$$A(g) \rightleftharpoons 2D(g) \qquad K_c = \frac{[D]^2}{[A]}$$

CD-ROM   Screen 16.14, Disturbing a Chemical Equilibrium (4): Volume Changes.

At constant temperature, a decrease in volume (increase in pressure) increases the concentrations of both A and D. In the expression for Q, the concentration of D is squared and the concentration of A is raised to the first power. As a result, the numerator of Q increases more than the denominator as pressure increases. Thus, $Q > K_c$, and this equilibrium shifts to the left. Conversely, an increase in volume (decrease in pressure) shifts this reaction to the right until equilibrium is reestablished, because $Q < K_c$. We can summarize the effect of pressure (volume) changes on *this* gas-phase system at equilibrium.

Study this tabulation carefully. How would these conclusions change for a reaction in which there are more moles of gaseous reactants than moles of gaseous products?

Stress	$Q*$	Direction of Shift of A(g) $\rightleftharpoons$ 2D(g)
Volume decrease, pressure increase	$Q > K_c$	Toward smaller number of moles of gas (left for *this* reaction)
Volume increase, pressure decrease	$Q < K_c$	Toward larger number of moles of gas (right for *this* reaction)

*In Q for this *reaction, there are more moles of gaseous product than gaseous reactant.*

In general, for reactions that involve gaseous reactants or products, LeChatelier's Principle allows us to predict the following results.

1. If there is *no change in the total number of moles of gases in the balanced chemical equation*, a volume (pressure) change does not affect the position of equilibrium; Q is unchanged by the volume (or pressure) change.

2. If a *balanced chemical equation involves a change in the total number of moles of gases*, changing the volume (or pressure) of a reaction mixture changes the value of Q; it *does not change* the value of K_c. For such a reaction:

   (a) A decrease in volume (increase in pressure) shifts a reaction in the direction that produces the *smaller total number of moles of gas*, until Q again equals K_c.

   (b) An increase in volume (decrease in pressure) shifts a reaction in the direction that produces the *larger total number of moles of gas*, until Q again equals K_c.

One practical application of these ideas is illustrated in the next section by the Haber process.

The foregoing argument applies only when pressure changes are due to volume changes. It *does not apply* if the total pressure of a gaseous system is raised by merely pumping in an inert gas, for example, He. If the gas that is introduced is not involved in the reaction, the *partial pressure* of each reacting gas remains constant, so the system remains at equilibrium.

Many students find it hard to believe that as a different gas is pumped into a container the partial pressures of any gases already present do not change.

EXAMPLE 17-9   *Disturbing a System at Equilibrium: Predictions*

(a) Given the following reaction at equilibrium in a closed container at 500°C, predict the effect of increasing the pressure by decreasing the volume.

$$N_2(g) + 3H_2(g) \rightleftharpoons 2NH_3(g)$$

(b) Make the same prediction for the following reaction at high temperature.

$$H_2(g) + I_2(g) \rightleftharpoons 2HI(g)$$

Plan

We apply LeChatelier's Principle to predict the effect on each reaction.

Solution

(a) Increasing the pressure favors the reaction that produces the smaller number of moles of gas (forward in this case).

More NH_3 is formed.

(b) This reaction involves the same number of moles of gas on both sides, so a pressure (volume) change does not disturb the equilibrium. There is no effect on the position of equilibrium.

Changes in Temperature

Consider the following exothermic reaction at equilibrium:

$$A + B \rightleftharpoons C + D + heat \qquad (\Delta H \text{ is negative})$$

Heat is produced by the forward (exothermic) reaction. Suppose we increase the temperature at constant pressure by adding heat to the system. This favors the reaction to the left, removing some of the extra heat. Lowering the temperature favors the reaction to the right as the system replaces some of the heat that was removed.

By contrast, for an endothermic reaction at equilibrium,

$$W + X + heat \rightleftharpoons Y + Z \qquad (\Delta H \text{ is positive})$$

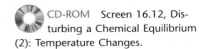

CD-ROM   Screen 16.12, Disturbing a Chemical Equilibrium (2): Temperature Changes.

Charles D. Winters

The gas-phase equilibrium for the *exothermic* reaction

$$2NO_2(g) \rightleftharpoons N_2O_4(g) \qquad \Delta H^0_{rxn} = -57.2 \text{ kJ/mol rxn}$$

The two flasks contain the same *total mass* of gas. NO_2 is brown, whereas N_2O_4 is colorless. The higher temperature (50°C) of the flask on the left favors the reverse reaction; this mixture is more highly colored because it contains more NO_2. The flask on the right, at the temperature of ice water, contains less brown NO_2 gas. The lower temperature favors the formation of the N_2O_4.

Charles Steele

Effect of temperature changes on the equilibrium of the *endothermic* reaction

$$[Co(OH_2)_6]^{2+} + 4Cl^- + heat \rightleftharpoons$$
$$[CoCl_4]^{2-} + 6H_2O$$

We begin with a purple equilibrium mixture of the pink and blue complexes at room temperature (not shown). In hot water the forward reaction (endothermic) is favored and K_c is higher, so the solution is blue (*right*). At 0°C, the reverse reaction (exothermic) is favored and K_c is lower, so the solution is pink (*left*). Each insert shows the structure of the cobalt complex species present in highest concentration; other ions and solvent molecules are not shown.

an increase in temperature at constant pressure favors the reaction to the right. A decrease in temperature favors the reaction to the left.

The *value of any equilibrium constant changes* as the temperature changes. Changing the temperature of a reaction at equilibrium thus causes Q to differ from K_c, but this is now because K_c has changed. The reaction then proceeds in the direction that moves Q toward the new value of K_c. As we will see in Section 17-13,

the K_c values of exothermic reactions decrease with increasing T, and the K_c values of endothermic reactions increase with increasing T. No other stresses affect the value of K_c.

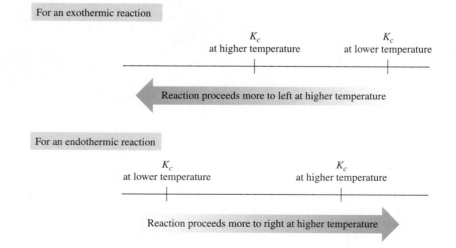

Addition of a Catalyst

Adding a catalyst to a system changes the rate of the reaction (Section 16-9), but this *cannot* shift the equilibrium in favor of either products or reactants. Because a catalyst affects the activation energy of *both* forward and reverse reactions equally, it changes both rate constants by the same factor, so their ratio, K_c, does not change.

Can you use the Arrhenius equation (Section 16-8) to show that lowering the activation energy barrier increases forward and reverse rates by the same factor?

> Adding a catalyst to a reaction at equilibrium has no effect; it changes neither Q nor K_c.

The same equilibrium mixture is achieved with or without the catalyst, but the equilibrium is established more quickly in the presence of a catalyst.

Not all reactions attain equilibrium; they may occur too slowly, or else products or reactants may be continually added or removed. Such is the case with most reactions in biological systems. On the other hand, some reactions, such as typical acid–base neutralizations, achieve equilibrium very rapidly.

EXAMPLE 17-10 *Disturbing a System at Equilibrium: Predictions*

Given the following reaction at equilibrium in a closed container at 500°C, predict the effect of each of the following changes on the amount of NH_3 present at equilibrium: (a) raising the temperature; (b) lowering the temperature; (c) introducing some platinum catalyst.

$$N_2(g) + 3H_2(g) \rightleftharpoons 2NH_3(g) \qquad \Delta H^0 = -92 \text{ kJ/mol rxn}$$

Plan

We apply LeChatelier's Principle to each part of the question individually.

Solution

(a) The negative value for ΔH tells us that the forward reaction is exothermic. Raising the temperature favors the endothermic reaction (reverse in this case).

> Some NH_3 is used up.

(b) Lowering the temperature favors the exothermic reaction (forward in this case).

> More NH_3 is formed.

(c) A catalyst does not favor either reaction.

> It would have no effect on the amount of NH_3.

You should now work Exercises 54, 57, and 58.

Now we shall illustrate the commercial importance of these changes.

17-7 THE HABER PROCESS: A PRACTICAL APPLICATION OF EQUILIBRIUM

As Germany prepared for World War I, Britain controlled the seas and thus the access to the natural nitrates in India and Chile that were needed to prepare explosives. Fritz Haber (1868–1934) developed the process to provide a cheaper and more reliable source of explosives. The current use of the process is more humanitarian: most NH_3 is used to produce fertilizers. In the United States, approximately 100 pounds of NH_3 is required per person per year.

Nitrogen, N_2, is very unreactive. The Haber process is the economically important industrial process by which atmospheric N_2 is converted to ammonia, NH_3, a soluble, reactive compound. Innumerable dyes, plastics, explosives, fertilizers, and synthetic fibers are made from ammonia. The Haber process provides insight into kinetic and thermodynamic factors that influence reaction rates and the positions of equilibria. In this process the reaction between N_2 and H_2 to produce NH_3 is never allowed to reach equilibrium, but moves toward it.

$$N_2(g) + 3H_2(g) \rightleftharpoons 2NH_3(g) \qquad \Delta H^0 = -92 \text{ kJ/mol}$$

$$K_c = \frac{[NH_3]^2}{[N_2][H_2]^3} = 3.6 \times 10^8 \qquad \text{(at 25°C)}$$

The process is diagrammed in Figure 17-3. The reaction is carried out at about 450°C under pressures ranging from 200 to 1000 atmospheres. Hydrogen is obtained from coal gas or petroleum refining and nitrogen from liquefied air.

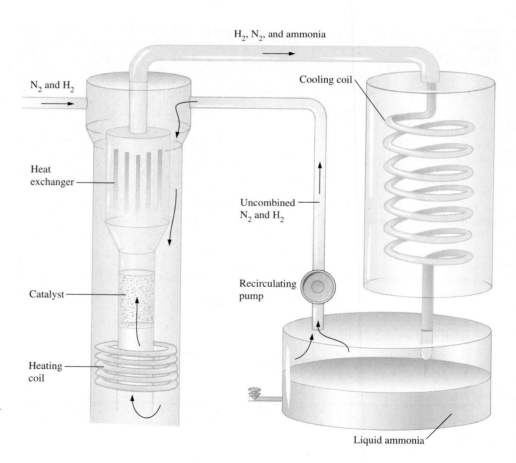

Figure 17-3 A simplified representation of the Haber process for synthesizing ammonia.

TABLE 17-1	*Effect of T and P on Yield of Ammonia*			
		Mole % NH_3 in Equilibrium Mixture		
°C	K_c	10 atm	100 atm	1000 atm
209	650	51	82	98
467	0.5	4	25	80
758	0.014	0.5	5	13

A nighttime photo of a large plant for the commercial production of ammonia, NH_3. Such an installation can produce up to 7000 metric tons of ammonia per day. There are nearly 100 such plants in the world.

The value of K_c is 3.6×10^8 at 25°C. This very large value of K_c indicates that *at equilibrium* virtually all of the N_2 and H_2 (mixed in a 1:3 mole ratio) would be converted into NH_3. At 25°C, the reaction occurs so slowly, however, that no measurable amount of NH_3 is produced within a reasonable time. Thus, the large equilibrium constant (a thermodynamic factor) indicates that the reaction proceeds toward the right almost completely. It tells us *nothing*, however, about how fast the reaction occurs (a kinetic factor).

There are four moles of gas on the left side of the equation and only two moles of gas on the right, so increasing the pressure favors the production of NH_3. The Haber process is therefore carried out at very high pressures, as high as the equipment will safely stand.

The reaction is exothermic (ΔH^0_{rxn} is negative), so increasing the temperature favors the *decomposition* of NH_3 (the reverse reaction). But, the rates of both forward and reverse reactions increase as the temperature increases.

The addition of a catalyst of finely divided iron and small amounts of selected oxides also speeds up both the forward and reverse reactions. This allows NH_3 to be produced not only faster but at a lower temperature, which increases the yield of NH_3 and extends the life of the equipment.

Table 17-1 shows the effects of increases in temperature and pressure on the equilibrium yield of NH_3, starting with 1:3 mole ratios of $N_2:H_2$. K_c decreases by more than ten orders of magnitude, from 3.6×10^8 at 25°C to only 1.4×10^{-2} at 758°C. This tells us that the reaction proceeds *very far to the left* at high temperatures. Casual examination of the data might suggest that the reaction should be run at lower temperatures, because a higher percentage of the N_2 and H_2 is converted into NH_3. The reaction occurs so slowly, however, even in the presence of a catalyst, that it cannot be run economically at temperatures below about 450°C.

The emerging reaction mixture is cooled down, and NH_3 (bp = −33.43°C) is removed as a liquid. This prevents the reaction from reaching equilibrium and favors the forward reaction. The unreacted N_2 and H_2 are recycled. Excess N_2 is used to favor the reaction to the right.

In practice, the mixed reactants are compressed by special pumps and injected into the heated reaction vessel.

Ten orders of magnitude is 10^{10}, that is, 10 billion.

$$1 \times 10^{10} = 10,000,000,000$$

SC*L*INKS.

TOPIC: Haber Process
GO TO: www.scilinks.org
*sci***LINKS CODE:** WCH1710

17-8 DISTURBING A SYSTEM AT EQUILIBRIUM: CALCULATIONS

We can use equilibrium constants to calculate new equilibrium concentrations that result from adding one or more species to, or removing one or more species from, a system at equilibrium.

EXAMPLE 17-11 *Disturbing a System at Equilibrium: Calculations*

Some hydrogen and iodine are mixed at 229°C in a 1.00-liter container. When equilibrium is established, the following concentrations are present: $[HI] = 0.490\ M$, $[H_2] = 0.080\ M$, and $[I_2] = 0.060\ M$. If an additional 0.300 mol of HI is then added, what concentrations will be present when the new equilibrium is established?

$$H_2(g) + I_2(g) \rightleftharpoons 2HI(g)$$

Plan

We use the initial equilibrium concentrations to calculate the value of K_c. Then we determine the new concentrations after some HI has been added and calculate Q. The value of Q tells us which reaction is favored. Then we can represent the new equilibrium concentrations. We substitute these representations into the K_c expression and solve for the new equilibrium concentrations.

Solution

Calculate the value of K_c from the first set of equilibrium concentrations.

$$K_c = \frac{[HI]^2}{[H_2][I_2]} = \frac{(0.490)^2}{(0.080)(0.060)} = 50$$

When we add 0.300 mol of HI to the 1.00-liter container, the [HI] instantaneously increases by 0.300 M.

	$H_2(g)$ +	$I_2(g)$ $\rightleftharpoons$	$2HI(g)$
equilibrium	0.080 M	0.060 M	0.490 M
mol/L added	0 M	0 M	+0.300 M
new initial conc'n	0.080 M	0.060 M	0.790 M

Substitution of these *new initial* concentrations into the reaction quotient gives

$$Q = \frac{[HI]^2}{[H_2][I_2]} = \frac{(0.790)^2}{(0.080)(0.060)} = 130$$

> It is obvious that adding some HI favors the reaction to the left. If more than one substance is added to the reaction mixture, it might not be obvious which reaction will be favored. Calculating Q always lets us make the decision.

Because $Q > K_c$, the reaction proceeds to the left to establish a new equilibrium. The new equilibrium concentrations can be determined as follows. Let $x = $ mol/L of H_2 formed; so $x = $ mol/L of I_2 formed, and $2x = $ mol/L of HI consumed.

> Equal concentrations of H_2 and I_2 must be formed by the *new progress* of the reaction.

	$H_2(g)$ +	$I_2(g)$ $\rightleftharpoons$	$2HI(g)$
new initial conc'n	0.080 M	0.060 M	0.790 M
change due to rxn	+x M	+x M	−2x M
new equilibrium	(0.080 + x) M	(0.060 + x) M	(0.790 − 2x) M

Substitution of these values into K_c allows us to evaluate x.

$$K_c = 50 = \frac{(0.790 - 2x)^2}{(0.080 + x)(0.060 + x)} = \frac{0.624 - 3.16x + 4x^2}{0.0048 + 0.14x + x^2}$$

$$0.24 + 7.0x + 50x^2 = 0.624 - 3.16x + 4x^2$$

$$46x^2 + 10.2x - 0.38 = 0$$

> To "consume a negative quantity of HI" would be to form HI. The value $x = -0.25$ would lead to $[H_2] = (0.080 + x)\ M = (0.080 - 0.25)\ M = -0.17\ M$. A negative concentration is impossible, so $x = -0.25$ is the extraneous root.

Solution by the quadratic formula gives $x = 0.032$ and -0.25.

Clearly, $x = -0.25$ is the extraneous root, because x cannot be less than zero in this case. This reaction does not consume a negative quantity of HI, because the reaction is proceeding toward the left. Thus, $x = 0.032$ is the root with physical meaning, so the new equilibrium concentrations are

$$[H_2] = (0.080 + x) M = (0.080 + 0.032) M = \boxed{0.112 \ M}$$

$$[I_2] = (0.060 + x) M = (0.060 + 0.032) M = \boxed{0.092 \ M}$$

$$[HI] = (0.790 - 2x) M = (0.790 - 0.064) M = \boxed{0.726 \ M}$$

In summary,

Original Equilibrium	Stress Applied	New Equilibrium
$[H_2] = 0.080 \ M$ $[I_2] = 0.060 \ M$ $[HI] = 0.490 \ M$	Add 0.300 M HI	$[H_2] = 0.112 \ M$ $[I_2] = 0.092 \ M$ $[HI] = 0.726 \ M$

We see that some of the additional HI is consumed, but not all of it. More HI remains after the new equilibrium is established than was present before the stress was imposed. The new equilibrium $[H_2]$ and $[I_2]$ are substantially greater than the original equilibrium concentrations, however.

You should now work Exercise 64.

We can also use the equilibrium constant to calculate new equilibrium concentrations that result from decreasing the volume (increasing the pressure) of a gaseous system that was initially at equilibrium.

EXAMPLE 17-12   *Disturbing a System at Equilibrium: Calculations*

At 22°C the equilibrium constant, K_c, for the following reaction is 4.66×10^{-3}. (a) If 0.800 mol of N_2O_4 were injected into a closed 1.00-liter container at 22°C, how many moles of each gas would be present at equilibrium? (b) If the volume were then halved (to 0.500 L) at constant temperature, how many moles of each gas would be present after the new equilibrium has been established?

$$N_2O_4(g) \rightleftharpoons 2NO_2(g) \qquad K_c = \frac{[NO_2]^2}{[N_2O_4]} = 4.66 \times 10^{-3}$$

Plan

(a) We are given the value of K_c and the initial concentration of N_2O_4. We write the reaction summary, which gives the representation of the equilibrium concentrations. Then we substitute these into the K_c expression and solve for the new equilibrium concentrations.

(b) We obtain the *new initial* concentrations by adjusting the equilibrium concentrations from Part (a) for the volume change. Then we solve for the *new* equilibrium concentrations as we did in Part (a).

Solution

(a) Let x = mol/L of N_2O_4 consumed and $2x$ = mol/L of NO_2 formed.

	$N_2O_4(g)$	$\rightleftharpoons$	$2NO_2(g)$
initial	0.800 M		0 M
change due to rxn	$-x \ M$		$+2x \ M$
equilibrium	$(0.800 - x) M$		$2x \ M$

CHAPTER 17: Chemical Equilibrium

$$K_c = \frac{[NO_2]^2}{[N_2O_4]} = 4.66 \times 10^{-3} = \frac{(2x)^2}{0.800 - x} = \frac{4x^2}{0.800 - x}$$

$$3.73 \times 10^{-3} - 4.66 \times 10^{-3}x = 4x^2$$

$$4x^2 + 4.66 \times 10^{-3}x - 3.73 \times 10^{-3} = 0$$

The value of x is the number of moles per liter of N_2O_4 that react. So x must be positive and cannot be greater than 0.800 M.

$$0 < x < 0.800\ M$$

Solving by the quadratic formula gives $x = 3.00 \times 10^{-2}$ and $x = -3.11 \times 10^{-2}$. We use $x = 3.00 \times 10^{-2}$.

The original equilibrium concentrations are

$$[NO_2] = 2x\ M = 6.00 \times 10^{-2}\ M$$

$$[N_2O_4] = (0.800 - x)\ M = (0.800 - 3.00 \times 10^{-2})\ M = 0.770\ M$$

$$\underline{?}\ mol\ NO_2 = 1.00\ L \times \frac{6.00 \times 10^{-2}\ mol\ NO_2}{L} = 6.00 \times 10^{-2}\ mol\ NO_2$$

$$\underline{?}\ mol\ N_2O_4 = 1.00\ L \times \frac{0.770\ mol\ N_2O_4}{L} = 0.770\ mol\ N_2O_4$$

LeChatelier's Principle tells us that a decrease in volume (increase in pressure) favors the production of N_2O_4.

$$Q = \frac{(0.120)^2}{1.54} = 9.35 \times 10^{-3}$$

$$Q > K$$

$$\therefore \xleftarrow{\text{shift left}}$$

(b) When the volume of the reaction vessel is halved, the concentrations are doubled, so the *new initial* concentrations of N_2O_4 and NO_2 are 2(0.770 M) = 1.54 M and 2(6.00 × 10⁻² M) = 0.120 M, respectively.

	$N_2O_4(g)$	$\rightleftharpoons$	$2NO_2(g)$
new initial	1.54 M		0.120 M
change due to rxn	$+x\ M$		$-2x\ M$
new equilibrium	$(1.54 + x)\ M$		$(0.120 - 2x)\ M$

$$K_c = \frac{[NO_2]^2}{[N_2O_4]} = 4.66 \times 10^{-3} = \frac{(0.120 - 2x)^2}{1.54 + x}$$

Rearranging into the standard form of a quadratic equation gives

$$x^2 - 0.121x + 1.81 \times 10^{-3} = 0$$

Solving as before gives $x = 0.104$ and $x = 0.017$.

The root $x = 0.104$ would give a *negative* concentration for NO_2, which is impossible.

The maximum value of x is 0.060 M, because $2x$ may not exceed the concentration of NO_2 that was present after the volume was halved. Thus, $x = 0.017\ M$ is the root with physical significance. The new equilibrium concentrations in the 0.500-liter container are

$$[NO_2] = (0.120 - 2x)\ M = (0.120 - 0.034)\ M = 0.086\ M$$

$$[N_2O_4] = (1.54 + x)\ M = (1.54 + 0.017)\ M = 1.56\ M$$

$$\underline{?}\ mol\ NO_2 = 0.500\ L \times \frac{0.086\ mol\ NO_2}{L} = 0.043\ mol\ NO_2$$

$$\underline{?}\ mol\ N_2O_4 = 0.500\ L \times \frac{1.56\ mol\ N_2O_4}{L} = 0.780\ mol\ N_2O_4$$

In summary,

First Equilibrium	Stress	New Equilibrium
0.770 mol of N_2O_4	Decrease volume from	0.780 mol of N_2O_4
0.0600 mol of NO_2	1.00 L to 0.500 L	0.043 mol of NO_2

The concentrations $[N_2O_4]$ and $[NO_2]$ *both increase* because of the large decrease in volume. However, the *number of moles* of N_2O_4 increases, while the *number of moles* of NO_2 decreases. We predict this from LeChatelier's Principle.

You should now work Exercise 66.

✓ **Problem-Solving Tip:** *There Are Several Ways to Solve Equilibrium Problems*

When a stress is applied to a system originally at equilibrium, it is no longer at equilibrium. As we did in Example 17-12(b), we can apply the stress to the *old equilibrium values* and then treat these as the new "initial values." Alternatively, we could adjust the *original concentration values* to reflect the stress, and then treat these as the new "initial values." We could consider $[N_2O_4]_{initial}$ in Example 17-12(b) as the *original* 0.800 mol of N_2O_4 from Part (a) in the *new* volume, 0.500 L. That is, $[N_2O_4]_{initial} = 0.800$ mol/0.500 L = 1.60 *M*, with no NO_2 having yet been formed. $[NO_2]_{initial} = 0$ *M*. From that starting point, the reaction would proceed to the *right*.

17-9 PARTIAL PRESSURES AND THE EQUILIBRIUM CONSTANT

It is often more convenient to measure pressures rather than concentrations of gases. Solving the ideal gas equation, $PV = nRT$, for pressure gives

$$P = \frac{n}{V}(RT) \qquad \text{or} \qquad P = M(RT)$$

$\left(\dfrac{n}{V}\right)$ is $\left(\dfrac{\text{no.mol}}{\text{L}}\right)$

The pressure of a gas is directly proportional to its concentration (n/V). For reactions in which all substances that appear in the equilibrium constant expression are gases, we sometimes prefer to express the equilibrium constant in terms of partial pressures *in atmospheres* (K_P) rather than in terms of concentrations (K_c).

In general for a reaction involving gases,

$$aA(g) + bB(g) \rightleftharpoons cC(g) + dD(g) \qquad K_P = \frac{(P_C)^c(P_D)^d}{(P_A)^a(P_B)^b}$$

K_P has no units for the same reasons that K_c has no units.

For instance, for the following reversible reaction,

$$N_2(g) + 3H_2(g) \rightleftharpoons 2NH_3(g) \qquad K_P = \frac{(P_{NH_3})^2}{(P_{N_2})(P_{H_2})^3}$$

EXAMPLE 17-13 *Calculation of K_P*

In an equilibrium mixture at 500°C, we find $P_{NH_3} = 0.147$ atm, $P_{N_2} = 6.00$ atm, and $P_{H_2} = 3.70$ atm. Evaluate K_P at 500°C for the following reaction.

$$N_2(g) + 3H_2(g) \rightleftharpoons 2NH_3(g)$$

One error that students sometimes make when solving K_P problems is to express pressures in torr. Remember that these pressures must be expressed in atmospheres.

Plan

We are given equilibrium partial pressures of all reactants and products. So we write the expression for K_P and substitute partial pressures in atmospheres into it.

Solution

$$K_P = \frac{(P_{NH_3})^2}{(P_{N_2})(P_{H_2})^3} = \frac{(0.147)^2}{(6.00)(3.70)^3} = \boxed{7.11 \times 10^{-5}}$$

You should now work Exercises 70 and 73.

17-10 RELATIONSHIP BETWEEN K_P AND K_c

If the ideal gas equation is rearranged, the molar concentration of a gas is

$$\left(\frac{n}{V}\right) = \frac{P}{RT} \quad \text{or} \quad M = \frac{P}{RT}$$

Substituting P/RT for n/V in the K_c expression for the N_2–H_2–NH_3 equilibrium gives the relationship between K_c and K_P for *this* reaction.

$$K_c = \frac{[NH_3]^2}{[N_2][H_2]^3} = \frac{\left(\dfrac{P_{NH_3}}{RT}\right)^2}{\left(\dfrac{P_{N_2}}{RT}\right)\left(\dfrac{P_{H_2}}{RT}\right)^3} = \frac{(P_{NH_3})^2}{(P_{N_2})(P_{H_2})^3} \times \frac{\left(\dfrac{1}{RT}\right)^2}{\left(\dfrac{1}{RT}\right)^4}$$

$$= K_P (RT)^2 \quad \text{and} \quad K_P = K_c (RT)^{-2}$$

In general the relationship between K_c and K_P is

Δn refers to the numbers of moles of gaseous substances in the balanced equation, *not* in the reaction vessel.

$$K_P = K_c(RT)^{\Delta n} \quad \text{or} \quad K_c = K_P(RT)^{-\Delta n} \qquad \Delta n = (n_{gas\ prod}) - (n_{gas\ react})$$

For reactions in which equal numbers of moles of gases appear on both sides of the equation, $\Delta n = 0$ and $K_P = K_c$.

In Example 17-1, we saw that for the ammonia reaction at 500°C (or 773 K), $K_c = 0.286$. We can describe this equilibrium in terms of partial pressures using K_P.

$$NH_3(g) + 3H_2(g) \rightleftharpoons 2NH_3(g) \qquad \Delta n = 2 - 4 = -2$$

$$K_P = K_c(RT)^{\Delta n} = (0.286)[(0.0821)(773)]^{-2} = 7.10 \times 10^{-5}$$

This is essentially the same value we obtained in Example 17-13.

✓ **Problem-Solving Tip:** *Be Careful About the Value of R*

To decide which value of R to use when you convert between K_c and K_P, you can reason as follows. K_c involves molar concentrations, for which the units are mol/L; K_P involves pressures expressed in atm. Thus the appropriate value of R to use for these conversions must include these units. We use $0.08206 \dfrac{L \cdot atm}{mol \cdot K}$, rounded to the number of places appropriate to the problem.

For *gas-phase reactions*, we can calculate the amounts of substances present at equilibrium using either K_P or K_c. The results are the same by either method (when they are both expressed in the same terms). To illustrate, let's solve the following problem by both methods.

EXAMPLE 17-14   *Calculations with K_c and K_P*

We place 10.0 grams of $SbCl_5$ in a 5.00-liter container at 448°C and allow the reaction to attain equilibrium. How many grams of $SbCl_5$ are present at equilibrium? Solve this problem (a) using K_c and molar concentrations and (b) using K_P and partial pressures.

$$SbCl_5(g) \rightleftharpoons SbCl_3(g) + Cl_2(g) \quad \text{at 448°C, } K_c = 2.51 \times 10^{-2} \quad \text{and} \quad K_P = 1.48$$

(a) Plan (using K_c)

We calculate the initial concentration of $SbCl_5$, write the reaction summary, and represent the equilibrium concentrations; then we substitute into the K_c expression to obtain the equilibrium concentrations.

(a) Solution (using K_c)

Because we are given K_c, we use concentrations. The initial concentration of $SbCl_5$ is

$$[SbCl_5] = \frac{10.0 \text{ g } SbCl_5}{5.00 \text{ L}} \times \frac{1 \text{ mol}}{299 \text{ g}} = 0.00669 \text{ } M \text{ } SbCl_5$$

Let x = mol/L of $SbCl_5$ that react. In terms of molar concentrations, the reaction summary is

	$SbCl_5$	$\rightleftharpoons$	$SbCl_3$	+	Cl_2
initial	0.00669 M		0		0
change due to rxn	$-x$ M		$+x$ M		$+x$ M
equilibrium	$(0.00669 - x)$ M		x M		x M

$$K_c = \frac{[SbCl_3][Cl_2]}{[SbCl_5]}$$

$$2.51 \times 10^{-2} = \frac{(x)(x)}{0.00669 - x}$$

$$x^2 = 1.68 \times 10^{-4} - 2.51 \times 10^{-2}x$$

$$x^2 + 2.51 \times 10^{-2}x - 1.68 \times 10^{-4} = 0$$

Solving by the quadratic formula gives

$$x = 5.49 \times 10^{-3} \quad \text{and} \quad -3.06 \times 10^{-2} \text{(extraneous root)}$$

$$[SbCl_5] = (0.00669 - x) \, M = (0.00669 - 0.00549) \, M = 1.20 \times 10^{-3} \, M$$

$$\underline{?} \text{ g SbCl}_5 = 5.00 \text{ L} \times \frac{1.20 \times 10^{-3} \text{ mol}}{L} \times \frac{299 \text{ g}}{\text{mol}} = \boxed{1.79 \text{ g SbCl}_5}$$

Let us now solve the same problem *using K_P and partial pressures.*

(b) Plan (using K_P)

Calculate the initial partial pressure of $SbCl_5$, and write the reaction summary. Substitution of the representations of the equilibrium partial pressures into K_P gives their values.

(b) Solution (using K_P)

We calculate the initial *pressure* of $SbCl_5$ in atmospheres, using $PV = nRT$.

$$P_{SbCl_5} = \frac{nRT}{V} = \frac{\left[(10.0 \text{ g})\left(\frac{1 \text{ mol}}{299 \text{ g}}\right)\right]\left(0.0821 \frac{L \cdot atm}{mol \cdot K}\right)(721 \text{ K})}{5.00 \text{ L}} = 0.396 \text{ atm}$$

The *partial pressure* of each substance is proportional to the *number of moles* of that substance.

Clearly, $P_{SbCl_3} = 0$ and $P_{Cl_2} = 0$ because only $SbCl_5$ is present initially. We write the reaction summary in terms of partial pressures in atmospheres, because K_P refers to pressures in atmospheres.

Let $y =$ decrease in pressure (atm) of $SbCl_5$ due to reaction. In terms of partial pressures, the reaction summary is

	$SbCl_5$	$\rightleftharpoons$	$SbCl_3$	$+$	Cl_2
initial	0.396 atm		0		0
change due to rxn	$-y$ atm		$+y$ atm		$+y$ atm
equilibrium	$(0.396 - y)$ atm		y atm		y atm

If we did not know the value of K_P, we could calculate it from the known value of K_c, using the relationship $K_P = K_c(RT)^{\Delta n}$.

$$K_P = \frac{(P_{SbCl_3})(P_{Cl_2})}{P_{SbCl_5}} = 1.48 = \frac{(y)(y)}{0.396 - y}$$

$$0.586 - 1.48y = y^2 \qquad y^2 + 1.48y - 0.586 = 0$$

Solving by the quadratic formula gives

$$y = 0.325 \quad \text{and} \quad -1.80 \text{ (extraneous root)}$$

$$P_{SbCl_5} = (0.396 - y) = (0.396 - 0.325) = 0.071 \text{ atm}$$

We use the ideal gas law, $PV = nRT$, to calculate the number of moles of $SbCl_5$.

$$n = \frac{PV}{RT} = \frac{(0.071 \text{ atm})(5.00 \text{ L})}{\left(0.0821 \frac{L \cdot atm}{mol \cdot K}\right)(721 \text{ K})} = 0.0060 \text{ mol SbCl}_5$$

We see that within roundoff range the same result is obtained by both methods.

$$\underline{?} \text{ g SbCl}_5 = 0.0060 \text{ mol} \times \frac{299 \text{ g}}{\text{mol}} = \boxed{1.8 \text{ g SbCl}_5}$$

You should now work Exercises 75 and 78.

17-11 HETEROGENEOUS EQUILIBRIA

Thus far, we have considered only equilibria involving species in a single phase, that is, **homogeneous equilibria. Heterogeneous equilibria** involve species in more than one phase. Consider the following reversible reaction at 25°C.

$$2HgO(s) \rightleftharpoons 2Hg(\ell) + O_2(g)$$

When equilibrium is established for this system, a solid, a liquid, and a gas are present. Neither solids nor liquids are significantly affected by changes in pressure. The fundamental definition of the equilibrium constant in thermodynamics is in terms of the activities of the substances involved.

> For any pure solid or pure liquid, the activity is taken as 1 (Section 17-2), so terms for pure liquids and pure solids *do not* appear in the K expressions for heterogeneous equilibria.

Thus, for the reaction

$$2HgO(s) \rightleftharpoons 2Hg(\ell) + O_2(g) \qquad K_c = [O_2] \qquad \text{and} \qquad K_P = P_{O_2}$$

These equilibrium constant expressions indicate that equilibrium exists at a given temperature for *one and only one* concentration and one partial pressure of oxygen in contact with liquid mercury and solid mercury(II) oxide.

Charles Steele

A photograph of the reaction

$$2HgO(s) \rightleftharpoons 2Hg(\ell) + O_2(g)$$

The reaction is not at equilibrium here, because O_2 gas has been allowed to escape.

EXAMPLE 17-15 K_c and K_P for Heterogeneous Equilibrium

Write both K_c and K_P for the following reversible reactions.
(a) $2SO_2(g) + O_2(g) \rightleftharpoons 2SO_3(g)$
(b) $2NH_3(g) + H_2SO_4(\ell) \rightleftharpoons (NH_4)_2SO_4(s)$
(c) $S(s) + H_2SO_3(aq) \rightleftharpoons H_2S_2O_3(aq)$

Plan

We apply the definitions of K_c and K_P to each reaction.

Solution

(a) $K_c = \dfrac{[SO_3]^2}{[SO_2]^2[O_2]}$ $K_P = \dfrac{(P_{SO_3})^2}{(P_{SO_2})^2(P_{O_2})}$

(b) $K_c = \dfrac{1}{[NH_3]^2} = [NH_3]^{-2}$ $K_P = \dfrac{1}{(P_{NH_3})^2} = (P_{NH_3})^{-2}$

(c) $K_c = \dfrac{[H_2S_2O_3]}{[H_2SO_3]}$ K_P undefined; no gases involved

EXAMPLE 17-16   Heterogeneous Equilibria

The value of K_P is 27 for the thermal decomposition of potassium chlorate at a given high temperature. What is the partial pressure of oxygen in a closed container in which the following system is at equilibrium at the given temperature? (This can be a dangerous reaction.)

$$2KClO_3(s) \xrightarrow{\text{heat}} 2KCl(s) + 3O_2(g)$$

Plan

Because two solids, $KClO_3$ and KCl, and only one gas, O_2, are involved, we see that K_P involves only the partial pressure of O_2, that is, $K_P = (P_{O_2})^3$.

Solution

We are given

$$K_P = (P_{O_2})^3 = 27$$

Let x atm $= P_{O_2}$ at equilibrium. Then we have

$$(P_{O_2})^3 = 27 = x^3 \qquad \boxed{x = 3.0 \text{ atm}}$$

This tells us that the partial pressure of oxygen at equilibrium is 3.0 atm.

You should now work Exercises 80 and 91.

17-12 RELATIONSHIP BETWEEN ΔG^0_{rxn} AND THE EQUILIBRIUM CONSTANT

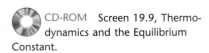
CD-ROM Screen 19.9, Thermo-dynamics and the Equilibrium Constant.

Let's consider in thermodynamic terms what may happen when two substances are mixed together at constant temperature and pressure. First, as a result of mixing, there is usually an increase in entropy due to the increase in disorder. If the two substances can react with each other, the chemical reaction begins, heat is released or absorbed, and the concentrations of the substances in the mixture change. An additional change in entropy, which depends on changes in the nature of the reactants and products, also begins to occur. The evolution or absorption of heat energy, the changes in entropy, and the changes in concentrations all continue until equilibrium is established. Equilibrium may be reached with large amounts of products formed, with virtually of all of the reactants remaining unchanged, or at *any* intermediate combination of concentrations.

The standard free energy change for a reaction is ΔG^0_{rxn}. This is the free energy change that would accompany *complete* conversion of *all* reactants initially present at standard conditions to *all* products at standard conditions—the standard reaction (Section 15-15). The free energy change for any other concentrations or pressures is ΔG_{rxn} (no superscript zero). The two quantities are related by the equation

> Thermodynamic standard states are (1) pure solids or pure liquids at 1 atm, (2) solutions of one-molar concentrations, and (3) gases at partial pressures of 1 atm.

$$\Delta G_{rxn} = \Delta G^0_{rxn} + RT \ln Q$$

R is the universal gas constant, T is the absolute temperature, and Q is the reaction quotient (Section 17-4). When a system is *at equilibrium*, $\Delta G_{rxn} = 0$ (Section 15-15) and $Q = K$ (Section 17-4). Recall that the reaction quotient may represent nonequilibrium concentrations (or partial pressures) of products and reactants. As reaction occurs, the free energy of the mixture and the concentrations change until at equilibrium $\Delta G_{rxn} = 0$, and the concentrations of reactants and products satisfy the equilibrium constant. At that point, Q becomes equal to K (Section 17-4). Then

$$0 = \Delta G^0_{rxn} + RT \ln K \qquad \text{(at equilibrium)}$$

Rearranging gives

$$\Delta G^0_{rxn} = -RT \ln K$$

> The energy units of R *must* match those of ΔG^0. We usually use
>
> $R = 8.314$ J/mol·K

This equation shows the relationship between the standard free energy change and the *thermodynamic equilibrium constant*.

For the following generalized reaction, the **thermodynamic equilibrium constant** is defined in terms of the activities of the species involved.

$$a\text{A} + b\text{B} \rightleftharpoons c\text{C} + d\text{D} \qquad K = \frac{(a_C)^c (a_D)^d}{(a_A)^a (a_B)^b}$$

where a_A is the activity of substance A, and so on. The mass action expression to which it is related involves concentration terms for species in solution and partial pressures for gases.

When the relationship $\Delta G^0_{rxn} = -RT \ln K$ is used with

1. all gaseous reactants and products, K represents K_P;
2. all solution reactants and products, K represents K_c;
3. a mixture of solution and gaseous reactants, K represents the thermodynamic equilibrium constant, and we do not make the distinction between K_P and K_c.

> We will encounter equilibrium calculations of solution and gaseous species in Chapter 21.

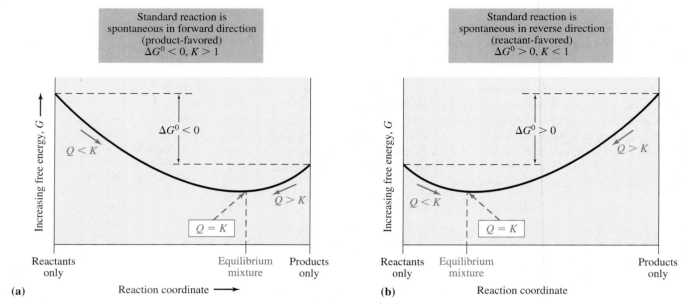

Figure 17-4 Variation in total free energy for a reversible reaction carried out at constant *T*. The *standard free energy change*, ΔG^0, represents the free energy change for the *standard reaction*—the *complete* conversion of reactants into products. In (a) this change is negative, indicating that the standard reaction is product-favored (spontaneous); the collection of just products would be more stable than the collection of just reactants. The mixture of reactants and products corresponding to the minimum of the curve is even more stable, however, and represents the equilibrium mixture. Because ΔG^0 is negative, $K > 1$, and the equilibrium mixture contains more products than reactants. At any point on the curve, comparing Q and K indicates the direction in which the reaction must proceed to approach equilibrium, that is, which way is "downhill" in free energy. The plot in (b) is for positive ΔG^0 (the standard reaction is reactant-favored). In this case $K < 1$, and the equilibrium mixture contains more reactants than products.

Figure 17-4 displays the relationships between free energy and equilibrium. The *left* end of each curve represents the total free energy of the reactants and the *right* end of each curve represents the total free energy of the products at standard state conditions. The difference between them is ΔG^0_{rxn}; like *K*, ΔG^0_{rxn} depends only on temperature and is a constant for any given reaction.

From the preceding equation that relates ΔG^0_{rxn} and *K*, we see that when ΔG^0_{rxn} is negative, $\ln K$ *must* be positive, and *K* is greater than 1. This tells us that products are favored over reactants at equilibrium. This case is illustrated in Figure 17-4a. When ΔG^0_{rxn} is positive, $\ln K$ *must* be negative, and *K* is less than 1. This tells us that reactants are favored over products at equilibrium (Figure 17-4b). In the *rare* case of a chemical reaction for which $\Delta G^0_{rxn} = 0$, then $K = 1$ and the numerator and the denominator must be equal in the equilibrium constant expression, (i.e., $[C]^c[D]^d \ldots = [A]^a[B]^b \ldots$). These relationships are summarized as follows.

ΔG^0_{rxn}	K	Product Formation
$\Delta G^0_{rxn} < 0$	$K > 1$	Products favored over reactants at equilibrium
$\Delta G^0_{rxn} = 0$	$K = 1$	At equilibrium when $[C]^c[D]^d \ldots = [A]^a[B]^b \ldots$ (very rare)
$\Delta G^0_{rxn} > 0$	$K < 1$	Reactants favored over products at equilibrium

The direction of approach to equilibrium and the actual free energy change (ΔG_{rxn}) are *not* constants. They vary with the conditions and the initial concentrations. If the initial concentrations correspond to $Q < K$, equilibrium is approached from left to right on the curves in Figure 17-4, and the forward reaction predominates. If $Q > K$, equilibrium is approached from right to left, and the reverse reaction predominates.

The magnitude of ΔG^0_{rxn} indicates the *extent* to which a chemical reaction occurs under standard state conditions, that is, how far the reaction goes toward the formation of products before equilibrium is reached. The more negative the ΔG^0_{rxn} value, the larger is the value of K and the more favorable is the formation of products. We think of some reactions as going to "completion." These generally have very negative ΔG^0_{rxn} values. The more positive the ΔG^0_{rxn} value, the smaller is the value of K and the less favorable is the formation of products.

EXAMPLE 17-17 K Versus ΔG^0_{rxn}

Use the data in Appendix K to calculate K_P for the following reaction at 25°C.

$$2N_2O(g) \rightleftharpoons 2N_2(g) + O_2(g)$$

Plan

The temperature is 25°C, so we evaluate ΔG^0_{rxn} for the reaction from ΔG^0_f values in Appendix K. The reaction involves only gases, so K is K_P. This means that $\Delta G^0_{rxn} = -RT \ln K_P$. We solve for K_P.

Solution

$$\Delta G^0_{rxn} = [2\,\Delta G^0_{fN_2(g)} + \Delta G^0_{fO_2(g)}] - [2\,\Delta G^0_{fN_2O(g)}]$$
$$= [2(0) + 0] - [2(104.2)] = -208.4 \text{ kJ/mol, or } -2.084 \times 10^5 \text{ J/mol}$$

This is a gas-phase reaction, so ΔG^0_{rxn} is related to K_P by

$$\Delta G^0_{rxn} = -RT \ln K_P$$

$$\ln K_P = \frac{\Delta G^0_{rxn}}{-RT} = \frac{-2.084 \times 10^5 \text{ J/mol}}{-(8.314 \text{ J/mol} \cdot \text{K})(298\text{K})} = 84.1$$

$$K_P = e^{84.1} = \boxed{3.3 \times 10^{36}}$$

Units cancel when we express ΔG^0 in joules per mole. We interpret this as meaning "per mole of reaction"— that is, for the number of moles of each substance shown in the balanced equation.

On some calculators, we evaluate e^x as follows: Enter the value of x, then press INV followed by ln x.

The very large value of K_P tells us that the equilibrium lies *very* far to the right. This reaction is so slow at 25°C, however, that very little N_2O decomposes to N_2 and O_2 at that temperature.

You should now work Exercise 84.

EXAMPLE 17-18 K Versus ΔG^0_{rxn}

In Examples 15-16 and 15-17 we evaluated ΔG^0_{rxn} for the following reaction at 25°C and found it to be +173.1 kJ/mol. Calculate K_P at 25°C for this reaction.

$$N_2(g) + O_2(g) \rightleftharpoons 2NO(g)$$

Plan

In this example we use $\Delta G^0_{rxn} = -RT \ln K_P$.

Solution

$$\Delta G^0_{rxn} = -RT \ln K_P$$

$$\ln K_P = \frac{\Delta G^0_{rxn}}{-RT} = \frac{1.731 \times 10^5 \text{ J/mol}}{-(8.314 \text{ J/mol} \cdot \text{K})(298 \text{ K})}$$

$$= -69.9$$

$$K_P = e^{-69.9} = \boxed{4.4 \times 10^{-31}}$$

 The energy units in ΔG, often given in kJ, must be converted to J to match those of R.

This very small number indicates that at equilibrium almost no N_2 and O_2 are converted to NO at 25°C. For all practical purposes, the reaction does not occur at 25°C.

You should now work Exercise 86.

A very important application of the relationships in this section is the use of measured K values to calculate ΔG^0_{rxn}.

EXAMPLE 17-19   K Versus ΔG^0_{rxn}

The equilibrium constant, K_P, for the following reaction is 5.04×10^{17} at 25°C. Calculate ΔG^0_{298} for the hydrogenation of ethylene to form ethane.

$$C_2H_4(g) + H_2(g) \rightleftharpoons C_2H_6(g)$$

Plan

We use the relationship between ΔG^0 and K_P.

Solution

$$\Delta G^0_{298} = -RT \ln K_P$$
$$= -(8.314 \, \text{J/mol} \cdot \text{K})(298 \, \text{K}) \ln (5.04 \times 10^{17})$$
$$= -1.01 \times 10^5 \, \text{J/mol}$$
$$= \boxed{-101 \, \text{kJ/mol}}$$

You should now work Exercise 88(a).

17-13   EVALUATION OF EQUILIBRIUM CONSTANTS AT DIFFERENT TEMPERATURES

Chemists have determined equilibrium constants for thousands of reactions. It would be an impossibly huge task to catalog such constants at every temperature of interest for each reaction. Fortunately, there is no need to do this. If we determine the equilibrium constant, K_{T_1}, for a reaction at one temperature, T_1, and also its ΔH^0, we can then *estimate* the equilibrium constant at a second temperature, T_2, using the **van't Hoff equation.**

Compare the form of this equation to the Arrhenius equation (Section 16-8) and to the Clausius–Clapeyron equation (Section 13-9).

$$\ln\left(\frac{K_{T_2}}{K_{T_1}}\right) = \frac{\Delta H^0}{R}\left(\frac{1}{T_1} - \frac{1}{T_2}\right)$$

Thus, if we know ΔH^0 for a reaction and K at a given temperature (say 298 K), we can use the van't Hoff equation to calculate the value of K at any other temperature.

EXAMPLE 17-20   Evaluation of K_P at Different Temperatures

We found in Example 17-18 that $K_P = 4.4 \times 10^{-31}$ at 25°C (298 K) for the following reaction. $\Delta H^0 = 180.5$ kJ/mol for this reaction. Evaluate K_P at 2400. K.

$$N_2(g) + O_2(g) \rightleftharpoons 2NO(g)$$

Plan

We are given K_P at one temperature, 25°C, and the value of ΔH^0. We are given the second temperature, 2400. K. These data allow us to evaluate the right side of the van't Hoff equation, which gives us $\ln(K_{T_2}/K_{T_1})$. Because we know K_{T_1}, we can find the value for K_{T_2}.

2400 K is a typical temperature inside the combustion chambers of automobile engines. Large quantities of N_2 and O_2 are present during gasoline combustion, because the gasoline is mixed with air.

Solution

Let $T_1 = 298$ K and $T_2 = 2400$. K. Then

$$\ln\left(\frac{K_{T_2}}{K_{T_1}}\right) = \frac{\Delta H^0}{R}\left(\frac{1}{T_1} - \frac{1}{T_2}\right)$$

Let us first evaluate the right side of the equation.

$$\ln\left(\frac{K_{T_2}}{K_{T_1}}\right) = \frac{1.805 \times 10^5 \text{ J/mol}}{8.314 \text{ J/mol} \cdot \text{K}}\left(\frac{1}{298 \text{ K}} - \frac{1}{2400. \text{ K}}\right) = 63.8$$

Now take the inverse logarithm of both sides.

$$\frac{K_{T_2}}{K_{T_1}} = e^{63.8} = 5.1 \times 10^{27}$$

Solving for K_{T_2} and substituting the known value of K_{T_1}, we obtain

$$K_{T_2} = (5.1 \times 10^{27})(K_{T_1}) = (5.1 \times 10^{27})(4.4 \times 10^{-31}) = \boxed{2.2 \times 10^{-3}} \text{ at 2400. K}$$

You should now work Exercise 88(b–f).

✓ **Problem-Solving Tip:** *Use the Correct K*

The K values that appear in the van't Hoff equation represent the *thermodynamic equilibrium constant* (see Section 17-12). For a gas-phase reaction (such as that in Example 17-20), K represents K_P; if the value of K_c were given, we would have to convert it to K_P (see Section 17-10) before using the van't Hoff equation.

In Example 17-20 we see that K_{T_2} (K_P at 2400. K) is quite small, which tells us that the equilibrium favors N_2 and O_2 rather than NO. Nevertheless, K_{T_2} is very much larger than K_{T_1}, which is 4.4×10^{-31}. At 2400. K, significantly more NO is present at equilibrium, relative to N_2 and O_2, than at 298 K. So automobiles emit small amounts of NO into the atmosphere, sufficient to cause severe air pollution problems. Catalytic converters (Section 16-9) are designed to catalyze the breakdown of NO into N_2 and O_2.

The K_P value could be converted to K_c using the relationship $K_c = K_P(RT)^{-\Delta n}$ (Section 17-10).

$$2NO(g) \rightleftharpoons N_2(g) + O_2(g)$$

This reaction is spontaneous. Catalysts do not shift the position of equilibrium. They favor neither consumption nor production of NO. They merely allow the system to reach equilibrium more rapidly. The time factor is very important because the NO stays in the automobile exhaust system for only a very short time.

Key Terms

Activity (of a component of an ideal mixture)   A dimensionless quantity whose magnitude is equal to molar concentration in an ideal solution, equal to partial pressure (in atmospheres) in an ideal gas mixture, and defined as 1 for pure solids or liquids.

Chemical equilibrium   A state of dynamic balance in which the rates of forward and reverse reactions are equal; there is no net change in concentrations of reactants or products while a system is at equilibrium.

Dynamic equilibrium   An equilibrium in which processes occur continuously, with no *net* change.

Equilibrium constant, K   A quantity that indicates the extent to which a reversible reaction occurs. K varies with temperature.

Heterogeneous equilibria   Equilibria involving species in more than one phase.

Homogeneous equilibria   Equilibria involving only species in a single phase, that is, all gases, all liquids, or all solids.

K_c   Equilibrium constant with amounts of reactants and products expressed as molar concentrations.

K_P   Equilibrium constant with amounts of reactants and products expressed as partial pressures.

LeChatelier's Principle   If a stress (change of conditions) is applied to a system at equilibrium, the system shifts in the direction that reduces the stress, to move toward a new state of equilibrium.

Mass action expression   For a reversible reaction,

$$a\text{A} + b\text{B} \rightleftharpoons c\text{C} + d\text{D}$$

the product of the molar concentrations of the products (species on the right), each raised to the power that corresponds to its coefficient in the balanced chemical equation, divided by the product of the concentrations of the reactants (species on the left), each raised to the power that corresponds to its coefficient in the balanced chemical equation. At equilibrium the mass action expression is equal to K; at other conditions, it is Q.

$$\frac{[\text{C}]^c[\text{D}]^d}{[\text{A}]^a[\text{B}]^b} = Q \text{ or, at equilibrium, } K_c$$

Reaction quotient, Q   The mass action expression under any set of conditions (not necessarily equilibrium); its magnitude relative to K determines the direction in which reaction must occur to establish equilibrium.

Reversible reactions   Reactions that do not go to completion and occur in both the forward and reverse directions.

van't Hoff equation   The relationship between ΔH^0 for a reaction and its equilibrium constants at two different temperatures.

Exercises

Basic Concepts

1. Define and illustrate the following terms: (a) reversible reaction, (b) static equilibrium, (c) equilibrium constant.
2. Equilibrium constants do not have units. Explain.
3. Distinguish between the terms "static equilibrium" and "dynamic equilibrium." Which kind does chemical equilibrium represent?
4. (a) Describe three examples of static equilibrium. (b) Describe three examples of dynamic equilibrium (besides chemical equilibrium).
5. Explain the significance of (a) a very large value of K, (b) a very small value of K, and (c) a value of K of about 1.
6. What can be said about the magnitude of the equilibrium constant in a reaction whose equilibrium lies far to the right? To the left?
7. What is the relationship between equilibrium and the rates of opposing processes?
8. What does the value of an equilibrium constant tell us about the time required for the reaction to reach equilibrium?
9. When giving the value of an equilibrium constant, it is necessary also to write the balanced chemical equation. Why? Give examples to illustrate your explanation.
10. (a) How is the equilibrium constant related to the forward and reverse rate constants? (b) Can the rate expressions for forward and reverse reactions be written from the balanced chemical equation? Explain. (c) Can the equilibrium constant expression be written from the balanced chemical equation? Explain.

11. (a) Sketch a set of curves similar to Figure 17-2 for concentration changes with time for a reaction

$$2\text{A(g)} + \text{B(g)} \rightleftharpoons \text{C(g)} + 2\text{D(g)}$$

assuming that K is much greater than 1. In each case, assume that A and B start at the same concentration and that no C or D are present.

(b) Repeat part (a) for the case that K is much less than 1.

12. (a) Sketch a set of curves similar to Figure 17-2 for concentration changes with time for a reaction

$$3\text{A(g)} + \text{B} \rightleftharpoons 2\text{C(g)}$$

assuming that K is much greater than 1. In each case, assume that A and B start at the same concentration and that no C is present.

(b) Repeat part (a) for the case that K is much less than 1.

13. At some temperature, the reaction

$$\text{N}_2\text{(g)} + 3\text{H}_2\text{(g)} \rightleftharpoons 2\text{NH}_3\text{(g)}$$

has an equilibrium constant K_c numerically equal to 1. State whether each of the following is true or false, and explain why.

(a) An equilibrium mixture must have the H_2 concentration three times that of N_2 and the NH_3 concentration twice that of H_2.

(b) An equilibrium mixture must have the H_2 concentration three times that of N_2.

(c) A mixture in which the H_2 concentration is three times that of N_2 *and* the NH_3 concentration is twice that of N_2 could be an equilibrium mixture.

(d) A mixture in which the concentration of each reactant and each product is 1 M is an equilibrium mixture.

(e) Any mixture in which the concentrations of all reactants and products are equal is an equilibrium mixture.

(f) An equilibrium mixture must have equal concentrations of all reactants and products.

14. Why do we omit concentrations of pure solids and pure liquids from equilibrium constant expressions?

15. Consider the following compounds, in the states indicated, as possible reactants or products in a chemical reaction. Which of these compounds would be omitted from the equilibrium constant expression? $H_2O(s)$, $H_2O(\ell)$, $H_2O(g)$, $HCl(g)$, $HCl(aq)$, $NaHCO_3(s)$, $CH_3OH(\ell)$, $Cl_2(aq)$, $N_2(g)$, $NH_3(\ell)$, $CO(g)$, and $Fe_2O_3(s)$.

16. Consider the following compounds, in the states indicated, as possible reactants or products in a chemical reaction. Which of these compounds would be omitted from the equilibrium constant expression? Explain. $CaCO_3(s)$, $H_2SO_4(\ell)$, $NaOH(s)$, $NaOH(aq)$, $O_2(g)$, $CH_3COOH(\ell)$, $HI(g)$, $HCl(aq)$, $I_2(s)$, $C(graphite)$, and $SO_3(g)$.

Equilibrium Constant Expression and Value of *K*

17. Write the expression for K_c for each of the following reactions:

(a) $CO_2(g) + H_2(g) \rightleftharpoons CO(g) + H_2O(g)$
(b) $2NO_2(g) \rightleftharpoons 2NO(g) + O_2(g)$
(c) $2CHCl_3(g) + 3H_2(g) \rightleftharpoons 2CH_4(g) + 3Cl_2(g)$
(d) $H_2(g) + I_2(g) \rightleftharpoons 2HI(g)$
(e) $2NOCl(g) \rightleftharpoons 2NO(g) + Cl_2(g)$

18. Write the expression for K_c for each of the following reactions:

(a) $2H_2O(g) + 2SO_2(g) \rightleftharpoons 2H_2S(g) + 3O_2(g)$
(b) $4NH_3(g) + 5O_2(g) \rightleftharpoons 4NO(g) + 6H_2O(g)$
(c) $PCl_3(g) + Cl_2(g) \rightleftharpoons PCl_5(g)$
(d) $N_2O_4(g) \rightleftharpoons 2NO_2(g)$
(e) $2SO_2(g) + O_2(g) \rightleftharpoons 2SO_3(g)$

19. Write the expression for K_c for each of the following reactions:

(a) $2CO(g) + O_2(g) \rightleftharpoons 2CO_2(g)$
(b) $SrCO_3(s) \rightleftharpoons SrO(s) + CO_2(g)$
(c) $2HBr(g) \rightleftharpoons H_2(g) + Br_2(\ell)$
(d) $P_4(g) + 3O_2(g) \rightleftharpoons P_4O_6(s)$
(e) $N_2(g) + O_2(g) \rightleftharpoons 2NO(g)$

20. Write the expression for K_c for each of the following reactions:

(a) $2H_2O_2(g) \rightleftharpoons 2H_2O(g) + O_2(g)$
(b) $2ZnS(s) + 3O_2(g) \rightleftharpoons 2ZnO(s) + 2SO_2(g)$
(c) $NH_3(g) + HCl(g) \rightleftharpoons NH_4Cl(s)$
(d) $NaF(s) + H_2SO_4(\ell) \rightleftharpoons NaHSO_4(s) + HF(g)$
(e) $2Cl_2(g) + 2H_2O(g) \rightleftharpoons 4HCl(g) + O_2(g)$

21. Write the expression for K_c for each of the following reactions:

(a) $TlCl_3(s) \rightleftharpoons TlCl(s) + Cl_2(g)$
(b) $CuCl_4{}^{2-}(aq) \rightleftharpoons Cu^{2+}(aq) + 4Cl^-(aq)$
(c) $3O_2(g) \rightleftharpoons 2O_3(g)$
(d) $4H_3O^+(aq) + 2Cl^-(aq) + MnO_2(s) \rightleftharpoons$
$$Mn^{2+}(aq) + 6H_2O(\ell) + Cl_2(aq)$$

22. On the basis of the equilibrium constant values, choose the reactions in which the *products* are favored.

(a) $NH_3(aq) + H_2O(\ell) \rightleftharpoons NH_4{}^+(aq) + OH^-(aq)$
$$K = 1.8 \times 10^{-5}$$
(b) $Au^+(aq) + 2CN^-(aq) \rightleftharpoons [Au(CN)_2]^-(aq)$
$$K = 2 \times 10^{38}$$
(c) $PbC_2O_4(s) \rightleftharpoons Pb^{2+}(aq) + C_2O_4{}^{2-}(aq)$ $K = 10^{-11}$
(d) $HS^-(aq) + H^+(aq) \rightleftharpoons H_2S(aq)$ $K = 1.0 \times 10^7$

23. On the basis of the equilibrium constant values, choose the reactions in which the *reactants* are favored.

(a) $H_2O(\ell) \rightleftharpoons H^+(aq) + OH^-(aq)$
$$K = 1.0 \times 10^{-14}$$
(b) $[AlF_6]^{3-}(aq) \rightleftharpoons Al^{3+}(aq) + 6F^-(aq)$
$$K = 2 \times 10^{-24}$$
(c) $Ca_3(PO_4)_2(s) \rightleftharpoons 3Ca^{2+}(aq) + 2PO_4{}^{3-}(aq)$
$$K = 1. \times 10^{-25}$$
(d) $2Fe^{3+}(aq) + 3S^{2-}(aq) \rightleftharpoons Fe_2S_3(s)$
$$K = 1. \times 10^{88}$$

Calculation of *K*

24. The reaction between nitrogen and oxygen to form $NO(g)$ is represented by the chemical equation

$$N_2(g) + O_2(g) \rightleftharpoons 2NO(g)$$

Equilibrium concentrations of the gases at 1500 K are 1.7×10^{-3} mol/L for O_2, 6.4×10^{-3} mol/L for N_2, and 1.1×10^{-5} mol/L for NO. Calculate the value of K_c at 1500 K from these data.

25. At elevated temperatures, BrF_5 establishes the following equilibrium.

$$2BrF_5(g) \rightleftharpoons Br_2(g) + 5F_2(g)$$

The equilibrium concentrations of the gases at 1500 K are 0.0064 mol/L for BrF_5, 0.0018 mol/L for Br_2, and 0.0090 mol/L for F_2. Calculate the value of K_c.

26. At some temperature the reaction

$$PCl_3(g) + Cl_2(g) \rightleftharpoons PCl_5(g)$$

is at equilibrium when the concentrations of PCl_3, Cl_2, and PCl_5 are 10, 9, and 12 mol/L, respectively. Calculate the value of K_c for this reaction at that temperature.

27. For the reaction

$$CO(g) + H_2O(g) \rightleftharpoons CO_2(g) + H_2(g)$$

the value of the equilibrium constant, K_c, is 1.845 at a given temperature. We place 0.500 mole CO and 0.500 mole H_2O in a 1.00-L container at this temperature, and allow the reaction to reach equilibrium. What will be the equilibrium concentrations of all substances present?

28. Given: $A(g) + B(g) \rightleftharpoons C(g) + 2D(g)$
One mole of A and one mole of B are placed in a 0.400-liter container. After equilibrium has been established, 0.20 mole of C is present in the container. Calculate the equilibrium constant, K_c, for the reaction.

29. Nitrogen reacts with hydrogen to form ammonia.

$$N_2(g) + 3H_2(g) \rightleftharpoons 2NH_3(g)$$

An equilibrium mixture at a given temperature is found to contain 0.31 mol/L N_2, 0.50 mol/L H_2, and 0.14 mol/L NH_3. Calculate the value of K_c at the given temperature.

30. For the following equation, $K_c = 7.9 \times 10^{11}$ at 500 K.

$$H_2(g) + Br_2(g) \rightleftharpoons 2HBr(g)$$

(a) $\frac{1}{2}H_2(g) + \frac{1}{2}Br_2(g) \rightleftharpoons HBr(g)$ $K_c = ?$
(b) $2HBr(g) \rightleftharpoons H_2(g) + Br_2(g)$ $K_c = ?$
(c) $4HBr(g) \rightleftharpoons 2H_2(g) + 2Br_2(g)$ $K_c = ?$

31. The equilibrium constant for the reaction

$$2SO_2 + O_2 \rightleftharpoons 2SO_3$$

is $K_c = 279$ at a given high temperature. What is the value of the equilibrium constant for each of the following reactions at this temperature?
(a) $2SO_3 \rightleftharpoons 2SO_2 + O_2$
(b) $SO_2 + \frac{1}{2}O_2 \rightleftharpoons SO_3$

32. A sealed tube initially contains 9.84×10^{-4} mol H_2 and 1.38×10^{-3} mol I_2. It is kept at 350°C until the reaction

$$H_2(g) + I_2(g) \rightleftharpoons 2HI(g)$$

comes to equilibrium. At equilibrium, 4.73×10^{-4} mol I_2 is present. Calculate (a) the numbers of moles of H_2 and HI present at equilibrium; (b) the equilibrium constant, K_c, for the reaction.

33. NO and O_2 are mixed in a container of fixed volume kept at 1000 K. Their initial concentrations are 0.0200 mol/L and 0.0300 mol/L, respectively. When the reaction

$$2NO(g) + O_2(g) \rightleftharpoons 2NO_2(g)$$

has come to equilibrium, the concentration of NO_2 is 2.2×10^{-3} mol/L. Calculate (a) the concentration of NO at equilibrium, (b) the concentration of O_2 at equilibrium, and (c) the equilibrium constant, K_c, for the reaction.

34. Antimony pentachloride decomposes in a gas-phase reaction at high temperatures.

$$SbCl_5(g) \rightleftharpoons SbCl_3(g) + Cl_2(g)$$

(a) At some temperature, an equilibrium mixture in a 5.00-L container is found to contain 6.91 g of $SbCl_5$, 16.45 g of $SbCl_3$, and 5.11 g of Cl_2. Evaluate K_c.
(b) If 15.0 grams of $SbCl_5$ is placed in the 5.00-liter container and allowed to establish equilibrium at the temperature in part (a), what will be the equilibrium concentrations of all species?

35. At standard temperature and pressure, the reaction indicated by the following equation has an equilibrium constant, K_c, equal to 0.021.

$$2HI(g) \rightleftharpoons H_2(g) + I_2(g)$$

Calculate the equilibrium constant, K_c, for the reverse equation.

36. The following reaction has an equilibrium constant, K_c, equal to 1538 at 1800°C.

$$2NO(g) + O_2(g) \rightleftharpoons 2NO_2(g)$$

Calculate the equilibrium constant, K_c, for the reverse equation.

The Reaction Quotient, Q

37. Define the reaction quotient, Q. Distinguish between Q and K.

38. Why is it useful to compare Q with K? What is the situation when (a) $Q = K$? (b) $Q < K$? (c) $Q > K$?

39. How does the form of the reaction quotient compare with that of the equilibrium constant? What is the difference between these two expressions?

40. If the reaction quotient is larger than the equilibrium constant, what will happen to the reaction? What will happen if $Q < K$?

41. $K_c = 19.9$ for the reaction

$$Cl_2(g) + F_2(g) \rightleftharpoons 2ClF(g)$$

What will happen in a reaction mixture originally containing $[Cl_2] = 0.5$ mol/L, $[F_2] = 0.2$ mol/L, and $[ClF] = 7.3$ mol/L?

42. The concentration equilibrium constant for the gas-phase reaction

$$H_2CO \rightleftharpoons H_2 + CO$$

has the numerical value 0.50 at a given temperature. A mixture of H_2CO, H_2, and CO is introduced into a flask at this temperature. After a short time, analysis of a small sample of the reaction mixture shows the concentrations to be $[H_2CO] = 0.50$ M, $[H_2] = 0.80$ M, and $[CO] = 0.25$ M. Classify each of the following statements about this reaction mixture as true or false.
(a) The reaction mixture is at equilibrium.
(b) The reaction mixture is not at equilibrium, but no further reaction will occur.
(c) The reaction mixture is not at equilibrium, but will move toward equilibrium by using up more H_2CO.

(d) The forward rate of this reaction is the same as the reverse rate.

43. The value of K_c at 25°C for

$$C(graphite) + CO_2(g) \rightleftharpoons 2CO(g)$$

is 3.7×10^{-23}. Describe what will happen if 3.0 mol of CO and 3.0 mol of CO_2 are mixed in a 1.0-L container with a suitable catalyst to make the reaction "go" at this temperature.

Uses of the Equilibrium Constant, K_c

44. For the reaction described by the equation

$$N_2(g) + C_2H_2(g) \rightleftharpoons 2HCN(g)$$

$K_c = 2.3 \times 10^{-4}$ at 300°C. What is the equilibrium concentration of hydrogen cyanide if the initial concentrations of N_2 and acetylene (C_2H_2) were 3.3 mol/L and 2.0 mol/L, respectively?

45. The equilibrium constant, K_c, for the reaction

$$Br_2(g) + F_2(g) \rightleftharpoons 2BrF(g)$$

is 55.3. What are the equilibrium concentrations of all these gases if the initial concentrations of bromine and fluorine were both 0.180 mol/L?

46. $K_c = 96.2$ at 400 K for the reaction

$$PCl_3(g) + Cl_2(g) \rightleftharpoons PCl_5(g)$$

What is the concentration of Cl_2 at equilibrium if the initial concentrations were 0.25 mol/L for PCl_3 and 6.0 mol/L for Cl_2?

47. $K_c = 5.85 \times 10^{-3}$ at 25°C for the reaction

$$N_2O_4(g) \rightleftharpoons 2NO_2(g)$$

Twenty (20.0) grams of N_2O_4 is confined in a 5.00-L flask at 25°C. Calculate (a) the number of moles of NO_2 present at equilibrium and (b) the percentage of the original N_2O_4 that is dissociated.

48. The reaction of iron and water vapor results in an equilibrium

$$3Fe(s) + 4H_2O(g) \rightleftharpoons Fe_3O_4(s) + 4H_2(g)$$

and an equilibrium constant, K_c, of 4.6 at 850°C. What is the concentration of hydrogen present at equilibrium if the reaction is initiated with 28 g of H_2O and excess Fe in a 10.0-liter container?

49. The reaction of iron and water vapor results in an equilibrium

$$3Fe(s) + 4H_2O(g) \rightleftharpoons Fe_3O_4(s) + 4H_2(g)$$

and an equilibrium constant, K_c, of 4.6 at 850°C. What is the concentration of water present at equilibrium if the reaction is initiated with 8.0 g of H_2 and excess iron oxide, Fe_3O_4, in a 16.0-liter container?

50. Carbon dioxide reacts with hot carbon in the form of graphite. The equilibrium constant, K_c, for the reaction is 10.0 at 850°C.

$$CO_2(g) + C(graphite) \rightleftharpoons 2CO(g)$$

If 25.0 g of carbon monoxide is placed in a 2.5-L reaction vessel and heated to 850°C, what is the mass of carbon dioxide at equilibrium?

51. Carbon dioxide reacts with hot carbon in the form of graphite. The equilibrium constant, K_c, for the reaction is 10.0 at 850°C.

$$CO_2(g) + C(graphite) \rightleftharpoons 2CO(g)$$

If 25 g of carbon dioxide and 55 grams of graphite are placed in a 2.5-L reaction vessel and heated to 850°C, what is the mass of carbon monoxide at equilibrium?

52. A 62.5-gram sample of HI was placed in a 1.50-L reaction vessel and allowed to come to equilibrium as illustrated in the following equation

$$2HI(g) \rightleftharpoons H_2(g) + I_2(g)$$

The equilibrium constant, K_c, is 0.830. Calculate the concentration of each species present at equilibrium.

Factors That Influence Equilibrium

53. State LeChatelier's Principle. Which factors have an effect on a system at equilibrium? How does the presence of a catalyst affect a system at chemical equilibrium? Explain your answer.

54. What will be the effect of increasing the total pressure on the equilibrium conditions for (a) a chemical equation that has more moles of gaseous products than gaseous reactants, (b) a chemical equation that has more moles of gaseous reactants than gaseous products, (c) a chemical equation that has the same number of moles of gaseous reactants and gaseous products, and (d) a chemical equation in which all reactants and products are pure solids, pure liquids, or in aqueous solution?

55. Suppose the following exothermic reaction is allowed to reach equilibrium.

$$A(g) + 3B(g) \rightleftharpoons 2C(g) + 3D(g)$$

Then we make each of the following changes, and allow the reaction to reestablish equilibrium. Tell whether the amount of B present at the new equilibrium will be (i) greater than, (ii) less than, or (iii) the same as the amount of B before the change was imposed.

(a) The temperature is increased while the volume is kept constant;

(b) more A is added;

(c) more C is added;

(d) a small amount of D is removed;

(e) the pressure is increased by decreasing the volume.

56. Suppose the following exothermic reaction is allowed to reach equilibrium.

$$A(g) + 3B(g) \rightleftharpoons 2C(g) + 3D(g)$$

Then we make each of the following changes and allow the reaction to reestablish equilibrium. For each change, tell whether the value of the equilibrium constant will be (i) greater than, (ii) less than, or (iii) the same as before the change was imposed.
 (a) The temperature is increased while the volume is kept constant;
 (b) more A is added;
 (c) more C is added;
 (d) a small amount of D is removed;
 (e) the pressure is increased by decreasing the volume.

57. What would be the effect on the equilibrium position of an equilibrium mixture of Br_2, F_2, and BrF_5 if the total pressure of the system were increased?

$$2BrF_5(g) \rightleftharpoons Br_2(g) + 5F_2(g)$$

58. What would be the effect on the equilibrium position of an equilibrium mixture of carbon, oxygen, and carbon monoxide if the total pressure of the system were increased?

$$2C(s) + O_2(g) \rightleftharpoons 2CO(g)$$

59. A weather indicator can be made with a hydrate of cobalt(II) chloride, which changes color as a result of the following reaction.

$$\underset{\text{pink}}{[Co(OH_2)_6]Cl_2(s)} \rightleftharpoons \underset{\text{blue}}{[Co(OH_2)_4]Cl_2(s)} + 2H_2O(g)$$

Does a blue color indicate "moist" or "dry" air? Explain.

60. Predict whether the equilibrium for the photosynthesis reaction described by the equation

$$6CO_2(g) + 6H_2O(\ell) \rightleftharpoons C_6H_{12}O_6(s) + 6O_2(g)$$
$$\Delta H^0 = 2801.69 \text{ kJ/mol}$$

would (i) shift to the right, (ii) shift to the left, or (iii) remain unchanged if (a) $[CO_2]$ were increased; (b) P_{O_2} were increased; (c) one half of the $C_6H_{12}O_6$ were removed; (d) the total pressure were decreased; (e) the temperature were increased; (f) a catalyst were added.

61. What would be the effect of decreasing the temperature on each of the following systems at equilibrium?
 (a) $H_2(g) + I_2(g) \rightleftharpoons 2HI(g)$; $\Delta H^0 = -9.45$ kJ/mol
 (b) $PCl_5(g) \rightleftharpoons PCl_3(g) + Cl_2(g)$; $\Delta H^0 = 92.5$ kJ/mol
 (c) $2SO_2(g) + O_2(g) \rightleftharpoons 2SO_3(g)$; $\Delta H^0 = -198$ kJ/mol
 (d) $2NOCl(g) \rightleftharpoons 2NO(g) + Cl_2(g)$; $\Delta H^0 = 75$ kJ/mol
 (e) $C(s) + H_2O(g) \rightleftharpoons CO(g) + H_2(g)$;
      $\Delta H^0 = 131$ kJ/mol

62. What would be the effect of decreasing the pressure by increasing the volume on each of the following systems at equilibrium?
 (a) $2CO(g) + O_2(g) \rightleftharpoons 2CO_2(g)$
 (b) $2NO(g) \rightleftharpoons N_2(g) + O_2(g)$

 (c) $N_2O_4(g) \rightleftharpoons 2NO_2(g)$
 (d) $Ni(s) + 4CO(g) \rightleftharpoons Ni(CO)_4(g)$
 (e) $N_2(g) + 3H_2(g) \rightleftharpoons 2NH_3(g)$

63. The value of K_c is 0.020 at 2870°C for the reaction

$$N_2(g) + O_2(g) \rightleftharpoons 2NO(g)$$

There are 0.800 mole of N_2, 0.500 mole of O_2, and 0.400 mole of NO in a 1.00-liter container at 2870°C. Is the system at equilibrium or must the forward or reverse reaction occur to a greater extent to bring the system to equilibrium?

64. Given: $A(g) + B(g) \rightleftharpoons C(g) + D(g)$
 (a) At equilibrium a 1.00-liter container was found to contain 1.60 moles of C, 1.60 moles of D, 0.40 mole of A, and 0.40 mole of B. Calculate the equilibrium constant for this reaction.
 (b) If 0.20 mole of B and 0.20 mole of C are added to this system, what will the new *equilibrium* concentration of A be?

65. Given: $A(g) + B(g) \rightleftharpoons C(g) + D(g)$
When one mole of A and one mole of B are mixed and allowed to reach equilibrium at room temperature, the mixture is found to contain $\frac{2}{3}$ mole of C.
 (a) Calculate the equilibrium constant.
 (b) If two moles of A were mixed with two moles of B and allowed to reach equilibrium, how many moles of C would be present at equilibrium?

66. Given: $A(g) \rightleftharpoons B(g) + C(g)$
 (a) When the system is at equilibrium at 200°C, the concentrations are found to be: $[A] = 0.30 \ M$, $[B] = [C] = 0.25 \ M$. Calculate K_c.
 (b) If the volume of the container in which the system is at equilibrium is suddenly doubled at 200°C, what will the new equilibrium concentrations be?
 (c) Refer back to part (a). If the volume of the container is suddenly halved at 200°C, what will the new equilibrium concentrations be?

*67. The equilibrium constant, K_c for the dissociation of phosphorus pentachloride is 9.3×10^{-2} at 252°C. How many moles and grams of PCl_5 must be added to a 2.0-liter flask to obtain a Cl_2 concentration of 0.17 M?

$$PCl_5(g) \rightleftharpoons PCl_3(g) + Cl_2(g)$$

68. At 25°C, K_c is 5.84×10^{-3} for the dissociation of dinitrogen tetroxide to nitrogen dioxide.

$$N_2O_4(g) \rightleftharpoons 2NO_2(g)$$

 (a) Calculate the equilibrium concentrations of both gases when 3.50 grams of N_2O_4 is placed in a 2.00-liter flask at 25°C.
 (b) What will be the new equilibrium concentrations if the volume of the system is suddenly increased to 3.00 liters at 25°C?
 (c) What will be the new equilibrium concentrations if the volume is decreased to 1.00 liter at 25°C?

K in Terms of Partial Pressures

69. Write the K_P expression for each reaction in Exercise 17.

70. Under what conditions are K_c and K_P for a reaction numerically equal? Are K_c and K_P numerically equal for any of the reactions in Exercises 17 and 18? Which ones?

71. 0.0100 mol of NH_4Cl and 0.0100 mol of NH_3 are placed in a closed 2.00-L container and heated to 603 K. At this temperature, all the NH_4Cl vaporizes. When the reaction

$$NH_4Cl(g) \rightleftharpoons NH_3(g) + HCl(g)$$

has come to equilibrium, 5.8×10^{-3} mol of HCl is present. Calculate (a) K_c and (b) K_P for this reaction at 603 K.

72. CO_2 is passed over graphite at 500 K. The emerging gas stream contains 4.0×10^{-3} mol percent CO. The total pressure is 1.00 atm. Assume that equilibrium is attained. Find K_P for the reaction

$$C(graphite) + CO_2(g) \rightleftharpoons 2CO(g)$$

73. At 425°C, the equilibrium partial pressures of H_2, I_2, and HI are 0.06443 atm, 0.06540 atm, and 0.4821 atm, respectively. Calculate K_P for the following reaction at this temperature.

$$2HI(g) \rightleftharpoons H_2(g) + I_2(g)$$

74. The equilibrium constant, K_P, for the reaction indicated by the following equation is 0.715 at 47°C.

$$N_2O_4(g) \rightleftharpoons 2NO_2(g)$$

Calculate the partial pressures of N_2O_4 and NO_2 in an experiment in which 2.5 moles of N_2O_4 is placed in a 5.0-L flask and allowed to establish equilibrium at 47°C.

75. The equilibrium constant, K_P, is 1.92 at 252°C for the decomposition reaction of phosphorus pentachloride indicated in the following equation.

$$PCl_5(g) \rightleftharpoons PCl_3(g) + Cl_2(g)$$

Calculate the partial pressures of all species present after 6.0 moles of PCl_5 is placed in an evacuated 4.0-liter container and equilibrium is reached at 252°C.

76. The following equilibrium partial pressures were measured at 750°C: $P_{H_2} = 0.387$ atm, $P_{CO_2} = 0.152$ atm, $P_{CO} = 0.180$ atm, and $P_{H_2O} = 0.252$ atm. What is the value of the equilibrium constant, K_P, for the reaction?

$$H_2 + CO_2 \rightleftharpoons CO + H_2O$$

77. For the reaction

$$H_2(g) + Cl_2(g) \rightleftharpoons 2HCl(g)$$

$K_c = 193$ at 2500 K. What is the value of K_P?

78. For the reaction

$$Br_2(g) \rightleftharpoons 2Br(g)$$

$K_P = 2550$ at 4000 K. What is the value of K_c?

79. A stream of gas containing H_2 at an initial partial pressure of 0.200 atm is passed through a tube in which CuO is kept at 500 K. The reaction

$$CuO(s) + H_2(g) \rightleftharpoons Cu(s) + H_2O(g)$$

comes to equilibrium. For this reaction, $K_P = 1.6 \times 10^9$. What is the partial pressure of H_2 in the gas leaving the tube? Assume that the total pressure of the stream is unchanged.

Relationships Among K, ΔG⁰, ΔH⁰, and T

80. In the distant future, when hydrogen may be cheaper than coal, steel mills may make iron by the reaction

$$Fe_2O_3(s) + 3H_2(g) \rightleftharpoons 2Fe(s) + 3H_2O(g)$$

For this reaction, $\Delta H = 96$ kJ/mol and $K_c = 8.11$ at 1000 K. (a) What percentage of the H_2 remains unreacted after the reaction has come to equilibrium at 1000 K? (b) Is this percentage greater or less if the temperature is decreased to below 1000 K?

81. What kind of equilibrium constant can be calculated from a ΔG^0 value for a reaction involving only gases?

82. What must be true of the value of ΔG^0 for a reaction if (a) $K \gg 1$; (b) $K = 1$; (c) $K \ll 1$?

83. A mixture of 3.00 mol of Cl_2 and 3.00 mol of CO is enclosed in a 5.00-L flask at 600°C. At equilibrium, 3.3% of the Cl_2 has been consumed.

$$CO(g) + Cl_2(g) \rightleftharpoons COCl_2(g)$$

(a) Calculate K_c for the reaction at 600°C.
(b) Calculate ΔG^0 for the reaction at this temperature.

84. (a) Use the tabulated thermodynamic values of ΔH_f^0 and S^0 to calculate the value of K_P at 25°C for the gas-phase reaction

$$CO + H_2O \rightleftharpoons CO_2 + H_2$$

(b) Calculate the value of K_P for this reaction at 200°C, by the same method as in part (a).
(c) Repeat the calculation of part (a), using tabulated values of ΔG_f^0.

85. The equilibrium constant K_c of the reaction

$$H_2(g) + Br_2(g) \rightleftharpoons 2HBr(g)$$

is 1.6×10^5 at 1297 K and 3.5×10^4 at 1495 K. (a) Is ΔH^0 for this reaction positive or negative? (b) Find K_c for the reaction

$$\tfrac{1}{2}H_2(g) + \tfrac{1}{2}Br_2(g) \rightleftharpoons HBr(g)$$

at 1297 K. (c) Pure HBr is placed in a container of constant volume and heated to 1297 K. What percentage of the HBr is decomposed to H_2 and Br_2 at equilibrium?

86. The air pollutant sulfur dioxide can be partially removed from stack gases in industrial processes and converted to sulfur trioxide, the acid anhydride of commercially important sulfuric acid. Write the equation for the reaction, using the smallest whole-number coefficients. Calculate the value of the equilibrium constant for this reaction at 25°C, from values of ΔG_f^0 in Appendix K.

87. The value of ΔH^0 for the reaction in Exercise 86 is -197.6 kJ/mol. (a) Predict qualitatively (i.e., without calculation) whether the value of K_P for this reaction at 500°C would be greater than, the same as, or less than the value at room temperature (25°C). (b) Now calculate the value of K_P at 500°C.

*88. The following is an example of an alkylation reaction that is important in the production of isooctane (2,2,4-trimethylpentane) from two components of crude oil: isobutane and isobutene. Isooctane is an antiknock additive for gasoline. The thermodynamic equilibrium constant, K, for this reaction at 25°C is 4.3×10^6, and ΔH^0 is -78.58 kJ/mol.
   (a) Calculate ΔG^0 at 25°C.
   (b) Calculate K at 800°C.
   (c) Calculate ΔG^0 at 800°C.
   (d) How does the spontaneity of the forward reaction at 800°C compare with that at 25°C?
   (e) Why do you think the reaction mixture is heated in the industrial preparation of isooctane?
   (f) What is the purpose of the catalyst? Does it affect the forward reaction more than the reverse reaction?

$$CH_3-\underset{\underset{CH_3}{|}}{\overset{\overset{CH_3}{|}}{C}}-H + CH_3-\underset{}{\overset{\overset{CH_3}{|}}{C}}=CH_2 \xrightarrow[\text{catalyst}]{\text{heat}}$$

   isobutane   isobutene

$$CH_3-\underset{\underset{CH_3}{|}}{\overset{\overset{CH_3}{|}}{C}}-CH_2-\underset{\underset{H}{|}}{\overset{\overset{CH_3}{|}}{C}}-CH_3$$

   isooctane

89. At sufficiently high temperatures, chlorine gas dissociates, according to

$$Cl_2(g) \rightleftharpoons 2Cl(g)$$

At 800°C, K_P for this reaction is 5.63×10^{-7}. (a) A sample originally contained Cl_2 at 1 atm and 800°C. Calculate the percentage dissociation of Cl_2 when this reaction has reached equilibrium. (b) At what temperature would Cl_2 (originally at 1 atm pressure) be 1% dissociated into Cl atoms?

Mixed Exercises

90. At 700°C, K_P is 1.50 for the reaction

$$C(s) + CO_2(g) \rightleftharpoons 2CO(g)$$

Suppose the total gas pressure at equilibrium is 1.00 atm. What are the partial pressures of CO and CO_2?

91. At -10°C, the solid compound $Cl_2(H_2O)_8$ is in equilibrium with gaseous chlorine, water vapor, and ice. The partial pressures of the two gases in equilibrium with a mixture of $Cl_2(H_2O)_8$ and ice are 0.20 atm for Cl_2 and 0.00262 atm for water vapor. Find the equilibrium constant K_P for each of these reactions.
   (a) $Cl_2(H_2O)_8(s) \rightleftharpoons Cl_2(g) + 8H_2O(g)$
   (b) $Cl_2(H_2O)_8(s) \rightleftharpoons Cl_2(g) + 8H_2O(s)$
   Why are your two answers so different?

92. A flask contains $NH_4Cl(s)$ in equilibrium with its decomposition products.

$$NH_4Cl(s) \rightleftharpoons NH_3(g) + HCl(g)$$

For this reaction, $\Delta H = 176$ kJ/mol. How is the mass of NH_3 in the flask affected by each of the following disturbances? (a) The temperature is decreased. (b) NH_3 is added. (c) HCl is added. (d) NH_4Cl is added, with no appreciable change in the gas volume. (e) A large amount of NH_4Cl is added, decreasing the volume available to the gases.

93. The equilibrium constant for the reaction

$$H_2(g) + Br_2(\ell) \rightleftharpoons 2HBr(g)$$

is $K_P = 4.5 \times 10^{18}$ at 25°C. The vapor pressure of liquid Br_2 at this temperature is 0.28 atm. (a) Find K_P at 25°C for the reaction

$$H_2(g) + Br_2(g) \rightleftharpoons 2HBr(g)$$

(b) How will the equilibrium in part (a) be shifted by an increase in the volume of the container if (1) liquid Br_2 is absent; (2) liquid Br_2 is present? Explain why the effect is different in these two cases.

94. Given that K_P is 4.6×10^{-14} at 25°C for the reaction

$$2Cl_2(g) + 2H_2O(g) \rightleftharpoons 4HCl(g) + O_2(g)$$
$$\Delta H^0 = +115 \text{ kJ/mol}$$

Calculate K_P and K_c for the reaction at 400°C and at 800°C.

95. $K_c = 19.9$ for the reaction

$$Cl_2(g) + F_2(g) \rightleftharpoons 2ClF(g)$$

What will happen in a reaction mixture originally containing $[Cl_2] = 0.200$ mol/L, $[F_2] = 0.300$ mol/L, and $[ClF] = 0.950$ mol/L?

96. A mixture of CO, H_2, CH_4, and H_2O is kept at 1133 K until the reaction

$$CO(g) + 3H_2(g) \rightleftharpoons CH_4(g) + H_2O(g)$$

has come to equilibrium. The volume of the container is 0.100 L. The equilibrium mixture contains 1.21×10^{-4} mol CO, 2.47×10^{-4} mol H_2, 1.21×10^{-4} mol CH_4, and 5.63×10^{-8} mol H_2O. Calculate K_P for this reaction at 1133 K.

97. What would the pressure of hydrogen be at equilibrium when $P_{WCl_6} = 0.012$ atm and $P_{HCl} = 0.10$ atm?

$$WCl_6(g) + 3H_2(g) \rightleftharpoons W(s) + 6HCl(g)$$
$$K_P = 1.37 \times 10^{21} \text{ at } 900 \text{ K.}$$

CONCEPTUAL EXERCISES

98. When the beam of the triple beam balance stops swinging has the balance reached a dynamic or a static equilibrium? Explain.

99. Theoretically, in closed systems all chemical reactions are reversible, and can be characterized with equilibrium constants. Students attempting to correlate the information from all the chapters of a chemistry text may wonder if the reactions described in the chemical equilibrium chapters are somehow different than those used in connection with stoichiometry, gas laws, etc. Speculate about the selection of reactions that the authors have used as illustrations in the various sections of this text.

100. What is the relationship between the forward and reverse reactions in an equilibrium?

101. The term "equilibrium" brings to mind the word "equal." What is the relationship between the two terms?

102. The masses of participants in a chemical equilibrium are not the same on both sides of the reaction. Does the equilibrium concept violate the Law of Conservation of Matter? Explain.

103. A sample of benzoic acid, a solid carbon-containing acid, is in equilibrium with an aqueous solution of benzoic acid. A tiny quantity of D_2O, water containing the isotope ^2H, deuterium, is added to the solution. The solution is allowed to stand at constant temperature for several hours, after which some of the solid benzoic acid is removed and analyzed. The benzoic acid is found to contain a tiny quantity of deuterium, D, and the formula of the deuterium containing molecules is C_6H_5COOD. Explain how this can happen.

104. Imagine yourself the size of atoms and molecules inside a beaker containing the following equilibrium mixture with a K greater than 1.

$$[Co(OH_2)_6]^{2+}(aq) + 4Cl^-(aq) \rightleftharpoons$$
$$\text{pink}$$
$$CoCl_4{}^{2-}(aq) + 6H_2O(\ell)$$
$$\text{blue}$$

Write a brief description of what you observe around you before and after additional water is added to the mixture.

BUILDING YOUR KNOWLEDGE

105. Hemoglobin, Hb, has four Fe atoms per molecule that, on the average, pick up roughly three molecules of O_2.

$$Hb(aq) + 3O_2(g) \rightleftharpoons Hb(O_2)_3(aq)$$

Discuss high altitude sickness in terms of this equilibrium.

Astronauts incur the risk of high altitude sickness.

106. At room temperature, the equilibrium constant for the reaction

$$2SO_2 + O_2 \rightleftharpoons 2SO_3$$

is 6.98×10^{24}. Calculate ΔG^0_{rxn} and $\Delta G^0_f(SO_3)$, given the additional information that $\Delta G^0_f(SO_2) = -300.194$ kJ/mol. Check your answer by looking up $\Delta G^0_f(SO_3)$ in Appendix K.

107. At 25°C, 550.0 g of deuterium oxide, D_2O (20.0 g/mol; density 1.10 g/mL), and 498.5 g of H_2O (18.0 g/mol; density 0.997 g/mL) are mixed. The volumes are additive. 47.0% of the H_2O reacts to form HDO. Calculate K_c at 25°C for the reaction

$$H_2O + D_2O \rightleftharpoons 2HDO$$

108. At its normal boiling point of 100°C, the heat of vaporization of water is 40.66 kJ/mol. What is the equilibrium vapor pressure of water at 65°C?

109. Use the data in the preceding exercise to calculate the temperature at which the vapor pressure of water is 1.25 atm.

BEYOND THE TEXTBOOK

Go to the textbook website at

http://www.brookscole.com/chemistry/whitten

for additional activities and exercises based on the General Chemistry Interactive CD-ROM, the World Wide Web, and library resources.

InfoTrac College Edition

For additional readings, go to InfoTrac College Edition, your online research library at:

http://infotrac.thomsonlearning.com

18 Ionic Equilibria I: Acids and Bases

Many foods contain weak acids. Citrus fruits contain citric acid and ascorbic acid (vitamin C).

Charles D. Winters

OBJECTIVES

After you have studied this chapter, you should be able to

- Identify strong electrolytes and calculate concentrations of their ions
- Understand the autoionization of water
- Understand the pH and pOH scales
- Use ionization constants for weak monoprotic acids and bases
- Discuss the concepts of solvolysis and hydrolysis
- Describe how polyprotic acids ionize in steps and how to calculate concentrations of all species in solutions of polyprotic acids
- Apply acid–base concepts to salts of strong acids and weak bases
- Apply acid–base equilibrium concepts to salts of strong bases and weak acids
- Apply acid–base equilibrium concepts to salts of weak bases and strong acids
- Apply acid–base equilibrium concepts to salts of weak bases and weak acids
- Apply acid–base equilibrium concepts to salts of small, highly charged cations

A queous solutions are very important. Nearly three fourths of the earth's surface is covered with water. Enormous numbers of chemical reactions occur in the oceans and smaller bodies of water. Plant and animal fluids are mostly water. Life processes (chemical reactions) of all plants and animals occur in aqueous solutions or in contact with water. Before we were born, all of us developed in sacs filled with aqueous solutions, which protected and helped nurture us until we had developed to the point that we could live in the atmosphere.

18-1 A REVIEW OF STRONG ELECTROLYTES

In previous discussions we have seen that water-soluble compounds may be classified as either electrolytes or nonelectrolytes. **Electrolytes** are compounds that ionize (or dissociate

 CD-ROM Screens 5.2, Solutions, and 17.5, Strong Acids and Bases.

TABLE 18-1	*Some Strong Acids and Strong Bases*

Strong Acids	
HCl	HNO_3
HBr	$HClO_4$
HI	$HClO_3$
	H_2SO_4

Strong Bases	
LiOH	
NaOH	
KOH	$Ca(OH)_2$
RbOH	$Sr(OH)_2$
CsOH	$Ba(OH)_2$

into their constituent ions) to produce aqueous solutions that conduct an electric current. **Nonelectrolytes** exist as molecules in aqueous solution, and such solutions do not conduct an electric current.

Strong electrolytes are ionized or dissociated completely, or very nearly completely, in dilute aqueous solutions. Strong electrolytes include strong acids, strong bases, and most soluble salts. You should review the discussions of these substances in Sections 4-2 and 10-8. Some strong acids and strong bases are listed again in Table 18-1. See Section 4-2, Part 5, for the solubility guidelines for ionic compounds.

Concentrations of ions in aqueous solutions of strong electrolytes can be calculated directly from the molarity of the strong electrolyte, as the following example illustrates.

EXAMPLE 18-1   *Calculation of Concentrations of Ions*

Calculate the molar concentrations of Ba^{2+} and OH^- ions in 0.030 M barium hydroxide.

Plan

Write the equation for the dissociation of $Ba(OH)_2$, and construct the reaction summary. $Ba(OH)_2$ is a strong base that is completely dissociated.

Solution

From the equation for the dissociation of barium hydroxide, we see that *one* mole of $Ba(OH)_2$ produces *one* mole of Ba^{2+} ions and *two* moles of OH^- ions.

	$Ba(OH)_2(s) \longrightarrow$	$Ba^{2+}(aq)$ +	$2OH^-(aq)$
(*strong base*)			
initial	0.030 M		
change due to rxn	$-0.030\ M$	$+0.030\ M$	$+2(0.030)\ M$
final	0 M	0.030 M	0.060 M

$$[Ba^{2+}] = 0.030\ M \quad \text{and} \quad [OH^-] = 0.060\ M$$

You should now work Exercises 4 and 6.

Recall that we use a single arrow ($\rightarrow$) to indicate that a reaction goes to completion, or nearly to completion, in the indicated direction.

Recall that we use a double arrow ($\rightleftharpoons$) to indicate that the reaction can go in either direction to reach equilibrium.

CD-ROM   Screen 17.3, The Acid–Base Properties of Water.

18-2   THE AUTOIONIZATION OF WATER

Careful experiments on its electrical conductivity have shown that pure water ionizes to a very slight extent.

$$H_2O(\ell) + H_2O(\ell) \rightleftharpoons H_3O^+(aq) + OH^-(aq)$$

Because the H_2O is pure, its activity is 1, so we do not include its concentration in the equilibrium constant expression. This equilibrium constant is known as the **ion product for water** and is usually represented as K_w.

$$K_w = [H_3O^+][OH^-]$$

The formation of an H_3O^+ ion by the ionization of water is always accompanied by the formation of an OH^- ion. Thus, in *pure* water the concentration of H_3O^+ is *always* equal to the concentration of OH^-. Careful measurements show that, in pure water at 25°C,

$$[H_3O^+] = [OH^-] = 1.0 \times 10^{-7}\ mol/L$$

Substituting these concentrations into the K_w expression gives

$$K_w = [H_3O^+][OH^-] = (1.0 \times 10^{-7})(1.0 \times 10^{-7})$$
$$= 1.0 \times 10^{-14} \quad \text{(at 25°C)}$$

Although the expression $K_w = [H_3O^+][OH^-] = 1.0 \times 10^{-14}$ was obtained for pure water, *it is also valid for dilute aqueous solutions at 25°C.* This is one of the most useful relationships chemists have discovered. It gives a simple relationship between H_3O^+ and OH^- concentrations in *all* dilute aqueous solutions.

The *value* of K_w is different at different temperatures (Table 18-2), but the *relationship* $K_w = [H_3O^+][OH^-]$ is still valid.

In this text, we shall assume a temperature of 25°C for all calculations involving aqueous solutions unless we specify another temperature.

Solutions in which the concentration of solute is less than about 1 mol/L are usually called dilute solutions.

TABLE 18-2	K_w *at Some Temperatures*
Temperature (°C)	K_w
0	1.1×10^{-15}
10	2.9×10^{-15}
25	1.0×10^{-14}
37*	2.4×10^{-14}
45	4.0×10^{-14}
60	9.6×10^{-14}

Normal human body temperature.

EXAMPLE 18-2 *Calculation of Ion Concentrations*

Calculate the concentrations of H_3O^+ and OH^- ions in a 0.050 *M* HNO_3 solution.

Plan

Write the equation for the ionization of HNO_3, a strong acid, and construct the reaction summary, which gives the concentrations of H_3O^+ (and NO_3^-) ions directly. Then use the relationship $K_w = [H_3O^+][OH^-] = 1.0 \times 10^{-14}$ to find the concentration of OH^- ions.

Solution

The reaction summary for the ionization of HNO_3, a strong acid, is

	HNO_3	$+ H_2O \longrightarrow$	H_3O^+	$+ NO_3^-$
(*strong acid*)				
initial	0.050 *M*		≈ 0 *M*	0 *M*
change due to rxn	-0.050 *M*		$+0.050$ *M*	$+0.050$ *M*
at equil	0 *M*		0.050 *M*	0.050 *M*

$$[H_3O^+] = [NO_3^-] = 0.050 \ M$$

The $[OH^-]$ is determined from the equation for the autoionization of water and its K_w.

	$2H_2O \rightleftharpoons$	$H_3O^+(aq)$	$+ OH^-$
initial		0.050 *M*	
change due to rxn	$-2x$ *M*	$+x$ *M*	$+x$ *M*
at equil		$(0.050 + x)$ *M*	x *M*

$$K_w = [H_3O^+][OH^-]$$
$$1.0 \times 10^{-14} = (0.050 + x)(x)$$

For the remainder of this textbook we will be working nearly exclusively with aqueous solutions, so we will often omit the "(aq)" and "(ℓ)."

Because the product $(0.050 + x)(x)$ is a very small number, we know that x must be very small. Thus, it will not matter whether we add x to 0.050; we can assume that $(0.050 + x) \approx 0.050$. We substitute this approximation into the equation and solve.

Recall that

$$[OH^-]_{\text{from } H_2O} = [H_3O^+]_{\text{from } H_2O}$$

in *all* aqueous solutions. So we know that $[H_3O^+]_{\text{from } H_2O}$ must also be 2.0×10^{-13} *M*. The 2.0×10^{-13} is only negligible when it is added to or subtracted from a number that is at least two orders of magnitude larger.

$$1.0 \times 10^{-14} = (0.050)(x) \quad \text{or} \quad x = \frac{1.0 \times 10^{-14}}{0.050} = 2.0 \times 10^{-13} \, M = [OH^-]$$

We see that the assumption that x is much smaller than 0.050 was a good one.

You should now work Exercise 14.

In solving Example 18-2 we assumed that *all* of the H_3O^+ (0.050 *M*) came from the ionization of HNO_3 and neglected the H_3O^+ formed by the ionization of H_2O. The ionization of H_2O produces only 2.0×10^{-13} *M* H_3O^+ and 2.0×10^{-13} *M* OH^- *in this solution*. Thus, we were justified in assuming that the $[H_3O^+]$ is derived solely from the strong acid. An easier way to carry out the calculation to find the $[OH^-]$ concentration is to write directly

$$K_w = [H_3O^+][OH^-] = 1.0 \times 10^{-14} \quad \text{or} \quad [OH^-] = \frac{1.0 \times 10^{-14}}{[H_3O^+]}$$

Then we substitute to obtain

$$[OH^-] = \frac{1.0 \times 10^{-14}}{0.050} = 2.0 \times 10^{-13} \, M$$

From now on, we shall use this more direct approach for such calculations.

When nitric acid is added to water, large numbers of H_3O^+ ions are produced. The large increase in $[H_3O^+]$ shifts the water equilibrium far to the left (LeChatelier's Principle), and the $[OH^-]$ decreases.

$$H_2O(\ell) + H_2O(\ell) \rightleftharpoons H_3O^+(aq) + OH^-(aq)$$

In acidic solutions the H_3O^+ concentration is always greater than the OH^- concentration. We should not conclude that acidic solutions contain no OH^- ions. Rather, the $[OH^-]$ is always less than 1.0×10^{-7} *M* in such solutions. The reverse is true for basic solutions, in which the $[OH^-]$ is always greater than 1.0×10^{-7} *M*. By definition, "neutral" aqueous solutions at 25°C are solutions in which $[H_3O^+] = [OH^-] = 1.0 \times 10^{-7}$ *M*.

Solution	General Condition	At 25°C	
acidic	$[H_3O^+] > [OH^-]$	$[H_3O^+] > 1.0 \times 10^{-7}$	$[OH^-] < 1.0 \times 10^{-7}$
neutral	$[H_3O^+] = [OH^-]$	$[H_3O^+] = 1.0 \times 10^{-7}$	$[OH^-] = 1.0 \times 10^{-7}$
basic	$[H_3O^+] < [OH^-]$	$[H_3O^+] < 1.0 \times 10^{-7}$	$[OH^-] > 1.0 \times 10^{-7}$

18-3 THE pH AND pOH SCALES

SCiLINKS.

TOPIC: Acids/Bases
GO TO: www.scilinks.org
*sci*LINKS CODE: WCH0430

The pH and pOH scales provide a convenient way to express the acidity and basicity of dilute aqueous solutions. The **pH** and **pOH** of a solution are defined as

$$pH = -\log [H_3O^+] \quad \text{or} \quad [H_3O^+] = 10^{-pH}$$

$$pOH = -\log [OH^-] \quad \text{or} \quad [OH^-] = 10^{-pOH}$$

Note that we use pH rather than pH_3O. At the time the pH concept was developed, H_3O^+ was represented as H^+. Various "p" terms are used. In general a lowercase "**p**" before a symbol means "negative logarithm of the symbol." Thus, pH is the negative logarithm of the H_3O^+ concentration, **pOH** is the negative logarithm of the OH^- concentration, and **pK** refers to the negative logarithm of an equilibrium constant. It is convenient to describe the autoionization of water in terms of **pK_w**.

> When dealing with pH, we always use the base-10 (common) logarithm, *not* the base-*e* (natural) logarithm. This is because pH is *defined* using base-10 logarithms.

$$pK_w = -\log K_w \qquad (= 14.0 \text{ at } 25°C)$$

EXAMPLE 18-3   *Calculation of pH*

Calculate the pH of a solution in which the H_3O^+ concentration is 0.050 mol/L.

Plan

We are given the value for $[H_3O^+]$; so we take the negative logarithm of this value.

Solution

$$[H_3O^+] = 0.050\ M = 5.0 \times 10^{-2}\ M$$
$$pH = -\log [H_3O^+] = -\log [5.0 \times 10^{-2}] = \boxed{1.30}$$

This answer contains only *two* significant figures. The "1" in 1.30 is *not* a significant figure; it comes from the power of ten.

You should now work Exercise 22.

 In a pH value, only the digits after the decimal point are significant figures.

EXAMPLE 18-4   *Calculation of H_3O^+ Concentration from pH*

The pH of a solution is 3.301. What is the concentration of H_3O^+ in this solution?

Plan

By definition, $pH = -\log [H_3O^+]$. We are given the pH, so we solve for $[H_3O^+]$.

Solution

From the definition of pH, we write

$$-\log [H_3O^+] = 3.301$$

Multiplying through by -1 gives

$$\log [H_3O^+] = -3.301$$

Taking the inverse logarithm (antilog) of both sides of the equation gives

$$[H_3O^+] = 10^{-3.301} \qquad \text{so} \qquad \boxed{[H_3O^+] = 5.00 \times 10^{-4}\ M}$$

You should now work Exercise 26.

A convenient relationship between pH and pOH in *all dilute solutions at 25°C* can be easily derived. We start with the K_w expression.

$$[H_3O^+][OH^-] = 1.0 \times 10^{-14}$$

Taking the logarithm of both sides of this equation gives

$$\log [H_3O^+] + \log [OH^-] = \log (1.0 \times 10^{-14})$$

Multiplying both sides of this equation by -1 gives

$$(-\log [H_3O^+]) + (-\log [OH^-]) = -\log (1.0 \times 10^{-14})$$

or

$$pH + pOH = 14.00$$

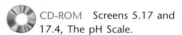

We can now relate $[H_3O^+]$ and $[OH^-]$ as well as pH and pOH.

$$[H_3O^+][OH^-] = 1.0 \times 10^{-14} \qquad \text{and} \qquad pH + pOH = 14.00 \qquad \text{(at 25°C)}$$

From this relationship, we see that pH and pOH can *both* be positive only if *both* are less than 14. If either pH or pOH is greater than 14, the other is obviously negative.

Please study carefully the following summary. It will be helpful.

Solution	General Condition	At 25°C
acidic	$[H_3O^+] > [OH^-]$ $pH < pOH$	$[H_3O^+] > 1.0 \times 10^{-7}\ M > [OH^-]$ $pH < \quad 7.00 \quad < pOH$
neutral	$[H_3O^+] = [OH^-]$ $pH = pOH$	$[H_3O^+] = 1.0 \times 10^{-7}\ M = [OH^-]$ $pH = \quad 7.00 \quad = pOH$
basic	$[H_3O^+] < [OH^-]$ $pH > pOH$	$[H_3O^+] < 1.0 \times 10^{-7}\ M < [OH^-]$ $pH > \quad 7.00 \quad > pOH$

To develop familiarity with the pH and pOH scales, consider a series of solutions in which $[H_3O^+]$ varies from $10\ M$ to $1.0 \times 10^{-15}\ M$. Obviously, $[OH^-]$ will vary from $1.0 \times 10^{-15}\ M$ to $10\ M$ in these solutions. Table 18-3 summarizes these scales.

TABLE 18-3 *Relationships Among $[H_3O^+]$, pH, pOH, and $[OH^-]$*

$[H_3O^+]$	pH			pOH	$[OH^-]$	
10^{-15}	15			-1	10^1	
10^{-14}	14			0	1	Increasing basicity
10^{-13}	13			1	10^{-1}	
10^{-12}	12			2	10^{-2}	
10^{-11}	11			3	10^{-3}	
10^{-10}	10		OH^- concentration	4	10^{-4}	
10^{-9}	9			5	10^{-5}	
10^{-8}	8			6	10^{-6}	
10^{-7}	7			7	10^{-7}	Neutral
10^{-6}	6			8	10^{-8}	
10^{-5}	5	H_3O^+ concentration		9	10^{-9}	
10^{-4}	4			10	10^{-10}	
10^{-3}	3			11	10^{-11}	
10^{-2}	2			12	10^{-12}	
10^{-1}	1			13	10^{-13}	
1	0			14	10^{-14}	Increasing acidity
10^1	-1			15	10^{-15}	

At any temperature,
$pH + pOH = pK_w$.

These are very important relationships. Remember them!

CD-ROM   Screens 5.17 and 17.4, The pH Scale.

Note that the pH can be 0 or a small negative number for concentrated strong acid solutions. A pH = -1.0 represents a $[H^+] = 10.0\ M$, which is certainly possible.

pH Range for a Few Common Substances

Substance	pH Range
Gastric contents (human)	1.6–3.0
Soft drinks	2.0–4.0
Lemons	2.2–2.4
Vinegar	2.4–3.4
Tomatoes	4.0–4.4
Beer	4.0–5.0
Urine (human)	4.8–8.4
Milk (cow's)	6.3–6.6
Saliva (human)	6.5–7.5
Blood plasma (human)	7.3–7.5
Egg white	7.6–8.0
Milk of magnesia	10.5
Household ammonia	11–12

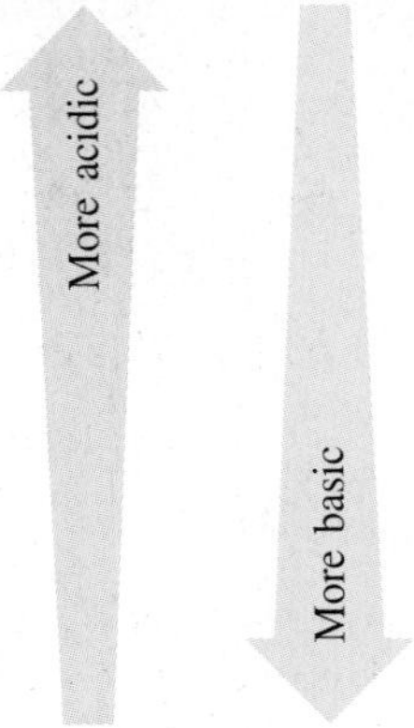

More acidic

More basic

EXAMPLE 18-5   pH and pOH of an Acidic Solution

Calculate $[H_3O^+]$, pH, $[OH^-]$, and pOH for a 0.015 M HNO_3 solution.

Plan

We write the equation for the ionization of the strong acid, HNO_3, which gives us $[H_3O^+]$. Then we calculate pH. We use the relationships $[H_3O^+][OH^-] = 1.0 \times 10^{-14}$ and pH + pOH = 14.00 to find pOH and $[OH^-]$.

Solution

$$HNO_3 + H_2O \longrightarrow H_3O^+ + NO_3^-$$

Because nitric acid is a strong acid (it ionizes completely), we know that

$$[H_3O^+] = \boxed{0.015\ M}$$

$$pH = -\log [H_3O^+] = -\log (0.015) = -(-1.82) = \boxed{1.82}$$

We also know that pH + pOH = 14.00. Therefore,

$$pOH = 14.00 - pH = 14.00 - 1.82 = \boxed{12.18}$$

Because $[H_3O^+][OH^-] = 1.0 \times 10^{-14}$, $[OH^-]$ is easily calculated.

$$[OH^-] = \frac{1.0 \times 10^{-14}}{[H_3O^+]} = \frac{1.0 \times 10^{-14}}{0.015} = \boxed{6.7 \times 10^{-13}\ M}$$

You should now work Exercise 27.

All ions are hydrated in aqueous solution. We often omit the designations (ℓ), (g), (s), (aq), and so on.

EXAMPLE 18-6   *Calculations Involving pH and pOH*

Calculate $[H_3O^+]$, pH, $[OH^-]$, and pOH for a 0.015 M $Ca(OH)_2$ solution.

Plan

We write the equation for the ionization of the strong base $Ca(OH)_2$, which gives us $[OH^-]$. Then we calculate pOH. We use the relationships $[H_3O^+][OH^-] = 1.0 \times 10^{-14}$ and pH + pOH = 14.00 to find pH and $[H_3O^+]$.

Solution

$$Ca(OH)_2 \xrightarrow{H_2O} Ca^{2+} + 2OH^-$$

Because calcium hydroxide is a strong base (it dissociates completely), we know that

$$[OH^-] = 2 \times 0.015\ M = \boxed{0.030\ M}$$

$$pOH = -\log[OH^-] = -\log(0.030) = -(-1.52) = \boxed{1.52}$$

We also know that pH + pOH = 14.00. Therefore,

$$pH = 14.00 - pOH = 14.00 - 1.52 = \boxed{12.48}$$

Because $[H_3O^+][OH^-] = 1.0 \times 10^{-14}$, $[H_3O^+]$ is easily calculated.

$$[H_3O^+] = \frac{1.0 \times 10^{-14}}{[OH^-]} = \frac{1.0 \times 10^{-14}}{0.030} = \boxed{3.3 \times 10^{-13}\ M}$$

You should now work Exercises 28 and 29.

> When asked for the pH of a basic solution, you should usually first work out the pOH value. It is very important to remember to then subtract the pOH value from 14.0 to calculate the pH. You should also remember the simple concept that the pH for a basic solution is greater than 7.0.

A pH meter is much more accurate than an indicator for obtaining pH values.

The pH of a solution can be determined using a pH meter (Figure 18-1) or by the indicator method. Acid–base **indicators** are intensely colored complex organic compounds that have different colors in solutions of different pH (Section 19-4). Many are weak acids or weak bases that are useful over rather narrow ranges of pH values. *Universal indicators* are mixtures of several indicators; they show several color changes over a wide range of pH values.

In the indicator method we prepare a series of solutions of known pH (standard solutions). We add a universal indicator to each; solutions with different pH have different

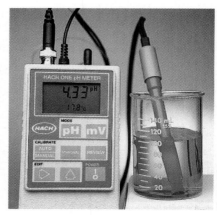

Figure 18-1   A pH meter gives the pH of a solution directly. When the electrode is dipped into a solution, the meter displays the pH. The pH meter is based on the glass electrode. This sensing device generates a voltage that is proportional to the pH of the solution in which the electrode is placed. The instrument has an electrical circuit to amplify the voltage from the electrode and a meter that relates the voltage to the pH of the solution. Before being used, a pH meter must be calibrated with a series of solutions of known pH.

Charles D. Winters

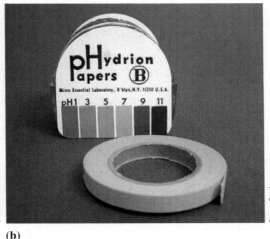

(a)   **(b)**

Figure 18-2   (a) Solutions containing a universal indicator. A universal indicator shows a wide range of colors as pH varies. The pH values are given by the black numbers. These solutions range from quite acidic (*upper left*) to quite basic (*lower right*). (b) Universal pH paper has been soaked in universal indicator allowing a wide range of pH's to be measured easily by comparing the test strip to the color scale on the container.

colors (Figure 18-2). We then add the same universal indicator to the unknown solution and compare its color to those of the standard solutions. Solutions with the same pH have the same color.

Universal indicator papers can also be used to determine pH. A drop of solution is placed on a piece of paper or a piece of the paper is dipped into a solution. The color of the paper is then compared with a color chart on the container to establish the pH of the solution.

18-4 IONIZATION CONSTANTS FOR WEAK MONOPROTIC ACIDS AND BASES

We have discussed strong acids and strong bases. There are relatively few of these. Weak acids are much more numerous than strong acids. For this reason you were asked to learn the list of strong acids (see Table 18-1). You may assume that other acids you encounter in this text will be weak acids. Table 18-4 contains names, formulas, ionization constants, and pK_a values for a few common weak acids; Appendix F contains a longer list of K_a values. *Weak* acids ionize only slightly in dilute aqueous solution. Our classification of acids as strong or weak is based on the *extent to which they ionize in dilute aqueous solution*.

Several weak acids are familiar to us. Vinegar is a 5% solution of acetic acid, CH_3COOH. Carbonated beverages are saturated solutions of carbon dioxide in water, which produces carbonic acid.

$$CO_2 + H_2O \rightleftharpoons H_2CO_3$$

Citrus fruits contain citric acid, $C_3H_5O(COOH)_3$. Some ointments and powders used for medicinal purposes contain boric acid, H_3BO_3. These everyday uses of weak acids suggest that there is a significant difference between strong and weak acids. The difference is that *strong acids ionize completely in dilute aqueous solution, whereas weak acids ionize only slightly.*

The pH of some common substances is shown by a universal indicator. Refer to Figure 18-2 to interpret the indicator colors.

How can you tell that carbonated beverages are *saturated* CO_2 solutions?

Would you think of using sulfuric or nitric acid for any of these purposes?

A **monoprotic acid** has only one ionizable H. Recall that **pK_a** means $-\log K_a$.

TABLE 18-4	Ionization Constants and pK_a Values for Some Weak Monoprotic Acids		

Acid	Ionization Reaction	K_a at 25°C	pK_a
hydrofluoric acid	$HF + H_2O \rightleftharpoons H_3O^+ + F^-$	7.2×10^{-4}	3.14
nitrous acid	$HNO_2 + H_2O \rightleftharpoons H_3O^+ + NO_2^-$	4.5×10^{-4}	3.35
acetic acid	$CH_3COOH + H_2O \rightleftharpoons H_3O^+ + CH_3COO^-$	1.8×10^{-5}	4.74
hypochlorous acid	$HOCl + H_2O \rightleftharpoons H_3O^+ + OCl^-$	3.5×10^{-8}	7.45
hydrocyanic acid	$HCN + H_2O \rightleftharpoons H_3O^+ + CN^-$	4.0×10^{-10}	9.40

 CD-ROM Screen 17.6, Weak Acids and Bases.

Let us consider the reaction that occurs when a weak acid, such as acetic acid, is dissolved in water. The equation for the ionization of acetic acid is

$$CH_3COOH(aq) + H_2O(\ell) \rightleftharpoons H_3O^+(aq) + CH_3COO^-(aq)$$

The equilibrium constant for this reaction could be represented as

$$K_c = \frac{[H_3O^+][CH_3COO^-]}{[CH_3COOH][H_2O]}$$

We should recall that the thermodynamic definition of K is in terms of activities. In dilute solutions, the activity of the (nearly) pure H_2O is essentially 1. The activity of each dissolved species is numerically equal to its molar concentration. Thus the **ionization constant** of a weak acid, K_a, does not include a term for the concentration of water.

We often use HA as a general representation for a monoprotic acid and A^- for its conjugate base. The equation for the ionization of a weak acid can be written as

$$HA \rightleftharpoons H^+ + A^-$$

For example, for acetic acid, we can write either

$$K_a = \frac{[H^+][A^-]}{[HA]}$$

or

The value of K_a can be found in Table 18.4.

$$K_a = \frac{[H_3O^+][CH_3COO^-]}{[CH_3COOH]} = 1.8 \times 10^{-5}$$

This expression tells us that in dilute aqueous solutions of acetic acid, the concentration of H_3O^+ multiplied by the concentration of CH_3COO^- and then divided by the concentration of *nonionized* acetic acid is equal to 1.8×10^{-5}.

Ionization constants for weak acids (and bases) must be calculated from *experimentally determined data*. Measurements of pH, conductivity, or depression of freezing point provide data from which these constants can be calculated.

EXAMPLE 18-7 Calculation of Ka and pKa from Equilibrium Concentrations

Nicotinic acid is a weak monoprotic organic acid that we can represent as H_A.

$$H_A + H_2O \rightleftharpoons H_3O^+ + A^-$$

A dilute solution of nicotinic acid was found to contain the following concentrations at equilibrium at 25°C. What are the K_a and pK_a values? $[HA] = 0.049$ M; $[H_3O^+] = [A^-] = 8.4 \times 10^{-4}$ M.

Plan

We are given *equilibrium* concentrations, and so we substitute these into the expression for K_a.

Solution

$$HA + H_2O \rightleftharpoons H_3O^+ + A^- \qquad K_a = \frac{[H_3O^+][A^-]}{[HA]}$$

$$K_a = \frac{(8.4 \times 10^{-4})(8.4 \times 10^{-4})}{0.049} = 1.4 \times 10^{-5}$$

The equilibrium constant expression is

$$K_a = \frac{[H_3O^+][A^-]}{[HA]} = 1.4 \times 10^{-5} \qquad pK_a = -\log(1.4 \times 10^{-5}) = 4.85$$

You should now work Exercise 32.

The structure of nicotinic acid is

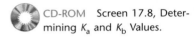

Nicotinic acid, also called niacin, is a necessary vitamin in our diets. It is not physiologically related to nicotine.

CD-ROM   Screen 17.8, Determining K_a and K_b Values.

EXAMPLE 18-8   *Calculation of K_a from Percent Ionization*

In 0.0100 M solution, acetic acid is 4.2% ionized. Calculate its ionization constant.

Plan

We write the equation for the ionization of acetic acid and its equilibrium constant expression. Next we use the percent ionization to complete the reaction summary and then substitute into the K_a expression.

Solution

The equations for the ionization of CH_3COOH and its ionization constant are

$$CH_3COOH + H_2O \rightleftharpoons H_3O^+ + CH_3COO^- \qquad \text{and} \qquad K_a = \frac{[H_3O^+][CH_3COO^-]}{[CH_3COOH]}$$

Because 4.2% of the CH_3COOH ionizes,

$$M_{CH_3COOH} \text{ that ionizes} = 0.042 \times 0.0100 \ M = 4.2 \times 10^{-4} \ M$$

Each mole of CH_3COOH that ionizes forms one mole of H_3O^+ and one mole of CH_3COO^-. We represent this in the reaction summary.

	CH_3COOH	$+ H_2O \rightleftharpoons$	H_3O^+	$+$	CH_3COO^-
initial	0.0100 M		≈ 0 M		0 M
change	-4.2×10^{-4} M		$+4.2 \times 10^{-4}$ M		$+4.2 \times 10^{-4}$ M
at equil	9.58×10^{-3} M		4.2×10^{-4} M		4.2×10^{-4} M

Substitution of these values into the K_a expression gives the value for K_a.

$$K_a = \frac{[H_3O^+][CH_3COO^-]}{[CH_3COOH]} = \frac{(4.2 \times 10^{-4})(4.2 \times 10^{-4})}{9.58 \times 10^{-3}} = 1.8 \times 10^{-5}$$

You should now work Exercise 38.

Some common household weak acids. A strip of paper impregnated with a universal indicator is convenient for estimating the pH of a solution. Refer to Figure 18-2(b) to interpret the colors of the indicator strips.

EXAMPLE 18-9   Calculation of K_a from pH

The pH of a 0.115 M solution of chloroacetic acid, $ClCH_2COOH$, is measured to be 1.92. Calculate K_a for this weak monoprotic acid.

Plan

For simplicity, we represent $ClCH_2COOH$ as HA. We write the ionization equation and the expression for K_a. Next we calculate $[H_3O^+]$ for the given pH and complete the reaction summary. Finally, we substitute into the K_a expression.

Solution

The ionization of this weak monoprotic acid and its ionization constant expression may be represented as

$$HA + H_2O \rightleftharpoons H_3O^+ + A^- \quad \text{and} \quad K_a = \frac{[H_3O^+][A^-]}{[HA]}$$

We calculate $[H_3O^+]$ from the definition of pH.

$$pH = -\log [H_3O^+]$$

$$[H_3O^+] = 10^{-pH} = 10^{-1.92} = 0.012 \ M$$

We use the usual reaction summary as follows. At this point, we know the *original* [HA] and the *equilibrium* $[H_3O^+]$. From this information, we fill out the "change" line and then deduce the other equilibrium values.

	HA	$+ \ H_2O \rightleftharpoons$	H_3O^+	$+$	A^-
initial	0.115 M		$\approx 0 \ M$		0 M
change due to rxn	$-0.012 \ M$		$+0.012 \ M$		0.012 M
at equil	0.103 M		0.012 M		0.012 M

Now that all concentrations are known, K_a can be calculated.

$$K_a = \frac{[H_3O^+][A^-]}{[HA]} = \frac{(0.012)(0.012)}{0.103} = \boxed{1.4 \times 10^{-3}}$$

You should now work Exercise 40.

✓ Problem-Solving Tip: *Filling in Reaction Summaries*

In Examples 18-8 and 18-9 the value of an equilibrium concentration was used to determine the change in concentration. You should become proficient at using a variety of data to determine values that are related via a chemical equation. Let's review what we did in Example 18-9. Only the equilibrium expression, initial concentrations, and the equilibrium concentration of H_3O^+ were known when we started the reaction summary. The following steps show how we filled in the remaining values in the order indicated by the numbered red arrows.

1. $[H_3O^+]_{equil} = 0.012 \ M$, so we record this value
2. $[H_3O^+]_{initial} \approx 0$, so change in $[H_3O^+]$ due to rxn must be $+0.012 \ M$
3. Formation of 0.012 M H_3O^+ consumes 0.012 M HA, so the change in [HA] $= -0.012 \ M$

4. $[HA]_{equil} = [HA]_{orig} + [HA]_{chg} = 0.115\ M + (-0.012\ M) = 0.103\ M$

5. Formation of $0.012\ M\ H_3O^+$ also gives $0.012\ M\ A^-$

6. $[A^-]_{equil} = [A^-]_{orig} + [A^-]_{chg} = 0\ M + 0.012\ M = 0.012\ M$

At equilibrium, $[H_3O^+] = 0.012\ M$ so

	HA	$+ H_2O \rightleftharpoons$	H_3O^+	$+$	A^-
initial	$0.115\ M$		$0\ M$		$0\ M$
change due to rxn	$-0.012\ M$		$+0.012\ M$		$0.012\ M$
at equil	$0.103\ M$		$0.012\ M$		$0.012\ M$

Ionization constants are equilibrium constants for ionization reactions, so their values indicate the extents to which weak electrolytes ionize. At the same concentrations, acids with larger ionization constants ionize to greater extents (and are stronger acids) than acids with smaller ionization constants. From Table 18-4, we see that the order of decreasing acid strength for these five weak acids is

$$HF > HNO_2 > CH_3COOH > HOCl > HCN$$

Conversely, in Brønsted–Lowry terminology (Section 10-4), the order of increasing base strength of the anions of these acids is

$$F^- < NO_2^- < CH_3COO^- < OCl^- < CN^-$$

Recall that in Brønsted–Lowry terminology, an acid forms its conjugate base by losing a H^+.

If we know the value of the ionization constant for a weak acid, we can calculate the concentrations of the species present in solutions of known initial concentrations.

EXAMPLE 18-10   *Calculation of Concentrations from K_a*

(a) Calculate the concentrations of the various species in $0.10\ M$ hypochlorous acid, HOCl. For HOCl, $K_a = 3.5 \times 10^{-8}$. (b) What is the pH of this solution?

Plan

We write the equation for the ionization of the weak acid and its K_a expression. Then we represent the *equilibrium* concentrations algebraically and substitute into the K_a expression. Then we solve for the concentrations and the pH.

We have written the formula for hypochlorous acid as HOCl rather than HClO to emphasize that its structure is H—O—Cl.

Solution

(a) The equation for the ionization of HOCl and its K_a expression are

$$HOCl + H_2O \rightleftharpoons H_3O^+ + OCl^- \quad \text{and} \quad K_a = \frac{[H_3O^+][OCl^-]}{[HOCl]} = 3.5 \times 10^{-8}$$

We would like to know the concentrations of H_3O^+, OCl^-, and nonionized HOCl in solution. An algebraic representation of concentrations is required, because there is no other obvious way to obtain the concentrations.

Let x = mol/L of HOCl that ionizes. Then, write the "change" line and complete the reaction summary.

	HOCl	$+ H_2O \rightleftharpoons$	H_3O^+	$+ OCl^-$
initial	$0.10\ M$		$\approx 0\ M$	$0\ M$
change due to rxn	$-x\ M$		$+x\ M$	$+x\ M$
at equil	$(0.10 - x)\ M$		$x\ M$	$x\ M$

We neglect the 1.0×10^{-7} mol/L of H_3O^+ produced by the ionization of *pure* water. Recall (see Section 18-2) that the addition of an acid to water suppresses the ionization of H_2O, so $[H_3O^+]$ from H_2O is even less than $1.0 \times 10^{-7}\ M$.

CD-ROM   Screen 17.9,
Estimating the pH of Weak
Acid Solutions.

Substituting these algebraic representations into the K_a expression gives

$$K_a = \frac{[H_3O^+][OCl^-]}{[HOCl]} = \frac{(x)(x)}{(0.10 - x)} = 3.5 \times 10^{-8}$$

This is a quadratic equation, but it is not necessary to solve it by the quadratic formula. The small value of the equilibrium constant, K_a, tells us that not very much of the original acid ionizes. Thus we can assume that $x \ll 0.10$. If x is small enough compared with 0.10, it will not matter (much) whether we subtract it, and we can assume that $(0.10 - x)$ is very nearly equal to 0.10. The equation then becomes

$$\frac{x^2}{0.10} \approx 3.5 \times 10^{-8} \qquad x^2 \approx 3.5 \times 10^{-9} \qquad so \qquad x \approx 5.9 \times 10^{-5}$$

In our algebraic representation we let

$$[H_3O^+] = x\,M = 5.9 \times 10^{-5}\,M \qquad [OCl^-] = x\,M = 5.9 \times 10^{-5}\,M$$

$$[HOCl] = (0.10 - x)\,M = (0.10 - 0.000059)\,M = 0.10\,M$$

$$[OH^-] = \frac{K_w}{[H_3O^+]} = \frac{1.0 \times 10^{-14}}{5.9 \times 10^{-5}} = 1.7 \times 10^{-10}\,M$$

(b) $pH = -\log(5.9 \times 10^{-5}) = 4.23$

You should now work Exercise 42.

✔ **Problem-Solving Tip:** *Simplifying Quadratic Equations*

We often encounter quadratic or higher-order equations in equilibrium calculations. With modern programmable calculators, solving such problems by iterative methods is often feasible. But frequently a problem can be made much simpler by using some mathematical common sense.

When the linear variable (x) in a *quadratic* equation is added to or subtracted from a much larger number, it can often be disregarded if it is sufficiently small. A reasonable rule of thumb for determining whether the variable can be disregarded in equilibrium calculations is this: If the exponent of 10 in the K value is -3 or less (-4, -5, -6, etc.), then the variable may be small enough to disregard when it is added to or subtracted from a number greater than 0.05. Solve the problem neglecting x; then compare the value of x with the number it would have been added to (or subtracted from). If x is more than 5% of that number, the assumption was *not* justified, and you should solve the equation using the quadratic formula.

Let's examine the assumption as it applies to Example 18-10. Our quadratic equation is

$$\frac{(x)(x)}{(0.10 - x)} = 3.5 \times 10^{-8}$$

Because 3.5×10^{-8} is a very small K_a value, we know that the acid ionizes only slightly. Thus, x must be very small compared with 0.10, so we can write $(0.10 - x) \approx 0.10$.

The equation then becomes $\frac{x^2}{0.10} \approx 3.5 \times 10^{-8}$. To solve this, we rearrange and take the square roots of both sides. To check, we see that the result, $x = 5.9 \times 10^{-5}$, is only 0.059% of 0.10. This error is much less than 5%, so our assumption is justified. You may also wish to use the quadratic formula to verify that the answer obtained this way is correct to within roundoff error.

The preceding argument is purely algebraic. We could use our chemical intuition to reach the same conclusion. A small K_a value (10^{-3} or less) tells us that the extent of ionization is very small; therefore, nearly all of the weak acid exists as nonionized molecules. The amount that ionizes is insignificant compared with the concentration of nonionized weak acid.

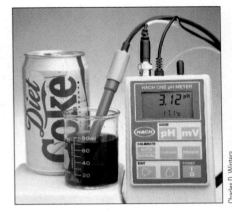

The pH of a soft drink is measured with a modern pH meter. Many cola drinks are quite acidic due to the dissolved CO_2 and phosphoric acid.

From our calculations in Example 18-10 we can draw some conclusions. In a solution containing *only a weak monoprotic acid*, the concentration of H_3O^+ is equal to the concentration of the anion of the acid. The concentration of the dilute nonionized acid is approximately equal to the molarity of the solution.

EXAMPLE 18-11 *Percent Ionization*

Calculate the percent ionization of a 0.10 M solution of acetic acid.

Plan

Write the ionization equation and the expression for K_a. Follow the procedure used in Example 18-10 to find the concentration of acid that ionized. Then, substitute the concentration of acid that ionized into the expression for percent ionization. Percentage is defined as (part/whole) $\times$ 100%, so the percent ionization is

$$\% \text{ ionization} = \frac{[CH_3COOH]_{\text{ionized}}}{[CH_3COOH]_{\text{initial}}} \times 100\%$$

Solution

The equations for the ionization of CH_3COOH and its K_a are

$$CH_3COOH + H_2O \rightleftharpoons H_3O^+ + CH_3COO^- \qquad K_a = \frac{[H_3O^+][A^-]}{[HA]} = 1.8 \times 10^{-5}$$

We proceed as we did in Example 18-10. Let $x = [CH_3COOH]_{\text{ionized}}$.

	$CH_3COOH + H_2O \rightleftharpoons$	H_3O^+	$+ CH_3COO^-$
initial	0.10 M	$\approx 0\,M$	0 M
change due to rxn	$-x\,M$	$+x\,M$	$+x\,M$
at equil	$(0.10 - x)\,M$	$x\,M$	$x\,M$

We could write the original $[H_3O^+]$ as 1.0×10^{-7} M. In very *dilute* solutions of weak acids, we might have to take this into account. In this acid solution, $(1.0 \times 10^{-7} + x) \approx x$.

Substituting into the ionization constant expression gives

$$K_a = \frac{[H_3O^+][A^-]}{[HA]} = \frac{(x)(x)}{(0.10 - x)} = 1.8 \times 10^{-5}$$

If we make the simplifying assumption that $(0.10 - x) \approx 0.10$, we have

$$\frac{x^2}{0.10} = 1.8 \times 10^{-5} \qquad x^2 = 1.8 \times 10^{-6} \qquad x = 1.3 \times 10^{-3}$$

This gives $[CH_3COOH]_{ionized} = x = 1.3 \times 10^{-3}\ M$. Now we can calculate the percent ionization for $0.10\ M\ CH_3COOH$ solution.

Note that we need not solve explicitly for the equilibrium concentrations $[H_3O^+]$ and $[CH_3COO^-]$ to answer the question. From the setup, we see that these are both $1.3 \times 10^{-3}\ M$.

$$\% \text{ ionization} = \frac{[CH_3COOH]_{ionized}}{[CH_3COOH]_{initial}} \times 100\% = \frac{1.3 \times 10^{-3}\ M}{0.10\ M} \times 100\% = \boxed{1.3\%}$$

Our assumption that $(0.10 - x) \approx 0.10$ is reasonable because $(0.10 - x) = (0.10 - 0.0013)$. This is only about 1% different than 0.10. When K_a for a weak acid is significantly greater than 10^{-3}, however, this assumption would introduce considerable error.

You should now work Exercise 48.

In dilute solutions, acetic acid exists primarily as nonionized molecules, as do all weak acids; there are relatively few hydronium and acetate ions. In $0.10\ M$ solution, CH_3COOH is 1.3% ionized; for each 1000 molecules of CH_3COOH originally placed in the solution, there are 13 H_3O^+ ions, 13 CH_3COO^- ions, and 987 nonionized CH_3COOH molecules. For weaker acids of the same concentration, the number of molecules of nonionized acid would be even larger.

By now we should have gained some "feel" for the strength of an acid by looking at its K_a value. Consider $0.10\ M$ solutions of HCl (a strong acid), CH_3COOH (see Example 18-11), and HOCl (see Example 18-10). If we calculate the percent ionization for $0.10\ M$ HOCl (as we did for $0.10\ M\ CH_3COOH$ in Example 18-11), we find that it is 0.059% ionized. In $0.10\ M$ solution, HCl is very nearly completely ionized. The data in Table 18-5 show that the $[H_3O^+]$ in $0.10\ M$ HCl is approximately 77 times greater than that in $0.10\ M\ CH_3COOH$ and approximately 1700 times greater than that in $0.10\ M$ HOCl.

Many scientists prefer to use pK_a values rather than K_a values for weak acids. Recall that in general, "p" terms refer to negative logarithms. The pK_a value for a weak acid is just the negative logarithm of its K_a value.

EXAMPLE 18-12 pK_a *Values*

The K_a values for acetic acid and hydrofluoric acid are 1.8×10^{-5} and 7.2×10^{-4}, respectively. What are their pK_a values?

Plan

pK_a is defined as the negative logarithm of K_a (i.e., $pK_a = -\log K_a$) so we take the negative logarithm of each K_a.

Solution

For CH_3COOH,

$$pK_a = -\log K_a = -\log (1.8 \times 10^{-5}) = -(-4.74) = \boxed{4.74}$$

For HF,

$$pK_a = -\log K_a = -\log (7.2 \times 10^{-4}) = -(-3.14) = \boxed{3.14}$$

You should now work Exercise 50.

TABLE 18-5	Comparison of Extents of Ionization of Some Acids			
Acid Solution	Ionization Constant	[H$_3$O$^+$]	pH	Percent Ionization
0.10 M HCl	very large	0.10 M	1.00	$\approx$100
0.10 M CH$_3$COOH	1.8 $\times$ 10^{-5}	0.0013 M	2.89	1.3
0.10 M HOCl	3.5 $\times$ 10^{-8}	0.000059 M	4.23	0.059

An inert solid has been suspended in the liquid to improve the quality of this photograph of a pH meter.

From Example 18-12, we see that the stronger acid (HF in this case) has the larger K_a value and the smaller pK_a value. Conversely, the weaker acid (CH$_3$COOH in this case) has the smaller K_a value and the larger pK_a value. The generalization is

the larger the value of K_a, the smaller is the value of pK_a, and the stronger is the acid.

A similar statement is true for weak bases; that is, a stronger base has the greater K_b value and the smaller pK_b value.

EXAMPLE 18-13 Acid Strengths and K_a Values

Given the following list of weak acids and their K_a values, arrange the acids in order of (a) increasing acid strength and (b) increasing pK_a values.

Acid	K_a
HOCl	3.5 $\times$ 10^{-8}
HCN	4.0 $\times$ 10^{-10}
HNO$_2$	4.5 $\times$ 10^{-4}

Plan

(a) We see that HNO$_2$ is the strongest acid in this group because it has the largest K_a value. HCN is the weakest because it has the smallest K_a value.
(b) We do not need to calculate pK_a values to answer the question. We recall that the weakest acid has the largest pK_a value and the strongest acid has the smallest pK_a value, so the order of increasing pK_a values is just the reverse of the order in Part (a).

Solution

(a) Increasing acid strength: HCN < HOCl < HNO$_2$
(b) Increasing pK_a values: HNO$_2$ < HOCl < HCN

You should now work Exercise 111.

You may know that hydrofluoric acid dissolves glass. But HF is *not* a strong acid. The reaction of glass with hydrofluoric acid occurs because silicates react with HF to produce silicon tetrafluoride, SiF$_4$, a very volatile compound. This reaction tells us nothing about the acid strength of hydrofluoric acid.

Do not confuse the strength of an acid with its reactivity. Acid strength refers to the extent of ionization of the acid, and not the reactions that it undergoes.

Beckman Instruments

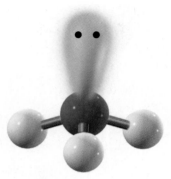

Ammonia, NH_3

Trimethylamine, $(CH_3)_3N$

The subscript "b" indicates that the substance ionizes as a base. We do not include $[H_2O]$ in the K_b expression for the same reasons described for K_a. Recall that **pK_b** means $-\log K_b$.

Charles D. Winters

The odor that we associate with fish is due to the presence of amines. This is one reason why lemon is often added to seafood. The citric acid (a weak acid) neutralizes the odor of the amines

Thus far we have focused our attention on acids. Very few common weak bases are soluble in water. Aqueous ammonia is the most frequently encountered example. From our earlier discussion of bonding in covalent compounds (Section 8-8), we recall that there is one unshared pair of electrons on the nitrogen atom in NH_3. When ammonia dissolves in water, it accepts H^+ from a water molecule in a reversible reaction (Section 10-4). We say that NH_3 ionizes slightly when it undergoes this reaction. Aqueous solutions of NH_3 are basic because OH^- ions are produced.

$$: NH_3 + H_2O \rightleftharpoons NH_4^+ + OH^-$$

Amines are derivatives of NH_3 in which one or more H atoms have been replaced by organic groups, as the following structures indicate.

$H-\overset{\cdot\cdot}{\underset{\underset{H}{|}}{N}}-H$	$H_3C-\overset{\cdot\cdot}{\underset{\underset{H}{|}}{N}}-H$	$H_3C-\overset{\cdot\cdot}{\underset{\underset{CH_3}{|}}{N}}-H$	$H_3C-\overset{\cdot\cdot}{\underset{\underset{CH_3}{|}}{N}}-CH_3$
ammonia NH_3	methylamine CH_3NH_2	dimethylamine $(CH_3)_2NH$	trimethylamine $(CH_3)_3N$

Thousands of amines are known, and many are very important in biochemical processes. Low-molecular-weight amines are soluble weak bases. The ionization of trimethylamine, for example, forms trimethylammonium ions and OH^- ions.

$$CH_3-\overset{\cdot\cdot}{\underset{\underset{CH_3}{|}}{N}}-CH_3 + \overset{H^+ \text{ transfer}}{\textcircled{H}}-OH \rightleftharpoons \left[CH_3-\overset{\overset{H}{|}}{\underset{\underset{CH_3}{|}}{N}}-CH_3 \right]^+ + OH^-$$

trimethylamine $(CH_3)_3N$ trimethylammonium ion $(CH_3)_3NH^+$

Now let us consider the behavior of ammonia in aqueous solutions. The reaction of ammonia with water and its ionization constant expression are

$$NH_3 + H_2O \rightleftharpoons NH_4^+ + OH^-$$

and

$$K_b = \frac{[NH_4^+][OH^-]}{[NH_3]} = 1.8 \times 10^{-5}$$

The fact that K_b for aqueous NH_3 has the same value as K_a for CH_3COOH is pure coincidence. It does tell us that in aqueous solutions of the same concentration, CH_3COOH and NH_3 are ionized to the same extent. Table 18-6 lists K_b and pK_b values for a few common weak bases. Appendix G includes a longer list of K_b values.

We use K_b's for weak bases in the same way we used K_a's for weak acids and pK_b values for weak bases in the same way we used pK_a values for weak acids.

EXAMPLE 18-14   *pH of a Weak Base Solution*

Calculate the $[OH^-]$, pH, and percent ionization for a 0.20 M aqueous NH_3 solution.

Plan

Write the equation for the ionization of aqueous NH_3 and represent the equilibrium concentrations algebraically. Then, substitute into the K_b expression and solve for $[OH^-]$ and $[NH_3]_{ionized}$.

TABLE 18-6	*Ionization Constants and* pK_b *Values for Some Weak Bases*		
Base	**Ionization Reaction**	K_b **at 25°C**	pK_b
ammonia	$NH_3 + H_2O \rightleftharpoons NH_4^+ + OH^-$	1.8×10^{-5}	4.74
methylamine	$(CH_3)NH_2 + H_2O \rightleftharpoons (CH_3)NH_3^+ + OH^-$	5.0×10^{-4}	3.30
dimethylamine	$(CH_3)_2NH + H_2O \rightleftharpoons (CH_3)_2NH_2^+ + OH^-$	7.4×10^{-4}	3.13
trimethylamine	$(CH_3)_3N + H_2O \rightleftharpoons (CH_3)_3NH^+ + OH^-$	7.4×10^{-5}	4.13
pyridine	$C_5H_5N + H_2O \rightleftharpoons C_5H_5NH^+ + OH^-$	1.5×10^{-9}	8.82
aniline	$C_6H_5NH_2 + H_2O \rightleftharpoons C_6H_5NH_3^+ + OH^-$	4.2×10^{-10}	9.38

aniline

pyridine

Solution

The equation for the ionization of aqueous ammonia and the algebraic representations of equilibrium concentrations follow. Let $x = [NH_3]_{ionized}$.

	NH_3	$+ H_2O \rightleftharpoons$	NH_4^+	$+ OH^-$
initial	$0.20\ M$		$0\ M$	$\approx 0\ M$
change due to rxn	$-x\ M$		$+x\ M$	$+x\ M$
at equil	$(0.20 - x)\ M$		$x\ M$	$x\ M$

Substitution into the ionization constant expression gives

$$K_b = \frac{[NH_4^+][OH^-]}{[NH_3]} = 1.8 \times 10^{-5} = \frac{(x)(x)}{(0.20 - x)}$$

Again, we can simplify this equation. The small value of K_b tells us that the base is only slightly ionized, so we can assume that $x \ll 0.20$, or $(0.20 - x) \approx 0.20$, and we have

$$\frac{x^2}{0.20} = 1.8 \times 10^{-5} \qquad x^2 = 3.6 \times 10^{-6} \qquad x = 1.9 \times 10^{-3}\ M$$

Then $[OH^-] = x = \boxed{1.9 \times 10^{-3}\ M,}$ pOH = 2.72, and pH = 11.28. $[NH_3]_{ionized} = x$, so the percent ionization may be calculated.

$$\% \text{ ionization} = \frac{[NH_3]_{ionized}}{[NH_3]_{initial}} \times 100\% = \frac{1.9 \times 10^{-3}}{0.20} \times 100\% = \boxed{0.95\% \text{ ionized}}$$

You should now work Exercises 54 and 110.

Since K_a and K_b values are usually known to only two significant figures, many answers in the following calculations will be limited to two significant figures. pH = 11.28, for example, has only two significant figures because the 11 relates to the exponent of ten.

The value of x is only about 1% of the original concentration, so the assumption is justified.

EXAMPLE 18-15   Household Ammonia

The pH of a household ammonia solution is 11.50. What is its molarity? Assume that ammonia is the only base (or acid) present.

Plan

We are given the pH of an aqueous NH_3 solution. Use pH + pOH = 14.00 to find pOH, which we can convert to $[OH^-]$. Then, complete the ionization reaction summary, and substitute the representations of equilibrium concentrations into the K_b expression.

Solution

At equilibrium pH = 11.50; we know that pOH = 2.50, so $[OH^-] = 10^{-2.50} = 3.2 \times 10^{-3}\ M$. This $[OH^-]$ results from the reaction, so we can write the change line. Then, letting x represent the *initial* concentration of NH_3, we can complete the reaction summary.

Measurement of the pH of a solution of household ammonia.

At equilibrium $[OH^-] = 3.2 \times 10^{-3}\,M$, so

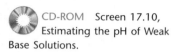

	NH_3	$+ H_2O \rightleftharpoons$	NH_4^+	$+$	OH^-
initial	$x\,M$		$0\,M$		$\approx 0\,M$
change	$-3.2 \times 10^{-3}\,M$		$+3.2 \times 10^{-3}\,M$		$+3.2 \times 10^{-3}\,M$
at equil	$(x - 3.2 \times 10^{-3})\,M$		$3.2 \times 10^{-3}\,M$		$3.2 \times 10^{-3}\,M$

Substituting these values into the K_b expression for aqueous NH_3 gives

$$K_b = \frac{[NH_4^+][OH^-]}{[NH_3]} = \frac{(3.2 \times 10^{-3})(3.2 \times 10^{-3})}{(x - 3.2 \times 10^{-3})} = 1.8 \times 10^{-5}$$

This suggests that $(x - 3.2 \times 10^{-3}) \approx x$. So we can approximate.

$$\frac{(3.2 \times 10^{-3})(3.2 \times 10^{-3})}{x} = 1.8 \times 10^{-5} \quad \text{and} \quad x = 0.57\,M\,NH_3$$

The solution is $0.57\,M\,NH_3$. Our assumption that $(x - 3.2 \times 10^{-3}) \approx x$ is justified.

You should now work Exercises 54, 56, and 58.

CD-ROM   Screen 17.10,
Estimating the pH of Weak
Base Solutions.

 We can *never* neglect an x term in multiplication or division!

✓ **Problem-Solving Tip:** *It Is Not Always x that Can Be Neglected*

Students sometimes wonder about the approximation in Example 18-15, thinking that only x can be neglected. We can consider neglecting one term in an expression only when the expression involves *addition* or *subtraction*. The judgment we must make is whether *either* of the terms is sufficiently smaller than the other that ignoring it would not significantly affect the result. In Example 18-15, x represents the *initial* concentration of NH_3; 3.2×10^{-3} represents the concentration that ionizes, which cannot be greater than the original x. We know that NH_3 is a *weak* base ($K_b = 1.8 \times 10^{-5}$), so only a small amount of the original ionizes. We can safely assume that $3.2 \times 10^{-3} \ll x$, so $(x - 3.2 \times 10^{-3}) \approx x$, the approximation that we used in solving the example.

Gaseous ammonia, NH_3, and gaseous hydrogen chloride, HCl, react to form solid NH_4Cl, the white smoke. In the reverse reaction, solid NH_4Cl decomposes when heated, to form gaseous NH_3 and HCl.

18-5 POLYPROTIC ACIDS

Thus far we have considered only *monoprotic* weak acids. Acids that can furnish *two* or more hydronium ions per molecule are called **polyprotic acids**. The ionizations of polyprotic acids occur stepwise, that is, one proton at a time. An ionization constant expression can be written for each step, as the following example illustrates. Consider phosphoric acid as a typical polyprotic acid. It contains three acidic hydrogen atoms and ionizes in three steps.

Each K_a expression includes $[H_3O^+]$, so each K_a expression must be satisfied by the *total* concentration of H_3O^+ in the solution.

$$H_3PO_4 + H_2O \rightleftharpoons H_3O^+ + H_2PO_4^- \qquad K_{a1} = \frac{[H_3O^+][H_2PO_4^-]}{[H_3PO_4]} = 7.5 \times 10^{-3}$$

$$H_2PO_4^- + H_2O \rightleftharpoons H_3O^+ + HPO_4^{2-} \qquad K_{a2} = \frac{[H_3O^+][HPO_4^{2-}]}{[H_2PO_4^-]} = 6.2 \times 10^{-8}$$

$$HPO_4^{2-} + H_2O \rightleftharpoons H_3O^+ + PO_4^{3-} \qquad K_{a3} = \frac{[H_3O^+][PO_4^{3-}]}{[HPO_4^{2-}]} = 3.6 \times 10^{-13}$$

We see that K_{a1} is much greater than K_{a2} and that K_{a2} is much greater than K_{a3}. This is generally true for polyprotic *inorganic* acids (Appendix F). Successive ionization constants often decrease by a factor of approximately 10^4 to 10^6, although some differences are outside this range. Large decreases in the values of successive ionization constants mean that each step in the ionization of a polyprotic acid occurs to a much lesser extent than the previous step. Thus, the $[H_3O^+]$ produced in the first step is very large compared with the $[H_3O^+]$ produced in the second and third steps. As we shall see, except in extremely dilute solutions of H_3PO_4, the concentration of H_3O^+ may be assumed to be that furnished by the first step in the ionization alone.

Citric acid is another example of a triprotic acid. It's structure is

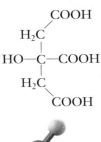

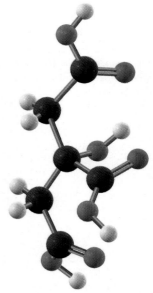

EXAMPLE 18-16 *Solutions of Weak Polyprotic Acid*

Calculate the concentrations of all species present in 0.100 M H_3PO_4.

Plan

Because H_3PO_4 contains three acidic hydrogens per formula unit, we show its ionization in three steps. For each step, write the appropriate ionization equation, with its K_a expression and value. Then, represent the equilibrium concentrations from the *first* ionization step, and substitute into the K_{a1} expression. Repeat the procedure for the second and third steps *in order*.

Solution

First we calculate the concentrations of all species that are formed in the first ionization step. Let x = mol/L of H_3PO_4 that ionize; then $x = [H_3O^+]_{1st} = [H_2PO_4^-]$.

$$H_3PO_4 \quad + H_2O \rightleftharpoons H_3O^+ + H_2PO_4^-$$
$$(0.100 - x)\ M \qquad\qquad x\ M \qquad x\ M$$

Substitution into the expression for K_{a1} gives

$$K_{a1} = \frac{[H_3O^+]_{1st}[H_2PO_4^-]}{[H_3PO_4]} = 7.5 \times 10^{-3} = \frac{(x)(x)}{(0.100 - x)}$$

This equation must be solved by the quadratic formula because K_{a1} is too large to neglect x relative to 0.100 M. Solving gives the *positive* root $x = 2.4 \times 10^{-2}$. Thus, from the first step in the ionization of H_3PO_4,

$x = -3.1 \times 10^{-2}$ is the extraneous root of the quadratic formula.

$$x\ M = \boxed{[H_3O^+]_{1st} = [H_2PO_4^-] = 2.4 \times 10^{-2}\ M}$$

$$(0.100 - x)\ M = \boxed{[H_3PO_4] = 7.6 \times 10^{-2}\ M}$$

For the second step, we use the $[H_3O^+]$ and $[H_2PO_4^-]$ from the first step. Let y = mol/L of $H_2PO_4^{2-}$ that ionize; then $y = [H_3O^+]_{2nd} = [HPO_4^{2-}]$.

$$H_2PO_4^- \quad + H_2O \rightleftharpoons \quad H_3O^+ \quad + HPO_4^{2-}$$
$$(2.4 \times 10^{-2} - y)\ M \qquad (2.4 \times 10^{-2} + y)M \quad y\ M$$
$$\text{from 1st step} \qquad \text{from 2nd step}$$

Substitution into the expression for K_{a2} gives

$$K_{a2} = \frac{[H_3O^+]_{1st\ \&\ 2nd}[HPO_4^{2-}]}{[H_2PO_4^-]} = 6.2 \times 10^{-8} = \frac{(2.4 \times 10^{-2} + y)(y)}{(2.4 \times 10^{-2} - y)}$$

Examination of this equation suggests that $y \ll 2.4 \times 10^{-2}$, so

$$\frac{(2.4 \times 10^{-2})(y)}{2.4 \times 10^{-2}} = 6.2 \times 10^{-8} \qquad y = \boxed{6.2 \times 10^{-8}\ M = [HPO_4^{2-}] = [H_3O^+]_{2nd}}$$

The pH of solutions of most polyprotic acids is governed by the first
ionization step.

We see that $[HPO_4^{2-}] = K_{a2}$ and $[H_3O^+]_{2nd} \ll [H_3O^+]_{1st}$. In general, in solutions of reasonable concentration of weak polyprotic acids for which $K_{a1} \gg K_{a2}$ and that contain no other
electrolytes, *the concentration of the anion produced in the second ionization step is always equal to K_{a2}*.

For the third step, we use $[H_3O^+]$ from the *first* step and $[HPO_4^{2-}]$ from the *second* step. Let
$z = mol/L$ of HPO_4^{2-} that ionize; then $z = [H_3O^+]_{3rd} = [PO_4^{3-}]$.

$$HPO_4^{2-} \;+\; H_2O \rightleftharpoons H_3O^+ \;+\; PO_4^{3-}$$
$$(6.2 \times 10^{-8} - z)\,M \qquad\qquad (2.4 \times 10^{-2} + z)\,M \quad z\,M$$

from 2nd step — from 3rd step   from 1st step   from 3rd step

$$K_{a3} = \frac{[H_3O^+][PO_4^{3-}]}{[HPO_4^{2-}]} = 3.6 \times 10^{-13} = \frac{(2.4 \times 10^{-2} + z)(z)}{(6.2 \times 10^{-8} - z)}$$

We make the usual simplifying assumption, and find that

$$z\,M = \boxed{9.3 \times 10^{-19}\,M = [PO_4^{3-}]} = [H_3O^+]_{3rd}$$

$[H_3O^+]_{2nd} = y = 6.2 \times 10^{-8}$ was disregarded in the second step and is
also disregarded here.

The $[H_3O^+]$ found in the three steps can be summarized:

$$
\begin{aligned}
[H_3O^+]_{1st} &= 2.4 \times 10^{-2}\,M = 0.024\,M \\
[H_3O^+]_{2nd} &= 6.2 \times 10^{-8}\,M = 0.000000062\,M \\
[H_3O^+]_{3rd} &= 9.3 \times 10^{-19}\,M = 0.00000000000000000093\,M \\
\hline
[H_3O^+]_{total} &= 2.4 \times 10^{-2}\,M = 0.024\,M
\end{aligned}
$$

We see that the H_3O^+ furnished by the second and third steps of ionization is negligible compared with that from the first step.

You should now work Exercise 60.

We have calculated the concentrations of the species formed by the ionization of
$0.100\,M\ H_3PO_4$. These concentrations are compared in Table 18-7. The concentration of
$[OH^-]$ in $0.100\,M\ H_3PO_4$ is included. It was calculated from the known $[H_3O^+]$ using the
ion product for water, $[H_3O^+][OH^-] = 1.0 \times 10^{-14}$.

Nonionized H_3PO_4 is present in greater concentration than any other species in
$0.100\,M\ H_3PO_4$ solution. The only other species present in significant concentrations are
H_3O^+ and $H_2PO_4^-$. Similar statements can be made for other weak polyprotic acids for
which the last K is very small.

TABLE 18-7	*Concentrations of the Species in 0.10 M H_3PO_4 (Example 18-16)*
Species	**Concentration (mol/L)**
H_3PO_4	7.6×10^{-2} = 0.076
H_3O^+	2.4×10^{-2} = 0.024
$H_2PO_4^-$	2.4×10^{-2} = 0.024
HPO_4^{2-}	6.2×10^{-8} = 0.000000062
OH^-	4.2×10^{-13} = 0.00000000000042
PO_4^{3-}	9.3×10^{-19} = 0.00000000000000000093

Phosphoric acid is a typical *weak* polyprotic acid. Let us now describe solutions of sulfuric acid, a *very strong* polyprotic acid.

EXAMPLE 18-17   *Solutions of Strong Polyprotic Acid*

Calculate concentrations of all species present in 0.10 M H_2SO_4. $K_{a2} = 1.2 \times 10^{-2}$.

Plan

Because the first ionization step of H_2SO_4 is complete, we read the concentrations for the first step from the balanced equation. The second ionization step is *not* complete, and so we write the ionization equation, the K_{a2} expression, and the algebraic representations of equilibrium concentrations. Then we substitute into K_{a2} for H_2SO_4.

Solution

As we pointed out, the first ionization step of H_2SO_4 is complete.

$$H_2SO_4 + H_2O \xrightarrow{100\%} H_3O^+ + HSO_4^-$$
$$0.10\ M \qquad\qquad 0.10\ M \quad 0.10\ M$$

The second ionization step is not complete, however.

$$HSO_4^- + H_2O \rightleftharpoons H_3O^+ + SO_4^{2-} \quad \text{and} \quad K_{a2} = \frac{[H_3O^+][SO_4^{2-}]}{[HSO_4^-]} = 1.2 \times 10^{-2}$$

Let $x = [HSO_4^-]$ that ionizes. $[H_3O^+]$ is the sum of the concentrations produced in the first and second steps. So we represent the equilibrium concentrations as

$$HSO_4^- + H_2O \rightleftharpoons H_3O^+ + SO_4^{2-}$$
$$(0.10 - x)\ M \qquad\qquad (0.10 + x)\ M \quad x\ M$$

from 1st step   from 2nd step

Substitution into the ionization constant expression for K_{a2} gives

$$K_{a2} = \frac{[H_3O^+][SO_4^{2-}]}{[HSO_4^-]} = 1.2 \times 10^{-2} = \frac{(0.10 + x)(x)}{0.10 - x}$$

Clearly, x cannot be disregarded because K_{a2} is too large. This equation must be solved by the quadratic formula, which gives $x = 0.010$ and $x = -0.12$ (extraneous). So $[H_3O^+]_{2nd} = [SO_4^{2-}] = 0.010\ M$. The concentrations of species in 0.10 M H_2SO_4 are

$$[H_2SO_4] \approx 0\ M \quad [HSO_4^-] = (0.10 - x)\ M = 0.09\ M \quad [SO_4^{2-}] = 0.010\ M$$
$$[H_3O^+] = (0.10 + x)\ M = 0.11\ M$$

$$[OH^-] = \frac{K_w}{[H_3O^+]} = \frac{1.0 \times 10^{-14}}{0.11} = 9.1 \times 10^{-14}\ M$$

In 0.10 M H_2SO_4 solution, the extent of the second ionization step is 10%.

You should now work Exercise 62.

In Table 18-8 we compare 0.10 M solutions of these two polyprotic acids. Their acidities are very different.

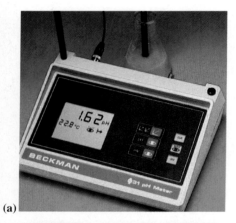

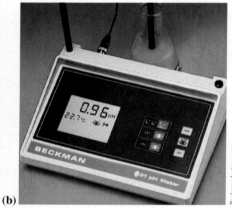

(a) pH of 0.10 M H_3PO_4. (b) pH of 0.10 M H_2SO_4.

	TABLE 18-8	Comparison of 0.10 M Solutions of Two Polyprotic Acids (Examples 18-16, 18-17)	
		0.10 M H_3PO_4	0.10 M H_2SO_4
K_{a1}		7.5×10^{-3}	very large
K_{a2}		6.2×10^{-8}	1.2×10^{-2}
K_{a3}		3.6×10^{-13}	—
$[H_3O^+]$		$2.4 \times 10^{-2} M$	$0.11 M$
pH		1.62	0.96
[nonionized molecules]		$7.6 \times 10^{-2} M$	$\approx 0 M$

18-6 SOLVOLYSIS

Solvolysis is the reaction of a substance with the solvent in which it is dissolved. The solvolysis reactions that we will consider in this chapter occur in aqueous solutions so they are called *hydrolysis* reactions. **Hydrolysis** is the reaction of a substance with water. Some hydrolysis reactions involve reaction with H_3O^+ or OH^- ions. One common kind of hydrolysis involves reaction of the anion of a *weak acid* with water to form *nonionized acid molecules* and OH^- ions. This upsets the H_3O^+/OH^- balance in water and produces basic solutions. This reaction is usually represented as

$$\underset{\substack{\text{anion of} \\ \text{weak acid}}}{A^-} + H_2O \rightleftharpoons \underset{\text{weak acid}}{HA} + OH^- \qquad \text{(excess } OH^-\text{, so solution is basic)}$$

Recall that in

basic solutions	$[H_3O^+] < [OH^-]$ or $[OH^-] > 1.0 \times 10^{-7} M$
neutral solutions	$[H_3O^+] = [OH^-] = 1.0 \times 10^{-7} M$
acidic solutions	$[H_3O^+] > [OH^-]$ or $[H_3O^+] > 1.0 \times 10^{-7} M$

Examples of conjugate acid–base pairs.

Acid		Conjugate Base
strong (HCl)	$\longrightarrow$	very weak (Cl^-)
weak (HCN)	$\longrightarrow$	stronger, but still weak (CN^-)

Base		Conjugate Acid
strong (OH^-)	$\longrightarrow$	very weak (H_2O)
weak (NH_3)	$\longrightarrow$	stronger, but still weak (NH_4^+)

In Brønsted–Lowry terminology anions of strong acids are extremely weak bases, whereas anions of weak acids are stronger bases (Section 10-4). To refresh your memory, consider the following examples.

Nitric acid, a common strong acid, is essentially completely ionized in dilute aqueous solution. *Dilute* aqueous solutions of HNO_3 contain equal concentrations of H_3O^+ and NO_3^- ions. In dilute aqueous solution nitrate ions show almost no tendency to react with H_3O^+ ions to form nonionized HNO_3; thus, NO_3^- is a very weak base.

$$HNO_3 + H_2O \xrightarrow{100\%} H_3O^+ + NO_3^-$$

On the other hand, acetic acid (a weak acid) is only slightly ionized in dilute aqueous solution. Acetate ions have a strong tendency to react with H_3O^+ to form CH_3COOH molecules.

$$CH_3COOH + H_2O \rightleftharpoons H_3O^+ + CH_3COO^-$$

Hence, the CH_3COO^- ion is a stronger base than the NO_3^- ion, but it is still weak.

In dilute solutions, strong acids and strong bases are completely ionized or dissociated. In the following sections we consider dilute aqueous solutions of salts. Based on our classification of acids and bases, we can identify four different kinds of salts.

1. Salts of strong bases and strong acids
2. Salts of strong bases and weak acids
3. Salts of weak bases and strong acids
4. Salts of weak bases and weak acids

$SC{\overset{i}{L}}INKS$.

TOPIC: Salts
GO TO: www.scilinks.org
*sci*LINKS CODE: WCH1820

CD-ROM   Screen 17.11, Acid–Base Properties of Salts: Hydrolysis.

18-7 SALTS OF STRONG BASES AND STRONG ACIDS

We could also describe these as salts that contain the cation of a strong base and the anion of a strong acid. Salts derived from strong bases and strong acids give *neutral* solutions because neither the cation nor the anion reacts appreciably with H_2O. Consider an aqueous solution of NaCl, which is the salt of the strong base NaOH and the strong acid HCl. Sodium chloride is ionic even in the solid state. It dissociates into hydrated ions in H_2O. H_2O ionizes slightly to produce equal concentrations of H_3O^+ and OH^- ions.

$$NaCl \text{ (solid)} \xrightarrow[100\%]{H_2O} Na^+ + Cl^-$$

$$H_2O + H_2O \rightleftharpoons OH^- + H_3O^+$$

We see that aqueous solutions of NaCl contain four ions, Na^+, Cl^-, H_3O^+ and OH^-. The cation of the salt, Na^+, is such a weak acid that it does not react appreciably with water. The anion of the salt, Cl^-, is such a weak base that it does not react appreciably with water. Solutions of salts of strong bases and strong acids are therefore *neutral* because neither ion of such a salt reacts to upset the H_3O^+/OH^- balance in water.

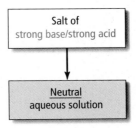

The salt of a strong acid and a strong base is often referred to as a *neutral salt*.

18-8 SALTS OF STRONG BASES AND WEAK ACIDS

When salts derived from strong bases and weak acids are dissolved in water, the resulting solutions are always basic. This is because anions of weak acids react with water to form hydroxide ions. Consider a solution of sodium acetate, $NaCH_3COO$, which is the salt of the strong base NaOH and the weak acid CH_3COOH. It is soluble and dissociates completely in water.

$$NaCH_3COO \text{(solid)} \xrightarrow[100\%]{H_2O} Na^+ + \boxed{CH_3COO^-}$$

$$\underset{\text{Equilibrium is shifted}}{H_2O + H_2O} \rightleftharpoons OH^- + \boxed{H_3O^+}$$

(result is excess OH^-)

$$\Updownarrow$$

$$CH_3COOH + H_2O$$

Acetate ion is the conjugate base of a *weak* acid, CH_3COOH. Thus, it combines to some extent with H_3O^+ to form CH_3COOH. As H_3O^+ is removed from the solution, causing

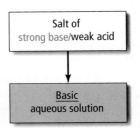

Though the salt of a strong base and a weak acid is not strictly included in the definition of a basic salt (section 10-9), many chemists refer to such a salt as a *weakly basic salt*.

This is like the reaction of a molecular weak base

$$NH_3 + H_2O \rightleftharpoons NH_4^+ + OH^-$$

(excess OH^- is produced; the solution becomes basic)

Just as we can write a K_b expression to describe the extent of ionization of NH_3, we can do the same for the acetate ion.

more H_2O to ionize, an excess of OH^- builds up. So the solution becomes basic. The net result of the preceding equations can be written as a single equation. This equation describes the *hydrolysis of acetate ions.*

$$CH_3COO^- + H_2O \rightleftharpoons CH_3COOH + OH^-$$

The equilibrium constant for this reaction is called a (base) hydrolysis constant, or K_b for CH_3COO^-.

$$K_b = \frac{[CH_3COOH][OH^-]}{[CH_3COO^-]} \qquad (K_b \text{ for } CH_3COO^-)$$

We can evaluate this equilibrium constant from other known expressions. We multiply the preceding expression by $[H_3O^+]/[H_3O^+]$ to give

$$K_b = \frac{[CH_3COOH][OH^-]}{[CH_3COO^-]} \times \frac{[H_3O^+]}{[H_3O^+]} = \frac{[CH_3COOH]}{[H_3O^+][CH_3COO^-]} \times \frac{[H_3O^+][OH^-]}{1}$$

We recognize that

$$K_b = \frac{1}{K_{a\,(CH_3COOH)}} \times \frac{K_w}{1} = \frac{K_w}{K_{a\,(CH_3COOH)}} = \frac{1.0 \times 10^{-14}}{1.8 \times 10^{-5}}$$

which gives

$$K_b = \frac{[CH_3COOH][OH^-]}{[CH_3COO^-]} = 5.6 \times 10^{-10}$$

We have calculated K_b, the hydrolysis constant for the acetate ion, CH_3COO^-.

Base hydrolysis constants, K_b's, for anions of weak acids can be determined experimentally. The values obtained from experiments agree with the calculated values. Please note that this K_b refers to a reaction in which the anion of a weak acid acts as a base.

We can do the same kind of calculations for the anion of any weak monoprotic acid and find that $K_b = K_w/K_a$, where K_a refers to the ionization constant for the weak monoprotic acid from which the anion is derived.

This equation can be rearranged to

$$K_w = K_a K_b \qquad \text{(valid for } any \text{ conjugate acid–base pair in aqueous solution)}$$

If either K_a or K_b is known, the other can be calculated.

For conjugate acid–base pairs

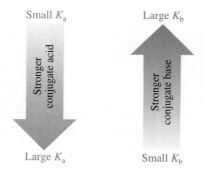

EXAMPLE 18-18 K_b *for the Anion of a Weak Acid*

(a) Write the equation for the reaction of the base CN^- with water. (b) The value of the ionization constant for hydrocyanic acid, HCN, is 4.0×10^{-10}. What is the value of K_b for the cyanide ion, CN^-?

Plan

(a) The base CN^- accepts H^+ from H_2O to form the weak acid HCN and OH^- ions. (b) We know that $K_a K_b = K_w$. So we solve for K_b and substitute into the equation.

Solution

(a) $CN^- + H_2O \rightleftharpoons HCN + OH^-$

(b) We are given $K_a = 4.0 \times 10^{-10}$ for HCN, and we know that $K_w = 1.0 \times 10^{-14}$.

$$K_{b(CN^-)} = \frac{K_w}{K_a} = \frac{1.0 \times 10^{-14}}{4.0 \times 10^{-10}} = 2.5 \times 10^{-5}$$

You should now work Exercises 78 and 80.

EXAMPLE 18-19 *Calculations Based on Hydrolysis*

Calculate $[OH^-]$, pH, and the percent hydrolysis for 0.10 M solutions of (a) sodium acetate, $NaCH_3COO$, and (b) sodium cyanide, $NaCN$.

Plan

We recall that all sodium salts are soluble and that all soluble ionic salts are completely dissociated in H_2O. We recognize that both $NaCH_3COO$ and $NaCN$ are salts of strong bases and weak acids. The anions in such salts hydrolyze to give basic solutions. In the preceding text we determined that K_b for $CH_3COO^- = 5.6 \times 10^{-10}$, and in Example 18-18 we determined that K_b for $CN^- = 2.5 \times 10^{-5}$. As we have done before, we first write the appropriate chemical equation and equilibrium constant expression. Then we complete the reaction summary, substitute the algebraic representations of equilibrium concentrations into the equilibrium constant expression, and solve for the desired concentration(s).

> If we did not know either K_b value, we could use $K_{a(CH_3COOH)}$ to find $K_{b(CH_3COO^-)}$ or $K_{a(HCN)}$ to find $K_{b(CN^-)}$.

Solution

(a) The overall equation for the reaction of CH_3COO^- with H_2O and its equilibrium constant expression are

$$CH_3COO^- + H_2O \rightleftharpoons CH_3COOH + OH^-$$

$$K_b = \frac{[CH_3COOH][OH^-]}{[CH_3COO^-]} = 5.6 \times 10^{-10}$$

Let x = mol/L of CH_3COO^- that hydrolyzes. Then $x = [CH_3COOH] = [OH^-]$.

	CH_3COO^- + H_2O $\longrightarrow$	CH_3COOH +	OH^-
initial	0.10 M	0 M	$\approx 0\ M$
change due to rxn	$-x\ M$	$+x\ M$	$+x\ M$
at equil	$(0.10 - x)\ M$	$x\ M$	$x\ M$

Because the value of K_b (5.6×10^{-10}) is quite small, we know that the reaction does not go very far. We can assume $x \ll 0.10$, so $(0.10 - x) \approx 0.10$; this lets us simplify the equation to

$$5.6 \times 10^{-10} = \frac{(x)(x)}{0.10} \quad \text{so} \quad x = 7.5 \times 10^{-6}$$

$$x = \boxed{7.5 \times 10^{-6}\ M = [OH^-]} \quad pOH = 5.12 \quad \text{and} \quad \boxed{pH = 8.88} \qquad pH = 14 - pOH$$

The 0.10 M $NaCH_3COO$ solution is distinctly basic.

$$\% \text{ hydrolysis} = \frac{[CH_3COO^-]_{\text{hydrolyzed}}}{[CH_3COO^-]_{\text{initial}}} \times 100\% = \frac{7.5 \times 10^{-6}\ M}{0.10\ M} \times 100\%$$

$$= \boxed{0.0075\% \text{ hydrolysis}}$$

(b) Perform the same kind of calculation for 0.10 M $NaCN$. Let y = mol/L of CN^- that hydrolyzes. Then $y = [HCN] = [OH^-]$.

(a)

(b)

The pH of 0.10 M NaCH$_3$COO is 8.88. The pH of 0.10 M NaCN is 11.20. An inert solid has been suspended in the liquids to improve the quality of these photographs.

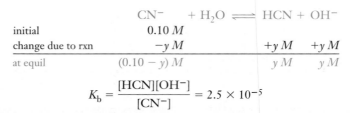

$$CN^- + H_2O \rightleftharpoons HCN + OH^-$$

	$CN^- + H_2O \rightleftharpoons$	HCN	$+ OH^-$
initial	0.10 M		
change due to rxn	$-y\,M$	$+y\,M$	$+y\,M$
at equil	$(0.10 - y)\,M$	$y\,M$	$y\,M$

$$K_b = \frac{[HCN][OH^-]}{[CN^-]} = 2.5 \times 10^{-5}$$

Substitution into this expression gives

$$\frac{(y)(y)}{(0.10 - y)} = 2.5 \times 10^{-5} \quad \text{so} \quad y = 1.6 \times 10^{-3}\,M$$

$$y = [OH^-] = 1.6 \times 10^{-3}\,M \qquad pOH = 2.80 \quad \text{and} \quad \boxed{pH = 11.20}$$

The 0.10 M NaCN solution is even more basic than the 0.10 M NaCH$_3$COO solution in Part (a).

$$\% \text{ hydrolysis} = \frac{[CN^-]_{\text{hydrolyzed}}}{[CN^-]_{\text{initial}}} \times 100\% = \frac{1.6 \times 10^{-3}\,M}{0.10\,M} \times 100\%$$

$$= 1.6\% \text{ hydrolysis}$$

You should now work Exercises 83 and 84.

The 0.10 M solution of NaCN is much more basic than the 0.10 M solution of NaCH$_3$COO because CN$^-$ is a much stronger base than CH$_3$COO$^-$. This is expected because HCN is a much weaker acid than CH$_3$COOH so that the K_b for CN$^-$ is much larger than the K_b for CH$_3$COO$^-$.

The percent hydrolysis for 0.10 M CN$^-$ (1.6%) is about 210 times greater than the percent hydrolysis for 0.10 M CH$_3$COO$^-$ (0.0075%). In Table 18-9 we compare 0.10 M solutions of CH$_3$COO$^-$, CN$^-$, and NH$_3$ (the familiar molecular weak base). We see that CH$_3$COO$^-$ is a much weaker base than NH$_3$, whereas CN$^-$ is a slightly stronger base than NH$_3$.

Beckman Instruments

TABLE 18-9	*Data for 0.10 M Solutions of NaCH$_3$COO, NaCN, and NH$_3$*		
	0.10 M NaCH$_3$COO	**0.10 M NaCN**	**0.10 M aq NH$_3$**
K_a for parent acid	1.8×10^{-5}	4.0×10^{-10}	
K_b for anion	5.6×10^{-10}	2.5×10^{-5}	K_b for NH$_3$ = 1.8×10^{-5}
$[OH^-]$	$7.5 \times 10^{-6}\,M$	$1.6 \times 10^{-3}\,M$	$1.3 \times 10^{-3}\,M$
% hydrolysis	0.0075%	1.6%	1.3% ionized
pH	8.88	11.20	11.11

18-9 SALTS OF WEAK BASES AND STRONG ACIDS

Ammonium chloride is an ionic salt that is soluble in water.

The second common kind of hydrolysis reaction involves the reaction of the cation of a weak base with water to form nonionized molecules of the weak base and H$_3$O$^+$ ions. This upsets the H$_3$O$^+$/OH$^-$ balance in water, giving an excess of H$_3$O$^+$, and making such solutions *acidic*. Consider a solution of ammonium chloride, NH$_4$Cl, the salt of aqueous NH$_3$ and HCl.

$$NH_4Cl(solid) \xrightarrow[100\%]{H_2O} \boxed{NH_4^+} + Cl^- \qquad \text{(result is excess } H_3O^+\text{)}$$

$$\underbrace{H_2O + H_2O \rightleftharpoons \boxed{OH^- } + H_3O^+}_{\text{Equilibrium is shifted}}$$

$$\Updownarrow$$

$$NH_3 + H_2O$$

Salt of
weak base/strong acid

↓

Acidic
aqueous solution

Ammonium ions from NH_4Cl react to some extent with OH^- to form nonionized NH_3 and H_2O molecules. This reaction removes OH^- from the system, so it causes more H_2O to ionize to produce an excess of H_3O^+. The net result of the preceding equations can be written as a single equation with its equilibrium constant expression.

$$NH_4^+ + H_2O \rightleftharpoons NH_3 + H_3O^+ \qquad K_a = \frac{[NH_3][H_3O^+]}{[NH_4^+]}$$

The expression $K_w = K_aK_b$ is valid for *any* conjugate acid–base pair in aqueous solution. We use it for the NH_4^+/NH_3 pair.

$$K_{a\,(NH_4^+)} = \frac{K_w}{K_{b\,(NH_3)}} = \frac{1.0 \times 10^{-14}}{1.8 \times 10^{-5}} = 5.6 \times 10^{-10} = \frac{[NH_3][H_3O^+]}{[NH_4^+]}$$

The fact that K_a for the ammonium ion, NH_4^+, is the same as K_b for the acetate ion should not be surprising. Recall that the ionization constants for CH_3COOH and aqueous NH_3 are equal (by coincidence). Thus, we expect CH_3COO^- to hydrolyze to the same extent as NH_4^+ does.

Many chemists refer to the salt produced by the reaction of a weak base and a strong acid as a *weakly acidic salt.*

Similar equations can be written for cations derived from other weak bases such as $CH_3NH_3^+$ and $(CH_3)_2NH_2^+$.

If you wish to derive $K_a = K_w/K_b$ for this case, multiply the K_a expression by $[OH^-]/[OH^-]$ and simplify.

EXAMPLE 18-20   *pH of a Soluble Salt of a Strong Acid and a Weak Base*

Calculate the pH of a 0.20 *M* solution of ammonium nitrate, NH_4NO_3.

Plan

We recognize that NH_4NO_3 is the salt of a weak base, NH_3, and a strong acid, HNO_3, and that the cations of such salts hydrolyze to give acidic solutions. From our earlier calculations, we know that K_a for $NH_4^+ = 5.6 \times 10^{-10}$. We proceed as we did in Example 18-19.

Solution

The cation of the weak base reacts with H_2O. Let x = mol/L of NH_4^+ that hydrolyzes. Then $x = [NH_3] = [H_3O^+]$.

	NH_4^+	$+ H_2O \rightleftharpoons$	NH_3	$+ H_3O^+$
initial	0.20 *M*		0 *M*	≈0 *M*
change due to rxn	$-x$ *M*		$+x$ *M*	$+x$ *M*
at equil	$(0.20 - x)$ *M*		x *M*	x *M*

Substituting into the K_a expression gives

$$K_a = \frac{[NH_3][H_3O^+]}{[NH_4^+]} = \frac{(x)(x)}{0.20 - x} = 5.6 \times 10^{-10}$$

Making the usual simplifying assumption gives $x = 1.1 \times 10^{-5}$ *M* $= [H_3O^+]$ and pH = 4.96. The 0.20 *M* NH_4NO_3 solution is distinctly acidic.

You should now work Exercise 90.

Ammonium nitrate is widely used as a fertilizer because of its high nitrogen content. It contributes to soil acidity.

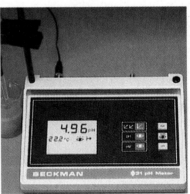

The pH of 0.20 *M* NH_4NO_3 solution is 4.96.

Taming Acids with Harmless Salts

From yellowing paper in old books and newsprint to heartburn and environmental spills, many Americans encounter unwanted or excessive amounts of acid. Neutralizing unwanted acids with hydroxide bases might appear to be a good way to combat these acids. But even more effective chemicals exist that can neutralize acids without the risks posed by hydroxide bases. These acid-neutralizing chemicals are salts of weak acids and strong bases. Such salts can neutralize acids because hydrolysis makes their aqueous solutions basic. More importantly, there is a significant advantage in using relatively harmless salts such as sodium hydrogen carbonate (baking soda) rather than stronger bases such as sodium hydroxide (lye). For example, if we used too much sodium hydroxide to neutralize sulfuric acid spilled from a car battery, any excess lye left behind would pose an environmental and human health threat about equal to that of the spilled sulfuric acid. (Lye is the major ingredient in such commercial products as oven cleaners and Drāno.) We would not be concerned, however, if a little baking soda were left on the ground after the sulfuric acid from the car battery had been neutralized.

The same principle applies to acid indigestion. Rather than swallow lye (ugh!) or some other strong base to neutralize excess stomach acid, most people take antacids. Antacids typically contain salts such as calcium carbonate, sodium hydrogen carbonate (sodium bicarbonate), and magnesium carbonate, all of which are salts of weak acids. These salts hydrolyze to form hydroxide ions that reduce the degree of acidity in the stomach. Physicians also prescribe these and similar salts to treat peptic ulcers. The repeated use of antacids should always be under the supervision of a physician.

Salts of weak acids and strong bases can be used effectively against a major acid spill in much the same way they are used against sulfuric acid from a car battery or excess stomach acid. In a recent major acid spill, a tank car filled with nitric acid was punctured by the coupling of another rail car, spilling 22,000 gallons of concentrated nitric acid onto the ground. Many thousand residents living near the spill were evacuated. There were no fatalities or serious injuries, and there was no major environmental damage; resident fire fighters neutralized the concentrated nitric acid by using airport snow blowers to spread *relatively* harmless sodium carbonate (washing soda) over the contaminated area.

Salts of weak acids and strong bases are also being used to combat the destructive aging process of paper. Think how serious this problem is for the Library of Congress, which loses 70,000 books each year to the decomposition of aging paper. Many of the twenty million books in the Library of Congress have a life expectancy of only 25 to 40 years. Paper ages because of the hydrolysis of aluminum sulfate. Aluminum sulfate has been used in the paper manufacturing process since the 1850s because it is an inexpensive sizing compound (it keeps ink from spreading out on paper). Aluminum sulfate is the salt of an insoluble weak base and a strong acid; it hydrolyzes in the water in paper (typically 4–7% H_2O) to give an acidic environment. The acid eats away at cellulose fibers, which causes the paper to turn yellow and eventually disintegrate. To combat this aging, the Library of Congress individually treats its collections with solutions of salts of weak acids and strong bases at great cost. Meanwhile, the paper industry is fighting this aging process by increasing its output of alkaline paper. Some alkaline paper contains calcium carbonate, the same salt found in several brands of antacids. Calcium carbonate increases the pH of paper to between 7.5 and 8.5. Special manufacturing techniques produce calcium carbonate that is very fine and that has uniform particle size. Alkaline papers are expected to last about 300 years, in contrast to the average 25- to 40-year life expectancy of standard acidic paper.

Salts that hydrolyze to produce basic solutions can settle upset stomachs, prevent yellowing pages, and neutralize major and minor acid spills. A knowledge of hydrolysis is very useful and has many applications.

Ronald DeLorenzo
Middle Georgia College

Science Museum Library

18-10 SALTS OF WEAK BASES AND WEAK ACIDS

Salts of weak bases and weak acids are the fourth class of salts. Most are soluble. Salts of weak bases and weak acids contain cations that would give acidic solutions and anions that would give basic solutions. Will solutions of such salts be neutral, basic, or acidic? They may be any one of the three depending on the relative strengths of the weak molecular acid and weak molecular base from which each salt is derived. Thus, salts of this class may be divided into three types that depend on the relative strengths of their parent weak bases and weak acids.

Salts of Weak Bases and Weak Acids for Which $K_b = K_a$

The common example of a salt of this type is ammonium acetate, NH_4CH_3COO, the salt of aqueous NH_3 and CH_3COOH. The ionization constants for both aqueous NH_3 and CH_3COOH are 1.8×10^{-5}. We know that ammonium ions react with water to produce H_3O^+.

$$NH_4^+ + H_2O \rightleftharpoons NH_3 + H_3O^+ \qquad K_a = \frac{[NH_3][H_3O^+]}{[NH_4^+]} = 5.6 \times 10^{-10}$$

We also recall that acetate ions react with water to produce OH^-.

$$CH_3COO^- + H_2O \rightleftharpoons CH_3COOH + OH^-$$

$$K_b = \frac{[CH_3COOH][OH^-]}{[CH_3COO^-]} = 5.6 \times 10^{-10}$$

Because these K values are equal, the NH_4^+ produces just as many H_3O^+ ions as the CH_3COO^- produces OH^- ions. Thus, we predict that ammonium acetate solutions are neutral, and they are. There are very few salts that have cations and anions with equal K values.

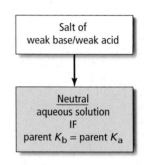

Salts of Weak Bases and Weak Acids for Which $K_b > K_a$

Salts of weak bases and weak acids for which K_b is greater than K_a are always basic because the anion of the weaker acid hydrolyzes to a greater extent than the cation of the stronger base.

Consider NH_4CN, ammonium cyanide. K_a for HCN (4.0×10^{-10}) is much smaller than K_b for NH_3 (1.8×10^{-5}), so K_b for CN^- (2.5×10^{-5}) is much larger than K_a for NH_4^+ (5.6×10^{-10}). This tells us that the CN^- ions hydrolyze to a much greater extent than do NH_4^+ ions, and so ammonium cyanide solutions are distinctly basic. Stated differently, CN^- is much stronger as a base than NH_4^+ is as an acid.

$$\left.\begin{array}{l} NH_4^+ + H_2O \rightleftharpoons NH_3 + H_3O^+ \\ CN^- + H_2O \rightleftharpoons HCN + OH^- \end{array}\right\} \rightarrow 2H_2O$$

The second reaction occurs to greater extent ∴ the solution is basic.

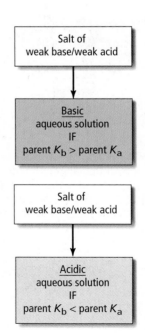

Salts of Weak Bases and Weak Acids for Which $K_b < K_a$

Salts of weak bases and weak acids for which K_b is less than K_a are acidic because the cation of the weaker base hydrolyzes to a greater extent than the anion of the stronger acid.

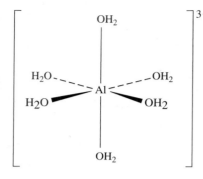

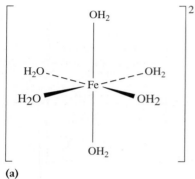

(a)

(b)

Figure 18-3 (a) Lewis structures of hydrated aluminum ions, $[Al(OH_2)_6]^{3+}$, and hydrated iron(II) ions, $[Fe(OH_2)_6]^{2+}$. (b) Ball-and-stick model of either of these ions.

Consider ammonium fluoride, NH_4F, the salt of aqueous ammonia and hydrofluoric acid. K_b for aqueous NH_3 is 1.8×10^{-5} and K_a for HF is 7.2×10^{-4}. So the K_a value for NH_4^+ (5.6×10^{-10}) is slightly larger than the K_b value for F^- (1.4×10^{-11}). This tells us that NH_4^+ ions hydrolyze to a slightly greater extent than F^- ions. In other words, NH_4^+ is slightly stronger as an acid than F^- is as a base. Ammonium fluoride solutions are slightly acidic.

$$NH_4^+ + H_2O \rightleftharpoons NH_3 + H_3O^+$$
$$F^- + H_2O \rightleftharpoons HF + OH^-$$
$$\longrightarrow 2H_2O$$

The first reaction occurs to greater extent ∴ the solution is acidic.

EXAMPLE 18-21   *Predicting Which Salts Are Acidic and Which Are Basic*

Classify each of the following soil additives as an acidic salt, a neutral salt, or a basic salt.
(a) $(NH_4)_2SO_4$ (b) NH_4NO_3 (c) Na_2CO_3

Plan

We recognize these salts as being the products of acid/base reactions. We can categorize each parent acid and base as being stronger and weaker. We then predict the properties of each salt based upon the strength of its parent acid and parent base.

Solution

(a) $(NH_4)_2SO_4$ is the salt formed by the reaction of NH_3 (a weak base) and H_2SO_4 (a strong acid). Because H_2SO_4 is stronger as an acid than NH_3 is as a base, $(NH_4)_2SO_4$ is an acidic salt.
(b) NH_4NO_3 is the salt formed by the reaction of NH_3 (a weak base) and HNO_3 (a strong acid). Because HNO_3 is stronger as an acid than NH_3 is as a base, NH_4NO_3 is an acidic salt.
(c) Na_2CO_3 is the salt formed by the reaction of NaOH (a strong base) and H_2CO_3 (a weak acid). Because NaOH is stronger as a base then H_2CO_3 is as an acid, Na_2CO_3 is a basic salt.

You should now work Exercise 106.

18-11 SALTS THAT CONTAIN SMALL, HIGHLY CHARGED CATIONS

Solutions of certain common salts of strong acids are acidic. For this reason, many homeowners apply iron(II) sulfate, $FeSO_4 \cdot 7H_2O$, or aluminum sulfate, $Al_2(SO_4)_3 \cdot 18H_2O$, to the soil around "acid-loving" plants such as azaleas, camelias, and hollies. You have probably tasted the sour, "acid" taste of alum, $KAl(SO_4)_2 \cdot 12H_2O$, a substance that is frequently added to pickles.

Each of these salts contains a small, highly charged cation and the anion of a strong acid. Solutions of such salts are acidic because these cations hydrolyze to produce excess hydronium ions. Consider aluminum chloride, $AlCl_3$, as a typical example. When solid anhydrous $AlCl_3$ is added to water, the water becomes very warm as the Al^{3+} ions become hydrated in solution. In many cases, the interaction between positively charged ions and the negative ends of polar water molecules is so strong that salts crystallized from aqueous solution contain definite numbers of water molecules. Salts containing Al^{3+}, Fe^{2+}, Fe^{3+}, and Cr^{3+} ions usually crystallize from aqueous solutions with six water molecules bonded (coordinated) to each metal ion. These salts contain the hydrated cations $[Al(OH_2)_6]^{3+}$, $[Fe(OH_2)_6]^{2+}$, $[Fe(OH_2)_6]^{3+}$, and $[Cr(OH_2)_6]^{3+}$, respectively, in the solid state. Such species also exist in aqueous solutions. Each of these species is octahedral, meaning that the metal ion (M^{n+}) is located at the center of a regular octahedron, and the O atoms in six H_2O molecules are

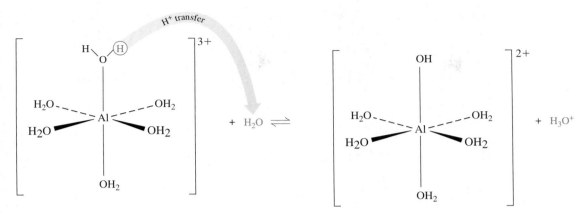

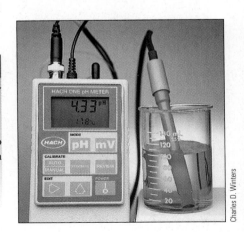

Figure 18-4 Hydrolysis of hydrated aluminum ions to produce H_3O^+—that is, the removal of a proton from a coordinated H_2O molecule by a noncoordinated one.

located at the corners (Figure 18-3). In the metal–oxygen bonds of the hydrated cation, electron density is decreased around the O end of each H_2O molecule by the positively charged metal ion. This weakens the H—O bonds in coordinated H_2O molecules relative to the H—O bonds in noncoordinated H_2O molecules. Consequently, the coordinated H_2O molecules can donate H^+ to solvent H_2O molecules to form H_3O^+ ions. This produces acidic solutions (Figure 18-4).

The equation for the hydrolysis of hydrated Al^{3+} is written as follows.

$$[Al(OH_2)_6]^{3+} + H_2O \rightleftharpoons [Al(OH)(OH_2)_5]^{2+} + H_3O^+$$

$$K_a = \frac{[[Al(OH)(OH_2)_5]^{2+}][H_3O^+]}{[[Al(OH_2)_6]^{3+}]} = 1.2 \times 10^{-5}$$

Removing an H^+ converts a coordinated water molecule to a coordinated hydroxide ion and decreases the positive charge on the hydrated species.

Hydrolysis of hydrated small, highly charged cations may occur beyond the first step. In many cases these reactions are quite complex. They may involve two or more cations reacting with each other to form large polymeric species. For most common hydrated cations, consideration of the first hydrolysis constant is adequate for our calculations.

EXAMPLE 18-22 *Percent Hydrolysis*

Calculate the pH and percent hydrolysis in 0.10 *M* $AlCl_3$ solution. $K_a = 1.2 \times 10^{-5}$ for $[Al(OH_2)_6]^{3+}$ (often abbreviated Al^{3+}).

Plan

We recognize that $AlCl_3$ produces a hydrated, small, highly charged cation that hydrolyzes to give an acidic solution. We represent the equilibrium concentrations and proceed as we did in earlier examples.

Solution

The equation for the reaction and its hydrolysis constant can be represented as

$$[Al(OH_2)_6]^{3+} + H_2O \rightleftharpoons [Al(OH)(OH_2)_5]^{2+} + H_3O^+$$

$$K_a = \frac{[[Al(OH)(OH_2)_5]^{2+}][H_3O^+]}{[[Al(OH_2)_6]^{3+}]} = 1.2 \times 10^{-5}$$

The blue color and the acidity of a 0.10 *M* $CuSO_4$ solution are both due to the hydrated Cu^{2+} ion.

Let x = mol/L of $[Al(OH_2)_6]^{3+}$ that hydrolyzes. Then x = $[Al(OH)(OH_2)_5]^{2+}$ = $[H_3O^+]$.

	$[Al(OH_2)_6]^{3+}$	+ H_2O ⇌	$[Al(OH)(OH_2)_5]^{2+}$	+ H_3O^+
initial	0.10 M			
change due to rxn	$-x\ M$		$+x\ M$	$+x\ M$
at equil	$(0.10 - x)\ M$		$x\ M$	$x\ M$

$$\frac{(x)(x)}{(0.10 - x)} = 1.2 \times 10^{-5} \quad \text{so} \quad x = 1.1 \times 10^{-3}$$

$[H_3O^+] = 1.1 \times 10^{-3}\ M$, pH = 2.96, and the solution is quite acidic.

$$\% \text{ hydrolysis} = \frac{[Al^{3+}]_{\text{hydrolyzed}}}{[Al^{3+}]_{\text{initial}}} \times 100\% = \frac{1.1 \times 10^{-3}\ M}{0.10\ M} \times 100\% = 1.1\% \text{ hydrolyzed}$$

$[Al(OH_2)_6]^{3+}$ is often abbreviated as Al^{3+}. Recall that x represents the concentration of Al^{3+} that hydrolyzes.

The pH of 0.10 M $AlCl_3$ is 2.96. The pH of 0.10 M CH_3COOH is 2.89.

As a reference point, CH_3COOH is 1.3% ionized in 0.10 M solution (see Example 18-11). In 0.10 M solution $AlCl_3$ is 1.1% hydrolyzed. The acidities of the two solutions are very similar.

You should now work Exercise 96.

Smaller, more highly charged cations are stronger acids than larger, less highly charged cations (Table 18-10). This is because the smaller, more highly charged cations interact with coordinated water molecules more strongly.

For isoelectronic cations in the same period in the periodic table, the smaller, more highly charged cation is the stronger acid. (Compare K_a values for hydrated Li^+ and Be^{2+} and for hydrated Na^+, Mg^{2+}, and Al^{3+}.) For cations with the same charge from the same group in the periodic table, the smaller cation hydrolyzes to a greater extent. (Compare K_a

Pepto-Bismol contains $BiO(HOC_4H_6COO)$, bismuth subsalicylate, a *hydrolyzed* bismuth salt. Such salts "coat" polar surfaces such as glass and the lining of the stomach.

TABLE 18-10	Ionic Radii and Hydrolysis Constants for Some Cations		
Cation	**Ionic Radius (Å)**	**Hydrated Cation**	K_a
Li^+	0.90	$[Li(OH_2)_4]^+$	1×10^{-14}
Be^{2+}	0.59	$[Be(OH_2)_4]^{2+}$	1.0×10^{-5}
Na^+	1.16	$[Na(OH_2)_6]^+$ (?)	10^{-14}
Mg^{2+}	0.85	$[Mg(OH_2)_6]^{2+}$	3.0×10^{-12}
Al^{3+}	0.68	$[Al(OH_2)_6]^{3+}$	1.2×10^{-5}
Fe^{2+}	0.76	$[Fe(OH_2)_6]^{2+}$	3.0×10^{-10}
Fe^{3+}	0.64	$[Fe(OH_2)_6]^{3+}$	4.0×10^{-3}
Co^{2+}	0.74	$[Co(OH_2)_6]^{2+}$	5.0×10^{-10}
Co^{3+}	0.63	$[Co(OH_2)_6]^{3+}$	1.7×10^{-2}
Cu^{2+}	0.96	$[Cu(OH_2)_6]^{2+}$	1.0×10^{-8}
Zn^{2+}	0.74	$[Zn(OH_2)_6]^{2+}$	2.5×10^{-10}
Hg^{2+}	1.10	$[Hg(OH_2)_6]^{2+}$	8.3×10^{-7}
Bi^{3+}	0.74	$[Bi(OH_2)_6]^{3+}$	1.0×10^{-2}

Charles Steele

values for hydrated Be^{2+} and Mg^{2+}.) If we compare cations of the same element in different oxidation states, the smaller, more highly charged cation is the stronger acid. (Compare K_a values for hydrated Fe^{2+} and Fe^{3+} and for hydrated Co^{2+} and Co^{3+}.)

Key Terms

Acidic salt A salt that contains an ionizable hydrogen atom, typically formed from the reaction of a weak base with a strong acid.

Amines Derivatives of ammonia in which one or more hydrogen atoms has been replaced by organic groups.

Basic salt A salt formed from the reaction of a weak acid and a strong base.

Hydrolysis The reaction of a substance with water.

Hydrolysis constant An equilibrium constant for a hydrolysis reaction.

Indicator An organic compound that exhibits different colors in solutions of different acidities.

Ion product for water An equilibrium constant for the ionization of water,

$$K_w = [H_3O^+][OH^-] = 1.0 \times 10^{-14} \text{ at } 25°C$$

Ionization constant An equilibrium constant for the ionization of a weak electrolyte.

Monoprotic acid An acid that can form only one hydronium ion per molecule; may be strong or weak.

Neutral salt A salt formed from the reaction of an acid and a base of equal strength.

pH The negative logarithm of the concentration (mol/L) of the H_3O^+ (or H^+) ion; the commonly used scale ranges from 0 to 14.

pK_a The negative logarithm of K_a, the ionization constant for a weak acid.

pK_b The negative logarithm of K_b, the ionization constant for a weak base.

pK_w The negative logarithm of the ion product for water.

pOH The negative logarithm of the concentration (mol/L) of the OH^- ion; the commonly used scale ranges from 14 to 0.

Polyprotic acid An acid that can form two or more hydronium ions per molecule.

Solvolysis The reaction of a substance with the solvent in which it is dissolved.

Exercises

NOTE *All exercises in this chapter assume a temperature of 25°C unless they specify otherwise. All logarithms are common (base 10).*

Review of Strong Electrolytes

1. List names and formulas for (a) seven strong acids; (b) six weak bases; (c) the common strong bases; (d) ten soluble ionic salts.

2. (a) How are a strong acid and a weak acid similar? How are they different? (b) How are a strong base and a weak base similar? How are they different?

3. Which of the following are strong electrolytes: a Group IA hydroxide, a Group IIIA hydroxide, $Cu(OH)_2$, $Be(OH)_2$, H_3AsO_3, HBr, H_3PO_4, salts of a Group IA metal?

4. Calculate the molarity of each of the following solutions. (a) 17.52 g of NaCl in 175 mL of solution; (b) 50.5 g of H_2SO_4 in 675 mL of solution; (c) 0.135 g of phenol, C_6H_5OH, in 1.5 L of solution.

5. Square brackets, [ ], are often used in some mathematical statements in chemistry. What is the meaning associated with square brackets?

6. Calculate the concentrations of the constituent ions in solutions of the following compounds in the indicated concentrations. (a) 0.45 M HBr; (b) 0.045 M KOH; (c) 0.0155 M $CaCl_2$.

7. Calculate the concentrations of the constituent ions in solutions of the following compounds in the indicated concentrations. (a) 0.025 M $Sr(OH)_2$; (b) 0.025 M $HClO_3$; (c) 0.025 M K_2SO_4.

8. Calculate the concentrations of the constituent ions in the following solutions. (a) 1.55 g of KOH in 1.50 L of solution; (b) 0.520 g of $Ba(OH)_2$ in 250. mL of solution; (c) 1.44 g of $Ca(NO_3)_2$ in 100. mL of solution.

9. Calculate the concentrations of the constituent ions in the following solutions. (a) 1.57 g of $Al_2(SO_4)_3$ in 425. mL of solution; (b) 45.8 g of $CaCl_2 \cdot 6H_2O$ in 5.00 L of solution; (c) 18.4 g of HBr in 675 mL of solution.

The Autoionization of Water

10. (a) Write a chemical equation showing the ionization of water. (b) Write the equilibrium constant expression for this equation. (c) What is the special symbol used for this equilibrium constant? (d) What is the relationship

between [H$^+$] and [OH$^-$] in pure water? (e) How can this relationship be used to define the terms "acidic" and "basic"?

11. Use K_w to explain the relationship between the hydronium ion concentration and the hydroxide ion concentration in aqueous solutions.

12. (a) Why is the concentration of OH$^-$ produced by the ionization of water neglected in calculating the concentration of OH$^-$ in a 0.080 M solution of NaOH? (b) Demonstrate that the concentration of OH$^-$ from H$_2$O can be neglected in the 0.080 M solution of NaOH.

13. Calculate the concentrations of OH$^-$ in the solutions described in Exercises 6(a), 7(b), and 9(c), and compare them with the OH$^-$ concentration in pure water.

14. Calculate the concentrations of H$_3$O$^+$ in the solutions described in Exercises 6(b), 7(a), and 8(b), and compare them with the H$_3$O$^+$ concentration in pure water.

15. Calculate [OH$^-$] that is in equilibrium with
(a) [H$_3$O$^+$] = 3.6 × 10^{-4} mol/L
(b) [H$_3$O$^+$] = 9.5 × 10^{-9} mol/L

16. Calculate [H$_3$O$^+$] that is in equilibrium with

$$[OH^-] = 8.87 \times 10^{-6} \text{ mol/L}$$

The pH and pOH Scales

17. Write mathematical definitions for pH and pOH. What is the relationship between pH and pOH? How can pH be used to define the terms "acidic" and "basic"?

18. What are the logarithms of the following numbers? (a) 0.00052; (b) 150.2; (c) 5.8 × 10^{-11}; (d) 2.9 × 10^{-7}.

19. A sample of milk is found to have a pH of 6.50. What are the concentrations of H$_3$O$^+$ and OH$^-$ ions in this sample?

Charles D. Winters

20. The normal pH of human blood ranges from 7.35 to 7.45. Calculate the concentrations of H$_3$O$^+$ and OH$^-$ ions in human blood that has a pH of 7.45.

Charles D. Winters

Blood analysis instrument.

21. Calculate the pH of a 1.55 × 10^{-4} M solution of HClO$_4$, a strong acid, at 25°C.

22. Calculate the pH of the following solutions. (a) 2.00 × 10^{-1} M HCl; (b) 0.050 M HNO$_3$; (c) 0.65 g · L^{-1} HClO$_4$.

23. Calculate the pH of the following solutions. (a) 6.5 × 10^{-2} M HBr; (b) 0.0062 M HI; (c) 2.84 g HNO$_3$ in 450. mL solution.

24. Calculate the pH of a 9.8 × 10^{-4} M solution of NaOH.

25. Calculate the pH of a 1.5 × 10^{-8} M solution of HCl.

26. A solution of HNO$_3$ has a pH of 2.52. What is the molarity of the solution?

27. Complete the following table. Is there an obvious relationship between pH and pOH? What is it?

Solution	[H$_3$O$^+$]	[OH$^-$]	pH	pOH
0.25 M HI	——	——	——	——
0.067 M RbOH	——	——	——	——
0.020 M Ba(OH)$_2$	——	——	——	——
0.00030 M HClO$_4$	——	——	——	——

28. Calculate the following values for each solution.

Solution	[H$_3$O$^+$]	[OH$^-$]	pH	pOH
(a) 0.065 M NaOH	——	——	——	——
(b) 0.065 M HCl	——	——	——	——
(c) 0.065 M Ca(OH)$_2$	——	——	——	——

29. Complete the following table by appropriate calculations.

[H$_3$O$^+$]	pH	[OH$^-$]	pOH
(a) ——	4.84	——	——
(b) ——	10.61	——	——
(c) ——	——	——	2.90
(d) ——	——	——	9.47

30. Predict which acid of each pair is the stronger acid. Briefly explain how you arrived at your answer. (a) H_3PO_4 or H_3AsO_4; (b) $HClO_3$ or HIO_3; (c) HCl or HI; (d) $HClO_2$ or HClO. (*Hint:* Review Chapter 10.)

31. Predict which acid of each pair is the stronger acid. Briefly explain how you arrived at your answer. (a) H_2O or H_2S; (b) H_2SO_3 or H_2SO_4; (c) H_2CO_3 or HCO_3^-; (d) $HBrO_3$ or HIO_3. (*Hint:* Review Chapter 10.)

32. Write a chemical equation that represents the ionization of a weak acid, HA. Write the equilibrium constant expression for this reaction. What is the special symbol used for this equilibrium constant?

33. What is the relationship between the strength of an acid and the numerical value of K_a? What is the relationship between the acid strength and the value of pK_a?

34. (a) What is the pH of pure water at body temperature, 37°C? Refer to Table 18-2. (b) Is this acidic, basic, or neutral? Why?

35. Fill in the blanks in this table for the given solutions. Refer to Table 18-2.

Sol'n	Temp. (°C)	Concentration (mol/L) [H₃O⁺]	[OH⁻]	pH
(a)	25	1.0×10^{-5}	_____	_____
(b)	0	_____	_____	2.75
(c)	60	_____	_____	7.00
(d)	25	_____	4.5×10^{-8}	_____

36. Write a chemical equation that represents the equilibrium between water and a weak base, B. Write the equilibrium constant expression for this reaction. What is the special symbol used for this equilibrium constant?

37. What is the relationship between base strength and the value of K_b? What is the relationship between base strength and the value of pK_b?

38. A 0.0730 M solution of a monoprotic acid is known to be 1.07% ionized. What is the pH of the solution? Calculate the value of K_a for this acid.

39. A 0.085 M aqueous solution of a weak, monoprotic acid is 0.85% ionized. Calculate the value of the ionization constant, K_a, for this acid.

40. The pH of a 0.025 M solution of butanoic acid, C_3H_7COOH, is 3.21. What is the value of the ionization constant, K_a, for butanoic acid?

41. The pH of a 0.35 M solution of uric acid is 2.17. What is the value of K_a for uric acid, a monoprotic acid?

42. Calculate the concentrations of all the species present in a 0.42 M benzoic acid solution. (See Appendix F.)

43. Find the concentrations of the various species present in a 0.35 M solution of HOBr. What is the pH of the solution? (See Appendix F.)

44. Hydrofluoric acid can be used to etch glass. Calculate the pH of a 0.44 M HF solution.

45. The ionization constant, K_a, for propanoic acid, C_2H_5COOH, is 1.3×10^{-5}. What is the percent ionization in a 0.75 M solution of propanoic acid? Propanoic acid is a monoprotic acid.

46. Calculate the pH and pOH of a household ammonia solution that contains 2.25 moles of NH_3 per liter of solution.

47. Calculate the percent ionization in a 0.45 M NH_3 solution.

48. What is the percent ionization in a 0.0450 M solution of formic acid, HCOOH?

49. What is the percent ionization in (a) a 0.150 M CH_3COOH solution and (b) a 0.0150 M CH_3COOH solution?

50. The K_a values for two weak acids are 7.8×10^{-5} and 4.5×10^{-10}, respectively. What are their pK_a values?

51. What is the concentration of OI^- in equilibrium with $[H_3O^+] = 0.045$ mol/L and [HOI] = 0.527 mol/L?

52. Pyridine is 0.053% ionized in 0.00500 M solution. What is the pK_b of this monobasic compound?

53. A 0.068 M solution of benzamide has a pOH of 2.91. What is the value of pK_b for this monobasic compound?

54. In a 0.0100 M aqueous solution of methylamine, CH_3NH_2, the equilibrium concentrations of the species are $[CH_3NH_2] = 0.0080$ mol/L and $[CH_3NH_3^+] = [OH^-] = 2.0 \times 10^{-3}$ mol/L. Calculate K_b for this weak base.

$$CH_3NH_2(aq) + H_2O(\ell) \rightleftharpoons CH_3NH_3^+ + OH^-$$

55. What is the concentration of NH_3 in equilibrium with $[NH_4^+] = 0.010$ mol/L and $[OH^-] = 1.2 \times 10^{-5}$ mol/L?

56. Calculate $[OH^-]$, percent ionization, and pH for (a) 0.15 M aqueous ammonia, and (b) 0.15 M methylamine solution.

57. Calculate $[H_3O^+]$, $[OH^-]$, pH, pOH, and percent ionization for 0.20 M aqueous ammonia.

58. Because K_b is larger for triethylamine

$$(C_2H_5)_3N(aq) + H_2O(\ell) \rightleftharpoons (C_2H_5)_3NH^+ + OH^-$$
$$K_b = 5.2 \times 10^{-4}$$

than for trimethylamine

$$(CH_3)_3N(aq) + H_2O(\ell) \rightleftharpoons (CH_3)_3NH^+ + OH^-$$
$$K_b = 7.4 \times 10^{-5}$$

an aqueous solution of triethylamine should have a larger concentration of OH^- ion than an aqueous solution of trimethylamine of the same concentration. Confirm this statement by calculating the $[OH^-]$ for 0.018 M solutions of both weak bases.

59. The equilibrium constant of the following reaction is 1.35×10^{-15}.

$$2D_2O \rightleftharpoons D_3O^+ + OD^-$$

D is deuterium, ^2H. Calculate the pD of pure deuterium oxide (heavy water). What is the relationship between $[D_3O^+]$ and $[OD^-]$ in pure D_2O? Is pure D_2O acidic, basic, or neutral?

Polyprotic Acids

60. Calculate the concentrations of the various species in a 0.100 M H_3AsO_4 solution. Compare the concentrations with those of the analogous species in 0.100 M H_3PO_4 solution (Example 18-16 and Table 18-7).

61. Citric acid, the acid in lemons and other citrus fruits, has the structure

$$CH_2COOH$$
$$|$$
$$HO-C-COOH$$
$$|$$
$$CH_2COOH$$

which we may abbreviate as $C_3H_5O(COOH)_3$ or H_3A. It is a triprotic acid. Write the chemical equations for the three stages in the ionization of citric acid with the appropriate K_a expressions.

Charles D. Winters

62. Calculate the concentrations of H_3O^+, OH^-, $HSeO_4^-$, and SeO_4^{2-} in 0.15 M H_2SeO_4, selenic acid, solution.

63. Some kidney stones are crystalline deposits of calcium oxalate, a salt of oxalic acid, $(COOH)_2$. Calculate the concentrations of H_3O^+, OH^-, $COOCOOH^-$, and $(COO^-)_2$ in 0.12 M $(COOH)_2$. Compare the concentrations with those obtained in Exercise 62. How can you explain the difference between the concentrations of $HSeO_4^-$ and $COOCOOH^-$? between SeO_4^{2-} and $(COO^-)_2$?

64. Rust stains can be removed from painted surfaces with a solution of oxalic acid, $(COOH)_2$. Calculate the pH of a 0.045 M oxalic acid solution.

65. Calculate the pH and pOH of a carbonated soft drink that

is 0.0035 M carbonic acid solution. Assume that there are no other acidic or basic components.

Charles D. Winters

Hydrolysis

66. Define and illustrate the following terms clearly and concisely: (a) solvolysis; (b) hydrolysis.

67. Predict which base of each pair is the stronger base. Briefly explain how you arrived at your answer. (a) NO_2^- or NO_3^-; (b) BrO_3^- or IO_3^-; (c) HSO_3^- or HSO_4^-; (d) F^- or Cl^-. (*Hint:* You may wish to review Chapter 10.)

68. Predict which base of each pair is the stronger base. Briefly explain how you arrived at your answer. (a) PH_3 or NH_3; (b) Br^- or F^-; (c) ClO_3^- or ClO_2^-; (d) HPO_4^{2-} or PO_4^{3-}. (*Hint:* You may wish to review Chapter 10.)

69. Some anions, when dissolved, undergo no significant reaction with water molecules. What is the relative base strength of such an anion compared with water? What effect will dissolution of such anions have on the pH of the solution?

70. Some cations in aqueous solution undergo no significant reactions with water molecules. What is the relative acid strength of such a cation compared with water? What effect will dissolution of such cations have on the pH of the solution?

71. How can salts be classified conveniently into four classes? For each class, write the name and formula of a salt that fits into that category. Use examples other than those used in illustrations in this chapter.

Salts of Strong Bases and Strong Acids

72. What determines whether the aqueous solution of a salt is acidic, basic, or neutral?

73. Why do salts of strong bases and strong acids give neutral aqueous solutions? Use KNO_3 to illustrate. Write names and formulas for three other salts of strong bases and strong acids.

74. Which of the following salts is the salt of a strong base and a strong acid? (a) Na_3PO_4; (b) K_2CO_3; (c) LiF; (d) $BaSO_4$; (e) $NaClO_3$.

75. Which of the following salts is the salt of a strong base and a strong acid? (a) $Ba_3(PO_4)_2$; (b) KNO_3; (c) NaI; (d) $CaCO_3$; (e) $LiClO_4$.

Salts of Strong Bases and Weak Acids

76. Why do salts of strong bases and weak acids give basic aqueous solutions? Use sodium hypochlorite, $NaOCl$, to illustrate. (Clorox, Purex, and other "chlorine bleaches" are 5% $NaOCl$.)

Charles D. Winters

77. Some anions react with water to upset the H_3O^+/OH^- balance. What is the relative base strength of such an anion compared with water? What effect will dissolution of such anions have on the pH of the solution?

78. Calculate the equilibrium constant for the reaction of azide ions, N_3^-, with water.

79. Write names and formulas for three salts of strong bases and weak acids other than those that appear in Section 18-8.

80. Calculate hydrolysis constants for the following anions of weak acids: (a) NO_2^-; (b) OBr^-; (c) $HCOO^-$. What is the relationship between K_a, the ionization constant for a weak acid, and K_b, the hydrolysis constant for the anion of the weak acid? (See Appendix F.)

81. Calculate the equilibrium constant for the reaction of hypoiodite ions (OI^-) with water.

82. Calculate the pH of 1.25 M solutions of the following salts: (a) $NaCH_3COO$; (b) $KOBr$; (c) $LiCN$.

83. Calculate the pH of 0.15 M solutions of the following salts: (a) $NaNO_2$; (b) $NaOCl$; (c) $NaHCOO$.

84. (a) What is the pH of a 0.15 M solution of KOI? (b) What is the pH of a 0.15 M solution of KF?

Salts of Weak Bases and Strong Acids

85. Why do salts of weak bases and strong acids give acidic aqueous solutions? Illustrate with NH_4NO_3, a common fertilizer.

86. Write names and formulas for four salts of weak bases and strong acids.

87. Use values found in Table 18-6 and in Appendix G to calculate hydrolysis constants for the following cations of weak bases: (a) NH_4^+; (b) $CH_3NH_3^+$, methylammonium ion; (c) $C_6H_5NH_3^+$, anilinium ion.

88. Use the values found in Table 18-6 and Appendix G to calculate hydrolysis constants for the following cations of weak bases: (a) $(CH_3)_2NH_2^+$, dimethylammonium ion; (b) $C_5H_5NH^+$, pyridinium ion; (c) $(CH_3)_3NH^+$, trimethylammonium ion.

89. Can you make a general statement relating parent base strength and extent of hydrolysis of the cations of Exercise 87 by using hydrolysis constants calculated in that exercise?

90. Calculate the pH of 0.26 M solutions of (a) NH_4NO_3; (b) $(CH_3)NH_3NO_3$; (c) $C_6H_5NH_3NO_3$.

Salts of Weak Bases and Weak Acids

91. Why are some aqueous solutions of salts of weak acids and weak bases neutral, whereas others are acidic and still others are basic?

92. Write the names and formulas for three salts of a weak acid and a weak base that give (a) neutral, (b) acidic, and (c) basic aqueous solutions.

93. If both the cation and anion of a salt react with water when dissolved, what determines whether the solution will be acidic, basic, or neutral? Classify aqueous solutions of the following salts as acidic, basic, or neutral. (a) $NH_4F(aq)$ and (b) $CH_3NH_3OI(aq)$.

Salts That Contain Small, Highly Charged Cations

94. Choose the hydrated cations that react with water to give acidic solutions. (a) $[Be(H_2O)_4]^{2+}$; (b) $[Al(H_2O)_6]^{3+}$; (c) $[Fe(H_2O)_6]^{3+}$; (d) $[Cu(H_2O)_6]^{2+}$. Write chemical equations for the reactions.

95. Why do some salts that contain cations related to insoluble bases (metal hydroxides) and anions related to strong acids give acidic aqueous solutions? Use $Fe(NO_3)_3$ to illustrate.

96. Calculate pH and percent hydrolysis for the following (Table 18-10). (a) 0.15 M $Al(NO_3)_3$, aluminum nitrate; (b) 0.075 M $Co(ClO_4)_2$, cobalt (II) perchlorate; (c) 0.15 M $MgCl_2$, magnesium chloride.

*97. Given pH values for solutions of the following concentrations, calculate hydrolysis constants for the hydrated cations: (a) 0.00050 M $CeCl_3$, cerium(III) chloride, pH = 5.99; (b) 0.10 M $Cu(NO_3)_2$, copper(II) nitrate, pH = 4.50; (c) 0.10 M $Sc(ClO_4)_3$, scandium perchlorate, pH = 3.44.

Mixed Exercises

98. A weak acid, HA, has a pK_a = 5.35. What is the concentration of the anion, A^-, in a 0.100 M solution?

99. Calculate the pH of the following solutions. (a) 0.0070 M $Ca(OH)_2$; (b) 0.25 M chloroacetic acid, $ClCH_2COOH$, K_a = 1.4 × 10⁻³; (c) 0.055 M pyridine, C_5H_5N.

100. Classify aqueous solutions of the following salts as acidic basic, or essentially neutral. Justify your choice. (a) $(NH_4)HSO_4$; (b) $(NH_4)_2SO_4$; (c) $LiCl$; (d) $LiBrO$; (e) $AlCl_3$.

101. Repeat Exercise 100 for (a) Na_2SO_4; (b) NH_4Cl; (c) KCl; (d) NH_4CN. (See Appendix F.)

102. In aqueous solution some cations react with water to upset the H_3O^+/OH^- balance. What is the relative acid strength of such a cation compared with water? What effect will dissolution of these cations have on the pH of the solution?

103. Some plants require acidic soils for healthy growth. Which of the following could be added to the soil around such plants to increase the acidity of the soil? Write equations to justify your answers. (a) $FeSO_4$; (b) Na_2SO_4; (c) $Al_2(SO_4)_3$; (d) $Fe_2(SO_4)_3$; (e) $BaSO_4$. Arrange the salts that give acidic solutions in order of increasing acidity. (Assume equal molarities of the salt solutions.)

104. Some of the following salts are used in detergents and other cleaning materials because they produce basic solutions. Which of the following could *not* be used for this purpose? Write equations to justify your answers. (a) Na_2CO_3; (b) Na_2SO_4; (c) $(NH_4)_2SO_4$; (d) Na_3PO_4.

105. Calculate the pH of each of the following solutions. (a) 0.038 g of barium hydroxide in 450. mL of solution; (b) 0.050 g of hydrogen iodide in 750. mL of solution; (c) 0.00075 g of HCl in 1.00 L of solution.

106. For each of the following pairs, tell which solution would have the lower pH. Tell how you arrive at each answer. (a) 0.015 M NH_4Br, ammonium bromide, and 0.015 M NH_4NO_3, ammonium nitrate; (b) 0.015 M ammonium perchlorate, NH_4ClO_4, and 0.010 M ammonium fluoride, NH_4F; (c) 0.010 M NH_4Cl and 0.050 M NH_4Cl. (*Hint:* Think before you calculate.)

CONCEPTUAL EXERCISES

107. How could we demonstrate that 0.010 M solutions of HCl and HNO_3 contain essentially no molecules of nonionized acid?

108. How could we demonstrate that 0.010 M solutions of HF and HNO_2 contain relatively few ions?

109. Carbonic acid, H_2CO_3, is diprotic and therefore has two ionization constants, K_{a1} = 4.2 × 10⁻⁷ and K_{a2} = 4.8 × 10⁻¹¹. The pH of a carbonic acid solution can be calculated without using K_{a2}. Explain using a 0.100 M solution of carbonic acid.

110. Answer the following questions for 0.15 M solutions of the weak bases listed in Table 18-6. (a) In which solution is (i) the pH highest; (ii) the pH lowest; (iii) the pOH highest; (iv) the pOH lowest? (b) Which solution contains (i) the highest concentration of the cation of the weak base; (ii) the lowest concentration of the cation of the weak base?

111. Answer the following questions for 0.15 M solutions of the weak acids listed in Table 18-4. Which solution contains (a) the highest concentration of H_3O^+; (b) the highest concentration of OH^-; (c) the lowest concentration of H_3O^+; (d) the lowest concentration of OH^-; (e) the highest concentration of nonionized acid molecules; (f) the lowest concentration of nonionized acid molecules?

BUILDING YOUR KNOWLEDGE

112. Ascorbic acid, $C_5H_7O_4COOH$, also known as vitamin C, is an essential vitamin for all mammals. Among mammals, only humans, monkeys, and guinea pigs cannot synthesize it in their bodies. K_a for ascorbic acid is 7.9 × 10⁻⁵. Calculate [H_3O^+] and pH in a 0.110 M solution of ascorbic acid.

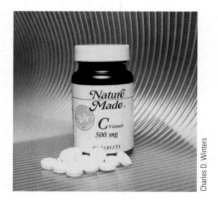

Charles D. Winters

113. Arrange the following common kitchen samples from most acidic to most basic.

carrot juice, pH 5.1	blackberry juice, pH 3.4
soap, pH 11.0	red wine, pH 3.7
egg white, pH 7.8	milk of magnesia, pH 10.5
sauerkraut, pH 3.5	lime juice, pH 2.0

*114. The buildup of lactic acid in muscles causes pain during extreme physical exertion. The K_a for lactic acid, C_2H_5OCOOH, is 8.4 × 10⁻⁴. Calculate the pH of a 0.110 M solution of lactic acid. Can you make a simplifying assumption in this case?

115. A 0.0100 molal solution of acetic acid freezes at −0.01938°C. Use this information to calculate the ionization constant for acetic acid. A 0.0100 molal solution is sufficiently dilute that it may be assumed to be 0.0100 molar without introducing a significant error.

116. The ion product for water, K_w, has the value 1.14×10^{-15} at 0°C and 5.47×10^{-14} at 50°C. Use the van't Hoff equation to estimate ΔH^0 for the ion product of water.

BEYOND THE TEXTBOOK

Go to the textbook website at

http://www.brookscole.com/chemistry/whitten

for additional activities and exercises based on the General Chemistry Interactive CD-ROM, the World Wide Web, and library resources.

InfoTrac College Edition

For additional readings, go to InfoTrac College Edition, your online research library at:

http://infotrac.thomsonlearning.com

19   Ionic Equilibria II: Buffers and Titration Curves

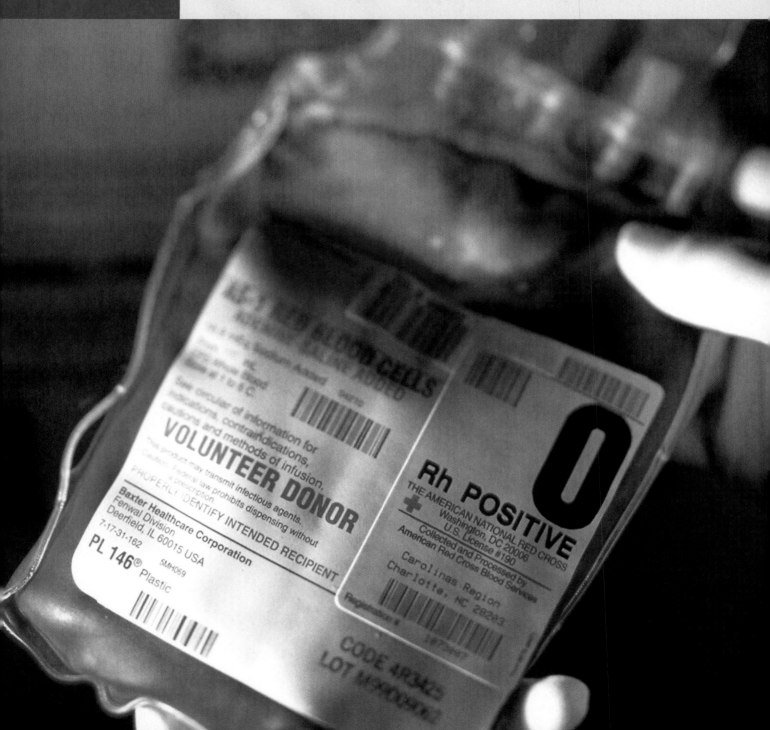

OBJECTIVES

After you have studied this chapter, you should be able to

- Explain the common ion effect and give illustrations of its operation
- Recognize buffer solutions and describe their chemistry
- Describe how to prepare a buffer solution of a specified pH
- Carry out calculations related to buffer solutions and their action
- Explain what acid–base indicators are and how they function
- Describe what species are present at various stages of titration curves for (a) strong acids and strong bases, (b) weak acids and strong bases, and (c) weak acids and weak bases
- Carry out calculations based on titration curves for (a) strong acids and strong bases and (b) weak acids and strong bases

It is critical that the acidity of our blood remain very near pH 7.4. This is accomplished with a physiological buffer system consisting primarily of a mixture of bicarbonate ion (HCO_3^-) and carbonic acid (H_2CO_3).

In previous chapters we calculated the acidity or basicity of aqueous solutions of strong acids, strong bases, weak acids, weak bases, and their salts. In this chapter we will study (1) solutions that have both weak acids and weak bases present, (2) indicators, and (3) titration curves.

It can be a challenge to recognize the type of solution present. This can be even more difficult if the solution is formed by a partial or total neutralization reaction. To help you to recognize the various solutions, a summary table (Table 19-7) is included in Section 19.8. We encourage you to look at the table often as you progress through this chapter and as you review this and previous chapters.

19-1 THE COMMON ION EFFECT AND BUFFER SOLUTIONS

In laboratory reactions, in industrial processes, and in the bodies of plants and animals, it is often necessary to keep the pH nearly constant despite the addition of acids or bases. The oxygen-carrying capacity of the hemoglobin in your blood and the activity of the enzymes in your cells are very sensitive to the pH of your body fluids. Our bodies use a combination of compounds known as a *buffer system* to keep the pH within a narrow range.

> A **buffer solution** contains a conjugate acid–base pair with both the acid and base in reasonable concentrations. The acidic component reacts with added strong bases. The basic component reacts with added strong acids.

The operation of a buffer solution depends on the *common ion effect*, a special case of LeChatelier's Principle.

> When a solution of a weak electrolyte is altered by adding one of its ions from an-other source, the ionization of the weak electrolyte is suppressed. This behavior is termed the **common ion effect**.

Many types of solutions exhibit this behavior. Two of the most frequently encountered kinds are

1. A solution of a weak acid *plus* a soluble ionic salt of the weak acid (e.g., CH_3COOH plus $NaCH_3COO$)
2. A solution of a weak base *plus* a soluble ionic salt of the weak base (e.g., NH_3 plus NH_4Cl)

Weak Acids Plus Salts of Weak Acids

Consider a solution that contains acetic acid *and* sodium acetate, a soluble ionic salt of CH_3COOH. The $NaCH_3COO$ is completely dissociated into its constituent ions, but CH_3COOH is only slightly ionized.

$$NaCH_3COO \xrightarrow{H_2O} Na^+ \; + \; \boxed{CH_3COO^-} \quad \text{(to completion)}$$

$$CH_3COOH + H_2O \rightleftharpoons \boxed{H_3O^+} + \boxed{CH_3COO^-} \quad \text{(reversible)}$$

Both CH_3COOH and $NaCH_3COO$ are sources of CH_3COO^- ions. The completely dissociated $NaCH_3COO$ provides a high $[CH_3COO^-]$. This shifts the ionization equilibrium of CH_3COOH far to the left as CH_3COO^- combines with H_3O^+ to form nonionized CH_3COOH and H_2O. The result is a drastic decrease in $[H_3O^+]$ in the solution.

> Solutions that contain a weak acid plus a salt of the weak acid are always less acidic than solutions that contain the same concentration of the weak acid alone.

In Example 18-11, we found that the H_3O^+ concentration in 0.10 M CH_3COOH is 1.3×10^{-3} mol/L (pH = 2.89). In Example 19-1, we will calculate the acidity of the same 0.10 M CH_3COOH solution, which was also made to be 0.20 M in the salt $NaCH_3COO$.

EXAMPLE 19-1 *Weak Acid/Salt of Weak Acid Buffer Solution*

Calculate the concentration of H_3O^+ and the pH of a buffer solution that is 0.10 M in CH_3COOH and 0.20 M in $NaCH_3COO$.

Plan

Write the appropriate equations for *both* $NaCH_3COO$ *and* CH_3COOH and the ionization constant expression for CH_3COOH. Then, represent the *equilibrium* concentrations algebraically and substitute into the K_a expression.

Solution

The appropriate equations and ionization constant expression are

$$\text{rxn 1} \quad NaCH_3COO \longrightarrow Na^+ + CH_3COO^- \quad \text{(to completion)}$$

$$\text{rxn 2} \quad CH_3COOH + H_2O \rightleftharpoons H_3O^+ + CH_3COO^- \quad \text{(reversible)}$$

$$K_a = \frac{[H_3O^+][CH_3COO^-]}{[CH_3COOH]} = 1.8 \times 10^{-5}$$

The general equations

$$NaA \longrightarrow Na^+ + A^-$$

and

$$HA \rightleftharpoons H^+ + A^-$$

may be used in place of these equations.

This K_a expression is valid for *all solutions* that contain CH_3COOH. In solutions that contain both CH_3COOH and $NaCH_3COO$, CH_3COO^- ions come from two sources. The ionization constant is satisfied by the *total* CH_3COO^- concentration.

Because $NaCH_3COO$ is completely dissociated, the $[CH_3COO^-]$ *from* $NaCH_3COO$ will be 0.20 mol/L. Let $x = [CH_3COOH]$ that ionizes; then x is also equal to $[H_3O^+]$ *and* equal to $[CH_3COO^-]$ *from* CH_3COOH. The *total* concentration of CH_3COO^- is $(0.20 + x)$ M. The concentration of nonionized CH_3COOH is $(0.10 - x)$ M.

$$
\begin{array}{llll}
\text{rxn 1} & NaCH_3COO & \longrightarrow Na^+ + & CH_3COO^- \\
& 0.20\ M & \Longrightarrow 0.20\ M & 0.20\ M \\
\text{rxn 2} & CH_3COOH + H_2O \rightleftharpoons & H_3O^+ + & CH_3COO^- \\
& (0.10 - x)\ M & x\ M & x\ M
\end{array}
$$

$$\longrightarrow \text{Total } [CH_3COO^-] = (0.20 + x)\ M$$

Substitution into the ionization constant expression for CH_3COOH gives

$$K_a = \frac{[H_3O^+][CH_3COO^-]}{[CH_3COOH]} = \frac{(x)(0.20 + x)}{0.10 - x} = 1.8 \times 10^{-5}$$

The small value of K_a suggests that x is very small. This leads to two assumptions.

Assumption	Implication
$x \ll 0.20$, so $(0.20 + x) \approx 0.20$	Most of the CH_3COO^- comes from $NaCH_3COO$ (rxn 1) and very little CH_3COO^- comes from ionization of CH_3COOH (rxn 2)
$x \ll 0.10$, so $(0.10 - x) \approx 0.10$	Very little of the CH_3COOH ionizes (rxn 2)

It is reasonable to assume that x (from the ionization of CH_3COOH) is small, because CH_3COOH is a weak acid (rxn 2), and its ionization is further suppressed by the high concentration of CH_3COO^- formed by the soluble salt, $NaCH_3COO$ (rxn 1).

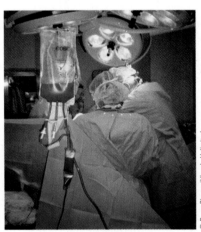

The pH of blood is controlled by buffers found in the blood.

You can verify the validity of the assumption by substituting the value $x = 9.0 \times 10^{-6}$ into the original equation.

Introducing these assumptions gives

$$\frac{0.20\,x}{0.10} = 1.8 \times 10^{-5} \quad \text{and} \quad x = 9.0 \times 10^{-6}$$

$$x\,M = \boxed{[H_3O^+] = 9.0 \times 10^{-6}\,M} \quad \text{so} \quad \boxed{pH = 5.05}$$

You should now work Exercise 8.

Let's calculate the percent ionization in the solution of Example 19-1.

$$\% \text{ ionization} = \frac{[CH_3COOH]_{ionized}}{[CH_3COOH]_{initial}} \times 100\%$$

$$= \frac{9.0 \times 10^{-6}\,M}{0.10\,M} \times 100\% = 0.0090\% \text{ ionized}$$

This compares with 1.3% ionization in 0.10 M CH_3COOH (Example 18-11). Table 19-1 compares these solutions. The third column shows that $[H_3O^+]$ is *140 times greater* in 0.10 M CH_3COOH than in the solution to which 0.20 mol/L $NaCH_3COO$ has been added (due to the common ion effect).

The calculation of the pH of any solution containing significant amounts of both a weak acid and the salt of the weak acid may be carried out as we have done in Example 19-1. Alternatively, one may proceed as follows. We can start by writing the equation for the ionization of the *weak monoprotic acid* and its K_a as we did previously.

HA and A$^-$ represent the weak acid and its conjugate base, respectively.

$$HA + H_2O \rightleftharpoons H_3O^+ + A^- \quad \text{and} \quad \frac{[H_3O^+][A^-]}{[HA]} = K_a$$

Solving this expression for $[H_3O^+]$ gives

$$[H_3O^+] = K_a \times \frac{[HA]}{[A^-]}$$

These are the kinds of assumptions we made in Example 19-1.

Consider a solution in which the concentrations of both the weak acid and its anion (from an added salt) are some reasonable values, such as greater than 0.050 M. Under these conditions the concentration of the anion, $[A^-]$, in the solution can be assumed to be entirely due to the dissolved salt. With these restrictions, the preceding expression becomes

$$[H_3O^+] = K_a \times \frac{[HA]}{[\text{conjugate base}]}$$

James W. Morgenthaler

The two solutions of Table 19-1, in the presence of universal indicator. The CH_3COOH solution is on the left.

TABLE 19-1	Comparison of $[H_3O^+]$ and pH in Acetic Acid and Sodium Acetate–Acetic Acid Solutions			
Solution	% CH_3COOH Ionized	$[H_3O^+]$		pH
0.10 M CH_3COOH	1.3%	$1.3 \times 10^{-3}\,M$	2.89	
0.10 M CH_3COOH and 0.20 M $NaCH_3COO$	0.0090%	$9.0 \times 10^{-6}\,M$	5.05	$\Delta pH = 2.16$

[HA] is the concentration of nonionized weak acid (in most cases this is the total acid concentration) and [conjugate base] is the concentration of the anion from the dissolved salt.

If we take the logarithm of both sides of the preceding equation, we obtain

$$\log [H_3O^+] = \log K_a + \log \frac{[acid]}{[conjugate\ base]}$$

Multiplying by -1 gives

$$-\log [H_3O^+] = -\log K_a - \log \frac{[acid]}{[conjugate\ base]}$$

Recalling that in Chapter 18 we defined $-\log [H_3O^+]$ (or for the general expression $-\log [H^+]$) as pH and $-\log K_a$ as pK_a, the preceding equation becomes

$$pH = pK_a + \log \frac{[conjugate\ base]}{[acid]} \qquad \text{(acid/salt buffer)}$$

 CD-ROM   Screen 18.4, pH of Buffer Solutions.

This equation is known as the **Henderson–Hasselbalch equation.** Workers in the biological sciences use it frequently.

EXAMPLE 19-2   *Weak Acid/Salt of Weak Acid Buffer Solution (via the Henderson–Hasselbalch Equation)*

Use the Henderson–Hasselbalch equation to calculate the pH of the buffer solution in Example 19-1.

Plan

The Henderson–Hasselbalch equation is $pH = pK_a + \log \frac{[conjugate\ base]}{[acid]}$. The value for pK_a for acetic acid can be calculated from the value of K_a found in Example 19-1 and many other places. The solution in Example 19-1 is 0.10 M in CH_3COOH and 0.20 M in $NaCH_3COO$. The values used for [conjugate base] and [acid] are their initial concentrations after mixing but before reaction.

Solution

The appropriate values needed for the Henderson–Hasselbalch equation are

$$pK_a = -\log K_a = -\log 1.8 \times 10^{-5} = 4.74$$

$$[conjugate\ base] = [CH_3COO^-] = [NaCH_3COO]_{initial} = 0.20\ M$$

$$[acid] = [CH_3COOH]_{initial} = 0.10\ M$$

$$pH = pK_a + \log \frac{[conjugate\ base]}{[acid]} = 4.74 + \log \frac{0.20}{0.10}$$

$$= 4.74 + \log 2.0 = 4.74 + 0.30 = \boxed{5.04}$$

The values $pK_a = 4.74$, log 2.0, and pH = 5.04 all have two significant figures.

You should now work Exercise 9.

Weak Bases Plus Salts of Weak Bases

Let us consider the second common kind of buffer solution, containing a weak base and its salt. A solution that contains aqueous NH_3 and ammonium chloride, NH_4Cl, a soluble ionic salt of NH_3, is typical. The NH_4Cl is completely dissociated, but aqueous NH_3 is only slightly ionized.

$$\text{rxn 1} \qquad NH_4Cl \xrightarrow{H_2O} \boxed{NH_4^+} + Cl^- \qquad \text{(to completion)}$$

$$\text{rxn 2} \qquad NH_3 + H_2O \rightleftharpoons \boxed{NH_4^+} + \boxed{OH^-} \qquad \text{(reversible)}$$

Both NH_4Cl and aqueous NH_3 are sources of NH_4^+ ions. The completely dissociated NH_4Cl provides a high $[NH_4^+]$. This shifts the ionization equilibrium of aqueous NH_3 far to the left, as NH_4^+ ions combine with OH^- ions to form nonionized NH_3 and H_2O. The result is that $[OH^-]$ is decreased significantly.

> Solutions that contain a weak base plus a salt of the weak base are always less basic than solutions that contain the same concentration of the weak base alone.

EXAMPLE 19-3   Weak Base/Salt of Weak Base Buffer Solution

Calculate the concentration of OH^- and the pH of a solution that is 0.20 M in aqueous NH_3 and 0.10 M in NH_4Cl.

Plan

Write the appropriate equations for *both* NH_4Cl and NH_3 and the ionization constant expression for NH_3. Then, represent the *equilibrium* concentrations algebraically, and substitute into the K_b expression.

Solution

The appropriate equations and algebraic representations of concentrations are

$$
\begin{array}{lllll}
\text{rxn 1} & NH_4Cl & \longrightarrow & \boxed{NH_4^+} + & Cl^- \\
 & 0.10\,M & & 0.10\,M & 0.10\,M \\
\text{rxn 2} & NH_3 + H_2O & \rightleftharpoons & \boxed{NH_4^+} + & OH^- \\
 & (0.20 - x)\,M & & x\,M & x\,M
\end{array}
$$

$$\text{Total } [NH_4^+] = (0.10 + x)\,M$$

Substitution into the K_b expression for aqueous NH_3 gives

$$K_b = \frac{[NH_4^+][OH^-]}{[NH_3]} = 1.8 \times 10^{-5} = \frac{(0.10 + x)(x)}{0.20 - x}$$

The small value of K_b suggests that x is very small. This leads to two assumptions.

Assumption	Implication
$x \ll 0.10$, so $(0.10 + x) \approx 0.10$	Most of the NH_4^+ comes from NH_4Cl (rxn 1) and very little NH_4^+ comes from ionization of NH_3 (rxn 2)
$x \ll 0.20$, so $(0.20 - x) \approx 0.20$	Very little of the NH_3 ionizes (rxn 2)

It is reasonable to assume that x (from the ionization of NH_3) is small, because NH_3 is a weak base (rxn 2), and its ionization is further suppressed by the high concentration of NH_4^+ formed by the soluble salt, NH_4Cl (rxn 1).

Introducing these assumptions gives

$$\frac{0.10x}{0.20} = 1.8 \times 10^{-5}\ M \quad \text{and} \quad x = 3.6 \times 10^{-5}\ M$$

$$x\ M = [OH^-] = 3.6 \times 10^{-5}\ M \quad \text{so} \quad pOH = 4.44 \quad \text{and} \quad pH = 9.56$$

You should now work Exercise 10.

The percent ionization of NH_3 in this solution is

$$\frac{3.6 \times 10^{-5}\ M_{ionized}}{0.20\ M_{original}} \times 100\% = 0.018\%.$$

In Example 18-14 we calculated $[OH^-]$ and pH for 0.20 M aqueous NH_3. Compare those results with the values obtained in Example 19-3 (Table 19-2). The concentration of OH^- is *53 times greater* in the solution containing only 0.20 M aqueous NH_3 than in the solution to which 0.10 mol/L NH_4Cl has been added. This is another demonstration of the common ion effect.

We can derive a relationship for $[OH^-]$ in a solution containing a weak base, B, *plus* a salt that contains the cation, BH^+, of the weak base, just as we did for weak acids. In general terms the equation for the ionization of a monoprotic weak base and its K_b expression are

$$B + H_2O \rightleftharpoons BH^+ + OH^- \quad \text{and} \quad \frac{[BH^+][OH^-]}{[B]} = K_b$$

Solving the K_b expression for $[OH^-]$ gives

$$[OH^-] = K_b \times \frac{[B]}{[BH^+]}$$

Taking the logarithm of both sides of the equation gives

$$\log\,[OH^-] = \log K_b + \log \frac{[B]}{[BH^+]}$$

Multiplying by -1, substituting pK_b for $-\log K_b$ and rearranging gives another form of the *Henderson–Hasselbalch equation* for solutions containing a weak base plus a salt of the weak base.

B and BH^+ represent the weak base and its conjugate acid, respectively— for example, NH_3 and NH_4^+.

$$pOH = pK_b + \log \frac{[BH^+]}{[B]} \quad \text{(base/salt buffer)}$$

James W. Morgenthaler

| TABLE 19-2 | *Comparison of $[OH^-]$ and pH in Ammonia and Ammonium Chloride—Ammonia Solutions* | | |

Solution	% NH_3 Ionized	$[OH^-]$	pH
0.20 M aq NH_3	0.95%	$1.9 \times 10^{-3}\ M$	11.28
0.20 M aq NH_3 and 0.10 M aq NH_4Cl	0.018%	$3.6 \times 10^{-5}\ M$	9.56

$\Delta pH = -1.72$

The two solutions in Table 19-2, in the presence of universal indicator. The NH_3 solution is on the left. Can you calculate the percentage of NH_3 that is ionized in these two solutions?

The Henderson–Hasselbalch equation is valid for solutions of weak bases plus salts of weak bases with univalent anions in reasonable concentrations. In general terms we can also write this equation as

$$pOH = pK_b + \log \frac{[\text{conjugate acid}]}{[\text{base}]} \qquad \text{(base/salt buffer)}$$

CD-ROM Screen 18.6, Adding Reagents to Buffer Solutions.

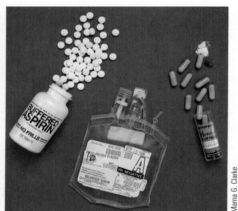
Marna G. Clarke

Three common examples of buffers. Many medications are buffered to minimize digestive upset. Most body fluids, including blood plasma, contain very efficient natural buffer systems. Buffer capsules are used in laboratories to prepare solutions of specified pH.

The net effect is to neutralize most of the H_3O^+ from HCl by forming nonionized CH_3COOH molecules. This slightly decreases the ratio $[CH_3COO^-]/[CH_3COOH]$, which governs the pH of the solution.

19-2 BUFFERING ACTION

A buffer solution is able to react with either H_3O^+ or OH^- ions, whichever is added.

Thus, a buffer solution resists changes in pH. When we add a modest amount of a strong base or a strong acid to a buffer solution, the pH changes very little.

The two common kinds of buffer solutions are the ones we have just discussed—namely, solutions containing (1) a weak acid plus a soluble ionic salt of the weak acid and (2) a weak base plus a soluble ionic salt of the weak base.

Solutions of a Weak Acid and a Salt of the Weak Acid

A solution containing acetic acid, CH_3COOH, and sodium acetate, $NaCH_3COO$, is an example of this kind of buffer solution. The acidic component is CH_3COOH. The basic component is $NaCH_3COO$ because the CH_3COO^- ion is the conjugate base of CH_3COOH. The operation of this buffer depends on the equilibrium

$$\underset{\substack{high \text{ conc} \\ \text{(from acid)}}}{CH_3COOH} + H_2O \rightleftharpoons H_3O^+ + \underset{\substack{high \text{ conc} \\ \text{(from salt)}}}{CH_3COO^-}$$

If we add a strong acid such as HCl to this solution, it produces H_3O^+. As a result of the added H_3O^+, the reaction occurs to the *left*, to use up most of the added H_3O^+ and reestablish equilibrium. Because the $[CH_3COO^-]$ in the buffer solution is high, this can occur to a great extent. The net reaction is

$$H_3O^+ + CH_3COO^- \longrightarrow CH_3COOH + H_2O \qquad (\approx 100\%)$$

or, as a formula unit equation,

$$\underset{\text{added acid}}{HCl} + \underset{\text{base}}{NaCH_3COO} \longrightarrow \underset{\text{weaker acid}}{CH_3COOH} + \underset{\text{salt}}{NaCl} \qquad (\approx 100\%)$$

This reaction goes nearly to completion because CH_3COOH is a *weaker* acid than HCl; even when mixed from separate sources, its ions have a strong tendency to form nonionized CH_3COOH molecules rather than remain separate.

When a strong base, such as NaOH, is added to the $CH_3COOH–NaCH_3COO$ buffer solution, it is consumed by the acidic component, CH_3COOH. This occurs in the following way. The additional OH^- causes the water autoionization reaction to proceed to the *left*.

$$2H_2O \rightleftharpoons H_3O^+ + OH^- \qquad \text{(shifts \textit{left})}$$

This uses up some H_3O^+, causing more CH_3COOH to ionize.

$$CH_3COOH + H_2O \rightleftharpoons CH_3COO^- + H_3O^+ \qquad \text{(shifts \textit{right})}$$

Because the $[CH_3COOH]$ is high, this can occur to a great extent. The net result is the neutralization of OH^- by CH_3COOH.

$$\underset{\text{added base}}{OH^-} + \underset{\text{acid}}{CH_3COOH} \longrightarrow \underset{\text{weaker base}}{CH_3COO^-} + \underset{\text{water}}{H_2O} \qquad (\approx 100\%)$$

The net effect is to neutralize most of the OH^- from NaOH. This slightly increases the ratio $[CH_3COO^-]/[CH_3COOH]$, which governs the pH of the solution.

EXAMPLE 19-4   *Buffered Solutions*

If we add 0.010 mol of solid NaOH to 1.00 liter of a buffer solution that is 0.100 M in CH_3COOH and 0.100 M in $NaCH_3COO$, how much will $[H_3O^+]$ and pH change? Assume that there is no volume change due to the addition of solid NaOH.

Plan

Calculate $[H_3O^+]$ and pH for the original buffer solution. Then, write the reaction summary that shows how much of the CH_3COOH is neutralized by NaOH. Calculate $[H_3O^+]$ and pH for the resulting buffer solution. Finally, calculate the change in pH.

Solution

For the 0.100 M CH_3COOH and 0.100 M $NaCH_3COO$ solution, we can write

$$[H_3O^+] = K_a \times \frac{[\text{acid}]}{[\text{salt}]} = 1.8 \times 10^{-5} \times \frac{0.100}{0.100} = 1.8 \times 10^{-5}\ M; \text{pH} = 4.74$$

We could have used the Henderson–Hasselbalch equation to solve for the pH in this example.

$$\text{pH} = pK_a + \log \frac{[\text{conjugate base}]}{[\text{acid}]}$$

When solid NaOH is added, it reacts with CH_3COOH to form more $NaCH_3COO$.

	NaOH	+ CH_3COOH	$\longrightarrow$ $NaCH_3COO$	+ H_2O
start	0.010 mol	0.100 mol	0.100 mol	
change due to rxn	-0.010 mol	-0.010 mol	$+0.010$ mol	
after rxn	0 mol	0.090 mol	0.110 mol	

The $[H_3O^+]$ can then be obtained from the pH value.

The volume of the solution is 1.00 liter, so we now have a solution that is 0.09 M in CH_3COOH and 0.110 M in $NaCH_3COO$. In this solution,

$$[H_3O^+] = K_a \times \frac{[\text{acid}]}{[\text{salt}]} = 1.8 \times 10^{-5} \times \frac{0.090}{0.110} = 1.5 \times 10^{-5}\ M; \text{pH} = 4.82$$

The addition of 0.010 mol of solid NaOH to 1.00 liter of this buffer solution decreases $[H_3O^+]$ from $1.8 \times 10^{-5}\ M$ to $1.5 \times 10^{-5}\ M$ and increases pH from 4.74 to 4.82, a change of 0.08 pH unit, which is a very slight change.

This is enough NaOH to neutralize 10% of the acid.

You should now work Exercise 22.

Let's describe what happens when we add 0.010 mole of solid NaOH to one liter of 0.100 M CH_3COOH (pH = 2.89) to give a solution that is 0.090 M in CH_3COOH and 0.010 M in $NaCH_3COO$. The pH of the solution is 3.78, which is only 0.89 pH unit higher than that of the 0.100 M CH_3COOH solution.

By contrast, adding 0.010 mole of NaOH to enough pure H_2O to give one liter produces a 0.010 M solution of NaOH: $[OH^-] = 1.0 \times 10^{-2}\ M$ and pOH = 2.00. The pH of this solution is 12.00, an increase of 5.00 pH units above that of pure H_2O.

$$\text{pH} + \text{pOH} = 14$$

See Table 19-3.

In summary, 0.010 mole of NaOH

added to 1.00 L of the $CH_3COOH/NaCH_3COO$ buffer, pH 4.74 $\longrightarrow$ 4.82

added to 1.00 L of 0.100 M CH_3COOH, pH 2.89 $\longrightarrow$ 3.78

added to 1.00 L of pure H_2O, pH 7.00 $\longrightarrow$ 12.00

In similar fashion we could calculate the effects of adding 0.010 mole of pure HCl(g) instead of pure NaOH to 1.00 liter of each of these three solutions. This would result in the following changes in pH:

See Table 19-3.

added to 1.00 L of the $CH_3COOH/NaCH_3COO$ buffer, pH 4.74 $\longrightarrow$ 4.66

added to 1.00 L of 0.100 M CH_3COOH, pH 2.89 $\longrightarrow$ 2.00

added to 1.00 L of pure H_2O, pH 7.00 $\longrightarrow$ 2.00

> ⚠ When a little strong acid is added to a solution of a weak acid, the $[H_3O]^+$ from the weak acid is negligible compared to the $[H_3O]^+$ from the strong acid.

The results of adding NaOH or HCl to these solutions (Table 19-3) demonstrate the efficiency of the buffer solution. We recall that each change of 1 pH unit means that the $[H_3O^+]$ and $[OH^-]$ change by a *factor* of 10. In these terms, the effectiveness of the buffer solution in controlling pH is even more dramatic.

Solutions of a Weak Base and a Salt of the Weak Base

An example of this type of buffer solution is one that contains the weak base ammonia, NH_3, and its soluble ionic salt ammonium chloride, NH_4Cl. The reactions responsible for the operation of this buffer are

$$NH_4Cl \xrightarrow{H_2O} NH_4^+ + Cl^- \quad \text{(to completion)}$$

$$\underset{\substack{high \text{ conc} \\ \text{(from base)}}}{NH_3} + H_2O \rightleftharpoons \underset{\substack{high \text{ conc} \\ \text{(from salt)}}}{NH_4^+} + OH^- \quad \text{(reversible)}$$

If a strong acid such as HCl is added to this buffer solution, the resulting H_3O^+ shifts the equilibrium reaction

$$2H_2O \rightleftharpoons H_3O^+ + OH^- \quad \text{(shifts } left\text{)}$$

TABLE 19-3	*Changes in pH Caused by Addition of Pure Acid or Base to One Liter of Solution*				
	When We Add 0.010 mol NaOH(s)		**When We Add 0.010 mol HCl(g)**		
We Have 1.00 L of Original Solution	pH *changes by*	$[H_3O^+]$ *decreases by a factor of*	pH *changes by*	$[H_3O^+]$ *increases by a factor of*	
buffer solution (0.10 M $NaCH_3COO$ and 0.10 M CH_3COOH)	+0.08 pH unit	1.2	−0.08 pH unit	1.2	
0.10 M CH_3COOH	+0.91	8.1	−0.89	7.8	
pure H_2O	+5.00	100,000	−5.00	100,000	

strongly to the *left*. As a result of the diminished OH^- concentration, the reaction

$$NH_3 + H_2O \rightleftharpoons NH_4^+ + OH^- \qquad \text{(shifts } right\text{)}$$

shifts markedly to the *right*. Because the $[NH_3]$ in the buffer solution is high, this can occur to a great extent. The net reaction is

$$\underset{\text{added acid}}{H_3O^+} + \underset{\text{base}}{NH_3} \longrightarrow \underset{\text{weak acid}}{NH_4^+} + \underset{\text{water}}{H_2O} \qquad (\approx 100\%)$$

When a strong base such as NaOH is added to the *original* buffer solution, it is neutralized by the more acidic component, NH_4Cl, or NH_4^+, the conjugate acid of ammonia.

$$NH_3 + H_2O \rightleftharpoons NH_4^+ + OH^- \qquad \text{(shifts } left\text{)}$$

Because the $[NH_4^+]$ is high, this can occur to a great extent. The result is the neutralization of OH^- by NH_4^+.

$$OH^- + NH_4^+ \longrightarrow NH_3 + H_2O \qquad (\approx 100\%)$$

or, as a formula unit equation,

$$\underset{\text{added base}}{NaOH} + \underset{\text{acid}}{NH_4Cl} \longrightarrow \underset{\text{weak base}}{NH_3} + \underset{\text{water}}{H_2O} + NaCl \qquad (\approx 100\%)$$

The net effect is to neutralize most of the H_3O^+ from HCl. This slightly increases the ratio $[NH_4^+]/[NH_3]$, which governs the pH of the solution.

The net effect is to neutralize most of the OH^- from NaOH. This slightly decreases the ratio $[NH_4^+]/[NH_3]$, which governs the pH of the solution.

Summary   Changes in pH are minimized in buffer solutions because the basic component can react with added H_3O^+ ions (producing additional weak acid), while the acidic component can react with added OH^- ions (producing additional weak base).

19-3   PREPARATION OF BUFFER SOLUTIONS

Buffer solutions can be prepared by mixing other solutions. When solutions are mixed, the volume in which each solute is contained increases, so solute concentrations change. These changes in concentration must be considered. If the solutions are dilute, we may assume that their volumes are additive.

EXAMPLE 19-5   *Preparation of Buffer Solutions*

Calculate the concentration of H_3O^+ and the pH of a buffer solution prepared by mixing 200. mL of 0.10 M NaF and 100. mL of 0.050 M HF. $K_a = 7.2 \times 10^{-4}$ for HF.

Plan

Calculate the number of millimoles (or moles) of NaF and HF and then the molarity of each solute in the solution after mixing. Write the appropriate equations for both NaF and HF, represent the equilibrium concentrations algebraically, and substitute into the K_a expression for HF.

Alternatively, one could substitute into the Henderson–Hasselbalch equation and solve for pH, then $[H_3O^+]$.

Solution

When two dilute solutions are mixed, we assume that their volumes are additive. The volume of the new solution will be 300. mL. Mixing a solution of a weak acid with a solution of its salt does not form any new species. So we have a straightforward buffer calculation. We calculate the number of millimoles (or moles) of each compound and the molarities in the new solution.

$$\text{? mmol NaF} = 200.\text{ mL} \times \frac{0.10\text{ mmol NaF}}{\text{mL}} = 20.\text{ mmol HF}$$

$$\text{? mmol HF} = 100.\text{ mL} \times \frac{0.050\text{ mmol HF}}{\text{mL}} = 5.0\text{ mmol HF}$$

in 300. mL

The molarities of NaF and HF in the solution are

$$\frac{20.\text{ mmol NaF}}{300.\text{ mL}} = 0.067\ M\text{ NaF} \quad \text{and} \quad \frac{5.0\text{ mmol HF}}{300.\text{ mL}} = 0.017\ M\text{ HF}$$

The appropriate equations and algebraic representations of concentrations are

$$\begin{array}{ccccc}
\text{NaF} & \longrightarrow & \text{Na}^+ & + & \text{F}^- \\
0.067\ M & & 0.067\ M & & 0.067\ M \\
\end{array}$$

$$\longrightarrow \text{Total [F}^-] = (0.067 + x)\ M$$

$$\begin{array}{ccccc}
\text{HF} & + \text{H}_2\text{O} \rightleftharpoons & \text{H}_3\text{O}^+ & + & \text{F}^- \\
(0.017 - x)\ M & & x\ M & & x\ M \\
\end{array}$$

Using the Henderson–Hasselbalch equation, we obtain:

$$\text{pH} = -\log(7.2 \times 10^{-4}) + \log\frac{0.067}{0.017}$$

$$\text{pH} = 3.14 - 0.596 = 3.74$$

Our assumption is valid.

Substituting into the K_a expression for hydrofluoric acid gives

$$K_a = \frac{[\text{H}_3\text{O}^+][\text{F}^-]}{[\text{HF}]} = \frac{(x)(0.067 + x)}{(0.017 - x)} = 7.2 \times 10^{-4}$$

Can we assume that x is negligible compared with 0.067 and 0.017 in this expression? When in doubt, solve the equation using the simplifying assumption. Then decide whether the assumption was valid. Assume that $(0.067 + x) \approx 0.067$ and $(0.017 - x) \approx 0.017$.

$$\frac{0.067x}{0.017} = 7.2 \times 10^{-4} \qquad x = \boxed{1.8 \times 10^{-4}\ M = [\text{H}_3\text{O}^+]}$$

$$\text{pH} = \boxed{3.74}$$

You should now work Exercise 34.

CD-ROM Screen 18.5, Preparing Buffer Solutions.

We often need a buffer solution of a given pH. One method by which such solutions can be prepared involves adding a salt of a weak base (or weak acid) to a solution of the weak base (or weak acid).

EXAMPLE 19-6   *Preparation of Buffer Solutions*

A solution is 0.10 M in aqueous NH_3. Calculate (a) the number of moles and (b) the number of grams of NH_4Cl that must be added to 500 mL of this solution to prepare a buffer solution with pH = 9.15. You may neglect the volume change due to addition of solid NH_4Cl.

Plan

Convert the given pH to the desired $[OH^-]$ by the usual procedure. Write the appropriate equations for the reactions of NH_4Cl and NH_3 and represent the equilibrium concentrations. Then, substitute into the K_b expression, and solve for the concentration of NH_4Cl required. This allows calculation of the amount of solid NH_4Cl to be added.

Solution

(a) Because the desired pH = 9.15, pOH = 14.00 − 9.15 = 4.85. So $[OH^-] = 10^{-pOH} = 10^{-4.85} = 1.4 \times 10^{-5}\ M\ OH^-$ desired. Let x mol/L be the necessary molarity of NH_4Cl. Because $[OH^-] = 1.4 \times 10^{-5}\ M$, this must be the $[OH^-]$ produced by ionization of NH_3. The equations and representations of equilibrium concentrations follow.

$$NH_4Cl \xrightarrow{100\%} \quad NH_4^+ \quad + \quad Cl^-$$
$$x\ M \quad \Longrightarrow \quad x\ M \qquad x\ M$$

$$NH_3 \quad + \quad H_2O \quad \rightleftharpoons \quad NH_4^+ \quad + \quad OH^-$$
$$(0.10 - 1.4 \times 10^{-5})\ M \qquad 1.4 \times 10^{-5}\ M \quad 1.4 \times 10^{-5}\ M$$

$$\text{Total } [NH_4^+] = (x + 1.4 \times 10^{-5})\ M$$

Substitution into the K_b expression for aqueous ammonia gives

$$K_b = \frac{[NH_4^+][OH^-]}{[NH_3]} = 1.8 \times 10^{-5} = \frac{(x + 1.4 \times 10^{-5})(1.4 \times 10^{-5})}{(0.10 - 1.4 \times 10^{-5})}$$

NH_4Cl is 100% dissociated, so $x \gg 1.4 \times 10^{-5}$. Then $(x + 1.4 \times 10^{-5}) \approx x$.

$$\frac{(x)(1.4 \times 10^{-5})}{0.10} = 1.8 \times 10^{-5} \qquad x = 0.13\ M = [NH_4^+] = M_{NH_4Cl}$$

Now we calculate the number of moles of NH_4Cl that must be added to prepare 500. mL (0.500 L) of buffer solution.

$$\underline{?}\ \text{mol } NH_4Cl = 0.500\ L \times \frac{0.13\ \text{mol } NH_4Cl}{L} = \boxed{0.065\ \text{mol } NH_4Cl}$$

(b) $$\underline{?}\ \text{g } NH_4Cl = 0.065\ \text{mol} \times \frac{53.5\ \text{g } NH_4Cl}{\text{mol } NH_4Cl} = \boxed{3.5\ \text{g } NH_4Cl}$$

You should now work Exercise 37.

Here x does *not* represent a *change* in concentration, but rather the initial concentration of NH_4Cl. We do *not* assume that $x \ll 1.4 \times 10^{-5}$, but rather the reverse.

$$pOH = pK_b + \log \frac{[\text{conjugate acid}]}{[\text{base}]}$$

$$4.85 = -\log(1.8 \times 10^{-5})$$
$$+ \log \frac{[\text{conjugate acid}]}{0.10}$$

$$4.85 = 4.74 + \log[\text{conjugate acid}]$$
$$- (-1.00)$$

$$[\text{conjugate acid}] = 0.13\ M = [NH_4^+]$$

We have used K_a or K_b expressions or the Henderson–Hasselbalch equation in its acid–salt or base–salt form to find the pH of buffered solutions. Either of these calculations involves a *ratio* of concentrations, for instance $\dfrac{[\text{conjugate base}]}{[\text{acid}]}$. Although the ratios are

Preparation of the buffer solution in Example 19-6. We add 3.5 grams of NH_4Cl to a 500-mL volumetric flask and dissolve it in a little of the 0.10 M NH_3 solution. We then dilute to 500 mL with the 0.10 M NH_3 solution.

Charles Steele

Fun with Carbonates

Carbonates react with acids to produce carbon dioxide. This property of carbonates has been exploited in many ways, both serious and silly.

One of the giddiest applications of this behavior of carbonates is in Mad Dawg, a foaming bubble gum developed in the early 1990s. If you chew a piece of this gum, large quantities of foam are produced so that it is difficult to keep the colorful lather from oozing out of your mouth. The froth begins to form as your teeth mix saliva with the gum's ingredients (sodium hydrogen carbonate, citric acid, malic acid, food coloring, and flavoring).

How is this foam produced? When citric acid and malic acid dissolve in saliva, they produce hydrogen ions which decompose the sodium hydrogen carbonate (baking soda) to produce carbon dioxide, a gas. These bubbles of carbon dioxide produce the foam. Large quantities of foam are produced because citric and malic acids taste sour, which stimulates salivation.

A common medical recipe for a similar combination of ingredients is found in Alka Seltzer tablets; these contain sodium hydrogen carbonate, citric acid, and aspirin. The acid and carbonate react in water to produce carbon dioxide, which gives the familiar fizz of Alka Seltzer.

Makeup artists add baking soda to cosmetics to produce monster-flesh makeup. When the hero throws acid (which is actually vinegar, a dilute solution of acetic acid) into the monster's face, the acetic acid reacts with sodium hydrogen carbonate to produce the disgustingly familiar scenes of "dissolving flesh" that we see in horror movies. The ability of baking soda to produce carbon dioxide delights children of all ages as it creates monsters in the movies.

Many early fire extinguishers utilized the reaction of sodium hydrogen carbonate with acids. A metal cylinder was filled with a solution of sodium hydrogen carbonate and water; a bottle filled with sulfuric acid was placed above the water layer. Inverting the extinguisher activated it by causing the

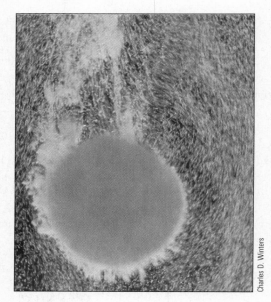

Alka Seltzer™.

acid to spill into the carbonate solution. The pressure produced by gaseous carbon dioxide gas pushed the liquid contents out through a small hose.

Kitchen oven fires can usually be extinguished by throwing baking soda onto the flame. When heated, carbonates decompose to produce carbon dioxide, which smothers fires by depriving them of oxygen.

Chefs frequently use the heat-sensitive nature of carbonates to test the freshness of a box of baking soda. Pouring some boiling water over a little fresh baking soda results in active bubbling. Less active bubbling means the baking soda is unlikely to work well in a baking recipe.

Ronald DeLorenzo
Middle Georgia College

written in terms of molarities, it is not always necessary to use concentrations in the calculation. Both reagents are present in *a single* buffer solution, so the solution volume cancels from the molarity ratio. For example:

$$\frac{[\text{conjugate base}]}{[\text{acid}]} = \frac{\dfrac{\text{mol conjugate base}}{\text{L soln}}}{\dfrac{\text{mol acid}}{\text{L soln}}} = \frac{\text{mol conjugate base}}{\text{mol acid}}$$

Thus we see that a molarity ratio in the acid–salt Henderson–Hasselbalch equation can be treated as a mole (or millimole) ratio. A similar conclusion can be reached for the base–salt version of the Henderson–Hasselbalch equation or for the K_a or K_b expressions that were used in previous buffer calculations.

In practice, a common method of preparing a buffer solution is by *partial* neutralization of a weak acid solution by adding a strong base solution. For example,

$$HA + NaOH \longrightarrow NaA + H_2O \qquad \text{(partial)}$$

If an appreciable amount of the weak acid remains unneutralized, then this solution contains significant concentrations of a *weak acid* and *its conjugate base*, just as though we had added the salt from a separate solution; thus, it is a buffer solution. Example 19-7 illustrates the preparation of a buffer by this method.

EXAMPLE 19-7 Buffer Preparation by Partial Neutralization

Calculate the pH of a solution obtained by mixing 400. mL of a 0.200 M acetic acid solution and 100. mL of a 0.300 M sodium hydroxide solution.

Plan

Sodium hydroxide, NaOH, is a strong base, so it reacts with acetic acid, CH_3COOH, to form sodium acetate, $NaCH_3COO$. If an appreciable amount of excess acetic acid is still present after the sodium hydroxide has reacted, the excess acetic acid and the newly formed sodium acetate solution form a buffered solution.

Solution

We first calculate how much of the weak acid has been neutralized. The numbers of millimoles of CH_3COOH and NaOH mixed are calculated as

$$\text{mmol } CH_3COOH = (0.200 \text{ mmol/mL}) \times 400. \text{ mL} = 80.0 \text{ mmol}$$

$$\text{mmol NaOH} = (0.300 \text{ mmol/mL}) \times 100. \text{ mL} = 30.0 \text{ mmol}$$

Not enough NaOH is present to neutralize all of the CH_3COOH, so NaOH is the *limiting reactant*.

	NaOH	+ CH_3COOH	$\longrightarrow$	$NaCH_3COO$	+ H_2O
start	30.0 mol	80.0 mmol		0	—
change	−30.0 mmol	−30.0 mmol		+30.0 mmol	—
after rxn	0.0 mmol	50.0 mmol		30.0 mmol	

Because $NaCH_3COO$ is a soluble salt, it provides 30.0 mmol CH_3COO^- to the solution. This solution contains a significant amount of CH_3COOH not yet neutralized *and* a significant amount of its conjugate base, CH_3COO^-. We recognize this as a buffer solution and can use the Henderson–Hasselbalch equation to find the pH.

$$pH = pK_a + \log \frac{\text{mmol conjugate base}}{\text{mmol acid}} = pK_a + \log \frac{\text{mmol } CH_3COO^-}{\text{mmol } CH_3COOH}$$

$$= 4.74 + \log \frac{30.0}{50.0} = 4.74 + \log(0.600) = 4.74 + (-0.222)$$

$$= \boxed{4.52}$$

You should now work Exercises 52b to 52f.

Recall that M is equal to

moles per liter

or

millimoles per milliliter.

We could solve this problem using the K_a expression as we did in Example 19-5.

19-4   ACID–BASE INDICATORS

In Section 11-2 we described acid–base titrations and the use of indicators to tell us when to stop a titration. Detection of the end point in an acid–base titration is only one of the important uses of indicators.

An **indicator** is an organic dye; its color depends on the concentration of H_3O^+ ions, or pH, in the solution. By the color an indicator displays, it "indicates" the acidity or basicity of a solution. Figure 19-1 displays solutions that contain three common indicators in solutions over the pH range 3 to 11. Carefully study Figure 19-1 and its legend.

The first indicators used were vegetable dyes. Litmus is a familiar example. Most of the indicators that we use in the laboratory today are synthetic compounds; that is, they have been made in laboratories by chemists. Phenolphthalein is the most common acid–base indicator. It is colorless in solutions of pH less than 8 ($[H_3O^+] > 10^{-8} \, M$) and turns bright pink as pH approaches 10.

Many acid–base indicators are weak organic acids, HIn, where "In" represents various complex organic groups. Bromthymol blue is such an indicator. Its ionization constant is

Phenolphthalein was the active component of the laxative Ex-Lax. It is sometimes added to laboratory ethyl alcohol to discourage consumption.

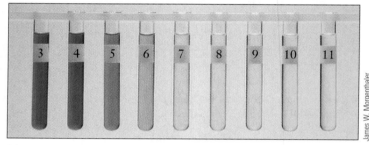

(a)

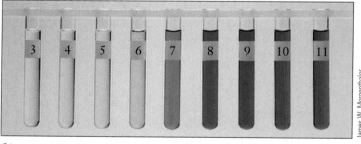

(b)

Figure 19-1 Three common indicators in solutions that cover the pH range 3 to 11 (the black numbers). (a) Methyl red is red at pH 4 and below; it is yellow at pH 7 and above. Between pH 4 and pH 7 it changes from red to red-orange, to orange, to yellow. (b) Bromthymol blue is yellow at pH 6 and below; it is blue at pH 8 and above. Between pH 6 and 8 it changes smoothly from yellow through green to blue. (c) Phenolphthalein is colorless below pH 8 and bright pink above pH 10. It changes smoothly from colorless to bright pink in the pH range 8 to 10.

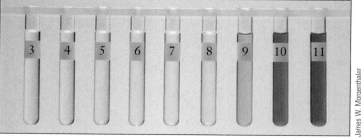

(c)

7.9×10^{-8}. We can represent its ionization in dilute aqueous solution and its ionization constant expression as

$$\text{HIn} + H_2O \rightleftharpoons H_3O^+ + \text{In}^-$$

$$K_a = \frac{[H_3O^+][\text{In}^-]}{[\text{HIn}]} = 7.9 \times 10^{-8}$$

color 1
yellow   ← for bromthymol blue →   color 2
blue

Bromthymol blue indicator is yellow in acidic solutions and blue in basic solutions.

HIn represents nonionized acid molecules, and In^- represents the anion (conjugate base) of HIn. The essential characteristic of an acid–base indicator is that HIn and In^- *must* have quite different colors. The relative amounts of the two species determine the color of the solution. Adding an acid favors the reaction to the left and gives more HIn molecules (color 1). Adding a base favors the reaction to the right and gives more In^- ions (color 2). The ionization constant expression can be rearranged.

$$\frac{[H_3O^+][\text{In}^-]}{[\text{HIn}]} = K_a \qquad \text{so} \qquad \frac{[\text{In}^-]}{[\text{HIn}]} = \frac{K_a}{[H_3O^+]}$$

This shows clearly how the $[\text{In}^-]/[\text{HIn}]$ ratio depends on $[H_3O^+]$ (or on pH) and the K_a value for the indicator. As a rule of thumb, when $[\text{In}^-]/[\text{HIn}] \geq 10$, color 2 is observed; conversely, when $[\text{In}^-]/[\text{HIn}] \leq \frac{1}{10}$, color 1 is observed.

Universal indicators are mixtures of several acid–base indicators that display a continuous range of colors over a wide range of pH values. Figure 18-2 shows concentrated solutions of a universal indicator in flat dishes so that the colors are very intense. The juice of red (purple) cabbage is a universal indicator. Figure 19-2 shows the color of red cabbage juice in solutions within the pH range 1 to 13.

One important use of universal indicators is in commercial indicator papers, which are small strips of paper impregnated with solutions of universal indicators. A strip of the paper is dipped into the solution of interest, and the color of the indicator on the paper indicates the pH of the solution. The photographs on page 753 illustrate the use of universal indicators to estimate pH. We shall describe the use of indicators in titrations more fully in Sections 19-5 and 19-6.

The red anthocyanin pigment in the common geranium is a naturally occurring acid–base indicator.

SC*i*LINKS.

TOPIC: Titration/
Indicators
GO TO: www.scilinks.org
*sci*LINKS CODE: WCH1920

Figure 19-2 The juice of the red (purple) cabbage is a naturally occurring universal indicator. From left to right are solutions of pH 1, 4, 7, 10, and 13.

TITRATION CURVES

19-5 STRONG ACID/STRONG BASE TITRATION CURVES

A **titration curve** is a plot of pH versus the amount (usually volume) of acid or base added. It displays graphically the change in pH as acid or base is added to a solution and shows how pH changes near the equivalence point.

The point at which the color of an indicator changes in a titration is known as the **end point**. It is determined by the K_a value for the indicator (Section 19-4). Table 19-4 shows a few acid–base indicators and the pH ranges over which their colors change. Typically, color changes occur over a range of 1.5 to 2.0 pH units.

The **equivalence point** is the point at which chemically equivalent amounts of acid and base have reacted.

> Ideally, the end point and the equivalence point in a titration should coincide.

In practice, we try to select an indicator whose range of color change includes the equivalence point. We use the same procedures in both standardization and analysis to minimize any error arising from a difference between end point and equivalence point.

Review Section 11-2 before you consider the titration of 100.0 mL of a 0.100 M solution of HCl with a 0.100 M solution of NaOH. As we know, NaOH and HCl react in a 1:1 ratio. We calculate the pH of the solution at several stages as NaOH is added.

Titrations are usually done with 50-mL or smaller burets. We have used 100. mL of solution in this example to simplify the arithmetic.

1. Before any NaOH is added to the 0.100 M HCl solution:

$$HCl + H_2O \xrightarrow{100\%} H_3O^+ + Cl^-$$

$$[H_3O^+] = 0.100\ M \quad \text{so} \quad pH = 1.00$$

If one prefers, these calculations can be done using moles and liters in the place of millimoles and milliliters.

2. After 20.0 mL of 0.100 M NaOH has been added:

	HCl	+	NaOH	$\longrightarrow$	NaCl	+ H$_2$O
start	10.0 mmol		2.0 mmol		0 mmol	
change	−2.0 mmol		−2.0 mmol		+2.0 mmol	
after rxn	8.0 mmol		0 mmol		2.0 mmol	

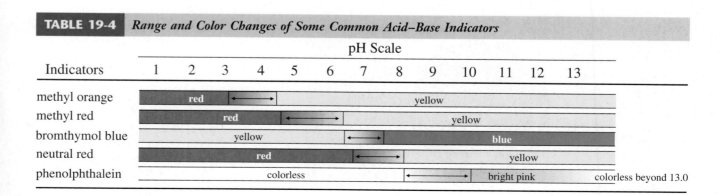

| TABLE 19-4 | *Range and Color Changes of Some Common Acid–Base Indicators* |

pH Scale

Indicators	1	2	3	4	5	6	7	8	9	10	11	12	13

methyl orange — red / yellow

methyl red — red / yellow

bromthymol blue — yellow / blue

neutral red — red / yellow

phenolphthalein — colorless / bright pink   colorless beyond 13.0

The concentration of unreacted HCl in the total volume of 120. mL is

$$M_{HCl} = \frac{8.00 \text{ mmol HCl}}{120. \text{ mL}} = 0.067 \text{ M HCl}$$

$$[H_3O^+] = 6.7 \times 10^{-2} \text{ M} \qquad \text{so} \qquad pH = 1.17$$

3. After 50.0 mL of 0.100 M NaOH has been added (midpoint of the titration):

	HCl	+	NaOH	⟶	NaCl	+ H₂O
start	10.0 mmol		5.0 mmol		0 mmol	
change	−5.0 mmol		−5.0 mmol		+5.0 mmol	
after rxn	5.0 mmol		0 mmol		5.0 mmol	

$$M_{HCl} = \frac{5.00 \text{ mmol HCl}}{150. \text{ mL}} = 0.033 \text{ M HCl}$$

$$[H_3O^+] = 3.3 \times 10^{-2} \text{ M} \qquad \text{so} \qquad pH = 1.48$$

4. After 100. mL of 0.100 M NaOH has been added:

	HCl	+	NaOH	⟶	NaCl	+ H₂O
start	10.0 mmol		10.0 mmol		0 mmol	
change	−10.0 mmol		−10.0 mmol		+10.0 mmol	
after rxn	0 mmol		0 mmol		10.0 mmol	

We have added enough NaOH to neutralize the HCl exactly so this is the equivalence point. A strong acid and a strong base react to give a neutral salt solution so pH = 7.00.

5. After 110.0 mL of 0.100 M NaOH has been added:

	HCl	+	NaOH	⟶	NaCl	+ H₂O
start	10.0 mmol		11.0 mmol		0 mmol	
change	−10.0 mmol		−10.0 mmol		+10.0 mmol	
after rxn	0 mmol		1.0 mmol		10.0 mmol	

The pH is determined by the excess NaOH.

Some household products. Each solution contains a few drops of a universal indicator. A color of yellow or red indicates a pH less than 7. A green to purple color indicates a pH greater than 7.

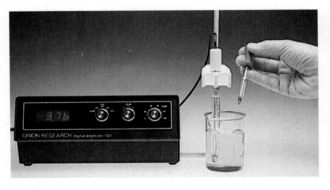
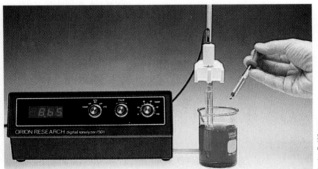

The end point of the titration of 0.100 M HCl with 0.100 M NaOH using another indicator, bromthymol blue.

$$M_{NaOH} = \frac{1.0 \text{ mmol NaOH}}{210. \text{ mL}} = 0.0048 \, M \text{ NaOH}$$

$$[OH^-] = 4.8 \times 10^{-3} \, M \quad \text{so} \quad pOH = 2.32 \quad \text{and} \quad pH = 11.68$$

Table 19-5 displays the data for the titration of 100.0 mL of 0.100 M HCl by 0.100 M NaOH solution. A few additional points have been included to show the shape of the curve better. These data are plotted in Figure 19-3a. This titration curve has a long "vertical section" over which the pH changes very rapidly with the addition of very small amounts of base. The pH changes from 3.60 (99.5 mL NaOH added) to 10.40 (100.5 mL of NaOH added) in the vicinity of the equivalence point (100.0 mL NaOH added). The midpoint of the vertical section (pH = 7.00) is the equivalence point. We can separate the calculations on this kind of titration into four distinct types that correspond to four regions of the titration curves.

CD-ROM   Screen 18.7, Titration Curves.

1. Before any strong base is added, the pH depends on the strong acid alone.
2. After some strong base has been added, but before the equivalence point, the remaining (excess) strong acid determines the pH.
3. At the equivalence point, the solution is neutral.
4. Beyond the equivalence point, excess strong base determines the pH.

Ideally, the indicator color change should occur at pH = 7.00. For practical purposes, indicators with color changes in the pH range 4 to 10 can be used in the titration of strong acids and strong bases because the vertical portion of the titration curve is so long. Figure 19-3 shows the ranges of color changes for methyl red and phenolphthalein, two widely used indicators. Both fall within the vertical section of the NaOH/HCl titration curve. When a strong acid is added to a solution of a strong base, the titration curve is inverted, but its essential characteristics are the same (Figure 19-3b).

In Figure 19-3a we see that the curve rises very slowly before the equivalence point. It then rises very rapidly near the equivalence point because there is no hydrolysis. The curve becomes almost flat beyond the equivalence point.

TABLE 19-5	*Titration Data for 100.0 mL of 0.100 M HCl Versus NaOH*		
mL of 0.100 M NaOH Added	mmol NaOH Added	mmol Excess Acid or Base	pH
0.0	0.00	10.0 H_3O^+	1.00
20.0	2.00	8.0	1.18
50.0	5.00	5.0	1.48
90.0	9.00	1.0	2.28
99.0	9.90	0.10	3.30
99.5	9.95	0.05	3.60
100.0	10.00	0.00 (eq. pt.)	7.00
100.5	10.05	0.05 OH^-	10.40
110.0	11.00	1.00	11.68
120.0	12.00	2.00	11.96

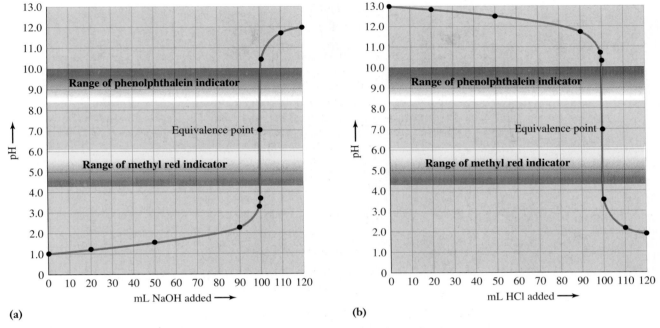

Figure 19-3 (a) The titration curve for 100. mL of 0.100 *M* HCl with 0.100 *M* NaOH. Note that the "vertical" section of the curve is quite long. The titration curves for other strong acids and bases are identical with this one *if* the same concentrations of acid and bases are used *and if* both are monoprotic. (b) The titration curve for 100. mL of 0.100 *M* NaOH with 0.100 *M* HCl. This curve is similar to that in Part (a), but inverted.

19-6 WEAK ACID/STRONG BASE TITRATION CURVES

When a weak acid is titrated with a strong base, the curve is quite different. The solution is buffered *before* the equivalence point. It is basic *at* the equivalence point because salts of weak acids and strong bases hydrolyze to give basic solutions. So, we can separate the calculations on this kind of titration into four distinct types, which correspond to four regions of the titration curves.

1. Before any base is added, the pH depends on the weak acid alone.

2. After some base has been added, but before the equivalence point, a series of weak acid/salt buffer solutions determines the pH.

3. At the equivalence point, hydrolysis of the anion of the weak acid determines the pH.

4. Beyond the equivalence point, excess strong base determines the pH.

Consider the titration of 100.0 mL of 0.100 *M* CH_3COOH with 0.100 *M* NaOH solution. (The strong electrolyte is added to the weak electrolyte.)

1. Before any base is added, the pH is 2.89 (Example 18-11 and Table 18-5).

2. As soon as some NaOH is added, but before the equivalence point, the solution is buffered because it contains both CH_3COOH and $NaCH_3COO$.

$$NaOH + CH_3COOH \longrightarrow NaCH_3COO + H_2O$$
$$\text{lim amt} \qquad \text{excess}$$

For instance, after 20.0 mL of 0.100 M NaOH solution has been added, we have

	NaOH	+ CH₃COOH	⟶ NaCH₃COO + H₂O
start	2.00 mmol	10.00 mmol	0 mmol
change	−2.00 mmol	−2.00 mmol	+2.00 mmol
after rxn	0 mmol	8.00 mmol	2.00 mmol

We recognize that this is a buffer solution, so we can use the ratio (mmol conjugate base)/(mmol acid) in the Henderson–Hasselbalch equation as we did in Example 19-7.

We could use the K_a expression for these calculations as we did in Example 19-5.

$$pH = pK_a + \log \frac{\text{mmol conjugate base}}{\text{mmol acid}} = pK_a + \log \frac{\text{mmol CH}_3\text{COO}^-}{\text{mmol CH}_3\text{COOH}}$$

$$= 4.74 + \log \frac{2.00}{8.00} = 4.74 + \log(0.250) = 4.74 + (-0.602) = 4.14$$

After some NaOH has been added, the solution contains both $NaCH_3COO$ and CH_3COOH, and so it is buffered until the equivalence point is reached. All points before the equivalence point are calculated in the same way.

Just before *the equivalence point, the solution contains relatively high concentrations of NaCH₃COO and relatively low concentrations of CH₃COOH. Just* after *the equivalence point, the solution contains relatively high concentrations of NaCH₃COO and relatively low concentrations of NaOH, both basic components. In both regions our calculations are only approximations. Exact calculations of pH in these regions are too complicated to discuss here.*

3. At the equivalence point, the solution is 0.0500 M in $NaCH_3COO$.

	NaOH	+ CH₃COOH	⟶ NaCH₃COO + H₂O
start	10.0 mmol	10.0 mmol	0 mmol
change	−10.0 mmol	−10.0 mmol	+10.0 mmol
after rxn	0 mmol	0 mmol	10.0 mmol

$$M_{\text{NaCH}_3\text{COO}} = \frac{10.0 \text{ mmol NaCH}_3\text{COO}}{200. \text{ mL}} = 0.0500 \ M \text{ NaCH}_3\text{COO}$$

The pH of a 0.0500 M solution of $NaCH_3COO$ is 8.72 (Example 18-19 shows a similar calculation). The solution is distinctly basic at the equivalence point because of the hydrolysis of the acetate ion.

4. Beyond the equivalence point, the concentration of the excess NaOH determines the pH of the solution just as it did in the titration of a strong acid.

Table 19-6 lists several points on the titration curve, and Figure 19-4 shows the titration curve for 100.0 mL of 0.100 M CH₃COOH titrated with a 0.100 M solution of NaOH. This titration curve has a short vertical section (pH ≈ 7 to 10), and the indicator range is limited. Phenolphthalein is the indicator commonly used to titrate weak acids with strong bases (see Table 19-4).

The titration curves for weak bases and strong acids are similar to those for weak acids and strong bases except that they are inverted (recall that strong is added to weak). Figure 19-5 displays the titration curve for 100.0 mL of 0.100 M aqueous ammonia titrated with 0.100 M HCl solution.

TABLE 19-6	Titration Data for 100.0 mL of 0.100 M CH₃COOH with 0.100 M NaOH			

mL 0.100 M NaOH Added		mmol Base Added	mmol Excess Acid or Base	pH
0.0 mL		0	10.0 CH₃COOH	2.89
20.0 mL		2.00	8.00	4.14
50.0 mL		5.00	5.00	4.74
75.0 mL	buffered region	7.50	2.50	5.22
90.0 mL		9.00	1.00	5.70
95.0 mL		9.50	0.50	6.02
99.0 mL		9.90	0.10	6.74
100.0 mL		10.0	0 (equivalence point)	8.72
101.0 mL		10.1	0.10 OH⁻	10.70
110.0 mL		11.0	1.0	11.68
120.0 mL		12.0	2.0	11.96

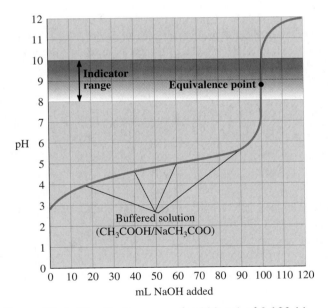

Figure 19-4 The titration curve for 100. mL of 0.100 M CH₃COOH with 0.100 M NaOH. The "vertical" section of this curve is much shorter than those in Figure 19-3 because the solution is buffered before the equivalence point.

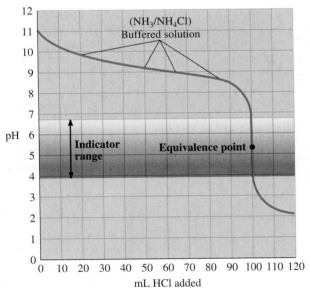

Figure 19-5 The titration curve for 100. mL of 0.100 M aqueous ammonia with 0.100 M HCl. The vertical section of the curve is relatively short because the solution is buffered before the equivalence point. The curve is very similar to that in Figure 19-4, but inverted.

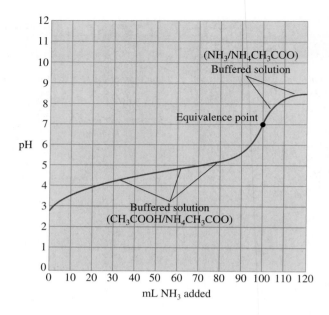

Figure 19-6 The titration curve for 100. mL of 0.100 M CH$_3$COOH with 0.100 M aqueous NH$_3$. Because the solution is buffered before and after the equivalence point, the vertical section of the curve is too short to be noticed. Color indicators cannot be used in such titrations. Instead, physical methods such as conductivity measurements can be used to detect the end point.

19-7 WEAK ACID/WEAK BASE TITRATION CURVES

In titration curves for weak acids and weak bases, pH changes near the equivalence point are too small for color indicators to be used. The solution is buffered both before and after the equivalence point. Figure 19-6 shows the titration curve for 100.0 mL of 0.100 M CH$_3$COOH solution titrated with 0.100 M aqueous NH$_3$. The calculation of values on the curve in Figure 19-6 other than the initial pH and the pH at the equivalence point is beyond the scope of this text.

> ✓ **Problem-Solving Tip:** *Titration Curves*
>
> You can consider a titration curve in four parts.
>
> 1. **Initial solution** (before any titrant is added).
> 2. **Region before the equivalence point.** This may or may not be buffered. The solution is buffered in this region if the substance being titrated is a weak acid or weak base.
> 3. **Equivalence point.** Its location depends on the concentrations of the acid and the base solutions; its pH depends on the strengths of the acid and base.
> 4. **Region beyond the equivalence point.** This becomes nearly flat as more and more excess reactant is added. We often calculate only one or two points in this region.
>
> Recognizing the four regions of a titration curve allows you to decide which kind of calculation is required.

19-8 SUMMARY OF ACID–BASE CALCULATIONS

In this and the previous chapter we have discussed several different types of acidic or alkaline solutions. Table 19-7 summarizes those many different types of solutions.

TABLE 19-7	*A Review of Our Understanding of Nonneutral Aqueous Solutions*		
Type of Aq. Solution	**Example(s)**	**Resulting Chemistry or Type of Solution; Calculation**	**Section**
strong acid (represented as HX)	HNO_3 or HCl	*Complete* ionization; reaction goes to completion to form H_3O^+ $$HX + H_2O \longrightarrow H_3O^+ + X^-$$	18-1
strong base [represented as MOH or $M(OH)_2$]	NaOH or $Ba(OH)_2$	*Complete* ionization; reaction goes to completion to form OH^- $$MOH \longrightarrow M^+ + OH^-$$ $$M(OH)_2 \longrightarrow M^{2+} + 2OH^-$$	18-1
weak acid (represented as HA)	CH_3COOH or HCN	*Partial* ionization; $HA + H_2O \rightleftharpoons H_3O^+ + A^-$ Must solve equilibrium expression involving K_a for the acid using *equilibrium* concentrations: $K_a = \dfrac{[H_3O^+][A^-]}{[HA]}$	18-4
weak base (represented as B)	NH_3	*Partial* ionization; $B + H_2O \rightleftharpoons BH^+ + OH^-$ Must solve equilibrium expression involving K_b for the base, using *equilibrium* concentrations: $K_b = \dfrac{[BH^+][OH^-]}{[B]}$	18-4
Salt of *strong base* & *strong acid* (represented as MX)	$NaNO_3$ (salt of NaOH and HNO_3)	No hydrolysis—*neutral solution*	18-7
Salt of *strong base* & *weak acid* (represented as MA)	NaCN (salt of NaOH and HCN)	*Hydrolysis* of conjugate base (A^-) of *weak acid* $\Longrightarrow$ *basic* solution $$A^- + H_2O \rightleftharpoons HA + OH^-$$ Solve weak base equilibrium for A^- using $K_{b\text{ for }A^-} = \dfrac{K_w}{K_{a\text{ for HA}}}$	18-8
Salt of *weak base* & *strong acid* (represented as BHX)	NH_4NO_3 (salt of NH_3 and HNO_3)	*Hydrolysis* of conjugate acid (BH^+) of *weak base* $\Longrightarrow$ *acidic* solution $$BH^+ + H_2O \rightleftharpoons B + H_3O^+$$ Solve weak acid equilibrium for BH^+ using $K_{a\text{ for }BH^+} = \dfrac{K_w}{K_{b\text{ for B}}}$	18-9
Salt of *weak base* & *weak acid* (represented as BHA)	NH_4CN (salt of NH_3 and HCN)	*Hydrolysis* of conjugate base (A^-) of *weak acid* $\Longrightarrow$ *basic* solution and *Hydrolysis* of conjugate acid (BH^+) of *weak base* $\Longrightarrow$ *acidic* solution. This solution can be *basic, neutral,* or *acidic,* depending on which hydrolysis occurs to a greater extent; use K_b and K_a values for the *hydrolysis reactions* to tell which ion will be the dominant factor. (We did not do calculations for this type of salt.)	18-10
Weak acid & *salt* of its conjugate base (represented as HA & MA)	CH_3COOH + $NaCH_3COO$ (or the product of the partial neutralization in a titration of a *weak acid* with a *strong base*)	Salt ionizes completely; $MA \longrightarrow M^+ + A^-$ Mixture of weak acid (HA) and its conjugate base (A^-) in significant concentrations gives a *buffer.* Use Henderson–Hasselbalch equation for acid–salt buffer (with initial concentrations or mole ratio): $$pH = pK_a + \log\frac{[\text{conjugate base}]}{[\text{acid}]}$$	19-1, 19-2, 19-3
Weak base & *salt* of its conjugate acid (represented as B & BHX)	NH_3 + NH_4Cl (or the product of the partial neutralization in a titration of a *weak base* with a *strong acid*)	Salt ionizes completely; $NH_4Cl \longrightarrow NH_4^+ + Cl^-$ Mixture of weak base (NH_3) and its conjugate acid (NH_4^+) in significant concentrations gives a *buffer.* Use Henderson–Hasselbalch equation for base–salt buffer (with initial concentrations or mole ratio): $$pOH = pK_b + \log\frac{[\text{conjugate acid}]}{[\text{base}]}$$	19-1, 19-2, 19-3

Key Terms

Buffer solution   A solution that resists changes in pH when strong acids or strong bases are added. A buffer solution contains an acid and its conjugate base, so it can react with added base or acid. Common buffer solutions contain either (1) a weak acid and a soluble ionic salt of the weak acid *or* (2) a weak base and a soluble ionic salt of the weak base.

Common ion effect   Suppression of ionization of a weak electrolyte by the presence in the same solution of a strong electrolyte containing one of the same ions as the weak electrolyte.

End point   The point at which an indicator changes color and a titration should be stopped.

Equivalence point   The point at which chemically equivalent amounts of reactants have reacted.

Henderson–Hasselbalch equation   An equation that enables us to calculate the pH or pOH of a buffer solution directly.

For acid–salt buffer $pH = pK_a + \log \dfrac{[\text{conj. base}]}{[\text{acid}]}$

For base–salt buffer $pOH = pK_b + \log \dfrac{[\text{conj. acid}]}{[\text{base}]}$

Indicator (for acid–base reactions)   An organic compound that exhibits different colors in solutions of different acidities; used to indicate the point at which reaction between an acid and a base is complete.

Titration   A procedure in which one solution is added to another solution until the chemical reaction between the two solutes is complete; usually the concentration of one solution is known and that of the other is unknown.

Titration curve (for acid–base titration)   A plot of pH versus volume of acid or base solution added.

Exercises

NOTE   *All exercises in this chapter assume a temperature of 25°C unless they specify otherwise. Values of K_a and K_b can be found in Appendix F or will be specified in the exercise.*

Basic Ideas

1. (a) What is the relationship between pH, pOH, and pK_w? (b) What is the relationship between K_a and pH? (c) What is the relationship between K_a and pOH?
2. Write the balanced equation for an acid–base reaction that would produce each of the following salts; predict whether an aqueous solution of each salt is acidic, basic, or neutral. (a) $NaNO_3$; (b) Na_2S; (c) $Al_2(SO_4)_3$; (d) $Mg(CH_3COO)_2$; (e) $(NH_4)_2SO_4$.
3. Write the balanced equation for an acid–base reaction that would produce each of the following salts; predict whether an aqueous solution of each salt is acidic, basic, or neutral. (a) CaF_2; (b) ZnS; (c) $ZnCl_2$; (d) K_3PO_4; (e) NH_4NO_3.
4. Under what circumstances can it be predicted that a neutral solution is produced by an acid–base reaction? (*Hint:* This question has more than one answer.)

The Common Ion Effect and Buffer Solutions

5. Which of the following solutions are buffers? Each solution was prepared by mixing and diluting appropriate quantities of the two solutes to yield the concentrations indicated. Explain your decision for each solution. (a) 0.10 M HCN and 0.10 M NaCN; (b) 0.10 M NaCN and 0.10 M NaCl; (c) 0.10 M NH_3 and 0.10 M NH_4Br; (d) 0.10 M NaOH and 0.90 M KOH.
6. Which of the following solutions are buffers? Each solution was prepared by mixing and diluting appropriate quantities of the two solutes to yield the concentra-

tions indicated. Explain your decision for each solution. (a) 1.0 M HCN and 0.20 M NaCN; (b) 0.10 M NaCN and 0.10 M HCN; (c) 0.10 M NH_4Cl and 0.90 M NH_4Br; (d) 0.10 M NaCl and 0.20 M HF.

7. Suppose that you have a solution that is 0.50 M in methylamine, CH_3NH_2, and 0.00050 M in the salt methylammonium chloride, CH_3NH_3Cl. Would you expect this to be an effective buffer solution? Why or why not?
8. Calculate pH for each of the following buffer solutions. (a) 0.10 M HF and 0.20 M KF; (b) 0.040 M CH_3COOH and 0.025 M $Ba(CH_3COO)_2$.
9. The pK_a of HOCl is 7.45. Calculate the pH of a solution that is 0.0222 M HOCl and 0.0444 M NaOCl.
10. Calculate the concentration of OH^- and the pH for the following buffer solutions. (a) 0.30 M NH_3(aq) and 0.20 M NH_4NO_3; (b) 0.15 M NH_3(aq) and 0.10 M $(NH_4)_2SO_4$.
11. Calculate the concentration of OH^- and the pH for the following solutions. (a) 0.45 M NH_3(aq) and 0.35 M NH_4NO_3; (b) 0.10 M aniline, $C_6H_5NH_2$, and 0.25 M anilinium chloride, $C_6H_5NH_3Cl$.
*12. Buffer solutions are especially important in our body fluids and metabolism. Write net ionic equations to illustrate the buffering action of (a) the $H_2CO_3/NaHCO_3$ buffer system in blood and (b) the NaH_2PO_4/Na_2HPO_4 buffer system inside cells.
13. Calculate the ratio of $[NH_3]/[NH_4^+]$ concentrations that gives (a) solutions of pH = 9.55 and (b) solutions of pH = 9.10.
14. We prepare two solutions as follows. In solution A 0.50 mole of potassium acetate is added to 0.25 mole of acetic acid and diluted to a final volume of 1.00 liter. In solution B 0.25 mole of potassium acetate is added to 0.50 mole of acetic acid and diluted to 1.00 liter. (a) Which solution is expected to have the lower pH? (b) Explain

how you can reach your conclusion without calculating the pH of each solution.

15. Compare the pH of a 0.25 M acetic acid solution to the pH of a solution composed of 0.25 M acetic acid to which 0.25 mole of sodium acetate per liter has been added.

16. Calculate the mass of sodium acetate that must be added to one liter of a 0.150 M acetic acid solution to produce a buffer with the pH of 4.75. Assume that there is no volume change.

17. A solution is produced by dissolving 0.075 mole of formic acid and 0.075 mole of sodium formate in sufficient water to produce 1.00 L of solution. (a) Calculate the pH of this buffer solution. (b) Calculate the pH of the solution after an additional 0.010 mole of sodium formate is dissolved.

Buffering Action

18. Consider the ionization of formic acid, HCOOH.

$$HCOOH + H_2O \rightleftharpoons HCOO^- + H_3O^+$$

What effect does the addition of sodium formate, NaHCOO, have on the fraction of formic acid molecules that undergo ionization in aqueous solution?

19. Briefly describe why the pH of a given buffer solution remains nearly constant when small amounts of acid or base are added. Over what pH range would a given buffer exhibit the best buffering action (nearly constant pH)?

20. What is the pH of a solution that is 0.15 M in $HClO_4$ and 0.15 M $KClO_4$? Is this a buffer solution?

21. (a) Find the pH of a solution that is 0.40 M in formic acid and 0.30 M in sodium formate. (b) Find the pH after 0.050 mol HCl has been added to 1.0 liter of the solution.

22. One liter of 0.400 M NH_3 solution also contains 12.78 g of NH_4Cl. How much will the pH of this solution change if 0.142 mole of gaseous HCl is bubbled into it?

23. (a) Find the pH of a solution that is 0.11 M in nitrous acid and 0.15 M in sodium nitrite. (b) Find the pH after 0.14 mol of NaOH has been added to 1.0 liter of the solution.

24. (a) Find the pH of a solution that is 1.00 M in NH_3 and 0.80 M in NH_4Cl. (b) Find the pH of the solution after 0.10 mol of HCl has been added to 1.0 liter of the solution. (c) A solution was prepared by adding NaOH to pure water to give 1.00 liter of solution whose pH = 9.34. Find the pH of this solution after 0.10 mol of HCl has been added to it.

25. (a) Calculate the concentrations of CH_3COOH and CH_3COO^- in a solution in which their total concentration is 0.200 mol/L and the pH is 4.50. (b) If 0.0100 mol of solid NaOH is added to 1.00 L of this solution, how much does the pH change?

26. A solution contains bromoacetic acid, $BrCH_2COOH$, and sodium bromoacetate, $NaBrCH_2COO$, with acid and salt concentrations that total 0.30 mol/L. If the pH is 3.10, what are the concentrations of the acid and the salt? $K_a = 2.0 \times 10^{-3}$ for $BrCH_2COOH$.

27. Calculate the concentration of propionate ion, $CH_3CH_2COO^-$, in equilibrium with 0.025 M CH_3CH_2COOH (propionic acid) and 0.10 M H^+ from hydrochloric acid. $K_a = 1.3 \times 10^{-5}$ for CH_3CH_2COOH.

28. Calculate the concentration of $C_2H_5NH_3^+$ in equilibrium with 0.012 M $C_2H_5NH_2$ (ethylamine) and 0.0011 M OH^- ion from sodium hydroxide. $K_b = 4.7 \times 10^{-4}$ for ethylamine.

29. When chlorine gas is dissolved in water to make "chlorine water," HCl (a strong acid) and HOCl (a weak acid) are produced in equal amounts.

$$Cl_2(g) + H_2O(\ell) \longrightarrow HCl(aq) + HOCl(aq)$$

What is the concentration of OCl^- ion in a solution containing 0.010 mol of each acid in 1.00 L of solution?

Preparation of Buffer Solutions

30. A buffer solution of pH 5.30 is to be prepared from propionic acid and sodium propionate. The concentration of sodium propionate must be 0.60 mol/L. What should be the concentration of the acid? $K_a = 1.3 \times 10^{-5}$ for CH_3CH_2COOH.

31. We need a buffer with pH 9.00. It can be prepared from NH_3 and NH_4Cl. What must be the $[NH_4^+]/[NH_3]$ ratio?

*32. What volumes of 0.150 M acetic acid and 0.100 M NaOH solutions must be mixed to prepare 1.00 L of a buffer solution of pH 4.50 at 25°C?

*33. One liter of a buffer solution is prepared by dissolving 0.150 mol of $NaNO_2$ and 0.070 mol of HCl in water. What is the pH of this solution? If the solution is diluted twofold with water, what is the pH?

*34. One liter of a buffer solution is made by mixing 500. mL of 1.25 M acetic acid and 500. mL of 0.600 M calcium acetate. What is the concentration of each of the following in the buffer solution? (a) CH_3COOH; (b) Ca^{2+}; (c) CH_3COO^-; (d) H^+. (e) What is the pH?

35. What must be the concentration of benzoate ion, $C_6H_5COO^-$, in a 0.055 M benzoic acid, C_6H_5COOH, solution so that the pH is 5.00?

36. What must be the concentration of chloroacetic acid, $ClCH_2COOH$, in a 0.015 M $NaCH_2ClCOO$ solution so that the pH is 3.00? $K_a = 1.4 \times 10^{-3}$ for $ClCH_2COOH$.

37. What must be the concentration of NH_4^+ in a 0.075 M NH_3 solution so that the pH is 8.80?

Acid–Base Indicators

38. (a) What are acid–base indicators? (b) What are the essential characteristics of acid–base indicators? (c) What determines the color of an acid–base indicator in an aqueous solution?

39. K_a is 7.9×10^{-8} for bromthymol blue, an indicator that can be represented as HIn. HIn molecules are yellow, and In^- ions are blue. What color will bromthymol blue be in a solution in which (a) $[H_3O^+] = 1.0 \times 10^{-4}$ M and (b) pH = 10.30?

*40. The indicator metacresol purple changes from yellow to purple at pH 8.2. At this point it exists in equal concentrations as the conjugate acid and the conjugate base. What are K_a and pK_a for metacresol purple, a weak acid represented as HIn?

41. A series of acid–base indicators can be used to estimate the pH of an unknown solution. Use the values given in Table 19-4 to determine the possible range of pH values of the following solution. The solution was colorless with phenolphthalein, yellow with methyl orange, and yellow with methyl red.

42. A series of acid–base indicators can be used to estimate the pH of an unknown solution. Use the values given in Table 19-4 to determine the possible range of pH values of the following solution. The solution was colorless in phenolphthalein, blue in bromthymol blue, and yellow in methyl orange.

43. Use Table 19-4 to choose one or more indicators that could be used to "signal" reaching a pH of (a) 3.5; (b) 7.0; (c) 10.3; (d) 8.0.

44. A solution of 0.020 M acetic acid is to be titrated with a 0.025 M NaOH solution. What is the pH at the equivalence point? Choose an appropriate indicator for the titration.

45. Demonstrate mathematically that neutral red is red in solutions of pH 3.00, whereas it is yellow in solutions of pH 10.00. HIn is red, and In$^-$ is yellow. K_a is 2.0×10^{-7}.

Strong Acid/Strong Base Titration Curves

46. Make a rough sketch of the titration curve expected for the titration of a strong acid with a strong base. What determines the pH of the solution at the following points? (a) No base added; (b) half-equivalence point; (c) equivalence point; (d) excess base added. Compare your curve with Figure 19-3.

For Exercises 47, 52, 57, and 58, calculate and tabulate $[H_3O^+]$, $[OH^-]$, pH, and pOH at the indicated points as we did in Table 19-4. In each case assume that pure acid (or base) is added to exactly 1 L of a 0.0100 molar solution of the indicated base (or acid). This simplifies the arithmetic because we may assume that the volume of each solution is constant throughout the titration. Plot each titration curve with pH on the vertical axis and moles of base (or acid) added on the horizontal axis.

47. Solid NaOH is added to 1 L of 0.0500 M HCl solution. Number of moles of NaOH added: (a) none; (b) 0.00500; (c) 0.01500; (d) 0.02500 (50% titrated); (e) 0.03500; (f) 0.04500; (g) 0.04750; (h) 0.0500 (100% titrated); (i) 0.0525; (j) 0.0600; (k) 0.0750 (50% excess NaOH). Consult Table 19-4, and list the indicators that could be used in this titration.

48. A 25.0-mL sample of 0.125 M HNO$_3$ is titrated with 0.100 M NaOH. Calculate the pH of the solution

(a) before the addition of NaOH and after the addition of (b) 5.0 mL; (c) 12.5 mL; (d) 25.0 mL; (e) 31.2 mL; (f) 37.5 mL of NaOH.

49. A 33.0-mL sample of 0.245 M HNO$_3$ solution is titrated with 0.213 M KOH. Calculate the pH of the solution (a) before the addition of KOH and after the addition of (b) 5.55 mL, (c) 12.0 mL, (d) 24.5 mL, (e) 35.2 mL, (f) 38.8 mL of KOH solution.

50. A 44.0-mL sample of 0.145 M HCl solution is titrated with 0.213 M NaOH. Calculate the pH of the solution (a) before the addition of NaOH and after the addition of (b) 5.55 mL, (c) 12.0 mL, (d) 20.5 mL, (e) 27.2 mL, (f) 31.8 mL of NaOH solution.

Weak Acid/Strong Base Titration Curves

51. Make a rough sketch of the titration curve expected for the titration of a weak monoprotic acid with a strong base. What determines the pH of the solution at the following points? (a) No base added; (b) half-equivalence point; (c) equivalence point; (d) excess base added. Compare your curve to Figure 19-4.

52. Solid NaOH is added to exactly 1 L of 0.0200 M CH$_3$COOH solution. Number of moles NaOH added: (a) none; (b) 0.00400; (c) 0.00800; (d) 0.01000 (50% titrated); (e) 0.01400; (f) 0.01800; (g) 0.01900; (h) 0.0200 (100% titrated); (i) 0.0210; (j) 0.0240; (k) 0.0300 (50% excess NaOH). Consult Table 19-4, and list the indicators that could be used in this titration.

53. A 44.0-mL sample of 0.202 M CH$_3$COOH solution is titrated with 0.185 M NaOH. Calculate the pH of the solution (a) before the addition of any NaOH solution and after the addition of (b) 15.5 mL, (c) 20.0 mL, (d) 24.0 mL, (e) 27.2 mL, (f) 48.0 mL, (g) 50.2 mL of NaOH solution.

54. A 32.44-mL sample of 0.182 M CH$_3$COOH solution is titrated with 0.185 M NaOH. Calculate the pH of the solution (a) before the addition of any NaOH solution and after the addition of (b) 15.55 mL, (c) 20.0 mL, (d) 24.02 mL, (e) 27.2 mL, (f) 31.91 mL, (g) 33.12 mL of NaOH solution.

55. A solution contains an unknown weak monoprotic acid, HA. It takes 46.24 mL of NaOH solution to titrate 50.00 mL of the HA solution to the equivalence point. To another 50.00-mL sample of the same HA solution, 23.12 mL of the same NaOH solution is added. The pH of the resulting solution in the second experiment is 5.14. What are K_a and pK_a of HA?

56. Calculate the pH at the equivalence point of the titration of 100.0 mL of each of the following with 0.150 M KOH: (a) 1.000 M acetic acid; (b) 0.100 M acetic acid; (c) 0.0100 M acetic acid.

Mixed Exercises

*57. Gaseous HCl is added to 1 L of 0.0100 M aqueous ammonia solution. Number of moles HCl added: (a) none; (b) 0.00100; (c) 0.00300; (d) 0.00500 (50% titrated);

(e) 0.00700; (f) 0.00900; (g) 0.00950; (h) 0.0100 (100% titrated); (i) 0.0105; (j) 0.0120; (k) 0.0150 (50% excess HCl). Consult Table 19-4, and list the indicators that could be used in this titration.

*58. Gaseous NH_3 is added to exactly 1 L of 0.0100 M HNO_3 solution. Number of moles NH_3 added: (a) none; (b) 0.00100; (c) 0.00400; (d) 0.00500 (50% titrated); (e) 0.00900; (f) 0.00950; (g) 0.0100 (100% titrated); (h) 0.0105; (i) 0.0130. What is the major difference between the titration curve for the reaction of HNO_3 and NH_3 and the other curves you have plotted? Consult Table 19-4. Can you suggest a satisfactory indicator for this titration?

59. Compare the pH of 0.54 M NaCl with the pH of 0.54 M NaCN.

60. Compare the pH of 0.54 M NaCl with the pH of 0.54 M NH_4Cl.

CONCEPTUAL EXERCISES

61. The pH of an equal molar acetic acid/sodium acetate buffer is 4.74. Draw a molecular representation of a small portion of this buffer solution. (You may omit the water molecules.) Draw another molecular representation of the solution after a very small amount of NaOH has been added.

62. Suppose you were asked on a laboratory test to outline a procedure to prepare a buffered solution of pH 8.0 using hydrocyanic acid, HCN. You realize that a pH of 8.0 is basic, and you find that the K_a of hydrocyanic acid is 4.0×10^{-10}. What is your response?

63. The odor of cooked fish is due to the presence of amines. This odor is lessened by adding lemon juice, which contains citric acid. Why does this work?

64. The *end point* of a titration is not the same as the *equivalence point* of a titration. Differentiate between these two concepts.

65. One function of our blood is to carry CO_2 from our tissues to our lungs. It is critical that the pH of our blood remains at 7.4 ± 0.05. Our blood is buffered to maintain that pH range and the primary buffer system is composed of HCO_3^-/H_2CO_3. (a) Calculate the ratio of HCO_3^- to H_2CO_3 in blood if there were no other buffers present. (b) Determine whether in the presence of the HCO_3^-/H_2CO_3 system alone, our blood would be more effective as a buffer against the addition of acidic or basic solutions.

BUILDING YOUR KNOWLEDGE

66. A 0.738-g sample of an unknown mixture composed only of NaCl and $KHSO_4$ required 36.8 mL of 0.115 M NaOH for titration. What is the percent by mass of $KHSO_4$ in the mixture?

67. Acetylsalicylic acid, the active ingredient in aspirin, has a K_a value of 3.0×10^{-4}. We dissolve 0.0100 mole of acetylsalicylic acid in sufficient water to make 1.00 L of solution and then titrate it with 0.500 M NaOH solution. What is the pH at each of these points in the titration? (a) before any of the NaOH solution is added; (b) at the equivalence point; (c) when a volume of NaOH solution has been added that is equal to half the amount required to reach the equivalence point.

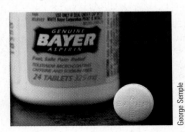

George Semple

*68. What is the pH of a solution that is a mixture of HOCl and HOI, each at 0.25 M concentration?

69. An unknown amount of water is mixed with 350. mL of a 6.0 M NaOH solution. A 75.0-mL sample of the resulting solution is titrated to neutrality with 52.5 mL of 6.00 M HCl. (a) Calculate the concentration of the diluted NaOH solution. (b) What was the concentration of the NaOH solution before it was diluted? (c) What volume of water was added? Assume that the volumes were additive.

70. A 3.5-L container of HCl had no concentration noted on its label. A 20.00-mL sample of this unknown HCl solution is titrated to a pH of 7.0 by 34.0 mL of 3.00 M NaOH solution. Determine the volume of this HCl solution required to prepare 1.5 L of 0.75 M HCl solution.

71. Calculate the pH at the equivalence point for the titration of a solution containing 150.0 mg of ethylamine, $C_2H_5NH_2$, with 0.1000 M HCl solution. The volume of the solution at the equivalence point is 250. mL. Select a suitable indicator. K_b for ethylamine appears in Exercise 28.

BEYOND THE TEXTBOOK

Go to the textbook website at

http://www.brookscole.com/chemistry/whitten

for additional activities and exercises based on the General Chemistry Interactive CD-ROM, the World Wide Web, and library resources.

InfoTrac College Edition

For additional readings, go to InfoTrac College Edition, your online research library at:

http://infotrac.thomsonlearning.com

OBJECTIVES

After you have studied this chapter, you should be able to

- Write solubility product constant expressions
- Explain how K_{sp} values are determined
- Use K_{sp} values in chemical calculations
- Recognize some common, slightly soluble compounds
- Describe fractional precipitation and how it can be used to separate ions
- Explain how simultaneous equilibria can be used to control solubility
- Describe some methods for dissolving precipitates

Some organisms produce solids that also occur naturally as minerals. This process is known as biomineralization. The calcium carbonate, $CaCO_3$, in an eggshell has the same crystal structure as in the mineral calcite.

So far we have discussed mainly compounds that are quite soluble in water. Although most compounds dissolve in water to some extent, many are so slightly soluble that they are called "insoluble compounds." We shall now consider those that are only very slightly soluble. As a rough rule of thumb, compounds that dissolve in water to the extent of 0.020 mole/liter or more are classified as soluble. Refer to the solubility guidelines (Table 4-8) as necessary.

Slightly soluble compounds are important in many natural phenomena. Our bones and teeth are mostly calcium phosphate, $Ca_3(PO_4)_2$, a slightly soluble compound. Also, many natural deposits of $Ca_3(PO_4)_2$ rock are mined and converted into agricultural fertilizer. Limestone caves have been formed by acidic water slowly dissolving away calcium carbonate, $CaCO_3$. Sinkholes are created when acidic water dissolves away most of the underlying $CaCO_3$. The remaining limestone can no longer support the weight above it, so it collapses, and a sinkhole is formed.

20-1 SOLUBILITY PRODUCT CONSTANTS

Suppose we add one gram of solid barium sulfate, $BaSO_4$, to 1.0 liter of water at 25°C and stir until the solution is *saturated*. Very little $BaSO_4$ dissolves; most of the $BaSO_4$ remains as undissolved solid. Careful measurements of conductivity show that one liter of a saturated solution of barium sulfate contains only 0.0025 gram of dissolved $BaSO_4$, no matter how much more solid $BaSO_4$ is added. The $BaSO_4$ that does dissolve is completely dissociated into its constituent ions.

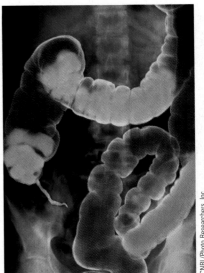

An X-ray photo of the gastrointestinal tract. The barium ions in $BaSO_4$ absorb X-radiation well.

Barium sulfate is the "insoluble" substance taken orally before stomach X-rays are made because the barium atoms absorb X-rays well. Even though barium ions are quite toxic, barium sulfate can still be taken orally without danger. The compound is so insoluble that it passes through the digestive system essentially unchanged.

$$BaSO_4(s) \underset{}{\overset{H_2O}{\rightleftharpoons}} Ba^{2+}(aq) + SO_4^{2-}(aq)$$

(We usually omit H_2O over the arrows in such equations.)

In equilibria that involve slightly soluble compounds in water, the equilibrium constant is called the **solubility product constant, K_{sp}.** The activity of the pure solid $BaSO_4$ is one (Section 17-11). Hence, the concentration of the solid is not included in the equilibrium constant expression. For a saturated solution of $BaSO_4$ in contact with solid $BaSO_4$, we write

$$BaSO_4(s) \rightleftharpoons Ba^{2+}(aq) + SO_4^{2-}(aq) \quad \text{and} \quad K_{sp} = [Ba^{2+}][SO_4^{2-}]$$

The solubility product constant for $BaSO_4$ is the product of the concentrations of its constituent ions in a saturated solution.

> The **solubility product expression** for a compound is the product of the concentrations of its constituent ions, each raised to the power that corresponds to the number of ions in one formula unit of the compound. The quantity is constant at constant temperature for a saturated solution of the compound. This statement is the **solubility product principle.**

The existence of a substance in the solid state can be indicated several ways. For example, $BaSO_4(s)$, $\overline{BaSO_4}$, and $BaSO_4\downarrow$ are sometimes used to represent solid $BaSO_4$. In this text we use the (s) notation for formulas of solid substances in equilibrium with their saturated aqueous solutions.

The dissolving process of ions going into solution. During precipitation, ions are producing more solid as the ions leave the solution. For a saturated solution, the dissolving and precipitation rates are equal. For the very slightly soluble salts discussed in this chapter, the ion concentrations are very low compared with the concentration in this illustration.

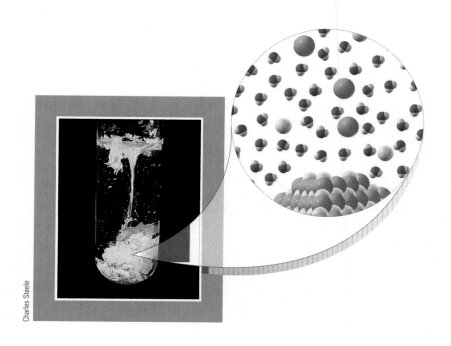

Charles Steele

Consider dissolving slightly soluble calcium fluoride, CaF_2, in H_2O.

$$CaF_2(s) \rightleftharpoons Ca^{2+}(aq) + 2F^-(aq) \qquad K_{sp} = [Ca^{2+}][F^-]^2 = 3.9 \times 10^{-11}$$

The very small amount of solid zinc phosphate, $Zn_3(PO_4)_2$, that dissolves in water gives three zinc ions and two phosphate ions per formula unit.

$$Zn_3(PO_4)_2(s) \rightleftharpoons 3Zn^{2+}(aq) + 2PO_4^{3-}(aq) \qquad K_{sp} = [Zn^{2+}]^3[PO_4^{3-}]^2 = 9.1 \times 10^{-33}$$

TOPIC: Solubility
GO TO: www.scilinks.org
sciLINKS CODE: WCH0440

Generally, we may represent the dissolution of a slightly soluble compound and its K_{sp} expression as

$$M_yX_z(s) \rightleftharpoons yM^{z+}(aq) + zX^{y-}(aq) \qquad \text{and} \qquad K_{sp} = [M^{z+}]^y[X^{y-}]^z$$

In some cases a compound contains more than two kinds of ions. Dissolution of the slightly soluble compound magnesium ammonium phosphate, $MgNH_4PO_4$, in water and its solubility product expression are represented as

$$MgNH_4PO_4(s) \rightleftharpoons Mg^{2+}(aq) + NH_4^+(aq) + PO_4^{3-}(aq)$$
$$K_{sp} = [Mg^{2+}][NH_4^+][PO_4^{3-}] = 2.5 \times 10^{-12}$$

We often shorten the term "solubility product constant" to "solubility product." Thus, the solubility products for barium sulfate, $BaSO_4$, and for calcium fluoride, CaF_2, are written as

$$K_{sp} = [Ba^{2+}][SO_4^{2-}] = 1.1 \times 10^{-10} \qquad K_{sp} = [Ca^{2+}][F^-]^2 = 3.9 \times 10^{-11}$$

The **molar solubility** of a compound is the number of moles that dissolve to give one liter of saturated solution.

CD-ROM Screens 18.8, Solubility and Precipitation; 18.9, Solubility Product Constant; and 18.10, Determining K_{sp}.

✓ Problem-Solving Tip: *Names and Formulas of Several Ions Must Be Recognized*

To write the required equations for the dissolution and dissociation of salts, you must recognize the names and formulas of common, stable ions that are often present. Table 2-3 contains the names and formulas of several ions; since then you have learned others. Review these names and formulas carefully (formulas of ions must include their charges). Many of the ions you need to know are present in the salts listed in Appendix H.

(a)

(b)

Barium sulfate, $BaSO_4$, occurs in the mineral barite (a). Calcium fluoride, CaF_2, occurs in the mineral fluorite (b). Both are clear, colorless crystals. Minerals are often discolored by impurities.

20-2 DETERMINATION OF SOLUBILITY PRODUCT CONSTANTS

Unless otherwise indicated, solubility product constants and solubility data are given for 25°C (Appendix H).

If the solubility of a compound is known, the value of its solubility product can be calculated.

EXAMPLE 20-1 *Molar Solubility and Product Constants*

We frequently use statements such as "The solution contains 0.0025 gram of $BaSO_4$." What we mean is that 0.0025 gram of solid $BaSO_4$ *dissolves* to give a solution that contains equal concentrations of Ba^{2+} and SO_4^{2-} ions.

One (1.0) liter of saturated barium sulfate solution contains 0.0025 gram of dissolved $BaSO_4$. (a) What is the molar solubility of $BaSO_4$? (b) Calculate the solubility product constant for $BaSO_4$.

Plan

(a) We write the equation for the dissolution of $BaSO_4$ and the expression for its solubility product constant, K_{sp}. From the solubility of $BaSO_4$ in H_2O, we calculate its molar solubility and the concentrations of the ions. (b) This lets us calculate K_{sp}.

Solution

(a) In saturated solutions, equilibrium exists between solid and dissolved solute. The equation for the dissolution of barium sulfate in water and its solubility product expression are

$$BaSO_4(s) \rightleftharpoons Ba^{2+}(aq) + SO_4^{2-}(aq) \qquad K_{sp} = [Ba^{2+}][SO_4^{2-}]$$

From the given solubility of $BaSO_4$ in H_2O we can calculate its *molar solubility*.

$$\frac{?\ mol\ BaSO_4}{L} = \frac{2.5 \times 10^{-3}\ g\ BaSO_4}{1.0\ L} \times \frac{1\ mol\ BaSO_4}{233\ g\ BaSO_4} = 1.1 \times 10^{-5}\ mol\ BaSO_4/L$$
(dissolved)

1.1×10^{-5} mole of solid $BaSO_4$ dissolves to give a liter of saturated solution.

(b) We know the molar solubility of $BaSO_4$. The dissolution equation shows that each formula unit of $BaSO_4$ that dissolves produces one Ba^{2+} ion and one SO_4^{2-} ion.

$$BaSO_4(s) \rightleftharpoons Ba^{2+}(aq) + SO_4^{2-}(aq)$$
$$1.1 \times 10^{-5}\ mol/L \Longrightarrow 1.1 \times 10^{-5}\ M \qquad 1.1 \times 10^{-5}\ M$$
(dissolved)

In a saturated solution $[Ba^{2+}] = [SO_4^{2-}] = 1.1 \times 10^{-5}\ M$. Substituting these values into the K_{sp} expression for $BaSO_4$ gives the calculated value of K_{sp}.

$$K_{sp} = [Ba^{2+}][SO_4^{2-}] = (1.1 \times 10^{-5})(1.1 \times 10^{-5}) = 1.2 \times 10^{-10}$$

Calculated values may differ slightly from those found in tables due to data or rounding differences.

You should now work Exercise 6.

EXAMPLE 20-2 *Molar Solubility and Solubility Product Constant*

One (1.00) liter of a saturated solution of silver chromate at 25°C contains 0.0435 gram of dissolved Ag_2CrO_4. Calculate (a) its molar solubility and (b) its solubility product constant.

Plan

We proceed as in Example 20-1.

Solution

(a) The equation for the dissolution of silver chromate in water and its solubility product expression are

$$Ag_2CrO_4(s) \rightleftharpoons 2Ag^+(aq) + CrO_4^{2-}(aq) \qquad and \qquad K_{sp} = [Ag^+]^2[CrO_4^{2-}]$$

The molar solubility of silver chromate is calculated first.

$$\frac{?\ mol\ Ag_2CrO_4}{L} = \frac{0.0435\ g\ Ag_2CrO_4}{1.00\ L} \times \frac{1\ mol\ Ag_2CrO_4}{332\ g\ Ag_2CrO_4} = \boxed{1.31 \times 10^{-4}\ mol/L}$$
(dissolved)

1.31×10^{-4} mole of solid Ag_2CrO_4 dissolves to give a liter of saturated solution.

(b) The equation for dissolution of Ag_2CrO_4 and its molar solubility give the concentrations of Ag^+ and CrO_4^{2-} ions in the saturated solution.

$$Ag_2CrO_4(s) \rightleftharpoons 2Ag^+(aq) + CrO_4^{2-}(aq)$$
$$1.31 \times 10^{-4}\ mol/L \Longrightarrow 2.62 \times 10^{-4}\ M \quad 1.31 \times 10^{-4}\ M$$
(dissolved)

Substitution into the K_{sp} expression for Ag_2CrO_4 gives the value of K_{sp}.

$$K_{sp} = [Ag^+]^2[CrO_4^{2-}] = (2.62 \times 10^{-4})^2(1.31 \times 10^{-4}) = \boxed{8.99 \times 10^{-12}}$$

The calculated K_{sp} of Ag_2CrO_4 is 8.99×10^{-12}.

You should now work Exercise 8.

⚠ Remember to raise $[Ag^+]$ to the power of its coefficient in the balanced chemical equation.

The molar solubility and K_{sp} values for $BaSO_4$ and Ag_2CrO_4 are compared in Table 20-1. These data show that the molar solubility of Ag_2CrO_4 is greater than that of $BaSO_4$. The K_{sp} for Ag_2CrO_4, however, is less than the K_{sp} for $BaSO_4$ because the expression for Ag_2CrO_4 contains a *squared* term, $[Ag^+]^2$.

If we compare K_{sp} values for two 1:1 compounds, for example, AgCl and $BaSO_4$, the compound with the larger K_{sp} value has the higher molar solubility. The same is true for any two compounds that have the same ion ratio, for example, the 1:2 compounds CaF_2 and $Mg(OH)_2$ and the 2:1 compound Ag_2CO_3.

$K_{sp(AgCl)} = 1.8 \times 10^{-10}$
$K_{sp(BaSO_4)} = 1.1 \times 10^{-10}$

The molar solubility of AgCl is only slightly higher than that of $BaSO_4$.

If two compounds have the *same ion ratio*, the one with the larger K_{sp} will have the higher *molar* solubility.

Appendix H lists some K_{sp} values. Refer to it as needed.

✓ **Problem-Solving Tip:** *It's Important to Know the Difference Between Solubility, Molar Solubility, and Solubility Product Expression*

Often it is necessary that you understand which of these is given or which is asked for. Recognition depends on a clear understanding of their definitions.

- The solubility of a compound is the amount of the compound that dissolves in a specified volume of solution. Solubility is usually expressed as either grams per liter or grams per 100. mL.
- The molar solubility of a compound is the number of moles of the compound that dissolve to give one liter of saturated solution. It is expressed in moles per liter.
- The solubility product expression for a compound is the product of the concentrations of its constituent ions, each raised to the power that corresponds to the number of ions in one formula unit of the compound. Like all equilibrium constants, it has no dimensions.

> Values of solubility product constants are usually tabulated to only two significant figures.

TABLE 20-1 *Comparison of Solubilities of $BaSO_4$ and Ag_2CrO_4*

Compound	Molar Solubility	K_{sp}
$BaSO_4$	1.1×10^{-5} mol/L	$[Ba^{2+}][SO_4^{2-}] = 1.1 \times 10^{-10}$
Ag_2CrO_4	1.3×10^{-4} mol/L	$[Ag^+]^2[CrO_4^{2-}] = 9.0 \times 10^{-12}$

20-3 USES OF SOLUBILITY PRODUCT CONSTANTS

> CD-ROM Screens 18.11, Estimating Salt Solubility, and 18.14, Using Solubility.

When the solubility product for a compound is known, the solubility of the compound in H_2O at 25°C can be calculated as Example 20-3 illustrates.

> The values of K_{sp} are obtained from Appendix H.

EXAMPLE 20-3 *Molar Solubilities from K_{sp} Values*

Calculate the molar solubilities, concentrations of the constituent ions, and solubilities in grams per liter for (a) silver chloride, AgCl ($K_{sp} = 1.8 \times 10^{-10}$), and (b) zinc hydroxide, $Zn(OH)_2$ ($K_{sp} = 4.5 \times 10^{-17}$).

Plan

We are given the value for each solubility product constant. In each case we write the appropriate balanced equation, represent the equilibrium concentrations, and then substitute into the K_{sp} expression.

Solution

(a) The equation for the dissolution of silver chloride and its solubility product expression are

$$AgCl(s) \rightleftharpoons Ag^+(aq) + Cl^-(aq) \qquad K_{sp} = [Ag^+][Cl^-] = 1.8 \times 10^{-10}$$

Each formula unit of AgCl that dissolves produces one Ag^+ and one Cl^-. We let x = mol/L of AgCl that dissolves, that is, the molar solubility.

$$\begin{array}{ccc} AgCl(s) & \rightleftharpoons & Ag^+(aq) + Cl^-(aq) \\ x \text{ mol/L} & \Longrightarrow & x\,M \qquad\quad x\,M \end{array}$$

Substitution into the solubility product expression gives

$$K_{sp} = [Ag^+][Cl^-] = (x)(x) = 1.8 \times 10^{-10} \qquad x^2 = 1.8 \times 10^{-10} \qquad x = 1.3 \times 10^{-5}$$

$$\boxed{x = \text{molar solubility of AgCl} = 1.3 \times 10^{-5} \text{ mol/L}}$$

One liter of saturated AgCl contains 1.3×10^{-5} mole of dissolved AgCl at 25°C. From the balanced equation we know the concentrations of the constituent ions.

$$\boxed{x = \text{molar solubility} = [Ag^+] = [Cl^-] = 1.3 \times 10^{-5} \text{ mol/L} = 1.3 \times 10^{-5}\,M}$$

Now we can calculate the mass of dissolved AgCl in one liter of saturated solution.

$$\frac{?\ g\ AgCl}{L} = \frac{1.3 \times 10^{-5} \text{ mol AgCl}}{L} \times \frac{143 \text{ g AgCl}}{1 \text{ mol AgCl}} = \boxed{1.9 \times 10^{-3} \text{ g AgCl/L}}$$

A liter of saturated AgCl solution contains only 0.0019 g of dissolved AgCl.

(b) The equation for the dissolution of zinc hydroxide, $Zn(OH)_2$, in water and its solubility product expression are

$$Zn(OH)_2(s) \rightleftharpoons Zn^{2+}(aq) + 2OH^-(aq) \qquad K_{sp} = [Zn^{2+}][OH^-]^2 = 4.5 \times 10^{-17}$$

We let x = molar solubility, so $[Zn^{2+}] = x$ and $[OH^-] = 2x$, and we have

$$\begin{array}{ccc} Zn(OH)_2(s) & \rightleftharpoons & Zn^{2+}(aq) + 2OH^-(aq) \\ x \text{ mol/L} & \Longrightarrow & x\,M \qquad 2x\,M \end{array}$$

Substitution into the solubility product expression gives

$$[Zn^{2+}][OH^-]^2 = (x)(2x)^2 = 4.5 \times 10^{-17}$$

$$4x^3 = 4.5 \times 10^{-17} \qquad x^3 = 11 \times 10^{-18} \qquad x = 2.2 \times 10^{-6}$$

$$x = \text{molar solubility of } Zn(OH)_2 = 2.2 \times 10^{-6} \text{ mol } Zn(OH)_2/L$$

$$x = [Zn^{2+}] = 2.2 \times 10^{-6}\,M \qquad \text{and} \qquad 2x = [OH^-] = 4.4 \times 10^{-6}\,M$$

The $[OH^-]$ is twice the molar solubility of $Zn(OH)_2$ because each formula unit of $Zn(OH)_2$ produces two OH^-.

We can now calculate the mass of dissolved $Zn(OH)_2$ in one liter of saturated solution.

$$\frac{?\text{ g } Zn(OH)_2}{L} = \frac{2.2 \times 10^{-6} \text{ mol } Zn(OH)_2}{L} \times \frac{99 \text{ g } Zn(OH)_2}{1 \text{ mol } Zn(OH)_2} = 2.2 \times 10^{-4} \text{ g } Zn(OH)_2/L$$

A liter of saturated $Zn(OH)_2$ solution contains only 0.00022 g of dissolved $Zn(OH)_2$.

You should now work Exercise 16.

✓ **Problem-Solving Tip:** *The Dissolution of a Slightly Soluble Base Is Not a K_b Problem*

The K_{sp} expression describes the equilibrium between a slightly soluble compound and its ions; in Example 20-3(b) one of those ions is OH^-. A K_b expression describes the equilibrium between a *soluble* basic species, for example, the ammonia molecule or the acetate ion, and the products it forms in solution, including OH^-. Do you see why the dissolution of $Zn(OH)_2$ is not a K_b problem? We found that $[OH^-] = 4.4 \times 10^{-6}\,M$ in a *saturated* $Zn(OH)_2$ solution. From this we find pOH = 5.36 and pH = 8.64. A saturated $Zn(OH)_2$ solution is not very basic because $Zn(OH)_2$ is not very soluble in H_2O. The $[OH^-]$ is 44 times greater than it is in pure water.

The Common Ion Effect in Solubility Calculations

The common ion effect applies to solubility equilibria just as it does to other ionic equilibria. Silver acetate, $AgCH_3COO$, is a slightly soluble salt:

$$AgCH_3COO \rightleftharpoons Ag^+(aq) + CH_3COO^-(aq)$$

If we add another salt that provides one of the product ions, say Ag^+ (from $AgNO_3$), this dissolution reaction is shifted to the *left* (LeChatelier's Principle). The common ion effect is thus a special case of LeChatelier's Principle. Example 20-4 illustrates this effect.

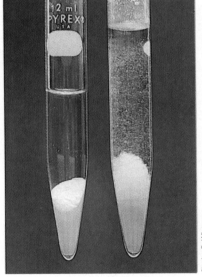

Charles D. Winters

The tube at the left contains a saturated solution of silver acetate, $AgCH_3COO$. When 1 M $AgNO_3$ is added to the tube, the equilibrium

$$AgCH_3COO(s) \rightleftharpoons$$
$$Ag^+(aq) + CH_3COO^-(aq)$$

shifts to the left and additional $AgCH_3COO$ precipitates. This demonstrates the common ion effect.

EXAMPLE 20-4 *Molar Solubilities and the Common Ion Effect*

For magnesium fluoride, MgF_2, $K_{sp} = 6.4 \times 10^{-9}$. (a) Calculate the molar solubility of magnesium fluoride in pure water. (b) Calculate the molar solubility of MgF_2 in 0.10 M sodium fluoride, NaF, solution. (c) Compare these molar solubilities.

CD-ROM   Screen 18.13, The
Common Ion Effect.

Plan

For Part (a), we write the appropriate chemical equations and solubility product expression, designate the equilibrium concentrations, and then substitute into the solubility product expression. For Part (b), we recognize that NaF is a soluble ionic compound that is completely dissociated into its ions. MgF_2 is a slightly soluble compound. Both compounds produce F^- ions so this is a common ion effect problem. We write the appropriate chemical equations and solubility product expression, represent the equilibrium concentrations, and substitute into the solubility product expression. For Part (c), we compare the molar solubilities by calculating their ratio.

Solution

(a) We let x = molar solubility for MgF_2, a slightly soluble salt.

$$MgF_2(s) \rightleftharpoons Mg^{2+}(aq) + 2F^-(aq) \quad \text{(reversible)}$$
$$x \text{ mol/L} \Longrightarrow \quad x\, M \qquad 2x\, M$$

$$K_{sp} = [Mg^{2+}][F^-]^2 = 6.4 \times 10^{-9}$$
$$(x)(2x)^2 = 6.4 \times 10^{-9}$$
$$x = 1.2 \times 10^{-3}$$

$1.2 \times 10^{-3}\, M$ = molar solubility of MgF_2 in pure water

(b) NaF is a soluble ionic salt and, therefore, 0.10 M F^- is produced by

$$NaF(s) \xrightarrow{H_2O} Na^+(aq) + F^-(aq) \quad \text{(complete)}$$
$$0.10\, M \Longrightarrow \quad 0.10\, M \qquad 0.10\, M$$

We let y = molar solubility for MgF_2, a slightly soluble salt.

$$MgF_2(s) \rightleftharpoons Mg^{2+}(aq) + 2F^-(aq) \quad \text{(reversible)}$$
$$y \text{ mol/L} \Longrightarrow \quad y\, M \qquad 2y\, M$$

The total $[F^-]$ is 0.10 M from NaF *plus* $2y$ M from MgF_2, or $(0.10 + 2y)$ M.

$$K_{sp} = [Mg^{2+}][F^-]^2 = 6.4 \times 10^{-9}$$
$$(y)(0.10 + 2y)^2 = 6.4 \times 10^{-9}$$

Very little MgF_2 dissolves, so y is small. This suggests that $2y \ll 0.10$, so $0.10 + 2y \approx 0.10$. Then

When the same ion is produced by a soluble and a slightly soluble salt, the concentration from the slightly soluble salt can be neglected.

$$(y)(0.10)^2 = 6.4 \times 10^{-9} \quad \text{and} \quad y = 6.4 \times 10^{-7}$$

$6.4 \times 10^{-7}\, M$ = molar solubility of MgF_2 in 0.10 M NaF

(c) The ratio of molar solubility in water to molar solubility in 0.10 M NaF solution is

$$\frac{\text{molar solubility (in } H_2O)}{\text{molar solubility (in NaF solution)}} = \frac{1.2 \times 10^{-3}\, M}{6.4 \times 10^{-7}\, M} = \frac{1900}{1}$$

The molar solubility of MgF_2 in 0.10 M NaF ($6.4 \times 10^{-7}\, M$) is about 1900 times less than it is in pure water ($1.2 \times 10^{-3}\, M$).

You should now work Exercises 20 and 26.

THE EFFECTS OF HYDROLYSIS ON SOLUBILITY

In Section 18-8 we discussed the hydrolysis of anions of weak acids. For example, we found that for CH_3COO^- and CN^- ions,

$$CH_3COO^- + H_2O \rightleftharpoons CH_3COOH + OH^-$$

$$K_b = \frac{[CH_3COOH][OH^-]}{[CH_3COO^-]} = 5.6 \times 10^{-10}$$

$$CN^- + H_2O \rightleftharpoons HCN + OH^- \qquad K_b = \frac{[HCN][OH^-]}{[CN^-]} = 2.5 \times 10^{-5}$$

We see that K_b for CN^-, the anion of a *very* weak acid, is much larger than K_b for CH_3COO^-, the anion of a much stronger acid. This tells us that in solutions of the same concentration, CN^- ions hydrolyze to a much greater extent than do CH_3COO^- ions. So we might expect that hydrolysis would have a much greater effect on the solubilities of cyanides such as AgCN than on the solubilities of acetates such as $AgCH_3COO$. It does.

Hydrolysis reduces the concentrations of anions of weak acids, such as F^-, $CO_3{}^{2-}$, CH_3COO^-, and CN^-, so its effect must be taken into account when we do very precise solubility calculations. Taking into account the effect of hydrolysis on solubilities of slightly soluble compounds is beyond the scope of this chapter, however.

CD-ROM Screen 18.12, Can a Precipitation Reaction Occur?

The Reaction Quotient in Precipitation Reactions

Another application of the solubility product principle is the calculation of the maximum concentrations of ions that can coexist in solution. From these calculations we can determine whether a **precipitate** will form in a given solution. The reaction quotient, Q (Section 17-4), is useful in such decisions. We compare Q_{sp} with K_{sp}.

If $Q_{sp} < K_{sp}$	Forward process is favored No precipitation occurs; if solid is present, more solid can dissolve
$Q_{sp} = K_{sp}$	Solution is *just* saturated Solid and solution are in equilibrium; neither forward nor reverse process is favored
$Q_{sp} > K_{sp}$	Reverse process is favored Precipitation occurs to form more solid

EXAMPLE 20-5 *Predicting Precipitate Formation*

If 100. mL of 0.00075 M sodium sulfate, Na_2SO_4, is mixed with 50. mL of 0.015 M barium chloride, $BaCl_2$, will a precipitate form?

Plan

We are mixing solutions of two soluble ionic salts. First we find the amount of each solute at the instant of mixing. Next we find the molarity of each solute *at the instant of mixing*. Then

Pouring ammonium sulfide solution into a solution of cadmium nitrate gives a precipitate of cadmium sulfide.

$$(NH_4)_2S + Cd(NO_3)_2 \longrightarrow CdS(s) + 2NH_4NO_3$$

Cadmium sulfide is used as a pigment in artists' oil-based paints.

TOPIC: Precipitation Reactions
GO TO: www.scilinks.org
*sci*LINKS CODE: WCH0480

James W. Morganthaler

we find the concentration of each ion in the *new* solution. Now we ask the question "*Could* any combination of the ions in this solution form a slightly soluble compound?" The answer is "Yes, Ba^{2+} and SO_4^{2-} *could* form $BaSO_4$," so we calculate Q_{sp} and compare it with K_{sp} to determine whether solid $BaSO_4$ is formed.

Solution

We find the *amount* of each solute at the instant of mixing.

$$\underline{?} \text{ mmol } Na_2SO_4 = 100. \text{ mL} \times \frac{0.00075 \text{ mmol } Na_2SO_4}{\text{mL}} = 0.075 \text{ mmol } Na_2SO_4$$

$$\underline{?} \text{ mmol } BaCl_2 = 50. \text{ mL} \times \frac{0.015 \text{ mmol } BaCl_2}{\text{mL}} = 0.75 \text{ mmol } BaCl_2$$

> Because the amounts of dissolved ions are so small, it is convenient to express molarity as mmol/mL. Review this use of mmol in Section 11-1.

When *dilute* aqueous solutions are mixed, their volumes can be added to give the volume of the resulting solution.

$$\text{volume of mixed solution} = 100. \text{ mL} + 50. \text{ mL} = 150. \text{ mL}$$

Then we find the *molarity of each solute at the instant of mixing.*

$$M_{Na_2SO_4} = \frac{0.075 \text{ mmol } Na_2SO_4}{150. \text{ mL}} = 0.00050 \ M \ Na_2SO_4$$

$$M_{BaCl_2} = \frac{0.75 \text{ mmol } BaCl_2}{150. \text{ mL}} = 0.0050 \ M \ BaCl_2$$

Now we find the *concentration of each ion* in the new solution.

$$Na_2SO_4(s) \xrightarrow{100\%} 2Na^+(aq) + SO_4^{2-}(aq) \qquad \text{(to completion)}$$
$$0.00050 \ M \Longrightarrow \quad 0.0010 \ M \qquad 0.00050 \ M$$

$$BaCl_2(s) \xrightarrow{100\%} Ba^{2+}(aq) + 2Cl^-(aq) \qquad \text{(to completion)}$$
$$0.0050 \ M \Longrightarrow \quad 0.0050 \ M \qquad 0.010 \ M$$

We consider the kinds of compounds mixed and determine whether a reaction could occur. Both Na_2SO_4 and $BaCl_2$ are soluble ionic salts. At the moment of mixing, the new solution contains a mixture of Na^+, SO_4^{2-}, Ba^{2+}, and Cl^- ions. We must consider the possibility of forming two new compounds, $NaCl$ and $BaSO_4$. Sodium chloride is a soluble ionic compound so Na^+ and Cl^- do not combine in dilute aqueous solutions. $BaSO_4$, however, is only very slightly soluble, and solid $BaSO_4$ will precipitate from the solution *if* $Q_{sp} > K_{sp}$ for $BaSO_4$. K_{sp} for $BaSO_4$ is 1.1×10^{-10}. Substituting $[Ba^{2+}] = 0.0050 \ M$ and $[SO_4^{2-}] = 0.00050 \ M$ into the Q_{sp} expression for $BaSO_4$, we get

> Recall that Q has the same form as the equilibrium constant, in this case K_{sp}, but the concentrations are not necessarily equilibrium concentrations.

$$Q_{sp} = [Ba^{2+}][SO_4^{2-}] = (5.0 \times 10^{-3})(5.0 \times 10^{-4}) = 2.5 \times 10^{-6} \qquad (Q_{sp} > K_{sp})$$

Because $Q_{sp} > K_{sp}$ solid $BaSO_4$ will precipitate until $[Ba^{2+}][SO_4^{2-}]$ just equals K_{sp} for $BaSO_4$.

You should now work Exercises 28 and 30.

When white solid potassium iodide, KI, and white solid lead(II) nitrate, $Pb(NO_3)_2$, are stirred together, some yellow lead(II) iodide, PbI_2, forms. This reaction occurs in the small amount of water present in these solids.

✓ **Problem-Solving Tip:** *Detection of Precipitates*

The human eye is not a very sensitive detection device. As a rule of thumb, a precipitate can be seen with the naked eye if $Q_{sp} > K_{sp}$ by a factor of 1000. In Example 20-5,

Q_{sp} exceeds K_{sp} by a factor of $\dfrac{2.5 \times 10^{-6}}{1.1 \times 10^{-10}} = 2.3 \times 10^4 = 23,000$. We expect to be able

to see the $BaSO_4$ precipitate that is formed. Modern techniques enable us to detect smaller amounts of precipitates.

EXAMPLE 20-6   *Initiation of Precipitation*

What $[Ba^{2+}]$ is necessary to start the precipitation of $BaSO_4$ in a solution that is 0.0015 M in Na_2SO_4? Assume that the Ba^{2+} comes from addition of a solid soluble ionic compound such as $BaCl_2$. For $BaSO_4$, $K_{sp} = 1.1 \times 10^{-10}$.

Plan

These are the compounds in Example 20-5. We recognize that Na_2SO_4 is a soluble ionic compound and that the molarity of SO_4^{2-} is equal to the molarity of the Na_2SO_4 solution. We are given K_{sp} for $BaSO_4$, so we solve for $[Ba^{2+}]$.

Solution

Because Na_2SO_4 is a soluble ionic compound, we know that $[SO_4^{2-}] = 0.0015\ M$. We can use K_{sp} for $BaSO_4$ to calculate the $[Ba^{2+}]$ required for Q_{sp} to just equal K_{sp}.

$$[Ba^{2+}][SO_4^{2-}] = 1.1 \times 10^{-10}$$

$$[Ba^{2+}] = \frac{1.1 \times 10^{-10}}{[SO_4^{2-}]} = \frac{1.1 \times 10^{-10}}{1.5 \times 10^{-3}} = 7.3 \times 10^{-8}\ M$$

Addition of enough $BaCl_2$ to give a barium ion concentration of $7.3 \times 10^{-8}\ M$ *just satisfies* K_{sp} for $BaSO_4$; that is, $Q_{sp} = K_{sp}$. Ever so slightly more $BaCl_2$ would be required for Q_{sp} to exceed K_{sp} and for precipitation of $BaSO_4$ to occur. Therefore

$$\boxed{[Ba^{2+}] > 7.3 \times 10^{-8}\ M}\qquad \text{(to initiate precipitation of } BaSO_4)$$

You should now work Exercise 29.

Often we wish to remove an ion from solution by forming an insoluble compound (as in water purification). We use K_{sp} values to calculate the concentrations of ions remaining in solution *after* precipitation has occurred.

The recovery of silver from the solutions used in developing and fixing photographic film and prints presents just such a problem. Silver is an expensive metal, and the recovery is profitable. Moreover, if not recovered, the silver ions would constitute an undesirable pollutant in water supplies.

EXAMPLE 20-7   *Concentration of Common Ion*

Suppose we wish to recover silver from an aqueous solution that contains a soluble silver compound such as $AgNO_3$ by precipitating insoluble silver chloride, $AgCl$. A soluble ionic compound such as $NaCl$ can be used as a source of Cl^-. What is the minimum concentration of chloride ion needed to reduce the dissolved silver ion concentration to a maximum of $1.0 \times 10^{-9}\ M$? For $AgCl$, $K_{sp} = 1.8 \times 10^{-10}$.

Plan

We are given K_{sp} for $AgCl$ and the required equilibrium $[Ag^+]$, so we solve for $[Cl^-]$.

Solution

The equation for the reaction of interest and the K_{sp} for AgCl are

$$AgCl(s) \rightleftharpoons Ag^+(aq) + Cl^-(aq) \quad \text{and} \quad [Ag^+][Cl^-] = 1.8 \times 10^{-10}$$

To determine the $[Cl^-]$ required to reduce the $[Ag^+]$ to 1.0×10^{-9} M, we solve the K_{sp} expression for $[Cl^-]$.

$$[Cl^-] = \frac{1.8 \times 10^{-10}}{[Ag^+]} = \frac{1.8 \times 10^{-10}}{1.0 \times 10^{-9}} = \boxed{0.18 \; M \; Cl^-}$$

To reduce the $[Ag^+]$ to 1.0×10^{-9} M (0.00000011 g Ag^+/L), NaCl would be added until $[Cl^-] = 0.18$ M in the solution.

You should now work Exercise 32.

Silver chloride precipitates when chloride ions are added to a solution containing silver ions.

Clearly cations such as Na^+ or K^+ must also be present in this solution.

20-4 FRACTIONAL PRECIPITATION

We sometimes wish to remove some ions from solution while leaving others with similar properties in solution. This separation process is called **fractional precipitation.** Consider a solution that contains Cl^-, Br^-, and I^- ions. These halide ions are anions of elements in the same family in the periodic table. We expect them to have similar properties. But we also expect some differences in properties, and that is what we find. Consider the solubility products for these silver halides.

Compound	Solubility Product
AgCl	1.8×10^{-10}
AgBr	3.3×10^{-13}
AgI	1.5×10^{-16}

These K_{sp} values show that AgI is less soluble than AgBr and that AgBr is less soluble than AgCl. Silver fluoride is quite soluble in water.

NaCl, NaBr, NaI, AgNO₃, and NaNO₃ are soluble compounds that are completely dissociated in dilute aqueous solution.

We ignore the extremely small change in volume caused by addition of solid AgNO₃.

EXAMPLE 20-8 *Concentration Required to Initiate Precipitation*

Solid silver nitrate is slowly added to a solution that is 0.0010 M each in NaCl, NaBr, and NaI. Calculate the $[Ag^+]$ required to initiate the precipitation of each of the silver halides. For AgI, $K_{sp} = 1.5 \times 10^{-16}$; for AgBr, $K_{sp} = 3.3 \times 10^{-13}$; and for AgCl, $K_{sp} = 1.8 \times 10^{-10}$.

Plan

We are given a solution that contains equal concentrations of Cl^-, Br^-, and I^- ions; all of which form insoluble silver salts. Then we slowly add Ag^+ ions. We use each K_{sp} to determine the $[Ag^+]$ that must be exceeded to initiate precipitation of each salt as we did in Example 20-6.

Solution

We calculate the $[Ag^+]$ necessary to begin to precipitate each of the silver halides. The solubility product for AgI is

$$[Ag^+][I^-] = 1.5 \times 10^{-16}$$

$[I^-] = 1.0 \times 10^{-3}$ M, so the $[Ag^+]$ that must be exceeded to start precipitation of AgI is

$$[Ag^+] = \frac{1.5 \times 10^{-16}}{[I^-]} = \frac{1.5 \times 10^{-16}}{1.0 \times 10^{-3}} = 1.5 \times 10^{-13} M$$

Therefore, AgI will begin to precipitate when $[Ag^+] > 1.5 \times 10^{-13} M$.

Repeating this kind of calculation for silver bromide gives

$$[Ag^+][Br^-] = 3.3 \times 10^{-13}$$

$$[Ag^+] = \frac{3.3 \times 10^{-13}}{[Br^-]} = \frac{3.3 \times 10^{-13}}{1.0 \times 10^{-3}} = 3.3 \times 10^{-10} M$$

Thus, $[Ag^+] > 3.3 \times 10^{-10} M$ is needed to start precipitation of AgBr.

For the precipitation of silver chloride to begin,

$$[Ag^+][Cl^-] = 1.8 \times 10^{-10}$$

$$[Ag^+] = \frac{1.8 \times 10^{-10}}{[Cl^-]} = \frac{1.8 \times 10^{-10}}{1.0 \times 10^{-3}} = 1.8 \times 10^{-7} M$$

To precipitate AgCl, we must have $[Ag^+] > 1.8 \times 10^{-7} M$.

We have shown that

to precipitate AgI,	$[Ag^+] > 1.5 \times 10^{-13} M$
to precipitate AgBr,	$[Ag^+] > 3.3 \times 10^{-10} M$
to precipitate AgCl,	$[Ag^+] > 1.8 \times 10^{-7} M$

> We are not suggesting that a $1.5 \times 10^{-13} M$ solution of AgNO$_3$ be added. We are pointing out the fact that when sufficient AgNO$_3$ has been added to the solution to make $[Ag^+] > 1.5 \times 10^{-13} M$, AgI begins to precipitate.

This calculation tells us that when AgNO$_3$ is added slowly to a solution that is 0.0010 M in each of NaI, NaBr, and NaCl, AgI precipitates first, AgBr precipitates second, and AgCl precipitates last. We can also calculate the amount of I$^-$ precipitated before Br$^-$ begins to precipitate and the amounts of I$^-$ and Br$^-$ precipitated before Cl$^-$ begins to precipitate (Example 20-9).

> As AgNO$_3$ is added to the solution containing Cl$^-$, Br$^-$, and I$^-$ ions, some AgBr and AgCl may precipitate *locally*. As the solution is stirred, AgBr and AgCl redissolve as long as $[Ag^+]$ is not large enough to exceed their K_{sp} values in the *bulk* of the solution.

EXAMPLE 20-9 *Fractional Precipitation*

Refer to Example 20-8. (a) Calculate the percentage of I$^-$ precipitated before AgBr precipitates. (b) Calculate the percentages of I$^-$ and Br$^-$ precipitated before Cl$^-$ precipitates.

Plan

From Example 20-8 we know the $[Ag^+]$ that must be exceeded to initiate precipitation of each of three silver halides, AgI, AgBr, and AgCl. We use each of these values of $[Ag^+]$ with the appropriate K_{sp} expression, in turn, to find the concentration of each halide ion that remains in solution (unprecipitated). We express these halide ion concentrations as percent unprecipitated. Then we subtract each from exactly 100% to find the percentage of each halide that precipitates.

Freshly precipitated AgCl is white (*left*), AgBr is very pale yellow (*center*), and AgI is yellow (*right*). Polarizabilities of these halide ions increase in the order Cl$^-$ < Br$^-$ < I$^-$. Colors of the silver halides become more intense in the same direction. Solubilities of the silver halides increase in the opposite direction.

Solution

In Example 20-7 we did a similar calculation, but we did not express the result in terms of the percentage of an ion precipitated.

(a) In Example 20-8 we found that AgBr begins to precipitate when $[Ag^+] > 3.3 \times 10^{-10}\ M$. This value for $[Ag^+]$ can be substituted into the K_{sp} expression for AgI to determine $[I^-]$ remaining *unprecipitated* when AgBr begins to precipitate.

$$[Ag^+][I^-] = 1.5 \times 10^{-16}$$

$$[I^-]_{unppt'd} = \frac{1.5 \times 10^{-16}}{[Ag^+]} = \frac{1.5 \times 10^{-16}}{3.3 \times 10^{-10}} = 4.5 \times 10^{-7}\ M$$

The percentage of I^- unprecipitated is

$$\%\ I^-_{unppt'd} = \frac{[I^-]_{unppt'd}}{[I^-]_{orig}} \times 100\% = \frac{4.5 \times 10^{-7}\ M}{1.0 \times 10^{-3}\ M} \times 100\%$$

$$= 0.045\%\ I^-\ unprecipitated$$

Therefore, 99.955% of the I^- precipitates *before* AgBr begins to precipitate.

We have subtracted 0.045% from *exactly* 100%. We have therefore *not* violated the rules for significant figures.

(b) Similar calculations show that *just before* AgCl begins to precipitate, $[Ag^+] = 1.8 \times 10^{-7}\ M$, and the $[I^-]$ unprecipitated is calculated as in Part (a).

$$[Ag^+][I^-] = 1.5 \times 10^{-16}$$

$$[I^-]_{unppt'd} = \frac{1.5 \times 10^{-16}}{[Ag^+]} = \frac{1.5 \times 10^{-16}}{1.8 \times 10^{-7}} = 8.3 \times 10^{-10}\ M$$

The percentage of I^- unprecipitated just before AgCl precipitates is

$$\%\ I^-_{unppt'd} = \frac{[I^-]_{unppt'd}}{[I^-]_{orig}} \times 100\% = \frac{8.3 \times 10^{-10}\ M}{1.0 \times 10^{-3}\ M} \times 100\%$$

$$= 0.000083\%\ I^-\ unprecipitated$$

Therefore, 99.999917% of the I^- precipitates before AgCl begins to precipitate.

A similar calculation for the amount of Br^- precipitated just before AgCl begins to precipitate gives

$$[Ag^+][Br^-] = 3.3 \times 10^{-13}$$

$$[Br^-]_{unppt'd} = \frac{3.3 \times 10^{-13}}{[Ag^+]} = \frac{3.3 \times 10^{-13}}{1.8 \times 10^{-7}} = 1.8 \times 10^{-6}\ M$$

$$\%\ Br^-_{unppt'd} = \frac{[Br^-]_{unppt'd}}{[Br^-]_{orig}} \times 100\% = \frac{1.8 \times 10^{-6}\ M}{1.0 \times 10^{-3}\ M} \times 100\%$$

$$= 0.18\%\ Br^-\ unprecipitated$$

Thus, 99.82% of the Br^- precipitates before AgCl begins to precipitate.

You should now work Exercises 36 and 38.

We have described the series of reactions that occurs when solid $AgNO_3$ is added slowly to a solution that is 0.0010 M in Cl^-, Br^-, and I^-. Silver iodide begins to precipitate first; 99.955% of the I^- precipitates before any solid AgBr is formed. Silver bromide begins to precipitate next; 99.82% of the Br^- and 99.999917% of the I^- precipitate before any solid AgCl forms. This shows that we can separate these ions very effectively by fractional precipitation.

20-5 SIMULTANEOUS EQUILIBRIA INVOLVING SLIGHTLY SOLUBLE COMPOUNDS

Many weak acids and bases react with many metal ions to form insoluble compounds. In such cases, we must take into account the weak acid or weak base equilibrium as well as the solubility equilibrium. The most common examples involve the reaction of metal ions with aqueous ammonia to form insoluble metal hydroxides.

 CD-ROM Screen 18.15, Simultaneous Equilibria.

EXAMPLE 20-10 *Simultaneous Equilibria*

We make a solution that is 0.10 M in magnesium nitrate, $Mg(NO_3)_2$, *and* 0.10 M in aqueous ammonia, a weak base. Will magnesium hydroxide, $Mg(OH)_2$, precipitate from this solution? K_{sp} for $Mg(OH)_2$ is 1.5×10^{-11}, and K_b for aqueous NH_3 is 1.8×10^{-5}.

Plan

We first write equations for the two *reversible* reactions and their equilibrium constant expressions. We note that $[OH^-]$ appears in both equilibrium constant expressions. From the statement of the problem we know the concentration of Mg^{2+}. We use the K_b expression for aqueous NH_3 to find $[OH^-]$. Then we calculate Q_{sp} for $Mg(OH)_2$ and compare it with its K_{sp}.

Solution

Two equilibria and their equilibrium constant expressions must be considered.

$$Mg(OH)_2(s) \rightleftharpoons Mg^{2+}(aq) + 2OH^-(aq) \qquad K_{sp} = [Mg^{2+}][OH^-]^2 = 1.5 \times 10^{-11}$$

$$NH_3(aq) + H_2O(\ell) \rightleftharpoons NH_4^+(aq) + OH^-(aq) \qquad K_b = \frac{[NH_4^+][OH^-]}{[NH_3]} = 1.8 \times 10^{-5}$$

The $[OH^-]$ in 0.10 M aqueous NH_3 is calculated as in Example 18.14.

$$\underset{(0.10 - x)M}{NH_3(aq)} + H_2O(\ell) \rightleftharpoons \underset{x\,M}{NH_4^+(aq)} + \underset{x\,M}{OH^-(aq)}$$

$$\frac{[NH_4^+][OH^-]}{[NH_3]} = 1.8 \times 10^{-5} = \frac{(x)(x)}{(0.10 - x)} \qquad x = 1.3 \times 10^{-3}\,M = [OH^-]$$

Magnesium nitrate is a soluble ionic compound, so $[Mg^{2+}] = 0.10\,M$. Now that both $[Mg^{2+}]$ and $[OH^-]$ are known, we calculate Q_{sp} for $Mg(OH)_2$.

$$[Mg^{2+}][OH^-]^2 = Q_{sp}$$

$$(0.10)(1.3 \times 10^{-3})^2 = 1.7 \times 10^{-7} = Q_{sp}$$

$K_{sp} = 1.5 \times 10^{-11}$, so we see that $Q_{sp} > K_{sp}$.

Therefore, $\boxed{Mg(OH)_2 \text{ would precipitate until } Q_{sp} = K_{sp}.}$

You should now work Exercise 46.

 We have calculated the $[OH^-]$ produced by the ionization of 0.10 M aqueous NH_3. *This is the equilibrium concentration of OH^- in this solution. There is no reason to double this value when we put it into the K_{sp} expression!*

Example 20-11 shows how we can calculate the concentration of weak base that is required to initiate precipitation of an insoluble metal hydroxide.

CD-ROM Screen 18.16, Solubility and pH.

EXAMPLE 20-11 *Simultaneous Equilibria*

What concentration of aqueous ammonia is necessary to just start precipitation of $Mg(OH)_2$ from a 0.10 M solution of $Mg(NO_3)_2$? Refer to Example 20-10.

Plan

We have the same reactions and equilibrium constant expressions as in Example 20-10. We are given $[Mg^{2+}]$, and so we use the K_{sp} expression of $Mg(OH)_2$ to calculate the $[OH^-]$ necessary to initiate precipitation. Then we find the molarity of aqueous NH_3 solution that would furnish the desired $[OH^-]$.

Solution

Two equilibria and their equilibrium constant expressions must be considered.

$$Mg(OH)_2(s) \rightleftharpoons Mg^{2+}(aq) + 2OH^-(aq) \qquad K_{sp} = 1.5 \times 10^{-11}$$

$$NH_3(aq) + H_2O(\ell) \rightleftharpoons NH_4^+(aq) + OH^-(aq) \qquad K_b = 1.8 \times 10^{-5}$$

We find the $[OH^-]$ necessary to initiate precipitation of $Mg(OH)_2$ when $[Mg^{2+}] = 0.10\ M$.

$$K_{sp} = [Mg^{2+}][OH^-]^2 = 1.5 \times 10^{-11}$$

$$[OH^-]^2 = \frac{1.5 \times 10^{-11}}{[Mg^{2+}]} = \frac{1.5 \times 10^{-11}}{0.10} = 1.5 \times 10^{-10} \qquad [OH^-] = 1.2 \times 10^{-5}\ M$$

Therefore, $[OH^-] > 1.2 \times 10^{-5}\ M$ to initiate precipitation of $Mg(OH)_2$.

Now we use the equilibrium of NH_3 as a weak base to find the $[NH_3]$ that will produce $1.2 \times 10^{-5}\ M\ OH^-$. Let x be the original $[NH_3]$.

$$NH_3(aq) + H_2O(\ell) \rightleftharpoons NH_4^+(aq) + OH^-(aq)$$
$$(x - 1.2 \times 10^{-5})\ M \qquad\qquad 1.2 \times 10^{-5}\ M \quad 1.2 \times 10^{-5}\ M$$

Because 1.2×10^{-5} and x are of comparable magnitude, neither can be disregarded in the term $(x - 1.2 \times 10^{-5})$.

$$K_b = \frac{[NH_4^+][OH^-]}{[NH_3]} = 1.8 \times 10^{-5} = \frac{(1.2 \times 10^{-5})(1.2 \times 10^{-5})}{x - 1.2 \times 10^{-5}}$$

$$1.8 \times 10^{-5}\ x - 2.16 \times 10^{-10} = 1.44 \times 10^{-10}$$

$$1.8 \times 10^{-5}\ x = 3.6 \times 10^{-10} \qquad \text{so} \qquad x = 2.0 \times 10^{-5}\ M = [NH_3]_{orig}$$

The solution must be ever so slightly greater than $2.0 \times 10^{-5}\ M$ in NH_3 to initiate precipitation of $Mg(OH)_2$ in a 0.10 M solution of $Mg(NO_3)_2$.

A solution that contains a weak base can be buffered (by addition of a salt of the weak base) to decrease its basicity. Significant concentrations of some metal ions that form insoluble hydroxides can be kept in such solutions.

EXAMPLE 20-12 *Simultaneous Equilibria*

What minimum number of moles of NH_4Cl must be added to 1.0 liter of solution that is 0.10 M in $Mg(NO_3)_2$ *and* 0.10 M in NH_3 to prevent precipitation of $Mg(OH)_2$?

Plan

These are the same compounds, in the same concentrations, that we used in Example 20-10. Because we know $[Mg^{2+}]$, we must find the maximum $[OH^-]$ that can exist in the solution *without exceeding* K_{sp} for $Mg(OH)_2$. Then we find the minimum concentration of NH_4Cl that is necessary to buffer the NH_3 solution to keep the $[OH^-]$ below the calculated value.

Solution

The buffering action of NH_4Cl in the presence of NH_3 decreases the concentration of OH^-. Again we have two equilibria.

$$Mg(OH)_2(s) \rightleftharpoons Mg^{2+}(aq) + 2OH^-(aq) \qquad K_{sp} = 1.5 \times 10^{-11}$$

$$NH_3(aq) + H_2O(\ell) \rightleftharpoons NH_4^+(aq) + OH^-(aq) \qquad K_b = 1.8 \times 10^{-5}$$

To find the *maximum* $[OH^-]$ *that can exist in solution without causing precipitation*, we substitute $[Mg^{2+}]$ into the K_{sp} for $Mg(OH)_2$.

$$[Mg^{2+}][OH^-]^2 = 1.5 \times 10^{-11}$$

$$[OH^-]^2 = \frac{1.5 \times 10^{-11}}{[Mg^{2+}]} = \frac{1.5 \times 10^{-11}}{0.10} = 1.5 \times 10^{-10}$$

$$[OH^-] = 1.2 \times 10^{-5}\ M \qquad \text{(maximum } [OH^-] \text{ possible)}$$

To prevent precipitation of $Mg(OH)_2$ in *this* solution, $[OH^-]$ must be equal to or less than $1.2 \times 10^{-5}\ M$. K_b for aqueous NH_3 is used to calculate the number of moles of NH_4Cl necessary to buffer 1.0 L of 0.10 M aqueous NH_3 so that $[OH^-] = 1.2 \times 10^{-5}\ M$. Let x = number of mol/L of NH_4Cl required.

$$
\begin{array}{cccc}
NH_4Cl(aq) & \longrightarrow & NH_4^+(aq) & + Cl^-(aq) \qquad \text{(to completion)} \\
x\ M & \rightleftharpoons & x\ M & x\ M \\
\end{array}
$$

$$
\begin{array}{cccc}
NH_3(aq) & + H_2O(\ell) \rightleftharpoons & NH_4^+(aq) & + OH^-(aq) \\
(0.10 - 1.2 \times 10^{-5})\ M & & 1.2 \times 10^{-5}\ M & 1.2 \times 10^{-5}\ M \\
\end{array}
$$

We can assume that $(x + 1.2 \times 10^{-5}) \approx x$ and $(0.10 - 1.2 \times 10^{-5}) \approx 0.10$.

$$\frac{(x)(1.2 \times 10^{-5})}{0.10} = 1.8 \times 10^{-5}$$

$$x = 0.15 \text{ mol of } NH_4^+ \text{ per liter of solution}$$

Addition of 0.15 mol of NH_4Cl to 1.0 L of 0.10 M aqueous NH_3 decreases $[OH^-]$ to $1.2 \times 10^{-5}\ M$. Then K_{sp} for $Mg(OH)_2$ is not exceeded in this solution, and so no precipitate would form.

You should now work Exercises 42 and 44.

Examples 20-10, 20-11, and 20-12 illustrate a very important point.

All relevant equilibria must be satisifed when more than one equilibrium is required to describe a solution.

In Example 20-10 we found that $Mg(OH)_2$ will precipitate from a solution that is 0.10 M in $Mg(NO_3)_2$ and 0.10 M in NH_3.

You may wish to refer to Example 19-6 to refresh your understanding of buffer solutions.

20-6   DISSOLVING PRECIPITATES

Solubility product constants, like other equilibrium constants, are thermodynamic quantities. They tell us about the extent to which a given reaction can occur, but nothing about how fast that reaction occurs.

A precipitate dissolves when the concentrations of its ions are reduced so that K_{sp} is no longer exceeded, that is, when $Q_{sp} < K_{sp}$. The precipitate then dissolves until $Q_{sp} = K_{sp}$. The precipitates can be dissolved by several types of reactions. All involve removing ions from solution.

Converting an Ion to a Weak Electrolyte

Three specific illustrations follow.

1. *Converting OH^- to H_2O.* Insoluble $Al(OH)_3$ dissolves in acids. H^+ ions react with OH^- ions [from the saturated $Al(OH)_3$ solution] to form the weak electrolyte H_2O. This makes $[Al^{3+}][OH^-]^3 < K_{sp}$, so that the dissolution equilibrium shifts to the right and $Al(OH)_3$ dissolves.

$$Al(OH)_3(s) \rightleftharpoons Al^{3+}(aq) + 3OH^-(aq)$$
$$3H^+(aq) + 3OH^-(aq) \longrightarrow 3H_2O(\ell)$$
$$\text{overall rxn:} \quad \overline{Al(OH)_3(s) + 3H^+(aq) \longrightarrow Al^{3+}(aq) + 3H_2O(\ell)}$$

2. *Converting NH_4^+ to NH_3.* Ammonium ions, from a salt such as NH_4Cl, dissolve insoluble $Mg(OH)_2$. The NH_4^+ ions combine with OH^- ions in the saturated $Mg(OH)_2$ solution. This forms the weak electrolytes NH_3 and H_2O. The result is $[Mg^{2+}][OH^-]^2 < K_{sp}$, and so the $Mg(OH)_2$ dissolves.

$$Mg(OH)_2(s) \rightleftharpoons Mg^{2+}(aq) + 2OH^-(aq)$$
$$2NH_4^+(aq) + 2OH^-(aq) \longrightarrow 2NH_3(aq) + 2H_2O(\ell)$$
$$\text{overall rxn:} \quad \overline{Mg(OH)_2(s) + 2NH_4^+(aq) \longrightarrow Mg^{2+}(aq) + 2NH_3(aq) + 2H_2O(\ell)}$$

This process, dissolution of $Mg(OH)_2$ in an NH_4Cl solution, is the reverse of the reaction we considered in Example 20-10. There, $Mg(OH)_2$ precipitated from a solution of aqueous NH_3.

3. *Converting S^{2-} to H_2S.* Nonoxidizing acids dissolve most insoluble metal sulfides. For example, 6 *M* HCl dissolves MnS. The H^+ ions combine with S^{2-} ions to form H_2S, a gas that bubbles out of the solution. The result is $[Mn^{2+}][S^{2-}] < K_{sp}$, and so the MnS dissolves.

$$MnS(s) \rightleftharpoons Mn^{2+}(aq) + S^{2-}(aq)$$
$$S^{2-}(aq) + H_2O(\ell) \longrightarrow HS^-(aq) + OH^-(aq)$$
$$2H^+(aq) + HS^-(aq) + OH^-(aq) \longrightarrow H_2S(g) + H_2O(\ell)$$
$$\text{overall rxn:} \quad \overline{MnS(s) + 2H^+(aq) \longrightarrow Mn^{2+}(aq) + H_2S(g)}$$

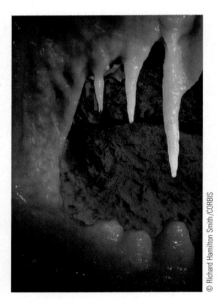

Stalactite and stalagmite formations in limestone caves result from the dissolution and reprecipitation of calcium carbonate, $CaCO_3$.

S^{2-} like O^{2-} does not exist in appreciable amounts in aqueous solutions.

Converting an Ion to Another Species by a Redox Reaction

Most insoluble metal sulfides dissolve in hot dilute HNO_3 because NO_3^- ions oxidize S^{2-} ions, or actually their hydrolysis product (HS^-), to elemental sulfur. This removes HS^- (and thus S^{2-}) ions from the solution and promotes the dissolving of more of the metal sulfide.

$$3HS^-(aq) + 2NO_3^-(aq) + 5H^+(aq) \longrightarrow 3S(s) + 2NO(g) + 4H_2O(\ell)$$

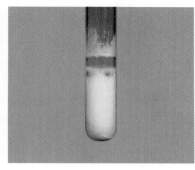

Manganese(II) sulfide, MnS, is salmon-colored. MnS dissolves in 6 *M* HCl. The resulting solution of $MnCl_2$ is pale pink.

Consider copper(II) sulfide, CuS, in equilibrium with its ions. This equilibrium lies far to the left. Removal of the S^{2-} or HS^- ions by oxidation to elemental sulfur favors the reaction to the right, and so CuS(s) dissolves in hot dilute HNO_3.

$$CuS(s) \rightleftharpoons Cu^{2+}(aq) + S^{2-}(aq)$$

$$S^{2-}(aq) + H_2O(\ell) \longrightarrow HS^-(aq) + OH^-(aq)$$

$$3HS^-(aq) + 3OH^-(aq) + 8H^+(aq) + 2NO_3^-(aq) \longrightarrow 3S(s) + 2NO(g) + 7H_2O(\ell)$$

We multiply the first two equations by 3, add the three equations, and cancel like terms. This gives the net ionic equation for dissolving CuS(s) in hot dilute HNO_3.

$$3CuS(s) + 2NO_3^-(aq) + 8H^+(aq) \longrightarrow 3Cu^{2+}(aq) + 3S(s) + 2NO(g) + 4H_2O(\ell)$$

It is sometimes convenient to ignore hydrolysis of S^{2-} ions in aqueous solutions. Leaving out the hydrolysis step may give the false impression that S^{2-} ions exist in solution. The overall net ionic equation for this redox reaction will be the same, however, with or without consideration of hydrolysis of the S^{2-} ions.

Complex Ion Formation

The cations in many slightly soluble compounds can form *complex ions*. This often results in dissolution of the slightly soluble compound. Some metal ions share electron pairs donated by molecules and ions such as NH_3, CN^-, OH^-, F^-, Cl^-, Br^-, and I^- form coordinate covalent bonds to metal ions. Coordinate covalent bonds are formed as these electron-donating groups (ligands) replace H_2O molecules from hydrated metal ions. The decrease in the concentration of the hydrated metal ion shifts the solubility equilibrium to the right.

"Ligand" is the name given to an atom or a group of atoms bonded to the central element in complex ions. Ligands are Lewis bases.

Copper(II) sulfide, CuS, is black. As CuS dissolves in 6 *M* HNO_3, some NO is oxidized to brown NO_2 by O_2 in the air. The resulting solution of $Cu(NO_3)_2$ is blue.

CD-ROM   Screen 18.17,
Complex Ion Formation and
Solubility.

Many copper(II) compounds react with excess aqueous NH_3 to form the deep-blue complex ion $[Cu(NH_3)_4]^{2+}$.

$$Cu^{2+}(aq) + 4NH_3(aq) \rightleftharpoons [Cu(NH_3)_4]^{2+}(aq)$$

The dissociation of this complex ion is represented as

$$[Cu(NH_3)_4]^{2+}(aq) \rightleftharpoons Cu^{2+}(aq) + 4NH_3(aq)$$

The equilibrium constant for the dissociation of a complex ion is called its **dissociation constant, K_d.**

As before, the outer brackets mean molar concentrations. The inner brackets are part of the formula of the complex ion.

$$K_d = \frac{[Cu^{2+}][NH_3]^4}{[[Cu(NH_3)_4]^{2+}]} = 8.5 \times 10^{-13}$$

Recall that $Cu^{2+}(aq)$ is really a hydrated ion, $[Cu(H_2O)_6]^{2+}$. The preceding reaction and its K_d expression are represented more accurately as

$$[Cu(NH_3)_4]^{2+} + 6H_2O \rightleftharpoons [Cu(H_2O)_6]^{2+} + 4NH_3$$

$$K_d = \frac{[[Cu(H_2O)_6]^{2+}][NH_3]^4}{[[Cu(NH_3)_4]^{2+}]} = 8.5 \times 10^{-13}$$

The more effectively a ligand competes with H_2O for a coordination site on the metal ions, the smaller K_d is. This tells us that in a comparison of complexes with the same number of ligands, the smaller the K_d value, the more stable the complex ion. Some complex ions and their dissociation constants, K_d, are listed in Appendix I.

For brevity we shall omit H_2O from formulas of hydrated ions. For example, we write $[Cu(OH_2)_6]^{2+}$ as Cu^{2+}.

Copper(II) hydroxide dissolves in an excess of aqueous NH_3 to form the deep-blue complex ion $[Cu(NH_3)_4]^{2+}$. This decreases the $[Cu^{2+}]$ so that $[Cu^{2+}][OH^-]^2 < K_{sp}$, and so the $Cu(OH)_2$ dissolves.

$$\begin{array}{rl} Cu(OH)_2(s) \rightleftharpoons & Cu^{2+}(aq) \quad\quad + 2OH^-(aq) \\ Cu^{2+}(aq) + 4NH_3(aq) \rightleftharpoons & [Cu(NH_3)_4]^{2+}(aq) \\ \hline \end{array}$$

overall rxn: $Cu(OH)_2(s) + 4NH_3(aq) \rightleftharpoons [Cu(NH_3)_4]^{2+}(aq) + 2OH^-(aq)$

Similarly, $Zn(OH)_2$ dissolves in excess NH_3 to form $[Zn(NH_3)_4]^{2+}$ ions.

$$Zn(OH)_2(s) + 4NH_3(aq) \rightleftharpoons [Zn(NH_3)_4]^{2+}(aq) + 2OH^-(aq)$$

Amphoteric hydroxides such as $Zn(OH)_2$ also dissolve in excess strong base by forming complex ions (Section 10-6).

$$Zn(OH)_2(s) + 2OH^-(aq) \rightleftharpoons [Zn(OH)_4]^{2-}(aq)$$

We see that we are able to shift equilibria (in this case, dissolve $Zn(OH)_2$) by taking advantage of complex ion formation.

Charles D. Winters

Concentrated aqueous NH_3 was added *slowly* to a solution of copper(II) sulfate, $CuSO_4$. Unreacted blue copper(II) sulfate solution remains in the bottom part of the test tube. The light-blue precipitate in the middle is copper(II) hydroxide, $Cu(OH)_2$. The top layer contains deep-blue $[Cu(NH_3)_4]^{2+}$ ions that were formed as some $Cu(OH)_2$ dissolved in excess aqueous NH_3.

Key Terms

Complex ions Ions resulting from the formation of coordinate covalent bonds between simple cations and other ions or molecules (ligands).

Dissociation constant K_d The equilibrium constant that applies to the dissociation of a complex ion into a simple ion and coordinating species (ligands).

Fractional precipitation Removal of some ions from solution by precipitation while leaving other ions, with similar properties, in solution.

Molar solubility The number of moles of a solute that dissolve to produce a liter of saturated solution.

Precipitate A solid formed by mixing in solution the constituent ions of a slightly soluble compound.

Solubility product constant, K_{sp} The equilibrium constant that applies to the dissolution of a slightly soluble compound.

Solubility product principle The solubility product constant expression for a slightly soluble compound is the product of the concentrations of the constituent ions, each raised to the power that corresponds to the number of ions in one formula unit.

Exercises

Consult Appendix H for solubility product constant values and Appendix I for complex ion dissociation constants, as needed.

Solubility Product

1. The solubility product constant values listed in Appendix H were determined at 25°C. How would those K_{sp} values change, if at all, with a change in temperature?
2. State the solubility product principle. What is its significance?
3. Why do we not include a term for the solid in a solubility product expression?
4. What do we mean when we refer to the molar solubility of a compound?
5. Write the solubility product expression for each of the following salts. (a) MgF_2; (b) $Bi_2(SO_4)_3$; (c) $CuCO_3$; (d) Ag_3PO_4.
6. Write the solubility product expression for each of the following salts. (a) $Co_3(AsO_4)_2$; (b) Hg_2I_2 [contains mercury(I) ions, Hg_2^{2+}]; (c) AuI_3; (d) Ag_2CO_3.
7. The K_{sp} value for $BaSO_4$ is calculated from the expression $K_{sp} = [Ba^{2+}][SO_4^{2-}]$, whereas the K_{sp} value for $Mg(OH)_2$ is calculated from the expression, $K_{sp} = [Mg^{2+}][OH^-]^2$. Explain why the hydroxide ion concentration is squared, but none of the other concentrations is squared.

Experimental Determination of K_{sp}

Values of K_{sp} calculated from the solubility data in these exercises may not agree exactly with the solubility products given in Appendix H because of rounding differences.

8. From the solubility data given for the following compounds, calculate their solubility product constants.
   (a) $SrCrO_4$, strontium chromate, 1.2 mg/mL
   (b) BiI_3, bismuth iodide, 7.7×10^{-3} g/L
   (c) $Fe(OH)_2$, iron(II) hydroxide, 1.1×10^{-3} g/L
   (d) SnI_2, tin(II) iodide, 10.9 g/L

9. From the solubility data given for the following compounds, calculate their solubility product constants.
   (a) CuBr, copper(I) bromide, 1.0×10^{-3} g/L
   (b) AgI, silver iodide, 2.8×10^{-8} g/10 mL
   (c) $Pb_3(PO_4)_2$, lead(II) phosphate, 6.2×10^{-7} g/L
   (d) Ag_2SO_4, silver sulfate, 5.0 mg/mL
10. Construct a table like Table 20-1 for the compounds listed in Exercise 8. Which compound has (a) the highest molar solubility; (b) the lowest molar solubility; (c) the largest K_{sp}; (d) the smallest K_{sp}?
11. Construct a table like Table 20-1 for the compounds listed in Exercise 9. Which compound has (a) the highest molar solubility; (b) the lowest molar solubility; (c) the largest K_{sp}; (d) the smallest K_{sp}?
12. A solution is produced by stirring 1 gram of calcium fluoride in 1.00 liter of water at 25°C. Careful analysis shows that 0.0163 grams of calcium fluoride has dissolved. Calculate the K_{sp} for calcium fluoride based on these data.
13. Calculate the K_{sp} for zinc phosphate if 1.18×10^{-4} grams of zinc phosphate dissolved to make 2.0 liters of a saturated solution.

Uses of Solubility Product Constants

14. Sea shells are calcium carbonate with traces of colored impurities. The solubility product for calcium carbonate is 4.8×10^{-9}. What is the solubility of calcium carbonate in grams per liter of solution?
15. Calculate molar solubilities, concentrations of constituent ions, and solubilities in grams per liter for the following compounds at 25°C: (a) $Cd(CN)_2$, cadmium cyanide; (b) PbI_2, lead iodide; (c) $Sr_3(AsO_4)_2$, strontium arsenate; (d) Hg_2CO_3, mercury(I) carbonate [the formula for mercury(I) ion is Hg_2^{2+}].
16. Calculate molar solubilities, concentrations of constituent ions, and solubilities in grams per liter for the following compounds at 25°C: (a) CuI, copper(I) iodide;

(b) $Ba_3(PO_4)_2$, barium phosphate; (c) PbF_2, lead(II) fluoride; (d) $Sr_3(PO_4)_2$, strontium phosphate.

17. What is the concentration of lead ions in one liter of saturated $PbCO_3$ solution?

18. Barium sulfate is used to produce distinct X-rays of the gastrointestinal tract. What is the maximum mass of barium sulfate that can dissolve in 5.00 liters of water, a volume much greater than the volume of the average gastrointestinal tract?

19. Calculate the molar solubility of CuBr in 0.012 M KBr solution.

20. Calculate the molar solubility of Ag_2SO_4 in 0.10 M K_2SO_4 solution.

21. Construct a table similar to Table 20-1 for the compounds listed in Exercise 15. Which compound has (a) the highest molar solubility; (b) the lowest molar solubility; (c) the highest solubility, expressed in grams per liter; (d) the lowest solubility, expressed in grams per liter?

22. Construct a table similar to Table 20-1 for the compounds listed in Exercise 16. Which compound has (a) the highest molar solubility; (b) the lowest molar solubility; (c) the highest solubility, expressed in grams per liter; (d) the lowest solubility, expressed in grams per liter?

23. Of the three compounds $CuCO_3$, $Ca(OH)_2$, and Ag_2CrO_4, which has (a) the highest molar solubility; (b) the lowest molar solubility; (c) the highest solubility, expressed in grams per liter; and (d) the lowest solubility, expressed in grams per liter?

24. Of the three compounds, Ag_2CO_3, AgCl, and $Pb(OH)_2$, which has (a) the highest molar solubility; (b) the lowest molar solubility; (c) the highest solubility, expressed in grams per liter; and (d) the lowest solubility, expressed in grams per liter?

25. What volume of water is required to dissolve 7.5 grams of copper(II) carbonate, $CuCO_3$?

26. Which has the greater molar solubility in 0.15 M K_2CrO_4 solution: $BaCrO_4$ or Ag_2CrO_4?

27. Will a precipitate form when 1.00 g of $AgNO_3$ is added to 50.0 mL of 0.050 M NaCl? If so, would you expect the precipitate to be visible?

28. Will a precipitate of $PbCl_2$ form when 5.0 g of solid $Pb(NO_3)_2$ is added to 1.00 L of 0.010 M NaCl? Assume that volume change is negligible.

29. Sodium bromide and lead nitrate are soluble in water. Will lead bromide precipitate when 1.03 g of NaBr and 0.332 g of $Pb(NO_3)_2$ are dissolved in sufficient water to make 1.00 L of solution?

30. Will a precipitate of $Cu(OH)_2$ form when 10.0 mL of 0.010 M NaOH is added to 1.00 L of 0.010 M $CuCl_2$?

31. A solution is 0.0100 M in Pb^{2+} ions. If 0.103 mol of solid Na_2SO_4 is added to 1.00 L of this solution (with negligible volume change), what percentage of the Pb^{2+} ions remain in solution?

32. A solution is 0.0100 M in Pb^{2+} ions. If 0.103 mol of solid NaI is added to 1.00 L of this solution (with negligible

volume change), what percentage of the Pb^{2+} ions remain in solution?

*33. A solution is 0.0100 M in $Ba(NO_3)_2$. If 0.103 mol of solid Na_3PO_4 is added to 1.00 L of this solution (with negligible volume change), what percentage of the Ba^{2+} ions remain in solution?

Fractional Precipitation

34. What is fractional precipitation?

35. Solid Na_2SO_4 is added slowly to a solution that is 0.10 M in $Pb(NO_3)_2$ and 0.10 M in $Ba(NO_3)_2$. In what order will solid $PbSO_4$ and $BaSO_4$ form? Calculate the percentage of Ba^{2+} that precipitates just before $PbSO_4$ begins to precipitate.

36. To a solution that is 0.010 M in Cu^+, 0.010 M in Ag^+, and 0.010 M in Au^+, *solid* NaBr is added slowly. Assume that there is no volume change due to the addition of solid NaBr. (a) Which compound will begin to precipitate first? (b) Calculate $[Au^+]$ when AgBr just begins to precipitate. What percentage of the Au^+ has precipitated at this point? (c) Calculate $[Au^+]$ and $[Ag^+]$ when CuBr just begins to precipitate.

37. A solution is 0.015 M in Pb^{2+} and 0.015 M in Ag^+. As Cl^- is introduced to the solution by the addition of solid NaCl, determine (a) which substance will precipitate first, AgCl or $PbCl_2$, and (b) the fraction of the metal ion in the first precipitate that remains in solution at the moment the precipitation of the second compound begins.

38. A solution is 0.050 M in K_2SO_4 and 0.050 M in K_2CrO_4. A solution of $Pb(NO_3)_2$ is added slowly without changing the volume appreciably. (a) Which salt, $PbSO_4$ or $PbCrO_4$, will precipitate first? (b) What is $[Pb^{2+}]$ when the salt in part (a) begins to precipitate? (c) What is $[Pb^{2+}]$ when the other lead salt begins to precipitate? (d) What are $[SO_4^{2-}]$ and $[CrO_4^{2-}]$ when the lead salt in part (c) begins to precipitate?

39. Solid $Pb(NO_3)_2$ is added slowly to a solution that is 0.015 M each in NaOH, K_2CO_3, and Na_2SO_4. (a) In what order will solid $Pb(OH)_2$, $PbCO_3$, and $PbSO_4$ begin to precipitate? (b) Calculate the percentages of OH^- and CO_3^{2-} that have precipitated when $PbSO_4$ begins to precipitate.

40. Suppose you have three beakers that contain, respectively, 100 mL of each of the following solutions: (i) 0.0015 M KOH; (ii) 0.0015 M K_2CO_3; (iii) 0.0015 M KCN. (a) If solid zinc nitrate, $Zn(NO_3)_2$, were added slowly to each beaker, what concentration of Zn^{2+} would be required to initiate precipitation? (b) If solid zinc nitrate were added to each beaker until $[Zn^{2+}] = 0.0015$ M, what concentrations of OH^-, CO_3^{2-}, and CN^- would remain in solution, that is, unprecipitated? Neglect any volume change when solid is added.

41. Suppose you have three beakers that contain, respectively, 100 mL each of the following solutions: (i) 0.0015 M KOH; (ii) 0.0015 M K_2CO_3; (iii) 0.0015 M KI. (a) If solid

lead nitrate, $Pb(NO_3)_2$, were added slowly to each beaker, what concentration of Pb^{2+} would be required to initiate precipitation? (b) If solid lead nitrate were added to each beaker until $[Pb^{2+}] = 0.0015\ M$, what concentrations of OH^-, CO_3^{2-}, and I^- would remain in solution, that is, unprecipitated? Neglect any volume change when solute is added.

Simultaneous Equilibria

42. If a solution is made 0.080 M in $Mg(NO_3)_2$, 0.075 M in aqueous ammonia, and 3.5 M in NH_4NO_3, will $Mg(OH)_2$ precipitate? What is the pH of this solution?

43. If a solution is made 0.090 M in $Mg(NO_3)_2$, 0.090 M in aqueous ammonia, and 0.080 M in NH_4NO_3, will $Mg(OH)_2$ precipitate? What is the pH of this solution?

*44. Calculate the solubility of CaF_2 in a solution that is buffered at $[H^+] = 0.0050\ M$ with $[HF] = 0.10\ M$.

*45. Calculate the solubility of $AgCN$ in a solution that is buffered at $[H^+] = 0.000200\ M$, with $[HCN] = 0.01\ M$.

46. If a solution is $2.0 \times 10^{-5}\ M$ in $Mn(NO_3)_2$ and $1.0 \times 10^{-3}\ M$ in aqueous ammonia, will $Mn(OH)_2$ precipitate?

47. If a solution is 0.040 M in manganese(II) nitrate, $Mn(NO_3)_2$, and 0.080 M in aqueous ammonia, will manganese(II) hydroxide, $Mn(OH)_2$, precipitate?

48. Milk of magnesia is a suspension of the slightly soluble compound $Mg(OH)_2$ in water. (a) What is the molar solubility of $Mg(OH)_2$ in a 0.015 M NaOH solution? (b) What is the molar solubility of $Mg(OH)_2$ in a 0.015 M $MgCl_2$ solution?

49. How many moles of $Cr(OH)_3$ will dissolve in 445 mL of a solution with a pH of 5.00?

50. Determine whether a precipitate forms when a 0.00050 M solution of magnesium nitrate is brought to a pH of 8.75.

*51. What concentration of NH_4NO_3 is necessary to prevent precipitation of $Mn(OH)_2$ in the solution of Exercise 47?

52. (a) What is the pH of a saturated solution of $Fe(OH)_2$? (b) What is the solubility in grams of $Fe(OH)_2$/100 mL of solution?

53. (a) What is the pH of a saturated solution of $Cu(OH)_2$? (b) What is the solubility in grams of $Cu(OH)_2$/100 mL of solution?

Dissolution of Precipitates and Complex Ion Formation

54. Explain, by writing appropriate equations, how the following insoluble compounds can be dissolved by the addition of a solution of nitric acid. (Carbonates dissolve in strong acids to form carbon dioxide, which is evolved as a gas and water.) What is the "driving force" for each reaction? (a) $Cu(OH)_2$; (b) $Sn(OH)_4$; (c) $ZnCO_3$; (d) $(PbOH)_2CO_3$.

55. Explain, by writing equations, how the following insoluble compounds can be dissolved by the addition of a solution of ammonium nitrate or ammonium chloride. (a) $Mg(OH)_2$; (b) $Mn(OH)_2$; (c) $Ni(OH)_2$.

56. The following insoluble sulfides can be dissolved in 3 M hydrochloric acid. Explain how this is possible, and write the appropriate equations. (a) MnS; (b) CuS.

57. The following sulfides are less soluble than those listed in Exercise 56 and can be dissolved in hot 6 M nitric acid, an oxidizing acid. Explain how, and write the appropriate balanced equations. (a) PbS; (b) CuS; (c) Bi_2S_3.

58. Why would MnS be expected to be more soluble in 0.10 M HCl solution than in water? Would the same be true for $Mn(NO_3)_2$?

*59. For each pair, choose the salt that would be expected to be more soluble in acidic solution than in pure water, and justify your choice: (a) $Hg_2(CH_3COO)_2$ or Hg_2Br_2; (b) $Pb(OH)_2$ or PbI_2; (c) AgI or $AgNO_2$.

Mixed Exercises

60. We mix 25.0 mL of a 0.0030 M solution of $BaCl_2$ and 50.0 mL of a 0.050 M solution of NaF. (a) Find $[Ba^{2+}]$ and $[F^-]$ in the mixed solution at the instant of mixing (before any possible reaction occurs). (b) Would BaF_2 precipitate?

*61. A concentrated, strong acid is added to a solid mixture of 0.015-mol samples of $Fe(OH)_2$ and $Cu(OH)_2$ placed in 1.0 L of water. At what values of pH will the dissolution of each hydroxide be complete? (Assume negligible volume change.)

62. A solution is 0.015 M in I^- ions and 0.015 M in Br^- ions. Ag^+ ions are introduced to the solution by the addition of solid $AgNO_3$. Determine (a) which compound will precipitate first, AgI or $AgBr$, and (b) the percentage of the halide ion in the first precipitate that is removed from solution before the precipitation of the second compound begins.

63. Calculate the molar solubility of Ag_2SO_4 (a) in pure water, (b) in 0.010 M $AgNO_3$, and (c) in 0.010 M K_2SO_4.

CONCEPTUAL EXERCISES

64. Suppose there was a deposit of gold(I) chloride in a river bed. What minimum volume of water, at 25°C, would be required to dissolve 1.0 gram of gold ions? Assume sufficient contact time for the water to become saturated with AuCl.

65. (a) Are "insoluble" substances really insoluble? (b) What do we mean when we refer to insoluble substances?

66. Solubility product calculations are actually based on heterogeneous equilibria. Why are pure solids and liquids exempted from these calculations?

67. Draw a picture of a portion of a saturated silver chloride solution at the molecular level. Show a small amount of solid plus some dissociated ions. You need not show water or waters of hydration. Prepare a second drawing that

includes the same volume of solution but twice as much solid. Should your drawing include more, fewer, or the same number of silver ions?

68. The solubility product constants of silver chloride, AgCl, and silver chromate, Ag_2CrO_4, are 1.8×10^{-10} and 9.0×10^{-12}, respectively. Suppose that the chloride, Cl^- (aq), and chromate, CrO_4^{2-} (aq), ions are both present in the same solution at concentrations of $0.010\ M$ each. A standard solution of silver ions, Ag^+ (aq), is dispensed slowly from a buret into this solution while it is stirred vigorously. Solid silver chloride is white, and silver chromate is red. What will be the concentration of Cl^- (aq) ions in the mixture when the first tint of red color appears in the mixture?

BUILDING YOUR KNOWLEDGE

69. How many moles of CO_3^{2-} must be added to 0.50 liter of a $0.10\ M\ Sr^{2+}$ solution to produce a solution that is $1.0 \times 10^{-6}\ M\ Sr^{2+}$? How many moles of CO_3^{2-} are in the final solution and how many moles of CO_3^{2-} are in the precipitate formed? Assume no volume change for the solution.

70. A fluoridated water supply contains 1 mg/L of F^-. What is the maximum amount of Ca^{2+}, expressed in grams per liter, that can exist in this water supply?

71. Many industrial operations require very large amounts of water as a coolant in heat exchange processes. Muddy or cloudy water is usually unsatisfactory because the dispersed solids may clog filters or deposit sediment in pipes and pumps. Murky water can be clarified on a large scale by adding agents to coagulate colloidal material, and then allowing the precipitate to settle out in holding tanks or ponds before the clarified water is sent to plant intakes. Recent methods employ the addition of both calcium hydroxide and magnesium carbonate. If 56 g of $Ca(OH)_2$ and 45 g of $MgCO_3$ were added to 520 liters of water, would these compounds form a precipitate of calcium carbonate?

72. Magnesium carbonate is used in the manufacture of a high density *magnesite* brick. This material is not well suited to general exterior use because the magnesium carbonate easily erodes. What percentage of 28 grams of surface-exposed $MgCO_3$ would be lost through the solvent action of 15 liters of water? Assume sufficient contact time for the water to become saturated with $MgCO_3$.

BEYOND THE TEXTBOOK

Go to the textbook website at

http://www.brookscole.com/chemistry/whitten

for additional activities and exercises based on the General Chemistry Interactive CD-ROM, the World Wide Web, and library resources.

InfoTrac College Edition

For additional readings, go to InfoTrac College Edition, your online research library at:

http://infotrac.thomsonlearning.com

Electrochemistry

Batteries are voltaic cells, which operate by electrochemical (redox) reactions.

Charles D. Winters

OUTLINE

OBJECTIVES

After you have studied this chapter, you should be able to

- Use the terminology of electrochemistry (terms such as "cell," "electrode," "cathode," "anode")

- Describe the differences between electrolytic cells and voltaic (galvanic) cells

- Recognize oxidation and reduction half-reactions, and know at which electrode each occurs for electrolytic cells and for voltaic cells

- Write half-reactions and overall cell reactions for electrolysis processes

- Use Faraday's Law of Electrolysis to calculate amounts of products formed, amounts of current passed, time elapsed, and oxidation state

- Describe the refining and plating of metals by electrolytic methods

- Describe the construction of simple voltaic cells from half-cells and a salt bridge, and understand the function of each component

- Write half-reactions and overall cell reactions for voltaic cells
- Compare various voltaic cells to determine the relative strengths of oxidizing and reducing agents
- Interpret standard reduction potentials
- Use standard reduction potentials, E^0, to calculate the potential of a standard voltaic cell, E^0_{cell}
- Use standard reduction potentials to identify the cathode and the anode in a standard cell
- Use standard reduction potentials to predict the spontaneity of a redox reaction
- Use standard reduction potentials to identify oxidizing and reducing agents in a cell or in a redox reaction
- Describe some corrosion processes and some methods for preventing corrosion
- Use the Nernst equation to relate electrode potentials and cell potentials to different concentrations and partial pressures
- Relate the standard cell potential (E^0_{cell}) to the standard Gibbs free energy change (ΔG^0) and the equilibrium constant (K)
- Distinguish between primary and secondary voltaic cells
- Describe the compositions and reactions of some useful primary and secondary cells (batteries)
- Describe the electrochemical processes involved in discharging and recharging a lead storage (automobile) battery

Electrochemistry deals with the chemical changes produced by electric current and with the production of electricity by chemical reactions. Many metals are purified or are plated onto jewelry by electrochemical methods. Digital watches, automobile starters, calculators, and pacemakers are just a few devices that depend on electrochemically produced power. Corrosion of metals is also an electrochemical process.

We learn much about chemical reactions from the study of electrochemistry. The amount of electrical energy consumed or produced can be measured quite accurately. All electrochemical reactions involve the transfer of electrons and are therefore *oxidation–reduction* reactions. The sites of oxidation and reduction are separated physically so that oxidation occurs at one location, and reduction occurs at the other. Electrochemical processes require some method of introducing a stream of electrons into a reacting chemical system and some means of withdrawing electrons. In most applications the reacting system is contained in a **cell,** and an electric current enters or exits by **electrodes.**

We classify electrochemical cells into two types.

1. **Electrolytic cells** are those in which electrical energy from an external source causes *nonspontaneous* chemical reactions to occur.

2. **Voltaic cells** are those in which *spontaneous* chemical reactions produce electricity and supply it to an external circuit.

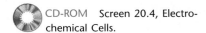

CD-ROM You should review oxidation–reduction (redox) reactions in text Section 4-7. Screens 20.2, Redox Reactions, and 20.3, Balancing Equations for Redox Reactions.

CD-ROM Screen 20.4, Electrochemical Cells.

We will discuss several kinds of electrochemical cells. From experimental observations we deduce what is happening at each electrode and for the overall reaction chemistry. We then construct simplified diagrams of the cells.

21-1 ELECTRICAL CONDUCTION

Electric current represents transfer of charge. Charge can be conducted through metals and through pure liquid electrolytes (that is, molten salts) or solutions containing electrolytes. The former type of conduction is called **metallic conduction.** It involves the flow of electrons with no similar movement of the atoms of the metal and no obvious changes in the metal (Section 13-17). **Ionic,** or **electrolytic, conduction** is the conduction of electric current by the motion of ions through a solution or a pure liquid. Positively charged ions migrate toward the negative electrode while negatively charged ions move toward the positive electrode. Both kinds of conduction, ionic and metallic, occur in electrochemical cells (Figure 21-1).

Ionic liquid is a newer term used for a room temperature molten salt.

21-2 ELECTRODES

Electrodes are surfaces on which oxidation or reduction half-reactions occur. They may or may not participate in the reactions. Those that do not react are called *inert electrodes.* Regardless of the kind of cell, electrolytic or voltaic, the electrodes are identified as follows.

The **cathode** is defined as the electrode at which *reduction* occurs as electrons are gained by some species. The **anode** is the electrode at which *oxidation* occurs as electrons are lost by some species.

Each of these can be either the positive or the negative electrode.

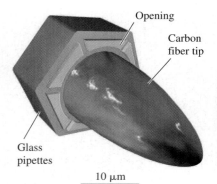

Opening

Carbon fiber tip

Glass pipettes

10 μm

Many reactions that take place in living cells are redox reactions. These reactions can be studied with miniature electrodes such as that shown here.

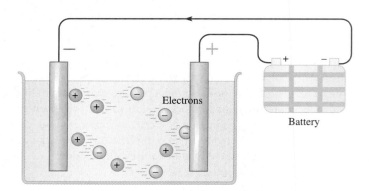

Electrons

Battery

Figure 21-1   The motion of ions through a solution is an electric current. This accounts for ionic (electrolytic) conduction. Positively charged ions migrate toward the negative electrode, and negatively charged ions migrate toward the positive electrode. Here the rate of migration is greatly exaggerated for clarity. The ionic velocities are actually only slightly greater than random molecular speeds.

SC*i*LINKS.

TOPIC: Electrochemical Cells
GO TO: www.scilinks.org
*sci***LINKS CODE:** WCH2110

ELECTROLYTIC CELLS

In some electrochemical cells *nonspontaneous* chemical reactions are forced to occur by the input of electrical energy. This process is called **electrolysis.** An electrolytic cell consists of a container for the reaction material with electrodes immersed in the reaction material and connected to a source of direct current. Inert electrodes are often used so that they do not become involved with the reaction.

Lysis means "splitting apart." In many electrolytic cells compounds are split into their constituent elements.

21-3 THE ELECTROLYSIS OF MOLTEN SODIUM CHLORIDE (THE DOWNS CELL)

Solid sodium chloride does not conduct electricity. Its ions vibrate about fixed positions, but they are not free to move throughout the crystal. Molten (melted) NaCl, however, is an excellent conductor because its ions are freely mobile. Consider a cell in which a source of direct current is connected by wires to two inert graphite electrodes (Figure 21-2a). They are immersed in a container of molten sodium chloride. When the current flows, we observe the following.

Molten NaCl, melting point 801°C, is a clear, colorless liquid that looks like water.

1. A pale green gas, which is chlorine, Cl_2, is liberated at one electrode.
2. Molten, silvery white metallic sodium, Na, forms at the other electrode and floats on top of the molten sodium chloride.

The metal remains liquid because its melting point is only 97.8°C. It floats because it is less dense than the molten NaCl.

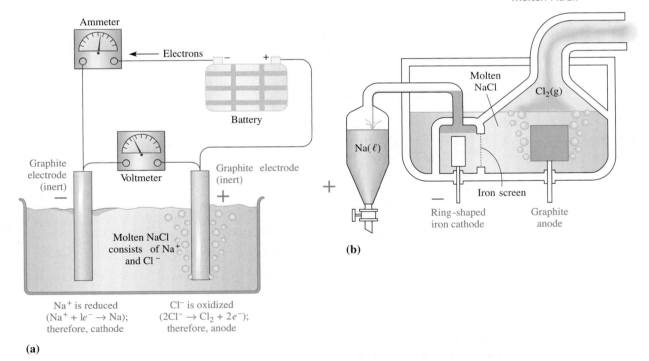

(a)

(b)

Figure 21-2 (a) Apparatus for electrolysis of molten sodium chloride. (b) The Downs cell, the apparatus in which molten sodium chloride is commercially electrolyzed to produce sodium metal and chlorine gas. The liquid Na floats on the denser molten NaCl.

From these observations we can deduce the processes of the cell. Chlorine must be produced by oxidation of Cl^- ions, and the electrode at which this happens must be the anode. Metallic sodium is produced by reduction of Na^+ ions at the cathode, where electrons are being forced into the cell.

$$2Cl^- \longrightarrow Cl_2(g) + 2e^- \qquad \text{(oxidation, anode half-reaction)}$$
$$2[Na^+ + e^- \longrightarrow Na(\ell)] \qquad \text{(reduction, cathode half-reaction)}$$
$$\underbrace{2Na^+ + 2Cl^-}_{2NaCl(\ell)} \longrightarrow 2Na(\ell) + Cl_2(g) \qquad \text{(overall cell reaction)}$$

In this chapter, as in Chapters 4 and 11, we often use red type to emphasize reduction and blue type to emphasize oxidation.

The formation of metallic Na and gaseous Cl_2 from NaCl is *nonspontaneous* except at temperatures very much higher than 801°C. The direct current (dc) source must supply electrical energy to force this reaction to occur. Electrons are used in the cathode half-reaction (reduction) and produced in the anode half-reaction (oxidation). They therefore travel through the wire from *anode* to *cathode*. The dc source forces electrons to flow non-spontaneously from the positive electrode to the negative electrode. The anode is the positive electrode and the cathode the negative electrode *in all electrolytic cells*. Figure 21-2a is a simplified diagram of the cell.

The direction of spontaneous *flow for negatively charged particles is from negative to positive.*

Sodium and chlorine must not be allowed to come in contact with each other because they react spontaneously, rapidly, and explosively to form sodium chloride. Figure 21-2b shows the **Downs cell** that is used for the industrial electrolysis of sodium chloride. The Downs cell is expensive to run, mainly because of the cost of construction, the cost of the electricity, and the cost of heating the NaCl to melt it. Nevertheless, electrolysis of a molten sodium salt is the most practical means by which metallic Na can be obtained, owing to its extremely high reactivity. Once liberated by the electrolysis, the liquid Na metal is drained off, cooled, and cast into blocks. These must be stored in an inert environment (e.g., in mineral oil) to prevent reaction with atmospheric O_2 or water.

Electrolysis of NaCl in the Downs cell is the main commercial method of producing metallic sodium. The Cl_2 gas produced in the Downs cell is cooled, compressed, and marketed. This partially offsets the expense of producing metallic sodium, but most chlorine is produced by the cheaper electrolysis of aqueous NaCl.

21-4 THE ELECTROLYSIS OF AQUEOUS SODIUM CHLORIDE

Consider the electrolysis of a moderately concentrated solution of NaCl in water, using inert electrodes. The following experimental observations are made when a sufficiently high voltage is applied across the electrodes of a suitable cell.

1. H_2 gas is liberated at one electrode. The solution becomes basic in that vicinity.
2. Cl_2 gas is liberated at the other electrode.

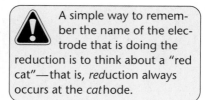

A simple way to remember the name of the electrode that is doing the reduction is to think about a "red cat"—that is, *reduction* always occurs at the *cat*hode.

Chloride ions are obviously being oxidized to Cl_2 in this cell, as they were in the electrolysis of molten NaCl. But Na^+ ions are not reduced to metallic Na. Instead, gaseous H_2 and aqueous OH^- ions are produced by reduction of H_2O molecules at the cathode. Water is more easily reduced than Na^+ ions. This is primarily because the reduction of Na^+ would produce the very active metal Na, whereas the reduction of H_2O produces the more stable products $H_2(g)$ and $OH^-(aq)$. The active metals Li, K, Ca, and Na (Table 4-14) displace H_2 from aqueous solutions, so we do not expect these metals to be produced in aqueous solution. Later in this chapter (Section 21-14) we learn the quantitative basis for predicting which of several possible oxidations or reductions is favored. The half-reactions and overall cell reaction for this electrolysis are

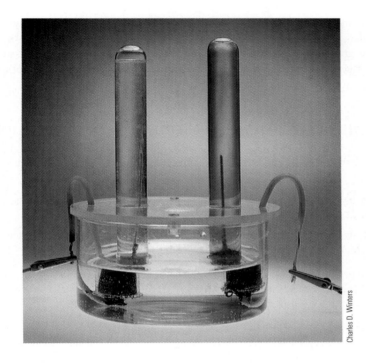

The electrolysis of the aqueous solution of KI, another Group IA – Group VIIA salt. At the cathode (*left*), water is reduced to $H_2(g)$ and OH^- ions, turning the phenolphthalein indicator pink. The characteristic brownish color of aqueous I_2 appears at the anode (*right*).

$$2Cl^- \longrightarrow Cl_2 + 2e^- \qquad \text{(oxidation, anode)}$$
$$2H_2O + 2e^- \longrightarrow 2OH^- + H_2 \qquad \text{(reduction, cathode)}$$
$$2H_2O + 2Cl^- \longrightarrow 2OH^- + H_2 + Cl_2 \qquad \text{(overall cell reaction as net ionic equation)}$$
$$\underbrace{+ 2Na^+} \longrightarrow \underbrace{+ 2Na^+} \qquad \text{(spectator ions)}$$
$$2H_2O + 2NaCl \longrightarrow 2NaOH + H_2 + Cl_2 \qquad \text{(overall cell reaction as formula unit equation)}$$

We will omit the notation that indicates states of substances—(s), (ℓ), (g), and (aq)—except where states are not obvious. This is a common shortcut when writing equations.

The cell is illustrated in Figure 21-3. As before, the battery forces the electrons to flow from the anode (+) through the wire to the cathode (−).

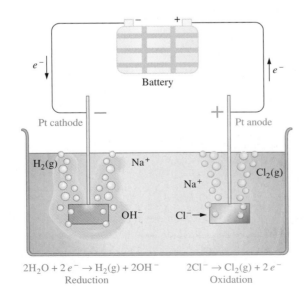

Figure 21-3 Electrolysis of aqueous NaCl solution. Although several reactions occur at both the anode and the cathode, the net result is the production of $H_2(g)$ and NaOH at the cathode and $Cl_2(g)$ at the anode. A few drops of phenolphthalein indicator were added to the solution. The solution turns pink at the cathode, where OH^- ions are formed and the solution becomes basic.

The overall cell reaction produces gaseous H_2 and Cl_2 and an aqueous solution of NaOH, called caustic soda. Solid NaOH is then obtained by evaporation of the residual solution. This is the most important commercial preparation of each of these substances. It is much less expensive than the electrolysis of molten NaCl, because it is not necessary to heat the solution.

Not surprisingly, the fluctuations in commercial prices of these widely used industrial products—H_2, Cl_2, and NaOH—have often paralleled one another.

21-5 THE ELECTROLYSIS OF AQUEOUS SODIUM SULFATE

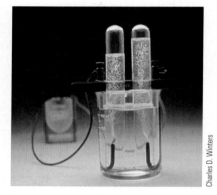

CD-ROM   Screen 20.11, Electrolysis.

In the electrolysis of aqueous sodium sulfate using inert electrodes, we observe the following.

1. Gaseous H_2 is produced at one electrode. The solution becomes basic around that electrode.

2. Gaseous O_2 is produced at the other electrode. The solution becomes acidic around that electrode.

As in the previous example, water is reduced in preference to Na^+ at the cathode. Observation 2 suggests that water is also preferentially oxidized relative to the sulfate ion, SO_4^{2-}, at the anode (Figure 21-4).

$$
\begin{array}{ll}
2(2H_2O + 2e^- \longrightarrow H_2 + 2OH^-) & \text{(reduction, cathode)} \\
\underline{2H_2O \longrightarrow O_2 + 4H^+ + 4e^-} & \text{(oxidation, anode)} \\
6H_2O \longrightarrow 2H_2 + O_2 + \underbrace{4H^+ + 4OH^-}_{4H_2O} & \text{(overall cell reaction)} \\
2H_2O \longrightarrow 2H_2 + O_2 & \text{(net reaction)}
\end{array}
$$

The net result is the electrolysis of water. This occurs because H_2O is more readily reduced than Na^+ and more readily oxidized than SO_4^{2-}. The ions of Na_2SO_4 conduct the current through the solution, but they take no part in the reaction.

Electrolysis of water to produce $O_2(g)$ and $H_2(g)$. Na_2SO_4 was dissolved in the water to improve the conductivity. Can you identify which tube in this picture is collecting the H_2 gas and the name of the electrode producing it (anode or cathode)?

Figure 21-4 The electrolysis of aqueous Na_2SO_4 produces $H_2(g)$ at the cathode and O_2 at the anode. Bromthymol blue indicator has been added to the solution. This indicator turns blue in the basic solution near the cathode (where OH^- is produced) and yellow in the acidic solution near the anode (where H^+ is formed).

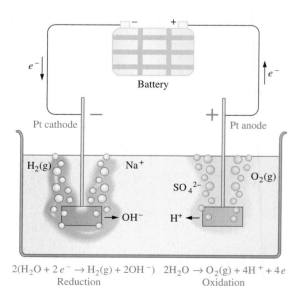

$2(H_2O + 2e^- \rightarrow H_2(g) + 2OH^-)$ $2H_2O \rightarrow O_2(g) + 4H^+ + 4e$
Reduction   Oxidation

TABLE 21-1	Amounts of Elements Produced at One Electrode in Electrolysis by 1 Faraday of Electricity		
Half-Reaction	Number of e^- in Half-Reaction	Product (electrode)	Amount Produced
$Ag^+(aq) + e^- \longrightarrow Ag(s)$	1	Ag (cathode)	1 mol = 107.868 g
$2H^+(aq) + 2e^- \longrightarrow H_2(g)$	2	H_2 (cathode)	$\frac{1}{2}$ mol = 1.008 g
$Cu^{2+}(aq) + 2e^- \longrightarrow Cu(s)$	2	Cu (cathode)	$\frac{1}{2}$ mol = 31.773 g
$Au^{3+}(aq) + 3e^- \longrightarrow Au(s)$	3	Au (cathode)	$\frac{1}{3}$ mol = 65.656 g
$2Cl^- \longrightarrow Cl_2(g) + 2e^-$	2	Cl_2 (anode)	$\frac{1}{2}$ mol = 35.453 g = 11.2 L_{STP}
$2H_2O(\ell) \longrightarrow O_2(g) + 4H^+(aq) + 4e^-$	4	O_2 (anode)	$\frac{1}{4}$ mol = 8.000 g = 5.60 L_{STP}

21-6 COUNTING ELECTRONS: COULOMETRY AND FARADAY'S LAW OF ELECTROLYSIS

In 1832–1833, Michael Faraday's studies of electrolysis led to this conclusion.

The amount of substance that undergoes oxidation or reduction at each electrode during electrolysis is directly proportional to the amount of electricity that passes through the cell.

This is **Faraday's Law of Electrolysis.** A quantitative unit of electricity is now called the faraday.

One **faraday** is the amount of electricity that corresponds to the gain or loss, and therefore the passage, of 6.022×10^{23} electrons, or *one mole* of electrons.

A smaller electrical unit commonly used in chemistry, physics, and electronics is the **coulomb (C).** One coulomb is defined as the amount of charge that passes a given point when 1 ampere (A) of electric current flows for 1 second. One **ampere** of current equals 1 coulomb per second. One faraday is equal to 96,485 coulombs of charge.

An ampere is usually called an "amp."

For comparison, a 100-watt household light bulb uses a current of about 0.8 ampere.

$$1 \text{ ampere} = 1 \frac{\text{coulomb}}{\text{second}} \quad \text{or} \quad 1 \text{ A} = 1 \text{ C/s}$$

$$1 \text{ faraday} = 6.022 \times 10^{23} \ e^- = 96,485 \text{ C}$$

Table 21-1 shows the amounts of several elements produced during electrolysis by the passage of 1 faraday of electricity. The use of electrochemical cells to relate the amount of reactant or product to the amount of current passed is called **coulometry.**

Michael Faraday (1791–1867) is considered the greatest experimental scientist of the 19th century. As a bookbinder's apprentice, he educated himself by extensive reading. Intrigued by his self-study of chemistry and by a lecture given by Sir Humphry Davy, the leading chemist of the day, Faraday applied for a position with Davy at the Royal Institution. He subsequently became director of that laboratory. His public lectures on science were very popular.

The amount of electricity in Examples 21-1 and 21-2 would be sufficient to light a 100-watt household light bulb for about 150 minutes, or 2.5 hours.

When the number of significant figures in the calculation warrants, the value 96,485 coulombs is usually rounded to 96,500 coulombs (9.65×10^4 C).

Remember that:

2.50 A $= 2.50$ C/s

Also convert time into seconds when doing these calculations!

EXAMPLE 21-1   *Electrolysis*

Calculate the mass of copper metal produced at the cathode during the passage of 2.50 amperes of current through a solution of copper(II) sulfate for 50.0 minutes.

Plan

The half-reaction that describes the reduction of copper(II) ions tells us the number of moles of electrons required to produce one mole of copper metal. Each mole of electrons corresponds to 1 faraday, or 9.65×10^4 coulombs, of charge. The product of current and time gives the number of coulombs.

$$\boxed{\begin{array}{c}\text{current}\\ \times \text{ time}\end{array}} \longrightarrow \boxed{\begin{array}{c}\text{no. of}\\ \text{coulombs}\end{array}} \longrightarrow \boxed{\begin{array}{c}\text{mol of } e^-\\ \text{passed}\end{array}} \longrightarrow \boxed{\begin{array}{c}\text{mass}\\ \text{of Cu}\end{array}}$$

Solution

The equation for the reduction of copper(II) ions to copper metal is

$$
\begin{array}{cccc}
Cu^{2+} + & 2e^- & \longrightarrow & Cu \quad \text{(reduction, cathode)}\\
\text{1 mol} & 2(6.02 \times 10^{23})e^- & & \text{1 mol}\\
\text{63.5 g} & 2(9.65 \times 10^4 \text{ C}) & & \text{63.5 g}
\end{array}
$$

We see that 63.5 grams of copper "plate out" for every 2 moles of electrons, or for every $2(9.65 \times 10^4$ coulombs) of charge. We first calculate the number of coulombs passing through the cell.

$$\underline{?}\ C = 50.0 \text{ min} \times \frac{60 \text{ s}}{1 \text{ min}} \times \frac{2.50 \text{ C}}{\text{s}} = 7.50 \times 10^3 \text{ C}$$

We calculate the mass of copper produced by the passage of 7.50×10^3 coulombs.

$$\underline{?}\ g\ Cu = 7.50 \times 10^3 \text{ C} \times \frac{1 \text{ mol } e^-}{9.65 \times 10^4 \text{ C}} \times \frac{63.5 \text{ g C u}}{2 \text{ mol } e^-} = \boxed{2.47 \text{ g Cu}} \quad \begin{array}{l}\text{(about the mass of}\\ \text{a copper penny)}\end{array}$$

Notice how little copper is deposited by this considerable current in 50 minutes.

You should now work Exercises 26 and 32.

EXAMPLE 21-2   *Electrolysis*

What volume of oxygen gas (measured at STP) is produced by the oxidation of water at the anode in the electrolysis of copper(II) sulfate in Example 21-1?

Plan

We use the same approach as in Example 21-1. Here we relate the amount of charge passed to the number of moles, and hence the volume of O_2 gas produced at STP.

$$\boxed{\begin{array}{c}\text{current}\\ \times \text{ time}\end{array}} \longrightarrow \boxed{\begin{array}{c}\text{no. of}\\ \text{coulombs}\end{array}} \longrightarrow \boxed{\begin{array}{c}\text{mol of } e^-\\ \text{passed}\end{array}} \longrightarrow \boxed{\begin{array}{c}L_{STP}\\ \text{of } O_2\end{array}}$$

Solution

The equation for the oxidation of water and the equivalence between the number of coulombs and the volume of oxygen produced at STP are

$$
\begin{array}{cccc}
2H_2O \longrightarrow & O_2 & + 4H^+ + & 4e^- \quad \text{(oxidation, anode)}\\
& \text{1 mol} & & 4(6.02 \times 10^{23})e^-\\
& 22.4\ L_{STP} & & 4(9.65 \times 10^4 \text{ C})
\end{array}
$$

The number of coulombs passing through the cell is 7.50×10^3 C. For every $4(9.65 \times 10^4$ coulombs) passing through the cell, 22.4 L of O_2 at STP is produced.

$$\underline{?}\ L_{STP}\ O_2 = 7.50 \times 10^3\ C \times \frac{1\ \text{mol}\ e^-}{9.65\ \times\ 10^4\ C} \times \frac{22.4\ L_{STP}\ O_2}{4\ \text{mol}\ e^-} = \boxed{0.435\ L_{STP}\ O_2}$$

You should now work Exercise 28.

Notice how little product is formed by what seems to be a lot of electricity. This suggests why electrolytic production of gases and metals is so costly.

CD-ROM   Screen 20.12, Coulometry.

21-7 COMMERCIAL APPLICATIONS OF ELECTROLYTIC CELLS

Several elements are produced commercially by electrolysis. In Sections 21-3 to 21-5, we described some electrolytic cells that produce sodium (the Downs cell), chlorine, hydrogen, and oxygen. Electrolysis of molten compounds is also the common method of obtaining other Group IA metals, IIA metals (except barium), and aluminum (Section 22-3). Impure metals can also be refined electrolytically, as we will describe for copper in Section 22-8.

Metal-plated articles are common in our society. Jewelry and tableware are often plated with silver. Gold is plated onto jewelry and electrical contacts. Copper is plated onto many objects for decorative purposes (Figure 21-5). Some automobiles have steel bumpers plated

A family memento that has been electroplated with copper. To aid in electroplating onto nonconductors, such as shoes, the material is first soaked in a concentrated electrolyte solution to make it conductive.

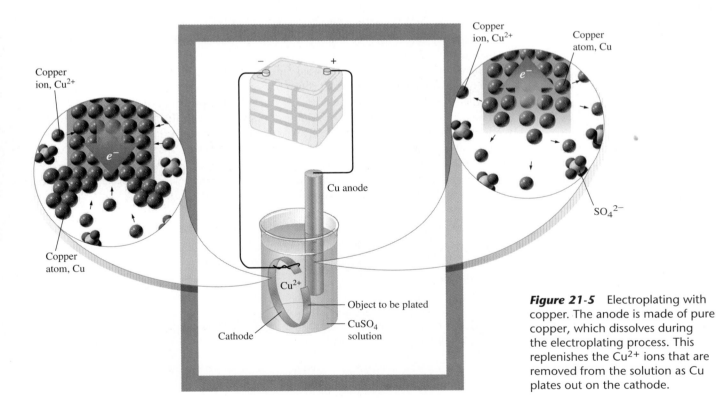

Figure 21-5 Electroplating with copper. The anode is made of pure copper, which dissolves during the electroplating process. This replenishes the Cu^{2+} ions that are removed from the solution as Cu plates out on the cathode.

Anodized aluminum is made by oxidizing a piece of aluminum in an electrolytic cell for a short period of time. The oxidation microscopically dissolves some of the Al surface and coats it with a protective and tough coating of aluminum oxide. This creates a matte finish that can be dyed different colors.

These are named for Allesandro Volta (1745–1827) and Luigi Galvani (1737–1798), two Italian physicists of the 18th century.

Neither a voltmeter nor an ammeter generates electrical energy. They are designed to measure electrical voltage or the amount of current flow.

Agar is a porous gelatinous material obtained from algae.

When we show a concentration in a *standard cell* as 1 *M*, it is assumed to be *exactly* 1 *M*.

with thin films of chromium. A chrome bumper requires approximately 3 seconds of electroplating to produce a smooth, shiny surface only 0.0002 mm thick. When the metal atoms are deposited too rapidly, they are not able to form extended lattices. Rapid plating of metal results in rough, grainy, black surfaces. Slower plating produces smooth surfaces. "Tin cans" are steel cans plated electrolytically with tin; these are sometimes replaced by cans plated in a fraction of a second with an extremely thin chromium film.

VOLTAIC OR GALVANIC CELLS

Voltaic, or **galvanic, cells** are electrochemical cells in which *spontaneous* oxidation–reduction reactions produce electrical energy. The two halves of the redox reaction are separated, requiring electron transfer to occur through an external circuit. In this way, useful electrical energy is obtained. Everyone is familiar with some voltaic cells. The batteries commonly used in flashlights, portable radios, photographic equipment, and many toys and appliances are voltaic cells. Automobile batteries consist of voltaic cells connected in series so that their voltages add. We will first consider some simple laboratory cells used to measure the potential difference, or voltage, of a reaction under study. We will then look at some common voltaic cells.

21-8 THE CONSTRUCTION OF SIMPLE VOLTAIC CELLS

A **half-cell** contains the oxidized and reduced forms of an element, or other more complex species, in contact with one another. A common kind of half-cell consists of a piece of metal (the electrode) immersed in a solution of its ions. Consider two such half-cells in separate beakers (Figure 21-6). The electrodes are connected by a wire. A voltmeter can be inserted into the circuit to measure the potential difference between the two electrodes, or an ammeter can be inserted to measure the current flow. The electric current is the result of the spontaneous redox reaction that occurs. We measure the potential of the cell.

The circuit between the two solutions is completed by a **salt bridge.** This can be any medium through which ions can slowly pass. A salt bridge can be made by bending a piece of glass tubing into the shape of a "U," filling it with a hot saturated salt/5% agar solution, and allowing it to cool. The cooled mixture "sets" to the consistency of firm gelatin. As a result, the solution does not run out when the tube is inverted (see Figure 21-6), but the ions in the gel are still able to move. A salt bridge serves three functions.

1. It allows electrical contact between the two solutions.
2. It prevents mixing of the electrode solutions.
3. It maintains the electrical neutrality in each half-cell as ions flow into and out of the salt bridge.

A cell in which all reactants and products are in their thermodynamic standard states (1 *M* for dissolved species and 1 atm partial pressure for gases) is called a **standard cell.**

21-9 THE ZINC–COPPER CELL

Consider a standard cell made up of two half-cells, one a strip of metallic Cu immersed in 1 M copper(II) sulfate solution and the other a strip of Zn immersed in 1 M zinc sulfate solution (see Figure 21-6). This cell is called the Daniell cell. The following experimental observations have been made about this cell.

1. The initial voltage is 1.100 volts.
2. The mass of the zinc electrode decreases. The concentration of Zn^{2+} increases in the solution around the zinc electrode as the cell operates.
3. The mass of the copper electrode increases. The concentration of Cu^{2+} decreases in the solution around this electrode as the cell operates.

The Zn electrode loses mass because some Zn metal is *oxidized* to Zn^{2+} ions, which go into solution. Thus the Zn electrode is the *anode*. At the *cathode*, Cu^{2+} ions are *reduced* to Cu metal. This plates out on the electrode, so its mass increases.

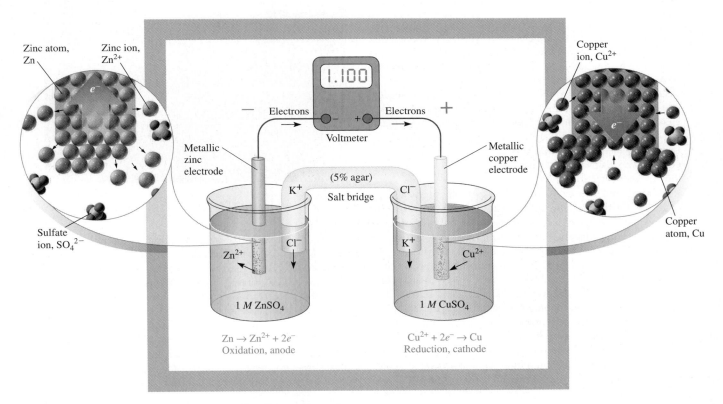

Figure 21-6   The zinc–copper voltaic cell utilizes the reaction

$$Zn(s) + Cu^{2+}(aq) \longrightarrow Zn^{2+}(aq) + Cu(s)$$

The standard potential of this cell is 1.10 volts.

$$\begin{aligned}
\text{Zn} &\longrightarrow \text{Zn}^{2+} + 2e^- &&\text{(oxidation, anode)}\\
\text{Cu}^{2+} + 2e^- &\longrightarrow \text{Cu} &&\text{(reduction, cathode)}\\
\hline
\text{Cu}^{2+} + \text{Zn} &\longrightarrow \text{Cu} + \text{Zn}^{2+} &&\text{(overall cell reaction)}
\end{aligned}$$

Electrons are released at the anode and consumed at the cathode. They therefore flow through the wire from anode to cathode, as in all electrochemical cells. In all *voltaic* cells the electrons flow spontaneously from the negative electrode to the positive electrode. So, in contrast with electrolytic cells, the anode is negative and the cathode is positive. To maintain electroneutrality and complete the circuit, two Cl^- ions from the salt bridge migrate into the anode solution for every Zn^{2+} ion formed. Two K^+ ions migrate into the cathode solution to replace every Cu^{2+} ion reduced. Some Zn^{2+} ions from the anode vessel and

Compare the $-/+$, anode–cathode, and oxidation–reduction labels and the directions of electron flow in Figures 21-2a and 21-6.

CHEMISTRY IN USE

Our Daily Lives

A Spectacular View of One Mole of Electrons

Early in our study of chemistry, we saw that atoms are made up of protons, neutrons, and electrons. We also discussed the incredibly large size of Avogadro's number, 6.022×10^{23}. Although individual atoms and molecules are invisible to the naked eye, one mole of atoms or molecules is easily detected. Because subatomic particles are even smaller than atoms and also invisible, you might never expect to see individual electrons. Let's consider the possibility, however, of seeing a faraday of charge. A faraday of charge contains Avogadro's number of electrons. Would this collection of 6.022×10^{23} electrons be visible? If so, what might it look like? It would look quite spectacular!

Throughout the 1980s, scientists carefully studied data collected during 5 million lightning flashes along the eastern United States. The data were collected by 36 instruments that were collectively known as the National Lightning Detection Network. The investigating scientists found that the electrical currents in lightning flashes over northern Florida measured about 45,000 amps, about double the 25,000-amp currents in lightning flashes over the New England states. This study showed that the amount of current flowing during lightning flashes was inversely proportional to the latitude (distance from the equator) of the storm.

One coulomb is the amount of charge that passes a point when a one-ampere current flows for one second (1 coulomb = 1 ampere · second). Thus, a current of 96,500 amps flowing for one second contains Avogadro's number of electrons, or one faraday of charge.

Measurements taken in northern Florida show that a typical two-second lightning strike over that section of the coun-

Gordon Garradd/Science Photo Library/Photo Researchers, Inc.

try would transfer approximately Avogadro's number of electrons between the clouds and the earth. So, for those living in northern Florida, a spectacular mental view of one mole of electrons can be obtained by visualizing a two-second lightning strike. Keep in mind that the average lightning strike lasts only a small fraction of a second, and that we can only have a mental view of a two-second lightning strike by extrapolation of what is seen in nature. Because New England lightning strikes produce only about half the current of lightning strikes over northern Florida, people in New England must try to imagine a four-second lightning strike.

Ronald DeLorenzo
Middle Georgia College

(a)   (b)

Charles D. Winters

James W. Morganthaler

(a) A strip of zinc was placed in a blue solution of copper(II) sulfate, $CuSO_4$ (left hand beaker). The Cu^{2+} in solution oxidizes the Zn metal and deposits Cu metal onto the strip of Zn, while the oxidized Zn^{2+} dissolved. The resulting zinc sulfate solution is colorless (right hand beaker). This is the same overall reaction as the one that occurs when the two half-reactions are separated in the zinc–copper cell (see Figure 21-6). (b) No reaction occurs when copper wire is placed in a colorless zinc sulfate solution. The reaction

$$Zn^{2+} + Cu(s) \longrightarrow Zn(s) + Cu^{2+}$$

is the *reverse* of the spontaneous reaction in Figure 21-6, so it is *nonspontaneous.*

some $SO_4{}^{2-}$ ions from the cathode vessel also migrate into the salt bridge. Neither Cl^- nor K^+ ions are oxidized or reduced in preference to the zinc metal or Cu^{2+} ions.

As the redox reaction proceeds consuming the reactants, the cell voltage decreases. When the cell voltage reaches zero, the reaction has reached equilibrium, and no further net reaction occurs. At this point, however, the Cu^{2+} ion concentration in the cathode cell is *not* zero. This description applies to any voltaic cell.

Voltaic cells can be represented as follows for the zinc–copper cell.

salt bridge
↓

$$Zn|Zn^{2+}(1\ M)\|Cu^{2+}\ (1\ M)|Cu$$

↖   ↗

species (and concentrations)
in contact with electrode surfaces

In this representation, a single line (|) represents an interface at which a potential develops, that is, an electrode. It is conventional to write the anode half-cell on the left in this notation.

The same reaction occurs when a piece of Zn is dropped into a solution of $CuSO_4$. The Zn dissolves and the blue color of Cu^{2+} ions disappears. Copper forms on the Zn and then settles to the bottom of the container. Because the oxidizing Cu^{2+} ions and reducing Zn metal are in direct contact with one another in this arrangement, the electron transfer is "short-circuited," so the spontaneous free energy available to perform work is wasted and simply heats the solution. Only by separating the two half-reactions into separate cells and connecting them with an external wire (circuit) and a salt bridge can one effectively use the electricity generated.

> ⚠ This is a shorthand notation for the cell reactions:
>
> $$Zn \longrightarrow Zn^{2+} + 2e^-$$
>
> and
>
> $$Cu^{2+} + 2e^- \longrightarrow Cu$$

✓ Problem-Solving Tip: *How to Tell the Anode from the Cathode*

The correspondence between the names *anode* and *cathode* and the charge on the electrode is *different* for electrolytic cells than for voltaic (galvanic) cells. Students sometimes get confused by trying to remember which is which. Check the definitions of these two terms in Section 21-2. The surest way to name these electrodes is to determine what process takes place at each one. Remember the "red cat" mnemonic—*red*uction occurs at the *cat*hode, regardless of the type of electrochemical cell one is dealing with.

21-10 THE COPPER–SILVER CELL

Now consider a similar standard voltaic cell consisting of a strip of Cu immersed in 1 M $CuSO_4$ solution and a strip of Ag immersed in 1 M $AgNO_3$ solution. A wire and a salt bridge complete the circuit. The following observations have been made.

1. The initial voltage of the cell is 0.462 volt.

2. The mass of the copper electrode decreases. The Cu^{2+} ion concentration increases in the solution around the copper electrode.

3. The mass of the silver electrode increases. The Ag^+ ion concentration decreases in the solution around the silver electrode.

In this cell the Cu electrode is the anode because Cu metal is oxidized to Cu^{2+} ions. The Ag electrode is the cathode because Ag^+ ions are reduced to metallic Ag (Figure 21-7).

$$
\begin{array}{ll}
Cu \longrightarrow Cu^{2+} + 2e^- & \text{(oxidation, anode)} \\
\underline{2(Ag^+ + e^- \longrightarrow Ag)} & \text{(reduction, cathode)} \\
Cu + 2Ag^+ \longrightarrow Cu^{2+} + 2Ag & \text{(overall cell reaction)}
\end{array}
$$

As before, ions from the salt bridge migrate to maintain electroneutrality. Some NO_3^- ions (from the cathode vessel) and some Cu^{2+} ions (from the anode vessel) also migrate into the salt bridge.

Recall that in the zinc–copper cell the copper electrode was the *cathode*; now in the copper–silver cell the copper electrode has become the *anode*.

KCl is not used in this salt bridge because the Cl^- ions would react with Ag^+ ions to form insoluble AgCl(s).

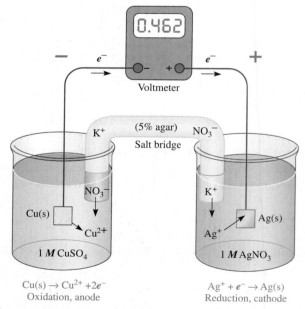

Figure 21-7 The copper–silver voltaic cell utilizes the reaction

$$Cu(s) + 2Ag^+(aq) \longrightarrow Cu^{2+}(aq) + 2Ag(s)$$

The standard potential of this cell is 0.462 volt. This standard cell can be represented as $Cu|Cu^{2+}(1\ M)||Ag^+(1\ M)|Ag$.

Copper atom, Cu

e^-

Copper ion, Cu^{2+}

Silver ion, Ag^+

e^-

Silver atom, Ag

Photo courtesy of Donald M. West

The standard $Cu|Cu^{2+}$(1 M)$\|Ag^+$(1 M)$|Ag$ cell.

Whether a particular electrode acts as an anode or a cathode in a voltaic cell depends on what the other electrode of the cell is.

The two cells we have described show that the Cu^{2+} ion is more easily reduced (is a stronger oxidizing agent) than Zn^{2+}, so Cu^{2+} oxidizes metallic zinc to Zn^{2+}. By contrast, Ag^+ ion is more easily reduced (is a stronger oxidizing agent) than Cu^{2+} ion, so Ag^+ oxidizes Cu atoms to Cu^{2+}. Conversely, metallic Zn is a stronger reducing agent than metallic Cu, and metallic Cu is a stronger reducing agent than metallic Ag. We can now arrange the species we have studied in order of increasing strength as oxidizing agents and as reducing agents.

$$Zn^{2+} < Cu^{2+} < Ag^+$$
Increasing strength as oxidizing agents

$$Ag < Cu < Zn$$
Increasing strength as reducing agents

James W. Morganthaler

James W. Morganthaler

(*Left*) A spiral of copper wire was placed in a colorless solution of silver nitrate, $AgNO_3$. The silver has been displaced from solution and adheres to the wire. The resulting copper nitrate solution is blue. The same reaction occurs when the two half-reactions are separated in the copper–silver cell (see Figure 21-7). (*Right*) No reaction occurs when silver wire is placed in a blue copper sulfate solution. The reaction

$$Cu^{2+} + 2Ag(s) \longrightarrow Cu(s) + 2Ag^+$$

is the reverse of the spontaneous reaction in Figure 21-7; it has a *negative* E^0_{cell} and is *nonspontaneous*.

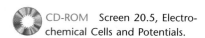
STANDARD ELECTRODE POTENTIALS

The potentials of the standard zinc–copper and copper–silver voltaic cells are 1.100 volts and 0.462 volts, respectively. The magnitude of a cell's potential measures the spontaneity of its redox reaction. *Higher (more positive) cell potentials indicate greater driving force for the reaction as written.* Under standard conditions, the oxidation of metallic Zn by Cu^{2+} ions has a greater tendency to go toward completion than does the oxidation of metallic Cu by Ag^+ ions. It is convenient to separate the total cell potential into the individual contributions of the two half-reactions. This lets us determine the relative tendencies of particular oxidation or reduction half-reactions to occur. Such information gives us a quantitative basis for specifying strengths of oxidizing and reducing agents. In the next several sections we shall see how this is done for standard half-cells.

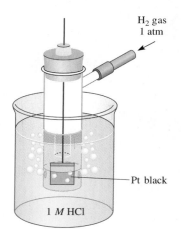

21-11 THE STANDARD HYDROGEN ELECTRODE

Every oxidation must be accompanied by a reduction (i.e., the electrons must have somewhere to go). So it is impossible to determine experimentally the potential of any *single* electrode. We therefore establish an arbitrary standard. The conventional reference electrode is the **standard hydrogen electrode (SHE).** This electrode contains a piece of metal electrolytically coated with a grainy black surface of inert platinum metal, immersed in a $1\,M\,H^+$ solution. Hydrogen, H_2, is bubbled at 1 atm pressure through a glass envelope over the platinized electrode (Figure 21-8).

Figure 21-8   The standard hydrogen electrode (SHE). A molecular-level view of the operation of the SHE as a cathode is shown in Figure 21-9 and as an anode in Figure 21-10.

> By international convention, the standard hydrogen electrode is arbitrarily assigned a potential of *exactly* 0.0000 . . . volt.

SHE Half-Reaction	E^0 (standard electrode potential)	
$H_2 \longrightarrow 2H^+ + 2e^-$	exactly 0.0000 . . . V	(SHE as anode)
$2H^+ + 2e^- \longrightarrow H_2$	exactly 0.0000 . . . V	(SHE as cathode)

The superscript in E^0 indicates thermodynamic standard-state conditions.

We then construct a standard cell consisting of a standard hydrogen electrode and some other standard electrode (half-cell). Because the defined electrode potential of the SHE contributes exactly 0 volt to the sum, the voltage of the overall cell then lets us determine the **standard electrode potential** of the other half-cell. This is its potential with respect to the standard hydrogen electrode, measured at 25°C when the concentration of each ion in the solution is $1\,M$ and the pressure of any gas involved is 1 atm.

> By agreement, we always present the standard cell potential for each half-cell as a *reduction* process.

21-12 THE ZINC–SHE CELL

This cell consists of an SHE in one beaker and a strip of zinc immersed in $1\,M$ zinc chloride solution in another beaker (Figure 21-9). A wire and a salt bridge complete the circuit. When the circuit is closed, the following observations can be made.

1. The initial potential of the cell is 0.763 volt.

2. As the cell operates, the mass of the zinc electrode decreases. The concentration of Zn^{2+} ions increases in the solution around the zinc electrode.

3. The H^+ concentration decreases in the SHE. Gaseous H_2 is produced.

We can conclude from these observations that the following half-reactions and cell reaction occur.

		E^0	
(oxidation, anode)	$Zn \longrightarrow Zn^{2+} + 2e^-$	0.763 V	
(reduction, cathode)	$2H^+ + 2e^- \longrightarrow H_2$	0.000 V	(by definition)
(cell reaction)	$Zn + 2H^+ \longrightarrow Zn^{2+} + H_2$	$E^0_{cell} = 0.763$ V	(measured)

The standard potential at the anode *plus* the standard potential at the cathode gives the standard cell potential. The potential of the SHE is 0.000 volt, and the standard cell potential is found to be 0.763 volt. So the standard potential of the zinc anode must be 0.763 volt. The $Zn|Zn^{2+}(1.0\ M)\|H^+(1.0\ M), H_2(1\ atm)|Pt$ cell is depicted in Figure 21-9.

Note that in *this* cell the SHE is the *cathode*, and metallic zinc reduces H^+ to H_2. The zinc electrode is the *anode* in this cell.

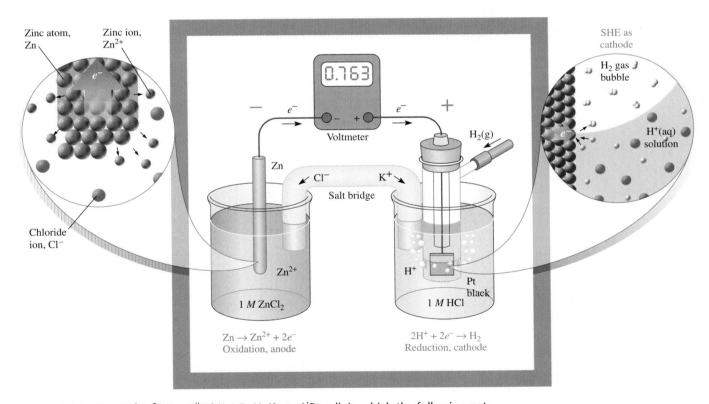

Figure 21-9 The $Zn|Zn^{2+}(1\ M)\|H^+(1\ M); H_2(1\ atm)|Pt$ cell, in which the following net reaction occurs.

$$Zn(s) + 2H^+(aq) \longrightarrow Zn^{2+}(aq) + H_2(g)$$

In this cell the standard hydrogen electrode functions as the cathode.

The idea of "electron pressure" helps to explain this process. The negative reduction potential for the half-reaction

$$Zn^{2+} + 2e^- \longrightarrow Zn \qquad E^0 = -0.763 \text{ V}$$

says that this reaction is *less favorable* than the corresponding reduction to H_2,

$$2H^+ + 2e^- \longrightarrow H_2 \qquad E^0 = 0.000 \text{ V}$$

Before they are connected, each half-cell builds up a supply of electrons waiting to be released, thus generating an electron pressure. Let's compare these electron pressures by reversing the two half-reactions to show production of electrons (and changing the signs of their E^0 values).

> What we have informally called "electron pressure" is the tendency to undergo oxidation. This is formally expressed as an *oxidation potential;* however, we usually tabulate reduction potentials (Table 21-2).

$$Zn \longrightarrow Zn^{2+} + 2e^- \qquad E^0_{oxidation} = +0.763 \text{ V}$$
$$H_2 \longrightarrow 2H^+ + 2e^- \qquad E^0_{oxidation} = 0.000 \text{ V}$$

The process with the more *positive* E^0 value is favored, so we reason that the electron pressure generated at the Zn electrode is greater than that at the H_2 electrode. As a result, when the cell is connected, the electrons released by the oxidation of Zn flow through the wire *from the Zn electrode to the H_2 electrode*, where they are consumed by the reduction of H^+ ions. Oxidation occurs at the zinc electrode (anode), and reduction occurs at the hydrogen electrode (cathode).

21-13   THE COPPER–SHE CELL

Another voltaic cell consists of an SHE in one beaker and a strip of Cu metal immersed in $1 \text{ } M$ copper(II) sulfate solution in another beaker. A wire and a salt bridge complete the circuit. For this cell, we observe the following (Figure 21-10).

1. The initial cell potential is 0.337 volt.

2. Gaseous hydrogen is consumed. The H^+ concentration increases in the solution of the SHE.

3. The mass of the copper electrode increases. The concentration of Cu^{2+} ions decreases in the solution around the copper electrode.

Thus, the following half-reactions and cell reaction occur.

		E^0	
(oxidation, anode)	$H_2 \longrightarrow 2H^+ + 2e^-$	0.000 V	(by definition)
(reduction, cathode)	$Cu^{2+} + 2e^- \longrightarrow Cu$	0.337 V	
(cell reaction)	$H_2 + Cu^{2+} \longrightarrow 2H^+ + Cu$	$E^0_{cell} = 0.337 \text{ V}$	(measured)

> Recall that in the Zn–SHE cell the SHE was the *cathode.*

The SHE functions as the *anode* in this cell, and Cu^{2+} ions oxidize H_2 to H^+ ions. The standard electrode potential of the copper half-cell is 0.337 volt as a *cathode* in the Cu–SHE cell.

Again, we can think of $E^0_{oxidation}$ in the two half-cells as "electron pressures."

$$Cu \longrightarrow Cu^{2+} + 2e^- \qquad E^0_{oxidation} = -0.337 \text{ V}$$
$$H_2 \longrightarrow 2H^+ + 2e^- \qquad E^0_{oxidation} = 0.000 \text{ V}$$

Now the hydrogen electrode has the higher electron pressure. When the cell is connected, electrons flow through the wire from the hydrogen electrode to the copper electrode. H_2 is oxidized to $2H^+$ (anode), and Cu^{2+} is reduced to Cu (cathode).

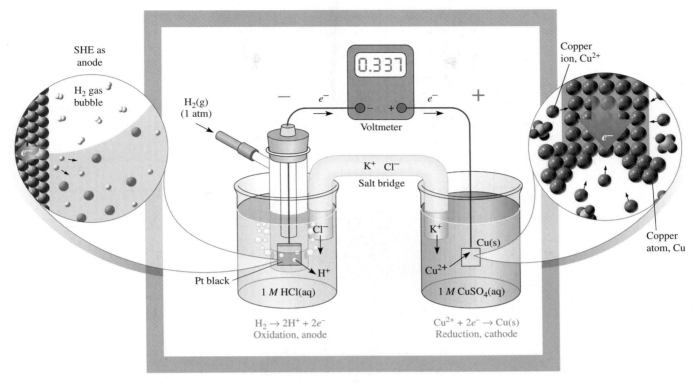

Figure 21-10   The standard copper–SHE cell,

$$Pt|H^+(1\ M);\ H_2(1\ atm)\|Cu^{2+}(1\ M)|Cu$$

In this cell, the standard hydrogen electrode functions as the anode. The net reaction is

$$H_2(g) + Cu^{2+}(aq) \longrightarrow 2H^+(aq) + Cu(s)$$

21-14   STANDARD ELECTRODE POTENTIALS

We can develop a series of standard electrode potentials by measuring the potentials of other standard electrodes versus the SHE in the way we described for the standard Zn–SHE and standard Cu–SHE voltaic cells. Many electrodes involve metals or non-metals in contact with their ions. We saw (Section 21-12) that the standard Zn electrode behaves as the anode versus the SHE and that the standard *oxidation* potential for the Zn half-cell is 0.763 volt.

The activity series (Table 4-14) is based on standard electrode potentials.

		$E^0_{oxidation}$
(as anode)	$Zn \longrightarrow Zn^{2+} + 2e^-$	$+0.763$ V
	reduced form $\longrightarrow$ oxidized form $+ ne^-$	(standard *oxidation* potential)

The *reduction* potential for the standard zinc electrode (to act as a *cathode* relative to the SHE) is therefore the negative of this, or -0.763 volt.

		$E^0_{oxidation}$
(as cathode)	$Zn^{2+} + 2e^- \longrightarrow Zn$	-0.763 V
	oxidized form $+ ne^- \longrightarrow$ reduced form	(standard *reduction* potential)

By international convention, the standard potentials of electrodes are tabulated for *reduction half-reactions*. These indicate the tendencies of the electrodes to behave as cathodes toward the SHE. Electrodes with positive E^0 values for reduction half-reactions act as *cathodes* versus the SHE. Those with negative E^0 values for reduction half-reactions act as anodes versus the SHE.

Electrodes with *Positive $E^0_{reduction}$*	Electrodes with *Negative $E^0_{reduction}$*
Reduction occurs *more readily* than the reduction of $2H^+$ to H_2. Electrode acts as a *cathode* versus the SHE.	Reduction is *more difficult* than the reduction of $2H^+$ to H_2. Electrode acts as an *anode* versus the SHE.

> The more positive the E^0 value for a half-reaction, the greater the tendency for the half-reaction to occur in the forward direction as written. Conversely, the more negative the E^0 value for a half-reaction, the greater the tendency for the half-reaction to occur in the reverse direction as written.

Table 21-2 lists standard reduction potentials for a few elements.

1. The species on the *left* side are all cations of metals, hydrogen ions, or elemental non-metals. These are all *oxidizing agents* (*oxidized forms* of the elements). Their strengths as oxidizing agents increase from top to bottom, that is, as the $E^0_{reduction}$ values become more positive. Fluorine, F_2, is the strongest oxidizing agent, and Li^+ is a very weak oxidizing agent.

2. The species on the *right* side are free metals, hydrogen, or anions of nonmetals. These are all *reducing agents* (*reduced forms* of the elements). Their strengths as reducing agents increase from bottom to top, that is, as the $E^0_{reduction}$ values become more negative. Metallic Li is a very strong reducing agent, and F^- is a very weak reducing agent.

> The more positive the reduction potential, the stronger the species on the left is as an oxidizing agent and the weaker the species on the right is as a reducing agent.

The oxidizing agent is reduced.

The reducing agent is oxidized.

21-15 USES OF STANDARD ELECTRODE POTENTIALS

The most important application of electrode potentials is the prediction of the spontaneity of redox reactions. Standard electrode potentials can be used to determine the spontaneity of redox reactions in general, whether or not the reactions can take place in electrochemical cells.

Suppose we ask the question: At standard conditions, will Cu^{2+} ions oxidize metallic Zn to Zn^{2+} ions, or will Zn^{2+} ions oxidize metallic copper to Cu^{2+}? One of the two possible reactions is spontaneous, and the reverse reaction is nonspontaneous. We must determine which one is spontaneous. We already know the answer to this question from experimental results (see Section 21-9), but let us demonstrate the procedure for predicting the spontaneous reaction.

TABLE 21-2 *Standard Aqueous Reduction Potentials in Aqueous Solution at 25°C*

Element	Reduction Half-Reaction	Standard Reduction Potential E^0 (volts)	
Li	$Li^+ + e^- \longrightarrow Li$	-3.045	
K	$K^+ + e^- \longrightarrow K$	-2.925	
Ca	$Ca^{2+} + 2e^- \longrightarrow Ca$	-2.87	
Na	$Na^+ + e^- \longrightarrow Na$	-2.714	
Mg	$Mg^{2+} + 2e^- \longrightarrow Mg$	-2.37	
Al	$Al^{3+} + 3e^- \longrightarrow Al$	-1.66	
Zn	$Zn^{2+} + 2e^- \longrightarrow Zn$	-0.763	
Cr	$Cr^{3+} + 3e^- \longrightarrow Cr$	-0.74	
Fe	$Fe^{2+} + 2e^- \longrightarrow Fe$	-0.44	
Cd	$Cd^{2+} + 2e^- \longrightarrow Cd$	-0.403	
Ni	$Ni^{2+} + 2e^- \longrightarrow Ni$	-0.25	
Sn	$Sn^{2+} + 2e^- \longrightarrow Sn$	-0.14	
Pb	$Pb^2 + 2e^- \longrightarrow Pb$	-0.126	
H_2	$2H^+ + 2e^- \longrightarrow H_2$	0.000	(reference electrode)
Cu	$Cu^{2+} + 2e^- \longrightarrow Cu$	$+0.337$	
I_2	$I_2 + 2e^- \longrightarrow 2I^-$	$+0.535$	
Hg	$Hg^{2+} + 2e^- \longrightarrow Hg$	$+0.789$	
Ag	$Ag^+ + e^- \longrightarrow Ag$	$+0.799$	
Br_2	$Br_2 + 2e^- \longrightarrow 2Br^-$	$+1.08$	
Cl_2	$Cl_2 + 2e^- \longrightarrow 2Cl^-$	$+1.360$	
Au	$Au^{3+} + 3e^- \longrightarrow Au$	$+1.50$	
F_2	$F_2 + 2e^- \longrightarrow 2F^-$	$+2.87$	

Left arrow (upward): Increasing strength as oxidizing agent; increasing ease of reduction.
Right arrow (upward): Increasing strength as reducing agent; increasing ease of oxidation.

1. Choose the appropriate half-reactions from a table of standard reduction potentials.

2. Write the equation for the half-reaction with the more positive (or less negative) E^0 value *for reduction* first, along with its potential.

3. Then write the equation for the other half-reaction *as an oxidation* and write its *oxidation potential*; to do this, reverse the tabulated reduction half-reaction, and change the sign of E^0. (Reversing a half-reaction or a complete reaction also changes the sign of its potential.)

4. Balance the electron transfer. *We do not multiply the potentials by the numbers used to balance the electron transfer!* The reason is that each potential represents a *tendency* for a reaction process to occur relative to the SHE; this does not depend on *how many times* it occurs. An electric potential is an *intensive property*.

5. Add the reduction and oxidation half-reactions, and add the reduction and oxidation potentials. E^0_{cell} will be *positive* for the resulting overall cell reaction. This indicates that the reaction as written is *product-favored (spontaneous)*. A *negative* E^0_{cell} value would indicate that the reaction is *reactant-favored (nonspontaneous)*.

For the cell described here, the Cu^{2+}/Cu couple has the more positive reduction potential, so we keep it as the reduction half-reaction and reverse the other half-reaction. Following the steps outlined, we obtain the equation for the spontaneous reaction.

$$Cu^{2+} + 2e^- \longrightarrow Cu \qquad\qquad +0.337\ V \longleftarrow \text{reduction potential}$$
$$\underline{\qquad\qquad Zn \longrightarrow Zn^{2+} + 2e^- \qquad\qquad +0.763\ V \longleftarrow \text{oxidation potential}}$$
$$Cu^{2+} + Zn \longrightarrow Cu + Zn^{2+} \qquad E^0_{cell} = +1.100\ V$$

The positive E^0_{cell} value tells us that the forward reaction is spontaneous at standard conditions. So we conclude that copper(II) ions oxidize metallic zinc to Zn^{2+} ions as they are reduced to metallic copper. (Section 21-9 shows that the potential of the standard zinc–copper voltaic cell is 1.100 volts. This is the spontaneous reaction that occurs.)

The reverse reaction has a negative E^0 and is nonspontaneous.

nonspontaneous
reaction: $\qquad\qquad Cu + Zn^{2+} \longrightarrow Cu^{2+} + Zn \qquad E^0_{cell} = -1.100\ \text{volts}$

To make it occur, we would have to supply electrical energy with a potential difference greater than 1.100 volts. That is, this nonspontaneous reaction would have to be carried out in an *electrolytic cell*.

EXAMPLE 21-3 *Predicting the Direction of Reactions*

At standard conditions, will chromium(III) ions, Cr^{3+}, oxidize metallic copper to copper(II) ions, Cu^{2+}, or will Cu^{2+} oxidize metallic chromium to Cr^{3+} ions? Write the cell reaction, and calculate E^0_{cell} for the spontaneous reaction.

Plan

We refer to the table of standard reduction potentials and choose the two appropriate half-reactions.

Solution

The copper half-reaction has the more positive reduction potential, so we write it first. Then we write the chromium half-reaction as an oxidation, balance the electron transfer, and add the two half-reactions and their potentials.

		E^0
$3(Cu^{2+} + 2e^- \longrightarrow Cu)$	(reduction)	+0.337 V
$\underline{2(Cr \longrightarrow Cr^{3+} + 3e^-)}$	(oxidation)	+0.74 V
$2Cr + 3Cu^{2+} \longrightarrow 2Cr^{3+} + 3Cu$		$E^0_{cell} = +1.08$ V

Because E^0_{cell} is positive, we know that the reaction is product-favored (spontaneous).

Cu^{2+} ions spontaneously oxidize metallic Cr to Cr^{3+} ions and are reduced to metallic Cu.

You should now work Exercise 54a.

 It is very important to remember to reverse the sign of the E^0 potential whenever you switch a half-reaction around, as with the chromium equation shown here. Also note that one does *not* multiply the E^0 potentials by the factors used to balance the number of electrons when adding the half-reactions together to give the overall balanced equation.

✓ Problem-Solving Tip: *The Sign of E^0 Indicates Spontaneity*

For a reaction that is spontaneous at *standard conditions*, E^0 must be positive. A negative value of E^0_{cell} indicates that the reverse of the reaction written would be spontaneous at standard conditions.

21-16 STANDARD ELECTRODE POTENTIALS FOR OTHER HALF-REACTIONS

In some half-cells, the oxidized and reduced species are both in solution as ions in contact with inert electrodes. For example, the standard iron(III) ion/iron(II) ion half-cell contains 1 M concentrations of the two ions. It involves the following half-reaction.

$$Fe^{3+} + e^- \longrightarrow Fe^{2+} \qquad E^0 = +0.771 \text{ V}$$

The standard dichromate ($Cr_2O_7^{2-}$) ion/chromium(III) ion half-cell consists of a 1 M concentration of each of the two ions in contact with an inert electrode. The balanced half-reaction in acidic solution (1.0 M H^+) is

$$Cr_2O_7^{2-} + 14H^+ + 6e^- \longrightarrow 2Cr^{3+} + 7H_2O \qquad E^0 = +1.33 \text{ V}$$

Standard reduction potentials for some other reactions are given in Table 21-3 and in Appendix J. These potentials can be used like those of Table 21-2.

Platinum metal is often used as the inert electrode material. These two standard half-cells could be shown in shorthand notation as

$Pt/Fe^{3+}(1 \ M), Fe^{2+}(1 \ M)$ and
$Pt/Cr_2O_7^{2-}(1 \ M), Cr^{3+}(1 \ M)$

 If H^+ appears anywhere in the half-reaction the solution is acidic. The presence of OH^- means that the solution is basic.

| **TABLE 21-3** | **Standard Reduction Potential for Selected Half-Cells** |

Reduction Half-Reaction	Standard Reduction Potential E^0 (volts)
$Zn(OH)_4^{2-} + 2e^- \longrightarrow Zn + 4OH^-$	−1.22
$Fe(OH)_2 + 2e^- \longrightarrow Fe + 2OH^-$	−0.877
$2H_2O + 2e^- \longrightarrow H_2 + 2OH^-$	−0.828
$PbSO_4 + 2e^- \longrightarrow Pb + SO_4^{2-}$	−0.356
$NO_3^- + H_2O + 2e^- \longrightarrow NO_2^- + 2OH^-$	+0.01
$Sn^{4+} + 2e^- \longrightarrow Sn^{2+}$	+0.15
$AgCl + e^- \longrightarrow Ag + Cl^-$	+0.222
$Hg_2Cl_2 + 2e^- \longrightarrow 2Hg + 2Cl^-$	+0.27
$O_2 + 2H_2O + 4e^- \longrightarrow 4OH^-$	+0.40
$NiO_2 + 2H_2O + 2e^- \longrightarrow Ni(OH)_2 + 2OH^-$	+0.49
$H_3AsO_4 + 2H^+ + 2e^- \longrightarrow H_3AsO_3 + H_2O$	+0.58
$Fe^{3+} + e^- \longrightarrow Fe^{2+}$	+0.771
$ClO^- + H_2O + 2e^- \longrightarrow Cl^- + 2OH^-$	+0.89
$NO_3^- + 4H^+ + 3e^- \longrightarrow NO + 2H_2O$	+0.96
$O_2 + 4H^+ + 4e^- \longrightarrow 2H_2O$	+1.229
$Cr_2O_7^{2-} + 14H^+ + 6e^- \longrightarrow 2Cr^{3+} + 7H_2O$	+1.33
$MnO_4^- + 8H^+ + 5e^- \longrightarrow Mn^{2+} + 4H_2O$	+1.507
$Cl_2 + 2e^- \longrightarrow 2Cl^-$	+1.360
$PbO_2 + HSO_4^{2-} + 3H^+ + 2e^- \longrightarrow PbSO_4 + 2H_2O$	+1.685

Increasing strength as oxidizing agent; increasing ease of reduction

Increasing strength as reducing agent; increasing ease of oxidation

EXAMPLE 21-4   *Predicting the Direction of Reactions*

In an acidic solution at standard conditions, will tin(IV) ions, Sn^{4+}, oxidize gaseous nitrogen oxide, NO, to nitrate ions, NO_3^-, or will NO_3^- oxidize Sn^{2+} to Sn^{4+} ions? Write the cell reaction and calculate E^0_{cell} for the spontaneous reaction.

Plan

We refer to the table of standard reduction potentials (see Table 21-3) and choose the appropriate half-reactions.

Solution

The NO_3^-/NO reduction half-reaction has the more positive E^0 value, so we write it first and write the Sn^{4+}/Sn^{2+} half-reaction as an oxidation. We balance the electron transfer and add the two half-reactions to obtain the equation for the *spontaneous* reaction. Then we add the half-reaction potentials to obtain the overall cell potential.

		E^0
$2(NO_3^- + 4H^+ + 3e^- \longrightarrow NO + 2H_2O)$	(reduction)	+0.96 V
$3(Sn^{2+} \longrightarrow Sn^{4+} + 2e^-)$	(oxidation)	−0.15 V
$2NO_3^- + 8H+ + 3Sn^{2+} \longrightarrow 2NO + 4H_2O + 3Sn^{4+}$		$E^0_{cell} = +0.81$ V

Because E^0_{cell} is positive for this reaction,

nitrate ions spontaneously oxidize tin(II) ions to tin(IV) ions and are reduced to nitrogen oxide in acidic solution.

You should now work Exercises 55 and 56.

 Pay careful attention to the conditions (acidic or basic) and the half-cell reactions mentioned in the problem. For example, there also is a half-reaction in Table 21-3 with NO_3^- being reduced to NO_2^-. This is not the correct half-reaction to use for this question.

✓ **Problem-Solving Tip:** *A Common Error in E^0_{cell} Calculations*

Remember the italicized warning in Step 4 of the procedure set out in Section 21-15. *We do not multiply the potentials by the numbers used to balance the electron transfer!* This is a very common error.

✓ **Problem-Solving Tip:** *Remember What We Mean by Standard Conditions*

When we say that a reaction takes place at *standard conditions*, we mean the following:

1. The temperature is the standard thermodynamic temperature, 25°C, unless stated otherwise.

2. All reactants and products are at *unit activity*. This means that

   a. Any solution species that takes part in the reaction is at a concentration of exactly 1 M;

   b. Any gas that takes part in the reaction is at a pressure of exactly 1 atm;

   c. Any other substance that takes part in the reaction is *pure*.

(When we say "takes part in the reaction" we mean either as a reactant or a product.) These are the same conditions that were described as *standard conditions* for thermodynamic purposes (Section 15-6). When one or more of these conditions is not satisfied, we must adjust our calculations for nonstandard conditions. We shall learn how to do this in Section 21-19.

Now that we know how to use standard reduction potentials, let's use them to explain the reaction that occurs in the electrolysis of aqueous NaCl. The first two electrolytic cells we considered involved *molten* NaCl and *aqueous* NaCl (see Sections 21-3 and 21-4). There was no doubt that in molten NaCl, metallic Na would be produced by reduction of Na$^+$, and gaseous Cl$_2$ would be produced by oxidation of Cl$^-$. But we found in aqueous NaCl, that H$_2$O, rather than Na$^+$, was reduced. This is consistent with the less negative reduction potential of H$_2$O, compared with Na$^+$.

$$E^0$$

$$2H_2O + 2e^- \longrightarrow H_2 + 2OH^- \qquad -0.828 \text{ V}$$
$$Na^+ + e^- \longrightarrow Na \qquad -2.714 \text{ V}$$

The more easily reduced species, H$_2$O, is reduced.

Electrode potentials measure only the relative *thermodynamic* likelihood for various half-reactions. In practice kinetic factors can complicate matters. For instance, sometimes the electrode process is limited by the rate of diffusion of dissolved species to or from the electrode surface. At some cathodes, the rate of electron transfer from the electrode to a reactant is the rate-limiting step, and a higher voltage (called *overvoltage*) must be applied to accomplish the reduction. As a result of these factors, a half-reaction that is *thermodynamically* more favorable than some other process still might not occur at a significant rate. In the electrolysis of NaCl(aq), Cl$^-$ is oxidized to Cl$_2$ gas (-1.360 V), instead of H$_2$O being oxidized to form O$_2$ gas (-1.229 V), because of the overvoltage of O$_2$ on Pt, the inert electrode.

21-17 CORROSION

Ordinary **corrosion** is the redox process by which metals are oxidized by oxygen, O$_2$, in the presence of moisture. There are other kinds, but this is the most common. The problem of corrosion and its prevention are of both theoretical and practical interest. Corrosion is responsible for the loss of billions of dollars annually in metal products. The mechanism of corrosion has been studied extensively. It is now known that the oxidation of metals occurs most readily at points of strain (where the metals are most "active"). Thus, a steel nail, which is mostly iron (Section 22-7), first corrodes at the tip and head (Figure 21-11). A bent nail corrodes most readily at the bend.

CD-ROM Screen 20.10, Corrosion.

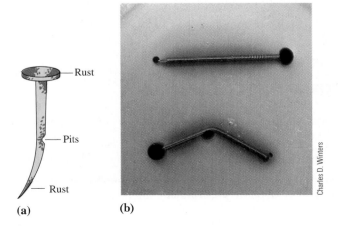

Rust

Pits

Rust

(a)   (b)

Figure 21-11 (a) A bent nail corrodes at points of strain and "active" metal atoms. (b) Two nails were placed in an agar gel that contained phenolphthalein and potassium ferricyanide, K$_3$[Fe(CN)$_6$]. As the nails corroded they produced Fe^{2+} ions at each end and at the bend. Fe^{2+} ions react with [Fe(CN)$_6$]$^{3-}$ ions to form Fe$_3$[Fe(CN)$_6$]$_2$, an intensely blue-colored compound. The rest of each nail is the cathode, at which water is reduced to H$_2$ and OH$^-$ ions. The OH$^-$ ions turn phenolphthalein pink.

Charles D. Winters

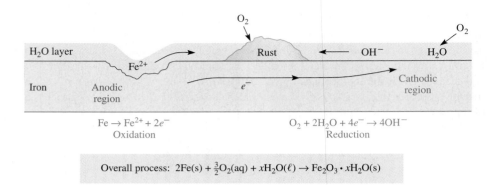

Overall process: $2Fe(s) + \frac{3}{2}O_2(aq) + xH_2O(\ell) \rightarrow Fe_2O_3 \cdot xH_2O(s)$

Figure 21-12   The corrosion of iron. Pitting appears at the anodic region, where iron metal is oxidized to Fe^{2+}. Rust appears at the cathodic region.

A point of strain in a steel object acts as an anode where the iron is oxidized to iron(II) ions, and pits are formed (Figure 21-12).

$$Fe \longrightarrow Fe^{2+} + 2e^- \qquad \text{(oxidation, anode)}$$

The electrons produced then flow through the nail to areas exposed to O_2. These act as cathodes where oxygen is reduced to hydroxide ions, OH^-.

$$O_2 + 2H_2O + 4e^- \longrightarrow 4OH^- \qquad \text{(reduction, cathode)}$$

At the same time, the Fe^{2+} ions migrate through the moisture on the surface. The overall reaction is obtained by balancing the electron transfer and adding the two half-reactions.

$2(Fe \longrightarrow Fe^{2+} + 2e^-)$	(oxidation, anode)
$O_2 + 2H_2O + 4e^- \longrightarrow 4OH^-$	(reduction, cathode)
$2Fe + O_2 + 2H_2O \longrightarrow 2Fe^{2+} + 4OH^-$	(net reaction)

The Fe^{2+} ions can migrate from the anode through the solution toward the cathode region, where they combine with OH^- ions to form iron(II) hydroxide. Iron is further oxidized by O_2 to the $+3$ oxidation state. The material we call rust is a complex hydrated form of iron(III) oxides and hydroxides with variable water composition; it can be represented as $Fe_2O_3 \cdot xH_2O$. The overall reaction for the rusting of iron is

$$2Fe(s) + \tfrac{3}{2}O_2(aq) + xH_2O(\ell) \longrightarrow Fe_2O_3 \cdot xH_2O(s)$$

21-18   CORROSION PROTECTION

There are several methods for protecting metals against corrosion. The most widely used are

1. Plating the metal with a thin layer of a less easily oxidized metal
2. Connecting the metal directly to a **sacrificial anode,** a piece of another metal that is more active and therefore preferentially oxidized

Salt (NaCl) accelerates the corrosion of most metals, especially steel, by making the water layer more electrically conductive. This makes it easier for the O_2 to oxidize the metal as it does not have to diffuse through the water layer to make physical contact with the metal surface. Salt does not, in itself, do any active corrosion.

Compare the potentials for the reduction half-reactions to see which metal is more easily oxidized. The more positive the reduction potential for a metal, the more stable the metal is as the free element and the harder it is to oxidize.

	$E^0_{reduction}$
$Mg^{2+} + 2e^- \longrightarrow Mg$	-2.37 V
$Zn^{2+} + 2e^- \longrightarrow Zn$	-0.763 V
$Fe^{2+} + 2e^- \longrightarrow Fe$	-0.44 V
$Sn^{2+} + 2e^- \longrightarrow Sn$	-0.14 V
$Cu^{2+} + 2e^- \longrightarrow Cu$	$+0.337$ V

Galvanized objects are steel coated with zinc to protect against corrosion.

Steel is plated with chromium for appearance as well as protection against corrosion.

Corrosion is an undesirable electrochemical reaction with very serious economic consequences.

3. Allowing a protective film, such as a metal oxide, to form naturally on the surface of the metal

4. Galvanizing, or coating steel with zinc, a more active metal

5. Applying a protective coating, such as paint

The thin layer of tin on tin-plated steel cans is less easily oxidized than iron, and it protects the steel underneath from corrosion. It is deposited either by dipping the can into molten tin or by electroplating. Copper is also less active than iron (see Table 21-2). It is sometimes deposited by electroplating to protect metals when food is not involved. Whenever the layer of tin or copper is breached, the iron beneath it corrodes even more rapidly than it would without the coating, because of the adverse electrochemical cell that is set up.

Figure 21-13a shows an iron pipe connected to a strip of magnesium, a more active metal, to protect the iron from oxidation. The magnesium is preferentially oxidized. It is called a "sacrificial anode." Similar methods are used to protect bridges and the hulls of ships from corrosion. Other inexpensive active metals, such as zinc, are also used as sacrificial anodes. Galvanizing (coating the iron with zinc), combines these two approaches. Even if the zinc coating is broken so the iron is exposed, the iron is not oxidized as long as it is in contact with the more easily oxidized zinc (Figure 21-13c).

Aluminum, a very active metal, reacts rapidly with O_2 from the air to form a surface layer of aluminum oxide, Al_2O_3, that is so thin that it is transparent. This very tough, hard substance is inert to oxygen, water, and most other corrosive agents in the environment. In this way, objects made of aluminum form their own protective layers and need not be treated further to inhibit corrosion.

SCi LINKS.

TOPIC: Corrosion
GO TO: www.scilinks.org
sciLINKS CODE: WCH2120

Acid rain endangers structural aluminum by dissolving this Al_2O_3 coating.

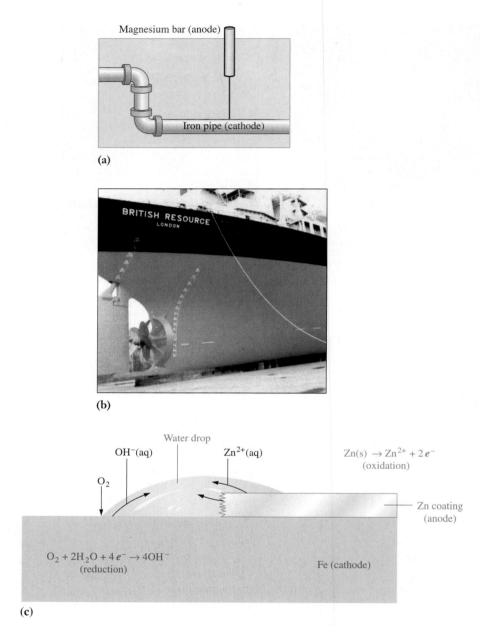

Figure 21-13   (a) Cathodic protection of buried iron pipe. A magnesium or zinc bar is oxidized instead of the iron. The "sacrificial" anode eventually must be replaced. (b) Cathodic protection of a ship's hull. The small yellow horizontal strips are blocks of titanium (coated with platinum) that are attached to the ship's hull. The hull is steel (mostly iron). When the ship is in salt water, the titanium blocks become the anode, and the hull the cathode, in a voltaic cell. Because oxidation always occurs at the anode, the ship's hull (the cathode) is protected from oxidation (corrosion). (c) Galvanizing is another method of corrosion protection. Even if the zinc coating is broken, it is still oxidized in preference to the less reactive iron as long as the two metals remain in contact.

EFFECT OF CONCENTRATIONS (OR PARTIAL PRESSURES) ON ELECTRODE POTENTIALS

21-19 THE NERNST EQUATION

Standard electrode potentials, designated E^0, refer to standard-state conditions. These standard-state conditions are one molar solutions for ions, one atmosphere pressure for gases, and all solids and liquids in their standard states at 25°C. (Remember that we refer to *thermodynamic* standard-state conditions, and not standard temperature and pressure as in gas law calculations.) As any of the standard cells described earlier operates, and concentrations or pressures of reactants change, the observed cell voltage drops. Similarly, cells constructed with solution concentrations different from one molar, or gas pressures different from one atmosphere, cause the corresponding potentials to deviate from standard electrode potentials.

The **Nernst equation** is used to calculate electrode potentials and cell potentials for concentrations and partial pressures other than standard-state values.

$$E = E^0 - \frac{2.303\, RT}{nF} \log Q$$

where

E = potential under the *nonstandard* conditions
E^0 = *standard* potential
R = gas constant, $8.314\ \text{J/mol} \cdot \text{K}$
T = absolute temperature in K
n = number of moles of electrons transferred in the reaction or half-reaction
F = faraday, $96{,}485\ \text{C/mol}\ e^- \times 1\ \text{J/(V} \cdot \text{C)} = 96{,}485\ \text{J/V} \cdot \text{mol}\ e^-$
Q = reaction quotient

In this equation the expression following the minus sign represents how much the *nonstandard* conditions cause the electrode potential to deviate from its standard value, E^0. The Nernst equation is normally presented in terms of base-10 logarithms, as we will do in this text.

The reaction quotient, Q, was introduced in Section 17-4. It involves a ratio of concentrations or pressures of products to those of reactants, each raised to the power indicated by the coefficient in the balanced equation. The Q expression that is used in the Nernst equation is the thermodynamic reaction quotient; it can include *both* concentrations and pressures. Substituting these values into the Nernst equation at 25°C gives

$$E = E^0 - \frac{0.0592}{n} \log Q \qquad \textit{(Note: in terms of base-10 log)}$$

At 25°C, the value of $\dfrac{2.303\, RT}{F}$ is 0.0592; at any other temperature, this term must be recalculated. Can you show that this term has the units $\text{V} \cdot \text{mol}$?

In general, half-reactions for standard reduction potentials are written

$$x\,\text{Ox} + ne^- \longrightarrow y\,\text{Red}$$

"Ox" refers to the oxidized species and "Red" to the reduced species; x and y are their coefficients, respectively, in the balanced equation. The Nernst equation for any *cathode* half-cell (*reduction* half-reaction) is

$$E = E^0 - \frac{0.0592}{n} \log \frac{[\text{Red}]^y}{[\text{Ox}]^x} \qquad \text{(reduction half-reaction)}$$

For the familiar half-reaction involving metallic zinc and zinc ions,

$$\text{Zn}^{2+} + 2e^- \longrightarrow \text{Zn} \qquad E^0 = -0.763 \text{ V}$$

the corresponding Nernst equation is

Metallic Zn is a pure solid, so its concentration does not appear in Q.

$$E = E^0 - \frac{0.0592}{2} \log \frac{1}{[\text{Zn}^{2+}]} \qquad \text{(for reduction)}$$

We substitute the E^0 value into the equation to obtain

$$E = -0.763 \text{ V} - \frac{0.0592}{2} \log \frac{1}{[\text{Zn}^{2+}]}$$

EXAMPLE 21-5　*The Nernst Equation*

Calculate the potential, E, for the $\text{Fe}^{3+}/\text{Fe}^{2+}$ electrode when the concentration of Fe^{2+} is exactly five times that of Fe^{3+}.

Plan

The Nernst equation lets us calculate potentials for concentrations other than one molar. The tabulation of standard reduction potentials gives us the value of E^0 for the reduction half-reaction. We use the balanced half-reaction and the given concentration ratio to calculate the value of Q. Then we substitute this into the Nernst equation with n equal to the number of moles of electrons involved in the half-reaction.

Solution

The reduction half-reaction is

$$\text{Fe}^{3+} + e^- \longrightarrow \text{Fe}^{2+} \qquad E^0 = +0.771 \text{ V}$$

This should make qualitative sense to you from LeChatelier's Principle. The more products that are present relative to reactants, the less the driving force (lower potential) to make more products.

We are told that the concentration of Fe^{2+} is five times that of Fe^{3+}, or $[\text{Fe}^{2+}] = 5[\text{Fe}^{3+}]$. Calculating the value of Q,

$$Q = \frac{[\text{Red}]^y}{[\text{Ox}]^x} = \frac{[\text{Fe}^{2+}]}{[\text{Fe}^{3+}]} = \frac{5[\text{Fe}^{3+}]}{[\text{Fe}^{3+}]} = 5$$

The balanced half-reaction shows one mole of electrons, or $n = 1$. Putting values into the Nernst equation,

$$E = E^0 - \frac{0.0592}{n} \log Q = +0.771 - \frac{0.0592}{1} \log 5 = (+0.771 - 0.041) \text{ V}$$

$$= \boxed{+0.730 \text{ V}}$$

You should now work Exercise 78.

The Nernst equation can be applied to balanced equations for redox reactions. One approach is to correct the reduction potential for each half-reaction to take into account the nonstandard concentrations or pressures.

EXAMPLE 21-6   *The Nernst Equation*

A cell is constructed at 25°C as follows. One half-cell consists of the Fe^{3+}/Fe^{2+} couple in which $[Fe^{3+}] = 1.00\ M$ and $[Fe^{2+}] = 0.100\ M$; the other involves the MnO_4^-/Mn^{2+} couple in acidic solution in which $[MnO_4^-] = 1.00 \times 10^{-2}\ M$, $[Mn^{2+}] = 1.00 \times 10^{-4}\ M$, and $[H^+] = 1.00 \times 10^{-3}\ M$. (a) Find the electrode potential for each half-cell with these concentrations, and (b) calculate the overall cell potential.

Plan

(a) We can apply the Nernst equation to find the reduction potential of each half-cell with the stated concentrations. (b) As in Section 21-15, we write the half-reaction with the more positive potential (*after* correction) along with its potential. We reverse the other half-reaction and change the sign of its E value. We balance the electron transfer and then add the half-reactions and their potentials to find the overall cell potential.

Solution

(a) For the MnO_4^-/Mn^{2+} half-cell *as a reduction*,

$$MnO_4^- + 8H^+ + 5e^- \longrightarrow Mn^{2+} + H_2O \qquad E^0 = +1.507\ V$$

$$E = E^0 - \frac{0.0592}{n} \log \frac{[Mn^{2+}]}{[MnO_4^-][H^+]^8}$$

$$= +1.507\ V - \frac{0.0592}{5} \log \frac{1.00 \times 10^{-4}}{(1.00 \times 10^{-2})(1.00 \times 10^{-3})^8}$$

$$= +1.507\ V - \frac{0.0592}{5} \log (1.00 \times 10^{22}) = +1.507\ V - \frac{0.0592}{5} (22.0)$$

$$= \boxed{+1.246\ V}$$

(b) For the Fe^{3+}/Fe^{2+} half-cell *as a reduction*,

$$Fe^{3+} + e^- \longrightarrow Fe^{2+} \qquad E^0 = +0.771\ V$$

$$E = E^0 - \frac{0.0592}{n} \log \frac{[Fe^{2+}]}{[Fe^{3+}]} = +0.771\ V - \frac{0.0592}{1} \log \frac{0.100}{1.00}$$

$$= +0.771\ V - \frac{0.0592}{1} \log (0.100) = +0.771\ V - \frac{0.0592}{1} (-1.00)$$

$$= \boxed{+0.830\ V}$$

The corrected potential for the MnO_4^-/Mn^{2+} half-cell is greater than that for the Fe^{3+}/Fe^{2+} half-cell, so we reverse the latter, balance the electron transfer, and add.

	E (corrected)
$MnO_4^- + 8H^+ + 5e^- \longrightarrow Mn^{2+} + 4H_2O$	+1.246 V
$5(Fe^{2+} \longrightarrow Fe^{3+} + e^-)$	−0.830 V
$MnO_4^- + 8H^+ + 5Fe^{2+} \longrightarrow Mn^{2+}\ 4H_2O + 5Fe^{3+}$	$E_{cell} = 0.416\ V$

The reaction in Example 21-6 is product-favored (spontaneous) under the stated conditions, with a potential of +0.416 volt *when the cell starts operation.* As the cell discharges and current flows, the product concentrations, $[Mn^{2+}]$ and $[Fe^{3+}]$, increase. At the same time, reactant concentrations, $[MnO_4^-]$, $[H^+]$, and $[Fe^{2+}]$, decrease. This increases

CD-ROM Screen 20.7, Electrochemical Cells at Nonstandard Conditions.

$\log Q_{cell}$, so the correction factor becomes more negative. Thus, the overall E_{cell} *decreases* (the reaction becomes less favorable). Eventually the cell potential approaches zero (equilibrium), and the cell "runs down." The cell is completely run down ($E_{cell} = 0$) when the term $\dfrac{0.0592}{n} \log Q_{cell}$ is equal in magnitude to E^0_{cell}.

We can also find the cell potential for a nonstandard cell by first finding E^0_{cell} or the overall standard cell reaction, and then using the Nernst equation to correct for nonstandard concentrations. The next example illustrates this approach.

EXAMPLE 21-7   *The Nernst Equation*

A cell is constructed at 25°C as follows. One half-cell consists of a chlorine/chloride, Cl_2/Cl^-, electrode with the partial pressure of $Cl_2 = 0.100$ atm and $[Cl^-] = 0.100$ *M*. The other half-cell involves the MnO_4^-/Mn^{2+} couple in acidic solution with $[MnO_4^-] = 0.100$ *M*, $[Mn^{2+}] = 0.100$ *M*, and $[H^+] = 0.100$ *M*. Apply the Nernst equation to the overall cell reaction to determine the cell potential for this cell.

Plan

First we determine the overall cell reaction and its *standard* cell potential, E^0_{cell}, as in Examples 21-3 and 21-4. Then we apply the Nernst equation to the overall cell.

Solution

The MnO_4^-/Mn^{2+} half-reaction has the more positive reduction potential, so we write it first. Then we write the Cl_2/Cl^- half-reaction as an oxidation, balance the electron transfer, and add the two half-reactions and their potentials to obtain the overall cell reaction and its E^0_{cell}.

	E^0
$2(MnO_4^- + 8H^+ + 5e^- \longrightarrow Mn^{2+} + 4H_2O)$	$+1.507$ V
$5(2Cl^- \longrightarrow Cl_2 + 2e^-)$	-1.360 V
$2MnO_4^- + 16H^+ + 10Cl^- \longrightarrow 2Mn^{2+} + 8H_2O + 5Cl_2$	$E^0_{cell} = +0.147$ V

In the overall reaction, $n = 10$. We then apply the Nernst equation to this overall reaction by substituting appropriate concentration and partial pressure values. Because Cl_2 is a gaseous component, its term in the Nernst equation involves its partial pressure, P_{Cl_2}, in atm.

$$E_{cell} = E^0_{cell} - \frac{0.0592}{n} \log \frac{[Mn^{2+}]^2 (P_{Cl_2})^5}{[MnO_4^-]^2 [H^+]^{16} [Cl^-]^{10}}$$

Remember that, in evaluating Q in the Nernst equation, (1) molar concentrations are used for dissolved species, and (2) partial pressures of gases are expressed in atmospheres.

$$= 0.147 \text{ V} - \frac{0.0592}{10} \log \frac{(0.100)^2 (0.100)^5}{(0.100)^2 (0.100)^{16} (0.100)^{10}}$$

$$= 0.147 \text{ V} - \frac{0.0592}{10} \log(1.00 \times 10^{21})$$

$$= 0.147 \text{ V} - \frac{0.0592}{10} (21.00) = \boxed{0.017 \text{ V}}$$

You should now work Exercises 80 and 84.

Now solve Example 21-6 by applying the Nernst equation to the *overall* cell reaction to determine the cell potential.

The method illustrated in Example 21-7, applying the Nernst equation to the *overall* cell reaction, usually involves less calculation than correcting the separate half-reactions as in Example 21-6. We interpret our results as follows: The positive cell potentials in Examples 21-6 and 21-7 tell us that each of these cell reactions is spontaneous *in the direction written,* for the concentrations given. If the resulting cell potential were negative, the *reverse* reaction would be favored at those concentrations. We could then reverse the equation for the overall cell reaction and change the sign of its potential to describe the spontaneous operation of the cell.

✓ Problem-Solving Tip: *Be Careful of the Value of n*

How do you know what value of n to use? Remember that n must be the number of moles of electrons transferred in the *balanced* equation for the process to which you apply the Nernst equation.

1. For a *half-reaction*, n represents the number of moles of electrons in that half-reaction. In Example 21-6 we applied the Nernst equation to each half-reaction separately, so we used $n = 5$ for the half-reaction

$$MnO_4^- + 8H^+ + 5e^- \longrightarrow Mn^{2+} + H_2O$$

and we used $n = 1$ for the half-reaction

$$Fe^{3+} + e^- \longrightarrow Fe^{2+}$$

2. For an *overall reaction*, n represents the total number of moles of electrons transferred. In Example 21-7 we applied the Nernst equation to an *overall* reaction in which 10 moles of electrons was transferred from 10 moles of Cl^- to 2 moles of MnO_4^-, so we used the value $n = 10$.

21-20 USING ELECTROCHEMICAL CELLS TO DETERMINE CONCENTRATIONS

We can apply the ideas of the preceding section to *measure* the voltage of a cell and then use the Nernst equation to solve for an *unknown* concentration. The following example illustrates such an application.

A pH meter uses the voltage of a cell to measure the H^+ concentration in a solution. Each change of one pH unit causes a voltage change of 0.0592 volts.

This cell is similar to the zinc–hydrogen cell that we discussed in Section 21-12, except that the hydrogen concentration is not (necessarily) 1.00 M.

EXAMPLE 21-8 *The Nernst Equation*

We construct an electrochemical cell at 25°C as follows. One half-cell is a standard Zn^{2+}/Zn cell, that is, a strip of zinc immersed in a 1.00 M Zn^{2+} solution; the other is a *nonstandard* hydrogen electrode in which a platinum electrode is immersed in a solution of *unknown* hydrogen ion concentration with gaseous hydrogen bubbling through it at a pressure of 1.000 atm. The observed cell voltage is 0.522 V. (a) Calculate the value of the reaction quotient Q. (b) Calculate $[H^+]$ in the second half-cell. (c) Determine the pH of the solution in the second half-cell.

Plan

We saw in Section 21-12 that the zinc–hydrogen cell operated with oxidation at the zinc electrode and reduction at the hydrogen electrode, with a *standard* cell potential of 0.763 V.

$$\text{overall:} \quad Zn + 2H^+ \longrightarrow Zn^{2+} + H_2 \qquad E^0_{cell} = 0.763 \text{ V}$$

(a) We rearrange the Nernst equation to solve for the reaction quotient, Q, from the measured cell voltage and $n = 2$. (b) We substitute concentrations and partial pressures in the expression for Q. Then we can solve for the only unknown, $[H^+]$. (c) The pH can be determined from the $[H^+]$ determined in Part (b).

Solution

(a)
$$E_{cell} = E_{cell}^0 - \frac{0.0592}{n} \log Q$$

Substituting and solving for Q,

$$0.522\ V = 0.763\ V - \frac{0.0592}{2} \log Q$$

$$\frac{0.0592\ V}{2} \log Q = (0.763 - 0.522)\ V = 0.241\ V$$

$$\log Q = \frac{(2)(0.241\ V)}{0.0592\ V} = 8.14$$

$$Q = 10^{8.14} = \boxed{1.4 \times 10^8}$$

(b) We write the expression for Q from the balanced overall equation, and solve for $[H^+]$.

$$Q = \frac{[Zn^{2+}]P_{H_2}}{[H^+]^2}$$

$$[H^+]^2 = \frac{[Zn^2]^+ P_{H_2}}{Q} = \frac{(1.00)(1.00)}{1.4 \times 10^8} = 7.1 \times 10^{-9}$$

$$[H^+] = \boxed{8.4 \times 10^{-5}\ M}$$

(c)
$$pH = -\log [H^+] = -\log (8.4 \times 10^{-5}) = \boxed{4.08}$$

You should now work Exercises 81 and 82.

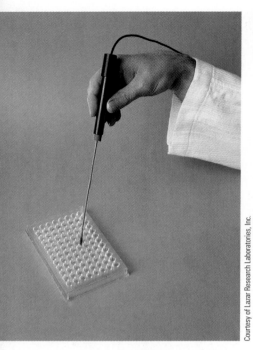

Microelectrodes have been developed to measure concentrations in very small volumes of solution.

Electrochemical procedures that use the principles illustrated here provide a convenient method for making many concentration measurements.

Enrichment   Concentration Cells

As we have seen, different concentrations of ions in a half-cell result in different half-cell potentials. We can use this idea to construct a **concentration cell,** in which both half-cells are composed of the same species, but in different ion concentrations. Suppose we set up such a cell using the Cu^{2+}/Cu half-cell that we introduced in Section 21-9. We put copper electrodes into two aqueous solutions, one that is 0.10 M $CuSO_4$ and another that is 1.00 M $CuSO_4$. To complete the cell construction, we connect the two electrodes with a wire and join the two solutions with a salt bridge as usual (Figure 21-14). Now the relevant reduction half-reaction in either half-cell is

$$Cu^{2+} + 2e^- \longrightarrow Cu \qquad E^0 = +0.337\ V$$

Thus the Cu^{2+} ions in the more concentrated half-cell can be considered as the reactant, and those in the more dilute cell as the product.

$$Cu^{2+}(1.00\ M) \longrightarrow Cu^{2+}(0.10\ M)$$

The overall cell potential can be calculated by applying the Nernst equation to the overall cell reaction. We must first find E^0, the standard cell potential *at standard concentrations;*

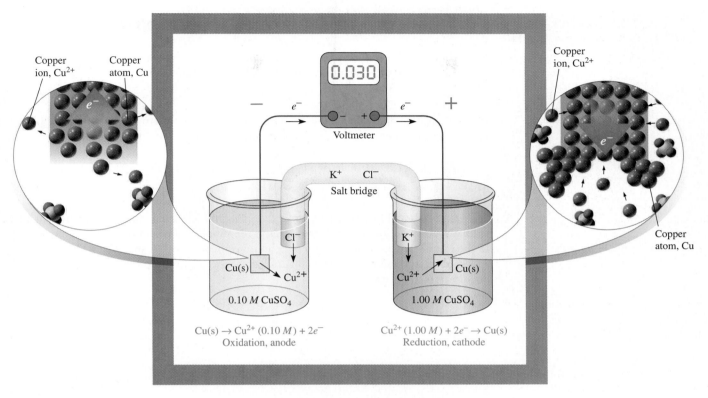

Figure 21-14 The concentration cell $Cu|Cu^{2+}(0.10\ M)\|Cu^{2+}(1.00\ M)|Cu$. The overall reaction lowers the $[Cu^{2+}]$ concentration in the more concentrated solution and increases it in the more dilute solution.

because the same electrode and the same type of ions are involved in both half-cells, this E^0 is always zero. Thus,

$$E_{\text{cell}} = E^0 - \frac{0.0592}{n}\log\frac{[\text{dilute solution}]}{[\text{concentrated solution}]}$$

$$= 0 - \frac{0.0592}{2}\log\frac{0.10}{1.00} = +0.030\ \text{V}$$

The overall cell potential is positive; the reaction is spontaneous as written.

   As the reaction proceeds, $[Cu^{2+}]$ decreases in the more concentrated half-cell and increases in the more dilute half-cell until the two concentrations are equal; at that point $E_{\text{cell}} = 0$, and equilibrium has been reached. This equilibrium $[Cu^{2+}]$ is the same concentration that would have been formed if we had just mixed the two solutions directly to obtain a solution of intermediate concentration.

> In any concentration cell, the spontaneous reaction is always in the direction that equalizes the concentrations.

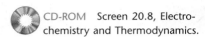

CD-ROM   Screen 20.8, Electrochemistry and Thermodynamics.

21-21 THE RELATIONSHIP OF E^0_{cell} TO ΔG^0 AND K

In Section 17-12 we studied the relationship between the standard Gibbs free energy change, ΔG^0, and the thermodynamic equilibrium constant, K.

$$\Delta G^0 = -RT \ln K$$

There is also a simple relationship between ΔG^0 and the standard cell potential, E^0_{cell}, for a redox reaction (reactants and products in standard states).

$$\Delta G^0 = -nFE^0_{cell}$$

ΔG^0 can be thought of as the *negative of the maximum electrical work* that can be obtained from a redox reaction. In this equation, n is the number of moles of electrons transferred in the overall process (mol e^-/mol rxn), and F is the faraday, 96,485 J/V·mol e^-.

Combining these relationships for ΔG^0 gives the relationship between E^0_{cell} values and equilibrium constants.

$$\underbrace{-nFE^0_{cell}}_{\Delta G^0} = \underbrace{-RT \ln K}_{\Delta G^0}$$

After multiplying by -1, we can rearrange.

$$nFE^0_{cell} = RT \ln K \quad \text{or} \quad E^0_{cell} = \frac{RT \ln K}{nF} \quad \text{or} \quad \ln K = \frac{nFE^0_{cell}}{RT}$$

If any one of the three quantities ΔG^0, K, and E^0_{cell} is known, the other two can be calculated using these equations. It is usually much easier to determine K for a redox reaction from electrochemical measurements than by measuring equilibrium concentrations directly, as described in Chapter 17. Keep in mind the following for all redox reactions at *standard conditions*.

Forward Reaction	ΔG^0	K	E^0_{cell}	
product-favored (spontaneous)	$-$	>1	$+$	
at equilibrium	0	1	0	(all substances at *standard conditions*)
reactant-favored (nonspontaneous)	$+$	<1	$-$	

EXAMPLE 21-9 *Calculation of ΔG^0 from Cell Potentials*

Calculate the standard Gibbs free energy change, ΔG^0, in J/mol at 25°C for the following reaction from standard electrode potentials.

$$3Sn^{4+} + 2Cr \longrightarrow 3Sn^{2+} + 2Cr^{3+}$$

Plan

We evaluate the standard cell potential as we have done before. Then we apply the relationship $\Delta G^0 = -nFE^0_{cell}$.

Recall from Chapter 15 that ΔG^0 can be expressed in joules per *mole of reaction*. Here we ask for the number of joules of free energy change that corresponds to the reaction of 2 moles of chromium with 3 moles of tin(IV) to give 3 moles of tin(II) ions and 2 moles of chromium(III) ions.

Solution

The standard reduction potential for the Sn^{4+}/Sn^{2+} couple is $+0.15$ volt; that for the Cr^{3+}/Cr couple is -0.74 volt. The equation for the reaction shows Cr being oxidized to Cr^{3+}, so the sign of the E^0 value for the Cr^{3+}/Cr couple is reversed. The overall reaction, the sum of the two half-reactions, has a cell potential equal to the sum of the two half-reaction potentials.

$$
\begin{array}{ll}
 & E^0 \\
\hline
3(Sn^{4+} + 2e^- \longrightarrow Sn^{2+}) & +0.15 \text{ V} \\
2(Cr \longrightarrow Cr^{3+} + 3e^-) & -(-0.74 \text{ V}) \\
\hline
3Sn^{4+} + 2Cr \longrightarrow 3Sn^{2+} + 2Cr^{3+} & E^0_{cell} = +0.89 \text{ V}
\end{array}
$$

The positive value of E^0_{cell} indicates that the forward reaction is spontaneous.

$$\Delta G^0 = -nFE^0_{cell} = -\left(\frac{6 \text{ mol } e^-}{\text{mol rxn}}\right)\left(\frac{9.65 \times 10^4 \text{J}}{\text{V} \cdot \text{mol } e^-}\right)(+0.89 \text{ V})$$

$$= -5.2 \times 10^5 \text{ J/mol rxn} \quad \text{or} \quad -5.2 \times 10^2 \text{ kJ/mol rxn}$$

You should now work Exercise 95.

The quite negative value of ΔG^0 tells us that the reaction is product-favored. This tells us nothing about the speed with which the reaction would occur.

EXAMPLE 21-10 *Calculation of K from Cell Potentials*

Use the standard cell potential to calculate the value of the equilibrium constant, K, at 25°C for the following reaction.

$$2Cu + PtCl_6^{2-} \longrightarrow 2Cu^+ + PtCl_4^{2-} + 2Cl^-$$

Plan

We calculate E^0_{cell} for the reaction as written. Then we use it to calculate K.

Solution

First we find the appropriate half-reactions. Cu is oxidized to Cu^+, so we write the Cu^+/Cu couple as an oxidation and reverse the sign of its tabulated E^0 value. We balance the electron transfer and then add the half-reactions. The resulting E^0_{cell} value can be used to calculate the equilibrium constant, K, for the reaction *as written*.

$$
\begin{array}{ll}
2(Cu \longrightarrow Cu^+ + e^-) & -(+0.521 \text{ V}) \\
PtCl_6^{2-} + 2e^- \longrightarrow PtCl_4^{2-} + 2Cl^- & +0.68 \text{ V} \\
\hline
2Cu + PtCl_6^{2-} \longrightarrow 2Cu^+ + PtCl_4^{2-} + 2Cl^- & E^0_{cell} = +0.16 \text{ V}
\end{array}
$$

As the problem is stated, we must keep the equation as written. We must therefore accept either a positive or a negative value of E^0_{cell}. A negative value of E^0_{cell} would lead to $K < 1$.

Then we calculate K.

$$\ln K = \frac{nFE^0_{cell}}{RT} = \frac{(2)(9.65 \times 10^4 \text{ J/V} \cdot \text{mol})(+0.16 \text{ V})}{(8.314 \text{ J/mol} \cdot \text{K})(298 \text{ K})} = 12.5$$

$$K = e^{12.5} = 2.7 \times 10^5$$

At equilibrium, $\quad K = \dfrac{[Cu^+]^2[PtCl_4^{2-}][Cl^-]^2}{[PtCl_6^{2-}]} = 2.7 \times 10^5.$

The forward reaction is product-favored (spontaneous), and the equilibrium lies far to the right.

You should now work Exercises 96 and 98.

PRIMARY VOLTAIC CELLS

CD-ROM   Screen 20.9,
Batteries.

As any voltaic cell produces current (*discharges*), chemicals are consumed. **Primary voltaic cells** cannot be "recharged." Once the chemicals have been consumed, further chemical action is not possible. The electrolytes or electrodes (or both) cannot be regenerated by reversing the current flow through the cell using an external direct current source. The most familiar examples of primary voltaic cells are the ordinary "dry" cells that are used as energy sources in flashlights and other small appliances.

21-22 DRY CELLS

The first dry cell was patented by Georges Leclanché (1839–1882) in 1866. The container of this dry cell, made of zinc, also serves as one of the electrodes (Figure 21-15). The other electrode is a carbon rod in the center of the cell. The zinc container is lined with porous paper to separate it from the other materials of the cell. The rest of the cell is filled with a moist mixture (the cell is *not* really dry) of ammonium chloride (NH_4Cl), manganese(IV) oxide (MnO_2), zinc chloride ($ZnCl_2$), and a porous, inert filler. Dry cells are sealed to keep the moisture from evaporating. As the cell operates (the electrodes must be connected externally), the metallic Zn is oxidized to Zn^{2+}, and the liberated electrons flow along the container to the external circuit. Thus, the zinc electrode is the anode (negative electrode).

TOPIC: Batteries
GO TO: www.scilinks.org
*sci***LINKS CODE:** WCH2130

$$Zn \longrightarrow Zn^{2+} + 2e^- \quad \text{(oxidation, anode)}$$

The carbon rod is the cathode, at which ammonium ions are reduced.

$$2NH_4^+ + 2e^- \longrightarrow 2NH_3 + H_2 \quad \text{(reduction, cathode)}$$

Addition of the half-reactions gives the overall cell reaction

$$Zn + 2NH_4^+ \longrightarrow Zn^{2+} + 2NH_3 + H_2 \qquad E_{cell} = 1.6 \text{ V}$$

(a)

(b)

Figure 21-15 (a) The Leclanché cell is a dry cell that generates a potential difference of about 1.6 volts. (b) Some commercial alkaline dry cells.

As H_2 is formed, it is oxidized by MnO_2 in the cell. This prevents collection of H_2 gas on the cathode, which would stop the reaction.

$$H_2 + 2MnO_2 \longrightarrow 2MnO(OH)$$

The ammonia produced at the cathode combines with zinc ions and forms a soluble compound containing the complex ions, $[Zn(NH_3)_4]^{2+}$.

$$Zn^{2+} + 4NH_3 \longrightarrow [Zn(NH_3)_4]^{2+}$$

The buildup of reaction products at an electrode is called *polarization* of the electrode.

This reaction reduces polarization due to the buildup of ammonia, and it keeps the concentration of Zn^{2+} from increasing substantially, which would decrease the cell potential by allowing it to diffuse away from the anode. Under heavy current conditions, NH_3 gas can build up in an ordinary dry cell slowing down the migration of the NH_4^+ ions to the cathode. This can cause a temporary reduction of the battery voltage and current making it appear as if the battery is prematurely dying. On sitting, however, the NH_3 gas has time to react with the Zn^{2+} ions to form $[Zn(NH_3)_4]^{2+}$, which can diffuse away more readily from the cathode. Thus, a regular battery can often regain full power after sitting for a while after a period of heavy use. Once the reactant chemicals are sufficiently depleted, however, the battery is "dead."

Alkaline dry cells are similar to Leclanché dry cells except that (1) the electrolyte is basic (alkaline) because it contains KOH, and (2) the interior surface of the Zn container is rough; this gives a larger surface area. Alkaline cells have a longer shelf life than ordinary dry cells, and they stand up better under heavy use. One gets considerably better performance from an alkaline battery by eliminating the NH_3 gas production that occurs in a regular NH_4Cl-based dry cell. The voltage of an alkaline cell is about 1.5 volts.

During discharge, the alkaline dry cell reactions are

$$\begin{array}{ll} Zn(s) + 2OH^-(aq) \longrightarrow Zn(OH)_2(s) + 2e^- & \text{(anode)} \\ 2MnO_2(s) + 2H_2O(\ell) + 2e^- \longrightarrow 2MnO(OH)(s) + 2OH^-(aq) & \text{(cathode)} \\ \hline Zn(s) + 2MnO_2(s) + 2H_2O(\ell) \longrightarrow Zn(OH)_2(s) + 2MnO(OH)(s) & \text{(overall)} \end{array}$$

Outer steel case

Steel outer top

Tin-plated inner top

Sealing and insulating gasket

Zn (anode)

KOH saturated with ZnO in absorbent material (electrolyte)

HgO mixed with graphite (cathode)

Gas vent

Barrier

Inner steel case

Courtesy of Eveready Battery Company

The mercury battery of the type frequently used in watches, calculators, and hearing aids is a primary cell. Although mercury in the water supply is known to cause health problems, no conclusive evidence has been found that the disposal of household batteries contributes to such problems. Nevertheless, manufacturers are working to decrease the amount of mercury in batteries. In recent years, the amount of mercury in alkaline batteries decreased markedly; at the same time, the life of such batteries has increased dramatically.

SECONDARY VOLTAIC CELLS

In **secondary voltaic cells,** or *reversible cells,* the original reactants can be regenerated. This is done by passing a direct current through the cell in the direction opposite of the discharge current flow. This process is referred to as *charging,* or recharging, a cell or battery. The most common example of a secondary voltaic cell is the lead storage battery, used in most automobiles.

21-23   THE LEAD STORAGE BATTERY

The lead storage battery is depicted in Figure 21-16. One group of lead plates contains compressed spongy lead. These alternate with a group of lead plates composed of lead(IV) oxide, PbO_2. The electrodes are immersed in a solution of about 40% sulfuric acid. When the cell discharges, the spongy lead is oxidized to lead ions, and the lead plates accumulate a negative charge.

$$Pb \longrightarrow Pb^{2+} + 2e^- \qquad \text{(oxidation)}$$

The lead ions then react with hydrogen sulfate ions from the sulfuric acid to form insoluble lead(II) sulfate. This begins to coat the lead electrode.

$$Pb^{2+} + HSO_4^- \longrightarrow PbSO_4(s) + H^+ \qquad \text{(precipitation)}$$

Thus, the net process at the anode *during discharge* is

$$Pb + HSO_4^- \longrightarrow PbSO_4(s) + H^+ + 2e^- \qquad \text{(anode during discharge)}$$

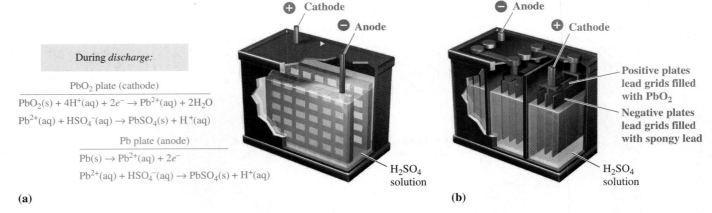

During *discharge:*

PbO$_2$ plate (cathode)

$PbO_2(s) + 4H^+(aq) + 2e^- \rightarrow Pb^{2+}(aq) + 2H_2O$
$Pb^{2+}(aq) + HSO_4^-(aq) \rightarrow PbSO_4(s) + H^+(aq)$

Pb plate (anode)

$Pb(s) \rightarrow Pb^{2+}(aq) + 2e^-$
$Pb^{2+}(aq) + HSO_4^-(aq) \rightarrow PbSO_4(s) + H^+(aq)$

(a)

(b)

Figure 21-16   (a) A schematic representation of one cell of a lead storage battery. The reactions shown are those taking place during the *discharge* of the cell. Alternate lead grids are packed with spongy lead and lead(IV) oxide. The grids are immersed in a solution of sulfuric acid, which serves as the electrolyte. To provide a large reacting surface, each cell contains several connected grids but for clarity, only one of each is shown in (a). Such a cell generates a voltage of about 2 volts. Six of these cells are connected together in series, although only three are shown in (b), so that their voltages add to make a 12-volt battery.

The electrons travel through the external circuit and re-enter the cell at the PbO_2 electrode, which is the cathode during discharge. Here, in the presence of hydrogen ions, the lead(IV) oxide is reduced to lead(II) ions, Pb^{2+}. These ions also react with HSO_4^- ions from the H_2SO_4 to form an insoluble $PbSO_4$ coating on the lead(IV) oxide electrode.

$$PbO_2 + 4H^+ + 2e^- \longrightarrow Pb^{2+} + 2H_2O \quad \text{(reduction)}$$
$$\underline{Pb^{2+} + HSO_4^- \longrightarrow PbSO_4(s) + H^+ \quad \text{(precipitation)}}$$
$$PbO_2 + 3H^+ + HSO_4^- + 2e^- \longrightarrow PbSO_4(s) + 2H_2O \quad \text{(cathode during discharge)}$$

The net cell reaction for discharge and its standard potential are obtained by adding the net anode and cathode half-reactions and their tabulated potentials. The tabulated E^0 value for the anode half-reaction is reversed in sign because it occurs as oxidation during discharge.

$$E^0$$

$$Pb + HSO_4^- \longrightarrow PbSO_4(s) + H^+ + 2e^- \quad -(-0.356 \text{ V})$$
$$\underline{PbO_2 + 3H^+ + HSO_4^- + 2e^- \longrightarrow PbSO_4(s) + 2H_2O \quad +1.685 \text{ V}}$$
$$Pb + PbO_2 + 2H^+ + 2HSO_4^- \longrightarrow 2PbSO_4(s) + 2H_2O \quad E^0_{cell} = +2.041 \text{ V}$$

$$\underbrace{2H_2SO_4}$$

One cell creates a potential of about 2 volts. Automobile 12-volt batteries have six cells connected in series. The potential declines only slightly during use, because solid reagents are being consumed. As the cell is used, some H_2SO_4 is consumed, lowering its concentration.

When a potential slightly greater than the potential the battery can generate is imposed across the electrodes, the current flow can be reversed. The battery can then be recharged by reversing all the reactions. The alternator or generator applies this potential when the engine is in operation. The reactions that occur in a lead storage battery are summarized as follows.

$$Pb + PbO_2 + 2H^+ + 2HSO_4^- \underset{\text{charge}}{\overset{\text{discharge}}{\rightleftharpoons}} 2PbSO_4(s) + 2H_2O$$

During many repeated charge–discharge cycles, some of the $PbSO_4$ falls to the bottom of the container and the H_2SO_4 concentration becomes correspondingly low. Eventually the battery cannot be recharged fully. It can be traded in for a new one, and the lead can be recovered and reused to make new batteries.

The decrease in the concentration of sulfuric acid provides an easy method for measuring the degree of discharge, because the density of the solution decreases accordingly. We simply measure the density of the solution with a hydrometer.

A *generator* supplies direct current (dc). An *alternator* supplies alternating current (ac), so an electronic rectifier is used to convert this to direct current for the battery.

This is one of the oldest and most successful examples of recycling.

21-24 THE NICKEL–CADMIUM (NICAD) CELL

The nickel–cadmium (nicad) cell has gained widespread popularity because it can be recharged. It thus has a much longer useful life than ordinary (Leclanché) dry cells. Nicad batteries are used in electronic toys, camcorders, and photographic equipment.

When the battery is delivering power the anode is cadmium, and the cathode is nickel(IV) oxide. The electrolytic solution is basic. The "discharge" reactions that occur in a nicad battery are

$$Cd(s) + 2OH^-(aq) \longrightarrow Cd(OH)_2(s) + 2e^- \quad \text{(anode)}$$
$$\underline{NiO_2(s) + 2H_2O(\ell) + 2e^- \longrightarrow Ni(OH)_2(s) + 2OH^-(aq) \quad \text{(cathode)}}$$
$$Cd(s) + NiO_2(s) + 2H_2O(\ell) \longrightarrow Cd(OH)_2(s) + Ni(OH)_2(s) \quad \text{(overall)}$$

To see why a nicad battery produces a constant voltage, write the Nernst equation for its reaction. Look at Q.

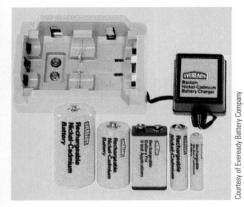

Courtesy of Eveready Battery Company

Rechargeable nicad batteries are used to operate many electrical devices.

SCi
LINKS.

TOPIC: Fuel Cells
GO TO: www.scilinks.org
*sci*LINKS CODE: WCH2140

The efficiency of energy conversion of the fuel cell operation is 60–70% of the theoretical maximum (based on ΔG). This represents about twice the efficiency that can be realized from burning hydrogen in a heat engine coupled to a generator.

Nathan D. Lewis/California Institute of Technology

Gaseous H_2 is produced from H_2O at an illuminated photoelectrode. Light from the sun may soon be used to produce hydrogen, the ultimate clean-burning fuel.

The solid reaction product at each electrode adheres to the electrode surface. Hence, a nicad battery can be recharged by an external source of electricity; that is, the electrode reactions can be reversed. Because no gases are produced by the reactions in a nicad battery, the unit can be sealed. The voltage of a nicad cell is about 1.4 volts, slightly less than that of a Leclanché cell. The toxicity of cadmium and the limited number of recharges that a nicad battery can handle before deactivating are problems that newer battery technologies are addressing. Nickel-metal-hydride and lithium–ion batteries are two newer generation cells that are environmentally friendlier, have longer lifetimes, and higher power to weight ratios.

21-25 THE HYDROGEN–OXYGEN FUEL CELL

Fuel cells are voltaic cells in which the reactants are continuously supplied to the cell and the products are continuously removed. The hydrogen–oxygen fuel cell (Figure 21-17) already has many applications. It is used in spacecraft to supplement the energy obtained from solar cells. Liquid H_2 is carried on board as a propellant. The boiled-off H_2 vapor that ordinarily would be lost is used in a fuel cell to generate electrical power.

Hydrogen (the fuel) is supplied to the anode compartment. Oxygen is fed into the cathode compartment. The diffusion rates of the gases into the cell are carefully regulated for maximum efficiency. Oxygen is reduced at the cathode, which consists of porous carbon impregnated with a finely divided Pt or Pd catalyst.

$$O_2 + 2H_2O + 4e^- \xrightarrow{\text{catalyst}} 4OH^- \qquad \text{(cathode)}$$

The OH^- ions migrate through the electrolyte to the anode, an aqueous solution of a base. The anode is also porous carbon containing a small amount of catalyst (Pt, Ag, or CoO). Here H_2 is oxidized to H_2O.

$$H_2 + 2OH^- \longrightarrow 2H_2O + 2e^- \qquad \text{(anode)}$$

The net reaction is obtained from the two half-reactions.

$$\begin{array}{ll} O_2 + 2H_2O + 4e^- \longrightarrow 4OH^- & \text{(cathode)} \\ 2(H_2 + 2OH^- \longrightarrow 2H_2O + 2e^-) & \text{(anode)} \\ \hline 2H_2 + O_2 \longrightarrow 2H_2O & \text{(net cell reaction)} \end{array}$$

The net reaction is the same as the burning of H_2 in O_2 to form H_2O, but combustion does not actually occur. Most of the chemical energy from the formation of H—O bonds is converted directly into electrical energy, rather than into heat energy as in combustion.

When the H_2/O_2 fuel cell is used aboard spacecraft, it is operated at a high enough temperature that the water evaporates at the same rate as it is produced. The vapor is then condensed to pure water.

Current research is aimed at modifying the design of fuel cells to lower their cost. Better catalysts would speed the reactions to allow more rapid generation of electricity and produce more power per unit volume. The H_2/O_2 cell is nonpolluting; the only substance released is H_2O. Catalysts have been developed that allow sunlight to decompose water into hydrogen and oxygen, which might be used to operate fuel cells, permitting the utilization of solar energy.

Fuel cells have also been constructed using fuels other than hydrogen, such as methane or methanol. Biomedical researchers envision the possibility of using tiny fuel cells to operate pacemakers. The disadvantage of other power supplies for pacemakers, which are pri-

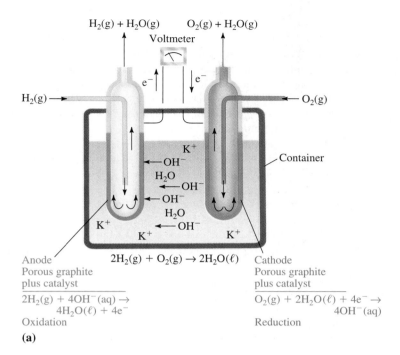

$$H_2(g) + H_2O(g) \qquad O_2(g) + H_2O(g)$$

Voltmeter

$H_2(g) \rightarrow$ $\qquad$ $\leftarrow O_2(g)$

Container

$$2H_2(g) + O_2(g) \rightarrow 2H_2O(\ell)$$

Anode
Porous graphite
plus catalyst
_____

$2H_2(g) + 4OH^-(aq) \rightarrow$
$\qquad 4H_2O(\ell) + 4e^-$

Oxidation

(a)

Cathode
Porous graphite
plus catalyst
_____

$O_2(g) + 2H_2O(\ell) + 4e^- \rightarrow$
$\qquad 4OH^-(aq)$

Reduction

Courtesy of UTC Fuel Cells

(b)

Figure 21-17 (a) Schematic drawing of a hydrogen–oxygen fuel cell. (b) Space shuttle fuel cells are self-contained units, each measuring 14 × 15 × 45 in. and weighing 260 lb. Three of these are installed under the payload bay, just aft of the crew compartment, and are fueled by H_2 and O_2 from cryogenic tanks. Each fuel cell is capable of providing 12 kW continuously, and up to 16 kW for short periods. A single unit can fully power the entire space shuttle in its day-to-day operation. The water produced is used for crew drinking and spacecraft cooling.

mary voltaic cells, is that their reactants are eventually consumed so that they require periodic surgical replacement. As long as the fuel and oxidizer are supplied, a fuel cell can—in theory, at least—operate forever. Eventually, tiny pacemaker fuel cells might be operated by the oxidation of blood sugar (the fuel) by the body's oxygen at a metal electrode implanted just below the skin.

Key Terms

Alkaline cell A dry cell in which the electrolyte contains KOH.

Ampere Unit of electric current; 1 ampere equals 1 coulomb per second.

Anode The electrode at which oxidation occurs.

Cathode The electrode at which reduction occurs.

Cathode protection Protection of a metal against corrosion by making it a cathode (attaching it to a sacrificial anode of a more easily oxidized metal).

Cell potential Potential difference, E_{cell}, between reduction and oxidation half-cells; may be at *nonstandard* conditions.

Concentration cell A voltaic cell in which the two half-cells are composed of the same species but contain different ion concentrations.

Corrosion Oxidation of metals in the presence of air and moisture.

Coulomb Unit of electric charge; the amount of charge that passes a given point when 1 ampere of electric current flows for 1 second.

Coulometry The use of electrochemical cells to relate the amount of reactant or product to the amount of current passed through the cell.

Downs cell An electrolytic cell for the commercial electrolysis of molten sodium chloride.

Dry cells Ordinary batteries (voltaic cells) for flashlights, radios, and so on; many are Leclanché cells.

Electrochemistry The study of the chemical changes produced by electric current and the production of electricity by chemical reactions.

Electrode potentials Potentials, E, of half-reactions as reductions versus the standard hydrogen electrode.

Electrodes Surfaces on which oxidation and reduction half-reactions occur in electrochemical cells.

Electrolysis The process that occurs in electrolytic cells.

Electrolytic cell An electrochemical cell in which electrical energy causes nonspontaneous redox reactions to occur.

Electrolytic conduction See *Ionic conduction*.

Electroplating Plating a metal onto a (cathodic) surface by electrolysis.

Faraday An amount of charge equal to 96,485 coulombs; corresponds to the charge on one mole of electrons, 6.022×10^{23} electrons.

Faraday's Law of Electrolysis The amount of substance that undergoes oxidation or reduction at each electrode during electrolysis is directly proportional to the amount of electricity that passes through the cell.

Fuel cell A voltaic cell in which the reactants (usually gases) are supplied continuously and products are removed continuously.

Galvanic cell See *Voltaic cell*.

Half-cell The compartment in a voltaic cell in which the oxidation or reduction half-reaction occurs.

Hydrogen–oxygen fuel cell A fuel cell in which hydrogen is the fuel (reducing agent) and oxygen is the oxidizing agent.

Ionic conduction Conduction of electric current by ions through a pure liquid or a solution; also called *Electrolytic conduction*.

Lead storage battery A secondary voltaic cell that is used in most automobiles.

Leclanché cell A common type of dry cell (battery).

Metallic conduction Conduction of electric current through a metal or along a metallic surface.

Nernst equation An equation that corrects standard electrode potentials for nonstandard conditions.

Nickel–cadmium cell (nicad battery) A dry cell in which the anode is Cd, the cathode is NiO_2, and the electrolyte is basic.

Polarization of an electrode Buildup of a product of oxidation or reduction at an electrode, preventing further reaction.

Primary voltaic cell A voltaic cell that cannot be recharged; no further chemical reaction is possible once the reactants are consumed.

Sacrificial anode A more active metal that is attached to a less active metal to protect the less active metal cathode against corrosion.

Salt bridge A U-shaped tube containing an electrolyte that connects two half-cells of a voltaic cell.

Secondary voltaic cell A voltaic cell that can be recharged; the original reactants can be regenerated by reversing the direction of current flow.

Standard cell A cell in which all reactants and products are in their thermodynamic standard states (1 M for solution species and 1 atm partial pressure for gases).

Standard cell potential The potential difference, E^0_{cell}, between standard reduction and oxidation half-cells.

Standard electrochemical conditions 1 M concentration for solution species, 1 atm partial pressure for gases, and pure solids and liquids.

Standard electrode A half-cell in which the oxidized and reduced forms of a species are present at unit activity: 1 M solutions of dissolved species, 1 atm partial pressure of gases, and pure solids and liquids.

Standard electrode potential By convention, the potential (E^0) of a half-reaction as a reduction relative to the standard hydrogen electrode, when all species are present at unit activity.

Standard hydrogen electrode (SHE) An electrode consisting of a platinum electrode that is immersed in a 1 M H^+ solution and that has H_2 gas bubbled over it at 1 atm pressure; defined as the reference electrode, with a potential of *exactly* 0.0000 . . . volt.

Voltage Potential difference between two electrodes; a measure of the chemical potential for a redox reaction to occur.

Voltaic cell An electrochemical cell in which spontaneous chemical reactions produce electricity; also called a galvanic cell.

Exercises

Redox Review and General Concepts

1. (a) Define oxidation and reduction in terms of electron gain or loss. (b) What is the relationship between the numbers of electrons gained and lost in a redox reaction? (c) Do all electrochemical cells involve redox reactions?

2. Define and illustrate (a) oxidizing agent and (b) reducing agent.

3. For each of the following unbalanced equations, (i) write the half-reactions for oxidation and for reduction, and (ii) balance the overall equation in acidic solution using the half-reaction method.
   (a) $Hg^{2+} + Pb \longrightarrow Hg + Pb^{2+}$
   (b) $MnO_2 + Cl^- \longrightarrow Mn^{2+} + Cl_2$
   (c) $Sn^{2+} + O_2 \longrightarrow Sn^{4+} + H_2O$

4. For each of the following unbalanced equations, (i) write the half-reactions for oxidation and reduction, and (ii) balance the overall equation using the half-reaction method.
   (a) $FeS + NO_3^- \longrightarrow$
       $NO + SO_4^{2-} + Fe^{2+}$ (acidic solution)
   (b) $Cr_2O_7^{2-} + Fe^{2+} \longrightarrow$
       $Cr^{3+} + Fe^{3+}$ (acidic solution)
   (c) $S^{2-} + Cl_2 + OH^- \longrightarrow$
       $SO_4^{2-} + Cl^- + H_2O$ (basic solution)

5. (a) Compare and contrast ionic conduction and metallic conduction. (b) What is an electrode? (c) What is an inert electrode?

6. Support or refute each of the following statements: (a) In any electrochemical cell the positive electrode is the one toward which the electrons flow through the wire. (b) The cathode is the negative electrode in any electrochemical cell.

7. For each of the following unbalanced equations, (i) write the half-reactions for oxidation and reduction, (ii) identify the species that lose and the species that gain electrons, and (iii) write the balanced net ionic equation for the overall reaction.
   (a) $Zn(s) + Au^{3+}(aq) \longrightarrow Zn^{2+}(aq) + Au(s)$
   (b) $NO_2^-(aq) + Cr_2O_7^{2-}(aq) \longrightarrow$
       $NO_3^-(aq) + Cr^{3+}(aq)$ (acidic solution)
   (c) $N_2O_4(aq) + Br^-(aq) \longrightarrow$
       $NO_2^-(aq) + Br_3^-(aq)$ (basic solution)

8. For each of the following unbalanced equations, (i) write the half-reactions for oxidation and reduction, (ii) identify the species that lose and the species that gain electrons, and (iii) write the balanced net ionic equation for the overall reaction.
   (a) $Ni(s) + Sn^{2+}(aq) \longrightarrow Ni^{2+}(aq) + Sn(s)$
   (b) $MnO_4^-(aq) + Cu(s) \longrightarrow$
       $Cu^{2+}(aq) + Mn^{2+}(aq)$ (acidic solution)
   (c) $Cr(OH)_3(s) + IO_3^-(aq) \longrightarrow$
       $I^-(aq) + CrO_4^{2-}(aq)$ (basic solution)

Electrolytic Cells: General Concepts

9. (a) Solids such as potassium bromide, KBr, and sodium nitrate, $NaNO_3$, do not conduct electric current even though they are ionic. Why? Can these substances be electrolyzed as solids? (b) Support or refute the statement that the Gibbs free energy change, ΔG, is positive for any electrolysis reaction.

10. (a) Metallic magnesium cannot be obtained by electrolysis of aqueous magnesium chloride, $MgCl_2$. Why? (b) There are no sodium ions in the overall cell reaction for the electrolysis of aqueous sodium chloride. Why?

11. Consider the electrolysis of molten aluminum oxide, Al_2O_3, dissolved in cryolite, Na_3AlF_6, with inert electrodes. This is the Hall–Héroult process for commercial production of aluminum (Section 22-6). The following experimental observations can be made when current is supplied: (i) Silvery metallic aluminum is produced at one electrode. (ii) Oxygen, O_2, bubbles off at the other electrode. Diagram the cell, indicating the anode, the cathode, the positive and negative electrodes, the half-reaction occurring at each electrode, the overall cell reaction, and the direction of electron flow through the wire.

12. Do the same as in Exercise 11 for the electrolysis of molten calcium chloride with inert electrodes. The observations are (i) Bubbles of pale green chlorine gas, Cl_2, are produced at one electrode. (ii) Silvery white molten metallic calcium is produced at the other electrode.

13. Do the same as in Exercise 11 for the electrolysis of aqueous potassium sulfate, K_2SO_4. The observations are (i) Bubbles of gaseous hydrogen are produced at one electrode, and the solution becomes more basic around that electrode. (ii) Bubbles of gaseous oxygen are produced at the other electrode, and the solution becomes more acidic around that electrode.

14. Do the same as in Exercise 11 for the electrolysis of an aqueous solution of copper(II) bromide, $CuBr_2$. The observations are (i) One electrode becomes coated with copper metal, and the color of the solution around that electrode fades. (ii) Around the other electrode, the solution turns brown, as bromine is formed and dissolves in water.

15. (a) Write the equation for the half-reaction when H_2O is reduced in an electrochemical cell. (b) Write the equation for the half-reaction when H_2O is oxidized in an electrochemical cell.

Faraday's Law

16. What are (a) a coulomb, (b) electric current, (c) an ampere, and (d) a faraday?

17. Calculate the number of electrons that have a total charge of 1 coulomb.

18. For each of the following cations, calculate (i) the number of faradays required to produce 1.00 mol of free metal and (ii) the number of coulombs required to produce 1.00 g of free metal. (a) Fe^{3+}, (b) Hg^{2+}, (c) Hg_2^{2+}.

19. For each of the following cations, calculate (i) the number of faradays required to produce 1.00 mol of free metal and (ii) the number of coulombs required to produce 1.00 g of free metal. (a) Fe^{2+}, (b) Cr^{3+}, (c) K^+.

20. In 400. min 2.40 g of copper is obtained by electrolysis of a copper(I) nitrate solution. (a) How many amperes is required for this experiment? (b) Using the same current and time, what mass of copper would be obtained from a copper(II) nitrate solution?

21. A mass of 1.25 g of silver is plated from a silver nitrate solution in 1.25 h. Calculate the (a) coulombs, (b) faradays, and (c) amperes necessary for this process.

22. Rhodium is an element that has the appearance of silver, but does not tarnish like silver and, because it is very hard, does not become worn or scratched like silver. What mass of rhodium could be plated by electrolysis of a $Rh(NO_3)_3$ solution with a 0.75-A current for 15.0 min?

23. Hydrogen may be the fuel of the future for automobiles according to some experts. Hydrogen can be isolated from water by electrolysis. Calculate the mass of hydrogen that is released when a 1.50-A current is passed through salt water for 5.00 h.

24. The mass of silver deposited on a spoon during electroplating was 0.915 mg. How much electric charge passed through the cell?

25. What mass of platinum could be plated onto a ring from the electrolysis of a platinum(II) salt with a 0.385-A current for 142 s?

26. What mass of silver could be plated onto a spoon from electrolysis of silver nitrate with a 3.50-A current for 45.0 min?

*27. We pass enough current through a solution to plate out *one* mole of nickel metal from a solution of $NiSO_4$. In other electrolysis cells, this same current plates out *two* moles of silver from $AgNO_3$ solution but liberates only *one-half* mole of O_2 gas. Explain these observations.

*28. A current is passed through 500. mL of a solution of CaI_2. The following electrode reactions occur:

anode: $2I^- \longrightarrow I_2 + 2e^-$
cathode: $2H_2O + 2e^- \longrightarrow H_2 + 2OH^-$

After some time, analysis of the solution shows that 35.7 mmol of I_2 has been formed. (a) How many faradays of charge have passed through the solution? (b) How many coulombs? (c) What volume of dry H_2 at STP has been formed? (d) What is the pH of the solution?

29. The cells in an automobile battery were charged at a steady current of 5.0 A for exactly 2 h. What masses of Pb and PbO_2 were formed in each cell? The overall reaction is

$2PbSO_4(s) + 2H_2O(\ell) \longrightarrow$
$\qquad Pb(s) + PbO_2(s) + 2H_2SO_4(aq)$

*30. The chemical equation for the electrolysis of a fairly concentrated brine solution is

$2NaCl(aq) + 2H_2O(\ell) \longrightarrow$
$\qquad Cl_2(g) + H_2(g) + 2NaOH(aq)$

What volume of gaseous chlorine would be generated at 752 torr and 15°C if the process were 83% efficient and if a current of 1.75 A flowed for 5.00 h?

31. An electrolytic cell contains 40.0 mL of a 0.152 M solution of $FeCl_3$. A current of 0.620 A is passed through the cell, causing deposition of Fe(s) at the cathode. What is the concentration of $Fe^{3+}(aq)$ in the cell after this current has run for 20.0 min?

32. Suppose 250. mL of a 0.433 M solution of $CuCl_2$ is electrolyzed. How long will a current of 0.75 A have to run in order to reduce the concentration of Cu^{2+} to 0.167 M? What mass of Cu(s) will be deposited on the cathode during this time?

33. One method for recovering metals from their ores is by electrodeposition. For example, after passing a current of 18.0 amps through a molten tantalum salt for 38 min, workers at a metals processing plant isolated 15.4 g of pure tantalum. What was the oxidation state of tantalum in the molten material from which the metal was isolated?

*34. Three electrolytic cells are connected in series; that is, the same current passes through all three, one after another. In the first cell, 1.20 g of Cd is oxidized to Cd^{2+}; in the second, Ag^+ is reduced to Ag; in the third, Fe^{2+} is oxidized to Fe^{3+}. (a) Find the number of faradays passed through the circuit. (b) What mass of Ag is deposited at the cathode in the second cell? (c) What mass of $Fe(NO_3)_3$ could be recovered from the solution in the third cell?

Voltaic Cells: General Concepts

35. What does voltage measure? How does it vary with time in a primary voltaic cell? Why?

36. (a) Why must the solutions in a voltaic cell be kept separate and not allowed to mix? (b) What are the functions of a salt bridge?

37. A voltaic cell containing a standard Fe^{3+}/Fe^{2+} electrode and a standard Ga^{3+}/Ga electrode is constructed, and the circuit is closed. Without consulting the table of standard reduction potentials, diagram and completely describe the cell from the following experimental observations. (i) The mass of the gallium electrode decreases, and the gallium ion concentration increases around that electrode. (ii) The ferrous ion, Fe^{2+}, concentration increases in the other electrode solution.

38. Repeat Exercise 37 for a voltaic cell that contains standard Co^{2+}/Co and Au^{3+}/Au electrodes. The observations are: (i) Metallic gold plates out on one electrode, and the gold ion concentration decreases around that electrode. (ii) The mass of the cobalt electrode decreases, and the cobalt(II) ion concentration increases around that electrode.

39. Appendix J lists selected reduction potentials in volts at 25°C. Why is it not necessary to list a mixture of reduction and oxidation potentials?

*40. In Section 4-10 we learned how to predict from the activity series (Table 4-14) which metals replace which others from aqueous solutions. From that table, we predict that zinc will displace silver. The equation for this process is

$$Zn(s) + 2Ag^+(aq) \longrightarrow Zn^{2+}(aq) + 2Ag(s)$$

Suppose we set up a voltaic cell based on this reaction. (a) What half-reaction would represent the reduction in this cell? (b) What half-reaction would represent the oxidation? (c) Which metal would be the anode? (d) Which metal would be the cathode? (e) Diagram this cell.

41. In a voltaic cell made with metal electrodes, is the more active metal more likely to be the anode or the cathode? Explain.

42. When metallic copper is placed into aqueous silver nitrate, a spontaneous redox reaction occurs. No electricity is produced. Why?

43. Assume that a voltaic cell utilizes the redox reaction

$$2Al(s) + 3Ni^{2+}(aq) \longrightarrow$$
$$2Al^{3+}(aq) + 3Ni(s) \text{ (acidic solution)}$$

Potassium and nitrate ions may also be present. Draw this voltaic cell, and label the anode, cathode, electron flow, and ion flow.

44. Assume that a voltaic cell, proposed as a method for the purification of uranium, utilizes the redox reaction

$$3Mg(s) + 2U^{3+}(aq) \longrightarrow$$
$$3Mg^{2+}(aq) + 2U(s) \text{ (acidic solution)}$$

Potassium and nitrate ions may also be present. Draw this voltaic cell, and label the anode, cathode, electron flow, and ion flow.

Standard Cell Potentials

45. (a) What are standard electrochemical conditions? (b) Why are we permitted to assign arbitrarily an electrode potential of exactly 0 V to the standard hydrogen electrode?

46. What does the sign of the standard reduction potential of a half-reaction indicate? What does the magnitude indicate?

47. (a) What are standard reduction potentials? (b) What information is contained in tables of standard reduction potentials (Tables 21-2 and 21-3, Appendix J)? How is the information in such tables arranged?

48. Standard reduction potentials are 1.36 V for $Cl_2(g)/Cl^-$, 0.799 V for $Ag^+/Ag(s)$, 0.521 V for $Cu^+/Cu(s)$, 0.337 V for $Cu^{2+}/Cu(s)$, −0.44 V for $Fe^{2+}/Fe(s)$, −2.71 V for $Na^+/Na(s)$, and −2.925 V for $K^+/K(s)$. (a) Arrange the oxidizing agents in order of increasing strength. (b) Which

of these oxidizing agents will oxidize Cu under standard-state conditions?

49. Standard reduction potentials are 1.455 V for the $PbO_2(s)/Pb(s)$ couple, 2.87 V for $F_2(g)/F^-$, 3.06 V for $F_2(g)/HF(aq)$, and 1.77 V for $H_2O_2(aq)/H_2O(\ell)$. Under standard-state conditions, (a) which is the strongest oxidizing agent, (b) which oxidizing agent(s) could oxidize lead to lead(IV) oxide, and (c) which oxidizing agent(s) could oxidize fluoride ion in an acidic solution?

50. Arrange the following less commonly encountered metals in an activity series from the most active to the least active: radium [$Ra^{2+}/Ra(s)$, $E^0 = -2.9$ V], rhodium [$Rh^{3+}/Rh(s)$, $E^0 = 0.80$ V], europium [$Eu^{2+}/Eu(s)$, $E^0 = -3.4$ V]. How do these metals compare in reducing ability with the active metal lithium [$Li^+/Li(s)$, $E^0 = -3.0$ V], with hydrogen, and with platinum [$Pt^{2+}/Pt(s)$, $E^0 = 1.2$ V], which is a noble metal and one of the least active of the metals?

51. Arrange the following metals in an activity series from the most active to the least active: nobelium [$No^{3+}/No(s)$, $E^0 = -2.5$ V], cobalt [$Co^{2+}/Co(s)$, $E^0 = -0.28$ V], chromium [$Cr^{3+}/Cr(s)$, $E^0 = -0.74$ V], thallium [$Tl^+/Tl(s)$, $E^0 = -0.34$ V], polonium [$Po^{2+}/Po(s)$, $E^0 = 0.65$ V].

52. Diagram the following cells. For each cell, write the balanced equation for the reaction that occurs spontaneously, and calculate the cell potential. Indicate the direction of electron flow, the anode, the cathode, and the polarity (+ or −) of each electrode. In each case, assume that the circuit is completed by a wire and a salt bridge. (a) A strip of magnesium is immersed in a solution that is 1.0 M in Mg^{2+}, and a strip of silver is immersed in a solution that is 1.0 M in Ag^+. (b) A strip of zinc is immersed in a solution that is 1.0 M in Zn^{2+}, and a strip of tin is immersed in a solution that is 1.0 M in Sn^{2+}.

53. Repeat Exercise 52 for the following cells. (a) A strip of chromium is immersed in a solution that is 1.0 M in Cr^{3+}, and a strip of gold is immersed in a solution that is 1.0 M in Au^{3+}. (b) A strip of aluminum is immersed in a solution that is 1.0 M in Al^{3+}, and a strip of lead is immersed in a solution that is 1.0 M in Pb^{2+}.

In answering Exercises 54-73, justify each answer by appropriate calculations. Assume that each reaction occurs at standard electrochemical conditions.

54. (a) Will Fe^{3+} oxidize Sn^{2+} to Sn^{4+} in acidic solution? (b) Will dichromate ions oxidize fluoride ions to free fluorine in acidic solution?

55. (a) Will dichromate ions oxidize arsenous acid, H_3AsO_3, to arsenic acid, H_3AsO_4, in acid solution? (b) Will dichromate ions oxidize hydrogen peroxide, H_2O_2, to free oxygen, O_2, in acidic solution?

56. (a) Will permanganate ions oxidize Cr^{3+} to $Cr_2O_7^{2-}$ in acidic solution? (b) Will sulfate ions oxidize arsenous acid, H_3AsO_3, to arsenic acid, H_3AsO_4, in acid solution?

57. Calculate the standard cell potential, E_{cell}^0, for the cell described in Exercise 37.

58. Calculate the standard cell potential, E_{cell}^0, for the cell described in Exercise 38.

59. (a) Write the equation for the oxidation of $Zn(s)$ by $Br_2(\ell)$. (b) Calculate the potential of this reaction under standard-state conditions. (c) Is this a spontaneous reaction?

60. For each of the following cells, (i) write the net reaction in the direction consistent with the way the cell is written; (ii) write the half-reactions for the anode and cathode processes; (iii) find the standard cell potential, E_{cell}^0, at 25°C; and (iv) tell whether the standard cell reaction actually occurs as given or in the reverse direction.
    (a) $Cr \,|\, Cr^{3+} \,\|\, Cu^{2+} \,|\, Cu$
    (b) $Ag \,|\, Ag^+ \,\|\, Cd^{2+} \,|\, Cd$

61. Repeat Exercise 60 for the following cells:
    (a) $Al \,|\, Al^{3+} \,\|\, Ce^{4+}, Ce^{3+} \,|\, Pt$
    (b) $Zn \,|\, Zn^{2+} \,\|\, Tl^+ \,|\, Tl$

62. Which of the following reactions are spontaneous in voltaic cells under standard conditions?
    (a) $H_2(g) \longrightarrow H^+(aq) + H^-(aq)$
    (b) $Zn(s) + 4CN^-(aq) + Ag_2CrO_4(s) \longrightarrow$
        $\qquad Zn(CN)_4{}^{2-}(aq) + 2Ag(s) + CrO_4{}^{2-}(aq)$
    (c) $MnO_2(s) + 4H^+(aq) + Sr(s) \longrightarrow$
        $\qquad Mn^{2+}(aq) + 2H_2O(\ell) + Sr^{2+}(aq)$
    (d) $Cl_2(g) + 2H_2O(\ell) + ZnS(s) \longrightarrow$
        $\qquad 2HOCl(aq) + H_2S(aq) + Zn(s)$

63. Consult a table of standard reduction potentials, and determine which of the following reactions are spontaneous under standard electrochemical conditions.
    (a) $Mn(s) + 2H^+(aq) \longrightarrow H_2(g) + Mn^{2+}(aq)$
    (b) $2Al^{3+}(aq) + 3H_2(g) \longrightarrow 2Al(s) + 6H^+(aq)$
    (c) $2Cr(OH)_3(s) + 6F^-(aq) \longrightarrow$
        $\qquad 2Cr(s) + 6OH^-(aq) + 3F_2(g)$
    (d) $Cl_2(g) + 2Br^-(aq) \longrightarrow Br_2(\ell) + 2Cl^-(aq)$

64. Which of each pair is the stronger reducing agent? (a) Ag or H_2, (b) Sn or Pb, (c) Hg or Au, (d) Cl^- in acidic solution or Cl^- in basic solution, (e) HCl or H_2S, (f) Ag or Au

65. Which of each pair is the stronger oxidizing agent? (a) Cu^+ or Ag^+, (b) Sn^{2+} or Sn^{4+}, (c) Fe^{2+} or Fe^{3+}, (d) I_2 or Br_2, (e) $MnO_4{}^-$ in acidic solution or $MnO_4{}^-$ in basic solution, (f) H^+ or Co^{2+}.

*66. The element ytterbium forms both 2+ and 3+ cations in aqueous solution. $E^0 = -2.797$ V for $Yb^{2+}/Yb(s)$, and -2.267 V for $Yb^{3+}/Yb(s)$. What is the standard-state reduction potential for the Yb^{3+}/Yb^{2+} couple?

*67. The standard reduction potential for Cu^+ to $Cu(s)$ is 0.521 V, and for Cu^{2+} to $Cu(s)$ it is 0.337 V. Calculate the E^0 value for the Cu^{2+}/Cu^+ couple.

68. Consider a suggestion for purifying uranium without an outside energy source by setting up a voltaic cell with the reaction

$3Mg(s) + 2U^{3+}(aq) \longrightarrow$
$\qquad 3Mg^{2+}(aq) + 2U(s) \qquad$ (acidic solution)

The standard reduction potentials are -1.798 for the uranium half-reaction and -2.714 for the magnesium half-reaction. (a) Will this setup work spontaneously? (b) Calculate the voltage produced by this cell as written.

69. A reaction is proposed for a nickel-cadmium battery as

$$Ni(s) + Cd^{2+}(aq) \longrightarrow Ni^{2+}(aq) + Cd(s)$$

(a) Is the reaction spontaneous as written? (b) Calculate the voltage produced by this voltaic cell as written.

70. Selecting from half-reactions involving the following species, write the spontaneous reaction that will produce the voltaic cell with the highest voltage in an acidic solution: K^+, Ca^{2+}, Ni^{2+}, H_2O_2, and F_2.

71. Propose the spontaneous reaction that will produce the voltaic cell with the highest voltage output by choosing from only reduction and oxidation potentials involving: Au^{3+} and Au, HgO and Hg, Ag_2O and Ag, S and S^{2-}, and $SO_4{}^{2-}$ and $SO_3{}^{2-}$.

72. Tarnished silver is coated with a layer of $Ag_2S(s)$. The coating can be removed by boiling the silverware in an aluminum pan, with some baking soda or salt added to make the solution conductive. Explain this from the point of view of electrochemistry.

*73. Describe the process of corrosion. How can corrosion of an easily oxidizable metal be prevented if the metal must be exposed to the weather?

Concentration Effects; Nernst Equation

For cell voltage calculations, assume that the temperature is 25°C unless stated otherwise.

*74. How is the Nernst equation of value in electrochemistry? How would the Nernst equation be modified if we wished to use natural logarithms, ln? What is the value of the constant in the following equation at 25°C?

$$E = E^0 - \frac{constant}{n} \ln Q$$

75. Identify all of the terms in the Nernst equation. What part of the Nernst equation represents the correction factor for nonstandard electrochemical conditions?

76. By putting the appropriate values into the Nernst equation, show that it predicts that the voltage of a standard half-cell is equal to E^0. Use the Zn^{2+}/Zn reduction half-cell as an illustration.

77. Calculate the potential associated with the following half-reaction when the concentration of the cobalt(II) ion is $1.0 \times 10^{-3} \, M$.

$$Co(s) \longrightarrow Co^{2+} + 2e^-$$

78. Calculate the reduction potential for hydrogen ion in a system having a perchloric acid concentration of $2.00 \times 10^{-4} \, M$ and a hydrogen pressure of 2.50 atm. (Recall that $HClO_4$ is a strong acid in aqueous solution.)

79. The standard reduction potentials for the $H^+/H_2(g)$ and $O_2(g)$, $H^+/H_2O(\ell)$ couples are 0.0000 V and 1.229 V, respectively. (a) Write the half-reactions and the overall reaction, and calculate E^0 for the reaction

$$2H_2(g) + O_2(g) \longrightarrow 2H_2O(\ell)$$

(b) Calculate E for the cell when the pressure of H_2 is 3.00 atm and that of O_2 is 1.30 atm.

80. Consider the cell represented by the notation

$$Zn(s)\,|\,ZnCl_2(aq)\,\|\,Cl_2(g,\,1\,atm);\,Cl^-(aq)\,|\,C$$

Calculate (a) E^0 and (b) E for the cell when the concentration of the $ZnCl_2$ is 0.21 mol/L.

81. What is the concentration of Ag^+ in a half-cell if the reduction potential of the Ag^+/Ag couple is observed to be 0.40 V?

82. What must be the pressure of fluorine gas in order to produce a half-cell reduction potential of the F_2/F^- couple of 2.70 V in a solution that contains 0.34 M F^-?

83. Calculate the cell potential of each of the following electrochemical cells at 25°C.
   (a) $Sn(s)\,|\,Sn^{2+}(6.5 \times 10^{-3}\,M)\,\|\,Ag^+(0.110\,M)\,|\,Ag(s)$
   (b) $Zn(s)\,|\,Zn^{2+}(0.500\,M)\,\|$
   $\qquad Fe^{3+}(7.2 \times 10^{-6}\,M),\,Fe^{2+}(0.15\,M)\,|\,Pt$
   (c) $Pt\,|\,H_2(1\,atm)\,|\,HCl(0.00880\,M)\,|\,Cl_2(1\,atm)\,|\,Pt$

84. Calculate the cell potential of each of the following electrochemical cells at 25°C.
   (a) $Pt\,|\,H_2(10.0\,atm),\,H^+(1.00 \times 10^{-3}\,M)\,\|$
   $\qquad Ag^+(0.00549\,M)\,|\,Ag(s)$
   (b) $Pt\,|\,H_2(1.00\,atm),\,H^+(pH = 5.97)\,\|$
   $\qquad H^+(pH = 3.47),\,H_2(1.00\,atm)\,|\,Pt$
   (c) $Pt\,|\,H_2(0.0361atm),\,H^+(0.0175\,M)\,\|$
   $\qquad H^+(0.0175\,M),\,H_2(5.98 \times 10^{-4}\,atm)\,|\,Pt$

*85. Find the potential of the cell in which identical iron electrodes are placed into solutions of $FeSO_4$ of concentration 1.5 mol/L and 0.15 mol/L.

*86. We construct a cell in which identical copper electrodes are placed in two solutions. Solution A contains 0.75 M Cu^{2+}. Solution B contains Cu^{2+} at some concentration known to be lower than that in solution A. The potential of the cell is observed to be 0.040 V. What is $[Cu^{2+}]$ in solution B?

*87. We construct a standard copper-cadmium cell, close the circuit, and allow the cell to operate. At some later time, the cell voltage reaches zero, and the cell is "run down." (a) What will be the ratio of $[Cd^{2+}]$ to $[Cu^{2+}]$ at that time? (b) What will be the concentrations?

*88. Repeat Exercise 87 for a standard zinc–nickel cell.

*89. The cell potential for the cell

$$Zn(s) + 2H^+(?M) \longrightarrow Zn^{2+}(2.5\,M) + H_2(g)\,(5.0\,atm)$$

is observed to be 0.445 V. What is the pH in the H^+/H_2 half-cell?

90. We wish to produce a 0.375-volt concentration cell using two hydrogen electrodes, both with hydrogen at a pressure of one atmosphere. One of the solutions has a pH of 1.5. Calculate the required pH of the other solution.

91. A concentration cell prepared using two hydrogen electrodes, both with the partial pressure of hydrogen being one atmosphere, produces 0.450 volts. The pH of one hydrogen electrode is 1.65; what is the pH of the other?

Relationships Among ΔG^0, E^0_{cell}, and K

92. How are the signs and magnitudes of E^0_{cell}, ΔG^0, and K related for a particular reaction? Why is the equilibrium constant K related only to E^0_{cell} and not to E_{cell}?

93. In light of your answer to Exercise 92, how do you explain the fact that ΔG^0 for a redox reaction *does* depend on the number of electrons transferred, according to the equation $\Delta G^0 = -nFE^0_{cell}$?

94. Calculate E^0_{cell} from the tabulated standard reduction potentials for each of the following reactions in aqueous solution. Then calculate ΔG^0 and K at 25°C from E^0_{cell}. Which reactions are spontaneous as written?
   (a) $MnO_4^- + 5Fe^{2+} \longrightarrow$
   $\qquad\qquad Mn^{2+} + 5Fe^{3+} \qquad$ (acidic solution)
   (b) $2Cu^+ \longrightarrow Cu^{2+} + Cu(s)$
   (c) $3Zn(s) + 2MnO_4^- + 4H_2O \longrightarrow$
   $\qquad\qquad 2MnO_2(s) + 3Zn(OH)_2(s) + 2OH^-$

95. Calculate ΔG^0 (overall) and ΔG^0 per mole of metal for each of the following reactions from E^0 values.
   (a) Zinc dissolves in dilute hydrochloric acid to produce a solution that contains Zn^{2+}, and hydrogen gas is evolved.
   (b) Chromium dissolves in dilute hydrochloric acid to produce a solution that contains Cr^{3+}, and hydrogen gas is evolved.
   (c) Silver dissolves in dilute nitric acid to form a solution that contains Ag^+, and NO is liberated as a gas.
   (d) Lead dissolves in dilute nitric acid to form a solution that contains Pb^{2+}, and NO is liberated as a gas.

96. Use tabulated reduction potentials to calculate the value of the equilibrium constant for the reaction

$$Sn^{4+} + 2Fe^{2+} \rightleftharpoons Sn^{2+} + 2Fe^{3+}$$

97. Use tabulated reduction potentials to calculate the equilibrium constant for the reaction

$$2I^- + Br_2(\ell) \rightleftharpoons I_2(s) + 2Br^-$$

98. Using the following half-reactions and E^0 data at 25°C:

$$PbSO_4(s) + 2e^- \longrightarrow Pb(s) + SO_4^{2-} \qquad E^0 = -0.356\,V$$
$$PbI_2(s) + 2e^- \longrightarrow Pb(s) + 2I^- \qquad E^0 = -0.365\,V$$

calculate the equilibrium constant for the reaction

$$PbSO_4(s) + 2I^- \rightleftharpoons PbI_2(s) + SO_4^{2-}$$

Practical Aspects of Electrochemistry

99. Distinguish among (a) primary voltaic cells, (b) secondary voltaic cells, and (c) fuel cells.

100. Sketch and describe the operation of (a) the Leclanché dry cell, (b) the lead storage battery, and (c) the hydrogen-oxygen fuel cell.

101. Why is the dry cell designed so that Zn and MnO_2 do not come into contact? What reaction might occur if they were in contact? How would this reaction affect the usefulness of the cell?

*102. People sometimes try to recharge dry cells, with limited success. (a) What reaction would you expect at the zinc electrode of a Leclanché cell in an attempt to recharge it? (b) What difficulties would arise from the attempt?

103. Briefly describe how a storage cell operates.

104. How does a fuel cell differ from a dry cell or a storage cell?

105. Does the physical size of a commercial cell govern the potential that it will deliver? What does the size affect?

Mixed Exercises

106. Consider the electrochemical cell represented by $Mg(s) \mid Mg^{2+} \parallel Fe^{3+} \mid Fe(s)$. (a) Write the ion-electron equations for the half-reactions and the overall cell equation. (b) The standard reduction potential for $Fe^{3+}/Fe(s)$ is -0.036 V at 25°C. Determine the standard potential for the reaction. (c) Determine E for the cell when the concentration of Fe^{3+} is 10.0 mol/L and that of Mg^{2+} is 1.00×10^{-3} mol/L. (d) If 150 mA is to be drawn from this cell for a period of 20.0 min, what is the minimum mass for the magnesium electrode?

107. A sample of Al_2O_3 dissolved in a molten fluoride bath is electrolyzed using a current of 1.20 A. (a) What is the rate of production of Al in g/h? (b) The oxygen liberated at the positive carbon electrode reacts with the carbon to form CO_2. What mass of CO_2 is produced per hour?

*108. The "life" of a certain voltaic cell is limited by the amount of Cu^{2+} in solution available to be reduced. If the cell contains 30.0 mL of 0.165 M $CuSO_4$, what is the maximum amount of electric charge this cell could generate?

109. A magnesium bar weighing 5.0 kg is attached to a buried iron pipe to protect the pipe from corrosion. An average current of 0.025 A flows between the bar and the pipe. (a) What reaction occurs at the surface of the bar? Of the pipe? In which direction do electrons flow? (b) How many years will be required for the Mg bar to be entirely consumed (1 year = 3.16×10^7 s)? (c) What reaction(s) will occur if the bar is not replaced after the time calculated in part (b)?

110. (a) Calculate the ratio of ion concentrations of Mn^{2+} and Fe^{2+} necessary to produce a voltaic cell of 1.56 volts. The electrodes are solid manganese and iron. (b) Draw

this voltaic cell. Indicate which electrode is the anode and which is the cathode, as well as the direction of electron flow.

111. (a) Calculate the ratio of ion concentrations of Mg^{2+} and Cu^{2+} necessary to produce a voltaic cell of 2.67 volts. The electrodes are solid magnesium and solid copper. (b) Draw this voltaic cell. Indicate which electrode is the anode and which is the cathode, as well as the direction of electron flow.

112. The production of uranium metal from purified uranium dioxide ore consists of the following steps:

$$UO_2(s) + 4HF(g) \longrightarrow UF_4(s) + 2H_2O(\ell)$$

$$UF_4(s) + 2Mg(s) \xrightarrow{heat} U(s) + 2MgF_2(s)$$

What is the oxidation number of U in (a) UO_2, (b) UF_4, and (c) U? Identify (d) the reducing agent and (e) the substance reduced. (f) What current could the second reaction produce if 0.500 g of UF_4 reacted each minute? (g) What volume of HF(g) at 25°C and 10.0 atm would be required to produce 0.500 g of U? (h) Would 0.500 g of Mg be enough to produce 0.500 g of U?

113. Which of each pair is the stronger oxidizing agent? (a) H^+ or Cl_2, (b) Zn^{2+} or Se in contact with acidic solution, (c) $Cr_2O_7{}^{2-}$ or Br_2 (acidic solution).

114. (a) Describe the process of electroplating. (b) Sketch and label an apparatus that a jeweler might use for electroplating silver onto jewelry. (c) A jeweler purchases highly purified silver to use as the anode in an electroplating operation. Is this a wise purchase? Why?

115. The same quantity of electric charge that deposited 0.612 g of silver was passed through a solution of a gold salt, and 0.373 g of gold was deposited. What is the oxidation state of gold in this salt?

CONCEPTUAL EXERCISES

116. Both I_2 and Br_2 are commercially prepared by oxidation of their halide binary salts from oceans or natural waters, and Cl_2 can be prepared by electrolyzing aqueous NaCl solutions. Why then is it not possible to prepare F_2 by electrolysis of an aqueous NaF solution?

117. Electroplating is performed by using a source of direct current. (a) Why can't an alternating current be used? (b) What would happen if an alternating current were used for electroplating?

118. Figure 21-5 is a schematic diagram of the electrolytic cell used to separate copper from the impurities, that is, zinc, iron, silver, gold, and platinum. Can this process be used to separate the metals left after the copper has been removed? Explain.

119. A zinc-copper cell like that shown in Figure 21-7 is constructed, except that an inert platinum wire is used instead of the salt bridge. Will the cell still produce a potential?

BUILDING YOUR KNOWLEDGE

120. An electrochemical cell was needed in which hydrogen and oxygen would react to form water. (a) Using the following standard reduction potentials for the couples given, determine which combination of half-reactions gives the maximum output potential:
$E^0 = -0.828$ V for $H_2O(\ell)/H_2(g)$, OH^-
$E^0 = 0.0000$ V for $H^+/H_2(g)$
$E^0 = 1.229$ V for $O_2(g)$, $H^+/H_2O(\ell)$
$E^0 = 0.401$ V for $O_2(g)$, $H_2O(\ell)/OH^-$
(b) Write the balanced equation for the overall reaction in (a).

121. (a) Given the following E^0 values at 25°C, calculate K_{sp} for cadmium sulfide, CdS.

$$Cd^{2+}(aq) + 2e^- \longrightarrow Cd(s) \qquad E^0 = -0.403 \text{ V}$$
$$CdS(s) + 2e^- \longrightarrow Cd(s) + S^{2-}(aq) \quad E^0 = -1.21 \text{ V}$$

(b) Evaluate ΔG^0 at 25°C for the process

$$CdS(s) \rightleftharpoons Cd^{2+}(aq) + S^{2-}(aq)$$

122. Refer to tabulated reduction potentials. (a) Calculate K_{sp} for AgBr(s). (b) Calculate ΔG^0 for the reaction

$$AgBr(s) \rightleftharpoons Ag^+(aq) + Br^-(aq)$$

*123. Under standard-state conditions, the following reaction is not spontaneous:

$$Br^- + 2MnO_4^- + H_2O(\ell) \longrightarrow$$
$$BrO_3^- + 2MnO_2(s) + 2OH^- \qquad E^0 = -0.022 \text{ V}$$

The reaction conditions are adjusted so that $E = 0.120$ V by making $[Br^-] = [MnO_4^-] = 1.60$ mol/L and $[BrO_3^-] = 0.60$ mol/L. (a) What is the concentration of hydroxide ions in this cell? (b) What is the pH of the solution in the cell?

124. Show by calculation that $E^0 = -1.662$ V for the reduction of Al^{3+} to Al(s), regardless of whether the equation for the reaction is written

(i) $\frac{1}{3}Al^{3+} + e^- \longrightarrow \frac{1}{3}Al(s) \qquad \Delta G^0 = 160.4$ kJ/mol

or

(ii) $Al^{3+} + 3e^- \longrightarrow Al(s) \qquad \Delta G^0 = 481.2$ kJ/mol

125. We wish to fill a balloon with H_2 at a pressure of 1.05 atm and a temperature of 25°C. The volume of the balloon when filled, is 750 mL. How long must a current of 2.75 A be passed through the cell in order to produce this amount of H_2 by electrolysis of water?

BEYOND THE TEXTBOOK

Go to the textbook website at

http://www.brookscole.com/chemistry/whitten

for additional activities and exercises based on the General Chemistry Interactive CD-ROM, the World Wide Web, and library resources

InfoTrac College Edition

For additional readings, go to InfoTrac College Edition, your online research library at:

http://infotrac.thomsonlearning.com

22 Metals I: Metallurgy

Molten steel being poured from a slag pot.

OBJECTIVES

After you have studied this chapter, you should be able to

- Describe major sources of metals
- Describe some pretreatment techniques for ores
- Describe some reduction processes that produce free metals
- Describe some techniques for refining (purifying) metals
- Describe the specific metallurgies of five metals: magnesium, aluminum, iron, copper, and gold

METALS

M etals are widely used for structural purposes in buildings, cars, railroads, ships, and aircraft. They also serve as conductors of heat and electricity. Medical and nutritional research during recent decades has provided much insight into important biological functions of metals. The metals Na, K, Ca, and Mg, as well as some nonmetals (C, H, O, N, P, and S), are present in the human body in substantial quantities. Many other metals are present in lesser quantities in the human body, but they are essential to our well-being (see the Chemistry in Use essay "Trace Elements and Life" in Chapter 23). In this chapter, we shall study the occurrence of metals and examine processes for obtaining metals from their ores.

22-1 OCCURRENCE OF THE METALS

In our study of periodicity we learned that metallic character increases toward the left and toward the bottom of the periodic table (Section 4-1) and that oxides of most metals are basic (Section 6-8, part 2). The oxides of some metals (and metalloids) are amphoteric (Section 10-6). In Section 13-17 we described metallic bonding and related the effectiveness of metallic bonding to the characteristic properties of metals.

The properties of metals influence the kinds of ores in which they are found, and the metallurgical processes used to extract them from their ores. Metals with negative standard reduction potentials (active metals) are found in nature in the combined state. Those with positive reduction potentials, the less active metals, may occur in the uncombined free state

SCiLINKS.

TOPIC: Minerals
GO TO: www.scilinks.org
*sci***LINKS CODE:** WCH2210

A cluster of 21 minerals that are present in ores, plus two samples of native ores (free metals, Ag and Bi). These are identified in the following list.

1. Bornite (iridescent)—COPPER
2. Dolomite (pink)—MAGNESIUM
3. Molybdenite (gray)—MOLYBDENUM
4. Skutterudite (gray)—COBALT, NICKEL
5. Zincite (mottled red)—ZINC
6. Chromite (gray)—CHROMIUM
7. Stibnite (*top right,* gray)—ANTIMONY
8. Gummite (yellow)—URANIUM
9. Cassiterite (rust, *bottom right*)—TIN
10. Vanadinite crystal on Goethite (red crystal)—VANADIUM
11. Cinnabar (red)—MERCURY
12. Galena (gray)—LEAD
13. Monazite (white)—RARE EARTHS: cerium, lanthanum, neodymium, thorium
14. Bauxite (gold)—ALUMINUM
15. Strontianite (white, spiny)—STRONTIUM
16. Cobaltite (gray cube)—COBALT
17. Pyrite (gold)—IRON
18. Columbinite (tan, gray stripe)—NIOBIUM, TANTALUM
19. Native BISMUTH (shiny)
20. Rhodochrosite (pink)—MANGANESE
21. Rutile (shiny twin crystal)—TITANIUM
22. Native SILVER (filigree on quartz)
23. Pyrolusite (black, powdery)—MANGANESE

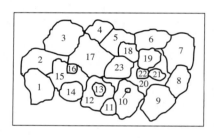

SCi LINKS
TOPIC: Mining Minerals
GO TO: www.scilinks.org
*sci*LINKS CODE: WCH2220

Pronounce "gangue" as one syllable with a soft final g.

as **native ores.** Examples of native ores are Cu, Ag, Au, and the less abundant Pt, Os, Ir, Ru, Rh, and Pd. Cu, Ag, and Au are also found in the combined state.

Many "insoluble" compounds of the metals are found in the earth's crust. Solids that contain these compounds are the **ores** from which metals are extracted. Ores contain **minerals,** comparatively pure compounds of the metals of interest, mixed with relatively large amounts of **gangue**—sand, soil, clay, rock, and other material. Soluble compounds are found dissolved in the sea or in salt beds in areas where large bodies of water have evaporated. Metal ores can be classified by the anions with which the metal ions are combined (Table 22-1 and Figure 22-1).

METALLURGY

Metallurgy is the commercial extraction of metals from their ores and the preparation of metals for use. It usually includes several steps: (1) mining the ore, (2) pretreatment of the ore, (3) reduction of the ore to the free metal, (4) refining or purifying the metal, and (5) alloying, if necessary.

TABLE 22-1	*Common Classes of Ores*
Anion	**Examples and Names of Minerals**
none (native ores)	Au, Ag, Pt, Os, Ir, Ru, Rh, Pd, As, Sb, Bi, Cu
oxide	hematite, Fe_2O_3; magnetite, Fe_3O_4; bauxite, Al_2O_3; cassiterite, SnO_2; periclase, MgO; silica, SiO_2
sulfide	chalcopyrite, $CuFeS_2$; chalcocite, Cu_2S; sphalerite, ZnS; galena, PbS; iron pyrites, FeS_2; cinnabar, HgS
chloride	rock salt, $NaCl$; sylvite, KCl; carnallite, $KCl \cdot MgCl_2 \cdot 6H_2O$
carbonate	limestone, $CaCO_3$; magnesite, $MgCO_3$; dolomite, $MgCO_3 \cdot CaCO_3$; rhodochrosite, $MnCO_3$
sulfate	gypsum, $CaSO_4 \cdot 2H_2O$; epsom salts, $MgSO_4 \cdot 7H_2O$; barite, $BaSO_4$
silicate	beryl, $Be_3Al_2Si_6O_{18}$; kaolinite, $Al_2(Si_2O_8)(OH)_4$; spodumene, $LiAl(SiO_3)_2$

Photos of several of these minerals appear in this chapter. How many can you find?

The most widespread minerals are silicates. But extraction of metals from silicates is very difficult. Metals are obtained from silicate minerals only when there is no other more economical alternative.

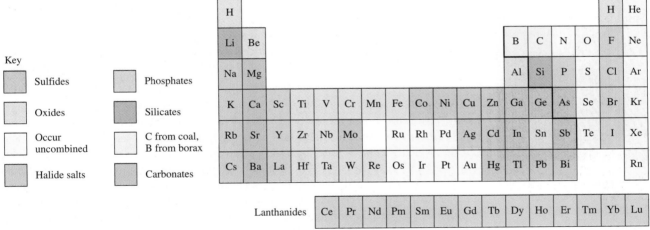

Figure 22-1 Major natural sources of the elements. The soluble halide salts are found in oceans, salt lakes, brine wells, and solid deposits. Most helium is obtained from wells in the United States and Russia. Most of the other noble gases are obtained from air.

22-2 PRETREATMENT OF ORES

After being mined, many ores must be concentrated by removal of most of the gangue. Most sulfides have relatively high densities and are denser than gangue. After pulverization, the lighter gangue particles are removed by a variety of methods. One involves blowing the lighter particles away via a cyclone separator (Figure 22-2a). These particles can be sifted out through layers of vibrating wire mesh or on inclined vibration tables.

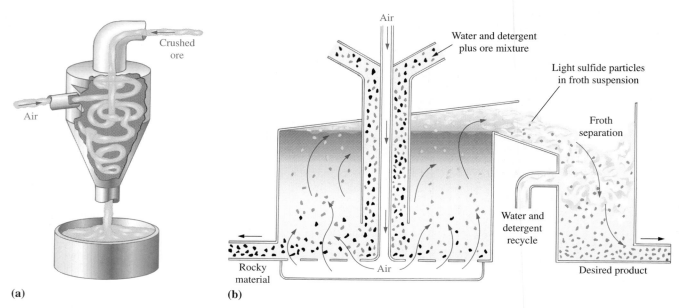

(a)   **(b)**

Some sulfide minerals: iron pyrite (fool's gold), FeS_2; black galena, PbS; yellow orpiment, As_2S_3.

Panning for gold is possible because of the high density of the native metal.

Figure 22-2 (a) The cyclone separator enriches metal ores. Crushed ore is blown in at high velocity. Centrifugal force takes the heavier particles, with higher percentages of metal, to the wall of the separator. These particles spiral down to the collection bin at the bottom. Lighter particles, not as rich in the metal, move into the center. They are carried out the top in the air stream. (b) A representation of the flotation process for enrichment of copper sulfide ore. The relatively light sulfide particles are suspended in the water–oil–detergent mixture and collected as a froth. The denser material sinks to the bottom of the container.

Hydrophilic ("water-loving") particles are made wet by water; hydrophobic ("water-hating") particles are not made wet by water. These two kinds of particles are separated in the **flotation process.** This method is particularly applicable to sulfides, carbonates, and silicates, which are not "wetted" by water or else can be made water-repellent by treatment. Their surfaces are easily covered by layers of oil or other flotation agents. A stream of air is blown through a swirled suspension of such an ore in water and oil (or other agent). Bubbles form in the oil on the mineral particles and cause them to rise to the surface. The bubbles are prevented from breaking and escaping by a layer of oil and emulsifying agent. A frothy ore concentrate forms at the surface. By varying the relative amounts of oil and water, the types of oil additive, the air pressure, and so on, it is even possible to separate one metal sulfide, carbonate, or silicate from another (Figure 22-2b).

Another pretreatment process involves chemical modification. This converts metal compounds to more easily reduced forms. Carbonates and hydroxides may be heated to drive off CO_2 and H_2O, respectively.

$$CaCO_3(s) \xrightarrow{\text{heat}} CaO(s) + CO_2(g)$$

$$Mg(OH)_2(s) \xrightarrow{\text{heat}} MgO(s) + H_2O(g)$$

Some sulfides are converted to oxides by **roasting,** that is, heating below their melting points in the presence of oxygen from air. For example,

$$2ZnS(s) + 3O_2(g) \xrightarrow{\text{heat}} 2ZnO(s) + 2SO_2(g)$$

Figure 22-3 Trees damaged by acid rain and air pollution in the southwestern United States.

Roasting sulfide ores causes air pollution. Enormous quantities of SO_2 escape into the atmosphere (Section 6-8, Part 4), where it causes great environmental damage (Figure 22-3). Federal regulations now require limitation of the amount of SO_2 that escapes with stack gases and fuel gases. Now most of the SO_2 is trapped and used in the manufacture of sulfuric acid (Section 24-12).

Some of these environmental problems are discussed in the Chemistry in Use essay "Acid Rain."

22-3 REDUCTION TO THE FREE METALS

The method used for reduction of metals from their compounds in ores to the free metals depends on how strongly the metal ions are bonded to anions. When the bonding is strong, more energy is required to reduce the metals. This makes reduction more expensive. The most active metals usually have the strongest bonding.

The least reactive metals occur in the free state and thus require no reduction. Examples include Au, Ag, and Pt. This is why gold and silver have been used as free metals since prehistoric times. Some less active metals, such as Hg, can be obtained directly from their sulfide ores by roasting. This reduces metal ions to the free metals by oxidation of the sulfide ions.

Rutile contains TiO_2.

$$HgS(s) + O_2(g) \xrightarrow{heat} SO_2(g) + Hg(g)$$

cinnabar   from air   obtained as vapor; later condensed

Roasting sulfides of the more active metals produces metal oxides, but no free metals.

$$2NiS(s) + 3O_2(g) \xrightarrow{heat} 2NiO(s) + 2SO_2(g)$$

In a process called **smelting,** the resulting metal oxides are then reduced to free metals with coke, CO, or another reducing agent, such as H_2, Fe, or Al.

$$SnO_2(s) + 2C(s) \xrightarrow{heat} Sn(\ell) + 2CO(g)$$

$$WO_3(s) + 3H_2(g) \xrightarrow{heat} W(s) + 3H_2O(g)$$

Coke is impure carbon.

(*Above*) The pink mineral rhodo-chrosite is manganese carbonate, $MnCO_3$. (*Below*) A pile of manganese metal. Manganese is an important element in steels that must withstand great shock, such as those used to construct rock crushers.

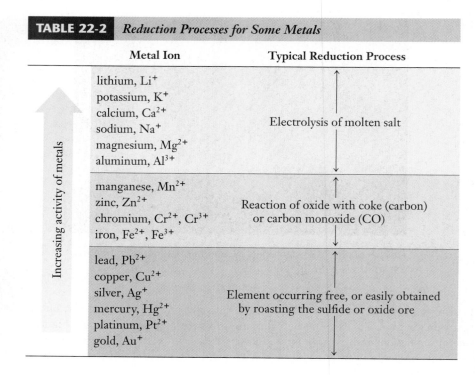

TABLE 22-2	*Reduction Processes for Some Metals*	
	Metal Ion	**Typical Reduction Process**
	lithium, Li^+ potassium, K^+ calcium, Ca^{2+} sodium, Na^+ magnesium, Mg^{2+} aluminum, Al^{3+}	Electrolysis of molten salt
	manganese, Mn^{2+} zinc, Zn^{2+} chromium, Cr^{2+}, Cr^{3+} iron, Fe^{2+}, Fe^{3+}	Reaction of oxide with coke (carbon) or carbon monoxide (CO)
	lead, Pb^{2+} copper, Cu^{2+} silver, Ag^+ mercury, Hg^{2+} platinum, Pt^{2+} gold, Au^+	Element occurring free, or easily obtained by roasting the sulfide or oxide ore

Increasing activity of metals →

The very active metals, such as Al and Na, are reduced electrochemically, usually from their anhydrous molten salts. If H_2O is present, it is reduced instead of the more active metals. Tables 22-2 and 22-3 summarize reduction processes for some metal ions.

22-4 REFINING OF METALS

Metals obtained from reduction processes are almost always impure. Further **refining** (purification) is usually required. This can be accomplished by distillation if the metal is more volatile than its impurities, as in the case of mercury. Among the metals purified by electrolysis (Sections 21-3 to 21-7) are Al (Section 22-6), Cu (Section 22-8), Ag, and Au (Section 22-9).

Zone refining is often used when extremely pure metals are desired for such applications as solar cells and semiconductors (Section 13-17). An induction heater surrounds a bar of the impure solid and passes slowly from one end to the other (Figure 22-4). As it passes, it melts portions of the bar, which slowly recrystallize as the heating element moves away. The impurity does not fit into the crystal as easily as the element of interest, so most of it is carried along in the molten portion until it reaches the end. Repeated passes of the heating element produce a bar of high purity. The end containing the impurities can be sliced off and recycled.

After refining, many metals are alloyed, or mixed with other elements, to change their physical and chemical characteristics. In some cases, certain impurities are allowed to remain during refining because their presence improves the properties of the metal. For example, a small amount of carbon in iron greatly enhances its hardness. Examples of alloys

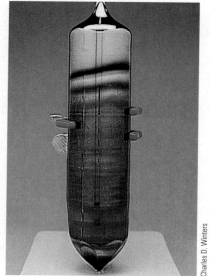

Ultrapure silicon is produced by zone refining.

TABLE 22-3	*Some Specific Reduction Processes*		
Metal	**Compound (ore)**	**Reduction Process**	**Comments**
mercury	HgS (cinnabar)	Roast reduction; heating of ore in air $HgS + O_2 \xrightarrow{heat} Hg + SO_2$	
copper	sulfides such as Cu_2S (chalcocite)	Blowing of oxygen through purified molten Cu_2S $Cu_2S + O_2 \xrightarrow{heat} 2Cu + SO_2$	Preliminary ore concentration and purification steps required to remove FeS impurities
zinc	ZnS (sphalerite)	Conversion to oxide and reduction with carbon $2ZnS + 3O_2 \xrightarrow{heat} 2ZnO + 2SO_2$ $ZnO + C \xrightarrow{heat} Zn + CO$	Process also used for the production of lead from galena, PbS
iron	Fe_2O_3 (hematite)	Reduction with carbon monoxide produced from coke $2C\text{ (coke)} + O_2 \xrightarrow{heat} 2CO$ $Fe_2O_3 + 3CO \xrightarrow{heat} 2Fe + 3CO_2$	
titanium	TiO_2 (rutile)	Conversion of oxide to halide salt and reduction with an active metal $TiO_2 + 2Cl_2 + 2C \xrightarrow{heat} TiCl_4 + 2CO$ $TiCl_4 + 2Mg \xrightarrow{heat} Ti + 2MgCl_2$	Also used for the reduction of UF_4 obtained from UO_2, pitchblende
tungsten	$FeWO_4$ (wolframite)	Reduction with hydrogen $WO_3 + 3H_2 \xrightarrow{heat} W + 3H_2O$	Used also for molybdenum
aluminum	$Al_2O_3 \cdot xH_2O$ (bauxite)	Electrolytic reduction (electrolysis) in molten cryolite, $Na_3[AlF_6]$, at 1000°C $2Al_2O_3 \longrightarrow 4Al + 3O_2$	
sodium	NaCl (sea water)	Electrolysis of molten chlorides $2NaCl \longrightarrow 2Na + Cl_2$	Also for calcium, magnesium, and other active metals in Groups IA and IIA

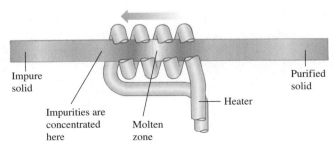

Figure 22-4 A representation of a zone refining apparatus.

include brass (Cu and Zn), bronze (Cu and Sn), duralumin (Al, Cu, Mg, and Mn), and stainless steel (Fe, Cu, Mn, plus others).

METALLURGIES OF SPECIFIC METALS

The metallurgies of Mg, Al, Fe, Cu, and Au will be discussed as specific examples. The order of increasing standard reduction potentials of these metals indicates the order of increasing ease of reduction to the free metals.

Reduction Half-Reaction	Standard Reduction Potential E^0, Volts
$Mg^{2+} + 2e^- \longrightarrow Mg$	-2.37
$Al^{3+} + 3e^- \longrightarrow Al$	-1.66
$Fe^{2+} + 2e^- \longrightarrow Fe$	-0.44
$Cu^{2+} + 2e^- \longrightarrow Cu$	$+0.337$
$Au^{3+} + 3e^- \longrightarrow Au$	$+1.50$

22-5 MAGNESIUM

Mg and Al are active metals, Fe and Cu are moderately active, and Au is relatively inactive.

Magnesium occurs widely in carbonate ores, but most Mg comes from salt brines and from the sea (Figure 22-5). Sea water is 0.13% Mg by mass. Because of its low density ($1.74\,g/cm^3$), Mg is used in lightweight structural alloys for such items as automobile and aircraft parts.

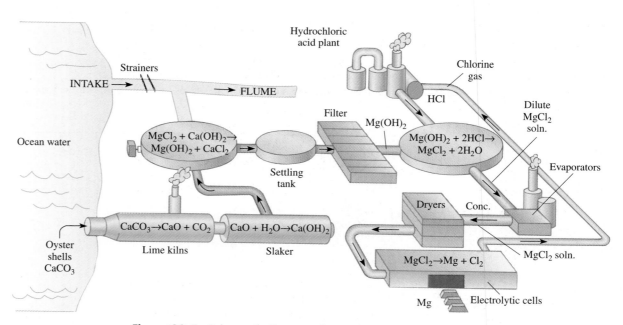

Figure 22-5 Schematic diagram of an industrial plant for the production of magnesium from the Mg^{2+} ions in sea water.

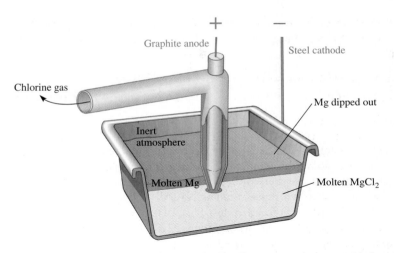

Figure 22-6 A cell for electrolyzing molten $MgCl_2$. The magnesium metal is formed on the steel cathode and rises to the top, where it is dipped off periodically. Chlorine gas is formed around the graphite anode and is piped off.

A bed of limestone, $CaCO_3$, along the Verde River, Arizona.

Magnesium ions are precipitated as $Mg(OH)_2$ by addition of $Ca(OH)_2$ (slaked lime) to sea water. The slaked lime is obtained by crushing oyster shells ($CaCO_3$), heating them to produce lime (CaO), and then adding a limited amount of water (slaking).

$$CaCO_3(s) \xrightarrow{heat} CaO(s) + CO_2(g) \qquad \text{lime production}$$

$$CaO(s) + H_2O(\ell) \longrightarrow Ca(OH)_2(s) \qquad \text{slaking lime}$$

$$Ca(OH)_2(s) + Mg^{2+}(aq) \longrightarrow Ca^{2+}(aq) + Mg(OH)_2(s) \qquad \text{precipitation}$$

The last reaction occurs because K_{sp} for $Mg(OH)_2$, 1.5×10^{-11}, is much smaller than that for $Ca(OH)_2$, 7.9×10^{-6}. The milky white suspension of $Mg(OH)_2$ is filtered, and the solid $Mg(OH)_2$ is then neutralized with HCl to produce a $MgCl_2$ solution. Evaporation of the H_2O leaves solid $MgCl_2$, which is then melted and electrolyzed (Figure 22-6) under an inert atmosphere to produce molten Mg and gaseous Cl_2. The products are separated as they are formed, to prevent recombination.

$$Mg(OH)_2(s) + 2[H^+(aq) + Cl^-(aq)] \longrightarrow [Mg^{2+}(aq) + 2Cl^-(aq)] + 2H_2O$$

$$\xrightarrow[\text{then melt solid}]{\text{evaporate solution,}} MgCl_2(\ell) \xrightarrow{\text{electrolysis}} Mg(\ell) + Cl_2(g)$$

Magnesium is cast into ingots or alloyed with other light metals. The Cl_2 by-product is used to produce more HCl for neutralization of $Mg(OH)_2$.

22-6 ALUMINUM

Aluminum is the most commercially important nonferrous metal. Its chemistry and uses will be discussed in Section 23-7. Aluminum is obtained from bauxite, or hydrated aluminum oxide, $Al_2O_3 \cdot xH_2O$. Aluminum ions can be reduced to Al by electrolysis only in the absence of H_2O. First the crushed bauxite is purified by dissolving it in a concentrated solution of NaOH to form soluble $Na[Al(OH)_4]$. Then $Al(OH)_3 \cdot xH_2O$ is precipitated from

Recall that Al_2O_3 is amphoteric. Impurities such as oxides of iron, which are not amphoteric, are left behind in the crude ore.

the filtered solution by blowing in carbon dioxide to neutralize the unreacted NaOH and one OH^- ion per formula unit of $Na[Al(OH)_4]$. Heating the hydrated product dehydrates it to Al_2O_3.

For clarity, we have not shown waters of hydration.

$$Al_2O_3(s) + 2NaOH(aq) + 3H_2O(\ell) \longrightarrow 2Na[Al(OH)_4](aq)$$

$$2Na[Al(OH)_4](aq) + CO_2(aq) \longrightarrow 2Al(OH)_3(s) + Na_2CO_3(aq) + H_2O(\ell)$$

$$2Al(OH)_3(s) \xrightarrow{\text{heat}} Al_2O_3(s) + 3H_2O(g)$$

A mixture of compounds typically has a lower melting point than any of the pure compounds (Chapter 14).

The melting point of Al_2O_3 is 2045°C; electrolysis of pure molten Al_2O_3 would have to be carried out at or above this temperature, with great expense. It can be done, however, at a much lower temperature when Al_2O_3 is mixed with much lower-melting cryolite, a mixture of NaF and AlF_3 often represented as $Na_3[AlF_6]$. The molten mixture can be electrolyzed at 1000°C with carbon electrodes. The cell used industrially for this process, called the **Hall–Héroult process**, is shown in Figure 22-7.

The inner surface of the cell is coated with carbon or carbonized iron, which functions as the cathode at which aluminum ions are reduced to the free metal. The graphite anode is oxidized to CO_2 gas and must be replaced frequently. This is one of the chief costs of aluminum production.

(cathode)	$4[Al^{3+} + 3e^- \longrightarrow Al(\ell)]$
(anode)	$3[C(s) + 2O^{2-} \longrightarrow CO_2(g) + 4e^-]$
(net reaction)	$4Al^{3+} + 3C(s) + 6O^{2-} \longrightarrow 4Al(\ell) + 3CO_2(g)$

Molten aluminum is denser than molten cryolite, so it collects in the bottom of the cell and is drawn off and cooled to a solid.

A more economical approach, the Alcoa chlorine process, has now been developed on a commercial scale. The anhydrous bauxite is first converted to $AlCl_3$ by reaction with Cl_2 in the presence of carbon. The $AlCl_3$ is then melted and electrolyzed to give aluminum, and the recovered chlorine is reused in the first step.

$$2Al_2O_3(s) + 3C(coke) + 6Cl_2(g) \longrightarrow 4AlCl_3(s) + 3CO_2(g)$$

$$2AlCl_3(\ell) \longrightarrow 2Al(\ell) + 3Cl_2(g)$$

This process uses only about 30% as much electrical energy as the Hall–Héroult process.

Charles D. Winters

Many consumer items are made of aluminum.

Figure 22-7 (a) Schematic drawing of a cell for producing aluminum by electrolysis of a melt of Al_2O_3 in $Na_3[AlF_6]$. The molten aluminum collects in the container, which acts as the cathode. (b) Casting molten aluminum. Electrolytic cells used in the Hall–Héroult process appear in the background.

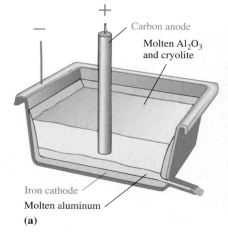

Carbon anode

Molten Al_2O_3 and cryolite

Iron cathode

Molten aluminum

(a)

Aluminum Association of America

(b)

The use of large amounts of electrical energy in electrolysis makes production of aluminum from ores an expensive metallurgy. Methods for recycling used Al use less than 10% of the energy required to make new metal from bauxite by the Hall–Héroult process. Processing of recycled Al now accounts for more than half of the production of this metal. Recycling of aluminum drink cans is now so successful that more than two thirds of this aluminum is used to produce new cans. The typical time elapsed from collection of cans until new ones made from the same aluminum appear on supermarket shelves is only a few weeks.

22-7 IRON

The most desirable iron ores contain hematite, Fe_2O_3, or magnetite, Fe_3O_4. As the available supplies of these high-grade ores have dwindled, taconite, which is magnetite in very hard silica rock, has become an important source of iron. The oxide is reduced in blast furnaces (Figure 22-8) by carbon monoxide. Coke mixed with limestone ($CaCO_3$) and crushed

In Fe_3O_4, the formal oxidation state of iron is $+\frac{8}{3}$. Two thirds of the Fe atoms are in the +3 state and the other one third are in the +2 state. The formula is sometimes written as $Fe_2O_3 \cdot FeO$.

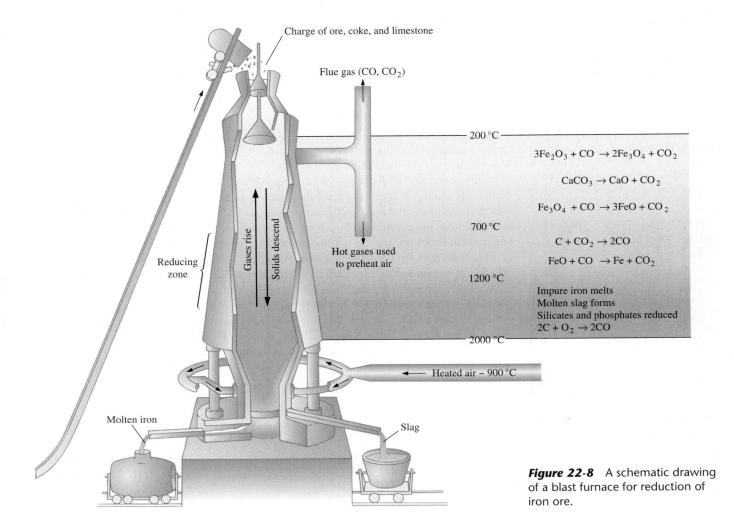

Charge of ore, coke, and limestone

Flue gas (CO, CO$_2$)

Gases rise

Solids descend

Reducing zone

Hot gases used to preheat air

200 °C

$$3Fe_2O_3 + CO \rightarrow 2Fe_3O_4 + CO_2$$

$$CaCO_3 \rightarrow CaO + CO_2$$

$$Fe_3O_4 + CO \rightarrow 3FeO + CO_2$$

700 °C

$$C + CO_2 \rightarrow 2CO$$

$$FeO + CO \rightarrow Fe + CO_2$$

1200 °C

Impure iron melts
Molten slag forms
Silicates and phosphates reduced
$$2C + O_2 \rightarrow 2CO$$

2000 °C

Heated air ~ 900 °C

Molten iron

Slag

Figure 22-8 A schematic drawing of a blast furnace for reduction of iron ore.

Iron ore is scooped in an open-pit mine.

Reaction of a metal oxide (basic) with a nonmetal oxide (acidic) forms a salt.

Considerable amounts of slag are also used to neutralize acidic soil. If there were no use for the slag, its disposal would be a serious economical and environmental problem.

Molten steel is poured from a basic oxygen furnace.

ore is admitted at the top of the furnace as the *charge*. A blast of hot air from the bottom burns the coke to carbon monoxide with the evolution of more heat.

$$2C(s) + O_2(g) \xrightarrow{\text{heat}} 2CO(g) + \text{heat}$$

Most of the oxide is reduced to molten iron by carbon monoxide, although some is reduced directly by coke. Several stepwise reductions occur (see Figure 22-8), but the main reactions for Fe_3O_4 can be summarized as follows:

$$\begin{array}{c} Fe_3O_4 + CO \longrightarrow 3FeO + CO_2 \\ 3(FeO + CO \longrightarrow Fe + CO_2) \\ \hline \text{overall:} \quad Fe_3O_4(s) + 4CO(g) \longrightarrow 3Fe(\ell) + 4CO_2(g) \end{array}$$

Much of the CO_2 reacts with excess coke to produce more CO to reduce the next incoming charge.

$$CO_2(g) + C(s) \xrightarrow{\text{heat}} 2CO(g)$$

The limestone, called a **flux,** reacts with the silica gangue in the ore to form a molten **slag** of calcium silicate.

$$\underset{\text{limestone}}{CaCO_3(s)} \xrightarrow{\text{heat}} CaO(s) + CO_2(g)$$

$$CaO(s) + \underset{\text{gangue}}{SiO_2(s)} \xrightarrow{\text{heat}} \underset{\text{slag}}{CaSiO_3(\ell)}$$

The slag is less dense than molten iron; it floats on the surface of the iron and protects it from atmospheric oxidation. Both are drawn off periodically. Some of the slag is subsequently used in the manufacture of cement.

The iron obtained from the blast furnace contains carbon, among other things. It is called **pig iron.** If it is remelted, run into molds, and cooled, it becomes **cast iron.** This is brittle because it contains much iron carbide, Fe_3C. If all the carbon is removed, nearly pure iron can be produced. It is silvery in appearance, quite soft, and of little use. If *some* of the carbon is removed and other metals such as Mn, Cr, Ni, W, Mo, and V are added, the mixture becomes stronger and is known as **steel.** There are many types of steel, containing alloyed metals and other elements in various controlled proportions. Stainless steels show high tensile strength and excellent resistance to corrosion. The most common kind contains 14–18% chromium and 7–9% nickel.

Pig iron can also be converted to steel by burning out most of the carbon with O_2 in a basic oxygen furnace (Figure 22-9). Oxygen is blown through a heat-resistant tube inserted below the surface of the *molten* iron. Carbon burns to CO, which subsequently escapes and burns to CO_2.

22-8 COPPER

Copper is widely used, especially in its alloys such as bronze (Cu and Sn) and brass (Cu and Zn). As a result, it is becoming very scarce. The U.S. Bureau of Mines estimates that the known worldwide reserves of copper ore will be exhausted during the first half of the

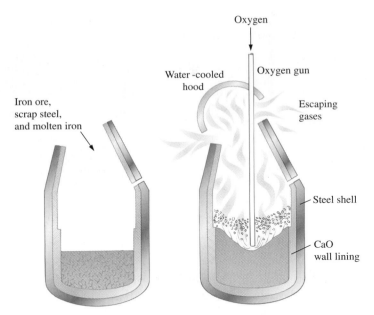

Figure 22-9 A representation of the basic oxygen process furnace. Much of the steel manufactured today is refined by blowing oxygen through a furnace that is charged with scrap and molten iron from a blast furnace. After the refined iron is withdrawn into a ladle, alloying elements are added to produce the desired steel. The steel industry is one of the nation's largest consumers of oxygen.

21st century (Figure 22-10). It is now profitable to mine ores containing as little as 0.25% copper. The increased use of fiber optics in place of copper in communications cables may help to lessen the demand for this metal. The use of superconducting materials in electricity transmission lines could eventually provide enormous savings.

The two main classes of copper ores are the mixed sulfides of copper and iron—such as chalcopyrite, $CuFeS_2$—and the basic carbonates—such as azurite, $Cu_3(CO_3)_2(OH)_2$, and malachite, $Cu_2CO_3(OH)_2$. Let's consider $CuFeS_2$ (or $CuS \cdot FeS$). The copper compound is separated from gangue by flotation (Figures 22-2b and 22-11) and then roasted to remove volatile impurities. Enough air is used to convert iron(II) sulfide, but not copper(II) sulfide, to the oxide.

$$2CuFeS_2(s) + 3O_2(g) \xrightarrow{heat} 2FeO(s) + 2CuS(s) + 2SO_2(g)$$

The roasted ore is then mixed with sand (SiO_2), crushed limestone ($CaCO_3$), and some unroasted ore that contains copper(II) sulfide in a reverberatory furnace at 1100°C. CuS is reduced to Cu_2S, which melts.

$$2CuS(s) + O_2(g) \longrightarrow Cu_2S(\ell) + SO_2(g)$$

The limestone and silica form a molten calcium silicate glass. This dissolves iron(II) oxide to form a slag less dense than the molten copper(I) sulfide, on which it floats.

$$CaCO_3(s) + SiO_2(s) \xrightarrow{heat} CaSiO_3(\ell) + CO_2(g)$$

$$CaSiO_3(\ell) + FeO(s) + SiO_2(s) \xrightarrow{heat} CaSiO_3 \cdot FeSiO_3(\ell)$$

James Cowlin

Figure 22-10 An open-pit copper mine near Bagdad, Arizona.

Modern Chemistry, Holt, Rinehart & Winston, © 1986

Figure 22-11 A copper ore being enriched by flotation.

(a)   **(b)**

Two copper-containing minerals. (a) Malachite, $Cu_3(CO_3)_2(OH)_2$ or $2CuCO_3 \cdot Cu(OH)_2$, is green. (b) azurite, $Cu_2CO_3(OH)_2$ or $CuCO_3 \cdot Cu(OH)_2$, is blue.

The steps of this process are described in Exercise 38 at the end of the chapter.

A sludge called anode mud collects under the anodes. It contains such valuable and difficult-to-oxidize elements as Au, Pt, Ag, Se, and Te. The separation, purification, and sale of these elements reduce the cost of refined copper.

The slag is periodically drained off. The molten copper(I) sulfide is drawn off into a Bessemer converter, where it is again heated and treated with air. This oxidizes sulfide ions to SO_2 and reduces copper(I) ions to metallic copper. The overall process is

$$Cu_2S(\ell) + O_2(g) \xrightarrow{\text{heat}} 2Cu(\ell) + SO_2(g)$$

The impure copper is refined using an electrolytic cell like the one shown in Figure 22-12. Thin sheets of very pure copper are made to act as cathodes by connecting them to the

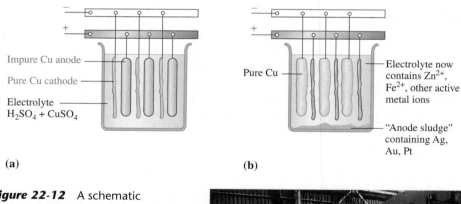

Impure Cu anode

Pure Cu cathode

Electrolyte
$H_2SO_4 + CuSO_4$

Pure Cu

Electrolyte now contains Zn^{2+}, Fe^{2+}, other active metal ions

"Anode sludge" containing Ag, Au, Pt

(a)   **(b)**

Figure 22-12 A schematic diagram of the electrolytic cell used for refining copper (a) before electrolysis and (b) after electrolysis. (c) Commercial electrolysis cells for refining copper.

(c)

negative terminal of a dc generator. Chunks of impure copper connected to the positive terminal function as anodes. The electrodes are immersed in a solution of copper(II) sulfate and sulfuric acid. When the cell operates, Cu from the impure anodes is oxidized and goes into solution as Cu^{2+} ions; Cu^{2+} ions from the solution are reduced and plate out as metallic Cu on the pure Cu cathodes. Other active metals from the impure bars also go into solution after oxidation. They do not plate out onto the cathode bars of pure Cu because of the far greater concentration of the more easily reduced Cu^{2+} ions that are already in solution. Overall, there is no net reaction, merely a simultaneous transfer of Cu from anode to solution and from solution to cathode.

$$Cu\ (impure) \longrightarrow Cu^{2+} + 2e^- \quad \text{(oxidation, anode)}$$
$$\underline{Cu^{2+} + 2e^- \longrightarrow Cu\ (pure) \quad \text{(reduction, cathode)}}$$
$$Cu\ (impure) \longrightarrow Cu\ (pure) \quad \text{(no net reaction)}$$

Nevertheless, the net effect is that small bars of very pure Cu and large bars of impure Cu are converted into large bars of very pure Cu and small bars of impure Cu.

The energy provided by the electric generator forces a decrease in the entropy of the system by separating the Cu from its impurities in the impure bars.

22-9 GOLD

Gold is an inactive metal (sometimes called a noble metal), so it occurs mostly in the native uncombined state. It is sometimes found as gold telluride. Because of its high density, metallic gold can be concentrated by panning. In this operation, gold-bearing sand and gravel are gently swirled with water in a pan. The lighter particles spill over the edge, and the denser nuggets of gold remain. Gold is concentrated by sifting crushed gravel in a stream of water on a slightly inclined shaking table that contains several low barriers. These impede the descent of the heavier gold particles but allow the lighter particles to pass over. In a modern extension to this centuries-old process, the gold is then alloyed with mercury and removed. The mercury is distilled away, leaving behind the pure gold.

Gold is also recovered from the anode sludge from electrolytic purification of copper Section 21-7. Gold is so rare that it is also obtained from very low-grade ores by the cyanide process. Air is bubbled through an agitated slurry of the ore mixed with a solution of NaCN. This causes slow oxidation of the metal and the formation of a soluble complex compound.

$$4Au(s) + 8CN^-(aq) + O_2(g) + 2H_2O(\ell) \longrightarrow 4[Au(CN)_2]^-(aq) + 4OH^-(aq)$$

After filtration, free gold can then be regenerated by electrolytic reduction or by reduction of $[Au(CN)_2]^-$ with zinc.

$$Zn(s) + 2[Au(CN)_2]^-(aq) \longrightarrow 2Au(s) + Zn(CN)_4^{2-}(aq)$$

E.I. DuPont de Nemours & Company

Mining low-gold ore in an open-pit mine.

Because of environmental concerns about mercury toxicity, the cyanide process is increasingly preferred. This is not to suggest that mercury is more toxic than cyanide. The problems due to mercury are greater in that it persists in the environment for a long time, and mercury poisoning is cumulative.

Key Terms

Alloying   Mixing of a metal with other substances (usually other metals) to modify its properties.

Cast iron   The brittle iron obtained when the pig iron from the blast furnace is remelted, run into molds, and cooled; contains much iron carbide, Fe_3C.

Charge (in metallurgy)   A sample of crushed ore as it is admitted to a furnace for smelting.

Flotation   A method by which hydrophobic (water-repelling) particles of an ore are separated from hydrophilic (water-attracting) particles in a metallurgical pretreatment process.

Flux   A substance added to react with the charge, or a product of its reduction, in metallurgy; usually added to lower a melting point.

Gangue   Sand, rock, and other impurities surrounding the mineral of interest in an ore.

Hall–Héroult process   A process in which a molten mixture of Al_2O_3, NaF, and AlF_3 is electrolyzed to obtain pure aluminum.

Metallurgy   The overall processes by which metals are extracted from ores.

Mineral   A naturally occurring inorganic solid substance having a definite chemical composition and characteristic crystalline structure, color, and hardness; contains a metal that can be extracted in a metallurgical process.

Native ore   A metal that occurs in an uncombined or free state in nature.

Ore   A natural deposit containing a mineral of an element to be extracted.

Pig iron   The iron obtained from the blast furnace.

Refining   Purifying of a substance.

Roasting   Heating a compound below its melting point in the presence of air.

Slag   Unwanted material produced during smelting.

Smelting   Chemical reduction of a metal from its ore by treating the molten ore with a reducing agent.

Steel   Iron alloyed with other metals, such as Mn, Cr, Ni, W, Mo, and V, and sometimes with C and Si.

Zone refining   A method of purifying a bar of metal by passing it through an induction heater; this causes impurities to move along in the melted portion.

Exercises

General Concepts

1. List the chemical and physical properties that we usually associate with metals.
2. Define the term "metallurgy." What does the study of metallurgy include?
3. What kinds of metals are least likely to occur in the uncombined (native) state in nature?
4. List the six anions (and their formulas) that are most often combined with metals in ores. Give at least one example of an ore of each kind. What anion is the most commonly encountered?
5. How does an ore differ from a mineral? Name the three general categories of procedures needed to produce pure metals from ores. Describe the purpose of each.
6. Briefly describe one method by which gangue can be separated from the desired mineral during the concentration of an ore.
7. Give the five general steps involved in extracting a metal from its ore and converting the metal to a useful form. Briefly describe the importance of each.
8. Describe the flotation method of ore pretreatment. Are any chemical changes involved?
9. Smelting is the process by which a metal ore is reduced. In this context, what does the term "reduced" mean?
10. Which metals are likely to be found in the free state in nature? Why are these metals in the free state in nature? Describe the location of these elements on the periodic table. Is there a pattern?

11. What kinds of ores are roasted? What kinds of compounds are converted to oxides by roasting? What kinds are converted directly to the free metals?
12. Of the following compounds, which would you expect to require electrolysis to obtain the free metals: KCl; $Cr_2(SO_4)_3$; Fe_2O_3; Al_2O_3; Ag_2S; $MgSO_4$? Why?
13. At which electrode is the free metal produced in the electrolysis of a metal compound? Why?
*14. Write the equation that describes the electrolysis of a brine solution to form $NaOH$, Cl_2, and H_2. What mass of each substance will be produced in an electrolysis cell for each mole of electrons passed through the cell? Assume 100% efficiency. (See Section 21-4)
15. The following equations represent reactions used in some important metallurgical processes.
    (a) $Fe_3O_4(s) + CO(g) \longrightarrow Fe(\ell) + CO_2(g)$
    (b) $MgCO_3(s) + SiO_2(s) \longrightarrow MgSiO_3(\ell) + CO_2(g)$
    (c) $Au(s) + CN^- + H_2O(\ell) + O_2(g) \longrightarrow$
    $$[Au(CN)_2]^- + OH^-$$
    Balance the equations. Which one(s) represent reduction to a free metal?
16. Repeat Exercise 15 for
    (a) Al_2O_3 (cryolite solution) $\xrightarrow{\text{electrolysis}}$ $Al(\ell) + O_2(g)$
    (b) $PbSO_4(s) + PbS(s) \longrightarrow Pb(\ell) + SO_2(g)$
    (c) $TaCl_5(g) + Mg(\ell) \longrightarrow Ta(s) + MgCl_2(\ell)$
17. Suggest a method of obtaining manganese from an ore containing manganese(III) oxide, Mn_2O_3. On what basis do you make the suggestion?

18. What is the purpose of utilizing the basic oxygen furnace after the blast furnace in the production of iron?

19. A blast furnace utilizes or produces five species that include carbon. (a) List the five species that contain carbon. (b) Explain the function of each step in the chain of events from heating the ore to the production of steel.

20. Magnesium hydroxide can be precipitated from sea water by adding slaked lime, $Ca(OH)_2$. (a) Why does this precipitation occur? (b) Could this process be used to remove sodium from sea water? Why?

21. Describe the metallurgy of (a) copper and (b) magnesium.

22. Describe the metallurgy of (a) iron and (b) gold.

Native gold.

23. Briefly describe the Hall–Héroult process for the commercial preparation of aluminum.

24. (a) Compare the refining of aluminum by the Hall–Héroult process and by the Alcoa chlorine process. (b) Which process is less expensive and why? (c) The Alcoa chlorine process is potentially more dangerous to the workers than is the Hall–Héroult process. Why?

25. A common iron ore is magnetite, Fe_3O_4. In this formula for magnetite the oxidation number of iron is not an integer. Write the formula in a form that includes only integer oxidation numbers.

26. Name some common minerals that contain iron. Write the chemical formula for the iron compound in each. What is the oxidation number of iron in each substance?

27. What is steel? How does the hardness of iron compare with that of steel?

28. (a) What is the difference in composition between pure iron and steel? (b) What is the name of the process by which steel is produced?

29. (a) What is an alloy? (b) Why are alloys produced? (c) How are alloys produced?

30. Describe and illustrate the electrolytic refining of Cu.

31. Name the undesirable gaseous product formed during the roasting of copper and other sulfide ores. Why is it undesirable?

CONCEPTUAL EXERCISES

32. One of the problems with roasting sulfide ores is the air pollution caused by the gas that is produced. Describe some consequences of releasing this gas into the atmosphere.

33. Gold can be obtained by the separation of gold from the anode sludge of the electrolytic purification of copper. (a) Considering the nature of gold and copper, is this source of gold to be expected? Explain. (b) Would finding silver in the sludge also be expected? Explain.

34. The iron formed in a blast furnace is called pig iron, which contains impurities that make the metal brittle. What are these impurities, and how did they come about?

35. Discuss how pig iron is refined to form steel.

BUILDING YOUR KNOWLEDGE

36. The reaction

$$FeO(s) + CO(g) \longrightarrow Fe(s) + CO_2(g)$$

takes place in the blast furnace at a temperature of 800 K. (a) Calculate ΔH^0_{800} for this reaction, using $\Delta H^0_{f,\,800} = -268$ kJ/mol for FeO, -111 kJ/mol for CO, and -394 kJ/mol for CO_2. Is this a favorable enthalpy change? (b) Calculate ΔG^0_{800} for this reaction, using $\Delta G^0_{f,\,800} = -219$ kJ/mol for FeO, -182 kJ/mol for CO, and -396 kJ/mol for CO_2. Is this a favorable free energy change? (c) Using your values of ΔH^0_{800} and ΔG^0_{800}, calculate ΔS^0_{800}.

37. During the operation of a blast furnace, coke reacts with the oxygen in air to produce carbon monoxide, which, in turn, serves as the reducing agent for the iron ore. Assuming the formula of the iron ore to be Fe_2O_3, calculate the mass of air needed for each ton of iron produced. Assume air to be 21% O_2 by mass, and assume that the process is 90.5% efficient.

38. The following reactions take place during the extraction of copper from copper ore.
   (a) $2Cu_2S(\ell) + 3O_2(g) \longrightarrow 2Cu_2O(\ell) + 2SO_2(g)$
   (b) $2Cu_2O(\ell) + Cu_2S(\ell) \longrightarrow 6Cu(\ell) + SO_2(g)$
   Identify the oxidizing and reducing agents. Show that each equation is correctly balanced by demonstrating that the increase and decrease in oxidation numbers are equal.

*39. Assuming complete recovery of metal, which of the following ores would yield the greater quantity of copper on a mass basis? (a) an ore containing 3.60 mass % azurite, $Cu(OH)_2 \cdot 2CuCO_3$, or (b) an ore containing 4.95 mass % chalcopyrite, $CuFeS_2$.

*40. What mass of copper could be electroplated from a solution of $CuSO_4$, using an electric current of 2.50 A flowing for 8.00 h? (Assume 100% efficiency.)

*41. (a) Calculate the weight, in pounds, of sulfur dioxide produced in the roasting of 1 ton of chalcocite ore containing 10.3% Cu_2S, 0.94% Ag_2S, and no other source of sulfur. (b) What weight of sulfuric acid can be prepared from the SO_2 generated, assuming 85% of it can be recovered from stack gases, and 88% of that recovered can be converted to sulfuric acid? (c) How many pounds of pure copper can be obtained, assuming 78% efficient extraction and purification? (d) How many pounds of silver can be produced, assuming 85% of it can be extracted and purified?

*42. Fifty-five pounds of Al_2O_3, obtained from bauxite, is mixed with cryolite and electrolyzed. How long would a 0.900-A current have to be passed to convert all the Al^{3+} (from Al_2O_3) to aluminum metal? What volume of oxygen, collected at 775 torr and 125°C, would be produced in the same period of time?

*43. Calculate the percentage of iron in hematite ore containing 60.5% Fe_2O_3 by mass. How many pounds of iron would be contained in 1 ton of the ore?

44. The roasting of cinnabar, HgS, is a process by which free mercury is obtained.

$$HgS + O_2 \overset{heat}{\rightleftharpoons} Hg + SO_2$$

Assuming that the roasting process is 100% efficient, what mass, in tons, of SO_2 is released into the air by the roasting of 400. tons of cinnabar?

45. The roasting of covellite, CuS, is a process by which free copper is obtained.

$$CuS + O_2 \overset{heat}{\rightleftharpoons} Cu + SO_2$$

Assuming that the roasting process is 90.5% efficient, what mass, in tons, of SO_2 is released into the air by the roasting of 400. tons of covellite?

46. Consider the separation of iron from hematite, Fe_2O_3, in a blast furnace which has an overall efficiency of 98.0% in order to leave a specific amount of carbon in the final product.

$$2C(coke) + O_2 \overset{heat}{\rightleftharpoons} 2CO$$

$$Fe_2O_3 + 3CO \overset{heat}{\rightleftharpoons} 2Fe + 3CO_2$$

Calculate the mass of coke, in tons, necessary for the ultimate reaction with 150. tons of hematite.

*47. Using data from Appendix K, calculate ΔG^0_{298} for the following reactions.
(a) $Al_2O_3(s) \longrightarrow 2Al(s) + \frac{3}{2}O_2(g)$
(b) $Fe_2O_3(s) \longrightarrow 2Fe(s) + \frac{3}{2}O_2(g)$
(c) $HgS(s) \longrightarrow Hg(\ell) + S(s)$
Are any of the reactions spontaneous? Are the ΔG^0_{298} values in line with what would be predicted based on the relative activities of the metal ions involved? Do increases in temperature favor these reactions?

*48. Find the standard molar enthalpies of formation of Al_2O_3, Fe_2O_3, and HgS in Appendix K. Are the values in line with what might be predicted, in view of the methods by which the metal ions are reduced in extractive metallurgy?

BEYOND THE TEXTBOOK

Go to the textbook website at

http://www.brookscole.com/chemistry/whitten

for additional activities and exercises based on the General Chemistry Interactive CD-ROM, the World Wide Web, and library resources.

InfoTrac College Edition

For additional readings, go to InfoTrac College Edition, your online research library at:

http://infotrac.thomsonlearning.com

Matthew McVay/Tony Stone Images

Metals and alloys have a wide variety of uses. Strength, lightness, and corrosion resistance are some properties that are required of metals used in aircraft construction.

OUTLINE

OBJECTIVES

After you have studied this chapter, you should be able to

- Describe the properties and occurrence of the Group IA metals
- Describe some important reactions of the Group IA metals
- Describe some important uses of the Group IA metals and their compounds
- Describe the properties and occurrence of the Group IIA metals
- Describe some important reactions of the Group IIA metals
- Describe some important uses of the Group IIA metals and their compounds
- Identify the post-transition metals
- Describe some important periodic trends in the properties of Group IIIA metals and some of their compounds
- Describe the important properties of aluminum and some of its compounds
- Identify the *d*-transition metals and describe some of their important properties and their typical compounds
- Describe typical oxidation states of the transition metals
- Describe some oxides, oxyanions, and hydroxides of chromium

I n this chapter we will discuss some **representative metals** and some *d*-**transition metals.** The representative elements are those in the A groups of the periodic table. They have valence electrons in their outermost *s* and *p* atomic orbitals. Metallic character increases from top to bottom within groups and from right to left within periods. All the elements in Groups IA (except H) and IIA are metals. The heavier members of Groups IIIA, IVA, and VA are called **post-transition metals.**

Hydrogen is included in Group IA in the periodic table, but it is *not* a metal.

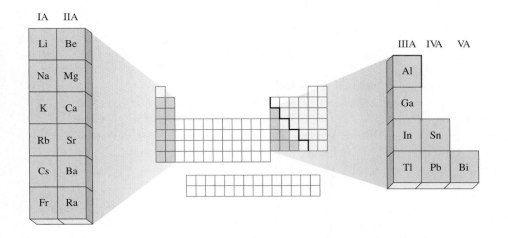

THE ALKALI METALS (GROUP IA)

23-1 GROUP IA METALS: PROPERTIES AND OCCURRENCE

The alkali metals are not found free in nature, because they are so easily oxidized. They are most economically produced by electrolysis of their molten salts. Sodium (2.6% abundance by mass) and potassium (2.4% abundance) are very common in the earth's crust. The other IA metals are quite rare. Francium consists only of short-lived radioactive isotopes formed by alpha-particle emission from actinium (Section 26-4). Both potassium and cesium also have natural radioisotopes. Potassium-40 is important in the potassium–argon radioactive decay method of dating ancient objects (Section 26-12). The properties of the alkali metals vary regularly as the group is descended (Table 23-1).

The free metals, except lithium, are soft, silvery, corrosive metals that can be cut with a knife; lithium is harder. Cesium is slightly golden and melts in the hand (wrapped in plastic because it is so corrosive). The relatively low melting and boiling points of the alkali metals result from their fairly weak bonding forces. Each atom can furnish only one electron for metallic bonding (Section 13-17). Because their outer electrons are so loosely held, the metals are excellent electrical and thermal conductors. They ionize when irradiated with low-energy light (the photoelectric effect). These effects become more pronounced with increasing atomic size. Cesium is used in photoelectric cells.

The low ionization energies of the IA metals show that the single electron in the outer shell is very easily removed. In all alkali metal compounds the metals exhibit the +1 oxidation state. Virtually all are ionic. The extremely high second ionization energies show that removal of an electron from a filled shell is impossible by chemical means.

See the discussion of electrolysis of sodium chloride in Section 21-3.

Sodium, like the other alkali metals, is an excellent conductor of electricity.

TABLE 23-1	*Properties of the Group IA Metals*					
Property	Li	Na	K	Rb	Cs	Fr
Outer electrons	$2s^1$	$3s^1$	$4s^1$	$5s^1$	$6s^1$	$7s^1$
Melting point (°C)	186	97.8	63.6	38.9	28.5	27
Boiling point (°C)	1347	904	774	688	678	677
Density (g/cm^3)	0.534	0.971	0.862	1.53	1.87	—
Atomic radius (Å)	1.52	1.86	2.27	2.48	2.65	—
Ionic radius, M$^+$ (Å)	0.90	1.16	1.52	1.66	1.81	—
Electronegativity	1.0	1.0	0.9	0.9	0.8	0.8
E^0 (volts): $M^+(aq) + e^- \longrightarrow M(s)$	−3.05	−2.71	−2.93	−2.93	−2.92	—
Ionization energies (kJ/mol)						
$M(g) \longrightarrow M^+(g) + e^-$	520	496	419	403	377	—
$M^+(g) \longrightarrow M^{2+}(g) + e^-$	7298	4562	3051	2632	2420	—
$\Delta H^0_{hydration}$ (kJ/mol): $M^+(g) + xH_2O \longrightarrow M^+(aq)$	−544	−435	−351	−293	−264	—

We might expect the standard reduction potentials of the metal ions to become less favorable (more negative) as the ionization energies for the metal atoms become more favorable (less endothermic). The magnitude of the standard reduction potential of Li, −3.05 volts, is unexpectedly large, however. The first ionization energy is the amount of energy absorbed when a mole of *gaseous* atoms ionize. The standard reduction potential, E^0, indicates the ease with which *aqueous* ions are reduced to the metal (Section 21-14). Thus, hydration energies must also be considered (Section 14-2). Because the Li$^+$ ion is so small, its charge density (ratio of charge to size) is very high. It therefore exerts a stronger attraction for polar H$_2$O molecules than do the other IA ions. These H$_2$O molecules must be stripped off during the reduction process in a very endothermic manner, making the E^0 for the Li$^+$/Li couple very negative (see Table 23-1).

The high charge density of Li$^+$ ion accounts for its ability to polarize large anions. This gives a higher degree of covalent character in Li compounds than in other corresponding alkali metal compounds. For example, LiCl is soluble in ethyl alcohol, a less polar solvent than water; NaCl is not. Salts of the alkali metals with small anions are very soluble in water, but salts with large and complex anions, such as silicates and aluminosilicates, are not very soluble.

Polarization of an anion refers to distortion of its electron cloud. The ability of a cation to polarize an anion increases with increasing charge density (ratio of charge to size) of the cation.

23-2 REACTIONS OF THE GROUP IA METALS

Many of the reactions of the alkali metals are summarized in Table 23-2. All are characterized by the loss of one electron per metal atom. These metals are very strong reducing agents. Reactions of the alkali metals with H$_2$ and O$_2$ were discussed in Sections 6-7 and 6-8, reactions with the halogens in Section 7-2; and reactions with water in Section 4-10, Part 2.

The high reactivities of the alkali metals are illustrated by their vigorous reactions with water. Lithium reacts readily; sodium reacts so vigorously that the hydrogen gas it forms may ignite; and potassium, rubidium, and cesium cause the H$_2$ to burst into flames when

TABLE 23-2	*Some Reactions of the Group IA Metals*

Reaction	Remarks
$4M + O_2 \longrightarrow 2M_2O$	Limited O_2
$4Li + O_2 \longrightarrow 2Li_2O$	Excess O_2 (lithium oxide)
$2Na + O_2 \longrightarrow Na_2O_2$	(sodium peroxide)
$M + O_2 \longrightarrow MO_2$	M = K, Rb, Cs; excess O_2 (superoxides)
$2M + H_2 \longrightarrow 2MH$	Molten metals
$6Li + N_2 \longrightarrow 2Li_3N$	At high temperature
$2M + X_2 \longrightarrow 2MX$	X = halogen (Group VIIA)
$2M + S \longrightarrow M_2S$	Also with Se, Te of Group VIA
$12M + P_4 \longrightarrow 4M_3P$	
$2M + 2H_2O \longrightarrow 2MOH + H_2$	K, Rb, and Cs react explosively
$2M + 2NH_3 \longrightarrow 2MNH_2 + H_2$	With $NH_3(\ell)$ in presence of catalyst; with $NH_3(g)$ at high temperature (solutions also contain M^+ + solvated e^-)

the metals are dropped into water. The large amounts of heat evolved provide the activation energy to ignite the hydrogen. The elements also react with water vapor in the air or with moisture from the skin.

$$2K(s) + 2H_2O(\ell) \longrightarrow 2[K^+(aq) + OH^-(aq)] + H_2(g) \qquad \Delta H^0 = -390.8 \text{ kJ/mol rxn}$$

Alkali metals are stored under anhydrous nonpolar liquids such as mineral oil.

As is often true for elements of the second period, Li differs in many ways from the other members of its family. Its ionic charge density and electronegativity are close to those of Mg, so Li compounds resemble those of Mg in some ways. This illustrates the **diagonal similarities** that exist between elements in successive groups near the top of the periodic table.

IA	IIA	IIIA	IVA
Li	Be	B	C
Na	Mg	Al	Si

Lithium is the only IA metal that combines with N_2 to form a nitride, Li_3N. Magnesium readily forms magnesium nitride, Mg_3N_2. Both metals readily combine with carbon to form carbides, whereas the other alkali metals do not react readily with carbon. The solubilities of Li compounds are closer to those of Mg compounds than to those of other IA compounds. The fluorides, phosphates, and carbonates of both Li and Mg are only slightly soluble, but their chlorides, bromides, and iodides are very soluble. Both Li and Mg form normal oxides, Li_2O and MgO, when burned in air at 1 atmosphere pressure. The other alkali metals form peroxides or superoxides.

The IA metal oxides are basic. They react with water to form strong bases.

$$Na_2O(s) + H_2O(\ell) \longrightarrow 2[Na^+(aq) + OH^-(aq)]$$
$$K_2O(s) + H_2O(\ell) \longrightarrow 2[K^+(aq) + OH^-(aq)]$$

Charles Steele

Sodium reacts vigorously with water.

$$2Na(s) + 2H_2O(\ell) \longrightarrow 2[Na^+(aq) + OH^-(aq)] + H_2(g)$$

The indicator, phenolphthalein, was added to the water. As NaOH forms, the solution turns pink.

23-3 USES OF GROUP IA METALS AND THEIR COMPOUNDS

Lithium, Li

Metallic lithium has the highest heat capacity of any element. It is used as a heat transfer medium in experimental nuclear reactors. Extremely lightweight lithium–aluminum alloys are used in aircraft construction. Lithium compounds are used in some lightweight dry cells and storage batteries because they have very long lives, even in extreme temperatures. LiCl and LiBr are very hygroscopic (they take up moisture readily) and are used in industrial drying processes and air conditioning. Lithium compounds are used for the treatment of some types of mental disorders (mainly manic depression).

Sodium, Na

Sodium is by far the most widely used alkali metal because it is so abundant. Its salts are essential for life. The metal itself is used as a reducing agent in the manufacture of drugs and dyes and in the metallurgy of such metals as titanium and zirconium.

$$TiCl_4(g) + 4Na(\ell) \xrightarrow{heat} 4NaCl(s) + Ti(s)$$

Highway lamps often incorporate Na arcs, which produce a bright yellow glow. A few examples of the uses of sodium compounds are NaOH, called caustic soda, lye, or soda lye (used for production of rayon, cleansers, textiles, soap, paper, and many polymers); Na_2CO_3, called soda or soda ash, and $Na_2CO_3 \cdot 10H_2O$, called washing soda (also used as a substitute for NaOH when a weaker base is acceptable); $NaHCO_3$, called baking soda or

(text continues on page 922)

The yellowish glow of some highway lamps is due to a sodium arc. Mercury lamps give a bluish glow.

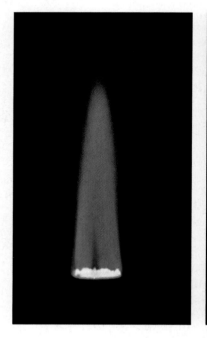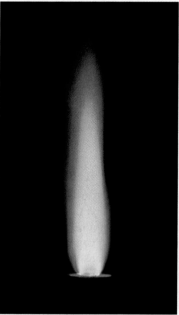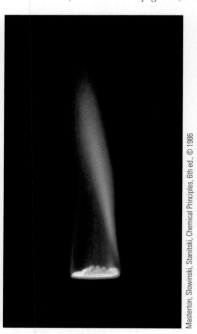

Spacings of energy levels are different for different alkali metals. The salts of the alkali metals impart characteristic colors to flames: lithium (*red*), sodium (*yellow*), and potassium (*violet*).

CHEMISTRY IN USE

The Development of Science

Trace Elements and Life

More than 300 years ago iron was the first trace element shown to be essential in the human diet. An English physician, Thomas Sedenham, soaked "iron and steel filings" in cold Rhenish wine. He used the resulting solution to treat patients suffering from chlorosis, now known to be an iron-deficiency anemia. Nearly 20 trace elements are now believed to be required by humans. The discovery of the biological functions of trace elements is an exciting and controversial area of human nutrition research.

The trace elements can be classified into several categories (see table). In 1989, the National Research Council recognized that iron, iodine, zinc, selenium, copper, chromium, manganese, and molybdenum were dietary essentials for humans. Fluorine is also considered to be valuable for human health, because of its benefits to the teeth and skeleton. These nine trace elements are required by humans and other animals because they are essential components in metalloenzymes and hormones or because they promote health in a specific tissue (such as fluorine in the teeth and skeleton). The trace elements required by the human body in milligram quantities include iron, zinc, copper, manganese, and fluorine. Trace elements required in microgram (μg) quantities include iodine, selenium, chromium, and molybdenum. Although probably required in μg quantities, no dietary recommendations have been made for arsenic, nickel, silicon, and boron even though there is evidence, primarily from animals, that they are essential. There is only weak evidence that cadmium, lead, lithium, tin, vanadium, and bromine are essential for humans.

Iron deficiency is one of the most common nutrient deficiencies in the world, occurring in up to 60% of the women, infants, and children of some countries. Anemia, characterized by a low concentration of hemoglobin in the blood or by a low volume of packed red blood cells, is the usual symptom of iron deficiency. Other symptoms include fatigue and cognitive disorders. Up to 1% of the population may have the genetic disease known as hereditary hemochromatosis, which results in excess absorption of dietary iron, and can lead to liver and heart damage. Concern about this disease may eventually lead to routine screening so that persons with this disorder can be treated before severe symptoms develop and can avoid foods and supplements with large amounts of iron. The recommended dietary allowance for iron for women aged 23 to 50 years is 50% higher than that for men in the same age group because of the iron lost in menstrual bleeding.

Iodine deficiency remains a major cause of mental retardation and infant mortality and morbidity worldwide—even

Many dietary supplements include essential trace elements.

Charles D. Winters

(Box continues on next page)

Dietary Trace Elements

Known to Be Essential	Known or Suspected Functions
iron	Hemoglobin, energy metabolism
iodine	Thyroid hormones
zinc	Enzymes, protein synthesis, cell division
copper	Hemoglobin, bone, nerves, vascular system
selenium	Enzymes, protect against oxidant stress
chromium	Insulin action
manganese	Enzymes, bone
molybdenum	Enzymes, sulfur metabolism
fluorine	Bones, teeth

Substantial Evidence for Essentiality	Known or Suspected Functions
arsenic	Amino acid metabolism
boron	Metabolism of calcium, magnesium, hormones
nickel	Not known, suspected in some enzymes
silicon	Bone and connective tissue

Weak Evidence for Essentiality	Known or Suspected Functions
bromine	Not yet known
cadmium	Not yet known
lead	Not yet known
lithium	Not yet known
tin	Not yet known
vanadium	Not yet known

though iodine was shown to be essential for human health nearly 100 years ago. More than 1 billion people are believed to be at risk for iodine deficiency. In 1986, the International Council for the Control of Iodine Deficiency Disorders was established in an effort to improve iodine nutrition and alleviate human suffering. This council works closely with the World Health Organization, the United Nations International Children's Fund, and the United Nations to alleviate iodine deficiency. Iodine is required for the thyroid hormones, thyroxine and triiodothyronine, that regulate the metabolic rate and O_2 consumption of cells. Iodine is also intimately involved in the control of growth and development, particularly during fetal and infant life.

In the 1930s, zinc was discovered to be a dietary essential in animals. Zinc deficiency was recognized as a potential public health problem in the 1960s in Iran, where endemic hypogonadism (delayed sexual development) and dwarfism were discovered in adolescents consuming insufficient dietary zinc.

More than 200 zinc enzymes have been discovered. Zinc is also important for the structure and function of biomembranes. Loss of zinc from these membranes results in increased susceptibility to oxidative damage, structural strains, and alterations in specific receptor sites and transport systems. Zinc also helps to stabilize the structures of RNA, DNA, and ribosomes. Several transcription factors contain "Zn fingers," which are needed for the binding of these transcription factors to the DNA. Thus, zinc is absolutely necessary for adequate growth, protein synthesis, and cell division. The best sources of zinc in the human diet are animal foods such as meat, fish, poultry, and dairy products.

Copper was also shown to be essential in the early 1900s. Copper is needed for the absorption and mobilization of iron, so a deficiency of copper causes a type of anemia that is difficult to distinguish from iron deficiency anemia. Copper is also needed for the cardiovascular system, bone, brain, and nervous system. Premature and malnourished infants are particularly susceptible to developing copper deficiency, in part because milk is a poor source of copper. Whole grains, legumes, and nuts are the major dietary sources of copper.

Selenium was first suspected of being a dietary essential in the 1950s. Selenium is considered to be an antioxidant nutrient because it is present in enzymes that help protect against toxic species of oxygen and free radicals. Selenium deficiency is a major public health problem in certain parts of China, where it increases the risk of heart disease, bone and joint disorders, and liver cancer. Selenium is currently under intensive investigation as a possible protector against cancer. The content of selenium in foods is highly variable and dependent on the selenium content of the soil. Generally the best sources of selenium are muscle meats, cereals, and grains.

The 1950s also saw the first evidence that chromium might be a dietary essential. Chromium is believed to promote the action of insulin and thus influences the metabolism of carbohydrates, fats, and proteins. Reports of severe human deficiency of chromium are rare and have been found primarily in people receiving only intravenous feedings for several months or years. Only a few laboratories in the world can accurately measure the amount of chromium in foods and body tissues because chromium is present in stainless steel, which is ubiquitous in analytical laboratories and easily contaminates biological samples.

Manganese and molybdenum are essential for enzymes in humans and other animals, but a dietary deficiency of these minerals is exceedingly rare in humans. Cobalt is essential for vitamin B_{12}, but the human body cannot make vitamin B_{12} from cobalt and thus requires the preformed vitamin from dietary sources. (It is possible to derive some vitamin B_{12} from bacterial synthesis in the digestive tract.)

Efforts to discover whether other elements might be essential intensified during the 1970s. Although it is believed that arsenic, nickel, silicon, and boron are probably essential to humans, it has been difficult to determine whether other minerals have specific biological functions in humans or other animals.

There are several reasons why it is difficult to establish the essentiality of trace elements. Some elements, such as arsenic and selenium, were first recognized for their extreme toxicity, so it has been difficult to convince many health specialists that a toxic element might also be a dietary essential at low levels. Also, most trace elements are present in extremely small amounts in diets and in tissues, and few laboratories are equipped to prevent contamination of samples and to measure these elements with the necessary precision.

Two factors have aided in the discovery of the roles of many trace elements. One is the availability of two highly sensitive analytical techniques, activation analysis and electrothermal atomic absorption spectroscopy, that allow detection of these elements in concentrations of only a few parts per billion. The other is the use of special isolation chambers that allow study of animals under carefully controlled conditions, free of unwanted contaminants. The diets fed to animals and their air supply must be carefully purified to keep out even traces of unwanted elements, and their cages must be made of plastics that contain no metals.

Our understanding of the biological functions of trace elements is changing the way scientists think about diet and health. For example, supplements of manganese, copper, and zinc in combination with calcium have recently been shown to improve human bone health to a greater extent than just calcium alone. Silicon and boron are also believed to be important for bone health. Deficiencies of selenium or copper are suspected by some scientists of increasing the risk of cancer or heart disease. Because chromium, copper, and zinc influence glucose metabolism, future prevention and treatment strategies for diabetes may involve these nutrients.

Mary Ann Johnson
College of Family and Consumer Sciences
University of Georgia

bicarbonate of soda (used for baking and other household uses); NaCl (used as table salt and as the source of all other compounds of Na and Cl); $NaNO_3$, called Chile saltpeter (a nitrogen fertilizer); Na_2SO_4, called salt cake, a by-product of HCl manufacture (used for production of brown wrapping paper and cardboard boxes); and NaH (used for synthesis of $NaBH_4$, which is used to recover silver and mercury from waste water).

Other Group IA Metals

Like salts of Na, those of potassium are essential for life. KNO_3, commonly known as niter or saltpeter, is used as a potassium and nitrogen fertilizer. Most other major industrial uses for K can be satisfied with the more abundant and less expensive Na.

There are very few practical uses for the rare metals rubidium, cesium, and francium. Cesium is used in some photoelectric cells (Section 5-11).

THE ALKALINE EARTH METALS (GROUP IIA)

sci LINKS.

TOPIC: Alkaline Earth Metals
GO TO: www.scilinks.org
sciLINKS CODE: WCH2320

23-4   GROUP IIA METALS: PROPERTIES AND OCCURRENCE

The alkaline earth metals are all silvery white, malleable, ductile, and somewhat harder than their neighbors in Group IA. Activity increases from top to bottom within the group, with Ca, Sr, and Ba being considered quite active. Each has two electrons in its highest occupied energy level. Both electrons are lost in ionic compound formation, though not as easily as the outer electron of an alkali metal. Compare the ionization energies in Tables 23-1 and 23-3. Most IIA compounds are ionic, but those of Be exhibit a great deal of covalent character. This is due to the extremely high charge density of Be^{2+}. Compounds of beryllium therefore resemble those of aluminum in Group IIIA (diagonal similarities). The IIA elements exhibit the $+2$ oxidation state in all their compounds. The tendency to form $2+$ ions increases from Be to Ra.

The alkaline earth metals show a wider range of chemical properties than the alkali metals. The IIA metals are not as reactive as the IA metals, but they are much too reactive to occur free in nature. They are obtained by electrolysis of their molten chlorides. Calcium and magnesium are abundant in the earth's crust, especially as carbonates and sulfates. Beryllium, strontium, and barium are less abundant. All known radium isotopes are radioactive and are extremely rare.

In Section 8-5 we found that gaseous $BeCl_2$ is linear. The Be atoms in $BeCl_2$ molecules, however, act as Lewis acids. In the solid state, the Cl atoms form coordinate covalent bonds to Be, resulting in a polymeric structure. In such compounds, Be follows the octet rule.

23-5   REACTIONS OF THE GROUP IIA METALS

Table 23-4 summarizes some reactions of the alkaline earth metals, which, except for stoichiometry, are similar to the corresponding reactions of the alkali metals. Reactions with hydrogen and oxygen were discussed in Sections 6-7 and 6-8.

Except for Be, all the alkaline earth metals are oxidized to oxides in air. The IIA oxides (except BeO) are basic and react with water to give hydroxides. Beryllium hydroxide, $Be(OH)_2$, is quite insoluble in water and is amphoteric. Magnesium hydroxide, $Mg(OH)_2$, is only slightly soluble in water. The hydroxides of Ca, Sr, and Ba are strong bases.

Beryllium is at the top of Group IIA. Its oxide is amphoteric, whereas oxides of the heavier members are basic. Metallic character increases from top to bottom within a group and

Amphoterism is the ability of a substance to react with both acids and bases (see Section 10-6).

TABLE 23-3	*Properties of the Group IIA Metals*					
Property	Be	Mg	Ca	Sr	Ba	Ra
Outer electrons	$2s^2$	$3s^2$	$4s^2$	$5s^2$	$6s^2$	$7s^2$
Melting point (°C)	1283	649	839	770	725	700
Boiling point (°C)	2484	1105	1484	1384	1640	1140
Density (g/cm³)	1.85	1.74	1.55	2.60	3.51	5
Atomic radius (Å)	1.12	1.60	1.97	2.15	2.22	2.20
Ionic radius, M^{2+} (Å)	0.59	0.85	1.14	1.32	1.49	—
Electronegativity	1.5	1.2	1.0	1.0	1.0	1.0
E^0 (volts): $M^{2+}(aq) + 2e^- \longrightarrow 2M(s)$	-1.85	-2.37	-2.87	-2.89	-2.90	-2.92
Ionization energies (kJ/mol)						
$M(g) \longrightarrow M^+(g) + e^-$	899	738	599	550	503	509
$M^+(g) \longrightarrow M^{2+}(g) + e^-$	1757	1451	1145	1064	965	(979)
$\Delta H^0_{hydration}$ (kJ/mol): $M^{2+}(g) \longrightarrow M^{2+}(aq)$	—	-1925	-1650	-1485	-1276	—

from right to left across a period. This results in increasing basicity and decreasing acidity of the oxides in the same directions, as shown in the following table.

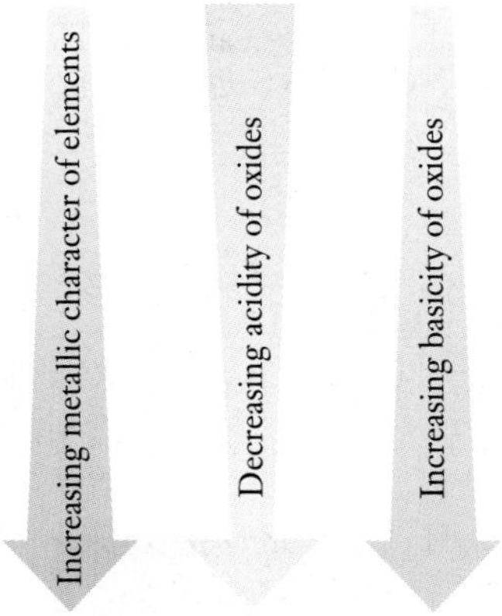

Group IA	Group IIA	Group IIIA
Li_2O (basic)	BeO (amphoteric)	B_2O_3 (amphoteric)
Na_2O (basic)	MgO (basic)	Al_2O_3 (amphoteric)
K_2O (basic)	CaO (basic)	Ga_2O_3 (amphoteric)
		In_2O_3 (basic)

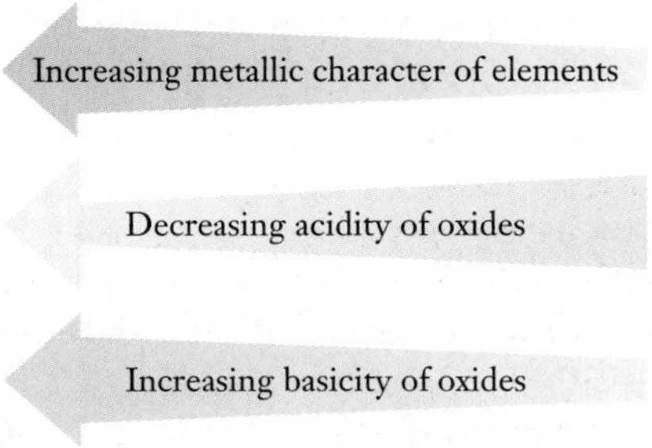

Increasing metallic character of elements

Decreasing acidity of oxides

Increasing basicity of oxides

Increasing metallic character of elements

Decreasing acidity of oxides

Increasing basicity of oxides

TABLE 23-4	*Some Reactions of the Group IIA Metals*
Reaction	Remarks
$2M + O_2 \longrightarrow 2MO$	Very exothermic (except Be)
$Ba + O_2 \longrightarrow BaO_2$	Almost exclusively
$M + H_2 \longrightarrow MH_2$	M = Ca, Sr, Ba at high temperatures
$3M + N_2 \longrightarrow M_3N_2$	At high temperatures
$6M + P_4 \longrightarrow 2M_3P_2$	At high temperatures
$M + X_2 \longrightarrow MX_2$	X = halogen (Group VIIA)
$M + S \longrightarrow MS$	Also with Se, Te of Group VIA
$M + 2H_2O \longrightarrow M(OH)_2 + H_2$	M = Ca, Sr, Ba at 25°C; Mg gives MgO at high temperatures
$M + 2NH_3 \longrightarrow M(NH_2)_2 + H_2$	M = Ca, Sr, Ba in $NH_3(\ell)$ in presence of catalyst; $NH_3(g)$ with heat

Magnesium burns in air.

The tires on the Hummer vehicles used in the Gulf War were made blowout-proof by strong, lightweight magnesium inserts. This is a photograph of a civilian model.

Beryllium does not react with pure water even at red heat. It reacts with solutions of strong bases to form the complex ion, $[Be(OH)_4]^{2-}$, and H_2. Magnesium reacts with steam to produce MgO and H_2. Ca, Sr, and Ba react with water at 25°C to form hydroxides and H_2 (see Table 23-4). Group IIA compounds are generally less soluble in water than corresponding IA compounds, but many are quite soluble.

23-6 USES OF GROUP IIA METALS AND THEIR COMPOUNDS

Beryllium, Be

Because of its rarity, beryllium has only a few practical uses. It occurs mainly as beryl, $Be_3Al_2Si_6O_{18}$, a gemstone which, with appropriate impurities, may be aquamarine (blue) or emerald (green). Because it is transparent to X-rays, "windows" for X-ray tubes are constructed of beryllium. Beryllium compounds are quite toxic.

Magnesium, Mg

Metallic magnesium burns in air with such a brilliant white light that it is used in photographic flash accessories and fireworks. It is very lightweight and is currently used in many alloys for building materials. Like aluminum, it forms an impervious coating of oxide that protects it from further oxidation. Given its inexhaustible supply in the oceans, it is likely that many more structural uses will be found for it as the reserves of iron ores dwindle.

Calcium, Ca

Calcium and its compounds are widely used commercially. The element is used as a reducing agent in the metallurgy of uranium, thorium, and other metals. It is also used as a scavenger to remove dissolved impurities such as oxygen, sulfur, and carbon in molten metals and to remove residual gases in vacuum tubes. It is a component of many alloys.

Limestone is mainly calcium carbonate.

A laboratory X-ray tube (*left*) and a close-up view of one of its windows (*right*). The windows are made of beryllium metal.

Heating limestone, $CaCO_3$, produces *quicklime*, CaO, which can then be treated with water to form *slaked lime*, $Ca(OH)_2$, an inexpensive base for which industry finds many uses. When slaked lime is mixed with sand and exposed to the CO_2 of the air, it hardens to form mortar. Heating gypsum, $CaSO_4 \cdot 2H_2O$, produces plaster of Paris, $2CaSO_4 \cdot H_2O$.

$$2[CaSO_4 \cdot 2H_2O(s)] \xrightarrow{\text{heat}} 2CaSO_4 \cdot H_2O(s) + 3H_2O(g)$$

Strontium, Sr

Strontium salts are used in fireworks and flares, which show the characteristic red glow of strontium in a flame. Strontium chloride is used in some toothpastes for persons with sensitive teeth. The metal itself has no practical uses.

Barium, Ba

Barium is a constituent of alloys that are used for spark plugs because of the ease with which it emits electrons when heated. It is used as a degassing agent for vacuum tubes. A slurry of finely divided barium sulfate, $BaSO_4$, is used to coat the gastrointestinal tract in preparation for X-ray photographs because it absorbs X-rays so well. It is so insoluble that it is not poisonous; all soluble barium salts are very toxic.

Calcium carbonate and calcium phosphate occur in seashells and animal bones.

THE POST-TRANSITION METALS

The metals below the stepwise division of the periodic table in Groups IIIA through VA are the **post-transition metals.** These include aluminum, gallium, indium, and thallium from Group IIIA; tin and lead from Group IVA; and bismuth from Group VA. Aluminum is the only post-transition metal that is considered very reactive. We will focus our discussion on the metals of Group IIIA.

There are no true metals in Groups VIA, VIIA, and VIIIA.

23-7 GROUP IIIA: PERIODIC TRENDS

The properties of the elements in Group IIIA (Table 23-5) vary less regularly down the groups than those of the IA and IIA metals. The Group IIIA elements are all solids. Boron, at the top of the group, is a nonmetal. Its melting point, 2300°C, is very high because it crystallizes as a covalent solid. The other elements, aluminum through thallium, form metallic crystals and have considerably lower melting points.

Aluminum, Al

Aluminum is the most reactive of the post-transition metals. It is the most abundant metal in the earth's crust (7.5%) and the third most abundant element. Aluminum is inexpensive compared with most other metals. It is soft and can be readily extruded into wires or rolled, pressed, or cast into shapes.

Because of its relatively low density, aluminum is often used as a lightweight structural metal. It is often alloyed with Mg and some Cu and Si to increase its strength. Many buildings are sheathed in aluminum, which resists corrosion by forming an oxide coating.

Pure aluminum conducts about two thirds as much electric current per unit volume as copper, but it is only one third as dense (Al, 2.70 g/cm^3; Cu, 8.92 g/cm^3). As a result, a mass of aluminum can conduct twice as much current as the same mass of copper. Aluminum is

This aluminum honeycomb material is made by bonding aluminum foil sheets to form hexagonal cells. It is used to make sandwich construction panels that have a very high strength-to-weight ratio.

TABLE 23-5   *Properties of the Group IIIA Elements*

Property	B	Al	Ga	In	Tl
Outer electrons	$2s^2 2p^1$	$3s^2 3p^1$	$4s^2 4p^1$	$5s^2 5p^1$	$6s^2 6p^1$
Physical state (25°C, 1 atm)	solid	solid	solid	solid	solid
Melting point (°C)	2300	660	29.8	156.6	303.5
Boiling point (°C)	2550	2367	2403	2080	1457
Density (g/cm³)	2.34	2.70	5.91	7.31	11.85
Atomic radius (Å)	0.85	1.43	1.35	1.67	1.70
Ionic radius, M^{3+} (Å)	(0.20)*	0.68	0.76	0.94	1.03
Electronegativity	2.0	1.5	1.7	1.6	1.6
E^0 (volts): $M^{3+}(aq) + 3e^- \longrightarrow M(s)$	(−0.90)*	−1.66	−0.53	−0.34	0.916
Oxidation states	−3 to +3	+3	+1, +3	+1, +3	+1, +3
Ionization energies (kJ/mol)					
$\quad M(g) \longrightarrow M^+(g) + e^-$	801	578	579	558	589
$\quad M^+(g) \longrightarrow M^{2+}(g) + e^-$	2427	1817	1971	1813	1961
$\quad M^{2+}(g) \longrightarrow M^{3+}(g) + e^-$	3660	2745	2952	2692	2867
$\Delta H^0_{hydration}$ (kJ/mol): $M^{3+}(g) + xH_2O \longrightarrow M^{3+}(aq)$	—	−4750	−4703	−4159	−4117

*For the covalent +3 oxidation state.

now used in electrical transmission lines and has been used in wiring in homes. The latter use has been implicated as a fire hazard, however, due to the heat that can be generated during high current flow at the junction of the aluminum wire and fixtures of other metals. As a result, aluminum wiring no longer meets building codes in many places.

Aluminum is a strong reducing agent.

$$Al^{3+}(aq) + 3e^- \longrightarrow Al(s) \qquad E^0 = -1.66 \text{ V}$$

Passive toward means "does not react with."

Aluminum is thus quite reactive, but a thin, very hard transparent film of Al_2O_3 forms when Al comes into contact with air. This protects it from further oxidation. For this reason it is even passive toward nitric acid, HNO_3, a strong oxidizing agent. When the oxide coating is sanded off, Al reacts vigorously with HNO_3.

$$Al(s) + 4HNO_3(aq) \longrightarrow Al(NO_3)_3(aq) + NO(g) + 2H_2O(\ell)$$

The very negative enthalpy of formation of aluminum oxide makes Al a very strong reducing agent for other metal oxides. The *thermite reaction* is a spectacular example (see Chapter 15 opening photo). It generates enough heat to produce molten iron for welding large steel parts such as shafts and train tracks.

$$2Al(s) + Fe_2O_3(s) \longrightarrow 2Fe(s) + Al_2O_3(s) \qquad \Delta H^0 = -852 \text{ kJ/mol}$$

Anhydrous Al_2O_3 occurs naturally as the extremely hard, high-melting mineral *corundum*, which has a network structure. It is colorless when pure, but becomes colored when transition metal ions replace a few Al^{3+} ions in the crystal. *Sapphire* is usually blue and

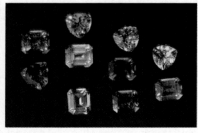

© Gemological Institute of America

Small amounts of different transition metal ions give different colors to sapphire, which is mostly aluminum oxide, Al_2O_3.

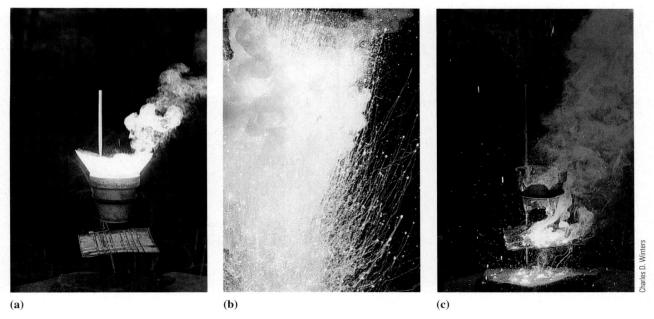

(a)   **(b)**   **(c)**

Figure 23-1 The thermite reaction. A mixture of Fe_2O_3 and aluminum powder was placed in a clay pot with a piece of magnesium ribbon as a fuse. (a) The reaction was initiated by lighting the magnesium fuse. (b) So much heat was produced by the reaction that the iron melted as it was produced. (c) The molten iron dropped out of the clay pot and burned through a sheet of iron that was placed under the pot.

contains some iron and titanium. *Ruby* is red due to the presence of small amounts of chromium.

Other Group IIIA Metals

Gallium is unusual in that it melts when held in the hand. It has the largest liquid state temperature range of any element (29.8° to 2403°C). It is used in transistors and high-temperature thermometers. Gallium-67 was one of the first artificially produced isotopes to be used in medicine. It concentrates in inflamed areas and in certain melanomas.

Indium is a soft, bluish metal that is used in some alloys with silver and lead to make good heat conductors. Most indium is used in electronics. Thallium is a soft, heavy metal that resembles lead. It is quite toxic and has no important practical uses as a free metal.

Periodic Trends

The atomic radii do not increase regularly as Group IIIA is descended (Figure 6-1). The atomic radius of Ga, 1.35 Å, is *less* than that of Al, 1.43 Å, which is directly above Ga. The transition elements are located between calcium (IIA) and gallium (IIIA), strontium (IIA) and indium (IIIA), and barium (IIA) and thallium (IIIA). The increase in nuclear charge that accompanies filling of the $(n-1)d$ subshell results in the contraction of the size of the atoms. This contraction is caused by the stronger attraction of the more highly charged

Gallium metal melts below body temperature.

CHEMISTRY IN USE   *The Development of Science*

The Most Valuable Metal in the World

Imagine paying $8 million for one pound of aluminum! Although aluminum currently costs less than $1.00 a pound, it was considered the most valuable metal in 1827. Aluminum was so cherished by royalty in the early to mid-1800s that they alone ate with aluminum spoons and forks while their lower-class guests dined with cheaper gold and silver service. Aluminum is the most abundant metal in the earth's crust (7.5%); why was it originally so expensive?

Aluminum was first prepared by the following displacement reaction.

$$AlCl_3 + 3K \longrightarrow Al + 3KCl$$

Charles D. Winters

Potassium was also expensive because it was made by passing an electric current (from a voltaic cell) through molten KCl. In addition to the great cost of energy required to melt large quantities of KCl, copper and zinc (used in voltaic cells) were also expensive metals in the early 1800s. Thus, the very small amount of aluminum produced by this displacement reaction was extremely expensive.

It was not practical to produce aluminum by passing an electric current through molten Al_2O_3 because it has a high melting point, 2000°C. This high temperature is difficult to achieve and maintain; the components of most voltaic cells melt below this temperature. Zinc melts at 420°C and copper at 1083°C.

The cost of aluminum began to drop as the result of two major advances in the late 1800s. The first came with the invention of the electric generator, which could produce electricity using steam or water. Electricity generated by steam or water was quite inexpensive compared with electricity generated by voltaic cells. Despite this cost reduction, aluminum still cost more than $100,000 a pound. The second advance

took place in 1886, when chemists discovered that they could lower the melting point of aluminum oxide by mixing it with complex salts, such as $Na_3[AlF_6]$. Since 1886, the price of aluminum has decreased markedly because of lower electricity costs, improved production techniques, and recycling of discarded aluminum products.

Although aluminum is no longer used in table services by royalty, it is of inestimable value in energy conservation. Around our homes we find energy-saving items such as aluminum storm doors and windows, insulation backed with aluminum foil, and aluminum siding. Because vehicle weight significantly affects gas mileage, substituting aluminum for heavier metals in cars, trucks, trains, and aircraft helps preserve our petroleum supplies. Although the cost of aluminum has decreased drastically, it is still a valuable metal because of its ability to help us conserve energy and to improve our standard of living at the same time.

Ronald DeLorenzo
Middle Georgia College

Compare the radii and densities of these elements with those of Group IA and IIA metals in the same rows.

nuclei for the outer electrons. This causes the radii of Ga, In, and Tl to be smaller than would be predicted from the radii of B and Al. Atomic radii strongly influence other properties. For example, Ga, In, and Tl are each much denser than the elements above them due to their unusually small atomic radii.

The Group IIIA elements have the ns^2np^1 outer electron configuration. Aluminum shows only the +3 oxidation state in its compounds. The heavier metals (Ga, In, Tl) can lose or share either the single p valence electron or the p and both s electrons to exhibit the +1 or

+3 oxidation state, respectively. In general the post-transition metals can exhibit oxidation states of $(g - 2)+$ and $g+$, where g = periodic group number. As examples, TlCl and $TlCl_3$ both exist, as do $SnCl_2$ and $SnCl_4$. The stability of the lower state increases as the groups are descended. This is called the **inert s-pair effect** because the two s electrons remain nonionized, or unshared, for the $(g - 2)+$ oxidation state. To illustrate, $AlCl_3$ exists but not AlCl; $TlCl_3$ is less stable than TlCl.

As is generally true, for each pair of compounds, covalent character is greater for the higher (more polarizing) oxidation state of the metal.

THE *d*-TRANSITION METALS

The term "transition elements" denotes elements in the middle of the periodic table. They provide a transition between the "base formers" on the left and the "acid formers" on the right. The term applies to both the *d*- and *f*-transition elements (*d* and *f* atomic orbitals are being filled across this part of the periodic table). All are metals. We commonly use the term "transition metals" to refer to the *d*-transition metals.

Oxides of most nonmetals are acidic, and oxides of most metals are basic (except those having high oxidation states).

IIIB	IVB	VB	VIB	VIIB	VIIIB			IB	IIB
Sc	Ti	V	Cr	Mn	Fe	Co	Ni	Cu	Zn
Y	Zr	Nb	Mo	Tc	Ru	Rh	Pd	Ag	Cd
La	Hf	Ta	W	Re	Os	Ir	Pt	Au	Hg
Ac	Rf	Db	Sg	Bh	Hs	Mt			

The *d*-transition metals are located between Groups IIA and IIIA in the periodic table. Strictly speaking, a *d*-transition metal must have a partially filled set of *d* orbitals. Zinc, cadmium, and mercury (Group IIB) and their cations have completely filled sets of *d* orbitals, but they are often discussed with *d*-transition metals because their properties are similar. All of the other elements in this region have partially filled sets of *d* orbitals, except the IB elements and palladium, which have completely filled sets. Some of the cations of these latter elements have partially filled sets of *d* orbitals.

Many transition metal ions are highly colored. These flasks contain (*left to right*), aqueous solutions of $Fe(NO_3)_3$, $Co(NO_3)_2$, $Ni(NO_3)_2$, $Cu(NO_3)_2$, and $Zn(NO_3)_2$. Colorless Zn^{2+} ions differ from the others by having completely filled $3d$ orbitals.

The early transition metals include Groups IIIB to VB, while the late transition metals are usually defined as Groups VIIIB, IB, and IIB. The middle Groups VIB and VIIB can often behave as either early or late transition metals in their chemistry.

In the "building" of electron configurations by the Aufbau Principle, the outer s orbitals are occupied before the inner d orbitals (Section 5-17).

23-8 GENERAL PROPERTIES

The following are general properties of transition elements.

1. All are metals.
2. Most are harder and more brittle and have higher melting points, boiling points, and heats of vaporization than nontransition metals.
3. Their ions and their compounds are usually colored.
4. They form many complex ions (Chapter 25).
5. With few exceptions, they exhibit multiple oxidation states.
6. Many of them are paramagnetic, as are many of their compounds.
7. Many of the metals and their compounds are effective catalysts.

Some properties of $3d$-transition metals are listed in Table 23-6.

23-9 OXIDATION STATES

Most transition metals exhibit more than one nonzero oxidation state. For the early transition metals, the *maximum* oxidation state is given by the metal's group number, but this is often not its most stable oxidation state (Table 23-7).

The outer s electrons lie outside the d electrons and are *always* the first ones lost in ionization. In the first transition series, scandium and zinc exhibit just one nonzero oxidation state. Scandium loses its two $4s$ electrons and its only $3d$ electron to form Sc^{3+}. Zinc loses its two $4s$ electrons to form Zn^{2+}.

$$_{21}Sc\ [Ar] \quad \uparrow\ \_\ \_\ \_\ \_ \quad \uparrow\downarrow \quad \xrightarrow{-3e^-} \quad _{21}Sc^{3+}\ [Ar]$$

$$_{30}Zn\ [Ar] \quad \uparrow\downarrow\ \uparrow\downarrow\ \uparrow\downarrow\ \uparrow\downarrow\ \uparrow\downarrow \quad \uparrow\downarrow \quad \xrightarrow{-2e^-} \quad _{30}Zn^{2+}\ [Ar]\ 3d^{10}$$

TABLE 23-6	*Properties of Metals in the First Transition Series*									
Properties	Sc	Ti	V	Cr	Mn	Fe	Co	Ni	Cu	Zn
Melting point (°C)	1541	1660	1890	1850	1244	1535	1495	1453	1083	420
Boiling point (°C)	2831	3287	3380	2672	1962	2750	2870	2732	2567	907
Density (g/cm³)	2.99	4.54	6.11	7.18	7.21	7.87	8.9	8.91	8.96	7.13
Atomic radius (Å)	1.62	1.47	1.34	1.25	1.29	1.26	1.25	1.24	1.28	1.34
Ionic radius, M^{2+} (Å)	—	0.94	0.88	0.89	0.80	0.74	0.72	0.69	0.70	0.74
Electronegativity	1.3	1.4	1.5	1.6	1.6	1.7	1.8	1.8	1.8	1.6
E^0 (V) for $M^{2+}(aq) + 2e^- \longrightarrow M(s)$	-2.08*	-1.63	-1.2	-0.91	-1.18	-0.44	-0.28	-0.25	$+0.34$	-0.76
IE (kJ/mol) first	631	658	650	652	717	759	758	757	745	906
second	1235	1310	1414	1592	1509	1561	1646	1753	1958	1733

*For $Sc^{3+}(aq) + 3e^- \longrightarrow Sc(s)$.

TABLE 23-7		*Nonzero Oxidation States of the 3d-Transition Metals**								
IIIB	IVB	VB	VIB	VIIB		VIIIB			IB	IIB
Sc	Ti	V	Cr	Mn	Fe	Co	Ni	Cu	Zn
								+1 r	
				+2	+2 r	+2	+2	+2	+2
+3		+3 r	+3		+3	+3 o			
	+4	+4		+4 o					
		+5 o							
			+6 o						
				+7 o					

Abbreviations: o = oxidizing agent; r = reducing agent.

Titanium and nickel also have only a single common nonzero oxidation state. All of the other 3d-transition metals exhibit at least two oxidation states in their compounds. For example, cobalt can form Co^{2+} and Co^{3+} ions.

$$Co\ [Ar]\quad \uparrow\downarrow\ \uparrow\downarrow\ \uparrow\ \uparrow\ \uparrow\quad \uparrow\downarrow\quad \xrightarrow{-2e^-}\quad Co^{2+}\ [Ar]\quad \uparrow\downarrow\ \uparrow\downarrow\ \uparrow\ \uparrow\ \uparrow\quad \_$$

$$Co\ [Ar]\quad \uparrow\downarrow\ \uparrow\downarrow\ \uparrow\ \uparrow\ \uparrow\quad \uparrow\downarrow\quad \xrightarrow{-3e^-}\quad Co^{3+}\ [Ar]\quad \uparrow\downarrow\ \uparrow\ \uparrow\ \uparrow\ \uparrow\quad \_$$

The most common oxidation states of the 3d-transition elements are +2 and +3. The elements in the middle of each series exhibit more oxidation states than those to the left or right. As one moves down a group, higher oxidation states become more stable and more common (opposite to the trend for representative elements). This is because the d electrons are more effectively shielded from the nucleus as the group is descended and are therefore more easily lost or more readily available for sharing. For example, cobalt commonly exhibits the +2 and +3 oxidation states. Rh and Ir are just below Co. Their common oxidation states are +3 and +4. The +4 state is slightly more stable for Ir than for the lighter Rh.

The oxides and hydroxides of lower oxidation states of a given transition metal are basic. Those containing intermediate oxidation states tend to be amphoteric, and those containing high oxidation states tend to be acidic. This is illustrated for the oxides and hydroxides of Cr in the next section.

23-10 CHROMIUM OXIDES, OXYANIONS, AND HYDROXIDES

Typical of the metals near the middle of a transition series, chromium shows several oxidation states. The most common are +2, +3, and +6 (Table 23-8).

Oxidation–Reduction

The most stable oxidation state of Cr is +3. Solutions of blue chromium(II) salts are easily air-oxidized to chromium(III).

$$Cr^{3+} + e^- \longrightarrow Cr^{2+} \qquad E^0 = -0.41\ V$$

Pentaamminechlorocobalt(III) chloride, [Co(NH₃)₅(Cl)]Cl₂, is a compound that contains cobalt in the +3 oxidation state (*left*). Hexaaquacobalt(II) chloride, [Co(OH₂)₆]Cl₂, contains cobalt in the +2 oxidation state (*right*).

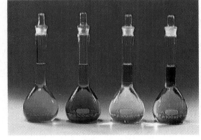

Aqueous solutions of some compounds that contain chromium. (*Left to right*) Chromium(II) chloride (CrCl₂) is blue; chromium(III) chloride (CrCl₃) is green; potassium chromate (K₂CrO₄) is yellow; potassium dichromate (K₂Cr₂O₇) is orange.

Ox. State	Oxide	Hydroxide	Name	Acidic/Basic	Related Salt	Name
+2	CrO black	$Cr(OH)_2$	chromium(II) hydroxide	basic	$CrCl_2$ anhydr. colorless aq. lt. blue	chromium(II) chloride
+3	Cr_2O_3 green	$Cr(OH)_3$	chromium(III) hydroxide	amphoteric	$CrCl_3$ anhydr. violet aq. green $KCrO_2$ green	chromium(II) chloride potassium chromite
+6	CrO_3 dk. red	H_2CrO_4 or $[CrO_2(OH)_2]$ $H_2Cr_2O_7$ or $[Cr_2O_5(OH)_2]$	chromic acid dichromic acid	weakly acidic acidic	K_2CrO_4 yellow $K_2Cr_2O_7$ orange	potassium chromate potassium dichromate

TABLE 23-8   Some Compounds of Chromium

Chromium(VI) species are oxidizing agents. Basic solutions containing chromate ions, CrO_4^{2-}, are weakly oxidizing. Acidification produces the dichromate ion, $Cr_2O_7^{2-}$, and chromium(VI) oxide, both powerful oxidizing agents.

$$Cr_2O_7^{2-} + 14H^+ + 6e^- \longrightarrow 2Cr^{3+} + 7H_2O \qquad E^0 = +1.33 \text{ V}$$

Chromate–Dichromate Equilibrium

Red chromium(VI) oxide, CrO_3, is the acid anhydride of two acids: chromic acid, H_2CrO_4, and dichromic acid, $H_2Cr_2O_7$. Neither acid has been isolated in pure form, although chromate and dichromate salts are common. CrO_3 reacts with H_2O to produce strongly acidic solutions containing hydrogen ions and (predominantly) dichromate ions.

$$2CrO_3 + H_2O \longrightarrow [2H^+ + Cr_2O_7^{2-}] \qquad \text{dichromic acid (red-orange)}$$

From such solutions orange dichromate salts can be crystallized after adding a stoichiometric amount of base. Addition of excess base produces yellow solutions from which only yellow chromate salts can be obtained. The two anions exist in solution in a pH-dependent equilibrium.

$$\underset{\text{yellow}}{\overset{+6}{2CrO_4^{2-}}} + 2H^+ \rightleftharpoons \underset{\text{orange}}{\overset{+6}{Cr_2O_7^{2-}}} + H_2O \qquad K_c = \frac{[Cr_2O_7^{2-}]}{[CrO_4^{2-}]^2[H^+]^2} = 4.2 \times 10^{14}$$

Adding a strong acid to a solution that contains $CrO_4^{2-}/Cr_2O_7^{2-}$ ions favors the reaction to the right and increases $[Cr_2O_7^{2-}]$. Adding a base favors the reaction to the left and increases $[CrO_4^{2-}]$.

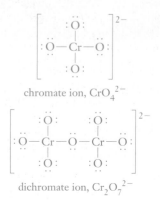

chromate ion, CrO_4^{2-}

dichromate ion, $Cr_2O_7^{2-}$

Chromate ion, CrO_4^{2-} Dichromate ion, $Cr_2O_7^{2-}$

Dehydration of chromate or dichromate salts with concentrated H_2SO_4 produces CrO_3. Chromium(VI) oxide is a strong oxidizing agent. A powerful "cleaning solution" once used for removing greasy stains and coatings from laboratory glassware was made by adding concentrated H_2SO_4 to a concentrated solution of $K_2Cr_2O_7$. The active ingredients are CrO_3, an oxidizing agent, and H_2SO_4, an excellent solvent.

This cleaning solution must be used with great caution because it is a strong oxidizing agent and is carcinogenic.

Chromium(III) hydroxide is amphoteric (Section 10-6).

$$Cr(OH)_3(s) + 3H^+ \longrightarrow Cr^{3+} + 3H_2O \qquad \text{(rxn. with acids)}$$

$$Cr(OH)_3(s) + OH^- \longrightarrow Cr(OH)_4^- \qquad \text{(rxn. with bases)}$$

Key Terms

Alkali metals   Group IA metals.

Alkaline earth metals   Group IIA metals.

d-Transition metals   Metals that have partially filled sets of d orbitals; the B groups of the periodic table.

Diagonal similarities   Chemical similarities of elements of Period 2 to elements of Period 3 one group to the right; especially evident toward the left of the periodic table.

Inert s-pair effect   The tendency of the two outermost s electrons to remain nonionized or unshared in compounds; characteristic of the post-transition metals.

Post-transition metals   Representative metals in the "p block."

Representative metals   Metals in the A groups in the periodic table; their outermost electrons are in s and p orbitals.

Exercises

1. How do the acidities or basicities of metal oxides vary with oxidation numbers of the same metal?

2. Discuss the general differences in electron configurations of representative elements and d-transition metals.

3. Compare the extents to which the properties of successive elements across the periodic table differ for representative elements and d-transition metals. Explain.

4. Compare the metals and nonmetals with respect to (a) number of outer-shell electrons, (b) electronegativities, (c) standard reduction potentials, and (d) ionization energies.

5. How do the physical properties of metals differ from those of nonmetals?

6. Define "malleable" and "ductile."

7. (a) What is the meaning of the statement, "alkali metals are corrosive?" (b) Are alkali metal ions corrosive? (c) Are alkali metal ions an important part of your diet?

8. Compare the alkali metals with the alkaline earth metals with respect to (a) atomic radii, (b) densities, (c) first ionization energies, and (d) second ionization energies. Explain the comparisons.

9. Summarize the chemical and physical properties of: (a) the alkali metals and (b) the alkaline earth metals.

10. (a) Compare and contrast the physical and chemical properties of the Group IA metals with those of the Group IIA metals. (b) Compare the periodic trends of the two groups.

11. Write the general outer-electron configurations for atoms of the IA and IIA metals. What oxidation state(s) would you predict for these elements? What types of bonding would you expect in most of the compounds of these elements? Why?

12. Write electron configurations ($\uparrow\downarrow$ notation) for (a) Ca, (b) Ca^{2+}, (c) Na, (d) Na^+, (e) Sn, (f) Sn^{2+}, and (g) Sn^{4+}.

13. Write electron configurations ($\uparrow\downarrow$ notation) for (a) Li, (b) Li^+, (c) Sr, (d) Sr^{2+}, (e) Al, (f) Al^{3+}, and (g) Ga^{3+}.

14. Are the elements in Groups IA and IIA found in the free state in nature? What are the primary sources for these elements?

15. Describe some uses for (a) lithium and its compounds and (b) sodium and its compounds.

16. Where do the metals of Groups IA and IIA fall with respect to H_2 in the activity series? What does this tell us about their reactivities with water and acids?

17. Write chemical equations describing the reactions of O_2 with each of the alkali and alkaline earth metals. Account for differences within each family.

18. Describe some uses for (a) calcium and its compounds and (b) magnesium and its compounds.

19. Write general equations for reactions of alkali metals with (a) hydrogen, (b) sulfur, and (c) ammonia. Represent the metal as M.

20. Write general equations for reactions of alkali metals with (a) water, (b) phosphorus, and (c) halogens. Represent the metal as M and the halogen as X.

21. Write general equations for reactions of alkaline earth metals with (a) hydrogen, (b) sulfur, and (c) ammonia. Represent the metal as M.

22. Write general equations for reactions of alkaline earth metals with (a) water, (b) phosphorus, and (c) chlorine. Represent the metal as M.

23. What is meant by the term "diagonal relationships"?

24. Give some illustrations of diagonal relationships in the periodic table, and explain each.

25. What is hydration energy? How does it vary for cations of the alkali metals?

26. How do hydration energies vary for cations of the alkaline earth metals?

27. How do the standard reduction potentials of the alkali metal cations vary? Why?

28. How do the standard reduction potentials of the alkaline earth metal cations vary? Why?

29. Why are the standard reduction potentials of lithium and beryllium out of line with respect to group trends?

*30. Calculate ΔH^0 values at 25°C for the reactions of 1 mol of each of the following metals with stoichiometric quantities of water to form metal hydroxides and hydrogen. (a) Li, (b) K, and (c) Ca. Rationalize the differences in these values.

31. How are the *d*-transition metals distinguished from other elements?

32. What are the general properties of the *d*-transition metals?

33. Why are trends in variations of properties of successive *d*-transition metals less regular than trends among successive representative elements?

34. Write out the electron configurations for the following species: (a) Sc; (b) Fe; (c) Cu; (d) Zn^{2+}; (e) Cr^{3+}; (f) Ni^{2+}; (g) Ag; (h) Ag^+.

35. Why do copper and chromium atoms have "unexpected" electron configurations?

36. Discuss the similarities and differences among elements of corresponding A and B groups of the periodic table, IIIA and IIIB for example.

37. Copper exists in the +1, +2, and +3 oxidation states. Which is the most stable? Which would be expected to be a strong oxidizing agent and which would be expected to be a strong reducing agent?

38. For a given transition metal in different oxidation states, how does the acidic character of its oxides increase? How do ionic and covalent character vary? Characterize a series of metal oxides as examples.

39. For different transition metals in the same oxidation state in the same group (vertical column) of the periodic table, how do covalent character and acidic character of their oxides vary? Why? Cite evidence for the trends.

40. Chromium(VI) oxide is the acid anhydride of which two acids? Write their formulas. What is the oxidation state of the chromium in these acids?

CONCEPTUAL EXERCISES

41. Some of the oxides of the transition metals are amphoteric. (a) What is meant by the term "amphoteric?" (b) Identify five oxides that are amphoteric, five oxides that are basic, and five oxides that are acidic.

42. Calcium carbonate is a very common compound on this planet. Name five common items that are composed primarily of calcium carbonate.

43. Magnesium and some of the magnesium alloys will burn if heated to a high enough temperature. Some magnesium alloys are used in construction and may have to be welded at a high temperature. Suggest techniques or requirements for welding these alloys so that they do not ignite.

44. Some mental disorders are treated with medications that contain lithium salts. Why is pure lithium unsuitable for such purposes?

45. Gallium thermometers are high-temperature thermometers. (a) Why can't they be used near room temperature? (b) Propose a reason why mercury thermometers cannot be used to measure temperatures in the upper range of gallium thermometers.

BUILDING YOUR KNOWLEDGE

*46. How many grams of Co_3O_4 (a mixed oxide, $CoO \cdot Co_2O_3$) must react with excess aluminum to produce 190. g of metallic cobalt, assuming 70.5% yield?

$$3Co_3O_4 + 8Al \xrightarrow{\text{heat}} 9Co + 4Al_2O_3$$

*47. Calculate ΔH^0, ΔS^0, and ΔG^0 for the reaction of 1 mol of Na with water to form aqueous NaOH and hydrogen.

*48. Calculate ΔH^0, ΔS^0, and ΔG^0 for the reaction of 1 mol of Rb with water to form aqueous RbOH and hydrogen. Compare the spontaneity of this reaction with that in Exercise 47.

*49. What is the ratio of $[Cr_2O_7^{2-}]$ to $[CrO_4^{2-}]$ at 25°C in a solution prepared by dissolving 1.5×10^{-3} mol of sodium chromate, Na_2CrO_4, in enough of an aqueous solution buffered at pH = 2.00 to produce 225 mL of solution?

*50. An unknown compound is known to be the carbonate of either a Group IA or Group IIA element. Repeated heating of 14.78 grams of the unknown produced a total of 4.48 L of carbon dioxide at 0°C and 1 atm pressure (101 kPa). Calculate the number of moles of CO_2 produced and identify the unknown compound.

BEYOND THE TEXTBOOK

Go to the textbook website at

http://www.brookscole.com/chemistry/whitten

for additional activities and exercises based on the General Chemistry Interactive CD-ROM, the World Wide Web, and library resources.

InfoTrac College Edition

For additional readings, go to InfoTrac College Edition, your online research library at:

http://infotrac.thomsonlearning.com

Some Nonmetals and Metalloids

24

The rose on the right is in an atmosphere of sulfur dioxide, SO$_2$. Gaseous SO$_2$ and aqueous solutions of HSO$_3^-$ and SO$_3^{2-}$ ions are used as bleaching agents. A similar process is used to bleach wood pulp before it is converted to paper.

Charles D. Winters

OUTLINE

OBJECTIVES

After you have studied this chapter, you should be able to

- Describe the occurrence and use of the noble gases
- Describe compounds of the noble gases
- Describe the occurrence and production of the halogens
- Describe some important reactions and compounds of the halogens
- Describe the occurrence and production of sulfur, selenium, and tellurium
- Describe some important reactions and compounds of the heavier Group VIA nonmetals
- Describe the occurrence and production of nitrogen and phosphorus
- Describe some important reactions of nitrogen and phosphorus
- Describe the occurrence and importance of silicon
- Describe a few important compounds of silicon

Only about 20% of the elements are classified as nonmetals. With the exception of H, they are in the upper right-hand corner of the periodic table. In this chapter we shall consider the chemistry and properties of nonmetals and metalloids that illustrate group trends and individuality of elements within groups of nonmetals.

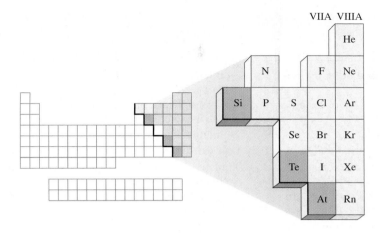

THE NOBLE GASES (GROUP VIIIA)

24-1 OCCURRENCE, USES, AND PROPERTIES

The noble gases are very low-boiling gases. Except for radon, they can be isolated by fractional distillation of liquefied air. Radon is collected from the radioactive disintegration of radium salts. Table 24-1 gives the percentage of each noble gas in the atmosphere.

Helium is produced in the United States from some natural gas fields. This source was discovered in 1905 by H. P. Cady and D. F. McFarland at the University of Kansas, when they were asked to analyze a nonflammable component of natural gas from a Kansas gas well. Uses of the noble gases are summarized in Table 24-2.

The noble gases are colorless, tasteless, and odorless. In the liquid and solid states the only forces of attraction among the atoms are very weak dispersion forces. Polarizability and interatomic interactions increase with increasing atomic size, and so melting and boiling points increase with increasing atomic number. The attractive forces among He atoms are so small that He remains liquid at 1 atmosphere pressure even at a temperature of 0.001 K.

SC*LINKS.*

TOPIC: Noble Gases
GO TO: www.scilinks.org
*sci*LINKS CODE: WCH2410

Radon is continually produced in small amounts in the uranium radioactive decay sequence (Section 26-11). Radon gas is so unreactive that it eventually escapes from the soil. Measurable concentrations of radon, a radioactive gas, have been observed in basements of many dwellings.

A pressure of about 26 atmospheres is required to solidify He at 0.001 K.

24-2 XENON COMPOUNDS

Until the early 1960s, chemists believed that the Group VIIIA elements would not combine chemically with any elements. In 1962, Neil Bartlett (1932–) and his research group at the University of British Columbia were studying the powerful oxidizing agent PtF_6.

The noble gases are often called the rare gases. They were formerly called the "inert gases" because it was incorrectly thought that they could not enter into chemical combination.

TABLE 24-1	*Percentages (by volume) of Noble Gases in the Atmosphere*				
He	**Ne**	**Ar**	**Kr**	**Xe**	**Rn**
0.0005%	0.015%	0.94%	0.00011%	0.000009%	≈0%

TABLE 24-2	*Uses of the Noble Gases*	
Noble Gas	**Use**	**Useful Properties or Reasons**
helium	1. Filling of observation balloons and other lighter-than-air craft	Nonflammable; 93% of lifting power of flammable hydrogen
	2. He/O_2 mixtures, rather than N_2/O_2, for deep-sea breathing	Low solubility in blood; prevents nitrogen narcosis and "bends"
	3. Diluent for gaseous anesthetics	Nonflammable, nonreactive
	4. He/O_2 mixtures for respiratory patients	Low density, flows easily through restricted passages
	5. Heat transfer medium for nuclear reactors	Transfers heat readily; does not become radioactive; chemically inert
	6. Industrial applications, such as inert atmosphere for welding easily oxidized metals	Chemically inert
	7. Liquid He used to maintain very low temperatures in research (cryogenics)	Extremely low boiling point
neon	Neon signs	Even at low Ne pressure, moderate electric current causes bright orange-red glow; can be modified by colored glass or mixing with Ar or Hg vapor
argon	1. Inert atmosphere for welding	Chemically inert
	2. Filling incandescent light bulbs	Inert; inhibits vaporization of tungsten filament and blackening of bulbs
krypton	Airport runway and approach lights	Gives longer life to incandescent lights than Ar, but more expensive
xenon	Xe and Kr mixture in high-intensity, short-exposure photographic flash tubes	Both have fast response to electric current
radon	Radiotherapy of cancerous tissues	Radioactive

They accidentally prepared and identified $O_2^+PtF_6^-$ by reaction of oxygen with PtF_6. Bartlett reasoned that xenon also should be oxidized by PtF_6 because the first ionization energy of O_2 (1.31×10^3 kJ/mol) is slightly larger than that of xenon (1.17×10^3 kJ/mol). He obtained a red crystalline solid initially believed to be $Xe^+PtF_6^-$ but now known to be a more complex compound.

Since Bartlett's discovery, many other noble gas compounds have been made. All involve very electronegative elements. Most are compounds of Xe, and the best characterized compounds are xenon fluorides. Oxygen compounds are also well known. Reaction of Xe with F_2, an extremely strong oxidizing agent, in different stoichiometric ratios produces xenon difluoride, XeF_2; xenon tetrafluoride, XeF_4; and xenon hexafluoride, XeF_6, all colorless crystals (Table 24-3).

All the xenon fluorides are formed in exothermic reactions. They are reasonably stable, with Xe—F bond energies of about 125 kJ/mol of bonds. For comparison, strong covalent bond energies range from about 170 to 500 kJ/mol, whereas bond energies of hydrogen bonds are typically less than 40 kJ/mol.

Oxygen is second only to fluorine in electronegativity.

Crystals of the noble gas compound xenon tetrafluoride, XeF_4.

Argonne National Laboratory

THE HALOGENS (GROUP VIIA)

The elements of Group VIIA are known as **halogens** (Greek, "salt formers"). The term **"halides"** is used to describe their binary compounds with less electronegative elements. The heaviest halogen, astatine, is an artificially produced element of which only short-lived radioactive isotopes are known.

TABLE 24-3	*Xenon Fluorides*				
Compound	Preparation (Molar ratio Xe:F₂)	Reaction Conditions	e^- Pairs Around Xe	Xe Hybridization, Molecular Geometry*	
XeF_2	1:1–3	400°C or irradiation or elec. discharge	5	sp^3d, linear	
XeF_4	1:5	Same as for XeF_2	6	sp^3d^2, square planar	
XeF_6	1:20	300°C and 60 atm or elec. discharge	7	sp^3d^3(?), exact geometry undetermined	

*See Tables 8-3, 8-4.

24-3 PROPERTIES

The elemental halogens exist as diatomic molecules containing single covalent bonds. Properties of the halogens show obvious trends (Table 24-4). Their high electronegativities indicate that they attract electrons strongly. Many binary compounds that contain a metal and a halogen are ionic.

The small fluoride ion (radius = 1.19 Å) is not easily polarized (distorted) by cations, whereas the large iodide ion (radius = 2.06 Å) is. As a result, compounds containing I^- ions show greater covalent character than those containing F^- ions. The properties of Cl^- and Br^- ions are intermediate between those of F^- and I^-.

The chemical properties of the halogens resemble one another more closely than do those of elements in any other periodic group, with the exception of the noble gases and possibly the Group IA metals. But their physical properties differ significantly. Melting and boiling points of the halogens increase from F_2 to I_2. This follows their increase in size and increase in ease of polarization of outer-shell electrons by adjacent nuclei, resulting in greater intermolecular attractive (dispersion) forces. All halogens except astatine are decidedly nonmetallic. They show the -1 oxidation number in most of their compounds. Except for fluorine, they also exhibit oxidation numbers of $+1$, $+3$, $+5$, and $+7$.

SCiLINKS

TOPIC: Halogens
GO TO: www.scilinks.org
sciLINKS CODE: WCH2420

TABLE 24-4	*Properties of the Halogens*				
Property	F	Cl	Br	I	At
Physical state (25°C, 1 atm)	gas	gas	liquid	solid	solid
Color	pale yellow	yellow-green	red-brown	violet (g); black (s)	—
Atomic radius (Å)	0.72	1.00	1.14	1.33	1.40
Ionic radius (X^-) (Å)	1.19	1.67	1.82	2.06	—
Outer shell e^-	$2s^22p^5$	$3s^23p^5$	$4s^24p^5$	$5s^25p^5$	$6s^26p^5$
First ionization energy (kJ/mol)	1681	1251	1140	1008	890
Electronegativity	4.0	3.0	2.8	2.5	2.1
Melting point (°C, 1 atm)	-220	-101	-7.1	114	—
Boiling point (°C, 1 atm)	-188	-35	59	184	—
X—X bond energy (kJ/mol)	155	242	193	151	—

We often represent a halogen atom as X, without specifying a particular halogen.

F$_2$ Cl$_2$ Br$_2$ I$_2$ At$_2$

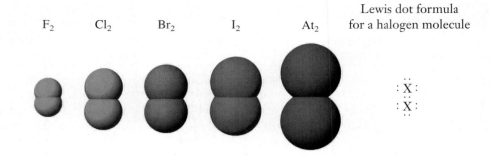

Lewis dot formula for a halogen molecule

$$:\overset{\cdot\cdot}{\underset{\cdot\cdot}{X}}:$$
$$:\overset{\cdot\cdot}{\underset{\cdot\cdot}{X}}:$$

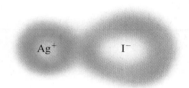

Ag$^+$ I$^-$

The diffuse cloud of the I$^-$ ion is easily polarized by the small Ag$^+$ ion.

24-4 OCCURRENCE, PRODUCTION, AND USES

The halogens are so reactive that they do not occur free in nature. The most abundant sources of halogens are halide salts. A primary source of iodine is NaIO$_3$. The halogens are obtained by oxidation of the halide ions.

$$2X^- \longrightarrow X_2 + 2e^-$$

The order of increasing ease of oxidation is F$^-$ < Cl$^-$ < Br$^-$ < I$^-$ < At$^-$.

Fluorine

Fluorine occurs in large quantities in the minerals *fluorspar* or *fluorite*, CaF$_2$; *cryolite*, Na$_3$AlF$_6$; and *fluorapatite*, Ca$_5$(PO$_4$)$_3$F. It also occurs in small amounts in sea water, teeth, bones, and blood. F$_2$ is such a strong oxidizing agent that it has not been produced by *direct* chemical oxidation of F$^-$ ions. The pale yellow gas is prepared by electrolysis of a molten mixture of KF + HF, or KHF$_2$, in a Monel metal cell. This must be done under anhydrous conditions because H$_2$O is more readily oxidized than F$^-$.

Monel metal is an alloy of Ni, Cu, Al, and Fe. It is resistant to attack by hydrogen fluoride.

$$2KHF_2 \xrightarrow[\text{melt}]{\text{electrolysis}} F_2(g) + H_2(g) + 2KF(s)$$

In 1986, Carl O. Christe discovered that the strong Lewis acid SbF$_5$ displaces the weaker Lewis acid MnF$_4$ from the hexafluoromanganate(IV) ion, [MnF$_6$]$^{2-}$. MnF$_4$ is thermodynamically unstable, and it decomposes into MnF$_3$ and F$_2$. Christe heated a mixture of potassium hexafluoromanganate(IV) and antimony(V) fluoride in a passivated Teflon–stainless steel container at 150°C for 1 hour. He obtained elemental F$_2$ in better than 40% yield. The overall reaction may be represented as

$$\overset{+1}{}\overset{+4}{}\overset{-1}{} \quad \overset{+5}{}\overset{-1}{} \quad \overset{+1}{}\overset{+5}{}\overset{-1}{} \quad \overset{+3}{}\overset{-1}{} \quad \overset{0}{}$$
$$2K_2MnF_6(s) + 4SbF_5(\ell) \longrightarrow 4KSbF_6(s) + 2MnF_3(s) + F_2(g)$$

The reactions of F$_2$ with other elements are dangerous because of the vigor with which F$_2$ oxidizes other substances. They must be carried out with *extreme caution!*

Fluorine is used as a fluorinating agent. Many fluorinated organic compounds are stable and nonflammable. They are used as refrigerants, lubricants, plastics (such as Teflon), insecticides, and, until recently, aerosol propellants. Stannous fluoride, SnF$_2$, is used in toothpaste.

Chlorine

Chlorine (Greek *chloros*, "green") occurs in abundance in NaCl, KCl, MgCl$_2$, and CaCl$_2$ in salt water and in salt beds. It is also present as HCl in gastric juices. The toxic, yellowish green gas is prepared commercially by electrolysis of concentrated aqueous NaCl, in which

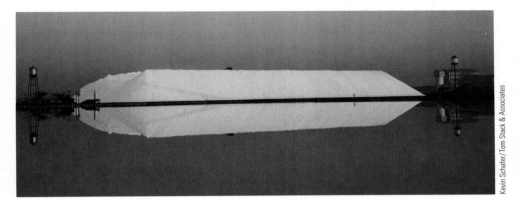

Chlorides occur in salt beds.

industrially important H_2 and caustic soda (NaOH) are also produced (Section 21-4). More than 26 billion pounds of chlorine was produced in the United States in 1997.

Chlorine is used to produce many commercially important products. Tremendous amounts of it are used in extractive metallurgy and in chlorinating hydrocarbons to produce a variety of compounds (such as polyvinyl chloride, a plastic). Chlorine is present as Cl_2, NaClO, $Ca(ClO)_2$, or Ca(ClO)Cl in household bleaches as well as in bleaches for wood pulp and textiles. Under carefully controlled conditions, Cl_2 is used to kill bacteria in public water supplies.

Bromine is a dark red, volatile liquid.

Bromine

Bromine (Greek *bromos*, "stench") is less abundant than fluorine and chlorine. In the elemental form it is a dense, freely flowing, corrosive, dark-red liquid with a brownish red vapor at 25°C. It occurs mainly in NaBr, KBr, $MgBr_2$, and $CaBr_2$ in salt water, underground salt brines, and salt beds. The major commercial source for bromine is deep brine wells in Arkansas that contain up to 5000 parts per million (0.5%) of bromide.

Bromine is used in the production of silver bromide for light-sensitive eyeglasses and photographic film; in the production of sodium bromide, a mild sedative; and in methyl bromide, CH_3Br, a soil fumigant that contributes to the destruction of the ozone layer.

Iodine reacts with starch (as in this potato) to form a deep-blue complex substance.

More recently available is an aqueous solution of an iodine complex of polyvinylpyrrolidone, or "povidone." It does not sting when applied to open wounds.

Iodine

Iodine (Greek *iodos*, "purple") is a violet-black crystalline solid with a metallic luster. It exists in equilibrium with a violet vapor at 25°C. The element can be obtained from dried seaweed or shellfish or from $NaIO_3$ impurities in Chilean nitrate ($NaNO_3$) deposits. It is contained in the growth-regulating hormone thyroxine, produced by the thyroid gland. "Iodized" table salt is about 0.02% KI, which helps prevent goiter, a condition in which the thyroid enlarges. Iodine has been used as an antiseptic and germicide in the form of tincture of iodine, a solution in alcohol.

The preparation of iodine involves reduction of iodate ion from $NaIO_3$ with sodium hydrogen sulfite, $NaHSO_3$.

$$2IO_3^-(aq) + 5HSO_3^-(aq) \longrightarrow 3HSO_4^-(aq) + 2SO_4^{2-}(aq) + H_2O(\ell) + I_2(s)$$

Iodine is then purified by sublimation (Figure 13-16).

24-5 REACTIONS OF THE FREE HALOGENS

Bromine reacts with powdered antimony so vigorously that the flask vibrates.

The free halogens react with most other elements and many compounds. For example, all the Group IA metals react with all the halogens to form simple binary ionic compounds (Section 7-2).

General Reaction	Remarks
$nX_2 + 2M \longrightarrow 2MX_n$	All X_2 with most metals (most vigorous reaction with F_2 and Group IA metals)
$X_2 + nX'_2 \longrightarrow 2XX'_n$	Formation of interhalogens (n = 1, 3, 5, or 7); X is larger than X'
$X_2 + H_2 \longrightarrow 2HX$	With all X_2.
$3X_2 + 2P \longrightarrow 2PX_3$	With all X_2, and with As, Sb, Bi replacing P
$5X_2 + 2P \longrightarrow 2PX_5$	Not with I_2; also Sb $\longrightarrow$ SbF_5, $SbCl_5$; As $\longrightarrow$ AsF_5; Bi $\longrightarrow$ BiF_5
$X_2 + H_2S \longrightarrow S + 2HX$	With all X_2
$X'_2 + 2X^- \longrightarrow 2X'^- + X_2$	$F_2 \longrightarrow Cl_2, Br_2, I_2$
	$Cl_2 \longrightarrow Br_2, I_2$
	$Br_2 \longrightarrow I_2$

The most vigorous reactions are those of F_2, which usually oxidizes other species to their highest possible oxidation states. Iodine is only a mild oxidizing agent (I^- is a mild reducing agent) and usually does not oxidize substances to high oxidation states. Consider the following reactions of halogens with two metals that exhibit variable oxidation numbers.

Iron and chlorine react to form iron(III) chloride, $FeCl_3$.

	With Fe			With Cu	
$2Fe + 3F_2$	$\longrightarrow$	$\overset{+3}{2FeF_3}$ (only)			
$2Fe + 3Cl_2$ (excess)	$\longrightarrow$	$\overset{+3}{2FeCl_3}$	$Cu + X_2$ $\longrightarrow$	$\overset{+2}{CuX_2}$	(X = F, Cl, Br)
$Fe + Cl_2$ (lim. amt.)	$\longrightarrow$	$\overset{+2}{FeCl_2}$			
$Fe + I_2$	$\longrightarrow$	$\overset{+2}{FeI_2}$ (only)	$2Cu + I_2$ $\longrightarrow$	$\overset{+1}{2CuI}$	(only)
$Fe^{3+} + I^-$	$\longrightarrow$	$Fe^{2+} + \frac{1}{2}I_2$	$Cu^{2+} + 2I^- \longrightarrow$	$\overset{+1}{CuI} + \frac{1}{2}I_2$	

24-6 THE HYDROGEN HALIDES AND HYDROHALIC ACIDS

For instance, aqueous solutions of hydrogen fluoride are called hydrofluoric acid.

The hydrogen halides are colorless gases that dissolve in water to give acidic solutions called hydrohalic acids. The gases have piercing, irritating odors. The abnormally high melting and boiling points of HF are due to its very strong hydrogen bonding (Figure 13-5).

Hydrogen halides can be prepared by combination of the elements.

$$H_2 + X_2 \longrightarrow 2HX(g) \qquad X = F, Cl, Br, I$$

The reaction with F_2 to produce HF is explosive and very dangerous. The reaction producing HCl does not occur significantly in the dark but occurs rapidly by a photochemical **chain reaction** when the mixture is exposed to light. Light energy is absorbed by Cl_2 molecules, which break apart into very reactive chlorine atoms, which have unpaired electrons (**radicals**). These subsequently attack H_2 molecules and produce HCl molecules, leaving hydrogen atoms (also radicals). The hydrogen radicals, in turn, attack Cl_2 molecules to form HCl molecules and chlorine radicals, and the process continues.

> A photochemical reaction is one in which a species (usually a molecule) interacts with radiant energy to produce very reactive species. These then undergo further reaction.

$$Cl_2 \xrightarrow{h\nu} 2 : \overset{..}{\underset{..}{Cl}} \cdot \qquad \text{chain initiation step}$$

$$\left. \begin{array}{l} : \overset{..}{\underset{..}{Cl}} \cdot + H_2 \longrightarrow HCl + H \cdot \\[2mm] H \cdot + Cl_2 \longrightarrow HCl + : \overset{..}{\underset{..}{Cl}} \cdot \end{array} \right\} \quad \text{chain propagation steps}$$

This chain reaction continues as long as there is a significant concentration of radicals. **Chain termination steps** eliminate two radicals and can eventually terminate the reaction.

$$\left. \begin{array}{l} H \cdot + H \cdot \longrightarrow H_2 \\[2mm] : \overset{..}{\underset{..}{Cl}} \cdot + : \overset{..}{\underset{..}{Cl}} \cdot \longrightarrow Cl_2 \\[2mm] H \cdot + : \overset{..}{\underset{..}{Cl}} \cdot \longrightarrow HCl \end{array} \right\} \quad \text{chain termination steps}$$

The reaction of H_2 with Br_2 is also a photochemical reaction. That of H_2 with I_2 is very slow, even at high temperatures and with illumination.

All hydrogen halides react with H_2O to produce *hydrohalic acids* that ionize.

$$H - \overset{..}{\underset{|}{O}} : + H - \overset{..}{\underset{..}{X}} : \rightleftharpoons H - \overset{..}{\underset{|}{O}} - H^+ + : \overset{..}{\underset{..}{X}} : ^-$$
$$\quad\quad H \qquad\qquad\qquad\qquad H$$

The reaction is essentially complete for dilute aqueous solutions of HCl, HBr, and HI. In dilute aqueous solutions; HF is a weak acid ($K_a = 7.2 \times 10^{-4}$). In concentrated solutions more acidic dimeric $(HF)_2$ units are present (Figure 24-1). They ionize as follows.

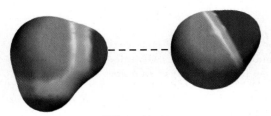

Dilute solution

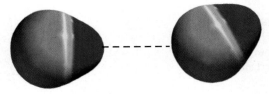

Concentrated solution $(HF)_2$

Figure 24-1 Hydrogen bonding (dashed lines) in dilute and concentrated aqueous solutions of hydrofluoric acid, HF.

$$(HF)_2(aq) + H_2O(\ell) \rightleftharpoons H_3O^+(aq) + HF_2^-(aq) \qquad K \approx 5$$

The order of increasing acid strengths of the aqueous hydrohalic acids is

$$HF \ll HCl < HBr < HI$$

The only acid used in industry to a greater extent than HCl is H_2SO_4. Hydrochloric acid is used in the production of metal chlorides, dyes, and many other commercially important products. It is also used on a large scale to dissolve metal oxide coatings from iron and steel prior to galvanizing or enameling.

Hydrofluoric acid is used in the production of fluorine-containing compounds and for etching glass. The acid reacts with silicates, such as calcium silicate, $CaSiO_3$, in the glass to produce a very volatile and thermodynamically stable compound, silicon tetrafluoride, SiF_4.

$$CaSiO_3(s) + 6HF(aq) \longrightarrow CaF_2(s) + SiF_4(g) + 3H_2O(\ell)$$

24-7   THE OXOACIDS (TERNARY ACIDS) OF THE HALOGENS

Table 24-5 lists the known oxoacids of the halogens, their sodium salts, and some trends in properties. Only three oxoacids, $HClO_4$, HIO_3, and H_5IO_6, have been isolated in anhydrous form. The others are known only in aqueous solution. In all these acids the H is bonded through an O.

The only oxoacid of fluorine that has been prepared is unstable hypofluorous acid, HOF. Aqueous *hypohalous acids* (except HOF) can be prepared by reaction of free halogens (Cl_2, Br_2, I_2) with cold water. The smaller the halogen, the farther to the right the equilibrium lies.

In HOF the oxidation states are F = −1, H = +1, O = 0.

These reactions all involve disproportionation of the halogen.

$$\overset{\scriptscriptstyle(0)}{X_2} + H_2O \rightleftharpoons \underset{\text{hydrohalic acid}}{\overset{\scriptscriptstyle(-1)}{HX}} + \underset{\text{hypohalous acid}}{\overset{\scriptscriptstyle(+1)}{HOX}} \qquad (X = Cl, Br, I)$$

Hypohalite salts can be prepared by reactions of the halogens with *cold* dilute bases.

TABLE 24-5   *Oxoacids of the Halogens and Their Salts*

Oxidation State	Acid	Name of Acid	Thermal Stability and Acid Strength	Sodium Salt	Name of Salt	Thermal Stability	Hydrolysis of Anion	Nature of Halogen
+1	HXO (HOX)	hypohalous acid		NaXO (NaOX)	sodium hypohalite			X = F, Cl, Br, I
+3	HXO_2	halous acid		$NaXO_2$	sodium halite			X = Cl, Br (?)
+5	HXO_3	halic acid	Increase	$NaXO_3$	sodium halate	Increases	Increase	X = Cl, Br, I
+7	HXO_4	perhalic acid		$NaXO_4$	sodium perhalate			X = Cl, Br, I
+7	H_5XO_6	paraperhalic acid		several types	sodium paraperhalates			X = I only

$$\overset{\scriptsize{0}}{X_2} + 2NaOH \longrightarrow \underset{\text{sodium halide}}{\overset{\scriptsize{-1}}{NaX}} + \underset{\text{sodium hypohalite}}{\overset{\scriptsize{+1}}{NaOX}} + H_2O \qquad (X = Cl, Br, I)$$

The hypohalites are used as bleaching agents. Sometimes Cl_2 is used as a bleach or as a disinfectant, as in public water supplies. It reacts slowly with H_2O to form HCl and HOCl. The hypochlorous acid then decomposes into HCl and O radicals, which kill bacteria.

$$Cl_2 + H_2O \rightleftharpoons HCl + HOCl$$

$$HOCl \longrightarrow HCl + \; : \overset{..}{\underset{.}{O}} \cdot$$

These oxygen radicals are very strong oxidizing agents. They are the effective bleaching and disinfecting agent in aqueous solutions of Cl_2 or hypochlorite salts.

Perchloric acid is the strongest of all common acids with respect to ionization. Hot, concentrated perchloric acid is a very strong oxidizing agent that can explode in the presence of reducing agents. Cold, dilute perchloric acid is only a weak oxidizing agent.

Solid household bleaches are usually $Ca(ClO)Cl$. This is prepared by reaction of Cl_2 with $Ca(OH)_2$.

$$Ca(OH)_2 + Cl_2 \longrightarrow Ca(ClO)Cl + H_2O$$

SULFUR, SELENIUM, AND TELLURIUM

24-8 OCCURRENCE, PROPERTIES, AND USES

Each Group VIA element is less electronegative than its neighboring halogen. Oxygen and sulfur are clearly nonmetallic, but selenium is less so. Tellurium is usually classified as a metalloid and forms metal-like crystals. Its chemistry is mostly that of a nonmetal. Polonium is a metal. All 29 isotopes of polonium are radioactive.

Irregularities in the properties of elements within a given family increase toward the middle of the periodic table. There are larger differences in the properties of the Group VIA elements than in the properties of the halogens. The properties of elements in the *second period* usually differ significantly from those of other elements in their families, because second-period elements have no low-energy d orbitals. So, the properties of oxygen are not very similar to those of the other Group VIA elements (Table 24-6). The maximum number of electrons that O can have in its valence shell is eight. The heavier Group VIA elements—S, Se, Te, and Po—have d orbitals in their valence shells; one or more of these

The d orbitals do not occur until the third energy level.

TABLE 24-6	Some Properties of Group VIA Elements				
Property	O	S	Se	Te	Po
Physical state (1 atm, 25°C)	gas	solid	solid	solid	solid
Color	colorless (very pale blue)	yellow	red-gray to black	brass-colored, metallic luster	—
Outermost electrons	$2s^2 2p^4$	$3s^2 3p^4$	$4s^2 4p^4$	$5s^2 5p^4$	$6s^2 6p^4$
Melting point (1 atm, °C)	-218	112	217	450	254
Boiling point (1 atm, °C)	-183	444	685	990	962
Electronegativity	3.5	2.5	2.4	2.1	1.9
First ionization energy (kJ/mol)	1314	1000	941	869	812
Atomic radius (Å)	0.73	1.03	1.19	1.42	1.68
Ionic ($2-$) radius (Å)	1.26	1.70	1.84	2.07	—
Common oxidation states	usually -2	$-2, +2, +4, +6$	$-2, +2, +4, +6$	$-2, +2, +4, +6$	$-2, +6$

Native sulfur. Elemental sulfur is deposited at the edges of some hot springs and geysers. This formation surrounds Emerald Lake in Yosemite National Park.

d orbitals can accommodate additional electrons to form up to six bonds. Thus each of the Group VIA elements except O can bond covalently to as many as six other atoms.

Sulfur

Sulfur makes up about 0.05% of the earth's crust. It was one of the elements known to the ancients. It was used by the Egyptians as a yellow coloring, and it was burned in some religious ceremonies because of the unusual odor it produced; it is the "brimstone" of the Bible. Alchemists tried to incorporate its "yellowness" into other substances to produce gold.

Sulfur occurs as the free element—predominantly S_8 molecules—and in metal sulfides such as galena, PbS; iron pyrite, FeS_2; and cinnabar, HgS. To a lesser extent, it occurs as metal sulfates such as barite, $BaSO_4$, and gypsum, $CaSO_4 \cdot 2H_2O$, and in volcanic gases as H_2S and SO_2.

Sulfur is found in much naturally occurring organic matter, such as petroleum and coal. Its presence in fossil fuels causes environmental and health problems because many sulfur-containing compounds burn to produce sulfur dioxide, an air pollutant.

Much of the sulfur used in the United States is recovered from sulfur compounds in natural gas and oil. During the oil-refining process, these compounds are reduced to hydrogen sulfide, which is then oxidized to sulfur in the Claus furnace.

$$8H_2S(g) + 4O_2(g) \longrightarrow S_8(\ell) + 8H_2O(g)$$

Elemental sulfur is mined along the U.S. Gulf Coast by the **Frasch process** (Figure 24-2). Most of it is used in the production of sulfuric acid, H_2SO_4, the most important of all industrial chemicals. Sulfur is used in the vulcanization of rubber and in the synthesis of many important sulfur-containing organic compounds.

In each of the three physical states, elemental sulfur exists in many forms. The two most stable forms of sulfur, the rhombic (mp 112°C) and monoclinic (mp 119°C) crystalline modifications, consist of different arrangements of S_8 molecules. These are puckered rings containing eight sulfur atoms (Figure 2-3) and all S—S single bonds. Above 150°C, sulfur becomes increasingly viscous and darkens as the S_8 rings break apart into chains that interlock with one another through S—S bonds. The viscosity reaches a maximum at 180°C, at which point sulfur is dark brown. Above 180°C, the liquid thins as the chains are broken down into smaller chains. At 444°C, sulfur boils to give a vapor containing S_8, S_6, S_4, and S_2 molecules.

Selenium

Selenium is quite rare (9×10^{-6}% of the earth's crust). It occurs mainly as an impurity in sulfur, sulfide, and sulfate deposits. It is obtained from the flue dusts that result from roasting sulfide ores and from the "anode mud" formed in the electrolytic refining of copper. It

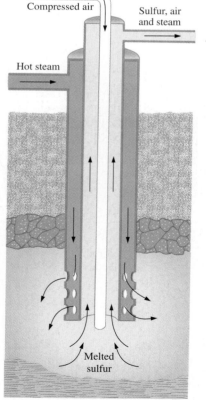

Figure 24-2 The Frasch process for mining sulfur. Three concentric pipes are used. Water at about 170°C and a pressure of 100 lb/in² (7 kg/cm²) is forced down the outermost pipe to melt the sulfur. Hot compressed air is pumped down the innermost pipe. It mixes with the molten sulfur to form a froth, which rises through the third pipe.

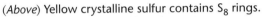

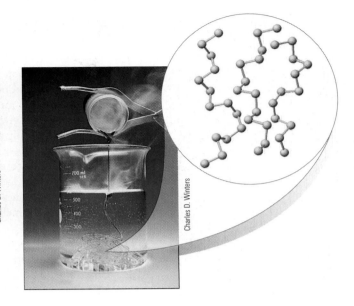

Charles D. Winters

(*Above*) Yellow crystalline sulfur contains S_8 rings.

(*Right*) When molten sulfur is heated above 150°C, an amorphous form of sulfur called "plastic sulfur" is formed as the S_8 rings break and the fragments link to form long chains.

is used as a red coloring in glass. The gray crystalline allotropic form of selenium has an electrical conductivity that is very light-sensitive, so it is used in photocopy machines and in solar cells.

Tellurium

Tellurium is even less abundant ($2 \times 10^{-7}\%$ of the earth's crust) than selenium. It occurs mainly in sulfide ores, especially with copper sulfide, and as the tellurides of gold and silver. It, too, is obtained from the "anode mud" from refining of copper. The element forms brass-colored, shiny, hexagonal crystals having low electrical conductivity. It is added to some metals, particularly lead, to increase electrical resistance and improve resistance to heat, corrosion, mechanical shock, and wear.

24-9 REACTIONS OF GROUP VIA ELEMENTS

Some reactions of the Group VIA elements are summarized in the following table.

General Equation	Remarks
$x\mathrm{E} + y\mathrm{M} \longrightarrow \mathrm{M}_y\mathrm{E}_x$	With many metals
$z\mathrm{E} + \mathrm{M}_x\mathrm{E}_y \longrightarrow \mathrm{M}_x\mathrm{E}_{y+z}$	Especially with S, Se
$\mathrm{E} + \mathrm{H}_2 \longrightarrow \mathrm{H}_2\mathrm{E}$	Decreasingly in the series O_2, S, Se, Te
$\mathrm{E} + 3\mathrm{F}_2 \longrightarrow \mathrm{EF}_6$	With S, Se, Te, and excess F_2
$2\mathrm{E} + \mathrm{Cl}_2 \longrightarrow \mathrm{E}_2\mathrm{Cl}_2$	With S, Se (Te gives $TeCl_2$); also with Br_2
$\mathrm{E}_2\mathrm{Cl}_2 + \mathrm{Cl}_2 \longrightarrow 2\mathrm{ECl}_2$	With S, Se; also with Br_2
$\mathrm{E} + 2\mathrm{Cl}_2 \longrightarrow \mathrm{ECl}_4$	With S, Se, Te, and excess Cl_2; also with Br_2
$\mathrm{E} + \mathrm{O}_2 \longrightarrow \mathrm{EO}_2$	With S (with Se, use $O_2 + NO_2$)

E represents a Group VIA element.

24-10 HYDRIDES OF GROUP VIA ELEMENTS

All the Group VIA elements form covalent compounds of the type H_2E (E = O, S, Se, Te, Po) in which the Group VIA element is in the -2 oxidation state. H_2O is a liquid that is essential for animal and plant life. H_2S, H_2Se, and H_2Te are colorless, noxious, poisonous gases. They are even more toxic than HCN. Egg protein contains sulfur, and its decomposition forms H_2S, giving off the odor of rotten eggs. H_2Se and H_2Te smell even worse. Their odors are usually ample warning of the presence of these poisonous gases.

Both the melting point and boiling point of water are very much higher than expected by comparison with those of the heavier hydrides (Figure 13-5). This is a consequence of hydrogen bonding in ice and liquid water (Section 13-2) caused by the strongly dipolar nature of water molecules. The electronegativity differences between H and the other VIA elements are much smaller than that between H and O, so no H-bonding occurs in H_2S, H_2Se, or H_2Te.

Aqueous solutions of hydrogen sulfide, selenide, and telluride are acidic; acid strength increases as the group is descended: $H_2S < H_2Se < H_2Te$. The same trend was observed for increasing acidity of the hydrogen halides. The acid ionization constants are

> H_2S is a stronger acid than H_2O. The solubility of H_2S in water is approximately 0.10 mol/L at 25°C.

> Here E represents S, Se, or Te.

		H_2S	H_2Se	H_2Te
$H_2E \rightleftharpoons H^+ + HE^-$	K_{a1}:	1.0×10^{-7}	1.9×10^{-4}	2.3×10^{-3}
$HE^- \rightleftharpoons H^+ + E^{2-}$	K_{a2}:	1.0×10^{-19}	$\approx 10^{-11}$	$\approx 1.6 \times 10^{-11}$

24-11 GROUP VIA OXIDES

Although others exist, the most important VIA oxides are the dioxides, which are acid anhydrides of H_2SO_3, H_2SeO_3, and H_2TeO_3; and the trioxides, which are anhydrides of H_2SO_4, H_2SeO_4, and H_6TeO_6. Let us consider SO_2 and SO_3 as examples.

Sulfur Dioxide, SO_2

Sulfur dioxide is a colorless, poisonous, corrosive gas with a very irritating odor. Even in small quantities, it causes coughing and nose, throat, and lung irritation. It is an angular molecule with trigonal planar electronic geometry, sp^2 hybridization at the S atom, and resonance stabilization.

Sulfur dioxide is produced in reactions such as the combustion of sulfur-containing fossil fuels and the roasting of sulfide ores.

$$2ZnS + 3O_2 \longrightarrow 2ZnO + 2SO_2$$

SO_2 is a waste product of these operations. In the past, it was released into the atmosphere along with some SO_3 produced by its reaction with O_2. A more environmentally friendly practice now is to trap SO_2 and SO_3 and use them to make H_2SO_4. Some coal contains up to 5% sulfur, so both SO_2 and SO_3 are present in the flue gases when coal is burned. No way has been found to remove all the SO_2 from flue gases of power plants. One way of removing most of the SO_2 involves the injection of limestone, $CaCO_3$, into the combustion zone of the furnace. Here $CaCO_3$ decomposes to lime, CaO. This then combines with SO_2

> If the SO_2 and SO_3 are allowed to escape into the atmosphere, they cause highly acidic rain.

+4 Oxidation State

Formula	Name
H_2SO_3	sulfurous acid
H_2SeO_3	selenous acid
H_2TeO_3	tellurous acid

+6 Oxidation State

Formula	Name
H_2SO_4	sulfuric acid
H_2SeO_4	selenic acid
H_6TeO_6	telluric acid

Large amounts of SO_2 and H_2S are released during volcanic eruptions.

to form calcium sulfite ($CaSO_3$), an ionic solid, which is collected and disposed of as solid waste

$$CaCO_3 \xrightarrow{\text{heat}} CaO + CO_2 \qquad \text{followed by} \qquad CaO + SO_2 \longrightarrow CaSO_3$$

This process is called scrubbing, and a disadvantage of it is the formation of huge quantities of solid waste ($CaSO_3$, unreacted CaO, and by-products).

Catalytic oxidation is now used by the smelting industry to convert SO_2 into SO_3. This is then dissolved in water to make solutions of H_2SO_4 (up to 80% by mass). The gases containing SO_2 are passed through a series of condensers containing catalysts to speed up the reaction. In some cases the impure H_2SO_4 can be used in other operations in the same plant.

Recently mined sulfur awaits shipment.

Sulfur Trioxide, SO_3

Sulfur trioxide is a liquid that boils at 44.8°C. It is the anhydride of H_2SO_4. It is formed by the reaction of SO_2 with O_2. The reaction is very exothermic, but ordinarily very slow. It is catalyzed commercially in the **contact process** by spongy Pt, SiO_2, or vanadium(V) oxide, V_2O_5, at high temperatures (400 to 700°C).

$$2SO_2(g) + O_2(g) \xrightleftharpoons{\text{catalyst}} 2SO_3(g) \qquad \Delta H^0 = -197.6 \text{ kJ/mol} \qquad \Delta S^0 = -188 \text{ J/K} \cdot \text{mol}$$

The high temperature favors SO_2 and O_2 but allows the reaction to proceed much more rapidly, so it is economically advantageous. The SO_3 is then removed from the gaseous reaction mixture by dissolving it in concentrated H_2SO_4 (95% H_2SO_4 by mass) to produce polysulfuric acids—mainly pyrosulfuric acid, $H_2S_2O_7$. This is called oleum, or fuming sulfuric acid. The addition of fuming sulfuric acid to water produces commercial H_2SO_4.

$$SO_3 + H_2SO_4 \longrightarrow H_2S_2O_7 \qquad \text{then} \qquad H_2S_2O_7 + H_2O \longrightarrow 2H_2SO_4$$

In the presence of certain catalysts, sulfur dioxide in polluted air reacts rapidly with O_2 to form SO_3. Particulate matter, or suspended microparticles, such as NH_4NO_3 and elemental S, act as efficient catalysts.

The prefix *pyro* means "heat" or "fire." Pyrosulfuric acid may also be obtained by heating concentrated sulfuric acid, which results in the elimination of one molecule of water from two molecules of sulfuric acid.

$$2H_2SO_4 \longrightarrow H_2S_2O_7 + H_2O$$

24-12 OXOACIDS OF SULFUR

Sulfurous Acid, H_2SO_3

Sulfur dioxide readily dissolves in water to produce solutions of sulfurous acid, H_2SO_3. The acid has not been isolated in anhydrous form.

$$H_2O + SO_2 \rightleftharpoons H_2SO_3$$

The acid ionizes in two steps in water

$$H_2SO_3 \rightleftharpoons H^+ + HSO_3^- \qquad K_{a1} = 1.2 \times 10^{-2}$$
$$HSO_3^- \rightleftharpoons H^+ + SO_3^{2-} \qquad K_{a2} = 6.2 \times 10^{-8}$$

When excess SO_2 is bubbled into aqueous NaOH, sodium hydrogen sulfite, $NaHSO_3$, is produced. This acid salt can be neutralized with additional NaOH or Na_2CO_3 to produce sodium sulfite.

$$NaOH + H_2SO_3 \longrightarrow NaHSO_3 + H_2O$$
$$NaOH + NaHSO_3 \longrightarrow Na_2SO_3 + H_2O$$

The sulfite ion is pyramidal and has tetrahedral electronic geometry as predicted by the VSEPR theory.

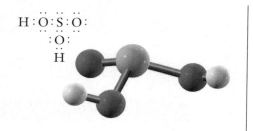

Sulfuric acid H_2SO_3

Hydrogen sulfite ion, HSO_3^-

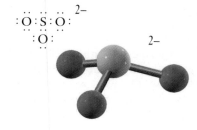

Sulfite ion, SO_3^{2-}

Sulfuric Acid, H_2SO_4

More than 40 million tons of sulfuric acid are produced annually worldwide. The contact process is used for the commercial production of most sulfuric acid. The solution sold commercially as "concentrated sulfuric acid" is 96–98% H_2SO_4 by mass and is about 18 molar H_2SO_4.

Pure H_2SO_4 is a colorless, oily liquid that freezes at 10.4°C and boils at 290 to 317°C while partially decomposing to SO_3 and water. There is some hydrogen bonding in solid and liquid H_2SO_4.

Tremendous amounts of heat are evolved when concentrated sulfuric acid is diluted. This illustrates the strong affinity of H_2SO_4 for water. H_2SO_4 is often used as a dehydrating agent. Dilutions should always be performed by adding the acid to water to avoid spattering the acid.

Pouring concentrated H_2SO_4 into an equal volume of H_2O liberates a lot of heat—enough to raise the temperature of the resulting solution by as much as 100°C.

Charles D. Winters

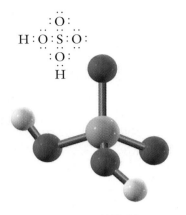

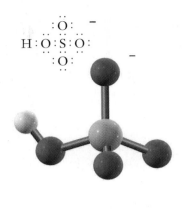

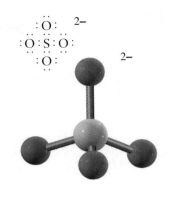

Sulfuric acid H_2SO_4 Hydrogen sulfate ion, HSO_4^- Sulfate ion, SO_4^{2-}

Sulfuric acid is a strong acid with respect to the first step of its ionization in water. The second ionization occurs to a lesser extent (Example 18-17).

$$H_2SO_4 \rightleftharpoons H^+ + HSO_4^- \qquad K_{a1} = \text{very large}$$
$$HSO_4^- \rightleftharpoons H^+ + SO_4^{2-} \qquad K_{a2} = 1.2 \times 10^{-2}$$

NITROGEN AND PHOSPHORUS

The Group VA elements provide a dramatic illustration of the vertical trends in metallic properties. In this family, nitrogen and phosphorus are nonmetals, arsenic is predominantly nonmetallic, antimony is more metallic, and bismuth is definitely metallic. Properties of the Group VA elements are listed in Table 24-7.

Oxidation states of the VA elements range from -3 to $+5$. Odd-numbered oxidation states are favored. The VA elements form very few monatomic ions. Ions with a charge of $3-$ occur for N and P, as in Mg_3N_2 and Ca_3P_2.

All of the Group VA elements show the -3 oxidation state in covalent compounds such as NH_3 and PH_3. The $+5$ oxidation state is found only in covalent compounds such as phosphorus pentafluoride, PF_5; nitric acid, HNO_3; and phosphoric acid, H_3PO_4; and in

SCI LINKS.

TOPIC: Nitrogen
GO TO: www.scilinks.org
sciLINKS CODE: WCH2430

TABLE 24-7	Properties of the Group VA Elements				
Property	**N**	**P**	**As**	**Sb**	**Bi**
Physical state (1 atm, 25°C)	gas	solid	solid	solid	solid
Color	colorless	red, white, black	yellow, gray	yellow, gray	gray
Outermost electrons	$2s^2 2p^3$	$3s^2 3p^3$	$4s^2 4p^3$	$5s^2 5p^3$	$6s^2 6p^3$
Melting point (°C)	-210	44 (white)	813 (gray, 28 atm)*	631 (gray)	271
Boiling point (°C)	-196	280 (white)	sublimes 613	1750	1560
Atomic radius (Å)	0.75	1.10	1.20	1.40	1.50
Electronegativity	3.0	2.1	2.1	1.9	1.8
First ionization energy (kJ/mol)	1402	1012	947	834	703
Oxidation states	-3 to $+5$	-3 to $+5$	-3 to $+5$	-3 to $+5$	-3 to $+5$

*Sublimes at lower pressures.

TABLE 24-8	*Oxidation States of Nitrogen and Examples*							
−3	**−2**	**−1**	**0**	**+1**	**+2**	**+3**	**+4**	**+5**
NH_3 ammonia	N_2H_4 hydrazine	NH_2OH hydroxylamine	N_2 nitrogen	N_2O dinitrogen oxide	NO nitrogen oxide	N_2O_3 dinitrogen trioxide	NO_2 nitrogen dioxide	N_2O_5 dinitrogen pentoxide
NH_4^+ ammonium ion		NH_2Cl chloramine		$H_2N_2O_2$ hyponitrous acid		HNO_2 nitrous acid	N_2O_4 dinitrogen tetroxide	HNO_3 nitric acid
NH_2^- amide ion						NO_2^- nitrite ion		NO_3^- nitrate ion

polyatomic ions such as NO_3^- and PO_4^{3-}. Each Group VA element exhibits the +3 oxidation state in one of its oxides, for instance, N_2O_3 and P_4O_6. These are acid anhydrides of nitrous acid, HNO_2, and phosphorous acid, H_3PO_3; both are weak acids. No other element exhibits more oxidation states than nitrogen (Table 24-8).

24-13 OCCURRENCE OF NITROGEN

Every protein contains nitrogen in each of its fundamental amino acid units.

Nitrogen, N_2, is a colorless, odorless, tasteless gas that makes up about 75% by mass and 78% by volume of the atmosphere. Nitrogen compounds form only a minor portion of the

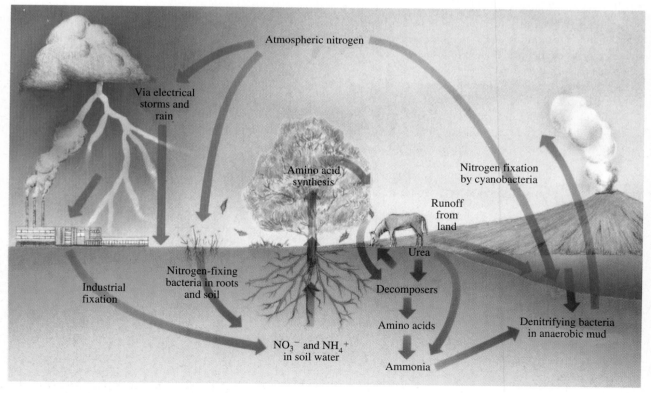

A schematic representation of the nitrogen cycle.

earth's crust, but all living matter contains nitrogen. The primary natural inorganic deposits of nitrogen are very localized. They consist mostly of KNO_3 and $NaNO_3$. Most sodium nitrate is mined in Chile.

The extreme abundance of N_2 in the atmosphere and the low relative abundance of nitrogen compounds elsewhere are due to the chemical inertness of N_2 molecules. This results from the very high bond energy of the $N{\equiv}N$ bond (945 kJ/mol).

Although N_2 molecules are relatively unreactive, nature provides mechanisms by which N atoms are incorporated into proteins, nucleic acids, and other nitrogenous compounds. The **nitrogen cycle** is the complex series of reactions by which nitrogen is slowly but continually recycled in the atmosphere (our nitrogen reservoir), lithosphere (earth), and hydrosphere (water).

When N_2 and O_2 molecules collide near a bolt of lightning, they can absorb enough electrical energy to produce molecules of NO. An NO molecule is quite reactive because it contains one unpaired electron. NO reacts readily with O_2 to form nitrogen dioxide, NO_2. Most NO_2 dissolves in rainwater and falls to the earth's surface. Bacterial enzymes reduce the nitrogen in a series of reactions in which amino acids and proteins are produced. These are then used by plants, eaten by animals, and metabolized. The metabolic products are excreted as nitrogenous compounds such as urea, $(NH_2)_2CO$, and ammonium salts such as $NaNH_4HPO_4$. These can also be enzymatically converted to ammonia, NH_3, and amino acids.

Nitrogen is converted directly into NH_3 in another way. Members of the class of plants called legumes (including soybeans, alfalfa, and clover) have nodules on their roots. Within the nodules live bacteria that produce an enzyme called nitrogenase. These bacteria extract N_2 directly from air trapped in the soil and convert it into NH_3. The ability of nitrogenase to catalyze this conversion, called *nitrogen fixation*, at ordinary temperatures and pressures with very high efficiency is a marvel to scientists. They must resort to very extreme and costly conditions to produce NH_3 from nitrogen and hydrogen (the Haber process, Section 17-7).

Ammonia is the source of nitrogen in many fertilizers. Unfortunately, nature does not produce NH_3 and related plant nutrient compounds rapidly enough to provide an adequate food supply for the world's growing population. Commercial synthetic fertilizers have helped to lessen this problem, but at great cost for the energy that is required to produce them.

Nitrogen is sold as compressed gas in cylinders. The boiling point of N_2 is $-195.8°C$ ($-320°F$). N_2 is obtained by fractional distillation of liquid air.

Root nodules on soybeans.

24-14 HYDROGEN COMPOUNDS OF NITROGEN

We have already described ammonia and some of its reactions. Please review Sections 17-7 and 18-4.

Liquid ammonia (bp $-33.4°C$) is used as a solvent for some chemical reactions. It is hydrogen bonded, just as H_2O is, but NH_3 is a much more basic solvent. Its weak *autoionization* produces the ammonium ion, NH_4^+, and the amide ion, NH_2^-. This is similar to H_2O, which ionizes to produce some H_3O^+ and OH^- ions.

$$NH_3(\ell) + NH_3(\ell) \rightleftharpoons NH_4^+ + NH_2^- \qquad K = 10^{-35}$$

$$\underset{\text{base}_2}{} \quad \underset{\text{acid}_1}{} \qquad \underset{\text{acid}_2}{} \quad \underset{\text{base}_1}{}$$

Many ammonium salts are known. Most are very soluble in water. They can be prepared by reactions of ammonia with acids. Reaction with nitric acid gives ammonium nitrate.

$$NH_3(aq) + [H^+(aq) + NO_3^-(aq)] \longrightarrow [NH_4^+ + NO_3^-(aq)] \qquad \text{ammonium nitrate}$$

Great advances have been made in cattle breeding in recent decades. Semen from superior bulls can be collected and stored in liquid nitrogen for 30 years or more.

Amines are organic compounds that are structurally related to ammonia. We think of them as being derived from NH_3 by the replacement of one or more hydrogens with organic groups (Sections 18-4 and 27-12). All involve sp^3-hybridized N. All are weak bases because of the unshared pair of electrons on N.

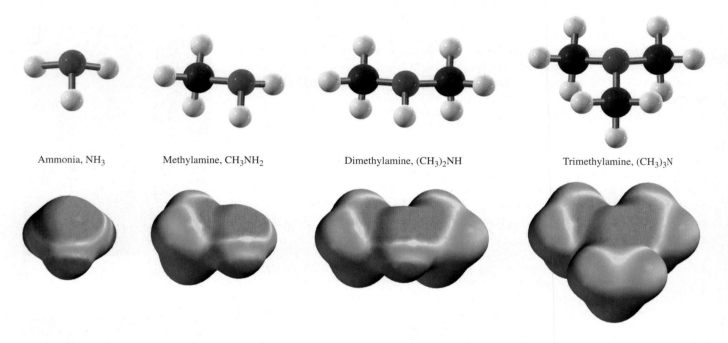

Ammonia, NH_3 Methylamine, CH_3NH_2 Dimethylamine, $(CH_3)_2NH$ Trimethylamine, $(CH_3)_3N$

24-15 NITROGEN OXIDES

Nitrogen forms several oxides, in which it exhibits positive oxidation states of 1 to 5 (Table 24-8). All have positive free energies of formation, owing to the high dissociation energy of N_2 and O_2 molecules. All are gases except N_2O_5, a solid that melts at 30.0°C.

Dinitrogen Oxide (+1 Oxidation State)

Molten ammonium nitrate undergoes autooxidation–reduction (decomposition) at 170 to 260°C to produce dinitrogen oxide, also called nitrous oxide.

$$\overset{\ominus 3}{N}H_4\overset{\oplus 5}{N}O_3(\ell) \xrightarrow{\text{heat}} \overset{\oplus 1}{N_2}O(g) + 2H_2O(g) + \text{heat}$$

At higher temperatures, explosions occur, producing N_2, O_2, and H_2O.

$$2NH_4NO_3(\ell) \xrightarrow{\text{heat}} 2N_2(g) + O_2(g) + 4H_2O(g) + \text{heat}$$

Dinitrogen oxide supports combustion because it produces O_2 when heated.

$$2N_2O(g) \xrightarrow{\text{heat}} 2N_2(g) + O_2(g) + \text{heat}$$

The molecule is linear but unsymmetrical, with a dipole moment of 0.17 D.

Some dentists use N_2O for its mild anesthetic properties. It is also known as laughing gas because of its side effects.

Dinitrogen oxide, or nitrous oxide, N_2O, mp −90.8°C, bp −88.8°C.

Nitrogen Oxide (+2 Oxidation State)

The first step of the Ostwald process (Section 24-16) for producing HNO_3 from NH_3 is used for the commercial preparation of nitrogen oxide, NO.

$$4NH_3(g) + 5O_2(g) \xrightarrow[\text{heat}]{\text{catalyst}} 4NO(g) + 6H_2O(g)$$

NO is not produced in large amounts in nature under usual conditions. It is formed by direct reaction of N_2 and O_2 in electrical storms.

NO is a colorless gas that condenses at $-152°C$ to a blue liquid. Gaseous NO is paramagnetic and contains one unpaired electron per molecule.

$$:N=\ddot{O}: \longleftrightarrow :N=\ddot{O}:$$

Its unpaired electron makes nitric oxide very reactive. Molecules that contain unpaired electrons are called *radicals*. NO reacts with O_2 to form NO_2, a brown, corrosive gas.

$$2NO(g) + O_2(g) \longrightarrow 2NO_2(g) + \text{heat}$$

In recent years, researchers have learned that very low concentrations of NO have important roles in a variety of physiological processes, not only in humans, but in such a variety of animals as barnacles, chickens, trout, and fruit flies. It has recently been found to be involved in transmission of signals by nerves, blood clotting, local control of blood flow, the immune system's ability to kill tumor cells and bacteria, and possibly even memory.

Nitrogen Dioxide and Dinitrogen Tetroxide (+4 Oxidation State)

Nitrogen dioxide is formed by reaction of NO with O_2. It is prepared in the laboratory by heating heavy metal nitrates.

$$2Pb(NO_3)_2(s) \xrightarrow{\text{heat}} 2PbO(s) + 4NO_2(g) + O_2(g)$$

The NO_2 molecule is angular. It is represented by resonance structures.

nitrogen dioxide

Because each NO_2 molecule contains one unpaired electron, NO_2 readily dimerizes to form colorless, diamagnetic dinitrogen tetroxide, N_2O_4, at low temperatures.

$$2NO_2(g) \rightleftharpoons N_2O_4(g) \qquad \Delta H^0 = -57.2 \text{ kJ/mol rxn}$$

brown   colorless

The N_2O_4 molecule is also represented by resonance structures.

Nitrogen oxide, or nitric oxide, NO, mp $-163.6°C$, bp $-151.8°C$, bond distance (1.15 Å) intermediate between N≡O (1.06 Å) and N=O (1.20 Å).

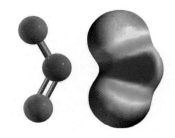

Nitrogen dioxide, NO_2, mp $-11.20°C$, one unpaired electron, bond length 1.197 Å; brown gas.

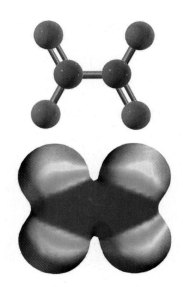

Dinitrogen tetroxide, N_2O_4, bp 21.2°C, no unpaired electrons, N—N bond length 1.64 Å, N—O bond length 1.17Å, O—N—O bond angle 126°.

CHEMISTRY IN USE

The Environment

Nitrogen Oxides and Photochemical Smog

Nitrogen oxides are produced in the atmosphere by natural processes. Human activities contribute only about 10% of all the oxides of nitrogen (collectively referred to as NO_x) in the atmosphere, but the human contribution occurs mostly in urban areas, where the oxides may be present in concentrations a hundred times greater than in rural areas.

Just as NO is produced naturally by the reaction of N_2 and O_2 in electrical storms, it is also produced by the same reaction at the high temperatures of internal combustion engines and furnaces.

$$N_2(g) + O_2(g) \rightleftharpoons 2NO(g) \qquad \Delta H^0 = +180 \text{ kJ/mol rxn}$$

At ordinary temperatures the reaction does not occur to a significant extent. Because it is endothermic, it is favored by high temperatures. Even in internal combustion engines and furnaces, the equilibrium still lies far to the left, so only small amounts of NO are produced and released into the atmosphere. Even very small concentrations of nitrogen oxides cause serious problems, however.

The NO radicals react with O_2 to produce NO_2 radicals. Both NO and NO_2 are quite reactive, and they do considerable damage to plants and animals. NO_2 reacts with H_2O in the air to produce corrosive droplets of HNO_3 and more NO.

$$\overset{+4}{3NO_2} + H_2O \longrightarrow \overset{+2}{NO} + 2\overset{+5}{HNO_3} \qquad \text{(nitric acid)}$$

The HNO_3 may be washed out of the air by rainwater (acid rain), or it may react with traces of NH_3 in the air to form solid NH_4NO_3, a *particulate* pollutant.

$$HNO_3 + NH_3 \longrightarrow NH_4NO_3$$

This situation occurs in all urban areas, but the problem is worse in warm, dry climates, which are conducive to light-induced (photochemical) reactions. Here ultraviolet (UV) radiation from the sun produces damaging oxidants. The brownish hazes that often hang over such cities as Los Angeles, Denver, and Mexico City are due to the presence of brown NO_2. Problems begin in the morning rush hour as NO is exhausted into the air. The NO combines with O_2 to form NO_2. Then, as the sun rises higher in the sky, NO_2 absorbs UV radiation and breaks down into NO and oxygen radicals.

$$NO_2 \overset{UV}{\longrightarrow} NO + O$$

The extremely reactive O radicals combine with O_2 to produce O_3 (ozone).

$$O + O_2 \longrightarrow O_3$$

Ozone is a powerful oxidizing agent that damages rubber, plastic materials, and all plant and animal life. It also reacts with hydrocarbons from automobile exhaust and evaporated gasoline to form secondary organic pollutants such as aldehydes and ketones (Section 27-11). The **peroxyacyl nitrates (PANs)**, perhaps the worst of the secondary pollutants, are especially damaging photochemical oxidants that are very irritating to the eyes and throat.

$$R - \overset{\overset{\displaystyle O}{\|}}{\underset{acyl}{C}} - \overbrace{O - O}^{peroxy} - \underbrace{N\overset{\displaystyle O}{\underset{\displaystyle O}{\diagdown}}}_{nitrate} \qquad R = \text{hydrocarbon chain or ring}$$

Catalytic converters in automobile exhaust systems reduce emissions of oxides of nitrogen.

Photochemical smog casts a haze over urban or industrial areas; its severity depends on the weather.

24-16 SOME OXOACIDS OF NITROGEN AND THEIR SALTS

The main oxoacids of nitrogen are nitrous acid, HNO_2, and nitric acid, HNO_3.

Nitrogen also forms hyponitrous acid, $H_2N_2O_2$, in which N is in the $+1$ oxidation state, as well as hyponitrite salts such as $Na_2N_2O_2$.

Nitrous Acid (+3 Oxidation State)

Nitrous acid, HNO_2, is unstable and cannot be isolated in pure form. It is prepared as a pale blue solution when H_2SO_4 reacts with cold aqueous sodium nitrite. Nitrous acid is a weak acid ($K_a = 4.5 \times 10^{-4}$). It acts as an oxidizing agent toward strong reducing agents and as a reducing agent toward very strong oxidizing agents.

Lewis formulas for nitrous acid and the nitrite ion follow.

nitrous acid   nitrite ion

Nitric Acid (+5 Oxidation State)

Pure nitric acid, HNO_3, is a colorless liquid that boils at 83°C. Light or heat causes it to decompose into NO_2, O_2, and H_2O. The presence of the NO_2 in partially decomposed aqueous HNO_3 causes its yellow or brown tinge. The Lewis structure of nitric acid is

$$H-\ddot{O}-N \overset{\ddot{O}}{\underset{\ddot{O}}{}} \longleftrightarrow H-\ddot{O}-N \overset{\ddot{O}}{\underset{O}{}}$$

HNO_3 is commercially prepared by the **Ostwald process.** At high temperatures, NH_3 is catalytically converted to NO, which is cooled and then air-oxidized to NO_2. Nitrogen dioxide reacts with H_2O to produce HNO_3 and some NO. The NO produced in the third step is then recycled into the second step. More than 18 billion pounds of HNO_3 was produced in the United States in 1997.

$$4NH_3(g) + 5O_2(g) \xrightarrow[1000°C]{Pt} 4NO(g) + 6H_2O(g)$$

$$2NO(g) + O_2(g) \xrightarrow{cool} 2NO_2(g)$$
$$3NO_2(g) + H_2O(\ell) \longrightarrow 2[H^+ + NO_3^-] + \boxed{NO(g)}$$
recycle

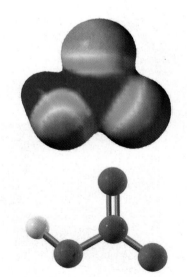

Nitric acid, HNO_3, mp −42°C; bp 83°C; bond lengths N—O (terminal) 1.22 Å, N—O (central) 1.41 Å.

Nitric acid is very soluble in water (≈ 16 mol/L). It is a strong acid and a strong oxidizing agent.

Nitric acid, HNO_3, reacts with protein-containing materials such as this feather, staining them yellow. Perhaps you have spilled nitric acid on your skin and seen it turn yellow.

Copper (*left beaker*) and zinc (*right beaker*) react with concentrated nitric acid.

NaNO₂ and NaNO₃ as Food Additives

The brown color of "old" meat is the result of oxidation of blood and is objectionable to many consumers. Nitrites and nitrates are added to food to retard this oxidation and also to prevent growth of botulism bacteria. Nitrate ions, NO_3^-, are reduced to NO_2^- ions, which are then converted to NO. This in turn reacts with the brown oxidized form of the heme in blood. This reaction keeps meat red longer. Controversy has arisen, however, concerning the possibility that nitrites combine with amines under the acidic conditions in the stomach to produce carcinogenic *nitrosoamines*.

$$\underbrace{\begin{array}{c} R' \\ \diagdown \\ N \\ \diagup \\ R \end{array}}_{\substack{\text{amine} \\ \text{group}}} - \underbrace{N=O}_{\substack{\text{nitroso} \\ \text{group}}} \qquad \text{(R and R}' = \text{organic groups)}$$

White phosphorus reacts with air, so it is stored under water. It contains tetrahedral P_4 molecules.

24-17 PHOSPHORUS

Phosphorus is always combined in nature. Phosphorus is present in all living organisms—as organophosphates and in calcium phosphates such as hydroxyapatite, $Ca_5(PO_4)_3(OH)$, and fluorapatite, $Ca_5(PO_4)_3F$, in bones and teeth. It also occurs in these and related compounds in phosphate minerals, which are mined mostly in Florida and North Africa.

Industrially, the element is obtained from phosphate minerals by heating them at 1200 to 1500°C in an electric arc furnace with sand (SiO_2) and coke.

The tips of "strike anywhere" matches contain tetraphosphorus trisulfide and red phosphorus. Friction converts kinetic energy into heat, which initiates a spontaneous reaction.

$$P_4S_3(s) + 8O_2(g) \longrightarrow P_4O_{10}(s) + 3SO_2(g)$$

$$2Ca_3(PO_4)_2 + 6SiO_2 + 10C \xrightarrow{\text{heat}} 6CaSiO_3 + 10CO + P_4$$

calcium phosphate   calcium silicate
(phosphate rock)   (slag)

Vaporized phosphorus is condensed to a white solid (mp = 44.2°C, bp = 280.3°C) under H_2O to prevent oxidation. Even when kept under H_2O, white phosphorus slowly converts to the more stable red phosphorus allotrope (mp = 597°C; sublimes at 431°C). Red phosphorus and tetraphosphorus trisulfide, P_4S_3, are used in matches. They do not burn spontaneously, yet they ignite easily when heated by friction. Both white and red phosphorus are insoluble in water.

The largest use of phosphorus is in fertilizers. Phosphorus is an essential nutrient, and nature's phosphorus cycle is very slow owing to the low solubility of most natural phosphates. Phosphate fertilizers are therefore essential. To increase the solubility of the natural phosphates, they are treated with H_2SO_4 to produce "superphosphate of lime," a mixture of two salts. This solid is pulverized and applied as a powder.

Superphosphate fertilizer.

$$Ca_3(PO_4)_2 + 2H_2SO_4 + 4H_2O \xrightarrow{\text{evaporate}} [Ca(H_2PO_4)_2 + 2(CaSO_4 \cdot 2H_2O)]$$

phosphate rock   calcium   calcium
                                   dihydrogen   sulfate
                                   phosphate   dihydrate

superphosphate of lime

SCiLINKS.

TOPIC: Solar Energy
GO TO: www.scilinks.org
sciLINKS CODE: WCH2440

This reaction represents the most popular use of sulfuric acid, the industrial chemical produced in largest quantity.

SILICON

Silicon is a shiny, blue-gray, high-melting, brittle metalloid. It looks like a metal, but it is chemically more like a nonmetal. It is second only to oxygen in abundance in the earth's crust, about 87% of which is composed of silica (SiO_2) and its derivatives, the silicate minerals. The crust is 26% Si, compared with 49.5% O. Silicon does not occur free in nature. Pure silicon crystallizes with a diamond-type structure, but the Si atoms are less closely packed than C atoms. Its density is 2.4 g/cm³ compared with 3.51 g/cm³ for diamond.

Pure silicon is used in solar cells to collect energy from the sun.

24-18   SILICON AND THE SILICATES

Elemental silicon is usually prepared by the high-temperature reduction of silica (sand) with coke. Excess SiO_2 prevents the formation of silicon carbide.

$$SiO_2(s, excess) + 2C(s) \xrightarrow{heat} Si(s) + 2CO(g)$$

Reduction of a mixture of silicon and iron oxides with coke produces an alloy of iron and silicon known as *ferrosilicon*. It is used in the production of acid-resistant steel alloys, such as "duriron," and in the "deoxidation" of steel. Aluminum alloys for aircraft are strengthened with silicon.

Elemental silicon is used to make silicone polymers. Its semiconducting properties (Section 13-17) are used in transistors and solar cells.

The biggest chemical differences between silicon and carbon are that (1) silicon does not form stable double bonds, (2) it does not form very stable Si—Si bonds unless the silicon atoms are bonded to very electronegative elements, and (3) it has vacant $3d$ orbitals in its valence shell into which it can accept electrons from donor atoms. The Si—O single bond is the strongest of all silicon bonds and accounts for the stability and prominence of silica and the silicates.

Silicon dioxide (silica) exists in two familiar forms in nature: quartz, small chips of which occur in sand; and flint (Latin *silex*), an uncrystallized amorphous type of silica. Silica is properly represented as $(SiO_2)_n$ because it is a polymeric solid of SiO_4 tetrahedra sharing all oxygens among surrounding tetrahedra (Figure 13-32c). For comparison, solid carbon dioxide (dry ice) consists of discrete O=C=O molecules, as does gaseous CO_2.

Some gems and semiprecious stones such as amethyst, opal, agate, and jasper are crystals of quartz with colored impurities.

Most of the crust of the earth is made up of silica and silicates. The natural silicates comprise a large variety of compounds. The structures of all these are based on SiO_4 tetrahedra, with metal ions occupying spaces between the tetrahedra. The extreme stability of the silicates is presumably due to the donation of extra electrons from O into vacant $3d$ orbitals of Si. In many common minerals, called aluminosilicates, Al atoms replace some Si atoms with very little structural change. Because an Al atom has one less positive charge in its nucleus than Si does, it is also necessary to introduce a univalent ion, such as K^+ or Na^+.

The physical characteristics of the silicates are often suggested by the arrangement of the SiO_4 tetrahedra. A single-chain silicate, diopside $[CaMg(SiO_3)_2]_n$, and a double-chain silicate, asbestos $[Ca_2Mg_5(Si_4O_{11})_2(OH_2)]_n$, occur as fibrous or needle-like crystals. Talc, $[Mg_3Si_4O_{10}(OH)_2]_n$, a silicate with a sheet-like structure, is flaky. Micas are sheet-like aluminosilicates with about one of every four Si atoms replaced by Al. Muscovite mica is $[KAl_2(AlSi_3O_{10})(OH)_2]_n$. Micas occur in thin sheets that are easily peeled away from one another.

The clay minerals are silicates and aluminosilicates with sheet-like structures. They result from the weathering of granite and other rocks. The layers have enormous "inner surfaces" that can absorb large amounts of H_2O. Clay mixtures often occur as minute platelets with a very large total surface area. When wet, the clays are easily shaped. When heated to high temperatures, they lose H_2O; when fired in a furnace, they become very rigid.

Fused sodium silicate, Na_2SiO_3, and calcium silicate, $CaSiO_3$, are the major components of the glass used in such things as drinking glasses, bottles, and window panes. *Glass* is a hard, brittle material that has no fixed composition or regular structure. Because it has no regular structure, it does not break evenly along crystal planes, but breaks to form

Si is much larger than C. As a result, Si—Si bonds are too long to permit the effective *pi* bonding that is necessary for multiple bonds.

Natural quartz crystals, SiO_2.

"Asbestos" refers to a group of impure magnesium silicate minerals. As you can see, asbestos is a fibrous material. When inhaled, these fibers are carcinogenic (cause cancer).

Charles D. Winters

Charles D. Winters

rounded surfaces and jagged edges. The basic ingredients are produced by heating a mixture of Na_2CO_3 and $CaCO_3$ with sand until it melts, at about 700°C.

$$[CaCO_3 + SiO_2](\ell) \xrightarrow{\text{heat}} CaSiO_3(\ell) + CO_2(g)$$

$$[Na_2CO_3 + SiO_2](\ell) \xrightarrow{\text{heat}} Na_2SiO_3(\ell) + CO_2(g)$$

The resulting "soda–lime" glass is clear and colorless (if all CO_2 bubbles escape and if the amounts of reactants are carefully controlled).

Key Terms

Amines   Organic compounds that are structurally related to ammonia by the replacement of one or more hydrogens.

Chain initiation step   The first step in a chain reaction; produces reactive species (such as radicals) that then propagate the reaction.

Chain propagation step   An intermediate step in a chain reaction; in such a step one or more reactive species is consumed, and another reactive species is produced.

Chain reaction   A reaction in which reactive species, such as radicals, are produced in more than one step. Consists of an initiation step, one or more propagation steps, and one or more termination steps.

Chain termination step   The combination of reactive species (such as radicals) which terminates the chain reaction.

Contact process   An industrial process by which sulfur trioxide and sulfuric acid are produced from sulfur dioxide.

Frasch process   A method by which elemental sulfur is mined or extracted. Sulfur is melted with superheated water (at 170°C under high pressure) and forced to the surface of the earth as a slurry.

Haber process   An industrial process for the catalyzed production of ammonia from N_2 and H_2 at high temperature and pressure; see Section 17-7.

Halides   Binary compounds of the halogens with less electronegative elements.

Halogens   Group VIIA elements; F, Cl, Br, I, and At. The free elements exist as diatomic molecules.

Nitrogen cycle   The complex series of reactions by which nitrogen is slowly but continually recycled in the atmosphere, lithosphere, and hydrosphere.

Noble gases   Group VIIIA elements; He, Ne, Ar, Kr, Xe, and Rn.

Ostwald process   An industrial process for the production of nitrogen oxide and nitric acid from ammonia and oxygen.

PANs   Abbreviation for peroxyacyl nitrates, photochemical oxidants in smog.

Particulate matter   Finely divided solid particles suspended in air.

Photochemical oxidants   Photochemically produced oxidizing agents capable of causing damage to plants and animals.

Photochemical smog   A brownish smog occurring in urban areas that receive large amounts of sunlight; caused by photochemical (light-induced) reactions among nitrogen oxides, hydrocarbons, and other components of polluted air that produce photochemical oxidants.

Radical   An atom or group of atoms that contains one or more unpaired electrons (usually very reactive species).

Exercises

The Noble Gases

1. (a) Write the Lewis dot representations of the noble gases. (b) Why are the noble gases so unreactive?
2. Why were the noble gases among the last elements to be discovered?
3. List some of the uses of the noble gases and reasons for the uses.
4. Arrange the noble gases in order of increasing (a) atomic radii, (b) melting points, (c) boiling points, (d) densities, and (e) first ionization energies.
5. Explain the order of increasing melting and boiling points of the noble gases in terms of polarizabilities of the atoms and forces of attraction between them.
6. What gave Neil Bartlett the idea that compounds of xenon could be synthesized? Which noble gases are known to form compounds? With which elements are the noble gas atoms bonded?
7. Describe the bonding and geometry in XeF_2, XeF_4, and XeF_6.
*8. Xenon(VI) fluoride can be produced by the combination of xenon(IV) fluoride with fluorine. Write a chemical equation for this reaction. What mass of XeF_6 could be produced from 2.75 g of XeF_4 and excess fluorine?

The Halogens

9. Write the electron configuration for each halogen atom. Write the Lewis symbol for a halogen atom, X. What is the

usual oxidation state of the halogens in binary compounds with metals, semiconducting elements, and most non-metals?

10. Write the Lewis structure of a halogen molecule, X_2. Describe the bonding in the molecule. What is the trend of bond length and strength going down the family from F_2 to I_2?

11. What types of intermolecular forces are found in molecular halogens? What is the trend in these forces going down the group from F_2 to I_2? Describe the physical state of each molecular halogen at room temperature and pressure.

12. List the halogens in order of increasing (a) atomic radii, (b) ionic radii, (c) electronegativities, (d) melting points, (e) boiling points, and (f) standard reduction potentials.

13. Write the equations describing the half-reactions and net reaction for the electrolysis of molten KF/HF mixtures. At which electrodes are the products formed? What is the purpose of the HF?

14. Carl O. Christe's preparation of F_2 did not involve direct chemical oxidation of fluoride ions. Explain.

15. Write equations describing general reactions of the free halogens, X_2, with (a) Group IA (alkali) metals, (b) Group IIA (alkaline earth) metals, and (c) Group IIIA metals. Represent the metals as M.

16. Write balanced equations for any reactions that occur in aqueous mixtures of (a) NaI and Cl_2, (b) NaCl and Br_2, (c) NaI and Br_2, (d) NaBr and Cl_2, and (e) NaF and I_2.

*17. An aqueous solution contains either NaBr or a mixture of NaBr and NaI. Using only aqueous solutions of I_2, Br_2, and Cl_2 and a small amount of CH_2Cl_2, describe how you might determine what is in the unknown solution.

*18. Write equations illustrating the tendency of F^- to stabilize high oxidation states of cations and the tendency of I^- to stabilize low oxidation states. Why is this the case?

19. Why are the free halogens more soluble in water than most nonpolar molecules?

20. Distinguish between hydrogen bromide and hydrobromic acid.

21. Refer to Figure 13-5. What is the order of decreasing boiling points of the hydrogen halides? Why is the HF "out of line?"

22. Describe the effect of hydrofluoric acid on glass.

23. What is the acid anhydride of perchloric acid?

24. Name the following compounds: (a) $KBrO_3$; (b) KOBr; (c) $NaClO_4$; (d) $NaClO_2$; (e) HOBr; (f) $HBrO_3$; (g) HIO_3; (h) $HClO_4$.

25. Write the Lewis formulas and structures of the four ternary acids of chlorine.

26. Write equations describing reactions by which the following compounds can be prepared: (a) hypohalous acids of Cl, Br, and I (in solution with hydrohalic acids); (b) hypohalite salts; (c) chlorous acid; (d) perchloric acid.

27. Most of the world's supply of bromine is obtained from sea water. The bromide content of sea water is about 1 part in 15,000. Discuss with the aid of equations, how bromine is produced from sea water.

28. Choose the strongest acid from each group: (a) HOCl, HOBr, HOI; (b) HOCl, $HClO_2$, $HClO_3$, $HClO_4$; (c) HOI, $HBrO_3$, $HClO_4$. Explain your choices.

Group IIIA Elements

29. Write abbreviated electron configurations for atomic oxygen, selenium, and polonium.

30. Write out the electron configurations of oxide, sulfide, and selenide ions.

31. Characterize the Group VIA elements with respect to color and physical state under normal conditions.

32. The Group VIA elements, except oxygen, can exhibit oxidation states ranging from -2 to $+6$, but not -3 or $+7$. Why?

33. Sulfur, selenium, and tellurium are all capable of forming six-coordinate compounds such as SF_6. Give two reasons why oxygen cannot be the central atom in such six coordinate molecules.

34. For the following species, draw (i) diagrams that show the hybridization of atomic orbitals and (ii) three-dimensional structures that show all hybridized orbitals and outermost electrons. (iii) Determine the oxidation state of the Group VIA element (other than oxygen) in each species. (a) H_2S; (b) SF_6; (c) SF_4; (d) SO_2; (e) SO_3

35. Repeat Exercise 34 for (a) SeF_6, (b) SO_3^{2-}, (c) SO_4^{2-}, (d) HSO_4^-, and (e) thiosulfate ion, $S_2O_3^{2-}$ (one S is central atom).

36. Write equations for the reactions of (a) S, Se, and Te with excess F_2; (b) O_2, S, Se, and Te with H_2; (c) S, Se, and Te with O_2.

37. Write equations for the reactions of (a) S and Te with HNO_3; (b) S and Se with excess Cl_2; (c) S and Se with Na, Ca, and Al.

38. Discuss the acidity of the aqueous Group VIA hydrides, including the relative values of acid ionization constants. What is primarily responsible for the order of increasing acidities in this series?

39. Compare the structures of the dioxides of sulfur, selenium, tellurium, and polonium. How do they relate to the metallic or nonmetallic character of these elements?

40. Write equations for reactions of
   (a) NaOH with sulfuric acid (1:1 mole ratio)
   (b) NaOH with sulfuric acid (2:1 mole ratio)
   (c) NaOH with sulfurous acid (1:1 mole ratio)
   (d) NaOH with sulfurous acid (2:1 mole ratio)
   (e) NaOH with selenic acid, H_2SeO_4 (1:1 mole ratio)
   (f) NaOH with selenic acid, (2:1 mole ratio)
   (g) NaOH with tellurium dioxide (1:1 mole ratio)
   (h) NaOH with tellurium dioxide (2:1 mole ratio)

*41. How much sulfur dioxide could be produced from complete combustion of 1.00 ton of coal containing 3.85% sulfur?

Industrial smokestack.

42. What mass of H_2SO_4 could be produced in the following process if 1.25 ton of FeS_2 is used? The *unbalanced* equations for the process are

$$FeS_2(s) + O_2(g) \longrightarrow Fe_2O_3(s) + SO_2(g)$$
$$SO_2(g) + O_2(g) \longrightarrow SO_3(g)$$
$$SO_3(g) + H_2SO_4(\ell) \longrightarrow H_2S_2O_7(\ell)$$
$$H_2S_2O_7(\ell) + H_2O(\ell) \longrightarrow H_2SO_4(aq)$$

43. Common copper ores in the western United States contain the mineral chalcopyrite, $CuFeS_2$. Assuming that a commercially useful ore contains 0.291 mass % Cu and that all the sulfur ultimately appears in the smelter stack gases as SO_2, calculate the mass of sulfur dioxide generated by the conversion of 1.00 ton of the ore.

*44. A gaseous mixture at some temperature in a 1.00-L vessel originally contained 1.00 mol SO_2 and 5.00 mol O_2. Once equilibrium conditions were attained, 77.8% of the SO_2 had been converted to SO_3. What is the value of the equilibrium constant (K_c) for this reaction at this temperature?

Nitrogen and Phosphorus

45. Characterize each of the Group VA elements with respect to normal physical state and color.

46. Write out complete electron configurations for the atoms of the Group VA elements; nitride ion, N^{3-}; and phosphide ion, P^{3-}.

47. Compare and contrast the properties of (a) N_2 and P_4; (b) HNO_3 and H_3PO_4; (c) N_2O_3 and P_4O_6.

48. Describe the natural nitrogen cycle.

49. List natural sources of nitrogen and phosphorus and at least two uses for each.

50. Discuss the effects of temperature, pressure, and catalysts on the Haber process for the production of ammonia. (You may wish to consult Section 17-7.)

51. Determine the oxidation states of nitrogen in the following: (a) NO_3^-; (b) NO; (c) N_2H_4; (d) NH_3; (e) NH_2^-.

52. Determine the oxidation states of nitrogen in the following: (a) N_2; (b) N_2O; (c) N_2O_4; (d) HNO_3; (e) HNO_2.

53. Draw three-dimensional structures showing all outer-shell electrons, describe molecular and ionic geometries, and indicate hybridization (except for N^{3-}) at the central ele-

ment, for the following species: (a) N_2, (b) N^{3-}; (c) NH_3, (d) NH_4^+; (e) NH_2^-, amide ion.

54. Draw three-dimensional structures showing all outer-shell electrons, describe molecular and ionic geometries, and indicate hybridization at the central element for the following species: (a) NH_2Br, bromamine; (b) HN_3, hydrazoic acid; (c) N_2O_2; (d) $NO_2^+NO_3^-$, solid nitronium nitrate; (e) HNO_3; (f) NO_2^-.

55. Draw three-dimensional structures showing all outer-shell electrons for the following species: (a) P_4; (b) P_4O_{10}; (c) As_4O_6; (d) H_3PO_4; (e) AsO_4^{3-}.

56. Write formula unit equations for the following: (a) thermal decomposition of potassium azide, KN_3; (b) reaction of gaseous ammonia with gaseous HCl; (c) reaction of aqueous ammonia with aqueous HCl; (d) thermal decomposition of ammonium nitrate at temperature above 260°C; (e) reaction of ammonia with oxygen in the presence of red hot platinum catalyst; (f) thermal decomposition of nitrous oxide (dinitrogen oxide), N_2O; (g) reaction of NO_2 with water.

57. Write the formula unit equation for the preparation of "superphosphate of lime."

58. Write two equations illustrating the ability of ammonia to function as a Lewis base.

59. In liquid ammonia would sodium amide, $NaNH_2$, be acidic, basic, or neutral? Would ammonium chloride, NH_4Cl, be acidic, basic, or neutral? Why?

60. Which of the following molecules have a nonzero dipole moment—that is, are polar molecules? (a) NH_3; (b) NH_2Cl; (c) NO_2; (d) NH_2OH; (e) HNO_3.

61. Describe with equations the Ostwald process for the production of nitrogen oxide, NO, and nitric acid.

62. Why is NO so reactive?

63. Write a Lewis formula for NO_2. Would you predict that it is very reactive? How about N_2O_4 (dimerized NO_2)?

64. At room temperature, a sample of NO_2 gas is brown. Explain why this sample loses its color as it is cooled.

65. Discuss the problem of NO_x emissions with respect to air pollution. Use equations to illustrate the important reactions.

66. What are the acid anhydrides of (a) nitric acid, HNO_3; (b) nitrous acid, HNO_2; (c) phosphoric acid, H_3PO_4; and (d) phosphorous acid, H_3PO_3?

67. Calcium phosphate (phosphate rock) is not applied directly as a phosphorus fertilizer. Why?

68. Discuss the use of sodium nitrite as a meat preservative.

CONCEPTUAL EXERCISES

69. Not much is known about astatine because it is very rare, is radioactive, and decays very quickly. Would you predict the chemical and physical properties of astatine to be more like

those of a metal or nonmetal? Defend your answer on the basis of astatine's location on the periodic table.

70. Elemental chlorine is obtained by the electrolysis of molten NaCl (Downs cell). Elemental fluorine is obtained by the electrolysis of KHF_2 in a cell made of Monel metal (a stainless steel alloy). Both of these processes are dangerous. Why?

71. Acid rain consists of acids formed by the reaction of acid anhydrides and water. List four acid anhydrides commonly involved in acid rain formation and the sources of each.

72. Sulfur deposits are mined by the Frasch process. Explain why free sulfur is not found on the earth's surface.

73. (a) On the basis of electronic structures, explain why most metal halides are ionic. (b) On the basis of your explanation, predict which halide compounds are likely to be covalent.

74. (a) Which of the halogens is the most active chemically and (b) which of the halogens is most likely to be reduced from the free state? (c) Which of the halogens is most likely to act as an oxidizing agent? (d) Which of the halogens is least likely to be an effective oxidizing agent? (e) Which of the halogens is likely to be a liquid under room conditions? (f) Which of the halogens is found free in nature?

75. The nitrogen cycle indicates that electrical storms (lightning) are a source of nitrogen compounds. (a) What is the function of the lightning? (b) What compounds of nitrogen are produced during electrical storms? (c) What is the effect of these compounds on the rain water?

BUILDING YOUR KNOWLEDGE

*76. Standard enthalpies of formation are -402 kJ/mol for $XeF_6(s)$ and -261.5 kJ/mol for $XeF_4(s)$. Calculate ΔH^0_{rxn} at 25°C for the preparation of XeF_6 from XeF_4 and $F_2(g)$.

77. The average atomic mass of N is 14.0067 amu. There are two isotopes which contribute to this average: ^{14}N (14.00307 amu) and ^{15}N (15.00011 amu). Calculate the percentage of ^{15}N atoms in a sample of naturally occurring nitrogen.

78. The $N \equiv N$ and the $N-N$ bond energies are listed in Tables 15-2 and 15-3. Predict whether four gaseous nitrogen atoms would form two gaseous nitrogen molecules or a gaseous tetrahedral molecule similar to P_4, basing your prediction on the amount of energy released as the molecules

are formed. Repeat the calculations for phosphorus using 485 kJ/mol for $P \equiv P$ and 201 kJ/mol for $P-P$.

*79. Commercial concentrated HNO_3 contains 69.5 mass % HNO_3 and has a density of 1.42 g/mL. What is the molarity of this solution? What volume of the concentrated acid should you use to prepare 5.00 L of dilute HNO_3 solution with a concentration of 4.00 M?

*80. What is the total mass of silicon in the crust of the earth? Assume that the radius of the earth is 6400 km, the crust is 50 km thick, the density of the crust is 3.5 g/cm^3, and 25.7 mass % of the crust is silicon.

*81. How many grams of xenon oxide tetrafluoride, $XeOF_4$, and how many liters of HF at STP could be prepared, assuming complete reaction of 15.5 g of xenon tetrafluoride, XeF_4, with a stoichiometric quantity of water according to the equation below?

$$6XeF_4(s) + 8H_2O(\ell) \longrightarrow$$
$$2XeOF_4(\ell) + 4Xe(g) + 16HF(g) + 3O_2(g)$$

*82. Argon crystallizes at $-235°C$ in a face-centered cubic unit cell with $a = 5.43$ Å. Determine the apparent radius of an argon atom in the solid.

*83. A reaction mixture contained 100. g of K_2MnF_6 and 174 g of SbF_5. Fluorine was produced in 35.3% yield. How many grams of F_2 were produced? What volume is this at STP?

BEYOND THE TEXTBOOK

Go to the textbook website at

http://www.brookscole.com/chemistry/whitten

for additional activities and exercises based on the General Chemistry Interactive CD-ROM, the World Wide Web, and library resources.

InfoTrac College Edition

For additional readings, go to InfoTrac College Edition, your online research library at:
http://infotrac.thomsonlearning.com

Coordination Compounds

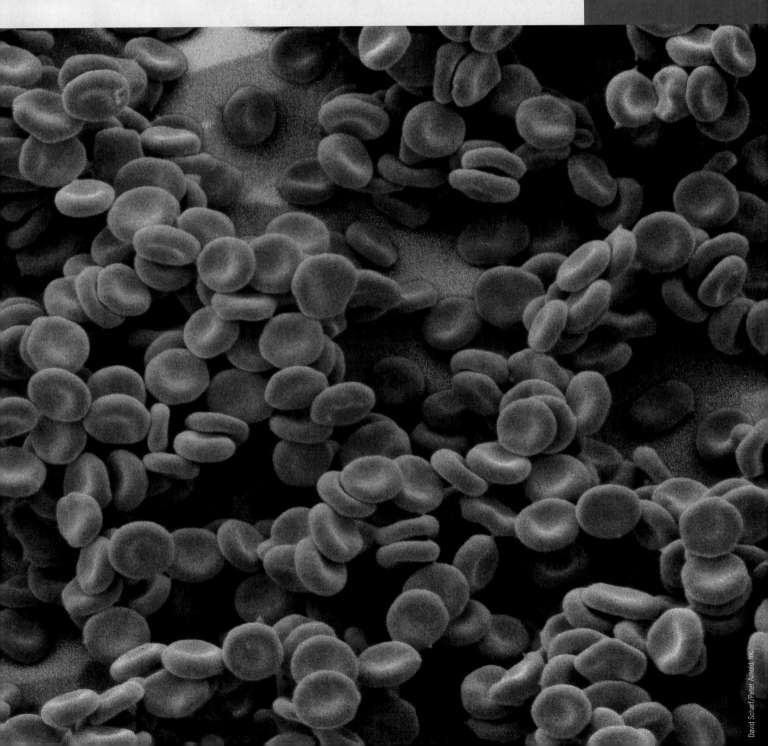

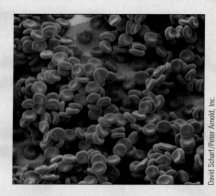

Red blood cells (1200x). The red blood cells that transport O_2 throughout our bodies contain hemoglobin, a coordination compound.

OUTLINE

OBJECTIVES

After you have studied this chapter, you should be able to

- Recognize coordination compounds
- Recognize metals that form soluble ammine complexes in aqueous solutions and write the formulas for common ammine complexes
- Use the terminology that describes coordination compounds
- Apply the rules for naming coordination compounds
- Recognize common structures of coordination compounds
- Describe various kinds of structural (constitutional) isomerism and distinguish among structural isomers
- Recognize stereoisomers
- Describe the crystal field theory of bonding in coordination compounds
- Explain the origin of color in complex species
- Use the spectrochemical series to explain colors of a series of complexes

Coordination compounds are found in many places on the earth's surface. Every living system includes many coordination compounds. They are also important components of everyday products as varied as cleaning materials, medicines, inks, and paints. A list of important coordination compounds appears to be endless because new ones are discovered every year.

25-1 COORDINATION COMPOUNDS

Covalent bonds in which the shared electron pair is provided by one atom are called *coordinate covalent or dative bonds.*

In Section 10-10 we discussed Lewis acid–base reactions. A *base* makes available a share in an electron pair, and an *acid* accepts a share in an electron pair, to form a **coordinate covalent bond** or **dative bond.** Such bonds are sometimes represented by arrows that point from the electron pair donor (Lewis base) to the acceptor (Lewis acid) when one wants to emphasize this type of bonding. But chemists usually just draw straight lines (or wedges or dashed lines) to indicate a bond.

ammonia,   boron trichloride,
a Lewis base   a Lewis acid

chloride ion,   tin(IV) chloride,   hexachlorostannate(IV) ion
a Lewis base   a Lewis acid

The red arrows represent coordinate covalent bonds. These arrows do not imply that two Sn—Cl bonds are different from the others. Once formed, all the Sn—Cl bonds in the $[SnCl_6]^{2-}$ ion are alike.

Common oxidation states and d-electron counts for the transition metals are given in Table 25-1. Most d-transition metal ions have vacant d orbitals that can accept shares in electron pairs. Many act as Lewis acids by forming coordinate covalent bonds in **coordination compounds (coordination complexes, or complex ions)**. Complexes of transition metal ions or molecules include cationic species (e.g., $[Cr(OH_2)_6]^{3+}$, $[Co(NH_3)_6]^{3+}$, $[Ag(NH_3)_2]^+$), anionic species (e.g., $[Ni(CN_4)]^{2-}$, $[MnCl_5]^{3-}$), and neutral species (e.g., $[Fe(CO)_5]$, $[Pt(NH_3)_2Cl_2]$). Many complexes are very stable, as indicated by their low dissociation constants, K_d (Section 20-6 and Appendix I).

We often write water as OH_2 rather than H_2O when we want to emphasize that oxygen is the donor atom.

TABLE 25-1	*Common Oxidation States and* d-*Electron Counts for the Transition Metals*								
Sc $+3$ d^0	**Ti** $+3$ d^1 $+4$ d^0	**V** $+3$ d^2 $+5$ d^0	**Cr** 0 d^6 $+2$ d^4 $+3$ d^3 $+4$ d^2 $+6$ d^0	**Mn** $+1$ d^6 $+2$ d^5 $+3$ d^4 $+5$ d^2 $+7$ d^0	**Fe** 0 d^8 $+2$ d^6 $+3$ d^5	**Co** 0 d^9 $+1$ d^8 $+2$ d^7 $+3$ d^6	**Ni** 0 d^{10} $+2$ d^8	**Cu** $+1$ d^{10} $+2$ d^9	**Zn** $+2$ d^{10}
Y $+3$ d^0	**Zr** $+4$ d^0	**Nb** $+3$ d^2 $+5$ d^0	**Mo** 0 d^6 $+2$ d^4 $+4$ d^2 $+6$ d^0	**Tc** $+1$ d^6 $+3$ d^4 $+5$ d^2 $+7$ d^0	**Ru** 0 d^8 $+2$ d^6 $+3$ d^5	**Rh** 0 d^9 $+1$ d^8 $+2$ d^7 $+3$ d^6	**Pd** 0 d^{10} $+2$ d^8 $+4$ d^6	**Ag** $+1$ d^{10}	**Cd** $+2$ d^{10}
La $+3$ d^0	**Hf** $+4$ d^0	**Ta** $+3$ d^2 $+5$ d^0	**W** 0 d^6 $+2$ d^4 $+4$ d^2 $+6$ d^0	**Re** $+1$ d^6 $+2$ d^5 $+3$ d^4 $+5$ d^2 $+7$ d^0	**Os** 0 d^8 $+2$ d^6 $+3$ d^5	**Ir** 0 d^9 $+1$ d^8 $+2$ d^7 $+3$ d^6	**Pt** 0 d^{10} $+2$ d^8 $+4$ d^6	**Au** $+1$ d^{10} $+2$ d^9 $+3$ d^8	**Hg** $+2$ d^{10}

When Lewis bases (also called *ligands*) bond to a transition metal, the d orbitals always fill before the next higher s orbital. The filled noble gas electron configuration associated with each transition metal is usually omitted and just the valence d-electron count is shown.

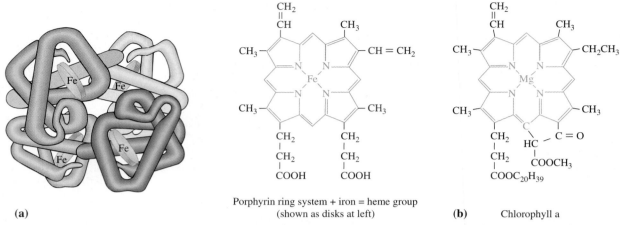

Porphyrin ring system + iron = heme group
(shown as disks at left)

(a)

(b)   Chlorophyll a

Figure 25-1 (a) A model of a hemoglobin molecule (MW = 64,500 amu). Individual atoms are not shown. The four heme groups in a hemoglobin molecule are represented by disks. Each heme group contains one Fe^{2+} ion and porphyrin rings. A single red blood cell contains more than 265 million hemoglobin molecules and more than 1 billion Fe^{2+} ions. (b) The structure of chlorophyll a, which also contains a porphyrin ring with a Mg^{2+} ion at its center. Chlorophyll is necessary for photosynthesis. The porphyrin ring is the part of the molecule that absorbs light. The structure of chlorophyll b is slightly different.

The hydrate usually written as $CuSO_4 \cdot 5H_2O$ is more properly shown by the coordination formula, $[Cu(OH_2)_4(SO_4)] \cdot H_2O$. The solid-state structure has four waters that are coordinated to the Cu^{2+} ion in a square planar arrangement, with two very weakly bonded SO_4^{2-} anions that bridge between different $[Cu(OH_2)_4]^{2+}$ complexes. The other water molecule is held in the solid by hydrogen-bonding.

Double salts are ionic solids resulting from the cocrystallization of two salts from the same solution into a single structure. In the example given, the solid is produced from an aqueous solution of iron(II) sulfate, $FeSO_4$, and ammonium sulfate, $(NH_4)_2SO_4$.

Many important biological substances contain coordinated metal atoms or ions. Hemoglobin and chlorophyll are two examples (Figure 25-1). Hemoglobin is a protein that carries O_2 in blood. It contains iron(II) ions bound to large porphyrin rings. The transport of oxygen by hemoglobin involves the coordination and subsequent release of O_2 molecules by the Fe(II) ions. Chlorophyll is necessary for photosynthesis in plants. It contains magnesium ions bound to porphyrin rings. Vitamin B_{12} is a large complex of cobalt. Coordination compounds have many important applications in such areas as medicine, water treatment, soil and plant treatment, protection of metal surfaces, analysis of trace amounts of metals, electroplating, catalysis, and textile dyeing.

Bonding in transition metal complexes was not understood until the pioneering research of Alfred Werner (1866–1919), a Swiss chemist who received the Nobel Prize in chemistry in 1913. Great advances have been made since in the field of coordination chemistry, but Werner's work remains the most important contribution by a single researcher.

Prior to Werner's work, the formulas of transition metal complexes were written with dots, $CrCl_3 \cdot 6H_2O$, $AgCl \cdot 2NH_3$, just like double salts such as iron(II) ammonium sulfate hexahydrate, $FeSO_4 \cdot (NH_4)_2SO_4 \cdot 6H_2O$. The properties of solutions of double salts are the properties expected for solutions made by mixing the individual salts. However, a solution of $AgCl \cdot 2NH_3$, or more properly $[Ag(NH_3)_2]Cl$, behaves quite differently from either a solution of (very insoluble) silver chloride or a solution of ammonia. The dots have been called "dots of ignorance," because they signified that the mode of bonding was unknown. Even today, many common coordination compounds such as $CrCl_3 \cdot 6H_2O$ are still written this way even though the most common coordination formula in the solid state is $[CrCl_2(OH_2)_4]Cl \cdot 2H_2O$. When this complex is dissolved in water, $[CrCl(OH_2)_5]^{2+}$ and eventually $[Cr(OH_2)_6]^{3+}$ are formed by chloride ligand dissociation and replacement with water molecules. Table 25-2 summarizes the types of experiments Werner performed and interpreted to lay the foundations for modern coordination theory.

TABLE 25-2 *Interpretation of Experimental Data by Werner*

Formula	Moles AgCl Precipitated per Formula Unit	Number of Ions per Formula Unit (based on conductance)	True Formula	Ions/Formula Unit	
$PtCl_4 \cdot 6NH_3$	4	5	$[Pt(NH_3)_6]Cl_4$	$[Pt(NH_3)_6]^{4+}$	$4\ Cl^-$
$PtCl_4 \cdot 5NH_3$	3	4	$[Pt(NH_3)_5Cl]Cl_3$	$[Pt(NH_3)_5Cl]^{3+}$	$3\ Cl^-$
$PtCl_4 \cdot 4NH_3$	2	3	$[Pt(NH_3)_4Cl_2]Cl_2$	$[Pt(NH_3)_4Cl_2]^{2+}$	$2\ Cl^-$
$PtCl_4 \cdot 3NH_3$	1	2	$[Pt(NH_3)_3Cl_3]Cl$	$[Pt(NH_3)_3Cl_3]^+$	Cl^-
$PtCl_4 \cdot 2NH_3$	0	0	$[Pt(NH_3)_2Cl_4]$	no ions	

Werner isolated platinum(IV) compounds with the formulas that appear in the first column of Table 25-2. He added excess $AgNO_3$ to solutions of carefully weighed amounts of the five salts. The precipitated AgCl was collected by filtration, dried, and weighed. He determined the number of moles of AgCl produced. This told him the number of Cl^- ions precipitated per formula unit. The results are in the second column. Werner reasoned that the precipitated Cl^- ions must be free (uncoordinated), whereas the unprecipitated Cl^- ions must be bonded to Pt so they could not be precipitated by Ag^+ ions. He also measured the conductances of solutions of these compounds of known concentrations. By comparing these with data on solutions of simple electrolytes, he found the number of ions per formula unit. The results are shown in the third column. Piecing the evidence together, he concluded that the correct formulas are the ones listed in the last two columns. The NH_3 and Cl^- within the brackets are bonded by dative bonds to the Lewis acid, Pt(IV) ion.

The conductance of a solution of an electrolyte is a measure of its ability to conduct electricity. It is related to the number of ions and the charges on ions in solution.

The charge on a complex is the sum of its constituent charges.

We can use this relationship to determine or confirm the charge on a complex species. For example, the charge on $[Pt(NH_3)_6]^{4+}$ can be calculated as

$$Charge = [\text{charge on Pt(IV)}] + 6 \times (\text{charge on } NH_3)$$
$$= (4+) + 6 \times (0) = 4+$$

The charge on $[Pt(NH_3)_4Cl_2]^{2+}$ is

$$Charge = [\text{charge on Pt(IV)}] + 4 \times (\text{charge on } NH_3) + 2 \times (\text{charge on } Cl^-)$$
$$= (4+) + 4 \times (0) + 2 \times (1-) = 2+$$

The oxidation state of the transition metal is very important and is indicated in a variety of ways. The formal notation is to use a Roman numeral in parentheses after the element symbol, as in Pt(IV). But the oxidation state can also be shown in the following ways, which are all equivalent and commonly used: Pt(+4), Pt^{4+}, and Pt^{IV}. The oxidation state of an atom is indicated whenever a metal atom appears by itself with a charge (Cu^{2+}, Rh^0, Co^{1-}). In order to figure out the oxidation state of a transition metal in a coordination complex, one must know the charges on the ligands (usually neutral or anionic) and the overall charge on the complex without any counterions present (or the charge on the counterion if present). You can practice this technique by verifying that all the Pt atoms in Table 25-2 are in the +4 oxidation state.

Colors of coordination compounds depend on which metals and ligands are present. From left: the $[Ni(NH_3)_6]^{2+}$ ion is purple; the $[Ni(H_2O)_6]^{2+}$ ion is green; the $[Cu(H_2O)_4]^{2+}$ ion is light blue; and the $[Cu(NH_3)_4]^{2+}$ ion is deep blue.

The careful addition of aqueous NH_3 to the top of a light blue solution of Cu^{2+} ions (bottom of beaker) produces a grayish-white $Cu(OH)_2$ precipitate (middle) and the soluble dark blue tetraammine complex, $[Cu(NH_3)_4]^{2+}$ (top).

CD-ROM   Screen 17.13, Cationic Lewis Acids.

25-2 AMMINE COMPLEXES

Ammine complexes contain NH_3 molecules bonded to metal ions. Because the ammine complexes are important compounds, we will describe some of them briefly.

Most metal hydroxides are insoluble in water, and so aqueous NH_3 initially reacts with nearly all metal ions to form insoluble metal hydroxides, or hydrated oxides.

$$2[(NH_3(aq) + H_2O(\ell) \rightleftharpoons NH_4^+(aq) + OH^-(aq)]$$
$$\underline{Cu^{2+}(aq) + 2OH^-(aq) \longrightarrow Cu(OH)_2(s)}$$
$$Cu^{2+}(aq) + 2NH_3(aq) + 2H_2O(\ell) \longrightarrow Cu(OH)_2(s) + 2NH_4^+(aq)$$

Similarly,

$$Cr^{3+} + 3NH_3 + 3H_2O \longrightarrow Cr(OH)_3(s) + 3NH_4^+$$

In general terms we can represent this reaction as

$$M^{n+} + nNH_3 + nH_2O \longrightarrow M(OH)_n(s) + nNH_4^+$$

where M^{n+} represents all of the common metal ions *except* the cations of the strong bases (Group IA cations and the heavier members of Group IIA: Ca^{2+}, Sr^{2+}, and Ba^{2+}).

The hydroxides of some metals and some metalloids are amphoteric (Section 10-6). Aqueous NH_3 is a weak base ($K_b = 1.8 \times 10^{-5}$), so the $[OH^-]$ is too low to dissolve amphoteric hydroxides to form hydroxo complexes.

However, a number of metal hydroxides do dissolve in an excess of aqueous NH_3 to form ammine complexes. For example, the hydroxides of copper and cobalt are readily soluble in an excess of *aqueous* ammonia solution. This occurs because of the reaction of NH_3 (a relatively good ligand) with the metal atoms to produce charged ammine complexes.

$$Cu(OH)_2(s) + 4NH_3 \rightleftharpoons [Cu(NH_3)_4]^{2+} + 2OH^-$$
$$Co(OH)_2(s) + 6NH_3 \rightleftharpoons [Co(NH_3)_6]^{2+} + 2OH^-$$

$Co(OH)_2$ (a blue compound that turns gray quickly) dissolves in excess aqueous NH_3 to form $[Co(NH_3)_6]^{2+}$ ions (*yellow-orange*).

	TABLE 25-3	*Common Metal Ions That Form Soluble Complexes with an Excess of Aqueous Ammonia*[a]	

Metal Ion	Insoluble Hydroxide Formed by Limited Aq. NH_3	Complex Ion Formed by Excess Aq. NH_3
Co^{2+}	$Co(OH)_2$	$[Co(NH_3)_6]^{2+}$
Co^{3+}	$Co(OH)_3$	$[Co(NH_3)_6]^{3+}$
Ni^{2+}	$Ni(OH)_2$	$[Ni(NH_3)_6]^{2+}$
Cu^+	$CuOH \longrightarrow \frac{1}{2}Cu_2O$[b]	$[Cu(NH_3)_2]^+$
Cu^{2+}	$Cu(OH)_2$	$[Cu(NH_3)_4]^{2+}$
Ag^+	$AgOH \longrightarrow \frac{1}{2}Ag_2O$[b]	$[Ag(NH_3)_2]^+$
Zn^{2+}	$Zn(OH)_2$	$[Zn(NH_3)_4]^{2+}$
Cd^{2+}	$Cd(OH)_2$	$[Cd(NH_3)_4]^{2+}$
Hg^{2+}	$Hg(OH)_2$	$[Hg(NH_3)_4]^{2+}$

[a] *The ions of Rh, Ir, Pd, Pt, and Au show similar behavior.*

[b] *CuOH and AgOH are unstable and decompose to the corresponding oxides.*

TABLE 25-4 *Typical Simple Ligands with Their Donor Atoms Shaded*

Molecule	Name	Name as Ligand	Ion	Name	Name as Ligand
$:NH_3$	ammonia	ammine	$:\overset{..}{Cl}:^-$	chloride	chloro
$:\overset{..}{O}H_2$	water	aqua	$:\overset{..}{F}:^-$	fluoride	fluoro
$:C\equiv O:$	carbon monoxide	carbonyl	$:C\equiv N:^-$	cyanide	cyano[a]
$:PH_3$	phosphine	phosphine	$:\overset{..}{O}H^-$	hydroxide	hydroxo
$:N=\overset{..}{O}$	nitrogen oxide	nitrosyl	$:N\overset{\overset{O.^-}{/\!/}}{\diagdown O:}$	nitrite	nitro[b]

[a]*Nitrogen atoms can also function as donor atoms, in which case the ligand name is "isocyano."*
[b]*Oxygen atoms can also function as donor atoms, in which case the ligand name is "nitrito."*

Interestingly, the metal hydroxides that exhibit this behavior are derived from the 12 metals of the cobalt, nickel, copper, and zinc families. All the common cations of these metals except Hg_2^{2+} (which disproportionates) form soluble complexes in the presence of excess aqueous ammonia (Table 25-3).

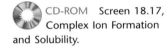 CD-ROM Screen 18.17, Complex Ion Formation and Solubility.

25-3 IMPORTANT TERMS

The Lewis bases in coordination compounds may be molecules, anions, or (rarely) cations. They are called **ligands** (Latin *ligare*, "to bind"). The **donor atoms** of the ligands are the atoms that donate shares in electron pairs to metals. In some cases it is not possible to identify donor atoms, because the bonding electrons are not localized on specific atoms. Some small organic molecules such as ethylene, $H_2C=CH_2$, bond to a transition metal through the electrons in their double bonds. Examples of typical simple ligands are listed in Table 25-4.

Ligands that can bond to a metal through only one donor atom at a time are **monodentate** (Latin *dent*, "tooth"). Ligands that can bond simultaneously through more than one donor atom are **polydentate**. Polydentate ligands that bond through two, three, four, five, or six donor atoms are called *bidentate, tridentate, quadridentate, quinquedentate*, (or *pentadentate*), and *hexadentate*, respectively. Complexes that consist of a metal atom or ion and polydentate ligands are called **chelate complexes** (Greek *chele*, "claw").

The **coordination number** of a metal atom or ion in a complex is the number of donor atoms to which it is coordinated, not necessarily the number of ligands. The **coordination sphere** includes the metal or metal ion (called the **central atom**) and its ligands, but no uncoordinated counterions. For example, the coordination sphere of hexaamminecobalt(III) chloride, $[Co(NH_3)_6]Cl_3$, is the hexaamminecobalt(III) ion, $[Co(NH_3)_6]^{3+}$. These terms are illustrated in Table 25-5.

The replacement of hydrogen atoms on PH_3 with hydrocarbon groups such as methyl (CH_3) generates phosphines (e.g., $P(CH_3)_3$), which are very strong ligands for middle to late transition metal atoms. In general, the stronger the donor ability of the lone pair on the ligand, the stronger the dative bond formed to the metal atom.

TABLE 25-5 Some Ligands and Coordination Spheres (complexes)

Ligand(s)	Classification	Coordination Sphere	Oxidation Number of M	Coordination Number of M
NH_3 ammine	monodentate	 $[Co(NH_3)_6]^{3+}$ hexaamminecobalt(III)	+3	6
$H_2N-CH_2-CH_2-NH_2$ (or N⌒N) ethylenediamine (en)	bidentate	 $[Co(en)_3]^{3+}$ tris(ethylenediamine)cobalt(III) ion	+3	6
Br^- bromo $H_2N-CH_2-CH_2-NH_2$ ethylenediamine (en)	monodentate bidentate	 $[Cu(en)Br_2]$ dibromoethylenediaminecopper(II)	+2	4
$H_2N-(CH_2)_2-\overset{\overset{\textstyle H}{\vert}}{N}-(CH_2)_2-NH_2$ (or N⌒N⌒N) diethylenetriamine (dien)	tridentate	 $[Fe(dien)_2]^{3+}$ bis(diethylenetriamine)iron(III) ion	+3	6
ethylenediaminetetraacetato (edta)	hexadentate	 $[Co(edta)]^-$ (ethylenediaminetetraacetato)-cobaltate(III) ion	+3	6

25-4 NOMENCLATURE

The International Union of Pure and Applied Chemistry (IUPAC) has adopted a set of rules for naming coordination compounds. The rules are based on those originally devised by Werner.

1. Cations are always named before anions, with a space between their names.

2. In naming the coordination sphere, ligands are named in alphabetical order. The prefixes di = 2, tri = 3, tetra = 4, penta = 5, hexa = 6, and so on specify the number of each kind of *simple* (monodentate) ligand. For example, in dichloro, the "di" indicates that two Cl^- ions act as ligands. For complicated ligands (polydentate chelating agents), other prefixes are used, including: bis = 2, tris = 3, tetrakis = 4, pentakis = 5, and hexakis = 6. The names of complicated ligands are enclosed in parentheses. The numeric prefixes are not used in alphabetizing. When a prefix denotes the number of substituents on a single ligand, as in dimethylamine, $NH(CH_3)_2$, it *is* used to alphabetize ligands.

3. The names of anionic ligands end in the suffix -o. Examples are F^-, fluoro; OH^-, hydroxo; O^{2-}, oxo; S^{2-}, sulfido; CO_3^{2-}, carbonato; CN^-, cyano; SO_4^{2-}, sulfato; NO_3^-, nitrato; $S_2O_3^{2-}$, thiosulfato.

4. The names of neutral ligands are usually unchanged. Four important exceptions are NH_3, ammine; H_2O, aqua; CO, carbonyl; and NO, nitrosyl.

5. Some metals exhibit variable oxidation states. The oxidation number of such a metal is designated by a Roman numeral in parentheses following the name of the complex ion or molecule.

6. The suffix "-ate" at the end of the name of the complex signifies that it is an anion. If the complex is neutral or cationic, no suffix is used. The English stem is usually used for the metal, but where the naming of an anion is awkward, the Latin stem is substituted. For example, "ferrate" is used rather than "ironate," and "plumbate" rather than "leadate" (Table 25-6).

TABLE 25-6 *Names for Some Metals in Complex Anions*

Metal	Name* of Metal in Complex Anions
aluminum	aluminate
antimony	antimonate
chromium	chromate
cobalt	cobaltate
copper	*cupr*ate
gold	*aur*ate
iron	*ferr*ate
lead	*plumb*ate
manganese	manganate
nickel	nickelate
platinum	platinate
silver	*argent*ate
tin	*stann*ate
zinc	zincate

Stems derived from Latin names for metals are shown in italics.

The term *ammine* (two m's) signifies the presence of ammonia as a ligand. It is different from the term *amine* (one m), which describes some organic compounds (Section 27-12) that are derived from ammonia.

The following examples illustrate these rules.

$K_2[Cu(CN)_4]$	potassium tetracyanocuprate(II)
$[Ag(NH_3)_2]Cl$	diamminesilver chloride
$[Cr(OH_2)_6](NO_3)_3$	hexaaquachromium(III) nitrate
$[Co(en)_2Br_2]Cl$	dibromobis(ethylenediamine)cobalt(III) chloride
$Ni(CO)_4$	tetracarbonylnickel(0)
$Na[Al(OH)_4]$	sodium tetrahydroxoaluminate
$[Pt(NH_3)_4][PtCl_6]$	tetraammineplatinum(II) hexachloroplatinate(IV)
$[Cu(NH_3)_2(en)]Br_2$	diammine(ethylenediamine)copper(II) bromide
$Na_2[Sn(OH)_6]$	sodium hexahydroxostannate(IV)
$[Co(en)_3](NO_3)_3$	tris(ethylenediamine)cobalt(III) nitrate
$K_4[Ni(CN)_2(ox)_2]$	potassium dicyanobis(oxalato)nickelate(II)
$[Co(NH_3)_4(OH_2)Cl]Cl_2$	tetraammineaquachlorocobalt(III) chloride

The oxidation state of aluminum is not given because it is always +3.

The abbreviation *ox* represents the oxalate ion $(COO)_2^{2-}$ or $C_2O_4^{2-}$.

TABLE 25-7   *Idealized Geometries for Various Coordination Numbers*

Coordination Number	Geometry	Examples	Model
2	L M L linear	$[Ag(NH_3)_2]^+$ $[Cu(CN)_2]^-$	$[Ag(NH_3)_2]^+$
4	L M L L L tetrahedral	$[Zn(CN)_4]^{2-}$ $[Cd(NH_3)_4]^{2+}$	$[Cd(NH_3)_4]^{2+}$
4	L L M L L square planar	$[Ni(CN)_4]^{2-}$ $[Cu(OH_2)_4]^{2+}$ $[Pt(NH_3)_2Cl_2]$	$[Cu(OH_2)_4]^{2+}$
5	L L M L L L trigonal bipyramidal	$Fe(CO)_5$ $[CuCl_5]^{3-}$	$Fe(CO)_5$
5	L L M L L L square pyramidal	$[Ni(CN)_5]^{3-}$ $[MnCl_5]^{3-}$	$[MnCl_5]^{3-}$
6	L L M L L L L octahedral	$[Fe(CN)_6]^{4-}$ $[Fe(OH_2)_6]^{2+}$	$[Fe(CN)_6]^{4-}$

25-5 STRUCTURES

The structures of coordination compounds are governed largely by the coordination number of the metal. Many have structures similar to the simple molecules and ions we studied in Chapter 8. Unshared pairs of electrons in d orbitals usually have only small influences on geometry because the sizes of the ligand atoms usually dominate. Table 25-7 summarizes the geometries for common coordination numbers.

Both tetrahedral and square planar geometries are common for complexes with coordination number 4. For coordination number 5, the trigonal bipyramidal structure and the square pyramidal structure are both common. Transition metal complexes with coordination numbers as high as 7, 8, and 9 are known. The geometries tabulated are ideal geometries. Actual structures are sometimes distorted, especially if the ligands are not all the same. The distortions are usually due to larger ligands bumping into one another (*steric effects*) and occasionally smaller anionic ligands repelling one another.

The d^8 electronic state for Rh, Ir, Pd, and Pt strongly favors a square planar coordination geometry for complexes of these metal ions.

ISOMERISM IN COORDINATION COMPOUNDS

Isomers are different compounds that have the same molecular formula; they have the same number and kinds of atoms arranged differently. The term *isomers* comes from the Greek word meaning "equal weights." *Because their structures are different, isomers have different physical and chemical properties.* Here we shall restrict our discussion of isomerism to that caused by different arrangements of ligands about central metal ions.

There are two major classes of isomers: structural (constitutional) isomers and stereoisomers. For coordination compounds, each can be further subdivided as follows.

Structural Isomers

1. ionization isomers
2. hydrate isomers
3. coordination isomers
4. linkage isomers

Stereoisomers

1. geometric (positional) isomers
2. optical isomers

Charles D. Winters

Some coordination compounds. Starting at the top left and moving clockwise:

$[Cr(CO)_6]$ (*white*), CO is the ligand.

$K_3\{Fe[(COO)_2]_3\}$ (*green*), $(COO)_2^{2-}$ (oxalate ion) is the ligand.

$[Co(H_2N—CH_2—CH_2—NH_2)_3]I_3$ (*yellow-orange*), ethylenediamine is the ligand.

$[Co(NH_3)_5(OH_2)]Cl_3$ (*red*), NH_3 and H_2O are ligands.

$K_3[Fe(CN)_6]$ (*red-orange*), CN^- is the ligand. (A drop of water fell on this sample causing the darker spot.)

Differences between **structural isomers** involve either more than one coordination sphere or different donor atoms on the same ligand. They contain *different atom-to-atom bonding sequences*. Simple stereoisomers of coordination compounds involve only one coordination sphere and the same ligands and donor atoms. Before considering stereoisomers, we will describe the four types of structural isomers.

25-6 STRUCTURAL (CONSTITUTIONAL) ISOMERS

Ionization (Ion–Ion Exchange) Isomers

These isomers result from the interchange of ions inside and outside the coordination sphere. For example, red-violet $[Co(NH_3)_5Br]SO_4$ and red $[Co(NH_3)_5SO_4]Br$ are ionization isomers.

> Isomers such as those shown here may or *may not* exist in the same solution in equilibrium. Such isomers are often formed by *different* reactions.

> We write the sulfato ligand as OSO_3 to emphasize that it is coordinated through a lone pair on one of the O atoms.

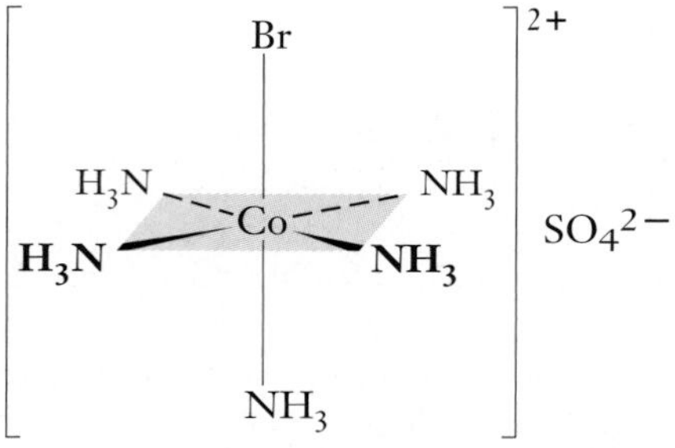

$[Co(NH_3)_5Br]SO_4$
pentaamminebromocobalt(III)
sulfate
A

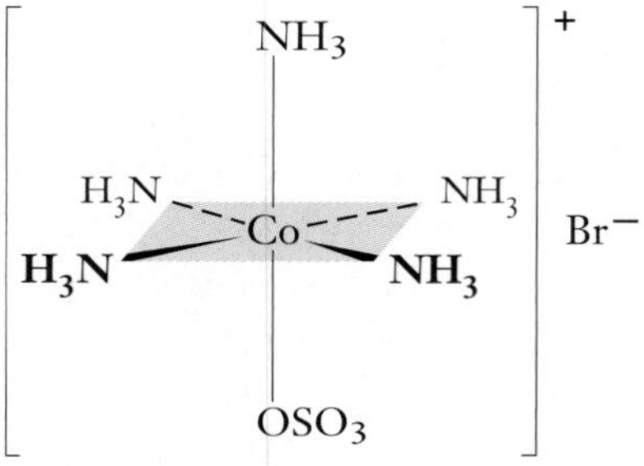

$[Co(NH_3)_5SO_4]Br$
pentaamminesulfatocobalt(III)
bromide
B

In structure A the SO_4^{2-} ion is free and is not bound to the cobalt(III) ion. A solution of A reacts readily with a solution of barium, $BaCl_2$, to precipitate $BaSO_4$ but does not react readily with $AgNO_3$. In structure B the SO_4^{2-} ion is bound to the cobalt(III) ion and so it does not react with $BaCl_2$ in aqueous solution. The Br^- ion is free, however, and a solution of B reacts with $AgNO_3$ to precipitate AgBr. *Equimolar* solutions of A and B also have different electrical conductivities. The sulfate solution, A, conducts electric current better because its ions have $2+$ and $2-$ charges rather than $1+$ and $1-$. Other examples of this type of isomerism include

> Recall that $BaSO_4$ ($K_{sp} = 1.1 \times 10^{-10}$) and AgBr ($K_{sp} = 3.3 \times 10^{-13}$) are classified as insoluble in H_2O, whereas $BaBr_2$ is soluble and Ag_2SO_4 ($K_{sp} = 1.7 \times 10^{-5}$) is moderately soluble.
>
> In A:
>
> $Ba^{2+} + SO_4^{2-} \longrightarrow BaSO_4(s)$
>
> $Ag^+ + SO_4^{2-} \longrightarrow$ no rxn
>
> In B:
>
> $Ag^+ + Br^- \longrightarrow AgBr(s)$
>
> $Ba^{2+} + Br^- \longrightarrow$ no rxn

$[Pt(NH_3)_4Cl_2]Br_2$	and	$[Pt(NH_3)_4Br_2]Cl_2$
$[Pt(NH_3)_4SO_4](OH)_2$	and	$[Pt(NH_3)_4(OH)_2]SO_4$
$[Co(NH_3)_5NO_2]SO_4$	and	$[Co(NH_3)_5SO_4]NO_2$
$[Cr(NH_3)_5SO_4]Br$	and	$[Cr(NH_3)_5Br]SO_4$

Hydrate Isomers

In some crystalline complexes, water can be *inside* and *outside* the coordination sphere. For example, when treated with excess $AgNO_3(aq)$, solutions of the following three hydrate isomers yield three, two, and one mole of AgCl precipitate, respectively, per mole of complex. The Ag cations react quickly only with the "free" chloride anions not coordinated to the Cr atom.

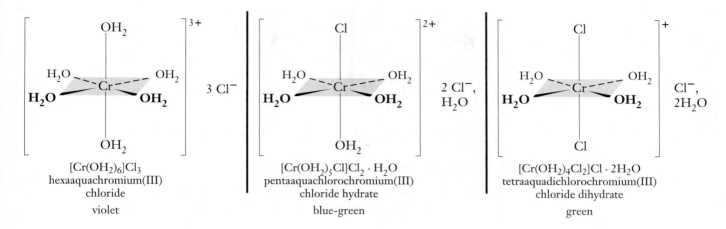

[Cr(OH₂)₆]Cl₃
hexaaquachromium(III)
chloride
violet

[Cr(OH₂)₅Cl]Cl₂ · H₂O
pentaaquachlorochromium(III)
chloride hydrate
blue-green

[Cr(OH₂)₄Cl₂]Cl · 2H₂O
tetraaquadichlorochromium(III)
chloride dihydrate
green

Coordination Isomers

Coordination isomerism can occur in compounds containing both complex cations and complex anions. Such isomers involve exchange of ligands between cation and anion, that is, between coordination spheres.

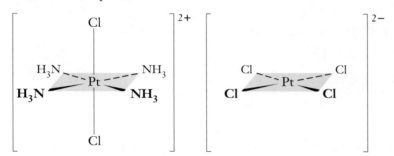

[Pt(NH₃)₄Cl₂][PtCl₄],
tetraamminedichloroplatinum(IV) tetrachloroplatinate(II)

[Pt(NH₃)₄][PtCl₆],
tetraamminedichloroplatinum(II) hexachloroplatinate(IV)

Linkage Isomers

Certain ligands can bind to metal ions in more than one way. One example of such a ligand is nitro, —NO₂⁻, and nitrito, —ONO⁻. The donor atoms are on the left in these representations.

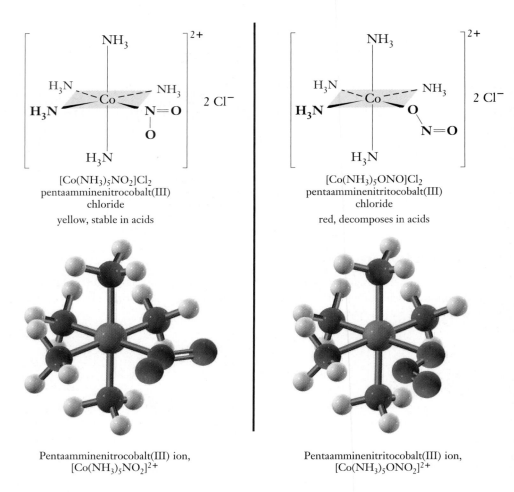

[Co(NH₃)₅NO₂]Cl₂
pentaamminenitrocobalt(III)
chloride
yellow, stable in acids

[Co(NH₃)₅ONO]Cl₂
pentaamminenitritocobalt(III)
chloride
red, decomposes in acids

Pentaamminenitrocobalt(III) ion,
[Co(NH₃)₅NO₂]²⁺

Pentaamminenitritocobalt(III) ion,
[Co(NH₃)₅ONO₂]²⁺

25-7 STEREOISOMERS

A complex with coordination number 2 or 3 that contains only *simple* ligands can have only one spatial arrangement. Try building models to see this.

Compounds that contain the same atoms and the same atom-to-atom bonding sequences, but that differ only in the spatial arrangements of the atoms relative to the central atom, are **stereoisomers.** Complexes with only *simple* ligands can exist as stereoisomers *only if* they have coordination number 4 or greater. The most common coordination numbers among coordination complexes are 4 and 6, and so they will be used to illustrate stereoisomerism.

Geometric (*cis–trans*) Isomers

In **geometric isomers,** or *cis-trans* **isomers,** of *coordination compounds,* the same ligands are arranged in different orders within the coordination sphere. Geometric isomerism occurs when atoms or groups of atoms can be arranged on two sides of a rigid structure. *Cis* means "adjacent to" and *trans* means "on the opposite side of." *Cis-* and *trans-* dimminedichloroplatinum(II) are shown here.

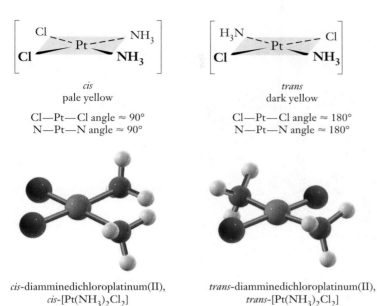

cis
pale yellow

Cl—Pt—Cl angle ≈ 90°
N—Pt—N angle ≈ 90°

trans
dark yellow

Cl—Pt—Cl angle ≈ 180°
N—Pt—N angle ≈ 180°

The *cis* isomer has been used successfully in cancer chemotherapy. The *trans* isomer has no such activity.

cis-diamminedichloroplatinum(II),
cis-[Pt(NH$_3$)$_2$Cl$_2$]

trans-diamminedichloroplatinum(II),
trans-[Pt(NH$_3$)$_2$Cl$_2$]

In the *cis* isomer, the chloro groups are closer to each other (on the same side of the square) than they are in the *trans* isomer. The ammine groups are also closer together in the *cis* complex.

Geometric isomerism is possible for octahedral complexes. For example, complexes of the type MA$_4$B$_2$ can exist in two isomeric forms. Consider as an example the complex ion [Co(NH$_3$)$_4$Cl$_2$]$^+$. The two like ligands (Cl$^-$) can be either *cis* or *trans* to each other. These two complexes are different colors: solutions and salts of the *cis* isomer are violet and those of the *trans* isomer are green.

Cis–trans isomerism is not possible for tetrahedral complexes, in which all angles are (ideally) 109.5°.

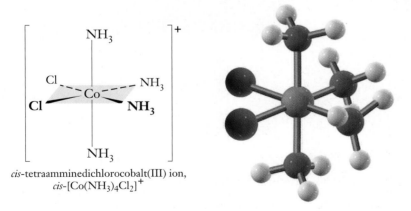

cis-tetraamminedichlorocobalt(III) ion,
cis-[Co(NH$_3$)$_4$Cl$_2$]$^+$

Ball-and-stick model of *cis*-[Co(NH$_3$)$_4$Cl$_2$]$^+$.
Note the positions of the green chloride ligands.

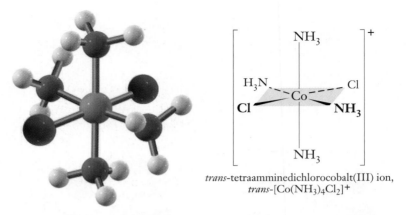

trans-tetraamminedichlorocobalt(III) ion,
trans-$[Co(NH_3)_4Cl_2]^+$

Ball-and-stick model of *trans*-$[Co(NH_3)_4Cl_2]^+$.
Note the positions of the green chloride ligands.

Complexes involving bidentate ligands, such as ethylenediamine, can also exhibit this kind of isomerism, as shown in Figure 25-2.

Octahedral complexes with the general formula MA_3B_3 can exhibit another type of geometric isomerism, called *mer–fac* isomerism. This can be illustrated with the complex ion $[Pt(NH_3)_3Cl_3]^+$ (see Table 25-2). In one isomer the three similar ligands (e.g., the Cl^- ligands) lie at the corners of a triangular face of the octahedron; this is called the *fac* isomer (for *facial*).

In the other isomer, the three similar ligands lie in the same plane; this is called the *mer* isomer (so called because the plane is analogous to a meridian on a globe).

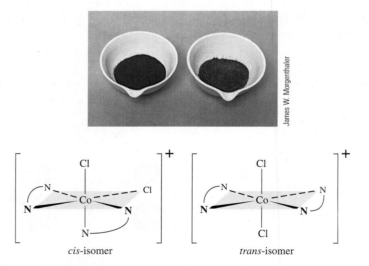

cis-isomer   *trans*-isomer

Figure 25-2   The dichlorobis(ethylenediamine)cobalt(III) ion, $[Co(en)_2Cl_2]^+$,

exists as a pair of *cis–trans* isomers. Ethylenediamine is represented as N⌒N.
Cis-dichlorobis(ethylenediamine)cobalt(III) perchlorate, $[Co(en)_2Cl_2]ClO_4$, is purple.
Trans-dichlorobis(ethylenediamine)cobalt(III) chloride, $[Co(en)_2Cl_2]Cl$, is green.

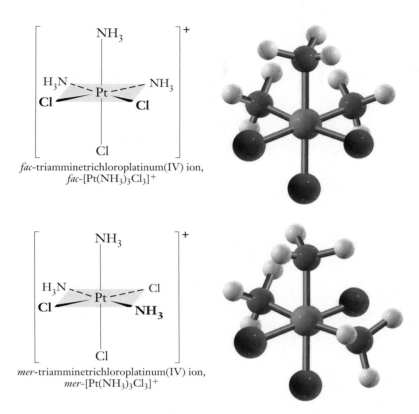

fac-triamminetrichloroplatinum(IV) ion,
fac-[Pt(NH$_3$)$_3$Cl$_3$]$^+$

mer-triamminetrichloroplatinum(IV) ion,
mer-[Pt(NH$_3$)$_3$Cl$_3$]$^+$

Ball-and-stick models of *fac*-[Pt(NH$_3$)$_3$Cl$_3$]$^+$ (*top*) and *mer*-[Pt(NH$_3$)$_3$Cl$_3$]$^+$ (*bottom*). The three chloro ligands in the *fac*-isomer are all on one triangular face of an octahedron, and the three ammine ligands are all on the opposite triangular face. The *mer* isomer has the chloro ligands in one meridian (in a plane), and the three ammine ligands in a perpendicular meridian.

Complexes of the type [MA$_2$B$_2$C$_2$] can exist in several isomeric forms. Consider as an example [Cr(OH$_2$)$_2$(NH$_3$)$_2$Br$_2$]$^+$. First, the members of all three pairs of like ligands may be either *trans* to each other (A) or *cis* to each other (B).

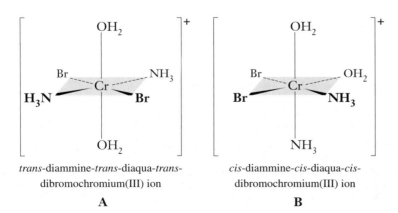

trans-diammine-*trans*-diaqua-*trans*-
dibromochromium(III) ion

A

cis-diammine-*cis*-diaqua-*cis*-
dibromochromium(III) ion

B

Then, members of one of the pairs may be *trans* to each other, but the members of the other two pairs are *cis*.

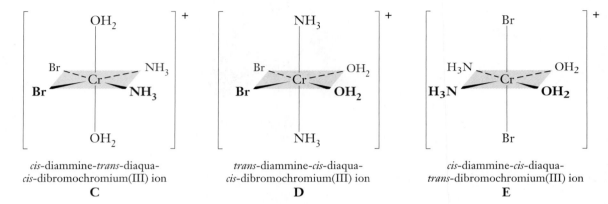

cis-diammine-*trans*-diaqua-
cis-dibromochromium(III) ion

C

trans-diammine-*cis*-diaqua-
cis-dibromochromium(III) ion

D

cis-diammine-*cis*-diaqua-
trans-dibromochromium(III) ion

E

In C, only the two H_2O ligands are in a *trans* arrangement; in D, only the NH_3 ligands are *trans*; and in E, only the two Br^- ligands are *trans*.

There is no trans–trans–cis isomer. Why?

Further interchange of the positions of the ligands produces no new geometric isomers. However, one of the five geometric isomers (B) can exist in two distinct forms called *optical isomers*.

Optical Isomers

An object that is not superimposable with its mirror image is said to be chiral.

The *cis*-diammine-*cis*-diaqua-*cis*-dibromochromium(III) geometrical isomer (B) exists in two forms that bear the same relationship to each other as left and right hands. They are *nonsuperimposable* mirror images of each other and are called **optical isomers** or **enantiomers.**

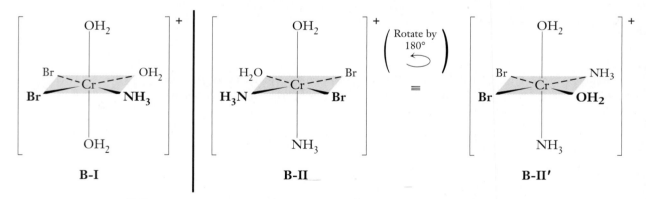

B-I

B-II

Rotate by 180°

$\equiv$

B-II'

Optical isomers of *cis*-diammine-*cis*-diaqua-*cis*-dibromochromium(III) ion

Complexes B-I and B-II are mirror images of each other. To see that they are not identical, imagine rotating B-II about its vertical axis by 180° (B-II'), so that the two Br^- ligands are at the left, as they are in B-I. Then B-II' has the OH_2 ligand at the right front position and the NH_3 ligand at the right rear position, which is not the same as in B-I. No rotation of B-II makes it identical to B-I. Thus these two arrangements are nonsuperimposable mirror images of each other. Ball-and-stick models of these two complexes appear in Figure 25-3.

Optical isomers interact with polarized light in different ways. Separate equimolar solutions of each rotate a *plane of polarized light* (see Figures 25-4 and 25-5) by equal amounts

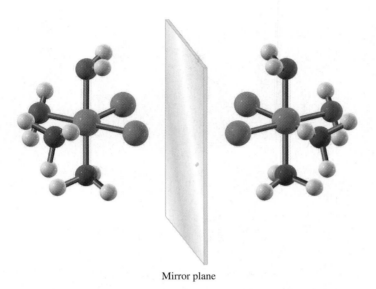

Figure 25-3 The optical isomers of the *cis*-diammine-*cis*-diaqua-*cis*-dibromochromium(III) ion.

Mirror plane

but in opposite directions. One solution is **dextrorotatory** (rotates to the *right*) and the other is **levorotatory** (rotates to the *left*). Optical isomers are called *dextro* and *levo* isomers. The phenomenon by which a plane of polarized light is rotated is called **optical activity.** It can be measured with a device called a polarimeter (Figure 25-5) or with more sophisticated instruments. A single solution containing equal amounts of the two isomers is a **racemic mixture.** This solution does not rotate a plane of polarized light. The equal and opposite effects of the two isomers exactly cancel. To exhibit optical activity, the *dextro* and *levo* isomers must be separated from each other. This is done by one of a number of chemical or physical processes broadly called *optical resolution.*

Another pair of optical isomers follows, each of which contains three molecules of ethylenediamine, a bidentate ligand.

Alfred Werner was also the first person to demonstrate optical activity in an inorganic compound (not of biological origin). This demonstration silenced critics of his theory of coordination compounds, and in his opinion, it was his greatest achievement. Louis Pasteur had demonstrated the phenomenon of optical activity many years earlier in organic compounds of biological origin.

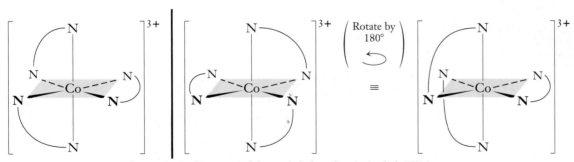

The two optical isomers of the *tris*(ethylenediamine)cobalt(III) ion

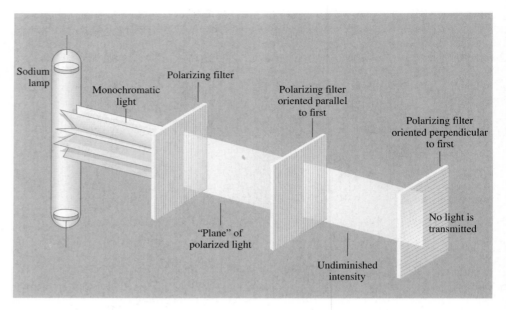

Figure 25-4   Light from a lamp or from the sun consists of electromagnetic waves that vibrate in all directions perpendicular to the direction of travel. Polarizing filters absorb all waves except those that vibrate in a single plane. The third polarizing filter, with a plane of polarization at right angles to the first, absorbs the polarized light completely.

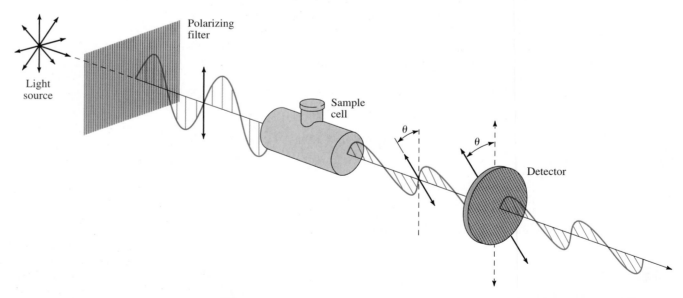

Figure 25-5   The principles of a **polarimeter.** The plane of polarization of plane-polarized light is rotated through an angle (θ) as it passes through an optically active medium. Species that rotate the plane to the right (clockwise) are dextrorotatory, and those that rotate it to the left are levorotatory.

BONDING IN COORDINATION COMPOUNDS

Bonding theories for coordination compounds should be able to account for structural features, colors, and magnetic properties. The earliest accepted theory was the valence bond theory (Chapter 8). It can account for structural and magnetic properties, but it offers no explanation for the wide range of colors of coordination compounds. The **crystal field theory** gives satisfactory explanations of color as well as of structure and magnetic properties for many coordination compounds. We will therefore discuss only this theory in the remainder of this chapter.

25-8 CRYSTAL FIELD THEORY

Hans Bethe (1906–   ) and J. H. van Vleck (1899–1980) developed the crystal field theory between 1919 and the early 1930s. It was not widely used, however, until the 1950s. In its original form, it assumed that the bonds between ligand and metal were completely ionic.

In a metal ion surrounded by ligands, the d orbitals are at higher energy than they are in an isolated metal ion. If the surrounding electrons were uniformly distributed about the metal ion, the energies of *all* five d orbitals would increase by the same amount (a *spherical crystal field*). Because the ligands approach the metal ion from different directions, they affect different d orbitals in different ways. Here we illustrate the application of these ideas to complexes with coordination number 6 (*octahedral crystal field*).

The $d_{x^2-y^2}$, and d_{z^2} orbitals are directed along a set of mutually perpendicular x, y, and z axes (Figure 5-24). As a group, these orbitals are called e_g **orbitals.** The d_{xy}, d_{yz}, and d_{xz} orbitals, collectively called t_{2g} **orbitals,** lie between the axes. The ligand donor atoms approach the metal ion along the axes to form octahedral complexes. Crystal field theory proposes that the approach of the six donor atoms (point charges) along the x, y, and z axes sets up an electric field (the crystal field). Electrons on the ligands repel electrons in e_g orbitals on the metal ion more strongly than they repel those in the t_{2g} orbitals (Figure 25-6). This removes the degeneracy of the set of d orbitals and splits them into two sets, the e_g set at higher energy and the t_{2g} set at lower energy.

Modern ligand field theory is based on crystal field theory. It attributes partial covalent character and partial ionic character to bonds. It is a more sophisticated theory, but too complicated to present here.

The labels e_g (d_{z^2}, $d_{x^2-y^2}$) and t_{2g} (d_{xz}, d_{yz}, d_{xy}) refer to the symmetry of those sets of d orbitals in an octahedral complex.

Recall that degenerate orbitals are orbitals of equal energy.

(a) e_g type orbitals (only $d_{x^2-y^2}$ shown here)

(b) t_{2g} type orbitals (only d_{xy} shown here)

Figure 25-6 Effects of the approach of ligands on the energies of d orbitals on the metal ion. In an octahedral complex, the ligands (L) approach the metal ion (M) along the x, y, and z axes, as indicated by the blue arrows. (a) The orbitals of the e_g type—$d_{x^2-y^2}$ (shown here) and d_{z^2}—point directly toward the incoming ligands, so electrons in these orbitals are strongly repelled. (b) The orbitals of the t_{2g} type—d_{xy} (shown here), d_{xz} and d_{yz}—do not point toward the incoming ligands, so electrons in these orbitals are less strongly repelled.

The t_{2g} orbitals were originally in a uniform spherical repulsive electric field. This changes because the incoming ligands localize the electric field and orient it directly toward the two e_g orbitals. As a result, the t_{2g} orbitals are now exposed to a less repulsive electric field, so they drop in energy (become more stable).

Typical values of Δ_{oct} are between 100 and 400 kJ/mol.

The energy separation between the two sets of orbitals is called the **crystal field splitting energy,** $\Delta_{octahedral}$, or Δ_{oct}. It is proportional to the *crystal field strength* of the ligands—that is, how strongly the ligand electrons repel the electrons on the metal ion. In modern bonding terms, the more strongly the ligands coordinate or bond to the metal atom, the larger the energy gap between the e_g and t_{2g} orbitals.

The d electrons on a metal ion occupy the t_{2g} set in preference to the higher energy e_g set. Electrons that occupy the e_g orbitals are strongly repelled by the relatively close approach of ligands. The occupancy of these orbitals tends to destabilize octahedral complexes.

As always, electrons occupy the orbitals in the arrangement that results in the lowest energy. The electron **pairing energy,** P, is the expenditure of energy that is necessary to pair electrons by bringing two negatively charged particles into the same region of space. We must compare this with the crystal field splitting energy, Δ_{oct}, the difference in energy between t_{2g} and e_g orbitals. Weak field ligands such as F^- cause only small splitting energies, whereas strong field ligands such as CN^- give large values of Δ_{oct}.

Weak field ligands bond weakly, while strong field ligands usually coordinate strongly.

If the splitting energy, Δ_{oct}, is smaller than the pairing energy, P, the electrons occupy all five nondegenerate orbitals singly before pairing. After all d orbitals are half-filled, additional electrons then pair with electrons in the lower energy t_{2g} set. Such a complex would have the same number of *unpaired* electrons on the metal atom or ion as when the metal is uncomplexed; this is called a **high spin complex.** But if the orbital splitting energy is greater than the pairing energy, the electrons will be at lower energy if they pair in the t_{2g} orbital before any electrons occupy the higher energy e_g orbitals. Such a complex could have fewer unpaired electrons on the metal atom than when the metal is uncomplexed, so it is called a **low spin complex.**

Let us now describe the hexafluorocobaltate(III) ion, $[CoF_6]^{3-}$, and the hexacyanocobalt(III) ion, $[Co(CN)_6]^{3-}$. Both contain the d^6 Co^{3+} ion. The $[CoF_6]^{3-}$ ion is a paramagnetic complex, whereas $[Co(CN)_6]^{3-}$ is a diamagnetic complex. We will focus our attention on the d electrons.

A diamagnetic species has no unpaired electrons. A paramagnetic species has one or more unpaired electrons, and the strength of the paramagnetism depends on the number of unpaired electrons. See Chapter 5.

The free Co^{3+} ion has six electrons (four unpaired) in its $3d$ orbitals.

Magnetic measurements indicate that $[CoF_6]^{3-}$ also has four unpaired electrons per ion. So there must be four electrons in t_{2g} orbitals and two in e_g orbitals. Fluoride, F^-, is a weak field ligand, so the crystal field splitting energy is very small ($\Delta_{oct} < P$) and electron pairing is unfavorable. Thus $[CoF_6]^{3-}$ is a *high spin* complex.

$[CoF_6]^{3-}$ is a *high spin* complex.

▶[Co(OH₂)₆]²⁺ ions are pink (*bottom*). A limited amount of aqueous ammonia produces $Co(OH)_2$, a blue compound that quickly turns gray (*middle*). $Co(OH)_2$ dissolves in excess aqueous ammonia to form [Co(NH₃)₆]²⁺ ions, which are yellow-orange (*top*).

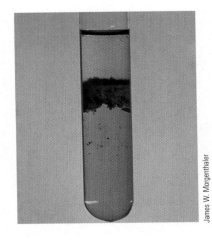

James W. Morgenthaler

On the other hand, [Co(CN)₆]³⁻ is diamagnetic, so all six d electrons must be paired in the t_{2g} orbitals. Cyanide ion, CN^-, is a strong field ligand, which generates a large crystal field splitting energy ($\Delta_{oct} > P$), making electron pairing more favorable; therefore, [Co(CN)₆]³⁻ is a low spin complex.

Δ_{oct} = large; strong crystal field due to CN^- (not to scale)

3d, free Co³⁺ ion

Co³⁺ ion in spherical crystal field

[Co(CN)₆]³⁻

[Co(CN)₆]³⁻ is a *low spin* complex.

The difference in electron configurations between [CoF₆]³⁻ and [Co(CN)₆]³⁻ is due to the relative magnitudes of the crystal field splitting, Δ_{oct}, caused by the different crystal field strengths of F^- and CN^-. The CN^- ion interacts with vacant metal orbitals more strongly than the F^- ion does. As a result, the crystal field splitting generated by the close approach of six CN^- ions (strong field ligands) to the metal ion is greater than that generated by the approach of six F^- ions (weak field ligands).

$$\Delta_{oct} \text{ for } [Co(CN)_6]^{3-} > \Delta_{oct} \text{ for } [CoF_6]^{3-}$$

Different low and high spin configurations exist only for octahedral complexes having metal ions with d^4, d^5, d^6, and d^7 configurations. For d^1-d^3 and d^8-d^{10} ions, only one possibility exists and is designated as high spin. All d^n possibilities are shown in Table 25-8.

A convenient way to describe d-transition metal ions is to indicate the number of d electrons, which is directly related to the oxidation state of the metal atom.

TABLE 25-8	*High and Low Spin Octahedral Configurations*						
d^n	Examples	High Spin	Low Spin	d^n	Examples	High Spin	Low Spin
d^1	Ti³⁺	— — e_g / ↑ — — t_{2g}	same as high spin	d^6	Fe²⁺, Ru²⁺, Pd⁴⁺, Rh³⁺, Co³⁺	↑ ↑ e_g / ⇅ ↑ ↑ t_{2g}	— — e_g / ⇅ ⇅ ⇅ t_{2g}
d^2	Ti²⁺, V³⁺	— — e_g / ↑ ↑ — t_{2g}	same as high spin	d^7	Co²⁺, Rh²⁺	↑ ↑ e_g / ⇅ ⇅ ↑ t_{2g}	↑ — e_g / ⇅ ⇅ ⇅ t_{2g}
d^3	V²⁺, Cr³⁺	— — e_g / ↑ ↑ ↑ t_{2g}	same as high spin	d^8	Ni²⁺, Pt²⁺	↑ ↑ e_g / ⇅ ⇅ ⇅ t_{2g}	same as high spin
d^4	Mn³⁺, Re³⁺	↑ — e_g / ↑ ↑ ↑ t_{2g}	— — e_g / ⇅ ↑ ↑ t_{2g}	d^9	Cu²⁺	⇅ ↑ e_g / ⇅ ⇅ ⇅ t_{2g}	same as high spin
d^5	Mn²⁺, Fe³⁺ Ru³⁺	↑ ↑ e_g / ↑ ↑ ↑ t_{2g}	— — e_g / ⇅ ⇅ ↑ t_{2g}	d^{10}	Zn²⁺, Ag⁺, Hg²⁺	⇅ ⇅ e_g / ⇅ ⇅ ⇅ t_{2g}	same as high spin

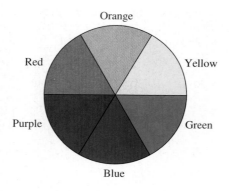

A color wheel shows colors and their complementary colors. For example, green is the complementary color of red. The data in Table 25-9 are given for specific wavelengths. Broad bands of wavelengths are shown in this color wheel.

TABLE 25-9 *Complementary Colors*

Wavelength Absorbed (Å)	Spectral Color (color absorbed)	Complementary Color (color observed)
4100	violet	lemon yellow
4300	indigo	yellow
4800	blue	orange
5000	blue-green	red
5300	green	purple
5600	lemon yellow	violet
5800	yellow	indigo
6100	orange	blue
6800	red	blue-green

25-9 COLOR AND THE SPECTROCHEMICAL SERIES

A substance appears colored because it absorbs light that corresponds to one or more of the wavelengths in the visible region of the electromagnetic spectrum (4000 to 7000 Å) and transmits or reflects the other wavelengths. Our eyes are detectors of light in the visible region, and so each wavelength in this region appears as a different color. A combination of all wavelengths in the visible region is called "white light"; sunlight is an example. The absence of all wavelengths in the visible region is blackness.

In Table 25-9 we show the relationships among colors absorbed and colors transmitted or reflected in the visible region. The first column displays the wavelengths absorbed. The **spectral color** is the color associated with the wavelengths of light absorbed by the sample. When certain visible wavelengths are absorbed from incoming "white" light, the light *not absorbed* remains visible to us as transmitted or reflected light. For instance, a sample that absorbs orange light appears blue. The **complementary color** is the color associated with the wavelengths that are not absorbed by the sample. The complementary color is seen when the spectral color is removed from white light.

Most transition metal compounds are colored, a characteristic that distinguishes them from most compounds of the representative elements. In transition metal *compounds* the d orbitals are often split into two sets of orbitals separated by energies, Δ_{oct} (for an octahedral complex), that correspond to wavelengths of light in the visible region. The absorption of visible light causes electronic transitions between orbitals in these sets. Table 25-10 gives the colors of some transition metal nitrates in aqueous solution. Solutions of representative (A group) metal nitrates are colorless.

One transition of a high spin octahedral Co(III) complex is depicted as follows.

TABLE 25-10 *Colors of Aqueous Solutions of Some Transition Metal Nitrates*

Transition Metal Ion	Color of Aq. Solution
Cr^{3+}	Deep blue
Mn^{2+}	Pale pink
Fe^{2+}	Pale green
Fe^{3+}	Orchid
Co^{2+}	Pink
Ni^{2+}	Green
Cu^{2+}	Blue

Planck's constant is $h = 6.63 \times 10^{-34}$ J·s.

Ground State		Excited State	Energy of Light Absorbed
e_g ↑ ↑	Absorption of light	↑↓ ↑	
t_{2g} ↑↓ ↑ ↑		↑ ↑ ↑	$\Delta E = h\nu$ depends on Δ_{oct}

The colors of complex compounds that contain a given metal depend on the ligands. The yellow compound at the left is a salt that contains $[Co(NH_3)_6]^{3+}$ ions. In the next three compounds, left to right, one NH_3 ligand in $[Co(NH_3)_6]^{3+}$ has been replaced by NCS^- (*orange*), H_2O (*red*), and Cl^- (*purple*). The green compound at the right is a salt that contains $[Co(NH_3)_4Cl_2]^+$ ions.

The frequency (ν), and therefore the wavelength and color, of the light absorbed are related to Δ_{oct}.[1] This, in turn, depends on the crystal field strength of the ligands. So the colors and visible absorption spectra of transition metal complexes, as well as their magnetic properties, provide information about the strengths of the ligand–metal interactions.

By interpreting the visible spectra of many complexes, it is possible to arrange common ligands in order of increasing crystal field strengths.

$$I^- < Br^- < Cl^- < F^- < OH^- < H_2O < (COO)_2^{2-} < NH_3 < en < NO_2^- < CN^-$$

Increasing crystal field strength

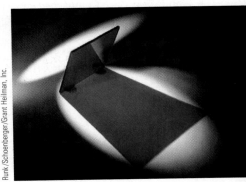

Adding a Co(II) salt to molten glass gives the glass a deep blue color. This blue glass is often used by glassblowers to block out the intense yellow light emitted by thermally excited sodium atoms present in hot borosilicate glass.

This arrangement is called the **spectrochemical series.** Strong field ligands, such as CN^-, produce large crystal field splittings and usually produce low spin complexes, where possible. Weak field ligands, such as Cl^-, produce small crystal field splittings and high spin complexes. Low spin complexes usually absorb higher-energy (shorter-wavelength) light than do high spin complexes. The colors of several six-coordinate Cr(III) complexes are listed in Table 25-11.

In $[Cr(NH_3)_6]Cl_3$, the Cr(III) is bonded to six ammonia ligands, which produce a relatively high value of Δ_{oct}. This causes the $[Cr(NH_3)_6]^{3+}$ ion to absorb relatively high energy visible light in the blue and violet regions. Thus, we see yellow-orange, the complementary color.

Water is a weaker field ligand than ammonia, and therefore Δ_{oct} is less for $[Cr(OH_2)_6]^{3+}$ than for $[Cr(NH_3)_6]^{3+}$. As a result, $[Cr(OH_2)_6]Br_3$ absorbs lower energy (longer wavelength) light. This causes the reflected and transmitted light to be higher energy bluish gray, the color that describes $[Cr(OH_2)_6]Br_3$.

We see the light that is transmitted (passes through the sample) or that is reflected by the sample.

TABLE 25-11	*Colors of Some Chromium(III) Complexes*		
$[Cr(OH_2)_6]Cl_3$	violet	$[Cr(NH_3)_6]Cl_3$	yellow
$[Cr(OH_2)_6]Br_3$	bluish gray	$[Cr(NH_3)_5Cl]Cl_2$	purple
$[Cr(OH_2)_4Cl_2]Cl$	green	$[Cr(NH_3)_4Cl_2]Cl$	violet
$[Cr(OH_2)_4Br_2]Br$	green	$[Cr(CON_2H_4)_6][SiF_6]_3$	green

Different anions sometimes cause compounds containing the same complex cation to have different colors. For example, two different-colored compounds in this table both contain the $[Cr(OH_2)_6]^{3+}$ ion.

[1]The numerical relationship between Δ_{oct} and the wavelength, λ, of the absorbed light is found by combining the expressions $E = h\nu$ and $\nu = c/\lambda$, where c is the speed of light. For *one* mole of a complex,

$$\Delta_{oct} = EN_A = \frac{hcN_A}{\lambda} \quad \text{where } N_A \text{ is Avogadro's number}$$

Key Terms

Ammine complexes Complex species that contain ammonia molecules bonded to metal ions.

Central atom The atom or ion to which the ligands are bonded in a complex species.

Chelate A ligand that utilizes two or more connected donor atoms in bonding to metals.

cis–trans isomerism See *Geometric isomerism.*

Complementary color The color associated with the wavelengths of light that are not absorbed—that is, the color transmitted or reflected.

Complex ion See *Coordination compound or complex.*

Coordinate covalent bond A covalent bond in which both shared electrons are donated by the same atom; a bond between a Lewis base and a Lewis acid.

Coordination compound or complex A compound containing coordinate covalent (dative) bonds between electron pair donors (ligands) and a metal.

Coordination isomers Isomers involving exchange of ligands between a complex cation and a complex anion of the same coordination compound.

Coordination number The number of donor atoms coordinated (bonded) to a metal.

Coordination sphere The metal ion and its coordinated ligands, but not any uncoordinated counterions.

Crystal field splitting energy The energy separation between the e_g and t_{2g} orbitals; a measure of the crystal field strength of the ligands (how strongly the ligand electrons repel the electrons on the metal ion).

Crystal field theory A theory of bonding in transition metal complexes in which ligands and metal ions are treated as point charges; a purely ionic model. Ligand point charges represent the crystal (electric) field perturbing the metal's d orbitals that contain nonbonding electrons.

Δ_{oct} The energy separation between e_g and t_{2g} sets of metal d orbitals caused by six ligands bonding in an octahedral geometry.

Dative bond See *Coordinate covalent bond* (also known as a coordination bond).

Dextrorotatory Describes an optically active substance that rotates the plane of plane-polarized light to the right; also called dextro.

Donor atom A ligand atom whose electrons are shared with a Lewis acid.

e_g orbitals A set of $d_{x^2-y^2}$ and d_{z^2} orbitals; those d orbitals within a set with lobes directed along the x, y, and z axes.

Enantiomers Stereoisomers that differ only by being nonsuperimposable mirror images of each other, like left and right hands; also called optical isomers.

Geometric isomerism Occurs when atoms or groups of atoms can be arranged in different ways on two sides of a rigid structure; also called *cis–trans* isomerism. In geometric isomers of coordination compounds, the same ligands are arranged in different orders within the coordination sphere.

High spin complex The crystal field designation for a complex in which all t_{2g} and e_g orbitals are singly occupied before any pairing occurs.

Hydrate isomers Isomers of crystalline complexes that differ in terms of the presence of water inside or outside the coordination sphere.

Ionization isomers Isomers that result from the interchange of ions inside and outside the coordination sphere.

Isomers Different compounds that have the same formula.

Levorotatory Refers to an optically active substance that rotates the plane of plane-polarized light to the left; also called levo.

Ligand A Lewis base in a coordination compound.

Linkage isomers Isomers in which a particular ligand bonds to a metal ion through different donor atoms.

Low spin complex The crystal field designation for a complex in which pairing occurs to fill the t_{2g} orbitals before any electrons occupy the e_g orbitals.

Monodentate Describes ligands with only one donor atom.

Optical activity The ability of one of a pair of optical isomers to rotate the plane of polarized light.

Optical isomers See *Enantiomers.*

Pairing energy The energy required to pair two electrons in the same orbital.

Plane-polarized light Light waves in which all the electric vectors are oscillating in one plane.

Polarimeter A device used to measure optical activity.

Polydentate Describes ligands with more than one donor atom.

Racemic mixture An equimolar mixture of optical isomers that is, therefore, optically inactive.

Spectral color The color associated with the wavelengths of light that are absorbed.

Spectrochemical series An arrangement of ligands in order of increasing ligand field strength.

Square planar complex A complex in which the metal atom or ion is in the center of a square plane, with a ligand donor atom at each of the four corners.

Stereoisomers Isomers that differ only in the way in which atoms are oriented in space; they include geometric and optical isomers.

Strong field ligand A ligand that exerts a strong crystal or ligand electric field and generally forms low spin complexes with metal ions when possible. A ligand that forms a strong dative bond.

Structural (constitutional) isomers (Applied to coordination compounds.) Isomers whose differences involve more than a single coordination sphere or else different donor atoms; they include ionization isomers, hydrate isomers, coordination isomers, and linkage isomers.

t_{2g} orbitals A set of d_{xy}, d_{yz}, and d_{xz} orbitals; those d orbitals within a set with lobes bisecting (midway between) the x, y, and z axes.

Weak field ligand A ligand that exerts a weak crystal or ligand field and generally forms high spin complexes with metals. A ligand that forms a weak dative bond.

Exercises

Basic Concepts

1. What property of transition metals allows them to form coordination compounds easily?
2. Suggest more appropriate formulas for $NiSO_4 \cdot 6H_2O$, $Cu(NO_3)_2 \cdot 4NH_3$, and $Co(NO_3)_2 \cdot 6NH_3$.
3. What are the two constituents of a coordination complex? What type of chemical bonding occurs between these constituents?
4. Define the term "coordination number" for the central atom or ion in a complex. What values of the coordination numbers for metal ions are most common?
5. Distinguish among the terms *ligands*, *donor atoms*, and *chelates*.
6. Identify the ligands and give the coordination number and the oxidation number for the central atom or ion in each of the following: (a) $[Co(NH_3)_2(NO_2)_4]^-$; (b) $[Cr(NH_3)_5Cl]Cl_2$; (c) $K_2[Cu(CN)_4]$; (d) $[Pd(NH_3)_4]^{2+}$.
7. Repeat Exercise 6 for (a) $Na_2[Pt(CN)_4]$; (b) $[Ag(NH_3)_2]^+$; (c) $[Pt(NH_3)_2Cl_4]$; (d) $[Co(en)_3]^{3+}$.
8. What is the term given to the phenomenon of ring formation by a ligand in a complex? Describe a specific example.
9. Describe the experiments of Alfred Werner on the compounds of the general formula $PtCl_4 \cdot nNH_3$, where $n = 2, 3, 4, 5, 6$. What was his interpretation of these experiments?
10. For each of the compounds of Exercise 9, write formulas indicating the species within the coordination sphere. Also indicate the charges on the complex ions.
11. Write a structural formula showing the ring(s) formed by a bidentate ligand such as ethylenediamine with a metal ion such as Fe^{3+}. How many atoms are in each ring? The formula for this complex ion is $[Fe(en)_3]^{3+}$.

Ammine Complexes

12. Write net ionic equations for reactions of solutions of the following transition metal salts in water with a *limited amount* of aqueous ammonia: (It is not necessary to show the ions as hydrated.) (a) $AgCl_2$; (b) $Zn(NO_3)_2$; (c) $Fe(NO_3)_3$; (d) $Co(NO_3)_2$; (e) $Ni(NO_3)_2$.

13. Write *net ionic* equations for the reactions of the insoluble products of Exercise 12 with an *excess* of aqueous ammonia, if a reaction occurs.
14. Which of the following insoluble metal hydroxides will dissolve in an excess of aqueous ammonia? (a) $Zn(OH)_2$; (b) $Cr(OH)_3$; (c) $Fe(OH)_2$; (d) $Ni(OH)_2$; (e) $Hg(OH)_2$.
15. Write net ionic equations for the reactions in Exercise 14.

Naming Coordination Compounds

16. Give systematic names for the following compounds.
    (a) $Fe(CO)_5$
    (b) $Na_2[Co(OH_2)_2(OH)_4]$
    (c) $[Ag(NH_3)_2]Br$
    (d) $[Cr(en)_3](NO_3)_3$
    (e) $[Co(NH_3)_4Cl]SO_4$
    (f) $K_2[Cu(CN)_4]$
    (g) $[Ni(NH_3)_4(H_2O)_2](NO_3)_2$
    (h) $Na[Al(H_2O)_2(OH)_4]$
    (i) $[Co(NH_3)_4Cl_2][Cr(C_2O_4)_2]$
17. Give systematic names for the following compounds.
    (a) $Na[Au(CN)_2]$
    (b) $Ni(CO)_4$
    (c) $[CoCl_6]^{3-}$
    (d) $[Co(H_2O)_6]^{3+}$
    (e) $Na_2[CuCl_4]$
    (f) $K[Cr(NH_3)_2(OH)_2Cl_2]$
    (g) $[Mo(NCS)_2(en)_2]ClO_4$
    (h) $[Mn(NH_3)_2(OH_2)_3(OH)]SO_4$
    (i) $[Co(en)_2Cl_2]I$
18. Give systematic names for the following compounds.
    (a) $[Ag(NH_3)_2]Cl$
    (b) $[Fe(en)_3]PO_4$
    (c) $[Co(NH_3)_6]SO_4$
    (d) $[Co(NH_3)_6]_2(SO_4)_3$
    (e) $[Pt(NH_3)_4]Cl_2$
    (f) $(NH_4)_2[PtCl_4]$
    (g) $[Co(NH_3)_5SO_4]NO_2$
    (h) $K_4[NiF_6]$
19. Give systematic names for the following compounds.
    (a) $K_2[PdCl_6]$
    (b) $(NH_4)_2[PtCl_6]$

(c) $[Co(en)_3]Cl_2$
(d) $Na_4[Fe(CN)_6]$
(e) $[Cr(en)_2(NH_3)_2](NO_3)_2$
(f) $[Co(NH_3)_4Cl_2]Cl$
(g) $(NH_4)_3[CuCl_5]$

20. Write formulas for the following.
    (a) diamminedichlorozinc
    (b) tin(IV) hexacyanoferrate(II)
    (c) tetracyanoplatinate(II) ion
    (d) potassium hexachlorostannate(IV)
    (e) tetraammineplatinum(II) ion
    (f) sodium hexachloronickelate(II)
    (g) tetraamminecopper(II)pentacyanohydroxoferrate(III)
    (h) diaquadicyanocopper(II)
    (i) potassium hexachloropalladate(IV)

21. Write formulas for the following compounds.
    (a) *trans*-diamminedinitroplatinum(II)
    (b) rubidium tetracyanozincate
    (c) triaqua-*cis*-dibromochlorochromium(III)
    (d) pentacarbonyliron(0)
    (e) sodium pentacyanohydroxocobaltate(III)
    (f) hexammineruthenium(III) tetrachloronickelate(II)
    (g) sodium tetracyanocadmate
    (h) hexaamminecobalt(III) chloride

Structures of Coordination Compounds

22. Write formulas, and provide names for three complex cations in each of the following categories.
    (a) cations coordinated to only monodentate ligands
    (b) cations coordinated to only bidentate ligands
    (c) cations coordinated to two bidentate and two monodentate ligands
    (d) cations coordinated to one tridentate ligand, one bidentate ligand, and one monodentate ligand
    (e) cations coordinated to one tridentate ligand and three monodentate ligands

23. Provide formulas and names for three complex anions that fit each description in Exercise 22.

24. How many geometric isomers can be formed by complexes that are (a) octahedral MA_2B_4 and (b) octahedral MA_3B_3? Name any geometric isomers that can exist. Is it possible for any of these isomers to show optical activity (exist as enantiomers)? Explain.

25. Write the structural formulas for (a) two isomers of $[Pt(NH_3)_2Cl_2]$, (b) four isomers (including linkage isomers) of $[Co(NH_3)_3(NO_2)_3]$, and (c) two isomers (including ionization isomers) of $[Pt(NH_3)_3Br]Cl$.

26. Determine the number and types of isomers that would be possible for each of the following complexes.
    (a) tetraamminediaquachromium(III) ion
    (b) triamminetrichlorochromium(III) ion
    (c) tris(ethylenediamine) chromium(III) ion
    (d) dichlorobis(ethylenediamine)platinum(IV) chloride
    (e) diamminedibromodichlorochromate(III) ion

*27. Indicate whether the complexes in each pair are identical or isomers.

(a)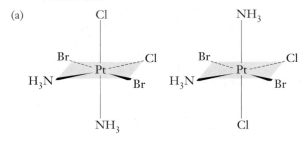

(b)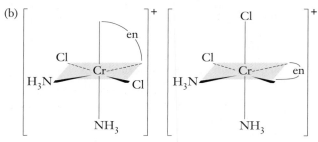

(c)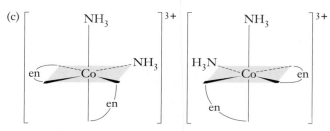

(d)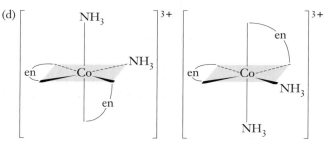

28. Distinguish between constitutional (structural) isomers and stereoisomers.

29. Distinguish between an optically active complex and a racemic mixture.

30. Write the formula for a potential ionization isomer of each of the following compounds. Name each one.
    (a) $[Co(NH_3)_4Br_2]Br$
    (b) $[Ni(en)_2(NO_2)_2]Cl_2$
    (c) $[Fe(NH_3)_5CN]SO_4$

31. Write the formula for a potential hydrate isomer of each of the following compounds. Name each one.
    (a) $[Cu(OH_2)_4]Cl_2$; (b) $[Ni(OH_2)_5Br]Br \cdot H_2O$

32. Write the formula for a potential coordination isomer of each of the following compounds. Name each one.
   (a) $[Co(NH_3)_6][Cr(CN)_6]$; (b) $[Ni(en)_3][Cu(CN)_4]$

33. Write the formula for a potential linkage isomer of each of the following compounds. Name each one.
   (a) $[Co(en)_2(NO_2)_2]Cl$; (b) $[Cr(NH_3)_5(CN)](CN)_2$

Crystal Field Theory

34. Describe clearly what Δ_{oct} is. How is Δ_{oct} actually measured experimentally? How is it related to the spectrochemical series?

35. On the basis of the spectrochemical series, determine whether the following complexes are low spin or high spin.
   (a) $[CrCl_4Br_2]^{3-}$
   (b) $[Co(en)_3]^{3+}$
   (c) $[Fe(OH_2)_6]^{3+}$
   (d) $[Fe(NO_2)_6]^{3-}$

36. On the basis of the spectrochemical series, determine whether the following complexes are low spin or high spin.
   (a) $[Cu(OH_2)_6]^{2+}$
   (b) $[MnF_6]^{3-}$
   (c) $[Co(CN)_6]^{3-}$
   (d) $[Cr(NH_3)_6]^{3+}$

37. Write out the electron distribution in t_{2g} and e_g orbitals for the following in an octahedral field.

Metal Ions	Ligand Field Strength
V^{2+}	Weak
Mn^{2+}	Strong
Mn^{2+}	Weak
Ni^{2+}	Weak
Cu^{2+}	Weak
Fe^{3+}	Strong
Cu^+	Weak
Ru^{3+}	Strong

38. Write formulas for two complex ions that would fit into each of the categories of Exercise 37. Name the complex ions you list.

39. Determine the electron distribution in (a) $[Co(CN)_6]^{3-}$, a low spin complex ion, and (b) $[CoF_6]^{3-}$, a high spin complex ion.

CONCEPTUAL EXERCISES

40. How can you predict whether a complex is high or low spin? How can you determine whether a complex is high or low spin?

41. Define the terms *diamagnetic* and *paramagnetic*. What feature of electronic structure is directly related to these properties?

42. What are three common metal coordination numbers found in coordination chemistry? Give the shape usually associated with each of the common coordination numbers.

43. Several isomers are possible for $[Co(en)(NH_3)(H_2O)_2Cl]^{2+}$. (Two of the isomers have optical isomers.) How many isomers can you draw? Draw the structure of an isomer that does not have an optical isomer. Draw the structure of an isomer that would have an optical isomer. (*Hint:* "en" is not large enough to bind at *trans* locations.)

BUILDING YOUR KNOWLEDGE

Values of equilibrium constants are listed in Appendix I.

44. Calculate (a) the molar solubility of $Zn(OH)_2$ in pure water, (b) the molar solubility of $Zn(OH)_2$ in 0.28 M NaOH solution, and (c) the concentration of $[Zn(OH)_4]^{2-}$ ions in the solution of (b).

45. The yellow complex oxalatorhodium compound $K_3[Rh(C_2O_4)_3]$ can be prepared from the wine-red complex compound $K_3[RhCl_6]$ by boiling a concentrated aqueous solution of $K_3[RhCl_6]$ and $K_2C_2O_4$ for two hours and then evaporating the solution until the product crystallizes.

$$K_3[RhCl_6](aq) + 3K_2C_2O_4(aq) \xrightarrow{\text{heat}} K_3[Rh(C_2O_4)_3](s) + 6KCl(aq)$$

What is the theoretical yield of the oxalato complex if 1.150 g of the chloro complex is heated with 4.95 g of $K_2C_2O_4$? In an experiment, the actual yield was 0.78 g. What is the percent yield?

46. Consider the formation of the triiodoargentate(I) ion.

$$Ag^+ + 3I^- \longrightarrow [AgI_3]^{2-}$$

Would you expect an increase or decrease in the entropy of the system as the complex is formed? The standard state absolute entropy at 25°C is 72.68 J/K mol for Ag^+, 111.3 J/K mol for I^-, and 253.1 J/K mol for $[AgI_3]^{2-}$. Calculate ΔS^0 for the reaction and confirm your prediction.

47. Molecular iodine reacts with I^- to form a complex ion.

$$I_2(aq) + I^- \rightleftharpoons [I_3]^-$$

Calculate the equilibrium constant for this reaction given the following data at 25°C.

$$I_2(aq) + 2e^- \longrightarrow 2I^- \quad E^0 = 0.535 \text{ V}$$
$$[I_3]^- + 2e^- \longrightarrow 3I^- \quad E^0 = 0.5338 \text{ V}$$

48. Calculate the pH of a solution prepared by dissolving 0.40 mol of tetraamminecopper(II) chloride, $[Cu(NH_3)_4]Cl_2$, in water to give 1.0 L of solution. Ignore hydrolysis of Cu^{2+}.

49. Use the following standard reduction potential data to answer the questions.

$$Co^{3+} + e^- \rightleftharpoons Co^{2+}$$
$$E^0 = 1.808 \text{ V}$$

$$Co(OH)_3(s) + e^- \rightleftharpoons Co(OH)_2(s) + OH^-$$
$$E^0 = 0.17 \text{ V}$$

$$[Co(NH_3)_6]^{3+} + e^- \rightleftharpoons [Co(NH_3)_6]^{2+}$$
$$E^0 = 0.108 \text{ V}$$

$$[Co(CN)_6]^{3-} + e^- \rightleftharpoons [Co(CN)_5]^{3-} + CN^-$$
$$E^0 = -0.83 \text{ V}$$

$$O_2(g) + 4H^+ (10^{-7} M) + 4e^- \rightleftharpoons 2H_2O(\ell)$$
$$E^0 = 0.815 \text{ V}$$

$$2H_2O(\ell) + 2e^- \rightleftharpoons H_2(g) + 2OH^-$$
$$E^0 = -0.828 \text{ V}$$

Which cobalt(III) species among those listed would oxidize water? Which cobalt(II) species among those listed would be oxidized by water? Explain your answers.

BEYOND THE TEXTBOOK

Go to the textbook website at

http://www.brookscole.com/chemistry/whitten

for additional activities and exercises based on the General Chemistry Interactive CD-ROM, the World Wide Web, and library resources.

InfoTrac College Edition

For additional readings, go to InfoTrac College Edition, your online research library at:

http://infotrac.thomsonlearning.com

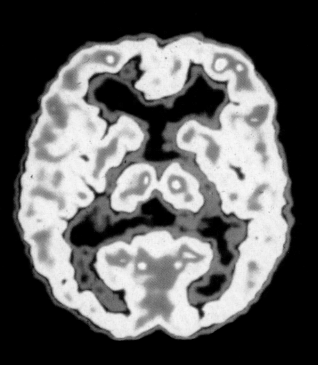
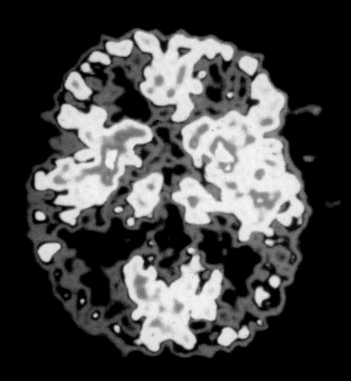

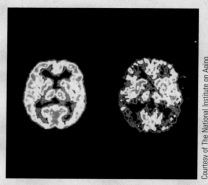

Courtesy of The National Institute on Aging

Positron Emission Tomography (PET) scans show differences in brain activity between a normal brain (left) and a brain affected by Alzheimer's disease (right). Glucose containing the radioactive isotope carbon-11 (^{11}C) was injected into the patients to obtain these images. When a carbon-11 isotope decays, it gives off a positron that immediately combines with an electron on another nearby atom. This matter/anti-matter reaction converts these two particles into two high energy gamma rays (180° apart) that are measured by a ring of detectors around the patient's head. Regions of the brain that are active consume more radioactive glucose and produce the brighter areas (red and yellow), while blue and black denote low or inactive portions.

OUTLINE

26-1	The Nucleus
26-2	Neutron–Proton Ratio and Nuclear Stability
26-3	Nuclear Stability and Binding Energy
26-4	Radioactive Decay
26-5	Equations for Nuclear Reactions
26-6	Neutron-Rich Nuclei (Above the Band of Stability)
26-7	Neutron-Poor Nuclei (Below the Band of Stability)
26-8	Nuclei with Atomic Number Greater Than 83
26-9	Detection of Radiation
26-10	Rates of Decay and Half-Life
26-11	Disintegration Series
26-12	Uses of Radionuclides
26-13	Artificial Transmutations of Elements
26-14	Nuclear Fission
26-15	Nuclear Fission Reactors
26-16	Nuclear Fusion

OBJECTIVES

After you have studied this chapter, you should be able to

- Describe the makeup of the nucleus
- Describe the relationships between neutron–proton ratio and nuclear stability
- Tell what is meant by the band of stability
- Calculate mass deficiency and nuclear binding energy
- Describe the common types of radiation emitted when nuclei undergo radio-active decay
- Write and balance equations that describe nuclear reactions
- Predict the different kinds of nuclear reactions undergone by nuclei, depending on their positions relative to the band of stability
- Describe methods for detecting radiation
- Understand half-lives of radioactive elements
- Carry out calculations associated with radioactive decay
- Interpret disintegration series
- Tell about some uses of radionuclides, including the use of radioactive elements for dating objects
- Describe some nuclear reactions that are induced by bombardment of nuclei with particles
- Tell about nuclear fission and some of its applications, including nuclear reactors
- Tell about nuclear fusion and some prospects for and barriers to its use for the production of energy

Chemical properties are determined by electron distributions and are only indirectly influenced by atomic nuclei. Until now, we have discussed ordinary chemical reactions, so we have focused on electron configurations. **Nuclear reactions** involve changes in the composition of nuclei. These extraordinary processes are often accompanied by the release of tremendous amounts of energy and by transmutations of elements. Some differences between nuclear reactions and ordinary chemical reactions follow.

Nuclear Reaction	Ordinary Chemical Reaction
1. Elements may be converted from one to another.	1. No new elements can be produced.
2. Particles within the nucleus are involved.	2. Only the electrons participate.
3. Tremendous amounts of energy are released or absorbed.	3. Relatively small amounts of energy are released or absorbed.
4. Rate of reaction is not influenced by external factors.	4. Rate of reaction depends on factors such as concentration, temperature, catalyst, and pressure.

Medieval alchemists spent years trying to convert other metals into gold without success. Years of failure and the acceptance of Dalton's atomic theory early in the 19th century convinced scientists that one element could not be converted into another. Then, in 1896 Henri Becquerel discovered "radioactive rays" (*natural* **radioactivity**) coming from a uranium compound. Ernest Rutherford's study of these rays showed that atoms of one element may indeed be converted into atoms of other elements by spontaneous nuclear disintegrations. Many years later it was shown that nuclear reactions initiated by bombardment of nuclei with accelerated subatomic particles or other nuclei can also transform one element into another—accompanied by the release of radiation (*induced radioactivity*).

Becquerel's discovery led other researchers, including Marie and Pierre Curie, to discover and study new radioactive elements. Many radioactive isotopes, or **radioisotopes**, now have important medical, agricultural, and industrial uses.

Nuclear fission is the splitting of a heavy nucleus into lighter nuclei. **Nuclear fusion** is the combination of light nuclei to produce a heavier nucleus. Huge amounts of energy are released when these processes occur. These processes could satisfy a large portion of our future energy demands. Current research is aimed at surmounting the technological problems associated with safe and efficient use of nuclear fission reactors and with the development of controlled fusion reactors.

Marie Sklodowska Curie (1867–1934) is the only person to have been honored with Nobel Prizes in both physics and chemistry. In 1903, Pierre (1859–1906) and Marie Curie and Henri Becquerel (1852–1908) shared the prize in physics for the discovery of natural radioactivity. Marie Curie also received the 1911 Nobel Prize in chemistry for her discovery of radium and polonium and the compounds of radium. She named polonium for her native Poland. Marie's daughter, Irene Joliot-Curie (1897–1956), and Irene's husband, Frederick Joliot (1900–1958), received the 1935 Nobel Prize in chemistry for the first synthesis of a new radioactive element.

In Chapter 5, we represented an atom of a particular isotope by its *nuclide symbol*. Radioisotopes are often called **radionuclides.**

26-1 THE NUCLEUS

In Chapter 5 we described the principal subatomic particles (Table 26-1). Recall that the neutrons and protons together constitute the nucleus, with the electrons occupying essentially empty space around the nucleus. The nucleus is only a minute fraction of the total volume of an atom, yet nearly all the mass of an atom resides in the nucleus. Thus, nuclei are extremely dense. It has been shown experimentally that nuclei of all elements have approximately the same density, 2.4×10^{14} g/cm^3.

If enough nuclei could be gathered together to occupy one cubic centimeter, the total weight would be about 250 million tons!

TABLE 26-1	*Fundamental Particles of Matter*	
Particle	Mass	Charge
Electron (e^-)	0.00054858 amu	1−
Proton (p or p^+)	1.0073 amu	1+
Neutron (n or n^0)	1.0087 amu	none

TOPIC: Basic Nuclear Science
GO TO: www.scilinks.org
*sci*LINKS **CODE:** WCH2610

From an electrostatic point of view, it is amazing that positively charged protons can be packed so closely together. Yet many nuclei do not spontaneously decompose, so they must be stable. In the early 20th century when Rutherford postulated the nuclear model of the atom, scientists were puzzled by such a situation. Physicists have since detected many very short-lived subatomic particles (in addition to protons, neutrons, and electrons) as products of nuclear reactions. Well over 100 have been identified. A discussion of these many particles is beyond the scope of a chemistry text. Furthermore their functions are not entirely understood, but it is now thought that they help to overcome the proton–proton repulsions and to bind nuclear particles (**nucleons**) together. The attractive forces among nucleons appear to be important over only extremely small distances, about 10^{-13} cm.

26-2 NEUTRON–PROTON RATIO AND NUCLEAR STABILITY

CD-ROM   Screen 2.17, Nuclear Stability: The Island of Stability.

The term **"nuclide"** is used to refer to different atomic forms of all elements. The term "isotope" applies only to different forms of the same element. Most naturally occurring nuclides have even numbers of protons and even numbers of neutrons; 157 nuclides fall into this category. Nuclides with odd numbers of both protons and neutrons are least common (there are only four), and those with odd–even combinations are intermediate in abundance (Table 26-2). Furthermore, nuclides with certain "magic numbers" of protons and neutrons seem to be especially stable. Nuclides with a number of protons *or* a number of neutrons *or* a sum of the two equal to 2, 8, 20, 28, 50, 82, or 126 have unusual stability. Examples are $^4_2$He, $^{16}_8$O, $^{42}_{20}$Ca, $^{88}_{38}$Sr, and $^{208}_{82}$Pb. This suggests an energy level (shell) model for the nucleus similar to the shell model of electron configurations.

The nuclide symbol for an element (Section 5-7) is

$$^A_Z E$$

where E is the chemical symbol for the element, Z is its atomic number, and A is its mass number.

Figure 26-1 is a plot of the number of neutrons (N) versus number of protons (Z) for the stable nuclides (the **band of stability**). For low atomic numbers, the most stable nuclides have equal numbers of protons and neutrons ($N = Z$). Above atomic number 20, the most stable nuclides have more neutrons than protons. Careful examination reveals an approximately stepwise shape to the plot due to the stability of nuclides with even numbers of nucleons.

TABLE 26-2	*Abundance of Naturally Occurring Nuclides*			
Number of protons	even	even	odd	odd
Number of neutrons	even	odd	even	odd
Number of such nuclides	157	52	50	4

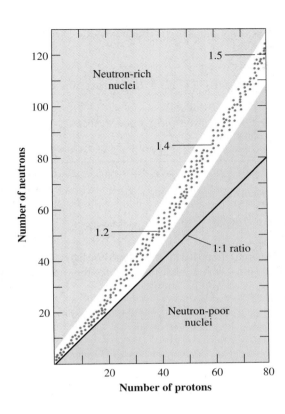

Figure 26-1 A plot of the number of neutrons versus the number of protons in stable nuclei. As atomic number increases, the *N/Z* ratio (the decimal fractions) of the stable nuclei increases. The stable nuclei are located in an area known as the band of stability. Most radioactive nuclei occur outside this band.

26-3 NUCLEAR STABILITY AND BINDING ENERGY

 **CD-ROM** Screen 2.18, Nuclear Stability: Binding Energy.

Experimentally, we observe that the masses of atoms other than $^1_1$H are always *less* than the sum of the masses of their constituent particles. We now know why this *mass deficiency* occurs. We also know that the mass deficiency is in the nucleus of the atom and has nothing to do with the electrons; however, *because tables of masses of isotopes include the electrons, we shall also include them.*

The **mass deficiency, Δm,** for a nucleus is the difference between the sum of the masses of electrons, protons, and neutrons in the atom (calculated mass) and the actual measured mass of the atom.

Do you remember how to find the numbers of protons, neutrons, and electrons in a specified atom? Review Section 5-7.

$$\Delta m = \text{(sum of masses of all } e^-, p^+, \text{ and } n^0) - \text{(actual mass of atom)}$$

For most naturally occurring isotopes, the mass deficiency is only about 0.15% or less of the calculated mass of an atom.

EXAMPLE 26-1 *Mass Deficiency*

Calculate the mass deficiency for chlorine-35 atoms in amu/atom and in g/mol atoms. The actual mass of a chlorine-35 atom is 34.9689 amu.

Plan

We first find the numbers of protons, electrons, and neutrons in one atom. Then we determine the "calculated" mass as the sum of the masses of these particles. The mass deficiency is the actual mass subtracted from the calculated mass. This deficiency is commonly expressed either as mass per atom or as mass per mole of atoms.

Solution

Each atom of $^{35}_{17}\text{Cl}$ contains 17 protons, 17 electrons, and $(35 - 17) = 18$ neutrons. First we sum the masses of these particles.

protons:	17×1.0073 amu	$= 17.124$ amu	(masses from Table 26-1)
electrons:	17×0.00054858 amu	$= 0.0093$ amu	
neutrons:	18×1.0087 amu	$= 18.157$ amu	

$$\text{sum} = 35.290 \text{ amu} \longleftarrow \text{calculated mass}$$

Then we subtract the actual mass from the "calculated" mass to obtain Δm.

$$\Delta m = 35.290 \text{ amu} - 34.9689 \text{ amu} = \boxed{0.321 \text{ amu}} \qquad \text{mass deficiency (in one atom)}$$

We have calculated the mass deficiency in amu/atom. Recall (Section 5-9) that 1 gram is 6.022×10^{23} amu. We can show that a number expressed in amu/atom is equal to the same number in g/mol of atoms.

$$\frac{?\text{ g}}{\text{mol}} = \frac{0.321 \text{ amu}}{\text{atom}} \times \frac{1 \text{ g}}{6.022 \times 10^{23} \text{ amu}} \times \frac{6.022 \times 10^{23} \text{ atoms}}{1 \text{ mol } ^{35}\text{Cl atoms}}$$

$$= \boxed{0.321 \text{ g/mol of } ^{35}\text{Cl atoms}} \longleftarrow \text{(mass } \textit{deficiency} \text{ in a mole of Cl atoms)}$$

You should now work Exercises 14a and 16a, b.

What has happened to the mass represented by the mass deficiency? In 1905, Einstein set forth the Theory of Relativity. He stated that matter and energy are equivalent. An obvious corollary is that matter can be transformed into energy and energy into matter. The transformation of matter into energy occurs in the sun and other stars. It happened on earth when controlled nuclear fission was achieved in 1939 (Section 26-14). The reverse transformation, energy into matter, has not yet been accomplished on a large scale. Einstein's equation, which we encountered in Chapter 1, is $E = mc^2$. E represents the amount of energy released, m the mass of matter transformed into energy, and c the speed of light in a vacuum, 2.997925×10^8 m/s (usually rounded off to 3.00×10^8 m/s).

A mass deficiency represents the amount of matter that would be converted into energy and released if the nucleus were formed from initially separate protons and neutrons. This energy is the **nuclear binding energy, *BE*.** It provides the powerful short-range force that holds the nuclear particles (protons and neutrons) together in a very small volume.

We can rewrite the Einstein relationship as

$$BE = (\Delta m)c^2$$

Specifically, if 1 mole of ^{35}Cl nuclei were to be formed from 17 moles of protons and 18 moles of neutrons, the resulting mole of nuclei would weigh 0.321 gram less than the original collection of protons and neutrons (Example 26-1).

Nuclear binding energies may be expressed in many different units, including kilojoules/mole of atoms, kilojoules/gram of atoms, and megaelectron volts/nucleon. Some useful equivalences are

1 megaelectron volt (MeV) = 1.60×10^{-13} J and 1 joule (J) = 1 kg·m²/s²

Let's use the value of Δm for ^{35}Cl atoms to calculate their nuclear binding energy.

EXAMPLE 26-2   Nuclear Binding Energy

Calculate the nuclear binding energy ^{35}Cl in (a) kilojoules per mole of Cl atoms, (b) kilojoules per gram of Cl atoms, and (c) megaelectron volts per nucleon.

Plan

The mass deficiency that we calculated in Example 26-1 is related to the binding energy by the Einstein equation.

Solution

The mass deficiency is 0.321 g/mol = 3.21×10^{-4} kg/mol.

(a) $BE = (\Delta m)c^2 = \dfrac{3.21 \times 10^{-4} \text{ kg}}{\text{mol } ^{35}\text{Cl atoms}} \times (3.00 \times 10^8 \text{ m/s})^2 = 2.89 \times 10^{13} \dfrac{\text{kg·m}^2/\text{s}^2}{\text{mol } ^{35}\text{Cl atoms}}$

$= 2.89 \times 10^{13}$ J/mol ^{35}Cl atoms = $\boxed{2.89 \times 10^{10} \text{ kJ/mol of } ^{35}\text{Cl atoms}}$

(b) From Example 26-1, the actual mass of ^{35}Cl is

$$\frac{34.9689 \text{ amu}}{^{35}\text{Cl atom}} \quad \text{or} \quad \frac{34.9689 \text{ g}}{\text{mol } ^{35}\text{Cl atoms}}$$

We use this mass to set up the needed conversion factor.

$$BE = \frac{2.89 \times 10^{10} \text{ kJ}}{\text{mol of } ^{35}\text{Cl atoms}} \times \frac{1 \text{ mol } ^{35}\text{Cl atoms}}{34.9689 \text{ g } ^{35}\text{Cl atoms}} = \boxed{8.26 \times 10^8 \text{ kJ/g } ^{35}\text{Cl atoms}}$$

(c) The number of nucleons in *one* atom of ^{35}Cl is 17 protons + 18 neutrons = 35 nucleons.

$BE = \dfrac{2.89 \times 10^{10} \text{ kJ}}{\text{mol of } ^{35}\text{Cl atoms}} \times \dfrac{1000 \text{ J}}{\text{kJ}} \times \dfrac{1 \text{ MeV}}{1.60 \times 10^{-13} \text{ J}} \times \dfrac{1 \text{ mol } ^{35}\text{Cl atoms}}{6.022 \times 10^{23} \,^{35}\text{Cl atoms}} \times \dfrac{1 \,^{35}\text{Cl atom}}{35 \text{ nucleons}}$

> The mass number, Z, is equal to the number of nucleons in one atom.

$= \boxed{8.57 \text{ MeV/nucleon}}$

You should now work Exercises 14 and 16.

The nuclear binding energy of a mole of ^{35}Cl nuclei, 2.89×10^{13} J/mol, is an enormous amount of energy—enough to heat 6.9×10^7 kg (≈ 76,000 tons) of water from 0°C to 100°C! Stated differently, this is also the amount of energy that would be required to separate 1 mole of ^{35}Cl nuclei into 17 moles of protons and 18 moles of neutrons. This has never been done.

Figure 26-2 is a plot of average binding energy per gram of nuclei versus mass number. It shows that nuclear binding energies (per gram) increase rapidly with increasing mass number, reach a maximum around mass number 50, and then decrease slowly. The nuclei with the highest binding energies (mass numbers 40 to 150) are the most stable. Large amounts of energy would be required to separate these nuclei into their component neutrons and

> Some unstable radioactive nuclei do emit a single proton, a single neutron, or other subatomic particles as they decay in the direction of greater stability. None decomposes entirely into elementary particles.

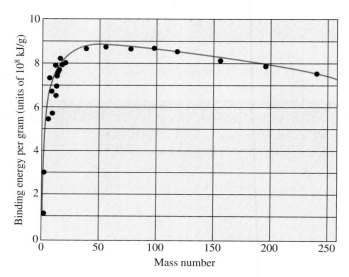

Figure 26-2 Plot of binding energy per gram versus mass number. Very light and very heavy nuclei are relatively unstable.

protons. Even though these nuclei are the most stable ones, *all* nuclei are stable with respect to complete decomposition into protons and neutrons because all (except ^1H) nuclei have mass deficiencies. In other words, the energy equivalent of the loss of mass represents an associative force that is present in all nuclei except ^1H. It must be overcome to separate the nuclei completely into their subatomic particles.

26-4 RADIOACTIVE DECAY

Nuclei whose neutron-to-proton ratios lie outside the stable region undergo spontaneous radioactive decay by emitting one or more particles or electromagnetic rays or both. The type of decay that occurs usually depends on whether the nucleus is above, below, or to the right of the band of stability (Figure 26-1). Common types of **radiation** emitted in decay processes are summarized in Table 26-3.

The particles can be emitted at different kinetic energies. In addition, radioactive decay often leaves a nucleus in an excited (high-energy) state. Then the decay is followed by gamma ray emission.

$$(\text{excited nucleus}) \longrightarrow {}^M_Z E^* \longrightarrow {}^M_Z E + {}^0_0 \gamma$$

The energy of the **gamma ray** ($h\nu$) is equal to the energy difference between the ground and excited nuclear states. This is like the emission of lower energy electromagnetic radiation that occurs as an atom in its excited electronic state returns to its ground state (Section 5-12). Studies of gamma ray energies strongly suggest that nuclear energy levels are quantized just as are electronic energy levels. This adds further support for a shell model for the nucleus.

The penetrating abilities of the particles and rays are proportional to their energies. Beta particles and positrons are about 100 times more penetrating than the heavier and slower-moving alpha particles. They can be stopped by a $\frac{1}{8}$-inch-thick (0.3 cm) aluminum

Recall that the energy of electromagnetic radiation is $E = h\nu$, where h is Planck's constant and ν is the frequency.

TABLE 26-3 *Common Types of Radioactive Emissions*

Type and Symbol[a]	Identity	Mass (amu)	Charge	Velocity	Penetration
beta (β^-, $_{-1}^{0}\beta$, $_{-1}^{0}e$)	electron	0.00055	1−	≤90% speed of light	low to moderate, depending on energy
positron[b] ($_{+1}^{0}\beta$, $_{+1}^{0}e$)	positively charged electron	0.00055	1+	≤90% speed of light	low to moderate, depending on energy
alpha (α, $_{2}^{4}\alpha$, $_{2}^{4}$He)	helium nucleus	4.0026	2+	≤10% speed of light	low
proton ($_{1}^{1}p$, $_{1}^{1}$H)	proton, hydrogen nucleus	1.0073	1+	≤10% speed of light	low to moderate, depending on energy
neutron ($_{0}^{1}n$)	neutron	1.0087	0	≤10% speed of light	very high
gamma ($_{0}^{0}\gamma$) ray	high-energy electro-magnetic radiation such as X-rays	0	0	speed of light	high

[a] *The number at the upper left of the symbol is the number of nucleons, and the number at the lower left is the number of positive charges.*

[b] *On the average, a positron exists for only about a nanosecond (1×10^{-9} second) before colliding with an electron and being converted into the corresponding amount of energy.*

plate. They can burn skin severely but cannot reach internal organs. Alpha particles have low penetrating ability and cannot damage or penetrate skin. They can damage sensitive internal tissue if inhaled, however. The high-energy gamma rays have great penetrating power and severely damage both skin and internal organs. They travel at the speed of light and can be stopped by thick layers of concrete or lead.

A technician cleans lead glass blocks that form part of the giant OPAL particle detector at CERN, the European center for particle physics near Geneva, Switzerland.

Robotics technology is used to manipulate highly radioactive samples safely.

26-5 EQUATIONS FOR NUCLEAR REACTIONS

In *chemical* reactions, atoms in molecules and ions are rearranged, but matter is neither created nor destroyed, and atoms are not changed into other atoms. In earlier chapters, we learned to write balanced chemical equations to represent chemical reactions. Such equations must show the same total number of atoms of each kind on both sides of the equation and the same total charge on both sides of the equation. In a nuclear reaction, a different kind of transformation occurs, one in which a proton can change into a neutron, or a neutron can change into a proton, but the total number of nucleons remains the same. This leads to two requirements for the equation for a nuclear reaction:

> 1. The sum of the mass numbers (the left superscript in the nuclide symbol) of the reactants must equal the sum of the mass numbers of the products.
> 2. The sum of the atomic numbers (the left subscript in the nuclide symbol) of the reactants must equal the sum of the atomic numbers of the products; this maintains charge balance.

Because such equations are intended to describe only the changes in the nucleus, they do not ordinarily include ionic charges (which are due to changes in the arrangements of electrons). In the following sections, we will see many examples of such equations for nuclear reactions.

26-6 NEUTRON-RICH NUCLEI (ABOVE THE BAND OF STABILITY)

Nuclei in this region have too high a ratio of neutrons to protons. They undergo decays that *decrease* the ratio. The most common such decay is **beta emission.** A **beta particle** is an electron ejected *from the nucleus* when a neutron is converted into a proton.

$$\substack{1\\0}n \longrightarrow \substack{1\\1}p + \substack{0\\-1}\beta$$

Thus, beta emission results in an increase of one in the number of protons (the atomic number) and a decrease of one in the number of neutrons, with no change in mass number. Examples of beta particle emission are

$$\substack{228\\88}Ra \longrightarrow \substack{228\\89}Ac + \substack{0\\-1}\beta \quad \text{and} \quad \substack{14\\6}C \longrightarrow \substack{14\\7}N + \substack{0\\-1}\beta$$

The sum of the mass numbers on each side of the first equation is 228, and the sum of the atomic numbers on each side is 88. The corresponding sums for the second equation are 14 and 6, respectively.

26-7 NEUTRON-POOR NUCLEI (BELOW THE BAND OF STABILITY)

Two types of decay for nuclei below the band of stability are **positron emission** or **electron capture** (*K* capture). Positron emission is most commonly encountered with artificially radioactive nuclei of the lighter elements. Electron capture occurs most often with heavier elements.

A **positron** has the mass of an electron but a positive charge. Positrons are emitted when protons are converted to neutrons.

$$\begin{smallmatrix}1\\1\end{smallmatrix}p \longrightarrow \begin{smallmatrix}1\\0\end{smallmatrix}n + \begin{smallmatrix}0\\+1\end{smallmatrix}\beta$$

Thus, positron emission results in a *decrease* by one in atomic number and an *increase* by one in the number of neutrons, with *no change* in mass number.

$$\begin{smallmatrix}38\\19\end{smallmatrix}K \longrightarrow \begin{smallmatrix}38\\18\end{smallmatrix}Ar + \begin{smallmatrix}0\\+1\end{smallmatrix}\beta \quad \text{and} \quad \begin{smallmatrix}15\\8\end{smallmatrix}O \longrightarrow \begin{smallmatrix}15\\7\end{smallmatrix}N + \begin{smallmatrix}0\\+1\end{smallmatrix}\beta$$

The same effect can be accomplished by electron capture (*K* capture), in which an electron from the *K* shell ($n = 1$) is captured by the nucleus.

$$\begin{smallmatrix}106\\47\end{smallmatrix}Ag + \begin{smallmatrix}0\\-1\end{smallmatrix}e \longrightarrow \begin{smallmatrix}106\\46\end{smallmatrix}Pd \quad \text{and} \quad \begin{smallmatrix}37\\18\end{smallmatrix}Ar + \begin{smallmatrix}0\\-1\end{smallmatrix}e \longrightarrow \begin{smallmatrix}37\\17\end{smallmatrix}Cl$$

Some nuclides, such as $\begin{smallmatrix}22\\11\end{smallmatrix}Na$, undergo both electron capture and positron emission.

$$\begin{smallmatrix}22\\11\end{smallmatrix}Na + \begin{smallmatrix}0\\-1\end{smallmatrix}e \longrightarrow \begin{smallmatrix}22\\10\end{smallmatrix}Ne \ (3\%) \quad \text{and} \quad \begin{smallmatrix}22\\11\end{smallmatrix}Na \longrightarrow \begin{smallmatrix}22\\10\end{smallmatrix}Ne + \begin{smallmatrix}0\\+1\end{smallmatrix}\beta \ (97\%)$$

Some of the neutron-poor nuclei, especially the heavier ones, *increase* their neutron-to-proton ratios by undergoing **alpha emission. Alpha particles** are helium nuclei, $\begin{smallmatrix}4\\2\end{smallmatrix}He$, consisting of two protons and two neutrons. Alpha emission also results in an increase of the neutron-to-proton ratio. An example is the alpha emission of lead-204.

$$\begin{smallmatrix}204\\82\end{smallmatrix}Pb \longrightarrow \begin{smallmatrix}200\\80\end{smallmatrix}Hg + \begin{smallmatrix}4\\2\end{smallmatrix}\alpha$$

Electron capture by the nucleus differs from an atom gaining an electron to form an ion.

α-particles carry a double positive charge, but charge is usually not shown in nuclear reactions.

26-8 NUCLEI WITH ATOMIC NUMBER GREATER THAN 83

All nuclides with atomic number greater than 83 are beyond the band of stability and are radioactive. Many of these decay by alpha emission.

$$\begin{smallmatrix}226\\88\end{smallmatrix}Ra \longrightarrow \begin{smallmatrix}222\\86\end{smallmatrix}Rn + \begin{smallmatrix}4\\2\end{smallmatrix}\alpha \quad \text{and} \quad \begin{smallmatrix}210\\84\end{smallmatrix}Po \longrightarrow \begin{smallmatrix}206\\82\end{smallmatrix}Pb + \begin{smallmatrix}4\\2\end{smallmatrix}\alpha$$

The decay of radium-226 was originally reported in 1902 by Rutherford and Soddy. It was the first transmutation of an element ever observed. A few heavy nuclides also decay by beta emission, positron emission, and electron capture.

Some isotopes of uranium ($Z = 92$) and elements of higher atomic number, the **transuranium elements,** also decay by spontaneous nuclear fission. In this process a heavy nuclide splits to form nuclides of intermediate mass and emits neutrons.

$$\begin{smallmatrix}252\\98\end{smallmatrix}Cf \longrightarrow \begin{smallmatrix}142\\56\end{smallmatrix}Ba + \begin{smallmatrix}106\\42\end{smallmatrix}Mo + 4\begin{smallmatrix}1\\0\end{smallmatrix}n$$

The only stable nuclide with atomic number 83 is $\begin{smallmatrix}209\\83\end{smallmatrix}Bi$.

26-9 DETECTION OF RADIATION

Photographic Detection

Emanations from radioactive substances affect photographic plates just as ordinary visible light does. Becquerel's discovery of radioactivity resulted from the unexpected exposure of such a plate, wrapped in black paper, by a nearby enclosed sample of a uranium-containing compound, potassium uranyl sulfate. After a photographic plate has been developed and fixed, the intensity of the exposed area is related to the amount of radiation that has struck the plate. Quantitative detection of radiation by this method is difficult and tedious.

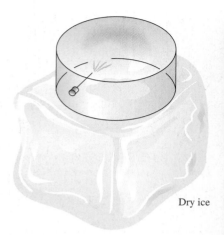

Figure 26-3 A cloud chamber. The emitter is glued onto a pin stuck into a stopper that is mounted on the chamber wall. The chamber has some volatile liquid in the bottom and rests on dry ice. The cool air near the bottom becomes supersaturated with vapor. When an emission speeds through this vapor, ions are produced. These ions serve as "seeds" about which the vapor condenses, forming tiny droplets, or fog.

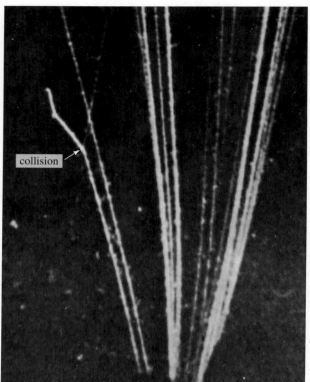

Figure 26-4 An historic cloud chamber photograph of alpha tracks in nitrogen gas. The forked track was shown to be due to a speeding proton (going off to the left) and an isotope of oxygen (going off to the right). It is assumed that the α-particle struck the nucleus of a nitrogen atom at the point where the track forks.

Detection by Fluorescence

Fluorescent substances can absorb high-energy radiation such as gamma rays and subsequently emit visible light. As the radiation is absorbed, the absorbing atoms jump to excited electronic states. The excited electrons return to their ground states through a series of transitions, some of which emit visible light. This method may be used for the quantitative detection of radiation, using an instrument called a **scintillation counter.**

Cloud Chambers

The original cloud chamber was devised by C. T. R. Wilson (1869–1959) in 1911. A chamber contains air saturated with vapor. Particles emitted from a radioactive substance ionize air molecules in the chamber. Cooling the chamber causes droplets of liquid to condense on these ions. The paths of the particles can be followed by observing the fog-like tracks produced. The tracks may be photographed and studied in detail. Figures 26-3 and 26-4 show a cloud chamber and a cloud chamber photograph, respectively.

Gas Ionization Counters

The Geiger counter can detect only β and γ radiation. The α-particles cannot penetrate the walls or window of the tube.

A common gas ionization counter is the **Geiger–Müller counter** (Figure 26-5). Radiation enters the tube through a thin window. Windows of different stopping powers can be used to admit only radiation of certain penetrating powers.

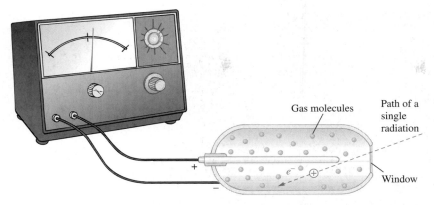

Figure 26-5 The principle of operation of a gas ionization counter. The center wire is positively charged, and the shell of the tube is negatively charged. When radiation enters through the window, it ionizes one or more gas atoms. The electrons are attracted to the central wire, and the positive ions are drawn to the shell. This constitutes a pulse of electric current, which is amplified and displayed on the meter or other readout.

A sample of carnotite, a uranium ore, shown with a Geiger–Müller counter.

26-10 RATES OF DECAY AND HALF-LIFE

Radionuclides have different stabilities and decay at different rates. Some decay nearly completely in a fraction of a second and others only after millions of years. The rates of all radioactive decays are independent of temperature and obey *first-order kinetics*. In Section 16-3 we saw that the rate of a first-order process is proportional only to the concentration of one substance. The rate law and the integrated rate equation for a first-order process (Section 16-4) are

$$\text{rate of decay} = k[\text{A}] \quad \text{and} \quad \ln\left(\frac{\text{A}_0}{\text{A}}\right) = akt$$

Here A represents the amount of decaying radionuclide of interest remaining after some time *t*, and A_0 is the amount present at the beginning of the observation. The *k* is the rate constant, which is different for each radionuclide. Each atom decays independently of the others, so the stoichiometric coefficient *a* is *always* 1 for radioactive decay. We can therefore drop it from the calculations in this chapter and write the integrated rate equation as

The reactant coefficient *a* is *always* 1 for radioactive decay.

$$\ln\left(\frac{\text{A}_0}{\text{A}}\right) = kt$$

Because A_0/A is a ratio, A_0 and A can represent either molar concentrations of a reactant or masses of a reactant. The rate of radioactive disintegrations follows first-order kinetics, so it is proportional to the amount of A present; we can write the integrated rate equation in terms of *N*, the number of disintegrations per unit time:

$$\ln\left(\frac{N_0}{N}\right) = kt$$

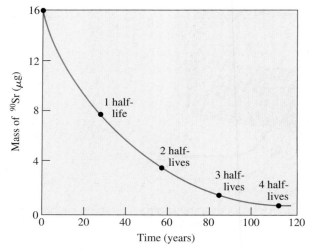

Figure 26-6   The decay of a 16-μg sample of $^{90}_{38}$Sr.

 CD-ROM   Screen 2.19, Half-Life.

In nuclear chemistry, the decay rate is usually expressed in terms of the **half-life, $t_{1/2}$,** of the process. This is the amount of time required for half of the original sample to react. For a first-order process, $t_{1/2}$ is given by the equation

$$t_{1/2} = \frac{\ln 2}{k} = \frac{0.693}{k}$$

The isotope strontium-90 was introduced into the atmosphere by the atmospheric testing of nuclear weapons. Because of the chemical similarity of strontium to calcium, it now occurs with Ca in measurable quantities in milk, bones, and teeth as a result of its presence in food and water supplies. It is a radionuclide that undergoes beta emission with a half-life of 28 years. It may cause leukemia, bone cancer, and other related disorders. If we begin with a 16-μg sample of $^{90}_{38}$Sr, 8 μg will remain after one half-life of 28 years. After 56 years, 4 μg will remain; after 84 years, 2 μg; and so on (Figure 26-6).

Because all radioactive decay follows first-order kinetics, a similar plot for any radionuclide shows the same shape of *exponential decay curve*. About ten half-lives (280 years for $^{90}_{38}$Sr) must pass for any radionuclide to lose 99.9% of its radioactivity.

In 1963, a treaty was signed by the United States, the Soviet Union, and the United Kingdom prohibiting the further testing of nuclear weapons in the atmosphere. Since then, strontium-90 has been disappearing from the air, water, and soil according to the curve in Figure 26-6. So the treaty has largely accomplished its aim up to the present.

Gamma rays destroy both cancerous and normal cells, so the beams of gamma rays must be directed as nearly as possible at only cancerous tissue.

EXAMPLE 26-3   *Rate of Radioactive Decay*

The "cobalt treatments" used in medicine to arrest certain types of cancer rely on the ability of gamma rays to destroy cancerous tissues. Cobalt-60 decays with the emission of beta particles and gamma rays, with a half-life of 5.27 years.

$$^{60}_{27}\text{Co} \longrightarrow {}^{60}_{28}\text{Ni} + {}^{0}_{-1}\beta + {}^{0}_{0}\gamma$$

How much of a 3.42-μg sample of cobalt-60 remains after 30.0 years?

▶ The γ-radiation from ^{60}Co is used to treat cancers near the surface of the body.

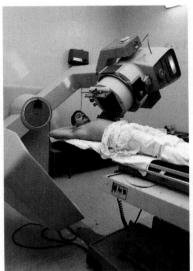

Plan

We determine the value of the specific rate constant, k, from the given half-life. This value is then used in the first-order integrated rate equation to calculate the amount of cobalt-60 remaining after the specified time.

Solution

We first determine the value of the specific rate constant.

$$t_{1/2} = \frac{0.693}{k} \qquad \text{so} \qquad k = \frac{0.693}{t_{1/2}} = \frac{0.693}{5.27 \text{ y}} = 0.131 \text{ y}^{-1}$$

This value can now be used to determine the ratio of A_0 to A after 30.0 years.

$$\ln\left(\frac{A_0}{A}\right) = kt = 0.131 \text{ y}^{-1}(30.0 \text{ y}) = 3.93$$

Taking the inverse ln of both sides, $\dfrac{A_0}{A} = 51$.

$$A_0 = 3.42 \ \mu\text{g, so}$$

$$A = \frac{A_0}{51} = \frac{3.42 \ \mu\text{g}}{51} = \boxed{0.067 \ \mu\text{g} \ ^{60}_{27}\text{Co}} \quad \text{remains after 30.0 years.}$$

You should now work Exercise 54.

26-11 DISINTEGRATION SERIES

Many radionuclides cannot attain nuclear stability by only one nuclear reaction. Instead, they decay in a series of disintegrations. A few such series are known to occur in nature. Two begin with isotopes of uranium, ^{238}U and ^{235}U, and one begins with ^{232}Th. All three of these end with a stable isotope of lead ($Z = 82$). Table 26-4 outlines in detail the ^{238}U, ^{235}U, and ^{232}Th disintegration series, showing half-lives. For any particular decay step, the decaying nuclide is called the **parent** nuclide, and the product nuclide is the **daughter.**

Uranium-238 decays by alpha emission to thorium-234 in the first step of one series. Thorium-234 subsequently emits a beta particle to produce protactinium-234 in the second step. The series can be summarized as shown in Table 26-4a. The *net* reaction for the ^{238}U series is

$$^{238}_{92}\text{U} \longrightarrow \ ^{206}_{82}\text{Pb} + 8 \ ^{4}_{2}\text{He} + 6 \ ^{0}_{-1}\beta$$

"Branchings" are possible at various points in the chain. That is, two successive decays may be replaced by alternative decays, but they always result in the same final product. There are also decay series of varying lengths starting with some of the artificially produced radionuclides (Section 26-13).

TABLE 26-4 | *Emissions and Half-Lives of Members of Natural Radioactive Series**

(a) ^{238}U Series	(b) ^{235}U Series	(c) ^{232}Th Series

(a) ^{238}U Series

$^{238}_{92}$U $\rightarrow \alpha$
$\downarrow$ 4.51×10^9 y

$^{234}_{90}$Th $\rightarrow \beta$
24.1 d

$^{234}_{91}$Pa $\rightarrow \beta$
6.75 h

$^{234}_{92}$U $\rightarrow \alpha$
$\downarrow$ 2.47×10^5 y

$^{230}_{90}$Th $\rightarrow \alpha$
$\downarrow$ 8.0×10^4 y

$^{226}_{88}$Ra $\rightarrow \alpha$
$\downarrow$ 1.60×10^3 y

$^{222}_{86}$Rn $\rightarrow \alpha$
3.82 d

$\beta \leftarrow$ $^{218}_{84}$Po $\rightarrow \alpha$
0.04%   3.05 m

$\alpha \leftarrow$ $^{218}_{85}$At $^{214}_{82}$Pb $\rightarrow \beta$
2 s   26.8 m

$\beta \leftarrow$ $^{214}_{83}$Bi $\rightarrow \alpha$
99.96%   19.7 m

$\alpha \leftarrow$ $^{214}_{84}$Po $^{210}_{81}$Tl $\rightarrow \beta$
1.6×10^{-4} s   1.32 m

$^{210}_{82}$Pb $\rightarrow \beta$
20.4 y

$\beta \leftarrow$ $^{210}_{83}$Bi $\rightarrow \alpha$
100%   5.01 d

$\alpha \leftarrow$ $^{210}_{84}$Po $^{206}_{81}$Tl $\rightarrow \beta$
138 d   4.19 m

$^{206}_{82}$Pb

(b) ^{235}U Series

$^{235}_{92}$U $\rightarrow \alpha$
$\downarrow$ 7.1×10^8 y

$^{231}_{90}$Th $\rightarrow \beta$
25.5 h

$^{231}_{91}$Pa $\rightarrow \alpha$
$\downarrow$ 3.25×10^4 y

$\beta \leftarrow$ $^{227}_{89}$Ac $\rightarrow \alpha$
98.8%   21.6 y

$\alpha \leftarrow$ $^{227}_{90}$Th $^{223}_{87}$Fr $\rightarrow \beta$
18.2 d   22 m

$^{223}_{88}$Ra $\rightarrow \alpha$
11.4 d

$^{219}_{86}$Rn $\rightarrow \alpha$
4.00 s

$\beta \leftarrow$ $^{215}_{84}$Po $\rightarrow \alpha$
5×10^{-4}% 1.78×10^{-3} s

$\alpha \leftarrow$ $^{215}_{85}$At $^{211}_{82}$Pb $\rightarrow \beta$
10^{-4} s   36.1 m

$\beta \leftarrow$ $^{211}_{83}$Bi $\rightarrow \alpha$
99.7%   2.16 m

$\alpha \leftarrow$ $^{211}_{84}$Po $^{207}_{81}$Tl $\rightarrow \beta$
0.52 s   4.79 m

$^{207}_{82}$Pb

(c) ^{232}Th Series

$^{232}_{90}$Th $\rightarrow \alpha$
$\downarrow$ 1.41×10^{10} y

$^{228}_{88}$Ra $\rightarrow \beta$
6.7 y

$^{228}_{89}$Ac $\rightarrow \beta$
6.13 h

$^{228}_{90}$Th $\rightarrow \alpha$
1.91 y

$^{224}_{88}$Ra $\rightarrow \alpha$
3.64 d

$^{220}_{86}$Rn $\rightarrow \alpha$
55.3 s

$\beta \leftarrow$ $^{216}_{84}$Po $\rightarrow \alpha$
0.014%   0.14 s

$\alpha \leftarrow$ $^{216}_{85}$At $^{212}_{82}$Pb $\rightarrow \beta$
3×10^{-4} s   10.6 h

$\beta \leftarrow$ $^{212}_{83}$Bi $\rightarrow \alpha$
66.3%   60.6 m

$\alpha \leftarrow$ $^{212}_{84}$Po $^{208}_{81}$Tl $\rightarrow \beta$
3.0×10^{-7} s   3.10 m

$^{208}_{82}$Pb

*Abbreviations are y, year; d, day; m, minute; and s, second. Less prevalent decay branches are shown in blue.

26-12 USES OF RADIONUCLIDES

Radionuclides have practical uses because they decay at known rates. Some applications make use of the radiation that is continuously emitted by radionuclides.

Radioactive Dating

The ages of articles of organic origin can be estimated by *radiocarbon dating*. The radio-isotope carbon-14 is produced continuously in the upper atmosphere as nitrogen atoms capture cosmic-ray neutrons.

$$^{14}_{7}N + ^{1}_{0}n \longrightarrow ^{14}_{6}C + ^{1}_{1}H$$

The carbon-14 atoms react with oxygen molecules to form $^{14}CO_2$. This process continually supplies the atmosphere with radioactive $^{14}CO_2$, which is removed from the atmosphere by photosynthesis. The intensity of cosmic rays is related to the sun's activity. As long as this remains constant, the amount of $^{14}CO_2$ in the atmosphere remains constant. $^{14}CO_2$ is incorporated into living organisms just as ordinary $^{12}CO_2$ is, so a certain fraction of all carbon atoms in living substances is carbon-14. This decays with a half-life of 5730 years.

$$^{14}_{6}C \longrightarrow ^{14}_{7}N + ^{0}_{-1}\beta$$

After death, the plant no longer carries out photosynthesis, so it no longer takes up $^{14}CO_2$. Other organisms that consume plants for food stop doing so at death. The emissions from the ^{14}C in dead tissue then decrease with the passage of time. The activity per gram of carbon is a measure of the length of time elapsed since death. Comparison of ages of ancient trees calculated from ^{14}C activity with those determined by counting rings indicates that cosmic ray intensity has varied somewhat throughout history. The calculated ages can be corrected for these variations. The carbon-14 technique is useful only for dating objects less than 50,000 years old. Older objects have too little activity to be dated accurately.

The *potassium–argon* and *uranium–lead* methods are used for dating older objects. Potassium-40 decays to argon-40 with a half-life of 1.3 billion years.

$$^{40}_{19}K + ^{0}_{-1}e \longrightarrow ^{40}_{18}Ar$$

Because of its long half-life, potassium-40 can be used to date objects up to 1 million years old by determination of the ratio of $^{40}_{19}K$ to $^{40}_{18}Ar$ in the sample. The uranium–lead method is based on the natural uranium-238 decay series, which ends with the production of stable lead-206. This method is used for dating uranium-containing minerals several billion years old because this series has an even longer half-life. All the ^{206}Pb in such minerals is assumed to have come from ^{238}U. Because of the very long half-life of $^{238}_{92}U$, 4.5 billion years, the amounts of intermediate nuclei can be neglected. A meteorite that was 4.6 billion years old fell in Mexico in 1969. Results of $^{238}U/^{206}Pb$ studies on such materials of extraterrestrial origin suggest that our solar system was formed several billion years ago.

In 1992, hikers in the Italian Alps found the remains of a man who had been frozen in a glacier for about 4000 years. This discovery was especially important because of the unusual preservation of tissues, garments, and personal belongings. Radiocarbon dating is used to estimate the ages of archaeological finds such as these.

Atmospheric testing of nuclear warheads has also caused minor fluctuations in the natural abundance of ^{14}C.

Gaseous argon is easily lost from minerals. Measurements based on the $^{40}K/^{40}Ar$ method may therefore not be as reliable as desired.

In Table 26-4(a) we see that the first step, the decay of ^{238}U, is the slowest step (longest half-life). We learned in Section 16-7 that the slowest step is the rate-determining step.

EXAMPLE 26-4 *Radiocarbon Dating*

A piece of wood taken from a cave dwelling in New Mexico is found to have a carbon-14 activity (per gram of carbon) only 0.636 times that of wood cut today. Estimate the age of the wood. The half-life of carbon-14 is 5730 years.

Plan

As we did in Example 26-3, we determine the specific rate constant k from the known half-life. The time required to reach the present fraction of the original activity is then calculated from the first-order decay equation.

Solution

First we find the first-order specific rate constant for ^{14}C.

$$t_{1/2} = \frac{0.693}{k} \quad \text{or} \quad k = \frac{0.693}{t_{1/2}} = \frac{0.693}{5730 \text{ y}} = 1.21 \times 10^{-4}\,\text{y}^{-1}$$

The present ^{14}C activity, N (disintegrations per unit time), is 0.636 times the original activity, N_0.

$$N = 0.636\,N_0$$

We substitute into the first-order decay equation

$$\ln\left(\frac{N_0}{N}\right) = kt$$

$$\ln\left(\frac{N_0}{0.636\,N_0}\right) = (1.21 \times 10^{-4}\,\text{y}^{-1})t$$

We cancel N_0 and solve for t.

$$\ln\left(\frac{1}{0.636}\right) = (1.21 \times 10^{-4}\,\text{y}^{-1})t$$

$$0.452 = (1.21 \times 10^{-4}\,\text{y}^{-1})t \quad \text{or} \quad t = \boxed{3.74 \times 10^3\text{y (or 3740 y)}}$$

You should now work Exercises 58 and 62.

EXAMPLE 26-5   *Uranium–Lead Dating*

A sample of uranium ore is found to contain 4.64 mg of ^{238}U and 1.22 mg of ^{206}Pb. Estimate the age of the ore. The half-life of ^{238}U is 4.51×10^9 years.

Plan

The original mass of ^{238}U is equal to the mass of ^{238}U remaining plus the mass of ^{238}U that decayed to produce the present mass of ^{206}Pb. We obtain the specific rate constant, k, from the known half-life. Then we use the ratio of original ^{238}U to remaining ^{238}U to calculate the time elapsed, with the aid of the first-order integrated rate equation.

Solution

First we calculate the amount of ^{238}U that must have decayed to produce 1.22 mg of ^{206}Pb, using the isotopic masses.

$$\underline{?} \text{ mg } ^{238}U = 1.22 \text{ mg } ^{206}Pb \times \frac{238 \text{ mg } ^{238}U}{206 \text{ mg } ^{206}Pb} = 1.41 \text{ mg } ^{238}U$$

Thus, the sample originally contained 4.64 mg + 1.41 mg = 6.05 mg of ^{238}U.

We next evaluate the specific rate (disintegration) constant, k.

$$t_{1/2} = \frac{0.693}{k} \quad \text{so} \quad k = \frac{0.693}{t_{1/2}} = \frac{0.693}{4.51 \times 10^9 \text{ y}} = 1.54 \times 10^{-10}\,\text{y}^{-1}$$

Now we calculate the age of the sample, t.

$$\ln\left(\frac{A_0}{A}\right) = kt$$

$$\ln\left(\frac{6.05 \text{ mg}}{4.64 \text{ mg}}\right) = (1.54 \times 10^{-10} \text{ y}^{-1})t$$

$$\ln 1.30 = (1.54 \times 10^{-10} \text{ y}^{-1})t$$

$$\frac{0.262}{1.54 \times 10^{-10} \text{ y}^{-1}} = t \quad \text{or} \quad t = 1.70 \times 10^9 \text{ years}$$

The ore is approximately 1.7 billion years old.

You should now work Exercise 76.

Medical Uses

The use of cobalt radiation treatments for cancerous tumors was described in Example 26-3. Several other nuclides are used as **radioactive tracers** in medicine. Radioisotopes of an element have the same chemical properties as stable isotopes of the same element, so they can be used to "label" the presence of an element in compounds. A radiation detector can be used to follow the path of the element throughout the body. Modern computer-based techniques allow construction of an image of the area of the body where the radioisotope is concentrated. Salt solutions containing ^{24}Na can be injected into the bloodstream to follow the flow of blood and locate obstructions in the circulatory system. Thallium-201 tends to concentrate in healthy heart tissue, whereas technetium-99 concentrates in abnormal heart tissue. The two can be used together to survey damage from heart disease.

Iodine-123 concentrates in the thyroid gland, liver, and certain parts of the brain. This radioisotope is used to monitor goiter and other thyroid problems, as well as liver and brain tumors. One of the most useful radioisotopes in medical applications in recent years is an isotope of technetium, an element that does not occur naturally on earth. This isotope, ^{99m}Tc, is produced by the decay of ^{99}Mo.

$$^{99}\text{Mo} \longrightarrow {}^{99m}\text{Tc} + {}_{-1}^{0}\beta$$

The "m" in the superscript of ^{99m}Tc stands for "metastable." This isotope is formed at a high energy, and then slowly decays with a half-life of 6.0 hours by emitting gamma radiation.

$$^{99m}\text{Tc} \longrightarrow {}^{99}\text{Tc} + \gamma$$

The "Chemistry in Use" feature on the textbook Web site describes some medical applications of ^{99m}Tc.

Another form of imaging that uses positron emitters (Section 26-6) is *positron emission tomography (PET)*. Isotopes commonly used in this technique are short-lived positron emitters such as ^{11}C ($t_{1/2} = 20.4$ min), ^{13}N ($t_{1/2} = 9.98$ min), ^{15}O ($t_{1/2} = 2.05$ min), and ^{18}F ($t_{1/2} = 110$ min). The appropriate isotope is incorporated into a chemical that is normally taken up by the tissues that are being investigated, for instance carbon dioxide or glucose including ^{11}C or water including ^{15}O. This radioactive chemical can then be administered by inhalation or injection. The patient is then placed into a cylindrical gamma ray detector. When these radioisotopes decay, the emitted positron quickly encounters an electron and reacts in a matter–antimatter annihilation, to give off two gamma rays in opposite directions.

$$_{+1}^{0}\beta + {}_{-1}^{0}\beta \longrightarrow 2\gamma$$

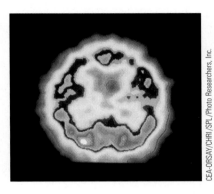

Positron emission tomography (PET) allows mapping of tissues. This PET scan shows the distribution of radioactive glucose in a healthy human brain.

This isotope of iodine is also used in the treatment of thyroid cancer. Because of its preferential absorption in the thyroid gland, it delivers radiation where it is needed.

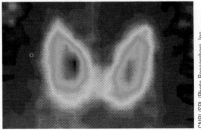

A scan of the radiation released by radioactive iodine concentrated in thyroid tissue gives an image of the thyroid gland.

The directions of emission of millions of such pairs of gamma rays detected over several minutes allow for a computer reconstruction of an image of the tissue containing the positron emitter.

The energy produced by the decay of plutonium-238 is converted into electrical energy in heart pacemakers. The relatively long half-life of the isotope allows the device to be used for 10 years before replacement.

Agricultural Uses

The pesticide DDT is toxic to humans and animals repeatedly exposed to it. DDT persists in the environment for a long time. It concentrates in fatty tissues. The DDT once used to control the screwworm fly was replaced by a radiological technique. Irradiating the male flies with gamma rays alters their reproductive cells, sterilizing them. When great numbers of sterilized males are released in an infested area, they mate with females, who, of course, produce no offspring. This results in the reduction and eventual disappearance of the population.

Labeled fertilizers can also be used to study nutrient uptake by plants and to study the growth of crops. Gamma irradiation of some foods allows them to be stored for longer periods without spoiling. For example, it retards the sprouting of potatoes and onions. In 1999, the FDA approved gamma irradiation of red meat as a way to curb food-borne illnesses. In addition to reducing levels of *Listeria, Salmonella*, and other bacteria significantly, such irradiation is currently the only known way to completely eliminate the dangerous strain of *Escherichia coli* bacteria in red meat. Absorption of gamma rays by matter produces no radioactive nuclides, so foods preserved in this way are *not* radioactive.

Industrial Uses

There are many applications of radiochemistry in industry and engineering. When great precision is required in the manufacture of strips or sheets of metal of definite thicknesses, the penetrating powers of various kinds of radioactive emissions are utilized. The thickness of the metal is correlated with the intensity of radiation passing through it. The flow of a liquid or gas through a pipeline can be monitored by injecting a sample containing a radioactive substance. Leaks in pipelines can also be detected in this way.

The procedure works because the female flies mate only once. In an area highly populated with sterile males, the probability of a "productive" mating is very small.

A weak radioactive source such as americium is used in some smoke detectors. Radiation from the source ionizes the air to produce a weak current. Smoke particles interrupt the current flow by attracting the ions. This decrease in current triggers the alarm.

PhotoDisc/GETTY

Courtesy of International Atomic Energy Agency

Moderate irradiation with gamma rays from radioactive isotopes has kept the strawberries at the right fresh for 15 days, while those at the left are moldy. Such irradiation kills mold spores but does no damage to the food. The food does *not* become radioactive.

In addition to the ^{238}Pu-based heart pacemaker already mentioned, lightweight, portable power packs that use radioactive isotopes as fuel have been developed for other uses. Polonium-210, californium-242, and californium-244 have been used in such generators to power instruments for space vehicles and in polar regions. These generators can operate for years with only a small loss of power.

Research Applications

The pathways of chemical reactions can be investigated using radioactive tracers. When radioactive $^{35}S^{2-}$ ions are added to a saturated solution of cobalt sulfide in equilibrium with solid cobalt sulfide, the solid becomes radioactive. This shows that sulfide ion exchange occurs between solid and solution in the solubility equilibrium.

$$CoS(s) \rightleftharpoons Co^{2+}(aq) + S^{2-}(aq) \qquad K_{sp} = 8.7 \times 10^{-23}$$

Photosynthesis is the process by which the carbon atoms in CO_2 are incorporated into glucose, $C_6H_{12}O_6$, in green plants.

$$6CO_2 + 6H_2O \xrightarrow[\text{chlorophyll}]{\text{sunlight}} C_6H_{12}O_6 + 6O_2$$

The process is more complex than the net equation implies; it actually occurs in many steps and produces a number of intermediate products. By using labeled $^{14}CO_2$, we can identify the intermediate molecules. They contain the radioactive ^{14}C atoms.

26-13 ARTIFICIAL TRANSMUTATIONS OF ELEMENTS

The first artificially induced nuclear reaction was carried out by Rutherford in 1915. He bombarded nitrogen-14 with alpha particles to produce an isotope of oxygen and a proton.

$$^{14}_{7}N + ^{4}_{2}\alpha \longrightarrow ^{1}_{1}H + ^{17}_{8}O$$

Such reactions are often indicated in abbreviated form, with the bombarding particle and emitted subsidiary particles shown parenthetically between the parent and daughter nuclei.

$$^{14}_{7}N \ (^{4}_{2}\alpha, \ ^{1}_{1}p) \ ^{17}_{8}O$$

Several thousand different artificially induced reactions have been carried out with bombarding particles such as neutrons, protons, deuterons ($^{2}_{1}H$), alpha particles, and other small nuclei.

SC*LINKS.*

TOPIC: Synthetic Elements
GO TO: www.scilinks.org
*sci*LINKS CODE: WCH2630

Bombardment with Positive Ions

A problem arises with the use of positively charged nuclei as projectiles. For a nuclear reaction to occur, the bombarding nuclei must actually collide with the target nuclei, which are also positively charged. Collisions cannot occur unless the projectiles have sufficient kinetic energy to overcome coulombic repulsion. The required kinetic energies increase with increasing atomic numbers of the target and of the bombarding particle.

Particle accelerators called **cyclotrons** (atom smashers) and **linear accelerators** have overcome the problem of repulsion. A cyclotron (Figure 26-7) consists of two hollow, D-shaped electrodes called "dees." Both dees are in an evacuated enclosure between the poles of an electromagnet. The particles to be accelerated are introduced at the center in the gap

The first cyclotron was constructed by E. O. Lawrence (1901–1958) and M. S. Livingston at the University of California in 1930.

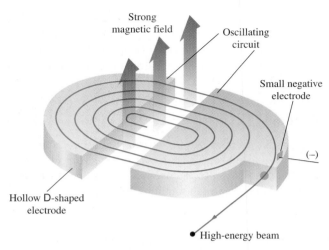

Strong
magnetic field

Oscillating
circuit

Small negative
electrode

(−)

Hollow D-shaped
electrode

High-energy beam

Figure 26-7   Schematic representation of a cyclotron.

The path of the particle is initially circular because of the interaction of the particle's charge with the electromagnet's field. As the particle gains energy, the radius of the path increases, and the particle spirals outward.

between the dees. The dees are connected to a source of high-frequency alternating current that keeps them oppositely charged. The positively charged particles are attracted toward the negative dee. The magnetic field causes the path of the charged particles to curve 180° to return to the space between the dees. Then the charges are reversed on the dees, so the particles are repelled by the first dee (now positive) and attracted to the second. This repeated process is synchronized with the motion of the particles. They accelerate along a spiral path and eventually emerge through an exit hole oriented so that the beam hits the target atoms (Figure 26-8).

Argonne National Laboratory

Figure 26-8   A beam of protons (bright blue stream) from a cyclotron at the Argonne National Laboratory. Nuclear reactions take place when protons and other atomic particles strike the nuclei of atoms.

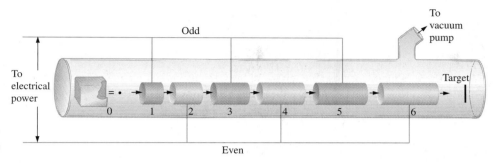

Figure 26-9 Diagram of an early type of linear accelerator. An alpha emitter is placed in the container at the left. Only those α-particles that happen to be emitted in line with the series of accelerating tubes can escape.

In a linear accelerator the particles are accelerated through a series of tubes within an evacuated chamber (Figure 26-9). The odd-numbered tubes are at first negatively charged and the even ones positively charged. A positively charged particle is attracted toward the first tube. As it passes through that tube, the charges on the tubes are reversed so that the particle is repelled out of the first tube (now positive) and toward the second (negative) tube. As the particle nears the end of the second tube, the charges are again reversed. As this process is repeated, the particle is accelerated to very high velocities. The polarity is changed at constant frequency, so subsequent tubes are longer to accommodate the increased distance traveled by the accelerating particle per unit time. The bombardment target is located outside the last tube. If the initial polarities are reversed, negatively charged particles can also be accelerated. The longest linear accelerator, completed in 1966 at Stanford University, is about 2 miles long. It is capable of accelerating electrons to energies of nearly 20 GeV.

Many nuclear reactions have been induced by such bombardment techniques. At the time of development of particle accelerators, there were a few gaps among the first 92 elements in the periodic table. Particle accelerators were used between 1937 and 1941 to synthesize three of the four "missing" elements: numbers 43 (technetium), 85 (astatine), and 87 (francium).

The first linear accelerator was built in 1928 by a German physicist, Rolf Wideroe.

One gigaelectron volt (GeV) = 1×10^9 eV = 1.60×10^{-10} J. This is sometimes called 1 billion electron volts (BeV) in the United States.

Courtesy of Fermilab

An aerial view of the particle accelerator dedicated in 1978 at the Fermi National Accelerator Laboratory (Fermilab), near Batavia, Illinois. This proton accelerator, 4 miles in circumference, accelerates protons to energies of 1 trillion electron volts.

$$\ce{^{96}_{42}Mo + ^{2}_{1}H -> ^{97}_{43}Tc + ^{1}_{0}}n$$

$$\ce{^{209}_{83}Bi + ^{4}_{2}\alpha -> ^{210}_{85}At + 3\ ^{1}_{0}}n$$

$$\ce{^{230}_{90}Th + ^{1}_{1}}p \longrightarrow \ce{^{223}_{87}Fr + 2\ ^{4}_{2}\alpha}$$

Many hitherto unknown, unstable, artificial isotopes of known elements have also been synthesized so that their nuclear structures and behavior could be studied.

Neutron Bombardment

Neutrons bear no charge, so they are not repelled by nuclei as positively charged projectiles are. They do not need to be accelerated to produce bombardment reactions. Neutron beams can be generated in several ways. A frequently used method involves bombardment of beryllium-9 with alpha particles.

$$\ce{^{9}_{4}Be + ^{4}_{2}\alpha -> ^{12}_{6}C + ^{1}_{0}}n$$

Nuclear reactors (Section 26-15) are also used as neutron sources. Neutrons ejected in nuclear reactions usually possess high kinetic energies and are called **fast neutrons.** When they are used as projectiles they cause reactions, such as (n, p) or (n, α) reactions, in which subsidiary particles are ejected. The fourth "missing" element, number 61 (promethium), was synthesized by fast neutron bombardment of neodymium-142.

$$\ce{^{142}_{60}Nd + ^{1}_{0}}n \longrightarrow \ce{^{143}_{61}Pm + ^{0}_{-1}\beta}$$

Fast neutrons move so rapidly that they are likely to pass right through a target nucleus without reacting. Hence, the probability of a reaction is low, even though the neutrons may be very energetic.

Slow neutrons ("thermal" neutrons) are produced when fast neutrons collide with **moderators** such as hydrogen, deuterium, oxygen, or the carbon atoms in paraffin. These neutrons are more likely to be captured by target nuclei. Bombardments with slow neutrons can cause neutron-capture (n, γ) reactions.

$$\ce{^{200}_{80}Hg + ^{1}_{0}}n \longrightarrow \ce{^{201}_{80}Hg + ^{0}_{0}\gamma}$$

Slow neutron bombardment also produces the ^3H isotope (tritium).

$$\ce{^{6}_{3}Li + ^{1}_{0}}n \longrightarrow \ce{^{3}_{1}H + ^{4}_{2}\alpha} \qquad (n,\ \alpha)\ \text{reaction}$$

E. M. McMillan (1907–1991) discovered the first transuranium element, neptunium, in 1940 by bombarding uranium-238 with slow neutrons.

$$\ce{^{238}_{92}U + ^{1}_{0}}n \longrightarrow \ce{^{239}_{92}U + ^{0}_{0}\gamma}$$

$$\ce{^{239}_{92}U} \longrightarrow \ce{^{239}_{93}Np + ^{0}_{-1}\beta}$$

Several additional elements have been prepared by neutron bombardment or by bombardment of the nuclei so produced with positively charged particles. Some examples are

$$\left. \begin{array}{l} \ce{^{238}_{92}U + ^{1}_{0}}n \longrightarrow \ce{^{239}_{92}U + ^{0}_{0}\gamma} \\[4pt] \ce{^{239}_{92}U} \longrightarrow \ce{^{239}_{93}Np + ^{0}_{-1}\beta} \\[4pt] \ce{^{239}_{93}Np} \longrightarrow \ce{^{239}_{94}Pu + ^{0}_{-1}\beta} \end{array} \right\} \quad \text{plutonium}$$

$$\ce{^{239}_{94}Pu + ^{4}_{2}\alpha} \longrightarrow \ce{^{242}_{96}Cm + ^{1}_{0}}n \qquad \text{curium}$$

$$\ce{^{246}_{96}Cm + ^{12}_{6}C} \longrightarrow \ce{^{254}_{102}No + 4\ ^{1}_{0}}n \qquad \text{nobelium}$$

$$\ce{^{252}_{98}Cf + ^{10}_{5}B} \longrightarrow \ce{^{257}_{103}Lr + 5\ ^{1}_{0}}n \qquad \text{lawrencium}$$

26-14 NUCLEAR FISSION

Isotopes of some elements with atomic numbers above 80 are capable of undergoing fission in which they split into nuclei of intermediate masses and emit one or more neutrons. Some fissions are spontaneous; others require that the activation energy be supplied by bombardment. A given nucleus can split in many different ways, liberating enormous amounts of energy. Some of the possible fissions that can result from bombardment of fissionable uranium-235 with fast neutrons follow. The uranium-236 is a short-lived intermediate.

$$^{235}_{92}\text{U} + ^{1}_{0}n \longrightarrow ^{236}_{92}\text{U} \begin{cases} \nearrow ^{160}_{62}\text{Sm} + ^{72}_{30}\text{Zn} + 4\,^{1}_{0}n + \text{energy} \\ \nearrow ^{146}_{57}\text{La} + ^{87}_{35}\text{Br} + 3\,^{1}_{0}n + \text{energy} \\ \longrightarrow ^{141}_{56}\text{Ba} + ^{92}_{36}\text{Kr} + 3\,^{1}_{0}n + \text{energy} \\ \searrow ^{144}_{55}\text{Cs} + ^{90}_{37}\text{Rb} + 2\,^{1}_{0}n + \text{energy} \\ \searrow ^{144}_{54}\text{Xe} + ^{90}_{38}\text{Sr} + 2\,^{1}_{0}n + \text{energy} \end{cases}$$

Recall that the binding energy is the amount of energy that must be supplied to the nucleus to break it apart into subatomic particles. Figure 26-10 is a plot of binding energy per nucleon versus mass number. It shows that atoms of intermediate mass number have the highest binding energies per nucleon; therefore, they are the most stable. Thus, fission is an energetically favorable process for heavy atoms, because atoms with intermediate masses and greater binding energies per nucleon are formed.

Which isotopes of which elements undergo fission? Experiments with particle accelerators have shown that every element with an atomic number of 80 or more has one or more isotopes capable of undergoing fission, provided they are bombarded at the right energy. Nuclei with atomic numbers between 89 and 98 fission spontaneously with long half-lives of 10^4 to 10^{17} years. Nuclei with atomic numbers of 98 or more fission spontaneously with shorter half-lives of a few milliseconds to 60.5 days. One of the *natural* decay modes of the transuranium elements is via spontaneous fission. In fact, all known nuclides with *mass numbers* greater than 250 do this because they are too big to be stable. Most nuclides with mass numbers between 225 and 250 do not undergo fission spontaneously (except for a few

The term "nucleon" refers to a nuclear particle, either a neutron or a proton.

1 MeV = 1.60×10^{-13} J

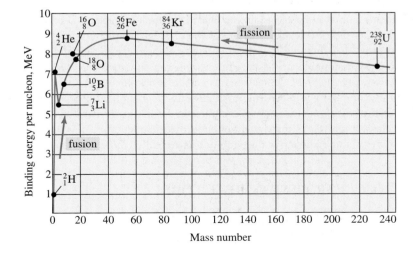

Figure 26-10 Variation in nuclear binding energy per nucleon with atomic mass. This plot shows the relative stability of the most stable isotopes of selected elements. The most stable nucleus is $^{56}_{26}\text{Fe}$, with a binding energy of 8.80 MeV per nucleon.

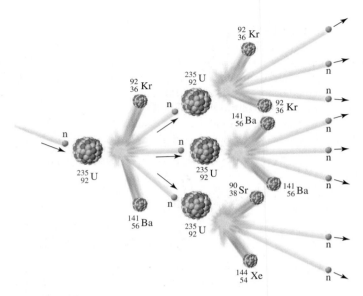

Figure 26-11 A self-propagating nuclear chain reaction. A stray neutron induces a single fission, liberating more neutrons. Each of them induces another fission, each of which is accompanied by release of two or three neutrons. The chain continues to branch in this way, very quickly resulting in an explosive rate of fission.

with extremely long half-lives). They can be induced to undergo fission when bombarded with particles of relatively low kinetic energies. Particles that can supply the required activation energy include neutrons, protons, alpha particles, and fast electrons. For nuclei lighter than mass 225, the activation energy required to induce fission rises very rapidly.

In Section 26-2 we discussed the stability of nuclei with even numbers of protons and even numbers of neutrons. We should not be surprised to learn that both ^{233}U and ^{235}U can be excited to fissionable states by slow neutrons much more easily than ^{238}U, because they are less stable. It is so difficult to cause fission in ^{238}U that this isotope is said to be "nonfissionable."

Typically, two or three neutrons are produced per fission reaction. These neutrons can collide with other fissionable atoms to repeat the process. If sufficient fissionable material, the **critical mass,** is contained in a small enough volume, a sustained **chain reaction** can result. If too few fissionable atoms are present, most of the neutrons escape and no chain reaction occurs. Figure 26-11 depicts a fission chain reaction.

In an atomic bomb, two subcritical portions of fissionable material are brought together to form a critical mass. A nuclear fission explosion results. A tremendous amount of heat energy is released, as well as many radionuclides whose effects are devastating to life and the environment. The radioactive dust and debris are called *fallout.*

26-15 NUCLEAR FISSION REACTORS

In a nuclear fission reactor, the fission reaction is controlled by inserting materials to absorb some of the neutrons so that the mixture does not explode. The energy that is produced can be safely used as a heat source in a power plant.

The launching of the nuclear submarine *Hyman G. Rickover* into the Thames River in Connecticut (August 27, 1983).

Courtesy of E. I. DuPont de Nemours Company

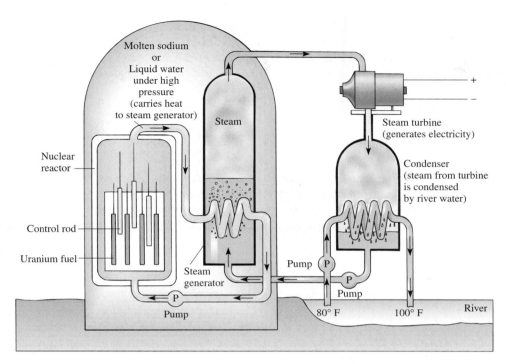

Figure 26-12 A schematic diagram of a light water reactor plant. This design includes two closed loops of water. The water that carries heat from the reactor to the steam generator is in a closed loop and is not released to the environment.

Light Water Reactors

Most commercial nuclear power plants in the United States are "light water" reactors, moderated and cooled by ordinary water. Figure 26-12 is a schematic diagram of a light water reactor plant. The reactor core at the left replaces the furnace in which coal, oil, or natural gas is burned in a fossil fuel plant. Such a fission reactor consists of five main components: (1) fuel, (2) moderator, (3) control rods, (4) cooling system, and (5) shielding.

Fuel

Rods of U_3O_8 enriched in uranium-235 serve as the fuel. Unfortunately, uranium ores contain only about 0.7% $^{235}_{92}U$. Most of the rest is nonfissionable $^{238}_{92}U$. The enrichment is done in processing and reprocessing plants by separating gaseous $^{235}UF_6$ from $^{238}UF_6$, prepared from the ore. Separation by diffusion is based on the slower rates of diffusion of heavier gas molecules (Section 12-14). Another separation procedure uses the ultracentrifuge.

A potentially more efficient method of enrichment would involve the use of sophisticated tunable lasers to ionize $^{235}_{92}U$ selectively and not $^{238}_{92}U$. The ionized $^{235}_{92}U$ could then be made to react with negative ions to form another compound, easily separated from the mixture. For this method to work, we must construct lasers capable of producing radiation monochromatic enough to excite one isotope and not the other—a difficult challenge.

Moderator

The fast neutrons ejected during fission are too energetic to be absorbed efficiently by other nuclei. Thus, they must be slowed by collisions with atoms of comparable mass that do not absorb them, called *moderators*. The most commonly used moderator is ordinary water, although graphite is sometimes used. The most efficient moderator is helium, which

Uranium is deposited on the negative electrode in the electrorefining phase of fuel reprocessing. The crystalline mass is about 97% LiCl and KCl. The remaining 3% uranium chloride is responsible for the amethyst color.

SC*LINKS*.

TOPIC: Nuclear Reactors
GO TO: www.scilinks.org
*sci*LINKS **CODE:** WCH2650

slows neutrons but does not absorb them all. The next most efficient is *heavy water* (deuterium oxide, $^2_1H_2O$ or $^2_1D_2O$). This is so expensive that it has been used chiefly in research reactors. A Canadian-designed power reactor that uses heavy water is more neutron-efficient than light water reactors. This design is the basis of the many reactors in Canada.

Control Rods

Cadmium and boron are good neutron absorbers.

$$^{10}_{5}B + ^1_0n \longrightarrow ^7_3Li + ^4_2\alpha$$

The rate of a fission reaction is controlled by the use of movable control rods, usually made of cadmium or boron steel. They are automatically inserted in or removed from spaces between the fuel rods. The more neutrons absorbed by the control rods, the fewer fissions occur and the less heat is produced. Hence, the heat output is governed by the control system that operates the rods.

Cooling System

Two cooling systems are needed. First, the moderator itself serves as a coolant for the reactor. It transfers fission-generated heat to a steam generator. This converts water to steam. The steam then goes to turbines that drive generators to produce electricity. Another coolant (river water, sea water, or recirculated water) condenses the steam from the turbine, and the condensate is then recycled into the steam generator.

The danger of meltdown arises if a reactor is shut down quickly. The disintegration of radioactive fission products still goes on at a furious rate, fast enough to overheat the fuel elements and to melt them. So it is not enough to shut down the fission reaction. Efficient cooling must be continued until the short-lived isotopes are gone and the heat from their disintegration is dissipated. Only then can the circulation of cooling water be stopped.

The 1979 accident at Three Mile Island, near Harrisburg, Pennsylvania, was due to stopping the water pumps too soon *and* the inoperability of the emergency pumps. A combination of mechanical malfunctions, errors, and carelessness produced the overheating that damaged the fuel assembly. It did not and *could not explode*, although melting of the core material did occur. The 1986 accident at Chernobyl, in the USSR, involved a reactor of a very different design and was far more serious. The effects of that disaster will continue for decades.

The neutrons are the worst problem of radiation. The human body contains a high percentage of H_2O, which absorbs neutrons very efficiently. A new weapon, the neutron bomb, produces massive amounts of neutrons and so is effective against people, but it does not produce the long-lasting radiation of the fission atomic bomb.

Shielding

It is essential that people and the surrounding countryside be adequately shielded from possible exposure to radioactive nuclides. The entire reactor is enclosed in a steel containment vessel. This is housed in a thick-walled concrete building. The operating personnel are further protected by a so-called biological shield, a thick layer of organic material made of compressed wood fibers. This absorbs the neutrons and beta and gamma rays that would otherwise be absorbed in the human body.

Breeder Reactors

The possibility of shortages in the known supply of $^{235}_{92}U$ has led to the development of **breeder reactors,** which can manufacture more fuel than they use. A breeder reactor is designed not only to generate electrical power but also to maximize neutron capture in the core by $^{238}_{92}U$. The fuel of a typical breeder reactor consists of the abundant but nonfissionable isotope $^{238}_{92}U$ mixed with $^{235}_{92}U$ or $^{239}_{94}Pu$, which produce neutrons when they undergo fission. Some of these neutrons are absorbed by $^{238}_{92}U$ to form $^{239}_{92}U$. This unstable uranium isotope soon leads, after two steps of beta emission, to $^{239}_{94}Pu$.

$$^{238}_{92}U + ^{1}_{0}n \longrightarrow {}^{239}_{92}U \xrightarrow[t_{1/2} = 23.4 \text{ min}]{\beta \text{ decay}} {}^{239}_{93}Np \xrightarrow[t_{1/2} = 2.35 \text{ days}]{\beta \text{ decay}} {}^{239}_{94}Pu$$

This fissionable $^{239}_{94}Pu$ can then be used as fuel in a reactor.

For every $^{235}_{92}U$ or $^{239}_{94}Pu$ nucleus that undergoes fission, more than one neutron is captured by $^{238}_{92}U$ to produce $^{239}_{94}Pu$. Thus, the breeder reactor can produce more fissionable material than it consumes. After about 7 years, enough $^{239}_{94}Pu$ can be collected to fuel a new reactor *and* to refuel the original one.

Nuclear Power: Hazards and Benefits

Controlled fission reactions in nuclear reactors are of great use and even greater potential. The fuel elements of a nuclear reactor have neither the composition nor the extremely compact arrangement of the critical mass of a bomb. Thus, no possibility of nuclear explosion exists. However, various dangers are associated with nuclear energy generation. The possibility of "meltdown" has been discussed with respect to cooling systems in light water reactors. Proper shielding precautions must be taken to ensure that the radionuclides produced are always contained within vessels from which neither they nor their radiations can escape. Long-lived radionuclides from spent fuel must be stored underground in heavy, shock-resistant containers until they have decayed to the point that they are no longer biologically harmful. As examples, strontium-90 ($t_{1/2}$ = 28 years) and plutonium-239 ($t_{1/2}$ = 24,000 years) must be stored for 280 years and 240,000 years, respectively, before they lose 99.9% of their activities. Critics of nuclear energy contend that the containers could corrode over such long periods, or burst as a result of earth tremors, and that transportation and reprocessing accidents could cause environmental contamination with radionuclides. They claim that river water used for cooling is returned to the rivers with too much heat (thermal pollution), thus disrupting marine life. (It should be noted, though, that fossil fuel electric power plants cause the same thermal pollution for the same amount of electricity generated.) The potential for theft also exists. Plutonium-239, a fissionable material, could be stolen from reprocessing plants and used to construct atomic weapons.

Proponents of the development of nuclear energy argue that the advantages far outweigh the risks. Nuclear energy plants do not pollute the air with oxides of sulfur, nitrogen, carbon, and particulate matter, as fossil fuel electric power plants do. The big advantage of nuclear fuels is the enormous amount of energy liberated per unit mass of fuel. At present, nuclear reactors provide about 22% of the electrical energy consumed in the United States. In some parts of Europe, where natural resources of fossil fuels are scarcer, the utilization of nuclear energy is higher. For instance, in France and Belgium, more than 80% of electrical energy is produced from nuclear reactors. With rapidly declining fossil fuel reserves, it appears likely that nuclear energy and solar energy will become increasingly important. Intensifying public concerns about nuclear power, however, may mean that further growth in energy production using nuclear power in the United States must await technological developments to overcome the remaining hazards.

Nuclear waste may take centuries to decompose, so we cannot afford to take risks in its disposal. Suggested approaches include casting it into ceramics, as shown here, to eliminate the possibility of the waste dissolving in ground water. The encapsulated waste could then be deposited in underground salt domes. Located in geologically stable areas, such salt domes have held petroleum and compressed natural gas trapped for millions of years. The political problems of nuclear waste disposal are at least as challenging as the technological ones. Refer to the Chemistry in Use box entitled "Managing Nuclear Wastes" in this chapter.

CHEMISTRY IN USE   *The Environment*

Managing Nuclear Wastes

Some may consider a career in managing nuclear waste as being just about the worst job anyone would ever want, but hundreds of technically trained people have spent years working to solve the problems associated with nuclear power. The major part of the continuing challenge is political. Nuclear power plants generate about 23% of the electricity in the United States. Most of the high-level nuclear waste (HLW) that is generated from nuclear power plants—in the form of spent nuclear fuel (SNF)—is generated where many people live, in the eastern half of the United States. The safest place for a repository is away from people, in a dry, remote location, probably in the western United States, where there are fewer people (and fewer votes!).

SNF constitutes about half of the HLW in the United States. The other half comes from the construction and existence of nuclear weapons. All HLW is a federal responsibility. About 90% of the radioactivity in nuclear waste is from HLW. The largest volume of nuclear waste is low-level waste (LLW) and that is mostly the responsibility of the state (or group of states) in which it is generated. LLW is rather awkwardly defined, being everything that is neither HLW nor defense waste and consists of wastes from hospitals; pharmaceutical labs; research labs; and the moon suits, tools, and the like from nuclear power plants. In the eastern United States, most of the LLW is in the form of the plastic beads that make up the ion-exchange resins used in nuclear power plants to clean various loops of water used in power production.

Plutonium wastes from the Los Alamos National Laboratory in northern New Mexico were trucked for the first time to the federal Waste Isolation Pilot Plant in Carlsbad in March 1999. The 600 pounds (270 kg) of waste consisted of plutonium-contaminated clothing and metal cans, packed in boxes and stainless steel containers. Most of the material was from the laboratory's manufacture of nuclear batteries used in NASA's deep space probes and will be buried in the depository carved out of ancient salt caverns about half a mile (0.8 km) below ground.

Most current attention is focused on SNF for two reasons. It is highly radioactive and it can be seen as a "local" problem because it is made where electric customers live. Europe has reprocessing plants to recover the unused fissionable material for new fuel, but the United States disallowed the practice in the 1970s. This partially explains why spent fuel rods have been piling up at U.S. nuclear plants.

Research focused on Yucca Mountain, Nevada, at the western edge of the National Test Site, for its suitability as a nuclear waste repository for SNF and some defense waste. In July 2002, after both houses of the US Congress voted to override a veto by the State of Nevada, President George W. Bush signed the bill making Yucca Mountain the central repository for the nation's nuclear waste. Many political leaders and residents of Nevada strongly opposed this plan, and they seriously question that nuclear waste can be safely kept out of the human environment for 10,000 years, as is required under the federal Nuclear Waste Policy Act.

The numbers describing SNF are barely comprehensible to most people. The volume of all existing SNF could cover a large football stadium to a depth of 4 or 5 feet, but no sensible person would want to confine that much heat and radioactivity to one place. Another description is the 70,000 metric tons of SNF generated to date in power plants, a figure that means little unless one understands thousand-kilogram quantities and knows the density of fission products. The plans for Yucca Mountain are that it will hold in its many miles of tunnels and caverns, all the SNF so far generated and expected to be generated in the next few years.

The SNF portion of HLW can be understood by chemists who see in it nearly every element on the periodic chart of the elements. After a ^{235}U nucleus undergoes fission and releases its excess nuclear binding energy, it leaves a pair of new atoms. These fission products are like newly born forms of the elements that are already well known and, like newborns, are unstable until they mature. There are about 1000 isotopes of about 100 different elements in SNF, and most are radioactive. They decay into stable elements at different rates, giving off alpha, beta, and gamma emissions. It will take about 7000 years until the SNF will be only as radioactive as the rocks and minerals that make up our planet.

These fission products are housed in long titanium rods, each about the diameter of a pencil, that constitute the fuel assembly in a nuclear power plant. Workers wearing gloves can handle fuel assemblies before fissioning occurs. But after removal from a nuclear reactor, the fuel assembly is stored in a cooling pool of water beside the reactor for at least 10 years. If the power plant has a small cooling pool, on-site storage of the oldest fuel assemblies occurs in specially constructed concrete casks until the federal government takes ownership and finds a suitable place for it. Fuel rod consolidation is sometimes practiced to save space because much of the space in a fuel assembly was present so power plant water could easily pass and pick up the heat generated by the fission process.

Other options that have been considered for HLW include outer space ejection and burial in deep ocean trenches. The consensus worldwide is that deep geological isolation is the best option. The United States leads in studying a specific site, Yucca Mountain. In other countries, even those generating a larger percentage of their power with nuclear power, the small volumes awaiting burial allow them more time to choose a location.

For the student who likes a challenge, nuclear waste management is a good one. From the cost-benefit analyses of all the ways we make and use electrical power to the way the wastes are handled, one can find an issue or a career. Here are a few key issues to study and discuss:

- *Transportation of the waste to its repository.* Should it be done by rail or by truck? Should there be public notification of the time of transport? Are there response measures in place in case of an accident?

- *The site's seismicity.* Will there be significant volcanic or seismic activity near the site in the next 10,000 years?

- *Hydrology.* Is there enough evidence to ensure that radionuclides will not seep into groundwater to any significant degree?

- *Public education.* Should conservation be taught, and should teachers promote or discourage the role of nuclear power in our nation's power mix?

- *Other technical options.* Should one investigate nuclear physics options that might transmute the long-lived radioisotopes into ones with shorter half-lives regardless of the costs?

- *Weapons disarmament.* Should the plutonium from "disarmed" nuclear weapons eventually be turned into nuclear fuel or made useless immediately and buried with other HLW?

For more information, see the following Web sites:
The Department of Energy, http://www.energy.gov
Yucca Mountain studies project, http://www.ymp.gov
The American Nuclear Society, http://www.ans.org
See also *Radwaste* magazine.

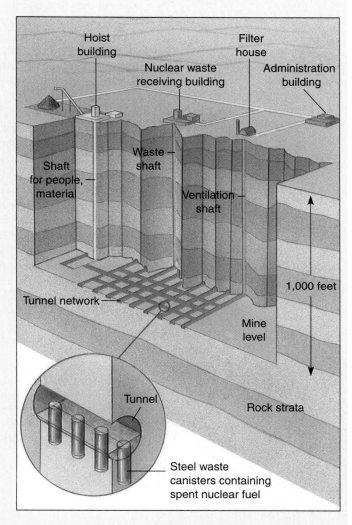

In the United States, permanent storage sites for high-level radioactive wastes will probably be deep underground in rock formations. Shown is the kind of nuclear waste facility designed for Yucca Mountain, which would be a three-square-mile complex of interconnected tunnels located in dense volcanic rock 305 meters (1000 feet) beneath the mountain.

Donald H. Williams
Professor of
Chemistry

26-16 NUCLEAR FUSION

The deuteron and triton are the nuclei of two isotopes of hydrogen, called deuterium and tritium. Deuterium occurs naturally in water. When the D_2O is purified as "heavy water," it can be used for several types of chemical analysis.

Fusion, the joining of light nuclei to form heavier nuclei, is favorable for the very light atoms. In both fission and fusion, the energy liberated is equivalent to the loss of mass that accompanies the reactions. Much greater amounts of energy per unit mass of *reacting atoms* are produced in fusion than in fission.

Spectroscopic evidence indicates that the sun is a tremendous fusion reactor consisting of 73% H, 26% He, and 1% other elements. Its major fusion reaction is thought to involve the combination of a deuteron, ^2_1H, and a triton ^3_1H, at tremendously high temperatures to form a helium nucleus and a neutron.

$$^2_1H + ^3_1H \longrightarrow ^4_2He + ^1_0n + \text{energy}$$

Thus, solar energy is actually a form of fusion energy.

Fusion reactions are accompanied by even greater energy production per unit mass of reacting atoms than are fission reactions. They can be initiated only by extremely high temperatures, however. The fusion of ^2_1H and ^3_1H occurs at the lowest temperature of any fusion reaction known, but even this is 40,000,000 K! Such temperatures exist in the sun and other stars, but they are nearly impossible to achieve and contain on earth. **Thermonuclear** bombs (called fusion bombs or hydrogen bombs) of incredible energy have been detonated in tests but, thankfully, never in war. In them the necessary activation energy is supplied by the explosion of a fission bomb.

It is hoped that fusion reactions can be harnessed to generate energy for domestic power. Because of the tremendously high temperatures required, no currently known structural material can confine these reactions. At such high temperatures all molecules dissociate and most atoms ionize, resulting in the formation of a new state of matter called a

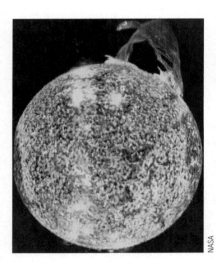

NASA

Our sun supplies energy to the earth from a distance of 93,000,000 miles. Like other stars, it is a giant nuclear fusion reactor. Much of its energy comes from the fusion of deuterium, ^2_1H, producing helium, ^4_2He.

The explosion of a thermonuclear (hydrogen) bomb releases tremendous amounts of energy. If we could learn how to control this process, we would have nearly limitless amounts of energy.

Courtesy of National Atomic Museum

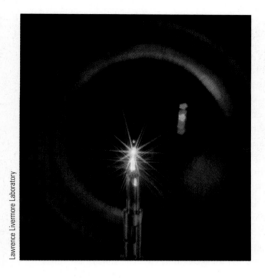

Lawrence Livermore Laboratory

Nuclear fusion provides the energy of our sun and other stars. Development of controlled fusion as a practical source of energy requires methods to initiate and contain the fusion process. Here a very powerful laser beam has initiated a fusion reaction in a 1-mm target capsule that contained deuterium and tritium. In a 0.5-picosecond burst, 10^{13} neutrons were produced by the reaction $^2_1H + ^3_1H \longrightarrow ^4_2He + ^1_0n$

plasma. A very high temperature plasma is so hot that it melts and decomposes anything it touches, including structural components of a reactor. The technological innovation required to build a workable fusion reactor probably represents the greatest challenge ever faced by the scientific and engineering community.

Plasmas have been called the fourth state of matter.

Recent attempts at the containment of lower-temperature plasmas by external magnetic fields have been successful, and they encourage our hopes. Fusion as a practical energy source, however, lies far in the future at best. The biggest advantages of its use would be that (1) the deuterium fuel can be found in virtually inexhaustible supply in the oceans; and (2) fusion reactions would produce only radionuclides of very short half-life, primarily tritium ($t_{1/2} = 12.3$ years), so there would be no long-term waste disposal problem. If controlled fusion could be brought about, it could liberate us from dependence on uranium and fossil fuels.

Courtesy of Princeton Physics Laboratory

The plasma in a fusion reactor must not touch the walls of its vacuum vessel, which would be vaporized. In the Tokamak fusion test reactor, the plasma is contained within a magnetic field shaped like a doughnut. The magnetic field is generated by D-shaped coils around the vacuum vessel.

Key Terms

Alpha emission   Radioactive decay in which an alpha particle is given off from the nucleus.

Alpha particle (α)   A particle that consists of two protons and two neutrons; identical to a helium nucleus.

Artificial transmutation   An artificially induced nuclear reaction caused by bombardment of a nucleus with subatomic particles or small nuclei.

Band of stability   A band containing stable (nonradioactive) nuclides in a plot of number of neutrons versus number of protons (atomic number).

Beta emission   Radioactive decay in which an electron is ejected *from the nucleus* to convert a neutron into a proton.

Beta particle (β)   An electron emitted from the nucleus when a neutron decays to a proton and an electron.

Binding energy (nuclear binding energy)   The energy equivalent ($E = mc^2$) of the mass deficiency of an atom.

Breeder reactor   A fission reactor that produces more fissionable material than it consumes.

Chain reaction   A reaction that, once initiated, sustains itself and expands.

Cloud chamber   A device for observing the paths of speeding particles as vapor molecules condense on the ionized air molecules in their tracks.

Control rods   Rods of materials such as cadmium or boron steel that act as neutron absorbers (not merely moderators), used in nuclear reactors to control neutron fluxes and therefore rates of fission.

Critical mass   The minimum mass of a particular fissionable nuclide, in a given volume, that is required to sustain a nuclear chain reaction.

Cyclotron   A device for accelerating charged particles along a spiral path.

Daughter nuclide   A nuclide that is produced in a nuclear decay.

Electron capture   Absorption of an electron from the first energy level (K shell) by a proton as it is converted to a neutron; also K capture; also positron emission.

Fast neutron   A neutron ejected at high kinetic energy in a nuclear reaction.

Fluorescence   Absorption of high-energy radiation by a substance and the subsequent emission of visible light.

Gamma ray (γ)   High-energy electromagnetic radiation.

Geiger-Müller counter   A type of gas ionization counter used to detect radiation.

Half-life of a radionuclide   The time required for half of a given sample to undergo radioactive decay.

Heavy water   Water containing deuterium, a heavy isotope of hydrogen, ^2_1H.

Linear accelerator   A device used for accelerating charged particles along a straight-line path.

Mass deficiency   The amount of matter that would be converted into energy if an atom were formed from constituent particles.

Moderator   A substance such as hydrogen, deuterium, oxygen, or paraffin capable of slowing fast neutrons upon collision.

Nuclear binding energy   The energy equivalent of the mass deficiency; energy released in the formation of an atom from subatomic particles.

Nuclear fission   The process in which a heavy nucleus splits into nuclei of intermediate masses and one or more neutrons are emitted.

Nuclear fusion   The combination of light nuclei to produce a heavier nucleus.

Nuclear reaction   A reaction involving a change in the composition of a nucleus; it can evolve or absorb an extraordinarily large amount of energy.

Nuclear reactor   A system in which controlled nuclear fission reactions generate heat energy on a large scale. The heat energy is subsequently converted into electrical energy.

Nucleons   Particles comprising the nucleus; protons and neutrons.

Nuclides   Different atomic forms of all elements (in contrast to isotopes, which are different atomic forms of a single element).

Parent nuclide   A nuclide that undergoes nuclear decay.

Plasma   A physical state of matter that exists at extremely high temperatures, in which all molecules are dissociated and most atoms are ionized.

Positron   A nuclear particle with the mass of an electron but opposite charge.

Positron emission   See *Electron capture*.

Radiation   High-energy particles or rays emitted in nuclear decay processes.

Radioactive dating   A method of dating ancient objects by determining the ratio of amounts of a parent nuclide and one of its decay products present in an object and relating the ratio to the object's age via half-life calculations.

Radioactive tracer   A small amount of radioisotope that replaces a nonradioactive isotope of the element in a compound whose path (e.g., in the body) or whose decomposition products are to be monitored by detection of radioactivity; also called a radioactive label.

Radioactivity   The spontaneous disintegration of atomic nuclei.

Radioisotope   A radioactive isotope of an element.

Radionuclide   A radioactive nuclide.

Scintillation counter   A device used for the quantitative detection of radiation.

Slow neutron   A fast neutron slowed by collision with a moderator.

Thermonuclear energy   Energy from nuclear fusion reactions.

Transuranium elements   The elements with atomic numbers greater than 92 (uranium); none occurs naturally and all must be prepared by nuclear bombardment of other elements.

Exercises

Nuclear Stability and Radioactivity

1. Define and compare nuclear fission and nuclear fusion. Briefly describe current uses of nuclear fission and fusion.

2. Differentiate between natural and induced radioactivity. Use the periodic table to identify the locations of those elements that are the result of induced radioactivity.

3. How do nuclear reactions differ from ordinary chemical reactions?

4. What is the equation that relates the equivalence of matter and energy? What does each term in this equation represent?

5. What is mass deficiency? What is binding energy? How are the two related?

6. What are nucleons? What is the relationship between the number of protons and the atomic number? What is the relationship among the number of protons, the number of neutrons, and the mass number?

7. Define the term "binding energy per nucleon." How can this quantity be used to compare the stabilities of nuclei?

8. Describe the general shape of the plot of binding energy per nucleon against mass number.

9. (a) Briefly describe a plot of the number of neutrons against the atomic number (for the stable nuclides). Interpret the observation that the plot shows a band with a somewhat step-like shape. (b) Describe what is meant by "magic numbers" of nucleons.

10. Potassium, $Z = 19$, has a series of naturally occurring isotopes: ^{39}K, ^{40}K, ^{41}K. Identify the isotope(s) of potassium that is/are most likely to be stable and which tend to decay.

11. Platinum, $Z = 78$, has a series of naturally occurring isotopes: ^{190}Pt, ^{192}Pt, ^{194}Pt, ^{195}Pt, ^{196}Pt, and ^{198}Pt. Identify the isotope(s) of platinum that is/are most likely to be stable and which tend to decay.

12. Indicate the type of emission and the decay product predicted for each unstable isotope listed in Exercise 10.

13. Indicate the type of emission and the decay product predicted for each unstable isotope listed in Exercise 11.

14. The actual mass of a ^{62}Ni atom is 61.9283 amu. (a) Calculate the mass deficiency in amu/atom and in g/mol for this isotope. (b) What is the nuclear binding energy in kJ/mol for this isotope?

15. The actual mass of a ^{108}Pd atom is 107.90389 amu. (a) Calculate the mass deficiency in amu/atom and in g/mol for this isotope. (b) What is the nuclear binding energy in kJ/mol for this isotope?

16. Calculate the following for ^{64}Zn (actual mass = 63.9291 amu). (a) mass deficiency in amu/atom; (b) mass deficiency in g/mol; (c) binding energy in J/atom; (d) binding energy in kJ/mol; (e) binding energy in MeV/nucleon.

17. Calculate the following for ^{49}Ti (actual mass = 48.94787 amu). (a) mass deficiency in amu/atom; (b) mass deficiency in g/mol; (c) binding energy in J/atom; (d) binding energy in kJ/mol; (e) binding energy in MeV/nucleon.

18. Calculate the nuclear binding energy in kJ/mol for each of the following (a) $^{127}_{53}$I; (b) $^{81}_{35}$Br; (c) $^{35}_{17}$Cl. The atomic masses are 126.9044 amu, 80.9163 amu, and 34.96885 amu, respectively.

19. Repeat Exercise 18 for (a) $^{36}_{16}$S, (b) $^{39}_{19}$K, and (c) $^{24}_{12}$Mg. Their respective atomic masses are 35.96709 amu, 38.96371 amu, and 23.98504 amu. Which of these nuclides has the greatest binding energy per nucleon?

20. Compare the behaviors of α, β, and γ radiation (a) in an electrical field, (b) in a magnetic field, and (c) with respect to ability to penetrate various shielding materials, such as a piece of paper and concrete. What is the composition of each type of radiation?

21. Why are α-particles that are absorbed internally by the body particularly dangerous?

22. Name some radionuclides that have medical uses, and list the uses.

23. Describe how radionuclides can be used in (a) research, (b) agriculture, and (c) industry.

24. Name and describe four methods for detection of radiation.

25. Describe how (a) nuclear fission and (b) nuclear fusion generate more stable nuclei.

26. What evidence exists to support the theory that nucleons are arranged in "shells" or energy levels within the nucleus?

Nuclear Reactions

27. Consider a radioactive nuclide with a neutron/proton ratio that is larger than those for the stable isotopes of that element. What mode(s) of decay might be expected for this nuclide, and why?

28. Repeat Exercise 27 for a nuclide with a neutron/proton ratio that is smaller than those for the stable isotopes.

29. Calculate the neutron/proton ratio for each of the following radioactive nuclides, and predict how each of the nuclides might decay: (a) $^{13}_{5}$B (stable mass numbers for B are 10 and 11); (b) $^{92}_{38}$Sr (stable mass numbers for Sr are between 84 and 88); (c) $^{192}_{82}$Pb (stable mass numbers for Pb are between 204 and 208).

30. Repeat Exercise 29 for (a) $^{193}_{79}$Au (stable mass number for Au is 197), (b) $^{189}_{75}$Re (stable mass numbers for Re are 185 and 187), and (c) $^{137}_{59}$Pr (stable mass number for Pr is 141).

31. Write the symbols for the daughter nuclei in the following radioactive decays (β refers to an e^-).

(a) ^{125}Sn $\xrightarrow{-\beta}$ (d) ^{147}Sm $\xrightarrow{-\alpha}$

(b) ^{13}C $\xrightarrow{-n}$ (e) ^{184}Ir $\xrightarrow{-\beta}$

(c) ^{11}B $\xrightarrow{-\gamma}$ (f) ^{40}K $\xrightarrow{+\beta}$

32. Predict the kind of decays you would expect for the following radionuclides. (a) $^{60}_{27}Co$ (n/p ratio too high); (b) $^{20}_{11}Na$ (n/p ratio too low); (c) $^{222}_{86}Rn$; (d) $^{67}_{29}Cu$; (e) $^{238}_{92}U$; (f) $^{11}_{6}C$.

33. What are nuclear bombardment reactions? Explain the shorthand notation used to describe bombardment reactions.

34. Fill in the missing symbols in the following nuclear bombardment reactions.

(a) $^{23}_{11}Na + ? \longrightarrow ^{23}_{12}Mg + ^{1}_{0}n$

(b) $^{59}_{27}Co + ^{1}_{0}n \longrightarrow ^{56}_{25}Mn + ?$

(c) $^{96}_{42}Mo + ^{4}_{2}He \longrightarrow ^{100}_{43}Tc + ?$

(d) $^{209}_{83}Bi + ? \longrightarrow ^{210}_{84}Po + ^{1}_{0}n$

(e) $^{238}_{92}U + ^{16}_{8}O \longrightarrow ? + 5 ^{1}_{0}n$

35. Fill in the missing symbols in the following nuclear bombardment reactions.

(a) $^{232}_{90}Th + ? \longrightarrow ^{240}_{96}Cm + 4 ^{1}_{0}n$

(b) $? + ^{1}_{1}H \longrightarrow ^{29}_{14}Si + ^{0}_{0}\gamma$

(c) $^{26}_{?}Mg + ? \longrightarrow ^{26}_{?}Al + ^{1}_{0}n$

(d) $^{40}_{18}Ar + ? \longrightarrow ^{43}_{?}K + ^{1}_{1}H$

36. Write the symbols for the daughter nuclei in the following nuclear bombardment reactions. (a) $^{60}_{28}Ni$ (n, p); (b) $^{98}_{42}Mp$ ($^{1}_{0}n, \beta$); (c) $^{35}_{17}Cl$ (p, α).

37. Write the symbols for the daughter nuclei in the following nuclear bombardment reactions. (a) $^{20}_{10}N$ (α, γ); (b) $^{15}_{7}Ne$ (p, α); (c) $^{10}_{5}B$ (n, α)

38. Write the nuclear equation for each of the following bombardment processes. (a) $^{14}_{7}N$ (α, p) $^{17}_{8}O$; (b) $^{106}_{46}Pd$ (n, p) $^{106}_{45}Rh$; (c) $^{23}_{11}Na$ (n, β^-)X. Identify X.

39. Repeat Exercise 38 for the following. (a) $^{113}_{48}Cd$ (n, α) $^{114}_{48}Cd$; (b) $^{6}_{3}Li$ (n, α) $^{3}_{1}H$; (c) $^{2}_{1}H$ (γ, p)X. Identify X.

40. Write the shorthand notation for each of the following nuclear reactions.

(a) $^{6}_{3}Li + ^{1}_{0}n \longrightarrow ^{4}_{2}He + ^{3}_{1}H$

(b) $^{31}_{15}P + ^{2}_{1}H \longrightarrow ^{32}_{15}P + ^{1}_{1}H$

(c) $^{238}_{92}U + ^{1}_{0}n \longrightarrow ^{239}_{93}Np + ^{0}_{-1}e$

41. Repeat Exercise 40 for the following.

(a) $^{253}_{99}Es + ^{4}_{2}He \longrightarrow ^{256}_{101}Md + ^{1}_{0}n$

(b) $^{27}_{13}Al + ^{1}_{0}n \longrightarrow ^{26}_{13}Al + 2 ^{1}_{0}n$

(c) $^{37}_{17}Cl + ^{1}_{1}H \longrightarrow ^{1}_{0}n + ^{37}_{18}Ar$

42. Write the nuclear equations for the following processes. (a) $^{63}_{28}Ni$ undergoing β^- emission; (b) two deuterium ions undergoing fusion to give $^{3}_{2}He$ and a neutron; (c) a nuclide being bombarded by a neutron to form $^{7}_{3}Li$ and an α-particle (identify the unknown nuclide); (d) $^{14}_{7}N$ being bombarded by a neutron to form three α-particles and an atom of tritium.

43. Write the nuclear equations for the following processes. (a) $^{220}_{86}Rn$ undergoing α decay; (b) $^{110}_{49}In$ undergoing positron emission; (c) $^{127}_{53}I$ being bombarded by a proton to form $^{121}_{54}Xe$ and seven neutrons; (d) tritium and deuterium undergoing fusion to form an α-particle and a neutron; (e) $^{95}_{42}Mo$ being bombarded by a proton to form $^{95}_{43}Tc$ and radiation (identify this radiation).

44. "Radioactinium" is produced in the actinium series from $^{235}_{92}U$ by the successive emission of an α-particle, a β^--particle, an α-particle, and a β^--particle. What are the symbol, atomic number, and mass number for "radioactinium?"

45. An alkaline earth element (Group IIA) is radioactive. It undergoes decay by emitting three α-particles in succession. In what periodic table group is the resulting element found?

46. A nuclide of element rutherfordium, $^{257}_{104}Rf$, is formed by the nuclear reaction of californium-249 and carbon-12, with the emission of four neutrons. This new nuclide rapidly decays by emitting an α-particle. Write the equation for each of these nuclear reactions.

47. Supply the missing information to each of the following equations.

(a) $^{53}_{24}Cr + ^{4}_{2}He \longrightarrow ^{1}_{0}n + ?$

(b) $^{187}_{75}Re + \beta \longrightarrow ?$

(c) $^{243}_{95}Am + ^{1}_{0}n \longrightarrow ^{244}_{96}Cm + ? + \gamma$

(d) $^{35}_{17}Cl + ? \longrightarrow ^{32}_{16}S + ^{4}_{2}H$

48. Supply the missing information to each of the following reactions.

(a) $^{14}_{?}N + ? \longrightarrow ^{17}_{?}O + p$

(b) $^{235}_{92}U + ^{1}_{0}n \longrightarrow ^{137}_{54}? + 2 ^{1}_{0}n + ?$

(c) $^{241}_{95}Am + ^{12}_{?}C \longrightarrow 4 ^{1}_{0}n + ?$

49. Describe how (a) cyclotrons and (b) linear accelerators work.

Rates of Decay

50. What does the half-life of a radionuclide represent? How do we compare the relative stabilities of radionuclides in terms of half-lives?

51. Why must all radioactive decays be first order?

52. Describe the process by which steady-state (constant) ratios of carbon-14 to (nonradioactive) carbon-12 are attained in living plants and organisms. Describe the method of radiocarbon dating. What factors limit the use of this method?

53. The half-life of $^{19}_{8}O$ is 29 s. What fraction of the isotope originally present would be left after 15.0 s?

54. The half-life of $^{11}_{6}C$ is 20.3 min. How long will it take for 92.5% of a sample to decay? How long will it take for 99.0% of the sample to decay?

55. The activity of a sample of tritium decreased by 5.5% over the period of a year. What is the half-life of $_1^3H$?

56. A very unstable isotope of beryllium, ^8Be, undergoes emission with a half-life of 0.07 fs. How long does it take for 99.90% of a 2.0-μg sample of ^8Be to undergo decay?

57. The $_6^{14}C$ activity of an artifact from a burial site was 8.6/min · g C. The half-life of $_6^{14}C$ is 5730 years, and the current $_6^{14}C$ activity is 12.3/min · g C (that is, 12.3 disintegrations per minute per gram of carbon). How old is the artifact?

58. A piece of wood from a burial site was analyzed using $_6^{14}C$ dating and was found to have an activity of 10.8/min · g C. Using the data given in Exercise 57 for $_6^{14}C$ dating, determine the age of this piece of wood.

59. Analysis of an ant found in a piece of amber provided 14.0 disintegrations of $_6^{14}C$/min · g C, whereas a living ant of the same species produces 16.0 disintegrations per minute. Calculate the approximate age of the fossilized ant. The half-life of carbon-14 is 5730 years.

60. A skeleton was found in the woods and the police would like to place the approximate time of death. A sample of bone from the skeleton produces 12.5 disintegrations per min per g of C. Bone of recent origin produces 14.0 disintegrations per min · per g C. Calculate the approximate age of the bone sample. The half-life of carbon-14 is 5730 years.

61. Strontium-90 is one of the harmful radionuclides that results from nuclear fission explosions. It decays by beta emission with a half-life of 28 years. How long would it take for 99.9% of a given sample released in an atmospheric test of an atomic bomb to disintegrate?

62. Carbon-14 decays by beta emission with a half-life of 5730 years. Assuming a particular object originally contained 8.50 μg of carbon-14 and now contains 0.80 μg of carbon-14, how old is the object?

Fission and Fusion

63. Briefly describe a nuclear fission process. What are the two most important fissionable materials?

64. What is a chain reaction? Why are nuclear fission processes considered chain reactions? What is the critical mass of a fissionable material?

65. Where have continuous nuclear fusion processes been observed? What is the main reaction that occurs in such sources?

66. The reaction that occurred in the first fusion bomb was $_3^7Li$ (p, α) X. (a) Write the complete equation for the process, and identify the product, X. (b) The atomic masses are 1.007825 amu for $_1^1H$, 4.00260 amu for α, and 7.01600 amu for $_3^7Li$. Find the energy for the reaction in kJ/mol.

67. Summarize how an atomic bomb works, including how the nuclear explosion is initiated.

68. Discuss the pros and cons of the use of nuclear energy instead of other, more conventional types of energy based on fossil fuels.

69. Describe and illustrate the essential features of a light water fission reactor.

70. How is fissionable uranium-235 separated from nonfissionable uranium-238?

71. Distinguish between moderators and control rods of nuclear reactors.

72. What are the major advantages and disadvantages of fusion as a potential energy source compared with fission? What is the major technological problem that must be solved to permit development of a fusion reactor?

Mixed Exercises

*73. Calculate the binding energy, in kJ/mol of nucleons, for the following isotopes. (a) $_8^{15}O$ with a mass of 15.00300 amu; (b) $_8^{16}O$ with a mass of 15.99491 amu; (c) $_8^{17}O$ with a mass of 16.99913 amu; (d) $_8^{18}O$ with a mass of 17.99915 amu; (e) $_8^{19}O$ with a mass of 19.0035 amu. Which of these would you expect to be most stable?

*74. The first nuclear transformation (discovered by Rutherford) can be represented by the shorthand notation $_7^{14}N$ (α, p) $_8^{17}O$. (a) Write the corresponding nuclear equation for this process. The respective atomic masses are 14.00307 amu for $_7^{14}N$, 4.00260 amu for $_2^4He$, 1.007825 amu for $_1^1H$, and 16.99913 amu for $_8^{17}O$. (b) Calculate the energy change of this reaction in kJ/mol.

75. A proposed series of reactions (known as the carbon-nitrogen cycle) that could be important in the very hottest region of the interior of the sun is

$$^{12}C + {}^1H \longrightarrow A + \gamma$$
$$A \longrightarrow B + {}_{+1}^0e$$
$$B + {}^1H \longrightarrow C + \gamma$$
$$C + {}^1H \longrightarrow D + \gamma$$
$$D \longrightarrow E + {}_{+1}^0e$$
$$E + {}^1H \longrightarrow {}^{12}C + F$$

Identify the species labeled A–F.

*76. The ultimate stable product of ^{238}U decay is ^{206}Pb. A certain sample of pitchblende ore was found to contain ^{238}U and ^{206}Pb in the ratio 67.8 atoms ^{238}U:32.2 atoms ^{206}Pb. Assuming that all the ^{206}Pb arose from uranium decay, and that no U or Pb has been lost by weathering, how old is the rock? The half-life of ^{238}U is 4.51 × 10⁹ years.

*77. Potassium-40 decays to ^{40}Ar with a half-life of 1.3 × 10⁹ years. What is the age of a lunar rock sample that contains these isotopes in the ratio 33.4 atoms ^{40}K:66.6 atoms ^{40}Ar? Assume that no argon was originally in the sample and that none has been lost by weathering.

CONCEPTUAL EXERCISES

78. Both nuclear and conventional power plants produce wastes to which the environment is sensitive. Discuss these wastes and the current problems created by the wastes.
79. If the earth is 4.5×10^9 years old and the amount of radioactivity in a sample becomes smaller with time, how is it possible for there to be any radioactive elements left on earth that have half-lives of less than a few million years?

BUILDING YOUR KNOWLEDGE

80. Show by calculation which reaction produces the larger amount of energy per atomic mass unit of material reacting.

    fission: $^{235}_{92}U + ^{1}_{0}n \longrightarrow ^{94}_{40}Zr + ^{140}_{58}Ce + 6\,^{0}_{+1}e + 2\,^{1}_{0}n$

    fusion: $2\,^{2}_{1}H \longrightarrow ^{3}_{1}H + ^{1}_{1}H$

    The atomic masses are 235.0439 amu for $^{235}_{92}U$; 93.9061 amu for $^{94}_{40}Zr$; 139.9053 amu for $^{140}_{58}Ce$; 3.01605 amu for $^{3}_{1}H$; 1.007825 amu for $^{1}_{1}H$; 2.0140 amu for $^{2}_{1}H$.

81. What would be the volume of helium, measured at STP, generated from the decay of the ^{8}Be sample in Exercise 56?
82. Why is carbon-14 dating not accurate for estimating the age of material more than 50,000 years old?

BEYOND THE TEXTBOOK

Go to the textbook website at

http://www.brookscole.com/chemistry/whitten

for additional activities and exercises based on the General Chemistry Interactive CD-ROM, the World Wide Web, and library resources.

InfoTrac College Edition

For additional readings, go to InfoTrac College Edition, your online research library at:

http://infotrac.thomsonlearning.com

Waxes are high-molecular-weight straight-chain alkanes, alcohols, and carboxylic acids.

OUTLINE

OBJECTIVES

After you have studied this chapter, you should be able to

- Describe saturated hydrocarbons (alkanes and cycloalkanes)—their structures and their nomenclature
- Describe unsaturated hydrocarbons (alkenes and alkynes)—their structures and their nomenclature
- Draw and name constitutional isomers of alkanes
- Draw and name constitutional and geometric isomers of alkenes
- Describe and name some aromatic hydrocarbons—benzene, condensed aromatics, and substituted aromatic compounds
- Describe some common functional groups and name compounds from the following classes

  halides
  alcohols and phenols
  ethers
  aldehydes and ketones
  amines
  carboxylic acids and some of their derivatives
- Recognize examples of the three fundamental classes of organic reactions: substitution, addition, and elimination
- Recall some uses of three fundamental classes of organic reactions and some compounds that are prepared by each class of reaction
- Recognize some common polymers and the reactions by which they are formed; identify the monomers from which they are formed

O rganic chemistry is the study of carbon-containing compounds that contain C—C or C—H bonds, and sometimes a few other elements such as oxygen, nitrogen, sulfur, phosphorus, and the halogens. Why is one entire branch of chemistry devoted to the behavior of the compounds of just one element? The answer is twofold: (1) There are many more compounds that contain carbon than there are compounds that do not (more than 11 million organic compounds have been identified), and (2) the molecules containing carbon can be so much larger and more complex (a methane molecule contains five atoms per molecule and DNA contains tens of billions of atoms per molecule).

Originally the term "organic" was used to describe compounds of plant or animal origin. "Inorganic" was used to describe compounds from minerals. In 1828, Friedrich Wöhler (1800–1882) synthesized urea by boiling ammonium cyanate with water.

$$\underset{\substack{\text{ammonium cyanate} \\ \text{(an inorganic compound)}}}{NH_4OCN} \xrightarrow[\text{boil}]{H_2O} \underset{\substack{\text{urea} \\ \text{(an organic compound)}}}{H_2N-\overset{\displaystyle O}{\overset{\|}{C}}-NH_2}$$

This disproved the theory that held that organic compounds could be made only by living things. Today many organic compounds are manufactured from inorganic materials.

We encounter organic chemistry in every aspect of our lives. All life is based on a complex interrelationship of thousands of organic substances—from simple compounds such as sugars, amino acids, and fats to vastly more complex ones such as the enzymes that catalyze life's chemical reactions and the huge DNA molecules that carry genetic information from one generation to the next. The food we eat (including many additives); the clothes we wear; the plastics and polymers that are everywhere; our life-saving medicines; the paper on which we write; our fuels; many of our poisons, pesticides, dyes, soaps and detergents—all involve organic chemistry.

We normally think of petroleum and natural gas as fuel sources, but most synthetic organic materials are also derived from these two sources. More than half of the top 50 commercial chemicals are organic compounds derived in this way. Indeed, petroleum and natural gas may one day be more valuable as raw materials for organic synthesis than as fuel sources. If so, we may greatly regret our delay in developing alternative energy sources while burning up vast amounts of our petroleum and natural gas deposits as fuels.

A carbon atom has four electrons in its outermost shell with ground state configuration $1s^2 2s^2 2p^2$. The C atom can attain a stable configuration by forming four covalent bonds. As we saw in Chapter 8, each C atom can form single, double, or triple bonds by utilizing various hybridizations. The bonding of carbon is summarized in Table 27-1 using examples that we saw in Chapters 7 and 8. Carbon is unique among the elements in the extent to which it forms bonds between like atoms and in the diversity of compounds that are formed. Bonds between atoms of the same element rarely occur for any element except carbon. Carbon atoms form long chains, branched chains, and rings that may also have chains attached to them. A tremendous variety of carbon-containing compounds is known.

Although millions of organic compounds are known, the elements they contain are very few: C and H; often N, O, S, P, or a halogen; and sometimes another element. The great number and variety of organic compounds are a result of the many different arrangements of atoms, or *structures*, that are possible. The chemical and physical properties of organic compounds are related to the structures of their molecules. Thus, the basis for organizing and understanding organic chemistry is our understanding of structure and bonding.

In this text we shall give only an introduction to organic chemistry. In this chapter we organize organic compounds into the most common classes or "families" according to their structural features and learn to name various types of compounds. In this chapter and the next, we will present a few typical reactions that organic substances undergo.

CD-ROM   Screen 11.1, Organic Chemistry: Chapter Opener.

Urea, $H_2N-CO-NH_2$, is the principal end product of metabolism of nitrogen-containing compounds in mammals. It is eliminated in the urine. An adult man excretes about 30 grams of urea in 24 hours.

CD-ROM   To see many molecular models that you can rotate on-screen, open any chapter on the CD-ROM; then click on "Molecular Models" under the "Tools" menu.

CD-ROM   Screen 11.2, Carbon–Carbon Bonds.

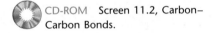

*SCi*LINKS.   **TOPIC:** Organic Compounds
GO TO: www.scilinks.org
*sci*LINKS CODE: WCH2710

TABLE 27-1	Hybridization of Carbon in Covalent Bond Formation			
Hybridization and Resulting Geometry	Orbitals Used by Each C Atom	Bonds Formed by Each C Atom		Example
sp^3, tetrahedral	four sp^3 hybrids	four σ bonds	ethane	H H | | H—C—C—H | | H H
sp^2, trigonal planar	three sp^2 hybrids, one p orbital	three σ bonds, one π bond	ethylene	H ⟍ ⟋ H C=C H ⟋ ⟍ H
sp, linear	two sp hybrids, two p orbitals	two σ bonds, two π bonds	acetylene	H—C≡C—H

CD-ROM   Screen 11.3,
Hydrocarbons.

Organic molecules are based on a framework of carbon–carbon and carbon–hydrogen bonds. Many compounds contain *only* the two elements C and H; they are called **hydrocarbons.** Hydrocarbons that contain a delocalized ring of pi bonds such as the benzene ring (Section 9-6) are called **aromatic hydrocarbons.** Those that do not contain such delocalized systems are called **aliphatic hydrocarbons.** Aliphatic hydrocarbons that contain only sigma (σ) bonds (i.e., only single bonds) are called **saturated hydrocarbons.** Those that contain both sigma and pi (π) bonds (i.e., double, triple, or delocalized bonds) are called **unsaturated hydrocarbons.** These classifications are diagrammed in Figure 27-1. The first seven sections of this chapter are devoted to the study of hydrocarbons.

A **functional group** is a special arrangement of atoms within an organic molecule that is responsible for some characteristic chemical behavior (reactions) of the compound. Different molecules that contain the same functional groups have similar chemical behavior. We shall follow the study of hydrocarbons with a presentation of some important characteristic functional groups.

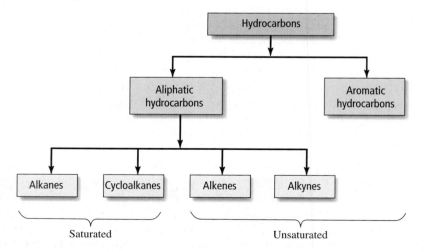

Figure 27-1   Classification of hydrocarbons.

SATURATED HYDROCARBONS

27-1 ALKANES AND CYCLOALKANES

The *saturated hydrocarbons,* or **alkanes,** are compounds in which each carbon atom is bonded to four other atoms. **Cycloalkanes** are saturated hydrocarbons that contain carbon atoms joined to form a ring.

The term "saturated" comes from early studies in which chemists tried to add hydrogen to various organic substances. Those to which no more hydrogen could be added were called saturated, by analogy with saturated solutions.

Alkanes

Each H atom in a saturated hydrocarbon is bonded to only one C atom. Saturated hydrocarbons contain only single bonds. Petroleum and natural gas are composed mostly of saturated hydrocarbons.

In Section 8-7 we examined the structure of the simplest alkane, *methane*, CH_4. We saw that methane molecules are tetrahedral with sp^3 hybridization at carbon (Figure 27-2).

Ethane, C_2H_6, is the next simplest saturated hydrocarbon. Its structure is quite similar to that of methane. Two carbon atoms share a pair of electrons. Each carbon atom also shares an electron pair with each of three hydrogen atoms. Both carbon atoms are sp^3 hybridized (Figure 27-3). *Propane*, C_3H_8, is the next member of the family (Figure 27-4).

$SC\overset{\bullet}{L}_{INKS.}$

TOPIC: Alkanes
GO TO: www.scilinks.org
*sci***LINKS CODE:** WCH2720

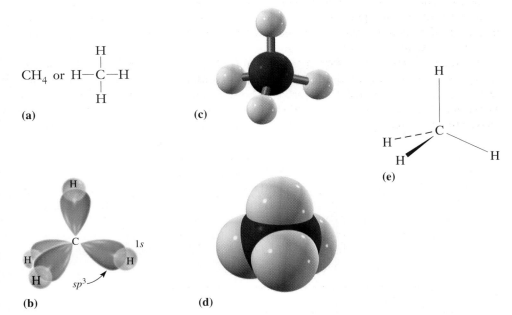

(a)

CH_4 or $H-\overset{\displaystyle H}{\underset{\displaystyle H}{\overset{|}{\underset{|}{C}}}}-H$

(c)

(e)

(b)

(d)

Figure 27-2 Representations of a molecule of methane, CH_4. (a) The condensed and Lewis formulas for methane. (b) The overlap of the four sp^3 carbon orbitals with the *s* orbitals of four hydrogen atoms forms a tetrahedral molecule. (c) A ball-and-stick model, (d) a space-filling model of methane, and (e) a three-dimensional representation that uses the wedged line to indicate a bond coming forward and a dashed line to represent a bond projecting backward.

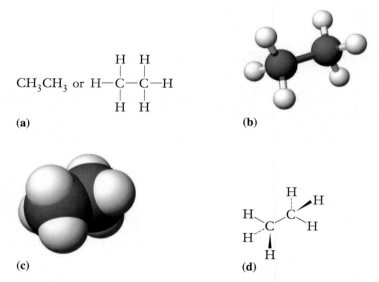

Figure 27-3   Models of ethane, C_2H_6. (a) The condensed and Lewis formulas for ethane, (b) a ball-and-stick model, (c) a space-filling model of ethane, and (d) a 3-D representation.

Iso = "same"; *mer* = "part." As we saw in Sections 25-6 and 25-7, isomers are substances that have the same numbers and kinds of atoms arranged differently. Isomerism in organic compounds is discussed more systematically in Chapter 28.

Two different compounds have the formula C_4H_{10} but different structures and, hence, different properties. Such *isomers* result when two molecules contain the same atoms bonded together in different orders. The structures of these two isomeric C_4H_{10} alkanes are shown in Figure 27-5. These two structures correspond to the two ways in which a hydrogen atom can be removed from a propane molecule and replaced by a —CH_3 group. If a —CH_3 replaces an H on either of the end carbon atoms, the result is butane. It is considered to be a straight-chain hydrocarbon because there is no branching. Though somewhat misleading, the term "straight chain" is widely used. The carbon chains are linear only in the structural formulas that we write. They are actually zigzag due to the tetrahedral bond angles at each carbon and are sometimes further kinked or twisted. Think of such a chain of carbon atoms as *continuous*. We can trace a single path from one terminal carbon to the other and pass through every other C atom *without backtracking*.

If the —CH_3 group replaces an H from the central carbon atom of propane, the result is the *branched-chain hydrocarbon* 2-methylpropane, or *isobutane*. This is the simplest branched-chain hydrocarbon.

The formulas of the alkanes can be written in general terms as C_nH_{2n+2}, where *n* is the number of carbon atoms per molecule. The first five members of the series are

Figure 27-4   Ball-and-stick and space-filling models and a three-dimensional representation of propane, C_3H_8.

	CH_4	C_2H_6	C_3H_8	C_4H_{10}	C_5H_{12}
Number of C atoms = n =	1	2	3	4	5
Number of H atoms = $2n + 2$ =	4	6	8	10	12

The formula of each alkane differs from the next by CH_2, a *methylene group*.

A series of compounds in which each member differs from the next by a specific number and kind of atoms is called a **homologous series**. The properties of members of such a series are closely related. The boiling points of the lighter members of the straight-chain saturated hydrocarbon series are shown in Figure 27-6. As the molecular weights of the straight-chain hydrocarbons increase, their boiling points also increase regularly. Properties such as boiling point depend on the forces between molecules (Chapter 13). Carbon–carbon and carbon–hydrogen bonds are essentially nonpolar and are arranged tetrahedrally around each C atom. As a result, saturated hydrocarbons are nonpolar molecules, and the only significant intermolecular forces are dispersion forces (Section 13-2). These forces, which are due to induced dipoles, become stronger as the sizes of the molecules and the number of electrons in each molecule increase. Thus, trends such as those depicted in Figure 27-6 are due to the increase in effectiveness of dispersion forces.

Some systematic method for naming compounds is necessary. The system in use today is prescribed by the International Union of Pure and Applied Chemistry (IUPAC). The names of the first 20 straight-chain alkanes are listed in Table 27-2. You should become familiar with at least the first ten. The names of the alkanes starting with pentane have prefixes (from Greek) that give the number of carbon atoms in the molecules. All alkane names have the *-ane* ending.

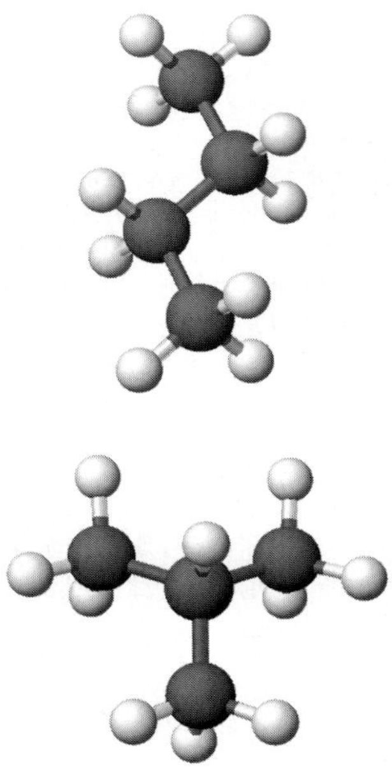

Constitutional isomers (structural isomers) have the same molecular formula but differ in the order in which their atoms are attached.

We have seen that there are two saturated C_4H_{10} hydrocarbons. For the C_5 hydrocarbons, there are three possible arrangements of the atoms. These three different C_5H_{12} alkanes are examples of constitutional isomers.

Figure 27-5 Ball-and-stick models of the two isomeric C_4H_{10} hydrocarbons, butane, $CH_3CH_2CH_2CH_3$, and methylpropane, CH_3CHCH_3.
$\quad\quad\quad\quad\quad\quad\quad\quad\quad$ |
$\quad\quad\quad\quad\quad\quad\quad\quad\quad CH_3$

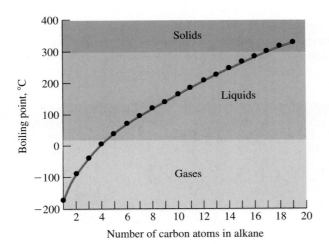

Figure 27-6 A plot of normal boiling point versus the number of carbon atoms in straight-chain, saturated hydrocarbons.

Rotation about a single bond does *not* generate a different isomer. For example, the following two drawings differ only by rotation about the bond shown in red; they both represent the same isomer of C_5H_{12} (see Example 27-1).

TABLE 27-2	*Some Straight-Chain Hydrocarbons (Alkanes)*			
Molecular Formula	IUPAC Name	Normal bp (°C)	Normal mp (°C)	State at Room Temperature
CH_4	methane	−161	−184	gas
C_2H_6	ethane	−88	−183	
C_3H_8	propane	−42	−188	
C_4H_{10}	butane	+0.6	−138	
C_5H_{12}	pentane	36	−130	liquid
C_6H_{14}	hexane	69	−94	
C_7H_{16}	heptane	98	−91	
C_8H_{18}	octane	126	−57	
C_9H_{20}	nonane	150	−54	
$C_{10}H_{22}$	decane	174	−30	
$C_{11}H_{24}$	undecane	194.5	−25.6	
$C_{12}H_{26}$	dodecane	214.5	−9.6	
$C_{13}H_{28}$	tridecane	234	−6.2	
$C_{14}H_{30}$	tetradecane	252.5	+5.5	
$C_{15}H_{32}$	pentadecane	270.5	10	
$C_{16}H_{34}$	hexadecane	287.5	18	
$C_{17}H_{36}$	heptadecane	303	22.5	solid
$C_{18}H_{38}$	octadecane	317	28	
$C_{19}H_{40}$	nonadecane	330	32	
$C_{20}H_{42}$	eicosane	205 (at 15 torr)	36.7	

The number of constitutional isomers increases rapidly as the number of carbon atoms in saturated hydrocarbons increases. There are five isomeric C_6H_{14} alkanes (Table 27-3). Table 27-4 displays the number of isomers of some saturated hydrocarbons (alkanes). Most of the isomers have not been prepared or isolated. They probably never will be.

As the degree of branching increases for a series of molecules of the same molecular weight, the molecules become more compact. A compact molecule can have fewer points of contact with its neighbors than more extended molecules do. As a result, the total induced dipole forces (dispersion forces) are weaker for branched molecules, and the boiling points of such compounds are lower. This pattern can be seen in the data in Table 27-3 and in Example 27-1.

EXAMPLE 27-1   *Isomeric Alkanes*

Write the Lewis formula of each of the three isomers having the molecular formula of C_5H_{12}. The boiling points of the three isomers are 9.5°C, 27.9°C, and 36.1°C. Match each compound with its boiling point.

Plan

We first write the carbon skeleton of the straight-chain isomer (no branching). Then we remove one carbon from the straight-chain structure and reposition it to form the skeleton for a branched isomer. We check to see whether branching is possible at other carbon atoms. Then we

TABLE 27-3	Isomeric C_6H_{14} Alkanes		
Normal IUPAC Name	**Formula**		**bp (°C)**
hexane	$CH_3CH_2CH_2CH_2CH_2CH_3$		68.7
2-methylpentane	$CH_3CH_2CH_2CHCH_3$ 　　　　　　 $\vert$ 　　　　　 CH_3		60.3
3-methylpentane	$CH_3CH_2CHCH_2CH_3$ 　　　　 $\vert$ 　　　 CH_3		63.3
2,2-dimethylbutane	CH_3 　　　 $\vert$ $CH_3CH_2CCH_3$ 　　　 $\vert$ 　　　 CH_3		49.7
2,3-dimethylbutane	$CH_3CH-CHCH_3$ 　　 $\vert$　　 $\vert$ 　 CH_3　 CH_3		58.0

TABLE 27-4	Numbers of Possible Constitutional Isomers of Alkanes

Formula	Isomers
C_7H_{16}	9
C_8H_{18}	18
C_9H_{20}	35
$C_{10}H_{22}$	75
$C_{11}H_{24}$	159
$C_{12}H_{26}$	355
$C_{13}H_{28}$	802
$C_{14}H_{30}$	1,858
$C_{15}H_{32}$	4,347
$C_{20}H_{42}$	366,319
$C_{25}H_{52}$	36,797,588
$C_{30}H_{62}$	4,111,846,763

remove a second carbon from the unbranched isomer and, if possible, form additional branching, and so on. Finally, we add hydrogen atoms to each skeleton to complete the structure.

Solution

The three compounds are:

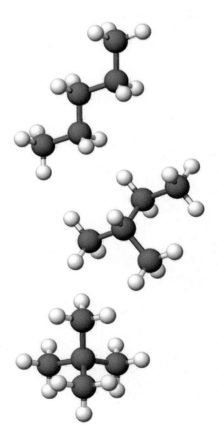

The compound with the most branching would be expected to be the one with the lowest boiling point, and the straight-chain isomer would be expected to have the highest boiling point.

$CH_3C(CH_3)_3$, bp = 9.5°C; $CH_3CH_2CH(CH_3)_2$, bp = 27.9°C; and $CH_3CH_2CH_2CH_2CH_3$, bp = 36.1°C.

(Models of these compounds are shown in the margin.)

You should now work Exercise 15.

$CH_3CH_2CH_2CH_2CH_3$, $CH_3CH_2CH(CH_3)_2$, and $CH_3C(CH_3)_3$.

Cycloalkanes

Noncyclic compounds are sometimes referred to as acyclic compounds.

The cyclic saturated hydrocarbons, or *cycloalkanes*, have the general formula C_nH_{2n}. The cycloalkanes (and other ring compounds that we will encounter later) are often shown in simplified skeletal form in which each intersection of two lines represents a C atom; we mentally add enough H atoms to give each carbon atom four bonds. The first four unsubstituted cycloalkanes and their simplified representations are

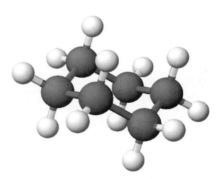

cyclopropane   cyclobutane   cyclopentane   cyclohexane

In skeletal representations, each intersection represents a carbon with any attached hydrogens. (See Section 27-3 for more details on writing skeletal representations.)

In some of these structures, the bond angles are somewhat distorted from the ideal tetrahedral angle of 109.5°, the most severe distortions being 60° in cyclopropane and 90° in cyclobutane. As a result, these rings are said to be "strained," and these two compounds are unusually reactive for saturated hydrocarbons. Cyclopentane is stable with a nearly flat ring, because the bond angles in a regular pentagon (108°) are near the tetrahedral angle (109.5°).

Cyclohexane is not stable as a flat ring. The bond angles of the flat ring would be 120°, not the 109.5°, of an sp^3-hybridized carbon. To avoid the strain, the ring "puckers" and becomes nonplanar (Figure 27-7).

Figure 27-7 A model showing a nonplanar arrangement of cyclohexane.

27-2 NAMING SATURATED HYDROCARBONS

It is important to realize that many compounds (and their names) were so familiar to chemists before the development of the IUPAC system (beginning about 1890) that they continued to be called by their common, or "trivial," names. In this and the next chapter, IUPAC names of compounds appear in blue type, and their common alternative names are shown in black type.

The IUPAC naming system is based on the names of the unbranched hydrocarbons given in Table 27-2 and their higher homologues. To name a branched-chain hydrocarbon, we first find the longest chain of carbon atoms and use the root name that corresponds to that hydrocarbon. We then indicate the positions and kinds of *substituents* attached to the chain. **Alkyl group** substituents attached to the longest chain are thought of as fragments of hydrocarbon molecules obtained by the removal of one hydrogen atom. We give them names related to the parent hydrocarbons from which they are derived. Other alkyl groups are named similarly (Table 27-5). We use the general symbol R to represent any alkyl group. The cycloalkyl groups derived from the first four cycloalkanes are called cyclopropyl, cyclobutyl, cyclopentyl, and cyclohexyl, respectively.

cyclopropyl　　　cyclobutyl　　　cyclopentyl　　　cyclohexyl

TABLE 27-5　*Some Alkanes and the Related Alkyl Groups*

Parent Hydrocarbon	Alkyl group, —R	
CH_4, methane	—CH_3, methyl	
C_2H_6, ethane	—C_2H_5, ethyl	
C_3H_8, propane	—C_3H_7, propyl	—C_3H_7, isopropyl
C_4H_{10}, butane	—C_4H_9, butyl	—C_4H_9, sec-butyl (read as "secondary butyl")
C_4H_{10}, methylpropane (common name, isobutane)	—C_4H_9, t-butyl (read as "tertiary butyl")	—C_4H_9, isobutyl

Summary of IUPAC Rules for Naming Alkanes

1. Find the longest chain of C atoms. Choose the base name that describes the number of C atoms in this chain, with the ending *-ane* (see Table 27-2). The longest chain may not be obvious at first if branches of different sizes are present.

2. Number the C atoms in this longest chain beginning at the end nearest the first branching. If necessary, go to the second branch closest to an end, and so on, until a difference is located. If there is branching at equal distances from both ends of the longest chain, begin numbering at the end nearest the branch that is first in alphabetical order.

3. Assign the name and position number to each substituent. Arrange the substituents in alphabetical order. Hyphenated prefixes, such as *tert-* and *sec-*, are not used in alphabetization of the substituents.

4. Use the appropriate prefix to group like substituents: *di-* = 2, *tri-* = 3, *tetra-* = 4, *penta-* = 5, and so on. Don't consider these prefixes when alphabetizing attached groups.

5. Write the name as a single word. Use hyphens to separate numbers and letters (plus some hyphenated prefixes) and commas to separate numbers. Don't leave any spaces.

Formerly the rule was to name the substituents in order of increasing complexity. This was sometimes difficult to determine. With the use of computers in literature searches, it became necessary to adopt the more definitive alphabetization of the names of substituents.

Let's name the following compound.

$$(CH_3)_2CHCH_2CH_2CH_2CH_3 \ or \ (CH_3)_2CH(CH_2)_3CH_3$$

We follow Rules 1 and 2 to number the carbon atoms in the longest chain.

$$\underset{1}{CH_3}-\underset{2}{CH}-\underset{3}{CH_2}-\underset{4}{CH_2}-\underset{5}{CH_2}-\underset{6}{CH_3}$$

with CH₃ branch on carbon 2

Parentheses are used to conserve space. Formulas written with parentheses must indicate unambiguously the structure of the compound. The parentheses here indicate that the CH_3 group is attached to the C next to it.

The methyl group is attached to the *second* carbon atom in a *six-carbon* chain, so the compound is named 2-methylhexane.

It is incorrect to name the compound 5-methylhexane because that violates Rule 2.

The following examples further illustrate the rules of nomenclature.

Remember that the (dash) formulas indicate which atoms are bonded to each other. They do *not* show molecular geometry. In this molecule, each C atom is tetrahedrally bonded to four other atoms.

EXAMPLE 27-2   Naming Alkanes

Give the IUPAC name of the compound represented by the structural formula

$$CH_3-\overset{\displaystyle CH_3}{\underset{\displaystyle CH_3}{C}}-CH_2-CH_2-CH_3 \quad or \quad (CH_3)_3C(CH_2)_2CH_3$$

Plan

We first find the longest carbon chain and number it to give the substituent encountered first the smallest possible number. Then we name the substituents as in Table 27-5 and specify the number of each as indicated in Rule 4.

Solution

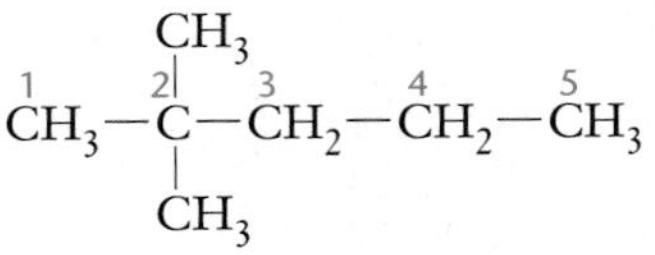

$$
\begin{array}{c}
\quad\quad\quad\;\; CH_3 \\
\quad\quad\quad\;\; \overset{2}{|} \\
\overset{1}{CH_3}-\overset{2}{C}-\overset{3}{CH_2}-\overset{4}{CH_2}-\overset{5}{CH_3} \\
\quad\quad\quad\;\; | \\
\quad\quad\quad\;\; CH_3
\end{array}
$$

The longest chain contains five carbons, so this compound is named as a derivative of pentane. There are two methyl groups, both at carbon number 2. The IUPAC name of this compound is 2,2-dimethylpentane.

EXAMPLE 27-3   *Naming Alkanes*

Give the IUPAC name of the compound represented by the structural formula

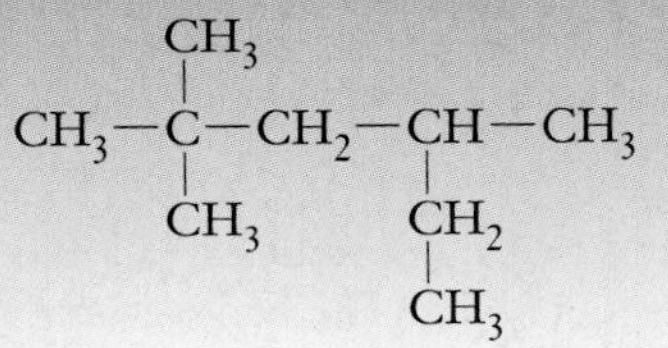

$$
\begin{array}{c}
\quad\quad\quad\;\; CH_3 \\
\quad\quad\quad\;\; | \\
CH_3-C-CH_2-CH-CH_3 \\
\quad\quad\;\; | \quad\quad\quad | \\
\quad\quad\;\; CH_3 \quad\quad CH_2 \\
\quad\quad\quad\quad\quad\quad\;\; | \\
\quad\quad\quad\quad\quad\quad\;\; CH_3
\end{array}
$$

Plan

The approach is the same as in Example 27-2. We should be aware that the longest carbon chain might not be *written* in a straight line.

Solution

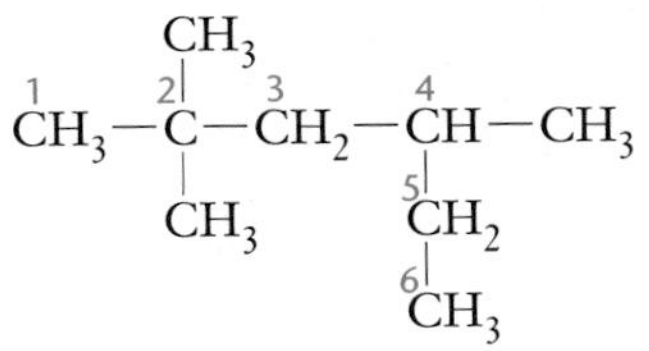

$$
\begin{array}{c}
\quad\; CH_3 \\
\quad\; \overset{2}{|} \\
\overset{1}{CH_3}-\overset{2}{C}-\overset{3}{CH_2}-\overset{4}{CH}-CH_3 \\
\quad\; | \quad\quad\quad \overset{5}{|} \\
\quad CH_3 \quad\quad CH_2 \\
\quad\quad\quad\quad\quad \overset{6}{|} \\
\quad\quad\quad\quad\quad CH_3
\end{array}
\quad or \quad
\begin{array}{c}
\quad\quad\; CH_3 \\
\quad\quad\; \overset{2}{|} \\
\overset{1}{CH_3}-\overset{2}{C}-\overset{3}{CH_2}-\overset{4}{CH}-\overset{5}{CH_2}-\overset{6}{CH_3} \\
\quad\quad\; | \quad\quad\quad | \\
\quad\; CH_3 \quad\quad CH_3
\end{array}
$$

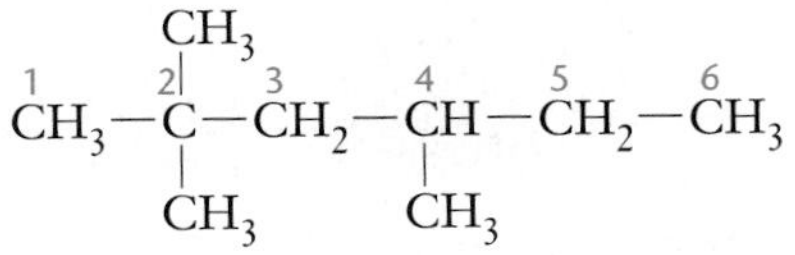

(rewritten to emphasize the six-C chain)

The longest chain contains six carbons, so this compound is named as a derivative of hexane. There are three methyl substituents: two at carbon number 2 and one at carbon number 4. The IUPAC name of this compound is 2,2,4-trimethylhexane.

You should now work Exercises 13 and 14.

EXAMPLE 27-4   *Writing Formulas from Names*

Write the structure for 4-*tert*-butyl-2,5-dimethylheptane.

Plan

The root name "heptane" indicates that there is a seven-carbon chain.

$$
C-C-C-C-C-C-C
$$

The names and numbers of the substituents tell us where to attach the alkyl groups.

Solution

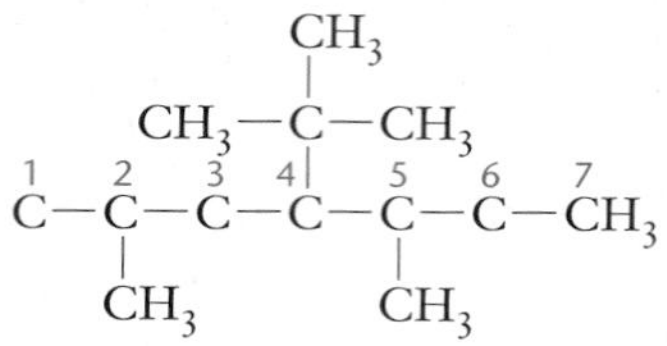

Then we fill in enough hydrogens to saturate each C atom and arrive at the structure

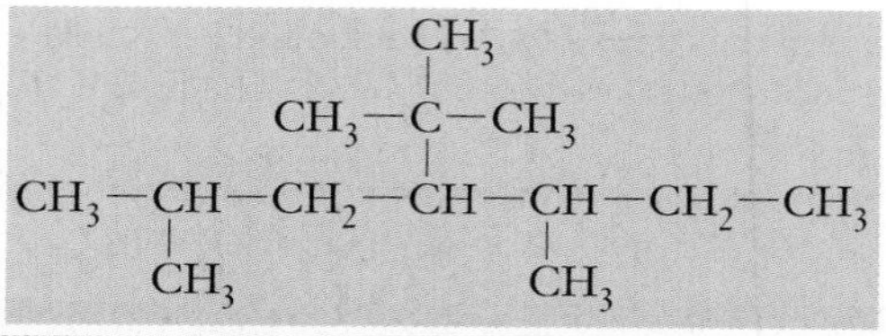

EXAMPLE 27-5   *Writing Formulas from Names*

Write the structure for the compound 2-cyclopropyl-3-ethylpentane.

Plan

The root name "pentane" tells us that the structure is based on a five-carbon chain. We place the substituents at the positions indicated by the numbers in the name.

Solution

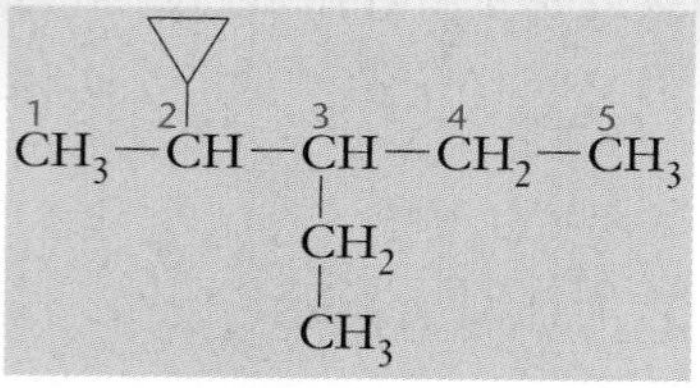

where the symbol represents the cyclopropyl group, $CH_2—CH_2$ CH

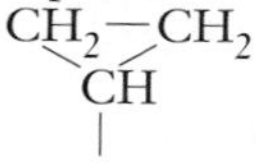

You should now work Exercise 16.

Recall that the cycloalkanes may also be represented by the general formula C_nH_{2n}.

The names of substituted cycloalkanes are derived analogously to those of alkanes. (1) The base name is determined by the number of carbon atoms in the ring using the same base name as the alkane with the addition of cyclo in front. (2) If only one substituent is attached to the ring, no "location number" is required because all positions in a cycloalkane are equivalent. (3) Two or more functional groups on the ring are identified by location numbers, which should be assigned sequentially to the ring carbons in the order that gives the *smallest sum* of location numbers.

1,3-dimethylcyclopentane can exist as two different isomers; one with both methyl groups on the same side of the ring (*cis*-), another with the methyl groups on opposite sides of the plane of the ring (*trans*-). (See Section 27-3.)

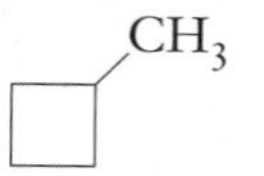

methylcyclobutane

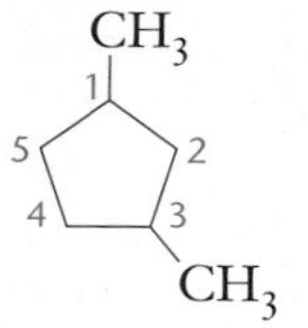

1,3-dimethylcyclopentane
(not 1,4-dimethylcyclopentane)

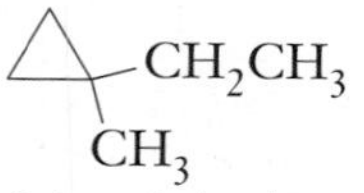

1-ethyl-1-methylcyclopropane

UNSATURATED HYDROCARBONS

There are three classes of unsaturated hydrocarbons: (1) the alkenes and their cyclic counterparts, the cycloalkenes; (2) the alkynes and the cycloalkynes; and (3) the aromatic hydrocarbons.

27-3 ALKENES

The simplest **alkenes** contain one carbon–carbon double bond, C=C, per molecule. The general formula for noncyclic alkenes is C_nH_{2n}. The simplest alkene is C_2H_4, ethene, which is usually called by its common name, ethylene.

The bonding in ethylene (Figure 27-8) was described in Section 8-13. The hybridization (sp^2) and bonding at other double-bonded carbon atoms are similar. Both carbon atoms in C_2H_4 are located at the centers of trigonal planes. Rotation about C=C double bonds does not occur significantly at room temperature. Compounds that have the general formula (XY)C=C(XY) therefore exist as a pair of *cis–trans* isomers. Figure 27-9 shows the *cis–trans* isomers of 1,2-dichloroethene. The existence of compounds with different arrangements of groups on the opposite sides of a bond with restricted rotation is called **geometric isomerism.** This *cis–trans* isomerism can occur across double bonds in alkenes and across single bonds in rings.

Two shared electron pairs draw the atoms closer together than a single electron pair does. Thus, carbon–carbon double bonds are shorter than C—C single bonds, 1.34 Å versus 1.54 Å. The physical properties of the alkenes are similar to those of the alkanes, but their chemical properties are quite different.

The root for the name of each alkene is derived from the alkane having the same number of C atoms as the longest chain containing the double bond. In the trivial (common) system of nomenclature, the suffix *-ylene* is added to the characteristic root. In systematic (IUPAC) nomenclature, the suffix *-ene* is added to the characteristic root.

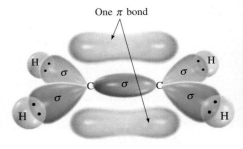

Figure 27-8 Four C—H σ bonds, one C—C σ bond (*green*), and one C—C π bond (*yellow*) in the planar C_2H_4 molecule.

SCiLINKS.

TOPIC: Alkenes
GO TO: www.scilinks.org
*sci***LINKS CODE:** WCH2730

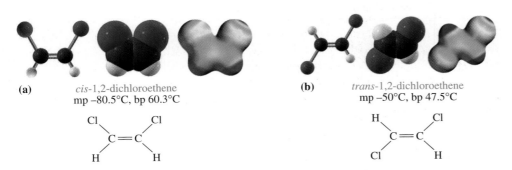

(a) *cis*-1,2-dichloroethene
mp −80.5°C, bp 60.3°C

(b) *trans*-1,2-dichloroethene
mp −50°C, bp 47.5°C

Figure 27-9 Two isomers of 1,2-dichloroethene are possible because rotation about the double bond is restricted. This is an example of *geometric* isomerism. A ball-and-stick model, space-filling model, and electrostatic charge potential plot are shown for each isomer. (a) The *cis* isomer. (b) The *trans* isomer.

Petroleum

Petroleum, or crude oil, was discovered in the United States (Pennsylvania) in 1859 and in the Middle East (Iran) in 1908. It has been found in many other locations since these initial discoveries and is now pumped from the ground in many parts of the world. Petroleum consists mainly of hydrocarbons. Small amounts of organic compounds containing nitrogen, sulfur, and oxygen are also present. Each oil field produces petroleum with a particular set of characteristics. Distillation of petroleum produces several fractions.

Gasoline is much in demand, so higher hydrocarbons (C_{12} and up) are "cracked" to increase the amount of gasoline that can be made from a barrel of petroleum. The hydrocarbons are heated, in the absence of air and in the presence of a catalyst, to produce a mixture of smaller alkanes that can be used in gasoline. This process is called *catalytic cracking*.

The **octane number** (rating) of a gasoline indicates how smoothly it burns and how much engine "knock" it produces. (Engine knock is caused by premature detonation of fuel in the combustion chamber.) 2,2,4-Trimethylpentane, isooctane, has excellent combustion properties and was arbitrarily assigned an octane number of 100. Heptane, $CH_3(CH_2)_5CH_3$, has very poor combustion properties and was assigned an octane number of zero.

$$CH_3-\underset{\underset{CH_3}{|}}{\overset{\overset{CH_3}{|}}{C}}-CH_2-\underset{CH_3}{\overset{\overset{CH_3}{|}}{CH}}-CH_3$$

isooctane (octane number = 100)

$$CH_3CH_2CH_2CH_2CH_2CH_2CH_3$$

heptane (octane number = 0)

Petroleum refinery towers.

Mixtures of these two were prepared and burned in test engines to establish the octane scale. The octane number of such a mixture is the percentage of isooctane in it. Gasolines burned in standard test engines are assigned octane numbers based on the compression ratio at which they begin to knock. A 90-octane fuel produces the same amount of knock as the 90% isooctane/10% heptane mixture. Branched-chain compounds produce less knock than straight-chain compounds. The octane numbers of two isomeric hexanes are

$$CH_3(CH_2)_4CH_3$$

hexane (octane number = 25)

$$(CH_3)_3CCH_2CH_3$$

2,2-dimethylbutane (octane number = 92)

Petroleum Fractions

Fraction*	Principal Composition	Distillation Range (°C)
natural gas	C_1-C_4	below 20
bottled gas	C_5-C_6	20–60
gasoline	C_4-C_{12}	40–200
kerosene	$C_{10}-C_{16}$	175–275
fuel oil, diesel oil	$C_{15}-C_{20}$	250–400
lubricating oils	$C_{18}-C_{22}$	above 300
paraffin	$C_{23}-C_{29}$	mp 50–60
asphalt	—	viscous liquid ("bottoms fraction")
coke	—	solid

Other descriptions and distillation ranges have been used, but all are similar.

Summary of IUPAC Rules for Naming Alkenes and Cycloalkenes

1. Locate the C atoms in the *longest* C chain *that contains the double bond*. Use the base name prefix with the ending *-ene*.

2. Number the C atoms of this chain sequentially *beginning at the end nearer the double bond*. Insert the number describing the position of the double bond (indicated by its *first* carbon location) before the base name. (This is necessary only for chains of four or more C atoms, because only one position is possible for a double bond in a chain of two or three carbon atoms.)

3. In naming alkenes, the double bond takes positional precedence over substituents on the carbon chain. The double bond is assigned the lowest possible number.

4. To name compounds with possible geometric isomers, consider the two largest groups within the carbon chain that contains the double bond—these are indicated as part of the base name. The isomer in which the largest groups at each end of the C=C are located on opposite sides is called *trans*. If the largest groups are on the same side, the molecule is referred to as *cis*. Insert the prefix *cis*- or *trans*- just before the number of the double bond to indicate whether the largest groups are on the same or opposite sides, respectively, of the double bond.

5. For cycloalkenes, the double bond is assumed to be between C atoms 1 and 2, so no position number is needed to describe it.

The prefix *trans*- means "across" or "on the other side of." As a reminder of this terminology, think of words such as "transatlantic."

Some illustrations of this naming system follow.

$$CH_2{=}CH_2 \qquad CH_3{-}CH{=}CH_2 \qquad \overset{4}{C}H_3\overset{3}{C}H_2{-}\overset{2}{C}H{=}\overset{1}{C}H_2$$

systematic: ethene   propene   1-butene
trivial: (ethylene)   (propylene)   (butylene)

$$\overset{1}{C}H_3\overset{2}{C}H{=}\overset{3}{C}H\overset{4}{C}H_3 \qquad CH_3{-}\underset{\underset{CH_3}{|}}{C}{=}CH_2$$

2-butene

methylpropene
(isobutylene)

There are two isomers of 2-butene, *cis*-2-butene and *trans*-2-butene.

The following two names illustrate the application of Rule 3.

$$\overset{4}{C}H_3{-}\overset{3}{C}H_2{-}\underset{\underset{CH_3}{|}}{\overset{2}{C}}{=}\overset{1}{C}H_2 \qquad \overset{1}{C}H_3{-}\overset{2}{C}H{=}\overset{3}{C}H{-}\underset{\underset{CH_3}{|}}{\overset{4}{C}}H{-}\overset{5}{C}H_3$$

2-methyl-1-butene   4-methyl-2-pentene

4-Methyl-2-pentene can exist as either of two isomers. Their names illustrate the application of Rule 4.

4-methyl-*cis*-2-pentene   4-methyl-*trans*-2-pentene

Some alkenes, called **polyenes,** have two or more carbon–carbon double bonds per molecule. The suffixes *-adiene, -atriene,* and so on are used to indicate the number of (C=C) double bonds in a molecule.

$$\overset{1}{C}H_2=\overset{2}{C}H-\overset{3}{C}H=\overset{4}{C}H_2 \qquad \overset{4}{C}H_3-\overset{3}{C}H=\overset{2}{C}=\overset{1}{C}H_2$$

1,3-butadiene   1,2-butadiene

1,3-Butadiene and similar molecules that contain *alternating* single and double bonds are described as having **conjugated double bonds.** Such compounds are of special interest because of their polymerization reactions (Section 27-19).

✓ **Problem-Solving Tip:** *When Is the Location of a Substituent or a Multiple Bond Not Included in an IUPAC Name?*

If only one location is possible for a substituent or double bond, the number that indicates the location can be left out of the compound's name. If there is doubt, put the number in the name. For example, "2-methyl-1-propene" is named simply methylpropene. There are no other locations possible for the substituent or the double bond. The name without the numbers is the correct name. Both numbers are necessary to name 3-methyl-1-butene, however.

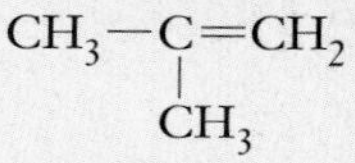

$$CH_3-\underset{\underset{CH_3}{|}}{C}=CH_2 \qquad CH_3-\underset{\underset{CH_3}{|}}{CH}-CH=CH_2$$

methylpropene 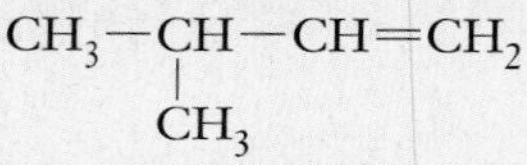   3-methyl-1-butene

---

EXAMPLE 27-6   *Naming Alkenes*

Give the IUPAC names of the following two alkenes.

(a) 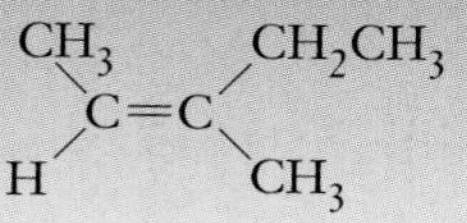 $\underset{H}{\overset{CH_3}{\diagdown}}C=\underset{CH_3}{\overset{CH_2CH_3}{\diagup}}$ (b) 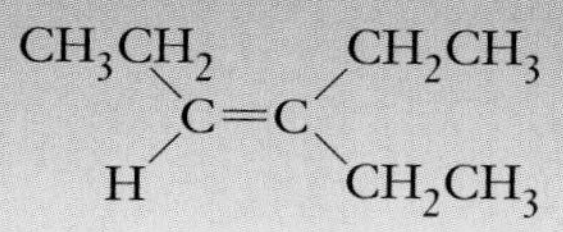 $\underset{H}{\overset{CH_3CH_2}{\diagdown}}C=\underset{CH_2CH_3}{\overset{CH_2CH_3}{\diagup}}$

Plan

For each compound, we first find the longest chain that includes the double bond, and then number it beginning at the end nearer the double bond (Rules 1 and 2). Then we specify the identities and positions of substituents in the same way we did for alkanes. In (a) we specify the geometric isomer by locating the two largest groups attached to the double-bonded carbons and then describing their relationship using the *cis–trans* terminology.

Solution

(a)

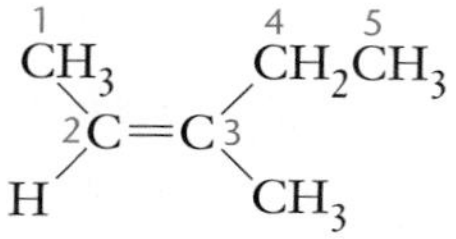

$$\underset{H}{\overset{\overset{1}{C}H_3}{\diagdown}}\underset{2}{C}=\underset{3}{C}\underset{CH_3}{\overset{\overset{4}{C}H_2\overset{5}{C}H_3}{\diagup}}$$

The longest such chain contains five atoms and has a double bond beginning at atom 2; thus, the compound is named as a derivative of 2-pentene. Now we must apply Rule 4. The two largest groups in the chain are the terminal —CH_3 (carbon 1) and the —CH_2CH_3 (carbons 4 and 5); these are on the *same side* of the double bond, so we name the compound as a derivative of *cis*-2-pentene. The only substituent is the methyl group at carbon 3. The full name of the compound is 3-methyl-*cis*-2-pentene.

(b) There are two choices for the longest chain, and either one would have an ethyl substituent.

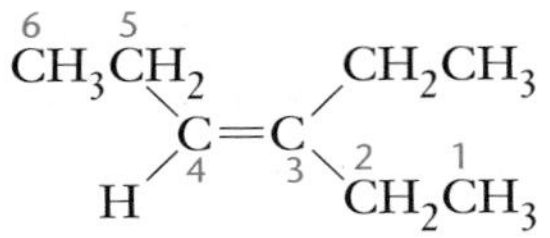

$$\underset{H}{\overset{\underset{6}{CH_3}\underset{5}{CH_2}}{}}\underset{4}{C}=\underset{3}{C}\underset{\underset{1}{CH_2CH_3}}{\overset{\underset{5}{CH_2}\underset{}{CH_3}}{}}\underset{2}{}$$ or $$CH_3CH_2 \quad CH_2CH_3 \\ C=C \\ H \quad CH_2CH_3$$

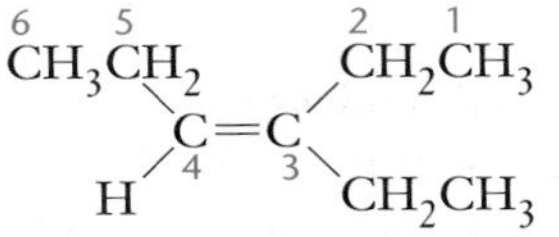

We could number either chain from the other end and still have the double bond starting at carbon 3; we number from the end that gives the carbon bearing the ethyl group the lowest possible position number, 3. Carbon 3 has two equivalent substituents, so geometric isomerism is not possible, and we do not use the *cis–trans* terminology. The name is 3-ethyl-3-hexene.

You should now work Exercise 20.

EXAMPLE 27-7   *Naming Alkenes*

Name the following alkene.

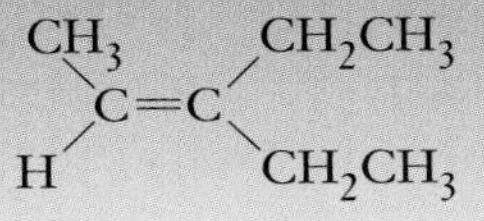

$$\underset{H}{\overset{CH_3}{}}\,C=C\,\underset{CH_2CH_3}{\overset{CH_2CH_3}{}}$$

Plan

As we did in Example 27-6(b), we first find the longest chain that includes the double bond, and then number it beginning at the end nearer the double bond (Rules 1 and 2). Then we specify the identities and positions of substituents in the same way we did for alkanes.

Solution

There are two choices for the longest chain, and either one would have an ethyl substituent.

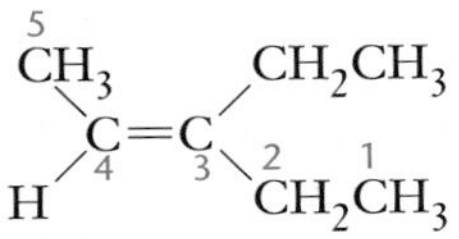

$$\overset{5}{CH_3}\quad CH_2CH_3 \\ C=C \\ H \quad CH_2CH_3$$ or $$CH_3 \quad CH_2CH_3 \\ C=C \\ H \quad CH_2CH_3$$

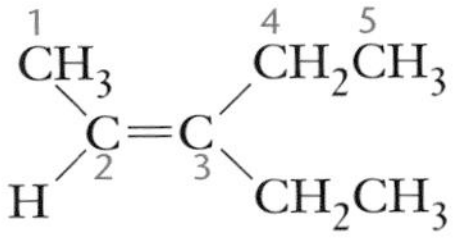

Carbon 3 has two equivalent substituents, so geometric isomerism is not possible, and we do not use the *cis-trans* terminology. The name is 3-ethyl-2-pentene.

You should now work Exercise 22.

One would not number the carbons as:

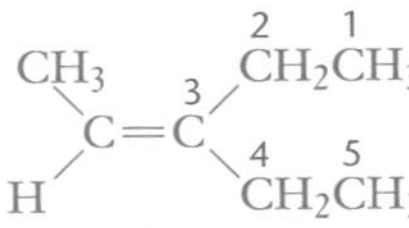

$$CH_3 \quad \overset{2}{CH_2}\overset{1}{CH_3} \\ \underset{3}{C}=\underset{4}{C} \\ H \quad \underset{5}{CH_2CH_3}$$

Rule 1 requires that we use the longest chain that includes both carbons of the double bond.

The cycloalkenes are represented by the general formula C_nH_{2n-2}. Two cycloalkenes and their skeletal representations are

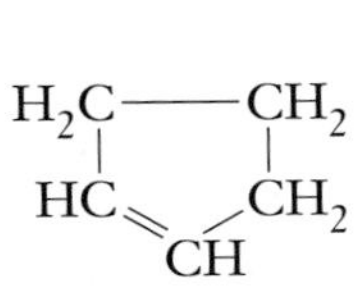

$$\underset{\underset{CH}{HC}}{\overset{H_2C}{}}\,\underset{\underset{CH_2}{CH_2}}{\overset{CH_2}{}}$$ or   ⬠

cyclopentene

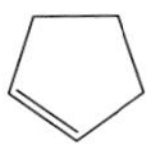

$$\underset{\underset{CH}{HC}}{\overset{H_2C}{}}\,\overset{CH_2}{\underset{CH_2}{}}$$ or   ⬡

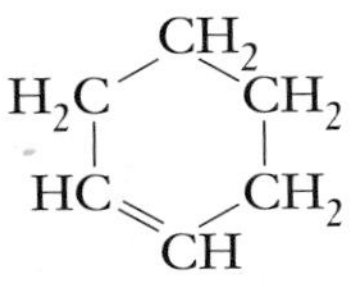

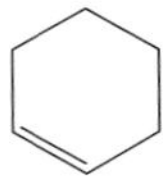

cyclohexene

> ✓ **Problem-Solving Tip:** *How to Draw Skeletal Representations*
>
> The rules for drawing skeletal representations follow.
>
> 1. A carbon atom is assumed to be at each intersection of two lines and at the end of each line. All carbon–carbon bonds are shown as lines. A carbon atom might be shown for clarity.
> 2. Because carbon has four bonds, we mentally supply hydrogens to give each carbon its four bonds.
> 3. Atoms other than hydrogen and carbon are shown. The following skeletal representation is that of 3-chloro-4-ethyl-*cis*-3-heptene. 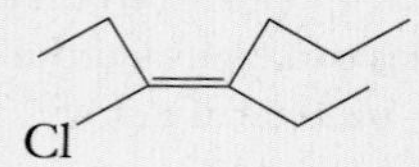
>
>
> and its formula is $C_9H_{17}Cl$.

EXAMPLE 27-8   *Cycloalkenes*

Write the skeletal representation of 3-methylcyclohexene.

Plan

We draw the ring of the specified size with one double bond in the ring. We number the ring so the double bond is between atoms 1 and 2. Then we add the designated substituents at the indicated positions.

Solution

We number the six-membered ring so that the double bond is between atoms 1 and 2.

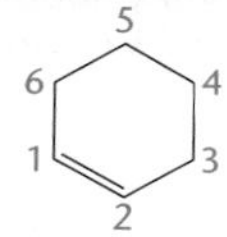

A methyl group is attached at carbon 3; the correct structure is

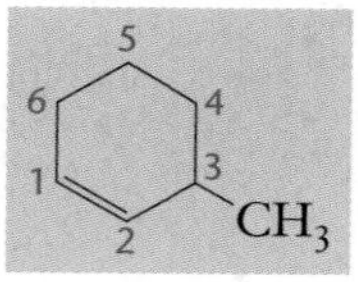

Remember that each intersection of two lines represents a carbon atom; there are enough H atoms at each C atom to make a total of four bonds to carbon.

You should now work Exercise 21.

27-4 ALKYNES

The **alkynes,** or acetylenic hydrocarbons, contain one or more carbon–carbon triple bonds, —C≡C—. The noncyclic alkynes with one triple bond per molecule have the general formula C_nH_{2n-2}. The bonding in all alkynes is similar to that in acetylene (Section 8-14).

Triply bonded carbon atoms are *sp*-hybridized. The triply bonded atoms and their adjacent atoms lie on a straight line (Figure 27-10).

Alkynes are named like the alkenes except that the suffix *-yne* is added to the characteristic root. The first member of the series is commonly called acetylene. Its molecular formula is C_2H_2. It is thermodynamically unstable, decomposing explosively to C(s) and H_2(g) at high pressures. It may be converted into ethene (also known as ethylene) and then to ethane by the addition of hydrogen. The condensed formulas for several small alkynes are

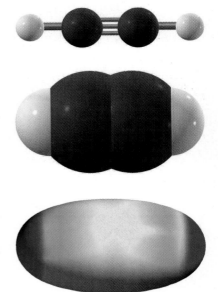

Figure 27-10 Models of acetylene, H—C≡C—H.

$$CH\equiv CH \qquad CH_3-C\equiv CH \qquad CH_3-CH_2-C\equiv CH$$

ethyne   propyne   1-butyne
(acetylene)

$$CH_3-C\equiv C-CH_3 \qquad CH_3-CH-C\equiv CH$$
$$\qquad\qquad\qquad\qquad\qquad\qquad\qquad\qquad\qquad | $$
$$\text{2-butyne} \qquad\qquad\qquad\qquad\qquad CH_3$$

3-methyl-1-butyne

The triple bond takes positional precedence over substituents on the carbon chain. It is assigned the lowest possible number in naming.

Summary of IUPAC Rules for Naming Alkynes

Alkynes are named like the alkenes except for the following two points.

1. The suffix *-yne* is added to the characteristic root.
2. Because the linear arrangement about the triple bond does not lead to geometric isomerism, the prefixes *cis-* and *trans-* are not used.

EXAMPLE 27-9   *Alkynes*

Write the structural formula of 5,5-dimethyl-2-heptyne.

Plan

The structure is based on a seven-carbon chain with a triple bond beginning at carbon 2. We add methyl groups at the positions indicated.

Solution

$$\overset{1}{C}-\overset{2}{C}\equiv\overset{3}{C}-\overset{4}{C}-\overset{5}{C}-\overset{6}{C}-\overset{7}{C}$$

Two methyl groups are attached to carbon 5, and sufficient hydrogens are added to complete the bonding at each C atom.

$$\begin{array}{c} \qquad\qquad\qquad CH_3 \\ \overset{1}{}\quad\overset{2}{}\quad\overset{3}{}\overset{4}{}\quad\overset{5}{}|\ \overset{6}{}\quad\overset{7}{} \\ CH_3C\equiv CCH_2CCH_2CH_3 \\ \qquad\qquad\qquad | \\ \qquad\qquad\qquad CH_3 \end{array}$$

The carbon chain is numbered to give the triple bond the lowest possible numbers even though that results in higher position numbers for the methyl substituents.

You should now work Exercise 28.

> ## EXAMPLE 27-10   *Naming Alkynes*
>
> Give the IUPAC name of the compound
>
> $$CH_3CHC{\equiv}CCH_2CH_3$$
> $$\overset{|}{CH_3}$$
>
> **Plan**
>
> We find the longest chain and number it to give the triple bond the lowest possible number. Then we specify the substituent(s) by name and position number.
>
> **Solution**
>
> The longest continuous carbon chain that includes the triple bond has six C atoms. There are two ways in which we could choose and number such a chain, and in both the triple bond would be between C atoms 3 and 4.
>
> $$\overset{1}{C}H_3\overset{2}{C}H\overset{3}{C}{\equiv}\overset{4}{C}\overset{5}{C}H_2\overset{6}{C}H_3 \qquad \overset{6}{C}H_3\overset{5}{C}H\overset{4}{C}{\equiv}\overset{3}{C}\overset{2}{C}H_2\overset{1}{C}H_3$$
> $$\overset{|}{CH_3} \qquad\qquad\qquad \overset{|}{CH_3}$$
>
> Because we also want the methyl substituent to have the lowest possible number, we choose the first possibility. The name of the compound is 2-methyl-3-hexyne.
>
> *You should now work Exercise 30.*

AROMATIC HYDROCARBONS

SC*i*LINKS.

TOPIC: Aromatic Compounds
GO TO: www.scilinks.org
*sci*LINKS CODE: WCH2740

The main components of coal gas are hydrogen ($\approx$ 50%) and methane ($\approx$ 30%).

Distillation of coal tar produces a variety of aromatic compounds.

Originally the word "aromatic" was applied to pleasant-smelling substances. The word now describes benzene, its derivatives, and certain other compounds that exhibit similar chemical properties. Some have very foul odors because of substituents on the benzene ring. On the other hand, many fragrant compounds do not contain benzene rings.

Steel production at one time required large amounts of coke. This was prepared by heating bituminous coal to high temperatures in the absence of air. This process also favors production of *coal gas* and *coal tar*. Because of the enormous amount of coal converted to coke, coal tar was produced in large quantities. It served as a source of aromatic compounds. For each ton of coal converted to coke, about 5 kg of aromatic compounds were obtained. Today petroleum refining is the major source of aromatic hydrocarbons.

Early research on the reactions of the aromatic hydrocarbons led to methods for preparing a great variety of dyes, drugs, flavors, perfumes, and explosives. More recently, large numbers of polymeric materials, such as plastics and fabrics, have been prepared from these compounds.

27-5 BENZENE

Benzene is the simplest aromatic hydrocarbon. By studying its reactions, we can learn a great deal about aromatic hydrocarbons. Benzene was discovered in 1825 by Michael Faraday when he fractionally distilled a by-product oil obtained in the manufacture of illuminating gas from whale oil.

Elemental analysis and determination of its molecular weight showed that the molecular formula for benzene is C_6H_6. The formula suggests that it is highly unsaturated. But its properties are quite different from those of alkenes and alkynes.

The facts that only one monosubstitution product is obtained in many reactions and that no additional products can be prepared show conclusively that benzene has a *symmetrical ring structure*. Stated differently, every H atom is equivalent to every other H atom, and this is possible only in a symmetrical ring structure (a).

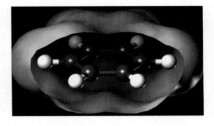

A computer-generated model of a molecule of benzene, C_6H_6. A ball-and-stick model is shown inside a representation of the molecular surface.

(a)

(skeleton only)

(b)

The debate over the structure and bonding in benzene raged for at least 30 years. In 1865, Friedrich Kekulé (1829–1896) suggested that the structure of benzene was intermediate between two structures [part (b) of the preceding structures] that we now call resonance structures. We often represent benzene as

or, more simply

The structure of benzene is described in detail in Section 9-6 in terms of MO theory.

All 12 atoms in a benzene molecule lie in a plane. This suggests sp^2 hybridization of each carbon. The six sp^2-hybridized C atoms lie in a plane, and the unhybridized p orbitals extend above and below the plane. Side-by-side overlap of the p orbitals forms pi orbitals (Figure 9-11). The electrons associated with the pi bonds are *delocalized* over the entire benzene ring (Figure 27-11a,b).

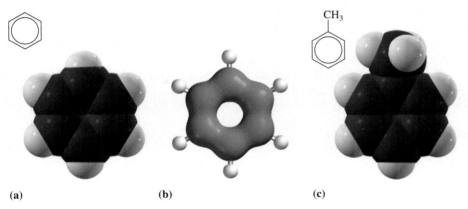

(a)   (b)   (c)

Figure 27-11 (a) A model of the benzene molecule, C_6H_6, and (b) its electron distribution. (c) A model of toluene, $C_6H_5CH_3$. This is a derivative of benzene in which one H atom has been replaced by an alkyl group.

27-6 OTHER AROMATIC HYDROCARBONS

Benzene molecules bearing alkyl substituents are called **alkylbenzenes.** The simplest of these is methylbenzene (common name, toluene), shown in Figure 27-11c. The dimethylbenzenes are called xylenes. Three different compounds (Table 27-6) have the formula $C_6H_4(CH_3)_2$ (see margin). These three xylenes are *constitutional isomers*. In naming these (as well as other disubstituted benzenes), we use prefixes *ortho-* (abbreviated *o-*), *meta-* (*m-*), or *para-* (*p-*) to refer to relative positions of substituents on the benzene ring. The *ortho-* prefix refers to two substituents located on *adjacent* carbon atoms; for example, 1,2-dimethylbenzene is *o*-xylene. The *meta-* prefix identifies substituents on C atoms 1 and 3, so 1,3-dimethylbenzene is *m*-xylene. The *para-* prefix refers to substituents on C atoms 1 and 4, so 1,4-dimethylbenzene is *p*-xylene.

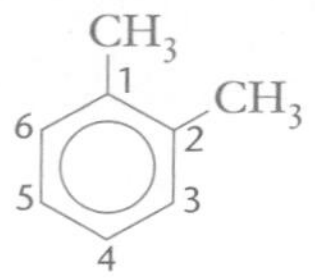

ortho-xylene
bp = 144°C
mp = −27°C

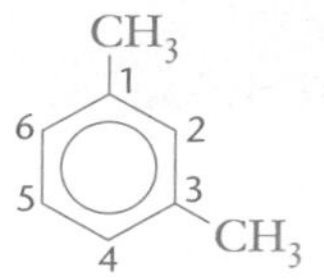

meta-xylene
bp = 139°C
mp = −54°C

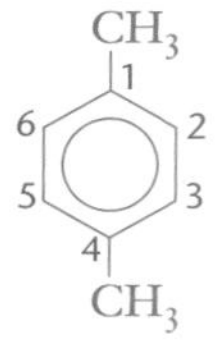

para-xylene
bp = 138°C
mp = 13°C

Summary of Rules for Naming Derivatives of Benzene

1. If there is only one group on the ring, no number is needed to designate its position.
2. If there are two groups on the ring, we use the traditional designations.

   ortho- or *o-* for 1,2-disubstitution
   meta- or *m-* for 1,3-disubstitution
   para- or *p-* for 1,4-disubstitution

3. If there are three or more groups on the ring, location numbers are assigned to give the *minimum sum* of numbers.

Examples are

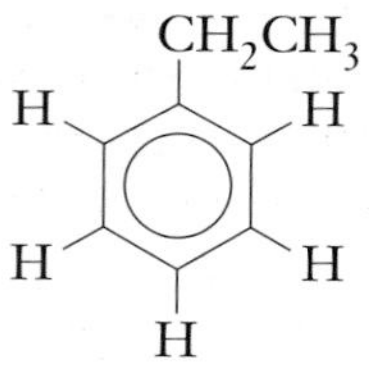

ethylbenzene

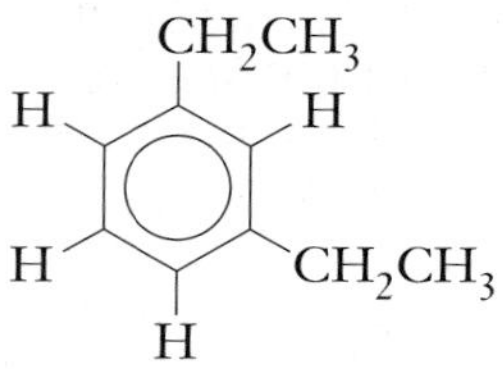

m-diethylbenzene, or 1,3-diethylbenzene

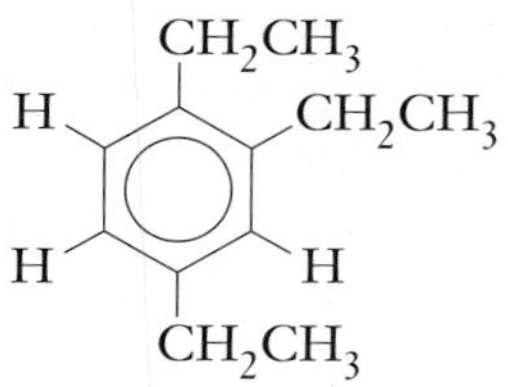

1,2,4-triethylbenzene

When an H atom is removed from a benzene molecule, C_6H_6, the resulting group, C_6H_5- or 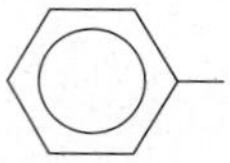 , is called "phenyl." Sometimes we name mixed alkyl-aromatic hydrocarbons on that basis.

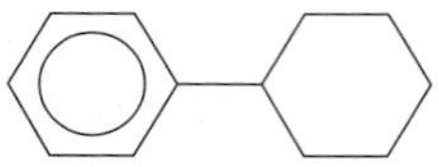

phenylcyclohexane

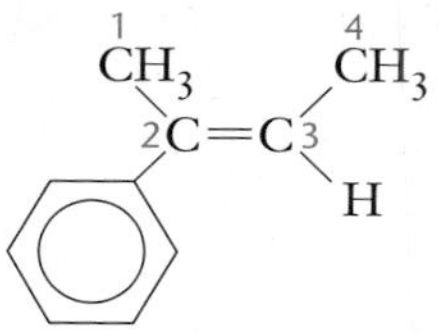

2-phenyl-*cis*-2-butene

TABLE 27-6	*Aromatic Hydrocarbons from Coal Tar*			
Name	**Formula**	**Normal bp (°C)**	**Normal mp (°C)**	**Solubility**
benzene	C_6H_6	80	+6	
toluene	$C_6H_5CH_3$	111	−95	
o-xylene	$C_6H_4(CH_3)_2$	144	−27	
m-xylene	$C_6H_4(CH_3)_2$	139	−54	All insoluble
p-xylene	$C_6H_4(CH_3)_2$	138	+13	in water
naphthalene	$C_{10}H_8$	218	+80	
anthracene	$C_{14}H_{10}$	342	+218	
phenanthrene	$C_{14}H_{10}$	340	+101	

Another class of aromatic hydrocarbons consists of "condensed" or "fused-ring" aromatic systems. The simplest of these are naphthalene, anthracene, and phenanthrene.

naphthalene, $C_{10}H_8$ anthracene, $C_{14}H_{10}$ phenanthrene, $C_{14}H_{10}$

No hydrogen atoms are attached to the carbon atoms that are involved in fusion of aromatic rings, that is, carbon atoms that are members of two or more aromatic rings.

The traditional name is often used as part of the base name in naming an aromatic hydrocarbon and its derivatives. You should know the names and structures of the aromatic hydrocarbons discussed thus far: benzene, toluene, the three xylenes, naphthalene, anthracene, and phenanthrene.

Distillation of coal tar provides four volatile fractions as well as the pitch that is used for surfacing roads and in the manufacture of "asphalt" roofing (Figure 27-12). Eight aromatic

Naphthalene

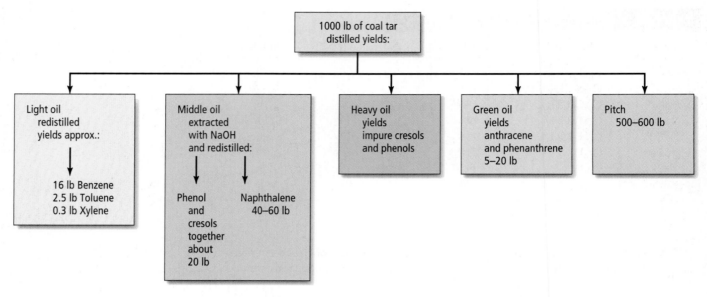

Figure 27-12 Fractions obtained from coal tar.

hydrocarbons obtained in significant amounts by efficient fractional distillation of the "light oil" fraction are those in Table 27-6.

27-7 HYDROCARBONS: A SUMMARY

Hydrocarbons contain only carbon and hydrogen. They can be subdivided into various groups. To assist you in organizing what you have studied so far in this chapter, take a few minutes and study Figure 27-13.

FUNCTIONAL GROUPS

 CD-ROM   Screen 11.5, Functional Groups.

As you study the following sections, you may wish to refer to the summary in Section 27-15.

An **alkyl** group is a group of atoms derived from an alkane by the removal of one hydrogen atom. An **aryl** group is a group of atoms derived from an aromatic system by the removal of one hydrogen atom.

The study of organic chemistry is greatly simplified by considering hydrocarbons as parent compounds and describing other compounds as derived from them. In general, an organic molecule consists of a skeleton of carbon atoms with special groups of atoms attached to that skeleton. These special groups of atoms are often called **functional groups** because they represent the most common sites of chemical reactivity (function). The only functional groups that are possible in hydrocarbons are double and triple (i.e., pi) bonds. Atoms other than C and H are called **heteroatoms,** the most common being O, N, S, P, and the halogens. Most functional groups contain one or more heteroatoms.

In the next several sections we will introduce some common functional groups that contain heteroatoms and learn a little about the resulting classes of compounds. We will continue to represent alkyl groups with the symbol R—. We commonly use that symbol to represent either an aliphatic (e.g., alkyl) or an aromatic (e.g., an aryl such as phenyl) group. When we specifically mean an aryl group, we will use the symbol Ar—.

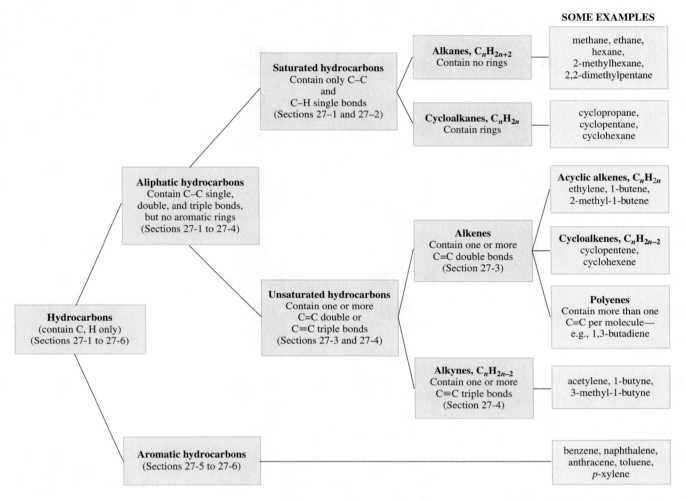

SOME EXAMPLES

Figure 27-13 A classification of hydrocarbons.

27-8 ORGANIC HALIDES

Almost any hydrogen atom in a hydrocarbon can be replaced by a halogen atom to give a stable compound. Table 27-7 shows some organic halides and their names.

In the IUPAC naming system, the organic halides are named as *halo-* derivatives of the parent hydrocarbons. The prefix *halo-* can be *fluoro-*, *chloro-*, *bromo-*, or *iodo-*. Simple alkyl chlorides are sometimes given common names as alkyl derivatives of the hydrogen halides. For instance, the IUPAC name for $CH_3CH_2—Cl$ is chloroethane; it is commonly called ethyl chloride by analogy to $H—Cl$, hydrogen chloride.

A carbon atom can be bonded to as many as four halogen atoms, so an enormous number of organic halides can exist. Completely fluorinated compounds are known as *fluorocarbons* or sometimes *perfluorocarbons*. The fluorocarbons are even less reactive than hydrocarbons. Saturated compounds in which all H atoms have been replaced by some combination of Cl and F atoms are called *chlorofluorocarbons* or sometimes *freons*. These compounds were widely used as refrigerants and as propellants in aerosol cans. However, the release of chlorofluorocarbons into the atmosphere has been shown to be quite damaging to the earth's ozone layer. Since January 1978, the use of chlorofluorocarbons in aerosol cans in the United States has been prohibited, and efforts to develop both controls for existing

Freon is a DuPont trademark for certain chlorofluorocarbons; other companies' related products are known by other names. Typical freons are trichlorofluoromethane, $CFCl_3$ (called Freon-11), and dichlorodifluoromethane, CF_2Cl_2 (called Freon-12).

TABLE 27-7 *Some Organic Halides*

Formula	Structural Formula	Ball-and-Stick Model	Electrostatic Charge Potential Plot	bp (°C)	IUPAC Name Common Name
CH_3Cl	H H—C—Cl H			−23.8	chloromethane methyl chloride
CH_2Cl_2	Cl H—C—Cl H			40.2	dichloromethane methylene chloride
$CHCl_3$	Cl H—C—Cl Cl			61	trichloromethane chloroform
CCl_4	Cl Cl—C—Cl Cl			76.8	tetrachloromethane carbon tetrachloride
$CHCl_2Br$	Cl H—C—Cl Br			90	bromodichloromethane —
$(CH_3)_2CHI$	H H H H—C—C—C—H H I H			89.5	2-iodopropane isopropyl iodide
C_5H_7Cl				25	3-chlorocyclopentene (other isomers possible) —
C_6H_5I	I			118	iodobenzene phenyl iodide
C_6H_4ClBr	Br Cl			204	1-bromo-2-chloro-benzene (other isomers possible) *o*-bromochlorobenzene
$C_{14}H_9Cl_5$	Cl Cl—C—Cl Cl——C——Cl H				1,1,1-trichloro-2,2-bis(*p*-chlorophenyl)-ethane DDT

chlorofluorocarbons and suitable replacements continue. The production and sale of freons have been banned in many countries.

27-9 ALCOHOLS AND PHENOLS

Alcohols and phenols contain the hydroxyl group (—O—H) as their functional group. **Alcohols** may be considered to be derived from saturated or unsaturated hydrocarbons by the replacement of at least one H atom by a hydroxyl group. The properties of alcohols result from a hydroxyl group attached to an *aliphatic* carbon atom, —C—O—H. Ethanol (ethyl alcohol) is the most common example (Figure 27-14).

When a hydrogen atom on an aromatic ring is replaced by a hydroxyl group (Figure 27-15), the resulting compound is known as a **phenol.** Such compounds behave more like acids than alcohols. Alternatively, we may view alcohols and phenols as derivatives of water in which one H atom has been replaced by an organic group.

water   ethanol   phenol

Indeed, this is a better view. The structure of water was discussed in Section 8-9. The hydroxyl group in an alcohol or a phenol is covalently bonded to a carbon atom, but the O—H

CD-ROM   Screen 11.6, Functional Groups (1): Reactions of Alcohols.

The simplest phenol is called phenol. The most common member of a class of compounds is often called by the class name. Salt, sugar, alcohol, and phenol are examples.

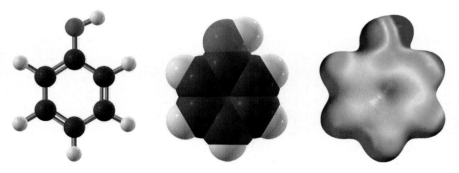

Figure 27-14   Models of ethanol (also called ethyl alcohol or grain alcohol), CH_3CH_2OH.

Figure 27-15   Models of phenol, C_6H_5OH.

bond is quite polar. The oxygen atom has two unshared electron pairs, and the C—O—H bond angle is nearly 104.5°.

The presence of a bonded alkyl or aryl group changes the properties of the —OH group. *Alcohols* are so very weakly acidic that they are thought of as neutral compounds. *Phenols* are weakly acidic.

Many properties of alcohols depend on whether the hydroxyl group is attached to a carbon that is bonded to *one*, *two*, or *three* other carbon atoms.

> **Primary alcohols** contain one R group; **secondary alcohols** contain two R groups; and **tertiary alcohols** contain three R groups bonded to the carbon atom to which the —OH group is attached.

Representing alkyl groups as R, we can illustrate the three classes of alcohols. The R groups may be the same or different.

In writing organic structures, we often use primes when we wish to specify that the alkyl groups might be different, for example, R, R′, R″.

$$
\begin{array}{ccc}
\overset{\displaystyle H}{\underset{\displaystyle H}{R-\overset{|}{\underset{|}{C}}-OH}} &
\overset{\displaystyle R'}{\underset{\displaystyle H}{R-\overset{|}{\underset{|}{C}}-OH}} &
\overset{\displaystyle R'}{\underset{\displaystyle R''}{R-\overset{|}{\underset{|}{C}}-OH}} \\[2ex]
\text{a primary (1°) alcohol} & \text{a secondary (2°) alcohol} & \text{a tertiary (3°) alcohol}
\end{array}
$$

Nomenclature

The systematic name of an alcohol consists of the root name plus an *-ol* ending. A numeric prefix indicates the position of the —OH group on a chain of three or more carbon atoms.

$$
CH_3-OH \qquad CH_3-CH_2-CH_2-OH \qquad \overset{\displaystyle OH}{CH_3-\overset{|}{CH}-CH_3} \qquad \overset{\displaystyle OH}{\underset{\displaystyle CH_3}{CH_3-\overset{|}{\underset{|}{C}}-CH_2CH_3}}
$$

methanol	1-propanol	2-propanol	2-methyl-2-butanol
methyl alcohol	propyl alcohol	isopropyl alcohol	*t*-pentyl alcohol
(wood alcohol)	(a primary alcohol)	(a secondary alcohol)	(a tertiary alcohol)

EXAMPLE 27-11   *Formulas of Alcohols*

Acyclic compounds contain no rings.

There are four constitutional isomers of the saturated acyclic four-carbon alcohols with one —OH per molecule. Write the structural formula of each, and identify each as primary, secondary, or tertiary. Name each isomer.

Plan

The carbon skeleton can be either a straight chain, C—C—C—C, or branched, $\overset{\displaystyle C-C-C}{\underset{\displaystyle C}{\overset{|}{}}}$.

Each skeleton has two different types of carbons to which the —OH group can be attached.

Solution

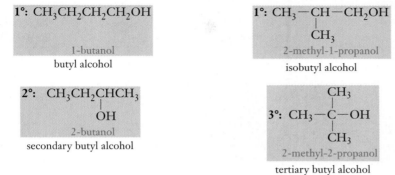

1°: CH₃CH₂CH₂CH₂OH

1-butanol
butyl alcohol

1°: CH₃—CH—CH₂OH
          |
          CH₃

2-methyl-1-propanol
isobutyl alcohol

2°: CH₃CH₂CHCH₃
              |
              OH

2-butanol
secondary butyl alcohol

3°: CH₃—C—OH
         |
         CH₃
with CH₃ above C

2-methyl-2-propanol
tertiary butyl alcohol

You should now work Exercise 46.

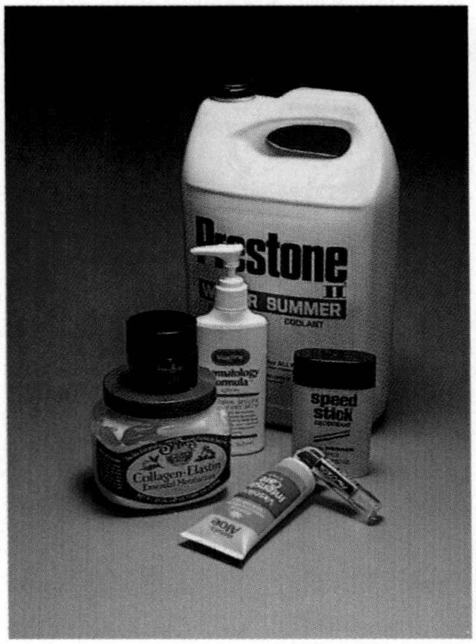

The **polyols** are alcohols that contain more than one —OH group per molecule. Those containing two OH groups per molecule are called **glycols.** Important examples of polyols include

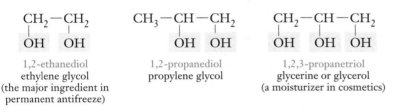

CH₂—CH₂
  | |
 OH   OH

1,2-ethanediol
ethylene glycol
(the major ingredient in
permanent antifreeze)

CH₃—CH—CH₂
          | |
         OH   OH

1,2-propanediol
propylene glycol

CH₂—CH—CH₂
  | | |
 OH   OH   OH

1,2,3-propanetriol
glycerine or glycerol
(a moisturizer in cosmetics)

Phenols are usually referred to by their common names. Examples are

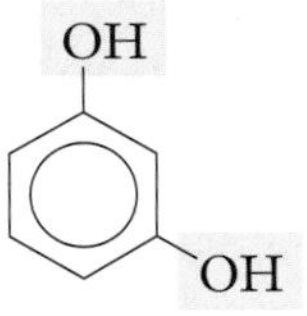

OH / OH   OH / OH   OH—CH₃   OH / CH₃   OH / CH₃

resorcinol 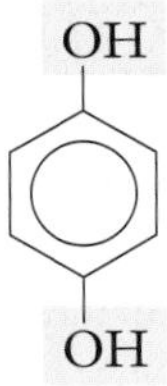   hydroquinone 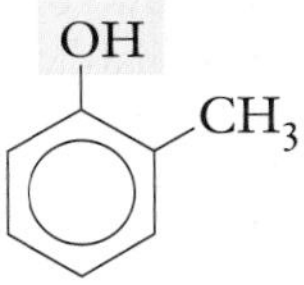   o-cresol 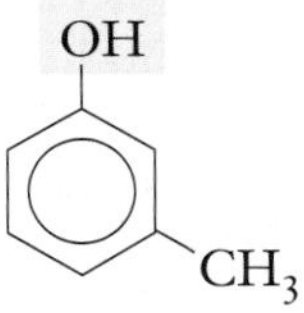   m-cresol 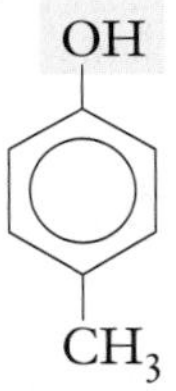   p-cresol

As you might guess, cresols occur in "creosote," a wood preservative.

Polyols are used in permanent antifreeze and in cosmetics.

Charles Steele

The *o-, m-,* and *p-* notation was introduced in Section 27-6.

Physical Properties

The hydroxyl group, —OH, is quite polar, whereas alkyl groups, R, are nonpolar. The properties of alcohols depend on two factors: (1) the number of hydroxyl groups per molecule and (2) the size of the nonpolar portion of the molecule.

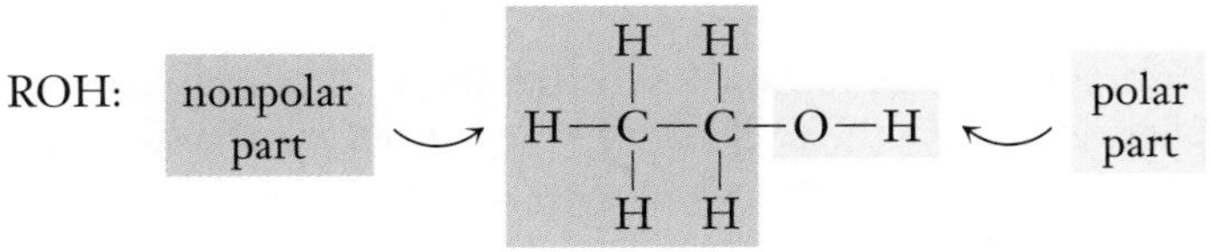

ROH: nonpolar part ⤳   H—C—C—O—H   ⤶ polar part
                                      | |
                              (with H H above and H H below the two carbons)

TABLE 27-8	*Physical Properties of Primary Alcohols*		
Name	Formula	Normal bp (°C)	Solubility in H_2O (g/100 g at 20°C)
methanol	CH_3OH	65	miscible
ethanol	CH_3CH_2OH	78.5	miscible
1-propanol	$CH_3CH_2CH_2OH$	97	miscible
1-butanol	$CH_3CH_2CH_2CH_2OH$	117.7	7.9
1-pentanol	$CH_3CH_2CH_2CH_2CH_2OH$	137.9	2.7
1-hexanol	$CH_3CH_2CH_2CH_2CH_2CH_2OH$	155.8	0.59

The low-molecular-weight alcohols are soluble in water in all proportions (miscible). Beginning with the four butyl alcohols, solubility in water decreases rapidly with increasing molecular weight. This is because the nonpolar parts of such molecules are much larger than the polar parts. Many polyols are very soluble in water because they contain two or more polar —OH groups per molecule.

Table 27-8 shows that the boiling points of unbranched primary alcohols increase, and their solubilities in water decrease, with increasing molecular weight. The boiling points of alcohols are much higher than those of the corresponding alkanes (see Table 27-2) because of the hydrogen bonding of the hydroxyl groups.

Most phenols are solids at 25°C. Phenols are only slightly soluble in water unless they contain other functional groups that interact with water.

Uses

Many alcohols and phenols have considerable commercial importance. Methanol, CH_3OH, was formerly produced by the destructive distillation of wood and is sometimes called wood alcohol. It is now produced in large quantities from carbon monoxide and hydrogen. It is very toxic and causes permanent blindness when taken internally.

Fermentation is an enzymatic process carried out by certain kinds of bacteria.

Ethanol, CH_3CH_2OH, also known as ethyl alcohol or grain alcohol, can be prepared by fermentation of sugar. The most ancient written literature refers to beverages that were obviously alcoholic!

Many simple alcohols are important raw materials in the industrial synthesis of polymers, fibers, explosives, plastics, and pharmaceutical products. Phenols are widely used in the preparation of plastics and dyes. Solutions of phenols are used as antiseptics and disinfectants. Some uses of polyols depend on their relatively high boiling points. For instance, glycerine is used as a wetting agent in cosmetic preparations. Ethylene glycol (bp 197°C), which is miscible with water, is used in commercial permanent antifreeze.

27-10 ETHERS

When the word "ether" is mentioned, most people think of the well-known anesthetic, diethyl ether. There are many ethers. Their uses range from artificial flavorings to refrigerants and solvents. An **ether** is a compound in which an O atom is bonded to two organic groups.

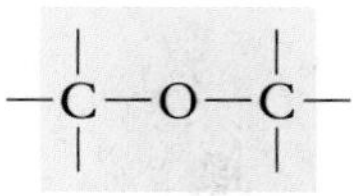

$$-\overset{\displaystyle |}{\underset{\displaystyle |}{C}}-O-\overset{\displaystyle |}{\underset{\displaystyle |}{C}}-$$

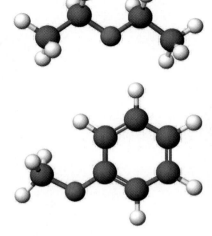

Models of diethyl ether (*top*) and methyl phenyl ether (*bottom*).

Alcohols are considered derivatives of water in which one H atom has been replaced by an organic group. Ethers may be considered derivatives of water in which both H atoms have been replaced by organic groups.

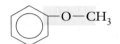

$$\underset{\text{water}}{\text{H—O—H}} \qquad \underset{\text{alcohol}}{\text{R—O—H}} \qquad \underset{\text{ether}}{\text{R—O—R}'}$$

The similarity is only structural, however, because ethers are only slightly polar and are chemically rather unreactive. In fact, their physical properties are similar to those of the corresponding alkanes; for example, CH_3OCH_3 is like $CH_3CH_2CH_3$.

Three kinds of ethers are known: (1) aliphatic, (2) aromatic, and (3) mixed. Common names are used for ethers in most cases.

$$H_3C-O-CH_3 \qquad H_3C-O-CH_2CH_3$$

methoxymethane
dimethyl ether
(an aliphatic ether)

methoxyethane
methyl ethyl ether
(an aliphatic ether)

methoxybenzene
methyl phenyl ether
anisole
(a mixed ether)

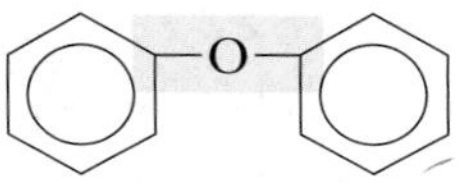

phenoxybenzene
diphenyl ether
(an aromatic ether)

When the benzene ring is a substituent, it is called a phenyl group ($-C_6H_5$).

Diethyl ether is a very low boiling liquid (bp 35°C). Dimethyl ether is a gas that is used as a refrigerant. The aliphatic ethers of higher molecular weights are liquids, and the aromatic ethers are liquids and solids.

Even ethers of low molecular weight are only slightly soluble in water. Diethyl ether is an excellent solvent for organic compounds. It is widely used to extract organic compounds from plants and other natural sources.

27-11 ALDEHYDES AND KETONES

Aldehydes and ketones contain the carbonyl group, $>\!C=O$. In **aldehydes,** at least one H atom is bonded to the carbonyl group. **Ketones** have two carbon atoms bonded to a carbonyl group. Models of formaldehyde (the simplest aldehyde) and acetone (the simplest ketone) are shown in Figure 27-16.

The simplest aldehyde, formaldehyde,

$$\overset{\displaystyle O}{\overset{\displaystyle \|}{H-C-H}}$$, has two H atoms and no alkyl or aryl groups.

$$\overset{O}{\overset{\|}{R-C-H}} \qquad \overset{O}{\overset{\|}{Ar-C-H}} \qquad \overset{O}{\overset{\|}{R-C-R}} \qquad \overset{O}{\overset{\|}{Ar-C-Ar}} \qquad \overset{O}{\overset{\|}{Ar-C-R}}$$

aliphatic
aldehyde

aromatic
aldehyde

aliphatic
ketone

aromatic
ketone

mixed
ketone

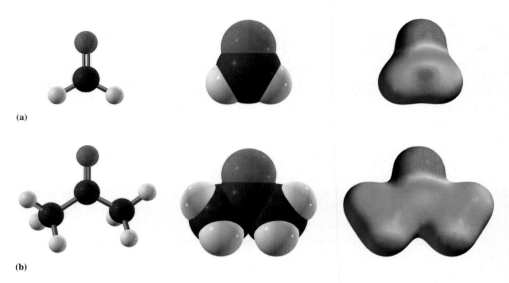

Figure 27-16　(a) Models of formaldehyde, HCHO, the simplest aldehyde. (b) Models of acetone, CH_3—CO—CH_3, the simplest ketone.

Aldehydes are usually called by their common names. These are derived from the name of the acid with the same number of C atoms (Table 27-9). The systematic (IUPAC) name is derived from the name of the parent hydrocarbon. The suffix *-al* is added to the characteristic stem. The carbonyl group takes positional precedence over most other substituents.

Formaldehyde has long been used as a disinfectant and as a preservative (including embalming fluid) for biological specimens. Its main use is in the production of certain plastics and in binders for plywood. Many important natural substances are aldehydes and ketones. Examples include sex hormones, some vitamins, camphor, and the flavorings extracted

TABLE 27-9	*Properties of Some Simple Aldehydes*	
Common Name	**Formula**	**Normal bp (°C)**
formaldehyde (methanal)	H—C—H with O double bonded to C	−21
acetaldehyde (ethanal)	CH_3—C—H with O double bonded to C	20.2
propionaldehyde (propanal)	CH_3CH_2C—H with O double bonded to C	48.8
benzaldehyde	C_6H_5—C—H with O double bonded to C	179.5

from almonds and cinnamon. Aldehydes contain a carbon–oxygen double bond, so they are very reactive compounds. As a result, they are valuable in organic synthesis, particularly in the construction of carbon chains.

The simplest ketone is called acetone. Other simple, commonly encountered ketones are usually called by their common names. These are derived by naming the alkyl or aryl groups attached to the carbonyl group.

$$CH_3-\overset{\overset{\displaystyle O}{\|}}{C}-CH_3 \qquad CH_3-\overset{\overset{\displaystyle O}{\|}}{C}-CH_2CH_3 \qquad CH_3CH_2-\overset{\overset{\displaystyle O}{\|}}{C}-CH_2CH_3$$

acetone   methyl ethyl ketone   diethyl ketone

cyclohexanone   acetophenone   benzophenone
                      (methyl phenyl ketone)   (diphenyl ketone)

The systematic names for ketones are derived from their parent hydrocarbons. The suffix -one is added to the characteristic stem.

$$\overset{1}{CH_3}-\overset{2}{\overset{\overset{\displaystyle O}{\|}}{C}}-\overset{3}{CH_2}-\overset{4}{CH_3} \qquad \overset{1}{CH_3}-\overset{2}{CH_2}-\overset{3}{\overset{\overset{\displaystyle O}{\|}}{C}}-\overset{4}{\overset{\overset{\displaystyle CH_3}{|}}{CH}}-\overset{5}{CH_2}-\overset{6}{CH_3}$$

butanone   4-methyl-3-hexanone

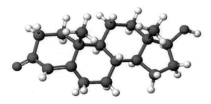

The ketones are excellent solvents. Acetone is very useful because it dissolves most organic compounds yet is miscible with water. Acetone is widely used as a solvent in the manufacture of lacquers, paint removers, explosives, plastics, drugs, and disinfectants. Some ketones of high molecular weight are used extensively in blending perfumes. Structures of some naturally occurring aldehydes and ketones are

Steroid molecules have similar molecular shapes but different biochemical functions. Progesterone (*top*), a female sex hormone, and testosterone (*bottom*), a male sex hormone. Both are ketones.

benzaldehyde   cinnamaldehyde   vanillin
(almonds)   (cinnamon)   (vanilla)

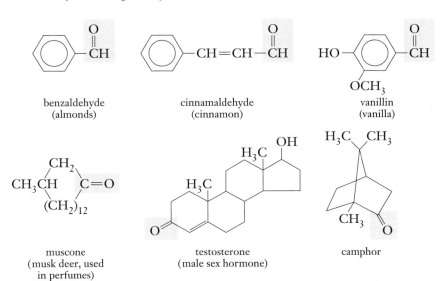

muscone   testosterone   camphor
(musk deer, used   (male sex hormone)
in perfumes)

TABLE 27-10   *Boiling Points of Ammonia and Some Amines*

Name	Formula	Boiling Point (°C)
ammonia	NH_3	−33.4
methylamine	CH_3NH_2	−6.5
dimethylamine	$(CH_3)_2NH$	7.4
trimethylamine	$(CH_3)_3N$	3.5
ethylamine	$CH_3CH_2NH_2$	16.6
aniline	$C_6H_5NH_2$	184
ethylenediamine	$H_2NCH_2CH_2NH_2$	116.5
pyridine	C_5H_5N	115.3
pyrrole	C_4H_5N	129.8

Ammonia acts as a Lewis base because there is one unshared pair of electrons on the N atom (Section 10-10).

27-12 AMINES

The **amines** are derivatives of ammonia in which one or more hydrogen atoms have been replaced by alkyl or aryl groups. Many low-molecular-weight amines are gases or low-boiling liquids (Table 27-10). Amines are basic compounds (Table 18-6, Section 28-4). The aliphatic amines of low molecular weight are soluble in water. Aliphatic diamines of fairly high molecular weight are soluble in water because each molecule contains two highly polar —NH_2 groups that form hydrogen bonds with water.

The odors of amines are quite unpleasant; many of the malodorous compounds that are released as fish decay are low-molecular-weight amines. Amines of high molecular weight are nonvolatile, so they have little odor. One of the materials used to manufacture nylon, hexamethylenediamine, is an aliphatic amine. Many aromatic amines are used to prepare organic dyes that are widely used in industrial societies. Amines are also used to produce many medicinal products, including local anesthetics and sulfa drugs.

Amines are widely distributed in nature in the form of amino acids and proteins, which are found in all higher animal forms, and in alkaloids, which are found in most plants. Some of these substances are fundamental building blocks of animal tissue, and minute amounts of others have dramatic physiological effects, both harmful and beneficial. Countless other biologically important substances, including many vitamins, antibiotics, and other drugs, contain **amino groups,** —NR_2 (where R can represent an H, alkyl, or aryl group).

Structure and Nomenclature

There are three classes of amines, depending on whether one, two, or three hydrogen atoms have been replaced by organic groups. They are called **primary, secondary,** and **tertiary amines,** respectively.

nicotine

strychnine

NH_3	RNH_2	R_2NH	R_3N
$H-\underset{H}{\overset{}{N}}-H$	$CH_3-\underset{H}{\overset{}{N}}-H$	$CH_3-\underset{H}{\overset{}{N}}-CH_3$	$CH_3-\underset{CH_3}{\overset{}{N}}-CH_3$
ammonia	methylamine (a primary amine)	dimethylamine (a secondary amine)	trimethylamine (a tertiary amine)

The systematic names of amines are based on consideration of the compounds as derivatives of ammonia. Amines of more complex structure are sometimes named as derivatives of the parent hydrocarbon with the term *amino-* used as a prefix to describe —NH_2.

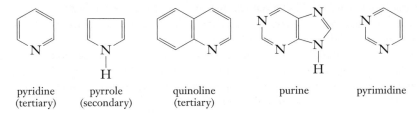

2-aminobutane or *sec*-butylamine   1-amino-3-ethylcyclohexane

Aniline is the simplest aromatic amine. Many aromatic amines are named as derivatives of aniline.

aniline   *p*-chloroaniline   3,4-dibromo-*N*-methyl-   *N,N*-dimethylaniline
(primary)   (primary)   aniline (secondary)   (tertiary)

Heterocyclic amines contain nitrogen as a part of the ring, bound to two carbon atoms. Many of these amines are found in coal tar and a variety of natural products. Some aromatic and heterocyclic amines are called by their common names.

pyridine   pyrrole   quinoline   purine   pyrimidine
(tertiary)   (secondary)   (tertiary)

Genes, the units of chromosomes that carry hereditary characteristics, are long stretches of double helical deoxyribonucleic acid, or DNA. DNA is composed of four *nucleotide bases:* adenine, guanine, cytosine, and thymine. The first two are modified purines, and the latter two are modified pyrimidines. The sequence of these building blocks in DNA acts as a code for the order of amino acids in the proteins of an organism. The DNA in each cell of an organism contains the instructions for making the complete organism (see Section 28-10).

adenine   guanine   cytosine   thymine

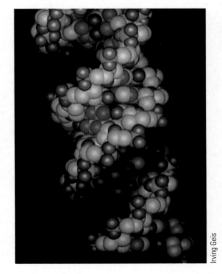

A space-filling model of a portion of the DNA double helical structure.

Irving Geis

TABLE 27-11	*Some Aliphatic Carboxylic Acids*		
Formula		**Common Name**	**IUPAC Name**
HCOOH		formic acid	methanoic acid
CH_3COOH		acetic acid	ethanoic acid
CH_3CH_2COOH		propionic acid	propanoic acid
$CH_3CH_2CH_2COOH$		butyric acid	butanoic acid
$CH_3CH_2CH_2CH_2CH_2COOH$		caproic acid	hexanoic acid
$CH_3(CH_2)_{10}COOH$		lauric acid	dodecanoic acid
$CH_3(CH_2)_{14}COOH$		palmitic acid	hexadecanoic acid
$CH_3(CH_2)_{16}COOH$		stearic acid	octadecanoic acid

Formic acid was obtained by distillation of ants (L. *formica,* "ant"); acetic acid occurs in vinegar (L. *acetum,* "vinegar"); butyric acid in rancid butter (L. *butyrum,* "butter"); and stearic acid in animal fats (Gr. *stear,* "beef suet"). Caproic acid is one of the so-called "goat acids." Its odor is responsible for the name.

27-13 CARBOXYLIC ACIDS

In the general formulas for alcohols and ethers, R cannot be H. In carboxylic acids, R can be H.

Compounds that contain the **carboxyl group,** $\overset{\displaystyle O}{\overset{\displaystyle \|}{-C-O-H}}$, are acidic. They are called **carboxylic acids.** Their general formula is R—COOH. Most are *weak acids.* They are much stronger acids than most phenols, however. Carboxylic acids are named systematically by dropping the terminal *-e* from the name of the parent hydrocarbon and adding *-oic acid* (Table 27-11). Many older names survive, however, and organic acids are often called by common names. In aromatic acids the carboxyl group is attached to an aromatic ring (Figure 27-17).

Organic acids occur widely in natural products, and many have been known since ancient times. Their common (trivial) names are often derived from a Greek or Latin word that indicates the original source (see Table 27-11).

Long-chain aliphatic carboxylic acids are often referred to as *fatty acids* because many have been obtained from animal fats.

The names of modified carboxylic acids are often derived from the trivial names of the acids. Positions of substituents are sometimes indicated by lowercase Greek letters, beginning with the carbon *adjacent* to the carboxyl carbon, rather than by numbering the carbon chain.

Derivatives of α-hydroxy carboxylic acids are currently used in some cosmetic preparations.

2-bromopropanoic acid
(α-bromopropionic acid)

4-hydroxybutanoic acid
(γ-hydroxybutyric acid)

EXAMPLE 27-12   *Isomeric Carboxylic Acids*

Write the structural formula of each of the four carboxylic acids having the molecular formula of $C_5H_{10}O_2$. Give the IUPAC name of each.

Plan

We first write the structural formula of the carboxylic acid that does not have branching. Then we remove one carbon from the straight-chain structure and reposition it to form branched isomers. We check to see whether branching is possible at more than one location. We find that branching is possible at two different locations. Remove a second carbon from the unbranched isomer and, if possible, from additional branching. The two removed carbons can be added as methyl branches. Each isomer is named by naming the longest chain that contains the carboxylic acid group. The final *-e* of the name is changed to *-oic acid*. Names for branches appear at the beginning of the name.

Solution

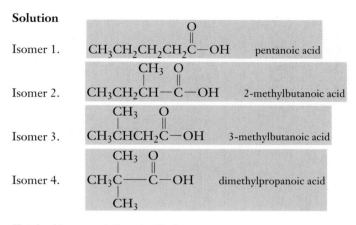

Isomer 1. CH₃CH₂CH₂CH₂C—OH pentanoic acid

Isomer 2. CH₃CH₂CH—C—OH 2-methylbutanoic acid

Isomer 3. CH₃CHCH₂C—OH 3-methylbutanoic acid

Isomer 4. CH₃C——C—OH dimethylpropanoic acid

You should now work Exercise 72a,b.

Some carboxylic acid molecules contain more than one —COOH group (Table 27-12). These acids are nearly always called by their common names. For example, oxalic acid, found in rhubarb, spinach, and tomatoes, is a dicarboxylic acid; citric acid, found in citrus fruits, is a tricarboxylic acid.

Aromatic acids are called by their common names or named as derivatives of benzoic acid, which is considered the "parent" aromatic acid.

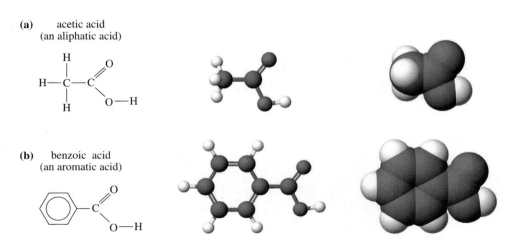

(a) acetic acid
(an aliphatic acid)

(b) benzoic acid
(an aromatic acid)

Figure 27-17 (a) Models of acetic acid. (b) Models of benzoic acid.

TABLE 27-12 *Some Naturally Occurring Organic Acids*

	Name	Source
acetic acid	CH_3-COOH	vinegar
citric acid	$HOOC-CH_2-\overset{\overset{\displaystyle OH}{|}}{\underset{\underset{\displaystyle COOH}{|}}{C}}-CH_2-COOH$	citrus fruits
lactic acid	$CH_3-\underset{\underset{\displaystyle OH}{|}}{CH}-COOH$	sour milk
malic acid	$HOOC-CH_2-\underset{\underset{\displaystyle OH}{|}}{CH}-COOH$	apples, watermelons, grape juice, wine
oxalic acid	$HOOC-COOH$	rhubarb, spinach, tomatoes
quinic acid	(cyclohexane ring structure)	cranberries
tartaric acid	$HOOC-\underset{\underset{\displaystyle OH}{|}}{CH}-\underset{\underset{\displaystyle OH}{|}}{CH}-COOH$	grape juice, wine

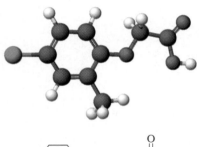

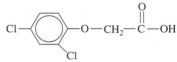

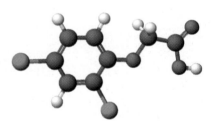

These two derivatives of acetic acids act as herbicides (weed killers) by overstimulating the plant's growth system.

benzoic acid   *p*-chlorobenzoic acid   2,4-dimethylbenzoic acid

Although many carboxylic acids occur in the free state in nature, many occur as amides or esters (Section 27-14). **α-Amino acids** are substituted carboxylic acids with the general structure

$$R-\overset{\overset{\displaystyle H}{|}}{\underset{\underset{\displaystyle NH_2}{|}}{C}}-\overset{\overset{\displaystyle O}{||}}{C}-OH$$

where R can be either an alkyl or an aryl group. α-Amino acids are the components of proteins, which make up the muscle and tissue of animals. Many other acids are important in the metabolism and synthesis of fats by enzyme systems. Acetic acid (the acid in vinegar) is the end product in the fermentation of most agricultural products. It is the fundamental unit used by living organisms in the biosynthesis of such widely diverse classes of natural products as long-chain fatty acids, natural rubber, and steroid hormones.

27-14 SOME DERIVATIVES OF CARBOXYLIC ACIDS

Three important classes of acid derivatives are formed by the replacement of the hydroxyl group by another atom or group of atoms. Each of these derivatives contains an acyl group (outlined in red in the following general structure types).

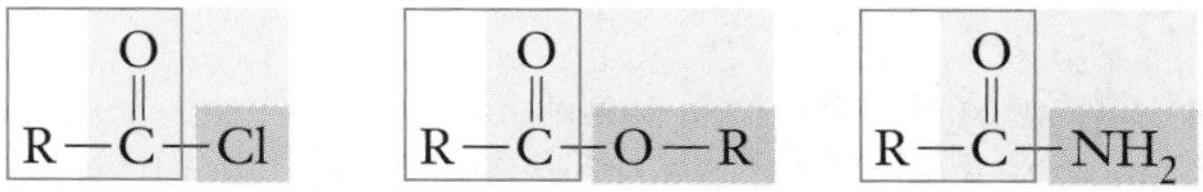

an acyl chloride   an ester   an amide
(an acid chloride)

Crystals of glycine viewed under polarized light. Glycine, the simplest amino acid, has the structure shown in the text, with R = H.

Aromatic compounds of these types (with R = aryl groups) are also encountered frequently.

Acyl Halides (Acid Halides)

The **acyl halides,** sometimes called **acid halides,** are structurally related to carboxylic acids by the replacement of the —OH group by a halogen, most often —Cl. They are usually named by combining the stems of the common names of the carboxylic acids with the suffix *-yl* and then adding the name of the halide ion. Examples are

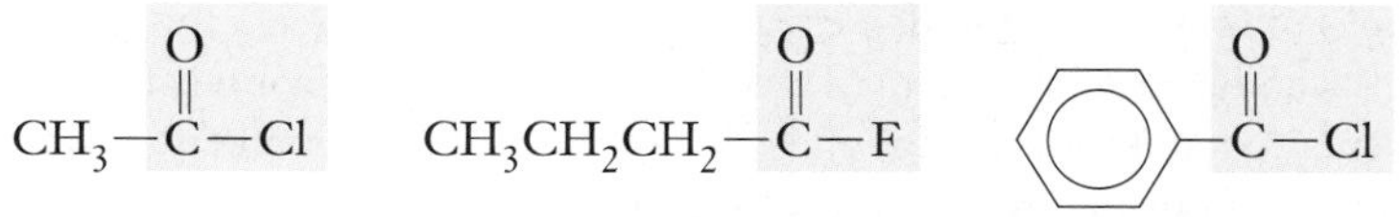

acetyl chloride   butyryl fluoride   benzoyl chloride

SCILINKS.

TOPIC: Acyl Halides
GO TO: www.scilinks.org
*sci*LINKS CODE: WCH2770

Acid halides are very reactive and have not been observed in nature.

Esters

An **ester** can be thought of as the result of removing one molecule of water from a carboxylic acid and an alcohol. Removing a molecule of water from acetic acid and ethyl alcohol gives ethyl acetate.

As we will see in Section 28-6, one method of forming esters involves acid-catalyzed reaction of an alcohol with a carboxylic acid.

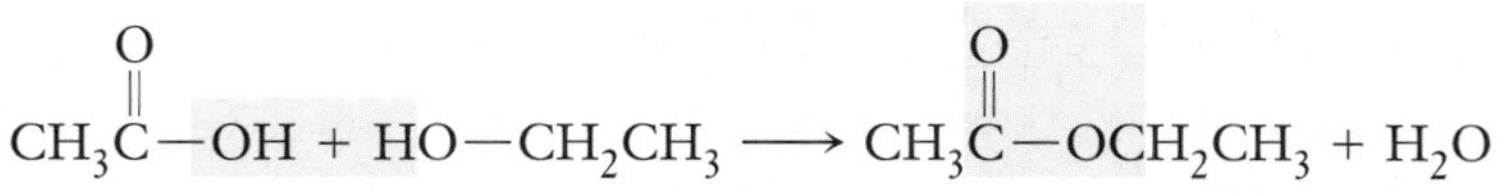

acetic acid   ethyl alcohol   ethyl acetate

The $CH_3-\overset{O}{\overset{\|}{C}}-$ fragment is derived from acetic acid, the parent acid; the $O-CH_2CH_3$ fragment is derived from ethanol, the parent alcohol. Models of ethyl acetate, an ester, are shown in Figure 27-18.

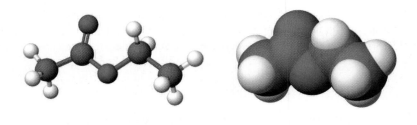

Figure 27-18   Models of ethyl acetate, $CH_3{-}\overset{\overset{\displaystyle O}{\|}}{C}{-}O{-}CH_2CH_3$, an ester.

Esters are nearly always called by their common names. These consist of, first, the name of the alkyl group in the alcohol, and then the name of the anion derived from the acid.

$$CH_3CH_2\overset{\overset{\displaystyle O}{\|}}{C}{-}OC(CH_3)_3 \qquad CH_3\overset{\overset{\displaystyle O}{\|}}{C}{-}O{-}\bigcirc \qquad \bigcirc{-}\overset{\overset{\displaystyle O}{\|}}{C}{-}OCH_3$$

       t-butyl propionate   phenyl acetate   methyl benzoate

Because of their inability to form hydrogen bonds, esters tend to be liquids with boiling points much lower than those of carboxylic acids of similar molecular weights.

Most simple esters are pleasant-smelling substances. They are responsible for the flavors and fragrances of most fruits and flowers and many of the artificial fruit flavorings that are used in cakes, candies, and ice cream (Table 27-13). Esters of low molecular weight are excellent solvents for nonpolar compounds. Ethyl acetate is an excellent solvent that gives many nail polish removers their characteristic odor.

Fats (solids) and **oils** (liquids) are esters of glycerol and aliphatic acids of high molecular weight. "Fatty acids" are all organic acids whose esters occur in fats and oils. Fats and oils have the general formula

CD-ROM   Screen 11.7, Functional Groups (2): Fats and Oils.

Biophoto Associates

Apple blossoms.

TABLE 27-13	*Some Common Esters*	
Ester	**Formula**	**Odor of**
isoamyl acetate	$CH_3COOC_5H_{11}$	bananas
ethyl butyrate	$C_3H_7COOC_2H_5$	pineapples
amyl butyrate	$C_3H_7COOC_5H_{11}$	apricots
octyl acetate	$CH_3COOC_8H_{17}$	oranges
isoamyl isovalerate	$C_4H_9COOC_5H_{11}$	apples
methyl salicylate	$C_6H_4(OH)(COOCH_3)$	oil of wintergreen
methyl anthranilate	$C_6H_4(NH_2)(COOCH_3)$	grapes

$$R-\overset{\overset{\displaystyle O}{\|}}{C}-O-CH_2$$
$$R-\overset{\overset{\displaystyle O}{\|}}{C}-O-CH$$
$$R-\overset{\overset{\displaystyle O}{\|}}{C}-O-CH_2$$

The fatty acid portions, $R-\overset{\overset{\displaystyle O}{\|}}{C}-$, may be saturated or unsaturated. The R's may be the same or different groups.

SCiLINKS.

TOPIC: Fats
GO TO: www.scilinks.org
sciLINKS CODE: WCH2780

Naturally occurring fats and oils are mixtures of many different esters. Milk fat, lard, and tallow are familiar, important fats. Soybean oil, cottonseed oil, linseed oil, palm oil, and coconut oil are examples of important oils.

The triesters of glycerol are called glycerides. *Simple glycerides* are esters in which all three R groups are identical. Two examples are

Glycerol is

$$HO-CH_2$$
$$HO-CH$$
$$HO-CH_2$$

$$H_2C-O-\overset{\overset{\displaystyle O}{\|}}{C}-(CH_2)_{14}CH_3$$
$$HC-O-\overset{\overset{\displaystyle O}{\|}}{C}-(CH_2)_{14}CH_3$$
$$H_2C-O-\overset{\overset{\displaystyle O}{\|}}{C}-(CH_2)_{14}CH_3$$

glyceryl tripalmitate
(palmitin)

$$H_2C-O-\overset{\overset{\displaystyle O}{\|}}{C}-(CH_2)_{16}CH_3$$
$$HC-O-\overset{\overset{\displaystyle O}{\|}}{C}-(CH_2)_{16}CH_3$$
$$H_2C-O-\overset{\overset{\displaystyle O}{\|}}{C}-(CH_2)_{16}CH_3$$

glyceryl tristearate
(stearin)

Most natural fatty acids contain even numbers of carbon atoms because they are synthesized in the body from two-carbon acetyl groups.

Waxes are esters of fatty acids and alcohols other than glycerol. Most are derived from long-chain fatty acids and long-chain monohydric alcohols. Both usually contain even numbers of carbon atoms. Beeswax is largely $C_{15}H_{31}COOC_{30}H_{61}$; carnauba wax contains $C_{25}H_{51}COOC_{30}H_{61}$. Both are esters of myricyl alcohol, $C_{30}H_{61}OH$.

Fats are solid esters of glycerol and (mostly) saturated acids. Oils are liquid esters that are derived primarily from unsaturated acids and glycerol. The acid portion of a fat usually contains an even number of carbon atoms, often 16 or 18. Some acids that occur frequently in fats and oils are

butyric	$CH_3CH_2CH_2COOH$
lauric	$CH_3(CH_2)_{10}COOH$
myristic	$CH_3(CH_2)_{12}COOH$
palmitic	$CH_3(CH_2)_{14}COOH$
stearic	$CH_3(CH_2)_{16}COOH$
oleic	$CH_3(CH_2)_7CH=CH(CH_2)_7COOH$
linolenic	$CH_3CH_2CH=CHCH_2CH=CHCH_2CH=CH(CH_2)_7COOH$
ricinoleic	$CH_3(CH_2)_5CHOHCH_2CH=CH(CH_2)_7COOH$

Figure 27-19a shows a model of stearic acid, a long-chain saturated fatty acid.

Honeybees produce the wax to build their honeycombs.

Amides

Amides are thought of as derivatives of organic acids and ammonia, primary amines, or secondary amines. Amides contain the $-\overset{\overset{\displaystyle O}{\|}}{C}-N\overset{\diagup}{\diagdown}$ functional group. They are named as

Figure 27-19 Models of long-chain fatty acids. The saturated fatty acids (a) are linear and tend to pack like sticks of wood to form solid masses in blood vessels, thereby constricting them. The *trans* unsaturated fatty acids have a slight Z-shaped kink in the chain, but are also essentially linear molecules. By contrast, *cis* unsaturated fatty acids (b) are bent and so do not pack as well as linear structures and do not collect in blood vessels as readily. Many natural vegetable fats and oils contain esters of *cis* unsaturated fatty acids, or polyunsaturated fatty acids. Health problems associated with saturated fatty acids can be decreased by eating less animal fat, butter, and lard. Problems due to *trans* fatty acids are reduced by avoiding processed vegetable fats.

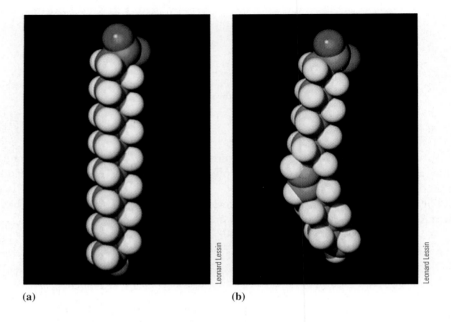

(a)   (b)

derivatives of the corresponding carboxylic acids, the suffix -*amide* being substituted for -*ic acid* or -*oic acid* in the name of the parent acid. The presence of alkyl or aryl substituents attached to nitrogen is designated by prefixing the letter *N* and the name of the substituent to the name of the unsubstituted amide.

| acetamide | benzamide | *N,N*-dimethylacetamide | *N*-phenylacetamide (acetanilide) |

CD-ROM   Screen 11.8, Functional Groups (3): Amino Acids and Proteins.

Proteins are complex amides of high molecular weight. Some synthetic fibers are also polyamides (Sections 27-19 and 28-9).

Crystals of the amide known as acetaminophen (Tylenol) viewed under polarized light. The structure of acetaminophen is

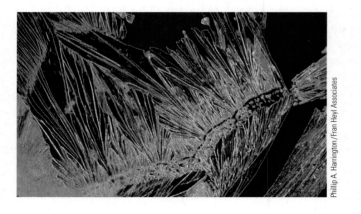

CHEMISTRY IN USE

Our Daily Lives

Butter, Margarine, and *trans* Fats

Humans had consumed butter for thousands of years before France introduced the world to margarine in the late 1860s; by 1874, margarine had reached the United States. Margarine consumption has increased rapidly, due in part to an increased risk of heart disease that has been associated with consumption of butter.

Both butter and margarine are primarily fats, but they contain different kinds of fats. The main ingredient in butter is cream, the concentrated fat from milk, whereas the main ingredient in margarine is vegetable oil, the concentrated fat from plants. Although it is widely accepted that animal fats pose a greater risk for heart disease than vegetable oils, most vegetable oils used in margarine and processed foods are modified by hydrogenation.

The hydrogenation process adds hydrogen atoms to unsaturated vegetable oils; this causes vegetable oils to solidify, which makes them creamier and prolongs their shelf life. Unfortunately, the hydrogenation process changes about 40% of the naturally occurring *cis* molecules of vegetable oil into *trans* isomers; the resulting fat molecules are referred to as *trans* fats. Oils produced in nature contain only *cis* isomers and are considered healthful, but the *trans* fats have been linked to many thousands of deaths due to heart disease. Diets high in hydrogenated vegetable oil may be as unhealthy as those high in saturated animal fats.

To understand the different effects produced by animal fats, vegetable oils, and *trans* fats, we need to know a little about blood cholesterol. Medical specialists recognize several types of cholesterol, two of which are HDL (high-density lipoproteins) and LDL (low-density lipoproteins). Because high levels of HDL reduce the risk of heart disease, HDL has become known as "good" cholesterol; because high levels of LDL increase the risk for heart disease, LDL has become known as "bad" cholesterol. In the average U.S. adult, the total amount of cholesterol (including HDL and LDL) circulating in the blood is about 200 mg per 100 mL of serum.

A diet high in animal fats usually increases blood cholesterol levels beyond 200 mg per 100 mL of serum, thereby increasing the risk for heart disease. When animal fats increase the total blood cholesterol level, however, they increase the levels of both HDL and LDL. In contrast, vegetable oils decrease the total blood cholesterol level, which makes them

Blood pressure check.

healthier foods than other fats. *Trans* fats (from hydrogenated vegetable oils) also decrease the total blood cholesterol level, giving the appearance of being healthy, but unfortunately they accomplish this reduction by simultaneously reducing the desirable HDL levels and increasing the undesirable LDL levels.

The damage done by hydrogenated vegetable oils is particularly disturbing because margarine is one of the largest sources of calories in the U.S. diet; approximately 40% of margarine fat is the *trans* isomer. Hydrogenated vegetable oils are widely used in shortening, cookies, crackers, chips, and other processed foods. Ironically, most of these foods are touted as being healthy because they contain no cholesterol or saturated fats.

What can we do to protect our health? For many years, nutritionists have been telling us what appears to be the "best solution": reduce your total fat intake to 20% of your daily calories, and reduce your consumption of animal fats by substituting vegetable oils. Whenever possible, use vegetable oils that have not been hydrogenated.

Ronald DeLorenzo
Middle Georgia College

FUNCTIONAL GROUP   CLASS OF COMPOUNDS   SOME EXAMPLES

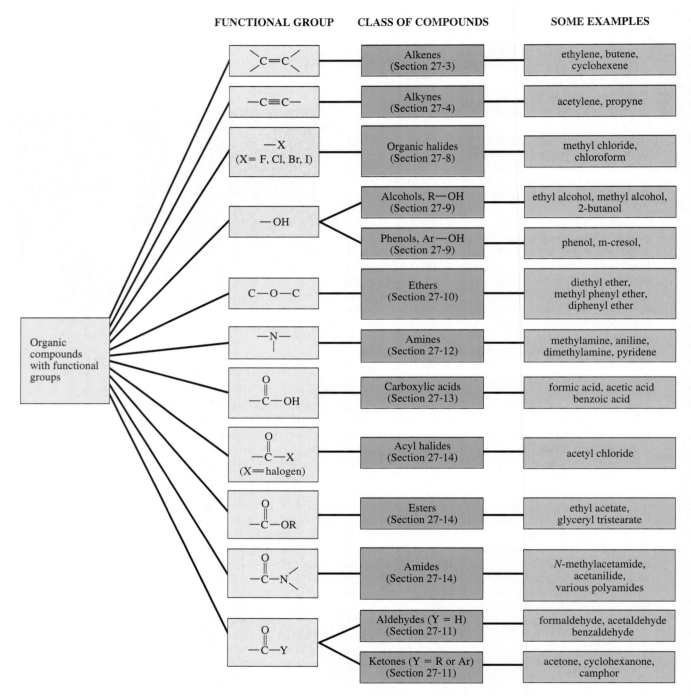

Figure 27-20 Summary of some functional groups and classes of organic compounds.

27-15 SUMMARY OF FUNCTIONAL GROUPS

Some important functional groups and the corresponding classes of related compounds are summarized in Figure 27-20.

FUNDAMENTAL CLASSES OF ORGANIC REACTIONS

Organic compounds display very different abilities to react, ranging from the limited reactivity of hydrocarbons and fluorocarbons to the great variety of reactions undergone by the millions of organic molecules that contain several functional groups. Reactivity depends on structure. We can usually predict the kinds of reactions a compound can undergo by identifying the functional groups it contains. But the electronic and structural features that are *near* a functional group can also affect its reactivity. One of the fascinations of organic chemistry is our ability to "fine-tune" both physical and chemical properties by making small changes in structure. The successes of this approach are innumerable, including the development of fuels and their additives or alternatives, the improvement of pharmaceuticals to enhance their effectiveness and minimize their ill effects, and the development of polymers and plastics with an incredible variety of properties and uses.

In the following sections of this chapter we will present a few of the kinds of reactions that organic compounds undergo. A topic of such vast scope as reactivity of organic compounds can be made manageable only if we divide our study of this field into subtopics. Nearly all organic transformations involve at least one of three fundamental classes of reactions. The following three sections will address substitution, addition, and elimination reactions. We will also look at some reaction sequences that combine reaction steps from more than one of the fundamental classes.

27-16 SUBSTITUTION REACTIONS

In a **substitution reaction** an atom or a group of atoms attached to a carbon atom is removed, and another atom or group of atoms takes its place. No change occurs in the degree of saturation at the reactive carbon atom.

The saturated hydrocarbons (alkanes and cycloalkanes) are chemically rather inert materials. For many years they were known as *paraffins* because they undergo few reactions. They do not react with such powerful oxidizing agents as potassium permanganate and potassium dichromate. They do react with the halogens, however, with oxygen when ignited, and with concentrated nitric acid. As expected, members of a homologous series (see Section 27-1) have similar chemical properties. If we study the chemistry of one of these compounds, we can make predictions about the others with a fair degree of certainty.

The saturated hydrocarbons can react without a big disruption of the molecular structure only by *displacement*, or *substitution of one atom for another.* At room temperature, chlorine and bromine react very slowly with saturated straight-chain hydrocarbons. At higher temperatures, or in the presence of sunlight or other sources of ultraviolet light, H atoms in the hydrocarbon can be replaced easily by halogen atoms. These substitution reactions are called **halogenation** reactions.

The reaction is represented as

Note that only one half of the chlorine atoms occur in the organic product. The other half form hydrogen chloride, a commercially valuable compound.

$$
\underset{\text{methane}}{H-\overset{\displaystyle H}{\underset{\displaystyle H}{\overset{|}{\underset{|}{C}}}}-H} \; + \; \underset{\text{chlorine}}{Cl-Cl} \; \xrightarrow[\text{UV}]{\text{heat or}} \; \underset{\substack{\text{chloromethane} \\ \text{(methyl chloride)} \\ \text{bp} = -23.8°C}}{H-\overset{\displaystyle H}{\underset{\displaystyle H}{\overset{|}{\underset{|}{C}}}}-Cl} \; + \; HCl
$$

Many organic reactions produce more than a single product. For example, the chlorination of CH_4 may produce several other products in addition to CH_3Cl, as the following equations show.

$$
H-\overset{\displaystyle H}{\underset{\displaystyle H}{\overset{|}{\underset{|}{C}}}}-Cl \; + \; Cl-Cl \; \longrightarrow \; \underset{\substack{\text{dichloromethane} \\ \text{(methylene chloride)} \\ \text{bp} = 40.2°C}}{Cl-\overset{\displaystyle H}{\underset{\displaystyle H}{\overset{|}{\underset{|}{C}}}}-Cl} \; + \; HCl
$$

$$
Cl-\overset{\displaystyle H}{\underset{\displaystyle H}{\overset{|}{\underset{|}{C}}}}-Cl \; + \; Cl-Cl \; \longrightarrow \; \underset{\substack{\text{trichloromethane} \\ \text{(chloroform)} \\ \text{bp} = 61°C}}{Cl-\overset{\displaystyle H}{\underset{\displaystyle Cl}{\overset{|}{\underset{|}{C}}}}-Cl} \; + \; HCl
$$

$$
Cl-\overset{\displaystyle H}{\underset{\displaystyle Cl}{\overset{|}{\underset{|}{C}}}}-Cl \; + \; Cl-Cl \; \longrightarrow \; \underset{\substack{\text{tetrachloromethane} \\ \text{(carbon tetrachloride)} \\ \text{bp} = 76.8°C}}{Cl-\overset{\displaystyle Cl}{\underset{\displaystyle Cl}{\overset{|}{\underset{|}{C}}}}-Cl} \; + \; HCl
$$

When a hydrocarbon has more than one C atom, its reaction with Cl_2 is more complex. The first step in the chlorination of ethane gives the product that contains one Cl atom per molecule.

$$
\underset{\text{ethane}}{H-\overset{\displaystyle H}{\underset{\displaystyle H}{\overset{|}{\underset{|}{C}}}}-\overset{\displaystyle H}{\underset{\displaystyle H}{\overset{|}{\underset{|}{C}}}}-H} \; + \; Cl-Cl \; \xrightarrow[\text{UV}]{\text{heat or}} \; \underset{\substack{\text{chloroethane} \\ \text{(ethyl chloride)} \\ \text{bp} = 13.1°C}}{H-\overset{\displaystyle H}{\underset{\displaystyle H}{\overset{|}{\underset{|}{C}}}}-\overset{\displaystyle H}{\underset{\displaystyle H}{\overset{|}{\underset{|}{C}}}}-Cl} \; + \; HCl
$$

When a second hydrogen atom is replaced, a mixture of the two possible products is obtained.

$$H-\underset{\underset{H}{|}}{\overset{\overset{H}{|}}{C}}-\underset{\underset{H}{|}}{\overset{\overset{H}{|}}{C}}-Cl + Cl-Cl \xrightarrow{\text{heat or UV}} \left\{\begin{array}{c} H-\underset{\underset{H}{|}}{\overset{\overset{H}{|}}{C}}-\underset{\underset{H}{|}}{\overset{\overset{Cl}{|}}{C}}-Cl \\ \text{1,1-dichlorethane} \\ \text{bp} = 57°C \\[1em] Cl-\underset{\underset{H}{|}}{\overset{\overset{H}{|}}{C}}-\underset{\underset{H}{|}}{\overset{\overset{H}{|}}{C}}-Cl \\ \text{1,2-dichloroethane} \\ \text{bp} = 84°C \end{array}\right\} + HCl$$

The product mixture does not contain equal numbers of moles of the dichloroethanes so we do not show a stoichiometrically balanced equation. Because reactions of saturated hydrocarbons with chlorine can produce many products, the reactions are not always as useful as might be desired.

Substitution is the most common kind of reaction of an aromatic ring. Halogenation, with chlorine or bromine, occurs readily in the presence of iron or anhydrous iron(III) chloride (a Lewis acid) catalyst.

$$\text{benzene} + Cl_2 \xrightarrow{FeCl_3} \text{chlorobenzene} + HCl$$

When iron is added as a catalyst, it reacts with chlorine to form iron(III) chloride, which is the true catalyst.

The equation is usually written in condensed form as

$$\text{⬡} + Cl_2 \xrightarrow{FeCl_3} \text{⬡-Cl} + HCl$$

Aromatic rings can undergo *nitration*, substitution of the *nitro* group $-NO_2$, in a mixture of concentrated nitric and sulfuric acids at moderate temperatures.

The H_2SO_4 is both a catalyst and a dehydrating agent. The H in the product H_2O comes from the hydrocarbon; the OH comes from HNO_3.

$$\text{⬡} + HNO_3 \xrightarrow[50°]{\text{conc. } H_2SO_4} \text{⬡-}NO_2 + H_2O$$

nitric acid   nitrobenzene

The explosive TNT (2,4,6-trinitrotoluene) is manufactured by the nitration of toluene in three steps.

$$\text{toluene} + 3HNO_3 \xrightarrow{H_2SO_4} \text{2,4,6-trinitrotoluene (TNT)} + 3H_2O$$

toluene   nitric acid   2,4,6-trinitrotoluene (TNT)

A model of a molecule of TNT.

Groups other than hydrogen can be substituted by other atoms or groups of atoms. For example, the bromine atom of an alkyl bromide can be replaced with a hydroxyl group to form an alcohol. The mechanism of this type of reaction is often studied in introductory organic chemistry courses.

$$CH_3-\underset{\underset{Br}{|}}{CH}-CH_3 + OH^- \longrightarrow CH_3-\underset{\underset{OH}{|}}{CH}-CH_3 + Br^-$$

Simple inorganic esters may be thought of as compounds that contain one or more alkyl groups covalently bonded to the anion of a *ternary* inorganic acid. Unless indicated, the term "ester" refers to organic esters.

Alcohols react with common inorganic oxyacids to produce **inorganic esters.** For instance, nitric acid reacts with alcohols to produce nitrates by substitution of nitrate, $-ONO_2$, for hydroxyl, $-OH$.

$$\underset{\text{ethanol}}{CH_3CH_2OH} + \underset{\text{nitric acid}}{HNO_3} \longrightarrow \underset{\text{ethyl nitrate}}{CH_3CH_2-ONO_2} + H_2O$$

Nobel's brother had been killed and his father physically disabled in a nitroglycerine explosion in 1864. Nobel willed $9.2 million to establish a fund for annual prizes in physics, chemistry, medicine, literature, and peace. The prizes were first awarded in 1901.

The substitution reaction of nitric acid with glycerol produces the explosive nitroglycerine. Alfred Nobel's (1833–1896) discovery in 1866 that this very sensitive material could be made into a "safe" explosive by absorbing it into diatomaceous earth or wood meal led to his development of dynamite.

$$\underset{\text{glycerol}}{\begin{array}{l} H_2C-OH \\ | \\ HC-OH \\ | \\ H_2C-OH \end{array}} + 3HNO_3 \xrightarrow{H_2SO_4} \underset{\substack{\text{glyceryl trinitrate} \\ \text{(nitroglycerine)}}}{\begin{array}{l} H_2C-ONO_2 \\ | \\ HC-ONO_2 \\ | \\ H_2C-ONO_2 \end{array}} + 3H_2O$$

Interestingly, nitroglycerine ("nitro") is taken by some who have heart disease. It acts as a vasodilator (dilates the blood vessels) to decrease arterial tension.

Cold, concentrated H_2SO_4 reacts with alcohols to form **alkyl hydrogen sulfates.** The reaction with lauryl alcohol is an important industrial reaction.

$$\underset{\text{1-dodecanol (lauryl alcohol)}}{CH_3(CH_2)_{10}CH_2-OH} + \underset{\substack{\text{sulfuric} \\ \text{acid}}}{HOSO_3H} \longrightarrow \underset{\text{lauryl hydrogen sulfate}}{CH_3(CH_2)_{10}CH_2-OSO_3H} + H_2O$$

The neutralization reaction of an alkyl hydrogen sulfate with NaOH then produces the sodium salt of the alkyl hydrogen sulfate.

$$CH_3(CH_2)_{10}CH_2-OSO_3H + Na^+OH^- \longrightarrow \underset{\text{sodium lauryl sulfate (a detergent)}}{CH_3(CH_2)_{10}CH_2-OSO_3^-Na^+} + H_2O$$

Sodium salts of the alkyl hydrogen sulfates that contain about 12 carbon atoms are excellent detergents. They are also biodegradable. (Soaps and detergents were discussed in Section 14-18.)

27-17 ADDITION REACTIONS

An **addition reaction** involves an *increase* in the number of atoms or groups attached to carbon. The molecule becomes more nearly saturated.

The principal reactions of alkenes and alkynes are addition reactions rather than substitution reactions. For example, contrast the reactions of ethane and ethylene with Cl_2.

CD-ROM Screen 11.4, Hydrocarbons and Addition Reactions.

ethane:
$$CH_3—CH_3 + Cl_2 \longrightarrow CH_3—CH_2Cl + HCl \quad \text{(substitution, slow)}$$

ethylene:
$$H_2C{=}CH_2 + Cl_2 \longrightarrow \underset{\underset{Cl}{|}}{CH_2}{-}\underset{\underset{Cl}{|}}{CH_2} \quad \text{(addition, rapid)}$$

Carbon–carbon double bonds are *reaction sites* and so represent *functional groups*. Most addition reactions involving alkenes and alkynes proceed rapidly at room temperature. By contrast, many substitution reactions of the alkanes require catalysts and high temperatures.

Bromine adds readily to the alkenes to give dibromides. The reaction with ethylene is

$$H_2C{=}CH_2 + Br_2 \longrightarrow \underset{\underset{Br}{|}}{CH_2}{-}\underset{\underset{Br}{|}}{CH_2} \quad \begin{array}{l}\text{1,2-dibromoethane}\\ \text{(ethylene dibromide)}\end{array}$$

The addition of Br_2 to alkenes is used as a simple qualitative test for unsaturation. Bromine, a dark red liquid, is dissolved in a nonpolar solvent. When an alkene is added, the solution becomes colorless as the Br_2 reacts with the alkene to form a colorless compound. This reaction may be used to distinguish between alkanes and alkenes.

Hydrogenation is an extremely important addition reaction of the alkenes. Hydrogen adds across double bonds at elevated temperatures, under high pressures, and in the presence of an appropriate catalyst (finely divided Pt, Pd, or Ni).

$$CH_2{=}CH_2 + H_2 \xrightarrow[\text{heat}]{\text{catalyst}} CH_3—CH_3$$

Some unsaturated hydrocarbons are converted to saturated hydrocarbons in the manufacture of high-octane gasoline and aviation fuels. Unsaturated vegetable oils can also be converted to solid cooking fats (shortening) by hydrogenation of most of the carbon–carbon double bonds present (Figure 27-21). Most of the double bonds that remain are the *trans*-isomer.

$$
\begin{array}{c}
\overset{\displaystyle O}{\overset{\displaystyle \|}{H_2COC}}(CH_2)_7CH{=}CH(CH_2)_7CH_3 \\
|\\
\overset{\displaystyle O}{\overset{\displaystyle \|}{HCOC}}(CH_2)_7CH{=}CH(CH_2)_7CH_3 \\
|\\
\overset{\displaystyle O}{\overset{\displaystyle \|}{H_2COC}}(CH_2)_7CH{=}CH(CH_2)_7CH_3 \\
\text{olein (an oil, liquid)}
\end{array}
\xrightarrow[\substack{\text{Ni catalyst}\\ \text{heat}}]{3H_2}
\begin{array}{c}
\overset{\displaystyle O}{\overset{\displaystyle \|}{H_2COC}}(CH_2)_{16}CH_3 \\
|\\
\overset{\displaystyle O}{\overset{\displaystyle \|}{HCOC}}(CH_2)_{16}CH_3 \\
|\\
\overset{\displaystyle O}{\overset{\displaystyle \|}{H_2COC}}(CH_2)_{16}CH_3 \\
\text{stearin (a fat, solid)}
\end{array}
$$

The *hydration reaction* (addition of water) is another very important addition reaction of alkenes. It is used commercially for the preparation of a wide variety of alcohols from petroleum by-products. Ethanol, the most important industrial alcohol, is produced industrially by the hydration of ethylene from petroleum, using H_2SO_4 as a catalyst.

$$H_2C{=}CH_2 + H_2O \xrightarrow{H_2SO_4} H_3C—CH_2OH$$

Figure 27-21 Hydrogenation of the olefinic double bonds in vegetable oil converts it to solid fat.

Think of H_2O as HOH.

One of the commercially most important addition reactions of the alkenes forms *polymers*. This reaction will be discussed in Section 27-19.

The alkynes contain two pi bonds, both of which are sources of electrons, and they are more reactive than the alkenes. The most common reaction of the alkynes is addition across the triple bond. The reactions with hydrogen and with bromine are typical.

$$H-C\equiv C-H \xrightarrow{H_2} \begin{array}{c} H \\ \diagdown \\ C=C \\ \diagup \qquad \diagdown \\ H \qquad\quad H \end{array}$$

$$H-C\equiv C-H \xrightarrow{Br_2} \begin{array}{c} H \qquad\quad Br \\ \diagdown \qquad \diagup \\ C=C \\ \diagup \qquad \diagdown \\ Br \qquad\quad H \end{array}$$

1,2-dibromoethene

Other unsaturated bonds can also undergo addition reactions. Probably the most important example is the carbonyl group, $-\overset{\displaystyle O}{\overset{\|}{C}}-$. Because of the availability of unshared pairs of electrons on the oxygen atom, the products can undergo a wide variety of subsequent reactions. For example, HCN adds to the C=O bond of acetone.

$$CH_3-\overset{\displaystyle O}{\overset{\|}{C}}-CH_3 + HCN \xrightarrow{NaOH(aq)} CH_3-\overset{\displaystyle OH}{\underset{\displaystyle CN}{\overset{|}{\underset{|}{C}}}}-CH_3$$

This reaction is a key early step in the production of the transparent plastic known as Plexiglas or Lucite.

27-18 ELIMINATION REACTIONS

An **elimination reaction** involves a *decrease* in the number of atoms or groups attached to carbon. The degree of unsaturation increases.

An elimination reaction, **dehydrohalogenation,** can occur for chloro-, bromo- and iodoalkanes. In such a reaction, the halogen, X, from one C atom and a hydrogen from an adjacent C atom are eliminated. A double bond between two carbon atoms is formed; the molecule becomes *more unsaturated*. The net reaction is the transformation of an alkyl halide (or haloalkane) into an alkene. Dehydrohalogenation reactions usually require a strong base such as sodium hydroxide, NaOH.

$$H-\overset{\displaystyle H}{\underset{\displaystyle H}{\overset{|}{\underset{|}{C}}}}-\overset{\displaystyle H}{\underset{\displaystyle Br}{\overset{|}{\underset{|}{C}}}}-H + NaOH \longrightarrow \begin{array}{c} H \qquad\quad H \\ \diagdown \qquad \diagup \\ C=C \\ \diagup \qquad \diagdown \\ H \qquad\quad H \end{array} + H_2O + NaBr$$

bromoethane　　　　　　　　　　　ethene
(ethylene)

Charles D. Winters

Concentrated sulfuric acid is an excellent dehydrating agent. Here it removes water from sucrose, a sugar with the formula $C_{12}H_{22}O_{11}$. Dehydration of sucrose produces (mostly) carbon.

A related reaction is **dehydration,** in which an alcohol is converted into an alkene and water by the elimination of H— and —OH from adjacent carbon atoms. The dehydration of an alcohol to form an alkene can be considered the reverse of the hydration of an alkene to form an alcohol (Section 27-17). Dehydration reactions are catalyzed by acids.

Such simple elimination reactions are relatively rare. Elimination reactions, however, frequently occur as individual steps in more complex reaction sequences. Many more elimination reactions are encountered in courses in organic chemistry.

27-19 POLYMERIZATION REACTIONS

A **polymer** is a large molecule that is a high-molecular-weight chain of small molecules. The small molecules that are linked to form polymers are called **monomers.** Typical polymers consist of hundreds or thousands of monomer units and have molecular weights up to thousands or millions.

> **Polymerization** is the combination of many small molecules (monomers) to form large molecules (polymers).

The word fragment -*mer* means "part." Recall that *isomers* are compounds that are composed of the same (*iso*) parts (*mers*). A *monomer* is a "single part"; a large number of monomers combine to form a *polymer,* "many parts."

Polymers are divided into two classes: natural and synthetic. Important biological molecules such as proteins, nucleic acids, and polysaccharides (starches and the cellulose in wood and cotton) are natural polymers. Natural rubber and natural fibers such as silk and wool are also natural polymers. Familiar examples of synthetic polymers include plastics such as polyethylene, Teflon, and Lucite (Plexiglas) and synthetic fibers such as nylon, Orlon, and Dacron. In this section we will describe some processes by which polymers are formed from organic compounds.

SC*i*LINKS.

TOPIC: Polymers
GO TO: www.scilinks.org
*sci*LINKS CODE: WCH2790

Addition Polymerization

CD-ROM   Screen 11.9, Synthetic Organic Polymers (1): Addition Polymerization.

Polymerization is an important addition reaction (Section 27-17) of the alkenes. Polymers formed by this kind of reaction are called **addition polymers.** The formation of polyethylene is an important example. In the presence of appropriate catalysts (a mixture of aluminum trialkyls, R_3Al, and titanium tetrachloride, $TiCl_4$), ethylene polymerizes into chains containing 800 or more carbon atoms.

$$nCH_2{=}CH_2 \xrightarrow{\text{catalyst}} +CH_2{-}CH_2{\cdot}_n$$
$$\text{ethylene} \qquad\qquad \text{polyethylene}$$

The polymer may be represented as $CH_3(CH_2{-}CH_2)_nCH_3$, where n is approximately 400. Polyethylene is a tough, flexible plastic. It is widely used as an electrical insulator and for the fabrication of such items as unbreakable refrigerator dishes, plastic cups, and squeeze bottles. Polypropylene is made by polymerizing propylene, $CH_3{-}CH{=}CH_2$, in much the same way. Teflon is made by polymerizing tetrafluoroethylene in a similar reaction.

Teflon is a trade name owned by Du-Pont, a company that has developed and manufactured many fluorinated polymers.

$$nCF_2{=}CF_2 \xrightarrow[\text{heat}]{\text{catalyst}} +CF_2{-}CF_2{\cdot}_n$$
$$\text{tetrafluoroethylene} \qquad\qquad \text{Teflon}$$

The molecular weight of Teflon is about 2×10^6. Approximately 20,000 $CF_2{=}CF_2$ molecules polymerize to form a single giant molecule. Teflon is a very useful polymer. It does *not* react with concentrated acids and bases or with most oxidizing agents. It does not dissolve in most organic solvents.

Natural rubber is obtained from the sap of the rubber tree, a sticky liquid called latex. Rubber is a polymeric hydrocarbon formed in the sap by the combination of about 2000 molecules of 2-methyl-1,3-butadiene, commonly called isoprene. The molecular weight of rubber is about 136,000.

Many cooking utensils with "non-stick" surfaces are coated with a polymer such as Teflon.

$$2nCH_2{=}\overset{\overset{\textstyle CH_3}{|}}{C}{-}CH{=}CH_2 \longrightarrow +(CH_2{-}\overset{\overset{\textstyle CH_3}{|}}{C}{=}CH{-}CH_2{-}CH_2{-}\overset{\overset{\textstyle CH_3}{|}}{C}{=}CH{-}CH_2{)_n}$$
$$\text{isoprene} \qquad\qquad\qquad\qquad\qquad \text{natural rubber}$$

When natural rubber is warmed, it flows and becomes sticky. To eliminate this problem, **vulcanization** is used. This is a process in which sulfur is added to rubber and the mixture is heated to approximately 140°C. Sulfur atoms combine with some of the double bonds in the linear polymer molecules to form bridges that bond one rubber molecule to another. This cross-linking by sulfur atoms converts the linear polymer into a three-dimensional polymer. Fillers and reinforcing agents are added during the mixing process to increase the durability of rubber and to form colored rubber. Carbon black is the most common reinforcing agent. Zinc oxide, barium sulfate, titanium dioxide, and antimony(V) sulfide are common fillers.

Numerous other polymers are elastic enough to be called by the generic name "rubber."

Some synthetic rubbers are superior to natural rubber in some ways. Neoprene is a synthetic elastomer (an elastic polymer) with properties quite similar to those of natural rubber. The basic structural unit is 2-chloro-1,3-butadiene, commonly called chloroprene, which differs from isoprene in having a chlorine atom rather than a methyl group at carbon 2 of the 1,3-butadiene chain.

$$nCH_2{=}CH{-}\overset{\overset{\textstyle Cl}{|}}{C}{=}CH_2 \xrightarrow{\text{polymerization}} +(CH_2{-}CH{=}\overset{\overset{\textstyle Cl}{|}}{C}{-}CH_2{)_n}$$
$$\text{chloroprene} \qquad\qquad\qquad\qquad \text{neoprene (a synthetic rubber)}$$

E. I. DuPont de Nemours & Company

Neoprene is less affected by gasoline and oil and is more elastic than natural rubber. It resists abrasion well and is not swollen or dissolved by hydrocarbons. It is widely used to make hoses for oil and gasoline, electrical insulation, and automobile and refrigerator parts.

When two different monomers are mixed and then polymerized, **copolymers** are formed. Depending on the ratio of the two monomers and the reaction conditions, the order of the units can range from quite regular (e.g., alternating) to completely random. In this way, polymers with a wide variety of properties can be produced. The most important rubber produced in the largest amount in the United States is styrene butadiene rubber (SBR), a polymer of styrene with butadiene in a $1:3$ molecular ratio.

The double bonds in SBR can be cross-linked by vulcanization as described for natural rubber. SBR is used primarily for making tires. Other copolymers are used to make car bumpers, body and chassis parts, wire insulation, sporting goods, sealants, and caulking compounds.

Some addition polymers and their uses are listed in Table 27-14.

Condensation Polymerization

Some polymerization reactions are based on **condensation reactions,** in which two molecules combine by splitting out or eliminating a small molecule. For such a polymer to be formed, each monomer must have two functional groups, one on each end. A polymer formed in this way is called a **condensation polymer.** There are many useful condensation polymers, based on a wide variety of bifunctional molecules.

Polyesters (short for "*polymeric esters*") are condensation polymers that are formed when *dihydric alcohols* react with *dicarboxylic acids*. An ester linkage is formed at each end of each monomer molecule to build up large molecules. A useful polyester is prepared from terephthalic acid and ethylene glycol.

CD-ROM   Screen 11.10, Synthetic Organic Polymers (2): Condensation Polymerization.

Dihydric alcohols contain two —OH groups per molecule. They are often called glycols.

polyethylene terephthalate (also known as PET)
(Dacron, Mylar)

TABLE 27-14 *Some Important Addition Polymers*

Polymer Name (some trade names)	Some Uses	Polymer Production, Tons/Year in United States	Monomer Name	Monomer Formula
polyethylene (Polythene)	electrical insulation; toys and molded objects; bags; squeeze bottles	14 million	ethylene	$\begin{array}{c} H \\ \diagdown \\ C=C \\ \diagup \\ H \end{array} \begin{array}{c} H \\ \diagup \\ \diagdown \\ H \end{array}$
polypropylene (Herculon, Vectra)	bottles; films; lab equipment; toys; packaging film; filament for rope, webbing, carpeting; molded auto and appliance parts	6.7 million	propylene	$\begin{array}{c} H \\ \diagdown \\ C=C \\ \diagup \\ H \end{array} \begin{array}{c} H \\ \diagup \\ \diagdown \\ CH_3 \end{array}$
polyvinyl chloride (PVC)	pipe, siding, gutters; floor tiles	7.4 million	vinyl chloride	$\begin{array}{c} H \\ \diagdown \\ C=C \\ \diagup \\ H \end{array} \begin{array}{c} H \\ \diagup \\ \diagdown \\ Cl \end{array}$
polyacrylonitrile (Orlon, Acrilan)	acrylic fibers for carpets, clothing, knitwear	95,000	acrylonitrile	$\begin{array}{c} H \\ \diagdown \\ C=C \\ \diagup \\ H \end{array} \begin{array}{c} H \\ \diagup \\ \diagdown \\ CN \end{array}$
polystyrene (Styrene, Styrofoam, Styron)	molded toys, dishes, kitchen equipment; insulating foam, e.g., ice chests; rigid foam packaging	3.3 million	styrene	$\begin{array}{c} H \\ \diagdown \\ C=C \\ \diagup \\ H \end{array} \begin{array}{c} H \\ \diagup \\ \diagdown \\ C_6H_5 \end{array}$
poly(methyl methacrylate) (Plexiglas, Lucite)	high-quality transparent objects; water-based paints; contact lenses	507,000	methyl methacrylate	$\begin{array}{c} H \\ \diagdown \\ C=C \\ \diagup \\ H \end{array} \begin{array}{c} CH_3 \\ \diagup \\ \diagdown \\ C-OCH_3 \\ \parallel \\ O \end{array}$
polybutadiene	automotive tire tread, hoses, belts; metal can coatings	610,000	1,3-butadiene	$\begin{array}{c} H \\ \diagdown \\ C=C \\ \diagup \\ H \end{array} \begin{array}{c} H \\ \diagup \\ \diagdown \\ C=C \\ \diagup \quad \diagdown \\ H \quad\quad H \end{array}$
ethylene-propylene copolymer	appliance parts; auto hoses, bumpers, body and chassis parts; coated fabrics	324,000	ethylene, propylene	formulas given previously
SBR copolymer	tires	1 million	styrene, 1,3-butadiene	formulas given previously

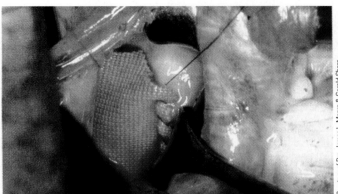

A patch made of Dacron polymer is used to close a defect in a human heart.

More than 2 million tons of this polymer is produced annually in the United States. Dacron, the fiber produced from this polyester, accounts for approximately 50% of all synthetic fibers. It absorbs very little moisture, and its properties are nearly the same whether it is wet or dry. Additionally, it possesses exceptional elastic recovery properties so it is used to make "permanent-press" fabrics. This polyester can also be made into films of great strength (e.g., Mylar), which can be rolled into sheets $\frac{1}{30}$ the thickness of a human hair. Such films can be magnetically coated to make audio- and videotapes.

The polymeric amides, **polyamides,** are an especially important class of condensation polymers. **Nylon** is the best known polyamide. It is prepared by heating anhydrous hexamethylenediamine with anhydrous adipic acid, a dicarboxylic acid. This substance is often called nylon 66 because the parent diamine and dicarboxylic acid each contain six carbon atoms.

$$
\left. \begin{array}{c} \underset{\text{adipic acid}}{\text{HO}-\overset{\overset{\text{O}}{\|}}{\text{C}}-(\text{CH}_2)_4-\overset{\overset{\text{O}}{\|}}{\text{C}}-\text{OH}} \\ + \\ \underset{\text{hexamethylenediamine}}{\text{H}_2\text{N}-(\text{CH}_2)_6-\text{NH}_2} \end{array} \right\} \xrightarrow[-\text{H}_2\text{O}]{\text{heat}} -\text{NH}\left(\overset{\overset{\text{O}}{\|}}{\text{C}}-(\text{CH}_2)_4-\overset{\overset{\text{O}}{\|}}{\text{C}}-\text{NH}-(\text{CH}_2)_6-\text{NH}\right)_n\overset{\overset{\text{O}}{\|}}{\text{C}}-
$$

nylon 66
(a polyamide)

(Top) This baby feeding spoon uses a thermochromic polymer that visibly and reversibly changes color at a given temperature. The temperature at which the color change occurs can be tailored by chemical modification of the polymer. (Bottom) Light-emitting plastics, such as polyphenylene vinylene (PPV), are polymers that emit light when a voltage is applied across them. These new plastics can be made to emit light of any color and could be used in many future applications, including light-emitting devices (LEDs) and thin-screen televisions.

The molecular weight of the polymer varies from about 10,000 to about 25,000. It melts at about 260–270°C.

Molten nylon is drawn into threads (Figure 27-22). After cooling to room temperature, these can be stretched to about four times their original length. The "cold drawing" process orients the polymer molecules so that their long axes are parallel to the fiber axis. At regular intervals there are N—H---O hydrogen bonds that *cross-link* adjacent chains to give strength to the fiber.

Petroleum is the ultimate source of both adipic acid and hexamethylenediamine. We do not mean that these compounds are present in petroleum, only that they are made from it. The same is true for many other industrial chemicals. The cost of petroleum is an important factor in our economy because so many products are derived from petroleum.

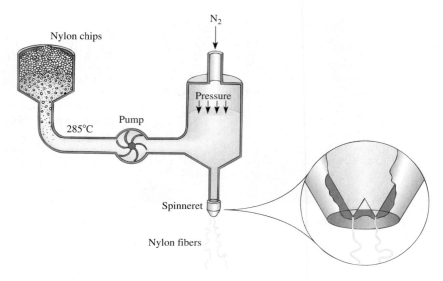

Figure 27-22 Fibers of synthetic polymers are made by extrusion of the molten material through tiny holes, called *spinnerets*. After cooling, nylon fibers are stretched to about four times their original length to orient the polymer molecules.

Nylon is formed at the interface where hexamethylenediamine (in the lower water layer) and adipyl chloride (a derivative of adipic acid, in the upper hexane layer) react. The nylon can be drawn out and wound on a stirring rod. In industry, adipic acid is used in place of adipyl chloride.

Some types of natural condensation polymers play crucial roles in living systems. **Proteins** (Section 28-9) are polymeric chains of L-*amino acids* linked by peptide bonds. A **peptide bond** is formed by the elimination of a molecule of water between the amino group of one amino acid and the carboxylic acid group of another.

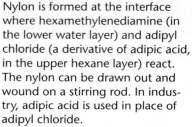

When this process is carried out repeatedly, a large molecule called a **polypeptide** is formed.

Proteins are discussed in greater detail in Section 28-9.

Key Terms

Acid halide   See *Acyl halide.*

Acyl group   The group of atoms remaining after removal of an —OH group of a carboxylic acid.

Acyl halide   A compound derived from a carboxylic acid by replacing the —OH group with a halogen (X), usually —Cl;

general formula is $R-\overset{\overset{\textstyle O}{\|}}{C}-X$; also called acid halide.

Addition polymer   A polymer formed by an addition reaction.

Addition reaction   A reaction in which two atoms or groups of atoms are added to a molecule, one on each side of a double or triple bond. The number of groups attached to carbon *increases*, and the molecule becomes more nearly saturated.

Alcohol   A hydrocarbon derivative in which an H attached to a carbon atom not part of an aromatic ring has been replaced by an —OH group.

Aldehyde   A compound in which an alkyl or aryl group and a hydrogen atom are attached to a carbonyl group; general

formula is $R-\overset{\overset{\textstyle O}{\|}}{C}-H$; R may be H.

Aliphatic hydrocarbons   Hydrocarbons that do not contain aromatic rings.

Alkanes   See *Saturated hydrocarbons.*

Alkenes   Unsaturated hydrocarbons that contain a carbon–carbon double bond.

Alkyl group   A group of atoms derived from an alkane by the removal of one hydrogen atom.

Alkylbenzene   A compound containing an alkyl group bonded to a benzene ring.

Alkynes   Unsaturated hydrocarbons that contain a carbon–carbon triple bond.

Amide   A compound containing the $-\overset{\overset{\textstyle O}{\|}}{C}-N\big\langle$ group.

Amine   A derivative of ammonia in which one or more hydrogen atoms have been replaced by alkyl or aryl groups.

Amino acid   A compound containing both an amino group and a carboxylic acid group.

Amino group   The —NR_2 group, where R can represent an H, alkyl, or aryl group.

Aromatic hydrocarbons   Benzene and similar condensed ring compounds; contain delocalized rings of electrons.

Aryl group   The group of atoms remaining after a hydrogen atom is removed from an aromatic system.

Carbonyl group   The $-\overset{\overset{\textstyle O}{\|}}{C}-$ group.

Carboxylic acid   A compound containing the carboxyl group

$-\overset{\overset{\textstyle O}{\|}}{C}-O-H$.

Condensation polymer   A polymer that is formed by a condensation reaction.

Condensation reaction   A reaction in which a small molecule, such as water or hydrogen chloride, is eliminated and two molecules are joined.

Conjugated double bonds   Double bonds that are separated from each other by one single bond, as in C=C—C=C.

Constitutional isomers   Compounds that contain the same numbers of the same kinds of atoms but that differ in the order in which their atoms are bonded together. Also known as structural isomers.

Copolymer   A polymer formed from two different compounds (monomers).

Cycloalkanes   Cyclic saturated hydrocarbons.

Dehydration   The reaction in which H— and —OH are eliminated from adjacent carbon atoms to form water and a more unsaturated bond.

Dehydrohalogenation   An elimination reaction in which a hydrogen halide, HX (X = Cl, Br, I), is eliminated from a haloalkane. A C=C double bond is formed.

Elimination reaction   A reaction in which the number of groups attached to carbon *decreases*. The degree of unsaturation in the molecule increases.

Ester   A compound of the general formula $R-\overset{\overset{\textstyle O}{\|}}{C}-O-R'$, where R and R' may be the same or different and may be either aliphatic or aromatic. R may be H. R' cannot be H.

Ether   A compound in which an oxygen atom is bonded to two alkyl or two aryl groups, or one alkyl and one aryl group.

Fat   A solid triester of glycerol and (mostly) saturated fatty acids.

Fatty acid   A long-chain aliphatic acid; many can be obtained from animal fats.

Functional group   A group of atoms that represents a potential reaction site in an organic compound.

Geometric isomers   Compounds with different arrangements of groups on the opposite sides of a bond with restricted rotation, such as a double bond; for example, *cis–trans* isomers of certain alkenes.

Glyceride   A triester of glycerol.

Halogenation reaction   A substitution reaction in which a hydrogen atom is replaced by a halogen atom.

Heteroatom   An atom other than C or H in an organic molecule; the most common heteroatoms are O, N, S, P, and the halogens.

Heterocyclic amine   An amine in which nitrogen is part of a ring.

Homologous series   A series of compounds in which each member differs from the next by a specific number and kind of atoms.

Hydration reaction   A reaction in which the elements of water, H and OH, add across a double or triple bond.

Hydrocarbon   A compound that contains only carbon and hydrogen.

Hydrogenation   The reaction in which hydrogen adds across a double or triple bond.

Isomers   Different compounds that have the same molecular formula.

Ketone   A compound in which a carbonyl group is bound to two alkyl or two aryl groups, or to one alkyl and one aryl group.

Monomers   The small molecules that are linked to form polymers.

Oil   A liquid triester of glycerol and unsaturated fatty acids.

Organic chemistry   The study of carbon-containing compounds that contain C—C or C—H bonds and sometimes a few other elements such as oxygen, nitrogen, sulfur, phosphorus, and the halogens.

Peptide bond   A bond formed by elimination of a molecule of water between the amino group of one amino acid and the carboxylic acid group of another.

Phenol   A hydrocarbon derivative that contains an —OH group bound to an aromatic ring.

Pi bond   A chemical bond formed by the side-to-side overlap of atomic orbitals.

Polyamide   A polymeric amide.

Polyene   A compound that contains more than one double bond per molecule.

Polyester   A polymeric ester.

Polymerization   The combination of many small molecules (monomers) to form large molecules (polymers).

Polymers   Large molecules formed by the combination of many small molecules (monomers).

Polyol   An alcohol that contains more than one —OH group; a polyol with two —OH groups is called a glycol.

Polypeptide   A polymer composed of amino acids linked by peptide bonds.

Primary alcohol   An alcohol with no or one R group bonded to the carbon bearing the —OH group.

Primary amine   An amine in which one H atom of ammonia has been replaced by an organic group.

Protein   A naturally occurring polymeric chain of L-amino acids linked together by peptide bonds.

Saturated hydrocarbons   Hydrocarbons that contain only single bonds. They are also called *alkanes* or *paraffin hydrocarbons*.

Secondary alcohol   An alcohol with two R groups bonded to the carbon bearing the —OH group.

Secondary amine   An amine in which two H atoms of ammonia have been replaced by organic groups.

Sigma bond   A chemical bond formed by the end-to-end overlap of atomic orbitals.

Structural isomers   See *Constitutional isomers.*

Substitution reaction   A reaction in which an atom or a group of atoms attached to a carbon atom is replaced by another atom or group of atoms. No change occurs in the degree of saturation at the reactive carbon.

Tertiary alcohol   An alcohol with three R groups bonded to the carbon bearing the —OH group.

Tertiary amine   An amine in which three H atoms of ammonia have been replaced by organic groups.

Unsaturated hydrocarbons   Hydrocarbons that contain double or triple carbon–carbon bonds.

Vulcanization   The process in which sulfur is added to rubber and heated to 140°C, to cross-link the linear rubber polymer into a three-dimensional polymer.

Exercises

Basic Ideas

1. (a) Give a brief definition of *organic chemistry*. (b) Can life, as currently known, exist without organic compounds? (c) What happened to the "vital force" theory?

2. How is carbon's tendency to bond to other carbon atoms unique among the elements?

3. (a) Why do many organic compounds have low melting and boiling points? (b) Some organic compounds, like table sugar, do not have a boiling point. Suggest a reason why.

4. (a) What are the principal sources of organic compounds? (b) Some chemists argue that the ultimate source of all naturally occurring organic compounds is carbon dioxide. Could this be possible? *Hint:* Think about the origins of coal, natural gas, and petroleum.

5. List 10 "everyday" uses of organic compounds. Choose examples from at least five different classes of compounds.

Aliphatic Hydrocarbons

6. (a) Define and compare the structures of alkanes, alkenes, and alkynes. (b) Write the general formula of each of these hydrocarbon families. (c) What is a saturated hydrocarbon?

7. (a) What are hydrocarbons? (b) What are the cycloalkanes? (c) Write the general formula for acyclic alkanes.

8. Describe the bonding in and the geometry of molecules of the following alkanes: (a) methane; (b) ethane; (c) propane; (d) butane. How are the formulas for these compounds similar? How are they different?

9. (a) What are straight-chain hydrocarbons? (b) What are branched-chain hydrocarbons? (c) Cite three examples of each.

10. (a) What is a homologous series? (b) Provide specific examples of compounds that are members of a homologous series. (c) What is a methylene group? (d) How does each

member of a homologous series differ from compounds that come before and after it in the series? (e) Name three homologous series that are also aliphatic hydrocarbons.

11. (a) What are alkyl groups? (b) Draw structures for and write the names of the first five straight-chain alkyl groups. (c) What is the origin of the names for alkyl groups?

12. Could a substance with the molecular formula C_3H_8 be a cycloalkane? Could C_3H_8 be a branched alkane having a methyl group as a substituent attached to the longest chain?

13. Name the following compound by the IUPAC system: $CH_3CH(CH_3)CH_2CH_2CH_3$. Draw a constitutional isomer of this compound, and give its correct IUPAC name.

14. Give the IUPAC name of each of the following compounds.
   (a) $CH_3CH_2CH(CH_3)CH_2CH_3$
   (b) $CH_3CH_2CH(CH_2CH_3)CH_2CH_3$
   (c) $CH_3CH(CH_2CH_2CH_3)CH_2CH_3$
   (d) $CH_3CH_2CH_2CH(CH_3)_2$
   (e) $CH_3CH(CH_3)CH_2CH_2CH_3$

15. Write structural formulas for the isomeric saturated hydrocarbons having the molecular formula C_6H_{14}. Which one would have the highest boiling point?

16. Write structural formulas for the constitutional isomers of C_6H_{12} that contain a cyclopropane ring. Name each by the IUPAC system.

17. Write the structural formula for 2,2-dimethylpentane.

18. Write the IUPAC name for each of the following.

(a) (b) (c)

19. Write structural formulas for 1,1,2-trimethylcyclohexane, isopropylcyclobutane, and *sec*-butylcyclohexane.

20. (a) How does the general formula for the alkenes differ from the general formula for the alkanes? (b) Why are the general formulas identical for alkenes and cycloalkanes that contain the same number of carbon atoms?

21. (a) What are cycloalkenes? (b) What is their general formula? (c) Provide the structural formulas and names of three examples.

22. Describe the bonding at each carbon atom in (a) ethene, (b) propene, (c) 1-butene, and (d) 2-methyl-2-butene.

23. (a) What are geometric isomers? (b) Why is rotation around a double bond not possible at room temperature? (c) What do *cis* and *trans* mean? (d) Draw structures for *cis*- and *trans*-3-methyl-3-hexene. How should their melting and boiling points compare?

24. How do carbon-carbon single bond lengths and carbon–carbon double bond lengths compare? Why?

25. (a) What are alkynes? (b) What other name is used to describe them? (c) What is the general formula for alkynes?

(d) How does the general formula for alkynes compare with the general formula for cycloalkenes? Why?

26. Describe the bonding and geometry associated with the triple bond of alkynes.

27. Write the structural formulas of the following compounds: (a) 1-butyne; (b) 2-methylpropene; (c) 2-ethyl-3-methyl-1-pentene; (d) 3-methyl-1-butyne.

28. Write the structural formulas of the following compounds: (a) 2-butyne; (b) 1,3-pentadiene; (c) 3,3-dimethylcyclobutene; (d) 3,4-diethyl-1-hexyne.

29. Write the IUPAC names for the following compounds.

(a)   (c)

(b) $CH_2{=}C(CH_3)_2$ (d)

30. Repeat Exercise 29 for

(a)   (d)

(b)   (e)

(c) $CH_3C{\equiv}CCH_3$ (f)

31. Each of the hydrocarbon families has members that are cyclic hydrocarbons. (a) What is a cyclic hydrocarbon? (b) Write the structural formula of one example of each of the following: (i) cycloalkene; (ii) cycloalkane; and (iii) cycloalkyne.

Aromatic Hydrocarbons

32. (a) What distinguishes an aromatic compound from other cyclic compounds? (b) What is the principal source of aromatic hydrocarbons?

33. (a) The structure for benzene is sometimes drawn as a hexagon enclosing a circle; what is the meaning of the circle? (b) Write the structural formulas of the three most common aromatic hydrocarbons.

34. (a) What is a phenyl group? (b) How many isomeric monophenylnaphthalenes are possible?

35. There are only three isomeric trimethylbenzenes. Write their structural formulas, and name them.

36. (a) How many isomeric dibromobenzenes are possible? (b) What names are used to designate these isomers?

37. Write the structural formulas for the following compounds: (a) *p*-chlorofluorobenzene; (b) ethylbenzene; (c) 1,3,5-tribromobenzene; (d) 1,3-diphenylbutane; (e) *p*-chlorotoluene.

38. Write the IUPAC names for the following compounds.

(a)   (b)   (c)   (d)

Alkyl and Aryl Halides

39. (a) Write the general representation for the formula of an alkyl halide. (b) How does this differ from the representation for the formula of an aryl halide?

40. Write the IUPAC names for the following halides.

(a) $-CH_2CH_2Cl$ (c) CH_2ClCH_2Cl

(b) $CH_3-CH-CH_2Cl$ (d) $Cl-C=C-Cl$

41. Write the structural formulas for the following: (a) 2,2-dichloropentane; (b) 3-bromo-1-butene; (c) 1,2-dichloro-2-fluoropropane; (d) *p*-dichlorobenzene.

42. Name the following.

(a)   (b)   (c)   (d)

43. The compound 1,2-dibromo-3-chloropropane (DBCP) was used as a pesticide in the 1970s. Recently, agricultural workers have claimed that exposure to DBCP made them sterile. Write the structural formula for this compound.

Alcohols and Phenols

44. (a) What are alcohols and phenols? (b) How do they differ? (c) Why can alcohols and phenols be viewed as derivatives of hydrocarbons? as derivatives of water?

45. (a) Distinguish among primary, secondary, and tertiary alcohols. (b) Write names and formulas for three alcohols of each type.

46. Write structural formulas for and write the IUPAC names of the eight (saturated) alcohols that contain five carbon atoms and one -OH group per molecule. Which ones may be classified as primary alcohols? secondary alcohols? tertiary alcohols?

47. (a) What are glycols? (b) Write the structural formula of three examples. (c) Why are glycols more soluble in water than monohydric alcohols that contain the same number of carbon atoms?

48. Refer to Table 27-8, and explain the trends in boiling points and solubilities of alcohols in water.

49. Why are methyl alcohol and ethyl alcohol called wood alcohol and grain alcohol, respectively?

50. Why are most phenols only slightly soluble in water?

51. Write the structural formula for each of the following compounds: (a) 2-butanol; (b) cyclohexanol; (c) 1,4-butanediol.

52. Give the IUPAC name for each of the following compounds.

(a) $CH_3CH_2CHCH_2OH$ (c) $CH_2-CH_2-CH_2$
     CH_3 OH   OH

(b) $CH_3-C-CH_2CH_2OH$ (d) CH_3-C-OH
with CH_3 groups

53. Which of the following compounds are phenols? Give the IUPAC name for each compound.

(a) $-CH_2CH_2OH$

(b) $-OH$ (d) $-OH$

(c) $-OH$ and $-OH$ (e) $-OH$

54. Write the structural formulas for the following: (a) *p*-chlorophenol; (b) 4-nitrophenol (the nitro group is -NO_2); (c) *m*-nitrophenol.

Ethers

55. Distinguish among aliphatic ethers, aromatic ethers, and mixed ethers.

56. Briefly describe the bonding around the oxygen atom in dimethyl ether. What intermolecular forces are found in this ether?

57. What determines whether an ether is "symmetrical" or "asymmetrical"?

*58. Write the structural formulas for the following: (a) methoxymethane; (b) 2-methoxypropane; (c) 1,3-diethoxybutane; (d) ethoxybenzene; (e) methoxycyclobutane.

59. Give the IUPAC name for the following ethers.

   (a) $CH_3-O-CH_2CH_2CH_3$

   (b) $CH_3-O-CH-CH_3$
        $\quad\quad\quad\quad\;|$
        $\quad\quad\quad\quad CH_3$

   (c) ─O─CH₂CH₃

   (d) ─O─CH₃

Amines

60. (a) What are amines? (b) Why are amines described as derivatives of ammonia?

61. Write the general representation for the formula for a compound that is (a) a primary amine, (b) a secondary amine, (c) a tertiary amine. Is $(CH_3)_2CHNH_2$ a secondary amine? Give a reason for your answer.

62. Name the following amines.

   (a) CH_3-CH_2
        $\quad\quad\quad\;\;|$
        $\quad\quad\quad\;\;NH$
        $\quad\quad\quad\;\;|$
        CH_3-CH_2

   (b) $O_2N-$$-NH_2$

   (The $-NO_2$ substituent is called "nitro-.")

   (c) $-NHCH_3$

   (d) $CH_3CH_2CH_2CH_2-N-CH_2CH_2CH_2CH_3$
        $\quad\quad\quad\quad\quad\quad\quad\;\;\;|$
        $\quad\quad\quad\quad\quad\quad\quad CH_2CH_2CH_2CH_3$

63. The stench of decaying proteins is due in part to two compounds whose structures and common names are

   $H_2NCH_2CH_2CH_2CH_2NH_2$ putrescine
   $H_2NCH_2CH_2CH_2CH_2CH_2NH_2$ cadaverine

   Name these compounds as amino-substituted alkanes.

Aldehydes and Ketones

64. (a) List several naturally occurring aldehydes and ketones. (b) What are their sources? (c) What are some uses of these compounds?

65. (a) Distinguish between aldehydes and ketones. (b) Cite three examples (each) of aliphatic and aromatic aldehydes and ketones by drawing structural formulas, and give the IUPAC names of the compounds.

*66. Write the structural formulas for the following: (a) 2-methylbutanal; (b) propanal; (c) *o*-ethoxybenzaldehyde; (d) butanone; (e) bromopropanone; (f) 3-heptanone.

*67. Name the following compounds.

   (a) $CH_3CHCH_2\overset{\displaystyle O}{\overset{\displaystyle \|}{C}}H$
        $\quad\;\;|$
        $\quad\;\;CH_3$

   (b) =O

   (c) $H-\underset{\displaystyle Br}{\underset{\displaystyle |}{\overset{\displaystyle Br}{\overset{\displaystyle |}{C}}}}-CH_2-\overset{\displaystyle O}{\overset{\displaystyle \|}{C}}H$

   (d) $-\overset{\displaystyle O}{\overset{\displaystyle \|}{C}}-CH_2-CH_3$

Carboxylic Acids and Their Derivatives

68. (a) What are carboxylic acids? (b) Write structural formulas for and write the IUPAC names of five carboxylic acids.

69. Why are aliphatic carboxylic acids sometimes called fatty acids? Cite two examples.

70. (a) What are acyl chlorides, or acid chlorides? (b) Write structural formulas for four acid chlorides, and give their IUPAC names.

71. (a) What are esters? (b) Write structural formulas for four esters, and give their IUPAC names.

72. Write the structural formulas for the following: (a) methylpropanoic acid; (b) 2-bromobutanoic acid; (c) *p*-nitrobenzoic acid; (d) potassium benzoate; (e) 2-aminopropanoic acid.

73. (a) What are fats? What are oils? (b) Write the general formulas for fats and oils.

74. (a) What are glycerides? Distinguish between simple glycerides and mixed glycerides. (b) Write the IUPAC names and structural formulas for three simple glycerides.

75. Write the IUPAC names and structural formulas for some acids that occur in fats and oils (as esters).

76. What are waxes?

77. Give the IUPAC names for the following esters.

   (a) $CH_3\overset{\displaystyle O}{\overset{\displaystyle \|}{C}}-OCH_2CH_3$ (b) $CH_3\overset{\displaystyle O}{\overset{\displaystyle \|}{C}}OCH_3$

78. Name the following esters.

   (a) $-\overset{\displaystyle O}{\overset{\displaystyle \|}{C}}-O-$

$$\overset{O}{\overset{\|}{}}$$

(b) $CH_3CH_2CH_2COCH_2CH_3$

Reactions of Organic Molecules

79. (a) When an addition reaction occurs, the molecule becomes more nearly saturated. Explain. (b) When an elimination reaction occurs, the molecule becomes more unsaturated. Explain.

80. (a) What is a substitution reaction? (b) What is a halogenation reaction? (c) What is an addition reaction?

81. (a) Describe the reaction of methane with chlorine in ultraviolet light. (b) Write equations that show structural formulas for all compounds that can be formed by reaction (a). (c) Name all compounds in these equations. (d) Why are the halogenation reactions of the larger alkanes of limited value?

82. (a) Describe the reaction of ethane with chlorine in ultraviolet light. (b) Write equations that show structural formulas for all compounds that can be formed by reaction (a). (c) Give the IUPAC names for all compounds in these equations.

83. Which of the following compounds could undergo addition reactions? (a) propene; (b) 1,3-butadiene; (c) cyclopentene; (d) acetone.

84. Most reactions of the alkanes that do not disrupt the carbon skeleton are substitution reactions, whereas the alkenes are characterized by addition to the double bond. What does this statement mean?

85. How can bromination be used to distinguish between alkenes and alkanes?

86. Write equations for two reactions in which alkenes undergo addition reactions with halogens. Give the IUPAC names for all compounds.

87. When a substitution reaction occurs, there is no change in saturation. Explain.

88. (a) What is hydrogenation? (b) Why is it important? (c) Write equations for two reactions that involve hydrogenation of alkenes. (d) Give the IUPAC names for all compounds in part (c).

89. Describe two qualitative tests that can be used to distinguish between alkenes and alkanes. (b) Cite some specific examples. (c) What difference in reactivity is the basis for the qualitative distinction between alkanes and alkenes?

90. (a) Why are alkynes more reactive than alkenes? (b) What is the most common kind of reaction that alkynes undergo? (c) Write equations for three such reactions. (d) Give the IUPAC names for all compounds in part (c).

91. (a) What is the most common kind of reaction that the benzene ring undergoes? (b) Write equations for the reaction of benzene with chlorine in the presence of an iron catalyst and for the analogous reaction with bromine.

92. Write equations to illustrate both aromatic and aliphatic substitution reactions of toluene using (a) chlorine, (b) bromine, and (c) nitric acid.

93. Classify each reaction as substitution, addition, or elimination.
(a) $CH_3CH_2Br + CN^- \longrightarrow CH_3CH_2CN + Br^-$
(b) $CH_3CHCH_2Br + Zn \longrightarrow CH_3CH{=}CH_2 + ZnBr_2$
      |
    Br
(c) $C_6H_6 + HNO_3 \xrightarrow{H_2SO_4} C_6H_5NO_2 + H_2O$

94. Describe a simple test to distinguish between the two isomers 2-pentene and cyclopentane.

Polymers

95. (a) What is a polymer? (b) What is the term for the smaller molecule that serves as the repeating unit making up a polymer? (c) What are typical molecular weights of polymers?

96. (a) What is polymerization? (b) Write equations for three polymerization reactions.

97. (a) Give an example of a condensation reaction. (b) What is the essential feature of monomers used in condensation polymerizations?

*98. The examples of condensation polymers given in the text are all copolymers, that is, they contain two different monomers. Is it possible for a single monomer to polymerize so as to form a condensation *homo*polymer? If so, suggest an example. If not, explain why not.

99. (a) What is a monomer? (b) Name three polymers commonly found in a classroom and their use or function.

100. (a) What is a copolymer? (b) What is a condensation polymer? (c) Name three polymers that are addition polymers.

101. Poly(vinyl alcohol) has a relatively high melting point, 258°C. How would you explain this behavior? A segment of the polymer is

$$\cdots CH_2CHCH_2CHCH_2CHCH_2CHCH_2CH \cdots$$
$$\quad\ \ |\quad\ \ |\quad\ \ |\quad\ \ |\quad\ \ |$$
$$\ \ OH\ \ \ OH\ \ \ OH\ \ \ OH\ \ \ OH$$

102. What changes could be made in the structures of polymer molecules that would increase the rigidity of a polymer and raise its melting point?

103. Methyl vinyl ketone, $CH_3\overset{O}{\overset{\|}{C}}CH{=}CH_2$, can be polymerized by addition polymerization. The addition reaction involves only the C=C bond. Write the structural formula of a four-unit segment of this polymer.

104. (a) What is rubber? (b) What is vulcanization? (c) What is the purpose of vulcanizing rubber? (d) What are fillers and reinforcing agents? (e) What is their purpose?

105. (a) What is an elastomer? (b) Cite a specific example. (c) What are some of the advantages of neoprene compared with natural rubber?

106. (a) What are polyamides? (b) In what kind of reaction are polyamides formed?

107. (a) What are polyesters? (b) What is Dacron? (c) How is Dacron prepared? (d) What is Mylar? (e) Is it reasonable to assume that a polyester can be made from propylene glycol and terephthalic acid? If so, sketch its structure.

108. Suppose the following glycol is used with terephthalic acid to form a polyester. Write the structure of the polymer, showing two repeating units.

$$HOCH_2 - \bigcirc - CH_2OH$$

109. Write the structural formula of the monomer used in preparation of each of the following polymers.

(a) $\cdots -\underset{\underset{CH_3}{|}}{C}-CH_2-\underset{\underset{CH_3}{|}}{C}-CH_2-\underset{\underset{CH_3}{|}}{C}-CH_2-\cdots$

(b) $\cdots -\underset{\underset{CH_3}{|}}{\overset{\overset{CH_3}{|}}{C}}-CH_2-\underset{\underset{CH_3}{|}}{\overset{\overset{CH_3}{|}}{C}}-CH_2-\underset{\underset{CH_3}{|}}{\overset{\overset{CH_3}{|}}{C}}-CH_2-\cdots$

110. Is it possible to produce a copolymer by addition polymerization? If so, give an example. If not, explain why not.

111. (a) What is nylon? (b) How is it prepared?

112. Common nylon is called nylon 66. (a) What does this mean? (b) Write formulas for two other possible nylons.

Mixed Exercises

*113. Identify the class of organic compounds (ester, ether, ketone, and so on) to which each of the following belongs.

(a) $\bigcirc - CH_2CH_2OH$

(b) $\bigcirc - O - \bigcirc$

(c) $\bigcirc - \bigcirc - OH$

(d) $CH_3\overset{\overset{O}{\|}}{C}CH_2 - \bigcirc$

*114. Identify the class of organic compounds (ester, ether, ketone, etc.) to which each of the following belongs.

(a) $\bigcirc - \overset{\overset{O}{\|}}{C}OH$ (d) $\bigcirc - \overset{\overset{O}{\|}}{C}NH_2$

(b) $\bigcirc - CH_2O - \overset{\overset{O}{\|}}{C}H$

(c) $\bigcirc - CH_2CH_2\overset{\overset{O}{\|}}{C}Cl$

*115. Identify and name the functional groups in each of the following.

(a) $HO-\overset{\overset{O}{\|}}{C}-\bigcirc-\overset{\overset{O}{\|}}{C}-OH$

(b) $CH_2=CH-\overset{\overset{O}{\|}}{C}-O-CH_2CH_2CH_2CH_3$

(c) $\bigcirc$ with OH, NO$_2$, O=C-CH$_2$CH$_2$OH

(d) $\bigcirc - \underset{\underset{CH_3}{|}}{N}-CH_2CH_3$

116. Identify and name the functional groups in the following.

(a) $CH_3CH_2-\overset{\overset{O}{\|}}{C}-CH_2-\overset{\overset{O}{\|}}{C}-CH_2CH_3$

(b) $HO-\bigcirc-\underset{\underset{OCH_3}{}}{\overset{\overset{OH}{|}}{C}}HCH_2-\underset{}{\overset{\overset{H}{|}}{N}}-CH_3$

(c) dioxane (also known as 1,4-dioxin)

(d) morphine

(e) epinephrine (adrenaline)

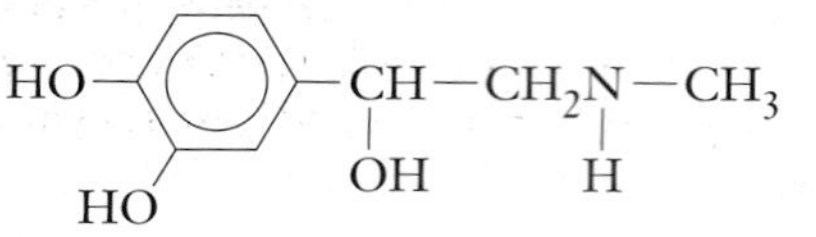

117. Identify and name the functional groups in each of the following.

(a) morpholine

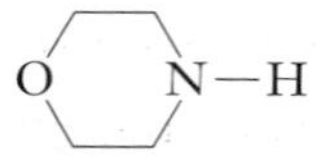

(b) citric acid

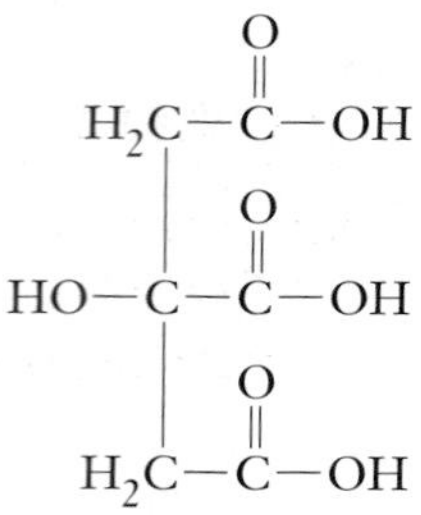

(c) coniine (from the hemlock plant; the poison that Socrates drank)

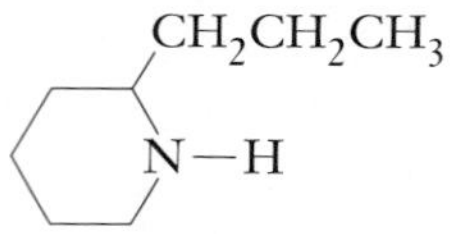

(d) glucose (a simple sugar, also known as dextrose)

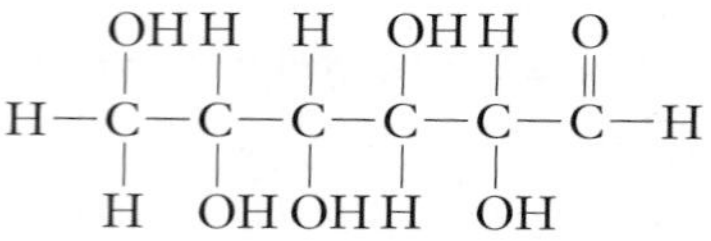

(e) vitamin C (also called ascorbic acid)

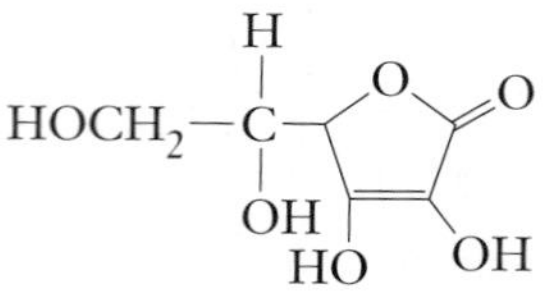

118. Name the following compounds.

(a) $CH_3CH(CH_3)CH_2CH_2OH$

(b)

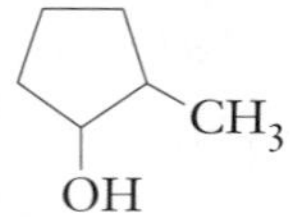

(c) 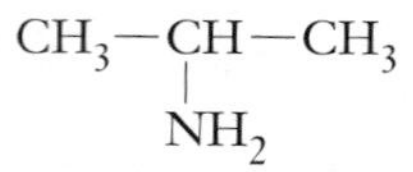 $CH_3-CH-CH_3$
      |
      NH_2

(d) 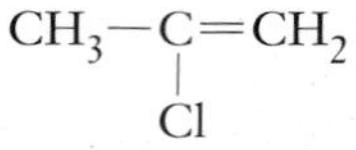 $CH_3-C=CH_2$
          |
          Cl

(e) 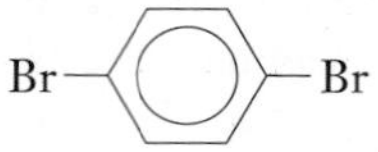 $Br-\bigcirc-Br$

(f) $(CH_3CH_2)_3N$

(g) 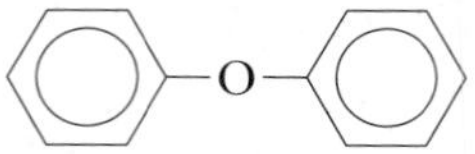 $\bigcirc-O-\bigcirc$

(h)

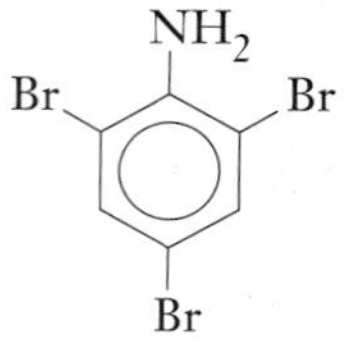

119. Write the structural formulas for the following compounds: (a) *p*-bromotoluene; (b) cyclohexanol; (c) 2-methoxy-3-methylbutane; (d) diethylamine; (e) *o*-chlorophenol; (f) 1,4-butanediol.

120. Give the IUPAC names for the following compounds.

(a) 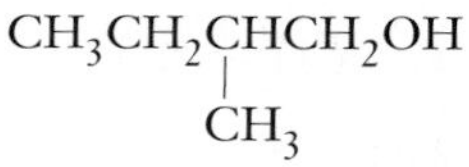 $CH_3CH_2CHCH_2OH$
              |
              CH_3

(b) $CH_3CH_2CH_2CH_2NH_2$

(c) 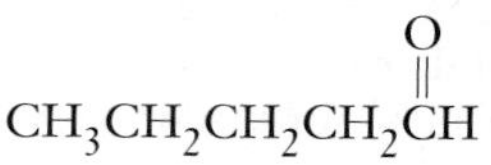 $CH_3CH_2CH_2CH_2\overset{O}{\overset{\|}{C}}H$

(d) 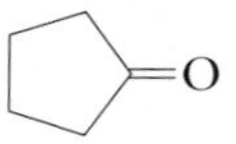 $\bigcirc=O$

(e) 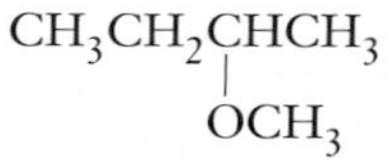 $CH_3CH_2CHCH_3$
              |
              OCH_3

(f) 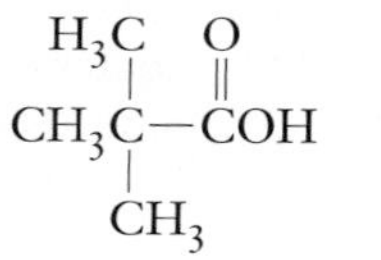 $CH_3\overset{H_3C}{\underset{CH_3}{C}}-\overset{O}{\overset{\|}{C}}OH$

CONCEPTUAL EXERCISES

121. Suggest why the C_6 ring in cyclohexane is not planar.

122. Can an aromatic compound also be a saturated compound?

BUILDING YOUR KNOWLEDGE

123. (a) How do the melting points and boiling points of the alkanes vary with molecular weight? (b) Do you expect them to vary in this order? (c) Why or why not? Use intermolecular forces to explain your answer.

124. Write a Lewis formula for butanone.

125. (a) What are resonance structures? (b) Draw resonance structures for benzene. (c) What do we mean when we say that the electrons associated with the π bonds in benzene are delocalized over the entire ring?

126. Lidocaine has replaced Novocain (procaine) as the favored anesthetic in dentistry. What functional group do the two compounds have in common?

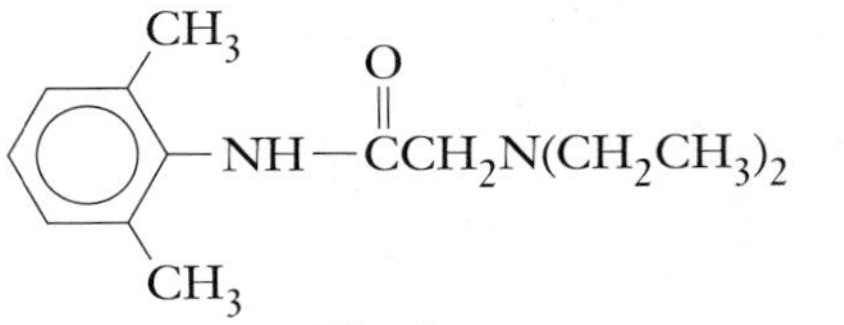

lidocaine

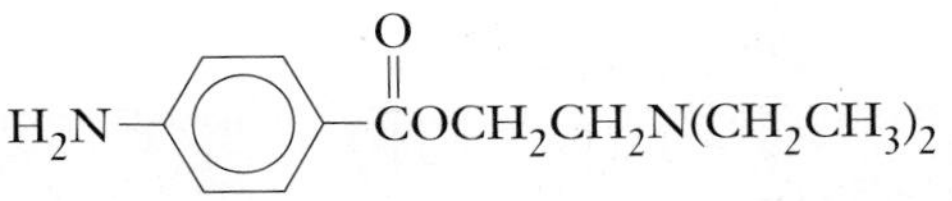

Novocain

127. PHB stands for poly(hydroxybutanoate). (a) Write the structural formula of 4-hydroxybutanoic acid. (b) Write the structural formula of an ester that can be formed from two molecules of 4-hydroxybutanoic acid. How is it formed? Would the ester be very soluble in water? Give a reason for your answer. (c) Write the repeating unit in the polymeric form of the polyester, hydroxybutanoate.

BEYOND THE TEXTBOOK

Go to the textbook website at

http://www.brookscole.com/chemistry/whitten

for additional activities and exercises based on the General Chemistry Interactive CD-ROM, the World Wide Web, and library resources.

InfoTrac College Edition

For additional readings, go to InfoTrac College Edition, your online research library at:

http://infotrac.thomsonlearning.com

28 Organic Chemistry II: Shapes, Selected Reactions, and Biopolymers

Capsaicin, the substance responsible for the "hot" in peppers, has phenol, ester, and amide functional groups. In addition to acting as an inflammatory agent, it can act as a pain reliever, presumably by blocking nerve endings involved in the transmission of pain.

OBJECTIVES

After you have studied this chapter, you should be able to

- Describe the differences between constitutional isomers and stereoisomers
- Distinguish between isomers and conformations
- Describe some common organic acids and bases and their relative strengths
- Recall some reactions that involve oxidation and reduction of organic compounds
- Recall reactions that form carboxylic acid derivatives
- Recall reactions in which esters and amides are formed
- Write equations in which carbonyl groups are used to produce other functional groups
- Write equations showing the hydrolysis of esters (saponification)
- Describe the structure of carbohydrates
- Describe the structure of proteins
- Describe the structure of nucleic acids

SHAPES OF ORGANIC MOLECULES

As we learned in Chapters 25 and 27, the chemical and physical properties of a substance depend on the arrangements, as well as the identities, of its atoms.

CD-ROM To see many molecular models that you can rotate on-screen, open any chapter on the CD-ROM; then click on "Molecular Models" under the "Tools" menu.

Isomers are substances that have the same number and kind of atoms—that is, the same *molecular formula*—but with different structural formulas. *Because their structures are different, isomers are different substances and have different properties.*

Isomers can be broadly divided into two major classes: constitutional isomers and stereoisomers. In Chapter 25 we discussed isomerism in coordination compounds, and in

1101

Chapter 27 we learned about some isomeric organic compounds. In this chapter we will take a more systematic look at some three-dimensional aspects of organic structures—a subject known as **stereochemistry** ("spatial chemistry").

28-1 CONSTITUTIONAL ISOMERS

> **Constitutional** (or **structural**) **isomers** differ in the *order* in which their atoms are bonded together.

In our studies of hydrocarbons in Sections 27-1 through 27-7, we saw some examples of constitutional isomerism. Recall that there are three isomers of C_5H_{12}.

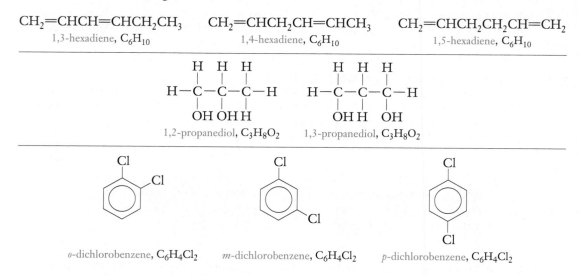

pentane, C_5H_{12} methylbutane
                             isopentane, C_5H_{12}

dimethylpropane
neopentane, C_5H_{12}

These three isomers differ in the lengths of their base chains but not in the functional groups present (i.e., only alkyl groups are present in this case). As a result, they differ somewhat in their melting and boiling points but differ only very slightly in the reactions they undergo.

In one kind of constitutional isomerism, the compounds have the same *number* and *kind* of functional groups on the *same base chain* or the *same ring* but in different positions. Isomers that vary only in the location of groups of atoms usually have very similar chemical and physical properties differing mainly in physical properties such as melting and boiling points. The following groups of isomers are examples in which the type of functional group does not change.

$CH_2{=}CHCH{=}CHCH_2CH_3$ $CH_2{=}CHCH_2CH{=}CHCH_3$ $CH_2{=}CHCH_2CH_2CH{=}CH_2$
1,3-hexadiene, C_6H_{10} 1,4-hexadiene, C_6H_{10} 1,5-hexadiene, C_6H_{10}

1,2-propanediol, $C_3H_8O_2$ 1,3-propanediol, $C_3H_8O_2$

o-dichlorobenzene, $C_6H_4Cl_2$ *m*-dichlorobenzene, $C_6H_4Cl_2$ *p*-dichlorobenzene, $C_6H_4Cl_2$

Sometimes the different order of arrangements of atoms results in different functional groups. Some examples of this type of constitutional isomerism follow.

An alcohol and an ether:

H   H
| |
H—C—C—OH   H—C—O—C—H
| |
H   H

ethanol, C_2H_6O methoxymethane, C_2H_6O
(dimethyl ether)

An aldehyde and a ketone:

H   H   H   O   H   H   O   H
| | | ‖ | | ‖ |
H—C—C—C—C—H   H—C—C—C—C—H
| | | | | |
H   H   H   H   H   H

butanal, C_4H_8O butanone, C_4H_8O
(butyraldehyde)   (methyl ethyl ketone)

28-2 STEREOISOMERS

> In **stereoisomers** the atoms are linked together in the same atom-to-atom order, but their arrangements in space are different.

There are two types of stereoisomers: geometric and optical.

Geometric Isomers

Geometric isomers (or *cis–trans* isomers) differ only in the spatial orientation of groups about a plane or direction. Two geometric isomers have the same molecular formula, the same functional groups, the same base chain or ring, and the same order of attachment of atoms; they differ in orientation either (1) around a double bond or (2) across the ring in a cyclic compound. If the larger groups are on opposite sides of the ring or the double bond, the designation *trans* appears in the name; if they are on the same side, the designation is *cis*. We learned in Section 27-3 about the geometric isomerism associated with the double bond in alkenes such as the 1,2-dichloroethenes (Figure 27-9). Similarly, two or more substituents can be either on the same side or on opposite sides of the ring, as shown in Figures 28-1 and 28-2. This kind of isomerism is possible when substituents have replaced an

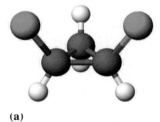

(a)

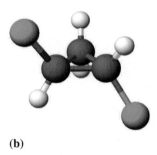

(b)

Figure 28-1 Models of (a) *cis*-dichlorocyclopropane and (b) *trans*-dichlorocyclopropane.

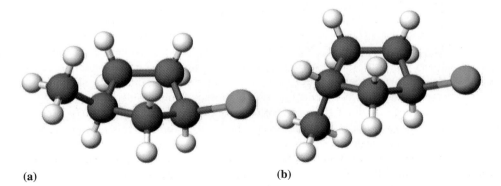

(a)   (b)

Figure 28-2 Models of (a) *cis*-1-chloro-3-methylcyclopentane and (b) *trans*-1-chloro-3-methylcyclopentane.

Figure 28-3 Mirror images. Place your left hand in front of a mirror; you will observe that it looks like your real right hand. We say that the two hands are mirror images of each other; each hand is in every way the "reverse" of the other. Now try placing one hand directly over the other; they are not identical. Hence, they are nonsuperimposable mirror images. Each hand is a *chiral* object.

H from a —CH$_2$— unit in a ring. Because substituents on an aromatic ring are bonded in the plane of the ring, such substitutions do not lead to geometric isomerism.

Optical Isomers

Many objects are mirror images of each other and cannot be superimposed. Your two hands are a familiar example of this; each hand is a nonsuperimposable mirror image of the other (Figure 28-3). "Superimposable" means that if one object is placed over the other, the positions of all parts will match.

An object that is *not* superimposable with its mirror image is said to be **chiral** (from the Greek word *cheir*, meaning "hand"); an object that *is* superimposable with its mirror image is said to be **achiral.** Examples of familiar objects that are chiral are a screw, a propeller, a foot, an ear, and a spiral staircase; examples of common objects that are achiral are a plain cup with no decoration, a pair of eyeglasses, and a sock.

> **Optical isomers** that are nonsuperimposable mirror images of each other (chiral) are called **enantiomers** of one another. Enantiomers can exist in two forms that bear the same relationship to each other as do left and right hands.

As an example of this, consider first the two models of bromochloromethane, CH$_2$BrCl, shown in Figure 28-4. They are mirror images of each other, and they can be superimposed. Thus, this molecule is *achiral* and is not capable of optical isomerism. Now consider bromochloroiodomethane, CHBrClI (Figure 28-5). This molecule is not superimposable with its mirror image, so it is *chiral*, and the two forms are said to be enantiomers of each other. Any compound that contains four different atoms or groups bonded to the same carbon atom is chiral; that is, it exhibits optical isomerism. Such a carbon is said to be *asymmetric* (meaning "without symmetry"). Most simple chiral molecules contain at least one asymmetric carbon atom, although there are other ways in which molecular chirality can occur.

Stereoisomers have the same type and number of atoms, connected in the same order, but arranged differently in space. Optical isomers (enantiomers) and geometric isomers are subgroups of stereoisomers. They differ, however, in that geometric isomers have *different* physical and chemical properties, whereas optical isomers have physical properties that are identical (e.g., melting point, boiling point, and density). Optical isomers also undergo the same chemical reactions, except when they interact with other chiral compounds. Consequently, their properties in biological systems may be very different. They also often exhibit different solubilities in solvents that are composed of chiral molecules.

Optical isomers also differ from each other in one notable physical property: They interact with polarized light in different ways. The main features of this subject were presented in Chapter 25. Separate equimolar solutions of two optical isomers rotate a plane of polarized light (Figures 25-4 and 25-5) by equal amounts but in opposite directions. One of the optical isomers is designated as the D-isomer and its nonsuperimposable mirror image is designated as the L-isomer. The phenomenon in which a plane of polarized light is

We speak of a screw or a propeller as being "right-handed" or "left-handed."

Ray Ellis/Photo Researchers, Inc.

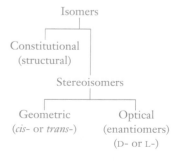

Isomers

Constitutional
(structural)

Stereoisomers

Geometric   Optical
(*cis-* or *trans-*)   (enantiomers)
(D- or L-)

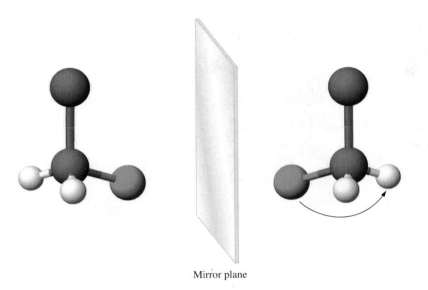

Figure 28-4 Models of two mirror-image forms of bromochloromethane, CH_2BrCl. The two models are the same (superimposable), so they are achiral. CH_2BrCl does *not* exhibit optical isomerism.

Mirror plane

rotated by samples of either isomer is called **optical activity.** It can be measured with a polarimeter (Figure 25-5). A **racemic mixture** is a single sample containing equal amounts of the two optical isomers of a compound. Such a solution does not rotate a plane of polarized light because the equal and opposite effects of the two isomers exactly cancel. The isomers must be separated from each other to exhibit optical activity.

One very important way in which optical isomers differ chemically from one another is in their biological activities. α-Amino acids have the general structure

$$H_2N-\underset{\underset{R}{|}}{\overset{\overset{H}{|}}{C}}-COOH$$

where R represents any of a number of common substituents. The central carbon atom has four atoms or groups bonded to it. In α-amino acids the four groups are all different, so each amino acid can exist as two optical isomers. The exception is glycine, in which R = H.

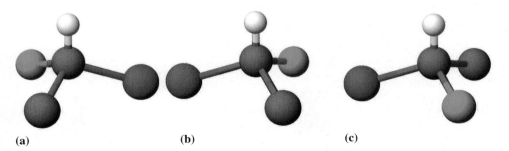

(a)   (b)   (c)

Figure 28-5 (a, b) Models of the two mirror-image forms of bromochloroiodomethane, CHBrClI. (Colors: Cl, *green;* Br, *red;* I, *purple.*) (c) The same model as in (a), turned so that H and I point the same as in (b); however, the Br and Cl atoms are not in the same positions in (b) and (c). The two models in (a) and (b) cannot be superimposed on each other no matter how we rotate them, so they are chiral. These two forms of CHBrClI represent *different compounds* that are optical isomers of each other.

Figure 28-6   Models of the two optical isomers of phenylalanine. The naturally occurring phenyl-alanine in all living systems is the form shown on the left.

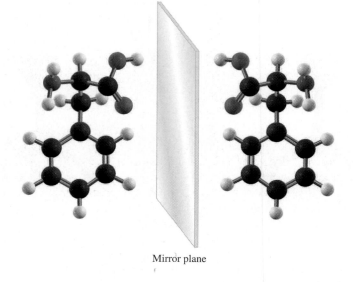

Mirror plane

Ball-and-stick model of L-phenylalanine.

Figure 28-6 shows this mirror-image relationship for optical isomers of phenylalanine, in which R = —CH₂C₆H₅. All naturally occurring phenylalanine in living systems is the L-form. In fact, only one isomer of each of the various optically active amino acids is found in proteins.

28-3   CONFORMATIONS

As we saw in Section 27-3, rotation does not occur around carbon–carbon double bonds at room temperature.

The **conformations** of a compound differ from one another in the *extent of rotation about one or more single bonds.* The C—C single bond length, 1.54 Å, is relatively independent of the structure of the rest of the molecule.

Rotation about a single C—C bond is possible; in fact, at room temperature it occurs very rapidly. In drawings and models, like those in Figure 28-7, it might at first appear that rotation about a single bond would be unrestricted. The electrons of the groups attached to the carbon atoms do repel one another somewhat, however. At room temperature ethane molecules possess sufficient energy to cause rapid rotation about the single carbon–carbon bond from one conformation to another. The staggered conformation of ethane is slightly more stable (of lower energy) than the eclipsed conformation (see Figure 28-7); in the eclipsed conformation there is a slightly stronger repulsion between H atoms on adjacent C atoms.

Figure 28-7   Two possible conformations of ethane. (a) Staggered. (b) Eclipsed. Rotation of one CH₃ group about the C—C single bond, as shown by the curved arrows, converts one conformation to the other.

Staggered conformation   Eclipsed conformation

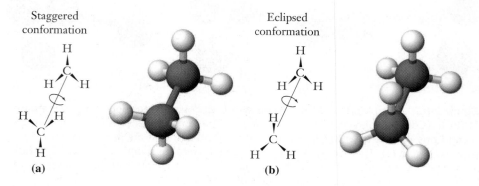

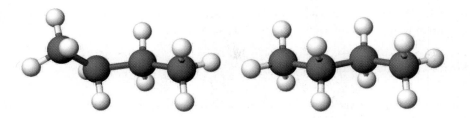

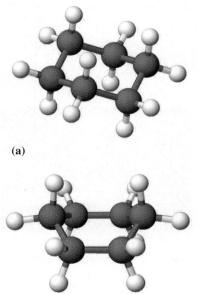

Figure 28-8 Two staggered conformations of butane, C_4H_{10}.

Consider the two conformations of butane shown in Figure 28-8. Again, staggered conformations are *slightly* more stable than eclipsed ones. At room temperature many conformations are present in a sample of butane.

Take care to distinguish between conformational differences and isomerism. The two forms of butane shown in Figure 28-8 are *not* isomers of each other. Either form can be converted to the other by rotation about a single bond, which is a very easy process that does not involve breaking any bonds. By contrast, at least one chemical bond would have to be broken and then re-formed to convert one isomer to another. This is most obvious with isomerism in which a conversion would change the order of attachment of the atoms. It is also true for geometric isomers that differ in orientation about a double bond. To convert such a *cis* isomer to a *trans* isomer, it would be necessary to rotate part of the molecule about the double bond. Such a rotation would move the *p* orbitals out of the parallel alignment that is necessary to form the pi component of the double bond (Section 8-13). The breaking of this pi bond is quite costly in terms of energy; it occurs only with the input of energy in the form of heat or light.

We saw in Section 27-1 that cyclohexane adopts a folded or "puckered" form. Cyclohexane actually exists in different stable forms, including the *chair* and *boat* forms (Figure 28-9). The chair form is the more stable of the two because, on the average, the hydrogens (or other substituents) are farther from one another than in the boat form. Chair and boat cyclohexane, however, *are not* different compounds. Either form is easily converted into the other by rotation around single bonds without breaking any bonds, and the two forms cannot be separated. Thus, they are different *conformations* of cyclohexane.

(a)

Figure 28-9 Two conformations of cyclohexane. (a) Chair and (b) boat.

(b)

SELECTED REACTIONS

Organic compounds can have quite different abilities to react, ranging from the limited reactivity of hydrocarbons and fluorocarbons to the great variety of reactions undergone by the thousands of organic molecules that contain several functional groups. Reactivity depends on structure. We can usually predict the kinds of reactions a compound can undergo by identifying the functional groups it contains.

In Chapter 27 we saw three fundamental classes of organic reactions and we associated the fundamental reactions with specific functional groups. In this chapter we will present some additional types of organic reactions.

 CD-ROM Screen 11.5, Functional Groups.

28-4 REACTIONS OF BRØNSTED–LOWRY ACIDS AND BASES

Many organic compounds can act as weak Brønsted–Lowry acids or bases. Their reactions involve the transfer of H^+ ions, or *protons* (Section 10-4). Like similar reactions of inorganic compounds, these acid–base reactions of organic acids and bases are usually fast and

In the Brønsted–Lowry description, an *acid* is a *proton donor* and a *base* is a *proton acceptor*. Review the terminology of conjugate acid–base pairs in Section 10-4.

reversible. Consequently, we can discuss the acidic or basic properties of organic compounds in terms of equilibrium constants (Section 18-4).

Some Organic Acids

The most important organic acids contain carboxyl groups, $-\overset{\overset{\displaystyle O}{\|}}{C}-O-H$. They are called carboxylic acids (Section 27-13). They ionize slightly when dissolved in water, as illustrated with acetic acid.

$$CH_3COOH + H_2O \rightleftharpoons CH_3COO^- + H_3O^+ \qquad K_a = \frac{[CH_3COO^-][H_3O^+]}{[CH_3COOH]} = 1.8 \times 10^{-5}$$

acid₁   base₂   base₁   acid₂

Acetic acid is 1.3% ionized in 0.10 M solution. Regardless of the lengths of the chains, the acid strengths of the monocarboxylic acids are approximately the same, with K_a values (in water) in the range 10^{-5} to 10^{-4}. Their acid strengths increase dramatically when electronegative substituents are present on the α-carbon atom (K_a values in water range from 10^{-3} to 10^{-1}). Compare acetic acid and the three substituted acetic acids in Table 28-1. There are two main reasons for this increase: (1) the electronegative substituents pull electron density from the carboxylic acid group, and (2) the more electronegative substituents help to stabilize the resulting carboxylate anion by spreading the negative charge over more atoms.

The alcohols are so *very weakly acidic* that they do not react with strong bases. They have about the same acid strength as water (see Table 28-1), and some of their reactions are analogous to those of water.

The reactive metals react with alcohols to form *alkoxides* with the liberation of hydrogen.

$$2CH_3CH_2-OH + 2Na \longrightarrow H_2 + 2[Na^+ + CH_3CH_2O^-]$$

ethanol   sodium ethoxide
(an alkoxide)

Sources of some naturally occurring carboxylic acids.

This is similar to the reaction of water with active metals.

$$2H-OH + 2Na \longrightarrow$$
$$H_2 + 2[Na^+ + OH^-]$$

In Section 18-3 we defined

$$pK_a = -\log K_a$$

When K_a goes up by a factor of 10, pK_a goes down by one unit. We see that the stronger an acid, the lower its pK_a value.

TABLE 28-1	K_a *and* pK_a *Values of Some Carboxylic Acids*		
Name	**Formula**	K_a	pK_a
formic acid	HCOOH	1.8×10^{-4}	3.74
acetic acid	CH_3COOH	1.8×10^{-5}	4.74
propanoic acid	CH_3CH_2COOH	1.4×10^{-5}	4.85
monochloroacetic acid	$ClCH_2COOH$	1.5×10^{-3}	2.82
dichloroacetic acid	$Cl_2CHCOOH$	5.0×10^{-2}	1.30
trichloroacetic acid	Cl_3CCOOH	2.0×10^{-1}	0.70
benzoic acid	C_6H_5COOH	6.3×10^{-5}	4.20
phenol*	C_6H_5OH	1.3×10^{-10}	9.89
ethanol*	CH_3CH_2OH	$\approx 10^{-18}$	≈ 18

Phenol and ethanol are not carboxylic acids. Phenol is weakly acidic compared with carboxylic acids, whereas ethanol is even weaker than water.

Alkoxides are strong bases that react with water (hydrolyze) to form the parent alcohol and a metal hydroxide.

$$[Na^+ + CH_3CH_2O^-] + H{-}OH \longrightarrow CH_3CH_2OH + [Na^+ + OH^-]$$

sodium ethoxide   ethanol

Phenols react with metallic sodium to produce *phenoxides*; the reactions are analogous to those of alcohols. Because phenols are more acidic than alcohols, their reactions are more vigorous.

Reaction of sodium metal with ethanol gives sodium ethoxide and hydrogen.

We saw in Section 18-9 that salts derived from strong acids and weak bases react with water (hydrolyze) to give acidic solutions. The example given there involved the acidic character of the ammonium ion.

$$NH_4^+ + H_2O \rightleftharpoons NH_3 + H_3O^+ \qquad K_a = \frac{[NH_3][H_3O^+]}{[NH_4^+]} = 5.6 \times 10^{-10}$$

acid$_1$ base$_2$ base$_1$ acid$_2$

The weaker a base, the stronger its conjugate acid (Section 10-4).

Similar hydrolysis reactions occur with organic ammonium salts.

$$RNH_3^+ + H_2O \rightleftharpoons RNH_2 + H_3O^+ \qquad K_a = \frac{[RNH_2][H_3O^+]}{[RNH_3^+]}$$

acid$_1$ base$_2$ base$_1$ acid$_2$

We recall from Chapter 18 that the relationship $K_w = K_a K_b$ describes the strengths of any conjugate acid–base pair in aqueous solution. For instance, we can use this relationship for the $CH_3NH_3^+/CH_3NH_2$ pair (we obtain K_b for methylamine, CH_3NH_2, from Appendix G).

$$K_{a(CH_3NH_3^+)} = \frac{K_w}{K_{b(CH_3NH_2)}} = \frac{1.0 \times 10^{-14}}{5.0 \times 10^{-4}} = 2.0 \times 10^{-11}$$

In summary, we can rank the acid strengths of these classes of organic species.

carboxylic acids > phenols > substituted ammonium ions > alcohols

Our discussion has emphasized water solutions of these acids. Many organic compounds are soluble in numerous other solvents. The properties of acids and bases in other solvents depend on solvent properties such as polarity, acidity or basicity, and polarizability.

Some Organic Bases

The most important organic bases are the amines. When dissolved in water, they are partially converted to substituted ammonium ions. This equilibrium is defined as shown in the following equations for the ionization of primary, secondary, and tertiary amines.

The formula of a quaternary ammonium salt is R_4NCl.

$pK_b = -\log K_b$

The weaker a base, the higher its pK_b value.

TABLE 28-2	Basicities of Ammonia and Some Amines in Water		
Name	**Formula**	K_b	pK_b
ammonia	NH_3	1.8×10^{-5}	4.74
methylamine	CH_3NH_2	5.0×10^{-4}	3.30
dimethylamine	$(CH_3)_2NH$	7.4×10^{-4}	3.13
trimethylamine	$(CH_3)_3N$	7.4×10^{-5}	4.13
ethylamine	$CH_3CH_2NH_2$	4.7×10^{-4}	3.33
aniline	$C_6H_5NH_2$	4.2×10^{-10}	9.38
ethylenediamine	$H_2NCH_2CH_2NH_2$	8.5×10^{-5} (K_{b1})	4.07
pyridine	C_5H_5N	1.5×10^{-9}	8.82

Several antibacterial mouthwashes that contain a quaternary pyridinium chloride salt.

$$\underset{\text{base}_1}{RNH_2} + H_2O \rightleftharpoons \underset{\text{acid}_2}{RNH_3^+} + \underset{\text{acid}_1}{OH^-} \qquad K_b = \frac{[RNH_3^+][OH^-]}{[RNH_2]}$$

primary, 1°

$$\underset{}{R_2NH} + H_2O \rightleftharpoons R_2NH_2^+ + OH^- \qquad K_b = \frac{[R_2NH_2^+][OH^-]}{[R_2NH]}$$

secondary, 2°

$$R_3N + H_2O \rightleftharpoons R_3NH^+ + OH^- \qquad K_b = \frac{[R_3NH^+][OH^-]}{[R_3N]}$$

tertiary, 3°

Most low-molecular-weight aliphatic amines are somewhat stronger bases than ammonia. Table 28-2 shows that aliphatic amines are much stronger bases than aromatic and heterocyclic amines. The basicities of amines often decrease roughly in the order tertiary > secondary > primary. Other structural factors and solvation effects, however, may outweigh this tendency, especially with tertiary amines.

We can rank the base strengths of common organic bases as

alkoxides > aliphatic amines ≈ phenoxides > carboxylates ≈ aromatic amines ≈ heterocyclic amines

We could describe the oxidation and reduction of organic compounds in terms of changes in oxidation numbers, just as we did for inorganic compounds in Sections 4-4 and 4-7. Formal application of oxidation number rules to organic compounds often leads to fractional oxidation numbers for carbon. For organic species, the descriptions in terms of increase or decrease of oxygen or hydrogen are usually easier to apply.

28-5 OXIDATION–REDUCTION REACTIONS

Oxidation of an organic molecule usually corresponds to *increasing* its *oxygen* content or *decreasing* its *hydrogen* content. **Reduction** of an organic molecule usually corresponds to *decreasing* its *oxygen* content or *increasing* its *hydrogen* content.

For example, the oxygen content increases when an aldehyde is converted to a carboxylic acid, so this process is an oxidation.

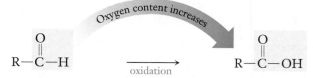

Converting a primary alcohol to an aldehyde or a secondary alcohol to a ketone is also an oxidation; the hydrogen content decreases.

CHEMISTRY IN USE

Research & Technology

Chemical Communication

The geometries of molecules play important roles in chemical reactivity. Molecular geometry is particularly important in a group of substances known as pheromones. *Pheromones* are chemicals used for communication between members of the same species. Pheromone activity has been observed in many forms of life, from insects to humans, and pheromone research is being done at many scholarly institutions.

If you've ever observed lines of ants moving in opposite directions, you have observed the influence of pheromones on insect behavior. When an ant finds food, it immediately heads toward its nest while secreting 9-oxy-2-decenoic acid from an abdominal gland. When other ants cross this acid trail, they compulsively follow it to the food source and carry the nourishment back to their nest. Soon, many ants will be following the acid trail and reinforcing it with their own 9-oxy-2-decenoic acid secretions. Eventually, the food source becomes exhausted, trail reinforcement stops, and the acid trail evaporates. Ants are so dependent on the acid trail that if a part of it were wiped away the ants following the trail in both directions would come to a complete stop. They wouldn't know where to go.

Perhaps an even more impressive example than the total dependence on chemical communication by ants is demonstrated with the so-called "death pheromone." Immediately upon the death of an ant, fellow ants continue to groom the dead ant and treat it as if it were still living. This attention continues until the dead ant's body produces the death pheromone, 10-octadecenoic acid. On sensing this pheromone, colleagues carry the dead ant to the nearest garbage site. Interestingly, if 10-octadecenoic acid is applied to a living ant, the living ant is similarly dumped into the garbage. The discarded ant will quickly return only to be carried off again, and this process continues until the death pheromone evaporates.

Because pheromones are used by female insects to indicate their state of fertile readiness, pheromones have proven to be an effective weapon in controlling some crop-damaging insects. For example, when a specific mating pheromone is applied to crops, male cotton bollworms and female tobacco budworms compulsively mate with one another. Because of physical incompatibilities, their bodies become interlocked and both insects eventually die. Less drastic uses of pheromones to control crop damage involve baiting traps with sex

© 1990 Trygve Steen

pheromones to lure and trap male insects. Trapping males eventually slows reproduction, and the insect population may decrease to controllable levels. Some of these sex pheromones are so powerful that a single drop has the potential of attracting millions of males. In fact, some male insects can detect a single molecule of female pheromone from a great distance and then successfully seek out and find the female.

Chemical communication is not confined to the insect world. Female dogs secrete the chemical *p*-hydroxybenzoate to attract males. Just like the ants and cotton bollworms, who are dependent on detecting chemicals for their actions, male dogs will attempt to mate with various objects to which *p*-hydroxybenzoate has been applied.

When we examine the molecular structures and functional groups of known pheromones, we find that they have little in common. Some pheromones contain stereoisomers, and some insects can distinguish between the stereoisomers. The structures of pheromones play vital roles in their activity. Part of the structure is an upper limit of about 20 carbon atoms, a limit probably imposed by Graham's Law. Most pheromones must travel through the air; those with low molecular weights are often more volatile. Scientists suspect that the physical motions of pheromone molecules, which are also a function of molecular structure, play an important role in the communication mechanism.

Ronald DeLorenzo
Middle Georgia College

The notation R′ emphasizes that the two R groups may be the same (R = R′, e.g., in the formation of acetone, $CH_3-\overset{\displaystyle O}{\overset{\|}{C}}-CH_3$) or different

(R ≠ R′, e.g., in the formation of methyl ethyl ketone, $CH_3-\overset{\displaystyle O}{\overset{\|}{C}}-CH_2CH_3$).

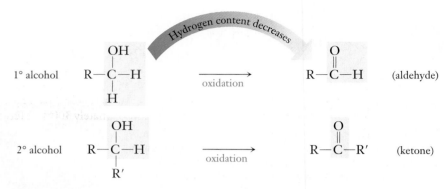

An aldehyde can be oxidized to a carboxylic acid.

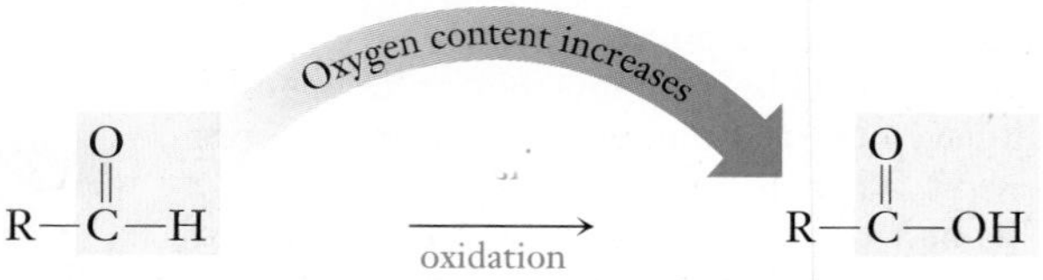

In each of these "oxidation" reactions, something else must act as the oxidizing agent (which is reduced). These oxidizing agents are often inorganic species such as dichromate ions, $Cr_2O_7^{2-}$. The reverse of each of the preceding reactions is a reduction of the organic molecule. In this reverse reaction, the reducing agent (the substance that is oxidized) is often an inorganic compound.

Let us look at a few important types of organic oxidations and reductions.

Oxidation of Alcohol

Aldehydes can be prepared by the oxidation of *primary* alcohols. The reaction mixture is heated to a temperature slightly above the boiling point of the aldehyde so that the aldehyde distills out as soon as it is formed. Potassium dichromate in the presence of dilute sulfuric acid is the common oxidizing agent.

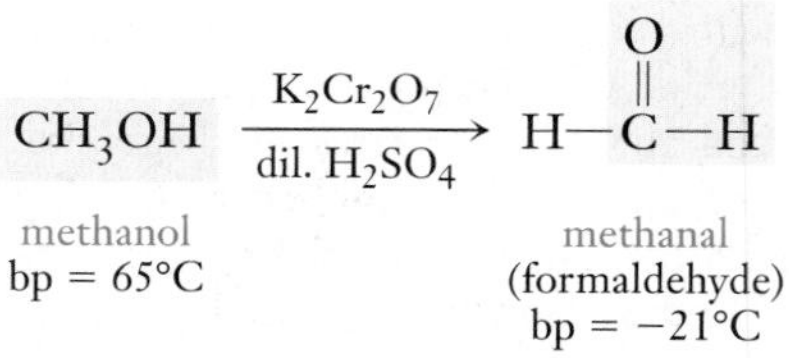

Aldehydes are easily oxidized to carboxylic acids. They must therefore be removed from the reaction mixture as soon as they are formed. Aldehydes have lower boiling points than the alcohols from which they are formed, so the removal of aldehydes is easily accomplished.

$$CH_3OH \xrightarrow[\text{dil. } H_2SO_4]{K_2Cr_2O_7} H-\overset{\displaystyle O}{\overset{\|}{C}}-H$$

methanol
bp = 65°C

methanal
(formaldehyde)
bp = −21°C

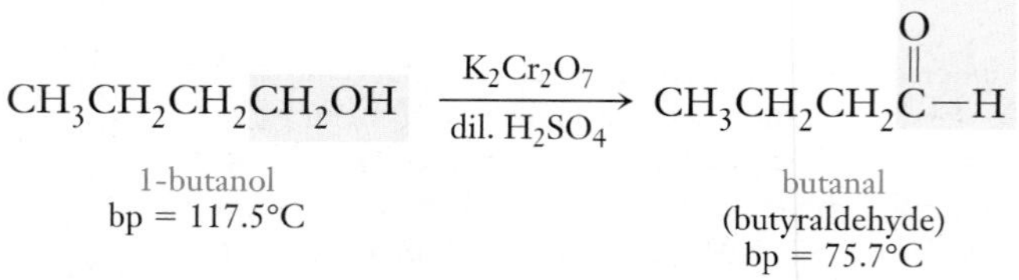

$$CH_3CH_2CH_2CH_2OH \xrightarrow[\text{dil. } H_2SO_4]{K_2Cr_2O_7} CH_3CH_2CH_2\overset{\displaystyle O}{\overset{\|}{C}}-H$$

1-butanol
bp = 117.5°C

butanal
(butyraldehyde)
bp = 75.7°C

Ketones can be prepared by the oxidation of *secondary* alcohols. Ketones are not as susceptible to oxidation as are aldehydes.

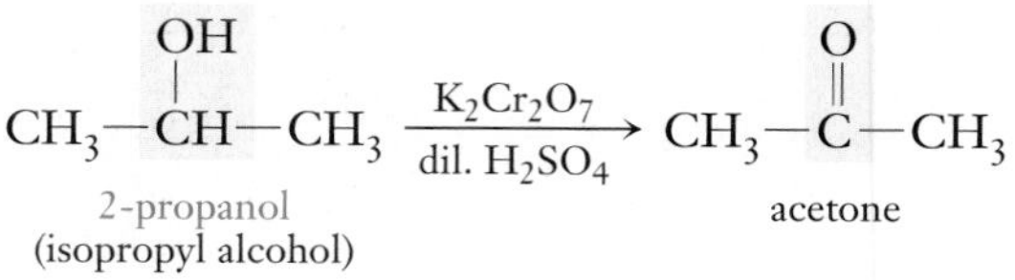

Ketones are not as easily oxidized as aldehydes, because oxidation of a ketone requires the breaking of a carbon–carbon bond. Thus, it is not as important that they be quickly removed from the reaction mixture.

$$CH_3-\overset{\displaystyle OH}{\overset{|}{C}H}-CH_3 \xrightarrow[\text{dil. } H_2SO_4]{K_2Cr_2O_7} CH_3-\overset{\displaystyle O}{\overset{\|}{C}}-CH_3$$

2-propanol
(isopropyl alcohol)

acetone

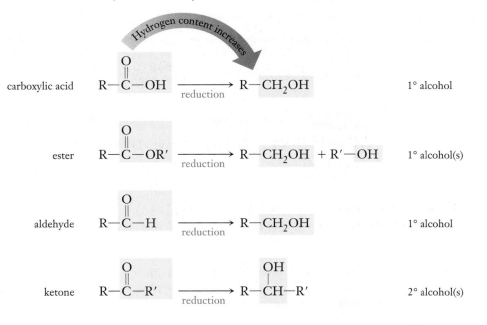

cyclooctanol → cyclooctanone

$K_2Cr_2O_7$ / dil. H_2SO_4

These two reactions can also be described as a type of *elimination* reaction (Section 27-18). Two hydrogen atoms are eliminated to form a C=O double bond.

Aldehydes and ketones can be prepared by a catalytic process that involves passing alcohol vapors and air over a copper gauze or powder catalyst at approximately 300°C. Here the oxidizing agent is O_2.

$$2CH_3OH + O_2 \xrightarrow[300°C]{Cu} 2H-\overset{\overset{O}{\|}}{C}-H + 2H_2O$$

methanol   formaldehyde

Formaldehyde is quite soluble in water; the gaseous compound can be dissolved in water to give a 40% solution.

Acetaldehyde can be prepared by the similar oxidation of ethanol.

$$2CH_3CH_2OH + O_2 \xrightarrow[300°C]{Cu} 2CH_3-\overset{\overset{O}{\|}}{C}-H + 2H_2O$$

ethanol   acetaldehyde

Oxidation of tertiary alcohols is difficult because the breaking of a carbon–carbon bond is required. Such oxidations are of little use in synthesis.

Reduction of Carbonyl Compounds

Reduction of a variety of compounds that contain the carbonyl group provides synthetic methods to produce primary and secondary alcohols. A common, very powerful reducing agent is lithium aluminum hydride, $LiAlH_4$; other reducing agents include sodium in alcohol and sodium borohydride, $NaBH_4$.

Hydrogen content increases

carboxylic acid $R-\overset{\overset{O}{\|}}{C}-OH \xrightarrow{\text{reduction}} R-CH_2OH$ 1° alcohol

Organic reactions are sometimes written in extremely abbreviated form. This is often the case when a variety of common oxidizing or reducing agents will accomplish the desired conversion.

ester $R-\overset{\overset{O}{\|}}{C}-OR' \xrightarrow{\text{reduction}} R-CH_2OH + R'-OH$ 1° alcohol(s)

aldehyde $R-\overset{\overset{O}{\|}}{C}-H \xrightarrow{\text{reduction}} R-CH_2OH$ 1° alcohol

ketone $R-\overset{\overset{O}{\|}}{C}-R' \xrightarrow{\text{reduction}} R-\overset{\overset{OH}{|}}{CH}-R'$ 2° alcohol(s)

Oxidation of Alkylbenzenes

Unsubstituted aromatic hydrocarbons (Sections 27-5 and 27-6) are quite resistant to oxidation by chemical oxidizing agents. The reactions of strong oxidizing agents with alkylbenzenes illustrate the stability of the benzene ring system. Heating toluene with a basic solution of $KMnO_4$ results in a nearly 100% yield of benzoic acid. The ring itself remains intact; only the nonaromatic portion of the molecule is oxidized.

$$\text{toluene} \xrightarrow[\text{(2) HCl(aq)}]{\text{(1) heat, OH}^-, \text{KMnO}_4} \text{benzoic acid}$$

Two such alkyl groups on an aromatic ring are oxidized to give a diprotic acid, as the following example illustrates.

$$p\text{-xylene} \xrightarrow{\text{oxidation}} \text{terephthalic acid}$$

Terephthalic acid is used to make "polyesters," an important class of polymers (Section 27-19).

Combustion of Organic Compounds

Recall that ΔH^0 is negative for an exothermic process.

The most extreme oxidation reactions of organic compounds occur when they burn in O_2. Such *combustion reactions* (Section 6-8) are highly exothermic. When the combustion takes place in excess O_2, the products are CO_2 and H_2O. Examples of alkane combustions are

$$CH_4(g) + 2O_2(g) \longrightarrow CO_2(g) + 2H_2O(\ell) \quad \Delta H^0 = -891 \text{ kJ/mol rxn}$$
$$2C_8H_{18}(\ell) + 25O_2(g) \longrightarrow 16CO_2(g) + 18H_2O(\ell) \quad \Delta H^0 = -1.090 \times 10^4 \text{ kJ/mol rxn}$$

The **heat of combustion** is the amount of energy *liberated* per mole of hydrocarbon burned. Heats of combustion are assigned positive values *by convention* (Table 28-3) and are therefore equal in magnitude, but opposite in sign, to ΔH^0 values for combustion reactions.

Acetylene is produced by the slow addition of water to calcium carbide.

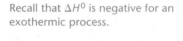

$$CaC_2(s) + 2H_2O(\ell) \longrightarrow$$
$$HC{\equiv}CH(g) + Ca(OH)_2(s)$$

The light of one kind of headlamp used by miners and cave explorers is given off by the combustion of acetylene.

TABLE 28-3 *Heats of Combustion of Some Alkanes*

		Heat of Combustion	
Hydrocarbon		*kJ/mol*	*J/g*
methane	CH_4	891	55.7
propane	C_3H_8	2220	50.5
pentane	C_5H_{12}	3507	48.7
octane	C_8H_{18}	5450	47.8
decane	$C_{10}H_{22}$	6737	47.4
ethanol*	C_2H_5OH	1372	29.8

Not an alkane; included for comparison only.

The combustion of hydrocarbons produces large volumes of CO_2 and H_2O in addition to large amounts of heat. The rapid formation of these gases at high temperature and pressure drives the pistons or turbine blades in internal combustion engines.

In the absence of sufficient oxygen, partial combustion of hydrocarbons occurs. The products may be carbon monoxide (a very poisonous gas) or carbon (which deposits on spark plugs, in the cylinder head, and on the pistons of automobile engines). Many modern automobile engines now use microcomputer chips and sensors to control the air supply and to optimize the fuel/O_2 ratio. The reactions of methane with insufficient oxygen are

$$2CH_4 + 3O_2 \longrightarrow 2CO + 4H_2O \quad \text{and} \quad CH_4 + O_2 \longrightarrow C + 2H_2O$$

All hydrocarbons undergo similar reactions.

The alkenes, like the alkanes, burn in *excess* oxygen to produce carbon dioxide and water in exothermic reactions.

$$CH_2{=}CH_2(g) + 3O_2 \text{ (g, excess)} \longrightarrow 2CO_2(g) + 2H_2O(\ell) \quad \Delta H^0 = -1387 \text{ kJ/mol rxn}$$

When an alkene (or any other unsaturated organic compound) is burned in air, a yellow, luminous flame is observed, and considerable soot (unburned carbon) is formed. This reaction provides a qualitative test for unsaturation. Saturated hydrocarbons burn in air without forming significant amounts of soot.

Acetylene lamps are charged with calcium carbide. Very slow addition of water produces acetylene, which is burned as it is produced. Acetylene is also used in the oxyacetylene torch for welding and cutting metals. When acetylene is burned with oxygen, the flame reaches temperatures of about 3000°C.

Like other hydrocarbons, the *complete combustion* of aromatic hydrocarbons, such as benzene, releases large amounts of energy.

$$2C_6H_6(\ell) + 15O_2(g) \longrightarrow 12CO_2(g) + 6H_2O(\ell) \quad \Delta H^0 = -6548 \text{ kJ/mol rxn}$$

Because they are so unsaturated, aromatic hydrocarbons burn *in air* with a yellow, sooty flame.

28-6 FORMATION OF CARBOXYLIC ACID DERIVATIVES

The carboxylic acid derivatives introduced in Section 27-14 can be formed by *substitution* of another group in place of —OH in the carboxyl group. The acyl halides (acid halides) are usually prepared by treating acids with PCl_3, PCl_5, or $SOCl_2$ (thionyl chloride). In general terms, the reaction of acids with PCl_5 may be represented as

$$\underset{\text{acid}}{R-\overset{\displaystyle O}{\overset{\|}{C}}-OH} + \underset{\substack{\text{phosphorus}\\\text{pentachloride}}}{PCl_5} \longrightarrow \underset{\substack{\text{an acyl chloride}\\\text{(an acid chloride)}}}{R-\overset{\displaystyle O}{\overset{\|}{C}}-Cl} + HCl(g) + \underset{\substack{\text{phosphorus}\\\text{oxychloride}}}{POCl_3}$$

$$\underset{\text{acetic acid}}{CH_3-\overset{\displaystyle O}{\overset{\|}{C}}-OH} + PCl_5 \longrightarrow \underset{\text{acetyl chloride}}{CH_3-\overset{\displaystyle O}{\overset{\|}{C}}-Cl} + HCl(g) + POCl_3$$

The acyl halides are much more reactive than their parent acids. Consequently, they are often used in reactions to introduce an acyl group into another molecule.

Charles Steele

Hexane, C_6H_{14}, an alkane, burns cleanly in air to give CO_2 and H_2O (*top*). 1-Hexene, C_6H_{12}, an alkene, burns with a flame that contains soot (*middle*). Burning *o*-xylene, an aromatic hydrocarbon, produces large amounts of soot (*bottom*).

When an organic acid is heated with an alcohol, an equilibrium is established with the resulting *ester* and water. The reaction is catalyzed by traces of strong inorganic acids, such as a few drops of concentrated H_2SO_4.

$$CH_3-\overset{\overset{\displaystyle O}{\|}}{C}-OH + CH_3CH_2-OH \underset{}{\overset{H^+,\ heat}{\rightleftharpoons}} CH_3-\overset{\overset{\displaystyle O}{\|}}{C}-O-CH_2CH_3 + H_2O$$

acetic acid　　　　ethyl alcohol　　　　　　　　　　ethyl acetate, an ester

Many experiments have shown conclusively that the OH group from the acid and the H from the alcohol are the atoms that form water molecules.

In general terms, the reaction of an organic acid and an alcohol may be represented as

$$R-\overset{\overset{\displaystyle O}{\|}}{C}-OH + R'-O-H \rightleftharpoons R-\overset{\overset{\displaystyle O}{\|}}{C}-O-R' + H_2O$$

acid　　　　alcohol　　　　　　　ester

(R and R' may be the same or different groups.)

Reactions between acids and alcohols are usually quite slow and require prolonged boiling (refluxing). The reactions between most acyl halides and most alcohols, however, occur very rapidly without requiring the presence of an acid catalyst.

$$CH_3-\overset{\overset{\displaystyle O}{\|}}{C}-Cl + CH_3CH_2-O-H \longrightarrow CH_3-\overset{\overset{\displaystyle O}{\|}}{C}-O-CH_2CH_3 + HCl$$

acetyl chloride　　　　ethyl alcohol　　　　　　　　ethyl acetate

Amides are usually *not* prepared by the reaction of an amine with an organic acid. Acyl halides react readily with primary and secondary amines to produce amides. The reaction of an acyl halide with a primary or secondary amine produces an amide and a salt of the amine.

In this preparation, one half of the amine is converted to an amide and the other half to a salt.

$$2CH_3NH_2 + CH_3-\overset{\overset{\displaystyle O}{\|}}{C}-Cl \longrightarrow CH_3-\overset{\overset{\displaystyle O}{\|}}{C}-\underset{\underset{\displaystyle CH_3}{|}}{N}-H + CH_3NH_3^+ + Cl^-$$

methylamine　　　acetyl chloride　　　N-methylacetamide　　　methylammonium
(a primary amine)　(an acyl halide)　　　(an amide)　　　　chloride (a salt)

28-7 HYDROLYSIS OF ESTERS

Because most esters are not very reactive, strong reagents are required for their reactions. Esters can be hydrolyzed by refluxing with solutions of strong bases.

$$CH_3-\overset{\overset{\displaystyle O}{\|}}{C}-O-CH_2CH_3 + Na^+OH^- \overset{heat}{\longrightarrow} CH_3\overset{\overset{\displaystyle O}{\|}}{C}-O^-Na^+ + CH_3CH_2OH$$

ethyl acetate　　　　　　　　　　　　sodium acetate　　　ethanol

The hydrolysis of esters in the presence of strong bases is called **saponification** (soapmaking).

In general terms, the hydrolysis of esters may be represented as

$$R-\overset{\overset{\displaystyle O}{\|}}{C}-O-R' + Na^+OH^- \overset{heat}{\longrightarrow} R-\overset{\overset{\displaystyle O}{\|}}{C}-O^-Na^+ + R'OH$$

ester　　　　　　　　　　　　　　salt of an acid　　　alcohol

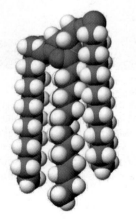

Tripalmitin, a triglyceride.

Soap can be made by boiling animal fat with an alkaline solution obtained from the ashes of hardwood. The resulting soap can be "salted out" by adding sodium chloride, making use of the fact that soap is less soluble in a salt solution than in water.

Like other esters, fats and oils (Section 27-14) can be hydrolyzed in strongly basic solution to produce salts of the acids and the trihydric alcohol glycerol. The resulting alkali metal salts (commonly sodium or potassium) of long-chain fatty acids are **soaps.** In Section 14-18 we described the cleansing action of soaps and detergents.

CD-ROM Screen 11.7, Functional Groups (2): Fats and Oils.

$$H_2C-O-\overset{\overset{\displaystyle O}{\|}}{C}-(CH_2)_{16}CH_3$$
$$HC-O-\overset{\overset{\displaystyle O}{\|}}{C}-(CH_2)_{16}CH_3 \; + \; 3Na^+OH^- \xrightarrow[\;H_2O\;]{\text{heat}} \; 3CH_3(CH_2)_{16}\overset{\overset{\displaystyle O}{\|}}{C}-O^-Na^+ \; + \; \begin{matrix} H_2C-OH \\ HC-OH \\ H_2C-OH \end{matrix}$$
$$H_2C-O-\overset{\overset{\displaystyle O}{\|}}{C}-(CH_2)_{16}CH_3$$

| glyceryl tristearate (a fat) | | sodium stearate (a soap) | glycerol (glycerine) |

BIOPOLYMERS

As we study the molecules and ions in living organisms such as humans, we learn that substances with a huge range of sizes and complexities are involved. These include small molecules and ions such as sodium ions, trace metals, ketones, alcohols, and carboxylic acids. Steroids, hormones, vitamins, fats, and oils are somewhat larger. As we continue up the scale of sizes of substances present in living systems, we encounter the mammoth polymers (biopolymers, or polymers found in biological systems). Concepts discussed in the first 27 chapters and the first half of this chapter can be applied to large molecules and ions as well as small ones. In the following sections we will look briefly at three classes of biopolymers (carbohydrates, proteins, and nucleic acids), keeping in mind that almost all of the chemical concepts used in these discussions were introduced earlier in this text.

28-8 CARBOHYDRATES

The word *saccharide* comes from the Latin *saccharum,* for "sugar" or "sweet."

Carbohydrates have the general formula $C_n(H_2O)_m$, thus the name hydrated carbons or, more commonly, carbohydrates. **Monosaccharides,** or simple sugars, are the monomers from which more complex carbohydrates are constructed. The most important simple sugars contain either five or six carbon atoms. Monosaccharides that contain an aldehyde group are called **aldoses,** and those that contain a ketone group are called **ketoses.** The most commonly encountered monosaccharide is glucose, $C_6H_{12}O_6$, also known as dextrose or blood sugar. Figure 28-10a shows glucose as it is often written to emphasize its aldehyde functional group. We can see from this formula that carbons 2, 3, 4, and 5 in glucose each are bonded to four *different* groups and are, therefore, chiral. Each different monosaccharide has a unique arrangement of atoms about its chiral atoms. Some important monosaccharides are shown in Table 28-4.

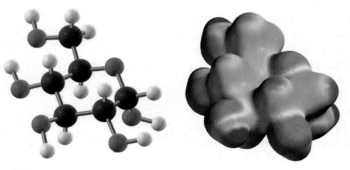

(a)

(b)

(c1) α-glucose

(c2) β-glucose

(d)

Figure 28-10 The cyclization of glucose to form either the α- or the β-cyclic form. (a) The linear representation of the aldehyde form of glucose. (b) A coiled representation that shows the aldehyde function group near the —OH group of carbon 5. (c1, c2) The two cyclic forms of glucose. (d) Models of the α-form of glucose.

TABLE 28-4 *Some Important Monosaccharides*

Five Carbon Atom Monosaccharides

CHO	CHO	CH_2OH
H—C—OH	HO—C—H	C=O
H—C—OH	H—C—OH	H—C—OH
H—C—OH	H—C—OH	H—C—OH
CH_2OH	CH_2OH	CH_2OH
ribose	arabinose	ribulose

Six Carbon Atom Monosaccharides

CHO	CHO	CHO	CH_2OH
H—C—OH	HO—C—H	H—C—OH	C=O
HO—C—H	HO—C—H	HO—C—H	HO—C—H
H—C—OH	H—C—OH	HO—C—H	H—C—OH
H—C—OH	H—C—OH	H—C—OH	H—C—OH
CH_2OH	CH_2OH	CH_2OH	CH_2OH
glucose	mannose	galactose	fructose

Besides their aldehyde or ketone form, saccharides can also exist in a cyclic form. Figure 28-10 shows the interconversion of glucose between its linear form and its cyclic form. Figure 28-10b shows the molecule rotated into a conformation that brings the hydroxyl group of carbon 5 near the aldehyde carbon. In this conformation, a reversible reaction occurs in which a hydrogen is moved from the hydroxyl group to the carbonyl oxygen, and an ether bond is formed between the two carbons (Figure 28-10c). During the formation of the ether bond, a ring is also formed and carbon atom 1 becomes chiral. This new chiral center can exist as either of two isomers; the resulting stereoisomers are referred to as α- or β-forms.

Most naturally occurring carbohydrates contain more than one monosaccharide unit. Complex carbohydrates are formed when two or more monosaccharides are linked. This linkage between two monosaccharides occurs by elimination of a molecule of water and the formation of an ether bond between two of the cyclic structures, as shown in Figure 28-11. The newly formed C—O bond linking the rings is called the **glycosidic bond.** Figure 28-11 illustrates the linkage of α-glucose and β-fructose to form sucrose. Fructose is a ketone-containing carbohydrate that exists in the cyclic form shown. A **disaccharide** is a molecule consisting of two monosaccharides joined together by a glycosidic bond. Sucrose is a disaccharide; two other disaccharides, each involving only glucose units, are shown in Figure 28-12. In such a molecule, the C—O—C linkage between the monosaccharide units involves carbon number 1 or 2 of one monosaccharide and carbon number 1, 4, or 6 of the other monosaccharide. The arrangement about the carbon number 1 or 2 is what is described as the α- or β-linkage. Using these notations, we describe maltose as consisting of two glucose units joined by an α-1,4-linkage. Lactose has a β-1,4-linkage. Table 28-5 lists some of the common sugars.

SCiLINKS.

TOPIC: Carbohydrates
GO TO: www.scilinks.org
*sci*LINKS CODE: WCH2830

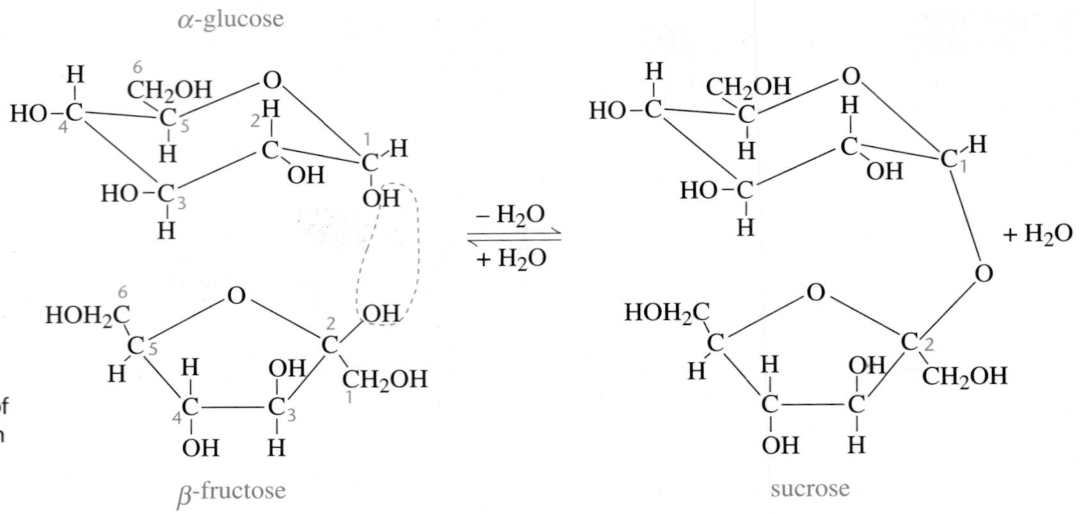

Figure 28-11 Formation of sucrose by the condensation of an α-glucose molecule with a β-fructose molecule.

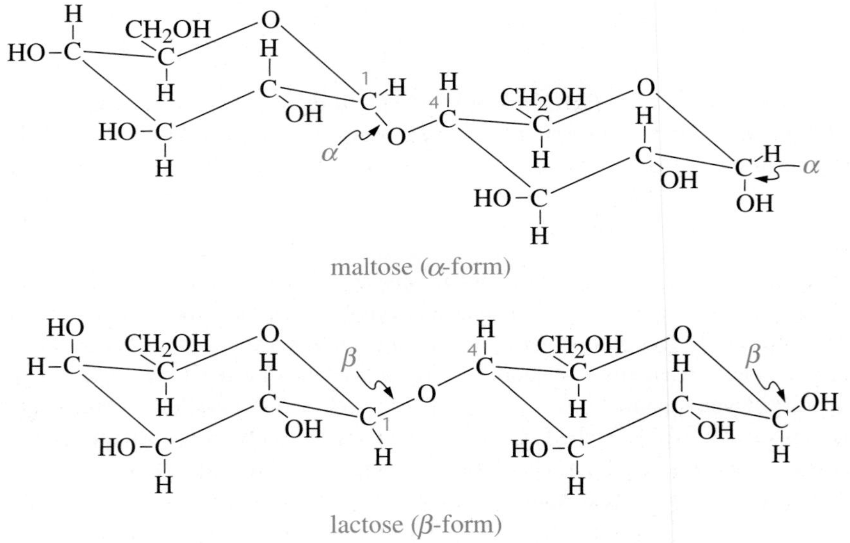

Figure 28-12 Two disaccharides: maltose and lactose. Each involves two monosaccharides. Maltose has an α-1,4-linkage and lactose has a β-1,4-linkage.

TABLE 28-5	Sweetness of Common Sugars Relative to Sucrose

Substance	Sweetness Relative to Sucrose at 1.00
lactose (milk sugar, a disaccharide)	0.16
galactose (a monosaccharide in milk sugar)	0.32
maltose (a disaccharide used in beer making)	0.33
glucose (dextrose, a common monosaccharide)	0.74
sucrose (table sugar, a disaccharide)	1.00
fructose (fruit sugar, a monosaccharide)	1.74

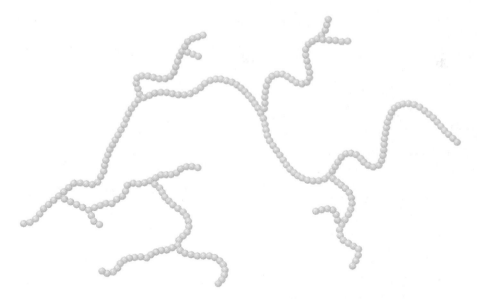

Figure 28-13 Amylopectin, or glycogen.

A carbohydrate that contains three monosaccharides is called a **trisaccharide,** and so on. Carbohydrates that contain from four to ten monosaccharides are usually called by the general term **oligosaccharides.** Carbohydrates that contain larger numbers of monosaccharide units are called **polysaccharides.**

Amylose, a type of starch found in plants, is a polysaccharide. It consists of a linear combination of several thousand glucose units joined by α-1,4-linkages. In our bodies we store glucose in a form known as amylopectin, or glycogen. Glycogen is similar to amylose; it typically contains about 30,000 glucose monomers with the majority linked via an α-1,4-linkage but, unlike amylose, it contains occasional branches via α-1,6-linkages. In Figure 28-13 each sphere represents a glucose unit. All links are α-1,4 except where branching takes place; branching utilizes an α-1,6-linkage. Our digestive systems contain enzymes (see Sections 16-9 and 28-9) that can break this biopolymer, glycogen, into its monomeric glucose units as we need energy to drive other biological processes.

Cellulose, the structural biopolymer found in plants, contains the β-1,4-linkage of glucose units. Our digestive systems are incapable of breaking the β-1,4-linkage, so we cannot utilize cellulose as a source of stored glucose. Some animals such as termites and cattle are able to digest cellulose.

Bacteria in the digestive tracts of termites and cattle produce an enzyme that allows them to digest the cellulose in wood and grass.

TABLE 28-6 *The 20 Common Amino Acids Found in Proteins*

Nonpolar R Groups

alanine (Ala)

$$CH_3\underset{\underset{NH_2}{|}}{C}HCOOH$$

glycine (Gly)

$$H\underset{\underset{NH_2}{|}}{C}HCOOH$$

isoleucine (Ile)

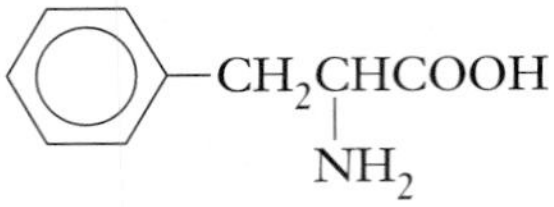

$$CH_3CH_2\underset{\underset{NH_2}{|}}{\overset{\overset{CH_3}{|}}{C}}HCHCOOH$$

leucine (Leu)

$$(CH_3)_2CHCH_2\underset{\underset{NH_2}{|}}{C}HCOOH$$

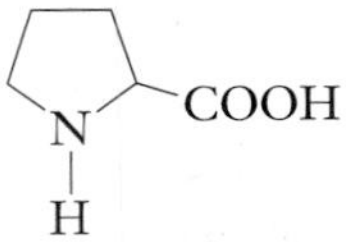

methionine (Met)

$$CH_3SCH_2CH_2\underset{\underset{NH_2}{|}}{C}HCOOH$$

phenylalanine (Phe)

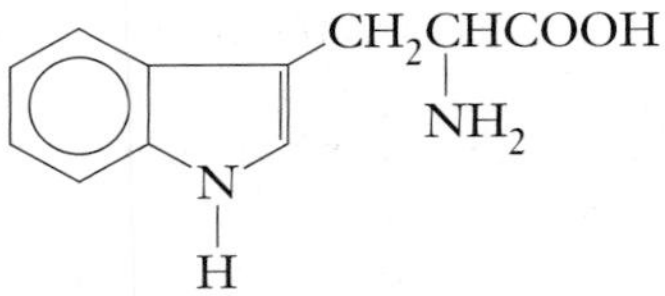

$$\text{⟨benzene ring⟩}-CH_2\underset{\underset{NH_2}{|}}{C}HCOOH$$

proline (Pro)

$$\text{⟨pyrrolidine ring⟩}-COOH$$

tryptophan (Trp)

$$\text{⟨indole ring⟩}-CH_2\underset{\underset{NH_2}{|}}{C}HCOOH$$

valine (Val)

$$(CH_3)_2CH\underset{\underset{NH_2}{|}}{C}HCOOH$$

Polar, Neutral R Groups

asparagine (Asn)

$$H_2N\overset{\overset{O}{\|}}{C}CH_2\underset{\underset{NH_2}{|}}{C}HCOOH$$

glutamine (Gln)

$$H_2N\overset{\overset{O}{\|}}{C}CH_2CH_2\underset{\underset{NH_2}{|}}{C}HCOOH$$

serine (Ser)

$$HOCH_2\underset{\underset{NH_2}{|}}{C}HCOOH$$

threonine (Thr)

$$CH_3\overset{\overset{OH}{|}}{C}H\underset{\underset{NH_2}{|}}{C}HCOOH$$

Polar, Acidic R Groups

aspartic acid (Asp)

$$HOOCCH_2\underset{\underset{NH_2}{|}}{C}HCOOH$$

glutamic acid (Glu)

$$HOOCCH_2CH_2\underset{\underset{NH_2}{|}}{C}HCOOH$$

cysteine (Cys)

$$HSCH_2\underset{\underset{NH_2}{|}}{C}HCOOH$$

tyrosine (Tyr)

$$HO-\text{⟨benzene ring⟩}-CH_2\underset{\underset{NH_2}{|}}{C}HCOOH$$

Polar, Basic R Groups

arginine (Arg)

$$H_2N\overset{\overset{NH}{\|}}{C}NHCH_2CH_2CH_2\underset{\underset{NH_2}{|}}{C}HCOOH$$

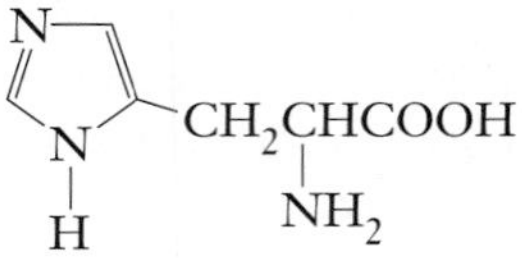

histidine (His)

$$\text{⟨imidazole ring⟩}-CH_2\underset{\underset{NH_2}{|}}{C}HCOOH$$

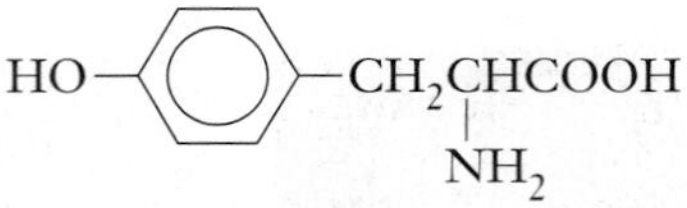

lysine (Lys)

$$H_2NCH_2CH_2CH_2CH_2\underset{\underset{NH_2}{|}}{C}HCOOH$$

28-9 POLYPEPTIDES AND PROTEINS

Two amino acids can react to form an amide bond (see Section 27-19). Compounds formed by the linking of small numbers of amino acids are called **peptides,** and the amide bond linking the amino acid monomers is called the **peptide bond.**

$$H-\underset{\underset{H}{|}}{\overset{\overset{H}{|}}{N}}-\underset{\underset{R}{|}}{\overset{\overset{H}{|}}{C}}-\underset{}{\overset{\overset{O}{\|}}{C}}-OH + H-\underset{\underset{H}{|}}{\overset{\overset{H}{|}}{N}}-\underset{\underset{R'}{|}}{\overset{\overset{H}{|}}{C}}-\underset{}{\overset{\overset{O}{\|}}{C}}-OH \longrightarrow H-\underset{\underset{H}{|}}{\overset{\overset{H}{|}}{N}}-\underset{\underset{R}{|}}{\overset{\overset{H}{|}}{C}}-\underset{}{\overset{\overset{O}{\|}}{C}}-\underset{\underset{H}{|}}{\overset{\overset{H}{|}}{N}}-\underset{\underset{R'}{|}}{\overset{\overset{H}{|}}{C}}-\underset{}{\overset{\overset{O}{\|}}{C}}-OH + H_2O$$

peptide bond

Prefixes di-, tri-, and so on are used to indicate the number of amino acid monomers that are joined to form the peptide. The compound indicated in the preceding formula is a **dipeptide.**

Twenty different α-amino acids (see Section 27-13) are found widely in nature. Each amino acid is commonly given a three-letter abbreviation. These acids differ in the R groups attached to the α-carbon (the carbon adjacent to the —COOH group). Differences in R groups lead to differences in the properties of the α-amino acids. The amino acids are usually classified by their R groups, or side chains, according to two criteria. The first classification depends on whether the R groups are polar or nonpolar, the second describes whether the R group contains an acidic or basic group. The 20 common amino acids are shown, according to these classifications, in Table 28-6.

Many small peptides are known to have physiological significance. Aspartame, a common artificial sweetener, is a dipeptide. Glutathione, a tripeptide, acts as a scavenging agent for harmful oxidizing agents believed to cause cancer. Enkephalins, which are naturally occurring analgesics (pain relievers) that occur in the brain, are pentapeptides. The antibiotics gramicidin-S and tyrocidine A are cyclic decapeptides (i.e., they contain ten amino acids linked into a ring by peptide bonds) produced by a bacterium.

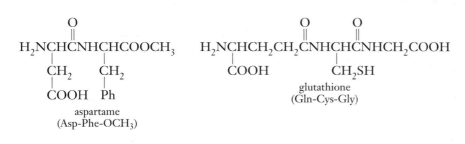

$$\underset{\underset{\underset{COOH}{|}}{\underset{CH_2}{|}}}{H_2NCHCNHCHCOOCH_3}\underset{\underset{Ph}{|}}{\underset{CH_2}{|}}$$

aspartame
(Asp-Phe-OCH₃)

$$\underset{\underset{COOH}{|}}{H_2NCHCH_2CH_2CNHCHCNHCH_2COOH}\underset{\underset{CH_2SH}{|}}{}$$

glutathione
(Gln-Cys-Gly)

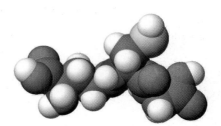

Models of aspartame (*top*) and glutathione (*bottom*).

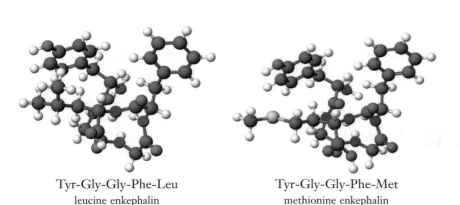

Tyr-Gly-Gly-Phe-Leu
leucine enkephalin

Tyr-Gly-Gly-Phe-Met
methionine enkephalin

Some organisms such as bacteria also produce D-amino acids.

A **protein** is a biopolymer, of molecular weight typically 5000 g/mol or greater, consisting of one or more polypeptide chains. The α-amino acids that occur in nearly all proteins are the L optical isomer.

An **enzyme** is a protein that catalyzes a biological reaction.

Each protein consists of one or more polypeptide chains with a unique sequence of amino acids. The order of these monomers is termed the **primary structure** of the protein. The many possible interactions between different amino acids in the polypeptide chain cause each protein molecule to adopt a characteristic shape, usually depending on its environment. The **secondary structure** of a protein is the arrangement in space of the polypeptide backbone, without reference to the conformations of the side chains (R groups). Myoglobin is the oxygen-storing protein in muscles. Its amino acid sequence (primary structure) and a three-dimensional representation of its secondary structure are shown in Figure 28-14. The **tertiary structure** of a protein describes the overall shape of the protein, including the side chains and any other nonpeptide components of the protein. Myoglobin is referred to as a globular protein. Some proteins consist of multiple polypeptide units called *subunits*. These subunits are held to one another by noncovalent interactions such as hydrogen bonding, dipole–dipole interactions, or ionic attractions, to form the **quaternary structure** of the protein.

SC*i*LINKS.

TOPIC: Proteins
GO TO: www.scilinks.org
*sci*LINKS CODE: WCH2840

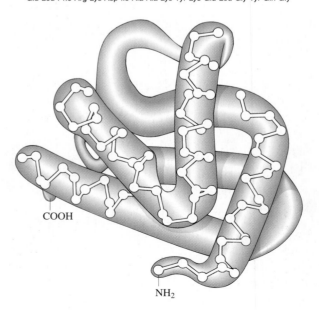

Val-Leu-Ser-Glu-Gly-Glu-Trp-Gln-Leu-Val-Leu-His-Val-Trp-Ala-Lys-Val-
Glu-Ala-Asp-Val-Ala-Gly-His-Gly-Gln-Asp-Ile-Leu-Ile-Arg-Leu-Phe-Lys-
Ser-His-Pro-Glu-Thr-Leu-Glu-Lys-Phe-Asp-Arg-Phe-Lys-His-Leu-Lys-
Thr-Glu-Ala-Glu-Met-Lys-Ala-Ser-Glu-Asp-Leu-Lys-Lys-His-Gly-Val-Thr-
Val-Leu-Thr-Ala-Leu-Gly-Ala-Ile-Leu-Lys-Lys-Lys-Gly-His-His-Glu-Ala-
Glu-Leu-Lys-Pro-Leu-Ala-Gln-Ser-His-Ala-Thr-Lys-His-Lys-Ile-Pro-Ile-Lys-
Tyr-Leu-Glu-Phe-Ile-Ser-Glu-Ala-Ile-Ile-His-Val-Leu-His-Ser-Arg-His-
Pro-Gly-Asn-Phe-Gly-Ala-Asp-Ala-Gln-Gly-Ala-Met-Asn-Lys-Ala-Leu-
Glu-Leu-Phe-Arg-Lys-Asp-Ile-Ala-Ala-Lys-Tyr-Lys-Glu-Leu-Gly-Tyr-Gln-Gly

COOH

NH$_2$

Figure 28-14 The amino acid sequence and the three-dimensional representation of myoglobin, the oxygen-storing protein in muscles.

A model of the peptide chain of the protein ribonuclease A. The different colors represent regions within the folding and coiling patterns.

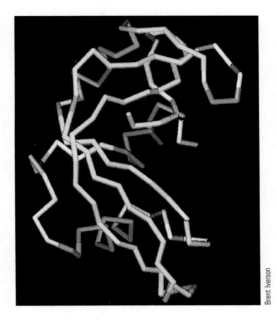

The folding and coiling patterns of a protein often affect its biological activity, for instance, its effectiveness as an enzyme. Forces such as heat and ionic environment can alter the pattern of folding and coiling, causing the enzyme to become inactive, or *denatured*.

Proteins make up more than 50% of the dry weight of animals and bacteria. They perform many important functions in living organisms, a few of which are indicated in Table 28-7. Each protein carries out a specific biochemical function. Each is a polypeptide with its own unique *sequence* of amino acids. The amino acid sequence of a protein determines exactly how it folds up in a three-dimensional conformation and how it performs its precise biochemical task.

A "ribbon" model of the protein calmodulin. In this type of model, the ribbon represents the polypeptide chain. This protein coordinates with Ca^{2+} ions (white spheres) and aids in transporting them in living systems.

TABLE 28-7	*Some Functions of Proteins*
Example	**Function**
Enzymes	
amylase	converts starch to glucose
DNA polymerase I	repairs DNA molecule
transaminase	transfers amino group from one amino acid to another
Structural Proteins	
viral coat proteins	outer covering of virus
keratin	hair, nails, horns, hoofs
collagen	tendons, cartilage
Hormones	
insulin, glucagon	regulate glucose metabolism
oxytocin	regulates milk production in female mammals
vasopressin	increases retention of water by kidney
Contractile Proteins	
actin	thin contractile filaments in muscle
myosin	thick filaments in muscle
Storage Proteins	
casein	a nutrient protein in milk
ferritin	stores iron in spleen and egg yolk
Transport Proteins	
hemoglobin	carries O_2 in blood
myoglobin	carries O_2 in muscle
serum albumin	carries fatty acids in blood
cytochrome *c*	transfers electrons
Immunological Proteins	
γ-globulins	form complexes with foreign proteins
Toxins	
neurotoxin	blocker of nerve function in cobra venom
ricin	nerve toxin in South American frog (most toxic substance known—0.000005 g is fatal to humans)

28-10   NUCLEIC ACIDS

We have seen that polysaccharides are polymers composed of a single type of monomer (carbohydrates), as are proteins (amino acids). The third type of biopolymer is more complex. Nucleic acids use three very different types of monomers: the phosphate group, one of two simple carbohydrate units (*deoxyribose* or *ribose*), and selected organic bases (Figure 28-15). A typical segment of the resulting polymer is shown in Figure 28-16.

    There are two types of nucleic acids: **ribonucleic acid (RNA)** and **deoxyribonucleic acid (DNA).** One of the ways in which they differ is in the carbohydrate that they contain.

sci*LINKS*

TOPIC: DNA
GO TO: www.scilinks.org
*sci*LINKS CODE: WCH2850

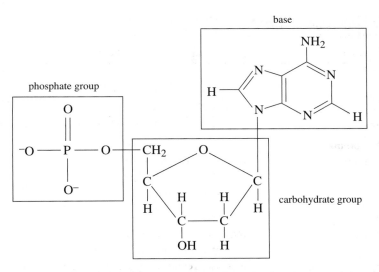

Figure 28-15 The three kinds of units that are polymerized to form nucleic acids.

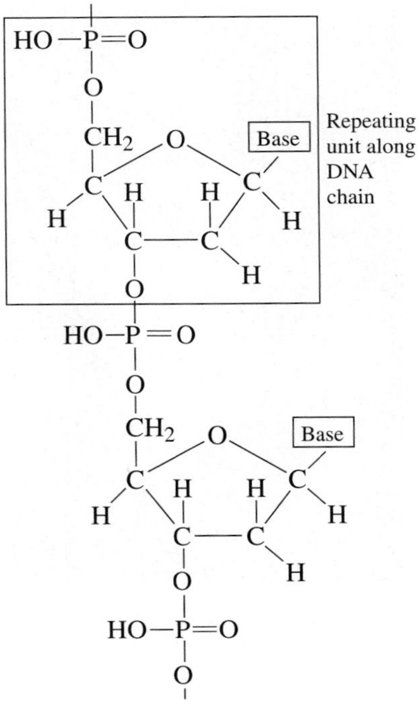

The two carbohydrates in nucleic acids are ribose (contained in RNA) and 2-deoxyribose (contained in DNA); these carbohydrates differ in the presence or absence of an oxygen atom on carbon 2.

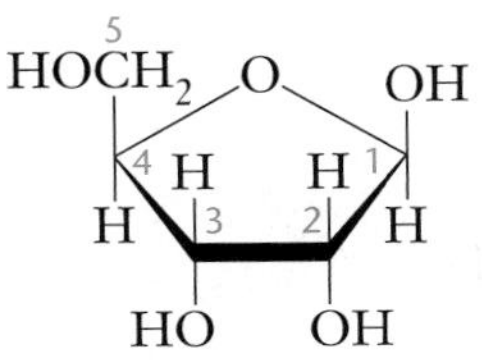

ribose

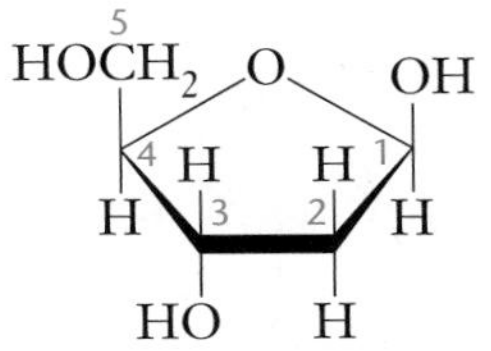

2-deoxyribose

Figure 28-16 A short segment of DNA (deoxyribonucleic acid).

Repeating unit along DNA chain

The bases found in nucleic acids are shown in Figure 28-17. Each type of nucleic acid contains two pyrimidine bases and two purine bases. The two purine bases adenine (A) and guanine (G) and the pyrimidine base cytosine (C) are found in both RNA and DNA. In

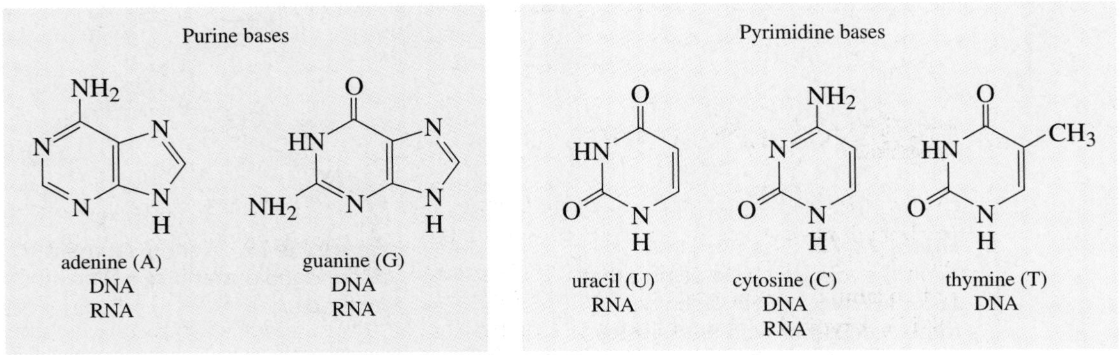

Purine bases

adenine (A)
DNA
RNA

guanine (G)
DNA
RNA

Pyrimidine bases

uracil (U)
RNA

cytosine (C)
DNA
RNA

thymine (T)
DNA

Figure 28-17 The bases found in DNA and RNA.

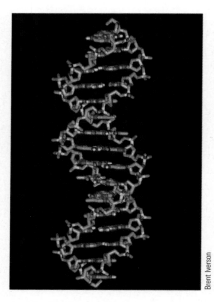

An idealized model of DNA.

RNA, the other pyrimidine base is uracil (U), whereas DNA contains thymine (T). Specific pairs of these bases hydrogen bond to each other in particularly favorable arrangements, so they always occur in distinctive hydrogen-bonded base pairs. Adenine and thymine (in DNA) or uracil (in RNA) each form two hydrogen bonds, so they are always paired together; similarly, guanine and cytosine each form three hydrogen bonds, so they always pair to each other (Figure 28-18). The resulting complementary **base-pairing** links two strands of the polymer in a very specific order to form a double helix. This order is the key to nucleic acid function. Figure 28-19 shows a short segment of the double strand of DNA. In that figure, we see that adenine is paired with thymine and guanine is paired with cytosine. In RNA, adenine forms base pairs with uracil instead of with thymine.

DNA is one of the largest molecules known. Human DNA is estimated to have up to three billion base pairs, resulting in a molecular weight in the tens of billions. RNA is smaller and more variable, with molecular weights of 20,000 to 40,000. Genetic information for each organism is stored in its DNA as a result of the locations and sequence of the base pairs. The information is replicated when the strands unravel and new complementary strands are formed. The genetic information of the DNA is used to guide the many syntheses that occur in a living cell. For example, for the synthesis of a specific protein, a specific portion of the DNA double helix unwinds. Then an RNA molecule is built using the sequence of base pairs of the DNA as a pattern. The RNA then migrates from the cell's nucleus to the location where the protein is to be synthesized. There the base sequence of the RNA determines the sequence of amino acids to be used in the protein construction. With its many possible sequences of the same base pairs to transmit information to control synthesis of other molecules, DNA contains a vast amount of genetic information.

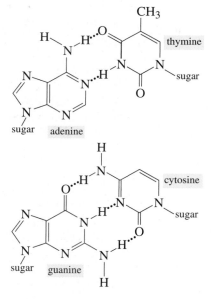

Figure 28-18 The two kinds of hydrogen-bonded base-pairing that occur in DNA. In RNA, adenine forms a base pair with uracil instead of with thymine.

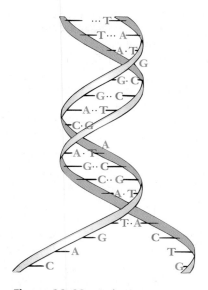

Figure 28-19 A short segment of the double strand of a DNA molecule.

Key Terms

Achiral Describes an object that *can* be superimposed with its mirror image.

Aldose A monosaccharide that contains an aldehyde group.

Base pairing The complementary hydrogen bonding of cytosine with guanine and of adenine with thymine (in DNA) or with uracil (in RNA).

Biopolymer Polymers found in biological systems.

Brønsted–Lowry acid A species that acts as a proton donor.

Brønsted–Lowry base A species that acts as a proton acceptor.

Carbohydrate A biopolymer made up of saccharide units.

Chiral Describes an object that *cannot* be superimposed with its mirror image.

Conformation One specific geometry of a molecule. The conformations of a compound differ from one another only by rotation about single bonds.

Constitutional isomers Compounds that contain the same number of the same kinds of atoms but that differ in the order in which their atoms are bonded together. Also known as structural isomers.

2-Deoxyribose The carbohydrate found in DNA.

Dipeptide A compound in which two amino acid monomers have joined to form the peptide.

Disaccharide A molecule consisting of two monosaccharides joined together by a glycosidic bond.

DNA Deoxyribonucleic acid; a nucleic acid consisting of phosphate, 2-deoxyribose, and the four bases adenine, cytosine, guanine, and thymine.

Enantiomers See *Optical isomers.*

Enzyme A protein that catalyzes a biological reaction.

Geometric isomers Compounds with different arrangements of groups on the opposite sides of a bond with restricted rotation, such as a double bond or a single bond in a ring; for example, *cis–trans* isomers of certain alkenes.

Glycosidic bond A bond linking monomers in a polysaccharide.

Ketose A monosaccharide that contains a ketone group.

Monomers The small molecules from which polymers are formed.

Monosaccharide Simple sugars (monomers) from which more complex carbohydrates are constructed.

Nucleic acid A biopolymer consisting of repeating units of ribose or deoxyribose, phosphate, and selected bases.

Oligosaccharide A molecule consisting of four to ten monosaccharides joined together by glycosidic bonds.

Optical activity The rotation of a plane of polarized light by chiral molecules.

Optical isomers Molecules that are nonsuperimposable mirror images of each other, that is, that bear the same relationship to each other as do left and right hands; also called *enantiomers.*

Oxidation (as applied to organic compounds) The increase of oxygen content or the decrease of hydrogen content of an organic molecule.

Peptide bond A bond formed by elimination of a molecule of water between the amino group of one amino acid and the carboxylic acid group of another.

Peptides Compounds formed by the linking of small numbers of amino acids.

Polymerization The combination of many small molecules (monomers) to form large molecules (polymers).

Polymers Large molecules formed by the combination of many small molecules (monomers).

Polypeptide A polymer composed of amino acids linked by peptide bonds.

Polysaccharide Carbohydrates that contain more than ten monosaccharide units.

Primary structure The sequence of the monomeric units in proteins, carbohydrates, and so on.

Protein A naturally occurring polymeric chain of L-amino acids linked together by peptide bonds.

Racemic mixture A single sample containing equal amounts of the two enantiomers (optical isomers) of a compound; does not rotate the plane of polarized light.

Reduction (as applied to organic compounds) The decrease of oxygen content or the increase of hydrogen content of an organic molecule.

Ribose The carbohydrate found in RNA.

RNA Ribonucleic acid, a nucleic acid consisting of phosphate, ribose, and the four bases adenine, cytosine, guanine, and uracil.

Saponification The hydrolysis of esters in the presence of strong bases.

Soap The alkali matal (Na or K) salt of a long-chain fatty acid.

Stereochemistry The study of the three-dimensional aspects of molecular structures.

Stereoisomers Isomers in which the atoms are linked together in the same atom-to-atom order, but with different arrangements in space. See *Geometric isomers, Optical isomers.*

Trisaccharide A molecule consisting of three monosaccharides joined together by glycosidic bonds.

Exercises

Geometries of Organic Molecules

1. Distinguish between two types of constitutional isomerism. Give an example of each type.
2. Distinguish between two types of stereoisomerism. Give an example of each type.
3. Which of the following compounds can exist as *cis* and *trans* isomers? Draw them. (a) 2,3-dimethyl-2-butene; (b) 2-chloro-2-butene; (c) dichlorobenzene; (d) 1,1-dichlorocyclobutane.
4. Which of the following compounds can exist as *cis* and *trans* isomers? Draw them. (a) 2-butene; (b) 2,3-dibromo-2-butene; (c) 2-bromo-2-butene; (d) 1,2-dichlorocyclopentane.
5. Distinguish between conformations and isomers.
6. What is optical isomerism? What are enantiomers?
7. Which of the following compounds would exhibit optical isomerism?

(a) $CH_3CHCH_2CH_3$ (c) $HO-$ ⬡ $-\overset{\overset{H}{|}}{\underset{\underset{Cl}{|}}{C}}-$ ◯
   |
   Br

(b) CH_3CHCH_3 (d) $CH_3CH{=}CHCH_3$
   |
   OH

8. Draw three-dimensional representations of the enantiomeric pairs in Exercise 7.
9. Write formulas and names for the isomers of (a) bromochlorobenzene, (b) trifluorobenzene, and (c) chlorotoluene. What kind of isomerism is illustrated by each of these sets of compounds?
10. Write formulas and names for the isomers of (a) butanol, (b) pentachloropropene, and (c) chlorofluoroethane.
11. Draw structural representations of two conformations of butane. Indicate the one having the lower potential energy.
12. Draw structural representations of two conformations of cyclohexane. Give the name commonly applied to each.

Selected Reactions

13. (a) List the four most acidic organic functional groups. (b) List the four common organic functional groups that are bases.
14. Why are aqueous solutions of amines basic? Show, with equations, how the dissolution of an amine in water is similar to the dissolution of ammonia in water.
15. Show that the reactions of amines with inorganic acids such as HCl are similar to the reactions of ammonia with inorganic acids.
*16. What are the equilibrium concentrations of the species present in a 0.15 M solution of aniline? $K_b = 4.2 \times 10^{-10}$
$$C_6H_5NH_2(aq) + H_2O(\ell) \rightleftharpoons C_6H_5NH_3^+ + OH^-$$

*17. Which solution would be more acidic: a 0.15 M solution of aniline hydrochloride, $C_6H_5NH_3Cl$ ($K_b = 4.2 \times 10^{-10}$ for aniline, $C_6H_5NH_2$), or a 0.15 M solution of methylamine hydrochloride, CH_3NH_3Cl ($K_b = 5.0 \times 10^{-4}$ for methylamine, CH_3NH_2)? Justify your choice.
18. Choose the compound that is the stronger acid in each set.

(a) $CH_3CH_2CH_2OH$ or   ⬡$-\overset{\overset{O}{\|}}{C}-OH$

(b) CH_3CH_2OH or   ◯$-OH$

(c) ◯$-OH$ or   ◯$-\overset{\overset{O}{\|}}{C}-OH$

(d) ⬡$-OH$ or   ◯$-OH$

19. (a) What are alkoxides? (b) What do we mean when we say that the low-molecular-weight alkoxides are strong bases?
20. (a) Write equations for the reactions of three alcohols with metallic sodium. (b) Name all compounds in these equations. (c) Show how these reactions are similar to the reaction of metallic sodium with water.
21. Which physical property of aldehydes is used to advantage in their production from alcohols?
22. The K_b for lidocaine (a monobasic organic compound, MW = 234.3) is 7.0×10^{-6}. What is the pH of a 1.5 percent solution of lidocaine? The density of the solution is 1.00 g/mL.
23. How are the terms "oxidation" and "reduction" often used in organic chemistry? Classify the following changes as either oxidation or reduction: (a) CH_4 to CH_3OH; (b) $CH_2{=}CH_2$ to $CH_3\text{-}CH_3$; (c) CH_3CH_2CHO to $CH_3CH_2CH_3$;

(d) ◯$-CH_3$ to   ◯$-\overset{\overset{O}{\|}}{C}-OH$

24. Classify the following changes as either oxidation or reduction: (a) CH_3OH to CO_2 and H_2O; (b) CH_2CH_2 to CH_3CHO; (c) CH_3COOH to CH_3CHO; (d) $CH_3CH{=}CH_2$ to $CH_3CH_2CH_3$.
25. Write equations to illustrate the oxidation of the following aromatic hydrocarbons by potassium permanganate in basic solution: (a) toluene; (b) ethylbenzene; (c) 1,2-dimethylbenzene.
26. (a) Do you expect toluene to produce soot as it burns? Why? (b) Would you expect the flames to be blue or yellow?

27. Describe the preparation of three aldehydes from alcohols, and write appropriate equations. Name all reactants and products.
28. Describe the preparation of three ketones from alcohols, and write appropriate equations. Name all reactants and products.
29. An ester is the product of the reaction of an alcohol with a carboxylic acid and is the organic version of an inorganic salt. What is an inorganic ester?
30. (a) Write equations for the formation of three inorganic esters. (b) Name the inorganic ester formed in each case.
31. (a) What is nitroglycerine? (b) Write the chemical equation that shows the preparation of nitroglycerine. (c) List two important uses for nitroglycerine. Are they similar?
32. Write equations for the formation of three different esters, starting with an acid and an alcohol in each case. Name all compounds.
33. Write equations for the formation of three different esters, starting with a different acid chloride and a different alcohol in each case. Name all compounds.
34. Write equations for the hydrolysis of (a) methyl acetate, (b) ethyl formate, (c) butyl acetate, and (d) octyl acetate. Name all products.
35. (a) What is a saponification reaction? (b) Why is this kind of reaction called saponification?

Biopolymers

36. What is necessary if a molecule is to be capable of polymerization? Name three types of molecules that can polymerize.
37. A cellulose polymer has a molecular weight of 750,000. Estimate the number of units of the monomer, β-glucose ($C_6H_{12}O_6$) in this polymer. This polymerization reaction can be represented as

$$x C_6H_{12}O_6 \longrightarrow \text{cellulose} + (x - 1)H_2O$$

38. Write structural formulas of two monosaccharides that contain six carbon atoms.
39. Write the structural formula of a trisaccharide containing only α-1,4-linked glucose units.
40. Write two structural formulas of fructose, one in its cyclic form and one in its straight-chain form.
41. Describe the structure of a natural amino acid molecule. What kind of isomerism do most amino acids exhibit? Why?
42. How are the amino acid units in a polypeptide joined together? What are the links called?
43. Consider only two amino acids:

$$NH_2-\underset{\underset{R'}{|}}{\overset{\overset{H}{|}}{C}}-COOH \quad \text{and} \quad NH_2-\underset{\underset{R}{|}}{\overset{\overset{H}{|}}{C}}-COOH$$

Write the structural formulas for the dipeptides that could be formed containing one molecule of each amino acid.
*44. How many different dipeptides can be formed from the three amino acids A, B, and C? Write the sequence of amino acids in each. Assume that an amino acid could occur more than once in each dipeptide.
*45. How many different tripeptides can be formed from the two amino acids A and B? Write the sequence of amino acids in each. Assume that an amino acid can occur no more than twice in each tripeptide.
46. Aspartame (trade name NutraSweet) is a methyl ester of a dipeptide:

Write the structural formulas of the two amino acids that are combined to make aspartame (ignoring optical isomerism).
47. Write the structural formula of the carbohydrate found in deoxyribonucleic acid (DNA). How does this carbohydrate differ from the one found in ribonucleic acid (RNA)?
48. Write structural formulas of the four bases found in DNA. Circle at least one functional group on each that gives the compound its base properties.
49. Draw a short segment of the DNA biopolymer that contains three each of the phosphate, carbohydrate, and base groups.
50. In the double-strand form of DNA the base groups are paired. What base is found paired (hydrogen bonded) with thymine?

Mixed Exercises

51. Identify the major products of each reaction.

52. Identify the major products of each reaction.

(a) —CH$_2$OH + Na $\longrightarrow$

(b) CH$_3$CH$_2\overset{\overset{\displaystyle O}{\|}}{C}OCH_3$ $\xrightarrow[\text{heat}]{\text{KOH(aq)}}$

(c) $\xrightarrow[\text{heat}]{\text{NaOH(aq)}}$

53. Write a structural formula for each of the following compounds:
   (a) CH$_3$CH$_2$OH
   (b) CH$_3$COOH
   (c) CH$_3$CHOHCH$_3$
   (d) CH$_3$OCH$_3$
   (e) CH$_3$CHCHCH$_2$CH$_3$

54. Write a structural formula for each of the following compounds:
   (a) CH$_3$CH$_2$ONO$_2$
   (b) CH$_3$CHCH$_2$
   (c) CH$_3$CH$_2$CH$_2$OH
   (d) ClCH$_2$CHCHCl
   (e) CH$_3$CH$_2$CH$_2$ONa

55. (a) Name three biopolymers found in animals. (b) Name at least one biopolymer that is composed of identical monomers.

56. Name the three types of monomers that are found in DNA. Give a structural formula of an example of each.

CONCEPTUAL EXERCISES

57. Nylon is decomposed by acids, but polyethylene is not. Suggest an explanation for this difference in behavior.

58. How does the heat of combustion of ethyl alcohol compare with the heats of combustion of low-molecular-weight saturated hydrocarbons on a per-mole basis and on a per-gram basis?

BUILDING YOUR KNOWLEDGE

59. A laboratory procedure calls for oxidizing 2-propanol to acetone using an *acidic* solution of K$_2$Cr$_2$O$_7$. An insufficient amount of K$_2$Cr$_2$O$_7$ is on hand, however, so the laboratory instructor decides to use an acidic solution of KMnO$_4$ instead. What mass of KMnO$_4$ is required to carry out the same amount of oxidation as 0.500 g of K$_2$Cr$_2$O$_7$?

60. The chemical equation for the water gas reaction is

$$C(s) + H_2O(g) \rightleftharpoons CO(g) + H_2(g)$$

At 1000 K, the value of K_p for this reaction is 3.2. When we treat carbon with steam and allow the reaction to reach equilibrium, the partial pressure of water vapor is observed to be 15.6 atm. What are the partial pressures of CO and H$_2$ under these conditions?

61. (a) In aqueous solution, acetic acid exists mainly in the molecular form ($K_a = 1.8 \times 10^{-5}$). Calculate the freezing point depression for a 0.10 molal aqueous solution of acetic acid, ignoring any ionization of the acid. $K_f = 1.86°C/molal$ for water. (b) In nonpolar solvents such as benzene, acetic acid exists mainly as dimers

as a result of hydrogen bonding. Calculate the freezing point depression for a 0.10 molal solution of acetic acid in benzene. $K_f = 5.12°C/molal$ for benzene. Assume complete dimer formation.

62. What is the pH of a 0.15 *M* solution of sodium benzoate? $K_a = 6.3 \times 10^{-5}$ for benzoic acid, C$_6$H$_5$COOH. Would this solution be more or less acidic than a 0.15 *M* solution of sodium acetate? $K_a = 1.8 \times 10^{-5}$ for acetic acid, CH$_3$COOH.

63. (a) A sample of an organic compound contains only carbon, hydrogen, and oxygen. Combustion of a 0.240 g sample of this compound produces 0.352 g of carbon dioxide and 0.144 g of water. What is the simplest formula of the compound? (b) A solution containing 0.20 g of the compound dissolves in 25.0 g of water freezes at −0.248°C. What is the molecular formula of the compound?

64. The compound in Exercise 63 reacted violently with sodium to give a colorless gas. It also reacted with ethanol in the presence of concentrated sulfuric acid to form a substance that had a fruity smell. From these observations, draw a structural formula of the compound and write balanced equations for the reactions.

BEYOND THE TEXTBOOK

Go to the textbook website at

http://www.brookscole.com/chemistry/whitten

for additional activities and exercises based on the General Chemistry Interactive CD-ROM, the World Wide Web, and library resources.

InfoTrac College Edition

For additional readings, go to InfoTrac College Edition, your online research library at:

http://infotrac.thomsonlearning.com

SOME MATHEMATICAL OPERATIONS

In chemistry we frequently use very large or very small numbers. Such numbers are conveniently expressed in *scientific*, or *exponential*, *notation*.

A-1 SCIENTIFIC NOTATION

In scientific notation, a number is expressed as the *product of two numbers*. By convention, the first number, called the digit term, is between 1 and 10. The second number, called the *exponential term*, is an integer power of 10. Some examples follow.

$$
\begin{aligned}
10000 &= 1 \times 10^4 & 24327 &= 2.4327 \times 10^4 \\
1000 &= 1 \times 10^3 & 7958 &= 7.958 & \times 10^3 \\
100 &= 1 \times 10^2 & 594 &= 5.94 & \times 10^2 \\
10 &= 1 \times 10^1 & 98 &= 9.8 & \times 10^1 \\
1 &= 1 \times 10^0 & & \\
1/10 = 0.1 &= 1 \times 10^{-1} & 0.32 &= 3.2 & \times 10^{-1} \\
1/100 = 0.01 &= 1 \times 10^{-2} & 0.067 &= 6.7 & \times 10^{-2} \\
1/1000 = 0.001 &= 1 \times 10^{-3} & 0.0049 &= 4.9 & \times 10^{-3} \\
1/10000 = 0.0001 &= 1 \times 10^{-4} & 0.00017 &= 1.7 & \times 10^{-4}
\end{aligned}
$$

Recall that, by definition, $(\text{any base})^0 = 1$.

The exponent of 10 is the number of places the decimal point must be shifted to give the number in long form. A *positive exponent* indicates that the decimal point is *shifted right* that number of places. A *negative exponent* indicates that the decimal point is *shifted left*. When numbers are written in *standard scientific notation*, there is one nonzero digit to the left of the decimal point.

$$
7.3 \times 10^3 = 73 \times 10^2 \quad = 730 \times 10^1 \quad = 7300
$$
$$
4.36 \times 10^{-2} = 0.436 \times 10^{-1} \quad = 0.0436
$$
$$
0.00862 = 0.0862 \times 10^{-1} = 0.862 \times 10^{-2} = 8.62 \times 10^{-3}
$$

In scientific notation the digit term indicates the number of significant figures in the number. The exponential term merely locates the decimal point and does not represent significant figures.

Addition and Subtraction

In addition and subtraction all numbers are converted to the same power of 10, and the digit terms are added or subtracted.

$$
(4.21 \times 10^{-3}) + (1.4 \times 10^{-4}) = (4.21 \times 10^{-3}) + (0.14 \times 10^{-3}) = \underline{4.35 \times 10^{-3}}
$$
$$
(8.97 \times 10^4) - (2.31 \times 10^3) = (8.97 \times 10^4) - (0.231 \times 10^4) = \underline{8.74 \times 10^4}
$$

Multiplication

The digit terms are multiplied in the usual way, the exponents are added algebraically, and the product is written with one nonzero digit to the left of the decimal.

Two significant figures in answer.

Two significant figures in answer.

$$(4.7 \times 10^7)(1.6 \times 10^2) = (4.7)(1.6) \times 10^{7+2} = 7.52 \times 10^9 = \underline{7.5 \times 10^9}$$
$$(8.3 \times 10^4)(9.3 \times 10^{-9}) = (8.3)(9.3) \times 10^{4-9} = 77.19 \times 10^{-5} = \underline{7.7 \times 10^{-4}}$$

Division

The digit term of the numerator is divided by the digit term of the denominator, the exponents are subtracted algebraically, and the quotient is written with one nonzero digit to the left of the decimal.

$$\frac{8.4 \times 10^7}{2.0 \times 10^3} = \frac{8.4}{2.0} \times 10^{7-3} = \underline{4.2 \times 10^4}$$

Three significant figures in answer.

$$\frac{3.81 \times 10^9}{8.412 \times 10^{-3}} = \frac{3.81}{8.412} \times 10^{[9-(-3)]} = 0.45292 \times 10^{12} = \underline{4.53 \times 10^{11}}$$

Powers of Exponentials

The digit term is raised to the indicated power, and the exponent is multiplied by the number that indicates the power.

$$(1.2 \times 10^3)^2 = (1.2)^2 \times 10^{3 \times 2} = 1.44 \times 10^6 = \underline{1.4 \times 10^6}$$
$$(3.0 \times 10^{-3})^4 = (3.0)^4 \times 10^{-3 \times 4} = 81 \times 10^{-12} = \underline{8.1 \times 10^{-11}}$$

These instructions are applicable to most calculators. If your calculator has other notation, consult your calculator's instruction booklet.

Electronic Calculators   To square a number: (1) enter the number and (2) touch the (x^2) button.

$$(7.3)^2 = 53.29 = \underline{53} \qquad \text{(two sig. figs.)}$$

To raise a number y to power x: (1) enter the number; (2) touch the (y^x) button; (3) enter the power; and (4) touch the (=) button.

$$(7.3)^4 = 2839.8241 = \underline{2.8 \times 10^3} \qquad \text{(two sig. figs.)}$$
$$(7.30 \times 10^2)^5 = 2.0730716 \times 10^{14} = \underline{2.07 \times 10^{14}} \qquad \text{(three sig. figs.)}$$

Roots of Exponentials

The exponent must be divisible by the desired root if a calculator is not used. The root of the digit term is extracted in the usual way, and the exponent is divided by the desired root.

$$\sqrt{2.5 \times 10^5} = \sqrt{25 \times 10^4} = \sqrt{25} \times \sqrt{10^4} = \underline{5.0 \times 10^2}$$
$$\sqrt[3]{2.7 \times 10^{-8}} = \sqrt[3]{27 \times 10^{-9}} = \sqrt[3]{27} \times \sqrt[3]{10^{-9}} = \underline{3.0 \times 10^{-3}}$$

Electronic Calculators   To extract the square root of a number: (1) enter the number and (2) touch the $(\sqrt{x})$ button.

$$\sqrt{23} = 4.7958315 = \underline{4.8} \qquad \text{(two sig. figs.)}$$

On some models, this function is performed by the $\sqrt[x]{y}$ button.

To extract some other root: (1) enter the number y; (2) touch the (INV) and then the (y^x) button; (3) enter the root to be extracted, x; and (4) touch the (=) button.

A-2 LOGARITHMS

The logarithm of a number is the power to which a base must be raised to obtain the number. Two types of logarithms are frequently used in chemistry: (1) common logarithms (abbreviated log), whose base is 10, and (2) natural logarithms (abbreviated ln), whose base is $e = 2.71828$. The general properties of logarithms are the same no matter what base is used. Many equations in science were derived by the use of calculus, and these often involve natural (base e) logarithms. The relationship between $\log x$ and $\ln x$ is as follows.

$$\ln x = 2.303 \log x$$

$\ln 10 = 2.303$

Finding Logarithms   The common logarithm of a number is the power to which 10 must be raised to obtain the number. The number 10 must be raised to the third power to equal 1000. Therefore, the logarithm of 1000 is 3, written $\log 1000 = 3$. Some examples follow.

Number	Exponential Expression	Logarithm
1000	10^3	3
100	10^2	2
10	10^1	1
1	10^0	0
1/10 = 0.1	10^{-1}	-1
1/100 = 0.01	10^{-2}	-2
1/1000 = 0.001	10^{-3}	-3

To obtain the logarithm of a number other than an integral power of 10, you must use either an electronic calculator or a logarithm table. On most calculators, you do this by (1) entering the number and then (2) pressing the (log) button.

$$\log 7.39 = 0.8686444 = \underline{0.869}$$
$$\log 7.39 \times 10^3 = 3.8686 \quad = \underline{3.869}$$
$$\log 7.39 \times 10^{-3} = -2.1314 \quad = \underline{-2.131}$$

The number to the left of the decimal point in a logarithm is called the *characteristic*, and the number to the right of the decimal point is called the *mantissa*. The characteristic only locates the decimal point of the number, so it is usually not included when counting significant figures. The mantissa has as many significant figures as the number whose log was found.

To obtain the natural logarithm of a number on an electronic calculator, (1) enter the number and (2) press the (ln) or (ln x) button.

$$\ln 4.45 = 1.4929041 = \underline{1.493}$$
$$\ln 1.27 \times 10^3 = 7.1468 \quad = \underline{7.147}$$

Finding Antilogarithms   Sometimes we know the logarithm of a number and must find the number. This is called finding the *antilogarithm* (or *inverse logarithm*). To do this on a calculator, (1) enter the value of the log; (2) press the (INV) button; and (3) press the (log) button.

$$\log x = 6.131; \quad \text{so } x = \text{inverse log of } 6.131 = \underline{1.352 \times 10^6}$$
$$\log x = -1.562; \quad \text{so } x = \text{inverse log of } -1.562 = \underline{2.74 \times 10^{-2}}$$

On some calculators, the inverse log is found as follows:
1. enter the value of the log
2. press the (2ndF) (second function) button
3. press (10^x)

On some calculators, the inverse natural logarithm is found as follows:
1. enter the value of the ln
2. press the (2ndF) (second function) button
3. press (e^x)

To find the inverse natural logarithm, (1) enter the value of the ln; (2) press the (INV) button; and (3) press the (ln) or (ln x) button.

$$\ln x = 3.552; \qquad \text{so } x = \text{inverse ln of } 3.552 = \underline{3.49 \times 10^1}$$

$$\ln x = -1.248; \qquad \text{so } x = \text{inverse ln of } -1.248 = \underline{2.87 \times 10^{-1}}$$

Calculations Involving Logarithms

Because logarithms are exponents, operations involving them follow the same rules as the use of exponents. The following relationships are useful.

$$\log xy = \log x + \log y \qquad \text{or} \qquad \ln xy = \ln x + \ln y$$

$$\log \frac{x}{y} = \log x - \log y \qquad \text{or} \qquad \ln \frac{x}{y} = \ln x - \ln y$$

$$\log x^y = y \log x \qquad \text{or} \qquad \ln x^y = y \ln x$$

$$\log \sqrt[y]{x} = \log x^{1/y} = \frac{1}{y} \log x \qquad \text{or} \qquad \ln \sqrt[y]{x} = \ln x^{1/y} = \frac{1}{y} \ln x$$

A-3 QUADRATIC EQUATIONS

Algebraic expressions of the form

$$ax^2 + bx + c = 0$$

are called **quadratic equations.** Each of the constant terms (a, b, and c) may be either positive or negative. All quadratic equations may be solved by the **quadratic formula.**

$$x = \frac{-b \pm \sqrt{b^2 - 4ac}}{2a}$$

If we wish to solve the quadratic equation $3x^2 - 4x - 8 = 0$, we use $a = 3$, $b = -4$, and $c = -8$. Substitution of these values into the quadratic formula gives

$$x = \frac{-(-4) \pm \sqrt{(-4)^2 - 4(3)(-8)}}{2(3)} = \frac{4 \pm \sqrt{16 + 96}}{6}$$

$$= \frac{4 \pm \sqrt{112}}{6} = \frac{4 \pm 10.6}{6}$$

The two roots of this quadratic equation are

$$x = 2.4 \qquad \text{and} \qquad x = -1.1$$

As you construct and solve quadratic equations based on the observed behavior of matter, you must decide which root has physical significance. Examination of the *equation that defines x* always gives clues about possible values for x. In this way you can tell which is extraneous (has no physical significance). Negative roots are often extraneous.

When you have solved a quadratic equation, you should always check the values you obtained by substitution into the original equation. In the preceding example we obtained $x = 2.4$ and $x = -1.1$. Substitution of these values into the original quadratic equation, $3x^2 - 4x - 8 = 0$, shows that both roots are correct. Such substitutions often do not give a perfect check because some round-off error has been introduced.

ELECTRON CONFIGURATIONS OF THE ATOMS OF THE ELEMENTS

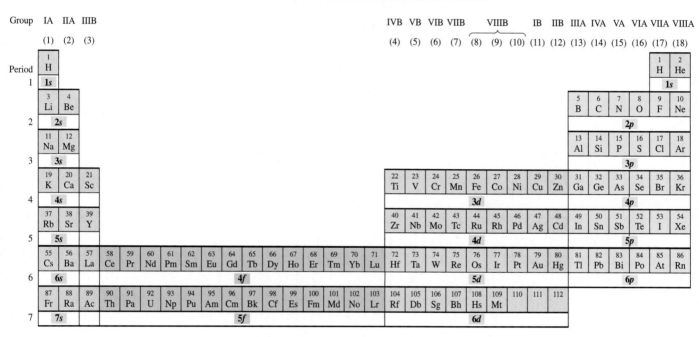

A periodic table colored to show the kinds of atomic orbitals (subshells) being filled in different parts of the periodic table. The atomic orbitals are given below the symbols of blocks of elements. The electronic structures of the A group elements are quite regular and can be predicted from their positions in the periodic table, but there are many exceptions in the *d* and *f* blocks. The populations of subshells are given in the table on pages A-6 and A-7.

Electron Configurations of the Atoms of the Elements

Element	Atomic Number	Populations of Subshells										
		1s	2s	2p	3s	3p	3d	4s	4p	4d	4f	5s
H	1	1										
He	2	2										
Li	3	2	1									
Be	4	2	2									
B	5	2	2	1								
C	6	2	2	2								
N	7	2	2	3								
O	8	2	2	4								
F	9	2	2	5								
Ne	10	2	2	6								
Na	11	Neon core			1							
Mg	12				2							
Al	13				2	1						
Si	14				2	2						
P	15				2	3						
S	16				2	4						
Cl	17				2	5						
Ar	18	2	2	6	2	6						
K	19	Argon core						1				
Ca	20							2				
Sc	21						1	2				
Ti	22						2	2				
V	23						3	2				
Cr	24						5	1				
Mn	25						5	2				
Fe	26						6	2				
Co	27						7	2				
Ni	28						8	2				
Cu	29						10	1				
Zn	30						10	2				
Ga	31						10	2	1			
Ge	32						10	2	2			
As	33						10	2	3			
Se	34						10	2	4			
Br	35						10	2	5			
Kr	36	2	2	6	2	6	10	2	6			
Rb	37	Krypton core										1
Sr	38											2
Y	39									1		2
Zr	40									2		2
Nb	41									4		1
Mo	42									5		1
Tc	43									5		2
Ru	44									7		1
Rh	45									8		1
Pd	46									10		
Ag	47									10		1
Cd	48									10		2

Element	Atomic Number	Krypton core	4d	4f	5s	5p	5d	5f	6s	6p	6d	7s
In	49		10		2	1						
Sn	50		10		2	2						
Sb	51		10		2	3						
Te	52		10		2	4						
I	53		10		2	5						
Xe	54		10		2	6						
Cs	55		10		2	6			1			
Ba	56		10		2	6			2			
La	57		10		2	6	1		2			
Ce	58		10	1	2	6	1		2			
Pr	59		10	3	2	6			2			
Nd	60		10	4	2	6			2			
Pm	61		10	5	2	6			2			
Sm	62		10	6	2	6			2			
Eu	63		10	7	2	6			2			
Gd	64		10	7	2	6	1		2			
Tb	65		10	9	2	6			2			
Dy	66		10	10	2	6			2			
Ho	67		10	11	2	6			2			
Er	68		10	12	2	6			2			
Tm	69		10	13	2	6			2			
Yb	70		10	14	2	6			2			
Lu	71		10	14	2	6	1		2			
Hf	72		10	14	2	6	2		2			
Ta	73		10	14	2	6	3		2			
W	74		10	14	2	6	4		2			
Re	75	Krypton core	10	14	2	6	5		2			
Os	76		10	14	2	6	6		2			
Ir	77		10	14	2	6	7		2			
Pt	78		10	14	2	6	9		1			
Au	79		10	14	2	6	10		1			
Hg	80		10	14	2	6	10		2			
Tl	81		10	14	2	6	10		2	1		
Pb	82		10	14	2	6	10		2	2		
Bi	83		10	14	2	6	10		2	3		
Po	84		10	14	2	6	10		2	4		
At	85		10	14	2	6	10		2	5		
Rn	86		10	14	2	6	10		2	6		
Fr	87		10	14	2	6	10		2	6		1
Ra	88		10	14	2	6	10		2	6		2
Ac	89		10	14	2	6	10		2	6	1	2
Th	90		10	14	2	6	10		2	6	2	2
Pa	91		10	14	2	6	10	2	2	6	1	2
U	92		10	14	2	6	10	3	2	6	1	2
Np	93		10	14	2	6	10	4	2	6	1	2
Pu	94		10	14	2	6	10	6	2	6		2
Am	95		10	14	2	6	10	7	2	6		2
Cm	96		10	14	2	6	10	7	2	6	1	2
Bk	97		10	14	2	6	10	9	2	6		2
Cf	98		10	14	2	6	10	10	2	6		2
Es	99		10	14	2	6	10	11	2	6		2
Fm	100		10	14	2	6	10	12	2	6		2
Md	101		10	14	2	6	10	13	2	6		2
No	102		10	14	2	6	10	14	2	6		2
Lr	103		10	14	2	6	10	14	2	6	1	2
Rf	104		10	14	2	6	10	14	2	6	2	2
Db	105		10	14	2	6	10	14	2	6	3	2
Sg	106		10	14	2	6	10	14	2	6	4	2
Bh	107		10	14	2	6	10	14	2	6	5	2
Hs	108		10	14	2	6	10	14	2	6	6	2
Mt	109		10	14	2	6	10	14	2	6	7	2

APPENDIX C

COMMON UNITS, EQUIVALENCES, AND CONVERSION FACTORS

C-1 FUNDAMENTAL UNITS OF THE SI SYSTEM

The metric system was implemented by the French National Assembly in 1790 and has been modified many times. The International System of Units, or *le Système International* (SI), represents an extension of the metric system. It was adopted by the eleventh General Conference of Weights and Measures in 1960 and has also been modified since. It is constructed from seven base units, each of which represents a particular physical quantity (Table I).

The first five units listed in Table I are particularly useful in general chemistry. They are defined as follows.

1. The *meter* is defined as the distance light travels in a vacuum in 1/299,792,458 second.

2. The *kilogram* represents the mass of a platinum–iridium block kept at the International Bureau of Weights and Measures at Sèvres, France.

3. The *second* was redefined in 1967 as the duration of 9,192,631,770 periods of a certain line in the microwave spectrum of cesium-133.

4. The *kelvin* is 1/273.16 of the temperature interval between absolute zero and the triple point of water.

5. The *mole* is the amount of substance that contains as many entities as there are atoms in exactly 0.012 kg of carbon-12 (12 g of ^{12}C atoms).

TABLE I	SI Fundamental Units	
Physical Quantity	**Name of Unit**	**Symbol**
length	meter	m
mass	kilogram	kg
time	second	s
temperature	kelvin	K
amount of substance	mole	mol
electric current	ampere	A
luminous intensity	candela	cd

Prefixes Used with Metric Units and SI Units

Decimal fractions and multiples of metric and SI units are designated by the prefixes listed in Table II. Those most commonly used in general chemistry are underlined.

TABLE II	Traditional Metric and SI Prefixes					
Factor	Prefix	Symbol	Factor	Prefix	Symbol	
10^{12}	tera	T	10^{-1}	deci	d	
10^9	giga	G	10^{-2}	centi	c	
10^6	mega	M	10^{-3}	milli	m	
10^3	kilo	k	10^{-6}	micro	μ	
10^2	hecto	h	10^{-9}	nano	n	
10^1	deka	da	10^{-12}	pico	p	
			10^{-15}	femto	f	
			10^{-18}	atto	a	

C-2 DERIVED SI UNITS

In the International System of Units all physical quantities are represented by appropriate combinations of the base units listed in Table I. A list of the derived units frequently used in general chemistry is given in Table III.

TABLE III	Derived SI Units		
Physical Quantity	Name of Unit	Symbol	Definition
area	square meter	m^2	
volume	cubic meter	m^3	
density	kilogram per cubic meter	kg/m^3	
force	newton	N	$kg \cdot m/s^2$
pressure	pascal	Pa	N/m^2
energy	joule	J	$kg \cdot m^2/s^2$
electric charge	coulomb	C	$A \cdot s$
electric potential difference	volt	V	$J/(A \cdot s)$

Common Units of Mass and Weight

1 pound = 453.59 grams

1 pound = 453.59 grams = 0.45359 kilogram
1 kilogram = 1000 grams = 2.205 pounds
1 gram = 10 decigrams = 100 centigrams
      = 1000 milligrams
1 gram = 6.022×10^{23} atomic mass units
1 atomic mass unit = 1.6606×10^{-24} gram
1 short ton = 2000 pounds = 907.2 kilograms
1 long ton = 2240 pounds
1 metric tonne = 1000 kilograms = 2205 pounds

Common Units of Length

1 inch = 2.54 centimeters (exactly)

1 mile = 5280 feet = 1.609 kilometers
1 yard = 36 inches = 0.9144 meter
1 meter = 100 centimeters = 39.37 inches = 3.281 feet
      = 1.094 yards
1 kilometer = 1000 meters = 1094 yards = 0.6215 mile
1 Ångstrom = 1.0×10^{-8} centimeter = 0.10 nanometer
      = 1.0×10^{-10} meter = 3.937×10^{-9} inch

Common Units of Volume

1 quart = 0.9463 liter
1 liter = 1.056 quarts

1 liter = 1 cubic decimeter = 1000 cubic centimeters
      = 0.001 cubic meter
1 milliliter = 1 cubic centimeter = 0.001 liter
      = 1.056×10^{-3} quart
1 cubic foot = 28.316 liters = 29.902 quarts
      = 7.475 gallons

Common Units of Force* and Pressure

1 atmosphere = 760 millimeters of mercury
      = 1.01325×10^5 pascals
      = 1.01325 bar
      = 14.70 pounds per square inch
1 bar = 10^5 pascals = 0.98692 atm
1 torr = 1 millimeter of mercury
1 pascal = $1 \ kg/m \cdot s^2 = 1 \ N/m^2$

*Force: 1 newton (N) = $1 \ kg \cdot m/s^2$, i.e., the force that, when applied for 1 second, gives a 1-kilogram mass a velocity of 1 meter per second.

Common Units of Energy

1 joule = 1×10^7 ergs

1 thermochemical calorie* = 4.184 joules = 4.184×10^7 ergs
      = 4.129×10^{-2} liter-atmospheres
      = 2.612×10^{19} electron volts
1 erg = 1×10^{-7} joule = 2.3901×10^{-8} calorie
1 electron volt = 1.6022×10^{-19} joule = 1.6022×10^{-12} erg = 96.487 kJ mol†
1 liter-atmosphere = 24.217 calories = 101.325 joules = 1.01325×10^9 ergs
1 British thermal unit = 1055.06 joules = 1.05506×10^{10} ergs = 252.2 calories

*The amount of heat required to raise the temperature of one gram of water from 14.5°C to 15.5°C.

†Note that the other units are per particle and must be multiplied by 6.022×10^{23} to be strictly comparable.

PHYSICAL CONSTANTS

Quantity	Symbol	Traditional Units	SI Units
Acceleration of gravity	g	980.6 cm/s	9.806 m/s
Atomic mass unit ($\frac{1}{12}$ the mass of ^{12}C atom)	amu or u	1.6606×10^{-24} g	1.6606×10^{-27} kg
Avogadro's number	N	6.0221367×10^{23} particles/mol	6.0221367×10^{23} particles/mol
Bohr radius	a_0	0.52918 Å 5.2918×10^{-9} cm	5.2918×10^{-11} m
Boltzmann constant	k	1.3807×10^{-16} erg/K	1.3807×10^{-23} J/K
Charge-to-mass ratio of electron	e/m	1.75882×10^{8} coulomb/g	1.75882×10^{11} C/kg
Electronic charge	e	1.60218×10^{-19} coulomb 4.8033×10^{-10} esu	1.60218×10^{-19} C
Electron rest mass	m_e	9.10940×10^{-28} g 0.00054858 amu	9.10940×10^{-31} kg
Faraday constant	F	96,485 coulombs/eq 23.06 kcal/volt·eq	96,485 C/mol e^- 96,485 J/V·mol e^-
Gas constant	R	$0.08206 \dfrac{\text{L} \cdot \text{atm}}{\text{mol} \cdot \text{K}}$ $1.987 \dfrac{\text{cal}}{\text{mol} \cdot \text{K}}$	$8.3145 \dfrac{\text{kPa} \cdot \text{dm}^3}{\text{mol} \cdot \text{K}}$ 8.3145 J/mol·K
Molar volume (STP)	V_m	22.414 L/mol	22.414×10^{-3} m³/mol 22.414 dm³/mol
Neutron rest mass	m_n	1.67495×10^{-24} g 1.008665 amu	1.67495×10^{-27} kg
Planck constant	h	6.6262×10^{-27} erg·s	6.6262×10^{-34} J·s
Proton rest mass	m_p	1.6726×10^{-24} g 1.007277 amu	1.6726×10^{-27} kg
Rydberg constant	R_∞	3.289×10^{15} cycles/s 2.1799×10^{-11} erg	1.0974×10^{7} m^{-1} 2.1799×10^{-18} J
Speed of light (in a vacuum)	c	$2.99792458 \times 10^{10}$ cm/s (186,281 miles/second)	2.99792458×10^{8} m/s

$\pi = 3.1416$ $2.303\,R = 4.576\ cal/mol \cdot K = 19.15\ J/mol \cdot K$

$e = 2.71828$ $2.303\,RT\ (at\ 25°C) = 1364\ cal/mol = 5709\ J/mol$

$ln\,X = 2.303\ log\,X$

APPENDIX E

SOME PHYSICAL CONSTANTS FOR A FEW COMMON SUBSTANCES

Specific Heats and Heat Capacities for Some Common Substances

Substance	Specific Heat (J/g · °C)	Molar Heat Capacity (J/mol · °C)
Al(s)	0.900	24.3
Ca(s)	0.653	26.2
Cu(s)	0.385	24.5
Fe(s)	0.444	24.8
Hg(ℓ)	0.138	27.7
H_2O(s), ice	2.09	37.7
H_2O(ℓ), water	4.18	75.3
H_2O(g), steam	2.03	36.4
C_6H_6(ℓ), benzene	1.74	136
C_6H_6(g), benzene	1.04	81.6
C_2H_5OH(ℓ), ethanol	2.46	113
C_2H_5OH(g), ethanol	0.954	420
$(C_2H_5)_2O$(ℓ), diethyl ether	3.74	172
$(C_2H_5)_2O$(g), diethyl ether	2.35	108

Heats of Transformation and Transformation Temperatures of Several Substances

Substance	mp (°C)	Heat of Fusion (J/g)	ΔH_{fus} (kJ/mol)	bp (°C)	Heat of Vaporization (J/g)	ΔH_{vap} (kJ/mol)
Al	658	395	10.6	2467	10520	284
Ca	851	233	9.33	1487	4030	162
Cu	1083	205	13.0	2595	4790	305
H_2O	0.0	334	6.02	100	2260	40.7
Fe	1530	267	14.9	2735	6340	354
Hg	−39	11	23.3	357	292	58.6
CH_4	−182	58.6	0.92	−164	—	—
C_2H_5OH	−117	109	5.02	78.3	855	39.3
C_6H_6	5.48	127	9.92	80.1	395	30.8
$(C_2H_5)_2O$	−116	97.9	7.66	35	351	26.0

Vapor Pressure of Water at Various Temperatures

Temperature (°C)	Vapor Pressure (torr)	Temperature (°C)	Vapor Pressure (torr)	Temperature (°C)	Vapor Pressure (torr)	Temperature (°C)	Vapor Pressure (torr)
−10	2.1	21	18.7	51	97.2	81	369.7
−9	2.3	22	19.8	52	102.1	82	384.9
−8	2.5	23	21.1	53	107.2	83	400.6
−7	2.7	24	22.4	54	112.5	84	416.8
−6	2.9	25	23.8	55	118.0	85	433.6
−5	3.2	26	25.2	56	123.8	86	450.9
−4	3.4	27	26.7	57	129.8	87	468.7
−3	3.7	28	28.3	58	136.1	88	487.1
−2	4.0	29	30.0	59	142.6	89	506.1
−1	4.3	30	31.8	60	149.4	90	525.8
0	4.6	31	33.7	61	156.4	91	546.1
1	4.9	32	35.7	62	163.8	92	567.0
2	5.3	33	37.7	63	171.4	93	588.6
3	5.7	34	39.9	64	179.3	94	610.9
4	6.1	35	42.2	65	187.5	95	633.9
5	6.5	36	44.6	66	196.1	96	657.6
6	7.0	37	47.1	67	205.0	97	682.1
7	7.5	38	49.7	68	214.2	98	707.3
8	8.0	39	52.4	69	223.7	99	733.2
9	8.6	40	55.3	70	233.7	100	760.0
10	9.2	41	58.3	71	243.9	101	787.6
11	9.8	42	61.5	72	254.6	102	815.9
12	10.5	43	64.8	73	265.7	103	845.1
13	11.2	44	68.3	74	277.2	104	875.1
14	12.0	45	71.9	75	289.1	105	906.1
15	12.8	46	75.7	76	301.4	106	937.9
16	13.6	47	79.6	77	314.1	107	970.6
17	14.5	48	83.7	78	327.3	108	1004.4
18	15.5	49	88.0	79	341.0	109	1038.9
19	16.5	50	92.5	80	355.1	110	1074.6
20	17.5						

APPENDIX F

IONIZATION CONSTANTS FOR WEAK ACIDS AT 25°C

Acid	Formula and Ionization Equation		K_a
Acetic	CH_3COOH	$\rightleftharpoons H^+ + CH_3COO^-$	1.8×10^{-5}
Arsenic	H_3AsO_4	$\rightleftharpoons H^+ + H_2AsO_4^-$	$2.5 \times 10^{-4} = K_{a1}$
	$H_2AsO_4^-$	$\rightleftharpoons H^+ + HAsO_4^{2-}$	$5.6 \times 10^{-8} = K_{a2}$
	$HAsO_4^{2-}$	$\rightleftharpoons H^+ + AsO_4^{3-}$	$3.0 \times 10^{-13} = K_{a3}$
Arsenous	H_3AsO_3	$\rightleftharpoons H^+ + H_2AsO_3^-$	$6.0 \times 10^{-10} = K_{a1}$
	$H_2AsO_3^-$	$\rightleftharpoons H^+ + HAsO_3^{2-}$	$3.0 \times 10^{-14} = K_{a2}$
Benzoic	C_6H_5COOH	$\rightleftharpoons H^+ + C_6H_5COO^-$	6.3×10^{-5}
Boric*	$B(OH)_3$	$\rightleftharpoons H^+ + BO(OH)_2^-$	$7.3 \times 10^{-10} = K_{a1}$
	$BO(OH)_2^-$	$\rightleftharpoons H^+ + BO_2(OH)^{2-}$	$1.8 \times 10^{-13} = K_{a2}$
	$BO_2(OH)^{2-}$	$\rightleftharpoons H^+ + BO_3^{3-}$	$1.6 \times 10^{-14} = K_{a3}$
Carbonic	H_2CO_3	$\rightleftharpoons H^+ + HCO_3^-$	$4.2 \times 10^{-7} = K_{a1}$
	HCO_3^-	$\rightleftharpoons H^+ + CO_3^{2-}$	$4.8 \times 10^{-11} = K_{a2}$
Citric	$C_3H_5O(COOH)_3$	$\rightleftharpoons H^+ + C_4H_5O_3(COOH)_2^-$	$7.4 \times 10^{-3} = K_{a1}$
	$C_4H_5O_3(COOH)_2^-$	$\rightleftharpoons H^+ + C_5H_5O_5COOH^{2-}$	$1.7 \times 10^{-5} = K_{a2}$
	$C_5H_5O_5COOH^{2-}$	$\rightleftharpoons H^+ + C_6H_5O_7^{3-}$	$7.4 \times 10^{-7} = K_{a3}$
Cyanic	$HOCN$	$\rightleftharpoons H^+ + OCN^-$	3.5×10^{-4}
Formic	$HCOOH$	$\rightleftharpoons H^+ + HCOO^-$	1.8×10^{-4}
Hydrazoic	HN_3	$\rightleftharpoons H^+ + N_3^-$	1.9×10^{-5}
Hydrocyanic	HCN	$\rightleftharpoons H^+ + CN^-$	4.0×10^{-10}
Hydrofluoric	HF	$\rightleftharpoons H^+ + F^-$	7.2×10^{-4}
Hydrogen peroxide	H_2O_2	$\rightleftharpoons H^+ + HO_2^-$	2.4×10^{-12}
Hydrosulfuric	H_2S	$\rightleftharpoons H^+ + HS^-$	$1.0 \times 10^{-7} = K_{a1}$
	HS^-	$\rightleftharpoons H^+ + S^{2-}$	$1.0 \times 10^{-19} = K_{a2}$
Hypobromous	$HOBr$	$\rightleftharpoons H^+ + OBr^-$	2.5×10^{-9}
Hypochlorous	$HOCl$	$\rightleftharpoons H^+ + OCl^-$	3.5×10^{-8}
Hypoiodous	HOI	$\rightleftharpoons H^+ + OI^-$	2.3×10^{-11}
Nitrous	HNO_2	$\rightleftharpoons H^+ + NO_2^-$	4.5×10^{-4}
Oxalic	$(COOH)_2$	$\rightleftharpoons H^+ + COOCOOH^-$	$5.9 \times 10^{-2} = K_{a1}$
	$COOCOOH^-$	$\rightleftharpoons H^+ + (COO)_2^{2-}$	$6.4 \times 10^{-5} = K_{a2}$
Phenol	HC_6H_5O	$\rightleftharpoons H^+ + C_6H_5O^-$	1.3×10^{-10}
Phosphoric	H_3PO_4	$\rightleftharpoons H^+ + H_2PO_4^-$	$7.5 \times 10^{-3} = K_{a1}$
	$H_2PO_4^-$	$\rightleftharpoons H^+ + HPO_4^{2-}$	$6.2 \times 10^{-8} = K_{a2}$
	HPO_4^{2-}	$\rightleftharpoons H^+ + PO_4^{3-}$	$3.6 \times 10^{-13} = K_{a3}$
Phosphorous	H_3PO_3	$\rightleftharpoons H^+ + H_2PO_3^-$	$1.6 \times 10^{-2} = K_{a1}$
	$H_2PO_3^-$	$\rightleftharpoons H^+ + HPO_3^{2-}$	$7.0 \times 10^{-7} = K_{a2}$
Selenic	H_2SeO_4	$\rightleftharpoons H^+ + HSeO_4^-$	Very large $= K_{a1}$
	$HSeO_4^-$	$\rightleftharpoons H^+ + SeO_4^{2-}$	$1.2 \times 10^{-2} = K_{a2}$
Selenous	H_2SeO_3	$\rightleftharpoons H^+ + HSeO_3^-$	$2.7 \times 10^{-3} = K_{a1}$
	$HSeO_3^-$	$\rightleftharpoons H^+ + SeO_3^{2-}$	$2.5 \times 10^{-7} = K_{a2}$
Sulfuric	H_2SO_4	$\rightleftharpoons H^+ + HSO_4^-$	Very large $= K_{a1}$
	HSO_4^-	$\rightleftharpoons H^+ + SO_4^{2-}$	$1.2 \times 10^{-2} = K_{a2}$
Sulfurous	H_2SO_3	$\rightleftharpoons H^+ + HSO_3^-$	$1.2 \times 10^{-2} = K_{a1}$
	HSO_3^-	$\rightleftharpoons H^+ + SO_3^{2-}$	$6.2 \times 10^{-8} = K_{a2}$
Tellurous	H_2TeO_3	$\rightleftharpoons H^+ + HTeO_3^-$	$2 \times 10^{-3} = K_{a1}$
	$HTeO_3^-$	$\rightleftharpoons H^+ + TeO_3^{2-}$	$1 \times 10^{-8} = K_{a2}$

Boric acid acts as a Lewis acid in aqueous solution.

IONIZATION CONSTANTS FOR WEAK BASES AT 25°C

Base	Formula and Ionization Equation				K_b
Ammonia	NH_3	$+ H_2O \rightleftharpoons NH_4^+$		$+ OH^-$	1.8×10^{-5}
Aniline	$C_6H_5NH_2$	$+ H_2O \rightleftharpoons C_6H_5NH_3^+$		$+ OH^-$	4.2×10^{-10}
Dimethylamine	$(CH_3)_2NH$	$+ H_2O \rightleftharpoons (CH_3)_2NH_2^+$		$+ OH^-$	7.4×10^{-4}
Ethylenediamine	$(CH_2)_2(NH_2)_2$	$+ H_2O \rightleftharpoons (CH_2)_2(NH_2)_2H^+$		$+ OH^-$	$8.5 \times 10^{-5} = K_{b1}$
	$(CH_2)_2(NH_2)_2H^+$	$+ H_2O \rightleftharpoons (CH_2)_2(NH_2)_2H_2^{2+}$		$+ OH^-$	$2.7 \times 10^{-8} = K_{b2}$
Hydrazine	N_2H_4	$+ H_2O \rightleftharpoons N_2H_5^+$		$+ OH^-$	$8.5 \times 10^{-7} = K_{b1}$
	$N_2H_5^+$	$+ H_2O \rightleftharpoons N_2H_6^{2+}$		$+ OH^-$	$8.9 \times 10^{-16} = K_{b2}$
Hydroxylamine	NH_2OH	$+ H_2O \rightleftharpoons NH_3OH^+$		$+ OH^-$	6.6×10^{-9}
Methylamine	CH_3NH_2	$+ H_2O \rightleftharpoons CH_3NH_3^+$		$+ OH^-$	5.0×10^{-4}
Pyridine	C_5H_5N	$+ H_2O \rightleftharpoons C_5H_5NH^+$		$+ OH^-$	1.5×10^{-9}
Trimethylamine	$(CH_3)_3N$	$+ H_2O \rightleftharpoons (CH_3)_3NH^+$		$+ OH^-$	7.4×10^{-5}

APPENDIX H

SOLUBILITY PRODUCT CONSTANTS FOR SOME INORGANIC COMPOUNDS AT 25°C

Substance	K_{sp}	Substance	K_{sp}
Aluminum compounds		*Chromium compounds*	
$AlAsO_4$	1.6×10^{-16}	$CrAsO_4$	7.8×10^{-21}
$Al(OH)_3$	1.9×10^{-33}	$Cr(OH)_3$	6.7×10^{-31}
$AlPO_4$	1.3×10^{-20}	$CrPO_4$	2.4×10^{-23}
Antimony compounds		*Cobalt compounds*	
Sb_2S_3	1.6×10^{-93}	$Co_3(AsO_4)_2$	7.6×10^{-29}
Barium compounds		$CoCO_3$	8.0×10^{-13}
$Ba_3(AsO_4)_2$	1.1×10^{-13}	$Co(OH)_2$	2.5×10^{-16}
$BaCO_3$	8.1×10^{-9}	$CoS\ (\alpha)$	5.9×10^{-21}
$BaC_2O_4 \cdot 2H_2O^*$	1.1×10^{-7}	$CoS\ (\beta)$	8.7×10^{-23}
$BaCrO_4$	2.0×10^{-10}	$Co(OH)_3$	4.0×10^{-45}
BaF_2	1.7×10^{-6}	Co_2S_3	2.6×10^{-124}
$Ba(OH)_2 \cdot 8H_2O^*$	5.0×10^{-3}	*Copper compounds*	
$Ba_3(PO_4)_2$	1.3×10^{-29}	$CuBr$	5.3×10^{-9}
$BaSeO_4$	2.8×10^{-11}	$CuCl$	1.9×10^{-7}
$BaSO_3$	8.0×10^{-7}	$CuCN$	3.2×10^{-20}
$BaSO_4$	1.1×10^{-10}	$Cu_2O\ (Cu^+ + OH^-)^\dagger$	1.0×10^{-14}
Bismuth compounds		CuI	5.1×10^{-12}
$BiOCl$	7.0×10^{-9}	Cu_2S	1.6×10^{-48}
$BiO(OH)$	1.0×10^{-12}	$CuSCN$	1.6×10^{-11}
$Bi(OH)_3$	3.2×10^{-40}	$Cu_3(AsO_4)_2$	7.6×10^{-36}
BiI_3	8.1×10^{-19}	$CuCO_3$	2.5×10^{-10}
$BiPO_4$	1.3×10^{-23}	$Cu_2[Fe(CN)_6]$	1.3×10^{-16}
Bi_2S_3	1.6×10^{-72}	$Cu(OH)_2$	1.6×10^{-19}
Cadmium compounds		CuS	8.7×10^{-36}
$Cd_3(AsO_4)_2$	2.2×10^{-32}	*Gold compounds*	
$CdCO_3$	2.5×10^{-14}	$AuBr$	5.0×10^{-17}
$Cd(CN)_2$	1.0×10^{-8}	$AuCl$	2.0×10^{-13}
$Cd_2[Fe(CN)_6]$	3.2×10^{-17}	AuI	1.6×10^{-23}
$Cd(OH)_2$	1.2×10^{-14}	$AuBr_3$	4.0×10^{-36}
CdS	3.6×10^{-29}	$AuCl_3$	3.2×10^{-25}
Calcium compounds		$Au(OH)_3$	1.0×10^{-53}
$Ca_3(AsO_4)_2$	6.8×10^{-19}	AuI_3	1.0×10^{-46}
$CaCO_3$	4.8×10^{-9}	*Iron compounds*	
$CaCrO_4$	7.1×10^{-4}	$FeCO_3$	3.5×10^{-11}
$CaC_2O_4 \cdot H_2O^*$	2.3×10^{-9}	$Fe(OH)_2$	7.9×10^{-15}
CaF_2	3.9×10^{-11}	FeS	4.9×10^{-18}
$Ca(OH)_2$	7.9×10^{-6}	$Fe_4[Fe(CN)_6]_3$	3.0×10^{-41}
$CaHPO_4$	2.7×10^{-7}	$Fe(OH)_3$	6.3×10^{-38}
$Ca(H_2PO_4)_2$	1.0×10^{-3}	Fe_2S_3	1.4×10^{-88}
$Ca_3(PO_4)_2$	1.0×10^{-25}	*Lead compounds*	
$CaSO_3 \cdot 2H_2O^*$	1.3×10^{-8}	$Pb_3(AsO_4)_2$	4.1×10^{-36}
$CaSO_4 \cdot 2H_2O^*$	2.4×10^{-5}	$PbBr_2$	6.3×10^{-6}

SOLUBILITY PRODUCT CONSTANTS FOR SOME INORGANIC COMPOUNDS AT 25°C (continued)

Substance	K_{sp}	Substance	K_{sp}
Lead compounds (cont.)		*Nickel compounds (cont.)*	
$PbCO_3$	1.5×10^{-13}	NiS (α)	3.0×10^{-21}
$PbCl_2$	1.7×10^{-5}	NiS (β)	1.0×10^{-26}
$PbCrO_4$	1.8×10^{-14}	NiS (γ)	2.0×10^{-28}
PbF_2	3.7×10^{-8}	*Silver compounds*	
$Pb(OH)_2$	2.8×10^{-16}	Ag_3AsO_4	1.1×10^{-20}
PbI_2	8.7×10^{-9}	$AgBr$	3.3×10^{-13}
$Pb_3(PO_4)_2$	3.0×10^{-44}	Ag_2CO_3	8.1×10^{-12}
$PbSeO_4$	1.5×10^{-7}	$AgCl$	1.8×10^{-10}
$PbSO_4$	1.8×10^{-8}	Ag_2CrO_4	9.0×10^{-12}
PbS	8.4×10^{-28}	$AgCN$	1.2×10^{-16}
Magnesium compounds		$Ag_4[Fe(CN)_6]$	1.6×10^{-41}
$Mg_3(AsO_4)_2$	2.1×10^{-20}	Ag_2O ($Ag^+ + OH^-$)†	2.0×10^{-8}
$MgCO_3 \cdot 3H_2O$*	4.0×10^{-5}	AgI	1.5×10^{-16}
MgC_2O_4	8.6×10^{-5}	Ag_3PO_4	1.3×10^{-20}
MgF_2	6.4×10^{-9}	Ag_2SO_3	1.5×10^{-14}
$Mg(OH)_2$	1.5×10^{-11}	Ag_2SO_4	1.7×10^{-5}
$MgNH_4PO_4$	2.5×10^{-12}	Ag_2S	1.0×10^{-49}
Manganese compounds		$AgSCN$	1.0×10^{-12}
$Mn_3(AsO_4)_2$	1.9×10^{-11}	*Strontium compounds*	
$MnCO_3$	1.8×10^{-11}	$Sr_3(AsO_4)_2$	1.3×10^{-18}
$Mn(OH)_2$	4.6×10^{-14}	$SrCO_3$	9.4×10^{-10}
MnS	5.1×10^{-15}	$SrC_2O_4 \cdot 2H_2O$*	5.6×10^{-8}
$Mn(OH)_3$	$\approx 1.0 \times 10^{-36}$	$SrCrO_4$	3.6×10^{-5}
Mercury compounds		$Sr(OH)_2 \cdot 8H_2O$*	3.2×10^{-4}
Hg_2Br_2	1.3×10^{-22}	$Sr_3(PO_4)_2$	1.0×10^{-31}
Hg_2CO_3	8.9×10^{-17}	$SrSO_3$	4.0×10^{-8}
Hg_2Cl_2	1.1×10^{-18}	$SrSO_4$	2.8×10^{-7}
Hg_2CrO_4	5.0×10^{-9}	*Tin compounds*	
Hg_2I_2	4.5×10^{-29}	$Sn(OH)_2$	2.0×10^{-26}
$Hg_2O \cdot H_2O$*		SnI_2	1.0×10^{-4}
($Hg_2^{2+} + 2OH^-$)†	1.6×10^{-23}	SnS	1.0×10^{-28}
Hg_2SO_4	6.8×10^{-7}	$Sn(OH)_4$	1.0×10^{-57}
Hg_2S	5.8×10^{-44}	SnS_2	1.0×10^{-70}
$Hg(CN)_2$	3.0×10^{-23}	*Zinc compounds*	
$Hg(OH)_2$	2.5×10^{-26}	$Zn_3(AsO_4)_2$	1.1×10^{-27}
HgI_2	4.0×10^{-29}	$ZnCO_3$	1.5×10^{-11}
HgS	3.0×10^{-53}	$Zn(CN)_2$	8.0×10^{-12}
Nickel compounds		$Zn_2[Fe(CN)_6]$	4.1×10^{-16}
$Ni_3(AsO_4)_2$	1.9×10^{-26}	$Zn(OH)_2$	4.5×10^{-17}
$NiCO_3$	6.6×10^{-9}	$Zn_3(PO_4)_2$	9.1×10^{-33}
$Ni(CN)_2$	3.0×10^{-23}	ZnS	1.1×10^{-21}
$Ni(OH)_2$	2.8×10^{-16}		

*$[H_2O]$ does not appear in equilibrium constants for equilibria in aqueous solution in general, so it does not appear in the K_{sp} expressions for hydrated solids.

†Very small amounts of oxides dissolve in water to give the ions indicated in parentheses. These solid hydroxides are unstable and decompose to oxides as rapidly as they are formed.

APPENDIX I

DISSOCIATION CONSTANTS FOR SOME COMPLEX IONS

Dissociation Equilibrium			K_d
$[AgBr_2]^-$	$\rightleftharpoons$	$Ag^+ + 2Br^-$	7.8×10^{-8}
$[AgCl_2]^-$	$\rightleftharpoons$	$Ag^+ + 2Cl^-$	4.0×10^{-6}
$[Ag(CN)_2]^-$	$\rightleftharpoons$	$Ag^+ + 2CN^-$	1.8×10^{-19}
$[Ag(S_2O_3)_2]^{3-}$	$\rightleftharpoons$	$Ag^+ + 2S_2O_3^{2-}$	5.0×10^{-14}
$[Ag(NH_3)_2]^+$	$\rightleftharpoons$	$Ag^+ + 2NH_3$	6.3×10^{-8}
$[Ag(en)]^+$	$\rightleftharpoons$	$Ag^+ + en^*$	1.0×10^{-5}
$[AlF_6]^{3-}$	$\rightleftharpoons$	$Al^{3+} + 6F^-$	2.0×10^{-24}
$[Al(OH)_4]^-$	$\rightleftharpoons$	$Al^{3+} + 4OH^-$	1.3×10^{-34}
$[Au(CN)_2]^-$	$\rightleftharpoons$	$Au^+ + 2CN^-$	5.0×10^{-39}
$[Cd(CN)_4]^{2-}$	$\rightleftharpoons$	$Cd^{2+} + 4CN^-$	7.8×10^{-18}
$[CdCl_4]^{2-}$	$\rightleftharpoons$	$Cd^{2+} + 4Cl^-$	1.0×10^{-4}
$[Cd(NH_3)_4]^{2+}$	$\rightleftharpoons$	$Cd^{2+} + 4NH_3$	1.0×10^{-7}
$[Co(NH_3)_6]^{2+}$	$\rightleftharpoons$	$Co^{2+} + 6NH_3$	1.3×10^{-5}
$[Co(NH_3)_6]^{3+}$	$\rightleftharpoons$	$Co^{3+} + 6NH_3$	2.2×10^{-34}
$[Co(en)_3]^{2+}$	$\rightleftharpoons$	$Co^{2+} + 3en^*$	1.5×10^{-14}
$[Co(en)_3]^{3+}$	$\rightleftharpoons$	$Co^{3+} + 3en^*$	2.0×10^{-49}
$[Cu(CN)_2]^-$	$\rightleftharpoons$	$Cu^+ + 2CN^-$	1.0×10^{-16}
$[CuCl_2]^-$	$\rightleftharpoons$	$Cu^+ + 2Cl^-$	1.0×10^{-5}
$[Cu(NH_3)_2]^+$	$\rightleftharpoons$	$Cu^+ + 2NH_3$	1.4×10^{-11}
$[Cu(NH_3)_4]^{2+}$	$\rightleftharpoons$	$Cu^{2+} + 4NH_3$	8.5×10^{-13}
$[Fe(CN)_6]^{4-}$	$\rightleftharpoons$	$Fe^{2+} + 6CN^-$	1.3×10^{-37}
$[Fe(CN)_6]^{3-}$	$\rightleftharpoons$	$Fe^{3+} + 6CN^-$	1.3×10^{-44}
$[HgCl_4]^{2-}$	$\rightleftharpoons$	$Hg^{2+} + 4Cl^-$	8.3×10^{-16}
$[Ni(CN)_4]^{2-}$	$\rightleftharpoons$	$Ni^{2+} + 4CN^-$	1.0×10^{-31}
$[Ni(NH_3)_6]^{2+}$	$\rightleftharpoons$	$Ni^{2+} + 6NH_3$	1.8×10^{-9}
$[Zn(OH)_4]^{2-}$	$\rightleftharpoons$	$Zn^{2+} + 4OH^-$	3.5×10^{-16}
$[Zn(NH_3)_4]^{2+}$	$\rightleftharpoons$	$Zn^{2+} + 4NH_3$	3.4×10^{-10}

*The abbreviation "en" represents ethylenediamine, $H_2NCH_2CH_2NH_2$.

STANDARD REDUCTION POTENTIALS IN AQUEOUS SOLUTION AT 25°C

Acidic Solution	Standard Reduction Potential, E^0 (volts)
$Li^+(aq) + e^- \longrightarrow Li(s)$	-3.045
$K^+(aq) + e^- \longrightarrow K(s)$	-2.925
$Rb^+(aq) + e^- \longrightarrow Rb(s)$	-2.925
$Ba^{2+}(aq) + 2e^- \longrightarrow Ba(s)$	-2.90
$Sr^{2+}(aq) + 2e^- \longrightarrow Sr(s)$	-2.89
$Ca^{2+}(aq) + 2e^- \longrightarrow Ca(s)$	-2.87
$Na^+(aq) + e^- \longrightarrow Na(s)$	-2.714
$Mg^{2+}(aq) + 2e^- \longrightarrow Mg(s)$	-2.37
$H_2(g) + 2e^- \longrightarrow 2H^-(aq)$	-2.25
$Al^{3+}(aq) + 3e^- \longrightarrow Al(s)$	-1.66
$Zr^{4+}(aq) + 4e^- \longrightarrow Zr(s)$	-1.53
$ZnS(s) + 2e^- \longrightarrow Zn(s) + S^{2-}(aq)$	-1.44
$CdS(s) + 2e^- \longrightarrow Cd(s) + S^{2-}(aq)$	-1.21
$V^{2+}(aq) + 2e^- \longrightarrow V(s)$	-1.18
$Mn^{2+}(aq) + 2e^- \longrightarrow Mn(s)$	-1.18
$FeS(s) + 2e^- \longrightarrow Fe(s) + S^{2-}(aq)$	-1.01
$Cr^{2+}(aq) + 2e^- \longrightarrow Cr(s)$	-0.91
$Zn^{2+}(aq) + 2e^- \longrightarrow Zn(s)$	-0.763
$Cr^{3+}(aq) + 3e^- \longrightarrow Cr(s)$	-0.74
$HgS(s) + 2H^+(aq) + 2e^- \longrightarrow Hg(\ell) + H_2S(g)$	-0.72
$Ga^{3+}(aq) + 3e^- \longrightarrow Ga(s)$	-0.53
$2CO_2(g) + 2H^+(aq) + 2e^- \longrightarrow (COOH)_2(aq)$	-0.49
$Fe^{2+}(aq) + 2e^- \longrightarrow Fe(s)$	-0.44
$Cr^{3+}(aq) + e^- \longrightarrow Cr^{2+}(aq)$	-0.41
$Cd^{2+}(aq) + 2e^- \longrightarrow Cd(s)$	-0.403
$Se(s) + 2H^+(aq) + 2e^- \longrightarrow H_2Se(aq)$	-0.40
$PbSO_4(s) + 2e^- \longrightarrow Pb(s) + SO_4^{2-}(aq)$	-0.356
$Tl^+(aq) + e^- \longrightarrow Tl(s)$	-0.34
$Co^{2+}(aq) + 2e^- \longrightarrow Co(s)$	-0.28
$Ni^{2+}(aq) + 2e^- \longrightarrow Ni(s)$	-0.25
$[SnF_6]^{2-}(aq) + 4e^- \longrightarrow Sn(s) + 6F^-(aq)$	-0.25
$AgI(s) + e^- \longrightarrow Ag(s) + I^-(aq)$	-0.15
$Sn^{2+}(aq) + 2e^- \longrightarrow Sn(s)$	-0.14
$Pb^{2+}(aq) + 2e^- \longrightarrow Pb(s)$	-0.126
$N_2O(g) + 6H^+(aq) + H_2O + 4e^- \longrightarrow 2NH_3OH^+(aq)$	-0.05
$2H^+(aq) + 2e^- \longrightarrow H_2(g)$ (reference electrode)	0.000
$AgBr(s) + e^- \longrightarrow Ag(s) + Br^-(aq)$	0.10
$S(s) + 2H^+(aq) + 2e^- \longrightarrow H_2S(aq)$	0.14
$Sn^{4+}(aq) + 2e^- \longrightarrow Sn^{2+}(aq)$	0.15
$Cu^{2+}(aq) + e^- \longrightarrow Cu^+(aq)$	0.153
$SO_4^{2-}(aq) + 4H^+(aq) + 2e^- \longrightarrow H_2SO_3(aq) + H_2O$	0.17

STANDARD REDUCTION POTENTIALS IN AQUEOUS SOLUTION AT 25°C (continued)

Acidic Solution	Standard Reduction Potential, E^0 (volts)
$SO_4^{2-}(aq) + 4H^+(aq) + 2e^- \longrightarrow SO_2(g) + 2H_2O$	0.20
$AgCl(s) + e^- \longrightarrow Ag(s) + Cl^-(aq)$	0.222
$Hg_2Cl_2(s) + 2e^- \longrightarrow 2Hg(\ell) + 2Cl^-(aq)$	0.27
$Cu^{2+}(aq) + 2e^- \longrightarrow Cu(s)$	0.337
$[RhCl_6]^{3-}(aq) + 3e^- \longrightarrow Rh(s) + 6Cl^-(aq)$	0.44
$Cu^+(aq) + e^- \longrightarrow Cu(s)$	0.521
$TeO_2(s) + 4H^+(aq) + 4e^- \longrightarrow Te(s) + 2H_2O$	0.529
$I_2(s) + 2e^- \longrightarrow 2I^-(aq)$	0.535
$H_3AsO_4(aq) + 2H^+(aq) + 2e^- \longrightarrow H_3AsO_3(aq) + H_2O$	0.58
$[PtCl_6]^{2-}(aq) + 2e^- \longrightarrow [PtCl_4]^{2-}(aq) + 2Cl^-(aq)$	0.68
$O_2(g) + 2H^+(aq) + 2e^- \longrightarrow H_2O_2(aq)$	0.682
$[PtCl_4]^{2-}(aq) + 2e^- \longrightarrow Pt(s) + 4Cl^-(aq)$	0.73
$SbCl_6^-(aq) + 2e^- \longrightarrow SbCl_4^-(aq) + 2Cl^-(aq)$	0.75
$Fe^{3+}(aq) + e^- \longrightarrow Fe^{2+}(aq)$	0.771
$Hg_2^{2+}(aq) + 2e^- \longrightarrow 2Hg(\ell)$	0.789
$Ag^+(aq) + e^- \longrightarrow Ag(s)$	0.7994
$Hg^{2+}(aq) + 2e^- \longrightarrow Hg(\ell)$	0.855
$2Hg^{2+}(aq) + 2e^- \longrightarrow Hg_2^{2+}(aq)$	0.920
$NO_3^-(aq) + 3H^+(aq) + 2e^- \longrightarrow HNO_2(aq) + H_2O$	0.94
$NO_3^-(aq) + 4H^+(aq) + 3e^- \longrightarrow NO(g) + 2H_2O$	0.96
$Pd^{2+}(aq) + 2e^- \longrightarrow Pd(s)$	0.987
$AuCl_4^-(aq) + 3e^- \longrightarrow Au(s) + 4Cl^-(aq)$	1.00
$Br_2(\ell) + 2e^- \longrightarrow 2Br^-(aq)$	1.08
$ClO_4^-(aq) + 2H^+(aq) + 2e^- \longrightarrow ClO_3^-(aq) + H_2O$	1.19
$IO_3^-(aq) + 6H^+(aq) + 5e^- \longrightarrow \frac{1}{2}I_2(aq) + 3H_2O$	1.195
$Pt^{2+}(aq) + 2e^- \longrightarrow Pt(s)$	1.2
$O_2(g) + 4H^+(aq) + 4e^- \longrightarrow 2H_2O$	1.229
$MnO_2(s) + 4H^+(aq) + 2e^- \longrightarrow Mn^{2+}(aq) + 2H_2O$	1.23
$N_2H_5^+(aq) + 3H^+(aq) + 2e^- \longrightarrow 2NH_4^+(aq)$	1.24
$Cr_2O_7^{2-}(aq) + 14H^+(aq) + 6e^- \longrightarrow 2Cr^{3+}(aq) + 7H_2O$	1.33
$Cl_2(g) + 2e^- \longrightarrow 2Cl^-(aq)$	1.360
$BrO_3^-(aq) + 6H^+(aq) + 6e^- \longrightarrow Br^-(aq) + 3H_2O$	1.44
$ClO_3^-(aq) + 6H^+(aq) + 5e^- \longrightarrow \frac{1}{2}Cl_2(g) + 3H_2O$	1.47
$Au^{3+}(aq) + 3e^- \longrightarrow Au(s)$	1.50
$MnO_4^-(aq) + 8H^+(aq) + 5e^- \longrightarrow Mn^{2+}(aq) + 4H_2O$	1.507
$NaBiO_3(s) + 6H^+(aq) + 2e^- \longrightarrow Bi^{3+}(aq) + Na^+(aq) + 3H_2O$	1.6
$Ce^{4+}(aq) + e^- \longrightarrow Ce^{3+}(aq)$	1.61
$2HOCl(aq) + 2H^+(aq) + 2e^- \longrightarrow Cl_2(g) + 2H_2O$	1.63
$Au^+(aq) + e^- \longrightarrow Au(s)$	1.68
$PbO_2(s) + SO_4^{2-}(aq) + 4H^+(aq) + 2e^- \longrightarrow PbSO_4(s) + 2H_2O$	1.685
$NiO_2(s) + 4H^+(aq) + 2e^- \longrightarrow Ni^{2+}(aq) + 2H_2O$	1.7
$H_2O_2(aq) + 2H^+(aq) + 2e^- \longrightarrow 2H_2O$	1.77
$Pb^{4+}(aq) + 2e^- \longrightarrow Pb^{2+}(aq)$	1.8
$Co^{3+}(aq) + e^- \longrightarrow Co^{2+}(aq)$	1.82
$F_2(g) + 2e^- \longrightarrow 2F^-(aq)$	2.87

STANDARD REDUCTION POTENTIALS IN AQUEOUS SOLUTION AT 25°C *(continued)*

Basic Solution	Standard Reduction Potential, E^0 (volts)
$SiO_3^{2-}(aq) + 3H_2O + 4e^- \longrightarrow Si(s) + 6OH^-(aq)$	−1.70
$Cr(OH)_3(s) + 3e^- \longrightarrow Cr(s) + 3OH^-(aq)$	−1.30
$[Zn(CN)_4]^{2-}(aq) + 2e^- \longrightarrow Zn(s) + 4CN^-(aq)$	−1.26
$Zn(OH)_2(s) + 2e^- \longrightarrow Zn(s) + 2OH^-(aq)$	−1.245
$[Zn(OH)_4]^{2-}(aq) + 2e^- \longrightarrow Zn(s) + 4OH^-(aq)$	−1.22
$N_2(g) + 4H_2O + 4e^- \longrightarrow N_2H_4(aq) + 4OH^-(aq)$	−1.15
$SO_4^{2-}(aq) + H_2O + 2e^- \longrightarrow SO_3^{2-}(aq) + 2OH^-(aq)$	−0.93
$Fe(OH)_2(s) + 2e^- \longrightarrow Fe(s) + 2OH^-(aq)$	−0.877
$2NO_3^-(aq) + 2H_2O + 2e^- \longrightarrow N_2O_4(g) + 4OH^-(aq)$	−0.85
$2H_2O + 2e^- \longrightarrow H_2(g) + 2OH^-(aq)$	−0.828
$Fe(OH)_3(s) + e^- \longrightarrow Fe(OH)_2(s) + OH^-(aq)$	−0.56
$S(s) + 2e^- \longrightarrow S^{2-}(aq)$	−0.48
$Cu(OH)_2(s) + 2e^- \longrightarrow Cu(s) + 2OH^-(aq)$	−0.36
$CrO_4^{2-}(aq) + 4H_2O + 3e^- \longrightarrow Cr(OH)_3(s) + 5OH^-(aq)$	−0.12
$MnO_2(s) + 2H_2O + 2e^- \longrightarrow Mn(OH)_2(s) + 2OH^-(aq)$	−0.05
$NO_3^-(aq) + H_2O + 2e^- \longrightarrow NO_2^-(aq) + 2OH^-(aq)$	0.01
$O_2(g) + H_2O + 2e^- \longrightarrow OOH^-(aq) + OH^-(aq)$	0.076
$HgO(s) + H_2O + 2e^- \longrightarrow Hg(\ell) + 2OH^-(aq)$	0.0984
$[Co(NH_3)_6]^{3+}(aq) + e^- \longrightarrow [Co(NH_3)_6]^{2+}(aq)$	0.10
$N_2H_4(aq) + 2H_2O + 2e^- \longrightarrow 2NH_3(aq) + 2OH^-(aq)$	0.10
$2NO_2^-(aq) + 3H_2O + 4e^- \longrightarrow N_2O(g) + 6OH^-(aq)$	0.15
$Ag_2O(s) + H_2O + 2e^- \longrightarrow 2Ag(s) + 2OH^-(aq)$	0.34
$ClO_4^-(aq) + H_2O + 2e^- \longrightarrow ClO_3^-(aq) + 2OH^-(aq)$	0.36
$O_2(g) + 2H_2O + 4e^- \longrightarrow 4OH^-(aq)$	0.40
$Ag_2CrO_4(s) + 2e^- \longrightarrow 2Ag(s) + CrO_4^{2-}(aq)$	0.446
$NiO_2(s) + 2H_2O + 2e^- \longrightarrow Ni(OH)_2(s) + 2OH^-(aq)$	0.49
$MnO_4^-(aq) + e^- \longrightarrow MnO_4^{2-}(aq)$	0.564
$MnO_4^-(aq) + 2H_2O + 3e^- \longrightarrow MnO_2(s) + 4OH^-(aq)$	0.588
$ClO_3^-(aq) + 3H_2O + 6e^- \longrightarrow Cl^-(aq) + 6OH^-(aq)$	0.62
$2NH_2OH(aq) + 2e^- \longrightarrow N_2H_4(aq) + 2OH^-(aq)$	0.74
$OOH^-(aq) + H_2O + 2e^- \longrightarrow 3OH^-(aq)$	0.88
$ClO^-(aq) + H_2O + 2e^- \longrightarrow Cl^-(aq) + 2OH^-(aq)$	0.89

APPENDIX K

SELECTED THERMODYNAMIC VALUES AT 298.15 K

Species	ΔH_f^0 (kJ/mol)	S^0 (J/mol·K)	ΔG_f^0 (kJ/mol)	Species	ΔH_f^0 (kJ/mol)	S^0 (J/mol·K)	ΔG_f^0 (kJ/mol)
Aluminum				*Cesium*			
Al(s)	0	28.3	0	Cs$^+$(aq)	−248	133	−282.0
AlCl$_3$(s)	−704.2	110.7	−628.9	CsF(aq)	−568.6	123	−558.5
Al$_2$O$_3$(s)	−1676	50.92	−1582	*Chlorine*			
Barium				Cl(g)	121.7	165.1	105.7
BaCl$_2$(s)	−860.1	126	−810.9	Cl$^-$(g)	−226	—	—
BaSO$_4$(s)	−1465	132	−1353	Cl$_2$(g)	0	223.0	0
Beryllium				HCl(g)	−92.31	186.8	−95.30
Be(s)	0	9.54	0	HCl(aq)	−167.4	55.10	−131.2
Be(OH)$_2$(s)	−907.1	—	—	*Chromium*			
Bromine				Cr(s)	0	23.8	0
Br(g)	111.8	174.9	82.4	(NH$_4$)$_2$Cr$_2$O$_7$(s)	−1807	—	—
Br$_2$(ℓ)	0	152.23	0	*Copper*			
Br$_2$(g)	30.91	245.4	3.14	Cu(s)	0	33.15	0
BrF$_3$(g)	−255.6	292.4	−229.5	CuO(s)	−157	42.63	−130
HBr(g)	−36.4	198.59	−53.43	*Fluorine*			
Calcium				F$^-$(g)	−322	—	—
Ca(s)	0	41.6	0	F$^-$(aq)	−332.6	—	−278.8
Ca(g)	192.6	154.8	158.9	F(g)	78.99	158.6	61.92
Ca^{2+}(g)	1920	—	—	F$_2$(g)	0	202.7	0
CaC$_2$(s)	−62.8	70.3	−67.8	HF(g)	−271	173.7	−273
CaCO$_3$(s)	−1207	92.9	−1129	HF(aq)	−320.8	—	−296.8
CaCl$_2$(s)	−795.0	114	−750.2	*Hydrogen*			
CaF$_2$(s)	−1215	68.87	−1162	H(g)	218.0	114.6	203.3
CaH$_2$(s)	−189	42	−150	H$_2$(g)	0	130.6	0
CaO(s)	−635.5	40	−604.2	H$_2$O(ℓ)	−285.8	69.91	−237.2
CaS(s)	−482.4	56.5	−477.4	H$_2$O(g)	−241.8	188.7	−228.6
Ca(OH)$_2$(s)	−986.6	76.1	−896.8	H$_2$O$_2$(ℓ)	−187.8	109.6	−120.4
Ca(OH)$_2$(aq)	−1002.8	76.15	−867.6	*Iodine*			
CaSO$_4$(s)	−1433	107	−1320	I(g)	106.6	180.66	70.16
Carbon				I$_2$(s)	0	116.1	0
C(s, graphite)	0	5.740	0	I$_2$(g)	62.44	260.6	19.36
C(s, diamond)	1.897	2.38	2.900	ICl(g)	17.78	247.4	−5.52
C(g)	716.7	158.0	671.3	HI(g)	26.5	206.5	1.72
CCl$_4$(ℓ)	−135.4	216.4	−65.27	*Iron*			
CCl$_4$(g)	−103	309.7	−60.63	Fe(s)	0	27.3	0
CHCl$_3$(ℓ)	−134.5	202	−73.72	FeO(s)	−272	—	—
CHCl$_3$(g)	−103.1	295.6	−70.37	Fe$_2$O$_3$(s, hematite)	−824.2	87.40	−742.2
CH$_4$(g)	−74.81	186.2	−50.75	Fe$_3$O$_4$(s, magnetite)	−1118	146	−1015
C$_2$H$_2$(g)	226.7	200.8	209.2	FeS$_2$(s)	−177.5	122.2	−166.7
C$_2$H$_4$(g)	52.26	219.5	68.12	Fe(CO)$_5$(ℓ)	−774.0	338	−705.4
C$_2$H$_6$(g)	−84.86	229.5	−32.9	Fe(CO)$_5$(g)	−733.8	445.2	−697.3
C$_3$H$_8$(g)	−103.8	269.9	−23.49	*Lead*			
C$_6$H$_6$(ℓ)	49.03	172.8	124.5	Pb(s)	0	64.81	0
C$_8$H$_{18}$(ℓ)	−268.8	—	—	PbCl$_2$(s)	−359.4	136	−314.1
C$_2$H$_5$OH(ℓ)	−277.7	161	−174.9	PbO(s, yellow)	−217.3	68.70	−187.9
C$_2$H$_5$OH(g)	−235.1	282.6	−168.6	Pb(OH)$_2$(s)	−515.9	88	−420.9
CO(g)	−110.5	197.6	−137.2	PbS(s)	−100.4	91.2	−98.7
CO$_2$(g)	−393.5	213.6	−394.4				
CS$_2$(g)	117.4	237.7	67.15				
COCl$_2$(g)	−223.0	289.2	−210.5				

SELECTED THERMODYNAMIC VALUES AT 298.15 K (continued)

Species	ΔH_f^0 (kJ/mol)	S^0 (J/mol·K)	ΔG_f^0 (kJ/mol)	Species	ΔH_f^0 (kJ/mol)	S^0 (J/mol·K)	ΔG_f^0 (kJ/mol)
Lithium				*Rubidium*			
Li(s)	0	28.0	0	Rb(s)	0	76.78	0
LiOH(s)	−487.23	50	−443.9	RbOH(aq)	−481.16	110.75	−441.24
LiOH(aq)	−508.4	4	−451.1	*Silicon*			
Magnesium				Si(s)	0	18.8	0
Mg(s)	0	32.5	0	SiBr$_4$(ℓ)	−457.3	277.8	−443.9
MgCl$_2$(s)	−641.8	89.5	−592.3	SiC(s)	−65.3	16.6	−62.8
MgO(s)	−601.8	27	−569.6	SiCl$_4$(g)	−657.0	330.6	−617.0
Mg(OH)$_2$(s)	−924.7	63.14	−833.7	SiH$_4$(g)	34.3	204.5	56.9
MgS(s)	−347	—	—	SiF$_4$(g)	−1615	282.4	−1573
Mercury				SiI$_4$(g)	−132	—	—
Hg(ℓ)	0	76.02	0	SiO$_2$(s)	−910.9	41.84	−856.7
HgCl$_2$(s)	−224	146	−179	H$_2$SiO$_3$(s)	−1189	134	−1092
HgO(s, red)	−90.83	70.29	−58.56	Na$_2$SiO$_3$(s)	−1079	—	—
HgS(s, red)	−58.2	82.4	−50.6	H$_2$SiF$_6$(aq)	−2331	—	—
Nickel				*Silver*			
Ni(s)	0	30.1	0	Ag(s)	0	42.55	0
Ni(CO)$_4$(g)	−602.9	410.4	−587.3	*Sodium*			
NiO(s)	−244	38.6	−216	Na(s)	0	51.0	0
Nitrogen				Na(g)	108.7	153.6	78.11
N$_2$(g)	0	191.5	0	Na$^+$(g)	601	—	—
N(g)	472.704	153.19	455.579	NaBr(s)	−359.9	—	—
NH$_3$(g)	−46.11	192.3	−16.5	NaCl(s)	−411.0	72.38	−384
N$_2$H$_4$(ℓ)	50.63	121.2	149.2	NaCl(aq)	−407.1	115.5	−393.0
(NH$_4$)$_3$AsO$_4$(aq)	−1268	—	—	Na$_2$CO$_3$(s)	−1131	136	−1048
NH$_4$Cl(s)	−314.4	94.6	−201.5	NaOH(s)	−426.7	—	—
NH$_4$Cl(aq)	−300.2	—	—	NaOH(aq)	−469.6	49.8	−419.2
NH$_4$I(s)	−201.4	117	−113	*Sulfur*			
NH$_4$NO$_3$(s)	−365.6	151.1	−184.0	S(s, rhombic)	0	31.8	0
NO(g)	90.25	210.7	86.57	S(g)	278.8	167.8	238.3
NO$_2$(g)	33.2	240.0	51.30	S$_2$Cl$_2$(g)	−18	331	−31.8
N$_2$O(g)	82.05	219.7	104.2	SF$_6$(g)	−1209	291.7	−1105
N$_2$O$_4$(g)	9.16	304.2	97.82	H$_2$S(g)	−20.6	205.7	−33.6
N$_2$O$_5$(g)	11	356	115	SO$_2$(g)	−296.8	248.1	−300.2
N$_2$O$_5$(s)	−43.1	178	114	SO$_3$(g)	−395.6	256.6	−371.1
NOCl(g)	52.59	264	66.36	SOCl$_2$(ℓ)	−206	—	—
HNO$_3$(ℓ)	−174.1	155.6	−80.79	SO$_2$Cl$_2$(ℓ)	−389	—	—
HNO$_3$(g)	−135.1	266.2	−74.77	H$_2$SO$_4$(ℓ)	−814.0	156.9	−690.1
HNO$_3$(aq)	−206.6	146	−110.5	H$_2$SO$_4$(aq)	−907.5	17	−742.0
Oxygen				*Tin*			
O(g)	249.2	161.0	231.8	Sn(s, white)	0	51.55	0
O$_2$(g)	0	205.0	0	Sn(s, grey)	−2.09	44.1	0.13
O$_3$(g)	143	238.8	163	SnCl$_2$(s)	−350	—	—
OF$_2$(g)	23	246.6	41	SnCl$_4$(ℓ)	−511.3	258.6	−440.2
Phosphorus				SnCl$_4$(g)	−471.5	366	−432.2
P(g)	314.6	163.1	278.3	SnO$_2$(s)	−580.7	52.3	−519.7
P$_4$(s, white)	0	177	0	*Titanium*			
P$_4$(s, red)	−73.6	91.2	−48.5	TiCl$_4$(ℓ)	−804.2	252.3	−737.2
PCl$_3$(g)	−306.4	311.7	−286.3	TiCl$_4$(g)	−763.2	354.8	−726.8
PCl$_5$(g)	−398.9	353	−324.6	*Tungsten*			
PH$_3$(g)	5.4	210.1	13	W(s)	0	32.6	0
P$_4$O$_{10}$(s)	−2984	228.9	−2698	WO$_3$(s)	−842.9	75.90	−764.1
H$_3$PO$_4$(s)	−1281	110.5	−1119	*Zinc*			
Potassium				ZnO(s)	−348.3	43.64	−318.3
K(s)	0	63.6	0	ZnS(s)	−205.6	57.7	−201.3
KCl(s)	−436.5	82.6	−408.8				
KClO$_3$(s)	−391.2	143.1	−289.9				
KI(s)	−327.9	106.4	−323.0				
KOH(s)	−424.7	78.91	−378.9				
KOH(aq)	−481.2	92.0	−439.6				

APPENDIX L

ANSWERS TO SELECTED EVEN-NUMBERED NUMERICAL EXERCISES

CHAPTER 1

28. (a) 423.$\underline{06}$ mL = 4.2306×10^2 mL (5 sig. fig.)
  (b) 0.0001$\underline{073040}$ g = 1.073040×10^{-4} g (7 sig. fig.)
  (c) 1$\underline{081.02}$ pounds = 1.08102×10^3 pounds (6 sig. fig.)
30. (a) 5060 **(b)** 0.0040010 **(c)** 16100 **(d)** 0.0000206 **(e)** 9000.
  (f) 0.009000
32. 3.2×10^4 cm^3
34. (a) 2.85×10^{-2} km **(b)** 3.63×10^4 m **(c)** 4.47×10^5 g
  (d) 1.32×10^3 mL **(e)** 5.59 L **(f)** 6.251×10^6 cm^3
36. (a) 28.3 L **(b)** 2.11 pt **(c)** 0.4252 km/L
38. 36.70 cents/L
40. 2.33 g/mL
42. 1.65 g/cm^3
44. (a) 58.6 cm^3 **(b)** 3.88 cm **(c)** 1.53 in
46. (a) 21.982 cm^3 **(b)** 7.644 g **(c)** 21.134 g **(d)** 21.134 cm^3
  (e) 0.848 cm^3 **(f)** 9.01 g/cm^3
48. 1.61×10^3 g
50. (a) -12°C **(b)** 274.6 K **(c)** 167°F **(d)** 50.5°F
52. (c) 285.3°R
54. For Al, 660.4°C and 1221°F; for Ag, 961.9°C and 1763°F
56. 38.6°C, 311.8 K
58. 865 J
60. (a) 1.16×10^6 J **(b)** 12.5°C
62. 0.49 J/g°C
64. (a) 90.9 tons of ore **(b)** 102 kg of ore
66. 7.77 m
68. 120 mg KCN (2 sig. fig.)
74. 18°C (65°F) or 65°C; 65°C is hotter
76. 300 mL (1 sig. fig.)
80. 4.85×10^7 atoms

CHAPTER 2

24. 2.411
26. (a) 159.808 amu **(b)** 34.0146 amu **(c)** 183.187 amu
  (d) 194.1903 amu
30. (a) 36.5 amu **(b)** 60.0 amu **(c)** 98.0 amu **(d)** 98.1 amu
32. 352.2 g Au, 0.3522 kg Au
34. 1.91×10^{25} H atoms
36. (a) 0.724 mol NH$_3$ **(b)** 0.0339 mol NH$_4$Br **(c)** 0.027 mol
  PCl$_5$ **(d)** 0.969 mol Sn
38. (a) 1.71×10^{23} molecules CO$_2$ **(b)** 2.69×10^{23} molecules N$_2$
  (c) 6.07×10^{22} molecules P$_4$ **(d)** 1.21×10^{23} molecules P$_2$
44. 2.66×10^{-16} g CH$_4$
48. 78.25% Ag

50. (a) C$_9$H$_9$N **(b)** 131.17 g/mol
52. (a) C$_{13}$H$_{24}$N$_4$O$_3$S (FW = 316 g/mol) **(b)** same
54. C$_8$H$_{11}$O$_3$N
56. NaHCO$_3$
58. C$_6$H$_{14}$N$_2$O$_2$
60. C$_2$O$_2$
62. (a) 54.82% C, 5.624% H, 7.104% N, 32.45% O **(b)** 80.87%
  C, 11.70% H, 7.430% O **(c)** 63.15% C, 5.300% H, 31.55% O
64. Cu$_3$(CO$_3$)$_2$(OH)$_2$ is 55.31% Cu, Cu$_2$S is 79.84% Cu, CuFeS$_2$
  is 34.63% Cu, CuS is 66.46% Cu, Cu$_2$O is 88.82% Cu,
  Cu$_2$CO$_3$(OH)$_2$ is 57.49% Cu
66. 0.1088 g C and 0.00647 g H giving 94.36% C and 5.61% H
68. CH$_2$
70. C$_2$H$_6$O
72. (a) 3.43 g O **(b)** 6.86 g O
74. (a) 5.53 g O **(b)** 8.29 g O
76. 797 g Hg
78. 63.3 g KMnO$_4$
80. 110 lb Cu$_2$S (to 2 sig. fig.)
82. (a) 556 g CuSO$_4$·H$_2$O **(b)** 500. g CuSO$_4$
84. 31.6 g Cr in ore, 19.0 g Cr recovered
86. (a) 100. lb MgCO$_3$ **(b)** 275 lb impurity **(c)** 28.9 lb Mg
88. (a) 63.92% **(b)** 47.6%
90. (a) 2.00 mol O$_3$ **(b)** 6.00 mol O **(c)** 96.0 g O$_2$ **(d)** 64.0 g O$_2$
92. C$_{14}$H$_{24}$O
96. C$_6$H$_{10}$O$_4$
98. CH$_3$CH$_2$OH and CH$_3$OCH$_3$ will produce the most water
100. 225 g NaCl
102. ZnSO$_4$ is 36.0% more economical
104. (a) 1.30 mol Ag **(b)** 1.30 mol Ag **(c)** 5.26×10^{-3} mol Ag
  (d) 0.217 mol Ag
106. C$_3$H$_6$O$_2$, C$_2$H$_6$O$_2$
108. 568 mL ethanol

CHAPTER 3

12. (b) 1200 molecules H$_2$ **(c)** 800 molecules NH$_3$
14. (b) 11 mol HCl **(c)** 5.7 mol H$_2$O
16. 1470 g NaHCO$_3$ (to 3 sig. fig.)
18. (a) 9.8 mol O$_2$ **(b)** 3.2 mol O$_2$ **(c)** 3.2 mol O$_2$ **(d)** 3.2 mol O$_2$
  (e) 13 mol O$_2$
20. (a) 15.00 mol O$_2$ **(b)** 12.00 mol NO **(c)** 18.00 mol H$_2$O
22. 128 g O$_2$
24. 90.43 g Fe$_3$O$_4$
26. 55.08 g NaI
28. 27.6 g C$_3$H$_8$

30. 102.3 g NH_3

32. 653 g superphosphate

34. 91.78 g K

36. 12.9 g $Ca_3(PO_4)_2$

38. (a) $AgNO_3$ (b) 9.9 g $BaCl_2$ (c) 35.1 g AgCl

40. 147 g PCl_5, 114 g PCl_5

42. (a) 1.47 g O_2 (b) 74.8%

44. 97.3%

46. 131 mg SO_2, 82.4%

48. (a) 350. g Fe   (b) 47.6%

50. 71.4 g H_2TeO_3

52. 677 g $KClO_3$

54. (b) 2 mol N atoms   (c) 265.9 g NH_3

56. 191 kg Zn

58. 89.1 g $(NH_4)_2SO_4$

60. 631 mL $(NH_4)_2SO_4$ soln

62. 0.619 M Na_3PO_4

64. (a) 18 g Na_2HPO_4 (b) same

66. (a) 20.0% $CaCl_2$ (b) 2.13 M $CaCl_2$

68. 0.0112 M $BaCl_2$ soln

70. 28.7 M HF soln

72. 0.469 L conc. HCl soln

74. 221 mL $Ba(OH)_2$ soln

76. 286 mL NaOH soln

78. 0.0285 L KOH soln

80. 8.23 mL HNO_3 soln

82. 0.00433 M $AlCl_3$ soln

84. 6.43% Fe_3O_4 in ore

86. 1.42 g Br_2

90. 1.0% $SnCl_2$ < 1.0% $AlCl_3$ < 1.0% NaCl in molarity

92. 68.0 mL HNO_3

94. A: 24 particles/L; B: 12 particles/L; C: 6 particles/L; D: 16 particles/L; E: 12 particles/L; F: 20 particles/L

96. (a) 62.7% yield

98. (a) CH_3CH_2OH (b) 81.6% yield

100. 0.536 M NaCl

102. 623 g H_3PO_4

104. 0.0867 g AgCl

106. for Zn: $65.4/mol H_2; for Al: $36.0/mol H_2

CHAPTER 4

8. ~90°C

42. (a) oxidation no. of P: in PCl_5, +5; in P_4O_6, +3; in P_4O_{10}, +5; in HPO_3, +5; in H_3PO_3, +3; in $POCl_3$, +5; in $H_4P_2O_7$, +5; in $Mg_3(PO_4)_2$, +5

(b) oxidation no. of Br: in Br^-, −1; in BrO^-, +1; in BrO_2^-, +3; in BrO_3^-, +5; in BrO_4^-, +7

(c) oxidation no. of Mn: in MnO, +2; in MnO_2, +4; in $Mn(OH)_2$, +2; in K_2MnO_4, +6; in $KMnO_4$, +7; in Mn_2O_7, +7

(d) oxidation no. of O: in OF_2, +2; in Na_2O, −2; in Na_2O_2, −1; in KO_2, −1/2

44. (a) oxidation no. of N: in N^{3-}, −3; in NO_2^-, +3; in NO_3^-, +5; in N_3^-, −1/3; in NH_4^+, −3

(b) oxidation no. of Cl: in Cl_2, 0; in HCl, −1; in HClO, +1; in $HClO_2$, +3; in $KClO_3$, +5; in Cl_2O_7, +7; in $Ca(ClO_4)_2$, +7

134. (a) 0.184 mol O_2 (b) 0.220 mol O_2 (c) 0.0346 mol O_2

CHAPTER 5

6. The charge on each droplet is a multiple of 1.63×10^{-19} coulombs

10. 2.3×10^{-14}

14. (a) mass % for e^- = 0.021816%   (b) mass % for p = 40.059%   (c) mass % for n = 59.918%

26. 7.5%

28. 79.90 amu

30. 55.85 amu

32. 69.17% ^{63}Cu

34. O, 15.999 amu; Cl, 35.453 amu

36. 73.3 amu

38. 52.0 amu

40. (a) 3.07×10^{14} s^{-1} (b) 6.79×10^{14} s^{-1} (c) 6.10×10^9 s^{-1} (d) 6.10×10^{18} s^{-1}

42. (a) 4.47×10^{14} s^{-1} (b) 2.96×10^{-19} J/photon   (c) red

44. 5.85×10^{-19} J/photon, 352 kJ/mol

46. 2.5×10^{13} miles

50. 320 nm (violet/near ultraviolet)

54. 216 kJ/mol

56. 3.20×10^{15} s^{-1}

60. 2.53×10^{18} photons

62. (a) 1.59×10^{-14} m   (b) 3.97×10^{-32} m

64. 1.88×10^3 m/s

102. (a) $n = 3, \ell = 1, m_\ell = -1, 0, +1$ (b) $n = 6, \ell = 3, m_\ell = -3, -2, -1, 0, +1, +2, +3$ (c) $n = 3, \ell = 1, m_\ell = -1, 0, +1$ (d) $n = 4, \ell = 3, m_\ell = -3, -2, -1, 0, +1, +2, +3$

114. 6 electrons

120. 3.76×10^{-36} kg

122. 216 kJ/mol

124. 3.35 m

CHAPTER 6

46. F_2, 0.71 ; Cl_2, 0.99 , Cl-F, 1.70

80. Ge, 122 pm

86. 1.05×10^{15} s^{-1}

88. 135 kJ

CHAPTERS 7 AND 8

No numerical exercises in these chapters.

CHAPTER 9

24. (a) X_2, bo = 3   (b) X_2, bo = 1.5   (c) X_2^-, bo = 2.5

28. (b) N_2, bo = 3, N_2^-, bo = 2.5   (b) N_2^+, bo = 2.5

30. NO^+, bo = 3

32. CN, bo = 2.5; CN^-, bo = 3; CN^{2-}, bo = 2.5

46. NO, bo = 2.5; NO^+, bo = 3; NO^-, bo = 2. Bond lengths are inversely related to bond order.

CHAPTER 10

No numerical exercises in this chapter.

CHAPTER 11

4. 0.157 M $MgSO_4$

6. 5.292 M H_2SO_4

8. 1.44 M NaCl

10. 4.50 M KI

12. 0.0229 M BaI_2

14. 1.24 M Na_3PO_4, 0.56 M NaOH

16. 26.2 mL CH_3COOH soln

18. 0.857 M CH_3COOH

20. (a) 1.33 L NaOH soln, 0.222 L H_3PO_4 soln   (b) 0.889 L NaOH soln, 0.222 L H_3PO_4 soln

26. 0.2134 M HNO_3

28. 0.07159 M NaOH

30. 2.0 mL HNO_3 soln

34. 23.9% $(COOH)_2 \cdot 2H_2O$

36. 0.104 g $CaCO_3$

38. 0.0354 M NaOH

40. (a) 0.55 N HCl   (b) 1.1 N H_2SO_4 (c) 1.6 N H_3PO_4 (d) 0.55 N NaOH

42. 1.62 N H_3PO_4

44. 0.249 M H_3AsO_4, 0.746 N H_3AsO_4

46. 0.217 N HCl, 0.217 M HCl

48. 32.7% $Mg(OH)_2$

54. oxidizing agents: (a) MnO_4^- (b) $Cr_2O_7^{2-}$ (c) MnO_4^- (d) $Cr_2O_7^{2-}$

64. 3.52 mL $KMnO_4$ soln

66. 4.21 mL $KMnO_4$ soln

68. 0.2960 g As_2O_3

70. 4.2 mL NO_3^- soln

72. 0.159 M $KMnO_4$

74. 0.2196 M HCl

76. 5.61 mmol HCl, 9.26 mL HCl

78. 430. mL HCl soln

80. 21.1 mL H_2SO_4 soln

82. 18.08 mL NaOH soln

84. (a) 11.1 mL HI soln   (b) 132 mL HI soln   (c) 34.6 mL HI soln

88. 59.36% Fe

90. 39 g AgCl

92. 80 mmol HCl in stomach; 9.28 mmol HCl

96. 1.2 g $NaHCO_3$

CHAPTER 12

6. (a) 14.5 psi   (b) 75.2 cm Hg   (c) 29.6 in Hg   (d) 100. kPa   (e) 0.989 atm   (f) 33.5 ft H_2O

10. 1.84×10^3 psi

14. 1.09 atm

16. (a) 26 atm   (b) 0.38 L

18. 890 balloons

24. 0.84 L

26. (a) 1.441 L   (b) increase by 41 cm

28. $-78.5°C$, 3.26 L; $-195.8°C$, 1.30 L; $-268.9°C$, 0.0713 L

32. 486 K or 213°C

38. 5.24 g/L

40. (a) #1—Kr, #2—He

44. 17.7 atm

46. (a) 1.80×10^8 L Cl_2, 6.34×10^6 ft^3 Cl_2 (b) 40.0 ft

48. 29.6 g/mol, 2% error

50. 30.0 g/mol, $C_2H_6(g)$

52. 46.7 amu/molecule

56. P_{total} = 132 atm, P_{CHCl_3} = 39.4 atm

58. X_{He} = 0.440, X_{Ar} = 0.299, X_{Xe} = 0.261

60. (a) 10.12 atm   (b) 3.38 atm   (c) 3.38 atm

62. 477 mL

64. (a) He, 3.43 atm; N_2, 1.29 atm   (b) 4.72 atm   (c) X_{He}, 0.727

66. 74.8 g NaN_3

68. 7.15 g SO_2

70. 4.45 g $KClO_3$

72. 3.33 L NH_3

74. 109 g KNO_3

76. 25.1% S by mass

78. (a) 2.83 L C_8H_{18} (b) 44.8 min

80. (a) 3.35×10^{-23} mL   (b) 0.0202 L   (c) 8.9×10^{-4} or 0.089%

82. 1.4

84. 1.17

92. compressibility factor, 0.695; ideal pressure, 43.2 atm

94. (a) 0.821 atm   (b) 0.805 atm

96. 0.11 g H_2O

98. 5000 L air (1 sig. fig.)

100. 121 g/mol

102. (a) 6130 g H_2O (b) 6150 mL H_2O

104. 2%

106. C_2N_2 (MW = 52.1 g/mol)

108. 1.106 g

118. 1.3 L O_2

120. 206 g $MgSiO_3$

CHAPTER 13

36. bp = $\sim34°C$.

38. 344.4 torr

40. (a) 43.4 kJ/mol

42. bp = $\sim1°C$

46. 6.38×10^3 J/mol

48. 2.28×10^5 J

50. 2.01×10^5 J

52. 54.1°C

54. 58.7°C

56. (a) 6.14×10^3 J (b) 19.7 g H_2O

58. For H_2O, 439.1 torr; for D_2O, 412.1 torr

60. (c) $\Delta H_{vap} = 3.72 \times 10^4$ J/mol (d) 3.48×10^2 K or 75°C

62. 4.34×10^4 J or 43.4 kJ

64. 2.94×10^{-3} torr

88. (a) a/2 (b) 6 Cl^- ions (c) $a(2^{1/2})/2$ (d) 6 Na^+ ions

90. 1.00 g/cm^3

92. 22.4 g/cm^3

94. (a) 8 (b) tetrahedron with 4 nearest neighbors (d) 1.545 Å
(e) 3.517 g/cm^3

96. AW = $(1.37 \times 10^{-21}$ g)/$(6.64 \times 10^{-24}$ mol) = 206 g/mol,
Pb (207.2 g/mol)

100. 1.542 Å

108. 0.988 atm, 0.982 atm, 0.6%

112. (a) true (b) false (c) false

114. CS_2 < acetone < ethanol

124. 64 g/mol, Cu

130. (a) 3.2×10^{18} atoms B (b) 5.8×10^{-5} g B

CHAPTER 14

18. X_{CH_4} (at 25°C) = 2.4×10^{-4}; X_{CH_4} (at 50°C) = 1.7×10^{-4};
decreases

26. Add 14.3 g NaCl to enough water to make exactly 1 L soln

28. 1.81 m C_6H_5COOH in C_2H_5OH

30. $X_{C_2H_5OH}$ = 0.540; X_{H_2O} = 0.460

32. 0.7767 M K_2SO_4, 0.8197 m K_2SO_4, 12.50% K_2SO_4,
X_{H_2O} = 0.9854

34. (a) 1.48 m $C_6H_{12}O_6$

38. (a) 7.01 torr (b) 67.6 torr

40. $P_{acetone}$ = 172 torr, $P_{chloroform}$ = 148 torr

42. P_{total} = 320. torr, $X_{acetone}$ = 0.538, $X_{chloroform}$ = 0.462

44. (a) $P_{chloroform}$ = 55 torr (b) $P_{acetone}$ = 210 torr
(c) P_{total} = 280 torr

46. 101.15°C

48. −4.18°C

50. (a) camphor (b) water (c) camphor (d) nitrobenzene

52. −18.6°C

54. 78.54°C

56. 1035°C

58. 30.4 g $C_{10}H_8$

60. $C_4H_8O_4$ (120.10 g/mol)

62. (a) 70.% $C_{10}H_8$, 30.% $C_{14}H_{10}$ (b) 80.4°C

66. (a) 3 (b) 2 (c) 5 (d) 2

68. $CaCl_2$ < $KClO_3$ < CH_3COOH < CH_3OH (based on i)

70. 751 torr

72. 0.15 m Na_2SO_4 has lower fp

76. 100.63°C

78. 1%

80. 79%

84. 0.431 atm

86. ΔT_f = 0.105°C, ΔT_b = 0.0290°C

88. 45.6 atm

90. (a) -1.02×10^{-4} °C (b) 1.35×10^{-3} atm or 1.02 torr
(c) 1000% error (d) 10% error

98. bp = 100.443°C; π = 20.4 atm

100. (a) 60.5 g/mol (b) 117 g/mol

104. 37% lactose by mass

118. 1.4 atm, 0.059 M

CHAPTER 15

14. (a) 2630 kJ (b) 61.3 g O_2

16. +98.9 kJ/mol rxn

28. −583 kJ/mol

30. −1015.4 kJ/mol rxn

32. +197.8 kJ/mol rxn

34. −289 kJ/mol rxn

36. (a) −36.0 kJ/mol rxn (b) −1656 kJ/mol rxn
(c) +624.6 kJ/mol rxn

38. +46.36 kJ/g C_3H_8; +44.27 kJ/g C_8H_{18}

40. +333 kJ heat released

42. −65.3 kJ/mol SiC(s)

46. (a) −121 kJ/mol rxn (b) −103 kJ/mol rxn

48. −379 kJ/mol rxn

50. −93 kJ/mol HCl; −270 kJ/mol HF

52. 329 kJ/mol

54. 264 kJ/mol

56. 288 kJ/mol

58. 381 J/°C

60. 0.66 J/g·°C

62. (a) $+2.1 \times 10^3$ J (b) -2.1×10^3 J

64. −41.9 kJ/g $C_6H_6(\ell)$; −3270 kJ/mol $C_6H_6(\ell)$

66. −23.1 kJ or −18.9 kJ/g Mg or −460. kJ/mol Mg

68. -2.08×10^3 J/g $C_{10}H_{22}(\ell)$; −297 kJ/mol $C_{10}H_{22}(\ell)$

74. −289 J

76. (a) by surroundings (b) by system (c) by surroundings

78. (a) +983 J (b) 0 J

92. (a) −128.8 J/(mol rxn)·K (b) −182 J/(mol rxn)·K
(c) +148.6 J/(mol rxn)·K

96. −4.78 kJ, spontaneous

98. −53.42 kJ/mol rxn

100. (a) ΔH^0 = −71.75 kJ/mol rxn, ΔS^0 = −268.0 J/(mol rxn)·K,
ΔG^0 = +8.15 kJ/mol rxn
(b) ΔH^0 = +14.7 kJ/mol rxn, ΔS^0 = +31.2 J/(mol rxn)·K,
ΔG^0 = +5.4 kJ/mol rxn
(c) ΔH^0 = −367.6 kJ/mol rxn, ΔS^0 = −11.62 J/(mol rxn)·K,
ΔG^0 = −364.1 kJ/mol rxn

104. (a) ΔH^0 = −196.0 kJ/mol rxn, ΔG^0 = −233.6 kJ/mol rxn,
ΔS^0 = +125.6 J/(mol rxn)·K

108. (a) spontaneous at all temperatures (b) spontaneous at
T > 1294 K (c) spontaneous at T > 980.7 K (d) spontaneous
at T > 3000. K

110. (a) 97°C

114. (a) −156 kJ/mol $C_6H_{12}(\ell)$ (b) −165 kJ/mol $C_6H_5OH(s)$

116. $q = +10700$ J; $w = -789$ J; $\Delta E = 9900$ J

122. (a) sp. ht. $= 0.13$ J/g °C, tungsten, W (0.135 J/g °C)
    (b) yes, sp. ht $= 0.26$ J/g °C, molybdenum, Mo

126. 2.4 hr walking

128. for Pb: sp. ht $= 0.128$ J/g °C; molar heat capacity, 26.5 J/mol °C

130. (a) 30.0 kJ/g   **(b)** 7.17 kcal/g   **(c)** 35.9 kcal

CHAPTER 16

10. O_2: 1.38 M/min, NO: 1.10 M/min, H_2O: 1.65 M/min

14. 1/2

18. rate $= (2.5 \times 10^{-2} \ M^{-1} \cdot min^{-1})[B][C]$

20. decrease by factor of 8

22. rate $= (1.2 \times 10^2 \ M^{-2} \cdot s^{-1})[ClO_2]^2[OH^-]$

24. rate $= (2.5 \times 10^{-3} \ M^{-2} \cdot s^{-1})[A][B]^2$

26. rate $= (20. \ M^{-4} \cdot s^{-1})[A]^2[B]^3$

28. 0.300 M/s, 1.20 M/s

30. (a) 4.52×10^{-29} s^{-1} **(b)** 4.04×10^{-10} s^{-1}

34. 5.29 s

36. (a) 2.5×10^6 s **(b)** 41 days **(c)** 0.723 g Cs **(d)** 0.67 g

38. (a) 2.4×10^9 s or 76 yrs **(b)** 0.59 M NO_2, 54 g NO_2
    (c) 0.91 M NO

40. 1680 s or 28.0 min

42. 0.024 min^{-1}

44. (a) zero order rxn, rate $= k = 0.00273 \ mM \cdot s^{-1}$,
    $[HI] = [HI]_0 - akt = 5.46 \ mM - (2)(0.00273 \ mM \cdot s^{-1})t$
    (b) 2.18 mmol/L

46. 4.4 s

52. 340 kJ/mol rxn

54. 84 kJ/mol rxn

56. 103 kJ/mol rxn

58. (a) 270. kJ/mol **(b)** 7.7×10^{-14} s^{-1} **(c)** 730. K

60. 2600 g CO_2/L·min

62. (a) 0.24 s^{-1} **(b)** 9°C

64. (a) no **(b)** yes **(c)** no

66. (a) rate $= k[A]^2[B]^2$ **(b)** rate $= k[A][B][D]$

68. rate $= k[O_3]^2/[O_2]$

70. (a) N_2Cl **(b)** yes

72. (b) yes

76. (a) 2.9×10^{-12} mol N_2O_5 **(b)** 19 s

78. (a) rate $= k[Hb][CO]$ **(b)** 0.280 L/μmol·s
    (c) 0.252 μmol/L·s

88. 64 kJ/mol

92. $-200.$ kJ/mol rxn

CHAPTER 17

24. $K_c = 1.1 \times 10^{-5}$

26. $K_c = 0.13$

28. $K_c = 0.12$

30. (a) $K_c' = 8.9 \times 10^5$ **(b)** $K_c'' = 1.3 \times 10^{-12}$
    (c) $K_c''' = 1.6 \times 10^{-24}$

32. (a) 7.4×10^{-5} mol H_2, 1.8×10^{-3} mol HI **(b)** $K_c = 93$

34. (a) $K_c = 4.49 \times 10^{-2}$ **(b)** $[SbCl_5] = 1.6 \times 10^{-3}$ M,
    $[SbCl_3] = [Cl_2] = 8.40 \times 10^{-3}$ M

36. $K_c' = 6.502 \times 10^{-4}$

42. (a) false **(b)** false **(c)** true **(d)** false

44. 0.038 M

46. 5.8 M

48. 0.093 M

50. 1 g (1 sig. fig.)

52. $[HI] = 0.116$ M, $[H_2] = [I_2] = 0.105$ M

64. 0.35 M

66. (a) $K_c = 0.21$ **(b)** $[A] = 0.12$ M, $[B] = [C] = 0.15$ M
    (c) $[A] = 0.71$ M, $[B] = [C] = 0.39$ M

68. (a) $[N_2O_4] = 0.0144$ M, $[NO_2] = 9.18 \times 10^{-3}$ M
    (b) $[N_2O_4] = 9.03 \times 10^{-3}$ M, $[NO_2] = 7.26 \times 10^{-3}$ M
    (c) $[N_2O_4] = 0.031$ M, $[NO_2] = 0.0136$ M

72. $K_p = 1.6 \times 10^{-9}$

74. $P_{N_2O_4} = 12$ atm, $P_{NO_2} = 2.9$ atm

76. $K_p = 0.771$

78. $K_c = 7.76$

80. (a) 33.2%

84. (a) $K_p = 1.1 \times 10^5$ **(b)** $K_p = 2.3 \times 10^2$ **(c)** $K_p = 1.0 \times 10^5$

86. $K_p = 2.3 \times 10^{24}$

88. (a) -37.9 kJ/mol rxn at 25°C **(b)** $K_{p \ 1073 \ K} = 4.9 \times 10^{-4}$
    (c) $\Delta G^0 = +68.0$ kJ/mol rxn at 800°C

90. $K_p = 1.50$, $P_{CO} = 0.685$ atm, $P_{CO_2} = 0.315$ atm

94. At 400°C: $K_p = 8.3 \times 10^{-3}$, $K_c = 1.5 \times 10^{-4}$; at 800°C:
    $K_p = 16$, $K_c = 0.18$

96. $K_p = 0.00432$

106. $\Delta G^0_{rxn} = -142$ kJ/mol, $\Delta G^0_{f \ SO_3(g)} = -371$ kJ/mol

108. $K_2 = 0.26$

CHAPTER 18

4. (a) 1.71 M NaCl **(b)** 0.763 M H_2SO_4
    (c) 9.6×10^{-4} M C_6H_5OH

6. (a) $[H^+] = [Br^-] = 0.45$ M **(b)** $[K^+] = [OH^-] = 0.045$ M
    (c) $[Ca^{2+}] = 0.0155$ M, $[Cl^-] = 0.310$ M

8. (a) $[K^+] = [OH^-] = 0.0184$ M **(b)** $[Ba^{2+}] = 0.0122$ M,
    $[OH^-] = 0.0243$ M **(c)** $[Ca^{2+}] = 0.0878$ M, $[NO_3^-] = 0.176$ M

16. 1.13×10^{-9} M

18. (a) -3.28 **(b)** 2.1767 **(c)** -10.24 **(d)** -6.54

20. 3.5×10^{-8} M, 2.9×10^{-7} M

22. (a) 0.699 **(b)** 1.30 **(c)** 2.19

24. 10.99

26. 3.0×10^{-3} M

28. (a) pH $= 12.81$ **(b)** pH $= 1.19$ **(c)** pH $= 13.11$

34. (a) pH $= 6.82$ **(b)** neutral

38. pH $= 3.107$, $K_a = 8.45 \times 10^{-6}$

40. $K_a = 1.6 \times 10^{-5}$

42. $[C_6H_5COOH] = 0.42$ M, $[H_3O^+] = [C_6H_5COO^-] =$
    5.1×10^{-3} M, $[OH^-] = 2.0 \times 10^{-12}$ M

44. pH $= 1.75$

46. pH $= 11.80$

48. 6.2% ionized

50. For weak acid #1: $pK_a = 4.11$; for weak acid #2: $pK_a = 9.35$

52. $pK_b = 8.82$

54. $K_b = 5.0 \times 10^{-4}$

56. (a) $[OH^-] = 1.6 \times 10^{-3}\ M$, 1.1% ionized, pH = 11.20
(b) $[OH^-] = 8.4 \times 10^{-3}\ M$, 5.6% ionized, pH = 11.92

58. For triethylamine: $[OH^-] = 2.8 \times 10^{-3}\ M$, for trimethylamine: $[OH^-] = 1.1 \times 10^{-3}\ M$

60.

0.100 M H$_3$AsO$_4$ Solution		0.100 M H$_3$PO$_4$ Solution	
Species	*Concentration (M)*	*Species*	*Concentration (M)*
H_3AsO_4	0.095	H_3PO_4	0.076
H_3O^+	0.0050	H_3O^+	0.024
$H_2AsO_4^-$	0.0050	$H_2PO_4^-$	0.024
$HAsO_4^{2-}$	5.6×10^{-8}	HPO_4^{2-}	6.2×10^{-8}
OH^-	2.0×10^{-12}	OH^-	4.2×10^{-13}
AsO_4^{3-}	3.4×10^{-18}	PO_4^{3-}	9.3×10^{-19}

62. $[H_3O^+] = 0.16\ M$, $[OH^-] = 6.2 \times 10^{-14}\ M$, $[HSeO_4^-] = 0.14\ M$, $[SeO_4^{2-}] = 0.01\ M$

64. pH = 1.5

78. $K = 5.3 \times 10^{-10}$

80. (a) $K_b = 2.2 \times 10^{-11}$ (b) $K_b = 4.0 \times 10^{-6}$
(c) $K_b = 5.6 \times 10^{-11}$

82. (a) pH = 9.42   (b) pH = 11.35   (c) pH = 11.75

84. (a) pH = 11.89   (b) pH = 8.15

88. (a) $K_a = 1.4 \times 10^{-11}$ (b) $K_a = 6.7 \times 10^{-6}$
(c) $K_a = 1.4 \times 10^{-10}$

90. (a) pH = 4.92   (b) pH = 5.64   (c) pH = 2.60

96. (a) pH = 2.89, 0.87% hydrolysis   (b) pH = 5.21, 8.2×10^{-3}% hydrolysis   (c) pH = 6.17, 4.5×10^{-4}% hydrolysis

98. (a) pH = 1.97   (b) $[NO_2^-] = 0.011\ M$

112. $[H_3O^+] = 2.9 \times 10^{-3}\ M$, pH = 2.54

114. pH = 2.04

CHAPTER 19

8. (a) pH = 3.44   (b) pH = 4.84

10. (a) $[OH^-] = 2.7 \times 10^{-5}\ M$, pH = 9.43
(b) $[OH^-] = 1.4 \times 10^{-5}\ M$, pH = 9.13

16. 12 g NaCH$_3$COO

20. pH = 0.82

22. pH decreases by 0.40 units (9.48 $\longrightarrow$ 9.08)

24. (a) pH = 9.34   (b) pH = 9.26   (c) pH = 1.00

26. $[BrCH_2COOH] = 8.6 \times 10^{-2}\ M$, $[NaBrCH_2COO] = 0.21\ M$

28. $[C_2H_5NH_3^+] = 0.0017\ M$

30. $[CH_3CH_2COOH] = 0.23\ M$

32. 0.34 L NaOH, 0.66 L CH$_3$COOH

34. (a) 0.625 M CH$_3$COOH   (b) 0.300 M Ca^{2+}
(c) 0.600 M CH$_3$COO$^-$ (d) $1.9 \times 10^{-5}\ M$ H$^+$ (d) pH = 4.72

36. $1.1 \times 10^{-2}\ M$ ClCH$_2$COOH

40. $K_a = 6 \times 10^{-9}$, $pK_a = 8.2$

42. pH between 7.6 and 8

44. pH = 8.40, phenolphthalein

48. (a) pH = 0.903   (b) pH = 1.059   (c) pH = 1.302
(d) pH = 1.91   (e) pH = 7.00   (f) pH = 12.00

50. (a) pH = 0.839   (b) pH = 0.979   (c) pH = 1.166
(d) pH = 1.506   (e) pH = 2.08   (f) pH = 11.71

52. (a) pH = 3.22   (b) pH = 4.14   (c) pH = 4.57   (d) pH = 4.74
(halfway to the equivalence pt.)   (e) pH = 5.11   (f) pH = 5.70
(g) pH = 6.0   (h) pH = 8.52 (at the equivalence pt.)
(i) pH = 11.00   (j) pH = 11.60   (k) pH = 12.00

54. (a) pH = 2.74   (b) pH = 4.72   (c) pH = 4.97   (d) pH = 5.23
(e) pH = 5.51   (f) pH = 8.86   (g) pH = 11.54

56. (a) pH = 8.93   (b) pH = 8.76   (c) pH = 8.36

58. (a) pH = 2.00   (b) pH = 2.05   (c) pH = 2.22   (d) pH = 2.30
(e) pH = 3.00   (f) pH = 3.3   (g) pH = 5.62   (h) pH = 8.0
(i) pH = 8.73

60. pH (NaCl) = 7.00, pH (NH$_4$Cl) = 4.76

66. 78.1% KHSO$_4$ by mass

68. pH = 4.03

70. 0.22 L HCl

CHAPTER 20

8. (a) $K_{sp} = 3.5 \times 10^{-5}$ (b) $K_{sp} = 7.7 \times 10^{-19}$
(c) $K_{sp} = 6.9 \times 10^{-15}$ (d) $K_{sp} = 1.01 \times 10^{-4}$

12. $K_{sp} = 3.65 \times 10^{-11}$

14. 6.9×10^{-3} g CaCO$_3$/L

16. (a) 2.3×10^{-6} mol CuI/L; $2.3 \times 10^{-6}\ M$ Cu$^+$; $2.3 \times 10^{-6}\ M$ I$^-$;
4.4×10^{-4} g CuI/L
(b) 6.5×10^{-7} mol Ba$_3$(PO$_4$)$_2$/L; $2.0 \times 10^{-6}\ M$ Ba^{2+};
$1.3 \times 10^{-6}\ M$ PO$_4^{3-}$; 3.9×10^{-4} g Ba$_3$(PO$_4$)$_2$/L
(c) 2.1×10^{-3} mol PbF$_2$/L; $2.1 \times 10^{-3}\ M$ Pb^{2+};
$4.2 \times 10^{-3}\ M$ F$^-$; 0.51 g PbF$_2$/L
(d) 2.5×10^{-7} mol Sr$_3$(PO$_4$)$_2$/L; $7.4 \times 10^{-7}\ M$ Sr^{2+};
$4.9 \times 10^{-7}\ M$ PO$_4^{3-}$; 1.1×10^{-4} g Sr$_3$(PO$_4$)$_2$/L

18. 0.012 g BaSO$_4$

20. 6.5×10^{-3} mol Ag$_2$SO$_4$/L of 0.10 M K$_2$SO$_4$

28. Precipitation does not occur

30. Precipitation does occur

32. 0.013% Pb^{2+} in soln

36. (a) AuBr   (b) $1.5 \times 10^{-6}\ M$ Au$^+$; 99.985% Au$^+$ precipitated
(c) $9.4 \times 10^{-11}\ M$ Au$^+$; $6.2 \times 10^{-7}\ M$ Ag$^+$

38. (a) PbCrO$_4$ (b) $3.6 \times 10^{-13}\ M$ Pb^{2+} (c) $3.6 \times 10^{-7}\ M$ Pb^{2+}
(d) $5.0 \times 10^{-8}\ M$ CrO$_4^{2-}$

40. (a) (i) $2.0 \times 10^{-11}\ M$ Zn^{2+} (ii) $1.0 \times 10^{-8}\ M$ Zn^{2+}
(iii) $3.6 \times 10^{-6}\ M$ Zn^{2+}
(b) (i) $1.7 \times 10^{-7}\ M$ OH$^-$ (ii) $1.0 \times 10^{-8}\ M$ CO$_3^{2-}$
(iii) $7.3 \times 10^{-5}\ M$ CN$^-$

42. pH = 7.59

44. 2.5×10^{-7} mol CaF$_2$/L

46. A precipitate will form but not be seen

48. (a) 6.7×10^{-8} mol Mg(OH)$_2$/L of 0.015 M NaOH
(b) 1.6×10^{-5} mol Mg(OH)$_2$/L of 0.015 M MgCl$_2$

50. Precipitation will not occur

52. (a) pH = 9.40   (b) 1.1×10^{-4} g Fe(OH)$_2$/100 mL

60. Precipitation will not occur

62. 99.955% I^- removed

64. 1.1×10^4 L H_2O

68. 6.0×10^{-6} M Cl^-

70. 0.6 g Ca^{2+}/L

72. 29% loss of $MgCO_3$

CHAPTER 21

18. **(i) (a)** 3 **(b)** 2 **(c)** 1
(ii) (a) 5.18×10^3 C **(b)** 962 C **(c)** 481 C

20. **(a)** 0.152 A **(b)** 1.20 g Cu

22. 0.24 g Rh

24. 0.818 C

26. 10.6 g Ag

28. **(a)** 0.0714 faradays **(b)** 6.89×10^3 C **(c)** 0.800 L_{STP} H_2
(d) pH = 13.155

30. 3.2 L Cl_2

32. 4.6 hr, 4.2 g Cu

34. **(a)** 0.0214 faradays **(b)** 2.31 g Ag **(c)** 5.18 g $Fe(NO_3)_3$

52. **(a)** +3.17 V **(b)** +1.05 V

54. **(a)** yes **(b)** no

56. **(a)** yes **(b)** no

58. +1.78 V

60. **(a)** 1.08 V **(b)** −1.202 V

62. **(a)** no **(b)** yes **(c)** yes **(d)** no

64. **(a)** H_2 **(b)** Sn **(c)** Hg **(d)** Cl^- in base **(e)** H_2S **(f)** Ag

66. −1.207 V

68. **(a)** yes **(b)** +0.916 V

70. +5.80 V (K^+/K with F_2/F^-)

78. −0.231 V

80. **(a)** +2.123 V **(b)** +2.165 V

82. $P_{F2} = 2.1 \times 10^{-7}$ atm

84. **(a)** +0.873 V **(b)** +0.148 V **(c)** +0.0527 V

86. 0.034 M Cu^{2+}

88. **(a)** $[Zn^{2+}]/[Ni^{2+}] = 2 \times 10^{17}$ **(b)** 1×10^{-17} M Ni^{2+};
2.00 M Zn^{2+}

90. pH = 7.83

94. **(a)** $E^0_{cell} = +0.74$V; $\Delta G^0 = -360$ kJ/mol rxn; $K = 10^{62}$
(b) $E^0_{cell} = +0.368$V; $\Delta G^0 = -35.5$ kJ/mol rxn; $K = 1.7 \times 10^6$
(c) $E^0_{cell} = +1.833$V; $\Delta G^0 = -1061$ kJ/mol rxn; $K = 8.6 \times 10^{185}$

96. $K = 10^{-21}$

98. $K = 2$ (1 sig. fig.)

106. **(b)** +2.33 V **(c)** +2.44 V **(d)** 0.0227 g Mg

108. 955 C

110. $[Mn^{2+}]/[Fe^{2+}] = 10^{-28}$

112. **(a)** +4 **(b)** +4 **(c)** 0 **(d)** Mg(s) **(e)** UF_4(s) **(f)** 10.2 A
(g) 0.0206 L HF(g) **(h)** yes, 2.45 g U

122. **(a)** $K_{sp} = 10^{-12}$ **(b)** $-G^0 = +68$ kJ/mol rxn

124. **(i)** −1.662 V **(ii)** −1.662 V

CHAPTER 22

14. 39.997 g NaOH, 1.008 g H_2, 35.45 g Cl_2

36. **(a)** −15 kJ/mol rxn **(b)** +5 kJ/mol rxn **(c)** −25 J/(mol rxn)·K

40. 23.7 g Cu

42. 4.99 yr; 1.18×10^4 L O_2

44. 110. tons SO_2

46. 34.5 tons C (coke)

CHAPTER 23

30. **(a)** −201.4 kJ/mol rxn **(b)** −138.9 kJ/mol rxn
(c) −415.0 kJ/mol rxn

46. 368 g Co_3O_4

48. $\Delta H^0_{rxn} = -195.4$ kJ/mol Rb(s), $\Delta S^0_{rxn} = 29.4$ J/K per 1 mol Rb(s),
$\Delta G^0_{rxn} = -204.0$ kJ/mol Rb(s)

50. Q = Li in Li_2CO_3

CHAPTER 24

8. 3.25 g XeF_6

42. 2.04 tons H_2SO_4

44. $K_c = 2.7$

76. −140 kJ/mol rxn

80. 2.2×10^{25} g Si

82. 1.92 Å

CHAPTER 25

44. **(a)** 2.2×10^{-6} mol $Zn(OH)_2$/L **(b)** 0.0087 mol $Zn(OH)_2$/L
(c) $[Zn(OH)_4^{2-}] = 0.0087$ M

46. $\Delta S^0_{rxn} = -153.5$ J/(mol rxn)·K

48. pH = 10.43

CHAPTER 26

14. **(a)** 0.587 g/mol **(b)** 5.28×10^{10} kJ/mol of ^{62}Ni atoms

16. **(a)** 0.602 amu/atom **(b)** 0.602 g/mol **(c)** 9.00×10^{-11} J/atom
(d) 5.42×10^{10} kJ/mol **(e)** 8.79 MeV/nucleon

18. **(a)** 1.04×10^{11} kJ/mol of ^{127}I atoms **(b)** 6.83×10^{10} kJ/mol of
^{81}Br atoms **(c)** 2.89×10^{10} kJ/mol of ^{35}Cl atoms

30. **(a)** $n/p = 1.49$ **(b)** $n/p = 1.47, 1.49$ **(c)** $n/p = 1.39$

54. 76.0 min (92.5%), 135 min (99.0%)

56. 0.7 fs (to 1 sig. fig.)

58. 2880 yr

60. 937 yr

62. 1.95×10^4 yr

66. **(b)** $\Delta E = -1.68 \times 10^9$ kJ/mol rxn

74. **(b)** $\Delta E = +1.15 \times 10^8$ kJ/mol rxn

76. 2.52×10^9 yr

80. fission = -1.40×10^{-13} J/amu ^{235}U; fusion = -1.53×10^{-13} J/amu ^{2}H

CHAPTER 27

No numerical exercises in this chapter.

CHAPTER 28

16. $[C_6H_5NH_2] = 0.15\ M$; $[C_6H_5NH_3^+] = [OH^-] = 7.9 \times 10^{-6}\ M$; $[H_3O^+] = 1.3 \times 10^{-9}\ M$

22. pH = 10.83

44. 9; A-A, A-B, A-C, B-A, B-B, B-C, C-A, C-B, C-C

60. $P_{CO} = P_{H_2} = 7.1$ atm

62. pH (sodium benzoate) = 8.69; more acidic

GLOSSARY/INDEX

Glossary terms, printed in **boldface,** are defined here as well as in the text (location indicated by boldface page numbers) and in Key Terms. Page numbers followed by *i* indicate illustrations or their captions; page numbers followed by *t* indicate tables.